MANY-BODY THEORY OF MOLECULES, CLUSTERS, AND CONDENSED PHASES

World Scientific Series in 20th Century Physics

Vol. 22 A Quest for Symmetry — Selected Works of Bunji Sakita
edited by K. Kikkawa, M. Virasoro and S. R. Wadia

Vol. 23 Selected Papers of Kun Huang (with Commentary)
edited by B.-F. Zhu

Vol. 24 Subnuclear Physics — The First 50 Years: Highlights from Erice to ELN
by A. Zichichi
edited by O. Barnabei, P. Pupillo and F. Roversi Monaco

Vol. 25 The Creation of Quantum Chromodynamics and the Effective Energy
by V. N. Gribov, G. 't Hooft, G. Veneziano and V. F. Weisskopf
edited by L. N. Lipatov

Vol. 26 A Quantum Legacy — Seminal Papers of Julian Schwinger
edited by K. A. Milton

Vol. 27 Selected Papers of Richard Feynman (with Commentary)
edited by L. M. Brown

Vol. 28 The Legacy of Léon Van Hove
edited by A. Giovannini

Vol. 29 Selected Works of Emil Wolf (with Commentary)
edited by E. Wolf

Vol. 30 Selected Papers of J. Robert Schrieffer — In Celebration of His 70th Birthday
edited by N. E. Bonesteel and L. P. Gor'kov

Vol. 31 From the Preshower to the New Technologies for Supercolliders — In Honour of Antonino Zichichi
edited by B. H. Wiik, A. Wagner and H. Wenninger

Vol. 32 In Conclusion — A Collection of Summary Talks in High Energy Physics
edited by J. D. Bjorken

Vol. 33 Formation and Evolution of Black Holes in the Galaxy
— Selected Papers with Commentary
edited by H. A. Bethe, G. E. Brown and C.-H. Lee

Vol. 35 A Career in Theoretical Physics, 2nd Edition
by P. W. Anderson

Vol. 36 Selected Papers (1945–1980) with Commentary
by Chen Ning Yang

Vol. 37 Adventures in Theoretical Physics
— Selected Papers with Commentaries
by Stephen L. Adler

Vol. 38 Matter Particled — Patterns, Structure and Dynamics
— Selected Research Papers of Yuval Ne'eman
edited by R. Ruffini and Y. Verbin

Vol. 39 Searching for the Superworld — A Volume in Honour of Antonino Zichichi on the Occasion of the Sixth Centenary Celebrations of the University of Turin, Italy
edited by S. Ferrara and R. M. Mössbauer

Vol. 40 Murray Gell-Mann — Selected papers
edited by H. Fritzsch

Vol. 41 Many-Body Theory of Molecules, Clusters, and Condensed Phases
edited by N. H. March and G. G. N. Angilella

For information on Vols. 1–21, please visit http://www.worldscibooks.com/series/wsscp_series.shtml

World Scientific Series in 20th Century Physics Vol. 41

MANY-BODY THEORY OF MOLECULES, CLUSTERS, AND CONDENSED PHASES

editors

N. H. March

Oxford University, UK & University of Antwerp, Belgium

G. G. N. Angilella

Università di Catania, CNISM & INFN, Italy

World Scientific

NEW JERSEY · LONDON · SINGAPORE · BEIJING · SHANGHAI · HONG KONG · TAIPEI · CHENNAI

Published by

World Scientific Publishing Co. Pte. Ltd.

5 Toh Tuck Link, Singapore 596224

USA office: 27 Warren Street, Suite 401-402, Hackensack, NJ 07601

UK office: 57 Shelton Street, Covent Garden, London WC2H 9HE

British Library Cataloguing-in-Publication Data
A catalogue record for this book is available from the British Library.

MANY-BODY THEORY OF MOLECULES, CLUSTERS, AND CONDENSED PHASES
World Scientific Series in 20th Century Physics — 41

ISBN-13 978-981-4271-77-6
ISBN-10 981-4271-77-2

Printed in Singapore by B & JO Enterprise

Contents

Part 1: Quantal electron crystals

Part 2: Structure, forces and electronic correlation functions in liquid metals

Part 2a. Nuclear structure factor and pair potentials in some *sp* liquid metals

Part 3: One-body potential theory of molecules and condensed matter

Partisc 3a. Thomas–Fermi semiclassical approximation

Part 3b. Transcending Thomas–Fermi theory

Part 1: Quantal electron crystals

We shall begin this commentary on Part 1 with a little historical background, including a brief meeting of one of us (NHM) with the creator of the field of quantal electron solids in zero magnetic field, namely Professor E. P. Wigner [Phys. Rev. **46**, 1002 (1934)].

NHM made his first contribution in this area through the virial theorem. In Ref. [1.1], written nearly a quarter of a century after Wigner's innovation, the many-electron total correlation energy per electron was partitioned thereby into kinetic and potential contributions. This was followed, almost immediately, by the variational calculation of Young and March (1960), using the two-electron (or equivalently second-order) correlated density matrix [1.3]. Their theory had explicitly built in the quantal electron crystal as the extreme low-density limit.

In 1968, NHM's former research student, John Durkan, together with Professor R. J. Elliott and NHM [1.4] were motivated by the early transport experiments by E. H. Putley on highly compensated InSb in an applied magnetic field, introducing in the Reviews of Modern Physics the concept of a magnetically induced Wigner solid (customarily now abbreviated in the very substantial existing literature as MIWS). Following a further key experiment by D. Somerford on the same three-dimensional InSb material [J. Phys. C **4**, 1570 (1971)], Care and March [1.5] discussed theoretically the electrical conduction near the Wigner transition, emphasizing however, besides MIWS, the need to treat carefully the random fields of the acceptor centres in the highly-compensated specimen of InSb. This was followed by the review, of Care and March in 1975, on *Electron crystallization* [1.6].

But the experimental breakthrough in the field of MIWS came just 20 years after the Durkan–Elliott–March proposal [1.4] through the utilization of the two-dimensional electron assembly in a high-quality GaAs-AlGaAs heterojunction by E. Y. Andrei *et al.* [Phys. Rev. Lett. **60**, 2765 (1988)]. Their experiment, in essence, tested for a defining property of a (now electron!) solid, namely its rigidity to shear. Soon afterwards, a different experiment on the luminescence spectrum of such a heterojunction was reported by H. Buhmann *et al.* [Phys. Rev. Lett. **66**, 926 (1991)], and a schematic mapping was proposed for the melting curve of the MIWS. This was followed by the thermodynamics of melting of a two-dimensional Wigner electron crystal by Lea, March, and Sung [1.13]. In their work, the change in magnetization along the MIWS melting curve was set out, as very reminiscent of the de Haas-van Alphen effect, but now occurring in the so-called Laughlin electron liquid in equilibrium with the MIWS. The relevance of the anyon model to the strongly correlated electron liquid was also stressed in [1.13] (see also [1.18]).

Two other very different areas in which Wigner's ideas in zero magnetic field played a salient role were (1) the hole liquid in a high-T_c cuprate, and whether this could undergo Wigner crystallization at suffciently low density, and (ii) Wigner molecules. As to (i), the matter was raised in a semiquantitative study in [1.14], and a quantitative theory was subsequently proposed by M. P. A. Fisher *et al.* [Phys. Rev. ..., ...(...)]. In area (ii), an exactly solvable model was put forward by P. Capuzzi *et al.* [1.24] for two electrons, and multi-electron assemblies were treated, for example, by Amovilli and March [Phys. Lett. A **324**, 46 (2004)].

The Editors, as mentioned above, wish to add here to this brief Commentary on Part I by referring first to a somewhat extraordinary meeting between E. P. Wigner and N. H. March, at Louisiana State University in Baton Rouge. NHM had been invited by (the late) Professor J. Callaway, and was giving a seminar on MIWS. As Joe Callaway (JC) was taking NHM into the Seminar to introduce him to the audience, JC whispered that Professor Wigner was visiting that day and would be there! Professor Wigner made no contribution until the Seminar was over. But then, he came forward to the lecture bench and simply asked "Had all that happened as a result of his 'elementary' research papers on zero magnetic field in 1934 and 1938?"!

As the concluding comments here, the Editors emphasize two further points. One is that this field has been brought to full fruition only by the interplay of fundamental many-body theory, key experiments referred to above, and by quantum computer simulation [see D. M. Ceperley and B. J. Alder, Phys. Rev. Lett. **45**, 566 (1980)]. Finally, the Editors emphasize that Part 1 of the present Volume could not possibly have materialized without the inspirational collaboration of Professors C. M. Care, R. J. Elliott, M. J. Lea, Dr D. Somerford and Professor W. H. Young with NHM's research group.

PHYSICAL REVIEW VOLUME 110, NUMBER 3 MAY 1, 1958

Kinetic and Potential Energies of an Electron Gas

N. H. March
Department of Physics, The University, Sheffield, England
(Received January 3, 1958)

It is pointed out that the kinetic and potential energies of an electron gas may be obtained exactly in the high-density limit by using the virial theorem in conjunction with the results of Gell-Mann and Brueckner. For intermediate and low densities, the results obtained from Wigner's formula are also referred to, although for the intermediate region these are likely to be at best semiquantitative.

1. INTRODUCTION

FOLLOWING earlier investigations by Wigner,[1] there has recently been considerable progress in dealing with correlation effects in an electron gas.[2–4] In particular, the high-density limit has been examined rather thoroughly using perturbation theory, and the two leading terms in the correlation energy are now known exactly from the work of Macke[2] and Gell-Mann and Brueckner.[4] Unfortunately, all the methods so far applied successfully to the correlation problem have been of considerable complexity, and it is difficult therefore to follow in detail the physical nature of the correlation effects as the density is varied. Indeed, to do so completely would require explicit knowledge of the many-body wave function or, alternatively, the second-order density matrix, over the entire range of electron densities. There is one further, and very simple, method, however, of gaining some limited insight into the way in which the correlation energy arises. This is to follow the separate kinetic and potential energy terms as the electron density changes. As far as we are aware, this has not been attempted previously, and the purpose of this article is to point out that once the exact correlation energy is known, the separate kinetic and potential energy contributions can be obtained almost immediately by suitable application of the many-body virial theorem. Thus, from the results of Gell-Mann and Brueckner, the high-density forms of these energy terms are obtained exactly. We shall also make a very rough calculation for intermediate and low densities using Wigner's result, although we should stress that the separate energy terms obtained by applying the virial theorem to an approximate correlation energy will be much more seriously in error in general than the original total energy.

2. FORM OF THE VIRIAL THEOREM

For the present problem, with Coulombic interactions, the virial theorem may be written in the form

$$2T+V=-r_s(dE/dr_s), \tag{1}$$

where T, V, and E are, respectively, kinetic, potential, and total energies per particle, and r_s is the radius of a sphere containing one electron.[5] Since $T+V=E$, we have

$$T=-E-r_s(dE/dr_s). \tag{2}$$

As is customary, we now write the total energy E as

$$E=\epsilon_F+\epsilon_x+\epsilon_c, \tag{3}$$

the three separate terms being, respectively, the Fermi energy, the exchange energy, and the correlation energy. Using the explicit results

$$\epsilon_F=\frac{3}{5}\left(\frac{9\pi}{4}\right)^{\frac{2}{3}}\frac{1}{r_s^2}\approx\frac{2.21}{r_s^2}, \tag{4}$$

and

$$\epsilon_x=-\frac{3}{2\pi}\left(\frac{9\pi}{4}\right)^{\frac{1}{3}}\frac{1}{r_s}\approx-\frac{0.916}{r_s}, \tag{5}$$

and introducing ΔT and ΔV as the changes in kinetic and potential energies due to correlation, we find

$$\Delta T=T-\epsilon_F=-\epsilon_c-r_s(d\epsilon_c/dr_s), \tag{6}$$

$$\Delta V=V-\epsilon_x=2\epsilon_c+r_s(d\epsilon_c/dr_s). \tag{7}$$

These results are quite general, and once ϵ_c is known as a function of r_s we may easily obtain ΔT and ΔV from them.

3. RESULTS FOR HIGH-DENSITY LIMIT

The correlation energy in the high-density limit may be written, following Gell-Mann and Brueckner, as

$$\epsilon_c=A\ln r_s+C+\text{terms that vanish as } r_s\rightarrow 0, \tag{8}$$

where $A=(2/\pi^2)(1-\ln 2)=0.0622$ and $C=-0.096\pm0.002$. Equations (6), (7), and (8) then yield immediately

$$\Delta T=-A\ln r_s-(A+C)+\text{terms that vanish as } r_s\rightarrow 0, \tag{9}$$

and

$$\Delta V=2A\ln r_s+(A+2C)+\text{terms that vanish as } r_s\rightarrow 0. \tag{10}$$

We note that for $r_s<1$, both the leading terms in the

[1] E. P. Wigner, Phys. Rev. **46**, 1002 (1934) and Trans. Faraday Soc. **34**, 678 (1938).
[2] W. Macke, Z. Naturforsch. **5a**, 192 (1950).
[3] D. Pines, Phys. Rev. **92**, 626 (1953) and in *Solid State Physics*, edited by F. Seitz and D. Turnbull (Academic Press, Inc., New York, 1955), Vol. 1, p. 367.
[4] M. Gell-Mann and K. A. Brueckner, Phys. Rev. **106**, 364 (1957).
[5] Throughout this article we shall measure lengths in atomic units and energies in Rydbergs.

kinetic energy are positive, while both the potential energy terms are negative, the decrease in the total energy of the electron gas coming from the gain in the potential energy due to the correlations keeping the electrons, on the average, further apart, this outweighing the associated increase in kinetic energy. Unfortunately, it seems necessary to have more information on the higher terms of the series (8) before deciding on the range of r_s over which the two available terms will yield reasonable numerical values for ϵ_c. For low densities, the results of Wigner are probably not seriously in error, and we therefore consider now whether we can obtain supplementary information on the kinetic and potential energies in this way.

4. USE OF WIGNER'S FORMULA

Wigner's formula may be written[6]

$$\epsilon_{cW} = -0.88/(r_s+7.8). \tag{11}$$

To show the relation with the Gell-Mann and Brueckner work for the high-density limit, we have plotted both Eqs. (8) and (11) in Fig. 1. We have, somewhat arbitrarily, used Eq. (8) out to $r_s=1$, and the Wigner formula over the range $1<r_s<2$, and we should emphasize that in doing so we have probably extended both formulas outside their proper range of validity. However, both formulas lead to practically the same energy at $r_s=1$, although, as is only to be expected, they do not join smoothly here.

Let us now use the approximate result (11) in the Eqs. (6) and (7) which we derived from the virial theorem, although, as we have already remarked, errors in Eq. (11) will be very appreciably magnified in the separate terms thus obtained. We find then

$$\Delta T_W = 6.9/(r_s+7.8)^2, \tag{12}$$

$$\Delta V_W = -(0.88r+13.8)/(r_s+7.8)^2, \tag{13}$$

and these results are again plotted in Fig. 1 for $1<r_s<2$, together with the values given by Eqs. (9) and (10) for $r_s<1$. From Fig. 1 it then appears very likely that Eqs. (12) and (13) are considerably overestimating the separate energy changes due to correlation in the density range represented by $1<r_s<2$. From a qualitative point of view, however, ΔT_W is always positive and ΔV_W always negative according to Eqs. (12) and (13),

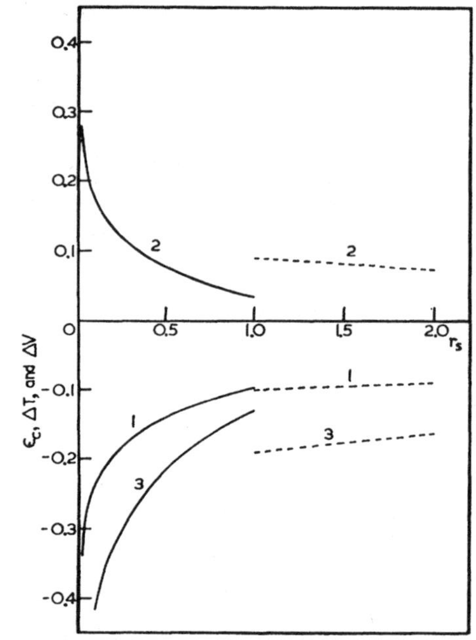

FIG. 1. Energy changes due to correlation effects, as functions of r_s. The solid curves, for $r_s<1$, are as follows: 1, ϵ_c from Eq. (8); 2, ΔT from Eq. (9); 3, ΔV from Eq. (10). The dashed curves, for $1<r_s<2$, are as follows: 1, ϵ_{cW} from Eq. (11); 2, ΔT_W from Eq. (12); 3, ΔV_W from Eq. (13).

agreeing with the conclusions drawn from Eqs. (9) and (10) for the high-density limit.

It is perhaps interesting to note also that in the low-density limit $r_s \to \infty$, the correlation energy is entirely potential energy in first order, but that according to Eqs. (4) and (12) the "correction" to the kinetic energy is about three times the Fermi energy. This kinetic energy "correction" may need quantitative modification later when a more accurate discussion of the low-density limit becomes available. However, it would appear that the present discussion establishes the general features of the separate energy curves and shows that over the entire range of densities it is the balance between an increased kinetic energy and a decreased potential energy of greater absolute value which yields the correlation energy, the ratio of kinetic to potential energy change going from $-\frac{1}{2}$, in the limit $r_s \to 0$, to 0 in the limit $r_s \to \infty$.

[6] See, for example, the second article cited in reference 3.

Probability Density of Electron Separation in a Uniform Electron Gas

By N. H. March and W. H. Young
Department of Physics, The University, Sheffield

[Received November 7, 1958]

We briefly report here a new method of approach to the correlation problem in a uniform electron gas, within the framework of density matrix theory. The earlier important paper of Mayer (1955) using this method has been shown independently by Tredgold (1957) and Koppe (1957) to lead to a variational scheme which is not complete, due to difficulties associated with subsidiary conditions. We propose here a uniform method of describing the electron correlations throughout the entire density range, and show that our method does indeed lead to accurate results in the high and low density limits. We hope to report detailed calculations for intermediate densities later, but at least we show in this note how a reasonably accurate picture of the correlation effects may be obtained over a wide density range.

Following Mayer (1955), we introduce the probability density $F_2(\xi_{12})$ of separation ξ_{12} between electrons; this being essentially the diagonal element of the second order density matrix. We use the variable $\boldsymbol{\xi}$, rather than $\mathbf{r}$, defined by

$$r_s\boldsymbol{\xi}=\left(\frac{3}{4\pi}\right)^{1/3}\mathbf{r} \quad . \quad . \quad . \quad . \quad . \quad . \quad . \quad . \quad (1)$$

where r_s is the radius of a sphere containing one electron. Then, as Mayer has shown, the potential energy per particle V may be written

$$V=-\frac{1}{2}\left(\frac{3}{4\pi}\right)^{1/3}\cdot\frac{1}{r_s}\int_0^\infty 4\pi\xi\{1-F_2(\xi, r_s)\}d\xi \quad . \quad . \quad . \quad (2)$$

and to emphasize the importance of F_2 we first prove a new result; that the total energy per particle E may be written exactly in terms of F_2. This may be achieved by using the many-body virial theorem for this problem in the form given recently by one of us (March 1958). Re-writing this theorem as a differential equation for the energy $E(r_s)$, given $V(r_s)$ from (2), it may be shown that

$$E=\frac{3}{10}\left(\frac{9\pi}{4}\right)^{2/3}\cdot\frac{1}{r_s^2}+\frac{1}{r_s^2}\int_0^{r_s} r_s V\,dr_s, \quad . \quad . \quad . \quad . \quad (3)$$

the constant of integration being simply the Fermi energy.

Thus we focus attention on the calculation of F_2, and for simplicity of presentation we shall at first neglect spin. With N particles, let $\Psi(\mathbf{r}_1, \mathbf{r}_2 \ldots \mathbf{r}_N)$ be any antisymmetric wave function such that it is normalized to unity. We now define corresponding density matrices (see for example Löwdin 1955)

$$\gamma(\mathbf{r}_1'|\mathbf{r}_1)=N\int \Psi^*(\mathbf{r}_1', \mathbf{r}_2\ldots\mathbf{r}_N)\Psi(\mathbf{r}_1, \mathbf{r}_2\ldots\mathbf{r}_N)d\mathbf{r}_2\ldots d\mathbf{r}_N, \quad . \quad (4)$$

$$\Gamma(\mathbf{r}_1'\mathbf{r}_2'|\mathbf{r}_1\mathbf{r}_2)=\frac{N(N-1)}{2}\int \Psi^*(\mathbf{r}_1', \mathbf{r}_2', \mathbf{r}_3\ldots\mathbf{r}_N)\Psi(\mathbf{r}_1, \mathbf{r}_2, \mathbf{r}_3\ldots\mathbf{r}_N)d\mathbf{r}_3\ldots d\mathbf{r}_N. \quad . \quad . \quad . \quad (5)$$

Then considering the function Φ defined by

$$\Phi(\mathbf{r}_1, \mathbf{r}_2\ldots\mathbf{r}_N)=\sqrt{\left(\frac{N}{\gamma(\mathbf{r}_1|\mathbf{r}_1)}\right)}\Psi(\mathbf{r}_1, \mathbf{r}_2\ldots\mathbf{r}_N) \quad . \quad . \quad . \quad (6)$$

it is easily shown that

$$\int \Phi^*(\mathbf{r}_1, \mathbf{r}_2\ldots\mathbf{r}_N)\Phi(\mathbf{r}_1, \mathbf{r}_2\ldots\mathbf{r}_N)\,d\mathbf{r}_2\ldots d\mathbf{r}_N=1. \quad . \quad . \quad . \quad (7)$$

Furthermore Φ is antisymmetrical in $\mathbf{r}_2, \mathbf{r}_3\ldots\mathbf{r}_N$, and can therefore be used to construct density matrices of various orders for $N-1$ particles, in coordinates $\mathbf{r}_2\ldots\mathbf{r}_N$. In particular

$$\frac{2}{\gamma(\mathbf{r}_1|\mathbf{r}_1)}\Gamma(\mathbf{r}_1\mathbf{r}_2'|\mathbf{r}_1\mathbf{r}_2)$$

is a first order density matrix for $N-1$ particles, for any arbitrary fixed $\mathbf{r}_1$.

We now apply the above general result to the electron gas problem. Here $\gamma(\mathbf{r}_1|\mathbf{r}_1)$, the mean electron density, is a constant, γ say, and the 'particle density' for $N-1$ particles is

$$\frac{2}{\gamma}\,\Gamma(\mathbf{r}_1\mathbf{r}_2|\mathbf{r}_1\mathbf{r}_2)\equiv\frac{2}{\gamma}\,\Gamma(\mathbf{r}_1\mathbf{r}_2)$$

for fixed $\mathbf{r}_1$. We may write in this case

$$\frac{2}{\gamma}\Gamma(\mathbf{r}_1\mathbf{r}_2)=f(\mathbf{r}_{12})\,;\quad \mathbf{r}_{12}=\mathbf{r}_2-\mathbf{r}_1 \qquad (8)$$

and it is sufficient to investigate $(2/\gamma)\Gamma(0\,\mathbf{r}_2)=f(\mathbf{r}_2)$, where we can think of axes as fixed at electron 1. Clearly to assume any first order density matrix for $N-1$ particles is not sufficient, for the antisymmetry property requires that

$$f(0)=0 \qquad (9)$$

for particles of like spin. Then $f(\mathbf{r})$ is intimately related to $F_2(\xi)$, the latter representing an angular average of $f(\mathbf{r})$.

Having seen how the second order density matrix may be re-expressed in a form related to a first order matrix for $N-1$ particles we shall now attempt to set up a model in keeping with the requirements put forward here. We restrict ourselves to a first order density matrix of the simplest kind, constructed from a set of orthonormal single particle functions $\psi_i(\mathbf{r})$, as

$$\sum_1^{N-1}\psi_i^*(\mathbf{r}')\psi_i(\mathbf{r}).$$

We were essentially helped in our considerations by seeking a unified treatment which includes both the Hartree–Fock solution, known to be correct in the high density limit, and the low density case (Wigner 1938). Wigner's low density calculation assumed that the potential energy was minimized when the electrons went on to the sites of a body-centred cubic lattice. Now, of course, this can only be interpreted as a lattice relative to a given electron we have singled out, as no real set of lattice points can occur preferentially in a uniform gas. Thus the lattice can be thought of as describing the configuration of all the other electrons, once axes are fixed on one particle. Of course, even then, all orientations of this lattice are equally probable, and so eventually we must average over angles to obtain physically significant results. Wigner finds that orbitals centred on the lattice sites can be represented by harmonic oscillator functions

$$\psi=\left(\frac{\alpha}{\pi}\right)^{3/4}\exp\left(\frac{-\alpha r^2}{2}\right)\,;\quad \alpha=\frac{1}{r_s^{3/2}}, \qquad (10)$$

for which ψ^2 tends to the delta function in the low density limit. If the density is very small, the functions (10) can be treated as orthogonal (effectively zero outside their own sphere) and hence are a valid set from which to form a first order density matrix of the kind we require. However, with N particles, we have N harmonic oscillator functions, and to obtain a suitable first order density for $N-1$ particles we must remove the density contributed by one of the orbitals; the one centred on the origin. We can then build up an acceptable function F_2 by averaging over angles. The result thus obtained for $r_s = 100$ is shown in curve 1 of the figure, and for such a low density the present method should provide a truly accurate representation of the real situation. We see that the probability density F_2 is violently oscillatory and of a highly complex functional form. In order to follow the change in F_2 as r_s is reduced, we have plotted in curve 2 the case $r_s = 4$. We should emphasize that extrapolation to such a density is not justified fundamentally, as the orbitals are now no longer

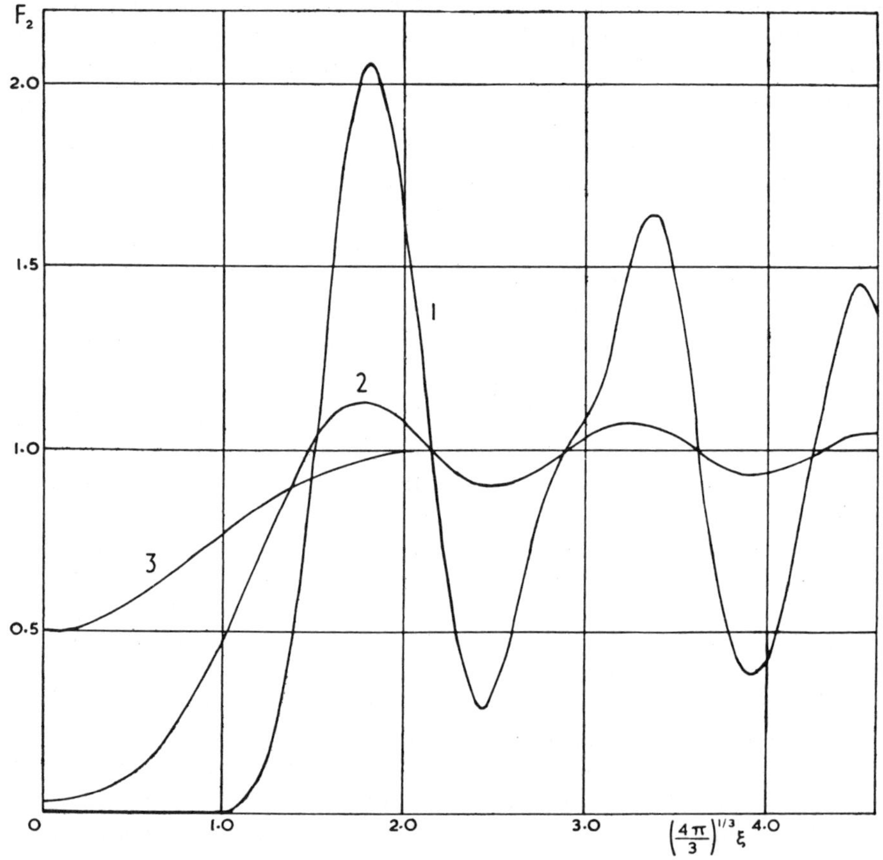

Probability density $F_2(\xi)$ as function of electron separation ξ. Curve 1. Low density form for $r_s = 100$. Curve 2. Extrapolated low density form for $r_s = 4$. Curve 3. Fermi hole, correct in the limit $r_s \to 0$.

orthogonal, and also do not make $f(0)$ zero for particles of like spin. Various estimates of the range of validity of the low density argument of Wigner give $r_s > \sim 10$–20. Nevertheless, whichever value we assume for the lower limit of approximate validity, it is clear that F_2 is still oscillatory about the line $F_2 = 1$, to a significant degree, and any acceptable probability density must contain such behaviour.

We see then that Wigner's arguments can be fitted into the present scheme, and give a good deal of information about the form of F_2. However, we wish to see how the high density form can be obtained from the present method. One thing which suggests itself is to extrapolate the low density F_2 into $r_s = 0$, but doing this we find no Fermi hole, due essentially to the failure of the Gaussian orbitals to (*a*) satisfy orthogonality requirements and (*b*) have nodes at the positions of like spin electrons. The question then arises as to what the localized functions centred on lattice sites must be in this limit. It seemed clear to us that since plane waves provide the correct solution in the high density limit, free-electron Wannier functions must be intimately related to this case. We conclude by showing that this is indeed so, and we demonstrate that for a one-dimensional Fermi gas, the exact solution may be obtained in this way.

If we consider a lattice spacing π/k_0, then the free electron Wannier functions may be written

$$\psi_m(x) = \frac{1}{(\pi k_0)^{1/2}} \frac{\sin k_0(x - m\pi/k_0)}{(x - m\pi/k_0)} = \frac{(-1)^m \sin k_0 x}{(\pi k_0)^{1/2}(x - m\pi/k_0)}. \quad (11)$$

Summing over the sites of an infinite lattice we can write

$$\sum_{-\infty}^{\infty} \psi_m^*(x')\psi_m(x) \simeq \frac{1}{\pi k_0} \int_{-\infty}^{\infty} \frac{\sin k_0(x' - m\pi/k_0)}{(x' - m\pi/k_0)} \frac{\sin k_0(x - m\pi/k_0)}{(x - m\pi/k_0)} \, dm$$

$$= \frac{k_0}{\pi} \frac{\sin k_0(x' - x)}{k_0(x' - x)}. \quad (12)$$

Hence, in particular

$$\sum_{-\infty}^{\infty} \psi_m^2(x) = k_0/\pi. \quad (13)$$

This latter result may be obtained by direct summation and is exact.

Now, according to our model, we form the density for $(N-1)$ particles by removing the density contributed by the localized orbital centred on the origin, in order to make $f(0) = 0$, and ensure normalization to $(N-1)$. Then we find

$$\sum_{-\infty}^{\infty} \psi_n^2(x) - \psi_0^2(x) = \frac{k_0}{\pi}\left[1 - \frac{\sin^2 k_0 x}{(k_0 x)^2}\right]. \quad (14)$$

In this case the exact second order density matrix can be written down from a two-by-two determinant (see for example Löwdin 1955) involving the first order matrix

$$\frac{k_0}{\pi} \frac{\sin k_0(x' - x)}{k_0(x' - x)},$$

the result being

$$\frac{1}{2} \begin{vmatrix} \dfrac{k_0}{\pi} & \dfrac{\sin k_0(x' - x)}{\pi(x' - x)} \\ \dfrac{\sin k_0(x' - x)}{\pi(x' - x)} & \dfrac{k_0}{\pi} \end{vmatrix} = \frac{1}{2}\left(\frac{k_0}{\pi}\right)^2 \left[1 - \left\{\frac{\sin k_0(x' - x)}{k_0(x' - x)}\right\}^2\right]. \quad (15)$$

From (8), we must multiply (14) by $\gamma/2=k_0/2\pi$ in order to compare with (15). The equivalence then evident proves that we obtain an exact description of the Fermi hole in this case, using localized orbitals on a lattice; the orbitals being free electron Wannier functions.

So far, we have not seen how to extend our high density arguments exactly to the real three-dimensional gas, with two spin directions. However, the following model will lead, at least, to a fairly accurate representation of the Fermi hole. In this high density limit we think of Wigner's body-centred-cubic lattice as two interpenetrating simple cubic lattices, and we centre on each lattice site a free electron Wannier function for a simple cubic lattice, the functions on one lattice being multiplied by a spin function α, those on the other lattice by a spin function β. Then we can again form a first order matrix for $N-1$ particles, and the probability density we obtain may be written

$$1-\frac{1}{2}\left[\frac{\sin c\xi_x}{c\xi_x}\frac{\sin c\xi_y}{c\xi_y}\frac{\sin c\xi_z}{c\xi_z}\right]^2;\quad c=2^{-1/3}\pi. \quad . \quad . \quad . \quad (16)$$

To obtain $F_2(\xi)$ we must average this over angles, and we have not yet achieved any closed result. However, we shall obtain a fair approximation to the Fermi hole, as the following argument shows. If we go back to the basic form for the simple cubic Wannier functions,

$$\psi(\mathbf{r})=\frac{1}{8(\pi k_0)^{3/2}}\int_{\text{unit cell}}\exp(i\mathbf{k}\,.\,\mathbf{r})dk_x\,dk_y\,dk_z \quad . \quad . \quad (17)$$

and replace the cube of volume $8k_0^3$ in $\mathbf{k}$ space by an equal volume sphere before performing the integration, then we obtain the exact Fermi hole.

Further work is now in progress to determine an acceptable form of localized orbital $\psi(\mathbf{r})$ in the intermediate density range, which goes into the Gaussian function (10) at low densities, and into functions at least intimately related to the Wannier functions we have described in the high density limit.

Acknowledgment

One of us (W. H. Y.) wishes to acknowledge the award of a Town Trustees Fellowship by the University of Sheffield.

References

Koppe, H., 1957, *Z. Phys.*, **148,** 135.
Löwdin, P. O., 1955, *Phys. Rev.*, **97,** 1474.
March, N. H., 1958, *Phys. Rev.*, **110,** 604.
Mayer, J. E., 1955, *Phys. Rev.*, **100,** 1579.
Tredgold, R. H., 1957, *Phys. Rev.*, **105,** 1421.
Wigner, E. P., 1938, *Trans. Faraday Soc.*, **34,** 678.

A density matrix approach to correlation in a uniform electron gas

By W. H. Young and N. H. March
Department of Physics, The University, Sheffield

(*Communicated by R. E. Peierls, F.R.S.—Received* 23 *December* 1959)

The energy of a uniform electron gas can be specified completely in terms of its second-order density matrix. Mayer has therefore suggested that trial matrices satisfying all the usual conditions might be employed to determine variationally correlation energies and pair functions. Unfortunately, the particular choice made by Mayer did not satisfy all the Pauli conditions on the second-order matrix. However, matrices satisfactory from this point of view are presented here and the consequences of assuming such forms are investigated. Since one of the matrices is a direct generalization of the Hartree–Fock expression, a description of correlation effects is assured and it appears that this should be adequate for all densities.

1. Introduction

Mayer (1955) has approached the correlation problem in a uniform electron gas by exhibiting the dependence of the total energy on the second-order density matrix. He then suggests a variational method based on trial second-order matrices satisfying all the normalization, Hermiticity, antisymmetry, Pauli and translational invariance properties. The Pauli conditions pertaining to the second-order matrix (see equations (2·12) and (2·13)) have previously proved difficult to satisfy beyond the Hartree–Fock (H.F.) approximation, and in this respect Mayer's method fails to some extent, as has been pointed out by Koppe (1957) and Tredgold (1957). The object of the present work is to exhibit second-order matrices of considerable generality satisfying automatically the usual conditions plus the Pauli conditions expressed in equations (2·12) and (2·13). One of the resulting schemes (method (A) of § 5) is then shown rigorously to satisfy the final necessary Pauli conditions on the first-order matrix (see equations (2·10) and (2·11)). A more powerful method ((B) of § 5) is arrived at by a direct generalization of the H.F. case, and while a rigorous proof that the conditions (2·10) and (2·11) are automatically satisfied has not as yet been found, plausible arguments are presented (see the appendix) for believing that this is again the case. But, in any event, it appears unlikely that this could constitute a major difficulty to method (B) as the correctness of the Pauli conditions on the first-order matrix can always be verified directly for any special variational form. This would be somewhat in the spirit of Mayer's original approach; our essential contribution then lies in ensuring that the second-order Pauli conditions are fulfilled. We show also in this paper that while method (A) leads to an energy somewhat in error in the high-density limit, and can therefore be useful only for intermediate and low densities, method (B) will yield energies which are lower than the corresponding H.F. figures, because the procedure consists of a direct generalization of the H.F. method. In particular, in the low-density limit, both methods contain the description of Wigner (1938; see also March & Young 1959) and this leads us to suppose that the present approach should be useful over the whole density range.

2. General properties of density matrices

As full discussions of density matrices are now available (Mayer 1955; Löwdin 1955; Koppe 1957; Ayres 1958) it will be sufficient to state briefly here our definitions and those general density matrix properties which will be of use later.

Let $\Phi(\mathbf{x}_1\mathbf{x}_2\ldots\mathbf{x}_N)$ be some wave function antisymmetric in each pair of its co-ordinates and normalized to unity. Here the $\mathbf{x}_i$ denote the space co-ordinates $\mathbf{r}_i$ and spin co-ordinates σ_i. Then we may define the following density matrices for an N particle system:

first-order (with spin)

$$\gamma_\sigma(\mathbf{x}_1'\,|\,\mathbf{x}_1) = N\int\Phi^*(\mathbf{x}_1'\mathbf{x}_2\ldots\mathbf{x}_N)\,\Phi(\mathbf{x}_1\mathbf{x}_2\ldots\mathbf{x}_N)\,\mathrm{d}\mathbf{x}_2\ldots\mathrm{d}\mathbf{x}_N; \tag{2·1}$$

second-order (with spin)

$$\Gamma_\sigma(\mathbf{x}_1'\mathbf{x}_2'\,|\,\mathbf{x}_1\mathbf{x}_2) = \tfrac{1}{2}N(N-1)\int\Phi^*(\mathbf{x}_1'\mathbf{x}_2'\mathbf{x}_3\ldots\mathbf{x}_N)\,\Phi(\mathbf{x}_1\mathbf{x}_2\ldots\mathbf{x}_N)\,\mathrm{d}\mathbf{x}_3\ldots\mathrm{d}\mathbf{x}_N; \tag{2·2}$$

first-order spinless $\qquad \gamma(\mathbf{r}_1'\,|\,\mathbf{r}_1) = \int\gamma_\sigma(\mathbf{r}_1'\sigma_1\,|\,\mathbf{r}_1\sigma_1)\,\mathrm{d}\sigma_1; \qquad$ (2·3)

second-order spinless

$$\Gamma(\mathbf{r}_1'\mathbf{r}_2'\,|\,\mathbf{r}_1\mathbf{r}_2) = \int\Gamma_\sigma(\mathbf{r}_1'\sigma_1\mathbf{r}_2'\sigma_2\,|\,\mathbf{r}_1\sigma_1\mathbf{r}_2\sigma_2)\,\mathrm{d}\sigma_1\,\mathrm{d}\sigma_2. \tag{2·4}$$

The diagonal parts of these matrices have direct physical interpretations. For example,

$$\rho(\mathbf{r}_1) \equiv \gamma(\mathbf{r}_1\,|\,\mathbf{r}_1) = \text{number of particles} \times \text{probability density of finding a particle at } \mathbf{r}_1, \tag{2·5}$$

and

$$P(\mathbf{r}_1\mathbf{r}_2) \equiv \Gamma(\mathbf{r}_1\mathbf{r}_2\,|\,\mathbf{r}_1\mathbf{r}_2) = \text{number of pairs} \times \text{probability density of finding simultaneously one particle at } \mathbf{r}_1 \text{ and another at } \mathbf{r}_2. \tag{2·6}$$

At this stage we can write down a number of simple properties of the matrices, namely

$$\gamma(\mathbf{r}_1'\,|\,\mathbf{r}_1) = \frac{2}{N-1}\int\Gamma(\mathbf{r}_1'\mathbf{r}_2\,|\,\mathbf{r}_1\mathbf{r}_2)\,\mathrm{d}\mathbf{r}_2\ \text{etc.,} \quad \int\rho(\mathbf{r}_1)\,\mathrm{d}\mathbf{r}_1 = N\ \text{etc.} \quad \text{(normalization),} \tag{2·7}$$

$$\Gamma_\sigma(\mathbf{x}_1\mathbf{x}_2\,|\,\mathbf{x}_1'\mathbf{x}_2') = \Gamma_\sigma^*(\mathbf{x}_1'\mathbf{x}_2'\,|\,\mathbf{x}_1\mathbf{x}_2) \quad \text{etc.,} \quad \text{(Hermiticity),} \tag{2·8}$$

and

$$\Gamma_\sigma(\mathbf{x}_1'\mathbf{x}_2'\,|\,\mathbf{x}_1\mathbf{x}_2) = -\Gamma_\sigma(\mathbf{x}_1'\mathbf{x}_2'\,|\,\mathbf{x}_2\mathbf{x}_1) \quad \text{etc.,} \quad \text{(antisymmetry).} \tag{2·9}$$

Finally, Löwdin (1955) and Koppe (1957) have shown that if $\{\psi_i(\mathbf{x})\}$ is any complete orthonormal set of spin-orbitals of suitable type, then

$$\gamma_\sigma(\mathbf{x}'\,|\,\mathbf{x}) = \Sigma a_{kl}\,\psi_k^*(\mathbf{x}')\,\psi_l(\mathbf{x}), \tag{2·10}$$

where the a_{kk} are real and

$$\Sigma a_{kk} = N, \quad 0 \leqslant a_{kk} \leqslant 1 \quad (\text{all } k). \tag{2·11}$$

Also,

$$\Gamma_\sigma(\mathbf{x}_1'\mathbf{x}_2'\,|\,\mathbf{x}_1\mathbf{x}_2) = \sum_{\substack{k_1<k_2\\ l_1<l_2}} a_{k_1k_2l_1l_2}\frac{1}{\sqrt{2}}\begin{vmatrix}\psi_{k_1}^*(\mathbf{x}_1') & \psi_{k_2}^*(\mathbf{x}_1')\\ \psi_{k_1}^*(\mathbf{x}_2') & \psi_{k_2}^*(\mathbf{x}_2')\end{vmatrix}\frac{1}{\sqrt{2}}\begin{vmatrix}\psi_{l_1}(\mathbf{x}_1) & \psi_{l_2}(\mathbf{x}_1)\\ \psi_{l_1}(\mathbf{x}_2) & \psi_{l_2}(\mathbf{x}_2)\end{vmatrix}, \tag{2·12}$$

where the $a_{k_1k_2k_1k_2}$ are real and

$$\sum_{k_1<k_2} a_{k_1k_2k_1k_2} = \tfrac{1}{2}N(N-1), \quad 0 \leqslant a_{k_1k_2k_1k_2} \leqslant 1 \quad (\text{all } k_1, k_2). \tag{2·13}$$

We refer to the results of equations (2·10) to (2·13) as the Pauli conditions.

We now go on to discuss the uniform electron gas problem and the special density matrix properties for this case.

3. Special properties of density matrices associated with the uniform electron gas

Following Mayer (1955), let us consider a uniform electron gas described by the Hamiltonian

$$H = -\frac{1}{2}\sum_1^N \nabla^2_{r_i} + \sum_{i<j}\frac{1}{|\mathbf{r}_i-\mathbf{r}_j|} - \frac{3}{4\pi r_s^3}\sum_1^N\int\frac{d\mathbf{r}_0}{|\mathbf{r}_i-\mathbf{r}_0|} + \frac{1}{2}\left(\frac{3}{4\pi r_s^3}\right)^2\iint\frac{d\mathbf{r}_0\,d\mathbf{r}_1}{|\mathbf{r}_0-\mathbf{r}_1|}. \tag{3·1}$$

Here the first two terms have an obvious physical interpretation. The last two terms represent respectively the electron–positive ion interaction energy and the self-energy of the continuous positive charge. As usual, r_s measures the mean interparticle spacing. We now make the simple linear change of variable

$$\boldsymbol{\xi} = \left(\frac{3}{4\pi}\right)^{\frac{1}{3}}\frac{\mathbf{r}}{r_s} \tag{3·2}$$

and the Hamiltonian becomes

$$H = \left(\frac{3}{4\pi}\right)^{\frac{2}{3}}\frac{1}{r_s^2}H_0, \tag{3·3}$$

where
$$H_0 = -\frac{1}{2}\sum_1^N \nabla^2_{\xi_i} + \left(\frac{4\pi}{3}\right)^{\frac{1}{3}} r_s\left\{\sum_{i<j}\frac{1}{|\boldsymbol{\xi}_i-\boldsymbol{\xi}_j|} - \sum_1^N\int\frac{d\boldsymbol{\xi}_0}{|\boldsymbol{\xi}_i-\boldsymbol{\xi}_0|} + \frac{1}{2}\iint\frac{d\boldsymbol{\xi}_0\,d\boldsymbol{\xi}_1}{|\boldsymbol{\xi}_0-\boldsymbol{\xi}_1|}\right\}. \tag{3·4}$$

Thus, as pointed out by Ferrell (1958), we are led to consider the essentially equivalent problem in $\boldsymbol{\xi}$-space of a uniform electron gas of unit density, modified to have a coupling parameter $\left(\frac{4\pi}{3}\right)^{\frac{1}{3}} r_s$.

Suppose now that $\gamma(\boldsymbol{\xi}' \mid \boldsymbol{\xi})$ is the first-order spinless density matrix and $P(\boldsymbol{\xi}'\boldsymbol{\xi})$ is the diagonal part of the second-order spinless matrix, both derived from some common wave function. Then, as has been shown by Mayer, the energy bound corresponding to (3·3) is given by

$$\mathscr{E} = -\frac{1}{2}\left(\frac{3}{4\pi}\right)^{\frac{2}{3}}\frac{1}{r_s^2}\int[\nabla^2_{\xi}\gamma(\boldsymbol{\xi}'\mid\boldsymbol{\xi})]_{\boldsymbol{\xi}'=\boldsymbol{\xi}}\,d\boldsymbol{\xi} - \frac{1}{2}\left(\frac{3}{4\pi}\right)^{\frac{1}{3}}\frac{1}{r_s}\iint\frac{[1-2P(\boldsymbol{\xi}'\boldsymbol{\xi})]}{|\boldsymbol{\xi}'-\boldsymbol{\xi}|}\,d\boldsymbol{\xi}'\,d\boldsymbol{\xi}. \tag{3·5}$$

By taking N large we can assume the usual translational invariance properties

$$\gamma(\boldsymbol{\xi}'\mid\boldsymbol{\xi}) = f(\boldsymbol{\xi}'-\boldsymbol{\xi}) \quad (f(0)=1), \tag{3·6}$$

$$2P(\boldsymbol{\xi}'\boldsymbol{\xi}) = F(\boldsymbol{\xi}'-\boldsymbol{\xi}). \tag{3·7}$$

The quantity $F(\boldsymbol{\xi})$ is of considerable physical interest, being the probability density that an electron (of either spin) is at a position $\boldsymbol{\xi}$ relative to some given electron, and is referred to as the pair function. Indeed the calculation of the exact F for all r_s can be regarded as the central problem for it has been shown (March & Young 1959; March 1958; Cowan & Kirkwood 1958; March, Young & Murray, to be published) that the total energy may be expressed in terms of this function only.

4. Generalized H.F. density matrices

It will be recalled that any first-order density matrix can be written in the form (2·10) where the conditions (2·11) apply. Furthermore, a choice of trial matrix of the Dirac form

$$\sum_{1}^{N} \phi_k^*(\mathbf{x}') \, \phi_k(\mathbf{x}), \tag{4·1}$$

where the ϕ's are orthonormal leads to a valid variational scheme. Now we also know that any second-order matrix can be written in the form (2·12) where the conditions (2·13) hold. It therefore seems worth while investigating whether we can choose a trial second-order matrix of the form

$$\sum_{k_1<k_2} \phi_{k_1k_2}^*(\mathbf{x}_1'\mathbf{x}_2') \, \phi_{k_1k_2}(\mathbf{x}_1\mathbf{x}_2) = \tfrac{1}{2}\sum_{k_1,k_2} \phi_{k_1k_2}^*(\mathbf{x}_1'\mathbf{x}_2') \, \phi_{k_1k_2}(\mathbf{x}_1\mathbf{x}_2), \tag{4·2}$$

where the $\phi_{k_1k_2}(\mathbf{x}_1\mathbf{x}_2)$ are antisymmetric in $\mathbf{x}_1$ and $\mathbf{x}_2$ and satisfy the orthonormality requirements

$$\iint \phi_{k_1k_2}^*(\mathbf{x}_1\mathbf{x}_2) \, \phi_{l_1l_2}(\mathbf{x}_1\mathbf{x}_2) \, d\mathbf{x}_1 \, d\mathbf{x}_2 = \delta_{k_1l_1}\delta_{k_2l_2}. \tag{4·3}$$

In this way equations (2·7), (2·8) and (2·9) are immediately satisfied.

It is readily verified that a single Slater determinant of orthonormal orbitals $\psi_k(\mathbf{x})$ corresponds to the choice

$$\frac{1}{\sqrt{2}}\begin{vmatrix} \psi_{k_1}(\mathbf{x}_1) & \psi_{k_2}(\mathbf{x}_1) \\ \psi_{k_1}(\mathbf{x}_2) & \psi_{k_2}(\mathbf{x}_2) \end{vmatrix} \tag{4·4}$$

of two-particle functions. We therefore generalize the above using the device of 'transformed orbitals' which has proved useful (Macke 1955; March & Young 1958) in studying the foundations of the statistical theories. If, for the moment, we consider the simpler ferromagnetic gas, then the spin factors enter only trivially into all calculations and may therefore be omitted. We now make the following choice of generalized H.F. two-particle orbitals.

$$\phi_{k_1k_2}(\mathbf{r}_1\mathbf{r}_2) = [J(\mathbf{r}_1-\mathbf{r}_2)]^{\frac{1}{2}} \frac{1}{\sqrt{2}}\begin{vmatrix} \psi_{k_1}[\frac{1}{2}(\mathbf{r}_1+\mathbf{r}_2)+\frac{1}{2}\mathbf{R}(\mathbf{r}_1-\mathbf{r}_2)] & \psi_{k_2}[\frac{1}{2}(\mathbf{r}_1+\mathbf{r}_2)+\frac{1}{2}\mathbf{R}(\mathbf{r}_1-\mathbf{r}_2)] \\ \psi_{k_1}[\frac{1}{2}(\mathbf{r}_1+\mathbf{r}_2)-\frac{1}{2}\mathbf{R}(\mathbf{r}_1-\mathbf{r}_2)] & \psi_{k_2}[\frac{1}{2}(\mathbf{r}_1+\mathbf{r}_2)-\frac{1}{2}\mathbf{R}(\mathbf{r}_1-\mathbf{r}_2)] \end{vmatrix}, \tag{4·5}$$

where J is the Jacobian of the transformation $\mathbf{R}$, and consider the trial matrix

$$\Gamma(\mathbf{r}_1'\mathbf{r}_2' \mid \mathbf{r}_1\mathbf{r}_2) = \tfrac{1}{2}\sum_{k_1,k_2} \phi_{k_1k_2}^*(\mathbf{r}_1'\mathbf{r}_2') \, \phi_{k_1k_2}(\mathbf{r}_1\mathbf{r}_2). \tag{4·6}$$

The range of k_1 and k_2 in (4·6) is not explicitly specified since more than one choice is considered below, though, as might be expected, particular emphasis is given to that choice which corresponds to the direct generalization of the H.F. procedure. The choice (4·5) satisfies (4·3) and is antisymmetric provided

$$\mathbf{R}(-\mathbf{r}) = -\mathbf{R}(\mathbf{r}). \tag{4·7}$$

It remains to show that the Pauli conditions are satisfied for H.F. plane wave ψ's.

5. The satisfaction of the Pauli conditions for generalized H.F. matrices

For notational convenience we will work in one-dimensional ξ-space, all arguments in this section being capable of immediate extension to three dimensions and indeed, along lines which will be indicated later, to two spins. We therefore take

$$\psi_k(\xi) = \frac{1}{N^{\frac{1}{2}}} \exp \frac{2\pi i k \xi}{N}, \tag{5·1}$$

equations (4·5) and (4·6) now being written

$$\phi_{k_1 k_2}(\xi_1 \xi_2) = \frac{\sqrt{2i}}{N} \exp i\pi \frac{(k_1+k_2)}{N} (\xi_1+\xi_2)\,[y'(\xi_1-\xi_2)]^{\frac{1}{2}} \sin \pi(k_1-k_2)\, y(\xi_1-\xi_2) \tag{5·2}$$

and

$$\Gamma(\xi_1'\xi_2' \mid \xi_1\xi_2) = \tfrac{1}{2} \sum_{k_1, k_2} \phi^*_{k_1 k_2}(\xi_1'\xi_2')\, \phi_{k_1 k_2}(\xi_1 \xi_2), \tag{5·3}$$

where we take y for **R** and y' for J. We begin by proving two lemmas.

Lemma 1: *Suppose a trial first-order matrix is written in terms of its 'natural orbitals' (see Löwdin 1955)*

$$\gamma(\xi' \mid \xi) = \sum_i a_i \phi_i^*(\xi')\, \phi_i(\xi), \tag{5·4}$$

where the a_i are real and $0 \leqslant a_i \leqslant 1$ for all i. *Then in any other representation*

$$\gamma(\xi' \mid \xi) = \sum_{r,s} b_{rs} \psi_r^*(\xi')\, \psi_s(\xi), \tag{5·5}$$

where the b_{rr} are real and $0 \leqslant b_{rr} \leqslant 1$ for all r.

To prove this we suppose $\qquad \phi_i = \sum_r l_{ir} \psi_r. \qquad$ (5·6)

Then substitution in (5·4) and comparison with (5·5) gives

$$b_{rs} = \sum_i a_i l_{ir}^* l_{is}. \tag{5·7}$$

Hence the b_{rr} are real and

$$0 \leqslant b_{rr} = \sum_i a_i \mid l_{ir} \mid^2 \leqslant \sum_i |l_{ir}|^2 = 1, \tag{5·8}$$

the latter step following from the conditions

$$(\phi_i, \phi_i) = (\psi_i, \psi_i) = 1.$$

Lemma 2: A precisely similar result holds for a trial second-order matrix, the proof being exactly as in lemma 1.

In view of lemma 2, the satisfaction of the Pauli conditions for the expansion of our Γ (defined by (5·3)) in terms of two-particle functions is immediate. It is by no means obvious, however, that the corresponding requirements on γ are satisfied. To verify that these hold it is necessary to identify, and establish the Pauli conditions for, the first-order natural orbital expansion. An appeal to lemma 1 then suffices.

To identify the first-order natural orbitals we observe that

$$\gamma(\xi' \mid \xi) = \frac{1}{N-1} \sum_{k_1, k_2} \int_{-\infty}^{\infty} \phi^*_{k_1 k_2}(\xi' \xi_2)\, \phi_{k_1 k_2}(\xi \xi_2)\, \mathrm{d}\xi_2$$

$$= \frac{1}{N-1} \sum_{k_1, k_2} \int_{-\infty}^{\infty} \frac{2}{N^2} \exp\left\{ i\pi \frac{(k_1+k_2)}{N} (\xi' - \xi) \right\} [y'(\xi' - \xi_2)\, y'(\xi - \xi_2)]^{\frac{1}{2}}$$

$$\times \sin \pi \frac{(k_1 - k_2)}{N} (\xi' - \xi_2) \sin \pi \frac{(k_1 - k_2)}{N} (\xi - \xi_2)\, \mathrm{d}\xi_2$$

$$= \frac{1}{N-1} \sum_{k_1, k_2} \int_{-\infty}^{\infty} \frac{2}{N^2} \exp\left\{ i\pi \frac{(k_1+k_2)}{N} (\xi' - \xi) \right\} \left[y'\left(\frac{\xi' - \xi}{2} - X \right) y'\left(\frac{\xi - \xi'}{2} - X \right) \right]^{\frac{1}{2}}$$

$$\times \sin \pi \frac{(k_1 - k_2)}{N} \left(\frac{\xi' - \xi}{2} - X \right) \sin \pi \frac{(k_1 - k_2)}{N} \left(\frac{\xi' - \xi}{2} - X \right) \mathrm{d}X.$$

The latter is a function of $\xi' - \xi$ only and so

$$\gamma(\xi' \mid \xi) = \frac{1}{N} \sum_{m=-\infty}^{\infty} \alpha_m \exp\left(-\frac{2\pi i m \xi'}{N} \right) \exp\left(\frac{2\pi i m \xi}{N} \right). \tag{5·9}$$

Our final task is therefore to show that for all m, α_m is real and $0 \leqslant \alpha_m \leqslant 1$.

Let us consider plane wave expansions of the kind

$$\phi_{k_1 k_2}(\xi \xi_2) = \sum_{r, s=-\infty}^{\infty} \psi_r(\xi)\, C^{k_1 k_2}_{r\ s} \psi_s(\xi_2), \tag{5·10}$$

the ψ's being defined by (5·1). Then, since y is odd, the $\phi_{k_1 k_2}(\xi \xi_2)$ are antisymmetric and so

$$C^{k_1 k_2}_{r\ s} + C^{k_1 k_2}_{s\ r} = 0. \tag{5·11}$$

Also, from (5·10)

$$\int \phi^*_{k_1' k_2'}(\xi' \xi_2)\, \phi_{k_1 k_2}(\xi \xi_2)\, \mathrm{d}\xi_2 = \sum_{r, i} \psi^*_r(\xi')\, d_{ri}(k_1' k_2', k_1 k_2)\, \psi_i(\xi), \tag{5·12}$$

where
$$d_{ri}(k_1' k_2', k_1 k_2) = \sum_{s=-\infty}^{\infty} (C^{k_1' k_2'}_{r\ s})^* C^{k_1 k_2}_{r\ s}. \tag{5·13}$$

In particular,
$$d^{k_1 k_2}_{r\ r} \equiv d_{rr}(k_1 k_2, k_1 k_2) = \sum_{s=-\infty}^{\infty} |C^{k_1 k_2}_{r\ s}|^2. \tag{5·14}$$

We note from (4·3), (5·12) and (5·13) the orthonormality conditions

$$\sum_{r, s=-\infty}^{\infty} (C^{k_1' k_2'}_{r\ s})^* C^{k_1 k_2}_{r\ s} = \delta_{k_1' k_1} \delta_{k_2' k_2}. \tag{5·15}$$

Also, by (5·12),
$$\gamma(\xi' \mid \xi) = \frac{1}{N-1} \sum_{k_1, k_2} \sum_{r, i=-\infty}^{\infty} \psi^*_r(\xi')\, d^{k_1 k_2}_{r\ i} \psi_i(\xi), \tag{5·16}$$

and so comparison with (5·9) and (5·13) gives

$$\alpha_r = \frac{1}{N-1} \sum_{k_1, k_2} d_{r\ r}^{k_1 k_2}. \tag{5·17}$$

By (5·14) the α_r are real and positive and it now remains to show that $\alpha_r \leqslant 1$ for all r.

At this stage it seems necessary to use the explicit form of the $C_{r\ s}^{k_1 k_2}$, equations (5·2) and (5·10) giving

$$C_{r\ s}^{k_1 k_2} = C_{r-\lambda\ s-\lambda}^{k_1-\lambda\ k_2-\lambda} \quad (\lambda, \text{ any integer}), \tag{5·18}$$

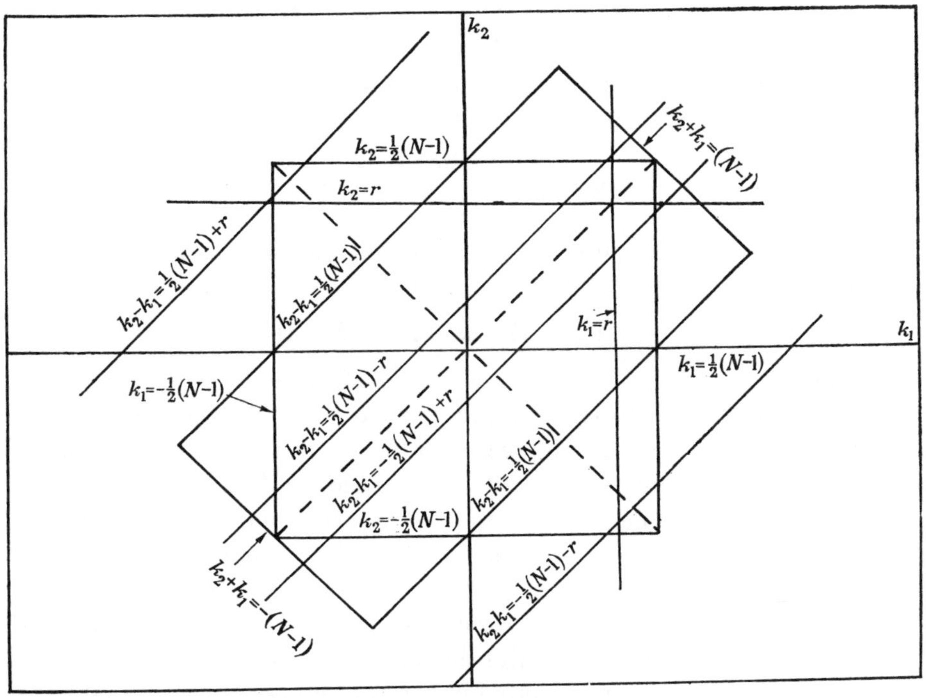

FIGURE 1. Diagram of (k_1, k_2) space. Rectangular configuration, method (A); square configuration, method (B).

whence, by (5·14),

$$\begin{aligned}
\sum_{m=-\infty}^{\infty} d_{r\ r}^{m\ m+n} &= \sum_{m=-\infty}^{\infty} \sum_{s=-\infty}^{\infty} |C_{r\ s}^{m\ m+n}|^2 \\
&= \sum_{m=-\infty}^{\infty} \sum_{s=-\infty}^{\infty} |C_{r-m\ s-m}^{0\ \ n}|^2 \\
&= \sum_{m=-\infty}^{\infty} \sum_{s=-\infty}^{\infty} |C_{r-m\ s}^{0\ \ n}|^2 \\
&= \sum_{m=-\infty}^{\infty} \sum_{s=-\infty}^{\infty} |C_{m\ s}^{0\ n}|^2 \\
&= 1 \quad \text{by equation (5·15).}
\end{aligned} \tag{5·19}$$

Hence for any integral m_0, m_1 we have

$$\sum_{m=m_0}^{m_1} d_r^{m}{}_r^{m+n} \leqslant 1. \tag{5·20}$$

We are now in the position to consider choices of the ranges of k_1 and k_2 in equation (5·3). Two cases will be dealt with.

(A) Suppose all non-equal k_1, k_2 are taken for which

$$-\tfrac{1}{2}(N-1) \leqslant k_1-k_2 \leqslant \tfrac{1}{2}(N-1), \quad -(N-1) \leqslant k_1+k_2 \leqslant (N-1), \tag{5·21}$$

this region of (k_1, k_2) space being shown in figure 1. Then, for all r,

$$\alpha_r = \frac{1}{N-1} \sum_{\lambda=-\frac{1}{2}(N-1)}^{\frac{1}{2}(N-1)}{}' \sum_{k_1=-(N-1)}^{N-1} d_r^{k_1}{}_r^{k_1+\lambda}, \tag{5·22}$$

where Σ' denotes the omission of $\lambda = 0$ from the summation. The term $d_r^{k_1}{}_r^{k_1}$ is in fact zero from (5·2), (5·10) and (5·14). Using (5·20),

$$\alpha_r \leqslant \frac{1}{N-1} \sum_{\lambda=-\frac{1}{2}(N-1)}^{\frac{1}{2}(N-1)}{}' 1 = 1. \tag{5·23}$$

Hence the Pauli conditions are completely satisfied in this case.

(B) Suppose we take all non-equal k_1, k_2 such that

$$-\tfrac{1}{2}(N-1) \leqslant k_1, k_2 \leqslant \tfrac{1}{2}(N-1). \tag{5·24}$$

This corresponds to a direct generalization of the H.F. case. Unfortunately we have not so far been able to give a rigorous proof of the result (5·23) for this case, but we believe it to be true again and we present in the appendix arguments which are at least suggestive of its correctness for all y. We should emphasize in any case that one has control over the Pauli conditions on the first-order matrix through the explicit form of variational function y, and it is therefore always open to proceed, as Mayer did with different trial functions, to impose these conditions directly. Unlike Mayer's approach, however, our trial forms ensure the more difficult Pauli conditions on the second-order matrix. In view of the arguments we present, we shall proceed on the assumption that a general proof of the correctness of these conditions on γ can eventually be given for all transformations y. In connexion with such a proof, it should be noted that so far the full power of the orthonormality requirements (5·15) does not appear to have been employed (at any rate, not when considering γ) and any rigorous proof that the Pauli conditions are satisfied by γ will probably use these equations.

6. One-dimensional, one-spin variational methods

The methods corresponding to (A) and (B) will now be examined in detail in one dimension for a two-body potential U_2 of such a form that all integrals below in which it appears are convergent. If x_s is the mean particle spacing, a bound to the total energy is (cf. equation (3·5))

$$\mathscr{E} = -\frac{1}{2x_s^2}\int_{\text{length } N}\left[\frac{\partial^2}{\partial \xi^2}\gamma(\xi' \mid \xi)\right]_{\xi'=\xi} d\xi - \frac{1}{2}\iint U_2(x_s(\xi'-\xi))\,[1-2P(\xi'\xi)]\,d\xi'\,d\xi. \tag{6·1}$$

(A) Here we have

$$2\Gamma(\xi'\xi_2' \mid \xi\xi_2)$$
$$= \sum_{\substack{-\frac{1}{2}(N-1)\leqslant k_1-k_2\leqslant\frac{1}{2}(N-1)\\ -(N-1)\leqslant k_1+k_2\leqslant(N-1)}} \frac{2}{N^2}\exp\left[\mathrm{i}\pi\frac{(k_1+k_2)}{N}\{-(\xi'+\xi_2')+(\xi+\xi_2)\}\right]$$
$$\times [y'(\xi'-\xi_2')\,y'(\xi-\xi_2)]^{\frac{1}{2}}\sin\pi\frac{(k_1-k_2)}{N}y(\xi'-\xi_2')\sin\pi\frac{(k_1-k_2)}{N}y(\xi-\xi_2)$$
$$= 2[y'(\xi'-\xi_2')\,y'(\xi-\xi_2)]^{\frac{1}{2}}\iint_{\substack{-\frac{1}{2}\leqslant k_1-k_2\leqslant\frac{1}{2}\\ -1\leqslant k_1+k_2\leqslant 1}}\exp\left[\mathrm{i}\pi(k_1+k_2)\{-(\xi'+\xi_2')+(\xi+\xi_2)\}\right]$$
$$\times \sin\pi(k_1-k_2)\,y(\xi'-\xi_2')\sin\pi(k_1-k_2)\,y(\xi-\xi_2)\,\mathrm{d}k_1\,\mathrm{d}k_2$$
$$= 2[y'(\xi'-\xi_2')\,y'(\xi-\xi_2)]^{\frac{1}{2}}\frac{\sin\pi\{-(\xi'+\xi_2')+(\xi+\xi_2)\}}{\pi\{-(\xi'+\xi_2')+(\xi+\xi_2)\}}$$
$$\times\left[\frac{\sin\frac{1}{2}\pi\{y(\xi'-\xi_2')-y(\xi-\xi_2)\}}{\pi\{y(\xi'-\xi_2')-y(\xi-\xi_2)\}}-\frac{\sin\frac{1}{2}\pi\{y(\xi'-\xi_2')+y(\xi-\xi_2)\}}{\pi\{y(\xi'-\xi_2')+y(\xi-\xi_2)\}}\right] \tag{6·2}$$

and in particular
$$2P(\xi\xi_2) = y'(\xi-\xi_2)\left[1-\frac{\sin\pi y(\xi-\xi_2)}{\pi y(\xi-\xi_2)}\right]. \tag{6·3}$$

The property of translational invariance

$$\gamma(\xi'-\mu \mid \xi-\mu) = \gamma(\xi' \mid \xi) \quad (\text{any real } \mu), \tag{6·4}$$

an expression of the boundary conditions or the compatability of (5·9) and (5·16) is now invoked. Using the formulae

$$(N-1)\,\gamma(\xi' \mid \xi) = \int_{\text{length } N} 2\Gamma(\xi'\xi_2 \mid \xi\xi_2)\,\mathrm{d}\xi_2 + \text{const.} + o\left(\frac{1}{N}\right), \tag{6·5}$$

$$N\gamma(\xi' \mid \xi) = \int_{\text{length } N+1} 2\Gamma(\xi'\xi_2 \mid \xi\xi_2)\,\mathrm{d}\xi_2 + \text{const.} + o\left(\frac{1}{N}\right), \tag{6·6}$$

we may write for very many particles

$$\gamma(\xi' \mid \xi) = \int_{t-\frac{1}{2}}^{t+\frac{1}{2}} 2\Gamma(\xi'\xi_2 \mid \xi\xi_2)\,\mathrm{d}\xi_2 \quad (\text{large } t). \tag{6·7}$$

Thus by (6·2) we have

$$\gamma(\xi' \mid \xi) = \frac{2\sin\pi(\xi'-\xi)}{\pi(\xi'-\xi)}\int_{t-\frac{1}{2}}^{t+\frac{1}{2}}[y'(\xi'-\xi_2)\,y'(\xi-\xi_2)]^{\frac{1}{2}}\frac{\sin\frac{1}{2}\pi\{y(\xi'-\xi_2)-y(\xi-\xi_2)\}}{\pi\{y(\xi'-\xi_2)-y(\xi-\xi_2)\}}\,\mathrm{d}\xi_2 \tag{6·8}$$

for large t and application of (6·4) extends (6·8) to all t. It is now readily demonstrated that the truth of (6·8) for all t is entirely equivalent to the two conditions

$$\gamma(\xi' \mid \xi) = \frac{2\sin\pi(\xi'-\xi)}{\pi(\xi'-\xi)}\int_{-\frac{1}{2}}^{\frac{1}{2}}[y'(\xi'-\xi_2)\,y'(\xi-\xi_2)]^{\frac{1}{2}}\frac{\sin\frac{1}{2}\pi\{y(\xi'-\xi_2)-y(\xi-\xi_2)\}}{\pi\{y(\xi'-\xi_2)-y(\xi-\xi_2)\}}\,\mathrm{d}\xi_2 \tag{6·9}$$

and
$$\int_{t-\frac{1}{2}}^{t+\frac{1}{2}} y'(\xi)\,\mathrm{d}\xi = y(t+\tfrac{1}{2})-y(t-\tfrac{1}{2}) = 1 \quad (\text{all } t). \tag{6·10}$$

Straightforward use of (6·3) and (6·9) in (6·1) now gives

$$\frac{\mathscr{E}}{N}=\frac{1}{x_s^2}\left\{\frac{\pi^2}{6}+\int_{-\frac{1}{2}}^{\frac{1}{2}}\left(\frac{\pi^2}{24}y'^3+\frac{1}{8}\frac{y''^2}{y'}\right)\mathrm{d}\xi\right\}-\frac{1}{2}\int_{-\infty}^{\infty}U_2(x_s\xi)\left[1-y'\left(1-\frac{\sin\pi y}{\pi y}\right)\right]\mathrm{d}\xi \tag{6·11}$$

and for optimum y we should require to minimize the latter, subject to the subsidiary conditions

$$\left.\begin{aligned}&\text{(i)}\ \ y'(\xi)\geqslant 0,\\&\text{(ii)}\ \ y(-\xi)=-y(\xi),\\&\text{(iii)}\ \ y(\xi+m)=y(\xi)+m\quad(\text{all integral } m).\end{aligned}\right\} \tag{6·12}$$

Condition (iii) is equivalent to (6·10) in view of (ii). Clearly if y is determined in $(0, \frac{1}{2})$ it is known everywhere.

(B) In this case it will be observed that the second-order density matrix is obtained from the corresponding H.F. matrix by expressing the latter in terms of $\xi'+\xi_2'$, $\xi+\xi_2$, $\xi'-\xi_2'$, $\xi-\xi_2$, replacing these by $\xi'+\xi_2'$, $\xi+\xi_2$, $y(\xi'-\xi_2')$, $y(\xi-\xi_2)$ respectively and multiplying the whole by $[y'(\xi'-\xi_2')\,y'(\xi-\xi_2)]^{\frac{1}{2}}$. We therefore have

$$2\Gamma(\xi'\xi_2'\,|\,\xi\xi_2)=[y'(\xi'-\xi_2')\,y'(\xi-\xi_2)]^{\frac{1}{2}}$$
$$\times\begin{vmatrix} l[\frac{1}{2}(\xi'+\xi_2')+\frac{1}{2}y(\xi'-\xi_2')-\frac{1}{2}(\xi+\xi_2)-\frac{1}{2}y(\xi-\xi_2)] & l[\frac{1}{2}(\xi'+\xi_2')-\frac{1}{2}y(\xi'-\xi_2')-\frac{1}{2}(\xi+\xi_2)-\frac{1}{2}y(\xi-\xi_2)] \\ l[\frac{1}{2}(\xi'+\xi_2')+\frac{1}{2}y(\xi'-\xi_2')-\frac{1}{2}(\xi+\xi_2)+\frac{1}{2}y(\xi-\xi_2)] & l[\frac{1}{2}(\xi'+\xi_2')-\frac{1}{2}y(\xi'-\xi_2')-\frac{1}{2}(\xi+\xi_2)+\frac{1}{2}y(\xi-\xi_2)]\end{vmatrix}, \tag{6·13}$$

where
$$l(\xi)=(\sin\pi\xi)/\pi\xi. \tag{6·14}$$

Here
$$2P(\xi\xi_2)=y'(\xi-\xi_2)\,[1-l^2\{y(\xi-\xi_2)\}], \tag{6·15}$$

and by arguments similar to those used in case (A) we obtain

$$\gamma(\xi'\,|\,\xi)=\int_{-\frac{1}{2}}^{\frac{1}{2}}[y'(\xi'-\xi_2)\,y'(\xi-\xi_2)]^{\frac{1}{2}}\,l[\tfrac{1}{2}(\xi'-\xi)+\tfrac{1}{2}y(\xi'-\xi_2)-\tfrac{1}{2}y(\xi-\xi_2)]\times l[\tfrac{1}{2}(\xi'-\xi)-\tfrac{1}{2}y(\xi'-\xi_2)+\tfrac{1}{2}y(\xi-\xi_2)]\,\mathrm{d}\xi_2 \tag{6·16}$$

and

$$\frac{\mathscr{E}}{N}=\frac{1}{x_s^2}\left\{\frac{\pi^2}{12}+\int_{-\frac{1}{2}}^{\frac{1}{2}}\left(\frac{\pi^2}{12}y'^3+\frac{1}{8}\frac{y''^2}{y'}\right)\mathrm{d}\xi\right\}-\frac{1}{2}\int_{-\infty}^{\infty}U_2(x_s\xi)\,[1-y'(1-l^2(y))]\,\mathrm{d}\xi, \tag{6·17}$$

the subsidiary conditions again being given by (6·12).

Let us now compare the methods (A) and (B). It will be noted that if as $x_s\to 0$ the potential energy becomes negligible then $y'\to$ const., $y''\to 0$ and method (A) yields an energy per particle of $5\pi^2/24x_s^2$, to be compared with the correct result $\pi^2/6x_s^2$ given by method (B). In practice, however, the main correlation problem exists at intermediate rather than high densities in an electron gas, and in this region it seems entirely possible that even method (A) could lead to energies lower than the H.F. results. Certainly, in the low density limit $x_s\to\infty$, the kinetic energy is negligible and an argument similar to Wigner's shows that we have particles on a lattice in this case, and both methods (A) and (B) should be satisfactory. Thus, for both cases, the limiting forms of y and y' are as shown in curves I and III of figures 2 and 3. From

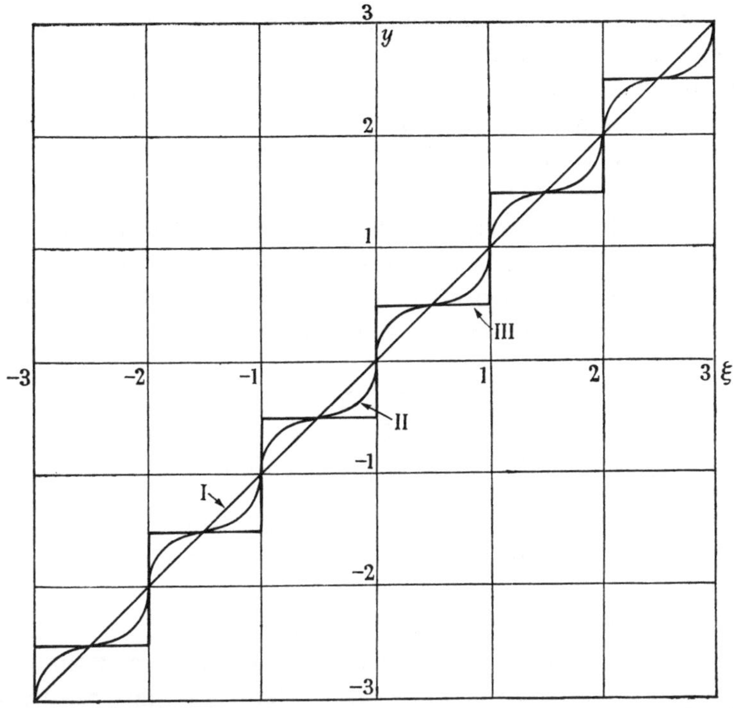

FIGURE 2. Form of $y(\xi)$ in one-dimensional, one-spin case. (I) High-density form; (II) inferred form at intermediate densities; (III) low-density form.

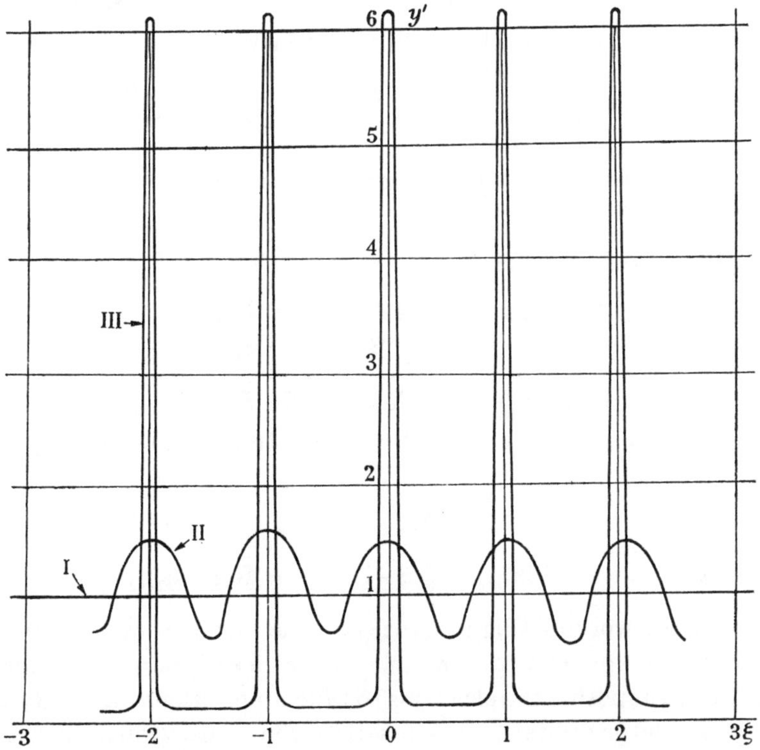

FIGURE 3. Form of y' in one-dimensional, one-spin case. (I) High-density form; (II) inferred form at intermediate densities; (III) low-density form.

these, the forms of y and y' at intermediate densities are inferred in curves II. Thus, with the reservation concerning method (A) in the high-density limit, it appears that both methods are flexible enough to cover a wide density range usefully.

While it is true that we have shown rigorously that the Pauli conditions on γ are automatically fulfilled only by (A) for all y, the method (B) has further attractions over (A), in addition to its advantage at high densities:

(i) It has a simple physical interpretation in terms of the localized orbital picture of March & Young (1959), who attempted to describe the pair function in terms of orbitals centred on the lattice sites, in such a way that if $\omega(\xi)$ denotes the orbital localized at the origin, then the pair function is

$$F(\xi) = \sum_{-\infty}^{\infty} \omega^2(\xi - m) - \omega^2(\xi). \tag{6·18}$$

In fact, the choice

$$\omega(\xi) = y'^{\frac{1}{2}}(\sin \pi y)/\pi y \tag{6·19}$$

leads to an F identical with that defined through (6·15).

(ii) The replacement of $(\sin \pi y)/\pi y$ in (6·3) by $(\sin^2 \pi y)/\pi^2 y^2$ in (6·15) is important from a numerical point of view, the latter being positive definite and more strongly convergent than the former for y satisfying (6·12).

For these reasons we will consider henceforth only generalizations of method (B), for which, as we have already remarked, it is felt that a general proof of the Pauli conditions on γ for all y may be found to establish (B) variationally as firmly as (A).

7. Generalized H.F. method in three dimensions and one spin

We now take as two-particle orbitals the functions

$$\phi_{\mathbf{k}_1\mathbf{k}_2}(\boldsymbol{\xi}\boldsymbol{\xi}_2) = \frac{\sqrt{2}\,\mathrm{i}}{N} \exp\left[\mathrm{i}\pi \frac{(\mathbf{k}_1+\mathbf{k}_2).(\boldsymbol{\xi}+\boldsymbol{\xi}_2)}{N^{\frac{1}{3}}}\right] [J(\boldsymbol{\xi}-\boldsymbol{\xi}_2)]^{\frac{1}{2}} \sin\pi \frac{(\mathbf{k}_1-\mathbf{k}_2).\mathbf{R}(\boldsymbol{\xi}-\boldsymbol{\xi}_2)}{N^{\frac{1}{3}}} \tag{7·1}$$

J being the Jacobian of $\mathbf{R}$. Following a procedure similar to that of (B) above we obtain

$$2\Gamma(\boldsymbol{\xi}'\boldsymbol{\xi}_2' \mid \boldsymbol{\xi}\boldsymbol{\xi}_2) = [J(\boldsymbol{\xi}'-\boldsymbol{\xi}_2')\,J(\boldsymbol{\xi}-\boldsymbol{\xi}_2)]^{\frac{1}{2}} \times \begin{vmatrix} l[\frac{1}{2}(\boldsymbol{\xi}'+\boldsymbol{\xi}_2')+\frac{1}{2}\mathbf{R}(\boldsymbol{\xi}'-\boldsymbol{\xi}_2')-\frac{1}{2}(\boldsymbol{\xi}+\boldsymbol{\xi}_2)-\frac{1}{2}\mathbf{R}(\boldsymbol{\xi}-\boldsymbol{\xi}_2)] & l[\frac{1}{2}(\boldsymbol{\xi}'+\boldsymbol{\xi}_2')-\frac{1}{2}\mathbf{R}(\boldsymbol{\xi}'-\boldsymbol{\xi}_2')-\frac{1}{2}(\boldsymbol{\xi}+\boldsymbol{\xi}_2)-\frac{1}{2}\mathbf{R}(\boldsymbol{\xi}-\boldsymbol{\xi}_2)] \\ l[\frac{1}{2}(\boldsymbol{\xi}'+\boldsymbol{\xi}_2')+\frac{1}{2}\mathbf{R}(\boldsymbol{\xi}'-\boldsymbol{\xi}_2')-\frac{1}{2}(\boldsymbol{\xi}+\boldsymbol{\xi}_2)+\frac{1}{2}\mathbf{R}(\boldsymbol{\xi}-\boldsymbol{\xi}_2)] & l[\frac{1}{2}(\boldsymbol{\xi}'+\boldsymbol{\xi}_2')-\frac{1}{2}\mathbf{R}(\boldsymbol{\xi}'-\boldsymbol{\xi}_2')-\frac{1}{2}(\boldsymbol{\xi}+\boldsymbol{\xi}_2)+\frac{1}{2}\mathbf{R}(\boldsymbol{\xi}-\boldsymbol{\xi}_2)] \end{vmatrix}, \tag{7·2}$$

where

$$l(\boldsymbol{\xi}) = 3\zeta^{-3}(\sin\zeta - \zeta\cos\zeta), \quad \zeta = (6\pi^2)^{\frac{1}{3}}\,\xi. \tag{7·3}$$

In particular,

$$2P(\boldsymbol{\xi}\boldsymbol{\xi}_2) = J(\boldsymbol{\xi}-\boldsymbol{\xi}_2)\,[1 - l^2\{\mathbf{R}(\boldsymbol{\xi}-\boldsymbol{\xi}_2)\}]. \tag{7·4}$$

In view of the work of Wigner (1938) on the low-density limit from which it is argued (March & Young 1959) that an accurate description of the pair function can be deduced by supposing the electrons to take up positions on a body-centred cubic lattice via oscillator orbitals, we will satisfy the translational invariance property by dividing $\boldsymbol{\xi}$-space up into unit volume Wigner–Seitz cells corresponding to such a body-centred cubic structure. Then, using arguments similar to those employed in one dimension in the previous section, we obtain

$$\gamma(\boldsymbol{\xi}' \mid \boldsymbol{\xi}) = \int_C [J(\boldsymbol{\xi}'-\boldsymbol{\xi}_2)\,J(\boldsymbol{\xi}-\boldsymbol{\xi}_2)]^{\frac{1}{2}}\, l[\tfrac{1}{2}(\boldsymbol{\xi}'-\boldsymbol{\xi})+\tfrac{1}{2}\mathbf{R}(\boldsymbol{\xi}'-\boldsymbol{\xi}_2)-\tfrac{1}{2}\mathbf{R}(\boldsymbol{\xi}-\boldsymbol{\xi}_2)] \times l[\tfrac{1}{2}(\boldsymbol{\xi}'-\boldsymbol{\xi})-\tfrac{1}{2}\mathbf{R}(\boldsymbol{\xi}'-\boldsymbol{\xi}_2)+\tfrac{1}{2}\mathbf{R}(\boldsymbol{\xi}-\boldsymbol{\xi}_2)]\,\mathrm{d}\boldsymbol{\xi}_2 \quad (7\cdot5)$$

and referring specifically to the electron gas potential

$$\frac{\mathscr{E}}{N} = \left(\frac{3}{4\pi}\right)^{\frac{2}{3}} \frac{1}{r_s^2}\left\{\frac{3}{20}(6\pi^2)^{\frac{2}{3}} + \frac{(6\pi^2)^{\frac{2}{3}}}{20}\int_C J\sum_1^3 (\nabla R_i)^2\,\mathrm{d}\boldsymbol{\xi} + \frac{1}{8}\int \frac{(\nabla J)^2}{J}\,\mathrm{d}\boldsymbol{\xi}\right\} - \frac{1}{2}\left(\frac{3}{4\pi}\right)^{\frac{1}{3}} \frac{1}{r_s}\int_{\text{all space}} \frac{[1-J(1-l^2(\mathbf{R}))]}{\xi}\,\mathrm{d}\boldsymbol{\xi}, \quad (7\cdot6)$$

where C is any cell. The subsidiary conditions are

$$\left.\begin{array}{ll}\text{(i)} & J(\boldsymbol{\xi}) \geqslant 0, \\ \text{(ii)} & \mathbf{R}(-\boldsymbol{\xi}) = -\mathbf{R}(\boldsymbol{\xi}), \\ \text{(iii)} & \mathbf{R}(\boldsymbol{\xi}+\boldsymbol{\mu}) = \mathbf{R}(\boldsymbol{\xi})+\boldsymbol{\mu},\end{array}\right\} \quad (7\cdot7)$$

where $\boldsymbol{\mu}$ is any translation vector between equivalent lattice points.

It will be noted that the pair function defined through (7·4) is in general angle-dependent and so corresponds to a non-physical situation since in a uniform electron gas there are no preferential directions. This, of course, is true in particular of the Wigner low-density lattice into which the former case degenerates as $r_s \to \infty$. We may, however, consider such functions as representatives of infinitely degenerate systems and regard the angular average of $F(\boldsymbol{\xi})$ as the physically meaningful quantity (see March & Young 1959).

Equation (7·6) is thus proposed for dealing with correlation in a uniform ferromagnetic electron gas. Clearly there are practical difficulties associated with the minimization of (7·6), largely because of the geometry of the Wigner–Seitz cells, but these do not appear to be insurmountable. A discussion of such practical aspects will, however, be postponed until later (§ 10) and meanwhile we will extend the results of §§ 6 and 7 to the case of two spins.

8. Generalized H.F. method in one dimension and two spins

In view of the form of y' (figure 3) for the ferromagnetic gas it was thought that the situation in the present case might be characterized by two functions, y' and z' say, corresponding to the relative density distributions of the two spins and having the properties (figure 4):

$$\left.\begin{array}{ll}\text{(i)} & y'(\xi) \geqslant 0, \\ \text{(ii)} & y(-\xi) = -y(\xi), \\ \text{(iii)} & y(\xi+1) = z(\xi)+1, \\ \text{(iv)} & y(\xi+2) = y(\xi)+2.\end{array}\right\} \quad (8\cdot1)$$

It turns out that this is in fact the case.

We therefore seek to generalize the procedure of §4 along the above lines and note that in the non-magnetic case the use of a single Slater determinant of orthonormal orbitals $\psi_k(\xi)\,\chi_{\pm\frac{1}{2}}(\sigma)$ is equivalent to a choice of two types of two-particle functions (cf. equation (4·4)), namely those of the kind

$$\frac{1}{\sqrt{2}}\begin{vmatrix}\psi_{k_1}(\xi_1)\,\chi_{\frac{1}{2}}(\sigma_1) & \psi_{k_2}(\xi_1)\,\chi_{\frac{1}{2}}(\sigma_1)\\ \psi_{k_1}(\xi_2)\,\chi_{\frac{1}{2}}(\sigma_2) & \psi_{k_2}(\xi_2)\,\chi_{\frac{1}{2}}(\sigma_2)\end{vmatrix}, \tag{8·2}$$

in which only one spin function appears, and those of the kind

$$\frac{1}{\sqrt{2}}\begin{vmatrix}\psi_{k_1}(\xi_1)\,\chi_{\frac{1}{2}}(\sigma_1) & \psi_{k_2}(\xi_1)\,\chi_{-\frac{1}{2}}(\sigma_1)\\ \psi_{k_1}(\xi_2)\,\chi_{\frac{1}{2}}(\sigma_2) & \psi_{k_2}(\xi_2)\,\chi_{-\frac{1}{2}}(\sigma_2)\end{vmatrix}, \tag{8·3}$$

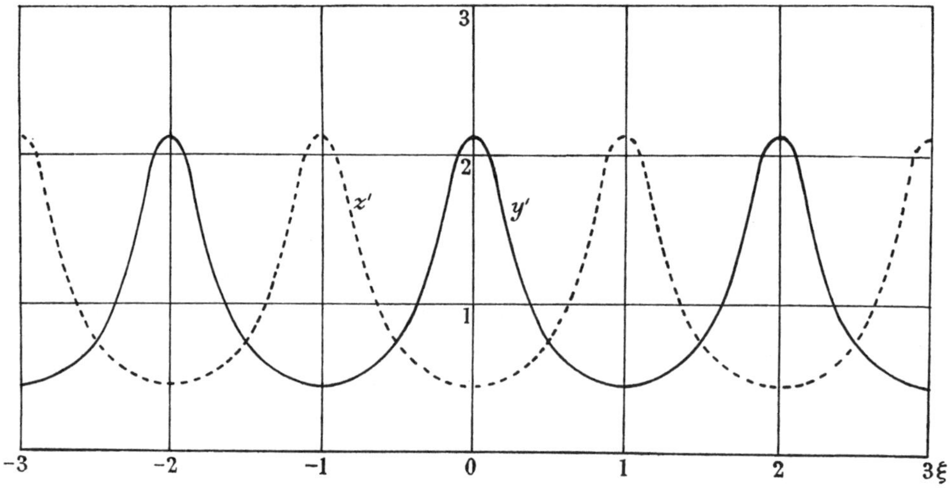

FIGURE 4. Form of y' and z' in one-dimensional, two-spin case (schematic).

in which both spin factors occur. We now replace these respectively by

$$[y'(\xi_1-\xi_2)]^{\frac{1}{2}}\frac{1}{\sqrt{2}}\begin{vmatrix}\psi_{k_1}(y_1)\,\chi_{\frac{1}{2}}(\sigma_1) & \psi_{k_2}(y_1)\,\chi_{\frac{1}{2}}(\sigma_1)\\ \psi_{k_1}(y_2)\,\chi_{\frac{1}{2}}(\sigma_2) & \psi_{k_2}(y_2)\,\chi_{\frac{1}{2}}(\sigma_2)\end{vmatrix} \tag{8·4}$$

and

$$[z'(\xi_1-\xi_2)]^{\frac{1}{2}}\frac{1}{\sqrt{2}}\begin{vmatrix}\psi_{k_1}(z_1)\,\chi_{\frac{1}{2}}(\sigma_1) & \psi_{k_2}(z_1)\,\chi_{-\frac{1}{2}}(\sigma_1)\\ \psi_{k_1}(z_2)\,\chi_{\frac{1}{2}}(\sigma_2) & \psi_{k_2}(z_2)\,\chi_{-\frac{1}{2}}(\sigma_2)\end{vmatrix}, \tag{8·5}$$

where

$$y_1 = \tfrac{1}{2}(\xi_1+\xi_2)+\tfrac{1}{2}y(\xi_1-\xi_2), \tag{8·6}$$

$$y_2 = \tfrac{1}{2}(\xi_1+\xi_2)-\tfrac{1}{2}y(\xi_1-\xi_2), \quad \text{etc.,} \tag{8·7}$$

y and z satisfying equations (8·1). It is directly verifiable that the aggregate of orbitals of the kind (8·4) and (8·5) is orthonormal.

For ψ's defined by (5·1), all the usual conditions are satisfied as in the ferromagnetic case and we have

$$2\Gamma(\xi_1'\xi_2'\,|\,\xi_1\xi_2) = \tfrac{1}{2}[y'(\xi_1'-\xi_2')\,y'(\xi_1-\xi_2)]^{\frac{1}{2}}\begin{vmatrix}l(y_1^{(1)}-y_1) & l(y_1^{(1)}-y_2)\\ l(y_2^{(1)}-y_1) & l(y_2^{(1)}-y_2)\end{vmatrix}$$
$$+\tfrac{1}{2}[z'(\xi_1'-\xi_2')\,z(\xi_1-\xi_2)]^{\frac{1}{2}}\,l(z_1^{(1)}-z_1)\,l(z_2^{(1)}-z_2), \tag{8·8}$$

where $$y_1^{(1)} = \tfrac{1}{2}(\xi_1'+\xi_2')+\tfrac{1}{2}y(\xi_1'-\xi_2'), \quad \text{etc.} \tag{8·9}$$

and $$l(\xi) = (\sin\tfrac{1}{2}\pi\xi)/\tfrac{1}{2}\pi\xi. \tag{8·10}$$

We note in particular that $$F(\xi) = \tfrac{1}{2}y'(1-l^2(y))+\tfrac{1}{2}z' \tag{8·11}$$

and it is of interest to note that, as in the one-spin case, the pair function can be interpreted in terms of localized orbitals (cf. equation (6·18)) since by using equations (8·1) we can write (8·11) in the form

$$F(\xi) = \frac{1}{2}\sum_{-\infty}^{\infty}\omega^2(\xi-m)-\tfrac{1}{2}\omega^2(\xi), \tag{8·12}$$

where $$\omega(\xi) = y'^{\frac{1}{2}}(\sin\tfrac{1}{2}\pi y)/\tfrac{1}{2}\pi y. \tag{8·13}$$

The kinetic energy per particle is evaluated by the technique used in the ferromagnetic case and we obtain

$$\frac{\mathscr{E}}{N} = \frac{1}{x_s^2}\left\{\frac{\pi^2}{48}+\int_{-\frac{1}{2}}^{\frac{1}{2}}\left(\frac{\pi^2}{96}y'^3+\frac{\pi^2}{96}z'^3+\frac{1}{16}\frac{y''^2}{y'}+\frac{1}{16}\frac{z''^2}{z'}\right)\mathrm{d}\xi\right\}$$
$$-\frac{1}{2}\int_{-\infty}^{\infty}U_2(x_s\xi)\,[1-\tfrac{1}{2}y'(1-l^2(y))-\tfrac{1}{2}z']\,\mathrm{d}\xi. \tag{8·14}$$

It would seem just possible from a variational point of view that a somewhat improved scheme might be obtained by relaxing condition (8·1 (iii)) but we shall not consider this point here. Instead we now turn to the full three-dimensional problem.

9. Generalized H.F. method for a non-magnetic electron gas

Here we are guided by the considerations of §8 and the desirability of reproducing Wigner's result in the low-density limit. We begin by defining the usual body-centred cubic structure in $\boldsymbol{\xi}$-space and regarding it as consisting of two interpenetrating simple cubic lattices, one associated with spin $\frac{1}{2}$ and the other with spin $-\frac{1}{2}$ electrons. Next we define two transformations **R** and **S** possessing Jacobians J and K corresponding to like and unlike spins respectively, having properties analogous to equations (8·1), namely

$$\left.\begin{array}{ll}\text{(i)} & J(\boldsymbol{\xi}) \geqslant 0,\\ \text{(ii)} & \mathbf{R}(-\boldsymbol{\xi}) = -\mathbf{R}(\boldsymbol{\xi}),\\ \text{(iii)} & \mathbf{R}(\boldsymbol{\xi}+\boldsymbol{\mu}_1) = \mathbf{S}(\boldsymbol{\xi})+\boldsymbol{\mu}_1,\\ \text{(iv)} & \mathbf{R}(\boldsymbol{\xi}+\boldsymbol{\mu}_2) = \mathbf{R}(\boldsymbol{\xi})+\boldsymbol{\mu}_2,\end{array}\right\} \tag{9·1}$$

where here $\boldsymbol{\mu}_1$ and $\boldsymbol{\mu}_2$ denote any vectors between equivalent lattice points of the body-centred lattices and of either simple cubic lattice respectively. We may thus view J and K as corresponding to two periodic charge distributions identical except for spin and the relative displacement of the two simple cubic lattices. In this way we generalize Wigner's low-density limit picture to all densities in a quantum mechanical way.

The analysis is now as before, $\boldsymbol{\xi}$, **R**, **S**, J and K replacing ξ, y, z, y' and z' in equations (8·2) to (8·11) but equation (8·10) being replaced by its three-dimensional equivalent $$l(\boldsymbol{\xi}) = 3\zeta^{-3}(\sin\zeta-\zeta\cos\zeta), \quad \zeta = (3\pi^2)^{\frac{1}{3}}\xi. \tag{9·2}$$

Thus, for example, $$F = \tfrac{1}{2}J(1-l^2(\mathbf{R}))+\tfrac{1}{2}K. \tag{9·3}$$

The kinetic energy is calculated in the usual way and we obtain

$$\frac{\mathscr{E}}{N}=\left(\frac{3}{4\pi}\right)^{\frac{2}{3}}\frac{1}{r_s^2}\left\{\frac{3}{20}(3\pi^2)^{\frac{2}{3}}+\frac{(3\pi^2)^{\frac{2}{3}}}{40}\int_C\left[J\sum_1^3(\nabla R_i)^2+K\sum_1^3(\nabla S_i)^2\right]d\boldsymbol{\xi}\right.$$
$$\left.+\frac{1}{16}\int_C\left[\frac{(\nabla J)^2}{J}+\frac{(\nabla K)^2}{K}\right]d\boldsymbol{\xi}\right\}-\frac{1}{2}\left(\frac{3}{4\pi}\right)^{\frac{1}{3}}\frac{1}{r_s}\int_{\text{all space}}\frac{[1-\frac{1}{2}J(1-l^2(\mathbf{R}))-\frac{1}{2}K]}{\xi}d\boldsymbol{\xi},\tag{9·4}$$

where C denotes a unit volume Wigner–Seitz cell. The expression (9·4) is now to be minimized for **R** defined over two adjacent cells (**R** and **S** thus being automatically determined over all space through (9·1)).

Finally, we indicate one practical method of applying the three-dimensiona formulae.

10. The practical application of the three-dimensional formulae

The main purpose of this paper is to exhibit the explicit density matrices and show that they describe the physical situation in a realistic way. The numerical variational calculations to which they lead must be regarded as a separate project. However, it is felt that it would be undesirable to end without indicating that accurate numerical use of our equations appears entirely possible. For the purposes of illustration we will consider the ferromagnetic electron gas and the restricted minimization of (7·6).

The method which suggests itself as a starting point is that of a spherical approximation in which each Wigner–Seitz cell is replaced by a sphere of equal (unit) volume of radius $\xi_0=(3/4\pi)^{\frac{1}{3}}$ and an explicit form for **R** chosen. This is in fact how we proceed in the following but in doing so it is necessary to recognize that fairly large errors are thereby introduced unless special precautions are taken. The source and correction of these will be indicated presently. Choosing, therefore, **R** spherically symmetrical in the sphere surrounding the origin, we write

$$\mathbf{R}=R(\xi)\,\boldsymbol{\xi}/\xi,\tag{10·1}$$

whence

$$J=\left|\frac{1}{3\xi^2}\frac{d}{d\xi}R^3\right|\tag{10·2}$$

and

$$\sum_1^3(\nabla R_i)^2=\frac{2R^2}{\xi^2}+R'^2.\tag{10·3}$$

Furthermore, denoting the sphere centred at the origin by S_0 and the aggregate of all other spheres by S', the potential energy may be written

$$\frac{V}{N}=-\frac{1}{2}\left(\frac{3}{4\pi}\right)^{\frac{1}{3}}\frac{1}{r_s}\left\{\int_{\text{all space}}\frac{d\boldsymbol{\xi}}{\xi}-\int\frac{J}{\xi}d\boldsymbol{\xi}\right\}$$
$$+\frac{1}{2}\left(\frac{3}{4\pi}\right)^{\frac{1}{3}}\frac{1}{r_s}\int_{S_0}\frac{J[1-l^2(\mathbf{R})]}{\xi}d\boldsymbol{\xi}-\frac{1}{2}\left(\frac{3}{4\pi}\right)^{\frac{1}{3}}\frac{1}{r_s}\int_{S'}\frac{Jl^2(\mathbf{R})}{\xi}d\boldsymbol{\xi}$$
$$=-\frac{0\cdot896}{r_s}+\frac{1}{2}\left(\frac{3}{4\pi}\right)\frac{1}{r_s}\int_{S_0}\frac{J[1-l^2(\mathbf{R})]}{\xi}d\boldsymbol{\xi}-\frac{1}{2}\left(\frac{3}{4\pi}\right)^{\frac{1}{3}}\frac{1}{r_s}\int_{S'}\frac{Jl^2(\mathbf{R})}{\xi}d\boldsymbol{\xi},\tag{10·4}$$

from simple electrostatic arguments and a result from Wigner (1938) that

$$-\frac{1}{2}\left(\frac{3}{4\pi}\right)^{\frac{1}{3}}\frac{1}{r_s}\left\{\int_{\text{all space}}\frac{d\boldsymbol{\xi}}{\xi}-\Sigma'\frac{1}{\xi_i}\right\} = -\frac{0\cdot896}{r_s}, \tag{10·5}$$

where the $\boldsymbol{\xi}_i$ denote all the points of the above body-centred cubic lattice and Σ' means that the point at the origin is excluded from the summation. The subsidiary conditions now become

$$\int_0^{\xi_0} J4\pi\xi^2\,d\xi = 1 \tag{10·6}$$

and

$$J'(\xi_0) = 0, \tag{10·7}$$

with equation (7·7 (iii)) serving to define $\mathbf{R}$ and so J in spheres other than S_0. The kinetic energy (defined in terms of integrals over S_0 only) and the first two terms of (10·4) are likely to account for an overwhelming part of the total energy and might be minimized with respect to the scalar R. A correction taking into account the third term of (10·4) could then be made.

Now it will be noted that, in the low-density limit, the replacement of cells by spheres in itself involves no further approximations since each cell (other than that containing the origin) possesses only a point charge at its centre. Also, at low densities, when the cells contain oscillator orbitals highly localized at the centres, we expect high accuracy. This is, of course, due to the fact that J values near the cell boundaries are very small and redistributions necessary in the spherical approximation are negligible. On the other hand, in the high-density limit $\mathbf{R} = \boldsymbol{\xi}$, $J = 1$ everywhere, J values near the boundaries are not small and one might expect serious errors in this case. The extent of these can be illustrated by evaluating the left-hand side of (10·5) by such an approximation, replacing the potential $\int_{\text{all space}}\frac{d\boldsymbol{\xi}}{\xi}$ by the potential due to unit volume spheres of unit charge density centred on the lattice points. Simple electrostatic arguments then give $-0\cdot75/r_s$ as the value of the left-hand side to be compared with the correct answer $-0\cdot896/r_s$. The former value is that used by Wigner (1934) in his earlier discussion of the low-density electron gas. It seems therefore that in the spherical approximation caution must be exercised at high densities and it might be desirable within this simplified framework to choose J to vanish at the boundary of S_0. Since, however, the spread of J is not restricted within the sphere the above somewhat artificial restriction should not prevent us from obtaining good energies up to quite high densities.

11. Conclusions

It thus appears that the H.F. theory of the uniform gas may be generalized in a powerful manner by utilizing transformed two-body orbitals, and improvement on the H.F. energies seems thereby assured for all densities. Indeed, the method looks particularly promising from a practical point of view, as it has been demonstrated directly that it yields highly accurate results in both the high- and the low-density limits. This is in marked contrast to the methods utilizing perturbation theory (see, for example, the review of Brueckner 1959) which seem restricted in application to a density range considerably higher than that encountered in metals under normal conditions. Clearly, however, detailed numerical calculations are required for intermediate densities before the accuracy which may be achieved with our variational approach can be finally established, and such work is now being planned.

From a fundamental point of view, two connected matters seem to deserve further attention. As Löwdin has shown, knowledge of the Dirac density matrix is sufficient to enable the total wave function in the H.F. approximation to be systematically built up. It would be very valuable if this result could be generalized to yield a correlated many-body wave function from second-order matrices of the form (4·2). For example, this would greatly facilitate comparison between the present method and that of Jastrow (1955). Secondly, unless this many-body wave function is forthcoming further work is obviously called for in order to place method (B) of §5 on a completely firm footing. At present, we believe that this will be rigorously possible, but should it not prove to be the case, method (A) is still available, for which all the Pauli conditions have been established rigorously.

Appendix

The equation (5·19) states that, for any fixed r, the summation of $d_{r\ r}^{k_1 k_2}$ over any line $k_2 - k_1 = \text{const.}$ is unity and this is fundamental in establishing the first-order Pauli conditions for method (A) specified by the equations (5·21). Unfortunately, it appears to be inadequate for discussing method (B) where the ranges of k_1 and k_2 are given by the square D: $-\frac{1}{2}(N-1) \leqslant k_1, k_2 \leqslant \frac{1}{2}(N-1)$. (Reference to figure 1 will prove helpful throughout the following argument.) Straightforward use of (5·19) in the latter case gives

$$\alpha_r \leqslant \frac{1}{N-1} \sum_{\lambda=-(N-1)}^{N-1}{}' \sum_{k_1=-\frac{1}{2}(N-1)}^{\frac{1}{2}(N-1)} d_{r\ r}^{k_1\, k_1+\lambda} \leqslant 2, \tag{A 1}$$

which is twice the desired bound. The losses in accuracy which arise from using (5·19) in this particular case can be illustrated by considering (k_1, k_2) space in the H.F. approximation when

$$\phi_{k_1 k_2} = \frac{1}{\sqrt{2}} \begin{vmatrix} \psi_{k_1}(\xi_1) & \psi_{k_2}(\xi_1) \\ \psi_{k_1}(\xi_2) & \psi_{k_2}(\xi_2) \end{vmatrix}. \tag{A 2}$$

From (5·10)
$$C_{k_1 k_2}^{k_1 k_2} = \frac{1}{\sqrt{2}}; \qquad C_{k_2 k_1}^{k_1 k_2} = -\frac{1}{\sqrt{2}} \quad (k_1 \neq k_2), \tag{A 3}$$

all other $C_{r\ s}^{k_1 k_2}$ being zero and so by (5·14)

$$d_{r\ r}^{k_1 k_2} = \begin{cases} \frac{1}{2} \text{ if } r = k_1 \text{ or } k_2 \quad (k_1 \neq k_2) \\ 0 \text{ otherwise.} \end{cases} \tag{A 4}$$

Thus $d_{r\ r}^{k_1 k_2}$ is zero everywhere except along the lines $k_1 = r$, $k_2 = r$, where (but at the exceptional point $k_1 = k_2 = r$ where it vanishes) it is $\frac{1}{2}$. We now see why accuracy is lost in method (B), by attempting finite sums along the diagonal lines $k_2 - k_1 = k$ for various k, the ranges of k_1 and k_2 being specified by (5·24). We consider the most important case when $|r| \leqslant \frac{1}{2}(N-1)$. For $0 < |k| \leqslant \frac{1}{2}(N-1) - |r|$, both $d_{r\ r}^{k_1\ r}$ and $d_{r\ r}^{r\ k_2}$ are within D in (k_1, k_2) space and so for this range of k (5·19) works well. Now suppose

$\frac{1}{2}(N-1)-|r| < |k| \leqslant \frac{1}{2}(N-1)+|r|$. Here only one of $d_{r\;r}^{k_1\,r}$ and $d_{r\;r}^{r\,k_2}$ lies in D and so summing over that part of a diagonal within D we obtain a sum of $\frac{1}{2}$, the remaining contribution of $\frac{1}{2}$ to the infinite diagonal sum coming from outside. When

$$\tfrac{1}{2}(N-1)-|r| < |k| \leqslant N-1,$$

the situation is even worse, all contributions to the infinite diagonal sums coming from outside D. We see therefore why in a naïve application of (5·19) we lose accuracy in the H.F. case.

The above analysis of the H.F. limit in which, for example, there are reflective symmetries about the lines $k_1 = k_2$ and $k_1+k_2 = 2r$ suggests that certain properties of this kind might be generally true for all densities and might therefore be used to obtain a sharper bound than that given by (A 1). Such appears to be the situation. For example, using (5·18)

$$\sum_{m=-\infty}^{r-\frac{1}{2}n}\sum_{s=-\infty}^{\infty}|C_{r\;\,s}^{m\;m+n}|^2 = \sum_{m=-\infty}^{r-\frac{1}{2}n}\sum_{s=-\infty}^{\infty}|C_{r-(m+\frac{1}{2}n)\;s-(m+\frac{1}{2}n)}^{-\frac{1}{2}n\;\;\frac{1}{2}n}|^2$$

$$= \sum_{m=-\infty}^{r-\frac{1}{2}n}\sum_{s=-\infty}^{\infty}|C_{r-(m+\frac{1}{2}n)\;s}^{-\frac{1}{2}n\;\;\frac{1}{2}n}|^2 = \sum_{\mu=0}^{\infty}\sum_{s=-\infty}^{\infty}|C_{\mu\;\;s}^{-\frac{1}{2}n\;\frac{1}{2}n}|^2, \quad \text{(A 5)}$$

and similarly
$$\sum_{m=r-\frac{1}{2}n}^{\infty}\sum_{s=-\infty}^{\infty}|C_{r\;\,s}^{m\;m+n}|^2 = \sum_{\mu=-\infty}^{0}\sum_{s=-\infty}^{\infty}|C_{r\;\;s}^{-\frac{1}{2}n\;\frac{1}{2}n}|^2. \quad \text{(A 6)}$$

Now it is readily seen that
$$|C_{-r\;\;-s}^{-\frac{1}{2}n\;\frac{1}{2}n}|^2 = |C_{r\;\;s}^{-\frac{1}{2}n\;\frac{1}{2}n}|^2 \quad \text{(A 7)}$$

and so the sums (A 5) and (A 6) are equal and by (5·19) both are equal to $\frac{1}{2}$.

The above result can be used at once to give bounds to the α_r. If $|r| \geqslant \frac{1}{2}(N-1)$, all the $d_{r\;r}^{k_1\,k_2}$ lie on the same side of $k_1+k_2 = 2r$. We have $2(N-1)$ diagonal sums each less than or equal to $\frac{1}{2}$ and so $\alpha_r \leqslant 1$. If $|r| \leqslant \frac{1}{2}(N-1)$ there are $2(N-1-2|r|)$ diagonals to which the above result does not apply and so relying on (5·19) for these we can only say for this range of r that

$$\alpha_r \leqslant \frac{1}{N-1}\{1.2(N-1-2|r|)+\tfrac{1}{2}.4|r|\} = 2-\frac{2|r|}{N-1}. \quad \text{(A 8)}$$

It seems likely that further progress should be possible along these lines, but so far we have been unable to achieve a stronger result in this latter case.

References

Ayres, R. U. 1958 *Phys. Rev.* **111**, 1453.
Brueckner, K. A. 1959 *The many body problem*, p. 175. London: Methuen.
Cowan, R. D. & Kirkwood, J. G. 1958 *Phys. Rev.* **111**, 1460.
Ferrell, R. A. 1958 *Phys. Rev. letters*, **1**, 443.
Jastrow, R. 1955 *Phys. Rev.* **98**, 1479.
Koppe, H. 1957 *Z. Phys.* **148**, 135.
Löwdin, P. O. 1955 *Phys. Rev.* **97**, 1474, 1490, 1509.
Macke, W. 1955 *Phys. Rev.* **100**, 992; *Ann. Phys., Lpz.*, **17**, 1.
March, N. H. 1958 *Phys. Rev.* **110**, 604.
March, N. H. & Young, W. H. 1958 *Proc. Phys. Soc.* A, **72**, 182.
March, N. H. & Young, W. H. 1959 *Phil. Mag.* **4**, 384.
Mayer, J. E. 1955 *Phys. Rev.* **100**, 1579.
Tredgold, R. H. 1957 *Phys. Rev.* **105**, 1421.
Wigner, E. P. 1934 *Phys. Rev.* **46**, 1002.
Wigner, E. P. 1938 *Trans. Faraday Soc.* **34**, 678.

PRINTED IN GREAT BRITAIN AT THE UNIVERSITY PRESS, CAMBRIDGE
(BROOKE CRUTCHLEY, UNIVERSITY PRINTER)

REVIEWS OF MODERN PHYSICS VOLUME 40, NUMBER 4 OCTOBER 1968

SESSION VI—HEAVILY DOPED SEMICONDUCTORS

Localization of Electrons in Impure Semiconductors by a Magnetic Field

J. DURKAN
The University, Sheffield, England
R. J. ELLIOTT
University of Oxford, Oxford, England
N. H. MARCH
The University, Sheffield, England

The magnetic-field-dependent activation energy required to explain the rapid increase in the Hall coefficient with decreasing temperature in *n*-type InSb from 2°–5°K and in fields of up to 15 kG, observed by Putley and other workers, is interpreted in terms of the energy gap between the two lowest donor levels. This gap appears to be rather insensitive to carrier screening. Further, the existence of a threshold magnetic field, below which no activation energy is found, is interpreted as evidence of a Wigner transition, in which the electrons in the lowest impurity band cease to conduct when the donor-wave-function overlap, or the effective mass for electron transport, reaches a critical value. Finally, the relation between the activation energy for conduction in the extreme high-field limit and the Yafet–Keyes–Adams ionization energy is briefly discussed. It is anticipated that this ionization energy will be more closely approached at high fields, but perhaps never quite attained in practice.

I. INTRODUCTION

Interest in the behavior of a hydrogenic impurity center in a semiconductor in a high magnetic field was first stimulated by the work of Yafet, Keyes, and Adams.[1] They demonstrated that the ionization energy of such a center increased with magnetic field, in the manner shown in the uppermost curve of Fig. 1. This curve is actually for *n*-type InSb, which is a particularly favorable case for discussing such magnetic-field-dependent effects. Here the ratio of the zero-point energy in a magnetic field, $eB\hbar/2m^*c$, to the effective Rydberg $m^*e^4/2K^2\hbar^2$, is unity when the magnetic field B is $\sim$2kG, since in this case the effective mass m^* is $0.013m$ and the dielectric constant K is 16. Fields much in excess of this value are easily achieved and we expect such high fields to markedly localize the wave functions around the donor impurities. However, since the effective Bohr radius in InSb is 5.7×10^{-6} cm, even in the purest specimens available substantial overlap of donor wave functions occurs at zero magnetic field, and isolated donor levels cannot therefore be studied when $B=0$.

Nevertheless, Hall-effect measurements of Putley[2] and Nad' and Oleinikov,[3] in the temperature range 2°–5°K, showed that, in magnetic fields up to 15 kG, there is a rapid increase in the Hall coefficient R with decreasing temperature. R was found to follow an exponential law and this established the existence of an activation energy for electrons to become conducting, provided the magnetic field exceeded some threshold value. Furthermore, for a fixed temperature, a marked rise in R was found to occur as the magnetic field was increased above this threshold value, indicating a magnetic-field dependence of the activation energy, in qualitative accord with the predictions of Yafet, Keyes, and Adams. However, Curves 1 to 4 of Fig. 1 show the derived activation energies for four compensated samples with the listed donor and acceptor concentrations N_D and N_A, respectively, and it is seen that the observed activation energies are smaller than the Yafet–Keyes–Adams energy by a large factor.

[1] Y. Yafet, R. W. Keyes, and E. N. Adams, J. Phys. Chem. Solids **1**, 137 (1956).

[2] E. H. Putley, Proc. Phys. Soc. (London) **76**, 802 (1960); J. Phys. Chem. Solids **22**, 241 (1961); *Semiconductors and Semimetals*, R. K. Willardson and A. C. Beer, Eds. (Academic Press Inc., New York, 1966), Vol. 1.

[3] F. Y. A. Nad' and A. Y. A. Oleinikov, Fiz. Tverd. Tela **6**, 2064 (1964) [Sov. Phys.—Solid State **6**, 1629 (1965)].

II. ISOLATED DONOR LEVELS AND CARRIER SCREENING

To understand these results over the range of magnetic fields shown in Fig. 1, Durkan and March[4] have recently given a theory of the screening of charged impurities by free carriers, in which the cylindrically symmetric screening about the field direction, taken as the z axis, is accounted for from the outset. This screened potential depends on the electron density n_0, temperature T, magnetic field B, as well as on K and m^*. In the high-field limit and for the nondegenerate case the Fourier components $V(q)$ of the screened potential are

$$\tilde{V}(q)=\frac{4\pi e}{(2\pi)^{\frac{3}{2}}}\left\{Kq^2+4\pi e^2 n_0\beta\int_0^1 dy\exp-\left[\frac{q_z^2\hbar^2\beta(1-y^2)}{8m^*}\right]\exp-\left[\frac{(q_x^2+q_y^2)\hbar^2}{4m^*\mu_0^*B}\left(\coth\mu_0^*B\beta-\frac{\cosh\mu_0^*B\beta y}{\sinh\mu_0^*B\beta}\right)\right]\right\}^{-1}$$

$$\left(\beta=(k_BT)^{-1};\ \mu_0^*=\frac{e\hbar}{2m^*c}\right),\qquad(1)$$

which reduce to the bare Coulomb potential when the carrier density n_0 is zero.

In this screened potential, variational calculations were carried out for the two lowest impurity levels, as a function of magnetic field, with trial wave functions given by

$$\psi_0=\exp(-r^2/a^2)\exp(-z^2/b^2),\qquad r^2=x^2+y^2,\quad(2)$$

for the ground state and

$$\psi_1=r\exp(-i\phi)\exp(-r^2/a^2)\exp(-z^2/b^2)\quad(3)$$

for the first excited state, ϕ measuring an angle around the magnetic field. Following earlier work, a^2 was taken equal to $4\hbar c/eB$, related to the classical magnetic radius, while b was varied to minimize the energy, and details are given by Durkan and March.[4]

The main conclusion is that while the individual levels are appreciably changed by screening, the energy difference between the ground state and the first excited state is quite insensitive to the detailed choice of n_0 over a wide range of densities. This energy gap (essentially the hydrogenic result) is shown in Fig. 1, and the magnitude of the observed activation energy accords semiquantitatively with this gap, over a substantial range of magnetic fields. We wish to stress that with the less-pure specimens studied by Sladek,[5] we expect the screening to be important and this is borne out by the experiments which give activation energies of magnitudes similar to those shown in Fig. 1, but at much higher magnetic fields. The threshold magnetic field is also much higher, and we return to this question in Sec. III.

The model then on which the Hall-effect measurements can be understood is one in which the impurity states higher than the ground state are broadened by overlap of the donor wave functions and eventually merge into a quasicontinuum with the InSb conduction band.

III. THRESHOLD MAGNETIC FIELD AND WIGNER TRANSITION

We wish now to comment on the interpretation of the threshold magnetic field required for an activation energy to be observed.

The first possibility suggested is that the carrier screening discussed in Sec. II is sufficient to suppress all bound states in the screened potential around a donor. This situation was discussed recently, with a rather less realistic screened potential than (1) by Fenton and Haering.[6] However, the criterion of Fenton and Haering needs some modification for, as the energy gap becomes very small, the donor wave functions become exceedingly diffuse and a great deal of overlap will occur. Thus, the impurity level has a bandwidth E_b say, and the criterion for conduction without an activation energy is that the energy gap E_g is $\sim\frac{1}{2}E_b$, rather than $E_g=0$. The criterion of Fenton and Haering for such conduction is rather too stringent.

However, in the case of the rather pure specimens considered in the present paper, the isolated donor levels do not appear to be greatly affected by screening. Our interpretation of the results of Putley[2] is that, below the threshold magnetic field, conduction is taking place in the lowest impurity band. Then, as the overlap of the donor wave functions is decreased by increasing magnetic field, we expect a transition to a nonconducting state, akin to the crystallization of electrons in a uniform background suggested long ago by Wigner,[7] as the electron density is lowered. This transition, to be distinguished from the Mott transition, can occur with much less than one electron/site, which is the case with the heavily compensated specimens

[4] J. Durkan and N. H. March, Proc. Phys. Soc. (London) **1**, 1118 (1968).

[5] R. J. Sladek, J. Phys. Chem. Solids **5**, 157 (1958); **8**, 515 (1959).

[6] F. W. Fenton and R. R. Haering, Phys. Rev. **159**, 593 (1967). Similar considerations due to one of us (N.H.M.) have been briefly referred to by Putley (see Ref. 2).

[7] E. P. Wigner, Phys. Rev. **46**, 1002 (1934); Trans. Faraday Soc. **34**, 678 (1938).

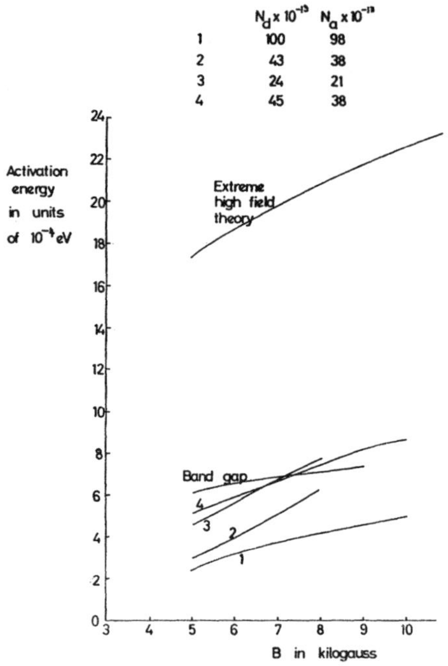

Fig. 1. Sample characteristics for specimens 1–4 are shown, along with the observed activation energies. The curve labeled "band gap" shows the separation between the ground and first excited isolated donor levels. The curve labeled "extreme high field theory" is the ionization energy of the hydrogenic impurity center as calculated by Yafet, Keyes, and Adams (Ref. 1).

considered here.[8] The long-range Coulomb interactions in such a low-density system of electrons are hardly screened and a primitive theory, extending that proposed by Wigner, would go as follows. Let the electron concentration be n, and define r_0 by

$$4\pi r_0^3/3 = 1/n. \tag{4}$$

According to Lindemann's criterion, used in his melting law, the transition will occur when the zero-point displacement is a fixed fraction of the inter-electronic spacing. If we assume, at high fields, that it is the magnetic field rather than the Coulomb repulsion which localizes the electrons, then we find

$$r_0^2 \sim a^2, \tag{5}$$

where, as we saw earlier, $a^2 \sim hc/eB$. This gives threshold fields of the order observed by both Putley[2] and Sladek,[5] but the approach has the drawback that it does not depend on the donor concentration N_D.

A second argument is that the Wigner transition occurs when the potential energy per electron is roughly equal to the kinetic energy associated with electron localization. This kinetic energy should reflect the width of the impurity band. Although it is difficult to estimate this for random impurities, it should be roughly proportional to N_D if overlap is appreciable. Suppose that the donor concentration which makes the bandwidth equal to the binding energy is N_0; then the criterion becomes

$$r_0 \sim N_D a_0/N_0, \tag{6}$$

where a_0 is the effective Bohr radius of the hydrogenic impurity center. A plausible value for N_0 is given by

$$1/N_0 \sim 4\pi a_0^3/3. \tag{7}$$

However, in a magnetic field, one should replace a_0^3 by $b(hc/eB)$, where b is the extension of the orbit along the field direction, which also decreases with increasing B. In fact, $hc/eB \sim 10^{-8}\text{cm}^2$ for $B \sim 10$ kG and the critical fields observed by both Putley and Sladek fit in roughly if we take

$$r_0 = 3N_D a_0/N_0. \tag{8}$$

The two estimates of the threshold fields are not very different in the present problem.

In spite of the disorder of the donors, an activation energy appears rather suddenly at a fairly well-defined threshold field according to Putley's measurements, and this seems to support our hypothesis that we are seeing here an example of a Wigner transition.

Finally, it is of interest to consider what will happen in the present model as the magnetic field is greatly increased beyond the range shown in Fig. 1. As the broadened first excited state is narrowed by the localization of the wave function (3) in the magnetic field, we expect an activation energy characteristic of the next excited state (or group of states) to be observed and so on. Thus, the Yafet–Keyes–Adams curve may be expected to become a rather better approximation to the observed activation energy in the extreme high-field limit, though, because of bunching of impurity levels below the bottom of the InSb conduction band, we do not expect the ionization energy ever quite to be reached in practice.

Discussion of March's Paper

R. W. Keyes (I.B.M.): I didn't understand why it was that you excluded the possibility that the difference between the observed ionization energy and the one Yafet, Adams, and I calculated isn't just due to screening.

N. H. March: No, there didn't seem to be a possibility of enough screening even if we said that the screening was due to all the available electrons. In the specimens which I talked about first, that never altered the ionization energy substantially. But now in the specimens of Dr. Sladek, where there are many more carriers, then of course, as you saw, the curves for the isolated levels began to decay away. Certainly there, the small activation energy shown in Fig. 1 for the impure specimens would be an ionization energy greatly reduced by screening. And so I think there are two mechanisms and I think that even for the threshold fields there could be two mechanisms. One I suggested was due to electron localization in the lowest impurity band, and the other one actually losing the bound state into the band by screening, but these are appropriate for very different conduction electron densities. The latter one is discussed in the recent paper by Fenton and Haering, as referred to in the

[8] No doubt, the random fields of the acceptor centers will eventually have to be considered carefully in a definitive theory.

text. However, these authors did not refer to the very low-density case where the Wigner transition takes place.

H. BROOKS (Harvard University): The real point is that you were dealing experimentally with very highly compensated samples so that the screening was minimized. I should point out one effect which would alter the screening although I don't think it will alter your explanation. That is that some years ago I showed that if you take into account in highly compensated samples the presence of both the donors and the acceptors and the fact that some of them can be populated and so on, there is an additional contribution to the screening due to the statistical population of the donors around another donor, so to speak, but this can double the screening but it can't change it in order of magnitude.

L. J. NEURINGER (Massachusetts Institute of Technology): Magnetic freeze-out *has* been studied in the high-field region to 200 kG. In a recent paper [Phys. Rev. Letters **18**, 773 (1967)] by Hanamura, Beckman, and myself, we found that both the magnitude and magnetic-field dependence ($\epsilon_i \sim H^{1/3}$) of the ionization energy, determined from Hall-coefficient measurements on uncompensated, heavily doped specimens ($n \sim 10^{16}$ cm^{-3}), obeyed the Yafet, Keyes, and Adams theory. It would appear that our results have serious consequences for your theory with regard to the importance of screening and with respect to the magnetic-field dependence of the ionization energy which your theory would predict at high magnetic fields. It also appears that in deducing the ionization energy from the Hall coefficient data you have neglected the fact that there is present two-band conduction. With regard to the threshold magnetic field for freeze-out H_0, we found good agreement with experiment by simply equating the volume occupied by the electronic wave function in this high magnetic field to the volume occupied by a single impurity, as a result $H_0 \sim N_{\text{imp}}^{6/7}$. I would venture to say that magnetic freeze-out can best be studied at high magnetic fields using heavily doped, uncompensated samples because (a) one is free of the complications introduced by two-band conduction, and (b) the fluctuation in the electric field at the various donor sites in the crystal, produced by the compensating acceptors, does not play a role as it does in the compensated samples.

N. H. MARCH: Well, I regret of course that I did not know about those results. They seem to agree satisfactorily with our prediction that the Yafet–Keyes–Adams ionization energy should be almost regained in very high fields.

Electrical conduction in the Wigner lattice in n type InSb in a magnetic field

Abstract. It is argued that in highly compensated n type InSb in a magnetic field exceeding a critical value, both electron correlations and disorder are involved in an essential way.

Wigner crystallization is then anticipated, and the electrical conductivity measurements of Somerford are interpreted as due to electrons diffusing via defects in the Wigner electronic lattice, and in particular via vacant electronic sites.

In view of the theoretical interest of such Wigner crystallization, further experiments are proposed on n type InSb and other materials (though higher magnetic fields are then needed), including ac transport measurements and Bragg scattering of neutrons.

Introduction

Earlier theoretical studies (Durkan and March 1968, Durkan *et al.* 1969) of highly compensated n type InSb in a magnetic field have concerned themselves with interpreting Hall measurements, particularly those of Putley (1960, 1961, 1966). Recently Somerford (1971) in this department made measurements of the conductivity at ^{3}He temperatures.

Briefly, the interpretation of the Hall effect is that until the critical magnetic field is reached, conduction is occurring in the lowest impurity band. Beyond the critical field, overlap in this impurity band has become so small, due to the shrinking of the donor orbits in the magnetic field, that the Hall conduction is taking place by activation of carriers to a higher energy band.

Two striking features of the experimental results are:
(i) the 'metal'–insulator transition at some critical magnetic field
(ii) the lower activation energy for conduction than for the Hall effect.

Arguments are presented here which lead us to conclude that (i) is substantially a many body transition in the lowest impurity band from an electron fluid to a 'Wigner electron crystal'. The observation (ii) is explained as due to electrons diffusing in this lattice, such motion giving a negligible contribution to the dc Hall effect, which therefore is dominated by the band mechanism referred to above.

Mechanism for localization

There is disorder present in this system and the possibility of electron localization due to such disorder clearly exists as we reduce the overlap between sites (compare Anderson 1958) by increasing the applied magnetic field. But such a transition to localization could not explain Somerford's experimental results, for, as Mott (1969) has shown, the activation energy for conduction would tend to zero as $T \to 0$ on this model.

In order to explain the relatively sharp metal–insulator transition, as well as to understand the nonzero activation energy of Somerford, we must therefore invoke, in addition to the disorder, the interelectronic forces. To show that the Coulomb correlations are indeed important in the present low electron density system, we have made approximate calculations of criteria for the onset of (*a*) localization due to disorder and (*b*) Wigner condensation; in the absence of a practicable approach to the real situation in which they exist together. To consider (*a*) and (*b*), we have assumed that the donors form a lattice and we have considered only the random fields of the acceptor centres, for the work of Herbert and Jones (1971) strongly indicates that the inclusion of structural disorder will alter the conclusions only quantitatively. We take the form of the probability distribution of the energies at a donor site to be that given by Mott and Twose (1961), which has a width $e^2/\epsilon r_a$ where r_a is the mean separation of the acceptors given by $(3/4\pi N_a)^{1/3}$, N_a being the acceptor density and ϵ the dielectric constant. Arguments paralleling those of Anderson then lead to the criterion

$$\frac{e^2}{\epsilon r_d} > c_D V \qquad \text{(disorder localization)} \tag{1}$$

where the donors with mean separation r_d interact through terms V_{ij}, these being matrix elements which we take to have the form $V_{ij} = V$ when i and j are nearest neighbours and zero otherwise. Similarly the Wigner transition, which takes place when the potential energy of each electron exceeds some multiple of its kinetic energy leads to

$$\frac{e^2}{\epsilon r_d} > c_W V \qquad \text{(Wigner localization)} \tag{2}$$

We remark at this point that, in the criteria (1) and (2), the effect of increasing the magnetic field is to reduce V.

The actual values of c_D and c_W remain somewhat uncertain, but we want to emphasize that $c_D \propto K^{-1/3}$ while c_W turns out to be proportional to $(1 - K)^{2/3}$ where K is the compensation ratio N_a/N_d. It is the high compensation $K \gtrsim 0{\cdot}8$ which is one essential feature causing the Wigner transition to be important in the present system, relative to a one electron transition induced by disorder. With admitted uncertainty in the numerical estimates, we find c_D and c_W to be of the same order of magnitude in the highly compensated specimens used by Somerford and this confirms that both disorder and Coulomb interaction are essential features in this regime.

Low density electron system in magnetic field

Thus, while we must clearly recognize the disorder in the donor and acceptor sites, we conclude that the experimental observations (i) and (ii) involve interelectronic correlation which we anticipate must lead to Wigner condensation. For this reason we now discuss the formation of such an electron crystal, ignoring the disorder in the zeroth order approximation.

Following Carr (1961), the Hamiltonian $T + V$ may be written, including a vector potential A and working in the effective mass approximation:

$$T = \sum_i \left(p_i + \frac{e}{c} A_i\right)^2 \Big/ 2m^* \tag{3}$$

and

$$V = \frac{1}{\epsilon}\left(\frac{e^2\rho^2}{2}\iint \frac{\mathrm{d}r\mathrm{d}r'}{|r - r'|} + \frac{e^2}{2}\sum\sum_{i\neq j}\frac{1}{|r_i - r_j|} - e^2\rho\sum_i \int \frac{\mathrm{d}r}{|r - r_i|}\right) \tag{4}$$

with $\rho = 3/4\pi r_s^3$, the sums being taken over $N_d - N_a$ electrons. Using arguments similar to those of Wigner, the ground state above a certain critical magnetic field has the form of a regular electron lattice, in which the electrons oscillate about their lattice sites, in cigar shaped orbitals elongated along the field direction. Taking the effective impurity Rydberg as the unit of energy, the harmonic oscillator Hamiltonian of Carr, namely

$$H = -\sum_i \nabla_i^2 + \sum_i \sum_j C(n_i - n_j)\, u_i u_j \tag{5}$$

is regained. The tensor $C(n_i - n_j)$, with n_i a lattice vector, is identical to Carr's except that

$$C_{xx}(0) = C_{yy}(0) = \frac{1}{r_s^3} + \frac{\gamma^2}{4} \tag{6}$$

where γ is the ratio of the zero point energy of a free carrier in the magnetic field to the impurity Rydberg. Typically, the effective Bohr radius is 6×10^{-6} cm and $N_d - N_a \sim 10^{13}$ cm^{-3}, yielding $r_s \sim 4$. Thus from equation (6) we expect the magnetic field to become important in $C_{xx}(0)$ and $C_{yy}(0)$ when $\gamma \gtrsim 1$, in agreement with the observed critical fields. In writing equation (6) we have assumed that the ground state orbitals have zero angular momentum. The form of the terms $C_{xx}(0)$ and $C_{yy}(0)$ confirms that cigar shaped orbitals elongated parallel to the magnetic field would indeed occur in this system.

Conduction in Wigner electron crystal at nonzero temperature

We must next discuss the basic conduction process in the Wigner electron crystal. A full treatment will require a quantum mechanical calculation of the current-carrying excited states and this is being undertaken. Our model for these states will be motivated by analogy with atomic diffusion in crystals. Thus, we describe these states in terms of an electron vacancy in the Wigner crystal: evidently such states will be separated from the ground state by an energy gap. Other excited states will be those in which the Wigner crystal structure itself changes somewhat from the ground state structure but these do not seem to afford an explanation of the observed conduction.

A major contribution to the activation energy for diffusion via such electron vacancies can then be ascribed to the vacancy formation energy. A model for estimating this can be set up by assuming that the electron crystal does not relax on forming the vacancy though such relaxation will no doubt be important in a quantitative theory. An estimate of the formation energy U can then be made and it is found that there are two terms U_1 and U_2, the first being a field independent quantity which is of the order of the Madelung energy $e^2/\epsilon r_e$ with r_e the mean interelectronic spacing. The second is magnetic field dependent and can be estimated at high fields by a Taylor expansion (see Carr's Appendix 1) and to first order the field dependence is given by the square of the displacement of an electron from its lattice site. This displacement is of the order of the classical magnetic radius $(e\hbar/cB)^{1/2}$, and hence the field dependence of U_2 is proportional to $|B|^{-1}$, which appears qualitatively to accord with Somerford's measurements. At this point we should remark that the conductivity results of Somerford show some relatively small anisotropy. This indicates to us that a picture of two dimensional localization in a plane perpendicular to the magnetic field is an oversimplification and that a genuine three dimensional structure is involved.

The cigar shaped orbitals on the lattice sites point, of course, along the magnetic field, but the orientation of the Wigner crystal is presumably determined by the impurity background, and it seems possible in this way to avoid a very anisotropic conductivity.

The model of conduction therefore can be described in elementary physical terms as the diffusion of electrons via vacancies in the Wigner (strained) crystal. A vacancy, which we take here to be localized, would not be of this form in 'jellium', where the hole would have a wavevector $\boldsymbol{k}$. It is the background of acceptors and donors that removes the degeneracy of the jellium model. As discussed above, a major term in the activation energy Q for diffusion, and hence via the Einstein relation, for the electrical conductivity, is the electron vacancy formation energy. If we write the usual diffusion relation

$$D = D_0 \exp(-Q/k_B T) \tag{7}$$

and take Q to be the activation energy measured by Somerford, we can make an estimate from his conductivity measurements of $D_0 \sim \nu r_e^2$ where ν is a characteristic frequency for electron–vacancy interchange. The frequency ν is then found to be $\sim 10^6\ s^{-1}$ which is much lower than a characteristic frequency (Debye frequency) of the Wigner crystal. However, we expect the overlap of wavefunctions on adjacent sites of the Wigner crystal to enter, with overlap integral S very much less than unity. Since S depends exponentially on the ratio of the classical magnetic radius to the spacing of electrons in the Wigner crystal, we expect a small value of ν on this model.

As remarked above, we are currently engaged in working out a detailed quantitative description motivated by this classical analogy. However, Somerford's activation energy for conduction is indeed showing a tendency to saturate with increasing magnetic field, at a value around the Madelung energy $e^2/\epsilon r_e$, and this prediction of the present model, that the high field saturation value of the activation energy should depend only on the electron density, is accessible to experimental test. Secondly, it is clear from our model that disorder effects should become more important relative to electron–electron correlations as the degree of compensation is reduced. However, there will be competition with carrier screening effects in this case.

In addition, it would be of considerable interest in attempting to confirm the existence of the Wigner crystallization in this material (other materials should be examined also, though they require higher fields) if neutron experiments could eventually be carried out to study (i) whether non-current-carrying low-lying excited states of the Wigner crystal, which are known to have phonon character, are observable and (ii) whether long-range magnetic order associated with the Wigner condensation could be detected. Because of the very low electron concentration, such neutron experiments, though involving measurement of Bragg scattering, will no doubt be very difficult at the present time. Further transport measurements are therefore clearly called for, particularly the ac conductivity at ^{3}He temperatures in a frequency range where the applied frequency passes through the frequency ν for electron–vacancy interchange.

Numerous colleagues have given us the benefit of their comments on this work and we are particularly grateful to Dr I G Austin, Dr R J Elliott, Dr D C Herbert, Dr R D Lowde and Dr D Somerford for their criticism. One of (CMC) wishes to acknowledge the award of a Postgraduate Studentship by the Science Research Council.

Department of Physics,
The University,
Sheffield

C. M. CARE
N. H. MARCH
18*th November* 1971

ANDERSON, P. W., 1958, *Phys. Rev.*, **109**, 1492–505.
CARR, W. J., 1961, *Phys. Rev.*, **122**, 1437–46.
DURKAN, J., and March, N. H., 1968, *J. Phys.* C: *Solid St. Phys.*, **1**, 1118–27.

DURKAN, J., ELLIOTT, R. J., and MARCH, N. H., 1968, *Rev. mod. Phys.*, **40**, 812–5.
HERBERT, D. C., and JONES, R., 1971, *J. Phys.* C: *Solid St. Phys.*, **4**, 1145–61.
MOTT, N. F., 1969, *Phil. Mag.*, **19**, 835.
MOTT, N. F., and TWOSE, W. D., 1961, *Adv. Phys.*, **10** 107–70.
PUTLEY, E. H., 1960, *Proc. Phys. Soc.*, **76**, 802–5.
—— 1961, *J. Phys. Chem. Solids*, **22**, 241–7.
—— 1966, *Semiconductors and Semi-Metals*, vol 1, eds R. K. Willardson and A. C. Beer (New York: Academic Press).
SOMERFORD, D. J., 1971, *J. Phys.* C: *Solid St. Phys.*, **4**, 1570–5.

[101]

Electron crystallization

By C. M. Care and N. H. March
Department of Physics, Imperial College,
South Kensington, London, S.W.7

[Received 20 August 1974]

Abstract

The arguments leading to the concept of an electron crystal in the low density regime of 'jellium' are reviewed. Ground-state properties discussed include estimates of the critical density at which the transition to a crystalline state takes place, and then in the low density regime the dielectric function, pair function and momentum distribution are dealt with. The magnetic character of the ground state as a function of density is also considered.

The low-lying excitations of the Wigner electron crystal are phonon-like and hence the low temperature specific heat obeys a T^3 law. Defect models are considered in order to throw further light on the character of excited states.

Finally experimental conditions favourable for electron crystallization are briefly considered.

CONTENTS

§ 1. Introduction

We consider a system of degenerate interacting electrons moving in a uniform background of neutralizing positive charge (jellium model). It is now generally accepted that the electronic assembly will crystallize out at sufficiently low densities. This idea is due to Wigner (1934, 1938) and the resulting electron lattice structure is termed the Wigner lattice.

The arguments for lattice formation, and the conditions under which it is expected to occur, are elaborated on below, in various directions. However, qualitatively the picture is simply that the high density situation, in which a well-defined spherical Fermi surface exists, becomes unstable because at sufficiently low density the potential energy/electron, A/r_s, r_s being the mean interelectronic distance, outweighs the Fermi kinetic energy/particle. This latter quantity scales as $1/r_s^2$ and hence its dominance at high densities.

Since, as remarked above, the state with a well-defined Fermi surface becomes unstable with respect to a state where electron localization occurs, a metal–non-metal transition will occur. For the purposes of this review, which is restricted almost entirely to the Wigner transition, we classify the four metal-insulator transitions most usually discussed as follows:

(i) The Wigner transition. As remarked, this occurs due to long-range Coulombic repulsions between electrons, in a uniform neutralizing background.

(ii) The Mott transition. This occurs in an array of atoms with one electron per atom, as the spacing between the atoms becomes sufficiently large.

(iii) The Anderson transition. A purely one-electron transition, in sharp contrast to the other three; this occurs due to electron localization induced by disorder.

(iv) The Hubbard transition. This occurs in narrow bands, due to short-range interactions between electrons of opposite spin: the crucial interaction taking place on a single site.

Most of this short article is concerned with summarizing the knowledge of the ground and the excited states that we now possess, on this jellium model, in the low density limit. However, in § 5 we briefly summarize the conditions under which it seems possible (at least in principle) to observe Wigner crystallization. Though circumstantial evidence exists in at least two areas for electron crystallization, there is probably as yet no final answer to the question raised more than forty years ago: " Does the Wigner lattice exist? "

§ 2. Ground-state properties

We turn then to discuss the ground-state properties, dealing first with the total energy, and then with space and momentum properties: namely, pair function, momentum density and dielectric constant.

2.1. *Energy*

The work of Wigner was extended by Carr (1961) who calculated a more accurate value for the energy per particle of the Wigner lattice by taking proper account of the lattice vibrations. He also considered the specific heat of the lattice and the problem of its magnetic properties, which we deal with in later sections.

Carr first showed that for sufficiently large r_s the effect of the electron exchange only contributed a term falling off exponentially with $r_s^{1/2}$ and thus that the electrons could be considered as being localized at specific lattice sites. He then improved Wigner's estimate of the contribution of the lattice vibrations by calculating the normal modes and finding their zero-point energy.

Coldwell-Horsfall and Maradudin (1960) independently estimated the zero-point energy of the electrons using a moment trace method (Montroll 1942). Subsequently the anharmonic contribution was calculated (Carr *et al.* 1961) and the ground-state energy per particle was expressed in the form

$$\frac{E}{N}=-\frac{1{\cdot}792}{r_s}+\frac{2{\cdot}66}{r_s^{3/2}}-\frac{0{\cdot}73}{r_s^{2}}+0(r_s^{-5/2})+E_{\rm exp} \tag{2.1}$$

where Carr (1961) quotes $E_{\rm exp}$ as

$$E_{\rm exp}=\left(\frac{21}{r_s}-\frac{4{\cdot}8}{r_s^{3/4}}-\frac{1{\cdot}16}{r_s^{5/4}}\right)\exp\left(-2{\cdot}06r_s^{1/2}\right)-\left(\frac{2{\cdot}06}{r_s^{5/4}}-\frac{0{\cdot}66}{r_s^{7/4}}\right)\exp\left(-1{\cdot}55r_s^{1/2}\right). \tag{2.2}$$

In eqn. (2.1) the first term is the Madelung energy of the appropriate b.c.c. lattice which was calculated by Fuchs (1935). The second term is the best estimate of the vibrational energy of the electrons. It is interesting to note that this is not much different from Wigner's estimate using the Einstein model for the lattice vibrations. The third term arises from the anharmonic contribution to the ground-state energy. The last term is due to exchange and is shown in more detail in eqn. (2.2). The above expressions represent the first terms in an expansion of the energy per particle in powers of $r_s^{-1/2}$. Actual numerical results for the energy per particle as estimated by Carr (1961) are shown in fig. 1. We note in passing that the frequency spectrum for the lattice was calculated by Clark (1958) (fig. 2) and, further, that there is a sum rule characteristic of the Coulomb lattice due to Kohn (see Clark 1958):

$$\sum_{j=1}^{3}\omega_{\mathbf{q}j}^{2}=\frac{4\pi Ne^2}{m}=\omega_p^{2}, \tag{2.3}$$

Fig. 1

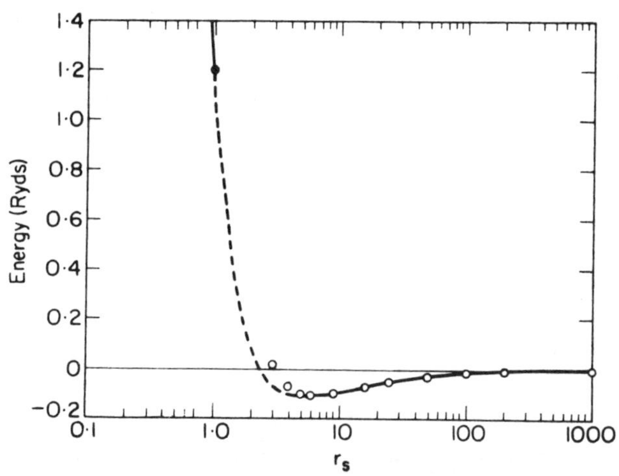

Total energy per particle versus r_s. ○ Result of eqn. (2.1) with neglect of anharmonic terms. ● Result of high density perturbation theory of Gell-Mann and Brueckner (1957). Dashed curve: constructed by Carr (1961) to connect end regions. (From Carr (1961).)

Fig. 2

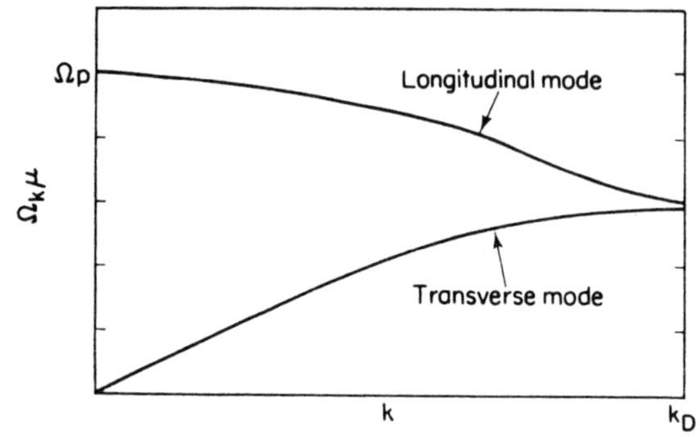

Dispersion curves for phonon modes in the Wigner lattice. Longitudinal mode in 100 direction.

where ω_p is the plasma frequency for the electrons, and ω_{qj} is the jth vibrational mode with frequency ω and wave vector $\mathbf{q}$. The relation (2.3) is true for all values of the vector $\mathbf{q}$, and is derived, for instance, in Pines (1963) and also by Brout (1959).

2.2. *Momentum Distribution and Pair Function*

To contrast with the Wigner lattice properties, let us first consider the probability $P(k)$ of occupation of a plane wave state k when the density is very high. Clearly the answer is the usual Fermi distribution

$$\begin{aligned} P(\mathbf{k}) &= 1 \qquad k < k_f \\ &\quad 0 \qquad k > k_f \end{aligned} \tag{2.4}$$

k_f being the Fermi momentum. Daniel and Vosko (1960) have discussed, using high density perturbation theory, the development of this momentum

Fig. 3

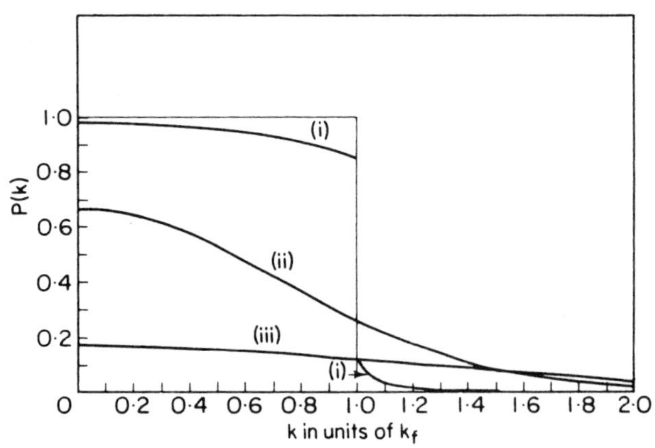

Momentum distribution function $P(k)$ versus k/k_f, k_f being the Fermi momentum. Curve i: Fermi distribution, as modified by effect of 'switching on' interactions at high density. Curve ii: Momentum distribution following electron crystallization. Curve iii: Same as (ii), except that density is much lower.

distribution as the density is decreased, or the interactions are ‘ switched on ’. The discontinuity at k_f is diminished, but remains, as shown in fig. 3.

Now we turn to the strong coupling or Wigner limit. In the most elementary terms, suppose the coupling is so strong that electrons vibrate but little about their (assumed body-centred-cubic) lattice sites. The field in which they vibrate can be estimated by constructing a Wigner–Seitz cell around one lattice site, arguing that it can be replaced, due to its high symmetry, by a sphere. Then, the only appreciable contribution to the potential acting on an electron is that due to the positive background charge within the sphere. Other cells, because of their high symmetry, contribute only multipole terms to the potential in the cell under discussion. Thus the potential energy in which the electron moves is simply

$$V(\mathbf{r}) = -\frac{e^2 r^2}{2 r_s{}^3} + \text{Const.}, \tag{2.5}$$

and the ground-state wave function is that of a three-dimensional isotropic harmonic oscillator, given by

$$\psi = \left(\frac{\alpha}{\pi}\right)^{3/4} \exp -\tfrac{1}{2}\alpha r^2, \tag{2.6}$$

where

$$\alpha = r_s{}^{-3/2}.$$

Following March and Sampanthar (1962), the momentum distribution is readily obtained from the Fourier transform of this Wigner orbital as

$$P(\mathbf{K}) = \frac{3\pi^{1/2}}{r_s{}^{3/4}} \exp\left\{-\left(\frac{9\pi}{4}\right)^{2/3} r_s{}^{-1/2} K^2\right\}, \tag{2.7}$$

where $K = k/k_f$. The range of validity of this result is restricted because the Wigner orbitals have been assumed orthogonal. A necessary, though not sufficient condition, for this result for $P(K)$ to be valid is that

$$3\pi^{1/2}/r_s{}^{3/4} < 1 \qquad \text{or} \qquad r_s > 9{\cdot}3 a_0. \tag{2.8}$$

The discontinuity characteristic of weakly interacting Fermions is seen to disappear in the strong coupling limit.

Turning to the electronic pair function $g(r)$, the usual Fermi hole result

$$g(r) = 1 - \frac{9}{2}\left\{\frac{j_1(k_f r)}{k_f r}\right\}^2, \tag{2.9}$$

where

$$j_1(\rho) = (\sin \rho - \rho \cos \rho)/\rho^2$$

is profoundly changed at low densities. Using the same Wigner orbitals (2.6), the pair function has the form shown in fig. 4. This, of course, reflects the lattice behaviour, the finite width of the peaks being due to the electrons oscillating about the Wigner lattice sites (March and Young, 1959).

Fig. 4

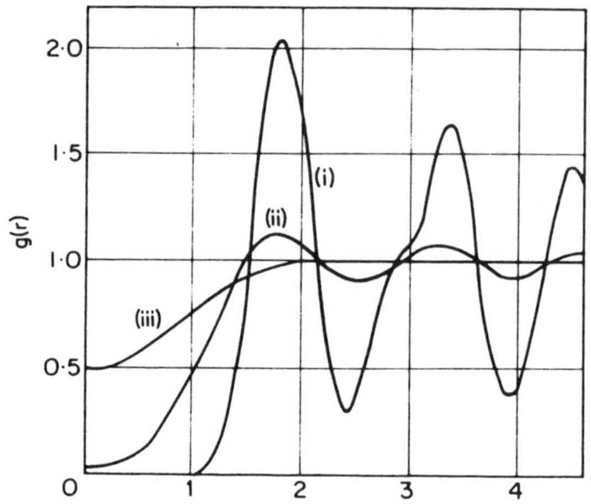

Pair functions $g(r)$ versus r/r_s, for different densities. Curve i: Wigner form for $r_s = 100$. Curve ii : Wigner form for $r_s = 10$. Curve iii : Fermi hole.

2.3. *Dielectric properties*

The space-averaged longitudinal dielectric constant $\epsilon(\mathbf{k},\omega)$ was derived by Bagchi (1969). Using classical arguments he showed that

$$\frac{1}{\epsilon(\mathbf{k},\omega)} = 1 - \frac{\omega_p{}^2}{k^2} \sum_\lambda \frac{(\mathbf{k} \cdot \hat{\epsilon}_{k\lambda})^2}{[\omega^2(\mathbf{k},\lambda) - \omega^2]}, \tag{2.10}$$

where $\hat{\epsilon}_{\mathbf{k}\lambda}$ is an eigenvector associated with the wave vector $\mathbf{k}$ and mode λ. As predicted by de Wette (1964) the long wavelength static dielectric constant is seen to be negative and very large. We note that this result is independent of whether $k/\omega \to 0$ or $\omega/k \to 0$ as k and ω tend to zero. The long wavelength ($k = 0$) dielectric constant is shown as a function of frequency in fig. 5. Bagchi

Fig. 5

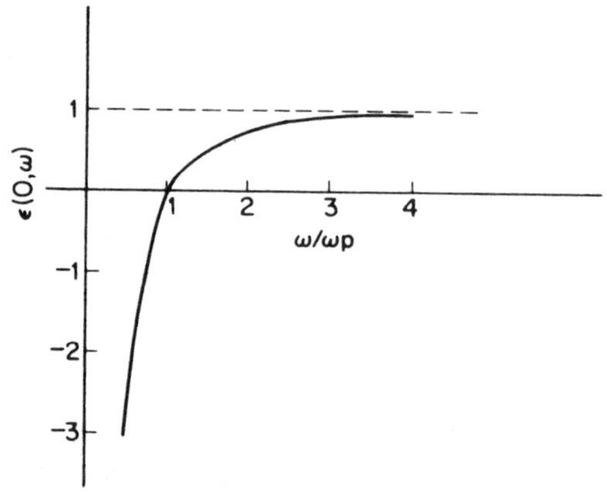

Long-wavelength limit ($k = 0$) of dielectric constant of Wigner lattice as function of frequency ω. (From Bagchi (1969).)

Estimates of the critical density for the Wigner transition ($T=0^{\circ}$K)
a = Critical value of r_s

Author	Critical value of r_s	Method
Wigner (1938)	$6 \lesssim a \lesssim 10$	Comparison of energy with other states
Nozières and Pines (1958)	$a=20$ $(\delta=\frac{1}{4})$	Lindemann criterion using longitudinal plasmon modes only $\delta=\langle u^2\rangle_{av}/R_0$
Coldwell-Horsfall and Maradudin (1960)	$\delta=\frac{1}{2}\Rightarrow a=6{\cdot}5$ $\delta=\frac{1}{4}\Rightarrow a=104$	Lindemann melting criterion
Carr (1961)	$a\approx 5$ $r_s<270$ Antiferromagnetic $r_s>270$ Ferromagnetic	Comparison of energy with other states
Mott (1961)	$a\approx 20$	Kinetic energy and potential energy of same order of magnitude
de Wette (1964)	$100>a>47$	Existence of bound states
Misawa (1965)	Ferromagnetic instability in electron gas for $(m^*/m)r_s$ between 7 and 10	Analytic continuation of high density results
van Horn (1967)	$a=27$	Improvement of de Wette's arguments
Piettrass (1967 b)	$a\gtrsim 5$	Soft transverse modes due to including anharmonic effects in a self-consistent field approximation
Edwards and Hillel (1968)	$a\approx 10$ $10\lesssim r_s\lesssim 40$ Antiferromagnetic $r_s>40$ Ferromagnetic	Comparison of energies of various states by variational calculation
Rajagopal and Mahenti (1967)	$r_s>9{\cdot}4$ Electron gas is ferromagnetic	Using 'plasma cut-off' screening
Osaka (1967)	$r_s\geqslant 19{\cdot}0$ Electron gas is ferromagnetic	RPA screening
Carmi (1968)	$r_s=7{\cdot}6$ Phase transition in electron gas	Behaviour of free energy
Kugler (1969)	(1) $a=21{\cdot}9$ (2) $a\simeq 700$	Inclusion of anharmonic effects in RHA. Estimate (1) from absence of RHA solutions. Estimate (2) includes higher-order anharmonic terms.

Table—continued

Author	Critical value of r_s	Method
Wiser and Cohen (1969)	$r_s \gtrsim 5$ Mechanical instability in electron gas	Use best estimate for correlation energy. Transition is to low density gas *not* Wigner lattice
Isihara (1972)	$a = 14{\cdot}4$	Padé approximant interpolation of ground-state energy

explains the anomalous behaviour of the dielectric constant for small k and ω by showing that the Wigner lattice is a limiting case of a classical Lorentz lattice. He also makes the interesting speculation that the negative dielectric constant in certain regions of k and ω could give rise to an effective attractive force between two conduction electrons placed in the lattice. This would lead to the possibility of superconductivity.

§ 3. The Wigner transition

3.1. *Estimates of the critical density*

We now consider the stability of the electron lattice and in particular estimates of the critical r_s at which the 'Wigner transition' is expected to occur at zero degrees Kelvin.

In the table we summarize the many different estimates for the critical value a of the density parameter r_s at which the electron gas will condense into the ordered Wigner lattice. We also include estimates of the values of r_s at which magnetic and mechanical instabilities occur. Carr (1961) and Wigner estimate a, by comparing the energy of the lattice state with that of the high density system. The most common method of estimating a is by using the Lindemann melting criterion which states that melting will occur when the mean square electron displacement $\langle u^2\rangle_{av}$ is some fraction δ of the intersite separation R_0 (cf. footnote on page 157 of Kugler 1969). Using this method Nozières and Pines (1958) found a to be about 20. However, the method is criticized by Coldwell-Horsfall and Maradudin on the grounds that a is very dependent upon the choice of δ. Kugler (1969) extends this criticism by showing that the result also depends strongly upon the way in which $\langle u^2\rangle_{av}$ is determined. He calculates the value of a by locating the point at which his 'renormalized harmonic approximation' (RHA) breaks down. After including higher-order anharmonic effects he finds a is approximately 700. De Wette estimated that a lay between 47 and 100 by establishing a criterion for the appearance of bound states in the potential experienced by each electron. This criterion was later improved by Van Horn (1967) to give an estimate of 27 for a.

Several authors, e.g. van Horn (1967), Wiser and Cohen (1969), have noted that the uniform background electron gas model is unstable above r_s about 5 to 7. Van Horn considers that stability can be maintained if the background is allowed to relax.

Wiser and Cohen assert that the low density phase of the electron system is a gas and the high density phase is a liquid. We also include in the table the estimates of Misawa (1965) and Rajogopal and Mahenti (1967) for the onset of a ferromagnetic instability in the electron gas, although this is not necessarily related to the onset of Wigner crystallization. We also note that de Llano and Tolmachev (1971) discuss the gas–liquid instability in a fermion system in terms of the breakdown of perturbation theory based on the plane wave Hartree–Fock solutions, although they give no estimate of the critical density. Foldy (1971) considers the phase transitions within the Wigner lattice from body-centred to face-centred cubic structure at various temperatures.

The arguments of Kugler strongly suggest that the harmonic approximation is not good for r_s less than 1000. The Lindemann criterion gives results which are very sensitive to the choice of δ and Bagchi argues that the Wigner lattice is on the limit of stability. For these reasons we feel it is just possible that the Wigner lattice may only be observed for r_s greater than a few hundred Bohr radii. Though we favour the smaller estimates, the whole subject of the transition density remains rather open.

However, even the largest estimates do not preclude the possibility of an intermediate ' liquid-like ' phase in which the electrons are highly correlated. These arguments are supported by those given at the end of § 4.

3.2. *Charge density waves*

The fact that a one-electron description with a total wave function which is a single Slater determinant may still be useful in the presence of strong Coulomb interactions was stressed by Overhauser (1968); see also Young (1960).

The idea is to allow a departure from plane wave orbitals by writing

$$\psi_{\mathbf{k}}(\mathbf{r}) = \exp(i\mathbf{k}\,.\,\mathbf{r}) + \lambda \exp(i(\mathbf{k}+\mathbf{K})\,.\,\mathbf{r}), \tag{3.1}$$

where $\mathbf{K}$ can be thought of, roughly speaking, as a reciprocal lattice vector of the Wigner lattice. Using λ as a variational parameter and calculating it as a function of r_s, it is found that λ is identically zero from $r_s = 0$ to a finite value r_c say, and subsequently λ begins to increase with r_s.

This is in the three-dimensional jellium model. In the more academic one-dimensional case (compare Kugler 1974, to be published), it is only necessary, at absolute zero, to demonstrate the existence of charge density waves at some chosen density. Then the theorem that phase transitions cannot occur in a one-dimensional system can be invoked to argue that such charge density waves must always be present†.

In the language used above, the function $\lambda(r_s)$ is non-zero for any non-zero r_s, even though $\lambda \to 0$ as $r_s \to 0$. This is, of course, in sharp contrast to the three-dimensional jellium model, where, as we saw, λ is identically zero out to a finite value of r_s.

3.3. *Magnetic ordering and spin density waves*

Several authors (Herring 1969, Carr 1961, Edwards and Hillel 1968) have discussed the magnetic properties of the Wigner lattice.

† We wish to acknowledge a stimulating discussion with Professor P. Choquard on this and other points relating to Wigner crystallization. Phase transitions occurring in the ground state, as the density is changed, may not be fully covered by the usual theorems however.

If one takes the Wigner lattice to have the body-centred cubic structure, then it is physically very plausible to assign upward spin electrons to one of the two interpenetrating simple cubic lattices that make up the b.c.c. lattice, and downward spins to the other. This would mean an antiferromagnetic electron crystal.

Unfortunately, it is quite difficult to decide, in the low density regime, whether there will be a range of r_s in which ferromagnetism will be favoured, but this seems likely for very large r_s, antiferromagnetic ordering being favoured for intermediate r_s.

Carr has estimated that the antiferromagnetic to ferromagnetic transition will occur at $r_s \sim 270$. However, Edwards and Hillel (1968) find that the Wigner lattice will be antiferromagnetic in the region $10 \lesssim r_s \lesssim 40$.

We note in passing that Shuster and Kazinskaya (1971) considered the possibility of an intermediate phase between the Wigner lattice and the electron gas in which more than one electron occupied each lattice site. They estimated the range of stability of such a system and the form of the expansion for the energy per particle.

§ 4. Excited states

4.1. *Phonons and plasmons*

All the work discussed above follows Wigner in that it is concerned with the ground state of the electron gas at low densities. In contrast to the high density case, the low-lying excitations of the Wigner lattice are the collective modes, the phonons of the lattice. We note that the plasmon modes now form an ' optical ' type of branch and the ' acoustic ' modes come from the transverse oscillations. This behaviour is a result of the rigid uniform background as explained by Clark (1958).

In the harmonic approximation it is possible to derive the specific heat and it is found to be of the Debye T^3 form for all normal temperatures, since the Debye temperature, θ, is very high. Thus Carr (1961) finds

$$\theta = 2{\cdot}45 \times 10^5 r_s^{-3/2} K. \tag{4.1}$$

The specific heat per particle is found to be

$$C_v = Akr_s^{9/2}[(kT)_{Ry}]^3, \tag{4.2}$$

where Carr finds to zeroth order that $A = 62$ and Coldwell-Horsfall and Maradudin estimate $A = 56{\cdot}21$. The large value of θ and low value of C are a consequence of the large velocity of sound in the Wigner lattice. The large velocity of sound arises because the electrons are only very loosely bound to the lattice sites (Bagchi 1969). We emphasize that the results (4.1) and (4.2) depend upon the validity of the harmonic approximation.

4.2. *Defect models*

It is generally assumed that the Wigner lattice is non-conducting at $T = 0°K$ in the sense of having a gap against single-particle excitations (Kohn 1967). We ignore the possibility that the lattice might move as a whole, and assume that it is ' pinned ' in some way. We now speculate on the nature of the excited states which would be current-carrying. It seems likely that these will be associated with defects in the lattice. One such possible excitation is the promotion of an electron from a lattice site to an interstitial position. The interstitial and associated vacancy are related to the excitation pair considered by de Wette in his discussion of lattice stability. For reasons of symmetry the 'interstitial' states and 'vacancy' states must be delocalized and form a narrow band. This is considered by van Horn (1967). The energy separating these bands will be due mainly to electrostatic interaction between the electrons and as a first approximation we may take de Wette's estimate for the band gap of $0{\cdot}2/r_s$ Ryds. At a finite temperature the concept of the lattice is only meaningful if there are few vacancies. Thus we should only expect the Wigner lattice to be a useful concept provided

$$(kT)_{\text{Ryd}} \lesssim \frac{1}{100}\left(\frac{0{\cdot}2}{r_s}\right), \quad r_s > a. \tag{4.3}$$

This is similar to the conclusion reached by Mott (1961). If kT exceeds the value in (4.3) we should anticipate that the long-range order of the Wigner lattice would break down, although the resulting 'fluid' would show extreme correlation. It is interesting to note that the major part of the Madelung energy in eqn. (2.1) may be derived simply by isolating each electron in a sphere of positive charge of radius r_s. Indeed, the energy associated with the long-range order is only $0{\cdot}45\%$ of the total Madelung energy (de Wette 1964).

§ 5. Experimental conditions favourable for electron crystallization

We isolate here a number of areas in which it seems possible to envisage electron crystallization occurring:

(i) Highly compensated semiconductors.
(ii) Inversion layers in silicon devices.
(iii) Magnetic insulating materials (e.g. magnetite).
(iv) Outside suitable metal surfaces.

In all these cases, of course, one has to recognize

(*a*) the granular density (as opposed to the uniform assumption in jellium) of the ions, and the effect of this on localization of electrons,
(*b*) other elements, such as disorder, in (i) and (ii) especially,
(*c*) the importance of assessing the effects of temperature, the Wigner lattice being very easy to disrupt by thermal agitation.

We shall briefly summarize below the evidence for and against electron crystallization in these four areas: there are doubtless some others, perhaps more definitive, that are going to come to light as interest in this field grows.

(i) *Doped semiconductors*

The point to be made here is that, for example, in a material like n-type InSb, which has been widely investigated because of its technological use in infra-red detectors (e.g. the Putley detector), one can control the free carrier density by compensating the material, and hence arrange say to have one electron per n donors, where n can easily be made ~ 100. Thus, in this way we can readily simulate a really low electron density system.

The second point that needs emphasizing is that the size of such ' impurity ' orbits, in a jellium model, is large; they spread over many lattice spacings. Hence, it seems likely that such electrons will not be crucially affected by the periodic lattice.

The disadvantages are

(i) The fluctuating electric field effects of the acceptors, which introduce an element of disorder into the system. One must *not* expect therefore that observations on this system will conform in fine detail with the picture which emerges for low density electrons in jellium.
(ii) The fact that, in the purest specimens of n-type InSb currently available, there is such a degree of overlap of the donor orbits that an impurity band is formed in which metallic-like conduction occurs.

However, from the work of Putley, and various Russian scientists, it is known that the application of modest magnetic fields can alter crucially the Hall conduction properties. These results were interpreted by Durkan *et al.* (1968) and Durkan and March (1970) in terms of a Wigner transition. Subsequent experiments on electrical conduction by Somerford (1971) are in general accord with such a description (Care and March 1971) and some high frequency experiments purport to reveal a frequency to be associated with the Wigner lattice (Biskupski *et al.* 1973).

Difficulties which may need full consideration are

(i) the very low melting temperature of such a lattice,
(ii) the treatment *together* of electron correlation and disorder-induced transitions (see Care and March 1971, Care 1974, to be published). The conclusions from such a discussion appear to be perhaps dependent on whether disorder is added to a Wigner picture, or whether correlation is taken account of in an Anderson model (Herbert and Jones 1971, see also Suprapto 1974).

However, this appears to be a fruitful area for further study, both experimentally and theoretically.

(ii) *Inversion layers*

Very beautiful experiments have been carried out at Cambridge (Pepper *et al.* 1974) on Si inversion layers, in MOS and MNOS devices.

Briefly, the MOS devices have led to a definite confirmation of the essential features of Anderson localization, due to disorder. But in the MNOS devices the disorder element is minimized. There seems some evidence for a Wigner-type transition, of a two-dimensional kind†, though some as yet unexplained effects (e.g. the increase in conductivity which takes place before the insulating phase takes over) occur.

Again, this looks at the present time a very promising area for much further work regarding electron crystallization.

(iii) *Magnetic insulating phases*

Here we can only be very brief and point out that proposals have been made for a Wigner transition in magnetite. Because of some uncertainty in the

† Chaplik (1972) appears to have first suggested that Wigner crystallization might occur in MOS transistors.

magnetic properties to be expected in an electron crystal, plus the fact that a Hubbard model can yield, with appeal only to short-range interactions, splitting of bands at an arbitrary electron atom ratio, it seems difficult in this system at present to isolate effects which depend *crucially* on the long-range Coulomb interactions (Sokoloff 1971).

(iv) *Region of conduction electron spill-out from metal surfaces*

The importance of the spill-out region of electron density outside a metal surface is becoming quite clear in a number of applications. By now, in an electron gas framework, density profiles exist which allow relatively successful calculation of a number of surface properties: work functions (Smith 1969), surface energies (Lang and Kohn 1970, Harris and Jones 1974), and, more tentatively, ion–ion interaction energies (Brown *et al.* 1974). In this latter paper it has been pointed out that in the tail of this distribution, where the electron density is really low, Coulomb repulsion effects might lead to a tendency to Wigner lattice formation (maybe two-dimensional). However, these electrons are not in a region where there is positive-ion compensation and the problem may be different in character from jellium.

Crandall and Williams (1972) and Crandall (1973) consider the possibility of a two-dimensional Wigner crystallization in the electrons trapped on the surface of liquid helium. The results of Sommer and Tanner (1971) support this hypothesis.

§ 6. Lattice formation in other systems

The problem of crystallization in Coulomb plasmas is related to the Wigner crystallization but involves the condensation of ions rather than electrons. There has been considerable interest in this area in stellar physics and for completeness we quote a selection of the relevant references: Salpeter (1961), Mestel and Ruderman (1967), Van Horn (1968) and Kovetz and Shaviv (1970) all consider the problem of the crystallization of *ions* in white dwarfs. The crystallization of the crust of neutron stars is considered, for instance, by Ruderman (1968, 1969) and plays an important part in the theory of pulsars.

A related problem to that of the electron gas is that of a charged Bose gas and this is discussed by Foldy (1961). In the low density limit the solutions become independent of the statistics and thus it is anticipated that the Bose system will also condense into a regular lattice at low densities.

§ 7. Summary and conclusion

There can be little doubt about the essential correctness of Wigner's ideas in the jellium model; namely, that electron crystallization must occur at sufficiently low density. Doubt remains as to

(*a*) the precise magnitude of the transition density,

(*b*) the precise magnetic ordering of the ground state, through antiferromagnetism seems likely for intermediate r_s. Otherwise, the understanding of ground-state properties has reached quite a sophisticated, if not yet fully quantitative level.

But a much less clear-cut situation exists with regard to excited states.

However, it is quite clear that

(*a*) there are low-lying phonon excitations. These may offer one way of probing electron lattices experimentally in the future; to try to establish the existence of lattice, and if successful, to attempt to find its structure (this is *not* necessarily the same, of course, in ground and excited states);

(*b*) the plasmons must be accounted for, at all densities;

(*c*) a picture of states in which, by analogy with classical diffusion, electrons can hop via defects in the Wigner lattice has some relevance (de Wette (1964); for some different, but related ideas connected with the experimental area (i) above, see Care and March (1971)).

It is probably premature to say that there is, as yet, definitive experimental evidence for electron crystallization in nature, though the circumstantial evidence in areas (i) and (ii) appears to us to be quite strong and we have little doubt that the next decade will see a great deal of further progress in this field.

Finally, we have contrasted electron crystallization with the Fermi sphere electron gas model. That there are metallic systems (in particular molten metals) in which an electron liquid is probably present has been proposed recently by Egelstaff *et al.* (1974) on the basis of diffraction data†. However, while the table might suggest that in the low density alkali metals we are not too far away from the density at which Wigner correlation effects lead to electrons avoiding one another, the effects in these diffraction experiments are also revealed in high electron density metals (e.g. Al). There is little doubt here that electron–ion interaction, as well as electron–electron repulsion, must be included in any realistic description of this electron liquid: it is *not* a consequence of the jellium model.

APPENDIX

Boundary conditions and the interpretation of the Wigner lattice

There are at least, two, somewhat different, ways of viewing the Wigner lattice. One is simply to argue that the symmetry is broken when $r_s > a$, the critical density, and that the electron density becomes periodic, with the periodicity of a definite crystal lattice‡. This is analogous then to the charge density wave picture of Overhauser, discussed in § 3.2. Clearly we have singled out an origin, and we have orientated the lattice in space.

An alternative point of view is to argue as follows. We 'sit' on one electron and view the surrounding electrons from this vantage point. Then the lattice is representing the (anisotropic) pair correlation functions $g(r)$ of the electrons. If we form a total wave function for this 'lattice', say Ψ_L and take the many electron density matrix

$$\Gamma_L(\mathbf{x}_1 \ldots \mathbf{x}_N;\ \mathbf{x}_1' \ldots \mathbf{x}_N') = \Psi_L^*(\mathbf{x}_1 \ldots \mathrm{x}_N)\Psi_L(\mathbf{x}_1' \ldots \mathbf{x}_N')$$

† *Note added in proof*—X-ray inelastic scattering data reported by Platzman and Eisenberger (1974, *Phys. Rev. Lett.*, **33**, 152) on crystalline Be and Al also cannot be accounted for on the basis of a weakly correlated electron gas. They refer to an incipient Wigner lattice; however electron–ion interaction is again important as well as in liquid metals (S. Cusack, March, Parrinello and Tosi, to be published).

‡ Often taken to be body-centred-cubic, following the work of Fuchs (1935).

where $\mathbf{x}_n$ denotes both space and spin coordinates of the nth electron, and N is the total number of electrons in volume Ω, then we form the density matrix to be used to describe the ground state by

(i) averaging over all orientations of the lattice, for a fixed origin;

(ii) averaging over all positions of the origin of the unit cell.

Such a procedure leads us to ground-state density matrices corresponding to

(*a*) a uniform particle density N/Ω;

(*b*) an isotropic pair function $g(r)$.

Working with the first description (i.e. charge density waves) leads to a direct characterization of one degenerate component Γ_L of the ground-state density matrix Γ of the second picture.

As remarked above, the matter of which description one adopts is one of boundary conditions. Charge density waves assume that both an origin is fixed in space, as well as the orientation of the Wigner lattice. The second picture insists that the particle density is uniform, i.e. no choice of origin is preferred over any other, in a uniform positive background.

Of course, the two pictures discussed here are academic, for when we turn to real systems, ' granular ' ions fix both an origin and an orientation, as well as boundary conditions being fixed by the presence of surfaces.

Acknowledgments

One of us (C.M.C.) held a Science Research Council Fellowship during the period when this review was written.

References

Bagchi, A., 1969, *Phys. Rev.*, **178,** 707.
Biskupski, G., Dubois, H., and Ferre, D., 1973, *C. r. hebd. Séanc. Acad. Sci., Paris*, B, **276,** 781.
Brout, R., 1959, *Phys. Rev.*, **113,** 43.
Brown, J. S., Brown, R. C., and March, N. H., 1974, *Physics Lett.* A, **47,** 489.
Care, C. M., and March, N. H., 1971, *J. Phys.* C, **4,** L372.
Carmi, G., 1968, *J. Math. Phys.*, **9,** 2120.
Carr, W. J., 1961, *Phys. Rev.*, **122,** 1437.
Carr, W. J., Coldwell-Horsfall, R. A., and Fein, A. E., 1961, *Phys. Rev.*, **124,** 747.
Chaplik, A. V., 1972, *Soviet Phys. JETP*, **35,** 395.
Clark, C. B., 1958, *Phys. Rev.*, **109,** 1133.
Coldwell-Horsfall, R. A., and Maradudin, A. A., 1960, *J. Math. Phys.*, **1,** 395.
Crandall, R. S., 1973, *Phys. Rev.* A, **8,** 2136.
Crandall, R. S., and Williams, R., 1972, *Phys. Rev.* A, **5,** 2183.
Daniel, E., and Vosko, S. H., 1960, *Phys. Rev.*, **120,** 2041.
de Llano, M., and Tolmachev, V. V., 1971, *Physics Lett.* B, **37,** 37.
de Wette, F. W., 1964, *Phys. Rev.* A, **135,** 287.
Durkan, J., Elliott, R. J., and March, N. H., 1968, *Rev. mod. Phys.*, **40,** 812.
Durkan, J., and March, N. H., 1968, *J. Phys.* C, **1,** 1118.
Edwards, S. F., and Hillel, A. J., 1968, *J. Phys.* C, **1,** 61.
Egelstaff, P. A., March, N. H., and McGill, N. C., 1974, *Can. J. Phys.*, 52, 1651.
Foldy, L. L., 1961, *Phys. Rev.*, **124,** 649; 1971, *Ibid.* B, **3,** 3472.
Fuchs, K., 1935, *Proc. R. Soc.*, **151,** 585.
Gell-Mann, M., and Brueckner, K. A., 1957, *Phys. Rev.*, **106,** 364.
Harris, J., and Jones, R., 1974, *Physics Lett.*, A, **46,** 407.

HERBERT, D. C., and JONES, R., 1971, *J. Phys.* C, **4,** 1145.
HERRING, C., 1966, *Magnetism*, Vol. 4 (Rado & Suhl).
ISIHARA, A., 1972, *Physics Lett.* A, **39, 313**.
KOHN, W., 1967, *Phys. Rev. Lett.*, **19,** 789.
KOVETZ, A., and SHAVIV, G., 1970, *Astron. Astrophys.*, **8,** 398.
KUGLER, A. A., 1969, *Ann. Phys.*, **53,** 133.
LANG, N. D., and KOHN, W., 1970, *Phys. Rev.* B, **1,** 4555.
MARCH, N. H., and SAMPANTHAR, S., 1962, *Acta physica hung.*, **14,** 67.
MARCH, N. H., and YOUNG, W. H., 1959, *Phil. Mag.*, **4,** 384.
MISAWA, S., 1965, *Phys. Rev.*, **140,** A1645.
MONTROLL, E. W., 1942, *J. chem. Phys.*, **10,** 218.
MESTEL, L., and RUDERMAN. M. A., 1967, *Mon. Not. R. astr. Soc.*, **136,** 27.
MOTT, N. F., 1961, *Phil. Mag.*, **6,** 287.
NOZIÈRES, P., and PINES, D., 1958, *Phys. Rev.*, **111,** 442.
OSAKA, Y., 1967, *J. phys. Soc. Japan*, **22,** 1513.
OVERHAUSER, A. W., 1968, *Phys. Rev.*, **167,** 691.
PEPPER, M., POLLITT, S. I., ADKINS, C. J. and OAKLEY, R. E., 1974, *Physics Lett.* A, **47,** 71.
PIETRASS, B., 1967 a, *Physics Lett.* A, **24,** 555; 1967 b, *Phys. Stat. Sol.*, **24,** 571.
PINES, D., 1963, *Elementary Excitations in Solids* (New York: Benjamin).
RAJAGOPAL, A. K., and MAHENTI, S. D., 1967, *Phys. Rev.* **158,** 353.
RUDERMAN, M. A., 1968, *Nature, Lond.*, **218,** 1128; 1969, *Ibid.*, **223,** 597.
SALPETER, E. E., 1961, *Astrophys. J.*, **134,** 669.
SHUSTER, G. V., and KAZINSKAYA A. I., 1971, *Soviet Phys. Solid State*, **13,** 1038.
SMITH, J. R., 1969, *Phys. Rev.*, **181,** 522.
SOKOLOFF, J. B., 1971, *Phys. Rev.* B, **3,** 3162, see also FAZEKAS, P, 1972, *Solid State Commun.*, **10,** 175.
SOMMER, W. T., and TANNER, D. J., 1971, *Phys. Rev. Letts.*, **27,** 1345.
SOMERFORD, D. J., 1971, *J. Phys.* C, **4,** 1570.
SUPRAPTO, B., 1974, *J. Phys.* C, **7,** 1822.
VAN HORN, H. M., 1967, *Phys. Rev.*, **157,** 342; 1968, *Astrophys. J.*, **151,** 227.
YOUNG, W. H., 1960, *Phil. Mag.*, **6,** 371.
WIGNER, E. P., 1934, *Phys. Rev.*, **46,** 1002; 1938, *Trans. Faraday Soc.*, **34,** 678.
WISER, N., and COHEN, M. H., 1969, *J. Phys.* C, **2,** 193.

[*The Editors do not hold themselves responsible for the views expressed by their correspondents.*]

J. Phys. C: Solid State Phys., Vol. 9, 1976. Printed in Great Britain.

LETTER TO THE EDITOR

Thermodynamics of Wigner crystallization

M Parrinello and N H March
Department of Physics, The Blackett Laboratory, Imperial College, London SW7 2BK, UK

Received 16 January 1976

Abstract. The thermodynamics of Wigner crystallization is considered. The melting curve in the classical regime is $p = AT^4$. In the $T = 0$ limit, the change in potential energy is related to the volume change. A dimensional argument is given to show that in four dimensions, the melting curve in the ρ–T plane is given by $T = K\rho^{1/2}$, a fully quantal result. Finally, a schematic plot of critical density versus dimensionality is achieved.

The Wigner transition is currently of considerable interest (Care and March 1975, Elliott and Kleppman 1975). It is the purpose of this letter to consider the thermodynamics of Wigner crystallization. We begin with the case of constant (p, T).

Gibbs free energy G (p, T constant)

The equilibrium condition for the fluid–crystal transition is

$$G_1 = G_2, \tag{1}$$

which for changes ΔE, ΔS and $\Delta\mathscr{V}$ in internal energy, entropy and volume, respectively, across the transition gives

$$\Delta E = T\Delta S - p\Delta\mathscr{V}. \tag{2}$$

Now for a Coulomb system, the virial theorem reads

$$2K + V = 3p\mathscr{V} \tag{3}$$

where K is the kinetic and V the potential energy. Hence

$$p\Delta\mathscr{V} = \tfrac{2}{3}\Delta K + \tfrac{1}{3}\Delta V \tag{4}$$

and from equations (4) and (2) it follows that

$$T\Delta S = \tfrac{5}{3}\Delta K + \tfrac{4}{3}\Delta V. \tag{5}$$

The following limiting results occur for p and T constant:

(*a*) *Classical.* The classical limit yields

$$\Delta S = \tfrac{4}{3}\Delta V/T \tag{6}$$

and

$$\Delta\mathscr{V} = \tfrac{1}{3}\Delta V/p. \tag{7}$$

Using the Clausius–Clapeyron equation

$$\frac{\mathrm{d}p}{\mathrm{d}T} = \frac{\Delta S}{\Delta \mathscr{V}}, \tag{8}$$

we find the melting curve in this classical regime is

$$p = AT^4. \tag{9}$$

(*b*) *Zero temperature.* From equation (5), putting $T\Delta S = 0$ we find

$$\Delta E = \Delta K + \Delta V = \tfrac{1}{5}\Delta V \tag{10}$$

and

$$\Delta \mathscr{V} = -\frac{1}{5p}\Delta V. \tag{11}$$

Helmholtz free energy F ($\mathscr{V}$, T constant)

Keeping $\mathscr{V}$ and T constant, we have from the virial result (3)

$$\mathscr{V}\Delta p = \tfrac{2}{3}\Delta E - \tfrac{1}{3}\Delta V. \tag{12}$$

From the equality of the Helmholtz free energy in the two phases we have

$$\Delta E = T\Delta S. \tag{13}$$

Equations (12) and (13) give the following limiting results:

(*a*) *Classical*

$$\Delta p = \frac{1}{3\mathscr{V}}\Delta V \tag{14}$$

and

$$\Delta S = \Delta V/T. \tag{15}$$

The melting curve which follows from equations (14) and (15) has already been given (see, for example, Pollock and Hansen 1973) as

$$\mathscr{V} = AT^{-3} \tag{16}$$

or

$$T = \chi\rho^{1/3} \tag{17}$$

where

$$\chi = (Ze)^2(\tfrac{4}{3}\pi)^{1/3}/158k_{\mathrm{B}}, \tag{18}$$

Ze being the charge on the point ions.

(*b*) *Zero temperature.* The $T = 0$ case becomes, in this case of constant $\mathscr{V}$,

$$\Delta E = 0 \tag{19}$$

$$\Delta p = -\frac{1}{3\mathscr{V}}\Delta V. \tag{20}$$

Results for d dimensions

These simple thermodynamic considerations can be extended to the case of a Coulomb system having d dimensions, in which case the virial theorem (compare Toulouse (1975) who does not consider non-zero pressure) reads

$$2T + (d - 2)V = dp\mathscr{V}. \tag{21}$$

The most interesting case (apart from $d = 3$!) is as usual $d = 4$, when equation (21) becomes

$$2E = 4p\mathscr{V}. \tag{22}$$

This allows one to calculate the melting curve in the density–temperature (ρ, T) plane as

$$T = C\rho^{1/2}. \tag{23}$$

It should be noted that equation (23) is *fully* quantal. From equation (23) the critical value of r_s for crystallization, r_s^* say, is infinite. This also follows from scaling considerations. In fact, for a change of scale

$$r \rightarrow r' = r/s, \tag{24}$$

the jellium Hamiltonian

$$H = -\frac{\hbar^2}{2m}\sum_i \nabla_i^2 + \frac{e^2}{2}\sum_{i \neq j}\frac{1}{|r_i - r_j|^{d-2}} \rightarrow H'. \tag{25}$$

The transformed Hamiltonian is evidently

$$H' = s^2\left(-\frac{\hbar^2}{2m}\sum_i \nabla_i^2 + \frac{e^2}{2}s^{-\epsilon}\sum_{i \neq j}\frac{1}{|r_i - r_j|^{d-2}}\right) \tag{26}$$

where $\epsilon = 4 - d$. Thus, for $s \rightarrow \infty$ (high density) we approach the free-particle case for $\epsilon > 0$ and the Wigner case for $\epsilon < 0$.

We must therefore have the qualitative plot for r_s^* as a function of dimensionality d shown in figure 1. The point $d = 1$ corresponds to $r_s^* = 0$. This is because phase transitions cannot occur in one dimension, and for strong repulsive (Coulomb) interactions

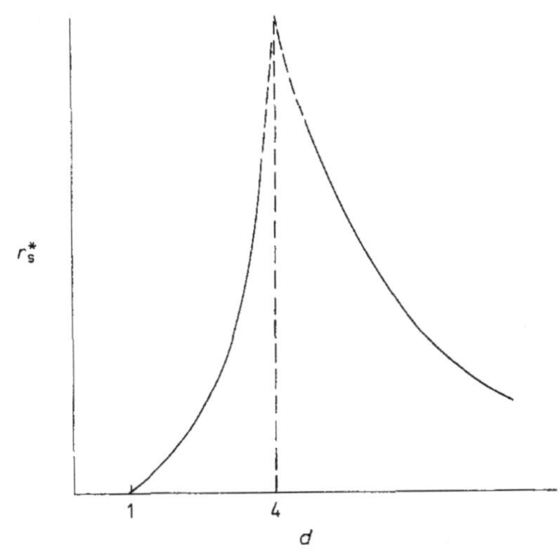

Figure 1. Critical r_s for Wigner crystallization as a function of dimensionality d (schematic only).

in this case an ordered state must form (compare Sutherland 1975). Work relating to ϵ-expansions is continuing.

One of us (MP) wishes to acknowledge support from the Science Research Council during the course of this work.

References

Care C M and March N H 1975 *Adv. Phys.* **24** 101
Elliott R J and Kleppmann W G 1975 *J. Phys. C: Solid St. Phys.* **8** 2729
Pollock E L and Hansen J P 1973 *Phys. Rev.* A **8** 3110
Sutherland B 1975 *Phys. Rev. Lett.* **35** 185
Toulouse G 1975 *J. Physique* **36** 1137

PHYSICAL REVIEW B VOLUME 19, NUMBER 4 15 FEBRUARY 1979

Phenomenological theory of first- and second-order metal-insulator transitions at absolute zero

N. H. March
Theoretical Chemistry Department, University of Oxford, Oxford, England

M. Suzuki
Department of Physics, University of Tokyo 113, Japan

M. Parrinello
Department of Physics, University of Messina, Messina, Italy
(Received 20 October 1977)

A phenomenological theory of metal-insulator transitions at $T = 0$ is set up in which the order parameter is the discontinuity q in the single-particle occupation probability at the Fermi surface. Applied to the case of the second-order metal-insulator transition in a half-filled Hubbard band, q is found to have a critical exponent of unity, and the relation to Gutzwiller's variational treatment is exposed. The enhancement of the spin susceptibility by the Hubbard interaction is also treated. The same phenomenology is then applied to jellium, where a first-order metal-insulator transition occurs when the conducting electron fluid crystallizes on to the Wigner lattice. The form of the spin susceptibility, chemical potential, and energy is given near to, but on the high-density side of, the transition.

I. INTRODUCTION

A good deal of interest has centered around metal-insulator transitions recently[1] but progress in purely microscopic terms has proved difficult. In this paper, therefore, we set up a phenomenological treatment at $T=0$, by first characterizing the order parameter q in the metallic phase, and then developing the ground-state energy in powers of q.

We discuss both first- and second-order metal-insulator transitions. Specifically, in Sec. II we treat the metal-insulator transition in a half-filled Hubbard band. The case of the metal-insulator transition in jellium, from a conducting electron fluid to an insulating Wigner electron crystal[2] is dealt with in Sec. III. As discussed earlier by Parrinello and March,[3] this is a first-order transition. We then immediately discuss a half-filled Hubbard band.

II. SECOND-ORDER METAL-INSULATOR TRANSITION IN HALF-FILLED HUBBARD BAND

A. Assumptions and discussion

(i) A suitable generalized order parameter is the discontinuity q at the Fermi surface in the metallic phase of the single-particle occupation number $n_{\vec{k}}$.
(ii) At a critical strength, for example, C_0, of the Hubbard interaction C, which is the energy it costs to put two electrons with antiparallel spin on the same site, a metal-insulator transition occurs, at which the discontinuity q is reduced to zero.
(iii) The average number of doubly occupied sites, for example, ν, is a function of q, i.e., the interaction energy is $C\nu(q)$.

The phenomenology follows Suzuki[4] in that the ground-state energy $E(q)$ is expanded about the singular point $q=0$ as

$$E(q)=E_0+E_1q+E_2q^2+\cdots. \tag{2.1}$$

The form of E_1 is assumed such that

$$E_1(C)=\alpha(C_0-C), \quad \text{i.e.,} \quad E_1(C_0)=0, \quad \alpha<0, \tag{2.2}$$

while $E_2(C_0)>0$.

The energy minimization

$$\frac{\partial E}{\partial q}=0, \tag{2.3}$$

evidently yields

$$E_1+2E_2q+\cdots=0. \tag{2.4}$$

Near the transition, i.e., for very small q, we find

$$q=-E_1/2E_2=Q(1-C/C_0); \quad Q=-\alpha C_0/2E_2. \tag{2.5}$$

Substituting q from (2.5) into (2.1), the minimum energy is found to be

$$E_{\min}=E_0+(\alpha C_0Q+E_2Q^2)(1-C/C_0)^2. \tag{2.6}$$

Thus, the difference in energy between the metallic and the insulating state, for example, ΔE, is given by

$$\Delta E=(\alpha C_0Q+E_2Q^2)(1-C/C_0)^2. \tag{2.7}$$

Evidently the number of doubly occupied sites ν is given in the Hubbard model, from Feynman's theorem by $d\Delta E/dC$ and, hence,

$$\nu=\frac{d\Delta E}{dC}=-2/C_0(\alpha C_0Q+E_2Q^2)(1-C/C_0). \tag{2.8}$$

The behavior exhibited in Eqs. (2.5), (2.7), and (2.8) is, with regard to the "critical indices," identical to that predicted by Gutzwiller's variational method.[5,6] This method, in fact, applies forms (2.5), (2.7), and (2.8), which are shown to be valid near $C = C_0$ by the above phenomenology, right into $C = 0$. In Gutzwiller's work, $Q = 1 + C/C_0$, while $\alpha C_0 Q + E_2 Q^2$ is chosen so that both ΔE and ν reduce to their Hartree-Fock values as $C \to 0$.

B. Spin susceptibility near metal-insulator transition

As pointed out by Brinkman and Rice,[6] there is some experimental interest in connection with V_2O_3 in the way the electron-spin susceptibility is enhanced by the interaction C in the metallic phase as the metal-insulator transition is approached.

Again, following Suzuki,[4] the energy expansion is generalized to allow for nonzero magnetization m. If h is the applied magnetic field, we therefore write

$$E(m, q) = E_0(m) + am^2 - hm + E_1 q + E_2 q^2 + \cdots + eqm^2 . \tag{2.9}$$

Minimizing with respect to m, the result is

$$2am + 2eqm = h , \tag{2.10}$$

and hence the spin susceptibility χ is given by

$$m = \{1/2[a(C) + eq]\} h \equiv \chi h . \tag{2.11}$$

From (2.11) it can be seen that since $q \to 0$ as $C \to C_0$ the question of the enhancement of the susceptibility as the metal-insulator transition is approached rests on the behavior of a as a function of interaction strength C. Provided $a \to 0$ as $C \to C_0$ as $(1 - C/C_0)$, or faster, then using (2.5) in (2.11) we obtain

$$\chi \propto 1/(1 - C/C_0) , \tag{2.12}$$

which is the form given by Brinkman and Rice.[6]

III. WIGNER CRYSTALLIZATION

A. Metallic versus crystal transitions

As discussed earlier by Parrinello and March,[3] in jellium we are dealing with a first-order transition from a metallic electron fluid to an insulating Wigner electron crystal, in contrast to the second-order transition discussed in Sec. II.

Evidently in jellium, if r_s is the usual interelectronic spacing, the ground state energy E is a function of r_s and a transition to an electron crystal occurs as r_s is increased to a critical value, for example, r_s^*. Again, the discontinuity q in the single particle occupation number (which in jellium is equivalent to the momentum distribution) at the Fermi surface is used as the order parameter. Allowing from the outset for nonzero magnetization m we expand $E(q, m)$ for the first-order transition as

$$E(m, q) = E_0 + am^2 - hm + E_1 q + E_2 q^2 + E_3 q^3 + em^2 q + \cdots \tag{3.1}$$

Minimizing with respect to m, we get again Eqs. (2.10) and (2.11). Since q does not now go to zero at the first-order transition, but suffers a discontinuity on passing into the insulating phase, we expect χ to be discontinuous, but not divergent.

If we repeat the minimization with respect to q, at $m = 0$ as in Sec. II, then we find

$$0 = \frac{\partial E}{\partial q} = E_1 + 2E_2 q + 3E_3 q^2 , \tag{3.2}$$

with solution

$$q_0 = -E_2 \pm (E_2^2 - 3E_1 E_3)^{1/2} / 3E_3 . \tag{3.3}$$

Here the positive sign in Eq. (3.3) corresponds to the minimum of energy, as shown in Fig. 1. The discontinuity q_c in the order parameter at the metal-insulator transition is also shown in Fig. 1. It is straightforward to show that

$$\Delta E \equiv E(q_0) - E_0 = \tfrac{2}{3} E_1 q_0 + \tfrac{1}{3} E_2 q_0^2 , \tag{3.4}$$

and using (3.4) one finds

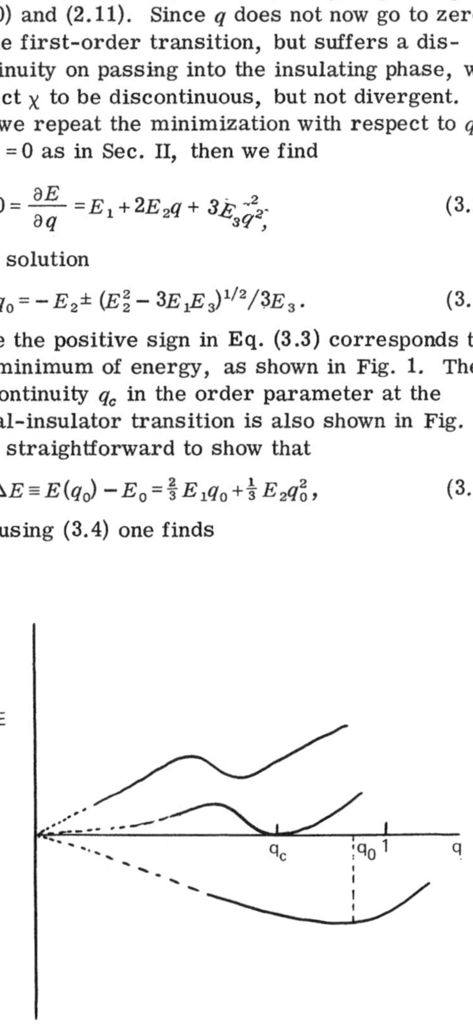

FIG. 1. Schematic form of ground-state energy E as a function of magnitude of discontinuity q in momentum distribution at the Fermi surface of jellium. q has maximum value of 1, when $r_s = 0$. As r_s increases from zero, the lowest curve shows discontinuity q_0 for an $r_s < r_s^*$. q_c is the smallest discontinuity which can exist in the metallic phase, i.e., the value of q_0 at r_s^* minus. The upper curve shows behavior in insulating Wigner crystal, i.e., for $r_s > r_s^*$. The minimum energy now occurs for $q = 0$.

$$\Delta E = -E_1E_2/9E_3 + (-\tfrac{2}{9}E_2^2/E_3 + \tfrac{2}{3}E_1)q_0$$
$$\equiv f(r_s) + g(r_s)q_0(r_s) . \qquad (3.5)$$

Now the critical interelectronic spacing r_s^* is such that $\Delta E(r_s^*) = 0$ since $E(q_c) = E_0$. The corresponding q_c is given by

$$q_c = q_0(r_s^*) , \qquad (3.6)$$

in (3.3) with the plus sign and we obtain

$$\Delta E = B(r_s^* - r_s) , \qquad (3.7)$$

by expanding (3.5) near the critical value r_s^*. Equation (3.7) is in accord with our expectations, if B is assumed negative, that the Wigner transition occurs at an r_s greater than the equilibrium distance, for example, r_e, at which $(dE/dr_s)|r_e = 0$. Thus, the phenomenological treatment of the Wigner transition leads to the required result

$$E(r_s) = E(r_s^*) + B(r_s^* - r_s) + \cdots, \qquad (3.8)$$

for $r_s^* - r_s$ small and positive.

B. Relation to virial theorem and to chemical potential

We conclude the discussion of the Wigner first-order transition by noting that for jellium at $T = 0$ we have the virial theorem[7]

$$2T + U = -r_s \frac{dE}{dr_s} , \qquad (3.9)$$

where T is the kinetic energy and U is the potential energy. Hence,

$$T = -E - r_s \frac{dE}{dr_s} = -E(r_s^*) + Br_s^* - 2B(r_s^* - r_s) , \qquad (3.10)$$

whereas

$$U = E - T = 2E(r_s^*) + 3B(r_s^* - r_s) - Br_s^* . \qquad (3.11)$$

Evidently, since dE/dr_s is discontinuous at the Wigner transition, it follows from (3.9) that T and U separately are discontinuous even though the sum $E = T + U$ is not.

Finally, we note that the chemical potential μ in jellium is given by[8]

$$\mu = \tfrac{5}{3}T + \tfrac{4}{3}U = E(r_s^*) + \tfrac{1}{3}Br_s^* + \tfrac{2}{3}B(r_s^* - r_s) + \cdots . \qquad (3.12)$$

IV. SUMMARY

The main results for the second-order metal-insulator transition in a half-filled Hubbard band are: (i) The generalized order parameter q has a critical exponent of unity. (ii) The difference in energy between the metallic state and the paramagnetic insulating state is $\alpha(1 - C/C_0)^2$, C_0 being the critical interaction strength for the transition. (iii) The susceptibility χ near the metal-insulator transition can be strongly enhanced as supposed by Brinkman and Rice.

For the first-order Wigner transition, the same basic phenomenology leads to: (iv) A form of the spin susceptibility near the transition which will lead to χ discontinuous, but not divergent, and (v) The expected form of expansion of $E(r_s)$ around the critical r_s for the transition. Though E is continuous, kinetic and potential energies separately must suffer discontinuities across the transition.

ACKNOWLEDGMENTS

Two of us (N.H.M. and M.P.) wish to thank Dr. David Wallace and Dr. Royce Zia for much help, and for continuing interest in this work. The contributions of two of us (N.H.M. and M.S.) were largely carried out while both were visitors at the Theoretical Physics Institute, University of Alberta, Edmonton. They wish to thank Professor D.D. Betts, Professor A. B. Bhatia, and Professor Y. Takahashi for generous hospitality during the visit.

[1]N. F. Mott, *The Metal-Insulator Transition* (Taylor and Francis, London, 1975).
[2]C. M. Care and N. H. March, Adv. Phys. 24, 101 (1973).
[3]M. Parrinello and N. H. March, J. Phys. C 9, LI47 (1976).
[4]M. Suzuki, Prog. Theor. Phys. 58, 1151 (1977).
[5]M. C. Gutzwiller, Phys. Rev. A 137, 1826 (1965).
[6]W. E. Brinkman and T. M. Rice, Phys. Rev. B 2, 4302 (1970).
[7]N. H. March, Phys. Rev. 110, 604 (1958).
[8]N. H. March, Phys. Lett. A 39, 150 (1972).

Solid State Communications, Vol.50,No.8, pp.725-728, 1984. Printed in Great Britain.

0038-1098/84 $3.00 + .00 Perganom Press Ltd.

COOPERATIVE MAGNETISM IN METALLIC JELLIUM AND IN THE INSULATING WIGNER ELECTRON CRYSTAL*

F. Herman and N. H. March**
IBM Research Laboratory, San Jose, California 95193 USA

(Received 6 March 1984 by S. Lundqvist)

Total energies for the paramagnetic (P) and ferromagnetic (F) metallic phases of jellium are calculated using interpolation formulae adjusted to the results of computer simulation studies. For the Wigner insulator (W), the total energy given by an early theoretical model is found to agree closely with recent computer simulation results. Using these total energies, the phase transitions from P to F and from F to W are estimated to take place at $r_s = 79 \pm 1$ and 84 ± 4, respectively, in units of the Bohr radius. Although consistent with earlier estimates based directly on the computer simulations at a limited number of r_s values, the present estimates are more precise. Since the P to F and F to W transitions come so close together on an r_s scale, the F phase may just barely emerge as the lowest energy phase or may not so emerge at all. Using the virial theorem, kinetic energy curves are constructed and kinetic energy discontinuities at the first-order phase transitions are thereby estimated. The momentum distributions for various phases of jellium are also discussed in terms of simple models. Finally, the phase boundary between antiferromagnetic and paramagnetic states of the Wigner crystal for $T \neq 0$ is briefly considered. It is shown that one is dealing with antiferromagnetism in the millidegree Kelvin range.

1. Ground State of Wigner Insulator

Wigner[1,2] drew attention a long time ago to the phenomenon of electron crystallization at sufficiently low densities in the so-called jellium model, in which interacting electrons move in a uniform, neutralizing, and unresponsive background of positive charge. Many properties associated with electron crystallization have been reviewed by Care and March.[3] More recent work, using a Monte Carlo computer method, by Ceperley and Alder,[4] has settled the fact that, if r_s is the mean inter-electronic spacing, related to the electron density $\rho = N/V$ for N electrons in volume V by $\rho = 3/4\pi r_s^3$, then the critical value of r_s at which electron crystallization occurs, r_c say, is $r_c = 100 \pm 20$, in units of the Bohr radius.

That antiferromagnetism in the Wigner crystal is favored at the transition density $\rho_c = 3/4\pi r_c^3$ is intuitively clear. This is because the Madelung energy favors the body-centered-cubic electron crystal, and a long-range ordered Néel antiferromagnetic insulator at absolute zero is quite natural for this structure, the upward spin electrons being located on the sites of one of the two interpenetrating simple cubic lattices, and the downward spins on the other.

Ceperley and Alder[4] also showed that the critical value of r_s for the transition from the paramagnetic to the ferromagnetic metallic phase of jellium is 75 ± 5. The large uncertainties in their critical r_s estimates arise primarily from the fact that their computer simulation studies were carried out at only a limited number of widely spaced r_s values.

In order to narrow down these estimates, as we wish to do, it is necessary to obtain an accurate representation for the total energy $E(r_s)$ as a continuous function of r_s over a wide range of r_s for the various jellium phases. In the case of the Wigner crystal, we had originally planned to develop an interpolation formula and fit this to the computer simulation results, but we were spared this effort when we discovered that the theoretical expression[5]

$$E(r_s) = -1.79186/r_s + 2.65/r_s^{3/2} - 0.73/r_s^2 \ \text{Ry}, \quad (1)$$

without any adjustment at all, reproduces the Ceperley-Alder energies at $r_s = 50$, 100, 130 and 200 to within 0.124, 0.003, 0.002 and 0.006 mRy, respectively. The corresponding computer simulation uncertainties are 0.010, 0.003, 0.002, and 0.001 mRy.

The above expression includes the electrostatic energy of the Wigner crystal, the energy of zero-point motion of the electrons about their lattice points, and the leading anharmonic correction to this motion, but neglects higher order anharmonic corrections as well as exchange and orbital overlap corrections. After exploring various modifications of Eq. (1), such as the inclusion of an exponentially decaying function of r_s which brings the fitting error at $r_s = 50$ below 0.010 mRy without affecting the fitting errors at the remaining values of r_s, we concluded that Eq. (1) — as it stands — is sufficiently accurate for our purposes in the range between $r_s = 75$ and 125, where the various phase transitions of interest take place. Using Eq. (1) and the virial theorem,[6]

* Supported in part by Office of Naval Research Contract Number N00014-79-C-0814.

** Permanent address: Theoretical Chemistry Department, University of Oxford, Oxford OX1 3TG, England

$$K + E = -r_s\, dE/dr_s \,, \tag{2}$$

the kinetic energy is found to be

$$K(r_s) = 1.325/\, r_s^{3/2} - 0.73/r_s^2 \,. \tag{3}$$

In the Wigner crystal, the electronic orbitals can be described by non-overlapping Gaussians centered on the BCC lattice sites at sufficiently low densities (r_s > 9.3).[3] The single-particle occupation probability of plane wave state k can then be approximated by the Fourier transform of the Wigner orbitals:

$$n(k/k_F) = \frac{3\pi^{1/2}}{r_s^{3/4}} \exp\,[-(9\pi/4)^{2/3} r_s^{-1/2} (k/k_F)^2] \,. \tag{4}$$

The corresponding kinetic energy is $K(r_s) = 1.5/r_s^{3/2}$, which may be regarded as a first approximation to Eq. (3).

2. Low-Density Metallic Phases

The ground-state energies for the paramagnetic (P) and ferromagnetic (F) metallic phases of jellium have been calculated by Ceperley and Alder[4] between $1.0 \le r_s \le 100$ and $2.0 \le r_s \le 100$, respectively, at selected values of r_s. These results have been carefully fitted by Vosko, Wilk, and Nusair (VWN)[7] and by Perdew and Zunger (PZ).[8] In spite of their drastically different analytical forms, the VWN and PZ interpolation formulae both provide good overall fits to the Ceperley-Alder data over the entire range of r_s. Looked at in greater detail, however, the PZ fit is closer than the VWN fit for $r_s \le 20$, while for $r_s \ge 50$, the PZ fitting errors (though quite small) are about 5 times larger than their VWN counterparts overall, and in some cases are of opposite sign. Accordingly, we will base our results on a weighted average of $E(r_s)$ given by the VWN and PZ interpolation formulae, with VWN being given 5 times as much weight as PZ.

In Fig. 1 we plot the kinetic energies $K(r_s)$ for the various phases, as well as the kinetic energy differences

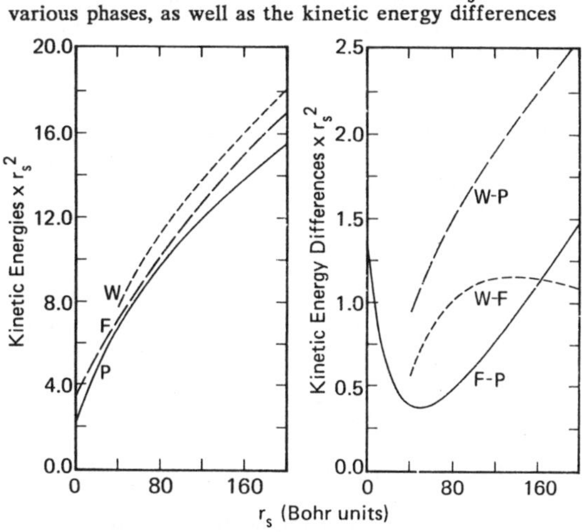

Fig. 1. Kinetic energies and kinetic energy differences for the paramagnetic (P) and ferromagnetic (F) metallic phases and for the Wigner insulating phase (W) of jellium, all × r_s^2. Energies are in Rydberg units.

between these phases. The kinetic energies for P and F were obtained from the weighted average of the VWN and PZ fits and the virial theorem, while $K(r_s)$ for the Wigner crystal (W) was obtained from Eq. (3). As expected, K is higher for the ferromagnetic than for the paramagnetic phase, since the Fermi sphere has radius $2^{1/3}$ times that in the doubly occupied spin-degenerate case of the paramagnetic. However, we note that at r_s = 79 ± 1, where the paramagnetic and ferromagnetic total ground state energies become equal,[9] as can be seen from Fig. 2, there is a discontinuity in K, ΔK = K(P) − K(F) = 0.08 mRy. If we use the full Fermi

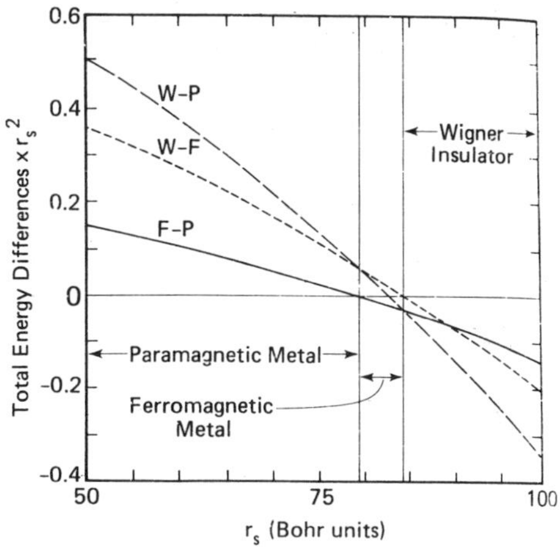

Fig. 2. Total energy differences (× r_s^2) between various phases of jellium.

sphere model, with r_s = 79, we would find ΔK(single-particle) = 0.22 mRy, showing that the contribution of electron-hole correlation excitations to the discontinuity is − 0.14 mRy. This discontinuity in K is another reflection of the fact that a first-order transition is occurring at about r_s = 79 between paramagnetic and ferromagnetic metallic phases. Bloch[10] in pioneering work using the Hartree-Fock approximation found a similar transition at much lower $r_s \sim 6$, but Wigner[1,2] has pointed out that electron correlation would decrease the stability of the ferromagnetic state from the Hartree-Fock prediction.

If we accept Eq. (1) as an accurate description of the total energy for the Wigner crystal in the range $75 \le r_s \le 100$, and make suitable allowance for computer simulation and interpolation uncertainties, we find that the cross-over between the paramagnetic metal and insulator phases occurs at r_s = 82 ± 3, while the ferromagnetic metal to insulator transition occurs at r_s = 84 ± 4 (cf. Fig. 2). Moreover, the discontinuity ΔK for these metal-insulator transitions is found to be somewhat larger than that for the paramagnetic to ferromagnetic transition discussed earlier (cf. Fig. 1).

With the aid of Eq. (1) and the PZ and VWN interpolation formulae, then, we have been able to pinpoint the values of r_s at which the various phase transitions occur. Our estimates are consistent with the earlier estimates of Ceperley and Alder, but the uncertainties in our estimates are substantially smaller. In fact, according to our estimates, the transition densities come so close together that the ferromagnetic phase may just barely emerge as the lowest energy phase or may not so emerge at all. This question deserves further theoretical study, as does the question of the possible occurrence of partially polarized metallic phases at intermediate densities.

3. Single-Particle Occupation Probabilities for Low-Density Metallic Phases

While n(k) is a continuous function of k for the Wigner crystal (cf. Eq. (4)), there is a discontinuity in n(k) at the Fermi surface in normal metals, as is well known, and also in low density metallic phases, as has been demonstrated, for example, by the computer simulation studies of Ceperley and Alder.[11] (See also Fig. 3 below.) March et al.[12] recently suggested that this discontinuity in n(k) could be used as an order parameter in metal-insulator transitions.

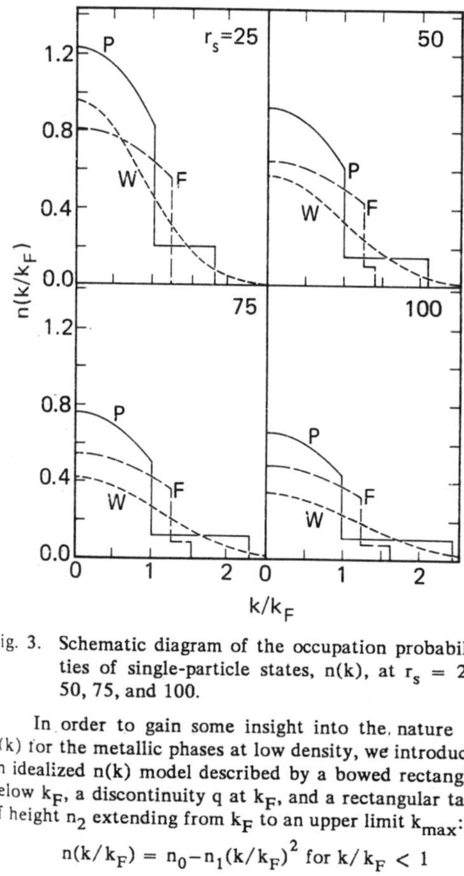

Fig. 3. Schematic diagram of the occupation probabilities of single-particle states, n(k), at r_s = 25, 50, 75, and 100.

In order to gain some insight into the nature of n(k) for the metallic phases at low density, we introduce an idealized n(k) model described by a bowed rectangle below k_F, a discontinuity q at k_F, and a rectangular tail of height n_2 extending from k_F to an upper limit k_{max}:

$$n(k/k_F) = n_0 - n_1(k/k_F)^2 \text{ for } k/k_F < 1$$

$$n_0 = n_1 + q + n_2$$

$$n(k/k_F) = n_2 \text{ for } 1 < k/k_F < k_{max}/k_F.$$

Using normalization and the kinetic energy:

$$N/V = \frac{3}{4\pi r_s^3} = \frac{4\pi}{(2\pi)^3}\int n(k)k^2dk = \frac{\lambda k_F^3}{6\pi^2},$$

$$K(r_s) = \int n(k)k^4dk \Big/ \int n(k)k^2dk,$$

where λ is 2 for phase P and 1 for phase F, we can solve for n_0, n_1, and n_2 in terms of the dimensionless ratios $M = k_{max}/k_F$ and $K_0 = \lambda\, K(r_s)/(3k_F^2/5)$, and an assumed ratio such as q/n_0. Although we would expect the tail to fall off with increasing k, rather than remain constant up to some cutoff, as we have assumed, modeling a decaying tail realistically would involve introducing still more parameters, and we already have two parameters too many for a self-contained solution.

In order to obtain a closed solution, we must impose two conditions on our model. Perhaps the simplest approach is to assume that n_1, q, and n_2 are proportional to one another, with fixed ratios, over the entire range of r_s of interest, say between r_s = 5 and 100. Using the computer simulation results for n(k) obtained earlier by Ceperley and Alder[11] for the metallic phases at r_s = 50 as a guide, we adopted the proportion $n_1{:}q{:}n_2$ = 2:3:1. We can now express K_0, as well as all the parameters, in terms of M:

$$K_0 = (M^5 + 25/7)/(M^3 + 19/5),$$

$$n_0 = 3n_1 = 2q = 6n_2 = 6\lambda/(M^3 + 19/5).$$

For a given r_s, we can find the value of M which corresponds to the calculated value of K_0, or $K(r_s)$, and thereby obtain the remaining parameters. The model remains valid as long as $n_0 \le \lambda$, which implies $M > 1.3$, and in practice imposes the limits $r_s > 4$ for phase P and $r_s > 9$ for phase F.

In Fig. 3 we display the n(k) given by the above model for the P and F metallic phases of jellium at r_s = 25, 50, 75, and 100. By design, our curves for P and F for r_s = 50 resemble the corresponding computer simulation results of Ceperley and Alder as closely as the simplicity of our model permits. For comparison, we also show n(k) for the insulating W phase as given by Eq. (4). All of these sketches should be viewed as semi-quantitative only, but they do describe some essential physical features of n(k). We have also calculated n(k) for r_s = 5 and 10 for the P phase using the above model, and these n(k) are consistent with unpublished computer simulation calculations by Ceperley (private communication), confirming the usefulness of our model even at densities just above the metallic range.

4. Wigner Crystal at Elevated Temperatures

While a good deal of attention has focused on the ground state of the Wigner electron crystal, less attention has been given to properties at elevated temperatures T. However, the thermodynamics of Wigner crystallization has been discussed[13] as has also the approximate form of the melting curve of the electron crystal[14,15] in the density-temperature plane. In this concluding section, we will consider briefly the phase boundary in the ρ-T plane of the electron crystal which separates the antiferromagnetic and paramagnetic phases.

In his pioneering work on the quantitative theory of the low-density Wigner crystal, Carr[16] recognized the antiferromagnetic nature of the ground state near ρ_c.

By calculating the exchange integral and then comparing the energy of antiferromagnetic and paramagnetic states, assuming little correlation between spins in the latter, he estimated the Néel temperature T_N for large r_s as

$$T_N \sim 1.6\times10^5\left[(13r_s^{-1}-3.2r_s^{-3/4})\exp(-1.55r_s^{1/2}) - (10.5r_s^{-1}-2.4r_s^{-3/4})\exp(-2.06r_s^{1/2})\right] \quad (5)$$

where r_s is measured in units of the Bohr radius and T_N in degrees Kelvin. Evaluating Eq. (5) at r_s = 80, 100, and 120, which span Ceperley and Alder's and our own estimate for the transition density, we obtain T_N = 6.47, 0.85, and 0.14 mK, respectively. In this range of densities, then, we are dealing with cooperative antiferromagnetism in the millidegree Kelvin range. In contrast, Carr[16] mentions $T_N \sim 15$ K, due to the lack of knowledge of the transition density ρ_c at that time. Carr noted, however, that Eq. (5) gives $T_N > 0$ for $r_s <$ 270 and thus the maximum range of density for the phase boundary under discussion is $100 < r_s < 270$.

It is worth noting here that in treating the order of magnetic phase transitions, the volume dependence of the transition temperature plays an essential role.[17,18] The derivative $dT_N/d(V/N)$ as a function of V/N $(= 4\pi r_s^3/3)$ is easily obtained from Eq. (5). Estimates we have made suggest that the antiferromagnetic-paramagnetic phase transition is second-order near $r_s \sim$ 270, but further work is called for on this point, over the entire phase boundary. We expect the sublattice magnetization M(T) to be a useful order parameter for the phase transition and that the usual phenomenological treatment of this[19] will still be useful.

We conclude by remarking that although jellium has the very obvious physical limitation that we are dealing, certainly in the Wigner crystal, with antiferromagnetism in the millidegree range, due to the extreme weakness of the exchange forces, yet it might provide a valuable prototype model to exhibit the effect of the long-range Coulomb force on magnetic phase diagrams. In this sense it might complement the widely used models based on the assumption of short-range intra-atomic electron-electron repulsion.

Acknowledgments – The authors are grateful to Dr. D. M. Ceperley for stimulating correspondence and to Drs. I. P. Batra, P. Lambin, R. K. Nesbet, J. C. Scott, and J. B. Torrance for fruitful discussions and helpful comments on the manuscript. One of us (N. H. M.) wishes to thank the IBM management for partial financial support during his visit to San Jose in the Winter of 1983.

REFERENCES

1. E. P. Wigner, Phys. Rev. **46**, 1002 (1934).
2. E. P. Wigner, Trans. Faraday Soc. **34**, 678 (1938).
3. C. M. Care and N. H. March, Adv. Phys. **24**, 101 (1975).
4. D. M. Ceperley and B. J. Alder, Phys. Rev. Lett. **45**, 566 (1980).
5. R. A. Coldwell-Horsfall and A. A. Maradudin, J. Math. Phys. **1**, 395 (1960); W. J. Carr, R. A. Coldwell-Horsfall, and A. E. Fein, Phys. Rev. **124**, 747 (1961).
6. N. H. March, Phys. Rev. **110**, 604 (1958).
7. S. H. Vosko, L. Wilk, and M. Nusair, Can. J. Phys. **58**, 1200 (1980); see also G. S. Painter, Phys. Rev. **B24**, 4264 (1981).
8. J. P. Perdew and A. Zunger, Phys. Rev. **B23**, 5048 (1981).
9. The energies become equal at r_s = 75.6 and 79.7 for the PZ and VWN fits, respectively.
10. F. Bloch, Z. Phys. **57**, 545 (1929).
11. D. M. Ceperley and B. J. Alder, J. Phys. (Paris) **41**, Colloque C7-295 (1980); and private communication.
12. N. H. March, M. Suzuki and M. Parrinello, Phys. Rev. **B19**, 2027 (1979).
13. M. Parrinello and N. H. March, J. Phys. **C9**, L147 (1976).
14. A. Ferraz, N. H. March and M. Suzuki, Phys. Chem. Liquids **8**, 153 (1978).
15. A. Ferraz, N. H. March and M. Suzuki, Phys. Chem. Liquids **9**, 59 (1979).
16. W. J. Carr, Phys. Rev. **122**, 1437 (1961).
17. C. P. Bean and D. S. Rodbell, Phys. Rev. **126**, 104 (1962).
18. N. P. Grazhdankina, Sov. Phys. Uspekhi **11**, 727 (1969).
19. C. Kittel, *Introduction to Solid State Physics* (Wiley, New York, 1956)

Quantum-Mechanical Wigner Electron Crystallization with and without Magnetic Fields

M. J. LEA
Department of Physics, Royal Holloway and Bedford New College, University of London, Egham, Surrey TW20 0EX, United Kingdom

N. H. MARCH
Theoretical Chemistry Department, University of Oxford, Oxford OX1 3UB, United Kingdom

Abstract

More than 50 years ago, Wigner argued that Coulombically interacting electrons moving in a uniform, nonresponsive, neutralizing background would exhibit a phase transition in the ground state, at sufficiently low density, to a localized electron crystal. For jellium, Ceperley and Alder demonstrated, in the '80s, by quantum computer simulation, that the Wigner electron crystal formed when the mean interelectronic spacing r_s was ~80–100 Bohr radii. The electron crystal is (a) an insulator, (b) magnetic, probably with long-range Néel-type antiferromagnetism, (c) phononlike in its low-lying non-current-carrying excited states, and (d) defective at elevated temperatures, with hopping conduction likely to occur. Durkan, Elliott, and March pointed out, again some 30 years after Wigner, that, under suitable conditions, Wigner electron crystallization could be aided by appropriate application of magnetic fields. These workers considered n-type InSb in a magnetic field, and stressed the importance of observing Bragg reflections in X-ray or neutron experiments. While their considerations were immediately relevant to 3-dimensional Wigner electron crystals, advances in semiconductor technology have now led Andrei et al. to use a high-quality GaAs/GaAlAs heterojunction to study 2-dimensional Wigner crystallization, as induced by application of a magnetic field. Here, after a brief review of these experiments, some discussion is given of the melting curve of a 2-dimensional Wigner crystal in a magnetic field.

Introduction

More than 50 years ago, Wigner [1,2] made the exciting proposal that, in jellium at $T = 0$, Coulombic repulsion e^2/r_{ij} between electrons i and j at separation r_{ij} would eventually localize electrons on the sites of a body-centered-cubic (bcc) lattice; the Wigner electron crystal, as the electron density was lowered sufficiently.

When this area was reviewed by Care and March [3], some 40 years after Wigner's work, the critical electron density, n_c say, for the electron liquid–electron crystal transition in the ground-state of jellium, was not known, their table (p. 107) showing that the corresponding critical interelectronic separation, $r_c = (3/4\pi)^{1/3} n_c^{-1/3}$, from different theoretical models ranged from a value $\sim 6a_0$, $a_0 = \hbar^2/me^2$, i.e., only slightly greater than the density in metallic cesium, to $r_c \sim 3000a_0$. The advent of quantum computer simulation led Ceperley and Alder [4] to pin down r_c to ~ 80–$100a_0$ in the jellium model.

International Journal of Quantum Chemistry: Quantum Chemistry Symposium 23, 717–729 (1989)

CCC 0020-7608/89/230717-13$04.00

Though it was suggested by Egelstaff, March, and McGill [5] that incipient Wigner electron ordering was already present in a variety of liquid metals [6], this proposal being supported by subsequent experimental analysis by Dobson [7] on Na, and Johnson [8] on K, it was, of course, obvious that this was a precursor to Wigner crystallization, as one was dealing with a manifestly metallic state, in contrast to the localization of electrons on the sites of the Wigner crystal. However, because of continuing interest in such "fingerprints" of the Wigner electron crystal, we briefly review the main physical properties of the Wigner electron crystal in the next section. Because of the low density ($r_c \sim 100a_0$) given by the computer experiment of Ceperley and Alder, even expanded metals (see, for example, Ref. [9]) do not afford the right medium for decisive study of Wigner crystallization.

Therefore, in the third section, Wigner crystallization induced by applied magnetic fields will be considered, starting with the proposal of Durkan, Elliott, and March [10], first in the context of highly compensated semiconductors (e.g., n-type InSb in a magnetic field). This work, of course, was carried out before the greatly increased understanding of the Hall effect, though it was designed to interpret the Hall measurements of Putley [11] on n-type InSb. Subsequent experimental work will be reviewed, culminating in the beautiful experiments of Andrei et al. [12] on a 2-dimensional Wigner solid induced by application of a magnetic field. In the fourth section, some brief discussion is given of the theory of the melting curve of the 2-dimensional Wigner crystal. The final section constitutes a summary, along with some proposed directions for future work.

Predicted Physical Properties of Wigner Electron Crystal in Ground and Low-Lying Excited States

Here, we briefly summarize the physical properties which one must expect the Wigner electron crystal to exhibit in its ground and low-lying excited states. These may be listed as follows:

(i) The ground state is insulating, electrons vibrating about the sites of the Wigner bcc lattice.

(ii) Closely related to (i), the Fermi surface discontinuity in the electronic momentum distribution disappears in the insulating phase (see Fig. 1).

(iii) Long-range magnetic order is to be expected in the ground state. Herman and March [13], following earlier work of Carr [14], used the Ceperley–Alder numerical results to argue that, most probably, Néel-type antiferromagnetism exists, upward spin electrons being located on one of the two interpenetrating simple cubic lattices forming the bcc electron crystal, downward spins being on the other.

(iv) Phononlike, noncurrent carrying, low-lying excited states exist.

(v) Defects (e.g., interstitials [15] and vacancies [16]) can occur at elevated temperatures and, via these, small nonzero electronic conductivity can be anticipated at $T \neq 0$.

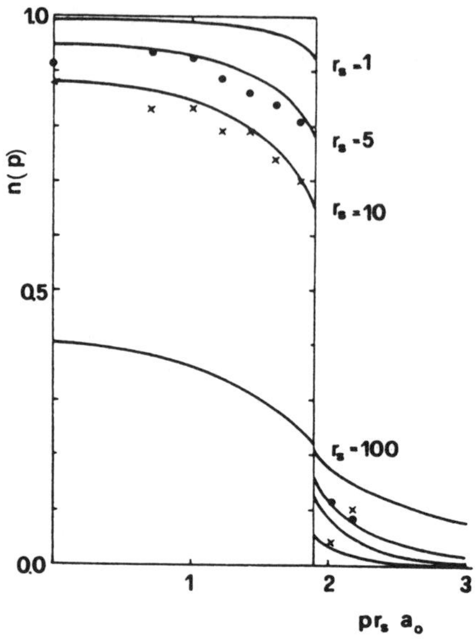

Figure 1. Electronic momentum distribution $n(p)$ in jellium model for various values of mean interelectronic spacing in jellium. Note that discontinuity in $n(p)$ of unity for r_s tends to zero has reduced to practically zero for $r_s = 100$; i.e., near to the Wigner transition to an electron crystal.

So far, only (i) has been tested by "experiment," and then, as discussed, on the computer rather than in the physics laboratory!

Because of the low critical density n_c, reflected in the Ceperley–Alder limits $r_c \sim$ 80–100a_0, it has been clear for a long time that the low density electrons in a suitable semiconductor provide the most promising medium in which to search for Wigner crystallization. This prompted Durkan and March [17] and Durkan, Elliott and March [10] to study magnetic-field dependent effects in semiconductors.

Magnetic-Field-Induced Wigner Crystallization

It is of some interest to review briefly the arguments presented in the work of Durkan, Elliott, and March [10] at the 1968 San Francisco Conference on the metal–insulator transition. Their work was motivated by the experiments of Putley [11] on n-type InSb in a magnetic field, in which a transition was observed which could be interpreted as "conduction in a metallic-like impurity band" formed by overlap of hydrogenic donor impurity wave functions, which, under sufficiently large applied magnetic field B, became so narrow because of reduced overlap in directions perpendicular to the magnetic field that metallic conduction ceased. Arguments by Durkan, Elliott, and March [10], which were order-of-magnitude in nature, strongly suggested that one could expect magnetic fields to aid Wigner electron crystallization.

Some attention was subsequently focussed on the way one might try to observe such Wigner electron crystallization experimentally. The work already cited of Egelstaff, March, and McGill focussed on comparison of X-ray and neutron scattering in (ideally) the lowest-electron density metals (alkalis) while Elliott and his co-workers [18,19] studied neutron scattering specifically from materials with favorable g values.

However, from different directions, evidence began to accumulate much more recently for quantum-mechanical Wigner crystallization in 2-dimensional systems {e.g., abnormal behaviour observed in dc transport for a (Landau level) filling factor $\nu < \frac{1}{5}$ [20]}, and therefore in the remainder of this section we shall focus on this area.

Andrei et al. test directly for a defining property of a solid, namely its rigidity to shear. An unbounded charged 2-dimensional liquid has only plasmon excitations in zero field, related to the number of electrons per unit area, n_s say, by

$$\omega_p^2 = 2\pi e^2 n_s q / \varepsilon m \tag{1}$$

The application of a perpendicular magnetic field H introduces a gap:

$$\omega_+^2 = \omega_p^2 + \omega_c^2 \tag{2}$$

There are no gapless bulk excitations, as previously discussed, for example, by Girvin, MacDonald, and Platzman [21,22].

As Andrei et al. [12] stress, a system capable of resisting static shear, characterized by modulus μ does, in sharp contrast, exhibit a gapless magnetophonon branch of frequency

$$\omega_- = c(2\pi\mu)^{1/2} q^{3/2} / H, \qquad \omega_c \gg \omega_p \tag{3}$$

Andrei et al. [12], entirely convincingly, take the presence of such a mode as positive evidence of a solid phase. In practice, their experiment consists of causing a longitudinal electric wave of defined wave vector q to interact with the electrons and sweeping its frequency. By this technique, they establish beyond reasonable doubt the existence of an electron solid phase.

Melting of 3- and 2-Dimensional Wigner Electron Crystals

Because of the now considerable experimental interest aroused by the study of resistance to shear by the French group [12], it is of obvious interest to consider in more detail the melting curve of the Wigner electron crystal.

Parrinello and March [23] discussed the thermodynamics of Wigner crystallization in d dimensions, and drew a (schematic) picture of the critical r_s for the phase transition versus dimensionality. It should be noted that the electron–electron interaction is there taken to satisfy Laplace's equation in d dimensions.

Though we are interested here in the quantum-mechanical Wigner transition (classical Wigner crystallization is simulated, according to Ferraz and March [24], by the freezing of liquid Na, with relatively uniform conduction electron density, into a bcc metal crystal) Parrinello and March [23] pointed out that the melting curve of the Wigner crystal in 3 dimensions approached the classical melting criterion ($\Gamma = e^2/r_s kT \sim 170$) in the extreme low density limit, because of the absence of any quantum-mechanical tunneling as $r_s \rightarrow \infty$. Ferraz, March, and Suzuki [25,26] interpolated between this limit and the $T = 0$ transition using the Clausius–Clapeyron equation for a first-order electron liquid–solid phase transition in 3 dimensions, and their main conclusions have been subsequently supported by the more quantitative work of Nagara, Nagata, and Nakamura [27] based on a partial summation to all orders of the Wigner–Kirkwood expansion, in 3 dimensions also. Subsequent to the work of Refs. [23–25], Fukuyama, Platzman and Anderson [28] have discussed melting of the 2-dimensional Wigner crystal.

Possible Experimental Evidence for Quantal 3-Dimensional Wigner Crystallization

The early experiments of Somerford [29] on n-type InSb were interpreted by Care and March [16] in terms of a Wigner transition in a magnetic field, and this interpretation was subsequently pressed by Kleppmann and Elliott [18]. In particular, these latter workers could interpret the anisotropy of the conductivity observed by Somerford, even though it is true that the experiments exhibited a large scatter. They concluded that although the interpretation of Somerford's experiment is consistent with the Wigner transition, this does not prove its existence (see also Mansfield [30]). Therefore, below, most attention will be focused on the interpretation of the recent experiments of Andrei et al. [12]. These studied transitions in a magnetic field in a 2-dimensional GaAs heterojunction, as already explained. However, before turning to this experiment, it is of some interest in the context of the work of Somerford [29] to consider first the melting of a 3-dimensional Wigner electron crystal in a magnetic field.

Melting in a Magnetic Field. To illustrate the form of phase diagram to be expected for the Wigner transition in a magnetic field, Figure 2 shows the thermal energy kT_m at melting, in units of e^2/r_s. This diagram has been constructed by making use of the data presented by Kleppmann and Elliott [18]. Adopting their proposal, the melting temperature has been estimated from the difference in ground-state energies of localized and delocalized electrons in an applied magnetic field.

While the indications from Figure 2 are that there could be some (approximate) convergence to a common limit in the high-field regime, the main merit is that the somewhat complicated crossover of curves in the intermediate range [18] is avoided by using the convenient variable $n^{2/3}/H$. At zero temperature, melting leads back to values of $n^{2/3}/H$, which are only relatively weakly dependent on the specimen electron density n, the value of which characterizes the curves labeled A–D in Figure 2.

In spite of some variation in shape associated with the change in electron density from curve A ($n = 10^{22}$ cm^{-3}) to curve D ($n = 5 \times 10^{23}$ cm^{-3}), Figure 2 demonstrates the general trend of a decrease from the limit as H tends to infinity to a "critical" value of $n^{2/3}/H$ at $T_m = 0$ in all four cases. It is interesting, in view of the apparently somewhat simpler results for 2-dimensional melting to be presented below, to comment here as follows:

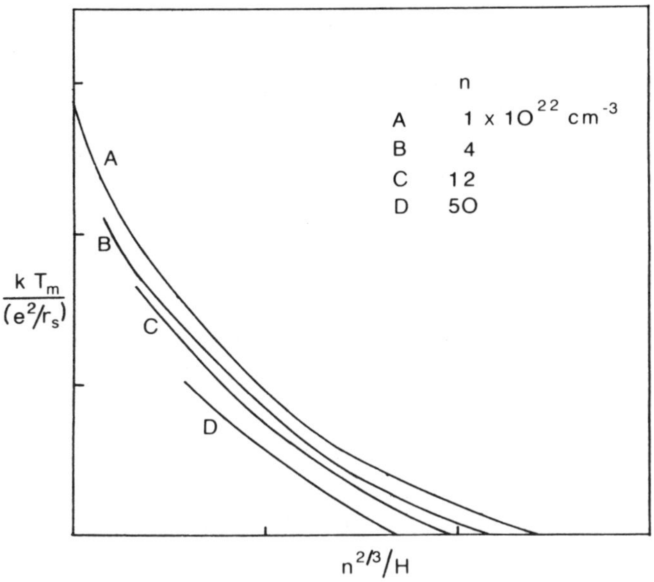

Figure 2. Melting curves of 3-dimensional Wigner crystal in magnetic field H, as constructed from numerical estimates of Kleppman and Elliott [18] of the difference in total energy of localized and delocalized electronic phases.

(i) The Kleppmann-Elliott calculations are variational in character and it is, of course, entirely possible that the difference of the ground-state energies of localized and delocalized phases are subject to some uncertainty.

(ii) More serious is the fact that the phase diagram in Figure 2 has been constructed from a knowledge only of the ground-state energy difference. Undoubtedly the nature of the phase diagram is sensitive to the detail of the elementary excitations in the different phases, as is made clear by the work of Ferraz, March, and Suzuki [26] for the zero-field melting curve.

(iii) The data used to construct Figure 2 above is available only over a relatively limited range of magnetic field, namely a factor of about 20 variation starting from around 10^9 gauss.

Having discussed what can be expected for Wigner electron crystals in 3 dimensions, let us turn now to the interpretation of the data of Andrei et al. [12] for 2-dimensional Wigner crystallization induced by a magnetic field.

General Form of Phase Diagram for Wigner Transition in 2 Dimensions

In the last decade there has been much interest in the Wigner transition in 2 dimensions. Experimentally, the two systems presently available are in semiconductors and at the surface of cryogenic substrates (compare Appendix A for classical Wigner crystallization). In GaAs heterostructures, where high electron mobilities can be achieved, a low effective mass and a high dielectric constant leads to a large Bohr radius ~ 10 nm. Consequently, the values of r_s which can be achieved are low; typically less than 3. Conversely, for electrons on helium, hydrogen, or neon (Appendix A), $r_s > 1000$, which takes one well into the classical regime. It will be instructive to include in the phase diagrams below the clues one has from the classical limit, even though the main focus of the present work is quantum-mechanical Wigner crystallization.

Bearing these points in mind, there are two ways in which it will prove useful to depict (somewhat schematically) the phase diagram associated with the 2-dimensional Wigner transition in a magnetic field. The first of these, shown in Figure 3, closely parallels the discussion of the 3-dimensional case in the subsection Melting in a Magnetic Field above. Here the phase diagram is represented by a plot of reduced temperature T_m/T_{mc}, where T_{mc} is the classical melting temperature vs. filling factor ν. The different curves shown then in Figure 3 correspond to different densities n_s, now related to r_s by $n_s = 1/\pi r_s^2$. If we denote in this 2-dimensional case the critical value of r_s at $T = 0$ and in zero magnetic field by r_W, then the points from the data of Andrei et al. [12], indicated by dots on Figure 3, are in the regime $r_s < r_W$, while the crosses are from the work of Stone et al. [31] in the classical regime. In Appendix B, a theoretical model is presented which will, at least in general terms, allow one to fill in curves also for $r_s \sim r_W$ in Figure 3.

Returning briefly to the work of Stone et al. [31] for 2-dimensional electrons on helium depicted in Figure 3, it is worth adding here some comments on the nature of the effects of a magnetic field.

Effects of a Magnetic Field. A perpendicular magnetic field H will strongly affect the dynamics of the electrons and modify the plasmon and transverse phonon

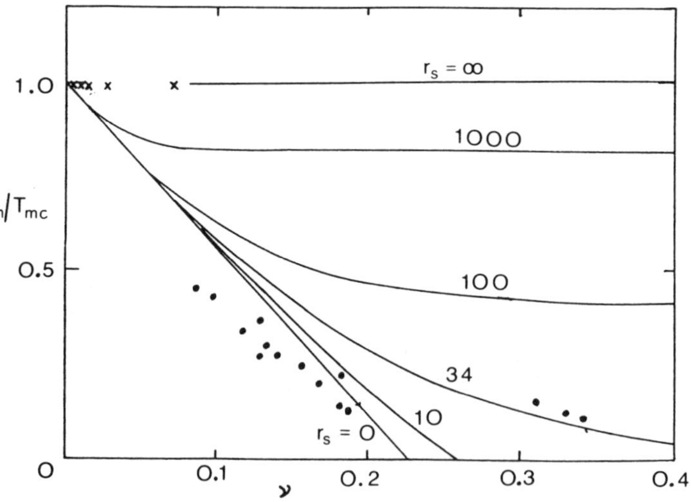

Figure 3. Schematic analogue of Figure 2 for 2-dimensional case. Crosses reflect the trends of the data of Stone et al. [31] in the near classical limit. The dots likewise show the near quantum limit data of Andrei et al. [12]. A precise phase diagram based on the model of Appendix B will be published elsewhere.

spectra in the Wigner crystal (see also Appendix B). However, in the low density region, the effect on the shear modulus [see Eq. (3)] and the stability of the crystal should be small, following the general principle that the thermodynamic variables of a classical system should be independent of H. Thus, the work of Stone et al. [31], depicted already in Figure 3, has demonstrated experimentally that the melting temperature of a 2-dimensional electron solid at $r_s = 3700$ is independent of field for $\nu > 0.005$. In the quantal region, the effect of H is to suppress the zero-point motion and to increase T_m. In the limit of high magnetic fields, $\nu \ll 1$, T_m should approach the classical value T_{mc}, though there may be a correction in real systems due to the finite spatial extent of the localization normal to the 2-dimensional plane, as discussed by Zhang and Das Sarma [32]. A schematic phase diagram in a field is shown in Figure 4; this alternative plot is of $1/r_s$ vs. T, for different values of H. It should be noted in connection with Figures 3 and 4 that for $r_s < r_W$ there should be a critical Landau level filling factor ν_c. The value ν_c^0 in the extreme quantum limit $r_s = 0$ has been estimated by Fukuyama and Yoshioka [33] as 0.104 and by Lozovik et al. [34] as 0.126. At lower fields the electrons are fluid and exhibit the fractional quantum Hall effect in the Laughlin state. Some of these points are dealt with in a little more detail in Appendices A and B.

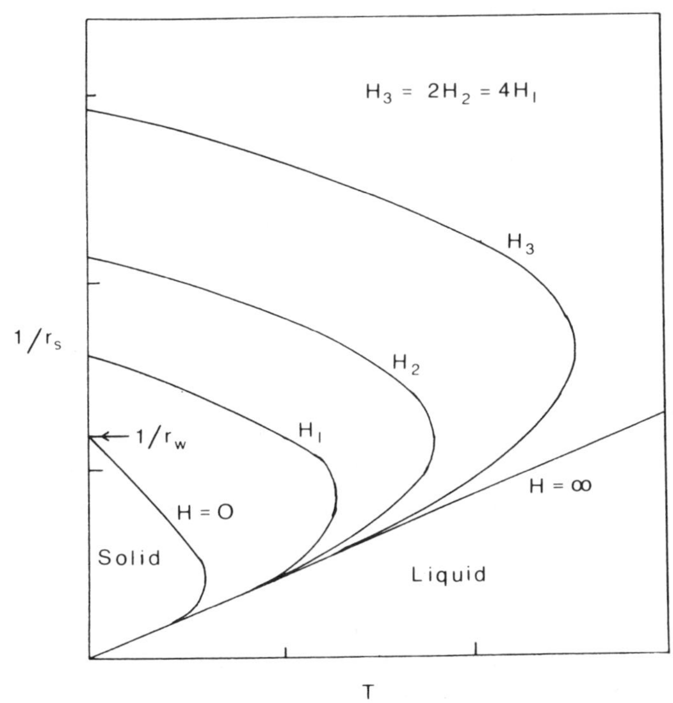

Figure 4. Alternative plot of phase diagram of 2-dimensional Wigner crystal in a magnetic field. The ordinate is now $1/r_s$, plotted against temperature T. Different curves correspond to different magnetic fields.

Summary and Directions for Further Work

As already mentioned above, though the observation of resistance to shear confirms the presence of an "electron solid" rather than an electron fluid, in the work of the French group [12], it does not demonstrate long-range order beyond reasonable doubt. This aspect therefore deserves further study, and appears to necessitate diffraction experiments (cf. Refs. [5], [18], and [19]).

In the light of the 2-dimensional electron solidification and the (at least provisional) mapping out of the melting curve of the Wigner solid, further theory developing that reported in the previous section is clearly called for. Though the French group have ingeniously exploited magnetophonon properties of the 2-dimensional electron solid, it also seems worthwhile, for the future, to reopen (a) the interpretation of the available experiments of Somerford [29] and of Mansfield [30], which may pertain to 3-dimensional Wigner electron solids, and (b) to extend these measurements and their interpretation, by further experimental studies, bearing in mind progress in understanding the Hall effect. As emphasized by the French group, this has resulted in a deeper understanding than was possible at the time of the work of Durkan, Elliott, and March [10] of the basic reasons why magnetic fields are so favorable in inducing Wigner electron solidification; for when the filling factor $\nu < 1$ there is spatial freedom for all the electrons for the same zero-point energy, and this has led to resurgence of theoretical studies [34–39]. In this context, the work of March and Tosi [40] on a localized oscillator in a magnetic field, used in Appendix B below, might fruitfully be combined with the quantum-chemical approach of Jones and Trickey [41], used so far only in zero magnetic field.

APPENDIX A: CLUES FROM CLASSICAL 2D MELTING

In the limit of high magnetic fields, $\nu \ll 1$, the Wigner melting temperature should approach the classical value T_{mc}, as the magnetic length becomes less than the interparticle spacing and the effects of zero-point motion are suppressed. In real systems there may be a correction because of the finite spatial dimension of the localization normal to the 2D plane (Zhang and Das Sarma [32]). It is therefore instructive to look briefly at the experimental and theoretical situation in the low density limit, $r_s > 300$.

The 2D classical solid has been observed experimentally for electrons on the surface of liquid helium by Grimes and Adams [42] and on solid neon by Kajita [43]. This hexagonal solid melts for $\Gamma = \Gamma_m = 127 \pm 3$ (Deville [44]). No significant quantum corrections have been found for $r_s > 1200$, the present experimental limit. It is now clear that this melting is a Kosterlitz–Thouless transition (of Nelson and Halperin [45]) due to the thermal unbinding of dislocation pairs which leads to a jump in shear modulus, μ_s at T_m to $\mu_{KT} = 4\pi k T_m / a^2$. The temperature dependence of μ_s has been measured both indirectly by Gallet et al. [46] and directly by Deville et al. [47] from the transverse branch of the magnetophonon spectrum. Below T_m the modulus is renormalized by thermally activated dislocation pairs and also by strong anharmonic effects. The experimental data have been beautifully reproduced in molecular dynamic computer simulations by Morf [48], which are also in good agreement with theoretical calculations by Fisher [49] and by Chang and Maki [50]. Recent measurements of the specific heat in this system by Glattli et al. [51] showed no latent heat and that the transition was continuous as expected for dislocation unbinding.

The effect of a perpendicular magnetic field in the low density regime should be small, following the general principle that the thermodynamic variables of a classical system should be independent of H. This has been confirmed recently by Stone et al. [31], who found experimentally that the melting temperature of a 2D classical electron solid ($r_s = 3700$) was independent of field for $\nu > 0.005$. Saitoh [52], however, has calculated the phase diagram in a magnetic field using a modified Lindemann criterion and predicted a substantial effect at this density. This perhaps shows the difficulty of relating the dynamics of the solid (which are strongly modified by a field) to a stability criterion such as $\mu_s = \mu_{KT}$.

It is worth mentioning here that the 2D classical solid should melt into an orientationally ordered, or hexatic, phase [44]. This was found in a computer simulation by Frenkel and McTague [53] on a Lennard-Jones solid but has not yet been observed experimentally.

APPENDIX B: AN ELEMENTARY MODEL OF THE MELTING CURVE OF A 2D ELECTRON CRYSTAL IN A MAGNETIC FIELD

Because of the difficulty of studying electron assemblies with strong long-range Coulombic interactions, we present an elementary model for the melting curve of a 2D electron solid, density n, in a perpendicular magnetic field, based on the Lindemann melting criterion. In a field the longitudinal plasmon mode $\omega_p(q)$ and the transverse phonon mode $\omega_t(q)$ are transformed to two magnetophonon modes $\omega_\pm(q)$, where

$$\omega_+^2 = \omega_c^2 + \omega_p^2 + \omega_t^2, \qquad \omega_- = \omega_p\omega_t/\omega_c \tag{4}$$

In high fields, $\omega_c \gg \omega_p$, ω_t the zero-point motion is contained in the ω_+ mode, close to the cyclotron frequency, ω_c. This can be described quantum-mechanically by the theory of March and Tosi [40] for a localized oscillator in a magnetic field with a vector potential $\mathbf{A} = (-\frac{1}{2}Hy, \frac{1}{2}Hx, 0)$. The canonical density matrix of such a Wigner electron oscillator is given by

$$C(\mathbf{r}, \mathbf{r}_0, \beta) = f(\beta)\exp\{-i(x_0y + y_0x)\phi(\beta) - [(x - x_0)^2 + (y - y_0)^2]g(\beta) - [(x + x_0)^2 + (y + y_0)^2]h(\beta)\} \tag{5}$$

where the quantities in Eq. (5) are given explicitly by March and Tosi and later workers. Defining the mean square displacement $\langle r^2 \rangle$ by

$$\langle r^2 \rangle = \frac{\displaystyle\int r^2 C(\mathbf{r}, \mathbf{r}, \beta)d\mathbf{r}}{\displaystyle\int C(\mathbf{r}, \mathbf{r}, \beta)d\mathbf{r}} \tag{6}$$

this is readily calculated to yield

$$\langle r^2 \rangle = \frac{\hbar}{m\omega_c b} \frac{1}{\{\coth(b\alpha) - \cosh\alpha/\sinh(b\alpha)\}} = \frac{\hbar}{m\omega_L bF} \tag{7}$$

where $\alpha = \hbar\omega_L/kT$. The parameter b is given by

$$b = (1 + \omega_W^2/\omega_L^2)^{1/2} \tag{8}$$

where the frequency ω_W of the localised Wigner oscillator can be taken from Bonsall and Maradudin [54] as $8e^2/ma^3$, where a is the lattice spacing. In terms of the parameters $r_s = (\pi n)^{-1/2}/a_0$ (a_0 is the Bohr radius) and the Landau filling factor $\nu = nhc/eH$, the melting criterion becomes, for $\alpha \gg 1$,

$$\frac{\langle r^2 \rangle}{a^2} = \frac{K\nu}{b} = \gamma_M, \qquad b = (1 + D\nu^2 r_s)^{1/2} \tag{9}$$

where $\gamma_M \simeq 0.08$ [38]. The numerical constants K and D (of order unity) should be calculated for a full theory but are used as fitting parameters in this simple model.

In the zero temperature, high field limit the electrons will form a classical 2D solid at all densities. In the extreme quantum limit, $r_s = 0$, there is a critical filling factor $\nu_c^0 = \gamma_M/K$ at which the crystal will melt as the field is reduced. However, for $r_s > r_W = K^2/D\gamma_M$, the crystal will remain stable even in zero field [55], as shown on the phase diagram in Figure 4.

At finite temperatures (but keeping $\alpha \gg 1$) the zero-point motion needs supplementing by contributions from phonon-like models. In zero field the uncoupled ω_p and ω_t modes contain both kinetic and potential energy and give a mean square displacement proportional to the temperature T. In a field the ω_- mode produces elliptical lattice vibrations at low frequency with mostly potential energy but the same longitudinal and transverse displacements as in the separate ω_p and ω_t modes in zero field. Hence the mean displacements due to thermal vibrations are unchanged by a magnetic field. In 2D the sum over all lattice modes diverges logarithmically and the Lindemann criterion is modified so that $\langle r^2 \rangle$ refers to the displacements from the mean position relative to the nearest neighbors. The effect of temperature is thus to add a term

$$\frac{\langle r^2 \rangle}{a^2} = \text{const}\,\frac{T}{T_{mc}}\,\gamma_M \tag{10}$$

to Eq. (9), where T_{mc} is the classical melting temperature given by $e^2/r_s K\Gamma_m$ and $\Gamma_m = 126$ is the ratio of potential to kinetic energy at the classical melting point. In the high field limit, all crystals will melt at T_{mc}.

At finite temperatures and fields, the melting temperature T_m is given by

$$\frac{T_c}{T_{mc}}\gamma_M + \frac{K}{bF(\alpha b)} = \gamma_M \tag{11}$$

This model can be fitted to the data of Andrei et al. [12] who found $\nu_c^0 = 0.23 \pm 0.04$. The resulting phase diagrams for several values of r_s are shown schematically in Figure 3 as T_m/T_{mc} vs. ν. The data of Andrei et al., for $1.6 < r_s < 2.5$, are shown but note that the extrapolation to high field gives $T_m < T_{mc}$ because of a correction due to the perpendicular extent of the electronic wave function. The data of Stone et al. [31] for $r_s = 3700$ confirms the lack of field dependence in the low density limit. This simple model overestimates the effects of a magnetic field for $r_s > 300$. Figure 4 shows schematic phase diagrams for different field strengths on a plot of $1/r_s$ vs. T.

Bibliography

[1] E. P. Wigner, Phys. Rev. **46**, 1002 (1934).
[2] E. P. Wigner, Trans. Faraday Soc. **34**, 678 (1938).
[3] C. M. Care and N. H. March, Adv. Phys. **24**, 101 (1975).
[4] D. M. Ceperley and B. J. Alder, Phys. Rev. Lett. **45**, 566 (1980).
[5] P. A. Egelstaff, N. H. March, and N. C. McGill, Can. J. Phys. **52**, 1651 (1974).
[6] See also S. Tamaki, Can. J. Phys. **65**, 286 (1987).
[7] P. J. Dobson, J. Phys. C**11**, L295 (1978).
[8] M. W. Johnson, private communication and to appear, 1988.
[9] R. G. Chapman and N. H. March, Phys. Rev. B. **38**, 792 (1988); R. Winter and F. Hensel, Phys. Chem. Liq. **20**, 1 (1989).
[10] J. Durkan, R. J. Elliott, and N. H. March, Rev. Mod. Phys. **40**, 812 (1968).
[11] E. H. Putley, Proc. Phys. Soc. (London) **76**, 802 (1960).
[12] E. Y. Andrei, G. Deville, D. C. Glattli, F. I. B. Williams, E. Paris, and B. Etienne, Phys. Rev. Lett. **60**, 2765 (1988).
[13] F. Herman and N. H. March, Solid State Commun. **50**, 725 (1984).
[14] W. J. Carr, Phys. Rev. **122**, 1437 (1961).
[15] F. W. de Wette, Phys. Rev. A**135**, 287 (1964).
[16] C. M. Care and N. H. March, J. Phys. C**4**, L372 (1971).
[17] J. Durkan and N. H. March, J. Phys. C**1**, 1118 (1968).
[18] W. G. Kleppmann and R. J. Elliott, J. Phys. C**8**, 2729 (1975).
[19] R. J. Elliott and W. G. Kleppmann, J. Phys. C**8**, 2737 (1975).
[20] E. Mendez, M. Heiblum, L. L. Chang, and L. Esaki, Phys. Rev. B**28**, 4886 (1983).
[21] S. M. Girvin, A. H. MacDonald, and P. M. Platzman, Phys. Rev. Lett. **54**, 581 (1985).
[22] S. M. Girvin, A. H. MacDonald, and P. M. Platzman, Phys. Rev. B**33**, 2481 (1986).
[23] M. Parrinello and N. H. March, J. Phys. C**9**, L147 (1976).
[24] A. Ferraz and N. H. March, Solid State Commun. **36**, 977 (1980).
[25] A. Ferraz, N. H. March, and M. Suzuki, Phys. Chem. Liq. **8**, 153 (1978).
[26] A. Ferraz, N. H. March, and M. Suzuki, Phys. Chem. Liq. **9**, 59 (1979).
[27] H. Nagara, Y. Nagata, and T. Nakamura, Phys. Rev. A**36**, 1859 (1987); N. H. March, *ibid.* A**37**, 4526 (1988).
[28] H. Fukuyama, P. M. Platzman, and P. W. Anderson, Phys. Rev. B**19**, 5211 (1979).
[29] D. J. Somerford, J. Phys. C**4**, 1570 (1971).
[30] R. Mansfield, J. Phys. C**4**, 2084 (1971).
[31] A. O. Stone, M. J. Lea, P. Fozooni, and J. Frost, 1989, to appear.
[32] F. C. Zhang and S. Das Sarma, Phys. Rev. B**33**, 2903 (1986).
[33] H. Fukuyama and D. Yoshioka, J. Phys. Soc. Jpn. **48**, 1853 (1980).
[34] Yu. E. Lozovik, V. M. Fartzdinov, and B. Abdullaev, J. Phys. C**18**, L807 (1985).
[35] K. Maki and X. Zotos, Phys. Rev. B**28**, 4349 (1983).
[36] P. K. Lam and S. M. Girvin, Phys. Rev. B**30**, 473 (1984).

[37] D. D. Levesque, J. J. Weis, and A. H. MacDonald, Phys. Rev. **B30**, 1056 (1984).
[38] S. T. Chui, T. M. Hakim, and K. B. Ma, Phys. Rev. **B33**, 7110 (1986).
[39] S. Kivelson, C. Kallin, D. P. Arovas, and J. R. Schrieffer, Phys. Rev. **B36**, 1620 (1987).
[40] N. H. March and M. P. Tosi, J. Phys. **A18**, L643 (1985).
[41] R. S. Jones and S. B. Trickey, J. Phys. **C18**, 6355 (1985).
[42] C. C. Grimes and G. Adams, Phys. Rev. Lett. **42**, 795 (1979).
[43] K. Kajita, J. Phys. Soc. Jpn. **54**, 4092 (1985).
[44] G. Deville, J. Low Temp. Phys. **72**, 135 (1988).
[45] D. R. Nelson and B. I. Halperin, Phys. Rev. **B19**, 2457 (1979).
[46] F. Gallet, G. Deville, A. Valdes, and F. I. B. Williams, Phys. Rev. Lett. **49**, 212 (1982).
[47] G. Deville, A. Valdes, E. Y. Andrei, and F. I. B. Williams, Phys. Rev. Lett. **53**, 588 (1984).
[48] R. H. Morf, Phys. Rev. Lett. **43**, 931 (1979).
[49] D. S. Fisher, Phys. Rev. **B26**, 5009 (1982).
[50] M. Chang and K. Maki, Phys. Rev. **B27**, 1646 (1983).
[51] D. C. Glattli, E. Y. Andrei, and F. I. B. Williams, Phys. Rev. Lett. **60**, 420 (1988).
[52] M. Saitoh, Surf. Sci. **196**, 8 (1988).
[53] D. Frenkel and J. P. McTague, Phys. Rev. Lett. **42**, 1632 (1979).
[54] L. Bonsall and A. A. Maradudin, Phys. Rev. **B15**, 1959 (1977).
[55] D. M. Ceperley, Phys. Rev. **B18**, 3126 (1978).

Received April 7, 1989

Phys. Chem. Liq., 1990, Vol. 21, pp. 183–193
Reprints available directly from the publisher
Photocopying permitted by licence only

THE ELECTRON LIQUID-SOLID PHASE TRANSITION IN TWO DIMENSIONS IN A MAGNETIC FIELD

M. J. LEA* and N. H. MARCH†

* *Department of Physics, Royal Holloway & Bedford New College, (University of London), Egham, Surrey, TW20 0EX. England.*

† *Theoretical Chemistry Department, University of Oxford, 5, South Parks Road, Oxford, OX1 3UB. England.*

(Received 27 October 1989)

We present a simple model for the melting curve of a two-dimensional electron solid in a magnetic field and compare it with recent experiments on the phase diagram of 2-D electrons on cryogenic substrates and in GaAs/GaAlAs heterojunctions.

KEY WORDS: Wigner crystallisation, Landau filling factor, melting temperature.

1 INTRODUCTION

Some 50 years ago Wigner[1] proposed that at sufficiently low density the delocalized electron liquid in jellium would give way to a localized electron crystal. The physical reason for this, as he clearly recognized, was that at low density where the potential energy dominates the kinetic contribution to the total energy E, the electrons would avoid each other maximally by going on to the sites of a lattice—the Wigner crystal. The lattice with the lowest Madelung energy is body-centered-cubic (bcc) and most probably this will exhibit long-range antiferromagnetic order[2,3], the upward spins occupying one of the two interpenetrating simple cubic lattices and the downward spins the other.

Unfortunately, it has, to date, not proved possible to simulate Wigner's electron crystal in the laboratory, though quantum Monte Carlo calculations[4] on the jellium model have fully vindicated Wigner's ideas on the electron liquid-electron crystal phase transition. However, in 1968, Durkan, Elliott and March[5] proposed an interpretation of the Hall effect data of Putley[6] on n-type InSb in a magnetic field in terms of Wigner crystallization taking place at a critical value of the applied magnetic field. Subsequent experiments by Somerford[7] on the same system were interpreted by Care and March[8] as a Wigner transition aided by the magnetic field. Later, Kleppmann and Elliott[9,10] pressed this interpretation and showed that the anisotropy of the conductivity measured by Somerford was consistent with the Wigner transition. More recently, McDonald and Bryant[11] have recalculated the energy of a 3-D electron plasma in a magnetic field including a precise treatment of the exchange energy. They find a phase diagram with several other types of ground state ordering besides the simple electron Wigner crystal, depending on the field strength and the density.

Very recently, interest in the magnetically induced Wigner solid (MIWS) has been revived by the beautiful experiments of Andrei *et al.*[12] and Glattli *et al.*[13] on a 2-D electron assembly in a GaAs/GaAlAs heterojunction. These workers pointed out that a crucial difference between an electron liquid (delocalized state) and an electron solid is that only the latter can sustain low frequency shear waves. Evidence was given by Andrei *et al.*[12] that their 2-D electrons, in a magnetic field, could support shear waves. Though an alternative interpretation of the experimental data of Andrei *et al.* has been suggested by Stormer and Willett,[14] the experiments have now been repeated and the MIWS provides the most probable explanation of the data. The sharp onset of a propagating mode was used to plot the phase diagram for Wigner crystallisation as a function of temperature and field.

The purpose of this paper is to present a simple model of two-dimensional Wigner crystallisation and its melting curve to an electron liquid phase in an applied magnetic field. This model will then be compared with experiments in both classical and quantum 2-D electron solids.

2 THE WIGNER OSCILLATOR MODEL

The model incorporates, in a phenomenological way, the quantum effects of zero-point motion, the Kosterlitz-Thouless melting criterion and some aspects of the anharmonicity of the electron lattice vibrations. The work of March and Tosi[15] on a localised harmonic oscillator in a magnetic field is taken as the starting point. They calculated the canonical matrix for this system which reduces to the result for free electrons of Sondheimer and Wilson[16] as the oscillator force constant tends to zero. The Landau energy levels for free electrons in a magnetic field of arbitrary strength are embodied fully in this limit. With non-zero force constant for the localised oscillator, the energy levels calculated by Darwin[17] are involved in the calculation of the canonical density matrix and therefore of its trace which is essentially the partition function.

The idea behind the present work is then simply stated. Each electron is taken as oscillating in a harmonic potential well produced by the other electrons in the Wigner crystal. This is an Einstein model including field dependent zero point motion. The mean square displacement $\langle r^2 \rangle$ in the plane will be field and temperature dependent and can be calculated from the canonical density matrix $C(\mathbf{r}, \beta)$ as

$$\langle r^2 \rangle = \frac{\int r^2 C(\mathbf{r}, \beta) \mathrm{d}\mathbf{r}}{\int C(\mathbf{r}, \beta) \mathrm{d}\mathbf{r}} \tag{1}$$

where $\beta = 1/kT$. This calculation[18] gives

$$\langle r^2 \rangle = \frac{\hbar}{m\omega_L b F(\alpha, b)} \tag{2}$$

where ω_L is the Larmor frequency $eH/2mc$,

$$F(a, b) = \coth(b\alpha) - \frac{\cosh(\alpha)}{\sinh(b\alpha)} \tag{3}$$

and $\alpha = \hbar\omega_L/kT$. The parameter b is given by

$$b = (1 + \omega_W^2/\omega_L^2)^{1/2} \tag{4}$$

where ω_W is the natural frequency of the localised Wigner oscillator, which will be close to the characteristic phonon frequency ω_0 given by Bonsall and Maradudin[19] as $\omega_0^2 = 8e^2/ma^3$.

It is convenient to work in terms of the reduced variables $t = T/T_{mc}$ where T_{mc} is the melting temperature of the classical crystal, the Landau level filling factor $\nu = nhc/eH$ and $r_s = r_0/a_B$ where $r_0 = (\pi n)^{-1/2}$ in 2 dimensions. Eq. (2) then becomes

$$\frac{\langle r^2 \rangle}{a^2} = \frac{(\sqrt{3}/2\pi)\nu}{bF(\alpha, b)} = \frac{0.276\nu}{bF(\alpha, b)}$$

$$\text{with } \alpha = \frac{\Gamma_m}{\nu r_s t}; \qquad b = (1 + D\nu^2 r_s)^{1/2} \tag{5}$$

and we have assumed a triangular lattice for the 2-D crystal. The only free parameter is the constant $D = 1.158\,(\omega_W/\omega_L)^2$ which defines the natural frequency of the Wigner oscillators. It is instructive to consider the limiting forms of eq. (5). For $r_s \gg 1$, in the classical limit

$$\frac{\langle r^2 \rangle}{a^2} = \frac{0.552t}{\Gamma_m D} \tag{6}$$

and this result is also obtained in the high field limit, $\nu = 0$, for all values of r_s. Thus, as is well known, a magnetic field suppresses the zero-point motion and produces a classical 2-D crystal. At $t = 0$, the only contribution to $\langle r^2 \rangle$ comes from the zero-point motion in which the quantum limit, $r_s = 0$, is given by

$$\begin{aligned} \langle r^2 \rangle/a^2 &= 0.276\nu \\ &= 2l_B^2/a^2 \end{aligned} \tag{7}$$

where $l_B = hc/eH$ is the magnetic length. For $r_s \gg 1$ at $t = 0$ we have the zero-point motion of the 2-D oscillator

$$\langle r^2 \rangle/a^2 = \frac{0.276}{(Dr_s)^{1/2}} \tag{8}$$

In this model $\langle r^2 \rangle$ is defined with respect to a lattice point in the crystal. But in 2-D $\langle r^2 \rangle$ diverges logarithmically[20] as the overall crystal size increases to infinity, due to low frequency acoustic modes. However, this does not prevent the formation of a crystal as the relevant parameter is the mean relative displacement of nearest neighbours defined as

$$\langle (r_i - r_{i+1})^2 \rangle/a^2 = \gamma \tag{9}$$

and called the mean melting parameter by Lozovik *et al.*[21] For our Einstein model we have

$$\gamma = \langle r_i^2 \rangle / a^2 + \langle r_{i+1}^2 \rangle / a^2 - \langle 2 r_i r_{i+1} \rangle / a^2$$
$$= 2 \langle r^2 \rangle / a^2 \tag{10}$$

since the motion of neighbouring electrons is uncorrelated.

There is strong experimental evidence[22] that the melting of a 2-D classical crystal is a Kosterlitz-Thouless (K-T) transition due to the thermal unbinding of dislocation pairs when the shear modulus μ is given by

$$\mu = 4\pi kT/a^2. \tag{11}$$

Hence the melting temperature T_{mc} of a 2-D classical crystal is

$$T_{mc} = \frac{e^2}{r_0 k \Gamma_m} \tag{12}$$

where $\Gamma_m = 127 \pm 3$, as determined experimentally for electrons on helium.[23]

In order to apply the K-T criterion to our general model we require a relationship between μ and the parameter γ. In the classical limit the shear modulus decreases with temperature (for $t \ll 1$) as

$$\mu^* = \mu/\mu_0 = 1 - Ct \tag{13}$$

where μ_0 is the zero temperature modulus and C has been measured experimentally[23] as 0.3 ± 0.1. Computer simulations by Morf[24] (which gave $C = 0.23$) and Bedanov *et al.*[25] and calculations by Lozovik *et al.*[21] strongly suggest that the decrease in μ is due to anharmonicity of the electron lattice. However, this anharmonicity seems to have much less effect on γ which increases almost linearly with temperature,[25] as eq. (6) until the K-T transition intervenes at $\gamma = \gamma_c = 0.03$. In this classical limit this Lindemann melting criterion and the K-T transition both lead to a melting temperature T_{mc} as given in eq. (12). We can use this result to determine the vibrational parameter $D = 1.104/\gamma_c \Gamma_m = 0.292$ in our model eq. (5). In the high density, quantum limit at $T = 0$ the K-T transition is suppressed (i.e. only occurs when μ falls to zero) and μ will continue to decrease as $\gamma = 0.552v$ increases beyond γ_c. Lozovik *et al.*[21] have calculated μ and γ versus v in this limit and found that μ decreased and γ increased almost linearly until the crystal became unstable at $v = v_q$ (0.126 in their calculations) due to anharmonicity.

Combining the classical and quantum results we therefore assume that, in general,

$$\begin{aligned} \mu^* &= 1 - A\gamma(r_s, H, T); \gamma \le \gamma_q \\ &= \mu_1 \qquad ; \gamma = \gamma_q \\ &= 0 \qquad ; \gamma \ge \gamma_q \end{aligned} \tag{14}$$

where γ is given by eqs. (10) and (5), until $\gamma = \gamma_q$ at $\mu^* = \mu_1$ when the crystal becomes unstable. Glattli *et al.*[13] found that the low temperature limit of their postulated Wigner crystal occurred for $v = v_q = 0.192$ which is equivalent to $\gamma_q = 0.106$ and we use this value. Hence the only free parameter in our model is μ_1. Even this is tightly constrained since, in eq. (13), $C = (1 - \mu_1)\gamma_c/\gamma_q$ and hence for $C = 0.23$[24], $\mu_1 = 0.19$ while for $C = 0.3 \pm 0.1$[22] we find $\mu_1 = -0.06 \pm 0.35$.

Finally, from eqs. (11) and (14) the 2-D melting temperature t_m is given by

$$t_m = \frac{1 - A\gamma(r_s, H, T)}{1 - A\gamma_c}; \gamma \leq \gamma_q$$

$$= 0 \qquad ; \gamma > \gamma_q \tag{15}$$

For $\gamma < \gamma_q$ this will be a K-T transition while for $\gamma = \gamma_q$ the transition could be driven by an anharmonic instability in the lattice vibrations, though other mechanisms could be invoked. Note that eq. (15) is not equivalent to a pure Lindemann criterion since γ_m, the value at the melting point, is

$$\gamma_m = t_m\gamma_c + (1 - t_m)\gamma_0 \tag{16}$$

where $\gamma_0 = 1/A$. Our model corresponds to a temperature dependent γ_m.

At finite temperatures we should also consider the thermally excited modes in the crystal. In a magnetic field the longitudinal plasmon mode $\omega_p(q)$ and the transverse phonon mode $\omega_t(q)$ are transformed to two magnetophonon modes $\omega_\pm(q)$ where

$$\omega_+^2 = \omega_c^2 + \omega_p^2 + \omega_t^2; \omega_- = \omega_p\omega_t/\omega_c. \tag{17}$$

In a complementary approach to the one taken in this paper Saitoh[26] calculated $\langle r^2 \rangle$ from the excitation of these collective ω_+ and ω_- modes, and used the Lindemann criterion to obtain the melting curve. The Einstein model gives qualitatively similar results, though Saitoh's calculations predict larger quantum effects at relatively low electron densities.

In high fields $\omega_c \gg \omega_p$, the zero-point energy is contained in the ω_+ mode, close to the cyclotron frequency. It is this motion that has been treated quantum-mechanically in the Einstein model above. The low frequency transverse ω_- mode will be thermally excited for

$$\frac{\hbar\omega_-}{kT} \approx \frac{\hbar\omega_c}{kT}\left(\frac{\omega_0}{\omega_c}\right)^2 \approx \frac{\Gamma_m \nu}{2t} < 1 \tag{18}$$

where ω_c is the cyclotron frequency. It was the propagation of the ω_- mode which Andrei *et al.*[12] used as the criterion for a solid. But this mode will only contribute thermally to $\langle r^2 \rangle$ for $\nu < 0.02t$ in the high field classical limit. It would then give a term proportional to T but since eq. (5) already provides such a term in this limit no extra contribution from the ω_- mode is required.

Figure 1 shows a schematic phase diagram in the t-ν plane for several values of r_s. In the classical low density limit, $r_s \to \infty$, the melting temperature is independent of magnetic field. As the density increases the zero field melting temperature decreases. In the quantum limit, $r_s \to 0$, the phase diagram is independent of density and the melting temperature has its classical value for $\nu = 0$. As ν increases t_m decreases, but remains a K-T transition until $\gamma = \gamma_q$ when some other instability occurs at $\nu = 0.192$

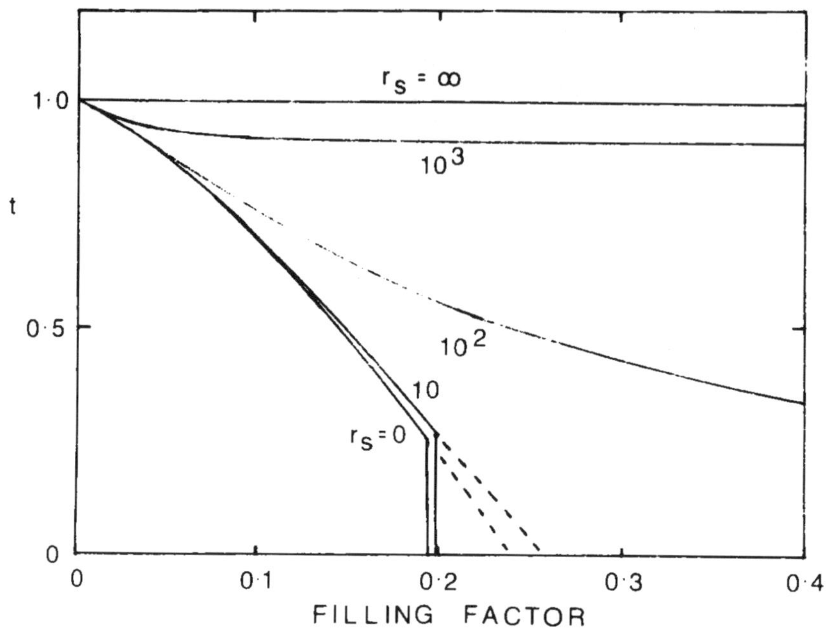

Figure 1 Schematic phase diagram of the 2-D electron solid for several values of r_s.

for the parameters used here. The dashed lines show the extrapolation of the K-T transition.

We will now compare this simple model with experiments on classical and quantum 2-D electron solids.

2.1. *The classical 2-D electron solid*

It is well established that in the classical, or low density limit, $r_s = \infty$, the melting temperature equals T_{mc}, eq.(12), for a K-T transition with $\Gamma_m = 127 \pm 3$ [23]. A magnetic field should have no effect on the thermodynamics of a classical solid and T_m should be independent of H.[28,29] However, at high densities T_m should be less than T_{mc}. These effects have been calculated by several authors,[26–31] notably by Chang and Maki,[29] Fisher[28] and Saitoh[26] though the results vary considerably.

Saitoh[26] calculated γ for the longitudinal and transverse magnetoplasmons in the 2-D crystal and used the modified Lindemann criterion $\gamma = \gamma_m = 0.03$ to obtain the phase diagram for all densities. His results predict relatively large increases in $T_m(B)$ with magnetic field at densities which are easily accessible for electrons on helium. Chang and Maki[29] calculated the shear modulus, allowing for anharmonicity, for all densities in zero field. In the classical region, $r_s > 300$ they found that the small quantum corrections increased μ, i.e. stiffened the crystal, and hence $T_m(0) > T_{mc}$ though the effect was small. Fisher[28] calculated the shear modulus for a high density crystal in the high-field, or classical, limit and also found an initial increase in $T_m > T_{mc}$ as the field is reduced from infinity.

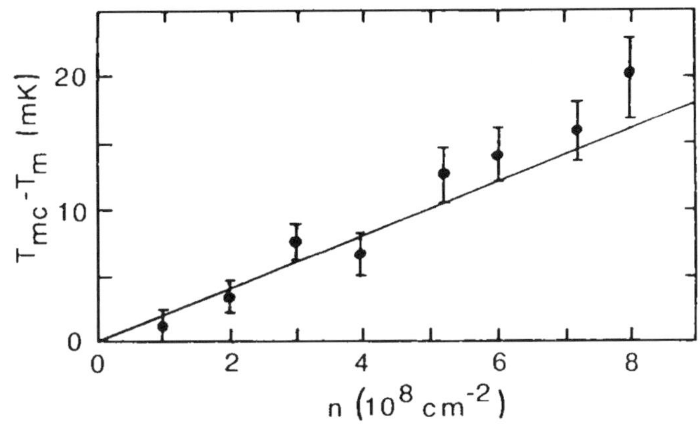

Figure 2 The experimental data of Mehrotra *et al.*[32] for 2-D electrons on helium showed a decrease in the melting temperature in zero field below the classical result Eq. (12). The solid line shows the model calculations.

Our model is equivalent to a temperature dependent Lindemann criterion, eq. (16), and hence will always give $T_m < T_{mc}$ though the magnitude of the quantum corrections will be less than the pure Lindemann criterion used by Saitoh.

Experimentally, the situation is unclear. In zero field Mehrotra, Guenin and Dahm[32] found a decrease in T_m below T_{mc} for electrons on helium for $n < 8 \times 10^8\ \mathrm{cm}^{-2}$ as shown in Figure 2. The solid line shows the quantum corrections calculated from our model with $\mu_1 = 0.2$, which are in good agreement with the experiments. No quantum corrections to T_m were reported by Kono[33] or Deville[22] for $n < 12 \times 10^8\ \mathrm{cm}^{-2}$ for electrons on helium or by Kajita[34] for $n < 2 \times 10^{10}\ \mathrm{cm}^{-2}$ for electrons on solid neon, though in each case the measured T_m and the predicted quantum corrections are within the error bars.

Recently, Stone *et al.*[35] have measured the field dependence of T_m for electrons on helium for $n = 8.1 \times 10^8\ \mathrm{cm}^{-2}$ as shown in Figure 3. They found that T_m was almost independent of field, possibly decreasing slightly to 7 T. These results are in better agreement, within the error bars, with the theories of Fisher[28] and Chang and Maki[29] than with the simple model presented here.

However, both the experiments and theory are difficult and subtle in this region. The main conclusion is that quantum effects are small and difficult to compute or observe in the experimentally accessible region for electrons on cryogenic substrates.

2.2. *The quantum 2-D crystal*

The 2-D electron assembly has been extensively studied in Si MOSFET's and in GaAl/GaAlAs heterostructures. In zero magnetic field these systems are in the high density limit and the electrons form a Fermi degenerate gas. As the magnetic field is increased, the Integer and Fractional Quantum Hall effects are observed.[36] The latter is characterized by a series of new ground states of the electron fluid, which is providing a wealth of new Physics. At the highest fields, for $\nu < 0.2$, Andrei *et al.*[12] and

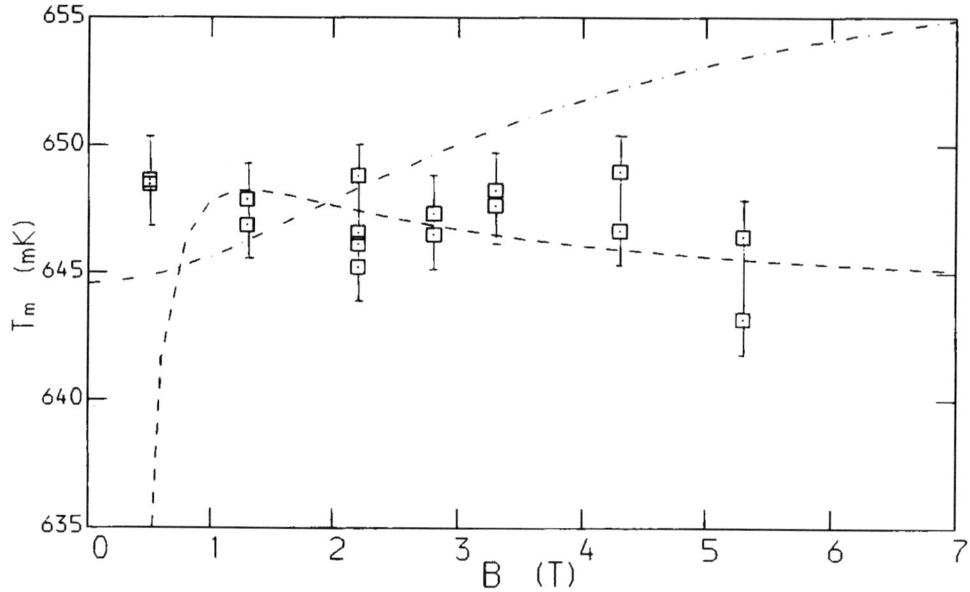

Figure 3 The experimental data of Stone *et al.*[35] for 2-D electrons on helium, $n = 8.1 \times 10^8$ cm^{-2}, showed that T_m was almost independent of magnetic field from 0.5 to 7*T*. The dashed line shows the calculation of Fisher[28] (with $T_{mc} = 643.5$ mK) while the dot-dashed line shows the result of the model used here (with $T_{mc} = 662$ mK).

Glattli *et al.*[13] have reported the probable occurrence of the MIWS for $\nu < \nu_q = 0.192 \pm 0.004$. The electron density for these experiments gives $1.6 < r_s < 2.5$. The experimental solid-fluid transitions t_m in the t-ν plane are shown in Figure 4. The data for many densities is shown and while the error bars are quite large, the general shape of the phase diagram is clear.

For $r_s \leq 5$, our model calculations suggest that, on a t-ν plot, the phase diagram corresponds to the quantum limit and is independent of density. The model phase diagram is shown in Figure 4 for $\mu_1 = 0.2$ (line a) and 0 (line b). For $\mu_1 = 0$ the transition remains of the K-T type down to the lowest temperatures. However, a better fit to the data is obtained for $\mu_1 = 0.2$ with a sharp drop in t_m at $\nu_q = 0.192$. Thus the transition would be of K-T type for $t_m > 0.24$, but some other form of crystalline instability would occur at lower temperatures. In a real system the detailed form of the phase diagram near ν_q would depend on the increasing anharmonicity and the nature of the excitations and ground state of the solid and fluid phases.

In the high field limit, $\nu \to 0$, the transition temperature of the ideal 2-D system should tend to T_{mc}. But at high electon densities the finite extent of the wave functions perpendicular to the 2-D plane will modify the Coulomb interaction[37] and should reduce t_m by a factor of about 0.87.[12] With this in mind the model phase diagram agrees reasonably well with the data.

This model can also be used to calculate the phase diagram for intermediate values of r_s, a region not yet accessible experimentally. A key parameter here is the limiting

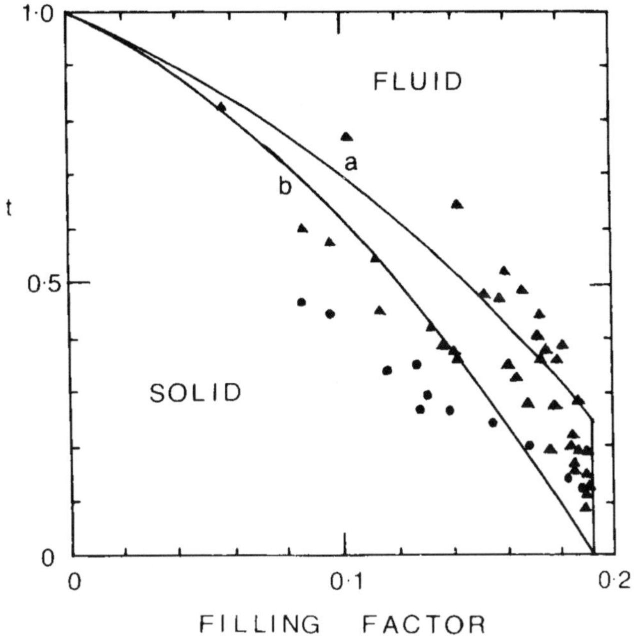

Figure 4 The experimental data of Andrei *et al.*[12] (o) and of Glattli *et al.*[13] (Δ) for 2-D electrons in GaAs/GaAlAs heterostructures. The points represent the onset of a new mode in the response of the electron gas which was interpreted as a phase transition to an MIWS. The solid lines show the model phase diagram for $\mu_1 = 0.2$ (line a) and $\mu = 0$ (line b).

value of $r_s = r_W$ at which the 2-D Wigner solid in stable in zero magnetic field. For our model that is given by

$$r_W = \frac{0.276\gamma_c \Gamma_m}{\gamma_q^2} \tag{20}$$

and has the value 93 for the parameters used. This has not yet been measured experimentally though a wide range of values have been found theoretically[27,25] from 4.5 to 1400. Probably the best estimate, $r_W = 33$, comes from the computer simulations of Ceperley[30] and Imada and Takahashi[31]. In this region an Einstein model may be less valid as the zero-point motion of the propagating plasmons will determine $\langle r^2 \rangle$ and the crystal could remain stable to higher densities.

It is interesting to speculate on the existence of the 2-D hexatic phase on this phase diagram. In the classical limit, Nelson and Halperin[38] argued that the K-T transition should be from a solid to an orientationally-ordered phase with the transition to a true fluid occurring at a temperature $t_1 \geq 1$. This hexatic phase has not yet been observed experimentally, though a transition in viscosity was found by Frankel and McTague[39] in a computer simulation, at $t_1 = 1.58$. Since the $\nu = 0$ phase diagram in the high density limit is the same as the classical one, this phase should also be present. Other phases of the fluid or solid could also exist.

SUMMARY

In 2-D a simple phenomenological model has been presented as a framework for discussion of the experiments on the 2-D classical electron solid on helium and neon and on the MIWS in GaAs/GaAlAs heterostructures. The parameters of the model are determined, within close limits, by results from computer simulations and experiments. While it is not expected to be fully quantitative it does give a guide to the phase diagram in both the classical and quantum regions.

The similarity between the results of Andrei *et al.*[12] and Glattli *et al.*[13] and the model developed in the present paper lends strong support to their claim that quantum-mechanical electron solidification has occurred in their experiment, aided by a magnetic field. Much more sophisticated theoretical approaches[5,13,27–31,40] have been used for aspects of this problem and may be generalised in the future. Many other types of ground state order could also be present in the phase diagram besides the simplest electron solid discussed here.

As for further experimental studies, propagation of shear waves establishes an electron solid, but not longer range order. Bragg diffraction studies are still needed to establish crystallinity for the Wigner solid in semiconductors, as has been done using ripplons for the 2D classical solids of electrons on helium[41] and of positive ions below the helium surface.[42]

References

1. E. P. Wigner (1934), *Phys. Rev.* **46**, 1002; *Trans. Faraday Soc.* **34**, 678 (1938).
2. W. J. Carr (1961), *Phys Rev.* **122**, 1437.
3. F. Herman and N. H. March (1984), *Solid State Commun.* **50**, 725.
4. D. M. Ceperley and B. J. Alder (1980), *Phys. Rev. Lett.* **45**, 566.
5. J. Durkan, R. J. Elliott and N. H. March (1968), *Rev. Mod. Phys.* **40**, 812.
6. E. H. Putley (1960), *Proc. Phys. Soc.* (London) **76**, 802.
7. D. J. Somerford (1971), *J. Phys.* **C4**, 1570.
8. C. M. Care and N. H. March (1971), *J. Phys.* **C4**, L372.
9. W. G. Kleppmann and R. J. Elliott (1975), *J. Phys.* **C8**, 2729.
10. R. J. Elliott and W. Kleppmann (1975), *J. Phys.* **C8**, 2737.
11. A. H. McDonald and G. W. Bryant (1987), *Phys. Rev. Letts.* **58**, 515.
12. E. Y. Andrei, G. Deville, D. C. Glattli, F. I. B. Williams, E. Paris and B. Etienne (1988), *Phys. Rev. Lett.* **60**, 2765.
13. D. C. Glattli, G. Deville, V. Duburcq, F. I. B. Williams, E. Paris, B. Etienne and E. Y. Andrei, Proc. 8th *Intern. conf. on Electron Prop. of 2-D Systems*, Surface, Science. In Press.
14. H. L. Stormer and R. L. Willett (1989), *Phys Rev. Lett.*, **62**, 972.
15. N. H. March and M. P. Tosi (1985), *J. Phys.* **A18**, L643.
16. E. H. Sondheimer and A. H. Wilson (1951), *Proc. Roy. Soc.* **A 210**, 173.
17. C. G. Darwin (1931), *Proc. Camb. Phil. Soc.* **27**, 86.
18. N. H. March and M. P. Tosi (1985), *Nuovo Cimento* **6D**, 521.
19. L. Bonsall and A. A. Maradudin (1977), *Phys. Rev. B.* **15**, 1959.
20. N. D. Mermin (1968), *Phys Rev.* **176**, 250.
21. Yu. E. Lozovik, V. M. Fartztdinov and B. Abdullaev (1985), *J. Phys. C.* **18**, L807.
22. G. Deville (1988), *J. Low Temp. Phys.* **72**, 135.
23. F. Gallet, G. Deville, A. Valdes and F. I. B. Williams (1982), *Phys Rev. Letts.* **48**, 212; G Deville, A. Valdes, E. Y. Andrei and F. I. B. Williams (1984), *Phys. Rev. Letts.* **53**, 588.
24. R. H. Morf (1979), *Phys. Rev. Letters* **43**, 931.
25. V. M. Bedanov, G. V. Gadiyak and Yu. E. Lozovik (1985), *Phys. Lett.* **109A**, 289.

26. M. Saitoh (1988), *Surf. Sci.* **196**, 8.
27. P. M. Platzman and H. Fukuyama (1974), *Phys. Rev. B* **10**, 3150.
28. D. S. Fisher (1982), *Phys. Rev. B* **26**, 5009.
29. M. Chang and K. Maki (1983), *Phys. Rev. B* **27**, 1646.
30. D. Ceperley (1978), *Phys. Rev. B* **18**, 3126.
31. M. Imada and M. Takahashi (1984), *J. Phys. Soc. Japan*, **53**, 3770.
32. R. Mehrotra, B. M. Guenin and A. J. Dahm (1982), *Phys. Rev. Letts.* **48**, 641.
33. K. Kono (1987), *J. Phys. Soc. Japan* **56**, 1111.
34. K. Kajita (1985), *J. Phys. Soc. Japan*, **54**, 4092.
35. A. O. Stone, M. J. Lea, P. Fozooni and J. Frost: *J. Phys. CM*. In Press.
36. K. von Klitzing, G. Dorda and M. Pepper (1980), *Phys Rev. Lett.* **45**, 494; D. C. Tsui, H. L. Stormer and A. C. Gossard (1982), *Phys. Rev. Lett.* **48**, 1559; R. B. Laughlin (1983), *Phys. Rev. Lett.* **50**, 1395.
37. F. C. Zhang and S. Das Sarma (1986), *Phys. Rev. B* **33**, 2903.
38. D. R. Nelson and B. I. Halperin (1979), *Phys. Rev.* **19**, 2457.
39. D. Frenkel and J. P. McTague (1979), *Phys. Rev. Letts.* **42**, 1632.
40. K. Maki and X. Zotos (1983), *Phys Rev. B* **28**, 4349; P. K. Lam and S. M. Girvin (1984), *Phys. Rev. B* **30**, 473; D. Levesque, J. J. Weis and A. H. MacDonald (1984), *Phys. Rev. B* **30** 1056; S. T. Chui, T. M. Hakim and K. B. Ma (1986), *Phys. Rev. B* **33**, 7110; S. Kivelson, C. Kallin, D. P. Arovas and J. R. Schrieffer (1987), *Phys. Rev. B.* **36**, 1620.
41. C. C. Grimes and G. Adams (1979), *Phys. Rev. Lett.* **42**, 795.
42. C. J. Mellor and W. F. Vinen, *Proc 8th Intern. Conf. on Electronic Properties of 2-D Systems*, Surface Science. In Press.

J. Phys.: Condens. Matter 3 (1991) 3493–3503. Printed in the UK

The shear modulus and the phase diagram for two-dimensional Wigner electron crystals

M J Lea† and N H March‡

† Department of Physics, Royal Holloway and Bedford New College, University of London, Egham Hill, Surrey TW20 0EX, UK

‡ Theoretical Chemistry Department, University of Oxford, 5 South Parks Road, Oxford OX1 3UB, UK

Received 31 December 1990, in final form 20 February 1991

Abstract. Defect energies in two- and three-dimensional classical crystals correlate with the shear modulus μ; in turn this relates the melting temperature T_m intimately to μ. Thus, a model is first proposed for the shear modulus for two-dimensional Wigner crystals. The melting temperature is then determined from the Kosterlitz–Thouless melting criterion or from an anharmonic instability inherent in the model. The relative positions of these transitions depend on the model parameters used. The calculation is generalized to include (a) zero-point motion which is dominant in the quantum limit and (b) the effect of a magnetic field. For the high field case, this modelling allows T_m to be plotted versus the Landau-level filling factor ν. The predictions of the model are thereby brought into contact with the experiments of Andrei *et al* and Glattli *et al* which have been interpreted as evidence for a magnetically induced Wigner solid (MIWS) in the electron assembly in a GaAs/AlGaAs heterojunction in strong magnetic fields. The model exhibits some of the features observed experimentally.

1. Introduction

In classical monatomic crystals, correlations have long been known to exist between melting temperature T_m, vacancy formation energy E_v and elastic moduli. Some understanding of these empirical correlations has been afforded by (a) statistical mechanical models at elevated temperatures, appropriate say to condensed argon (Bhatia and March 1984) and (b) current models of force fields including many-body contributions in metals like Cu (Johnson 1988, see also March 1989). In case (b), Johnson has stressed that the highest correlation among elastic constants and E_v is via the shear modulus μ. While the above relates to three dimensions, the well-known Kosterlitz–Thouless (1973, 1978) transition in two-dimensional classical crystals is driven by the shear modulus and the thermal unbinding of dislocation pairs.

Here the focus is on 2D Wigner electron crystallization, with and without magnetic fields. In zero magnetic field B there is a large body of work to date: we note in particular two very relevant studies. One is the classical limit of the phonon spectrum, worked out by Bonsall and Maradudin (1977). This provides a first-principles basis for scaling the model calculations made in the present paper for the shear modulus. The second is the quantum Monte Carlo study of Ceperley (1978), in which he calculated the mean interelectronic separation r_W, relative to the Bohr radius, at which the transition from electron liquid to 2D Wigner electron crystal occurs, in the quantal limit at $T = 0$. He found the value $r_W = 33$ which again is a most valuable piece of information in the present modelling. Turning to non-zero magnetic fields, which is in fact the prime motivation for the present study, the magnetic-field-assisted Wigner electron crystallization proposed by Durkan *et al* (1968) has been invoked by Andrei *et al* (1988) and Glattli *et al* (1990) as an explanation of their experiments on 2D electrons. These workers demonstrated the sudden onset of a new electron-phonon mode at low frequencies in the electron assembly in a GaAs/AlGaAs heterojunction in a high magnetic field that they suggested was a fingerprint of the magnetically induced Wigner solid (MIWS). They were thus able to map out the melting curve of the proposed Wigner solid as a function of the Landau-level filling factor $\nu = nh/eB$. Once such a plot was made there remained only a rather weak residual dependence on the carrier density, n, related to the mean interelectronic separation, r_0, by

$$n = (\pi r_0^2)^{-1}. \tag{1.1}$$

While there have been a number of attempts to calculate the melting curve of the Wigner crystal for $B = 0$ (Ferraz *et al* 1978, 1979; Nagara *et al* 1987, see also March 1988), it is only recently that much attention has been focused on melting in the presence of an applied magnetic field (Elliott and Kleppmann 1975, Saitoh 1988, Lea and March 1990). This is therefore one focal point of the present work: to propose a model to predict the melting curve in a magnetic field.

The outline of the paper is then as follows. In section 2, a model calculation is proposed of the shear modulus μ of a 2D Wigner electron crystal. The idea behind the calculation is motivated by extending the approach of Bonsall and Maradudin (1977), which applied in the limit of complete electron localization. Section 3 is then concerned with self-consistent solutions for three limiting cases of the model: (i) the classical limit, (ii) the quantum limit where zero-point motion dominates and (iii) the high magnetic field regime. The relationship to experiment, both the heterojunction data referred to above for the high magnetic field case and the quantum Monte Carlo result for zero magnetic field, is also given in section 3. A summary is given in section 4, where some proposals for further experiments that would be of interest are put forward.

2. Half-width of the electron density profile related to shear modulus

March and Tosi (1985) have considered the effective electron density of a localized Wigner oscillator in a magnetic field B of arbitrary strength. This can be viewed as the Einstein model of a Wigner crystal in an applied magnetic field. As well as for the ground state, the effects of harmonic restoring force, magnetic field and temperature, T, still contrive to leave a Gaussian profile for the electron density $\gamma(r)$ at each site, which we shall simply write in the form

$$\gamma(r) = C \exp(-r^2/\lambda^2). \tag{2.1}$$

Here it is evident that the quantity λ measures the half-width of the localized electron density profile; this is related to a convenient dimensionless parameter σ which we shall take as the root-mean-square shear strain below. It is clear that these two quantities are intimately linked. For zero-point Debye shear waves the mean-square shear strain, or differential shear displacement, $\langle\sigma_0^2\rangle = 4\langle u_0^2\rangle/3r_0^2$, where $\langle u_0^2\rangle$ is the mean-square displacement, resolved along any crystal axis. However, for thermal phonons, $\langle u_T^2\rangle$ diverges logarithmically as the sample size increases, while σ_T^2 does not, and is a natural vibrational parameter for a 2D crystal.

What we emphasize is that the distribution in space of the electron density in the Wigner crystal can be characterized through the whole parameter range of magnetic field B, temperature T and carrier density n by this half-width λ. Though this can be calculated within the framework of the Einstein model, we shall in fact use the shear strain parameter σ, and model its dependence on the above parameters in the present approach. The Einstein results for λ then can be regarded as providing one useful guideline in achieving satisfactory modelling.

2.1. *Modelling of the shear modulus as function of shear strain parameter*

An essential first step in setting up a model to calculate the phase diagram of the 2D Wigner crystal for a wide selection of parameter values is to relate the shear modulus μ to the dimensionless parameter σ.

As starting point, we note that Bonsall and Maradudin (1977) gave the Madelung energy of any 2D crystalline array of localized electrons. Their results were later confirmed by Borwein *et al* (1988). The ground state is a triangular crystal (one electron per lattice point on a 2D hexagonal Bravais lattice) with energy $-1.1061\, e^2/r_0$ per electron. Figure 1 displays the energies $E_1(\alpha)$ and $E_2(\alpha)$ for a simple shear strain α applied along the $\langle 11\bar{2}\rangle$ and $\langle 10\bar{1}\rangle$ directions, respectively. For small strains, E_1 and E_2 are close to the harmonic energy

$$E_0(\alpha) = -1.1061e^2/r_0 + \tfrac{1}{2}\mu_0\pi r_0^2\alpha^2 \qquad (2.2)$$

where $\mu_0 = 0.044\, e^2/r_0^3$ is the shear modulus at $T = 0$ for a classical electron crystal. For large strains, there is both anisotropy and anharmonicity. To a very good approximation

$$E_1(\alpha) = [-1.1013 - 0.0048\cos(\sqrt{3}\pi\alpha)]e^2/r_0 \qquad (2.3)$$

with $E_2(\alpha) > E_0(\alpha)$ with

$$E_1(\alpha) + E_2(\alpha) = 2E_0(\alpha). \qquad (2.4)$$

The effect of this anharmonicity is to cause the shear modulus to depend on the thermal motion and zero-point motion of the electrons. We assume that this gives a Gaussian distribution† $\exp(-\alpha^2/2\sigma^2)$ of resolved shear strain along any direction with variance σ^2. Hence the mean energy for an additional infinitesimal strain α_1 can be written as

$$\overline{E(\alpha_1)} = \int_{-\infty}^{\infty} E(\alpha + \alpha_1)\exp(-\alpha^2/2\sigma^2)\, \mathrm{d}\alpha \qquad (2.5)$$

The effective shear modulus is then given by $\partial^2\bar{E}/\partial\alpha_1^2$, as α_1 tends to zero. Substituting the expressions for E_1 and E_2 in (2.5) we obtain

$$\mu_1^*(\sigma) = \mu_1(\sigma)/\mu_0 = \exp(-D\sigma^2) \qquad \mu_2^*(\sigma) = 2 - \mu_1^*(\sigma) \qquad (2.6)$$

where $D = 3\pi^2/2 = 14.8$ while μ_1^* and μ_2^* are the normalized shear moduli for simple

† In the appendix, some motivation for this assumption is provided by taking the site density γ to move rigidly as the sites move under shear. This argument relates σ to the half-width of the site density.

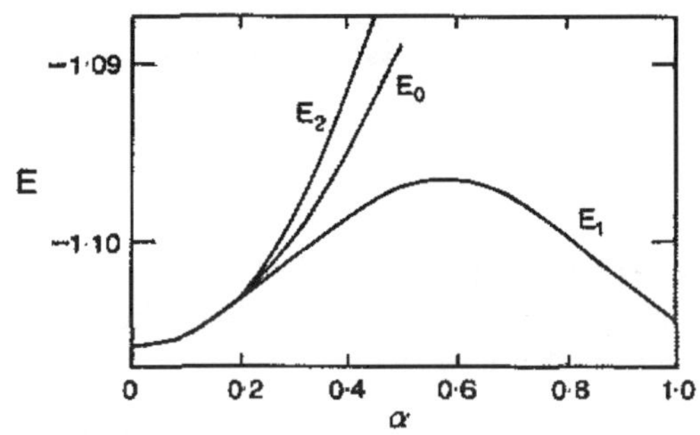

Figure 1. The Coulomb energy per electron, in units of e^2/r_0, for a 2D triangular electron crystal as a function of a static shear strain α parallel to the $\langle 11\bar{2}\rangle$ direction (energy E_1) and the $\langle 10\bar{1}\rangle$ direction (energy E_2). The parabola E_0 shows the harmonic energy variation for the zero-temperature shear modulus. The calculations were made using the expressions given by Bonsall and Maradudin (1977).

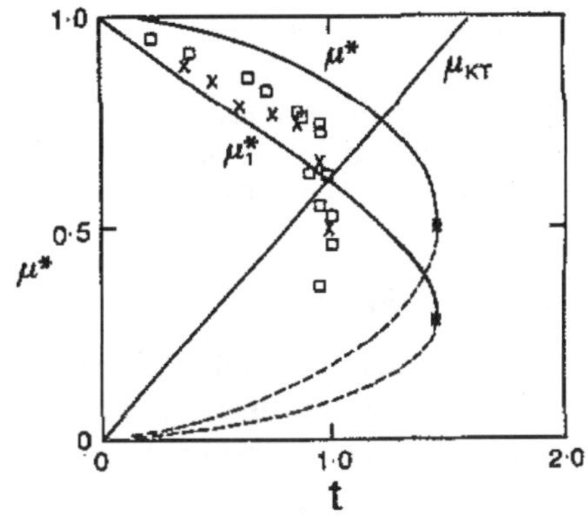

Figure 2. The reduced shear moduli μ^* and μ_1^*, calculated from (3.2) as a function of reduced temperature t. The points on the lines show the anharmonic stability limit. The broken curves show the unstable solutions of (3.2). The full line μ_{KT} shows the theoretical locus of the Kosterlitz–Thouless transition. The crosses are the results from computer simulations by Morf (1979), while the squares are the experimental data of Deville *et al* (1984) for electrons on liquid helium.

shear strains along the $\langle 11\bar{2}\rangle$ and $\langle 10\bar{1}\rangle$ directions, respectively. A static shear stress on the crystal produces both types of strain and a resultant modulus μ^*, isotropic by symmetry, is given by

$$1/\mu^* = 1/2\mu_1^* + 1/2\mu_2^* \qquad \mu^*(\sigma) = \mu_1^*(\sigma)\mu_2^*(\sigma). \tag{2.7}$$

The next stage is to calculate the mean-square shear produced by the propagating shear modes, angular frequency $\omega(q)$ and amplitude U_q, in the crystal. The mean-square component of displacement of each electron, mass m, along any direction is given by

$$\langle u^2\rangle = \frac{1}{2}\sum_q |U_q|^2 = \frac{1}{2}\sum_q \frac{\varepsilon}{mn\omega^2} \tag{2.8}$$

where

$$\varepsilon = (\hbar\omega_q/2)\coth(\hbar\omega_q/2k_B T) \tag{2.9}$$

is the mean energy per mode. The mean-square shear strain, resolved along any direction, is

$$\sigma^2 = \langle\alpha^2\rangle = \frac{1}{2}\sum_q q^2|U_q|^2 = \frac{1}{2}\sum_q \frac{q^2\varepsilon}{mn\omega^2} \tag{2.10}$$

for pure shear modes, and will be isotropic by symmetry in a triangular crystal with contributions from all shear modes.

3. Instabilities and melting

The purpose of this section is to convert the result (2.6) relating reduced shear modulus μ to half-width σ into potentially observable predictions. First, let us consider the classical limit in zero field.

3.1. Classical limit

Specifically, the above calculation of the scaled shear modulus will first be employed to construct the temperature dependence of the shear modulus μ and to exhibit the instability of the lattice. A phonon model leads to a relation of the form

$$\sigma^2 = Ak_B T/\mu(\sigma) \tag{3.1}$$

A being a constant and of course, model dependent.

A value of A can be obtained from a simplified Debye model. An unscreened 2D plasma is incompressible in the long-wavelength limit, so we include only the shear modes (which have propagation and polarization directions in the plane of the electrons). The shear modes propagating along the two symmetry directions are: $\omega_1 = c_1 q$ where $c_1^2 = \mu_1/mn$ for q_1 parallel to $\langle 10\bar{1}\rangle$ and polarization along $\langle 11\bar{2}\rangle$, and $\omega_2 = c_2 q$ with $c_2^2 = \mu_2/mn$ for q_2 parallel to $\langle 11\bar{2}\rangle$ and polarization along $\langle 10\bar{1}\rangle$. Integrating (2.10) up to the Debye wave vector $q_D = 2/r_0$ leads to

$$\sigma^2 = (k_B T/2\pi r_0^2)(1/2\mu_1 + 1/2\mu_2) = k_B T/2\pi r_0^2 \mu = 0.0285t/\mu^*(\sigma) \tag{3.2}$$

where t is the temperature normalized to the Kosterlitz–Thouless melting temperature $T_{mc} = e^2/r_0\Gamma_m k_B$, where Γ_m has been determined experimentally to be 127 (Deville 1988) for electrons on liquid helium.

Equation (3.2) has been solved self-consistently to obtain the shear modulus $\mu^*(t)$ as a function of temperature. The results are shown, along with plots of $\mu_1^*(t)$ and $\mu^*(t)$ in figure 2. Several interesting points emerge from this admittedly simple model. First, the shear mode with q_1 parallel to $\langle 10\bar{1}\rangle$, which corresponds to the close-packed lines of electrons sliding past each other, softens as t increases while the q_2 mode stiffens. It is to be noted that $\mu_1^*(t)$ decreases linearly with t at low temperatures. The total shear modulus $\mu^*(t)$ also decreases to an anharmonic instability at $\mu^* = 0.5$, $t = 1.46$ and $\sigma = 0.29$. For $\sigma > 0.29$, (3.2) still has a solution (shown as a broken curve in figure 2) but the crystal will then be unstable. The Kosterlitz–Thouless transition occurs at a temperature T_m such that

$$T_m = \mu a^2/4\pi k_B \tag{3.3}$$

where a is the lattice spacing. This transition will therefore occur when the reduced shear modulus $\mu_{KT} = 0.62t$ as shown by the full line in figure 2. This intersects the $\mu^*(t)$ graph for this model at $t = 1.22$. Also shown in figure 2 are measurements of μ^* by Deville *et al* (1984), together with the computer simulation results of Morf (1979). The shear modulus is found to decrease linearly at low temperature and this has been shown to be due to anharmonicity in detailed calculations by Chang and Maki (1983). The rapid decrease in μ near the transition found by Morf has been ascribed to renormalization due to thermally excited dislocations.

Note that the shear modulus and the anharmonic instability in the present model can be scaled in temperature by adjusting the absolute value of A. Hence the relative positions of the Kosterlitz–Thouless transition and the anharmonic instability can be

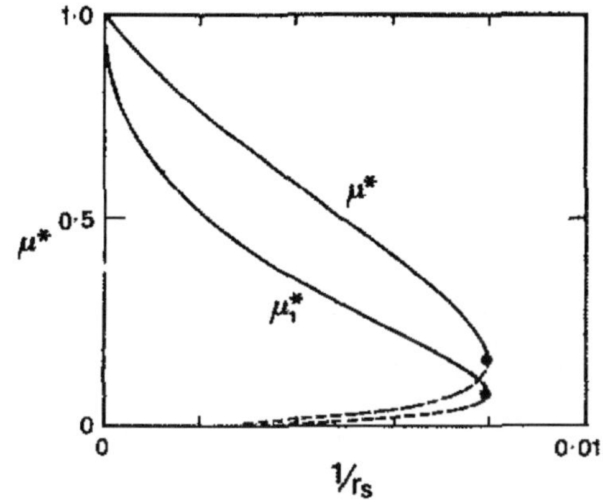

Figure 3. The reduced shear moduli μ^* and μ_1^*, calculated from (3.4) as a function of $1/r_s$ at $T = 0$. The points on the lines show the anharmonic stability limit. The broken curves show the unstable solutions of (3.2).

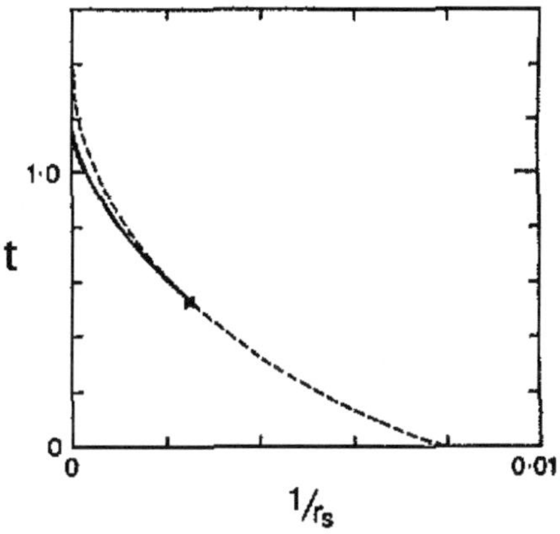

Figure 4. The model phase diagram of the 2D electron crystal in zero field on a t versus $1/r_s$ plot. The full curve shows the locus of the Kosterlitz–Thouless transition while the broken curve is the locus of the anharmonic instability. Note the predicted change in the nature of the transition as the electron density increases.

varied and can occur very close together: with reference to figure 2 an increase in σ by only 20% above the Debye model would bring the upper curve into accord with the available data. The two transitions are closely linked in that dislocations are produced by slip along the $\langle 11\bar{2}\rangle$ direction which corresponds to the softened q_1 shear mode. It is tempting to associate the region between the Kosterlitz–Thouless transition and the anharmonic instability with the postulated hexatic phase (Nelsen and Halperin 1979).

3.2. *Quantum limit*

Equation (2.6) can also be applied to the quantum crystal at $T = 0$ in zero field. In this case σ^2 is due to zero-point shear strain and the two-mode Debye model gives

$$\sigma_0^2 = (1/3\sqrt{\pi r_s})(1/2\sqrt{\mu_1} + 1/2\sqrt{\mu_2}) = (0.45/\sqrt{r_s})(1/\sqrt{\mu_1^*} + 1/\sqrt{\mu_2^*}) \qquad (3.4)$$

where $r_s = r_0/a_B$ and $a_B = \hbar^2/me^2$ is the Bohr radius. This can be solved for μ^* and μ_1^* as functions of $1/r_s$, the results being shown in figure 3. As zero-point motion increases, the shear modulus falls until the crystal becomes unstable at $\mu^* = 0.16$ and $r_s = r_W = 125$. There are no experimental results on this transition, which is the transition Wigner originally proposed (Wigner 1934, 1938), though in two dimensions, but Ceperley has shown by computer simulation that $r_W = 33$. Hence the quantum crystal is more stable than our simple model suggests. The notorious sensitivity of r_W to the model chosen is clear from the table of Care and March (1975). The Debye model probably overestimates the zero-point motion. Siringo *et al* (1991) have also shown that the force constant between disks of electronic charge is greater than for point charges. Finally this model does not consider any specifically quantum effects which may result from the overlap of the individual electronic wave functions. Nonetheless a possible mechanism for 2D Wigner quantum melting is clearly indicated, as an anharmonic instability.

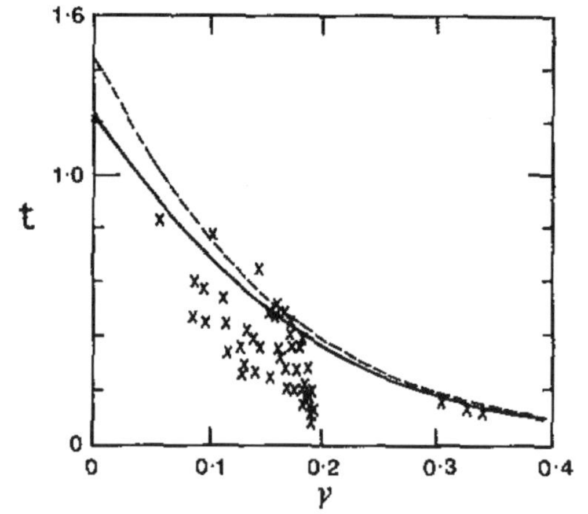

Figure 5. The model phase diagram of the 2D electron crystal in a magnetic field on a t versus ν plot. The full curve shows the locus of the Kosterlitz–Thouless transition while the broken curve is the locus of the anharmonic instability. The data points are from Andrei *et al* (1988) and Glattli *et al* (1990).

It is interesting to follow this instability and the K–T transition at finite temperature, as r_s decreases from infinity in the classical limit, using the expression for σ^2, supplemented by thermal contributions. The loci of these transitions on the t–$1/r_s$ plane is shown in figure 4. At $r_s = \infty$ the K–T transition occurs below the instability, as already discussed. But as r_s decreases the two transitions merge until, for $r_s < 400$, the K–T transition no longer occurs. In the present model the transition is an anharmonic instability for $125 < r_s < 400$.

3.3. *The magnetically induced Wigner solid (MIWS)*

In the quantum limit of high density, when $r_s = 0$, it is well established theoretically that an infinite magnetic field can suppress the zero-point motion and induce a classical 2D electron crystal. As the field is reduced, or as the Landau-level filling factor ν increases, the cyclotron motion of individual electrons increases with a mean-square displacement along any direction $\langle u^2 \rangle = l_B^2$, where $l_B = (\hbar/eB)^{1/2}$ is the magnetic length. If we assume that this displacement produces both longitudinal and shear strains then the resolved shear strain can be written as

$$\sigma_B^2 = 0.0285t/\mu^*(\sigma) + G\nu \tag{3.5}$$

where the first term is taken to be the same as for the classical crystal in zero field. Fourier transforming the cyclotron motion into components of longitudinal and shear displacements and integrating to find the mean-square shear strain gives $G = 0.5$ in the limit $l_B < r_0$. Taking $G = 0.5$, (3.5) is solved self-consistently. It is found that as ν increases the K–T transition and the instability temperature decrease as shown in figure 5. For the model parameters used here, the instability lies above the K–T transition at all fields, but the relative position could well be field dependent.

This phase diagram has some of the features of the experimental data of Andrei *et al* (1988) and Glattli *et al* (1990). However, it now seems possible that there may be regions of 2D electron liquid phases interspersed with solid phases (Jiang *et al* 1990, Buhmann *et al* 1991) and that the phase diagram calculated here could form an 'envelope' for the solid regions. On this interpretation such a liquid phase is seen in the data close to a filling factor $\nu = 0.192$. Also the three points at $\nu > 0.3$, originally excluded by

Andrei *et al* from their analysis, could represent a solid phase in this region. It is known from the zero-magnetic-field treatments of melting, already referred to above, that near $T = 0$ the phase diagram will depend sensitively on the nature of the low-lying excitations in the two phases and these are not carefully treated in the present model. It is to be noted that Lozovik *et al* (1985) have exposed an anharmonic instability as the Landau level factor increases beyond a critical value.

4. Discussion and summary

That there is an intimate relation between the dimensionless parameter σ characterizing the shear strain and the electron density profile (2.1) has already been emphasized. Some work on deviations of this profile from Gaussian form is available from the study of Gann *et al* (1979). Related to this, and for the ground-state in zero magnetic field, the 3D density functional calculations of Senatore and Pastore (1990) offer a way of checking the Gaussian structure factor $\rho(k)$, by extracting the Fourier components in the periodic density in the Wigner crystal at the reciprocal lattice vectors $\boldsymbol{K}_i$. Equally important would be a study of the way that shear affected the localized density profile. Underlying the present work is the model in which the 'localized Gaussian blobs' move rigidly as the lattice sites are shifted by a 'frozen phonon'. This could be tested, at least in principle, by the approach of Senatore and Pastore (1990), who refer to both BCC and FCC lattices. As pointed out by Perrin *et al* (1985), it is possible by continuous deformation via a body-centred tetragonal phase to pass from BCC to FCC; each periodic lattice could be explored to check the way in which the 'localized' blobs on the Wigner sites have to be deformed as a crystal is sheared. In connection with atomic crystals, the model of localized blobs moving without deformation is the analogue of the rigid-ion model, which in turn is equivalent to a pair force field.

Such refinements will no doubt mean that the simple modelling of μ in terms of σ in (2.6) will have to be transcended. Nevertheless, this modelling suggests that it will be of considerable interest to explore whether Bragg reflection studies of some sort are feasible on a heterojunction. If the phonon mode found by Andrei *et al* (1988) and by Glattli *et al* (1990) is the lower hybrid magnetophonon mode derived from the shear mode, this would confirm that one is dealing with a Wigner solid; not necessarily with long-range crystalline order. So this would be the first objective of Bragg reflection studies: to demonstrate crystallinity. The second would be to test the prediction of the present study of an intimate connection between shear modulus and scattering factors.

A further point of considerable interest is to recognize that a more general theoretical model will not merely cover a range of filling factors to $\nu > 0.4$, but will need to relate Wigner crystal theory to Laughlin-like electron liquid states as well as, possibly, the Hall crystal postulated by Halperin *et al* (1986). This should then reveal what will no doubt be a close connection between the region of parameter space treated in the present work and integral and fractional quantum Hall effects. Already, a re-entrant phase diagram is emerging from experimental studies (Jiang *et al* 1990, Buhmann *et al* 1991) with a series of interspersed liquid and solid phases as ν increases. The 'envelope' of these phases seems to be close to the melting line in figure 5. The host material may also have a strong influence on the phase diagram of the 2D electrons. For instance, Kohler *et al* (1986) have suggested that a phonon-mediated transverse charge-density wave state in a quantizing magnetic field may lead to Wigner crystallization over the whole range of magnetic quantization.

Finally, to summarize, the main conclusions of the model presented here are:

(i) There are two possible instabilities: one arising directly from anharmonicity in the simple model presented; the other is a Kosterlitz–Thouless transition. As the physical parameters are varied, it is possible to find 'cross-over' between the instabilities.

(ii) The quantum limit $T = 0$ in zero field is clearly sensitive to (*a*) tunnelling, which is not incorporated in the present model; it is related to Wigner oscillator wave-function overlap and (*b*) possible ring exchange, relating to magnetism, as discussed by March and Tosi (1980).

(iii) At the present stage of development of magnetic-field-assisted Wigner crystallization, the simple model presented here seems particularly useful. The main features of the phase diagram established experimentally by Andrei *et al* (1988) and by Glattli *et al* (1990) are compatible with the present predictions of Wigner crystallization in the high-field regime, except near $T = 0$ (see figure 5) where more careful treatment of the low-lying excitations in the liquid and the crystal phases is clearly called for (for $B = 0$ see Ferraz *et al* 1978, 1979). Subsequent experiments on non-linear electrical conductivity provide further strong evidence in support of pinned Wigner crystals over a substantial range of electric field, followed eventually by 'sliding' of the Wigner crystallites.

Appendix: Energy stored in electric field as a function of strain

The purpose of this appendix is to provide motivation from an approximate electronic theory for the phenomenological treatment of the shear modulus given in section 2.1. For convenience of presentation, we specifically treat a 3D Wigner electron crystal in zero magnetic field.

The starting point is an equilibrium crystal with localized electron densities $\gamma(r)$ centred on lattice sites l, the unit cell being denoted by Ω_l. When shear is applied, we denote the new sites by $l + \Delta l$ and the volume of integration is through a unit cell denoted by $\Omega_{l+\Delta l}$, though the density is held constant in the shear.

If the total ground-state electron density in the crystal without shear is denoted by $\rho(r, l)$ then we assume

$$\rho(r, l) = \sum_l \gamma(|r - l|). \tag{A1}$$

Evidently, in a 'rigid blobs' model, analogous to the lattice dynamical rigid-ion model of normal metal crystals built from electrons and granular ions, the density in the sheared lattice is given by

$$\rho(r, l + \Delta l) = \sum \gamma(|r - l - \Delta l|) \tag{A2}$$

where Δl is not necessarily constant through the unit cell. The density change $\Delta\rho(r)$ can then be written, to first-order in the (assumed small) site displacement, Δl,

$$\Delta\rho = \rho(r, l + \Delta l) - \rho(r, l) = \Delta l \cdot \operatorname{grad} \rho(r, l). \tag{A3}$$

Next let us consider the energy stored in the electric field $E(r)$. By a similar argument, the change in energy per electron due to shear is approximately

$$8\pi\Delta\mathscr{E}_{\text{shear}} = \int [E^2(r, l + \Delta l) - E^2(r, l)]\, dr \tag{A4}$$

where the integration is taken over the common volume of the two unit cells involved.

Writing the integrand as $(E_{l+\Delta l}+E_l)\,(E_{l+\Delta l}-E_l)$, one has again for Δl small the approximate form $2E_l(\Delta \boldsymbol{l}\cdot \mathrm{grad}\,\bar{E})$.

At this stage we invoke Poisson's equation

$$\mathrm{div}\,\boldsymbol{E} = 4\pi\Delta\rho \tag{A5}$$

and after some manipulation, one has for the interesting term for present purposes of $O(\Delta l^2)$, the form

$$\Delta\mathscr{E}_{\mathrm{shear}} \sim \int E_i \Delta \boldsymbol{l}\cdot(\boldsymbol{r}/r)\Delta\rho\,\mathrm{d}\boldsymbol{r}. \tag{A6}$$

The important conclusion here is that, when the localized density $\gamma(r) = N\exp(-r^2/\lambda^2)$, the grad ρ term involved in $\Delta\rho$ contains a term of the form $\exp(-r^2/\lambda^2)2r/\lambda^2$. This is plainly the term which is strongly dependent on the half-width, λ, of the density profile; therefore in the remainder of the integrand in (A6) one use that the localized limit λ tends to zero. Essentially then, this approximate microscopic theory provides a basis in electron theory for the phenomenological assumption embodied in (2.5) in the limit $\alpha_1 \to 0$. The argument used there is to write, for a harmonic potential $\frac{1}{2}x^2$, the result

$$\text{Energy} = (\tfrac{1}{2}x^2) \times \text{distribution in } x,$$

or for a general energy such as calculated by Bonsall and Maradudin (1977) and displayed in figure 1, $\frac{1}{2}x^2$ is replaced by $E(x)$. The distribution in x written above has, in (2.5), Gaussian form with half-width σ. The above treatment therefore crucially links electron density and the shear strain dimensionless parameter, σ.

References

Andrei E Y, Deville G, Glattli D C, Williams F I B, Paris E and Etienne B 1988 *Phys. Rev. Lett.* **60** 2765

Bhatia A B and March N H 1984 *J. Chem. Phys.* **80** 2076

Bonsall L and Maradudin A A 1977 *Phys. Rev.* B **15** 1959

Borwein D, Borwein J M, Shail R and Zucker I J 1988 *J. Phys. A: Math. Gen.* **21** 1519

Buhmann H, Joss W, van Klitzing K, Kukuskim I V, Plant A S, Martinez G, Ploog K and Timofeev V B 1991 *Phys. Rev. Lett.* **6** 926

Care C M and March N H 1975 *Adv. Phys.* **24** 101

Ceperley D M 1978 *Phys. Rev.* B **18** 3126

Chang M and Maki K 1983 *Phys. Rev.* B **18** 1646

Deville G 1988 *J. Low Temp. Phys.* **72** 135

Deville G, Valdes A, Andrei E Y and Williams F I B 1984 *Phys. Rev. Lett.* **53** 588

Durkan J, Elliott R J and March N H 1968 *Rev. Mod. Phys.* **40** 812

Elliott R J and Kleppmann W G 1975 *J. Phys. C: Solid State Phys.* **8** 2729, 2737

Ferraz, A, March N H and Suzuki M 1978 *Phys. Chem. Liquids* **8** 153; 1979 *Phys. Chem. Liquids* **9** 59

Gann R C, Chakravarty S and Chester G V 1979 *Phys. Rev.* B **20** 326

Glattli D C, Deville G, Duburcq G, Williams F I B, Paris E and Etienne B 1990 *Surf. Sci.* **229** 334

Halperin B I, Tesanovic Z and Axel F 1986 *Phys. Rev. Lett.* **57** 922

Jiang H W, Willett R L, Stormer H L, Tsui D C, Pfeiffer L N and West K W 1990 *Phys. Rev. Lett.* **65** 633

Johnson R A 1988 *Phys. Rev.* B **37** 3924

Kohler H, Roos M and Dato P 1986 *J. Phys. C: Solid State Phys.* **19** 5215

Kosterlitz J M and Thouless D J 1973 *J. Phys. C: Solid State Phys.* **6** 1181; 1978 *Prog. Low Temp. Phys.* B **7** 371

Lea M J and March N H 1990 *Phys. Chem. Liq.* **21** 183

Lozovik Yu E, Fartztdinov V M and Abdullaev B 1985 *J. Phys. C: Solid State Phys.* **18** L807

March N H 1989 *Phys. Rev.* B **40** 3356; 1988 *Phys. Rev.* A **37** 4526

March N H and Tosi M P 1985 *J. Phys.* A **18** L643

March N H and Tosi M P 1980 *Phys. Chem. Liquids* **10** 113
Morf R H 1979 *Phys. Rev. Lett.* **43** 931
Nagara H, Nagata Y and Nakamura T 1987 *Phys. Rev.* A **36** 1859
Nelson D R and Halperin B I 1979 *Phys. Rev.* B **19** 2457
Perrin R C, March N H and Taylor R 1985 *Solid State Commun.* **53** 287
Ploog K 1990 *Physica* B + C at press
Saitoh M 1988 *Surf. Sci.* **196** 8
Senatore G and Pastore G 1990 *Phys. Rev. Lett.* **64** 303
Siringo F, Lea M J and March N H 1991 *Phys. Chem. Liquids* at press
Wigner E P 1934 *Phys. Rev.* **46** 1002; 1938 *Trans. Faraday Soc.* **34** 678

J. Phys.: Condens. Matter 3 (1991) 4301–4306. Printed in the UK

LETTER TO THE EDITOR

Thermodynamics of melting of a two-dimensional Wigner electron crystal

M J Lea†, N H March‡ and W Sung‡§

† Department of Physics, Royal Holloway and Bedford New College, University of London, Egham, Surrey TW20 0EX, UK

‡ Theoretical Chemistry Department, University of Oxford, South Parks Rd., Oxford OX1 3UB, UK

Received 28 March 1991

Abstract. The thermodynamics of the Wigner crystallization of two-dimensional electrons in a magnetic field is set out. The results relate the slope of the melting curve as a function of the Landau level filling factor ν, to changes in magnetization, ΔM, and entropy, ΔS, on melting, and identify points where ΔM, ΔS and $\Delta E = 0$, with E being the internal energy. The phase diagram as revealed by recent experiments on GaAs/AlGaAs heterojunctions is thereby analysed. Remarkable features are inferred in the magnetization of the strongly correlated electron liquid.

The proposal made by Durkan *et al* (1968) that localization of electrons in impure semiconductors by strong magnetic fields could assist Wigner crystallization has been brought to fruition by Andrei *et al* (1988), Glattli *et al* (1990) and Jiang *et al* (1990). These workers used the two-dimensional electron gas (2DEG) in GaAs/AlGaAs heterojunctions in strong magnetic fields to observe the magnetically-induced Wigner solid (MIWS). Their major findings have been confirmed by Buhmann *et al* (1991) who observed specifically an additional line in the luminescence spectrum of such a heterojunction below a critical Landau level filling factor ν_c and a critical temperature ($T_c = 1.4$ K at 26 T). The lack of correlation between ν_c and the disorder-related properties of the system testifies to the intrinsic nature of the line and its appearance, they conclude, signals the formation of a Wigner solid.

In the above experiments, it proved possible to map out approximately the melting curve of the Wigner solid as a function of the Landau level filling factor, ν, given in terms of the (areal) electron density n and the magnetic field H applied perpendicular to the electron layer, by

$$\nu = nhc/eH. \tag{1}$$

The schematic phase diagram proposed by Buhmann *et al* (1991) is shown in figure 1. There are four crystal phases, marked C1, C2, C3 and C4, interspersed with the liquid phase at filling factors corresponding to the fractional quantum Hall effect (FQHE) at $\nu = \nu_q = 1/q$ with $q = 5$, 7 and 9. The final solid phase ends at a critical filling factor of $\nu_c = 0.28 \pm 0.02$. These experimental results have prompted us to generalise the

§ Permanent address: Department of Physics, Pohang Institute of Science and Technology, Korea.

0953-8984/91/234301 + 06 $03.50 © 1991 IOP Publishing Ltd

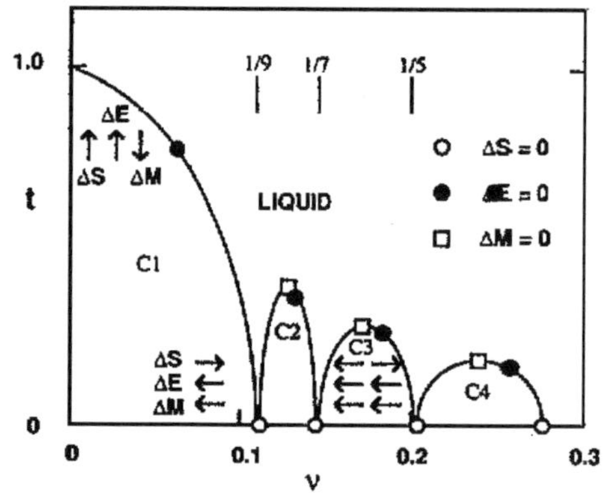

Figure 1. Schematic phase diagram as proposed by Buhmann *et al* (1991), showing the four crystal phases, C1 to C4, and the reentrant liquid phase at $\nu = \frac{1}{9}$, $\frac{1}{7}$ and $\frac{1}{5}$. The symbols mark the points ΔM, ΔS and $\Delta E = 0$ as deduced from thermodynamics. The arrows show the direction of ΔM, ΔS and ΔE across the phase boundary nearest to the arrows. The arrows in phases C2 and C4 match those shown in phase C3. It should be noted that, as alternatives, the possibilities of a finite liquid range at $T = 0$ or a solid always stable at the lowest temperatures, are not excluded.

thermodynamics of Wigner crystallization, already considered by Parrinello and March (1976) in zero magnetic field. The starting point of the present study of the thermodynamics of an electron crystal to electron liquid first-order melting transition is the result (see, for example, Pippard 1966) for the melting temperature, T_m, as a function of magnetic field, H. At constant area Ω, this can be written

$$(\partial T_m/\partial H)_\Omega = -\Delta M/\Delta S. \tag{2}$$

If the subscript C denotes the crystal phase and the subscript L the liquid phase, then $\Delta M = M_L - M_C$ is the change of magnetization on melting, while $\Delta S = S_L - S_C$ is the corresponding entropy change. This equation is readily cast into a form directly useful in analysing the measurements referred to above by using the relation between ν and H given in (1) to find

$$(\partial T_m/\partial \nu)_\Omega = (H/\nu)\,\Delta M/\Delta S. \tag{3}$$

Let us comment immediately on some properties which follow from (3):

(i) Turning points in the melting curve in the (ν, T) plane correspond to $\Delta M = 0$, provided the entropy change, ΔS, does not simultaneously go to zero. This result means that maxima (see figure 1) in the melting curve immediately locate points where the magnetization in the strongly correlated liquid equals that in the Wigner solid.

(ii) Infinite slopes on the melting curve, at specific values of ν away from $\nu = 0$, imply $\Delta S = 0$ for non-zero ΔM.

(iii) (3) can also be used to locate points where $\Delta E = 0$, where E is the internal energy. If we write the magnetic Helmholtz free energy, F, relevant to phase equilibrium at constant T, field H and area Ω, as (Pippard 1966)

$$F = E - TS - HM \tag{4}$$

the equilibrium condition $F_L = F_C$ evidently then yields

$$\Delta E - T\,\Delta S - H\,\Delta M = 0. \tag{5}$$

As envisaged in (iii) above, the existence of a point, or points, on the melting curve where $\Delta E = 0$ yields at such points

$$\Delta M/\Delta S = -T_m/H \tag{6}$$

the slope of the melting curve is then given by (from (3))

$$(\partial T_m/\partial \nu)_\Omega = -T_m/\nu. \tag{7}$$

This equation can only be satisfied for $\partial T_m/\partial \nu$ negative and then whenever the melting curve lies at a tangent to a hyperbola of the form $T_m \nu$ = constant. Hence we now have a prescription for locating points where ΔM, ΔS and ΔE are zero, and these have been marked on the phase diagram in figure 1. At each point the relative signs of the non-zero differentials are then fixed by the relations in (5). Also, equation (5) requires that ΔS and ΔM cannot simultaneously have opposite signs to ΔE.

Following Parrinello and March (1976), it is also useful to invoke the virial theorem in this treatment of Wigner crystallization. For non-zero magnetic field this reads

$$2K + 2HM + U = 2p\Omega \tag{8}$$

where K is the kinetic energy, U potential energy ($E = U + K$) and p the pressure. Equation (8) is for a two-dimensional electron assembly with a three-dimensional Coulomb interaction (Tao 1990). For incompressible phases not affected by pressure we have†

$$2\Delta K + 2H\,\Delta M + \Delta U = 0. \tag{9}$$

Combining this with (5) we identify two additional points $\Delta U = 0$ and $\Delta K = 0$, which occur when $\partial T_m/\partial \nu = -T_m/2\nu$ and $-T_m/3\nu$, respectively. In summary we have

(i) When $\Delta M = 0$,

$$\Delta E = T_m\,\Delta S = \tfrac{1}{2}\Delta U = -\Delta K. \tag{10}$$

(ii) When $\Delta S = 0$ or $T = 0$,

$$\Delta E = H\,\Delta M = \tfrac{1}{4}\Delta U = -\tfrac{1}{3}\Delta K. \tag{11}$$

(iii) When $\Delta E = 0$

$$T_m\,\Delta S = -H\,\Delta M = -\tfrac{1}{2}\Delta U = \tfrac{1}{2}\Delta K. \tag{12}$$

(iv) When $\Delta U = 0$

$$\Delta E = -H\,\Delta M = \tfrac{1}{2}T_m\,\Delta S = \Delta K. \tag{13}$$

(v) When $\Delta K = 0$,

$$\Delta E = -2H\,\Delta M = \tfrac{2}{3}T_m\,\Delta S = \Delta U. \tag{14}$$

Let us now consider some of these results in the light of the schematic phase diagram (figure 1) from Buhmann *et al* (1991), into which they have subsumed the main features compatible with microscopic theory and experiment. It is clear (see Andrei *et al* 1988) that, as ν tends to zero, the melting temperature, T_m, should tend to that of a classical one-component plasma. This has a melting temperature, T_{mc}, given by

$$T_{mc} = e^2(\pi n)^{1/2}/\kappa k_B \Gamma_m \tag{15}$$

where k_B is Boltzmann's constant while $\Gamma_m = 127 \pm 3$ (Deville 1988) and κ is the dielectric constant of the host material. Hence figure 1 shows $t_m = T_m/T_{mc}$ versus ν with $t_m = 1$ at $\nu = 0$. As this point is approached, we anticipate on physical grounds that the entropy S_L of the liquid will be greater than that of the solid S_C‡. Since there is no physical reason for an infinite slope of the melting curve at $\nu = 0$, it is clear from (1) and (3) that, with

† Fully quantitative work may require an estimate of the electron pressure contribution relative to the magnetic field term.

‡ Should this expectation not be borne out by experiment, the changes to be made in figure 1 are straightforward.

ΔS not equal to zero, ΔM must approach zero at least as rapidly as ν^2 along the melting curve, at constant density.

We have labelled figure 1 with the points corresponding to $\Delta M = 0$, $\Delta E = 0$ and $\Delta S = 0$. The arrows show the directions in which S, M and E increase across the melting curve. For phase C1, the arrows are completely determined by the condition $\Delta S > 0$ for the classical crystal at $\nu = 0$, which also implies $\Delta M < 0$ and $\Delta E > 0$ in the same limit. In this picture $\Delta S = 0$ occurs only at $T = 0$ as required by the third law of thermodynamics, which implies that the melting curve approaches the axis with a vertical slope. The arrows in the other regions C2, C3 and C4 follow from either (i) immediately above, taking $\Delta S > 0$ at the highest temperatures in each solid phase, or (ii) assuming that the thermodynamic functions behave in a similar way each time the phase diagram approaches $\nu = 0$. Figure 2 shows the points at which $\Delta E = 0$, $\Delta U = 0$ and $\Delta K = 0$ that occur for $\partial t_m/\partial \nu = -t_m/\nu$, $-t_m/2\nu$ and $-t_m/3\nu$, respectively.

The simplest conceptual picture that emerges from this phase diagram is that the four separate solid-phase regions are essentially the same Wigner solid (presumably a triangular crystal with one electron per lattice point on a 2D hexagonal Bravais lattice) with regions near $\nu = \frac{1}{5}$, $\frac{1}{7}$ and $\frac{1}{9}$ where the liquid phase has a lower free energy. It is conceivable that the four solid phases are different but this seems somewhat unlikely, particularly since the ν values indicate that the change in ordering is occurring in the liquid. There have been many theoretical calculations of the ground state energies of the Wigner solid and the fractional quantum Hall effect liquid states (Levesque *et al* 1984, Lam and Girvin 1984, Yoshioka *et al* 1983, Halperin 1984, Egorov 1986, Markiewicz 1986, Thugman and Kivelson 1985, Kivelson *et al* 1987, Isihara 1989). There is a consensus that the solid has lower energy at very small values of ν but that the FQHE states with $\nu = 1/q$, as introduced by Laughlin (1983), have lower energies above some critical $\nu_c \sim 1/6.5$ (Lam and Girvin 1984). The energy of the FQHE states also has cusps at $\nu = 1/q$ with the limiting slopes being related to the energy gap of the quasiparticle excitations in the liquid (Halperin *et al* 1984).

Figure 1 shows that the entropy change on melting, ΔS, is greater than zero on either side of the liquid state at ν_q. We assume that, at low temperatures,

$$\Delta S = A(\nu)T^{\alpha} \tag{16}$$

where $A(\nu)$ and α are positive constants which may vary strongly around ν_q. (The entropy from the thermal excitations from the ground state at ν_q of the form $\exp(-\Delta/kT)$, where Δ is the quasiparticle energy gap, will be negligible near $T = 0$.) This entropy could come from low-lying excitations in the liquid or from quasiparticle excitations related to $|\nu - \nu_q|$ (which might give $\alpha = 0$). Then, on the melting curve†,

$$T_m \Delta S = \Delta E - H\Delta M = \Delta \mathscr{E} = A(\nu)T_m^{\alpha+1}. \tag{17}$$

The magnetization of the liquid, M_L, will be discontinuous at a cusp in the energy at $\nu = 1/q$, and could change sign (see Isihara 1989). If the change in magnetization on melting ΔM is approximately equal to M_L, then the relative signs of ΔM agree with the conclusions from figure 1, as sketched in figure 3, assuming that the transitions occur each side of ν_q. This field dependence of ΔM is very reminiscent of the de Haas–van Alphen effect at integral ν values, suggesting that the magnetism of the electron liquid phase is intimately connected with the exotic variation of ΔM shown here. The conclusion here

† $\Delta \mathscr{E}$, the free energy at $T = 0$, has the re-entrant cusp minima following the melting curve (figure 1). Since $\mathscr{E}_s$ is a monotonically increasing function of ν, this feature is entirely due to the electron liquid, consistent with the theoretical results mentioned earlier.

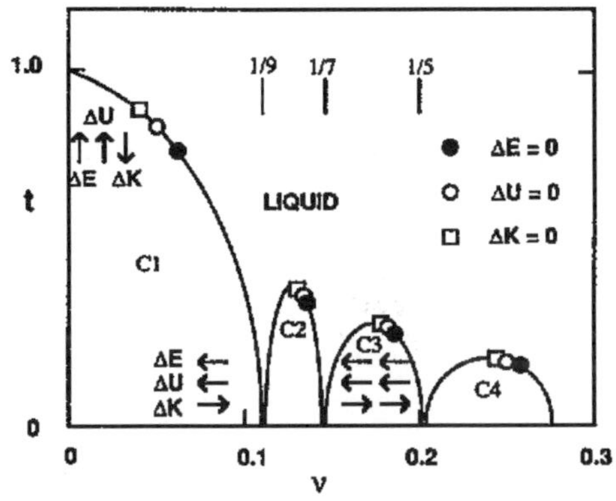

Figure 2. Schematic phase diagram showing the points ΔE, ΔU and $\Delta K = 0$ as deduced from thermodynamics. The arrows show the direction of ΔE, ΔU and ΔK across the phase boundary nearest to the arrows. The arrows in phases C2 and C4 match those shown in phase C3.

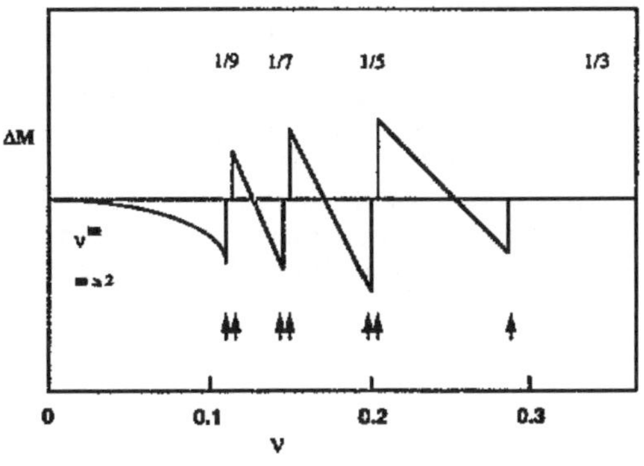

Figure 3. Schematic diagram of the change in magnetization on melting along the melting curve shown in figure 1. The arrows mark the liquid–solid transitions.

is that, for a finite width of reentrant liquid phase at finite temperature, the magnetization and entropy are properties of the liquid, connected by (17) on the melting line. However as T decreases towards zero, the entropy must go to zero. If this occurs by some further ordering among the excitations, then ΔM may also go to zero at the same time. If $\Delta M/\Delta S$ remained a constant in this process then $\partial t_m/\partial \nu$ would approach the axis with a finite slope, from (3). While it is not inconceivable that the orbital magnetism of the strongly correlated electron liquid could be consistent with the anyon model (Wilczek 1990), we are currently studying simpler microscopic models. The presence of minima in the free energy could also lead to the possibility of other phenomena, such as phase separation. However the relative energies of the different phases are small compared with the total Coulomb energy of the electron sheet so that the free energy versus density plot will probably exhibit only a small cusp with no minima at ν_q.

One of us (WS) wishes to acknowledge support from Pohang Institute of Science and Technology and the Ministry of Education in Korea during his sabbatical leave in Oxford.

References

Andrei E Y, Deville G, Glattli D C, Williams F I B, Paris E and Etienne B 1988 *Phys. Rev. Lett.* **60** 2765

Buhmann H, Joss W, von Klitzing K, Kukuskin I V, Plaut A S, Martinez G, Ploog K and Timofeev V B 1991 *Phys. Rev. Lett.* **66** 926

Deville G 1988 *J. Low Temp. Phys.* **72** 135

Durkan J, Elliott R J and March N H 1968 *Rev. Mod. Phys.* **40** 812

Glattli D C, Deville G, Duburcq G, Williams F I B, Paris E and Etienne B 1990 *Surf. Sci.* **229** 334

Egorov V S 1986 *Sov. Phys.–Solid State* **28** 177

Isihara A 1989 *Solid State Physics* vol 42 (New York: Academic) p 271

Jiang H W, Willett R L, Stormer H L, Tsui D C, Pfeiffer L N and West K W 1990 *Phys. Rev. Lett.* **65** 633

Kivelson S, Kallin C, Arovas D A and Schrieffer J R 1987 *Phys. Rev.* B **36** 1620

Halperin B I 1984 *Phys. Rev. Lett.* **52** 1583

Lam P K and Girvin S M 1984 *Phys. Rev.* B **30** 493

Laughlin R B 1983 *Phys. Rev. Lett.* **50** 1395
Levesque D, Weis J J and McDonald A H 1984 *Phys. Rev.* B **30** 1056
Markiewicz R S 1986 *Phys. Rev.* B **34** 4172; *Phys. Rev.* B **34** 4177
Parrinello M and March N H 1976 *J. Phys. C: Solid State Phys.* **9** L147
Pippard A B 1966 *Elements of Classical Thermodynamics* (Cambridge: Cambridge University Press)
Tao Z C 1990 *Phys. Lett.* **151A** 172
Thugman S A and Kivelson S 1985 *Phys. Rev.* B **31** 5280
Wilczek F 1990 *Anyons* (Singapore: World Scientific) p 87
Yoshioka D, Halperin B I and Lee P A 1983 *Phys. Rev. Lett.* **50** 1219

Phys. Chem. Liq., 1991, Vol. 24, pp. 131–135
Reprints available directly from the publisher
Photocopying permitted by license only

LETTER

Can the Hole Liquid Undergo Wigner Crystallization in High-T_c $La_{2-x}Sr_xCuO_4$ at Low Density?

N. H. MARCH*, B. V. PARANJAPE† and F. SIRINGO*

**Theoretical Chemistry Department, University of Oxford, 5 South Parks Road, Oxford OX1 3UB, UK*

†*Department of Physics, University of Alberta, Edmonton, Alberta, Canada*

(Received 4 April 1991)

A microscopic interpretation is proposed for the behaviour of the superconducting transition temperature T_c as a function of hole concentration p in $La_{2-x}Sr_xCuO_4$ when oxygen vacancies are suppressed. At high $p \sim 0.32$ a BCS-type formula for T_c is assumed, whereas at low $p \sim 0.06$ it is proposed that the holes undergo Wigner crystallization. Near the Wigner transition, one has a strongly correlated hole liquid, and Cooper pair binding will become increasingly difficult as $p \to 0.06$ from above, because the Fermi distribution is strongly changed from the unit step function, appropriate near $p \sim 0.32$. Some experiments are proposed to test the microscopic model put forward here.

KEY WORDS: Cooper pairs, strongly correlated hole liquid.

The hole concentration p in $La_{2-x}Sr_xCuO_4$ has been used as a variable in mapping out the superconducting phase of this high-T_c material, especially by Torrance *et al*[1]. The phase boundaries of the superconducting state that these and other workers[2,3] observed are reproduced in Figure 1. It is to be noted that no oxygen vacancies were allowed in their experiment, since these vacancies were filled by annealing samples in 100 bars of oxygen pressure.

The purpose of this work is to propose a microscopic interpretation of the boundaries to the superconducting phase shown in Figure 1. We note first with Torrance *et al.*[1] that superconductivity disappears in metallic $La_{2-x}Sr_xCuO_4$ at high hole concentrations, the critical p, denoted by p_c below, being $\simeq 0.32$.

Since in Ref. 1 it is emphasized that at low temperatures and around p_c one has a normal metal–superconductor transition, we shall assume in this regime a BCS-like formula[4]

$$k_B T_c = E_M \exp\left(-\frac{2}{g(E_f)V}\right) \tag{1}$$

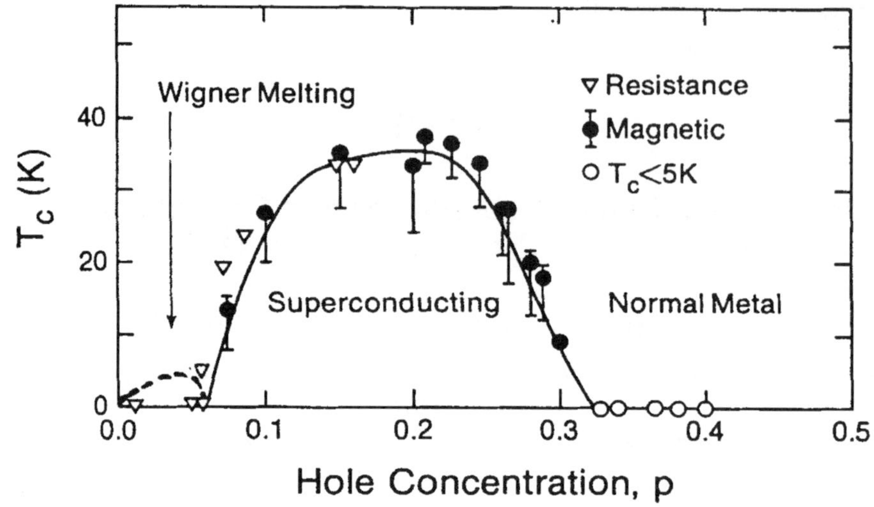

(a)

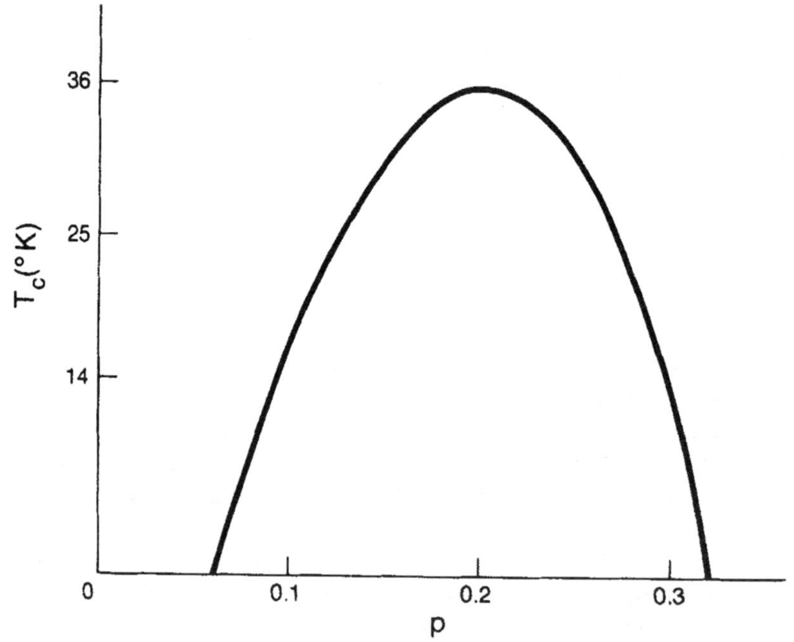

(b)

Figure 1 (*a*) The superconducting transition temperature T_c as a function of hole concentration p, after Torrance *et al*[1]. Use has been made of resistance measurements[2] and magnetic measurements[1,3] in establishing this curve. The schematic form of the melting curve of the Wigner hole crystal proposed here is shown by the dashed line. A strongly correlated hole liquid will then exist between these phase boundaries.

(*b*) Schematic form of T_c vs. p. To right of maximum, Eq. (3) plus Eq. (4) extended to include quadratic term in $(p - p_c)$ were employed. To left of maximum, Eqn. (6) was similarly extended with quadratic term.

where E_m is some characteristic energy, the nature of which is to be discussed below, $g(E_f)$ is the density of states at the Fermi-energy E_f while V is a suitable measure of the attractive interaction between holes, leading to Cooper pairs.

In equation (1), which we shall assume can characterize the right-hand part only of the $T_c - p$ plot in Figure 1 when T_c is sufficiently low, i.e. near p_c, let us now utilize a further simple thermodynamic result[5], namely that

$$T_c = (3\gamma/(a-b))^{1/2} \tag{2}$$

where it is assumed that the specific heat per unit volume of a metal has the form $C_s = aT^3$ in the superconducting state while $C_n = bT^3 + \gamma T$ in the normal state. Assuming in the normal state that the electronic specific heat γ is directly proportional to $g(E_f)$, one can use Eq. (2) in Eq. (1) to write

$$\frac{k_B T_c}{E_m} \equiv t_c = \exp\left(-\frac{F(p)}{t_c^2}\right) \tag{3}$$

where one has lumped into the p-dependent function $F(p)$ the dependences of a, b and possibly also E_m on the hole concentration. Equation (3) is the first result of importance for the present argument.

To make a preliminary test of Eq. (3), we have taken E_m to correspond, a little arbitrarily, with the maximum $T_c \sim 36$ K. When this is done, we find the approximate result, over a limited range of p, that

$$F(p) \doteqdot 2.6(p - p_c) \tag{4}$$

or

$$t_c^2 \ln t_c = 2.6(p - p_c) \tag{5}$$

This, we propose, determines, albeit approximately, the shape of the $T_c - p$ boundary in Figure 1 near to the highest hole concentration p_c for which superconductivity exists in this material.

At this point, it is useful to switch attention to the left-hand side of the phase boundary in Figure 1, but again near $T_c = 0$, which will be assumed once more to correspond to a specific hole concentration, denoted below by p_w. Specifically at $T = 0$, one proceeds, by lowering the hole concentration p, to find a 'phase transition' from a state with itinerant holes to a state in which the holes are localized for $p < p_w$.

It seems to us natural to assume that Coulomb repulsion between the holes, which becomes the more important compared to their kinetic (Fermi) energy as p is lowered, will eventually order the holes into a quantal Wigner crystal state[6]. Such strong hole correlations would lead to the filled Fermi sphere picture, so basic to the usual argument for Cooper pairing, which is assumed correct near $p = p_c$, being modified drastically near p_w. In particular, the discontinuity in the momentum distribution

$n(k)$ at $k = k_f$, with k_f the Fermi wave number, which is unity in the usual Cooper pairing theory, will be decreased by hole correlations to a value $q(p)$ say, which reduces to zero as one passes into the Wigner state of localized holes. Of course, whether $q(p)$ falls smoothly to zero at p_w depends on the order of the transition. We expect the hole liquid-solid transition to be first-order but that the discontinuity in q will be small anyway in the strongly correlated hole liquid near $p = p_w$. Then, it is known from heavy Fermion theory[7] that one should compare thermal energies $k_B T$ with qE_f, not E_f, which plainly means that thermal excitation can be expected to play a much greater role near $p = p_w$ than around $p = p_c$, where a BCS-like picture still seems appropriate.

An argument based on the Clausius–Clapeyron equation then leads us to conclude that since, near $p = p_w$, the entropy of the superconductor at small $T \simeq 0$ will be greater than that of the Wigner crystal, then $(dT_c/dp)_{p=p_w}$ will be finite. This appears not to be inconsistent with the phase diagram given by Torrance *et al*[1], though it would be too strong a statement to say that the experimental data established this result. However, writing

$$T_c = a_w(p - p_w) \tag{6}$$

one can form $t_c^2 \ln t_c$ appearing in Eq. (5), but now near $p = p_w$.

In conclusion, we want to reiterate the following points which we propose must be incorporated in any microscopic model of the superconducting phase boundary found in Ref. 1 and depicted in the upper half of Figure 1:

i) A BCS-like formula for T_c should work at low T_c near $p = p_c$ (see lower part of Figure 1).

ii) The transition at $T = 0$ from itinerant holes to a Wigner crystal of holes as p is lowered through p_w should profoundly influence the left-hand side of Figure 1 near $p = p_w$. Thermodynamics plus experiment are consistent with the proposed form (6), though this is not established by the data available.

iii) Near p_w, the hole liquid is strongly correlated, and the Fermi sphere picture is a poor starting point for discussing Cooper pairing, the discontinuity q in $n(k)$ near $k = k_f$ departing strongly from unity and indeed being zero in the Wigner crystal of holes. This may well explain why Cooper pairs no longer bind for $p \sim p_w$, because the role of the Exclusion Principle is weakened.

Outstanding questions which remain are as follows:

i) What is the nature of the elementary excitations in the superconducting state near $p = p_w$? In particular, is there a linear term in the specific heat in the superconductor in this regime?

ii) Is there an additional phase boundary, starting out at $p = p_w$ and going to the left of Figure 1 representing the melting of the Wigner crystal of holes? If so, what is the behaviour of the holes between these two phase boundaries at low T_c?

iii) In addition to the Coulomb correlations leading to hole localization at $T = 0$

for $p < p_w$, does the local variation in Sr concentration play any role through Anderson localization?

iv) Is the maximum value of T_c related to qE_f where q is the discontinuity in the Fermi momentum distribution at k_f? Measurements of the magnetic susceptibility of the holes in the region between the phase boundaries representing (a) the melting of the Wigner crystal and (b) the transition $T_c - p$ line may be useful in clarifying this point.

Acknowledgements

One of us (NHM) wishes to acknowledge that his contribution to this work was made during a visit to the Physics Department of the University of Alberta, Edmonton during 1989. In particular, Professors H. R. Glyde and F. C. Khanna are thanked for generous hospitality. Another of us (FS) acknowledges a Fellowship award from the Fondazione Angelo Della Riccia, Italy, which made possible his stay in Oxford, U.K. Finally, NHM wishes to acknowledge valuable discussions with Dr. A. Bezinge during a visit to IBM, San José in 1988.

References

1. J. B. Torrance, Y. Tokura, A. I. Nazzal, A. Bezinge, T. C. Huang and S. S. P. Parkin, *Phys. Rev. Lett.*, **61**, 1127 (1988).
2. M. W. Shafer, T. Penney and B. L. Olson, *Phys. Rev.* **B36**, 4047 (1987).
3. R. B. van Dover, R. J. Cava, B. Batlogg and E. A. Reitman, *Phys. Rev.* **B35**, 5337 (1987).
4. See, for example, N. H. March, W. H. Young and S. Sampanthar, *The Many-Body Problem in Quantum Mechanics* (Cambridge: University Press 1967).
5. See, for instance, A. B. Pippard, *Elements of Classical Thermodynamics*, (Cambridge: University Press) 1966.
6. For a review of quantal Wigner crystallization, see C. M. Care and N. H. March, *Advances in Physics* **24**, 101 (1975).
7. See, for example, T. M. Rice, K. Ueda, H. Ott and H. Rudigier, *Phys. Rev.* **B31**, 594 (1985); also R. G. Chapman and N. H. March, *Phys. Rev.* **B38**, 792 (1988).

Phys. Chem. Liq., 1991, Vol. 23, pp. 115–121
Reprints available directly from the publisher
Photocopying permitted by license only

LETTER

Self-consistent Force Constant Calculation for a Two-dimensional Wigner Electron Crystal in High Magnetic Fields, and Limitations of Lindemann's Law of Melting

F. SIRINGO*, M. J. LEA† and N. H. MARCH*

**Theoretical Chemistry Department, University of Oxford, 5 South Parks Road, Oxford OX1 3UB, England.*

and

†Department of Physics, Royal Holloway and Bedford New College, University of London, Egham Hill, Egham, Surrey TW20 0EX, England.

(Received October 9, 1990)

The Einstein oscillator model of the Wigner electron crystal takes a specified force constant, which is an explicit function of the electronic separation r_s. Here a method is devised for its self-consistent calculation. For the gaussian density of the Wigner oscillator, with width λ in terms of r_s, experimental data on the melting temperature T_m versus Landau filling factor ν is employed to plot the critical overlap parameter, λ_c say, versus T_m. This shows that λ_c increases as T_m decreases, which is in contrast to Lindemann's Law.

KEY WORDS: Landau filling factor, two-dimensional electron crystal, melting.

A body of work now exists on the melting of the Wigner electron crystal in zero magnetic field[1–4]. Unfortunately, no experiment is presently known in which the quantal regime of the melting curve can be explored in the laboratory. Therefore, attention has been focussing on magnetic field assisted Wigner crystallization[5]; the motivation being the recent experimental work of Andrei *et al.*[6] and Glattli *et al.*[7]. This was carried out on a GaAs–GaAlAs heterojunction in a magnetic field; the appearance of a new low frequency propagating mode is taken as a fingerprint of an electron solid. The conductivity in this new phase has the hallmarks of pinned Wigner crystallites: Bragg reflection studies have not proved feasible to date, but will eventually be required to confirm long-range order.

Because of the above motivation, we have been studying models of the melting curves of Wigner crystals in magnetic fields in 2 and 3 dimensions[8,9]. The present work focusses on the two-dimensional case. The new aspect of the investigation is to refine the harmonic oscillator model[10] and in particular to devise a method to make it self consistent, in a sense to be set out below. This treatment is then applied to an analysis of the results of Glattli *et al.*[7], whose experiment gives data on the melting temperature T_m of the two-dimensional electron solid as a function of the Landau filling factor $\nu = nh/eB$, with B the magnetic field strength and n the density of electrons per unit area.

Turning immediately to the self-consistent approach to the Einstein model, we note first that such self-consistency is lacking in the normal procedure. This is because the spring constant $m\Omega^2$ in the harmonic potential $\frac{1}{2}m\Omega^2r^2$ is usually written, with units of length the appropriate Bohr radius $a_b = h^2\varepsilon\varepsilon_0/\pi e^2 m$ (see below for numerical values):

$$m\Omega^2 = g/r_s^2(e^2/4\pi\varepsilon\varepsilon_0), \qquad n = (\pi r_s^2)^{-1} \tag{1}$$

where g has previously been taken as a constant which characterizes a given model. In fact Eq. (1) is valid only in the extreme low density limit, r_s tends to infinity. In general the scaling of Eq. (1) with r_s is correct but g is a somewhat complicated function of r_s and T, in addition to the magnetic field B if this is used, as here, to assist the Wigner transition.

Below, the question is raised, and answered in the affirmative, as to whether one can proceed self-consistently to a calculation of the oscillator force constant in such an Einstein description of the Wigner crystal. As already mentioned, the results are then combined with experimental data[7] on the melting temperature T_m vs. Landau filling factor ν to plot the overlap parameter λ_c (cf Eq. (13) below) corresponding to the melting transition, versus T_m.

The complete Hamiltonian for the single oscillator in 2 dimensions is

$$H = \frac{(\mathbf{p} - e\mathbf{A}/c)^2}{2m} + \tfrac{1}{2}m\Omega^2(x^2 + y^2) \tag{2}$$

Here the vector potential $\mathbf{A}$ is taken in terms of the magnitude of the field B in the z direction as $\mathbf{A} = (-By/2, Bx/2, Bx, 0)$ and we define $\omega = eB/2m$. Below we shall regard Ω^2 as a mean-field parameter which depends on the behaviour of the oscillators surrounding the one under discussion. Knowledge of Ω^2 allows the true potential felt by each oscillator to be evaluated in the harmonic approximation. The theory of March and Tosi[11] can be used in order to get the density matrix from the Hamiltonian (2). The electron density can then be evaluated, and in atomic units this can be written as

$$\rho(r, \beta) = \frac{1}{\pi\sigma^2}\exp(-r^2/\sigma^2) \tag{3}$$

where

$$\sigma^2 = \tilde{\omega}^{-1}\frac{\sinh(\beta\tilde{\omega})}{\cosh(\beta\tilde{\omega}) - \cosh(\beta\omega)}. \tag{4}$$

Here $\tilde{\omega} = [\omega^2 + \Omega^2]^{1/2}$ while $\beta = (k_B T)^{-1}$. Assuming a given geometrical structure for the crystal, and summing up over all the oscillator sites $i \neq j$ we find the density surrounding the site j as

$$\rho_j(\mathbf{r}, \beta) = \sum_{i \neq j} \rho(\mathbf{r} - \mathbf{R}_i, \beta) \tag{5}$$

where $\mathbf{R}_i$ denote the lattice vectors.

The Coulomb repulsion energy for an electron on site j is then

$$\phi_j(\mathbf{r}) = \int d\mathbf{r}' \frac{1}{|\mathbf{r} - \mathbf{r}'|} \sum_{i \neq j} \rho(\mathbf{r} - \mathbf{R}_i, \beta). \tag{6}$$

From the rotational invariance symmetry of the crystal and in particular if the crystal is invariant under rotations through an angle $\Theta \neq 0, \pi$, then in general

$$\left.\begin{aligned} \left(\frac{\partial \phi_j}{\partial x}\right)_0 &= \left(\frac{\partial \phi_j}{\partial y}\right)_0 = 0 \\ \left(\frac{\partial^2 \phi_j}{\partial x\, \partial y}\right)_0 &= 0 \\ \left(\frac{\partial^2 \phi_j}{\partial x^2}\right)_0 = \left(\frac{\partial^2 \phi_j}{\partial y^2}\right)_0 &= \tfrac{1}{2}(\nabla^2 \phi_j)_0 \end{aligned}\right\} \tag{7}$$

Then the harmonic expansion of Eq. (6) reads

$$\phi_j(\mathbf{r}) \simeq \phi_j(0) + \tfrac{1}{4}(\nabla^2 \phi)_0 (x^2 + y^2). \tag{8}$$

This is the point now at which the model can be made self-consistent by imposing

$$\Omega^2 = \tfrac{1}{2}(\nabla^2 \phi_j)_0 = \int d\mathbf{r}' \frac{1}{|\mathbf{r} - \mathbf{r}'|^3} \sum_{i \neq j} \rho(\mathbf{r} - \mathbf{R}_i, \beta). \tag{9}$$

Since for a uniformly charged 2-D layer $(\nabla^2 \phi) = 0$, we can invoke the superposition principle and we can add any constant uniform charge density to ρ:

$$\Omega^2 = \int d\mathbf{r}' \frac{1}{|\mathbf{r} - \mathbf{r}'|^3} \sum_{i \neq j} \{\rho(\mathbf{r} - \mathbf{R}_i, \beta) - \rho_0\} \tag{10}$$

where ρ_0 is an arbitrary constant. In particular, the choice $\rho_0 = \sum_{i \neq j} \rho(-\mathbf{R}_i, \beta)$ removes divergence difficulties from the integration. Eq. (10) then becomes, after using Eqs. (3) and (4):

$$\Omega^2 = \sigma^{-3} \sum_i F\left(\left|\frac{\mathbf{R}_i}{\sigma}\right|\right) \tag{11}$$

where the function $F(x)$ is defined as

$$F(x) = \exp(-x^2) \int_0^\infty \frac{dr}{r^2} [\exp(-r^2) I_0(2rx) - 1]. \tag{12}$$

Here $I_0(y) = J_0(iy)$ is the Bessel function of imaginary argument.

Equations (11) and (4) provide a self-consistent way of calculating the spring constant Ω^2 and the density $\rho(r, \beta)$ for the single oscillator. From Eq. (4) it is to be noted that the mean square displacement $\langle r^2 \rangle = \sigma^2$, and it is convenient to introduce the overlap parameter

$$\lambda = \sigma/r_s. \tag{13}$$

In Eq. (11), the quantity $\sum_i F(|\mathbf{R}_i|r_s/\lambda)$ is a function of λ only if the geometry of the crystal is fixed. If we define

$$g(\lambda) = \lambda^{-3} \sum_i F\left(\frac{|\mathbf{R}_i|r_s}{\lambda}\right) \tag{14}$$

then it follows that

$$\Omega^2 = g(\lambda)/r_s^3 \tag{15}$$

which is to be compared with Eq. (1). Evidently for a different set of variables r_s, T, B we find a different self-consistent solution and a different value of λ. The function $g(\lambda)$ is plotted in Figure 1 for the hexagonal structure, the functional dependence on λ subsuming the above variables. The calculations have also been carried out for the square lattice, and it is found that the curve in Figure 1 is relatively insensitive to structure. However the square lattice results are not plotted as this structure is known to be unstable to shear deformations in the classical limit[12]. The overlap parameter increases with increase of density and temperature, and Figure 2 shows plots of constant λ in the density-temperature plane at $B = 0$. In the classical limit r_s tends to infinity

$$\lambda^2 g(\lambda) \sim 2r_s/\beta \tag{16}$$

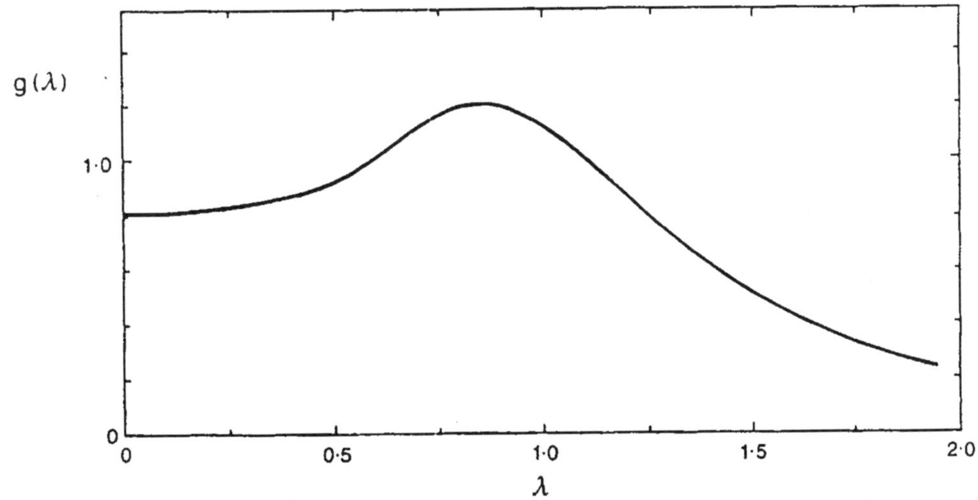

Figure 1 Spring constant characterized by $g(\lambda)$ in Eq. (1) as a function of the overlap parameter λ. Note that λ tends to zero is the classical limit corresponding to $g(0) = 0.81$ for the hexagonal lattice. The corresponding square lattice value differs by only about 1%. The large λ limit corresponds to the regime where tunnelling is important.

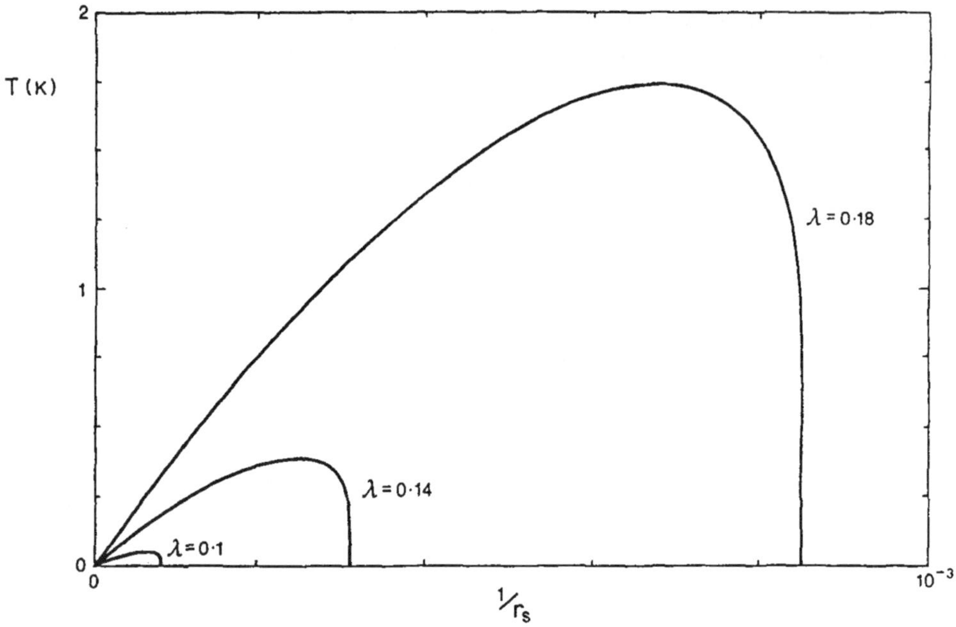

Figure 2 Shows curves of constant λ in the density-temperature plane for the case of zero magnetic field. (Actual plot is T vs. r_s^{-1} where r_s is related to the density n by Eq. (1)).

while in the limit T tends to zero the result is

$$\lambda^4 = \frac{v^2}{(1 + gr_s v^2)}. \tag{17}$$

In the case of zero field in fact one can write the implicit equation

$$\beta = \frac{r_s^{3/2}}{\{g(\lambda)\}^{1/2}} \ln\left[\frac{\lambda^2\{g(\lambda)\}^{1/2} r_s^{1/2} + 1}{\lambda^2\{g(\lambda)\}^{1/2} r_s^{1/2} - 1}\right]. \tag{18}$$

This is now the point to turn to the main focus of the paper; the influence of the Landau filling factor in relation to melting in high magnetic fields. There are various possible approaches; the one we have settled on after investigation is to utilize the experimental results of Glattli *et al.*[7] for the measured melting temperature T_m of the two-dimensional electron solid versus v. Their melting temperature is normalized to the classical melting temperature T_{mc} given by

$$T_{mc} = e^2/\varepsilon r_s k_B \Gamma_m \tag{19}$$

where $\Gamma_m = 127 \pm 3$ has been found experimentally[13] for electrons on the surface of liquid helium. For GaAs/GaAlAs $\varepsilon = 13$ and $m = 0.067m_e$ where m_e is the free electron mass. The electron separation r_s is normalized to the Bohr radius $a_b = h^2\varepsilon\varepsilon_0/\pi e^2 m = 1.03 \times 10^{-8}$ m in this material. For a density of 10^{11}cm^{-2} this gives $r_s = 1.74$. Using the result of the present model for λ as a function of v, T/T_{mc} and r_s,

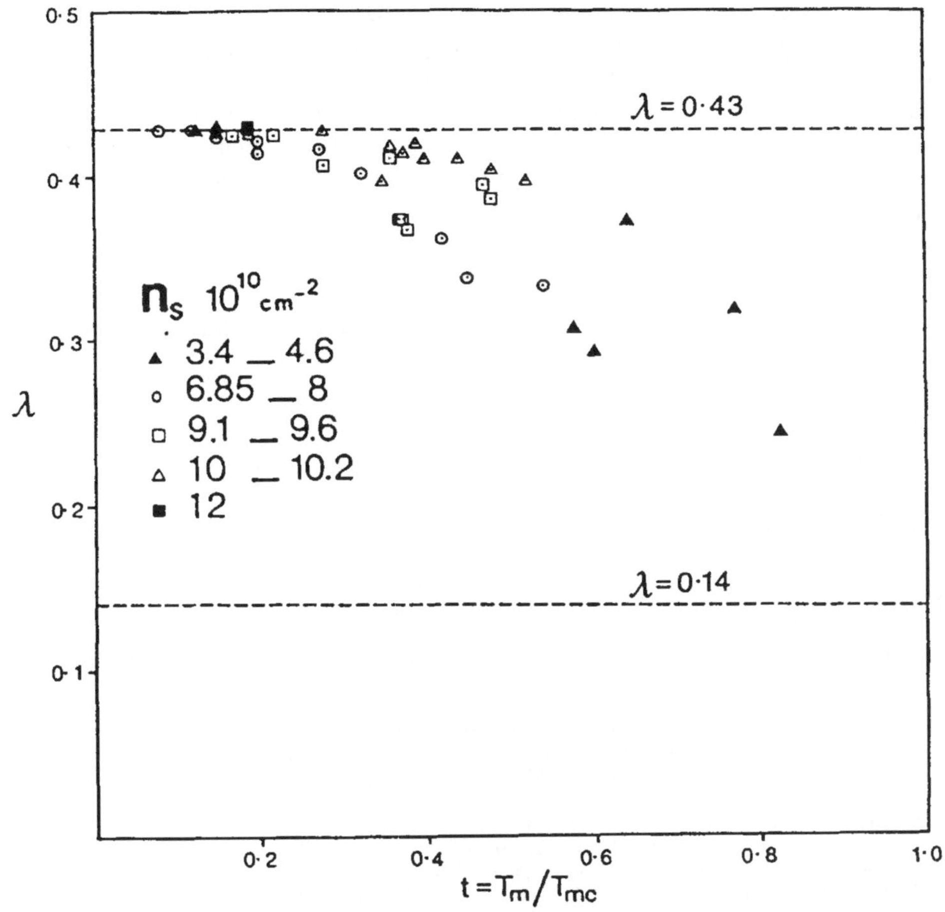

Figure 3 Shows critical overlap parameter λ_c versus reduced temperature T/T_{mc} as extracted from T versus Landau filling factor ν according to Glattli *et al*[7]. Different densities are recorded on the inset. The dashed line corresponds to the classical limiting value of 0.14 shown at $T/T_{mc} = 1$.

the critical value λ_c at melting can be derived from the phase diagram as determined by the experimental data. The classical limit, $\nu = 0$, corresponds to $\lambda_c = 0.14$; a rather low value. The definition of λ used here avoids the logarithmic singularity in the vibrational amplitude of two-dimensional crystals. Other values of λ_c derived from the work of Bedanov *et al.*[14] and Gann *et al.*[15] are 0.22 and 0.30 respectively, though the latter value may be enhanced by the logarithmic effects in their computer simulation. The melting here is thought to be a Kosterlitz–Thouless[16] transition, driven by the shear modulus and the thermal unbinding of dislocation pairs. The T_m tends to zero limit in a field occurs for $\nu = \nu_c = 0.192$ which corresponds to $\lambda_c = 0.43$. Evidently the critical λ is different for the melting of a two-dimensional classical crystal and its quantal counterpart where zero-point motion dominates. A similar conclusion was also reached by Lozovik *et al.*[17]. It is certainly worthy of note that

Ceperley's result[18] $r_W = 33$ for the melting of a two-dimensional quantal crystal in zero field also gives $\lambda_c = 0.44$ from Eq. (17). Figure 3 also shows the λ_c values obtained from the entire set of data of Glattli *et al*[7]. It can be seen that λ_c drops from the quantal value of 0.44 to the classical estimate of 0.14 as the magnetic field is increased. Hence Lindemann's Law must be transcended for two-dimensional melting in a magnetic field. It is probable that anharmonicity, not included here, plays an important role.

Acknowledgement

One of us (F.S.) wishes to acknowledge a grant from the Italian CNR which made possible his visit to Oxford, and thereby this collaboration.

References

1. A. Ferraz, N. H. March and M. Suzuki, *Phys. Chem. Liquids*, **8**, 153 (1978).
2. A. Ferraz, N. H. March and M. Suzuki, *Phys. Chem. Liquids*, **9**, 59 (1979).
3. M. Parrinello and N. H. March, *J. Phys.* **C9**, L147 (1976).
4. H. Nagara, Y. Nagata and T. Nakamura, *Phys. Rev.*, **A36**, 1859 (1987); N. H. March ibid, **A37**, 4526 (1988).
5. J. Durkan, R. J. Elliott and N. H. March, *Rev. Mod. Phys.*, **40**, 812 (1968).
6. E. Y. Andrei, G. Deville, D. C. Glattli, F. I. B. Williams, E. Paris and B. Etienne, *Phys. Rev. Lett.*, **60**, 2765 (1988).
7. D. C. Glattli, G. Deville, V. Duburcq, F. I. B. Williams, E. Paris and B. Etienne, *Surf. Sci.*, **229**, 344 (1990).
8. M. J. Lea and N. H. March, *Int. J. Quantum Chem. Symp.*, **23**, 717 (1989).
9. M. J. Lea and N. H. March, *Phys. Chem. Liquids*, **21**, 183 (1990).
10. See, for example, the review by C. M. Care and N. H. March, *Adv. Phys.*, **24**, 101 (1975).
11. N. H. March and M. P. Tosi, *J. Phys.*, **A18**, L643 (1985).
12. L. Bonsall and A. A. Maradudin, *Phys. Rev.*, **B15**, 1959 (1977).
13. G. Deville, *J. Low Temp. Phys.*, **72**, 135 (1988).
14. V. M. Bedanov, G. V. Gadiyak and Yu. E. Lozovik, *Phys. Lett.*, **109A**, 289 (1985).
15. R. C. Gann, S. Chakravarty and G. V. Chester, *Phys. Rev.*, **B20**, 326 (1979).
16. J. M. Kosterlitz and D. J. Thouless, *J. Phys.*, **C6**, 1181 (1973); *Prog. Low Temp. Phys.*, **7B**, 371 (1978).
17. Yu. E. Lozovik, V. M. Farztdinov and B. Abdullaev, *J. Phys.*, **C18**, L807 (1985).
18. D. M. Ceperley, *Phys. Rev.*, **B18**, 3126 (1978).

Volume 157, number 2,3 PHYSICS LETTERS A 22 July 1991

Remnants of the Fermi surface in the Wigner electron crystal phase of a strongly interacting one-dimensional system

A. Holas

Institute of Physical Chemistry of the Polish Academy of Sciences, Kasprzaka 44, 01-224 Warsaw, Poland

and

N.H. March

Theoretical Chemistry Department, University of Oxford, 5 South Parks Road, Oxford OX1 3TG, UK

Received 18 April 1991; accepted for publication 13 May 1991
Communicated by A.A. Maradudin

The work of Ovchinnikov and Zabrodin is used to study the non-analytic behaviour in the electronic momentum distribution in the Wigner crystalline phase. For a jellium, a comparison is made with our earlier three-dimensional study of the electronic pair function and the momentum distribution.

As is well known, a D-dimensional system of non-interacting (ni) fermions forms an isotropic paramagnetic (P) liquid. Its momentum distribution $n_P^{ni}(\boldsymbol{k})$ is the step function

$$n_P^{ni}(\boldsymbol{k}) = \theta(k_F - |\boldsymbol{k}|) , \tag{1}$$

having unit discontinuity at the Fermi momentum k_F. As a function of the uniform electron density ρ, this latter quantity is given by

$$\begin{aligned} k_F &= (3\pi^2\rho)^{1/3}, \quad D=3 , \\ &= (2\pi\rho)^{1/2}, \quad D=2 , \\ &= (\tfrac{1}{2}\pi\rho), \quad D=1 , \end{aligned} \tag{2}$$

where double occupancy of all states within the D-dimensional Fermi sphere has been assumed.

Switching on a repulsive interaction between the electrons (corresponding to the mean interelectronic spacing $r_s > 0$ for jellium models), modifies the momentum distribution (1) to a form $n_P(\boldsymbol{k})$. This becomes less than unity for $|\boldsymbol{k}| < k_F$ and develops a tail for $|\boldsymbol{k}| > k_F$. Such a behaviour in a three-dimensional jellium can be seen, for example, in fig. 24 of the review by Hedin and Lundqvist [1]. A discontinuity of magnitude z_F say, at $|\boldsymbol{k}| = k_F$, less than unity is found in this case. It should be stressed here that the characteristic momentum k_F is still given by eq. (2) as for the non-interacting case. With increasing strength of the interaction, z_F diminishes and again at $D=3$ this behaviour is well demonstrated by fig. 26 of ref. [1]. In particular, $z_F \approx 0.5$ at $r_s = 5$. Another example worthy of note in this context is the model $n_P(\boldsymbol{k})$ obtained by Herman and March [2] for $r_s \geqslant 25$. A similar behaviour is observed for $D=2$ in the Monte Carlo simulations of Tanatar and Ceperley [3]; e.g. the discontinuity is estimated as $z_F = 0.2$ at $r_s = 10$. With further increase in the strength of the repulsive interaction, phase transitions occur (in the region of r_s around 80 or 40 for $D=3$ or 2, respectively), leading ultimately to the Wigner (W) crystalline phase, which exhibits some spin structure. The question then arises as to whether the W phase has "memory" of the Fermi surface which existed in the paramagnetic liquid phase. In modelling various phases of a jellium for $D=3$, Herman and March [2] assumed that the momentum distribution $n_W(k)$ for the Wigner crystal phase is a continuous and smooth function of $|\boldsymbol{k}|$, without any structure in the region of k_F. But this assumption reflects a lack of information on this function $n_W(\boldsymbol{k})$ rather than an established fact.

A recent investigation by Ovchinnikov and Zabrodin [4] on the $D=1$ fermion system with a strong repulsion in the Wigner crystal phase, having antiferromagnetic spin arrangement, will be utilized below to shed light on the question raised above. Specifically, these authors calculated the leading term of the large R expansion of the one-particle density matrix to be [4]

$$G(R)=(\rho R)^{-\nu}\cos(\tfrac{1}{2}\pi\rho R+\phi_0)\,, \tag{3}$$

to within a normalization factor omitted here. In eq. (3) the phase ϕ_0 depends only weakly on R, while the exponent ν can assume non-integral positive values; e.g. $\nu=\frac{9}{8}$ for a particular limiting case corresponding to very strong interaction. The possibility of using $G(R)$ to obtain information on $n_W(\boldsymbol{k})$ is connected with the fact that these quantities are related by Fourier transform (FT). To be precise, the FT relation is, as discussed for example by Thakkar et al. [5],

$$B(\boldsymbol{R})=\int d^D\boldsymbol{k}\; n(\boldsymbol{k})\exp(-i\boldsymbol{k}\cdot\boldsymbol{R})\,, \tag{4}$$

where $B(\boldsymbol{R})$, the so-called folded density used for Compton profile analysis, is defined from the one-particle spinless density matrix $\Gamma^{(1)}(\boldsymbol{r},\boldsymbol{r}')$ as

$$B(\boldsymbol{R})=\int d^D\boldsymbol{r}\,\Gamma^{(1)}(\boldsymbol{r},\boldsymbol{r}+\boldsymbol{R})\,. \tag{5}$$

While the definition adopted by Ovchinnikov and Zabrodin [4] corresponds to $G(\boldsymbol{R})=\Gamma^{(1)}(\boldsymbol{0},\boldsymbol{R})$, nevertheless, since their result (3) was obtained in the limit of large $R=|\boldsymbol{R}|$, $R\gg\rho^{-1}$, the dependence on $\boldsymbol{r}$ in eq. (5) becomes irrelevant in this limit. Hence one can write

$$B(\boldsymbol{R})\simeq CG(\boldsymbol{R})\,, \tag{6}$$

where the constant factor C will subsequently be omitted since eq. (3) is only defined to within a normalization factor as already mentioned. Taking into account the one-dimensionality and isotropy of the result (3), relation (5) with (6) can be rewritten as

$$G(R)=2\int_0^\infty dk\cos(kR)n_W(k)$$

$$=\mathrm{Re}\left(2\int_{-\infty}^{\infty} dk\exp(ikR)n_W(k)\theta(k)\right). \tag{7}$$

The second form of eq. (7) above allows us to use a powerful tool developed by Lighthill [6] for investigating the asymptotic properties of FT. We have previously used this technique to obtain a large r expansion of the pair correlation function $g(r)$ for $D=3$ of the dense electron liquid in the jellium model [7] from knowledge of the non-analyticities occurring in its FT, the liquid structure factor $S(k)$. All Lighthill's results relevant in the present context may be found quoted in this previous study [7]. Here, the Lighthill theorem will be applied in the opposite manner: knowing the asymptotic form of $G(R)$ in eq. (3), we wish to extract the non-analyticities occurring in $n_W(k)$. To this end, let us rewrite eq. (3) in the form

$$G(R)=\mathrm{Re}[(\rho R)^{-\nu}\exp(ik_F R+i\phi_0)]\,. \tag{8}$$

It should again be stressed that k_F appearing in eq. (8) is just the same paramagnetic Fermi momentum as in eq. (2). If we now assume that the momentum distribution $n_W(k)$ is non-analytic at certain k points, say $k=k_m$, $m=1, 2, ..., M$, then according to Lighthill [6] it follows from eq. (7) that the asymptotic large R expansion of $G(R)$ must be a sum of terms like $\mathrm{Re}[\exp(ik_m R)F_m(R)]$, where $F_m(R)$ is a non-analytic function of $1/R$; see eq. (2.3) of ref. [7] for further details. Since, in fact, the leading term of the asymptotic expansion is known to have the form (8), the strongest non-analyticity (na) of $n_W(k)$ must occur at $k=k_F$. This is then found to have the form

$$n_W^{na}(k)=C_F|k-k_F|^{(\nu-1)}$$

$$\times[\tan(\tfrac{1}{2}\pi\nu)+\tan(\phi_0)\,\mathrm{sgn}(k-k_F)]\,, \tag{9}$$

where C_F is a constant; the, as yet unknown, weak dependence of ϕ_0 on R having here been neglected. The correctness of eq. (9) can be verified by re-inserting it into eq. (7) and then applying the Lighthill procedure described by eqs. (2.1)–(2.3) with (A9.2) and (A9.3) of ref. [7]. Then the result (8) is recovered with some real constant factor multiplying it.

As seen from eq. (9), the one-dimensional Fermi system in the Wigner crystalline phase "remembers" the Fermi surface at $k=k_F$ which existed in the weakly interacting liquid phase as a point where $n_P(k)$ was discontinuous. Now the same Fermi surface is connected with the non-analyticity of $n_W(k)$, which has a power-law character with a "fractional" exponent equal to $\nu-1$. The function recorded in eq. (9) represents (for $1<\nu<2$) either an extremely sharp kink of finite height, or depth, at $k=k_F$ when $|\tan(\frac{1}{2}\pi\nu)|>|\tan(\phi_0)|$ or, in the opposite case, a monotonic function with infinite slope at $k=k_F$.

Presently we know of no evidence for similar "remnants" of the Fermi surface in the Wigner electron crystal phase for $D=2$ and 3, although the present study leads us to expect that such a phenomenon should occur for $D>1$ also. However, in these cases, a deformation of the "Fermi surface", from a spherical (circular) in the liquid phase, to a "warped" form in accord with the lattice symmetry in the Wigner phase W is to be anticipated.

In connection with the above-mentioned investigations [7,4], we conclude by pointing out some interesting similarities and differences between our own earlier work [7] on three-dimensional and that of Ovchinnikov and Zabrodin [4] on one-dimensional cases. In both studies, the asymptotic, large R expressions are obtained for the chosen characteristics of many-fermion systems with interaction: in ref. [7] for the pair correlation function $g(R)$, i.e. the diagonal element of the two-particle density matrix for the $D=3$ liquid paramagnetic phase, and in ref. [4] for $G(R)$, i.e. the off-diagonal one-particle density matrix for $D=1$ in the antiferromagnetic Wigner crystal phase. The leading terms of these expansions turn out to have the form of products of an oscillatory factor, with a phase $2k_FR$ and k_FR respectively, and a power-law factor, $1/R^{4+\mu}$ and $1/R^{\nu}$ respectively, where the "fractional" character of the exponents is directly related to the presence of the electron–electron interaction. The difference in the phase of the oscillatory factors is due to the fact that in the case of $g(R)$ it reflects the Kohn-like anomaly of $S(k)$, corresponding to the diameter $2k_F$ of the Fermi sphere, while in the case of $G(R)$ it reflects the anomaly (9) at the Fermi surface, thereby corresponding to the radius k_F. In both cases, the same "paramagnetic" k_F, eq. (2), occurs.

References

[1] L. Hedin and S. Lundqvist, in: Solid state physics, Vol. 23 (Academic Press, New York, 1969).

[2] F. Herman and N.H. March, Solid State Commun. 50 (1984) 725.

[3] B. Tanatar and D.M. Ceperley, Phys. Rev. B 39 (1989) 5005.

[4] A.A. Ovchinnikov and A.V. Zabrodin, Phys. Lett. A 151 (1990) 420.

[5] A.J. Thakkar, A.M. Simas and V.H. Smith, Chem. Phys. 63 (1981) 175.

[6] M.J. Lighthill, Introduction to Fourier analysis and generalized functions (Cambridge Univ. Press, Cambridge, 1959).

[7] A. Holas and N.H. March, Phys. Chem. Liq. 17 (1987) 215.

J. Phys.: Condens. Matter 4 (1992) 5263–5272. Printed in the UK

Melting of Wigner electron crystals: phenomenology and anyon magnetism

M J Lea†, N H March‡ and W Sung§

† Royal Holloway and Bedford New College, University of London, Egham, Surrey TW20 0EX, UK

‡ Theoretical Chemistry Department, University of Oxford, 5 South Parks Road, Oxford OX1 3UB, UK

§ Department of Physics, Pohang Institute of Technology, Korea

Received 19 December 1991

Abstract. The appropriate generalization of the Clausius–Clapeyron equation to describe the first-order melting transition of two-dimensional Wigner electron crystals is used (a) within a phenomenological framework and (b) in conjunction with microscopic theories relating to anyons and composite fermions, to represent the main features of the melting curve as a function of Landau level filling factor ν.

1. Introduction

The proposal of Durkan *et al* (1968) that Wigner electron crystallization should be aided by an applied magnetic field H has now been brought to fruition, starting with the experiment of Andrei *et al* (1988) on two-dimensional GaAs/AlGaAs heterojunctions. This work, in which preliminary data on the melting curve of the Wigner crystal in two dimensions as a function of field were presented, has subsequently been confirmed and extended (see Glattli *et al* 1991). Structure in the phase diagram was reported by Glattli *et al* (1991), Plaut *et al* (1991) and Li *et al* (1991). In particular Buhmann *et al* (1991), using luminescence measurements on the same system as that studied by Andrei *et al* (1988) have suggested a schematic phase diagram, giving the melting temperature T_m versus the Landau level filling factor $\nu = nhc/eH$, where n is the electron density, as shown in figure 1(*a*). The most notable features of this phase diagram are the re-entrant liquid phases. The latter work prompted us to give the thermodynamics of the Wigner crystal melting curve in a magnetic field H (Lea *et al* 1991). The main result of that study was for the slope of the melting curve in the (H, T_m) plane:

$$\partial T_m/\partial H = -\Delta M/\Delta S \tag{1}$$

where ΔM and ΔS represent the changes across the melting curve of the magnetization M and the entropy S respectively. The re-entrant aspects of the phase diagram plotted by Buhmann *et al* (1991) having maxima in the (H, T_m) plane, define points where $\Delta M = 0$. Rewriting (1) in terms of ν readily yields

$$\partial T_m/\partial \nu = (H/\nu)\Delta M/\Delta S. \tag{2}$$

0953-8984/92/235263+10$04.50

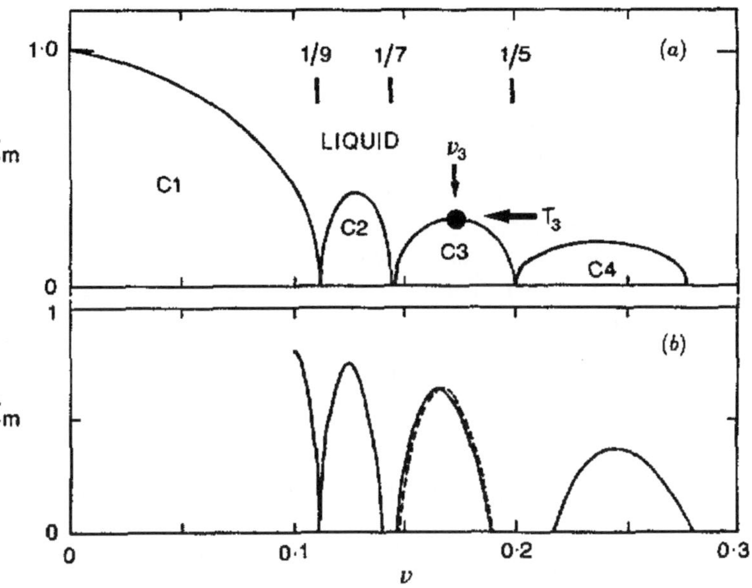

Figure 1. (*a*) Schematic phase diagram showing Wigner crystals in regions C_1–C_4 as proposed by Buhmann *et al* (1991). (*b*) Model calculations of the phase diagram. The solid lines show a model based on (10), (12) and (13) as described in the text. The dashed line in phase C3 only shows the schematic form given by the phenomenological equation (4).

It is clear from (2) that, to model the Wigner crystal melting curve, one must study the changes in magnetization ΔM, and the entropy ΔS.

The purpose of the present paper is twofold:

(i) to provide a phenomenological model that will allow the integration of (2);

(ii) to study the magnetism of the anyon model in the present context, bearing in mind the remarkable variation of the magnetism of the (Laughlin) electron liquid as a function of ν, deduced in our earlier work (Lea *et al* 1991). This also leads to an analysis based on composite fermions.

2. Phenomenological model

It seems natural enough, in view of the zeros in ΔM already referred to, at temperature T_3 and $\nu = \nu_3$, say, in region C_3 (see figure 1(*a*)), to express ΔM as an expansion in $\nu^{-1} - \nu_3^{-1}$. Assuming that the entropy difference along the melting curve is a maximum near ν_3 and falls to zero as T_m goes to zero, we can write (2) as

$$\partial T_m/\partial \nu \propto (H/\nu)(\nu^{-1} - \nu_3^{-1})/[1 - a(\nu^{-1} - \nu_3^{-1})^2] \tag{3}$$

where a is a constant close to unity, which can be integrated to give

$$T_m(\nu) = T_3[1 + C \ln(1 - a(\nu^{-1} - \nu_3^{-1})^2)] \tag{4}$$

where the parameters C and a must be determined experimentally. The form of $T_m(\nu)$ given by (4) is shown in figure 1(*b*), for the C_3 phase. Further terms in the expansion in $\nu^{-1} - \nu_3^{-1}$ could be included that would change the detailed shape of the phase boundaries. This analysis should apply between any two re-entrant phases.

3. Anyon model: classical limit

Having established a primitive phenomenological form appropriate to represent regions C_2–C_4 in figure 1, let us now turn to a model which can give further insight into the microscopic origin of the phase diagram. This is the anyon model, in which one has fractional statistics (Wilczek 1990) characterized by a parameter γ, which has the value 0 for fermions and the value $\pm\frac{1}{2}$ for bosons.

A dilute (non-degenerate) gas of non-interacting two-dimensional anyons (mass m) in a magnetic field has been studied by Johnson and Canright (1990) and also by Dowker and Chang (1990). The second virial coefficient B_2 in the equation of state for the pressure P,

$$P/nkT = 1 + nB_2(T) + O(n^2) \tag{5}$$

is given by

$$B_2(T) = (\lambda^2/x)[\gamma - \exp(4\gamma x)/(2\sinh(2x)) + 1/(4\tanh(x))] \tag{6}$$

where λ is the de Broglie thermal wavelength given by $\lambda^2 = h^2/2\pi mkT$, $x = \hbar\omega_c/2kT$ and ω_c is the cyclotron frequency. Hence the differential magnetization per unit area of the anyons relative to a classical gas is

$$\Delta M_a(\gamma) = -(n^2/2m)\,\partial B_2/\partial x. \tag{7}$$

This is shown in figure 2 as a function of γ for $x = 1$, 2, 5 and 10. Note that the magnetization is not symmetric about $\gamma = 0$, but does have the same value for $\gamma = \pm\frac{1}{2}$.

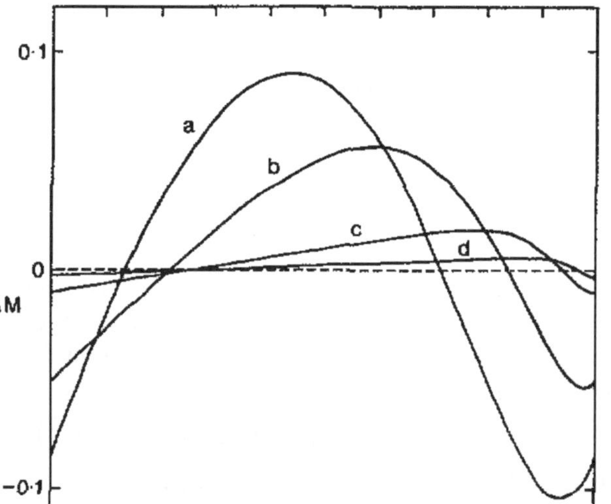

Figure 2. Anyon magnetism ΔM_a of an anyon gas according to (7), versus the fractional statistics parameter γ. Different curves (a–d) correspond to values of the parameter $x = 1$, 2, 5 and 10 respectively.

In order to relate (2) to a field-dependent magnetization it is necessary to relate the statistics parameter γ to ν. To gain orientation, let us refer back to figure 1. Taking the admittedly schematic form there literally, one notes that $T_m = 0$ at values $\nu = 1/q$, where $q = 9$, 7 and 5. At these very points, microscopic theory predicts a Bose condensate (Lee 1991), although this is immediately unstable against raising the temperature. However, this motivates the assumption that at $\nu = 1/q$ the value of γ corresponds to the boson value. Hence as the number of flux quanta per electron, $1/\nu$, increases, the particles alternate between fermions and bosons. Turning to the maxima in the melting curve in the (ν, T_m) plane, one has $\Delta M = 0$, and since the Wigner electron crystal is built from fermions, we expect that the first-order melting transition described by the analogue, equation (1), of the Clausius–Clapeyron equation will degenerate at these points to a second-order transition, since $\Delta M = 0$ there. Hence one assumes that the anyons are fermions at the maxima in the melting curve. The assumption that γ is a continuous variable between these limits is related to a mean-field approximation (Johnson and Canright 1990). If we assume that for non-integral values of q the particles are anyons with fractional statistics, then the magnetization will be dependent on the field, from (2).

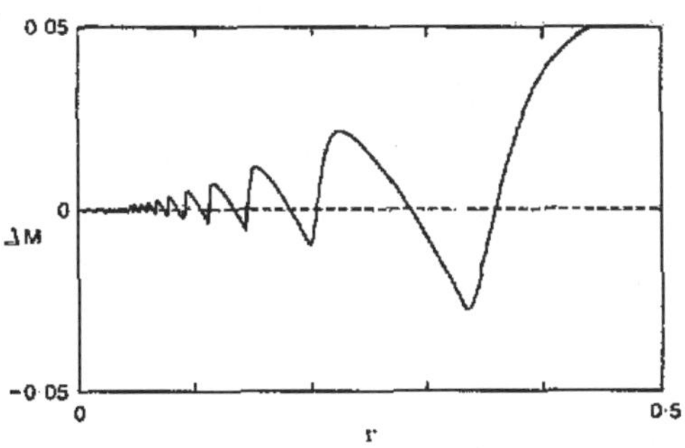

Figure 3. ΔM_a as a function of Landau level filling factor ν. This plot is obtained from figure 2 by using the γ–ν relation given in (8).

These ideas can be subsumed in the relation

$$\gamma = 1/2\nu - j \tag{8}$$

where the integer j is chosen in such a way that γ varies in the range from $-\frac{1}{2}$ to $+\frac{1}{2}$ as ν decreases. This assumption allows plots of the field-dependent ΔM_a versus ν of an anyon gas for various parameter values, from (5) and (6), as shown in figure 3. The plot is calculated for $x = A/\nu$, which corresponds to a variable field at a fixed temperature, for the arbitrary value $A = 1$. It can be seen that the magnetization of this non-interacting anyon gas has many of the features required to explain qualitatively the phase diagram of the 2D Wigner crystal. In particular, ΔM_a changes sign at $\nu = 1/q$ as required to explain the re-entrant nature of the solid regions C_2, C_3 and C_4. It also predicts that there may be further structure in the phase diagram of the region C_1, at $\nu = 1/q$, where q is an odd integer. It appears likely that the purely statistical effects described here will persist in the presence of interactions and that similar magnetization changes will also occur at the FQHE states, at integral fractions.

While, in view of the simplified nature of the assumption†, equation (7), we do not feel it appropriate to press the detail quantitatively, it seems remarkable that the anyon model does indeed extend the de-Haas–van-Alphen-type singularities referred to by Lea *et al* (1991) into the region $\nu \ll 1$.

4. Low-temperature treatment: composite fermion model

A closely related approach at low temperatures, in contrast to the classical limit discussed above, is to use the idea of 'composite' fermions (Jain 1989, 1990, 1992, Jain and Goldman 1992) in which a fermion plus an even number of flux quanta is also a fermion. These composite fermions are then regarded as separate entities in the magnetic field of the remaining flux lines, with a new filling factor ν_i given by

$$\nu_i = \nu/(1 - 2i\nu) = p/(q - 2ip) \tag{9}$$

where $2i$ is the number of flux quanta associated with each composite fermion. The second part of (9) gives ν_i for the fractional quantum Hall states with $\nu = p/q$. For each of these states a value of i can be chosen such that ν_i is an integer. The sign of ν_i depends on whether the flux in the composite fermion is parallel or anti-parallel to the applied field. For the FQHE states with $p = 1$, $\nu_i = 1$ with $i = 1, 2, 3, 4, 5$ for $q = 3, 5, 7, 9, 11$ and so on. This idea has been successfully used by Jain and Goldman (1992) to map the fractional quantum Hall effect states to equivalent integer quantum Hall effect states.

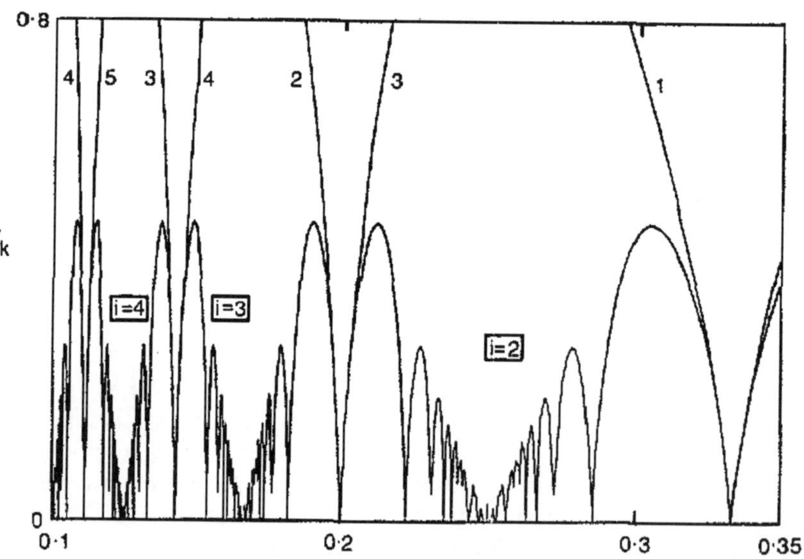

Figure 4. The entropy $S(\nu)/k$ per electron for the composite fermion model, calculated using (9). The various lines are marked by the value of i, the number of flux quanta pairs in each composite fermion. Note that $S(\nu)$ goes to zero at each ordered state $\nu = p/q$ of the fractional quantum Hall effect.

† We reiterate that the virial expansion used is only strictly valid in the classical limit.

4.1. Entropy of the composite fermions

If the entropy $S(\nu)$ of the liquid goes to zero at the fractional filling factors, then we can relate the entropy to the equivalent integer state based on ν_i. Using the results for the de Haas–van Alphen effect we have the entropy per electron $S(\nu)$ for a partially full level:

$$S(\nu) = (-k/\nu_i)[\nu_i \ln \nu_i + (1-\nu_i)\ln(1-\nu_i)] \tag{10}$$

for $\nu_i \leqslant 1$, and equivalent forms for $\nu_i > 1$, where ν_i is given by (9). The entropy of these fermions for $i = 1$, 2 and 3 is shown in figure 5 for $0.1 < \nu < 0.35$ and goes to zero at each ordered state of the fractional quantum Hall effect. This entropy expression will only be valid as long as the system behaves as an assembly of weakly interacting composite fermions. Note, however, that the entropy close to the $\nu = \frac{1}{5}$ state, for instance, has the same form for $i = 2$ and $i = 3$. The functional form of $S(\nu)$ between two neighbouring states is close to the phenomenological expression used in section 2. According to the third law of thermodynamics, the entropy $S(\nu)$ away from the ordered states must also go to zero in the very-low-temperature limit. This could occur via further ordering among the excitations or, in real systems, via some form of localization due to random potentials, as in the integer quantum Hall effect, which could also lead to the observed Hall plateaus. The effect of these localizing potentials would presumably also lead to a pinned Wigner crystal as the solid ground state, as also observed experimentally (Glattli *et al* 1991).

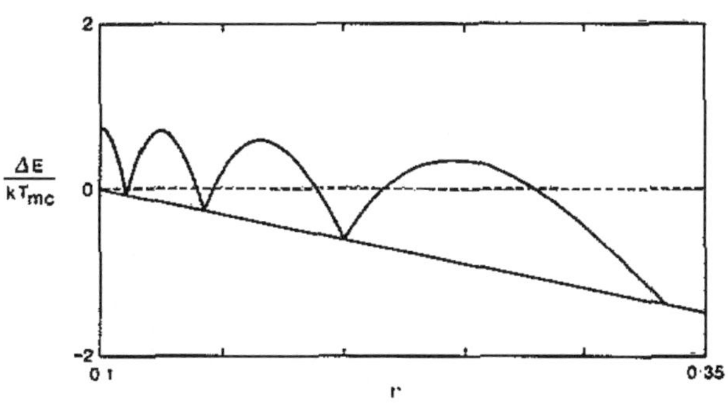

Figure 5. Form of internal energy along the melting curve, expressed as $\Delta E/kT_{\rm mc}$. Gradients at the displayed cusps in ΔE are known in terms of excitation energies and the energy gaps Δ_ν. This information is used asymptotically in this figure at $\nu = 1/q$. The turning points shown in ΔE versus ν are as required by the thermodynamic arguments presented by Lea *et al* (1991).

4.2. Energy of the electron liquid

The internal energies (including the magnetic potential energy term $-HM$, which is different from the definition employed by Lea *et al* (1991) of the electron liquid and solid have been calculated by several authors (see Isihara (1989) for a review). For $\nu \leqslant 0.1$ the solid is thought to have the lower free energy at $T = 0$ and is then the ground state. The Laughlin liquid states have the lower ground-state energies for $\nu = 1/q$ where $q = 3,5,7$ and probably 9. If the temperature T of one of these states is increased, then quasi-particle excitations are thermally excited across an energy gap

Δ. Tao (1990) has shown that this energy gap is field dependent and must have the form

$$\Delta = \Delta_\nu e^2/l \qquad \Delta/kT_{mc} = \Gamma_m \Delta_\nu \sqrt{2/\nu} \tag{11}$$

where $l = (\hbar c/eH)^{1/2}$ is the magnetic length, and Δ_ν is a dimensionless parameter to be calculated or found from experiment. For $\nu = 1/q = \frac{1}{3}$. $\Delta_\nu = 0.018$ (Isihara 1989). The second part of the equation gives the energy gap in units of the transition temperature of the classical electron gas $T_{mc} = e^2(\pi n)^{1/2}/k\Gamma_m$ where $\Gamma_m = 127$ (Deville 1988). If the magnetic field is increased or decreased then quasi-particle or quasi-hole excitations will be generated with energies ϵ_+ and ϵ_- respectively where $\varepsilon_+ + \varepsilon_- = \Delta$, since an increase in temperature creates neutral pairs of excitations. If an excitation is equivalent to an extra or missing flux line then we can write the energy per electron near $\nu = \nu_q = 1/q$ as

$$E(\nu) = E_q + [|(\nu - \nu_q)|/\nu_q] q\varepsilon_\pm \tag{12}$$

where E_q is the ground-state energy at $\nu = 1/q$. Hence the energy $E(\nu)$ will have cusps at $\nu = 1/q$, as shown in figure 5.

4.3. *Magnetization of the electron liquid*

The magnetization $M = -(\partial F/\partial H)_T$ where F is the free energy $F = E - TS$. At zero temperature, a cusp in the energy near ν_q, for instance, will give a jump in the magnetization per electron of magnitude $q(\varepsilon_+ + \varepsilon_-)/H = q\Delta/H$ as ν passes through the ordered state. Hence the magnetization may change sign as ν passes through each of the ordered states indicated by $S = 0$ in figure 5. The schematic variation of M with ν will therefore be very similar in structure to the diagram for the anyon gas in figure 3, although many more states may be revealed.

4.4. *The liquid–solid phase diagram*

We can now return to the discussion of the field dependence of the electron–solid phase boundary. If the liquid phases at $\nu = \frac{1}{3}, \frac{1}{5}, \frac{1}{7}$ and $\frac{1}{9}$ do have lower internal energy than the solid they will form the ground state at zero temperature. For $\nu < \nu_q$ the energy will increase as excitations are created until $E_S = E_L$ (the subscripts refer to the solid and liquid phases). At this point there will be a phase transition to the solid phase. Hence the phase diagram will indeed contain re-entrant liquid phases as proposed by Buhmann *et al* (1991). The experimental evidence indicates that these occur at $\nu = \frac{1}{5}, \frac{1}{7}$ and $\frac{1}{9}$ while Plaut *et al* (1991) found that the highest value of ν for which the solid was present was $\nu = 0.28 \pm 0.02$. However, it must be noted that in the original results of Andrei *et al* (1988) a possible solid phase was found for $\nu \simeq 0.34$. Glattli *et al* (1991) have also reported a liquid phase close to $\nu = \frac{2}{9}$ which would correspond to $i = 2$, $\nu_i = 1$ in (9). We can use the experimental value of Plaut *et al* (1991) to determine $\Delta E = E_L - E_S$ at $\nu = \frac{1}{3}$ by assuming $\epsilon_+ = \epsilon_- = \Delta/2$ with $\Delta_\nu = 0.009$ (Isihara 1989) in (12), so that $\Delta E = 0$ at $\nu = 0.28$. If we also assume that ΔE at the ground states $\nu = \nu_q$ decreases in magnitude until the solid phase becomes the only stable phase for $\nu > 0.1$ (see Isihara 1989), we can then sketch $\Delta E(\nu)$ along the melting curve, expressed in units of kT_{mc} as shown in figure 6. The gradients at the cusps in ΔE are determined by the excitation energies and the

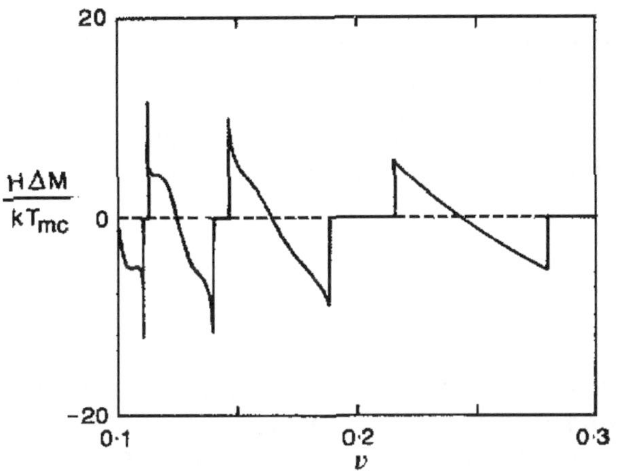

Figure 6. The magnetization change ΔM along the melting curve, expressed as $H\,\Delta M/kT_{mc}$, for comparison with the magnetization deduced thermodynamically (Lea *et al* 1991).

energy gaps Δ_ν. Ordered states at other fractional values of ν may also produce cusps in ΔE, but the energy gaps may well be smaller. Only if $\Delta E = E_L - E_S < 0$ will a re-entrant phase occur.

The phase boundary for $T > 0$ can, in principle, be found by equating the free energies $F_S = F_L$ and hence

$$T_m = (E_L - E_S)/(S_L - S_S) = \Delta E/\Delta S \tag{13}$$

where ΔE and ΔS are evaluated along the melting curve. For the purposes of this discussion we will assume that the entropy of the solid is small compared with the liquid and hence that $\Delta S \simeq S_L$, the entropy of the liquid. If $\Delta E > 0$, but S_L decreases near a fractional state, then T_m would increase rapidly and the neglect of the solid entropy would no longer be a good approximation. Indeed if the solid entropy becomes greater than S_L while $\Delta E > 0$, an increase in temperature will stabilize the solid phase. However, along the melting curve, which is the region of interest here, thermal excitations will probably smooth out the variations in S_L shown in figure 4, and also any small cusps in ΔE. To demonstrate the origin of the observed melting curve, we have therefore used the internal energy shown in figure 5, and an entropy variation that asymptotically approaches the calculated entropy, equation (10), for values of ν close to the liquid ground states. We have taken the energy gaps to increase with ν as $\Delta_\nu = 0.054\nu$. The resultant phase diagram, as calculated from (13) is shown in figure 1(*b*) for $\nu > 0.1$. This does indeed have some of the features of the melting curve postulated from the experimental results, although further structure may also occur in the phase diagram at other ordered states.

We can also assume that the magnetization difference between the two phases is dominated by the field-dependent magnetization discussed above and that $\Delta M \simeq M_L$, the magnetization of the liquid phase. Introducing $t_m = T_m/T_{mc}$ we can then write (2) as

$$\partial t_m/\partial\nu = \pm\sqrt{2}\Gamma_m\Delta_\nu/\nu^{3/2}S(\nu) \tag{14}$$

close to the liquid ground state. Note the important property that $\partial t_{\rm m}/\partial\nu$ is a function only of ν, as required by scaling arguments (Tao 1990). All the thermodynamic functions should depend only on ν and $t = T/T_{\rm mc}$ in this region of the (t, ν) plane. The magnetization change along the melting curve $\Delta M = -(\partial\Delta F/\partial H)_T$ as calculated from our simple model is shown in figure 6 and shows the features that were derived from thermodynamic arguments by Lea *et al* (1991).

5. Discussion and summary

In summary we have

(i) proposed a phenomenological model for which the analogue, equation (1), of the Clausius–Clapeyron equation for the melting curve of a Wigner crystal can be integrated—the simplest model of the phase boundaries of the phases C_2–C_4 in figure 1 gives a re-entrant phase diagram;

(ii) calculated the magnetization of two closely related microscopic models: (a) an anyon gas model in the classical limit and (b) a composite fermion model, which complements (a) in that the magnetization and entropy are determined at low temperatures.

While the present models do not allow a fully quantitative prediction of the melting curve of the two-dimensional Wigner crystal, the main features of the magnetization of the electron liquid are reproduced by the two microscopic models (a) and (b) above. It is our distinct impression that the present work supports the usefulness of the anyon concept in magnetic fields, although further work remains to be done to relate the model to the composite fermion approach. Currently, our view is that the two models are complementary—one being readily calculable in the classical limit and the other being most tractable at low temperatures. Both models, in the end, relate to an electron associated with an integral number of flux lines, and the decomposition into 'fermions plus additional flux lines', as opposed to 'anyons' may be a matter of semantics, although we have at present no decisive conclusion on this point†. Our final comment is that while we feel that our work supports the model for anyons in magnetic fields, no conclusions are to be drawn from the present study as to the usefulness, or otherwise, of the model of field-free anyons in the context of high-temperature superconductivity.

Acknowledgments

This collaboration began during a visit by one of us (WS) to the Theoretical Chemistry Department, University of Oxford, and we gratefully acknowledge support from the Pohang Institute of Science and Technology and the Ministry of Education in Korea.

References

Andrei E Y, Deville G, Glattli D C, Williams F I B, Paris E and Etienne B 1988 *Phys. Rev. Lett.* **60** 2765

† Since this work was completed, the work of Cho and Rim (1992 *Ann. Phys.* **213** 295) has appeared and is relevant to dealing with Bose–Einstein condensation anyons.

Buhmann H, Joss W, von Klitzing K, Kukuskin I V, Plaut A S, Martinez G, Ploog K and Timofeev V B 1991 *Phys. Rev. Lett.* **66** 926

Deville G 1988 *J. Low Temp. Phys.* **72** 135

Dowker J S and Chang M 1990 *ICTP Report* ICTP, Trieste

Durkan J, Elliott R J and March N H 1968 *Rev. Mod. Phys.* **40** 812

Glattli D C, Andrei E Y, Clarke R S, Deville S, Dorin C, Etienne B E, Foxon C T, Harris J J, Paris E, Probst O, Williams F I B and Wright P A 1991 *Physica* B **169** 328

Isihara A 1989 *Solid State Physics* vol 42 (New York: Academic) p 272

Jain J K 1989 *Phys. Rev. Lett.* **63** 199

—— 1990 *Phys. Rev.* B **41** 7653

—— 1992 *Proc. 9th Int. Conf. on Electronic Properties of Two-Dimensional Systems: Surf. Sci.* **263** at press

Jain J K and Goldman V J 1992 *Phys. Rev.* B **45** 1255

Johnson M D and Canright G S 1990 *Phys. Rev.* B **41** 6870

Lea M J, March N H and Sung W J 1991 *J. Phys.: Condens. Matter* **3** 4301

Lee D-H 1991 *Physica* B **169** 37

Li Y P, Sajoto T, Engel L W, Tsui D S and Shagegan M 1991 *Phys. Rev. Lett.* **67** 1630

Plaut A S, Buhmann H, Joss W, von Klitzing K, Kukuskin I V, Martinez G, Ploog K and Timofeev V B 1991 *Physica* B **169** 557

Tao Z C 1990 *Phys. Lett.* A **151** 172

Wilczek F 1990 *Anyons* (Singapore: World Scientific)

J. Phys.: Condens. Matter 5 (1993) B149–B156. Printed in the UK

Melting of a magnetically induced Wigner electron solid and anyon properties

N H March

Theoretical Chemistry Department, University of Oxford, 5 South Parks Road, Oxford OX1 3UB, UK

Received 5 October 1992

Abstract. The early proposal that magnetic fields applied to impure semiconductors would assist Wigner electron crystallization has been verified experimentally in GaAs/AlGaAs heterojunctions. These experimental studies have established some of the basic features of the melting curve of the two-dimensional Wigner electron solid as a function of Landau level filling factor. Using thermodynamics plus the anyon model, remarkable magnetic behaviour of the electron liquid in equilibrium with the Wigner electron solid is exhibited and interpreted. In addition to anyon magnetism, the momentum and statistical distribution functions for anyons are treated.

1. Introduction

The localization of electrons in impure semiconductors was discussed by Durkan *et al* (1968) in relation to transport properties of highly compensated n-type InSb in an applied magnetic field. In this work, the proposal was made that Wigner electron crystallization could be aided by localization due to strong applied magnetic fields. This phenomenon of a magnetically induced Wigner solid (MIWS) has subsequently been observed by Andrei *et al* (1988) in a GaAs/AlGaAs heterojunction. Their findings have been confirmed and extended by the photoluminescence study of Buhmann *et al* (1991).

Motivated by these studies, and in particular by the desire to understand the melting curve of the MIWS, Lea *et al* (1991) have given the thermodynamics of such melting in a magnetic field. Their results are utilized in section 2 immediately below to treat the equilibrium between the Laughlin (1988) liquid and the Wigner solid. Remarkable magnetic properties of the electron liquid can then be deduced, following Lea *et al* (1991), by combining the thermodynamic treatment with the 'experimental' phase diagram of Buhmann *et al* (1991).

Following this, in section 3 a mechanism of melting based on the study of the shear modulus is discussed: this seems most appropriate for the Landau level filling factor $\nu <\sim 0.2$.

Then, in section 4, the model of anyon magnetism is employed to give a microscopic interpretation of the results of section 2. This is followed in section 5 by a discussion of statistical and momentum distribution functions for anyons. Section 6 constitutes a summary, together with some directions which seem to be potentially fruitful for further work.

0953-8984/93/SB0149+08$07.50

2. Equilibrium between Laughlin liquid and Wigner solid

The analogue of the Clausius–Clapeyron equation for melting in a magnetic field was given by Lea *et al* (1991) as

$$(\partial T_m/\partial \nu) = (H/\nu)\Delta M/\Delta S \tag{2.1}$$

where T_m is the melting temperature, while ν is the Landau level filling factor given in terms of the (areal) electron density n and the magnetic field H applied perpendicular to the electron layer by

$$\nu = nhc/eH. \tag{2.2}$$

Finally in equation (2.1) $\Delta M = M_L - M_S$ is the change of magnetization on melting, while $\Delta S = S_L - S_S$ is the corresponding entropy change.

The schematic phase diagram proposed by Buhmann *et al* (1991) is shown in figure 1. The four solid phases, marked C1–C4, are interspersed with the liquid phase at filling factors corresponding to the fractional quantum Hall effect at $\nu = \nu_q = 1/q$ with $q = 5$, 7 and 9. The final solid phase terminates at a critical filling factor $\nu_c = 0.28 \pm 0.02$.

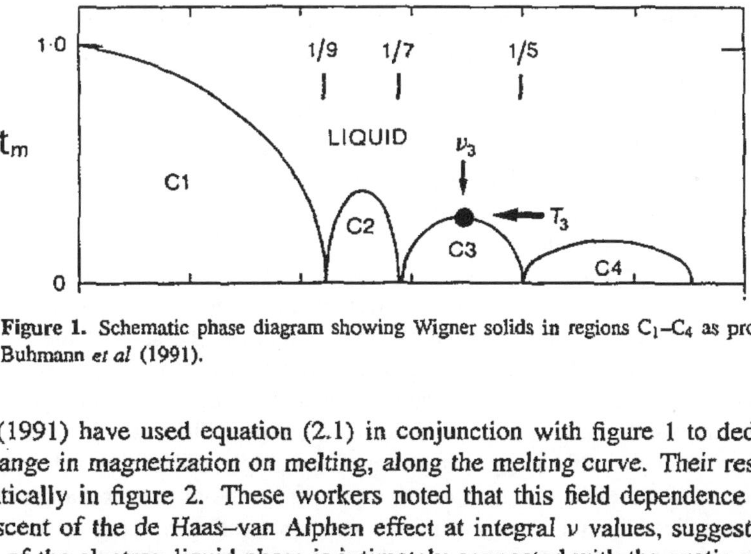

Figure 1. Schematic phase diagram showing Wigner solids in regions C_1–C_4 as proposed by Buhmann *et al* (1991).

Lea *et al* (1991) have used equation (2.1) in conjunction with figure 1 to deduce the form of the change in magnetization on melting, along the melting curve. Their results are shown schematically in figure 2. These workers noted that this field dependence of ΔM is very reminiscent of the de Haas–van Alphen effect at integral ν values, suggesting that the magnetism of the electron liquid phase is intimately connected with the exotic variation of ΔM shown here. The relation of their proposal to the anyon model will be taken up in section 4. However, before turning to that, it is relevant here to discuss the relation between the above phase diagram and the shear modulus.

3. Shear modulus and phase diagram

A further property, the shear modulus, will be discussed in this section, in relation to the phase diagram of two-dimensional Wigner crystals. The arguments below are based on the work of Lea and March (1991; see also March and Tosi, 1985) and the later study of Chui

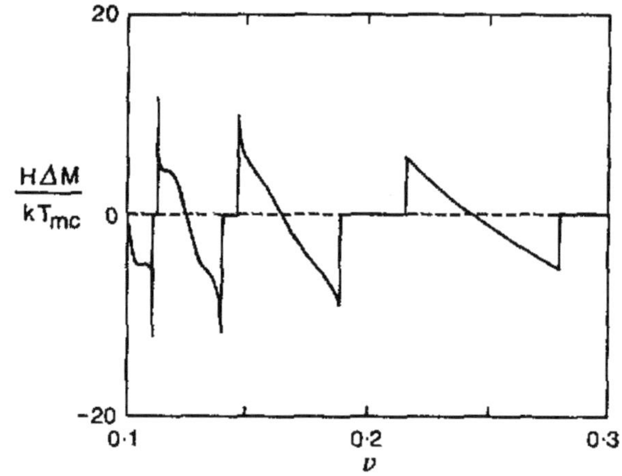

Figure 2. Schematic form of the magnetization of the electron liquid, as a function of the Landau level filling factor ν, along the melting curve (after Lea *et al* 1992).

and Esfarjani (1991). Both these treatments bear a close relationship to the Kosterlitz–Thouless mechanism, which is a classical treatment in which the focus is the role of dislocations. Depending on the magnitude of the core energy of a dislocation, melting occurs discontinuously when grain boundaries are spontaneously generated (see Chui 1983) or continuously when dislocations unbind (Kosterlitz and Thouless 1973, Young 1979, Nelson and Halperin 1979).

Whereas the work of Lea and March (1991) was based on anharmonicity in the two-dimensional Wigner crystal, which was then discussed in relation to the Kosterlitz–Thouless transition, Chui and Esfarjani (1991, and references therein) treated the quantal Wigner transition within the framework of dislocation generation. Specifically, these workers calculated the change in energy of a single dislocation as the virtual emission and reabsorption of a pair of phonons by the dislocation. Their finding was that the elastic energy and the quantum correction are both of the order of $\ln A$, where A is the area of the system. For the two-dimensional electron assembly, the sum of the two contributions becomes zero close to the experimental and simulation results. This leads to the proposal of a mechanism of quantum melting due to the creation of dislocations (Chui and Esfarjani 1991). In the solid phase and at elevated temperatures, dislocations occur as bound pairs. In analogy with the self-energy calculation for a single dislocation referred to above, the interaction between dislocations involves a term from the virtual exchange of two phonons. This renormalized interaction can then be used in treating melting at finite temperatures. Employing the Kosterlitz–Thouless criterion then leads to a solid–liquid phase boundary in the high-field limit which is in fair agreement with experiment out to a Landau level filling factor $\nu = 0.22$. The effect of anharmonicity, considered also in some detail by Lea and March, is to cause a lowering of the melting temperature for $\nu < 0.22$ according to Chui and Esfarjani (1991; Bonsall and Maradudin 1977).

4. Anyon magnetism and melting curve

Let us return at this point to discuss a model of anyon magnetism which is related closely to the work described in section 2 above. Such a model, even though we shall use only the classical limit below, can give further insight into the microscopic origin (Lea *et al* 1992) of the phase diagram sketched in figure 1. In the anyon model, one has fractional statistics (Wilczek 1990) characterized by a parameter γ, which has the value zero for fermions and the value $\pm\frac{1}{2}$ for bosons.

A dilute (non-degenerate) gas of non-interacting two-dimensional anyons (mass m) in a magnetic field has been studied by Johnson and Canright (1990) and also by Dowker and Chang (1990). The second virial coefficient B_2 in the equation of state for the pressure P,

$$P/nkT = 1 + nB_2(T) + O(n^2) \tag{4.1}$$

is given by

$$B_2(T) = (\lambda^2/x)\{\gamma - \exp(4\gamma x)/[2\sinh(2x)] + 1/[4\tanh(x)]\} \tag{4.2}$$

where λ is the de Broglie thermal wavelength given by $\lambda^2 = h^2/2\pi mkT$, $x = \hbar\omega_c/2kT$ and ω_c is the cyclotron frequency. Hence the differential magnetization per unit area of the anyons relative to a classical gas is

$$\Delta M_a(\gamma) = -(n^2/2m)\partial B_2/\partial x. \tag{4.3}$$

This is shown in figure 3 as a function of γ for $x = 1, 2, 5$ and 10. It should be noted that the magnetization is not symmetric about $\gamma = 0$, but does have the same value for $\gamma = \pm\frac{1}{2}$.

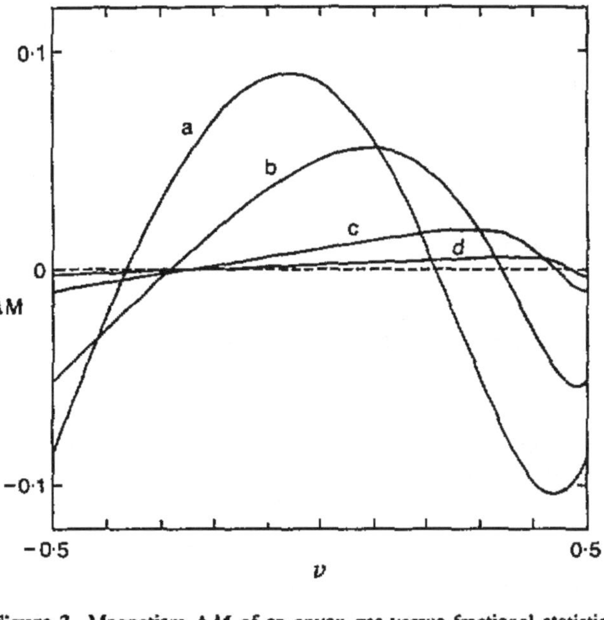

Figure 3. Magnetism ΔM of an anyon gas versus fractional statistics parameter γ. Different curves (a–d) correspond to values of parameter $x = 1, 2, 5$ and 10 respectively (after Lea *et al* 1992).

In order to relate equation (4.3) to a field-dependent magnetization it is necessary to connect the statistics parameter γ to ν. To gain orientation, let us refer back to figure 1.

Taking the admittedly schematic form there literally, one notes that $T_m = 0$ at values $\nu = 1/q$, where $q = 9$, 7 and 5. At these very points, microscopic theory predicts a Bose condensate (Lee 1991), although this is immediately unstable against raising the temperature. However, this motivates the assumption that at $\nu = 1/q$ the value of γ corresponds to the boson value. Hence as the number of flux quanta per electron, $1/\nu$, increases, the particles alternate between fermions and bosons. Turning to the maxima in the melting curve in the (ν, T_m) plane, one has $\Delta M = 0$, and since the Wigner electron crystal is built from fermions, we expect that the first-order melting transition described by the analogue, equation (2.1), of the Clausius–Clapeyron equation will degenerate at these points to a second-order transition, since $\Delta M = 0$ there. Hence one assumes that the anyons are fermions at the maxima in the melting curve. The assumption that γ is a continuous variable between these limits is related to a mean-field approximation (Johnson and Canright 1991). If we assume that for non-integral values of q the particles are anyons with fractional statistics, then the magnetization will be dependent on the field, from equation (4.3).

These ideas can be subsumed in the relation

$$\gamma = 1/2\nu - j \tag{4.4}$$

where the integer j is chosen in such a way that γ varies in the range from $-\frac{1}{2}$ to $+\frac{1}{2}$ as ν decreases. This assumption allows plots of the field-dependent ΔM_a versus ν of an anyon gas for various parameter values, from equations (4.1) and (4.2) (see Lea *et al* 1992). The magnetization of this non-interacting anyon gas has many of the features required to explain qualitatively the phase diagram of the 2D Wigner crystal (compare figure 2). In particular, ΔM_a changes sign at $\nu = 1/q$ as required to explain the re-entrant nature of the solid regions C_2, C_3 and C_4. It also predicts that there may be further structure in the phase diagram of the region C_1, at $\nu = 1/q$, where q is an odd integer. It appears likely that the purely statistical effects described here will persist in the presence of interactions and that similar magnetization changes will also occur at the FQHE states, at integral fractions.

While, in view of the simplified nature of the assumption in equation (4.4), we do not feel it appropriate to press the detail quantitatively, it seems remarkable that the anyon model does indeed extend the de Haas–van Alphen-type singularities referred to by Lea *et al* (1991) into the region $\nu \ll 1$.

5. Statistical and momentum distribution function for anyons

From the above discussion of anyon magnetism, we turn to treat the statistical distribution function $f(\epsilon)$ for anyons, and the closely related momentum distribution function $n(k)$. The work below follows the treatment of March *et al* (1992).

5.1. Chemical model generalizing Bose and Fermi distribution functions

Let us start out from the treatment of collisions in a gas of fermions following, for example, Ma (1985). Suppose we have a collision $1 + 2 \leftrightarrow 3 + 4$, i.e. the states 1 and 2 interact to change to states 3 and 4. We note that collisions are necessary to allow thermodynamic equilibrium to be attained, even in an ideal gas. Then, following Ma, this reaction has the rate

$$f_1 f_2 (1 - f_3)(1 - f_4) R$$

where $1 - f_3$ and $1 - f_4$ are the probabilities that there are no particles in states 3 and 4. These factors must be present for fermions because if states 3 and 4 are occupied then the reaction cannot occur. Now one invokes the fact that this rate must equal that in the reverse direction, i.e.

$$f_1 f_2(1 - f_3)(1 - f_4)R = f_3 f_4(1 - f_1)(1 - f_2)R'. \tag{5.1}$$

Ma next uses time-reversal symmetry of quantum mechanics to show that $R = R'$ and this leads to the form (see Ma 1985, equation (3.21))

$$f(\epsilon)/1 - f(\epsilon) = \exp(-\alpha - \beta\epsilon). \tag{5.2}$$

We next note that, for bosons, the factors $(1 - f_3)$ and $(1 - f_4)$ etc. in equation (5.1) become $(1 + f_3)$ and $(1 + f_4)$: it is as though, if states f_3 and f_4 are already occupied, other states would be more inclined to go to them.

Turning to fractional statistics, then interchange of two particles introduces a phase factor of the form $\exp(2\pi i\gamma)$, where we choose the sign such that $\gamma = 0$ for fermions and $\gamma = \frac{1}{2}$ for bosons. The main assumption on which the generalized collision theory presented here is based is that this reduces the tendency $(1 + f(\epsilon))$ for particles to like the same state for bosons to $(1 - a(\gamma)f(\epsilon))$ where the anyon factor $a(\gamma)$ will be taken to have the properties

$$a(0) = 1 \qquad a(\tfrac{1}{2}) = -1 \tag{5.3}$$

$$-1 \leqslant a(\gamma) \leqslant 1. \tag{5.4}$$

Obviously the 'boundary conditions' specified in equation (5.3) for $\gamma = 0$ and $\frac{1}{2}$ ensure the correct Fermi–Dirac and Bose–Einstein limits.

Then, whereas Ma rewrites equation (5.2) for bosons and fermions as

$$f(\epsilon)/1 \pm f(\epsilon) = \exp(-\alpha - \beta\epsilon) \tag{5.5}$$

one now has for anyons

$$f(\epsilon)/1 - a(\gamma)f(\epsilon) = \exp(-\alpha - \beta\epsilon) \tag{5.6}$$

or

$$1/f(\epsilon) = \exp(\alpha + \beta\epsilon) + a(\gamma). \tag{5.7}$$

Equation (5.7) represents the proposal of March *et al* (1992) for the shape of the statistical distribution function for anyons. While the argument of this section (see, however, section 5.2 below) does not contain within itself a unique procedure for determining the anyon shape factor, the fractional statistics phase factor $\exp(2\pi i\gamma)$ prompts us to assume that $a(\gamma)$ will be expressible as an appropriate Fourier series compatible with the condition (5.4). Then in the absence of further information at this stage, use of Occam's razor leads us to propose

$$a(\gamma) = \cos(2\pi\gamma). \tag{5.8}$$

Thus, by combining equations (5.7) and (5.8) one has the result

$$f(\epsilon) = 1/[\exp(\alpha + \beta\epsilon) + \cos(2\pi\gamma)]. \tag{5.9}$$

One notes immediately that with the special form (5.9) arising from the shape of $f(\epsilon)$ in equation (5.7), the particular case $\gamma = \frac{1}{4}$ leads back precisely to Boltzmann statistics.

5.2. Derivation of occupation number from the Hamiltonian of the free anyon gas

From the above physical argument, we now proceed to summarize the derivation by March *et al* (1992) of results using the Hamiltonian of the anyon gas with statistical parameter $n = 1/(2\gamma)$ and transformations similar to those employed by Mori (1991). Explicitly the Hamiltonian H takes the form

$$H = (1/2m)\int d^2r\psi^+(r)[p - A(r)]^2\psi(r). \tag{5.10}$$

Here $\psi(r)$, $\psi^+(r)$ are the Fermion field operators. $A(r)$ represents the vector potential

$$A(r) = \int d^2r'K(r - r')\rho(r') \tag{5.11}$$

where $K(r) = (1/n)\,\text{grad}[\theta(r)]$ and $\theta(r)$ is the azimuthal angle of the two-dimensional vector r when $r > a$ and zero otherwise. After integration the limit as a tends to zero is to be taken.

The next step is to employ the operator

$$U(r) = \exp[(i/n)\int d^2r'\theta(r - r')\rho(r')] \tag{5.12}$$

to define the anyon field operators. Here it is to be noted that if $U(r')$ is obtained from $U(r)$ after rotation by 2π about the z axis then $\theta(r') = \theta(r)+2\pi$ and $U(r') = \exp(2\pi iN/n)U(r)$ where $\hat{N} = \int d^2r\psi^+(r)\psi(r)$. Following Mori (1991), one then defines the anyon field operators by (March *et al* 1992):

$$\tilde{\psi}^+(r) = \psi^+(r)U(r) : \tilde{\psi}(r) = U^+(r)\psi(r). \tag{5.13}$$

Using standard quantum mechanical methods (Kittel 1963), the Hamiltonian can be expressed in terms of these operators as a free anyon Hamiltonian

$$H = -\frac{1}{2M}\int d^2r\tilde{\psi}^+(r)\nabla^2\tilde{\psi}(r). \tag{5.14}$$

The occupation numbers n_k, or equivalently the momentum distribution function, can be calculated, the result being

$$\langle n_k\rangle_T = \{\cos(\pi/n) + \exp[\beta(\epsilon_k - \mu)]\}^{-1}. \tag{5.15}$$

Thus in this (perturbative) approximation, the occupation numbers have the same form as the ones obtained by using detailed balance considerations. For higher-order perturbation theory this form is essentially the same because of the same operator in all higher-order terms.

In summary, for the case of plane waves, corresponding to ϵ in equation (5.9) equal to $k^2/2m$, it is clear that equation (5.15) leads back to the form (5.9), with the factor $a(\gamma) = \cos(2\pi\gamma)$, which was obtained by physical arguments. In connection with the argument of section 5.1, it is to be noted that, although time-reversal symmetry is not valid for anyons, when combined with parity it leaves the Hamiltonian of anyons invariant. Returning to equation (5.1), it then follows that $R = R'$.

B156 *N H March*

6. Summary and proposed directions for future work

Both phenomenological and microscopic models can be employed to usefully interpret the experimentally determined melting curve of the two-dimensional Wigner solid (Andrei *et al* 1988, Buhmann *et al* 1991). From a microscopic standpoint, the anyon model of the electron liquid in equilibrium with the Wigner solid affords at least a qualitative interpretation of the remarkable variation of the magnetism of the electron liquid with Landau level filling factor.

It is evident that further experiments to extend and deepen the present knowledge of the melting of the Wigner solid are of first-rank importance. Can, for example, direct magnetic measurements be made on the Laughlin electron liquid for $\nu <\sim 0.28$?

The theory is still at an early stage. However, some progress might well come from a study of multiple quantum well structures. This could lead to a better understanding of the 'cross-over' from two-dimensional to three-dimensional Wigner behaviour. In this context, the extension of the photoluminescence experiment to embrace the n-type InSb three-dimensional material discussed by Durkan *et al* (1968; see also Care and March 1975) would be of considerable interest for the future.

Acknowledgments

The work reported in this paper has been carried out with numerous co-workers (see references for a complete list). Especial thanks are due to Dr Lea of RHBNC, London University, and to Drs Gidopoulos and Theophilou (Demokritos, Attikis, Greece) for the collaborative work (March *et al* 1992) summarized in section 5.

References

Andrei E Y, Deville G, Glattli D C, Williams F I B, Paris E and Etienne B 1988 *Phys. Rev. Lett.* **60** 2765
Bonsall L and Maradudin A A 1977 *Phys. Rev.* B **15** 1959
Buhmann H, Joss W, von Klitzing K, Kukuskim I V, Plant A S, Martinez G, Ploog K and Timofeev V B 1991 *Phys. Rev. Lett.* **6** 926
Care C M and March N H 1975 *Adv. Phys.* **24** 101
Chui S T 1983 *Phys. Rev.* B **28** 178
Chui S T and Esfarjani K 1991 *Phys. Rev.* B **44** 11 498
Dowker J S and Chang M 1990 *ICTP, Trieste, Report*
Durkan J, Elliott R J and March N H 1968 *Rev. Mod. Phys.* **40** 812
Johnson M D and Canright G S 1990 *Phys. Rev.* B **41** 6870
—— 1991 *Comment. Solid State Phys.* **15** 77
Kittel C 1963 *Quantum Theory of Solids* (New York: Wiley) p 407
Kosterlitz J M and Thouless D J 1973 *J. Phys. C: Solid State Phys.* **6** 1181
Laughlin R B 1988 *Phys. Rev. Lett.* **60** 2577
Lea M J and March N H 1991 *J. Phys.: Condens. Matter* **3** 3493
Lea M J, March N H and Sung W 1991 *J. Phys.: Condens. Matter* **3** L4301
—— 1992 *J. Phys.: Condens. Matter* **4** 5263
Lee D-H 1991 *Physica* B **169** 37
Ma S K 1985 *Statistical Mechanics* (Singapore: World Scientific)
March N H, Gidopoulos N, Theophilou A K, Lea M J and Sung W 1993 *Phys. Chem. Liq.* at press
March N H and Tosi M P 1985 *J. Phys. A: Math. Gen.* **18** L643
Mori H 1991 *Phys. Rev.* B **43** 5474
Nelson D R and Halperin B I 1979 *Phys. Rev.* B **19** 2457
Wilczek F 1990 *Fractional Statistics and Anyons* (Singapore: World Scientific)
Young A P 1979 *Phys. Rev.* B **19** 1855

Recent progress in the field of electron correlation

G. Senatore

Dipartimento di Fisica Teorica, Università di Trieste, I-34014 Grignano (TS), Italy

N. H. March

Department of Theoretical Chemistry, University of Oxford, Oxford, OX1 3UB, England

Electron correlation plays an important role in determining the properties of physical systems, from atoms and molecules to condensed phases. Recent theoretical progress in the field has proven possible using both analytical methods and numerical many-body treatments, for realistic systems as well as for simplified models. Within the models, one may mention the jellium and the Hubbard. The jellium model, while providing a simple, rough approximation to conduction electrons in metals, also constitutes a key ingredient in the treatment of electrons in condensed phases within density-functional formalism. The Hubbard- and the related Heisenberg-model Hamiltonians, on the other hand, are designed to treat situations in which very strong correlations tend to bring about site localization of electrons. The character of the interactions in these lattice models allows for a local treatment of correlations. This is achieved by the use of projection techniques that were first proposed by Gutzwiller for the multicenter problem, being the natural extension of the Coulson-Fischer treatment of the H_2 molecule. Much work in this area is analytic or semianalytic and requires approximations. However, a full many-body treatment of both realistic and simplified models is possible by resorting to numerical simulations, i.e., to the so-called quantum Monte Carlo method. This method, which can be implemented in a number of ways, has been applied to atoms, molecules, and solids. In spite of continuing progress, technical problems still remain. Thus one may mention the fermion sign problem and the increase in computational time with the nuclear charge in atomic and related situations. Still, this method provides, to date, one of the most accurate ways to calculate correlation energies, both in atomic and in multicenter problems.

CONTENTS

I. INTRODUCTION

In this review, we shall consider some aspects of electron correlation in molecules and solids on which recent progress has proven possible. It is natural enough to start with the homogeneous electron fluid: i.e., the so-called jellium model of a metal, in which electrons interacting Coulombically move in a nonresponsive uniform, positive neutralizing background charge. Here, the ground-state energy is simply a function of the mean density $\rho_0 = 3/4\pi r_s^3 a_0^3$; $a_0 = \hbar^2/me^2$. The pair correlations in this system are by now relatively well understood, as is the transition from a strongly correlated electron liquid to an insulating Wigner electron crystal at a suitable low value of the density, i.e., at large enough r_s. This is a spectacular manifestation of the effects of correlation. Only in the last decade have quantum Monte Carlo (QMC) simulations largely settled the critical value of r_s for this transition, placing it at about 100. Some attention will be given to such QMC results in Sec. IV of the present article, where correlation energies for small molecules, as well as numerical solutions of Hubbard and Heisenberg Hamiltonians, will also be referred to.

The jellium model affords the basis for the so-called theory of the inhomogeneous electron gas, which had its origin in the Thomas-Fermi method. This latter theory was formally completed by the Hohenberg-Kohn theorem (1964), which states that the ground-state energy of an inhomogeneous electron gas in a unique functional of its electron density $\rho(\mathbf{r})$. However, the functional is not yet known. There are fairly recent reviews both on

 0034-6861/94/66(2)/445(35)/$08.50

the electron gas (Singwi and Tosi, 1981; Ichimaru, 1982) and on the so-called density-functional theory (Bamzai and Deb, 1981; Ghosh and Deb, 1982; Lundqvist and March, 1983; Callaway and March, 1984; Jones and Gunnarsson, 1989; Parr and Yang, 1989). Thus we shall keep this discussion relatively short, focusing on the way Wigner electron crystallization can be treated within the density-functional framework. It is relevant in this context to stress at the outset that, while density-functional theory reveals the way in which electron correlation enters the calculation of the single-particle density, there is no means within the formal framework for calculating the required functional describing correlation.

Section III is concerned with the multicenter problems of molecules and solids, and the qualitative aspects of electron correlation are first stressed by summarizing the idea behind the Coulson-Fischer (1949) wave function for the ground state of the H_2 molecule. This idea is then taken up in relation to the work of Gutzwiller (1963, 1964, 1965) on strongly correlated electrons in narrow energy bands. In this treatment, electron interactions are treated by means of the Hubbard U, which, by definition, is the energy required to place two electrons with antiparallel spin on the same site. As in the Coulson-Fischer treatment, one reduces the weight of the ionic configurations that are obtained by expanding out a single Slater determinant of Bloch wave functions.

Because of the current excitement concerning superconductivity, the relevance of the Hubbard Hamiltonian in providing a quantitative basis for the development of Pauling's ideas on the resonant valence-bond theory of metals is briefly summarized also within the context of Gutzwiller's method.

To this stage in the article, the main theoretical development is via largely analytical methods. However, as mentioned already, important progress in the calculation of correlation energy in multicenter problems has resulted from the development of quantum computer simulation, pioneered in general terms by Anderson (1975, 1976, 1980) and applied to jellium by Ceperley and Alder (1980). Important here is to input a trial wave function, transcending wherever possible a single Slater determinant. Briefly, one form of the QMC method starts from the time-dependent Schrödinger equation or the equivalent Bloch equation in imaginary time, in contrast to the methods used elsewhere in the article. The alternative approach works directly with the many-particle Green's function, depending on all electronic coordinates and on an energy parameter. The strengths and the weaknesses of these methods are assessed. As well as the jellium results already mentioned, the parallel development in calculating correlation energies in small molecules will also be described. At the time of writing, though there are still some technical problems, this computer simulation approach provides a very accurate way of calculating electronic correlation energies in multicenter problems. Some promising directions for future work are suggested in concluding the article.

II. HOMOGENEOUS ELECTRON ASSEMBLY

We now turn to the discussion of correlation in a model appropriate to simple metals. It will be useful to focus first on the electron assembly formed by the conduction electrons in a metal like *Na* or *K*. In these cases, there is a whole body of evidence that demonstrates the weakness of the electron-ion interaction. Therefore the jellium model, introduced initially by Sommerfeld long ago, where the ionic lattice is smeared out into a uniform neutralizing background of positive charge in which the correlated electronic motion takes place, affords a valuable starting point.

Electron correlation can be treated quantitatively in this jellium model by numerical means (see, e.g., Sec. IV.D) at all densities, including metallic ones. Approximate treatments, on the other hand, are simpler in the two limiting cases of very high and very low densities. Though at first the problem looks totally different from the molecular case (see especially Sec. III.A), there is, in fact, a close parallel in the sense that in the high-density limit a delocalized picture (cf. molecular orbitals) is correct while, in the low-density limit, electron localization (cf. valence bond theory) is again induced by strong Coulomb repulsion between electrons.

In the following we shall focus especially on the regime of extreme low densities, where the effects of correlation are dominant, leaving aside the range of high and metallic densities (see, e.g., Singwi and Tosi, 1981). Here we shall just recall that at high densities, i.e., for $r_s \to 0$, the kinetic energy dominates the potential energy and an independent-particle description provides a reasonable starting point to describe the electron assembly. In fact, a single plane-wave determinant of spin orbitals is a Hartree-Fock solution for the jellium model, yielding a homogeneous density distribution (see, e.g., March *et al.*, 1967) and an energy

$$\frac{E_{\mathrm{HF}}}{N} = \left[\frac{2.21}{r_s^2} - \frac{0.916}{r_s} \right] Ry \; . \tag{2.1}$$

Many-body perturbation on such a state, in which electrons are fully delocalized, yields an approximate estimate of the correlation energy E_c (Gell-Mann and Brueckner, 1957), defined as the difference between the true ground-state energy E_0 and its Hartree-Fock approximation E_{HF}, i.e., $E_c = E_0 - E_{\mathrm{HF}}$.

A. Low-density Wigner electron crystal

Turning to the extreme low-density limit $r_s \to \infty$, Wigner (1934, 1938) pointed out that the above delocalized picture broke down completely, and once the potential energy became large compared with the kinetic energy, the electrons would then want to avoid each other maximally. He stressed that this situation would be achieved by electrons becoming localized on the sites of a lattice. He argued that one must find the stable lattice by minimizing the Madelung energy. Of the lattices so far examined, the bcc lattice has the lowest Madelung term. This yields, for the electron bcc Wigner crystal, the energy

$$\lim_{r_s\to\infty}\frac{E}{N}=-\frac{1.792}{r_s}Ry\ , \tag{2.2}$$

which shows, by comparison with the Hartree-Fock plane-wave result (2.1), that this latter approximation is no longer of any physical utility, the energy being too high by a factor of about 2.

As the repulsive coupling between the electrons is relaxed, that is, r_s reduced below the range of validity of Eq. (2.2), the electrons will vibrate about the bcc lattice sites. Since the bcc lattice has a Wigner-Seitz cell of high symmetry, it is a useful first approximation to neglect the (multipole) fields of the other cells in considering the vibration of an electron in its own cell. Then the potential energy $V(r)$, in which the electron vibrates, is created solely by the uniform positive background in its own cell, which, in the spherical approximation, gives

$$V(r)=\frac{e^2r^2}{2r_s^3a_0^3}+\text{const} \tag{2.3}$$

with the ground-state isotropic harmonic-oscillator wave function as

$$\psi(r)=\left[\frac{\alpha}{\pi}\right]^{3/4}\exp(-\tfrac{1}{2}\alpha r^2),\quad \alpha=(r_sa_0)^{-3/2}\ . \tag{2.4}$$

This Wigner oscillator leads to a kinetic energy per electron in the low-density limit as

$$\lim_{r_s\to\infty}\frac{T}{N}=\frac{3}{2r_s^{3/2}}Ry\ . \tag{2.5}$$

This is qualitatively different from the independent electron result $(2.21/r_s^2)Ry$. In fact, an obvious effect of switching on the electron-electron interaction away from the $r_s\to 0$ limit is to promote electrons outside the Fermi sphere of radius k_F, leaving holes inside (cf. Fig. 3). This creation of electron-hole pairs obviously increases the kinetic energy. Thus, if one writes $T/N=K/r_s^2$, the increase in kinetic energy appears as an increase in $K(r_s)$. In the limit $r_s\to\infty$, the creation of particle-hole pairs is so prolific that K diverges as $r_s^{1/2}$ and the Fermi sphere picture breaks down completely. The calculation yielding Eq. (2.5) is an Einstein-type model, whereas one should, of course, treat the vibrational modes of the bcc Wigner crystal by collective phonon theory. The coefficient 3 in Eq. (2.5) is then reduced to 2.66 (Carr, 1961; Coldwell-Horsfall and Maradudin, 1963).

The Fermi distribution is then profoundly altered by the creation of particle-hole pairs, and another way of seeing this is to take the Fourier transform of the Gaussian orbital (2.4), which, of course, yields also a Gaussian momentum distribution. At most, remnants of a Fermi surface remain in the low-density limit [in one dimension, this statement is made quantitative by Holas and March (1991)].

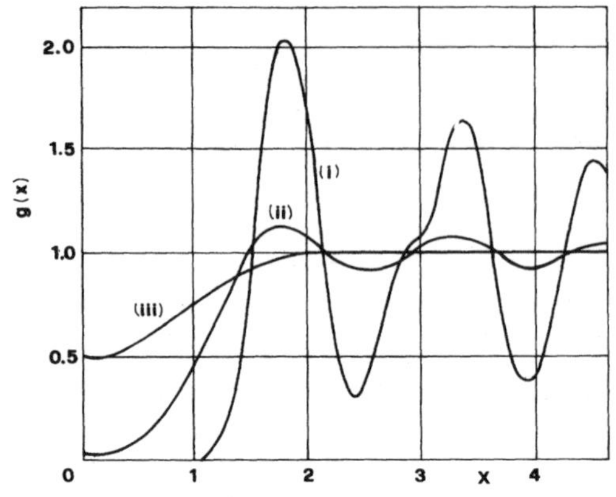

FIG. 1. Pair-correlation function $g(r)$ vs $x=r/r_sa_0$, for different densities: (i) Wigner form for $r_s=100$; (ii) Wigner form for $r_s=10$; (iii) Fermi hole, correct in the limit $r_s\to 0$.

A measure of the order present in a many-body system is afforded by the so-called pair-correlation function $g(r)$. This gives the probability of finding pairs of electrons at distance r. In the Wigner-crystal regime, and within the approximate framework of electrons behaving like Einstein oscillators, it is a straightforward matter to construct $g(r)$. Results for different values of r_s are shown in Fig. 1, taken from the work of March and Young (1959). The tendency of electrons to avoid each other with increasing r_s is apparent, as well as the greater amount of order present in the state in which the electrons are localized in Gaussian orbitals in contrast to that in which they are in plane waves.

B. Density-functional theory of many-electron system

One approach to incorporate, approximately, of course, electron correlation is afforded by the density-functional theory. However, because of related recent reviews (Lundqvist and March, 1983; Callaway and March, 1984; Jones and Gunnarsson, 1989) and books (see, e.g., Parr and Yang, 1989), our treatment will be very brief, merely to establish the notation and recall a few facts.

The theory has developed from the pioneering work of Thomas (1926) and Fermi (1928) and Dirac (1930), with a later paper by Slater (1951) also being very important in developing the present form of the density-functional theory. The step that was lacking, namely, the proof that the ground-state energy of a many-electron system is indeed a unique functional of the electron density, was taken by Hohenberg and Kohn (1964), who supplied the proof for a nondegenerate ground state.

It has proven helpful in developing the theory to separate the energy functional into a number of parts, closely paralleling the energy principle of the original Thomas-Fermi theory (cf. March, 1975).

Thus one writes the total energy $E[\rho]$ as a sum of an independent-particle kinetic-energy functional $T_s[\rho]$, electrostatic potential-energy terms, and then a contribution $E_{xc}[\rho]$ from exchange and correlation which are usefully considered together. Explicitly, we write

$$E[\rho]=T_s[\rho]+\int \rho(\mathbf{r})v(\mathbf{r})d\mathbf{r} + \frac{e^2}{2}\int \frac{\rho(\mathbf{r})\rho(\mathbf{r}')}{|\mathbf{r}-\mathbf{r}'|}d\mathbf{r}d\mathbf{r}'+E_{xc}[\rho] , \quad (2.6)$$

where $v(\mathbf{r})$ is the external potential acting on the electrons ($-Ze^2/r$ in an atom).

According to Hohenberg and Kohn (1964), $E[\rho]$ is a minimum at the equilibrium density distribution in the given external potential $v(\mathbf{r})$. Thus one minimizes this total energy with respect to the ground-state electron density $\rho(\mathbf{r})$, subject only to the condition that the electron density satisfy the normalization condition

$$\int \rho(\mathbf{r})d\mathbf{r}=N , \quad (2.7)$$

for a system with N electrons.

With the introduction of a Lagrange multipler μ, which has the significance of the chemical potential of the electron cloud of the atom, molecule, or solid being considered, the Euler equation of the variational problem reads

$$\mu=\frac{\delta T_s}{\delta\rho(\mathbf{r})}+v(\mathbf{r})+e^2\int\frac{\rho(\mathbf{r}')}{|\mathbf{r}-\mathbf{r}'|}d\mathbf{r}'+\frac{\delta E_{xc}}{\delta\rho(\mathbf{r})} , \quad (2.8)$$

which evidently expresses the constant chemical potential μ throughout the charge distribution as a sum of various contributions which vary from point to point, arising from kinetic, electrostatic, and exchange-plus-correlation contributions. In this form, as Kohn and Sham (1965) emphasized, thereby formally completing the treatment of Slater (1951), one can interpret the problem as posed in terms of equivalent independent-particle equations, which then bypasses the fact that even the independent-particle kinetic-energy functional is still not known in closed form [as a functional of $\rho(\mathbf{r})$]. Thus by solving single-particle Schrödinger equations, with a total one-body potential energy given by

$$v_{\mathrm{eff}}(\mathbf{r})=v(\mathbf{r})+e^2\int\frac{\rho(\mathbf{r}')}{|\mathbf{r}-\mathbf{r}'|}d\mathbf{r}'+\frac{\delta E_{xc}}{\delta\rho(\mathbf{r})} , \quad (2.9)$$

one can avoid any approximation in the independent-particle kinetic energy. Furthermore, the above argument suggests that, in principle, the exact many-electron problem of calculating the ground-state electron density $\rho(\mathbf{r})$ can be reduced to a one-body problem (but see Parr and Yang, 1989, for a discussion of the subtleties of the so-called v representability). Naturally, the many-electron effects in the correlation energy functional are now subsumed in the one-body potential energy $v_{\mathrm{eff}}(\mathbf{r})$, through the functional derivative $\delta E_{xc}/\delta\rho(\mathbf{r})$. Of course, exact knowledge of this quantity would require exact solution of the many-electron problem, which is currently not feasible. Extraction of the potential $v_{\mathrm{eff}}(\mathbf{r})$ directly from the ground-state density has proved possible for the Be atom, however (Hunter and March, 1989).

C. Local-density approximations to exchange and correlation

As the simplest example of the use of Eq. (2.9), let us derive the so-called Dirac-Slater exchange potential. Here one neglects correlation and approximates the exchange energy density by its value in a uniform electron gas with its local density inserted at the point in question. This leads to the result that the total exchange energy, A, say, in Dirac-Slater approximation is given by

$$A=-c_e\int\rho(\mathbf{r})^{4/3}d\mathbf{r},\quad c_e=\frac{3e^2}{4}\left[\frac{3}{\pi}\right]^{1/3} . \quad (2.10)$$

Taking the functional derivative required by Eq. (2.9) leads immediately to the Dirac-Slater exchange potential

$$v_{\text{Dirac-Slater}}(\mathbf{r})=-\tfrac{4}{3}c_e\rho(\mathbf{r})^{1/3} . \quad (2.11)$$

This local-density approximation (LDA) has proved very valuable. In fact, one can extend it to treat the full exchange-correlation functional by setting

$$E_{xc}^{\mathrm{LDA}}[\rho]=\int\epsilon_{xc}(\rho(\mathbf{r}))d\mathbf{r} . \quad (2.12)$$

Above, $\epsilon_{xc}(\rho_0)$ is the exchange-correlation energy per particle of the electron fluid at uniform density ρ_0, which is accurately known (cf., Sec. IV.D) and suitably parametrized (Vosko *et al.*, 1980; Perdew and Zunger, 1981).

It turns out that in the few examples that can be solved for the exchange energy, the Dirac-Slater form is a very useful approximation (see, however, Overhauser, 1985). For example, Miglio *et al.* (1981) have shown, in the case of the infinite barrier model of a metal surface, that although the electron density varies strongly at the surface, nevertheless the local-exchange theory discussed above remains a remarkably useful approximation.

The local-density approximation and its local spin-density extension have been widely applied, going all the way from atoms and molecules to clusters and solids. There are successes (many) and failures (not negligible ones), depending both on the physical quantity under consideration and on the class of systems. An up-to-date survey of the overall situation was given recently by Jones and Gunnarsson (1989). Modifications of the LDA schemes that go beyond the local dependence in Eq. (2.12), seek to include in the exchange and correlation functional additional information on the behavior of the single-particle density and/or to enforce exact limiting behaviors of the resulting exchange-correlation potential (see, e.g., Becke, 1992 and Johnson *et al.*, 1993). Of the many situations where the above theory, based as it is on using uniform electron-gas relations locally, is too crude, the electron Wigner crystal at zero temperature, i.e., in the fully degenerate limit, constitutes a good example.

D. Density-functional treatment of Wigner crystallization

Underlying the LDA of Eq. (2.12) is the assumption that one is concerned with systems with modulations of the electronic density which are neither too large nor too rapidly varying. In fact, LDA is commonly used in situations where the density is far from satisfying such requirements. Nonetheless, it is not surprising that LDA should perform poorly, especially in the treatment of the Wigner crystal mentioned in Sec. II.A, since this is characterized by both a strongly localized density and a crucial role of correlations. Below, we summarize in some detail the application to the study of the Wigner crystal of an approximation scheme that transcends LDA. However, before doing so, we anticipate that the most accurate assessment of the freezing density in jellium, $r_s=100\pm20$, comes from quantum Monte Carlo simulations (cf. Sec. IV.D), which also establish that the transition to a bcc regular structure takes place from a fully spin-polarized liquid, this phase being lower in energy with respect to the unpolarized liquid for $r_s>75$. Therefore in the following we shall content ourselves with a discussion of the coexistence between the spin-polarized liquid and a regular crystalline phase.

Senatore and Pastore (1990) find that the application of the LDA to the freezing of the spin-polarized electron fluid into a bcc crystal yields an exceedingly low value for the freezing density, $r_s=22$. Hence they propose an approximation for exchange and correlation in which, rather than on the full functional $E_{xc}[\rho]$, one focuses on the difference $\Delta=E_{xc}[\rho_s]-E_{xc}[\rho_0]$ between the solid and the liquid phase and then resorts to an expansion of Δ about the liquid, in powers of the density difference $\rho_Q(\mathbf{r})=\rho_s(\mathbf{r})-\rho_0$. To second order, such an expansion yields

$$E_{xc}[\rho_s]=E_{xc}[\rho_0]+\frac{1}{2}\int\int d\mathbf{r}\,d\mathbf{r}'[-\chi^{-1}(\mathbf{r}-\mathbf{r}')+\chi_0^{-1}(\mathbf{r}-\mathbf{r}')]\rho_Q(\mathbf{r})\rho_Q(\mathbf{r}') \,, \tag{2.13}$$

with $\chi(r)$ and $\chi_0(r)$ being, respectively, the static response function of the homogeneous liquid and the response function of the noninteracting electrons (i.e., the Lindhard function). One of the motivations for using such a quadratic approximation is that, for classical liquids, it is known to work surprisingly well (for a recent review, see Baus, 1990).

As we mentioned in Sec. II.B, the Euler-Lagrange problem for the ground-state energy functional of interacting particles can be conveniently recast into the so-called Kohn-Sham problem (i.e., the self-consistent problem of noninteracting particles in a density-dependent one-body potential). Once the exchange-correlation functional is known, it is a simple matter to obtain at once the density-dependent Kohn-Sham potential. From Eqs. (2.9) and (2.13) one gets, in Fourier transform,

$$v_{\text{eff}}(\mathbf{q})=\rho_Q(\mathbf{q})[-\chi^{-1}(q)+\chi_0^{-1}(q)] \,. \tag{2.14}$$

It may be worth stressing at this point that, whereas the LDA borrows only a thermodynamic property of the homogeneous liquid (i.e., the exchange-correlation energy), the quadratic approximation involves—in principle—structural information of the liquid at all the wave vectors which are relevant in the modulated phase. Thus, for a regular solid, one finds that such a region of wave vectors extends from about $2q_F$ onward.

To perform calculations, one needs the response function of the homogeneous interacting-electron liquid. This quantity, however, is not exactly known. Senatore and Pastore have therefore employed the so-called STLS (Singwi, Tosi, Land, Sjölander) decoupling scheme (Singwi *et al.*, 1968) to construct $\chi^{-1}(q)$ from the structure factor $S(k)$ obtained from the QMC simulations.

The calculation of the ground-state energy of the Wigner crystal requires the self-consistent solution of Kohn-Sham equations for the Bloch orbitals of a single fully occupied energy band, since there is one electron per unit cell and one is considering the spin-polarized state. This can be accomplished by using standard computational techniques for band-structure calculations. The results that are obtained can be summarized as follows.

The quadratic approximation predicts freezing into the bcc lattice at $r_s=102$, a value which compares extremely well with the QMC prediction of $r_s=100\pm20$. In addition, a Lindemann ratio γ (rms deviation about the lattice site divided by the nearest-neighbor distance) of 0.34 is obtained, whereas QMC suggests $\gamma=0.30\pm0.02$ for all quantum systems studied to date. The calculated density still turns out to be well localized, even if considerably less so than in classical freezing. This, together with the high symmetry of the periodic structure, suggests the possibility of a tight-binding approximation in which Bloch orbitals are built from one Gaussian orbital per site with a variational width. Using this approximation, calculations simplify considerably, whereas the results for the freezing r_s and γ change only slightly. In fact, one gets $r_s=107$ and $\gamma=0.29$.

Senatore and Pastore have also investigated the stability of the fcc electron crystal. They find, within the fuller calculations, that the fcc solid is, in fact, the stable phase between $r_s=97$ and $r_s=108$, being in this range lower in energy than the bcc crystal (and, of course, also lower in energy than the homogeneous liquid). For higher values of the coupling r_s, the bcc remains the stable phase, in agreement with the findings of harmonic lattice calculations (see, e.g., Foldy, 1971).

An investigation of the importance of higher-order

terms in the expansion of Eq. (2.13), when applied to the study of Wigner crystallization, was subsequently conducted by Moroni and Senatore (1991), who resorted to density-functional-theory schemes of the weighted-density type. While there are quantitative changes in the details of the freezing transition, the agreement with the QMC results remains very good. One should note that the freezing theory summarized above crucially relies on the knowledge of the static response function of the quantum liquid. The sensitivity of the results to the accuracy with which $\chi(q)$ is known is thus an important issue. At present there is little knowledge about the precise form of static response in quantum fluids. However, some progress has recently been made for the two-dimensional electron gas and for ^{4}He (Moroni *et al.*, 1992), using QMC techniques, and, in fact, work using similar means is in progress on the three-dimensional electron fluid (Moroni *et al.*, 1993).

III. LOCALIZED VERSUS MOLECULAR-ORBITAL THEORIES OF ELECTRONS IN MULTICENTER PROBLEMS: GUTZWILLER VARIATIONAL METHOD

We begin by stressing that one-center (i.e., atomic) correlation effects have to be treated by quantitative examination of the problem, except for isolated instances of collective effects within specific shells (see, for example, Wendin, 1986). In contrast, as already mentioned, we have qualitative consequences in multicenter problems, which are worthy of full consideration. Let us start by reviewing the situation in the H_2 molecule, going back to the pioneering work on the chemical bond by Heitler and London (1927).

This Heitler-London description merely asserted that a useful ground-state symmetric space wave function could be built up from the atomic orbitals (1s functions) centered on nuclei a and b, namely, ϕ_a and ϕ_b. After symmetrization, one is led to the wave function

$$\Psi_{\rm HL}(1,2)=\phi_a(1)\phi_b(2)+\phi_b(1)\phi_a(2) \; . \qquad (3.1)$$

The first term on the right-hand side of Eq. (3.1) evidently corresponds to electron 1 on nucleus a and electron 2 on nucleus b. The second part is added because of the indistinguishability of electrons.

Turning to the delocalized description, one introduces a molecular orbital $\psi_{\rm MO}$, and, in the singlet ground state, one puts two electrons into it with opposed spins. Then the MO total space wave function is written in the form

$$\Psi_{\rm MO}(1,2)=\psi_{\rm MO}(1)\psi_{\rm MO}(2) \qquad (3.2)$$

and, in terms of the 1s atomic orbitals, in the approximation in which the molecular orbital is built up as a linear combination of atomic orbitals,

$$\Psi_{\rm LCAO\text{-}MO}(1,2)=[\phi_a(1)+\phi_b(1)][\phi_a(2)+\phi_b(2)]$$
$$=\Psi_{\rm HL}(1,2)+\phi_a(1)\phi_a(2)+\phi_b(1)\phi_b(2) \; . \qquad (3.3)$$

In the second part of Eq. (3.3), we have noted explicitly that the LCAO-MO wave function can be viewed as a linear superposition of the Heitler-London covalent terms and an equally weighted admixture of ionic terms, $\phi_a(1)\phi_a(2)$ evidently representing both electrons on nucleus a, etc. That Eq. (3.3) is incorrect as the internuclear distance R gets large compared with the size a_0 of the 1s hydrogen orbitals is quite clear; the molecule dissociates into two neutral H atoms, just as described by the original Heitler-London wave function (3.1).

A. Coulson-Fischer wave function with asymmetric orbitals, for H_2 molecule

An important clarification of the role of electron correlation in molecules came with the work of Coulson and Fischer (1949) on H_2. They asked the question as to what was the best admixture of covalent and ionic states at each internuclear distance R, by contemplating asymmetric molecular orbitals $\phi_a+\lambda\phi_b$, $\lambda\leq 1$, and $\phi_b+\lambda\phi_a$, the former representing, with $\lambda<1$, the electron primarily but not wholly belonging to nucleus a, etc. Then they formed the (unsymmetrized) variational wave function

$$\Psi_{\rm Coulson\text{-}Fischer}=[\phi_a(1)+\lambda\phi_b(1)][\phi_b(2)+\lambda\phi_a(2)] \; . \qquad (3.4)$$

Determining λ as a function of R by minimization of $\langle H\rangle$ with respect to the wave function (3.4), H being the total Hamiltonian of the H_2 molecule, they found the following situation: (a) for $R<1.6R_{\rm equilibrium}$, $\lambda=1$; (b) for $R>1.6R_{\rm equilibrium}$, λ falls quite rapidly to zero as R is increased. $R_{\rm equilibrium}$ being the equilibrium internuclear separation. For $\lambda=1$, Eq. (3.4) becomes identical with Eq. (3.3), whereas for $R>1.6R_{\rm equilibrium}$ we see that electrons quickly "go back on to their own atoms."

This idea, that one decreases the weight of the ionic configurations in a molecular-orbital treatment, has been taken up in the work of Gutzwiller (1963, 1964, 1965) for treating strong correlations in narrow energy bands, and we shall discuss the results of his method in some detail below.

The important point to be stressed from the above is that electron correlations can have the qualitative effect of driving electrons back on to their own atoms when the internuclear spacing becomes substantially larger than the size of the atomic orbitals involved.

It is convenient at this point to follow Falicov and Harris (1969) and refer to a model for a two-electron homopolar molecule. These workers discuss the eigenstates of the one-band Hamiltonian for such a two-electron system, and their results are summarized in Appendix A. They then use the exact solution for the ground state as a standard to assess the validity of the MO, Heitler-London and other states having either spin- or charge-density waves.

By definition, of course, the MO approximation is undercorrelated, the Heitler-London states being always overcorrelated. The spin-density and charge-density waves are less easily classified, with the question of under- or overcorrelated depending on the strength of the interaction.

Falicov and Harris construct from spin- and charge-density-wave states, which have broken symmetry, symmetrized versions. These symmetrical states were always found in their work to be slightly undercorrelated. This work has been extended by Huang *et al.* (1976).

B. Gutzwiller's variational method

As mentioned above, a possible way to account for correlation of antiparallel electrons in a multicenter problem is to partly project out from a given uncorrelated wave function those components corresponding to double occupied (ionic) sites. In the simple case of H_2, this was most simply done by Coulson and Fischer (1949) as set out in Sec. III.A. The generalization of such an approach to situations with an arbitrary number of centers is due to Gutzwiller (1963, 1965). Here we shall just outline the key points of Gutzwiller's approach, which has been reviewed by Vollhardt (1984). In the next two sections, we shall discuss in some detail the application of the Gutzwiller variational method to treat the correlation in molecules, on the one hand, and the Hubbard Hamiltonian, on the other. The latter has received renewed attention in connection with the exciting discovery of high-T_c superconductivity (Bednorz and Müller, 1986).

The starting point of Gutzwiller's variational approach is the uncorrelated wave function, for the problem under consideration. This is constructed, for a regular lattice with N sites and one Wannier orbital ϕ per site, from the Bloch waves $\Psi_{\mathbf{k}}(\mathbf{r})$,

$$\Psi_{\mathbf{k}}(\mathbf{r})=\frac{1}{\sqrt{N}}\sum_i \exp(i\mathbf{k}\mathbf{R}_i)\phi(\mathbf{r}-\mathbf{R}_i)\ . \tag{3.5}$$

Using the second-quantization formalism, the uncorrelated ground-state wave function can be written as

$$|\Phi_0\rangle=\prod_{\mathbf{k}\in K} a^\dagger_{\mathbf{k}\uparrow}\prod_{\mathbf{q}\in Q} a^\dagger_{\mathbf{q}\downarrow}|0\rangle\ , \tag{3.6}$$

where $a^\dagger_{\mathbf{k}\sigma}$ is the creation operator of an electron in the Bloch wave $\Psi_{\mathbf{k}}$ and with spin projection σ, $|0\rangle$ is the vacuum state, and K and Q are sets of points in reciprocal space, which in general may be delimited by different Fermi surfaces. If one denotes the creation operator of an electron in the Wannier orbital ϕ at the site i by $a^\dagger_{i\sigma}$,

$$a^\dagger_{i\sigma}=\frac{1}{\sqrt{N}}\sum_{\mathbf{k}}\exp(-i\mathbf{k}\mathbf{R}_i)a^\dagger_{\mathbf{k}\sigma}\ , \tag{3.7}$$

and the corresponding number operator by $n_{i\sigma}$, the Gutzwiller wave function can be written as

$$|\Phi\rangle=\prod_i[1-(1-g)n_{i\uparrow}n_{i\downarrow}]|\Phi_0\rangle=g^D|\Phi_0\rangle\ . \tag{3.8}$$

Above, $D=\sum_i n_{i\uparrow}n_{i\downarrow}$ is the operator that counts the number of double occupied sites. Clearly, in the wave function Φ, the components containing double occupied sites are reduced by a fractional amount $1-g$ $(0\le g\le 1)$ with respect to their value in the uncorrelated wave function Φ_0, thus reducing the repulsive interaction energy among antiparallel electrons. For a given Hamiltonian H, the variational parameter g has to be determined by minimizing the ground-state energy

$$E_0(g)=\frac{\langle\Phi|H|\Phi\rangle}{\langle\Phi|\Phi\rangle}\ . \tag{3.9}$$

To fix ideas, let us briefly consider the case originally investigated by Gutzwiller, that of the so-called Hubbard Hamiltonian,

$$H=\sum_{ij}\sum_\sigma t_{ij}a^\dagger_{i\sigma}a_{j\sigma}+U\sum_i n_{i\uparrow}n_{i\downarrow}\ , \tag{3.10}$$

which is designed to describe fermions on a lattice in a narrow-band system. The first term on the right-hand side of Eq. (3.10) is the kinetic energy due to the hopping of electrons between sites, and the second one crudely describes the on-site repulsion of antiparallel-spin electrons. It is straightforward to show that for $g=1$ one simply regains the uncorrelated state, which of course is the exact ground state for zero on-site repulsion, $U=0$. In this case $D=D_0\equiv N_\uparrow N_\downarrow/N$. On the other hand, $g=0$ corresponds to a fully correlated wave function in which the components containing double occupied sites are suppressed and $D=0$. This is the exact ground state for $U=\infty$. Therefore, for finite repulsion, one will have $0<g<1$ and $0<D<D_0$, since the effect of correlation is precisely to reduce the number of doubly occupied sites present in the uncorrelated wave function.

Thus far, even for the Hubbard Hamiltonian, the evaluation of the averages in Eq. (3.9) has proved too difficult to carry out analytically in the general case, with the notable exception of the one-dimensional lattice (Gebhard and Vollhardt, 1987; Metzner and Vollhardt, 1987). Hence one has either to employ numerical methods (Horsch and Kaplan, 1983; Gros *et al.*, 1987a, 1987b; Yokoyama and Shiba, 1987a, 1987b) or to resort to approximate treatments, as originally done by Gutzwiller (1963, 1965). It has subsequently been shown that the approximations originally introduced by Gutzwiller to evaluate the averages in Eq. (3.9) are just equivalent to the neglect of the spin-configuration dependence of the various terms appearing in the expansion of both $\langle\phi|H|\phi\rangle$ and $\langle\phi|\phi\rangle$ (Ogawa *et al.*, 1975; see also Vollhardt, 1984). While the interested reader is urged to consult the original papers for the details of Gutzwiller's approximation (GA), it is convenient to report here at least the main results of GA as applied to the Hubbard Hamiltonian.

Within the GA, the average of Eq. (3.9) with the Hamiltonian given in (3.10) can be written as

$$E_0/N=n_\uparrow q_\uparrow\bar{\epsilon}_\uparrow+n_\downarrow q_\downarrow\bar{\epsilon}_\downarrow+Ud\ . \tag{3.11}$$

Here, $n_\sigma=N_\sigma/N$; q_σ is the discontinuity of the momentum distribution $n_{\mathbf{k}\sigma}$ at the Fermi surface; $d=D/N$, with D the average number of double occupied sites; and

$$\bar{\epsilon}_\sigma=\frac{1}{N_\sigma}\sum_{|\mathbf{k}|<k_{F\sigma}}\epsilon(\mathbf{k}) . \quad (3.12)$$

In Eq. (3.12), $\epsilon(\mathbf{k})$ is the energy eigenvalue of the Bloch wave $\Psi_\mathbf{k}$,

$$\epsilon(\mathbf{k})=\frac{1}{N}\sum_{ij} t_{ij}\exp[-i\mathbf{k}(\mathbf{R}_i-\mathbf{R}_j)] , \quad (3.13)$$

and the zero of energy has been chosen so that $\sum_\mathbf{k}\epsilon(\mathbf{k})=0$; i.e., $t_{ii}=0$. The GA simply yields

$$q_\sigma=\frac{\{[(n_\sigma-d)(1-n_\sigma-n_{-\sigma}+d)]^{1/2}+[(n_{-\sigma}-d)d]^{1/2}\}^2}{n_\sigma(1-n_{-\sigma})} , \quad (3.14)$$

where g has been eliminated in favor of d, and d has to be determined variationally, by minimizing the ground-state energy $E_0(d)$ as given in Eq. (3.11). In the special case of a paramagnetic half-filled band, for which $n_\uparrow=n_\downarrow=1/2$, $q_\uparrow=q_\downarrow=q$, and $\bar{\epsilon}_\uparrow=\bar{\epsilon}_\downarrow=\bar{\epsilon}_0<0$, one obtains, after minimization,

$$d=\frac{1}{4}\left[1-\frac{U}{U_c}\right] , \quad (3.15)$$

$$q=1-\left[\frac{U}{U_c}\right]^2 , \quad (3.16)$$

$$E_0/N=|\bar{\epsilon}_0|\left[1-\frac{U}{U_c}\right]^2 , \quad (3.17)$$

with $U_c=8|\bar{\epsilon}_0|$. This shows that, within the GA, at a finite critical value of the interaction $U=U_c$, d, q, and E_0 all vanish. The vanishing of the discontinuity in the momentum distribution at the Fermi surface, and consequently of the kinetic energy, would signal a metal-insulator transition (Brinkman and Rice, 1970; see also March *et al.*, 1979 and March and Parrinello, 1982). In fact, at the critical strength U_c, all sites would be singly occupied and the particles fully localized. We shall return to this point below.

Before turning to the discussion of how the Gutzwiller variational method has been generalized to treat interatomic and intra-atomic correlation in molecules, we have to mention more recent work by Kotliar and Ruckenstein (1986). These authors have proposed a new treatment of the Hubbard Hamiltonian that makes use of auxiliary boson fields, in analogy with the so-called *slave-boson* approach first proposed by Barnes (1976). They have shown that, within the functional-integral formalism, a particular choice of the auxiliary fields yields exactly the result of GA, as a mean-field or, more precisely, a saddle-point approximation to the exact functional integral. The merit of such an approach is that in principle it would allow for systematic improvement upon GA, in that one could study the effect of Gaussian fluctuations about the saddle point. However, nothing has been done in this direction, to our knowledge, at the time of writing.

C. Local approach to correlation in molecules

As we have discussed above, the local approach to correlation was introduced by Coulson and Fischer in the qualitative treatment of the H_2 molecule. Subsequently, Gutzwiller extended such an idea to solids to deal with strong correlation in narrow bands originating from very localized orbitals such as d orbitals, within the model description afforded by the Hubbard Hamiltonian. Starting from Gutzwiller's systematic scheme, a further step toward the quantitative treatment of correlation in small molecules was taken by Stollhoff and Fulde (1977, 1978).

Their main point is as follows. One still wants to set up a variational scheme in which those components of the wave functions that correspond to two electrons being too close are reduced. To this end, the wave function may be described, for instance, by assigning its value on the points of a spatial mesh (the grid) in which the system of interest, say, the molecule, is embedded. In practice, one may as well give at such points the amplitudes of basis-set functions, in terms of which the uncorrelated wave function is expanded. Within this picture, the wave function of Eq. (3.8) corresponds to a grid whose points coincide with the locations of atoms in the molecule. A quantitative treatment clearly requires that one improve upon the ansatz of Eq. (3.8) in at least two ways. One should (a) choose a finer grid, and (b) correlate the electronic motion not only on-site but also between different sites. The latter point can be easily realized by generalizing the variational wave function to read (Stollhoff and Fulde, 1977)

$$|\Phi\rangle=\prod_{i=0}^{I}P_i|\Phi_0\rangle , \quad (3.18)$$

where the P_i's are projection operators defined by

$$P_i=\prod_j(1-\eta_i 0_{ij}) , \quad (3.19)$$

and

$$0_{0j}=n_{j\uparrow}n_{j\downarrow} , \quad (3.20)$$

$$0_{ij}=\sum_{\sigma\sigma'}n_{j\sigma}n_{j+i\sigma'} . \quad (3.21)$$

For $I=0$, Eq. (3.18) yields the original Gutzwiller ansatz. However, for nonzero I, density correlations between different grid points are also taken into account, up to the Ith neighboring points. Notice that, in the latter case, one is correlating also the motion of parallel-spin electrons on different grid points. The reason for this is that, though the uncorrelated wave function $|\Phi_0\rangle$ is antisymmetrized, the Pauli exclusion principle is completely effective only at very short distances.

The problem of a finer grid can also be dealt with in quite a simple way (Stollhoff and Fulde, 1978). Instead of directly making the grid finer, one can alternatively choose to keep as grid points the atomic position in the molecule, while using at each site a number of basis-set functions with varying degrees of localization, possibly also off-center. Thus the operator $a_{i\sigma}^{\dagger}$ creates an electron in the basis-set function i rather than in one of the self-consistent one-particle orbitals in terms of which $|\Phi_0\rangle$ is constructed. Here, the index i labeling the creation operator stands, in fact, for the pair (i,α), where α indicates either one of the original basis-set functions or a suitable new combination; it might be, for instance, an s-p hybrid.

The effect of taking into account off-site correlation through the use of the variational wave function given in Eq. (3.18) with $I \neq 0$ has been investigated by Stollhoff and Fulde (1977) by studying the H_6 model of Mattheiss (1961). This model has the advantage that one knows the exact solution, which was obtained numerically. The model assumes six sites equally spaced on a ring, with one s atomic orbital at each site. From these, one constructs a Wannier orbital, in terms of which the Hamiltonian reads

$$H = \sum_{ij\sigma} t_{ij} a_{i\sigma}^{\dagger} a_{j\sigma}^{\dagger} + \sum_{\substack{ijkl \\ \sigma\sigma'}} V_{ijkl} a_{i\sigma}^{\dagger} a_{j\sigma'}^{\dagger} a_{l\sigma'} a_{k\sigma} \, . \tag{3.22}$$

Bloch waves are then built from the Wannier orbital, and from these the uncorrelated Hartree-Fock ground state $|\Phi_0\rangle$ is obtained. The correlation energy of the system was studied by Stollhoff and Fulde for various values of the first-neighbor distance R, with the variational wave function of Eq. (3.18). They found that the simpler Gutzwiller ansatz, i.e., $I=0$, only yields about 50% of the exact correlation energy at the exact equilibrium distance, $R=1.8$ a.u. On the other hand, the situation improves substantially by going to $I=1$, and even more for $I=2$. In the latter case, one recovers more than 95% of the correlation energy for any distance. The situation is illustrated in some detail in Fig. 2. It should also be noticed that a further increase of I from 2 to 3 does not produce anything new, since $I=3$ corresponds to the largest distance of the atoms in the ring.

Stollhoff and Fulde also investigated the accuracy of a simplified version of Eq. (3.18) obtained by linearization,

$$|\Phi\rangle = \left[1 - \sum_{i=0}^{I} \eta_i \sum_{j} 0_{ij}\right] |\Phi_0\rangle \, , \tag{3.23}$$

in order to reduce computational complexity. With this simplified variational ansatz, they found that at the equilibrium distance about 90% of the correlation energy was still recovered with $I=2$. However, the agreement with the exact results becomes worse at larger distances, in contrast with the systematic improvement found with the full variational wave function of Eq. (3.18).

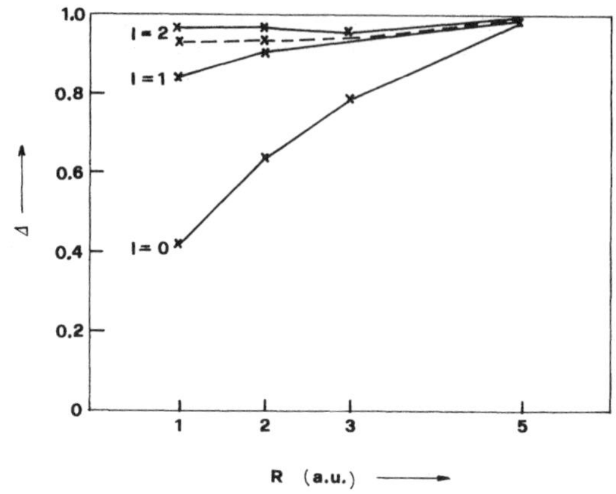

FIG. 2. Gain in correlation energy $\Delta = (E_{HF} - E_{Gutz.})/(E_{HF} - E_{exact})$ for the H_6 model by using the ansatz of Eq. (3.18) to evaluate the ground-state energy $E_{Gutz.}$. Results are shown for $I=0$ to 2. The dashed line is the approximation given by Eq. (3.23). From Stollhoff and Fulde (1977).

It should be mentioned that, in general, using the linearized wave function of Eq. (3.23) to variationally calculate the correlation energy yields the so-called size-consistency problem. In practice the correlation energy does not turn out to be proportional to the electron number; i.e., it is not extensive as it should be in the limit of a large system. A way to restore the correct number dependence is to expand the variational ground-state energy of Eq. (3.9) to a given order in the variational parameters, say, the second, starting from the full wave function of Eq. (3.18). In the case of small molecules, the difference that one finds in correlation energy, with respect to the use of the linearized wave function (3.23), is only of a few percent (Stollhoff and Fulde, 1980). Another possible way to tackle the size-consistency problem begins with the rewriting of the product of projection operators appearing in Eq. (3.18) in an exponential form. This makes it possible to prove a linked-cluster theorem (see, for instance, Horsch and Fulde, 1979), which in turn allows a systematic expansion of the correlation energy in terms of connected diagrams. Again, the correct number dependence of the energy is automatically preserved to any order in the expansion in powers of the variational parameters. It should also be mentioned that spin-spin correlation can be dealt with as well, within the local approach. To this purpose one can just enlarge the class of projection operators appearing in Eq. (3.18) by defining further operators $0_{ij} = s_i \cdot s_j$, where s_i is the spin operator at the site i (Stollhoff and Fulde, 1978).

Clearly, in the study of the H_6 model only interatomic correlations could be taken into account, and admittedly within an oversimplified model. Therefore, for a more stringent test, Stollhoff and Fulde (1978, 1980) studied, within the simplified variational ansatz of Eq. (3.23), some small atoms and molecules, taking into account intra-atomic correlations as well. To this end, the radial correlation was treated by considering basis-set functions of s type with different degrees of localization. On the other hand, angular correlations were introduced by using appropriate hybrid functions. In particular, s-p mixing was used to get sets of tetragonal hybrids, and d and f functions were used to obtain hybrids with hexagonal and octagonal symmetry, respectively. Within their simple scheme they were able to recover 93% of the experimental correlation energy for He and 90% for H_2. Thus, starting from the Hartree-Fock uncorrelated ground state, in both cases the results of the local approach based on a Gutzwiller-like ansatz yielded a fraction of the correlation energy, very close (a few percent of difference) to the one obtained in configuration-interaction (CI) calculations with the same basis sets. Similar results were also obtained for He_2 and Be.

A very detailed study of Ne and CH_4 along these lines (Stollhoff and Fulde, 1980) shows that also in more complex situations the local approach is capable of accounting for the correlation energy obtainable with CI calculations, within a few percent. One should notice, though, that the agreement with experiment of the correlation energy yielded by CI calculations with reasonable basis sets tends to worsen with increasing complexity of the system studied.

More recently, Oles *et al.* (1986) proposed another computational scheme for the treatment of correlations in more complex molecules containing C, N, and H atoms. While the interatomic correlations are still treated within a local approach based on semiempirical self-consistent-field calculations, intra-atomic correlations are dealt with by means of a different and simpler scheme. Good agreement is found with experimental results, discrepancies being within a few percent. They have also shown that simple algebraic parametrizations are possible for the various contributions to the correlation energy. In particular, interatomic correlations are found to depend only on bond lengths, whereas intra-atomic correlations are found to be determined for a given atom by its total charge and fraction of p electrons. A further study, on the determination of optimal local functions for the calculation of the correlation energy within the local approach, has also recently appeared (Dieterich and Fulde, 1987).

In the foregoing, we have been concerned with the application of a local approach to the calculation of correlation in molecules. Yet we should like to mention here an important development along these lines concerning extended systems. The above local approach, in fact, has been suitably modified (Horsch and Fulde, 1979) to treat also short-range and long-range correlation in the ground state of solids. A first attempt to calculate in this manner the correlation contribution to the ground-state energy of diamond (Kiel *et al.*, 1982) suffered the limitation of a poorly converged uncorrelated ground state. When an uncorrelated ground state of good quality is used, however, the local approach performs remarkably well, reproducing the electronic contribution to the binding energy of diamond with an accuracy of about 2% (Stollhoff and Bohnen, 1988).

D. Gutzwiller's variational treatment of the Hubbard model

As we have mentioned above, the Gutzwiller variational method was originally applied to the study of the so-called Hubbard model, characterized by the Hamiltonian of Eq. (3.10). In spite of the apparent simplicity of such a Hamiltonian, the relevant variational calculation is of considerable difficulty. Therefore Gutzwiller solved the problem in an approximate manner. It has been only recently, after some decades, that a renewed interest in this problem has brought about a certain amount of work leading to the exact solution of the variational problem, by direct numerical evaluation, by analytical means, or by the Monte Carlo method.

Kaplan, Horsch, and Fulde began a study in 1982 to assess the accuracy of the Gutzwiller variational wave function (GVW) in describing the ground state of the single-band Hubbard model with only nearest-neighbor hopping, in the atomic limit, i.e., the limit in which $U \rightarrow \infty$. In this case there is no variation to be taken. In fact, $g=0$, and the Gutzwiller wave function reads

$$|\Phi_\infty\rangle = \prod_i [1 - n_{i\uparrow} n_{i\downarrow}] |\Phi_0\rangle \ . \tag{3.24}$$

They were able to perform, for the half-filled-band situation, the direct numerical evaluation of the spin-spin correlation function,

$$q_l = \langle s_i^z s_{i+l}^z \rangle \ , \tag{3.25}$$

for a number of small regular rings. The actual calculations were done for rings of 6, 10, 14, and 18 sites. By extrapolating from the results for finite rings, they found that in the thermodynamic limit (number of sites N going to infinity) $q_1 = -0.1474$, thus being within 0.2% from the exact result, i.e., within the error bars of their calculation. For q_2, instead, the discrepancy with the exact result was of about 7%. They also noticed that the spin-spin correlations that they obtained reproduced as well qualitative features of the exact ones. Hence for large U, at least in one dimension, Gutzwiller's wave function describes the short-range spin-spin correlations in an accurate manner. We should notice here that, in the atomic limit and for the half-filled-band case, the exact ground state of the 1D Hubbard model is known to be the same as for the antiferromagnetic Heisenberg chain, for which q_1 was exactly evaluated by Bonner and Fisher (1964), and q_2 by Takahashi (1977). In fact, in the large-U limit, the Hubbard Hamiltonian goes, to leading order in t/U, into the Heisenberg Hamiltonian with antiferromagnetic coupling,

$$H = J \sum_{\langle ij \rangle} (\mathbf{s}_i \cdot \mathbf{s}_j - \tfrac{1}{4}) \ , \tag{3.26}$$

where J is proportional to t^2/U and t is the hopping energy. We shall return to this point in some detail.

Kaplan, Horsch, and Fulde also studied the energy for large but finite U. In this case one has to start from the full variational wave function (3.8) and consider an expansion of the energy in powers of the parameter g, to be determined variationally. They found that the leading term of such an expansion was of the form $E=-N\alpha t^2/U$, with α depending on the kinetic-energy operator and on the zero- and first-order wave functions obtained from the expansion of the variational wave function (3.8) in powers of g. Note that the zero-order wave function, which was given in Eq. (3.24), has no sites with double occupancy, as is clear by inspection. Similarly, the first-order one has on average only one site doubly occupied. They found that, at variance with the spin-spin correlations, the coefficient α yielded by the GVW was in gross disagreement with the known exact value. It is now known, from the exact solution of the Hubbard model in one dimension with the GVW (Metzner and Vollhardt, 1987), that α is not a constant, but vanishes as $1/\ln(U/t)$ in the limit considered by Kaplan, Horsch, and Fulde. These authors argued that the unsatisfactory result for the energy yielded by the GVW was due to the incorrect description of correlations between doubly occupied sites and empty sites, or holes, as we shall call them in the present context. Therefore they considered a modification of the GVW containing a second variational parameter to improve the treatment of such correlations. In this manner they were able to reproduce the exact value of α within 1%.

A further step toward the assessment of the reliability of the GVW was then taken by Horsch and Kaplan (1983). They extended the calculation for finite rings to other values of N, corresponding to open-shell systems, and also performed calculations for much larger systems with N up to 100 by using the Monte Carlo method to evaluate the relevant averages. This second step was particularly important. First, it fully confirmed the conclusions previously obtained from the exact calculations for small rings. Secondly, it was the first application of the Monte Carlo method to this problem. Before turning to the presentation of more recent studies on the Hubbard model, with a variational Monte Carlo technique based on the GVW, we shall summarize below the exact results that have been recently obtained in one dimension for arbitrary band filling and interaction strengths.

1. Exact analytic results in one dimension

Analytic results for the ground-state energy and momentum distribution function for the Hubbard model with on-site repulsion have been recently obtained by Metzner and Vollhardt (1987) for the paramagnetic situation. The key point of their attack on the problem at hand lies in a new approach to the calculation of the expectation values of relevant operators on the GVW.

The calculation of the ground-state energy appears to require the evaluation of the expectation values of both kinetic and potential energy for arbitrary values of the variational parameter g, $0\le g\le 1$. Let us start from the potential energy. It is clear from Eq. (3.10) that, apart from the proportionality constant U, the potential-energy operator is the same as the operator D that counts the number of doubly occupied sites. Thus one has to calculate $\langle D\rangle$ on the wave function of Eq. (3.8). Let us indicate the product $n_{i\uparrow}n_{i\downarrow}$ by D_i. It is then not difficult to show that

$$d=\frac{\langle D\rangle}{N}=g^2\sum_{m=1}^{\infty}(g^2-1)^{m-1}c_m\ , \tag{3.27}$$

where N is the number of sites, and the coefficients $c_m=x_m/N(m-1)!$ are suitable expectation values, which, if one puts $x_m=y_m/\langle\Phi|\Phi\rangle$, are explicitly given by

$$y_m=\sum_{f_1,\ldots,f_m}{}'\ \langle\Phi_0|D_{f_1}\cdots D_{f_m}|\Phi_0\rangle,\quad m\le N\ , \tag{3.28}$$

with $f_i\ne f_j$ (prime on sum). The expectation value appearing above can be transformed into the sum over all the possible pairs of contractions, $P_{ij}\equiv\langle a_{i\sigma}^{\dagger}a_{j\sigma}\rangle_0$, using Wick's theorem. We shall indicate such a sum by $\{\cdots\}_0$. The important point to realize is that, because of the prime on the sum in Eq. (3.28), one can choose to put $\langle a_{i\sigma}a_{j\sigma}^{\dagger}\rangle\equiv-P_{ji}$, since in this second kind of contraction one has always $i\ne j$ and therefore there is no δ_{ij} contribution. Equation (3.28) can then be rewritten as

$$y_m=\sum_{f_1,\ldots,f_m}{}'\ \{D_{f_1}\cdots D_{f_m}\}_0\ . \tag{3.29}$$

One can show that the sum $\{\cdots\}_0$ appearing above is, in fact, the determinant of a matrix having the P_{ij} as elements. Hence, for any $f_i=f_j$, the sum $\{\cdots\}_0$ vanishes, since the relative determinant has two identical rows. As a result, one has the freedom to remove the prime from the sum in Eq. (3.29). Having done so, one can next observe that in a diagrammatic analysis of x_m with lines corresponding to the P_{ij}, the contributions to $\{\cdots\}_0$ arising from disconnected diagrams just cancel the norm $\langle\Phi|\Phi\rangle$. Therefore one is left with

$$x_m=\sum_{f_1,\ldots,f_m}\{D_{f_1}\cdots D_{f_m}\}_0^c\ . \tag{3.30}$$

We notice that thus far no reference was made to interaction strength or to dimensionality.

While in the above the analysis was quite general, to date further progress has proved possible only in one dimension. In particular, in one dimension and for zero total spin ($n_\uparrow=n_\downarrow=n/2$), it is possible to show that c_m is proportional to n^{m+1}. One can restrict attention to $0\le n\le 1$, since, due to particle-hole symmetry for $1\le n\le 2$, the relation $d(n)=d(2-n)+n-1$ holds. By using the dependence of d on n and imposing the continuity of its first derivative, the proportionality constant in c_m can be calculated. One obtains

$$c_m=\tfrac{1}{2}(-1)^{m+1}n^{m+1}/(m+1)\ . \tag{3.31}$$

With the above equation for the coefficients c_m, the series in Eq. (3.27) is exactly summed to

$$d=\frac{n^2}{2}\left[\frac{g}{1-G^2}\right]^2\left[\ln\frac{1}{G^2}+G^2-1\right]\ , \tag{3.32}$$

with $G^2=1-n+ng^2$. We stress that, as is clear from Eq. (3.32), for the half-filled band—for which $n=1$—the correlation energy Ud is found to be nonanalytic in g, because of the presence of the term $\ln(1/g)$. For strong correlations ($g\rightarrow 0$), one finds $d=g^2\ln(1/g)$. Moreover, the double occupancy d is never vanishing for finite correlations, in contrast with the GA result of Eq. (3.15). This seems to be the case also in two and three dimensions when the GVW is exactly handled (Yokoyama and Shiba, 1987a).

A similar, though more complicated, analysis allows the calculation of the momentum distribution function for the same situation considered immediately above, i.e., $n_\uparrow=n_\downarrow=n/2$. We shall content ourselves here with quoting just the result for the discontinuity q in the momentum distribution function $n_{k\sigma}$,

$$q=G^{-1}[(G+g)/(1+g)]^2\ . \tag{3.33}$$

The overall shape of $n_{k\sigma}$ for half filling and at different values of g is illustrated in Fig. 3. It should be noted that for $n=1$ the above equation reduces to $q=4g/(1+g)^2$, which is the result of Gutzwiller's approximation (GA). This, however, does not imply that, for this particular value of filling, the kinetic energy coincides with that of GA, since g has to be determined by minimizing the total energy, and the equations for the potential energy are different in the two cases.

By combining Eq. (3.32) with the result for the momentum distribution function, one may calculate $E_0(g)$ in one dimension for the paramagnetic ground state. Minimization with respect to g yields the energy for given values of t, U, n. For $n=1$, the comparison with the result of Gutzwiller's approximate treatment and the exact result of Lieb and Wu (1968) shows that the GVW, when exactly handled, gives an energy that is intermediate between the other two. This is illustrated in Fig. 4. In particular, specializing to the case $n=1$ and $U/t\rightarrow\infty$, with only nearest-neighbor hopping, one finds $\bar{\epsilon}_0=-4t/\pi$, and

$$E/N=-(4/\pi)^2(t^2/U)(\ln\bar{U})^{-1}\ , \tag{3.34}$$

where $\bar{U}=U/|\bar{\epsilon}_0|$. We note, with reference to the numerical results of Kaplan *et al.* discussed earlier, that the energy is certainly proportional to t^2/U, but the proportionality factor goes logarithmically to zero with U/t. Thus, for the half-filled-band case, the GVW gives a result that is qualitatively different from the exact result. From Eq. (3.33) it is clear that a discontinuity in the momentum distribution remains at any finite U, whereas the exact solution of the Hubbard model of Lieb and Wu (1968) gives an insulating system for any nonzero U (see also Ferraz *et al.*, 1978). It would seem from these considerations that the GVW is not a particularly good ansatz for the ground state of the Hubbard model. However, we have already seen from the results of Kaplan *et al.* that it performs much better in the calculation of spin-spin correlations than for the energy.

Using techniques similar to those employed in the energy calculations, Gebhard and Vollhardt (1987) have calculated, for the paramagnetic ground state, a number of correlation functions,

$$C_j^{XY}=N^{-1}\sum_i\langle X_iY_{i+j}\rangle-\langle X\rangle\langle Y\rangle\ , \tag{3.35}$$

where $X_i,Y_i=S_i^z,n_i,D_i,H_i$. Here $n_i=n_{i\uparrow}+n_{i\downarrow}$, and $H_i=(1-n_{i\uparrow})(1-n_{i\downarrow})$ is the number operator for the holes. In addition, $X=N^{-1}\sum_i X_i$, and the Fourier transforms of the functions defined in Eq. (3.35) are sim-

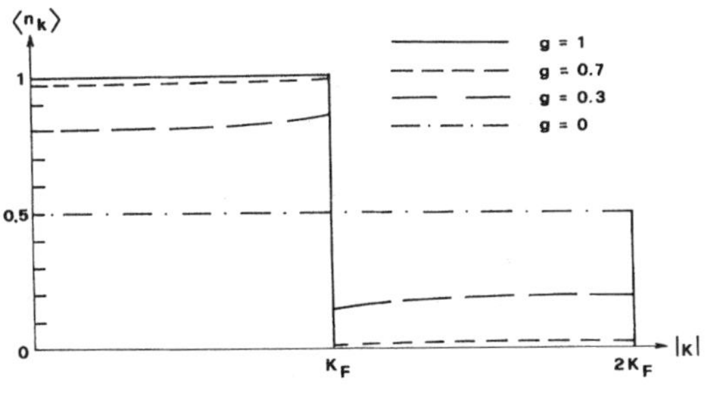

FIG. 3. Momentum distribution $\langle n_k\rangle$ for the one-dimensional Hubbard model. The results obtained by using the Gutzwiller variational wave function are shown for several values of the correlation parameter g in the case of a half-filled band ($n_\uparrow=n_\downarrow=\frac{1}{2}$). From Metzner and Vollhardt (1987).

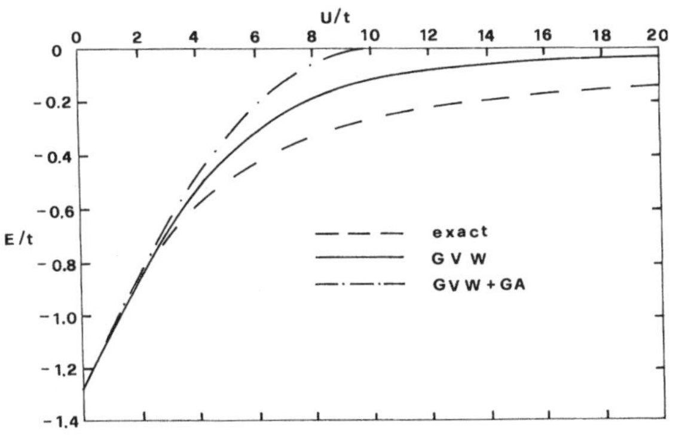

FIG. 4. Ground-state energy E for the one-dimensional Hubbard model with $n_\uparrow=n_\downarrow=\frac{1}{2}$ as a function of U. The results for E, as calculated with the Gutzwiller variational wave function (GVW), are compared with the result of the Gutzwiller approximation (GA). From Metzner and Vollhardt (1987).

ply denoted by $C^{XY}(q)$. The results for the spin-spin correlation function $q_l \equiv C_l^{SS}$ obtained by Gebhardt and Vollhardt fully confirm the numerical calculations of q_1 by Kaplan *et al.* However, with the analytical solution it is possible to establish that there is, in fact, a difference of 0.2% between the exact and the GVW results in the atomic limit, and that such a difference is not due to numerical uncertainties. In addition, the features of the exact q_l found in numerical investigation (Betsuyaku and Yokota, 1986; Kaplan *et al.*, 1987) on the Heisenberg antiferromagnetic chain are reasonably reproduced, though the agreement worsens at larger distances, as already suggested by Kaplan *et al.* (1982). The comparison between the GVW result and the exact result for the hole-hole correlation function $C^{HH}(q)$ in the atomic limit also shows an overall agreement, which tends to improve as the number of holes tends to zero. Finally, we notice that Gebhardt and Vollhardt also comment on the correlations between holes and doubly occupied sites. Their conclusion is that these kinds of correlations are not described particularly well by the GVW and that this might well be the reason for the logarithmic singularity found in the energy for $g \to 0$.

2. Numerical results

Recently, a number of numerical investigations on the Hubbard model with the GVW were conducted with the help of the Monte Carlo method. One should distinguish between two kinds of investigations. In one case the original Gutzwiller program is implemented; i.e., ground-state properties of the Hubbard Hamiltonian are variationally calculated with the wave function of Eq. (3.8), for arbitrary n, t, U and dimensionality. In the other case, following the observation of Kaplan *et al.* (1982) that the GVW for $g=0$ gives an accurate description of spin correlation in the Heisenberg antiferromagnetic chain, one works on the effective Hamiltonian to which the Hubbard one reduces in the limit $U \to \infty$, thus restricting one to the strong-coupling situation.

Yokoyama and Shiba (1987a) have systematically studied the Hubbard model, following the first of the above-mentioned approaches. Thus they have performed calculations in one, two, and three dimensions over the full range of coupling U/t and for both half-filled and non-half-filled bands. Here we shall not enter into the technical details of their variational Monte Carlo technique, with which they performed calculations with up to 216 sites ($6\times6\times6$ in three dimensions). We shall merely give a brief review of their main results.

In one dimension and for $n=1$, they have calculated $E_{\rm kin}/t$ and $d=E_{\rm pot}/U$ as functions of g between 0 and 1. For a given value of U, one can then construct from such functions the total energy as a function of g and find its minimum by inspection. They have also calculated the momentum distribution. We merely note that in this case the exact solution is available and that according to Metzner and Vollhardt (1987) the agreement with their analytic results is excellent. In the limit in which $U \to \infty$, however, they have not been able to isolate the logarithmic correction to the quadratic dependence t^2/U of the energy, but it would have been surprising otherwise. Similar calculations were performed for the test case $n=0.84$, taken as representative of a band less than half filled. The results for the energy are in fair agreement with the exact result, even if they observe that improvements upon the GVW are called for if one wishes to obtain better agreement. The need to take into account intersite correlation is also mentioned, a point that has already been stressed by Stollhoff and Fulde (1977), as we discussed at some length earlier.

In two dimensions only the case $n=1$ on a square lattice was considered. They found, for the total energy, that the variational results and those of the Gutzwiller approximation are much closer than they were in one dimension, as can be seen by comparing the 2D case reported in Fig. 5 with the 1D case in Fig. 4. This is in accord with the expectation that the GA should become more accurate in higher dimensions. However, comparison with the Hartree-Fock antiferromagnetic energy shows that the discrepancies with this are sizable for large U, whereas they were much smaller in one dimension.

The calculations in three dimensions were performed for a cubic lattice and again for a half-filled band. The results for the total energy as a function of the interaction strength U/t are similar to those found in two dimensions. Furthermore, in going from two to three dimensions, changes qualitatively similar to those observed in going from one to two dimensions are apparent. Thus the agreement between variational and approximate treatments, i.e., with the GA, improve further, while for large U the discrepancy with the antiferromagnetic Hartree-Fock energy, which is lower, remains sizable.

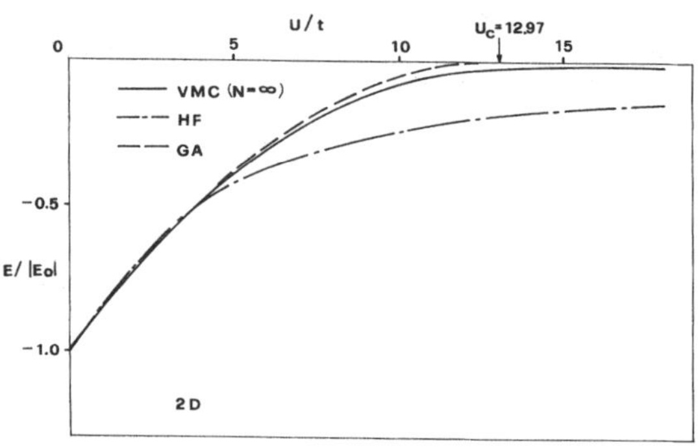

FIG. 5. Normalized ground-state energy for the two-dimensional Hubbard model as a function of U: VMC, variational Monte Carlo with the Gutzwiller variational wave function (extrapolated at an infinite number of sites, $N=\infty$); HF, antiferromagnetic Hartree-Fock; GA, Gutzwiller approximation. The half-filled-band case ($n_\uparrow = n_\downarrow = \frac{1}{2}$) is shown. From Yokoyama and Shiba (1987a).

This suggests that, in two and three dimensions, one should perform the variational calculations using a GVW based on the Hartree-Fock ground state of the antiferromagnetic system, rather than on that of the paramagnetic one.

To introduce the second of the approaches mentioned above, we shall briefly summarize the derivation of the effective Hamiltonian which can be obtained from the Hubbard one for strong couplings U/t. We shall follow the very straightforward method given by Gros *et al.* (1987a), but reference should also be made to Castellani *et al.* (1979) and to Hirsch (1985). One can start from the Hubbard Hamiltonian given in Eq. (3.10) and rearrange the kinetic energy T, specializing to the case of nearest-neighbor hopping, to read

$$H=T_h+T_d+T_{\text{mix}}+V , \tag{3.36}$$

with

$$T_h=-t\sum_{\langle ij\rangle,\sigma}(1-n_{i,-\sigma})a_{i\sigma}^{\dagger}a_{j\sigma}(1-n_{j,-\sigma}) , \tag{3.37}$$

$$T_d=-t\sum_{\langle ij\rangle,\sigma}n_{i,-\sigma}a_{i\sigma}^{\dagger}a_{j\sigma}n_{j,-\sigma} , \tag{3.38}$$

$$T_{\text{mix}}=-t\sum_{\langle ij\rangle,\sigma}n_{i,-\sigma}a_{i\sigma}^{\dagger}a_{j\sigma}(1-n_{j,-\sigma}) -t\sum_{\langle ij\rangle,\sigma}(1-n_{i,-\sigma})a_{i\sigma}^{\dagger}a_{j\sigma}n_{j,-\sigma} , \tag{3.39}$$

where T_h and T_d are the kinetic energies describing the propagation of holes and doubly occupied sites, respectively, in their Hubbard bands, and T_{mix} is clearly a mixing term that couples such bands. V is, of course, the usual on-site repulsion energy, $V=U\sum_i n_{i\uparrow}n_{i\downarrow}$. The next step is to apply to the Hamiltonian a suitable unitary transformation,

$$H_{\text{eff}}=e^{iS}He^{-iS}=H+i[S,H]+\cdots , \tag{3.40}$$

such that, in lowest order in t/U, T_{mix} vanishes on the right-hand side of Eq. (3.40). This can be accomplished by choosing S such that $i[S,T_h+T_d+V]=-T_{\text{mix}}$, i.e.,

$$S=\sum_{n,m}|n\rangle\frac{\langle n|T_{\text{mix}}|m\rangle}{i(\epsilon_n-\epsilon_m)}\langle m| , \tag{3.41}$$

with $|n\rangle$ and $|m\rangle$ being eigenstates of T_h+T_d+V. Even if these eigenstates are not known in the general case, for very large U it must be $\epsilon_n-\epsilon_m=\pm U+O(t)$. Thus one gets

$$S=-\frac{it}{U}\sum_{\langle ij\rangle,\sigma}n_{i,-\sigma}a_{i\sigma}^{\dagger}a_{j\sigma}(1-n_{j,-\sigma}) +\frac{it}{U}\sum_{\langle ij\rangle,\sigma}(1-n_{i,-\sigma})a_{i\sigma}^{\dagger}a_{j\sigma}n_{j,-\sigma} . \tag{3.42}$$

If, as is the case here, one is interested in taking matrix elements of H_{eff} between states with no doubly occupied sites, such as the infinite-U Gutzwiller wave function of Eq. (3.24), it can be easily shown that H_{eff} can be written as

$$H_{\text{eff}}=T_h+i[S,T_{\text{mix}}]+SVS =T_h+\frac{2t^2}{U}\sum_i\sum_{\tau,\tau'}[(a_{i+\tau}^{\dagger}\mathbf{s}a_{i+\tau'})\cdot(a_i^{\dagger}\mathbf{s}a_i) -\tfrac{1}{4}(a_{i+\tau}^{\dagger}a_{i+\tau'})\cdot(a_i^{\dagger}a_i)] . \tag{3.43}$$

Above, $a_i\mathbf{s}a_j\equiv\sum_{\sigma\sigma'}a_{i\sigma}^{\dagger}(\mathbf{s})_{\sigma\sigma'}a_{j\sigma'}$ and $a_ia_j\equiv\sum_\sigma a_{i\sigma}^{\dagger}a_{j\sigma}$, with the vector $\mathbf{s}$ being the spin operator, and the indices τ and τ' running over the first neighbors of i. We should note two things. In Eq. (3.43) we have also considered the term SVS, which would appear as of higher order in S. But the expansion parameter is t/U, and such a term turns out to be of the same order of $i[S,T_{\text{mix}}]$. In addition, if three-site terms (i.e., $\tau\neq\tau'$) can be neglected, then Eq. (3.43) simplifies to

$$H_{\text{eff}}=T_h+\frac{2t^2}{U}\sum_{\langle ij\rangle}(\mathbf{s}_i\cdot\mathbf{s}_j-\tfrac{1}{4}n_in_j) , \tag{3.44}$$

with $n_i=n_{i\uparrow}+n_{i\downarrow}$ being the site number operator. This is exactly true for the half-filled-band case, for which one also has $n_i=1$. However, for $n\neq1$, three-site contributions may become important, as has recently been discussed by Yokoyama and Shiba (1987b). It should be noted that calculating averages of H_{eff} on $|\Phi_\infty\rangle$ is equivalent to calculating averages of H on a modified wave function $|\tilde{\Phi}_\infty\rangle=(1-iS)|\Phi_\infty\rangle$, that is, $\langle|\Phi_\infty|H_{\text{eff}}|\Phi_\infty\rangle=\langle|\tilde{\Phi}_\infty|H|\tilde{\Phi}_\infty\rangle$, provided terms of order t^2/U^2 have been neglected. In particular, the number of doubly occupied sites is nonzero on this wave function, or, more precisely, of order t^2/U^2. For the half-filled band, if one puts $H_{\text{eff}}=T_h+V_{\text{corr}}$, then

$$D=\langle\tilde{\Phi}_\infty|\frac{1}{U}V|\tilde{\Phi}_\infty\rangle=-\frac{1}{2U}\langle\Phi_\infty|V_{\text{corr}}|\Phi_\infty\rangle . \tag{3.45}$$

Before turning to the numerical results that have recently been obtained by employing the strong-coupling Hamiltonian of Eq. (3.43) or (3.44), we should like to stress one point. As observed first by Kaplan *et al.* (1982), $|\Phi_\infty\rangle$, while giving a good description of the spin-spin correlations, yields poor results, if used to calculate the energy directly from the Hubbard Hamiltonian. One should have clearly in mind therefore that $|\Phi_\infty\rangle$ is a good ground state for H_{eff}, but not for H. Thus the combined use of H_{eff} and $|\Phi_\infty\rangle$ is equivalent, as we have already noted, to working with the Hubbard Hamiltonian, but using an improved Gutzwiller-like wave function.

Using the approach mentioned above, i.e., working in terms of H_{eff} and $|\Phi_\infty\rangle$, Gros *et al.* (1987a) have conducted an extensive Monte Carlo study of the Hubbard model for strong coupling. They restrict their investigation to one dimension to obtain good numerical accuracy, even if their interest is, of course, in the three-dimensional situation. On the other hand, they consider variable band filling, magnetization, and degeneracy N_f of the local state. Their results for the spin-spin correlation function q_1 at half-filling have established, before the exact solution with the GVW appeared (Gebhard and

Vollhardt, 1987), that the small difference between the exact result for this quantity and that obtained from the infinite-U Gutzwiller wave function is beyond the numerical uncertainty of the Monte Carlo calculation. The investigation of the kinetic energy as a function of the filling, on the other hand, shows good agreement with the exact result of both Monte Carlo and GA results. Gros, Joynt, and Rice also discuss the fact that for large U the hole-hole correlation function can be put into correspondence with that of a system of spinless noninteracting fermions. This allows for an assessment of the quality of their results. They find fair agreement, even if some qualitative features of the exact result, related to the presence of a sharp Fermi surface, are missing. Moreover, an enhancement with respect to the exact result of correlations at short distances is found. Regarding the total energy, for large U and values of the filling very close to 1, the accuracy of the Monte Carlo result is dominated by the kinetic-energy contribution, with an accuracy of about 6%. In fact, the potential energy, being determined by q_1, has a much better accuracy, i.e., 0.2%. Finally, they find that when excited-state Gutzwiller wave functions are considered, the accuracy of GA is much reduced. For the cases in which direct comparison with the analytical results of Metzner and Vollhardt (1987) and Gebhard and Vollhardt (1987) was possible, excellent agreement was found.

One of the reasons for the revived interest in the Hubbard Hamiltonian is its possible relevance in the understanding of the mechanism underlying high-temperature superconductivity, as suggested by Anderson (1987). In particular, investigations in the strong correlation regime are called for. The importance of the scheme, briefly outlined above, for dealing with the large coupling situation is then apparent. An investigation in this direction was made recently by Yokoyama and Shima (1987b), within the effective Hamiltonian approach. They used the effective Hamiltonian of Eq. (3.39) and considered both a paramagnetic and an antiferromagnetic ground state. This can be done by simply changing the type of uncorrelated wave function $|\Phi_0\rangle$, from which the Gutzwiller wave function of Eq. (3.24) is built. One and two dimensions were considered. The purpose of the study was to characterize the competition between the two types of ground states. Before briefly summarizing their findings, it should be mentioned that they have also shown that the simple infinite-U Gutzwiller wave function constitutes one of the many possible realizations of Anderson's resonating valence-bond state or singlet state. In one dimension, they find that for both half-filled and non-half-filled bands, the singlet state is stable against the Néel (antiferromagnetic) state. The situation is quite different in two dimensions. For the half-filled case the singlet state is found to be unstable against the Néel state. Whereas the results for the energy and magnetization are in reasonable accord with estimates obtained with different approaches, they question whether the same situation would be found if next-nearest hopping were to be included. Moving from the half-filled band ($n=1$), at very large U, the ferromagnetic state becomes favorable in energy. However, by decreasing n at some point ($n \cong 0.61$) the singlet state takes over again. On the other hand, if U is decreased, the Néel state appears to again decrease in energy. From these results they draw a qualitative phase diagram as shown in Fig. 6. This should be considered meaningful only in the small t/U region, where the effective Hamiltonian approach is appropriate.

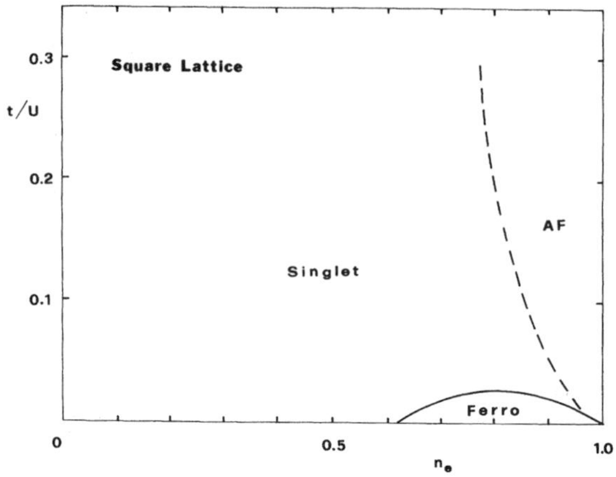

FIG. 6. Phase diagram of the two-dimensional square-lattice Hubbard model, as inferred from variational Monte Carlo using a Gutzwiller-type wave function (Yokoyama and Shiba, 1987b). Here $n_e = n_\uparrow + n_\downarrow$ is the band filling.

Finally, we should mention here the work of Gros, Joynt, and Rice (1987b). Using the effective Hamiltonian approach, they investigate the stability of generalized Gutzwiller wave functions against Cooper pairing, in the large-U limit and in two dimensions. This is done by evaluating the binding energy of two holes in the variational wave function. As they note, no real attempt was made to optimize their wave function. They find that the paramagnetic or singlet state is stable against s-wave pairing but unstable against d-wave pairing. On the other hand, the antiferromagnetic state is stable against both kinds of pairings. They discuss a possible pairing mechanism and the relevance of their results for high-T_c superconductors.

E. Resonating valence-bond states

In the foregoing we have seen how the Hubbard Hamiltonian—for large values of the coupling U—can be transformed into an antiferromagnetic Heisenberg Hamiltonian by means of a suitable unitary transformation. In this connection and also in relation to the high-T_c superconductivity, it seems appropriate here to briefly review the mathematical formulation due to Anderson (1973) of the concept of the resonating valence-bond (RVB) states first put forward by Pauling (1949).

In discussing the ground-state properties of the triangular two-dimensional Heisenberg antiferromagnet for $S=\frac{1}{2}$, Anderson (1973; see, also, Fazekas and Anderson, 1974) proposed that at least for this system, and perhaps also in other cases, the ground state might be the analog of the precise singlet in the Bethe (1931) solution of the linear antiferromagnetic chain. In fact, the zero-order energy of a state consisting purely of nearest-neighbor singlet pairs is more nearly realistic than that of the Néel state.

For electrons on a lattice, one can think of a singlet bond or pair as the state formed when two electrons with opposite spin are localized on two distinct sites. A resonating valence-bond state is a coherent superposition of such singlet bonds; its energy is further lowered as a result of the matrix elements connecting the different valence-bond configurations.

Heuristically, valence bonds can be viewed as real-space Cooper pairs that repel one another, a joint effect of the Pauli principle and the Coulomb interaction. When there is one electron per site, charge fluctuations are suppressed, leading to an insulating state. However, as one moves away from half filling, current can flow; the system becomes superconducting as the valence bonds Bose condense.

Anderson (1987), while stressing the difficulty of making quantitative calculations with RVB states, in fact gives a suggestive representation of them by exploiting the Gutzwiller-type projection technique.

Clearly, a delocalized or mobile valence bond can be written as

$$b_\tau^\dagger|\Phi_0\rangle = \frac{1}{\sqrt{N}}\left[\sum_i a_{i\uparrow}^\dagger a_{i+\tau\downarrow}^\dagger\right]|0\rangle = \frac{1}{\sqrt{N}}\left[\sum_k a_{k\uparrow}^\dagger a_{-k\downarrow}^\dagger \exp i(k\cdot\tau)\right]|0\rangle , \quad (3.46)$$

where $b_\tau^\dagger$ is the creation operator for a valence-bond state with lattice vector τ; $a_{i\sigma}^\dagger$, the single-electron creation operator; and N, the total number of sites. A distribution of bond lengths can be obtained by summing $b_\tau^\dagger$ over τ with appropriate weights. One then gets a new creation operator,

$$b^\dagger = \sum_k c_k a_{k\uparrow}^\dagger a_{-k\downarrow}^\dagger , \quad (3.47)$$

with the restriction

$$\sum_k c_k = 0 , \quad (3.48)$$

if double occupancy is to be avoided.

Anderson proceeds by (a) Bose condensating such mobile valence bonds,

$$|\Phi\rangle = (b^\dagger)^{N/2}|0\rangle , \quad (3.49)$$

and by (b) projecting out the double occupancy—which would otherwise be present—with an infinite-U Gutzwiller projection operator $P_d=\prod_i(1-n_{i\uparrow}n_{i\downarrow})$,

$$|\Phi_{\rm RVB}\rangle = P_d(b^\dagger)^{N/2}|0\rangle . \quad (3.50)$$

One can then show that the RVB state written above can be obtained with simple manipulations from a standard BCS state by projecting on the state with N particles and projecting out, at the same time, the double occupancy:

$$|\Phi_{\rm RVB}\rangle = P_{N/2}P_d\left[\frac{1}{\sqrt{1+c_k}} + \frac{c_k}{\sqrt{1+c_k}} a_{k\uparrow}^\dagger a_{-k\downarrow}^\dagger\right]|0\rangle . \quad (3.51)$$

Baskaran *et al.* (1987) have subsequently shown that by treating the large-U Hamiltonian of Eq. (3.44) with a mean-field (Hartree-Fock) approximation, one obtains precisely a BCS-type Hamiltonian which—for half filling—yields the RVB state heuristically introduced above. In fact, one also finds that $\sum_k c_k=0, |c_k|=1$ and the c_k change sign across what they call a *pseudo-Fermi* surface. We shall not deal here with the nature of excitations from such a RVB state, referring the reader instead to the original papers.

IV. QUANTUM MONTE CARLO CALCULATION OF CORRELATION ENERGY

Until quite recently, two systematic methods have provided the main routes to the calculation of correlation energies for many-electron systems at zero temperature, namely, configuration interaction (CI) and many-body perturbation theory. Relatively recent overviews of these two approaches are available [see, e.g., Shavitt *et al.* (1977) and Wilson (1981)], and therefore we shall not go over that ground in the present article. In the last ten years, however, progress in the calculation of electronic correlation has also been made by using a completely different approach, the so-called quantum Monte Carlo (QMC) method.

The goal of QMC [see, e.g., Ceperley (1981), Reynolds *et al.* (1982), Ceperley and Alder (1984), and Kalos (1984)] is to obtain the exact ground-state wave function of a many-body system by numerically solving the Schrödinger equation in one of its equivalent forms. In practice, this is achieved by means of iterative algorithms, which propagate the wave function from a suitable starting guess to the exact ground-state value. Thus, in the diffusion Monte Carlo (DMC) method, one is concerned directly with the evolution in imaginary time of the wave function, which corresponds to a diffusion process in configuration space. In the Green's-function Monte Carlo (GFMC) technique, on the other hand, a time-integrated form of the Green's function or resolvent is used to propagate the wave function. In fact, in either case the sampling of an appropriate Green's function is required, and this is accomplished by means of suitable random-walk algorithms.

Another implementation of the QMC (Blankenbecler *et al.*, 1981; Scalapino and Sugar, 1981; Koonin *et al.*, 1982) emphasizes the role of the imaginary-time propagator $e^{-\beta H}$ more directly. For long times β (small temperatures $T=1/\beta$), the imaginary-time propagator is dominated by the ground-state energy E_0. Thus E_0 can be suitably extracted from the knowledge of the partition function $Z=\mathrm{Tr}e^{-\beta H}$ or, equally well, from that of the expectation value of the imaginary-time propagator on a state which is not orthogonal to the ground state. In either case, the key simplification in the calculations comes from the mapping of the problem of interacting particles onto that of independent particles. This is accomplished at the expense of introducing *auxiliary* external fields (Koonin *et al.*, 1982; Hirsch, 1983; Sugiyama and Koonin, 1986) having their own probability density.

In the following, we first review the main features of the QMC method for a many-body system in the DMC and GFMC implementations, discussing also the complications that arise from antisymmetry when dealing with systems of fermions, which is the case of interest here. Some applications of the QMC method—to the study of the homogeneous electron assembly and small molecules—are then briefly summarized. Finally, we give the basic ideas underlying the auxiliary external field technique and review some of the recent results on the two-dimensional Hubbard model.

A. Diffusion Monte Carlo method

The Schrödinger equation in imaginary time for an assembly of N identical particles of mass m interacting with a potential $V(\mathbf{R})$ reads

$$-\frac{\partial\phi(\mathbf{R},t)}{\partial t}=(H-E_T)\phi(\mathbf{R},t)$$
$$=[-D\nabla^2+V(\mathbf{R})-E_T]\phi(\mathbf{R},t)\ , \qquad (4.1)$$

where $D=\hbar/2m$, $\mathbf{R}$ is the $3N$-dimensional vector specifying the coordinates of the particles, t is the imaginary time in units of $\hbar$, and the constant E_T represents a suitable shift of the zero of energy. The (imaginary) time evolution of an arbitrary *trial* wave function is easily obtained from its expansion in terms of the eigenfunctions $\phi_i(\mathbf{R})$ of the Hamiltonian H as

$$\phi(\mathbf{R},t)=\sum_i N_i\exp[-(E_i-E_T)t]\phi_i(\mathbf{R})\ . \qquad (4.2)$$

Here, E_i is the energy eigenvalue corresponding to $\phi_i(\mathbf{R})$, and the coefficients N_i are fixed by the initial conditions, i.e., by the chosen trial wave function. Clearly, for long times one finds

$$\phi(\mathbf{R},t)=N_0\exp[-(E_0-E_T)t]\phi_0(\mathbf{R})\ , \qquad (4.3)$$

provided $N_0\neq 0$. Moreover, if E_T is adjusted to be the true ground-state energy E_0, asymptotically one obtains a steady-state solution, corresponding to the ground-state wave function ϕ_0. Thus the problem of determining the ground-state eigenfunction of the Hamiltonian H is equivalent to that of solving Eq. (4.1) with the appropriate boundary conditions.

It is not difficult to recognize in Eq. (4.1) two equations that are well known in physics, though combined together. In fact, if only the term with the Laplacian were present on the right-hand side of Eq. (4.1), one would have a diffusion equation, such as the equation describing Brownian motion (see, for instance, van Kampen, 1981). On the other hand, by retaining only the term $[V(\mathbf{R})-E_T]\phi(\mathbf{R},t)$, one would obtain a rate equation, that is, an equation describing branching processes such as radioactive decay or birth and death processes in a population.

A convenient manner of *simulating* Eq. (4.1) is as follows. Let $\psi(\mathbf{R})$ be the wave function at $t=0$. One can generate an ensemble of systems (points in the $3N$-dimensional space representing electronic configuration) distributed with density $\psi_T(\mathbf{R})$. Hence the time evolution of the wave function will correspond to the motion in configuration space of such systems or *walkers*, as determined by Eq. (4.1). In particular, the Laplacian and energy terms in Eq. (4.1) will cause, respectively, random diffusion and branching (deletion or duplication) of the walkers describing the wave function. This schematization of the Schrödinger equation is particularly appealing from the practical point of view. However, a density is non-negative. Thus the same should hold for the wave function. This would seem to limit the applicability of such a scheme to the ground state of a bosonic system. For the sake of simplicity, we shall accept this restriction for a while and deal with the complications associated with Fermi statistics later.

Solving the Schrödinger equation in its form (4.1) by random-walk processes with branching is not particularly efficient. In fact, the branching term can become very large whenever the interaction potential $V(\mathbf{R})$ does so, causing large fluctuations in the number of walkers. This slows down the convergence toward the ground state. A more efficient computational scheme is obtained by introducing the so-called *importance sampling*. This amounts to considering the evolution equation for the probability distribution $f(\mathbf{R},t)=\phi(\mathbf{R},t)\psi_T(\mathbf{R})$, rather than directly for the wave function ϕ. By using Eq. (4.1) and the definition of $f(\mathbf{R},t)$, one obtains with a little algebra

$$-\frac{\partial f(\mathbf{R},t)}{\partial t}=-D\nabla^2 f+[E_L(\mathbf{R})-E_T]f+D\nabla[fF_Q(\mathbf{R})]\ . \qquad (4.4)$$

Above, $E_L(\mathbf{R})\equiv H\psi_T/\psi_T$ defines the *local energy* associated with the trial wave function, and

$$F_Q(\mathbf{R})\equiv\nabla\ln|\psi_T(\mathbf{R})|^2=2\nabla\psi_T(\mathbf{R})/\psi_T(\mathbf{R}) \qquad (4.5)$$

is the *quantum trial force*, which drifts the walkers away from regions where $\psi(\mathbf{R})$ is small. Obviously, when dealing with Eq. (4.4), the walkers are to be drawn from the probability distribution f. With a judicious choice of ψ_T, one can make $E_L(\mathbf{R})$ a smooth function close to E_T throughout the configuration space and thus reduce branching. In this respect it is essential that $\psi_T(\mathbf{R})$

reproduce the correct cusp behavior as any two particles approach each other, so as to exactly cancel the infinities originating from $V(\mathbf{R})$. That $F_Q(\mathbf{R})$ is a force acting on the walkers becomes clear by comparing Eq. (4.4) with the Fokker-Planck (FP) equation (see, e.g., van Kampen, 1981). It turns out, with the rate term neglected, that the term containing F_Q has to be identified with the so-called *drift term* of the FP equation, and consequently F_Q must be identified with the external force acting on the walkers. We stress again that an important feature of this force is to drift the walkers away from regions of low probability. In fact, from its definition it is apparent that the force becomes highly repulsive in regions where the trial wave function becomes small, diverging where the latter vanishes. Thus the trial wave function determines the probability with which different regions of configuration space are sampled. Therefore it is important that ψ_T be a good approximation to ϕ_0 in order to keep the walkers mainly in regions of configuration space which are really significant in the statistical averages. We finally observe that the asymptotic solution of Eq. (4.4) is

$$f(\mathbf{R},T)=\psi_T(\mathbf{R})\phi_0(\mathbf{R})\exp[-(E_0-E_T)t] , \tag{4.6}$$

which becomes a steady-state solution when E_T is adjusted to E_0.

The differential equation (4.4) for the probability distribution $f(\mathbf{R},t)$ can be recast in integral form as

$$f(\mathbf{R},t+t')=\int d\mathbf{R}' f(\mathbf{R}',t')K(\mathbf{R}',\mathbf{R},t) , \tag{4.7}$$

where the Green's function $K(\mathbf{R}',\mathbf{R},t)$ is a solution of Eq. (4.4) with the boundary condition $K(\mathbf{R}',\mathbf{R},0) =\delta(\mathbf{R}-\mathbf{R}')$ and is simply related to the Green's function $G(\mathbf{R}',\mathbf{R},t)$ for Eq. (4.1), $K(\mathbf{R}',\mathbf{R},t) =\psi_T(\mathbf{R}')G(\mathbf{R}',\mathbf{R},t)\psi_T^{-1}(\mathbf{R})$. The advantage of the above equation is that for short times t it is possible to write approximate simple expressions for $K(\mathbf{R}',\mathbf{R},t)$ (Moskowitz *et al.*, 1982b; Reynolds *et al.*, 1982). The time evolution of $f(\mathbf{R},t)$ for finite t can then be obtained by successive iterations of Eq. (4.7), starting from the initial distribution $f(\mathbf{R},0)$ and using a small time step. In each time iteration, advantage can be taken of the positivity of K for short times so as to interpret it as a transition density. A suitable modified random-walk algorithm can then be devised, which allows the integration of Eq. (4.7) during a small time interval. The essential steps of such an algorithm are as follows. The walkers representing the initial distribution are allowed to diffuse randomly and drift under the action of the quantum force F_Q. After the new positions have been reached, each walker is deleted or placed in the *new generation* in an appropriate number of copies, depending on the size of the local energy at the old and at the new position relative to the reference energy E_T. Finally, the number of walkers in the new generation is renormalized to the initial population. The interested reader may find a detailed description of such a random-walk algorithm in Reynolds *et al.* (1982).

Once the long-time probability distribution has been obtained and made stationary by a suitable shift of the constant E_T, one can estimate equilibrium quantities. For the ground-state energy E_0, in particular, by using the fact that H is Hermitian, one finds

$$E_0=\frac{\int \psi_T H\phi_0}{\int \psi_T\phi_0}=\frac{\int E_L f}{\int f}=\frac{\sum_i E_L(\mathbf{R}_i)}{\sum_i} . \tag{4.8}$$

The sum on the right-hand side of the above equation runs over the positions of all the walkers representing the equilibrium probability distribution.

It should be stressed that the short-time approximation to the Green's function introduces a systematic error in the computations. However, this error should decrease with a decrease in the time step. Moreover, in the calculation by Reynolds *et al.* (1982), an acceptance/rejection step was used which, within the calculational scheme outlined above, should correct for the approximate nature of G.

B. Green's-function Monte Carlo technique

We have seen that in the diffusion Monte Carlo method one starts from the Schrödinger equation in imaginary time in order to construct an integral equation which can then be solved iteratively. Such an equation contains a time-dependent Green's function. In the Green's-function Monte Carlo technique, one proceeds in a similar fashion, starting, however, from the Schrödinger eigenvalue equation. The resulting integral equation, as we shall see, differs from Eq. (4.7) mainly through the absence of time. In fact, it could be directly obtained from Eq. (4.7) by a straightforward time integration (Ceperley and Alder, 1984). Here, however, we choose to follow another route (Ceperley and Kalos, 1979; Kalos, 1984) which in the present context is somewhat more instructive.

The Schrödinger eigenvalue equation for an N-body system, described by the Hamiltonian H introduced in Eq. (4.1), is

$$H\phi(\mathbf{R})=E\phi(\mathbf{R}) . \tag{4.9}$$

Let us assume that the potential $V(\mathbf{R})$ appearing in H is such that the spectrum of H is bounded from below, E_0 being the lowest eigenvalue, i.e., the ground-state energy. This is a natural assumption for a physical system in the nonrelativistic limit. One can choose a positive constant V_0 such that $E_0+V_0>0$, and rewrite Eq. (4.9) as

$$(H+V_0)\phi(\mathbf{R})=(E+V_0)\phi(\mathbf{R}) . \tag{4.10}$$

The resolvent $G(\mathbf{R},\mathbf{R}')$, i.e., the Green's function for Eq. (4.10), is then defined by

$$(H+V_0)G(\mathbf{R},\mathbf{R}')=\delta(\mathbf{R}-\mathbf{R}') . \tag{4.11}$$

Due to the manner in which V_0 has been chosen, it is

clear that the operator $H+V_0$ is positive definite. This is immediate in the energy representation. The same, of course, holds for the inverse operator $G=1/(H+V_0)$. Actually, a much more strict inequality involving G can be established, namely,

$$G(\mathbf{R},\mathbf{R}')=\langle \mathbf{R}|G|\mathbf{R}'\rangle \geq 0 \quad \text{for all } \mathbf{R},\mathbf{R}' , \tag{4.12}$$

which is of central importance for the calculational scheme that we are about to describe. A brief sketch of how one can prove the inequality (4.12) is given in Appendix B. Here, we merely note for future reference that G is integrable. Therefore, because of the above inequality, it can be regarded as a probability, or more precisely as a transition density.

A space integration of Eq. (4.10), after multiplying by $G(\mathbf{R},\mathbf{R}')$ and utilizing Eq. (4.11), yields the desired integral equation,

$$\phi(\mathbf{R})=(E+V_0)\int d\mathbf{R}' G(\mathbf{R},\mathbf{R}')\phi(\mathbf{R}') . \tag{4.13}$$

In the above equation one has to solve simultaneously for E and $\phi(\mathbf{R})$. However, if one has a trial wave function $\psi_T(\mathbf{R})$ and, consequently, a trial energy E_T, it is tempting to try solving Eq. (4.13) by iteration. One would generate a sequence of wave functions, according to

$$\Phi_n(\mathbf{R})=(E_T+V_0)\int d\mathbf{R}' G(\mathbf{R},\mathbf{R}')\Phi_{n-1}(\mathbf{R}') \tag{4.14}$$

and with $\Phi_0(\mathbf{R})=\psi_T(\mathbf{R})$. In fact, this series converges precisely to the ground-state wave function, and it does so exponentially fast. This can be easily seen by expanding both $G(\mathbf{R},\mathbf{R}')$ and $\Phi_0(\mathbf{R})$ in eigenfunctions of H,

$$G(\mathbf{R},\mathbf{R}')=\sum_i \frac{\phi_i(\mathbf{R})\phi_i(\mathbf{R}')}{E_i+V_0} , \tag{4.15}$$

$$\Phi_0(\mathbf{R})=\sum_i c_i\phi_i(\mathbf{R}) . \tag{4.16}$$

One immediately finds

$$\Phi_n(\mathbf{R})=\sum_i \left[\frac{E_T+V_0}{E_i+V_0}\right]^n c_i\phi_i(\mathbf{R}) . \tag{4.17}$$

So, provided $c_0\neq 0$, one obtains

$$\lim_{n\to\infty} \Phi_0(\mathbf{R}) \propto c_0\phi_0(\mathbf{R}) , \tag{4.18}$$

which is the desired result.

As in the case of the diffusion Monte Carlo method, one has cast the problem of finding the ground-state wave function and energy of a given Hamiltonian in integral form, suitable for iterative techniques. For the sake of simplicity, we shall temporarily restrict the discussion to bosonic systems, for which both $G(\mathbf{R},\mathbf{R}')$ and the ground-state wave function are non-negative. If one assumes for a while that $G(\mathbf{R},\mathbf{R}')$ is known, a random-walk algorithm similar to the one described for the diffusion Monte Carlo technique is readily constructed. One can proceed as follows. First, at random, an initial population of walkers is drawn from the probability distribution $\Phi_0(\mathbf{R})$, say, at the positions $\{\mathbf{R}_i\}$. Then, a new set of configurations $\{\mathbf{R}'_j\}$ is generated at random, with each one having a conditional density $\sum_i (E+V_0)G(\mathbf{R}_i,\mathbf{R}'_j)$. This density determines the number of walkers corresponding to each new configuration. Finally, a renormalization of the new walker population to the initial size yields the new generation. The above series of steps corresponds to the first iteration of Eq. (4.14). The successive iterations can be performed in the same manner, by only changing the first step. In the nth iteration, in fact, one has to take as the initial population of walkers the new generation of the previous iteration.

It is clear from Eq. (4.17) that the change in size of the walker population is determined by the trial energy E_T. Depending on whether this is larger or smaller than the true ground-state energy E_0, the population will asymptotically grow or decline. This suggests a way of estimating the ground-state energy (Ceperley and Kalos, 1979). In fact, from the space integration of Eq. (4.17) it follows that asymptotically

$$E_0=E_T+V_0\left[\frac{N_n}{N_{n+1}}-1\right] , \tag{4.19}$$

with N_n the initial walker population in the nth iteration and N_{n+1} the new walker population but before size renormalization. The above energy estimator, known as the *growth* estimator, is unfortunately biased even in the limit in which Φ_n is converging to ϕ_0, as discussed by Ceperley and Kalos (1979). A better energy estimator is the one introduced with Eq. (4.8), when describing the diffusion Monte Carlo method. However, in order to use that estimator one has to work with the density $f(\mathbf{R})=\phi(\mathbf{R})\psi_T(\mathbf{R})$ rather than directly with the wave function. This can be easily arranged by multiplying Eq. (4.13) by $\psi_T(\mathbf{R})$ on either side, so as to obtain the new integral equation

$$f(\mathbf{R})=(E+V_0)\int d\mathbf{R}' K(\mathbf{R},\mathbf{R}')f(\mathbf{R}') , \tag{4.20}$$

where

$$f(\mathbf{R})=\phi(\mathbf{R})\psi_T(\mathbf{R}) \tag{4.21}$$

and

$$K(\mathbf{R},\mathbf{R}')=\psi_T(\mathbf{R})G(\mathbf{R},\mathbf{R}')\psi_T^{-1}(\mathbf{R}') . \tag{4.22}$$

Clearly, Eq. (4.20) can still be simulated by means of the random-walk algorithm already described, with $f(\mathbf{R})$ and $K(\mathbf{R},\mathbf{R}')$ replacing, respectively, $\phi(\mathbf{R})$ and $G(\mathbf{R},\mathbf{R}')$. Within such a calculation scheme, successive iteration of Eq. (4.20) will produce a walker population that will asymptotically tend to be distributed with density $f=\phi_0\psi_T$. Consequently, the *local* energy estimator of Eq. (4.8) will also tend to the exact ground-state energy.

In the foregoing we have assumed knowledge of $G(\mathbf{R},\mathbf{R}')$. In practice, for a system of interacting particles, G is not known and must also be sampled by means of suitable techniques. This can be achieved by relating the exact Green's function G to some trial or reference Green's function G_T, which is known analytically or is numerically calculable. In the domain Green's-Function method of Kalos *et al.* (1974; see also Ceperley and Kalos, 1979 and Moskowitz and Schmidt, 1986), the trial Green's function is taken to be that appropriate to independent particles in a constant potential, with the motion restricted to a subdomain of the whole space. In another implementation of GFMC due to Ceperley (1983; see also Ceperley and Alder, 1984), a better trial Green's function is introduced, which is much closer to the exact one and is defined over the whole space. In either case, the integral equation relating G_T to G is such that a random-walk algorithm, somehow similar to the one outlined above for the calculation of f, can be used. In fact, a global random-walk algorithm can be devised that combines the iterations of Eq. (4.20) with those needed to correct for the difference between G and G_T. Details as to how this is done in practice are to be found in the references quoted above. Here, we shall simply observe again that, in the approach due to Ceperley and Alder (1984), Eq. (4.20) is obtained as a time average or Laplace transform of Eq. (4.7). In addition, the sampling of the time-integrated Green's function [$K(\mathbf{R},\mathbf{R}')$ in the present notation] is obtained also in practical calculations by summing its time-dependent counterpart. This makes such an approach very similar to the diffusion Monte Carlo method, while removing the truncation error due to the use of a finite time step.

C. Quantum Monte Carlo technique with fermions: Fixed-node approximation and nodal relaxation

Here we shall discuss the modifications that one can make to the calculational schemes described above when the restriction of a non-negative wave function is relaxed. Of course, this is necessary if one is to be able at all to deal with Fermi systems and, in general, with excited states of the many-body Hamiltonian H, introduced in Eq. (4.1).

For the sake of clarity, let us briefly recall the relation between the eigenfunctions of H and those of a many-body system described by H but also obeying Bose or Fermi statistics. For bosons (with zero spin), only those wave functions of H that are symmetric under the exchange of any two particle coordinates are admissible solutions. On the other hand, the ground-state wave function of H is completely symmetric, since the ground-state energy is nondegenerate and H commutes with any particle permutation. So, the ground state of H, which of course is characterized by a non-negative wave function, is also the bosonic ground state. Needless to say, the bosonic excited states will have wave functions with positive and negative regions. For fermions, the symmetry of the wave functions is determined by the chosen spin configuration that we shall denote here by s. In general, for many fermions it will always be true that a certain number of particles, say, M, will have the same spin projection. The admissible wave functions of H will then be only those which are antisymmetric with respect to the permutation of any two particle coordinates, within the group of particles having the same spin projections. This is equivalent to saying that only excited states of H can be considered, since, as we have already observed, its ground-state wave function is completely symmetric. It follows at once that the fermionic wave functions must have positive and negative regions, separated by nodal surfaces in the $3N$-dimensional space of the particle coordinates.

The need for dealing with wave functions $\phi(\mathbf{R})$ that change sign would seem to preclude the use of random-walk algorithms such as those discussed above, in that they require a positive wave function. In fact, if one knew the location of the nodal surfaces and consequently of the connected domains in which they break the whole space, the problem would not be a real one. Since the equations involved in the evolution of the wave function are linear, one could just consider in each domain the evolution of the modulus of ϕ. However, such nodal surfaces are not known in more than one dimension, and the problem remains. A simple way of remedying it is to take as nodal surfaces those of the trial wave function $\psi_T(\mathbf{R})$, assumed to be a good approximation to the exact wave function at least in the location of such surfaces. This approximation corresponds to a solution of the Schrödinger equation within a restricted class of functions, and so it should have a variational character. As such it should yield energies that are upper bounds to the exact ones, in the absence of other approximations or sources of error.

A practical way of realizing this so-called fixed-node approximation, within calculations based on random-walk algorithms, is to delete those walkers that, during the calculation, cross a nodal surface of the trial wave function (see, for instance, Reynolds *et al.*, 1982). An alternative and better choice, however, is merely to reject those moves which correspond to such crossings (see the recent discussion on this point by Umrigar *et al.*, 1991). These are simple procedures for enforcing the vanishing of the wave function at the nodal surfaces. In fact, both procedures correspond to solving the Schrödinger equation separately in each domain, the evolution of the walkers in one domain having become independent from that in the other domains. More precisely, if $\{V_\alpha\}$ denotes the domains bounded by the nodal surfaces of ψ_T, one is solving for the ground state of the Hamiltonian H in each domain V_α, according to

$$H\phi_\alpha(\mathbf{R},s)=\varepsilon_\alpha\phi_\alpha(\mathbf{R},s) \;, \tag{4.23}$$

and

$$\phi_\alpha(\mathbf{R},s)\psi_T(\mathbf{R},s)>0, \quad \mathbf{R}\in V_\alpha \;, \tag{4.24}$$

$$\phi_\alpha(\mathbf{R},s)=0, \quad \mathbf{R}\notin V_\alpha \;. \tag{4.25}$$

Above, we have also indicated the spin configuration s.

It should be clear that within the random-walk approach, the asymptotic stationarity of the walker population can only be attained if $E_T=\varepsilon_m$, with ε_m being the minimum among all the ε_α corresponding to volumes V_α which were populated at the beginning of the random walk. Therefore, if the trial energy E_T is suitably adjusted so as to yield stationarity, the asymptotic walker population will be distributed with density

$$f_\infty(\mathbf{R})=f(\mathbf{R},t\to\infty)=c_m\phi_m(\mathbf{R},s)\psi_T(\mathbf{R},s) \ . \quad (4.26)$$

We wish now to make our statement about the variational nature of the fixed-node approximation more precise. First of all, we notice that since H is completely symmetric under the exchange of any two particle coordinates, the symmetry of wave functions is preserved during the evolution, i.e., the random walk. Thus from any one of the ϕ_α, which is defined in a given domain, a wave function with the antisymmetry dictated by the chosen spin configuration can be formally constructed by summing the given ϕ_α over all the permutations P of the electrons, according to

$$\hat{\phi}_\alpha(\mathbf{R},s)=\sum_P(-)^P\phi_\alpha(P\mathbf{R},s) \ . \quad (4.27)$$

From the variational principle applied to the Hamiltonian H it follows that the variational energy associated with the above wave function satisfies

$$\frac{\int d\mathbf{R}\hat{\phi}_\alpha^* H\hat{\phi}_\alpha}{\int d\mathbf{R}\hat{\phi}_\alpha^*\hat{\phi}_\alpha}=\varepsilon_\alpha>E_0 \ , \quad (4.28)$$

where E_0 is the minimum eigenvalue among those relative to the eigenfunctions of H with the given symmetry, i.e., is the exact energy of the fermionic ground state, for the given spin configuration. On the other hand, from Eq. (4.26) and the Hermiticity of H it also follows that the local energy estimator, which was explicitly given in Eq. (4.8), tends asymptotically to ε_m. This completes the proof that the fixed-node approximation yields a variational upper bound to the exact ground-state energy of a fermionic system.

Before discussing to what extent it is possible to improve on the fixed-node approximation, it is worth noting a detail which may appear technical at this point, but will prove useful in what follows. In applying the above scheme, one is solving the Schrödinger equation by means of random walks separately in each of the domains fixed by the trial antisymmetric function ψ_T. In practice, the sign of ψ_T is needed so as to delete those walkers that change domains, or, better, to reject those moves that imply such crossings. It should be clear to the reader that, with regard to importance sampling, nothing changes if one defines a *guidance* function ψ_G to be positive everywhere and takes as distribution $f=\phi\psi_G$, provided that the sign of ψ_T is still used to locate the nodal surfaces. In fact, the fixed-node approximation, as presented above, corresponds to taking $\psi_G=|\psi_T|$. However, nothing changes in the foregoing discussion if one makes a different choice for ψ_G. We stress that, to yield an efficient sampling, ψ_G should, in any case, be a good approximation to the modulus of the exact ground state. This is equivalent to saying that ψ_G should not differ too much from $|\psi_T|$, since ψ_T was assumed from the start to be a good approximation to the exact ground state.

It was mentioned that the difficulties in applying a random-walk algorithm to the calculation of the Fermionic ground state arise from the fact that the corresponding wave function possesses positive and negative regions and therefore cannot be regarded as a walker's density. However, this difficulty is easily remedied, at least formally. The manner in which this can be done is as follows. One can simply regard positive and negative regions of the trial wave function as determining the positive densities of different objects, the white and black walkers. In fact, an antisymmetric wave function can always be written as the difference between two positive functions. The partitioning in white and black walkers mentioned above corresponds to a particular choice of such functions. It is clear from the linearity of Eqs. (4.1) and (4.14) that one can consider the evolution of each one of these functions separately. At each step of the iteration process yielding the evolution, the difference between the two functions will give an antisymmetric wave function (Kalos, 1984).

The approach outlined above to the problem of Fermi statistics is the essence of the so-called nodal relaxation method (Ceperley and Alder, 1980, 1984; Ceperley, 1981). However, while it seems relatively simple and straightforward, such a method turns out to be unstable when numerically implemented. The reason for this is clear. The two positive functions into which the antisymmetric wave function is divided are no longer antisymmetric. As such, they acquire a projection onto the symmetric Bose ground state, which has a lower energy than the Fermi ground state. As the evolution proceeds, the Bose component of each one of the functions grows exponentially with respect to the antisymmetric component. Thus, in principle, in taking the difference between the two functions, the Bose component should cancel out, leaving the antisymmetric Fermi component. In practice, because of the finite numerical precision, after a sufficient number of iterations the latter component is completely lost, and the difference between the two functions is just noise. This is a manifestation of the so-called *fermion sign problem.* In order to avoid the loss of the fermionic signal, one has to suitably restrict the evolution time. Therefore it is important that the trial wave function be very close to the exact Fermi ground state, since the closer it is the shorter the evolution time can be made. For the above reasons, the estimates of the Fermi ground state based on the nodal relaxation method have also been termed "transient estimates" (Kalos, 1984).

Before closing this necessarily incomplete presentation of the quantum Monte Carlo technique, we intend to give a very brief idea of the way the nodal relaxation method has been practically implemented (see, for instance,

Ceperley and Alder, 1984). To fix ideas, we shall consider the application of the method within the Green's-function formalism with importance sampling. Thus the starting point is Eq. (4.20), which gives the evolution of the distribution $f(\mathbf{R})=\phi(\mathbf{R})\psi_T(\mathbf{R})$. We shall, however, distinguish between a trial wave function, which provides the initial nodal surfaces, and a positive guidance function, to be used in the importance sampling. In this manner one can deal with the evolution of a positive distribution, for which a random-walk algorithm can still be used, and attach to each walker the appropriate sign at the end, in a way which depends on the number of crossings of the walker through the nodal surfaces of ψ_T.

Since the importance sampling is made according to ψ_G, it is useful to introduce a corresponding Green's function,

$$D(\mathbf{R},\mathbf{R}')=\psi_G(\mathbf{R})G(\mathbf{R},\mathbf{R}')\psi_G^{-1}(\mathbf{R}') , \tag{4.29}$$

and an appropriate weight function,

$$W(\mathbf{R})=\psi_T(\mathbf{R})\psi_G^{-1}(\mathbf{R}) . \tag{4.30}$$

The relation between the above Green's function and the one defined in Eq. (4.22) is

$$K(\mathbf{R},\mathbf{R}')=W(\mathbf{R})D(\mathbf{R},\mathbf{R}')W^{-1}(\mathbf{R}') . \tag{4.31}$$

If we maintain for f the definition $f=\phi\psi_T$ and introduce correspondingly a new distribution $g=\phi\psi_G$, relative to the guidance wave function, one can rewrite Eq. (4.20) as

$$g(\mathbf{R})=(E+V_0)\int d\mathbf{R}'D(\mathbf{R},\mathbf{R}')g(\mathbf{R}') \tag{4.32}$$

and, of course,

$$f(\mathbf{R})=W(\mathbf{R})g(\mathbf{R}) . \tag{4.33}$$

Equation (4.32) must be iterated, starting from a suitable initial distribution. The best that one can choose is to start from the fixed-node distribution $g_{FN}=\phi_{FN}\psi_G$, obtained, for instance, from a previous calculation with guidance function ψ_G. To this fixed-node distribution corresponds an initial g,

$$g_0(\mathbf{R})=\sigma_0(\mathbf{R})|g_{FN}(\mathbf{R})|\equiv\sigma_0(\mathbf{R})|g(\mathbf{R})| \tag{4.34}$$

where

$$\sigma_0(\mathbf{R})=W(\mathbf{R})/|W(\mathbf{R})|=\mathrm{sgn}[\psi_T(\mathbf{R})] . \tag{4.35}$$

From the computational point of view, one can just begin iterating Eq. (4.32) starting from $g=|g_0|$, and assigning to each new walker a *counter* in which is stored the sign σ_0 of the initial fixed-node parent. In this manner, only positive distributions enter the random walk. The final distribution $f=\phi\psi_T$ is then easily obtained from the walker population corresponding to the final g by assigning to each walker a weight and a sign according to $W(\mathbf{R})\sigma_0$, where $\mathbf{R}$ is the position of the walker and σ_0 is the sign associated with the position of the initial fixed-node parent, at the beginning of the random walk. This immediately follows from Eqs. (4.32)–(4.34).

Once the nodal relaxation has been performed, the fermionic ground-state energy can be evaluated via the local energy estimator, introduced with Eq. (4.8), which in the present case can be written as

$$E_0=\frac{\sum_i W(\mathbf{R}_i)\sigma_0 E_L(\mathbf{R}_i)}{\sum_i W(\mathbf{R}_i)\sigma_0} . \tag{4.36}$$

We stress once again that, in applying the nodal relaxation method to practical calculation, it is essential to start from a very good guess for the fermionic ground state, in order to minimize the length of the random walk associated with the relaxation of the nodes. In fact, during the evolution of the modulus of the wave function, the fermionic signal decreases in favor of the bosonic component. This leads, for instance, to an exponential growth in the variance of the energy with respect to its average value, as can be seen from the definition of average implied by the local energy estimator of Eq. (4.36) (Kalos, 1984; Schmidt and Kalos, 1984).

Finally, we wish to comment on the choice of the guidance wave function. ψ_G should be sufficiently close to $|\psi_T|$ to ensure an efficient sampling of configuration space, ψ_T having been chosen so as to be a good approximation to the exact ground state. However, in the case in which nodal relaxation is performed, other considerations come into play. In particular, within the framework of the diffusion Monte Carlo method, it appears from Eqs. (4.4) and (4.5) that a zero of the guidance function corresponds to an infinitely repulsive force acting on the walkers. This means that the walkers are pushed away from the nodal surfaces of ψ_T and cannot cross them. In other words, the statistical system associated with the walkers is nonergodic, in that the walkers cannot redistribute themselves among the various domains determined by the nodal surfaces. On the other hand, within the Green's-function Monte Carlo procedure, the vanishing of ψ_G yields an excessive branching in the walker generation (Ceperley and Alder, 1984). In either case, a simple way to remedy this inconvenience is to take $\psi_G(\mathbf{R})=|\psi_T(\mathbf{R})|[1+s(\mathbf{R})]$, where $s(\mathbf{R})$ is a nonnegative function which is very small away from nodal surfaces and vanishes at infinity. Near a nodal surface, however, $s(\mathbf{R})$ behaves in such a way as to make $\psi_G(\mathbf{R})$ small but finite. A reasonable choice of $s(\mathbf{R})$ is obtained by compromising between reduction of branching and efficiency of the sampling (Ceperley and Alder, 1984).

D. Monte Carlo computer experiments on phase transitions in uniform interacting-electron assembly

The ground-state energy of a uniform interacting-electron assembly has been computed by Ceperley and Alder (1980) with the diffusion Monte Carlo method described in Sec. IV.A. In particular, these authors have calculated the ground-state energy for four distinct phases of this system of charged particles, at various densities. They have considered (a) the unpolarized Fermi fluid, (b) the fully polarized Fermi fluid, (c) the Bose fluid, and (d) the Bose crystal on a bcc lattice. With this information, which can be regarded as the most accurate to date, they were able to predict the transitions between the various phases with reasonable accuracy.

For each phase, the calculations were performed by first generating fixed-node wave functions, or more precisely the corresponding random-walker populations, and then applying to such fixed-node distributions the nodal relaxation scheme discussed in Sec. IV.C. The trial wave functions for the Fermi phases were chosen as a product of Slater determinants, one for each spin-projection population, times a Jastrow factor ensuring the cusp condition as any two electrons approach each other. The Slater determinants were constructed from plane waves with the wave vector lying within the Fermi sphere. Of course, at given density the Fermi wave vector of the fully polarized electron fluid is $2^{1/3}$ larger than that of the unpolarized system. For the crystalline phase, the one-particle orbitals were chosen as Gaussians centered on the lattice sites with a width chosen variationally.

The analysis of the convergence of the nodal relaxation shows that in the present case the Hartree-Fock nodes, which were employed in the calculations, constitute a good approximation to the nodes of the exact wave function. In fact, it was found that the convergence of the relaxation process was relatively quick. The effect of the finite number of particles and finite time step on the results of the calculations was also systematically studied, and extrapolations to infinite number of particles and zero time step were performed. The systematic error originating from the finiteness of the *sample* was found to be one order of magnitude larger than the statistical error, in spite of the fact that the interactions between the particles and their images in the periodically extended space were taken into account with an Ewald summation procedure (see, e.g., Ceperley, 1978) to eliminate the major surface effect.

The results of Ceperley and Alder for the charged Fermi and Bose systems are fully summarized in Table I. A more direct interpretation of their findings is obtained

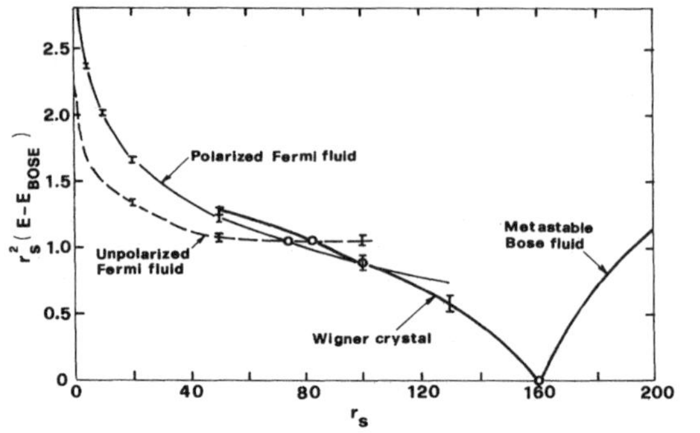

FIG. 7. Energy of the four phases studied relative to that of the lowest boson state times r_s^2 in rydbergs vs r_s in Bohr radii. Below $r_s=160$, the Bose fluid is the most stable phase; above, the Wigner crystal is most stable. The energies of the polarized and unpolarized Fermi fluid are seen to intersect at $r_s=75$. The polarized (ferromagnetic) Fermi fluid is stable between $r_s=75$ and $r_s=100$, the Fermi Wigner crystal above $r_s=100$, and the normal paramagnetic Fermi fluid below $r_s=75$. From Ceperley and Alder (1980).

from Fig. 7 where the quantity $r_s^2(E-E_{\mathrm{Bose}})$ is plotted against r_s. Thus the energy of each phase is referred to the Bose ground state. The two curves corresponding, respectively, to the paramagnetic ground state and to the Wigner crystal intersect when $r_s \approx 80$ Bohr radii. Interest in the fully polarized or ferromagnetic state was indicated by Bloch's work within the Hartree-Fock approximation (Bloch, 1928). Bloch's theory represents the simplest example of spin-density-functional theory. Of course, it is now well known that Hartree-Fock theory predicts too readily the existence of ferromagnetism. This is because it correlates parallel-spin electrons essentially correctly through the Fermi hole, whereas antiparallel-spin electrons are uncorrelated. Thus the energy of the fully ferromagnetic state is predicted more accurately than that of the paramagnetic state. In particular, Bloch's theory leads to ferromagnetism for $r_s>6$. This is only a slightly larger r_s than for metallic cesium,

TABLE I. Ground-state energy of the charged Fermi and Bose systems. r_s is the Wigner-sphere radius in units of Bohr radii. Energies are in rydbergs, and the digits in parentheses represent the error bar in the last decimal place. The four phases are the paramagnetic or unpolarized Fermi fluid (PMF); the ferromagnetic or polarized Fermi fluid (FMF); the Bose fluid (BF); and the Bose crystal with a bcc lattice (bcc).

r_s	PMF	FMF	BF	bcc
1.0	1.174(1)	· · ·	· · ·	· · ·
2.0	0.0041(4)	0.2517(6)	−0.4531(1)	· · ·
5.0	−0.1512(1)	−0.1214(2)	−0.21663(6)	· · ·
10.0	−0.10675(5)	−0.1013(1)	−0.12150(3)	· · ·
20.0	−0.06329(3)	−0.06251(3)	−0.06666(2)	· · ·
50.0	−0.02884(1)	−0.02878(2)	−0.02927(1)	−0.02876(1)
100.0	−0.015321(5)	−0.015340(5)	−0.015427(4)	−0.015339(3)
130.0	· · ·	· · ·	−0.012072(4)	−0.0123037(2)
200.0	· · ·	· · ·	−0.008007(3)	−0.008035(1)

the lowest electron-density metal. There is no sign of a tendency to ferromagnetism in the physical properties of this metal.

In fact, as a comparison of the curves from computer experiments displayed in Fig. 7 shows, the ferromagnetic state does not become stable with respect to the paramagnetic state until $r_s > 70$, indicating the vital importance of electron correlation in discussing this magnetic transition. This ferromagnetic state intersects the Wigner crystal at $r_s \approx 100$. Around such values of r_s the Bose and Fermi crystals differ in energy by an amount which is less than 1.0×10^{-6} rydbergs.

While Ceperley and Alder point out that their computer studies need refinement, there can be little doubt that the Wigner crystal becomes stable in the range $70 < r_s < 100$, and this, of course, is very valuable information. We notice that one area of obvious importance, if the Wigner electron crystal is to be unambiguously identified in three dimensions, is the question of the temperature at which melting of the electron crystal will occur.

E. Calculation of correlation energy for small molecules

Configuration-interaction calculations have been able, in the past, to account typically for about 80% of the correlation energy of molecules such as water (see, for example, Meyer, 1971 and Rosenberg and Shavitt, 1975). However, interesting chemistry occurs on an energy scale of only a fraction of the correlation energy. For example, the O-H bond strength in water is about 50% of the correlation energy. Thus the correlation energy computed using large CI wave functions differs from the exact (nonrelativistic, Born-Oppenheimer) energy by an amount of this same order of magnitude. Improving the CI results can be difficult, since convergence to the exact result is slow and can be nonuniform. Nevertheless, present-day state-of-the-art CI calculations yield whenever possible very accurate energy estimates (Mårtensson-Pendrill *et al.*, 1991) and to date still provide the yardstick against which all other calculations are measured. The numerical effort involved in CI calculations for a system of N electrons increases with a power of N which is between 4 and 5.

The quantum Monte Carlo method, at least in principle, appears to be free of the limitations inherent to an expansion procedure. In practice, in the absence of a stable algorithm to implement nodal relaxation, optimizing the nodes of the trial wave function becomes a crucial issue. Moreover, treating molecules with large nuclear charges Z by QMC may require very large computational times, even larger than in CI. In fact, though Reynolds *et al.* (1982) optimistically estimated that the computational effort needed in QMC would increase only as the third power of the number of electrons, subsequent and more careful estimates yield, for an atom with nuclear charge Z, a computational effort that increases either as $Z^{5.5}$ (Ceperley, 1986) or as $Z^{6.5}$ (Hammond *et al.*, 1987).

The quantum Monte Carlo method was developed and used primarily in the fields of nuclear and condensed-matter theory (cf. Sec. IV.D on jellium). However, subsequent chemical calculations have been performed (Anderson, 1975, 1976, 1980; Mentch and Anderson, 1981; Moskowitz and Kalos, 1981; Alder *et al.*, 1982; Moskowitz *et al.*, 1982a, 1982b; Ceperley and Alder, 1984; Moskowitz and Schmidt, 1986). In the following we shall briefly review some of these calculations. In particular, we shall illustrate the quality of calculations performed (a) with diffusion Monte Carlo, (b) with the Green's-function Monte Carlo, (c) within the fixed-node approximation, and (d) with allowance for nodal relaxation. We shall also comment again on limitations and possible developments of QMC.

The DMC method has been applied by Reynolds *et al.* (1982) to the calculation of the ground-state energy of some small molecules, H_2, LiH, Li_2, H_2O, within the fixed-node approximation. Trial wave functions of different sophistication were considered by these authors so as to show the importance of ψ_T on the efficiency of the sampling. All their importance functions were in the form of a product: a Slater determinant for each of the two groups of electrons with given spin projection times a correlation factor of Jastrow type. Of course, the Slater determinants were taken in such a way as to ensure the symmetry associated with the particular choice of the total spin projection, whereas the Jastrow factor was such as to reproduce the correct cusp behavior of the wave function as the electrons approach each other. They have considered three kinds of trial wave functions corresponding to Slater determinants constructed, respectively, from (a) a minimal basis set of Slater-type atomic orbitals, (b) a somewhat enhanced basis set and/or an optimized version of (a) and (c) localized Gaussian orbitals. In case (c) the Jastrow factor contained additional terms to reproduce also the cusp behavior associated with the electron-nuclear Coulomb attraction: this should have the effect of making the local energy associated with ψ_T even smoother.

In Table II, the fixed-node (FN) quantum Monte Carlo ground-state energy for some molecules is reported for the three choices of ψ_T listed above. Reported also for comparison are the energies obtained with the Hartree-Fock (HF) approximation, the best CI calculations, and the *exact* clamped nuclei or Born-Oppenheimer approximation, in the usual nonrelativistic framework afforded by the many-electron Schrödinger equation. All the energies are in hartrees. It is clear that, with the exception of the water molecule, in all cases the fixed-node energy accounts for most of the correlation energy. There are improvements in going from simpler to more sophisticated wave functions, so that with the best trial wave function (III) the fixed-node energy accounts for 95% or more of the correlation energy. It should be noted that in the case of the hydrogen molecule the ground-state wave function has no nodes. Therefore differences between fixed-node and exact energies of H_2, which are beyond

TABLE II. Total ground-state energy (in hartrees) of some small molecules (from Reynolds *et al.*, 1982). The figure in parentheses is the statistical error. The various symbols are explained in the text.

	H_2	LiH	Li_2	H_2O
HF	−1.1336	−7.987	−14.872	−76.0675
FN-I	−1.1745(8)	−8.047(5)	−14.985(5)	−76.23(2)
FN-III		−8.059(4)	−14.991(7)	−76.377(7)
FN-III	−1.174(1)	−8.067(2)	−14.990(2)	
Best CI	−1.1737	−8.0647	−14.903	−76.3683
Exact	−1.17447	−8.0699	−14.9967	−76.4376

the statistical error, give a measure of the numerical error associated with the use of a finite time step in the diffusion Monte Carlo calculations. In the case of the water molecule, it seems that none of the trial functions that were used is of especially good quality. In fact, in such a case only about 80% of the correlation energy is accounted for by the fixed-node calculations.

Reynolds *et al.* (1982) have also performed fixed-node calculations for the ground-state energy of the lithium diatomic molecule at various internuclear distances around the equilibrium one. The results of such calculations, based on a ψ_T of type II, are reported in Table III, together with the Hartree-Fock and the exact energy. It is found that 90% or more of the correlation energy is obtained also in this case.

As noted by Reynolds *et al.* (1982), the Born-Oppenheimer approximation, adopted throughout their work, can also be relaxed. This is achieved by allowing the nuclei, as well as the electrons, to diffuse. The diffusion constant for each nucleus is then $\hbar/2M$, where M denotes the nuclear mass. Thus the nuclei diffuse considerably more slowly than the electrons, and this makes the calculation longer.

Thus it is found that by using relatively simple trial functions ψ_T and making only a modest computational effort, one can obtain with the fixed-node QMC at least as much, and often more, of the correlation energy than proves possible by CI calculations to date, for simple molecules.

Fixed-node calculations for small molecules using the domain Green's-function (DGF) method have been performed by Moskowitz and Schmidt (1986). As these authors stress, the use of the DGF method is free of the systematic error introduced in the DMC by the finite time step. Here we shall just briefly comment on their results.

TABLE III. Ground-state energies (in hartrees) at selected nuclear separations for Li_2. The symbols are as in Table I. Typical statistical uncertainty on the fixed-node results is 0.005 hartrees.

r (bohr)	HF	FN-II	Exact
3	−14.786	−14.905	−14.915
4	−14.853	−14.968	−14.983
5.05 (equil.)	−14.872	−14.991	−14.997
6	−14.869	−14.985	−14.992
7	−14.859	−14.976	−14.982

Among the systems that they consider are the LiH molecule, the Be atom, and the BeH_2 molecule. With their calculations they are able to reproduce ground-state correlation energies within chemical accuracy, i.e., a few percent, when using the best trial functions. The same degree of accuracy is also found in the prediction of excitation energies for the Be atom and the energy barrier for insertion of Be in H_2.

It should be stressed once again that the quality of fixed-node energies depends on the capability of the chosen trial wave function to reproduce the nodal surfaces of the exact ground state. Thus, while the fixed-node approximation yields upper bounds to the exact energy, the systematic optimization of such results does not appear to be easy. In this respect the nodal relaxation technique, which was developed by Ceperley and Alder (1980) for the electron gas in the first instance, may prove valuable. One has to keep clearly in mind, though, that such a technique only gives transient estimates for the energy. In other words, if the relaxation is performed for times too long, it becomes intrinsically unstable, with the interesting signal that decreases exponentially whereas the noise remains constant. Thus the nodal relaxation technique crucially depends on the availability of a good starting guess for the ground-state wave function, and this may be provided by a preliminary (or, in actual practice, contemporary) fixed-node calculation.

Ceperley and Alder (1984) have performed further calculations on some of the small molecules previously considered with the DMC method, improving on them in two ways. They have changed to Green's-function Monte Carlo, utilizing a technique introduced by Ceperley (1983), to sample the exact Green's function, thus eliminating any error associated with the use of a finite time step. Further, they have implemented the nodal relaxation. In these new calculations they have also considered the H_3 molecule. Below, we shall briefly review their study as an example of how nodal relaxation can work. Since we have already described in some detail the main points of the nodal relaxation method, we shall merely present their results for the ground-state energy of the molecules considered. We shall also try to exemplify a bit the problem of the *convergence* of the nodal relaxation. For these calculations, with the exception of the H_2O molecule, trial functions of type III as used in the work of Reynolds *et al.* (1982) were employed. For H_2O a more sophisticated ψ_T was used. We refer the interested reader to the original paper by Ceperley and Alder (1984) for the details of their trial functions.

In Table IV we report the ground-state energy for LiH, Li_2, H_2O, and for three different configurations of H_3 denoted by H_3(I), H_3(II), and H_3(III). At first glance, the nodal relaxation appears to be able in all cases to bring down the fixed-node energy so as to agree with the exact energy. However, a few comments are necessary. It is true that the relaxed-node energies coincide with the exact ones within the statistical error. Nevertheless, the statistical error increases with increasing total energy. In addition, as has already been mentioned, the nodal relaxation can only provide transient estimates of the fermionic ground-state energy. Therefore it appears necessary to examine the convergence of the nodal relaxation in detail. To this end one can consider the total energy as a function of the generation number, starting from the first generation after the beginning of nodal release. Such curves for two typical cases are shown in Figs. 8–10. In Fig. 8 the relaxing of the total energy for LiH is shown. It is clear that in this case one can confidently speak of convergence. We stress that in the figure are also reported the results of runs in which the trial wave function was deoptimized to show how this destroys the convergence. The case of H_2O is illustrated in Fig. 9. Although the relaxed energy at the end of the run is in agreement with the exact energy, within the statistical uncertainty, there is no indication that convergence was reached, in that the slope of the energy curve does not seem to diminish.

A better way of studying the convergence of the nodal relaxation is to look at the energy difference between successive generations. Of course, a sign of convergence should be the vanishing of such a quantity after a sufficient number of generations. This quantity can also be evaluated more accurately than the total energy itself, as discussed by Ceperley and Alder (1984). In Fig. 10 the difference in release-node energy is shown for LiH. It is clear that with a good trial wave function the node relaxation can be considered to be converged. However, in the case of H_2O, reported in Fig. 11, it is not possible to say much, since the error bars increase too fast with the generation number even in the difference calculation.

The conclusion that can be drawn from the above discussion is that whereas the nodal relaxation appears to work well with light molecules, there are still problems

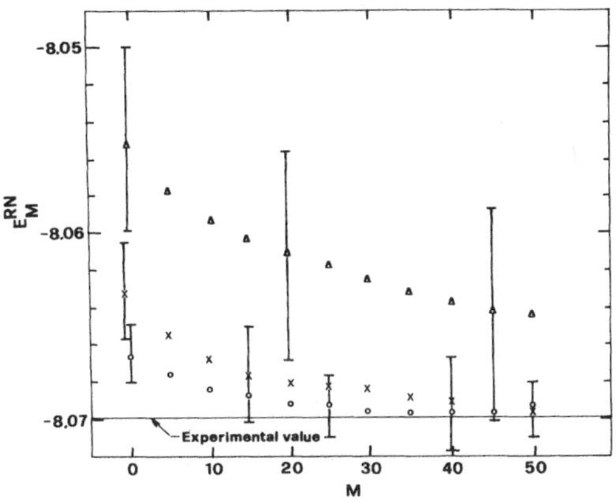

FIG. 8. Energy (in hartrees) vs the number of generations since node release for the molecule LiH. The results for three different trial functions are shown. (○) indicates the results obtained with the best trial function. The parameters in the other two trial functions were deliberately deoptimized to raise the fixed-node energy. From Ceperley and Alder (1984).

with the heavier ones. This is related to the fact that in the heavier molecules, because of the larger nuclear charges Z, the total energies are also larger and hence so are the error bars. Moreover, with increasing Z, the difference between the Bose and Fermi ground-state energy increases and, correspondingly, the rate (exponential) at which the Bose component obscures the fermionic one.

A number of ways to partially improve the QMC computations for heavier molecules are listed by Ceperley and Alder (1984). One way is based on a different treatment of inner electrons, possibly by means of pseudopotentials, so as to deal, in practice, with an equivalent problem with lower Z. Another way considers the possibility of deleting all the random walks that frequently cross the nodes. They also cite the possibility of directly calculating energy differences by means of correlated random walks. To date, practical calculations of heavier atoms and molecules, within the fixed-node approximation, have mainly exploited the separation of electrons in core and valence, followed by a different treatment of the two kinds of electrons (Hammond *et al.*, 1987; Hurley and Christiansen, 1987; Hammond *et al.*, 1988; Yoshida and Iguchi, 1988; Bachelet *et al.*, 1989). The use of pseu-

TABLE IV. Comparison of fixed-node and relaxed-node ground-state energies (in hartrees) with CI and exact results. RN indicates the relaxed-node energy, and the other symbols are as in Tables I and II. Δ is the difference $E_{FN}-E_{RN}$.

Molecule	FN	RN	Δ	Exact	CI
H_3(I)	−1.6581(3)	−1.6591(1)	0.0009(2)	−1.65919	−1.65876
H_3(II)	−1.6239(3)	−1.6244(3)	0.0005(2)	−1.62451	−1.62337
H_3(III)	−1.6606(2)	−1.6617(2)	0.0011(2)	−1.66194	−1.66027
LiH	−8.067(1)	−8.071(1)	0.004(1)	−8.0705	−8.0690
Li_2	−14.990(2)	−14.994(2)	0.004(1)	−14.9967	−14.903
H_2O	−76.39(1)	−76.43(2)	0.04(1)	−76.437	−76.368

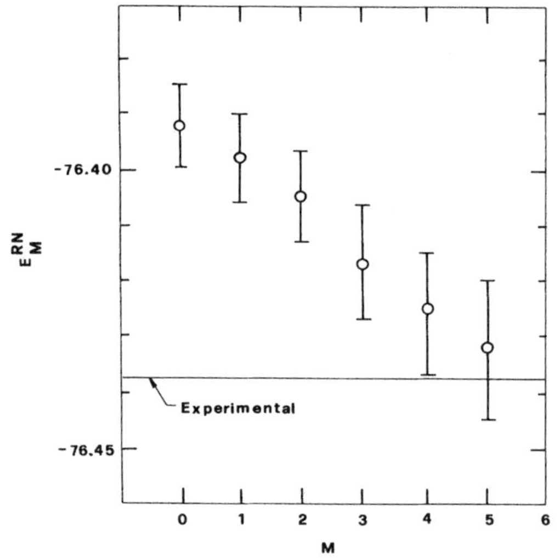

FIG. 9. Energy (in hartrees) vs the number of generations since node release for the molecule H_2O. From Ceperley and Alder (1984).

dopotentials to achieve the core-valence separation introduces, however, a new complication: good pseudopotentials are nonlocal, and this is in conflict with key aspects of QMC. To overcome this complication, workers have resorted to two tricks: either the nonlocal potentials are *made local* by suitable approximations (Hammond *et al.*, 1987; Hurley and Christiansen, 1987; Yoshida and Iguchi, 1988), or they are transformed into local ones by introducing appropriate pseudo-Hamiltonians (Bachelet *et al.*, 1989). A different approach has also been proposed and tested, in which core and valence electrons are both described in terms of wave functions, with the core, however, being treated variationally while the important valence electrons are treated by QMC (Hammond *et al.*, 1988).

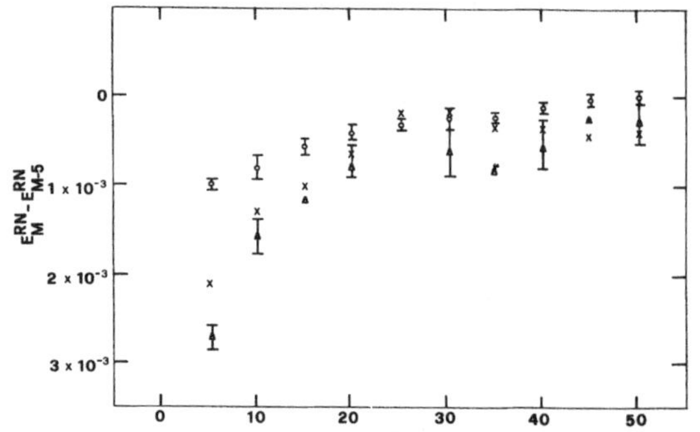

FIG. 10. Change in release-node energy (in hartrees) every five generations since node release for the molecule LiH. From Ceperley and Alder (1984).

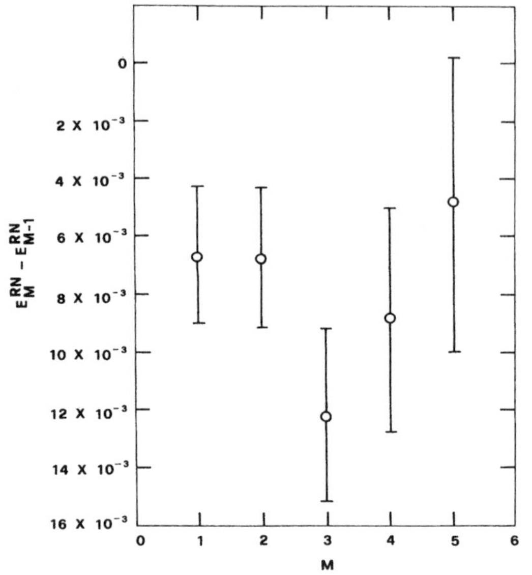

FIG. 11. Change in release-node energy (in hartrees) every five generations since node release for the molecule H_2O. From Ceperley and Alder (1984).

The severe limitation imposed on QMC calculations by the big increase with Z of the computation time remains an open problem, even though improved diffusion Monte Carlo algorithms (Umrigar *et al.*, 1991) may allow time steps considerably larger than in the past, thus partially alleviating the slowing down of calculations. In fact, the attempts mentioned above to overcome such a slowing down with a core-valence separation still appear far from developing into standard, transferable algorithms. On the other hand, even though the use of sophisticated, specifically tailored wave functions may allow very accurate predictions of the correlation energy in fixed-node calculations (Umrigar *et al.*, 1991), nodal relaxation remains very difficult to implement, because of the lack of a stable algorithm. Thus the fermion sign problem and the computational slowing down for species with large Z remain open challenges for the quantum Monte Carlo treatment of many-electron systems.

F. Auxiliary-field quantum Monte Carlo method and Hubbard model

The central idea of the auxiliary-field quantum Monte Carlo (AFQMC) method, as we have already anticipated, is to exactly rewrite the propagator of a many-particle system—with two-body interactions—in terms of a propagator for independent particles interacting with auxiliary external fields. While this procedure introduces the need for averaging over the values taken by such extra fields, it permits the use of well-established techniques to treat the propagator for independent particles. The strategy is relatively simple and contains three main steps: (a) find the appropriate transformation that replaces the coupling between the particles with a coupling to suitably chosen classical fields; (b) solve the new independent particle problem; i.e., calculate the trace (partition function) or a suitable matrix element of the propagator; (c) perform the average over the auxiliary fields, usually with techniques borrowed from classical statistical mechanics. For the sake of simplicity, rather than considering the general case of an arbitrary Hamiltonian with quadratic interactions (see, e.g., Sugiyama and Koonin, 1986; Negele and Orland, 1988), here we shall concentrate on the specific case of the Hubbard model, for which we shall then discuss some applications.

The basic relation that allows for the introduction of the auxiliary fields is the Hubbard-Stratonovich (HS) identity (Stratonovich, 1957; Hubbard, 1959) valid for any Hermitian operator $\hat{O}$,

$$e^{(1/2)\alpha^2\hat{O}^2}=\frac{1}{\sqrt{2\pi}}\int_{-\infty}^{\infty}dx\, e^{-(1/2)x^2}e^{-\alpha x\hat{O}}\ , \tag{4.37}$$

or appropriate extension of it (see below). The Hubbard Hamiltonian of Eq. (4.10) contains a one-body term, $K=\sum_{ij\sigma}t_{ij}a^{+}_{i\sigma}a_{j\sigma}$, and a two-body interaction $V=U\sum_i n_{i\uparrow}n_{i\downarrow}$. Therefore it appears to be a good candidate for the HS transformation. However, the propagator $e^{-\beta H}$ contains the sum H_0+V, rather than merely V. Here $H_0=K-\mu\sum_{i\sigma}a^{+}_{i\sigma}a_{i\sigma}$. The trick for circumventing this difficulty is to resort to the Trotter formula,

$$\begin{aligned}e^{-\beta(H_0+V)}&=\prod_{r=1}^{L}e^{-\varepsilon(H_0+V)}\\&=\prod_{r=1}^{L}e^{-\varepsilon H_0}e^{-\varepsilon V}+O(\varepsilon)\ ,\end{aligned} \tag{4.38}$$

with $\varepsilon=\beta/L$, so that the exponential of V appears explicitly. In fact, V still needs some rearranging. One possibility is to rewrite $e^{-\varepsilon V}$—in each of the time slices generated by the Trotter formula—using the relation (Hirsch, 1983)

$$n_{i\uparrow}n_{i\downarrow}=-\tfrac{1}{2}(n_{i\uparrow}-n_{i\downarrow})^2+\tfrac{1}{2}(n_{i\uparrow}+n_{i\downarrow})\ , \tag{4.39}$$

which is valid for fermions ($n^2_{i\sigma}=n_{i\sigma}$). Then, use of the HS transformation in Eq. (4.38) and elementary manipulations yield the propagator in a new form,

$$\begin{aligned}U(\beta)&=e^{-\beta H}\\&=\left[\frac{\varepsilon}{2\pi}\right]^{NL/2}\\&\quad\times\int\prod_{r=1}^{L}\prod_{i=1}^{N}dx_{ri}e^{-(\varepsilon/2)x^2_{ri}}\prod_{r'=1}^{L}e^{-\varepsilon h_{r'}(x)}\ ,\end{aligned} \tag{4.40}$$

valid to order ε, with

$$h_r(x)=H_{0r}+\sum_i\left[\sqrt{U}x_{ri}(n_{ri\uparrow}-n_{ri\downarrow})+\frac{U}{2}(n_{ri\uparrow}+n_{ri\downarrow})\right] \tag{4.41}$$

and N the number of sites of the finite lattice on which the Hubbard model is considered. It is evident that the Hamiltonian $h_r(x)$ contains only one-body operators: in particular, it describes independent particles moving in an external field $\sqrt{U}x_{ri}$. Thus, through a Gaussian averaging, the imaginary-time propagator $U(\beta)$ is related to a new propagator for independent particles,

$$U_\varepsilon(\beta,x)=\prod_{r=1}^{L=\beta/\varepsilon}e^{-\varepsilon h_r(x)}\ . \tag{4.42}$$

A number of comments are in order. The accuracy of Eq. (4.40) increases with increasing $L=\beta/\varepsilon$, and this alternative representation of the propagator only becomes exact in the limit $\varepsilon\to 0$ ($L\to\infty$). However, one can study systematically such a convergence and accordingly estimate the systematic error introduced by a finite Trotter time ε. The reader will have noted the presence in Eqs. (4.40) of a time index denoted by r. This new labeling simply keeps track of the Trotter time slice.

As we have anticipated, quantities that one typically wants to calculate are the partition function $Z=\mathrm{Tr}U(\beta)$ or a *projected* partition function $Z_T=\langle\Phi_T|U(\beta)|\Phi_T\rangle$, where Φ_T is a trial wave function nonorthogonal to the ground state. In either case, for $\beta\to\infty$ one obtains a limiting behavior $Z(Z_T)\propto e^{-\beta E_0}$. Thus one can obtain E_0 as

$$E_0=-\lim_{\beta\to\infty}\frac{1}{\beta}Z(\beta)=-\lim_{\beta\to\infty}\frac{1}{\beta}Z_T(\beta)\ . \tag{4.43}$$

From Eqs. (4.40) and (4.41) it is clear that the evaluation of Z requires that of

$$\tilde{Z}(x)=\mathrm{Tr}U_\varepsilon(\beta,x)\ . \tag{4.44}$$

This can be evaluated (Blankenbecler *et al.*, 1981) in terms of a determinant involving the matrices associated with the Hamiltonians $h_r(x)$ of Eq. (4.41). Once $\tilde{Z}(x)$ is known, the problem of calculating Z is reduced to that of evaluating a multidimensional integral over the variables $\{x_{ri}\}$. In fact, one readily obtains

$$Z=\int dx\, e^{-\beta\tilde{U}(x)}\mathrm{sgn}(\tilde{Z}(x))\ , \tag{4.45}$$

where

$$\int dx\equiv\left[\frac{\varepsilon}{2\pi}\right]^{NL/2}\int\prod_{r=1}^{L}\prod_{i=1}^{N}dx_{ri} \tag{4.46}$$

and

$$\tilde{U}(x)\equiv\frac{1}{2L}\sum_{r=1}^{L}x^2_{ri}-\frac{1}{\beta}\ln|\tilde{Z}(x)|\ . \tag{4.47}$$

Thus the problem of calculating the partition function for a quantum system is reduced to the classical problem of calculating a configurational integral. We shall merely note that numerical problems may, however, arise from the possible presence of nodal surfaces in the function $\tilde{Z}(x)$. These can give (a) problems of ergodicity in sampling the integral of Eq. (4.44) as well as (b) problems of signal-to-noise ratio if Z results from the cancellation of large contributions of opposite sign. This is a manifestation of the so-called fermion sign problem.

The considerations above, and, in particular, Eqs. (4.44)–(4.46), remain valid if one calculates a projected partition function—the only difference being that Z and $\tilde{Z}(x)$ are replaced by Z_T and $\tilde{Z}_T(x) = \langle \Phi_T | U_\varepsilon(\beta,x) | \Phi_T \rangle$, respectively. Appropriate techniques are available (Sugiyama and Koonin, 1986; see also Sorella, 1989) to evaluate $Z_T(x)$. With manipulations similar to those reviewed above, it is also possible to reduce the calculation of relevant correlation functions to that of pseudoclassical averages. It seems proper at this point to add that, particularly for the Hubbard Hamiltonian, it has proved possible (Hirsch, 1983) to write a discrete transformation of the HS type, i.e.,

$$e^{-\varepsilon U n_{i\uparrow} n_{i\uparrow}} = \frac{1}{2} \sum_{\sigma=\pm 1} e^{-\varepsilon[J\sigma(n_{i\uparrow}-n_{i\downarrow})+U(n_{i\uparrow}+n_{i\downarrow})/2]} , \tag{4.48}$$

with $\cosh(\varepsilon J) = e^{\varepsilon U/2}$. This transformation maps the quantum Hubbard problem onto an equivalent classical problem where one has to sample over Ising variables, rather than continuous ones. According to Hirsch (1986), in calculations using the transformation of Eq. (4.47), the sign problem is not severe.

To give a practical illustration of the AFQMC, we shall discuss recent work on the 2D Hubbard model, stimulated by its possible relevance to the high-T_c superconductors. We shall not attempt here to give an exhaustive review, and therefore the interested reader is urged to consult the original papers mentioned below for a detailed discussion and further references.

Hirsch and Tang (1989) have studied the 2D Hubbard model with nearest-neighbor hopping for lattices as big as 8×8 and for imaginary times up to $\beta=20$, in units in which the hopping energy $-t=-1$. They have performed AFQMC simulations based on the sampling of the partition function Z for different values of the coupling U and of the filling ρ—defined as the ratio between the number of electrons N_e and the number of sites in the lattice N, $\rho = N_e/N$. Because of electron-hole symmetry, one can take $\rho \le 1$. The doping can be suitably defined as $\delta = 1-\rho$. Since the lattice can accommodate $2N$ electrons, $\rho=1$ corresponds to half filling and, similarly, $\rho=0.5$ to a quarter filling. Hirsch and Tang have looked especially into the magnetic properties of the system, which can be characterized by the magnetic structure factor

$$S(\mathbf{q}) = \frac{1}{N} \sum_{ij} e^{i\mathbf{q}\cdot(\mathbf{l}_i - \mathbf{l}_j)} \langle (n_{i\uparrow} - n_{i\downarrow})(n_{j\uparrow} - n_{j\downarrow}) \rangle \tag{4.49}$$

and, in particular, by its value at the wave vector $\mathbf{q}_m = (\pi,\pi) \equiv \pi$. They find that at half filling the magnetic structure factor exhibits a sharp peak at $\mathbf{q}_m$. Such a peak, which is shown in Fig. 12, tends to become more and more pronounced with increasing lattice size or coupling U. Also apparent—from the same figure—is the saturation of $S(\pi)$ with increasing simulation time β. In fact, as the time becomes shorter, the drop in $S(\pi)$ becomes more and more pronounced for larger lattices and couplings. The presence of such a peak in $S(\mathbf{q})$ is interpreted as a sign of antiferromagnetic long-range order. As they move from $\rho=1$, they find a rapid suppression of such an ordering, as can be seen from Fig. 13, and they conjecture that the antiferromagnetic order disappears immediately away from half filling. Hirsch and Tang also find that at low temperatures (long imaginary times) the properties of the 2D Hubbard model at half filling are well described by a spin-wave theory with renormalized local moment and spin-wave velocity.

The results of Hirsch and Tang are consistent with those of recent MC simulations (Reger and Young, 1988) of the two-dimensional antiferromagnetic Heisenberg model with which the two-dimensional Hubbard model at half filling is equivalent for large U. Moreover, they are fully confirmed by independent investigations on the Hubbard model. In Fig. 14 we show the results of White *et al.* (1989) for the spin-spin correlation function

$$c(l_x, l_y) = \frac{1}{N} \sum_i \langle m(\mathbf{l}_i + \mathbf{l}) m(\mathbf{l}_i) \rangle , \tag{4.50}$$

where $m(\mathbf{l}_i)$ is the local magnetic operator along the z axis. Clearly, while staggered spin correlations are clearly visible for half filling, these are nearly absent for the quarter-filled case. In fact, results of Sorella (1989), obtained using the projected partition function technique and Langevin dynamics to sample the pseudoclassical partition function, show that, even with a doping δ of only 12%, such correlations are practically absent. Thus

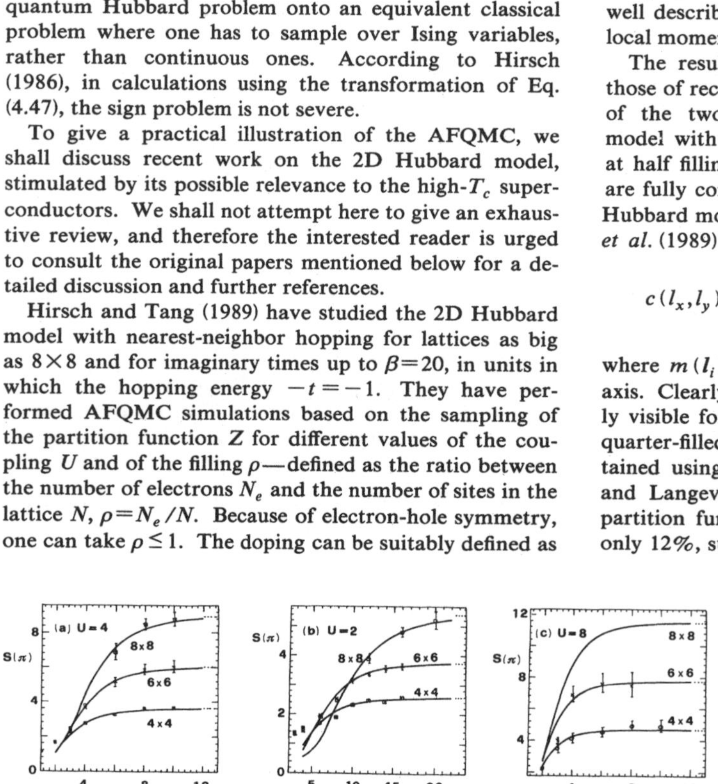

FIG. 12. Magnetic structure factor for $S(\pi) \equiv S[\mathbf{q}=(\pi,\pi)]$ vs imaginary time β for lattices of size 4×4, 6×6, and 8×8, at half filling. The solid lines are results of spin-wave theory (Hirsch and Tang, 1989). The dashed lines at the right side indicate the $T=0$ limits of the spin-wave results. From Hirsch and Tang (1989).

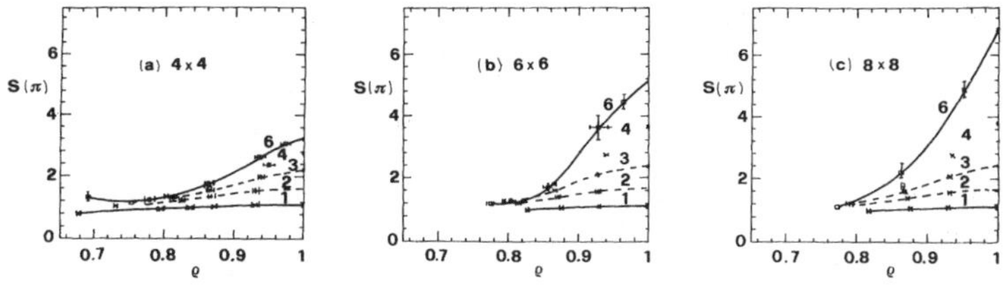

FIG. 13. Magnetic structure factor for $S(\pi)\equiv S[\mathbf{q}=(\pi,\pi)]$ vs band filling for $U=4$, with 4×4, 6×6, and 8×8 lattices at various imaginary times. The numbers next to the curves indicate β; the curves are drawn through the points to guide the eye. From Hirsch and Tang (1989).

the conjecture of Hirsch and Tang on the suppression of antiferromagnetic order away from half filling seems fully confirmed.

White *et al.* have also presented results for the single-particle momentum occupation $n(\mathbf{k})\equiv\langle n_{\mathbf{k}\sigma}\rangle=\langle a^{\dagger}_{\mathbf{k}\sigma}a_{\mathbf{k}\sigma}\rangle$. As can be seen from Fig. 15, the one-electron momentum distribution for the interacting electrons at half filling is broadened more than at a quarter filling, with respect to the noninteracting-electron case. This is also confirmed by a subsequent study by Moreo *et al.* (1990), which in addition, from an investigation of the compressibility and of the electron self-energy, provides evidence for a one-electron gap at half filling. Of course, this is consistent with the presence of antiferromagnetic long-range order. On the contrary, there is no evidence of a gap at a quarter filling where the system appears to be in a Fermi-liquid state—as is also suggested by the sharpness of the one-electron momentum distribution which more closely resembles that for noninteracting electrons. Thus one can conclude that in two dimensions the Hubbard model predicts antiferromagnetic long-range order for any finite U at half filling behaving like an insulator, whereas such order disappears on moving away from $\rho=1$ and the system behaves rather like a paramagnetic Fermi liquid.

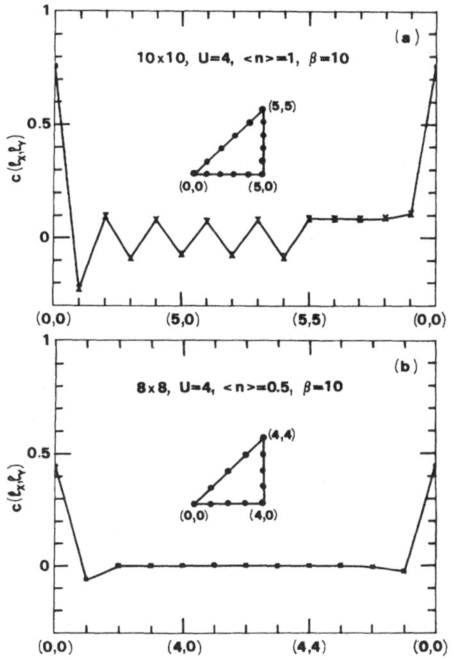

FIG. 14. Spin-spin correlation function $c(l_x,l_y)$. The horizontal axis traces out the triangular path seen in the center of the figure. Strong antiferromagnetic correlations are visible in (a), which is for a half-filled band ($\langle n\rangle\equiv\rho=1.0$), but are nearly absent in (b), which is at quarter filling ($\langle n\rangle\equiv\rho=0.5$). From White *et al.* (1989).

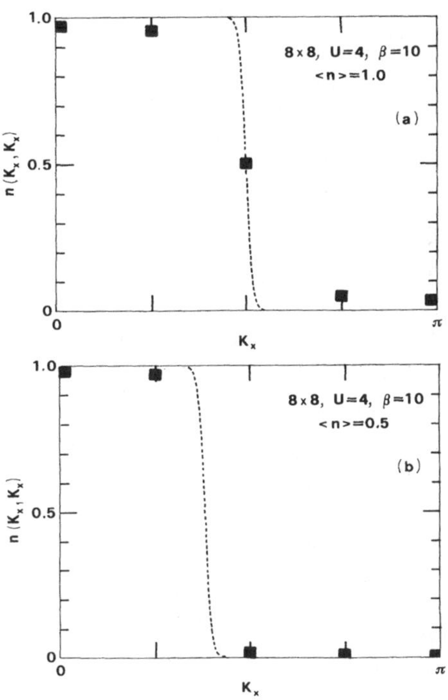

FIG. 15. Momentum distribution $n(\mathbf{k})$ for an 8×8 lattice with $U=4$ and $\beta=10$ at (a) half filling ($\langle n\rangle\equiv\rho=1.0$) and at (b) quarter filling ($\langle n\rangle\equiv\rho=0.5$). The dashed curves are the $U=0$ results. From White *et al.* (1989).

V. SUMMARY AND FUTURE DIRECTIONS

The major topics discussed in this article fall into two categories: (a) full many-body treatments of molecules and solids, and (b) simplified models, in which, however, strong correlations can be accommodated, at least in principle.

These two categories overlap in important areas. Thus (a) includes the jellium model of a metal which, in spite of its simplicity, has provided the basis for the development of the original Thomas-Fermi-Dirac method (see Gombás, 1949; March, 1957) into the theory of the inhomogeneous electron gas or, more formally, density-functional theory (Hohenberg and Kohn, 1964). Within this framework, fairly realistic calculations can now be performed on properties determined by the ground-state charge density alone for metallic, semiconducting, and insulating crystals. These calculations build the jellium results into the exchange and correlation contribution to the one-body crystal potential. While this treatment, which follows closely the pioneering work of Slater (1951) and was formalized into the current approach by Kohn and Sham (1965), is by now widely used in extensive systems, it has obvious limitations due to the forcing of the many-electron problem into a one-electron mold. One of these, the gap problem in insulators and semiconductors, is so severe that the correction to the one-body potential band gap is the order of the band gap itself (see, for example, Sham and Schlüter, 1983). To correct this, one must go back to a fully many-electron approach involving the nonlocal mass operator (Pickett, 1986; Godby *et al.*, 1988; Fiorentini and Baldereschi, 1992).

Having dealt with the way the results of the jellium model can be built into realistic calculations on molecules and solids, we turn to the second class, under (b), which is provided by Hubbard- and related Heisenberg-model Hamiltonians. These are designed to treat situations in which very strong electron-electron correlations tend to bring about site localization of electrons. Here the basic idea, at the simplest level, is to keep antiparallel electrons apart by imposing an energy penalty, U, for allowing two electrons with antiparallel spins simultaneously on a given atomic site. It is interesting that the variational wave-function approaches of Coulson and Fischer for H_2 from the quantum-chemical angle, and subsequently Gutzwiller from the standpoint of low-order density-matrix theory, address what amounts to essentially the same basic point: how to reduce the weight of ionic configurations in a molecular-orbital or band-theory approach. The strong interplay between chemical and physical points of view should be clear from the account of Sec. III, where a local approach to correlation in molecules is also given some prominence. We shall return to this interplay below.

But before doing so, we must stress here, following Sec. IV, that the very accurate correlation energy calculations have come from quantum Monte Carlo computer techniques. However, crucial input into such calculations is a trial wave function. For the jellium model, which was simulated by the above technique by Ceperley and Alder (1980), the trial function was a Slater determinant of plane waves, multiplied by a product of pairs (Bijl-Jastrow) wave function. The analytical efforts expended earlier on such wave functions, which unfortunately almost always had to invoke relatively uncontrolled approximations to allow the calculation of the appropriate low-order density matrices determining the energy, have been brought to fruition through the addition of the quantum Monte Carlo technique. But as well as the jellium results, correlation energies of small molecules are also reported in Sec. IV, together with some numerical solutions of the two-dimensional Hubbard model.

This prompts us to return to the theme raised above as to the fruitful interplay between physical and chemical ideas in treating realistic many-electron systems. This was already apparent in the early and originally largely parallel developments in the quantum chemistry of the polyenes, which anticipated, through the recognition of alternating single and double bonds (see, for example, the survey for Murrell, 1971), what solid-state physicists today call "Peierls's theorem," namely, that one-dimensional metals cannot exist. Later the exact solution of the one-dimensional Hubbard Hamiltonian by Lieb and Wu (1968; see also Kotrla, 1990) was a significant step in the study of strong correlations, though it did not allow for bond alternation. However, this has now been done in the quantitative, though admittedly approximate, work of Kajzar and Friedel (1987). In this same context, the high-T_c superconductors (see, for example, Micnas *et al.*, 1990) have had, as one by-product, to bring Pauling's theory of resonating valence bonds (RVB) back into fashion, as briefly summarized in Sec. III.E.

But, in more general terms, one question that surely remains at the heart of the theory of the high-T_c ceramic oxide superconductors is whether the chemistry of these very specific systems can in fact be subsumed within the parameter space of the currently popular two-dimensional Hubbard model. If this proved to be the case, then the system specificity already referred to must mean that high-T_c superconductivity can only be obtained in a very tiny portion of this parameter space. Certainly, the recent discussion of Anderson (1990) suggests that, already, considerable new insight is coming from low-dimensional Hubbard models, when solved in a highly accurate manner via Tomonaga-Luttinger-liquids theory. This approach, going back to Luther (1979) and taken forward in a major way by Haldane (1981), promises to have a lot to say that will be, at least, highly relevant to the normal state of high-T_c superconductors. The main point to stress here is that careful many-body analysis of interacting fermion systems reveals the possibility of two fixed points. One is the well-established Landau-Fermi-liquid theory. In this theory the interaction parameters are marginal operators around a single fixed point: essentially the free Fermi liquid [Anderson (1990), who refers to the work of Benfatto and Gallavotti

(1990)]. The Luttinger liquid was defined by Haldane (1981), who showed that a large variety of one-dimensional quantum fluids could all be solved by common techniques based on transforming to phase and phase-shift variables for the Fermi-surface excitations. As Anderson (1990) emphasizes, these systems are characterized by fractionation of quantum numbers and, often, a Fermi surface with nonclassical exponents. He argues that the Luttinger liquid is a fixed point, of the same renormalization group that "usually" yields the Landau-Fermi liquid as a unique fixed point. This Luttinger-liquid state, according to Haldane (1981), already embraces a large class of interacting one-dimensional systems and, Anderson has argued, should also include some two-dimensional systems in which the band spectrum is bounded above, i.e., systems with Mott-Hubbard gaps and an upper Hubbard band. Anderson makes this approach the basis for a theory that appears to be useful in calculating normal-state, and some superconducting, properties of high-T_c superconductors.

The fact that charge and spin acquire distinct spectra in the Haldane-Anderson approach, plus the excitement surrounding the possible role of particles with fractional statistics in two dimensions, the anyons of Wilczek, following the demonstration by Kalmeyer and Laughlin (1987) that a gas of anyons has a superconducting ground state, mean (see also Halperin *et al.*, 1989) that many-electron theorists have truly major challenges ahead. Since the early books by Thouless (1961) and by March *et al.* (1967) on many-body theory, the area of application of many-electron techniques has expanded hugely. The correlation problem was important in the early days (see Wigner, 1934, 1938), but now it has moved to the center of the stage. Whether present models are rich enough to embrace much chemistry underlying the attractive interactions between holes in the ceramic oxide superconductors [see, however, the work of Callaway *et al.* (1990) on small cluster calculations using the Hubbard model], the framework existing now for many-body studies surveyed in the present article should be flexible enough to increasingly embrace realistic systems and, in the high-T_c materials, to eventually incorporate band-structure effects in the specific ceramic oxides, as well as the long-range Coulomb repulsion between holes, which seems to be missing from current studies of the Cooper pair binding.

APPENDIX A: MODEL OF TWO-ELECTRON HOMOPOLAR MOLECULE

Here we shall summarize the results of the model of Falicov and Harris (1969) of a two-electron homopolar molecule.

To define their one-band Hamiltonian, it is convenient to use second quantization operators and to restrict the model to four orbitals, one of each spin in each of the two centers. The Hamiltonian then takes the form

$$H=H_\alpha+H_\beta+H_u+H_K \tag{A1}$$

where

$$H_\alpha=\alpha(n_{1\uparrow}+n_{1\downarrow}+n_{2\uparrow}+n_{2\downarrow}) , \tag{A2}$$

$$H_\beta=\beta(c^\dagger_{1\uparrow}c^\dagger_{2\uparrow}+c^\dagger_{1\downarrow}c_{2\downarrow}+c^\dagger_{2\uparrow}c_{1\uparrow}+c^\dagger_{2\downarrow}c_{1\downarrow}) , \tag{A3}$$

$$H_u=U(n_{1\uparrow}n_{1\downarrow}+n_{2\uparrow}n_{2\downarrow}) , \tag{A4}$$

$$H_K=K(n_{1\uparrow}n_{2\downarrow}+n_{1\uparrow}n_{2\downarrow}+n_{1\downarrow}n_{2\uparrow}+n_{1\downarrow}n_{2\downarrow}) . \tag{A5}$$

Here $c^\dagger_{i\sigma}$, $c_{i\sigma}$, and $n_{i\sigma}=c^\dagger_{i\sigma}c_{i\sigma}$ are, respectively, creation, annihilation, and number operators for the orbital of spin centered on nucleus i. H_α and H_β are, respectively, the single-particle diagonal and off-diagonal terms; H_u is the intra-atomic Coulomb repulsion, while H_k is the corresponding interatomic term. The parameters β, U, and K are positive-definite quantities, such that $U>K$.

Since one is considering only two-electron states, the following results can be exploited:

(a) The electron-number operator

$$N=\sum_{i\sigma} n_i\sigma \tag{A6}$$

is completely diagonal and can be replaced everywhere by the number 2; e.g.,

$$H_\alpha=2\alpha . \tag{A7}$$

(b) The operator

$$N^2=\sum_{ij\sigma\sigma'} n_{i\sigma}n_{j\sigma'} \tag{A8}$$

is also completely diagonal and can be replaced everywhere by the number 4.

(c) Recalling that

$$n_{i\sigma}n_{i\sigma}\equiv n_{i\sigma} \tag{A9}$$

for any i and σ, we can utilize (a) and (b) immediately above to write the identity

$$\begin{aligned}4=2&+2(n_{1\uparrow}n_{1\downarrow}+n_{2\uparrow}n_{2\downarrow})\\&+2(n_{1\uparrow}n_{2\uparrow}+n_{1\uparrow}n_{2\downarrow}+n_{1\downarrow}n_{2\uparrow}+n_{1\downarrow}n_{2\downarrow}) .\end{aligned} \tag{A10}$$

(d) The Hamiltonian can be rewritten

$$H=2\alpha+K+\mathcal{H} \tag{A11}$$

with

$$\mathcal{H}=H_\beta+H_u+(H_k-K) , \tag{A12}$$

where the final term can be written

$$H_K-K=-K(n_{1\uparrow}n_{1\downarrow}+n_{2\uparrow}n_{2\downarrow})=-(K/U)H_u . \tag{A13}$$

From the above expressions, or alternatively from simple physical arguments, it can be readily seen that the eigenvalues of H can depend on U and K only through the difference $U-K$. This permits one to write the ground-state energy E as

TABLE V. Matrix elements of H.

	$\vert 1\uparrow 2\uparrow\rangle$	$\vert 1\downarrow 2\downarrow\rangle$	$\vert 1\uparrow 1\downarrow\rangle$	$\vert 2\uparrow 2\downarrow\rangle$	$\vert 1\uparrow 2\downarrow\rangle$	$\vert 2\uparrow 1\downarrow\rangle$
$\langle 1\uparrow 2\uparrow\vert$	0	0	0	0	0	0
$\langle 1\downarrow 2\downarrow\vert$	0	0	0	0	0	0
$\langle 1\uparrow 1\downarrow\vert$	0	0	$U-K$	0	$-\beta$	$-\beta$
$\langle 2\uparrow 2\downarrow\vert$	0	0	0	$U-K$	$-\beta$	$-\beta$
$\langle 1\uparrow 2\downarrow\vert$	0	0	$-\beta$	$-\beta$	0	0
$\langle 2\uparrow 1\downarrow\vert$	0	0	$-\beta$	$-\beta$	0	0

$$E=\langle H\rangle=2\alpha+K+\epsilon\ , \tag{A14}$$

where

$$\epsilon=\langle\mathcal{H}\rangle=\beta\epsilon(x) \tag{A15}$$

with ϵ only dependent on the variable $x=(U/K)/\beta$.

We now discuss the exact ground state.

Any eigenstate of Eq. (A12) with two electrons should be a linear combination of the six states

$$|1\uparrow 2\uparrow\rangle,\ \ |1\downarrow 2\downarrow\rangle,\ \ |1\uparrow 1\downarrow\rangle\ ,$$

$$|2\uparrow 2\downarrow\rangle,\ \ |1\uparrow 2\downarrow\rangle,\ \ |2\uparrow 1\downarrow\rangle,$$

where

$$|i\sigma j\sigma'\rangle=c_{i\sigma}^{+}c_{j\sigma'}^{+}|0\rangle \tag{A16}$$

with $|0\rangle$ as the vacuum state. The matrix elements of H in this manifold are given in Table V, taken from Falicov and Harris (1969).

An exact diagonalization of the Hamiltonian matrix yields from the ground state $|G\rangle$

$$\begin{aligned}|G\rangle=&2[16+x^2(x^2+16)^{1/2}]^{-1/2}(|1\uparrow 1\downarrow\rangle+|2\uparrow 2\downarrow\rangle)\\&+0.5[x+(x^2+16)^{1/2}]\\&\times[16+x^2+x(x^2+16)^{1/2}]^{-1/2}\\&\times(|1\uparrow 2\downarrow\rangle+|2\uparrow 1\downarrow\rangle)\ ,\end{aligned} \tag{A17}$$

while the energy, expressed in the variables discussed above, is characterized by

$$\epsilon_G=0.5[x-(x^2+16)^{1/2}]\ . \tag{A18}$$

Falicov and Harris have used these exact results to assess the accuracy of various approximate solutions such as the Heitler-London and molecular-orbital solutions discussed in Sec. III.A, but we shall not go into detail here, except to say that the most successful approximate solutions of their model are of the form of symmetrized spin-density-wave trial functions.

The generalization of this model to more complicated, many-electron chains has also been discussed (Fenton, 1968; Harris and Falicov, 1969), and we refer the interested reader to these papers for details.

APPENDIX B: POSITIVITY OF THE STATIC GREEN'S FUNCTION $G(\mathbf{R},\mathbf{R}')$

Here we shall prove that the Green's function $G(\mathbf{R},\mathbf{R}')$, obtained as a solution of Eq. (4.11) with homogeneous boundary conditions, is non-negative. We start by remarking that, with a suitable choice of the constant V_0, all the eigenvalues of the Schrödinger operator $H+V_0$ with homogeneous boundary conditions are positive and in particular the lowest one, $\varepsilon_0\equiv E_0+V_0>0$. Of course the ground-state wave function $\phi_0(\mathbf{R})$ has the same sign everywhere.

From Eq. (4.15), it is evident that $G(\mathbf{R},\mathbf{R}')$ is positive and vanishes only at points $\mathbf{R}'$ where all wave functions are vanishing. Therefore, $G(\mathbf{R}',\mathbf{R}')=0$ also implies $G(\mathbf{R},\mathbf{R}')=0$ for all values of $\mathbf{R}$.

On the other hand, $\mathbf{G}(\mathbf{R}',\mathbf{R}')>0$ implies that, for given $\mathbf{R}'$, the function $\tilde{\phi}_0(\mathbf{R})$ defined by

$$\tilde{\phi}_0(\mathbf{R})\equiv G(\mathbf{R},\mathbf{R}'),\ \ \text{fixed } \mathbf{R}'\ , \tag{B1}$$

is positive in a region around $\mathbf{R}'$ for reasons of continuity. Let us assume *per absurdum* that regions exist where $\tilde{\phi}_0(\mathbf{R})$ is negative. Let D_2 be the domain where $\tilde{\phi}_0(\mathbf{R})<0$, the point $\mathbf{R}'$ being in its complement D_1, and $D=D_1+D_2$ being the $3N$-dimensional domain of interest. Clearly, nodal surfaces must exist that separate negative and positive regions. It is not difficult to convince oneself that it is always possible to find a subdomain of D_2 (denoted by D'), possibly coinciding with D_2, such that $\tilde{\phi}_0(\mathbf{R})$ does not change sign in D' and vanishes on its boundary. It follows that in the closed domain D' the function $\tilde{\phi}_0(\mathbf{R})$ satisfies

$$(H+V_0)\tilde{\phi}_0(\mathbf{R})=0\ , \tag{B2}$$

with homogeneous boundary conditions. In other words, the Schrödinger operator $H+V_0$, restricted to D' and with the condition that the eigenfunctions vanish at the boundary, possesses the ground-state eigenfunction $\tilde{\phi}_0$ with eigenvalue $\tilde{\varepsilon}_0=0$. Clearly, one has $\varepsilon_0>\tilde{\varepsilon}_0$. However, it can be shown using the calculus of variations (see, for instance, Courant and Hilbert, 1953, particularly Chap. VI,2, theorem 3) that under the present circumstances, i.e., D' is a proper subdomain of D, it must instead hold the condition $\tilde{\varepsilon}_0>\varepsilon_0$. Thus one is led to a contradiction. Therefore $G(\mathbf{R},\mathbf{R}')$ is non-negative.

REFERENCES

Alder, B. J., D. M. Ceperley, and P. J. Reynolds, 1982, J. Chem. Phys. **86**, 1200.

Anderson, J. B., 1975, J. Chem. Phys. **63**, 1499.

Anderson, J. B., 1976, J. Chem. Phys. **65**, 4121.
Anderson, J. B., 1980, J. Chem. Phys. **73**, 3897.
Anderson, P. W., 1973, Mater. Res. Bull. **8**, 153.
Anderson, P. W., 1987, Science **235**, 1196.
Anderson, P. W., 1990, Phys. Rev. Lett. **64**, 1839.
Bachelet, G. B., D. M. Ceperley, and M. G. B. Chiocchetti, 1989, Phys. Rev. Lett. **62**, 2088.
Bamzai, A. S., and B. M. Deb, 1981, Rev. Mod. Phys. **53**, 95.
Barnes, S. E., 1976, J. Phys. F **6**, 1375.
Baskaran, G., Z. Zou, and P. W. Anderson, 1987, Solid State Commun. **63**, 973.
Baus, M., 1990, J. Phys. Condens. Matter **2**, 2111.
Becke, A. D., 1992, J. Chem. Phys. **96**, 2155.
Bednorz, J. G., and K. A. Müller, 1986, Z. Phys. B **64**, 189.
Benfatto, G., and G. Gallavotti, 1990, Phys. Rev. B **42**, 9967.
Bethe, H. A., 1931, Z. Phys. **71**, 205.
Betsuyaku, M., and I. Yokota, 1986, Phys. Rev. B **33**, 6505.
Blankenbecler, R., D. J. Scalapino, and R. L. Sugar, 1981, Phys. Rev. D **24**, 2278.
Bloch, F., 1928, Z. Phys. **57**, 545.
Bonner, J. C., and M. E. Fisher, 1964, Phys. Rev. **135A**, 640.
Brinkman, W. F., and T. M. Rice, 1970, Phys. Rev. B **2**, 4302.
Callaway, J., and N. H. March, 1984, Solid State Phys. **38**, 136.
Callaway, J., D. P. Chen, D. G. Kanhere, and Qiming-Li, 1990, Phys. Rev. B **42**, 465.
Carr, W. J., 1961, Phys. Rev. **122**, 1437.
Castellani, C., C. Di Castro, D. Feinberg, and J. Ranninger, 1979, Phys. Rev. Lett. **43**, 1957.
Ceperley, D. M., 1978, Phys. Rev. B **18**, 3126.
Ceperley, D. M., 1981, in *Recent Progress in Many-Body Theories,* edited by J. G. Zabolitzky, M. de Llano, M. Fortes, and J. W. Clarck (Springer-Verlag, Berlin), p. 262.
Ceperley, D. M., 1983, J. Comput. Phys. **51**, 404.
Ceperley, D. M., 1986, J. Stat. Phys. **43**, 815.
Ceperley, D. M., and B. J. Alder, 1980, Phys. Rev. Lett. **45**, 566.
Ceperley, D. M., and B. J. Alder, 1984, J. Chem. Phys. **81**, 5833.
Ceperley, D. M., and M. H. Kalos, 1979, in *Monte Carlo Methods in Statistical Physics,* edited by K. Binder (Springer-Verlag, Berlin), p. 145.
Coldwell-Horsfall, R. A., and A. A. Maradudin, 1963, J. Math. Phys. **4**, 582.
Coulson, C. A., and I. Fisher, 1949, Philos. Mag. **40**, 386.
Courant, R., and D. Hilbert, 1953, *Methods of Mathematical Physics, Vol. 1* (Interscience, New York), Chap. 6.
Dieterich, K., and P. Fulde, 1987, J. Chem. Phys. **87**, 2976.
Dirac, P. A. M., 1930, Proc. Cambridge Philos. Soc. **26**, 376.
Falicov, L. M., and R. A. Harris, 1969, J. Chem. Phys. **51**, 3153.
Fazekas, P., and P. W. Anderson, 1974, Philos. Mag. **30**, 432.
Fenton, E. W., 1968, Phys. Rev. Lett. **21**, 1427.
Fermi, E., 1928, Z. Phys. **48**, 73.
Ferraz, A., P. J. Grout, and N. H. March, 1978, Phys. Lett. A **66**, 155.
Fiorentini, V., and A. Baldereschi, 1992, J. Phys. Condens. Matter **4**, 5967.
Foldy, L. L., 1971, Phys. Rev. B **3**, 3472.
Gebhard, F., and D. Vollhardt, 1987, Phys. Rev. Lett. **59**, 1472.
Gell-Mann, M., and K. A. Brueckner, 1957, Phys. Rev. **106**, 364.
Ghosh, S. K., and B. M. Deb, 1982, Phys. Rep. **92**, 1.
Godby, R. W., M. Schlüter, and L. J. Sham, 1988, Phys. Rev. B **37**, 10 159.
Gombás, P., 1949, *Die Statistische Theorie des Atoms und ihre Andwendungen* (Springer, Vienna).
Gros, G., R. Joynt, and T. M. Rice, 1987a, Phys. Rev. B **36**, 381.
Gros, G., R. Joynt, and T. M. Rice, 1987b, Z. Phys. B **68**, 425.
Gutzwiller, M. C., 1963, Phys. Rev. Lett. **10**, 159.
Gutzwiller, M. C., 1964, Phys. Rev. **134**, A923.
Gutzwiller, M. C., 1965, Phys. Rev. **137**, A1726.
Haldane, F. D. M., 1981, J. Phys. C **14**, 2585.
Halperin, B. I., J. March-Russell, and F. Wilczek, 1989, Phys. Rev. B **40**, 8726.
Hammond, B. L., P. J. Reynolds, and W. A. Lester, Jr., 1987, J. Chem. Phys. **87**, 1130.
Hammond, B. L., P. J. Reynolds, and W. A. Lester, Jr., 1988, Phys. Rev. Lett. **61**, 2312.
Harris, R. A., and L. M. Falicov, 1969, J. Chem. Phys. **50**, 4590.
Heitler, W., and F. London, 1927, Z. Phys. **44**, 455.
Hirsch, J. E., 1983, Phys. Rev. B **28**, 4059.
Hirsch, J. E., 1985, Phys. Rev. Lett. **54**, 1317.
Hirsch, J. E., 1986, J. Stat. Phys. **43**, 841.
Hirsch, J. E., and S. Tang, 1989, Phys. Rev. Lett. **62**, 591.
Hohenberg, P., and W. Kohn, 1964, Phys. Rev. **136**, B864
Holas, A., and N. H. March, 1991, Phys. Lett. **157**, 160.
Horsch, P., and P. Fulde, 1979, Z. Phys. B **36**, 23.
Horsch, P., and T. Kaplan, 1983, J. Phys. C **10**, L1203.
Huang, C., J. A. Moriarty, and A. Sher, 1976, Phys. Rev. B **14**, 2539.
Hubbard, J., 1959, Phys. Lett. **3**, 77.
Hunter, G., and N. H. March, 1989, Int. J. Quantum Chem. **35**, 649.
Hurley, M. M., and P. A. Christiansen, 1987, J. Chem. Phys. **86**, 1069.
Ichimaru, S., 1982, Rev. Mod. Phys. **54**, 1017.
Johnson, B. G., P. M. W. Gill, and J. Pople, 1993, J. Chem. Phys. **98**, 5612.
Jones, R. O., and O. Gunnarsson, 1989, Rev. Mod. Phys. **61**, 689.
Kajzar, F., and J. Friedel, 1987, Phys. Rev. B **35**, 9614.
Kalmeyer, V., and R. B. Laughlin, 1987, Phys. Rev. Lett. **59**, 2095.
Kalos, M. H., 1984, in *Monte Carlo Method in Quantum Physics,* NATO ASI Series, edited by M. H. Kalos (Reidel, Dordrecht), p. 19.
Kalos, M. H., D. Levesque, and L. Verlet, 1974, Phys. Rev. A **9**, 2178.
Kaplan, T. A., P. Horsch, and J. Borysowicz, 1987, Phys. Rev. B **35**, 1877.
Kaplan, T. A., P. Horsch, and P. Fulde, 1982, Phys. Rev. Lett. **49**, 889.
Kiel, B., G. Stollhoff, C. Weigel, P. Fulde, and H. Stoll, 1982, Z. Phys. B **46**, 1.
Kohn, W., and L. J. Sham, 1965, Phys. Rev. **140**, A1133.
Koonin, S. E., G. Sugiyama, and H. Friedrich, 1982, in *Time Dependent Hartree-Fock and Beyond,* edited by K. Goeke and P. G. Reinhard (Springer-Verlag, Berlin), p. 214.
Kotliar, G., and A. E. Ruckenstein, 1986, Phys. Rev. Lett. **57**, 1362.
Kotrla, M., 1990, Phys. Lett. A **145**, 33.
Lieb, E. H., and F. Y. Wu, 1968, Phys. Rev. Lett. **20**, 1445.
Lundqvist, S., and N. H. March, 1983, Eds., *Theory of Inhomogeneous Electron Gas* (Plenum, New York).
Luther, A. M., 1979, Phys. Rev. B **19**, 320.
March, N. H., 1957, Adv. Phys. **6**, 1.
March, N. H., 1975, *Self-Consistent Fields in Atoms* (Pergamon, Oxford).
March, N. H., and M. Parrinello, 1982, *Collective Effects in Solids and Liquids* (Hilger, Bristol).

March, N. H., M. Suzuki, and M. Parrinello, 1979, Phys. Rev. B **19**, 2027.
March, N. H., and W. H. Young, 1959, Philos. Mag. **4**, 384.
March, N. H., W. H. Young, and S. Sampanthar, 1967, *The Many-Body Problem in Quantum Mechanics* (Cambridge University, Cambridge, England).
Mårtensson-Pendrill, A., S. A. Alexander, L. Adamowicz, N. Oliphant, J. Olsen, P. Öster, H. M. Quiney, S. Salomonson, and D. Sundholm, 1991, Phys. Rev. A **43**, 3355.
Mattheiss, L. F., 1961, Phys. Rev. **123**, 1209.
Mentch, F., and J. B. Anderson, 1981, J. Chem. Phys. **74**, 6307.
Metzner, W., and D. Vollhardt, 1987, Phys. Rev. Lett. **59**, 121.
Meyer, W., 1971, Int. J. Quantum Chem. Quantum Chem. Symp. No. 5, 341.
Micnas, R., J. Ranninger, and S. Robaszkiewicz, 1990, Rev. Mod. Phys. **62**, 113.
Miglio, L., M. P. Tosi, and N. H. March, 1981, Surf. Sci. **111**, 119.
Moreo, A., D. J. Scalapino, R. L. Sugar, S. R. White, and N. E. Bickers, 1990, Phys. Rev. B **41**, 2313.
Moroni, S., D. M. Ceperley, and G. Senatore, 1992, Phys. Rev. Lett. **69**, 1837.
Moroni, S., and G. Senatore, 1991, Phys. Rev. B **44**, 9864.
Moroni, S., G. Senatore, and D. M. Ceperley, 1993, unpublished.
Moskowitz, J. W., and M. H. Kalos, 1981, Int. J. Quantum Chem. **20**, 1107.
Moskowitz, J. W., and K. E. Schmidt, 1986, J. Chem. Phys. **85**, 2868.
Moskowitz, J. W., K. E. Schmidt, M. A. Lee, and M. H. Kalos, 1982a, J. Chem. Phys. **76**, 1064.
Moskowitz, J. W., K. E. Schmidt, M. A. Lee, and M. H. Kalos, 1982b, J. Chem. Phys. **77**, 349.
Murrell, J. N., 1971, *The Theory of the Electronic Spectra of Organic Molecules* (Chapman and Hall, London).
Negele, J. W., and H. Orland, 1988, *Quantum Many-Particle Systems* (Addison-Wesley, Redwood City, CA), p. 332.
Ogawa, T., K. Kanda, and T. Matsubara, 1975, Prog. Theor. Phys. **53**, 614.
Oles, A. M., F. Pfirsch, and P. Fulde, 1986, J. Chem. Phys. **85**, 5183.
Overhauser, A. W., 1985, Int. J. Quantum Chem. **S14**.
Parr, G., and W. Yang, 1989, *Density Functional Theory of Atoms and Molecules* (Oxford University, Oxford).
Pauling, L., 1949, *The Nature of the Chemical Bond* (Cornell University, Ithaca).
Perdew, J. P., and A. Zunger, 1981, Phys. Rev. B **23**, 5048.
Pickett, W. E., 1986, Comments Solid State Phys. **12**, 1 (Part I) and 57 (Part II).
Reger, J. D., and A. P. Young, 1988, Phys. Rev. B **37**, 5978.
Reynolds, P. J., D. M. Ceperley, B. J. Alder, and W. A. Lester, 1982, J. Chem. Phys. **77**, 5593.
Rosenberg, B. J., and I. Shavitt, 1975, J. Chem. Phys. **63**, 2162.
Scalapino, D. J., and R. L. Sugar, 1981, Phys. Rev. B **24**, 4295.
Schmidt, K. E., and M. H. Kalos, 1984, in *Application of the Monte Carlo Method in Statistical Physics,* edited by K. Binder (Springer-Verlag, Berlin), p. 125.
Senatore, G., and G. Pastore, 1990, Phys. Rev. Lett. **64**, 303.
Sham, L. J., and M. Schlüter, 1983, Phys. Rev. Lett. **51**, 1888.
Shavitt, I., B. Ross, and B. O. Siegbahn, 1977, in *Modern Theoretical Chemistry, Vol. 3,* edited by H. F. Schaefer, III (Plenum, New York), p. 189.
Singwi, K. S., and M. P. Tosi, 1981, Solid State Phys. **36**, 177.
Singwi, K. S., M. P. Tosi, R. H. Land, and A. Sjölander, 1968, Phys. Rev. **176**, 589.
Slater, J. C., 1951, Phys. Rev. **81**, 385.
Sorella, S., 1989, Ph.D. thesis (ISAS, Trieste).
Stollhoff, G., and K. P. Bohnen, 1988, Phys. Rev. B **37**, 4678.
Stollhoff, G., and P. Fulde, 1977, Z. Phys. B **26**, 257.
Stollhoff, G., and P. Fulde, 1978, Z. Phys. B **29**, 231.
Stollhoff, G., and P. Fulde, 1980, J. Chem. Phys. **73**, 4548.
Stratonovich, R. D., 1957, Dokl. Akad. Nauk SSSR **115**, 1097 [Sov. Phys. Dokl. **2**, 416 (1958)].
Sugiyama, G., and S. E. Koonin, 1986, Ann. Phys. (N.Y.) **168**, 1.
Takahashi, M., 1977, J. Phys. C **10**, 1289.
Tan, L., Q. Li, and J. Callaway, 1991, Phys. Rev. B **44**, 341.
Thomas, L. H., 1926, Proc. Cambridge Philos. Soc. **23**, 542.
Thouless, D. J., 1961, *The Quantum Mechanics of Many-Body Systems* (Academic, New York).
Umrigar, C. J., K. J. Runge, and M. P. Nightingale, 1991, in *Monte Carlo Methods in Theoretical Physics,* edited by S. Caracciolo and A. Fabrocini (ETS Editrice, Pisa), p. 161.
van Kampen, N. G., 1981, *Stochastic Processes in Physics and Chemistry* (North-Holland, Amsterdam).
Vollhardt, D., 1984, Rev. Mod. Phys. **56**, 99.
Vosko, S. H., L. Wilk, and M. Nusair, 1980, Can. J. Phys. **58**, 1200.
Wendin, G., 1986, Comments Mod. Phys. D **17**, 115.
White, S. R., D. J. Scalapino, R. L. Sugar, E. Y. Loh, J. E. Gubernatis, and R. T. Scalettar, 1989, Phys. Rev. B **40**, 506.
Wigner, E. P., 1934, Phys. Rev. **46**, 1002.
Wigner, E. P., 1938, Trans. Faraday Soc. **34**, 678.
Wilson, S., 1981, in *Theoretical Chemistry, Vol. 4* (Royal Society of Chemistry, London), p. 1.
Yokoyama, H., and H. Shiba, 1987a, J. Phys. Soc. Jpn. **56**, 1490.
Yokoyama, H., and H. Shiba, 1987b, J. Phys. Soc. Jpn. **56**, 3570.
Yoshida, T., and K. Iguchi, 1988, J. Chem. Phys. **88**, 1032.

Phys. Chem. Liq., 2000, Vol. 38, pp. 151–154
Reprints available directly from the publisher
Photocopying permitted by license only

Letter

THERMODYNAMICS OF THE EQUILIBRIUM BETWEEN A FRACTIONAL QUANTUM HALL LIQUID AND A WIGNER ELECTRON SOLID

N. H. MARCH*

Oxford University, Oxford, OX1 3QZ, England and
Department of Physics, University of Antwerp,
RUCA, Antwerp, Belgium

(Received 17 September 1998)

A recent paper of Wu *et al.*, discusses thermodynamic observables in a fractional quantum Hall (FQH) liquid. In particular the de Haas-van Alphen effect is considered theoretically. Here, it is pointed out that laboratory experiments on GaAs/AℓGaAs heterojunctions, plus the analogue of the Clausius–Clapeyron equation in an applied magnetic field, allow schematic analysis of the orbital magnetism of the FQH liquid. There is general agreement with the theoretical results of Wu *et al.*

Keywords: Quantum hall effect; orbital magnetism; Wigner solid

In a recent article, Wu *et al.* [1] have given a very detailed theoretical treatment of the thermal activation of quasiparticles and the thermodynamic observables in fractional quantum Hall (FQH) liquids. Their important conclusion is that what are usually thought of as different but equivalent pictures of the FQH effect (anyon, composite Fermion, and composite Boson) can exhibit significant differences at finite

*Private address for correspondence: 6 Northcroft Road, Egham, Surrey, TW20 ODU.

temperature, though they are all equivalent at $T = 0$. Then their final sentence reads 'In particular, it is more desirable that these theoretical predictions would be put to experimental tests, if the tremendous difficulties in measuring thermodynamic quantities of a thin layer of electron gas could be overcome someday'.

It is therefore the more remarkable that some existing experiments on GaAs/AℓGaAs heterojunctions in magnetic fields **B** applied perpendicular to the two-dimensional electron assembly can provide insight into some of the thermodynamic properties of the FQH liquid studied theoretically by Wu *et al.* [1]. The two laboratory experiments focussed on below are those of Andrei *et al.* [2] and the subsequent, quite different, type of investigation by Buhmann *et al.* [3]. Both experiments, in spite of their very different techniques, allow information to be extracted on the equilibrium between a FQH liquid and a Wigner electron solid, in which electron localization driven purely by Coulomb interaction in Wigner's original papers [4, 5] is now magnetically induced [6]. We wish to add to these two laboratory experiments the quantum computer simulation results of Ortiz, Ceperley and Martin [7] (OCM).

Starting with the computer simulation data, OCM for $r_s = 20$, r_s as usual measuring the mean interelectronic spacing, have by means of a stochastic method appropriate for systems with broken time-reversal symmetry, studied the transition between an incompressible $\nu = 1/m$ FQH liquid and a Wigner solid. For $m = 5$, the work of OCM demonstrates that 'further work must be done to show definitively which phase is stable'.

Returning to the laboratory experiments, Buhmann *et al.*, as a result of their measured luminescence spectra, have proposed a qualitative form of the phase diagram showing 4 Wigner solid phases, with the FQH liquid assumed as the ground-state at filling factors $\nu = 1/5$, $1/7$ and $1/9$ (see their Fig. 4(c)). The earlier radio spectroscopic data of Andrei *et al.* [2] are generally compatible with the Buhmann *et al.*, phase diagram, though the features shown in this diagram differ most from Andrei *et al.*, around $\nu = 1/5$, $1/7$ and $1/9$. However, the computer simulation results are not incompatible with the Buhmann *et al.* (T, ν) phase diagram in their Figure 4(c).

We now turn to the thermodynamic interpretation [8] of the Buhmann *et al.*, phase diagram, in relation to the calculation of thermodynamic observables, and in particular the de Haas-van Alphen effect, by Wu *et al.* [1]. Lea, March and Sung [8] use, in particular, to discuss the thermodynamics of an electron solid to electron liquid first-order melting transition, the result for the melting temperature T_m as a function of magnetic field H, at constant area Ω:

$$\left(\frac{\partial T_m}{\partial H}\right)_\Omega = -\frac{\Delta M}{\Delta S}. \tag{1}$$

If the subscript s denotes the solid phase and ℓ the liquid phase, then ΔM in Eq. (1) is $M_\ell - M_s$ which is the change in magnetization on melting, while $\Delta S = S_l - S_s$ is the corresponding entropy change. Using the Landau level filling factor ν, given in terms of the (areal) electron density n and the magnetic field H applied perpendicular to the electron layer in the heterojunction:

$$\nu = \frac{nhc}{eH}, \tag{2}$$

one immediately recasts Eq. (1) into a relation used for interpreting the Buhmann *et al.* (T, ν) phase diagram, namely

$$\left(\frac{\partial T_m}{\partial \nu}\right)_\Omega = \left(\frac{H}{\nu}\right)\frac{\Delta M}{\Delta S}. \tag{3}$$

Using this Eq. (3), and making plausible assumptions about the entropy change on melting, Lea *et al.* [8] in their Figure 3 draw a schematic diagram of the change in magnetization ΔM on melting along the melting curve of the Buhmann *et al.*, proposed phase diagram. Lea *et al.*, conclude that 'this field dependence of ΔM is very reminiscent of the de Haas-van Alphen effect at integral ν values, suggesting that the magnetism of the electron liquid phase is intimately connected with the exotic variation of ΔM ...'. Though the range of ν values in Figure 3 of Lea *et al.*, is somewhat different from that in the very recent discussion of Wu *et al.* [1], the gist of the conclusions is the same in the two cases.

In later work, Lea *et al.* [9] use both anyon and composite Fermion models to represent the main features of the melting of such Wigner electron solids as a function of the Landau level filling factor. Of course, the later studies of Wu *et al.*, transcend in their detailed treatment the results in references 8 and 9.

In summary, the detailed theoretical study of Wu *et al.* [1] of thermodynamic observables in FQH liquids makes it now the more urgent to bring such theoretical predictions into quantitative contact with (a) laboratory experiments such as reported in Refs. [2, 3] and (b) quantal computer simulations of the kind discussed in Ref. [7]. In the latter area, it would seem of great interest to study the orbital magnetism of the FQH liquid in the range of ν values discussed in Refs. [8, 9].

Acknowledgements

The author wishes to thank Professor D. M. Ceperley for a most valuable discussion on the melting of Wigner solids after the paper by NHM at the Heraklion International Workshop on 'Electron correlation and materials properties'. Thanks are due also to Professor M. J. Lea for his continuing support, to Dr. P. Schmidt of ONR for much stimulation and encouragement in the area of matter in intense external fields, and the ONR for partial financial support. This study was brought to fruition at ICTP Trieste and thanks are due to Professor Yu Lu and his colleagues for their interest, and for generous hospitality.

References

[1] Wu, Y. S., Yu, Y., Hatsugai, Y. and Kohmoto, M. (1998). *Phys. Rev.*, **B57**, 9907.
[2] Andrei, E. Y., Deville, G., Glattli, D. C., Williams, F. I. B., Paris, E. and Etienne, B. (1988). *Phys. Rev. Lett.*, **60**, 2765.
[3] Buhmann, H., Joss, W., von Klitzing, K., Kukuskim, I. V., Plant, A. S., Martinez, G., Ploog, K. and Timofeev, V. B. (1991). *Phys. Rev Lett.*, **B6**, 926.
[4] Wigner, E. P. (1934). *Phys. Rev.*, **46**, 1002.
[5] Wigner, E. P. (1938). *Trans. Faraday Soc.*, **34**, 678.
[6] Durkan, J., Elliott, R. J. and March, N. H. (1968). *Rev. Mod. Phys.*, **40**, 812.
[7] Ortiz, G., Ceperley, D. M. and Martin, R. M. (1993). *Phys. Rev. Lett.*, **71**, 2777.
[8] Lea, M. J., March, N. H. and Sung, W. (1991). *J. Phys. Condens. Matter*, **3**, CL4301.
[9] Lea, M. J., March, N. H. and Sung, W. (1992). *J. Phys. Condens. Matter*, **4**, 5263.

Fulleride Superconductivity Compared and Contrasted with RVB Theory of High T_c Cuprates

RICHARD H. SQUIRE,[1] NORMAN H. MARCH[2,3]
[1]Department of Natural Science, Marshall University, 918 Woodland Ave., South Charleston, West Virginia 25303, USA
[2]Department of Physics, University of Antwerp (RUCA), Groenborgerlaan, Antwerp, Belgium
[3]Oxford University, Oxford, England

Received 19 March 2004; accepted 19 March 2004
Published online 29 July 2004 in Wiley InterScience (www.interscience.wiley.com).
DOI 10.1002/qua.20171

ABSTRACT: Recently we proposed a microscopic mechanism for alkali-metal doped Fulleride superconductivity. The aim of the current study is to compare and contrast such Fulleride superconductivity with that of high T_c cuprates. We focused earlier on "topological" superconductivity in the Fulleride case and so it seemed natural to make contact with the resonating valence bond treatment of Anderson et al. for the high T_c cuprates. Finally, some experimental points of contact involve the product of electrical resistivity and nuclear spin lattice relaxation times in the normal state of both classes of superconductors considered here. Int J Quantum Chem 100: 1092–1103, 2004

Key words: boson–fermion model; fulleride superconductivity; high-temperature superconductivity; resonating valence bond; topological doping; pairing mechanism

Introduction

It seems that there is general agreement that fullerene superconductivity begins with almost complete transfer of electrons from the dopant to the C_{60} molecule [1, 2]. There is also agreement that two and four electrons transferred onto C_{60} produce an insulator, whereas three transferred electrons generate a superconductor under suitable conditions. These statements about the widely varying properties of the various charged C_{60} molecules must be properly answered for a viable microscopic theory [3].

In the current work we sharpen and refine our earlier model by presenting more-detailed microscopic and mesoscopic mechanisms for the observations of the various charged fullerenes mentioned above including superconductivity. In addition, an interesting extrapolation of our superconductivity mechanisms to high-temperature superconductivity (HTSC) naturally exists in two

Correspondence to: R. H. Squire; e-mail: rhsquire@charter.net ways: first, through use of a similar Hamilton; and second, via the route of decompactification, which explicitly maps our final model onto HTSC.

The paper begins with a discussion of different-looking Hamiltonians to uncover the seemingly mysterious nature of topological doping. The third section contains the original premises with some extensions. The relevancy of preformed bosons in terms of specific features (such as a pseudospin gap) contained in the general boson–fermion model is discussed in the fourth section. The heart of the mechanism is the "intermolecular" pairing due to resonant interaction resulting from approach of the boson and fermion energy levels as proposed in the fifth section. The sixth section contains a brief discussion of extensions of the model to HTSC, followed by a proposed method of experimentally determining the existence of a precursor 2e boson in the seventh section. Conclusions and future work follow.

Hamiltonians and Topological Doping

In the 1960s several chemists recognized the importance of a new type of excitation termed a "misfit" [4]. In an extended conjugated system such as polyacetylene (PA), $(CH)_n$, an added electron or hole created a different bonding pattern. The region of the misfit is confined to a small number of atoms, costing only a modest amount of energy. This misfit can be described by a continuum model, would be essentially free to move through defect-free material, and would have a small effective mass so it would be a quantum particle. Some physicists have called this particle a soliton; however, it truly is not, as it cannot pass through another identical particle without dissipation. Nevertheless, we continue this established naming practice. The chemists understood that a single electronic state would appear near the center of the energy gap; some additional reasoning as to why this should be so is presented below.

Doped PA can be described by the Su, Schrieffer, Heeger (SSH) Hamiltonian [4],

$$H = -\sum_{n,s} t_0(c^{\dagger}_{n+1,s}c_{ns} + c^{\dagger}_{ns}c_{n+1,s}) + \alpha \sum_{n,s} (u_{n+1} - u_n) \times (c^{\dagger}_{n+1,s}c_{n,s} + c^{\dagger}_{n,s}c_{n+1,s}) + \frac{1}{2}\sum_n K(u_{n+1} - u_n)^2 + \frac{1}{2}\sum_n M\dot{u}_n^2,$$

where $c^{\dagger}_{ns}$ creates and $c_{n,s}$ destroys a π electron of spin $s(\pm 1/2)$ on the n-th CH group, t_0 is the electron-hopping matrix element, and α is the electron–phonon coupling constant. Constant K is the effective spring constant of the bond length change due to the Peierls distortion, and the last term is the kinetic energy of nuclear motion with M equal the mass of the CH group. Parallel PA polymer chains have very little hopping from chain to chain; one has essentially a one-dimensional conductor. In addition, there are two distinct bonding patterns that give rise to two degenerate ground states (or "vacua") related by a reflection symmetry in the coordinate displacement that we used as a parity symmetry (a φ^4 theory is more appropriate than Sine–Gordon) (Ref. [1], Appendix A). The electron hopping along the quasi–one-dimensional chain of CH groups is described by taking $\varphi_n = (-1)u_n$ with $n = \pm 1$. The electron–phonon or electron lattice displacement interaction that is unstable with respect to the spontaneous symmetry breaking of the reflection symmetry (Peierls distortion) so $u_n \rightarrow -u_n$, which gives rise to kink (soliton) and antikink (antisoliton) solution of φ^4 theory. The kink in PA is sometimes called a domain wall, because it "interpolates" between two vacua [4].

A second Hamiltonian of interest is the short-range resonating valence bond (SR-RVB) Hamiltonian [2b], which represents the Heisenberg Hamiltonian (the Hubbard Hamiltonian in the large U limit) with the effects of lattice distortion to examine the competition between the Ne'el and RVB states. This Hamiltonian has a superficial similarity to the essentially one-dimensional SSH Hamiltonian,

$$H = \sum_{\langle i,j \rangle} \left\{ \frac{4(t_0 - \alpha u_{ij})^2}{U} \bar{S}_i \cdot \bar{S}_j + \frac{K}{2}(\bar{u}_i - \bar{u}_j)^2 \right\},$$

and most of the symbols are defined in the same manner (U is the repulsion energy between electrons on the same site). However, this model is deceptive in its appearance, as it has a complex, rich phase structure [5]. The SR-RVB Hamiltonian represents a short-range form of RVB as opposed to Anderson's RVB [6], which captures the Mott insulator physics of the parent compounds of HTSC, for example, La_2CuO_4. However, this model has not yet been solved; one reason is that its holes and spins are highly entangled, with no obvious method of separating them. A brief description is that the RVB is a coherent superposition of states.

Each site is strongly bonded to only one neighbor in a singlet pair. A resonance between two valence bond configurations lowers the energy by

$$H_{res} = -t \sum_{plaquette} \{|\,\|\,\rangle\langle = | + h.c.\} + V \sum_{plaquette} \{|\,\|\,\rangle\langle\,\|\,| + | = \rangle\langle = |\}$$

for a square lattice. Using this standard graphic notation, the t term runs over $|\alpha\rangle\langle\beta|$ for every possible appropriate pair of configurations so that β differs from α only by the replacement of a "vertical" pair by a horizontal pair of dimers on one plaquette. Rokshar and Kivelson found that when $V = t$ (the "RK point"), the ground-state wavefunction is

$$|\Psi_0\rangle = \frac{1}{\sqrt{N_s}} \sum_{\alpha} |\alpha\rangle,$$

the sum running over all dimer configurations. The probability weight is the same for all the N_s states, just as in the classical ensemble. This quantum dimer model captures the low-energy dynamics of valence bond–dominated phases of a quantum Heisenberg antiferromagnetic. From this model emerge two fundamentally different lattices: a bipartite and a nonbipartite. The latter is a spin liquid with topological order $\mathbb{Z}_2$, whereas the former has only crystalline phases. Fradkin further discusses the attributes of both Kiverson's and Anderson's RVB states [5]; we return to this discussion when we examine dimer liquids as opposed to VB crystals.

In our initial discussion of C_{60} superconductivity [1], we used the linear sigma model (Appendix A of Ref. [1]; Ref. [7]) and wrote it in terms of the Dirac equation,

$$L = \frac{|\dot{\Delta}|^2}{g^2\bar{\omega}^2} - \frac{|\Delta|^2}{g^2} + \bar{\psi}(i\hat{D} - |\Delta|e^{i\gamma_5\varphi})\psi$$

to emphasize topological doping. Here, $\hat{D} = \tilde{\gamma}(i\partial_\mu - A_\mu^{ext})$; $\tilde{\gamma}_0$, $\tilde{\gamma}_1$, $\tilde{\gamma}_5$ are 2D Dirac matrices, g is a coupling constant, Δ is a gap parameter, and $\bar{\omega}$ is a characteristic phonon frequency. This is just another way of looking at the Peierls instability that opens up a gap, always at the Fermi level, so the spectrum depends strongly on the number of particles (or filling factor). Because there is a gap in the electronic spectrum and the only gapless mode is the sliding charge density wave (CDW), Fröhlich long ago concluded this system is superconducting [8] (this is incorrect; it is an ideal conductor, as there is no Meissner effect in a strictly one-dimensional system). With a redefinition of terms, this model is also exactly the same as the TLM (Takayama, Lin-Liu, Maki) model, [9] and both can be derived from the discrete SSH Hamiltonian, as well. This interconnection of Hamiltonians provides us with different insights into the same physical phenomena.

This soliton verification in PA assisted a thought process that evolved a very deep relation: Whenever there is a solution to a conjugation symmetry Dirac equation with a topologically nontrivial background field like a soliton, there always are self-conjugate, normalizable, zero-energy solutions in addition to the positive and negative solutions related by conjugation. This conclusion is reached not only physically but also mathematically, because index theorems count the zero eigenvalues (or spectral flows) to insure the number is nonvanishing whenever the topology of the background is nontrivial.

The so-called zero modes are wavefunctions at zero energy (usually the Fermi level), and they can be present when a quantum particle is present in a background with a nontrivial topological structure. The connectivity of C_{60} leads to a ground state with just such a nontrivial topology [10]. Dramatic physical consequences can be associated with quantum fluctuations of these zero modes, as they are related to chiral and parity anomalies and fractional fermion numbers. Although there seems to be some questions as to the intrinsic critical properties in terms of critical zero modes in the two-dimensional random hopping model, we assume that they are formed in fullerenes, based on the nontrivial connectivity and the evidence associated with the SSH model. In dimensions higher than one, say, for a two-leg ladder, soliton states also exist [11a]. There are examples where the solitons are bound into pairs, and a phase diagram of a generalized SSH model was presented for a square lattice [11b].

Extended Micromechanism

In the theory of (HTSC) [6], the focus is principally on an electronic liquid where bosonic fields modulate the interactions between electrons. In this study we apply certain ideas from this theory, supplemented by our modifications of the SSH/Kivelson model to C_{60} superconductivity. The following four key concepts and assumptions are crucial:

(1) Carbon "chain" C_{60} has two types of boson (soliton) solutions. Because C_{60} is a non–simply connected carbon molecule, there are always soliton solutions. Where are they located in the spectrum? Because a simple energetic argument supports a soliton wave lower in energy than a plane wave [4], standard theory would strongly suggest they are in the middle of the energy gap. Calculations and experiment locate solitons in PA at 0.42 eV and 0.65–0.75 eV, respectively, well within the gap of 1.4 eV. The stabilization due to the Jahn–Teller (JT) effect in C_{60}^{3-} would suggest their location at about −0.3 eV relative to the lowest unoccupied molecular orbital (LUMO). Gunnarsson found independent evidence that suggest a local intramolecular pairing "surprisingly resistant to coulomb repulsion" [12]. In addition to energy considerations, there is an inherent topological stability due to a "collective" vibrational mode following the electronic wavefunction. From the work of Herzberg and Longuet-Higgins [13] and others, a sign change in the electronic wavefunction as one traverses a closed path around a potential surface intersection in nuclear configuration space induces a compensating sign change on the part of the nuclear wavefunction if the full wavefunction is to be continuous and single-valued. This is equivalent to a Chern number for a collective vibrational mode; hence, the modes discussed below are topologically active and in some ways may be regarded as "phasons." In PA it is estimated that a soliton occupies about 30 —$(CH)_n$— units [4]. Thus, a soliton/antisoliton pair could be easily created on a molecule of C_{60}. This first type of soliton is localized (or confined) on a C_{60}^{3-} molecule and contains two paired electrons, so it is a charge 2 boson. A general argument leads to the conclusion that the midgap state always has an even degeneracy; we assume it to be twofold [14]. Because this soliton is "confined," there is no macroscopic charge flow. This has been called a "loop" soliton [4e] and leads to a pseudospin gap, as we show below. The second type of boson is formed from two itinerant electrons in an intermolecular manner and leads to "deconfinement" of all bosons.

(2) The molecular (confined) soliton solution has a microscopic description in terms of a JT distortion as an extension of a one-dimensional Peierls–Frolich distortion. There are three different solutions for the JT interaction as a function of the number of electrons doped onto C_{60} (i.e., as $n = 1 \rightarrow 6$). The Schrödinger equation and energy for $n = 1, 5$ is [15]

$$-\frac{1}{2}\left[\frac{1}{q^2}\frac{\partial}{\partial q}\left(q^2\frac{\partial}{\partial q}\right)+\frac{\partial}{\partial\alpha}\left(\sin\alpha\frac{\partial}{\partial\alpha}\right)+\frac{1}{q^2\sin^2\alpha}\frac{\partial^2}{\partial\gamma^2}-q^2+kq\right]\psi$$

$$-\frac{1}{6q^2}\left[\frac{1}{\sin\theta}\frac{\partial}{\partial\theta}\left(\sin\theta\frac{\partial}{\partial\theta}\right)+\frac{1}{\sin^2\theta}\frac{\partial}{\partial\phi^2}\right]\psi = E_{1,5}\psi$$

$$E_{1,5}=\left(-\frac{1}{2}k^2+\frac{3}{2}\right)+\left(\frac{1}{6k^2}\right)L(L+1),$$

where k is the coupling coefficient and L is an integer required by phase invariance to be odd. So, the ground state is $L = 1$, and there is a Berry phase for these solutions. The first line above is a three-dimensional harmonic oscillator with a displace origin, whereas the second part is the Hamiltonian for a rotator. For $n = 2, 4$, the solution is very similar; however, there is no Berry phase in these cases.

For $n = 3$, the Schrödinger equation is

$$-\frac{1}{2}\left[\frac{1}{q}\frac{\partial}{\partial q}\left(q\frac{\partial}{\partial q}\right)+\frac{1}{q^2}\frac{\partial^2}{\partial\alpha^2}-q^2+2\sqrt{3}\,kq\right]\psi$$

$$-\frac{1}{8q^2}[4\lambda_x^2+4\lambda_y^2+\lambda_z^2]\psi=E_3\psi.$$

The first line is a two-dimensional harmonic oscillator, again with a displaced origin ($2\sqrt{3}\,kq$), and line two is a symmetric top. The energy minimum is at $q = \sqrt{3}\,k$, so, substituting, we get the energy for $n = 3$ as

$$E_3=-\frac{3}{2}(k^2+1)+\frac{1}{24k^2}+\frac{1}{6k^2}\left[L(L+1)-\frac{3}{4}K^2\right].$$

There are eight fivefold degenerate H_g coordinates participating in these relations, and a simple approximation allows us to visualize the C_{60} surface distortions. For $n = 1, 2, 4, 5$, there is a symmetric ellipsoidal distortion $(3z^2 - r^2)$. For $n = 3$, all three angles (γ, θ, ϕ) are required as the distortion in a directional $(x^2 - y^2)$ type.

(3) Crucial to the intermolecular pairing is the symmetry of the different types of JT distortions: no intermolecular pairing for the $(3z^2 - r^2)$ unimodal modes for $n = 1, 2, 4, 5$ electrons, and pairing for the $(x^2 - y^2)$ bimodal mode for three electrons. The $(x^2 - y^2)$ pairing is the same as in HTSC. Recently, Manini et al. [16] showed that electron hopping in a lattice of molecules possessing a Berry phase can lead to pairing. In their model, the electron hopping needs to be accompanied by orbital operators that switch on and off the Berry phase as the electron number changes. With the benefit of hindsight, their result does not appear to be completely correct, as there is yet no superconductivity known for C_{60}^{1-} or C_{60}^{5-}. However, they showed a pairing mechanism exists for a Berry phase that is inherent to Wiegmann's "topological instability." Our comment is that topological instability is necessary but not sufficient; there are additional factors, such as symmetry and energetics, that are also important (see the fifth section).

(4) The soliton is a weak coupling dual of the strong coupling fermion interaction. Duality between fermions and bosons has been discussed for some time. The classical example has been the correspondence between the massive Thirring model for fermions and the sine-Gordon equation for solitons (bosons) [17]. Using the linear sigma model we discussed earlier,

$$L = \frac{|\dot{\Delta}|^2}{g^2\omega^2} - \frac{|\Delta|^2}{g^2} - \bar{\psi}(i\hat{D} - |\Delta| e^{i\gamma_5\varphi})\psi,$$

we follow others and bosonize this Lagrangian [17]

$$L = \frac{\pi}{2}[(\partial\phi_\uparrow)^2 + (\partial\phi_\downarrow)^2] + \bar{\Delta}[\cos(2\pi\phi_\uparrow + \varphi) + \cos(2\pi\phi_\downarrow + \varphi)] + \frac{1}{\pi}\varepsilon_{\mu\upsilon}A_\mu^{external}\partial_\upsilon(\phi_\uparrow + \phi_\downarrow).$$

Here, $A_\mu^{external}$ is an external field, so we can keep track of response functions, and introducing spin and charge densities,

$$\phi_c = \frac{\phi_\uparrow + \varphi_\downarrow}{\sqrt{2}} \quad \phi_s = \frac{\phi_\uparrow - \varphi_\downarrow}{\sqrt{2}},$$

what results is spin and charge separation in one dimension, as shown,

$$L = \frac{\pi}{2}[(\partial\phi_c)^2 + (\partial\phi_s)^2] + \bar{\Delta}\cos(\sqrt{2}\,\pi\phi_c)\cos(\sqrt{2}\,\pi\phi_s) + \sqrt{2}\,\varepsilon_{\mu\upsilon}A_\mu^{ext}\partial_\upsilon\phi_c.$$

If the energy is less than $\bar{\Delta}$, a singlet current adiabatically follows the density of the solitons of the phonon filed $j_\mu = (2/\pi)\varepsilon_{\mu\upsilon}\partial_\upsilon\varphi$; put differently, the φ field follows the field $-\sqrt{2}\,\pi\phi_c$ so that the argument of the first cos term is fixed. This provides the dynamics of an independent gapped spin density wave and a gapless charge density wave, viz.,

$$L = L_s + L_c$$

$$L_s = \frac{\pi}{2}(\partial\phi_s)^2 + \bar{\Delta}\cos(\sqrt{2}\,\pi\phi_s)$$

$$L_c = \frac{\pi}{2}(\partial\phi_c)^2 + \sqrt{2}\,\varepsilon_{\mu\upsilon}A_\mu^{ext}\partial_\upsilon\phi_c$$

$$-\sqrt{2}\,\pi\phi_c = \varphi.$$

Returning to Fröhlich's picture [8]; if a system has topologically distinct degenerate vacua, a configuration $\varphi(x, t)$ that transforms one vacuum into another (from $[\varphi(x = \pm\infty), t = 0]$ to $[\varphi(x = \pm\infty, t = \infty)]$) is a hydrodynamic mode, irrespective of whether the particle spectrum is gapped or not. As Fröhlich put it, "the whole system, electrons and solitons can move through the lattice without being disturbed." In our case, the confined boson, providing an approximation to a charge density wave, moves through the C_{60}^{3-} molecules as a collective surface wave.

The solution of the Schrödinger equation at a static kink $\Delta(x \to \pm\infty) \to \pm|\Delta_0|$ with minimal energy is the

$$i\partial_x\psi_L - i\Delta\psi_R = E\psi_L$$

$$-i\partial_x\psi_R + i\Delta\psi_L = E\psi_R$$

zero mode located in the core of the kink,

$$\psi_L = \psi_R = e^{-\int_0^x \Delta(x)}.$$

Each time we add an electron to a midgap state, we add a twist that rearranges the period of the CDW. This constitutes a level crossing that can create a nonconservation of the axial current (axial anomaly); bosonization keeps track of this: two electrons in our soliton (boson) state have no topological instability. An odd electron creates the instability that is related to the Berry phase and the Chern number [7, 13].

We also may connect with the SR-RVB theory. An electron in a strongly correlated system can be thought of as a composite object made up of two quasiparticles, a spinon that carries spin half and no charge, and an eon that carries charge e but no spin [18]. Anderson strongly suggested that spin-charge separation is essential to HTSC [6]; performing the spin-charge separation adds considerably to the number of states and seems to complicate matters in our model.

Overview of Possible C_{60} Superconducting Transition

Our most basic premise is that a soliton is quantized as a boson and stabilized by a JT interaction in a topological way. The boson now interacts with the remaining "itinerant electron" as has been suggested by the boson–fermion model [19] used to describe HTSC as well as Bose–Einstein condensation (BEC). We think these enhancements to our C_{60} superconductor model not only are necessary but also offer some interesting features.

It seems generally accepted that the onset of HTSC is controlled by phase fluctuations resulting in an exceptionally rich phase structure containing a pseudophase. Our model of superconductivity in fullerenes contains some (surprising) similarities. We show a finite-order parameter expected to persist well above T_c (the superconducting transition temperature) to some temperature T^p, the pairing temperature. Between T_c and T^p is our pseudogap phase, where we have electron pairing, and it is the pair density, n_p, as opposed to coherence length, that is the dominant factor. The pseudogap opens up driven by amplitude fluctuations of dynamical electron-pair formation uncorrelated in space and "confined" to a C_{60} molecule. Pair formation in some cases is only indirectly related to incipient superconductivity, as they might also be formed in C_{60}^{2-} and C_{60}^{4-}, which are insulators, but the pairs are essential to inducing superconductivity when they are in resonance with unpaired electrons and become "unconfined" in C_{60}^{3-}. Here, as the temperature is lowered to T_c, intermolecular coupling of the unpaired electron to an unpaired electron on an adjacent C_{60} molecule becomes favored, driving the system to a superconducting state where all three electron pairs become deconfined.

Boson–Fermion Model

We modify an existing boson–fermion (BF) model to fit the proposed soliton (boson) and electron in C_{60}^{3-} [19]. This adaptation in the so-called "atomic limit" provides a familiar framework of local bonding, antibonding, and nonbonding states between bosons and fermions [20, 21]. The properties of the BF model include the opening of a pseudogap between T^p, where the incoherent pairs are first formed, and T_c, where superconductivity begins. The impact of this pseudogap should be noticeable in the spectral features.

In the Wannier representation the BF Hamiltonian is

$$H = (D - \mu) \sum_{j,\sigma} c_{j,\sigma}^{\dagger} c_{j,\sigma} + t \sum_{j,\delta,\sigma} c_{j+\delta,\sigma}^{\dagger} c_{j,\sigma} + (\Delta_B - 2\mu) \sum_{j} b_j^{\dagger} b_j + g \sum_{j} [c_{j,\uparrow}^{\dagger} c_{j,\downarrow}^{\dagger} b_j + b_j^{+} c_{j,\downarrow} c_{j,\uparrow}],$$

where μ denotes the common chemical potential to insure total charge conservation, $n_{tot} = n_{F\uparrow} + n_{F\downarrow} + 2n_B$ (n_B and $n_{F\uparrow,\downarrow}$, respectively, are the total number of hard-core bosons and electrons with up/down spin states confined to a C_{60} molecule) or, stated differently,

$$n = \frac{1}{N}\left(\sum_{j,\sigma} \langle c_{j\sigma}^{\dagger} c_{j\sigma} \rangle + 2 \sum_{j} \langle b_j^{\dagger} b_j \rangle \right).$$

The intermolecular hopping integral for the itinerant electrons is t, with a half bandwidth given by $D = zt$, z denoting the coordination number of the underlying lattice; Δ_B is the energy level of the hard-core bosons relative to electron level, a very important feature. This model has been scrutinized relative to BEC for several years. The main theme we intend to utilize is that below a certain temperature, T^p, local electron pairing, for us confined to a single C_{60} molecule, begins to form and exist over a finite time, as well as space, interval. As a consequence, a pseudogap opens up in the density of states (DOS) of the itinerant electrons centered on the Fermi level.

For each site the Hilbert space is the direct product of bosonic and fermionic states,

$$\{|0\rangle, |\uparrow\rangle, |\downarrow\rangle, |\uparrow\downarrow\rangle\} \otimes \{[0), [1)\}.$$

The local interaction term, v, mixes the states, and the resulting states are:

$$\text{Bonding } |B^-\rangle = U|\uparrow\downarrow\rangle \otimes [0) - V|0\rangle \otimes [1)$$

$$\text{Antibonding } |B^+\rangle = V|\uparrow\downarrow\rangle \otimes [0) + U|0\rangle \otimes [1).$$

The eigenvalues are

$$\varepsilon_\pm = (2\varepsilon_0 + E_0)/2 \pm \gamma$$

with

$$U^2 = \frac{1}{2}\left(1 - \frac{2\varepsilon_0 - E_0}{2\gamma}\right)$$

$$V^2 = \frac{1}{2}\left(1 + \frac{2\varepsilon_0 - E_0}{2\gamma}\right) \quad UV = \frac{v}{2\gamma}$$

with $\varepsilon_0 = zt - \mu$, $E_0 = \Delta_B - 2\mu$, and

$$\gamma = \sqrt{(\varepsilon_0 - E_0/2)^2 + v^2}.$$

The local one-particle fermionic Green's function,

$$G_{Loc}(\tau) = -\langle T[c_{j,\uparrow}(\tau)c^\dagger_{j,\uparrow}]\rangle$$

is

$$G_{Loc}(i\omega_n) = \frac{Z_f}{i\omega_n - \varepsilon_0} + \frac{1 - Z_f}{i\omega_n - v^2/(i\omega_n + \varepsilon_0 - E_0)},$$

with Z_f a temperature-dependent factor weighting the population of each state. The Green's function contains three poles with spectral weights Z_f, $U^2(1 - Z_f)$, and $V^2(1 - Z_f)$. The first term is the contribution from the nonbonding one-particle fermionic state, and the second term is the hybridized two-particle state (bonding $|B^-\rangle$ and antibonding $|B^+\rangle$). For simplicity, we consider only the symmetric case where $\langle c^\dagger_{j,\sigma}c_{j,\sigma}\rangle = \langle b^\dagger_j b_j\rangle = 1/2$ and, thus, ε_0, $E_0 = 0$. Then, $Z_f = 2/(3 + \cosh(\beta v))$ and βv is the only parameter ($\beta = 1/k_BT$). As T goes to zero, the nonbonding component's spectral weight goes to zero [21, 22].

If we generalize the above "atomic" Green's function to a lattice and add a kinetic term from the Hubbard model (diagonal in $\bar{k}$ space),

$$G(\bar{k}, i\omega_n) = \frac{1}{[G_{Loc}(i\omega_n)]^{-1} - t_k}$$

(t_k is the Fourier transform of the hopping term.) The three-pole structure in the atomic Green's function is still present in the lattice Green's function. With a parabolic density of states, $\rho_0(\varepsilon)$ and a fermion bandwidth, $D = 2zt$, and $-D/2 \le \varepsilon \le D/2$,

$$\rho_0(\varepsilon) = \frac{8}{\pi D}\sqrt{(\varepsilon/D + 1/2)(\varepsilon/D - 1/2)},$$

provided the temperature is zero, we now have a two-pole Green's function,

$$G(\bar{k}, i\omega_n) = \frac{1}{2}\left(1 + \frac{t_{\bar{k}}}{4\gamma_{\bar{k}}}\right)\frac{1}{i\omega_n - t_{\bar{k}}/2 - \gamma_{\bar{k}}} + \frac{1}{2}\left(1 - \frac{t_{\bar{k}}}{4\gamma_{\bar{k}}}\right)\frac{1}{i\omega_n - t_{\bar{k}}/2 + \gamma_{\bar{k}}}.$$

With $\gamma_{\bar{k}} = \sqrt{t_{\bar{k}}^2/4 + v^2}$, the density of states

$$\rho(\varepsilon) = \rho_0(\varepsilon - v^2/\varepsilon)[\theta(\varepsilon - E_1^-)\theta(E_1^+ - \varepsilon) + \theta(\varepsilon - E_2^-)\theta(E_2^+ - \varepsilon)]$$

where

$$E_1^\pm = \pm\frac{D}{4} + \frac{1}{2}\sqrt{\left(\frac{D}{2}\right)^2 + 4v^2}$$

$$\text{and } E_2^\pm = \pm\frac{D}{4} - \frac{1}{2}\sqrt{\left(\frac{D}{2}\right)^2 + 4v^2}$$

at $T = 0$ shows a gap at the Fermi level of size $\Delta_G = 2E_1^-$. Several different groups with various approaches obtained this conclusion [19, 21, 22]. All reached the same conclusion for such a gap in the density of states that deepens as the temperature is lowered. (There are other approaches that generate a pseudogap, i.e., Fisher's $\mathbb{Z}_2$ formulation [23], and Schrieffer's "spin bag," or spin fluctuation [24]).

As the temperature is decreased, spectral weight is transferred from the Fermi energy to the bands with contributions from the bonding and antibonding two-electron states. The reduction of the Fermi energy spectral density below the characteristic temperature T^p creates the opening of a pseudogap in the density of states. So, below T^p, the electrons get heavily damped, while the electron pairs continue to become better defined quasiparticles. The resistivity just before the superconducting state resembles that of an insulator, and it is determined by electrons with a pseudogap. As the superconducting state is approached, the transport is increasingly controlled by the phase fluctuations of the electron pairs.

Mesoscopic Coupling Leading to Superconductivity

There are a number of theories that begin with "precondensed pairs" of electrons and diverge from one another for a number of reasons. The original concept of preformed pairs, described by Nozieres and Schmitt-Rink [25], examined the free energy of pairs and the effects on a superfluid phase transition as Bardeen, Cooper, Schrieffer theory crosses over to Bose-Einstein condensation. More recently, the BF model of Ranninger [26] was conceived as a "bipolaron" theory that postulated preformed pairs with strong electron–phonon coupling that bose-condensed at T_c. Micnas et al. [27] considered a model of a two-component superconductor consisting of coexisting preformed local electron pairs and itinerant fermions. The authors examined a number of possibilities including BCS–mean field theory, Kosterlitz-Thouless theory, and BCS-to-BEC crossover. We propose a preformed (above T_c) boson topologically stabilized on a C_{60} "mesostructure." As mentioned above, one method of attack on the long-distance RVB's complexity has been the creation of mesostructures and examining their relationships. In our case, C_{60} is a natural case in itself, as it contains a rich internal structure, as evidenced by the various phases upon doping, especially the $n = 2$, 3, and 4 electrons, insulating, superconducting, and insulating, respectively, at low temperatures. Last, and in a somewhat different vein, Ohashi and Griffin [28] discussed BCS–BEC crossover of Fermi atoms with preformed Cooper pairs in which the attractive interaction contains a Feshbach resonance. We refer to the works mentioned below, because the analysis performed provides a framework of support for the pairing mechanism we propose.

The complete Hamiltonian for the superconducting C_{60}^{3-} is

$$H = H_{kef} + H_{keb}(local) + H_{JT} + H_{\rm int},$$

where the first two terms are the kinetic terms for itinerant fermions and localized (confined) bosons containing terms discussed in the SSH and BF models. The term H_{JT} is the JT interaction described in the third section, and we have modified

$$H_{\rm int} = -\sum_{i,j} J_{ij} b_i^\dagger b_j + \frac{1}{\sqrt{N}} \sum_q I\phi_k(B_q^\dagger b_q + b_q^\dagger B_q).$$

Here, the first term is a boson-hopping term, effectively zero, until the superconducting state is obtained; I is the coupling constant; and $B_q^\dagger = \sum_k \phi_k c_{k+(q/2),\uparrow}^\dagger c_{-k+(q/2),\downarrow}^\dagger$ is the singlet pair creation operator of c-electrons. The pairing symmetry is ϕ_k, which is important in the intermolecular boson formation.

As discussed, the JT bosons allow the formation of a pseudogap at T_p through their interaction with itinerant electrons, and we call this the microscopic portion of the theory. The detailed mesoscopic coupling proposal begins with any two adjacent fullerene molecules. As the temperature is further lowered, the electrons get heavily dampened and the boson energy level is lowered. As this level gets close to the bottom of the fermionic band (Figure 1), the effective attraction varies as $I^2/2(\Delta - \mu)$ with $\Delta \to \mu$ (remember that I is a coupling constant and negative); put in terms of an overall effective force,

$$U_{eff} = U_{el} + I^2/2(\Delta - \mu),$$

that is, the effective attraction mediated by the electronic bosons overcomes the strong electronic repulsion, U_{el}, and U_{eff} becomes negative and allows coupling between the itinerant electrons at T_c, allowing an intermolecular boson to be formed. To compare with a simplified BCS interaction, we use

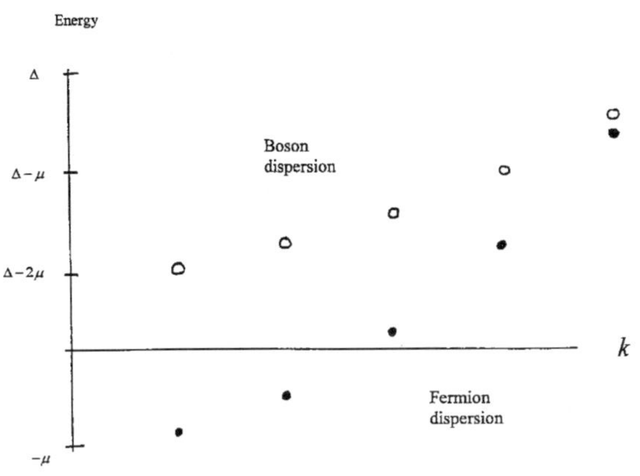

FIGURE 1. Fermion–boson Feshbach resonance.

a "jellium" model, with the solid approximated by a fluid of electrons and point ions developed by Pines [29]. As elaborated by de Gennes [30], the model in a certain approximation gives

$$U_{eff} = U_{coul} + U_{electron\text{-}phonon}$$

or

$$U_{eff} = U_{coul} + \frac{4\pi e^2}{q^2 + k_s^2} \cdot \frac{\omega_q^2}{\omega^2 - \omega_q^2},$$

where ω_q is a characteristic frequency. If $\omega < \omega_q$, the second term becomes attractive. The physical idea is that one electron polarizes the medium by attracting positive ions, such that these positive ions in turn attract the second electron, giving an effective attraction strong enough to overcome the repulsive coulomb interaction. Then, superconductivity results. In our model we have replaced the phonons by an electronic mechanism of attraction. The attraction is strongest with $(x^2 - y^2)$ symmetry. Excess energy from the new type of boson causes wild fluctuations, with dissipation most likely due to formation of quasiparticles by a fluctuating condensate particle number [31]. This excess energy is transferred to the confined bosons, enabling them to gain energy and allowing all bosons to become equivalent. (This is not a Bose-Einstein condensation of preformed pairs.) A phase diagram is shown in Figure 2. Many of the "free" bosons phase lock and participate in the newly formed resonating valence bonds, thereby resulting in superconductivity. Note that the superconductivity described by U_{eff} is entirely electronic in nature, with a second-order "assist" by JT vibrons in establishing the preformed pair. Stated differently, at T_c there is predominantly one kind of Cooper pair, as the preformed molecular bosons have been deconfined. The coupling between the intermolecular boson (Cooper pair) and the hard-core boson b_q is through the center of mass of the two subsystems. The following expression now holds for N, the number of electrons for the C_{60} dimer,

$$N = N_{Cooper} + N_{preformed} = 6,$$

and the two fullerene molecules can contribute up to three Cooper pairs to the "resonating spin liquid". It is interesting to note that the transition to superconductivity is a dimensional increase from a confined two-dimensional C_{60} "crystal" to a three-dimensional "spin liquid" with resonating valence bonds.

A connection at this point can be made with current phase-transition theories of Senthil et al. [32a] in the context of Moessner et al. [32b] and Fradkin and Shenker [32c]. A C_{60} molecules seems to behave as an SR-RVB plaquette. The addition of three electrons permits a charge-2 scalar (soliton) to exist on the plaquette. As the temperature is lowered below T_c, the opening of a deconfining phase is allowed as the dimension is raised from two to three. Based on the short experimental coherence length, it would seem as though a boson band is not formed, but a spin liquid fluctuating from plaquette to plaquette is.

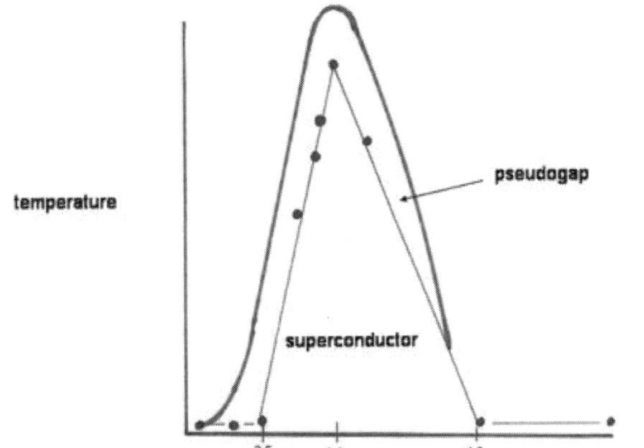

FIGURE 2. Phase diagram for Fullerides.

It is interesting to discuss the mechanism here in the context of the "exciton" superconducting mechanism that has not been found to date [33]. The logic was as follows: Consider a polymer or sandwich-type compound where electronic polarizability could be transferred from the side chains or layers. Physically, one electron polarizes the lattice and a second electron takes advantage of it to give an attractive interaction, an "electronic" BCS-type mechanism. The superconducting transition temperatures can be estimated from a BCS-type equation:

$$T_c(BCS) = \hbar\omega_D \exp\left(\frac{1}{\lambda}\right) \quad T_c(exciton) = E_{ex}\exp\left(\frac{1}{\lambda}\right).$$

Now the difference between prefactors $\hbar\omega_D$ and E_{ex}, where the former is the Debye temperature and the latter is the electronic equivalent, is two to three orders of magnitude (larger for E_{ex}). This has led to speculation that much higher T_c values (approaching room temperature) might be obtainable with an electronic mechanism, hence, the original interest.

Fullerene and HTSC Microscopic and Mesoscopic Structures

Through the course of this paper we discuss HTSC and fullerene superconductivity as though they were one in the same (and this seemed natural given that we were unaware of the RVB and BF models' connection). We originally embarked on the study of fullerenes as a two-dimensional PA, C_nH_n, that was compactified. If we had a planar two-dimensional C_nH_n structure, we could draw a circle at some large radius, R, centered at the origin, and create a boundary based on complete rings of C_nH_n. (Admittedly this is somewhat naive, as the five-membered rings in C_{60} are crucial (but see Fig. 1(a) in Ref. [10]). We could then fit the two-dimensional C_nH_n structure onto the surface of a sphere with the origin at the south pole, and the circle at the north pole with R contracting to a point. This allows us to "compactify" C_nH_n, say, $n = 60$, into a sphere. A good feature of this construction is that we have a finite (molecular) structure as opposed to an essentially infinite plane.

We can do the opposite and allow the C_nH_n fullerene sphere to be opened onto plane and use many pieces of the C_{60} mesoscopic structure to describe what could be similar features in planar HTSC. For example, the loop solitons might become planar stripes and could possibly interact with a passing, slow-moving (low-temperature) itinerant electron to create an attraction to an adjacent copper oxide plane in a similar attractive condition. Then, what is intermolecular pairing in C_{60} could be viewed as interplanar pairing in HTSC. This is not as strange as it might seem, with the knowledge that the BFM and the RVB theories have been mapped into each other [34].

Possible Experimental Determination of Precursor $2e$ Bosons

Here we discuss similarities and differences in relation between normal-state resistivity R and nuclear spin lattice relaxation time T_1 in high T_c cuprates and in fullerides.

We begin with a brief summary of the arguments given by Egorov and March [35] based on carriers in high T_c cuprates moving through an assembly of Cu ions exhibiting antiferromagnetic spin fluctuations. Employing the results of two-dimensional Fermi liquid theory for this system [36], the in-plane resistivity R is related to the magnetic susceptibility $\chi(\underset{\sim}{Q})$ at the antiferromagnetic wave vector $\underset{\sim}{Q}$ by

$$R \propto T^2\chi(\underset{\sim}{Q}). \quad \text{(VII.1)}$$

If $\chi(\underset{\sim}{Q})$ has Curie-Weiss behavior $\propto 1/T$, with T the absolute temperature, then one recovers the characteristic $R \propto T$ form. Combining Eq. (VII.1) with the corresponding result

$$(TT_1)^{-1} \propto \chi(\underset{\sim}{Q}). \quad \text{(VII.2)}$$

For the temperature dependence of the nuclear spin lattice relaxation time T_1, Egorov and March [35] eliminated the susceptibility χ between Eqs. (VII.1) and (VII.2), to find

$$RT_1 \propto T \quad \text{(VII.3)}$$

from the two-dimensional Fermi liquid assumption. Their use of experimental data for R and T_1 led to the result shown in Figure 2 of [35]. Equation (VII.3), reproduced well above T_c, nevertheless deviates from the experimental curve as T_c is approached from above. This was interpreted as a fingerprint of the formation of precursor 2*e* bosons [37, 38]. The evidence for such pairing above T_c in the cuprates was subsequently reviewed by Timusk and Statt [39].

This situation is next compared and contrasted with that in the fullerides, having been reviewed by Gunnarsson [3].

Some Facts Concerning Alkali-Doped C_{60} Compounds

Below we are concerned mainly with two classes of the superconducting fullerides: (1) A_3C_{60}, where A = K, Rb, Cs, or some combination of these alkali elements; and (2) $Na_2A_xC_{60}$ ($x \leq 1$), though Gunnarsson [3] also reviewed three other classes.

In classes (1) and (2) with $x = 1$, three electrons are doped into the t_{1u} [3], whereas in class (2) with $x < 1$, the valence is less than three.

To make contact with the "model" prediction of Egorov and March [35] for the high T_c cuprates, we refer immediately to the nuclear magnetic resonance studies of Maniwa et al. [40] on $NaAC_{60}$. Because we desire to construct the simplest possible model for fullerides having a degree of realism, we single out the case when $A = Cs$ or the ^{13}C spin-lattice relaxation time T_1 as measured by Maniwa et al. [40]. As their Figure 2 shows for the fulleride Na_2CsC_{60}, one has over a temperature range extending up to 100 K the experimental finding

$$T_1T = \text{constant}. \tag{VII.4}$$

We combine this result [Eq. (VII.4)] with a form of resistivity for doped fullerenes emerging from various experimental articles [41–45]

$$R = a_1 + a_2T^2. \tag{VII.5}$$

To compare with Eq. (VII.3), we use the model relations Eqs. (VII.4) and (VII.5) to obtain a further simplistic model prediction but now for the alkali-doped C_{60} compounds as

$$RT_1 = \frac{b_1}{T} + b_2T. \tag{VII.6}$$

Thus, Eq. (VII.6) would predict a linear behavior of RT_1 at high temperature but an increase in RT_1 at lower temperature as the term b_1/T begins to play a role. This is qualitatively compatible with the high T_c cuprate behavior shown in Figure 2.

It is tempting to suggest therefore that the constant b_1 in Eq. (VII.6) could have a fingerprint of a precursor 2*e* boson in the doped fullerenes and further experiments on, for example, Na_2CsC_{60} to examine this effect may prove to be of considerable interest.

Conclusions and Future Work

There are several obvious conclusions: (1) The model set out in the third section has a microscopic and a mesoscopic character for fullerene superconductivity, illustrating confining and deconfining transitions; (2) it predicts a pseudogap for fullerenes; and (3) we believe that this model has similarities with the RVB model for HTSC. To our knowledge, a pseudogap has not been observed in fullerenes, but Eqs. (VII.5) and (VII.6) above are consistent with available experiments and may have fingerprints of precursor bosons.

In most cases, the next step would be to better define the other interactions that participate in the superconducting state. However, we recently became aware of the relation between the BF model and RVB. So, rather than attempt to construct a superconducting theory from first principles, it seems far more appropriate to establish firmer connections with RVB. This seems especially appropriate because a criticism of the BF model has been the ambiguity of model parameters; we know how many electrons and bosons we have and we have two-order parameters and a superfluid stiffness. Reconciliation with RVB is needed here.

Anderson [46] predicted that superconductivity breaks down once the single-particle mean level spacing *d* becomes larger than the (bulk) superconducting gap Δ. Anderson's theory, based on an assumption as to where the fluctuation domination region begins, has been tested with metal grain superconductors. This prediction was made before HTSC and Fullerides were discovered. If the pairing mechanism proposed in this paper is correct, Fullerides and HTSC compounds would provide a much more controlled test and in principle, answer the question, "How many molecules does it take to have a superconductor?"

References

1. March, N. H.; Squire, R. H. Int J Quantum Chem 2003, 92, 261.
2. (a) Chakravarty, S.; Kivelson, S. e-print cond-mat/0012305; (b) Kivelson, S. A.; Rokhsar, D. S.; Sethna, J. P. Phys Rev 1987, B35, 8865; (c) Emery, V. J. Phys Rev Lett 1990, 65, 1076.
3. Gunnarsson, O. Rev Mod Phys 1997, 69, 575.
4. (a) Longuet-Higgins, H. C.; Salem, L. Proc R Soc London Series A 1959, 251, 172; Pople, J. A.; Walmsley, S. H. Mol Phys 1962, 5, 15; Su, W.-P.; Schrieffer, J. R.; Heeger, A. Phys Rev Lett 1970, 42, 1698; Jaciw, R.; Schieffer, J. R. Nucl Phys 1981, B 190 [FS3], 253; (b) Heeger, A. J.; Kivelson, S.; Schrieffer, J. R.; Su, W.-P. Rev Mod Phys 1988, 60, 781; (c) for a very readable account, see Jackiw, R. Diverse Topics in Theoretical and Mathematical Physics; World Scientific: Singapore, 1995; 79 and 449; (d) Tolbert, L. M. Acc Chem Res 1992, 25, 561; (e) Matsutani, S. e-print math-ph/0308027.
5. Fradkin, E. Field Theory of Condensed Matter Systems; Addison-Wesley: Redwood City, 1991.
6. Anderson, P. W. The Theory of Superconductivity in the High T_c Cuprates; Princeton University Press: Princeton, 1997.
7. (a) Wiegmann, P. B. Prog Theor Phys 1992, 107, 243; Phys Rev 1999, B59, 15705; Abanov, A. G.; Wiegmann, P. B. hep-th/0006157; hep-th/9911025; hep-th/0105213; (b) for a perspective, see Angilella, G. G. N.; March, H. H.; Pucci, R.; Squire, R. H. Phys Chem Liquids 2002, 40, 353.
8. Fröhlich, H. Proc R Soc 1954, A223, 296.
9. Takayama, M.; Lin-Liu, Y. R.; Maki, K. Phys Rev B 1980, 21, 2388.
10. Coffey, D.; Trugman, S. A. Phys Rev Lett 1992, 69, 176.
11. (a) Brouwer, P. W.; Racine, E.; Furusaki, A.; Hatsugai, Y.; Morita, Y.; Mudry, C. e-print cond-may/0201580; (b) Figge, M. T.; Mostovoy, M.; Knoester, J. Phys Rev Lett 2001, 86, 4572; Voo, K.-K.; Mou, C.-Y. e-print/0309298.
12. Han, J. E.; Gunnarsson, O.; Crespi, V. H. e-print cont-mat/0208454.
13. Herzberg, G.; Longuet-Higgins, H. C. Disc Farad Soc 1963, 35, 77.
14. Witten, E. Nucl Phys 1983, B223, 422.
15. (a) Auerbach, A.; Manini, N.; Tosatti, E. Phys Rev 1994, B49, 12998, 13008; (b) O'Brien, M. C. M. J Phys C 1971, 4, 2524; J Phys C 1972, 5, 2045; J Phys C 1976, 9, 3153; J Phys A: Math Gen 1989, 22, 1779; Phys Rev 1996, B53, 3775; Phys Rev B 1996, 53, 3775; (c) a nice summary is Chancey, C. C.; O'Brien, M. C. M. The Jahn–Teller Effect in C_{60} and Other Icosahedral Complexes; Princeton University Press: Princeton, 1997; Varma, C. M.; Zaanen, J.; Raghavachari, K. Science, 1991, 254, 989; Gunnarsson, O. Phys Rev 1995, B51, 3493.
16. Manini, N.; Tosatti, E.; Doniach, S. e-print cond-mat/9407080; Santoro, G.; Manini, N.; Parola, A.; Tosatti, E. e-print cond-mat/9511093.
17. Stone, M. Bosonization; World Scientific: Singapore, 1994.
18. Kivelson, S.; Rokhsar, D. S. Physica 1988, C 153, 531; Rokhsar, D. S.; Kivelson, S. Phys Rev Lett 1988, 61, 2376.
19. Cuoco, M.; Noce, C.; Ranninger, J.; Romano, A. e-print cond-mat/0305075.
20. Doniach, S.; Sondheimer, E. H. Green's Functions for Solid State Physicists; Addison-Wesley: Redwood City, 1974; p 184.
21. Domanski, T.; Ranninger, J.; Robin, J. M. e-print cond-mat/9702056.
22. Devillard, P.; Ranninger, J. e-print cond-mat/0003378; Micnas, R.; Robaszkiewicz, S.; Bussman-Holder, A. e-print cond-mat/0212367.
23. Senthil, T.; Fisher, M. P. A. Phys Rev 2000, B62, 7850; J Phy A: Math Gen 2001, 34, L119.
24. Kampf, A.; Schrieffer, J. R. Physica 1990, B163, 267.
25. Nozieres, P.; Schmitt-Rink, J. J Low Temp Phys 1985, 59, 159.
26. Ranninger, J.; Tripodi, L. e-print cond-mat/0212332.
27. Micnas, R.; Robaszkiewicz, S.; Bussmann-Holder, A. e-print cond-mat/0309434.
28. Ohashi, Y.; Griffin, A. e-print cond-mat/0210185.
29. Pines, D. Phys Rev 1958, 109, 280.
30. de Gennes, P. G. Superconductivity in Metals and Alloys; Benjamin, W. A. New York, 1966, Reprinted by Addison-Wesley: Reading, 1989; p. 102.
31. Zapata, I.; Sols, F.; Leggett, A. J. e-print cond-mat/0212604.
32. (a) Senthil, T.; Balents, L.; Sachdev, S.; Vishwanath, A.; Fisher, M. P. A. e-print cond-mat/0312617; (b) Moessner, R.; Sondhi, S. L.; Fradkin, E. e-print condt-mat/0103396; (c) Fradkin, E.; Shenker, S. H. Phys Rev 1979, D19, 3682.
33. Little, W. A. Phys Rev 1964, 134, A1416; Ginzburg, V. L. Sov Phys JETP 1965, 20, 1549; Allender, D.; Bray, J.; Bardeen, J. Phys Rev 1973, B7, 1020.
34. Kochetov, E.; Mierzejewski, M. e-print cond-mat/0204420.
35. Egorov, S. A.; March, N. H. Phys Chem Liquids 1994, 28, 141.
36. Kohno, H.; Yamada, K. Prog Theor Phys 1991, 85, 13.
37. Moriya, T. Spin Fluctuations in Itinerant Electron Magnetism; Springer: Berlin, 1985; see also Moriya, T. Acta Phys Polonica 2003, B34, 287.
38. Egorov, S. A.; March, N. H.; Pucci, R. Phys Chem Liquids; see also March, N. H. Electron Correlation in Molecules and Condensed Phases; Plenum: New York, 1996; p 354.
39. Timusk, T.; Statt, B. Rep Prog Phys 1999, 62, 61.
40. Maniwa, Y.; Saito, T.; Kurne, K.; Kikuchi, K.; Ikemoto, I.; Suzuki, S.; Achiba, Y.; Hirosawa, I.; Kosaka, M.; Tanigaki, K. Phys Rev 1995, B52, R7054.
41. Xiang, X.-D., Hou, J. G.; Briceño, G.; Vareka, W. A.; Mostovoy, R.; Zettl, A.; Crespi, V. H.; Cohen, M. L. Science 1992, 56, 1190.
42. Klein, O.; Gruner, G.; Huang, S.-M.; Wiley, J. B.; Kaner, R. B. Phys Rev 1992, B46, 11247.
43. Hori, J. G.; Xiang, X.-D.; Zettl, A.; Cohen, M. L. referred to by Vareka, W. A.; Zettl, A. Phys Rev Lett 1994, 72, 4121.
44. Zhao, W. B.; Zhang, X. D.; Ye, Z. Y.; Zhang, J. L.; Li, C. Y.; Yin, D. L. Solid State Comm 1993, 85, 945.
45. Hebard, A. F.; Palstra, T. T. M.; Haddon, R. C.; Fleming, R. M. Phys Rev 1993, B48, 9945.
46. Anderson, P. W. J Phys Chem Solids 1959, 11, 28.

Available online at www.sciencedirect.com
SCIENCE DIRECT
ELSEVIER
PHYSICS LETTERS A
Physics Letters A 346 (2005) 355–358
www.elsevier.com/locate/pla

Fingerprints of quantal Wigner solid-like correlations in D-dimensional assemblies

Norman H. March [a,b], Richard H. Squire [c,*]

[a] *Donostia International Physics Center, 20018 San Sebastian, Spain*
[b] *Oxford University, Oxford, England, UK*
[c] *Department of Chemistry, West Virginia University, Institute of Technology, 1102 Orndorff Hall, Montgomery, WV 25136, USA*

Received 3 June 2005; accepted 11 June 2005

Available online 24 June 2005

Communicated by V.M. Agranovich

Abstract

After a brief summary of the mechanism of quantal Wigner electron crystallization, we emphasize experimental situations in three, two, and one dimensions in which fingerprints of Wigner solid-like correlations are in evidence. Special stress is placed on the way applied magnetic fields can aid Wigner localization, as first pointed out in relation to n-type InSb by Durkan et al. [Rev. Mod. Phys. 40 (1968) 812]. Semiconductor heterojunctions with a two-dimensional electron fluid in the presence of a transverse magnetic field are given some prominence in relation to melting of the Wigner solid and its connection with magnetic properties. It is also proposed that a Wigner "hole" solid may be induced by a magnetic field of 60 tesla in a specific high T_c cuprate and that the fulleride phase diagram may also contain Wigner solids.

Wigner's concept of a quantum electron crystal [1,2] goes back some seven decades. He was interested in the correlations between electrons induced by the Coulombic repulsion energy e^2/r_{ij}, where electrons i and j are separated by a distance, r_{ij}, originally in a simple monovalent metal like Na. Already it was known that in the ground state, provided the system was modeled by smearing out the granular sodium ions into a uniform, neutralizing, non-responsive background in which the Coulombic repelling electrons moved, the energy per electron, E/N, had the form

$$\frac{E}{N}=\frac{2.21}{r_s^2}-\frac{0.916}{r_s}, \qquad \rho=\frac{3}{4\pi r_s^3}. \tag{1}$$

Here, ρ denotes the uniform electron density and r_s measures the mean interelectronic spacing in the limit of small r_s or high density ρ. In this limit the ground-state wave function is usefully approximated as a Hartree–Fock determinant of plane waves $\exp(i\vec{k}\cdot\vec{r})$, where a sphere in $\vec{k}$ space is filled with electrons for

* Corresponding author.
E-mail address: rhsquire@charter.net (R.H. Squire).

0375-9601/$ – see front matter © 2005 Elsevier B.V. All rights reserved.
doi:10.1016/j.physleta.2005.06.050

$|\vec{k}| < k_F$, the Fermi wave number. The first term on the RHS of Eq. (1) is kinetic or Fermi energy, while the second term is exchange potential energy. What Wigner recognized from Eq. (1) was that as the density ρ was lowered, or equivalently r_s was increased, eventually in this "jellium" model the potential energy would dominate the kinetic energy. So, Wigner [1,2] examined the Madelung energy of various lattices towards the sites at which the electrons would go, he argued, to minimize their mutual interaction potential energy. His finding was dramatic: namely, that as $r_s \to \infty$ the Madelung energy of a body-centered cubic lattice that had the lowest value of this electrostatic energy, would yield instead of Eq. (1),

$$\frac{E}{N} \to \frac{-1.8}{r_s} \tag{2}$$

which is twice as low as the Hartree–Fock formula (1) yields as $r_s \to \infty$. Of course, this meant that a Slater determinant of delocalized electrons was a poor approximation to the wave function in the low density limit: the electrons were "localized".

Thus, at some value of the interelectronic spacing r_s, as the density is lowered, there was a phase transition in the ground state of "jellium" from an electronic fluid to an electronic crystal. Care and March [3] in their review already converged on the fact that the critical value of r_s for the phase transition was very much greater than that in the lowest density simple monovalent metal Cs, where r_s is 5.5 in atomic units $a_0 = \hbar^2/(me^2)$. Thus, these workers recognized that to find Wigner crystals in nature, the electron fluids in simple metals like the alkalis did not provide a useful medium in which to explore Wigner electron solids experimentally. Rather, it would be in the much lower electron densities found in suitable semiconducting systems, such as highly compensated InSb in three dimensions (3D), or much later, in heterojunctions such as GaAs/AlGaAs, in which low (areal) density 2D electron fluids could be created. For 2D electron assemblies, quantum computer simulations were carried out to complement the earlier three-dimensional study of Ceperley and Alder [4] who had found the critical r_s for 3D jellium to be around 80–100 atomic units, thus confirming that Wigner solids were to be sought experimentally in semiconducting media rather than in metals. The r_s in 2D jellium, measuring now areal density, was lower (around 27 a.u., see, e.g., [5]).

With this brief theoretical background, we turn to experiments on 3D, 2D, and 1D assemblies and enquire as to the strength of the evidence regarding fingerprints of Wigner solid-like correlations at low temperatures induced by Coulombic repulsion in quantum mechanical regimes. We begin with 3D and the early work of Durkan, Elliott, and March [6] who were interested in a metal–insulator transition induced in a highly compensated n-type InSb subjected to an applied magnetic field, $\vec{H}$. The conclusion of these authors was that, as the field strength $\vec{H}$ was increased, a Wigner transition was occurring at a critical magnetic field. Measurements of electrical resistivity on 3D n-type InSb by Somerford [7] (see also Care and March [8]) showed that a metal–insulator transition occurred at a critical magnetic field. This proposal of magnetically assisted Wigner transition has proved fruitful, particularly, and much more recently, on 2D electron fluids. In a GaAs/AlGaAs heterojunction subjected to a magnetic field perpendicular to the plane of the 2D electron assembly, the phenomenon now known as magnetically induced Wigner solid (MIWS) was seen. Indeed, just 20 years after the proposal by Durkan et al. [6], Andrei et al. [9] conducted a beautiful experiment that leads us from the 3D n-type InSb to the 2D heterojunction. One fingerprint of MIWS in 3D was pointed out by Lea et al. [10] using data from Kleppmann and Elliott [11]. In the two-dimensional heterojunction, the melting curve of the Wigner solid had reentrant form (see also Buhmann et al. [12], and Lea et al. [13]). In addition, the melting took place into a Laughlin liquid—of course, the medium for the fractional quantum Hall effect. Thermodynamic analysis of the reentrant melting curve as a function of magnetic field led Lea et al. [13] to a remarkable behavior of the magnetic susceptibility in the Laughlin liquid.

For the future it is important to consider more carefully the sliding of Wigner lattices and also the magnetism. The latter has been discussed from the standpoint of both experiment and theory by Okamoto and Kawasi [14]. Pinning effects prevent the Wigner lattice from sliding at low electric fields as discussed by Pudalov et al. [15]. The magnetic properties of the Wigner crystal are assumed to be determined by ring exchange interaction by analogy with bcc ^{3}He [16]. Okamoto and Kawasi present a calculation taking into account the Aharonov–Bohm effect on two-, three-, and four-electron ring exchange interactions, and are led to conclude that this calculation is in accord with their experimental findings. The ferromagnetic ground state they predict for the Wigner solid becomes unstable when the magnetic flux through the exchange path changes the three-electron ring exchange interaction to antiferromagnetic. This phase transition is thought to be magnetic field induced. All these aspects may eventually need consideration when more quantitative measurements become available on the reentrant melting curve that separates the Wigner solid in a 2D-heterojunction with a magnetic field perpendicular to the 2D electron liquid from the Laughlin electron liquid which is the seat of the fractional quantum Hall effect. It is also possible that magnetoquantum oscillations of tunneling conductance between two quasi-two-dimensional layers of electron liquids in equilibrium in tilted magnetic fields [17] will, under appropriate conditions, carry fingerprints of a tendency towards strong Wigner-like correlations.

Recent HTSC experimental studies of the copper oxide planes in cuprates by Ando et al. [18] show signatures of a quantum phase transition at 1/8-hole doping in an applied magnetic field of 58 T. We propose that at sufficiently low temperatures, the observed characteristics of the charge confinement may well be due to a MIWS of holes (see also [19]). This is a variant of the idea that applied magnetic fields would aid charge carrier repulsions in bringing about a MIWS as was pointed out by Durkan et al. [6] earlier. It would be of great interest if measurements prove possible on the specimens of Ando et al. as the magnetic field is varied of either the MIWS melting curve as a function of magnetic field, or of orbital magnetic susceptibility.

Very recently, Bylander et al. [20] have perfected a technique to observe individual electrons by sending a direct current through a microelectronic circuit with a one-dimensional chain of small "islands" connected by conducting or superconducting tunnel junctions. Small charges tunneling through this array are time-correlated thereby allowing Bylander et al. to measure single-electron tunneling oscillations. One experiment described uses a superconducting array with 50 junctions, each with a capacitance, C_A, with the stray capacitance of an electrode inside the array as C_0. An excess localized charge on one part of an array repels neighboring "islands" such that the resulting potential (called a charged soliton) is spread over several junctions; the soliton length, L, is $L = \sqrt{C_A/C_0}$. When the array is biased above a threshold, V_t, the solitons repel each other and a *moving* quasi-Wigner lattice of charged solitons is formed as shown both theoretically and experimentally (indirectly). In general both electron and Cooper pair tunneling can take place with the threshold for electron injection lower than that of the Cooper pairs.

There is a connection here with the work in which we have proposed that HTSC and fulleride superconductivity might have similar mechanisms based on their position on the "Uemura plot" [19,21]. Using our interpretation of Ando et al.'s work, Bylander's study becomes increasingly interesting as a method to test a superconducting model proposed by us for the fullerides [19]. In the fulleride phase diagram at two and four doped electrons on a fullerene crystal, the model suggests these doping levels produce two different Wigner crystals of charge carriers with a superconducting phase between them. Using a quantum rotor analogy [19,22,23], this portion of the phase diagram is suggestive of that of Buhmann et al. [12]. Might a one-dimensional array of either of these species produce a moving Wigner lattice? Can a 1D array of fullerene molecules be used in place of the superconducting "islands" such that doping might occur by a bias. Would it be possible to see a transition from a moving Wigner lattice to a pure superconducting current (and back)? One additional future direction is to describe this proposed Wigner solid of charge carriers more quantitatively in anticipation that it might be revealed spectroscopically [24].

Acknowledgements

N.H.M. wishes to acknowledge that his contribution to this study was brought to fruition during a visit to DIPC, San Sebastian. N.H.M. thanks Professors P.M. Echenique and A. Rubio for their very generous hospitality.

References

[1] E.P. Wigner, Phys. Rev. 46 (1934) 1002.
[2] E.P. Wigner, Trans. Faraday Soc. 34 (1938) 678.
[3] C.M. Care, N.H. March, Adv. Phys. 24 (1974) 101.
[4] D.M. Ceperley, B.J. Alder, Phys. Rev. Lett. 45 (1980) 567.

[5] N.H. March, M.P. Tosi, Coulomb Liquids, Academic Press, New York, 1984.
[6] J. Durkan, R.J. Elliott, N.H. March, Rev. Mod. Phys. 40 (1968) 812.
[7] D.J. Somerford, J. Phys. C 4 (1971) 1570.
[8] C.M. Care, N.H. March, J. Phys. C 4 (1971) L372.
[9] E.Y. Andrei, G. Deville, D.C. Glattli, F.I.B. Williams, E. Paris, B. Etienne, Phys. Rev. Lett. 60 (1988) 2765.
[10] M.J. Lea, N.H. March, Int. J. Quantum Chem. Symp. 23 (1989) 717;
N.H. March, Phys. Chem. Liq. 38 (2000) 151.
[11] H. Kleppmann, R.J. Elliott, J. Phys. C 8 (1975) 2747.
[12] H. Buhmann, W. Joss, K. von Klitzing, I.V. Kukuskhim, A.S. Plart, G. Martinez, K. Ploog, V.B. Tunofeev, Phys. Rev. Lett. 86 (1991) 926.
[13] M.J. Lea, N.H. March, W. Sung, J. Phys.: Condens. Matter 3 (1991) L4301;
M.J. Lea, N.H. March, W. Sung, J. Phys.: Condens. Matter 4 (1992) 5263.
[14] T. Okamoto, S. Kawasi, Phys. Rev. B 57 (1998) 9097.
[15] V.M. Pudalov, M. D'Iorio, S.V. Kravchenko, J.W. Campbell, Phys. Rev. Lett. 70 (1993) 1866.
[16] D.D. Osheroff, J. Low Temp. Phys. 87 (1992) 297, and references therein.
[17] S.K. Lyo, Phys. Rev. B 57 (1998) 9114.
[18] Y. Ando, S. Ono, X.F. Sub, J. Takeya, F.F. Balakirev, J.B. Betts, G.S. Boebinger, Phys. Rev. Lett. 92 (2004) 247004.
[19] R.H. Squire, N.H. March, in: Proc. Sanibel Symp., Int. J. Quantum Chem. (2005), submitted for publication.
[20] J. Bylander, T. Duty, P. Delsing, Nature 434 (2005) 361.
[21] Y.J. Uemura, L.P. Le, G.M. Luke, B.J. Sternlieb, W.D. Wu, J.H. Brewer, T.M. Riseman, C.L. Seaman, M.B. Maple, M. Ishikawa, D.G. Hinks, J.D. Jorgensen, G. Saito, H. Yamochi, Phys. Rev. Lett. 66 (1991) 2665;
Y.J. Uemura, cond-mat/0406301.
[22] M.P.A. Fisher, P.B. Weichman, G. Grinstein, D.S. Fisher, Phys. Rev. B 40 (1989) 546.
[23] A. Paramekanti, L. Balents, M.P.A. Fisher, cond-mat/0203171.
[24] H.A. Fertig, in: S.D. Sarma, A. Pinczuk (Eds.), Perspectives in Quantum Hall Effects, Wiley, New York, 1977.

phys. stat. sol. (b) **242**, No. 9, 1817–1819 (2005) / **DOI** 10.1002/pssb.200461735

On the phase diagram of a two-dimensional electron gas near integer fillings and fractions such as 1/5 and 1/7

Francisco Claro[*, 1], **Alejandro Cabo**[**, 2], and **Norman H. March**[3]

[1] Facultad de Física, Pontificia Universidad Católica de Chile, Santiago de Chile, Chile
[2] Grupo de Física Teórica, Instituto de Cibernética y Física, La Habana, Cuba
[3] Department of Physics, University of Antwerp, Antwerp, Belgium and Oxford University, Oxford, England

Received 11 October 2004, revised 5 May 2005, accepted 5 May 2005
Published online 29 June 2005

PACS 73.43.Cd

The phase diagram of the 2DEG at high magnetic fields is discussed in the light of recent experiments showing the presence of crystalline phases around integer as well as certain fractional fillings. We suggest that a Hall crystal is a proper competing phase compatible with the odd denominator rule that characterizes the fractional quantum Hall effect, and that near integer filling a pinned crystal coexists with an incompressible fluid with no longitudinal resistance.

The boundaries between different phases of the 2DEG at high magnetic fields has been the subject of much interest recently. Although originally it was believed that a Wigner Crystal (WC) would only form in the lower Landau level and below filling 1/7 [1], microwave resonance experiments showed that this state appeared around higher fractions such as 1/5 [2], and integer fillings like 1 and 2 [3]. Numerical diagonalization of the hamiltonian over a unit cell containing few particles with periodic boundary conditions also showed the presence of rich phases at a range of filling factors [4]. A schematic diagram of the melting curve characterizing the equilibrium between a Wigner solid and a Laughlin liquid for the thermodynamic limit was proposed more than a decade ago by Buhmann et al. [5, 6]. This was followed by a discussion based on a first order phase transition by Lea, March and Sung [7]. These authors subsequently applied microscopic theory, for both anyon and composite Fermion models, to demonstrate that the main features of the Buhmann et al. diagram could be explained by invoking the magnetization, of the de Haas–van Alphen character, of the Laughlin/CF liquid [8–10].

In this work we refine and extend somewhat the phase diagram of Buhmann et al. In the course of such refinement, we shall want to consider one further phase besides the Wigner solid: a 'Hall'crystal (HC) [11]. This state was found in early work by one of us (F.C.) [12, 13], using the Hartree–Fock (HF) approximation. It is a charge density wave with a fraction of an electron per unit cell that is conveniently taken to be 1/2. As ν is varied, the energy per particle presents cusps at all fillings with odd denominator, revealing a gap, while it forms a Fermi liquid at all even denominators in accordance with experiment. The cusps are hierarchically deployed, in such a way that fractions of smaller denominators are associated with larger gaps. Thus, the HC state at 1/5 presents a large gap while that at 2/11 has a smaller gap, and 3/19 is smaller still. Cabo and Claro later argued that the HC shows strong cohesive energy determined by cooperative ring exchange effects, making the HC a suitable candidate for the ground state

[*] Corresponding author: e-mail: fclaro@puc.cl
[**] e-mail: cabo@cidet.icmf.inf.cu

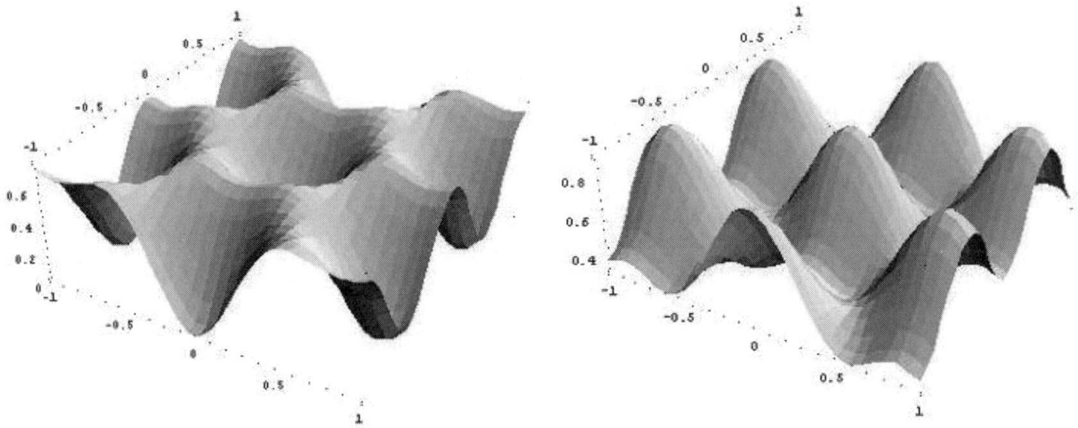

Fig. 1 (online colour at: www.pss-b.com) Charge density wave profile of the Hartree–Fock approximation for the electron distribution in the Hall crystal (left). For comparison, the Wigner crystal density is shown in the right.

[14]. Its second order correlation correction turns out to be an order of magnitude larger than that for the Wigner solid [15]. This strong enhancement may be understood on the basis of the electron densities of the competing states, shown in Fig. 1. As can be observed in the figure, the HC presents marked percolating channels, suggesting that the Wannier like localized states of the problem have appreciable overlap. This is in contrast with the WC structure that shows isolated gaussian-shaped peaks with no connecting channels between them. To be fully operative, cooperative rings of exchange effects require long paths with large exchange probabilities along them and these are more easily found in the HC state than in the WC.

The first experimental evidence for a Wigner crystal at low fillings was the finding of a sharp increase in the longitudinal resistance at the edges of the Hall plateau around filling 1/5, understood to arise from a pinned state with space charge fluctuations [16]. In a recent experiment, Pan et al. found that upon increasing the temperature this wing resistance decreased, developing structure consistent with the formation of further fractional quantum Hall effect (FQHE) at fractions such as 2/11, 3/17, 2/13 and 3/19 [17]. One interpretation of this finding is that the pinned WC formed at low temperatures melts down into a Laughlin liquid as the temperature is increased. Another interpretation is that the low temperature phase is a pinned HC exhibiting the characteristic FQHE odd-denominator rule, which becomes depinned as the sample is heated. Further heating finally destroys the FQHE with a temperature boundary profile consistent with the gap hierarchy exhibited by the HC state.

From the above description the following picture emerges. Fractions of small denominator such as 1/1, 1/3, 1/5, 1/7 and others form a liquid state of null resistance down to zero temperature. If a Hall crystal were to form at such fractions, pinning and the ensuing rise in longitudinal resistance would occur at lower temperatures than presently reported, possibly below a few mK. As the filling departs from such selective fractions a pinned crystal is formed, that could be a WC or a HC. Both would give a microwave signal and either model is consistent with the data of Refs. [2] and [3]. Both are also consistent with the data of Pan et al. [17], although the physics is quite different: in the WC model there is a high temperature FQHE liquid phase, as originally suggested by Buhmann et al. [5], while in the HC model the crystal is itself a FQHE state that becomes depinned as the temperature rises.

In connection with $\nu = 1$ and 2 we would like to add some comments related to the recent detection of crystalline states by Chen et al. [3], around such integral fillings. They find that the magnetic field region in which the microwave absorption shows peaks is well inside the plateaus [18]. Then, the question about the compatibility of the crystalline nature of the ground state and the quantized value of the Hall conductivity over a range of filling fractions emerges. We believe that the occurrence of such an effect is consistent with the results of Cabo and Martinez [19]. It is argued there that the linear response of a

2DEG in the integral quantum Hall regime can be roughly described as a tendency of the impurities to act as effective charge reservoirs, receiving or releasing the exact amount of electrons to ensure the local integral filling condition in large sample areas. The mechanism is triggered by the presence of a fairly sharp cusp in the energy per particle, making preferable energy-wise for the extra electrons or holes to occupy other available states than to join the main integral Hall state, which is preserved across de plateau. The following picture then seems a plausible way to explain the findings of Ref. [3]. First, when the filling factor is near the values $\nu = 1, 2$ for a very clean sample, it can be expected that the charge reservoirs receiving the excess or defect charges are the small number of imperfections present. However, as the filling deviates more and more from the integer value moving towards fractions such as $\nu = 10/9$ or 8/9, it could happen that the limited number of localization centers in the high quality sample are saturated to their capacity. Excess or defect charges would then cluster around impurities forming pinned crystalline regions which could be viewed as a system of dynamically generated localization centers. It is possible that such mechanism is also present near major fractions, such as 1/5, 1/7 and others.

Acknowlegdements N.H.M. and A.C. acknowledge the kind hospitality of the AS ICTP allowing this collaborative work and wish to thank Prof. V. E. Kravtsov for the stimulating atmosphere provided. F.C. thanks the support of FONDECYT through grants 1020829 and 7020829.

References

[1] P. K. Lam and S. M. Girvin, Phys. Rev. B **30**, 473 (1984).
[2] T. Sajoto, Y. P. Li, L. W. Engel, D. C. Tsui, and M. Shayegan, Phys. Rev. Lett. **70**, 2321 (1993).
[3] Y. P. Chen, R. M. Lewis, L. W. Engel, D. C. Tsui, P. D. Ye, L. N. Pfeiffer, and K. W. West, Phys. Rev. Lett. **91**, 016801 (2003).
[4] N. Shibata and D. Yoshioka, Phys. Rev. Lett. **86**, 5755 (2001).
[5] H. Buhmann, W. Joss, K. von Klitzing, I. V. Kukushkin, G. Martinez, A. S. Plaut, K. Ploog, and V. B. Timofeev, Phys. Rev. Lett. **65**, 1056 (1990).
[6] R. Price, X. Zhu, and P. M. Platzman, Phys. Rev. B **48**, 11473 (1993).
[7] M. J. Lea, N. H. March, and W. Sung, J. Phys.: Condens. Matter **3**, 6810 (1991).
[8] M. J. Lea, N. H. March, and W. Sung, J. Phys.: Condens. Matter **4**, 5263 (1992).
[9] Y. S. Wu, Y. Yu, Y. Hatsugai, and M. Kohmoto, Phys. Rev. B **57**, 9907 (1998).
[10] N. H. March, Phys. Chem. Liquids **38**, 151 (2000).
[11] B. I. Halperin, Z. Tesanovic, and F. Axel, Phys. Rev. Lett. **57**, 922 (1986).
[12] F. Claro, Solid State Commun. **53**, 27 (1985).
[13] F. Claro, Phys. Rev. B **35**, 7980 (1987).
[14] A. Cabo and F. Claro, Phys. Rev. B **70**, 235320 (2004).
[15] D. Yoshioka and P. A. Lee, Phys. Rev. B **27**, 4986 (1983).
[16] H. W. Jiang, R. L. Willett, H. L. Stormer, D. C. Tsui, L. N. Pfeiffer, K. W. Baldwin, and K. W. West, Phys. Rev. Lett. **65**, 633 (1990); Phys. Rev. B **44**, 8107 (1991).
[17] W. Pan, H. L. Stormer, D. C. Tsui, L. N. Pfeiffer, K. W. Baldwin, and K. W. West, Phys. Rev. Lett. **88**, 176802 (2002).
[18] D. C. Tsui, private communication.
[19] A. Cabo Montes de Oca and D. Martinez-Pedrera, Phys. Rev. B **67**, 245310 (2003).

Available online at www.sciencedirect.com
SCIENCE @ DIRECT®
ELSEVIER
Physics Letters A 339 (2005) 207–211
PHYSICS LETTERS A
www.elsevier.com/locate/pla

Wigner bosonic molecules with repulsive interactions and harmonic confinement

P. Capuzzi [a,*], N.H. March [b,c], M.P. Tosi [a]

[a] NEST-INFM and Classe di Scienze, Scuola Normale Superiore, Piazza dei Cavalieri 7, I-56126 Pisa, Italy
[b] Department of Physics, University of Antwerp, 171 Groenenborgerlaan, B-2020 Antwerp, Belgium
[c] Oxford University, Oxford, England, UK

Received 16 February 2005; accepted 22 February 2005

Available online 21 March 2005

Communicated by V.M. Agranovich

Abstract

In earlier work, the momentum distribution of harmonically confined bosons has been calculated for interparticle interactions which are also harmonic for an arbitrary number N of repelling Bose particles. After a brief discussion of the pair function in this model, attention is also given, but now for $N = 2$, to the single-particle density for the repulsive interaction potential λ/r_{12}^2, where r_{12} is the interboson distance and λ the repulsive coupling strength. The single-particle density is displayed for specific values of λ, with attention focusing on the limit of strong repulsive coupling. Throughout, the focus is on the fingerprints of Wigner bosonic molecules.

Keywords: Wigner bosonic molecules; Exactly solvable models; Harmonic confinement; Harmonic interactions

In earlier work [1], attention has been focused on the momentum distribution of harmonically confined bosons for interparticle interactions which also have harmonic form for an arbitrary number N of repelling (and also attracting) Bose particles. In this Letter, following a brief discussion of the pair correlation function $P(\mathbf{r}_1, \mathbf{r}_2)$ in this model, attention is also given, but now for $N = 2$, to the single-particle density $\rho(r)$ for the repulsive interaction potential changed to the form λ/r_{12}^2 where $r_{12} = |\mathbf{r}_1 - \mathbf{r}_2|$ denotes the interboson separation while λ measures the repulsive coupling strength. In particular, the ground-state particle density $\rho(r)$ is displayed for specific values of λ, with attention focusing on the change from the Gaussian form of $\rho(r)$ as λ goes from zero to large values of this repulsive coupling strength. It will be shown that $\rho(r)$ for large λ already contains clear fingerprints of the formation of a Wigner bosonic molecule.

Following the culture of the background to the present study, let us utilize the early work of Cohen

* Corresponding author.
E-mail address: capuzzi@sns.it (P. Capuzzi).

0375-9601/$ – see front matter © 2005 Elsevier B.V. All rights reserved.
doi:10.1016/j.physleta.2005.02.078

and Lee [2] on the low-order density matrices for N bosons with harmonic confinement and also harmonic interparticle repulsion. In particular, from Eq. (2.35) of [2] for the second-order density matrix, we can write the diagonal pair function $P(\mathbf{r}_1, \mathbf{r}_2)$ already referred to above as

$$P(\mathbf{r}_1, \mathbf{r}_2) = \frac{N(N-1)}{2}\left\{\frac{N\omega\delta_N^2/\pi^2}{(N-2)\omega + 2\delta_N}\right\}^{3/2} \times \exp\{-2b_1(r_1^2 + r_2^2) - 2b_2\mathbf{r}_1 \cdot \mathbf{r}_2 + b_3(2\mathbf{r}_1 \cdot \mathbf{r}_2 + r_1^2 + r_2^2)\}, \tag{1}$$

where b_1–b_3 are recorded explicitly in Eqs. (2.38)–(2.40) of [2]. Finally, $\delta_N^2 = \omega^2 \pm N\gamma^2$, where the confinement potential V is $\frac{1}{2}\omega^2 r^2$ and the interaction potential is $\pm\frac{1}{2}\gamma^2 r_{ij}^2$. The limit of strong repulsions is achieved for $\delta_N \simeq 0$ and a further increase of γ would break the external confinement. The important point here is the appearance in Eq. (1), via the scalar product $\mathbf{r}_1 \cdot \mathbf{r}_2$, of the angle between the position vectors $\mathbf{r}_1$ and $\mathbf{r}_2$, superposed of course on 'confinement' terms of the expected Gaussian form. If we fix particle 1 say at position $(0, 0, a)$ in Cartesian coordinates, then the top panel in Fig. 1 displays $P([0, 0, a], \mathbf{r}_2)$ to demonstrate this conditional probability distribution of boson separation for two bosons. For large bosonic repulsion, the tendency towards 'particle localization' is becoming evident. The effect of repulsions is more clearly seen in the behavior of the pair function $g([0, 0, a], \mathbf{r}_2)$, defined as

$$g(\mathbf{r}_1, \mathbf{r}_2) = \frac{2N}{N-1}\frac{P(\mathbf{r}_1, \mathbf{r}_2)}{\rho(\mathbf{r}_1)\rho(\mathbf{r}_2)}. \tag{2}$$

For non-interacting particles $g(\mathbf{r}_1, \mathbf{r}_2) \equiv 1$ reflecting the absence of correlations, while on switching on the interactions g develops a peak at $\mathbf{r}_2 = (\delta_N + \omega)/(\delta_N - \omega)\mathbf{r}_1$, which tends to $\mathbf{r}_2 = -\mathbf{r}_1$ in the limit of strong interactions. This is accompanied with a diminution of the width of g along the direction of $\mathbf{r}_1 - \mathbf{r}_2$. The location of the peaks is expected since, given a particle at $\mathbf{r}_1$, in order to minimize the energy at large interparticle repulsions, r_{12} must be maximal and thus $\mathbf{r}_2 = -\mathbf{r}_1$. These effects are illustrated in the bottom panel of Fig. 1.

We have also considered the distributions of 4 and 6 bosons with interparticle harmonic repulsions. In this case g has a maximum when

$$\mathbf{r}_2 = \frac{\delta_N + (N-1)\omega}{\delta_N - \omega}\mathbf{r}_1, \tag{3}$$

however, its width is non-monotonic on increasing the strength of the interaction, diverging for extremely low values of δ_N. This suggests that, in contrast to the two-boson problem, the spatial distribution of 4 or 6 bosons does not possess a structure at large couplings. In Fig. 2 we show contour plots of P for $N = 2, 4$ and 6 bosons and two values of the interparticle interaction close to the strong repulsion limit. The contour plots evince the strongly elongated structure of the pair function for two bosons at $\delta_N \simeq 0$ and illustrate the regular behavior of the distributions of 4 and 6 particles on increasing δ_N. However, we stress here that the ground-state density $\rho(r)$ retains its spherical Gaussian form for all N, though naturally the half-width of the Gaussian depends on the strength of the interparticle interactions. These comments parallel those pertaining to the momentum distribution in [1].

This is the point to consider the alternative choice of interparticle repulsion energy λ/r_{12}^2. Here we can draw on the work of Crandall et al. [3]. While these authors were concerned exclusively with a two-electron atomic model, with therefore no reference to bosons, we can utilize the fact that they have an exact, spatially symmetric solution of the Schrödinger equation for confinement potential $V(r)$ given by

$$V(r) = \frac{1}{2}m\omega^2 r^2, \tag{4}$$

and repulsive interaction potential energy $W(r_{12})$ given by

$$W(r_{12}) = \lambda/r_{12}^2, \quad \lambda > 0. \tag{5}$$

The nodeless solution which they utilize for their two-electron atomic model, times of course a singlet spin function to give an overall antisymmetric fermion wave function, has the unnormalized form [3, Eq. (3.14)]

$$\Psi(\mathbf{r}_1, \mathbf{r}_2) = e^{-m\omega r_1^2/2\hbar}e^{-m\omega r_2^2/2\hbar}r_{12}^{\alpha}, \tag{6}$$

where $\alpha = [(1 + 4\lambda m/\hbar^2)^{1/2} - 1]/2$. Clearly, when the interparticle repulsion is switched off in the present bosonic application, the $r_{12} \equiv |\mathbf{r}_1 - \mathbf{r}_2|$ term in (6) is replaced by unity and one has an elementary product wave function in which the two bosons under

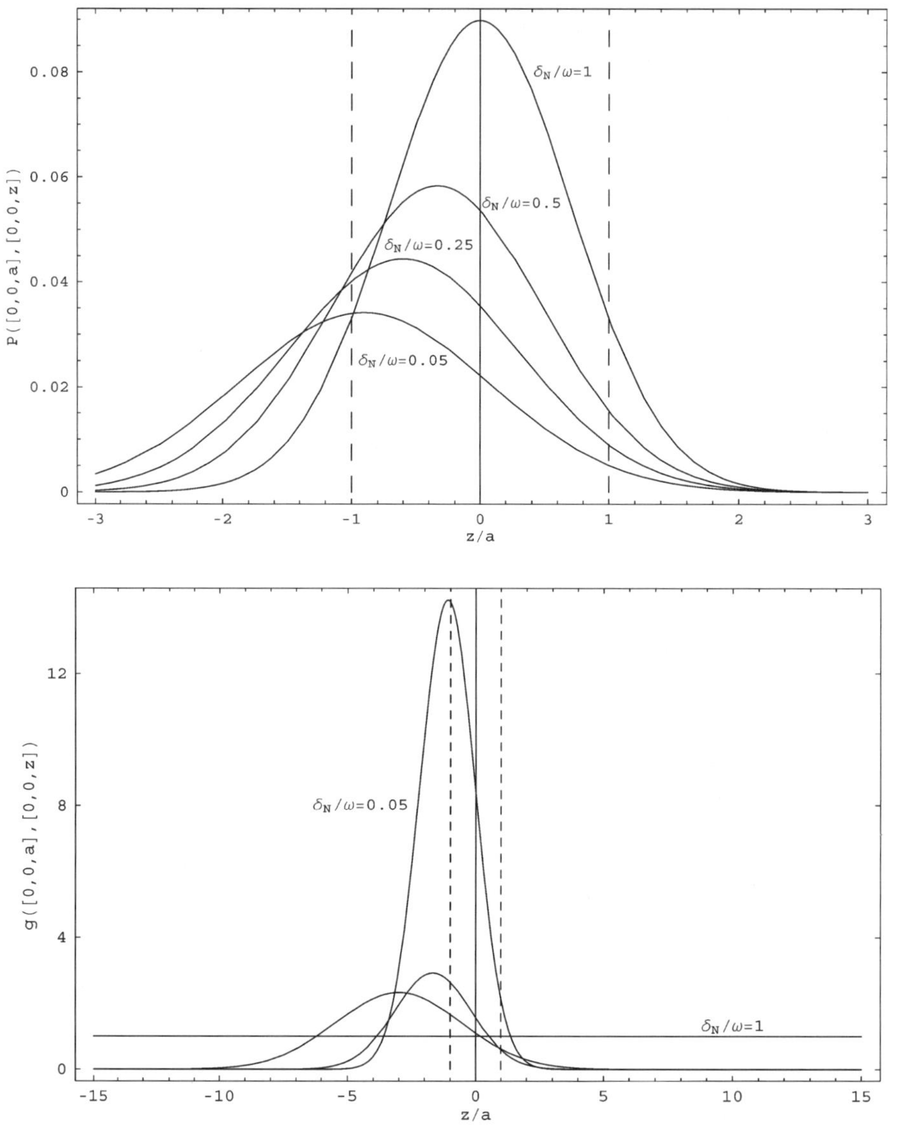

Fig. 1. Pair functions $P(\mathbf{r}_1, \mathbf{r}_2)$ (top panel) and $g(\mathbf{r}_1, \mathbf{r}_2)$ (bottom panel) for $\mathbf{r}_1 = a\hat{z}$ and $\mathbf{r}_2 = z\hat{z}$ of two bosons interacting with harmonic repulsions as functions of z/a. The different curves correspond to the values of $\delta_N/\omega = 1$, 0.5, 0.25 and 0.05. The dashed lines locate $|z_1| = a$.

discussion occupy the same Gaussian orbital. While, of course, we can immediately construct $P(\mathbf{r}_1, \mathbf{r}_2)$, after appropriate normalization, as $|\Psi(\mathbf{r}_1, \mathbf{r}_2)|^2$ from Eq. (6), we shall rather start displaying below the single-particle density given by

$$\rho(r) = \mathcal{N} \int \Psi^2(\mathbf{r}_1, \mathbf{r}_2)\, d\mathbf{r}_2, \tag{7}$$

where $\mathcal{N}$ is to be chosen such that

$$\int \rho(r)\, d\mathbf{r} = 2 \tag{8}$$

in the present two-boson model. Explicitly, using Eq. (6) in Eq. (7), we have

$$\rho(r) = \mathcal{N} e^{-m\omega r^2/\hbar} \int e^{-m\omega r_2^2/\hbar} |\mathbf{r} - \mathbf{r}_2|^{2\alpha}\, d\mathbf{r}_2. \tag{9}$$

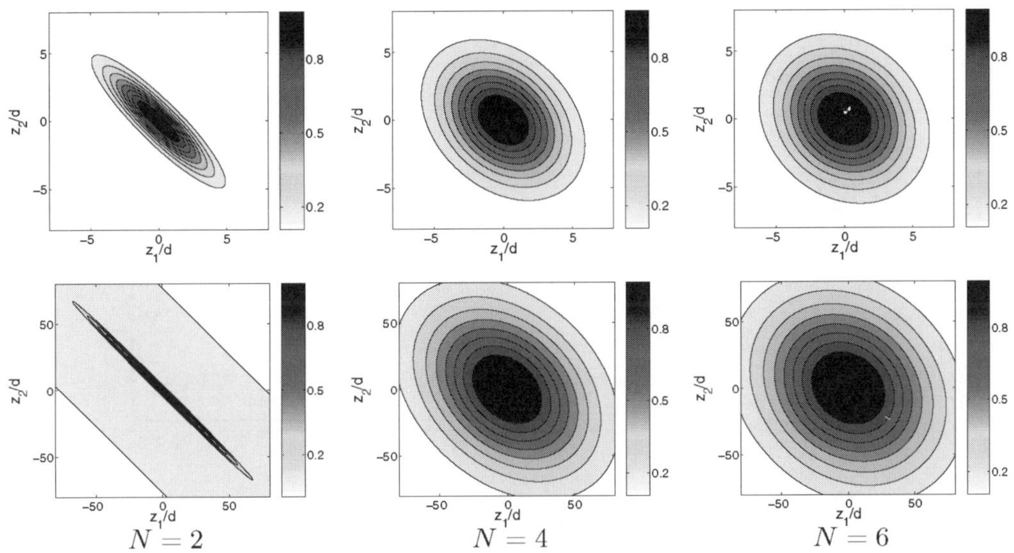

Fig. 2. Contour plots of the pair functions $P([0,0,z_1],[0,0,z_2])$ as functions of z_1/d and z_2/d with $d=\sqrt{\hbar/(m\omega)}$ for N bosons interacting with the harmonic potential $\frac{1}{2}\omega^2 r^2$. The left, middle and right columns correspond to $N=2$, 4 and 6 bosons respectively, and the top and bottom rows to $\delta_N/\omega=0.05$ and 10^{-3}.

The integral entering Eq. (9) gives

$$\rho(r)=\frac{2^{1-\alpha}}{\pi^{3/2}}\left(\frac{m\omega}{\hbar}\right)^{3/2} \times e^{-2m\omega r^2/\hbar}\,{}_1F_1\left(\tfrac{3}{2}+\alpha,\tfrac{3}{2},\tfrac{m\omega r^2}{\hbar}\right), \tag{10}$$

where ${}_1F_1$ is the confluent hypergeometric function [4]. For discrete values of λ given by

$$\left[\left(1+\frac{4\lambda m}{\hbar^2}\right)^{1/2}-1\right]=2n, \tag{11}$$

where n is an integer, $\rho(r)$ can be explicitly written as $\exp(-m\omega r^2/\hbar)$ times a polynomial of order $2n$. In practice, $\rho(r)$ has to be evaluated numerically and sample results for $\rho(r)$ are displayed in Fig. 3. The repulsive interaction between the two bosons, first widens the density profile and then leads to the formation of a maximum displaced from the trap center. This can be viewed as the first signature of the formation of a Wigner molecule of bosonic particles.

Let us now examine the pair function for two bosons with λ/r_{12}^2 repulsion. In Fig. 4 we show the contour plots of $|\Psi(\mathbf{r}_1,\mathbf{r}_2)|^2$ for $\alpha=0.1$ and 10. We observe the appearance of two well separated regions where the conditional probability is maximum. From these plots it is clear that the probability distribution for the distance between the two particles is a highly peaked function around

$$r_{12}^0/d=\sqrt{2+2\alpha}, \tag{12}$$

where $d=\sqrt{\hbar/(m\omega)}$. Furthermore, the differences with the harmonic repulsions are notable. Even though we found that the probability distribution of two bosons is maximum when they are in opposite positions, their repulsion is not as strong as to separate them from each other at a minimum distance as two bosons interacting with λ/r_{12}^2 are. In this sense, the lack of strong localization can be attributed to the weak interactions present in the exact model of harmonic interactions.

To conclude, it is of interest to make contact with the numerical study of Romanovsky et al. [5]. Though these workers considered different repulsive interactions, namely, contact and Coulomb, from those studied in the present Letter, some similarities are evident. However, we have only $N=2$ for the inverse square interaction potential λ/r_{12}^2, whereas the numerical studies of [5] considered $N=6$. For Coulomb repulsions, they noted, in two dimensions, in contrast to the present three-dimensional study, the tendency to

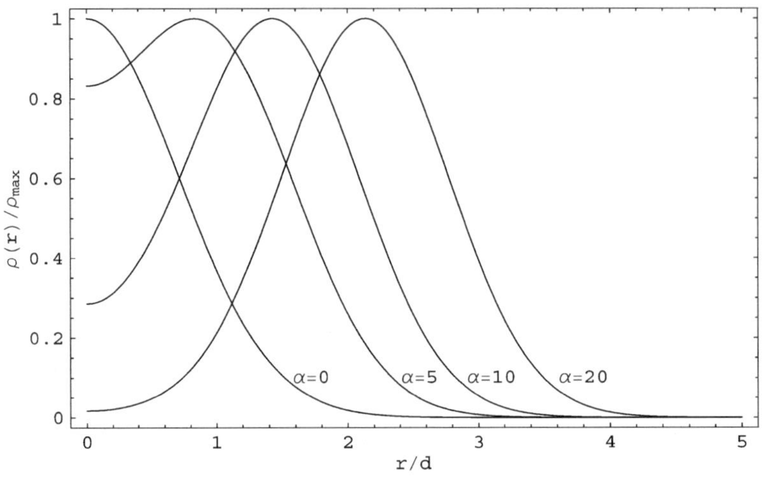

Fig. 3. Boson densities $\rho(r)/\rho_{\max}$ for two bosons with the λ/r_{12}^2 interaction potential as functions of r/d with $\rho_{\max}$ the maximum density for given α, and $d=\sqrt{\hbar/(m\omega)}$. The different lines correspond to different values of α (measuring the strength of the repulsive interaction: see Eq. (6)) as indicated in the plot.

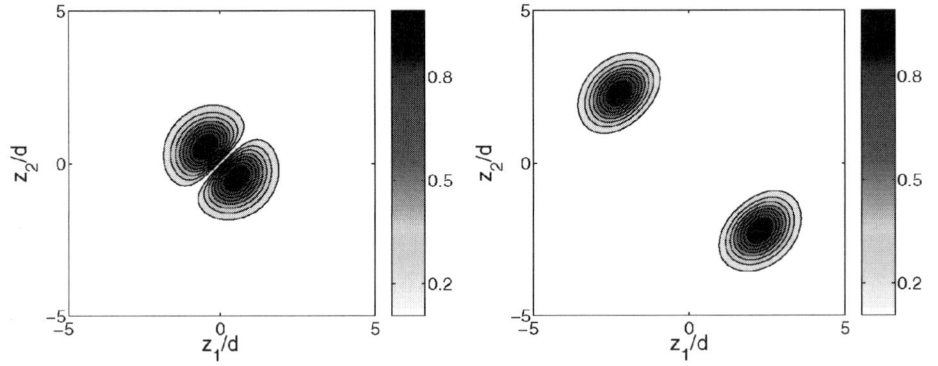

Fig. 4. Contour plots of the pair functions $P([0,0,z_1],[0,0,z_2])$ as functions of z_1/d and z_2/d for two bosons interacting with the λ/r_{12}^2 potential. Left and right panels correspond to $\alpha=0.1$ and 10, respectively.

ring formation of the bosons, with, as they term it, a breaking of rotational symmetry.

Acknowledgements

This work has been partially supported by an Advanced Research Initiative of Scuola Normale Superiore di Pisa (SNS) and by the Istituto Nazionale di Fisica della Materia (INFM). N.H. March acknowledges that his contribution to this Letter was largely made during a visit to SNS in 2005. He wishes to thank SNS for the very stimulating atmosphere where this study was carried out and for generous support.

References

[1] N.H. March, M.P. Tosi, Phys. Chem. Liq. 39 (2001) 183.

[2] L. Cohen, C. Lee, J. Math. Phys. 26 (1985) 3105.

[3] R. Crandall, R. Whitnell, R. Bettega, Am. J. Phys. 52 (1984) 438.

[4] M. Abramowitz, I.A. Stegun (Eds.), Handbook of Mathematical Functions with Formulas, Graphs, and Mathematical Tables, Dover, New York, 1972.

[5] I. Romanovsky, C. Yannouleas, U. Landman, Phys. Rev. Lett. 93 (2004) 230405.

Available online at www.sciencedirect.com

SCIENCE @ DIRECT®

ELSEVIER

Physics Letters A 349 (2006) 271–275

PHYSICS LETTERS A

www.elsevier.com/locate/pla

Phase diagram of two-dimensional electron gas in a perpendicular magnetic field around Landau level filling factors $\nu = 1$ and 3

Norman H. March [a,b,c], Alejandro Cabo [a,d,*], Francisco Claro [e]

[a] *Abdus Salam International Center for Theoretical Physics, Trieste, Italy*
[b] *Department of Physics, University of Antwerp, Antwerp, Belgium*
[c] *Oxford University, Oxford, England, UK*
[d] *Grupo de Física Teórica, Instituto de Cibernética, Matemática y Física, La Habana, Cuba*
[e] *Facultad de Física, Pontificia Universidad Católica de Chile, Santiago de Chile, Chile*

Received 30 August 2005; accepted 7 September 2005

Available online 16 September 2005

Communicated by V.M. Agranovich

Abstract

The measured melting curve $T_m(\nu)$ between the crystal and liquid phases is analyzed using thermodynamics to extract the change of magnetization ΔM as a function of the Landau level filling factor ν, near $\nu = 1$. An explanation of $\Delta M(\nu)$ is proposed in terms of skyrmions. Near $\nu = 3$, a Wigner crystal is the most probable solid phase, experiments excluding skyrmions.

In previous works [1–3] the authors have proposed the possible existence of a Hall crystal and/or skyrmions at very low temperatures near the Landau filling factor $\nu = 1$. The recent NMR relaxation experiments of Gervais et al. [4] strengthen the earlier claim of Barrett et al. [5] that skyrmions exist around $T \sim 40$ mK near $\nu = 1$, and a possible transition from a liquid to a lattice of skyrmions could occur there. We point out here the thermodynamical result [6] for the slope of the melting curve $T_m(\nu)$:

$$\frac{\partial}{\partial \nu} T_m(\nu) = \frac{B}{\nu} \frac{\Delta M}{\Delta S}, \tag{1}$$

where ΔM and ΔS represent magnetization and entropy changes respectively at the transition temperature T_m of skyrmion melting and B and ν are the magnetic field and the filling factor, respectively. The three estimated values of the solid–liquid transition temperature T_m, according to the measurements of Gervais et al., are illustrated by the black points in Fig. 1. The continuous curve in the figure is determined by a fit to the measured points after also adding two physical conditions: (a) the transition temperature in the

* Corresponding author.
E-mail address: cabo@cidet.icmf.inf.cu (A. Cabo).

0375-9601/$ – see front matter
doi:10.1016/j.physleta.2005.09.011

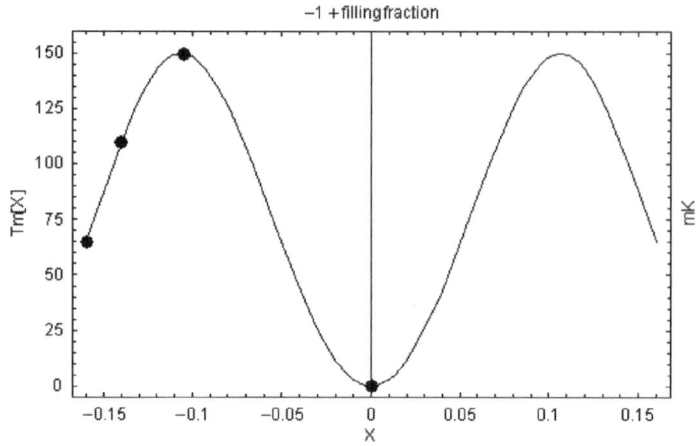

Fig. 1. The experimental measurements of Gervais et al. for the solid to liquid transition temperatures T_m for $\nu = 0.84, 0.86, 0.895$ are shown as black points. The solid curve is a fit to an even polynomial curve of $X = \nu - 1$, which is also satisfying $T_m = 0$ at $X = 0$ as required by the zero density of skyrmions at $\nu = 1$.

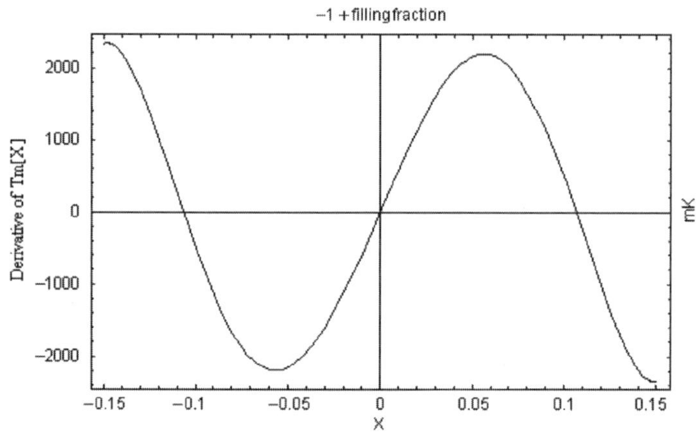

Fig. 2. The derivative of the estimated curve for $T_m(\nu)$ showing the various change of signs of $\Delta M/\Delta S$ as indicated by Eq. (1).

limit $\nu \to 1$ should vanish because the density of the skyrmion gas tends to zero [5], and (b) to assume, in a first approximation, a skyrmion–antiskyrmion symmetry around $\nu = 1$. The even polynomial of the filling factor deviation $X = \nu - 1$, employed for the fit, included up to the sixth powers of X. It should be noticed that the curves for the liquid to solid phase of the skyrmions estimated from the specific heat measurement of Bayot et al. [8] do not show the particle and antiparticle symmetry. However, taking into account that the vanishing of the critical temperature T_m for $\nu = 1$ should be valid in any case (see, e.g., Ref. [7]), the qualitative argument below will not be affected by this simplifying assumption in the first stage of the analysis. The qualitative behavior shown in Fig. 1 (apart from the above mentioned asymmetry) is also shown by the measurements in Ref. [8]. It will be seen that the analysis given here will supply a possible mechanism for the lack of symmetry around $\nu = 1$.

The behavior of T_m in Fig. 1 can be studied by considering Eq. (1). The derivative of T_m is depicted in Fig. 2. Firstly, it may be noted that for $X < 0$ (the region in which the experimental points of Gervais et al. were measured) there exists a negative value of $X = X_c$ for which ΔM should be zero. This is so, because we expect physically that $S_{\text{solid}} < S_{\text{liquid}}$. Defining now $\Delta S = S_{\text{solid}} - S_{\text{liquid}} = S_s - S_l$ (and also $\Delta M = M_s - M_l$) it follows that ΔS is nonvanishing and negative. Then, the magnetization of the liquid at $X = X_c$, is identical to that of the solid at the zero of T'_m for $X = X_c$. Over X_c up to $X = 0$ the quantity ΔM should become positive, since from thermodynamics $\Delta M/\Delta S < 0$, hence ΔM and ΔS must have the opposite sign for $X > X_c$. Similarly, for the region $X < X_c$, the magnetization change should acquire negative values.

The filling factor values for which the internal energies of the crystal and the liquid coincide can also be estimated by considering the relation

$$\frac{\partial}{\partial \nu} T_m(\nu) = -\frac{T_m(\nu)}{\nu}, \tag{2}$$

which should be obeyed at those points [6]. This equation has the following solutions: $X = -0.104$ and $X = 0$, lying in the interval defined by $X = -0.16$ (corresponding to the lowest measured filling factor) and $X = 0$. It is of interest that the equal energy filling factors are very close to the values for which the magnetization of the crystal and the liquid are also coinciding. The equality of the internal energy at $X = 0$, that is at $\nu = 1$, is an expected result since the skyrmion density vanishes at $\nu = 1$.

In addition, the inset of Fig. 3 of Gervais et al., as these workers stress, suggests that there is a critical

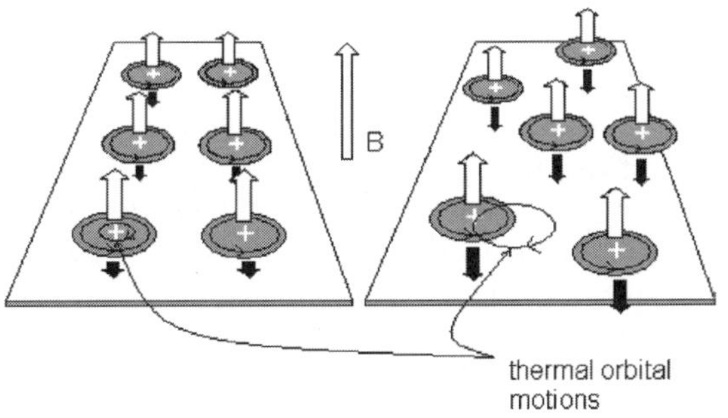

Fig. 3. Diagrams representing the skyrmion crystal (left-hand side) and the skyrmion liquid (right-hand side). The white vertical arrows represent the individual magnetic moments of each particle due to its internal (Hall) motion. The black vertical black ones indicate the contribution to the magnetic moments coming from the thermally induced orbital motion of the center of mass. The less constrained orbital movement in the liquid phase can be expected to increase the thermally induced orbital components. This is illustrated by the longer black arrows in the liquid diagram.

filling factor in the range $\nu \approx 0.80$–0.83 where a quantum phase transition exists. Following the treatment of reference March, Suzuki and Parrinello [9], it is then relevant to ask what is the order parameter of this transition. We propose the low frequency shear modulus, say G, of the skyrmion crystal, as a candidate. If the quantum phase transition remains first order, in analogy with (1) for the melting curve, then it follows from this equation that, since the third law of thermodynamics requires $\Delta S \to 0$, that the melting curve has a positively infinite slope at the critical filling factor near 0.8. This assumes the unlikely event that $\Delta M \to 0$ such that $\Delta M/\Delta S$ remains finite does not occur. Also, such a first order transition would imply that the low frequency shear modulus of the skyrmion crystal goes discontinuously to zero at the quantum phase transition.

We next advance an explanation for the form of the estimated solid–liquid phase transition curve of an assembly of skyrmions. The picture to be exposed incorporates concepts coming from the "anyon" type of description of the QHE state near $\nu = 1$ [6,10,11]. We expect that the qualitative explanation given can help to make contact between "anyon" descriptions and the transition between liquid and crystal skyrmion phases.

Let us consider the region $X < 0$ near $\nu = 1$ and first assume that X is close to zero. That is, the density of skyrmion excitations N is low and naturally they will tend to crystallize [7]. This situation is illustrated at the left-hand side in Fig. 3, which shows a crystalline arrangement of skyrmions. For $\nu < 1$ the skyrmions should really be of the antiskyrmion kind having a positive net charge. This charge distribution is indicated in the figure by the plus sign. The curly arrows lying inside each skyrmion picture signal the flow of diamagnetic Hall currents, giving rise to the magnetic moment of the assumed almost isolated skyrmion. This magnetic moment is depicted as the white vertical arrow over each skyrmion picture. Since, the temperature is nonvanishing and is also close to the critical value T_m, the excitations will also have a kinetic motion. This should produce a mean increment of the magnetic moment per particle. It is clear that this increment will be always a diamagnetic one: that is, opposite to the external magnetic field B. This contribution is shown as a black arrow below each of the particle symbols. In the crystal phase (left-hand side) it seems that the crystal potential could drastically restrain the realization of large circular motion of the center of mass of each skyrmion at thermal velocities, around the crystal position. Then, the diamagnetic orbital moment addition to each skyrmion $m_{S,o\text{–}s}$, can be expected to be reduced in absolute value by the crystal potential.

The liquid phase state of the skyrmion gas is illustrated in the right-hand side of Fig. 3. The same conventions for the symbols are used and the temperature is also assumed to be equal to T_m but slightly above the phase transition. In this case, since the gas is very diluted in the considered situation, the skyrmion proper magnetic moments can be expected to be coincident in value with the ones in the crystal phase. The situation seems to be different for the orbital component of the magnetic moment created by the thermal fluctuations. In this case, the constraining action of the crystal is not present and the skyrmions have the chance of orbiting with similar thermal velocities but in greater circles with appreciably more degrees of freedom than their solid counterparts. Thus, an increase in magnitude of the orbital magnetic moment contribution $m_{S,o\text{–}l}$ with respect to its solid counterpart $m_{S,o\text{–}s}$, should be expected in the liquid phase.

Therefore, in the region $X < 0$ down to X_c it is to be expected that

$$\begin{aligned} M_s - M_l &= N(m_S + m_{S,o\text{–}s}) - N(m_S + m_{S,o\text{–}l}) \\ &= N(-m_{S,o\text{–}l} + m_{S,o\text{–}s}), \end{aligned}$$

$$\Delta M = M_s - M_l > 0.$$

This is the same conclusion that the estimated phase diagram predicts, according to Eq. (1).

As the density of antiskyrmions N grows as X diminishes approaching to X_c in the liquid phase, the scattering among the skyrmions becomes more intense. This effect can be expected to inhibit the realization of large circular motions of the center of mass of orbiting antiskyrmions, leading to a decrease of the magnitude of $m_{S,o\text{–}l}$ down to equalize it to $m_{S,o\text{–}s}$ at $X = X_c$. Further, as the density increases even more, the scattering plus overlapping effects of the skyrmions in the liquid phase makes natural a further reduction of the net magnetization of the liquid with respect to the value in the more organized solid arrangement, allowing steady currents and magnetic moments. This situation is in accord with the property

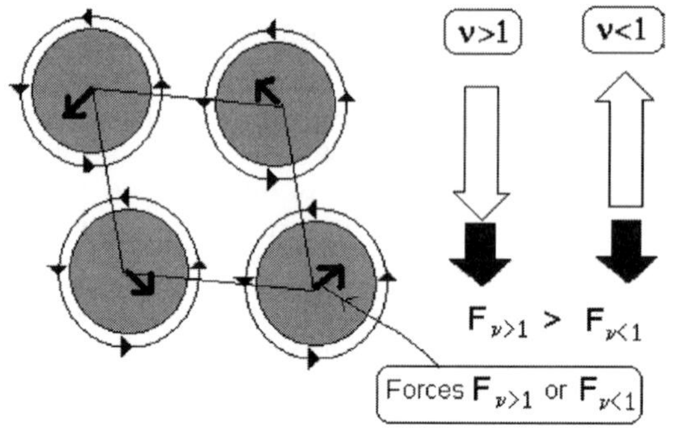

Fig. 4. Diagrams illustrating the appearance of shear forces in a crystal formed by identical rotating bodies (left-hand side). The right hand side shows the two ways of combinations of the internal rotations due to Hall (Chern–Simons) current and charge densities constituting the skyrmions, and the thermally induced diamagnetic rotation of the center of mass of the excitation. For $\nu > 1$ they are in the same sense, and for $\nu < 1$ they are opposed. Thus, for $\nu > 1$ the crystal should be expected to be more deformed.

$\Delta M < 0$ following from the estimated critical curve for $X < X_c < 0$.

Let us propose below a qualitative explanation for the skyrmion–antiskyrmion observed asymmetry in the heat capacity measurements in Ref. [8]. For this purpose we notice that the orbital magnetic moment generated by the thermal motion is not changing its sign when the filling factor passes across the $\nu = 1$ value. In addition it can be also underlined that the skyrmion has a rotating internal motion. These circumstances are illustrated in Fig. 4. Therefore, such rotational movements inside the crystal should generate shear forces tending to distort it in some measure, as illustrated in the left-hand side of Fig. 4. But, after taking into account that for $\nu > 1$, both, the internal skyrmion rotational motion and the thermal one coincide in sense, and that, on the contrary, for $\nu < 1$ these motions are opposed, it follows that the shear forces distorting the skyrmion crystal should be higher for $\nu > 1$. But, the crystal melting is strongly determined by the resistance to shear deformations, and then, the $\nu > 1$ crystal should be in a state closer to an instability under shear deformations. Thus, it seems natural to expect that the melting in the $\nu < 1$ region will require higher temperatures to be attained, as experimentally observed [8].

Let us remark that in contrast to the behavior near $\nu = 1$, where experiment strongly points to the ground-state being a skyrmion crystal [4,5], there is accord between experimental groups that a skyrmion crystal does not form the ground state of the two-dimensional electron assembly near $\nu = 3$, where a Wigner crystal seems the most likely candidate from existing experiments. Therefore, the whole situation around integral filling factors seems to reinforce the view advanced in [2,10] that the main ground state at exactly integer filling factors shows an electromagnetic Chern–Simons (equivalent to a Hall one [10]) response that either:

(a) Transforms the set of impurities, when they exist in dirty samples, in a kind of "charge reservoir", which accepts or releases electrons above or below $\nu = 1$, respectively.
(b) Or, in rough terms, generates "artificial impurities" localizing charges, as skyrmions or antiskyrmion excitations near $\nu = 1$ and "*electron*" or "*hole*" like Wigner crystals near $\nu = 3$, again when ν is above or below the corresponding filling factor, respectively.

The coexistence of all these crystal or liquid structures for magnetic fields well inside plateaus is a supporting experimental fact for this view, as stated in [2,10]. The demonstration of the validity of this property for realistic planar samples is under consideration. In Ref. [10] it was only shown for the simpler case of a superlattice of planar samples.

To conclude, near $\nu = 1$ the measured phase diagram given by Gervais et al. has been interpreted using the thermodynamic relation (1) and making the plausible assumption that $S_s < S_l$. The measured melting curve requires various change of signs of the magnetization difference ΔM in the region $\nu = 0.8$–1.2. A qualitative interpretation of the nature of these changes is advanced by considering the internal structure of the skyrmions and their thermal motions. The observed asymmetry of the melting into a skyrmion liquid phase above and below $\nu = 1$ is proposed to be produced by the diamagnetic moment created by the thermal motion, which reduces or increases the magnitude of the internal magnetic moments of the skyrmions and antiskyrmions, respectively. Therefore, an explanation of the observed properties of the Skyrme excitations near $\nu = 1$ is proposed and based on an electromagnetic and "anyon" description [6,10,11]. However, further work, both experiment and theory, is required on this potentially important matter.

Acknowledgements

N.H.M. and A.C.M. wish to acknowledge that their contribution to this Letter was brought to fruition during a stay at the AS ICTP, Trieste in 2005. This stay was made possible by Prof. V.E. Kravtsov, to whom they are grateful for generous hospitality and the very stimulating environment. The support of FONDECYT, Grants 1020829 and 7020829, for the activity of F.C. is also very much recognized.

References

[1] A. Cabo, F. Claro, Phys. Rev. B 70 (2004) 235320.
[2] F. Claro, A. Cabo, N.H. March, Phys. Status Solidi B 242 (2005) 1817.
[3] A. Cabo, F. Claro, N.H. March, cond-mat/0309166.
[4] G. Gervais, H.L. Stormer, D.C. Tsui, P.L. Kuhns, W.G. Moulton, A.P. Reyes, L.N. Pfeiffer, K.W. Baldwin, K.W. West, Phys. Rev. Lett. 94 (2005) 196803.
[5] S.E. Barrett, G. Dabbagh, L.N. Pfeiffer, K.W. West, R. Tycko, Phys. Rev. Lett. 74 (1995) 5112.
[6] M.J. Lea, N.H. March, W. Sung, J. Phys.: Condens. Matter 4 (1992) 5263;
M.J. Lea, N.H. March, W. Sung, J. Phys.: Condens. Matter 3 (1991) 6810.
[7] L. Brey, H.A. Fertig, R. Côte, A.H. MacDonald, Phys. Rev. Lett. 75 (1995) 2562.
[8] V. Bayot, E. Grivei, S. Melinte, M.B. Santos, M. Shayegan, Phys. Rev. Lett. 76 (1996) 4584.
[9] N.H. March, M. Suzuki, M. Parrinello, Phys. Rev. B 19 (1979) 2027.
[10] A. Cabo, D. Martinez-Pedrera, Phys. Rev. B 67 (2003) 245310.
[11] S.L. Sondi, A. Karlhede, S. Kivelson, E.H. Rezayi, Phys. Rev. B 47 (1993) 16419.

Quantum Statistics of Charged Particles and Fingerprints of Wigner Crystallization in *D* Dimensions

N. H. MARCH
Oxford University, Oxford, England; Department of Physics, University of Antwerp, Groenenborgerlaan 171, B-2020 Antwerp, Belgium

Received 25 February 2006; accepted 22 March 2006
Published online 1 August 2006 in Wiley InterScience (www.interscience.wiley.com).
DOI 10.1002/qua.21145

ABSTRACT: After a brief summary of the physical arguments underlying Wigner's original concept in 1934 of a quantal electron crystal, theoretical interpretation of a number of experimental findings are presented. These include (i) low-density carriers in semiconductors in applied magnetic fields in both three, and recently two, dimensions; and (ii) low-temperature phase diagram of underdoped high T_c cuprates; fullerides with relatively low T_c are also referred to in a related context. Interpretation of areas (i) and (ii) focuses on the relevance of both Fermi–Dirac and anyonic (fractional) statistics, the latter in relation to the proposed melting curve of the two-dimensional (2D) magnetically induced Wigner solid into the Laughlin liquid phase, which is the seat of the fractional quantum Hall effect. A brief discussion follows of crystalline phases additional to the Wigner solid, namely Skyrmion and Hall crystals. Bose–Einstein statistics is then referred to, but now in relation to finite-size confined quantal assemblies, with fingerprints of Wigner molecules the focus. Finally, quasi-1D lattices are considered, both in Bechgaard salts and in the very recent single-electron counting experiment of Bylander et al. Int J Quantum Chem 106: 3032–3042, 2006

Key words: quantal electron crystals; melting; Laughlin liquid

1. Background and Outline

The focus of the present article is on the formation of Wigner crystals of charged particles in D-dimensional assemblies. For Fermions, the idea that a quantal crystal phase of electrons could form at sufficiently low densities goes back to Wigner [1, 2]. Section 2 begins with a summary of the physical arguments underlying Wigner's original concept. Experimentally, to observe fingerprints of quantal electron solids, it has proved very important to generalize Wigner's original proposal to consider applied magnetic fields to aid Wigner localization, as advocated in the work of Durkan et al. [3] in interpreting electrical transport measurements in 3-dimensional (3D) highly compensated n-type InSb. These investigators stressed that the experimental observations could be interpreted in terms of a magnetically induced Wigner solid (MIWS). Twenty years later, Andrei et al. [4] made use of major advances in semiconductor technology to subject the 2D electron assembly in a GaAs/Al-GaAs heterojunction to an applied magnetic field perpendicular to the plane of the electrons. Andrei et al. [4] observed that, at a critical value of the magnetic field, a solid phase of the $D = 2$ electron assembly was created. Their experiment was followed by photoluminescence measurements by Buhmann et al. [5]. The resulting melting curve of the 2D Wigner solid in a magnetic field was analyzed by Lea et al. [6, 7], using thermodynamics plus an anyon model of the orbital magnetization of the Laughlin (FQHE) liquid. This work [6, 7] is summarized in Section 2 in the present work.

Part of the content of Section 3 is to examine, as a function of the 2D Landau level filling factor ν, which is proportional to the ratio of the areal carrier density n to the magnetic field strength H, two solid phases additional to the Wigner solid: (i) the Hall crystal proposed by Halperin et al. [8]; and (ii) the Skyrmion (Ref. [9]; see also Ref. [10]) or quantum soliton, crystal.

Section 4 then reviews, more briefly because of the absence to date of confirmation (or otherwise) from experiment, theoretical proposals for Wigner crystallization in high T_c cuprates as well as in doped fullerides. In Section 5, finite-size systems with Bose–Einstein statistics are considered, but now for $D = 3$, both from computer studies [11] and from a model analytical study with repulsive interactions of inverse square law form [12]. Section 6 consists of a summary, together with some suggestions for future directions that should prove fruitful to study Wigner lattices in quasi 1D systems.

2. Summary of Wigner's Concept of a 3-Dimensional Quantal Electron Crystal and Some Further Consequences in *D* Dimensions

In 1934, Wigner [1] was interested in studying the many-electron correlation effects in simple monovalent metals, such as the alkalis Na and K. To deal with the electron–electron repulsion energy e^2/r_{ij} between each pair of valence electrons at separation r_{ij}, he analyzed the so-called jellium model. This is such that the granularity of the Na^+ ions in sodium metal, say, is removed by smearing out their positive charges into a nonresponsive neutralizing background in which Coulombically repelling electrons moved. Given the uniform density of the electrons, n say, which in the 3D model Wigner considered is related to the mean interelectronic spacing r_s by

$$n = 3/4\pi r_s^3, \tag{1}$$

Wigner already knew that in the limit of high-density n, or equivalently from Eq. (1), in the limit $r_s \to 0$, the ground-state energy per electron, E/N say, was given by

$$\frac{E}{N} = \frac{2 \cdot 21}{r_s^2} - \frac{0 \cdot 916}{r_s}: \qquad r_s \to 0, \tag{2}$$

where the right-hand side of Eq. (2) is in Rydbergs, if r_s is given in units of the Bohr radius $a_0 = \hbar^2/me^2$, with m the electron mass. The first term on the right-hand side is the kinetic energy of a Fermi gas at $T = 0$, in which a sphere in momentum space of radius $p_f = \hbar k_f$ with the Fermi wave number k_f related to density n by

$$n = \frac{k_f^3}{3\pi^2}, \tag{3}$$

is completely filled with electrons.

The second term in Eq. (2), of order e^2/r_s, is the exchange energy per particle, due to the so-called Fermi hole created around an electron, with upward spin say, for electrons with parallel spin, due to the Pauli exclusion principle for fermions.

What Wigner [1, 2] already recognized was that the Hartree–Fock approximation (2), based on a determinant of plane waves, became inappropriate in the low-density limit $r_s \to \infty$, where the potential energy dominated the kinetic contribution. He argued that in this low-density limit, therefore, one must minimize the potential energy by making electrons avoid one another maximally by going on to the sites of a regular lattice. He then noted that the body-centered cubic (bcc) lattice had the lowest Madelung energy, yielding as $r_s \to \infty$, instead of Eq. (2):

$$\frac{E}{N} \to -\frac{1 \cdot 8}{r_s} \tag{4}$$

Comparing Eqs. (2) and (4), we see immediately that the correlation energy is essentially equal to the exchange energy. We have passed from the itinerant electron state as $r_s \to 0$, dominated by kinetic Fermi energy, to a localized quantal electron crystal.

2.1. FURTHER CONSEQUENCES OF 3D WIGNER ELECTRON CRYSTALLIZATION

Following the analysis given above of the high- and low-density limits of the ground-state energy of the 3D jellium model, we want to summarize briefly four further consequences of Wigner's ideas:

1. Ceperley and Alder [13] reported quantum computer simulations of the ground-state energy of different phases in the jellium model. Their major conclusion was that the critical r_s for formation of the 3D quantal electron crystal was approximately $80a_0$, confirming the earlier conclusion in the review by Care and March [14] that one must seek Wigner crystallization in the much lower carrier densities in semiconducting materials, the lowest-density alkali metal, Cs, having an r_s value of $5.5a_0$.
2. The Wigner low-density limit (3) needs to be "corrected" as r_s is reduced somewhat, by a term due to electrons vibrating around the lattice sites of the bcc Wigner lattice. This then yields

$$\frac{E}{N} \to -\frac{1\cdot 8}{r_s} + 0\left(\frac{1}{r_s^{3/2}}\right), \tag{5}$$

and especially Coldwell–Horsfall and Maradudin [15] made a thorough quantitative study of the energy in this low-density expansion. This expansion was related to the Ceperley–Alder quantum computer simulation by Herman and March [16], who also discussed the momentum distribution $n(p)$ in 3D. Appendix A compares this 3D $n(p)$ with results for a 1D model [17], and points out uncertainties that remain in the "borderline," but physically very important, 2D case (see, e.g., Ref. [18]).
3. The magnetic properties of the quantal Wigner electron crystal were also considered, albeit briefly, by Herman and March [16]. The simplest picture in the extreme low-density limit $r_s \to \infty$ is to suppose that on the bcc lattice, which consists of two interpenetrating simple cubic structures, the upward spin electrons sit on one of these simple cubic arrays and the downward spins on the other. This yields a Néel antiferromagnet. However, as Carr [19] and later Herman and March [16] stressed, the exchange interaction between the Wigner localized orbitals, of Gaussian form (see, e.g., Ref. [20]) becomes very weak, and the energy difference for example, between ferro- and antiferro ordering is very tiny. Magnetism, but taken as the difference between the Wigner solid and the Laughlin liquid phases across the melting curve, is a crucial ingredient in analyzing theoretically the measured melting properties, but now in an applied magnetic field and in 2D.
4. Relating to melting, in classical crystals there is an intimate correlation between the temperature T_m at melting and the vacancy formation energy [21]. Therefore, we note that some attention has been given to the energies of "electron defects" [14] in the Wigner lattice.

2.2. CRITICAL INTERELECTRONIC SPACING r_s IN THE GROUND-STATE OF WIGNER ELECTRON CRYSTALS IN D DIMENSIONS

In this section, we take examples in dimensions 2 and 1, to complement the above discussion of Wigner's treatment in 3D, motivated by electron correlation in the alkali metals; therefore, it is relevant to draw attention to the work of Parrinello and March [22]. These investigators produced a schematic plot, redrawn in Figure 1, of the critical interelectronic spacing r_s in the ground state of Wigner electron crystals in D dimensions, assuming that the electron–electron interaction satisfied Laplace's equation. But for interaction e^2/r_{ij}, which is only thereby reproduced at $D = 3$, the same qualitive features will be reproduced for $D = 1$ and 2; therefore, we stress in Figure 1 that in 2D, the critical radius r_s must be significantly smaller, and that Wigner crystallization should be easier to search for experimentally than in 3D with the Ceperley–Alder [13] critical radius of $80a_0$. Furthermore, for $D = 1$, the critical $r_s \to 0$, which is a consequence of a theorem that the writer suspects goes back to Peierls, but is also associated correctly with the names of Mermin and Wagner. Thus, 1D is qualitatively different in this respect from $D = 2$ and 3, and this situation is expanded on in Appendix A.

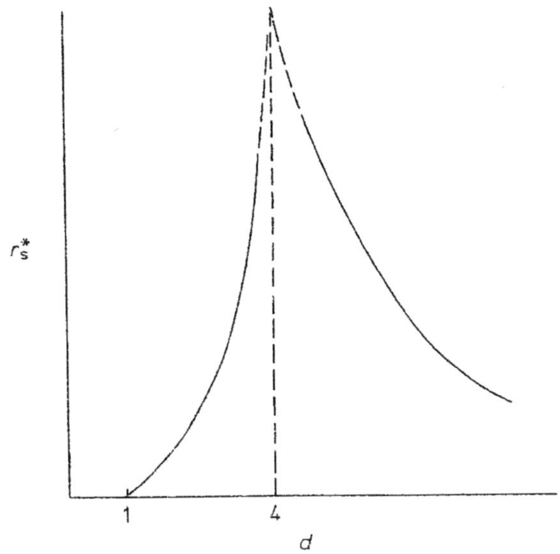

FIGURE 1. Schematic plot of critical interelectronic spacing r_s versus dimensionality. (Redrawn from Parrinello and March [22].)

With this summary of results stemming from Wigner's original work at zero magnetic field, we turn in Section 3 to magnetically induced Wigner solids, first in 3D, but then in the 2D electron assembly in semiconducting heterojunctions.

3. Quantal Electron Crystal Formation Caused by Applying Magnetic Fields

3.1. THREE-DIMENSIONAL *n*-TYPE INSB

Following Hall effect experiments by Putley [23] on highly compensated *n*-type InSb, Durkan et al. [3] proposed as a theoretical interpretation of these measurements that impurity band conduction was suppressed at a critical value of the applied magnetic field by formation of a quantal electron crystal. Support for their theory came from the later experiment of Somerford [24] on the electrical resistivity at ^{3}He temperatures in this same material at this critical magnetic field. His finding was that a sudden increase of resistivity, to be likened to an itinerant electron system becoming localized by application of the field, was occurring, the resistivity increasing by a factor $\sim 10^7$. Care and March [25] gave a theoretical explanation of Somerford's findings in terms of quantal Wigner electron crystal formation aided by the magnetic field. The subsequent theoretical study by Elliott and Kleppmann [26] showed that a quantitative calculation of the resistivity based on Wigner crystal formation was entirely consistent with Somerford's measurements.

There was then a lull in this area, until the pioneering experiment of Andrei et al. [4], to which we now turn.

3.2. TWO-DIMENSIONAL ELECTRON LIQUID IN GaAs/AlGaAs HETEROJUNCTION IN A TRANSVERSE MAGNETIC FIELD

While it was already clear from the study by Durkan et al. [3] that a Bragg diffraction experiment was desirable, this has so far not proved practicable. Thus, it was of considerable interest when Andrei et al. [4], in essence, exploited the fact that an electron liquid phase could not support the propagation of low-frequency shear waves, whereas an electron solid could. This is a gross statement; in fact, the experiment carried out by Andrei et al. involved magnetophonons. But it demonstrated already, beyond reasonable doubt, that when the 2D electron liquid in GaAs/AlGaAs was subjected to a transverse magnetic field beyond a critical value, such "shear" waves could propagate, proving the existence of a quantal electron solid. The subsequent photoluminescence experiment by Buhmann et al. [5] supported the conclusions of Andrei et al. [4] and furthermore led to the drawing of a schematic phase diagram.

The theory of such as phase diagram was then given in the study of Lea et al. [6, 7], starting from the analogue of the Clausius–Clapeyron equation, but now in the presence of an applied magnetic field. These investigators wrote the thermodynamic relation determining the slope $\partial T_m/\partial H$ of the melting curve of the quantal electron solid into the Laughlin electron liquid, the seat of the quantum Hall effect, as

$$\frac{\partial T_m}{\partial H} = -\frac{\Delta M}{\Delta S}. \tag{3.1}$$

Hence, T_m is the melting temperature, H the magnetic field strength, M the magnetization, and S entropy. Lea et al. [6] then introduced the Landau level filling factor ν defined in terms of the areal electron density n as

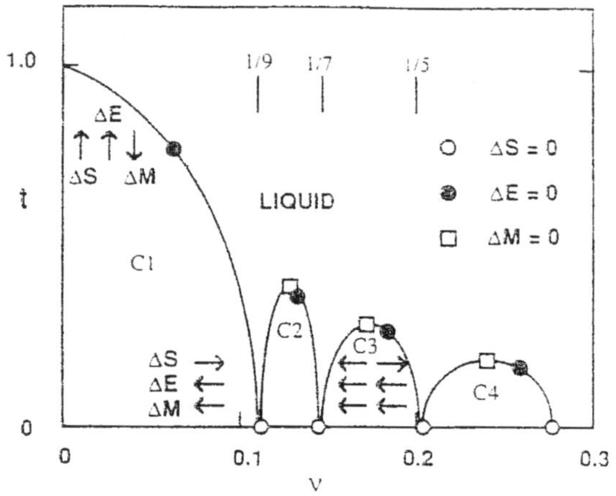

FIGURE 2. Schematic phase diagram showing crystal phases C1 to C4. At Landau level filling factors ν = 1/9, 1/7, and 1/5, diagram assumes Laughlin electron–liquid is the ground state. Open circles show when entropy change ΔS from liquid to solid is zero, while squares denote when magnetization change ΔM is equal to zero. ΔE, the internal energy change, is given in terms of ΔS and ΔM by $T\Delta S + H\Delta M$, with H the field strength specifically $\Delta M = M_L - M_S$, etc. where subscripts L and S denote liquid and solid, respectively. Arrows show the direction of ΔM, ΔS, and ΔE across the phase boundary nearest to the arrows. (Redrawn from Lea et al. [6].)

$$\nu = \frac{nhc}{eH} \tag{3.2}$$

to reach the result

$$\frac{\partial T_m(\nu)}{\partial \nu} = \frac{H}{\nu}\frac{\Delta M}{\Delta S}. \tag{3.3}$$

The schematic phase diagram as analyzed in Ref. [7] is shown in Figure 2. In their study, Lea et al. [6] had shown that such a "reentrant" phase diagram could be explained by combining the result (3.3) with an anyon model (see also Appendix B) of the orbital magnetism of the Laughlin electron liquid, into which state the quantal electron solid formed by applying the transverse magnetic field melts at milliKelvin temperatures (Fig. 2).

3.3. OTHER SOLID PHASES IN COMPETITION WITH QUANTAL WIGNER SOLID

To complete this section, we discuss briefly below two other solid phases that, at least in principle, are in competition with the Wigner solid as the ground-state at favorable values of ν of the 2D electron assembly in a transverse magnetic field. In early work, Halperin et al. [8] proposed the Hall crystal, which was conceived to have electron ordering but also to retain transport properties, and in particular the Hall effect, akin to those of the Laughlin liquid. At the time of writing, we know of no experimental confirmation of the existence of this phase. Therefore, we turn directly to a further solid state, i.e., that comprising quantum solitons, or Skyrmions [24, 25]. The very recent experimental work of Gervais et al. [27] following Barrett et al [28] had been discussed, and a proposed theoretical interpretation given by March et al. [29]. Experiment [27] and theory [29] both focus on a Landau level filling factor ν near unity. In [29], the thermodynamic Eq. (3.3) was again used, but this time to discuss the melting curve of the Skyrmion crystal observed in Refs. [28] and [27]. An explanation of $\Delta M(\nu)$ around $\nu = 1$ may be found in Ref. [29] to follow naturally from a Skyrmion picture (the reader is referred especially to Fig. 3 of Ref. [29]).

4. Theoretical Proposals for Wigner Crystallization in High T_c Cuprates

Having discussed at some length Wigner solids in 3D and 2D semiconducting materials, i.e., n-type InSb and GaAs/AlGaAs heterojunctions, respectively, where extensive experimental studies are available, we turn now to consider the phase diagram of high T_c cuprates. We shall set out a number of theoretical proposals that have been put forward to date, but we shall be brief because of the absence of experimental confirmation, at the time of writing.

The upper part of Figure 3 shows the proposal made by Siringo et al. [30] for Wigner hole crystallization at low doping concentration: ie well below the superconducting phase of the high T_c cuprates. These investigators anticipated that such a hole crystal would melt at very low temperatures, as indicated also schematically by the dashed curve in Figure 3. The independent proposal subsequently put forward by Balents et al. [31] is shown in the lower part of Figure 3. These investigators develop a specific many-body theory to support their proposal (the interested reader is referred to their article for details). Experimental studies, if possible at milliKelvin temperatures, would, of course, be of considerable interest for further development and if necessary refinement of the above ideas.

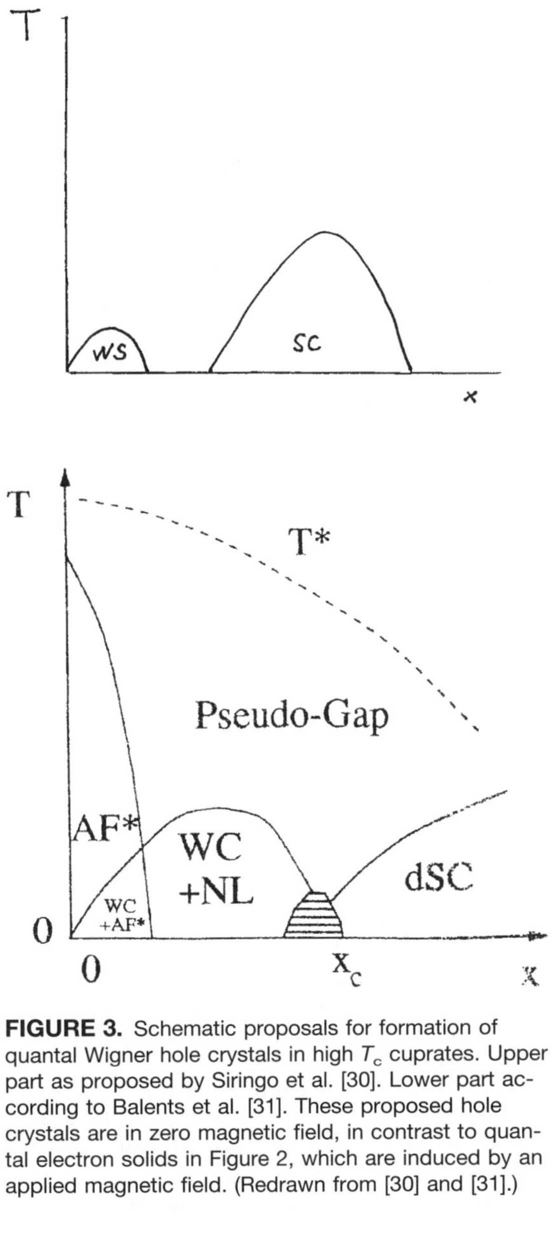

FIGURE 3. Schematic proposals for formation of quantal Wigner hole crystals in high T_c cuprates. Upper part as proposed by Siringo et al. [30]. Lower part according to Balents et al. [31]. These proposed hole crystals are in zero magnetic field, in contrast to quantal electron solids in Figure 2, which are induced by an applied magnetic field. (Redrawn from [30] and [31].)

4.1. SUPPRESSION OF SUPERCONDUCTIVITY BY AN APPLIED MAGNETIC FIELD OF 58 TESLA IN A PARTICULAR HIGH T_C CUPRATE

Very recently, Squire and March [32, 33] presented a study on superconductivity both in a high T_c cuprate and in doped fullerides. In both cases, theoretically motivated proposals were put forward involving Wigner solid-like correlations.

The first case was motivated by the experimental studies of Ando et al. [34] on a specific high T_c cuprate, which show signatures of a phase transition under hole-doping in an applied magnetic field at 58 T. By drawing an analogy with the early work of Durkan et al. [3], Squire and March [32, 33], invoked above in Section 3, propose that the magnetic field suppresses superconductivity at the phase transition, resulting in a quantal Wigner solid of holes.

In doped fullerides, but now without an applied magnetic field, Squire and March [32] argue that Coulombic repulsion between charged carriers, at specific doping considerations, can lead to a localized structure for these charges, due to the occurrence of Wigner solid-like correlations.

5. Bose–Einstein States With Strong Interparticle Repulsion and Wigner Molecule Formation

Bose–Einstein condensates (BECs) in harmonic confinement were first studied experimentally with weakly interacting neutral atoms. Subsequently, however, control of the interaction strength has proved feasible, and thereby bosonic states can be explored in circumstances corresponding to strongly repelling particles. This is then the motivation for this section, following Wigner solidification in Fermi–Dirac and in anyonic (fractional) statistical assemblies (see also Appendix B).

However, the recent work we summarize below is not now solely on Coulombic repulsions, nor is it on extended assemblies, but rather on small bosonic clusters in 2D, and very briefly, also in 3D.

5.1. TRAPPED LOCALIZED BOSON CLUSTERS IN TWO DIMENSIONS

Quite recently, Romanovsky et al. [35] carried out numerical studies of small numbers of bosons in 2D harmonic traps. These investigators have treated such clusters for both a repulsive contact potential and Coulombic repulsion by numerical calculations. In essence, their procedure is to describe the strongly repelling bosons through symmetry breaking at the Hartree–Fock (HF) level, followed by "post-HF" symmetry restoration. To exemplify their method, they consider N bosons with N = 2–7. They regard their method as appropriate to reflect the transition from a BEC state to a crystal array. In the case of 2D confinement, which they study, the trapped localized bosons form "crystalline patterns," which have ring character both for the repulsive contact potential and for pure Coulombic repulsion (Wigner molecule formation). From the HF calculation, these patterns are reflected immediately in the single-particle densities. Following the symmetry restoration referred to above, the single-particle densities become rotationally symmetric. However, the "localized symmetry" can be exposed in the so-called conditional-probability distribution $P(\mathbf{r}, \mathbf{r}_0)$. This is defined as the probability of finding a boson at $\mathbf{r}$, given that the observer "sits" on another boson at $\mathbf{r}_0$.

To summarize the main findings of Romanovsky et al. [35], the strongly interacting bosons form polygonal-ring-like patterns, for both interactions considered. In particular, their Figure 2(b) gives ring radii for Coulombic repulsion for $N = 6$ harmonically confined 2D bosons. Their qualitative findings should encourage experimental work which aims to study such strongly repelling bosons in two dimensions.

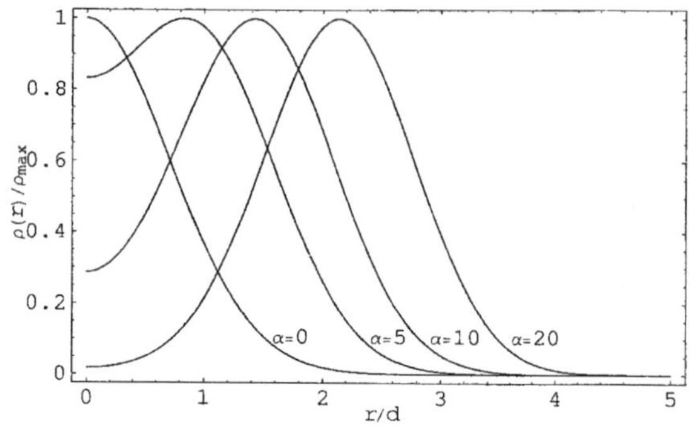

FIGURE 4. Boson densities $\rho(r)/\rho_{max}$ for two bosons with repulsive interaction λ/r_{12}^2, plotted as functions of r/d, with ρ_{max} the maximum value of ρ for given interaction strength, while $d = (\hbar/m\omega)^{1/2}$, where ω defines the strength of harmonic confinement. Curves are characterized by different values of the interaction strength. Note that above a certain λ, maximum of $\rho(r)$ is no longer at $r = 0$, heralding Wigner molecule formation.

5.2. TWO BOSONS IN THREE DIMENSIONS, BUT WITH INVERSE SQUARE REPULSION: WIGNER MOLECULE FORMATION

Subsequent to the above study [35], Capuzzi et al. [36] have used an analytically solvable model [37] to describe two bosons in 3D that interact via an inverse square law, i.e., the potential energy of repulsion is λ/r_{12}^2, with r_{12} the interparticle separation. Capuzzi et al. [36] then depict the interplay between the harmonic confining potential $(1/2)m\omega^2r^2$, and this repulsion for (i) the single-particle state density $\rho(r)$, and (ii) the pair function analogous to $P(\mathbf{r}, \mathbf{r}_0)$ above. To illustrate Wigner molecule formation found by Capuzzi et al. we restrict ourselves here to the single-particle density, which is shown in Figure 4. For weak repulsion λ the ground-state density has its maximum at the origin $r = 0$. But as λ is increased, the maximum moves to, say, $r = a$, and hence $2a$ is an estimate of the bond length of the bosonic Wigner molecule (the reader is referred to Fig. 2 of Capuzzi et al. [36] for plots depicting the pair correlation function associated with this "Wigner ordering").

6. Summary and Future Directions Involving Especially One-Dimensional Wigner Lattices

Experimental evidence is now compelling that the magnetically induced Wigner solid (MIWS) proposed in 1968 is observed at milliKelvin temperatures in a GaAs/AlGaAs heterojunction in a transverse magnetic field.

The melting curve of this Wigner solid is reentrant, with three confirmed regions $0 < \nu \leq 1/9$, $1/9 \leq \nu \leq 1/7$, and $1/7 \leq \nu \leq 1/5$ [37], where the Landau level filling factor $\nu = nhc/eH$. The areal density n is related in this 2D situation to the mean interelectronic spacing r_s by $n = 1/\pi r_s^2$, replacing the 3D form in Eq. (1). The thermodynamic interpretation of the above melting curve, combined with the consequences of an anyon model, shows that the general reentrant features are semiquantitatively given in terms of the orbital magnetism of the Laughlin electron liquid, this magnetism being de Haas–van Alphen-like.

Returning to the starting point for the proposal of MIWS, 3D highly compensated n-type InSb should be studied further experimentally with a view to finally understanding the dramatic metal–insulator transition induced by an applied magnetic field.

Following this brief proposal on a 3D Wigner solid, we conclude the article by commenting on two areas in which quasi-1D Wigner lattices might well figure prominently. The first area involves quasi-1D Bechgaard salts. Early theoretical work by Hubbard [38] (see also Torrance [39]) is relevant here, as is the recent study by Demler et al. [40]. As these investigators note, these Bechgaard salts have an important feature, which is the proximity of a superconducting state and a magnetically ordered insulating state.

Let us give a little more detail on the second area we single out as worthy of much further study within the context of 1D Wigner ordering. The motivation is the technique reported very recently by Bylander et al. [41] to observe individual electrons. This is done by passing a direct current through a microelectronic circuit with a 1D chain of small "islands" connected by conducting or superconducting tunnel junctions [33]. Charges tunneling through this array are time-correlated, permitting measurements of single-electron tunneling oscillations by Bylander et al. [41].

One experiment reported by these investigators employed a superconducting array with fifty junctions. Each of these had a capacitance C_A with the stray capacitance of an electrode inside the array denoted by C_0. It is then noted that an excess localized charge on one part of an array repels neighboring islands. The result, leading to a type of charged soliton, is spread over several junctions. Grossly, the extent of this soliton is given by $(C_A/C_0)^{1/2}$. Biasing the array above a threshold V, such charged solitons repel each other and a moving quasi-Wigner lattice of these "charged solitons" is formed as shown theoretically, and also, although indirectly, by experiment. Both electron and Cooper pair tunneling can occur, with the threshold for electron injection lower than that for Cooper pairs. Elsewhere, Squire and March [33] have suggested a possible connection with the work reported above on a high T_c cuprate, with superconductivity suppressed by a sufficiently high applied magnetic field.

It seems to this writer that the two areas referred to above offer exciting prospects for further study, both experimentally and theoretically, of quantal Wigner crystallization in low-dimensional geometries.

ACKNOWLEDGMENTS

N. H. M. thanks Professors F. Claro, M. J. Lea, and R. H. Squire and Dr. A. Cabo for numerous valuable discussions on the area embraced by the present study. Thanks are also due to Professor D. Van Dyck and Professor D. Lamoen for making possible my continuing affiliation with the University of Antwerp.

References

1. Wigner, E. P. Phys Rev 1934, 46, 1002.
2. Wigner, E. P. Trans Faraday Soc 1938, 34, 678.
3. Durkan, J.; Elliott, R. J.; March, N. H. Rev Mod Phys 1968, 40, 812.
4. Andrei, E. Y.; Deville, G.; Glattli, D. C.; Williams, F. I. B.; Paris, E.; Etienne, B. Phys Rev Lett 1988, 60, 2765.
5. Buhmann, H.; Joss, W.; von Klitzing, K.; Kukushkin, I. V.; Plaut, A. S.; Martinez, G.; Ploog, K.; Timofeev, V. B. Phys Rev Lett 1991, 66, 926.
6. Lea, M. J.; March, N. H.; Sung, W. J Phys Condens Matter 1991, 3, 4301.
7. Lea, M. J.; March, N. H.; Sung, W. J Phys Condens Matter 1992, 4, 5263.
8. Halperin, B. I.; Tesanovic, Z.; Axel, F. Phys Rev Lett 1986, 57, 922.
9. Skyrme, T. H. R. Nuclear Phys 1962, 31, 556.
10. Sondi, S. L.; Karlhede, A.; Kivelson, S.; Rezayi, E. H. Phys Rev B 1993, 47, 16419.
11. Yannouleas, C.; Landman, U. J Phys A Solid State 2006, 203, 1160.
12. Capuzzi, P.; March, N. H.; Tosi, M. P. Phys Lett A 2005, 339, 207.
13. Ceperley, D. M.; Alder, B. J. Phys Rev Lett 1980, 45, 566.
14. Care, C. M.; March, N. H. Adv Phys 1975, 24, 101.
15. Coldwell-Horsfall, R.; Maradudin, A. A. J Math Phys 4, 582.
16. Herman, F.; March, N. H. Solid State Commun 1984, 50, 725.
17. Holas, A.; March, N. H. Phys Lett A 1991, 157, 160.
18. March, N. H., Ed. Electron Correlation in Solids; Imperial College Press: London, 1999.
19. Carr, W. J. Phys Rev 1961, 122, 1437.
20. March, N. H.; Young, W. H. Philos Mag 1959, 4, 384.
21. March, N. H. Liquid Metals: Concepts and Theory; Cambridge University Press: New York, 2005.
22. Parrinello, M.; March, N. H. J Phys C Solid State 1976, 9, L147.
23. Putley, E. Proc Phys Soc 1960, 76, 802.
24. Somerford, D. J. J Phys C 1971, 4, 1570.
25. Care, C. M.; March, N. H. J Phys C 1971, 4, L372.
26. Elliott, R. J.; Kleppmann, O. J Phys C 1975, 8, 2737.
27. Gervais, G.; Stormer, H. L.; Tsui, D. C.; Kuhns, P. L.; Moulton, W. G.; Reyes, A. P.; Pfeiffer, L. N.; Baldwin, K. W.; West, K. W. Phys Rev Lett 2005, 94, 196803.
28. Barrett, S. E.; Dabbash, G.; Pfeiffer, L. N.; West, K. W.; Tycko, R. Phys Rev Lett 1995, 74, 5112.
29. March, N. H.; Cabo, A.; Claro, F. Phys Lett A 2006, 349, 271.
30. Siringo, F.; March, N. H.; Paranjape, B. V. Phys Chem Liquids 1991, 24, 131.
31. Balents, L.; Fisher, M. P. A.; Nayak, C. Phys Rev B 1999, 60, 1654.
32. Squire, R. H.; March, N. H. Int J Quantum Chem 2005, 105, 883.
33. March, N. H.; Squire, R. H. Phys Lett A 2005, 344, 383.

34. Ando, Y.; Ono, S.; Sub, X. F.; Takeya, J.; Balakirev, F. F.; Betts, J. B.; Boebinger, G. S. Phys Rev Lett 2004, 92, 247004.
35. Romanovsky, I.; Yannouleas, C.; Landman, U. Phys Rev Lett 2004, 93, 230405.
36. Capuzzi, P.; March, N. H.; Tosi, M. P. Phys Lett A 2005, 339, 207.
37. Claro, F.; Cabo, A.; March, N. H. Physica Status Solidi B 2005, 242, 1817.
38. Hubbard, J. Phys Rev B 1979, 20, 4584.
39. Torrance, J. B. Phys Rev B 1978, 17, 3099.
40. Demler, E.; Lopatnikova, A.; Simon, S. H. Phys Rev B 2004, 70, 115326.
41. Bylander, J.; Duty, T.; Delsing, P. Nature 2005, 434, 361.
42. Holas, A.; March, N. H. Phys Lett A 1991, A157, 160.
43. Smith, V. H. In Electron Spin and Momentum Densities and Chemical Reactivity; Mezey, P. G.; Robertson, B., Eds.; Kluwer: Dordrecht, The Netherlands, 1998.
44. Ovchinnikov, A. A.; Zabrodin, A. V. Phys Lett A 1990, 151, 420.
45. Angilella, G. G. N.; March, N. H.; Pucci, R.; Squire, R. H. Phys Chem Liquids 2002, 40, 353.
46. Baskaran, G. In Electron Correlation in the Solid State; March, N. H., Ed.; Imperial College Press: London, 1999; p 263.
47. March, N. H. J Phys Condens Matter 1993, 5, B149.
48. Leinaas, J. M.; Myrheim, J. Nuovo Cimento 1977, B37, 1.
49. March, N. H.; Gidopoulos, N.; Theophilou, A. K.; Lea, M. J.; Sung, W. Phys Chem Liquids 1993, 26, 135.
50. Wu, Y. Phys Rev Lett 1995, 73, 922; 1995, 74, 3906.
51. Angilella, G. G. N.; March, N. H.; Pucci, R. Phys Chem Liquids 2006, (in press).

Appendix A: Momentum Distribution $n(p)$ in Jellium Model in D Dimensions

In the main text, it has been stressed that in 3D there is a transition from an itinerant electron fluid in the high-density regime to a quantal electron crystal in the low-density limit in which the mean interelectronic separation $r_s \to \infty$. As Ceperley and Alder [13] demonstrated numerically in this 3D case, the critical r_s for this transition to a broken symmetry state was $\sim 80a_0$.

In Figure 1, this situation has been contrasted schematically with the 1D case, where it can be seen that the critical r_s tends to zero. Of course, when the electron–electron interactions are switched off, there is a discontinuity at the Fermi momentum p_f in any dimension D. What will be emphasized in this Appendix is that there exists a solvable analytical model in the 1D case, which Holas and March [42] have used to show that in spite of the disappearance of the discontinuity when even arbitrarily weak interactions are "switched on" in 1D, memory of the sharp Fermi "surface" remains through nonanalytic behavior in $n(p)$ around the Fermi momentum p_f.

To obtain the nonanalytic features of the momentum distribution $n_w(p)$ in the 1D Wigner (W) crystal, Holas and March [42] use the so-called folded density employed in Compton profile analysis [43] and denoted below by $B(\mathbf{R})$ as

$$B(\mathbf{R}) = \int d^D\mathbf{p}n(p)\exp(-i\mathbf{p}\cdot\mathbf{R}) \qquad (A.1)$$

in D dimensions. But $B(\mathbf{R})$ is related to the first-order spinless density matrix $\gamma(\mathbf{r}, \mathbf{r}')$ by

$$B(\mathbf{R}) = \int d^D\mathbf{r}\gamma(\mathbf{r}, \mathbf{r} + \mathbf{R}). \qquad (A.2)$$

Matrix γ was studied earlier by Ovchinnikov and Zabrodin [44]; Holas and March [42] use their model to extract, via Eqs. (A.1) and (A.2), the nonanalytic behavior of the momentum distribution $n_w(p)$ in this 1D model with strong repulsive interaction. This nonanalytic part of $n_w(p)$ has the form near the Fermi momentum p_f [42]:

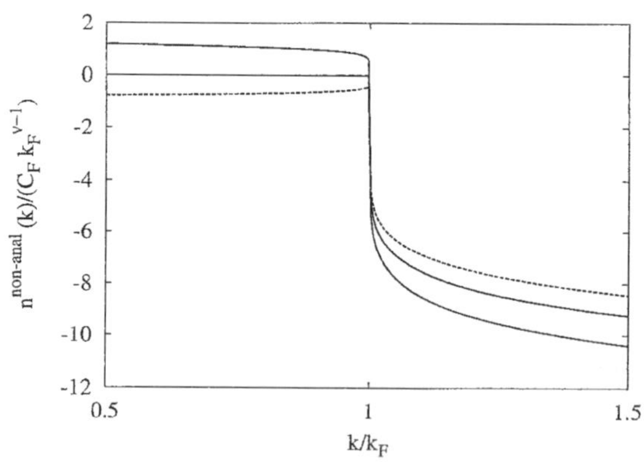

FIGURE A1. Nonanalytic contribution to momentum distribution of one-dimensional electron liquid, defined in Eq. (A.3) plotted versus k/k_F, where $p = \hbar k$ and $\hbar k_F$ is the Fermi momentum. As the Fermi momentum is traversed, behavior of plots changes from monotonic to nonmonotonic, according to whether, in Eq. (A.3), $|\tan(\phi_0)| \gtrless |\tan((1/2)\pi\nu)|$. Plots are for $\nu = 9/8$ and $2\phi_0/\pi = 1.1$, 1.125, and 1.15. (Redrawn from Angilella et al. [45].)

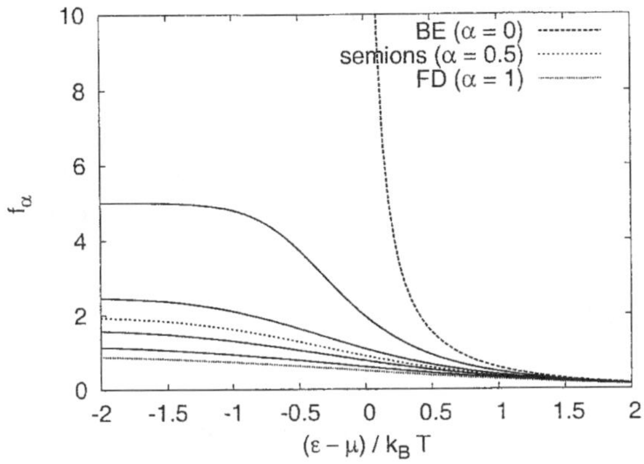

FIGURE A2. Distribution function for anyon statistics for different values of the anyon parameter α. The dashed curves are for $\alpha = 1/2$ (semions) and the limiting cases $\alpha = 0$ (Bose–Einstein) and $\alpha = 1$ (Fermi–Dirac). Solid curves correspond to the values $\alpha =$ 0.2, 0.4, 0.6, and 0.8. (Redrawn from Angilella et al. [51].)

$$n_w^{na}(p) = C_F|p - p_f|^{(\nu-1)} \times [\tan(\tfrac{1}{2}\pi\nu) + \tan(\phi_0)sgn(p - p_f)], \quad \text{(A.3)}$$

where C_F is a constant. Although this contribution clearly tends to zero as $p \rightarrow p_f$, and thus there is no discontinuity in $n_w(p)$ and $p = p_f$, Eq. (A.3) shows that fingerprints of the originally sharp Fermi "surface" with the repulsive interactions "switched off," remain in the presence of strong interparticle repulsion (see also Fig. A1, redrawn from [45]).

What is known in 2D is not decisive! In this "borderline" case, Baskaran [46] has presented arguments leading to the conclusion that, in the jellium model, repulsive interactions remove the discontinuity at $r_s =$ 0 in the momentum distribution $n(p)$. However, the true situation in 2D remains highly controversial.

Appendix B: Simple Collision Model Bringing About a (Partial) Unification of Fermi–Dirac, Bose–Einstein, and Anyonic Quantum Statistical Distribution Functions

Three types of quantum statistics of charged particles are referred to in the present article. Therefore, it seemed of some importance in this Appendix to present an elementary collision model (see also Ref. [47]) which, in the limit of quantum gases (with the Coulomb interactions treated rather fully in the body of the text) brings about at least a partial unification of the statistical distributions of these three types.

B1. COLLISIONS IN A GAS OF FERMIONS

Let us denote the basic collision as $1 + 2 \leftrightarrow 3 + 4$; i.e., states 1 and 2 interact to change to states 3 and 4. This simple reaction involving Fermions has a rate of the form

$$f_1f_2(1 - f_3)(1 - f_4)R$$

where "generalized Pauli exclusion" requires the inclusion of factors $(1 - f_3)$ and $(1 - f_4)$, which are the probabilities that there are no particles in states 3 and 4.

Next, this rate must be equated to that in the reverse direction. Therefore, introducing R' instead of "rate" R above, this results in the equality

$$f_1f_2(1 - f_3)(1 - f_4)R = f_3f_4(1 - f_1)(1 - f_2)R'. \quad \text{(B.1)}$$

The next step is to invoke time-reversal symmetry of quantum mechanics, which readily yields $R' = R$ and it is then easily verified that the Fermi–Dirac (FD) statistical distribution function $f(\epsilon)$, with ϵ the energy satisfies

$$\frac{f(\epsilon)}{1 - f(\epsilon)} = \exp(-\alpha - \beta\epsilon), \quad \text{(B.2)}$$

where β is recognized as the reciprocal of the thermal energy k_BT. Elementary solution of Eq. (B.2) for $f(\epsilon)$ yields the well-known structure of FD statistics.

B2. BOSE-EINSTEIN STATISTICAL DISTRIBUTION FUNCTION

This can be very brief, having established the result of FD statistics via Eqs. (B.1) and (B.2). In essence, in Eq. (B.1) the factors of the form $(1 - f_i)$ with $i = 1$–4 need replacing by $(1 + f_i)$. To take a specific example, it is as though states f_3 and f_4, say, are already occupied, and that other states, in the case of BE statistics under consideration, are then more inclined to go to them. Hence, in Eq. (B.2) the denominator $1 - f(\epsilon)$ in FD statistics becomes $1 + f(\epsilon)$ and by taking the reciprocal of the resulting equation one immediately has the shape of the BE statistical distribution function as

$$\frac{1}{f(\epsilon)} + 1 = \exp(\alpha + \beta\epsilon), \quad \text{(B.3)}$$

the well-established result once again.

B3. RESTRICTED TO TWO DIMENSIONS ONLY: ANYONS WITH FRACTIONAL STATISTICS

The simple collision model has led to a rather unified presentation of the statistical distribution function of FD and BE statistics. In early work on the 2D case only, where it was already known that anyonic fractional statistics apply [48], March et al. [49] made early use of the collision model. The shape of the BE result [Eq. (B.3)] then prompts the assertion that the corresponding result for anyon fractional states should have the form

$$\frac{1}{f(\epsilon)} = \exp\left(\frac{\epsilon - \mu}{k_B T}\right) + a. \quad \text{(B.4)}$$

Unfortunately, while the ansatz [Eq. (B.4)] readily recovers the FD distribution function with the choice $a = 1$ and the BE result with $a = -1$, the above collision model does not contain within itself a unique procedure designed to fix the quantity a in terms of a fractional statistics parameter, conventionally denoted by α, where α now lies between 0 and 1. This is why the collision model provides what we term a "partial" unification only of the three types of statistics under discussion.

Because, as noted above, the as yet unknown function $a(\alpha)$ reduces to 1 and -1 for FD and BE cases, respectively, it was the natural use of Occam's razor to assume $a(\alpha)$ to be independent of the energy ϵ. In the subsequent important work of Wu [50], it was demonstrated that away from the limits $\alpha = 0$ and $\alpha = 1$, $a(\alpha)$ does indeed involve the energy ϵ, albeit usually somewhat weakly. Appealing to his theory, let us write, for convenience of presentation, that $a(\alpha, \epsilon)$ takes the form

$$a(\alpha, \epsilon) = W(\zeta) + \alpha - \zeta. \quad \text{(B.5)}$$

Wu's beautiful result is then that the function $W(\zeta)$ in Eq. (B.5) satisfies the following equation:

$$W^{\alpha}[1 + W]^{1-\alpha} = \zeta. \quad \text{(B.6)}$$

To illustrate Eq. (B.6) in limiting cases, FD statistics corresponds to the "endpoint" $\alpha = 1$ when Eq. (B.6) immediately yields $W = \zeta$, and hence from Eq. (B.5), $a = 1$, a result already noted above. Likewise, BE statistics is recovered by taking the other endpoint of the range $0 \leq \alpha \leq 1$ specified for the fractional statistics parameter α, i.e., $\alpha = 0$. Inserting this BE limit into Eq. (B.6) yields $[1 + W] = \zeta$, and hence $a = -1$, a result already noted above for BE statistics.

In their analysis of the orbital magnetism of the Laughlin liquid, which in turn allowed the melting curve $T_m(\nu)$ to be analyzed using the thermodynamic result (3.3), Lea et al. [7], exploited the above fractional statistics parameter (for fuller details, see Ref. [7]).

Part 2: Structure, forces and electronic correlation functions in liquid metals

The theme of structure and forces in liquid metals was set out in the early work of Johnson and March (JM below) [2a.2]. The liquid structure factor $S(\mathbf{q})$, directly accessible via neutron diffraction experiments, and the pair correlation function $g(\mathbf{r})$ which is, in essence, the Fourier transform of $S(\mathbf{q})$, were stressed as providing a route to the extraction of an (assumed) pair potential $\phi(\mathbf{r}$. However, besides $g(\mathbf{r})$ and $\phi(\mathbf{r}$ being involved in the so-called force equaion, the triplet correlation g_3 also entered. Whereas this was 'decoupled' in terms of solely $g(\mathbf{r})$ in [2a.2], the proposal of JM was brought to full fruition by Reatto *et al.* by means of computer simulation. All the salient features of their pair potential for liquid Na near freezing, obtained actually from X-ray diffraction data as input, were regained from first-principles electron theory by Perrot and March [2a.12].

The theme of Part 2b relating to electronic correlation functions in liquid metals was also directly connected to experiment by Egelstaff, March, and McGill [2b.3] after earlier theoretical study by Tosi and March [2b.2]. In a pure liquid metal like Na, consisting of Na^+ ions (i) and valence electrons e) $S(\mathbf{q})$ has to be supplemented by two further partial structure factors $S_{ie}(\mathbf{q})$ and $S_{ee}(\mathbf{q})$, in analogy with molten salt NaCl, involving $S_{\mathrm{NaNa}}(\mathbf{q})$ and $S_{\mathrm{ClCl}}(\mathbf{q})$. Cusack *et al.* [2b.4] followed by giving the theory of $S_{ee}(\mathbf{q})$ in molten nearly-free-electron metals (see also the further study of Leys and March [2b.10]). This latter work was prompted by the important use of computer simulation to obtain $S_{ie}(\mathbf{q})$ in liquid Mg by de Wijs *et al.* [Phys. Rev. Lett. **75**, 4480 (1995)].

Part 2, in summary, has emerged as a result of initial theoretical proposals which have subsequently been complemented by relevant diffraction studies (e.g. [2b.5] on electron diffraction and [2b.6] using neutrons). In both 2a and 2b, the interplay involving many-body theory (both N-electron Schröödinger wave mechanics and classical statistical mechanics), diffraction experiments and computer simulation has been an outstanding feature.

We believe that both themes reported here are worthy of fuller investigation in the future. One focal point should then be, for pure liquid metals first of all, neutron diffraction measurements of $S(\mathbf{q})$ of sufficiently high accuracy to allow the so-called Ornstein–Zernicke direct correlation function $c(r)$ to be extracted, especially at large r. Following JM [2a.1], Johnson, Hutchinson, and March [Proc. Roy. Soc. A (London) **282**, 283 (1964)] emphasized the justification of the pair potential $\phi(\mathbf{r}$ concept for s–p bonded liquid metals (see [2a.3] following the electron theory of [3b.3]) having been established, that the direct correlation function $c(r)$ at large r [2a.5] should satisfy the asymptotic relation

$$c(r) = -\phi(r)/kBT, \qquad r \to \infty\,. \tag{1}$$

An analytic generalization of this result appropriate for hard-sphere-liquid s–p metals [Bernasconi and March, Phys. Chem. Liq. **15**, 169 (1986)] was proposed by Woodhead-Galloway, Gaskell, and March [J. Phys. C **1**, 271 (1968)].

Turning to two-component systems, we wish to note here, in concluding this Commentary on Part 2, the importance of the area of supercooled multicomponent metallic liquids and bulk metallic glasses for both pure science and for technological applications [see, e.g., Alonso and March, in press (2009)]. In particular, Cowlam *et al.* [J. Non-Crystalline Solids **205–207**, 567 (1996), and earlier refs. there] have emphasized the use of neutron diffraction data with the technique of isotopic enrichment, to extract partial directly correlations $c_{ij}(r)$ and corresponding pair potential $\phi_{ij}(r)$ generalizing Eq. (1) above in this area of binary metallic glasses.

Volume 7, number 3 PHYSICS LETTERS 15 November 1963

ELECTRONIC MOMENTUM DISTRIBUTION IN LIQUID METALS AND LONG-RANGE OSCILLATORY INTERACTIONS BETWEEN IONS

T. GASKELL and N. H. MARCH
Department of Physics, University of Sheffield

Received 19 October 1963

Gustafson et al. [1] have recently reported positron annihilation measurements in liquid and solid mercury, and have interpreted their results in terms of an electronic momentum distribution $P(k)$, which, in the liquid, has then considerable diffuseness, in contrast to the Fermi distribution with a sharp edge at k_f.

This is especially interesting in the light of the long-range oscillatory inter-ionic potentials in liquid metals which were derived by Johnson and March [2] from measured radial distribution function data. These oscillations can be understood in terms of the incomplete screening of a charge centre by the conduction electrons. However, it is known that the oscillations induced by the perturbation are closely connected with the sharpness of the Fermi surface [3-5]. Further work on the inter-ionic potential in liquid Hg [6] reveals damping of the oscillations, and since Gustafson et al. also determined $P(k)$ for liquid Hg it seemed of considerable interest to work out the screening theory for a general momentum distribution. The purpose of this communication is to express the screened potential round a test charge, or equivalently the dielectric constant, in a liquid metal, in terms of $P(k)$.

To do so, we start out from the Green function formulation of Edwards [7], and we take, following his work, the simple forms

$$G_0{\pm}(k,E) = \frac{1}{E-\frac{1}{2}k^2 \pm i\Gamma(k,E)} \qquad (1)$$

as describing our model of the liquid metal, where Γ^{-1} is playing the role of a relaxation time. For more general dispersion laws, we could replace $\frac{1}{2}k^2$ by $E(k)$ and the arguments would be modified in an obvious way. In the presence of a test charge, with screened potential $V(\boldsymbol{r})$, we have the perturbed Green's functions in configuration space, to first order in V as

$$G^{\pm}(\boldsymbol{r},\boldsymbol{r}_0) = G_0^{\pm}(\boldsymbol{r},\boldsymbol{r}_0) + \int d\boldsymbol{r}_1\, G_0^{\pm}(\boldsymbol{r},\boldsymbol{r}_1)\, V(\boldsymbol{r}_1)\, G_0^{\pm}(\boldsymbol{r}_1,\boldsymbol{r}_0). \qquad (2)$$

We now form the electron density $\rho(r)$ by taking the diagonal of $(1/2\pi i)[G^+ - G^-]$ from 0 to the Fermi energy E_f, and we find

$$\rho(r) = \rho_0 + \int_0^{E_f} dE \int d\boldsymbol{r}_1\, V(\boldsymbol{r}_1)[G_0^{+2}(\boldsymbol{r},\boldsymbol{r}_1) - G_0^{-2}(\boldsymbol{r},\boldsymbol{r}_1)], \qquad (3)$$

where ρ_0 is the mean electron density in the liquid metal. By Fourier transformation, and using the fact that the momentum distribution $P(q)$ is given by

$$P(q) = \frac{1}{2\pi i}\int_0^{E_f} dE[G_0^-(q,E) - G_0^+(q,E)], \qquad (4)$$

we readily obtain

$$\rho(q) = (2\pi)^3 \rho_0\, \delta(q) - \frac{V(q)}{2\pi^3}\int dk\, \frac{1}{k^2 - |\boldsymbol{k}-\boldsymbol{q}|^2} \times [P(|\boldsymbol{k}|) - P(|\boldsymbol{k}-\boldsymbol{q}|)]. \qquad (5)$$

It is of interest to remark that the result (5) may be shown to be of the same form as for temperature excitation.

By combining (5) with Poisson's equation, we obtain the Fourier components $V(q)$ of the screening potential round a charge Z

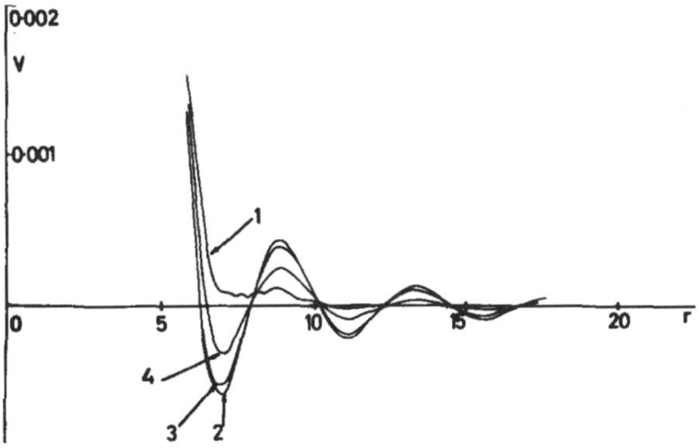

Fig. 1. Screened potential in liquid metal. Energy units are Rydbergs. Distances are given in atomic units.
Curve 1 Potential for Δ of eq. (9) = 0.2, k_f = 0.711.
Curve 2 Potential for Fermi distribution with k_f = 0.711 and for unit charge.
Curve 3 Potential for Δ of eq. (9) = 0.04.
Curve 4 Potential for Δ of eq. (9) = 0.1.

Volume 7, number 3 PHYSICS LETTERS 15 November 1963

$$V(q) = \frac{-4\pi Z}{q^2 - (4/\pi q)\int_0^\infty \mathrm{d}k\, kP(k)\ln\left|\frac{q-2k}{q+2k}\right|}, \tag{6}$$

where $P(k)$ is normalised so that

$$3\int_0^\infty \mathrm{d}k\, P(k)\, k^2 = k_\mathrm{f}^3 . \tag{7}$$

Eq. (5) immediately yields the usual result for a charge in a Fermi gas when we take $P(k)$ to have a sharp edge at k_f: namely

$$V(q) = \frac{-4\pi Z}{q^2 + (k_\mathrm{f}/\pi)\, g(q/2k_\mathrm{f})} \tag{8}$$

where $g(x) = 2 + [(x^2-1)/x]\ln|(1-x)/(1+x)|$.

Gustafson et al. have interpreted their experiments for liquid Hg in terms of a $P(k)$ given by

$$P(k) = \frac{A}{\exp((k^2-1)/\Delta)+1}, \tag{9}$$

where k is in units of k_f and $\Delta \approx 0.2$. For normalisation in accordance with (7), we find $A \approx 0.951$. However, inserting this into (6) and performing the Fourier transform of the resulting $V(q)$ numerically we obtain the surprising result shown in curve 1 of fig. 1. The diffuseness of the Fermi surface is so large that the character of the oscillations for a sharp surface, shown in curve 2, is completely destroyed. We have therefore examined the influence of varying Δ in (9) while preserving normalisation, and we show in curves 3 and 4 results for $\Delta = 0.04$ and 0.1 respectively.

Our conclusion then, is that if the momentum distribution in liquid Hg is indeed as diffuse as Gustafson et al. have claimed, the long-range oscillations round an ion would not occur. This is in sharp contrast to the conclusions from the analysis of measured radial distribution function data for 8 liquid metals including Hg, for a variety of temperatures which has been carried out by Johnson, Hutchinson and March [6]. In every case, long-range oscillations are found. A broadening of the Fermi surface of approximately 6%, which as Gustafson et al. remarked, has earlier been estimated by Knight et al. [8] from the electrical resistivity of liquid Hg, can readily be accomodated however, and the oscillations remain.

We therefore feel that the direct interpretation of positron annihilation experiments in a liquid in terms of electronic momentum distribution should be carefully re-examined, in the light of the present findings, and this problem is at present being investigated in this Department.

One of us (N.H.M.) wishes to ackowledge a most valuable discussion on this problem with Dr. J. Hubbard and Mr. J. Beeby during a brief visit to A.E.R.E., Harwell às a Vacation Consultant in the summer of this year. He also wishes to thank Dr. A. B. Lidiard and Dr. W. Marshall for their hospitality during this visit to the Theoretical Physics Division. Thanks are also due to Dr. J. E. Enderby and Dr. S. Misawa for some interesting comments on this work.

References

1) D.R. Gustafson, A.R. Mackintosh and D.J. Zaffarano, Phys Rev. 130 (1963) 1455.
2) M.D. Johnson and N.H. March, Physics Letters 3 (1963) 313.
3) W. Kohn and S.H. Vosko, Phys. Rev. 119 (1960) 912.
4) N.H. March and A.M. Murray, Proc. Phys. Soc. 79 (1962) 1001.
5) R.L. Odle and C.P. Flynn, Proc. Phys. Soc. 81 (1963) 412.
6) M.D. Johnson, P. Hutchinson and N.H. March, to be published.
7) S.F. Edwards, Proc. Roy. Soc. A 267 (1962) 518.
8) W.D. Knight, A.G. Berger and V. Heine, Ann. Phys. 8 (1959) 173.

* * * * *

Volume 3, number 7 PHYSICS LETTERS 15 February 1963

LONG-RANGE OSCILLATORY INTERACTION BETWEEN IONS IN LIQUID METALS

M. D. JOHNSON and N. H. MARCH
Department of Physics, The University, Sheffield

Received 14 January 1963

The purpose of this communication is to present an analysis of measured radial distribution functions for liquid metals, which reveals long-range oscillatory interactions between the ions.

In order to proceed with this analysis, we make the assumption that the total potential energy function in a liquid may be expressed as a sum of pair potentials $\varphi(r_{ij})$. This assumption has sometimes been criticised for liquid metals, but for reasons which we discuss briefly below, we believe that the criticisms are not valid. In addition, for the monatomic liquids we consider, φ is also assumed to be central. We now follow Born and Green [1] and note first that there is an exact relation between the pair correlation function n_2, the third order distribution function n_3, and the pair potential φ. To express this in a form convenient for our purpose, we first write

$$n_2(r) = n_0^2\, e^{-\beta U(r)}\,, \qquad \beta = 1/kT\,, \tag{1}$$

where n_0 is the mean density in the liquid, and $U(r)$ is a potential of mean force defined by (1).

Then we have [2]

$$\frac{\partial U(\boldsymbol{r}_2-\boldsymbol{r}_1)}{\partial \boldsymbol{r}_1} = \frac{\partial \varphi(\boldsymbol{r}_2-\boldsymbol{r}_1)}{\partial \boldsymbol{r}_1} + \int \frac{n_3(\boldsymbol{r}_1\boldsymbol{r}_2\boldsymbol{r}_3)}{n_2(\boldsymbol{r}_1\boldsymbol{r}_2)}\, \frac{\partial \varphi(\boldsymbol{r}_3-\boldsymbol{r}_1)}{\partial \boldsymbol{r}_1}\, d\boldsymbol{r}_3\,. \tag{2}$$

However, we can only derive $\varphi(r)$ from $U(r)$ using (2) if n_3 is known. This compels us to make the most serious approximation of our treatment: namely to write

$$n_0^3 n_3(\boldsymbol{r}_1\boldsymbol{r}_2\boldsymbol{r}_3) = n_2(\boldsymbol{r}_1\boldsymbol{r}_2)\, n_2(\boldsymbol{r}_1\boldsymbol{r}_3)\, n_2(\boldsymbol{r}_2\boldsymbol{r}_3)\,, \tag{3}$$

which is the well known superposition approximation of Kirkwood [3]. We then obtain an integral equation for $\varphi(r)$: namely [2]

$$\varphi(r) = U(r) + \frac{\pi}{n_0^3 r} \int_0^\infty ds\, n_2(s)\, \frac{\partial \varphi(s)}{\partial s} \times \int_{-s}^{s} dt\, (s^2-t^2)\,(t+r)\,[n_2(|t+r|) - n_0^2]\,. \tag{4}$$

This equation has been solved for sodium at 203°C, using the distribution function $n_2(r)$ as given by the neutron diffraction study of Ginrich and Heaton [4]. The procedure adopted was to use $U(r)$ as a first approximation to $\varphi(r)$, inside the integral, and then to iterate until $\varphi(r)$ reproduced itself. Curve 1 of fig. 1 shows $U(r)$ as obtained from (1) while curve 2 shows the result for $\varphi(r)$ as given by solution of (4). The essential point we wish to stress is that the oscillations in $U(r)$ are enhanced by the integral term in (4). As further confirmation of our findings, we have also obtained results for Na at 114°C, and for this case $U(r)$ is given by curve 3 of fig. 1. For this temperature, however, we used $\varphi(r)$ as given in curve 2 inside the integral term of (4), and the new $\varphi(r)$ thus obtained is shown in

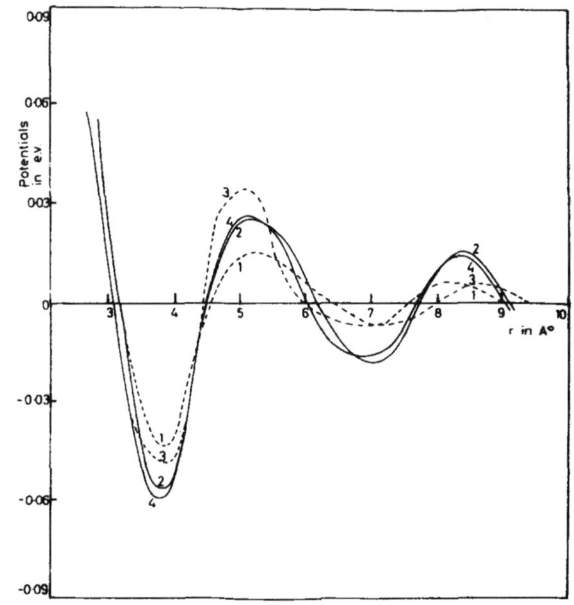

Fig. 1. Potentials for liquid Na.
Curve 1. Potential $U(r)$ of mean force at 203°C.
Curve 2. Pair potential $\varphi(r)$, derived from $U(r)$ of curve 1.
Curve 3. Potential $U(r)$ of mean force at 114°C.
Curve 4. Approximate pair potential $\varphi(r)$, derived from $U(r)$ of curve 3.

curve 4. It will be seen that $\varphi(r)$ has almost reproduced itself, and that the temperature dependence of $U(r)$ has been almost completely removed by the "correction" term in (4). With the present procedure, it would be unreasonable to attach significance to the small differences between curves 2 and 4.

Finally, it may perhaps be objected that the oscillations in $\varphi(r)$ are already present in $U(r)$ and that a similar situation would exist for liquid insulators. Therefore we have again employed (4) for the case of liquid argon, using the data of Henshaw [5] at 84°K and those of Eisenstein and Gingrich [6] at 149°K, and the corresponding results for $U(r)$ are shown in curves 1 and 2 of fig. 2. It rapidly became clear from numerical solution of (4) that

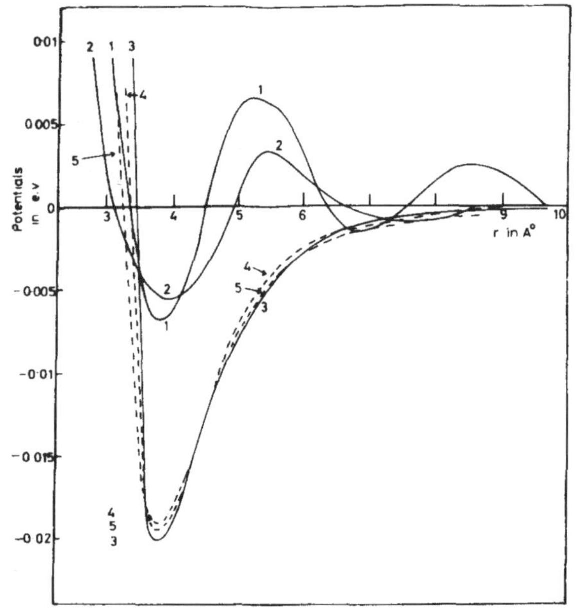

Fig. 2. Potentials for liquid Ar.
Curve 1. Potential $U(r)$ of mean force at 84°K.
Curve 2. Potential $U(r)$ of mean force at 149°K.
Curve 3. Pair potential $\varphi(r)$ as given by Dobbs and Jones.
Curve 4. Pair potential $\varphi(r)$, derived from $U(r)$ of curve 1.
Curve 5. Pair potential for $\varphi(r)$, derived from $U(r)$ of curve 2.

$U(r)$ was a very bad approximation to $\varphi(r)$ for liquid argon. Therefore, we took the pair potential as given by Dobbs and Jones [7], and shown in curve 3 of fig. 2 as a first approximation, and the final solutions $\varphi(r)$ of (4) are presented in curves 4 and 5 of fig. 2. It is then clear that whereas in Na the oscillations in $U(r)$ are considerably enhanced by the term involving n_3, in Ar the oscillations are annulled. This seems to us to give a convincing demonstration that whereas in liquid argon, the long-range forces are essentially of the Van der Waals type, in liquid sodium they arise from the polarisation of the conduction electron gas by the metal ions. Indeed, Corless and March [8] have recently given the theory of the interaction between charged centers in a dense electron gas, and while this theory, designed for dealing with interactions between defects in metal crystal, needs quantitative modification in our case, it is gratifying that the amplitude and wavelength of the long-range oscillations which we find are in general accord with this treatment. Furthermore, the work of Corless and March, based on a first-order account of the conduction electron polarisation, affords some theoretical basis for the use of pair potentials in liquid metals, as may be seen from eq. (A1.4) of their paper [8].

It is a pleasure to thank Dr. J. E. Enderby for stimulating our interest in the problem of liquid metals and for valuable discussions in which the relevance to this field of earlier point defect studies became apparent. The award of a D.S.I.R. Research Studentship is gratefully acknowledged by one of us (M.D.J.).

References

1) M. Born and H. S. Green, Proc. Roy. Soc. A 188 (1946) 10.
2) H. S. Green, Molecular theory of fluids (North-Holland Publishing Company, 1952) p. 75.
3) J. G. Kirkwood, J. Chem. Phys. 3 (1935) 300.
4) N. S. Gingrich and L. Heaton, J. Chem. Phys. 34 (1961) 873.
5) D. G. Henshaw, Phys. Rev. 105 (1957) 976.
6) A. Eisenstein and N. S. Gingrich, Phys. Rev. 58 (1940) 307.
7) E. R. Dobbs and G. O. Jones, Rep. Progr. Phys. 20 (1957) 516.
8) G. K. Corless and N. H. March, Phil. Mag. 6 (1961) 1285.

* * * * *

J. Phys. Chem. Solids Pergamon Press 1963. Vol. 24, pp. 1305–1308. Printed in Great Britain.

INTERACTION ENERGIES BETWEEN DEFECTS IN METALS

J. WORSTER and N. H. MARCH

Department of Physics, The University, Sheffield

(*Received* 24 *April* 1963)

Abstract—The predictions of perturbation theory regarding the interaction between charged impurities in metals are presented. The unperturbed conduction electrons are described by plane waves and the defects, treated as point charges, are screened to first order in the Hartree approximation. In the case of monovalent and divalent matrices, it is shown that at the nearest-neighbour distance the qualitative results are the same as would be predicted electrostatically. For trivalent matrices, the sign of the interaction is changed.

1. INTRODUCTION

IN TWO recent papers, CORLESS and MARCH[1, 2] have given a theory of the interaction between defects in metals using a perturbative treatment based on the Dirac density matrix,[3, 4] while, independently, BLANDIN, DÉPLANTÉ and FRIEDEL[5] have developed an essentially equivalent method. Indeed, some of the basic results of these theories are at least implicit in an early paper on electron-phonon interaction by BARDEEN[6] as well as in later work of NOZIÈRES and PINES.[7]

2. CALCULATION OF INTERACTION ENERGIES

Computations yielding the explicit numerical results predicted by the above mentioned theories have recently been carried out and it is the purpose of this note to report the main results for impurities in some monovalent, divalent, and trivalent metals.* In terms of the screened potential $V(r)$ due to unit excess positive charge embedded in the metal, the interaction energy between two impurity centres, distance a apart, and carrying excess charges Z_1 and Z_2 is given by [1, 2]

$$-Z_1 Z_2 V(a). \tag{1}$$

$V(a)$ may be obtained directly by solution of the equation, due to MARCH and MURRAY[3]

$$\nabla^2 V = \frac{2k_f^2}{\pi^2} \int \mathrm{d}\boldsymbol{r}_1 V(\boldsymbol{r}_1) \frac{j_1(2k_f|\boldsymbol{r}-\boldsymbol{r}_1|)}{|\boldsymbol{r}-\boldsymbol{r}_1|^2}, \tag{2}$$

where k_f is the magnitude of the Fermi wave vector, and

$$j_1(\rho) = \frac{\sin\rho - \rho\cos\rho}{\rho^2}, \tag{3}$$

or equivalently from the Fourier components of $V(\boldsymbol{r})$ given by[6, 8, 9]

$$V(\boldsymbol{k}) = \frac{-4\pi}{k^2 + \dfrac{k_f}{\pi} g\left(\dfrac{k}{2k_f}\right)} \tag{4}$$

where

$$g(x) = 2 + \frac{x^2-1}{x} \ln\left|\frac{1-x}{1+x}\right|. \tag{5}$$

Whereas some direct numerical solutions of (2) are already available,[4] the present results have been found by employing equation (4) as follows. We may write

$$rV(r) = -\frac{1}{\pi i} \int_{-\infty}^{\infty} \frac{x e^{2iRx}}{x^2 + \dfrac{1}{4\pi k_f} g(x)} \mathrm{d}x, \tag{6}$$

* Professor J. Friedel has kindly informed us that similar calculations have been carried out independently at Orsay by Drs Blandin and Déplanté, are in agreement with our results.

where $R = k_f r$. Integrating

$$\frac{1}{\pi i}\,\frac{z e^{2iRz}}{z^2+\dfrac{1}{\pi k_f}\left[\dfrac{1}{2}+\dfrac{z^2-1}{4z}\ln\dfrac{1-z}{1+z}\right]}$$

round a semicircle in the upper half of the complex plane gives

$$-rV = \alpha e^{-2\beta R}+\frac{2}{\pi}Im\int_1^\infty dx$$

$$\times\left[\frac{x e^{2iRx}}{x^2+\dfrac{1}{4\pi k_f}g(x)}-\frac{x e^{2iR\,x}}{x^2+\dfrac{1}{4\pi k_f}\left[g(x)-i\pi\dfrac{(x^2-1)}{x}\right]}\right]$$

where $2\pi k_f\beta^3 = \beta+(1+\beta^2)\tan^{-1}\beta$ (7)

and

$$\alpha = \frac{4\pi k_f\beta^3}{4\pi k_f\beta^3-\beta+(1-\beta^2)\tan^{-1}\beta}.$$

The integrand is now expanded in powers of $1/\pi k_f$, and $g(x)$ is expanded in powers of $1/x$. The integral becomes

$$\sum_{n=1}^{\infty}\left(-\frac{1}{\pi k_f}\right)^n F_n(R) \quad (8)$$

where

$$F_n(R) = Re\frac{1}{2}\sum_{r=1}^{n}\binom{n}{r}\left(-\frac{i\pi}{4}\right)^r$$

$$\times\sum_{s=0}^{\infty} C_{s,n-r,r}I_{4n+1-3r+2s}(R),$$

$$I_t(R) = \int_1^\infty dx\, x^{-t}\, e^{2iRx}$$

$$C_{s,u,r} = \sum_{p=0}^{r}(-1)^p\binom{r}{p}C_{s-p,u,0}$$

and

$$\sum_{s=0}^{\infty} C_{s,u,0}x^s \equiv \left[\sum_{s=0}^{\infty}\frac{x^s}{(2s+1)(2s+3)}\right]^u$$

The series converges for $2\pi k_f > 1$, and is quite suitable for accurate computation.

3. DISCUSSION OF RESULTS

We show in Fig. 1 our results for $V(r)$ for an impurity in Cu and in Mg, and similar results are

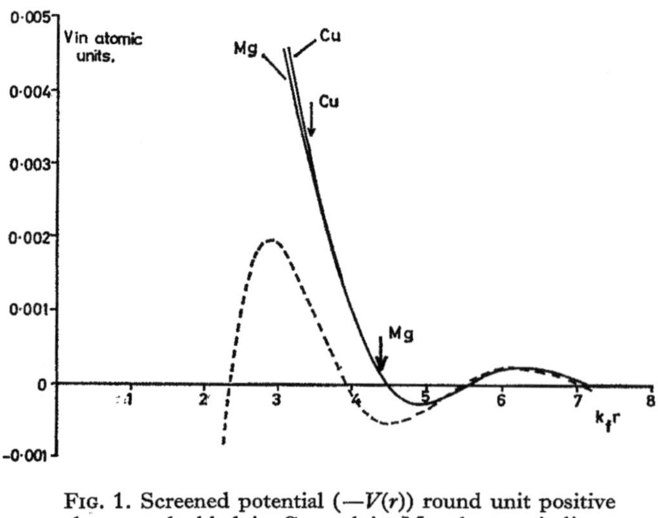

FIG. 1. Screened potential ($-V(r)$) round unit positive charge embedded in Cu and in Mg. Arrows indicate near-neighbour distance in the two cases. Broken line shows asymptotic form.

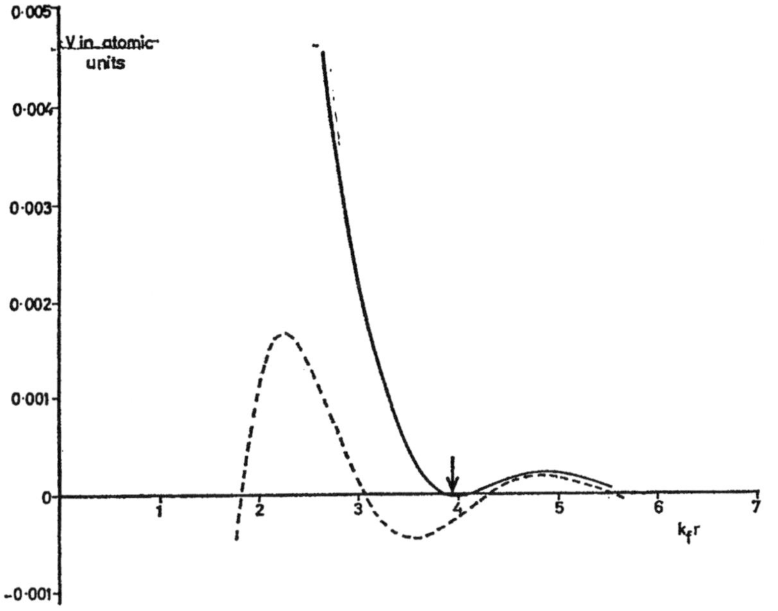

FIG. 2. Screened potential ($-V(r)$) round unit positive charge embedded in Al. Broken line shows asymptotic form.

plotted in Fig. 2 for Al as the matrix metal. For comparison, we have plotted also the asymptotic forms, given by

$$V(r) = -\frac{\pi}{k_f(1+2\pi k_f)^2}\frac{\cos 2k_f r}{r^3} \tag{9}$$

It will be seen that, at the near-neighbour distances in the three metals, $V(a)$ has a sign determined electrostatically by the sign of the charge on the impurity centre for Cu and Mg, while the sign is changed for Al. This is different from the predictions using the asymptotic form, which would group Mg and Al together,[5] with Cu behaving differently.

Furthermore, the magnitudes of the interaction energies are very small at the near-neighbour distance for Mg and Al, whereas this is not so when the asymptotic form is used. Thus, for impurities in Mg and Al, interactions other than between near-neighbours may be the most important. In this connection, it may be of interest to observe that the first node in V disappears at $k_f = 0{\cdot}939$ and that it occurs actually at the nearest-neighbour distance of trivalent close-packed metals for $k_f = 0{\cdot}9325$.

We realize, of course, that the present form of the theory has a number of shortcomings:

(i) Relaxation effects are ignored.

(ii) Bloch wave effects are not properly accounted for although CORLESS and MARCH[2] have shown that equation (1) is still valid to a first approximation in a point-ion model.

(iii) Low order perturbation theory is not quantitatively accurate, and a phase shift in the oscillatory potential would be introduced by higher order terms. Unfortunately, the use of equation (1), if $V(a)$ is calculated to higher order in perturbation theory, is not rigorously justified.

(iv) Electron exchange and correlation effects have been neglected, and a Hartree treatment employed throughout. Preliminary calculations performed at Orsay[10] show that exchange, included without correlation, can cause significant changes, as also noted by CORLESS and MARCH.[1] Such a procedure, however, is known in other problems such as electronic specific heat and free electron diamagnetism to lead to non-physical results unless correlation effects are also accounted for.

While we feel that with the numerous approximations involved, it is unreasonable to suppose that the results presented here can be entirely quantitative, they point to the interest of further studies, both experimental and theoretical, of interaction energies between defects especially in multivalent matrices.

Acknowledgements—One of us (N.H.M.) wishes to thank Professor J. Friedel, both for communicating unpublished results, and for a fruitful correspondence over a long period on the present problem. We wish to acknowledge that this work was performed under contract with the United Kingdom Atomic Energy Authority, A.E.R.E., Harwell.

REFERENCES

1. Corless G. K. and March N. H., *Phil. Mag.* **6,** 1285 (1961).
2. Corless G. K. and March N. H., *Phil. Mag.* **7,** 1765 (1962).
3. March N. H. and Murray A. M., *Phys. Rev.* **120,** 830 (1960).
4. March N. H. and Murray A. M., *Proc. Roy. Soc.* **A261,** 119 (1961).
5. Blandin A., Déplanté J. L. and Friedel J., *Proc. Kyoto Conference on Crystal Lattice Defects, J. phys. Soc. Japan* **18,** Suppl. II, 85 (1963).
6. Bardeen J., *Phys. Rev.* **52,** 688 (1937).
7. Nozières P. and Pines D., *Phys. Rev.* **109,** 762 (1958).
8. Langer J. S. and Vosko S. H., *J. Phys. Chem. Solids* **12,** 196 (1960).
9. March N. H. and Murray A. M., *Proc. Roy. Soc.* **A266,** 559 (1962).
10. Friedel J., Private communication.

[453]

Interatomic Forces and the Structure of Liquids

By J. E. Enderby and N. H. March†
Department of Physics, The University, Sheffield

CONTENTS

§ 1. Introduction

We wish in this article to assemble the evidence relating to the dependence of the properties of liquids on the detailed form of the interatomic forces. It has, of course, for a long time, been evident that the distribution functions required to describe classical fluids must depend in detail on the forces, but, more recently, evidence has been accumulating that hitherto unsuspected differences of a qualitative kind exist between the types of forces in conducting and insulating fluids.

We shall be primarily concerned here with arguments which start from the simplest possible assumptions about classical fluids:

(i) That the total potential energy of the fluid can be expressed as a sum of pair potentials $\phi(r_{ij})$ between atoms or ions i and j.

† On leave of absence for session 1965–6 at Laboratory of Atomic and Solid State Physics, Cornell University, Ithaca, New York, U.S.A.

(ii) That, away from melting points and critical points, these potentials are largely temperature independent.

To simplify the problem as far as possible, we restrict our consideration to monatomic fluids (e.g. liquid Ar or liquid Na), where central forces may be assumed. Then in order to lay down some general theoretical concepts to guide us, we begin by considering the nature of the force law, described by the potential energy $\phi(r)$ for

(*a*) insulating fluids,

(*b*) conducting fluids.

Because most workers agree concerning the type of forces in (*a*), we begin by discussing fluids like liquid Ar.

1.1. *Interatomic Forces in Insulating Fluids* (*e.g. Liquid Argon*)

The general form of the force law is as shown in fig. 1. There are two regions which can usefully be distinguished:

(i) The inner region of strong repulsion. This is due to quantum mechanical overlap forces coming into play as soon as the charge clouds of the separate atoms overlap.

(ii) The (in general rather weak) van der Waals attraction, such that, at large r, $\phi(r) = -Ar^{-6}$. This law may have to be modified somewhat at extremely large r, due to retardation effects, but we shall assume in this article that such effects can be ignored.

Now it has been shown that similar force curves for Ar can be used to explain data in both the gaseous and the solid phase (see, for example, Dobbs and Jones 1957). We arrive therefore at the first conclusion we wish to stress, namely that the forces between Ar atoms appear to be, to a first approximation, the same in condensed matter as in free space.

Fig. 1

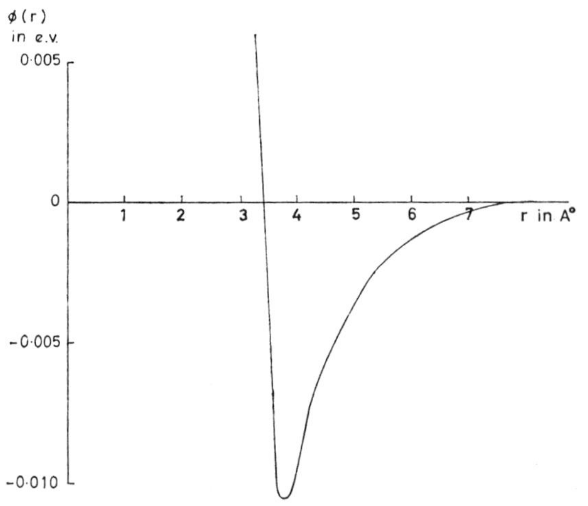

Dobbs–Jones potential for Ar.

There is a reservation to be made here, namely that there is evidence in solid Ar that many-body forces are playing a significant role. This appears to be important in considering the energy of formation of a vacancy in Ar (Foreman 1963), but, in the present state of the theory of liquids, we shall assume that away from transitions, the effect is of second order.

All that we have said above, we believe, will find general acceptance among workers in the field. We turn now to a more recent development; the nature of the forces between ions in conducting fluids.

1.2. *Interionic Forces in Conducting Fluids (e.g. Liquid Sodium)*

In the case of liquid Na, for example, which is highly conducting, measurements of the Hall coefficient strongly suggest that the number of conduction electrons per atom is the same as the valence. This time, therefore, in contrast to the situation for Ar, where to a first approximation all electrons can be thought of as perturbed only slightly from their free atom configurations, we can say that some electrons must be described by totally different wave functions from those appropriate to free atoms. Thus, the long-range forces in liquid Na can have little to do with the force law appropriate to Na atoms in free space. Naturally, as we approach the gaseous phase, this statement must become less precise, and this has to do with our earlier qualification that we are in a region away from transition points.

This all suggests that the interionic forces in liquid metals must have properties dominated by the high density of conduction electrons. That the forces can still be usefully described, as a first approximation, by effective pairwise interactions as in assumption (i), is less certain, but a detailed argument for an assembly of point charges in a Fermi gas which leads to this result has been given by Corless and March (1961); see also Ziman (1964). The extent to which this description yields a consistent picture for liquid metals is of considerable current interest. In this article, we focus attention primarily on the evidence available from the measured structures of liquid metals although reference is also made to other properties.

Our conclusion is that assumptions (i) and (ii) are well justified and lead to a form of $\phi(r_{ij})$ in liquid metals which has quite characteristic features.

§ 2. Radial Distribution Function $g(r)$

Macroscopic properties such as surface tension, viscosity and the temperature range over which the liquid phase exists, differ markedly between van der Waals insulators and liquid metals. What is less well known is the way in which the liquid structure is influenced by the forces. By analogy with crystalline solids, where, for example, van der Waals forces tend to produce close-packed structures, and covalent bonds tend to favour open structures like the diamond lattice, it seems that different forces ought to lead to characteristically different fluid structures. We would expect this effect to require rather careful consideration since the work of Bernal and his colleagues has shown that the gross features of the radial distribution function (the quantitative measure of the liquid structure defined in § 2.1) can be understood in terms of the random packing of hard spheres (Bernal 1959). Nevertheless, we shall now show that the experimental structure data contain clear evidence as to differences between the forces in insulating and conducting fluids.

2.1. *Scattering of X-rays and Neutrons*

It will be convenient to define the radial distribution function $g(r)$ via scattering theory. We will consider an assembly of N identical atoms irradiated by monochromatic radiation of wavelength λ. If A is the appropriate atomic scattering factor, then the angular distribution of the coherent scattered radiation is given by:

$$I_c(K)=A^2\left\langle\left|\sum_{i=1}^{N}\exp i\mathbf{K}.\mathbf{r}_i\right|^2\right\rangle, \quad \ldots \quad (2.1)$$

where

$$K=|\mathbf{K}|=\frac{4\pi\sin\theta}{\lambda},$$

$\mathbf{r}_i$ is the position of the ith atom, θ is one half of the scattering angle and the brackets indicate a time average. We now define a structure factor, $S(K)$, by:

$$S(K)=\frac{I_c}{NA^2}=\frac{1}{N}\left\langle\left|\sum_{i=1}^{N}\exp(\mathbf{K}.\mathbf{r}_i)\right|^2\right\rangle. \quad \ldots \quad (2.2)$$

The lack of long-range order means that the concept of a unit cell is no longer applicable and so, following Zernike and Prins (1927), we introduce the radial distribution function $g(r)$. This is defined by setting $n_0 g(r)4\pi r^2\,dr$ equal to the number of atoms in the spherical shell of radius r and thickness dr, n_0 being N/V, the number of atoms per unit volume. Allowing $g(r)$ to be a continuous function of r we find from (2.2) that

$$\begin{aligned}S(\mathbf{K})&=1+n_0\int_V g(\mathbf{r})\exp(i\mathbf{K}.\mathbf{r})\,d\mathbf{r}\\&=1+n_0\int_V(g(\mathbf{r})-1)\exp(i\mathbf{K}.\mathbf{r})\,d\mathbf{r}+n_0\int_V\exp(i\mathbf{K}.\mathbf{r})\,d\mathbf{r}. \quad (2.3)\end{aligned}$$

We now let $V\rightarrow\infty$ so that the last term becomes a δ function at $\mathbf{K}=0$ (the 'zero-order scattering') and need not be considered further. Moreover since for liquids $g(\mathbf{r})\equiv g(|\mathbf{r}|)$ we can carry out the angular integrations with the result:

$$S(K)=1+\frac{4\pi n_0}{K}\int_0^\infty(g(r)-1)r\sin Kr\,dr. \quad \ldots \quad (2.4)$$

Evidently (2.4) can be inverted to yield:

$$h(r) \equiv (g(r)-1) = \frac{1}{2\pi^2 n_0 r}\int_0^\infty (S(K)-1)K \sin Kr\, dK. \qquad (2.5)$$

$h(r)$, which is usually referred to as the total correlation function, can thus be obtained directly from an experimental determination of $S(K)$. Of course, the calculation of $h(r)$ from the measured $S(K)$ is not without its difficulties and the sort of errors which can arise have been fully discussed by Paalman and Pings (1963).

§ 3. Three-body and Ornstein–Zernike Correlation Functions

We will now introduce the three-body correlation function $n_3(\mathbf{r}_1\,\mathbf{r}_2\,\mathbf{r}_3)$, defined such that the probability that volume elements $d\mathbf{r}_1$, $d\mathbf{r}_2$, $d\mathbf{r}_3$ around $\mathbf{r}_1$, $\mathbf{r}_2$, $\mathbf{r}_3$ are occupied by molecules is given by:

$$n_3(\mathbf{r}_1\,\mathbf{r}_2\,\mathbf{r}_3)\, d\mathbf{r}_1\, d\mathbf{r}_2\, d\mathbf{r}_3.$$

Then, as has been known for a long time (cf. Green 1952), we can obtain an exact relation, for classical fluids, between the pair function $g(r)$, the pair force $\phi(r)$ and the triplet correlation function n_3. This relation, which can be derived rigorously from the classical partition function, is readily obtained by the following physical argument. We first express the radial distribution function $g(r)$ in the Boltzmann form:

$$g(r_{12}) = \exp\left[-\frac{U(r_{12})}{kT}\right], \quad . \quad . \quad . \quad . \quad . \quad . \quad . \qquad (3.1)$$

and then we recognize that $U(r_{12})$ is playing the role of a potential of mean force. The total force acting on atom 1 is therefore minus the gradient of U with respect to $\mathbf{r}_1$, and this can be split up into two parts, due to the pair force between atoms 1 and 2, and due to the other atoms. Thus, remembering that the probability of finding atom 3 in the volume element $d\mathbf{r}_3$, when atoms 1 and 2 are certainly in volume elements $d\mathbf{r}_1$ and $d\mathbf{r}_2$ around $\mathbf{r}_1$ and $\mathbf{r}_2$ is:

$$\frac{n_3(\mathbf{r}_1\,\mathbf{r}_2\,\mathbf{r}_3)\, d\mathbf{r}_3}{n_0^{\,2} g(r_{12})}$$

we may write:

$$-\frac{\partial U(r_{12})}{\partial \mathbf{r}_1} = -\frac{\partial \phi(r_{12})}{\partial \mathbf{r}_1} - \int \frac{n_3(\mathbf{r}_1\,\mathbf{r}_2\,\mathbf{r}_3)}{n_0^{\,2} g(r_{12})}\frac{\partial \phi(r_{13})}{\partial \mathbf{r}_1}\, d\mathbf{r}_3. \quad . \quad . \qquad (3.2)$$

This equation, as it stands, is exact, but to use it to derive the forces from the structure it will clearly be essential to make some assumption about n_3. This function is therefore at the very heart of the liquids problem, and we have only a rudimentary understanding of it.

Before proceeding to this next stage, we shall find it convenient to introduce the so-called direct correlation function, due to Ornstein and Zernike. The idea is again to split the total correlation function into a pair term, and that due to the remaining atoms. Thus, if f is the direct correlation function, we write:

$$h(r)=f(r)+n_0\int f(|\mathbf{r}_1-\mathbf{r}|)h(\mathbf{r}_1)\,d\mathbf{r}_1. \quad . \quad . \quad . \quad . \quad (3.3)$$

Clearly, if this decomposition were to have a rigorous interpretation, some three-body correlation function should have entered the definition. Since, however, we might expect simplifications when any two atoms are far apart, in the sense that the three-body correlation function can be expressed in terms of pair correlations, $f(r)$ may perhaps have asymptotic significance. This we examine below and show to be the case (cf. Johnson *et al.* 1964). Before doing so, we notice from the definition of $f(r)$ given in (3.3) that, by Fourier transform, $f(r)$ is determined by the structure factor $S(K)$ through the relation:

$$f(r)=\frac{1}{2\pi^2 n_0 r}\int_0^\infty dK\,K\left\{\frac{S(K)-1}{S(K)}\right\}\sin Kr, \quad . \quad . \quad . \quad (3.4)$$

which is to be compared with eqn. (2.5) for $h(r)$. Finally, we add to these equations the well-known thermodynamic relation:

$$S(0)=kT\,n_0\,K_{\mathrm{T}}, \quad . \quad . \quad . \quad . \quad . \quad . \quad (3.5)$$

where K_{T} is the isothermal compressibility. This will prove of central importance for the arguments which follow.

At this point, we must return to the basic force eqn. (3.2) and consider how we can put it into a useful, albeit approximate, form.

3.1. *Born–Green Assumption for Triplet Correlation Function*

Born and Green (1947), following earlier work by Kirkwood, proposed the approximation in which n_3 is built up as a product of pair correlation functions. We want to emphasize that such a description may not be adequate for some purposes, and we shall examine other approaches subsequently. It seems to us the most useful starting point at present, provided that the best knowledge (that is experimentally measured!) of $g(r)$ is employed in constructing n_3. Thus we write:

$$\frac{n_3(\mathbf{r}_1\,\mathbf{r}_2\,\mathbf{r}_3)}{n_0{}^3}=g(r_{12})g(r_{23})g(r_{31})$$

and, inserting this result into the force eqn. (3.2), we find a relation between the liquid structure $g(r)$ and the interatomic potential $\phi(r)$. After some manipulation (Rushbrooke 1960) the equation takes the form:

$$\ln g(r)+\frac{\phi(r)}{kT}=-n_0\int E(|\mathbf{r}_1-\mathbf{r}|)h(\mathbf{r}_1)\,d\mathbf{r}_1, \quad . \quad . \quad . \quad (3.6)$$

where

$$E(t)=\int_\infty^t g(r)\phi'(r)/kT\,dr. \quad . \quad . \quad . \quad . \quad . \quad . \quad (3.7)$$

This equation can be solved for $\phi(r)$ in principle, knowing $g(r)$ experimentally.

3.2. *Other Methods*

The Born–Green equation involves the function $E(r)$, which, for large r, goes like $\phi(r)/kT$, since $g(r) \to 1$ for large r. Since we introduced $f(r)$ as a direct correlation function, the roughest assumption would be (cf. eqn. (3.1)) that it behaves like:

$$f(r) \sim \exp\left[-\frac{\phi(r)}{kT}\right] - 1. \qquad (3.8)$$

If this were validated, then at large r, we would clearly obtain:

$$f(r) \sim -\frac{\phi(r)}{kT}. \qquad (3.9)$$

Thus, at least at large r, $f(r) = -E(r)$, and if, in eqn. (3.6), we replace $E(r)$ by $-f(r)$ we obtain a second approximate equation, known as the hyper-chain equation. This has the form:

$$\ln g(r) + \frac{\phi(r)}{kT} = n_0 \int f(|\mathbf{r}_1 - \mathbf{r}|) h(\mathbf{r}_1)\, d\mathbf{r}_1, \qquad (3.10)$$

and using the definition of $f(r)$ we find immediately:

$$\begin{aligned} h - f &= \ln g(r) + \frac{\phi(r)}{kT} \\ &= \ln(1+h) + \frac{\phi(r)}{kT}. \end{aligned} \qquad (3.11)$$

Yet a third equation, much in fashion, is the Percus–Yevick equation, which differs from (3.11) above in that $(h-f)$ is replaced by $\ln(1+h-f)$, which is clearly all right only if $h-f$ is small. Thus we have:

$$\ln(1+h-f) = \ln(1+h) + \frac{\phi(r)}{kT}. \qquad (3.12)$$

From eqns. (3.11) and (3.12) we can now check the consistency of our assumptions concerning the asymptotic relation between f and ϕ. Since for large r, $h \ll 1$, it is readily verified that eqn. (3.9) is regained. In these arguments, two reservations are to be made. The first is that $h^2(r) \ll \phi(r)/kT$ and the second that we must not approach the critical point at all closely.

Unfortunately, if we return to the Born–Green theory, we find a different multiplying constant in the relation between f and $-\phi/kT$ at least for van der Waals fluids (Gaskell 1965).

The common ground then is that all three approximate equations indicate that the direct correlation function behaves asymptotically like the potential $\phi(r)$. The diagrammatic methods (hyper-chain and Percus–Yevick) appear to us to be yielding the correct asymptotic form, but we are not able to give a rigorous proof. Notwithstanding, they appear to make very non-physical assumptions about n_3, when analysed from the standpoint of the force equation.

§ 4. Empirical Determinations of Pair Potential $\phi(r)$

4.1. *Analysis of Measurements of* $g(r)$

In the work of Johnson *et al.* (1964; see also Johnson and March 1963) the experimentally determined $g(r)$ was used in the Born–Green and Percus–Yevick equations, and results for $\phi(r)$ thereby obtained. Naturally, because of the different assumptions for n_3 which are at least implicit in these theories, the results for ϕ differ between the two theories.

Nevertheless, in the original work, there appeared to be common features which may be summarized as follows:

(i) $\phi(r)$ for liquid metals had oscillatory form, for all metals considered, from either theory.

(ii) For the one insulator considered, namely liquid argon, both theories gave a potential which was monotonic at large r. These potentials were not quantitatively in complete accord with that given in § 1, but the general form of that interaction was reproduced.

The oscillations were interpreted by Johnson *et al.* as reflecting the dielectric screening of the ions by the conduction electrons. Then, in the simplest model (point ion) of a liquid metal, $\phi(r) \sim B \cos 2k_f r/r^3$ (see, for example, Langer and Vosko (1959) for a many-body treatment and March and Murray (1960, 1962) for a Hartree calculation) where k_f is the magnitude of the Fermi wave vector. However, it has subsequently emerged that the wavelengths, which were only roughly given by π/k_f, are more accurately correlated with the positions of the first peak in the structure factor $S(K)$. Putting this another way, the wavelengths of the oscillations in $\phi(r)$ follow closely those in $U(r)$ (or $g(r)$).

This observation therefore suggests that the detailed structure of the ion-core is crucial in influencing the pair interaction for intermediate values of r. This shifts the emphasis somewhat from the point $K = 2k_f$ in the structure factor (see § 5). Nevertheless, on the theoretical grounds considered below, we expect the region around $2k_f$ to influence the asymptotic behaviour of ϕ. The problem of the precise form of the dielectric screening of a complex ion-core is clearly of the utmost interest, but so far has only been tackled in a relatively crude pseudo-potential framework (Harrison 1964, Ziman 1964).

With regard to the method used in the calculations by Johnson *et al.*, the following objections can be raised:

(i) Some of the basic structure data on which the analysis was performed leaves a good deal to be desired. For example, it appears that for sodium, two recent radial distribution function curves available at 100°c and 114°c (Gingrich and Heaton 1961, Orton *et al* 1960), differ markedly in the heights of the first and second peaks by a factor of 1·5, though their positions agree.

(ii) The results for $\phi(r)$, as was clear in the original paper, depend on the assumption made concerning the three-body correlation function n_3. Nevertheless, the relative temperature independence of the potentials obtained from the Born–Green theory suggests that this is more accurate than the Percus–Yevick theory for intermediate distances. However, if $f(r)$ behaves like $-\phi(r)/kT$, as we are at present inclined to believe, then this suggests that the Percus–Yevick and hyper-chain theories are preferable asymptotically.

4.2. *Nature of Direct Correlation Function $f(r)$ in Liquids*

The importance of the direct correlation function $f(r)$ was emphasized by Johnson *et al.* (1964) and we now wish to develop further their arguments as to the fundamentally different character of f in insulating and conducting fluids. Indeed, the common element in the interatomic force calculations we have discussed above may be thought of as the direct correlation function. What happens subsequently depends on the approximate theory used. For example, the Percus–Yevick theory simply yields:

$$\phi_{\mathrm{PY}} = +kT \ln\left(1 - \frac{f}{g}\right). \qquad (4.1)$$

Fig. 2

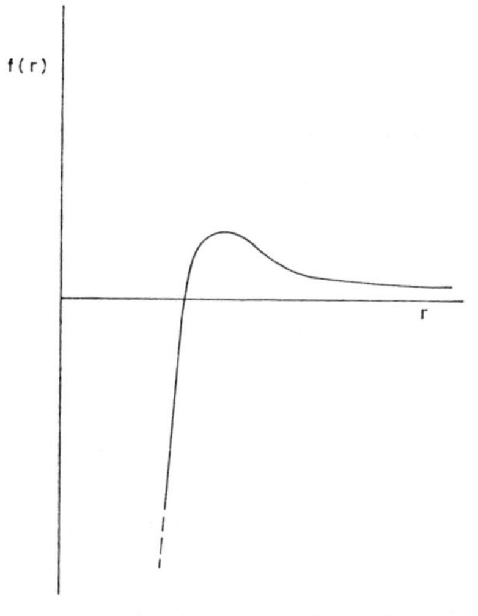

Form of $f(r)$ for liquid insulators.

From ϕ_{PY} calculated by Johnson *et al.*, and the observed $g(r)$, $f(r)$ can be immediately obtained and the general forms appropriate to Ar and Pb are shown in figs. 2 and 3 respectively.

It is now possible to enquire in detail just where the characteristic differences between Ar and Pb which are seen in figs. 2 and 3 arise in the experimental structure factors, since, in contrast to ϕ, f can in principle be derived directly from experiment. The comparison can immediately be made in **K** space, and at first sight the form of $\tilde{f}(K)$, which is simply

$1-1/S$ from eqn. (3.4), again appears rather similar in all cases, down to the lowest K value (~ 1 Å^{-1}) normally achieved in the past. This is demonstrated in fig. 4.

However, these plots of $\tilde{f}(K)$ take no account of the wide variation of $\tilde{f}(0)$ between metals and insulators, revealed in table 1, $\tilde{f}(0)$ being calculable, of course, from the thermodynamic result for $S(0)$ given in eqn. (3.5).

Fig. 3

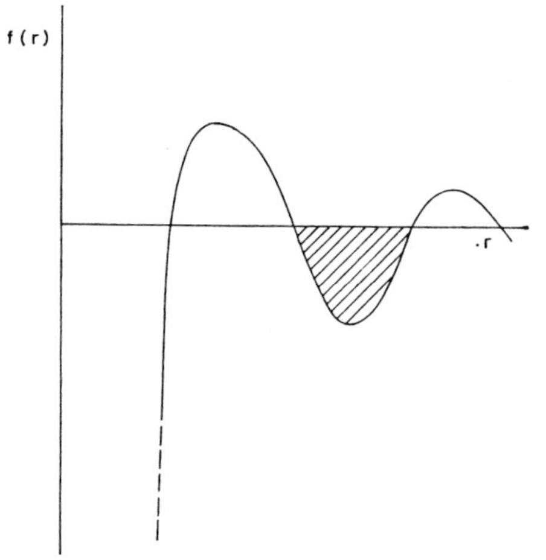

Form of $f(r)$ for liquid metals.

Figure 5 shows, therefore, the data of fig. 4, reduced by $\tilde{f}(0)$. The striking difference is that $\tilde{f}(K)/\tilde{f}(0)$ is much more localized for Pb than for Ar. To press this point further, we show in fig. 6, results in the range $0 < K < 2$ Å^{-1} for Pb (Egelstaff, Enderby, North and Rainey 1965, to be published), Ga (Ascarelli 1965, private communication), and Ar (Henshaw 1957). For Ar and later for Ne, whilst the lowest value of K at which $S(K)$ was measured was greater than for Pb and Ga, Henshaw has reported that for both cases, at the lowest angle, the compressibility value has almost been achieved. Again, the increased localization in the metals is clearly shown.

Table 1. Values of $\tilde{f}(0)$ for liquids

	Ar	Ne	Na	K	Hg	Ga	Pb	Sn	Bi
$\tilde{f}(0)$	−14·6	−17·6†	−37·5	−40·7	−134	−106‡	−109	−124	−99
T°K	84	24	371	337	323	323	600	505	544

† Estimated from diffraction data of Henshaw (1958).
‡ Taking isothermal compressibility $K_T = 4{\cdot}0 \times 10^{-12}$ c.g.s. units.

Fig. 4

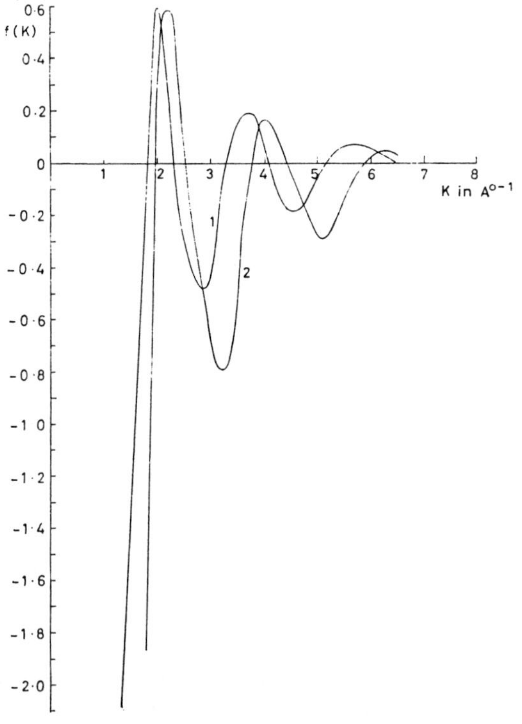

$\bar{f}(K)$ for Ar (curve 1) at 84°K and Pb (curve 2) at 600°K.

Fig. 5

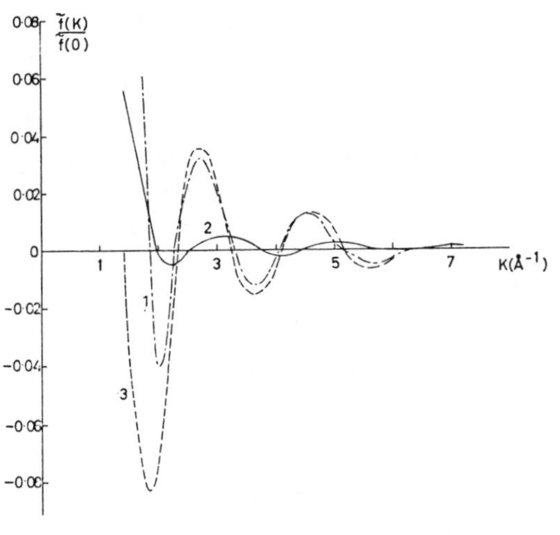

$\bar{f}(K)/\bar{f}(0)$ for Ar and Pb. Curve 1 : Ar, 2 : Pb, 3 : Model defined by eqn. (4.2).

Fig. 6

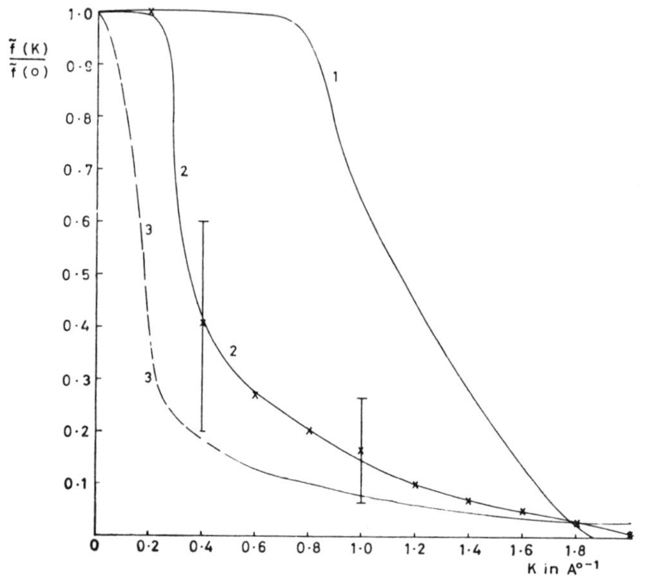

$\bar{f}(K)/\bar{f}(0)$ for small K. Curve 1 : Ar, 2 : Pb, 3 : Ga.

We turn now to discuss the way in which this localization affects the r space behaviour of f. Clearly a crucial region is that shaded in fig. 3, which for Pb, from the work of Johnson *et al.*, occurs between 4 and 6 Å. We thus construct

$$\int K \frac{\bar{f}(K)}{\bar{f}(0)} \sin Kr\, dK$$

for Pb and Ar at $r = 4{\cdot}5$ Å, as an example. The integrands are shown in figs. 7 and 8 and it is immediately clear that the sign for Pb is determined by the behaviour out to 2 Å^{-1}, and that $f(r)$ is negative at this distance, as anticipated in fig. 3. On the other hand, for Ar, the cancellation between the inner region $0 < K < 2$ Å^{-1} and the large K contribution is severe, and while $f(r)$ at $r = 4{\cdot}5$ Å is much smaller than for Pb, high accuracy is clearly required out to $K \sim 10$ Å^{-1} to obtain completely quantitative results.

This discussion, therefore, suggests that it is in **r** space that localization occurs for Ar, in contrast to the **K** space localization for metals.

To pursue this a little further, choose a crude $f(r)$ as shown in fig. 9. The transform may be written as:

$$\frac{\bar{f}(K)}{\bar{f}(0)} = 3\frac{j_1(Ka)}{Ka}, \quad . \quad . \quad . \quad . \quad . \quad . \quad . \quad . \quad (4.2)$$

where $j_1(\rho)$ is the first-order spherical Bessel function, given by $(\sin\rho - \rho\cos\rho)/\rho^2$. In order to obtain approximate agreement with the experimentally determined wavelength in $\bar{f}(K)$, we must choose a for Ar as $\sim 3{\cdot}3$ Å. This corresponds closely to the hard core radius.

Fig. 7

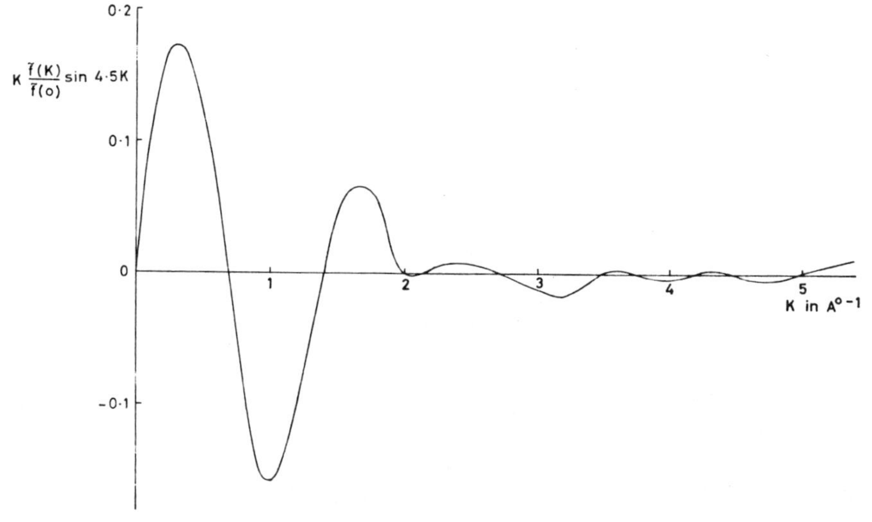

$K\frac{\bar{f}(K)}{\bar{f}(0)}\sin Kr$ for Pb at $r=4{\cdot}5$ Å.

Fig. 8

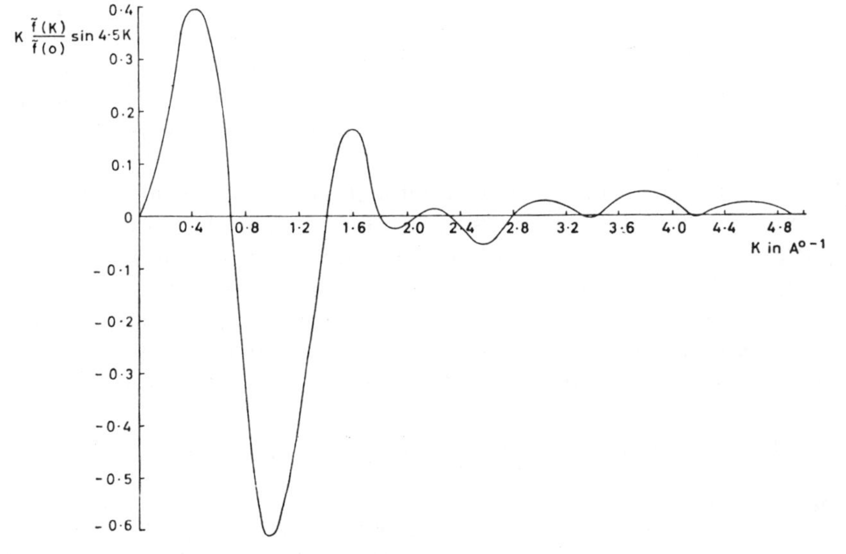

$K\frac{\bar{f}(K)}{\bar{f}(0)}\sin Kr$ for Ar at $r=4{\cdot}5$ Å.

Fig. 9

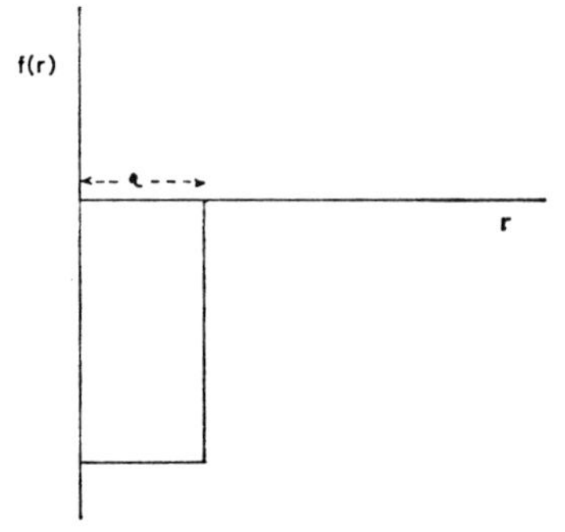

Model of localized $f(r)$ for Ar.

The results thereby obtained are shown in curve 3 of fig. 5, and we see that a good account is obtained of the oscillatory features beyond $2{\cdot}5\,\text{Å}^{-1}$. The major defect of the model is that the first minimum in $\tilde{f}(K)/\tilde{f}(0)$ is too deep by a factor of 2, and this leads to a slight violation of the necessary condition that $\tilde{f}(K) < 1$. But the basic point is that such a model correctly relates the size of $\tilde{f}(0)$ to the amplitude of the oscillations beyond the first minimum.

This argument has primarily focused attention on the hard core. The detail of the interaction $\phi(r)$ is determined, of course, around its minimum, by the height of the maximum of $f(r)$†.

More precise measurements of $S(K)$, and the inversion to obtain $f(r)$ for a range of temperatures will undoubtedly be of great value in clarifying the details of the forces and the applicability of the various approximations to the liquid state discussed above. There is little doubt, however, that the main features of the force curve are already exposed by our present knowledge of $f(r)$.

But, when we turn to consider liquid metals, the situation is completely different. If we consider Pb as an example, then the wavelength and amplitude of the oscillations in $\tilde{f}(K)$ are similar to the case of Ar, but $\tilde{f}(0)$ is a factor of more than 7 greater. Thus, the model discussed above is wholly inappropriate, for as eqn. (4.2) shows, the amplitude of the oscillations is simply proportional to $\tilde{f}(0)$. The conclusion is clearly that no short-range force model is at all appropriate for the metal.

† In this connection, we should stress that the hard core model is not to be regarded as an accurate description of $f(r)$. To obtain good results, it is necessary to account quantitatively for the first minimum in $\tilde{f}(K)/\tilde{f}(0)$, and this can be done in a crude way by subtracting a term $K^2 \exp(-K^2)$ from (4.2). When this is done, the essential features of $f(r)$ as shown in fig. 2 are reproduced, although its asymptotic form (see § 5) is not proportional to r^{-6}. As Enderby *et al.* (1965) have shown, all that is necessary to give this feature is a K^3 term in the model at small K.

It should be added that Ascarelli, for liquid Ga, has used the Percus–Yevick theory to calculate $\phi(r)$ directly from the K space data. In agreement with the earlier work (Johnson *et al.* 1964) on another trivalent metal, Al, he found a well-defined region where ϕ was positive beyond its first minimum. However, no further well-defined oscillations were obtained beyond 5 Å. This difference could arise because of the well-known structural peculiarities which have been reported in liquid Ga from time to time, and in particular the evidence for pairing (Faber 1964, Gustafson *et al.* 1963).

Thus, in summary, we can note four aspects of the curve for $f(K)$ in liquid metals:

(i) The small angle scattering, which leads to negative $f(r)$ around 4–7 Å, and which demonstrates the long-range character of the forces.

(ii) The large $\mathbf{K}$ behaviour of $f(\mathbf{K})$, which appears to derive primarily from the hard core (as for Ar).

(iii) The first peak, which appears to come partly from the hard core, and partly from the long-range forces. In view of the rather different character of the results of Johnson *et al.* on the one hand, and Ascarelli's work on Ga on the other, this aspect of the problem is evidently worth much more careful study.

(iv) The further region already referred to, around $K=2k_f$, which, on theoretical grounds, appears to be important. This, as we shall now discuss, would appear to determine the asymptotic behaviour, but has probably not begun to dominate the results of Johnson *et al.* out to the values of r considered in their work.

§ 5. Asymptotic Behaviour of $f(r)$ and $\phi(r)$

5.1. *Classical Insulating Fluids*

Before considering the influence of the region around $K=2k_f$ in metals, we briefly outline the arguments for van der Waals forces. There we have:

$$\phi(r) \sim -Ar^{-6}, \qquad (5.1)$$

and hence:

$$f(r) \sim \frac{A}{kT} r^{-6}. \qquad (5.2)$$

We now make use of the result (see, for example, Lighthill 1958) that

$$\int_0^\infty F(x) \sin xr\, dx \sim \frac{F(0)}{r} - \frac{F''(0)}{r^3} + \frac{F''''(0)}{r^5} + \dots \qquad (5.3)$$

This result is valid if $F(k)$ is well behaved, with its derivatives, and this appears to be true physically for insulating fluids. Using this result, we see immediately that to obtain an r^{-6} dependence in $f(r)$, we must have $F''(0)=0$, $F''''(0)\neq 0$, where $F(K)\equiv K(1-S^{-1})$. This implies an expansion for small k of the form:

$$S(K) = S(0) + a_2K^2 + a_3K^3 + \dots, \qquad (5.4)$$

where a_3 is given by:

$$a_3 = \pi^2 n_0 \{S(0)\}^2 A/12kT. \qquad (5.5)$$

It then follows after some calculation that

$$h(r) \sim \frac{1}{kT}\{S(0)\}^2 A r^{-6}. \qquad (5.6)$$

5.2. *Liquid Metals*

We assume, as discussed in § 4, that the forces between ions reflect the dielectric screening of the ions by the conduction electrons. Then as we saw $\phi(r) \sim B\cos(2k_f r)/r^3$, and this behaviour is known to arise from a term in the Fourier transform of $\phi(r)$ which, near $K = 2k_f$, behaves as:

$$(K - 2k_f)\ln|K - 2k_f| \qquad (5.7)$$

(see, for example, Langer and Vosko (1959)).

Thus, in this case, because of the discontinuity in the gradient of the function in (5.7) at $K = 2k_f$, the result (5.3) is not applicable. The form (5.7) implies, using the asymptotic relation $f(r) \pm -\phi(r)/kT$ and that the Fourier transform is simply related via (3.4) to $S(K)$, the following expansion around $2k_f$:

$$S(K) = S(2k_f) + S_1, \qquad (5.8)$$

where S_1, related directly to (5.7), dominates the asymptotic behaviour in r space. $h(r)$ is then immediately related to $f(r)$ since, from (2.5) and (3.4),

$$\frac{S-1}{S} \sim \text{const.} + \frac{S_1}{\{S(2k_f)\}^2} \qquad (5.9)$$

and

$$S - 1 \sim \text{const.} + S_1 \qquad (5.10)$$

near $K = 2k_f$. Since S_1, as we have seen, leads to the asymptotic form $r^{-3}\cos 2k_f r$, we may evidently write from (5.9) and (5.10):

$$f(r) \sim -\frac{B\cos 2k_f r}{kTr^3} \qquad (5.11)$$

and

$$h(r) \sim -\frac{B\{S(2k_f)\}^2 \cos 2k_f r}{kTr^3}. \qquad (5.12)$$

The contrast with the classical liquid insulators is then immediately clear when we compare (5.6) and (5.12). $\{S(0)\}^2$, which is usually very small, no longer appears in $h(r)$, but is replaced by $\{S(2k_f)\}^2$, and from experimental structure data this is always of the order of unity.

We should stress that some modification of our simple model will occur at sufficiently large r owing to the damping of the $r^{-3}\cos 2k_f r$ behaviour by slight Fermi surface blurring (Gaskell and March 1963). However, we do not expect this to modify our essential conclusion that, whereas $h(r)$ and $f(r)$ approach zero as an inverse power of r for classical insulating fluids, for liquid metals $f(r)$ and $h(r)$ both oscillate about zero asymptotically.

Finally, these results are in marked contrast to the old result of Ornstein and Zernike, which suggested that $g(r)$ approached its asymptotic value as $\exp(-qr)/r$. This appears to be a consequence of assuming that $S(K)$ had no singularities, and was a function solely of K^2. However, this work suggests that the origin is a singular point for classical insulating fluids, and $2k_f$ is a singular point for liquid metals. These appear to lead to the conclusion that, in either case, the range of $h(r)$ and $f(r)$ is the range of $\phi(r)$. It would be dangerous to infer that this is a general conclusion, independent of the form of ϕ. However, it should be mentioned that Widom (1964), before the work of Enderby *et al.* (1965), had suggested that $h(r) \sim -\phi(r)/kT$ †.

§ 6. Evidence Bearing on Existence of Sharp Fermi Surface in Liquid Metals

In the original work of Johnson *et al.* the oscillatory nature of the ion–ion interactions was interpreted as dielectric screening of the ions by the conduction electrons. This effect can be expected only if the Fermi surface is still a meaningful concept. These workers stressed that the decay of the oscillations ought to give information on the Fermi surface blurring, but, in view of the conclusion discussed above that the asymptotic region where the Kohn anomaly should dominate was probably not reached in the calculations of these authors, quantitative estimates of the blurring must await further accurate structure measurements, around small K, the first peak, and the Kohn region around $2k_f$.

It is therefore of interest, at this stage, to consider what other evidence we have which bears on the existence of a sharp Fermi surface. We discuss below two other observations which seem relevant to this point.

6.1. *Electronic Mean Free Path*

If one accepts at its face value the evidence of measurements of the Hall coefficient for liquid metals, one concludes that the relaxation time τ and the dispersion law $E(\mathbf{k})$ are independent of the direction of $\mathbf{k}$. Under these circumstances the conductivity can be expressed as:

$$\sigma = \frac{e^2 \lambda k_f^2}{3\pi^2 h}, \qquad (6.1)$$

where k_f is the radius of the Fermi surface which is calculated from the electron density n as:

$$k_f = (3\pi^2 n)^{1/3}. \qquad (6.2)$$

Thus it is possible to calculate the mean free path, λ, and the results are shown in table 2. The mean free path can be regarded as the range over

† Dr. Widom has kindly informed us (private communication) that he now believes the evidence available to be strongly in favour of the result $f(r) \sim -\phi(r)/kT$.

which an electronic wave packet loses its coherence so that the spread in k values for electrons taking part in the conduction process (i.e. around k_f) is given by $\Delta k \pm 1/\lambda$. Thus $\Delta k/k_f$ can in fact be crudely estimated as shown in table 2.

Table 2

	λ (Å)	Δk (Å^{-1})	k_f (Å^{-1})	$\Delta k/k_f$
Na	157	0·0064	0·89	0·007
Hg	7	0·14	1·34	0·10
Pb	6	0·17	1·54	0·11

6.2. *Positron Annihilation*

But the evidence which could, in principle, decide the question we are asking about the existence of a sharp edge at the Fermi surface most directly, is that available from positron annihilation experiments. In particular, Gustafson *et al.* (1963) have reported recently on positron annihilation measurements in both solid and liquid Hg, and their results have been interpreted in terms of an electronic momentum distribution $P(k)$, which, in the liquid, is considerably more diffuse than in the solid. Adopting a form for $P(k)$ given by:

$$P(k)=\frac{A}{\exp\left(\frac{k^2-1}{\Delta}\right)+1}, \quad . \quad . \quad . \quad . \quad . \quad . \quad (6.3)$$

with k in units of k_f, then $\Delta \approx 0{\cdot}2$.

While this diffuseness might have been supposed to arise from the disorder scattering of the electrons, there are two points which can be made in connection with it:

(i) The diffuseness may be due to the thermalization process, or the effect of disorder on the positron wave function.

(ii) While the diffuseness is roughly compatible with the mean free path discussed in § 6.1 above, Mott (to be published) has argued that the mean free path in liquid Hg is longer than obtained from the elementary conductivity argument of § 6.1.

If the diffuseness of $P(k)$ in Hg were indeed as large as that supposed by Gustafson *et al.*, then the considerations above would be in jeopardy. For Gaskell and March (1963) have shown that such a blurring of around 20% would damp out the oscillatory effects in the ion–ion interaction.

While the theory is rough, it shows us that oscillatory interactions are very sensitively dependent on the sharpness of the Fermi surface. It should be stressed that a blurring of around 10% would present no difficulty of this kind.

If the Fermi surface is not sharp, then we suggest that the stability of metallic liquid Hg itself presents some difficulty, and we therefore must assume that the positron annihilation results in the liquid require some degree of re-interpretation.

§ 7. Alternative Treatments of $\phi(r)$ from Solid State

The method described in § 4 is the only one so far proposed for the liquid state. However, several treatments of forces between ions in metallic crystals have been forthcoming recently, for Zn (Harrison 1963), Na (Cochran 1965), Al and Pb (Harrison 1964, 1965). Since direct comparison with the potentials obtained by Johnson *et al.* is possible for Na, Al and Pb, it is useful to summarize these results below.

Of course, this comparison between effective forces in the liquid and the solid state may not turn out to be wholly justified. Obvious factors which come in are:

(i) Density changes on melting.

(ii) Assumption of two-body forces, which is equivalent to Born approximation in the electron scattering by the (pseudo-) ions.

(iii) Non-central forces may arise in the solid, via dielectric screening of ions by electrons within a non-spherical Fermi surface.

We expect these effects to be important in certain groups of metals, for example (i) is comparatively large in Al, (ii) will be more important in light trivalent metals, again such as Al, while (iii) might be expected to play a role in the noble metals. However, for Na these effects are presumably small, and we therefore consider first Cochran's study of this metal.

7.1. *Sodium*

Cochran's work is based on the harmonic approximation, and employs the measured interatomic force constants. Since, however, only the first and second derivatives of ϕ at the equilibrium interatomic spacings are accessible in this way, some theoretical model must be invoked to complete the force curve, in contrast to the semi-empirical method for liquids.

The comparison made in fig. 10 shows that, while there are some general similarities between the force curves, the oscillations that Cochran obtains are much smaller (the amplitude at the first maximum is down by about a factor of four) and the oscillations are rapidly damped. Cochran also remarks that the force constants obtained from the potential of Johnson *et al.* are much too large beyond the first and second neighbours to explain the phonon results in the solid.

Two points should be made here. The first is that Johnson *et al.* noted that the oscillations in Na fall off less rapidly than for the other seven

Fig. 10

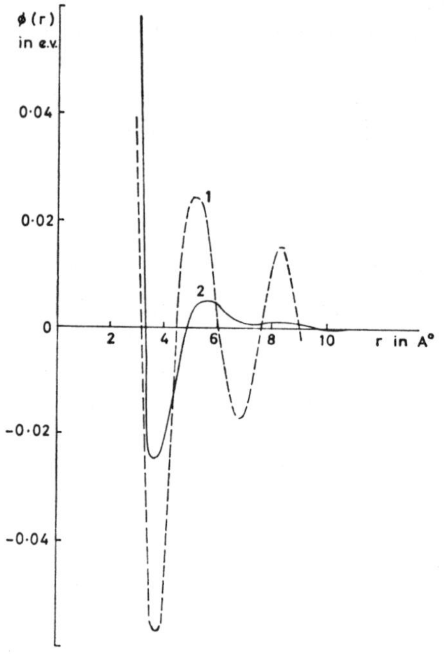

Liquid and solid $\phi(r)$'s for Na. Curve 1 : liquid. Curve 2 : solid.

metals considered. This may be connected with our earlier remark concerning the inaccuracies of the experimental data for this metal. The second is that Koenig (1964) has shown that the interplanar force constants for Na can be explained by dielectric screening theory.

Thus, there appears to be some ambiguity in the interpretation of the phonon data, but it does not seem possible to reconcile this quantitatively with the liquid analysis.

We are therefore forced to conclude that, even for such a simple metal as Na, the effective interionic interaction is not yet established with any certainty.

7.2. *Aluminium and Lead*

No work of comparable accuracy to that of Cochran exists for polyvalent metals. However, Harrison (1964, 1965) has attempted a calculation of ϕ from first principles in the framework of a pseudo-potential approximation for Al and Pb. The comparison is made in figs. 11 and 12.

At first sight, the agreement for Al appears impressive. Nevertheless, we should stress that the oscillations which Harrison obtains over the range of r shown in the figure are intimately bound up with the Kohn anomaly, which, as we have seen, is probably not the case for the liquid curve. It should also be added that Harrison's curve for Al correctly

Fig. 11

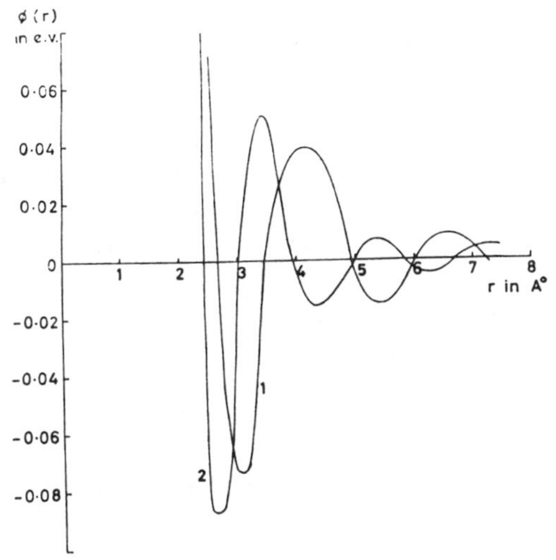

Liquid and solid $\phi(r)$'s for Al. Curve 1 : liquid. Curve 2 : solid.

Fig. 12

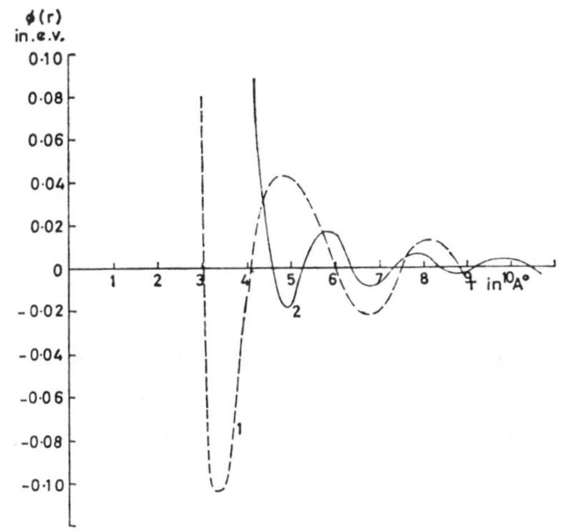

Liquid and solid $\phi(r)$'s for Pb. Curve 1 : liquid. Curve 2 : solid.

predicts the face-centred cubic structure as the stable one, in contrast to the potential of Johnson *et al.* (Worster and March 1964). On the other hand, the reverse is true for Pb. However, as Harrison points out, a cruder approximation was employed in deriving ϕ, which yields rather poor results around the principal minimum. It is also possible that in the heavier metals, spin–orbit coupling begins to play a significant role.

§ 8. Higher-order and Time-dependent Correlation Functions

In § 2 we saw how it was possible to interpret the scattering of radiation in terms of the pair distribution function. Actually, the treatment was based on the first Born approximation and if scattering effects are to be a source of information about n_3, it is necessary to consider the higher order terms. The treatment of Ferziger and Leonard (1962), concerned specifically with neutron scattering (although their treatment has wider applicability) is particularly relevant here. Thus for a set of identical scatterers, eqn. (2.1) becomes, in the second Born approximation,

$$I(\theta)=\left|\sum_{n=1}^{N} A\exp(i\mathbf{K}.\mathbf{r}_n)+\sum_{n,m}{}' A^2\exp(i\mathbf{k}.\mathbf{r}_m)\exp(-i\mathbf{k}'.\mathbf{r}_n)\frac{\exp(ik|\mathbf{r}_n-\mathbf{r}_m|)}{|\mathbf{r}_n-\mathbf{r}_m|}\right|^2$$

$$=A^2\left|\sum_{n=1}^{N}\exp(i\mathbf{K}.\mathbf{r}_n)\right|^2+A^3\left[\sum_{l}\exp(-i\mathbf{K}.\mathbf{r}_l)\sum_{n,m}\exp(i\mathbf{k}.\mathbf{r}_m)\right.$$

$$\left.\times\exp(-i\mathbf{k}'.\mathbf{r}_n)\frac{\exp(ik|\mathbf{r}_n-\mathbf{r}_m|)}{|\mathbf{r}_n-\mathbf{r}_m|}+\text{c.c.}\right]+0(A^4). \quad (8.1)$$

The first term is, as we have seen, related to $g(r)$.

The terms involving A^3 can be split up into a single integral involving g and a triple integral involving n_3 which, unfortunately, cannot be evaluated in closed form. It appears therefore that even if A could be made large enough so that the A^3 terms were significant, scattering experiments can only be used to test model n_3's; for example, of the Born–Green type.

A further interesting contribution to this field has appeared very recently (Frisch and McKenna 1965). Here, the emphasis was on the double scattering of electromagnetic radiation and although the same general conclusions on the possibilities of gaining information on n_3 emerge, two new points were made. The first of these is the need to use linearly polarized incident radiation, for this enables the singly and doubly scattered light to be separated by their polarization properties. The second point is that, because of the necessity of highly intense primary beams of radiation of suitable wavelength, the use of lasers might have advantages over other sources. Unfortunately, this would appear, at the moment, to restrict the method to the neighbourhood of the critical point.

Finally, we note that, even with consideration of pair correlations alone, $g(r)$ is insufficient to describe the dynamics of the fluid. Thus, it will be recalled that $g(r)$ had two components, a delta function at the origin, and the part we have discussed in detail. A complete theory of atomic motions must discuss how the delta function, and the pair correlation function we have considered, develop with time. Analogous to the structure factor $S(\mathbf{K})$, the quantity which is accessible experimentally is the scattering law $S(\mathbf{K}\omega)$. This gives the probability that the neutron absorbs energy $\hbar\omega$ from the liquid, and imparts momentum $\hbar\mathbf{K}$ to it.

A variety of experiments on the inelastic scattering of neutrons from liquids have been carried out, and substantial programmes are in hand (see, for example, Cocking and Egelstaff 1965). The conclusions bearing on the forces between ions, are, as yet, rather limited and qualitative. It does appear, however, that the structure in the inelastic region is far more pronounced for Na than for Ar (Dasannacharya and Rao 1965). This could be regarded as supporting evidence for the rather simple type of force (and collective motion) operating in insulators as opposed to metals. Clearly, a final satisfactory theory of liquids will have to relate $S(\mathbf{K}\omega)$, or its Fourier transform, the time dependent generalization of $g(r)$, to the forces $\phi(r)$. At present, we have only some basic sum rules to guide us (de Gennes 1959). In this connection, it should be noted that Randolph (1964) has evaluated the second moment of $S(\mathbf{K}\omega)$ from his neutron measurements for liquid Na. His surprising conclusion is that the moment theorem appears not to be satisfied, and that this may imply the need to invoke velocity dependent forces. If his findings are confirmed by subsequent experimental work, then, of necessity, some fairly substantial modifications to the picture presented here will be called for.

§ 9. Summary

The major conclusions of this article are therefore:

(i) By describing the scattering of x-rays and neutrons from liquids by $\tilde{f}(\mathbf{K})$ rather than $S(\mathbf{K})$, striking differences clearly emerge between insulating and metallic fluids. Roughly, these may be summed up as localization in $\mathbf{r}$ space for insulators, and in $\mathbf{K}$ space for conductors.

(ii) Existing approximate theories reveal an intimate connection between $f(r)$ and $\phi(r)$.

(iii) To advance beyond approximate theories would imply accurate knowledge of n_3 for use in the basic force eqn. (3.2). It should be stressed, of course, that precise measurements of $S(\mathbf{K})$ over a wide temperature range would be an essential prerequisite for any further developments.

(iv) While there seems strong evidence that the forces in liquid metals arise from the ion-core screening by conduction electrons with a relatively sharp Fermi surface, no well-defined correlation has as yet emerged between forces and electronic mean free paths. The nature of the electron states in conducting fluids will have to be greatly developed beyond our present understanding before this problem can be tackled successfully.

(v) Finally, and more tentatively, the common features between force laws which are revealed from discussions starting from the solid and liquid phases respectively hold out possibilities for a rather closely integrated description of condensed matter. In particular, it will be of great interest to explore fully the problem of the stability of crystal structures in terms of interionic forces having the characteristic features discussed here.

Notes added in proof

(i) Since this article was written, Dr. T. Gaskell has drawn our attention to earlier calculations of $f(r)$ for Ar (Reetz, A., and Lund, L. H., 1957, *J. chem. Phys.*, **26,** 518). Except for one case, $f(r)$ was found to have one node for $r > 1$ Å. Also, $f(r)$ was positive for $r > 4$ Å. In the exceptional case referred to, $f(r)$ went weakly negative around $r = 5$ Å, although subsequently it was positive, in agreement with Johnson *et al.*

Fortunately, the situation has been clarified greatly by a private communication just received from Professor Pings. Here again, results for $f(r)$ have been calculated for fluid Ar from 13 sets of data. In every case, $f(r)$ has the general form shown in fig. 2 and is short-ranged (Mikolaj and Pings, to be published).

(ii) Results for the ion–ion interaction in Li metal have been obtained recently by Meyer, Nestor and Young (1965; *Phys. Letters*, **18,** 10) using a pseudo-potential approach. The results agree rather well in amplitude, wavelength and phase with the liquid force curve of Johnson *et al.*

(iii) Dr. N. Ashcroft has informed us that successful calculations of electrical properties of liquid metals have been made recently (Ashcroft and Lekner, to be published) in which an exact analytic solution of the Percus–Yevick eqn. (3.12) for hard spheres (Thiele, E., 1963, *J. chem. Phys.* **39,** 474; Wertheim, M. S., 1963, *Phys. Rev. Letters*, **10,** 321) has been used to represent the structure factor $S(K)$. While the overall representation of $S(K)$ is good, particularly in the region of the first peak, there are generally appreciable departures from the experimentally determined $S(0)$; cf. eqn. (3.5).

(iv) It is perhaps worth stressing that for liquid metals, the potential of mean force $U(r)$, defined in eqn. (3.1), and the pair potential $\phi(r)$ have similar qualitative forms according to the calculations of Johnson *et al.*, in complete contrast to the situation in Ar. This seems particularly true for the polyvalent metals. This, combined with the fact that the Born–Green theory appeared to yield relatively temperature independent potentials, suggests that the virial expansion may perhaps be useful with oscillatory potentials, even for the high densities appropriate to liquid metals. In this case also, the form (3.8), the so-called Mayer function of cluster expansion theory, ought to afford a useful starting approximation for the direct correlation function. Such calculations are at present being carried out by Mr. E. R. Muller and one of us (N.H.M.).

References

Bernal, J. D., 1959, *Nature, Lond.*, **183,** 141.

Born, M., and Green, H. S., 1947, *Proc. roy. Soc.* A, **190,** 455.

Cochran, W., 1965, *Inelastic Scattering of Neutrons*, **1** (Vienna: Int. Atomic Energy Agency), p. 3.

Cocking, S., and Egelstaff, P. A., 1965, *Phys. Letters*, **16,** 130.

Corless, G. K., and March, N. H., 1961, *Phil. Mag.*, **6,** 1285.

Dasannacharya, B. A., and Rao, K. R., 1965, *Phys. Rev.*, **137,** A517.

de Gennes, 1959, *Physica*, **25,** 825.

Dobbs, E. R., and Jones, G. O., 1957, *Rep. Progr. Phys.*, **20,** 516.

Enderby, J. E., Gaskell, T., and March, N. H., 1965, *Proc. phys. Soc., Lond.*, **85,** 217.

Faber, T., 1964, *Solid State Communications*, **1,** 41.

Ferziger, J. H., and Leonard, A., 1962, *Phys. Rev.*, **128,** 2188.

Foreman, A. J., 1963, *Phil. Mag.*, **8,** 1211.

Frisch, H. L., and McKenna, J., 1965, *Phys. Rev.*, **139,** A68.
Gaskell, T., 1965, *Proc. phys. Soc., Lond.*, **86,** 693.
Gaskell, T., and March, N. H., 1963, *Phys. Letters*, **7,** 169.
Gingrich, N. S., and Heaton, L., 1961, *J. chem. Phys.*, **34,** 873.
Green, H. S., 1952, *Molecular Theory of Fluids* (Amsterdam : North-Holland Publishing Company).
Gustafson, D. R., and Mackintosh, A. R., 1963, *Phys. Letters*, **5,** 234.
Gustafson, D. R., Mackintosh, A. R., and Zaffarano, D. J., 1963, *Phys. Rev.*, **130,** 1455.
Harrison, W. A., 1963, *Phys. Rev.*, **129,** 2512; 1964, *Ibid.*, **136,** A1107; 1965, *Ibid.*, **139,** A179.
Henshaw, D. G., 1957, *Phys. Rev.*, **105,** 976 ; 1958, *Ibid.*, **111,** 1470.
Johnson, M. D., and March, N. H., 1963, *Phys. Letters*, **3,** 313.
Johnson, M. D., Hutchinson, P., and March, N. H., 1964, *Proc. roy. Soc.* A, **282,** 283.
Koenig, S. H., 1964, *Phys. Rev.*, **135,** A1693.
Langer, J. S., and Vosko, S. H., 1959, *J. Phys. Chem. Solids*, **12,** 196.
Lighthill, M. J., 1958, *An Introduction to Fourier Analysis and Generalized Functions* (Cambridge University Press).
March, N. H., and Murray, A. M., 1960, *Phys. Rev.*, **120,** 830 ; 1962, *Proc. roy. Soc.* A, **266,** 559.
Orton, B. R., Shaw, B. A., and Williams, G. I., 1960, *Acta Met.*, **8,** 177.
Paalman, H. H., and Pings, C. J., 1963, *Rev. mod. Phys.*, **35,** 389.
Randolph, 1964, *Phys. Rev.*, **134,** A1238.
Rushbrooke, G. S., 1960, *Physica*, **26,** 259.
Widom, B., 1964, *J. chem. Phys.*, **41,** 74.
Worster, J., and March, N. H., 1964, *Solid State Communications*, **2,** 245.
Zernike, F., and Prins, J. A., 1927, *Z. Phys.*, **41,** 184.
Ziman, J. M., 1964, *Advanc. Phys.* **13,** 89.

Phys. Chem. Liq., 1982, Vol. 11, pp. 207–217
0031-9104/82/1103-0207$06.50/0

Small Angle Scattering from Liquids: van der Waals Forces in Argon and Collective Mode in Na

C. C. MATTHAI and N. H. MARCH

Theoretical Chemistry Department, University of Oxford, 1 South Parks Road, Oxford OX1 3TG, England.

(*Received May 14, 1981*)

Because of discrepancies between recent estimates of van der Waals forces in liquid metals, we have examined available experimental data on the small angle scattering of X-rays from liquid Na, and have compared it with neutron scattering data on liquid argon. In the latter case, it follows from first principles that the liquid structure factor $S(k)$ has the small k expansion

$$S(k) = S(0) + a_2k^2 + a_3k^3 + \cdots \tag{i}$$

Therefore we have plotted $(S(k) - S(0))/k^2$ from available data for both Na and Ar, and they are immediately found to be qualitatively different. For a given state of argon, a_2 and a_3 can be extracted and are quite consistent with theoretical estimates.

For Na, the small angle scattering does not have the form (i) and it is demonstrated that there is a term proportional to k at small k. This means that one of the leading terms determining the approach of the radial distribution function $g(r)$ to its asymptotic value of unity at large r is proportional to r^{-4}. It is shown that this behaviour arises from the dispersion of a collective mode, in which the density fluctuations oscillate independently with a well defined dispersion relation $\omega(k)$. This relation has the form, with v_s the velocity of sound,

$$\omega(k) = v_sk + dk^2 + \cdots \tag{ii}$$

We have estimated the dispersion for Na, K and Rb from an approximate theory and have compared it with the experimentally observed collective mode for liquid Rb.

Finally, after extracting the k term discussed above, the magnitude of the k^3 term is estimated from experimental data on Na. It is quite consistent with the data to take the coefficient of k^3 as zero and possible reasons for this are discussed.

1 INTRODUCTION

Recently, there has been interest in the area of van der Waals forces in liquid metals. Chatterjee[1] has used some thermodynamic properties to establish approximate relations of the van der Waals constant c_6 in the form c_6/r^6, in

TABLE I

Estimates of the van der Waals coefficient c_6

	c_6 in units of 10^{-78} Jm^6		
Metal	Chatterjee	Upadhyaya *et al.*	Mahanty and Taylor
K	369.10	1.46	
Rb		2.99	
Cs	744.60	6.77	
Cu	36.30		1.00
Ag	63.74		4.39
Au	84.20		6.36

terms of the critical temperature and the van der Waals gas constants. His values for the alkali and noble metals are compared in Table I with the estimates made by Mahanty and Taylor[2] and Upadhyaya, Wang and Moore[3] from dielectric screening theory. They are seen to differ by a large factor, which is obviously disturbing.

We have therefore examined whether we could test these estimates by examining the small angle scattering data from liquid Na. In order to do this, it is instructive to compare and contrast the small angle scattering in liquid sodium with that in liquid argon. In this latter case, there is a precise theory of the small angle scattering, first given by Enderby, Gaskell and March[4] and leading to the form for the small k expansion of the liquid structure factor $S(k)$:

$$S(k) = S(0) + a_2 k^2 + a_3 k^3 + \cdots \tag{1.1}$$

The form of a_3 is given in terms of the van der Waals constant c_6 by

$$a_3 = \pi^2 \rho \{S(0)\}^2 c_6 / 12 k_B T \tag{1.2}$$

where $S(0)$ is given by the well-known result of fluctuation theory as

$$S(0) = \rho k_B T K_T \tag{1.3}$$

K_T being the isothermal compressibility and ρ the number density of atoms.

Because of the well-defined theory of the small-angle scattering from, say, liquid argon, we have plotted in Figure 1 the measured small-angle scattering from the data of Yarnell, Katz and Wenzel,[5] in the form suggested by Eq. (1.1), namely a plot of $(S(k) - S(0))/k^2$. This plot allows us to estimate a_2 and a_3. It is true that Yarnell *et al.*[5] have had to extrapolate their measured data to join on to the compressibility data for $S(0)$ in Eq. (1.3). We indicate in Figure 1 the points that are measured and those that are the extrapolated values of Yarnell *et al.*[5] While, ideally, the measured points need extending down to smaller angles, we have extracted from the work of Yarnell *et al.*

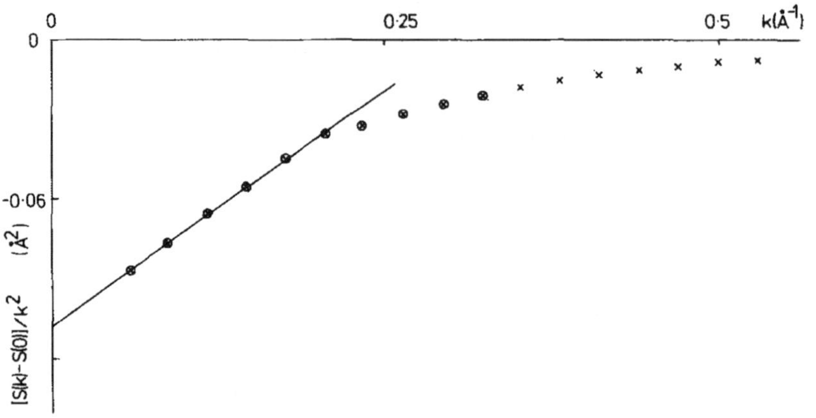

FIGURE 1 $(S(k) - S(0))/k^2$ against k for argon from data of Yarnell *et al.*, × denote measured points while ⊗ are extrapolated values given by Yarnell *et al.* The straight line is the fit to the form (1.1).

the coefficients a_2 and a_3 and these are recorded in Table II. Though these values depend on the extrapolated data, one can have confidence in this when the values are compared with the theoretical estimate of a_3 given in Eq. (1.2). As Robinson and March[6] have shown, the constant c_6 in liquid argon is expected to be reduced from its free space value by some few per cent only, due to the van der Waals interactions occurring in the condensed dielectric liquid medium. As for a_2, Woodhead–Galloway, Gaskell and March[7] have estimated the coefficient for various thermodynamic states. These theoretical estimates of a_2 and a_3 are also recorded in Table II. We conclude from Table II that the data of Yarnell *et al.*[5] for argon is in accord with the small angle scattering theory embodied in Eqs. (1.1) and (1.2).

At this point, we turn to compare and contrast the small angle X-ray scattering data for Na, as observed by Greenfield, Wellendorf and Wiser[8] with that discussed above for argon. Figure 2 immediately shows then that the situation is qualitatively different and that it makes no sense to try to

TABLE II

Constants a_1, a_2 and a_3 appearing in expansions (1.1) and (1.4)

	a_1 (Å)	a_2 (Å)^2	a_3 (Å^3)
Ar (calc.)	—	−0.08	0.375
Ar (expt)	—	−0.12	0.35
Na	0.002	0.0184	0
K	0.0095	0.025	0

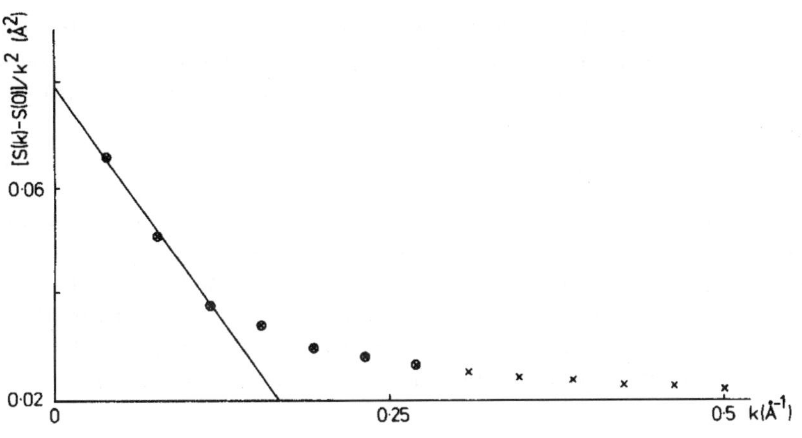

FIGURE 2 Same as Figure 1 but for liquid Na, from data of Greenfield *et al.*

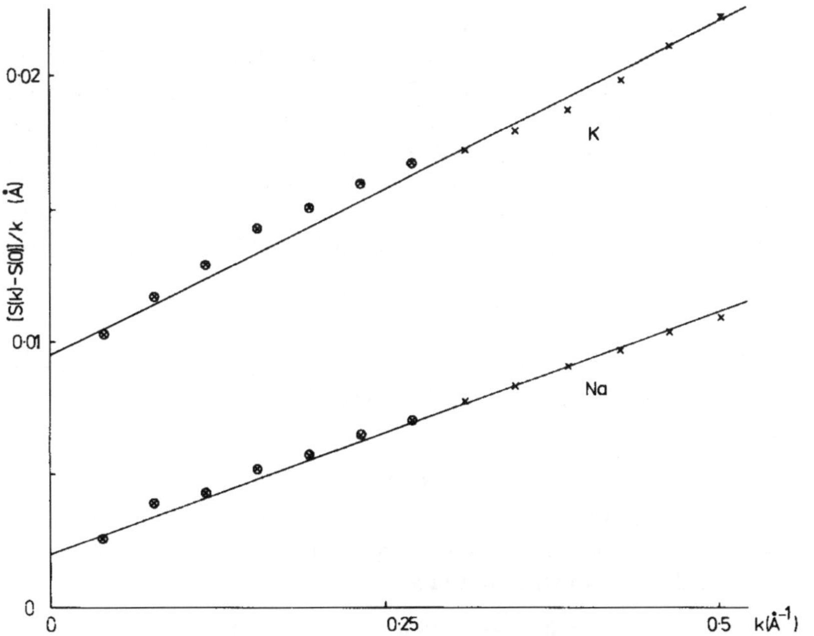

FIGURE 3 $(S(k) - S(0))/k$ against k for liquid Na and K. Straight lines show best fit to Eq. (1.4).

interpret this data on Na with the formula (1.1), because the overall slope is not constant as for Ar and is of the opposite sign. Replotting the data, it emerges that what is omitted from Eq. (1.1) as applied to Na is a term linear in k. Therefore, we shall write, to $O(k^2)$,

$$S(k) = S(0) + a_1 k + a_2 k^2 + \cdots \tag{1.4}$$

The question then arises as to the magnitude of the linear term in k required to fit the experimental data, and equally important, how to interpret it. We have extracted from the plot of $(S(k) - S(0))/k$ in Figure 3 the value $a_1 = 2 \times 10^{-3}$ Å, and we turn immediately to discuss the interpretation in Section 2 below. However, it is of interest to note from Figure 3 that X-ray measurements on liquid potassium[8] also reveal a linear term in k.

2 SMALL ANGLE SCATTERING AND COLLECTIVE OSCILLATIONS IN LIQUID METALS

As has been emphasized by Bratby, Gaskell and March,[9] and more recently by Young and coworkers,[10] the specific heat properties of simple liquid metals like Na near their freezing points are very simple, namely that the specific heat at constant volume c_v is near to $3R$, R being as usual the gas constant, while the ratio of the specific heats γ is near to unity. The simplest way to understand this is to liken the liquid metal to a collection of independent harmonic oscillators. These oscillators are to be thought of as associated with independent oscillations of the density fluctuations $\rho_{\mathbf{k}}$, defined by

$$\rho_{\mathbf{k}}(t) = \sum_i \exp(i\mathbf{k} \cdot \mathbf{r}_i(t)) \tag{2.1}$$

the sum being over the positions $\mathbf{r}_i(t)$ of all the ions at time t. Roughly, the above experiments on specific heats prompt one to write that $\rho_{\mathbf{k}}(t)$ oscillates according to

$$\rho_{\mathbf{k}}(t) = \rho_{\mathbf{k}}(0) \exp(i\omega(k)t) \tag{2.2}$$

where $\omega(k)$ is the frequency of the collective mode as a function of k. The simplest theory,[11] based essentially on Eq. (2.2), is to write the van Hove function, or dynamical structure factor, $S(k, \omega)$ as

$$S(k, \omega) = 2\pi S(k)\delta(\omega - \omega(k)). \tag{2.3}$$

Assuming that the collective mode exhausts the sum rule, with m being the ionic mass,

$$\int \omega^2 S(k, \omega) \frac{d\omega}{2\pi} = \frac{k^2 k_B T}{m}, \tag{2.4}$$

an assumption we test below by comparing its consequences with experiment, then it follows that

$$\omega^2(k) = \frac{k^2 k_B T}{mS(k)} \tag{2.5}$$

Using the form (1.4) for small k in Eq. (2.5), it follows that, with the ratio of the specific heats taken as unity in accord with the above discussion

$$\omega(k) = v_s k + dk^2 + \cdots \tag{2.6}$$

Clearly, the dispersion of the collective mode at small k is then the origin of the k term in Eq. (1.4) and in the above simplest theory of the collective mode one notes that

$$d = -\frac{v_s a_1}{2S(0)}. \tag{2.7}$$

3 ESTIMATE OF DISPERSION OF COLLECTIVE MODE FOR LIQUID ALKALI METALS

Though we have given the above specific heat evidence in favour of collective modes in small core metals like liquid Na, to our knowledge the only simple metal in which collective modes have been directly demonstrated by measur-

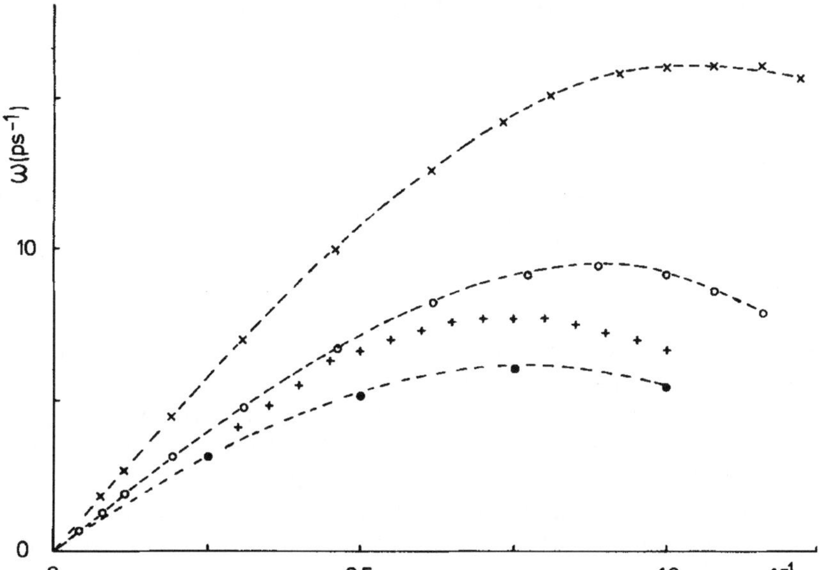

FIGURE 4 Dispersion relation $\omega(k)$ against k. × Na from Eq. (2.5); ○ K from Eq. (2.5); ● Rb from Eq. (2.5); + Rb from data of Copley and Rowe.

ing the peaks in $S(k, \omega)$, and plotting out the dispersion relation, is in liquid Rb. In Figure 4, we have plotted $\omega(k)$ against k from the data of Copley and and Rowe[12] for Rb. Also plotted in Figure 4 are the results obtained using Eq. (2.5) for Na, K and Rb. While we have used the $S(k)$ data of Greenfield *et al.*[8] for Na and K, for the structure factor of Rb we have used the results obtained in the molecular dynamics study of Rahman.[13]

We see for Rb that the simple theory of the collective mode used above is successful in reproducing the position of the maximum in $\omega(k)$. The quantitative discrepancy between theory and experiment in the numerical values of $\omega(k)$ for Rb shows that the approximations (2.3) and (2.4), while useful to get the main features of the dispersion relation, are not fully quantitative. From Figure 4, it is worth noting that the position of the maximum in $\omega(k)$, which occurs at 0.75 Å^{-1} for Rb, shifts to around 0.9 Å^{-1} for K and 1.1 Å^{-1} for Na.

4 IS THERE A VAN DER WAALS INTERACTION IN SIMPLE LIQUID METALS?

In addition to the small-angle scattering, one expects that the so-called Kohn anomaly around the diameter of the Fermi sphere $k = 2k_f$, will also affect the long range behaviour of $g(r)$ in liquid metals. As was suggested earlier,[14] this would give the leading asymptotic term in $g(r)$ to be of the form

$$g(r) \sim \frac{\cos(2k_f r)}{r^3} \tag{4.1}$$

though it was recognized that the finite mean-free path will eventually require a damping factor of the form $\exp(-r/1)$ where 1 is of the order of the electronic mean-free path, multiplying Eq. (4.1). The suggestion was also made[14] that the van der Waals forces would be screened out by the electron gas.

Having seen above that there is a long-range term in $g(r) \sim r^{-4}$ in liquid Na, K and Rb, arising from the dispersion of the collective mode, we finally turn to enquire, in the light of the evidence presented in the Introduction, whether there is a van der Waals term in these liquid alkalis. This would mean a k^3 term to be added in Eq. (1.4). To examine this point, we have extracted the k term from Eq. (1.4), and then in liquid Na and K we have plotted the data in Figure 5 in a suitable form to reveal the k^3 term. From this figure, it can be clearly seen that it is consistent with the experimental data to argue that there is no k^3 term in liquid Na, though we have demonstrated its presence conclusively for liquid argon by similar methods of analysis of

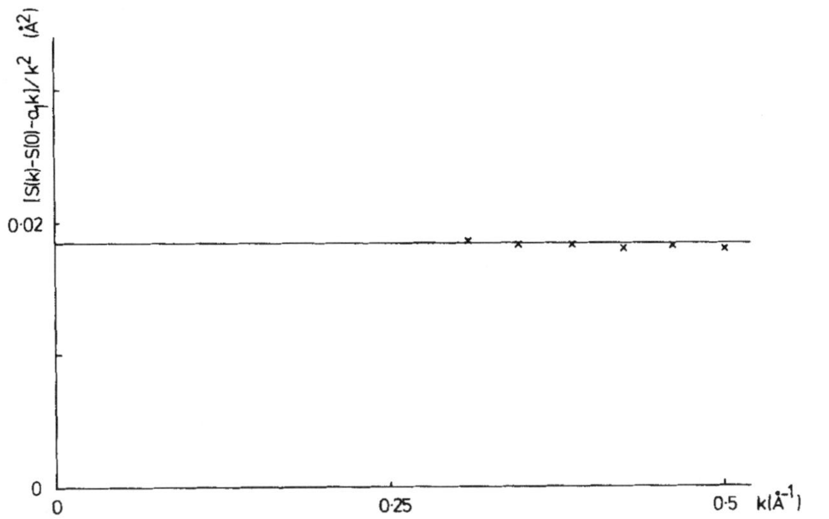

FIGURE 5 $(S(k) - S(0) - a_1 k)/k^2$ against k from data on Na. The straight line which is the best fit is made to have an intercept consistent with slope in Figure 3.

the small-angle data. In Figure 6, we demonstrate that the low k measurements on liquid Na can be fitted very well with an expansion of the type given in Eq. (1.4); that is without a k^3 term.

In the light of this conclusion as to the absence of the k^3 term in liquid Na, we want finally to re-examine the dielectric screening theory to see how such a conclusion might be compatible with that theory. Rehr, Zaremba and Kohn[15] examined the forces between ions immersed in an electron gas and found that the interaction could be expressed in terms of the ionic polarizability $\alpha(\omega)$ as

$$U(r) = -\frac{\hbar}{2\pi}\int_0^\infty \mathrm{d}u\alpha^2(iu)\left[\left(\frac{\partial^2 V(r, iu)}{\partial r^2}\right)^2 + \frac{2}{r^2}\left(\frac{\partial V(r, iu)}{\partial r}\right)^2\right]. \tag{4.2}$$

The screened coulomb interaction, $V(r, iu)$ is given by

$$V(r, iu) = \frac{2}{\pi}\int_0^\infty \mathrm{d}k\,\frac{\sin kr}{kr}\,\frac{1}{\varepsilon(k, iu)}, \tag{4.3}$$

where the frequency dependent dielectric function $\varepsilon(k, iu)$ describes the dynamic response of the electron gas. Mahanty and Taylor,[2] using a linear response theory, also obtained Eq. (4.2).

We see that for $\omega = 0$, the screened coulomb interaction between ions gives rise to the long-range behaviour of $g(r)$ as given by Eq. (4.1). Although the asymptotic limit of $\omega = 0$ in the random phase approximation may be

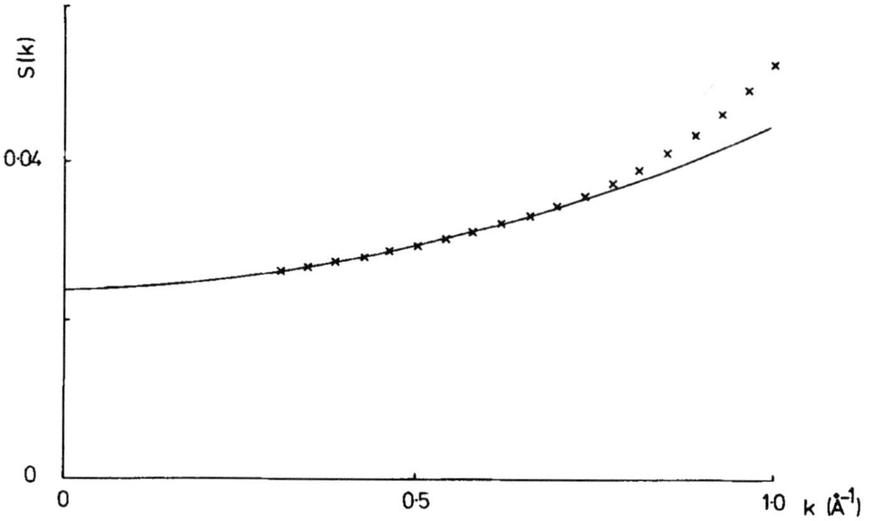

FIGURE 6 $S(k)$ against k for Na from Eq. (1.4), with coefficients from Table II. × indicate experimental points.

obtained analytically, this is not the case when the frequency dependence is included. Mahanty and Taylor, in an effort to obtain the asymptotic form of $U(r)$ from Eq. (4.2), used the hydrodynamic approximation to $\varepsilon(k, \omega)$:

$$\varepsilon_{//}(k, \omega) = 1 - \frac{5\omega_p^2}{(5\omega^2 - 3\hbar^2 k_f^2 k^2)}, \tag{4.4}$$

with ω_p the usual plasma frequency, which then, in the limit k tends to zero, gives the van der Waals form

$$U(r) = \frac{c_6}{r^6} \tag{4.5}$$

where

$$c_6 = \frac{3}{2\pi} \int_0^\infty \left(\frac{u^2}{\omega_p^2 + \mu^2} \right)^2 \alpha^2(iu)\, du. \tag{4.6}$$

However, to obtain the form (4.5), there is the assumption that the long-range behaviour of $U(r)$ arises from $\varepsilon(0, iu)$. Since we find that, for liquid Na and K, the analysis of the $S(k)$ measurements for small k indicate the absence of a term in the interaction potential between ions which falls off as r^{-6}, this assumption needs to be investigated further.

We add that in the case of Ar, where there is no electronic screening, i.e. $\varepsilon(k, \omega) = 1$, the van der Waals coefficient c_6 extracted from the $S(k)$ measurements agrees well with that obtained from thermodynamic arguments, giving

TABLE III

Dielectric function $\varepsilon(q, iu)$ in the random phase approximation (RPA) and in the hydrodynamic approximation in the limit $q \to 0$ for a typical metal (Al)

$\frac{u}{\omega_p}$	$\varepsilon_{RPA}(q, iu)$ as $q \to 0$	$\varepsilon_H(0, iu)$
2	1.20	1.25
1	1.47	2
0.5	2.9	5
0.1	48	101

confidence in our analysis of the $S(k)$ measurements. Also, it is relevant to mention that we have calculated $\varepsilon(q, iu)$ in the random phase approximation for some values of the arguments and these are compared in Table III with the hydrodynamic approximation (4.4). There are substantial differences, and this is enough to point to the fact that Eq. (4.4) should be transcended in future theoretical work on long-range forces in liquid metals.

5 CONCLUSION

Using the experimental scattering data of Yarnell *et al.*[5] on liquid argon the small-angle scattering theory embodied in Eqs. (1.1) and (1.2) is fully confirmed. But turning to liquid Na just above the freezing point, the data of Greenfield *et al.*[8] exclude the small k expansion (1.1) for Na and are in favour instead of the form (1.4). This form is shown to follow from the simplest theory of the collective oscillations of the density fluctuations. The order of magnitude of the dispersion observed, admittedly for liquid Rb, and not presently directly for Na, is consistent with the order of magnitude required to explain the small angle scattering.

It is clear from the present work that there is now a very worthwhile experimental program on small angle scattering that should clarify our understanding of the way collective modes, when they exist, will modify the small angle scattering. We have emphasized here that only when the collective modes can be extracted from the small angle scattering will it be possible to answer the question of whether van der Waals terms of the form c_6/r^6 really exist in simple liquid metals, or whether they are wholly screened out by the response of the electron gas to the oscillating dipoles on the liquid metal ions.

Present indications, which we have argued are not inconsistent with dielectric screening theory, are that in Na there is no term in c_6/r^6, but neither theory nor experiment is completely conclusive on this point and further work is needed.

References

1. B. Chatterjee, *J. Chem. Phys.*, **72**, 2050 (1980).
2. J. Mahanty and R. Taylor, *Phys. Rev.*, **B17**, 554 (1978).
3. C. Upadyaya, S. Wang, and R. A. Moore, *Can. J. Phys.*, **58**, 905 (1980).
4. J. E. Enderby, T. Gaskell, and N. H. March, *Proc. Phys. Soc.*, **85**, 217 (1965).
5. J. L. Yarnell, M. J. Katz, and R. G. Wenzel, *Phys. Rev.*, **A7**, 2130 (1973).
6. G. Robinson and N. H. March, *J. Phys.*, **C5**, 2553 (1972).
7. J. Woodhead-Galloway, T. Gaskell, and N. H. March, *J. Phys.*, **C1**, 1 (1968).
8. A. J. Greenfield, J. Wellendorf, and N. Wiser, *Phys. Rev.*, **A4**, 1607 (1971).
9. P. Bratby, T. Gaskell, and N. H. March, *Phys. Chem. Liquids*, **2**, 53 (1970).
10. W. H. Young, I. Yokoyama and I. Ohkoshi, *Phys. Chem. Liquids*, **10**, 273 (1981).
11. See, for example, N. H. March and M. P. Tosi, *Atomic Dynamics in Liquids* (Macmillan: London) 1976.
12. J. R. D. Copley and J. M. Rowe, *Phys. Rev. Letts.*, **32**, 49 (1974).
13. A. Rahman, *Phys. Rev.*, **A9**, 1667 (1974).
14. See, for example, N. H. March, *Liquid Metals* (Pergamon: Oxford) 1968.
15. J. J. Rehr, E. Zaremba, and W. Kohn, *Phys. Rev.*, **B12**, 2062 (1975).

Phys. Chem. Liq., 1983, Vol. 13, pp. 65–74
0031-9104/83/1301-0065$18.50/0

Structure and Forces in Simple Liquid Metals

I. EBBSJÖ
Studsvik Science Research Laboratory, S-611 82 Nyköping, Sweden.

G. G. ROBINSON
Department of Physics, University of Sheffield, England.

and

N. H. MARCH
Theoretical Chemistry Department, University of Oxford, Oxford OX1 3TG, England.

(*Received February 9, 1983*)

This paper is based on the exact force equation for a classical monatomic liquid with pairwise interactions.

This is handled by expanding the triplet correlation function $g^{(3)}$ in Legendre polynomials to yield

$$g'(r) + g(r)\frac{\varphi'(r)}{k_B T} = -\frac{4\pi\rho}{3k_B T}\int_0^\infty s^2\varphi'(s)Q_1(r, s)\,\mathrm{d}s \tag{i}$$

where $g(r)$ is the radial distribution function, $\varphi(r)$ the pair potential, ρ is the number density, while Q_1 is the $l = 1$ term in the Legendre polynomial expansion of $g^{(3)}$.

From molecular dynamics data for some simple liquids, we have calculated the left-hand-side of Eq. (i). This has been used to draw some conclusions about the usefulness of the Kirkwood superposition approximation in extracting forces from structure. Used in the right way, the Kirkwood approximation yields a quantitative estimate of the three-body term in the force equation for Na, but is much less useful for argon.

1 INTRODUCTION

Since the work of Johnson and March,[1,2] there has been a good deal of interest in whether it is possible, in a monatomic liquid, to invert the customary procedure of classical statistical mechanics, in which a pair potential $\varphi(r)$ is specified, and a radial distribution function $g(r)$ calculated from it, and to extract $\varphi(r)$ from the measured structure factors. This area was reviewed some time ago by Enderby and March.[3] Work was done on Lennard-Jones fluids by Krumhansl and Wang,[4] employing molecular dynamics, and recent relevant work is that of Haymet *et al.*[5] These workers devised a self consistent approach to the calculation of pair and triplet functions, but, in common with the work of Krumhansl and Wang, they have recently tested the above treatment at liquid state densities against new molecular dynamical results of the Lennard-Jones (12, 6) fluid. For high temperature, they find, by using a topological reduction of the diagrammatic expansion of the triplet correlation function appearing in the basic force equation (2.2) below, good agreement with the triplet function found from molecular dynamics.[6] However, for lower densities, their approach is not in quantitative agreement with the molecular dynamics approach.

Although, independently, as a part of the programme of work reported in our present paper, we also studied the case of Lennard-Jones fluids, we shall only briefly refer to our own findings at the end of the paper, in view of the recent extensive work on this class of fluids. Rather, we shall focus attention on a simple liquid metal, Na, here, though it will be useful subsequently to compare and contrast the results with those for liquid argon.

Returning to the proposal of Johnson and March for extracting force laws from structure data, two points are now generally agreed by workers in the area:

i) It is essential that structure data is available over a wide range of momentum transfer q, the small angle scattering data being of primary importance in pinning down the more interesting long-range features of the pair potentials. This is still a difficulty, though Greenfield *et al.*[7] have shown that the X-ray method on Na and K can make accurate small angle scattering data available; see also the very recent work of Waseda.[8]

ii) Related to the discussion of Lennard-Jones fluids above, but more generally now, great care is needed with the treatment of the three-particle distribution function. Thus, it is clear that, while the type of approximation made for this central quantity must take cognizance of the range of the forces, and is therefore certain to be different for liquid argon on the one hand, and liquid Na on the other, the non-critical use of, for instance, the Kirkwood superposition as closure, leading to the Born-Green theory of structure, must be transcended, at the present level of sophistication of liquid structure theories. Thus, for hard core-like interactions, with short-range tails, as in liquids like argon, the Born-Green theory of structure is poor, and we have now considerably more accurate structural theories.

The present work is a contribution essentially in the area (ii) above. Since we have the problem (i), we shall in the present paper mainly make use of computer experiments on simple liquids.

The outline of the paper is as follows. In Section 2 we outline the method of attack on the problem of what information can be gained about the triplet correlation function from molecular dynamics. As pointed out by Hutchinson,[9] the force equation, which is our main tool here, involves only one term (the p or $l = 1$ component) in the expansion of the triplet function in Legendre polynomials (Eq. (2.3) below). Therefore it must not be expected that full information on the triplet function $g^{(3)}$ can be extracted by our procedure, which uses only $g(r)$ and $\varphi(r)$ as described below.

2 FORCE EQUATION IN TERMS OF p HARMONIC OF TRIPLET CORRELATION FUNCTION

If we introduce the potential of mean force $U(r)$ in a classical monatomic liquid through the relation

$$g(r) = \exp\left(\frac{-U(r)}{k_B T}\right) \tag{2.1}$$

then it is easy to write down the force equation as

$$-\frac{\partial U(r_{12})}{\partial \mathbf{r}_1} = -\frac{\partial \varphi(r_{12})}{\partial \mathbf{r}_1} - \rho \int \frac{\partial \varphi(r_{13})}{\partial \mathbf{r}_1} \frac{g^{(3)}(\mathbf{r}_1, \mathbf{r}_2, \mathbf{r}_3)}{g(r_{12})} \mathbf{dr}_3 \tag{2.2}$$

where we can see that the first term on the right-hand side is the direct force between particles 1 and 2 separated by distance r_{12} while in the remaining term the force on particle 1 due to a particle at $\mathbf{r}_3$ is weighted by the probability that a particle is at $\mathbf{r}_3$ if there are certainly particles at $\mathbf{r}_1$ and $\mathbf{r}_2$, namely $g^{(3)}/g(r_{12})$.

We stress that Eq. (2.2) is an exact member of the Born–Green–Yvon hierarchy, only dependent on the assumption of pairwise interactions. Following Hutchinson,[9] we now write, with $r_{13} = s, r_{23} = t, r_{12} = r$, and with θ the angle between the vectors r_{13} and r_{12},

$$g^{(3)}(r, s, \theta) = \sum_{l=0}^{\infty} Q_l(r, s) P_l(\cos \theta) \tag{2.3}$$

and evidently, using the orthogonality properties of the Legendre polynomials, we obtain

$$Q_1(r, s) = \frac{3}{2} \int_0^{\pi} g^{(3)}(r, s, \theta) \cos \theta \sin \theta \, d\theta. \tag{2.4}$$

But it is now easy to show that the same orthogonality properties cut out all the terms in the triplet correlation function except the term with $l = 1$, to yield

$$g'(r) + \frac{g(r)}{k_B T} \varphi'(r) = -\frac{4\pi\rho}{3k_B T} \int_0^{\infty} s^2 \varphi'(s) Q_1(r, s) \, ds \tag{2.5}$$

This equation is the focal point of the present work; for convenience we define the left-hand side as $f(r)$, i.e.

$$f(r) = g'(r) + \frac{g(r)}{k_B T} \varphi'(r). \tag{2.6}$$

This is as far as we can get without either (a) analytical approximation to $g^{(3)}$, or in the form (2.5) to Q_1, or (b) use of experimental data. As already mentioned, we shall use here not real experimental data but the results of computer experiments. The way we do this, and some results, are set out immediately below in Section 3.

3 USE OF MOLECULAR DYNAMICS DATA

The point we exploit here is that, whereas in an X-ray or neutron experiment we have available merely the structure factor $S(q)$, over a still limited range of q, and therefore, in principle, $g(r)$ by Fourier transform, in molecular dynamical simulation we have both the input force $-\varphi'(r)$ and by computer calculation the corresponding $g(r)$. Thus we can calculate the left-hand-side of Eq. (2.5), or using the definition (2.6) we can calculate the quantity $f_{MD}(r)$. Clearly, this information is giving us, at very least, from Eq. (2.5), a precise if limited way of testing approximations to $Q_1(r, s)$ in Eq. (2.5), and this is the point of view adopted below.

3.1 Use of molecular dynamical simulation results on liquid Na[10]

The potential of Taylor[11] has been used to calculate the curve in Figure 1 for the force $-\varphi'(r)$ along the connection line between two particles. For comparison, we have plotted the mean force $-U'(r)$ derived from the calculation of $g(r)$ with the Taylor potential, in the manner described below. The different points plotted for $-U'(r)$ are for two different truncation distances of the Taylor potential, the dots corresponding to a truncation of the force at 7.3 Å and the + signs to a distance of 10.03 Å. The curve for $\varphi'(r)$ and the results for $U'(r)$ are at first sight pretty similar, but there is more structure in the mean force than in the pair force. We emphasize that this situation is strictly for the model potential adopted; it is not proven that this is precisely the situation in liquid Na. The data are, specifically, for a density $\rho = 2.41 \cdot 10^{22}\ \text{cm}^{-3}$ and for $T \simeq 400$ K. The mean force $-U'(r)$ plotted is simply $k_B T \cdot g'(r)/g(r)$, the pair function $g(r)$ being that generated by the Taylor potential inserted into the molecular dynamics programme.

Figure 2 shows $f(r)$ defined in Eq. (2.6), calculated from the molecular dynamics data, the dots corresponding to data from the system with the

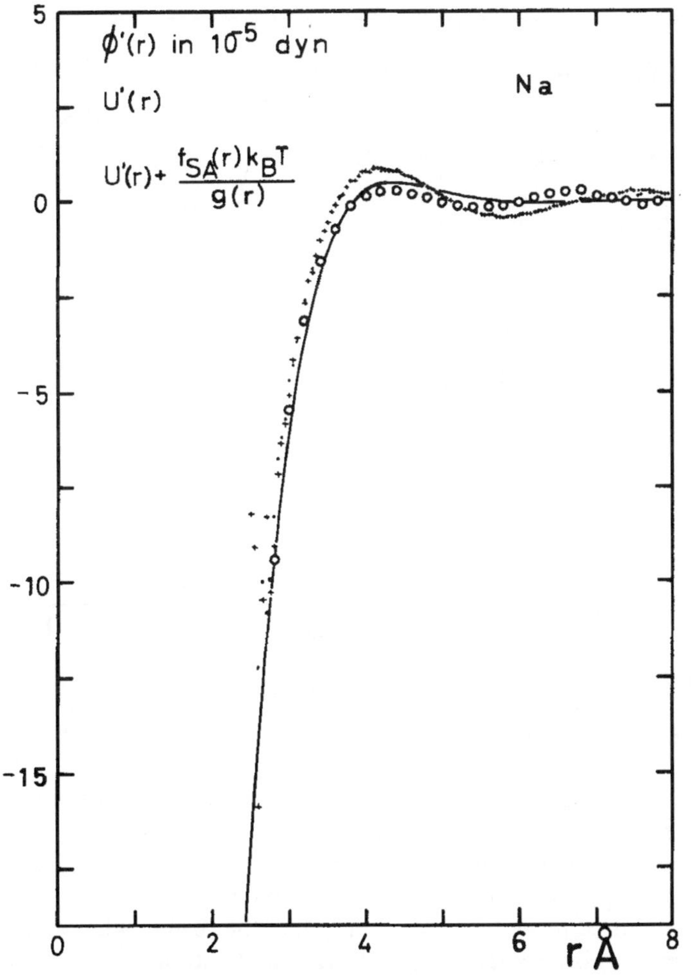

FIGURE 1 The pair force ϕ' (r)(——), the mean force U' (r)(· and +) and the 'recovered' pair force (○) by the use of the Kirkwood approximation as described in the text for liquid 'computer sodium'. The pair distribution function has its main maximum at $r \simeq 3.63$ Å and the pair potential its main minimum at $r \simeq 3.85$ Å. The data for U' (r) show that the evaluation procedure for g (r) and its first derivative in the small r-region could be improved.

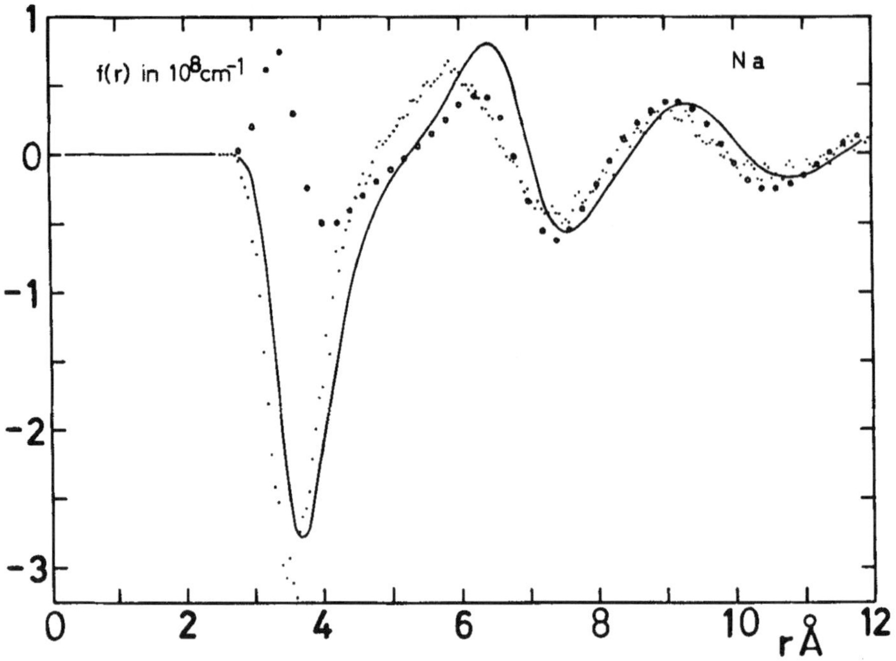

FIGURE 2 Comparison between $f_{MD}(r)(\cdots)$ and $f_{SA}(r)$(——) for liquid 'computer sodium'. The curve indicated by the open circles is obtained by the use of the mean force $U'(r)$ instead of the pair force in the Kirkwood approximation!

truncation distance $r_c = 7.3$ Å. Beyond this distance $f(r)$ is obtained from the slope of the pair distribution function.

Evidently, there are pronounced features in this curve; a very deep minimum around 3.6 Å followed by subsidiary maxima and minima.

For comparison, we have now computed the results when the Kirkwood approximation is adopted for $g^{(3)}$, it being understood however that the $g(r)$ used here is the 'exact' $g(r)$ of the molecular dynamics simulation.

The continous line is calculated then from the approximation

$$f_{SA}(r) = -\frac{\pi\rho g(r)}{k_B T}\int_0^{r_c} g(s)\varphi'(s)\left\{\int_{|r-s|}^{r+s} tg(t)\frac{r^2+s^2-t^2}{r^2}\,dt\right\} ds. \quad (3.1)$$

The main features of the $f(r)$ curve are clearly given by this approximation. The most serious error introduced is a phase shift, which causes all the maxima and minima to lie at rather too large r, the discrepancies in position being one or two tenths of an Å. The first minimum is a little too shallow, and the first maximum a little too pronounced, so that the Kirkwood approximation, used in the manner described above, overemphasizes the structure a little in the $f(r)$ curve.

The other point we feel is significant is the following one. In an actual calculation to try to extract the force law, as proposed in Refs. 1 and 2, one would use the 'exact', i.e. measured $g(r)$ in the Kirkwood decoupled $g^{(3)}$, exactly as we have done here. But one would, in the absence of exact knowledge of the pair potential have to put some trial form into the expression for $f(r)$. One obvious procedure is to put in the mean force derived directly from the experimental $g(r)$.

But the effect of doing this can be gauged from the present model of liquid Na. Thus in the curve marked by open circles in Figure 2 we have calculated $f_{SA}(r)$ with this one modification. The effect is dramatic. One finds now a totally different form of $f(r)$, which has practically a node at the position of the principal minimum of the true $f(r)$.

The final curve we have added to Figure 1 is the plot of $f_{SA}(r)k_B T/g(r) + U'(r)$, with $f_{SA}(r)$ calculated with exact model $g(r)$ and force law.

The conclusion is clearly that it is of vital importance in the three-body term to calculate it with a consistent $g(r)$ and force. If we take the potential of mean force, and the (inconsistent) model $g(r)$, huge errors are introduced.

Of course, we can ask the question: do we really know, in liquid Na metal, from the experimental facts, that the true pair force and the mean force differ by as much as is shown in Figure 1, with the Taylor potential. If we assume, for a moment, that the real force is significantly closer to the potential of mean force than in Figure 1, then Figure 2 makes it tempting to conclude that $f(r)$ may be substantially smaller in magnitude near 3.6 Å in real molten Na than with the Taylor potential.

So the question which remains would be: for real molten Na, do we know that the potential of mean force is *not* an almost self-consistent choice of pair potential. Of course, the potential of mean force, extracted from measured $g(r)$ data, will be temperature dependent. But we know that the liquid metal potential will depend on density, so is this effect merely reflecting the reduced density with increasing temperature?

What seems clear for the model we have studied is that, actually at the precise point of solution, the Kirkwood superposition for $g^{(3)}$, with the exact $g(r)$ as input data, and combined in the term in the force equation with the 'exactly consistent' pair potential, is leading for model liquid Na to a pretty useful result for the three-body term needed to extract the force. In order to see whether this is a situation which is specific to a simple liquid metal like Na, we have also carried out some calculations on liquid Ar. We shall not report these in the same detail as done for Na, as a large body of accurate work already exists for liquid argon, as summarized in the Introduction.

3.2 Summary of results for liquid argon

In the case of liquid argon, we made no attempt to use a particularly refined pair potential. The Lennard-Jones potential (12, 6) has been employed by Schofield and Hutchinson (private communication) to obtain the pair function $g(r)$ from molecular dynamics. The results correspond to a density of 0.0214 atoms/Å^3 and a temperature of 87.2 K. The value of $f(r)$ given by the superposition approximation is represented by the continous line in Figure 3, and is to be compared with the molecular dynamics results given by the crosses. It is immediately clear that the result of the superposition approximation is of much poorer quality for Ar than for Na, the molecular dynamics data revealing much more structure in $f(r)$ than is found for the model of liquid Na presented here.

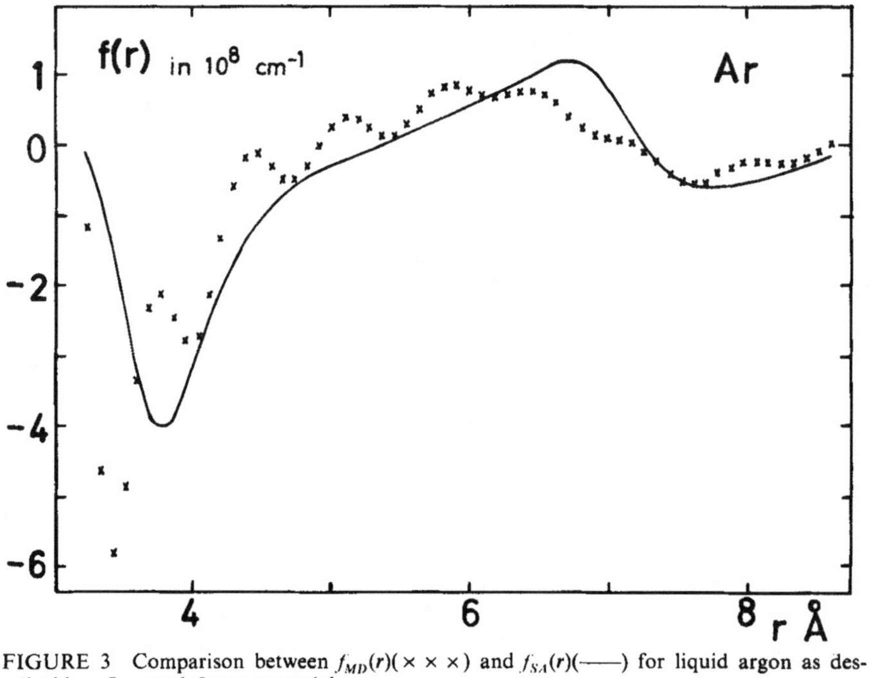

FIGURE 3 Comparison between $f_{MD}(r)(\times\times\times)$ and $f_{SA}(r)$(——) for liquid argon as described by a Lennard-Jones potential.

4 CONCLUSION

What is clear from the above discussion is that, provided one always uses the superposition approximation with the exact $g(r)$ corresponding to the chosen input potential, and treates the three-body term in the force equation with a consistent $g(r)$ and $\varphi(r)$, then the Kirkwood superposition approximation can be of quantitative value for potentials having the character of liquid Na. On the other hand, this procedure is poor for liquid Ar. This illustrates, quite clearly, that as has been stressed on a number of occasions,[12] it is essential, before asking for an appropriate decoupling procedure for the three-atom correlation function, to ask about the nature of the force law. The present paper leaves little doubt that in simple liquid metals the Kirkwood superposition approximation, when carefully employed, has quantitative merit, in sharp contrast to almost 'billiard ball' liquids like Ar.

References

1. M. D. Johnson and N. H. March, *Phys. Lett.*, **3**, 313 (1963).
2. See also M. D. Johnson, P. Hutchinson and N. H. March, Proc.Roy. Soc., **A282**, 283 (1964).
3. J. E. Enderby and N. H. March, *Adv. Phys.*, **14**, 453 (1965).
4. J. A. Krumhansl and S. Wang, *J. Chem. Phys.*, **56**, 2034, 2179 and 4287 (1972).
5. A. D. J. Haymet, S. A. Rice and W. G. Madden, *J. Chem. Phys.*, **74**, 3033 (1981); **75**, 4696(1981).
6. W. J. McNeil, W. G. Madden, A. D. J. Haymet and S. A. Rice, *J. Chem. Phys.*, **78**, 388 (1983).
7. A. J. Greenfield, J. Wellendorf and N. Wiser, *Phys. Rev.*, **A4**, 1607 (1971).
8. Y. Waseda, 1983, private communication and to be published.
9. P. Hutchinson, *Proc. Phys. Soc.*, **91**, 506 (1967).
10. I. Ebbsjö and T. Kinell, to be published.
11. R. Taylor, 1982, private communication, see L. Dagens, M. Rasolt and R. Taylor, *Phys. Rev.*, **B11**, 2726 (1975).
12. See, for example, N. H. March and M. P. Tosi *Atomic Dynamics in Liquids* (Macmillan: London), 1976, p34.

Properties of liquid direct correlation function at melting temperature related to vacancy formation energy in condensed phases of rare gases

A. B. Bhatia
Department of Physics, University of Alberta, Edmonton, Alberta, Canada T6G 2J1
N. H. March
Theoretical Chemistry Department, University of Oxford, 1 South Parks Road, Oxford OX1 3TG, England

(Received 24 October 1983; accepted 15 November 1983)

For condensed phases of rare gases, the approximate relation

$$[E_V + \tfrac{1}{2}B\Omega] = -\left(\frac{k_B T\,[c(r=0)+2]}{2}\right)_{T_m} \tag{I}$$

is derived, E_V being the vacancy formation energy, B^{-1} the isothermal compressibility, Ω the atomic volume, $c(r)$ the liquid direct correlation function, and T_m the melting temperature.

I. INTRODUCTION

Empirical correlations between melting temperature T_m vacancy formation energy E_V and isothermal compressibility $\kappa_T \equiv B^{-1}$ have been known for a long time. Recent work by Varotsos and co-workers[1] has pursued one of these, embodied in the empirical form $E_V \simeq CB\Omega$, where Ω is the atomic volume and C is supposedly constant for a wide variety of solids; but in these studies of Varotsos *et al.*,[1] no first principles derivation of the above empirical relation was attempted. Additionally, we note here that experimentalists have often commented on the correlation between E_V and $k_B T_m$, the thermal energy corresponding to the melting temperature T_m.

As to microscopic theory, a pair force description has frequently been applied to calculate the vacancy formation energy in, say solid argon. If, in addition to the assumption of a solely pairwise potential $\phi(r)$, atomic relaxation round the vacancy is neglected in these closepacked solids, then the vacancy formation energy E_V in the cold solid is related directly to the depth of the pair potential $|\phi(a)|$, a being the near-neighbor distance, by $E_V = 6|\phi(a)|$. Under these assumptions, E_V is identical to the cohesive energy per atom W, although if one allows for small contributions from three-body forces then Foreman and Lidiard[2] point out that E_V is no longer equal to W, the ratio of the three-body contributions to these quantities being a factor of 2.

Our aim in the present work is to examine the microscopic basis for the empirical correlations noted above, in an admittedly restricted framework. Thus we shall confine our attention solely to condensed phases of rare gases, excluding helium. Furthermore, we shall retain the assumptions that we can neglect the small three-body force contributions, as well as atomic rearrangement around a vacancy in the crystal. Though, as already mentioned, pair force calculations on these materials are numerous, we shall examine the basis for correlations involving E_V, B, and $k_B T_m$ by working a pair force theory of E_V which is posed in a form analogous to liquid state descriptions.[3,4]

II. VACANCY FORMATION ENERGY AND LIQUID STATE STRUCTURE AT MELTING TEMPERATURE

Though the work was done with the vacancy formation energy E_V in simple metals specifically in mind, Faber,[3] in his book on liquid metals, has expressed E_V for a pair force model and in the absence of relaxation in terms of the pair potential, assumed to possess a Fourier transform as in the pseudopotential description of simple metals, and the structure factor $S(q)$. More conveniently for present purposes, Minchin *et al.*[4] have given an **r** space description of Faber's result, their formula for E_V being

$$E_V + p\Omega = -\frac{\rho}{2}\int g\phi\, d\mathbf{r} - \frac{\rho}{6}\int r\frac{\partial\phi}{\partial r} g\, d\mathbf{r}, \tag{2.1}$$

$\rho \equiv \Omega^{-1}$ being the atomic number density. Equation (2.1) reduces to the Faber formula when the pressure p is put equal to zero.

The relation of Eq. (2.1) with $p = 0$ to the result $E_V = 6|\phi(a)|$ quoted above, with $g(r)$ interpreted as the probability density of finding an atom at r if there is an atom at the origin, is readily seen by using Eq. (2.1) in the cold solid with short-range pair forces. Then the first term on the right-hand-side of Eq. (2.1) becomes $-(z/2)\phi(a)$, where z is the coordination number, while, since a is the equilibrium near-neighbor spacing, $(\partial\phi/\partial r)_{r=a} = 0$ and the second term vanishes.

A. Evaluation of E_V for hot solid using liquid state structure at melting temperature

However, because of the appearance of the melting temperature T_m in the empirical correlations which motivated the present study, let us turn our attention to the approximate evaluation of Eq. (2.1) in the hot solid near T_m. Of course, the determination of g in this case is a major problem necessitating a full treatment of large effects due to anharmonicity. To make progress in understanding the status of the empirical correlations involving E_V, T_m, and B we shall add the further basic assumption of the present treatment: That in the condensed phases of rare gases, where there are known to be no major changes of local coordination on melting, in sharp contrast say to tetrahedrally bonded covalent solids like Ge and Si which melt into a dense metallic phase, we can usefully evaluate E_V in the hot solid by inserting the appropriate liquid structure, described by the pair function $g(r)$ at the melting temperature T_m.

Making this assumption, let us next consider the right-hand side of Eq. (2.1) when we insert $g(r)|_{T_m}$. From the argu-

 0021-9606/84/052076-03$02.10

ment for the cold solid outlined above, we anticipate that the term involving $\partial\phi/\partial r$ will make a minor contribution. This is immediately confirmed by invoking the virial equation expression for the pressure p, namely

$$p = \rho k_B T - \frac{\rho^2}{6}\int r\frac{\partial\phi}{\partial r} g(\mathbf{r})\, d\mathbf{r}. \tag{2.2}$$

Putting $p = 0$ yields the second term on the right-hand side at T_m as $-k_B T_m$ and since E_V is known to be substantially larger than this, the minor correction from the $\partial\phi/\partial r$ term in Eq. (2.1) for condensed rare gases is confirmed. Incidentally, the formula written by Minchin *et al.*[4] for metals neglects the free gas pressure $\rho k_B T$ relative to the electron gas pressure.

When we turn to the first term on the right-hand side of Eq. (2.1), it should first be noted from the work of Parrinello *et al.*[5] and Johnson *et al.*[6] on the liquid partial structure factors of a binary conformal solution, the thermodynamic properties of which were discussed originally by Longuet-Higgins,[7] that the basic interchange energy of the mixture model involves just the volume integral of $g\phi$ appearing in Eq. (2.1), g and ϕ referring however to the reference liquid of that model. Johnson *et al.*[6] in fact evaluated $\int g\phi\, dr$ in the course of their work using data on liquid argon from the computer simulation results of Verlet.[8] Reading off the value of the integral from their Fig. 1, we find for the particular thermodynamic state they employ

$$-\frac{\rho}{2}\int g\phi\, d\mathbf{r} \simeq 0.08 \text{ eV}, \tag{2.3}$$

which is in semiquantitative agreement with the measured vacancy formation energy of ~ 0.05 eV for solid argon.

B. Relation of E_V to liquid direct correlation function at T_m

Encouraged by the above numerical evaluation of the formula (2.1) using computer simulation data, we shall next show how that equation can be approximately related to measured liquid structure data at the melting temperature T_m. The appropriate tool required to describe the structure turns out to be the Ornstein–Zernike direct correlation function $c(r)$, related to $g(r) \equiv h(r) + 1$ by

$$h(r) = c(r) + \rho\int c(|\mathbf{r} - \mathbf{r}'|)h(\mathbf{r}')d\mathbf{r}'. \tag{2.4}$$

At this point, the work of Woodhead-Galloway *et al.*[9] on $c(r)$ for liquid argon will be invoked. They calculate $c(r)$ as

$$c(r) \simeq c_{\text{hard sphere}}(r) - \frac{\phi_{\text{long range}}(r)}{k_B T}, \tag{2.5}$$

where $\phi_{\text{long range}}(r) \equiv \phi(\mathbf{r})$ outside the hard core of diameter σ. Using the Percus–Yevick solution[10,11] for $c_{\text{hard sphere}}(r) \equiv c_{hs}(r)$ and noting that it is identically zero outside $r = \sigma$ in this approximate theory, it follows from Eq. (2.5) that

$$c(r) \simeq -\frac{\phi(r)}{k_B T}, \quad r > \sigma, \tag{2.6}$$

but in the integral $\int g\phi\, d\mathbf{r}$ appearing in Eq. (2.1), $g \simeq 0$ for $r < 0$, and hence, putting the $\partial\phi/\partial r$ term as $-k_B T_m$ as discussed above, we find

$$E_V \simeq \tfrac{1}{2}\left(\rho k_B T\left\{\int g(r)c(r)\, d\mathbf{r} - \frac{2}{\rho}\right\}\right)_{T_m}, \tag{2.7}$$

or putting $g = 1 + h$:

$$\frac{E_V}{k_B T_m} \simeq \tfrac{1}{2}\left(\rho\int c(\mathbf{r})\, d\mathbf{r} + \rho\int h(r)c(r)\, d\mathbf{r} - 2\right). \tag{2.8}$$

Defining $\tilde{c}(q)$ as the Fourier transform of $c(r)$ through

$$\tilde{c}(q) = \rho\int c(r)\exp(i\mathbf{q}\cdot\mathbf{r})\, d\mathbf{r}, \tag{2.9}$$

noting from Eq. (2.4) that (putting $r = 0$ and recalling that $g(r = 0) = 0$ or $h(r = 0) = -1$)

$$\rho\int h(r)c(r)\, d\mathbf{r} = -[1 + c(r = 0)], \tag{2.10}$$

Eq. (2.8) becomes

$$\frac{E_V}{k_B T_m} \simeq \tfrac{1}{2}(\tilde{c}(0) - c(r = 0) - 3)_{T_m}. \tag{2.11}$$

This is the desired relation between the vacancy formation energy E_V, measured in units of $k_B T_m$, and the liquid direct correlation function $c(r)$ at the melting temperature T_m Equation (2.11) is the main result of this paper. We briefly discuss below how contact can be established with the known empirical correlations referred to in the Introduction.

III. MICROSCOPIC THEORY AND EMPIRICAL CORRELATIONS

Recalling that the empirical correlations involve the isothermal compressibility $\kappa_T \equiv B^{-1}$, the isothermal bulk modulus being denoted by B, and relating the liquid structure factor $S(q)$ to $\tilde{c}(q)$ through the Fourier transform of Eq. (2.4), namely

$$\tilde{c}(q) = \frac{S(q) - 1}{S(q)}, \tag{3.1}$$

Eq. (2.11) becomes

$$\frac{E_V}{k_B T_m} \simeq \tfrac{1}{2}\left(-\frac{1}{S(0)} - c(r = 0) - 2\right)_{T_m}. \tag{3.2}$$

However, $S(0)$ is related to κ_T through the fluctuation theory result

$$S(0) = \rho k_B T\kappa_T \equiv k_B T/B\Omega \tag{3.3}$$

and hence

$$\frac{E_V}{k_B T_m} + \tfrac{1}{2}\frac{B\Omega}{k_B T_m} \simeq -\tfrac{1}{2}c(r = 0)|_{T_m} - 1, \tag{3.4}$$

which is equivalent to Eq. (I) of the Abstract.

A. Use of hard sphere approximation for $c(r = 0)$

If, following Woodhead-Galloway *et al.*,[9] we employ the approximation (2.5) and note that, by definition, $\phi_{\text{long range}}(\mathbf{r})$ is zero inside the core, then

$$c(r = 0) \simeq c_{hs}(r = 0) \tag{3.5}$$

and using the exact Percus–Yevick solution of Wertheim[10] and of Thiele[11] as an approximation to $c_{hs}(r = 0)$ in E. (3.5), we obtain

$$c(r = 0) \simeq -(1 + 2\eta)^2/(1 - \eta)^4, \tag{3.6}$$

where η is the packing fraction defined by

$$\eta = (\pi/6)\rho\sigma^3. \tag{3.7}$$

For many simple liquids near the melting temperature, $\eta \simeq 0.45$ and Eq. (3.4) becomes

$$\frac{E_V}{k_B T_m} + \frac{1}{2}\frac{B\Omega}{k_B T_m} \simeq 19. \tag{3.8}$$

Since $S_{T_m}(0)$ for liquid argon[12] is $\simeq 0.052$, this yields

$$\tfrac{1}{2}B\Omega / k_B T_m = \frac{1}{2S_{T_m}(0)} \simeq 9.6$$

and hence $E_V/k_B T_m \simeq 9.4$ in reasonable accord with 0.08 eV/$k_B T_m \simeq 11$ if T_m is taken as 85 K. For krypton, using the data on B and Ω from Street and Staveley[13] and $T_m = 117$ K in Eq. (3.8), one finds also $E_V/k_B T_m \simeq 9.4$. Experimental values of E_V are quoted[1] to be in the range 0.05–0.069 eV for argon and 0.077–0.086 eV for krypton.

IV. DISCUSSION AND SUMMARY

The main achievement of this work is to relate the vacancy formation energy E_V in the hot solid, measured in units of the thermal energy $k_B T_m$, to two properties of the direct correlation function $c(r)$ at the melting point, namely $c(r=0)$ and $\rho \int c(r)\, dr$. The latter is a bulk property, expressible through the isothermal compressibility. The former quantity $c(r=0)$ is, at least in principle, an "observable" via structure data for $S(q)$, since

$$c(r=0) = \frac{1}{8\pi^3 \rho}\int \left(\frac{S(q)-1}{S(q)}\right) dq. \tag{4.1}$$

In the model adopted in the present discussion, the relative constancy of the packing fraction η at T_m leads to the relation between observable macroscopic quantities embodied in Eq. (3.8).

To reduce Eqs. (3.8) or (2.11) further so that they yield $E_V \propto B\Omega$, as proposed, for example, in the recent studies of Varotsos *et al.*,[1] would require [through Eq. (2.11)] that $c(r=0)$ is directly related to $\tilde{c}(0)$. This is certainly the case in hard spheres Percus–Yevick solution, namely

$$\tilde{c}_{hs}^{\mathrm{PY}}(0) = 1 + c_{hs}(r=0), \tag{4.2}$$

but this just reduces Eq. (2.11) to $E_V/k_B T_m \simeq -1$, showing that the attractive long range part of the interatomic potential ϕ, which contributes singificantly to $\tilde{c}(0)$, is crucial to the understanding of E_V. Since a direct relation between $\tilde{c}(0)$ and $c(r=0)$ in presence of a ϕ is not available, it does not seem possible at present to provide a full microscopic basis for the relation $E_V \propto B\Omega$. However, Eq. (3.8) supplemented with a dimensionality argument does provide an explanation for it as follows:

If one recalls the dimensionality arguments leading up to the Einstein and Lindemann relations which connect a characteristic vibration frequency of a solid to the bulk modulus B and to the melting temperature T_m, respectively,[14] one easily sees that

$$T_m = c' B\Omega, \tag{4.3}$$

where the constant of proportionality c' is expected roughly to be the same for all solids of a given class (e.g., inert gas solids except He) and B and Ω are to be taken close to the melting point. Use of Eq. (4.3) in Eq. (3.8) immediately leads to the empirical relations $E_V \propto B\Omega$ or $E_V \propto T_m$ reported in the literature.

The final point we wish to make brings us back to the cold solid, for which the short-range pair force model yields

$$E_V \simeq \tfrac{1}{2}z|\phi(a)|. \tag{4.4}$$

If we use the obvious fact that $a > \sigma$, then Eq. (2.5) permits us to write

$$\frac{E_V}{k_B T_m} \simeq \tfrac{1}{2}zc(a)|_{T_m}. \tag{4.5}$$

Of course, it is known experimentally that there is some variation of E_V with temperature T from the cold solid value in Eq. (4.4) to the hot solid value at $T = T_m$ given in Eq. (2.11). Although to the accuracy of the evaluations in the present paper this variation is unimportant, for reasons which will become clear presently, we allow for it by adding (-1) to the right-hand side of Eq. (4.5). Then using Eq. (2.11), one obtains a potentially useful though of course approximate relation

$$zc(a)|_{T_m} \simeq [\tilde{c}(0) - c(r=0) - 1], \tag{4.6}$$

where z is the coordination number. Note that Eq. (4.6) is exact for the hard sphere Percus–Yevick solution. Since it is still quite difficult from experimental diffraction studies to plot $c(r)$ vs r for all r, the relation at T_m between the maximum value $c(a)$ of $c(r)$, the $r=0$ value, and the volume integral of $c(r)$ may be valuable as an additional constraint in liquid structural studies at the melting point.

ACKNOWLEDGMENTS

One of us (NHM) wishes to thank the University of Alberta for the generous award of a Visiting Professorship in the Physics Department, which made possible his contribution to this work. ABB wishes to acknowledge partial supporty by the Natural Sciences and Engineering Research Council of Canada.

[1]P. Varotsos and K. Alexopoulos, Phys. Rev. B **15**, 4111 (1977); P. Varotsos, W. Ludwig and K. Alexopoulos, *ibid.* B **18**, 2683 (1978); P. Varotsos, J. Phys. F **10**, 571 (1980).
[2]A. J. E. Foreman and A. B. Lidiard, Philos. Mag. **8**, 97 (1963).
[3]T. E. Faber, *Introduction to the Theory of Liquid Metals* (Cambridge University, Cambridge, 1972).
[4]P. Minchin, A. Meyer, and W. H. Young, J. Phys. F **4**, 2117 (1974).
[5]M. Parrinello, M. P. Tosi, and N. H. March, Proc. R. Soc. London Ser. A **341**, 91 (1974).
[6]M. W. Johnson, N. H. March, D. I. Page, M. Parrinello, and M. P. Tosi, J. Phys. C **8**, 751 (1975).
[7]H. C. Longuet-Higgins, Proc. R. Soc. London **205**, 247 (1951).
[8]L. Verlet, Phys. Rev. **165**, 201 (1968).
[9]J. Woodhead-Galloway, T. Gaskell, and N. H. March, J. Phys. C **1**, 271 (1968).
[10]M. S. Wertheim, Phys. Rev. Lett. **10**, 321 (1963).
[11]E. J. Thiele, J. Chem. Phys. **38**, 1959 (1963).
[12]Y. L. Yarnell, M. J. Katz, R. G. Wenzel, and S. H. Koenig, Phys. Rev. A **7**, 2130 (1973).
[13]N. B. Street and L. A. K. Staveley, J. Chem. Phys. **55**, 2495 (1971).
[14]See for example, Fritz Reiche, *The Quantum Theory* (Methuen, London, 1922).

Phys. Chem. Liq., 1984, Vol. 13, pp. 313–316
0031-9104/84/1304–0313$18.50/0

Letter

Relation between Principal Peak Height, Position and Width of Structure Factor in Dense Monatomic Liquids

A. B. BHATIA

Department of Physics, University of Alberta, Edmonton, Alberta T6G 2J1 Canada

and

N. H. MARCH

Theoretical Chemistry Department, University of Oxford, 1 South Parks Road, Oxford OX1 3TG, England

(*Received September 16, 1983*)

In monatomic dense liquids, the condition that the pair function $g(r)$ vanishes at $r = 0$ is shown to lead to the approximate relation

$$S(q_m) = \text{constant}\,\frac{q_m}{\Delta q}\,;\ \text{constant} \simeq 0.3, \tag{I}$$

for the principal peak height, at position q_m, of the structure factor $S(q)$ in terms of the peak width Δq. In (I), $2\Delta q$ is defined precisely as the distance between the two adjacent nodes of $S(q) - 1$ which embrace the peak position q_m. Examples for liquid argon, sodium and potassium using measured diffraction data confirm the form (I), with constant $\simeq 3/8$.

First principles theory by which the pair function $g(r)$, or the structure factor $S(q)$, of a dense monatomic liquid, related by

$$g(r) - 1 = \frac{1}{(2\pi)^3\rho}\int [S(q) - 1]e^{i\mathbf{q}\cdot\mathbf{r}}d^3q \tag{1}$$

can be calculated from a specified force law continues to prove somewhat intractable. Therefore, it remains of interest to examine what regularities are exhibited by structure factors $S(q)$ which are available for a variety of dense liquids from diffraction experiments. Here, we shall concern ourselves exclusively with data on argon at 85 K, and on liquid metallic sodium and potassium, each in two thermodynamic states.

In classical liquids, the pair function must satisfy $g(0) = 0$, which then yields according to eqn. (1)

$$-2\pi^2 \frac{N}{V} \equiv -2\pi^2 \rho = \int_0^{\infty} [S(q) - 1] q^2 dq, \tag{2}$$

with N atoms in volume V. An approximate evaluation of the integral (2) in such dense liquids as cited above can be carried out as follows. Let q_m denote the position of the principal peak of $S(q)$. Suppose the peak width, $2\Delta q$ say, is to be measured by the distance between the two adjacent nodes of $S(q) - 1$ which embrace q_m, and that any asymmetry of the peak about q_m is neglected. If we now write eqn. (2) as

$$-2\pi^2 \rho = \int_0^{q_m - \Delta q} [S(q) - 1] q^2 dq + \int_{q_m - \Delta q}^{q_m + \Delta q} [S(q) - 1] q^2 dq + \int_{q_m + \Delta q}^{\infty} [S(q) - 1] q^2 dq, \tag{3}$$

then for an $S(q)$ appropriate to dense fluids, such as argon near the triple point, or Na and K near the melting point, the following approximations are reasonable:

(i) To replace $S(q) - 1$ by -1 over the range of the first integral in eqn. (3).

(ii) To neglect the third integral in eqn. (3), because of the oscillations about zero of $S(q) - 1$.

(iii) To estimate the second integral by the triangular area

$$[S(q_m) - 1] q_m^2 \Delta q.$$

Using these simplifications and introducing the mean interatomic separation R_A through $\rho = 3/4\pi R_A^3$ it is readily shown that

$$S(q_m) q_m^2 \Delta q \doteqdot \frac{1}{3} q_m^3 \left(1 - \frac{9\pi}{2} \frac{1}{(R_A q_m)^3}\right). \tag{4}$$

Empirically, in dense liquids, $R_A q_M \simeq 4.4$ and the second term in the round bracket in eqn. (4) contributes 0.15 compared with unity. Thus, one is left with the result

$$S(q_m) \sim 0.3 \frac{q_m}{\Delta q}. \tag{5}$$

When we confront the approximate prediction (5) with the accurate diffraction data of Yarnell *et al*[1] on liquid argon at 85 K we find $S(q_m) = 2.70$, $q_m = 2.00$ and $\Delta q = 0.275$ (all q_m, Δq in $A^{\circ -1}$), yielding $S(q_m)/(q_m/\Delta q) = 0.37$, which is nearer to 3/8 than the predicted 0.3 in eqn. (5). It is satisfying that the data of Greenfield *et al*[2] on liquid potassium at 65 C yields $S(q_m) = 2.73$, $q_m = 1.62$, $\Delta q = 0.225$ and hence a constant of 0.38, while for the experiment at 135 C, $S(q_m) = 2.51$, $q_m = 1.62$, $\Delta q = 0.24$, and the constant in eqn. (5) is 0.37. For Na at 100 C, $S(q_m) = 2.80$, $q_m = 2.02$, $\Delta q = 0.27$, the constant being 0.37 while at 200 C, $S(q_m) = 2.46$, $q_m = 2.00$, $\Delta q = 0.29$, and the constant is 0.36. Thus, for the five experiments we find eqn. (5) to be quantitative when 0.3 is replaced by 3/8. The fact $3/8 > 0.3$ seems to indicate that the third integral actually has a non-zero negative value.

Turning to $g(r)$, it is worth noting that an argument in which $S(0)$ is evaluated via $\int_0^\infty [g(r) - 1]r^2\, dr$ can be carried out with assumptions made paralleling (i) to (iii) above for calculating $g(0)$. We merely record here the result:

$$g(r_m)r_m^2 \Delta r = \tfrac{1}{3}r_m^3 - \tfrac{1}{3}R_A^3(1 - S(0)), \tag{6}$$

with definitions paralleling precisely those for $S(q)$. Since $g(r)$ is less readily accessible, we shall not comment further on eqn. (6) as it stands. However, for the data of Yarnell *et al* on liquid argon at 85 K, we find

$$g(r_m) = 3.05,\ \Delta r = 0.545\ A^\circ,\ r_m = 3.68\ A^\circ \tag{7}$$

and if we form $r_m/\Delta r$ we find 6.7, to be compared with $q_m/\Delta q = 7.2$. Thus very approximately

$$\frac{q_m}{\Delta q} \simeq \frac{r_m}{\Delta r}. \tag{8}$$

Though the main aim of this letter has now been achieved, eqns. (5) and (8) together prompt us finally to comment on the criterion, which it seems was first pointed out by Verlet, that simple liquids like argon freeze when $S(q_m) \simeq 2.8$. Ferraz and March,[3] using the one-component plasma model as starting point, drew attention to the fact that simple liquid metals, and in particular Na and K, also freeze when $S(q_m) \simeq 2.7$. In the cases of Ar, Na and K where freezing involves only minor changes in local coordination, then, at the melting temperature T_m, use of $S(q_m)|_{T_m} = 2.8$ yields from eqns. (5) and (8) the estimate $(\Delta r/r_m)_{T_m} \sim 0.11$. But Lindemann's law of melting, according to Faber,[4] gives $(\Delta r/R_A)_{T_m} \sim 0.2$ if we identify here Δr as the root mean square displacement of the atoms. Since $r_m \sim 1.8\, R_A$ these results are seen to be roughly consistent. Thus, there is no conflict between freezing criteria based on $S(q_m)|_{T_m} = 2.8$ on the one hand and Lindemann's law on the other, even though the latter is sometimes interpreted, for simple liquid metals say, in terms of $S(0)|_{T_m}$ = constant. Somewhat more generally, any acceptable microscopic theory of dense fluids, not far from the triple point, will have to be in accord with the above approximate relation between peak height, position and width of the structure factor $S(q)$.

Acknowledgement

One of us (NHM) wishes to thank the University of Alberta for the generous award of a Visiting Professorship in the Physics Department, which made possible his contribution to this work. ABB wishes to acknowledge partial support by the Natural Sciences and Engineering Research Council of Canada.

References

1. J. L. Yarnell, M. J. Katz, R. G. Wenzel and S. H. Koenig, *Phys. Rev.*, **A7**, 2130 (1973).
2. A. J. Greenfield, J. Wellendorf and N. Wiser, *Phys. Rev.*, **A4**, 1607 (1971).
3. A. Ferraz and N. H. March, *Solid State Commun.*, **36**, 977 (1980).
4. T. E. Faber, *Introduction to the Theory of Liquid Metals*, (Cambridge: University Press) (1972).

Phys. Chem. Liq., 1989, Vol. 20, pp. 241–245
Reprints available directly from the publisher
Photocopying permitted by license only

ELECTRON CORRELATION, CHEMICAL BONDING AND THE METAL-INSULATOR TRANSITION IN EXPANDED FLUID ALKALIS

N. H. MARCH

Theoretical Chemistry Department, University of Oxford, 5 South Parks Road, Oxford OX1 3UB, England.

(Received 7 February 1989)

Experimental data which demands that one transcends the 'method of neutral pseudoatoms' for expanded alkali metals is first summarized. A chemical 'bond' provides a more basic building block than a pseudoatom. This 'bond' can be characterised by potential curves $^1\Sigma_u^+$ and $^3\Sigma_u^+$ of the appropriate free-space diatom (e.g. Na_2) plus local coordination properties. The relation to the metal-insulator transition is thereby discussed as the coordination number reduces to 2.

KEY WORDS: Coordination number, Peierls transition.

A simple picture, on which the 'nearly free electron' (NFE) theory of electrical conductivity of *s*-*p* liquid metals is based[1,2], views the conduction electrons in plane wave states as scattered by an array of weak spherical ionic pseudopotentials arranged in accord with the known short-range ionic order. While in no way denying the successes of this viewpoint, work on expanded fluid alkali metals has shown that its range of validity is severely limited. Thus departures from this picture can be demonstrated as follows:

i) No spherical pseudoatom model[3] can explain the diffraction data[4] on liquid Na[5] and K[6].

ii) No explanation is afforded by such a model of the results obtained by combining magnetic susceptibility[7] and Knight shift[8] data. This shows that the probability density at the (Cs) nucleus of Fermi level electrons relative to the free atom decreases as the density is lowered[9].

iii) The NFE electrical conductivity theory[1,2] breaks down severely as one moves along the liquid-vapour coexistence curves of the heavy alkalis Rb and Cs[10], towards the critical density d_c.

Chapman and March[11] have utilized the discontinuity, q say, at the Fermi surface of the momentum distribution $n(p)$ to interpret the maximum in the susceptibility of Cs, showing that at the corresponding density, q was 0.5 in the electron gas (jellium) model and about 0.2 in Cs at the maximum of the magnetic susceptibility, this being estimated from

$$k_B T_{\max} \sim q E_f \tag{1}$$

with E_f the Fermi energy. This small value of q is consistent with strong electron-electron correlations as the critical density d_c is approached, while the major difference from the jellium value testifies to the simultaneous importance of electron-ion interaction. In their work, Chapman and March[10,12] write the conductivity in terms of the first-order density matrix, and thereby develop an expansion of the data in a power series in q, around the metal-insulator transition. This approach seems to give the gist of the experimental data.

Because of the factors (i)–(iii) above, it is obviously necessary to adopt another approach than pseudoatoms to describe the expanded alkalis. The treatment below will be motivated by the model of March and Tosi[13], who showed that the use of sp^3 hybridization for normal liquid Na and K, combined with Pauling's resonating valence bond[14] approach, could allow an interpretation of the diffraction data discussed in Refs 3–5.

This chemical bonding point of view, as will be argued below, is now strongly supported by the extensive ionic structural data on the heavy alkalis along the coexistence curve, and especially on fluid Cs which will be a focal point below. The neutron scattering experiments at high temperatures and pressures thereby required have been reviewed very recently by Winter and Bodensteiner[15] as well as by Winter and Hensel[16]. The former workers write 'in comparison to the strong density dependence of the number of nearest neighbours, the distance of nearest neighbours remains almost constant during the expansion'.

An explanation of such behaviour is afforded, as partly anticipated by Warren and Mattheiss[9], by employing a quasicrystalline model of the liquid structure. Then they point out that the coordination number z varies almost linearly with density d and that useful models are given by $z = 8$ body-centred cubic (BCC), $z = 6$, simple cubic (SC) whereas for $z = 4$ both simple tetragonal and diamond structures were considered. Their data relating z and d has been extrapolated linearly to the critical density $d_c = 0.38 \text{g/cm}^3$, with a resulting coordination number near to $z = 2$. This will be argued to motivate a model of the metal-insulator transition below, but for the moment let us note that d_c at $z = 2$, plus the Warren-Mattheiss data[9] for z vs. d is well fitted (to $< 2\%$) by

$$d_{\text{liquid}} = az + b \tag{2}$$

where $a = 0.23$ and $b = -0.08$, both in g cm^{-3}.

The constancy of the near-neighbour distance along the Cs liquid–vapour coexistence curve suggests, when coupled with the work of March and Tosi[13] on liquid Na and K, a chemical 'bond' as a basic building block and this is the point now to be pressed below.

Of course, in classical chemical arguments, one starts from an arrangement of nuclei, and then hybridizes appropriately: in Ref 13 one adopted a BCC arrangement of the ions and tetrahedral directionality plus resonance. This has prompted us to adopt the study of Malrieu *et al.*[17] of crystalline Na to the alkali fluids. This is not as drastic as it might sound, since the arguments of Ref. 17 rest, in fact, purely on consideration of first and second neighbours to a chosen Na ion. These workers, following ideas of Poshusta and Klein[18] on hydrogen, set up a Heisenberg Hamiltonian for Na which was characterized by the $^1\Sigma_g^+$ and $^3\Sigma_u^+$ potential curves of the free space diatom Na_2. These were taken, in fact from the semi-empirical study of Konawalow *et al.*[19] and were then used to calculate the energy as a function of near-neighbour distance r for both BCC and SC structures. Their results[17] can be summarized as follows:

$$E_{BCC}(r) = 4R(r) + 3R\left(\frac{2}{\sqrt{3}}r\right) - 4g(r)[1 + f(r) - 15f(r)^3] \tag{3}$$

where[18]

$$R(r) = E(^3\Sigma_u^+),\ g(r) = [E(^3\Sigma_u^+) - E(^1\Sigma_g^+)]/2,$$

with

$$f_{BCC}(r) = \frac{g(r)}{14g(r) - 12g\left(\frac{2}{\sqrt{3}}r\right)} \tag{4}$$

while

$$E_{SC}(r) = 3R(r) + 4R(\sqrt{2}r) - 3g(r)[1 + f(r) - 11f(r)^3] \tag{5}$$

with

$$f_{SC}(r) = \frac{g(r)}{[10g(r) - 16g(\sqrt{2}r)]} \tag{6}$$

It is noteworthy that when $E_{BCC}(r)$ and $E_{SC}(r)$ above are used to calculate the equilibrium near-neighbour distance one obtains for Na the values[17] $r_{BCC} = 7.07$a.u and $r_{SC} = 6.82$a.u which are remarkably close in spite of the very different densities. The experimental value for normal BCC Na is 6.99a.u.

Equations (3) and (5) respectively can be viewed as characterizing the 'bonds' in BCC and SC structures by the properties of the diatom Na_2 determining $R(r)$ and $g(r)$, plus the near-neighbour and second neighbour coordination, the equilibrium r_{BCC} being a factor of 1.2 larger than the free-space molecular bond length.

This characterization in terms of the free-space diatom Na_2 now prompts one to return to the result (2) for fluid Cs, and to discuss its relation to the metal-insulator transition. First, it is helpful, in contemplating the way that the two-component liquid metal built from Cs^+ ions and electrons e^- could eventually have structure with coordination number $z = 2$, to refer to the classical counterpart of a mixture of charged hard spheres and neutralizing point ions studied by Gillan *et al.*[20] by Monte Carlo simulation. For strong coupling, it is demonstrated there that polymerization can occur.

Returning to the quantum-mechanical case of a linear chain of, say, Cs ions, one has from Peierls theorem that there will be a bond length distortion, since a one-dimensional metal cannot exist. Here then is a mechanism for the metal-insulator transition, which can be expected to coincide with the critical point. This seems to be consistent with presently available data.

In summary, a metallic 'bond', defined by Eqs. (3) and (5) for BCC and SC coordination, emerges as a valuable building block for treating the expanded alkali metals. It relates these directly to properties of the free-space diatoms through the potential curves of the $^3\Sigma_u^+$ and $^1\Sigma_g^+$ states, as well as to local coordination. It emerges that, through Eq. (2), the metal-insulator transition can be understood as occurring at coordination number 2, representing a linear chain. It is then argued that a Peierls distortion will occur, leading to alkali diatoms at a free-space equilibrium bond length which is reduced from that of the metallic 'bond'. This is consistent with what is known about dense Cs vapor which consists of atoms and diatoms[7] whose ground states are spin singlets.

Evidently future structural studies using X-rays and electrons[4] to complement existing neutron data along the coexistence curves of the fluid alkalis will be of considerable interest. If the present viewpoint is supported, one can expect in the difference diffraction data a changing pattern as the ionic coordination is altered by moving towards the critical point. But with presently available neutron data, kindly supplied by Professor F. Hensel in advance of publication, J. Ascough and the writer are presently extracting ionic interaction potentials by inverting the neutron structure data along the coexistence curve for Cs, employing the method proposed by Johnson and March[21], which has recently been brought to full fruition for normal liquid Na by Reatto[22] and coworkers.

Acknowledgements

This work was started during the latter part of a visit to IBM, San Jose, in the Winter of 1988. Thanks are due to Drs F. Herman and J. B. Torrance for their kind hospitality during this visit. The work was completed at UCSD, La Jolla and grateful thanks are due to Professors H. T. Hammel and A. A. Yayanos of the Scripps Institution of Oceanography for making this stay possible. Finally, the writer is most grateful to Richard Chapman and John Ascough for many valuable discussions on the area embraced by this work.

References

1. J. M. Ziman, *Phil. Mag.*, **6**, 1013 (1961).
2. K. S. Krishnan and A. B. Bhatia, *Nature (London)* **156**, 503 (1945).
3. J. M. Ziman, *Adv. Phys.*, **13**, 89 (1964).
4. P. A. Egelstaff, N. H. March and N. C. McGill, *Can. Jour. Phys.*, **52**, 1651 (1974).
5. P. J. Dobson, *J. Phys.*, **C11**, L295 (1978).
6. M. W. Johnson, private communication and to be published (1989).
7. W. Freyland, *Phys. Rev.*, **B20**, 5104 (1979).
8. W. W. Warren, U.El-Hanany and G. F. Brennert in Proc 5th Int. Conf. Liquids and Amorphous Metals, *J. Non-Cryst. Solids*, **61-62**, 23 (1984).
9. See also W. W. Warren and L. F. Mattheiss, *Phys. Rev.*, **B30**, 3103 (1984).
10. R. G. Chapman and N. H. March, to be published (1989).
11. R. G. Chapman and N. H. March, *Phys. Rev.*, **B38**, 792 (1988).
12. See also N. H. March, *Phys. Chem. Liquids* (in press) 1989.
13. N. H. March and M. P. Tosi, *Phys. Chem. Liquids*, **10**, 113 (1981).
14. L. Pauling, *Proc. Nat. Acad. Sci.* **39**, 551 (1953).
15. R. Winter and T. Bodensteiner, *High Pressure Research*, **1**, 23 (1988).
16. R. Winter and F. Hensel, *Phys. Chem. Liquids* **20**, 1 (1989).

17. J. P. Malrieu, D. Maynau and J. P. Daudey, *Phys. Rev.*, **B30**, 1817 (1984).
18. R. D. Poshusta and D. J. Klein, *Phys. Rev. Lett.*, **48**, 1555 (1982).
19. D. D. Konowalow, M. E. Rosenkrantz and M. L. Olson, *J. Chem. Phys.*, **72**, 2612 (1980).
20. M. J. Gillan, B. Larsen, M. P. Tosi and N. H. March, *J. Phys.*, **C9**, 889 (1976).
21. M. D. Johnson and N. H. March, *Phys. Letters*, **3**, 314 (1963).
22. L. Reatto, *Phil. Mag.*, **58**, 37 (1988).

PHYSICAL REVIEW B VOLUME 38, NUMBER 1 1 JULY 1988

Magnetic susceptibility of expanded fluid alkali metals

R. G. Chapman and N. H. March
Theoretical Chemistry Department, University of Oxford, 5 South Parks Road, Oxford OX1 3UB, England
(Received 18 November 1987; revised manuscript received 17 February 1988)

From a phenomenological treatment of a correlation-induced metal-insulator transition at elevated temperatures, an interpretation of the behavior of the magnetic susceptibility of expanded fluid alkali metals is proposed. Renormalization of the Fermi temperature due to correlation is shown to play an essential role. An important conclusion is that the momentum distribution at the Fermi surface is quantitatively very different from that in jellium at the same density.

There has been much interest, both theoretical and experimental, in the magnetic properties of highly correlated systems. In particular, expanded alkali metals provide excellent subjects for the investigation of the influence of electron correlation on electrical and magnetic properties, and its role in the metal-insulator transition, which, for these materials, is believed to occur near to the liquid-vapor critical region.[1] The experiments of Freyland and co-workers[2-4] on the magnetic susceptibilities of alkali metals along the liquid-vapor coexistence curve have stimulated much interest, and it is on these that this note is focused. Warren[5,6] extracted the paramagnetic contribution to the total susceptibility due to the conduction electrons, and pointed out that the behavior in the low-density liquid region is consistent with a correlation-enhanced Pauli paramagnetism, limited by the free-spin Curie value. This is illustrated in Fig. 1, which incorporates the recent determination of the liquid-vapor coexistence curve for cesium by Jüngst, Knuth, and Hensel.[7] In what follows, we consider the behavior of the magnetic susceptibility in the vicinity of the metal-insulator transition, initially by means of a phenomenological description, and then by utilizing a microscopic theory developed for heavy-electron systems.

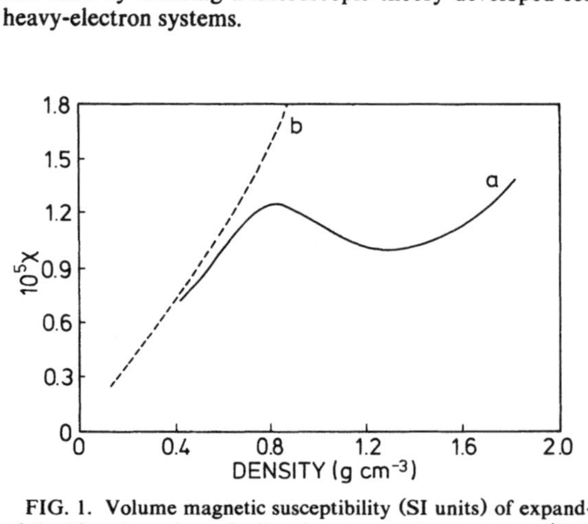

FIG. 1. Volume magnetic susceptibility (SI units) of expanded liquid cesium along the liquid-vapor coexistence curve (Ref. 6). Full curve a is based on the experimental data of Freyland (Ref. 2), and dashed curve b indicates the Curie limit for free spins. The critical density is 0.379 g cm^{-3} (Ref. 7).

The starting point of the present description is the phenomenological treatment of a correlation-induced metal-insulator transition at absolute zero $T=0$, due to March, Suzuki, and Parrinello.[8] Following Suzuki,[9] these authors wrote the ground-state energy per atom of a half-filled band near to the metal-insulator transition (in the metallic phase) as an expansion in the magnetization per atom m and the quasiparticle renormalization factor q. The latter is the discontinuity in the single-particle occupation number at the Fermi surface, which decreases continuously to zero at the transition. Thus

$$E(m,q)=E_0+am^2+\cdots+bq+cq^2+\cdots+eqm^2+\cdots . \tag{1}$$

The volume magnetic susceptibility χ is then

$$\chi=\frac{n_0\mu_0\mu_B^2}{2(a+eq)} , \tag{2}$$

where n_0 is the number density. Hence if a is zero, or vanishes at least as fast as q on approaching the transition, the susceptibility is enhanced, as in the Brinkman-Rice model[10] for such a system. For that particular case, we may readily identify the coefficients in the expansion (1), and find $a=0$, $b=(U_c-U)/8$, $c=U/32$, and $e=[1+\frac{3}{2}\bar{\varepsilon}N(\varepsilon_F)]/2N(\varepsilon_F)$, giving $\chi\sim q^{-1}$ as the authors originally noted. Here U is the Hubbard onsite interaction, $N(\varepsilon_F)$ is the electronic density of states per atom at the Fermi energy, U_c is the value of U at the metal-insulator transition, and $\bar{\varepsilon}$ is the band energy in absence of correlation.

Let us consider next the situation for $T\neq 0$. A natural extension of Eq. (2) writes the free energy per atom at temperature T as

$$F(m,q,T)=E_0+a(T)m^2+\cdots+b(T)q+c(T)q^2+\cdots+e(T)qm^2+\cdots . \tag{3}$$

Here, of course, the interpretation of q in terms of the average number of doubly occupied sites (given explicitly in the Brinkman-Rice model) is more appropriate, since the discontinuity in the single-particle occupation number will not be such a well-defined quantity. Furthermore, as $T\to\infty$, we expect a reversion to Curie-like behavior in the susceptibility, as the degeneracy temperature of the

Fermi fluid is exceeded. This is achieved if the coefficient a in Eq. (3) is proportional to T, with the coefficient e much less strongly dependent on T, and remaining finite as $T \to 0$. Thus if $a = \alpha T$, then as $T \to \infty$, $\chi \to n_0\mu_0\mu_B^2/2\alpha T$, and for lower temperatures χ is always less than this limiting value, for all densities. If $\alpha = \frac{1}{2}k_B$, this is simply the Curie law for the electrons. As $T \to 0$ we have $\chi \to n_0\mu_0\mu_B^2/2eq$, and require $1/2e \to N(\varepsilon_F)$ in the limit of high density ($q \to 1$), if we are to regain the Pauli susceptibility. Thus we expect a gradual transition between these limiting behaviors, when $a \approx eq$, i.e., $2eq \approx k_BT$.

To proceed further, we draw upon the work of Rice, Ueda, Ott, and Rudigier,[11] who extended the original Brinkman-Rice model to finite temperatures, in order to describe the normal-state properties of heavy-electron systems. For a half-filled band, they wrote the free energy per atom as

$$F = \sum_{\mathbf{k}\sigma} q_\sigma \varepsilon_{\mathbf{k}} n_{\mathbf{k}\sigma} + Ud + k_BT \sum_{\mathbf{k}\sigma} w_{\mathbf{k}} [n_{\mathbf{k}\sigma} \ln n_{\mathbf{k}\sigma} + (1-n_{\mathbf{k}\sigma})\ln(1-n_{\mathbf{k}\sigma})] , \tag{4}$$

where $n_{\mathbf{k}\sigma} = \{1+\exp[q(\varepsilon_{\mathbf{k}}-\mu)/w_{\mathbf{k}}k_BT]\}^{-1}$, $w_{\mathbf{k}}$ is a renormalization factor in $\mathbf{k}$ space, introduced to account for the nonorthogonality of the quasiparticle states $n_{\mathbf{k}\sigma}$, and d is the average number of doubly occupied sites. Although clearly the details of $w_{\mathbf{k}}$ are not known, the constraints

$$\bar{w} = \sum_{\mathbf{k}} w_{\mathbf{k}} = [(\tfrac{1}{2}-d)\ln(\tfrac{1}{2}-d) + d\ln 2]/\ln 2 , \tag{5}$$

$$\bar{w}_{-1} = \sum_{\mathbf{k}} w_{\mathbf{k}}^{-1} = 2 , \tag{6}$$

$$w \to 1 \text{ as } k \to k_f , \tag{7}$$

hold for the situation considered here. With

$$q = \{[d(n_\uparrow - d)]^{1/2} + [d(n_\downarrow - d)]^{1/2}\}^2/n_\uparrow n_\downarrow , \tag{8}$$

the magnetic susceptibility may be obtained by expanding the free energy to $O(m^2)$:

$$\frac{n_0\mu_0\mu_B^2}{\chi} = 2q\left[1 - \frac{1}{4(1-2d)^2}\right]\sum_{\mathbf{k},\sigma}\varepsilon_{\mathbf{k}} n_{\mathbf{k}\sigma} + k_BT\left[\sum_{\mathbf{k},\sigma} w_{\mathbf{k}}^{-1} n_{\mathbf{k}\sigma}(1-n_{\mathbf{k}\sigma})\right]^{-1} . \tag{9}$$

Now as $T \to 0$, the second term becomes $q/N(\varepsilon_f)$, and writing $\bar{\varepsilon} = \sum_{\mathbf{k}\sigma} \varepsilon_{\mathbf{k}} n_{\mathbf{k}\sigma}$ in this limit, then

$$\chi = \frac{n_0\mu_0\mu_B^2 N(\varepsilon_f)}{q}\left[1 + 2\bar{\varepsilon}N(\varepsilon_f)\left[1 - \frac{1}{4(1-2d)^2}\right]\right]^{-1} , \tag{10}$$

regaining the Brinkman-Rice result.[10] For $k_BT \gg q\varepsilon_f$, then $n_{\mathbf{k}\sigma} \approx \frac{1}{2}$, so that

$$\chi = \frac{n_0\mu_0\mu_B^2}{k_BT} , \tag{11}$$

using the value for $\bar{w}_{-1}$ from Eq. (6). Hence there is a crossover between enhanced Pauli paramagnetism and Curie behavior, for $k_BT \approx q\varepsilon_f$, and the structure of the theory is as suggested earlier in the phenomenological description. We may also note that for the low-temperature, high-density region of Fig. 1, Eq. (10) provides a mechanism for the observed enhancement of the susceptibility over the Pauli value and thus the upturn in the curve. Here, we may assume that the reduction of d from its uncorrelated value of $\frac{1}{4}$ is small. Writing $d = \frac{1}{4} - \delta$, where δ is small and positive, we find from Eq. (8) that $q = 1$ to $O(\delta)$, and on expanding the bracket in Eq. (10) similarly, we obtain

$$\chi = \frac{n_0\mu_0\mu_B^2 N(\varepsilon_F)}{1 + 16\bar{\varepsilon}N(\varepsilon_F)\delta} . \tag{12}$$

Since $\bar{\varepsilon} < 0$, χ shows a Stoner-like enhancement, the effect of which will tend to diminish as the density falls, while at the same time the $1/q$ factor becomes more important.

Referring to Fig. 1, then at the maximum in the susceptibility curve, as Warren[5] pointed out, the enhanced Pauli susceptibility appears to be restricted by the Curie limit. At this point, $\rho \approx 0.8$ g cm^{-3} and $T \approx 1780$ K, giving $q \approx 0.18$ (i.e., $d \approx 0.02$). Thus the renormalization of the electron degeneracy temperature due to the effects of correlation is strongly marked. More generally, we will expect a crossover region in the behavior of the susceptibility to the high-density side of the metal-insulator transition, as indicated schematically in Fig. 2. The experimental data for rubidium and sodium,[3,4] which do not show a maximum like those for cesium,[2] do not extend to low enough densities for this to be seen, although we should expect them to before the metal-insulator transition is reached. We note, of course, that near to the liquid-vapor critical region, the susceptibility is likely to be subject to diamagnetic corrections, arising from the presence of aggregate species.

In summary, it is shown that the magnetic-

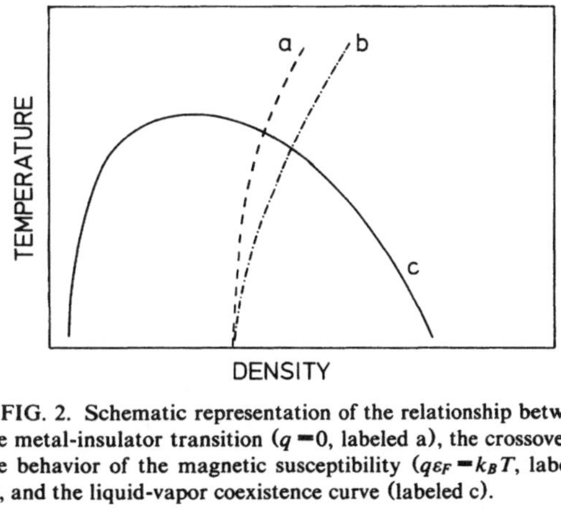

FIG. 2. Schematic representation of the relationship between the metal-insulator transition ($q = 0$, labeled a), the crossover in the behavior of the magnetic susceptibility ($q\varepsilon_F = k_BT$, labeled b), and the liquid-vapor coexistence curve (labeled c).

susceptibility data on expanded fluid cesium can be interpreted by means of a phenomenological treatment, developed from that given by March, Suzuki, and Parrinello[8] for $T=0$. This is supported, at a microscopic level, by adapting the work of Rice *et al.*[11] on heavy-electron systems. If the peak in the susceptibility-density plot for cesium along the coexistence curve is interpreted as the limitation of enhanced Pauli behavior and consequent crossover to a Curie-like regime, then q at this point is ≈ 0.18. This figure constitutes an upper bound for the value of q at the metal-insulator transition, and it is of interest here to note the marked difference in behavior from the jellium model. There, $q \approx 0.53$ at the corresponding density,[12] and the metal-insulator transition occurs at a much lower density.[13] It is evident that the influence of electron-ion interaction in the real system is substantial.

One of us (R.G.C.) wishes to acknowledge partial support by United Kingdom Science and Engineering Research Council (SERC) during the course of this work. We are grateful to Dr. W. W. Warren, Jr., for correspondence, and for kindly providing the data for Fig. 1.

[1]See, for example, F. Yonezawa and T. Ogawa, Prog. Theor. Phys. Suppl. **72**, 1 (1982).

[2]W. Freyland, Phys. Rev. B **20**, 5104 (1979).

[3]W. Freyland, J. Phys. (Paris) Colloq. **41**, C8-74 (1980).

[4]L. Bottyan, R. Dupree, and W. Freyland, J. Phys. F **13**, L173 (1983).

[5]W. W. Warren, Phys. Rev. B **29**, 7012 (1984).

[6]W. W. Warren, in *Amorphous and Liquid Metals,* edited by E. Lücher, G. Fritsch, and G. Jacucci, NATO Advanced Study Institute, Series E118 (Martinus Nijhoff, Dordrecht, 1987), p. 304.

[7]S. Jüngst, B. Knuth, and F. Hensel, Phys. Rev. Lett. **55**, 2160 (1985).

[8]N. H. March, M. Suzuki, and M. Parrinello, Phys. Rev. B **19**, 2027 (1979).

[9]M. Suzuki, Prog. Theor. Phys. **58**, 1151 (1977).

[10]W. F. Brinkman and T. M. Rice, Phys. Rev. B **2**, 4302 (1970).

[11]T. M. Rice, K. Ueda, H. R. Ott, and H. Rudigier, Phys. Rev. B **31**, 594 (1985).

[12]L. J. Lantto, Phys. Rev. B **22**, 1380 (1980).

[13]D. M. Ceperley and B. J. Alder, Phys. Rev. Lett. **45**, 566 (1980).

Phys. Chem. Liq., 1990, Vol. 21, pp. 251–255
Reprints available directly from the publisher
Photocopying permitted by license only

LETTER
Structure Inversion: Pair Potentials with Common Characteristics from Three Theories at Low Density on Liquid–Vapour Coexistence Curve of Cs

J. A. ASCOUGH and N. H. MARCH

Theoretical Chemistry Department, University of Oxford, 5 South Parks Road, Oxford OX1 3UB, UK.

(Received 2 January 1990)

Pair potentials with common features are shown to follow from three approximate theories for a low density state studied experimentally by diffraction in the work of Hensel *et al.* along the liquid–vapour coexistence curve of metallic fluid Cs. The reasons for this are exposed.

KEY WORDS: Ornstein–Zernike direct correlation function. Inverse problem.

The proposal of Johnson and March[1] to extract the pair potential $\phi(r)$ from diffraction data yielding the liquid structure factor $S(k)$ has been recently brought to full fruition for Na at the melting point by Reatto and coworkers[2–4]. Their potential has all the characteristics found in a subsequent, all electron, calculation on liquid Na by Perrot and March[5]. The work of Refs 2–4 demonstrates clearly that one approximate structural theory, the modified hypernetted chain (HNC) method, is not quantitative at the melting point of Na, and therefore must be refined by iteration using computer simulation.

Our interest here is in another, but heavier, alkali, Cs, and we have no reason to doubt that the same conclusion will obtain: namely that available approximate structural theories will not be sufficiently accurate for quantitative work at the melting point. However, Hensel and coworkers[6] have measured $S(k)$ at various thermodynamic states along the liquid–vapour coexistence curve. What will be shown below is that three approximate structural theories, applied to a low density state of Cs examined by Hensel *et al.*[6], all yield pair potentials $\phi(r)$ with common features, and the reasons for this will be exposed.

Let us take as starting point the approximate method proposed by de Angelis and March[7] for extracting the pair potential $\phi(r)$ from the measured structure factor $S(k)$ and the Fourier transform of $(S(k) - 1)$, namely $g(r) - 1$, with $g(r)$ the usual liquid pair function. Then, writing $g(r)$ for the ions, which may be treated classically, in the Boltzmann form

$$g(r) = \exp(-U(r)/k_B T) \tag{1}$$

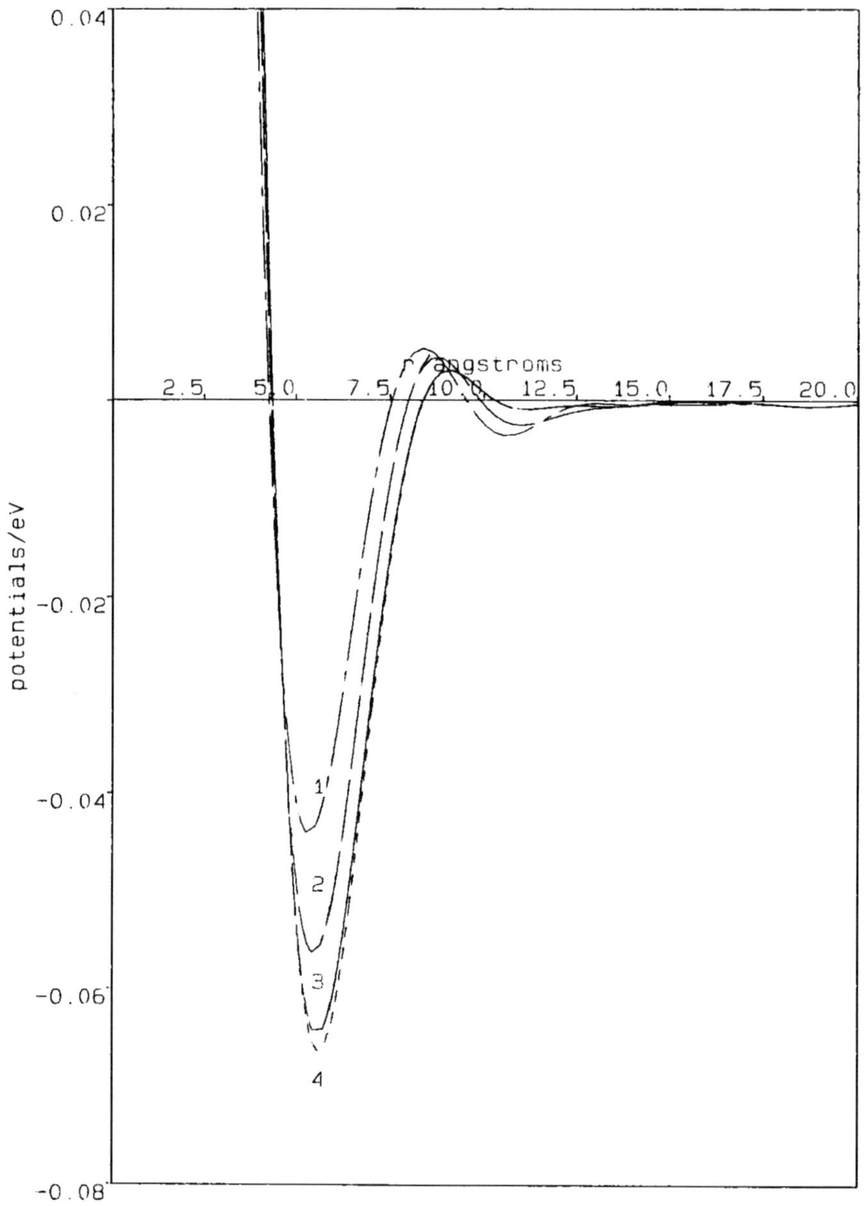

Figure 1 Curve 1. Potential of mean force $U(r)$ from experiment. This is for the thermodynamic state of fluid metallic Cs corresponding to density $\rho = 0.00416$ Å^{-3} and to temperature $T = 1923$ K. Curve 2. Pair potential $\phi(r)$ extracted from experiment via Eq. (2). Curve 3. $\phi(r)$ obtained from same data as used in curve 1, but via Eq. (3). Curve 4. $\phi(r)$ obtained using Percus–Yevick approximation[8].

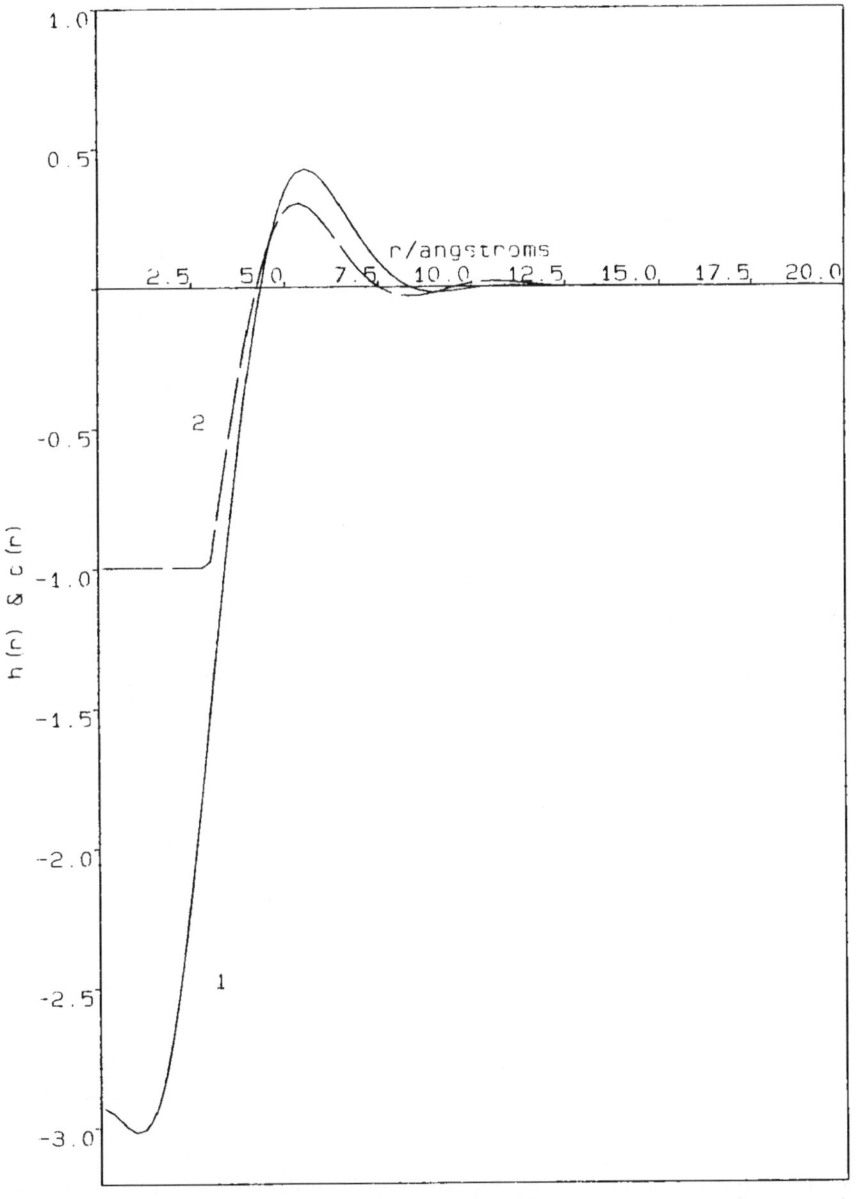

Figure 2 Curve 1. Ornstein–Zernike function for same state of fluid Cs as in Figure 1. This was calculated from diffraction data of Hensel and coworkers[6] by Fourier transform of $[S(k) - 1]/S(k)$. Curve 2. Shows total correlation function $h(r) = g(r) - 1$. This was calculated by Fourier transform of $[S(k) - 1]$, using again experimental data for the same thermodynamic state as in Figure 1.

the pair potential is related[7] to the potential of mean force $U(r)$ and the structure factor $S(k)$ by

$$\frac{\phi(r)}{k_B T} = \frac{U(r)}{k_B T} + \frac{1}{8\pi^3 \rho} \int [S(k) - 1]^2 \exp(i\mathbf{k} \cdot \mathbf{r}) \, d\mathbf{k}, \tag{2}$$

where ρ is the number density of ions. For the thermodynamic state of liquid Cs specified, $U(r)$ in curve 1 of Figure 1 has been derived from $-k_B T \ln g(r)$ according to Eq. (1), with $g(r)$ obtained by Fourier transform of the data of Hensel *et al.*[6]. With this same data, $U(r)$, according to Eq. (2), has to be "corrected" by the Fourier transform of $[S(k) - 1]^2$. The resulting pair potential $\phi(r)$ is displayed in curve 2 of Figure 1.

The hypernetted chain (HNC) theory in its original form gives

$$\frac{\phi(r) - U(r)}{k_B T} = h(r) - c(r) \tag{3}$$

where the total correlation function $h(r)$ is simply $g(r) - 1$ while $c(r)$ is the usual Ornstein–Zernike direct correlation function. This latter function $c(r)$ is plotted in curve 1 of Figure 2, while the corresponding $h(r)$ is shown in curve 2. Using these results in Eq. (3), with the $U(r)$ already utilized in curve 1 of Figure 1, the corresponding pair potential is drawn in curve 3 of that Figure. There is seen to be good overall agreement between the potentials extracted using Eqs (2) and (3). Thirdly, we have studied the inversion of the Percus-Yevick (PY) approximation[8]; the result being shown in curve 4 of Figure 1.

While none of the three approximate theories used above can be expected to lead to a quantitatively reliable pair potential at the melting point of Cs, all of them evidently yield pair potentials $\phi(r)$ at the low density thermodynamic state considered here with common characteristics. After studying the details of the calculations for the three cases, the reason for the agreement is found to lie in the fact that, in the regions of $\mathbf{r}$ space important in determining the main characteristics of $\phi(r)$, the quantity $h - c$ is small compared with $h(r)$*, except of course very near the nodes of this latter quantity.

Finally, it needs stressing that, although this work has focussed only on a low density metallic state of Cs, closer approach to the critical point will lead eventually to a situation where the form of $g(r)$ becomes dominated by critical fluctuations rather than by the pair force law. The approximate theories employed above are not then rich enough to embrace this regime. Therefore the limited range of validity of the procedure proposed here must be borne in mind. We are currently engaged, with Mr. C. M. Allison, in setting up computer simulation to fully test the pair potentials $\phi(r)$ in Figure 1. When this simulation has been completed, a full account of this work will be published.

* The weaker condition $|h - c|$ considerably less than unity leads to $\phi_{HNC}(r)$ in agreement with $\phi_{PY}(r)$.

Acknowledgment

We are most grateful to Professor F. Hensel for generously supplying us with the structural data in advance of its full publication. One of us (N.H.M) wishes to express his thanks to Drs F. Perrot and M.W.C. Dharma-wardana for numerous stimulating discussions on liquid metal pair potentials. Finally, the other of us (J.A.A) acknowledges the award of a postgraduate studentship by the Science and Engineering Research Council.

References

1. M. D. Johnson and N. H. March, *Phys. Lett.*, **3**, 313 (1963).
2. D. Levesque, J. J. Weis and L. Reatto, *Phys. Rev. Lett.*, **54**, 451 (1985).
3. L. Reatto, D. Levesque and J. J. Weis, *Phys. Rev.*, **A33**, 3451 (1986).
4. L. Reatto, *Phil. Mag.*, **A58**, 37 (1988).
5. F. Perrot and N. H. March, to be published (1990).
6. See, for example, R. Winter and F. Hensel, *Phys. Chem. Liquids*, **20**, 1 (1989).
7. U. de Angelis and N. H. March, *Phys. Lett.*, **56A**, 287 (1976).
8. J. K. Percus and G. J. Yevick, *Phys. Rev.*, **110**, 1 (1958).

PHYSICAL REVIEW A VOLUME 41, NUMBER 8 15 APRIL 1990

Pair potentials for liquid sodium near freezing from electron theory and from inversion of the measured structure factor

F. Perrot
Centre d'Etudes de Limeil-Valenton, 94195 Villeneuve St. Georges CEDEX, France

N. H. March
Theoretical Chemistry Department, University of Oxford, 5 South Parks Road, Oxford OX1 3UB, England

(Received 5 October 1989)

An effective pair potential for liquid sodium near freezing has been calculated from electron theory using the density-functional method. The main features of the potential extracted by Reatto, Levesque, and Weis [Phys. Rev. A **33**, 3451 (1986)] by inverting the measured structure factor of Greenfield, Wellendorf, and Wiser [Phys. Rev. A **4**, 1607 (1971)] are faithfully reflected by electron theory. To obtain precise agreement between the two methods will evidently require further progress in setting up nonlocal exchange and correlation functionals.

The proposal of Johnson and March[1] to invert the measured structure factor $S(k)$ of liquid metals to extract a pair potential $\phi(r)$ has recently been brought to full fruition for a particular thermodynamic state of liquid Na. Specifically, the x-ray diffraction data of Greenfield, Wellendorf, and Wiser[2] at $T=100\,°\mathrm{C}$ and density equal to 0.929 $\mathrm{g\,cm^{-3}}$ have been inverted by Reatto, Levesque, and Weis[3] to obtain $\phi(r)$ for this state. These workers employed an iterative predictor-corrector method in which the predictor was the modified hypernetted-chain approximation of liquid state structural theory, while the corrector was simulation. Reatto, Levesque, and Weis[3] verified the convergence of their method for a Lennard-Jones fluid and for a model potential for Al.

In parallel with this progress in extracting a pair force from structural data has been the development of methods for obtaining $\phi(r)$ based on electron theory. While work was carried out on the lightest alkali metal some time ago, both for a single Li ion embedded in a electron liquid of appropriate metallic density by Dagens, Rasolt, and Taylor[4] as well as for crystalline bcc Li by Perrin and co-workers,[5,6] which was of first-principles character, heavier alkali metals have hitherto been tackled using pseudopotentials, as in the work of Price, Singwi, and Tosi[7] on Rb.

In the case of Na, a detailed analysis of the charge density was reported by Rasolt and Taylor. These authors compared the density displaced by a Na ion in an electron gas, as obtained using nonlocal model potentials[8] or in a full nonlinear density-functional calculation done by Dagens.[9] Excellent agreement between these two methods was noticed when the parameters of the model potential were conveniently chosen.

The present calculations on liquid Na were based on a first-principles approach known as the neutral pseudo atom method.[9] The details will be published elsewhere.[10] Briefly, the calculations reported here consist of two distinct parts. First, a Kohn-Sham density functional calculation was performed for a Na nucleus, charge $11e$, embedded in the center of a cavity in a jellium background, of mean conduction electron density $\bar{n}$ corresponding to the mass density of 0.929 $\mathrm{g\,cm^{-3}}$. Naturally, an approximate form $V_{xc}(\mathbf{r})$ of the exchange-correlation contribution to the one-body potential $V(\mathbf{r})$, given by

$$V(\mathbf{r})=V_{\mathrm{Hartree}}(\mathbf{r})+V_{xc}(\mathbf{r})\ , \qquad (1)$$

must be adopted at the outset. Two such forms of $V_{xc}(\mathbf{r})$ were used in the present calculation, the first due to

TABLE I. Self-consistent density-functional calculation of the density displaced by an Na atom embedded in a cavity in jellium at electron density $\bar{n}=0.003\,604\,6$ a.u. The radius R_a of the cavity is such that $\frac{4}{3}\pi R_a^3\bar{n}=Z^*=1$.

	Exchange correlation	
	Ichimaru[a]	Hedin-Lundqvist[b]
Eigenvalues (Ry)		
E_{1s}	−74.9628	−74.9314
E_{2s}	−3.6336	−3.6293
E_{2p}	−1.6281	−1.6244
Average of r: $\langle r\rangle$ (a.u.)		
1s	0.1439	0.1439
2s	0.7916	0.7919
2p	0.8090	0.8096
Average of r^2: $\langle r^2\rangle$ (a.u.)		
1s	0.0279	0.0279
2s	0.7600	0.7609
2p	0.8578	0.8591
Total energy of embedding (Ry)		
ΔE	−322.8526	−322.8237
Friedel sum		
$\dot{Z}_F$	−0.000 495	−0.000 488
Total displaced charge (integrated in r space)		
Q	10.000 508	10.000 512

[a]Reference 11.
[b]Reference 12.

TABLE II. Characteristics of diffraction potential compared with electron theory form.

Positions of turning points and nodes (Å)	Principal minimum	First maximum	Second minimum	First node
Diffraction[a] $\phi(r)$	3.9	5.76	7.44	3.3
Electron theory $\phi(r)$	3.73	5.67	7.37	3.20

[a]Reference 13.

Ichimaru[11] and the second, the older form, due to Hedin and Lundqvist.[12]

Denoting the displaced charge around the Na nucleus in the jellium background with cavity by $\Delta n(r)$, Fig. 1 shows the radial density $D(r)=4\pi r^2\Delta n(r)$ for the two choices of V_{xc} in Eq. (1.1). Table I shows relevant data for one-electron eigenvalues, etc., in the two cases.

The second stage of the calculations involves the construction of the appropriate **q** space quantities $\overline{\Delta n}(q)$ and $w(q)$ needed to insert in the usual second-order perturbation formula for the pair interaction

$$\phi(r)=\frac{(Z^*e)^2}{r}+\int w(q)\overline{\Delta n}(q)\exp(-i\mathbf{q}\cdot\mathbf{r})d\mathbf{q} \ . \tag{2}$$

One of us describes elsewhere[10] in full detail the way $\overline{\Delta n}(q)$ is to be calculated from $\Delta n(r)$: care is needed to ensure that $\overline{\Delta n}(q)$ normalizes to the valence Z^*, which is unity for Na, to ensure precise cancellation with the Coulomb term in Eq. (1.2) at sufficiently large r. Given $\overline{\Delta n}(q)$, the (weak) potential $w(q)$ is calculated, by linear response theory,

$$w(q)=\frac{1}{\chi(q)}\overline{\Delta n}(q) \ , \tag{3}$$

$$\frac{1}{\chi(q)}=\frac{1}{\pi_0(q)}+\left[\frac{4\pi}{q^2}+X\right] , \tag{4}$$

where π_0 is the random-phase approximation (RPA) density response function and X the local field correction in the local density approximation for exchange and correlation effects. X is q independent and given by

$$X=\frac{dV_{xc}(\bar{n})}{d\bar{n}} \ . \tag{5}$$

The calculations reported here differ from those of Ref. 9 for the following reasons.

(i) We have used the Ichimaru (or Hedin-Lundqvist) exchange and correlation functional, instead of a pure Kohn-Sham exchange,

(ii) The density is that of the liquid ($\rho=0.929$ g cm^{-3}) instead of the solid density ($\rho=1.0067$ g cm^{-3}).

Nevertheless, our pair potential and those of Rasolt and Taylor are in qualitative agreement. The position of the first minimum is approximately the same in both theories. The main difference resides in the depth of the well: $\beta\phi$ ranges between -0.72 and -0.87 in Ref. 8 (depending on which model potential is used), and ours is $\beta\phi=-1.22$.

With the simplest nontrivial form of $\chi(q)$ involving $dV_{xc}(\bar{n})/d\bar{n}$, curve 1 of Fig. 2 was obtained. The choice of V_{xc} did not affect this curve to graphical accuracy. Table II records the characteristic features of this potential, where it is compared with the same features of the potential extracted by inverting the measured structure factor $S(k)$. All the trends of the potential obtained from the diffraction measurements are clear in the potential

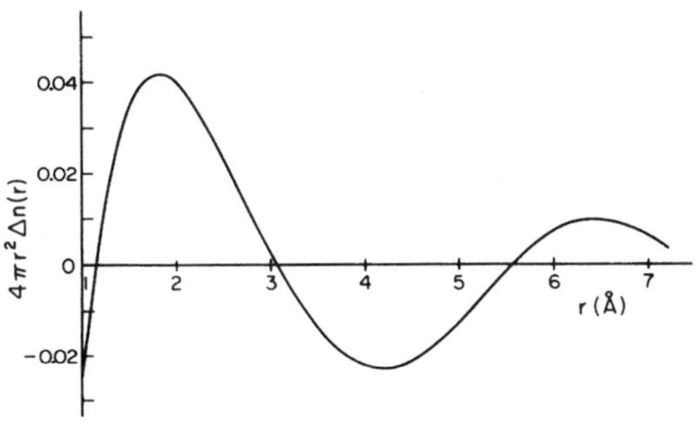

FIG. 1. Radial valence density $4\pi r^2\Delta n(r)$ for Na nucleus embedded in the center of a cavity in originally uniform electron liquid. For graphical accuracy, this is the same for either choice of $V_{xc}(r)$ in Eq. (1.1) for Ref. 11 or 12.

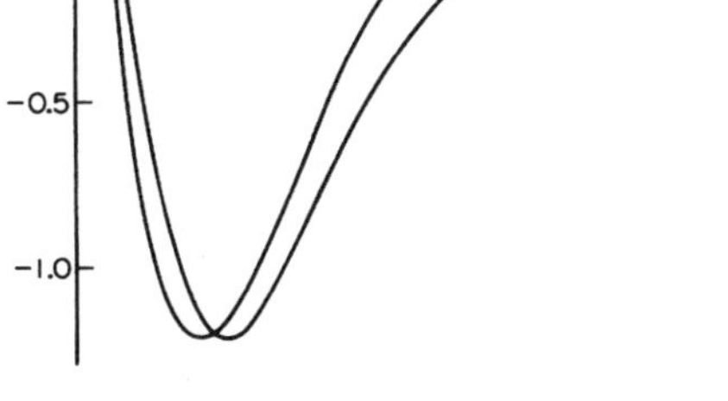

FIG. 2. Pair potentials for liquid Na at $T=100$ °C and density of 0.929 g cm^{-3}. Electron theory: upper curve for large r. Inversion of liquid structure factor, taken from Refs. 3 and 13: lower curve for larger r.

$\phi(r)$ calculated here from electron theory.

In conclusion, without any adjustable parameters electron theory gives a pair potential $\phi(r)$ at freezing for liquid Na in quite good general agreement with that extracted from the diffraction data of Greenfield, Wellendorf, and Wiser,[2] and by Reatto, Levesque, and Weis.[3] As for detailed comparison, the depth of $\phi(r)$ at the principal minimum is almost the same for the two potentials, though, as Table II shows, the theoretical $\phi(r)$ has its minimum a few tenths of an angstrom inside the diffraction form.[13] But the maxima and minima are practically at the same positions, though the diffraction $\phi(r)$ remains negative at larger r while the electron theory $\phi(r)$ just crosses the axis. On this latter point, we have checked that a modified $\chi(q)$ with a Hubbard-type correction would lead to a negative $\phi(r)$ at larger r, but at the expense of substantially deepening the principal minimum, this moving now to a position slightly outside the diffraction potential minimum. Thus the $\phi(r)$ extracted from $S(k)$ for liquid Na, while not fully quantitatively reproduced in detail by electron theory, nevertheless reflects all the trends predicted by that theory. Evidently, increasing effort to measure the structure factor $S(k)$ for liquid metals over the widest possible range of momentum transfer is now a matter of considerable importance.

This collaboration was made possible by our attendance at the workshop on atoms and ions in strongly coupled plasmas at the Institute for Theoretical Physics, University of California at Santa Barbara. It is a pleasure to thank the coordinators, Dr. J. Callaway, Dr. H. E. DeWitt, and Dr. M. H. Van Horn for the stimulating atmosphere they created and for numerous discussions, and Professor J. S. Langer and Professor J. R. Schrieffer for their generous hospitality.

[1]M. D. Johnson and N. H. March, Phys. Lett. **3**, 313 (1963).
[2]A. J. Greenfield, J. Wellendorf, and N. Wiser, Phys. Rev. A **4**, 1607 (1971).
[3]L. Reatto, D. Levesque, and J. J. Weis, Phys. Rev. A **33**, 3451 (1986).
[4]L. Dagens, M. Rasolt, and R. Taylor, Phys. Rev. B **11**, 2726 (1975).
[5]R. C. Perrin, R. Taylor, and N. H. March, J. Phys. F **5**, 1490 (1975).
[6]See also N. H. March, R. C. Perrin, and R. Taylor, Solid State Commun. **53**, 287 (1985).
[7]D. L. Price, K. S. Singwi, and M. P. Tosi, Phys. Rev. B **2**, 2983 (1970).
[8]M. Rasolt and R. Taylor, Phys. Rev. B **11**, 2717 (1975).
[9]L. Dagens, J. Phys. C **5**, 2333 (1972); J. Phys. (Paris) **34**, 879 (1973); **36**, 521 (1975).
[10]F. Perrot (unpublished).
[11]S. Ichimaru, Rev. Mod. Phys. **54**, 1017 (1982).
[12]L. Hedin and S. Lundqvist, Solid State Phys. **23**, 2 (1969).
[13]L. Reatto, Philos. Mag. A **58**, 37 (1988).

Phys. Chem. Liq., 1990, Vol. 22, pp. 133–148
Reprints available directly from the publisher
Photocopying permitted by license only

REVIEW ARTICLE
Information Content of Diffraction Experiments on Liquids and Amorphous Solids

N. H. MARCH

Theoretical Chemistry Department University of Oxford, 5 South Parks Road, Oxford OX1 3UB, England.

(Received 4 July 1990)

Starting with simple monatomic liquids such as condensed rare gases, the information that can be extracted from diffraction experiments is discussed. The rich interplay between atomic and electronic structure in liquid metals such as Na and K is emphasized: expanded Rb and Cs taken up the liquid–vapour coexistence curve towards the critical point are also mentioned. A liquid metal like Na is, in fact, a two-component system, Na^+ and electrons. Classical two-component systems are also considered, and specifically $BaCl_2$ which freezes into a fast-ion conducting phase. The structural theory of two-component liquids is extended to apply to amorphous Si, the 'second component' now being the bond charge. The way a combination of X-ray and electron diffraction can elucidate the nature of the local ordering, as well as the directional bonding, is stressed. There is clear evidence here for the bonding electrons, directly from the X-ray and electron scattering intensity.

KEY WORDS: Partial structure factors, directional bonding

1 INTRODUCTION

That the short-range order existing in liquids such as condensed rare gases could be extracted from X-ray diffraction experiments has been known for a long time. If we write the intensity I of X-rays of wavelength λ scattered through an angle 2Θ by a liquid sample containing N atoms:

$$I(k) = Nf^2(k)S(k):k = \frac{4\pi \sin \Theta}{\lambda} \tag{1.1}$$

then $S(k)$ is the liquid structure factor, related to the pair correlation function $g(r)$ by

$$S(k) = 1 + \rho \int [g(r) - 1] \exp(i\mathbf{k} \cdot \mathbf{r}) d\mathbf{r}, \tag{1.2}$$

with ρ the atomic number density. If one considers neutron rather than X-ray diffraction, then the atomic scattering factor $f(k)$ in Eq. (1.1) is replaced by a k independent neutron scattering length.

It is fair to say that all simple liquids near their triple point have remarkably similar forms of $S(k)$ and $g(r)$. In one sense, this is disappointing, as a hard sphere model can account for a lot of features. However, on closer inspection, departures from hard spheres become apparent and, we shall argue, it is these departures that are the matters of most interest for the chemical physics of condensed matter.

In Section 2 therefore, we shall summarize what we can learn from $S(k)$ measurements, taking as prototypes Ar and Na. Here theory is now relatively well developed, for such one-component liquids. Actually, in Section 5 we shall return to liquid Na, with some surprises to follow!

Section 3 then takes two examples of ionic melts: RbCl and $BaCl_2$, the latter, in the crystalline phase, being a fast-ion conductor: these also serve to introduce the structural description of two-component systems.

The other examples both involve electrons in disordered systems: probed directly by X-rays, and also usefully by electron diffraction. Cases treated are SiO_2 and amorphous Si in Section 4, and also in Section 5, as already remarked, liquid Na. Section 6 discusses some promising future directions for experiment and theory.

2 STRUCTURE AND FORCES IN SIMPLE MONATOMIC LIQUIDS NEAR TRIPLE POINT: e.g. Ar and Na

One can usefully group properties that are expected to arise in $S(k)$ and $g(r)$ according to first-principles liquid-state theory into three areas.

2.1 Small angle scattering

For van der Waals liquids such as Ar, the pair potential $\phi(r)$ at large r is expected to take the (non-retarded) form

$$\phi(r) = -\frac{c_6}{r^6} \tag{2.1}$$

This was shown by Enderby, Gaskell and March[1] to lead to a small k expansion of the structure factor $S(k)$ of the form

$$S(k) = S(0) + a_2 k^2 + a_3 k^3 + \cdots \tag{2.2}$$

From the theoretical result (true far from the critical point) that at sufficiently large r the Ornstein–Zernike direct correlation function $c(r)$ reflects the pair potential $\phi(r)$ through

$$c(r) = -\frac{\phi(r)}{k_B T}, \; r \text{ large} \tag{2.3}$$

one derives, using (2.1) the result for the coefficient of the k^3 term in Eq. (2.2):

$$a_3 = \pi^2 \rho \{S(0)\}^2 c_6 / 12 k_B T \tag{2.4}$$

and this result (2.4) has been tested by confrontation with the diffraction experiments of Yarnell[2] *et al.* for argon. The theoretical estimate of c_6 is in good accord with that extracted from Eq. (2.2): the work of Robinson and March[3] (see also Renne and Nijboer[4]) showing that a few percent reduction only in c_6 is to be expected because of the van der Waals interactions taking place in a dielectric liquid. The situation in liquid metals is more complex and is not accounted for by Eq. (2.2), as discussed by Matthai and March[5].

2.2 *Freezing and melting criteria*

The second feature we focus on is the so-called Verlet rule[6], that $S(k)$, as we cool down the liquid, approaches a maximum value at its principal peak of about 2.8, before freezing occurs. This, found by Verlet for Lennard-Jones liquids, also holds for simple metals, which, as pointed out by Ferraz and March[7], are usefully described around the principle peak by the classical one-component plasma model (OCP). This OCP freezes when $S(k)$ reaches a maximum of about 2.7, in accord with the X-ray experiments of Greenfield, Wellendorf and Wiser[8] on liquid Na and K near the melting temperature T_m. Bhatia and March[9] have shown that:

i) There is no conflict between the Verlet rule for freezing and the Lindemann law for melting, which in turn relates to

ii) The result, derived from the condition $g(r = 0) = 0$ in dense classical liquids:

$$S(k)_{\max} \doteqdot \frac{3}{8}\frac{k_{\max}}{\Delta k} \tag{2.5}$$

where $2\Delta k$ is the distance between the two adjacent nodes of $S(k) - 1$ embracing the principal peak. Silbert (private communication) has verified the approximate validity of Eq. (2.5) above from diffraction data on wide range of liquids.

2.3 *Inverse problem of extracting forces from measured structures*

An attractive possibility, pointed out some 25 years ago by Johnson and March[10], is to extract the entire pair potential $\phi(r)$ directly from diffraction measurements of $S(k)$. Their work is taken, nowadays, to define the 'inverse problem' of extracting forces from structure. This has recently been brought to fruition, especially by Levesque, Reatto and Weis[11] and also by Dharma-wardana and Aers[12] (for a fuller review, see March[13]). The idea is simply expressed. If we write, for a classical liquid $g(r)$ in the Boltzmann form

$$g(r_{12}) = \exp\left(-\frac{U(r_{12})}{k_B T}\right), \tag{2.6}$$

then the potential of mean force $U(r_{12})$ can be used to construct the total force $-\partial U(r_{12})/\partial \mathbf{r}_1$ acting for atom 1. In turn, this arises from the sum of the pair force $-\partial\phi(\mathbf{r}_{12})/\partial\mathbf{r}_1$ exerted by atom 2 at separation r_{12}, and the contribution from the other atoms. Thus, we can write

$$-\frac{\partial U(r_{12})}{\partial \mathbf{r}_1} = -\frac{\partial\phi(r_{12})}{\partial\mathbf{r}_1} - \rho\int \frac{g^{(3)}(\mathbf{r}_1\mathbf{r}_2\mathbf{r}_3)}{g(r_{12})}\frac{\partial\phi(r_{13})}{\partial\mathbf{r}_1}\,d\mathbf{r}_3, \tag{2.7}$$

the so-called force equation where the final term in Eq. (2.7) evidently involves the probability $g^{(3)}(\mathbf{r}_1\mathbf{r}_2\mathbf{r}_3)/g(r_{12})$ that a third atom is at $\mathbf{r}_3$ given atoms 1 and 2 at

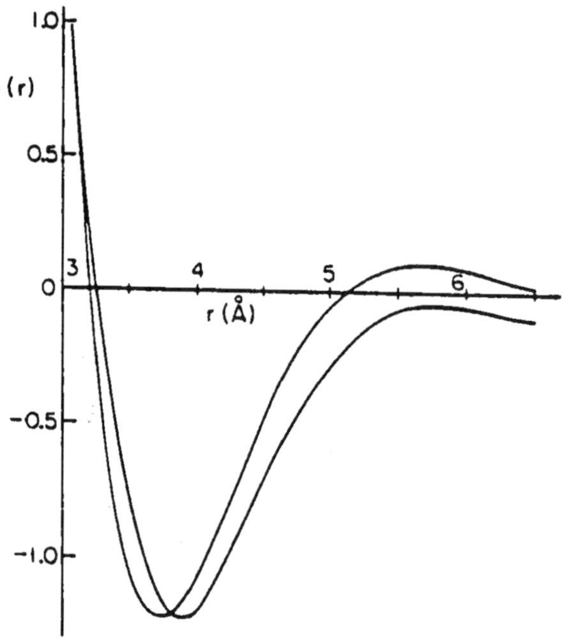

Figure 2.1 Pair potential $\phi(r)$ for liquid Na at melting point T_m in units of $\beta^{-1} = k_B T_m$. Lower curve for larger r is obtained by inversion of structure factor $S(k)$ and is taken from Refs. 11 and 14. Upper curve for large r is electron theory calculation of Perrot and March[14].

$\mathbf{r}_1$ and $\mathbf{r}_2$. Eq. (2.7) is an exact consequence of classical statistical mechanics for a liquid with pairwise additive forces. Evidently, since U is known from $g(r)$ through Eq. (2.6), $\phi(r)$ can be extracted only if knowledge of $g^{(3)}$ is assumed. By an iterative predictor-corrector technique in which the predictor is the modified hypernetted-chain method, while the corrector is simulation, Levesque, Reatto and Weis have adopted the proposal of Johnson and March and have successfully extracted $\phi(r)$ for Ar from simulation data and Na from diffraction measurements. This procedure is fully reviewed by Reatto[14]: the potential $\phi(r)$ thereby obtained from $S(k)$ for Na liquid near freezing is shown in Figure 2.1

It is worth pointing out that approximate analytical theories are still of considerable interest and these have taken two main directions:

i) Theories leading to

$$U(r) - \phi(r) = \int h(|\mathbf{r} - \mathbf{r}'|)F(\mathbf{r}')d\mathbf{r}', \tag{2.8}$$

which were proposed by the writer[15], and have been developed recently by Gaskell[16], and

ii) A theory, due to Barrat, Hansen and Pastore[17], based not on the Kirkwood decoupling of $g^{(3)}$ as a product of pair functions $g(r_{12})g(r_{23})g(r_{31})$ but of the three-body Ornstein–Zernike function

$$c^{(3)}(\mathbf{r}_1\mathbf{r}_2\mathbf{r}_3) = t(r_{12})t(r_{23})t(r_{31}) \tag{2.9}$$

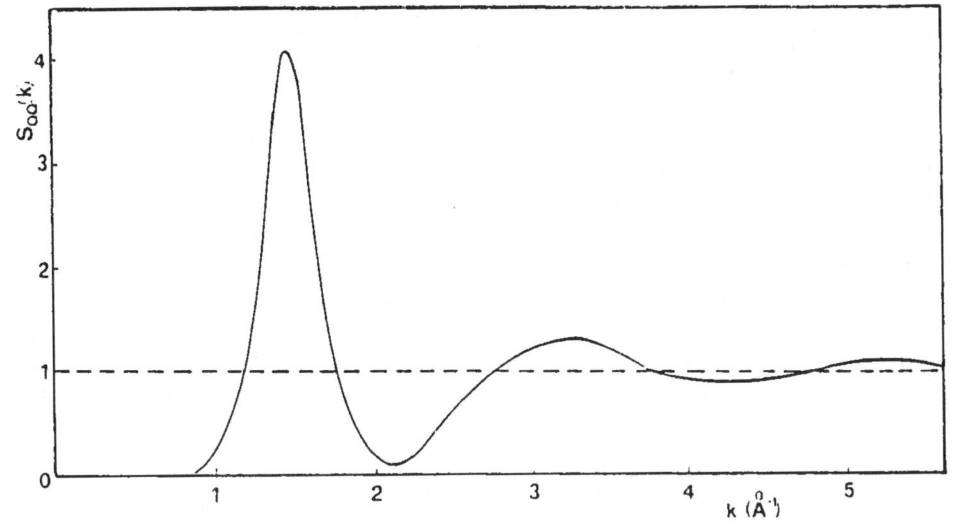

Figure 3.1 Charge–charge structure factor for molten RbCl, constructed from neutron diffraction data on partial structure factors (after March and Tosi[23]).

Here t is chosen empirically to yield correctly $c^{(3)}$ from the density derivative $\partial c/\partial \rho$. This assumption (2.9) has been considered recently in the context of the inverse problem by March and Senatore[18], though, to date, no numerical results of this approximate theory of the inverse problem are available to the writer's knowledge.

In summary, a great deal of valuable information is contained in measurements of $S(k)$ for liquids near the triple point. In view of the very fine experiments of Jüngst, Knuth and Hensel[19] on the structure factor of Rb and Cs along the coexistence curve of these expanded fluid metals, an extension of the above analysis for extracting $\phi(r)$ will be of considerable interest to study the variation of $\phi(r)$ with thermodynamic state as the metal-insulator transition is approached*. This transition, especially in expanded fluid Na, has technological[20] as well as fundamental interest[21].

3 IONIC MELTS: ESPECIALLY RbCl AND $BaCl_2$

Page and Mika[22] used the technique of isotopic substitution on CuCl to determine the partial structure factors S_{11}, S_{22} and S_{12} of an ionic melt. This same technique is illuminating in showing, in molten alkali halides, the charge ordering known to exist in crystalline NaCl say. The appropriate combination of the corresponding pair functions of g_{11}, g_{22} and g_{12} reflecting the charge–charge (QQ) correlation function is (see, for example, March and Tosi[23])

$$g_{QQ}(r) = g_{11} + g_{22} - 2g_{12} \tag{3.1}$$

and the Fourier transform $S_{QQ}(k)$ for molten RbCl is shown in Figure 3.1.

---

* Note added in proof. Preliminary results have now been reported (J. A. Ascough and N. H. March (1990) Phys. Chem. Liquids 21, 251) for a low density metallic state of Cs.

It is interesting that before the diffraction data on ionic melts became available, many of the important features of the structure factors of the molten alkali halides had been determined by Singer and co-workers[24], by computer simulation, reflecting the better understanding of force fields in these melts than in, say, liquid metals. Biggin and Enderby[25] have discussed at length the meaning of coordination numbers extracted from diffraction experiments. The coordination number is, though slightly ambiguous as these workers emphasize, reduced somewhat on melting.

After this brief introduction to the structure of a two-component liquid, we shall comment in a little more detail on what can be gleaned from partial structure factor measurements on $BaCl_2$. This material is in the class of fast-ion conductors having, in the solid state, the fluorite structure. As March and Tosi[26] have emphasized, the partial structure factors in the liquids already reflect the fact that, on freezing, $BaCl_2$ will have dominant Frenkel defects, in contrast to RbCl with dominant Schottky defects. In other words, liquid structure already, just above the melting temperature, has fingerprints of the defect structure to be expected on freezing.

March and Tosi[27] have given a zero-order theory, related to the work of Ramakrishnan and Yussouff[28] for the freezing of simple liquids such as Ar and Na, for the freezing of $BaCl_2$ into a fast-ion conducting state. Just below the melting point, making admittedly drastic assumptions about the order parameters in **k** space, they can characterize the lattice liquid formed by the anions (i.e. liquid in the periodic potential created by the frozen cation sub-lattice) by Fourier components of the density at the first reciprocal lattice vector. This can be related to structure through the admittedly approximate linear response theory to yield

$$\rho_{2G} = \frac{(\rho_{21}/\rho_{11})^{1/2}\tilde{c}_{12}(G)\rho_{1G}}{(\rho_{21}/\rho_{20}) - \tilde{c}_{22}(G)}. \tag{3.2}$$

As was to be expected, the coupling of the cation sub-lattice 1 to the liquid density modulations ρ_{2G} is determined by the Ornstein–Zernike cross-correlation function $\tilde{c}_{12}(G)$. The denominator in Eq. (3.2) represents, in essence, the response function of the system 2, i.e. the anions in $BaCl_2$. ρ_{21} denotes the mean anion singlet density in the liquid phase etc. D'Aguanno *et al.*[29] have subsequently studied this problem numerically and later work of Rovere and Tosi[30] has been concerned with the fast-ion transition itself. However, for some cautionary remarks about the truncation of the sums over reciprocal lattice vectors, see Haymet *et al.*[31], and further work is clearly called for in this area.

4 X-RAY AND ELECTRON SCATTERING FROM DISORDERED SOLIDS

Having introduced the partial structure factors of two-component liquids, with reference to RbCl and $BaCl_2$, we turn to consider diffraction from disordered solids. Again, it will be helpful to do it by examples: we mention here SiO_2[32], amorphous Si[33] and amorphous C[33,34]. While, we believe, the scattering theory outlined below will have more general usefulness, it has been applied most extensively so far to

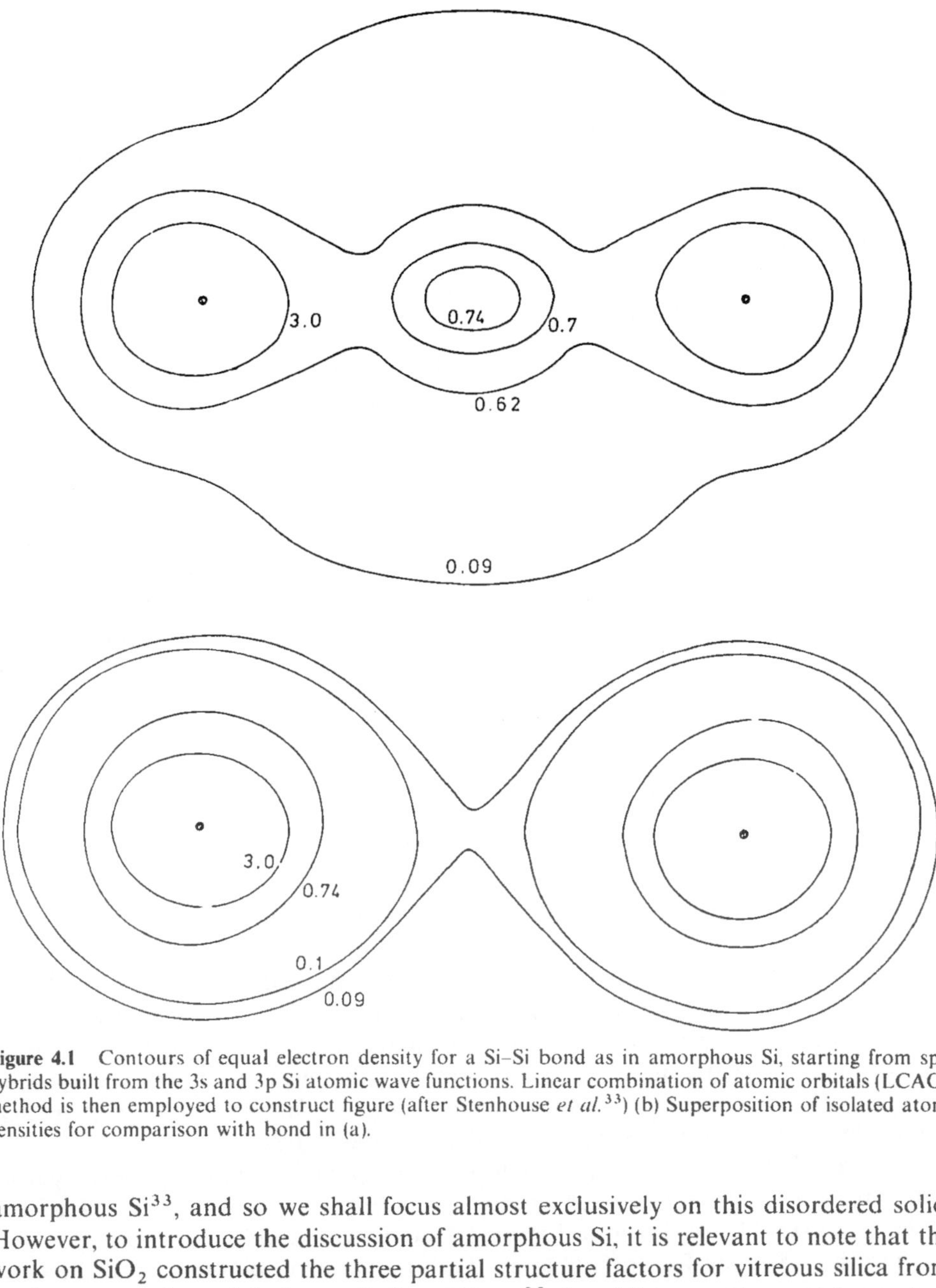

Figure 4.1 Contours of equal electron density for a Si–Si bond as in amorphous Si, starting from sp^3 hybrids built from the 3s and 3p Si atomic wave functions. Linear combination of atomic orbitals (LCAO) method is then employed to construct figure (after Stenhouse *et al.*[33]) (b) Superposition of isolated atom densities for comparison with bond in (a).

amorphous Si[33], and so we shall focus almost exclusively on this disordered solid. However, to introduce the discussion of amorphous Si, it is relevant to note that the work on SiO_2 constructed the three partial structure factors for vitreous silica from random network models due to Bell and Dean[35]. In addition to the model structure factors thereby obtained, X-ray scattering factors must be specified in order to calculate the X-ray intensity (cf Eq. (4.2) below). In Ref. 32, though it was recognised that chemical bonding should eventually be included explicitly, neutral Si and O atomic scattering factors were employed. It was noted from the work of King[36] that the height of the first peak in the X-ray scattering is significantly altered by using instead the scattering factors for Si^{4+} and O^{2-}. Following this brief comment on vitreous silica, we turn to the main example of the section, namely amorphous Si.

Given the nuclear–nuclear pair correlation function $g(r)$ for amorphous Si, one could attempt to use Eq. (1.1) with $f(k)$ replaced by a neutral atomic scattering factor for Si (see Figure 4 of Ref. 33). With a suitable continuous random network (CRN) model for $g(r)$, one calculates $S(k)$ from Eq. (1.2) and hence the X-ray intensity $I_x(k)$. The conclusion is, unambiguously, that the first peak in the X-ray intensity to too low to agree with experiment. When the same CRN model is used to calculate the electron diffraction intensity using[33]

$$I_e(k) = \text{const } k^{-4} S(k)[Z - f_{\text{atomic}}(k)]^2, \tag{4.1}$$

then the first peak is too high to agree with experiment.

The reason for the disagreement is not far to seek in qualitative terms. Figure 4.1(a) shows contours of equal electron density for a Si–Si bond as in amorphous Si, starting from atomic sp^3 hybrids built from the 3s and 3p wave functions. Then the linear combination of atomic orbitals (LCAO) method is employed to construct Figure 4.1(a). If we compare these electron density contours with those of Figure 4.1(b), which is merely a superposition of isolated atom densities[33], the marked difference between this superposition density and that of the LCAO density shown in Figure 4.1(a) is the absence in (b) of the closed, almost spherical, contours round the bond centre.

4.1 Effects of chemical bonding on diffraction intensities

The LCAO contours indicate that to model the covalent bond one should include, in addition to the superposition of spherical charge clouds on each Si nucleus, a spherical charge distribution at the bond centre, an idea exploited by Phillips[37] and other workers. Whereas Phillips was concerned with dielectric properties and lattice dynamics, and was able to model the bond charge as a point charge, the concern here is modelling diffraction intensities which will evidently require an extended spatial bond charge distribution. Before discussing the form of this, let us emphasize next that this immediately leads to the introduction of a further structural characterization, the bond centre, to supplement $S(k)$ discussed already.

The same CRN model was used therefore to calculate the nuclei-bond centre partial structure factor $S_{nb}(k)$ and the bond centre-bond centre structure factor $S_{bb}(k)$: these, taken from Ref. 33, are shown in Figure 4.2, along with $S(k)$.

Once the scattering factors $f_a(k)$ and $f_b(k)$ of the 'atomic' blobs (f_a) and the bond charge (f_b) are known, then the scattering is as from a two-component liquid and we can write for the X-ray intensity with suitable normalization[33]

$$I_x(k) = f_c^2(k)S(k) + 2f_b^2(k)S_{bb}(k) + 2(S_{nb}(k) - 1)f_c(k)f_b(k) \tag{4.2}$$

with $f_c = 4f_a$.

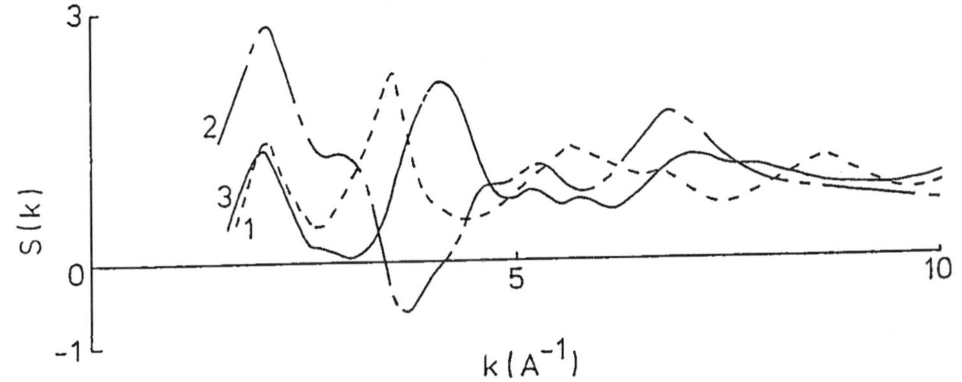

Figure 4.2 Partial structure factors for amorphous Si. (1) $S(k)$ is the nuclear–nuclear structure factor for amorphous Si (2) $S_{nb}(k)$ is nuclei–bond structure factor while (3) $S_{bb}(k)$ is that for bond-centre structure.

Similarly, the electron scattering intensity is given by

$$I_e(k) = \text{const } k^{-4}\{[Z - f_c(k)]^2 S(k) + 2f_b(k)^2 S_{bb}(k) - 2[S_{nb}(k) - 1][Z - f_c(k)] f_b(k)\}. \tag{4.3}$$

Putting $f_b = 0$ and $f_c = f_{\text{atomic}}$ in Eq. (4.3), Eq. (4.1) is regained. Equations (4.2) and (4.3) are the basic equations for modelling the diffraction intensities, to allow for the effects of chemical bonding.

The choice of f_a and f_b is fully discussed in Ref. 33, and these form factors are simply plotted in Figure 4.3. These, in turn, lead back to the equal electron density

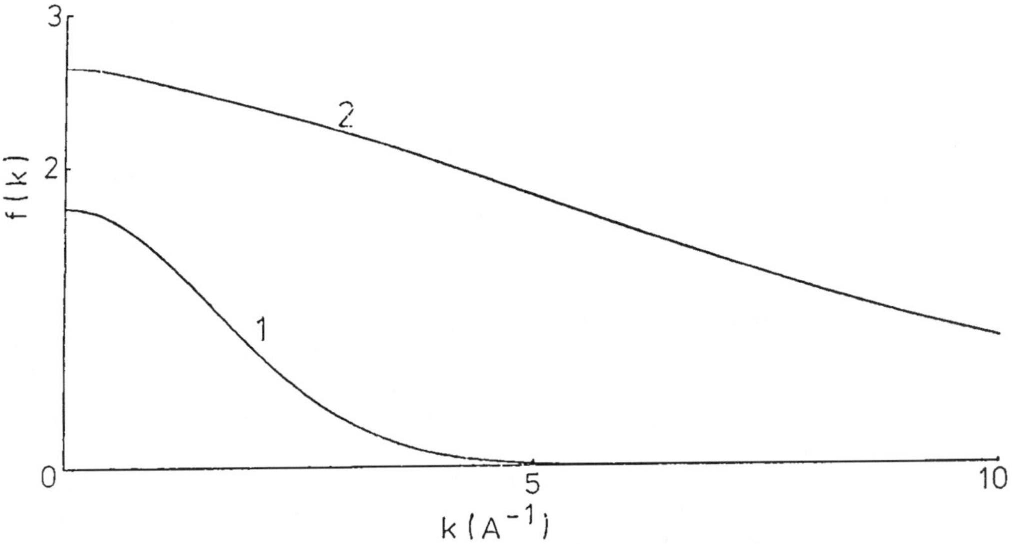

Figure 4.3 Form factors f_a and f_b appearing in intensity Eq. (4.2). $f_a(k)$ is upper curve.

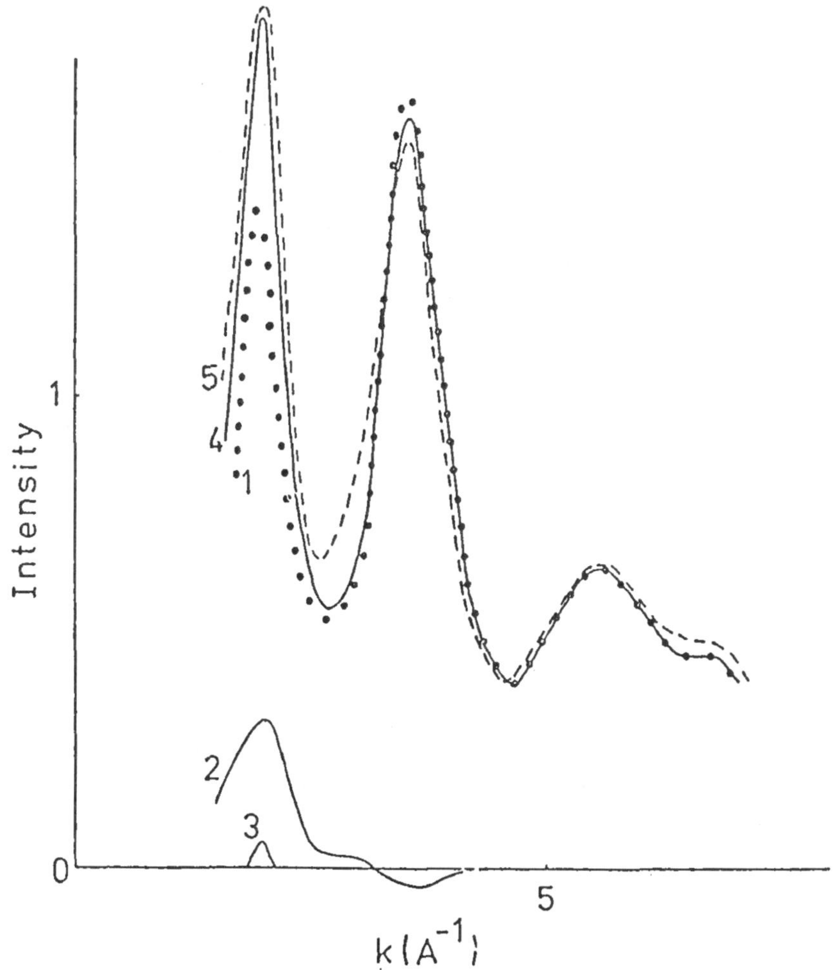

Figure 4.4 X-ray scattering intensity for amorphous silicon, using continuous random network model. (1) Contribution from Si–Si correlations (2) From Si-bond-centre correlations (3) Bond-centre contribution (4) Total predicted intensity (5) Experiment[33].

contours which are practically indistinguishable from those in Figure 4.1(a). Reference to Figures 4.4 and 4.5 show how the inclusion of the bond charge brings theory and experiment into substantial agreement for both X-ray and electron diffraction.

Though, no doubt, some refinements of structure and of electron distribution could now be made, we emphasize in the present context that one can gain confidence in the CRN structural model (with odd membered rings[33]) and in the detailed treatment of chemical bonding. A model built from ordered units turns out to be clearly inferior to the CRN model, while a superposition of spherical blobs on the Si nuclei fails to account for the diffraction intensities. As shown, in fact, in Ref. 33, use of the bond

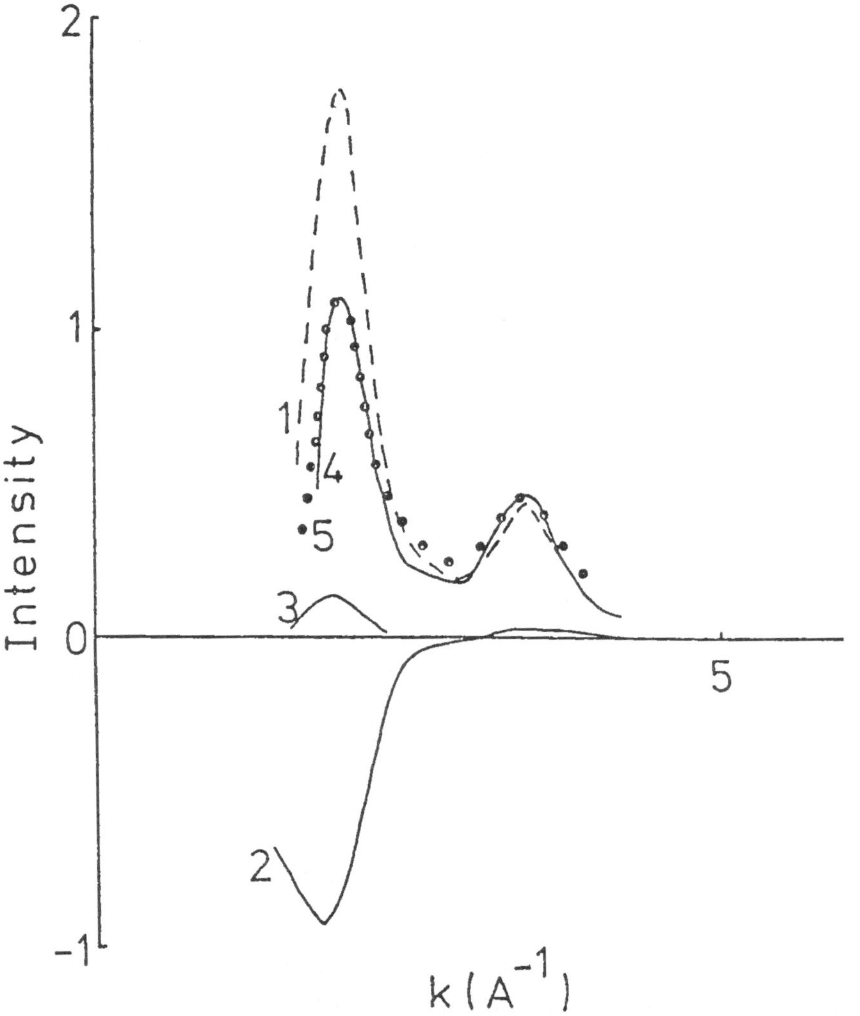

Figure 4.5 Electron scattering intensity for amorphous silicon using continuous random network model. Labelling (1) to (4) as in Figure 4.4. (5) Experiment[33].

distribution in Figure 4.1(c) accounts rather well for the 'forbidden' 222 reflection in crystalline Si. Naturally in the crystal, however, modern density functional theory combined with Bloch's theorem, leads to fully first principle electron densities, which can again be used to calculate Bragg reflection intensities[38].

Having emphasized that there is important information about chemical bonding in X-ray and electron diffraction intensities, let us return to liquid metals, and the important problem of electron–electron correlations among the valence electrons. Again, we shall find a (this time 'resonating') bond picture emerging, following earlier ideas of Pauling[39] on bonding in metals.

5 ELECTRON–ELECTRON PAIR CORRELATIONS IN LIQUID METALS

Returning to simple liquid metals like Na, these are really two-component liquids, Na^+ ions and valence electrons[40]. Thus, again, one wants to describe the structure in terms of three structure factors: the nuclear–nuclear structure factor $S(k)$, the valence electron–ion term $S_{iv}(k)$ and, of prime interest here, the electron–electron structure factor $S_{vv}(k)$[41].

Following Ref. 41, one writes the differential cross sections for X-ray and electron scattering in the forms

$$\left(\frac{d\sigma}{d\Omega}\right)_x = \text{const } S_x(k) \tag{5.1}$$

and

$$\left(\frac{d\sigma}{d\Omega}\right)_e = \frac{\text{const}}{k_4} S_e(k). \tag{5.2}$$

Then, $S_{vv}(k)$ can be written[41]

$$S_{vv}(k) = (1/Z)\{f_e(k)S_x(k) + f(k)S_e(k) - Zf(k)f_e(k)S(k)\} \tag{5.3}$$

Here $f(k)$ is the Fourier transform of the electron density in the core, while

$$f_e(k) = Z - f(k) \tag{5.4}$$

with Z the atomic number.

Thus, at least in principle, measurement via Eqs. (5.1) and (5.2) of $S_x(k)$ and $S_e(k)$, plus use of $S(k)$ determined by neutrons as already discussed in Section 2, will allow the electron–electron structure factor $S_{vv}(k)$ to be determined from Eq. (5.3), given a core scattering factor $f(k)$.

In practice, a number of difficulties remain[42,43,13]. However, by studying the main peaks of $S_x(k)$ and $S(k)$, it was originally argued in Ref. 41 that the X-rays were showing behaviour more like a 'Bragg peak' than the neutrons; the conclusion being that there was spatially more extensive short-range order of the electrons than the ions.

More recent work of Dobson[44] and of Johnson[45] has pressed the idea of Ref. 41 further for liquid Na and K respectively near their melting points. Both workers present evidence for short-range ordering over substantial differences and local coordination like that of a face-centered cubic (fcc) lattice. While in Ref. 41, Wigner electron correlations were referred to, these would lead to body-centered cubic (bcc) ordering.

Therefore, subsequent to the work of Dobson[44], March and Tosi[46] have presented an interpretation of the above fcc-like ordering in terms of a combination of both electron–electron and electron–ion interactions. They point out that a model which can explain the observed reflections characteristic of a fcc lattice in liquid Na and K can be built up by:

a) Using sp^3 type bonding charges, with Pauling resonance invoked between occupied and unoccupied bonds and

b) Assuming, once local electron coordination characteristic of such bonds is formed, Wigner-type lattice ordering[47] can propagate the fcc lattice over distances of 30 to 40 Å.

In a little more detail, March and Tosi propose their model, after examining the X-ray diffraction pattern to be expected from a number of different models for the arrangement of bond charges. Models in which bond charges are placed between (a) all near-neighbour ions or (b) all next near-neighbour ions can be excluded. This is because, unless one breaks the correlations between first neighbour ions, one will necessarily obtain a diffraction pattern characteristic of a bcc structure, given that local ion ordering in the liquid just above the freezing point resembles that in the (bcc) crystal. Such a breaking of correlations has in fact been examined but still does not lead to an fcc diffraction pattern.

However, a model that does lead to fcc reflections is to consider bond charges between each ion and four of its near-neighbours, in a tetrahedral configuration. Again, in the language of chemical hybridization used in Section 4, we could think of mixing 3s and 3p atomic orbitals in Na say, with sp^3 hybridization once again. It is cautioned in Ref. 46 that atomic orbitals are poor as a starting point for Na, Wannier orbitals being more appropriate.

The tetrahedral configuration proposed[46] is shown in Figure 5.1. The electron 'bond charges' add up to a fcc lattice. The 'unoccupied' bonds to the remaining 4 corners of the cube from the body centre would have to resonate with those drawn in Figure 5.1, in the manner of Pauling[39].

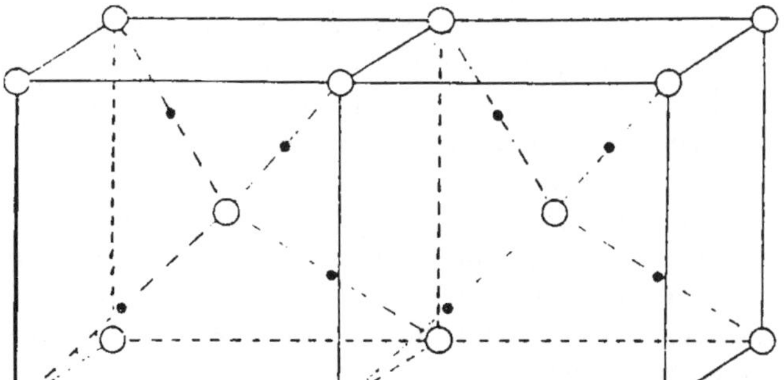

Figure 5.1 Electronic model proposed by March and Tosi[46] to interpret diffraction experiments from liquid Na near freezing. Bond charges (solid circles) are depicted schematically between each ion (open circles) and four of its near-neighbours, in tetrahedral array. 'Unoccupied' bonds to remaining 4 corners of cube would have to resonate[39] with those drawn in the Figure.

6 SUMMARY AND FUTURE DIRECTIONS

Having surveyed, by specific examples mainly, the information that can be derived from diffraction measurements on disordered systems, we wish to conclude by emphasizing areas that seem potentially fruitful for the future.

First of all, regarding the 'inverse problem' of Section 2.3, we have already stressed for a liquid metal like Na the interest in studying the variation of the force law with thermodynamic state along the coexistence curve. Of course, one has to recognise here that, eventually, as the critical point is approached, critical fluctuations 'take over' and numerous, though by no means all, properties become insensitive to the force law. More obviously, in the same area, accurate neutron diffraction data over a wide range of monentum transfer k should now allow effective pair interactions to be extracted from liquid state experiments for the technologically important transition metals.

Still on liquid metals, further understanding of supercooled liquids is clearly needed. A particularly relevant example in the present context is the work of Mountain[48] on supercooled liquid Rb. Though some diffraction experiments have been reported on liquid Ga[49], Mountain points out the difficulties to date in performing clean experiments of this type on simple supercooled liquids, and therefore the interest in computer simulation of this problem. By obtaining $g(r)$ for Rb, Mountain identifies two types of change as the temperature is lowered:

i) 'Sharpening' of the structure, which occurs in both equilibrium and supercooled liquid states:

ii) Evolution in the structure in the form of a shift in the position of the first maximum of $g(r)$ to larger r. This occurs in the equilibrium liquid but is not perceptible in the supercooled liquid[48].

Mountain notes in relation to the above that $g(r)$ represents the average environment of a particle. In the supercooled region, the near-neighbour environment becomes more sharply defined in space but does not shift its position as it does in the equilibrium liquid. That is to say, the local environment stops evolving, in the sense of a peak shift, once the freezing temperature is reached and in the supercooled region this fully developed liquid structure can only become more sharply defined until crystallization intervenes. Also in Ref. 48, following earlier studies[50,51], Mountain reiterates (cf. Section 2.2 above) that another empirical freezing criterion comes back in terms of the ratio R of the magnitude of $g(r)$ at the first minimum to the magnitude at the first maximum as $R = 0.20 \pm 0.01$. Further, R is linear with temperature for liquid states. Clearly, this work prompts further studies, both scattering experiments and approximate liquid state theory, on both the liquid-crystal phase transition and on the less well explored area of supercooling.

Related to this, there has been a resurgence of interest in the glassy transition. The work of Bernu *et al.*[52], again by computer experiment, is notable; however, so far the generalization to the dynamic structure factor has been the most fruitful approach here and it would take us too far from our main theme to give more than a few central references here[53,54].

Finally, let us return to chemical bonding in disordered systems. While in materials like amorphous silicon and carbon, the existence of well defined covalent bonds has been recognised for decades, it is, of course, extremely interesting that evidence for chemical bonding in low electron density metals like the alkalis is building up strongly (see Section 5 for liquid Na: and Refs. 55 and 56 for the crystalline phase of this sp metal).

Some stimulation for investigating sp metals further in the solid state has come from a desire to find a theory of superconductivity which would embrace both the recently discovered high T_c ceramic oxides[57] and the simple metals. As an example, the work of Messmer[58] on crystalline Be metal can be sited; here, as for the liquid alkalis discussed in Section 5, both chemical- and Wigner-lattice-type arguments are invoked for strong spatial correlations between valence electrons; the present writer[59] has drawn attention recently to the possible link between Dobson's analysis[42] of diffraction data on the trivalent liquid metal Al and Messmer's ideas[58] on divalent crystalline hexagonal-close-packed Be. There seems no doubt that there is currently a great deal more to be done in exploiting fully, by a combination of diffraction experiments and first-principles theory, just what we can learn as to the combined role of electron-ion and electron-electron interactions in both ionic structural correlations and pair correlations between valence electrons. Chemical insight is surely going to be of great importance in these future studies, even in metals like Na and Al which were thought, for a very long time, to be almost 'ideal' examples of electron delocalization.

Acknowledgements

It is a pleasure to thank R. G. Chapman, P. J. Grout, M. W. Johnson and M. P. Tosi for numerous valuable discussions relating to the general area of this article.

References

1. J. E. Enderby, T. Gaskell and N. H. March (1965), *Proc. Phys. Soc.*, **85**, 217.
2. J. L. Yarnell, M. J. Katz, R. G. Wenzel and S. H. Koenig (1973), *Phys. Rev.*, **A7**, 2130.
3. G. Robinson and N. H. March (1972), *Jour. Phys.*, **C5**, 2553.
4. M. J. Renne and B. R. A. Nijboer (1973), *Jour. Phys.*, **C6**, L10.
5. C. C. Matthai and N. H. March (1982), *Phys. Chem. Liquids*, **11**, 207.
6. L. Verlet (1968), *Phys. Rev.*, 165,201; see also S. H. Galam and J. P. Hansen (1976), *Phys. Rev.*, **A14**, 816.
7. A. Ferraz and N. H. March (1980), *Sol. State Commun.*, **36**, 977.
8. A. J. Greenfield, J. Wellendorf and N. Wiser (1971), *Phys. Rev.*, **A4**, 1607.
9. A. B. Bhatia and N. H. March (1984), *Phys. Chem. Liquids*, **13**, 313.
10. M. D. Johnson and N. H. March (1963), *Phys. Lett.*, **3**, 313.
11. D. Levesque, J. J. Weis and L. Reatto (1985), *Phys. Rev. Lett.*, **54**, 451; (1986) *Phys. Rev.*, **A33**, 3451.
12. M. W. C. Dharma-wardana and G. C. Aers (1983), *Phys. Rev.*, **B28**, 1701.
13. N. H. March (1987), *Can. J. Phys.*, **65**, 219.
14. L. Reatto (1988), *Phil. Mag.*, **A58**, 37; F. Perrot and N. H. March (1990), *Phys. Rev.*, **A41**, 4521.
15. N. H. March (1987), *Phys. Chem. Liquids*, **16**, 205; Liquid metals (Pergamon: Oxford) 1967.
16. T. Gaskell (1988), *J. Phys.*, **C21**, 1.
17. J. L. Barrat, J. P. Hansen and G. Pastore (1987), *Phys. Rev. Lett.*, **58**, 2075.
18. N. H. March and G. Senatore (1988), *Phys. Chem. Liquids*, **17**, 331.
19. S. Jungst, B. Knuth and F. Hensel (1985), *Phys. Rev. Lett.*, **55**, 2160, and references there.
20. F. Hensel (1988), *Chemistry in Britain*, **24**, 457.
21. See, for example, R. G. Chapman and N. H. March (1988), *Phys. Rev.*, **B38**, 792; also N. H. March (1990) *J. Math. Chem.* in press.

22. D. I. Page and K. Mika (1971), *J. Phys.*, **C4**, 3034.
23. N. H. March and M. P. Tosi (1984), Coulomb liquids (Academic: London).
24. See, for example, L. V. Woodcock and K. Singer (1971), *Trans Faraday Soc.*, **67**, 12.
25. S. Biggin and J. E. Enderby (1982), *Jour. Phys.*, **C15**, L305.
26. N. H. March and M. P. Tosi, see Ref. 23; also N. H. March, Chemical Physics of Liquids (Gordon and Breach: London) 1990.
27. N. H. March and M. P. Tosi (1981), *Phys. Chem. Liquids*, **11**, 79, 89, 129.
28. T. V. Ramakrishnan and M. Yussouff (1977), *Sol. State Commun.*, **21**, 389; (1979), *Phys. Rev.*, **B19**, 2775.
29. B. D'Aguanno, M. Rovere, M. P. Tosi and N. H. March (1983), *Phys. Chem. Liquids*, **13**, 113.
30. M. Rovere and M. P. Tosi (1986), *Repts. Prog. Phys.*, **49**, 1001; Sol. State Commun. **55**, 1109.
31. A. D. J. Haymet *et al.*, to be published.
32. B. Stenhouse, P. J. Grout and N. H. March (1976), *Phys. Lett.*, **A57**, 99.
33. B. Stenhouse, P. J. Grout, N. H. March and J. Wenzel (1977), *Phil. Mag.*, **36**, 129.
34. B. Stenhouse and P. J. Grout (1978), *J. Non-Crystalline Solids*, **27**, 247.
35. R. J. Bell and P. Dean (1972), *Phil. Mag*, **25**, 1381.
36. S. V. King (1967), *Nature*, 215,615.
37. J. C. Phillips (1968), *Phys. Rev*, **166**, 832.
38. L. C. Balbas, A. Rubio, J. A. Alonso, N. H. March and G. Borstel (1988), *J. Phys. Chem. Solids*, **49**, 1013.
39. L. Pauling (1949), The Nature of the Chemical Bond (Cornell University Press: Ithaca).
40. See, for example, N. H. March and M. P. Tosi (1973), *Annals of Physics*, **81**, 414.
41. P. A. Egelstaff, N. H. March and N. C. McGill (1974), *Can. J. Phys.*, **52**, 1651.
42. J. Chihara (1987), *J. Phys.*, **F17**, 295.
43. See also S. Tamaki (1987), *Can. J. Phys.*, **65**, 286.
44. P. J. Dobson (1978), *Jour. Phys.*, **C11**, L295.
45. M. W. Johnson (1990), private communication.
46. N. H. March and M. P. Tosi (1980), *Phys. Chem. Liquids*, **10**, 113.
47. E. P. Wigner (1934), *Phys. Rev.*, **46**, 1002; (1938) *Trans. Faraday Soc.*, **34**, 678; C. M. Care and N. H. March (1975), *Adv. Phys.*, **24**, 101; M. J. Lea and N. H. March (1989), *Int. J. Quantum Chem. Symp.*, **23**, 717.
48. R. D. Mountain (1982), *Phys. Rev.*, **A26**, 2859.
49. C. G. Windsor, private communication.
50. R. D. Mountain, H. J. Raveche and W. B. Streett (1974), *J. Chem. Phys.*, **61**, 1970.
51. F. F. Abraham (1980), *J. Chem. Phys.*, **72**, 359.
52. B. Bernu, Y. Hiwatari and J. P. Hansen (1985), *Jour. Phys.*, **C18**, L371.
53. E. Leutheusser (1984), *Phys. Rev.*, **A29**, 2765.
54. U. Bengtzelius, W. Götze and A. Sjölander (1984) *Jour. Phys.*, **C17**, 5915; U. Bengtzelius (1986), *Phys. Rev.*, **A33**, 3433.
55. J. P. Malrieu, J. P. Daudey and D. Maynau (1984), *Phys. Rev.*, **B30**, 1817.
56. N. H. March (1990), *Chem. Soc. Far. Trans.* **86**, 1203.
57. J. G. Bednorz and K. A. Muller (1986), *Z. Phys.*, **B64**, 189; see also P. W. Anderson (1987), *Science*, **235**, 1196.
58. R. P. Messmer (1987), *Sol. State Commun.*, **63**, 405.
59. N. H. March (1989), *Phys. Chem. Liquids*, **20**, 241.

PHYSICAL REVIEW A VOLUME 42, NUMBER 8 15 OCTOBER 1990

Binding in pair potentials of liquid simple metals from nonlocality in electronic kinetic energy

F. Perrot
Centre d'Etudes de Limeil-Valenton, 94195 Villeneuve St. Georges CEDEX, France

N. H. March
Theoretical Chemistry Department, University of Oxford, 5 South Parks Road, Oxford OX1 3UB, England
(Received 13 November 1989)

An explicit expression is obtained for the pair potential $\phi(R)$ in liquid simple metals from low-order density-gradient theory when the superposition of single-center displaced charges is employed. Corrections are thereby explicitly exhibited to the local Thomas-Fermi result $\phi_{\rm TF}(R)$ from second- and fourth-order inhomogeneity corrections $\Delta T_2(R)$ and $\Delta T_4(R)$ to the Thomas-Fermi electronic kinetic-energy change $\Delta T_{\rm TF}(R)$. The important point to emphasize is that, in each order, the potential $V(R/2)$ and displaced charge $\Delta(R/2)$ of a single screened ion at the midpoint of the metallic bond completely determines the pair potential in the density-gradient theory. These approximate results are illustrated by explicit calculations on liquid Na and liquid Be near their respective freezing points. While the pair potentials obtained by including $\Delta T_2(R)+\Delta T_4(R)$ are major improvements over the linear response result, the remaining nonlocal corrections to the total kinetic-energy change $\Delta T(R)$ are substantial, as again demonstrated by explicit results for Na and Be. The importance of further study of a possible functional relation between $\Delta T(R)-\Delta T_{\rm TFG}(R)$ and the potential $V(R/2)$ at the midpoint of the metallic bond is finally emphasized.

I. INTRODUCTION

The interaction between a pair of test charges Ze in a metallic medium was studied by means of the linearized Thomas-Fermi (TF) method, the latter being the forerunner of density-functional theory, in the early work of Alfred and March.[1] Their result was, from the linear equation for the screened potential V,

$$\nabla^2 V=q^2 V, \quad q^2=\frac{4k_f}{\pi a_0}, \quad \bar{n}=\frac{k_f^3}{3\pi^2} \tag{1.1}$$

with q^{-1} the TF screening length and k_f the Fermi wave number, that the two-center potential is the superposition

$$V=V_1+V_2 \tag{1.2}$$

of the screened potentials centered on the charges 1 and 2, and the total displaced charge $\Delta=n-\bar{n}$ was similarly obtained by superposition of one-center densities. This superposition property lies at the heart also of the present study. The result for the pair interaction $\phi(R)$ for the two test charges Z at separation R was

$$\phi(R)=\frac{Z^2}{R}\exp(-qR)\ . \tag{1.3}$$

This is simply the interaction of the second charge Z with the screened potential $V(R)=-Z\exp(-qR)/R$ of the first charge.

Corless and March[2] subsequently generalized the linear TF theory[1] and demonstrated that the pair interaction $\phi(R)$ was still

$$\phi(R)=-ZV(R)\ , \tag{1.4}$$

but that the screened potential $V(R)$ must now be calculated from the equation[3]

$$\nabla^2 V(r)=\frac{2k_f^2}{\pi^2}\int\frac{V(\mathbf{r}')j_1(2k_f|\mathbf{r}-\mathbf{r}'|)d\mathbf{r}'}{|\mathbf{r}-\mathbf{r}'|^2}\ ; \tag{1.5}$$

$j_1(x)=(\sin x-x\cos x)/x^2$. This reduces to Eq. (1.1) only for such slowly varying potentials that $V(\mathbf{r}')$ inside the integral in Eq. (1.5) can be replaced by $V(r)$, an approximation which is too primitive to apply to liquid simple metals. Unlike the semiclassical theory based on Eq. (1.1), Eq. (1.5) leads to binding. The **k**-space solution is equivalent to the use of the Lindhard[4] dielectric function, discussed also by Langer and Vosko.[5]

In the present paper, generalization of the above results will be affected in a number of directions, namely, by the following.

(i) Corrections of the TF result (1.1) by adding the lowest-order density-gradient correction $t_2=(\lambda/8)(\nabla n)^2/n$ to the TF kinetic-energy density. The next-order correction t_4 will also be included.

(ii) Incorporating exchange and correlation, albeit approximately, into the theory at local-density level.

(iii) Demonstrating that same incorporation of single-center nonlinearity can be made in the local-density functional theory.

(iv) Interpretation of $\phi(R)$ in all cases as a sum of local plus nonlocal pieces,

$$\phi(R)=\phi_{\rm loc}(R)+\phi_{\rm nloc}(R)\ . \tag{1.6}$$

Anticipating one of the important results to be presented in the present paper, let us merely note at this stage from Eqs. (1.3) and (1.4) that, with $V(R)=-Z\exp(-qR)/R$,

$$\phi_{\rm loc}(R)=\frac{R}{4}\left[V\left(\frac{R}{2}\right)\right]^2 . \tag{1.7}$$

This formula will be shown to be more general than the linear TF method from which it has just been derived. It evidently expresses the local form of the pair potential solely in terms of the electrostatic potential of one screened charge at the midpoint of the metallic bond. Further, Eq. (1.7) shows that the local form is always repulsive, which is related to what is nowadays referred to as Teller's theorem,[6] namely, that binding is not possible in a local (TF) theory.

Given the result (1.7), it is then obvious that the difficult part of the problem of constructing realistic pair potentials $\phi(R)$ is to judiciously approximate the nonlocal contribution in Eq. (1.6) and the main aim of the present work is to discuss this point.

Therefore, to gain initial insight, let us turn immediately to correct the linear TF theory based on Eqs. (1.1)–(1.3) by adding the lowest-order gradient correction to the TF kinetic-energy density.

II. PAIR POTENTIALS IN LINEAR RESPONSE THEORY WITH LOWEST-ORDER GRADIENT CORRECTION

Following the pioneering work of von Weizäcker,[7] one writes the most elementary nonlocal correction to the TF kinetic-energy density in terms of the electron density $n(\mathbf{r})$ in the form

$$t_{\rm TFG}^{(1)}=t_{\rm TF}+\frac{\lambda}{8}\frac{(\nabla n)^2}{n} , \tag{2.1}$$

where $t_{\rm TF}=\frac{3}{10}(3\pi^2)^{2/3}n^{5/3}$. Then, for von Weizsäcker's choice of λ, namely unity, March and Murray[8] converted the resulting density-functional Euler equation to a Schrödinger equation by the substitution

$$\psi=n^{1/2} . \tag{2.2}$$

After linearization of the Euler equation, they were able to solve it self-consistently together with the Poisson equation. As shown in the Appendix, a completely general solution can be obtained for arbitrary values of λ, and the final result is

$$\phi_{\rm TFG}^{(1)}(R)=\frac{1}{4}R\left\{\left[V\left(\frac{R}{2}\right)\right]^2-\gamma^2\left[\Delta\left(\frac{R}{2}\right)\right]^2\right\} , \tag{2.3}$$

where

$$\Delta=n(r)-\bar{n} \tag{2.4}$$

is the displaced electron density and γ^2 is given by

$$\gamma^2=\lambda\frac{\pi}{\bar{n}} . \tag{2.5}$$

The derivation of Eq. (2.3) is carried out in reciprocal space, using the Fourier transform of the response function associated with the kinetic-energy functional in Eq. (2.1).

In the linear response formalism invoked above, $\phi=\phi_{\rm LR}=-ZV$, V and Δ are intrinsic single-center quantities, independent of any two-center model. It is surprising to see that Eq. (2.3), a relation between these quantities, anticipates on the physical picture of binding, where the pair interaction is determined by properties at the midpoint of the bond.

The important points to stress about the nonlocal contribution proportional to $R[\Delta(R/2)]^2$ to the pair potential (2.3) are the following.

(i) It is always attractive; i.e., the nonlocality in the kinetic energy tends to bind the charged centers, though for some choices of λ it is easy to show from the Appendix that the pair potential remains repulsive everywhere.

(ii) As with the local contribution to $\phi_{\rm TFG}(R)$ in Eq. (1.7), the nonlocal contribution $R[\Delta(R/2)]^2$ depends only on a property at the midpoint of the metallic bond.

Naturally, one must not assume that Eq. (2.3) is precise: it is valid within the lowest-order density-gradient treatment. Much of the ensuing discussion will focus on the way Eq. (2.3), plus its extension to include the next-order density-gradient correction T_4 to the kinetic energy (see Sec. IV below), can most fruitfully be converted into realistic information about the true pair interaction $\phi(R)$ in liquid simple metals. With this as the main focal point, it is worthwhile next to introduce the total valence screening charge, $Q(R)$ say, inside a sphere of radius R, and to relate this quantity to $\phi(R)$.

III. TOTAL SCREENING CHARGE $Q(R)$ RELATED TO PAIR INTERACTION $\phi(R)$ IN LINEAR RESPONSE THEORY

The definition of the total screening charge $Q(R)$ is evidently given in terms of the displaced charge $\Delta(r)$ by

$$Q(r)=\int_0^r\Delta(r)4\pi r^2dr . \tag{3.1}$$

From Poisson's equation, the potential $V_e(r)$ treated by the valence displaced charge $\Delta(r)$ around say Na^+ or Be^{++} in their own liquid metals is related to $Q(r)$ since

$$\frac{1}{r^2}\frac{d}{dr}\left[r^2\frac{dV_e}{dr}\right]=-4\pi\Delta(r) \tag{3.2}$$

and hence

$$\frac{dV_e}{dr}=-\frac{Q(r)}{r^2} . \tag{3.3}$$

But as already discussed in Sec. I, in linear response theory $\phi(r)=-ZV(r)$ and hence the turning points of the pair potential occur at positions R_t, i.e., where

$$\frac{dV}{dr}=\frac{dV_e}{dr}-\frac{Z}{r^2}=0 \tag{3.4}$$

or

$$Q(R_t)=Z . \tag{3.5}$$

Evidently a necessary, though not sufficient, condition for Eq. (3.5) to be satisfied is that $\Delta(r)$ oscillates. To illustrate Eq. (3.5), we have plotted in Fig. 1 $Q(R)$ versus R for liquid Na near freezing using our earlier results.[9] Evi-

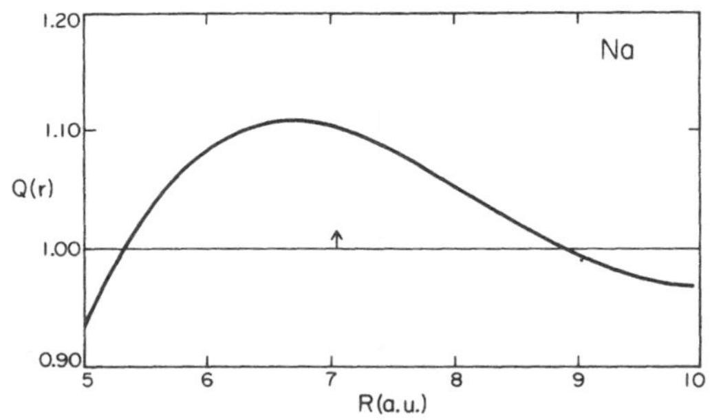

FIG. 1. Total valence screening charge $Q(R)$ defined in Eq. (3.1), for liquid Na near freezing. $Q(R)$ determines the pair interaction $Q(R)$ through Eq. (3.6) in linear response theory. Arrow denotes position of principal minimum in "exact" pair potential ϕ.

dently since the positions of the insertions of $Q(R)/Z$ with unity do not agree with the turning points of the pair potential $\phi(R)$ that we calculated for liquid Na near freezing, linear response theory must be transcended to faithfully reproduce the Na pair potential. Therefore, we shall return to the TF theory corrected by low-order gradients in a more general discussion of pair binding in liquid metals below. However, to summarize the results of this section; when linear response theory is valid (i.e., when $\phi=\phi_{\rm LR}=-ZV$) the total valence screening charge $Q(R)$ uniquely determines the pair potential $\phi(R)$, defined to tend to zero at infinite separation, by integration of the equation obtained from Eqs. (3.3) and (3.4), namely,

$$\frac{d\phi(R)}{dR}=Z\left[\frac{Q(R)-Z}{R^2}\right], \tag{3.6}$$

where Z is the valence.

IV. SUPERPOSITION OF SINGLE-ION DENSITIES AND THE PAIR INTERACTION FROM DENSITY-FUNCTIONAL THEORY

Having noted the inadequacy of the linear response formula for ϕ in liquid Na near freezing, we now return to a more fundamental discussion, still based on density-functional theory (DFT), of the basis of Eq. (2.3), which so far has been derived within the framework of linear response theory for the single-ion screening.

Here, we still retain the superposition of single-ion displaced charge, but focus then directly on the pair problem, utilizing the symmetry about a plane through the midpoint of the metallic bond, and perpendicular to that bond. By doing so, we shall provide a more fundamental derivation of Eq. (2.3), and shall also be able to extend it to include the higher density-gradient correction T_4, to be defined precisely below, to the kinetic energy in local TF theory.

As before, let us start from the DFT equation for a single ion in the electron gas of density $\bar{n}$. This can be written formally as

$$G'[\Delta+\bar{n}]-G'[\bar{n}]=-V(r)\ , \tag{4.1}$$

where G' is simply a shorthand notation for the functional derivative $\delta G/\delta n(r)$, which in TF theory is simply proportional to $[n(r)]^{2/3}$. Here, the functional $G=T+E_{\rm xc}$ includes exchange and correlation effects in the local density approximation. V is the usual electrostatic potential created by the displaced charge $\Delta(r)$, namely,

$$V=-\frac{Z}{r}+\frac{1}{r}*\Delta\ , \tag{4.2}$$

where the $*$ is simply shorthand for the convolution product.

A. Two-center energy calculation with superposition density

Let us consider two ions at positions $\mathbf{R}_i,\mathbf{R}_j$ and, let us also write $\Delta_i=\Delta(\mathbf{r}-\mathbf{R}_i)$ and $\Delta G(\Delta+\bar{n})=G[\Delta+\bar{n}]-G[\bar{n}]$. Then for the pair interaction ϕ between ions i and j we have

$$\begin{aligned}\phi=&\Delta G[\Delta_i+\Delta_j+\bar{n}]-\Delta G[\Delta_i+\bar{n}]-\Delta G[\Delta_j+\bar{n}]\\&+\left[-\frac{Z}{r_i}\right]\cdot\Delta_j+\left[-\frac{Z}{r_j}\right]\cdot\Delta_i\\&+\tfrac{1}{2}(\Delta_i+\Delta_j)\cdot\frac{1}{r}*(\Delta_i+\Delta_j)\\&-\tfrac{1}{2}\Delta_i\cdot\frac{1}{r}*\Delta_i-\tfrac{1}{2}\Delta_j\cdot\frac{1}{r}*\Delta_j+\frac{Z^2}{R}\ .\end{aligned} \tag{4.3}$$

Next it is to be noted from Eq. (2) that

$$\begin{aligned}ZV(R)&=-\frac{Z^2}{R}+\int\frac{Z}{|\mathbf{r}'-\mathbf{R}|}\Delta(\mathbf{r}')d\mathbf{r}'\\&=-\frac{Z^2}{R}+\int\frac{Z}{r''}\Delta(\mathbf{r}''+\mathbf{R})d\mathbf{r}''\\&=-\frac{Z^2}{R}+\left[\frac{Z}{r_i}\right]*\Delta_j\ .\end{aligned} \tag{4.4}$$

Utilizing Eq. (4.4) in Eq. (4.3), one finds

$$\begin{aligned}\phi=&\Delta G[\Delta_i+\Delta_j+\bar{n}]-\Delta G[\Delta_i+\bar{n}]-\Delta G[\Delta_j+\bar{n}]\\&+\left[-\frac{Z}{r_j}\cdot\Delta_i\right]+\tfrac{1}{2}\Delta_i\cdot\frac{1}{r}*\Delta_j+\tfrac{1}{2}\Delta_j\cdot\frac{1}{r}*\Delta_i-ZV(R)\ ,\end{aligned} \tag{4.5}$$

where the dot means the integral of the product of functions through the whole of space; technically the scalar product.

It is important to note at this point that, if one works with the exact density of the "molecule," written as $\Delta_i+\Delta_j+\rho$, instead of with $\Delta_i+\Delta_j$ in the superposition approximation, then Eq. (4.5) is correct to $O(\rho^2)$, due to the stationary properties of the energy functional. Let us now turn to examine several particular cases.

1. Linear response solid

This means $\Delta_i \ll \bar{n}$ and $\Delta_j \ll \bar{n}$ everywhere. Then one can expand the functionals in the form

$$\Delta G[\Delta_i+\Delta_j+\bar{n}]-\Delta G[\Delta_i+\bar{n}]-\Delta G[\Delta_j+\bar{n}]$$
$$=(\Delta_i+\Delta_j)\cdot G'[\bar{n}]-\Delta_i\cdot G'[\bar{n}]-\Delta_j G'[\bar{n}]$$
$$+O(\Delta_i^2,\Delta_i\Delta_j,\Delta_j^2)=0 \quad (4.6)$$

in first order. Thus Eq. (4.5) then yields

$$\phi=-ZV(R)+\cdots , \quad (4.7)$$

where the ellipsis represents second-order terms, which are

$$\tfrac{1}{2}G''[\bar{n}]\cdot[(\Delta_i+\Delta_j)^2-\Delta_i^2-\Delta_j^2]+\tfrac{1}{2}\Delta_i\cdot V_j+\tfrac{1}{2}\Delta_j\cdot V_i ,$$

but from Eq. (4.1)

$$\Delta_i\cdot G''[\bar{n}]=-V_i ,$$

so that the second-order terms vanish to leave the result

$$\phi_{\mathrm{LR}}=-ZV(R)+O(\Delta^3) . \quad (4.8)$$

2. Local density approximation for G

Inside the ion core i, Δ_i is much larger than $\bar{n}$ so that expansion in Δ_i is not permissible. But in this same region, Δ_j is very small compared with $\Delta_i+\bar{n}$ so that one can expand in Δ_j. Next, let us exploit the symmetry about the midpoint of the metallic bond in Eq. (4.5) and replace the integration over the whole of space by twice the integration in the half-space Ω_i containing i, as in Fig. 2.

Then we have

$$\phi+ZV(R)=2(\Delta G[\Delta_i+\Delta_j+\bar{n}]-\Delta G[\Delta_i+\bar{n}]$$
$$-\Delta G[\Delta_j+\bar{n}])+\Delta_i\circ V_j+\Delta_j\circ V_i . \quad (4.9)$$

In Eq. (4.9), the volume integrals appearing in ΔG are now restricted to the half-space Ω_i. Now given the assumption that G is local, one can expand for small Δ_j in the half-space Ω_i, to obtain

$$\phi+ZV(R)=2\{\Delta_j\circ G'[\Delta_i+\bar{n}]-\Delta_j\circ G'[\bar{n}]+O(\Delta_j^2)\}$$
$$+\Delta_i\circ V_j+\Delta_j\circ V_i$$
$$=-\Delta_j\circ V_i+\Delta i\circ V_j+O(\Delta_j^2) \quad (4.10)$$

with the help of Eq. (4.1). It is worth noting that Eq. (4.10) retains the full nonlocality of G around the center i and thus transcends linear response. The integrals in the scalar products of Eq. (4.10) are now restricted to the half-space Ω_i and are denoted by an open circle $\circ$. In the second step of Eq. (4.10), use has again been made of the basic Euler Eq. (4.1). Using $\nabla^2 V=-4\pi\{\Delta-Z\delta\}$ one finds

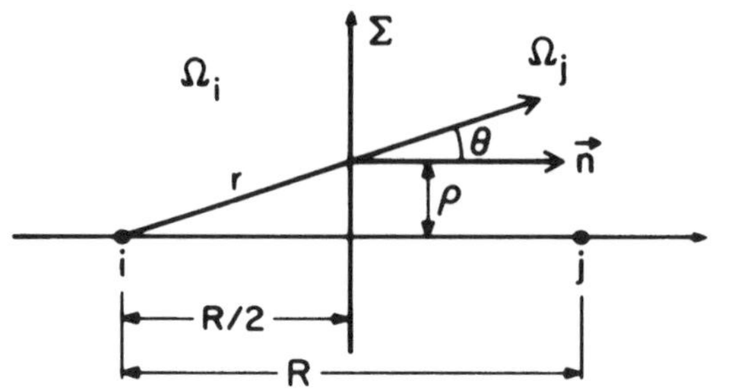

FIG. 2. This defines coordinate system used in calculating pair interaction in local density theory, plus low-order gradients, via surface integrals over the plane Σ, utilizing the symmetry of the homonuclear pair of ions.

$$\phi+ZV(R)=V_j\circ\left[-\frac{\nabla^2 V_i}{4\pi}+Z\delta(\mathbf{r}-\mathbf{R}_i)\right]$$
$$-V_i\circ\left[-\frac{\nabla^2 V_j}{4\pi}+Z\delta(\mathbf{r}-\mathbf{R}_j)\right] . \quad (4.11)$$

Because $\mathbf{R}_j$ is outside Ω_i, the term in $V_i\circ Z\delta(\mathbf{r}-\mathbf{R}_j)$ does not contribute. Therefore

$$\phi+ZV(R)=V(\mathbf{r}-\mathbf{R}_j)\circ Z\delta(\mathbf{r}-\mathbf{R}_i)$$
$$+\frac{1}{4\pi}(V_i\circ\nabla^2 V_j-V_j\circ\nabla^2 V_i) , \quad (4.12)$$

and one is left with an integral over the plane Σ:

$$\phi=\frac{1}{4\pi}\int_\Sigma (V_i\nabla_{\mathrm{n}}V_j-V_j\nabla_{\mathrm{n}}V_i)ds . \quad (4.13)$$

On this plane one evidently has

$$V_i=V_j ,$$
$$\nabla_{\mathrm{n}}V_i=-\nabla_{\mathrm{n}}V_j . \quad (4.14)$$

Using the coordinate system of Fig. 2,

$$\phi(R)=\frac{1}{4\pi}\int_0^\infty d\rho\, 2\pi\rho 2V(r)\left[-\frac{dV}{dr}\cos\theta\right] . \quad (4.15)$$

But $\cos\theta=R/2r$, $r^2=R^2/4+\rho^2$ or $r\,dr=\rho\,d\rho$ and hence

$$\phi(R)=-\int_{R/2}^\infty dr\, V(r)\frac{dV}{dr}\frac{R}{2}=\frac{R}{4}\left[V\left[\frac{R}{2}\right]\right]^2+O(\Delta_j^2) . \quad (4.16)$$

Thus, in a considerably more general framework than that employed in Secs. I and II, one regains the local form given in Eqs. (1.7) and (2.3). This establishes then on a more general basis the use of the form (1.7) for the first contribution to the pair potential in Eq. (1.6).

Having discussed fully the local contribution to $\phi(R)$ in Eq. (1.6), it is natural to examine further the status of the nonlocal form in Eq. (2.3). This is done immediately below.

B. Gradient corrections to $\phi_{\rm loc}(R)$

Following the procedure of Sec. II, let us consider next the case when G is given by the usual local leading term, plus a gradient correction

$$T_2=\frac{\lambda}{8}\int\frac{(\nabla n)^2}{n}d\mathbf{r}\ . \tag{4.17}$$

In the pair interaction one then obtains the additional contribution

$$T_2[\Delta_i+\Delta_j+\bar{n}]-T_2[\Delta_i+\bar{n}]-T_2[\Delta_j+\bar{n}]\ .$$

Expanding for small Δ_j the previous line reads

$$-\frac{\lambda}{8}\int_{\Omega_i}\frac{|\nabla\Delta_i|^2}{(\Delta_i+\bar{n})^2}\Delta_j d\mathbf{r}-\frac{\lambda}{8}\int_{\Omega_i}\frac{2\nabla\Delta_i\cdot\nabla\Delta_j}{\Delta_i+\bar{n}}d\mathbf{r}+O(\Delta_j^2)$$
$$=\frac{\lambda}{8}\int_{\Omega_i}\frac{|\nabla\Delta_i|^2}{(\Delta_i+\bar{n})^2}\Delta_j d\mathbf{r}-\frac{\lambda}{4}\int_{\Omega_i}\frac{\nabla^2\Delta_i}{\Delta_i+\bar{n}}\Delta_j d\mathbf{r}$$
$$+\frac{\lambda}{4}\int\frac{\nabla\Delta_i}{\Delta_i+\bar{n}}\Delta_j ds\ . \tag{4.18}$$

The first two terms are those which appear in the DFT Euler equation together with the TF term, to give $-\Delta_j\circ V_i$.

The remainder of Eq. (4.18) can be handled again as for the local case, to allow its evaluation on the plane Σ in Fig. 2. The outcome of this is to recover precisely the nonlocal term proportional to $R[\Delta(R/2)]^2$ in Eq. (2.3), which is thereby established on a more general footing.

1. Inclusion of fourth-order gradient correction

It will prove important in calculating the well depth of the first minima in the pair potentials of liquid Na (Ref. 9) and Be near freezing to include in the nonlocal part of $\phi(R)$ in Eq. (1.6) the next-order gradient correction t_4 for the kinetic energy.

Following the previous approach, the contribution is that which is associated with the surface integrals in

$$\Delta T_4=T_4[\Delta_i+\Delta_j+\bar{n}]-T_4[\Delta_i+\bar{n}]\ , \tag{4.19}$$

when expanded to first order in Δ_j. $T_4[\Delta_j+\bar{n}]$ does contribute to first order in Δ_j. Writing

$$d=\frac{1}{540}\frac{1}{(2\pi)^{3/2}}\ , \tag{4.20}$$

the form of t_4 calculated by Hodges[10] then yields

$$\Delta T_4=2d\int_{\Omega_i}\nabla^2\Delta_j\left[\frac{2\nabla^2\Delta_i}{(\Delta_i+\bar{n})^{5/3}}-\frac{(\frac{9}{8})(\nabla\Delta_i)^2}{(\Delta_i+\bar{n})^{8/3}}\right]d\mathbf{r}$$
$$+2d\int_{\Omega_i}(\nabla\Delta_j\cdot\nabla\Delta_i)\left[-\frac{(\frac{9}{4})\nabla^2\Delta_i}{(\Delta_i+\bar{n})^{8/3}}+\frac{(\frac{4}{3})(\nabla\Delta_i)^2}{(\Delta_i+\bar{n})^{11/3}}\right]d\mathbf{r}\ , \tag{4.21}$$

where terms in $\nabla^2\Delta_j$ have been collected together in the first integral along with those involving $\nabla\Delta_j$ in the second. The first integral may be transformed using Green's theorem into a volume integral and a surface integral. The volume integral is exactly that which is required in $-\Delta_j\circ V_i$. The surface integral contributes to the pair interaction for

$$2d\int_\Sigma\mathcal{F}[\Delta_i]\nabla\Delta_j\cdot d\mathbf{s}-2d\int_\Sigma\nabla\mathcal{F}[\Delta_i]\cdot d\mathbf{s}\Delta_j \tag{4.22}$$

on Σ, $\Delta_i=\Delta_j$, and $\nabla_n\Delta_j=-\nabla_n\Delta_i$, so that one gets immediately

$$2d\int_\Sigma-\nabla(\Delta_i\mathcal{F}[\Delta_i])\cdot d\mathbf{s} \tag{4.23}$$

or

$$2d\int_{R/2}^{\infty}-\frac{d}{dr}(\Delta\mathcal{F}[\Delta])2\pi r\,dr\frac{R}{2r} \tag{4.24}$$
$$=2\pi dR(\Delta\mathcal{F}[\Delta])_{r=R/2}\ . \tag{4.25}$$

In Eq. (4.25) we have used

$$\mathcal{F}[\Delta]=\frac{2\nabla^2\Delta}{(\Delta+\bar{n})^{5/3}}-\frac{\frac{9}{8}(\nabla\Delta)^2}{(\Delta+\bar{n})^{8/3}}\ . \tag{4.26}$$

Let us recall that the above derivation implies $\Delta_i=\Delta_j\ll\bar{n}$ on the Σ plane, though not everywhere. Let us now assume that the gradient of the density Δ may be measured in terms of a characteristic length l:

$$\nabla\Delta\sim\frac{\Delta}{l}\ . \tag{4.27}$$

One has for $\mathcal{F}[\Delta]$

$$\mathcal{F}[\Delta]\sim\frac{2\Delta}{l^2(\Delta+\bar{n})^{5/3}}-\frac{\frac{9}{8}\Delta^2}{l^2(\Delta+\bar{n})^{8/3}}$$
$$\sim\frac{\Delta}{l^2}\frac{2}{\bar{n}^{5/3}}\left[1-\frac{5}{3\bar{n}}-\frac{9}{8\bar{n}}\right],$$

so that, if one works to lower order in $\Delta/\bar{n}$, one is left with

$$\mathcal{F}[\Delta]\sim\frac{2\nabla^2\Delta}{\bar{n}^{5/3}} \tag{4.28}$$

and Eq. (4.25) becomes

$$\frac{4\pi dR}{(\bar{n})^{5/3}}(\Delta\nabla^2\Delta)_{r=R/2}\ . \tag{4.29}$$

Let us return to the second integral in Eq. (4.22). It gives a surface contribution which is

$$2d\int_\Sigma\Delta_j\left[\frac{-\frac{9}{4}\nabla^2\Delta_i}{(\Delta_i+\bar{n})^{8/3}}+\frac{\frac{4}{3}(\nabla\Delta_i)^2}{(\Delta_i+\bar{n})^{11/3}}\right]\nabla\Delta_i\cdot d\mathbf{s}$$

which, according to Eq. (4.23), is of the form

$$2\pi dR\int_\Sigma\frac{-\frac{9}{4}\Delta^3}{(\bar{n})^{8/3}l^3}dr\ . \tag{4.30}$$

The integrand is of order $[1/(\bar{n})^{8/3}]\Delta^3/l^3$ to be compared with order $[1/(\bar{n})^{5/3}](\Delta^2/l^3)$ for the previous contribution in Eq. (4.25). Equation (4.30) is of higher order in $\Delta/\bar{n}$ so that we can neglect it. Finally, the contribution of fourth-order gradients to the pair interaction is given by Eq. (4.29) only.

V. NUMERICAL RESULTS FOR THE GRADIENT EXPANSION PAIR INTERACTION IN LIQUID Na and Be

Let us first collect together the local TF plus gradient correction contributions to the pair potential. The result is

$$\phi^{(2)}_{\rm TFG}(R)=\tfrac{1}{4}R\left\{\left[V\left(\frac{R}{2}\right)\right]^2-\gamma^2\left[\Delta\left(\frac{R}{2}\right)\right]^2+\eta^2[\Delta\nabla^2\Delta]_{R/2}\right\}, \qquad (5.1)$$

where η^2 is taken from Eq. (4.29).

We have performed the calculation of this pair potential for liquid Na and Be near freezing. Let us stress the fact that we always use in Eq. (5.1) the V and Δ provided by the DFT calculation,[9,11] never the potential and electron density which would result from the self-consistent solution of the TF + gradient correction problem.

A. Pair potential for liquid Na

Figure 3 has been constructed for Na. The curve labeled $\phi(R)$ is our electron theory pair potential for liquid Na near freezing. The curves labeled $\phi^{(1)}_{\rm TFG}$ and $\phi^{(2)}_{\rm TFG}$ add T_2 with $\lambda=\frac{1}{9}$ given by Kirznits[12] and T_2+T_4, respectively, to the curve labeled $\phi_{\rm TF}(R)$. These curves were calculated as follows: (i) $\phi_{\rm TF}(R)$ from Eq. (1.7), (ii) $\phi^{(1)}_{\rm TFG}(R)$ from Eq. (2.3), and $\phi^{(2)}_{\rm TFG}(R)$ from Eq. (5.1).

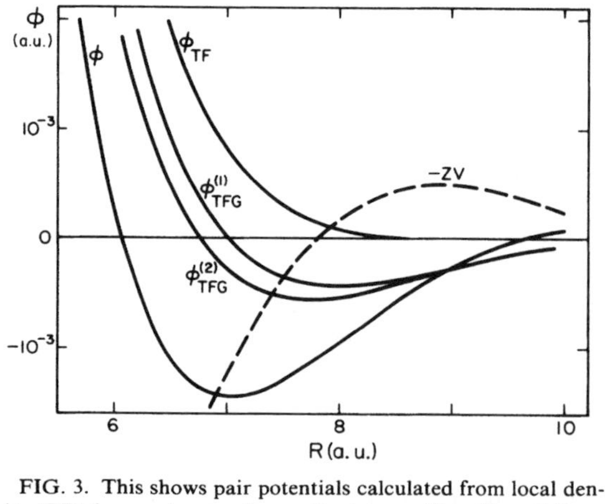

FIG. 3. This shows pair potentials calculated from local density (TF) theory plus density gradient corrections T_2 and T_4 to kinetic energy, for liquid Na near freezing. Various curves were obtained using $Q(R)$ in Fig. 1 from different degrees of approximation as follows. $\phi_{\rm TF}$, calculated from Eq. (1.7); $\phi^{(1)}_{\rm TFG}$ from Eq. (2.3) with $\lambda=\frac{1}{9}$ in which T_2 only is included; $\phi^{(2)}_{\rm TFG}$, from Eq. (5.1) containing both T_2 and T_4; $\phi_{\rm LR}=-ZV$; $\phi(R)$ is the pair potential obtained in our earlier work.

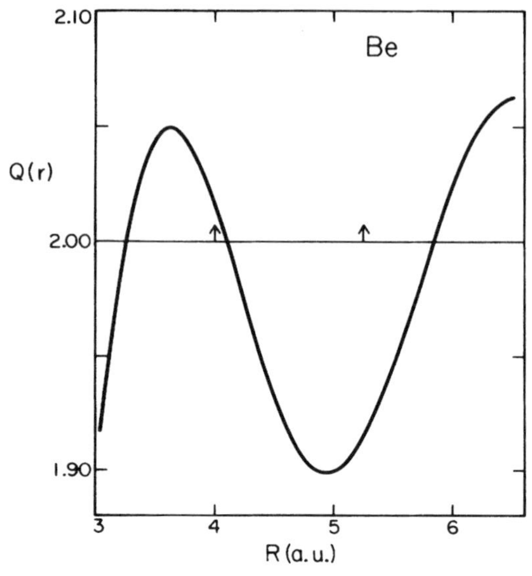

FIG. 4. Same as Fig. 1, but for liquid Be at a density equal to the solid density. Arrows denote first minimum and following maximum in "exact" pair potential $\phi(R)$.

B. Pair potential for liquid Be

It seemed of interest to perform similar calculations for the divalent liquid metal Be. First, the total valence screening charge $Q(R)$ was calculated using the DFT method.[11] It should be stressed that there are no adjust-

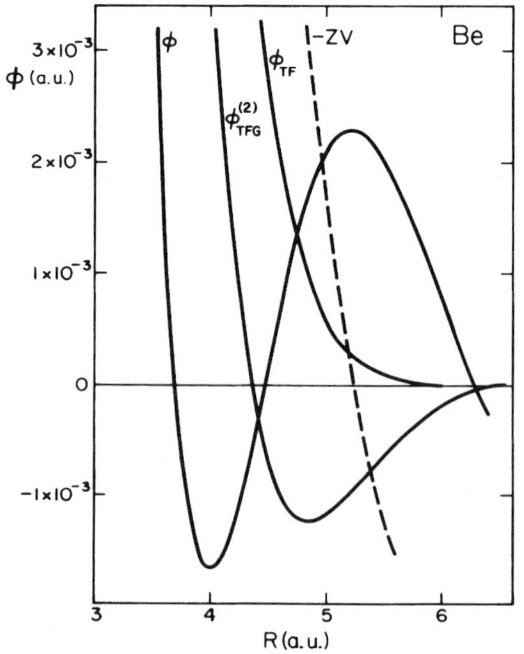

FIG. 5. Same as Fig. 3 but for Be.

able parameters in this calculation. The input into the full density functional calculation of $Q(R)$ is (a) the atomic number, (b) the mass density at freezing, and (c) an exchange-correlation potential.

Figure 4 shows $Q(R)$ versus R, which yields in linear response the pair interaction $\phi_{\rm LR}=-ZV$ in Fig. 5. This is shown compared with the "exact" $\phi(R)$ in the curve labeled ϕ. Again $\phi_{\rm TF}(R)$ and $\phi_{\rm TFG}^{(2)}$ have been constructed from $Q(R)$ in Fig. 4. The agreement for the well depth is quite comparable with that in liquid Na. The result is, at best, semiquantitative. However, it is more apparent for Be in Fig. 5 than for Na that the gradient expansion result is already a substantial improvement over the linear response curve $-ZV(R)$ around the principal minimum of $\phi(R)$.

These numerical results obviously indicate that the remaining nonlocality in the electronic kinetic energy is crucial to a quantitative calculation of the pair interaction in simple liquid metals. The discussion of these remaining contributions is the object of Sec. VI.

VI. REMAINING NONLOCALITY IN THE KINETIC ENERGY CONTRIBUTION TO THE PAIR INTERACTION

In the linear response approximation, the fully nonlocal kinetic energy change (plus exchange and correlation energy change) in bringing an ionic pair together from infinity, is given by (see Sec. IV A 1)

$$\begin{aligned}\Delta G(R)&=\tfrac{1}{2}G''[\bar{n}]\cdot[(\Delta_i+\Delta_j)^2-\Delta_i^2-\Delta_j^2]\\&=G''[\bar{n}]\cdot\Delta_i\cdot\Delta_j\\&=-\tfrac{1}{2}(V_i\cdot\Delta_j+V_j\cdot\Delta_i)\ .\end{aligned}\tag{6.1}$$

This expression may also be derived from the infinite order perturbation formalism of Stoddart and March,[13] which gives the changes in electron density and in kinetic energy density as

$$\Delta=n-\bar{n}=\sum_{p=1}^{\infty}n_p(r)\ ,\tag{6.2}$$

$$t-t_0=\mu\Delta-\sum_{p=1}^{\infty}\frac{p}{p+1}n_p(r)V(r)\ ,\tag{6.3}$$

with $V(r)$ the one-body potential in the Schrödinger equation. In this theory, exchange and correlation effects are neglected. The first-order term on the right-hand side of Eq. (6.3) does not contribute to the integrated kinetic energy. In linear response, the terms to be retained are only the $p=1$ terms, so that

$$\begin{aligned}\Delta T(R)&=-\tfrac{1}{2}(\Delta_i+\Delta_j)\cdot(V_i+V_j)+\tfrac{1}{2}\Delta_i\cdot V_i+\tfrac{1}{2}\Delta_j\cdot V_j\\&=-\tfrac{1}{2}(\Delta_i\cdot V_j+\Delta_j\cdot V_i)\ ,\end{aligned}$$

which is exactly Eq. (6.1) when exchange and correlation effects are neglected.

That Eq. (6.1) has achieved the summation of the gradient expansion of the kinetic energy terms in the pair interaction to all orders is now clear by rewriting the density-potential relation utilized in Eq. (1.5) as

$$V_1(\mathbf{r})=\int K(\mathbf{r}-s)\Delta_1(s)ds\tag{6.4}$$

with a similar expression for $V_2(\mathbf{r})$ in Eq. (6.1). Here K is a known kernel, given in Fourier transform by Stoddart and March[13] as

$$K(\mathbf{r}-\mathbf{s})=-(2\pi)^{-3}\int\frac{d\mathbf{q}\exp[i\mathbf{q}\cdot(\mathbf{r}-\mathbf{s})]}{J(q,k_f)}\ ,\tag{6.5}$$

where

$$J(q,k_f)=\frac{k_f}{\pi^2}\left[\frac{1}{2}+\frac{k_f}{2q}\left[1-\frac{q^2}{4k_f^2}\right]\ln\left|\frac{q+2k_f}{q-2k_f}\right|\right]\ .\tag{6.6}$$

By means of the small q expansion of $J(q,k_f)$, one gets

$$V_1(\mathbf{r})=-\frac{\pi^2}{k_f}[\Delta_1(\mathbf{r})-c_2\nabla_{\mathbf{r}}^2\Delta_1(\mathbf{r})+c_4\nabla_{\mathbf{r}}^2\nabla_{\mathbf{r}}^2\Delta_1(\mathbf{r})+\cdots]\ .\tag{6.7}$$

Substitution of this result (6.7) into Eq. (6.1), together with a similar equation for $V_2(\mathbf{r})$ in terms of $\Delta_2(\mathbf{r})$ gives back the early results of the linearized gradient expansion. Unfortunately, we are not able to achieve a similar resummation beyond the linear response approximation. To get a better feeling of the numerical importance of the terms beyond T_4, we can display the sum of the missing contribution as a function of R.

All the pair potentials plotted in Fig. 3 for liquid Na and Fig. 5 for liquid Be near freezing can be regarded as derivable from a single equation:

$$\phi(R)=\Delta G(R)-ZV(R)+\Delta U_1(R)\ .\tag{6.8}$$

Then the common element, for Na, for example, is the total potential-energy change

$$\Delta U(R)=-ZV(R)+\Delta U_1(R)\ ,\tag{6.9}$$

which is determined solely by the total valence screening charge $Q(R)$, plotted in Fig. 1 for Na and Fig. 4 for Be.

Thus, the kinetic plus exchange and correlation energy change $\Delta G(R)$ is solely responsible for the differences of the various curves for $\phi(R)$ in Figs. 3 and 5. We have now extracted the fully nonlocal energy change $\Delta G(R)$ from Eq. (6.8), using $\phi(R)$ and $-ZV(R)$ already plotted in Figs. 3 and 5, plus a calculation of $\Delta U_1(R)$ from

$$\Delta U_1(R)=\tfrac{1}{2}(\Delta_1\cdot V_2+\Delta_2\cdot V_1)\ ,\tag{6.10}$$

which is evidently the interaction potential energy between the displaced charge around one ionic center with the screened potential due to the other. The convolution involved in Eq. (6.10) is readily calculated in Fourier transform and the results are displayed for liquid Na and Be in Figs. 6 and 7, respectively. The difference between $\Delta T_{\rm TFG}^{(2)}(R)$ and $\Delta T(R)$ (equal to $\Delta G_{\rm TFG}^{(2)}-\Delta G$) is clearly the remaining nonlocality in the electronic kinetic-energy charge. Thus the difference $\Delta T(R)-\Delta T_{\rm TFG}^{(2)}(R)$ can be regarded as the effective sum of the remaining terms in the gradient expansion beyond T_2 and T_4.

The remaining question is whether what is embodied in

this difference is determined by the size and the shape of the total screening charge $Q(R)$ around the midpoint $R/2$ of the metallic bond.

VII. DISCUSSION AND SUMMARY

The main results of the present study are the low-order density-gradient equation (5.1) for the pair potential, and the calculation of the fully nonlocal electronic kinetic-energy charge $\Delta T(R)$ plotted in Figs. 6 and 7, for liquid Na and Be. The comparison of Eq. (5.1) with the pair potentials $\phi(R)$ including the fully nonlocal kinetic energy shows that Eq. (5.1) is not more than semiquantitative. However, its major merit is that it focuses on the importance of the form of the total valence screening charge $Q(R)$ around the midpoint of the metallic bond.

We think it possible that $\Delta T(R)-\Delta T^{(2)}_{\mathrm{TFG}}(R)$ may also be expressible, at least approximately, in terms of $Q(R/2)$ [or $V(R/2)$ or $\Delta(R/2)$]. Our argument is the following. The gradient expansion has dealt with the small-q components of the pair interaction, but cannot treat the components $q\sim 2k_F$. These components are responsible for the asymptotic behavior (Friedel oscillations):

$$\Delta(R)=\frac{A}{R^3}\cos(2k_F R)+O\left(\frac{1}{R^4}\right) \tag{7.1}$$

in linear response. In this asymptotic regime, the potential is

$$V(R)=\frac{\pi}{k_F^2}\Delta(R)+O\left(\frac{1}{R^4}\right) \tag{7.2}$$

so that it may be written as

$$V(R)=BR^3\left\{\left[\Delta\left(\frac{R}{2}\right)\right]^2-\frac{1}{4k_F^2}\left[\left.\frac{d\Delta}{dR}\right|_{R/2}\right]^{+2}\right\}, \tag{7.3}$$

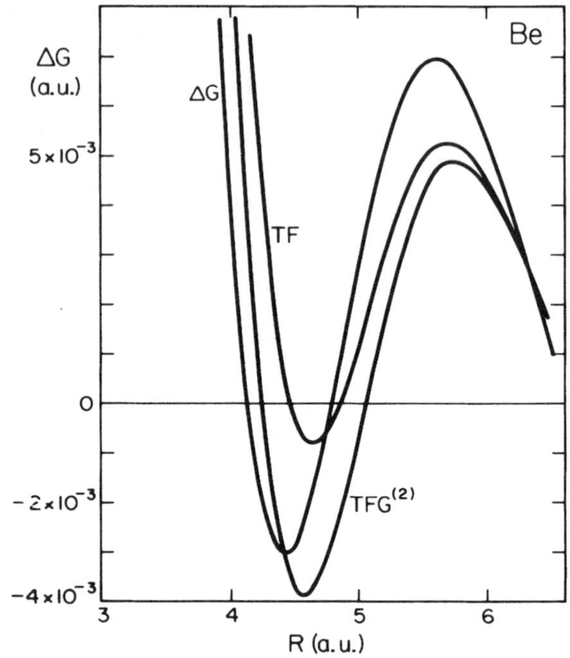

FIG. 7. Same as Fig. 6, but for Be.

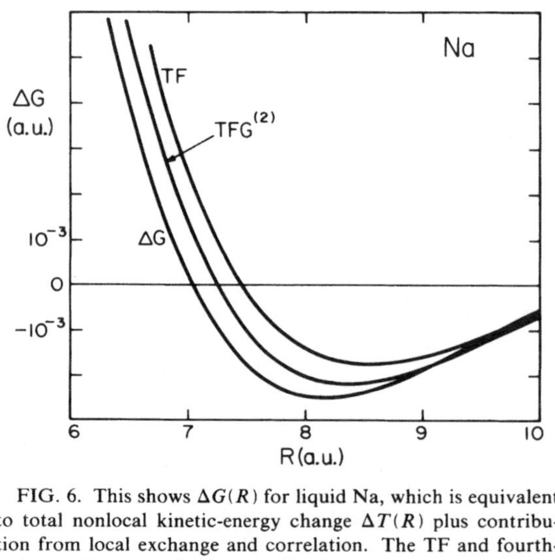

FIG. 6. This shows $\Delta G(R)$ for liquid Na, which is equivalent to total nonlocal kinetic-energy change $\Delta T(R)$ plus contribution from local exchange and correlation. The TF and fourth-order gradient contribution are also shown.

suggesting that, here again, it may be expressed in terms of Δ at $R/2$. It is not obvious whether this argument holds beyond the linear approximation, when a phase shift has to be introduced into Eq. (7.1). Clearly, further work is called for before a definitive conclusion can be reached on this point. A convenient approach could be the use of the formulation of ϕ in terms of the pressure tensor in the symmetry plane of the "molecule," as proposed by More.[14]

The final remarks concern the potential interest of the present study for the theory of bonding in free space molecules. Though, of course, a different derivation must be found (due to the lack of uniform background density $\bar{n}$), some of the present ideas should be relevant. In the case of "molecules" in high-temperature plasmas, where Friedel oscillations are damped by temperature effects, the TF and low-order gradient expansion results should become increasingly quantitative.

ACKNOWLEDGMENTS

The collaborative work reported here was made possible by the workshop at Institute for Theoretical Physics (ITP), University of California at Santa Barbara, on atoms and ions in strongly coupled plasmas in which both authors participated. We are very grateful to the coordinators, Dr. J. Callaway, Dr. H. E. DeWitt, and Dr. H. van Horn for the stimulating atmosphere they created. Thanks are also due to Professor J. S. Langer and Professor J. R. Schrieffer for the hospitality of ITP. Finally, financial support from the National Science Foundation under Grant No. PHY82-17853, supplemented by funds from the National Aeronautics and Space Administration, is acknowledged.

APPENDIX: VON WEIZSÄCKER'S GENERALIZATION OF TF THEORY IN LINEAR RESPONSE FOR ARBITRARY λ IN Eq. (2.1)

The purpose of this appendix is to effect the general demonstration of Eq. (2.3) with arbitrary value of λ in the first gradient correction to the *TF* kinetic energy. The value $\lambda=1$ corresponds to the original choice of von Weizsäcker.[7] The later work of Kirznits[12] and subsequently of Jones and Young[15] showed that the consistent generalization of TF theory for its range of validity of potentials which vary by but a small fraction of themselves over a characteristic de Broglie wavelength of the system yielded the value $\lambda=\frac{1}{9}$, used in constructing Figs. 3 and 5.

Taking the results of Jones and Young[15] as a convenient starting point, the screened ionic potential V in linear response theory can be written in Fourier transform for a general value of λ as

$$(2\pi)^3 V(r) = -4\pi Z \int e^{i\mathbf{q}\cdot\mathbf{r}} d\mathbf{q} \frac{1}{q^2+4\pi\dfrac{k_F}{\pi^2}\dfrac{1}{1+\dfrac{3\lambda}{4k_F^2}q^2}}$$

$$= -4\pi Z \int e^{i\mathbf{q}\cdot\mathbf{r}} d\mathbf{q} \frac{1+\alpha^2 q^2}{\alpha^2 q^4 + q^2 + q_{\mathrm{TF}}^2} , \tag{A1}$$

where

$$\alpha^2 = \frac{3\lambda}{4k_F^2} . \tag{A2}$$

The set of the denominator $\alpha^2 q^4 + q^2 + q_{\mathrm{TF}}^2$ may be conveniently classified by the parameter

$$\Delta = 1 - 4\alpha^2 q_{\mathrm{TF}}^2 , \tag{A3}$$

where Δ, being independent of position, is not to be confused with the displaced charge $\Delta(r)$.

1. $\Delta > 0$

The roots are $(-1-\sqrt{\Delta})/2\alpha^2$ and $(-1+\sqrt{\Delta})/2\alpha^2$ real, negative. Let us define

$$q_0^2 = \frac{1+\sqrt{\Delta}}{2\alpha^2}, \quad q_1^2 = \frac{1-\sqrt{\Delta}}{2\alpha^2} . \tag{A4}$$

After a straightforward decomposition of the integrand in Eq. (A1), one obtains the screened ionic potential V in the form

$$V(r) = -\frac{Z}{r}\frac{1}{2}\left[\frac{(\sqrt{\Delta}-1)}{\sqrt{\Delta}}e^{-q_0 r} + \frac{(\sqrt{\Delta}+1)}{\sqrt{\Delta}}e^{-q_1 r}\right] . \tag{A5}$$

From Poisson's equation, one finds the displaced charge Δ as

$$\Delta(r) = \frac{1}{4\pi r}\left[+\frac{Z}{2\sqrt{\Delta}}\right][(\sqrt{\Delta}-1)q_0^2 e^{-q_0 r} + (\sqrt{\Delta}+1)q_1^2 e^{-q_1 r}] . \tag{A6}$$

Using Eq. (A5) one obtains

$$\frac{r}{4}\left[V\left(\frac{r}{2}\right)\right]^2 = \frac{Z^2}{4\Delta r}\left[(\sqrt{\Delta}-1)^2 e^{-q_0 r} + (\sqrt{\Delta+1})^2 e^{-q_1 r} + 2(\Delta-1)e^{-(q_0+q_1)(r/2)}\right] . \tag{A7}$$

Thus it follows that

$$-ZV(r) - \frac{r}{4}\left[V\left(\frac{r}{2}\right)\right]^2 = -\frac{Z^2}{r}\frac{\alpha^4 q_0^2 q_1^2}{\Delta}(e^{-q_1 r} + e^{-q_0 r} - 2e^{-(q_0+q_1)(r/2)}) . \tag{A8}$$

Similarly the nonlocal contribution to ϕ is directly related to

$$4\pi r\left[\Delta\left(\frac{r}{2}\right)\right]^2 = \frac{Z^2}{\pi\Delta r}q_{\mathrm{TF}}^4(e^{-q_0 r} + e^{-q_1 r} - 2e^{-(q_0+q_1)(r/2)}) . \tag{A9}$$

Hence it follows that

$$-ZV(r) - \frac{r}{4}\left[V\left(\frac{r}{2}\right)\right]^2 = -\frac{\pi\alpha^2}{q_{\mathrm{TF}}^2}4\pi r\left[\Delta\left(\frac{r}{2}\right)\right]^2 , \tag{A10}$$

where

$$\frac{4\pi^2\alpha^2}{q_{\mathrm{TF}}^2} = \frac{\pi\lambda}{4\bar{n}} , \tag{A11}$$

which agrees with Eq. (2.3) in linear response theory.

2. $\Delta < 0$

Here $\Delta = 1-4\alpha^2 q_{\mathrm{TF}}^2 < 0$ and the roots are $(-1-i\sqrt{|\Delta|})/2\alpha^2$ and $(-1+i\sqrt{|\Delta|})/2\alpha^2$, with the real part negative. Let us define

$$q_a^2 = \frac{1+i\sqrt{|\Delta|}}{2\alpha^2}, \quad (q_a^*)^2 = \frac{1-i\sqrt{|\Delta|}}{2\alpha^2} , \tag{A12}$$

and retain for q_a and q_a^* the solutions with positive real part.

Hence the screened potential is now

$$V(r) = -\frac{Z}{r}\frac{1}{2}\left[\frac{i\sqrt{|\Delta|}-1}{i\sqrt{|\Delta|}}e^{-q_a r} + \frac{i\sqrt{|\Delta|}+1}{i\sqrt{|\Delta|}}e^{-q_a^* r}\right] , \tag{A13}$$

where

$$q_a^2=\rho^2 e^{i\theta},\quad \rho^2=\frac{\sqrt{1+|\Delta|}}{2\alpha^2}=\frac{q_{\rm TF}}{\alpha} \tag{A14}$$

with

$$\tan\theta=\sqrt{|\Delta|},\quad 0\le\theta\le\frac{\pi}{2}\ .$$

Thus

$$q_a=\rho e^{i(\theta/2)}\ ,$$
$$q_a^*=\rho e^{-i(\theta/2)}\ .$$

The potential becomes

$$V(r)=-\frac{Z}{2r}e^{-\rho\cos(\theta/2)r}\left[2\cos\left[\rho\sin\frac{\theta}{2}r\right]-\frac{i}{\sqrt{|\Delta|}}2i\sin\left[\rho\sin\frac{\theta}{2}r\right]\right]. \tag{A15}$$

Writing

$$\rho\cos\frac{\theta}{2}=\mu,\quad \rho\sin\frac{\theta}{2}=\nu$$

one finds

$$V(r)=-\frac{Z}{r}e^{-\mu r}\left[\cos\nu r+\frac{1}{\sqrt{|\Delta|}}\sin\nu r\right]. \tag{A16}$$

Using Poisson's equation with V in the form

$$V(r)=-\frac{Z}{r}\frac{1}{\sin\theta}\sin(\nu r+\theta)e^{-\mu r} \tag{A17}$$

yields the displaced charge $\Delta(r)$ as

$$\Delta(r)=\frac{Z}{r}\frac{1}{\sin\theta}\left[\frac{1}{4\pi}\frac{q_{\rm TF}}{\alpha}\right]\sin\nu r\, re^{-\mu r}\ . \tag{A18}$$

Hence

$$\frac{r}{4}\left[V\left[\frac{r}{2}\right]\right]^2=\frac{Z^2}{r}\frac{1}{\sin^2\theta}\sin^2\left[\frac{\nu r}{2}+\theta\right]e^{-\mu r}\ . \tag{A19}$$

It follows that

$$-ZV(r)-\frac{r}{4}\left[V\left[\frac{r}{2}\right]\right]^2=\frac{Z^2}{r}\frac{1}{\sin^2\theta}e^{-\mu r}\left[\sin\theta\sin(\nu r+\theta)-\sin^2\left[\frac{\nu r}{2}+\theta\right]\right] \tag{A20}$$

and

$$r\left[\Delta\left[\frac{r}{2}\right]\right]^2=\frac{Z^2}{r}\frac{1}{4\pi^2}\frac{q_{\rm TF}^2}{\alpha^2}\frac{1}{\sin^2\theta}\sin^2\left[\frac{\nu r}{2}\right]e^{-\mu r}\ . \tag{A21}$$

The transformation

$$\sin\theta\sin(\nu r+\theta)-\sin^2\left[\frac{\nu r}{2}+\theta\right]=-\sin^2\frac{\nu r}{2} \tag{A22}$$

then yields

$$-ZV(r)-\frac{r}{4}\left[V\left[\frac{r}{2}\right]\right]^2=-\frac{Z^2}{r}\frac{1}{\sin^2\theta}e^{-\mu r}\sin^2\frac{\nu r}{2} \tag{A23}$$

and comparison of Eqs. (A21) and (A23) leads to

$$-ZV(r)-\frac{r}{4}\left[V\left[\frac{r}{2}\right]\right]^2=-\frac{2\pi}{\bar n}\frac{\lambda}{8}r\left[\Delta\left[\frac{r}{2}\right]\right]^2\ . \tag{A24}$$

This is again equivalent to Eq. (2.3) in the linear response regime.

Hence the conclusion is that both for $\Delta>0$ and $\Delta<0$ in Eq. (A13), the result in linear response for the pair potential $\phi(r)$ is

$$\phi(r)=-ZV(r)=\frac{r}{4}\left[V\left[\frac{r}{2}\right]\right]^2-\frac{\pi\lambda}{4\bar n}r\left[\Delta\left[\frac{r}{2}\right]\right]^2\ . \tag{A25}$$

[1]L. C. R. Alfred and N. H. March, Philos. Mag. **2**, 985 (1957).
[2]G. K. Corless and N. H. March, Philos. Mag. **6**, 1285 (1961).
[3]N. H. March and A. M. Murray, Phys. Rev. **120**, 830 (1960); Proc. R. Soc. London, Ser. A **261**, 119 (1961).
[4]J. Lindhard, K. Dan. Selsk. Mat.-Fys. Medd. **28**, 8 (1954).
[5]J. S. Langer and S. H. Vosko, J. Phys. Chem. Solids **12**, 196 (1959).
[6]E. Teller, Rev. Mod. Phys. **34**, 627 (1962).
[7]C. F. von Weizäcker, Z. Phys. **96**, 431 (1935).
[8]N. H. March and A. M. Murray, Proc. R. Soc. London, Ser. A **256**, 600 (1960).
[9]F. Perrot and N. H. March, Phys. Rev. A **41**, 4521 (1990).
[10]C. H. Hodges, Can. J. Phys. **51**, 1428 (1973).
[11]F. Perrot, Phys. Rev. A (to be published).
[12]Zh. Eksp. Teor. Fiz. **32**, 115 (1957) [Sov. Phys.—JETP **5**, 64 (1957)].
[13]J. C. Stoddart and N. H. March, Proc. R. Soc. London, Ser. A **299**, 279 (1967).
[14]R. M. More, Adv. At. Mol. Phys. **21**, 305 (1985).
[15]W. Jones and W. H. Young, J. Phys. C **4**, 1322 (1971).

Phys. Chem. Liq., 1992, Vol. 25, pp. 59–64
Reprints available directly from the publisher
Photocopying permitted by license only

LETTER

The Deviation of the Pair Potential from the Potential of Mean Force in Molten Na Near Freezing

K. TANKESHWAR

International Centre for Theoretical Physics, 34100 Trieste, Italy

N. H. MARCH

Theoretical Chemistry Department, University of Oxford, 5 South Parks Road, Oxford OX1 3UB, UK

(Received 20 September 1991)

The difference between the pair potential $\phi(r)$ and the potential of mean force $U(r)$ is studied carefully in molten Na near freezing. Beyond some 4–5 Å the three-body term strongly cancels with $U(r)$, leaving $\phi(r)$ pretty small. Current approximate theories of liquid structure do not handle this cancellation well. Some improvements are therefore proposed.

KEY WORDS: Inverse problem, diffraction potential, force correlation function

The proposal of Johnson and March[1] to invert the liquid structure factor $S(k)$ obtained by diffraction experiments to obtain a pair potential $\phi(r)$ for liquid metals has been brought to fruition by Reatto and coworkers[2] for Na. It is then very encouraging that all the main features of this "diffraction" potential are found in electron theory study of Perrot and March[3]. However, it remains of interest to see whether the diffraction potential could be extracted from largely analytical liquid structural theories, thereby by-passing the still lengthy computer simulation studies required in the inversion procedure of Reatto *et al.* The present letter is aimed at this aspect of the pair potential inverse problem.

From the diffraction study of Greenfield *et al.*[4], we have plotted in curve 1 of Figure 1 the potential of mean force $U(r)$, in units of thermal energy k_BT, near the freezing point of liquid metallic Na. $U(r)$ is simply related to pair correlation function $g(r) = h(r) + 1$, with $h(r)$ the so-called total correlation function, by

$$g(r) = \exp(-U(r)\beta)\colon \beta = (k_BT)^{-1}. \tag{1}$$

Curve 2 shows the pair potential $\phi(r)$ of Reatto *et al.*[2]. The point we wish to emphasize at the outset is the smallness of $\phi(r)$ in the tail. There is plainly strong cancellation between the three-body term in the force equation[5] and $U(r)$, in arriving at

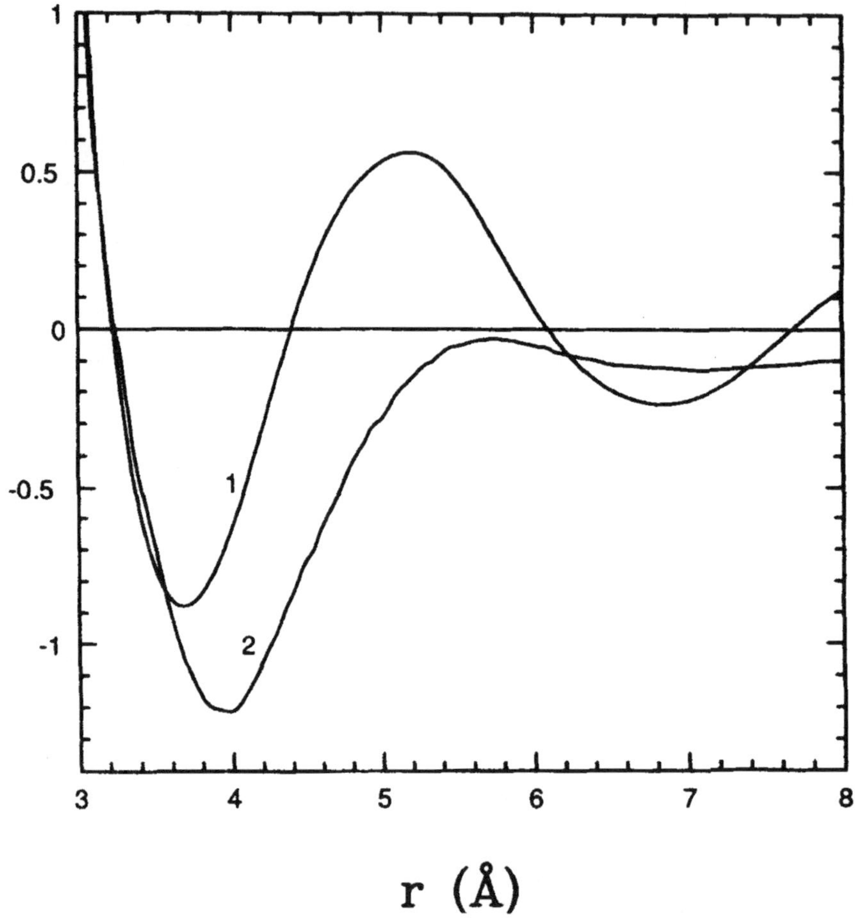

Figure 1 Curve 1: Potential of mean force $U(r)$ in units of $\beta = (k_B T)^{-1}$ for liquid Na near its freezing point. Curve 2: pair potential $\phi(r)$ of Reatto *et al.*[2].

this form of $\phi(r)$. Thus the use of approximate structural theories, in which g_3 is inevitably approximated, is delicate in this region.

To make this point quite clear, we show in Figure 2 a variety of such theories, and in particular their predictions of $\phi(r)$ when $U(r)$ from experiment shown in curve 1 of Figure 1 is used as input data. The dashed curve in Figure 2 is the result of approximate treatment proposed by Angelis and March[6], where $\phi(r)$ is given by

$$\beta\phi(r) = \beta U(r) + (8\pi^3\rho)^{-1} \int (S(k) - 1)^2 \exp(i\mathbf{k}.\mathbf{r})d\mathbf{k}. \tag{2}$$

The potential is a poor quantitative representation of that in curve 2 of Figure 1, and in particular, as stressed above, the approximation is too crude to annul in $\phi(r)$ the oscillation in $U(r)$ at large r. We have investigated possible improvements of

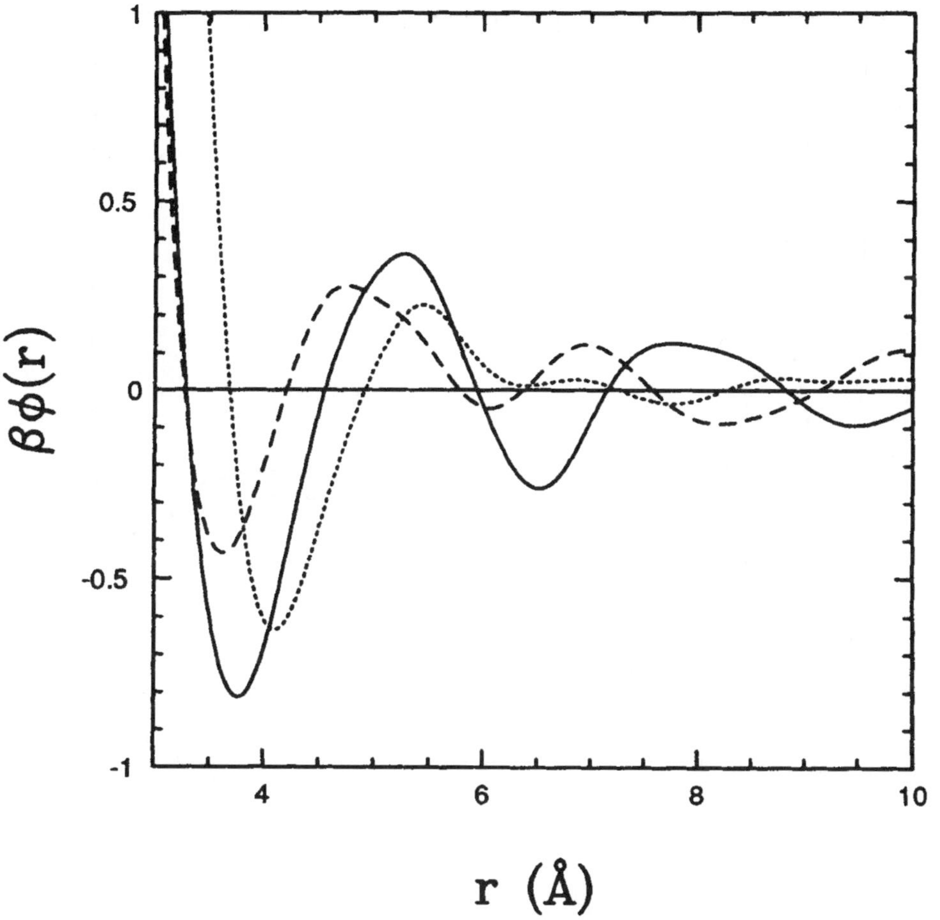

Figure 2 Effective pair potential $\phi(r)$ for liquid Na. Full, dashed and dotted curves are the results obtained from Eqs. (3), (2) and (4), respectively.

Eq. (2). One of these is that in which the Kirkwood approximation for g_3 in terms of $g(r)$ is combined in the force equation with the replacement in the three-body term of $\phi(r)$ by the "effective potential" $-k_BTc(r)$, where $c(r)$ is the Ornstein–Zernke direct correlation function. With neglect of term involving $c(r)dg(r)/dr$, one then finds

$$\beta\phi(r) = \beta U(r) + (8\pi^3\rho)^{-1} \int FT(g(r)c(r))(S(k) - 1) \exp(-i\mathbf{k}.\mathbf{r})d\mathbf{k} \tag{3}$$

with

$$FT(g(r)c(r)) = \rho \int g(r)c(r) \exp(i\mathbf{k}.\mathbf{r})d\mathbf{r}.$$

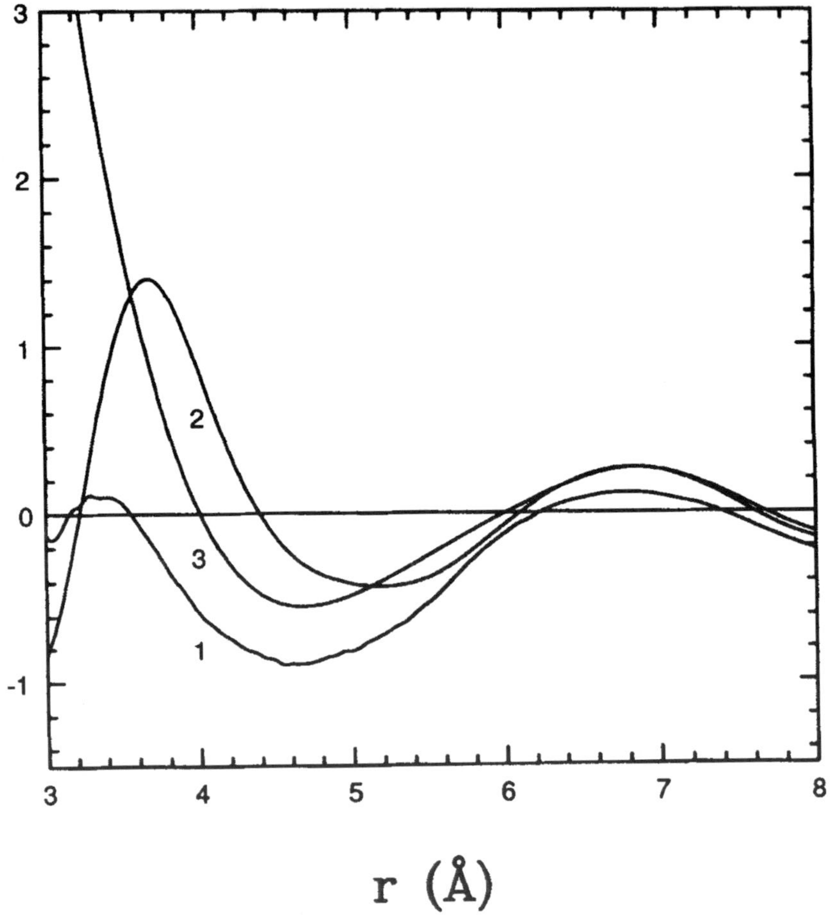

Figure 3 Curve 1: $\beta[\phi(r) - U(r)]$ obtained using data from Figure 1. Curve 2: $h(r)$. Curve 3: $h(r) - c(r)$.

The result of Eq. (3) is shown by full curve in Figure 2. Finally in Figure 2 the dotted curve shows the hypernetted chain (HNC) results:

$$\phi(r) = U(r) + k_B T(h - c). \tag{4}$$

As anticipated above, none of these three approximate structural theories is sufficiently refined to exhibit the cancellation between $U(r)$ and the three-body term revealed at large r in Figure 1.

Turning to possible refinements, we note that when $U(r)/k_B T$ is substantially less than unity, then Eq. (1) yields $h(r) = -U(r)/k_B T$. Thus after some 5 Å one can write from Figure 1 the result

$$\phi(r) - U(r) = h(r) k_B T \tag{5}$$

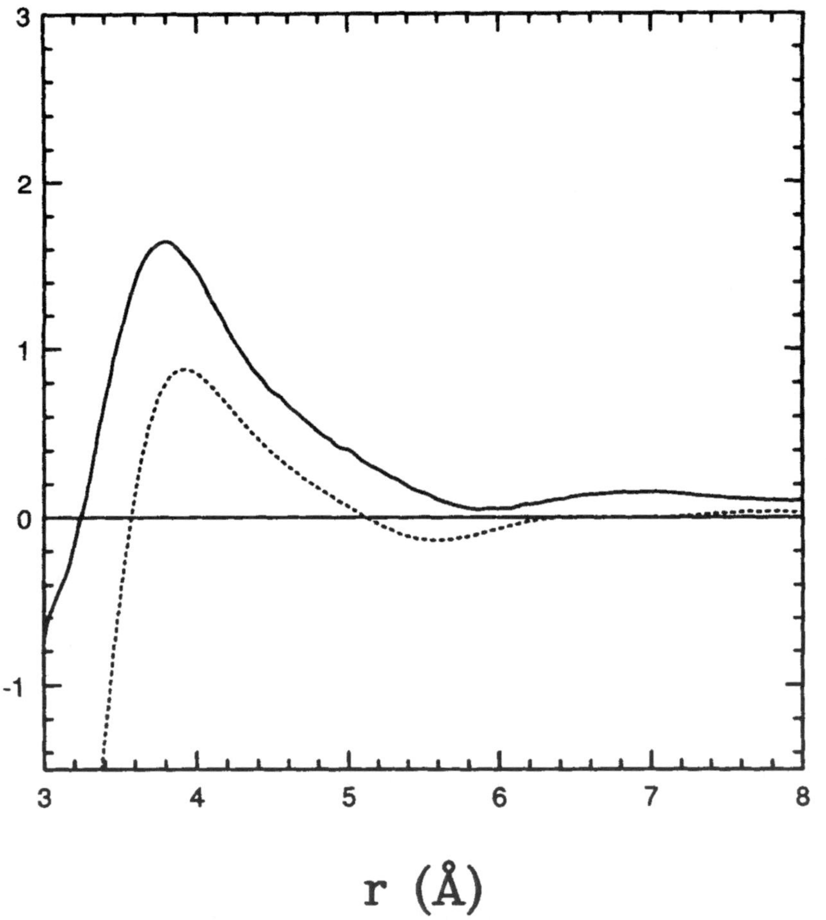

Figure 4 Full curve is $f(r)$ obtained from Eq. (7). Dashed curve represents direct correlation function $c(r)$.

which immediately builds in the large r cancellation emphasized above. Thus, in Figure 3, we plot the difference $\beta[\phi(r) - U(r)]$ obtained from Figure 1, in curve 1, while curve 2 shows $h(r)$. Curve 3 shows the HNC result h–c for comparison. For r greater than 4 Å, curves 1 and 3 have the similar shapes. Therefore, $\phi(r) - U(r) = h(r) - c(r) - B(r)$, where $B(r)$ is slow decaying function of r, may be the right modification. On the other hand for small r, it might be necessary to make an admixture of $h(r) - vc(r)$, where v is a constant hopefully very much less than unity.

In summary, while Eqs. (2), (3) and (4) lead to general shapes which are appropriate for pair potentials in conducting liquid like molten Na near freezing, they are far from quantitative at the first minimum, severely underestimating the depth of potential; Eq. (2) gives magnitude 0.43 for $\beta\phi(r)$, Eq. (3) 0.80, Eq. (4) 0.63 while the value of Perrot and March is 1.2, quite near to that of Reatto *et al.*[2]. Even more significant, the cancellation between $U(r)$ and the three-body term at large r is not adequate in any of Eqs. 2–4. In this respect, Eq. (5) is a definite improvement, though presently without fundamental status. However, it is tempting, in conclusion, to return to an observation made long ago by one of us[7] that a class of simple theories has the shape:

$$\phi(r) - U(r) = k_B T\rho \int G(\mathbf{r} - \mathbf{r}')h(\mathbf{r}')d\mathbf{r}' \tag{6}$$

Of course, if one retained the convolution from Eq. (6) for the deviation of $\phi(r)$ from $U(r)$, then Eq. (5) is regained by choice of the so-called force correlation function $G(r)$ as a δ function. Obviously this is too primitive, but it does suggest that $G(k)$, the FT of $G(r)$, is long-ranged in k space. Of course, an alternative is to give up the convolution from Eq. (6) and write

$$\beta[\phi(r) - U(r)] = h(r) - f(r) \tag{7}$$

To get an admittedly rough estimate of $f(r)$, we have taken $\phi(r) - U(r)$ from Figure 1 and the experimental $h(r)$. Then from Eq. (7), we have plotted $f(r)$ as curve 1 in Figure 4. For comparison, the HNC $c(r)$ is also shown as curve 2. It is found that $f(r)$ is always larger than $c(r)$.

Acknowledgement

We wish to acknowledge that this collaboration was made possible by the attendance of both authors at the summer 1991 Workshop in Condensed Matter at I.C.T.P., Trieste.

References

1. M. D. Johnson and N. H. March, *Phys. Lett.* **3**, 313 (1963).
2. See L. Reatto, *Phil. Mag.* **A58**, 37; and other references given there (1988).
3. F. Perrot and N. H. March, *Phys. Rev.* **A41**, 4521 (1990).
4. A. J. Greenfield, J. Wellendorf and N. Wiser, *Phys. Rev.* **A4**, 1607 (1971).
5. See, for example, N. H. March (1990). *Chemical Physics of Liquids* (Gordon and Breach: London), see also I. Ebbsjo, G. G. Robinson and N. H. March, *Phys. Chem. Liq.* **13**, 65 (1983).
6. U. de Angelis and N. H. March, *Phys. Lett.* **56A**, 287 (1976); see also D. I. Page, U. de Angelis and N. H. March, *Phys. Chem. Liq.* **12**, 53 (1982).
7. See N. H. March (1968). *'Liquid Metals'* (Pergamon: Oxford).

MOLECULAR PHYSICS, 1993, VOL. 80, NO. 4, 915–924

Three-particle correlation function and structural theories of dense metallic liquids

By K. I. GOLDEN†, N. H. MARCH‡ and A. K. RAY‡

†Department of Computer Science and Electrical Engineering, University of Vermont, Burlington, Vermont 05405, USA

‡Theoretical Chemistry Department, University of Oxford, 5 South Parks Road, Oxford OX1 3UB, England

(Received 5 April 1993; accepted 21 June 1993)

The first member of the Born–Green–Yvon hierarchy relates the pair correlation function $g(r)$ and the three-particle correlation function g_3 via an (assumed) pair potential $\phi(r)$. With the further assumption that the pair potential $\phi(r)$ and the potential of mean force separately possess their own Fourier transforms, one can derive the result that $S(k)\tilde{\phi}(k)/k_BT = -\tilde{E}(k)$, where $S(k)$ is the liquid structure factor, $\tilde{\phi}(k)$ the Fourier transform of $\phi(r)$, while $\tilde{E}(k)$ can be written in terms of the Fourier components of g_3 and $\phi(r)$. It is then shown that the precise form of $\tilde{E}(k)$ can be calculated for the one-component plasma in two dimensions with a $\ln r$ interaction for one particular coupling strength using the analytical pair function of Jancovici. This exact result is compared with results from two approximate dense liquid structural theories. Finally, some results are also presented for the liquid alkali metals Na and K near their freezing points, invoking electron theory for their pair potentials and X-ray measurements of their structure factors.

1. Introduction

Dense classical liquid structural theories start out from the first member of the Born–Green–Yvon (BGY) hierarchy which relates the pair function $g(r)$, the three-particle correlation function $g_3(\mathbf{r}_1, \mathbf{r}_2, \mathbf{r}_3)$ and the (assumed) pair potential $\phi(r)$. This equation is conveniently expressed in terms of the potential of mean force $U(r)$, defined by

$$g(r) = \exp\left[-U(r)/k_BT\right], \tag{1}$$

with

$$\frac{-\partial U(r_{12})}{\partial \mathbf{r}_1} = \frac{-\partial \phi(r_{12})}{\partial \mathbf{r}_1} - \rho \int \frac{g_3(\mathbf{r}_1, \mathbf{r}_2, \mathbf{r}_3)}{g(r_{12})} \frac{\partial \phi(r_{13})}{\partial \mathbf{r}_1} \, d\mathbf{r}_3. \tag{2}$$

For a given pair potential, this equation provides a direct route to the calculation of $g(r)$, provided an asumption can be made in which g_3 is written, but now approximately, in terms of the pair correlation function $g(r)$.

Born and Green [1] in very early work utilized the Kirkwood [2] approximation that g_3 can be written as the product $g(r_{12})\, g(r_{23}) g(r_{31})$, but severe thermodynamic inconsistencies are now known to arise, related to the violation of the asymptotic result for the Ornstein–Zernike direct correlation function $c(r)$ at large r. Far from the critical point, this relation is known to take the form [3]

$$c(r) = -\phi(r)/k_B T, \qquad r \text{ large}. \tag{3}$$

However, despite the violation of the large r result (3), the general 'shape' of the Born–Green approximation, first clearly set out by Rushbrooke [4], is of interest for what follows. Thus, inserting the Kirkwood assumption for g_3 into equation (2), this can be directly integrated to yield

$$\frac{U(r)}{k_B T} = \frac{\phi(r)}{k_B T} - \rho \int \mathscr{E}(|\mathbf{r} - \mathbf{r}'|) h(\mathbf{r}')\, \mathrm{d}\mathbf{r}' : \quad \mathscr{E}(r) = \frac{1}{k_B T} \int_r^\infty g(s) \frac{\partial \phi(s)}{\partial s}\, \mathrm{d}s. \tag{4}$$

Though equation (4) is no longer used for quantitative work, it will be brought into direct contact below with a formally exact development of equation (2) which is set out by two of the present authors elsewhere [5]. The purpose of the present paper is to bring this formally exact treatment into direct contact with various dense liquid structural theories, and hence to assess the accuracy of their treatment of the three-particle correlation function g_3.

2. Formally exact solution of equation (2) when $\phi(r)$ and $U(r)$ possess Fourier transforms

Below, we shall present first a brief summary of those results of [5] which we shall need in the present study. We shall restrict all further considerations to the special case when both the pair potential $\phi(r)$ and the potential of mean force $U(r)$ separately possess Fourier transforms $\tilde{\phi}(k)$ and $\tilde{U}(k)$ respectively. While this is undoubtedly a restrictive assumption, excluding, for example, Lennard-Jones 6–12 potentials from our further considerations, we shall consider two examples in sections 3 and 4 below for which these assumptions are valid. One is admittedly a simple model problem, namely a one-component plasma with an assumed $\ln r$ interaction [6], but the other is the case of simple s–p liquid metals such as Na or K, for which there is an effective pair potential $\phi(r)$ possessing a Fourier transform, for reasons set out by Corless and March [7].

2.1. *Fourier analysis of three-particle correlation function g_3*

The formal development of equation (2) in [5] hinges on the analysis of $g_3(\mathbf{r}_1, \mathbf{r}_2, \mathbf{r}_3)$ into the following form. First, one uses the vectors $\mathbf{r}_{12}$ and $\mathbf{r}_{32}$ as the independent variables, and separates out from g_3 the sum of the so-called total correlation functions $h(r_{ij})$, where

$$h(r) = g(r) - 1. \tag{5}$$

Explicitly therefore we write for the three-particle correlation function entering equation (2):

$$g_3(\mathbf{r}_{12}, \mathbf{r}_{32}) = 1 + h(r_{12}) + h(r_{32}) + h(r_{13}) + t(\mathbf{r}_{12}, \mathbf{r}_{32}). \tag{6}$$

The essential step required in what follows is then to analyse the remaining three-particle piece $t(\mathbf{r}_{12}, \mathbf{r}_{32})$ in equation (6) into its Fourier components, $\tilde{t}$ say. Then, given the assumption already set out that $\phi(r)$ and $U(r)$ separately possess Fourier transforms, we have solved equation (2) to yield the form:

$$\frac{\tilde{\phi}(k)}{k_B T} = -\frac{\tilde{E}(k)}{S(k)}, \tag{7}$$

with $S(k)$ the usual liquid structure factor, defined such that $S(k) - 1$ is simply the Fourier transform of $h(r)$ in equation (5). It only then remains to define the function $\tilde{E}(k)$ in equation (7), and this is conveniently done by separating it into two parts, namely [5]

$$\tilde{E}(k) = \tilde{h}(k) + \sum(k). \tag{8}$$

Then in equation (8) the summation $\sum$ can be written explicitly in terms of g_3 and ϕ as [5]:

$$\sum(k) = \frac{1}{V}\sum_{\mathbf{q}} \frac{\mathbf{k}\cdot\mathbf{q}}{k^2} \beta\rho\tilde{\phi}(q)\{\tilde{h}(|\mathbf{k}-\mathbf{q}|) + \rho\tilde{t}(\mathbf{k}-\mathbf{q},\mathbf{q})\}; \qquad \beta = (k_B T)^{-1}. \tag{9}$$

We must emphasize that under the assumptions made of the existence of Fourier transforms, equations (7)–(9) represent an exact solution of the BGY equation (2). We turn immediately to the first application of the above treatment, namely the two-dimensional one-component plasma (2D OCP) with $\ln r$ as the interaction potential.

3. 2D OCP; use of Jancovici pair function $g(r)$ to obtain three-particle term $\sum(k)$ in equation (8)

As already indicated in section 1 by reference to the (very approximate) treatment of Born and Green in equation (4), in applications of the BGY equation (2) it is usually necessary to decouple g_3 in terms of the two-particle correlation function $g(r)$, to make further progress in treating dense liquids. Fortunately, there is one model problem, namely the 2D OCP with $\ln r$ potential, where one can establish the precise relationship between $\tilde{E}(k)$, $\tilde{\phi}(k)$ and $S(k)$ in equation (7) in exact form, and hence compare this result for $\tilde{E}(k)$ with various approximations. This is therefore the aim of the present section.

To implement the above programme, the key point is to note that for the 2D OCP the pair function $g(r)$ has been calculated exactly by Jancovici [6] for a particular coupling strength $\Gamma = 2$. Here Γ is defined precisely as

$$\Gamma = \bar{\beta}\bar{e}^2, \tag{10}$$

where $\bar{\beta}^{-1}$ and $\bar{e}$ are respectively the thermal energy and charge per unit length. Thus $S(k)$ appearing in equation (7) can also be calculated exactly for this specific case $\Gamma = 2$. Since $\tilde{E}(k)$ in equation (7) is determined by $S(k)$ and $\tilde{\phi}(k)$, and since the 2D Fourier transform of $\ln r$ is known to be given by

$$\tilde{\phi}(k) = \frac{2\pi\bar{e}^2}{k^2}, \tag{11}$$

one can clearly plot the exact $\tilde{E}(k)$ against k for this model (see figure 1 below).

To achieve this, we note that the Jancovici pair function [6] has the form

$$g(r) = 1 - \exp(-r^2/a^2), \qquad a^2 = (\pi\rho)^{-1}. \tag{12}$$

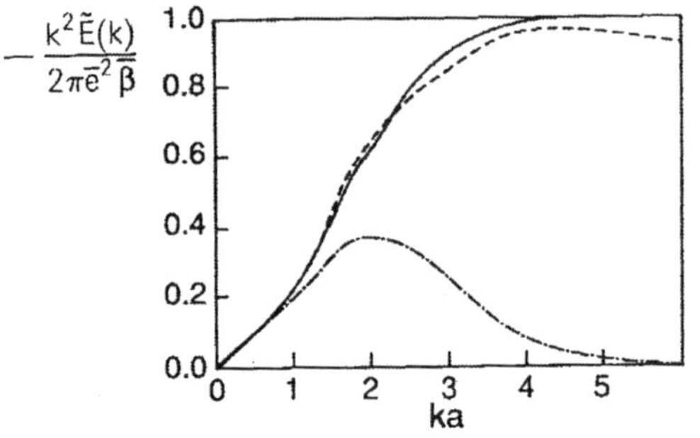

Figure 1. Plot of value of $k^2\tilde{E}(k)$, suitably scaled, against k in dimensionless units for one-component-plasma in two dimensions with $\ln r$ interaction at $\Gamma = 2$. Input into all three curves is (*i*) $S(k)$ from equation (13) and (*ii*) $\tilde{\phi}(k)$ from equation (11). Full curve — exact form of three-particle piece $\tilde{E}(k)$ in equations (7–9). Dot–dash curve — random phase approximation: equation (14). Dashed curve — hypernetted chain approximation: equation (20).

Taking the 2D Fourier transform of equation (12) to obtain the structure factor gives

$$S(k) = 1 - \exp(-k^2a^2/4). \tag{13}$$

Evidently, from $\tilde{\phi}(k)$ in equation (11) and $S(k)$ in equation (13) the exact $\tilde{E}(k)$ for the 2D OCP at $\Gamma = 2$ can be constructed and the result is plotted in figure 1. We turn immediately to a comparison of this exact curve with the results of approximate theories.

3.1. *Random phase approximation*

It is a simple matter to convince oneself that the random phase approximation (RPA), by now well known to be correct only in the long wavelength (small k) region, is equivalent to putting the summation $\sum(k)$ in equation (9) equal to zero. This very drastic assumption means that for $\tilde{E}(k)$, the quantity appearing in the central equation (7) of the present study, one has the approximation

$$\rho\tilde{E}_{\text{RPA}}(k) = \rho\tilde{h}(k) = S(k) - 1. \tag{14}$$

Inserting this into equation (7), and utilizing the relation between $S(k)$ and the Fourier transform of the direct correlation function $c(r)$ which results from the **r** space convolution definition of Ornstein and Zernike for $c(r)$:

$$\tilde{c}(k) = \frac{S(k) - 1}{S(k)}, \tag{15}$$

one finds for the RPA the results analogous to equation (3) but now in **k** space:

$$\tilde{c}_{\text{RPA}}(k) = -\frac{\tilde{\phi}(k)}{k_B T}. \tag{16}$$

We have plotted the RPA result in equation (14) in figure 1, where the exact form of $\tilde{E}(k)$ is given. It is immediately obvious that equation (14) is a very poor approximation for large k for the 2D OCP, falling off exponentially according to equation (13), whereas the exact $\tilde{E}(k)$ decays only slowly as k^{-2}. This is because, as already noted, the RPA is a long wavelength treatment only.

3.2. *Hypernetted chain approximation*

A much better approximation for dense liquids than the RPA is the hypernetted chain (HNC) treatment. This leads in **r** space to the result

$$\frac{U(r) - \phi(r)}{k_B T} = c(r) - h(r). \tag{17}$$

Inserting into the right-hand side of equation (17) the Ornstein–Zernike definition of $c(r)$, namely

$$h(r) = c(r) + \rho \int c(|\mathbf{r} - \mathbf{r}'|)h(r')\,\mathrm{d}\mathbf{r}', \tag{18}$$

one finds immediately a result having the same general 'shape' as the Born–Green (BG) approximation in equation (4), but with the function $\mathscr{E}(r)$ of that BG theory now replaced by the direct correlation function $c(r)$. In fact, $\mathscr{E}(r)$ from equation (4) evidently tends at sufficiently large r to $c(r)$ when one replaces $g(r)$ by its asymptotic large r limit of unity in equation (4) and then utilizes the result (16), though this latter step is, as noted above, not internally consistent within the BG framework. But what is important here is that both these theories have the same 'shape' in expressing $U(r) - \phi(r)$ as a convolution.

For the case of interest in the present study when both $U(r)$ and $\phi(r)$ have Fourier transforms, one can combine equations (17) and (18) to rewrite the HNC theory in **k** space as

$$\frac{\tilde{U}(k) - \tilde{\phi}(k)}{k_B T} = -\tilde{c}(k)[S(k) - 1]. \tag{19}$$

Hence one gets from equations (7) and (19) the HNC result for $\tilde{E}(k)$ as

$$\tilde{E}_{\mathrm{HNC}}(k) = -[S(k) - 1]^2 - \frac{\tilde{U}(k)}{k_B T} S(k). \tag{20}$$

Without going into detail here [8], one can utilize equations (1) and (12) to calculate an infinite series for $\tilde{U}(k)$ for the 2D OCP, and this can be used in conjunction with equations (20) and (13) to compute $\tilde{E}_{\mathrm{HNC}}(k)$. This later quantity is plotted for comparison with the exact result in figure 1. The result, as expected, is a great improvement on the RPA but still does not go to the correct limit at large k.

3.3. *Born–Green approximation and its refinement*

As already noted, the BG approximation (4) has identical shape to that of the HNC theory, but with $\mathscr{E}(r)$ replacing $c(r)$. Thus one can readily show that

$$\tilde{E}_{\mathrm{BG}}(k) = -[S(k) - 1]S(k)\tilde{\mathscr{E}}(k) - \frac{\tilde{U}(k)}{k_B T} S(k). \tag{21}$$

Since this, in general, violates the asymptotic result (3), we have not considered it further. But the later work of Gaskell [9] has proposed a modification of the BG theory which is consistent with equation (3). Utilizing his work, we have obtained a modified BG approximation which turns out to be

$$\tilde{E}_{\mathrm{MBG}}(k) = -\tilde{c}(k)[S(k)]^2\tilde{\mathscr{F}}(k) - \frac{\tilde{U}(k)}{k_B T} S(k), \tag{22}$$

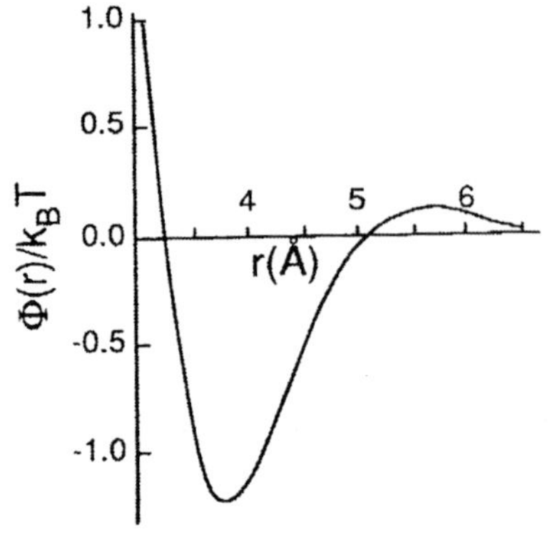

Figure 2. Electron theory pair potential $\phi(r)$ against r in liquid Na at freezing point. $\phi(r)$ is in units of $k_B T$ while r is in Å. Number density is 0·003 604 6 a.u. Fourier transform of $\phi(r)$ is used in figure 4 below.

where $\tilde{\mathscr{F}}(k)$ is the Fourier transform of the function $\mathscr{F}(r)$ in the Appendix. Again we have not pressed the detail numerically, but equation (22) transcends equation (21) in the sense that the correct long wavelength limit is now contained in equation (22).

From this problem, which provides a stringent test of approximations made in the BGY equation (2) in this admittedly restricted model, we turn finally to consider the treatment of section 2 in relation to work on the alkali metals (especially Na and K) near their freezing points.

4. Estimate of three-particle term $\tilde{E}(k)$ in equation (7) using electron theory of pair potentials in liquid alkali metals Na and K near freezing

As emphasized in [7], one can use effective pair potentials $\phi(r)$ to describe the structure of dense s–p liquid metals such as Na and K near their respective freezing points. This has culminated in rather precise calculations of the pair potential $\phi(r)$ from electron theory, by means of the density functional formalism [10]. These electron theory potentials do indeed possess Fourier transforms, as anticipated in [7]. Such potentials for Na [11] and also for K [12] near their freezing points (100°C for Na and 65°C for K) are plotted in figures 2 and 3 respectively. What is needed in the present work, to utilize equation (7), is $\tilde{\phi}(k)$ corresponding to these **r** space results and these have been calculated.†

Paralleling the arguments in section 3, in order to estimate the previously unknown three-particle term $\tilde{E}(k)$, we require not only $\tilde{\phi}(k)$ as calculated above, but also the structure factor $S(k)$. Fortunately, for Na and K near freezing these have been measured very accurately in the X-ray study of Greenfield *et al.* [13], and also, just for K, in the very recent neutron study of Johnson *et al.* [12]. Using these two quantities $\tilde{\phi}(k)$ and $S(k)$ as input into equation (7), the three-particle term has been obtained as follows. First, as is convenient in practice, let us subtract out the

† Dr F. Perrot has kindly calculated $\tilde{\phi}(k)$ for liquid potassium near freezing. Tables will be supplied by him, on request, from the following address: Centre d'Etudes de Limeil Valenton, 94195 Villeneuve St Georges Cedex, France.

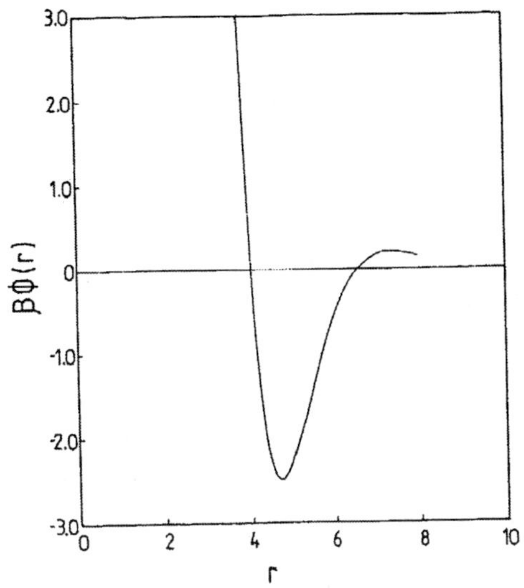

Figure 3. As in figure 2 but now for liquid K at freezing point. Density in this plot is 0·001 882 5.

bare monovalent ion Coulomb repulsion from $\tilde{\phi}(k)$ by writing

$$\tilde{\phi}(k) = \frac{4\pi}{k^2} + F(k), \tag{23}$$

where $F(k)$ is the precise quantity tabulated by inverting by Fourier transform the **r** space potentials in figures 2 and 3.

Substituting this result into equation (7), one can readily rearrange to yield

$$-\frac{k^2}{4\pi} k_{\mathrm{B}} T\tilde{E}(k) = S(k)\left[1 + \frac{k^2}{4\pi} F(k)\right]. \tag{24}$$

Hence the three-particle term is obtained in the form shown on the left-hand side of equation (24) and this is plotted against k in figure 4 for Na at freezing. The marked structure exhibited there is coming predominantly from the structure factor $S(k)$ as measured by Greenfield *et al.* [13]. We have also carried out the calculations for K using the $S(k)$'s from both [13] and [12]. As no new features are revealed, the same curve essentially being shifted bodily to smaller k in figure 4, we have not plotted the K results.

Again, it is immediately clear that the RPA result (14) is only useful in the long-wavelength limit, tending to zero at large k in contrast to the result shown in figure 4. Furthermore, the studies of Reatto *et al.* [14] using computer simulation to bypass the need for decoupling g_3, have already revealed the deficiencies of the HNC structural theory for liquid Na at freezing. Even with a 'bridge correction' calculated in the hard spheres approximation added to the HNC result (17), the pair potential extracted from the modified HNC procedure is only about half as deep as the potential shown in figure 2 for Na. We shall therefore not go into further detail here. But, in the future, when accurate experimental results become available for Na

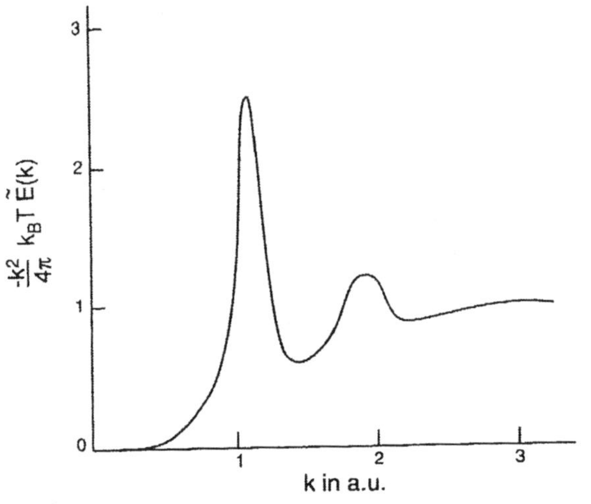

Figure 4. Plot of three-particle term $\tilde{E}(k)$ times k^2, suitably scaled, against k in a.u. for liquid Na at freezing point. Input data is $\tilde{\phi}(k)$ constructed from figure 2 by means of equation (23), plus experimental structure factor $S(k)$ from [13].

or K on the density dependence of the pair function, it will be of obvious interest to compute $\tilde{E}(k)$ from the modified BG result (22), to see if this transcends the modified HNC approximation.

5. Summary

To summarize, equation (7) is entirely equivalent to the BGY equation (2) in the admittedly restrictive case when both the pair potential $\phi(r)$ and the potential of mean force $U(r)$ possess Fourier transforms. Though this assumption is not valid for a Lennard-Jones 6–12 potential, it is true for (*a*) the two-dimensional one-component plasma model treated in section 3 and also (*b*) for the simple s–p liquid metals Na and K near freezing.

In these examples (*a*) and (*b*) therefore, knowledge of $S(k)$ and the Fourier transform $\tilde{\phi}(k)$ of the pair potential $\phi(r)$ allow the three-particle term $\tilde{E}(k)$ to be constructed. In example (*a*), the solution of Janovici [6] for the pair function $g(r)$ at coupling strength $\Gamma = 2$ allows a precise plot to be made as in figure 1. This affords a direct test of two approximate theories shown there, namely RPA and HNC. While the latter is much to be preferred, it is in error quantitatively at short wavelengths, though much less drastically than the RPA result, which is specifically a long wavelength approximation.

While the precision of example (*a*) is not yet possible for the liquid alkali metals, the electron theory (density functional) pair potentials shown in figures 2 and 3 allow one to construct again the three-particle contribution $\tilde{E}(k)$, which is shown explicitly for liquid Na in figure 4. This curve also utilized the measured liquid structure factor $S(k)$ for liquid Na at freezing, as obtained by Greenfield *et al.* [13], in conjunction with the Fourier components of the pair potential recorded in figure 3. The interest in having measurements on the density dependence of the pair function near freezing for liquid Na or K is pointed out, as this would permit one to test stringently the modified Born–Green treatment shown in equation (22).

We are deeply indebted to Dr F. Perrot for calculating $F(k)$ in equation (23) for us. We also wish to thank Dr M. W. Johnson of the Rutherford Laboratory for his continuing interest in this work and for valuable discussions.

Appendix

Generalization of $\mathscr{E}(r)$ in liquid structural theory based on the Kirkwood form of g_3 to include density derivative of the pair function $g(r)$

The ideas of Raveche and Kayser [15] (see also Gaskell [9]), suggest that a more appropriate approximation to the three-particle distribution function g_3 than the Kirkwood form [2] is given by

$$g_3(r_1,r_2,r_3) = g(r_{12})g(r_{23})g(r_{31})\{1+[\rho/2g(r_{13})]\,\partial g(r_{13})/\partial\rho[h(r_{12})+h(r_{32})]\}, \quad (\text{A }1)$$

where the density dependence of $g(r)$ now appears explicitly.

Inserting equation (A 1) in the BGY equation (2), making an additional approximation which however still preserves the asymptotic form (3) of $c(r)$, which was in fact the major motivation for adding a density derivative term, leads back to equation (4), provided $\mathscr{E}(r)$ there is replaced by $\mathscr{F}(r)$. In turn, this is defined to be given by

$$\mathscr{F}(r) = \frac{1}{k_\mathrm{B}T}\int_r^\infty \mathrm{d}s\left[g(s)+\frac{\rho}{2}\frac{\partial g(s)}{\partial\rho}\right]\frac{\partial\phi(s)}{\partial s}. \quad (\text{A }2)$$

Assuming that $\phi(r)$ and $U(r)$ have again separate Fourier transforms, the generalized form of equation (4) then yields

$$\frac{\tilde{\phi}(k)-\tilde{U}(k)}{k_\mathrm{B}T} = \tilde{\mathscr{F}}(k)\tilde{h}(k). \quad (\text{A }3)$$

Returning now to the formally exact equation (7), and replacing $1/S$ by $(1-\tilde{c})$, one has

$$\frac{\tilde{\phi}(k)}{k_\mathrm{B}T} = -\tilde{E}(k)[1-\tilde{c}(k)]. \quad (\text{A }4)$$

Hence it follows from equation (A 3) that the modified BG (MBG) result is

$$-\tilde{E}_\mathrm{MBG}(k)[1-\tilde{c}(k)] = \frac{\tilde{U}(k)}{k_\mathrm{B}T}+\tilde{\mathscr{F}}(k)\tilde{h}(k). \quad (\text{A }5)$$

On rearrangement, this is equivalent to equation (22) of the main text.

When density derivative data become available, it would be of interest to insert experimental information for $S(k)$ and $\tilde{U}(k)$ say for liquid metal Na near freezing, plus the electron theory data for $\tilde{\phi}(k)$ from figure 2, into equations (A 2) and (A 6) to compare $\tilde{E}_\mathrm{MBG}(k)$ with the form of $\tilde{E}(k)$ plotted in figure 4.

References

[1] Born, M., and Green, H. S., 1946, *Proc. R. Soc.* A, **188**, 10.

[2] KIRKWOOD, J. G., 1935, *J. chem. Phys.*, **3**, 300.
[3] See, for example, JOHNSON, M. D., HUTCHINSON, P., and MARCH, N. H., 1964, *Proc. R. Soc.* A, **282**, 283; ENDERBY, J. E., GASKELL, T., and MARCH, N. H., 1965, *Proc. phys. Soc.*, **85**, 217.
[4] RUSHBROOKE, G. S., 1960, *Physica*, **26**, 259.
[5] GOLDEN, K. I., and MARCH, N. H., 1993, *Phys. Chem. Liquids*, (to be published).
[6] JANCOVICI, B., 1981, *Phys. Rev. Lett.*, **46**, 386.
[7] CORLESS, G. K., and MARCH, N. H., 1961, *Phil. Mag.*, **6**, 1285.
[8] GOLDEN, K. I., and MARCH, N. H., 1993 (to be published).
[9] GASKELL, T., 1988, *J. Phys.* C, **21**, 1.
[10] PERROT, F., and MARCH, N. H., 1990, *Phys. Rev.* A, **42**, 4884.
[11] PERROT, F., and MARCH, N. H., 1990, *Phys. Rev.* A, **41**, 4521.
[12] JOHNSON, M. W., MARCH, N. H., PERROT, F., and RAY, A. K., 1993, *Phil. Mag.*, (to be published).
[13] GREENFIELD, A. J., WELLENDORF, J., and WISER, N., 1971, *Phys. Rev.* A, **4**, 1607.
[14] REATTO, L., 1988, *Phil. Mag.* A, **58**, 37, and other references therein.
[15] RAVECHE, H. J., and KAYSER, R. F., 1984, *Phys. Rev.* A, **29**, 1003.

PHYSICAL REVIEW E VOLUME 48, NUMBER 3 SEPTEMBER 1993

Long-range polarization interaction in simple liquid metals

M. Blazej and N. H. March
Theoretical Chemistry Department, University of Oxford, 5, South Parks Road, Oxford OX1 3UB, England
(Received 30 December 1992)

In earlier work [F. Perrot and N. H. March, Phys. Rev. A **41**, 4521 (1990); **42**, 4884 (1990)], the electron theory of pair potentials in simple s-p liquid metals, such as Na and K near freezing, was set out. The potential thus derived for Na is in good agreement in its major features with that obtained by inverting diffraction measurements of the liquid structure factor $S(k)$. The present study addresses polarization effects in the interatomic interaction in such s-p metals, a simple physical model being introduced. By a lower-bound argument, this is shown to lead to an attractive contribution to the effective pair potential which decays as R^{-4} at large interionic separation R. Since it is recognized that some difficulties still remain in justifying the simple model employed in a dense system, it is important to test experimentally whether such a long-range interaction is actually present in liquid Na or K near freezing; it is proposed that highly accurate small-angle neutron scattering could resolve this question precisely. Though less decisive for detection of the long-range polarization effect, the possible relevance to (i) long-wavelength phonon properties in hot (and cold) crystalline Na metal and (ii) the interaction between charged defects in metals (e.g., a divacancy) is also noted.

PACS number(s): 61.25.Mv, 61.20.−p

I. INTRODUCTION

In previous studies by Perrot and March [1,2; referred to below as PM], the density-functional theory of pair potentials in simple metals such as Na and K near freezing was set out, and brought into contact with results from liquid-structure inversion. As shown in PM for liquid Na just above its freezing point, all the salient features of the "diffraction" pair potential as extracted from the liquid structure x-ray measurements of Greenfield, Wellendorf, and Wiser [3] by Reatto [4] are reflected in the "density functional" potential.

However, some quantitative differences remain between these two pair potentials. This has prompted us, especially because the diffraction potential at large interionic separation lies below the zero potential axis, whereas the electron-theory potential lies above this axis out to 6.5 Å, to enquire whether there may be another physical contribution to "correct" the electron-theory potential. We have been led thereby to consider the possibility of a long-range polarization interaction between a pair of screened ions embedded in a metallic medium.

The issue of polarization effects in the interatomic interaction in metals has already attracted some interest in recent years, notable work being that of Maggs and Ashcroft [5]. These workers had in mind rather specifically metals with d electrons (e.g., silver), whereas in the present study we focus exclusively on a model that should have relevance to "nearly-free-electron" metals such as Na and K above. However, we must stress at the outset that all our present results are based on a static computation of (a lower bound to) the polarization interaction between two ions and their screening electron clouds, on the basis of which (model) the centers of the blobs of electrons are found to be displaced from the positions of the ions.

It needs to be said, at the outset, that it remains to give a full theoretical justification of this approach in a dense system, and in particular to use the polarization interaction calculated here as a pairwise additive potential. It is a problem for the future to build a bridge between the present static calculation and a fluctuation description. Nevertheless, we believe the consequences of our admittedly oversimplified static computation are of sufficient interest to prompt experimental studies of the long-range polarization interaction that arises in the present work.

Returning to the original motivation of our static calculation, early work by Gurney and Magee [6] on the free-space H_2 molecule allowed the 1s hydrogen orbitals to "float" away from the protons. The distance along the H_2 molecular axis by which these orbitals move was then treated as a variational parameter. Such a treatment immediately gave a substantial improvement in the ground-state energy as a function of internuclear separation.

Since the earlier work of PM was built directly on the superposition density of single ions, it seemed clear that the simplest approach was to allow such spherical blobs of screening charge to float off the ionic centers, without, however, any deformation of the single-ion spherical blobs. This is the picture adopted in the present work. Remarkably, it then turns out that the new potential energy curve can be largely characterized by the original pair potential of PM, plus the (assumed vector) displacements of the two blobs of screening charge from the ionic centers. Since, at large interionic separations, such displacements must become small, the "correction" to the standard pair potential $\phi(R)$ of PM can be expressed in terms of this potential and its low-order derivatives, plus the vector displacements.

Sections II and III are concerned with reporting a calculation based on this simple floating blobs model of the polarization interaction. Thus Sec. II sets out the correction to $\phi(R)$, say $\Delta\phi(R)$, the essential asymptotic result being expressed in Eq. (2.16) below. The vector displacements, however, are input information into this equation.

Then, in Sec. III, we determine the lower bound of this correction $\Delta\phi(R)$ to the pair potential. This leads, in what seems a rather natural way, to a polarization interaction (3.7), which falls off as the inverse fourth power of the interionic pair separation. Section IV is concerned with the possible experimental implications that such an inverse power law would have. Section V constitutes a summary, with some proposals for further work.

II. POLARIZATION-EFFECTS-CORRECTED PAIR INTERACTION

The model of "valence blobs of charge floating off the nuclei" offers an approach to take into account long-range correlation effects. This is already evident from the free-space study of London dispersion forces in homonuclear diatomic molecules by Egorov and March [7].

Within the framework of the derivation of pair interactions from density functional theory, this would correspond to a change in the functional $G[n]$, where n is the electron density, which describes kinetic plus exchange and correlation effects, when two ions are present. In the case of a single ion, $G[n]$ is that of the uniform electron gas $G_0[n]$, with the corresponding response function related to the functional second derivative. For two ions, we assume that the second derivative, and thus the response function, are changed due to polarization effects. The displaced density then becomes

$$\Delta n(\mathbf{r}+\mathbf{R}/2-\delta\mathbf{R}_1)+\Delta n(\mathbf{r}-\mathbf{R}/2-\delta\mathbf{R}_2) .$$

For any system containing one or two ions screened in a bath of conduction electrons, the change in total energy, with respect to that of the originally uniform electron background (i.e., without the ions embedded) can be written, as commonly done in second-order perturbation theory:

$$\Delta E=\int v_{\text{ext}}\bar{n}\,d\mathbf{r}+\tfrac{1}{2}\int v_{\text{ext}}\delta n\,d\mathbf{r} + (\text{change in ion-ion interaction}) . \tag{2.1}$$

In this equation, $\bar{n}$ denotes the density of the unperturbed homogeneous electron gas, v_{ext} the perturbing external potential, while δn is the total displaced electron density due to embedding the ions in the background of conduction electrons.

It will be assumed that the system with a pair of ions embedded in the electron gas, and the system with but a single ion, both satisfy Eq. (2.1). Thus the change in the pair interaction results from the change in the second term of ΔE, associated with floating the electron clouds, all the other quantities (v_{ext} and R, the interionic separation) being fixed.

We note here that Eq. (2.1) is applicable to weak perturbations, i.e., with v_{ext} taken as a pseudopotential. In what follows, it will be used with the true Coulomb potential of the nuclei. It is easy to verify that this does not change the form of the results at large R, except for the values of the constants.

A. Change in pair interaction due to floating of electronic clouds off the ionic centers

With the ions at positions separated by the vector $\mathbf{R}$, let us now allow the (assumed spherical) blob of screening charge to float off ion 1 by an amount determined by the vector $\delta\mathbf{R}_1$. Similarly, the electron valence cloud round ion 2 floats off by amount $\delta\mathbf{R}_2$.

The change $\Delta\phi(R)$ in the pair potential $\phi(R)$ is then calculable, with Δn denoting the density displaced by one ion acting alone. With the notation that a circle ($\circ$) means integration through the whole of space, $\Delta\phi(R)$ can be written, with Z denoting the valence ($Z=1$ for Na and K, 2 for Be),

$$\Delta\phi(R)=-\tfrac{1}{2}Z\left[\frac{1}{|\mathbf{r}+\delta\mathbf{R}_2|}+\frac{1}{|\mathbf{r}+\mathbf{R}+\delta\mathbf{R}_2|}+\frac{1}{|\mathbf{r}-\mathbf{R}+\delta\mathbf{R}_1|}+\frac{1}{|\mathbf{r}+\delta\mathbf{R}_1|}-\frac{1}{|\mathbf{r}|}-\frac{1}{|\mathbf{r}+\mathbf{R}|}-\frac{1}{|\mathbf{r}-\mathbf{R}|}-\frac{1}{|\mathbf{r}|}\right]\circ\Delta n(r) . \tag{2.2}$$

Recalling next that the "standard" pair interaction is

$$\phi(R)=\frac{Z^2}{R}-\frac{Z}{|\mathbf{r}-\mathbf{R}|}\circ\Delta n(r) , \tag{2.3}$$

one can transform the expression (2.2) to read

$$2\Delta\phi(R)=\phi(|\mathbf{R}+\delta\mathbf{R}_2|)+\phi(|\mathbf{R}-\delta\mathbf{R}_1|)-2\phi(R)-\frac{Z^2}{|\mathbf{R}+\delta\mathbf{R}_2|}-\frac{Z^2}{|\mathbf{R}-\delta\mathbf{R}_1|}+\frac{2Z^2}{R}+N(\delta R_1)+N(\delta R_2) . \tag{2.4}$$

In Eq. (2.4), $N(\delta\mathbf{R})$ denotes the quantity

$$N(\delta\mathbf{R})=-Z\left[\frac{1}{|\mathbf{r}+\delta\mathbf{R}|}-\frac{1}{\mathbf{r}}\right]\circ\Delta n(r) . \tag{2.5}$$

B. Approximate treatment for small floating distances

In what follows, we shall now develop the above results in a form which will lead to the asymptotic behavior of $\Delta\phi(R)$: the change in the standard $\phi(R)$ due to the long-range polarization interaction. From Eq. (2.5), one can first write

$$N(\delta\mathbf{R})=\frac{-Z}{\delta R}\int_0^{\delta R}\Delta n(x)4\pi x^2dx+Z\int_0^{\delta R}\Delta n(x)4\pi x\,dx . \tag{2.6}$$

Assuming that $\Delta n(r)$ is almost constant and equal to $\Delta n(0)$ in the range $0\rightarrow\delta R$, one readily obtains

$$N(\delta\mathbf{R})=\frac{2\pi}{3}Z\Delta n(0)(\delta R)^2 \tag{2.7}$$

(see also the Appendix, where the cruder Thomas-Fermi

approximation is employed).

This is the point at which to return to Eq. (2.4) for $\Delta\phi(R)$. One can expand this equation in terms of $s=\delta R/R$ and θ, the angle between $\delta\mathbf{R}$ and $\mathbf{R}$ for the first term on the right-hand side of Eq. (2.4). One can then develop such an expansion by noting that

$$\phi(|\mathbf{R}+\delta\mathbf{R}|)=\phi(R)+\phi'_R[|\mathbf{R}+\delta\mathbf{R}|-R] + \frac{\phi''_R}{2}[|\mathbf{R}+\delta\mathbf{R}|-R]^2+\cdots . \quad (2.8)$$

In terms of s and θ introduced above:

$$|\mathbf{R}+\delta\mathbf{R}|-R=R[s\cos\theta+\tfrac{1}{2}s^2(1-\cos^2\theta)+O(s^3)] \quad (2.9)$$

and

$$(|\mathbf{R}+\delta\mathbf{R}|-R)^2 \doteq R^2[s^2\cos^2\theta+O(s^3)] . \quad (2.10)$$

We must stress here the considerable advantage that has been gained in replacing Δn in Eq. (2.2) by ϕ to reach Eq. (2.4). Since, as discussed in PM, at large R, ϕ decreases as $R^{-3}\cos(2k_fR+\alpha)$, all its derivatives fall off with distance as R^{-3}, and in particular ϕ' is negligible compared to Z^2/R^2. Specifically, Eq. (2.4) becomes

$$2\Delta\phi(R)=R\left[\phi'_R+\frac{Z^2}{R^2}\right][-s_1\cos\theta_1+\tfrac{1}{2}s_1^2(1-\cos^2\theta_1)] + \frac{R^2}{2}\left[\phi''_R-\frac{2Z^2}{R^3}\right]s_1^2\cos^2\theta_1+Ms_1^2+O(s_1^3) + (\text{similar expression in } s_2, -\cos\theta_2) . \quad (2.11)$$

Here, the quantity M follows from Eq. (2.7) as

$$M=\frac{2\pi}{3}Z\Delta n(0)R^2\equiv mR^2 . \quad (2.12)$$

It is now useful to define three quantities A, B, and C and to note, as we do this, their asymptotic forms as $R\to\infty$. The definitions adopted are

$$A=R\left[\phi'_R+\frac{Z^2}{R^2}\right]\underset{R\to\infty}{\sim}\frac{Z^2}{R} , \quad (2.13)$$

$$B=\frac{R^2}{2}\left[\phi''_R-\frac{2Z^2}{R^3}\right]-\frac{A}{2}\underset{R\to\infty}{\sim}-\frac{3Z^2}{2R} , \quad (2.14)$$

and

$$C=M+\frac{A}{2}\underset{R\to\infty}{\sim}mR^2 . \quad (2.15)$$

Returning to Eq. (2.11), one then finds

$$2\Delta\phi(R)=-As_1\cos\theta_1+Bs_1^2\cos^2\theta_1+Cs_1^2 + (\text{similar terms with } s_1\to s_2,\ \theta_1\to\pi-\theta_2) . \quad (2.16)$$

Equation (2.16) is then the basic consequence of the "floating-valence-blobs" model to represent the long-range polarization interaction in simple liquid metals such as Na and K near freezing.

Of course, the above expression (2.16) is valid within that model, for given displacements of the screening charges from the ionic centers 1 and 2, described by the variables (s_i,θ_i), $i=1$ and 2. But to complete the calculation, one must determine these variations by a first-principles method.

III. LOWER BOUND TO THE CHANGE $\Delta\phi(R)$ IN THE PAIR INTERACTION

The major step still to be taken is to turn the form (2.16), which follows very generally for $\Delta\phi(R)$, the correction to the standard pair interaction $\phi(R)$ at large R, into an explicit asymptotic result for $\Delta\phi(R)$. We shall tackle this by determining the maximum correction that this floating-blobs model can make to the original pair potential $\phi(R)$. In order to find this lower bound to $\Delta\phi$, we must minimize Eq. (2.16) with respect to (s_i,θ_i), $i=1$ and 2. One is then led to the equations

$$As_1\sin\theta_1-2Bs_1^2\sin\theta_1\cos\theta_1=0 \quad (3.1)$$

and

$$-A\cos\theta_1+2Bs_1\cos^2\theta_1+2Cs_1=0 . \quad (3.2)$$

To satisfy Eq. (3.1), one must take the solution $\sin\theta_1=0$. The other solution $s_1\cos\theta_1=A/2B$ is not physically acceptable because, at large R, using Eq. (2.13) and (2.14), $A/B\to\frac{2}{3}$ and thus is not small. With $\sin\theta_1=0$, one has

$$\cos\theta_1\equiv\epsilon_1=\pm1 \quad (3.3)$$

and inserting this into Eq. (3.2) yields

$$-A\varepsilon_1+2(B+C)s_1=0 . \quad (3.4)$$

Inserting the asymptotic forms of A, B, and C from Eqs. (2.13)–(2.15), one finds

$$s_1\equiv\frac{\delta R_1}{R}=\frac{\epsilon_1Z^2}{2m}\frac{1}{R^3} \quad (3.5)$$

or

$$\delta R_1\equiv\frac{\epsilon_1Z^2}{2m}\frac{1}{R^2} \quad \text{for } \epsilon_1=1 . \quad (3.6)$$

The plus sign in Eq. (3.3) follows because the magnitude δR_1 must be positive.

The expression in $(s_2,-\cos\theta_2)$ is exactly symmetric, so that one finds $\epsilon_2=-1$. Hence, returning to Eq. (2.16) and inserting these results, together with the asymptotic forms for large R of A, B, and C, one is led to the final form of the lower bound:

$$\Delta\phi(R)=\frac{-A^2}{4(B+C)}\to\frac{-Z^4}{4m}\frac{1}{R^4} . \quad (3.7)$$

The displacements of the floating blobs of valence screening from the ionic centers are in opposite directions along the axis joining the ionic centers. The distance between the centers of the electronic charge is less than R.

It must be stressed that the inverse-fourth-power dependence on the interionic separation R given for the change in the standard pair potential $\phi(R)$ in Eq. (3.7) does not behave like the London dispersion force. It is specifically a polarization interaction in a simple metallic medium such as liquid Na or K near freezing.

IV. POSSIBLE EXPERIMENTAL IMPLICATIONS OF INVERSE-FOURTH-POWER POLARIZATION INTERACTION

The first comment to make in relation to Eq. (3.7) is that it is qualitatively in accord with the differences between the diffraction potential for Na obtained by Reatto [4] and the density functional standard pair potential calculated by Perrot and March [1]. The lowering of the standard electron theory potential predicted in Eq. (3.7) qualitatively brings the two potentials together at large R. However, the asymptotic form (3.7) could not be expected to work inside some 4.5 Å, as the electron theory and diffraction potentials cross one another at just less than 4 Å.

The second point relating to Eq. (3.7) is that, guided by pseudopotential representations of $\phi(R)$, which yield most directly its Fourier transform $\tilde{\phi}(k)$, the leading asymptotic form

$$\phi(R)\propto\frac{\cos(2k_fR+\alpha)}{R^3} \tag{4.1}$$

leads to singularities at $k=2k_f$ in $\tilde{\phi}(k)$. However, the correction $\Delta\phi(R)$ in Eq. (3.7) leads to $\Delta\tilde{\phi}(k)$ having a term proportional to k as $k\to 0$, i.e., the total Fourier transform now has two singular points: $k=0$ and $k=2k_f$, in contrast to the standard $\phi(R)$ in Eq. (4.1) for a sharp Fermi surface of diameter $2k_f$.

Such behavior in the Fourier transform of the effective potential is accessible to diffraction experiments on simple liquid metals through the Ornstein-Zernike direct correlation function $c(r)$. If $g(r)$ denotes the pair function, then the total correlation function $h(r)=g(r)-1$ is related to $c(r)$ via the convolution equation

$$h(r)=c(r)+n_a\int c(r')h(\mathbf{r}-\mathbf{r}')d\mathbf{r}' , \tag{4.2}$$

where n_a is the atomic-number density. With the usual liquid structure factor $S(k)$ related to the Fourier transform $\tilde{h}(k)$ through

$$\tilde{h}(k)=S(k)-1 , \tag{4.3}$$

Eq. (4.2) can be solved for $\tilde{c}(k)$, the Fourier transform of $c(r)$, as

$$\tilde{c}(k)=\frac{S(k)-1}{S(k)} . \tag{4.4}$$

Since $S(k=0)$ is known from the fluctuation theory result

$$S(0)=n_ak_BTK_T , \tag{4.5}$$

where K_T is the isothermal compressibility, it follows from Eqs. (4.5) and (4.4) that $\tilde{c}(k=0)$ is determined by thermodynamic measurements. Then it is widely accepted by workers in classical liquid-structure theory that, at sufficiently large r,

$$c(r)\sim\frac{-\phi(r)}{k_BT} . \tag{4.6}$$

Hence one is led, since there is an inverse-fourth-power "tail" according to the bound (3.7) on the corrected pair potential, to the small-angle form of $\tilde{c}(k)$:

$$\tilde{c}(k)=\tilde{c}(k=0)+c_1k+c_2k^2+\cdots . \tag{4.7}$$

Equation (4.7), a consequence of the R^{-4} form in Eq. (3.7) for liquid metals, is crucially different from the small-angle scattering from the insulating liquid argon, where $c_1=0$ and the London dispersion force referred to in Sec. I introduces a term c_3k^3 into Eq. (4.7). This term is verified to be present from the neutron diffraction measurements on liquid argon near the triple point by Yarnell *et al.* [8]. The magnitude of c_3 can be obtained in terms of the London dispersion coefficient C_6, and theory and experiment agree well for argon.

While there are indications [9] from the x-ray experiments of Greenfield, Wellendorf, and Wiser [3] that there is a k term in the expansion of $\tilde{c}(k)$ at small k as in Eq. (4.7), very careful small-angle scattering experiments will be required before contact can be made with the magnitude of the R^{-4} term given in Eq. (3.7). It does, however, depend explicitly on Z^4, so that a comparison of a trivalent liquid metal such as Al with monovalent Na might be useful experimentally.

V. SUMMARY AND PROPOSALS FOR FURTHER WORK

The lower bound (3.7) to the polarization interaction between screened ions in simple liquid metals such as Na and K falls off according to the floating-blobs model like the inverse fourth power of the interionic separation. Though the calculation of the coefficient of this R^{-4} interaction energy is from a lower-bound result, it does suggest a strong dependence on valence Z.

Already, such a term does tend to lessen the (already modest) discrepancy at large R between diffraction and electron theory standard pair potentials for liquid Na. However, it is pointed out that a direct test of the presence of an R^{-4} interaction at large R is possible by studying the small-angle scattering of neutrons from liquid Na or K, and Dr. M. W. Johnson of the Rutherford Laboratory, U.K., is currently assessing the feasibility of such an experiment. The aim would be to measure the liquid-structure factor $S(k)$ down to very small k, and then to construct the Ornstein-Zernike function $\tilde{c}(k)$ from Eq. (4.4). The question to be answered then is whether the small-angle scattering can be explained by the expansion (4.7) with $c_1\neq 0$. The magnitude of c_1 is directly connected with the magnitude of the R^{-4} polarization interaction in Eq. (3.7) through the asymptotic relation (4.5), which is valid in classical liquids provided one is far from the critical point.

Harker [10] has raised the question as to whether such a polarization interaction might also be accessible to experiment via phonon dispersion relations in a Na or K metallic crystal. In this context, it is highly relevant that recent studies of the phonon-dispersion relations in hot crystals, i.e., near the melting point, have expressed these dispersion relations in terms of the direct correlation function $c(r)$ in the liquid at melting [11,12]. The result may be written

$$\omega_q^2 = \frac{n_a}{m} \sum_{\mathbf{K}} \left\{ \left[\frac{\mathbf{q}}{q} \cdot \mathbf{q} + \mathbf{K} \right]^2 \tilde{\phi}(\mathbf{q}+\mathbf{K}) - \left[\frac{\mathbf{q}-\mathbf{K}}{q} \right]^2 \tilde{\phi}^2(\mathbf{K}) \right\}, \quad (5.1)$$

where the $\mathbf{K}$'s denote reciprocal lattice vectors, provided (i) the low-temperature result quoted in Eq. (5.1) has a Debye-Waller factor incorporated following Ferconi and Tosi [11] and (ii) the Fourier transform $\tilde{\phi}(k)$ of the pair interaction is replaced by

$$-\tilde{\phi}(k) \rightarrow (k_B T c(k))_{T=T_m} , \quad (5.2)$$

where T_m is the melting temperature. We emphasize below the hot crystal because of the link (5.2) to the direct correlation function in the liquid metal at freezing. Of course, the phonons in the cold crystal will also reflect any polarization interaction in the long-wavelength limit. Since, with polarization interaction as in Eq. (3.7), $\tilde{c}(k)$ has the small k expansion (4.7), we must expect from Eq. (5.1) that, at least in principle, $c_1 \neq 0$ will be reflected in the phonon dispersion ω_q in Eq. (5.1) when the replacement (5.2) is made. However, until a decisive theoretical prediction (or liquid diffraction measurement) of c_1 in Eq. (4.7), or equivalently of the coefficient of R^{-4} in Eq. (3.7), can be made, which may well require transcending the floating-blobs model employed throughout this paper (see Sec. I) no quantitative assessment will be possible of the effect of polarization interaction on the phonons.

In a different area, the work in Refs. [13] and [14] cited in the Appendix prompts us to add that the long-range polarization effect predicted in the present investigation may also have relevance to the interaction between charged defects in simple metals (e.g., a divacancy).

ACKNOWLEDGMENTS

We are deeply indebted to Dr. F. Perrot for his invaluable help and advice at every stage of this investigation. One of us (M.B.) has been supported by the ICI Soros Fund during his stay in Oxford. N.H.M. wishes to thank Dr. A. Harker and his colleagues in the Theoretical Studies Department, AEA Industrial Tech. Harwell, for much support and numerous valuable discussions.

APPENDIX: THOMAS-FERMI ENERGY FUNCTIONAL AND LINEARIZED DISPLACED CHARGE

To investigate the dependence of the results of the main text on the form of the energy functional, we have extended the treatment of March and co-workers [13,14] within the Thomas-Fermi approximation. These workers expanded the Thomas-Fermi kinetic energy to second order in the displaced charge, when measured from the unperturbed homogeneous electron gas, as in the main text. Within this framework, we have also used the floating-blobs model to repeat the calculation of the correction to the pair interaction, which in the Thomas-Fermi linearized approximation is $\phi(R) = [(Ze)^2/R]\exp(-qR)$, with q the inverse Thomas-Fermi screening length [see Eq. (A1) below]. We have studied the energy as a function of the amount of floating δR from the ionic centers, and have performed numerical calculations which show that there is no minimum for $\Delta\phi$, except at $\delta R = 0$. These calculations have also been repeated using the next-highest-order approximation in the displaced charge in the kinetic energy, with the same conclusion.

Looked at in a manner entirely paralleling the approach in the main text, one can use alternatively the linearized Thomas-Fermi approximation for the displaced charge, namely,

$$\Delta n = \frac{K}{r} \exp(-qr) , \quad q^2 = \frac{4k_f}{\pi a_0} \quad \text{where } a_0 = \frac{\hbar^2}{me^2} . \quad (A1)$$

Then the contribution (2.5) of the text can be readily evaluated and the central point is that, because of the semiclassical form (A1), the quantity $N(\delta R)$ now has a term linear in δR and a quadratic term with a negative coefficient. The conclusion is again the same as in the calculation outlined immediately above; there is no minimum in $\Delta\phi$ except at $\delta R = 0$. So one concludes that the polarization interaction in the text only follows provided a refined density-functional theory is used which transcends the semiclassical approximations inherent in the Thomas-Fermi theory.

[1] F. Perrot and N. H. March, Phys. Rev. A **41**, 4521 (1990).
[2] F. Perrot and N. H. March, Phys. Rev. A **42**, 4884 (1990).
[3] A. J. Greenfield, J. Wellendorf, and N. Wiser, Phys. Rev. A **4**, 1607 (1971).
[4] L. Reatto, Philos. Mag. A **58**, 37 (1988).
[5] A. C. Maggs and N. W. Ashcroft, Phys. Rev. Lett. **59**, 113 (1987).
[6] E. F. Gurney and J. L. Magee, J. Chem. Phys. **18**, 142 (1950).
[7] S. A. Egorov and N. H. March, Phys. Lett. A **157**, 57 (1991).
[8] J. L. Yarnell, M. J. Katz, R. G. Wenzel, and S. H. Koenig, Phys. Rev. A **7**, 2130 (1973).
[9] C. C. Matthai and N. H. March, Phys. Chem. Liq. **11**, 207 (1982).
[10] A. H. Harker (private communication).
[11] M. Ferconi and M. P. Tosi, Europhys. Lett. **14**, 797 (1991).
[12] N. H. March and B. V. Paranjape, Phys. Chem. Liq. **24**, 223 (1992).
[13] L. C. R. Alfred and N. H. March, Philos. Mag. **2**, 985 (1957).
[14] G. K. Corless and N. H. March, Philos. Mag. **6**, 1285 (1961).

Phys. Chem. Liq., 1994, Vol. 27, pp. 187–193
Reprints available directly from the publisher
Photocopying permitted by license only

LETTER

Liquid Structural Theories of Two- and Three-Dimensional Plasmas

K. I. GOLDEN and N. H. MARCH

Department of Theoretical Physics, Research School of Physical Sciences and Engineering, The Australian National University, Canberra, A.C.T. 2601, Australia*

Theoretical Chemistry Department, University of Oxford, 5 South Parks Road, Oxford OX1 3UB, UK.

(Received 1 May 1993)

Following a treatment of a two-dimensional and one-component plasma (OCP) with a ln r interaction which possesses a Fourier transform, proposals are made for a generalized structural theory in three dimensions, based on a separation of the direct correlation function into a "long" and a "short" range part.

KEY WORDS: Three-atom correlation function, Liquid metals, One-component plasma.

In this Letter, previous progress[1-3] in formally solving the lowest-order member of the Born–Green–Yvon (BGY) hierarchy for an assumed pair potential $\phi(r)$ possessing a Fourier transform $\tilde{\phi}(k)$ will be applied and also generalized. The main application considered is a two-dimensional (2D) one-component plasma (OCP) with a ln r interaction, though an outline is also given of a proposed approach to a liquid structural theory for a three-dimensional plasma such as liquid Na or K near freezing. The generalization effected is to exhibit the r space 'shape' of a general structural theory, avoiding now the need for $\phi(r)$ to possess a Fourier transform.

We first carry out k-space calculations which assume that the pair potential $\phi(r)$ possesses the Fourier transform $\tilde{\phi}(k)$. This will set the stage for the calculations involving the more general non-Fourier transformable $\phi(r)$ below.

It can be shown[3] that $\tilde{\phi}(k)$ can be written in terms of the liquid structure factor $S(k)$, where $(1/n)[S(k) - 1] = \tilde{h}(k)$ is the Fourier transform of $g(r) - 1$: $g(r)$ is the pair correlation function. The somewhat formal result is[3]

$$\frac{\tilde{\phi}(k)}{k_B T} = -\frac{\tilde{E}(k)}{S(k)}, \tag{1}$$

* Present Address: Department of Computer Science and Electrical Engineering, University of Vermont, Burlington, VT 05405, USA.

where $\tilde{E}(k)$ involves both the three-particle correlation function g_3 and the force $-\partial\phi(r)/\partial\mathbf{r}$. The **k**-space analysis of the BGY equation provides

$$\tilde{E}(k) = \tilde{h}(k) + \sum(\mathbf{k}), \tag{2}$$

where

$$\sum(\mathbf{k}) = \frac{1}{V}\sum_q \frac{\mathbf{k}\cdot\mathbf{q}}{k^2}\beta n\tilde{\phi}(q)\{\tilde{h}(|\mathbf{k}-\mathbf{q}|) + n\tilde{t}(\mathbf{k}-\mathbf{q},\mathbf{q})\}: \tag{3}$$

$\beta^{-1} \equiv k_B T$. In Eq. (3), $\tilde{t}$ is the Fourier transform of t in the **r**-space expression

$$g_3(\mathbf{r}_{12}, \mathbf{r}_{32}) = 1 + h(r_{12}) + h(r_{32}) + h(r_{13}) + t(\mathbf{r}_{12}, \mathbf{r}_{32}).$$

Generally to make progress in calculating $\sum(\mathbf{k})$, one must decouple g_3. In the case of the 2D OCP with $\ln r$ interaction, however, the pair function $g(r)$ has been calculated exactly by Jancovici[4] for a particular coupling strength $\Gamma = \bar{\beta}\bar{e}^2 = 2(\bar{\beta}^{-1}$ and $\bar{e}$ are thermal energy per unit length and charge per unit length); Jancovici obtained $g(r) = 1 - \exp(-r^2/a^2)$, whence $S(k) = 1 - \exp(-k^2a^2/4)$; a is the 2D Wigner–Seitz radius. Consequently, for this particular reference liquid, $\tilde{E}(k)$ can be exactly calculated from Eq. (1) since $\tilde{\phi}(k) = 2\pi\bar{e}/k^2$. One obtains

$$-\frac{k^2}{2\pi\bar{e}^2\bar{\beta}}\tilde{E}(k) = S(k). \tag{4}$$

The left-hand side of Eq. (4) versus wavenumber ka is plotted in Figure 1 (see also Golden *et al.*[5]) The continuous curve shows the exact value of $\tilde{E}(k)$ given by Eq. (4) with $S(k) = 1 - \exp(k^2a^2/4)$. This can now be used as a standard for comparison in assessing the accuracy of two well known approximate theories.

The simplest approximation for $\tilde{E}(k)$ is to put $\sum(\mathbf{k}) = 0$ in Eq. (2) Then $n\tilde{E}(k) = S(k) - 1$, whence Eq. (1) simplifies to the well-known Debye–Hückel (DH) formula.

$$\bar{\beta}n\tilde{\phi}(k) \approx -\tilde{c}(k); \tag{5}$$

$\tilde{c}(k) = 1 - 1/S(k)$ is the Fourier transform of the direct correlation function $c(r)$. This approximation evidently uses the asymptotic form for $c(r)$, namely $c(r) = -\beta\phi(r)$, outside its proper range of validity. Nevertheless, it is of interest to note the result

$$-\frac{k^2}{2\pi\bar{e}^2\bar{\beta}}\tilde{E}_{DH}(k) = \frac{k^2a^2}{4}\exp\left(-\frac{k^2a^2}{4}\right) \tag{6}$$

as the dash-dot-dash curve of Figure 1 (see also Ref. [5]).

Transcending the DH approximation, the hypernetted chain (HNC) formula is

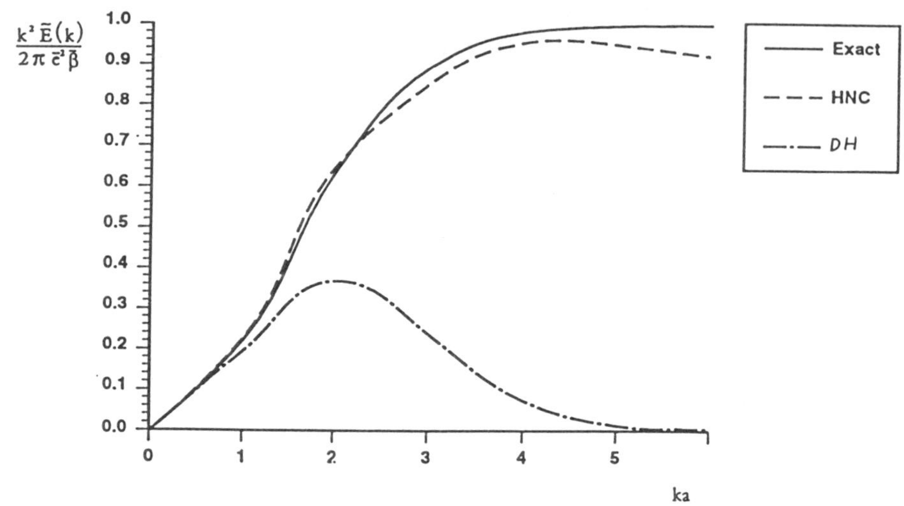

Figure 1 The function $[k^2/(2\pi\bar{e}^2\bar{\beta})]\tilde{E}(k)$ versus ka for the 2D OCP with ln r interaction potential. The solid curve is exact; the dash-dot-dash curve represents the Debye-Hückel approximation; and the dash curve represents the hypernetted chain approximation.

given by[3]

$$\tilde{E}_{HNC}(k) = -\frac{1}{n}\,\bar{\beta}\tilde{U}(k)S(k) - \frac{1}{n}\left[S(k) - 1\right]^2, \tag{7}$$

where

$$\begin{aligned}\bar{\beta}\tilde{U}(k) &= 2\pi n \int_0^\infty dr r J_0(kr) \ln\left[1 - \exp\left(-\frac{r^2}{a^2}\right)\right] \\ &= \sum_{p=1}^{\infty} \frac{1}{p^2} \exp\left(-\frac{k^2a^2}{4p}\right)\end{aligned} \tag{8}$$

is calculated via the potential of mean force, $U(r)$, derived from the definition $g(r) = \exp[-\bar{\beta}U(r)]$ and the exact pair correlation function, $g(r) = 1 - \exp(-r^2/a^2)$, for the 2D OCP at $\Gamma = 2$. The formula for $-k^2\tilde{E}_{HNC}(k)/(2\pi\bar{e}^2\bar{\beta})$ which follows from Eqs. (7) and (8) is displayed as the dash curve in Figure 1.

Not surprisingly, the HNC result (7) fits better with the exact form (4) than the DH result (6) though neither tends at large k to a finite constant given by the exact theory. It is, of course, of considerable interest that for the value of $\Gamma = 2$ of the coupling strength, one has an exact test of the accuracy of approximate structural theories.

Up to this point, the structure of the BGY equation for potentials $\phi(r)$ which possess a Fourier transform has been exploited. But for a liquid like argon, the

Lennard–Jones 6–12 potential is a reasonable zeroth order approximation and this does not have a Fourier transform. Our purpose below is therefore to recast the theory into an **r** space form applicable to more general pair potentials.

To do so, let us start by subtracting $\beta\tilde{U}(k)$ from both sides of Eq. (1) and introducing the function $\tilde{F}(k) = n\tilde{E}(k) + \beta\tilde{U}(k)$. Taking the inverse transform of the result, one then obtains

$$\frac{\phi(r) - U(r)}{k_B T} \equiv -\frac{\Delta(r)}{k_B T} = -F(r) + n\int F(|\mathbf{r} - \mathbf{r}'|)c(r')d\mathbf{r}', \tag{9}$$

where $F(r)$ is evidently the inverse transform of $\tilde{F}(k)$. Below, Eq. (9) will be solved on the assumption that though $\phi(r)$ and $U(r)$ do not separately have Fourier transforms, the difference $\Delta(r)$ does.

With this assumed, the Fourier transform of Eq. (9) gives

$$\frac{\tilde{\Delta}(k)}{k_B T} = -\tilde{F}(k) + \tilde{F}(k)\tilde{c}(k). \tag{10}$$

The simplest application of Eq. (10) is now to relate to the often useful, though clearly approximate, HNC approximation. To motivate such a relation, we recall that at large r, $\phi(r) \to -k_B T c(r)$ and $U(r) \to -k_B T h(r)$, whence

$$\frac{\phi(r) - U(r)}{k_B T} \equiv -\frac{\Delta(r)}{k_B T} \to h(r) - c(r); \ \underline{\text{large-r limit}}, \tag{11}$$

Guided by Eq. (11), we are led to write its small-k counterpart in the form

$$\frac{\tilde{\Delta}(k)}{k_B T} = [S(k) - 1] - \tilde{c}(k) + \delta(\Gamma); \ \underline{\text{small-}k\text{ limit}}, \tag{12}$$

where

$$\delta(\Gamma) = n\int d\mathbf{r}\ \ln g(r) + 1 - \beta\left(\frac{\partial p}{\partial n}\right)_T - S(0)\tilde{c}(0).$$

The k-independent $\delta(\Gamma)$ correction guarantees the *exactness* of Eq. (12) through $O(k^0)$; the coupling parameter, Γ, which was introduced above for the 2D OCP with $\ln r$ interaction, is generally defined to be the ratio of the average interaction energy to the thermal energy; e.g., $\Gamma = \beta(Ze)^2/a$ for the 3D ionic OCP with charge Ze. We note that in the weak coupling limit ($\Gamma \ll 1$), $\delta(\Gamma) = -\Gamma/4$ for the 2D OCP with ln r interaction. Eq. (12) is a valuable result for what follows.

We turn now to the "inverse" problem of classical liquid structure theory. The so-called "inverse" problem, namely the extraction of the pair potential $\phi(r)$ from diffraction measurements, was proposed by Johnson and March[6]. They recognized that the force (BGY) equation again provided the fundamental route. But then, decoupling of g_3 is required[1] which is often uncontrolled and violates the correct limiting result $c(r) \to -\beta\phi(r)$ at large r, and this does not permit the required accuracy to extract $\phi(r)$. Though the decoupling of g_3 has been bypassed in the work of Reatto[7] and co-workers, the price paid is to invoke computer simulation as the "corrector" in an iterative predictor-corrector technique. It remains of considerable importance to have a largely analytical solution of this inverse problem.

To this end, we outline a proposed solution by returning to the thermodynamically consistent decomposition of Kumar *et al.*[2] of the direct correlation function: they write $c(r) = c_p(r) + c_c(r)$, where $c_p(r)$ and $c_c(r)$ are the "potential" and "cooperative" parts. Keeping this decomposition in mind, Eq. (10) can be rewritten as

$$\tilde{c}_p(k) + \tilde{c}_c(k) = \frac{\beta\tilde{\Delta}(k)}{\tilde{F}(k)} + 1. \tag{14}$$

Kumar *et al.*[2] emphasized that the virial and compressibility routes to the equation of state were thermodynamically consistent, provided only that the cooperative part $c_c(k) \to 0$ as $k \to 0$; thus $\tilde{c}_p(k) \to \tilde{c}(k) = 1 - 1/S(k)$ in this limit. This property and the structure of Eq. (12) suggest rewriting the right-hand side of Eq. (14) as follows:

$$\begin{aligned}\tilde{c}_p(k) + \tilde{c}_c(k) = 1 &+ \frac{[S(k) - 1 - \tilde{c}(k)] + \delta(\Gamma)}{\tilde{F}(k)} \\ &+ \frac{(\beta\tilde{\Delta}(k) - [S(k) - 1 - \tilde{c}(k)]) - \delta(\Gamma)}{\tilde{F}(k)}.\end{aligned} \tag{15}$$

Since we have now arranged the last term on the right-hand side of Eq. (15) to tend to zero as $k \to 0$ from Eq. (12), we propose the "closure" of the present theory by identifying $\tilde{c}_p(k)$ as the first two right-hand side terms of (15), whence

$$[\tilde{c}(k)]^2 = -\frac{\tilde{F}(k)}{S(k)}[1 - c_p(k)] - \frac{\delta(\Gamma)}{S(k)}. \tag{16}$$

The final step in reaching the proposed solution of the inverse problem is to combine Eqs. (16) and (10). One obtains

$$\beta\tilde{\Delta}(k) = \frac{[\tilde{c}(k)]^2 + \delta(\Gamma)[1 - \tilde{c}(k)]}{1 - \tilde{c}_p(k)} \tag{17}$$

with its **r**-space counterpart

$$\begin{aligned}\Delta(r) - n\int d\mathbf{r}'\Delta(r')c_p(|\mathbf{r} - \mathbf{r}'|) \\ = k_B T n \int d\mathbf{r}' c(r')c(|\mathbf{r} - \mathbf{r}'|) - k_B T\delta(\Gamma)c(r); \; \mathbf{r} \neq 0.\end{aligned} \tag{18}$$

Eq. (9), which is formally exact, and the approximate Eq. (18) represent the principal results of this Letter. Eq. (18) is expected to be especially accurate at large r because of the underlying requirement of thermodynamic consistency; at small r, it is less satisfactory. By contrast, the complementary HNC structure

$$\Delta(r) - n \int d\mathbf{r}'\Delta(r')c(|\mathbf{r} - \mathbf{r}'|) = k_B T n \int d\mathbf{r}' c(r')c(|\mathbf{r} - \mathbf{r}'|) \tag{19}$$

is more satisfactory at small r, but is less accurate at large r.

Of course, in applying Eq. (18) to extract $\phi(r)$, one has immediately to recall from Ref. [2] that $c_p(r)$ involves $\phi(r)$ itself as well as T, n and $g(r)$. Therefore, for a given thermodynamic state, $c_p(r)$ is uniquely determined in terms of $\phi(r)$ provided diffraction measurements on $S(k)$ and also its density derivative $\partial S(k)/\partial n$ are made. Thus with a starting approximation to $\phi(r)$, say $\phi_0(r)$, experimental data then determine completely the right-hand side of Eq. (18), and hence a first approximation $\beta\phi_1(r)$ since $-\beta U(r) = \ln g(r)$ is also known from the diffraction measurements. One then recalculates $c_p(r)$ with this reset approximation to $\phi(r)$, and iterates to self consistency.

The essential points of this Letter can be summarized as follows. The shape of classical liquid structure theory is determined by the solution of the force equation (1) [with (2)]. This leads to an $\mathbf{r}$ space theory in which the difference between the pair potential $\phi(r)$ and the potential of mean force $U(r)$, denoted throughout this Letter by $\Delta(r)$, is a central quantity for the theory. Provided that $\Delta(r)$ has a Fourier transform (FT) $\tilde{\Delta}(k)$, even when such FTs do not exist for $\phi(r)$ and $U(r)$ separately, one can write $\Delta(r)$ precisely as the difference between a function $F(r)$ and the convolution of this same function with the direct correlation function $c(r)$ as in Eq. (9).

If one makes contact, for example, with HNC approximation of classical liquid structure theory, then the FT of $F(r)$, is $\tilde{F}_{HNC}(k) = -[S(k) - 1]^2$, where $S(k)$ is the liquid structure factor. Guided by the HNC structure for $\Delta(r)$ at large r, we formulate its small-k counterpart $\tilde{\Delta}(k)$ in a way which guarantees thermodynamic consistency at $k \to 0$. To set up a theory of $\tilde{F}(k)$ for all k, the thermodynamically consistent decomposition of $c(r)$ into the sum $c_p(r) + c_c(r)$ is finally invoked. A form of $\tilde{F}(k)$ and ultimately $\Delta(r)$ [Eq. (18)] is then proposed solely in terms of $c(r)$ and $c_p(r)$, the latter depending in an explicitly known way on $\phi(r)$, $g(r)$, and the density derivative of the pair function.

In summary, the 2D OCP has been considered within an exact liquid structural theory for one coupling strength $\Gamma = 2$. Comparison has been made with the Debye–Hückel and HNC approximations. Then, the shape of an $\mathbf{r}$ space theory avoiding the assumption that the pair potential $\phi(r)$ has a Fourier transform has been exposed. Finally, an outline proposal is presented, based on a separation of the direct correlation function $c(r)$ into "potential" and "cooperative" parts, which may well have application to three-dimensional plasmas such as liquid Na or K.

K.I.G. would like to thank Professor N. H. March and the Theoretical Chemistry Department at the University of Oxford for hospitality. K.I.G.'s contributions to this research have been partially supported by U.S. National Science foundation Grants

PHY-9115695 and INT-9215300, and by a visiting fellowship from the Institute of Advanced Studies at the Australian National University.

References

1. J. G. Kirkwood, *J. Chem. Phys.* **3**, 300 (1935).
2. N. Kumar, N. H. March, and A. Wasserman, *Phys. Chem. Liquids* **11**, 271 (1982).
3. K. I. Golden and N. H. March, "Thermodynamic consistency and classical liquid structure theory", to appear in *Phys. Chem. Liquids.*
4. J. Jancovici, *Phys. Rev. Lett.* **46**, 386 (1981).
5. K. I. Golden, N. H. March, and A. K. Ray, THREE-PARTICLE CORRELATION FUNCTION AND STRUCTURAL THEORIES OF DENSE METALLIC LIQUIDS, *Molecular Physics* **80**, 915 (1993).
6. M. D. Johnson and N. H. March, *Phys. Lett.* **3**, 313 (1963).
7. L. Reatto, *Phil. Mag.* **A58**, 37 (1988).

9486 J. Phys. Chem. 1994, 98, 9486–9487

Nature of Chemical Bonding in Highly Expanded Heavy Alkalis: Especially Cs and Rb

G. R. Freeman* and N. H. March†

Chemistry Department, University of Alberta, Edmonton, Alberta, Canada T6G 2G2, and Theoretical Chemistry Department, University of Oxford, 5 South Parks Road, Oxford OX1 3UB, England

Received: May 27, 1994®

Experimental data are now available on the structure of heavy alkalis, and especially Cs, in highly expanded forms. First, the neutron diffraction experiments of Hensel and co-workers on expanded fluid Cs taken up the liquid–vapor coexistence curve toward the critical point show a dramatic decrease in coordination number but a relatively minor increase in the nearest-neighbor Cs–Cs bond length. Less extensive data for liquid Rb exhibit similar features. Subsequently, by deposition on semiconductor substrates, and in particular GaAs and InSb, one- and two-dimensional ordered structures of Cs have been established, with near-neighbor Cs distances substantially longer than in bulk Cs and being governed by the geometry of the substrate. These experimental results and their relation to electrical conductivity are here analyzed in terms of a localized chemical bonding picture.

I. Introduction

The pioneering experiments of Hensel and co-workers[1] have led to remarkable progress in understanding the structure of expanded liquid rubidium and cesium. In particular, neutron diffraction data are available from these studies on the static liquid structure factor $S(q)$ for these two heavy alkalis, expanded by heating toward their critical points. For convenience we note here that the critical data of Rb and Cs are respectively T_c = 2017 K, P_c = 12.45 MPa, and d_c = 290 kg/m^3 and 1924 K, 9.25 MPa, and 380 kg/m^3. The law of rectilinear diameters breaks down for these liquid metals and their vapors.[1a] The data on the static structure factor $S(q)$ and its Fourier transform $g(r)$, the pair distribution function, were used by Hensel et al.[1b] to derive the characteristic changes of the microscopic structure and in particular the near-neighbor distance and the coordination number as a function of density.

We shall return to the interpretation of these experimental data in the dense fluid state characterized by short-range order (SRO) below. However, from a different type of study, namely deposition of cesium on semiconductor crystal surfaces, data have subsequently become available on expanded Cs structures with long-range order (LRO). While this later study is partially about Cs in interaction with the semiconducting substrate, there is again important information about chemical bonding in expanded Cs, but now with LRO. We find it useful to discuss these data immediately below, before returning to the data on expanded alkalis in the fluid state with only SRO.

II. Low-Dimensional Expanded Cs with LRO, through Deposition on Semiconducting Substrates

Whitman et al.[2] have reported the structural properties of Cs adsorbed on room temperature GaAs(110) and InSb(110) surfaces as observed with scanning tunneling microscopy. What these workers demonstrate is that Cs initially forms long, one-dimensional (1-D) zigzag chains on both surfaces. In particular, their Figure 1(a) shows specifically a large-area image of Cs chains on GaAs(110), including chains that are over 100 nm long. What their observations demonstrate in addition, that is important in the present context, is that the chains tend to be separated by tens of nanometers and have no long-range order along the [001] direction, thereby establishing that they are truly 1-D structures. The higher-resolution image (their Figure 1(b)) reveals that the Cs structures are zigzag chains of single atoms in registry with the GaAs(110) surface.

What needs especial emphasis in the present context is that the Cs–Cs nearest-neighbor (nn) distance in this structure is 0.69 nm, which is therefore considerably longer than the Cs–Cs nm distance of 0.52 nm in bulk Cs.

Whitman et al.[2] also adsorbed Cs on the InSb(110) surface, InSb being the III–V semiconductor with the largest lattice constant: about 15% larger than that of GaAs. Again, the formation of Cs zigzag chains was observed, but now with increase of the Cs–Cs nn distance to 0.80 nm.

Though we shall return below in section IV to results found when additional Cs adsorption occurs on GaAs(110), we want immediately to compare and contrast the above structures with deductions made from the diffraction experiments of Hansel et al.[1] on dense fluid Cs.

III. Modeling of Short-Range Order Using Chains (Coordination Number 2) in Dense Fluid Cs Near the Critical Density d_c = 380 kg/m^3

As already noted in section I, Hensel et al.[1b] derived both the nn bond length and the coordination number as a function of density d from neutron diffraction measurements of the dense fluid structure factor $S(q)$. One of us[3] earlier fitted their data for coordination number z by

$$d = az + b \tag{1}$$

where a = 230 and b = −80, both in kg/m^3. The relative constancy[1b] of the nn distance as d is lowered lends strong support to the view that a chemical bond is the basic building block in these expanded fluid states.

One of us has earlier noted,[3] but for the lighter alkali Na, that the study of Malrieu et al.[4] can be used to characterize this chemical bond. While their study[3] was on crystalline Na, it is not as drastic as it might sound to adapt their study to the alkali fluids, since the arguments of Malrieu et al.[4] rest purely on consideration of first and second neighbors to a chosen Na ion. These workers, following the ideas of Poshusta and Klein[5] on

* Author to whom correspondence should be addressed at the University of Alberta.
† University of Oxford.
® Abstract published in *Advance ACS Abstracts*, August 15, 1994.

0022-3654/94/2098-9486$04.50/0

hydrogen, set up a Heisenberg Hamiltonian for Na, which was characterized by the $^1\Sigma_g^+$ and $^3\Sigma_u^+$ potential energy curves of the free-space diatom Na_2. These were taken from the semiempirical study of Konawalow et al.[6] and were then used[4] to calculate the energy as a function of nn distance for both a body-centered cubic structure and a structure with lower coordination, namely simple cubic. We assume that this treatment[3,4] can be appropriately adapted to characterize the Cs chemical bond in terms of the corresponding potential energy curves of the Cs diatom.

Returning now to eq 1, which yields $z = 2$ for the coordination number of fluid Cs at the critical density, for reasons discussed in refs 3 and 7, consistent with the results extracted from experiment by Hensel and co-workers,[1b] it has already been proposed[3] that the coordination number is 2 (consistent with the existence of chains) in fluid Cs at density $d_c = 380$ kg/m^3. Returning to the data for Cs adsorbed on the semiconducting surfaces, it is natural again to assume zigzag chains, though the angle in these chains must clearly be $>\pi/3$ to yield a coordination number of 2, consistent with eq 1. Of course, for states with higher density, the coordination number increases and we shall return to discuss such states in comparison with the findings of Whitman et al.[2] for denser Cs adsorption in the summary below. Though the nn distance in zigzag chains found by Whitman et al.[2] is considerably longer than the chemical bond in fluid Cs, their 'expansion' of Cs is, of course, brought about the the substrate geometry. However, it is of interest, notwithstanding the marked differences between nn distances in fluid Cs (~0.56 nm) and Cs on GaAs (0.69 nm) and on InSb (0.80 nm), to consider the nature of the electrical conductivity associated with the 1-D Cs chain structures discussed above. Whitman et al.[2] refer, in support of their experimental approach, to the work of Ferraz et al.[7] in relation to the proximity of bulk Cs to the metal–insulator transition. To probe for metallic characteristics, Whitman et al.[2] measured the tunneling conductivity (reduction of the band gap) at zero bias, which is proportional to the density of states at the Fermi level.[8] Current versus voltage curves showed that the band gap of 1.45 eV of GaAs was reduced to 1.1 eV when 1-D chains were on the surface, which means that the 1-D Cs chain structures are insulating on GaAs(110).[2] Earlier, one of us[3] had suggested the same conclusion from a chain model for the SRO in fluid Cs at the critical density, though there the Peierls transition was invoked to propose the identity of the metal–insulator transition and the critical point in fluid Cs (see also section IV below for the reasons behind such attempts to 'mimic' SRO by models with LRO character).

IV. Discussion and Summary

It seems to be established beyond reasonable doubt that the basic building block for the structure of the expanded liquid metals Cs and Rb is a chemical bond with a rather constant length, ~0.54–0.57 nm.

If one follows the approach of Warren and Mattheiss[9] in using LRO models to mimic the local coordination in the fluid, then in conjunction with eq 1, which gives $z = 2$ at d_c, one would naturally, in view of the discussion of the semiconductor substrate adsorption in section II, adopt a largely 1-D model of zigzag chains,[3] with a nn distance ~0.57 nm. The findings of Whitman et al.[2] would then suggest that the fluid at this density would be insulating even though their nn distances are somewhat larger. This would be consistent with experiment, which currently does not distinguish clearly the metal–insulator transition and the critical point in Cs.[1b]

Raising the density and using the structure data of Hensel et al.[1b] at $T = 1923$ K, with number density $\varrho = 2.7 \times 10^{27}$ atom/m^3, Ascough and March[10] have extracted an effective pair potential $\phi(r)$, using the procedure proposed by Johnson and March.[11] This potential has a sharp minimum at the nn distance 0.56 nm followed by a repulsive region out to ~0.9 nm. Clusters with roughly equal bond lengths would be favored by this pair interaction, separated by substantial distances. Equation 1 predicts that this state has a coordination number near 3, and we now return to the study of Whitman et al.[2]

On allowing additional adsorption on GaAs(110) beyond that leading to the zigzag chains described in section II, Whitman et al. observe the formation of a 2-D overlayer consisting of five-atom Cs trapezoids arranged in a c(4 × 4) superlattice. It is true that the nn distance in their work is now shorter than that in bulk Cs, but there is space between these trapezoids. These configurations have an average coordination of ~3. As Whitman et al.[2] point out, nearly identical planar Cs clusters are also stable in the vapor phase.[12,13]

To summarize, the main purpose of the present work is to emphasize considerable similarities of the chemical bonding of the heavy alkalis (particularly Cs) in situations with SRO only (dense fluids) and in adsorption on semiconducting substrates. Local coordination appears to exhibit similar trends with density. In the liquid state, the bond length remains remarkably constant as the density is varied from the normal melting point of Cs to about twice the critical density. Of course, with Cs on the substrate, the nn distance is imposed on the Cs configuration by the substrate geometry. Nevertheless, in both InSb and GaAs, at low coverage, the Cs atoms take up zigzag chain configurations with low coordination number, just as found in the expanded liquid phase. The nn distances here are substantially larger than those in bulk Cs, so that again we are dealing with highly expanded states. Even at monolayer coverage, with a shorter Cs–Cs bond length now than in bulk Cs, there seems still a close relation between local coordination behavior and that found for Cs clusters in the vapor phase. It will clearly be of interest to extract pair potentials[10] at other densities from liquid structure data, as these potentials, in the metallic fluid phase, are known to have essential density dependence.[14,15]

References and Notes

(1) (a) Jüngst, S.; Knuth, B.; Hensel, F. *Phys. Rev. Lett.* **1985**, *55*, 2160. (b) Winter, R.; Hensel, F. *Phys. Chem. Liq.* **1989**, *20*, 1.

(2) (a) Whitman, L. J.; Stroscio, J. A.; Dragoset, R. A.; Celotta, R. J. *Phys. Rev. Lett.* **1991**, *66*, 1338. (b) First, P. N.; Dragoset, R. A.; Stroscio, J. A.; Celotta, R. J.; Feenstra, R. M. *J. Vac. Sci. Technol.* **1989**, *A7*, 2868.

(3) March, N. H. (a) *Phys. Chem. Liq.* **1989**, *20*, 241; (b) *J. Math. Chem.* **1990**, *4*, 271.

(4) Malrieu, J. P.; Maynau, D.; Daudey, J. P. *Phys. Rev.* **1984**, *B30*, 1817.

(5) Poshusta, R. D.; Klein, D. J. *Phys. Rev. Lett.* **1982**, *48*, 1555.

(6) Konowalow, D. D.; Rosenkrantz, M. E.; Olson, M. L. *J. Chem. Phys.* **1980**, *72*, 9612.

(7) Ferraz, A.; March, N. H.; Flores, F. *J. Phys. Chem. Solids* **1984**, *45*, 627.

(8) Stroscio, J. A.; Feenstra, R. M.; Fein, A. P. *Phys. Rev. Lett.* **1986**, *57*, 2579.

(9) Warren, W. W.; Mattheiss, L. F. *Phys. Rev.* **1984**, *B30*, 3103.

(10) Ascough, J. A.; March, N. H. *Phys. Chem. Liq.* **1990**, *21*, 251.

(11) Johnson, M. D.; March, N. H. *Phys. Lett.* **1963**, *3*, 313.

(12) Krauss, M.; Stevens, W. J. *J. Chem. Phys.* **1990**, *93*, 8915.

(13) Bonacic-Koutecky, V.; Fantucci, P.; Boustani, I.; Koutecky, J. In *Studies in Physical and Theoretical Chemistry*; Elsevier: Amsterdam, The Netherlands, 1989; Vol. 62, p 429.

(14) See, for example: March, N. H. In *Liquid Metals: Concepts and Theory*; Cambridge University Press: Oxford, U.K., 1990.

(15) (a) Hafner, J.; Heine, V. *J. Phys. F: Met. Phys.* **1983**, *13*, 2479. (b) Young, W. H. *Rep. Prog. Phys.* **1992**, *55*, 1.

Laser and Particle Beams (1998), *vol.* 16, *no.* 1, *pp.* 71–81
Printed in the United States of America

Diffraction and transport in dense plasmas: Especially liquid metals

By N.H. MARCH* AND M.P. TOSI**
*Oxford University, Oxford, England
**Istituto Nazionale di Fisica della Materia and Classe di Scienze, Scuola Normale Superiore, Pisa, Italy

(Received 1 July 1996; Accepted 17 September 1997)

Recent computer experiments on liquid Mg and Bi (and also on dense hydrogen) have focussed anew on issues involving static and dynamical structure in plasmas. In Mg and Bi, under normal liquid metal conditions, separation of core and valence electrons is valuable both for thermodynamics and in interpreting diffraction experiments. Mg is considered in some detail as a specific example where there is weak electron–ion interaction. Finally, dynamical structure is considered. After a brief summary relating back to the electron–electron pair correlation contribution in X-ray scattering, attention is next focussed on the (longitudinal) viscosity of alkali metals via the Kubo formula. This viscosity is shown to be dominated by ion–ion interactions. Nevertheless, an intimate relation at the melting point is exposed between shear viscosity, thermal conductivity, and electrical resistivity, the latter two transport coefficients being dominated by electrons.

1. Background and outline

The aim of the present study is to treat a liquid metal (e.g., Na, Mg, or Bi) as a two-component plasma and thereby to investigate both static and dynamical structure.

One important simplification will be made at the outset in pursuing this objective. Under normal conditions, liquid metals such as referred to above can be viewed as a collection of ions plus neutralizing electrons belonging to the liquid metal as a whole. Throughout, the ions (e.g., for Na^+ and Mg^{++} having a closed shell Ne-like electronic structure) will be treated as having core (c) electrons rigidly attached to their own nuclei.

Then, to describe the static structure of the liquid metal, one needs three partial structure factors:

1. The nucleus–nucleus structure factor: written conventionally as $S(q)$, and directly accessible via neutron scattering experiments,
2. The ion ($i \equiv$ nucleus): valence electron (e) structure factor, $S_{ie}(q)$ and
3. The valence electron: valence electron structure factor, denoted by $S_{ee}(q)$.

Egelstaff *et al.* (1974) proposed, by means of X-ray and electron diffraction experiments in addition to the neutron scattering measurement of $S(q)$, a route to find $S_{ie}(q)$ and $S_{ee}(q)$ and a substantial programme has developed along these lines (see, for example, Tamaki 1987).

However, very recently, computer simulation studies have appeared by de Wijs *et al.* (1995), and in particular $S_{ie}(q)$ has been determined thereby for liquid Mg and liquid Bi. Also, computer simulation work on dense hydrogen by Magro *et al.* (1996) has appeared. We have

 0263-0346/98 $12.50

discussed this elsewhere in relation to cage models of H and H_2 (March & Tosi 1996) and shall therefore restrict our discussion here to Mg and Bi.

Turning to dynamical structure, this is intimately related to transport, especially in the small q and low energy transfer limit, via so-called Kubo–Green formulae. To be more specific, consider as a simpler example than a liquid metal, largely to establish a notation, an insulator like argon (i.e., without itinerant electrons). Then, the static structure factor $S(q)$ has as its natural generalization the van Hove function $S(q, \omega)$. This is, essentially, the probability that a neutron incident on the liquid will transfer momentum $\hbar q$ and energy $\hbar\omega$ to the liquid. Light scattering also is important for insulators in the small q and small ω regime. The integral of $S(q, \omega)$ over all energy transfers leads back to the static (nucleus–nucleus) structure. Of course, for the two-component plasma one needs to supplement this by generalizations of $S_{ie}(q)$ and $S_{ee}(q)$, namely $S_{ie}(q, \omega)$ and $S_{ee}(q, \omega)$ respectively.

With this as general background, the outline of the paper is as follows. In section 2 immediately below, a brief summary of aspects of thermodynamics, plus requirements of perfect screening in a two-component metallic plasma will be given. This description will be related to an effective ion–ion interaction, for the specific example of liquid Na. Section 3 will then be concerned with the information which, in principle, is contained in an X-ray diffraction experiment on a two-component plasma. After setting out the basic theory, the structure factor $S_{ie}(q)$ will be modelled in the simplest realistic fashion for the case of weak electron–ion interaction (see, for example, March & Tosi 1984; Rasolt 1985).

Section 4 deals with dynamical structure and transport. Thus, we begin by noting that Baym (1964) has exploited the close similarities between neutron and electron scattering to give the theory of electrical conductivity in terms of $S(q, \omega)$ introduced above. He notes the conditions under which his treatment reduces to the customary nearly-free electron (NFE) theory associated with the names of Bhatia, Krishnan, and Ziman. This theory will then be used, in conjunction with the Wiedemann-Franz Law to explain some aspects of the thermal conductivity of the liquid alkali metals. The Kubo–Green formula has been used earlier by one of us (Tosi 1992) to demonstrate that the longitudinal viscosity $\eta_l = \frac{4}{3}\eta + \zeta$, with η shear and ζ bulk viscosities respectively, is dominated by ions again in the alkalis, the electrons making only percentage 'corrections.' This treatment is then related to Andrade's formula for $\eta(T_m)$ in liquid metals near the melting temperature T_m. This all enables a formula to be exposed relating thermal conductivity, electrical resistivity, and shear viscosity at the melting point. The above considerations have been restricted to liquid metals with well-defined valence. We add a brief comment in relation to recent experiments on W plasma (Kloss *et al.* 1996) where the valence is no longer constant through the range of experimental conditions.

2. Thermodynamics and screening in pure liquid metals

Let us begin by quoting the formula for the internal energy E in two-component theory, in terms of the **r** space correlation functions $g(r)$ from $S(q)$, $g_{ie}(r)$ from $S_{ie}(q)$ and $g_{ee}(r)$ from $S_{ee}(q)$. This formula reads

$$E = T_i + T_e + \frac{1}{2n_i}\int d\mathbf{r}[n_e^2 g_{ee}(r)\frac{e^2}{r} + n_i^2 g(r) v_{ii}(r) + 2n_i n_e g_{ie}(r)\bar{v}(r)], \tag{2.1}$$

T_i and T_e being the kinetic energies of ions and electrons in the metal at average number densities n_i and n_e respectively, while $v_{ii}(r)$ and $\bar{v}(r)$ are the bare ion–ion and ion–electron interaction potentials.

But, we can also write E in terms of the effective ion–ion interaction $\phi(r)$ mediated by the electrons. As proposed by Johnson and March (1963) and brought to full fruition by Reatto and coworkers (see Reatto 1988 and other references given there), this interaction $\phi(r)$, in appropriate cases like liquid Na (see below), can be obtained by inverting the (nucleus–nucleus) structure factor $S(q)$. The result (called 'diffraction' potential in figure 1) is compared with an electron theory calculation (Perrot & March 1990) based on the density functional method, in figure 1. The agreement is excellent, and prompts us to write the alternative form for the internal energy E as

$$E = T_i + \frac{1}{2} n_i \int d\mathbf{r} g(r)\phi(r). \tag{2.2}$$

We shall say a little more about electronic kinetic energy T_e when we treat dense hydrogen in section 4. Both equations (2.1) and (2.2) are highly relevant in the present context, as will become clearer when diffraction is discussed in some detail in the following section.

However, before that, let us record the results of perfect screening arguments (Watabe & Hasegawa 1973; Chihara 1973) for the long wavelength limits of the partial structure factors. These relations are

$$S_{ie}(q = 0) = z^{1/2} S(q = 0) \tag{2.3}$$

and

$$S_{ee}(q = 0) = zS(q = 0). \tag{2.4}$$

Since we have the result of fluctuation theory that $S(q = 0)$ is determined solely by thermodynamic properties through

$$S(0) = n_i k_B T K_T \tag{2.5}$$

with K_T the isothermal compressibility, it is clear that $S_{ie}(0)$ and $S_{ee}(0)$ are determined uniquely by equations (2.3) and (2.4). We note here that March *et al.* (1973) have generalized these equations to binary mixtures.

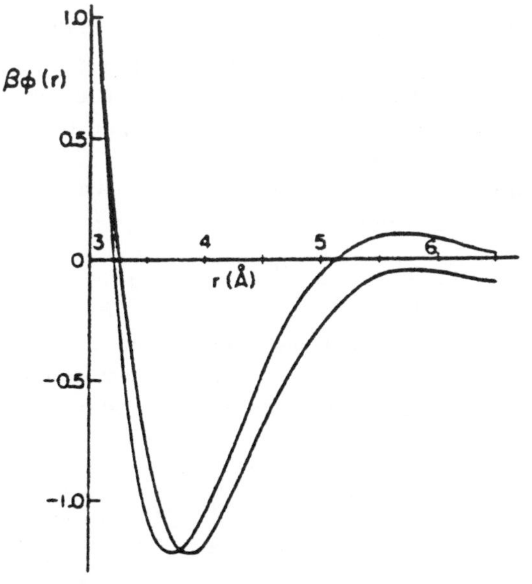

FIGURE 1. Shows 'diffraction' effective pair interaction $\phi(r)$ (in units of $k_B T$) obtained by inverting structure factor $S(q)$ near freezing point of liquid Na: this is lower curve at large r (see Reatto 1988 and other references given there). For comparison, electron theory potential (Perrot & March 1990) is also shown (upper curve at large r).

3. X-ray scattering on two-component plasmas

Following Egelstaff *et al.* (1974), the intensity of X rays scattered from a liquid metal, $I_X(q)$ say with $q=(4\pi/\lambda)\sin(\theta/2)$ where θ is the scattering angle and λ is the X-ray wavelength, can be written in terms of the total electron density $\rho(\mathbf{r})$ as

$$I_X(q) = \frac{1}{V}\int d\mathbf{r}\, d\mathbf{r}'\langle\rho(\mathbf{r})\rho(\mathbf{r}')\rangle\exp[i\mathbf{q}\cdot(\mathbf{r}-\mathbf{r}')] \\ \equiv F(\langle\rho(\mathbf{r})\rho(\mathbf{r}')\rangle). \tag{3.1}$$

In equation (3.1), V is the volume of the sample and F is evidently denoting the Fourier transform with respect to $\mathbf{r}-\mathbf{r}'$.

Making the core-valence electron separation emphasized in section 1,

$$\rho(\mathbf{r}) = \rho_c(\mathbf{r}) + \rho_v(\mathbf{r}), \tag{3.2}$$

and inserting equation (3.2) into (3.1) yields

$$I_X(q) = F(\langle\rho_c(\mathbf{r})\rho_c(\mathbf{r}')\rangle) + 2F(\langle\rho_c(\mathbf{r})\rho_v(\mathbf{r}')\rangle) + F(\langle\rho_v(\mathbf{r})\rho_v(\mathbf{r}')\rangle). \tag{3.3}$$

But $\rho_c(\mathbf{r})$ has the same 'structure' as the nuclei. We introduce the (rigid) core scattering factor $f_c(q)$ satisfying

$$\lim_{q\to 0} f_c(q) = Z_c = Z - z, \tag{3.4}$$

where Z_c is the number of core electrons (10 in liquid Mg), related to atomic number Z and valence z as shown in equation (3.4). Then, the formula (3.3) can be written in terms of $S(q)$ and $S_{ie}(q)$ as

$$I_X(q) = f_c^2(q)S(q) + 2f_c(q)S_{ie}(q) + F(\langle\rho_v(\mathbf{r})\rho_v(\mathbf{r}')\rangle). \tag{3.5}$$

We shall turn immediately below to relate equation (3.5) to both theoretical models and to recent computer studies of de Wijs *et al.* (1995).

3.1. *Modelling of $S_{ie}(q)$ for weak electron–ion interaction*

March and Tosi (1984; see also Rasolt 1985) give a treatment of $S_{ie}(q)$ in terms of $S(q)$ for weak electron–ion interaction. We shall evaluate this theoretical approach below, with the simplest possible modelling of its ingredients, which are (i) the bare electron–ion interaction $v(q)$ and (ii) the dielectric function $\epsilon(q)$.

The weak electron–ion result (see equation (6.22) of March and Tosi, 1984) can be conveniently expressed as the ratio $S_{ie}(q)/S(q)$:

$$\frac{S_{ie}(q)}{S(q)} = \frac{z^{1/2}v(q)}{(4\pi e^2/q^2)}\left[\frac{1}{\varepsilon(q)} - 1\right]. \tag{3.6}$$

We have normalized the ratio such that the perfect screening condition $S_{ie}(0)=z^{1/2}S(0)$, with z the valence, is satisfied. Now we choose (i) and (ii) above from the Ashcroft empty core model (Ashcroft 1968) and the Thomas-Fermi model respectively. In $\mathbf{r}$ space, the Ashcroft potential is simply

$$\tilde{v}(r) = -\frac{ze^2}{r}: \quad r > R_c \\ = 0 \quad r \le R_c \tag{3.7}$$

with Fourier transform

$$v(q) = -\frac{4\pi z e^2}{q^2}\cos(qR_c). \tag{3.8}$$

For Mg, where we have the computer data of de Wijs *et al.* (1995) for comparison, the core radius R_c is $R_c = 1.39a_o$ and the electron-sphere radius r_s is $r_s = 2.66a_o$, with a_o the Bohr radius $\hbar^2/me^2$, equal to 0.529 Å. Inserting the Thomas–Fermi dielectric function corresponding to the screened potential $(e^2/r)\exp(-k_{TF}r)$ with k_{TF}^{-1} the Thomas–Fermi screening length, related to the Fermi wave number k_f by $k_{TF}^2 = 4k_f/(\pi a_o)$ which is equivalent to $k_{TF} = 2.95(a_o/r_s)^{1/2}$ Å^{-1}, we find in the present modelling

$$\frac{S_{ie}(q)}{S(q)} = z^{1/2}[1 + (q/k_{TF})^2]^{-1}\cos(qR_c). \tag{3.9}$$

This result is plotted in figure 2 with values of R_c and k_{TF} appropriate to Mg ($R_c = 0.735$ Å and $k_{TF} = 1.81$ Å^{-1}).

For comparison, we note from the computer data of de Wijs *et al.* (1995) that the innermost node of $S_{ie}(q)$ occurs at about 2.1 Å^{-1}, near to that from the model (3.9). The deepest value of the ratio from the computer data is about -0.08 at $q_{\min} \approx 2.5$ Å^{-1}, whereas the model yields ≈ -0.13. For such a simple model, this agreement is pleasing. The important point to emphasize is that the innermost node of both the computer data and the model (3.9) lies inside the q value corresponding to the principal peak of the nucleus–nucleus structure factor $S(q)$. The deep negative region in $S_{ie}(q)$ is then in (anti)phase with the main peak of $S(q)$. The model overemphasizes structure in $S_{ie}(q)$ at larger q and probably needs 'smoothing' of the Ashcroft pseudopotential. However, the Gaussian smoothing factor employed by Rasolt (1985) has only a minor effect on the curve in figure 2 out to about 5 Å^{-1}. We emphasize that we have made no attempt to optimize the model (3.9), merely taking the simplest possible choices of $v(q)$ and $\varepsilon(q)$ to illustrate the main features in the computer data.

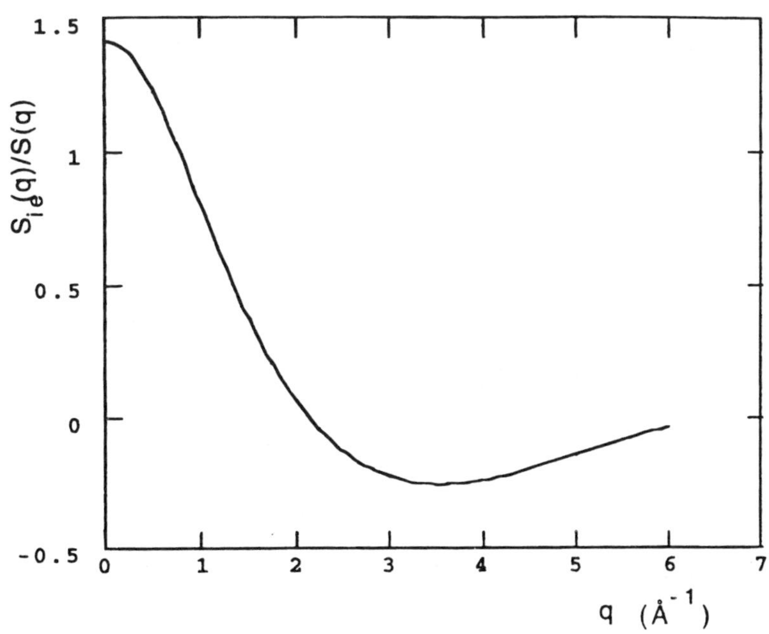

FIGURE 2. Result of model in equation (3.9) for ratio $S_{ie}(q)/S(q)$ in liquid Mg. Input into model is (i) core radius R_c, (ii) Thomas–Fermi inverse screening length k_{TF} and (iii) valency of Mg, $z = 2$. There is at least semi-quantitative agreement with the computer data of de Wijs *et al.* (1995), even though no attempt has been made to optimize model used here.

Below, we shall consider also the Bi data, though we expect that this is a case of strong electron–ion interaction (for example, liquid Bi on freezing becomes a semi-metal).

3.2. *Ratio of $S_{ie}(q)$ to nucleus–nucleus structure factor $S(q)$ for liquid Bi from computer data*

We show in figure 3 some points read from the figures of de Wijs *et al.* (1995) for liquid Bi, again the plot being conveniently made in terms of the ratio $S_{ie}(q)/S(q)$. This becomes, from the screening requirement at $q=0$, $S_{ie}(0)/S(0)=5^{1/2}=2.24$ for the Bi^{5+} ionic state. The data, as in figure 2, fall on a reasonably smooth curve, the important difference from Mg being that the first node in $S_{ie}(q)$ lies now at higher q than the principal maximum in $S(q)$. Thus $S_{ie}(q)$ now has a positive peak in phase with the main peak in $S(q)$ at $q \approx 2$ Å^{-1}.

3.3. *Nodes of $S_{ie}(q)$ and significance for electron–electron correlations as evidenced in X-ray diffraction*

It is quite apparent from the above that the precise knowledge of the nodal structure of $S_{ie}(q)$ is essential for quantitative work. For Bi, there are (at least) two nodes, at q_{oi} say with $i=1$ and 2, as evidenced in figure 3. Returning to equation (3.5) for the X-ray intensity $I_X(q)$, it is clear that this 'exposes' at q_{oi} the valence–valence electron density correlations, provided the accuracy of the X-ray measurement can distinguish $I_X(q_{oi})$ from $f_c^2(q_{oi})S(q_{oi})$. Quantum chemical calculations yield $f_c(q)$ to high accuracy, whereas $S(q)$ is accessible directly from neutron diffraction experiments.

For weak electron-ion interaction, we expect $F(\langle \rho_v(\mathbf{r})\rho_v(\mathbf{r}')\rangle)$ in equation (3.5) to be well approximated by the electron structure factor of the jellium model, say $S_{ee}^o(q)$. So, for liquid

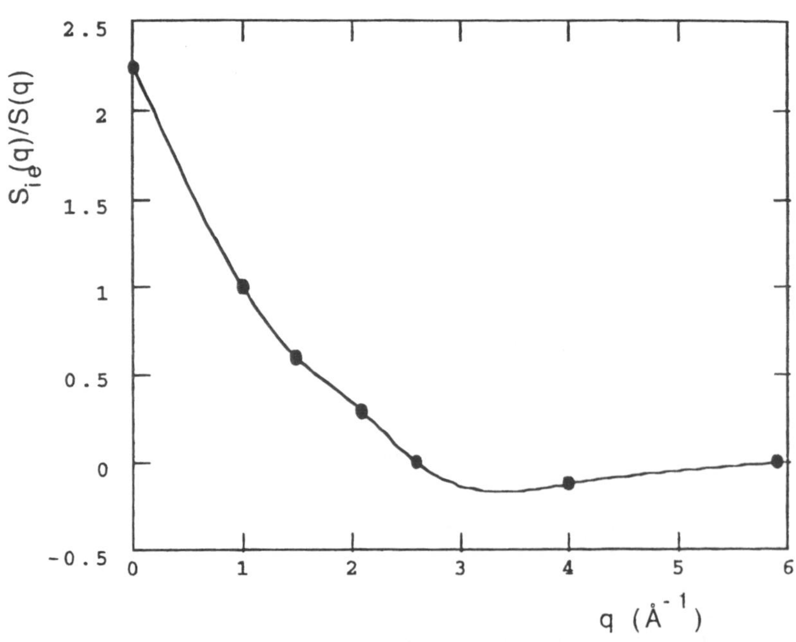

FIGURE 3. $S_{ie}(q)/S(q)$ for liquid Bi, but now plotted directly from data of de Wijs *et al.* (1995) plus perfect screening condition $S_{ie}(0)/S(0)=z^{1/2}$, with valence $z=5$ for Bi. The curve is a cubic spline drawn through the points.

metals like Mg or K there is, in principle at least, the possibility of measuring $S^o_{ee}(q_{oi})$. Samples of theoretical results for the electron structure factor in jellium are shown in figure 4 (from Chiofalo *et al.* 1994). However, for liquid Bi we expect

$$F(\langle\rho_v(\mathbf{r})\rho_v(\mathbf{r}')\rangle) = I^o_{ee}(q) + \Delta I_{ee}(q) \tag{3.10}$$

where $\Delta I_{ee}(q)$ arises from electron–ion interaction. Aspects of dynamical structure theory are needed even for the simplest treatment of $\Delta I_{ee}(q)$, so that we defer discussion to a later section.

4. Dynamical structure: especially transport coefficients of liquid metals

We turn from static to dynamical structure. Some attention will first be devoted to the thermal conductivity λ of the alkali metals. This will be approached via the Wiedemann–Franz Law, relating thermal and electrical conductivities of metals. This reads

$$\frac{\lambda\rho}{T} = L \tag{4.1}$$

where ρ is the electrical resistivity and L the Lorenz number.

In regard to ρ, we note first that Baym (1964) has exploited the relation between neutron and electron scattering, in the regime of validity of the Born approximation, to express the electrical resistivity in terms of the (nucleus–nucleus) dynamical structure factor $S(q,\omega)$. His result has the form

$$\rho = \frac{m^2}{12\pi^3\hbar^3 n_e e^2}\int_0^{2k_f} q^3|V_{ie}(\mathbf{q})|^2\,dq\int_{-\infty}^{\infty} S(q,\omega)\,\frac{\hbar\beta\omega}{\exp(\hbar\beta\omega)-1}\,\frac{d\omega}{2\pi}, \tag{4.2}$$

where $V_{ie}(\mathbf{q})$ is the screened electron–ion potential and $\beta = (k_BT)^{-1}$. In the region near and above the melting temperature, it can be shown that it is valid to re-express this in terms of the static $S(q)$, i.e., the integral of $S(q,\omega)$ over all energy transfers. This yields the nearly-free-electron formula

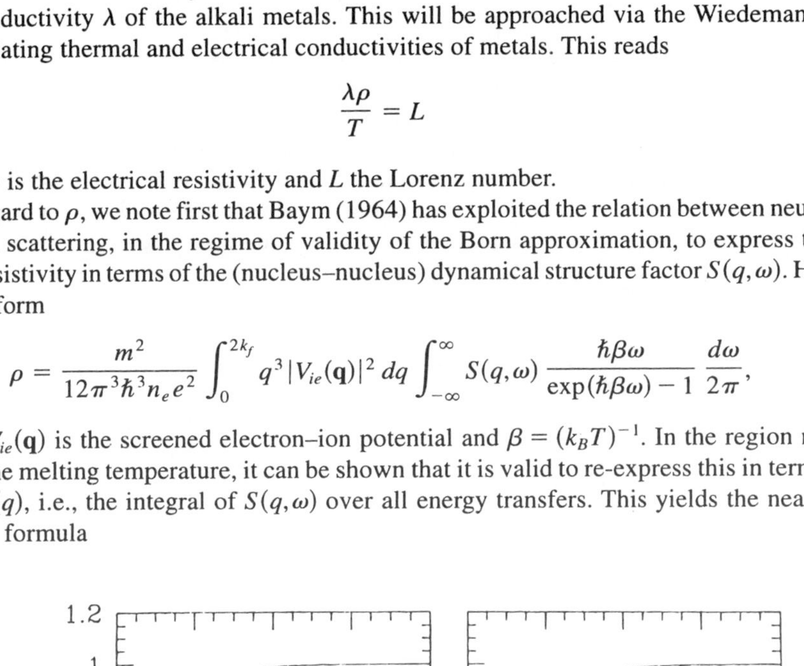

FIGURE 4. Samples of jellium results for the electron–electron structure factor at $r_s/a_o = 1$ and $r_s/a_o = 5$ (from Chiofalo *et al.* 1994). Notice that at $r_s/a_o = 5$ the effect of statistics is already minor, as shown by the comparison of electrons (fermions) with bosons. Liquid Mg corresponds to $r_s = 2.66a_o$. Evidently, one is not to expect structural detail in the jellium electron–electron structure factor for this liquid metal. However (see Egelstaff *et al.* 1974, Rasolt 1985), corrections to be made for the electron–ion interaction will introduce some structural detail through $\Delta I_{ee}(q)$ (see section 4.3).

$$\rho = \frac{12\pi m^2}{\hbar^3 n_e e^2 k_f^2} \int_0^1 |V_{ie}(\mathbf{q})|^2 S(q) \left(\frac{q}{2k_f}\right)^3 d\left(\frac{q}{2k_f}\right) \tag{4.3}$$

going back to Krishnan and Bhatia (1945) and brought to full fruition using pseudopotentials by Ziman (1961).

This formula has been used by Pastore *et al.* (1981) to evaluate ρ for the alkali metals as a function of temperature at constant density and their findings are shown in figure 5.

It will be seen, in relation to equation (4.1) that ρ/T is rather weakly dependent on T, with a negative temperature coefficient. Therefore, equation (4.1) tells us (see also, March & Tosi 1996) that the thermal conductivity λ is a weak function of T, with positive temperature coefficient, in agreement with experiment for liquid Na (see, for example, Sittig 1956).

4.1. *Relation to viscosity*

One of us (Tosi 1992) has used the Kubo formula to evaluate the longitudinal viscosity $\eta_l = \frac{4}{3}\eta + \zeta$, with η and ζ the shear and bulk viscosities respectively, for the liquid alkali metals from the (nucleus–nucleus) dynamical structure factor $S(q,\omega)$. This is evaluated in a two-component electron–ion plasma approach as the dynamical structure factor of the ionic assembly screened by the conduction electrons, the basic ingredients thus being (i) the dynamical structure factor of the classical one-component ionic plasma (OCP) and (ii) the dynamical electronic dielectric function $\varepsilon(q,\omega)$. In the Green-Kubo limit this yields

$$\eta_l = n_i M f_i + \tfrac{4}{15}\, n_e m v_f l_e \tag{4.4}$$

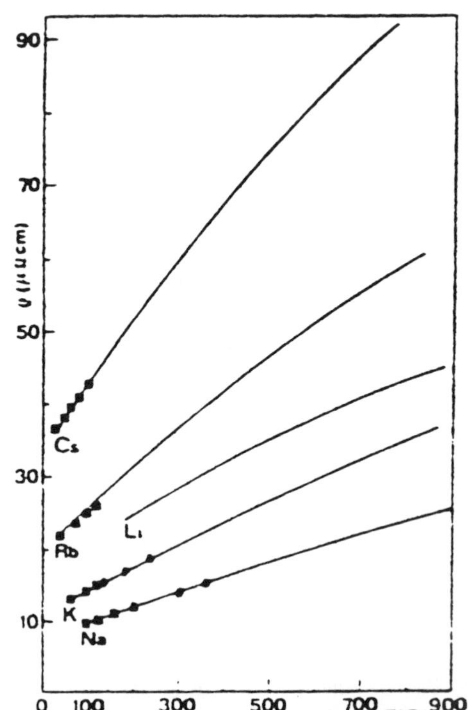

FIGURE 5. Electrical resistivity ρ of liquid alkali metals as a function of temperature at constant density from nearly-free-electron theory [see equation (4.3)], compared with experimental data (squares and dots). After Pastore *et al.* 1981.

where M is the ionic mass, f_i is a friction coefficient characteristic of the OCP, v_f is the Fermi velocity and l_e the electronic mean free path. Numerical evaluation of equation (4.4) shows that the first term on the right-hand side is dominant, the electronic contribution being only a percentage 'correction' to the ionic friction term (for details, see Tosi 1992).

From equation (4.4), with $\eta_l = \frac{4}{3}\eta + \zeta$ and $\zeta \ll \frac{4}{3}\eta$, these results lend support to the Andrade formula for the shear viscosity η_m of a liquid metal at the freezing temperature T_m,

$$\eta_m = A(MT_m)^{1/2} n_i^{2/3} \tag{4.5}$$

where A is a numerical constant given by $A = 5.1 \times 10^{-4}$ poise (Andrade 1934; see also Brown & March 1968; Faber 1972). Combining equation (4.5) with equation (4.1) then yields the relation

$$\lambda_m = \frac{L\eta_m^2}{A^2 \rho_m M n_i^{4/3}} \tag{4.6}$$

between the three transport coefficients at the freezing temperature (March & Tosi 1996).

4.2. *W plasma experiments and effective valence*

In Mg and Bi discussed above, there is a quite clear separation of bound and itinerant electrons, leading to the well defined valences of 2 and 5, respectively.

But in plasmas in general, the situation is often less clearcut. Thus, we report briefly here on the very recent measurements of Kloss *et al.* (1996). These authors studied the electrical conductivity of fluid W for subcritical and supercritical thermodynamic states. The ionic density in these experiments ranged from 4×10^{21} to 6×10^{22} cm^{-3}, while the temperature range embraced was from 3700 to 13000 K.

Briefly, we have utilized subsequently the Nernst–Einstein relation to throw light on the effective valence of *W* as the thermodynamic state is varied, based on the above transport measurements (March & Tosi 1997). The values of valence we extract are consistent with $z = 6$ near the melting point. Likalter has proposed in a private communication to Kloss *et al.* (1996) that $z \approx 3$ at the critical point and for the maximum observed expansion in their experiments Kloss *et al.* propose $z \approx 1$. These values are generally consistent with those extracted using the Nernst-Einstein relation.

4.3. *Diffraction revisited*

To complete this section on dynamical structure we must return to equation (3.10), where the valence–valence electron correlation function contribution to the X-ray scattering intensity $I_X(q)$ has been split into a jellium part already discussed plus an additional part $\Delta I_{ee}(q)$ due to the effect of the ion cores on the valence–valence electron correlations.

As Rasolt (1985) has discussed, the calculation of $\Delta I_{ee}(q)$ involves dynamical considerations. Essentially, he sets up the dynamical response functions $\chi_{\alpha\beta}(q,\omega)$ for the ion–electron mixture. Then he calculates, of course approximately, the electron–electron χ to second order in the electron–ion interaction. His figure 2 plots $\Delta I_{ee}(q)$ vs. q for the examples of liquid Na and Al. We want here to stress only two points. Both for Na and for Al, $\Delta I_{ee}(q)$ has an innermost node (at $\approx$0.95 Å^{-1} for Al and $\approx$1.57 Å^{-1} for Na). Therefore, at these q values the jellium contribution is being exposed. The second point to be noted is that his calculation leads to discontinuities in $\Delta I_{ee}(q)$ at $q = 2k_f$. Earlier examples exist (Stoddart *et al.* 1969) where non-linear effects in the electron–ion interaction change the singularity structure at $2k_f$, but we believe that further work on this issue is called for.

5. Conclusion and directions for further work

The two-component theory of a pure liquid metal has here prompted the analysis of computer studies. Thus de Wijs *et al.* (1995) focussed on the electron–ion structure factor $S_{ie}(q)$ in both liquid Mg and liquid Bi. Here, we have stressed that for liquid Mg a very simple model based on (i) weak electron–ion interaction, (ii) the Ashcroft pseudopotential (3.8), and (iii) the Thomas–Fermi dielectric screening approximation is adequate to reproduce the salient features of $S_{ie}(q)/S(q)$. The main feature of the electron–ion correlations in this liquid metal is the alternation in space between valence electrons and ions, as revealed by the (anti)phase relationship between the deep negative region in $S_{ie}(q)$ and the principal peak in $S(q)$. For liquid Bi, with much stronger electron–ion interaction, $S_{ie}(q)$ has instead a positive peak in phase with the principal peak in $S(q)$, followed by a broad negative region. These features indicate a state of intermediate-range order in the ionic and electronic components of liquid Bi, superposed onto alternation in space. The ratio $S_{ie}(q)/S(q)$ will undoubtedly be harder to model in this liquid metal than in liquid Mg. Screening conditions at $q = 0$ have an important role in both cases.

Turning to dynamical structure, aspects considered include some understanding of the weak temperature dependence, and of the sign of the temperature coefficient, of the thermal conductivity λ of liquid Na starting from the Wiedemann–Franz Law, plus the nearly-free-electron theory of electrical resistivity ρ. After a discussion of longitudinal viscosity of the alkali metals from the Kubo formula used on the dynamical structure factor $S(q,\omega)$, and its support for Andrade's formula for the shear viscosity η at the freezing temperature T_m, it is shown that a formula can be written relating λ, ρ, and η at T_m, even though λ and ρ are dominated by electrons while η has its major contribution, at least for the alkali metals, from the ions.

As an example different from the liquid metals considered above, with well defined valence, we have commented briefly on recent experiments on a W plasma, and in particular on the behavior of the effective valence over the range of states sampled in the measurements of electrical resistivity.

As for future directions, we have made proposals by which one can, at least in principle, explore the magnitude of the contribution of the (Fourier transform of the) valence–valence electron correlation function $\langle \rho_v(\mathbf{r})\rho_v(\mathbf{r}')\rangle$ to the X-ray scattering from a liquid metal. All important is to work at or near the nodes of the valence electron–ion structure factor as predicted by computer simulation studies. Liquid Mg seems an attractive case for further study here. But in longer terms, to obtain both electron–ion and electron–electron correlation functions for valence electrons, one will need to greatly improve electron diffraction techniques, to complement already good X-ray studies such as those of Greenfield *et al.* (1971) on liquid Na and K (see also Johnson *et al.* 1994).

Acknowledgments

One of the authors (N.H.M.) wishes to acknowledge generous hospitality from the Scuola Normale Superiore, during his visit in the Spring of 1996, where his contribution to the present work was brought to fruition. He also wishes to acknowledge partial financial support for this work from the EC Network on 'High energy density matter,' coordinated by Professor J.M. Martinez-Val in Madrid. It is a pleasure to thank Professor Martinez-Val for valuable suggestions which have had an important influence on the investigation reported here.

REFERENCES

Andrade, E.N. Da C. 1934 *Philos. Mag.* **17**, 497.

Ashcroft, N.W. 1968 *J. Phys. C* **1**, 232.

Baym, G. 1964 *Phys. Rev.* **135**, A1691.

Brown, R.C. & March, N.H. 1968 *Phys. Chem. Liquids* **1**, 141.

Chihara, J. 1973 *The Properties of Liquid Metals*, S. Takeuchi, ed. (Taylor and Francis: London), p. 137.

Chiofalo, M.L. *et al.* 1994 *Mod. Phys. Lett. B* **8**, 1207.

de Wijs, G.A. *et al.* 1995 *Phys. Rev. Lett.* **75**, 4480.

Egelstaff, P.A. *et al.* 1974 *Canad. J. Phys.* **52**, 1651.

Faber, T.E. 1972 *An Introduction to the Theory of Liquid Metals* (Cambridge University Press: Cambridge, UK).

Greenfield, A.J. *et al.* 1971 *Phys. Rev. A* **4**, 1607.

Johnson, M.D. & March, N.H. 1963 *Phys. Lett.* **3**, 313.

Johnson, M.W. *et al.* 1994 *Philos. Mag.* **B69**, 965.

Kloss, A. *et al.* 1996 *Phys. Rev. E* **54**, 5851.

Krishnan, K.S. & Bhatia, A.B. 1945 *Nature* **156**, 503.

Magro, W.R. *et al.* 1996 *Phys. Rev. Lett.* **76**, 1240.

March, N.H. & Tosi, M.P. 1984 *Coulomb Liquids* (Academic Press: New York).

March, N.H. & Tosi, M.P. 1996 *N. Cimento D* **18**, 1061.

March, N.H. & Tosi, M.P. 1997 *Phys. Chem. Liquids* **35**, 127.

March, N.H. *et al.* 1973 *J. Phys. C* **6**, L59.

Pastore, G. *et al.* 1981 *Physica* **111 B**, 283.

Perrot, F. & March, N.H. 1990 *Phys. Rev. A* **41**, 4521.

Rasolt, M. 1985 *Phys. Rev. B* **31**, 1615.

Reatto, L. 1988 *Philos. Mag. A* **58**, 37.

Sittig, M. 1956 *Sodium: Its Manufacture, Properties and Uses* (American Chemical Society Monograph Series, Reinhold Publishing: New York).

Stoddart, J.C. *et al.* 1969 *Phys. Rev.* **186**, 683.

Tamaki, S. 1987 *Canad. J. Phys.* **65**, 288.

Tosi, M.P. 1992 *N. Cimento* **14 D**, 559.

Watabe, M. & Hasegawa, M. 1973 *The Properties of Liquid Metals*, S. Takeuchi, ed. (Taylor and Francis: London), p. 133.

Ziman, J.M. 1961 *Philos. Mag.* **6**, 1013.

PHYSICAL REVIEW E **73**, 032201 (2006)

Structural corrections to Stokes-Einstein relation for liquid metals near freezing

N. H. March[1] and J. A. Alonso[2,*]

[1]*Department of Physics, University of Antwerp, Groenenborgerlaan 171, B-2020 Antwerp, Belgium and Oxford University, Oxford, England*

[2]*Departamento de Física Teórica, Atómica y Óptica, Universidad de Valladolid, 47011 Valladolid, Spain*

(Received 23 November 2005; published 1 March 2006)

Using recent progress relating the self-diffusion coefficient to excess entropy, structural corrections to the Stokes-Einstein relation are proposed for liquid metals near the melting temperature.

DOI: 10.1103/PhysRevE.73.032201 PACS number(s): 66.20.+d, 61.25.Mv, 68.35.Md

In an earlier work [1,2] we have been concerned with a correlation between a transport property, namely the shear viscosity η of liquid metals and a thermodynamic quantity, the surface tension γ. If, as for the *s-p* liquid metals, a pair potential $\Phi(r)$ represents a useful approximation to the interionic force field (see Ref. [3] for monovalent Na and Ref. [4] for the divalent metal Be), then an early formula of Fowler [5] expresses γ in terms of $\Phi(r)$ and the pair correlation function $g(r)$ of the bulk liquid

$$\gamma = \frac{\pi}{8}\rho^2 \int_0^\infty r^4 g(r) \frac{\partial \Phi(r)}{\partial r} dr, \tag{1}$$

where ρ is the particle number density. But for Na and Be, knowledge of the experimental $g(r)$ for a given thermodynamic state is enough to determine $\Phi(r)$ for that state, as demonstrated explicitly by Reatto *et al.* [6] for the case of sodium, using x-ray measurements [7] of $g(r)$. Thus, Eq. (1) then shows that, at least approximately, γ is a functional of $g(r)$ and the number density ρ, for a specific thermodynamic state. While this should occasion no surprise for such a thermodynamic property, the correlation between γ and η we exhibited in an earlier work [1,2] leads us to conclude that the transport property η is also, to a useful approximation, a functional of number density ρ and pair correlation function $g(r)$, at least for simple *s-p* metals like Na.

It is our purpose here to invoke now the important study of Dzugutov [8], who proposed, on the basis of computer simulation using pair potentials of various types, a relation between a further transport property, namely the self-diffusion coefficient D, and the excess entropy, this later quantity being the total entropy minus the ideal gas contribution. The result of Dzugutov, followed up in later work [9,10], reads

$$D = 0.194 a^4 g(a) \rho \left[\frac{\pi k_B T}{M}\right]^{1/2} e^{S_2/k_B}, \tag{2}$$

where k_B is the Boltzmann constant, T the temperature, and M is the atomic mass. Here S_2 denotes the two particle approximation to the excess entropy [11,12], given in terms of the radial distribution function $g(r)$ by

$$\frac{S_2}{k_B} = -2\pi\rho \int_0^\infty \{g(r)\ln g(r) - [g(r) - 1]\} r^2 dr \tag{3}$$

while a in Eq. (2) is the principal peak position of $g(r)$. For completeness we notice at this point that (a) some aspects of the Dzugutov result in Eq. (2) have appeared in the literature previously [13], and (b) corrections to the two-particle entropy approximation in Eq. (3) have been discussed by Laird and Haymet [14].

Since our main purpose here is to refine the Stokes-Einstein relation by introducing structural corrections via $g(r)$, we first stress that, to date, there is no direct analog of Eqs. (2) and (3) for the shear viscosity η. But fortunately, in a very early study, Brown and March [15] used physical arguments appropriate to liquid metals which were based fundamentally on the Green-Kubo transport theory, to evaluate η at the melting point T_m. They derived via this route the shear viscosity at the melting temperature as

$$\eta_{T_m} = c[(k_B T)^{1/2} M^{1/2} \rho^{2/3}]_{T=T_m}, \tag{4}$$

where c is a constant. Actually, in studies in the 1930s, Andrade [16] proposed a similar formula using kinetic theory but, as Faber [17] points out in his book, such a treatment is no longer adequate.

Evidently, then, by combining Eq. (4) with Eqs. (2) and (3), but now evaluated specifically at $T=T_m$, we can form the Stokes-Einstein combination, denoted x_m following March and Tosi [18]

$$x_m = \left[\frac{D\eta}{k_B T \rho^{1/3}}\right]_{T_m}. \tag{5}$$

The result for x_m then takes the form, after utilizing Eqs. (4) and (2)

$$x_m = c_1 [a^4 \rho^{4/3} g(a) e^{S_2/k_B}]_{T_m}, \tag{6}$$

where c_1 is a constant. This expression hence gives structural corrections to the Stokes-Einstein combination at the melting temperature, appearing on the right hand side of Eq. (5). While Eq. (6) constitutes the main result of the present study, we shall attempt in what follows to gain insight into likely reasons why x_m defined in Eq. (5) is rather constant for *s-p* electron liquid metals. To quantify this later statement, Table 1 in the paper of March and Tosi [18] shows that x_m^{-1} has an

*Electronic address: jaalonso@fta.uva.es

average value of 47 for five *s*-*p* metals, with a scatter of ±5.

Let us begin with the combination $a^4\rho^{4/3}g(a)$ appearing in Eq. (6) at the melting temperature T_m. We invoke then Eq. (6) of Bhatia and March [19] to write

$$a^2 g(a)\Delta r = \tfrac{1}{3}a^3 - \tfrac{1}{3}R_A^3[1 - S(0)] \tag{7}$$

to be utilized here at $T=T_m$. Δr in Eq. (7) is to be identified with the root mean square displacement of the ions, which in turn is a measure of the half width of the principal peak of $g(r)$, at position a. But Lindemann's law of melting, according to Faber [17], gives $(\Delta r/R_A)_{T_m} \approx 0.2$, where the mean interatomic separation R_A is related to the number density, $\rho = 3/4\pi R_A^3$. Since $a \approx 1.8R_A$, $\Delta r/a \approx 0.11$ at T_m, and hence from Eq. (7) it follows that

$$g(a) \approx 3.0 - 0.06[1 - S(0)]_{T_m}. \tag{8}$$

But the long wavelength limit $q \to 0$ of the structure factor $S(q)$ at T_m for *s*-*p* liquid metals is very much less than unity. Hence, since $a^3 \propto R_A^3 \propto \rho^{-1}$, it follows that $a^4\rho^{4/3}g(a)$ appearing in Eq. (6) at T_m is a constant from Eq. (8). It is worth remarking here that Eq. (8) is the counterpart in r space of the so-called Verlet rule that freezing occurs when the structure factor $S(q)$ has its principal maximum at a height of 2.8. No claim is made here that the r space criterion is (a) of comparable accuracy to the Verlet rule or (b) of easy access to experimental check (Reatto *et al.* [6] get $[g(a)]_{T_m}=2.43$ for liquid metal Na, which is substantially lower than the "prediction" (8)).

But naturally, given the proposal of Eq. (6), the most likely variation of x_m through the *s*-*p* metals will arise from the factor e^{S_2/k_B}. To seek a little further insight here, it is natural enough to write Eq. (3) as the sum of two parts

$$\left(\frac{S_2}{k_B}\right)_{T_m} = \left(\frac{S_u}{k_B}\right)_{T_m} - \frac{1}{2}[1 - S(0)]_{T_m}. \tag{9}$$

Here, S_u denotes the contribution to S_2 from the product $g(r)U(r)$, where $U(r)$ denotes the potential of mean force defined by the Boltzmann-like relation

$$g(r) = e^{-U(r)/k_BT}. \tag{10}$$

While one form of Lindemann's law gives $[S(0)]_{T_m}=\text{const}$ for liquid metals, this is quite approximate. More importantly, at T_m it is known from the fluctuation theory result that $S(0)=\rho k_B T K_T$, where K_T is the isothermal compressibility, is such that $[S(0)]_{T_m} \ll 1$, and hence the second contribution on the right hand side of Eq. (9) is essentially constant.

To date, the only evaluation of S_u/k_B near to the melting point from diffraction experiments is that made by Wallace [20] for Na, who used Waseda's data to yield $(S_u/k_B) = -2.6(\pm 0.2)$. For Na, this is therefore the dominant contribution to S_2/k_B appearing in the basic Eq. (6). But why might this be rather constant among *s*-*p* electron metals? We conjecture, but of course, experiment will be needed eventually to allow confirmation or refinement, that $\int g(r)U(r)d^3r$ might correlate closely with $\int g(r)\Phi(r)d^3r$. Thus, Tankeshwar and March [21] have compared graphically $U(r)$ and $\Phi(r)$ for Na near freezing, using $U(r)$ given by Eq. (10) from $g(r)$ taken from x-ray diffraction studies [7]. It is then clear from Fig. 1 of Tankeshwar and March [21] that $|\int g(r)U(r)d^3r| < |\int g(r)\Phi(r)d^3r|$ but that they may well be related to one another via a numerical factor, as their overall shapes are quite similar. From the work of Bhatia and March [22], $[\int g(r)\Phi(r)d^3r]_{T_m}$ is related, for Na say, to the vacancy formation energy E_v in the hot solid near T_m, E_v being scaled with the thermal energy of melting k_BT_m. This ratio is ≈ 9 from experiment for close-packed *s*-*p* metals, as can be seen, for example, from Rashid and March [23], with again a scatter, however, $\approx \pm 2$. These considerations show how it may be possible still to understand using Eq. (6) the relative constancy of x_m^{-1} of around 50 for five *s*-*p* metals, with a scatter of ±10%. Of course, the true test of the practical utility of Eq. (6) must await further "experimental" data on the pair correlation function $g(r)$ near melting: we trust the present proposal (6) will stimulate further careful diffraction measurements near freezing on other *s*-*p* electron metals.

Our final comment concerns materials with force fields transcending pair potential representations, and in particular *d*-electron liquid metals such as Ni or W. It should not be assumed that the above arguments take over simply to such cases: appeal being made to pair potential simplifications frequently in the present study.

This work was supported by MCYT of Spain (Grant No. MAT2002-04499-C02-01) and Junta de Castilla y Leon (Grant No. VA039A05). N.H.M. wishes to acknowledge the generous hospitality of L. M. Nieto and J. A. Alonso during a visit to the University of Valladolid.

[1] N. H. March and J. A. Alonso, Mol. Phys. **95**, 353 (1998).
[2] J. A. Alonso and N. H. March, Phys. Rev. E **60**, 4125 (1999).
[3] F. Perrot and N. H. March, Phys. Rev. A **41**, 4521 (1990).
[4] F. Perrot and N. H. March, Phys. Rev. A **42**, 4884 (1990).
[5] R. H. Fowler, Proc. R. Soc. London, Ser. A **159**, 229 (1937).
[6] L. Reatto, D. Levesque, and J. J. Weis, Phys. Rev. A **33**, 3451 (1986).
[7] A. J. Greenfield, J. Wellendorf, and N. Wiser, Phys. Rev. A **4**, 1607 (1971).
[8] M. Dzugutov, Nature (London) **381**, 137 (1996).
[9] J. J. Hoyt, M. Asta, and B. Sadigh, Phys. Rev. Lett. **85**, 594 (2000).
[10] C. Kaur, U. Harbola, and S. P. Das, J. Chem. Phys. **123**, 034501 (2005).
[11] R. D. Mountain and H. J. Raveché, J. Chem. Phys. **55**, 2250 (1971).
[12] A. Baranyai and D. J. Evans, Phys. Rev. A **40**, 3817 (1989).
[13] Y. Rosenfeld, Phys. Rev. E **62**, 7524 (2000).
[14] B. B. Laird and A. D. J. Haymet, Phys. Rev. A **45**, 5680 (1992).

[15] R. C. Brown and N. H. March, Phys. Chem. Liq. **1**, 141 (1968).
[16] E. N. da C. Andrade, Philos. Mag. **17**, 497 (1934).
[17] T. E. Faber, *Introduction to the Theory of Liquid Metals* (Cambridge University Press, Cambridge, 1972).
[18] N. H. March and M. P. Tosi, Phys. Rev. E **60**, 2402 (1999).
[19] A. B. Bhatia and N. H. March, Phys. Chem. Liq. **13**, 313 (1984).
[20] D. C. Wallace, J. Chem. Phys. **87**, 2282 (1987).
[21] K. Tankeshwar and N. H. March, Phys. Chem. Liq. **25**, 59 (1992).
[22] A. B. Bhatia and N. H. March, J. Chem. Phys. **80**, 2076 (1984).
[23] R. I. M. A. Rashid and N. H. March, Phys. Chem. Liq. **19**, 41 (1989).

ANNALS OF PHYSICS **81**, 414–437 (1973)

Quantum Theory of Pure Liquid Metals as Two-Component Systems

N. H. MARCH*

Department of Physics, Imperial College of Science and Technology, South Kensington, London S.W. 7, England

AND

M. P. TOSI

Istituto di Fisica dell' Universita' e GNSM del CNR, Messina, Italy

Received December 29, 1972

With the ultimate aims of clarifying the interpretation and the utility of effective ion–ion interactions in liquid metals, and of understanding the unusual isotopic mass dependence of the shear viscosity of liquid metal Li, a fully quantum statistical mechanical theory is developed from the many-body Hamiltonian of the conduction electron-positive ion assembly.

We have set up quantum equations of motion which are analogs of classical continuity and conservation equations by expanding the equation for the Wigner distribution function about its diagonal. The most important of these equations for our present purposes relates the time derivative of the current density $\mathbf{j}(\mathbf{r}, t)$ to the flux of current and to density–density correlation functions for electrons, electron–ions, and ions.

This theory is then applied to neutron scattering by liquid metals. While the theory is sufficiently general in principle to treat electron–ion interaction of arbitrary strength, it is shown that when the interaction is weak, the usual results are recovered, along with the effective ion–ion interaction. In this latter connection, it is also demonstrated how the effective Ornstein–Zernike function $C(q)$ in a liquid metal is related to bare ion and bare electron direct correlation functions and to the bare electron partial structure factor. Combining $C(q)$ with one of the classical equations of liquid structure such as Born–Green or Percus–Yevick then relates the effective ion–ion interaction to the partial correlation functions of the bare ions and electrons.

It is further shown how gradient expansions of the correlation functions lead to equations of motion for the density, current, and energy density which are simply the hydrodynamic equations of the present quantum theory of two-component systems. It is pointed out that the analog of the Navier–Stokes equation for the two-component system may be used to identify the quantity $\frac{4}{3}\eta + \zeta$ for the liquid metal, η and ζ being

* The work was largely carried out during a visit to the Department of Physics, University of Messina, Messina, Italy.

respectively the shear and bulk viscosities. Finally, it is demonstrated that $\frac{4}{3}\eta + \zeta$ depends explicitly on functional derivatives of the nonequilibrium pair distribution functions of ion–ion, electron–ion, and electron–electron correlations.

1. Introduction

The concept of effective ion–ion interactions has proved a fruitful one in the theory of liquid metals. As is well known, the conduction electrons play a major role in determining the form of this interaction. However, the precise interpretation of such effective ion–ion interactions is not well understood, and, hence, uncertainties still exist as to the range of physical properties which can be usefully calculated from effective interactions.

Another area where uncertainties remain is that of atomic transport in liquid metals. Thus, the shear viscosities of liquid metals Li^6 and Li^7 have been separately measured by Ban, Montgomery, and Randall [1] and reveal the surprising feature that they do not scale as the square root of the isotopic mass M. Indeed, at the melting temperature, these workers found that the viscosity of Li^7 was greater than that of Li^6 by a factor 1.44, whereas $(\frac{7}{6})^{1/2}$ is only 1.08. In contrast, a very recent experiment by Bratby and Harris [2] on the shear viscosity of isotopes of liquid Ne demonstrates excellent agreement with the root mass law, after a tiny quantum correction is made.

Though it is not possible to exclude quantum-mechanical tunneling effects through potential barriers for Li^+ ions at this stage, these experiments suggest that the differences arise because of the presence of conduction electrons in liquid Li. Thus, it seems clear that one must treat liquid Li as a two-component system; Li^+ ions and conduction electrons. While it is plausible to treat Li^+ ions classically at and above the melting temperature of Li, in fact we shall carry through the basic formulation of the two-component theory entirely quantum-mechanically. In this way, we can be certain that we have a theoretical framework general enough to allow us to deal, at least in principle, with the shear viscosity, even if quantum-mechanical tunneling of the Li^+ ions should turn out to be a significant feature.

At this point, it should be noted that Cowan and Kirkwood [3] had already recognized the need to treat liquid metals as two-component systems. Their pioneering work treated the ions classically, but unfortunately involved semi-classical Thomas–Fermi approximations in dealing with the electrons. Nevertheless, these workers were able to calculate approximate self-consistent pair distribution functions $g_{ii}(r)$, $g_{ie}(r)$, and $g_{ee}(r)$ expressing ion–ion, ion–electron, and electron–electron correlations, respectively. Their work, therefore, could be used to obtain static structure factors $S_{ii}(q)$, etc., defined as Fourier transforms of the radial distribution functions. Work in the same spirit, but relaxing the semiclassical approximations, has been carried through for a proton–electron assembly by Broyles *et al.* [4].

Very recently, Watabe and Hasegawa [5] and also Chihara [6] have discussed the dynamic generalizations of these structure factors, namely $S_{ii}(q, \omega)$, $S_{ie}(q, \omega)$, and $S_{ee}(q, \omega)$. Their theories are truly microscopic and, as such, inevitably involve approximations. However, these workers agree that when the above dynamical structure factors are integrated over all energy transfers, then with the ions treated classically

$$\lim_{q\to 0} [S_{ii}(q) = S_{ie}(q) = S_{ee}(q)] \tag{1.1}$$

for a monovalent liquid metal. This long wave length limit is already of interest because the electron–electron structure factor $S_{ee}(q)$ as $q \to 0$ tends to zero for the jellium model in which the positive ions are smeared out uniformly, in contrast to the result (1.1). Equation (1.1) is generalized by Watabe and Hasegawa to polyvalent metals with valency Z, when they find

$$\lim_{q\to 0} [S_{ee}(q) = ZS_{ii}(q)] \tag{1.2}$$

and

$$\lim_{q\to 0} [S_{ie}(q) = Z^{1/2}S_{ii}(q)]. \tag{1.3}$$

Evidently the usual thermodynamic result that

$$S_{ii}(0) = n_i k_B T K_T \tag{1.4}$$

must also hold, n_i being the number density of the ions and K_T the isothermal compressibility of the liquid metal. We should also refer here to the work of Takahashi [7] who develops a theory of liquid metals in the zero sound wave range, using the double-time Green function method.

Our aims in the present paper are different from those of the above workers. Thus, as indicated above, one of our ultimate interests is atomic transport, and it is, therefore, natural that we attempt to set up a quantum treatment in which analogs of classical continuity and conservation equations play a central role. Whereas, in a one-component theory of a classical liquid, the hydrodynamic regime has been fully studied by many workers (see, for instance, [8]), here in contrast we take as our starting point a quantum-mechanical theory of the coupled electron–ion system and a principal aim is to demonstrate that our microscopic equations contain quantum hydrodynamical equations for the two-component system. The only discussion we are aware of with a similar objective is that of Fröhlich [9]. His theory, however, is for a one-component system and is couched in general terms. Our present treatment is made quite explicit for a charged two-component fluid, namely a liquid metal.

2. Electron–Ion Hamiltonian and Density Fluctuation Operators

To implement the above program, let us set up a description of the assembly in terms of the local number densities $\rho_e(\mathbf{r}, t)$ and $\rho_i(\mathbf{r}, t)$ of electrons and ions, respectively, at position $\mathbf{r}$ at time t. We then introduce the density fluctuation operators $\rho_{\mathbf{k}}(t)$ for electrons and ions through

$$\rho_e(\mathbf{r}, t) = \sum_i \delta(\mathbf{r} - \mathbf{r}_i(t)) = \sum_{\mathbf{k}} \rho_{e\mathbf{k}}(t)\, e^{i\mathbf{k}\cdot\mathbf{r}} \tag{2.1}$$

and

$$\rho_i(\mathbf{r}, t) = \sum_j \delta(\mathbf{r} - \mathbf{R}_j(t)) = \sum_{\mathbf{k}} \rho_{i\mathbf{k}}(t)\, e^{i\mathbf{k}\cdot\mathbf{r}}. \tag{2.2}$$

Inverting these relations yields

$$\rho_{e\mathbf{k}}(t) = \mathscr{V}^{-1} \sum_i e^{-i\mathbf{k}\cdot\mathbf{r}_i(t)}; \qquad \rho_{i\mathbf{k}}(t) = \mathscr{V}^{-1} \sum_j e^{-i\mathbf{k}\cdot\mathbf{R}_j(t)}, \tag{2.3}$$

$\mathbf{r}_i$ denoting the position of the ith electron, $\mathbf{R}_j$ that of the jth ion, and $\mathscr{V}$ the total volume of the liquid metal.

Denoting electron–electron, ion–ion, and electron–ion interactions by v, V, and V_{ie}, respectively, and with ions of mass M, the total Hamiltonian reads

$$\begin{aligned} H &= \sum_i p_i^2/2m + \sum_j p_j^2/2M + \tfrac{1}{2} \sum_{i\neq i'} v(|\mathbf{r}_i - \mathbf{r}_{i'}|) \\ &\quad + \tfrac{1}{2} \sum_{j\neq j'} V(\mathbf{R}_j - \mathbf{R}_{j'}) + \sum_{ij} V_{ie}(\mathbf{r}_i - \mathbf{R}_j) \\ &= \sum_i p_i^2/2m + \sum_j p_j^2/2M + \tfrac{1}{2} \sum_{\mathbf{k}} v_{\mathbf{k}}(\rho_{e\mathbf{k}}^+ \rho_{e\mathbf{k}} - n_e) \\ &\quad + \tfrac{1}{2} \sum_{\mathbf{k}} V_{\mathbf{k}}(\rho_{i\mathbf{k}}^+ \rho_{i\mathbf{k}} - n_i) + \sum_{\mathbf{k}} V_{ie}(\mathbf{k})\, \rho_{e\mathbf{k}}^+ \rho_{i\mathbf{k}}, \end{aligned} \tag{2.4}$$

the density of electrons n_e and the ionic density n_i being related by $n_e = Zn_i$. It will be convenient subsequently to work with unit volume, $\mathscr{V} = 1$.

2.1. *Response Functions*

For use later in the discussion, we define the response functions

$$\begin{aligned} \chi_{ee}(\mathbf{q}, \omega) &= \int_0^\infty dt\, e^{-i\omega t}(1/i\hbar)\langle[\rho_{e\mathbf{q}}^+(0), \rho_{e\mathbf{q}}(t)]\rangle \\ &= \beta \int_0^\infty dt\, e^{-i\omega t}\langle\dot{\rho}_{e\mathbf{q}}^+(0); \rho_{e\mathbf{q}}(t)\rangle, \end{aligned} \tag{2.5}$$

$$\chi_{ie}(\mathbf{q}, \omega) = \int_0^\infty dt\ e^{-i\omega t}(1/i\hbar)\langle[\rho_{e\mathbf{q}}^+(0), \rho_{i\mathbf{q}}(t)]\rangle$$
$$= \beta \int_0^\infty dt\ e^{-i\omega t}\langle\dot{\rho}_{e\mathbf{q}}^+(0); \rho_{i\mathbf{q}}(t)\rangle, \tag{2.6}$$

and

$$\chi_{ii}(\mathbf{q}, \omega) = \int_0^\infty dt\ e^{-i\omega t}(1/i\hbar)\langle[\rho_{i\mathbf{q}}^+(0), \rho_{i\mathbf{q}}(t)]\rangle$$
$$= \beta \int_0^\infty dt\ e^{-i\omega t}\langle\dot{\rho}_{i\mathbf{q}}^+(0); \rho_{i\mathbf{q}}(t)\rangle. \tag{2.7}$$

Here, $\beta = 1/k_BT$, while $\langle X; Y\rangle$ is the canonical correlation function of Kubo, namely

$$\langle X; Y\rangle = (1/\beta) \int_0^\beta d\lambda\ \langle e^{\lambda H}Xe^{-\lambda H}Y\rangle \tag{2.8}$$

with

$$\langle[X(0), Y(t)]\rangle = i\hbar\beta\langle\dot{X}(0); Y(t)\rangle. \tag{2.9}$$

It is sometimes useful to define response functions for mass density. This is then related to the discussion of correlation functions in a classical binary alloy given by Bhatia, Thornton, and March [10] and we deal with these quantities in Appendix 1, along with a few of their properties.

2.2. *Equations of Motion for Density Matrices*

We next introduce field operators $\psi(\mathbf{r}, t)$ and $\Psi(\mathbf{r}, t)$ for electrons and ions, respectively, and then form the equations of motion for the density matrices defined as $\langle\psi^+(\mathbf{x}, t)\ \psi(\mathbf{x}', t)\rangle$ and $\langle\Psi^+(\mathbf{x}, t)\ \Psi(\mathbf{x}', t)\rangle$. These equations are

$$\{i\hbar(\partial/\partial t) - (\hbar^2/2m)(\nabla_{\mathbf{x}}^2 - \nabla_{\mathbf{x}'}^2)\}\langle\psi^+(\mathbf{x}, t)\ \psi(\mathbf{x}', t)\rangle$$
$$= -\int d\mathbf{x}''\ [v(\mathbf{x} - \mathbf{x}'') - v(\mathbf{x}' - \mathbf{x}'')]\langle\psi^+(\mathbf{x}, t)\ \psi(\mathbf{x}', t)\ \rho_e(\mathbf{x}'', t)\rangle$$
$$- \int d\mathbf{x}''\ [V_{ie}(\mathbf{x} - \mathbf{x}'') - V_{ie}(\mathbf{x}' - \mathbf{x}'')]\langle\psi^+(\mathbf{x}, t)\ \psi(\mathbf{x}', t)\ \rho_i(\mathbf{x}'', t)\rangle \tag{2.10}$$

and

$$\{i\hbar(\partial/\partial t) - (\hbar^2/2M)(\nabla_{\mathbf{x}}^2 - \nabla_{\mathbf{x}'}^2)\}\langle\Psi^+(\mathbf{x}, t)\ \Psi(\mathbf{x}', t')\rangle$$
$$= -\int d\mathbf{x}''\ [V(\mathbf{x} - \mathbf{x}'') - V(\mathbf{x}' - \mathbf{x}'')]\langle\Psi^+(\mathbf{x}, t)\ \Psi(\mathbf{x}', t)\ \rho_i(\mathbf{x}'', t)\rangle$$
$$- \int d\mathbf{x}''\ [V_{ie}(\mathbf{x} - \mathbf{x}'') - V_{ie}(\mathbf{x}' - \mathbf{x}'')]\langle\Psi^+(\mathbf{x}, t)\ \Psi(\mathbf{x}', t)\ \rho_e(\mathbf{x}'', t)\rangle. \tag{2.11}$$

Since these two equations have basically the same form, the calculation will be carried out in detail only for the electron equation of motion (2.10), and the final result will then be written down for Eq. (2.11).

3. Wigner Distribution Functions

Recalling our aim to set up the theory in terms of current and energy density, we next introduce the mixed density matrix or Wigner distribution function $f(\mathbf{p}, \mathbf{R}, t)$ defined by

$$f(\mathbf{p}, \mathbf{R}, t) = \int d\mathbf{r}\, e^{i\mathbf{p}\cdot\mathbf{r}/\hbar}\langle\psi^{+}(\mathbf{R} + \tfrac{1}{2}\mathbf{r}, t)\, \psi(\mathbf{R} - \tfrac{1}{2}\mathbf{r}, t)\rangle \tag{3.1}$$

with an entirely similar function $F(\mathbf{p}, \mathbf{R}, t)$ for the ions.

From this Wigner distribution function, we can now construct the density $n(\mathbf{R}, t)$, the current density $\mathbf{j}(\mathbf{R}, t)$ and the "kinetic energy tensor" $\pi(\mathbf{R}, t)$ as

$$n(\mathbf{R}, t) = \sum_{\mathbf{p}} f(\mathbf{p}, \mathbf{R}, t), \tag{3.2}$$

$$\mathbf{j}(\mathbf{R}, t) = (1/m) \sum_{\mathbf{p}} \mathbf{p} f(\mathbf{p}, \mathbf{R}, t), \tag{3.3}$$

and

$$\pi_{\alpha\beta}(\mathbf{R}, t) = (1/m) \sum_{\mathbf{p}} p_\alpha p_\beta f(\mathbf{p}, \mathbf{R}, t), \tag{3.4}$$

with analogous definitions of N, $\mathbf{J}$, and Π for the ions in terms of F and the ion mass M.

The equation of motion (2.10) can then be expanded about the diagonal to yield

$$\begin{aligned}\{i\hbar(\partial/\partial t) - (\hbar^2/m)\, \nabla_{\mathbf{R}} \cdot \nabla_{\mathbf{r}}\} \sum_{\mathbf{p}} e^{-i\mathbf{p}\cdot\mathbf{r}/\hbar} f(\mathbf{p}, \mathbf{R}, t) \\ = -\int dx''\, [\mathbf{r} \cdot \nabla_{\mathbf{R}} v(\mathbf{R} - \mathbf{x}'') + O(r^3)] \\ \times [\langle\rho_e(\mathbf{R}, t)\, \rho_e(\mathbf{x}'', t)\rangle + (m/i\hbar)\, \mathbf{r} \cdot \langle\mathbf{j}(\mathbf{R}, t)\, \rho_e(\mathbf{x}'', t)\rangle + O(r^2)] \\ -\int dx''\, [\mathbf{r} \cdot \nabla_{\mathbf{R}} V_{ie}(\mathbf{R} - \mathbf{x}'') + O(r^3)] \\ \times [\langle\rho_e(\mathbf{R}, t)\, \rho_i(\mathbf{x}'', t)\rangle + (m/i\hbar)\, \mathbf{r} \cdot \langle\mathbf{j}(\mathbf{R}, t)\, \rho_i(\mathbf{x}'', t)\rangle + O(r^2)],\end{aligned} \tag{3.5}$$

the current density operator $\mathbf{j}(\mathbf{R}, t)$ being given by

$$\mathbf{j}(\mathbf{R}, t) = (i\hbar/2m)\{[\nabla_{\mathbf{R}}\psi^{+}(\mathbf{R}, t)]\, \psi(\mathbf{R}, t) - \psi^{+}(\mathbf{R}, t)[\nabla_{\mathbf{R}}\psi(\mathbf{R}, t)]\}. \tag{3.6}$$

Equation (3.5) is the central equation from which continuity and conservation equations can now be obtained.

3.1. *Continuity Equation and Current Conservation*

Equating terms independent of $\mathbf{r}$ on both sides of Eq. (3.5) we find almost immediately that

$$\partial n(\mathbf{R}, t)/\partial t + \nabla_{\mathbf{R}} \cdot \mathbf{j}(\mathbf{R}, t) = 0, \tag{3.7}$$

the usual continuity equation.

The terms of $O(r)$ yield similarly

$$m(\partial \mathbf{j}(\mathbf{R}, t)/\partial t) + \nabla_{\mathbf{R}} \cdot \pi(\mathbf{R}, t) = -\int d\mathbf{x}'' \, \nabla_{\mathbf{R}} v(\mathbf{R} - \mathbf{x}'') \langle \rho_e(\mathbf{R}, t) \, \rho_e(\mathbf{x}'', t) \rangle - \int d\mathbf{x}'' \, \nabla_{\mathbf{R}} V_{ie}(\mathbf{R} - \mathbf{x}'') \langle \rho_e(\mathbf{R}, t) \, \rho_i(\mathbf{x}'', t) \rangle. \tag{3.8}$$

When one passes over to the jellium model, Eqs. (3.7) and (3.8) form the basis of the electron gas theory of Singwi *et al.* [11].

If we further express the current density in terms of a velocity field $\mathbf{v}(\mathbf{R}, t)$ by

$$\mathbf{j}(\mathbf{R}, t) = n(\mathbf{R}, t) \, \mathbf{v}(\mathbf{R}, t) \tag{3.9}$$

and separate the momentum flux tensor into two parts:

$$\pi_{\alpha\beta}(\mathbf{R}, t) = m j_\alpha(\mathbf{R}, t) \, v_\beta(\mathbf{R}, t) + \pi^0_{\alpha\beta}(\mathbf{R}, t), \tag{3.10}$$

then Eq. (3.8) can be rewritten as

$$mn(\mathbf{R}, t)(D\mathbf{v}(\mathbf{R}, t)/Dt) = -\nabla_{\mathbf{R}} \cdot \pi^0(\mathbf{R}, t) - \int d\mathbf{x}'' \, \nabla_{\mathbf{R}} v(\mathbf{R} - \mathbf{x}'') \langle \rho_e(\mathbf{R}, t) \, \rho_e(\mathbf{x}'', t) \rangle - \int d\mathbf{x}'' \, \nabla_{\mathbf{R}} V_{ie}(\mathbf{R} - \mathbf{x}'') \langle \rho_e(\mathbf{R}, t) \, \rho_i(\mathbf{x}'', t) \rangle, \tag{3.11}$$

where

$$D\mathbf{v}/Dt = \partial \mathbf{v}/\partial t + (\mathbf{v} \cdot \nabla_{\mathbf{R}})\mathbf{v}. \tag{3.12}$$

We shall see later that the Navier–Stokes equation for the liquid metal derives from a combination of Eq. (3.11) with the corresponding equation for the ions.

3.2. *Energy Transport Equation*

We turn to the final equation we wish to derive from Eq. (3.5). To $O(r^2)$, the diagonal parts then yield the result

$$\frac{i\hbar}{2}\frac{\partial}{\partial t}\left(\frac{-i}{\hbar}\right)^2 \sum_{\mathbf{p}} p_\alpha^2 f(\mathbf{p}, \mathbf{R}, t) - \frac{\hbar^2}{2m}\left(\frac{-i}{\hbar}\right)^3 \sum_{\mathbf{p}} \mathbf{p}\cdot\nabla_{\mathbf{R}} p_\alpha^2 f(\mathbf{p}, \mathbf{R}, t)$$
$$= -\frac{m}{i\hbar}\int d\mathbf{x}''\, \nabla_{R_\alpha} v(\mathbf{R}-\mathbf{x}'')\langle j_\alpha(\mathbf{R}, t)\, \rho_e(\mathbf{x}'', t)\rangle$$
$$-\frac{m}{i\hbar}\int d\mathbf{x}''\, \nabla_{R_\alpha} V_{ie}(\mathbf{R}-\mathbf{x}'')\langle j_\alpha(\mathbf{R}, t)\, \rho_i(\mathbf{x}'', t)\rangle. \tag{3.13}$$

Taking the trace, the following equation is obtained:

$$(\partial/\partial t)\, k(\mathbf{R}, t) + \nabla_{\mathbf{R}}\cdot \mathbf{j}_k(\mathbf{R}, t) = -\int d\mathbf{x}''\, \nabla_{\mathbf{R}} v(\mathbf{R}-\mathbf{x}'')\cdot\langle \mathbf{j}(\mathbf{R}, t)\, \rho_e(\mathbf{x}'', t)\rangle$$
$$-\int d\mathbf{x}''\, \nabla_{\mathbf{R}} V_{ie}(\mathbf{R}-\mathbf{x}'')\cdot\langle \mathbf{j}(\mathbf{R}, t)\, \rho_i(\mathbf{x}'', t)\rangle, \tag{3.14}$$

where the kinetic energy density $k(\mathbf{R}, t)$ is given by

$$k(\mathbf{R}, t) = \sum_{\mathbf{p}} (p^2/2m)\, f(\mathbf{p}, \mathbf{R}, t) = \tfrac{1}{2}\operatorname{Tr}\pi(\mathbf{R}, t) \tag{3.15}$$

and the kinetic energy flux $\mathbf{j}_k$ is simply

$$\mathbf{j}_k(\mathbf{R}, t) = \sum_{\mathbf{p}} (\mathbf{p}/m)(p^2/2m)\, f(\mathbf{p}, \mathbf{R}, t). \tag{3.16}$$

To see that Eq. (3.14) is, in fact, an energy transport equation, consider the identity below for the first term on the right side of Eq. (3.14):

$$-\int d\mathbf{x}''\, \nabla_{\mathbf{R}} v(\mathbf{R}-\mathbf{x}'')\cdot\langle \mathbf{j}(\mathbf{R}, t)\, \rho_e(\mathbf{x}'', t)\rangle$$
$$= -\nabla_{\mathbf{R}}\cdot\int d\mathbf{x}''\, v(\mathbf{R}-\mathbf{x}'')\langle \mathbf{j}(\mathbf{R}, t)\, \rho_e(\mathbf{x}'', t)\rangle$$
$$+\int d\mathbf{x}''\, v(\mathbf{R}-\mathbf{x}'')\langle \nabla_{\mathbf{R}}\cdot \mathbf{j}(\mathbf{R}, t)\, \rho_e(\mathbf{x}'', t)\rangle. \tag{3.17}$$

The continuity equation (3.7), written in operator form, gives

$$\langle \nabla_{\mathbf{R}}\cdot \mathbf{j}(\mathbf{R}, t)\, \rho_e(\mathbf{x}'', t)\rangle = -[(\partial/\partial t)\langle \rho_e(\mathbf{R}, t)\, \rho_e(\mathbf{x}'', t')\rangle]_{t'=t}\,, \tag{3.18}$$

and, hence, the last terms on the right side of Eq. (3.17) may be evaluated as

$$\int d\mathbf{x}''\, v(\mathbf{R}-\mathbf{x}'')\langle \nabla_{\mathbf{R}}\cdot \mathbf{j}(\mathbf{R},t)\,\rho_e(\mathbf{x}'',t)\rangle$$

$$= -\tfrac{1}{2}\int d\mathbf{x}''\, v(\mathbf{R}-\mathbf{x}'')(\partial/\partial t)\langle \rho_e(\mathbf{R},t)\,\rho_e(\mathbf{x}'',t)\rangle, \tag{3.19}$$

the last step following from Eq. (3.18) because of the symmetry of the correlation function $\langle \rho_e(\mathbf{R},t)\,\rho_e(\mathbf{x}'',t)\rangle$. But the electron–electron potential energy density $v_{ee}(\mathbf{R},t)$ is simply

$$v_{ee}(\mathbf{R},t) = \tfrac{1}{2}\int d\mathbf{x}''\, v(\mathbf{R}-\mathbf{x}'')\langle \rho_e(\mathbf{R},t)\,\rho_e(\mathbf{x}'',t)\rangle, \tag{3.20}$$

and so the right side of Eq. (3.19) becomes $-(\partial/\partial t)\, v_{ee}(\mathbf{R},t)$.

Similarly the second term on the right side of Eq. (3.14) can be evaluated to be

$$-\int d\mathbf{x}''\, \nabla_{\mathbf{R}} V_{ie}(\mathbf{R}-\mathbf{x}'')\cdot\langle \mathbf{j}(\mathbf{R},t)\,\rho_i(\mathbf{x}'',t)\rangle$$

$$= -\nabla_{\mathbf{R}}\cdot\int d\mathbf{x}''\, V_{ie}(\mathbf{R}-\mathbf{x}'')\langle \mathbf{j}(\mathbf{R},t)\,\rho_i(\mathbf{x}'',t)\rangle - (\partial/\partial t)\, V_{ie}(\mathbf{R},t), \tag{3.21}$$

where

$$V_{ie}(\mathbf{R},t) = \int d\mathbf{x}''\, V_{ie}(\mathbf{R}-\mathbf{x}'')\langle \rho_e(\mathbf{R},t)\,\rho_i(\mathbf{x}'',t^-)\rangle. \tag{3.22}$$

Thus, we can rewrite Eq. (3.14) as an energy transport equation for the electrons, namely

$$(\partial/\partial t)\,\epsilon(\mathbf{R},t) + \nabla_{\mathbf{R}}\cdot\mathbf{j}_\epsilon(\mathbf{R},t) = 0, \tag{3.23}$$

while for the ions we have

$$(\partial/\partial t)\,\mathscr{E}(\mathbf{R},t) + \nabla_{\mathbf{R}}\cdot\mathbf{J}_{\mathscr{E}}(\mathbf{R},t) = 0. \tag{3.24}$$

Here, the explicit forms of the energy densities for electrons and ions respectively are

$$\epsilon(\mathbf{R},t) = \sum_{\mathbf{p}} (p^2/2m)\, f(\mathbf{p},\mathbf{R},t) + \int d\mathbf{x}\, v(\mathbf{R}-\mathbf{x})\langle \rho_e(\mathbf{R},t)\,\rho_e(\mathbf{x},t^-)\rangle$$

$$+ \int d\mathbf{x}\, V_{ie}(\mathbf{R}-\mathbf{x})\langle \rho_e(\mathbf{R},t)\,\rho_i(\mathbf{x},t^-)\rangle \tag{3.25}$$

and

$$\mathscr{E}(\mathbf{R},t) = \sum_{\mathbf{p}} (p^2/2M)\, F(\mathbf{p},\mathbf{R},t) + \int d\mathbf{x}\, V(\mathbf{R}-\mathbf{x})\langle \rho_i(\mathbf{R},t)\,\rho_i(\mathbf{x},t^-)\rangle$$

$$+ \int d\mathbf{x}\, V_{ie}(\mathbf{R}-\mathbf{x})\langle \rho_i(\mathbf{R},t)\,\rho_e(\mathbf{x},t^-)\rangle. \tag{3.26}$$

Similarly, for the energy fluxes we have

$$\mathbf{j}_\varepsilon(\mathbf{R}, t) = \sum_{\mathbf{p}} (\mathbf{p}/m)(p^2/2m) f(\mathbf{p}, \mathbf{R}, t) + \int d\mathbf{x}\; v(\mathbf{R} - \mathbf{x})\langle \mathbf{j}(\mathbf{R}, t)\, \rho_e(\mathbf{x}, t)\rangle$$
$$+ \int d\mathbf{x}\; V_{ie}(\mathbf{R} - \mathbf{x})\langle \mathbf{j}(\mathbf{R}, t)\, \rho_i(\mathbf{x}, t)\rangle \tag{3.27}$$

and

$$\mathbf{J}_{\mathscr{E}}(\mathbf{R}, t) = \sum_{\mathbf{p}} (\mathbf{p}/M)(p^2/2M) F(\mathbf{p}, \mathbf{R}, t) + \int d\mathbf{x}\; V(\mathbf{R} - \mathbf{x})\langle \mathbf{J}(\mathbf{R}, t)\, \rho_i(\mathbf{x}, t)\rangle$$
$$+ \int d\mathbf{x}\; V_{ie}(\mathbf{R} - \mathbf{x})\langle \mathbf{J}(\mathbf{R}, t)\, \rho_e(\mathbf{x}, t)\rangle. \tag{3.28}$$

This completes then the set of continuity and conservation equations we set out to derive. These equations are, of course, fully microscopic, and the hydrodynamic equations will be obtained from them by gradient expansions.

Before indicating how these equations can be developed to obtain the quantum hydrodynamic limit, we shall demonstrate their usefulness as well as their generality by applying them to deal with neutron scattering from liquid metals.

4. Neutron Scattering from Liquid Metals

It is sometimes convenient to separate out short-range parts V^s and V^s_{ie} from the bare ion–ion and bare electron–ion interactions. In this case we can rewrite the basic equations for the electron and ion currents in terms of the electric field $\mathbf{E}(\mathbf{R}, t)$ at point $(\mathbf{R}, t)$

$$\mathbf{E}(\mathbf{R}, t) = \int d\mathbf{x}\; \nabla_{\mathbf{R}} v(\mathbf{R} - \mathbf{x})[n_e(\mathbf{x}, t) - Zn_i(\mathbf{x}, t)]. \tag{4.1}$$

We then find

$$m(\partial \mathbf{j}(\mathbf{R}, t)/\partial t) + \nabla_{\mathbf{R}} \cdot \pi(\mathbf{R}, t)$$
$$= -n_e\mathbf{E}(\mathbf{R}, t) - \int d\mathbf{x}\; \nabla_{\mathbf{R}} v(\mathbf{R} - \mathbf{x})\{\langle \rho_e(\mathbf{R}, t)\, \rho_e(\mathbf{x}, t)\rangle_c - Z\langle \rho_e(\mathbf{R}, t)\, \rho_i(\mathbf{x}, t)\rangle_c\}$$
$$- \int d\mathbf{x}\; \nabla_{\mathbf{R}} V^s_{ie}(\mathbf{R} - \mathbf{x})\langle \rho_e(\mathbf{R}, t)\, \rho_i(\mathbf{x}, t)\rangle \tag{4.2}$$

for the electrons, and the corresponding equation

$$M(\partial \mathbf{J}(\mathbf{R}, t)/\partial t) + \nabla_{\mathbf{R}} \cdot \Pi(\mathbf{R}, t)$$
$$= n_e \mathbf{E}(\mathbf{R}, t) - Z \int d\mathbf{x}\, \nabla_{\mathbf{R}} v(\mathbf{R} - \mathbf{x}) \{ Z \langle \rho_i(\mathbf{R}, t)\, \rho_i(\mathbf{x}, t) \rangle_c - \langle \rho_i(\mathbf{R}, t)\, \rho_e(\mathbf{x}, t) \rangle_c \}$$
$$- \int d\mathbf{x}\, \nabla_{\mathbf{R}} V^s(\mathbf{R} - \mathbf{x}) \langle \rho_i(\mathbf{R}, t)\, \rho_i(\mathbf{x}, t) \rangle - \int d\mathbf{x}\, \nabla_{\mathbf{R}} V^s_{ie}(\mathbf{R} - \mathbf{x}) \langle \rho_i(\mathbf{R}, t)\, \rho_e(\mathbf{x}, t) \rangle \tag{4.3}$$

for the ions. In these equations the suffix c denotes, as usual, the cumulant part of the correlation function.

Neutron scattering is described by the basic equations given above, namely the continuity equation (3.7) and its counterpart for the ions, Eqs. (4.2) and (4.3), and the energy transport equations (3.23) and (3.24), provided the Hamiltonian is supplemented by a neutron–ion interaction term represented by an external potential $V_{\text{ext}}(\mathbf{R}, t)$. This leads to the following modifications:

(i) on the right side of Eq. (4.3), a term $-n_i \nabla_{\mathbf{R}} V_{\text{ext}}(\mathbf{R}, t)$ now appears.

(ii) on the right side of Eq. (3.24), a term $-\mathbf{J}(\mathbf{R}, t) \cdot \nabla_{\mathbf{R}} V_{\text{ext}}(\mathbf{R}, t)$ is introduced.

Next, consider Eq. (4.2) which describes how the electrons behave. The electron–ion terms in this equation are represented by a field which is explicitly

$$-(1/n_e) \int d\mathbf{x}\, \nabla_{\mathbf{R}} V_{ie}(\mathbf{R} - \mathbf{x})\, \Delta \langle \rho_e(\mathbf{R}, t)\, \rho_i(\mathbf{x}, t) \rangle,$$

which can be developed as far as terms linear in V_{ext} to yield

$$\frac{1}{n_e} \Delta \langle \rho_e(\mathbf{R}, t)\, \rho_i(\mathbf{x}, t) \rangle$$
$$= \frac{1}{n_e} \int d\mathbf{x}'\, dt' \left. \frac{\delta \langle \rho_e(\mathbf{R}, t)\, \rho_i(\mathbf{x}, t) \rangle}{\delta V_{\text{ext}}(\mathbf{x}', t')} \right|_0 V_{\text{ext}}(\mathbf{x}', t')$$
$$= \frac{-i}{\hbar n_e} \int d\mathbf{x}'\, dt' \{ \langle T[\rho_i(\mathbf{x}', t')\, \rho_e(\mathbf{R}, t)\, \rho_i(\mathbf{x}, t)] \rangle - n_i(\mathbf{x}', t')\, g_{ei}(\mathbf{R} - \mathbf{x}) \}\, V_{\text{ext}}(\mathbf{x}', t'). \tag{4.4}$$

The bracket { } can be rewritten exactly as

$$\{ \; \} = n_e(\mathbf{R}, t) \langle T[\rho_i(\mathbf{x}', t')\, \rho_i(\mathbf{x}, t)] \rangle_c$$
$$+ n_i(\mathbf{x}, t) \langle T[\rho_i(\mathbf{x}', t')\, \rho_e(\mathbf{R}, t)] \rangle_c$$
$$+ \langle T[\rho_i(\mathbf{x}', t')\, \rho_e(\mathbf{R}, t)\, \rho_i(\mathbf{x}, t)] \rangle_c \,. \tag{4.5}$$

At this point, we can make contact with earlier treatments by noting that if the electron–ion interaction is assumed to be weak, then

$$\{\ \} \simeq n_e \langle T[\rho_i(\mathbf{x}', t')\, \rho_i(\mathbf{x}, t)]\rangle_c \tag{4.6}$$

and, hence, from Eq. (4.4)

$$(1/n_e)\, \Delta\langle \rho_e(\mathbf{R}, t)\, \rho_i(\mathbf{x}, t)\rangle \simeq -(i/\hbar) \int d\mathbf{x}'\, dt'\, \langle T[\rho_i(\mathbf{x}', t')\, \rho_i(\mathbf{x}, t)]\rangle_c\, V_{\text{ext}}(\mathbf{x}', t')$$
$$+ \text{ higher order terms in the electron–ion interaction.} \tag{4.7}$$

In this approximation we can, thus, treat the field of the ionic-density fluctuations as an external field acting on the electrons, which is evidently given explicitly by

$$-\nabla_{\mathbf{R}} \int d\mathbf{x}\, V_{ie}(\mathbf{R} - \mathbf{x})(-i/\hbar) \int d\mathbf{x}'\, dt'\, \langle T[\rho_i(\mathbf{x}', t')\, \rho_i(\mathbf{x}, t)]\rangle_c\, V_{\text{ext}}(\mathbf{x}', t')$$
$$= -\nabla_{\mathbf{R}} \int d\mathbf{x}\, V_{ie}(\mathbf{R} - \mathbf{x})\, n_i(\mathbf{x}, t). \tag{4.8}$$

In the last step, we have made use of the fact that $-(i/\hbar)\langle T[\rho_i(\mathbf{x}', t')\, \rho_i(\mathbf{x}, t)]\rangle_c$ is the (time-ordered) response function of the ions, that is $\chi_{ii}(\mathbf{x} - \mathbf{x}', t - t')$, to an external field coupled with the ion density. Comparing this result with Eq. (4.2) shows that it is equivalent to decoupling directly the term $\langle \rho_e(\mathbf{R}, t)\, \rho_i(\mathbf{x}, t)\rangle$ into $n_e(\mathbf{R}, t)\, n_i(\mathbf{x}, t)$.

With this approximation, a simple solution of the problem is readily obtained. The electron equations yield immediately

$$n_e(\mathbf{R}, t) = \int d\mathbf{x}\, dt'\, \chi_{ee}(\mathbf{R} - \mathbf{x}, t - t') \int d\mathbf{x}'\, V_{ie}(\mathbf{x} - \mathbf{x}')\, n_i(\mathbf{x}', t'). \tag{4.9}$$

This equation tells us that if we create an ionic density fluctuation $n_i(\mathbf{R}, t)$, then this is accompanied by the electron density fluctuation $n_e(\mathbf{R}, t)$ acting to screen the ionic fluctuation. We can write Eq. (4.9) in Fourier transform as

$$n_e(\mathbf{k}, \omega) = \chi_{ee}(\mathbf{k}, \omega)\, V_{ie}(\mathbf{k})\, n_i(\mathbf{k}, \omega). \tag{4.10}$$

It is good approximation because of the differences in excitation energies of electrons and ions to take the static limit of the response function in Eq. (4.10), when we can write

$$n_e(\mathbf{k}, \omega) \simeq \chi_{ee}(\mathbf{k}, 0)\, V_{ie}(\mathbf{k})\, n_i(\mathbf{k}, \omega) = \frac{V_{ie}(\mathbf{k})}{v(\mathbf{k})} \left[\frac{1}{\epsilon(\mathbf{k}, 0)} - 1\right] n_i(\mathbf{k}, \omega). \tag{4.11}$$

In the same decoupling approximation, we can then write the right side of Eq. (4.3) as

$$-n_i \nabla_{\mathbf{R}} V_{\text{ext}}(\mathbf{R}, t) - \int d\mathbf{x}\, \nabla_{\mathbf{R}} V_s(\mathbf{R} - \mathbf{x}) \langle \rho_i(\mathbf{R}, t)\, \rho_i(\mathbf{x}, t)\rangle,$$

where the effective, or screened, ion–ion interaction V_s is given by

$$V_s(\mathbf{R} - \mathbf{x}) = V(\mathbf{R} - \mathbf{x}) + \int d\mathbf{x}'\, d\mathbf{x}''\, dt' \times V_{ie}(\mathbf{R} - \mathbf{x}')\, \chi_{ee}(\mathbf{x}' - \mathbf{x}'', t - t')\, V_{ie}(\mathbf{x}'' - \mathbf{x}), \tag{4.12}$$

which in Fourier transform reads

$$\begin{aligned} V_s(\mathbf{k}) &= V(\mathbf{k}) + |V_{ie}(\mathbf{k})|^2\, \chi_{ee}(\mathbf{k}, 0) \\ &= V(\mathbf{k}) + \frac{|V_{ie}(\mathbf{k})|^2}{v(\mathbf{k})} \left[\frac{1}{\epsilon(\mathbf{k}, 0)} - 1\right]. \end{aligned} \tag{4.13}$$

Here and in Eq. (4.11) we have used the relation between the electron density response and the dielectric function $\epsilon(\mathbf{q}, \omega)$. Notice that this is the dielectric function of the electron gas in the presence of electron–ion interactions, since the decoupling of the electron–ion interaction has been used only to treat the effect of the external disturbance.

4.1. *Effective Ornstein–Zernike Function for a Liquid Metal*

The above argument is only valid for a weak electron–ion interaction. Because of the very obvious interest in the definition and the utility of the effective ion–ion interaction, we want here to supplement the above argument, by an operational approach to the effective ion–ion interaction which is valid when the electron–ion interaction is not weak. The argument involves the direct correlation function c_{ii}, c_{ie}, and c_{ee} which, following Pearson and Rushbrooke [12], we define in terms of the radial distribution functions $g_{ii}(r)$, etc. as

$$g_{ii}(r) - 1 = c_{ii}(r) + \int d\mathbf{r}'\, [g_{ii}(\mathbf{r} - \mathbf{r}') - 1]\, n_i c_{ii}(\mathbf{r}') + \int d\mathbf{r}'\, [g_{ie}(\mathbf{r} - \mathbf{r}') - 1]\, n_e c_{ei}(\mathbf{r}'), \tag{4.14}$$

$$g_{ie}(r) - 1 = c_{ie}(r) + \int d\mathbf{r}'\, [g_{ii}(\mathbf{r} - \mathbf{r}') - 1]\, n_i c_{ie}(\mathbf{r}') + \int d\mathbf{r}'\, [g_{ie}(\mathbf{r} - \mathbf{r}') - 1]\, n_e c_{ee}(\mathbf{r}'), \tag{4.15}$$

and

$$g_{ee}(r) - 1 = c_{ee}(r) + \int d\mathbf{r}' \, [g_{ei}(\mathbf{r} - \mathbf{r}') - 1] \, n_i c_{ie}(\mathbf{r}')$$
$$+ \int d\mathbf{r}' \, [g_{ee}(\mathbf{r} - \mathbf{r}') - 1] \, n_e c_{ee}(\mathbf{r}'). \tag{4.16}$$

We expect from these definitions of the direct correlation functions that they will have the range of the interactions, that is at large r, $c_{ii}(r) \propto V(r)$, $c_{ie} \propto V_{ie}(r)$, and $c_{ee} \propto e^2/r$, the proportionality constants being characteristic energies.

An effective direct correlation function, $C(r)$ say, was defined by Johnson *et al.* [13] as though the ions interacted through an effective interaction, and the liquid metal was treated then as a one-component system. The definition adopted for $C(r)$ was then evidently

$$g_{ii}(r) - 1 = C(r) + n_i \int d\mathbf{r}' \, [g_{ii}(\mathbf{r} - \mathbf{r}') - 1] \, C(\mathbf{r}'), \tag{4.17}$$

which yields, of course, the usual result

$$S_{ii}(q) = 1/(1 - C(q)), \tag{4.18}$$

where

$$C(q) = n_i \int C(r) \, e^{i\mathbf{q}\cdot\mathbf{r}} \, d\mathbf{r}. \tag{4.19}$$

If we now take the Fourier transforms of Eqs. (4.14) and (4.16) and note that we can eliminate the terms involving g_{ie} and c_{ie} between them, then we find

$$1 - C(q) = \frac{1 - c_{ii}(q)}{S_{ee}(q)[1 - c_{ee}(q)]}, \tag{4.20}$$

with

$$c_{ii}(q) = n_i \int d\mathbf{r} \, c_{ii}(r) \, e^{i\mathbf{q}\cdot\mathbf{r}}, \qquad c_{ee}(q) = n_e \int d\mathbf{r} \, c_{ee}(r) \, e^{i\mathbf{q}\cdot\mathbf{r}}. \tag{4.21}$$

Since $c_{ii}(q) \propto Z^2/q^2$ as $q \to 0$, and $c_{ee}(q) \propto 1/q^2$, while $S_{ee}(0)$ is finite, we regain the desired result $1 - C(0) = 1/S_{ii}(0)$ when we use Eq. (1.2). Because of the usefulness of this result, we give a simple proof of it based on screening in Appendix 2. The above argument shows us that $C(q)$, and, hence, the effective interaction is completely screened.

One way of viewing the effective Ornstein–Zernike function $C(q)$ is evidently, from Eq. (4.20) to say that it is related to the "bare ion" direct correlation function, but screened by $S_{ee}(q)[1 - c_{ee}(q)]$. Here the difference between the jellium result $S_{ee}(q) \to 0$ as $q \to 0$ and the nonzero result for $S_{ee}(0)$ in the liquid metal, pointed out by Chihara and by Watabe and Hasegawa, is seen to be crucial for screening.

If we now treat the ions classically, we can employ one of the equations of liquid structure theory, for example the Born–Green or Percus–Yevick theories, to relate $C(q)$ to the effective ion–ion interaction, as proposed by Johnson *et al.* [13]. In this way, the effective interaction is related directly and explicitly to the bare-ion and bare-electron correlation functions c_{ii}, S_{ee}, and c_{ee} through Eq. (4.20), without any necessity to assume a weak electron–ion interaction. Of course, such scheme will be affected by the inaccuracies implicit in the liquid structure equations, but should be useful to at least estimate the nonlinear effects of the electron–ion interaction.

4.2. *Connection with Bohm–Staver Formula for Velocity of Sound and with Electrical Conductivity*

We want to point out here two further consequences of our basic equations for the coupled electron–ion system in a liquid metal.

First, the argument above for the ion–ion response $\chi_{ii}(q, \omega)$ can be written in terms of the response χ_{ii}^0 in the absence of the electron–ion coupling term as

$$\chi_{ii}(q, \omega) = \frac{\chi_{ii}^0(q, \omega)}{1 - \chi_{ii}^0(q, \omega)\, F(q, \omega)}, \tag{4.22}$$

where

$$F(q, \omega) = |\, V_{ie}(q)|^2\, \chi_{ee}(q, \omega). \tag{4.23}$$

If we adopt a random phase approximation for $\chi_{ii}^0(q, \omega)$, namely

$$\chi_{ii}^0(q, \omega) = \frac{\chi_{ii}^{00}(q, \omega)}{1 - V(q)\, \chi_{ii}^{00}(q, \omega)}, \tag{4.24}$$

where χ_{ii}^{00} is the response function for free ions, then we find

$$\chi_{ii}(q, \omega) = \frac{\chi_{ii}^{00}(q, \omega)}{1 - \chi_{ii}^{00}(q, \omega)[V(q) + F(q, \omega)]}. \tag{4.25}$$

Calculating χ_{ii}^{00} from its imaginary part given in Appendix 3, the Bohm–Staver formula for structureless ions is readily recovered from the present treatment. The reader is referred to Appendix 3 for the details.

Secondly, we wish to comment quite briefly at this stage that the electrical conductivity of liquid metals can also be discussed on the basis of Eq. (4.2). In contrast to the treatment of neutron scattering, we have now to couple the electrons, of course, directly to an external field, when a formally exact expression for the mobility results. This transcends the customary Kubo–Greenwood independent electron formula in that electron–electron interactions are now included.

If we make the most elementary decoupling approximation in the mobility formula, the theory is readily seen to be related to the treatment of Baym [14], which considers the electron scattering by ionic density fluctuations. By generalizing the set of Eqs. (4.14)–(4.16) to include time-dependence, preliminary calculations indicate that $S_{ee}(k, \omega)$ now appears rather directly in the theory, in addition to the quantities $S_{ii}(k, \omega)$, and a screened electron–ion potential which are characteristic of Baym's formula. We are continuing the study of more refined decoupling procedures and hope to report the details of this work at a later stage.

5. Quantum Hydrodynamic Limit of Basic Equations

Because of the interest we emphasized in the isotopic mass dependence of the shear viscosity of liquid Li, we shall make a start here on the program of formulating a theory of atomic transport coefficients from the basic conservation equations given in this paper.

Let us first define the following quantities characteristic of the liquid metal as a whole:

$$\mathscr{M}(\mathbf{R}, t) = mn_e(\mathbf{R}, t) + Mn_i(\mathbf{R}, t), \tag{5.1}$$

$$\mathbf{J}_M(\mathbf{R}, t) = m\mathbf{j}(\mathbf{R}, t) + M\mathbf{J}(\mathbf{R}, t), \tag{5.2}$$

and

$$Q(\mathbf{R}, t) = Zn_i(\mathbf{R}, t) - n_e(\mathbf{R}, t). \tag{5.3}$$

Adding the continuity equations for electrons and ions we evidently find

$$\partial\mathscr{M}(\mathbf{R}, t)/\partial t + \nabla_{\mathbf{R}} \cdot \mathbf{J}_M(\mathbf{R}, t) = 0. \tag{5.4}$$

From the more interesting equations for $\partial\mathbf{j}/\partial t$ and $\partial\mathbf{J}/\partial t$, we find

$$\begin{aligned}\partial\mathbf{J}_M(\mathbf{R}, t)/\partial t &+ \nabla_{\mathbf{R}} \cdot [\pi(\mathbf{R}, t) + \Pi(\mathbf{R}, t)] \\ &= -\int d\mathbf{x}\; \nabla_{\mathbf{R}} v(\mathbf{R} - \mathbf{x})\langle Q(\mathbf{R}, t)\, Q(\mathbf{x}, t)\rangle_c \\ &\quad - \int d\mathbf{x}\; \nabla_{\mathbf{R}} V^s(\mathbf{R} - \mathbf{x})\langle \rho_i(\mathbf{R}, t)\, \rho_i(\mathbf{x}, t)\rangle \\ &\quad - \int d\mathbf{x}\; \nabla_{\mathbf{R}} V_{ie}^s(\mathbf{R} - \mathbf{x})\langle \rho_e(\mathbf{R}, t)\, \rho_i(\mathbf{x}, t) + \rho_i(\mathbf{R}, t)\, \rho_e(\mathbf{x}, t)\rangle.\end{aligned} \tag{5.5}$$

It is evidently Eq. (5.5) which will lead to the analog of the Navier–Stokes equation for the liquid metal as a whole. In particular, the correlation function

$\langle \rho_e(\mathbf{R}, t)\, \rho_i(\mathbf{x}, t) + \rho_i(\mathbf{R}, t)\, \rho_e(\mathbf{x}, t) \rangle$ has precisely the symmetry required to yield a "stress tensor" in the hydrodynamic limit.

We wish to comment that the right side of Eq. (5.5), giving the internal force driving the current has a rather direct physical interpretation. For structureless ions, the force is due only to correlations in the charge fluctuations, that is the first term on the right side. For ions with core electrons, Eq. (5.5) shows that there are additional forces due to short range ion–ion and short-range electron–ion interactions.

Having interpreted Eq. (5.5) it is now useful to cast it into linearized form and to introduce nonequilibrium pair distribution functions. Then we can show that the equation takes the form

$$\begin{aligned} D\mathbf{J}_M(\mathbf{R}, t)/Dt = & -\nabla_{\mathbf{R}} \cdot [\pi^0 + \Pi^0] - \int d\mathbf{x}''\, \nabla_{\mathbf{R}} v(\mathbf{R} - \mathbf{x}'') \langle \rho_e(\mathbf{R}, t)\, \rho_e(\mathbf{x}'', t) \rangle \\ & - \int d\mathbf{x}''\, \nabla_{\mathbf{R}} V(\mathbf{R} - \mathbf{x}'') \langle \rho_i(\mathbf{R}, t)\, \rho_i(\mathbf{x}'', t) \rangle \\ & - \int d\mathbf{x}''\, \nabla_{\mathbf{R}} V_{ie}(\mathbf{R} - \mathbf{x}'') \langle \rho_i(\mathbf{R}, t)\, \rho_e(\mathbf{x}'', t) + \rho_e(\mathbf{R}, t)\, \rho_i(\mathbf{x}'', t) \rangle \\ = & -\nabla_{\mathbf{R}} \cdot [\pi^0 + \Pi^0] - \int d\mathbf{x}''\, \nabla_{\mathbf{R}} v(\mathbf{R} - \mathbf{x}'')\, g_{ee}(\mathbf{R}, \mathbf{x}'', t) \\ & - \int d\mathbf{x}''\, \nabla_{\mathbf{R}} V(\mathbf{R} - \mathbf{x}'')\, g_{ii}(\mathbf{R}, \mathbf{x}'', t) \\ & - 2 \int d\mathbf{x}''\, \nabla_{\mathbf{R}} V_{ie}(\mathbf{R} - \mathbf{x}'')\, g_{ie}(\mathbf{R}, \mathbf{x}'', t), \end{aligned} \tag{5.6}$$

where g_{ee}, g_{ii}, and g_{ie} are the nonequilibrium pair distribution functions of electron–electron, ion–ion, and ion–electron correlations. No divergence difficulties arise in this equation, as is apparent from Eq. (5.5).

The final step in converting Eq. (5.6) into a hydrodynamic equation is to expand all the g's around their equilibrium values and then to carry out gradient expansions. Thus, for example, we write the deviation of g_{ii} from equilibrium as δg_{ii} and then split this into two parts, one δg_{ii}^{ρ} dependent on density and the second δg_{ii}^{v} dependent on velocity. We can then write for the change in $g_{ii}(r) \equiv g_{ii}(|\,\mathbf{x} - \mathbf{y}\,|)$ from equilibrium

$$\delta g_{ii} = \delta g_{ii}^{\rho} + \delta g_{ii}^{v}, \tag{5.7}$$

where

$$\delta g_{ii}^{\rho} = \tfrac{1}{2}[\delta n_i(\mathbf{x}, t) + \delta n_i(\mathbf{y}, t)]\, \Gamma_1^{\,i}(\mathbf{r}) + \tfrac{1}{2}[\delta n_e(\mathbf{x}, t) + \delta n_e(\mathbf{y}, t)]\, \Gamma_1^{\,e}(\mathbf{r}) + \cdots \tag{5.8}$$

with a similar expansion for δg_{ii}^{v}, following Fröhlich [9].

With analogous expansions of g_{ie} and g_{ee}, the analog of the Navier–Stokes equation can be written down, and the quantity $\frac{4}{3}\eta + \zeta$ related to combinations of quantities like $\int d\mathbf{r}(\partial V/\partial r)\, \Gamma_1{}^i(\mathbf{r})$, etc., as is clear from Eqs. (5.6) and (5.8). In view of the formidable number of functional derivatives (the Γ's, etc.) of the pair functions entering the theory, it would not be fruitful to write down the complete result. However, the fact that part of δg^ρ may be combined with the kinetic terms to yield a pressure term in analogy with that in the classical Navier–Stokes equation is demonstrated in Appendix 4.

All this would be very discouraging if our aim were to calculate the Γ's explicitly and, hence, make numerical estimates of $\frac{4}{3}\eta + \zeta$. Fortunately, we are interested in the way $\frac{4}{3}\eta + \zeta$ changes as we change the ionic mass from M to $M + \delta M$. In a one-component classical liquid, this problem can be very simply solved [15]. We are, therefore, currently developing approximate methods to study the variation of the nonequilibrium pair distribution functions as M changes by δM. Apart from this problem, it is better simply to regard $\frac{4}{3}\eta + \zeta$ as a parameter in the Navier–Stokes equation and to express the response functions in the hydrodynamic regime in terms of it, in analogy with the classical case.

6. Summary and Discussion

The basic microscopic quantum theory of liquid metals developed in this paper is embodied in a total of ten continuity and conservation equations for electrons and for ions. The main emphasis in the present work has centred, however, on the current conservation equations (4.2) and (4.3). These equations have been used to discuss the neutron scattering from liquid metals, and offered a proper basis for calculating the ionic response function $\chi_{ii}(k, \omega)$ from first principles. At this stage, contact has been established with $\chi_{ii}(k, \omega)$ as constructed from a one-component theory with an effective ion–ion interaction, when the electron–ion interaction is weak. When it is not weak, it is pointed out that the effective Ornstein–Zernike correlation function $C(r)$ is related to the partial direct correlation functions for bare ions and electrons, and to the structure factor for bare electrons. Combining this equation for $C(r)$ with approximate liquid structure theories appropriate for classical ions, such as the Born–Green and Percus–Yevick theories, relates the effective ion–ion interaction to the above partial correlation function, without assumptions being made about the strength of the electron–ion interactions.

However, in any of the current theories of liquid structure, except direct computer studies by molecular dynamics, etc., a three-body correlation function for the ions is decoupled and expressed in terms of the ionic pair function. The relation of such decoupling schemes in liquid metals to the form of the electron–ion interaction is not well understood and is currently being further investigated.

Finally, the hydrodynamic limit of the basic equations has been briefly discussed and a route to discuss the ionic mass dependence of $(\frac{4}{3}\eta + \zeta)$ outlined, with the nonequilibrium pair distribution functions of the electron–electron, ion–electron, and ion–ion correlations.

APPENDIX 1. MASS DENSITY AND MASS CONCENTRATION CORRELATION FUNCTIONS

An alternative description of the two-component system to that derived in the main text is offered by defining correlation functions for mass and charge (compare Bhatia, Thornton, and March [10]). Thus, we write

$$M_{\mathbf{k}}(t) = m\rho_{e\mathbf{k}}(t) + M\rho_{i\mathbf{k}}(t) \tag{A1.1}$$

and

$$X_{\mathbf{k}}(t) = \tfrac{1}{2}[M\rho_{i\mathbf{k}}(t) - m\rho_{e\mathbf{k}}(t)]. \tag{A1.2}$$

Now define correlation functions

$$S_{MM}(q, \omega) = (1/\mathscr{M}) \int_{-\infty}^{\infty} dt\, e^{-i\omega t} \langle M_{\mathbf{k}}^{+}(0); M_{\mathbf{k}}(t)\rangle \tag{A1.3}$$

and

$$S_{XX}(q, \omega) = (1/\mathscr{M}) \int_{-\infty}^{\infty} dt\, e^{-i\omega t} \langle X_{\mathbf{k}}^{+}(0); X_{\mathbf{k}}(t)\rangle, \tag{A1.4}$$

etc. In Eqs. (A1.3) and (A1.4), $\mathscr{M}$ is simply the total mass $mn_e + Mn_i$. Similarly, we define corresponding response functions

$$\begin{aligned} \chi_{MM}(q, \omega) &= \int_{0}^{\infty} dt\, e^{-i\omega t} (1/i\hbar) \langle [M_{\mathbf{q}}^{+}(0), M_{\mathbf{q}}(t)]\rangle \\ &= \beta \int_{0}^{\infty} dt\, e^{-i\omega t} \langle \dot{M}_{\mathbf{q}}^{+}(0); M_{\mathbf{q}}(t)\rangle, \end{aligned} \tag{A1.5}$$

etc. We can then calculate moments using the following relations

$$i\hbar\dot{\rho}_{e\mathbf{q}}^{+} = [\rho_{e\mathbf{q}}^{+}, H] = -(\hbar/m) \sum_{i} e^{i\mathbf{q}\cdot\mathbf{r}_i} (\mathbf{q}\cdot\mathbf{p}_i + \tfrac{1}{2}\hbar q^2) \tag{A1.6}$$

and

$$i\hbar\dot{\rho}_{i\mathbf{q}}^{+} = [\rho_{i\mathbf{q}}^{+}, H] = -(\hbar/M) \sum_{j} e^{i\mathbf{q}\cdot\mathbf{R}_j} (\mathbf{q}\cdot\mathbf{P}_j + \tfrac{1}{2}\hbar q^2). \tag{A1.7}$$

We then find

$$(1/\pi)\int_{-\infty}^{\infty} d\omega\, \omega\chi''_{MM}(q,\omega) = (1/i\hbar)\langle[M_{\mathbf{q}}(0), \dot{M}_{\mathbf{q}}^{+}(0)]\rangle = q^2\mathscr{M}, \tag{A1.8}$$

$$(1/\pi)\int_{-\infty}^{\infty} d\omega\, \omega\chi''_{MX}(q,\omega) = (1/i\hbar)\langle[X_{\mathbf{q}}(0), \dot{M}_{\mathbf{q}}^{+}(0)]\rangle = 0, \tag{A1.9}$$

and

$$(1/\pi)\int_{-\infty}^{\infty} d\omega\, \omega\chi''_{XX}(q,\omega) = (1/i\hbar)\langle[X_{\mathbf{q}}(0), \dot{X}_{\mathbf{q}}^{+}(0)]\rangle = q^2\mathscr{M}/4. \tag{A1.10}$$

To pass to the classical limit, we use the relations

$$\begin{aligned}\tilde{S}_{\rho\rho}(q,\omega) &\equiv (1/\mathscr{M})\int_{-\infty}^{\infty} dt\, e^{-i\omega t}\langle\tfrac{1}{2}\{\rho_{\mathbf{q}}^{+}(0), \rho_{\mathbf{q}}(t)\}\rangle \\ &\to (2/\beta\omega)(\chi''_{\rho\rho}(q,\omega)/\mathscr{M}),\end{aligned} \tag{A1.11}$$

etc., when we find

$$(1/2\pi)\int_{-\infty}^{\infty} \tilde{S}_{MM}(q,\omega)\,\omega^2\, d\omega = q^2/\beta \tag{A1.12}$$

and

$$(1/2\pi)\int_{-\infty}^{\infty} \tilde{S}_{XX}(q,\omega)\,\omega^2\, d\omega = \tfrac{1}{4}(q^2/\beta). \tag{A1.13}$$

Equations (A1.12) and (A1.13) are precisely the results given previously by Bhatia, Thornton, and March [10] for a classical binary alloy.

APPENDIX 2. Screening and Thermodynamics

We define pair distribution functions $g_{\alpha\beta}(r)$ and from these the partial structure factors

$$S_{ii}(q) = 1 + n_i \int [g_{ii}(r) - 1]\, e^{i\mathbf{q}\cdot\mathbf{r}}\, d\mathbf{r}, \tag{A2.1}$$

$$S_{ee}(q) = 1 + n_e \int [g_{ee}(r) - 1]\, e^{i\mathbf{q}\cdot\mathbf{r}}\, d\mathbf{r}, \tag{A2.2}$$

and

$$S_{ie}(q) = n_i^{1/2} n_e^{1/2} \int [g_{ie}(r) - 1]\, e^{i\mathbf{q}\cdot\mathbf{r}}\, d\mathbf{r}. \tag{A2.3}$$

Then we find from the formula of Kirkwood and Buff [16] for the isothermal compressibility K_T

$$K_T k_B T = \frac{1 + G_{ii}n_i + G_{ee}n_e + (G_{ii}G_{ee} - G_{ie}^2)\, n_i n_e}{n_i + n_e + (G_{ii} + G_{ee} - 2G_{ie})\, n_i n_e}, \tag{A2.4}$$

where

$$G_{\alpha\beta} = \int [g_{\alpha\beta}(r) - 1]\, d\mathbf{r}. \tag{A2.5}$$

The requirement that when we sit on a nucleus, of charge Ze, we screen it perfectly means that the density distribution of ions, namely $n_i g_{ii}(r)$ plus that of electrons, namely $n_{ie} = n_i g_{ie}(r)$ must correspond to a total charge

$$\begin{aligned} -Ze &= \int [-en_e g_{ie}(r) + Zen_i g_{ii}(r)]\, d\mathbf{r} \\ &= \int [-eZn_i g_{ie}(r) + Zen_i g_{ii}(r)]\, d\mathbf{r}, \end{aligned} \tag{A2.6}$$

which becomes, using Eq. (A2.5),

$$n_i G_{ie} = n_i G_{ii} + 1.$$

Evidently, from Eqs. (A2.1) at $q = 0$, we have then the relation

$$(n_i/n_e)^{1/2}\, S_{ie}(0) = S_{ii}(0), \tag{A2.8}$$

which is the result (1.3) given by Watabe and Hasegawa.

Similarly, sitting on an electron we can write for perfect screening

$$e = \int [-en_e g_{ee}(r) + Zen_i g_{ie}(r)]\, d\mathbf{r}, \tag{A2.9}$$

which yields from Eqs. (A2.2) and (A2.3) and the relation $n_e = Zn_i$

$$1 = -[S_{ee}(0) - 1] + (n_e/n_i)^{1/2}\, S_{ie}(0) \tag{A2.10}$$

$$S_{ee}(0) = Z^{1/2} S_{ie}(0) = Z S_{ii}(0) \tag{A2.11}$$

leading again to Eq. (1.2). We expect that Eq. (1.4) is correct for thermodynamic reasons, and if we substitute Eqs. (A2.11) and (A2.8) into (A2.4) we regain Eq. (1.4) as expected.

APPENDIX 3: ION–ION RESPONSE AND THE VELOCITY OF SOUND IN LIQUID METALS

As a further simple example of the present treatment of the coupled electron–ion system, we note that in the absence of electron–ion interactions the high frequency response of the ion–ion response function has an imaginary part

$$\operatorname{Im} \chi_{ii}^{0}(q, \omega) = -(\pi/2)(n_i q^2/M\omega_p{}^i)[\delta(\omega - \omega_p{}^i) - \delta(\omega + \omega_p{}^i)] \quad \text{(A3.1)}$$

in the long wave length limit, $\omega_p{}^i$ being the ionic plasma frequency

$$\omega_p{}^i = [4\pi n_i (Ze)^2/M]^{1/2}. \quad \text{(A3.2)}$$

We then find from Eq. (4.22) that

$$\operatorname{Re} \chi_{ii}^{-1} = \operatorname{Re}(\chi_{ii}^{0})^{-1} - \operatorname{Re} F(q, \omega). \quad \text{(A3.3)}$$

From Eq. (A3.1) the Kramers–Krönig relations yield

$$\operatorname{Re}(\chi_{ii}^{0})^{-1} = (M/n_i q^2)(\omega^2 - \omega_p^{i2}) \quad \text{(A3.4)}$$

while for point ions we have from Eq. (4.23) that

$$\operatorname{Re} F(q, \omega) = (4\pi Z^2 e^2/q^2) \operatorname{Re}[1/\epsilon(q, \omega) - 1]. \quad \text{(A3.5)}$$

Thus, from the zero of $\operatorname{Re}[\chi_{ii}^{-1}(q, \omega)]$ we find

$$\omega_q{}^2 = \omega_p^{i2} + (4\pi n_i Z^2 e^2/M) \operatorname{Re}[1/\epsilon(q, \omega_q) - 1]. \quad \text{(A3.6)}$$

Now $\epsilon(q, \omega)$ in the *low* frequency limit behaves in the random phase approximation as

$$\epsilon(q, \omega) \to 1 + 3\omega_p^{e2}/v_F{}^2 q^2, \quad \text{(A3.7)}$$

where $\omega_p{}^e$ is the electron plasma frequency and v_F is the Fermi velocity. Hence, we find for $q \to 0$

$$\omega_q{}^2 = (Zm/3M)\, v_F{}^2 q^2, \quad \text{(A3.8)}$$

which is simply the Bohm–Staver formula.

APPENDIX 4: FORM OF PRESSURE TERM IN NAVIER–STOKES EQUATION

We consider further here the hydrodynamic forms associated with the equations

$$\begin{aligned} &mn_e(D\mathbf{v}_e(\mathbf{R}, t)/Dt) \\ &\quad = -\nabla_{\mathbf{R}} \cdot \pi^0(\mathbf{R}, t) - n_e \int d\mathbf{x}'' \, \nabla_{\mathbf{R}} v(\mathbf{R} - \mathbf{x}'')[n_e(\mathbf{x}'', t) - Zn_i(\mathbf{x}'', t)] \\ &\quad\quad - \int d\mathbf{x}'' \, \nabla_{\mathbf{R}} v(\mathbf{R} - \mathbf{x}'')[\langle \rho_e(\mathbf{R}, t)\, \rho_e(\mathbf{x}'', t)\rangle_c - Z\langle \rho_e(\mathbf{R}, t)\, \rho_i(\mathbf{x}'', t)\rangle_c] \\ &\quad\quad - \int d\mathbf{x}'' \, \nabla_{\mathbf{R}} V_{ie}^s(\mathbf{R} - \mathbf{x}'')\langle \rho_e(\mathbf{R}, t)\, \rho_i(\mathbf{x}'', t)\rangle \end{aligned} \tag{A5.1}$$

and

$$\begin{aligned} &Mn_i(D\mathbf{v}_i(\mathbf{R}, t)/Dt) \\ &\quad = -\nabla_{\mathbf{R}} \cdot \Pi^0(\mathbf{R}, t) - Zn_i \int d\mathbf{x}'' \, \nabla_{\mathbf{R}} v(\mathbf{R} - \mathbf{x}'')[Zn_i(\mathbf{x}'', t) - n_e(\mathbf{x}'', t)] \\ &\quad\quad - \int d\mathbf{x}'' \, \nabla_{\mathbf{R}} V^s(\mathbf{R} - \mathbf{x}'')\langle \rho_i(\mathbf{R}, t)\, \rho_i(\mathbf{x}'', t)\rangle \\ &\quad\quad - \int d\mathbf{x}'' \, \nabla_{\mathbf{R}} V_{ie}^s(\mathbf{R} - \mathbf{x}'')\langle \rho_i(\mathbf{R}, t)\, \rho_e(\mathbf{x}'', t)\rangle, \end{aligned} \tag{A5.2}$$

where we have again separated out the short-range parts of the ion–ion and ion–electron potentials.

Let us now expand the cumulant correlation functions in terms of the density fluctuations of electrons and ions and their gradients. The analog of Eq. (5.8) gives for the change in the electron–electron correlations due to the density change

$$\begin{aligned} \Delta\langle \rho_e(\mathbf{R}, t)\, \rho_e(\mathbf{x}, t)\rangle_c &= \tfrac{1}{2}[2n_e(\mathbf{R}, t) + \mathbf{r} \cdot \nabla_{\mathbf{R}} n_e(\mathbf{R}, t)]\, \Gamma_2^e(\mathbf{r}) \\ &\quad + \tfrac{1}{2}[2n_i(\mathbf{R}, t) + \mathbf{r} \cdot \nabla_{\mathbf{R}} n_i(\mathbf{R}, t)]\, \Gamma_2^i(\mathbf{r}) + \cdots . \end{aligned} \tag{A5.3}$$

Here $\mathbf{r} = \mathbf{x} - \mathbf{R}$, while $\Gamma_2^e(\mathbf{r})$ is the partial derivative of the electron–electron correlation function with respect to the electron density, evaluated at equilibrium. We then find for the density fluctuation term

$$\begin{aligned} &- \int d\mathbf{x} \, \nabla_{\mathbf{R}} v(\mathbf{R} - \mathbf{x})\langle \rho_e(\mathbf{R}, t)\, \rho_e(\mathbf{x}, t)\rangle_c \\ &\quad = -\tfrac{1}{2} \int d\mathbf{x} \, \nabla_{\mathbf{R}} v(\mathbf{R} - \mathbf{x})[\mathbf{r} \cdot \nabla_{\mathbf{R}} n_e(\mathbf{R}, t)\, \Gamma_2^e(\mathbf{r}) + \mathbf{r} \cdot \nabla_{\mathbf{R}} n_i(\mathbf{R}, t)\, \Gamma_2^i(\mathbf{r})] \\ &\quad = \tfrac{1}{2} \left[\int d\mathbf{r}\, \Gamma_2^e(\mathbf{r})\, \nabla_{\mathbf{r}} v(\mathbf{r})\, \mathbf{r}\right] \cdot \nabla_{\mathbf{R}} n_e(\mathbf{R}, t) + \tfrac{1}{2} \left[\int d\mathbf{r}\, \Gamma_2^i(\mathbf{r})\, \nabla_{\mathbf{r}} v(\mathbf{r})\, \mathbf{r}\right] \cdot \nabla_{\mathbf{R}} n_i(\mathbf{R}, t) \\ &\quad = \tfrac{1}{6} \nabla_{\mathbf{R}} \left[n_e(\mathbf{R}, t) \int d\mathbf{r}\, \mathbf{r} \cdot \nabla_{\mathbf{r}} v(\mathbf{r})\, \Gamma_2^e(\mathbf{r}) + n_i(\mathbf{R}, t) \int d\mathbf{r}\, \mathbf{r} \cdot \nabla_{\mathbf{r}} v(\mathbf{r})\, \Gamma_2^i(\mathbf{r})\right] \\ &\quad = -\nabla_{\mathbf{R}} P_{ee}[n_e(\mathbf{R}, t), n_i(\mathbf{R}, t)]. \end{aligned} \tag{A5.4}$$

Here we have used the isotropy of the liquid metal at equilibrium, while

$$P_{ee}[n_e(\mathbf{R}, t), n_i(\mathbf{R}, t)] = -\tfrac{1}{6} \int d\mathbf{r}\, \mathbf{r} \cdot \nabla_\mathbf{r} v(\mathbf{r}) \langle \rho_e(\mathbf{R}, t)\, \rho_e(\mathbf{x}, t) \rangle_c . \qquad \text{(A5.5)}$$

The quantity on the left side of Eq. (A5.5) can be interpreted as the "partial pressure" due to the electron–electron interactions, in the sense that in the one-component neutral fluid, the right side of Eq. (A5.5) gives the pressure due to the interactions. The pressure term shown in Eq. (A5.4) is, of course, typical of the way pressure enters the classical Navier–Stokes equation.

References

1. N. T. Ban, C. M. Randall, and D. J. Montgomery, *Phys. Rev.* **128** (1962), 6.
2. P. A. W. Bratby and E. A. Harris, *Phys. Lett.* **40A** (1972), 25.
3. R. D. Cowan and J. G. Kirkwood, *J. Chem. Phys.* **29** (1958), 264.
4. A. A. Broyles, A. A. Barker, T. Dunn, and M. A. Pokrant, *International Journal of Quantum Chemistry* **3S** (1969), 293.
5. M. Watabe and M. Hasegawa, Second International Conference on Liquid Metals (Taylor and Francis, 1973), 133.
6. J. Chihara, Second International Conference on Liquid Metals (Taylor and Francis, 1973), 137.
7. H. Takahashi, *Physica*, **51** (1971), 333.
8. L. P. Kadanoff and P. C. Martin, *Ann. Phys.* **24** (1963), 419.
9. H. Fröhlich, *Physica* **37** (1967), 215.
10. A. B. Bhatia, D. E. Thornton, and N. H. March, *J. Phys. Chem. Liquids* (in press).
11. K. S. Singwi, A. Sjölander, M. P. Tosi, and R. H. Land, *Phys. Rev.* **B1** (1970), 1044.
12. F. J. Pearson and G. S. Rushbrooke, *Proc. Roy. Soc. Edinburgh* **A64** (1957), 305.
13. M. D. Johnson, P. Hutchinson, and N. H. March, *Proc. Roy. Soc.* **A282** (1964), 283.
14. G. Baym, *Phys. Rev.* **135** (1964), A1691.
15. R. C. Brown and N. H. March, *J. Phys. Chem. Liquids*, **1** (1968), 141.
16. J. G. Kirkwood and F. P. Buff, *J. Chem. Phys.* **19** (1951), 774.

IL NUOVO CIMENTO VOL. 15 B, N. 2 11 Giugno 1973

Small-Angle Scattering from Liquid Metals and Alloys and Electronic Correlation Functions.

M. P. TOSI (*)

Istituto di Fisica dell'Università - Messina
Gruppo Nazionale di Struttura della Materia del C.N.R. - Messina

N. H. MARCH

Department of Physics, Imperial College - London

(ricevuto l'1 Marzo 1973)

Summary. — The partial structure factors S_{ii}, S_{ie} and S_{ee} in pure liquid metals are considered at small wave vector q. Charge neutrality arguments relate the three functions through the valency Z at $q=0$. It is shown that the partial Ornstein-Zernike functions are proportional to q^{-2} in the long-wavelength limit, and that they are related, not only through Z, but through a common characteristic energy E. By setting up a model for the dynamic structure factors, and by assuming well-defined plasmon and phonon modes at small q, the characteristic energy is shown to be dominated by the zero-point energy of the electronic plasma oscillations. In turn, this same energy is shown to dominate the curvature of S_{ee} at $q=0$, the curvature being pronounced and S_{ee} therefore rising rapidly away from $q=0$. In contrast, because ion masses enter, the coefficients of q^2 in S_{ie} and S_{ii} are shown to be orders of magnitude smaller and these structure factors are therefore quite flat near $q=0$. However, the magnitudes of the coefficients of q^2 in S_{ie} and S_{ii} are shown to be influenced by the phonon dispersion, though insensitive to phonon damping. For weak electron-ion interaction, $S_{ie}(q)$ and $S(q)$ can be related through an electronic response function, and the resulting expression satisfies the exact limiting relation between S_{ie} and S at $q=0$. Finally, the q^2-term in S_{ee} in an alloy is discussed and is shown to be directly related to the average electronic plasma frequency, involving therefore the valencies of the components and the mean atomic volume. The partial structure factors S_{Ae} and S_{Be} can also be related to the (measurable) partial ionic structure factors for weak electron-ion interaction.

(*) Most of this work was carried out during a visit to Imperial College, London.

1. – Introduction.

Recent progress has been made in treating pure liquid metals as two-component systems, following earlier pioneering work by COWAN and KIRKWOOD [1].

As shown by WATABE and HASEGAWA [2] and by CHIHARA [3], the partial structure factors $S_{ii}(q) \equiv S(q)$, $S_{ie}(q)$ and $S_{ee}(q)$, for ion-ion, ion-electron and electron-electron pair correlations are simply related, via the valency Z at $q = 0$. These relations arise from the requirement that the metal as a whole be neutral.

In this paper we are concerned not with the extreme long-wavelength limit, that is, not with thermodynamics directly, but rather with the small-angle scattering problem. In particular we shall discuss the magnitudes of the q^2-terms in the small-q expansion of $S(q)$, S_{ie} and S_{ee}. We show, from a model in which we set up dynamical structure factors $S(q, \omega)$, $S_{ie}(q, \omega)$, $S_{ee}(q, \omega)$ assuming well-defined collective modes of *a*) phonon type and *b*) plasmon character, that the behaviour at small q is very different for the electronic correlation function S_{ee} from that of S_{ie} and S. We show that S_{ee} at long wavelengths is almost entirely determined by the electron plasma frequency and has strong curvature at the origin. In contrast, our arguments indicate that both plasmon and phonon contributions to S_{ie} and S should be small, and that $S(q)$ should be very flat, as observed for quite a number of liquid metals.

It is useful also to discuss partial Ornstein-Zernike correlation functions c_{ii}, c_{ie} and c_{ee}, following the definitions of PEARSON and RUSHBROOKE [4]. It is shown that these behave like q^{-2} as $q \to 0$, and, in addition to the valency, they are related by an energy E, which is determined in the above model in terms of plasmon energies and phonon dispersion. Because of the high energy of the electron plasma excitation, it is clear that E is dominated by this energy.

2. – Partial Ornstein-Zernike correlation functions and characteristic energy.

We shall adopt the definitions of partial Ornstein-Zernike correlation functions given earlier by PEARSON and RUSHBROOKE [4]. Solving their equa-

---

(1) R. D. COWAN and J. G. KIRKWOOD: *Journ. Chem. Phys.*, **29**, 264 (1958).

(2) M. WATABE and M. HASEGAWA: paper presented at *Second International Conference on Liquid Metals, Tokyo, September 1972*, to appear in *Advances in Physics* (1973).

(3) J. CHIHARA: paper presented at *Second International Conference on Liquid Metals, Tokyo, September 1972*, to appear in *Advances in Physics* (1973).

(4) F. J. PEARSON and G. S. RUSHBROOKE: *Proc. Roy. Soc. Edinburgh*, A **64**, 305 (1957).

tions for the partial structure factors, we find

$$1-c_{ii}(q)=\frac{S_{ee}(q)}{S(q)S_{ee}(q)-S_{ie}^2(q)}\,, \tag{1}$$

$$c_{ie}(q)\quad=\frac{S_{ie}(q)}{S(q)S_{ee}(q)-S_{ie}^2(q)}\,, \tag{2}$$

$$1-c_{ee}(q)=\frac{S(q)}{S(q)S_{ee}(q)-S_{ie}^2(q)}\,. \tag{3}$$

We notice that all three equations have a common denominator:

$$D(q)=S(q)S_{ee}(q)-S_{ie}^2(q)=\left[\frac{S(q)}{1-c_{ee}(q)}\right]^2(1-c_{ii}-c_{ee}+c_{ii}c_{ee}-c_{ie}^2)\,. \tag{4}$$

The extreme asymptotic forms of c can be written in the extreme long wavelength limit as

$$c_{ii}(q)\to-\frac{4\pi n_i Z^2 e^2}{q^2E}+C_i+O(q^2)\,, \tag{5}$$

$$c_{ie}(q)\to\frac{4\pi n_i n_e Ze^2}{q^2E}+C+O(q^2) \tag{6}$$

and

$$c_{ee}(q)\to-\frac{4\pi n_e e^2}{q^2E}+C_e+O(q^2)\,, \tag{7}$$

where E is a characteristic energy which we discuss further below.

We know, also, from the earlier work of WATABE and HASEGAWA and of CHIHARA, that

$$S(0)=A\,,\qquad S_{ie}(0)=Z^{\frac{1}{2}}A\,,\qquad S_{ee}(0)=ZA\,, \tag{8}$$

where, of course, from thermodynamics

$$S(0)=n_i k_B T K_T\,, \tag{9}$$

K_T being the isothermal compressibility of the liquid metal.

The partial structure factors at small q can be expanded in the forms

$$S(q)\quad=A+s_i q^2+\dots\,, \tag{10}$$

$$S_{ie}(q)=Z^{\frac{1}{2}}(A+sq^2+\dots) \tag{11}$$

and

$$S_{ee}(q) = Z(A + s_e q^2 + \ldots) . \tag{12}$$

Then we find for the denominator $D(q)$ the result at long wavelengths

$$D(q) = ZA(s_i + s_e - 2s)q^2 \equiv dq^2 . \tag{13}$$

Thus, in the extreme limit of long wavelengths, it is readily shown that the characteristic energy E is related to d by

$$E = 4\pi n_e e^2 d/S(0) . \tag{14}$$

Having established a somewhat general framework for discussion of the small-angle scattering, we turn immediately to the task of setting up the simplest model which reflects the salient characteristics of the coupled electron-ion system that a liquid metal represents.

3. – Plasmon-phonon model.

Under the assumption that in the long-wavelength limit single-particle excitation and multiparticle excitations are less important than collective modes, which we take to be

i) plasmonlike, with dispersion relation $\omega = \omega_p(q)$,

ii) phononlike, with $\omega = v_s(q)q$,

we can write approximate expressions for the partial dissipation functions as

$$\chi''_{ij}(q, \omega) = -\frac{\pi}{2}\sqrt{n_i n_j}\{\alpha_{ij}(q)[\delta(\omega - \omega_p(q)) - \delta(\omega + \omega_p(q))] + \beta_{ij}(q)[\delta(\omega - qv_s(q)) - \delta(\omega + qv_s(q))]\} . \tag{15}$$

We note at this point that

$$S_{ij}(q) = -\int \frac{d\omega}{2\pi}\chi''_{ij}(q, \omega)\,\text{ctgh}\left(\frac{\hbar\omega}{2k_B T}\right)(n_i n_j)^{-\frac{1}{2}} . \tag{16}$$

To satisfy the first moment theorem on χ''_{ij} in the long-wavelength limit, we must choose $\alpha_{ij}(q) \propto q^2$ and $\beta_{ij}(q) \propto q$. It is then easy to see that the static structure factor $S_{ij}(q)$ at $q = 0$ is entirely dominated by the phonon mode, and the compressibility sum rule (9) gives the coefficients β_{ij} in $\beta_{ij}(q) \to \beta_{ij}q$, as

$$\beta_{ii} = Z^{-1}\beta_{ee} = Z^{-\frac{1}{2}}\beta_{ie} = \hbar n_i K_T v_s , \tag{17}$$

v_s being the velocity of sound.

By treating the damping of the phonons, and the coupling of density fluctuations with entropy fluctuations, by hydrodynamic theory the sound velocity entering the above equation is identified with the isothermal sound velocity. Since, in liquid metals at the melting point, the ratio of specific heats $\gamma \sim 1.2 \div 1.3$, this distinction is not very important in practice.

Next, we consider the coefficients s, s_i and s_e introduced above. These coefficients involve both the plasmon mode and the phonons. More precisely, we find

$$Zs_e^{\text{plasmon}} = \frac{\hbar}{2m\omega_p}\frac{n_i M}{\mathcal{M}}, \tag{18}$$

$$s_i^{\text{plasmon}} = \frac{\hbar}{2M\omega_p}\frac{n_e m}{\mathcal{M}} \tag{19}$$

and

$$s^{\text{plasmon}} = -\frac{\hbar}{2m\omega_p}\frac{n_i m}{\mathcal{M}}. \tag{20}$$

Here m and M are electron and ion masses while $\mathcal{M}$ is defined by

$$\mathcal{M} = n_i M + n_e m \tag{21}$$

and ω_p is the reduced plasmon frequency

$$\omega_p^2 = 4\pi n_i Z^2 e^2 \left(\frac{1}{M} + \frac{1}{Zm}\right). \tag{22}$$

In practice, because of the large difference between electron and ion masses, ω_p can be taken equal to the electronic plasma frequency in the homogeneous electron gas and $Zs_e^{\text{plasmon}} \sim \hbar/2m\omega_p$, whereas s_i^{plasmon} and s^{plasmon} are extremely small. All the contributions (18)-(20) will be important, instead, in a degenerate electron-hole liquid.

On the other hand, the phonon contribution to the same coefficients depends on differences between coefficients of q^3-terms in $\beta_{ij}(q)$ and those in $qv_s(q)$. The suggestion from this calculation is that these terms are usually small in magnitude and therefore negligible compared with the plasmon contribution in s_e. While the phonon terms are likely to be important relative to the plasmon terms in s_i and s, the conclusion is that these latter coefficients are very small compared with s_e. These conclusions can be shown, by means of hydrodynamic forms, to be insensitive to phonon and to plasmon damping.

A few comments are called for on single- and multiparticle excitations. In the homogeneous electron gas, these excitations do not contribute to order q^2

as discussed for example by PINES and NOZIÈRES [5] since the pertinent matrix elements are of order q^2. Screening is responsible for this for the single-particle excitations. For multiparticle excitations, on the other hand, this result follows from the requirement of current conservation. In the two-component liquid metal, however, current is not conserved separately by the electrons and by the ions and so the matrix elements can be of order q, leading to contributions of order q^2 to the structure factors. Because of the large electronic plasmon energy, such contributions can hardly affect s_e, but for s and s_i possible contributions can arise from these mechanisms.

From this line of reasoning, we conclude that the coefficient s_e is essentially settled by the plasmon contribution, and is much larger than either s or s_i. This automatically settles the characteristic energy E as $\frac{1}{2}\hbar\omega_p$ in the partial Ornstein-Zernike functions. It is quite clear, then, that S_{ee} will show strong curvature at the origin, and will rise quickly from its (small) value $ZS(0)$ (typically 1/30 to 1/10 at the melting temperature), whereas, in comparison, $S(q)$ and $S_{ie}(q)$ will be flat for small q.

Because of the difficulty in estimating s_i and s, as seen from the above plasmon-phonon model, we shall turn next to relate $S_{ie}(q)$ to $S(q)$, which, of course, can be measured by X-ray or neutron experiments, in the case when the electron-ion interaction can be assumed weak.

4. – Relation between S_{ii} and S_{ie} for weak electron-ion interaction.

The Cowan-Kirkwood [1] model referred to above, and discussed in some detail in Appendix B, shows that

i) $S_{ie}(q)$ is a function(al) of $S(q)$,

ii) the asymptotic behaviour of $g_{ie}(r)$ and $g(r)$ can be related correctly (in that model) by a linear theory, valid at large r.

We therefore begin the investigation of an approximate relation between S_{ie} and S by examining the case of weak electron-ion interaction. In addition to this assumption, it is clear that the electron-electron correlations in the presence of the ions cannot be incorporated exactly and we should eventually have recourse to homogeneous electron gas theory.

Under a weak external potential, $V_{ext}(q, \omega)$, which is coupled to the electron density fluctuations, the polarizations of the electrons and of the ions are

$$\delta\varrho_e(q, \omega) = \chi_{ee}(q, \omega)\, V_{ext}(q, \omega) \tag{23}$$

---

[5] D. PINES and P. NOZIÈRES: *The Theory of Quantum Liquids* (New York, 1966).

and

$$\delta\varrho_i(q,\omega) = \chi_{ie}(q,\omega)\,V_{ext}(q,\omega)\,, \tag{24}$$

respectively. If the electron-ion interaction is weak, we can also write the polarization of the ions as

$$\delta\,\varrho_i(q,\omega) = \chi_{ii}(q,\omega)\,V_{ie}(q)\,\delta\varrho_e(q,\omega)\,, \tag{25}$$

whence

$$\chi_{ie}(q,\omega) = V_{ie}(q)\,\chi_{ee}(q,\omega)\,\chi_{ii}(q,\omega)\,. \tag{26}$$

It is a good approximation because of the difference in excitation energies of electrons and ions to take the static limit for the electronic response function in eq. (26), when we find

$$\chi_{ie}(q,\omega) = V_{ie}(q)\,\chi_{ee}(q,0)\,\chi_{ii}(q,\omega)\,, \tag{27}$$

which then gives

$$S_{ie}(q) = V_{ie}(q)\,\chi_{ee}(q,0)\,S(q)/\sqrt{Z}\,. \tag{28}$$

In the long-wavelength limit, $V_{ie}(q)\to -4\pi Ze^2/q^2$ and $\chi_{ee}(q,0)\to -q^2/4\pi e^2$, leading to

$$\lim_{q\to 0}\,[S_{ie}(q)/S(q)] = \sqrt{Z}\,, \tag{29}$$

which is the result of WATABE and HASEGAWA (2) and of CHIHARA (3).

We wish to emphasize that equation (28) for S_{ie} essentially multiplies $S(q)/\sqrt{Z}$ by the product of a bare electron-ion interaction and the electron-electron response function $\chi_{ee}(q,0)$. This equals the product of the screened electron-ion potential, which becomes usually quite small already for $q\sim k_F$, and the screened electron-electron response function which decreases rapidly for $q\sim 2k_F$, k_F being the Fermi wave number. Thus, we are led directly to the result that S_{ie} falls off from the very small value $Z^{\frac{1}{2}}S(0)$ at $q=0$ to about zero in the range 0 to k_F, or at most $2k_F$. Its behaviour may reflect the peak of $S(q)$ only in polyvalent metals, if $2k_F$ is sufficiently large.

The theory then leads immediately to the result that, if $S(q)$ increases away from $q=0$, then the curvature of $S_{ie}(q)$ is smaller than that of $S(q)$, or of opposite sign. In Sect. **5** we shall generalize the result (28) to the case of a binary alloy. Here $S_{ie}(q)$ can, for some concentrations, be much larger in the long-wavelength limit than for pure metals.

4·1. *Alternative derivation of characteristic energy.* – We now use eq. (28) to calculate the denominator $D(q)$ which appears in the partial Ornstein-Zernike functions given in eqs. (1)-(3). Then we find

$$D(q) = S(q)[S_{ee}(q) - \{V_{ie}(q)\chi_{ee}(q, 0)\}^2 S(q)/Z]\,. \tag{30}$$

By means of the property referred to above that $\lim_{q\to 0} V_{ie}(q)\chi_{ee}(q, 0) = Z$, it is readily shown that the square bracket in eq. (30) is of order q^2, and thus $S(q)$ outside the bracket can be replaced by $S(0) \equiv A$. From the small-q expansions (10)-(12) and writing

$$V_{ie}(q)\chi_{ee}(q, 0) = Z + fq^2 + O(q^4)\,, \tag{31}$$

it is readily shown that

$$D(q) = Aq^2[Zs_e - Zs_i - 2fA]\,. \tag{32}$$

We notice from the previous arguments that $s_e \gg s_i$, and that f is multiplied by $A(\equiv S(0))$ which is a small number $\sim 1/30$. Thus, to a good approximation

$$D(q) \equiv dq^2 = ZS(0)\, s_e q^2\,. \tag{33}$$

But from the plasmon contribution $Zs_e = \hbar/2m\omega_p$ and hence from eq. (14)

$$E = \tfrac{1}{2}\hbar\omega_p\,. \tag{34}$$

5. – Binary alloy.

The interest in this problem arises from the strong concentration dependence of the electronic correlation functions (S_{Ae}, S_{Be}, S_{ee}) in an A-B alloy. This has been demonstrated previously at $q = 0$ in Na-K and K-Hg by MARCH, TOSI and BHATIA ([6]).

The plasmon contribution to s_e in S_{ee} is rather clearly to be calculated for the mean electron density for the particular alloy in question. There can be little doubt that this will remain the dominant term in s_e.

In the light of the difficulties of estimating s_i and s in pure liquid metals, we restrict ourselves to some comments on the relation between S_{Ae}, S_{Be} and the partial ionic structure factors S_{AA}, S_{BB} and S_{AB}, which are again accessible through scattering experiments.

Thus, the $q = 0$ result ([6])

$$S_{Ae} = \left(\frac{n_A}{n_e}\right)^{\frac{1}{2}} \left\{ Z_A S_{AA} + Z_B S_{AB} \left(\frac{n_B}{n_A}\right)^{\frac{1}{2}} \right\} \tag{35}$$

([6]) N. H. MARCH, M. P. TOSI and A. B. BHATIA: *Journ. Phys. C*, **6**, L59 (1973).

is modified by replacing Z_A by $V_{Ae}(q)\chi_{ee}(q, 0)$, $V_{Ae}(q)$ being the electron-ion A interaction, and Z_B by $V_{Be}(q)\chi_{ee}(q, 0)$. The detailed argument is given in Appendix A.

We do not expect $V_{Be}(q)$ and $\chi_{ee}(q, 0)$ to depend strongly on concentration at small q, but we know from the work of McAlister and Turner (7) that in Na-K and K-Hg the partial ionic structure factors do indeed vary strongly with concentration.

6. – Summary.

We have discussed in this paper the low-q behaviour of the three correlation functions S_{ee}, S_{ie} and S in a pure liquid metal, and reached the conclusions that:

i) $S_{ee}(q)$ has strong curvature at the origin, rising rapidly from its value $ZS(0)$ at $q = 0$. The curvature is dominated by the electronic plasma frequency.

ii) S_{ie} and S have much smaller curvature at $q = 0$, the plasmon contributions being negligible, with phonon dispersion and multiparticle excitations playing a role.

iii) For weak electron-ion interaction, S_{ie} has either smaller curvature or curvature of opposite sign to $S(q)$, and is generally small for all q.

We have also shown how the above results can be generalized to the case of a binary alloy.

Appendix A

Relation between electron-ion and ion-ion structure factors in a binary alloy.

The argument is a generalization of that given in Sect. **4** for a pure metal. We first apply an external potential V_e to the electrons. The corresponding density changes for the three components can then be written as

$$\delta\varrho_e(q, \omega) = \chi_{ee}(q, \omega) V_e(q, \omega) , \tag{A.1}$$

$$\delta\varrho_A(q, \omega) = \chi_{Ae}(q, \omega) V_e(q, \omega) \tag{A.2}$$

and

$$\delta\varrho_B(q, \omega) = \chi_{Be}(q, \omega) V_e(q, \omega) . \tag{A.3}$$

(7) S. P. McAlister and R. Turner: *Journ. Phys. F*, **2**, L51 (1972).

The density changes of the ionic components are created by the potentials that they see due to the electronic polarization, which are given by

$$\text{potential seen by } A = V_{Ae}(q)\,\delta\varrho_e(q,\omega)\,, \tag{A.4}$$

while

$$\text{potential seen by } B = V_{Be}(q)\,\delta\varrho_e(q,\omega)\,. \tag{A.5}$$

Therefore it follows, by linear response theory, that

$$\delta\varrho_A(q,\omega) = \chi_{AA}(q,\omega)\,V_{Ae}(q)\,\delta\varrho_e(q,\omega) + \chi_{AB}(q,\omega)\,V_{Be}(q)\,\delta\varrho_e(q,\omega) \tag{A.6}$$

and

$$\delta\varrho_B(q,\omega) = \chi_{BA}(q,\omega)\,V_{Ae}(q)\,\delta\varrho_e(q,\omega) + \chi_{BB}(q,\omega)\,V_{Be}(q,\omega)\,\delta\varrho_e(q,\omega)\,. \tag{A.7}$$

These equations yield

$$\chi_{Ae}(q,\omega) = V_{Ae}(q)\,\chi_{ee}(q,\omega)\,\chi_{AA}(q,\omega) + V_{Be}(q)\,\chi_{ee}(q,\omega)\,\chi_{AB}(q,\omega)\,, \tag{A.8}$$

$\chi_{Be}(q,\omega)$ being obtained from this result by interchanging A and B.

If we neglect the frequency dependence of χ_{ee}, *i.e.* put $\chi_{ee}(q,\omega)=\chi_{ee}(q,0)$, then we find

$$S_{Ae}(q) = \left(\frac{n_A}{n_e}\right)^{\frac{1}{2}}\left\{V_{Ae}(q)\,\chi_{ee}(q,0)\,S_{AA}(q) + V_{Be}(q)\,\chi_{ee}(q,0)\,S_{AB}(q)\left(\frac{n_B}{n_A}\right)^{\frac{1}{2}}\right\}, \tag{A.9}$$

which is the desired result. S_{Be} follows by interchanging A and B.

Appendix B

Cowan-Kirkwood theory of liquid metals and plasmas.

Because of its role in motivating the theory of Sect. **4**, as well as its possible relevance to really high-temperature assemblies of ions and electrons, we shall summarize briefly here the relevant consequences of the Cowan-Kirkwood model for our purposes. The model gives no discrete bound states and therefore could only be quantitatively accurate when complete ionization has occurred. Furthermore, the electrons are treated by temperature-dependent Thomas-Fermi theory and therefore any effects depending on q's around $2k_F$ are lost. This probably means that the model will only work well at such high temperature that $k_B T \gg \frac{1}{2}\hbar\omega_p$.

The theory is expressible most directly in terms of the Fermi-Dirac function $I_{\frac{1}{2}}(\eta)$ defined by

$$I_{\frac{1}{2}}(\eta) = \int_0^\infty \mathrm{d}y\, y^{\frac{1}{2}}[1+\exp[y-\eta]]^{-1}\,. \tag{B.1}$$

Then we can write for the three pair functions

(B.2) $$n_e g_{ee}(r) = 4\pi(2mk_B Th^{-2})^{\frac{3}{2}} I_{\frac{1}{2}}(\eta_-) \,,$$

(B.3) $$n_e g_{ie}(r) = 4\pi(2mk_B Th^{-2})^{\frac{3}{2}} I_{\frac{1}{2}}(\eta_+)$$

and

(B.4) $$g(r) = \exp[-Z(\eta_+ - \eta_\infty)] \,.$$

Here, the η's are chemical potentials in units of $k_B T$, η_∞ being given in terms of the homogeneous electron gas density n_e as

(B.5) $$n_e = 4\pi(2mk_B Th^{-2})^{\frac{3}{2}} I_{\frac{1}{2}}(\eta_\infty) \,.$$

η_+ and η_- are related to solutions of the self-consistent Thomas-Fermi problem, obtained numerically by COWAN and KIRKWOOD.

However, for our purposes, we want to focus only on two points. First, we see from eq. (B.4) that

(B.6) $$\eta_+ = \eta_\infty - \frac{1}{Z} \ln g(r)$$

and hence from eqs. (B.3), (B.5) and (B.6) we can write

(B.7) $$g_{ie}(r) = \frac{I_{\frac{1}{2}}(\eta_+)}{I_{\frac{1}{2}}(\eta_\infty)} = \frac{I_{\frac{1}{2}}(\eta_\infty - Z^{-1} \ln g(r))}{I_{\frac{1}{2}}(\eta_\infty)} \,.$$

This relation, which holds for all r in the Cowan-Kirkwood theory, shows that, in this case, $g_{ie}(r)$ is an explicit function of the ion-ion radial distribution function $g(r)$.

The second, and related, point concerns the asymptotic behaviour of $g_{ie}(r)$ and $g(r)$, which is related to the small-q behaviour of $S_{ie}(q)$ and $S(q)$ discussed at length in the main text. We can show from the results of COWAN and KIRKWOOD that $g(r)-1$ has the asymptotic form

(B.8) $$g(r) - 1 = \frac{B}{r} \exp[-\lambda_+ r] \,,$$

where λ_+^{-1} is an appropriate screening length, while B must be determined by matching the asymptotic form (B.8) onto the numerical solution for small r. However, eq. (B.7) combined with eq. (B.8) then allows the asymptotic form of $g_{ie}(r)-1$ to be expressed explicitly as

(B.9) $$g_{ie}(r) - 1 = -\frac{1}{Z} \left.\frac{\mathrm{d}I_{\frac{1}{2}}(\eta)}{\mathrm{d}\eta}\right|_{\eta_\infty} \frac{B}{r} \exp[-\lambda_+ r] \,.$$

The electron-electron correlation function $g_{ee}(r)-1$, in contrast, involves, besides a term like eq. (B.8), a further screening length λ_-^{-1}, which is always the dominant term at large r.

It is, of course, possible to calculate completely S_{ee}, S_{ie} and S for all q in this model from eqs. (B.2)-(B.4), but since these results are not applicable to real liquid metals, for the reasons stated above and principally because of the absence of a sufficiently hard repulsive core arising from bound electrons in calculating $S(q)$, we only reiterate that the model may come into its own in really high-temperature plasmas.

● RIASSUNTO

I fattori di struttura parziali S_{ii}, S_{ie} e S_{ee} in metalli liquidi puri sono studiati a piccoli vettori d'onda. I tre fattori sono collegati tramite la valenza a $q = 0$. Si mostra anzitutto che le funzioni parziali di Ornstein-Zernike sono proporzionali a q^{-2} nel limite di grandi lunghezze d'onda, e sono collegate non solo tramite la valenza ma anche tramite un'energia caratteristica E. Con un modello per i fattori dinamici di struttura che assume modi fononici e plasmonici ben definiti a piccoli vettori d'onda, si mostra che l'energia caratteristica è dominata dall'energia di punto zero delle oscillazioni di plasma elettronico. La stessa energia domina la curvatura di S_{ee} a $q = 0$. Tale curvatura è forte e perciò S_{ee} cresce rapidamente al crescere del vettore d'onda. Al contrario, le curvature di S_{ie} ed S sono più piccole per ordini di grandezza, e perciò questi fattori di struttura sono piatti a piccoli vettori d'onda. Tuttavia, tali curvature sono affette dalla dispersione dei fononi, seppur insensibili allo smorzamento dei fononi. Se l'interazione elettrone-ione è debole, $S_{ie}(q)$ e $S(q)$ sono collegati tramite una funzione di risposta statica degli elettroni, in generalizzazione della relazione esatta valida per $q = 0$. Infine, il termine q^2 in S_{ee} in una lega è determinato dalla frequenza di plasma media, caraterizzata dalle valenze dei componenti e dal volume atomico medio. I fattori di struttura parziali S_{Ae} ed S_{Be} possono pure essere collegati ai fattori di struttura ionici parziali, se l'interazione elettrone-ione è debole.

Рессеяние на малые углы в жидких металлах и сплавах и электронные корреляционные функции.

Резюме (*). — Рассматриваются парциальные структурные факторы S_{ii}, S_{ie} и S_{ee} в чистых жидких металлах при малых волновых векторах q. Аргументы зарядовой нейтральности связывают эти три функции через валентность Z при $q=0$. Показывается, что парциальные функции Орнстейна-Зернике оказываются пропорциональными q^{-2} в длинноволновом пределе и что они связаны не только через Z, но и через общую характеристическую энергию E. Определяя модель для динамических структурных факторов и предполагая хорошо определенные моды плазмонов и фононов при малых q, показывается, что характеристическая энергия в основном определяется энергией нулевых колебаний электронной плазмы. Отмечается, что эта же энергия определяет кривизну S_{ee} при $q=0$, причем, кривизна является резко выраженной и, следовательно, S_{ee} быстро возрастает вдали от $q=0$. В противоположность этому, показывается, что коэффициемты q^2 в S_{ie} и S_{ii}, так как в них входят массы ионов, оказываются по порядку величины меньше и эти структурные факторы, таким образом, являются совсем плоскими вблизи $q=0$. Однако показывается, что коэффициенты q^2 в S_{ie} и S_{ii} зависят от дисперсии фононов, хотя являются нечувствительными к затуханию фононов. Для слабого электрон-ионного взаимодействия $S_{ie}(q)$ и $S(q)$ мпгут быть связаны через функцию электронного отклика. Полученное выражение удовлетворяет точному предельному соотношению между S_{ie} и S при $q=0$. Наконец, обсуждается член q^2 в S_{ee} в сплаве и показывается, что он непосредственно связан со средней частотой электронной плазмы, включая, таким образом, валентности компонент и средний размер атомов. Парциальные структурные факторы S_{Ae} и S_{Be} могут быть также связаны с (измеряемыми) парциальными ионными структурными факторами для слабого электрон-ионного взаимодействия.

(*) *Переведено редакцией.*

1651

Electron Correlation Functions in Liquids from Scattering Data

P. A. EGELSTAFF, N. H. MARCH[1] AND N. C. MCGILL[2]
Department of Physics, University of Guelph, Guelph, Ontario, N1G 2W1
Received October 12, 1973[3]

Recent work on the theory of liquid metals has involved correlation functions for the ion–electron density and the electron–electron density. The experimental determination of these functions is discussed for a general homonuclear fluid. It is shown that the electronic correlation functions may be extracted, in principle, by combining X-ray, neutron, and electron scattering data, though the smallness of the differences between the normalized data makes this difficult to do at present.

After reviewing the published scattering data, we conclude that the most useful procedure is to compare scattered intensities at a significant reference point, namely the principal maximum of the liquid structure factor. X-ray and neutron data are presented for liquified rare gases, molecular liquids, and liquid metals, and their averages considered. Systematic differences are found between these groups, which prompt the conclusion that electrons in molecular liquids are delocalized by chemical bonding, but that the conduction electrons in a liquid metal have a higher degree of short range order than the nuclei. In this event an electron liquid, rather than an electron gas, exists in liquid metals.

Des travaux récents sur la théorie des métaux liquides font intervenir les fonctions de corrélation pour la densité ion–électron et la densité électron–électron. La détermination expérimentale de ces fonctions est discutée pour le cas général d'un fluide homonucléaire quelconque. On montre que les fonctions de corrélation électronique peuvent être obtenues, en principe, en combinant les données de diffusion des rayons X, des neutrons et des électrons, bien que la chose soit difficile présentement, à cause de la petitesse des différences entre les données normalisées.

Après avoir passé en revue les données de diffusion publiées, nous concluons que la procédure la plus utile consiste à comparer les intensités diffusées en un point significatif, à savoir le maximum principal du facteur de structure du liquide. Des données de diffusion de rayons X et de neutrons sont présentées pour les gaz rares liquéfiés, les liquides moléculaires, les métaux liquides, et leurs moyennes sont considérées. On trouve des différences systématiques entre ces groupes, ce qui amène à conclure que des électrons, dans les liquides moléculaires, sont délocalisés par la liaison chimique, mais que les électrons de conduction dans un métal liquide, ont un degré d'ordre à courte portée plus élevé que les noyaux. Dans ces conditions, c'est un liquide d'électrons, plutôt qu'un gaz d'électrons, qui existe dans les métaux liquides.

[Traduit par le journal]

Can. J. Phys., **52**, 1651 (1974)

Introduction

For many years, it has been supposed that the ionic structure factor of a liquid could be measured by either neutron or X-ray scattering. It is evident that neutrons, being scattered by nuclei, directly determine the ionic structure factor. In contrast X rays are scattered by electrons, and so X-ray cross sections are related primarily to electronic correlation functions. Fast electrons are scattered by all charges present, and so electron scattering cross sections are related to both nuclear and electronic correlations. Electronic functions enter basic theories of liquid metals (Cowan and Kirkwood 1958; Chihara 1973; Watabe and Hasegawa 1973; March and Tosi 1973; Tosi and March 1973), and theories of the structure of molecules (*e.g.*, Herzberg 1939) may be reinterpreted in this way. It is the purpose of this paper to demonstrate, in a wide context, that electronic correlation functions in liquids can be extracted by combining neutron, X-ray, and electron scattering data, and that the best experimental data now available (although frequently presented in an unsuitable form) are only just accurate enough for this purpose. The differences between these three types of data have been emphasized in the past (*e.g.*, on liquids

[1]Permanent address: Department of Physics, Imperial College of Science and Technology, University of London, United Kingdom.
[2]Permanent address: Department of Theoretical Physics, University of St. Andrews, St. Andrews, Fife, United Kingdom.
[3]Revision received May 7, 1974.

by Pirenne (1946) and Egelstaff (1967*a*)), and also utilized in the experimental study of solids. However, we have adopted a different approach, and have expressed scattering cross sections in terms of separate ion–ion, ion–electron, and electron–electron correlation functions.

Since suitable electron scattering data are not known to us, the X-ray and neutron scattered intensities at the principal maximum of the liquid structure factor have been compared, after suitable normalization. Whereas for the liquified rare gases Ne and Ar the X-ray and neutron peaks have the same height, for molecular fluids O_2 and Br_2 the neutron peaks are higher than the X-ray peaks. This is most probably due to the intramolecular delocalization of electrons arising from chemical bonding. For liquid metals, the X-ray peak is higher than the neutron peak, in all nine cases for which we have been able to find suitable data. We conclude from this that there is probably short range order in the conduction electrons in liquid metals, and that the range of this order is greater than that for the ions.

After briefly reviewing the general scattering theory, we reformulate scattering cross sections in terms of ion and valence electron density fluctuations.

Scattering Theory

We first summarize the results of the theory of neutron, X-ray, and electron scattering.

In neutron scattering the scattered intensity is determined by the nuclear density–density correlation function (*e.g.*, Egelstaff 1967*b*)

$$[1] \quad \langle \rho_n(\boldsymbol{R})\rho_n(\boldsymbol{R}+\boldsymbol{r})\rangle = \frac{1}{V}\int_V \mathrm{d}\boldsymbol{r}' \,\langle \rho_n(\boldsymbol{r}')\rho_n(\boldsymbol{r}'+\boldsymbol{r})\rangle$$

where V is the volume, $\boldsymbol{R}$ is an arbitrary origin, and $\rho_n(\boldsymbol{r})$ is the nuclear density at position $\boldsymbol{r}$. The reduced neutron scattering cross section is then given by

$$[2] \quad \left(\frac{\mathrm{d}\sigma}{\mathrm{d}\Omega}\right)_n = (\text{const.}) \times \left\{\frac{1}{\rho_n}\int_V e^{i\boldsymbol{Q}\cdot\boldsymbol{r}}\,\langle \rho_n(\boldsymbol{R})\rho_n(\boldsymbol{R}+\boldsymbol{r})\rangle\,\mathrm{d}\boldsymbol{r}\right\}$$

where $\hbar\boldsymbol{Q}$ is the momentum transferred in the scattering event and ρ_n is the average value of $\rho_n(\boldsymbol{r})$. Denoting the quantity in the curly brackets by $S_n(Q)$, it is easily shown that (neglecting the forward scattering) it is related to the familiar nuclear pair correlation function $g_{nn}(r)$ by

$$[3] \quad S_n(Q) = 1 + \rho_n \int e^{i\boldsymbol{Q}\cdot\boldsymbol{r}}[g_{nn}(r) - 1]\,\mathrm{d}\boldsymbol{r}$$

For X rays, the scattered intensity is determined by the electron density–density correlation function $\langle \rho_e(\boldsymbol{R})\rho_e(\boldsymbol{R}+\boldsymbol{r})\rangle$ defined by an equation analogous to [1], and the X-ray scattering cross section is proportional to its Fourier transform, in correspondence with [2]. Fast electrons, however, are scattered by the potential they feel, which is related to the total charge density (nuclei plus electrons) by Poisson's equation. The total charge density (in units of e) is

$$[4] \quad q(\boldsymbol{r}) = Z\rho_n(\boldsymbol{r}) - \rho_e(\boldsymbol{r})$$

where Z is the atomic number. Thus, electron scattering is determined by the charge density–density correlation function $\langle q(\boldsymbol{R})q(\boldsymbol{R}+\boldsymbol{r})\rangle$ and the electron scattering cross section is proportional to its Fourier transform. If we substitute [4] into this last expression and recall that $\langle \rho_n\rho_n\rangle$ is determined by neutron scattering and $\langle \rho_e\rho_e\rangle$ by X-ray scattering, it is possible, at least in principle, to extract $\langle \rho_n\rho_e\rangle$ from the electron scattering data. These three correlation functions yield complete information about the pair correlations of nuclei and electrons in any homonuclear fluid.

Correlation Functions in the Rigid Core Model

It is often useful to divide the electrons in the neighborhood of each nucleus into Z_c 'core' electrons and Z_v 'valence' electrons, and to define the former as those attached rigidly to the nucleus, the spatial electronic distribution in the core being unaffected to a good approximation by the liquid environment. (We stress that this demarcation is not necessarily the same as that relevant to chemical bonding). Thus the total electron density is the sum of parts due to core and valence electrons: $\rho_e(\boldsymbol{r}) = \rho_c(\boldsymbol{r}) + \rho_v(\boldsymbol{r})$. This separation generates various terms in the cross sections such as

$$[5] \quad \int_V e^{i\boldsymbol{Q}\cdot\boldsymbol{r}}\langle \rho_c(\boldsymbol{R})\rho_v(\boldsymbol{R}+\boldsymbol{r})\rangle\,\mathrm{d}\boldsymbol{r}$$

It is readily shown that, with the above assumptions about the core, this term reduces to

$\rho_n f(Q) S_{iv}(Q)$, where $f(Q)$ is the Fourier transform of the electron density in the core (assumed spherically symmetric) and

$$[6] \quad S_{iv}(Q) = \frac{1}{\rho_n} \int_V e^{iQ\cdot r} \langle \rho_n(\boldsymbol{R}) \rho_v(\boldsymbol{R} + \boldsymbol{r}) \rangle \, d\boldsymbol{r}$$

subscripts i and v denoting ion and valence electron respectively. Clearly $f(0) = Z_c$, and we assume that the function $f(Q)$ is known by calculation. The corresponding form factor for electron scattering is

$$[7] \quad f_e(Q) = Z - f(Q)$$

satisfying $f_e(0) = Z_v$.

With these assumptions and manipulations, the cross sections for X-ray and electron scattering may be reduced to

$$[8] \quad \left(\frac{d\sigma}{d\Omega}\right)_x = (\text{const.}) S_x(Q) \qquad \left(\frac{d\sigma}{d\Omega}\right)_e = \frac{(\text{const.})}{Q^4} S_e(Q)$$

with

$$[9] \quad S_x(Q) = f^2(Q) S_{ii}(Q) + 2f(Q) S_{iv}(Q) + S_{vv}(Q)$$

$$S_e(Q) = f_e^2(Q) S_{ii}(Q) - 2f_e(Q) S_{iv}(Q) + S_{vv}(Q)$$

where

$$[10] \quad S_{ii}(Q) = \frac{1}{\rho_n} \int_V e^{iQ\cdot r} \langle \rho_n(\boldsymbol{R}) \rho_n(\boldsymbol{R} + \boldsymbol{r}) \rangle \, d\boldsymbol{r}$$

$$S_{vv}(Q) = \frac{1}{\rho_n} \int_V e^{iQ\cdot r} \langle \rho_v(\boldsymbol{R}) \rho_v(\boldsymbol{R} + \boldsymbol{r}) \rangle \, d\boldsymbol{r}$$

Equations 9, together with the neutron scattering result

$$[11] \quad (d\sigma/d\Omega)_n = (\text{const.}) S_n(Q) \qquad S_n(Q) \equiv S_{ii}(Q)$$

yield the following solutions for S_{iv} and S_{vv}:

$$[12] \quad S_{iv}(Q) = (1/2Z)\{S_x(Q) - S_e(Q) - Z(f(Q) - f_e(Q)) S_n(Q)\}$$

$$S_{vv}(Q) = (1/Z)\{f_e(Q) S_x(Q) + f(Q) S_e(Q) - Z f(Q) f_e(Q) S_n(Q)\}$$

These two equations, [12], form the basis of our proposal to extract electronic correlation functions from X-ray, neutron, and electron scattering and calculated values of $f(Q)$, when suitable data become available. These data need to be accurate and reliable since the final two terms on the right-hand side of [9] are expected to be a few percent of the ion–ion term, because $Z_v \ll Z$.

By analogy with [3], S_{iv} and S_{vv} may be readily expressed (again neglecting forward scattering) in terms of pair functions g_{iv} and g_{vv}. It is interesting to note the limiting behavior of S_{iv} and S_{vv} as $Q \to 0$, due to charge neutrality:

$$[13] \quad S_{iv}(0) \to Z_v S_{ii}(0) \qquad S_{vv}(0) \to Z_v^2 S_{ii}(0)$$

These results have been obtained previously (Chihara 1973; Watabe and Hasegawa 1973; March and Tosi 1973)

A further, simple assumption often made to reduce the above equations is that the valence electron density may be decomposed into a number of identical and spherically symmetrical localized parts which are rigidly attached to (and centered on) the ions. The Fourier transform of this 'rigid atom' distribution with $Z = (Z_c + Z_v)$ electrons will be written $F(Q)$, and this quantity will normally be comparable with $F_a(Q)$, the Fourier transform of the electron density in the isolated atom. But in general, environmental effects lead to some redistribution of valence electrons so that $F(Q) \neq F_a(Q)$, except at $Q = 0$ where both equal Z, and at sufficiently large Q where the core electrons dominate. In this model, which is widely used in the analysis of X-ray and electron scattering intensities, we can write

$$[14] \quad S_x(Q) = F^2(Q) S_{ii}(Q) \qquad S_e(Q) = [Z - F(Q)]^2 S_{ii}(Q)$$

Then from [12] we obtain

$$[15] \quad S_{iv}(Q) = [F(Q) - f(Q)] S_{ii}(Q) \qquad S_{vv}(Q) = [F(Q) - f(Q)]^2 S_{ii}(Q)$$

as would be expected if the valence electrons were a shell rigidly coupled to the ion, and had a form factor $f_v(Q) \equiv F(Q) - f(Q)$. In those cases for which $F(Q)$ is difficult to calculate, a test of this model would be (given S_{ii}) to deduce $F(Q)$ in [14] separately from the X-ray and electron scattering intensities, and compare the results.

In contrast to the conventional approach outlined above, it is interesting to consider also a more general model incorporating some degree of nonrigidity. Suppose that the valence electrons cannot be divided at any instant into identical localized parts centered on the ions, and that (in particular) they are free to form a structure of longer range order than that of the ions. In this case the principal peaks (at $Q = Q_0$ say) in S_{iv} and S_{vv} will have greater relative heights than in S_{ii}. A first approximation to this situation might be represented by

$$[16] \quad \begin{aligned} S_{iv}(Q_0) &= f_v(Q_0)(1 + \varepsilon)S_{ii}(Q_0) \\ S_{vv}(Q_0) &= f_v^2(Q_0)(1 + 2\varepsilon)S_{ii}(Q_0) \end{aligned}$$

where ε is a small positive number and $f_v(Q)$ is the form factor for the valence electrons in the simpler rigid atom approximation. In this event substitution in [9] yields

$$[17] \quad \begin{aligned} \frac{S_x(Q_0)}{F^2(Q_0)} &= S_{ii}(Q_0)\left[1 + 2\varepsilon\frac{f_v(Q_0)}{F(Q_0)}\right] \\ \frac{S_e(Q_0)}{[Z - F(Q_0)]^2} &= S_{ii}(Q_0)\left[1 - \frac{2\varepsilon f_v(Q_0)}{Z - F(Q_0)}\right] \end{aligned}$$

so that, in contrast to equation [14],

$$[18] \quad \frac{S_x(Q_0)}{F^2(Q_0)} > S_{ii}(Q_0) > \frac{S_e(Q_0)}{[Z - F(Q_0)]^2}$$

Thus, this type of nonrigid behavior may be investigated by comparing relative peak heights at Q_0 for the three kinds of data.

The Published Experimental Data

Unfortunately the published experimental data known to the authors are not suitable for the full form of analysis proposed here. The principal reasons for this are two-fold: (*a*) the accuracy is usually ~4% and the data are presented in small graphs involving larger reading errors, and (*b*) the X-ray and electron data are usually reduced to $S_{ii}(Q)$ via a form factor which is not specified explicitly (it would be preferable if $S_x(Q)$ and $S_e(Q)$ were tabulated separately). In order to use our methods it is necessary to find three sets of high quality data on the same element with the same state conditions. This is certainly not available over a wide range of Q values.

Thus in considering the experimental data we have decided to restrict attention to a single value of $S(Q)$ which may be measured fairly accurately and which is physically significant. This is the value at $Q = Q_0$. From [17] we see that

$$[19] \quad \begin{aligned} S_{xm} &\equiv \frac{S_x}{F_a^{\ 2}} = S_{ii}\frac{F^2}{F_a^{\ 2}}\left(1 + 2\varepsilon\frac{f_v}{F}\right) \\ S_{em} &\equiv \frac{S_e}{(Z - F_a)^2} \\ &= S_{ii}\frac{(Z - F)^2}{(Z - F_a)^2}\left(1 - \frac{2\varepsilon f_v}{Z - F}\right) \end{aligned}$$

where all functions of Q are evaluated at Q_0. In a number of previously published papers it has been customary to assume that S_{xm} and S_{em} are exactly equal to S_{ii} (indeed this is usually assumed for all Q, not just for $Q = Q_0$), enabling the calculation of S_{ii} to be carried out from the X-ray or electron scattering data. As we will show later, however, this is not generally the case experimentally (at least for $Q = Q_0$) and we require some theoretical account of the factors omitted.

We have been able to find only one case (Bi) in which all three types of diffraction data may be quoted with any confidence. The quantities shown in Table 1 for this case are $S_{ii}(Q_0)$, $S_{xm}(Q_0)$, and $S_{em}(Q_0)$. Noting that for bismuth $F \simeq F_a$ (and that pseudopotential theory indicates that F/F_a is much more likely to be just less than unity rather than just greater) we see from [19] that the data are consistent with the hypothesis of a nonrigid atom. Although this tentative conclusion may be undermined by unreliability in the figures quoted, we attempt later to reach firmer conclusions by analyzing the corresponding figures for a number of liquids and considering the average behavior. In doing this we are necessarily restricted to neutron and X-ray data, since we have been unable to find other satisfactory cases of electron data.

Although our primary interest is in liquid metals, it is useful to examine data on other homonuclear liquids also. This is done in the next two sections.

Liquified Rare Gases

The electronic density distribution for atoms in a liquified rare gas is usually assumed to be closely similar to that in the vapor. In this case the rigid atom model should be satisfactory *i.e.* $F(Q) \simeq F_a(Q)$ and $\varepsilon = 0$. Comparing $S_{xm}(Q_0)$

TABLE 1. A comparison of electron,[a] neutron,[b] and X-ray[b] data for bismuth

Electrons		Neutrons		X-rays	
Temp. (K)	$S_{em}(Q_0)$	Temp. (K)	$S_{ii}(Q_0)$	Temp. (K)	$S_{xm}(Q_0)$
~544[c]	1.75[d]	573	1.90	576	2.07

[a]Leonhardt R., Richter, H., and Rossteutscher, W. 1961. Z. Phys. 165, 121.
[b]Data from Table 2.
[c]The temperature is not clearly stated in [a], but is close to the melting temperature (544 K).
[d]The value of $S_{em}(Q_0)$ may need to be corrected to ~575 K and for the fact that electron diffraction experiments are made with a thin film rather than bulk liquid. Both of these corrections should reduce $S_{em}(Q_0)$.

with $S_{ii}(Q_0)$ for Ne and Ar in Table 2, it is seen that the two quantities agree to a few per cent, which is about the same as the experimental accuracy. The result may be more significant for argon than for neon because of the higher accuracy of the neutron data.

Homonuclear Molecular Liquids

We next discuss the relation between the X-ray and neutron data for homonuclear diatomic molecular liquids. Although we expect the electrons in such liquids to be fairly well localized within the various molecules, chemical bonding is almost certain to modify the electronic charge distribution from the free atom configuration, with consequent alteration to the atomic form factor. For example, molecular orbital theory and the Heitler–London theory of the H_2 molecule move charge into the center of the bond. Thus, as several authors (*e.g.*, Pirenne 1946) have pointed out, we would expect $F(Q) \neq F_a(Q)$ and that in some molecules $F(Q)$ will not even be spherically symmetric.

A guide to the effect on $F(Q)$ in the intermediate Q region (*e.g.*, near the principal peak in the structure factor) would be provided by a single parameter — the charge density at the nucleus — since by the inverse Fourier transform this equals

$$(8\pi^3)^{-1}\int F(Q)\,d\mathbf{Q}$$

while $F(0) = Z$, and $F(Q)|_{Q\to\infty}$ is dominated by the core electrons and is therefore unaffected by the chemical bonding. If, for example, the chemical bond moves electronic charge from the vicinity of the nucleus to the region between the two nuclei (or to anywhere else), then it would almost certainly follow that a symmetrized (atomic) $F(Q)$ would be less than $F_a(Q)$ in the intermediate Q region, the contribution to $F(Q)$ from the displaced charge being shorter ranged in Q than before.

Wahl (1966) has calculated by molecular orbital theory the electronic charge density in homonuclear diatomic molecules for elements in the periodic table between hydrogen and fluorine. So far the results have been presented only as charge density contour diagrams, and calculation of $F(Q)$ would require more detailed information. It is interesting, however, to compare the molecular charge distribution with that obtained by the superposition of two free atom charge distributions. Using oxygen as an example (one of the two molecular liquids listed in Table 2) and computing the free atom charge density from the appropriate Hartree–Fock wave functions (Hartree *et al.* 1939), it is found that the molecular charge density is, to a good approximation, equal to the superposition of two free atoms, but that some small redistribution of the charge in the molecule has definitely occurred. The charge density at the nucleus is not specified however, so preventing quantitative implementation of the above argument.

So far only the influence of the chemical bond has been discussed; we should also consider the change in $F(Q)$ due to placing the molecule in the liquid environment. We know of no calculations for this, although experiments suggest a small change for some molecules which is manifested in $S(Q)$ at very large Q.

Table 2 includes the data for O_2 and Br_2, and it is found that the neutron peaks are higher than the X-ray peaks. If these data are reliable, this points to the conclusion that $F(Q_0) < F_a(Q_0)$ and implies that chemical bonding causes electronic delocalization. Because the proportion of valence electrons (in the usual sense of chemical bonding) is greater in oxygen than for bromine, one might have expected a correspondingly larger

TABLE 2. Heights of the principal peak of the liquid structure factor for elements (near the triple point)

Element	X rays			Neutrons		
	Ref.	Temp. (K)	$S_{xm}(Q_0)$	Ref.	Temp. (K)	$S_n(Q_0)$
Rare gas fluids						
Neon	a	25.5	2.43	b	26	2.50
Argon	c	84.3	2.82	a	84.5	2.71
Molecular fluids						
Oxygen	e	64	3.11	f	62.4	3.33
Bromine	g	267	1.25	h	300	1.45
Liquid metals						
monovalent						
Na	k	373	2.80	l	373	2.45
K	k	338	2.73	l	338	2.35
Cu	m	1373	2.75	n	1423	2.41
polyvalent						
Al	p	938	2.50	q	950	2.30
Ga	r	293	2.50	r	293	2.35
Tl	s	623	2.66	t	593	2.30
Sn	m	523	2.50	t	523	2.35
Pb	u	602	2.80	t	613	2.60
Bi	v	576	2.07	t	573	1.90

NOTES: (*i*) The data selected are the best comparative (neutron and X-ray) data known to the authors. (*ii*) The quantities listed are: $S_n(Q_o)$ for neutrons and $S_x(Q_o)/F_a^2 \equiv S_{xm}(Q_o)$ for X-rays, where Q_o is the peak position. (*iii*) For references [b], [f], [h], the neutron counting rate curve has been corrected to $S(Q)$ using the methods described in reference [t]. (*iv*) The error in each entry is such that the neutron and X-ray data for each element could overlap in some cases: we draw attention, however, to the systematic behavior of each group.

[a]STIRPE, D. and THOMSON, C. W. 1962. J. Chem. Phys. **36**, 392. (Reported by SCHMIDT, P. W. and THOMPSON, C. W. 1968. *In* Simple dense fluids, *ed. by* H. L. Frisch and Z. W. Saltzberg (Academic Press, New York).)
[b]HENSHAW, D. G. 1958. Phys. Rev. **111**, 1470.
[c]GRINGRICH, N. S. and THOMSON, C. W. 1962. J. Chem. Phys. **36**, 2398. (Reported by SCHMIDT, P. W. and THOMPSON, C. W. 1968. *In* Simple dense fluids, *ed. by* H. L. Frisch and Z. W. Saltzberg (Academic Press, New York).)
[d]PAGE, D. I. 1972. Report AERE R6828, H.M.S.O. (Lond.).
[e]FURUMOTO, H. W. and SHAW, C. H. 1964. Phys. Fluids, **7**, 1026. (Reported by SCHMIDT, P. W. and THOMPSON, C. W. 1968. *In* Simple dense fluids, *ed. by* H. L. Frisch and Z. W. Saltzberg (Academic Press, New York).)
[f]HENSHAW, D. G. 1960. Phys. Rev. **119**, 22.
[g]GRUEBEL, R. W. and CLAYTON, G. T. 1967. J. Chem. Phys. **47**, 175. (Reported by SCHMIDT, P. W. and THOMPSON, C. W. 1968. *In* Simple dense fluids, *ed. by* H. L. Frisch and Z. W. Saltzberg (Academic Press, New York).)
[h]CAGLIOTTI, G. and RICCI, F. P. 1962. NUOVO CIMENTO, **24**, 103. ASCARELLI, P. and CAGLIOTTI, G. 1966. **Nuovo Cimento, 43, 375.**
[k]GREENFIELD, A. J., WELLENDORF, J., and WISER, N. 1971. Phys. Rev. A, **4**, 1607.
[l]GRINGRICH, N. S. and HEATON, L. 1961. J. Chem. Phys. **34** , 873. (Reported by ASHCROFT, N. W. and LEKNER, J. 1966. Phys. Rev. **145**, 83, from a table supplied by L. HEATON.)
[m]NORTH, D. M. and WAGNER, C. J. N. 1969. J. Appl. Cryst. **2**, 149. ——— 1970. Phys. Chem. Liquids, **2**, 87.
[n]BREUIL, M. and TOURAND, G. 1970. J. Phys. Chem. Solids, **31**, 549.
[p]FESSLER, R. R., KAPLOW, R., and AVERBACH, B. L. 1966. Phys. Rev. **150**, 34.
[q]LARSSON, K. E., DAHLBORG, U., and JOVIC, D. 1965. *In* Inelastic scattering of neutrons, Vol. II. (I.A.E.A., Vienna), p. 117.
[r]NARTEN, A. H. 1972. J. Chem. Phys. **56**, 1185.
[s]HALDER, N. C. and WAGNER, C. N. J. 1966. J. Chem. Phys. **45**, 482.
[t]NORTH, D. M., ENDERBY, J. E., and EGELSTAFF, P. A. 1968. J. Phys. C, **1**, 1075.
[u]KAPLOW, R., STRONG, S. L., and AVERBACH, B. L. 1965. Phys. Rev. **138**, A1336. (A comparison for lead is reported by EGELSTAFF, P. A. 1967. Adv. Phys. **16**, 147.)
[v]ISHERWOOD, S. P. and ORTON, B. R. 1968. Philos. Mag. **17**, 561.

difference between F and F_a. However, Wahl's charge density diagrams, which include displays of individual electron shells, make it clear that in the process of molecular formation modifications occur to the charge density, not just for the bonding electrons but also for some of the other, more tightly bound electrons, making any simple correlation less probable. Of course errors in the data could also be exaggerating the actual difference between oxygen and bromine.

Liquid Metals

In liquid metals the model of a rigid 'pseudo-atom' has sometimes been used, whose implication is that the valence electrons, though peaked about nuclei, will be more spread out than in the free atom. This will reduce the value of $F(Q_0)$ below that of $F_a(Q_0)$. Some theoretical and experimental information exists on this point for solids. Walter *et al.* (1973) discuss aluminum and show that at the [200] reciprocal lattice point the value of F calculated by pseudopotential theory is 2% less than F_a, compared to ~3% observed experimentally. Since the [200] point is close to the principal peak of the liquid $S(Q)$, we may expect a corresponding reduction for liquid aluminum. Similar data are available for other solid metals.

Thus, for a rigid ($\varepsilon = 0$) pseudoatom we would anticipate

$$[20] \quad S_{xm}(Q_0) = \frac{F^2(Q_0)}{F_a{}^2(Q_0)} S_{ii}(Q_0) < S_{ii}(Q_0)$$

Table 2 collects data for nine liquid metals, and in each case the X-ray peak is *higher* than the neutron peak. Due to errors in the data, any one of these by itself may not be significant, but we believe that the uniform trend displayed by the group as a whole is significant (see Table 3). Unless there is an unknown systematic error to one technique, this comparison suggests that the rigid pseudoatom model is not valid, and the data seem to concur more readily with the hypothesis of a nonrigid atom, as described by [19].

It is tempting to discuss the differences in Table 2 as a function of the proportion of valence electrons in each metal. We do not attempt this, partly due to the errors in the data mentioned above, and partly because we use the term 'valence electrons' here to denote those electrons for which the electron–electron interaction is comparatively strong ('core electrons' being those for which the electron–ion interaction is comparatively strong). In any attempted correlation of this type it would therefore be questionable as to which figure should be taken as the percentage of valence electrons, and we prefer simply to treat all the data on a common basis and emphasize the differences in the mean values of S_{xm} and S_{ii}.

Thus, our conclusion is that the differences between the X-ray and neutron scattering intensities at the principal peak for liquid metals imply that the valence electron density correlation function has a longer range order than the ion–ion correlation function. Because the long range component of g_{vv} is most likely to be due to the conduction electrons, it follows that the conduction electrons in liquid metals form an 'electron liquid' rather than an electron gas as has been customarily assumed. Due to different instantaneous correlations, the natural density fluctuations of this electron liquid can have longer range order than the ionic density fluctuations, but their principal period will be the same in order to preserve charge neutrality. When experimental data of sufficient accuracy become available to allow [12] to be used over a range of Q, we infer that this order will be revealed explicitly in the form of $S_{iv}(Q)$ and $S_{vv}(Q)$, which should both exhibit 'sharper' peaks than those in $S_{ii}(Q)$.

We now argue that this is a reasonable expectation for a liquid metal. The ionic pair correlation function $g_{ii}(r)$ will have a range of order determined roughly by the interatomic distance. The electron pair function $g_{vv}(r)$ is expected to oscillate with relatively small amplitude about unity, after rising from a low value through the usual Fermi and correlation hole. The range of the oscillations ought to correlate with the electronic mean free path, which suggests that the oscillations continue to larger r than those of g_{ii}. On the other hand, their amplitude should correlate with the conduction electron density and with the strength of the valence electron–ion interactions and thus might be expected to be smaller than that of g_{ii} except at large r.

Qualitative Model of Electronic Pair Function in Alkali Metals

To explain the experimental results to which attention is drawn here, a full theory of the two-component system, with proper inclusion of both electron–ion and electron–electron interactions, will eventually be required for the calculation of g_{vv} and g_{iv}. In a high density metal like Al for example, there can be little doubt that the electron–ion interaction dominates the Coulomb repulsions between electrons.

In the absence of a complete theory valid at all densities, we shall present below a crude model for the low density limit, in which electron–electron correlations become strong. This is of some interest, since the low density liquid metals are the alkalis, in which the electron–ion interaction is known to be especially weak (though even for the lowest density metal Cs, it is not clear to us whether the electron–electron interaction actually dominates).

We note first that 'electron gas' theories lead to a form of S_{vv} which does not reveal short range order. Thus, Fig. 1 shows the Fourier transform of $g_{vv}(r)$ as calculated by Chihara

(1973) for a model appropriate to Na. The result, as he points out, is quite similar to the results of homogeneous electron gas theory (Singwi *et al.* 1970) except at very small Q. Such a form of S_{vv} is too simple to account for the liquid metal scattering data. Thus, let us consider the influence of electron–electron repulsion as the electron density is lowered, in a homogeneous electron gas. The model used here is that of 'jellium': even though it is not susceptible to quantitative analysis, it may be useful for qualitative discussion. We know for this case that as the mean interelectronic distance r_s increases, to a value considerably greater than that in the lowest density alkali Cs, electron crystallization (the Wigner transition) eventually takes place. (We immediately stress, however, that we do not expect a Wigner lattice in any real metal). In r space, March and Young (1959) have plotted schematically the effect to be expected in the pair function as we interpolate between the electron gas and the Wigner crystal, and their curves are shown in Fig. 2. A more realistic calculation for finite r_s by Gaskell (1962) indicates that the picture of oscillations about unity is broadly as suggested by Fig. 2, but that this figure exaggerates the magnitude of the oscillations for (in the case of Na) r_s around 4.

In summary this model gives a qualitative picture in accord with the sign of the inequalities observed experimentally.

Conclusions

We have shown in this paper that electronic correlation functions in liquids can be extracted, in principle, by combining neutron, X-ray, and electron scattering intensities. The lack of precise experimental data has prevented quantitative implementation of our proposal on any liquid, but we hope there will now be interest in making improvements to all diffraction experiments and especially in producing scattering data for electrons. At present the most significant comparative criteria seem to be the peak heights of the liquid structure factors for neutrons and X-rays displayed in Table 2. To emphasize the conclusions we have drawn from this table, we give in Table 3 the average of the ratio of neutron to X-ray peak height for each group.

The errors in this table are difficult to estimate properly; we have assumed that the error on each entry in Table 2 may be considered as a random quantity and have, in many cases, estimated this error from details given in the publications cited. The most striking result in Table 3 is that metals give a ratio of less than unity, and this is probably not an artifact of the data in view of the values found for the other groups.

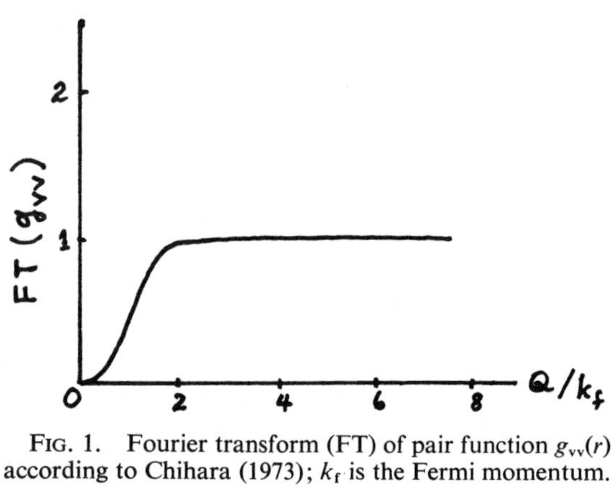

FIG. 1. Fourier transform (FT) of pair function $g_{vv}(r)$ according to Chihara (1973); k_f is the Fermi momentum.

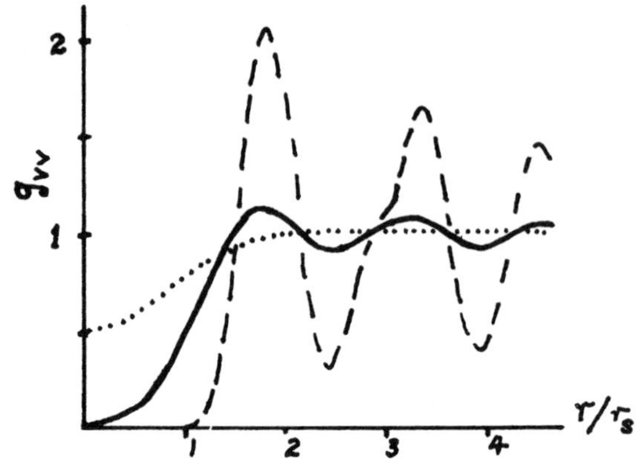

FIG. 2. Schematic form of electronic pair function $g_{vv}(r)$ according to March and Young (1959) in 'jellium'. The dotted curve is the Fermi hole function, and the dashed curve is the Wigner lattice function in the extreme low density limit (r_s = 100 Bohr radii). The solid curve is an interpolation suggested by the Wigner lattice model, for r_s = 4 Bohr radii. When damped somewhat at large r, to lose the long range order, this should resemble the pair function of an 'electron liquid' in alkali metals.

TABLE 3. Summary of intensity ratios

Group	Neutron peak height / X-ray peak height
Liquified rare gases	0.98 ± 0.03
Molecular fluids	1.12 ± 0.04
Monovalent metals	0.87 ± 0.04
Polyvalent metals	0.92 ± 0.03

With suitable models and the experimental data summarized in Table 3, we have attempted to demonstrate (*i*) the close approximation of atoms in rare gas liquids to free atoms in the vapor, (*ii*) the delocalization of electrons by chemical bonding in molecular liquids, and (*iii*) the existence of short range order in the conduction electrons in liquid metals, which appears to have a range greater than that for the ions.

Many experiments may be suggested in order to test these conclusions further and to provide detail concerning electronic correlations. Of these perhaps the most urgent are improved neutron experiments on sodium and potassium to confirm the data on these elements in Table 2. Electron scattering experiments on these would also be fruitful in allowing S_{vv} to be extracted and the structure of the electron liquid to be studied (although we recognize the inherent difficulties of electron diffraction experiments). Further work on molecular systems using all three techniques, especially where dissociation is occurring, may be of great interest too.

The ideas and results discussed in this paper have reconciled several outstanding experimental questions concerning the meaning of (apparently) discordant diffraction data, and have treated three different types of liquids on a common basis.

Acknowledgments

Two of us (N.H.M. and N.C.M.) would like to thank the Department of Physics, University of Guelph, for their hospitality during the period when this work was carried out. This research was supported by a grant from the National Research Council of Canada.

CHIHARA, J. 1973. *In* The properties of liquid metals, *ed. by* S. Takeuchi. Proceedings of Second International Conference, Tokyo (Taylor and Francis, London), p. 137.

COWAN, R. D. and KIRKWOOD, J. G. 1958. J. Chem. Phys. **29**, 264.

EGELSTAFF, P. A. 1967*a*. *In* The properties of liquid metals, *ed. by* P. D. Adams, H. A. Davies, and S. G. Epstein, Proceedings of the International Conference, Upton, New York (Taylor and Francis, London), p. 147.

——— 1967*b*. An introduction to the liquid state (Academic Press, London), chap. 6.

GASKELL, T. 1962. Proc. Phys. Soc. **80**, 1091.

HARTREE, D. R., HARTREE, W., and SWIRLES, B. 1939. Phil. Trans. R. Soc. **238**, 229.

HERZBERG, G. 1939. Molecular spectra and molecular structure (Prentice Hall, New York).

MARCH, N. H. and YOUNG, W. H. 1959. Philos. Mag. **4**, 384.

MARCH, N. H. and TOSI, M. P. 1973. Ann. Phys. **81**, 414.

PIRENNE, M. H. 1946. The diffraction of X-rays and electrons by free molecules (Cambridge University Press, London), chap. 9.

SINGWI, K. S., SJÖLANDER, A., TOSI, M. P., and LAND, R. H. 1970. Phys. Rev. B, **1**, 1044.

TOSI, M. P. and MARCH, N. H. 1973. Nuovo Cimento, B, **15**, 308

WAHL, A. C. 1966. Science, **151**, 961.

WALTER, J. P., FONG, C. Y., and COHEN, M. L. 1973. Solid State Commun. **12**, 303.

WATABE, M. and HASEGAWA, M. 1973. *In* The properties of liquid metals, *ed. by* S. Takeuchi, Proceedings of Second International Conference, Tokyo (Taylor and Francis, London), p. 133.

J. Phys. F: Metal Phys., Vol. 6, No. 5, 1976. Printed in Great Britain.

Electron–electron pair correlation function in solid and molten nearly-free electron metals

S Cusack, N H March, M Parrinello† and M P Tosi‡
The Blackett Laboratory, Department of Physics, Imperial College, London SW7 2B7

Received 22 September 1975 in final form 15 December 1975

Abstract. Scattering experiments on molten metals now afford the possibility of extracting valence electron correlation functions. Therefore, the theory of the pair correlation function between valence electrons in the presence of a weak electron–ion interaction v_{ie} is developed in this paper. To obtain explicit results to second-order in v_{ie}, we have used two different approaches: (i) diagrammatic analysis and (ii) density functional theory, in which the change in the electron–electron pair correlation function from its value in jellium is expressed in terms of the electron density change, for a fixed configuration of ions.

While the theory is applicable both to crystalline and molten nearly-free electron metals, the theory has been pushed through further for liquid metals, and in particular the most extensive numerical results have been obtained for liquid Na. The corrections from v_{ie} to the jellium pair function show some structure out to $k \sim k_f$, the Fermi momentum, in both the approximations (i) and (ii) employed. But for Na, the most important conclusion is that the corrections are small. Therefore, scattering experiments on liquid Na will, essentially, measure the electronic pair function in jellium at this density. Calculations for liquid Al show, however, that the corrections due to electron–ion interaction are substantially greater than for Na. Some comments are made relating to the x ray inelastic scattering experiments of Eisenberger and Platzman on crystalline metals.

1. Introduction

Egelstaff *et al* (1974) have recently proposed a method by which valence electron correlation functions in a pure liquid metal can be extracted from experiment. Given a model for the core electrons, what is required is to combine scattering data from x ray, neutron and electron diffraction experiments.

Therefore, in this paper, we give the theory of the valence electron pair correlation function in a nearly-free electron metal to second-order in the electron–ion interaction v_{ie}. While the theory is applicable to both crystalline and molten metals, our main concern in this paper is to press the theory through to numerical evaluation of the pair correlation function for the valence electrons in liquid Na. Some numerical results for molten Al are more briefly discussed.

In addition, the second-order term in the electron–ion correlation function is also given, the theory to first-order having been worked out earlier by two of us (Tosi and March 1973).

† Now at Department of Physics, University of Messina, Italy.
‡ Now at Department of Physics, University of Rome, Italy.

It would clearly also be of interest to complete the numerical calculations for crystalline metals. Such results could then be compared with those derived from scattering experiments on crystals, using the same principle as that proposed by Egelstaff *et al* (1974) for liquid metals. However, as will be seen below, the calculations for liquid metals are simpler, as only the diagonal elements of the response functions entering the theory are required. This is a consequence of the *constant* ensemble averaged valence electron density in liquid metals. In crystalline metals, the periodic electron density means that off-diagonal elements would have to be considered in order to obtain a complete picture of the pair function.

Inevitably, approximations have to be made in order to obtain numerical estimates from the theory presented here. But for the two approaches we have explored: (i) diagrammatic analysis and (ii) density functional theory, our main conclusions turn out to be rather independent of the detailed approximations made.

2. Theory of electronic pair correlation function

Throughout the present discussion, we shall assume knowledge of the pair function $g(r)$ of interacting electrons in a uniform background of positive charge; the so-called jellium model. If ρ_0 is the uniform electron density in jellium, it is convenient to work with the quantity

$$P_c^0(\boldsymbol{r}_1\boldsymbol{r}_2) = \rho_0^2[g(|\boldsymbol{r}_1 - \boldsymbol{r}_2|) - 1] \tag{2.1}$$

as the unperturbed pair function.

Then, in the presence of granular ions, there will be a spatially varying electron density, $\rho(\boldsymbol{r})$ say at position $\boldsymbol{r}$, and we shall work with the generalization of equation (2.1), namely

$$P_c(\boldsymbol{r}_1\boldsymbol{r}_2) = \langle\rho(\boldsymbol{r}_1)\,\rho(\boldsymbol{r}_2)\rangle - \langle\rho(\boldsymbol{r}_1)\rangle\,\langle\rho(\boldsymbol{r}_2)\rangle \equiv \langle\rho(\boldsymbol{r}_1)\,\rho(\boldsymbol{r}_2)\rangle_c \tag{2.2}$$

which is the density–density correlation function, minus the product of the one-electron densities. In equation (2.2) and elsewhere in the paper, $\langle\ \rangle_c$ denotes the cumulant average.

The aim of this work is now to calculate the change $\delta P_c(\boldsymbol{r}_1\boldsymbol{r}_2)$ from the jellium value (2.1) as we switch on a weak electron–ion interaction

$$U(\boldsymbol{r}) = \sum_l v_{ie}(\boldsymbol{r} - \boldsymbol{R}_l). \tag{2.3}$$

In writing down equation (2.3), it is assumed that we are dealing with a 'frozen-ion' configuration, defined by position vectors $\boldsymbol{R}_l$, the electron–ion interaction v_{ie} being a localized potential centred on each ionic site. Of course, for a crystalline metal the $\{\boldsymbol{R}_l\}$ will define the appropriate periodic lattice, but we take the $\{\boldsymbol{R}_l\}$ to be a general configuration in order to include the molten metal case. Naturally, in this latter case, we shall eventually have to carry out an ensemble average.

For the frozen-ion configuration represented by $U(\boldsymbol{r})$ in equation (2.3), it is clear that we can write formally, to second-order in $U(r)$, that the change in the electronic pair function from its jellium value is given by

$$\delta P_c(\boldsymbol{r}_1\boldsymbol{r}_2) = \int \chi_3(\boldsymbol{r}_1\boldsymbol{r}_2;\boldsymbol{r}_3)\,U(\boldsymbol{r}_3)\,d\boldsymbol{r}_3 + \tfrac{1}{2}\int \chi_4(\boldsymbol{r}_1\boldsymbol{r}_2;\boldsymbol{r}_3\boldsymbol{r}_4)\,U(\boldsymbol{r}_3)\,U(\boldsymbol{r}_4)\,d\boldsymbol{r}_3\,d\boldsymbol{r}_4. \tag{2.4}$$

The basic response functions χ_3 and χ_4 are at the heart of the theory and, it must be emphasized, they are properties of jellium. It is shown in Appendix A that these response functions are time-ordered density correlation functions, being respectively third and fourth order in the time-dependent density $\rho(\boldsymbol{r}t)$. Quite explicitly, they are given by

$$\chi_3(\boldsymbol{r}_1\boldsymbol{r}_2;\boldsymbol{r}_3) = (-\mathrm{i}/\hbar)\int \langle T[\rho(\boldsymbol{r}_1 0)\,\rho(\boldsymbol{r}_2 0)\,\rho(\boldsymbol{r}_3 t_3)]\rangle_c\,\mathrm{d}t_3 \tag{2.5}$$

and

$$\chi_4(\boldsymbol{r}_1\boldsymbol{r}_2:\boldsymbol{r}_3\boldsymbol{r}_4) = (-\mathrm{i}/\hbar)^2\iint \langle T[\rho(\boldsymbol{r}_1 0)\,\rho(\boldsymbol{r}_2 0)\,\rho(\boldsymbol{r}_3 t_3)\,\rho(\boldsymbol{r}_4 t_4)]\rangle_c\,\mathrm{d}t_3\,\mathrm{d}t_4. \tag{2.6}$$

Even though these response functions are properties of jellium, it is clearly not possible to calculate them exactly. Therefore, to motivate the first of the approximations we shall use in this paper, we re-express the change in the pair function, $\delta P_c(\boldsymbol{r}_1\boldsymbol{r}_2)$, due to the electron–ion interaction, as a functional of the corresponding change in the electron density, $\delta\rho(\boldsymbol{r}) = \rho(\boldsymbol{r}) - \rho_0$.

3. Electronic pair function as functional of charge density

That ground-state properties of a many-electron system could be expressed as functionals of the charge density was proposed long ago by Thomas and Fermi, and their theory was formally completed by the work of Hohenberg and Kohn (1964, referred to below as HK). Furthermore, for a system of N electrons, the exact ground-state density $\rho(\boldsymbol{r})$ can be expressed as the sum of the squares of the lowest N normalized eigenfunctions of a one-body potential $V(\boldsymbol{r})$. Because of this latter circumstance, we can write two equations for δP_c, namely

$$\delta P_c(\boldsymbol{r}_1\boldsymbol{r}_2) = \int \chi_3^{\mathrm{HK}}(\boldsymbol{r}_1\boldsymbol{r}_2;\boldsymbol{r}_3)\,V(\boldsymbol{r}_3)\,\mathrm{d}\boldsymbol{r}_3 + \tfrac{1}{2}\int \chi_4^{\mathrm{HK}}(\boldsymbol{r}_1\boldsymbol{r}_2;\boldsymbol{r}_3\boldsymbol{r}_4)\,V(\boldsymbol{r}_3)\,V(\boldsymbol{r}_4)\,\mathrm{d}\boldsymbol{r}_3\,\mathrm{d}\boldsymbol{r}_4 \tag{3.1}$$

and

$$\delta P_c(\boldsymbol{r}_1\boldsymbol{r}_2) = \int H_3(\boldsymbol{r}_1\boldsymbol{r}_2;\boldsymbol{r}_3)\,\delta\rho(\boldsymbol{r}_3)\,\mathrm{d}\boldsymbol{r}_3 + \tfrac{1}{2}\int H_4(\boldsymbol{r}_1\boldsymbol{r}_2;\boldsymbol{r}_3\boldsymbol{r}_4)\,\delta\rho(\boldsymbol{r}_3)\,\delta\rho(\boldsymbol{r}_4)\,\mathrm{d}\boldsymbol{r}_3\,\mathrm{d}\boldsymbol{r}_4. \tag{3.2}$$

But now the next step is to express the density change $\delta\rho(\boldsymbol{r})$ in terms of V and of U by the two further expansions

$$\delta\rho(\boldsymbol{r}) = \int F_1(\boldsymbol{r}\boldsymbol{r}_1)\,V(\boldsymbol{r}_1)\,\mathrm{d}\boldsymbol{r}_1 + \tfrac{1}{2}\int F_2(\boldsymbol{r}\boldsymbol{r}_1\boldsymbol{r}_2)\,V(\boldsymbol{r}_1)\,V(\boldsymbol{r}_2)\,\mathrm{d}\boldsymbol{r}_1\,\mathrm{d}\boldsymbol{r}_2 \tag{3.3}$$

where F_1 and F_2 are the independent particle linear and quadratic density response functions, and

$$\delta\rho(\boldsymbol{r}) = \int \mathscr{F}_1(\boldsymbol{r}\boldsymbol{r}_1)\,U(\boldsymbol{r}_1)\,\mathrm{d}\boldsymbol{r}_1 + \tfrac{1}{2}\int \mathscr{F}_2(\boldsymbol{r}\boldsymbol{r}_1\boldsymbol{r}_2)\,U(\boldsymbol{r}_1)\,U(\boldsymbol{r}_2)\,\mathrm{d}\boldsymbol{r}_1\,\mathrm{d}\boldsymbol{r}_2 \tag{3.4}$$

where $\mathscr{F}_1$ and $\mathscr{F}_2$ are the interacting jellium density response functions. Equations (3.1) to (3.4) are formally exact to second order. Equations (3.3) and (3.4) can be inverted to yield $V(\boldsymbol{r})$ and $U(\boldsymbol{r})$ respectively in terms of the density change $\delta\rho(\boldsymbol{r})$. Hence, the basic response functions χ_3 and χ_4 introduced in equation (2.4) can be related to the χ^{HK}'s via the functions H_3 and H_4. In particular

$$H_3(\boldsymbol{r}_1\boldsymbol{r}_2;\boldsymbol{r}_3) = \int \chi_3(\boldsymbol{r}_1\boldsymbol{r}_2;\boldsymbol{r}_4)\mathscr{F}_1^{-1}(\boldsymbol{r}_4\boldsymbol{r}_3)\,\mathrm{d}\boldsymbol{r}_4 \tag{3.5}$$

or alternatively

$$H_3(\boldsymbol{r}_1\boldsymbol{r}_2;\boldsymbol{r}_3) = \int \chi_3^{HK}(\boldsymbol{r}_1\boldsymbol{r}_2;\boldsymbol{r}_4)F_1^{-1}(\boldsymbol{r}_4\boldsymbol{r}_3)\,\mathrm{d}\boldsymbol{r}_4. \tag{3.6}$$

Similarly for H_4 we have

$$H_4(\boldsymbol{r}_1\boldsymbol{r}_2;\boldsymbol{r}_3\boldsymbol{r}_4) = \int \chi_4^{HK}(\boldsymbol{r}_1\boldsymbol{r}_2;\boldsymbol{r}_5\boldsymbol{r}_4)F_1^{-1}(\boldsymbol{r}_5\boldsymbol{r}_3)F_1^{-1}(\boldsymbol{r}_6\boldsymbol{r}_4)\,\mathrm{d}\boldsymbol{r}_5\,\mathrm{d}\boldsymbol{r}_6 - \int \chi_3^{HK}(\boldsymbol{r}_1\boldsymbol{r}_2;\boldsymbol{r}_5)$$

$$\times F_1^{-1}(\boldsymbol{r}_5\boldsymbol{r}_6)F_2(\boldsymbol{r}_6\boldsymbol{r}_7\boldsymbol{r}_8)F_1^{-1}(\boldsymbol{r}_7\boldsymbol{r}_3)F_1^{-1}(\boldsymbol{r}_8\boldsymbol{r}_4)\,\mathrm{d}\boldsymbol{r}_5\,\mathrm{d}\boldsymbol{r}_6\,\mathrm{d}\boldsymbol{r}_7\,\mathrm{d}\boldsymbol{r}_8 \tag{3.7}$$

along with the corresponding expression involving the χ's and $\mathscr{F}$'s.

At this point, it is helpful to transform to Fourier space, and to define

$$H_4(\boldsymbol{k}, -(\boldsymbol{k}+\boldsymbol{G}); \boldsymbol{G}-\boldsymbol{G}', \boldsymbol{G}') = \int \exp[\mathrm{i}\boldsymbol{k}.(\boldsymbol{r}_1-\boldsymbol{r}_2)]\exp[\mathrm{i}\boldsymbol{G}.(\boldsymbol{r}_3-\boldsymbol{r}_2)]$$

$$\times \exp[\mathrm{i}\boldsymbol{G}'.(\boldsymbol{r}_4-\boldsymbol{r}_3)]\,H_4(\boldsymbol{r}_1\boldsymbol{r}_2;\boldsymbol{r}_3\boldsymbol{r}_4)\,\mathrm{d}(\boldsymbol{r}_1-\boldsymbol{r}_2)\,\mathrm{d}(\boldsymbol{r}_3-\boldsymbol{r}_2)\,\mathrm{d}(\boldsymbol{r}_4-\boldsymbol{r}_3) \tag{3.8}$$

and similarly for F_2, χ_3^{HK} and χ_4^{HK}. Then we can derive for χ_4 the expression

$$\chi_4(\boldsymbol{k}, -(\boldsymbol{k}+\boldsymbol{G}); \boldsymbol{G}-\boldsymbol{G}', \boldsymbol{G}') = H_3(\boldsymbol{k}, -(\boldsymbol{k}+\boldsymbol{G}); \boldsymbol{G})\mathscr{F}_2(\boldsymbol{G}, \boldsymbol{G}'-\boldsymbol{G})$$

$$+ H_4(\boldsymbol{k}, -(\boldsymbol{k}+\boldsymbol{G}); \boldsymbol{G}-\boldsymbol{G}', \boldsymbol{G}')\mathscr{F}_1(\boldsymbol{G}-\boldsymbol{G}')\mathscr{F}_1(\boldsymbol{G}'). \tag{3.9}$$

So far the development is formal and exact to second-order in the electron–ion perturbation. However, to proceed further, some approximation must now be introduced.

4. Use of independent-particle response functions

An approximation central to the later development on molten metals (see §5 below) is now to replace χ_3^{HK} and χ_4^{HK} by equations (2.5) and (2.6) evaluated for independent particles. We denote these approximations by χ_3^0 and χ_4^0 respectively. Equations (2.5) and (2.6) can then be calculated directly using Wick's theorem, or alternatively by using the perturbation series given by March and Murray (1961) for the Dirac density matrix $\gamma(\boldsymbol{r}_1\boldsymbol{r}_2)$. This latter approach is possible because

$$P_c^0(\boldsymbol{r}_1\boldsymbol{r}_2) = -\tfrac{1}{2}[\gamma(\boldsymbol{r}_1\boldsymbol{r}_2)]^2 \tag{4.1}$$

in the independent particle model and evidently therefore

$$\delta P_c^0(\boldsymbol{r}_1\boldsymbol{r}_2) = -\gamma^0(\boldsymbol{r}_1\boldsymbol{r}_2)\,\delta\gamma(\boldsymbol{r}_1\boldsymbol{r}_2) - \tfrac{1}{2}\delta\gamma(\boldsymbol{r}_1\boldsymbol{r}_2)\,\delta\gamma(\boldsymbol{r}_1\boldsymbol{r}_2). \tag{4.2}$$

In this way, explicit expressions have been obtained for the quantities χ_3^0, χ_4^0, F_1 and F_2 and these are recorded in Appendix C.

Having established these independent particle approximations to χ_3^0 etc, we have reached the point at which we can set out the two basic approximations to χ_4 which we shall use in our numerical calculations on liquid metals.

The independent particle results just discussed can be used in equations (3.6) and (3.7) to obtain approximations to H_3 and H_4. Inserting these results into equation (3.9) an approximation to χ_4 can thus be generated. This is the density functional approximation to χ_4 which we shall apply explicitly to the case of liquid metals below. Before doing this, we consider a second, alternative approximation, based on the diagrammatic analysis initiated by Brovman and Kagan (1967). This is discussed more fully in Appendix B and again leads to an approximation for the response function χ_4 in terms of the independent particle response function χ_4^0. Explicit forms will be written down below for a restricted part (diagonal element) of χ_4, which is, as we shall see, all that is involved in calculating the pair function for molten metals. We must now turn to discuss this case, which, of course, will involve carrying out an ensemble average.

5. Ensemble average for molten metals

As shown in Appendix A, when allowance is made for ionic motions, additional terms involving ion–ion correlations enter equation (2.4). Using the adiabatic approximation, the time dependence in these terms can be simplified to a manageable form. In the case of the liquid metal it is then necessary to ensemble average over all possible ionic configurations. As a result, the first-order term vanishes and the second order terms that remain, involve the ion–ion structure factor $S_{ii}(q)$, directly accessible to experiment through neutron scattering. The resultant expression for the change in the valence electron pair function to second order in v_{ie} in a liquid metal is:

$$\delta P_e(\boldsymbol{r}_1, \boldsymbol{r}_2) = \tfrac{1}{2}\int \chi_4(\boldsymbol{r}_1\boldsymbol{r}_2; \boldsymbol{r}_3\boldsymbol{r}_4)\,\rho_i^2\, S_{ii}(\boldsymbol{x}-\boldsymbol{y})\, v_{ie}(\boldsymbol{r}_3-\boldsymbol{x})\, v_{ie}(\boldsymbol{r}_4-\boldsymbol{y})\, d\boldsymbol{r}_3\, d\boldsymbol{r}_4\, d\boldsymbol{x}\, d\boldsymbol{y}$$

$$+\int \mathscr{F}_1(\boldsymbol{r}_1\boldsymbol{r}_3)\, v_{ie}(\boldsymbol{r}_3-\boldsymbol{x})\,\rho_i^2\, S_{ii}(\boldsymbol{x}-\boldsymbol{y})\,\mathscr{F}_1(\boldsymbol{r}_3\boldsymbol{r}_4)\, v_{ie}(\boldsymbol{r}_4-\boldsymbol{y})\, d\boldsymbol{r}_3\, d\boldsymbol{r}_4\, d\boldsymbol{x}\, d\boldsymbol{y} \tag{5.1}$$

where ρ_i is the mean ionic density.

In Fourier transform, equation (5.1) becomes, in the case of a molten metal with valency z

$$\delta S(k) = (1/2z)\int S_{ii}(G')|v_{ie}(G')|^2\,\chi_4(\boldsymbol{k}, -\boldsymbol{k}; -\boldsymbol{G}', \boldsymbol{G}')\,[d\boldsymbol{G}'/(2\pi)^3]$$

$$+ (1/z)\, S_{ii}(k)|v_{ie}(k)|^2\,|\mathscr{F}_1(\boldsymbol{k})|^2. \tag{5.2}$$

Several authors (Watabe and Hasegawa 1973, Chihara 1973: see also March and Tosi 1973) have shown that charge neutrality in a liquid metal implies the limit

$$\lim_{k\to 0} \delta S(k) = z S_{ii}(0) \tag{5.3}$$

since the electron–electron structure factor in the long wavelength limit, $S(0)$, is zero for jellium. But it is straightforward to show that

$$\lim_{k\to 0} |\mathscr{F}_1(k)\, v_{ic}(k)|^2 = z^2 \tag{5.4}$$

when $v_{ic}(k) \sim -4\pi z/k^2$ for small k. Hence the second term in equation (5.2) yields the charge neutrality condition (5.3) which clearly must imply that

$$\lim_{k\to 0} \chi_4(\boldsymbol{k}, -\boldsymbol{k}, -\boldsymbol{G}', \boldsymbol{G}') = 0 \tag{5.5}$$

for the exact χ_4†.

Equation (5.2) shows, as we stressed earlier, that the whole of the response function χ_4 is *not* required to get the change in the pair function in a molten nearly-free electron metal; only the diagonal part $\boldsymbol{G} = 0$ of the full χ_4 defined in equation (3.9)‡. This is the reason we have deferred writing down our two different approximations to χ_4 until this point.

5.1. Approximate forms of χ_4 on diagonal

Returning to the discussion of §4.1, we notice that getting χ_4 for $\boldsymbol{G} = 0$ is equivalent to evaluating $\delta P_c(\boldsymbol{k}, -\boldsymbol{k})$. The first term on the right-hand side of equation (3.9) is zero in this case since

$$\lim_{G\to 0} \mathscr{F}_2(\boldsymbol{G}, \boldsymbol{G} - \boldsymbol{G}') = 0.$$

This latter condition is simply a statement that there is no second-order change in the average density. Collecting the results together we arrive at one of the two basic approximations for χ_4 employed here: the density functional approximation given by

$$\chi_4(\boldsymbol{k}, -\boldsymbol{k}; -\boldsymbol{G}', \boldsymbol{G}') \simeq \left(\frac{\mathscr{F}_1(\boldsymbol{G}')}{F_1(\boldsymbol{G}')}\right)^2 \left(\chi_4^0(\boldsymbol{k}, -\boldsymbol{k}, -\boldsymbol{G}', \boldsymbol{G}') - \chi_3^0(\boldsymbol{k}, -\boldsymbol{k}; 0)\,\frac{F_2(0, \boldsymbol{G}')}{F_1(0)}\right). \tag{5.6}$$

Evidently, all the quantities on the right-hand side of equation (5.6) are known, being independent particle objects; except $\mathscr{F}_1$. But this is related to the static dielectric function $\epsilon(G')$ and the proper polarization $\pi(G')$ for jellium by

$$\mathscr{F}_1(G') = \pi(G')/\epsilon(G') \tag{5.7}$$

with

$$\epsilon(G') = 1 - (4\pi/G'^2)\,\pi(G'). \tag{5.8}$$

At this point, we shall also write down the alternative approximation to the diagonal part of χ_4, as given by diagrammatic analysis.

As discussed in Appendix B, this can be written

$$\chi_4(\boldsymbol{k}, -\boldsymbol{k}; -\boldsymbol{G}', \boldsymbol{G}') \simeq \epsilon^{-2}(k)\epsilon^{-2}(G')\,\chi_4^0(\boldsymbol{k}, -\boldsymbol{k}, -\boldsymbol{G}', -\boldsymbol{G}'). \tag{5.9}$$

Both (5.6) and (5.9) have been used in the numerical work reported below on liquid Na and Al. One property of equations (5.6) and (5.9) that is important for our present

† However, we have not yet seen how to give a direct proof of this from equation (2.6).

‡ Though this is all that is required to calculate the change in the structure factor in molten metals, in crystalline problems one can expect off-diagonal terms to play a significant role.

purposes is that they obey the limiting condition (5.5). This follows directly from equation (5.9) because of the small k behaviour of $\epsilon(k)$. For the approximation (5.6) the argument is less direct but we shall now demonstrate that the same limit is obeyed. Because of the idempotency of the Dirac density matrix $\gamma(\boldsymbol{r}_1\boldsymbol{r}_2)$ it follows from equation (4.1) that

$$\int \delta P_{\rm c}^{0}(\boldsymbol{r}_1\boldsymbol{r}_2)\,\mathrm{d}\boldsymbol{r}_1 = -\delta\rho(\boldsymbol{r}_2). \tag{5.10}$$

Using equation (3.2) this implies that

$$\int H_4^{0}(\boldsymbol{r}_1\boldsymbol{r}_2\,;\boldsymbol{r}_3\boldsymbol{r}_4)\,\mathrm{d}\boldsymbol{r}_1 = 0 \tag{5.11}$$

or

$$\lim_{k\to 0} H_4^{0}(\boldsymbol{k}, -(\boldsymbol{k}+\boldsymbol{G}); \boldsymbol{G}-\boldsymbol{G}', \boldsymbol{G}') = 0. \tag{5.12}$$

Thus from equation (3.9), when $\boldsymbol{G}$ is zero, we are led to equation (5.5) also for the density functional approximation to χ_4. The importance of this limit for the case of liquid metals should be clear from the discussion at the beginning of this section.

6. Change in $S(k)$ for molten Na, due to electron–ion interaction

Figure 1 shows numerical results for $\delta S(k)$ for Na, obtained from equation (5.2). As input data we have used:

(i) The Ashcroft (1966) empty-core pseudopotential with core radius 1·67 atomic units.

(ii) The dielectric function of Singwi *et al* (1968).

The upper curve in figure 1 shows the second term on the RHS of equation (5.2) as a function of the wavenumber k (in units of the Fermi wavenumber $k_{\rm F}$). It is clear that this contribution to $\delta S(k)$ is local in k space, and the structure in it is easily associated with structure in $S_{\rm ii}(k)$ and $v_{\rm ie}(k)$. The position of the principal peak in $S_{\rm ii}(k)$ is marked, just beyond $2k_{\rm F}$. But $|v_{\rm ie}(k)|^2$ has become very small at this wavenumber, so that only a small contribution to $\delta S(k)$ remains ($\sim 0{\cdot}004$, compared to the jellium contribution ~ 1). The peak at $k \sim 1{\cdot}6\,k_{\rm F}$ reflects the low k side of the principal peak in $S_{\rm ii}(k)$, but is rapidly reduced to zero by the node in the pseudopotential at $k \sim 2k_{\rm F}$.

The two curves entirely below the axis in figure 1 show the calculated results for the other term in equation (5.2) using both the density functional and diagrammatic approximations to χ_4†. The non-locality of this term leads to a rather structureless contribution to $\delta S(k)$, with reasonable agreement between the two approximations except at small k. Here the $\epsilon^{-2}(k)$ factor in the diagrammatic analysis approximation gives rise to a much faster (k^4) approach to zero, a screening feature not present in the density functional approximation. However at the end of Appendix C it is shown that in other respects the density functional approximation is more satisfactory. The most important fact to emerge from calculation of this term however, is that it is small.

† The computer programme using the diagrammatic approximation was also run for Na using the Lindhard dielectric function instead of that of Singwi *et al*, this perhaps being a more internally consistent procedure. The results in this case, though sometimes as much as a factor of two larger than the results using the Singwi *et al* $\epsilon(k)$, for certain k values $\sim k_{\rm F}$, still give a very small contribution to $\delta S_{\rm ee}(k)$.

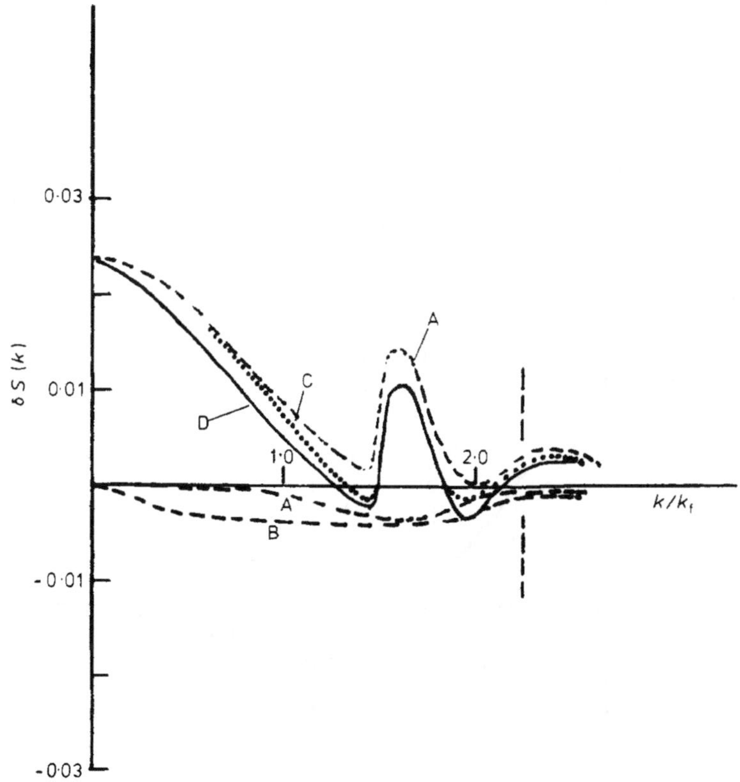

Figure 1. Change in $S(k)$ for liquid Na. Input data is Ashcroft pseudopotential, ionic structure factor and Singwi *et al* dielectric function. Upper curve A, contribution from charge neutrality term; lower curve A, contribution from χ_4 term using diagrammatic approximation; lower curve B, contribution from χ_4 term using density functional approximation; curve C, $\delta S(k)$ from diagrammatic approximation and curve D, $\delta S(k)$ from diagrammatic approximation. The most important point to note is that these contributions are never greater than about 0·03. The vertical dashed line denotes the position of principal peak in ion–ion structure factor S_{ii}.

Also in figure 1 are shown the two approximations to the total change in $S(k)$; that is, the sum of the local charge neutrality term and the non-local χ_4 term.

The conclusion seems quite clear for liquid Na. Except for small k, where the major contribution to $\delta S(k)$ comes from the charge neutrality term, the corrections due to ion–electron interaction outside $k \sim k_f$ are less than 2% for Na. Thus, if the proposal of Egelstaff *et al* (1974) can be implemented for liquid Na, then we shall essentially have a measure of the electronic structure factor of *jellium* at the Na density.

In view of the dependence of our results on χ_4, it is clear that some uncertainties must exist as soon as the electron–ion contributions become comparable with the jellium structure factor. It is also true that there must be corrections arising from the use of pseudopotentials and pseudo-wavefunctions. While such corrections are difficult to make in the present context, there is a whole body of evidence pointing to the smallness of such corrections in Na.

However, both sources of error referred to above, those in approximating χ_4, and those involved in pseudizing the problem, will be more serious in Al than in Na.

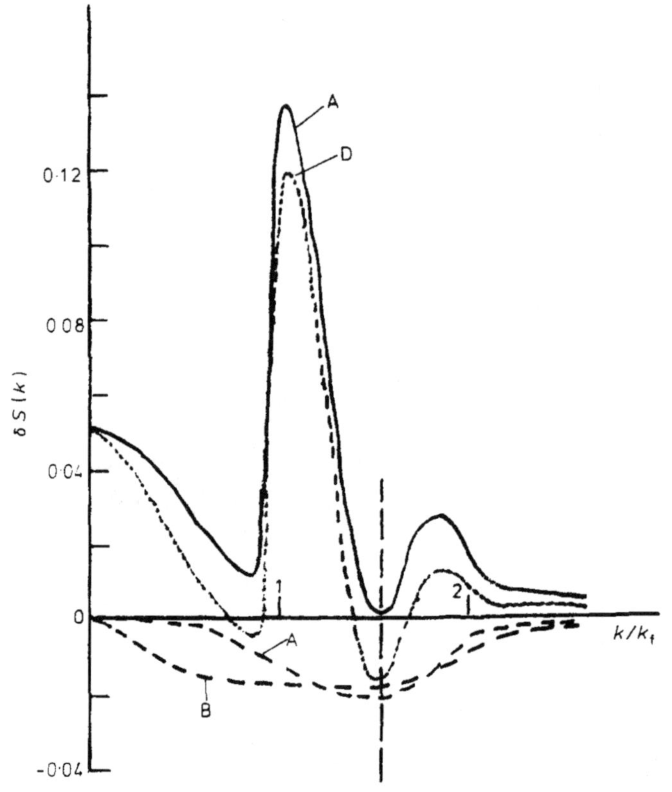

Figure 2. Change in $S(k)$ for liquid Al. Ashcroft pseudopotential with core radius 1·12 atomic units was used. Curves as in figure 1, except that the total diagrammatic approximation curve C has been omitted for clarity of diagram.

Therefore, though we think it of interest to report calculations in Al (see figure 2), we urge caution in any quantitative use of them. For Al, the pseudopotential core radius used was 1·12 atomic units. The general features of the results can be explained in a manner similar to those for Na as above. However, in the case of Al, the corrections to the jellium structure factor are much larger (up to 15%). The strong peak at $1{\cdot}1k_f$, which we stress is arising from the principal peak in S_{ii}, before this is annulled by the pseudopotential, should be detectable by experiment.

It is interesting to note that the exact four-body response function χ_4 will have a complicated singularity structure. Therefore, on taking the Fourier transform of $\delta S(k)$ to obtain the change in the electronic pair function in r space, we can expect these singularities to give rise to long-range oscillatory behaviour. We have not pursued this in detail, though the work of Brovman and Kagan (1974) on the singularity structure of closely related response functions will be useful in any future work on this point. Simple arguments, based on the form of χ_4^0, suggest that long-range oscillatory behaviour with wavelength π/k_f is to be expected in the electronic pair function. However, a complete treatment for liquid metals would have to take account of the electronic mean free path. This might be anticipated to introduce an exponential damping factor of the form $\exp(-\Delta kr)$, where Δk is a measure of the blurring of the Fermi surface and will be inversely proportional to the electronic mean free path.

7. Second-order theory of electron–ion correlation function

With a view to the analysis of the scattering experiments on molten metals when they become available, one will require not only $S(k)$ for electrons but $S_{ie}(k)$ for electron–ion correlations. Therefore we shall briefly summarize here the theory of S_{ie} to second-order in v_{ie}, the first-order theory being already available (Tosi and March 1973).

Using the functional derivative methods of Appendix A, one can show that to second-order in v_{ie} the quantity $\langle\rho_e(\mathbf{r}_1)\,\rho_i(\mathbf{r}_2)\rangle_c$ is given for liquid metals by

$$\langle\rho_e(\mathbf{r}_1)\,\rho_i(\mathbf{r}_1)\rangle = -(\mathrm{i}/\hbar)\int\langle T[\rho_e(\mathbf{r}_1 0)\,\rho_e(\mathbf{x}\,t_1)]\rangle_c\langle T[\rho_i(\mathbf{y}t_1)\,\rho_i(\mathbf{r}_2 0)]\rangle_c$$
$$\times\, v_{ie}(\mathbf{x}-\mathbf{y})\,\mathrm{d}\mathbf{x}\,\mathrm{d}\mathbf{y}\,\mathrm{d}t_1 + \tfrac{1}{2}(-\mathrm{i}/\hbar)^2\iint\langle T[\rho_e(\mathbf{r}_1 0)\,\rho_e(\mathbf{x}_1 t_1)\,\rho_e(\mathbf{x}_2 t_2)]\rangle_c$$
$$\times\,\langle T[\rho_i(\mathbf{r}_2 0)\rho_i(\mathbf{y}_1 t_1)\rho_e(\mathbf{y}_2 t_2)]\rangle_c v_{ie}(\mathbf{x}_1-\mathbf{y}_1)v_{ie}(\mathbf{x}_2-\mathbf{y}_2)$$
$$\times\,\mathrm{d}\mathbf{x}_1\,\mathrm{d}\mathbf{x}_2\,\mathrm{d}\mathbf{y}_1\,\mathrm{d}\mathbf{y}_2\,\mathrm{d}t_1\,\mathrm{d}t_2. \tag{7.1}$$

If we make the adiabatic approximation, by putting all times in the ionic correlation functions the same and then integrating over the electronic times we obtain

$$\langle\rho_e(\mathbf{r}_1)\,\rho_i(\mathbf{r}_2)\rangle = \int\mathscr{F}_1(\mathbf{r}_1,\mathbf{x})\,\rho_i^2 S_{ii}(\mathbf{y}-\mathbf{r}_2)\,v_{ie}(\mathbf{x}-\mathbf{y})\,\mathrm{d}\mathbf{x}\,\mathrm{d}\mathbf{y} + \tfrac{1}{2}\iint\mathscr{F}_2(\mathbf{r}_1,\mathbf{x}_1,\mathbf{x}_2)$$
$$\times\,\rho_i^3 S_{iii}(\mathbf{r}_2,\mathbf{y}_1,\mathbf{y}_2)\,v_{ie}(\mathbf{x}_1-\mathbf{y}_1)\,v_{ie}(\mathbf{x}_2-\mathbf{y}_2)\,\mathrm{d}\mathbf{x}_1\,\mathrm{d}\mathbf{x}_2\,\mathrm{d}\mathbf{y}_1\,\mathrm{d}\mathbf{y}_2 \tag{7.2}$$

where $\mathscr{F}_1$ and $\mathscr{F}_2$ are the linear and quadratic density response functions for jellium.

Taking the Fourier transform of this equation gives

$$S_{ie}(k) = \frac{\mathscr{F}_1(k)\,S_{ii}(k)\,v_{ie}(k)}{z^{1/2}} + \tfrac{1}{2}\int S_{iii}(\mathbf{k},\mathbf{G}-\mathbf{k},-\mathbf{G})\,\mathscr{F}_2(-\mathbf{k},-\mathbf{G}+\mathbf{k})$$
$$\times\, v_{ie}(\mathbf{G})\,v_{ie}(\mathbf{k}-\mathbf{G})\,[\mathrm{d}\mathbf{G}/(2\pi)^3]. \tag{7.3}$$

It is clear that the first-order term gives the correct charge neutrality limit

$$\lim_{k\to 0}\Delta S_{ie}(k) = \lim\frac{\mathscr{F}_1(k)\,v_{ie}(k)\,S_{ii}(k)}{z^{1/2}} = z^{1/2}S_{ii}(0) \tag{7.4}$$

provided the second-order term is zero in this limit. That this is true follows from Appendix C, equation (C.8) where we have

$$\lim_{k\to 0}\mathscr{F}_2(\mathbf{k},\mathbf{G}-\mathbf{k}) = \lim_{k\to 0}\mathscr{F}_1(k)\,\frac{\partial\,\mathscr{F}_1(G)}{\partial\rho} = 0. \tag{7.5}$$

The simplest way to make a calculation of the second-order term is

(i) use the Kirkwood approximation for S_{iii},
(ii) use the diagrammatic approximation for $\mathscr{F}_2$, that is

$$\mathscr{F}_2(\mathbf{k},\mathbf{G}-\mathbf{k}) = \frac{F_2(\mathbf{k},\mathbf{G}-\mathbf{k})}{\epsilon(k)\epsilon(\mathbf{G}-\mathbf{k})} \tag{7.6}$$

where F_2 is given by the Fourier transform of equation (C.18).

8. X ray scattering experiments

Though the detailed corrections to $S_{ee}(k)$ depend to a small extent on the way in which we approximate, the calculations reported here can leave little doubt that the electron–ion interaction in liquid Na is leading to corrections to $S_{ee}(k)$ of at most 2 or 3% at any wavenumber k.

In the light of this conclusion, we have therefore calculated the x ray scattering from Na to compare with the experiments of Greenfield and Wiser (1971). The procedure adopted is that discussed by Egelstaff *et al* (1974). The ionic form factor $f(k)$ has been taken from the work of Freeman (1959). For $S(\equiv S_{ee})$ and S_{ie} the data of Singwi *et al* (1968) and Chihara (1973) have been used. With these assumptions, the x ray scattering comes out to be very close to the neutron scattering.

The experiments indicate a sharper and narrower peak than the theory. The x ray scattering experiments are quite recent and of high accuracy. Clearly the neutron data needs refining, but unless substantial changes occur in this process, the conclusion drawn by Egelstaff *et al* is confirmed, namely that there appears to be somewhat more ordering of the electrons than the ions in liquid Na.

Furthermore, the present theory of the electronic pair function leaves little doubt that such ordering would have to be present in jellium; the contribution of the electron–ion interaction being but a few per cent in liquid Na.

This last conclusion, though drawn specifically for molten metals, is pointing in the same direction as the independent studies of Eisenberger and Platzman (1974) on crystalline metals. They appear to observe substantial departures from the Singwi *et al* structure factors for jellium, which are, it seems, arising from electron–electron interactions.

9. Summary

In the presence of weak electron–ion interactions, a theory of the electronic structure factor has been constructed. In a crystalline metal, equation (2.4) is the fundamental result, to second-order in v_{ie}. For liquid metals, the central result is equation (5.2).

Practical evaluation of equation (5.2) has been restricted to molten Na in detail, though some calculations have also been made for liquid Al. Two approximate forms of the jellium response function entering equation (5.2) have been studied, one from the diagrammatic analysis (equation 5.9) and one from density functional theory (equation 5.6). Both results lead back precisely to the charge neutrality conditions in the long wavelength limit. They do, however, lead to somewhat different numerical values away from $k = 0$. But for Na, the important conclusion is that the corrections from the electron–ion interaction are small. Therefore combining x ray, neutron and electron data† for molten Na ought to lead directly to the *jellium* electronic pair function at this density.

From preliminary calculations using neutron data, to estimate the x ray scattering, it seems that current approximations to the jellium pair function are not adequate in Na. But it must be stressed that this latter conclusion depends on the accuracy of the neutron data for Na, and it is obviously of importance that a really accurate neutron measurement on molten Na is now carried out.

† Dr P Dobson in this Department, in conjunction with Dr B Unvala in Metallurgy, is making an electron diffraction study currently on various liquid metals.

Acknowledgments

MPT wishes to acknowledge the award of a Senior Visiting Fellowship by the Science Research Council (SRC).

SC held a SRC postgraduate studentship during the period in which this work was carried out.

Finally thanks are due to the referees for detailed comments which have improved the presentation of the paper.

Appendix A

A.1. Functional derivative method to obtain response functions χ_3 and χ_4

Here the equations for χ_3 and χ_4 will be derived using the functional derivative technique of Kadanoff and Baym (1962). We switch on to the unperturbed jellium system, an electron–ion interaction of the form

$$H_{\text{int}}(t) = \int v_{\text{ie}}(\boldsymbol{x} - \boldsymbol{y})\, \rho_{\text{e}}(\boldsymbol{x}t)\, \rho_{\text{i}}(\boldsymbol{y}t)\, \mathrm{d}\boldsymbol{x}\, \mathrm{d}\boldsymbol{y}. \tag{A.1}$$

The aim is to generate a perturbation expansion for $\langle T[\rho_{\text{e}}(\boldsymbol{r}_1 t_1)\, \rho_{\text{e}}(\boldsymbol{r}_2 t_2)]\rangle_{\text{c}}$, from which the pair function $P_{\text{e}}(\boldsymbol{r}_1 \boldsymbol{r}_2)$ can be derived by putting $t_1 = t_2$. In equation (A.1), we can regard the electron–ion interaction $v_{\text{ie}}(\boldsymbol{x} - \boldsymbol{y})$ as an external potential coupling to $\rho_{\text{e}}(\boldsymbol{x}t)\, \rho_{\text{i}}(\boldsymbol{y}t)$ and we can then apply the functional derivative method to give, to first order

$$\delta^{(1)}\langle T[\rho_{\text{e}}(\boldsymbol{r}_1 t_1)\, \rho_{\text{e}}(\boldsymbol{r}_2 t_2)]\rangle_{\text{c}} = -(\mathrm{i}/\hbar) \int \langle T[\rho_{\text{e}}(\boldsymbol{r}_1 t_1)\, \rho_{\text{e}}(\boldsymbol{r}_2 t_2)\, \overline{\rho_{\text{e}}(\boldsymbol{x}t)\, \rho_{\text{i}}(\boldsymbol{y}t)}]\rangle_{\text{c}}\, v_{\text{ie}}(\boldsymbol{x} - \boldsymbol{y})\, \mathrm{d}\boldsymbol{x}\, \mathrm{d}\boldsymbol{y}\, \mathrm{d}t \tag{A.2}$$

where, in taking a cumulant average, $\overline{\rho_{\text{e}} \rho_{\text{i}}}$ must *not* be separated.

In the unperturbed system, the ions and electrons do not interact, and therefore the response function in equation (A.2) may be decoupled to yield

$$\delta^{(1)}\langle T[\rho_{\text{e}}(\boldsymbol{r}_1 t_1)\, \rho_{\text{e}}(\boldsymbol{r}_2 t_2)]\rangle_{\text{c}} = -(\mathrm{i}/\hbar) \int \langle T[\rho_{\text{e}}(\boldsymbol{r}_1 t_1)\, \rho_{\text{e}}(\boldsymbol{r}_2 t_2)\, \rho_{\text{e}}(\boldsymbol{x}t)]\rangle_{\text{c}} \langle \rho_{\text{i}}(\boldsymbol{y}t)\rangle\, v_{\text{ie}}(\boldsymbol{x} - \boldsymbol{y})\, \mathrm{d}\boldsymbol{x}\, \mathrm{d}\boldsymbol{y}\, \mathrm{d}t. \tag{A.3}$$

In second order, we obtain in a similar manner the result

$$\delta^{(2)}\langle T[\rho_{\text{e}}(\boldsymbol{r}_1 t_1)\, \rho_{\text{e}}(\boldsymbol{r}_2 t_2)]\rangle_{\text{c}} = \tfrac{1}{2}[-(\mathrm{i}/\hbar)]^2 \int \langle T[\rho_{\text{e}}(\boldsymbol{r}_1 t_1)\, \rho_{\text{e}}(\boldsymbol{r}_2 t_2)\, \overline{\rho_{\text{e}}(\boldsymbol{x}t)\, \rho_{\text{i}}(\boldsymbol{y}t)} \times \overline{\rho_{\text{e}}(\boldsymbol{x}'t')\, \rho_{\text{i}}(\boldsymbol{y}'t')}]\rangle_{\text{c}}\, v_{\text{ie}}(\boldsymbol{x} - \boldsymbol{y})\, v_{\text{ie}}(\boldsymbol{x}' - \boldsymbol{y}')\, \mathrm{d}\boldsymbol{x}\, \mathrm{d}\boldsymbol{y}\, \mathrm{d}\boldsymbol{x}'\, \mathrm{d}\boldsymbol{y}'\, \mathrm{d}t\, \mathrm{d}t'. \tag{A.4}$$

This may again be decoupled to yield

$$\delta^{(2)}\langle T[\rho_e(\boldsymbol{r}_1t_1)\,\rho_e(\boldsymbol{r}_2t_2)]\rangle_c$$

$$= \tfrac{1}{2}(-\mathrm{i}/\hbar)^2 \int \{\langle T[\rho_e(\boldsymbol{r}_1t_1)\,\rho_e(\boldsymbol{r}_2t_2)\,\rho_e(\boldsymbol{x}t)\,\rho_e(\boldsymbol{x}'t')]\rangle_c$$

$$\times \langle T[\rho_i(\boldsymbol{y}t)\,\rho_i(\boldsymbol{y}'t')]\rangle_c + \langle T[\rho_e(\boldsymbol{r}_1t_1)\,\rho_e(\boldsymbol{r}_2t_2)\,\rho_e(\boldsymbol{x}t)\,\rho_e(\boldsymbol{x}'t')]\rangle_c$$

$$\times \langle\rho_i(\boldsymbol{y}t)\rangle\,\langle\rho_i(\boldsymbol{y}'t')\rangle + 2\langle T[\rho_e(\boldsymbol{r}_1t_1)\,\rho_e(\boldsymbol{r}_2t_2)\,\rho_e(\boldsymbol{x}t)]\rangle_c$$

$$\times \langle T[\rho_i(\boldsymbol{y}t)\,\rho_i(\boldsymbol{y}'t')]\rangle_c\,\langle\rho_e(\boldsymbol{x}'t')\rangle + 2\langle T[\rho_e(\boldsymbol{r}_1t_1)\,\rho_e(\boldsymbol{x}t)]\rangle_c$$

$$\times \langle T[\rho_e(\boldsymbol{r}_2t_2)\,\rho_e(\boldsymbol{x}'t')]\rangle_c\,\langle T[\rho_i(\boldsymbol{y}t)\,\rho_i(\boldsymbol{y}'t')]\rangle_c\}\; v_{ie}(\boldsymbol{x}-\boldsymbol{y})\,v_{ie}(\boldsymbol{x}'-\boldsymbol{y}')$$

$$\times \mathrm{d}\boldsymbol{x}\,\mathrm{d}\boldsymbol{y}\,\mathrm{d}t\,\mathrm{d}\boldsymbol{x}'\,\mathrm{d}\boldsymbol{y}'\,\mathrm{d}t'. \tag{A.5}$$

A.2. Rigid ionic lattice

We first specialize to a rigid lattice, with sites $\boldsymbol{R}_l$. We have then

$$\langle\rho_i(\boldsymbol{y}t)\rangle = \sum_l \delta(\boldsymbol{y}-\boldsymbol{R}_l). \tag{A.6}$$

Furthermore, in this case $\langle T[\rho_i(\boldsymbol{y}t)\,\rho_i(\boldsymbol{y}'t')]\rangle_c = 0$ and we are left with the first-order term and only one of the second-order contributions. This then gives

$$\delta\langle T[\rho_e(\boldsymbol{r}_1t_1)\,\rho_e(\boldsymbol{r}_2t_2)]\rangle_c$$

$$= -(\mathrm{i}/\hbar)\int \langle T[\rho_e(\boldsymbol{r}_1t_1)\,\rho_e(\boldsymbol{r}_2t_2)\,\rho_e(\boldsymbol{x}t)]\rangle_c\; U(\boldsymbol{x})\,\mathrm{d}\boldsymbol{x}\,\mathrm{d}t$$

$$+ \tfrac{1}{2}(-\mathrm{i}/\hbar)^2 \int \langle T[\rho_e(\boldsymbol{r}_1t_1)\,\rho_e(\boldsymbol{r}_2t_2)\,\rho_e(\boldsymbol{x}t)\,\rho_e(\boldsymbol{x}'t')]\rangle$$

$$\times U(\boldsymbol{x})\,U(\boldsymbol{x}')\,\mathrm{d}\boldsymbol{x}\,\mathrm{d}\boldsymbol{x}'\,\mathrm{d}t\,\mathrm{d}t'. \tag{A.7}$$

Putting $t_1 = t_2$, we obtain the desired results for χ_3 and χ_4 (cf equations 2.4–2.6).

A.3. Molten metal

In the case of a liquid metal, on the other hand, we can write

$$\langle\rho_i(\boldsymbol{y}t)\rangle = \rho_i \tag{A.8}$$

and

$$\langle\rho_e(\boldsymbol{x}t)\rangle = \rho_e = z\,\rho_i \tag{A.9}$$

z being the valency. Therefore, if we choose the zero of energy such that

$$\int v_{ie}(\boldsymbol{x}-\boldsymbol{y})\,\mathrm{d}\boldsymbol{x} = 0 \tag{A.10}$$

we see that the first-order term vanishes, together with two of the second-order contributions. This leaves for the liquid metal the result

$$\delta\langle T[\rho_e(\boldsymbol{r}_1t_1)\,\rho_e(\boldsymbol{r}_2t_2)]\rangle_c$$

$$= \tfrac{1}{2}(-\mathrm{i}/\hbar)^2 \int \{\langle T[\rho_e(\boldsymbol{r}_1t_1)\,\rho_e(\boldsymbol{r}_2t_2)\,\rho_e(\boldsymbol{x}t)\,\rho_e(\boldsymbol{x}'t')]\rangle_c$$

$$\times \langle T[\rho_i(\boldsymbol{y}t)\,\rho_i(\boldsymbol{y}'t')]\rangle_c + 2\langle T[\rho_e(\boldsymbol{r}_1t_1)\,\rho_e(\boldsymbol{x}t)]\rangle_c\,\langle T[\rho_e(\boldsymbol{r}_2t_2)\,\rho_e(\boldsymbol{x}'t')]\rangle_c$$

$$\times \langle T[\rho_i(\boldsymbol{y}t)\,\rho_i(\boldsymbol{y}'t')]\rangle_c\}\; v_{ie}(\boldsymbol{x}-\boldsymbol{y})\,v_{ie}(\boldsymbol{x}'-\boldsymbol{y}')\,\mathrm{d}\boldsymbol{x}\,\mathrm{d}\boldsymbol{y}\,\mathrm{d}t\,\mathrm{d}\boldsymbol{x}'\,\mathrm{d}\boldsymbol{y}'\,\mathrm{d}t'. \tag{A.11}$$

Because of the very large ratio of ionic to electronic mass, we may use the adiabatic approximation and assume that, during the electronic motion, the ions are static. This means that we may put $t = t'$ in the ionic parts of the RHS of equation (A.11), while keeping $t \neq t'$ in the electronic parts. Also putting $t_1 = t_2$, we obtain

$$\delta P_e(\boldsymbol{r}_1\boldsymbol{r}_2) = \tfrac{1}{2}(-i/\hbar)^2 \int \langle T[\rho_e(\boldsymbol{r}_1 t_1)\,\rho_e(\boldsymbol{r}_2 t_1)\,\rho_e(\boldsymbol{x}t)\,\rho_e(\boldsymbol{x}'t')]\rangle_e \langle \rho_i(\boldsymbol{y})\,\rho_i(\boldsymbol{y}')\rangle_e$$

$$\times\, v_{ie}(\boldsymbol{x}-\boldsymbol{y})\,v_{ie}(\boldsymbol{x}'-\boldsymbol{y}')\,d\boldsymbol{x}\,d\boldsymbol{y}\,dt\,d\boldsymbol{x}'\,d\boldsymbol{y}'\,dt' + \int \mathscr{F}_1(\boldsymbol{r}_1-\boldsymbol{x}, t_1-t)$$

$$\times\, \mathscr{F}_1(\boldsymbol{r}_2-\boldsymbol{x}', t-t')\,\langle \rho_i(\boldsymbol{y})\,\rho_i(\boldsymbol{y}')\rangle_e\, v_{ie}(\boldsymbol{x}-\boldsymbol{y})\,v_{ie}(\boldsymbol{x}'-\boldsymbol{y}')$$

$$\times\, d\boldsymbol{x}\,d\boldsymbol{y}\,dt\,d\boldsymbol{x}'\,d\boldsymbol{y}'\,dt' \tag{A.12}$$

where we have used

$$\mathscr{F}_1(\boldsymbol{x}-\boldsymbol{y}, t_1-t_2) = -(i/\hbar)\langle T[\rho_e(\boldsymbol{x}t_1)\,\rho_e(\boldsymbol{y}t_2)]\rangle_e. \tag{A.13}$$

This is the same as (5.1).

Appendix B

B.1. Diagrammatic analysis of response function χ_4

Following the work of Brovman and Kagan (1967, see also Pethick 1970), it is possible to analyse the structure of χ_4 in terms of the proper four-body polarizability Γ_4. Their result is equivalent to the relation between χ_4, Γ_4 and the dielectric function $\epsilon(k, \omega)$

$$\chi_4(\boldsymbol{k}, -\boldsymbol{k}; -\boldsymbol{G}', \boldsymbol{G}') = \int (d\omega/2\pi)\,|\epsilon(k,\omega)|^{-2}\,\Gamma_4(\boldsymbol{k}\omega, -\boldsymbol{k}-\omega; -\boldsymbol{G}'0, \boldsymbol{G}'0)\,\epsilon^{-2}(G'). \tag{B.1}$$

If we now insert equation (B.1) into equation (5.2) we see that the role of the static dielectric functions appearing in (B.1) is simply to screen the electron–ion potential; however to proceed further, some approximation for Γ_4 is necessary. The simplest one, corresponding to a time-dependent Hartree approximation, is to take the free-particle value for Γ_4. But even in this approximation, the problem is not easy to handle, because the integration over the frequency has still to be carried out. Therefore if, as the final approximation, we neglect the frequency dependence of the dynamic dielectric function appearing in equation (B.1), then we find

$$\chi_4(\boldsymbol{k}, -\boldsymbol{k}; -\boldsymbol{G}', \boldsymbol{G}') = \epsilon^{-2}(k)\,\epsilon^{-2}(G')\,\chi_4^0(\boldsymbol{k}, -\boldsymbol{k}; -\boldsymbol{G}', \boldsymbol{G}') \tag{B.2}$$

where we have made use of the fact that

$$\int (d\omega/2\pi)\,\Gamma^{\text{free-particle}}(\boldsymbol{k}\omega, -\boldsymbol{k}-\omega; -\boldsymbol{G}'0, \boldsymbol{G}'0) = \chi_4^0(\boldsymbol{k}, -\boldsymbol{k}; -\boldsymbol{G}', \boldsymbol{G}'). \tag{B.3}$$

We finally note that equation (B.2) will automatically satisfy the long-wavelength limit (5.5) as in the limit $k \to 0$ the $\epsilon^{-2}(k)$ factor goes to zero as k^4 because of the perfect screening sum rule.

Appendix C

C.1. Evaluation of χ_4^0

The methods outlined in §4 lead to the closed result for χ_4^0 in $\boldsymbol{r}$ space:

$$\chi_4^0(\boldsymbol{r}_1\boldsymbol{r}_2;\boldsymbol{r}_3\boldsymbol{r}_4) = -\frac{k_f^4}{2\pi^6}\left(\frac{j_1(k_f|\boldsymbol{r}_1-\boldsymbol{r}_2|)}{|\boldsymbol{r}_1-\boldsymbol{r}_2|}\frac{j_1(k_f|\boldsymbol{r}_1-\boldsymbol{r}_3|+k_f|\boldsymbol{r}_3-\boldsymbol{r}_4|+k_f|\boldsymbol{r}_4-\boldsymbol{r}_2|)}{|\boldsymbol{r}_1-\boldsymbol{r}_3|\,|\boldsymbol{r}_3-\boldsymbol{r}_4|\,|\boldsymbol{r}_4-\boldsymbol{r}_2|}\right.$$
$$\left.+\frac{1}{2}\frac{j_1(k_f|\boldsymbol{r}_1-\boldsymbol{r}_3|+k_f|\boldsymbol{r}_3-\boldsymbol{r}_2|)}{|\boldsymbol{r}_1-\boldsymbol{r}_3|\,|\boldsymbol{r}_3-\boldsymbol{r}_2|}\frac{j_1(k_f|\boldsymbol{r}_1-\boldsymbol{r}_4|+k_f|\boldsymbol{r}_4-\boldsymbol{r}_2|)}{|\boldsymbol{r}_1-\boldsymbol{r}_4|\,|\boldsymbol{r}_4-\boldsymbol{r}_2|}\right) \tag{C.1}$$

where $j_1(x)$ is the first order spherical Bessel function $(\sin x - x\cos x/x^2)$. Fourier transforming this equation gives:

$$\chi_4^0(\boldsymbol{k},-(\boldsymbol{k}+\boldsymbol{G});\boldsymbol{G}-\boldsymbol{G}',\boldsymbol{G}') = -16\left(\int\frac{\mathrm{d}\boldsymbol{p}}{8\pi^3}\,[2\theta(k_f-|\boldsymbol{k}-\boldsymbol{p}|)\,J_3(\boldsymbol{p},\boldsymbol{p}+\boldsymbol{G}-\boldsymbol{G}',\boldsymbol{p}+\boldsymbol{G})\right.$$
$$\left.+J_2(\boldsymbol{p}-\boldsymbol{k},\boldsymbol{p}-\boldsymbol{k}+\boldsymbol{G}')\,J_2(\boldsymbol{p}+\boldsymbol{G}',\boldsymbol{p}+\boldsymbol{G})]\right) \tag{C.2}$$

where

$$\left.\begin{aligned}J_3(\boldsymbol{p},\boldsymbol{q},\boldsymbol{k}) &= \frac{\theta(k_f-|\boldsymbol{p}|)}{(p^2-q^2)(p^2-k^2)}+\frac{\theta(k_f-|\boldsymbol{q}|)}{(q^2-p^2)(q^2-k^2)}+\frac{\theta(k_f-|\boldsymbol{k}|)}{(k^2-p^2)(k^2-q^2)}\\ &\text{and}\\ J_2(\boldsymbol{p},\boldsymbol{q}) &= \frac{\theta(k_f-|\boldsymbol{p}|)-\theta(k_f-|\boldsymbol{q}|)}{q^2-p^2}.\end{aligned}\right\} \tag{C.3}$$

After some manipulation, the $\boldsymbol{G}=0$ limit of (C.2) may be written:

$$\begin{aligned}\chi_4^0(\boldsymbol{k},-\boldsymbol{k};-\boldsymbol{G}',\boldsymbol{G}') &= -\int\frac{\mathrm{d}\boldsymbol{p}}{\pi^3}\frac{2}{k_f}\,\theta(k_f-|\boldsymbol{k}-\boldsymbol{p}|)\frac{\delta(p-k_f)}{(2\boldsymbol{G}'.\boldsymbol{p}+G^2)}\\ &+\int\frac{\mathrm{d}\boldsymbol{p}}{\pi^3}\frac{8\theta(k_f-|\boldsymbol{k}-\boldsymbol{p}|)\,[\theta(k_f-|\boldsymbol{p}+\boldsymbol{G}'|)-\theta(k_f-p)]\,\boldsymbol{k}.\boldsymbol{G}'}{(G'^2+2\boldsymbol{G}'.\boldsymbol{p})^2\,(G'^2+2\boldsymbol{G}'.(\boldsymbol{p}-\boldsymbol{k}))}.\end{aligned} \tag{C.4}$$

C.2. Evaluation of χ_4^{HK} from equation (5.6) using long wavelength sum rules

Equation (5.6) for χ_4^{HK} is:

$$\chi_4^{\mathrm{HK}}(\boldsymbol{k},-\boldsymbol{k};-\boldsymbol{G}',\boldsymbol{G}') = \left(\frac{\mathscr{F}_1(\boldsymbol{G}')}{F_1(\boldsymbol{G}')}\right)^2\left(\chi_4^0(\boldsymbol{k},-\boldsymbol{k};-\boldsymbol{G}',\boldsymbol{G}')-\frac{\chi_3^0(\boldsymbol{k},-\boldsymbol{k};0)\,F_2(0,\boldsymbol{G}')}{F_1(0)}\right). \tag{C.5}$$

For the liquid metal calculations, only the spherical average over $\boldsymbol{G}'$ of χ_4^{HK} is required and in this case it can be shown that the second term on the RHS of (C.5) is exactly cancelled by the first term on the RHS of (C.4). Direct evaluation of this term in (C.4) gives

$$-4\theta(2k_f-k)\left(1-\frac{k}{2k_f}\right)\frac{1}{\pi G'}\ln\left|\frac{G'+2k_f}{G'-2k_f}\right|. \tag{C.6}$$

The corresponding term in (C.5) can be conveniently calculated using the following long wavelength sum rules for the various response functions. In general

$$\lim_{G\to 0} \chi_3[\boldsymbol{k}, -(\boldsymbol{k}+\boldsymbol{G}); \boldsymbol{G}] = \lim_{G\to 0} \mathscr{F}_1(\boldsymbol{G})(\partial/\partial\rho)\{\rho[S(k)-1]\} \tag{C.7}$$

where $S(k)$ is the exact jellium structure factor and

$$\lim_{G\to 0} \mathscr{F}_2(\boldsymbol{G}, \boldsymbol{G}'-\boldsymbol{G}) = \lim_{G\to 0} \mathscr{F}_2(\boldsymbol{G}', \boldsymbol{G}-\boldsymbol{G}') = \lim_{G\to 0} \mathscr{F}_1(\boldsymbol{G})(\partial/\partial\rho)\mathscr{F}_1(\boldsymbol{G}'). \tag{C.8†}$$

Applied to the non-interacting case these give

$$\chi_3^0(\boldsymbol{k}, -\boldsymbol{k}; 0) = F_1(0)(\partial/\partial\rho)\{\rho[S^{\mathrm{HF}}(k)-1]\} \tag{C.9}$$

and

$$F_2(0, \boldsymbol{G}') = F_1(0)(\partial/\partial\rho)F_1(\boldsymbol{G}') \tag{C.10}$$

with $S^{\mathrm{HF}}(k)$, the non-interacting Fermion structure factor, i.e.

$$S^{\mathrm{HF}}(k) - 1 = \theta(2k_{\mathrm{f}} - k)\left(\frac{3}{4}\frac{k}{k_{\mathrm{f}}} - \frac{1}{16}\frac{k^3}{k_{\mathrm{f}}^3}\right) \tag{C.11}$$

and

$$F_1(G) = -\frac{k_{\mathrm{f}}}{\pi^2}\left[\frac{1}{2} + \frac{k_{\mathrm{f}}}{2G}\left(1 - \frac{G^2}{4k_{\mathrm{f}}^2}\right)\ln\left|\frac{G+2k_{\mathrm{f}}}{G-2k_{\mathrm{f}}}\right|\right]. \tag{C.12}$$

Use of (C.9) and (C.10) in the second term of (C.5) (and including a factor of 4π arising from spherical averaging) gives the same expression as (C.6) thus demonstrating the exact cancellation. We are left with the following expression for $\chi_4^{\mathrm{HK}}(\boldsymbol{k}, -\boldsymbol{k}; -\boldsymbol{G}', \boldsymbol{G}')$

$$\chi_4^{\mathrm{HK}}(\boldsymbol{k}, -\boldsymbol{k}; -\boldsymbol{G}', \boldsymbol{G}')$$

$$= \left(\frac{\mathscr{F}_1(\boldsymbol{G}')}{F_1(\boldsymbol{G}')}\right)^2 \int \frac{\mathrm{d}\boldsymbol{p}}{\pi^3} \frac{8\theta(k_{\mathrm{f}} - |\boldsymbol{k}-\boldsymbol{p}|)[\theta(k_{\mathrm{f}} - |\boldsymbol{p}+\boldsymbol{G}'|) - \theta(k_{\mathrm{f}} - p)]\,\boldsymbol{k}.\boldsymbol{G}'}{(G'^2 + 2\boldsymbol{G}'.\boldsymbol{p})^2(G'^2 + 2\boldsymbol{G}'.(\boldsymbol{p}-\boldsymbol{k}))} \tag{C.13}$$

which clearly is zero in the $\boldsymbol{k} \to 0$ limit, as required.

C.3. Exact and approximate long wavelength limits of χ_4

A generalized Ward identity for the response function χ_4 as $\boldsymbol{G}' \to 0$ gives

$$\lim_{G'\to 0} [\epsilon(\boldsymbol{G}')]^2 \chi_4(\boldsymbol{k}, -\boldsymbol{k}; -\boldsymbol{G}', \boldsymbol{G}') = \pi^2(0)(\partial^2/\partial\rho^2)\{\rho[S(k)-1]\}. \tag{C.14}$$

The corresponding limit for the approximate form (C.13) may be shown to be

$$\lim_{G'\to 0} [\epsilon(\boldsymbol{G}')]^2 \chi_4^{\mathrm{HK}}(\boldsymbol{k}, -\boldsymbol{k}; -\boldsymbol{G}', \boldsymbol{G}') = \pi^2(0)(\partial^2/\partial\rho^2)\{\rho[S^{\mathrm{HF}}(k)-1]\}. \tag{C.15}$$

† Since $\lim_{G\to 0} \mathscr{F}_1(G) = 0$, this justifies the comment made above equation (5.6).

This clearly demonstrates the limitations of the approximation χ_4^{HK}, but at least the form (C.15) is closely related to the exact limit (C.14), which is less obvious for the alternative approximation based on diagrammatic analysis. It may be fruitful, in the future, to generalize equation (C.14) to non-zero $\boldsymbol{G}'$.

C.4. Summary of other independent particle results

We conclude this Appendix by collecting the independent particle forms for χ_3^0, F_1 and F_2. These are obtained by the methods of §4 and the $\boldsymbol{r}$ space forms are

$$\chi_3^0(\boldsymbol{r}_1\boldsymbol{r}_2;\boldsymbol{r}_3) = \frac{k_{\mathrm{f}}^4}{2\pi^5}\frac{j_1(k_{\mathrm{f}}|\boldsymbol{r}_1-\boldsymbol{r}_2|)}{|\boldsymbol{r}_1-\boldsymbol{r}_2|}\frac{j_1(k_{\mathrm{f}}|\boldsymbol{r}_1-\boldsymbol{r}_3|+k_{\mathrm{f}}|\boldsymbol{r}_3-\boldsymbol{r}_2|)}{|\boldsymbol{r}_1-\boldsymbol{r}_3||\boldsymbol{r}_3-\boldsymbol{r}_2|} \tag{C.16}$$

$$F_1(\boldsymbol{r}_1\boldsymbol{r}_2) = -\frac{k_{\mathrm{f}}^2}{2\pi^3}\frac{j_1(2k_{\mathrm{f}}|\boldsymbol{r}_1-\boldsymbol{r}_2|)}{|\boldsymbol{r}_1-\boldsymbol{r}_2|^2} \tag{C.17}$$

and

$$F_2(\boldsymbol{r}_1\boldsymbol{r}_2\boldsymbol{r}_3) = \frac{k_{\mathrm{f}}^2}{2\pi^4}\frac{j_1(k_{\mathrm{f}}|\boldsymbol{r}_1-\boldsymbol{r}_2|+k_{\mathrm{f}}|\boldsymbol{r}_2-\boldsymbol{r}_3|+k_{\mathrm{f}}|\boldsymbol{r}_3-\boldsymbol{r}_1|)}{|\boldsymbol{r}_1-\boldsymbol{r}_2||\boldsymbol{r}_2-\boldsymbol{r}_3||\boldsymbol{r}_3-\boldsymbol{r}_1|}. \tag{C.18}$$

References

Ashcroft N W 1966 *Phys. Lett.* **23** 48
Brovman E G and Kagan Yu 1967 *Sov. Phys.—JETP* **25** 365
——1974 *Sov. Phys.—Usp.* **17** 125
Chihara J 1973 *2nd Int. Conf. on Liquid Metals (Tokyo)* (London: Taylor and Francis) p 137
Egelstaff P A, March N H and McGill N C 1974 *Can. J. Phys.* **52** 1651
Eisenberger P and Platzman P L 1974 *Phys. Rev. Lett.* **33** 152
Freeman A J 1959 *Acta Cryst.* **12** 261
Greenfield A, Wiser N and Zellendorf 1971 *Phys. Rev.* A **4** 1607
Hohenberg P and Kohn W 1964 *Phys. Rev.* B **136** 864
Kadanoff L P and Baym G 1962 *Quantum Statistical Mechanics* (New York: Benjamin)
March N H and Murray A M 1961 *Proc. R. Soc.* **A261** 219
March N H and Tosi M P 1973 *Ann. Phys., NY* **81** 414
Pethick C J 1970 *Phys. Rev.* B **2** 1789
Singwi K S, Tosi M P, Land R H and Sjolander A 1968 *Phys. Rev.* **176** 589
Tosi M P and March N H 1973 *Nuovo Cim.* B **15** 308
Watabe M and Hasegawa M 1973 *2nd Int. Conf. on Liquid Metals (Tokyo)* (London: Taylor and Francis) p 439

Philosophical Magazine B, 1980, Vol. 42, No. 4, 511–515

Electron scattering by molten aluminium

By A. S. Brah and L. Virdhee
Department of Materials Science and Metallurgy, Imperial College, London SW7, England

and N. H. March
Department of Theoretical Chemistry, South Parks Road, Oxford, England

[Received 7 August 1979 and accepted 14 February 1980]

Abstract

Thin liquid aluminium films (150–500 Å) grown *in situ* on alumina substrates have been studied under U.H.V. conditions in a scanning transmission high-energy electron diffraction system incorporating inelastic filtering. The liquid structure factor was determined (after applying appropriate corrections) by using high-angle fitting and the Krogh-Moe methods. The reliability of the normalization was assessed by the Rahman criterion, giving an accuracy of 1% for the structure factor between $Q = 0{\cdot}5$ and $12{\cdot}0$ Å^{-1}.

§ 1. Introduction

Some results are presented on electron scattering from molten aluminium obtained by using refined experimental (e.g. U.H.V. and inelastic filtering) and theoretical techniques. Previous work on liquid aluminium by Bublik and Buntar (1958) and subsequently by Levin, Geld and Yakubchik (1975) suffered from the following drawbacks. (1) Experiments were carried out in relatively poor vacuum conditions without any facilities for inelastic energy filtering. (2) Semi-empirical techniques were used to eliminate inelastic effects. (3) No detailed account was given for the absorption losses or multiple scattering, nor were their normalization procedures sufficiently rigorous. (4) Bublik and Buntar (1958) reported a temperature dependence for liquid aluminium which to date has never been confirmed by any X-ray or neutron work. The aim of the present work is to rectify as far as possible these shortcomings.

§ 2. Experimental

A U.H.V. scanning high-energy electron diffraction (SHEED) system incorporating velocity filtering was used in the transmission mode to study the liquid aluminium films. This is fully described in previous work (Brah, Dobson, March, Unvala and Virdhee 1978, Brah, Unvala and Virdhee 1979). Owing to the highly reactive nature of molten aluminium it was found necessary to carry out all experimental work in a U.H.V. ($\leqslant 10^{-10}$ torr) environment. It also proved quite difficult to choose a chemically compatible support for the liquid film; after many trials with various substrate materials (e.g. MgO, NaCl, mica), alumina was found to be the most suitable. The alumina

0141–8637/80/4204 0511 $02·00

films were prepared using a number of methods including r.f. sputtering, and thermal and anodic oxidation. The latter technique was found to yield the best films which were sufficiently thin (50–100 Å) yet mechanically strong. Such films were collected onto molybdenum and stainless steel electron microscope grids (200 mesh) and inserted into the specially designed heating stage.

Thin films of liquid aluminium (SDC-nines purity) were prepared by *in situ* evaporation from a well outgassed, triple-strand tungsten conical basket by two alternative methods. The evaporation was carried out onto an already heated substrate (at 660°C) and films deposited at room temperature were subsequently annealed to obtain the molten phase.

§ 3. Results and discussion

Figure 1 (curve (*a*)) shows the intensity profile of the combined liquid aluminium and the alumina substrate at 660°C. The first peak of liquid aluminium is seen to be clearly displaced from the main peak of amorphous alumina (curve (*b*)), thus simplifying the matching along the x axis and the subsequent subtraction. The liquid profile is obtained by subtracting a suitably attenuated substrate profile from the combined profile as necessitated

Fig. 1

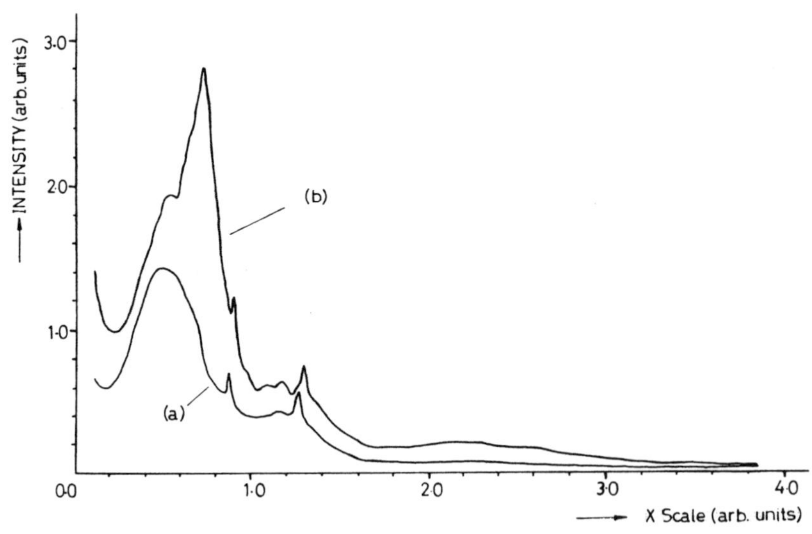

Intensity profiles of (*a*) the combined aluminium and alumina films, and (*b*) the alumina substrate.

by the absorption correction. The degree of attenuation is inferred from the low-angle region. It should, however, be mentioned that the intensity measurements a few milliradians away from zero-angle scattering are difficult on account of a large intensity contribution arising from the main beam ;

hence extrapolation is carried out in this region taking into account the isothermal compressibility limit. For the normalization constant, only the large-angle scattering intensity is used but, owing to multiple scattering, the intensity in the very-high-angle region ($Q > 10$ Å^{-1}, where $Q = 4\pi \sin(\theta/\lambda)$) was found to be too large by 10%.

Fig. 2

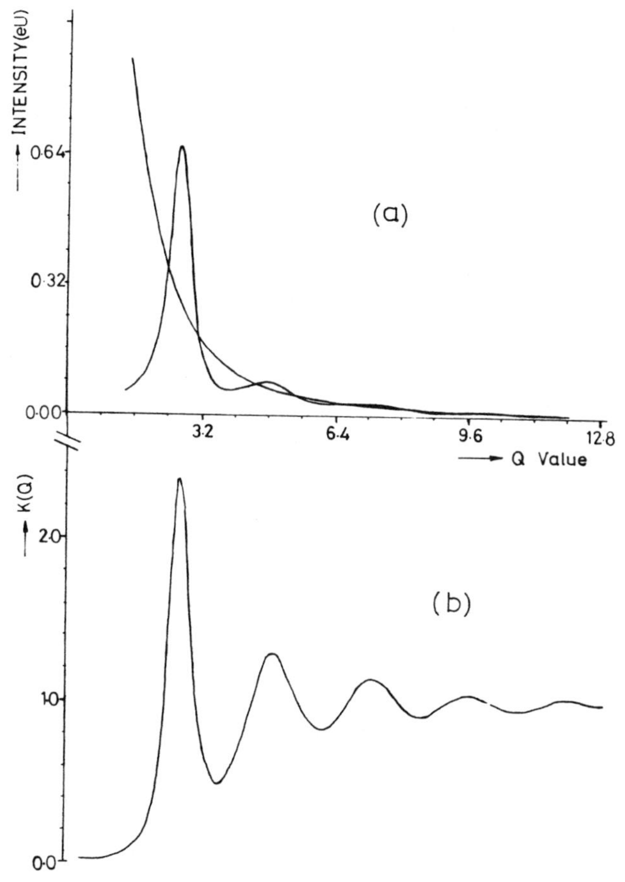

(*a*) Normalized intensity profile for liquid aluminium. (*b*) The structure factor $K(Q)$ for liquid aluminium.

After applying these corrections, the resulting liquid profile obtained is displayed in fig. 2 (*a*) where the data has been normalized by using the high-angle fitting criterion and the Krogh-Moe (1956) methods. The normalization constants obtained by these two methods agreed to within $\pm 1\%$. The atomic form factors used are those for electrons tabulated by Doyle and Turner (1968) based on the relativistic Hartree–Fock atomic fields. Before taking their squares these were multiplied by 1·05 to correct for relativistic effects for 25 keV incident electrons. Any further correction to the form factors similar to the dispersion corrections for X-rays cannot arise in the present case as the energy of the incident electrons is well clear of the lowest bound state ($\sim 1{\cdot}6$ keV) for the neutral aluminium atom. As shown previously (Brah *et al.* 1978, 1979), thermal corrections to the form factors for liquid metals are not applied. The resulting fit to the corrected liquid profile is also shown in fig. 2.

To obtain the structure factor $K(Q)$ the intensity is divided by the square of the atomic form factor. To assess the reliability, an independent test of the normalization is carried out according to the method suggested by Rahman (1965). This requires that the structure factor $K(Q)$ must satisfy the relationship

$$4\pi^2\rho_0 L j_1(\mu L) = \int_0^\infty Q[K(Q)-1][j_0(L(Q+\mu)) - j_0(L(Q-\mu))]\, dQ,$$

where $j_n(y)$ is the nth spherical Bessel function, μ is an arbitrary parameted L must be less than the effective hard-core diameter of the screened ion, anr, ρ_0 is the density of atoms.

Table 1. Typical results.

L (Å)	μ (Å^{-1})	Theoretical result	Adjustment factor for exact agreement
1·00	1·0	0·6278	0·9936
1·00	2·0	0·9076	0·9943
2·00	1·0	1·8151	0·9965
2·00	2·0	0·4841	0·9929
3·00	0·5	2·4774	0·9980
3·00	1·5	−0·0089	0·9917

Typical results obtained after performing the final integration for liquid aluminium at 660°C are shown in table 1. The value used for ρ_0 was 0·0528 atoms/Å—the experimental values determined by Waseda and Suzuki (1972) and Batalin, Kazimirov and Dmitruk (1972). It can be seen that a change of 1% in the structure factor is required to obtain exact agreement between the experimental and theoretical value. But, as the integration does not extend to infinity and the region between $Q=0$ and $0{\cdot}5$ Å^{-1} is missing, it is estimated that this might contribute an additional 1% error, thus giving an overall accuracy of 2%. The structure factor $K(Q)$ corresponding to this is shown graphically in fig. 2 (*b*).

Comparison with the electron scattering data of Bublik and Buntar (1958) and Levin *et al.* (1975) is given in table 2. It can be seen that there is fair agreement with positions of the peaks but the values for the structure factor are different, especially the value at the principal peak. Also the auxiliary peaks reported by these workers were not observed. These differences could be attributed to the refined experimental techniques used (such as U.H.V. conditions or energy filtering) and also the more rigorous correction procedure, hence justifying the present work. Furthermore, the present results are remarkably similar to those obtained by recent X-ray and neutron techniques (although on points of detail there are small differences), whereas the previous investigations do not share this similarity.

Table 2. Comparison of electron scattering data for aluminium at the melting point (665°C).

Source	Bublik and Buntar (1958)		Levin, Geld and Yakubchik (1975)		Present work	
	sin θ/λ	$K(Q)$	sin θ/λ	$K(Q)$†	sin θ/λ	$K(Q)$
First maxima	0·22	—	0·21 ~0·217	1·78	0·213	2·364
Auxiliary maxima	0·255	—	0·25 ~0·26	1·72	Not observed	—
Auxiliary maxima	0·363	—	Not observed	—	Not observed	—
Second maxima	0·422	—	0·38 ~0·39	1·25	0·397	1·300
Third maxima	—	—	0·64	~1·20	0·581	1·144
Fourth maxima	—	—	—	—	0·767	1·072
Fifth maxima	—	—	—	—	0·932	1·05

† Calculated from their graph.

A further detailed exposition of this work and comparison with X-ray and neutron results will follow.

Acknowledgments

We wish to thank Drs. B. A. Unvala and P. J. Dobson for their advice and encouragement and the S.R.C. for financial support.

References

Batalin, G. I., Kazimirov, V. P., and Dmitruk, B. F., 1972, *Izv. Akad. Nauk SSSR, Metall.*, **1,** 88.

Brah, A. S., Dobson, P. J., March, N. H., Unvala, B. A., and Virdhee, L., 1978, *Electron Diffraction 1927–1977*, edited by P. J. Dobson, J. B. Pendry and C. J. Humphreys, Inst. Phys. Conf. Ser. No. 41 (London, Bristol : The Institute of Physics), p. 430.

Brah, A. S., Unvala, B. A., and Virdhee, L., 1979, *Phil. Mag.* B, **40,** 335.

Bublik, A. I., and Buntar, A. G., 1958, *Kristallografiya*, **3,** 32.

Doyle, P. A., and Turner, P. S., 1968, *Acta crystallogr.* A, **24,** 290.

Krogh-Moe, J., 1956, *Acta crystallogr.*, **9,** 951.

Levin, E. S., Geld, P. V., and Yakubchik, V. P., 1975, *Izv. Akad. Nauk SSSR, Metall.*, **5,** 80.

Rahman, A., 1965, *J. chem. Phys.*, **42,** 3540.

Waseda, Y., and Suzuki, K., 1972, *Phys. Stat. Sol.* (b), **49,** 339.

Philosophical Magazine **B**, 1994, Vol. 69, No. 5, 965–977

A diffraction study of the structure of liquid potassium near freezing and density functional theory of pair potentials

By M. W. Johnson†, N. H. March‡, F. Perrot§ and A. K. Ray‡

† Rutherford Appleton Laboratory, Chilton, Didcot, Oxfordshire OX11 0QX, England
‡ Department of Theoretical Chemistry, University of Oxford, 5 South Parks Road, Oxford OX1 3UB, England
§ Centre d'Études de Limeil Valenton, 94195 Villeneuve St Georges Cedex, France

[*Received 25 September 1992*‖ *and accepted 29 January 1993*]

Abstract

New results will be reported from a neutron diffraction experiment to measure the liquid structure factor $S(Q)$ of potassium metal just above its freezing point. This experiment will then be analysed in relation to

(*a*) the X-ray measurements of Greenfield, Wellendorf and Wiser, and
(*b*) a density functional calculation of a potassium potential.

With respect to (*a*) the possible sources of experimental error will be examined and the confidence limits on the difference $S_n(Q)-S_X(Q)$ evaluated. Conclusions about electron correlation in liquid potassium will also be drawn. The pair potential calculated with respect to (*b*) will finally be discussed in relation to liquid structure theories of $S(Q)$.

§ 1. Introduction

In this paper we shall first discuss the difference between the structure factors measured by X-rays and neutrons for liquid metals. This question was first raised by Egelstaff, March and McGill (EMM) (1974) who pointed out some systematic differences between the height of the principal peak of $S(Q)_X$ and $S(Q)_n$ in the published data then available.

Subsequently Dobson (1978) has used previously published data on both sodium and aluminium to plot the function $S(Q)_X-S(Q)_n$, which for both metals appeared to reveal a 'solid-like' structure indicative of some form of medium range (30–40 Å) valence electron ordering as inferred by EMM. The sodium data for $S(Q)_n$ used by Dobson were taken from a small line drawing in the original publication by Gingrich and Heaton (1961). Moreover the data were not corrected for multiple scattering and no details were given of the method used for making the Placzek correction. Overall, the error estimate on this procedure could have produced errors of about 0·2 in $S(Q)_X-S(Q)_n$, equivalent to the observed effects. Dobson therefore suggested the numerical publication of good $S(Q)_n$ for liquid metals so that this comparison can be made accurately. In this paper we report new measurements on liquid potassium and examine the function $S(Q)_X-S(Q)_n$. This metal was chosen because of the excellent numerical results for $S(Q)_X$ previously published by Greenfield, Wellendorf and Wiser (1971).

‖ Received in final form 31 October 1992.

§ 2. Experimental details

The experimental results on liquid potassium were obtained using the diffractometer D2 at the Institute Laue–Langevin (Grenoble). This diffractometer has four detectors each separated by 6°. Data were recorded in 0·2° steps from 3·0° (in 2θ) to 115° with a beam height of 30 mm and collimation of 60′ 60′ 10′. Data at corresponding angles from the four detectors were summed together with a resulting intensity of approximately 10^4 counts per step. Two different samples were measured on two separate occasions using slightly different experimental conditions. In the first set of runs a neutron wavelength λ of 1·2213 Å was used and the sample located in a cloche filled with helium. On the second occasion a wavelength of 0·8780 Å was used and the cloche evacuated. In both experiments the data were recorded at 70°C.

§ 3. Analysis

Since the purpose of these experiments is to try to examine small differences between the X-ray and neutron measurements of $S(Q)$ the analysis of the data is of central importance. The aim is to determine the liquid structure factor $S(Q)$ defined by

$$S(Q)=\int_{-\infty}^{\infty} S(Q,\omega)\,\mathrm{d}\omega, \tag{1}$$

where $S(Q,\omega)$ is related to the double-differential cross-section by the expression

$$\frac{\mathrm{d}^2\sigma_{\mathrm{coh}}}{\mathrm{d}\Omega\,\mathrm{d}\omega}=b_{\mathrm{coh}}^2\frac{k}{k_0}S(Q,\omega). \tag{2}$$

Experimentally the function that is measured is the integral

$$\frac{\mathrm{d}\sigma_{\mathrm{coh}}}{\mathrm{d}\Omega}=b^2\int_{-\infty}^{\omega_0}\varepsilon(k)\frac{k}{k_0}S(Q,\omega)\,\mathrm{d}\omega, \tag{3}$$

where the integral is over the curved locus in (Q,ω) space, satisfying the kinematic scattering constraint. A method for deriving $S(Q)$ from the measured cross-section was first undertaken by Placzek (1952), although the method used here follows that of Yarnell, Katz, Wenzel and Koenig (1973), and their notation is used below.

By use of a Taylor-series expansion of $S(Q,\omega)$ along a path of constant ω, they derived the expression

$$(\varepsilon_0 b_{\mathrm{coh}}^2)^{-1}\frac{\mathrm{d}\sigma_{\mathrm{coh}}}{\mathrm{d}\Omega}=S(Q)-C_1Z+C_2Z^2-C_3Z\tau+\frac{m}{2M}(Z+\tau), \tag{4}$$

where

$$Z=\frac{\hbar^2 Q}{2ME},$$

$$\tau=\frac{k_{\mathrm{B}}T}{E_0},$$

$$C_1=1+\frac{k_0\varepsilon_1}{2\varepsilon_0},$$

$$C_2=\frac{3}{8}+\frac{\frac{5}{4}k_0\varepsilon_1}{2\varepsilon_0}+\frac{\frac{1}{4}k_0^2\varepsilon_2}{2\varepsilon_0},$$

$$C_3=-\frac{\frac{3}{2}k_0\varepsilon_1}{2\varepsilon_0}-\frac{\frac{1}{2}k_0^2\varepsilon_2}{2\varepsilon_0},$$

and the detector efficiency was approximated by the function

$$\varepsilon(\lambda)=1-\exp(-1{\cdot}8\lambda), \qquad \varepsilon_n(k)=\left.\frac{d^n\varepsilon}{dk^n}\right|_{k=k_0}.$$

The correction term derived from eqn. (4) was incorporated into the final expression for $S(Q)$ (eqn. (6)) via the term a_5. This correction is small and smoothly varying, having a value of less than 0·01 for $0 \leqslant Q \leqslant 4$ and reaching a value of 0·05 at $Q = 8\,\text{Å}^{-1}$.

The Placzek correction is not the only correction required, however. Experimental corrections are also required for

(*a*) background scattering from the container and surrounding materials,
(*b*) absorption and multiple scattering in the sample,
(*c*) instrumental resolution, and
(*d*) spectrometer efficiency.

The most important of these numerically is the background correction both for the can and the surrounding medium (helium or vacuum). These corrections were made using the 'empty-can' and 'no-cans' runs, the former being corrected to compensate for the absorption by the sample (via the term a_1 in eqn. (6)). The size of the background contribution is illustrated in fig. 1 which shows the raw data for 'sample', 'empty-can' and 'no-can' runs.

The correction due to absorption was calculated using the values for cylindrical samples in the *International Tables for X-ray Crystallography* (1969) and incorporated via the parameter a_2 in eqn. (6). The correction for multiple scattering was calculated

Fig. 1

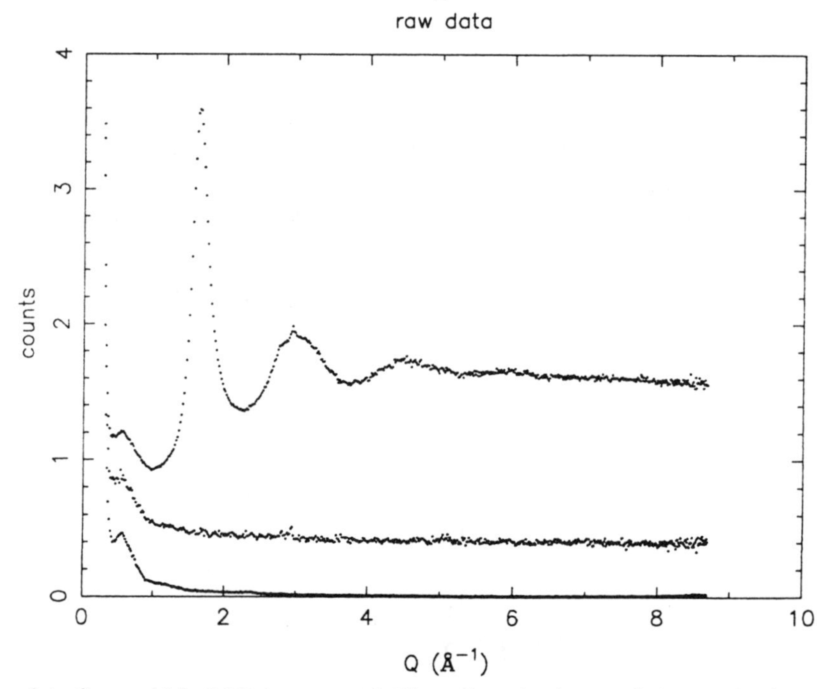

Raw data from which $S_n(Q)$ is extracted. Plots show background (lower), background + can (middle) and sample + can + background (top).

Table 1.

Q	S_{n1}	S_{n2}	S_X	Q	S_{n1}	S_{n2}	S_X	Q	S_{n1}	S_{n2}	S_X
1.000	0.132	0.147	0.086	3.500	0.966	0.987	0.930	6.000	1.024	1.014	1.015
1.025	0.138	0.156	0.093	3.525	0.953	0.973	0.921	6.025	1.026	1.014	1.015
1.050	0.145	0.167	0.102	3.550	0.940	0.961	0.912	6.050	1.026	1.015	1.014
1.075	0.154	0.178	0.112	3.575	0.929	0.950	0.904	6.075	1.026	1.015	1.014
1.100	0.165	0.190	0.124	3.600	0.919	0.941	0.897	6.100	1.026	1.015	1.013
1.125	0.178	0.204	0.138	3.625	0.910	0.933	0.892	6.125	1.026	1.015	1.013
1.150	0.194	0.221	0.153	3.650	0.903	0.927	0.887	6.150	1.024	1.015	1.012
1.175	0.212	0.240	0.172	3.675	0.898	0.923	0.883	6.175	1.022	1.015	1.012
1.200	0.231	0.262	0.193	3.700	0.894	0.920	0.880	6.200	1.020	1.014	1.011
1.225	0.255	0.289	0.218	3.725	0.893	0.919	0.878	6.225	1.018	1.014	1.010
1.250	0.282	0.321	0.246	3.750	0.892	0.920	0.877	6.250	1.015	1.014	1.009
1.275	0.316	0.361	0.280	3.775	0.894	0.923	0.878	6.275	1.011	1.013	1.008
1.300	0.358	0.411	0.324	3.800	0.896	0.926	0.880	6.300	1.007	1.013	1.008
1.325	0.409	0.474	0.383	3.825	0.900	0.930	0.882	6.325	1.002	1.013	1.007
1.350	0.472	0.550	0.461	3.850	0.905	0.936	0.887	6.350	0.997	1.012	1.006
1.375	0.553	0.643	0.564	3.875	0.911	0.942	0.892	6.375	0.994	1.011	1.005
1.400	0.661	0.756	0.695	3.900	0.918	0.948	0.899	6.400	0.993	1.011	1.005
1.425	0.803	0.896	0.860	3.925	0.925	0.955	0.906	6.425	0.994	1.010	1.004
1.450	0.984	1.068	1.062	3.950	0.933	0.961	0.915	6.450	0.995	1.009	1.003
1.475	1.213	1.276	1.306	3.975	0.941	0.968	0.924	6.475	0.997	1.008	1.002
1.500	1.492	1.525	1.588	4.000	0.950	0.975	0.934	6.500	0.997	1.008	1.002
1.525	1.823	1.816	1.904	4.025	0.959	0.981	0.945	6.525	0.996	1.007	1.001
1.550	2.186	2.119	2.232	4.050	0.968	0.988	0.955	6.550	0.994	1.006	1.000
1.575	2.517	2.387	2.512	4.075	0.977	0.994	0.965	6.575	0.993	1.005	1.000
1.600	2.752	2.576	2.682	4.100	0.986	1.000	0.976	6.600	0.993	1.004	0.999
1.625	2.858	2.666	2.718	4.125	0.995	1.005	0.986	6.625	0.994	1.003	0.998
1.650	2.824	2.653	2.643	4.150	1.004	1.011	0.996	6.650	0.996	1.002	0.998
1.675	2.650	2.533	2.484	4.175	1.012	1.016	1.004	6.675	0.998	1.001	0.998
1.700	2.395	2.331	2.268	4.200	1.021	1.021	1.013	6.700	1.000	1.000	0.997
1.725	2.123	2.091	2.026	4.225	1.028	1.026	1.020	6.725	1.001	1.000	0.997
1.750	1.875	1.856	1.789	4.250	1.036	1.031	1.027	6.750	1.001	0.999	0.997
1.775	1.657	1.648	1.571	4.275	1.043	1.035	1.033	6.775	1.000	0.998	0.997
1.800	1.471	1.467	1.388	4.300	1.049	1.039	1.039	6.800	0.998	0.997	0.996
1.825	1.315	1.313	1.245	4.325	1.056	1.042	1.043	6.825	0.997	0.996	0.996
1.850	1.185	1.186	1.133	4.350	1.061	1.046	1.047	6.850	0.995	0.996	0.996
1.875	1.078	1.082	1.042	4.375	1.066	1.049	1.051	6.875	0.993	0.995	0.996
1.900	0.991	0.998	0.965	4.400	1.069	1.051	1.053	6.900	0.992	0.994	0.996
1.925	0.920	0.930	0.897	4.425	1.073	1.054	1.054	6.925	0.991	0.994	0.996
1.950	0.863	0.875	0.837	4.450	1.075	1.056	1.055	6.950	0.991	0.993	0.996
1.975	0.817	0.829	0.784	4.475	1.077	1.057	1.055	6.975	0.991	0.993	0.996
2.000	0.779	0.790	0.738	4.500	1.078	1.058	1.055	7.000	0.992	0.993	0.997
2.025	0.748	0.757	0.701	4.525	1.078	1.059	1.053	7.025	0.994	0.992	0.997
2.050	0.723	0.730	0.671	4.550	1.078	1.059	1.051	7.050	0.996	0.992	0.997
2.075	0.702	0.708	0.647	4.575	1.077	1.059	1.049	7.075	0.997	0.992	0.997
2.100	0.687	0.692	0.630	4.600	1.075	1.059	1.046	7.100	0.999	0.992	0.997
2.125	0.676	0.679	0.617	4.625	1.072	1.058	1.042	7.125	1.000	0.992	0.998
2.150	0.668	0.671	0.610	4.650	1.069	1.057	1.038	7.150	1.001	0.992	0.998
2.175	0.662	0.666	0.606	4.675	1.066	1.055	1.034	7.175	1.002	0.992	0.998
2.200	0.659	0.664	0.605	4.700	1.062	1.053	1.029	7.200	1.002	0.993	0.998
2.225	0.658	0.664	0.606	4.725	1.058	1.051	1.024	7.225	1.003	0.993	0.998
2.250	0.659	0.666	0.610	4.750	1.054	1.049	1.019	7.250	1.003	0.993	0.999
2.275	0.661	0.669	0.615	4.775	1.049	1.046	1.014	7.275	1.003	0.994	0.999
2.300	0.665	0.674	0.623	4.800	1.044	1.043	1.008	7.300	1.003	0.994	0.999

Q	S_{n1}	S_{n2}	S_X	Q	S_{n1}	S_{n2}	S_X	Q	S_{n1}	S_{n2}	S_X
2.325	0.671	0.679	0.633	4.825	1.039	1.040	1.003	7.325	1.003	0.995	1.000
2.350	0.679	0.686	0.646	4.850	1.034	1.037	0.998	7.350	1.003	0.995	1.000
2.375	0.689	0.695	0.661	4.875	1.029	1.034	0.992	7.375	1.003	0.996	1.000
2.400	0.702	0.707	0.678	4.900	1.024	1.030	0.987	7.400	1.004	0.996	1.000
2.425	0.718	0.723	0.697	4.925	1.019	1.027	0.982	7.425	1.005	0.996	1.001
2.450	0.737	0.742	0.717	4.950	1.015	1.023	0.977	7.450	1.005	0.997	1.001
2.475	0.758	0.766	0.739	4.975	1.011	1.020	0.973	7.475	1.006	0.997	1.001
2.500	0.783	0.792	0.762	5.000	1.007	1.017	0.969	7.500	1.006	0.998	1.001
2.525	0.810	0.820	0.786	5.025	1.004	1.013	0.966	7.525	1.007	0.998	1.002
2.550	0.839	0.851	0.811	5.050	1.001	1.010	0.962	7.550	1.008	0.998	1.002
2.575	0.870	0.883	0.838	5.075	0.998	1.007	0.960	7.575	1.008	0.998	1.002
2.600	0.903	0.916	0.866	5.100	0.996	1.005	0.957	7.600	1.008	0.999	1.002
2.625	0.938	0.950	0.896	5.125	0.994	1.002	0.956	7.625	1.009	0.999	1.002
2.650	0.972	0.983	0.927	5.150	0.993	0.999	0.954	7.650	1.009	0.999	1.002
2.675	1.007	1.015	0.960	5.175	0.992	0.997	0.954	7.675	1.008	0.999	1.003
2.700	1.041	1.047	0.992	5.200	0.991	0.996	0.953	7.700	1.008	0.999	1.003
2.725	1.074	1.076	1.025	5.225	0.991	0.994	0.954	7.725	1.008	0.999	1.003
2.750	1.103	1.103	1.056	5.250	0.991	0.993	0.955	7.750	1.007	0.998	1.003
2.775	1.128	1.128	1.086	5.275	0.991	0.992	0.956	7.775	1.007	0.998	1.003
2.800	1.150	1.149	1.113	5.300	0.991	0.991	0.958	7.800	1.006	0.998	1.003
2.825	1.168	1.169	1.138	5.325	0.992	0.991	0.960	7.825	1.005	0.998	1.003
2.850	1.182	1.186	1.159	5.350	0.993	0.991	0.963	7.850	1.004	0.997	1.003
2.875	1.194	1.200	1.176	5.375	0.994	0.991	0.966	7.875	1.003	0.997	1.003
2.900	1.203	1.212	1.188	5.400	0.995	0.991	0.969	7.900	1.003	0.996	1.002
2.925	1.210	1.221	1.196	5.425	0.996	0.991	0.972	7.925	1.002	0.996	1.002
2.950	1.216	1.228	1.198	5.450	0.998	0.992	0.976	7.950	1.001	0.995	1.002
2.975	1.220	1.233	1.197	5.475	1.000	0.993	0.979	7.975	1.001	0.994	1.002
3.000	1.222	1.235	1.193	5.500	1.002	0.994	0.983	8.000	1.000	0.994	1.002
3.025	1.222	1.235	1.187	5.525	1.003	0.995	0.986	8.025	1.000	0.993	1.002
3.050	1.220	1.232	1.179	5.550	1.006	0.996	0.990	8.050	0.999	0.993	1.002
3.075	1.217	1.227	1.170	5.575	1.008	0.997	0.993	8.075	0.999	0.992	1.001
3.100	1.211	1.221	1.160	5.600	1.010	0.998	0.996	8.100	0.999	0.991	1.001
3.125	1.203	1.212	1.148	5.625	1.012	0.999	0.999	8.125	0.998	0.991	1.001
3.150	1.194	1.202	1.136	5.650	1.014	1.001	1.001	8.150	0.998	0.990	1.001
3.175	1.183	1.190	1.122	5.675	1.016	1.002	1.003	8.175	0.998	0.990	1.001
3.200	1.170	1.177	1.106	5.700	1.018	1.003	1.006	8.200	0.999	0.989	1.000
3.225	1.155	1.163	1.090	5.725	1.019	1.005	1.008	8.225	0.999	0.989	1.000
3.250	1.139	1.148	1.073	5.750	1.020	1.006	1.009	8.250	0.999	0.988	1.000
3.275	1.122	1.132	1.056	5.775	1.021	1.007	1.011	8.275	0.999	0.988	1.000
3.300	1.105	1.115	1.038	5.800	1.021	1.008	1.012	8.300	1.000	0.987	1.000
3.325	1.086	1.099	1.022	5.825	1.021	1.009	1.013	8.325	1.000	0.987	1.000
3.350	1.068	1.082	1.006	5.850	1.021	1.010	1.014	8.350	1.000	0.987	1.000
3.375	1.050	1.065	0.991	5.875	1.021	1.011	1.015	8.375	1.000	0.986	0.999
3.400	1.032	1.048	0.977	5.900	1.020	1.012	1.015	8.400	0.999	0.986	0.999
3.425	1.014	1.032	0.964	5.925	1.021	1.013	1.015	8.425	0.999	0.986	0.999
3.450	0.997	1.016	0.952	5.950	1.022	1.013	1.015	8.450	0.998	0.985	0.999
3.475	0.981	1.001	0.941	5.975	1.023	1.014	1.015	8.475	0.997	0.985	0.999

using the Monte Carlo program DISCUS (Johnson 1974) and is incorporated using the term a_4 in eqn. (6). This correction was also small and approximately linear varying from 0·020 at $Q=0\,\text{Å}^{-1}$ to 0·013 at $Q=8\,\text{Å}^{-1}$.

Corrections for the finite resolution of the diffractometer are difficult, and in this experiment they have been largely avoided by using a powder diffractometer with a high resolution. From the half-widths of the Bragg peaks used in the wavelength calibrations, the value of dQ/Q was estimated to be about 0·025 for Q in the range 1–3 Å. Thus, over the major peak $S(Q)$ where the pattern is changing most rapidly, the resolution width $dQ \approx 0{\cdot}04$ is small compared with the width of the peak in $S(Q)$ and is comparable with the step size with which the data were recorded. It is estimated that the largest correction necessary, which occurs where the second derivative is highest (Johnson 1977), is at the peak of $S(Q)$ and is only 1·1% (for the $\lambda = 1{\cdot}22$ data). Over the rest of the Q range the correction does not exceed 0·01%, and consequently none has been applied in this analysis.

Spectrometer efficiency is taken into account by comparing the potassium data with those obtained from a known vanadium standard. It is incorporated using the term a_3 in eqn. (6), which is derived from the expression

$$a_3 = \Sigma_V r_V^2 \Big/ \Sigma_K r_K^2. \tag{5}$$

Calculating a_3 to high precision is not possible owing to the uncertainty in the cross-sections (Σ_V, Σ_K) and radii (r_V, r_K), of both the potassium and the vanadium samples. Fortunately a_3 may be determined experimentally from the requirement that $S(Q) \rightarrow 1{\cdot}0$ at large Q, and this constraint was used in this analysis.

Fig. 2

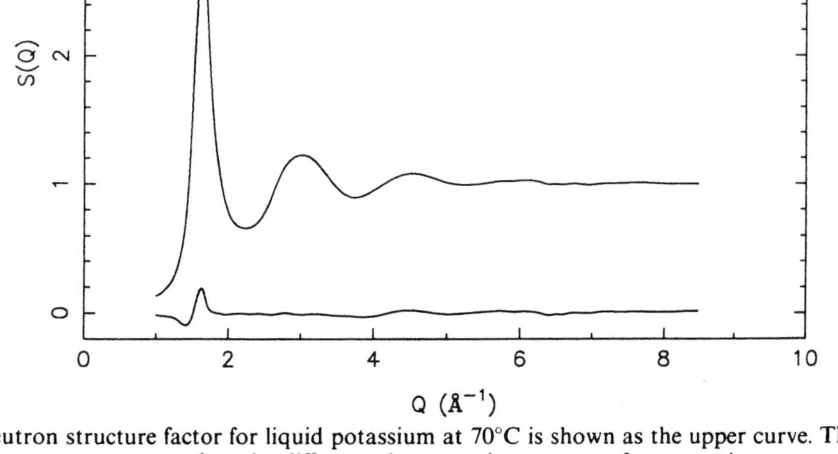

Neutron structure factor for liquid potassium at 70°C is shown as the upper curve. The lower curve plots the difference between the two sets of neutron data.

The expression from which $S(Q)$ was finally calculated was therefore

$$S(Q)=a_3\{a_2[c_1-c_2-a_1(c_3-c_2)]\}-a_4-\frac{\sigma_{\rm inc}}{\sigma_{\rm coh}}S_{\rm inc}(Q)+\left(1+\frac{\sigma_{\rm inc}}{\sigma_{\rm coh}}\right)a_5, \qquad (6)$$

where c_1 is the raw datum for sample + can + background, c_2 is the raw datum for background and c_3 is the raw datum for can + background.

In order to compare the neutron results, both between themselves and with the X-ray results, the data for $S(Q)$ had to be tabulated at the same set of Q values. This was achieved by fitting both the neutron and X-ray data with cubic splines (using the HARWELL subroutine library VC03A) and interpolating the result at a series of constant-Q intervals. The interpolated results for $S(Q)$ from both neutron experiments are listed in table 1, and the results for the first set shown graphically in fig. 2 (upper curve). The differences between the values of $S(Q)$ determined by the two sets of neutron runs made at $\lambda = 1{\cdot}22$ and $0{\cdot}878$ Å are plotted as the lower curve in fig. 2. The error at the first peak in $S(Q)$ is due to the different resolutions of the two sets of data. The variance of the difference between the two sets of data for $2{\cdot}0\,\text{Å}^{-1} < Q < 8{\cdot}52\,\text{Å}^{-1}$ is only 0·012. This error of about 1% between the two sets of neutron data is also consistent with the statistical error expected from the data.

§4. Comparison with X-ray data

The differences $\Delta S_1 = S_X - S_{n1}$ and $\Delta S_2 = S_X - S_{n2}$, between the X-ray and neutron results for $S(Q)$ for liquid potassium are illustrated graphically in fig. 3. The first point to be noted is that the differences are much smaller than those reported earlier by Dobson in his comparison of the published data on liquid sodium and aluminium. This is likely to be due to the increased precision with which these measurements have been made and the fact that the numerical values for the data did not have to be inferred from a small line drawing in a publication.

The values of ΔS below a Q value of $2\,\text{Å}^{-1}$ are comparable with the difference between the two neutron experiments and are probably due to resolution differences between the different measurements. Between 2·0 and $4\,\text{Å}^{-1}$, however, there appear to be a series of peaks apparent in both ΔS_1 and ΔS_2. The peaks are not apparent in the neutron difference plot (fig. 2) and do not appear to be correlated with any correction function. Their height, about 0·04, would appear to be significantly above the statistical error level of about 0·01, but only just. Interestingly, their positions would appear to be reasonable well described by a f.c.c. powder diffraction pattern, where the unit cell has a dimension of 7·1 Å. The positions of the first six peaks (*hkl*: 111, 002, 022, 113, 222, 004) for such a lattice are indicated in fig. 3 as vertical lines. Even more remarkable is the fact that the size of this f.c.c. unit cell is very close to the value expected (6·8 Å) if the single valence electron of each potassium atom were arranged in a f.c.c. lattice. This idea was first suggested by Dobson (1978) who found that ΔS for liquid sodium was fitted by a f.c.c. lattice ($a = 5{\cdot}38$ Å) again very close to the value (4·47 Å) expected if a single valence electron per atom is placed on a f.c.c. lattice. The half-width of the peaks appears to be in the range 0·15–0·25 Å^{-1}, which would indicate that the f.c.c. ordering extends over a range 25–40 Å, again similar to the result found by Dobson for liquid sodium.

There would appear to be some evidence, therefore, for the existence of electron crystallization in the two alkali metals studies so far, although the data from which this conclusion is drawn are still far from ideal. Clearly further data, especially from X-ray and neutron experiments performed on the same sample, would be most valuable.

Fig. 3

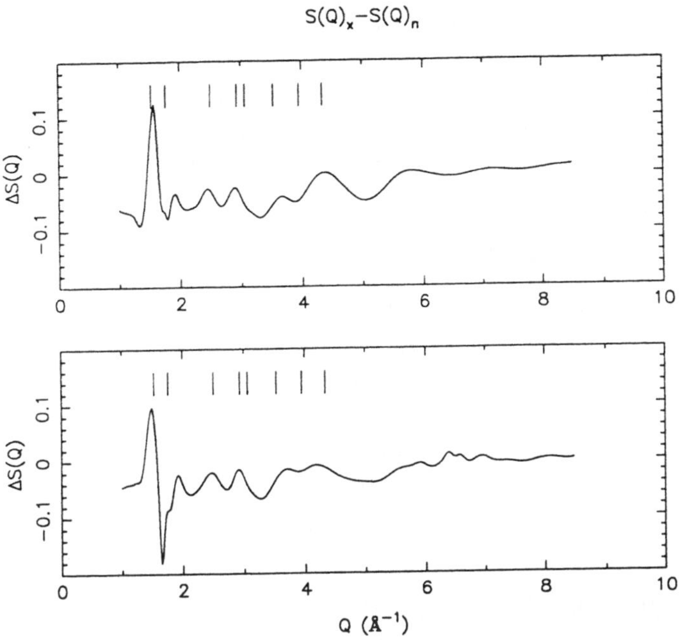

Difference between present neutron structure factors and X-ray structure factor of Greenfield *et al.* (1971).

§ 5. Theory

5.1. *Superposition of single-ion densities and pair potentials from density functional theory*

Earlier, density functional theory (DFT) has been used to calculate a pair potential in liquid sodium. Here we extend the arguments to liquid potassium at freezing. As in the earlier work on sodium (Perrot and March 1990a, b), let us still retain the superposition of single-ion displaced charges but focus then directly on the pair problem, utilizing the symmetry about a plane through the midpoint of the metallic bond, and perpendicular to that bond.

Let us start from the DFT for a single ion in the conduction electron liquid of density $\bar{\rho}$. This can be written formally as (for example Lundqvist and March (1983))

$$G'[\Delta+\bar{\rho}]-G'[\bar{\rho}]=-V(r), \tag{7}$$

where G' is simply shorthand notation for the functional derivative $\delta G/\delta\rho(r)$. Here, the functional $G=T+E_{xc}$ includes exchange–correlation effects in the local density approximation. V is the usual electrostatic potential created by the displaced charge $\Delta(r)$, namely

$$V=-\frac{Z}{r}+\frac{1}{r}*\Delta \tag{8}$$

where $*$ is simply shorthand for the convolution product.

With the two ions at positions R_i and R_j let us also write the displaced charge $\Delta_i=\Delta[r-R_i]$ and $\Delta G[\Delta+\bar{\rho}]=G[\Delta+\bar{\rho}]-G[\bar{\rho}]$. Then, for the pair interaction ϕ between ions i and j,

$$\phi=\Delta G[\Delta_i+\Delta_j+\bar{\rho}]-\Delta G[\Delta_i+\bar{\rho}]-\Delta G[\Delta_j+\bar{\rho}]$$
$$+\frac{Z}{r_i}\cdot\Delta_j+\frac{Z}{r_j}\cdot\Delta_i+\tfrac{1}{2}(\Delta_i+\Delta_j)\cdot\frac{1}{r}*(\Delta_i+\Delta_j)$$
$$-\tfrac{1}{2}\Delta_i\cdot\frac{1}{r}*\Delta_i-\tfrac{1}{2}\Delta_j\cdot\frac{1}{r}*\Delta_j+\frac{Z^2}{R}. \tag{9}$$

Next it is to be noted from eqn. (8) that

$$ZV(R)=-\frac{Z^2}{R}+\int\frac{Z}{|r'-R|}\Delta(r')\,\mathrm{d}r'$$
$$=-\frac{Z^2}{R}+\int\frac{Z}{r''}\Delta(r''+R)\,\mathrm{d}r''$$
$$=-\frac{Z^2}{R}+\frac{Z}{r_i}*\Delta_j. \tag{10}$$

Utilizing eqn. (10) in eqn. (9), one finds that

$$\phi=\Delta G[\Delta_i+\Delta_j+\bar{\rho}]-\Delta G[\Delta_i+\bar{\rho}]-\Delta G[\Delta_j+\bar{\rho}]$$
$$+\frac{-Z}{r_j}\cdot\Delta_i+\tfrac{1}{2}\Delta_i\cdot\frac{1}{r}*\Delta_j+\tfrac{1}{2}\Delta_j\cdot\frac{1}{r}*\Delta_i-ZV(R), \tag{11}$$

where the dots mean the integral of the product of functions through the whole of space; technically the scalar product.

The pair potential plotted in fig. 4 for liquid potassium can be regarded as derived from the equation (Perrot and March 1990b)

$$\phi(R)=\Delta G(R)-ZV(R)+\Delta U_1(R), \tag{12}$$

where the total potential energy change $\Delta U(R)$ is given by

$$\Delta U(R)=-ZV(R)+\Delta U_1(R), \tag{13}$$

which is determined solely by the total valence screening charge $Q(R)$; this is plotted in fig. 5 for potassium and integrates to the valency $Z=1$ as $R\rightarrow\infty$. The electron theory used to calculate $Q(R)$ in fig. 5 parallels that employed for sodium by Perrot and March (1990a, b). The corresponding electronic energy levels are recorded in table 2 and prominent features of $\phi(R)$ are collected in table 3. The features of the pair potential reflect, rather directly, the structure of $Q(R)$.

§6. Relation of electron theory to liquid structure

One of the problems underlying the inversion of the liquid structure factor $S(q)$ to extract $\phi(R)$ is the extreme cancellation known to occur in sodium between the potential mean force $U(R)$ and the three-body contribution to $\phi(R)$, outside a region of about 5 Å. Thus accurate input of the three-particle correlation function is necessary to reproduce the tail of the potential.

Fig. 4

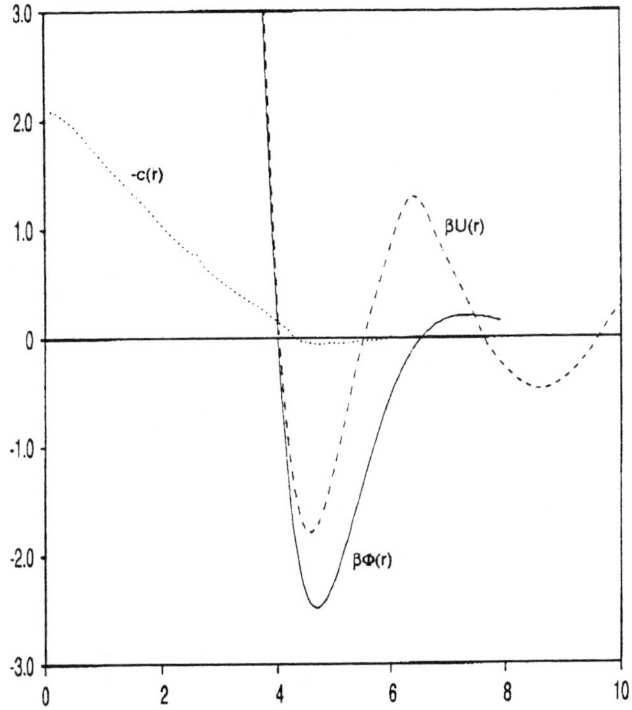

Electron theory pair potential $\phi(R)$ for liquid potassium at freezing. For comparison we have also plotted the potential $U(R)$ of mean force and the direct correlation function $c(r)$.

Fig. 5

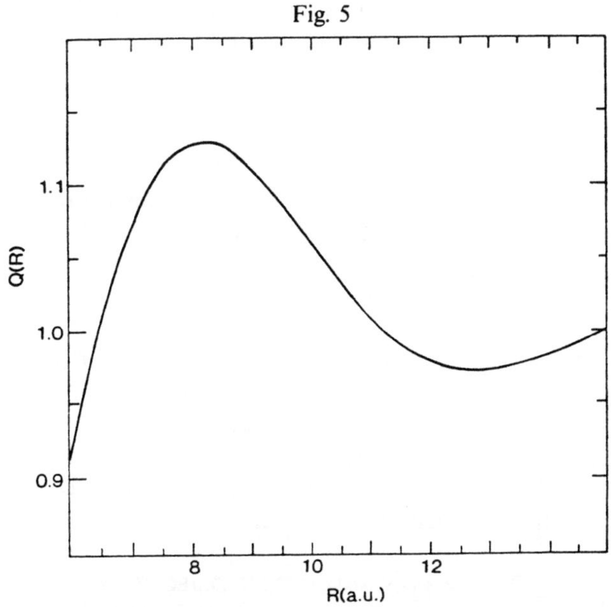

Total valence screening charge round a K^+ ion in a bath of electrons equal to valence electron density in liquid potassium. Note that $Q(R)$ tends to the valence $Z+1$ at sufficiently R.

Table 2. One-electron eigenvalues and properties of single screened K^+ ion in bath of valence electrons of density $\bar{\rho}=0{\cdot}001\,882\,5$ ($T=65°C$). $\langle r \rangle$ is the average of r, and $\langle r^2 \rangle$ is the average of r^2. Calculations were also made for a density corresponding to 135°C with $\bar{\rho}=0{\cdot}001\,845\,2$. The values obtained were essentially unchanged from those below except for the Friedel sum which was $-0{\cdot}002\,35$.

Quantum symbol	Eigenvalue (Ryd)	$\langle r \rangle$ (a.u.)	$\langle r^2 \rangle$ ((a.u.)^2)
1s	−256·49	0·0812	0·0090
2s	−25·31	0·3914	0·1815
2p	−20·20	0·3519	0·1536
3s	−2·19	1·2773	1·8935
3p	−1·01	1·4414	2·4813
Total energy E (Ryd)	−1196·44		
Friedel sum	−0·00247		

Table 3. Positions of prominent features in pair potential $\phi(R)$ for liquid potassium near freezing, from electron theory.

Average density $\bar{\rho}$	Principal minimum	First maximum	First node	Second node
0·001 882 6 ($T=65°C$)	8·857	13·86	7·588	12·39

Nevertheless, as shown by Tankeshwar and March (1992), the difference between $U(R)$ and $\phi(R)$ does reflect the direct correlation function $c(R)$ (fig. 6), which is known to be intimately related to $\phi(R)$ at large R through

$$c(R) \sim -\frac{\phi(R)}{k_B T}. \tag{14}$$

Therefore in fig. 4 we have constructed $U(R)$ from the 'measured' pair correlation function $g(R)$ for liquid potassium at freezing through

$$g(R) = \exp\left(\frac{-U(R)}{k_B T}\right) \tag{15}$$

for comparison with the electron theory $\phi(R)$. It can be seen that $U(R)$ and $\phi(R)$ are in excellent agreement at small R. We expect, from eqn. (14), that, at sufficiently large R, $\phi(R) = -k_B T c(R)$, but that this asymptotic value is only quantitative at larger R than in the figures reported in the present work. We are currently engaged in inverting the diffraction data on $S(Q)$ to determine a pair potential for direct comparison with the electron theory potential reported in the present work.

§ 7. Summary and future directions

Measurements of the neutron structure factor $S_n(Q)$ for liquid potassium just above freezing have been reported. The errors involved have been carefully analysed and discussed. The results have then been brought into direct contact with the X-ray

Fig. 6

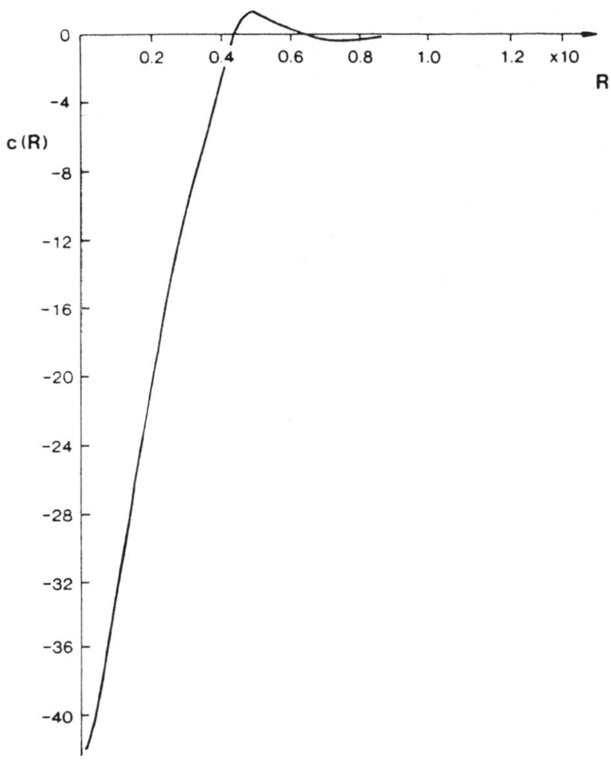

Direct correlation function $c(R)$ against R as calculated from X-ray data of Greenfield *et al.* (1971).

structure factor $S_X(Q)$ as measured by Greenfield *et al.* (1971). The differences, while considerably smaller than those reported for liquid sodium by Dobson (1978), do reveal medium-range order among the valence electrons, as inferred in the work of EMM. However, it is stressed that in future work it would be a great advantage to have X-ray and neutron measurements made together in any laboratory experiments.

The structure factor has then been brought into preliminary contact with a density functional calculation of the pair potential in liquid potassium at freezing. As is clear from fig. 4, there is already excellent accord for small separations between the potential of mean force $U(R)$, extracted from $g(R)$ derived by Fourier transform from $S(Q)$, and the electron theory pair potential $\phi(R)$. The experimental data of Greenfield *et al.* (1971) have also been used to calculate the direct correlation function $c(R)$, which when multiplied by the thermal energy $k_B T$ should give $\phi(R)$ at sufficiently large R. This asymptotic region has not been reached in the range of R presently accessible. As already mentioned, we are currently engaged in inverting the structure factor $S(Q)$ to derive a potential for direct comparison with the electron theory $\phi(R)$.

References

Dobson, P. J., 1978, *J. Phys.* C, **11,** L295.
Egelstaff, P. A., March, N. H., and McGill, N. C., 1974, *Can. J. Phys.*, **52,** 1651.
Gingrich, N. S., and Heaton, L., 1961, *J. chem. Phys.*, **34,** 873.

GREENFIELD, A. J., WELLENDORF, J., and WISER, N., 1971, *Phys. Rev.* A, **4**, 1607.
International Tables for X-ray Crystallography, 1969, Vol. II (Birmingham: Kynoch).
JOHNSON, M. W., 1974, Atomic Energy Research Establishment Report No. AERE R7682; 1977, Rutherford Appleton Laboratory Report No. RL-77-095.
LUNDQVIST, S., and MARCH, N. H. (editors), 1983, *Theory of the Inhomogeneous Electron Gas*, (New York: Plenum).
PERROT, F., and MARCH, N. H., 1990a, *Phys. Rev.* A, **41**, 4521; 1990b, *Ibid.*, **42**, 4884.
PLACZEK, G., 1952, *Phys. Rev.*, **86**, 377.
TANKESHWAR, K., and MARCH, N. H., 1992, *Phys. Chem. Liquids*, **25**, 59.
YARNELL, J. L., KATZ, M. J., WENZEL, R. G., and KOENIG, S. H., 1973, *Phys. Rev.* A, **7**, 2130.

Phys. Chem. Liq., 1999, Vol. 37, pp. 479–492
Reprints available directly from the publisher
Photocopying permitted by license only

Review Article

ELECTRONIC CORRELATION FUNCTIONS IN LIQUID METALS

N. H. MARCH*

Oxford University, Oxford, England

(Received 29 December 1997)

To determine experimentally the three pair correlation functions $g_{ii}(r)$, $g_{ie}(r)$ and $g_{ee}(r)$ in a pure liquid metal, i denoting ions and e electrons, requires three independent diffraction measurements. A brief review will be given in this difficult area, but progress is quite slow. One can make headway by confronting available experimental diffraction data with the results of computer experiments, and in particular on $g_{ie}(r)$. This will be illustrated with specific reference to recent computer simulations on liquid Mg and liquid Bi. For Mg, analytic modelling is also possible and this will be discussed.

Quite independently, computer experiments have recently appeared which describe the effects of isochoric heating on dense fluid hydrogen over a wide temperature range. This prompts again reference to analytic models, both caged atomic and molecular hydrogen being considered. Finally, though the electrical conductivity of the H plasma above has not yet been studied, a brief discussion of a possible mechanism of electronic transport in strongly coupled plasma will be presented.

Keywords: Two-component plasma; electron–ion correlation function; dense fluid hydrogen

1. INTRODUCTION

Substantially more than two decades ago, Egelstaff *et al.* [1] drew attention to the importance of extracting electronic correlation functions in liquid metals by combining X-ray, electron and neutron diffraction experiments. Then, for example, on liquid Mg, to be discussed at some length below, which can be considered as a two-

* Address for correspondence: 6, Northcroft Road, Egham, Surrey TW20 0DU, U.K.

component system (March and Tosi [2]) of ions Mg^{++} and electrons e^-, one could, at least in principle, extract the three partial structure factors $S_{nn}(q)$, $S_{nv}(q)$ and $S_{vv}(q)$, where n denotes nucleus and v is short for valence electrons. In early work (Watabe and Hasegawa [3], Chihara [4]), it was demonstrated that these three partial structure factors were related to the valence, say z, of the liquid metal by

$$S_{nv}(0) = z^{1/2} S_{nn}(0), \tag{1}$$

$$S_{vv}(0) = z S_{nn}(0), \tag{2}$$

while, from fluctuation theory, the nuclear structure factor $S_{nn}(0)$ at $q = 0$ is given by (see *e.g.*, Faber [5]):

$$S_{nn}(0) = n_i k_B T K_T, \tag{3}$$

n_i being the number of ions per unit volume, $k_B T$ the thermal energy and K_T the isothermal compressibility. These relations (1)–(3) will be important in some of the models to be discussed below (March and Tosi [6]).

In Section 2, the relation of the above three partial structure factors to X-ray scattering will be briefly summarized, following Egelstaff *et al.* [1]. However, it is important to note here that it has already been assumed in the above structural description that the electrons in a liquid metal such as Mg or Bi can be usefully classified into core and valence categories. In what follows, the core electrons will be assumed rigidly attached to the nuclei. Only the valence electrons therefore are described by the partial structure factors $S_{nv}(q)$ and $S_{vv}(q)$, and these two quantities will clearly enter the intensity of X-ray (and electron) diffraction from such liquid metals. However, the third partial structure factor, $S_{nn}(q)$, is directly accessible *via* neutron scattering. One recent neutron experiment which can be cited in the above context is on liquid K just above its freezing point (Johnson *et al.* [7]).

2. X-RAY SCATTERING FROM A LIQUID METAL

The X-ray scattering intensity $I_x(q)/N$ per atom is essentially the Fourier transform of the density–density correlation function

$\langle \rho(\mathbf{r}) \rho(\mathbf{r}') \rangle$:

$$\frac{I_x(q)}{N} = \int \langle \rho(\mathbf{r})\rho(\mathbf{r}') \rangle \exp(i\mathbf{q} \cdot \mathbf{r} - \mathbf{r}') d\mathbf{r}\, d\mathbf{r}' \tag{4}$$

Effecting the core-valence separation discussed above by writing $\rho = \rho_c + \rho_v$, one can rewrite Eq. (4) in the form

$$\frac{I_x(q)}{N} = f_c^2(q)\, S_{nn}(q) + 2\tilde{F}\langle \rho_c(\mathbf{r})\rho_v(\mathbf{r}') \rangle + \tilde{F}\langle \rho_v(\mathbf{r})\rho_v(\mathbf{r}') \rangle \tag{5}$$

where $\tilde{F}$ denotes Fourier transform. But as mentioned the core electrons are assumed 'tied' to nuclei and hence one has

$$\frac{I_x(q)}{N} = f_c^2(q)\, S_{nn}(q) + 2f_c(q)\, S_{nv}(q) + \tilde{F}\langle \rho_v(\mathbf{r})\rho_v(\mathbf{r}') \rangle \tag{6}$$

In Eqs. (5) and (6), $f_c(q)$ denotes the scattering factor of the core, *i.e.*, of Mg^{++} or Bi^{5+} for these two liquid metals to be treated further below. A similar expression for electron scattering has been written down by Egelstaff *et al.* [1].

2.1. Diffraction and Computer Experiments Relating to Electron–Ion Correlations

Direct experimental extraction of $S_{nv}(q)$ from diffraction measurements has so far proved difficult and progress has been therefore slow. The interested reader may refer to the review by Tamaki [8]; see however Steeb *et al.* [9]. We shall refer to such efforts again below, when we report, and analyze further, computer studies of electron–ion correlations.

We note, however, that for real progress using diffraction experiments it is essential to have high accuracy measurements using neutrons, X-rays and electrons on the same liquid metal under similar thermodynamic conditions. A full comparison of X-ray and neutron measurements on liquid K near freezing is presented by Johnson *et al.* [7] but unfortunately, to our knowledge, electron diffraction data is not available on this liquid metal near freezing at the time of writing.

3. COMPUTER SIMULATION ON ELECTRON – ION INTERACTION IN LIQUID MG

Though experimental diffraction studies to extract the electron – ion partial structure factor $S_{n\nu}(q)$ remain very difficult quantitatively, a major step forward in this general area has been taken by de Wijs *et al.* [10]. These workers follow Car and Parrinello [11] in combining molecular dynamics and density functional theory. They have, essentially, studied the valence electron density in liquid Mg (and also in liquid Bi, to be discussed also briefly below).

First of all, their calculated nuclear – nuclear structure factor $S_{nn}(q)$ can be compared with the measured $S_{nn}(q)$ for liquid Mg under comparable thermodynamic conditions. Experiment and theory are in good agreement.

3.1. Analytic Modelling to Expose Origin of Main Features in Electron – Ion Structure Factor $S_{n\nu}(q)$

On inspection of the computer data of de Wijs *et al.* [10], one sees that there is antiphase behaviour between $S_{nn}(q)$ and $S_{n\nu}(q)$ at the principal peak of the structure factor. In the modelling of March and Tosi [6], referred to as MT, it becomes quite clear what is the origin of this antiphase characteristic. Thus, MT give a formula for $S_{n\nu}(q)$ which they have used to analytically model the ratio $S_{n\nu}(q)/S_{nn}(q)$. The study of this ratio is motivated by Eq. (1), which relates the long wavelength limit of this quantity to the valence z(2 for liquid Mg near its freezing point). MT have then modelled $S_{n\nu}(q)/S_{nn}(q)$ using the simplest possible input. This is (i) a pseudopotential $v(k)$ representing the electron – ion interaction in Mg and (ii) a dielectric function $\varepsilon(q)$ accounting for metallic screening. No attempt was made to use refined choices. MT employ an Ashcroft pseudopotential for $v(q)$ characterized by a valence $z = 2$ and an Ashcroft radius already available in the literature. These workers simply used the Thomas-Fermi dielectric function

$$\varepsilon_{\mathrm{TF}}(q) = \frac{q^2 + q_{\mathrm{TF}}^2}{q^2} \tag{7}$$

where the Thomas-Fermi inverse screening length q_{TF} is determined from the Fermi wave number k_f by

$$q_{TF}^2 = \frac{4k_f}{\pi a_0} : a_0 = \frac{\hbar^2}{me^2}. \tag{8}$$

The main features of the computer data of de Wijs *et al.*, are then to be understood on the basis of the MT model summarised below. Their result employed is

$$\frac{S_{nv}(q)}{S_{nn}(q)} = \frac{z^{1/2}v(q)}{(4\pi e^2/q^2)}\left[\frac{1}{\varepsilon(q)} - 1\right] \tag{9}$$

The Ashcroft empty core form of $v(q)$ is

$$v(q) = \frac{-4\pi ze^2}{q^2}\cos(qR_c) \tag{10}$$

The literature value of $R_c = 1.39a_0$ was adopted for Mg, with $z = 2$ near freezing. The antiphase behaviour of $S_{nv}(q)$ with respect to the nuclear structure factor $S_{nn}(q)$ is due to the fact that the node in the pseudopotential occurs at a q value inside that characterizing the first peak in $S(q)$. Though no attempt was made to adjust the modelling to be fully quantitative, in fact all the essential features $S_{nv}(q)$ are already clear. The ratio $S_{nv}(q)/S_{nn}(q)$ according to the modelling of MT is displayed in Figure 1. We want to stress that in liquid Mg, the use of weak electron–ion interaction is valid.

To add a few comments relating to the comparison of the modelling shown in Figure 1 with the computer investigations on Mg, the node is at 2.1 Å^{-1} in the computer results: nearly predicted correctly by the MT model. The deepest value of the ratio shown in -0.13 whereas from the computer study of de Wijs *et al.*, it is -0.08.

MT also plot the same ratio, but now directly from the computer results for liquid Bi, and their results are shown in Figure 2. The extrapolated ratio from the data of de Wijs *et al.* [10] passes satisfactorily through the long wavelength limit $z^{1/2}$ with $z = 5$ for liquid Bi, as it must from the theoretical result (1) representing perfect screening, or equivalently the fact that long-range electric fields cannot exist in a metallic medium.

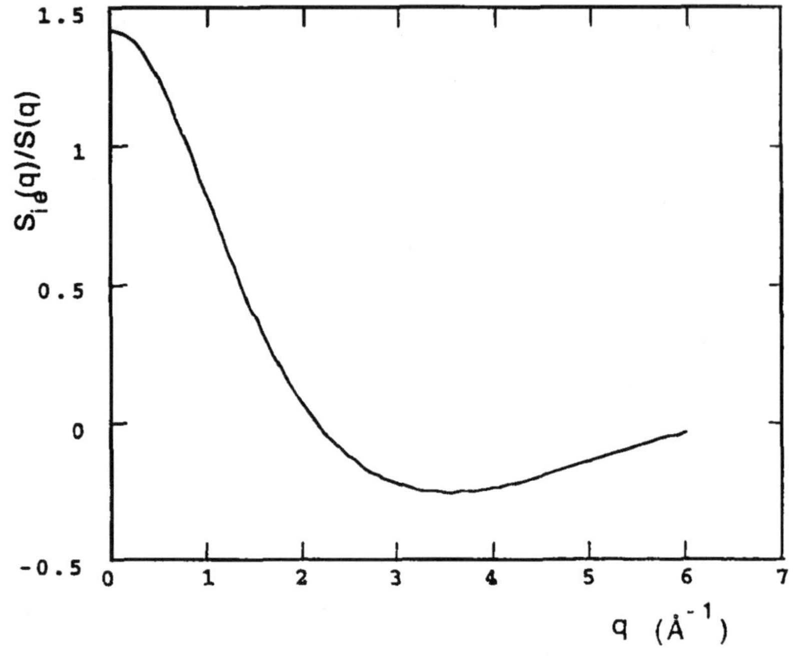

FIGURE 1 Ratio of electron–ion to nuclear–nuclear partial structure factors for liquid Mg near freezing, following March and Tosi [6]. Note that ratio at $q = 0$ is $z^{1/2}$, with valence $z = 2$ for Mg. Curve shown is constructed from Eq. (9) with $v(q)$, the electron–ion pseudopotential as in Eq. (10), with the Ashcroft radius $R_c = 1.39a_0 : a_0 = \hbar^2/me^2$. The Thomas-Fermi dielectric function in Eq. (7) was used in constructing curve in Figure. All the main features of the computer studies of de Wijs *et al.* [10] are apparent in the modelling displayed in this figure, with no adjustable parameters, as the literature value of R_c was adopted.

To sum up briefly, the assumption of weak electron–ion interaction, plus the simplest possible choices of electron–ion pseudopotential $v(q)$ in Eq. (10) and dielectric function $\varepsilon(q)$ in Eq. (7) satisfactorily account for the main features of the computer data of de Wijs *et al.* [10] on liquid Mg near freezing, as demonstrated in the work of March and Tosi [6]. However, such a simple treatment cannot be expected to work for liquid Bi, for though Hall effect measurements show that $z = 5$, crystalline Bi is a semimetal and this testifies to the fact that there the electron–ion interaction cannot be assumed weak.

de Wijs *et al.* [10] in their important study make contact with the diffraction data of Tamaki *et al.* [12]. Especially for Bi, it is a challenge

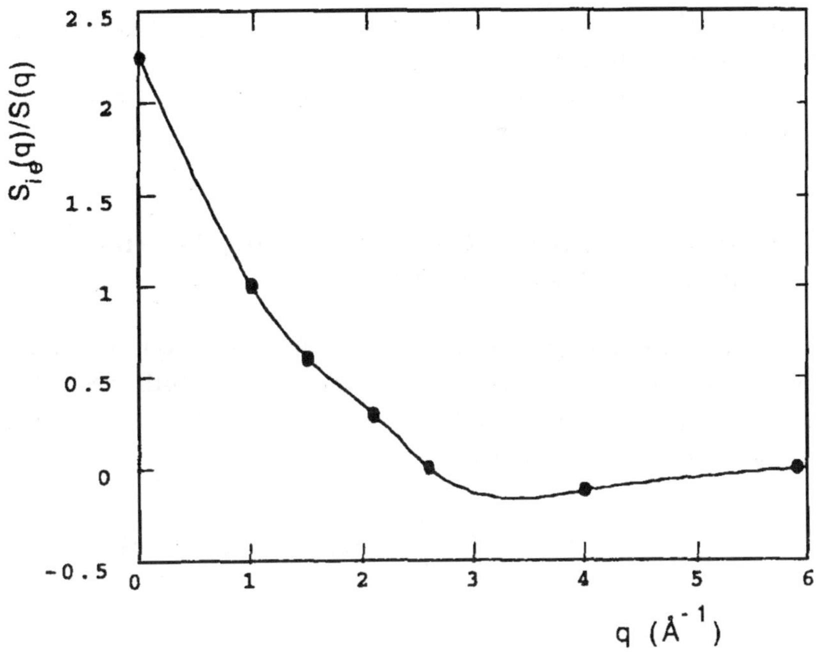

FIGURE 2 Similar to Figure 1 except that no model was invoked, the ratio of structure factors shown for liquid Bi being constructed by March and Tosi [6] directly from the computer data of de Wijs *et al.* [10]. There is no difficulty in extrapolating the computer results to pass through the theoretical limit $z^{1/2}$, given in Eq. (1) with $z = 5$.

for the experimentalist to see whether a quantitative account of Figure 2 can be obtained directly from experiment.

We conclude this section with a few brief comments on electron–electron correlations. It is tempting, from Eq. (6) for the X-ray intensity, to try to exploit the fact that the node in $S_{nv}(q)$ for Bi at around 2.1 Å^{-1} should 'expose' the electron–electron term at this value of q. Again, it is a challenge to experimentalists to see whether there one can establish with certainty a difference between $I_x(q)/N$ and $f_c^2(q)\ S_{nn}(q)$, as required by Eq. (6). This is presumably going to require considerable accuracy not only in the neutron and X-ray experiments needed for $S_{nn}(q)$ and $I_x(q)$ respectively but also in the quantum chemical calculation of the 'core' scattering factors $f_c(q)$. It remains of considerable interest to see if electron–electron correlations can be exposed by such diffraction studies. Refinement of electron diffraction measurements (Brah *et al.* [13, 14]) seems also a matter of high priority in this area.

We turn next to discuss electronic properties in dense fluid hydrogen [15].

4. COMPUTER SIMULATION OF DENSE FLUID HYDROGEN UNDER ISOCHORIC HEATING

To briefly introduce this section on dense fluid hydrogen, we begin by writing the internal energy E, which involves both electronic kinetic energy K and the kinetic energy of the ions K_{ion}. The result is (see *e.g.*, March and Tosi [16]):

$$E = K_{\text{ion}} + K + \frac{1}{2n_i}\int d\mathbf{r}\left[n_e^2 g_{ee}(r)\frac{e^2}{r} + n_i^2 g_{ii}(r)\, v_{ii}(r) + 2n_i n_e g_{ie}(r)\, \tilde{v}(r)\right] \tag{11}$$

In Eq. (11), the three partial radial distributions are evidently involved. The bare ion–ion potential is denoted by $v_{ii}(r)$, while $\tilde{v}(r)$ is the electron–ion interaction potential.

The computer simulation experiment of Magro *et al.* [15] has yielded the electronic kinetic energy K over a wide range of temperature T, for three densities. With the usual plasma parameter r_s measured in units of the Bohr radius a_0, related to the mean electron density ρ by

$$\rho = \frac{3}{4\pi r_s^3 a_0^3} \tag{12}$$

Figure 3 of the article by Magro *et al.* plots K *vs* T for r_s = 1.86, 2.0 and 2.2.

In the low temperature regime, one has molecular H_2 fluid whereas in the very hot assembly one has an atomic fluid. March and Tosi [17] attempted to gain insight, *via* admittedly somewhat crude models, into the variation of K with r_s (at constant temperature T say) in both the atomic and the molecular fluid regions.

4.1. H Atom in Cage

March and Tosi [17], in the high T regime of the computer results of Magro *et al.*, have analyzed the ground-state electronic kinetic energy K by using the model of a H atom enclosed by a spherical boundary of radius a with the proton at the centre. The wave function of the electron is put to zero at a. Using $2s$ and $3s$ wavefunctions ($\psi_{no}(r)$), each truncated at their innermost node, March and Tosi have plotted $K(a)$.

Subsequently, the writer has noted the 'boundary perturbation theory' developed by Barton *et al.* [18]. This allows one to calculate analytically the change $\Delta K = K(a) - K(\infty)$, where $K(\infty)$ is the usual H atom result

$$K(\infty) = -E(\infty) = \frac{e^2}{2a_0} \tag{13}$$

as follows from the usual hydrogenic $1s$ level plus the virial theorem.

Writing

$$u(r) = r\psi_{10}(r, a = \infty) \tag{14}$$

Barton *et al.* show first that one can write the energy change $\Delta E(a)$ entirely in terms of quadratures on the free space ($a = \infty$) quantity $u(r)$. Their result is given by

$$\frac{2m}{\hbar^2}\,\Delta E(a)\int_0^a \frac{dr_1}{u^2(r_1)}\int_0^{r_1} dr_2\, u^2(r_2) = 1. \tag{15}$$

Inserting $\psi_{10}(r)$ into Eqs. (14) and (15), one finds

$$\frac{\Delta E(a)}{|E(\infty)|} = 8\left(\frac{a}{a_0}\right)^2 \exp\left(\frac{-2a}{a_0}\right)\left[1 + O\left(\frac{a_0}{a}\right)\right]. \tag{16}$$

Now one invokes the virial theorem in the form

$$\Delta K + \Delta E = -\frac{ad\Delta E}{da} \tag{17}$$

which relates $\Delta K(a)$ and $\Delta E(a)$. Combining therefore Eqs. (16) and (17) yields the desired result

$$\Delta K(a) = 16|E(\infty)|\left(\frac{a}{a_0}\right)^3 \exp\left(\frac{-2a}{a_0}\right)\left[1 + O\left(\frac{a_0}{a}\right)\right] \tag{18}$$

This form (18) at small (a_0/a) fits well between the exact results of March and Tosi [17] from the 2*s* and 3*s* truncated wave functions and the value zero at $a = \infty$. The truncated 2*s* wave function at its node $2a_0$ corresponds to the Magro calculation for $r_s = 2$. The cage model H gives $K(2a_0) = 2.8$ Ryd which is substantially larger than the computer value of 1.6 Ryd. Similarly for $r_s = 1.86a_0$, the model gives ~ 0.2 Ryd, while the computer value is nearer 0.05 Ryd. The cage model therefore substantially overemphasizes the effect of compression on fluid hydrogen. This is confirmed by March and Tosi [17] by results obtained by softening the (infinite) barrier of the cage model.

4.2. H_2 Molecule With Electron – Electron Interactions, in Appropriate Cage

Using the careful study for LeSar and Herschbach [19], including electron – electron correlations, March and Tosi have repeated the calculations of K *vs* volume for the H_2 molecule in a spheroidal cage. Again, the gist of the variation found in the computer experiment is given by the model, but again the effect of compression is overemphasized by the infinite barrier model. As well as the kinetic energy, the internuclear distance of H_2 as a function of r_s is given in semiquantitative form by the cage model.

We now turn to the question of electrical conductivity, which was not reported on by Magro *et al.* [15] for dense fluid hydrogen in the final section.

5. MODEL OF ELECTRICAL CONDUCTIVITY OF STRONGLY COUPLED PLASMA

The idea presented in an earlier study (March and Tosi [20]) of the electrical conductivity of Cu plasma was to start from the Nernst-Einstein relation between mobility μ and electronic diffusion coefficient, D, namely

$$\mu = \frac{D}{k_B T} \tag{19}$$

where k_BT is the thermal energy associated with temperature T. Writing the conductivity in terms of the electron density n_e through

$$\sigma = n_e e^2 \mu \tag{20}$$

one finds immediately using Eq. (19)

$$\sigma = \frac{n_e e^2 D}{k_B T} \tag{21}$$

Rearranging Eq. (21) so that $Dn_e^{2/3} \propto D/(\text{length})^2$ appears, as in the previous work, (March and Tosi [20]) Eq. (21) becomes

$$\sigma n_e^{1/3} T = (Dn_e^{2/3})\left(\frac{e^2}{k_B}\right) \tag{22}$$

For monovalent Cu, $Dn_e^{2/3} = Dn_i^{2/3}$ turned out to be independent of the thermodynamic state in the strongly coupled plasma regime (March and Tosi [20], DeSilva and Kunze [21]).

This has therefore motivated a plot for W plasma of the quantity Y defined by

$$Y = \sigma n_i^{-1/3} T \tag{23}$$

as a function of temperature T. Note that Y differs from the left-hand side of Eq. (22) by a factor proportional to the cube root of the effective valence Z.

While σ and n_i are well specified in the paper of Kloss *et al.* [22] on tungsten plasma there is some question as the choice of data for the temperature. Thus, in the upper part (a) of their Figure 5, they plot measured temperature of the wire *vs* n_i for a wire explosion with lowest and highest energy input. Data has been used from both these curves. The product $Y = \sigma n_i^{-1/3} T$ turns out in both cases, even though there is considerable scatter at least partly coming from reading off of the data, to decrease substantially with increasing temperature T. The above argument, based on Eqs. (19)–(23), would interpret this decrease as reflecting the variation of $Z^{1/3}$ with temperature.

Under 'normal' conditions, Slater [23], for example, records the outer electronic configuration of the W atom as sd^5, suggesting, as noted also by Kloss *et al.*, that near the melting point $Z = 6$. Likalter in a private communication to Kloss *et al.*, has proposed $Z \approx 3$ at the critical point, and for the maximal observed expansion in their work Kloss *et al.*, propose $Z \approx 1$. These values are not inconsistent with the curves given in Figures 1 and 2 of March and Tosi [24] bearing in mind the substantial scatter among the points plotted.

To sum up, the earlier interpretation offered from the results on Cu plasma obtained in the conductivity measurements of DeSilva and Kunze [21] has been shown by March and Tosi [24] to permit the extraction of the effective valence Z as a function of temperature for W plasma from conductivity measurements. The results are quite compatible with the proposals made by Kloss *et al.* [22].

Subsequently, and in a somewhat different context, the same model as above has been used to interpret conductivity data on double-walled C nanotubes (March, Alonso and Rubio [25]).

6. SUMMARY AND FUTURE DIRECTIONS

The advent of computer studies such as those of de Wijs *et al.* [10] on liquid Mg and Bi near freezing and Magro *et al.* [15] on isochoric heating of dense fluid H over a vast range of temperature has greatly aided progress in the two-component theory of liquid metals. While Mg and Bi in the liquid state studied by de Wijs *et al.*, are undoubtedly good metals, it remains to be seen what is the electrical conductivity behaviour of dense fluid hydrogen over the range of thermodynamic states studied by Magro *et al.* [15].

However, it is highly relevant in the above context to record that shock wave measurements of Weir *et al.* [26] on liquid H and liquid D have unambiguously established that an insulator-metal transition occurs at high temperatures and at a pressure ~ 1.5 megabar. The point we need to stress here is that in the metallic phase observed to date there is coexistence with chemical bonding. Though the actual electrical resistivity is comparable with that observed in the heavy alkali liquid metal Rb (see Weir *et al.* [26]), the behaviour of liquid H and D seems more akin to that found in solid iodine, though in a much more accessible thermodynamic range in this latter material. Here, beyond question, the experimental electrical resistivity measurements of Drickamer [27] show a metallic phase at ~ 150 kbar, while at the same time diffraction experiments clearly reveal a 'molecular' metal with a well defined I_2 bond length. These two observations have been reconciled in the study of Siringo *et al.* ([28]; see also [29]). The mechanism of the metal-insulator transition here is energy band overlap, the gap in the insulating/semiconducting phase being gradually closed by application of laboratory pressures.

However, it must be emphasized that in the experiments of Weir *et al.* [26] the high temperatures are playing an important role in the insulator-metal transition in dense fluids H and D. These authors, in fact, conjecture that in the cold solids the applied pressure will probably need to be increased to the order of 3.5 megabar.

The work on H and D certainly needs following with both experiment and theory, particularly on electrical conductivity. The Nernst-Einstein approach of Section 5 would suggest electronic hopping as a possible mechanism here. One important consequence of that would be that the high-frequency conductivity $\sigma(\omega)$ should have power law behaviour, of the form σ proportional to ω^n with the exponent $n \sim 0.8$. This is to be contrasted with Drude-Zener type behaviour of the form $\sigma(\omega) = \sigma(0)/[1 + \omega^2\tau^2]$, with τ the relaxation time. For $\omega\tau \gg 1$ this leads to $\sigma(\omega) \alpha \omega^{-2}$.

To return finally to the matters raised in the Introduction, systematic work on one specific liquid metal is much needed. Since accurate X-ray and neutron experiments now exist on liquid K near freezing, this seems an obvious liquid metal on which to study electron diffraction. In parallel, it would be of interest if the computer simulation study paralleling those on liquid Mg and Bi could also be carried out on liquid K near freezing.

Acknowledgement

This article could not possibly have come to fruition without the great help of Professor Mario Tosi. It is a pleasure to acknowledge that the recent work reported was carried out during the tenure of visiting appointments in his group at the Scuola Normale Superiore in Pisa. Thanks are also due to Professor Renato Pucci and his colleagues at the University of Catania for invaluable discussions on the high-pressure behaviour of condensed iodine and condensed hydrogen.

Partial financial support for the work on liquid metals is acknowledged from the EC Network on 'High energy density matter', coordinated by Professor J. M. Martinez-Val (Madrid). Finally the writer is indebted to Professor Gabor Kalman for inviting him to deliver a Plenary Lecture at the Boston (1997) Conference on 'Strongly Coupled Plasma'. The present review resulted from that lecture.

References

[1] Egelstaff, P. A., March, N. H. and McGill, N. C. (1974). *Can. J. Phys.*, **52**, 1651.
[2] March, N. H. and Tosi, M. P. (1973). *Ann. Phys.* (NY) **81**, 414.
[3] Watabe, M. and Hasegawa, M. (1973). 2nd *Int. Conf. Liquid Metals* (Taylor and Francis), p. 133.
[4] Chihara, J., Ref. 3, p. 137.
[5] Faber, T. E. (1972). Theory of Liquid Metals (Cambridge University Press, UK).
[6] March, N. H. and Tosi, M. P. (1998). Laser and particle beams, **16**, 71.
[7] Johnson, M. W., March, N. H., Perrot, F. and Ray, A. K. (1994). *Phil. Mag.*, **B69**, 965.
[8] Tamaki, S. (1987). *Can. J. Phys.*, **65**, 286.
[9] Steeb, S., Knoll, W. and Lamparter, P. Z. (1983). *Naturforsch.*, **38a**, 395.
[10] de Wijs, G. A., Pastore, G., Selloni, A. and van der Lugt, W. (1995). *Phys. Rev. Lett.*, **75**, 4480.
[11] Car, R. and Parrinello, M. (1985). *Phys. Rev. Lett.*, **55**, 2471.
[12] Tamaki, S., Takeda, S. and Waseda, Y. (1994). *J. Phys. Soc. Japan*, **63**, 1794.
[13] Brah, A. S., Virdhee, L., Unvala, B. and March, N. H. (1979). Inst. of Phys. Proc. (UK).
[14] Brah, A. S., Virdhee, L. and March, N. H. (1980). *Phil. Mag.*, **42**, 511.
[15] Magro, W. R., Ceperley, D. M., PieRLeoni, C. and Bernu, B. (1996). *Phys. Rev. Lett.*, **76**, 1240.
[16] March, N. H. and Tosi, M. P. (1984). Coulomb Liquids (Academic: New York).
[17] March, N. H. and Tosi, M. P. (1996). *Nuovo Cimento*, **18D**, 1061.
[18] Barton, G., Bray, A. J. and McKane, A. J. (1990). *Am. J. Phys.*, **58**, 751.
[19] LeSar, R. and Herschbach, D. R. (1981). *J. Phys. Chem.*, **85**, 2798.
[20] March, N. H. and Tosi, M. P. (1995). *Phys. Chem. Liquids*, **29**, 273.
[21] DeSilva, A. W. and Kunze, H.-J. (1994). *Phys. Rev.*, **E49**, 4448.
[22] Kloss, A., Motzke, T., Grossjohann, R. and Hess, H. (1996). *Phys. Rev.*, **E54**, 5851.
[23] Slater, J. C. (1960). Quantum theory of atomic structure, Vol. I (McGraw-Hill, New York), p. 194.
[24] March, N. H. and Tosi, M. P. (1997). *Phys. Chem. Liquids*, in press.
[25] March, N. H., Alonso, J. A. and Rubio, A. (1997). *Phys. Stat. Solidi*, in press.
[26] Weir, S. T., Mitchell, A. C. and Nellis, W. J. (1996). *Phys. Rev. Lett.*, **76**, 1860.
[27] Drickamer, H. G. (1965). *Solid State Physics*, **17**, 1.
[28] Siringo, F., Pucci, R. and March, N. H. (1989). *High Pressure Research*, **8**, 1.
[29] March, N. H. (1992). *J. Math. Chem.*, **4**, 271.

ELSEVIER

Journal of Non-Crystalline Solids 250–252 (1999) 1–8

JOURNAL OF NON-CRYSTALLINE SOLIDS

www.elsevier.com/locate/jnoncrysol

Section 1. Liquid metals – electronic properties and structure

Local coordination, electronic correlations and relation between thermodynamic and transport properties of sp liquid metals

N.H. March [a,b]

[a] *University of Oxford, Oxford, UK*

[b] *Department of Physics, University of Antwerp, RUCA, Antwerp, Belgium*

Abstract

A brief discussion on local coordination in expanded alkalis will be followed by a short review of recent progress in determining electronic correlation functions by combining experiment using diffraction techniques with computer simulation. Then both critical point properties and melting temperatures of liquid sp metals will be discussed. In the latter case, the main topic will be on a metal model which will be used to correlate the ratio of surface tension to shear viscosity with a characteristic velocity. Conventional choice is to take this velocity as the thermal value $(k_B T/M)^{1/2}$. This choice has some merit. However an alternative is to use the velocity of sound, and a different formula then emerges which depends on the valency Z. Reference to some experimental support for such a formula is given. Finally, connection with diffusion, and with bulk viscosity, is discussed with some involvement of a collective mode model, such as prompted by early neutron inelastic scattering results on Rb near its melting point.

1. Introduction

In this article, attention will be given to the structure, thermodynamics and transport properties of liquid sp metals. Special attention will be paid to properties near (a) the critical point and (b) the melting temperature T_m. The critical point discussion is brief, but points to the interest in studying from first principles the compressibility ratio of the alkalis especially in relation to insulating liquids such as the condensed rare gases and diatomic liquids such as N_2 and O_2. Here is a topic which, in spite of the considerable advances using renormalization group methods, has to be fully integrated with the well-known statistical mechanical hierarchy. This area is considered in Section 2. Then in Section 3 recent progress in using electronic correlation functions in the structural description of sp liquid metals is reviewed.

Section 4 deals with dynamical structure and transport, but now using admittedly simplified models to describe some salient features discovered in an early work using neutron inelastic scattering. This description leads to a brief discussion of (i) the possible relation between shear and bulk viscosities in relation to the long wavelength structure factor and (ii) the inter-relation between a thermodynamic property, surface tension and a transport property such as shear viscosity. The fact that the (collective) sound velocity may affect this correlation for sp metals as the valence is increased (say for Sn) is proposed, by using an electron gas model i.e. a low-order

0022-3093/99/$ – see front matter
PII: S0022-3093(99)00336-1

density gradient framework. Finally, a summary will be given, together with some suggestions for future work, both experimental and theoretical.

2. Critical point properties: comparison of expanded alkali metals with other insulating liquids

I wish here to summarize some empirical and semi-empirical results on critical point properties, with attention again given mainly to expanded alkali metals. I introduce this topic by referring briefly to the recent study on the condensed rare gas systems in relation to corresponding facts on the expanded alkali metals. This discussion is prompted by the pioneering experiments of Hensel and coworkers on liquid Cs and Rb taken along the liquid–vapour coexistence curve towards the critical point.

Hensel et al. [1] concluded that the main decrease of the density that occurred was indicated in their structure factors obtained by neutron scattering via a decrease of the coordination number, z, with only a modest increase in the near-neighbour distance. This increase was, at the time, a most surprising result in that theorists had regarded the alkali metals, whether solid or liquid, as prime examples of the nearly free electron model. The experiments of Hensel et al., on the contrary, testified to the value of a 'chemical' model, in which the bond was the basic building block for the smaller density alkalis. I showed [2] that a useful approximation to the mass density d could be written

$$d = az + b, \tag{2.1}$$

where a and b are known constants.

Some further discussion of the dependence of metallic properties on local coordination number z is given in Appendix A, where reference is made to both solid and liquid properties, with the possibility that 'solid' might include both crystalline and amorphous materials. Here I will restrict attention to comparing and contrasting the effect on one combination of critical constants, p_c, ρ_c and T_c, namely the so-called compressibility ratio Z_c:

$$Z_c = p_c/\rho_c k_B T_c. \tag{2.2}$$

Table 1 then compares the Z_cs for insulating liquids, and in particular the rare gases with some homonuclear liquids added to assess the possible effect of bonding (e.g. N_2 and O_2), with known values for the liquid alkali metals.

March et al. [3] have used recently the BBGKY hierarchy to extract information on structural integrals from the constancy of Z_c at 0.29 for the rare gas elements such as Ar, Kr and Xe, within an assumed pair potential framework. The analysis points to the quantitative failure of the Van der Waals equation of state, but to the almost quantitative value of the Dieterici equation of state [4]. What is, at first sight, surprising is that the chemical bond in N_2 and O_2 does not alter the compressibility ratio from $Z_c = 0.29$.

But as Chapman and March [5] initially emphasized (see also the works of Likalter [6], Goldstein et al. [7], and Hess [8]), the situation is completely different in the alkali metals, as can again be seen by referring to the lower part of Table 1. Chapman and March [5] concluded that Coulomb interaction was crucial in understanding the Z_cs in the alkalis. So that, in spite of the bond being a building block in Cs and Rb near criticality, one is far from dealing with a system of

Table 1
Compressibility ratio, Z_c, at critical point for some insulating liquids compared with corresponding ratios in liquid alkali metals

Substance (insulating fluids)								
	H_2	He	N_2	O_2	Ar	Kr	Xe	CO
Z_c^{-1}	3.05	3.08	3.42	3.42	3.4	3.4	3.4	3.54
Substance (expanded liquid alkali metals) [a]								
	Li	Na	K	Rb	Cs			
Z_c	0.064	0.132	0.175	0.217	0.203			

[a] After Chapman and March [5]. These workers also pointed out that the critical constants T_c and V_c are related by $T_c V_c^n$ = constant. For the rare gas fluids, $n = -2$, while in complete contrast for the alkalis $n = 0.28$.

'neutral' chemically bonded species. To describe such a scenario structurally, the proposal by Egelstaff et al. [9] to describe the structural properties of a pure liquid metal such as Cs as a two-component mixture of ions (i) and of valence electrons (v) appears relevant, and therefore a summary of recent progress in this topic will be given below.

3. Local coordination and electronic correlation functions

We consider here, following the brief discussion of critical point data for the alkali metals, the possible description for the future of the expanded liquid metals Cs and Rb in terms of electronic correlation functions. That a liquid metal such as an alkali was a two-component system consisting of ions and electrons, and that therefore it required, as with molten NaCl, three pair correlation functions, to describe its structure was emphasized especially in the early studies of Chihara [10], Watabe et al. [11] and March and Tosi [12]. The nuclear (or ion) structure factor, $g(r)$, is already known from the neutron studies of Hensel et al. [1] for expanded Cs at a number of thermodynamic states along the liquid-vapour coexistence curve.

Egelstaff et al. [9] proposed a combination of electron and X-ray data as a route to the extraction of the two remaining correlation functions, valence electron-ion, g_{vi}, and valence–valence electron correlation, $g_{vv}(r)$. Unfortunately, progress has been slow by purely diffraction studies. However, a breakthrough has occurred recently in obtaining $g_{vi}(r)$ in liquid metals Mg and Bi [13]. March et al. [14] have reviewed these studies elsewhere [14], so that I shall consider below the computations of Ceperey et al. [15] on hydrogen plasma. Here, one can see the transition in isochoric studies from a molecular assembly of H_2 species at lower temperatures to an atomic system at other temperatures. In the course of this work, Ceperley et al. calculated the kinetic energy of the electrons. March and Tosi [16] have interpreted the gist of their results as arising from the kinetic energy of localization, by treating H and then H_2 enclosed in a suitably shaped 'box'. But what I want to emphasize here is the illuminating character of $g_{vv}(r)$ – of course there is only one electron per atom here and so problems that may arise in plasmas of distinguishing core and valence electrons as in Cs and Rb are absent. So it would clearly be very interesting if the work on Mg and Bi, in which the main point was $g_{vi}(r)$, could be extended to Cs and Rb in expanded states, and the question of $g_{vv}(r)$ could then be addressed by X-ray experiments to complement the available neutron studies of Hensel et al. [1].

With this rather brief discussion of (a) local coordination in expanded alkalis, plus the work of Ceperley et al. on hydrogen [15], and (b) possible further experiments, both laboratory and computer, we turn to the final topic to be treated, namely the relation between thermodynamic and transport coefficients in liquid metals near melting temperature T_m.

4. Relation between surface tension and shear viscosity at melting point

Several workers have noted that in the very simplest discussion of liquid surface tension, γ, by Fowler [17], who writes in terms of the radial distribution function, $g(r)$, and (assumed) pair potential, $\phi(r)$, as

$$\gamma = \frac{\pi}{8}\rho^2 \int_0^\infty \mathrm{d}r\, r^4\, \frac{\partial \phi(r)}{\partial r} g(r), \tag{4.1}$$

where ρ is the atomic number density, there is a direct relation to the shear viscosity formula of Born and Green [18], which involves precisely the same structural integral as in Eq. (4.1). Elimination of this integral then yields (see especially Egry [19]):

$$\frac{\gamma}{\eta}\left(\frac{M}{k_B T}\right)^{1/2} = \text{constant}, \tag{4.2}$$

where $k_B T$ measures the thermal energy, and one has in Eq. (4.2) introduced the measure $(k_B T)^{1/2}/M^{1/2}$ of the thermal velocity of an atom of mass M in liquid metal at temperature T. We shall be concerned dominantly (though not exclusively) with the choice $T = T_m$.

Alonso and March [20] have evaluated the ratio (4.2) for a number of liquid metals at the melting point. The constant in Eq. (4.2) is found to be 1.0 ± 0.2, except for some polyvalent metals including In and Sn for which the constant is 1.7. We shall return to the discussion of these exceptions in the penultimate section of the present paper.

Prompted by work on organic liquids [21] and on Lennard–Jones liquids [22], we shall attempt to introduce in place of the thermal velocity entering the dimensionless ratio in Eq. (4.2), a velocity reflecting collective behaviour of the liquid metal. The collective mode demonstrated in liquid Rb, [23], immediately motivates the use of the velocity of sound, c, in the present context. We turn therefore, in the following section, to consider the ratio $\gamma/\eta c$, at the melting point T_m of simple sp liquid metals.

5. Introduction of velocity of sound through the dimensionless ratio $\gamma/\eta c$ at the melting temperature T_m

Let us obtain orientation on the magnitude of the new dimensionless ratio proposed here, namely $\gamma/\eta c$, at the melting temperature T_m, by inserting the very simplest formulae currently available. For shear viscosity, one has the Andrade formula [24,25]:

$$\eta = A\, T_m^{1/2} \rho^{2/3} M^{1/2}. \tag{5.1}$$

For the surface tension, Brown and March [26] considered in an early work the simple model of a liquid metal surface in which the conduction electrons are assumed to move in their own Dirac–Slater exchange potential, proportional to the electron density $n(z)$ to the one-third power. In this model, with z taken perpendicular to an assumed planar surface, they then used the simplest form of density functional theory (due to Thomas [27], Fermi [28] and Dirac [29]), plus the lowest order Weizsacker–Kirznits density gradient correction [30] to the local electron gas kinetic energy, to explicitly calculate the density profile $n(z)$ through the liquid–vapour interface as

$$n(z) = \frac{n_{\text{liquid}}}{(1 + a\, \exp(-z/l))^3} : \; a \sim 1. \tag{5.2}$$

Since the surface tension is then determined solely by the density gradient contribution to the kinetic energy, one finds the result

$$l \propto E_f^{-1/2} : \; K_T \propto (n_{\text{liquid}}\, E_f)^{-1} \tag{5.3}$$

which in the model leads back to the correlation $\gamma K_T = \text{constant}\ l$ with l, evidently from Eq. (5.2), measuring the thickness of the liquid metal surface. In Eq. (5.3), E_f is the Fermi energy of the conduction electron gas.

Finally, I briefly summarize the theory of the velocity of sound c in a liquid metal of valency Z. To see the essence of the formula, actually going back to the earlier work of Bohm and Staver [31], I write down first the result for the ionic plasma frequency of ions of mass, M, in such an electronic assembly with bulk electron density, n_0, corresponding to ionic density, $n_i = n_0/Z$. Then I have from the usual Langmuir formula

$$\omega_p^{\text{ions}} = \left(\frac{4\pi (Ze)^2 n_i}{M}\right)^{1/2}. \tag{5.4}$$

But this equation is not the desired acoustic mode of the ions, and this fact is because of the neglect of screening of the ions by the electron gas. Using first in Eq. (5.4) the result, $Zn_i = n_0$, given above, I have the modified relation

$$\omega_p^{\text{ions}} = \left(\frac{4\pi e^2 Z n_0}{M}\right)^{1/2}. \tag{5.5}$$

But now I introduce the simplest form of (Thomas–Fermi) screening in which the ionic potential energy $-Ze^2/r$ goes into the screened Coulomb potential $-(Ze^2/r)\exp(-qr)$, where the inverse Thomas–Fermi screening length, q, is given in terms of the Fermi wave number, k_f, of the electron gas by

$$q^2 = 4k_f/\pi a_0 : \; a_0 = h^2/4\pi^2 m e^2 \tag{5.6a}$$

with m the electron mass in the assumed free electron gas of conduction electrons. Thus, in Fourier transform (k) space, I have the effect of screening represented by

$$\frac{Z}{k^2} \rightarrow \frac{Z}{q^2+k^2}, \tag{5.6b}$$

where the LHS of Eq. (5.6) is the Fourier transform of the bare ion potential, while the RHS is that of the screened Coulomb form. Thus in Eq. (5.5), I must make the change

$$Z \rightarrow \frac{Zk^2}{q^2}, \tag{5.7}$$

where I have taken the long wavelength limit $k \rightarrow 0$ to facilitate the extraction of the desired velocity of sound. Then I convert the so-called 'optic' mode form (5.5) into the acoustic mode

$$\omega = ck \tag{5.8}$$

and obtain the velocity of sound in the liquid metal as

$$c = \left(\frac{Zm}{3M}\right)^{1/2} v_{\mathrm{f}}, \tag{5.9}$$

where v_{f} is the Fermi velocity of the conduction electron cloud in the bulk liquid metal. Thus, I put together the formulae (5.1), (5.3) and (5.8) where we have specifically used the Andrade formula (5.1) at the melting temperature, T_{m}, to obtain the dimensionless ratio $\gamma/\eta c$ as given in Appendix B.

6. Surface tension–viscosity relation and liquid structure in long wave limit

I shall conclude this discussion of the relation between thermodynamic quantity, γ, and transport coefficient, η, by returning to the liquid structure, but now in the long wave limit, $k \rightarrow 0$. As Alonso and March [20] have noted, it is of considerable interest, in the light of Egry's earlier conclusion [19] that the form (4.2) relating γ/η to the thermal velocity has its justification in 'hard-sphere-like' assumptions, to consider the situation in real liquid metals in the long wave limit. This has been done, using experimental diffraction data, in the work of Bernasconi and March [32]. They have considered, in particular, the Ornstein–Zernika direct correlation function, $c(r)$, for a variety of liquid metals near melting temperatures and their results are reproduced in Table 2. The definition of $c(r)$ is given immediately below for completeness:

Table 2
Values of Ornstein–Zernike direct correlation function defined in Eq. (6.1) at $r=0$ and $k=0$ for some liquid metals near their freezing points. The ratio, $c(r=0)/\tilde{c}(k=0)$, is near unity for hard spheres according to the Percus–Yevick approximation (after Bernasconi and March [32])

Liquid metal	$-c(r=0)$	$-\tilde{c}(k=0)$	Ratio $c(r=0)/\tilde{c}(k=0)$
Na	43	41	1.0
K	42	40	1.0
Rb	45	42	1.1
Cs	50	38	1.3
Cu	60	47	1.3
Ag	51	53	1.0
Au	35	38	0.9
Mg	31	39	0.8
Al	45	54	0.8
Ca	34	200	0.2
Pb	44	110	0.4
Sn	40	140	0.3
Fe	46	48	1.0
Ni	41	50	0.8
Co	35	50	0.7

For purposes of comparison, liquid argon values are 33, 17 and 1.9.

$$r - 1 = h(r) = c(r) + \rho \int h(|\mathbf{r}-\mathbf{r}'|)c(\mathbf{r}')\,\mathrm{d}\mathbf{r}'. \tag{6.1}$$

What is actually tabulated there is $-c(r=0)$, and also $-\tilde{c}(k=0)$, where $\tilde{c}(k)$ is the Fourier transform of $c(r)$, and is related to the structure factor, $S(k)$, of the liquid metal by

$$\tilde{c}(k) = \frac{S(k)-1}{S(k)}. \tag{6.2}$$

As was shown by Bhatia and March [33] using the hard-sphere Percus–Yevick results for $c(r)$, there is an analytic relation in the above framework between these two quantities $c(r=0)$ and $c(k=0)$ given by

$$\tilde{c}_{\mathrm{HS}}^{\mathrm{PY}}(k=0) = 1 + c_{\mathrm{HS}}^{\mathrm{PY}}(r=0). \tag{6.3}$$

But as seen from Table 2, both $c(r=0)$ and $\tilde{c}(k=0)$ are very much greater in absolute magnitude than unity, and hence from Eq. (6.3) for hard spheres the ratio is predicted to be near unity. The results of Table 2 show that for the alkalis, the ratio is near unity, and indeed the only truly exceptional cases from the 15 liquid metals in

Table 2 (for comparison we have included 3 of the *d* electron liquid metals) are Ga, Pb and Sn.

7. Summary and proposed future directions

In the first part of the paper, the importance of local coordination change as the heavy alkalis are taken up the liquid-vapour coexistence curve towards the critical is the stressed point. Extrapolation of Eq. (2.1) to the critical density of Cs leads to the conclusion that the coordination number, z, reduces to $\sim$2 at the critical point, and this number has led me, elsewhere, to conjecture that zig–zag chains should be present and will undergo a Peierls distortion, leading to the coincidence of the critical point and the metal-insulator transition in this liquid system. The work of Likalter is also important in the present context, though Hess has pointed out that there are differences between Likalter [6] and me on the one hand and Goldstein et al. [7] on the other. A present view is that these differences may turn out to be of detail rather than of principle but more work is required, both theoretical and experimental.

This discussion led into a brief analysis of the electronic correlation functions as objects to treat metallic liquids in which bonding is pronounced and some emphasis was given to the computer calculations of Ceperley et al. [15] on molecular hydrogen. The valence–valence electron correlation function is especially illuminating in this context, and work on the alkalis using diffraction experiments plus computer simulation as carried out already on liquid Mg and liquid Bi near their melting points [13] is clearly called for (also private communication from Dr Price at Argonne National Lab).

A further topic treated was concerned with connecting a thermodynamic property, surface tension, γ, with a transport property, the shear viscosity, η. The thermal velocity connecting the two in the simplest statistical mechanical treatments has some quantitative difficulties for a few polyvalent metals, and therefore the possibility that the ratio, γ/η, may relate to a collective quantity, the velocity of sound, has been briefly explored. Dependence on the valence, Z, of the liquid metal is to be expected from a nearly free electron model, but this model is no doubt an oversimplification. The interest in further experiments relating to the connection between thermodynamics and transport is already clear. I note, in conclusion, that an earlier work on such a correlation was made by me [34] in relating the magnetic susceptibility of Cs along the liquid–vapour coexistence curve to the electrical conductivity of the liquid metal. In this application, there was no doubt that in expanded Cs both electron–electron and electron–ion correlations were important, which testifies further to the interest in exploring further the two-component view of a liquid metal, with three pair correlation functions.

Acknowledgements

The writer is especially indebted to much collaboration with Professors Alonso, Tosi, Freeman and Egty on the area discussed in the present article.

Appendix A. More general observations on local coordination in condensed metallic phases

I thought it of interest in the present context to add some brief and more general observations on local coordination in condensed phases. First of all, Freeman and March [34] drew attention to the local coordination of Cs atoms adsorbed on semiconducting substrates. This analysis started from the experimental studies of Whitman et al. [35,36], who used both InSb and GaAs as semiconducting substrates on which to adsorb the metal atoms. This work is reviewed in an article entitled 'The metallicity of metal thin-film overlayers' by Dowben [37].

Second, I want to point out the potential interest in generalizing the chemical model of crystalline solids, put forward by Klein et al. [38], in which the internal energy E of metallic crystals was studied as a function of lattice parameter. In particular, for K, results were presented [39] for bcc ($z=8$), simple cubic ($z=6$) diamond ($z=4$) and chain structures ($z=2$). Of course, the long-range order must be relaxed in the present form of the

Table 3
Specific heats of some sp liquid metals at their melting temperatures

	Na	K	Rb	Zn	Cd	Ga	Tl	Sn	Pb	Bi
$\gamma = c_p/c_v$	1.12	1.11	1.15	1.25	1.23	1.08	1.21	1.11	1.20	1.15
c_v/Nk_B	3.4	3.5	3.4	3.1	3.1	3.2	3.0	3.0	2.9	3.1

Note: A fully harmonic theory, based on independent density fluctuations, would yield $\gamma = 1$ and $c_v = 3$ in the above units.

quantum-chemical model in which the basic building block is the free space dimer (e.g. K_2). Use is made directly of chemical input of potential energy curves for singlet and triplet states as a function of dimer bond length, and should be generalized to embrace entropic considerations.

Subsequent to the work of March and Rubio [39] on metallic *K*, a comparison of low coordination materials has been made by the same authors, embracing also crystalline metallic hydrogen [40]. Nellis et al. [41] have produced liquid metallic hydrogen using shock wave pressures upward of 1.5 megabar, and have conjectured that even higher pressures will be required to form solid metallic hydrogen. Arguments based on the Clausius–Clapeyron equation, and utilizing the theory of Bhatia and March [42] for the way low temperature phase transitions depend on elementary excitations, one can show qualitatively that the slope of the melting line is consistent with the conjecture of Nellis et al. Some of the above arguments may well carry through to low temperature amorphous metals but we cannot go into details here.

Appendix B. Simple metal model for ratio, $\gamma/\eta c$, where γ is the surface tension, η the shear viscosity and c the velocity of sound, all taken at melting temperature, T_m

The deviations from the correlation proposed in Eq. (4.2), and stressed especially by Egry [19], for polyvalent metals, In and Sn, to which reference had been made, has prompted the question whether there is a more fundamental velocity connecting γ and η at the melting point of metals. In the work of Freeman and March [21], organic liquids were studied, but for want of data, room temperature had to be used for the comparison. One quantity explored in the organics, but presently without decisive conclusions, was the velocity of sound, *c*. Thus, below, I shall attempt, within a simple metal model, to estimate the ratio, $\gamma/\eta c$.

Following the arguments leading to Eqs. (5.2) and (5.3), we shall write γ in the form

$$\gamma K_T = l. \tag{A2.1}$$

Combining this next with Andrade's formula (5.1), we find the ratio γ/η. We next utilize the formula (5.9) for the velocity of sound. Putting these three formulae together, and invoking the result shown in Fig. 19(b) of Ref. [43], I reach the conclusion that

$$\frac{\gamma}{\eta_c} = (c_p/c_v)^{-1/2} G(Z), \tag{A2.2}$$

where the form of $G(Z)$ is linked with the curve drawn in Fig. 19(b) of Ref. [43]. The ratio (c_p/c_v) is given in Table 3 and does not vary widely at the melting point.

I want only to add one further comment. It is that for the organics, but now at room temperature rather than the melting temperature, T_m, considered here, $\gamma/\eta c$ appeared to correlate with the quantity, $D\rho^{2/3}$, with D the self-diffusion coefficient, and ρ the number density of molecules [21]. Further work is required on the theoretical basis for such a correlation, though plainly the accuracy of the Stokes–Einstein relation between D and η will be a factor.

References

[1] See, for example, R. Winter, F. Noll, T. Bondensteiner, W. Glaser, P. Chieux, F. Hensel, Z. Phys. Chem. 156 (1988) 145.
[2] N.H. March, J. Math. Chem. 4 (1990) 271.
[3] N.H. March, F. Perrot, M.P. Tosi, Molec. Phys. 93 (1998) 355.
[4] See, e.g., E.A. Guggenheim, Thermodynamics, North Holland, Amsterdam, 1949.

[5] R.G. Chapman, N.H. March, Phys. Chem. Liquids 16 (1986) 77.
[6] A.A. Likalter, Uspekhi Fiz. Nauk 162 (1992) 119.
[7] See, e.g., R.E. Goldstein, A. Parola, A.P. Smith, J. Chem. Phys. 91 (1989) 1843.
[8] H. Hess, Phys. Chem. Liquids 30 (1995) 251.
[9] P.A. Egelstaff, N.H. March, N.C. McGill, Can. J. Phys. 52 (1974) 1651.
[10] J. Chihara, J. Phys. F 17 (1987) 295.
[11] M. Watabe, in: Proceedings of the Conference on Liquid Metals, Japan, 1973.
[12] N.H. March, M.P. Tosi, Ann. Phys. 81 (1973) 414.
[13] G.A. de Wijs, G. Pastore, A. Selloni, W. van der Lugt, Phys. Rev. Lett. 19.
[14] N.H. March, M.P. Tosi, Plasmas and Ions 1 (1996) 13.
[15] D.M. Ceperley, B.J. Alder, Phys. Rev. B 36 (1987) 2092; W.R. Masto, D.M. Ceperley, C. Pierleoni, B. Bernu, Phys. Rev. Lett. 76 (1996) 1240.
[16] N.H. March, M.P. Tosi, Nuovo Cim. D 18 (1996) 1061.
[17] R.H. Fowler, Proc. Roy. Soc. A 159 (1937) 229.
[18] M. Born, H.S. Green, Proc. Roy. Soc. A 190 (1947) 455.
[19] I. Egry, Scripta Metallurgica et Materialia 26 (1992) 1349.
[20] J.A. Alonso, N.H. March, Molec. Phys. (1998).
[21] G.R. Freeman, N.H. March, J. Chem. Phys. 109 (1998) 10521.
[22] D.M. Heyes, N.H. March, to be published.
[23] J.R.D. Copley, J.M. Rowe, Phys. Rev. Lett. 32 (1974) 49.
[24] E.N. Andrade, C. da, Phil. Mag. 17 (1934) 497.
[25] See also, R.C. Brown, N.H. March, Phys. Chem. Liquids 1 (1968) 141.
[26] R.C. Brown, N.H. March, J. Phys. C 6 (1973) L363; see also Phys. Reports 24 (1976) 77.
[27] L.H. Thomas, Proc. Camb. Phil. Soc. 23 (1926) 542.
[28] E. Fermi, Z. Phys. 48 (1928) 73.
[29] P.A.M. Dirac, Proc. Camb. Phil. Soc. 26 (1930) 376.
[30] C.F. von Weizsacker, Z. Phys. 96 (1935) 431; D.A. Kirznits, Sov. Phys. JETP 5 (1957) 64.
[31] D. Bohm, T. Staver, Phys. Rev. 84 (1952) 836.
[32] J. Bernasconi, N.H. March, Phys. Chem. Liquids.
[33] A.B. Bhatia, N.H. March, J. Chem. Phys.
[34] G.R. Freeman, N.H. March, J. Phys. Chem. 98 (1994) 9486.
[35] L.J. Whitman, J.A. Stroscio, R.A. Dragoset, R.J. Celotta, Phys. Rev. Lett. 66 (1991) 1338.
[36] L.J. Whitman, J.A. Stroscio, R.A. Dragoset, R.J. Celotta, Science 251 (1991) 1206.
[37] P.A. Dowben, accepted for publication.
[38] N.H. March, M.P. Tosi, D.J. Klein, Phys. Rev. B 52 (1995) 9115.
[39] N.H. March, A. Rubio, Phys. Rev. B 56 (1997) 13865.
[40] N.H. March, A. Rubio, Phys. Stat. Solidi. in press.
[41] S.T. Weir, A.C. Mitchell, W.J. Nellis, Phys. Rev. Lett. 76 (1996) 1860.
[42] W.B. Leung, A.B. Bhatia, N.H. March, Phys. Lett. 58A (1976) 205.
[43] N.H. March, Liquid Metals, Pergamon, Oxford, UK, 1968.

JOURNAL OF CHEMICAL PHYSICS VOLUME 117, NUMBER 23 15 DECEMBER 2002

Thermodynamic consistency and integral equations for the liquid structure

F. E. Leys
University of Antwerp Groenenborgerlaan 171, B-2020 Antwerpen, Belgium

N. H. March
University of Antwerp Groenenborgerlaan 171, B-2020 Antwerpen, Belgium and Oxford University, Oxford, England

D. Lamoen
University of Antwerp Groenenborgerlaan 171, B-2020 Antwerpen, Belgium

(Received 20 May 2002; accepted 25 September 2002)

Within an assumed pair potential framework, it has been generally accepted for a long time that far from the critical point the asymptotic form of the direct correlation function $c(r)$ at large r is given by $[-\phi(r)/k_BT]$. Here $\phi(r)$ is the pair potential and k_BT the thermal energy. Subsequently, Kumar, March, and Wasserman [Phys. Chem. Liquids **11**, 271 (1982)] examined the condition for thermodynamic consistency between virial and compressibility equations of state. Their study, together with later work by Senatore, Rashid, and March [Phys. Chem. Liquids **16**, 1 (1986)], resulted in a decomposition of $c(r)$ into a potential part $c_p(r)$ given by Kumar *et al.* for all r and involving the pair function $g(r)$ and its density derivative, plus a "collective" part $c_c(r)$, which must obey a simple sum rule to satisfy thermodynamic consistency. The more recent study of B. C. Eu and K. Rah [J. Chem. Phys. **3**, 3327 (1999)] prompts us to bring their results into direct contact with the study of Kumar *et al.* The work of Eu and Rah gives a prominent place to the Mayer function $f(r)=e^{(-[\phi(r)/k_BT]}-1$ which tends to $-[\phi(r)/k_BT]$ as $r\to\infty$ for potentials tending to zero at infinity. [DOI: 10.1063/1.1522377]

I. INTRODUCTION

Analytic theories of liquid structure, which essentially relate the pair function $g(r)$ to an assumed pair potential $\phi(r)$ continue to provide a source of fundamental work in classical statistical mechanics. Probably the most notable of these theories, when account is taken of a balance between accuracy and relative simplicity, remain those of Percus and Yevick[1] (PY), and the later hypernetted chain[2] (HNC) procedure. The latter is now often supplemented by a so-called bridge function.[3]

Both PY and HNC theories give a central place to the Ornstein–Zernike (OZ) direct correlation function $c(r)$, defined in terms of the total correlation function $h(r)=g(r)-1$ by[4]

$$h(r)=c(r)+\rho\int h(|\mathbf{r}-\mathbf{r}'|)c(r')d\mathbf{r}' \tag{1}$$

with ρ denoting the atomic number density of the liquid.

It has been widely accepted for a long time that, providing one is far from the critical point, the range of $c(r)$ is determined by the range of the assumed pair potential $\phi(r)$. This general statement is made precise in that, for sufficiently large r,

$$c(r)\underset{r\to\infty}{\to}\frac{-\phi(r)/k_BT}{k_BT}, \tag{2}$$

where k_BT is the thermal energy.[5] Though, to date, there appears no rigorous general proof of Eq. (2), there is now a whole body of evidence supporting its validity.[6]

Kumar *et al.*[7] considered the implications for the direct correlation function $c(r)$ of imposing thermodynamic consistency between virial and compressibility equations of state, a matter pursued subsequently by Senatore, Rashid, and March.[8,9] The outcome of these studies was to divide $c(r)$ into a sum of two parts: a "potential" contribution $c_p(r)$ and a "collective" piece $c_e(r)$. Kumar *et al.* supplied an explicit form for $c_p(r)$ for all r, which contained the limiting form (2) for sufficiently large r, but which involved $g(r)$ and its density derivative. Quite explicitly, this form of $c_p(r)$ was

$$c_p(r)=-\frac{\phi(r)}{k_BT}\left(\frac{1}{6\rho r^2}\right)\frac{\partial^2}{\partial\rho\partial r}[\rho^2r^3g(r)]. \tag{3}$$

Thermodynamic consistency was then shown by Kumar *et al.* to be equivalent to the requirement that the collective piece $c_c(r)$ integrates to zero

$$\int c_c(r)d\mathbf{r}=0, \tag{4}$$

which implies that the potential part $c_p(r)$ contributes the whole of the isothermal compressibility calculated from $c(r)$.

II. THE CLOSURE OF EU AND RAH

A. Relation to the work of Kumar *et al.*

We turn immediately to confront the result (3) of Kumar *et al.* with the much more recent study of Eu and Rah.[10] The interested reader must, of course, refer to their paper for full details and also to the two studies of Martynov.[11,12] Below, it will suffice for the present purpose to quote the thermodynamically consistent closure proposed by Eu and Rah for the direct correlation function $c(r)$, which they denote by $c_E(r)$,

$$c_E(r)=\frac{1}{r^2\rho}\frac{\partial}{\partial\rho}\left[\frac{1}{6}\rho^2 f(r)\frac{\partial}{\partial r}r^3 y(r)\right]. \tag{5}$$

In their form (5), $f(r)$ is the so-called Mayer function, defined in terms of the pair potential $\phi(r)$ by

$$f(r)=e^{[-\phi(r)/k_BT]}-1, \tag{6}$$

which evidently tends to $-[\phi(r)/k_BT]$ as $r\to\infty$ provided $\phi(r)$ tends to zero at infinity. The function $y(r)$ in Eq. (5) is such that

$$y(r)=g(r)e^{\phi(r)/k_BT}. \tag{7}$$

If we assume the pair potential $\phi(r)$ to be independent of thermodynamic state, the expression (5) immediately reduces to

$$c_E(r)=f(r)\left(\frac{1}{6r^2\rho}\right)\frac{\partial^2}{\partial\rho\partial r}[\rho^2 r^3 y(r)], \tag{8}$$

and if we then retain only the first term in the expansions,

$$f(r)=\sum_{n=1}^{\infty}\frac{1}{n!}\left(-\frac{\phi(r)}{k_BT}\right)^n=-\frac{\phi(r)}{k_BT}+\cdots, \tag{9}$$

$$e^{\phi(r)/k_BT}=\sum_{n=0}^{\infty}\frac{1}{n!}\left(\frac{\phi(r)}{k_BT}\right)^n=1+\cdots \tag{10}$$

and insert them into the expression (8) one immediately regains the form (3) of Kumar *et al.* for the potential contribution $c_p(r)$. At finite temperatures, both expressions are thus seen to be identical only for r-values where the potential $\phi(r)$ is sufficiently small.

Establishing this relation immediately allows us to write down the approximation of Eu and Rah for the collective part of the direct correlation function, say $c_{Ec}(r)$,

$$c_{Ec}(r)=\frac{1}{6r^2\rho}\left[f(r)\frac{\partial^2}{\partial\rho\partial r}[\rho^2 r^3 y(r)]+\frac{\phi(r)}{k_BT}\frac{\partial^2}{\partial\rho\partial r}[\rho^2 r^3 g(r)]\right] \tag{11}$$

and since the form (5) satisfies thermodynamic consistency, it is clear from the study of Kumar *et al.* that this expression for the collective part should integrate to zero,

$$\int_0^\infty r^2 c_{Ec}(r)dr=0 \tag{12}$$

which is for the sake of completeness trivially shown to be correct in Appendix A.

Eu and Rah give an application to hard spheres the results of which exhibit considerable similarity with the PY approximation in liquid structure. However, care is needed, we feel, in both of these approaches for the case of longer ranged forces. As an admittedly extreme example, we consider in Appendix B the Eu–Rah closure, where one puts $c(r)=c_E(r)$, in relation to the two dimensional one-component plasma (2D OCP) having a $\phi(r)=-e^2\ln(r/L)$ interaction, which has been exactly solved by Jancovici for $g(r)$ for a single value of the coupling parameter $\Gamma=e^2/k_BT$ specified in Appendix B. While the Eu–Rah closure had been designed for three dimensions, and therefore the result quoted in Appendix B, as remarked, is an extreme example, the replacement of $-[\phi(r)/k_BT]$ in the Kumar *et al.* form of the "potential" contribution $c_p(r)$ by the Mayer function $f(r)$, as in the Eu–Rah closure is not appropriate at large r for the 2D OCP.

B. Relation to the force equation

The pair function $g(r)$ is related to the triplet correlation function $g^{(3)}(\mathbf{r}_1,\mathbf{r}_2,\mathbf{r}_3)$ and an assumed pair potential $\phi(r)$ through the first member of the so-called BBGKY hierarchy[4] often termed "the Force Equation."

Defining the potential of mean force $U(r)$ through

$$g(r)=e^{-[U(r)/k_BT]} \tag{13}$$

this equation reads

$$-\nabla_{r_1}U(r_{12})=-\nabla_{r_1}\phi(r_{12})-\rho\int\frac{g^{(3)}\mathbf{r}_1,\mathbf{r}_2,\mathbf{r}_3)}{g(r_{12})}\times\nabla_{r_1}\phi(r_{13})d\mathbf{r}_3. \tag{14}$$

In principle, if one inserts the closure of Eu and Rah for $c(r)$, into the OZ equation, one would obtain an expression relating $g(r)$ to the pair potential $\phi(r)$ and its derivative $\partial\phi(r)/\partial r$ and the density derivative of $g(r)$. From the work of Schofield[13] the density dependence of $g(r)$ (again for state-independent pair potentials) is known to be closely related to the triplet correlation function $g^{(3)}(\mathbf{r}_1,\mathbf{r}_2,\mathbf{r}_3)$ through the relation[4]

$$\left(\frac{\partial g(r)}{\partial\rho}\right)_T=\frac{1}{S(0)}\left[\int g^{(3)}(\mathbf{r}_1,\mathbf{r}_2,\mathbf{r}_3)d\mathbf{r}_3+\text{terms containing solely } g(r)\right], \tag{15}$$

where $S(0)$ is the long wavelength limit of the liquid structure factor $S(k)$. Through the OZ equation, the Eu–Rah closure thus implies an alternative expression for the Force Equation containing essentially the same ingredients.

We will not examine the relation between both expressions in detail here but merely note that the Force Equation and the alternative expression implied by the Eu–Rah closure use different parts of the triplet correlation function $g^{(3)}(\mathbf{r}_1,\mathbf{r}_2,\mathbf{r}_3)$. Specifically if we use the fact that, for a homogeneous isotropic liquid, $g^{(3)}(\mathbf{r}_1,\mathbf{r}_2,\mathbf{r}_3)$ cannot depend on the choice of the origin nor of the orientation of the triplet, we can decompose $g^{(3)}$ in the form[4]

$$g^{(3)}(\mathbf{r}_1,\mathbf{r}_2,\mathbf{r}_3)=\sum_{l=0}^{\infty}Q_l(s,t)P_l(\cos\theta) \tag{16}$$

with $s=|\mathbf{r}_1-\mathbf{r}_2|$, $t=|\mathbf{r}_2-\mathbf{r}_3|$, θ the angle between $\mathbf{r}_1-\mathbf{r}_2$ and $\mathbf{r}_2-\mathbf{r}_3$ and $P_l(\cos\theta)$ the Legendre polynomial of order

l. Using the orthogonality property of these polynomials it is straightforward to show that the Force Equation uses only the $l=1$ term of the expansion (16) while $(\partial g(r)/\partial\rho)_T$ uses only the $l=0$ term.

III. CONCLUSION AND FUTURE DIRECTIONS

The focus of this investigation has been on thermodynamic consistency in some of the simpler proposals for liquid structural theories. Such thermodynamic consistency, while obviously highly desirable, involves a space integral of $g(r)$ on the one hand and both space and density integrals of $c(r)$ on the other hand. Evidently therefore, local forms of thermodynamically consistent closure are not unique, as is evident from the two closures discussed here.

It is hence important to stress, in concluding, that the thermodynamically consistent closures of Eu and Rah and of Kumar *et al.* are related, provided that the potentials are non-Coulombic. Both treatments embody the density dependence of the pair function $g(r)$ in the expression proposed for the OZ direct correlation function. This is satisfactory for some purposes, as the triplet function is known to be related precisely to $(\partial g(r)/\partial\rho)_T$ by Eq. (15) valid for potentials independent of thermodynamic state. One can expect therefore that the proposals of Kumar *et al.* (when the potential part is taken as full closure) and of Eu and Rah should in general transcend the simpler structural theories such as HNC and Born–Green, which relate $c(r)$ to two-atom quantities at the same thermodynamic state. Of course, it is only fair to say that presently both for experiment and theory, accurate information on this density derivative still poses problems and therefore is an area to be investigated further in the future.

Finally, both referees have asked us to stress that, although considerable common ground has been demonstrated here to exist between the two closures under discussion, the Eu–Rah closure is the one appropriate for short range forces, and in particular for hard spheres, whereas the Kumar *et al.* closure $c_p(r)$ is to be preferred for long range forces.

APPENDIX A: PROOF OF THERMODYNAMIC CONSISTENCY

In the formalism of Kumar *et al.* proof of consistency is particularly simple. Since the integration in the sum rule is only over r, we can bring out the operator $(\partial/\partial\rho)\rho^2$ in both terms on the right-hand side of Eq. (11). For the integral over the remainder of the first term we then obtain

$$\int_0^\infty f(r)\frac{\partial}{\partial r}(r^3y(r))dr=f(r)r^3y(r)|_0^\infty - \int_0^\infty r^3y(r)\frac{\partial}{\partial r}f(r)dr \quad \text{(A1)}$$

and similarly for the integral over the second term

$$\int_0^\infty \frac{\phi(r)}{k_BT}\frac{\partial}{\partial r}(r^3g(r))dr=\frac{\phi(r)}{k_BT}r^3g(r)|_0^\infty - \frac{1}{k_BT}\int_0^\infty r^3g(r)\frac{\partial}{\partial r}\phi(r)dr. \quad \text{(A2)}$$

Using the fact that

$$y(r)\frac{\partial}{\partial r}f(r)=g(r)e^{[\phi(r)/k_BT]}e^{-[\phi(r)/k_BT]}\times\left[-\frac{1}{k_BT}\frac{\partial}{\partial r}\phi(r)\right] \quad \text{(A3)}$$

$$=-\frac{1}{k_BT}g(r)\frac{\partial}{\partial r}\phi(r) \quad \text{(A4)}$$

and that, given the assumptions made by Kumar *et al.*, we also have

$$f(r)r^3y(r)|_0^\infty=\frac{\phi(r)}{k_BT}r^3g(r)|_0^\infty=0. \quad \text{(A5)}$$

We immediately see that the RHS of Eq. (11) reduces to zero as required by the work of Kumar *et al.*

APPENDIX B: THE TWO-DIMENSIONAL ONE COMPONENT PLASMA

Following Senatore *et al.*[8] we first generalize the expressions for the direct correlation function to d dimensions yielding for the Kumar *et al.* potential contribution,

$$c_p(r)=-\frac{\phi(r)}{k_BT}\left(\frac{1}{2d\rho r^{d-1}}\right)\frac{\partial^2}{\partial\rho\partial r}[\rho^2r^3g(r)] \quad \text{(B1)}$$

and similarly for the Eu–Rah closure for state independent pair potentials in d dimensions,

$$c_E(r)=f(r)\left(\frac{1}{2d\rho r^{d-1}}\right)\frac{\partial^2}{\partial\rho\partial r}[\rho^2r^3y(r)]. \quad \text{(B2)}$$

We now turn to the 2D OCP. This is a two-dimensional electron gas immersed in a positive nonresponsive uniform background ensuring charge neutrality. The charges are assumed to interact via the solution of the two-dimensional Poisson's equation which yields $\phi(r)=-e^2\ln(r/L)$, with L an arbitrary scaling length. The coupling constant

$$\Gamma=\frac{e^2}{k_BT} \quad \text{(B3)}$$

determines an equilibrium state of the 2D OCP completely, independent of density.

For one particular coupling strength $\Gamma=2$, Jancovici[14] obtained an exact expression for the pair correlation function $g(r)$ in the form,

$$g(r)=1-e^{(-r^2/a^2)} \quad a^2=(\pi\rho)^{-1}, \quad \text{(B4)}$$

with ρ now denoting the area number density.

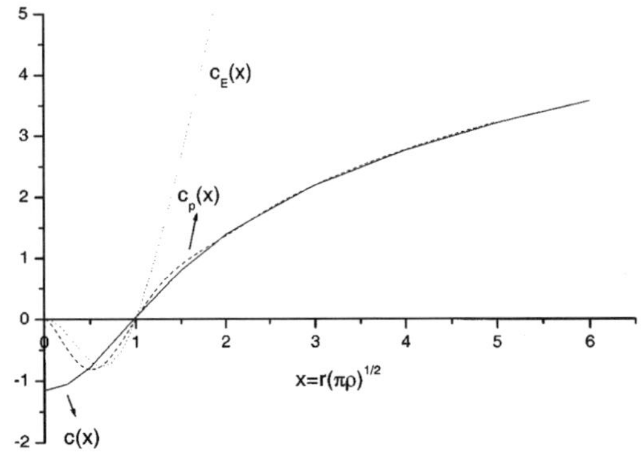

FIG. 1. Results for the direct correlation function for the classical two-dimensional one-component plasma for a value of the coupling parameter $\Gamma=2$. Solid line: exact result $c(x)$ as obtained by Jancovici (Ref. 15). The continuous curve was obtained by connecting a number of discrete points calculated from Eq. (B5) with straight lines. Dotted line: thermodynamicallly consistent direct correlation function $c_E(x)$ proposed by Eu and Rah (Ref. 10). Dashed line: potential part of the result proposed by Kumar *et al.* (Ref. 7).

The exact direct correlation function $c(r)$ can be obtained in the form[15]

$$c(r)=-2\gamma-\sum_{j=1}^{\infty}\frac{1}{j}(e^{-x^2/j}-1)\quad x=\frac{r}{a} \tag{B5}$$

with γ Euler's constant. Figure 1 depicts the exact expression for $c(r)$ from Eq. (B5) together with the approximate Eu–Rah closure $c_E(r)$ given by Eq. (B2) in two dimensions ($d=2$) and the potential contribution $c_p(r)$ given by Eq. (B1).

Evidently $c_p(r)$ begins to usefully approximate $c(r)$ by the time x reaches the value 2. Because of the presence of the Mayer function in $c_E(r)$, this model is not appropriate for a description by the Eu–Rah closure. However, $c_p(r)$ and $c_E(r)$ are in fair accord inside $x=1$, but both need "correcting" inside this value, $c_p(r)$ by the collective part $c_c(r)$. Of course, this example with ultra long-ranged interactions in no way denies the merit of the Eu–Rah closure for an extreme short-range force law.

[1] J. K. Percus and G. Yevick, Phys. Rev. **110**, 1 (1958).
[2] See, for example, G. S. Rushbrooke, Physica (Amsterdam) **26**, 259 (1960).
[3] See, e.g., J. P. Hansen and I. R. McDonald, *Theory of Simple Liquids*, 2nd ed. (Academic, New York, 1986); See also, for example, Y. Rosenfeld, D. Levesque, and J. J. Weis, Phys. Rev. A **39**, 3079 (1989).
[4] See, for instance, N. H. March and M. P. Tosi, *Atomic Dynamics in Liquids* (Dover, New York, 1991).
[5] See, for example, M. D. Johnson, P. Hutchinson, and N. H. March, Proc. R. Soc. London, Ser. A **282**, 283 (1964).
[6] See, for instance, F. E. Leys, N. H. March, F. Perrot, G. Straub, and V. E. Van Doren, High Pressure Physics and Technology **10**, 4 (2000); **10**, 84 (2000).
[7] N. Kumar, N. H. March, and A. Wasserman, Phys. Chem. Liq. **11**, 271 (1982).
[8] G. Senatore, R. I. M. A. Rashid, and N. H. March, Phys. Chem. Liq. **16**, 1 (1986).
[9] See also, N. H. March, *Liquid Metals Concepts and Theory* (Cambridge University Press, Cambridge, 1990).
[10] B. C. Eu and K. Rah, J. Chem. Phys. **111**, 3327 (1999).
[11] G. A. Martynov and G. N. Sarkisov, Mol. Phys. **49**, 1495 (1983).
[12] G. A. Martynov, Mol. Phys. **42**, 329 (1981).
[13] P. Schofield, Proc. Phys. Soc. London **88**, 149 (1966); see also N. H. March and M. P. Tosi, *Atomic Dynamics in Liquids* (Dover, New York, 1991).
[14] B. Jancovici, Phys. Rev. Lett. **46**, 386 (1981).
[15] See, for example, J. M. Caillol, D. Levesque, J. J. Weiss, and J. P. Hansen, J. Stat. Phys. **28**, 325 (1982).

INSTITUTE OF PHYSICS PUBLISHING JOURNAL OF PHYSICS A: MATHEMATICAL AND GENERAL

J. Phys. A: Math. Gen. **36** (2003) 5893–5898 PII: S0305-4470(03)54735-7

Electron–electron correlations in liquid s–p metals

F E Leys[1] and N H March[1,2]

[1] Physics Department, University of Antwerp, Groenenborgerlaan 171, B-2020 Antwerpen, Belgium
[2] University of Oxford, Oxford, UK

E-mail: Frederik.Leys@ua.ac.be

Received 9 October 2002
Published 22 May 2003
Online at stacks.iop.org/JPhysA/36/5893

Abstract
We present calculations for the valence electron–electron structure factor in liquid Mg near freezing, assuming knowledge of the jellium result. On the basis of this, we predict significant corrections to jellium short-range correlations in liquid s–p metals and in particular an increase in the electron–electron contact probability.

PACS numbers: 71.22.+I, 71.10.Ca, 71.10.+I, 71.45.Gm

1. Background and outline

Egelstaff *et al* [1] pointed out that a pure liquid metal with a well-defined valency could be fruitfully investigated as a two-component system of ions (i) and valence electrons (v). Thus by analogy, say with the classical molten salt, NaCl, three pair correlation functions are needed to characterize the structure. The first of these, $S_{ii}(k)$, is directly accessible by neutron scattering. While in principle, the remaining correlation functions $S_{iv}(k)$ and $S_{vv}(k)$ are also experimentally accessible by combining knowledge of $S_{ii}(k)$ with x-ray and electron diffraction studies, for Mg $S_{iv}(k)$ has become available recently by the computer simulation studies of de Wijs *et al* [2]. Thus, we present calculations for the remaining partial structure factor $S_{vv}(k)$.

The outline of the paper is then as follows. In section 2 the relevant theoretical results are briefly summarized, followed by numerical results on Mg in section 3. Section 4 treats corrections to jellium short-range correlations with results for Na and Mg. We conclude in section 5 with a summary of the main results and some proposals for future directions.

2. Theory: perturbation result for $S_{vv}(k)$ to second order in the electron–ion pseudopotential v_{iv}

Cusack *et al* [3] obtained the result that to second order in the (weak) electron–ion pseudopotential $v_{iv}(k)$, the correction $\delta S(k)$ to the (assumed known) jellium structure factor

0305-4470/03/225893+06$30.00

$S_{vv}^{jell}(k)$, because of the ionic background, consists of two terms which we shall simply label by 1 and 2

$$\delta S(k) = \delta S_1(k) + \delta S_2(k). \tag{1}$$

The first correction $\delta S_1(k)$ is local in the ionic structure factor and is given by

$$\delta S_1(k) = \frac{1}{Z_v} \left| \pi(k) \frac{v_{iv}(k)}{\epsilon(k)} \right|^2 S_{ii}(k) \tag{2}$$

where Z_v denotes the valency, $\pi(k)$ the proper jellium polarization and $\epsilon(k)$ the static dielectric function given by $\epsilon(k) = 1 - v(k)\pi(k)$ with $v(k) = \frac{4\pi}{k^2}$ representing the bare Coulombic interaction. As was pointed out by, e.g., March and Murray [4], the resultant electron density profile when one disturbs a uniform electron gas of density ρ_v^0 with a weak spherically symmetric pseudopotential is given in linear response by

$$\rho_v(r) = \rho_v^0 + \int F(|\mathbf{r} - \mathbf{r}'|) V(r) \, d\mathbf{r} \tag{3}$$

where the response function $F(r)$ is simply the Fourier transform of the proper polarization $\pi(k)$ and $V(r)$ is the screened pseudopotential. Hence, the correction $\delta S_1(k)$ can be rewritten as

$$\delta S_1(k) = \frac{|\tilde{\rho}_v(k)|^2}{Z_v} S_{ii}(k) \tag{4}$$

with $\tilde{\rho}_v(k)$ being the linear response approximation to the Fourier transform of the displaced density profile around every ion, commonly called the 'form factor'. The term $\delta S_1(k)$ can then be interpreted as the correlation that results between electrons because they form neutralizing shells around the correlated ions.

The second correction term $\delta S_2(k)$ is given by

$$\delta S_2(k) = \frac{1}{2Z_v} \frac{1}{(2\pi)^3} \int S_{ii}(q) |v_{iv}(q)|^2 \chi_4(\mathbf{k}, -\mathbf{k}; -\mathbf{q}, \mathbf{q}) \, d\mathbf{q} \tag{5}$$

where $\chi_4(\mathbf{k}, -\mathbf{k}; -\mathbf{q}, \mathbf{q})$ is the interacting four-body jellium response function which we calculated using the expression derived by Cusack *et al* [3] from the density functional theory.

3. Numerical results

In order to test the applicability of some of the main functions involved, we first calculated a quantity closely related to $\delta S_1(k)$, namely the ion–valence structure factor $S_{iv}(k)$, which to first order is given by [5]

$$S_{iv}(k) = \frac{1}{\sqrt{Z_v}} \left(\pi(k) \frac{v_{iv}(k)}{\epsilon(k)} \right) S_{ii}(k). \tag{6}$$

Figure 1 compares results for liquid Mg near freezing using equation (6) to the computer simulation result of de Wijs *et al* [2]. For the dielectric function a local field correction in the Hartree–Fock approximation [6] was used. The pseudopotential was calculated using the Ashcroft [7] empty core approximation with the standard Mg core radius value $R_c = 1.39a_0$, and the ion–ion structure factor $S_{ii}(k)$ was approximated by x-ray data at $T = 680$ K of the IAMP database [8]. Very good agreement over a substantial range of k-values is obtained.

The two constituents $\delta S_1(k)$ and $\delta S_2(k)$ are shown for liquid Mg in figure 2. The overall magnitude of both corrections is considerably increased compared to Na, but the main features are the same as those obtained by Cusack *et al* [3]. In particular we note that $\delta S_2(k \to 0) = 0$

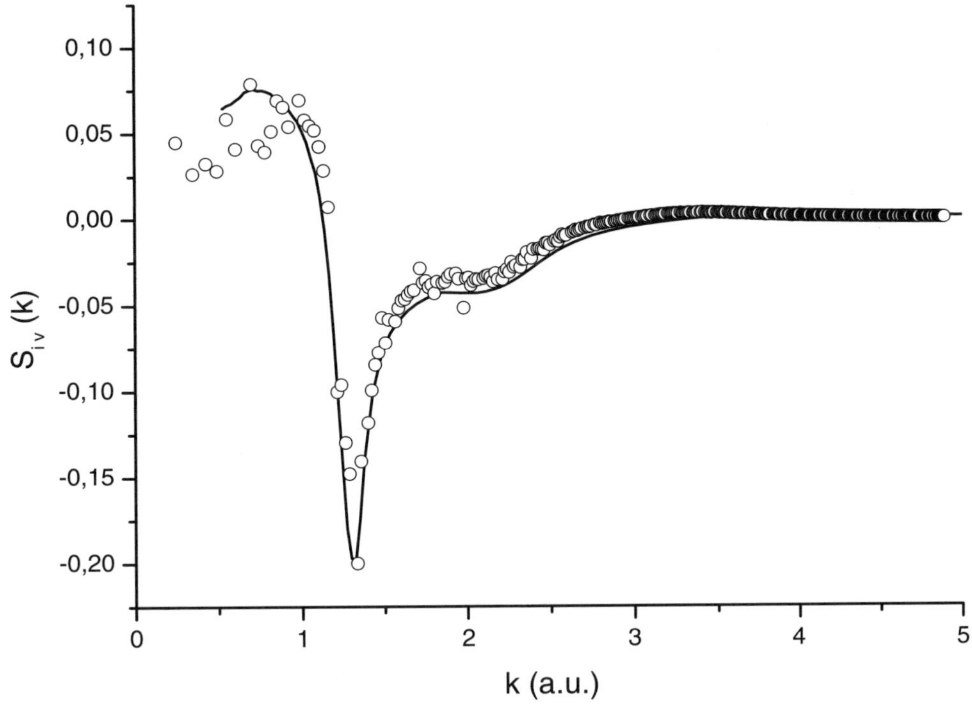

Figure 1. Ion–valence structure factor $S_{iv}(k)$ for liquid Mg. Circles are result from computer simulation of de Wijs *et al* [2], and solid line is linear response result.

and $\delta S_1(k \to 0) = Z_v S_{ii}(0)$ so that the normalization condition discussed in equation (10) is fulfilled. The pronounced peak in $S_{iv}(k)$ and in the first correction $\delta S_1(k)$ stems directly from the principal peak in $S_{ii}(k)$. The features of both peaks depend crucially on the position of the first node in the pseudopotential relative to the position of the principal peak in $S_{ii}(k)$. Because the pseudopotential appears squared in the expression for $\delta S_1(k)$, the antiphase behaviour of $S_{iv}(k)$ with respect to $S_{ii}(k)$ is not reproduced in $\delta S_1(k)$.

4. Corrections to jellium short-range electron–electron correlations

4.1. The electron–electron contact probability in a liquid metal

From the relation between the electronic pair correlation function $g_{vv}(r)$ and the structure factor $S_{vv}(k)$

$$g_{vv}(r) = 1 + \frac{1}{(2\pi)^3 \rho_v^0} \int (S_{vv}(k) - 1)\, e^{-i\mathbf{k}\cdot\mathbf{r}}\, d\mathbf{k} \tag{7}$$

it is straightforward to calculate the change $\delta g_{vv}(0)$ in the pair correlation function at the origin from its jellium value. We performed calculations for liquid Na ($r_s \approx 4$) and Mg ($r_s \approx 2.8$) near freezing and the results are listed in table 1. The corrections are significant and were found to be positive; the effect of the ionic background tends to 'fill up' the jellium exchange-correlation hole at the origin to some extent. With decreasing r_s, the magnitude of the first correction $\delta g_{vv}^1(r)$ was found to increase significantly while the second correction $\delta g_{vv}^2(r)$ remained nearly the same.

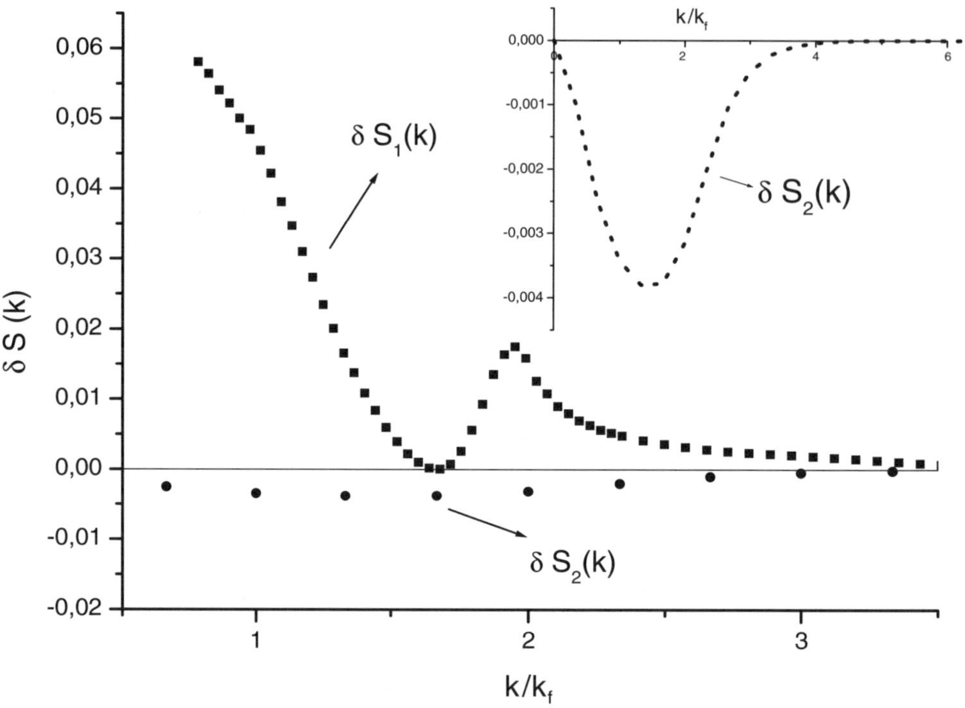

Figure 2. Corrections $\delta S_1(k)$ and $\delta S_2(k)$ to the jellium structure factor for liquid Mg. Inset shows $\delta S_2(k)$ on a more appropriate scale.

Table 1. Corrections to $g_{vv}^{jell}(0)$ in liquid Na and Mg near freezing. Contributions from δS_1 and δS_2 are given by $\delta g_{vv}^1(0)$ and $\delta g_{vv}^2(0)$ respectively. For comparison, the last row presents the contribution of correlation to $g_{vv}^{jell}(0)$ in jellium as obtained from equation (8). Total correction $\delta g_{vv}(0)$ adds up to 5% and 20% of $g_{vv}^{corr}(0)$ for Na and Mg respectively.

	Na ($r_s \approx 4$)	Mg ($r_s \approx 2.8$)
$\delta g_{vv}^1(0)$	+0.058	+1.102
$\delta g_{vv}^2(0)$	−0.040	−0.030
$\delta g_{vv}(0) = \delta g_{vv}^1(0) + \delta g_{vv}^2(0)$	+0.018	+0.072
$\delta g_{vv}^{corr}(0)$	−0.420	−0.381

It is useful to compare the magnitude of the corrections obtained above to those when taking into account electron correlation in jellium. From phase-shift analysis, Overhauser [9] obtained the following approximate result for the correction to $g_{vv}^{jell}(0)$ due to correlation

$$\delta g_{vv}^{corr}(0) = \frac{32}{(8+3r_s)^2} - \frac{1}{2}. \tag{8}$$

The results from this formula are also listed in table 1. For Na we see that the correction amounts to only 5% of that due to correlation, whereas for Mg almost 20% of the correlation hole at the origin is 'filled up' because of the ionic background. Although this does not follow straightforwardly from the theory of Cusack *et al* [3], obviously the obtained corrections $\delta g_{vv}(0)$ can apply only between electrons with opposite spins and cannot exceed the correction $\delta g_{vv}^{corr}(0)$ in order not to violate the Pauli exclusion principle.

Whereas the magnitude of the corrections is clearly surprising, the sign of the correction is not. This is because the 'exchange-correlation' hole in a liquid metal is on average less deep compared to jellium. This can easily be seen from the normalization condition for $g_{vv}(r)$ which is determined by the request that any perturbing charge inserted in a conducting medium must be completely screened. In jellium this leads to the condition

$$\rho_v^0 \int \left(g_{vv}^{jell}(r) - 1\right) d\mathbf{r} = -1 \tag{9}$$

whereas in a liquid metal, where the positive ions contribute to the screening, this implies

$$\rho_v^0 \int (g_{vv}(r) - 1)\, d\mathbf{r} = -1 + Z_v S_{ii}(0) \tag{10}$$

where $S_{ii}(0)$ is typically of the order of 0.02.

4.2. The electron–electron cusp condition

Having indicated directly above the way in which significant changes in $g_{vv}(0)$ result from the corrections to the structure factor, it is natural to investigate if the jellium cusp relation as obtained by Kimball [10]

$$\left.\frac{\partial g_{vv}^{jell}(r)}{\partial r}\right|_{r=0} = -\frac{1}{a_0} g_{vv}^{jell}(0) \tag{11}$$

is modified when the above corrections are taken into account. We therefore turn to the limiting behaviour of the correction terms $\delta S_1(k)$ and $\delta S_2(k)$ when k tends to infinity. Regarding the first correction $\delta S_1(k)$, we can make analytic progress assuming that $\tilde{\rho}_v(k)$ represents the *exact* valence density profile. The result of Carlsson and Ashcroft [11] generalizing Kato's theorem [12] to continuum states then gives the nuclear cusp relation

$$\left.\frac{\partial \rho_v(r)}{\partial r}\right|_{r=0} = -\frac{2Z_v}{a_0}\rho_v(0) \tag{12}$$

where the origin $r = 0$ now refers to an ionic centre. This directly implies from equation (4)

$$\delta S_1(k)|_{k\to\infty} = \frac{1}{Z_v}\left(8\pi\rho_v^0\left[\frac{2Z_v}{a_0}\rho_v(0)\right]\right)^2 \frac{1}{k^8}. \tag{13}$$

From our numerical calculations on Na and Mg, the second correction $\delta S_2(k)$ was found to tend to zero slightly faster than the predicted k^{-8} behaviour of $\delta S_1(k)$, namely proportional to k^{-10}, and so the first correction $\delta S_1(k)$ will dominate at large k. This leads to

$$\delta g_{vv}(r)|_{r\to 0} = \delta g_{vv}(0) - \frac{8\pi}{(2\pi)^3\rho_v^0 Z_v}\left(8\pi\rho_v^0\left[\frac{2Z_v}{a_0}\rho_v(0)\right]\right)^2 r^5 \tag{14}$$

and we conclude that the coefficient of the first-order term in the small r expansion of $g_{vv}(r)$ in a liquid s–p metal remains determined by the jellium value $g_{vv}^{jell}(0)$.

5. Summary and possible future directions

Calculations were performed for the valence electron–electron structure factor $S_{vv}(k)$ in s–p liquid metals, especially Mg, assuming knowledge of the jellium result. On the basis of these results a significant increase in $g_{vv}(0)$ was found. We stress that this arises from a partial cancellation of two terms of opposite sign and that our quantitative results may change somewhat depending on the approximation to χ_4. However, since both terms tend to differ by over a factor 3 with decreasing r_s, we are confident that our conclusion will remain unchanged at least qualitatively. The jellium cusp relation remains valid in a liquid metal, the first correction appearing only to fifth order in r.

As a natural continuation of the present work, a detailed study of intermediate- and long-range corrections to $g_{vv}^{jell}(r)$ is of interest for future work. In contrast to the short-range behaviour, these are expected to depend sensitively on the detailed and possibly singular structure of χ_4.

Finally, experimental determination of $S_{vv}(k)$ following the pioneering proposal of Egelstaff *et al* [1] is obviously of great interest and remains an important challenge for future work.

Acknowledgments

We would like to express our gratitude to Dr G A de Wijs and Professor W van der Lugt for supplying us with the numerical data for the ion–valence structure factor for Mg. It is also our pleasure to thank Professors N W Ashcroft and I A Howard, Dr D Lamoen and Dr G G N Angilella for stimulating and helpful discussions.

References

[1] Egelstaff P A, March N H and McGill N C 1974 *Can. J. Phys.* **52** 1651
[2] de Wijs G A, Pastore G, Selloni A and van der Lugt W 1995 *Phys. Rev. Lett.* **75** 4480
[3] Cusack S, March N H, Parrinello M and Tosi M P 1976 *J. Phys. F: Met. Phys.* **6** 749
[4] March N H and Murray A M 1960 *Phys. Rev.* **120** 830
[5] Tosi M P and March N H 1973 *Nuovo Cimento* B **15** 308
[6] Singwi K S, Tosi M P and Land R H 1968 *Phys. Rev.* **176** 589
[7] Ashcroft N W 1966 *Phys. Lett.* **23** 48
[8] IAMP database of [SCM-LIQ], webpage: http://www.iamp.tohoku.ac.jp
[9] Overhauser A W 1995 *Can. J. Phys.* **72** 683
[10] Kimball J C 1973 *Phys. Rev.* A **7** 1648
[11] Carlsson A E and Ashcroft N W 1982 *Phys. Rev* B **25** 3474
[12] Kato T 1957 *Pure Appl. Math.* **10** 151

Physics and Chemistry of Liquids
Vol. 43, No. 1, February 2005, 111–114

Taylor & Francis
Taylor & Francis Group

LETTER

Low density observations of Rb and Cs chains along the liquid–vapour coexistence curves to the critical point in relation to quantum-chemical predictions on the metal-insulator transitions in Li and Na rings

G.G.N. ANGILELLA†*, N.H. MARCH‡,§ and R. PUCCI†

†Dipartimento di Fisica e Astronomia, Università di Catania, and
Istituto Nazionale per la Fisica della Materia, UdR Catania,
Via S. Sofia, 64, I-95123 Catania, Italy,
‡Oxford University, Oxford, UK, §Department of Physics,
University of Antwerp, Groenenborgerlaan 171, B-2020 Antwerp, Belgium

(Received 5 October 2004)

Recent *ab initio* predictions concerning the metal-insulator (MI) transition in rings of the light alkali atoms, Li and Na, are compared and contrasted with experimental facts concerning diluted Rb and Cs alkalis. The main focus here is on the local coordination number as a function of density as these two heavy alkali metallic fluids are taken along the liquid–vapour coexistence curve towards the critical point, which in these cases coincides with the MI transition. Also recorded are the results of experiments in which Cs chains are observed at large interatomic spacing outside semiconducting substrates of InSb and GaAs.

Keywords: Alkalis; Critical point; Metal-insulator transition

1. Introduction

In a recent work, Paulus *et al.* [1] have reported *ab initio* calculations relating to the metal-insulator (MI) transition in Li rings. Subsequent studies by Alsheimer and Paulus [2] have embraced both the Na rings, as well as some mixed Na–Li systems.

Our purpose here is to note that in the heavier alkalis Rb and Cs, chains are observed under two very different physical areas. The first area is when one takes these two alkali metal fluids along the liquid–vapour (LV) coexistence curve towards the critical point. From neutron diffraction studies of Jüngst *et al.* [3], it is known that high coordination number z in these metallic liquids near the freezing point is reduced greatly as the

*Corresponding author. E-mail: giuseppe.angilella@ct.infn.it

Physics and Chemistry of Liquids
ISSN 0031-9104 print: ISSN 1029-0451 online © 2005 Taylor & Francis Ltd
http://www.tandf.co.uk/journals
DOI: 10.1080/00319100512331327117

Table 1. Bulk alkali metal equilibrium nearest-neighbour separation, R_e, compared with some results at separations larger than those required to induce the metal-insulator transition.

Alkali	R_e for bcc (Å)	Non-metallic separations (Å)	
Li rings	3.03	~4.6	[1]
Na rings	3.66	~4.8	[2]
K linear chains	4.52		see figure 4 of Ref. [8]
Rb linear chains	4.83		see figure 5a of Ref. [8]
Cs linear chains	5.2		see figure 5b of Ref. [8]
Cs zig-zag chains on InSb(110) surface	5.2	8.0	[6]
Cs zig-zag chains on GaAs(110) surface	5.2	6.9	[6]

density d is lowered along the LV coexistence curve. One of us [4] has fitted the experimental data by

$$d = az + b \tag{1}$$

where $a = 230$ and $b = -80$, both in kg/m^3 (see also [5]). Cs having a critical density of $380\,kg/m^3$, the conclusion is that $z \to 2$ as the critical point is approached. To present experimental accuracy, in these two heavy alkalis, Cs and Rb, the MI transition and the critical point coincide for these fluids.

The second area we wish to draw attention to in the present study is that of Cs outside semiconducting surfaces, and in particular, InSb(110) and GaAs(110) [6]. Here, the atom spacings in the Cs chains, which were observed experimentally, were dictated dominantly by the geometry of the semiconducting substrates. The spacing of Cs atoms is then substantially larger than for Cs body-centred cubic metal at atmospheric pressure (see table 1).

2. *Ab initio* studies of Paulus *et al.* on Li and Na rings

Recently, Paulus *et al.* [1] have applied *ab initio* quantum-chemical techniques to Li rings, with the aim of studying the analogue of a metal-insulator (MI) transition. By varying the interatomic distance, these authors have analyzed the character of the many-electron wave function. In particular, the importance of the s–p orbital quasi-degeneracy within the metallic regime has been emphasized. Parallel work embracing now both Na rings and mixed Na–Li systems has subsequently been reported [2].

3. Experimental studies on the heavy alkalis Rb and Cs: in relation to the formation of chains

Though the *ab initio* studies above [1,2] were on rings of some 10 alkali atoms, reference was also made to less accurate mean field studies of finite linear chains. It is this aspect that has prompted us to reopen older studies, dominantly but not wholly experimental, on chains of the heavy alkalis Rb and Cs.

Let us begin with a brief summary of the experimental findings of Jüngst *et al.* [3]. These authors used neutron diffraction to study the structure of several thermodynamic states of fluid Cs, lying along the liquid–vapour coexistence curve towards the critical point.

One important conclusion was that the near-neighbour 'bond' distance remained relatively constant ($\sim 5.4 - 5.7\,\text{Å}$) as the density was lowered towards the value at the critical point. The obvious implication is that this density reduction must come about for Rb and Cs largely by the lowering of the local coordination number z [see equation (1)] as the critical point is approached [see again equation (1)]. The other important point to be stressed here is that whereas in a divalent metal like Hg, there is a MI transition in the originally metallic fluid well before the critical point is reached, in the monovalent heavy alkalis Rb and Cs, experimental accuracy does not presently allow the critical point and the MI transition to be separated.

Therefore, it is of interest to turn to a second, very different, area, in which chains of Cs atoms have been detected experimentally. Thus Cs has been adsorbed on room temperature semiconducting substrates. This is then found, under suitable experimental conditions, to lead to the formation of quite long Cs chains outside, both InSb and GaAs [6], as observed with scanning tunneling microscopy. Taking the example of GaAs, one-dimensional zig-zag chains, more than 1000 Å in length, were observed by Whitman *et al.* [6].

4. Summary and future directions

We have compared and contrasted (see also table 1) the behaviour of configurations consisting of a (relatively small) number of light alkali atoms having low local coordination number z (e.g., a ring-like Li_{10} studied by *ab initio* quantum-chemical methods in [1]) with some experimental observations where z is small (~ 2) on the heavy alkalis Rb and Cs: especially as these metallic liquids approach the critical point, along the LV coexistence curve. Paulus *et al.* [1] stress for Li metal the substantial contribution of p electrons to the conduction band. They point out that with increasing interatomic separation the ratio of p and s electrons alters and in particular in the atomic limit, one is dealing with a $1s^2 2s$ atomic state.

Of course, in the heavy alkalis Rb and Cs, one must expect to deal with more angular momentum quantum numbers, and this will most probably affect the character of the MI transition. However, as experiments employing neutron scattering show, Rb and Cs retain a rather constant near-neighbour distance along the LV coexistence curve out to the critical density, and the MI transition to present experimental accuracy coincides with the critical point, where the local coordination number $z \sim 2$. It was proposed therefore in [4] that a Peierls transition occurred in Cs and Rb at the critical point. Also invoked is data for Cs adsorbed on semiconducting substrates, where chains are observed on both InSb and GaAs. Whitman *et al.* [6] invoke, at a semiquantitative level, the model of Ferraz *et al.* [7] in which, in general terms, the dissociation of H_2 embedded in an electron liquid is observed to occur as the electron density of the background fluid is increased. This model has some features in common with the observations of Whitman *et al.* [6].

Obviously, for the future, it would be of considerable interest if the *ab initio* quantum-chemical studies of [1] and [2] could be extended to the heavy alkali Rb, though already K would be of interest (see also the density functional work of March and Rubio [8]: this could be extended, as it was under compression to date, to lower conduction electron density). We note also the likely relevance of cluster studies, such as that recently reported on Li_8 by Grassi *et al.* [9], motivated by experimental determination of the geometry of the lowest isomers of such light alkali atom clusters [10]. Finally, it seems of interest to return briefly to table 1 above. There it is recorded first for Na rings that the interatomic separation has to be increased by a factor of ~ 1.3 from the equilibrium distance in bcc Na to induce a MI transition in the ring. Moving to the final entry in table 1, the non-metallic separation recorded for GaAs(110) is also 1.3 times the equilibrium spacing in bcc Cs metal. It is tempting therefore to conjecture that, in the case of GaAs, one is quite near the MI transition. In contrast, the formation of the same one-dimensional Cs zig-zag chains observed in InSb(110), the III-V semiconductor with the largest lattice constant, seems well into the insulating chain regime.

Acknowledgements

NHM wishes to acknowledge that his contribution to this study was brought to fruition during a visit to the University of Catania in 2004. He wishes to thank the Department of Physics and Astronomy for the stimulating atmosphere and generous hospitality.

References

[1] B. Paulus, K. Rosciszewski, P. Fulde, H. Stoll. *Phys. Rev. B*, **68**, 235115 (2003).
[2] W. Alsheimer, B. Paulus. *Eur. Phys. J. B*, **40**, 243 (2004).
[3] S. Jüngst, B. Knuth, F. Hensel. *Phys. Rev. Lett.*, **55**, 2160 (1985). See also R. Winter, F. Hensel. *Phys. Chem. Liq.*, **20**, 1 (1981).
[4] N.H. March. *J. Math. Chem.*, **4**, 271 (1990). See also N.H. March, *Phys. Chem. Liq.*, **20**, 241 (1989).
[5] G.R. Freeman, N.H. March. *J. Phys. Chem.*, **98**, 9486 (1994).
[6] L.J. Whitman, J.A. Stroscio, R.A. Dragoset, R.J. Celotta. *Phys. Rev. Lett.*, **66**, 1338 (1991).
[7] A. Ferraz, N.H. March, F. Flores. *J. Phys. Chem. Solids*, **45**, 627 (1984).
[8] N.H. March, A. Rubio. *Phys. Rev. B*, **56**, 13865 (1997).
[9] A. Grassi, G.M. Lombardo, G.G.N. Angilella, N.H. March, R. Pucci. *J. Chem. Phys.*, **120**, 11615 (2004).
[10] A. Kornath, A. Kaufmann, A. Zoermer, R. Ludwig. *J. Chem. Phys.*, **118**, 6957 (2003).

Part 3: One-body potential theory of molecules and condensed matter

The conceptual origins underlying Part 3 can be traced back, at very least, to the pioneering work of Hartree [reproduced in N. H. March, *Self-consistent fields* in atoms (Pergamon, Oxford, 1975)]. However, in Hartree's original procedure, each electron in the atom (Ar, say) under consideration moved in a somewhat different potential. But almost immediately following Hartree's pioneering work, Thomas and independently Fermi [see reprints in March, *op. cit.* (1975)] laid the foundations for a many-electron theory of atoms based on a common one-body potential $V(\mathbf{r})$. Within the Thomas–Fermi semiclassical framework, the ground-state electron density $\rho(\mathbf{r})$ was related directly to $V(\mathbf{r})$ by the relation

$$\phi(\mathbf{r}) = \frac{8\pi}{3h^3}(2m)^{3/2}[\mu - V(\mathbf{r})]^{3/2}\,, \tag{2}$$

μ being the chemical potential, which is constant at every point in the inhomogeneous electron cloud. The work of Latter [Phys. Rev. **99**, 510 (1955)] exploited the atomic potential $V(\mathbf{r})$ derived self-consistently by means of Eq. (2). Other early authors who contributed importantly to one-body potential theory were Herman and Skillman [*Atomic structure calculations* (Prentice-Hall, Englewood Cliffs, 1963)] and Green *et al.* [see, e.g., Phys. Rev. **184**, 1 (1969)].

However, directly relevant to Part 3 is the classic paper by Slater [Phys. Rev. **81**, 385 (1951); see also March, *op. cit.* (1975), for a reprint] and related work by Gaspar [see Theochem. article edited by A. Nagy]. Slater's work was formally completed through the study by Kohn and Sham [Phys. Rev. **140**, A1133 (1965)], whose work lies at the heart of current usage in density functional theory (DFT).

With this brief historical background, we wish to comment on a number of salient articles reprinted in Part 3. Thus Eq. (2) was used self-consistently in early work by March [Acta Cryst. **5**, 187 (1952)] on the ground-state electron density in the benzene molecule [3a.1]. Following this study, a central field model rather directly applicable to the polyatomic molecules CH_4, SiH_4, GeH_4, and SnH_4, again based on Eq. (2), was worked out fully in [3a.2]. This study was, much later, generously referred to, after the much later experimental discovery of C_{60}, as the March model of fullerene (see D. P. Clougherty and X. Zhu, Phys. Rev. A **56**, 632 (1997); F. Despa, Phys. Rev. B **57**, 7335 (1998); F. Siringo, G. Piccitto, and R. Pucci, Phys. Rev. A **46**, 4048 (1992)]. The reader is also referred in the above context to [3c.11] and [3c.16].

Important in Part 3b, in which the semiclassical result Eq. (2) is transcended, is the so-called differential virial theorem established in [3b.1] in one dimension, and generalized by Holas and March [3b.10] to three dimensions, and the formulation of an exact Thomas–Fermi method by perturbation theory in powers of $V(\mathbf{r})$ entering Eq. (2). The key articles are then [3b.2], [3c.2], and [3b.4]. At the time of writing, analytic sums of this infinite perturbation series are only available for special forms of $V(\mathbf{r})$ (e.g. harmonic confinement).

To single out another useful concept, we wish to refer here to the Pauli potential in [3b.1] [see also M. Levy and A. Görling, Phil. Mag. B **69**, 763 (1994) for a short review of this area, and also C. Herring and M. Chopra, Phys. Rev. A **37**, 31 (1988)]. Continuing interest in this concept is reflected in the study of March and Nagy [Phys. Rev. A, in press (2009)].

Turning finally to non-local potentials, these are, of course, typified by Hartree–Fock theory, and in particular the presence of the Fock operator instead of a local, Slater-like, potential $V(\mathbf{r})$. Fundamental theory on such non-local potentials is set out in [3c.5], [3c.7], and [3c.8]. Two applications which may be singled out are [3c.12] and [3c.18].

But, for the future, Hartree–Fock (HF) non-local potentials have now been transcended in the studies of Cordero, March, and Alonso [Phys. Rev. A **75**, 052502 (2007)], in which semi-empirical tuning of the ground-state HF electron densities of four light spherical atoms has led to results of quantum Monte Carlo quality, and of Amovilli, March, and Tolman [Phys. Rev. A **77**, 032503 (2008)], who constructs an off-diagonal idempotent Dirac density matrix from generalized HF equations. The extension of these two studies to multicentre molecules and clusters is clearly of considerable interest for the future.

Acta Cryst. (1952). **5**, 187

Theoretical Determination of the Electron Distribution in Benzene by the Thomas–Fermi and the Molecular-Orbital Methods

By N. H. March*

Wheatstone Physics Laboratory, King's College, London W.C. 2, England

(*Received* 23 *August* 1951)

The results of theoretical calculations of the electron distribution in benzene are reported. Density contours, both in the plane of the benzene ring and in a parallel plane at a height of 0·35 Å above the ring, are given for the Thomas–Fermi and the molecular-orbital methods. It is shown that in the molecular-orbital method the π electrons have only a small influence on the density in the parallel plane, except immediately above the carbon atoms, where they contribute about half of the total density. As far as comparison is valid the results appear to be in reasonable agreement with X-ray results of Robertson and his co-workers for naphthalene.

1. Introduction.

Interest in the problem of calculating theoretically the electron distribution in organic molecules has been aroused by recent accurate X-ray results (Abrahams, Robertson & White 1949 *a*, *b*; Mathieson, Robertson & Sinclair 1950 *a*, *b*) giving the density distributions in naphthalene and anthracene.

In view of the central position occupied by benzene, both in organic chemistry and in molecular theory, it was decided at the outset to make a detailed investigation of the electron distribution in benzene, rather than tackle at first the necessarily lengthier calculations for naphthalene and anthracene. Whilst no results exist as yet for benzene, a significant comparison can already be made with the experimental results quoted above. The extension of the work described here to the cases of naphthalene and anthracene would be straightforward but very laborious.

The electron distribution in benzene reported here has been calculated by two quite distinct methods. First the statistical method of Thomas and Fermi has been used, and secondly a full calculation has been carried out using the molecular-orbital method. As the Thomas–Fermi (T.F.) method does not make the usual distinction of molecular theory between π and σ electrons, it seemed that this method might be of value as a pointer to the validity of other treatments.

2. Use of the Thomas–Fermi method

Besides the intrinsic interest attached to the electron distribution in benzene the application of the T.F. method is of some interest from the point of view of the statistical theory and the method will thus be described in some detail.

As is well known, the T.F. method in atomic theory gives a useful overall representation of the electron density in heavy atoms. However the method has been restricted in its application to molecules by the mathematical difficulties presented by the non-linear character of the fundamental equation. The only direct attempt to solve the T.F. equation for a molecule has been made by Hund (1932), who has shown how a fairly good approximate solution can be obtained in the case of diatomic molecules. At the outset it was not of course clear whether Hund's method could be extended to systems other than those with axial symmetry; this work seems to show that the method can be successfully extended to our case.

3. Simplification of the problem

In the treatment of any neutral molecule by this method the fundamental problem is to solve the T.F. equation (see for example, Mott & Sneddon, 1948, p. 156)

$$\nabla^2 V = \mu V^{\frac{3}{2}}, \tag{1}$$

where

$$\mu = \frac{32\pi^2 e}{3h^3}(2me)^{\frac{3}{2}},$$

for the electrostatic potential V, subject to the conditions that as any nucleus is approached, V tends to the Coulomb potential due to that nucleus and that V tends to zero at infinity.

It must be admitted that to some extent benzene is not well suited to a T.F. treatment owing to the presence of hydrogen atoms where the small concentration of electrons renders any statistical method unreliable. Thus, even in the event of an exact solution of equation (1) being possible for benzene, the results would have to be regarded with due caution and probably the only regions in which any useful conclusions could be drawn would be in the regions between adjoining carbon nuclei, and in

* Present address: Department of Physics, The University, Sheffield, England.

the interior of the ring, where the effect of the hydrogen atoms would be exceedingly small. In view of this, and of a lack of knowledge as to the applicability of Hund's method of solution except in diatomic cases, it was decided at the outset to simplify the problem as follows. A neutral system of thirty-six electrons and six carbon nuclei arranged as in benzene is to be considered and the distribution of charge resulting from the T.F. equation investigated. The assumption is made that when the distribution in this system is known, then by simply adding to it the charge density contributions from six hydrogen atoms placed in their appropriate positions, an approximate distribution is obtained for benzene. Physically one would expect this approximation to have a negligible effect on the electron distribution in the interior of the ring and in the region between the carbon atoms, but that the C–H bonds would not be adequately described in the sense of the T.F. method (which at any rate, as we shall see later, does not cause enough charge to be pushed into the centre of a bond to give a completely adequate description).

4. Method of solution

By working in units of R, the half distance between neighbouring carbon nuclei, and in terms of the dimensionless function u defined by

$$V = \frac{Ze}{R} u,$$

equation (1) transforms to

$$\nabla^2 u = \gamma u^{\frac{3}{2}}, \tag{2}$$

where

$$\gamma = \mu (ZeR^3)^{\frac{1}{2}}.$$

In our case $2R$ is taken to be 1·39 Å, and then $\gamma = 4{\cdot}44$.

Following Hund, an approximate solution is sought of the form

$$u = v(r_1) + v(r_2) + \ldots + v(r_6), \tag{3}$$

where $r_1, \ldots, r_6$ denote the distances of a point from nuclei 1, ..., 6 respectively. In the neighbourhood of any one nucleus an approximate solution would be given by taking $v(r)$ simply as the solution for an atom, that is $(1/r)\varphi(\gamma^{\frac{2}{3}} r)$, where φ is the function calculated by Fermi, Bush & Caldwell, and Miranda, and conveniently tabulated by Gombás (1948, p. 358). However, with this form for $v(r)$, u would not satisfy the equation (2) at infinity for there the system behaves as a single atom of nuclear charge 36, and an asymptotic solution is

$$u = \frac{6}{r} \varphi(6^{\frac{1}{3}} \gamma^{\frac{2}{3}} r).$$

A solution is therefore assumed to be of the form

$$u = \frac{1}{r_1} \varphi(\gamma^{\frac{2}{3}} r_1 f(r_1)) + \frac{1}{r_2} \varphi(\gamma^{\frac{2}{3}} r_2 f(r_2)) + \ldots + \frac{1}{r_6} \varphi(\gamma^{\frac{2}{3}} r_6 f(r_6)), \tag{4}$$

where

$$f(r) = \frac{\lambda^2 + 6^{\frac{1}{3}} r^2}{\lambda^2 + r^2}, \tag{5}$$

and the question to be answered is whether, for any value of λ, it can be made a reasonable approximate solution of the differential equation. Equation (5) corresponds to Hund's expression for $f(r)$ for diatomic molecules, and is one of the simplest expressions which will give the necessary asymptotic forms at the six nuclei and at infinity. With u of the form (3),

$$\begin{aligned} \nabla^2 u &= \nabla^2 v(r_1) + \nabla^2 v(r_2) + \ldots + \nabla^2 v(r_6) \\ &= h(r_1) + h(r_2) + \ldots + h(r_6). \end{aligned} \tag{6}$$

When

$$v(r) = \frac{1}{r} \varphi(\gamma^{\frac{2}{3}} r f(r)),$$

$$h(r) = \frac{\gamma [f(r) + r f'(r)]^2 v(r)^{\frac{3}{2}}}{f(r)^{\frac{1}{2}}} + \gamma^{\frac{2}{3}} \left[\frac{2}{r} f'(r) + f''(r)\right] \varphi'(\gamma^{\frac{2}{3}} r f(r)). \tag{7}$$

With $f(r)$ defined by (5),

$$f(r) + r f'(r) = \frac{\lambda^2 [\lambda^2 + (6^{\frac{1}{3}} - 1) r^2] + 6^{\frac{1}{3}} r^4}{(\lambda^2 + r^2)^2}, \tag{8}$$

and

$$\frac{2}{r} f'(r) + f''(r) = \frac{2\lambda^2 (6^{\frac{1}{3}} - 1)(3\lambda^2 - r^2)}{(\lambda^2 + r^2)^3}. \tag{9}$$

This enables us to calculate $\nabla^2 u$.

Next, as in Hund's work, the quantity B, defined by

$$B = \gamma u^{\frac{3}{2}} - \nabla^2 u, \tag{10}$$

was calculated at a number of suitable points for various values of λ, and the value of λ was sought for which B was as small as possible over the region considered. As the axial symmetry of Hund's problem was lacking here, a choice of planes in which to calculate B had to be made. It was decided to consider points in the plane of the benzene ring and also in a plane parallel to the plane of the ring. This choice was made, first to ascertain whether the best value of λ in one plane was the best value in the other, and secondly to facilitate comparison with the results of the molecular-orbital method. The distance of the parallel plane from the plane of the ring was taken as $0{\cdot}5R$, (0·35 Å), of the order of the height of the plane of maximum density of the π electrons above the plane of the ring in the molecular-orbital method. Calculations were made for $\lambda^2 = 4, 5, 6$, and it was found that $\lambda^2 = 5$ was the best compromise value in both planes. The value of λ thus obtained completely defines the approximate solution (3) and hence the approximate density distribution. Unfortunately any estimate of the accuracy of the solution would be difficult.

5. Choice of λ^2

As the value of λ completely defines the approximate solution, a little more will be said here about the choice. Without detailed tables or elaborate contours, which are clearly not justified here, it is difficult to display the various results for B. The following qualitative summary should however suffice to show that a reasonable choice can be made.

Dividing the regions of space somewhat artificially into that inside the ring and that outside the ring, the following conclusions could be drawn:

(1) At 'internal' points in the plane of the ring a value 5 for λ^2 was slightly better than 4, 6 being certainly the worst value.

(2) At 'internal' points in the parallel plane 5 was clearly better than 4 or 6.

(3) At 'external' points in both planes $\lambda^2 = 6$ was found to be the best value, 4 being certainly the worst.

It is not difficult to see from these results that a change for the worse results in going from 5 to either 4 or 6, and that $\lambda^2 = 5$ is the best compromise value that can be obtained by this method. To give some indication of the magnitudes of the differences B, the mean values of $|B|/\gamma u^{\frac{3}{2}}$ have been evaluated for the various points considered for $\lambda^2 = 5$. In the order in which the regions are referred to above, the values are 0·032, 0·022, 0·123 and 0·159. These values suggest that the solution will be most reliable in the interior of the ring; which is satisfactory in view of the method of accounting for the hydrogen atoms adopted here. Also it is satisfactory that as far as it is possible to compare them, the trends in the values of B agree well with those found by Hund for the case of diatomic molecules.

Fig. 1. T. F. electron-density contours in plane of ring. Outer contour is 0·15 e.Å^{-3}, interval is 0·2 e.Å^{-3}. Density is infinite at all nuclei and to avoid complicating the figure contours finish at 1·35 e.Å^{-3} round the hydrogen atoms and at 1·75 e.Å^{-3} round the carbon atoms.

6. The approximate distribution in benzene

Remembering the assumption made at the outset, the charge density due to six hydrogen atoms, with their nuclei placed as in benzene, must be added to the density already obtained. The T. F. method, applicable to heavy atoms, obviously has little justification for hydrogen. A comparison of the radial charge density for hydrogen given by wave mechanics and by the T. F. method shows, however, that the two are considerably but not wildly different; and to avoid introducing functions foreign to the T. F. method the T. F. density was used. In any case, any conclusions drawn will be to a large extent independent of this, since the only regions that will be discussed are between the carbon nuclei and inside the ring. The charge contours in the plane of the ring are shown in Fig. 1, and those for the parallel plane in Fig. 2.

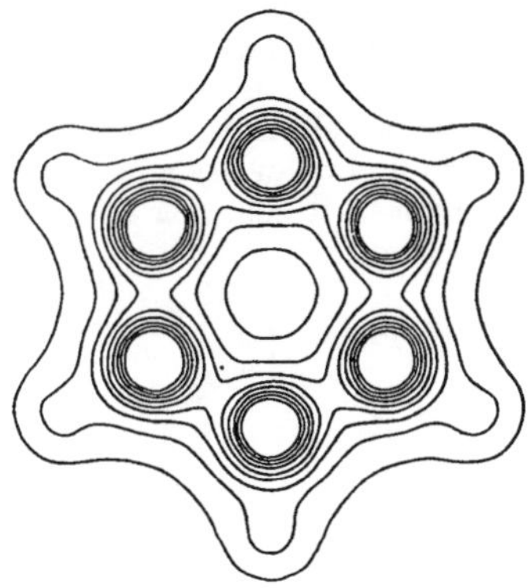

Fig. 2. T. F. electron-density contours in a plane parallel to the ring and at a height of 0·35 Å above the plane of the ring. Outer contour is 0·15 e.Å^{-3}, interval is 0·2 e.Å^{-3}. Density is 0·5 e.Å^{-3} immediately above the hydrogen atoms, and above the carbon atoms 3·8 e.Å^{-3}. Last contour shown is 1·75 e.Å^{-3}.

7. General outline of the molecular-orbital method

The principle of the second method used here, namely, the molecular-orbital method, can be described very simply as follows. Each individual electron is assigned a wave function and the total wave function is constructed as a determinant, in the usual way, from the one-electron wave functions, the total wave function thus constructed being antisymmetrical in the electrons. If the total wave function is denoted by $\Psi(r_1 \ldots r_n)$, where r_j represents all the coordinates of the jth electron, then the charge density due to electron 1 will be given by

$$\varrho(1) = \int |\Psi|^2 d\tau_2 \ldots d\tau_n \,, \tag{11}$$

with suitable normalisation of Ψ such that

$$\int \varrho(1) d\tau_1 = 1 \,. \tag{12}$$

From the form of the total wave function it follows that the charge density due to any other electron will be given by an exactly similar expression, so that the total density can be obtained.

8. Construction of one-electron functions

The $1s$ electrons of the carbon atoms are assumed to be unchanged by the formation of the molecule, and hence supply the first one-electron functions. To obtain the other one-electron functions needed, the $2s$ and $2p$ carbon atomic orbitals are first hybridized in the usual manner to form trigonal orbits. It is

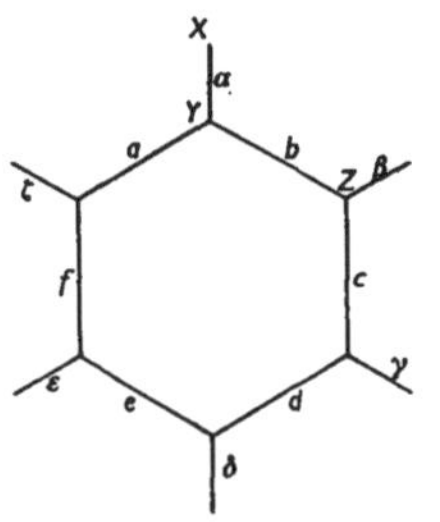

Fig. 3. Notation for the bond orbitals.

now assumed that 'molecular orbitals' (m.o.'s) which serve as the one-electron wave functions in the total wave function, can be formed in the usual way by linear combination of atomic orbitals.

Thus, for example, the m.o. based on bond Y–Z (see Fig. 3) and designated by b, can be written

$$b = \psi_Y(tr_{YZ}) + \psi_Z(tr_{ZY}) \,, \tag{13}$$

where $\psi_Y(tr_{YZ})$ denotes the trigonal orbit based on nucleus Y and directed along YZ. The form of $\psi_Y(tr)$ is given by

$$\psi_Y(tr_{YZ}) = \frac{1}{\sqrt{3}}\{\psi_Y(2s) + (\sqrt{2})\psi_Y(2p_{YZ})\} \,, \tag{14}$$

where $\psi_Y(2s)$ and $\psi_Y(2p_{YZ})$ are the $2s$ and $2p$ atomic orbitals of carbon. In a similar way the m.o.'s based on the C–H bonds can be formed. These are assumed to be of the form

$$\alpha = \psi_Y(tr_{YX}) + \psi_X(1s) \,, \tag{15}$$

where $\psi_X(1s)$ is the $1s$ hydrogen wave function. If the coefficient of $\psi_X(1s)$ in this orbital were changed slightly from unity, to allow a possible small polarity in the C–H bonds, the distribution in the C–H bonds would be slightly altered but the distribution between the carbon atoms and in the interior of the ring would almost certainly be negligibly affected. Each of these orbitals has axial symmetry about the particular bond on which it is based; they are the σ orbitals. From the $2p_Z$ atomic wave functions (the z axis being perpendicular to the plane of the ring) additional orbitals, the π orbitals, are formed. These can be written in the form (Mayer & Sklar, 1938)

$$\left.\begin{aligned}
\psi_0 &= (p_1+p_2+p_3+p_4+p_5+p_6)\\
\psi_1 &= (p_1+\omega p_2+\omega^2 p_3-p_4-\omega p_5-\omega^2 p_6)\\
\psi_2 &= (p_1+\omega^2 p_2-\omega p_3+p_4+\omega^2 p_5-\omega p_6)\\
\psi_3 &= (p_1-p_2+p_3-p_4+p_5-p_6)\\
\psi_4 &= (p_1-\omega p_2+\omega^2 p_3+p_4-\omega p_5+\omega^2 p_6)\\
\psi_5 &= (p_1-\omega^2 p_2-\omega p_3-p_4+\omega^2 p_5+\omega p_6)
\end{aligned}\right\} \tag{16}$$

where $p_1 \ldots p_6$ denote the $2p_Z$ orbitals based on Y, Z, etc., and $\omega = \exp(i\pi/3)$.

The electrons are then allocated as follows: two in each of the $1s$ orbits, a total of twelve being thus accommodated; two in each of the σ orbitals, twenty-four electrons being accommodated in these. Six electrons remain and in the ground state of the molecule, which we are considering here, the three π orbitals of lowest energy (ψ_0, ψ_1, ψ_5) are each doubly filled.

9. Atomic wave functions

For carbon, Slater wave functions have been used in the following calculations (Slater, 1930). A defect of these wave functions is that $\psi(2s)$ and $\psi(1s)$ are not orthogonal. In the present work this is remedied in part by writing for b, instead of (13)

$$b = n[\psi_Y(tr_{YZ}) + \psi_Z(tr_{ZY}) + \lambda\psi_Y(1s) + \lambda\psi_Z(1s)] \,, \tag{17}$$

where λ is chosen so that b is orthogonal to $\psi_Y(1s)$ and $\psi_Z(1s)$, and n is the normalizing factor. Using Slater wave functions,

$$n = 0{\cdot}542, \quad \lambda = -0{\cdot}212 \,.$$

Similarly, α is written in the form

$$\alpha = m[\psi_Y(tr_{YX}) + \psi_X(1s) + \mu\psi_Y(1s)] \,, \tag{18}$$

where μ is chosen so that α is orthogonal to $\psi_Y(1s)$. It is found that

$$m = 0{\cdot}543, \quad \mu = -0{\cdot}205 \,.$$

The $1s$ orbitals are now orthogonal to all the orbitals used in describing the molecule except that the $1s$ carbon orbital on atom r (say) is not strictly orthogonal to the $1s$ orbital on atom s, nor to the σ-bond orbitals which do not terminate on atom r. Such lack of orthogonalization will be small and can certainly be neglected with safety.

10. Expression for charge density

Let us denote the $1s$ functions by $s_1 \ldots s_6$ and the π wave functions ψ_0, ψ_1, ψ_5, by π_1, π_2, and π_3. The total wave function is now written in the usual determinantal form, in terms of $s_1 \ldots s_6$, $\pi_1 \ldots \pi_3$, $a \ldots f$ and $\alpha \ldots \zeta$. However $a \ldots f$, $\alpha \ldots \zeta$ are not orthogonal and in this form the problem cannot be easily handled. From a property of determinants, however, linear combinations of $a \ldots f$, $\alpha \ldots \zeta$ may be taken, and apart from a numerical factor multiplying it the determinant will be unchanged. Thus, still unnormalized Ψ can be written in a determinantal form with $\varphi_1 \ldots \varphi_{12}$ replacing $a \ldots \zeta$, where $\varphi_1 \ldots \varphi_{12}$ are suitable linear combinations of the $a \ldots \zeta$. A convenient choice is

$$\left.\begin{aligned} \varphi_1 &= a+b+c+d+e+f \\ \varphi_2 &= a+\omega b+\omega^2 c-d-\omega e-\omega^2 f \\ \varphi_3 &= a+\omega^2 b-\omega c+d+\omega^2 e-\omega f \\ \varphi_4 &= a-b+c-d+e-f \\ \varphi_5 &= a-\omega b+\omega^2 c+d-\omega e+\omega^2 f \\ \varphi_6 &= a-\omega^2 b-\omega c-d+\omega^2 e+\omega f \end{aligned}\right\} \tag{19}$$

and similarly φ_7 to φ_{12} are formed with $\alpha \ldots \zeta$ replacing $a \ldots f$. It is then easily shown that the only non-orthogonalities occurring are between φ_j and φ_{j+6}, where $j = 1 \ldots 6$, and a convenient notation is

$$\int \varphi_j^* \varphi_k d\tau = a_{j,k}\,. \tag{20}$$

From this point two procedures are possible. One is to use a method developed by Lennard-Jones (1931) for treating determinantal forms in which the one-electron functions are not orthogonal; the other is to proceed to form two new wave functions which are orthogonal, from φ_j and φ_{j+6}. The second procedure is the more convenient here though of course both lead to the same result. Thus, from the wave functions φ_j and φ_{j+6} a wave function

$$\chi_j = n_j(\varphi_j+\delta_j\varphi_{j+6}) \tag{21}$$

is constructed such that

$$\int (\varphi_j^*+\delta_j^*\varphi_{j+6}^*)\varphi_{j+6}d\tau = 0;$$

that is

$$a_{j,j+6}+\delta_j^* = 0\,,$$

or

$$\delta_j^* = -a_{j,j+6}, \quad \delta_j = -a_{j+6,j}\,, \tag{22}$$

and

$$1/|n_j|^2 = 1+\delta_j a_{j,j+6}+\delta_j^* a_{j+6,j}+\delta_j^*\delta_j\,. \tag{23}$$

We introduce now the convenient notation

$$a_{j,j+6}a_{j+6,j} = A_{j,j+6}\,. \tag{24}$$

Then

$$1/|n_j|^2 = 1-A_{j,j+6}\,. \tag{25}$$

Thus the wave functions χ_j, φ_{j+6} $(j = 1 \ldots 6)$ are mutually orthogonal, and the complete set of one-electron wave functions are now completely orthogonal (remembering the assumption about the 1*s* wave functions). Then the total charge density is given by

$$\varrho = 2\left[\sum_{j=1}^{6}|s_j|^2+\sum_{j=1}^{3}|\pi_j|^2+\sum_{j=1}^{6}|\chi_j|^2+\sum_{j=1}^{6}|\varphi_{j+6}|^2\right], \tag{26}$$

the factor 2 arising because all the orbits are doubly occupied. In this expression all the one-electron functions are assumed to be normalized. It is clear that if ϱ is integrated over all space the value of the integral will then be 42, the total number of electrons in the system. From (21)

$$\sum_{j=1}^{6}|\chi_j|^2 = \sum_{j=1}^{6}\frac{\varphi_j\varphi_j^*+\delta_j\varphi_j^*\varphi_{j+6}+\delta_j^*\varphi_{j+6}^*\varphi_j+\delta_j^*\delta_j\varphi_{j+6}^*\varphi_{j+6}}{1-A_{j,j+6}} \tag{27}$$

Introducing $\Phi_{j,k} = \varphi_j^*\varphi_k$, ϱ can be written

$$\varrho = 2\left[\sum_{j=1}^{6}|s_j|^2+\sum_{j=1}^{3}|\pi_j|^2+\sum_{j=1}^{6}\frac{\Phi_{j,j}+\Phi_{j+6,j+6}-(a_{j+6,j}\Phi_{j,j+6}+a_{j,j+6}\Phi_{j+6,j})}{1-A_{j,j+6}}\right]. \tag{28}$$

11. Calculation of ϱ

In order to calculate the overlaps $a_{j,j+6}$, the values of the overlaps between the original m.o.'s $a \ldots f$, $\alpha \ldots \zeta$, were needed. These were calculated using the wave functions described previously and are given below, $S_{a\beta}$ denoting $\int a\beta d\tau$; a and β being defined by equations (17) and (18).

$S_{ab} = 0{\cdot}1191$	$S_{a\alpha} = 0{\cdot}1307$	$S_{\alpha\beta} = 0{\cdot}1196$
$S_{ac} = 0{\cdot}0917$	$S_{a\beta} = -0{\cdot}0453$	$S_{\alpha\gamma} = 0{\cdot}0045$
$S_{ad} = 0{\cdot}0591$	$S_{a\gamma} = -0{\cdot}0100$	$S_{\alpha\delta} = 0{\cdot}0019$

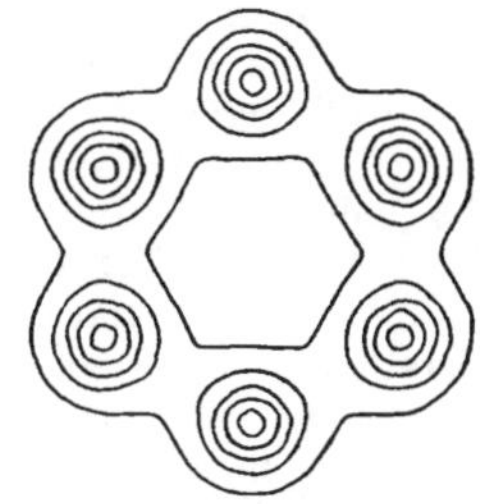

Fig. 4. π electron-density contours in a plane at a height of 0·35 Å above the plane of the ring. Outer contour is 0·15 e.Å^{-3}, interval is 0·2 e.Å^{-3}. All contours are shown.

Fig. 5. Molecular orbital electron-density contours in the plane of the ring. Outer contour is 0·15 e.Å^{-3}, interval is 0·2 e.Å^{-3}. To facilitate comparison with the T. F. contours, last contour shown is 1·75 e.Å^{-3}. The next contour, 1·95 e.Å^{-3}, defines the separate carbon atoms.

The actual calculation was carried out in terms of $a \ldots f$, $\alpha \ldots \zeta$, so that ϱ was written in terms of these wave functions. ϱ was then evaluated on a mesh of points, first in the plane of the ring and then in a plane parallel to it. The π electrons make no contribution to the density in the plane of the ring since the wave functions describing them have nodes in this plane. The parallel plane was taken at a distance of $0{\cdot}5R$

Fig. 6. Molecular orbital electron-density contours in a plane parallel to the plane of the ring and at a height of 0·35 Å above the plane of the ring. Outer contour is 0·15 e.Å^{-3}, interval is 0·2 e.Å^{-3}. All contours are shown.

(0·35 Å) from the plane of the ring; this is the same distance as in the T.F. calculation already described. The rather surprising feature is that even in this plane the density due to the σ and $1s$ electrons considerably outweighs that due to the π electrons, except immediately above the carbon atoms where the π electrons contribute about half of the total density. For interest the density contours due solely to the π electrons are shown in Fig. 4; the total density contours at intervals of 0·2 e.Å^{-3} are shown, for the plane of the ring in Fig. 5 and for the parallel plane in Fig. 6.

12. Comparison and discussion of results

If the contours for the total density as given by the two methods are compared the following conclusions can be drawn.

(i) *Plane of ring*

In the interior of the ring the variation of density given by the T.F. method is less rapid than that given by the m.o. method, as can be seen by the closer packing of the m.o. contours. The value at the centre of the ring is 0·29 e.Å^{-3} by the T.F. method and 0·20 e.Å^{-3} by the m.o. method. The positions of the 0·55 e.Å^{-3} and the 0·95 e.Å^{-3} contours inside the ring agree reasonably well with each other and also with the results obtained by Robertson for naphthalene. The bridge values, that is the values at the centre of a line joining adjacent carbon atoms, are of some interest. The values given by the two different theoretical methods differ quite considerably, the T.F. method giving 1·2 e.Å^{-3} and the m.o. method giving 1·9 e.Å^{-3}. Whilst it is not strictly permissible to compare these results with Robertson's values, one might suppose that the bond in naphthalene which would approximate most closely to a bond in benzene would be an end bond; this has a length of about 1·395 Å, that is about the same length as that of a C–C bond in benzene, and it is farthest from the effect of the excess carbons in naphthalene over those in benzene. For these end bonds Robertson gives a value slightly in excess of 1·5 e.Å^{-3}, a value which lies between the two theoretical values. One other point of some interest concerns the peak values at the carbon nuclei. With the T.F. model the density is infinite at the nuclei, a well known defect of the model. With the m.o. model the density is very large, around 830 e.Å^{-3}. However, in these calculations it has been assumed that the atoms are fixed. Higgs (1951) has, however, recently shown that the effect of vibrations is to reduce greatly the peak value of the density, so that comparison of our results near the carbon nuclei with Robertson's is meaningless.

Not much can be said at the moment about the distribution external to the ring, except that the T.F. method will be unreliable in the region of the hydrogen atoms and will over-emphasize their presence. The work of Robertson gives the 0·5 e.Å^{-3} contour in rough outline only and comparisons will not be very significant. But the general agreement appears to be reasonable.

(ii) *Parallel plane*

The distributions here are quite similar to those in the plane of the ring. The peak values in this plane differ very considerably, 3·8 e.Å^{-3} by the T.F. method, 1·8 e.Å^{-3} by the m.o. method. The 'atoms' are much more pronounced again in the T.F. method than in the m.o. method, and the general variation of density is more rapid for the m.o. method. The 'bridge' values are for the T.F. method 0·81 e.Å^{-3} and for the m.o. method 1·5 e.Å^{-3} whilst the values at the centre are 0·25 e.Å^{-3} and 0·17 e.Å^{-3} respectively. A surprising feature of the m.o. results is the small contribution due to the π electrons, and in view of the close interpenetration of σ and π electrons it seems to us quite remarkable that precise calculations can be made of the properties of organic molecules in which only the π electrons are considered.

13. Conclusion

Considering the nature of the approximations involved in both the T.F. and the m.o. methods, and the difference in view point of the two methods, the distributions are in reasonable agreement with each other, and with the experimental results for naphthalene as far as comparison is valid. The m.o. method seems to over-emphasize the amount of charge between the atoms, whilst the T.F. method tends to exaggerate the presence of the atoms at the expense of the 'bond' character. Also, as always, it should be remembered that the T.F. method gives poor results near the nuclei and at large distances, and that to some extent therefore charge contours 'flatter' its results. The results must await experimental work on the actual distribution in benzene before any more detailed discussion can be given of the relative merits of the two methods employed here.

It is a pleasure to thank Prof. C. A. Coulson for his most generous help and advice throughout the course of this work. The writer also wishes to acknowledge a grant from the Department of Scientific and Industrial Research.

References

Abrahams, S. C., Robertson, J. M. & White, J. G. (1949*a*). *Acta Cryst.* **2**, 233.

Abrahams, S. C., Robertson, J. M. & White, J. G. (1949*b*). *Acta Cryst.* **2**, 238.

Gombás, P. (1948). *Die Statistische Theorie des Atoms und ihre Anwendungen.* Vienna: Springer.

Higgs, P. W. (1951). Unpublished calculations.

Hund, F. (1932). *Z. Phys.* **77**, 12.

Lennard-Jones, J. E. (1931). *Proc. Camb. Phil. Soc.* **27**, 480.

Mayer, M. G. & Sklar, A. L. (1938). *J. Chem. Phys.* **6**, 645.

Mott, N. F. & Sneddon, I. N. (1948). *Wave Mechanics and its Applications.* Oxford: University Press.

Sinclair, V. C., Robertson, J. M. & Mathieson, A. M. (1950*a*). *Acta Cryst.* **3**, 245.

Sinclair, V. C., Robertson, J. M. & Mathieson, A. M. (1950*b*). *Acta Cryst.* **3**, 251.

Slater, J. C. (1930). *Phys. Rev.* **36**, 57.

[665]

THOMAS-FERMI FIELDS FOR MOLECULES WITH TETRAHEDRAL AND OCTAHEDRAL SYMMETRY

By N. H. MARCH

Communicated by C. A. COULSON

Received 15 April 1952

ABSTRACT. The Thomas-Fermi method has been applied to molecules with tetrahedral or octahedral symmetry in the approximation in which the nuclei of the outer atoms are smeared out over the surface of a sphere. Solutions are given for the molecules CH_4, CF_4, CCl_4, SiH_4, SiF_4 and SF_6, and using the results presented here it will not be a difficult matter to obtain the solutions for other molecules if they should be required.

A comparison of the results for CH_4 with those of Buckingham, Massey and Tibbs shows that whilst for such a light molecule the charge distribution is somewhat poor, the potential is already in good agreement with the results of the Hartree self-consistent field method. For CCl_4, however, the approximation of smoothing out all the protons appears to break down. An alternative approximation in which the K and L shell electrons are compressed into the nucleus and the modified nuclear charges smoothed, has been investigated and appears to give more reasonable results.

A discussion of the energy has been given, but the method is not sufficiently accurate to be of value for energetic considerations.

1. *Introduction.* In a recent paper Buckingham, Massey and Tibbs (3) (referred to hereafter as B.M.T.) have made a detailed investigation of the methane molecule by averaging the four protons over the surface of a sphere with the carbon atom as centre, and then applying the self-consistent field (s.c.f.) method of Hartree. The results were decidedly encouraging, and led them to suggest that results of interest would be obtained by applying the method to other molecules with tetrahedral or octahedral (t. or o.) symmetry. However, the labour involved in such calculations is considerable, and is particularly so unless a good first approximation is known. In the case of methane use was made of the treatment of Coulson (4) to obtain a starting field, which in fact proved to be a very good approximation to the true s.c. field. Similar calculations are not available, however, for other molecules, and, furthermore, results from atomic theory are not much help, as the nuclear field with which we are concerned here is different from that occurring in atoms.

As is well known, Dirac (6) has shown that by means of a semi-classical approximation one can pass from the Hartree s.c.f. method to the Thomas-Fermi method, and by doing so has provided the latter with a sound theoretical basis. One might therefore expect that the results of the T.F. method should be useful first approximations to the results of the s.c.f. method, and, indeed, it is well known in atomic theory that the T.F. potential is a useful starting point for s.c.f. calculations. Furthermore, as a theory in its own right, the method has not been without its successes, and it seemed a natural step therefore to investigate whether the T.F. method could be extended to molecules with t. and o. symmetry. The purpose of this paper is to present the results of this investigation.

2. *Application of the T.F. method.* Let us consider the molecules XY_n where $n = 4$ or 6, and following B.M.T. we suppose that the positive charges of the Y nuclei are smoothed out over the surface of a sphere with the X atom at the centre.

Let Ze be the positive charge of the central atom, and let N be the total number of remaining positive charges. Then the surface density over the sphere is given by

$$\sigma = Ne/4\pi R^2, \tag{1}$$

where R is the radius of the sphere. The problem is now clearly one of spherical symmetry. If we restrict ourselves for the moment to neutral molecules (for the case of positive ions see the appendix), then the essential problem is to solve the T.F. equation

$$\frac{1}{r}\frac{d^2}{dr^2}(rV) = \mu V^{\frac{3}{2}}, \tag{2}$$

where r is the distance measured from the central atom, V is the electrostatic potential and μ is given by

$$\mu = \frac{32\pi^2 e}{3h^3}(2me)^{\frac{3}{2}}, \tag{3}$$

subject to the appropriate boundary conditions. Let us define V_1 to be the electrostatic potential for $r \leqslant R$ and V_2 to be the electrostatic potential for $r > R$. Then the boundary conditions are

$$\left.\begin{aligned} &(a) \quad V_1 \to \frac{Ze}{r} \quad \text{as} \quad r \to 0, \\ &(b) \quad V_2 \to 0 \text{ at infinity}, \\ &(c) \quad (V_1)_R = (V_2)_R, \\ &(d) \quad -\left(\frac{dV_1}{dr}\right)_R + \left(\frac{dV_2}{dr}\right)_R = -4\pi\sigma, \end{aligned}\right\} \tag{4}$$

condition (d) following from a well-known result in electrostatics for the discontinuity in the field at a surface distribution of charge. It may easily be shown that with these boundary conditions the total number of electrons will be $Z + N$ as is of course required for a neutral molecule.

It is convenient to transform the T.F. equation into its usual dimensionless form

$$\frac{d^2\phi}{dx^2} = \frac{\phi^{\frac{3}{2}}}{x^{\frac{1}{2}}}, \tag{5}$$

where

$$V = \frac{Ze}{r}\phi, \quad r = bx \quad \text{and} \quad b = \left(\frac{3}{32\pi^2}\right)^{\frac{2}{3}} \frac{h^2}{2me^2 Z^{\frac{1}{3}}}. \tag{6}$$

The boundary conditions (4) then become

$$\left.\begin{aligned} &(a) \quad \phi_1(0) = 1, \\ &(b) \quad \phi_2 = 0 \text{ at infinity}, \\ &(c) \quad \phi_1(X) = \phi_2(X), \\ &(d) \quad \left(\frac{d\phi_1}{dx}\right)_X - \left(\frac{d\phi_2}{dx}\right)_X = \frac{N}{XZ}, \end{aligned}\right\} \tag{7}$$

where ϕ_1 and ϕ_2 are defined by

$$V_1 = \frac{Ze}{r}\phi_1, \quad V_2 = \frac{Ze}{r}\phi_2, \tag{8}$$

and X is the radius of the sphere in dimensionless units.

3. *Discussion of the solutions required.* The general form of the solutions of the T.F. equation is well known. However, it will clarify the position with regard to our particular problem if we consider the various forms of solution on a diagram of ϕ against x. The solutions with $\phi(0) = 1$ and with different initial slopes have already been interpreted physically. The solution I in Fig. 1 is the well-known T.F. free neutral atom solution. Solutions of type II with $\phi(0) = 1$ and which lie below the solution I have been interpreted physically as corresponding to positive ions. Solutions of type III, with $\phi(0) = 1$ and which lie above the solution I provide a T.F. description of the metallic state. These solutions with $\phi(0) = 1$ are clearly needed in our problem, but cannot provide the whole solution, since ϕ must tend to zero at infinity. Now there is only one solution amongst those we have mentioned so far which tends to zero at infinity and that is the T.F. atom solution, which is a solution of our problem for the trivial case $N = 0$. However, we need a solution which tends to zero at infinity and hence we must next look for solutions of this form.

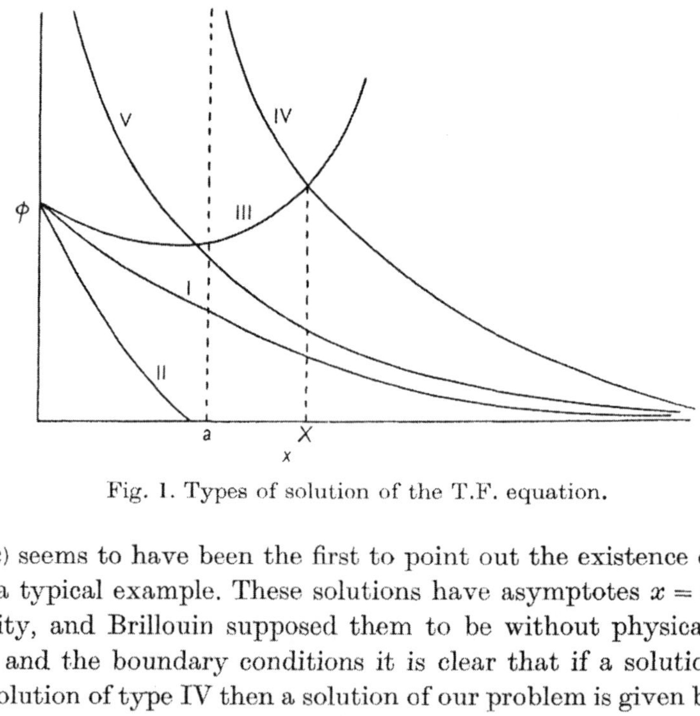

Fig. 1. Types of solution of the T.F. equation.

Brillouin (2) seems to have been the first to point out the existence of solutions of which IV is a typical example. These solutions have asymptotes $x = a$ and tend to zero at infinity, and Brillouin supposed them to be without physical significance. From Fig. 1 and the boundary conditions it is clear that if a solution of type III intersects a solution of type IV then a solution of our problem is given by solution III out to the point of intersection, and thereafter by the solution IV. The radius X is of course defined by the point of intersection, and from the slopes at X, the value of N/Z can be calculated for this particular solution. However, we have not dealt with all the solutions which tend to zero at infinity. As is well known a particular integral of the T.F. equation is

$$\phi = 144/x^3. \tag{9}$$

This solution divides the (ϕ, x) plane for the solutions which tend to zero at infinity into two parts. Above the solution $144/x^3$, marked V in Fig. 1, are the solutions of type IV that we have already discussed. Below the dividing curve are other solutions which tend to zero at infinity and which are finite at the origin, the usual T.F. atom

Table 1. *Solutions with* $\phi(0) = 1$

$-a_2$ / x_0 / $\phi(x_0)$	≈1·55 / ≈2·800 / ≈0·4767	1·56 / 3·043 / 0·4226	1·567 / 3·285 / 0·3775	1·5748 / 3·704 / 0·3148	1·581 / 4·330 / 0·2463	1·5849 / 5·229 / 0·1806	1·5853 / 5·401 / 0·1710	1·5866 / 6·206 / 0·1343	1·58756 / 7·790 / 0·0888	1·58764 / 8·015 / 0·0842	$-a_2$ / x_0 / $\phi(x_0)$
x											x
0·00	1·0000	1·0000	1·0000	1·0000	1·0000	1·0000	1·0000	1·0000	1·0000	1·0000	0·00
0·02	0·9727	0·9725	0·9724	0·9722	0·9721	0·9720	0·9720	0·9720	0·9720	0·9720	0·02
0·04	0·9485	0·9481	0·9478	0·9475	0·9472	0·9471	0·9471	0·9470	0·9470	0·9470	0·04
0·06	0·9261	0·9255	0·9251	0·9246	0·9243	0·9240	0·9240	0·9239	0·9239	0·9239	0·06
0·08	0·9052	0·9044	0·9039	0·9032	0·9027	0·9024	0·9024	0·9023	0·9022	0·9022	0·08
0·10	0·8856	0·8845	0·8838	0·8830	0·8824	0·8820	0·8820	0·8818	0·8817	0·8817	0·10
0·12	0·8669	0·8657	0·8649	0·8639	0·8632	0·8627	0·8626	0·8625	0·8624	0·8623	0·12
0·14	0·8493	0·8478	0·8468	0·8457	0·8448	0·8443	0·8442	0·8440	0·8439	0·8439	0·14
0·16	0·8324	0·8308	0·8296	0·8283	0·8273	0·8267	0·8266	0·8264	0·8262	0·8262	0·16
0·18	0·8163	0·8145	0·8132	0·8117	0·8106	0·8099	0·8098	0·8095	0·8094	0·8093	0·18
0·20	0·8009	0·7989	0·7974	0·7958	0·7945	0·7937	0·7936	0·7934	0·7932	0·7931	0·20
0·22	0·7862	0·7839	0·7823	0·7805	0·7791	0·7782	0·7781	0·7778	0·7776	0·7776	0·22
0·24	0·7720	0·7695	0·7678	0·7658	0·7643	0·7633	0·7632	0·7629	0·7626	0·7626	0·24
0·26	0·7584	0·7557	0·7538	0·7517	0·7500	0·7489	0·7488	0·7484	0·7482	0·7482	0·26
0·30	0·7328	0·7296	0·7274	0·7249	0·7229	0·7217	0·7215	0·7211	0·7208	0·7208	0·30
0·34	0·7089	0·7053	0·7027	0·6999	0·6976	0·6962	0·6960	0·6956	0·6952	0·6952	0·34
0·38	0·6867	0·6826	0·6797	0·6765	0·6739	0·6723	0·6721	0·6716	0·6712	0·6712	0·38
0·42	0·6660	0·6614	0·6582	0·6546	0·6517	0·6499	0·6497	0·6491	0·6486	0·6486	0·42
0·46	0·6467	0·6415	0·6379	0·6339	0·6307	0·6287	0·6285	0·6279	0·6274	0·6273	0·46
0·50	0·6285	0·6229	0·6189	0·6145	0·6110	0·6088	0·6085	0·6078	0·6073	0·6072	0·50
0·54	0·6115	0·6053	0·6010	0·5961	0·5923	0·5899	0·5896	0·5888	0·5882	0·5882	0·54
0·58	0·5955	0·5887	0·5841	0·5788	0·5746	0·5720	0·5717	0·5709	0·5702	0·5702	0·58
0·62	0·5805	0·5732	0·5681	0·5624	0·5579	0·5550	0·5547	0·5538	0·5531	0·5530	0·62
0·66	0·5664	0·5585	0·5530	0·5468	0·5419	0·5389	0·5385	0·5375	0·5368	0·5367	0·66
0·74	0·5407	0·5316	0·5252	0·5180	0·5124	0·5088	0·5085	0·5073	0·5064	0·5064	0·74
0·82	0·5179	0·5075	0·5002	0·4920	0·4856	0·4815	0·4811	0·4798	0·4788	0·4787	0·82
0·90	0·4978	0·4860	0·4777	0·4685	0·4612	0·4566	0·4561	0·4546	0·4535	0·4534	0·90
0·98	0·4800	0·4667	0·4575	0·4471	0·4389	0·4337	0·4332	0·4315	0·4302	0·4301	0·98
1·06	0·4644	0·4496	0·4392	0·4277	0·4185	0·4127	0·4121	0·4102	0·4088	0·4087	1·06
1·14	0·4507	0·4343	0·4228	0·4099	0·3998	0·3934	0·3927	0·3906	0·3890	0·3889	1·14
1·22	0·4389	0·4207	0·4080	0·3938	0·3826	0·3755	0·3748	0·3724	0·3707	0·3706	1·22
1·30	0·4288	0·4087	0·3947	0·3791	0·3667	0·3590	0·3581	0·3556	0·3537	0·3535	1·30
1·38	0·4202	0·3982	0·3828	0·3657	0·3522	0·3437	0·3427	0·3399	0·3378	0·3377	1·38
1·46	0·4131	0·3890	0·3722	0·3535	0·3387	0·3294	0·3284	0·3254	0·3231	0·3229	1·46
1·62	0·4031	0·3745	0·3546	0·3324	0·3149	0·3039	0·3027	0·2991	0·2963	0·2962	1·62
1·78	0·3983	0·3646	0·3412	0·3152	0·2946	0·2817	0·2803	0·2761	0·2729	0·2727	1·78
1·94	0·3982	0·3590	0·3317	0·3013	0·2774	0·2625	0·2608	0·2559	0·2522	0·2519	1·94
2·10	0·4029	0·3572	0·3256	0·2905	0·2629	0·2457	0·2438	0·2381	0·2338	0·2335	2·10
2·26	0·4120	0·3593	0·3229	0·2825	0·2508	0·2310	0·2289	0·2224	0·2175	0·2171	2·26
2·42	0·4256	0·3651	0·3233	0·2771	0·2409	0·2183	0·2158	0·2084	0·2028	0·2024	2·42
2·58	0·4439	0·3745	0·3267	0·2740	0·2328	0·2072	0·2044	0·1961	0·1897	0·1893	2·58
2·74	0·4668	0·3875	0·3331	0·2732	0·2266	0·1976	0·1945	0·1851	0·1779	0·1774	2·74
2·90	0·4947	0·4043	0·3424	0·2747	0·2221	0·1894	0·1859	0·1753	0·1672	0·1667	2·90
3·06	0·5278	0·4249	0·3548	0·2783	0·2191	0·1825	0·1785	0·1666	0·1576	0·1570	3·06
3·38	0·6114	0·4786	0·3890	0·2920	0·2176	0·1719	0·1670	0·1522	0·1410	0·1403	3·38
3·70	0·7217	0·5509	0·4368	0·3145	0·2217	0·1653	0·1593	0·1411	0·1274	0·1265	3·70
4·02	0·8649	0·6450	0·5000	0·3464	0·2315	0·1623	0·1550	0·1329	0·1163	0·1151	4·02
4·34	1·049	0·7658	0·5813	0·3888	0·2469	0·1626	0·1538	0·1271	0·1071	0·1058	4·34
4·66	1·287	0·9196	0·6845	0·4432	0·2684	0·1662	0·1556	0·1236	0·0997	0·0981	4·66
4·98	1·595	1·116	0·8148	0·5116	0·2965	0·1730	0·1603	0·1221	0·0938	0·0919	4·98
5·30	1·996	1·366	0·9790	0·5969	0·3321	0·1831	0·1679	0·1226	0·0892	0·0869	5·30
5·62	2·523	1·688	1·187	0·7028	0·3762	0·1967	0·1786	0·1250	0·0857	0·0831	5·62
5·94	—	—	1·450	0·8342	0·4303	0·2141	0·1926	0·1293	0·0834	0·0804	5·94
6·26	—	—	1·788	0·9979	0·4963	0·2356	0·2102	0·1355	0·0821	0·0786	6·26
6·58	—	—	2·225	1·203	0·5766	0·2619	0·2317	0·1439	0·0817	0·0777	6·58
6·90	—	—	2·795	1·460	0·6746	0·2935	0·2576	0·1543	0·0823	0·0777	6·90
7·22	—	—	3·549	1·787	0·7942	0·3313	0·2887	0·1672	0·0838	0·0785	7·22
7·54	—	—	—	—	—	0·3765	0·3257	0·1827	0·0863	0·0802	7·54
7·86	—	—	—	—	—	0·4302	0·3696	0·2011	0·0897	0·0827	7·86
8·18	—	—	—	—	—	—	0·4218	0·2228	0·0940	0·0860	8·18
8·50	—	—	—	—	—	—	0·4839	0·2482	0·0994	0·0903	8·50
8·82	—	—	—	—	—	—	0·5578	—	0·1059	0·0956	8·82
9·14	—	—	—	—	—	—	0·6461	—	0·1136	0·1018	9·14
9·46	—	—	—	—	—	—	—	—	0·1226	0·1092	9·46
9·78	—	—	—	—	—	—	—	—	0·1330	0·1177	9·78
10·10	—	—	—	—	—	—	—	—	0·1450	0·1276	10·10
10·42	—	—	—	—	—	—	—	—	0·1588	0·1390	10·42
10·74	—	—	—	—	—	—	—	—	0·1746	0·1520	10·74

solution being a particular case of this type of solution. It is clear that these solutions will intersect solutions with $\phi(0) = 1$, and together give us other acceptable solutions of our problem. We can see then that solutions of the T.F. equation of the form we require can be found, and that we are able to interpret solutions previously regarded as redundant from a physical viewpoint.

4. *Actual numerical solutions.* We have already seen that two types of solutions are required, solutions satisfying $\phi(0) = 1$ and with various initial slopes, and solutions tending to zero at infinity.

A number of solutions of the first type have already been obtained by Slater and Krutter (**12**) and recently by Feynman, Metropolis and Teller (F.M.T.) (**7**). However, as has been pointed out by Umeda (**13**), the F.M.T. results appear to be inconsistent with Slater and Krutter's, and furthermore, the F.M.T. solutions are not in a very useful region for our particular problem. Moreover, Slater and Krutter's solutions do not extend outwards far enough for our purpose. For these reasons, and because the solutions will be needed in a forthcoming discussion of the T.F. method for metals, a number of solutions have been calculated and are presented in Table 1. The method used to integrate the T.F. equation numerically was that described by Hartree (**10**) for second order equations with the first derivative absent. The integration was carried out in the usual manner by choosing the initial slope, and then obtaining starting values from the series expansion

$$\phi = 1 + a_2 x + a_3 x^{\frac{3}{2}} + a_4 x^2 + \ldots, \tag{10}$$

which was first given by Baker (**1**), and extended by F.M.T. The coefficients are collected in Table 2.

In the course of these calculations it has become clear that the slope at the origin, a_2, is by no means a good parameter to use in classifying the solutions, for a very small change in ϕ near the origin leads to a large change some distance out from the origin. Therefore, though we have generated our solutions by means of the initial slopes given in Table 1, a more reliable classification of the solutions is given by $\phi(x_0)$ and x_0, x_0 being the point at which a straight line passing through the origin is a tangent to the curve. (This is referred to again later.) An integral check was devised (in addition to the step-by-step check described by Hartree (**10**) which was applied at every stage of the calculation) and our solutions checked satisfactorily. As a final check one of our solutions was chosen, and two end-points of this solution were taken as defining boundary conditions. An integration was then started inwards towards the origin using the method which Fox and Goodwin (**8**), at the suggestion of Hartree, mention as being applicable to non-linear equations. The equation was used in the form

$$\frac{1}{w}\frac{d^2\phi}{dw^2} - \frac{1}{w^2}\frac{d\phi}{dw} = 2^{\frac{1}{2}}\phi^{\frac{3}{2}}, \quad \text{where } x = \frac{w^2}{2}, \tag{11}$$

to avoid difficulties at the origin. The final solution had the value 1 at the origin, with an error of one in the fifth decimal place, which means that our solutions should be entirely reliable to the four figures which we have tabulated. Our finding that the slope at the origin is not a very suitable parameter by means of which to classify solutions incidentally throws light on the discrepancy between Slater and Krutter's results and those of F.M.T. We have come to the conclusion that the results are inconsistent only with regard to the initial slopes, and that if the initial slopes of F.M.T. are discarded (we believe they are definitely incorrect, as our slopes are consistent with those of Slater and Krutter) and their solutions are classified by $\phi(x_0)$ and x_0, which they tabulate (and which are the important quantities for their applications), then we believe that their solutions are meaningful, if not entirely accurate. To graphical accuracy their values of $\phi(x_0)$ and x_0 lie on a smooth curve with our results.

Finally, it may be mentioned that an alternative method of integration would be to assume as boundary conditions two values of ϕ at chosen values of x, and then integrate inwards to the origin. Then a scale factor (defined later in equation (13)) could be applied to make this solution pass through the point $\phi(0) = 1$.

Table 2. *Coefficients of series expansion for small x*

$a_3 = \frac{4}{3}$	$a_6 = \frac{1}{3}$	$a_9 = \frac{2}{27} - \frac{1}{252}a_2^3$
$a_4 = 0$	$a_7 = \frac{3}{70}a_2^2$	$a_{10} = \frac{1}{175}a_2^2$
$a_5 = \frac{2}{5}a_2$	$a_8 = \frac{2}{15}a_2$	$a_{11} = \frac{31}{1485}a_2 + \frac{1}{1056}a_2^4$

Table 3. *Coefficients of series expansion for large x*

Coefficient	Value	Coefficient	Value	Coefficient	Value
F_2	$0{\cdot}62570F_1^2$	F_5	$0{\cdot}05508F_1^5$	F_8	$0{\cdot}00256F_1^8$
F_3	$0{\cdot}31339F_1^3$	F_6	$0{\cdot}02071F_1^6$	F_9	$0{\cdot}00085F_1^9$
F_4	$0{\cdot}13739F_1^4$	F_7	$0{\cdot}00741F_1^7$	F_{10}	$0{\cdot}00028F_1^{10}$

Let us pass on then to discuss the numerical determination of the solutions which tend to zero at infinity. It is a fortunate circumstance that all solutions of this type can be obtained rather simply. First, Coulson and March (5) have given an expansion valid for large x of the form

$$\phi = \frac{144}{x^3}\left(1 - \frac{F_1}{x^c} + \frac{F_2}{x^{2c}} - \frac{F_3}{x^{3c}} + \dots\right) \tag{12}$$

where the coefficients are given in Table 3*. In principle this will provide us with all solutions which tend to zero at infinity when we allow F_1 to take all possible values, both positive and negative. Fortunately, however, we can make use of the invariance property of the T.F. equation. This tells us that if $\phi(x)$ is a solution of the T.F. equation, then another solution is $\chi(y)$, where

$$\phi = a\chi, \quad x = by, \tag{13}$$

provided that $ab^3 = 1$. This property enables us to tabulate two master solutions, from which all solutions which are zero at infinity can be obtained. The reason why two master solutions are required is that, as we have already pointed out, the (ϕ, x) plane is divided in two by the solution $\phi = 144/x^3$ corresponding to $F_1 = 0$ in equation (12), and the scale-factor method can never transform a solution across this line. For

* Prof. Umeda (private communication) has pointed out that there are slight numerical errors in the last two decimals of F_2 and the last decimal of the remaining coefficients given in (5). The corrected values given in Table 3 were supplied by him and we have verified them.

convenience the two master solutions have been chosen to be defined by $F_1 = \pm 1$. With these values the series expansion has been used to calculate ϕ and ϕ' for large x, in fact as far as $x = 3$. After this, numerical integration inwards was begun. The master solutions are given in Tables 4 and 5. Actually, we already have a solution for

Table 4. *Master solution with* $F_1 = +1$

x	ϕ	$-\phi'$	x	ϕ	$-\phi'$	x	ϕ	$-\phi'$
0·1375	2576	25300	0·675	145·9	479·9	2·6	5·215	5·356
0·1500	2285	21350	0·700	134·6	429·8	2·8	4·276	4·103
0·1625	2039	18180	0·725	124·4	386·2	3·0	3·551	3·198
0·1750	1828	15590	0·750	115·2	348·1	3·2	2·982	2·530
0·1875	1647	13470	0·775	106·9	314·6	3·4	2·528	2·028
0·2000	1490	11710	0·800	99·44	285·1	3·6	2·163	1·645
0·2125	1353	10230	0·825	92·64	259·0	3·8	1·865	1·348
0·2250	1233	8991	0·850	86·46	235·9	4·0	1·619	1·116
0·2375	1128	7938	0·875	80·83	215·3	4·2	1·415	0·9316
0·2500	1034	7039	0·900	75·68	196·9			
0·2625	951·1	6267	0·925	70·96	180·5	4·6	1·100	0·6645
0·2750	877·0	5601	0·950	66·64	165·8	5·0	0·8718	0·4869
0·2875	810·7	5024	0·975	62·66	152·5	5·4	0·7029	0·3649
0·3000	751·1	4521	1·000	59·00	140·6	5·8	0·5751	0·2790
0·3125	697·4	4080	1·025	55·62	129·8	6·2	0·4766	0·2169
0·3250	648·8	3694	1·050	52·50	120·1	6·6	0·3994	0·1713
0·3375	604·8	3353				7·0	0·3381	0·1370
0·3500	564·8	3052	1·10	46·93	103·2	7·4	0·2887	0·1109
0·3625	528·4	2784	1·15	42·13	89·20	7·8	0·2485	0·09079
0·3750	495·1	2546	1·20	37·98	77·52	8·2	0·2154	0·07501
0·3875	464·6	2334	1·25	34·35	67·70	8·6	0·1881	0·06254
0·4000	436·7	2143	1·30	31·18	59·41	9·0	0·1651	0·05256
0·4125	411·0	1972	1·35	28·39	52·35			
0·4250	387·3	1818	1·40	25·93	46·32	9·8	0·1293	0·03791
0·4375	365·5	1679	1·45	23·74	41·14	10·6	0·1032	0·02804
0·4500	345·3	1553	1·50	21·80	36·67	11·4	0·08366	0·02118
0·4625	326·6	1439				12·2	0·06878	0·01630
0·4750	309·3	1336	1·6	18·51	29·42	13·0	0·05723	0·01275
0·4875	293·2	1242	1·7	15·86	23·89	13·8	0·04813	0·01012
0·5000	278·2	1156	1·8	13·69	19·61			
			1·9	11·91	16·25	15·4	0·03499	0·006611
0·525	251·2	1006	2·0	10·42	13·58	17·0	0·02624	0·004500
0·550	227·7	879·5	2·1	9·175	11·44	18·6	0·02018	0·003170
0·575	207·1	772·8	2·2	8·121	9·710	20·2	0·01585	0·002297
0·600	188·9	682·0	2·3	7·223	8·295			
0·625	172·9	604·3	2·4	6·453	7·130			
0·650	158·6	537·5						

positive F_1, the ordinary T.F. atom solution, but this is not known with accuracy for large distances, and hence we have calculated a new solution. A check of our solution is provided by the fact that by applying a suitable scale factor our solution is transformed to the existing solution, conveniently tabulated by Gombás (9). The values of ϕ' given in his table are, however, somewhat in error. With regard to our solutions, it is felt that at times there may be errors of 1 or 2 in the fourth place of ϕ and slightly greater errors in ϕ'.

5. *Discussion of the energy of the molecules.* Before we describe the results that we have obtained for particular molecules, the total energy of the systems that we are dealing with will be discussed. The total energy may be set up in the usual way, as the sum of the kinetic energy

$$T = c_k \int n^{\frac{5}{3}} d\tau, \quad \text{where} \quad c_k = \frac{3h^2}{10m}\left(\frac{3}{8\pi}\right)^{\frac{2}{3}}, \tag{14}$$

Table 5. *Master solution with* $F_1 = -1$

x	ϕ	$-\phi'$	x	ϕ	$-\phi'$	x	ϕ	$-\phi'$
0·5375	7360	69810	1·25	189·8	580·8	4·0	3·221	2·649
0·5500	6551	59990	1·30	163·5	477·1	4·2	2·744	2·141
0·5625	5854	51810	1·35	141·8	395·4	4·4	2·356	1·750
0·5750	5250	44970	1·40	123·7	330·3	4·6	2·038	1·443
0·5875	4725	39200	1·45	108·5	278·0	4·8	1·775	1·201
0·6000	4267	34310	1·50	95·73	235·6	5·0	1·555	1·007
0·6125	3864	30160	1·55	84·84	200·9	5·2	1·369	0·8512
0·6250	3510	26600	1·60	75·53	172·4	5·4	1·212	0·7240
0·6375	3197	23550	1·65	67·52	148·7			
0·6500	2920	20920	1·70	60·60	128·9	5·8	0·9631	0·5334
0·6625	2673	18640	1·75	54·58	112·3	6·2	0·7776	0·4015
0·6750	2452	16660	1·80	49·33	98·22	6·6	0·6368	0·3079
			1·85	44·73	86·29	7·0	0·5279	0·2400
0·700	2078	13420	1·90	40·67	76·11	7·4	0·4425	0·1898
0·725	1775	10920	1·95	37·09	67·38	7·8	0·3744	0·1520
0·750	1528	8972	2·00	33·92	59·86	8·2	0·3197	0·1232
0·775	1323	7435	2·05	31·09	53·36	8·6	0·2750	0·1009
0·800	1153	6210	2·10	28·57	47·71	9·0	0·2383	0·08337
0·825	1011	5224	2·15	26·31	42·78			
0·850	890·5	4425	2·20	24·28	38·48	9·8	0·1824	0·05787
0·875	788·3	3772				10·6	0·1426	0·04211
0·900	700·9	3233	2·3	20·80	31·36	11·4	0·1136	0·03111
0·925	625·9	2787	2·4	17·95	25·81	12·2	0·09192	0·02348
0·950	561·0	2414	2·5	15·60	21·43	13·0	0·07543	0·01805
0·975	504·7	2101	2·6	13·64	17·94			
1·000	455·5	1836	2·7	11·99	15·13	14·6	0·05260	0·01117
1·025	412·5	1612	2·8	10·59	12·84	16·2	0·03812	0·007277
1·050	374·7	1420	2·9	9·407	10·97	17·8	0·02850	0·004940
1·075	341·3	1256	3·0	8·390	9·426	19·4	0·02185	0·003470
1·100	311·7	1115						
1·125	285·4	992·6	3·2	6·754	7·072			
1·150	261·9	886·6	3·4	5·516	5·406			
1·175	240·9	794·4	3·6	4·561	4·202			
1·200	222·1	713·9	3·8	3·814	3·315			

and the potential energy

$$U = -\tfrac{1}{2}e\int nV_e \, d\tau - e\int nV_N \, d\tau + U_N, \tag{15}$$

where V_N is the potential due to the nuclei, V_e is the potential due to the electrons, and U_N is the nuclear-nuclear potential energy which we will discuss later. If the region between 0 and R is region 1 and between R and infinity is region 2, then let us introduce the following notation:

Let n_1 and n_2 be the electron densities in regions 1 and 2.
Let V_{e1} and V_{e2} be the electron potentials in regions 1 and 2.
Let V_{N1} and V_{N2} be the nuclear potentials in regions 1 and 2.
Then the total energy E in our spherically symmetrical problem can be written

$$E = c_k \int_0^R n_1^{\frac{5}{3}} 4\pi r^2 dr + c_k \int_R^\infty n_2^{\frac{5}{3}} 4\pi r^2 dr - \tfrac{1}{2}e \int_0^R n_1 V_{e1} 4\pi r^2 dr - \tfrac{1}{2}e \int_R^\infty n_2 V_{e2} 4\pi r^2 dr - e \int_0^R n_1 V_{N1} 4\pi r^2 dr - e \int_R^\infty n_2 V_{N2} 4\pi r^2 dr + U_N. \quad (16)$$

Remembering that we have the relations

$$\left.\begin{aligned} 4\pi n_1 e = \mu V_1^{\frac{3}{2}}, \quad 4\pi n_2 e = \mu V_2^{\frac{3}{2}}, \\ V_1 = V_{e1} + V_{N1}, \quad V_2 = V_{e2} + V_{N2}, \end{aligned}\right\} \quad (17)$$

and

then E can be written

$$E = 4\pi c_k \left(\frac{\mu}{4\pi e}\right)^{\frac{5}{3}} \int_0^R V_1^{\frac{5}{2}} r^2 dr + 4\pi c_k \left(\frac{\mu}{4\pi e}\right)^{\frac{5}{3}} \int_R^\infty V_2^{\frac{5}{2}} r^2 dr - \tfrac{1}{2}e \int_0^R n_1 V_1 4\pi r^2 dr - \tfrac{1}{2}e \int_0^R n_1 V_{N1} 4\pi r^2 dr - \tfrac{1}{2}e \int_R^\infty n_2 V_2 4\pi r^2 dr - \tfrac{1}{2}e \int_R^\infty n_2 V_{N2} 4\pi r^2 dr + U_N. \quad (18)$$

Let us first discuss the integrals appearing in the expression for the kinetic energy. Define

$$I_a^b = \int_a^b \mu\, V^{\frac{5}{2}} r^2 dr. \quad (19)$$

Then Milne (11) has shown that if $\dfrac{1}{r^2}\dfrac{d}{dr}\left(r^2 \dfrac{dV}{dr}\right) = \mu V^{\frac{3}{2}}$,

then
$$-\tfrac{7}{5} I_a^b = \left[r^2 V \frac{dV}{dr} + r^3 \left(\frac{dV}{dr}\right)^2 - \tfrac{4}{5} r^3 \mu V^{\frac{5}{2}}\right]_a^b. \quad (20)$$

This result can be applied immediately to the integrals appearing in the kinetic energy, and hence they can be evaluated in terms of V and its first derivative at the origin, R and infinity. The third and fifth integrals can immediately be expressed in terms of I_a^b and hence can also be obtained. The two integrals remaining are $-\tfrac{1}{2}e \int_0^R n_1 V_{N1} 4\pi r^2 dr$ and $-\tfrac{1}{2}e \int_R^\infty n_2 V_{N2} 4\pi r^2 dr$. To evaluate these integrals we must now introduce the explicit expressions for V_N. In region 1

$$V_{N1} = \frac{Ze}{r} + \frac{Ne}{R}, \quad (21)$$

and in region 2

$$V_{N2} = \frac{Ze}{r} + \frac{Ne}{r}. \quad (22)$$

Thus

$$\begin{aligned} -\tfrac{1}{2}e \int_0^R n_1 V_{N1} 4\pi r^2 dr &= -\tfrac{1}{2} Ze^2 4\pi \int_0^R n_1 r\, dr - \frac{1}{2}\frac{Ne^2}{R} \int_0^R n_1 4\pi r^2 dr \\ &= -\tfrac{1}{2} Ze \int_0^R \frac{1}{r}\frac{d^2}{dr^2}(r V_1)\, r\, dr - \frac{1}{2}\frac{Ne}{R} \int_0^R \frac{1}{r^2}\frac{d}{dr}\left(r^2 \frac{dV_1}{dr}\right) r^2 dr \\ &= -\tfrac{1}{2} Ze \left[\frac{d}{dr}(rV_1)\right]_0^R - \frac{1}{2}\frac{Ne}{R}\left[r^2 \frac{dV_1}{dr}\right]_0^R. \end{aligned} \quad (23)$$

Similarly $$-\tfrac{1}{2}e\int_R^\infty n_2 V_{N2} 4\pi r^2 dr = -\tfrac{1}{2}(Z+N)e\left[\frac{d}{dr}(rV_2)\right]_R^\infty. \tag{24}$$

Hence we have evaluated all the integrals appearing in the energy expression and all that remains is to express the result in a convenient form for evaluation.

First, the quantity $4\pi c_k(\mu/4\pi e)^{\frac{5}{3}}$ can be shown by substituting for c_k to be equal to $3\mu/5$. Using this result together with (17) and (18) we can write

$$E = \tfrac{1}{10}I_0^R + \tfrac{1}{10}I_R^\infty - \tfrac{1}{2}Ze\left[\frac{d}{dr}(rV_1)\right]_0^R - \frac{1}{2}\frac{Ne}{R}\left[r^2\frac{dV_1}{dr}\right]_0^R - \tfrac{1}{2}(Z+N)e\left[\frac{d}{dr}(rV_2)\right]_R^\infty. \tag{25}$$

But we can write

$$I_0^R + I_R^\infty = -\frac{5}{7}\frac{Z^2e^2a_2}{b} - \frac{5}{7}\left[\left(V_1 + r\frac{dV_1}{dr}\right)\frac{dV_1}{dr}r^2 - \left(V_2 + r\frac{dV_2}{dr}\right)\frac{dV_2}{dr}r^2\right]_R, \tag{26}$$

where a_2 is the slope of ϕ_1 at the origin. Finally, after some manipulation we obtain

$$E = \frac{3}{7}\frac{Z^2e^2a_2}{b} + \tfrac{3}{7}NeV_1(R) - \tfrac{1}{7}NeR\left(\frac{dV_1}{dr}\right)_R - \frac{3}{7}\frac{N^2e^2}{R} - \frac{ZNe^2}{R} + U_N. \tag{27}$$

We calculate U_N for the distribution of nuclei which actually occurs in the molecule and not for our hypothetical distribution. It can then be very easily shown that

$$U_N = \frac{Ne^2}{R}[Z + cN], \tag{28}$$

where $c = \dfrac{3\sqrt{6}}{32}$ for tetrahedral molecules

$= \dfrac{1+4\sqrt{2}}{24}$ for octahedral molecules.

Substituting (28) in the expression (27) for the energy, and writing the energy in terms of ϕ_1 instead of V_1, we have

$$E = \frac{3}{7}\frac{Z^2e^2}{b}\left[a_2 + \frac{4}{3}\frac{N}{Z}\frac{\phi_1(X)}{X} - \frac{1}{3}\frac{N}{Z}\left(\frac{d\phi_1}{dx}\right)_X - \frac{N^2}{XZ^2} + \frac{7c}{3}\frac{N^2}{XZ^2}\right]. \tag{29}$$

We also find after some calculation, starting again from the initial energy equation (19) and differentiating it with respect to R, that

$$\frac{dE}{dR} = \frac{ZNe^2}{R^2}\left[X\left(\frac{d\phi_1}{dx}\right)_X - \phi_1(X) - c\frac{N}{Z}\right]. \tag{30}$$

Using this expression for dE/dR, and writing down separately the expressions for the kinetic and potential energy, it can be shown that

$$2T + U = -R\frac{dE}{dR}, \tag{31}$$

which is the form the virial theorem takes in our problem. The equation (30) demonstrates immediately the possibility of a minimum in the energy versus radius-of-sphere curve. Using the solutions which we have obtained, the position of this minimum has been determined, and the results may be demonstrated by plotting N/Z against X_E, where X_E is the equilibrium radius of the sphere in dimensionless units. The graphs are shown in Fig. 2, curve I referring to tetrahedral, and curve II to octahedral molecules. These values of X_E can be compared immediately with the experimental bond lengths in molecules with t. and o. symmetry. However, it is immediately clear, on making the comparison, that there is little agreement between the theoretical and experimental values. For example, for CH_4, the experimental bond length is about a half of the theoretical value, whilst for SiF_4 and SF_6 it is about twice as great as the theoretical value.

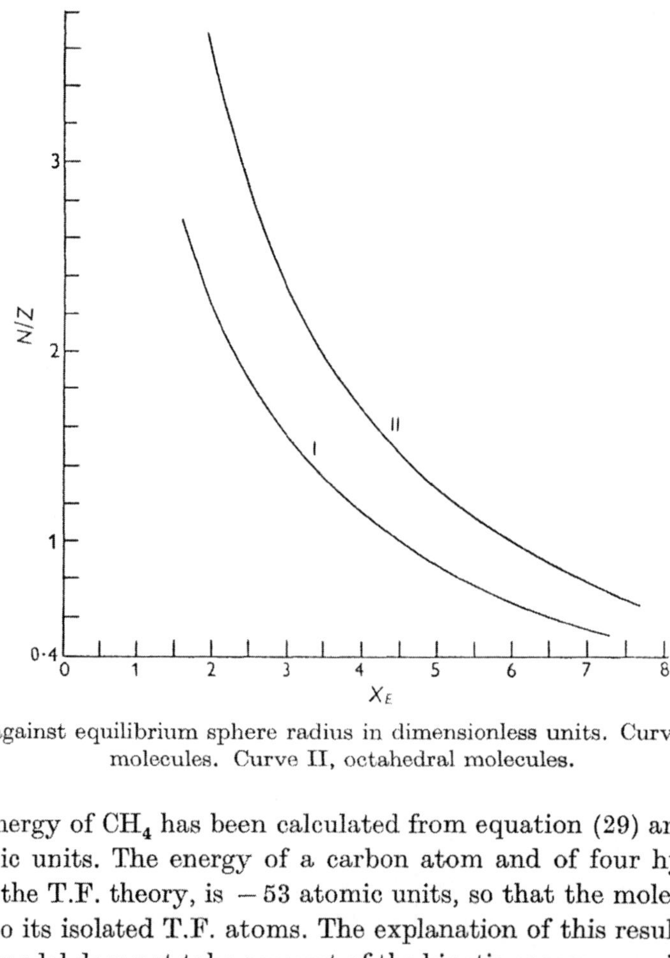

Fig. 2. N/Z against equilibrium sphere radius in dimensionless units. Curve I, tetrahedral molecules. Curve II, octahedral molecules.

The total energy of CH_4 has been calculated from equation (29) and comes out to be -51 atomic units. The energy of a carbon atom and of four hydrogen atoms, calculated on the T.F. theory, is -53 atomic units, so that the molecule is unstable with respect to its isolated T.F. atoms. The explanation of this result appears to be that the T.F. model does not take account of the kinetic energy associated with rapid changes of electron density, and in the T.F. atom this leads to an energy which is much too low. Now, in the molecular problem, we have replaced the four hydrogen nuclei by a surface distribution of charge and the electron density varies less violently near this distribution. B.M.T. give -40 atomic units for the total energy of the methane molecule so that our energy is still much too low.

It must be concluded that the method is therefore not adequate for energetic considerations. In the remainder of the work the bond lengths of the molecules have been taken from experiment, and the charge distribution and potential fields investigated in this case.

6. *Discussion of solutions for particular molecules.* After a few solutions had been obtained it proved possible, by judicious choice of the initial slopes by means of which the various solutions were generated, to obtain directly solutions which had a given N/Z (determined by the molecule under consideration) and a sphere radius equal to the experimental bond length. In all cases we have considered, the value of X corresponding to the particular solution we give to describe a molecule is different from the experimental value by less than 2 %. There seems little point in attempting to fix X to agree entirely with experiment, and indeed in some cases the experimental

Table 6. *Parameters defining solutions for particular molecules*

Molecules	N/Z	$-a_2$	F_1	b
CH_4	0·6667	1·581	+1	0·6690
CF_4	6·000	1·567	—	—
CCl_4	11·33	1·5748	−1	0·1520
SiH_4	0·2857	1·58756	+1	0·08062
SiF_4	2·571	1·5853	−1	0·550
SF_6	3·375	1·5853	−1	0·2240

accuracy would not warrant it. We have obtained explicit solutions for the molecules CH_4, CF_4, CCl_4, SiH_4, SiF_4 and SF_6, and the necessary information needed to define these solutions is given in Table 6. The solution inside the sphere is determined uniquely by giving the initial slope a_2, the solution corresponding to this slope being found in Table 1. To define the solution in the region external to the surface distribution it is necessary to give the value of F_1, that is, to define which master solution is to be used, and to give the appropriate scale factor b, which must be used to transform from the master solution to the particular solution we require. To obtain the solution in region 2 from the master solution it is only necessary to multiply the argument x of the master solution by $1/b$ to obtain the new argument, and to multiply the value ϕ of the master solution by b^3 to obtain the new value of ϕ. If the derivatives are required, the derivative ϕ' of the master solution has only to be multiplied by b^4. In Table 6 it will be noted first of all that no value of b is given for CF_4. The reason for this is that the solution in region 2 is not very different from $144/x^3$, and we have not tabulated the required range for this molecule. However, the solution for CF_4 in region 2 may be calculated immediately from the expression (12) with $F_1 = -0{\cdot}04681$. In the case of SiF_4 the solution in region 2 is insensitive to changes in b, and hence b should be used as if it were accurate to four figures. Table 6 therefore defines the solutions for the six molecules investigated here. It is clear that the method is quite general and that all that is required to obtain the solution for any other molecule is the appropriate solution in region 1. With the information given here it will be a relatively simple task to obtain the appropriate solution for any other molecule, if it should be desired.

7. *Comparison of results for* CH_4 *with B.M.T. results.* Particular attention has been paid to the methane molecule in this investigation, because at the moment this is the only molecule for which a Hartree self-consistent field treatment has been given. It provides us therefore with the only direct means of assessing the accuracy of our results.

It should be pointed out that a comparison of the results for this molecule with those obtained by the Hartree method does not provide an entirely fair test of the method, for we are dealing here with a system of only ten electrons, and we already know from the results of the T.F. method in atomic theory that the agreement between the charge distribution given by the T.F. and the Hartree methods is rather poor for light atoms, but that for heavy atoms the T.F. method gives a quite good overall representation of the distribution of charge.

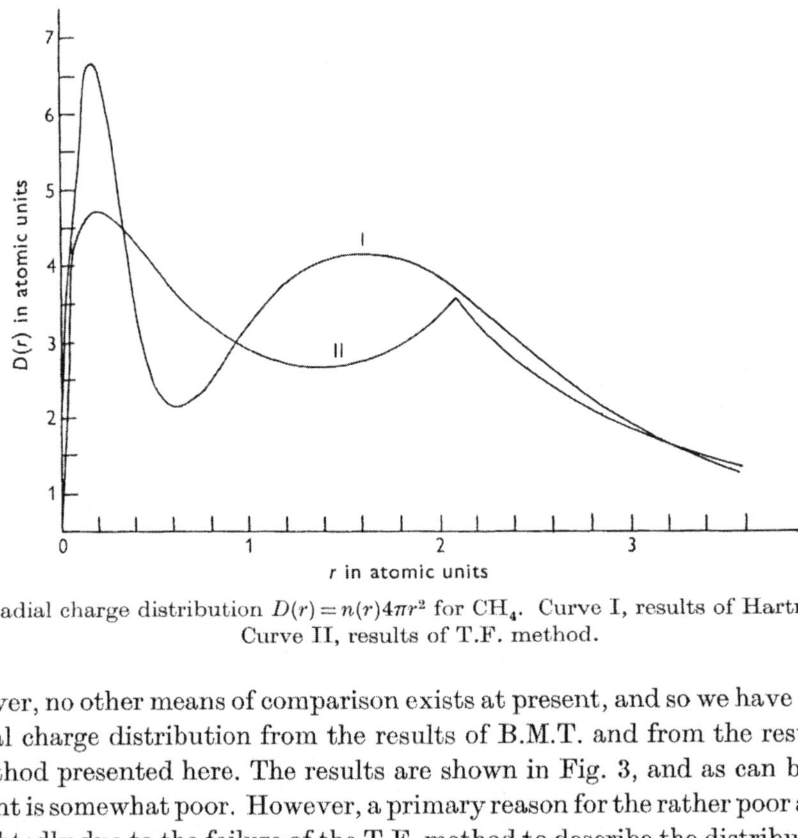

Fig. 3. Radial charge distribution $D(r) = n(r)4\pi r^2$ for CH_4. Curve I, results of Hartree method. Curve II, results of T.F. method.

However, no other means of comparison exists at present, and so we have calculated the radial charge distribution from the results of B.M.T. and from the results of the T.F. method presented here. The results are shown in Fig. 3, and as can be seen the agreement is somewhat poor. However, a primary reason for the rather poor agreement is undoubtedly due to the failure of the T.F. method to describe the distribution in the carbon atom, rather than a failure of the method in the molecular application we have made. One defect, however, does arise as a consequence of the application to molecules, and that is the non-smoothness of the curve at the position of the sphere of surface charge. This is an inevitable consequence of the relation between the charge density and the potential in the T.F. theory. However, it is quite possible that the defect will be less serious with a surface distribution than with point charges since the potential variation is less violent near a surface distribution than near point charges. It also seems reasonable to suppose that for heavier molecules the agreement between the T.F. and Hartree methods as regards the charge distribution will be much improved, assuming of course that the smoothing approximation is applied in both cases.

However, a much more gratifying result emerges when one attempts to compare the potential as given by the T.F. method for methane with that given by the Hartree treatment. The potential in the T.F. method is the potential which would act on an infinitesimal test charge, and to compare with this we have added the potentials arising from all the electrons in the Hartree treatment. The nuclear potential, which is the same of course in both methods, may conveniently be subtracted from the total potential, and the potential arising from the electrons calculated. The results are shown in Fig. 4 and are quite gratifying. They lead us to suppose that for heavier molecules the T.F. field will be an exceedingly useful first approximation to the true s.c. field, except in a small region near the central nucleus, in which region, however, the nuclear potential is very large, and V_e for most purposes is relatively unimportant.

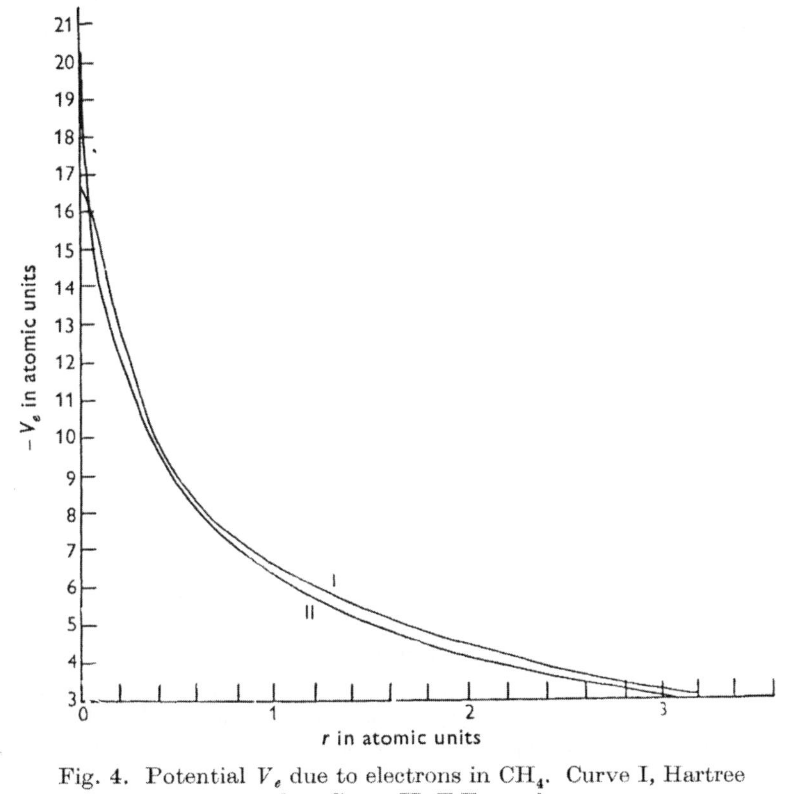

Fig. 4. Potential V_e due to electrons in CH_4. Curve I, Hartree results. Curve II, T.F. results.

8. *Discussion of results for* CCl_4*. Of the six molecules which we have discussed here, the one for which the smoothing approximation is the most questionable is CCl_4. This is due in the main to the presence of four groups of ten inner-shell electrons which are rather tightly bound to their own particular chlorine nucleus.

As a possible alternative approximation the K and L shell electrons around a particular chlorine nucleus have been compressed into the nucleus, thus reducing the nuclear charge to seven. These modified nuclear charges have been smoothed and the problem is then that of a molecule defined by $N/Z = 28/6$. The appropriate solution has been found by interpolation and the radial charge distributions obtained

* Thanks are due to Prof. C. A. Coulson for the suggestions from which this section arose.

by this and the previous method are displayed in Fig. 5. The curves differ greatly near the surface charge, due of course to the fact that in curve II the K- and L-shell electrons are missing. One might have expected that the curves would agree outside the region where the K and L shell electrons are important (indicated by the vertical lines in Fig. 5). However, we see that we must go very much further than this from the surface charge before there is any agreement. To attempt to determine which of the approximations is better, we have proceeded as follows. The number of electrons within a spherical shell of thickness 0·64 atomic units and of mean radius 3·32 atomic units has been found for both approximations. In the approximation in which all the electrons are considered this number is 25, and in the second case 9.

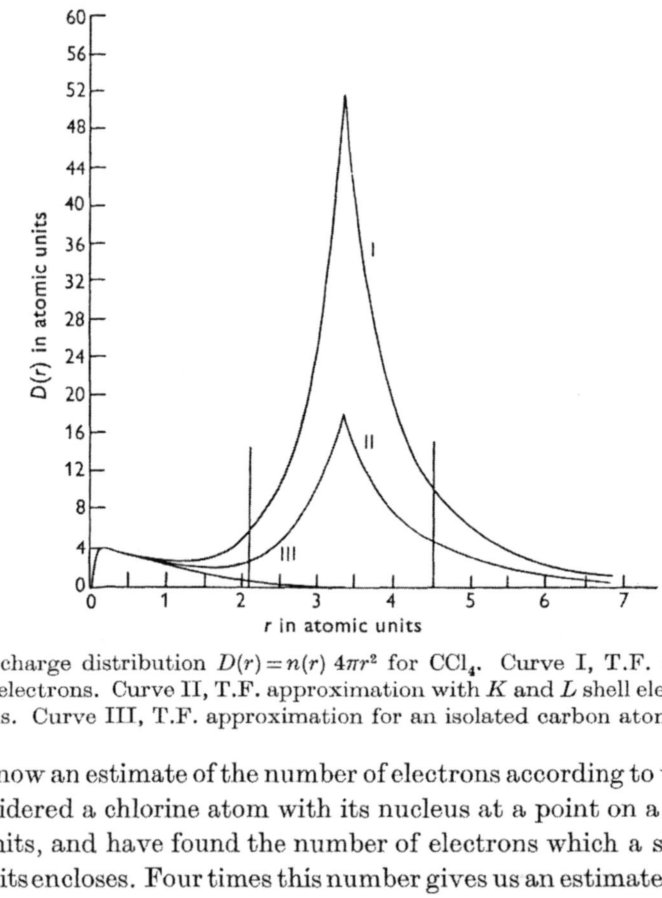

Fig. 5. Radial charge distribution $D(r)=n(r)\ 4\pi r^2$ for CCl_4. Curve I, T.F. approximation involving all electrons. Curve II, T.F. approximation with K and L shell electrons compressed into nucleus. Curve III, T.F. approximation for an isolated carbon atom.

We require now an estimate of the number of electrons according to wave mechanics. We have considered a chlorine atom with its nucleus at a point on a sphere of radius 3·32 atomic units, and have found the number of electrons which a shell of thickness 0·64 atomic units encloses. Four times this number gives us an estimate to compare with our T.F. value. To compare with our first approximation we must consider all electrons; to compare with the second method we must restrict ourselves to the $3s$ and $3p$ electrons. In the first case the number of electrons is about 40 (the T.F. atomic model gives 38); in the second case the number is 7.

Thus the number of electrons given by our first approximation (25) is considerably less than our estimate (40) for four chlorine atoms. If our result were to be believed, it would involve a considerable disturbance in the inner shells, and a marked shift of charge away from the chlorine atoms. Such a large movement of charge would seem to be unlikely, and it is thought that much of the effect must arise from our smoothing approximation in which we have implicitly treated the inner electrons rather violently. For just this reason, the second approximation of compressing the K and L electrons into the nucleus has been investigated.

In this case, the numbers of electrons inside the spherical shell considered are in reasonable agreement. The fact that the number given by our approximation is higher than the wave mechanical estimate may be due in part to neglect of the exclusion principle, for there is nothing to prevent outer electrons from falling inwards and taking the places of the K and L electrons which have been combined with the nuclei. However, neglecting the effect of the exclusion principle, it appears that the general disposition of the remaining electrons is sensibly described by this approximation. Hence, outside the region marked by the vertical lines in Fig. 5 we believe that curve II will give a better description of the distribution of electrons than curve I.

Possibly the most interesting region in our charge distributions is that midway between the carbon and the chlorine atoms. In this region, Fig. 5 shows how the charge cloud of the carbon atom merges into that originating from the chlorine atoms. The radial charge distribution due to an isolated carbon atom is shown in curve III of Fig. 5 for comparison. As we have already remarked, we believe curve II, which indicates less charge than curve I in the centre of the C—Cl bonds, to be nearer to the truth.

9. *Physical properties using the T.F. method.* For the molecules which we have discussed here, a calculation of various physical properties could now be made. For example, to mention three such properties, the diamagnetic susceptibilities, polarizabilities, and the shape of the Compton profiles in X-ray scattering could be found. However, in the light of experience in atomic theory there is little point in calculating any one of these properties. For each one is influenced very markedly by the form of the charge distribution, and our charge distributions fall off very slowly with distance exactly as in the atomic case, where the method has already been shown to be inadequate to describe any of these properties.

The only detailed discussion we have therefore made is that already described, in which the predictions of the method concerning the bond lengths in molecules were discussed. Here again it was found that the results were poor, and it must be concluded that the method is too inaccurate for energetic considerations.

10. *Conclusions.* It must be concluded that as a theory in its own right the T.F. method is inadequate to give a detailed description of the physical properties of molecules of the type considered here. This should not be regarded, however, as the failure of the method in the specific application made here, as an identical situation exists in atomic theory.

However, a comparison of the charge distribution and potential given by the T.F. method and by the Hartree method for the case of methane is more encouraging. The charge distributions are not in good agreement, but much of the discrepancy can be traced back to the failure of the T.F. method to describe the distribution of charge in an atom as light as carbon, and it is believed that for heavier molecules the agreement between the T.F. results and the Hartree results will be much improved. But the most important result which has emerged is the agreement between the potential calculated by the two methods. This would seem to indicate that in an extension of the work of B.M.T. to other molecules one could start confidently with the potentials presented in this paper as a first approximation. In addition, the potentials as given here may be useful in some problems; for example, in the calculation of electron scattering by molecules.

Finally, it may be remarked that a possible reason for part of the disagreement between the predicted bond lengths and those found experimentally may be due to the neglect of exchange. If one attempts to take account of exchange, unfortunately one finds that there is no longer a satisfactory free neutral atom. The defect would presumably carry over into the case discussed here, and thus it does not appear at present to be worth attempting such a generalization of the method.

The writer wishes to thank Prof. C. A. Coulson, F.R.S., for his valuable comments on the manuscript, and Mr J. F. Scott for useful advice on various aspects of the numerical work. The use of a calculating machine in the Department of Mathematics is gratefully acknowledged.

REFERENCES

(1) BAKER, E. B. *Phys. Rev.* (2), 36 (1930), 630.
(2) BRILLOUIN, L. *L'atome de Thomas-Fermi* (Actualités sci. industr. no. 160, Paris, 1934).
(3) BUCKINGHAM, R. A., MASSEY, H. S. W. and TIBBS, S. R. *Proc. roy. Soc.* A, 178 (1941), 119.
(4) COULSON, C. A. *Trans. Faraday Soc.* 33 (1937), 388.
(5) COULSON, C. A. and MARCH, N. H. *Proc. phys. Soc. Lond.* A, 63 (1950), 367.
(6) DIRAC, P. A. M. *Proc. Camb. phil. Soc.* 26 (1930), 376.
(7) FEYNMAN, R. P., METROPOLIS, N. and TELLER, E. *Phys. Rev.* (2), 75 (1949), 1561.
(8) FOX, L. and GOODWIN, E. T. *Proc. Camb. phil. Soc.* 45 (1949), 373.
(9) GOMBÁS, P. *Die statistische Theorie des Atoms und ihre Anwendungen* (Vienna, 1948), p. 358.
(10) HARTREE, D. R. *Mem. Manchr lit. phil. Soc.* 77 (1933), 91.
(11) MILNE, E. A. *Proc. Camb. phil. Soc.* 23 (1927), 794.
(12) SLATER, J. C. and KRUTTER, H. M. *Phys. Rev.* (2), 47 (1935), 559*.
(13) UMEDA, K. *Phys. Rev.* (2), 83 (1951), 651.

APPENDIX

The foregoing considerations have been confined to the case of neutral molecules. The extension of the method to deal with positive ions is considered here. We must now use the T.F. equation in its general form

$$\nabla^2(V-V_0) = \mu(V-V_0)^{\frac{3}{2}}, \tag{32}$$

where $-eV_0$ is the maximum energy of an electron and is only zero in the case of neutral molecules. If z represents the number of excess positive charges of the ion then we must satisfy the condition that

$$\int \nabla^2(V-V_0)\,d\tau = 4\pi(Z+N-z)\,e. \tag{33}$$

* For tables, see (9), p. 357.

Exactly as in the case of atomic ions, the molecule ions will have finite radii, which we will denote by R_b. We introduce as before the dimensionless quantities x, ϕ_1 and ϕ_2, defined by

$$r = bx, \quad V_1 - V_0 = \frac{Ze}{r}\phi_1, \quad V_2 - V_0 = \frac{Ze}{r}\phi_2. \tag{34}$$

Then, after some manipulation, equation (33) becomes

$$z/Z = -\phi_2'(X_b)\,X_b, \tag{35}$$

when we remember that condition (7*d*) has still to be satisfied in this case, and that the solutions that we shall require are solutions for which $\phi_2(X_b) = 0$. Similar solutions to those already discussed are still applicable in region 1, but in region 2 new solutions are required. The solutions needed are of the form shown in Fig. 6. The value of N/Z for the molecule described by a given solution is determined as in the case of neutral molecules by the difference in the slopes of ϕ_1 and ϕ_2 at X, whilst z/Z is determined by the slope of ϕ_2 at X_b. In order that the charge density shall vanish on the boundary R_b it is necessary to choose

$$V_0 = ze/R_b. \tag{36}$$

Thus the solution for a positive ion is completely defined.

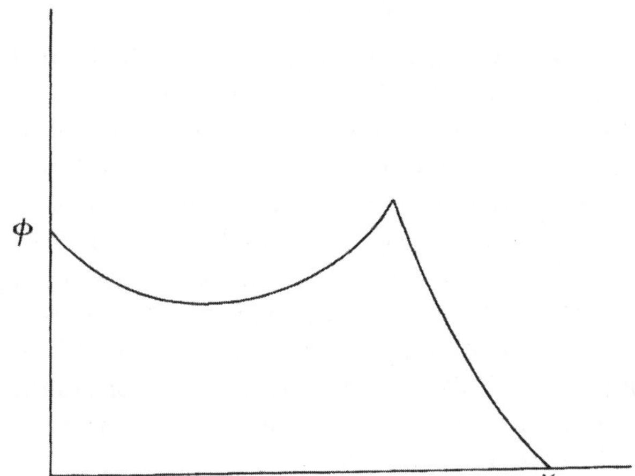

Fig. 6. Typical solution for a molecule ion.

DEPARTMENT OF PHYSICS
UNIVERSITY OF SHEFFIELD

The relation between the Wentzel–Kramers–Brillouin and the Thomas–Fermi approximations

By N. H. March
Department of Physics, The University, Sheffield

and J. S. Plaskett
Department of Physics, University of Virginia, U.S.A.

(*Communicated by C. A. Coulson, F.R.S.—Received* 19 *January* 1956)

It is shown that two essential approximations are made in using the customary Thomas–Fermi formula for the sum of the eigenvalues in any one-dimensional problem. The first is to start from the Wentzel–Kramers–Brillouin formula for the individual eigenvalues, and the second is to replace the summation by an integration.

The three-dimensional central field problem is then considered, and by similar arguments, though with an additional approximation, the usual Thomas–Fermi energy formula is again obtained. Possible ways of correcting the errors introduced by integrating instead of summing are discussed and illustrative examples given. In the three-dimensional case particular attention is given to the Coulomb field problem. Finally, brief reference is made to the possibility of correcting for the errors of the Wentzel–Kramers–Brillouin formula.

1. Introduction

A number of attempts have already been made to elucidate the nature of the approximations underlying the Thomas–Fermi (TF) method which has found wide application in dealing with many-body problems in quantum mechanics. Of outstanding importance are the accounts of Dirac (1930) and Brillouin (1934), and from the point of view of the present approach the more recent paper by Fényes (1949) is also relevant. In the last two papers explicit use is made of the Wentzel–Kramers–Brillouin (WKB) approximation, an approach we again adopt here. However, we believe that the present work leads to a much more clear-cut demonstration of the basic approximations underlying the usual TF energy formula than any that has been published hitherto. We restrict our considerations in this paper entirely to the energy; for a recent discussion of the TF particle density reference can be made to a paper by Plaskett (1953).

2. One-dimensional case

We take as our starting point the usual WKB formula for the eigenvalues in a potential $V(x)$ represented schematically in figure 1. According to this, the eigenvalues are given by the solutions of the equation

$$2\int_{x_1(E)}^{x_2(E)} p(E,x)\,\mathrm{d}x = (n+\tfrac{1}{2})\,h \quad (n = 0, 1, 2, \ldots), \tag{1}$$

where $$p(E,x) = \{2m(E-V(x))\}^{\frac{1}{2}},$$

and x_1 and x_2 are the classical turning points given by the solutions of

$$V(x) = E.$$

Since x_1 is a decreasing and x_2 an increasing function of E, the left-hand side of (1) is an increasing function of E that is zero for E equal to the minimum value of $V(x)$. It is convenient to regard n as a continuous variable, which is the function of E given by (1). The eigenvalues are then given by

$$n(E_0) = 0, \quad n(E_1) = 1, \quad \ldots, \quad n(E_k) = k, \ldots,$$

and we wish to calculate the sum of the first N eigenvalues, say E, given by

$$\mathrm{E} = \sum_{k=0}^{N-1} E_k. \tag{2}$$

Suppose now that $E(n)$ is the function inverse to $n(E)$ (for example, $E(k) = E_k$). We represent a possible form of this function in figure 2, and then of course we find the

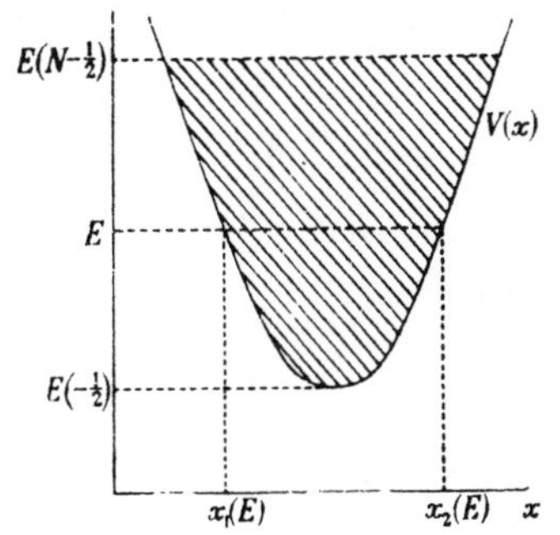

FIGURE 1. (E, x) plane for one-dimensional case.

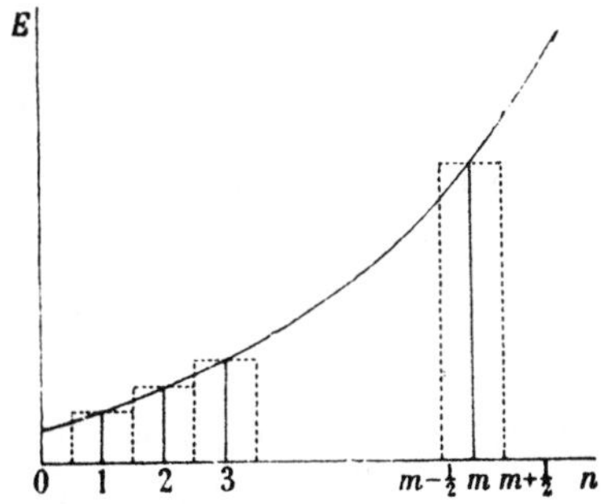

FIGURE 2. $E(n)$ represented schematically.

eigenvalues from the ordinates corresponding to integral values of n and finally sum these to obtain E. From a geometrical point of view this is equivalent to summing the areas of the rectangles shown in figure 2. This now suggests that the following simplification will often be quite accurate. We replace the area of each rectangle, say the one centred on $n = m$, by the area enclosed between the curve $E = E(n)$ and the ordinates $n = m - \frac{1}{2}$ and $m + \frac{1}{2}$. The approximation to the sum E thus obtained is

$$\mathrm{E} = \int_{-\frac{1}{2}}^{N-\frac{1}{2}} E(n)\,\mathrm{d}n, \tag{3}$$

where the limits of integration are obviously to be taken as shown. It is worth noting in passing that there is no approximation involved in using (3) if the function $E(n)$ is linear.

The expression (3) for the sum of the eigenvalues can be very easily simplified. We change the variable of integration to E ($E(n)$ is monotonic), differentiate (1) to obtain dn/dE, and then by substitution in (3) we have

$$\mathrm{E} = \int_{E(-\frac{1}{2})}^{E(N-\frac{1}{2})} \int_{x_1(E)}^{x_2(E)} \frac{2m}{h} p^{-1}(E,x)\,dx\,dE. \tag{4}$$

It should be noted in passing that in calculating dn/dE by differentiation of the integral in (1) with respect to E, no contribution arises from the fact that the end-points of the range of integration are functions of E, because the integrand is zero at both points. We now change the order of integration, and as we are evidently integrating over the shaded portion of the (E, x) plane shown in figure 1, we have

$$\mathrm{E} = \frac{2m}{h} \int_{x_1(E')}^{x_2(E')} \int_{E=V(x)}^{E'} E p^{-1}(E,x)\,dE\,dx, \tag{5}$$

where for brevity we have replaced $E(N-\frac{1}{2})$ by E'. The integration over E can now be performed (this was of course the point of reversing the order of integration) and we find with a little manipulation

$$\mathrm{E} = \int_{x_1(E')}^{x_2(E')} \left[\frac{1}{3}\frac{p^2(E',x)}{2m} + V(x)\right] \frac{2}{h} p(E',x)\,dx, \tag{6}$$

where E' is to be determined from the formula (cf. (1))

$$\int_{x_1(E')}^{x_2(E')} \frac{2}{h} p(E',x)\,dx = N. \tag{7}$$

This is the usual TF formula for the sum of the eigenvalues; to complete the TF scheme it is only necessary to identify $\frac{2}{h}p(E',x)$ with the particle density. In concluding this discussion of the one-dimensional case it is worth noting that the energy of the highest state, $E(N-1)$, is less than $E' = E(N-\frac{1}{2})$.

3. Three-dimensional case with spherical symmetry

The WKB approximation to the eigenvalue E_{kl} in a spherically symmetrical potential $V(r)$ is given by the solution of the equation

$$2\int_{r_1^l(E)}^{r_2^l(E)} \left\{2m\left[E - V(r) - \frac{\hbar^2}{2m}\frac{(l+\frac{1}{2})^2}{r^2}\right]\right\}^{\frac{1}{2}} dr = (k+\tfrac{1}{2})\,h, \tag{8}$$

where $r_1^l(E)$ and $r_2^l(E)$ are the roots of the equation

$$V(r) + \frac{\hbar^2}{2m}\frac{(l+\frac{1}{2})^2}{r^2} = E.$$

It should be noted that k is not the total quantum number; this is $k+l+1$. Also in (8) we have written $(l+\frac{1}{2})^2$ instead of $l(l+1)$ for the basic reasons considered in detail by Langer (1937). As before let us consider (8) as giving E as a function of the continuous variables k and l. The sum of the eigenvalues now required is

$$\mathrm{E} = \sum_{k,\, l \text{ integers } \geqslant 0} (2l+1)\, E(k,l) \tag{9}$$

for all k, l satisfying the condition $E(k,l) \leqslant E''$, say, where E'' is determined in any given problem by the number of levels we are considering.

Let us first of all leave the sum over l alone. Then we can immediately make use of the results given in (6) and (7) of §2. We find

$$\sum_{k=0}^{N_l-1} E(k,l) = \frac{2}{h}\int_{r_1^l(E_l')}^{r_2^l(E_l')} \left\{\frac{1}{3}\frac{p_3^l(r,E_l')}{2m} + \left(V + \frac{(l+\frac{1}{2})^2}{r^2}\frac{\hbar^2}{2m}\right) p_l(r,E_l')\right\} \mathrm{d}r, \tag{10}$$

where E_l' is determined by

$$\frac{2}{h}\int_{r_1^l(E_l')}^{r_2^l(E_l')} p_l(r,E_l')\,\mathrm{d}r = N_l. \tag{11}$$

N_l is determined by $\quad E(N_l-1,l) \leqslant E'', \quad$ but $\quad E(N_l,l) > E''$

and $p_l(E,r)$ is defined by

$$p_l(E,r) = \left\{2m\left[E - V(r) - \frac{(l+\frac{1}{2})^2}{r^2}\frac{\hbar^2}{2m}\right]\right\}^{\frac{1}{2}}. \tag{12}$$

However, in this formulation we have still to perform the sum over l.

In order to obtain the three-dimensional TF formula it is necessary to integrate over l and to replace E_l' by an energy independent of l. If the labour involved in using (10) and (11) and then summing over l is prohibitive and the further approximation of the three-dimensional TF formula is permissible, then that formula can more easily be obtained directly from (8) and (9).

To do this, we replace the sum (9) by an integral and write

$$\mathrm{E} = \sum_{k,\,l} (2l+1)\, E(k,l) \doteqdot \iint (2l+1)\, E(k,l)\,\mathrm{d}k\,\mathrm{d}l, \tag{13}$$

and we replace the number of states over which we are summing by an integral over the same region, that is, we write

$$\sum_{k,\,l} (2l+1) \doteqdot \iint (2l+1)\,\mathrm{d}k\,\mathrm{d}l. \tag{14}$$

The region of integration in both cases is bounded by

$$k = -\tfrac{1}{2}, \quad l = -\tfrac{1}{2} \quad \text{and} \quad E(k,l) = E', \tag{15}$$

where E' is to be chosen so that the integral in (14) is equal to N, the total number of occupied states. Unlike the one-dimensional case there are two approximations involved in passing from (9) to (13); as before, a sum is replaced by an integral, but in addition the region of integration as defined by (15) is incorrect. Instead of the curve $E(k,l) = E'$ we should have a boundary as shown in figure 3 and composed only of the lines k and l equal to half odd integers. If it were not for this incorrect region of integration, (14) would be exact. We shall now show that the integral in (13) is the usual TF expression for the sum of the eigenvalues.

Let us integrate with respect to k first. We have

$$\int_{-\frac{1}{2}}^{k'(l,E')} E(k,l)\,\mathrm{d}k = \frac{2}{h}\int_{r_1^l(E')}^{r_2^l(E')} \left\{\frac{1}{3}\frac{p_l^3(E',r)}{2m} + \left(V + \frac{(l+\frac{1}{2})^2}{r^2}\frac{\hbar^2}{2m}\right) p_l(E',r)\right\}\mathrm{d}r. \tag{16}$$

where $k'(l,E')$ is the solution of $E(k,l) = E'$ for k. Next we must integrate with respect to l, and reference to figure 4 will be helpful here. We have first to integrate

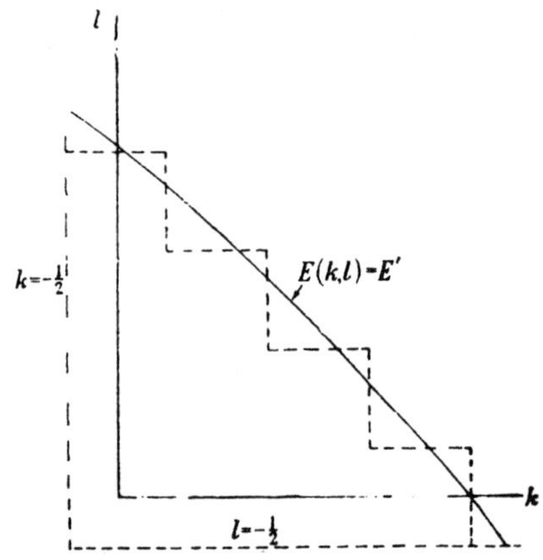

FIGURE 3. (k, l) plane illustrating regions of integration in central field problem.

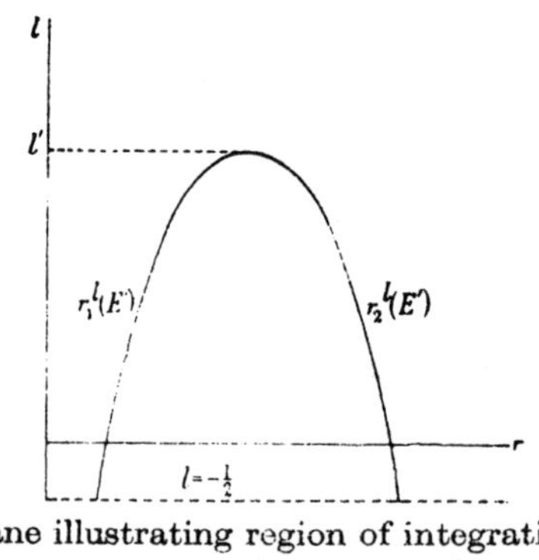

FIGURE 4. (l, r) plane illustrating region of integration in equation (18).

with respect to r from $r_1^l(E')$ to $r_2^l(E')$ and then with respect to l from $l = -\frac{1}{2}$ to the maximum value of l. Reference to figure 3 will show this to be l', the solution of $E(-\frac{1}{2}, l) = E'$. But from (8), $E(-\frac{1}{2}, l)\,(= E')$ is such that

$$2\int_{r_1^{l'}(E')}^{r_2^{l'}(E')} \left[E' - V(r) - \frac{(l'+\frac{1}{2})^2}{r^2}\frac{\hbar^2}{2m}\right]^{\frac{1}{2}}\mathrm{d}r = 0,$$

and this is only possible if $r_1^{l'}(E') = r_2^{l'}(E')$. The double integral over the (r, l) plane is, then, bounded by

$$l = -\tfrac{1}{2} \quad \text{and} \quad V(r) + \frac{(l+\frac{1}{2})^2}{r^2}\frac{\hbar^2}{2m} = E'. \tag{17}$$

We proceed now to perform the integration over l first. We find

$$\iint (2l+1)\,E(k,l)\,\mathrm{d}k\,\mathrm{d}l = \frac{2}{h}\int_{r_1^{-\frac{1}{2}}(E')}^{r_2^{-\frac{1}{2}}(E')}\int_{-\frac{1}{2}}^{l_{\max.}(r)}(2l+1)\left\{-\frac{p_l^3}{3m}+E'p_l\right\}\mathrm{d}l\,\mathrm{d}r, \tag{18}$$

where we have eliminated V from (16) using (12). $l_{\max.}(r)$ is of course obtained by solving (17). If we change the variable of integration to $(l+\frac{1}{2})^2$ we can perform the integration, and after a little manipulation we find

$$\iint (2l+1)\,E(k,l)\,\mathrm{d}k\,\mathrm{d}l$$
$$= \frac{2(2\pi)^2}{h^3}\int_{r^{-\frac{1}{2}}(E')}^{r_2^{-\frac{1}{2}}(E')}\left\{-\frac{2}{15m}[2m(E'-V)]^{\frac{5}{2}}+\tfrac{2}{3}E'[2m(E'-V)]^{\frac{3}{2}}\right\}r^2\,\mathrm{d}r$$
$$= \frac{4\pi}{3h^3}\int_{r_1^{-\frac{1}{2}}(E')}^{r_2^{-\frac{1}{2}}(E')}p'^3\left[\frac{3}{5}\frac{p'^2}{2m}+V\right]4\pi r^2\,\mathrm{d}r, \tag{19}$$

where we have written $p'^2 = 2m(E'-V)$, and $r_1^{-\frac{1}{2}}(E')$ and $r_2^{-\frac{1}{2}}(E')$ are the roots of $E' = V(r)$.

We must now calculate E' from (14). We have

$$\iint (2l+1)\,\mathrm{d}k\,\mathrm{d}l = \int (2l+1)\frac{2}{h}\int_{r_1^{\frac{1}{2}}(E')}^{r_2^{\frac{1}{2}}(E')}\left\{2m\left[E'-V-\frac{(l+\frac{1}{2})^2}{r^2}\frac{\hbar^2}{2m}\right]\right\}^{\frac{1}{2}}\mathrm{d}r\,\mathrm{d}l$$
$$= \frac{4\pi}{3h^3}\int_{r_1^{-\frac{1}{2}}(E')}^{r_2^{-\frac{1}{2}}(E')}p'^3 4\pi r^2\,\mathrm{d}r = N. \tag{20}$$

It will be recognized that (19) and (20) are the usual TF formulae giving the sum of the eigenvalues; the TF scheme is completed by identifying the particle density ρ with $\dfrac{4\pi}{3h^3}p'^3$.

4. Correction for error due to curvature in $E(n)$

We have noted already in the one-dimensional case that there is no error involved in replacing the sum of the WKB levels by an integral when $E(n)$ is linear. But in general there are errors arising from the curvature in $E(n)$, and the obvious method of correcting for these errors is to add a few terms of the Euler–Maclaurin formula as we see below. For (3) should be more correctly written

$$\sum_{k=0}^{N-1} E_k = \int_{-\frac{1}{2}}^{(N-1)+\frac{1}{2}} E(n)\,\mathrm{d}n + \sum_{s=1}^{m-1}\frac{B_{2s}(\frac{1}{2})}{(2s)!}[E^{2s-1}(N-\tfrac{1}{2})-E^{2s-1}(-\tfrac{1}{2})]$$
$$+\frac{NB_{2m}(\frac{1}{2})}{(2m)!}E^{2m}(-\tfrac{1}{2}+\theta N)\quad(0<\theta<1) \tag{21}$$

(see Milne-Thomson 1933, §7·5, (7)). The values of the first few Bernoulli polynomials, $B_{2s}(x)$ at $x=\frac{1}{2}$, are

$$B_2(\tfrac{1}{2}) = -\tfrac{1}{12},\quad B_4(\tfrac{1}{2}) = \tfrac{7}{240},\quad B_6(\tfrac{1}{2}) = -\tfrac{31}{1344}.$$

First of all we observe that with $m=1$ we have the remainder to (2) and a criterion for the sign of the error, showing it to be determined by the sign of $\mathrm{d}^2E/\mathrm{d}n^2$. Sometimes the remainder will be useful to estimate the magnitude of the error.

If we add on the first term in the sum, $s = 1$, we have instead of (6)

$$\sum_{k=0}^{N-1} E_k = \int_{x_1(E')}^{x_2(E')} \left(\frac{1}{3}\frac{p^2(E',x)}{2m} + V\right)\frac{2}{h}p(E',x)\,\mathrm{d}x - \frac{1}{24}\left[\left(\frac{\mathrm{d}E}{\mathrm{d}n}\right)_{N-\frac{1}{2}} - \left(\frac{\mathrm{d}E}{\mathrm{d}n}\right)_{-\frac{1}{2}}\right]$$

$$= \int_{x_1(E')}^{x_2(E')} \left(\frac{1}{3}\frac{p^2(E',x)}{2m} + V\right)\frac{2}{h}p(E',x)\,\mathrm{d}x$$

$$- \frac{1}{24}\left\{\frac{h}{2m}\frac{1}{\int_{x_1(E')}^{x_2(E')}[p(E',x)]^{-1}\,\mathrm{d}x} - \frac{h}{(2m)^{\frac{1}{2}}}\frac{1}{\pi}\left(\frac{V_0''}{2}\right)^{\frac{1}{2}}\right\}, \tag{22}$$

where V_0'' is the second derivative of $V(x)$ at its minimum. Here we have assumed in evaluating $(\mathrm{d}E/\mathrm{d}n)_{-\frac{1}{2}}$ that we can write

$$V(x) = V_0 + \frac{x^2}{2}V_0'' + O(x^3)$$

and

$$V'(x) = xV_0'' + O(x^2).$$

It is of some interest to compare the first term of the Euler–Maclaurin formula with the exact error of the TF energy expression. We shall now do this in three simple examples that can be worked out exactly without difficulty.

(1) *The linear harmonic oscillator*

In this case it is well known that the application of the WKB formula of (1) leads to the exact levels given by

$$E_n = (n+\tfrac{1}{2})\,h\nu_0 \quad (n = 0, 1, 2, \ldots), \tag{23}$$

where ν_0 is the classical frequency of the oscillator. Equation (23), regarded as defining $E(n)$, shows that this is linear, and therefore, as we have already noted, there is no approximation in replacing the sum of the eigenvalues by an integral. The first term in the Euler–Maclaurin formula obviously vanishes and all higher terms are clearly zero. Thus in this particular case the exact sum, the WKB sum and the TF expression for the sum are identical.

(2) *The square well with infinite walls*

Suppose that the length of the well is l and the zero of energy is taken on the bottom of the well. The WKB eigenvalues are then given by

$$2l(2mE)^{\frac{1}{2}} = (n+\tfrac{1}{2})\,h \quad (n = 0, 1, 2 \ldots),$$

and the sum of the first N eigenvalues is

$$\frac{h^2}{8ml^2}\sum_{n=0}^{N-1}(n+\tfrac{1}{2})^2 = \frac{h^2}{8ml^2}\frac{N(2N-1)(2N+1)}{12}.$$

The TF formula, on the other hand, is

$$\frac{h^2}{8ml^2}\int_{-\frac{1}{2}}^{N-\frac{1}{2}}(n+\tfrac{1}{2})^2\,\mathrm{d}n = \frac{h^2}{8ml^2}\frac{N^3}{3}.$$

Evaluating the first term of the Euler–Maclaurin correction we find the value

$$-\frac{\hbar^2}{8ml^2}\frac{N}{12},$$

and adding this to the TF energy is seen to lead to the correct result. This is of course because all other terms in the Euler–Maclaurin series are zero.

(3) Hydrogenlike potential with fixed l

The TF formula in this case is given by (10), whilst the WKB eigenvalues are given by

$$E_{kl} = -\frac{Z^2e^4m}{2\hbar^2}\frac{1}{(k+l+1)^2},$$

again the exact result. The first term in the Euler–Maclaurin correction is

$$-\frac{Z^2e^4m}{24\hbar^2}\left[\frac{1}{(N_l+l+\frac{1}{2})^3}-\frac{1}{(l+\frac{1}{2})^3}\right].$$

For large N_l this can be written very accurately as

$$\frac{Z^2e^4m}{24\hbar^2}(l+\tfrac{1}{2})^{-3}.$$

It should be noted, however, that the curvature of the $E(k, l)$ surface is so high for small l, especially for $l = 0$, that this Euler–Maclaurin correction is considerably too large. Omitting the factor $Z^2e^4m/\hbar^2$ for convenience and making $N_l = \infty$ we have drawn up table 1 to illustrate the position.

TABLE 1

l	0	1	2
first Euler–Maclaurin term	0·333	0·0123	0·00267
actual error	0·178	0·0109	0·00253

5. THREE-DIMENSIONAL FORMULA-ANALYSIS OF ERRORS IN THE COULOMB FIELD CASE

Let us now turn to the three-dimensional TF formula. Clearly, if the potential $V(r)$ behaves as r^{-1} as $r \to 0$, it will not be worth while to consider corrections of the Euler–Maclaurin type. (The situation is even worse when we now allow an l of $-\frac{1}{2}$, not just $l = 0$.) It is, nevertheless, instructive to analyze rather carefully the errors involved in the three-dimensional formula for a Coulomb potential. For this is very relevant to the physical problem of calculating atomic binding energies as we shall discuss in § 6.

From the previous analysis it is clear that the TF method can be regarded as first of all replacing

$$-\frac{Z^2}{2}\frac{2l+1}{(k+l+1)^2}$$

by $$-\frac{Z^2}{2}\int_{l-\frac{1}{2}}^{l+\frac{1}{2}}\int_{k-\frac{1}{2}}^{k+\frac{1}{2}}\frac{2y+1}{(x+y+1)^2}\,\mathrm{d}x\,\mathrm{d}y = -Z^2\left[(k+1)\ln\frac{k+l+2}{k+l+1}-k\ln\frac{k+l+1}{k+l}\right].$$

The errors for a number of small values of k and l are interesting and are shown in table 2 (apart from the factor Z^2). It is worth noting that the errors are not all of one sign; this is due to the fact that the surface $(2l+1)/(k+l+1)^2$ is not convex upwards everywhere. Actually it is neither convex nor concave for $k, l > -\frac{1}{2}$, that is, it lies both above and below its tangent plane at every point.

TABLE 2

l					
3	0·00439	—	—	—	
2	0·00990	0·00236	—	—	
1	0·03047	0·00323	0·00032	—	
0	0·19315	−0·00722	−0·00344	−0·00172	
	0	1	2	3	k

If we sum the errors over k with a fixed value of l and divide by $2l+1$ the result is very close to the error made in using (10), the greatest difference being for $l = 0$ as we should expect, although even here the difference is not sufficiently large to suggest that it will be worth while to use the more complicated formula (10) rather than the simpler equation (19).

Let us now compute the sum of the errors from $n = 1$ to N. To this end we put $k+l+1 = n$ and sum along the diagonal with constant n first. We have

$$-\frac{1}{2}\sum_{\text{diagonal}}\frac{2l+1}{n^2} = -\frac{1}{2}.$$

Similarly we sum the integrals over this diagonal and obtain

$$-\tfrac{1}{2}n(n+1)\ln\frac{n+1}{n}+\tfrac{1}{2}(n-1)\,n\ln\frac{n}{n-1}.$$

The sum of the errors from $n = 1$ to N is

$$-\frac{N}{2}+\sum_{n=1}^{N}\left[\tfrac{1}{2}n(n+1)\ln\frac{n+1}{n}-\tfrac{1}{2}(n-1)\,n\ln\frac{n}{n-1}\right] = \frac{1}{4}-\frac{1}{12N}+O\left(\frac{1}{N^2}\right).$$

However, it should be noted at this stage that the correction we have obtained is not the final correction to be applied to the TF formula for a Coulomb field. For as we emphasized in §3 the TF formula in three dimensions is subject to two errors. One is due to the curvature of the $(2l+1)E(k,l)$ surface, of which account has now been taken; the other is due to the replacement of the proper boundary, shown diagrammatically in figure 3, by the curve $E(k,l) = E'$. We shall now calculate this error directly in the Coulomb field case. Thus for the proper boundary, and when the levels from $n = 1$ to N are filled the energy in atomic units is

$$-\frac{Z^2}{2}\sum_{n=1}^{N}\left[n(n+1)\ln\frac{n+1}{n}-(n-1)\,n\ln\frac{n}{n-1}\right] = -Z^2\left[\frac{N}{2}+\frac{1}{4}-\frac{1}{12N}+O\left(\frac{1}{N^2}\right)\right].$$

This means now that $\frac{1}{6}N(N+1)(2N+1)$ levels are filled which fixes the E' of the TF method, through the equation

$$\int_{-\frac{1}{2}}\int_{-\frac{1}{2}}^{k+l+1<Z(-2E')^{-\frac{1}{2}}}(2l+1)\,\mathrm{d}k\,\mathrm{d}l = \tfrac{1}{3}Z^3(-2E')^{-\frac{3}{2}} = \frac{N(N+1)(2N+1)}{6}.$$

Having fixed E' the energy by the TF formula is

$$-\frac{Z^2}{2}\int_{-\frac{1}{2}}\int_{-\frac{1}{2}}^{k+l+1<Z(-2E')^{-\frac{1}{2}}} \frac{2l+1}{(k+l+1)^2}\,\mathrm{d}k\,\mathrm{d}l$$

$$= -\frac{Z^3}{2}(-2E')^{-\frac{1}{2}} = -Z^2\left[\frac{N}{2}+\frac{1}{4}-\frac{1}{24N}+O\left(\frac{1}{N^2}\right)\right].$$

So we see that this error is $Z^2/24N$. If this error is added on to the 'curvature' error the total error is

$$Z^2\left[\frac{1}{4}-\frac{1}{24N}+O\left(\frac{1}{N^2}\right)\right].$$

6. Correction of TF formula for atomic binding energies

We shall now go on to discuss the relevance of the above results to the calculation of atomic binding energies. The fact that the usual TF expression for the binding energy of atoms, namely,

$$E_{\mathrm{TF}} = -0\cdot7687Z^{\frac{7}{3}} \text{ atomic unit,} \tag{24}$$

is not in good agreement with experiment has been known for some time. Quite recently Scott (1952) has analyzed the reason for this, and has suggested that two terms should be added, one due the failure of the TF method in the neighbourhood of the nucleus, and the other due to exchange. The corrected result can then be written

$$-0\cdot7687Z^{\frac{7}{3}}+\tfrac{1}{2}Z^2-0\cdot221Z^{\frac{5}{3}},$$

where the second term arises from the shortcomings of the TF method near the nucleus, whilst the third is the exchange energy. More recently, Ballinger & March (1955) have pointed out that Scott's argument is not completely consistent. The essential point made by them is that if the TF formula is corrected for exchange, that is, if a term of order $Z^{\frac{5}{3}}$ is included, then a term of this order arising from the failure of the TF formula must also be considered. They showed that in the case of a Coulomb field there is in fact a term of order $Z^{\frac{5}{3}}$ of magnitude approximately one-third of the exchange energy. When this was included the already satisfactory agreement with experiment was improved. However, as these authors emphasized, their work threw no additional light on the validity of using the Coulomb field case to estimate the correction terms to be applied to the TF formula for atoms. As we shall now show, the more fundamental analysis we have been able to give in this paper casts further light on this point.

Let us go back to the final formula for the error in the Coulomb field case. This is

$$Z^2\left[\frac{1}{4}-\frac{1}{24N}+O\left(\frac{1}{N^2}\right)\right],$$

where we have filled all the levels up to N. We note first of all that letting N tend to infinity we reproduce Scott's result (an additional factor of two being required when the levels are doubly filled). On the other hand, choosing N to give $\frac{1}{2}Z$ filled levels we reproduce the correction of Ballinger & March, as expected.

One further point is of interest concerning the connexion of the present work with Scott's. He gave values for the errors due to the K, $K+L$, $K+L+M$, etc., shells, and these were

$$0{\cdot}221Z^2,\quad 0{\cdot}233Z^2,\quad 0{\cdot}238Z^2,\ldots,$$

to be compared with our values of

$$0{\cdot}193Z^2,\quad 0{\cdot}216Z^2,\quad 0{\cdot}226Z^2,\ldots.$$

It will be seen that the results are not identical, although the sum to infinity is the same. The reason is not far to seek. Scott has integrated over triangles (if we refer again to figure 3) whereas we have integrated over squares. The present procedure appears to be the more fundamental one.

We see now, however, that part of the $Z^{\frac{5}{3}}$ correction arises from the incorrect boundary that we take in the TF approximation. Thus, this would appear to depend to some extent on the higher energy levels, and it is clear that whilst the essential point made by Ballinger & March is correct, namely, that if the exchange energy is included then an additional term of order $Z^{\frac{5}{3}}$ must be added, it is necessary to estimate the size of this more carefully.

We have been able to see no simple analytical argument which will enable us to do this reliably. However, the following method allows us to achieve our object without a prohibitive amount of labour. What is effectively required is an estimate of the difference between the TF sum of the eigenvalues and the WKB sum for the proper screened potential of the TF theory for atoms. Now, of course, the TF energy expression (24) is not simply the sum of the eigenvalues for the self-consistent TF potential. We must take account of the fact that the sum of the eigenvalues includes twice the mutual potential energy of the electronic cloud. Fortunately, this can be calculated very easily. For, according to the virial theorem, the total potential energy U_{TF} is simply $2E_{\mathrm{TF}}$ and, furthermore (see, for example, Gombás 1949, p. 61), there is a simple relation between the electron-electron interaction energy U_{ee} and the electron-nuclear interaction energy U_{eN} in the TF atom. This is simply that

$$U_{ee} = -\tfrac{1}{7}U_{eN},$$

and thus we find

$$U_{ee} = -\tfrac{1}{7}[U_{\mathrm{TF}} - U_{ee}] \quad \text{or} \quad U_{ee} = -\tfrac{1}{6}U_{\mathrm{TF}} = -\tfrac{1}{3}E_{\mathrm{TF}}.$$

This leads to the result that the sum of the eigenvalues is given by

$$\mathrm{E} = E_{\mathrm{TF}} + U_{ee} = \tfrac{2}{3}E_{\mathrm{TF}} = -0{\cdot}5125Z^{\frac{7}{3}} \text{ atomic units.}$$

The method we propose to use is now as follows. For a particular case (we have chosen $Z = 18$) we have calculated the WKB levels for the proper TF potential by purely numerical methods, and the results are recorded in table 3. Adding the eigenvalues thus obtained for the $1s$, $2s$, $2p$, $3s$ and $3p$ levels with suitable weights, we then subtract from this sum the TF sum of the eigenvalues, modified by the correction $\frac{1}{2}Z^2$. The remainder should provide an estimate of the term in $Z^{\frac{5}{3}}$. Adding the exchange energy we thus obtain for the coefficient of the $Z^{\frac{5}{3}}$ term in the formula for atomic binding energies a value of 0·266 atomic unit (due to the possible errors in the WKB eigenvalues shown in the table; this might be in error by $\pm 0{\cdot}003$; see below, however), to be compared with 0·294 obtained by Ballinger & March and 0·221 given by Scott. The present result has of course been obtained by assuming that the effect of the higher order terms is small; a plausible assumption, for in the Coulomb field case the next three terms, of order $Z^{\frac{4}{3}}$, Z and $Z^{\frac{2}{3}}$, are all identically zero as was shown by Ballinger & March. The revised formula we propose is therefore

$$-0{\cdot}7687Z^{\frac{7}{3}}+\tfrac{1}{2}Z^2-0{\cdot}266Z^{\frac{5}{3}}.$$

TABLE 3. WKB EIGENVALUES FOR $Z = 18$, IN ATOMIC UNITS

1s	$-(107{\cdot}8\pm 0{\cdot}1)$
2s	$-(9{\cdot}41\pm 0{\cdot}02)$
2p	$-(7{\cdot}07\pm 0{\cdot}01)$
3s	$-(0{\cdot}56\pm 0{\cdot}01)$
3p	$-(0{\cdot}14\pm 0{\cdot}01)$

7. ATTEMPTS TO CORRECT FOR ERROR DUE TO APPROXIMATE NATURE OF WKB FORMULA

So far we have concerned ourselves with analyzing the fundamental approximations made in the TF method and then with possible methods of correcting for the error introduced by one of these approximations, namely, the replacement of a sum by an integral. It is finally of interest to inquire whether it is possible to correct for errors introduced by the use of the WKB formula should this be desirable in a given problem. This formula, as we have seen, forms the basis of the customary TF expression for the sum of the eigenvalues. We shall confine ourselves to a few remarks concerned with the one-dimensional case.

One very attractive method immediately suggests itself. It is to start from an expression giving the eigenvalues due to Milne (1930), of a somewhat similar form to the WKB result, but having the great virtue of being exact. This is

$$2\int_{-\infty}^{\infty} P(E,x)\,\mathrm{d}x = nh, \tag{25}$$

where P satisfies the second-order differential equation

$$\frac{P^2}{2m}-(E-V)-\frac{\hbar^2}{2m}P^{\frac{1}{2}}\frac{\mathrm{d}^2}{\mathrm{d}x^2}(P^{-\frac{1}{2}}) = 0.$$

Unfortunately, it does not appear to be possible to make significant progress from equation (25), for even when the sum over the eigenvalues is replaced by an integral the integration over E does not seem tractable in general, and without this simplification the method would seem to have little or no practical worth.

The remaining alternative seems to be to add further terms to the WKB formula (1), in the manner discussed, for example, by Dunham (1932). Including an additional term the calculation can be handled, and the final expression for the total energy can be written in a manner in which there is a formal resemblance with the expression given by Weizsäcker (see, for example, Plaskett 1953). However, the term involving $(\mathrm{d}\rho/\mathrm{d}x)^2/\rho$ in Weizsäcker's method has a coefficient $+\hbar^2/8m$, whereas in our case it is $-\hbar^2/24m$. But in any event the resemblance is only superficial, since the density ρ is not determined by minimizing the energy in our case, and our integral for the energy must be taken as a contour integration since it would diverge on approaching the turning points. As our result is unlikely to have much practical value we omit the details.

8. Conclusion

We have been able to analyze the essential approximations in the TF energy formula more precisely than has been done hitherto. We have discussed ways of correcting for the replacement of sums by integrals and have given illustrative examples. These show that in some cases it might be profitable to add some correction terms from the Euler–Maclaurin formula; in other cases it will clearly not be very useful to do this. A modified formula for atomic binding energies has been obtained as a consequence of the more fundamental investigation made in this paper; this formula seems to have a better theoretical foundation than any suggested so far. We have been able to see no very practicable method of correcting for the error introduced by the use of the WKB formula for the eigenvalues. Fortunately, it seems that in many cases of physical interest this will be a less important approximation than the replacement of the sum over the eigenvalues by an integral. Why this should be so is made more clear by the recent paper of Langer (1949), who has established that the formula (1) has approximate validity down to low values of n for a fairly wide range of potentials.

Finally, we ought to emphasize that the considerations of this paper have been confined to one-dimensional problems or to three-dimensional problems with spherically symmetrical potentials. There seems little hope at present that the detailed analysis given here and the rather thorough discussion of the errors involved in the TF energy formula can be extended to the general three-dimensional problem. Some progress has been made previously in this direction, however (Courant & Hilbert 1931, chapter VI; de Wet & Mandl 1950), and it would be of great interest, particularly in molecular problems, if reliable methods of correcting for the errors could be found in this general case.

References

Ballinger, R. A. & March, N. H. 1955 *Phil. Mag.* **46**, 246.

Brillouin, L. 1934 L'atome de Thomas-Fermi (*Actualités sci. industr.*, no. 160. Paris: Hermann.)

Courant, R. & Hilbert, D. 1931 *Methoden der mathematischen physik*, **1**, 2nd ed. Berlin: Springer.

Dirac, P. A. M. 1930 *Proc. Camb. Phil. Soc.* **26**, 376.

Dunham, J. L. 1932 *Phys. Rev.* **41**, 713.

Fényes, I. 1949 *Z. Phys.* **125**, 336.

Gombás, P. 1949 *Die statistische theorie des atoms und ihre anwendungen*. Vienna: Springer.

Langer, R. E. 1937 *Phys. Rev.* **51**, 669.

Langer, R. E. 1949 *Phys. Rev.* **75**, 1573.

Milne, W. E. 1930 *Phys. Rev.* **35**, 863.

Milne-Thomson, L. M. 1933 *The calculus of finite differences*. London: Macmillan.

Plaskett, J. S. 1953 *Proc. Phys. Soc.* A, **66**, 178.

Scott, J. M. C. 1952 *Phil. Mag.* **43**, 859.

Wet, J. S. de & Mandl, F. 1950 *Proc. Roy. Soc.* A, **200**, 572.

J. Phys. B: Atom. Molec. Phys., Vol. 5, March 1972. Printed in Great Britain

Non-relativistic theory of atomic and ionic binding energies for large atomic number

N H MARCH† and R J WHITE‡

† Department of Physics, The University of Sheffield
‡ Department of Mathematics, University of Southwestern Louisiana, Lafayette, Louisiana 70501, USA

MS received 16 November 1971

Abstract. The Z^{-1} expansion of the total energy of an ion with N electrons has by now been worked out carefully for systems with from 2 to 10 electrons, up to third order.

This paper is concerned with the behaviour of the energy coefficients $\epsilon_n(N)$ in the Z^{-1} expansion. The principal results are:

(i) To establish $\epsilon_n(N) \sim N^{n+1/3}$ for sufficiently large N, by appeal to the Thomas–Fermi theory of positive ions.

(ii) To obtain the first two terms in $\epsilon_0(N)$ exactly in such an asymptotic expansion.

(iii) To give approximate forms for large N for $\epsilon_1(N)$ and $\epsilon_2(N)$.

(iv) To point out that the total energy $E(Z, N)$ can be plotted, as Z becomes large, in a universal form by considering $E(Z, N)/Z^{7/3}$ against N/Z.

It is emphasized that the Thomas–Fermi theory of the binding energies of neutral atoms is giving us an approximate summation to all orders in Z^{-1} of the Z^{-1} expansion. The relation of the present theory to the result of Foldy that atomic binding energies in the range $Z = 2$ to 90 are approximately proportional to $Z^{12/5}$ is also discussed.

1. Introduction

Atomic binding energies have posed an interesting challenge to theorists for some time. Milne (1927) first tackled the problem of the Z (atomic number) dependence of binding energies for heavy atoms, using the theory of Thomas (1926) (which was also developed later and independently by Fermi (1928)). The binding energy was shown, in this theory, to be proportional to $Z^{7/3}$. Later, correction terms of order Z^2 and $Z^{5/3}, \ldots$ were added to this formula. With these correction terms, and correcting for relativistic effects which obviously become important in heavy atoms, this theory agrees reasonably well with experiment.

Subsequently the binding energy as a function of Z was considered on the basis of Hartree theory by Foldy (1951), who showed that this method, worked out for atomic numbers from $Z = 2$ to 90 say, led to a binding energy proportional more nearly to $Z^{12/5}$ in this range of atomic numbers.

An independent development of binding energies for atoms and ions was introduced by Hylleraas (1930) and has been used mainly for systems with small N (number of electrons). This perturbative approach, called Z^{-1} theory, has become more popular in the last few years and it is interesting that the statistical approach and the perturbative approach can be connected in a simple way. In Z^{-1} theory the problem is to generate the energy coefficients of the terms of different orders in Z^{-1} and Layzer (1959) has generated the first two such coefficients for $N = 2$ to 10. Layzer *et al* (1964) discuss the third coefficient and Knight (1969) calculates this coefficient for the $N = 2$ to 10 range. Higher order coefficients are not available over an appreciable range of N.

The purpose of this paper is to endeavour to estimate the asymptotic behaviour of the coefficients $\epsilon_n(N)$ of the Z^{-1} expansion for large N, from the Thomas–Fermi theory of positive ions. It will then be shown that the Thomas–Fermi theory yields an approximate summation of the terms in the Z^{-1} expansion, and the range of usefulness of this approximate sum is discussed.

2. Thomas–Fermi binding energies for positive ions

The total energy of a positive ion of atomic number Z and with N electrons can be calculated from the usual expression for the energy in Thomas–Fermi theory (see eg March 1957)

$$E(Z, N) = c_k \int (\rho(\boldsymbol{r}))^{5/3}\,\mathrm{d}\boldsymbol{r} + \int \rho(\boldsymbol{r})V_N(\boldsymbol{r})\,\mathrm{d}\boldsymbol{r} + \frac{1}{2}e^2 \int \frac{\rho(\boldsymbol{r}')\rho(\boldsymbol{r})}{|\boldsymbol{r}-\boldsymbol{r}'|}\,\mathrm{d}\boldsymbol{r}\,\mathrm{d}\boldsymbol{r}' \tag{2.1}$$

where $\rho(\boldsymbol{r})$ is the electron density in the ion, satisfying the normalization condition

$$\int \rho(\boldsymbol{r})\,\mathrm{d}\boldsymbol{r} = N \tag{2.2}$$

while V_N is the potential energy of an electron in the field of the nucleus, that is $-Ze^2/r$.

The electron density $\rho(\boldsymbol{r})$ and the total potential energy $V(\boldsymbol{r}) = V_N(\boldsymbol{r}) + V_e(\boldsymbol{r})$ where $V_e(\boldsymbol{r})$ is the electrostatic potential of the electron cloud, namely

$$V_e(\boldsymbol{r}) = e^2 \int \frac{\rho(\boldsymbol{r}')\,\mathrm{d}\boldsymbol{r}'}{|\boldsymbol{r}-\boldsymbol{r}'|} \tag{2.3}$$

are related by the usual Thomas–Fermi result

$$\rho(\boldsymbol{r}) = \frac{8\pi}{3h^3}(2m)^{3/2}(E - V(\boldsymbol{r}))^{3/2} \tag{2.4}$$

where E is the maximum energy of an electron in the ion, to be found from the normalization condition (2.2). For self-consistency we simply combine (2.4) with Poisson's equation to obtain

$$-\nabla^2 V = \frac{32\pi^2 e^2}{3h^3}(2m)^{3/2}(E - V(\boldsymbol{r}))^{3/2}. \tag{2.5}$$

Putting

$$E - V(\boldsymbol{r}) = \frac{Ze^2}{r}\phi(x) = \frac{Ze^2}{bx}\phi(x) \tag{2.6}$$

where x is the usual dimensionless length in the Thomas–Fermi (TF) theory, with

$$b = \frac{0{\cdot}88534}{a_0}Z^{1/3} \tag{2.7}$$

the potential field is determined by solving the equation

$$\frac{d^2\phi}{dx^2} = \frac{\phi^{3/2}}{x^{1/2}} \qquad \phi(0) = 1. \tag{2.8}$$

The nature of the solutions of equation (2.8) is sketched in figure 1; curve A is the neutral atom solution, while curve B is a positive ion solution, the construction shown giving the value of N/Z corresponding to this particular solution.

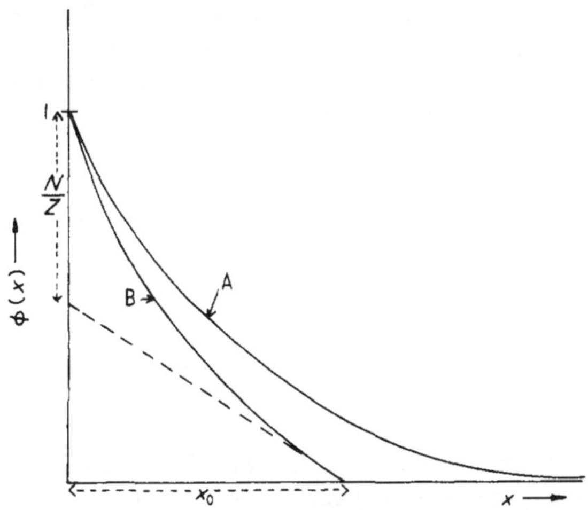

Figure 1. Showing schematically the types of solution of the Thomas–Fermi equation representing positive ions. The construction to determine N/Z for a particular solution should be noted. The isolated neutral atom solution, with infinite radius, is also shown. A neutral atom ($x_0 = \infty$); B positive ion.

Then after some manipulation the total electronic energy of a positive ion is given by (see eg Gombás 1950 p 237)

$$\begin{aligned} E(Z, N) &= \frac{12}{7}\left(\frac{2}{9\pi^2}\right)^{1/3}\left\{a_2 + \left(1-\frac{N}{Z}\right)^2 \frac{1}{x_0}\right\} Z^{7/3}\frac{e^2}{a_0} \\ &= 0{\cdot}4841\left\{a_2 + \left(1-\frac{N}{Z}\right)^2 \frac{1}{x_0}\right\} Z^{7/3}\frac{e^2}{a_0} \end{aligned} \tag{2.9}$$

in atomic units e^2/a_0. Here a_2 is the slope of $\phi(x)$ at the origin, while x_0 is the radius of the positive ion in dimensionless TF units. It should be noted that when $N = Z$, we know that $a_2 = -1{\cdot}58805$ and $x_0 = \infty$, and we get the usual TF result for atomic binding energies

$$E(Z, Z) = -0{\cdot}7687 Z^{7/3}\frac{e^2}{a_0}. \tag{2.10}$$

A little thought about the dimensionless TF method now tells us that

$$a_2 = a_2\left(\frac{N}{Z}\right) \quad \text{and} \quad x_0 = x_0\left(\frac{N}{Z}\right). \tag{2.11}$$

Thus, we conclude that for $Z \to \infty$, and $N \to \infty$ such that $(N/Z) \lesssim 1$, we must have

$$E(Z, N) = Z^{7/3} f\left(\frac{N}{Z}\right)\frac{e^2}{a_0}. \tag{2.12}$$

We have used available solutions of the TF equation to construct $f(N/Z)$. The work of Miranda (1934) allows us to obtain a_2 and x_0 for different values of N/Z and the results are collected in table 1. It is a pity that only three solutions shown there, plus the neutral atom ($x_0 = \infty$), are available in convenient form. However, with the addition of an analytical result as $x_0 \to 0$, we can get the form of $f(N/Z)$ in equation (2.12) sufficiently well for our present purposes. Should it prove necessary later, it will be a straightforward matter to get some more solutions of the TF equation (2.8) corresponding to smaller radii x_0 than are shown in table 1.

Table 1. Values of initial slope a_2 and positive ion radius x_0 for different values of N/Z

x_0	2·795	6·423	9·755	∞
a_2	−1.60	−1·589	−1·5882	−1·58805
N/Z	0·478	0·777	0·879	1

To study the limit $x_0 \to 0$, the use of an expansion of $\phi(x)$ around $x = 0$ suggests itself. The appropriate expansion has been given by Baker (1930) and takes the form

$$\phi(x) = 1 + a_2 x + \tfrac{4}{3}x^{3/2} + \tfrac{2}{5}a_2 x^{5/2} + \tfrac{1}{3}x^3 + \dots . \tag{2.13}$$

For sufficiently small x, the first two terms suffice and evidently since, for positive ions, $\phi(x_0) = 0$ we must have as $x_0 \to 0$, $a_2 x_0 = -1$. Since $x_0 \to 0$ corresponds to $N/Z \to 0$, this result used in equations (2.9) and (2.12) shows us that $f(N/Z)$ tends to zero as N/Z tends to zero. Actually, from Baker's expansion (2.13) we can show that $x_0 \sim A(N/Z)^{2/3}$ and $a_2 \sim -A^{-1}(Z/N)^{2/3}$, where A is a positive constant, in the limit $(N/Z) \to 0$. Unfortunately to determine A from Baker's series requires the sum to infinity of a sub-series which we have not yet achieved exactly. If we merely used the first three terms in equation (2.13) we would find $A = (3/2)^{3/2}$ but this is no more than a rough estimate†.

It is convenient to plot $(N/Z)^{2/3} f(N/Z)$ against N/Z using the results of table 1 and we show the form of this in curve A of figure 2. The crosses on this curve show the values from table 1, and the dashed curve is simply drawn in to join these values smoothly with the point at the origin. We expect this universal curve to be a valid representation of $E(Z, N)$ for sufficiently large Z and with $N/Z \lesssim 1$.

We can trivially rewrite equation (2.12) in the form

$$E(Z, N) = Z^2\left(Z^{1/3} f\left(\frac{N}{Z}\right)\right)\frac{e^2}{a_0} \tag{2.14}$$

and we now turn to examine how to connect this result with the Z^{-1} expansion.

† A better estimate from numerical summation is $A \simeq 1{\cdot}6$.

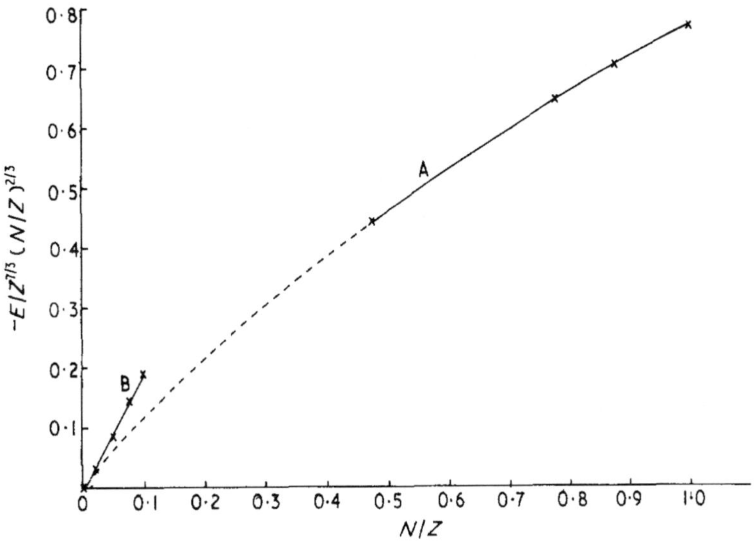

Figure 2. Binding energy $E(Z, N)$ plotted in reduced form against N/Z. Energy is in units e^2/a_0 (atomic units).

Curve A represents the Thomas–Fermi results. Crosses shown are from solution of Miranda (1934). Broken curve is simply graphical interpolation between crosses and point at the origin.

Curve B is obtained from Z^{-1} expansion for $Z = 100$, using results of table 2 for $N/Z \leqslant 0{\cdot}1$.

3. Energies in terms of Z^{-1} theory

If perturbation theory based on a hydrogenic zeroth-order representation is applied to an atom or ion of atomic number Z with N electrons, the energy may be written as

$$E(Z, N) = Z^2\left(\epsilon_0(N)+\frac{\epsilon_1(N)}{Z}+\frac{\epsilon_2(N)}{Z^2}+\ldots\right) \tag{3.1}$$

with convergence guaranteed for large enough Z, as shown by Kato (1951).

The coefficients ϵ_0, ϵ_1, and ϵ_2 are summarized in table 2 for the $N = 2$ to 10 ground states. Exchange and correlation effects can both occur in ϵ_1, but the presence of correlation effects in ϵ_1 indicates a degenerate zeroth-order state.

Table 2. Energy coefficients for Z^{-1} theory (energies in atomic units)

N	Configuration		$-\epsilon_0$	ϵ_1	$-\epsilon_2$	ϵ_3
2	^{1}S	1s^2	1	0·625	0·15766	0·008699
3	^{2}S	2s	1·125	1·0228	0·40837	−0·0230
4	^{1}S	2s^2	1·25	1·5593	0·8819	
5	^{2}P	2p	1·375	2·3275	1·8567	
6	^{3}P	2p^2	1·50	3·2589	3·2880	
7	^{4}S	2p^3	1·625	4·3535	5·2640	
8	^{3}P	2p^4	1·75	5·6619	8·1319	
9	^{2}P	2p^5	1·875	7·1343	11·7551	
10	^{1}S	2p^6	2·00	8·7708	16·2729	

The coefficient $\epsilon_0(N)$ is particularly simple and is easily tabulated for any N (given the configuration). Of more importance is the fact that the asymptotic large N dependence of ϵ_0 is readily obtained. Thus we show in the Appendix that

$$\epsilon_0(N) \simeq -2(1-\ln 2)6^{1/3}N^{1/3}+\tfrac{1}{2}+\mathrm{O}(N^{-1/3})$$
$$\simeq -1{\cdot}115N^{1/3}+\tfrac{1}{2}+\mathrm{O}(N^{-1/3}) \tag{3.2}$$

for large N (in atomic units e^2/a_0). In table 3 we give a comparison of $\epsilon_0(N)$ and the approximation obtained by retaining only the first two terms of equation (3.2) for a variety of N. It should be noted that the approximation is quite good even for small N.

Table 3. Comparison of $\epsilon_0(N)$ and asymptotic form of $\epsilon_0(N)$ for various N (energies in atomic units)

N	$\epsilon_0(N)$	$\tilde{\epsilon}_0(N)$†
2	1·00	0·91
3	1·13	1·11
4	1·25	1·28
5	1·38	1·41
6	1·50	1·53
7	1·63	1·63
8	1·75	1·73
9	1·88	1·82
10	2·00	1·90
15	2·28	2·25
20	2·51	2·53
30	3·06	2·97
40	3·35	3·31
50	3·64	3·61
60	3·88	3·87
70	4·19	4·10
88	4·49	4·46

† From equation 3.2.

We have no such exact asymptotic forms for ϵ_1 and ϵ_2. We do know that the TF theory discussed above should hold for large N and Z and thus we can compare equation (2.14) with equation (3.1) to obtain the large N asymptotic form of $\epsilon_n(N)$. We find the prediction that

$$\epsilon_n(N) \simeq A_n N^{n+1/3}+\dots \qquad \text{large } N. \tag{3.3}$$

It seems that if we plot ln ϵ_n versus ln N we might verify equation (3.3) if the asymptotic region of N is included. When we do this for ϵ_0, in which case equation (3.3) is, of course, proved exactly in an analytic fashion in equation (3.2) (see also the Appendix for details), we find a rather surprising result. The exponent predicted by the ln–ln plot is very close to $\frac{2}{5}$ instead of the expected $\frac{1}{3}$. This same result obtains if we plot the logarithm of the asymptotic approximation afforded by the first two terms of equation (3.2). Thus, it seems that N is never large enough (in the 2 to 90 range) to yield the characteristic asymptotic dependence graphically. This surely explains Foldy's result that in this range of atomic number the binding energy of neutral atoms goes like $Z^{12/5}$. Indeed,

ϵ_0 can be represented adequately in the $N = 20$ to 90 region by a formula of the type

$$\epsilon_0(N) \sim cN^{2/5} \tag{3.4}$$

instead of equation (3.2).

It is now of interest to see what we can learn about ϵ_1 and ϵ_2, where we have only values for rather low N ($N = 2$ to 10). If ϵ_0 is plotted against $N^{1/3}$ for $N = 2$ to 10 and this is compared with the linear prediction of equation (3.2) we find that the linear prediction seems visually close to a least squares result (see figure 3). This is readily verified. In fact

$$\tilde{\epsilon}_0 \text{ (least squares)} = -1{\cdot}13N^{1/3} + 0{\cdot}511 \tag{3.5}$$

using the $N = 2$ to 10 data.

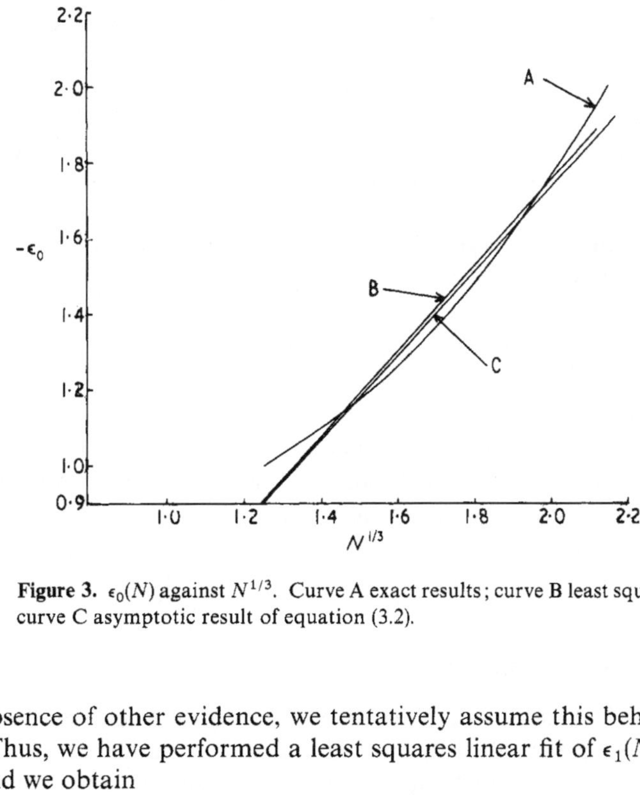

Figure 3. $\epsilon_0(N)$ against $N^{1/3}$. Curve A exact results; curve B least squares fit (equation (3.5)); curve C asymptotic result of equation (3.2).

In the absence of other evidence, we tentatively assume this behaviour persists for ϵ_1 and ϵ_2. Thus, we have performed a least squares linear fit of $\epsilon_1(N)/N$ and $\epsilon_2(N)/N^2$ with $N^{1/3}$ and we obtain

$$\tilde{\epsilon}_1 \text{ (least squares)} = 0{\cdot}661N^{4/3} - 0{\cdot}611N^{1/3} \tag{3.6}$$

and

$$\tilde{\epsilon}_2 \text{ (least squares)} = -0{\cdot}142N^{7/3} + 0{\cdot}158N^2. \tag{3.7}$$

Combining equations (3.2), (3.6) and (3.7) for the case of the neutral atom ($Z = N$) we obtain

$$E(Z, Z) \simeq -0{\cdot}596Z^{7/3} + 0{\cdot}047Z^2 + \ldots \tag{3.8}$$

the first term of which may be compared with the TF result (2.10).

A much more meaningful comparison with TF theory results if we calculate $E(Z, N)$ for $Z = 100$, $N = 2$ to 10 from the coefficients in table 1 and plot $E(Z, N)/Z^{7/3}$ versus N/Z. Results obtained for $N/Z \leqslant 0{\cdot}1$ are plotted in curve B of figure 2. Though the dashed curve representing the TF results is only approximate, it seems clear again that $Z = 100$ is too small for comparison with the universal curve which represents the asymptotic limit.

We want to stress that as $N/Z \to 1$ and for Z large the TF theory is indeed giving us an approximate summation to all orders of the Z^{-1} expansion.

4. Conclusion

This paper has succeeded in relating the Z^{-1} expansion and the Thomas–Fermi asymptotic formula for the binding energies of heavy ions. This latter theory predicts that the coefficients $\epsilon_n(N)$ in the Z^{-1} expansion (3.1) behave asymptotically as $N^{n+1/3}$ for large N, and this result is established exactly for $n = 0$. Though we have not been able to obtain exact results for $n \neq 0$, we give numerical estimates of the next two terms $\epsilon_1(N)$ and $\epsilon_2(N)$ for large N in equations (3.6) and (3.7). Considerable simplification in representing non-relativistic binding energies of atoms and ions should occur by plotting $(E(Z, N)/Z^{7/3})(N/Z)^{2/3}$ against N/Z, since this should become a universal curve for sufficiently large Z.

A final point worth noting is that the reciprocal of the radius of convergence of the Z^{-1} expansion is given by

$$\lim_{n\to\infty} \frac{\epsilon_{n+1}(N)}{\epsilon_n(N)} \sim \text{const } N \qquad \text{for large } N.$$

We anticipate, since there is no stable negative ion in the TF theory, that the constant will be unity, since the theory works for the neutral atom $N = Z$. The behaviour of the low order terms in equations (3.2), (3.6) and (3.7) is unfortunately very irregular and throws no light on the radius of convergence.

Appendix

Asymptotic form of $\epsilon_0(N)$ for large N

Consider atomic ground states with N electrons. In zeroth order (based on hydrogenic orbitals) the orbitals are described by a set of four quantum numbers (n, l, m_l, m_s) and the corresponding orbital energies are $-1/2n^2$. For our purpose it is sufficient to examine the case where m_l and m_s take on all possible values and where $k = n+l$ lies in the range $1 \leqslant k \leqslant K$ with n and l assuming all values consistent with a given k. There are some exceptions to this *aufbau*, but these are unimportant for our analysis since we will use a continuous model of the actual discrete process. The continuous model will be exact only for $N = 2, 4, 12, 20, 38, 56, 88, \ldots$ but the asymptotic form should hold for all N large enough.

Thus we have

$$N = \sum_{k=1}^{K} \sum_{n=n_0}^{k} 2(2l+1) \tag{A1}$$

where

$$l = k - n$$

$$n_0(k) = \frac{k}{2} + \frac{3}{4} + \frac{(-1)^k}{4}.$$

The sum (A1) is easily performed yielding

$$N = \tfrac{1}{6}K^3 + \tfrac{1}{2}K^2 + \tfrac{1}{3}K \tag{A2}$$

for K even, and

$$N = \tfrac{1}{6}K^3 + \tfrac{1}{2}K^2 + \tfrac{5}{6}K + \tfrac{1}{2} \tag{A3}$$

for K odd.

For large N both (A2) and (A3) lead to the result

$$K \sim (6N)^{1/3} - 1 + O(N^{-1/3}) \tag{A4}$$

which we will use later.

With this model ϵ_0 has the form

$$-\epsilon_0 = \sum_{k=1}^{K} \sum_{n=n_0}^{k} \frac{2(2l+1)}{2n^2}. \tag{A5}$$

This sum is somewhat more difficult to perform but we have succeeded in obtaining a closed form for the result. In fact

$$-\epsilon_0 = K - \frac{(K+3)(K+1)}{K^2} + \frac{\mathrm{d}}{\mathrm{d}K}\left[(K+1)^2\left\{\psi\left(\frac{K}{2}\right) - \psi(K)\right\}\right] \tag{A6}$$

for K even, and

$$-\epsilon_0 = K + \frac{\mathrm{d}}{\mathrm{d}K}\left[(K+1)^2\left\{\psi\left(\frac{K+1}{2}\right) - \psi(K+1)\right\}\right] \tag{A7}$$

for K odd. $\psi(x)$ is the digamma function defined as

$$\psi(x) = \frac{\Gamma'(x)}{\Gamma(x)}$$

where $\Gamma(x)$ is the usual gamma function.

Using either (A6) or (A7) we find that for large K, ϵ_0 has the form

$$\epsilon_0 \sim -2(1 - \ln 2)K + 2\ln 2 - \tfrac{3}{2} + O(K^{-1}). \tag{A8}$$

Since large K implies large N we use (A4) to obtain

$$\epsilon_0 \sim -2(1 - \ln 2)6^{1/3}N^{1/3} + \tfrac{1}{2} + O(N^{-1/3}). \tag{A9}$$

This represents the basic asymptotic form sought.

References

Baker E A 1930 *Phys. Rev.* **36** 630
Fermi E 1928 *Z. Phys.* **48** 73
Foldy L L 1951 *Phys. Rev.* **83** 397

Gombás P 1950 *Theories and Losungs methodem des Mehrteilchen problems der Wellenmechanik* (Basel: Verlag Birkhauser)
Hylleraas E A 1930 *Z. Phys.* **65** 209
Kato T 1951 *J. Fac. Sci., Tokyo Univ.* **16** 145
Knight R E 1969 *Phys. Rev.* **183** 45
Layzer D 1959 *Ann. Phys. NY* **8** 271
Layzer D, Horák Z, Lewis M and Thompson D 1964 *Ann. Phys. New York* **29** 101
March N H 1957 *Adv. Phys.* **6** 1
Milne E A 1927 *Proc. Camb. Phil. Soc.* **23** 794
Miranda C 1934 *Mem. Acad. Italia* **5** 283
Thomas L H 1926 *Proc. Camb. Phil. Soc.* **23** 542

Relation between the total energy and eigenvalue sum for neutral atoms and molecules

N. H. March

Theoretical Chemistry Department, University of Oxford, 1 South Parks Road, Oxford OX1 3TG, England
(Received 24 February 1977)

Ruedenberg has recently given arguments that the relation $E_{\text{total}} = (3/2)E_{\text{eigenvalue sum}}$ should hold approximately in molecules at their equilibrium configuration in self-consistent field theory. A first principles justification of this result is given here. The result has already been used in earlier work on atoms. The generalization to molecules is discussed.

I. INTRODUCTION

In a recent note[1] Ruedenberg has drawn attention to an approximate relation between the total energy E of a molecule in its equilibrium configuration in self-consistent field treatments and the sum of the one-electron eigenvalues $E_{\text{eigenvalue sum}} \equiv E_s$. The purpose of this paper is to provide some justification for Ruedenberg's result from first principles, on the basis of the Thomas–Fermi statistical theory, which can be regarded as the simplest possible self-consistent field method, valid, of course, for large numbers of electrons.[2]

Indeed, if one applies Ruedenberg's result to atoms, the fact that the total energy E is $(3/2)\ E_s$ was already used by March and Plaskett earlier.[3] However, in view of the obvious interest of Ruedenberg's result, it seems worthwhile summarizing the argument for atoms first, before going on to the molecular case. It will be useful to present the statistical theory in what follows as a special case of the density functional theory of many-electron systems.[4-6]

II. TOTAL ENERGY OF ATOMS AND MOLECULES IN TERMS OF ELECTRON DENSITY

The work of Hohenberg and Kohn[4] completes the Thomas–Fermi theory by demonstrating that the ground state energy E is a unique functional of the electron density $\rho(\mathbf{r})$. This allows the total energy to be written as

$$E = \int t_r(\rho)\, d\mathbf{r} + \int \rho V_N\, d\mathbf{r} + \frac{1}{2}\int \rho V_e\, d\mathbf{r} + V_{nn} + \int \epsilon_{xc}(\rho)\, d\mathbf{r}\ . \tag{2.1}$$

In Eq. (2.1), $t_r(\rho)$ is the single-particle kinetic energy density, V_N is the potential energy of the nuclear field, V_e is the electrostatic potential energy created by the ground state electron density $\rho(\mathbf{r})$, V_{nn} is the nuclear–nuclear interaction energy, while $\epsilon_{xc}(\rho)$ is the exchange and correlation energy density. Minimizing E with respect to ρ subject to

$$\int \rho\, d\mathbf{r} = N \tag{2.2}$$

where N is the total number of electrons in the molecule, yields the Euler equation of the variation problem as

$$\mu = \frac{\delta t_r(\rho)}{\delta\rho} + V_N + V_e + \frac{\delta\epsilon_{xc}(\rho)}{\delta\rho}\ . \tag{2.3}$$

As is well known,[5] this equation for the chemical potential μ, which has entered the theory as the Lagrange multiplier taking care of the subsidiary condition (2.2), shows that a one-body potential $V(\mathbf{r})$ defined by

$$V(\mathbf{r}) = V_N + V_e + \frac{\delta\epsilon_{xc}(\rho)}{\delta\rho} \tag{2.4}$$

can be introduced which, in principle, will allow the exact ground state density to be obtained from the sum of the squares of the one-electron wavefunctions ψ_i, with corresponding eigenvalues ϵ_i, over the lowest N states. Of course, in practice, ϵ_{xc} is not known exactly; this would be equivalent to the exact solution of the many-electron problem.

Nevertheless, it is now quite clear that the eigenvalue sum E_s given by

$$E_s = \sum_{i=1}^{N} \epsilon_i \tag{2.5}$$

is simply

$$E_s = \int t_r(\rho)\, d\mathbf{r} + \int \rho V(\mathbf{r})\, d\mathbf{r}\ , \tag{2.6}$$

which using Eq. (2.4) gives

$$E_s = \int t_r(\rho)\, d\mathbf{r} + \int \rho V_N\, d\mathbf{r} + \int \rho V_e\, d\mathbf{r} + \int \rho\, \frac{\delta\epsilon_{xc}(\rho)}{\delta\rho}\, d\mathbf{r}\ . \tag{2.7}$$

Combining Eqs. (2.1) and (2.7) it is clear that we have the exact relation

$$E = E_s - \frac{1}{2}\int \rho V_e\, d\mathbf{r} + V_{nn} + \int \left[\epsilon_{xc}(\rho) - \rho\, \frac{\delta\epsilon_{xc}(\rho)}{\delta\rho}\right] d\mathbf{r}\ . \tag{2.8}$$

If we denote the electron–electron electrostatic potential energy contribution $\frac{1}{2}\int \rho V_e\, d\mathbf{r}$ by V_{ee} then we have from Eq. (2.8),

$$E = E_s + (V_{nn} - V_{ee}) + \int \left[\epsilon_{xc}(\rho) - \rho\, \frac{\delta\epsilon_{xc}(\rho)}{\delta\rho}\right] d\mathbf{r}\ , \tag{2.9}$$

which we can regard as an exact generalization of Ruedenberg's result.[1]

Having reached this point by formally exact methods, we must now turn to approximations to make further progress. We deal first with atoms.

III. RELATION BETWEEN E AND E_s FOR NEUTRAL ATOMS

For an atom we drop V_{nn}. In the simplest self-consistent field theory, namely the Thomas–Fermi (TF) statistical theory, the Euler equation (2.3) is simplified by assuming

$$t_T \doteq c_k \rho^{5/3}, \quad \epsilon_{xc} \doteq 0 , \tag{3.1}$$

and it is well known for the neutral TF atom that $\mu = 0$. Thus, we have

$$0 = \tfrac{5}{3} c_k \rho^{2/3} + V_N + V_e . \tag{3.2}$$

Multiplying this equation by ρ and integrating over the whole of space we find immediately

$$0 = \tfrac{5}{3} T + V_{en} + 2V_{ee} , \tag{3.3}$$

where V_{en} is the potential energy of the electron–nuclear interaction and T is the total kinetic energy. The virial theorem, which Fock[7] first showed to be valid in the TF theory of atoms, yields

$$2T + V_{en} + V_{ee} = 0 . \tag{3.4}$$

Hence, from Eqs. (3.3) and (3.4) we obtain

$$V_{ee} = \tfrac{1}{3} T = -\tfrac{1}{3} E . \tag{3.5}$$

Returning to Eq. (2.9), with $\epsilon_{xc} = 0$, we find

$$\tfrac{2}{3} E = E_s , \tag{3.6}$$

a result already used by March and Plaskett.[3] This is the result given by Ruedenberg between the total energy and the eigenvalue sum for molecules in their equilibrium configuration; we turn next to this case. Whereas the above argument is rigorously valid within TF theory for atoms, there is a deficiency when this theory is applied to molecules, which means that this case is somewhat less clearcut. Nevertheless, the argument set out below for molecules throws light on the origin of the Ruedenberg result so we shall present it in some detail.

IV. ENERGY RELATIONS FOR MOLECULES IN EQUILIBRIUM CONFIGURATION

In equilibrium under purely Coulomb forces we know again for a molecule that, with $\bar{V}$ the average potential energy,

$$2T + \bar{V} = 0 , \tag{4.1}$$

which, for example in a diatomic molecule with $E = E(R)$, R being the internuclear distance, is a special case of

$$2T + \bar{V} = -R \frac{dE}{dR} , \tag{4.2}$$

as was first observed by Slater.[8] The right-hand side of Eq. (4.2) is simply the virial of the forces required to hold fast the nuclei at separation R, and this term is only zero at the equilibrium separation.

However, for molecules there is a difficulty with the TF statistical theory, as Teller[9] has shown, in that the theory does not lead to a minimum in the energy, that is, dE/dR is not zero for finite R.

Therefore, the argument given below will make use of the relation between V_{nn}, V_{en}, V_{ee}, and T from the statistical theory, but will use the equilibrium form (4.1) of the virial theorem, which we know, of course, to be exactly true for all molecules in equilibrium.

The argument now goes through as before, the essential equation being Eq. (3.3). Again with $\epsilon_{xc} = 0$ in Eq. (2.9), we now form $V_{nn} - V_{ee}$ using Eq. (4.1) to get

$$V_{nn} - V_{ee} = -2T - V_{en} - 2V_{ee} . \tag{4.3}$$

But clearly we can write from Eq. (2.7) with $\epsilon_{xc} = 0$,

$$\tfrac{1}{2} E_s = \tfrac{1}{2} T + \tfrac{1}{2} V_{en} + V_{ee} , \tag{4.4}$$

and hence for homonuclear diatomic molecules

$$V_{nn} - V_{ee} - \tfrac{1}{2} E_s = -\tfrac{5}{2} T - \tfrac{3}{2} V_{en} - 3V_{ee} = 0 \tag{4.5}$$

from Eq. (3.3). Thus, we regain the result desired

$$E = \tfrac{3}{2} E_s . \tag{4.6}$$

We reiterate that, in this derivation, we have combined the approximate TF relation (3.3) with the exact virial theorem at equilibrium at the observed R to get the result, the TF theory itself giving no stable molecule. Clearly the result (4.6) is approximate, but the results of self-consistent field calculations well fit the modification $E \doteq 1.55\, E_s$ as Ruedenberg points out.

We can make a modification to include Dirac–Slater exchange for which

$$\epsilon_{xc}(\rho) = -c_e \rho^{4/3} \tag{4.7}$$

and this adds a term equal to $\frac{4}{3}$ of the total exchange energy onto Eq. (3.3), plus a nonzero chemical potential, but we shall not pursue that further.

V. CONCLUSION

For neutral atoms the statistical theory leads to the result that the total energy is 3/2 times the eigenvalue sum without any approximations save those implicit in the statistical method.

For neutral molecules one regains this result if one takes the statistical relation between kinetic and potential energy contributions, in combination with the exact virial theorem. Exchange can usefully be included in the theory via Dirac–Slater theory if required later.

The arguments of Ruedenberg, given some basis in first principles theory in the present paper, should encourage a more careful study of the relation between different energy contributions in the equilibrium configuration of molecules. For heavy molecules we expect the result (3.3) to become increasingly accurate, especially if supplemented by Dirac–Slater exchange. Finally, of course, the entire considerations of this paper are based on nonrelativistic theory, whereas for molecules containing heavy atoms relativistic corrections will obviously be called for.

[1]K. Ruedenberg, J. Chem. Phys. **66**, 375 (1977).
[2]N. H. March, Adv. Phys. **7**, 1 (1957).
[3]N. H. March and J. S. Plaskett, Proc. Soc. (London) A **235**, 419 (1956).
[4]P. Hohenberg and W. Kohn, Phys. Rev. B **136**, 864 (1964).
[5]W. Kohn and L. J. Sham, Phys. Rev. A **140**, 1133 (1965).
[6]N. H. March, *Orbital Theories of Molecules and Solids* (Clarendon, Oxford, 1974), p. 95.
[7]V. Fock, Phys. Z. Sowjetunion **1**, 747 (1932).
[8]J. C. Slater, J. Chem. Phys. **1**, 687 (1933).
[9]E. Teller, Rev. Mod. Phys. **34**, 627 (1962).

Exact local density method for linear harmonic oscillator

G. P. Lawes and N. H. March

Theoretical Chemistry Department, University of Oxford, 1 South Parks Rd., Oxford, OX1 3TG, England
(Received 5 October 1978)

Using an exact form for the particle density for N particles moving independently in a linear harmonic oscillator potential, a differential equation is derived for the particle density. This is then shown to be exactly that which is obtained from a local density assumption for the kinetic energy density t, in the form $-1/2\Sigma\psi_i(\delta^2/\delta x^2_i)\psi_i$. The form of the kinetic energy density functional is then discussed in this example. The implications for wider classes of potential are briefly referred to.

I. INTRODUCTION

Modern density functional theory has, as a major aim, the direct calculation of the particle density ρ in a many-body system without recourse to wave function calculations. While a lot of progress has been made, wide use is made of local density approximations.

The purpose of this note is to demonstrate, by first setting up an exact differential equation for the density $\rho(x)$ for N particles moving in a linear harmonic oscillator potential, that the local density assumption made on the kinetic energy density,

$$t=-\frac{1}{2}\sum\psi_i\frac{\partial^2}{\partial x^2}\psi_i\ , \tag{1.1}$$

is in this case exact. This allows some progress to be made in discussing the form of t in this simple example. What will be shown below is that if we write the local density assumption in the form,

$$\frac{\delta t}{\delta\rho}=\left(\frac{\partial\rho}{\partial x}\right)^{-1}\frac{\partial t}{\partial x}\ , \tag{1.2}$$

then the Euler equation of the exact variation problem for the linear harmonic oscillator is exactly that derived from an exact expression for the density.

II. EXACT DIFFERENTIAL EQUATION FOR PARTICLE DENSITY FOR THE LINEAR HARMONIC OSCILLATOR

As Husimi[1] was the first to show, in units in which the Hamiltonian H takes the form,

$$H=\tfrac{1}{2}p^2+\tfrac{1}{2}x^2\ , \tag{2.1}$$

the density $\rho(x)$ can be written in terms of $\psi_N(x)$, the normalized wave function for the Nth state, as

$$\rho(x)=\tfrac{1}{2}\psi_N'^2+(N-\tfrac{1}{2}x^2)\psi_N^2\ . \tag{2.2}$$

Using the Schrödinger equation for ψ_N, namely

$$\psi_N''+2(N-\tfrac{1}{2}-\tfrac{1}{2}x^2)\psi_N=0\ , \tag{2.3}$$

one readily obtains the following derivatives:

$$\rho'(x)=-x\psi_N^2+\psi_N\psi_N' \tag{2.4}$$

and

$$\rho'''(x)=4x^2\psi_N\psi_N'-8N\psi_N\psi_N'+4Nx\psi_N^2-2x^3\psi_N^2-2x\psi_N'^2\ . \tag{2.5}$$

Following March,[2] if t is the kinetic energy density defined in Eq. (1.1), we use the relation;

$$\frac{\partial t}{\partial x}=-\tfrac{1}{2}\rho\frac{\partial V}{\partial x}-\tfrac{1}{8}\rho'''\ , \tag{2.6}$$

in our example, $\partial V/\partial x$ being simply x. Then we obtain for the right hand side of Eq. (2.6), after a short calculation using Eqs. (2.2) and (2.5),

$$-\tfrac{1}{2}x\rho-\tfrac{1}{8}\rho'''=(N-\tfrac{1}{2}x^2)(-x\psi_N^2+\psi_N\psi_N') \tag{2.7}$$

which yields, when use is made of Eq. (2.4), the desired differential equation for $\rho(x)$ for the linear harmonic oscillator,

$$(-\tfrac{1}{2}\rho\,\partial V/\partial x-\tfrac{1}{8}\rho''')=(N-\tfrac{1}{2}x^2)\rho'\ . \tag{2.8}$$

We stress that this is an exact differential equation for $\rho(x)$ for the linear harmonic oscillator, the ground state corresponding to the case $N=1$.

III. RELATION TO THE LOCAL DENSITY METHOD

Let us now examine the relation between the exact form (2.8) and the local density assumption embodied in writing Eq. (1.2). Then it follows by combining Eqs. (1.2) with the exact Euler equation of density functional theory, namely,

$$\delta t/\delta\rho=\mu-V\ , \tag{3.1}$$

where μ is the chemical potential, that for the harmonic oscillator the local density assumption implies the result,

$$\frac{\partial t}{\partial x}=(\mu-\tfrac{1}{2}x^2)\frac{\partial\rho}{\partial x}\ , \tag{3.2}$$

or making use of Eq. (2.6),

$$[-\tfrac{1}{2}\rho\frac{\partial V}{\partial x}-\tfrac{1}{8}\rho''']=(\mu-\tfrac{1}{2}x^2)\rho'\ . \tag{3.3}$$

It can be seen immediately that this is identical with Eq. (2.8) if $\mu=N$.

Thus, we have demonstrated here the remarkable result that the local density assumption embodied in Eq. (1.2) leads to the exact differential equation for $\rho(x)$ for the case of the harmonic oscillator.

IV. FORM OF THE KINETIC ENERGY DENSITY FOR THE LINEAR HARMONIC OSCILLATOR

The above conclusion strongly points to the fact that it should then be possible to demonstrate, for the linear harmonic oscillator, that the kinetic energy density t in Eq. (1.1) is indeed a function only of the local density $\rho(x)$. To do so, we next utilize the fact, that in addition to the diagonal density (2.2), Musimi[1] was able also to

 0021-9606/79/021007-03$01.00

exhibit the off-diagonal elements of the Dirac density matrix for the linear harmonic oscillator. Explicitly, the density matrix is given by

$$\rho(x', x) = \tfrac{1}{2}\psi_N(x')\psi_N(x) + \frac{1}{2(x'-x)}\left[\psi_N(x')\psi_N'(x) - \psi_N(x)\psi_N'(x)\right]. \tag{4.1}$$

We now form the kinetic energy density t from the result equivalent to Eq. (1.1), namely,

$$t = -\frac{1}{2}\frac{\partial^2}{\partial x^2}\rho(x', x)\bigg|_{x'=x} . \tag{4.2}$$

After a short calculation, we find the explicit expression for the kinetic energy density as

$$t = \tfrac{1}{12}\psi_N^2[x^4 - x^2(4N+1) + 2N + 4N^2] - \tfrac{1}{12}\psi_N'^2(x^2+1-2N) + \tfrac{1}{6}x\psi_N\psi_N' . \tag{4.3}$$

In order to illustrate the relation between t and ρ, we turn immediately to two specific examples.

A. Explicit form of $t(\rho)$ for the lowest state

Let us consider first the simplest case when only the lowest state is occupied. The normalized wave function is

$$\psi_1 = \frac{1}{\pi^{1/4}}\exp\left(\frac{-x^2}{2}\right) \tag{4.4}$$

Direct calculation of t from

$$t = -\tfrac{1}{2}\psi_1\frac{\partial^2\psi_1}{\partial x^2} \tag{4.5}$$

then yields

$$t = \frac{1}{2\pi^{1/2}}\exp(-x^2)(1-x^2) . \tag{4.6}$$

But from Eq. (4.4)

$$\rho = \psi_1^2 = \frac{1}{\pi^{1/2}}\exp(-x^2) , \tag{4.7}$$

hence it follows by combining (4.6) and (4.7) that

$$t = \tfrac{1}{2}\rho + \tfrac{1}{2}\rho(\ln\rho + \tfrac{1}{2}\ln\pi) . \tag{4.8}$$

Forming $\delta t/\delta\rho$, one finds

$$\frac{\delta t}{\delta\rho} = 1 + \tfrac{1}{2}(\ln\rho + \tfrac{1}{2}\ln\pi) , \tag{4.9}$$

and if use is made of Eq. (4.7), then one regains the Euler equation

$$\delta t/\delta\rho = 1 - \tfrac{1}{2}x^2 , \tag{4.10}$$

for this case of one level only occupied.

B. Form of the $\delta t/\delta\rho - \rho$ relation for the first two states occupied

As the second example, let us take the first two states only to be occupied. Explicit use of the first excited state wave function in addition to ψ_1 in Eq. (4.4) yields for the density

$$\rho(x) = \frac{1}{\pi^{1/2}}\exp(-x^2)(1+2x^2) \tag{4.11}$$

and for the kinetic energy density

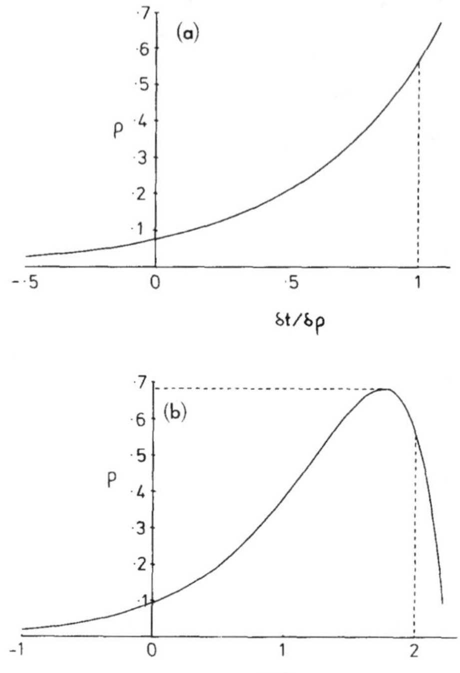

FIG. 1. (a) Shows ρ versus $\delta t/\delta\rho$ for lowest state only occupied ($N=1$). Dashed line shows maximum value of $\delta t/\delta\rho$. (b) Shows the same plot for the first two states occupied. The dashed lines indicate the maximum values of ρ and of $\delta t/\delta\rho$ from the exact results.

$$t = \frac{\exp(-x^2)}{2\pi^{1/2}}(-2x^4 + 5x^2 + 1) . \tag{4.12}$$

Evidently, Eqs. (4.11) and (4.12) could alternatively be obtained from Eqs. (2.2) and (4.3). Forming the right-hand side of Eq. (1.2) yields $2 - x^2/2$, which is from the general Euler equation equal to $\delta t/\delta\rho$. Using this result for $\delta t/\delta\rho$ in Eq. (4.11) one finds

$$\rho = \frac{1}{\pi^{1/2}}\exp\left(2\frac{\delta t}{\delta\rho} - 4\right)\left[1 + 2\left(4 - 2\frac{\delta t}{\delta\rho}\right)\right] . \tag{4.13}$$

This is an implicit relation for $\delta t/\delta\rho$, in terms of ρ and confirms the local nature of the relation in this example. Equations (4.13) and (4.9) are plotted in Fig. 1.

For general N, the relation between ρ and $\delta t/\delta\rho$ is summarized in the Appendix.

V. SUMMARY

Eq. (2.8) constitutes an exact differential equation for determining the density $\rho(x)$ for the linear harmonic oscillator. Remarkably, it turns out to be identical with that obtained from the local density assumption (1.2).

The local nature of the $\delta t/\delta\rho - \rho$ relation is confirmed by the two examples of Sec. IV. Though we have written down a formally exact relation between ρ and $\delta t/\delta\rho$ for arbitrary N in the Appendix, this is very complicated. It must, however, simplify again in the limit of very large N where the Thomas–Fermi statistical theory tells us that $\delta t/\delta\rho$ must asymptotically become proportional to ρ^2.

For a more general potential, in which, however, the higher derivatives are not too large, it seems from the present discussion that the local density differential equation given earlier by one of us[2] may still be a useful approximation, being exact for a harmonic potential as demonstrated.

APPENDIX: $\rho - \delta t/\delta\rho$ relation for general N

Let us write the wave function ψ_N in Eq. (2.2) for $\rho(x)$ explicitly in terms of the Hermite polynomials. Then we have[3]

$$\psi_N = \mathcal{N}_{N-1} \exp(-x^2/2)\, H_{N-1}(x) \; . \qquad \text{(A1)}$$

Thus, using the properties of the Hermite functions we find

$$\psi'_N = \mathcal{N}_{N-1} \exp(-x^2/2)[-xH_{N-1}(x) + 2(N-1)H_{N-2}(x)] \; , \qquad \text{(A2)}$$

where

$$\mathcal{N}_{N-1} = \frac{1}{\pi^{1/4}}\left(\frac{1}{2^{N-1}(N-1)!}\right)^{1/2} \; . \qquad \text{(A3)}$$

Hence, using the Euler equation again, as in the examples of Sec. IV.

$$\begin{aligned}\rho = \mathcal{N}^2_{N-1} \exp(2\delta t/\delta\rho - 2N)\{&NH^2_{N-1}([2N - 2\delta t/\delta\rho]^{1/2})\} \\ &+ 2(N-1)^2 H^2_{N-2}([2N - 2\delta t/\delta\rho]^{1/2}) \\ &- 2(N-1)(2N - 2\delta t/\delta\rho)^{1/2} H_{N-1}([2N - 2\delta t/\delta\rho]^{1/2}) \\ &\times H_{N-2}([2N - 2\delta t/\delta\rho]^{1/2}) \; . \qquad \text{(A4)}\end{aligned}$$

This includes the result (4.13) as the special case $N=2$. For intermediate N, the $\rho - \delta t/\delta\rho$ relation is evidently highly complicated but as remarked in the main text, it must simplify again for large N to yield asymptotically $\delta t/\delta\rho$ proportional to ρ^2.

[1]K. Husimi, Proc. Phys. Math. Soc. Jpn. **22**, 264 (1940).
[2]N. H. March, J. Chem. Phys. **70**, 587 (1979).
[3]See, for example, L. Pauling and E. B. Wilson, *Introduction to Quantum Mechanics* (McGraw-Hill, New York, 1935), p. 74.

Proc. Natl. Acad. Sci. USA
Vol. 77, No. 11, pp. 6285–6288, November 1980
Chemistry

Chemical potential, Teller's theorem, and the scaling of atomic and molecular energies

(Thomas–Fermi theory/electronegativity)

NORMAN H. MARCH AND ROBERT G. PARR*

Theoretical Chemistry Department, University of Oxford, Oxford 0X1, 3TG, England

Contributed by Robert G. Parr, July 18, 1980

ABSTRACT For atoms and homonuclear diatomic molecules, it is argued that the electronic energies have the forms

$$E(Z,N) = Z^{7/3}f_1(N/Z) + Z^{6/3}f_2(N/Z) + Z^{5/3}f_3(N/Z) + \cdots$$

and

$$E(Z,N,R) = Z^{7/3}F_1(N/Z,N^{2/3}/ZR) + Z^{6/3}F_2(N/Z,N^{2/3}/ZR) + \cdots, \text{ respectively,}$$

where Z is the atomic number, N is the number of electrons, and R is the internuclear distance. By using the Lieb–Simon theorem that the Thomas–Fermi theory is exact in the limit of large atomic number and the Teller theorem that molecules are not bound in the Thomas–Fermi theory, it is then shown, among other results, that the electron–electron repulsion energy for neutral systems has no term in Z^2 and that the nucleus–nucleus repulsion energy for neutral molecules is probably $\mathcal{O}(Z^{5/3})$. For neutral atoms, it is predicted and verified that the chemical potential (electronegativity) is $\mathcal{O}(Z^{-1/3})$ for large Z. Tetrahedral and octahedral molecules are briefly discussed.

Density descriptions of atoms and molecules have led to progress in establishing relationships between energy components in molecules. For example, Mucci and March (1) have shown that, in the simplest (Thomas–Fermi) density description, scaling relationships hold between various components, and also they have exhibited specific regularities in the nucleus–nucleus potential energy V_{nn}. Somewhat differently, Parr and Gadre (2) have proposed homogeneity properties of atomic and molecular energies, again in order to relate energy components.

In this paper, we shall first establish the form of the total energy $E(Z,N)$ for atomic ions with N electrons and nuclear charge Ze, for large Z and N. This is done by expanding the coefficient of $1/Z^n$ in the 1/Z expansion in powers of $N^{-1/3}$ for large N and formally summing the resulting series. Adding to this expansion the necessary requirement that the chemical potential μ of neutral atoms becomes 0 for large enough Z, we then obtain simple relationships between the separate energy components. We also show that for neutral atoms the chemical potential (electronegativity) must vary grossly with Z as $Z^{-1/3}$, and we verify this prediction from actual data.

The argument from the 1/Z expansion is then extended to the case of homonuclear diatomic molecular ions. We make use of a scaling result for the bare nuclear potential established from Thomas–Fermi statistical theory by Dreizler and March (3). This enables us again to write an expansion in reduced variables which allows relationships to be established between different energy components at equilibrium. By applying the theorem of Teller (4) that there is no molecular binding in the Thomas–Fermi theory, we argue that the nucleus–nucleus potential energy of neutral molecules at equilibrium is $\mathcal{O}(Z^{5/3})$ whereas the other major energy components involve the 7/3 power.

Atoms and atomic ions

The 1/Z expansion for the total energy $E(Z,N)$ may be written

$$\frac{E(Z,N)}{Z^2} = \epsilon_0(N) + \frac{1}{Z}\epsilon_1(N) + \cdots + \frac{1}{Z^n}\epsilon_n(N) + \cdots \quad [1]$$

As March and White (5) showed, for large N the coefficients ϵ_n can be expanded as

$$\epsilon_n(N) = a_nN^{n+1/3} + b_nN^n + c_nN^{n1/3} + \cdots, \quad [2]$$

which is a power series in $N^{-1/3}$. Forming $\epsilon_n(N)/Z^n$ and summing over n formally leads immediately to

$$E(Z,N) = Z^{7/3}f_1(N/Z) + Z^2f_2(N/Z) + Z^{5/3}f_3(N/Z) + \cdots. \quad [3]$$

As emphasized by March and White (5), the first term is the result of the Thomas–Fermi statistical theory and the function $f_1(N/Z)$ is already known from numerical solution of the Thomas–Fermi equation. Although the validity of Eq. 3 has not been strictly proved, we know of no evidence that it is not correct.

We now note that the chemical potential of the ion can be calculated from the formula

$$\mu = (\partial E/\partial N)_Z. \quad [4]$$

For the Thomas–Fermi neutral atom, it is well known that $\mu = 0$, and it follows by differentiating Eq. 3 that this is equivalent to requiring that $f_1'(N/Z) = 0$ at $N/Z = 1$. Indeed, a stronger result holds: Since the Thomas–Fermi theory is the exact theory as Z approaches infinity (6), the exact expression for μ, as calculated from Eqs. 3 and 4 for a neutral atom, must give $\mu = 0$ in the $Z = \infty$ limit. This implies that

$$-f_n'(N/Z)|_{N/Z=1} = 0, \quad n \leqslant 5. \quad [5]$$

This property of the functions $f_n(N/Z)$ has remarkable consequences for the energy components and for the chemical potential, as we shall see.

We can obtain the electron–nucleus potential energy V_{ne} by applying Feynman's theorem in the form

$$V_{ne} = Z(\partial E/\partial Z)_N. \quad [6]$$

Using Eq. 5, we find, for neutral atoms,

$$V_{ne} = (7/3)Z^{7/3}f_1(1) + 2Z^2f_2(1) + (5/3)Z^{5/3}f_3(1) + \cdots. \quad [7]$$

We note that, in higher terms, derivatives of f_n enter in accord with Eq. 5.

* To whom reprint requests should be addressed, at permanent address: Department of Chemistry, University of North Carolina, Chapel Hill, NC 27514.

But the virial theorem tells us that

$$2T + V_{ne} + V_{ee} = 0, \quad T = -E, V_{ee} = 2E - V_{ne}, \quad [8]$$

where T is the kinetic energy and V_{ee} is the electron–electron potential energy. We therefore have (for neutral atoms)

$$V_{ee} = -\frac{1}{3}Z^{7/3}f_1(1) + \frac{1}{3}Z^{5/3}f_3(1) + \cdots \quad [9]$$

We deduce that there is no term of $\mathcal{O}(Z^2)$ in the electron–electron repulsion energy.

For the chemical potential itself, we obtain, for neutral atoms,

$$\mu = Z^{-1/3}f_0'(1) + Z^{-2/3}f_7'(1) + \cdots. \quad [10]$$

Since the chemical potential is the negative of the electronegativity (7), the prediction is that the electronegativity of the elements goes like $Z^{-1/3}$ for large Z.

Homonuclear diatomic molecules

Having established the approach to the scaled form, Eq. 3, of the energy of atomic ions, via the $1/Z$ expansion, we turn now to homonuclear diatomic ions, with N electrons and each nucleus carrying charge Ze, at internuclear separation $2R$. We seek the form of the energy $E(Z,N,R)$, again from a $1/Z$ expansion.

We first summarize the electronic Hamiltonian $\mathcal{H}$ and its scaling properties. We have

$$\mathcal{H} = -\frac{1}{2}\sum_1^N \nabla^2_{x_i} + \sum_{i<j}\frac{e^2}{r_{ij}} - Ze^2\sum_{i=1}^N\left[\frac{1}{r_{\alpha i}} + \frac{1}{r_{\beta i}}\right], \quad [11]$$

where $r_{\alpha i}$ denotes the position of the ith electron from nucleus α, etc. If we measure distances in units of $1/Z$ and divide by Z^2, we obtain

$$\frac{\mathcal{H}}{Z^2} = -\frac{1}{2}\sum_1^N \nabla^2_{x_i} + \sum_{i<j}\frac{e^2}{Zx_{ij}} - e^2\sum_{i=1}^N\left[\frac{1}{x_{\alpha i}} + \frac{1}{x_{\beta i}}\right]. \quad [12]$$

This expression shows that the total electronic energy E of homonuclear diatomic molecular ion has the form

$$E(Z,N,R) = Z^2\left[\epsilon_0(N,X) + \frac{1}{Z}\epsilon_1(N,X) + \cdots + \frac{1}{Z^n}\epsilon_n(N,X) + \cdots\right], \quad [13]$$

where $X = ZR$.

We note, from the Hamiltonian of Eq. 12, that the electron–electron term does not affect the quantity $\epsilon_0(N,X)$, which is therefore a property of the eigenvalues in the bare Coulomb field of the two nuclei. For this problem, the Thomas–Fermi energy was first obtained by Dreizler and March (3). It is convenient to express their result in terms of the chemical potential for the bare nuclei as

$$\mu_{\text{bare nuclei}} = \frac{Ze^2}{R}G(N^{2/3}/X). \quad [14]$$

Since, for the bare nuclei Coulomb problem, the total energy (sum of one-electron energies) is of order $N\mu$, we have the scaling property of the total energy as

$$E_{\text{Coulomb}}(Z,N,R) \sim \frac{Z^2Ne^2}{X}Q(N^{2/3}/X). \quad [15]$$

Hence it follows that the scaling of the Coulomb term $\epsilon_0(N,X)$ in Eq. 13 is given by

$$\epsilon_0(N,X) = N^{1/3}S(N^{2/3}/X). \quad [16]$$

This result now motivates the assumed form of expansion for $\epsilon_n(N,X)$, which by analogy with atoms we take to be

$$\epsilon_n(N,X) = N^{n+1/3}A_n(N^{2/3}/X) + N^nB_n(N^{2/3}/X) + N^{n-1/3}C_n(N^{2/3}/X) + \cdots. \quad [17]$$

Substituting, in Eq. 13, of this large N expansion of the coefficients yields

$$E(Z,N,R) = Z^{7/3}\sum_n (N/Z)^{n+1/3}A_n(N^{2/3}/X) + Z^2\sum_n (N/Z)^nB_n(N^{2/3}/X) + \cdots, \quad [18]$$

which can be formally summed to yield

$$E(Z,N,R) = Z^{7/3}F_1(N/Z,N^{2/3}/X) + Z^2F_2(N/Z,N^{2/3}/X) + Z^{5/3}F_3(N/Z,N^{2/3}/X) + \cdots. \quad [19]$$

Our main interest being in neutral molecules, for them we can express this in the form

$$E(Z,Z,R) = Z^{7/3}F_1(Z^{1/3}R) + Z^2F_2(Z^{1/3}R) + Z^{5/3}F_3(Z^{1/3}R) + \mathcal{O}(Z^{4/3}), \quad [20]$$

which is our major result on the scaling of the homonuclear diatomic energy. As for atoms, the leading term is the self-consistent Thomas–Fermi answer; that this scales in accord with Eq. 20 is known from the work of Hund (8) and of Townsend and Handler (9). From the latter paper, approximate numerical values of the function F_1 in Eq. 20 can be extracted.

Again the electron-nucleus potential energy can be derived from Feynman's theorem. We find

$$V_{ne} = \frac{7}{3}Z^{7/3}F_1(Z^{1/3}R) + 2Z^2F_2(Z^{1/3}R) + \frac{5}{3}Z^{5/3}F_3(Z^{1/3}R) + \mathcal{O}(Z^{4/3}) + Z^{10/3}\frac{\partial F_1}{\partial Z} + Z^3\frac{\partial F_2}{\partial Z} + Z^{8/3}\frac{\partial F_3}{\partial Z} + \cdots. \quad [21]$$

At this stage we invoke the virial theorem in the form

$$2T + V_{ne} + V_{ee} + V_{nn} = -R\left(\frac{\partial W}{\partial R}\right)_{Z,N} \quad [22]$$

where W is the total molecular energy $E + V_{nn}$, including the nucleus–nucleus repulsion V_{nn}. But it is now clear that, because of the scaling property of Eq. 20, derivatives of E with respect to R are related to the Z derivatives in Eq. 21. Restricting ourselves to equilibrium, when $\partial W/\partial R$ equals zero, we can eliminate the derivative terms. We find

$$V_{ne}^{\text{equil}} = \frac{7}{3}Z^{7/3}F_1(Z^{1/3}R_e) + 2Z^2F_2(Z^{1/3}R_e) + \frac{5}{3}Z^{5/3}F_3(Z^{1/3}R_e) + \mathcal{O}(Z^{4/3}) + \frac{1}{3}V_{nn}^{\text{equil}}. \quad [23]$$

Also, we can obtain an expression involving V_{ee} from the virial theorem at equilibrium, and we find

$$V_{ee}^{\text{equil}} = -\frac{1}{3}Z^{7/3}F_1(Z^{1/3}R_e) + \frac{1}{3}Z^{5/3}F_3(Z^{1/3}R_e) + \mathcal{O}(Z^{4/3}) + \frac{2}{3}V_{nn}^{\text{equil}}. \quad [24]$$

We now argue that one can interpret Teller's theorem, that molecules are not bound in the Thomas–Fermi theory (4), such that as one takes the formal limit Z tends to infinity in the foregoing equation, we must find the equilibrium bond length

R_e becomes infinite, and this in turn means that V_{nn}^{equil} is a lower-order term in Z than V_{ee}, etc. Thus, it follows then that the term of $\mathcal{O}(Z^{7/3})$ in Eq. **24** arises entirely from V_{ee}. Also, there is no $\mathcal{O}(Z^2)$ term in V_{ee} for atoms; this comes from V_{ne} and we therefore make a formal separation of V_{ee} and V_{nn} from Eq. **24** as

$$V_{ee}^{\text{equil}} = -\frac{1}{3}Z^{7/3}F_1(Z^{1/3}R_e) + \frac{\beta}{3}Z^{5/3}F_3(Z^{1/3}R_e) + \mathcal{O}(Z^{4/3}) \quad [25]$$

and

$$V_{nn}^{\text{equil}} = -\frac{\gamma Z^{5/3}}{3}F_3(Z^{1/3}R_e) + \mathcal{O}(Z^{4/3}). \quad [26]$$

The sum of β and γ must be consistent with Eq. **24** and also there are inequalities since the sign of V_{nn} must be positive; these are easily written down. For present purposes, suffice it to remark that there is no restriction implied on the magnitude of the ratio β/γ in these expressions for V_{ee} and V_{nn}. This is relevant to the discussion in the following section. In principle, of course, both V_{ee} and V_{nn} could contain terms $\mathcal{O}(Z^2)$ if they cancelled to give Eq. **24**, but this seems highly unlikely. Our conclusion is that the theorem of Teller establishes for neutral molecules at equilibrium that the nucleus–nucleus potential energy is a smaller term of $\mathcal{O}(Z^{5/3})$ than the major energy components, of order $Z^{7/3}$.

We also note that we expect that the equilibrium bond length will take the form

$$\frac{1}{R_e} = \frac{a_1}{Z^{1/3}} + \frac{a_2}{Z^{2/3}} + \cdots, \quad [27]$$

for neutral homonuclear diatomic molecules, a result also embodying Teller's theorem. The first term, yielding $R_e \alpha Z^{1/3}$, appears to reflect gross trends in empirical data correctly.

A final useful result is obtained if we combine this analysis with a formula from Hartree–Fock theory, $E = \Sigma\epsilon_i - V_{ee}$, where $\Sigma\epsilon_i$ is the sum of occupied orbital energies. Namely, one finds

$$W = \frac{3}{2}\sum\epsilon_i - R(\partial W/\partial R)_{Z,N} - \frac{1}{2}Z^2F_2 - Z^{5/3}F_e + \cdots - \frac{3}{2}N\mu, \quad [28]$$

from which follows

$$W(R = R_e) = \frac{3}{2}(\sum\epsilon_i)^{\text{equil}} + \text{corrections}, \quad [29]$$

with the corrections identically zero in the Thomas–Fermi model but not otherwise. Deviations from the 3/2 ratio between total energy and sum of orbital energies are thereby ascribable to deviations from Thomas–Fermi behavior, as has already been pointed out (1, 7).

Tetrahedral and octahedral molecules

We pause here to deal with some results for tetrahedral and octahedral molecules, for which Mucci and March (1) have pointed point out regularities in the nucleus–nucleus potential energy V_{nn}^{equil}.

The nucleus–nucleus potential energy, for XY_m molecules, with $m = 4$ or 6, can be written in terms of the charge ze on the central X nucleus, the total charge ne on the outer nuclei (e.g., for CCl_4, $z = 6$ and $n = 68$), and the X—Y bond length R as

$$V_{nn} = n(z + cn)e^2/R \quad [30]$$

where $c = 0.22$ for tetrahedral and 0.27 for octahedral molecules. Mucci and March (1) drew attention to the central field model of such molecules where the charge ne was smeared uniformly over the surface of a sphere centered at X, and in particular to the result of March (10, 11) that the equilibrium bond length given by the Thomas–Fermi solution of this model is

$$R_e = z^{-1/3}g(n/z). \quad [31]$$

The existence of binding from the Thomas–Fermi theory is due to the conditions of Teller's theorem not being satisfied by this model, since one cannot dissociate into X and Y atoms in any natural way. Thus, one should be cautious of the prediction of V_{nn}^{equil} from such a model and the results of Mucci and March confirm that the model is not quantitatively applicable to XY_m, molecules when Y is heavy. Smearing the nuclear charges is a very poor way to treat heavy atoms in the outer positions.

However, it is attractive to use the form of the scaling in Eq. **31** for neutral tetrahedral and octahedral molecules and to write

$$R_e = z^a h(n/z), \quad [32]$$

where the exponent a is to be determined. We next note that if we inspect the plot of V_{nn}^{equil} against total number of electrons N(equal to $n + z$ for the neutral molecules under consideration), it can be represented by

$$V_{nn}^{\text{equil}} = AN^{5/3}, \quad A = 3.55 \text{ Hartree}. \quad [33]$$

We comment here that the ln–ln plot of the data of Mucci and March (1) gave a slope of 1.6; in the spirit of the expansions of the present paper we have therefore written the exponent in Eq. **33** as 5/3. Second, no sophisticated fitting procedure has been used and thus Eq. **33** may eventually need refinement. Third, Eq. **33** is a zero-order result in that the scatter about the average curve has been neglected. Evidently by inserting Eq. **33** into Eq. **30** we obtain the empirical form

$$R_e^{\text{empirical}} = \frac{nz\left(1 + \frac{cn}{z}\right)e^2}{Az^{5/3}\left(1 + \frac{n}{z}\right)^{5/3}} \quad [34]$$

or

$$\frac{R_e^{\text{empirical}}}{\left(1 + \frac{cn}{z}\right)} = z^{1/3}H(n/z) \quad [35]$$

where

$$H(x) = \frac{1.49x}{(1 + x)^{5/3}}\,\text{Å}. \quad [36]$$

By comparing Eqs. **32** and **35** we see that the exponent a in Eq. **32** is 1/3, in contrast to the value −1/3 in Eq. **31** for the central field model. The difference is another manifestation of Teller's theorem, V_{nn} being of order $z^{5/3}$ in the limit in which z becomes large with n/z tending to a constant. The central field model, in contrast, in the same limit yields V_{nn}^{equil} of order $z^{7/3}$. The constant A in Eq. **33** is analogous to γF_3 in Eq. **26**; estimates suggest γ is an order of magnitude bigger than β for these molecules.

For more general polyatomic molecules, it is clear that there are again some scaling properties in which the powers $Z^{7/3}$, Z^2, $Z^{5/3}$, etc. in Eq. **19** are replaced by homogeneous functions of the nuclear charges Z_α, Z_β, etc. of order 7/3, 6/3, 5/3, etc.

Numerical validation

While it is not the purpose of the present paper to apply comprehensively the formulas found, we here test the validity of three propositions: first, that Eq. **3** is a viable way to represent total energies of atoms; second, that Eq. **9** is a viable way to represent electron–electron repulsion energies of atoms; and third, that Eq. **10** is a useful way to describe atomic electronegativities.

We first examine the formula obtained from Eq. **3** by setting $N = Z$, the neutral atom energy expansion

$$E(Z,Z) = Z^{7/3}f_1(1) + Z^2 f_2(1) + Z^{5/3}f_3(1) + \cdots. \quad [37]$$

As already remarked, the term $f_1(1)$ is known from the self-consistent Thomas–Fermi theory, and corrections to that theory yield the approximate form, in hartrees (10),

$$E(Z,Z) = -0.7687Z^{7/3} + \frac{1}{2}Z^2 - 0.26Z^{5/3} + \mathcal{O}(Z^{4/3}). \quad [38]$$

Thus, we have already the approximate estimates $f_2(1) = 1/2$, $f_3(1) = -0.26$. More accurately, we may test Eq. **37** by fitting a formula of this form to Hartree–Fock data (12). The best three-term least-squares fit for the atoms from $Z = 2$ to $Z = 54$ is given by the formula

$$E = -0.7741Z^{7/3} + 0.5263Z^2 - 0.3073Z^{5/3}. \quad [39]$$

The general closeness of this to Eq. **38** is satisfactory and encouraging, as is the strikingly small root mean square (r.m.s.) error, 1.33. If we constrain the first coefficient to be exactly −0.7687, we obtain

$$E = -0.7687Z^{7/3} + 0.4904Z^2 - 0.2482Z^{5/3}, \quad [40]$$

also with r.m.s. error of 1.33, again highly satisfactory.

Similarly, we may test Eq. **9** by fitting Hartree–Fock data without assuming the Z^2 term to be missing. The best two-term least-square fit for the neutral atoms $Z = 2$ to $Z = 54$ is the expansion (r.m.s. error, 12.34)

$$V_{ee} = 0.2441Z^{7/3} + 0.0125Z^2. \quad [41]$$

Constraining the first coefficient to have the Thomas-Fermi value, 0.2562, the result is (r.m.s. error, 12.76)

$$V_{ee} = 0.2562Z^{7/3} + 0.0307Z^2, \quad [42]$$

The closeness of the first coefficient in Eq. **41** to the value 0.2562 and the closeness of the second coefficients in both Eqs. **41** and **42** to the value 0 together indicate that Eq. **9** is a valid way to express the quantity V_{ee}.

Testing of Eq. **10** is more difficult, because agreement on actual quantitative values of electronegativities is lacking, and electronegativities exhibit significant changes as one goes through a given row of the periodic table. But all tabulations appear to us to agree with the prediction from Eq. **10**, that the decrease with Z as one goes down a column in the periodic table will be as $Z^{-1/3}$. If, for example, one takes electronegativity values from Sanderson (13) for third, fourth, and fifth row elements, one obtains values of electronegativity times $Z^{1/3}$ as follows:

K	1.12	Ca	3.31	Ge	11.40
Rb	1.20	Sr	3.58	Sn	11.38
Cs	1.06	Ba	2.98		
As	12.51	Se	13.64	Br	14.82
Sb	12.39	Te	13.40	I	14.42.

Constancy within a column is clear. Variation within a row is considerable and is a matter for further study. What it means is that Eq. **10** factors into $Z^{-1/3}$ times an almost periodic function of some constant times $Z^{-1/3}$.

Conclusions

i. The $1/Z$ expansion for atomic ions, for large N, can be formally summed to yield Eq. **3**. The leading term is known from Thomas–Fermi theory. Estimates of $f_2(1)$ and $f_3(1)$ are available.

ii. The separate energy terms V_{ne} and V_{ee} for neutral atoms take the forms of Eqs. **7** and **9**, respectively. There is no term of $\mathcal{O}(Z^2)$ in V_{ee}.

iii. For homonuclear diatomic molecular ions, the scaling of the energy in Eq. **19** is proposed, the leading term being known again from Thomas–Fermi theory.

iv. The theorem of Teller, that there is no molecular binding in the Thomas–Fermi theory, is interpreted as proving that the nucleus–nucleus potential energy term is a smaller term in powers of the number of electrons N than for the other energy terms. Theory and empirical data strongly suggest that the term V_{nn} is $\mathcal{O}(Z^{5/3})$, whereas leading energy terms are $\mathcal{O}(Z^{7/3})$.

v. The theorem of Lieb and Simon, that Thomas–Fermi theory is exact for infinite Z, implies that the chemical potential or electronegativity of neutral atoms goes to zero as $Z^{-1/3}$ for large Z, which leads to the rule, verified from empirical data: Electronegativity decreases as $Z^{-1/3}$ as one goes down a column in the periodic table.

We thank Dr. Libero Bartolotti for performing the calculations leading to Eqs. 39 to 42 of the text. This work has been aided in part by a grant to the University of North Carolina from the National Science Foundation.

1. Mucci, J. F. & March, N. H. (1979) *J. Chem. Phys.* **71**, 1495–1497.
2. Parr, R. G. & Gadre, S. R. (1980) *J. Chem. Phys.* **72**, 3669–3673.
3. Dreizler, R. & March, N. H. (1980) *Z. Phys.* **A294**, 203–205.
4. Teller, E. (1962) *Rev. Mod. Phys.* **34**, 627–631.
5. March, N. H. & White, R. J. (1972) *J. Phys. B* **5**, 466–475.
6. Lieb, E. & Simon, B. (1977) *Adv. Math.* **23**, 22–116.
7. Parr, R. G., Donnelly, R. A., Levy, M. & Palke, W. F. (1978) *J. Chem. Phys.* **68**, 3801–3807.
8. Hund, F. (1932) *Z. Phys.* **77**, 12–25.
9. Townsend, J. R. & Handler, G. S. (1962) *J. Chem. Phys.* **36**, 3325–3329.
10. March, N. H. (1952) *Proc. Camb. Phil. Soc.* **48**, 665–682.
11. March, N. H. (1957) *Advances in Physics* **6**, 1–101.
12. Clementi, E. & Roetti, C. (1974) *At. Data Nucl. Data Tables* **14**, 177–478.
13. Sanderson, R. T. (1971) *Chemical Bonds and Bond Energy* (Academic, New York), p. 57.

J. Phys. B: At. Mol. Opt. Phys. **24** (1991) 4123–4128. Printed in the UK

Inhomogeneous electron gas theory of molecular dissociation energies

N H March

Theoretical Chemistry Department, University of Oxford, 5, South Parks Road, Oxford OX1 3UB, UK

Received 12 June 1991

Abstract. By separating the total energy of atoms and diatomic molecules into the sum of Thomas–Fermi, density gradient and exchange energies, the dissociation energy D, divided by the square of the total number of electrons in the molecule, is related by a simple analytic formula to the inhomogeneity kinetic energy of electron gas theory, for the equilibrium molecule. The shape of the resulting relation has the features of a semi-empirical correlation previously established.

1. Introduction

Following the pioneering studies of Thomas (1926) and Fermi (1928) on the electron density theory of atoms, early attempts were made to use such a treatment to deal with the electronic structure of molecules (e.g. Hund (1932) for N_2; March (1952) for benzene; for a review of this early phase see March (1957)). However, Teller (1962) demonstrated that, in self-consistent local density theories such as Thomas–Fermi plus the extension by Dirac (1930) to include exchange, homonuclear molecules cannot bind. That is to say: the energy of the two-centre problem of a homonuclear diatomic molecule like N_2 is always higher than that of the separated atoms. This seemed to present a severe problem for the theory of the inhomogeneous electron gas in molecules until Mucci and March (1983) proposed to make a merit out of Teller's theorem, by stressing that it implied that density gradients were crucial for molecular binding. This work was made more quantitative by Allan *et al* (1985) for diatomics, and subsequently the same result they obtained was confirmed to also embrace simple polyatomic molecules by Lee and Ghosh (1986). Specifically, it was shown that D/N^2, with D the dissociation energy and N the total number of electrons in the molecule, correlated with the inhomogeneity kinetic energy depending on the volume integral through the space of $(\nabla\rho)^2/\rho$. This was proposed by von Weizsäcker (1935) as a route to correct the Thomas–Fermi electron gas kinetic energy, T_0 say, which depended on the electron density $\rho(\boldsymbol{r})$ to the five thirds power:

$$T_0 = c_k \int \{\rho(\boldsymbol{r})\}^{5/3}\,\mathrm{d}\boldsymbol{r} \qquad c_k = \frac{3h^2}{10m}\left(\frac{3}{8\pi}\right)^{2/3}. \tag{1.1}$$

von Weizsäcker's inhomogeneity correction, T_W say, was explicitly

$$T_W = \frac{\hbar^2}{8m}\int \frac{(\nabla\rho)^2}{\rho}\,\mathrm{d}\boldsymbol{r}. \tag{1.2}$$

0953-4075/91/194123+06\$03.50

Kirzhnits (1957) later demonstrated that in the spirit of the TF method, in which use is made of a weakly inhomogeneous electron gas theory, this von Weizsäcker kinetic energy term T_W should be reduced by a factor of $\frac{1}{9}$, to yield the lowest-order gradient correction T_2 to the TF result (1.1) as

$$T_2 = \frac{\hbar^2}{72m} \int \frac{(\nabla \rho)^2}{\rho} \mathrm{d}\boldsymbol{r}. \tag{1.3}$$

To complete this introduction, let us record also the Dirac (1930) formula for exchange energy analogous to the TF kinetic energy T_0 in equation (1.1). The exchange energy density of a weakly inhomogeneous electron gas is then proportional to the density $\rho(\boldsymbol{r})$ to the four thirds power, and the total exchange energy A is given by

$$A = -c_e \int (\rho(\boldsymbol{r}))^{4/3} \mathrm{d}\boldsymbol{r} \qquad c_e = \frac{3}{4}\left(\frac{3}{\pi}\right)^{1/3} e^2. \tag{1.4}$$

The combination of the TF energy, the density gradient corrections and the exchange energy A is crucial in what follows.

2. Dissociation energy as a balance between TF energy, density gradient terms and exchange

The idea underlying the present work can be briefly stated. It is to write the total energy of the inhomogeneous electron gas in the one-centre problem of atoms and the multi-centre problem of molecules as a sum of three terms:

$$E = E_{\mathrm{TF}} + E_{\text{density gradient}} + E_{\text{exchange}}. \tag{2.1}$$

Let us immediately illustrate the merits of such a physical decomposition by considering the total ground-state energy of a neutral atom having atomic number Z. This may be written in the form

$$\frac{E_1(Z)}{(e^2/a_0)} = -0.7687 Z^{7/3} + \tfrac{1}{2}Z^2 - 0.26 Z^{5/3} + \dots \qquad a_0 = \hbar^2/me^2. \tag{2.2}$$

In this formula, the leading term is the energy of self-consistent TF theory, due to Milne (1927), the second term corrects for the rapid variation of the electron density in the K shell of atoms (see Scott (1952) and Ballinger and March (1955); the treatment of this term by Schwinger and Englert (1985) and co-workers is also relevant here), while the third term is dominated by exchange, but also has corrections from WKB levels replacing the continuous spectrum of the TF atom (March and Plaskett 1956).

Since the formula (2.2) gives a central focus to the method of the present paper, though admittedly restricted as it stands to one-centre (subscript 1 on E) atomic problems, let us next note that March and Parr (1980) considered the three-term formula for such neutral atoms displayed in equation (2.2) explicitly in relation to Hartree–Fock energies. They constrained the first term to the TF value in equation (2.2) but then made a least-squares fit of the terms of $O(Z^2)$ and $O(Z^{5/3})$ with the numerical Hartree–Fock ground-state energies. Their coefficients were 0.4904 and -0.2482 respectively, to be compared with the values in equation (2.2) from the theory of the inhomogeneous electron gas. Obviously the agreement is highly satisfactory. But what is more important for the present work is to note that when one separates the one-centre energy into the sum (2.1), the three terms, as given by electron density theory, have individually simple scaling properties with atomic number Z (here, for neutral atoms, obviously equal to the number of electrons N to be focused on below). This scaling of the three terms in equation (2.1), shown explicitly for neutral atoms in equation (2.2), is in spite of the fact that evidently, over the range of the Periodic Table, the total energy does not have simple scaling. This has motivated the search for a workable extension of the formula (2.2) for the one-centre energy E_1 to, at first, homonuclear diatomic molecules having nuclear charges Z and evidently, for the neutral molecules under discussion, the number of electrons $N=2Z$.

3. Approximate scaling properties for N_2, O_2 and F_2 at equilibrium, from electron gas theory

This then is the point to consider the two-centre energy, E_2 say, and in this section we shall consider specifically the three admittedly light homonuclear diatomic molecules N_2, O_2 and F_2 at the measured equilibrium bond lengths. Then, as Pucci and March (1987) have demonstrated explicitly, approximate scaling properties again hold if the energy is divided up into the three contributions shown in equation (2.1).

Because, from the work of Mucci and March (1983), $T_2=\frac{1}{9}T_W$ is crucial for molecular binding, we have collected in table 1 the values of T_2/Z^2 for these three molecules following Pucci and March (1987). What is important is to note the accurate scaling with Z^2. Similarly, we note (Pucci and March 1987) that the TF energy scales as $Z^{7/3}$ and the exchange energy as $Z^{5/3}$.

Next let us define the dissociation energy D in the usual way as

$$-D=E_2-2E_1. \tag{3.1}$$

But we have seen that both E_1 and E_2 can be separated into three terms with distinctive scaling properties and hence we can write

$$D=\gamma_1 Z^{7/3}+\gamma_2 Z^2+\gamma_3 Z^{5/3}+\dots \tag{3.2}$$

Recognizing that it will be difficult, because of the large quantities being subtracted in equation (3.1) to yield the usually small quantity D, to calculate precisely the coefficients γ_i in equation (3.2), let us next recall the explicit correlation exhibited, following Mucci and March (1983), by Allan *et al* (1985) and confirmed by Lee and Ghosh (1986):

D/N^2 correlates with T_2 or $\frac{1}{9}T_W$.

Since $N=2Z$ as above, let us rewrite equation (3.2) in the form

$$\frac{D}{Z^2}=\frac{D}{4N^2}=\gamma_1 Z^{1/3}+\gamma_2+\gamma_3 Z^{-1/3}+\dots \tag{3.3}$$

Table 1. Inhomogeneity kinetic energy T_2 defined by equation (1.3) for diatomic molecules N_2, O_2 and F_2 at their equilibrium separation.

Molecule	T_2/N^2 in atomic units[a]
N_2	0.049
O_2	0.050
F_2	0.050

[a] Values taken from Pucci and March (1987).

While equation (3.3) represents the 'shape' of the proposed inhomogeneous electron gas theory for the dissociation energies of homonuclear diatomic molecules built from atoms having atomic number Z, we next note that Lee and Ghosh showed that the correlation of D/N^2 with T_2 (or equivalently T_W which they used) extended to embrace also polyatomic molecules. Hence the next important step in the present work is to recast equation (3.3) in terms of T_2. To do so, we need only return to table 1 and note with Pucci and March that, for N_2, O_2 and F_2 under discussion, the inhomogeneity kinetic energy T_2 in the equilibrium molecules is such that T_2/Z^2 is accurately constant. Thus, we can now rewrite equation (3.3) in terms of $T_2^{1/6}$ instead of $Z^{1/3}$, to obtain

$$\frac{D}{N^2} = -\Gamma_1 T_2^{1/6} + \Gamma_2 - \Gamma_3 T_2^{-1/6} + \ldots . \tag{3.4}$$

What is now remarkable about the formula (3.4) is that, with the signs written deliberately as above, one can see that the density gradient term reflected by Γ_2 is just an additive constant to the Thomas–Fermi–Dirac (TFD) 'dissociation energy', which Teller's theorem asserts is negative. Evidently, for molecular binding, Γ_2 must be sufficiently positive to bring D/N^2 above zero. We turn immediately to relate the proposed formula (3.4) to the correlation exhibited by Allan *et al* (1985), and subsequently extended by Lee and Ghosh (1986) to embrace polyatomic molecules.

4. Connection of dissociation energy formula (3.4) with semi-empirical correlation

In figure 1, we have plotted essentially the quantity $(D/N^2 - \Gamma_2)$ against the inhomogeneity kinetic energy T_2, for two different values of the ratio Γ_3/Γ_1, though

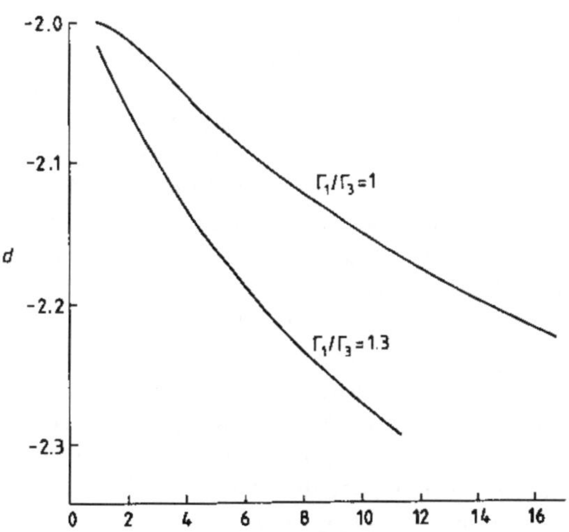

Figure 1. This figure shows the shape of the predicted dissociation energy D according to equation (3.4), plotted against the inhomogeneity kinetic energy T_2 of equation (1.3) found from the equilibrium molecular electron density $\rho(r)$. T_2 is in atomic units. The precise definition of the quantity d in the ordinate is $d = (\Gamma_1\Gamma_3)^{-1/2}(D/N^2 - \Gamma_2)$. The two curves correspond to different values of the ratio Γ_1/Γ_3 in equation (3.4). The value of d at the maximum, from equation (4.2), is -2, independent of Γ_1/Γ_3.

taken of order unity for reasons given below. This plot has been made over the range of T_2 covered in figure 1 of Allan *et al* (1985). If, however, we look at the form of equation (3.4) mathematically over a wider range of T_2 than plotted, it is obvious that the TF term in $T_2^{1/6}$ tends to zero as T_2 becomes small; this term comes into its own as T_2 becomes large. In contrast, the exchange term in equation (3.4), involving the coefficient Γ_3, becomes large as T_2 becomes small. With both Γ_1 and Γ_3 positive, to ensure that Teller's theorem is satisfied, it is clear that D/N^2 as a function of T_2 will have a maximum. Putting $(\partial/\partial T_2)(D/N^2)=0$, equation (3.4) yields at this turning point:

$$T_2^{1/3}=\Gamma_3/\Gamma_1. \tag{4.1}$$

From figure 1 of Lee and Ghosh (1986), such a maximum must occur for $T_W \leqslant 20$. At the turning point given by equation (4.1), one finds by inserting this equation into equation (3.4) that

$$[(D/N^2)_{\max}-\Gamma_2]=-2(\Gamma_1\Gamma_3)^{1/2}. \tag{4.2}$$

Evidently, for binding at this favourable point, one must have

$$\Gamma_2>2(\Gamma_1\Gamma_3)^{1/2}. \tag{4.3}$$

The general shape of the plots in figure 1 is encouraging in the light of the semi-empirical correlations exposed by Allan *et al* and by Lee and Ghosh.

5. Discussion and summary

For the future, it is evident that a complete, first principles theory of dissociation energies, based on electron density theory, will have to:

(i) predict the coefficients $\Gamma_1-\Gamma_3$ in equation (3.4);

(ii) assess the importance of the higher-order terms not displayed explicitly there.

We will return to (i) briefly below. As to (ii), the work of Clementi (1963), on correlation energies in atoms, might suggest that this enters at $O(Z)$ in the formula (2.2) for atomic binding energies. If the same proved to be true in the two-centre result (3.2), then equation (3.3) would evidently have a higher-order term of $O(Z^{-1})$. In turn this would introduce a contribution of $O(T_2^{-1/2})$ in equation (3.4). When D/N^2 is near to zero, then it seems clear that such a correlation term could well be significant†; near the (predicted) maximum in D/N^2 against T_2 we would not anticipate that correlation will play an essential role however.

Finally we return to point (i) above. The work of Townsend and Handler (1962) allows an estimate of Γ_1 from their TF self-consistent field calculations on N_2. Sheldon (1955) included exchange but not density gradients; his work enables Γ_3 to be estimated. These estimates are compatible with our findings in the present work, but clearly a full theory of the coefficients is for the future. This is especially so because Γ_2 is difficult to estimate presently. It is reassuring therefore that it represents a constant shift of the D/N^2 against T_2 curve, and does not affect its shape.

† Because of the comparison with Hartree-Fock energies of atoms in section 2, and particularly the comment of one of the referees, it should be pointed out in relation to the molecular dissociation problem that electron correlation is often necessary to correct the dissociation limit of Hartree-Fock theory. However, the present treatment is more directly analogous to Slater's (1951) simplification of the Hartree-Fock method, because of the use of the Dirac exchange energy (1.4); with of course the corresponding value derived from that formula of the Slater parameter α.

References

Allan N L, West C G, Cooper D L, Grout P J and March N H 1985 *J. Chem. Phys.* **83** 4562
Ballinger R A and March N H 1955 *Phil. Mag.* **46** 246
Clementi E 1963 *J. Chem. Phys.* **38** 2248
Dirac P A M 1930 *Proc. Camb. Phil. Soc.* **26** 376
Fermi E 1928 *Z. Phys.* **48** 73
Hund F 1932 *Z. Phys.* **77** 12
Kirzhnits D A 1957 *Sov. Phys.-JETP* **5** 64
Lee C and Ghosh S K 1986 *Phys. Rev.* A **33** 3506
March N H 1952 *Acta Crystallogr.* **5** 187
—— 1957 *Adv. Phys.* **6** 1
March N H and Parr R G 1980 *Proc. Natl. Acad. Sci., USA* **77** 6285
March N H and Plaskett J S 1956 *Proc. R. Soc.* A **235** 419
Milne E A 1927 *Proc. Camb. Phil. Soc.* **23** 794
Mucci J F and March N H 1983 *J. Chem. Phys.* **78** 6187
Pucci R and March N H 1987 *Phys. Rev.* A **35** 4428
Scott J M C 1952 *Phil. Mag.* **43** 859
Schwinger J and Englert B-G 1985 *Phys. Rev.* A **32** 26, 36, 47
Sheldon J W 1955 *Phys. Rev.* **99** 1291
Slater J C 1951 *Phys. Rev.* **81** 385
Teller E 1962 *Rev. Mod. Phys.* **34** 627
Thomas L H 1926 *Proc. Camb. Phil. Soc.* **23** 542
Townsend J R and Handler G S 1962 *J. Chem. Phys.* **36** 3325
von Weizsäcker C F 1935 *Z. Phys.* **96** 431

On the adiabatic connection method, and scaling of electron–electron interactions in the Thomas–Fermi limit

Mel Levy
Department of Chemistry, Tulane University, New Orleans, Louisiana 70118

Norman H. March
Inorganic Chemistry Department, University of Oxford, OX1 3QZ, United Kingdom

Nicholas C. Handy
Department of Chemistry, University of Cambridge, CB2 1EW, United Kingdom

(Received 6 September 1995; accepted 11 October 1995)

In this paper we examine three aspects of electron–electron scaling: (i) the electron–electron repulsions are only scaled in Thomas–Fermi theory; (ii) the electron–electron repulsions are scaled, and the one electron potential is adjusted to give a prescribed density, in Thomas–Fermi–Dirac theory; and (iii) new approaches to the adiabatic connection formulas are presented to help improve the exchange–correlation functional. A new generalized two-point expression is presented. Models (i) and (ii) are solved exactly. © *1996 American Institute of Physics.* [S0021-9606(96)02203-0]

I. INTRODUCTION

In modern density functional theory, the idea of an adiabatic connection to determine the exchange–correlation functional is taking on new importance. The early references include those of Harris and Jones,[1] Langreth and Perdew,[2] Gunnarsson and Lundqvist,[3] Levy,[4] and Harris.[5] Following Refs. 2 and 3 and the review in Parr and Yang,[6] the exchange–correlation functional $E_{xc}[\rho]$ may be written

$$E_{xc}[\rho]=\int_0^1 d\lambda\, U_{xc}^{\lambda}[\rho], \tag{1}$$

where

$$U_{xc}^{\lambda}[\rho]=\langle\Psi_\rho^\lambda|\hat{V}_{ee}|\Psi_\rho^\lambda\rangle-J[\rho]. \tag{2}$$

and $J[\rho]$ is the Coulomb energy. $\hat{V}_{ee}$ is the electron–electron repulsion operator, and Ψ_ρ^λ is the ground state wave function of a Hamiltonian $\hat{H}_\lambda$ which represents a system in which the electron–electron interaction is scaled

$$\hat{H}_\lambda=\hat{T}+\lambda\hat{V}_{ee}+\sum_i v_\lambda(\mathbf{r}_i). \tag{3}$$

It also minimizes, for fixed ρ, the constrained search expression[7]

$$\langle\Psi_\rho^\lambda|\hat{T}+\lambda\hat{V}_{ee}|\Psi_\rho^\lambda\rangle. \tag{4}$$

In the above $\rho(\mathbf{r})$ is to be interpreted as the exact ground state density of $\hat{H}_\lambda$. The above expression (4) was anticipated by an earlier expression for the noninteracting kinetic energy functional.[8]

In particular Becke[9,10] has used the above ideas in the context of a two point integration formula. He approximates[9] Eq. (1) as

$$E_{xc}[\rho]\sim\tfrac{1}{2}(U_{xc}^0+U_{xc}^1), \tag{5}$$

where U_{xc}^0 is the exchange energy of the Kohn–Sham determinant, and U_{xc}^1 is the potential energy contribution to the exchange–correlation energy of the fully interacting system. This led Becke to his half-and-half DFT functional, which was the forerunner of his fruitful three-parameter functional.[11] However, as we shall discuss in Sec. III Becke's three-parameter functional, unlike the half-and-half functional, really results from a partial abandonment of the adiabatic connection idea. [Actually, a very simple proof of Eq. (1), as well as a study of the implications of a linear approximation within the adiabatic connection, was given earlier.][4]

It is therefore this renaissance of interest in adiabatic scaling which has led us to investigate whether we can add anything of a precise nature to the discussion. In Sec. II we present two precise results, the first in the Thomas–Fermi limit for which the electron–electron interaction is scaled, and the second for which adiabatic scaling is examined in the Thomas–Fermi-Dirac limit. In Sec. III our analysis commences with an exact expression for U_{xc}^1 in terms of E_{xc}, and we shall provide a formula which enables one to utilize *nonlocal* exchange–correlation *potential energies*, such as those from gradient approximations, for U_{xc}^1 in Eq. (5).

II. ELECTRON–ELECTRON INTERACTION SCALING

In this section we shall examine two aspects of scaling the electron–electron interactions with a suitable atomic Hamiltonian. To achieve quite explicit results, we shall focus on the Thomas–Fermi (TF) limit. In Sec. II A we shall determine an exact result when the electron–electron repulsion term is scaled alone. In the Sec. II B we shall also scale the same term, but allow a variation in the one electron potential so that a prescribed density is obtained which is independent of the scaling.

A. Electron–electron interaction scaling alone

First we shall consider the Hamiltonian H_λ^Z, defined by

$$H_\lambda^Z=-\frac{\hbar^2}{2m}\sum_i^N\nabla_i^2-Ze^2\sum_i^N\frac{1}{r_i}+\lambda e^2\sum_{i<j}\frac{1}{r_{ij}}. \tag{6}$$

We shall consider $N \leqslant Z$, and both N and Z will be taken to be large.[12] We divide both sides of Eq. (6) by λ, to obtain a new Hamiltonian $H_\lambda^{Z'} = H_\lambda^Z/\lambda$, with an effective mass $m^* = m\lambda$ and an effective nuclear charge $Z^*e = Ze/\lambda$:

$$H_\lambda^{Z'} = -\frac{\hbar^2}{2m^*}\sum_i^N \nabla_i^2 - Z^*e^2\sum_i^N \frac{1}{r_i} + e^2\sum_{i<j}\frac{1}{r_{ij}}. \tag{7}$$

We immediately apply the TF approximation to this Hamiltonian, and focus on the ground-state energy $E'_\lambda(Z,N)$. Following Grout, Tal, and March,[13] this energy can be expressed as

$$E'_\lambda(Z,N) = Z^{*7/3} f\left(\frac{N}{Z^*}\right)\frac{e^2}{a_0^*}, \tag{8}$$

where $a_0^* = \hbar^2/m^*e^2$. The real energy $E_\lambda(Z,N)$ is $E'_\lambda(Z,N)\lambda$, and therefore

$$E_\lambda(Z,N) = \lambda^2 Z^{*7/3} f\left(\frac{N}{Z^*}\right)\frac{e^2}{a_0}. \tag{9}$$

On further reorganization the final result is

$$E_\lambda(Z,N) = \frac{Z^{7/3}}{\lambda^{1/3}} f\left(\lambda\frac{N}{Z}\right)\frac{e^2}{a_0}. \tag{10}$$

$f(N/Z)$ is known from the work of Grout *et al.* for the range $0 < N/Z \leqslant 1$, and so $f(\lambda N/Z)$ is also known for $\lambda \leqslant 1$. It is given in analytical form near $N/Z=1$ and near $N/Z=0$, and numerical form elsewhere. In particular one may write $E_\lambda(Z,N) = F(q^*)N^{1/3}Z^2$, where $q^* = 1 - N/Z^*$. From Table I of Ref. 13, one has $F(0) = -0.768$, $F(0.5) = -0.929$, and $F(1.0) = -1.145$.

Following March and White,[14] and later rigorous work by Tal and Levy,[15] we note that contact can be made between such a formula and the $1/Z$ expansion of atomic theory. Further, given a number of electrons N, it would be interesting to know, using an electronic structure *ab initio* program, how far the TF scaling exhibited above carries through to the full many-electron case for fixed (and usually small) N and variable Z.

B. Electron–electron scaling with constant density

For this discussion we introduce the Hamiltonian $\hat{H}_\lambda$,

$$\hat{H}_\lambda = \hat{T} + \lambda\hat{V}_{\mathrm{ee}} + \sum_i^N V_\lambda([\rho];\mathbf{r}_i). \tag{11}$$

Here the ground state density ρ is fixed independent of λ, and in atomic units

$$\hat{T} = -\sum_i^N \frac{1}{2}\nabla_i^2, \quad \hat{V}_{\mathrm{ee}} = \sum_{i<j}\frac{1}{r_{ij}}. \tag{12}$$

We shall now introduce the Thomas–Fermi–Dirac (TFD) ground state (GS) energy corresponding to different λ and a definite density $\rho_0(\mathbf{r})$ as the minimum obtained from

$$E_{\mathrm{GS},\lambda,\rho_0}^{\mathrm{TFD}} = \min_\rho\left[\int V_\lambda([\rho_0];\mathbf{r})\rho(\mathbf{r})d\mathbf{r} + F_\lambda^{\mathrm{TFD}}[\rho]\right] \tag{13}$$

$$= \int V_\lambda([\rho_0];\mathbf{r})\rho_0(\mathbf{r})d\mathbf{r} + F_\lambda^{\mathrm{TFD}}[\rho_0], \tag{14}$$

where

$$F_\lambda^{\mathrm{TFD}}[\rho] = c_k\int \rho^{5/3}(\mathbf{r})d\mathbf{r} + \lambda J[\rho] - \lambda c_X\int \rho^{4/3}(\mathbf{r})d\mathbf{r} \tag{15}$$

with the standard values $c_k = (3/10)(3/8\pi)^{2/3}$; $c_X = (3/4)(3/\pi)^{1/3}$ and $J[\rho]$ being the Coulomb energy

$$J[\rho] = \frac{1}{2}\int\int \frac{\rho(\mathbf{r}_1)\rho(\mathbf{r}_2)}{r_{12}}d\mathbf{r}_1 d\mathbf{r}_2. \tag{16}$$

The stationary principle gives

$$V_\lambda([\rho_0];\mathbf{r}) = -\frac{\delta F_\lambda^{\mathrm{TFD}}[\rho]}{\delta\rho(\mathbf{r})} \tag{17}$$

evaluated at $\rho = \rho_0$. It thus follows that

$$V_\lambda([\rho_0];\mathbf{r}) = -\frac{5}{3}c_k\{\rho_0(\mathbf{r})\}^{2/3} - \lambda\int\frac{\rho_0(\mathbf{r}')}{|\mathbf{r}-\mathbf{r}'|}d\mathbf{r}' + \frac{4}{3}\lambda c_X\{\rho_0(\mathbf{r})\}^{1/3}. \tag{18}$$

We immediately observe that in the TFD limit the scaling of the one-body potential V_λ is such that V_λ is linear in λ. We can write the ground state energy explicitly as

$$E_{\mathrm{GS},\lambda,\rho_0}^{\mathrm{TFD}} = -\frac{2}{3}c_k\int\rho^{5/3}(\mathbf{r})d\mathbf{r} - \frac{\lambda}{2}\int\frac{\rho_0(\mathbf{r}_1)\rho_0(\mathbf{r}_2)}{r_{12}} \times d\mathbf{r}_1 d\mathbf{r}_2 + \frac{1}{3}\lambda c_X\int\rho_0^{4/3}(\mathbf{r})d\mathbf{r}. \tag{19}$$

Evidently again, as with V_λ, $E_{\mathrm{GS},\lambda,\rho_0}^{\mathrm{TFD}}$, the ground state energy corresponding to electron density $\rho_0(\mathbf{r})$ at a coupling constant λ, is linear in λ.

It is possible to write two alternative forms of this result, the first relating the energy to its value $E_{\mathrm{GS},\lambda=0,\rho_0}^{\mathrm{TFD}}$ when the electron–electron interactions are "switched off."

$$E_{\mathrm{GS},\lambda,\rho_0}^{\mathrm{TFD}} = E_{\mathrm{GS},\lambda=0,\rho_0}^{\mathrm{TFD}} - \frac{\lambda}{2}\int\frac{\rho_0(\mathbf{r}_1)\rho_0(\mathbf{r}_2)}{r_{12}}d\mathbf{r}_1 d\mathbf{r}_2 + \frac{1}{3}\lambda c_X\int\rho^{4/3}(\mathbf{r})d\mathbf{r} \tag{20}$$

and the second works from the full coupling strength $\lambda=1$, in which case the result is

$$E_{\mathrm{GS},\lambda,\rho_0}^{\mathrm{TFD}} = E_{\mathrm{GS},\lambda=1,\rho_0}^{\mathrm{TFD}} + \frac{(1-\lambda)}{2}\int\frac{\rho_0(\mathbf{r}_1)\rho_0(\mathbf{r}_2)}{r_{12}}d\mathbf{r}_1 d\mathbf{r}_2 + \frac{1}{3}(\lambda-1)c_X\int\rho^{4/3}(\mathbf{r})d\mathbf{r}. \tag{21}$$

This completes our analysis of the Thomas–Fermi–Dirac problem.

III. A NEW LOOK AT THE TWO-POINT ADIABATIC CONNECTION FORMULA

A. New generalized linear formulas

In order to utilize U^1_{xc} in Eq. (5), one must have an expression for U^1_{xc}. With this in mind we note here that[16,17]

$$U^1_{xc}[\rho]=2E_{xc}[\rho]-\left.\frac{\partial E_{xc}[\rho_\lambda]}{\partial\lambda}\right|_{\lambda=1}, \tag{22}$$

where

$$\rho_\lambda(\mathbf{r})=\lambda^3\rho(\lambda\mathbf{r}). \tag{23}$$

Equation (22) is a special case of

$$U^\lambda_{xc}=2\lambda E_{xc}[\rho_{1/\lambda}]+\lambda^2\frac{\partial E_{xc}[\rho_{1/\lambda}]}{\partial\lambda}. \tag{24}$$

Now, in Eq. (5) one could use the exact expression for U^0_{xc} and one could approximate U^1_{xc} by some model, through employment of the analogue of expression (22). The resultant approximation to E_{xc} is denoted as $\tilde{E}_{xc}$, and is given by

$$\tilde{E}_{xc}=\tfrac{1}{2}U^0_{xc}[\rho]+\tfrac{1}{2}U^{\mathrm{mod},1}_{xc}[\rho], \tag{25}$$

where "mod" refers to some approximate model for U^1_{xc}, and where

$$U^{\mathrm{mod},1}_{xc}[\rho]=2E^{\mathrm{mod}}_{xc}[\rho]-\left.\frac{\partial E^{\mathrm{mod}}_{xc}[\rho_\lambda]}{\partial\lambda}\right|_{\lambda=1}. \tag{26}$$

The combination of Eqs. (25) and (26) yields

$$\tilde{E}_{xc}[\rho]=\frac{1}{2}U^0_{xc}[\rho]+E^{\mathrm{mod}}_{xc}[\rho]-\frac{1}{2}\left.\frac{\partial E^{\mathrm{mod}}_{xc}[\rho_\lambda]}{\partial\lambda}\right|_{\lambda=1} \tag{27}$$

or

$$\tilde{E}_{xc}[\rho]=\frac{1}{2}E_x[\rho]+\frac{1}{2}E^{\mathrm{mod}}_x[\rho]+E^{\mathrm{mod}}_c[\rho]-\frac{1}{2}\left.\frac{\partial E^{\mathrm{mod}}_c[\rho_\lambda]}{\partial\lambda}\right|_{\lambda=1}, \tag{28}$$

where we have imposed the requirement that

$$E^{\mathrm{mod}}_x[\rho_\lambda]=\lambda E^{\mathrm{mod}}_x[\rho] \tag{29}$$

which is the correct scaling requirement for exchange, and we have made use of the fact that $U^0_{xc}[\rho]$ equals the exact exchange functional, $E_x[\rho]$. Finally, one may, of course, decide to choose $E_x[\rho]$ itself for $E^{\mathrm{mod}}_x[\rho]$ on the right-hand of expression (28). In any case, formula (27) and (28) is our desired generalized linear (two-point) formula.

In Becke's first[9] paper concerning the two-point formula, the local density approximation (LDA) was employed for $U^{\mathrm{mod},1}_{xc}$. However, a two-point formula has never been written down with a local $U^{\mathrm{mod},1}_{xc}$ that is not the LDA, and a two-point formula has certainly never been written down with a *nonlocal* $U^{\mathrm{mod},1}_{xc}$, because it is far from widely known that an expression exists. With this in mind, we here emphasize that the appropriate expression does indeed exist, and it is our Eq. (27) with a nonlocal E^{mod}_{xc} on the right-hand side. Consequently we recommend that Eq. (27) be studied actively. (Becke's three parameter formula, in his subsequent paper,[11] contains nonlocal exchange and correlation approximations, *but does not* contain a corresponding $U^{\mathrm{mod},1}_{xc}$ (there is no $\partial E^{\mathrm{mod}}_{xc}[\rho]/\partial\lambda$), perhaps because our Eq. (27) was not yet known. As a result Becke abandons the adiabatic connection in his development of his semiempirical three parameter formula; this three parameter formula has nevertheless proven to be quite fruitful.

B. On the extent of the validity of the two-point formula (linear approximation)

Let us now look at the extent of the validity of the linear approximation when the *exact* U^λ_{xc} is employed in Eq. (5). First of all, it is known that $U^\lambda_{xc}[\rho]$ monotonically decreases[16] as λ increases, and it has already been shown that[17]

$$U^\lambda_{xc}[\rho]=\lambda U^1_{xc}[\rho_{1/\lambda}] \tag{30}$$

and that[17,18]

$$U^1_{xc}[\rho_{1/\lambda}]\sim\text{constant}+\lambda^{-1}U^0_{xc}[\rho] \tag{31}$$

as $\lambda\to 0$. Thus, we infer here, from Eqs. (30) and (31) that $U^\lambda_{xc}[\rho]$ is linear at small enough λ.

Next, in Eq. (22), we use

$$E_{xc}[\rho]=E_x[\rho]+E_c[\rho] \tag{32}$$

and recognize that

$$E_x[\rho]=U^0_{xc}[\rho] \tag{33}$$

and that

$$U^1_{xc}[\rho]=E_x[\rho]+E_c[\rho]-T_c[\rho]. \tag{34}$$

We consequently deduce that Eq. (5) implies

$$E_c[\rho]\simeq\tfrac{1}{2}(E_c[\rho]-T_c[\rho]) \tag{35}$$

which would be exact if

$$E_c[\rho]=-T_c[\rho]. \tag{36}$$

In other words, Eq. (5) would be exact if the density–functional definition of the correlation energy, $E_c[\rho]$, and the density–functional definition of the kinetic contribution to the correlation energy, $T_c[\rho]$, were to satisfy the same virial relation, for atoms, as are satisfied by their more familiar standard quantum chemistry counterparts. Here

$$E_c[\rho]=\langle\Psi^1_\rho|\hat{T}+\hat{V}_{ee}|\Psi^1_\rho\rangle-\langle\Psi^0_\rho|\hat{T}+\hat{V}_{ee}|\Psi^0_\rho\rangle \tag{37}$$

and

$$T_c[\rho]=\langle\Psi^1_\rho|\hat{T}|\Psi^1_\rho\rangle-\langle\Psi^0_\rho|\hat{T}|\Psi^0_\rho\rangle. \tag{38}$$

From the above definitions of $E_c[\rho]$ and $T_c[\rho]$, it follows that Eq. (5) is a reasonable approximation because Eq. (36) should generally be closely satisfied.

It turns out that the exact relation connecting $E_c[\rho]$ and $T_c[\rho]$ is[16]

$$E_c[\rho]=-T_c[\rho]+\left.\frac{\partial E_c[\rho_\lambda]}{\partial\lambda}\right|_{\lambda=1}. \tag{39}$$

Hence, it follows that another necessary and sufficient condition for the two-point formula to be exact, when the exact U_{xc}^{λ} is employed in Eq. (5), is for

$$\left.\frac{\partial E_c[\rho_\lambda]}{\partial\lambda}\right|_{\lambda=1}=0. \tag{40}$$

In support of the two-point formula, the magnitude of $\partial E_c[\rho_\lambda]/\partial\lambda|_{\lambda=1}$ has been found to be typically only about 10% of the magnitude of $E_c[\rho]$ in atoms.

We are now in a position to provide insights into why Eq. (27) improves upon the LDA when the LDA is "mod." In other words, with

$$\tilde{E}_{xc}^{\text{LDA}}[\rho]=\frac{1}{2}E_x[\rho]+\frac{1}{2}E_x^{\text{LDA}}[\rho]+E_c^{\text{LDA}}[\rho]-\frac{1}{2}\left.\frac{\partial E_c^{\text{LDA}}[\rho_\lambda]}{\partial\lambda}\right|_{\lambda=1} \tag{41}$$

we are now able to help explain why $\tilde{E}_{xc}^{\text{LDA}}[\rho]$ is generally superior to

$$E_{xc}^{\text{LDA}}[\rho]=E_x^{\text{LDA}}[\rho]+E_c^{\text{LDA}}[\rho]. \tag{42}$$

[Observe that the right-hand side of Eq. (41) is equivalent to Becke's original two-point formula.] First, since it is generally the case that $E_x^{\text{LDA}}[\rho]>E_x[\rho]$, the replacement of $\frac{1}{2}E_x^{\text{LDA}}[\rho]$, in Eq. (41), by $\frac{1}{2}E_x[\rho]$ helps. Moreover, the fact that $\partial E_c^{\text{LDA}}[\rho_\lambda]/\partial\lambda|_{\lambda=1}$, in Eq. (41), is too negative greatly offsets the fact $E_c^{\text{LDA}}[\rho]$ is too negative in Eq. (41). This may be explained in terms of the depth and the range of the exchange–correlation hole.[19,20]

IV. DISCUSSION

The continuing interest in the adiabatic connection of density functional theory makes it of interest to contemplate the way in which both the total ground state energy, and also the one-body potential, scale with the strength of the electron–electron interactions. Here explicit results are presented in the Thomas–Fermi and Thomas–Fermi–Dirac limits, which can be expected to be achieved in the atomic context considered here for large Z and N with $0<(N/Z)\leqslant 1$: i.e., for positive atomic ions. While the major part of the present note is concerned with such scaling of e^2/r_{ij} by λ, but keeping the electron density constant as λ is varied, we have also considered variations of λ at constant external potential. In the former case, it has been demonstrated that both one-body potential V_λ and the ground state energy are linear in λ, as exemplified by Eqs. (18) and (19), respectively. This linearity in λ for the total energy allows alternative forms (20) and (21) to be explicitly constructed. In Sec. III we have looked again at the two-point adiabatic connection formula, and commenced our discussion with an exact representation of the exchange–correlation potential energy for the fully interacting system. By introducing a model, we have given a new "two-point" expression for the exchange–correlation energy Eq. (28). In the future we feel that it would be worthwhile to investigate various models for E_x^{mod} and E_c^{mod} in expression (28). For instance, one might use E_x for E_x^{mod}, in Eq. (28), so that the *whole* exact exchange is utilized, or one might employ[21,22] CAM-LYP, PW-91,[23] or other nonlocal mixtures of E_x^{mod} and E_c^{mod}. There are countless possibilities. What is important is the fact that our new two-point formula should generally *improve upon* a given approximation for an exchange–correlation functional when $U_{xc}^{\text{mod},1}$ is superior to the corresponding E_{xc}^{mod}.

Finally, by knowing about relation (24), one is now able to extend the adiabatic connection approach to functionals that involve three or more points.

Note added in proof. A very interesting article has just appeared by Proynov, Ruiz, Vela, and Salahub[24] that contains a pertinent analysis of the two-point adiabatic connection formula. Our analysis overlaps with theirs to a certain extent.

ACKNOWLEDGMENTS

Two of us (M.L. and N.H.M.) wish to acknowledge that their collaboration in the present study was made possible by their participation in the International Workshop on Many-Body Theory which took place in Caracas, Venezuela in 1995. They wish to thank Professor E. V. Ludena and his colleagues for the stimulating atmosphere created at this workshop. M. L. also wishes to acknowledge partial support from the National Institute of Standards and Technology and N.H.M. also wishes to acknowledge partial support from the Office of Naval Research and in particular much stimulation and encouragement from Dr. P. Schmidt.

[1] J. Harris and R. O. Jones, J. Phys. F **4**, 1170 (1974).
[2] D. C. Langreth and J. P. Perdew, Phys. Rev. B **21**, 5469 (1980).
[3] O. Gunnarsson and B. I. Lundqvist, Phys. Rev. B **13**, 4274 (1976).
[4] M. Levy in *Density Functional Theory*, edited by J. Keller and J. L. Gasquez (Springer, New York, 1983).
[5] J. Harris, Phys. Rev. A **29**, 1648 (1984).
[6] R. G. Parr and W. Yang, *Density Functional Theory of Atoms and Molecules* (Oxford University Press, Oxford, 1989).
[7] M. Levy, Proc. Natl. Acad. Sci. USA **76**, 6062 (1979); Phys. Rev. A **26**, 1200 (1982).
[8] J. K. Percus, Int. J. Quantum Chem. **13**, 89 (1978).
[9] A. D. Becke, J. Chem. Phys. **98**, 1372 (1993).
[10] A. D. Becke, J. Chem. Phys. **88**, 1053 (1988).
[11] A. D. Becke, J. Chem. Phys. **98**, 5648 (1993).
[12] See, for example, N. H. March, *Electron Density Theory of Atoms and Molecules* (Academic, New York, 1992).
[13] P. J. Grout, Y. Tal, and N. H. March, J. Chem. Phys. **79**, 331 (1983).
[14] N. H. March and R. J. White, J. Phys. B **5**, 466 (1972).
[15] Y. Tal and M. Levy, Phys. Rev. A **23**, 408 (1981).
[16] M. Levy and J. P. Perdew, Phys. Rev. A **32**, 2010 (1985).
[17] M. Levy, Phys. Rev. A **43**, 4637 (1991).
[18] A. Gorling and M. Levy, Phys. Rev. A **45**, 1509 (1992).
[19] K. Burke, M. Erhzerhot, and J. P. Perdew (unpublished).
[20] K. Burke, J. P. Perdew, and M. Levy, Phys. Rev. A (submitted).
[21] G. J. Laming, V. Termath, and N. C. Handy, J. Chem. Phys. **99**, 8765 (1993).
[22] C. Lee, W. Yang, and R. G. Parr, Phys. Rev. B **37**, 785 (1988).
[23] J. P. Perdew, in *Electronic Structure of Solids '91*, edited by P. Ziesche and H. Eschrig (Academie, Berlin, 1991).
[24] E. I. Proynov, E. Ruiz, A. Vela, and D. R. Salahub, Int. J. Quantum Chem. **S29**, 61 (1995).

PHYSICAL REVIEW A **73**, 063205 (2006)

Two-dimensional electrostatic analog of the March model of C_{60} with a semiquantitative application to planar ring clusters

Claudio Amovilli*
Dipartimento di Chimica e Chimica Industriale, Università di Pisa, Via Risorgimento 35, I-56126 Pisa, Italy

Norman H. March†
Department of Physics, University of Antwerp, Groenenborgerlaan 171, B-2020 Antwerpen, Belgium
(Received 20 March 2006; published 22 June 2006)

The so-called March model of fullerene uses the one-center approximation of quantum chemistry to calculate the spherical π-electron density ρ by the Thomas-Fermi (TF) statistical method. Then the variational principle of the TF theory is employed by combining the electronic energy determined entirely by ρ with the discrete C_{60} nuclear potential energy to predict, semiquantitatively, the equilibrium radius. Here the same approach is worked out, but now largely analytically, in two dimensions, as a possible model of planar ring clusters. Again the equilibrium radii are calculated and consistency with realistic clusters of hydrogen and carbon rings is discussed in detail. Finally, an estimate of the "breathing" frequency of C_{14}, C_{18}, and C_{22} is given.

DOI: 10.1103/PhysRevA.73.063205 PACS number(s): 36.40.−c, 31.15.Bs

I. INTRODUCTION

The so-called March model [1,2] of fullerene approximates the 60 π electrons peaked around the surface of the buckyball by a spherical distribution of electron density ρ. This then leads, by a variational calculation based on the Thomas-Fermi (TF) energy functional $E_{TF}[\rho]$, but retaining the nuclear-nuclear potential energy for the correct discrete C nuclei positions [1], to an equilibrium radius of the buckminster fullerene somewhat larger than experiment, but already useful. Similar considerations, but with Hartree-Fock method instead of TF, have been successfully used in quantitative calculations on spherical boron cages [3].

The present study has been motivated by the cluster studies on C reported by Jones and Seifert [4]. In particular, they report by the density functional study (DFT), that the likely lowest isomers of C_N with N between 14 and 22 are planar rings with "cumulenic" spacing between the nuclei on the rings. This has prompted us to consider the analog of the three-dimensional March model of C_{60} to apply to such quasi two-dimensional planar ring clusters.

The outline of the paper is then as follows. In Sec. II below, for completeness, we briefly summarize the TF statistical model for such a two-dimensional (2D) planar electronic assembly. Then we focus on the use of electrostatics and charge interactions satisfying the corresponding 2D Poisson equation. Analogous to the March model of C_{60}, ρ, now the number of electrons per unit area, is taken to have circular symmetry. This is, in the spirit of the one-center expansion, equivalent to smearing N point charges f into a uniform line charge with total charge Nf. Whereas in C_{60}, Nf was, naturally enough, taken to be 60, i.e., one π electron per C atom [5,6], in cluster rings we shall study, now largely analytically, the equilibrium radius R_e of the rings as a function of N. The details of this are set out in Sec. III. In Sec. IV we analyze the consistency of the present model with real *ab initio* and DFT calculations on ring clusters. The paper concludes with a summary plus some proposals directions for further study in Sec. V.

II. BRIEF SUMMARY OF THE TWO-DIMENSIONAL THOMAS-FERMI THEORY

The semiclassical TF statistical method is based on two assumptions: (i) a chemical potential equation for the fastest electron with Fermi momentum $p_F(\mathbf{r})$ and (ii) a phase-space argument relating $p_F(\mathbf{r})$ to the electron density $\rho(\mathbf{r})$. Assumption (i) corresponding to writing the chemical potential μ as the sum of the kinetic term $p_F^2/2m$, with m as the electron mass, and a one-body potential energy denoted below by $V(\mathbf{r})$ yields explicitly

$$\mu = \frac{p_F^2}{2m} + V(\mathbf{r}) \tag{1}$$

which is valid for general dimensionality D. However, D enters the phase-space considerations in (ii) above, since a cell of volume h^D can hold two electrons with opposed spin in the spin compensated ground state we consider throughout the present study. Thus in 2D, one has for an area Ω of coordinate space and a momentum space area πp_F^2, that the number of electrons per unit area, $\rho(\mathbf{r})$, is given by

$$\rho(\mathbf{r}) = \left(\frac{n_{el}}{\Omega}\right)_{\mathbf{r}} = \frac{2\pi p_F^2(\mathbf{r})}{h^2}, \tag{2}$$

the numerator of the final term being the spin occupancy two times the area of occupied momentum space, while the denominator is the size of the basic phase-space cell in 2D. Removing $p_F^2(\mathbf{r})$ from Eq. (1), in which μ is constant throughout the whole of the inhomogeneous electron distribution under discussion, using Eq. (2) yields

*Electronic address: amovilli@dcci.unipi.it
†Also at: Oxford University, Oxford, England.

$$\mu = \frac{h^2}{4\pi m}\rho(\mathbf{r}) + V(\mathbf{r}). \quad (3)$$

This is the basic TF equation yielding the areal density $\rho(\mathbf{r})$ in 2D from a given one body potential $V(\mathbf{r})$. As usual μ is determined by the normalization condition on $\rho(\mathbf{r})$.

We turn in the following section to apply this Euler Eq. (3) of the TF statistical method to planar ring clusters.

III. SELF CONSISTENT FIELD AND ITS CONSEQUENCES IN QUASI-2D RING CLUSTERS

As in the TF neutral atom, and in C_{60} in the March model [1,2,5], the chemical potential μ can be put to zero in the neutral ring clusters under discussion. Equation (3) can be combined with the 2D form of Poisson's equation of electrostatics to yield, in atomic units ($\hbar=1$, $m=1$, $q_e=-1$, and $4\pi\epsilon_0=1$), for the electrostatic potential V_{es},

$$V_{es}(r) = -\pi\rho(r), \quad (4)$$

where V_{es} satisfies

$$\frac{\partial^2 V_{es}}{\partial r^2} + \frac{1}{r}\frac{\partial V_{es}}{\partial r} = 4V_{es}. \quad (5)$$

Scaling r to s, via $s=2r$, the electrostatic potential can be obtained by solution of Eq. (5), for a line charge of radius R, and with $t=2R$ as

$$V_{es}(r) = -\lambda[K_0(t)\Theta(t-s)I_0(s) + I_0(t)\Theta(s-t)K_0(s)], \quad (6)$$

where the quantity $\lambda(>0)$ is to be obtained from the density normalization condition ($n_{el}=Nf$),

$$Nf = 2\pi\int_0^\infty \rho(r)rdr, \quad (7)$$

while $\Theta(x)$ is the Heaviside step function and $K_0(x)$ and $I_0(x)$ are modified Bessel functions [7]. From the Gauss theorem, we have additionally that

$$q(s) = \frac{s}{2}V'_{es}(s), \quad (8)$$

which means that $q(s)$, the total charge, must tend to 0 as $s\to\infty$ for a neutral system.

A basic function of the present self consistent TF method for 2D ring clusters is then defined by $Q(s)=2q(s)/\lambda$. The merit of this definition is that $Q(s)$ no longer depends on normalization. It can be written, using Eqs. (6) and (8) as

$$Q(s) = -s[K_0(t)\Theta(t-s)I'_0(s) + I_0(t)\Theta(s-t)K'_0(s)]. \quad (9)$$

The jump (discontinuity due to line charge) at $s=t$ of Q, say ΔQ, is then given by

$$\Delta Q = t|I_0(t)K'_0(t) - K_0(t)I'_0(t)|, \quad (10)$$

where the right-hand side (rhs) of Eq. (10) contains the modulus of the Wronskian evaluated at t. Owing to the special form of the second order differential equation (5), the Wronskian is proportional to t^{-1} and, hence, ΔQ is constant. From (10) one finds $\Delta Q=1$. By the definition of Q this result leads to $\lambda=2Nf$, the jump of the function $q(s)$ being related to the total ring charge Nf. It is important to remark that, when t tends to 0, one recovers the two-dimensional TF model atom already discussed by Kventsel and Katriel [8].

Plots of scaled potential $v(s)=V_{es}(s)/\lambda$ and total charge $Q(s)$ are presented in Fig. 1 for various choices of scaled ring radii t.

A. Variational TF calculation of ring radii as a function of N

As already mentioned we follow the March model of C_{60} by using 2D TF theory variationally, based on the "line charge" circular density $\rho(r)$ given by Eq. (4). Then the energy $E_{TF}[\rho]$ in 2D is explicitly

$$E_{TF}[\rho] = \pi^2\int_0^\infty \rho(r)^2 rdr + 2\pi\int_0^\infty \rho(r)V_{lc}(r)rdr - \int \rho(r_1)\rho(r_2)\ln r_{12}d\mathbf{r}_1 d\mathbf{r}_2 + U_{nn}. \quad (11)$$

Here, V_{lc} is the confining potential due to the nuclear effective charges smeared on a circumference of radius R, namely

$$V_{lc}(r) = \begin{cases} 2Nf\ln R & (r\leq R) \\ 2Nf\ln r & (r>R) \end{cases}. \quad (12)$$

As already emphasized, U_{nn} is to be treated using the discrete nuclei on rings, in order to calculate the equilibrium radius R_e given by the equilibrium condition

$$\left(\frac{\partial E_{TF}}{\partial R}\right)_{R_e} = 0. \quad (13)$$

Following the Hellmann-Feynman theorem, Eq. (13) is equivalent to the balance of the forces acting on nuclei if the electron density entering the energy has been obtained variationally. Using the TF density coming from the potential given by Eq. (6), the radial component of the electric field due to electrons and acting on the ring is

$$\mathcal{E}_e = -\frac{2Nf\chi(2R)}{R}, \quad (14)$$

where

$$\chi(t) = tK_0(t)I'_0(t) \quad (15)$$

is the fraction of electrons inside the circle of radius $t=2R$. Then, for a set of N equispaced point charges f lying on a circumference of radius R, the resultant electric field acting on one of such charges and due to the other $N-1$ has the radial component (still in 2D)

$$\mathcal{E}_n = \frac{(N-1)f}{R} \quad (16)$$

which can be used to establish the force balance equation

$$\mathcal{E}_e + \mathcal{E}_n = 0. \quad (17)$$

Combining Eqs. (14), (16), and (17) we get

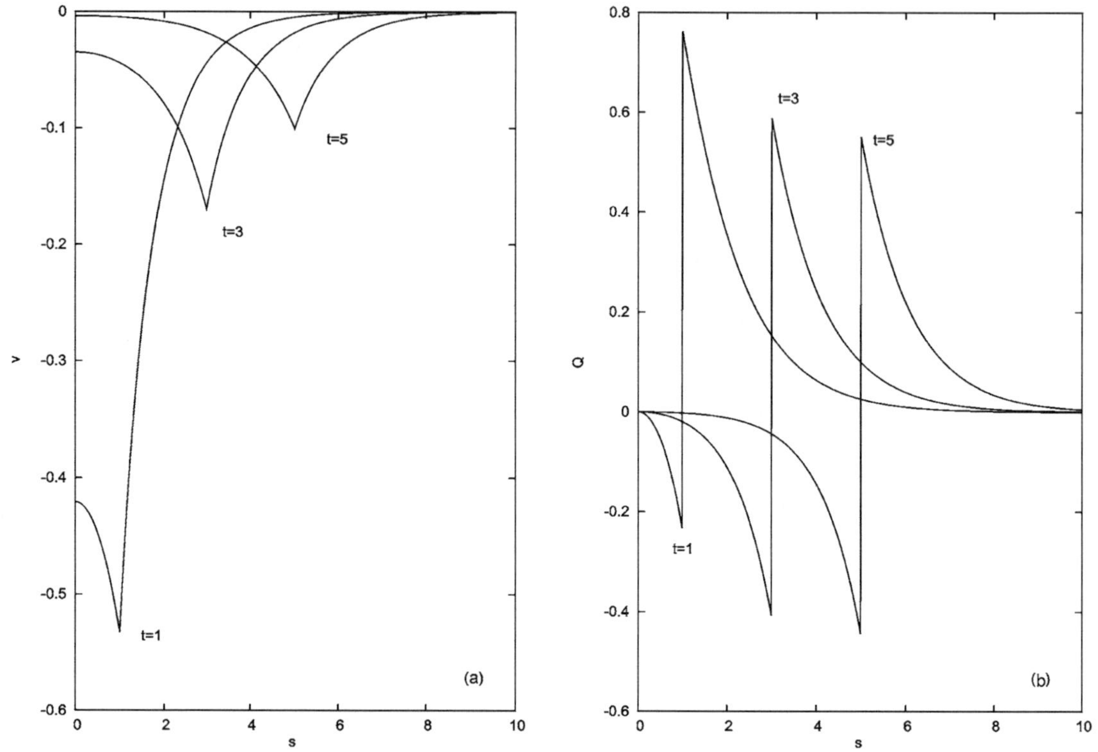

FIG. 1. Thomas-Fermi self consistent reduced potential (a) and charge (b) for a two-dimensional electron system, confined by a positive ring line charge, as a function of a properly scaled distance from the centre. Lengths (s and t) are in units of $2a_0$, Q is in units of Nf, the number of electrons in the ring, and v is in hartrees divided by $2Nf$ (see text for more details).

$$\chi(2R_e) = \frac{1}{2} - \frac{1}{2N}, \tag{18}$$

which can be solved to obtain R_e, the equilibrium radius in this model, as a function only of N, the number of atoms of the ring cluster.

In Fig. 2 we show the variation of the equilibrium radius with the number of ring atoms as it results from Eq. (18). From this plot one can see that R_e, in this model, tends rapidly to the limiting behavior, determined by the asymptotic forms of the modified Bessel functions, of about $N/4$ in atomic units. In the same figure we compare the equilibrium radii of this 2D model with realistic radii obtained from the DFT calculation on almost 2D ring clusters. This comparison will be discussed in the next section.

IV. CONSISTENCY OF THE 2D MARCH MODEL WITH REALISTIC PLANAR RING CLUSTERS

It is of interest now to compare the information that can be obtained from the model developed in the previous section with that arising from *ab initio* or DFT calculations on realistic planar ring clusters.

The prototype of such clusters could be rings of equispaced hydrogen atoms, $f=1$, constrained to lie on a circumference, while, in 2D, the analog of three-dimensional (3D) fullerenes are planar rings of N carbon atoms with N ranging from 14 to 22. The C rings of this size have been really produced in molecular beam experiments [9] and their stability has been discussed by Jones and Seifert [4]. Instead, the rings of hydrogen atoms considered here are only a theoretical construction that can be studied by standard methods for the electronic structure calculation.

In Fig. 2 we report the equilibrium radii of H and C clusters obtained from calculations performed at the B3LYP DFT level with basis sets flexible enough to give good constrained (circular) equilibrium geometries. For H rings we used even numbers of atoms ranging from 6 to 18, while for C rings we followed the study of Jones and Seifert [4] limiting attention to the three clusters C_{14}, C_{18}, and C_{22}. For C rings we made calculations on singlet states. For H clusters we considered singlet ground states when $N=4n+2$, n integer, and triplet states in the other cases.

From Fig. 2 is quite evident that in 2D the agreement between model equilibrium radii and radii of C ring clusters is not good as, instead, it happens in 3D for fullerenes. Some better consistency is instead observed between the model and our hypothetical H rings.

The existence of σ bonds in C rings modifies substantially the effective nuclear interaction from that expected in two dimensions. In order to overcome this inconsistency one can

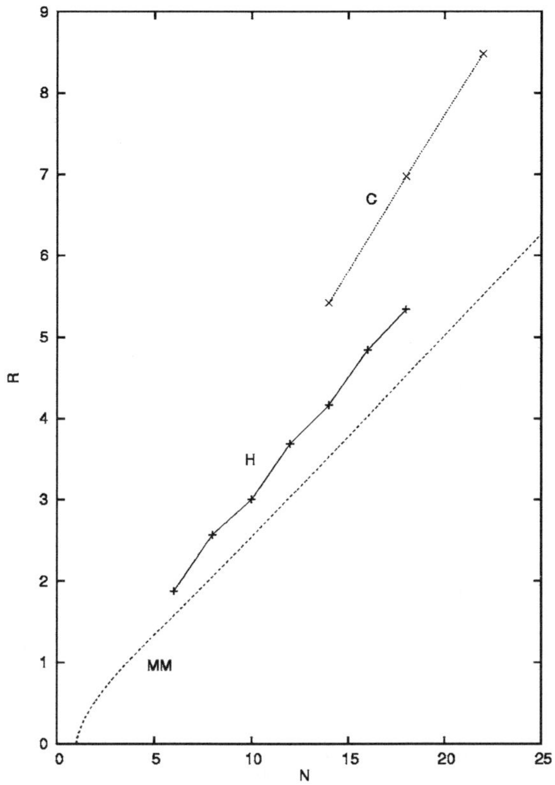

FIG. 2. Equilibrium radius as a function of the number of atoms of a ring cluster for the two dimensional March model of this work (MM) and for realistic H-ring (H) and C-ring (C) planar clusters. The radii are in atomic units.

modify the U_{nn} term in the energy functional (11). Equation (11) with the variational density (4) becomes

$$E = \pi \int_0^{\infty} \rho(r) V_{lc}(r) r dr + U_{nn} \tag{19}$$

in which we have

$$\pi \int_0^{\infty} \rho(r) V_{lc}(r) r dr = N^2 f^2 \ln R + N^2 f^2 K_0(2R) I_0(2R) \tag{20}$$

and

$$U_{nn} = - N(N-1) f^2 \ln R + \frac{N^2 f^2}{2} A_N, \tag{21}$$

where $A_N = (1/N) \Sigma_{j=2}^{N} \ln 2(1 - \cos \theta_{1j})$, θ_{1j} being the angle between nucleus 1, the ring centre and nucleus j. A_N does not depend on R. Now, to introduce some effect due to σ bonds, we propose to modify U_{nn} in the terms of order N, N being also the number of such bonds in C rings. σ bonds can be considered by including appropriate short ranged interactions acting between pairs of neighbor C atoms. Thus, we can write

$$U_{nn}^{(+\sigma)} = U_{nn} + \sum_{i=1}^{N} u(d_{i,i+1}) \tag{22}$$

where $u(d_{i,i+1})$ is the bond pair potential which depends on the distance $d_{i,i+1}$ between centres i and $i+1$ ($d_{N,N+1} = d_{N,1}$). In cumulenelike C rings the bond distances are all the same and, with the assumption $d_{i,i+1} \approx 2\pi R/N$, we have

$$U_{nn}^{(+\sigma)} = U_{nn} + N u\left(\frac{2\pi R}{N}\right). \tag{23}$$

At this point, by writing

$$U_{nn}^{(+\sigma)} = - N^2 f^2 \ln R + N g(R) \tag{24}$$

in which

$$g(R) = f^2 \ln R + \frac{f^2}{2} \sum_{j=2}^{N} \ln 2(1 - \cos \theta_{1j}) + u\left(\frac{2\pi R}{N}\right), \tag{25}$$

the energy E takes the simple form

$$E = N^2 f^2 K_0(2R) I_0(2R) + N g(R). \tag{26}$$

In the range of radii of C_N rings considered in this work, the modified Bessel function K_0 and I_0 can be substituted by their asymptotic expression. The product $K_0(t) I_0(t)$ in this regime is simply $1/2t$ with $t=2R$ in the present case. This fact, combined with the observation that the equilibrium radii are proportional to the number of atoms in the ring, suggests to maintain the $\ln R$ functional dependence in $g(R)$ of Eq. (26) in the region of the equilibrium geometries. Under these considerations we have

$$E(R) = \frac{N^2 f^2}{4R} + N(\alpha \ln R + \beta_N), \tag{27}$$

where β_N is a constant introduced to bring $E(R_{eq})$ to coincidence with that of calculated minimum. Here, we remark that the second term in the rhs of Eq. (27) corresponds to $Ng(R)$ of Eqs. (24) and (26). Equation (27) suggests that E/N could be written as a function of R/N in the more general form

$$\frac{E(R)}{N} = a \frac{N}{R} + b \ln\left(\frac{R}{N}\right) + c, \tag{28}$$

with a, b, and c related to f, α, and β_N through Eq. (27) above. Equation (28) is the analog of Eq. (9) of our previous work [6] on fullerenes in 3D.

In Fig. 3 we show results for fitted scaled energies E/N obtained from Hartree-Fock (HF) and B3LYP DFT calculations performed on C_{14}, C_{18}, and C_{22} ring clusters. The three curves, in both the two plots of Fig. 3, look roughly identical apart for a small energy shift. Fitted parameters a, b, and c are given in Table I. These numerical results confirm some universality of Eq. (28). The value of a can also be used to estimate the number of electrons of each atom, namely f, involved in the TF model density (4). From Eq. (27) we can

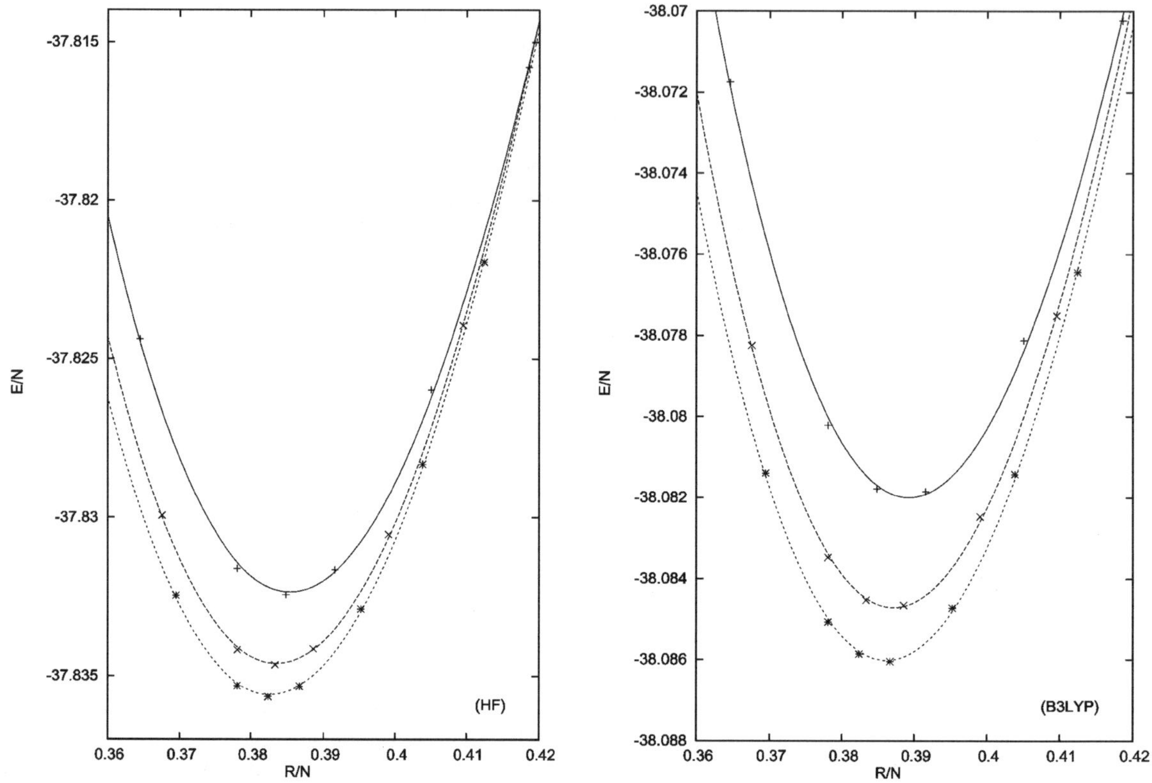

FIG. 3. Curves of scaled energies E/N against the scaled radius R/N, obtained at HF and DFT (B3LYP) level of calculation, for the three carbon rings C_{14} (upper curves), C_{18} (middle), and C_{22} (lower curves). Data are in atomic units.

write $f=2\sqrt{a}$, which tells us that this number is about 2.77 if the HF data are used or about 2.68 with B3LYP results.

Finally, with the parameters of Table I, we can calculate the frequency of the "breathing" symmetrical vibrational mode. The reduced mass for this motion is M/N where M is the mass of the carbon nucleus. By definition, the square of such angular frequency ω is given by

$$\omega^2 = \frac{N}{M}\left(\frac{\partial^2 E}{\partial R^2}\right)_{R_{eq}}, \tag{29}$$

where the energy E is the same as in Eq. (28) and plays the role of potential energy in the harmonic approximation for the study of the motion of nuclei in this case [10,11]. Combining Eqs. (28) and (29), we obtain

$$\omega = \sqrt{\frac{b^3}{a^2 M}}, \tag{30}$$

which is essentially a constant not depending on N. From the data of Table I we obtain $\bar{\nu}=\omega/2\pi c=1163\ \text{cm}^{-1}$ with B3LYP DFT and $1378\ \text{cm}^{-1}$ at the HF level.

We conclude this section on connection with realistic systems by mentioning quantum rings. Quantum rings are artificial systems of confined electrons observed at the nanoscale in semiconductors. They originate from 2D quantum dots by

TABLE I. Parameters entering the definition of the scaled energy defined in Eq. (28) obtained from fitting of HF and B3LYP results of calculations performed on C_{14}, C_{18} and C_{22} ring clusters.

Method	Parameter	C_{14}	C_{18}	C_{22}
HF	a	1.93±0.03	1.94±0.02	1.90±0.02
	b	5.02±0.08	5.06±0.06	4.97±0.07
	c	−38.065±0.005	−38.045±0.004	−38.030±0.004
B3LYP	a	1.80±0.03	1.81±0.02	1.77±0.03
	b	4.62±0.08	4.67±0.07	4.57±0.07
	c	−38.343±0.005	−38.323±0.004	−38.308±0.004

depleting the central region of the disk and thereby forming a ring (see, for example, Ref. [12]). We do not enter into the details of the methods used to study such systems, but we say that the model developed in the present work in Secs. II and III can also be formulated in a context of a quantum ring. In order to do this, we need to introduce the effective mass of electrons and the relative dielectric constant of the medium in all expressions of the model. The effective mass m^* enters through the definition of the Fermi energy, while the medium dielectric constant ϵ through the Poisson equation. For this case, we get the following coupled equations in atomic units:

$$\mu = V + \frac{\pi\rho}{m^*} \qquad -\nabla^2 V = \frac{4\pi}{\epsilon}\rho. \tag{31}$$

Of course, some attention will need to be paid to the confinement. The choice of a background positive line charge potential is still valid but the total charge and the radius should be differently defined in terms of some other external parameter. One more important point concerns the effect of an applied magnetic field, but this needs an extension of the present model.

V. SUMMARY AND PROPOSED FUTURE DIRECTIONS

The selfconsistent potential $V_{es}(r)$ in the analog of the March model in 2D planar ring clusters has first been calculated analytically. It is given by Eq. (6) in terms of modified Bessel functions. The 2D TF energy functional is then shown to lead to the potential energy function in Eq. (28), where the energy per atom of the ring cluster of radius R depends on R/N. This is the 2D counterpart of our early scaling of potential energy in the 3D fullerenes. The breathing frequency of the planar ring clusters C_{14}, C_{18}, and C_{22} is thereby estimated.

As to future directions, we already referred at the end of Sec. IV to a possible application of the present model to nanoscale quantum rings. Therefore, to conclude, we want to sketch briefly below how to eventually refine the 2D March model discussed fully above to take account of exchange, in what is essentially a local density (Dirac-Slater like) approximation. In 3D, the Slater $X\alpha$ method works as follows:

$$U_x = \frac{1}{2}\int d\mathbf{r}_1 \int d\mathbf{r}_2 \frac{\rho(r_1)\rho(r_2)f_x(r_{12})}{r_{12}}$$

$$-1 = \int d\mathbf{r}_2 \rho(r_2) f_x(r_{12}),$$

$$f_x(r_{12}) \approx -1(r_{12} \leqslant R) \approx 0(r_{12} > R),$$

$$R = \left(\frac{3}{4\pi\rho}\right)^{1/3},$$

$$U_x \approx -2\pi \int d\mathbf{r}_1 \rho(r_1)^2 \int_0^R r_{12} dr_{12}$$

$$= -\pi \int d\mathbf{r}_1 \rho(r_1)^2 R^2$$

$$\approx -\left(\frac{3\pi^{1/2}}{4}\right)^{2/3} \int d\mathbf{r}\rho(r)^{4/3}. \tag{32}$$

The counterpart of the above argument in 2D goes as follows:

$$U_x = -\int d\mathbf{r}_1 \int d\mathbf{r}_2 \rho(r_1)\rho(r_2) f_x(r_{12}) \ln r_{12},$$

$$R = \left(\frac{1}{\pi\rho}\right)^{1/2},$$

$$U_x \approx 2\pi \int d\mathbf{r}_1 \rho(r_1)^2 \int_0^R r_{12} \ln r_{12} dr_{12}$$

$$= \frac{\pi}{2} \int d\mathbf{r}_1 \rho(r_1)^2 R^2 (2 \ln R - 1)$$

$$\approx -\frac{1}{2} \int d\mathbf{r}\rho(r)[1 + \ln(\pi\rho)]. \tag{33}$$

The final line of Eq. (33) prompt us to add that the exchange potential energy U_x in 2D contains a term which is reminiscent of the Shannon entropy, the latter having an integrand of the form $\rho \ln \rho$.

ACKNOWLEDGMENTS

C.A. acknowledges financial support from MIUR (PRIN 2004) and from the University of Pisa (Fondi di Ateneo 2005). N.H.M. wishes to acknowledge that his contribution to this work was made during a visit to Scuola Normale Superiore (Pisa) in 2006. He thanks Professor M. P. Tosi for very generous hospitality.

[1] D. P. Clougherty and X. Zhu, Phys. Rev. A **56**, 632 (1997).
[2] F. Despa, Phys. Rev. B **57**, 7335 (1998).
[3] C. Amovilli and N. H. March, Chem. Phys. Lett. **347**, 459 (2001).
[4] R. O. Jones and G. Seifert, Phys. Rev. Lett. **79**, 443 (1997).
[5] F. Siringo, G. Piccitto, and R. Pucci, Phys. Rev. A **46**, 4048 (1992).
[6] C. Amovilli, I. A. Howard, D. J. Klein, and N. H. March, Phys. Rev. A **66**, 013210 (2002).
[7] M. Abramowitz and I. A. Stegun, *Handbook of Mathematical Functions* (National Bureau of Standards, Washington D.C., 1964).
[8] G. F. Kventsel and J. Katriel, Phys. Rev. A **24**, 2299 (1981).
[9] S. J. La Placa, P. A. Roland, and J. Wynne, Chem. Phys. Lett. **190**, 163 (1992).
[10] N. H. March, Proc. Cambridge Philos. Soc. **43**, 665 (1952).
[11] H. C. Longuet-Higgins and D. A. Brown, J. Inorg. Nucl. Chem. **1**, 60 (1955).
[12] B. C. Lee and C. P. Lee, Nanotechnology **15**, 848 (2004).

1.C Nuclear Physics 12 (1959) 237—240; © North-Holland Publishing Co., Amsterdam

APPROXIMATE SOLUTIONS OF THE DENSITY MATRIX EQUATION FOR A LOCAL AVERAGE FIELD

N. H. MARCH and W. H. YOUNG
Department of Physics, The University, Sheffield

Received 13 April 1959

Abstract: It is shown that a condition derived by Naqvi [1]) from the density matrix equation, and used by him to generalize the statistical theory, can be regarded as constituting a differential form of the virial theorem. This suggests that Naqvi's condition is satisfied by the customary treatments, and this is demonstrated directly.

A recent paper by Naqvi [1]) in which surface corrections to the statistical theory are examined by means of the density matrix method prompts us to record briefly some results of earlier unpublished work which have a bearing on Naqvi's investigation. We also comment on the relation between our earlier variational approach [2]) and Naqvi's procedure.

For brevity, we restrict our considerations to a one-dimensional case with a local average field, although various generalizations can be made to three dimensions, as described by Naqvi. Then, for particles moving in a common local potential $V(x)$, we wish to solve the equation

$$\frac{\partial^2 \rho}{\partial x'^2} - \frac{\partial^2 \rho}{\partial x^2} = \frac{2m}{\hbar^2} [V(x') - V(x)]\rho \tag{1}$$

with normalization, Hermiticity and idempotency conditions on the density matrix $\rho(x', x)$, such that the eigenvalue sum E given by

$$E = -\frac{\hbar^2}{2m} \int \left[\frac{\partial^2 \rho(x', x)}{\partial x^2}\right]_{x'=x} \mathrm{d}x + \int \rho(x, x) V \mathrm{d}x \tag{2}$$

is an absolute minimum. Following exactly Naqvi's procedure we use new coordinates

$$\xi = \tfrac{1}{2}(x' + x), \qquad \eta = \tfrac{1}{2}(x' - x)$$

and seek a solution of the resulting equation

$$\frac{\partial^2 \rho}{\partial \xi \partial \eta} = \frac{2m}{\hbar^2} [V(\xi + \eta) - V(\xi - \eta)]\rho \tag{3}$$

by expanding about the point $\eta = 0$. Thus we write

$$\rho(\xi, \eta) = n(\xi) + \sum_{j=1}^{\infty} \eta^{2j} a_{2j}(\xi), \tag{4}$$

where we have written $n(\xi)$ for the particle density $\rho(\xi, 0)$. By substitution in (3) it is easy to verify that

$$a'_2 = \frac{2m}{\hbar^2} nV', \tag{5}$$

$$a'_4 = \frac{m}{\hbar^2} [a_2 V' + \tfrac{1}{6} nV'''], \tag{6}$$

and higher terms may easily be obtained if required. We note that if the diagonal element were known, the entire matrix could be built up by this method. We should emphasize here that (5) is the essential result in Naqvi's attempt to generalize the statistical theory. Earlier unpublished work by us led to (5) being viewed as constituting a considerable generalization of the virial theorem, a kind of differential form of the customary expression; and the justification for this is outlined below.

Using (2) and (4) the kinetic energy T may immediately be written as

$$T = -\frac{\hbar^2}{4m} \int [\tfrac{1}{2} n''(x) + a_2(x)] \mathrm{d}x \tag{7}$$

and for most cases of interest we can drop the term in n'', although for boundary conditions other than the usual ones its retention may prove necessary. Assuming however that n'' integrates to zero, eq. (7) is easily rewritten as

$$T = \frac{\hbar^2}{4m} \int x a'_2 \mathrm{d}x - \frac{\hbar^2}{4m} (x a_2)_{\text{limits}}. \tag{8}$$

With the reservation again that no contribution arises from the boundaries, the use of (5) in (8) leads to the usual form of the virial theorem

$$2T = \int n x V' \mathrm{d}x. \tag{9}$$

Incidentally, this proof appears to be much more direct than the customary method based on wave functions.

Since the usual statistical theories are known to satisfy the virial theorem, it appeared likely to us that they would also satisfy (5) and this can be verified, as discussed below. This point does not appear to be made clear in Naqvi's work.

In the process of outlining the proof of (5) first for the original Thomas-Fermi theory, it seems worthwhile to note the way in which this approximation fits into the present framework, as we are not aware that this has been pointed out previously. If we go back to (3), and replace $V(\xi+\eta)-V(\xi-\eta)$ by $2\eta V'(\xi)$, corresponding to the neglect of the third and higher derivatives of the potential, the Thomas-Fermi density matrix [3])

$$\rho(\xi, \eta) = \frac{\sin\{2\pi\eta n(\xi)\}}{2\pi\eta} \tag{10}$$

is an exact solution of the resulting differential equation, provided

$$-nn' = \frac{4m}{h^2} V' \tag{11}$$

or, by integration,

$$n^2 = \frac{8m}{h^2} (E'-V), \tag{12}$$

giving the usual Thomas-Fermi density. Unfortunately, of course, this matrix usually corresponds to a cut-off in the particle density and does not satisfy the idempotency condition exactly. Nevertheless, comparing (4) and (10) we find

$$a_2(\xi) = -\tfrac{2}{3}\pi^3 n^3(\xi) \tag{13}$$

and from (11) and (13), eq. (5) is immediately verified.

Naqvi's basic aim in his proposed generalization may now be said to consist in modifying the Thomas-Fermi density matrix in such a way that (5) is preserved while the idempotency condition is satisfied with improved accuracy. It is therefore of interest to demonstrate that the variational density matrix for N particles which we proposed earlier [2]) entirely fulfils Naqvi's requirements. This matrix, which leads to a method intimately connected with von Weizsäcker's scheme, is defined as follows:

$$\rho(\xi, \eta) = \{y_1(\xi+\eta)y_1(\xi-\eta)\}^{\frac{1}{2}}\rho_0(\tfrac{1}{2}(y'+y),\ \tfrac{1}{2}(y'-y)). \tag{14}$$

Here we have used the notation $y' = y(\xi+\eta)$, $y = y(\xi-\eta)$,

$$\rho_0(\xi, \eta) = \frac{\sin 2\pi N\eta}{\sin 2\pi\eta}$$

and y is the variational function satisfying the Euler equation

$$\frac{h^2}{8m}(N^2-1)y_1{}^2+V-\frac{\hbar^2}{4m}\frac{y_3}{y_1}+\frac{\hbar^2}{8m}\frac{y_2{}^2}{y_1{}^2} = E', \tag{15}$$

where y_n denotes the nth derivative of y. We then find the following results for the matrix (14), by comparison with (4):

$$n = Ny_1 \tag{16}$$

$$a_2 = -\tfrac{2}{3}\pi^2 N(N^2-1)y_1{}^3+\tfrac{1}{2}N\left(y_3-\frac{y_2{}^2}{y_1}\right). \tag{17}$$

Again, using (15) and (17), it is a straightforward matter to verify (5). The exact idempotency of this matrix follows immediately as a consequence of its construction from N orthonormal transformed plane waves [2]).

We conclude that the 'differential form of the virial theorem' embodied in (5) is satisfied by existing statistical density matrices, and that it would appear necessary to go to (6) in order that the knowledge obtained by solution of the differential equation (3) may be used to improve the customary treatments. In this connection, a'_4 may be derived from the matrix (14) above and then compared with the right hand side of (6). While (6) is not exactly satisfied, remarkable similarity between a'_4 and the right hand side is evident, and in particular the terms of highest order in N are identical. Thus, in any given application, this would provide one possible method of testing the accuracy of the matrix (14). In this connection, we note finally that while (14) is satisfactory from a fundamental theoretical point of view, the possibility of obtaining higher numerical accuracy for a given potential by alternative approaches such as that suggested by Naqvi is certainly not excluded.

References

) M. A. Naqvi, Nuclear Physics **10** (1959) 256
2) N. H. March and W. H. Young, Proc. Phys. Soc. **72** (1958) 182
3) N. H. March, Proc. Phys. Soc. A **70** (1957) 839

PHYSICAL REVIEW VOLUME 120, NUMBER 3 NOVEMBER 1, 1960

Relation Between Dirac and Canonical Density Matrices, with Applications to Imperfections in Metals*

N. H. MARCH AND A. M. MURRAY
Department of Physics, The University, Sheffield, England

(Received June 17, 1960)

It is shown that the canonical density matrix in a single-particle framework may be related directly to the generalized canonical density matrix, containing the Fermi-Dirac function, and to the Dirac density matrix.

A study is then made of density matrices in central field problems. A new differential equation is derived, from the Bloch equation, for the diagonal element of the canonical density matrix. In the case of a continuum of energy levels, this is shown to lead directly to a differential equation for the diagonal element of the Dirac matrix, that is, the particle density. Free-electron density matrices are fully worked out and a perturbation theory based on these free-electron forms is presented.

It is further shown that for a nonspherical potential energy $V(\mathbf{r})$, the work of Green on the quantum-mechanical partition function may be utilized to yield a perturbation theory for the Dirac matrix. In this way, the correct formulation to replace Mott's well-known first-order approximation for dealing with imperfections in metals is obtained. A brief discussion of the way in which this removes qualitatively the difficulties of the Mott treatment is given and the possibility of direct numerical application in a self-consistent framework is pointed out.

1. INTRODUCTION

RECENTLY there has been a considerable revival of interest in the use of the density matrix in quantum-mechanical many-particle problems. If we consider a single-particle approximation to the solution of the many-body problem, then the essential tool is the matrix introduced long ago by Dirac.[1] Only recently have serious attempts been made to calculate this quantity directly.[2,3]

There are, unfortunately, a number of rather severe practical difficulties associated with the direct calculation of the Dirac density matrix. Chief among these is the awkward nature of the idempotency condition, but other points also are somewhat troublesome. Thus, while a differential equation may be obtained for the Dirac density matrix, this determines only very general properties of the solutions and must be used in conjunction with a variational principle.

The question naturally arises therefore whether some alternative formulation may be found, from which the Dirac density matrix may be obtained, but in which the calculations are eased by allowing the subsidiary conditions to be relaxed. The purpose of the present work is to show that the Dirac density matrix is rather directly related to the canonical density matrix, defined by Eq. (2.3), and since this latter quantity satisfies the Bloch equation (2.7) with a well-defined initial condition, it seems that a useful alternative approach for the calculation of the Dirac matrix is now available.

We remark at this stage that the need to develop the methods described here became apparent as a result of some earlier work on electronic wave functions around imperfections in metals[4] and we shall consider therefore examples particularly relevant to this field.

2. RELATION BETWEEN DIRAC AND CANONICAL DENSITY MATRICES

We consider the complete set of solutions of the Schrödinger equation

$$H\psi=E\psi, \tag{2.1}$$

and we denote the wave functions and energy levels by ψ_i and E_i, respectively. We now form the quantity

$$\sum_i \omega_i e^{-E_i/kT}\psi_i^*(\mathbf{r}')\psi_i(\mathbf{r}), \tag{2.2}$$

where the summation extends over all the energy levels. It will be convenient to consider the following choices of ω_i:

(a) $$\omega_i=1,$$

leading to the usual definition of the canonical density matrix

$$C(\mathbf{r}',\mathbf{r},1/kT)=\sum_i e^{-E_i/kT}\psi_i^*(\mathbf{r}')\psi_i(\mathbf{r}), \tag{2.3}$$

(b) $$\omega_i=\frac{e^{\zeta/kT}}{1+e^{(\zeta-E_i)/kT}},$$

leading to the generalized canonical density matrix

$$D(\mathbf{r}',\mathbf{r},1/kT,\zeta)=\sum_i \frac{e^{(\zeta-E_i)/kT}}{1+e^{(\zeta-E_i)/kT}}\psi_i^*(\mathbf{r}')\psi_i(\mathbf{r}). \tag{2.4}$$

The definition (2.4) contains the Dirac zero-temperature density matrix $\rho(\mathbf{r}',\mathbf{r},\zeta)$ as a special case, namely

$$\rho(\mathbf{r}',\mathbf{r},\zeta)=\lim_{T\to 0}D(\mathbf{r}',\mathbf{r},1/kT,\zeta), \tag{2.5}$$

where ζ must be determined from the number of occu-

* We wish to acknowledge that this work was performed under partial contractual support with the U. S. Army, through its European Research Office.

[1] P. A. M. Dirac, Proc. Cambridge Phil. Soc. **26**, 376 (1930).
[2] R. McWeeny, Proc. Roy. Soc. (London) **A235**, 496 (1956).
[3] N. H. March and W. H. Young, Proc. Phys. Soc. (London) **72**, 182 (1958).
[4] N. H. March and A. M. Murray, Proc. Roy. Soc. (London) **A256**, 400 (1960).

pied states N through the relation

$$\int \rho(\mathbf{r},\mathbf{r})d\tau = N. \tag{2.6}$$

Now it is well known that the canonical density matrix satisfies the Bloch equation

$$-\partial C/\partial\beta = HC; \quad \beta = 1/kT, \tag{2.7}$$

with the initial condition

$$C(\mathbf{r}',\mathbf{r},0) = \delta(\mathbf{r}'-\mathbf{r}), \tag{2.8}$$

and in Eqs. (2.7) and (2.8) is summarized the practical merit of the canonical matrix: it may be obtained without detailed knowledge of the individual wave functions and energy levels.

We shall now show directly that knowledge of C enables the Dirac matrix ρ and the generalized canonical matrix D to be found. We introduce the function $Q(\mathbf{r}',\mathbf{r},E)$ defined by

$$Q(\mathbf{r}',\mathbf{r},E) = \sum_i \psi_i^*(\mathbf{r}')\psi_i(\mathbf{r})\delta(E-E_i). \tag{2.9}$$

Hence

$$\rho(\mathbf{r}',\mathbf{r},\zeta) = \int_0^\zeta Q(\mathbf{r}',\mathbf{r},E)dE, \tag{2.10}$$

the Fermi level ζ being defined such that it lies just above the highest occupied level. From (2.3) and (2.9) we have

$$\begin{aligned} C(\mathbf{r}',\mathbf{r},\beta) &= \int_0^\infty Q(\mathbf{r}',\mathbf{r},E)e^{-\beta E}dE \\ &= \beta \int_0^\infty \left\{ \int_0^E Q(\mathbf{r}',\mathbf{r},E)dE \right\} e^{-\beta E}dE \\ &= \beta \int_0^\infty \rho(\mathbf{r}',\mathbf{r},E)e^{-\beta E}dE, \end{aligned} \tag{2.11}$$

using (2.10). From (2.4) and (2.9) we have

$$\begin{aligned} D(\mathbf{r}',\mathbf{r},\beta,\zeta) &= \int_0^\infty Q(\mathbf{r}',\mathbf{r},E)\frac{1}{1+e^{\beta(E-\zeta)}}dE \\ &= \int_0^\infty \rho(\mathbf{r}',\mathbf{r},E)\frac{\partial}{\partial E}\left\{\frac{-1}{1+e^{\beta(E-\zeta)}}\right\}dE. \end{aligned} \tag{2.12}$$

Equations (2.11) and (2.12) reveal immediately that knowledge of the canonical matrix C is sufficient to determine both ρ and D. We shall not consider the generalized canonical matrix D further in the present paper, as our main interest is in the Dirac matrix, but shall content ourselves with the remark that (2.12) will allow direct calculation of physical properties of Fermi-Dirac assemblies from a knowledge of the canonical density matrix.

Focusing attention therefore on ρ, in what follows, it is evident from (2.11) that $C(\mathbf{r}',\mathbf{r},\beta)/\beta$ is simply the Laplace transform of $\rho(\mathbf{r}',\mathbf{r},E)$. The inverse relation can be written in the form

$$\rho(\mathbf{r}',\mathbf{r},E) = \frac{1}{2\pi i}\int_{\sigma-i\infty}^{\sigma+i\infty} \frac{1}{\beta} e^{\beta E} C(\mathbf{r}',\mathbf{r},\beta)d\beta, \tag{2.13}$$

where σ is chosen such that the integrand has no poles in that part of the complex plane for which $R(\beta) \geqslant \sigma$.

3. EXAMPLE: A UNIFORM ELECTRON GAS

As a first example of the use of the relation (2.11), and because the results will later be made the basis of the perturbation treatment of Sec. 7, let us consider a uniform electron gas, neglecting interactions. As Sondheimer and Wilson[5] have shown, the Bloch equation (2.7) is readily solved, subject to the initial condition (2.8) and the result, in atomic units, is

$$C(\mathbf{r}',\mathbf{r},\beta) = \exp(-R^2/2\beta)/(2\pi\beta)^{\frac{3}{2}}, \tag{3.1}$$

where

$$R^2 = (x-x')^2 + (y-y')^2 + (z-z')^2.$$

In obtaining ρ, we start from the result[6]

$$\frac{a}{2\pi^{\frac{1}{2}}s^{\frac{3}{2}}}\exp\left(-\frac{a^2}{4s}\right) = \int_0^\infty \frac{e^{-st}}{\pi}\sin a(t)^{\frac{1}{2}}dt, \tag{3.2}$$

and by differentiating (3.2) with respect to a we find almost immediately

$$\begin{aligned} &\frac{a^3}{4\pi^{\frac{1}{2}}s^{\frac{5}{2}}}\exp\left(-\frac{a^2}{4s}\right) \\ &\qquad = \int_0^\infty \frac{e^{-st}}{\pi}[\sin a(t)^{\frac{1}{2}} - a(t)^{\frac{1}{2}}\cos a\,(t)^{\frac{1}{2}}]dt. \end{aligned} \tag{3.3}$$

Using (3.3) we obtain directly from (3.1) the result

$$\begin{aligned} &\frac{C(\mathbf{r}',\mathbf{r},\beta)}{\beta} \\ &\qquad = \frac{1}{2\pi^2R^3}\int_0^\infty e^{-\beta E}[\sin a(E)^{\frac{1}{2}} - a(E)^{\frac{1}{2}}\cos a(E)^{\frac{1}{2}}]dE, \end{aligned} \tag{3.4}$$

where $a = 2^{\frac{1}{2}}R$.

From (3.4) and (2.11), a simple closed form for the Dirac matrix results, namely

$$\rho(\mathbf{r}',\mathbf{r},\zeta) = \frac{1}{2\pi^2R^3}[\sin a(\zeta)^{\frac{1}{2}} - a(\zeta)^{\frac{1}{2}}\cos a(\zeta)^{\frac{1}{2}}]. \tag{3.5}$$

From (2.6), ζ is to be determined from the equation

$$\rho(\mathbf{r},\mathbf{r},\zeta) = N/V = (2\zeta)^{\frac{3}{2}}/6\pi^2, \tag{3.6}$$

---

[5] E. H. Sondheimer and A. H. Wilson, Proc. Roy. Soc. (London) **A210**, 173 (1951).

[6] See, for example, G. Doetsch, *Theorie und Anwendung der Laplace Transform* (Dover Publications, New York, 1943), p. 24.

where V is the volume, and this is the familiar relation between the electron density and the Fermi energy.

The result (3.5) has, of course, been obtained previously by direct use of plane waves, but the present argument shows how it may be found without recourse to the wave functions, and by a method which nevertheless circumvents consideration of the awkward idempotency condition.

4. DENSITY MATRICES IN CENTRAL FIELD PROBLEMS

We turn now to a case of particular importance in the imperfections field, namely that corresponding to a spherically symmetric potential energy $V(r)$. Here it is often convenient to exploit the fact that the wave equation separates in spherical polar coordinates r, θ, ϕ, as was done, for example, in our previous computations of electronic wave functions around a vacancy in a finite metal.[4] Then the solutions of Schrödinger's equation may be written in the form

$$\chi_{il}(r)Y_{lm}(\theta,\phi),$$

where

$$Y_{lm}(\theta,\phi)=\left[\frac{2l+1}{4\pi}\frac{(l-|m|)!}{(l+|m|)!}\right]^{\frac{1}{2}}P_l^m(\cos\theta)e^{im\phi},$$

and χ_{il} satisfies the radial Schrödinger equation

$$-\frac{1}{2}\frac{d^2}{dr^2}(r\chi_{il})=\left\{\frac{l(l+1)}{2r^2}+V-E_{il}\right\}r\chi_{il}. \qquad (4.1)$$

The eigenvalue E_{il} and the function χ_{il} are, of course, both independent of m. Hence from (2.3) we have

$$C(\mathbf{r}',\mathbf{r},\beta)=\sum_l\{\sum_i e^{-\beta E_{il}}\chi_{il}(r')\chi_{il}(r)\}\times\{\sum_m Y_{lm}^*(\theta',\phi')Y_{lm}(\theta,\phi)\}$$

$$=\sum_l\{\sum_i e^{-\beta E_{il}}\chi_{il}(r')\chi_{il}(r)\}\times\frac{(2l+1)}{4\pi}P_l(\cos\gamma), \qquad (4.2)$$

where γ is the angle between $\mathbf{r}'$ and $\mathbf{r}$ and the addition theorem for spherical harmonics has been used.

We now define C_l through the equation

$$4\pi C_l(r',r,\beta)=\sum_i e^{-\beta E_{il}}\chi_{il}(r')\chi_{il}(r), \qquad (4.3)$$

and thus (4.2) can be rewritten

$$C(\mathbf{r}',\mathbf{r},\beta)=\sum_l(2l+1)C_l(r',r,\beta)P_l(\cos\gamma). \qquad (4.4)$$

The initial condition on $C_l(r',r,\beta)$ may be obtained by viewing (4.1) as a one-dimensional wave equation, with wave function $r\chi_{il}(r)$. From (4.3) it then follows that $4\pi rr'C_l(r',r,\beta)$ is the corresponding one-dimensional canonical density matrix and hence

$$4\pi rr'C_l(r',r,0)=\delta(r'-r). \qquad (4.5)$$

This is easily shown to be consistent with the condition (2.8) on $C(\mathbf{r}',\mathbf{r},\beta)$. From (4.1) and (4.3), the equation satisfied by $C_l(r',r,\beta)$ is

$$\frac{1}{2}\frac{\partial^2}{\partial r^2}(rC_l)-\left\{\frac{l(l+1)}{2r^2}+V\right\}rC_l-\frac{\partial}{\partial\beta}(rC_l)=0. \qquad (4.6)$$

Often it would be very convenient in applications if the important diagonal element $C_l(r,r,\beta)$ could be calculated directly, without the necessity of obtaining the complete matrix $C_l(r',r,\beta)$. In fact, as is shown in Appendix I, the diagonal element, which we write for convenience as $Z_l(r,\beta)$, satisfies the differential equation

$$-\frac{1}{8}\frac{\partial^3}{\partial r^3}(r^2Z_l)-\frac{l(l+1)}{2r}\frac{\partial}{\partial r}(rZ_l)-\frac{1}{2}\frac{\partial V}{\partial r}r^2Z_l$$
$$-V\frac{\partial}{\partial r}(r^2Z_l)-\frac{\partial^2}{\partial\beta\partial r}(r^2Z_l)=0. \qquad (4.7)$$

Turning our attention now to the corresponding Dirac matrices, we define ρ_l through the equation

$$4\pi\rho_l(r',r,E)=\sum_i\chi_{il}(r')\chi_{il}(r), \qquad (4.8)$$

and then we have

$$\rho(\mathbf{r}',\mathbf{r},E)=\sum_l(2l+1)\rho_l(r',r,E)P_l(\cos\gamma), \qquad (4.9)$$

which is seen to be entirely analogous to (4.4). From (2.11), (4.4) and (4.9), and using the orthogonality of the Legendre polynomials, we have

$$C_l(r',r,\beta)=\beta\int_0^\infty \rho_l(r',r,E)e^{-\beta E}dE. \qquad (4.10)$$

Denoting the diagonal element of the matrix $\rho_l(r',r,E)$ as $n_l(r,E)$, it follows that

$$Z_l(r,\beta)=\beta\int_0^\infty n_l(r,E)e^{-\beta E}dE. \qquad (4.11)$$

When the eigenvalues E_{il} form a continuous set, it is shown in Appendix I that the differential equation (4.7) for Z_l leads to the equation

$$-\frac{1}{8}\frac{\partial^3}{\partial r^3}(r^2n_l)-\frac{l(l+1)}{2r}\frac{\partial}{\partial r}(rn_l)-\frac{1}{2}\frac{\partial V}{\partial r}r^2n_l$$
$$-V\frac{\partial}{\partial r}(r^2n_l)+\int_0^E E\frac{\partial^2}{\partial E\partial r}(r^2n_l)dE=0, \qquad (4.12)$$

for the "particle density" $n_l(r,E)$.

5. CENTRAL FIELD RESULTS FOR $V(r)=0$

As an application of the above relations, as well as the fact that the results will form the basis of the perturbation theory outlined in Sec. 6, we consider now the case when $V(r)=0$. Then Eq. (4.6) satisfied by $C_l(r',r,\beta)$ reduces to

$$-\frac{1}{2}\frac{\partial^2}{\partial r^2}(rC_l)-\frac{l(l+1)}{2r^2}(rC_l)-\frac{\partial}{\partial\beta}(rC_l)=0. \qquad (5.1)$$

The solutions of (5.1) satisfying the boundary condition (4.5) are, for the first two l values,

$$C_0(r',r,\beta)=\frac{1}{2^{\frac{3}{2}}\pi^{\frac{1}{2}}\beta^{\frac{1}{2}}r'r}\left\{\exp\left[-\frac{(r'-r)^2}{2\beta}\right]-\exp\left[-\frac{(r'+r)^2}{2\beta}\right]\right\},\quad(5.2a)$$

$$C_1(r',r,\beta)=\frac{1}{2^{\frac{3}{2}}\pi^{\frac{1}{2}}\beta^{\frac{1}{2}}r'r}\left\{\left(1-\frac{\beta}{r'r}\right)\exp\left[-\frac{(r'-r)^2}{2\beta}\right]+\left(1+\frac{\beta}{r'r}\right)\exp\left[-\frac{(r'+r)^2}{2\beta}\right]\right\}.\quad(5.2b)$$

Using the modified Bessel function

$$I_n(x)\equiv(-i)^nJ_n(ix)$$

the solution of (5.1) satisfying (4.5) may be obtained for all l in the form

$$C_l(r',r,\beta)=\frac{1}{4\pi\beta(r'r)^{\frac{1}{2}}}\exp\left(-\frac{r'^2+r^2}{2\beta}\right)I_{l+\frac{1}{2}}\left(\frac{r'r}{\beta}\right).\quad(5.3)$$

Setting $r'=r$, the corresponding solution of (4.7) with $V=0$ is

$$Z_l(r,\beta)=\frac{1}{4\pi\beta r}\exp\left(-\frac{r^2}{\beta}\right)I_{l+\frac{1}{2}}\left(\frac{r^2}{\beta}\right).\quad(5.4)$$

From the Laplace transform relation (4.10) we can obtain $\rho_l(r',r,E)$ from (5.2) and (5.3). Thus

$$\rho_0(r',r,E)=\frac{1}{4\pi^2r'r}\left\{\frac{\sin(r'-r)(2E)^{\frac{1}{2}}}{r'-r}-\frac{\sin(r'+r)(2E)^{\frac{1}{2}}}{r'+r}\right\},\quad(5.5a)$$

$$\rho_1(r',r,E)=\frac{1}{4\pi^2r'r}\left\{\left(\frac{\sin(r'-r)(2E)^{\frac{1}{2}}}{r'-r}-\frac{\cos(r'-r)(2E)^{\frac{1}{2}}}{r'r(2E)^{\frac{1}{2}}}\right)+\left(\frac{\sin(r'+r)(2E)^{\frac{1}{2}}}{r'+r}+\frac{\cos(r'+r)(2E)^{\frac{1}{2}}}{r'r(2E)^{\frac{1}{2}}}\right)\right\}.\quad(5.5b)$$

Equations (5.5) are particular cases of the general formula

$$\rho_l(r',r,E)=\frac{1}{2\pi^2}\int_0^k k^2j_l(kr')j_l(kr)dk,\quad(5.6)$$

where $k=(2E)^{\frac{1}{2}}$. Setting $r'=r$ we obtain from (5.5)

$$n_0(r,E)=\frac{1}{4\pi^2r^2}\left\{k-\frac{\sin2kr}{2r}\right\},\quad(5.7a)$$

$$n_1(r,E)=\frac{1}{4\pi^2r^2}\left\{k+\frac{\sin2kr}{2r}+\frac{\cos2kr-1}{kr^2}\right\},\quad(5.7b)$$

the general formula

$$n_l(r,E)=\frac{k^3}{4\pi^2}\{j_l^2(kr)-j_{l-1}(kr)j_{l+1}(kr)\}\quad(5.8)$$

being obtained from (5.6).

6. PERTURBATION TREATMENT FOR CENTRAL FIELDS

It appears possible for a spherical potential V to obtain a perturbation series solution of (4.12) which reduces to (5.7) or (5.8) when V vanishes. Writing

$$f=\frac{\partial}{\partial k}(r^2n_l),\quad\text{or}\quad r^2n_l=\int_0^k fdk,\quad(6.1)$$

Eq. (4.12) may be expressed as

$$\frac{\partial^3f}{\partial r^3}+4k^2\frac{\partial f}{\partial r}-\frac{4l(l+1)}{r^2}\frac{\partial f}{\partial r}+\frac{4l(l+1)}{r^3}f-8V\frac{\partial f}{\partial r}-4\frac{\partial V}{\partial r}f=0.\quad(6.2)$$

The three independent solutions of (6.2) when $V=0$ may be written

$$f_1=k^2r^2j_l^2(kr),\quad(6.3a)$$

$$f_2=k^2r^2j_l(kr)\bar{n}_l(kr),\quad(6.3b)$$

$$f_3=k^2r^2\bar{n}_l^2(kr),\quad(6.3c)$$

where, to avoid confusion with the particle density n_l, we have written $\bar{n}_l$ for the spherical Neumann function. The free-electron solution given in Sec. 5 is obtained from (6.1) by writing

$$f=f_1/2\pi^2.\quad(6.4)$$

To second order in V, it may be verified that the solution of (6.2) is then

$$\begin{aligned}f=\frac{f_1(r)}{2\pi^2}&+\frac{2}{\pi^2k}\times\left\{f_1(r)\int_r^\infty V(s)f_2(s)ds+f_2(r)\int_0^r V(s)f_1(s)ds\right\}\\&+\frac{8}{\pi^2k^2}\left\{f_2(r)\int_0^r V(s)f_1(s)\int_s^\infty V(t)f_2(t)dtds\right.\\&+f_1(r)\int_r^\infty V(s)f_2(s)\int_s^\infty V(t)f_2(t)dtds\\&-\tfrac{1}{2}f_1(r)\int_0^r V(s)f_3(s)\int_0^s V(t)f_1(t)dtds+\tfrac{1}{2}f_3(r)\\&\left.\times\int_0^r V(s)f_1(s)\int_0^s V(t)f_1(t)dtds\right\}.\end{aligned}\quad(6.5)$$

Thus combining (6.5) and (6.1), the perturbed density may be obtained to the same order of approximation.

7. PERTURBATION TREATMENT FOR NONSPHERICAL $V(r)$

After developing the methods described in the preceding sections, we became aware of a paper by Green[7] in which a very general perturbation treatment yielding the canonical density matrix was developed, in order to calculate the quantum mechanical partition function. We shall restrict ourselves to a first-order calculation here, although, in principle, higher terms may be obtained using Green's work. Equation (15) of Green's paper may be expressed almost immediately in our present notation as

$$C(\mathbf{r},\mathbf{r},\beta)=\frac{1}{(2\pi\beta)^{\frac{3}{2}}}\left[1-\int\frac{V(\mathbf{r}_1)\exp(-2R_1^2)/\beta}{\pi R_1}d\mathbf{r}_1\right], \quad (7.1)$$

where

$$R_1=\{(x-x_1)^2+(y-y_1)^2+(z-z_1)^2\}^{\frac{1}{2}}.$$

All that remains is to transform this to give the diagonal element of the Dirac matrix to the same order, using (2.11), the zeroth-order result being the free-electron density given in (3.6). The perturbed density is then found to be given by

$$n(\mathbf{r},\zeta)=\frac{k^3}{6\pi^2}-\frac{k^2}{4\pi^3}\int\frac{V(\mathbf{r}_1)j_1(2kR_1)}{R_1^2}d\mathbf{r}_1, \quad (7.2)$$

where $k=(2\zeta)^{\frac{1}{2}}$. This formula is new, and appears likely to be very useful in the imperfections field, as we shall now discuss.

8. APPLICATIONS TO IMPERFECTIONS IN METALS

Mott[8] developed a first-order approximation to deal with imperfections in metals, the result being expressed in the form

$$n(\mathbf{r},\zeta)=(2\zeta)^{\frac{3}{2}}/3\pi^2-q^2V(\mathbf{r})/4\pi, \quad (8.1)$$

where now we consider doubly filled levels. Here the screening radius $1/q$ is given in terms of the Fermi energy ζ by

$$q^2=2^{\frac{5}{2}}\zeta^{\frac{1}{2}}/\pi. \quad (8.2)$$

Various shortcomings of (8.1) are now well established. Thus, the density follows the potential too closely and at a point singularity, the density becomes infinite with the potential, an incorrect result. This is unfortunate, because in physically interesting problems involving positron annihilation in metals, the positron lifetime depends on the electron density at the positron and the Mott treatment is inadequate. Also, it is known from the work of Blandin *et al.*[9] and from previous computations by the present writers[4] that besides the localized screening predicted by (8.1) there are also long-range effects which Mott's treatment cannot account for.

We see now that the work presented in Sec. 7 allows the exact formulation of the first-order treatment, in which (8.1) must be replaced by

$$n(\mathbf{r},\zeta)=\frac{(2\zeta)^{\frac{3}{2}}}{3\pi^2}-\frac{k^2}{2\pi^3}\int\frac{V(\mathbf{r}_1)j_1(2kR_1)}{R_1^2}d\mathbf{r}_1. \quad (8.3)$$

The connection with Mott's treatment is seen directly when we make the assumption that the potential is slowly varying and replace $V(\mathbf{r}_1)$ by $V(\mathbf{r})$ in (8.3). Equation (8.1) then follows after a straightforward integration. Clearly, for a potential which is singular as r^{-1} the density given by (8.3) remains finite at the origin so that the qualitative defect of the Mott form is removed in this case. Also it seems that by combining (8.3) with Poisson's equation to yield

$$\nabla^2V(\mathbf{r})=\frac{2k^2}{\pi^2}\int\frac{V(\mathbf{r}_1)j_1(2kR_1)}{R_1^2}d\mathbf{r}_1, \quad (8.4)$$

we have a self-consistent field problem which should give some account of long-range effects in both the density and potential. Calculations are now being planned to enable the solutions of (8.4) to be obtained and we hope to report on this problem at a later stage.

We shall conclude by indicating the connection between the perturbation theory of Sec. 7 and that given for central field problems in Sec. 6. We have not seen at present a way to connect the two treatments completely generally in the case of a spherical potential,[9a] but we show here that they yield the same expression for the density at the origin to first order for central fields. This comparison is easily achieved, because only the partial density corresponding to $l=0$ contributes to the density at the origin when the separation in spherical harmonics is carried out. It follows from (6.5) and (6.1) that the density difference is given by

$$\frac{1}{\pi^2}\int_0^k dk\int_0^\infty V(s)\sin 2ks ds, \quad (8.5)$$

and the integration over k may be performed. The final result is

$$\frac{k^2}{\pi^2}\int_0^\infty j_1(2ks)V(s)ds, \quad (8.6)$$

and this is easily seen to be equivalent to that given by Eq. (7.2).

9. CONCLUSION

By exploiting the relation between the canonical and Dirac density matrices embodied in Eq. (2.11), we have

[7] H. S. Green, J. Chem. Phys. **20**, 1274 (1952).
[8] N. F. Mott, Proc. Cambridge Phil. Soc. **32**, 281 (1936).
[9] A. Blandin, E. Daniel, and J. Friedel, Phil. Mag. **4**, 180 (1959).

[9a] *Footnote added in proof.* This connection has now been found, and the proof is given in Appendix II.

shown how some progress may be made in the calculation of the Dirac matrix in certain cases. Suitable perturbation treatments based on free electrons may be developed, and the most important practical consequence would seem to be the exact first-order formulation which supersedes Mott's well-known treatment of imperfections in metals.

We also stress that the generalized canonical matrix D may be obtained from C, should it be required to calculate Fermi-Dirac physical properties at elevated temperatures. Furthermore we suggest that the new equations (4.7) and (4.12) for the diagonal elements of C and ρ, respectively, may have computational merit when perturbation theory fails. Generalizations of these equations when the potential is not spherical may also be obtained, but the results are complicated and will not therefore be recorded here.

APPENDIX I

In this appendix we derive (4.7), the equation satisfied by $Z_l(r,\beta)\equiv C_l(r,r,\beta)$, from (4.6) which is satisfied by $C_l(r',r,\beta)$. Multiplying (4.6) through by r' we obtain

$$\frac{1}{2}\frac{\partial^2}{\partial r^2}\{r'rC_l(r',r,\beta)\}-\left\{\frac{l(l+1)}{2r^2}+V(r)\right\}r'rC_l(r',r,\beta)-\frac{\partial}{\partial\beta}\{r'rC_l(r',r,\beta)\}=0. \quad \text{(A1.1)}$$

We now introduce new variables ξ, η defined by the relations

$$\xi=(r'+r')/2,\quad \eta=(r'-r)/2,$$

and hence $r=\xi+\eta$, $r'=\xi-\eta$. Since $r'rC_l(r',r,\beta)$ is symmetrical with respect to interchange of r and r' we may expand it in the form

$$r'rC_l(r',r,\beta)=\textstyle\sum_n \eta^{2n}c_n(\xi), \quad \text{(A1.2)}$$

where

$$r^{-2}c_0(r)=C_l(r,r,\beta)=Z_l(r,\beta).$$

We also expand $l(l+1)/2r^2+V(r)$ in the form

$$l(l+1)/2r^2+V(r)=\textstyle\sum_m \eta^m v_m(\xi), \quad \text{(A1.3)}$$

where

$$v_0(r)=l(l+1)/r^2+V(r),$$

and

$$v_1(r)=dv_0(r)/dr.$$

Writing (A1.1) in terms of ξ and η we obtain

$$\left\{\frac{1}{8}\frac{\partial^2}{\partial\eta^2}+\frac{1}{4}\frac{\partial^2}{\partial\eta\partial\xi}+\frac{1}{8}\frac{\partial^2}{\partial\xi^2}-\frac{\partial}{\partial\beta}\right\}\sum_n \eta^{2n}c_n(\xi)-\sum_m\sum_n \eta^{m+2n}c_n(\xi)v_m(\xi)=0.$$

Therefore it follows that

$$\sum_n\left\{\frac{n(2n-1)}{4}\eta^{2n-2}c_n+\frac{n}{2}\eta^{2n-1}c_n'+\tfrac{1}{8}\eta^{2n}c_n''-\eta^{2n}\frac{\partial c_n}{\partial\beta}\right\}-\sum_m\sum_n \eta^{m+2n}c_nv_m=0,$$

where primes on c_n denote derivatives with respect to ξ. We now equate to zero the coefficient of each power of η. The coefficient of η^0 gives then

$$\tfrac{1}{4}c_1+\tfrac{1}{8}c_0''-\frac{\partial c_0}{\partial\beta}-c_0v_0=0, \quad \text{(A1.4)}$$

and the coefficient of η,

$$\tfrac{1}{2}c_1'-c_0v_1=0. \quad \text{(A1.5)}$$

Hence, using the fact that v_1 is the derivative of v_0 and eliminating c_1 between (A1.4) and (A1.5) we obtain

$$\tfrac{1}{8}c_0'''-\tfrac{1}{2}v_0'c_0-v_0c_0'-\partial c_0'/\partial\beta=0. \quad \text{(A1.6)}$$

Replacing ξ by r, introducing the explicit form for v_0, and substituting $c_0(r)=r^2Z_l(r,\beta)$, (A1.6) is readily shown to be equivalent to (4.7).

Finally we show that (4.12) may be obtained directly from (4.7). Substituting (4.11) into (4.7), we obtain

$$\beta\int_0^\infty\left\{\frac{1}{8}\frac{\partial^3}{\partial r^3}(r^2n_l)-\frac{l(l+1)}{2r}\frac{\partial}{\partial r}(rn_l)-\frac{1}{2}\frac{\partial V}{\partial r}r^2n_l-V\frac{\partial}{\partial r}(r^2n_l)\right\}e^{-\beta E}dE-\frac{\partial}{\partial\beta}\left[\beta\int_0^\infty e^{-\beta E}\frac{\partial}{\partial r}(r^2n_l)dE\right]=0. \quad \text{(A1.7)}$$

Partial integration of the last term in the above yields

$$\beta\int_0^\infty\left\{\int_0^E E\frac{\partial^2}{\partial E\partial r}(r^2n_l)\right\}e^{-\beta E}dE,$$

and since (A1.7) must hold for all β, (4.12) follows.

APPENDIX II

We demonstrate in this Appendix the equivalence, to first order in V, of the methods of paragraphs 6 and 7, for the central field problem. The first-order terms of Eq. (6.5) may be written

$$\frac{1}{k}\frac{\partial}{\partial k}\{r^2(n_0-n)\}=-\frac{2k^2}{\pi^2}\sum_l(2l+1)\left[\int_0^r dsV(s)r^2s^2j_l(kr)\bar{n}_l(kr)j_l^2(ks)+\int_r^\infty dsV(s)r^2s^2j_l^2(kr)j_l(ks)\bar{n}_l(ks)\right] \quad \text{(A2.1)}$$

and we first focus attention on the quantity S defined by

$$S=-4k^2rs\sum_l(2l+1)j_l(kr)\bar{n}_l(kr)j_l^2(ks). \qquad \text{(A2.2)}$$

To proceed, we wish to differentiate S with respect to k, using the results of Schiff.[10] Strictly, terms corresponding to $l=0$ should be considered separately, but for convenience they may be included if we identify $j_{-1}(\rho)$ and $\bar{n}_{-1}(\rho)$ with $-\bar{n}_0(\rho)$ and $j_0(\rho)$, respectively. It then follows after some manipulation that

$$\frac{\partial S}{\partial k}=-4\sum_l(2l+1)k^2rs[r\{j_{l-1}(kr)\bar{n}_l(kr)-j_l(kr)\bar{n}_{l+1}(kr)\}\times j_l^2(ks)+sj_l(kr)\bar{n}_l(kr)\times\{j_{l-1}(ks)j_l(ks)-j_l(ks)j_{l+1}(ks)\}].$$

Using Schiff's Eq. (15.10) we obtain

$$\begin{aligned}\frac{\partial S}{\partial k}=&-4k^3r^2s^2\sum_l[j_{l-1}(kr)\bar{n}_l(kr)j_l(ks)j_{l-1}(ks)\\&+j_l(kr)\bar{n}_{l-1}(kr)j_{l-1}(ks)j_l(ks)\\&-j_l(kr)\bar{n}_{l+1}(kr)j_{l+1}(ks)j_l(ks)\\&-j_{l+1}(kr)\bar{n}_l(kr)j_l(ks)j_{l+1}(ks)]\\=&-4k^3r^2s^2[j_{-1}(kr)\bar{n}_0(kr)j_0(ks)j_{-1}(ks)\\&+j_0(kr)\bar{n}_{-1}(kr)j_{-1}(ks)j_0(ks)],\end{aligned}$$

since all other terms in the summation cancel. Interpreting $j_{-1}(\rho)$ and $\bar{n}_{-1}(\rho)$ in the manner discussed above we find

$$\begin{aligned}\frac{\partial S}{\partial k}&=\frac{4}{k}(\cos^2kr-\sin^2kr)\sin ks\cos ks\\&=\frac{1}{k}\{\sin 2k(r+s)-\sin 2k(r-s)\}. \qquad \text{(A2.3)}\end{aligned}$$

[10] L. I. Schiff, *Quantum Mechanics* (McGraw-Hill Book Company, Inc., New York, 1949), p. 78, Eq. (15.10).

Since $S(0)=0$, we have on integration, for $s\leqslant r$,

$$S(k)=\int_{2k(r-s)}^{2k(r+s)}du\frac{\sin u}{u}. \qquad \text{(A2.4)}$$

Thus, in (A2.1) we may write

$$\begin{aligned}&\frac{1}{k}\frac{\partial}{\partial k}\{r^2(n_0-n)\}\\&=\frac{1}{2\pi^2}\left\{\int_0^r dsV(s)rs\int_{2k(r-s)}^{2k(r+s)}du\frac{\sin u}{u}+\int_r^\infty dsV(s)rs\int_{2k(s-r)}^{2k(s+r)}du\frac{\sin u}{u}\right\}\\&=\frac{1}{2\pi^2}\int_0^\infty dsV(s)rs\int_{2k|r-s|}^{2k(r+s)}du\frac{\sin u}{u}. \qquad \text{(A2.5)}\end{aligned}$$

As may be verified by differentiation, the integral of (A2.5) is

$$r^2(n_0-n)=\frac{k^2}{2\pi^2}\int_0^\infty dsV(s)rs\times\int_{2k|r-s|}^{2k(r+s)}du\left(\frac{\sin u}{u^3}-\frac{\cos u}{u^2}\right). \qquad \text{(A2.6)}$$

Now when V has spherical symmetry, Eq. (7.2) may be written

$$n_0-n=\frac{k^2}{2\pi^2}\int_0^\pi d\theta\int_0^\infty dsV(s)\frac{s^2\sin\theta}{R_1^2}\times\left\{\frac{\sin 2kR_1}{(2kR_1)^2}-\frac{\cos 2kR_1}{2kR_1}\right\},$$

where $R_1^2=r^2+s^2-2rs\cos\theta$. Substituting $u=2kR_1$, it follows that

$$n_0-n=\frac{k^2}{2\pi^2}\int_0^\infty dsV(s)\frac{s}{r}\int_{2k|r-s|}^{2k(r+s)}du\left(\frac{\sin u}{u^3}-\frac{\cos u}{u^2}\right),$$

which is evidently equivalent to (A2.6).

[1285]

Electron Theory of Interaction between Point Defects in Metals

By G. K. Corless and N. H. March
Department of Physics, The University, Sheffield

[Received March 24, 1961]

Abstract

By employing a simple model of point charges in a uniform electron gas, it is shown how interaction energies between imperfections (for example, impurity–vacancy or vacancy–vacancy interactions) may be evaluated within the framework of low-order perturbation theory. Close contact may be established throughout with the earlier investigation of Alfred and March (1957), in which, however, a restrictive assumption that the perturbing potential was slowly varying had to be made. Our work now removes this assumption completely and provides a firm basis for the so-called 'electrostatic model' of the interaction between imperfections. In view of the long-range oscillations in the electrostatic potential around a defect, the interaction energy also has an oscillatory behaviour as a function of distance.

The relevance of our work to the Lazarus theory of impurity diffusion in metals is discussed. In this case, the numerical results are in fair agreement with those of the earlier treatment, and lend further support to the Lazarus theory. Also the investigation of Seeger and Bross (1956) on the di-vacancy may be critically examined by comparison with the present findings. It then appears that the terms in the total interaction energy which these earlier workers have omitted may be large. Indeed, in our calculations they are sufficient to change the sign of the interaction. It is finally pointed out that the present theory may have relevance in the interpretation of the results of Biondi and Rayne (1959, Rayne 1961) on optical absorption in Cu–Zn and Cu–Ge alloys.

§ 1. Introduction

We develop here a theory of the interaction between defects in metals, based on careful study of an admittedly over-simplified model. Thus we adopt the uniform gas approximation for the matrix metal, and assume that the point defects may be simulated by embedding appropriate point charges in the gas. Departures from such a model, particularly the use of Bloch waves rather than plane waves, and hence a non-uniform periodic density, for the conduction electrons in the unperturbed metal certainly require full investigation, but at present appear to lead to major complications.

We shall refer briefly here to a few of the numerous applications to which our interaction energy calculations are highly relevant. Lazarus (1954) developed a theory of impurity diffusion in metals based on a calculation of the interaction energy between an impurity and a vacancy (for a review of diffusion in metals, see Lazarus 1960). While Lazarus's earlier treatment was open to certain objections, particularly with regard to the neglect of correlation effects in diffusion (see, for example, Bardeen and Herring 1952, Manning 1958) and to the calculation of the energy of movement, the recent reformulation of the theory by Le Claire (1961) appears to go some way towards removing these unsatisfactory features. The other application of our work which we consider in some detail is to discuss the stability of the di-vacancy and to make an analysis of the earlier calculation of Seeger and Bross (1956).

The present work has been made possible by recent progress in the general theory of defects in metals (March and Murray 1960, 1961). While, in this earlier work, explicit results were obtained only for a single imperfection in an infinite metal, a full perturbation development of the Dirac density matrix was presented, and this is applicable to a general perturbing potential. We show here how considerable progress can be made in the two-centre case† (for example, impurity–vacancy or vacancy–vacancy complexes), using a first-order approximation. Our results reduce to the first-order treatment of Alfred and March (1957) when the assumption of a slowly varying potential is made. While these workers obtained the same final result as Lazarus for the interaction energy, it was clear that his original argument was valid only because of a major cancellation of the kinetic energy terms with part of the potential energy. We shall show by a quite general first-order argument that this cancellation occurs, independently of any assumption concerning the spatial variation of the potential.

§ 2. Density Matrices and Kinetic Energy Density

Following March and Murray (1961), we write for the Dirac density matrix $\rho(\mathbf{r}, \mathbf{r}_0, k)$, where $k^2 = 2E$, with E denoting the Fermi energy:

$$\rho(\mathbf{r}, \mathbf{r}_0, k) = \rho_0 + \rho_1 + \rho_2 + 0(V^3) \qquad (2.1)$$

where V is the perturbing potential energy. We then have the equations defining ρ_0, ρ_1 and ρ_2 as

$$\rho_0 = \frac{k^2}{\pi^2}\frac{j_1(k|\mathbf{r}-\mathbf{r}_0|)}{|\mathbf{r}-\mathbf{r}_0|}, \qquad (2.2)$$

$$\rho_1 = -\frac{k^2}{2\pi^3}\int d\mathbf{r}_1 \frac{V(\mathbf{r}_1) j_1(k|\mathbf{r}_1-\mathbf{r}_0| + k|\mathbf{r}-\mathbf{r}_1|)}{|\mathbf{r}_1-\mathbf{r}_0||\mathbf{r}-\mathbf{r}_1|}, \qquad (2.3)$$

$$\rho_2 = \frac{k^2}{4\pi^4}\int\int d\mathbf{r}_1 d\mathbf{r}_2 \frac{V(\mathbf{r}_1)V(\mathbf{r}_2) j_1(k|\mathbf{r}_1-\mathbf{r}_0| + k|\mathbf{r}_2-\mathbf{r}_1| + k|\mathbf{r}-\mathbf{r}_2|)}{|\mathbf{r}_1-\mathbf{r}_0||\mathbf{r}_2-\mathbf{r}_1||\mathbf{r}-\mathbf{r}_2|}. \qquad (2.4)$$

We note first that the kinetic energy T is given by

$$T = -\tfrac{1}{2}\int d\mathbf{r}[\nabla_{\mathbf{r}}^2 \rho]_{\mathbf{r}_0=\mathbf{r}}. \qquad (2.5)$$

† The generalization to M centres is readily achieved, as shown in Appendix 1.

Using (2.2)–(2.4) we then find by straightforward calculation:

$$\nabla_r^2\rho_0 = \frac{k^4}{\pi^2}\frac{j_1''(k|\mathbf{r}-\mathbf{r}_0|)}{|\mathbf{r}-\mathbf{r}_0|}, \qquad (2.6)$$

$$\nabla_r^2\rho_1 = \frac{2k^2}{\pi^2}\frac{V(\mathbf{r})j_1(k|\mathbf{r}-\mathbf{r}_0|)}{|\mathbf{r}-\mathbf{r}_0|}$$

$$-\frac{k^4}{2\pi^3}\int d\mathbf{r}_1\frac{V(\mathbf{r}_1)j_1''(k|\mathbf{r}-\mathbf{r}_1|+k|\mathbf{r}_1-\mathbf{r}_0|)}{|\mathbf{r}-\mathbf{r}_1||\mathbf{r}_1-\mathbf{r}_0|}, \qquad (2.7)$$

$$\nabla_r^2\rho_2 = \frac{k^4}{4\pi^4}\iint d\mathbf{r}_1 d\mathbf{r}_2\frac{V(\mathbf{r}_1)V(\mathbf{r}_2)j_1''(k|\mathbf{r}_1-\mathbf{r}_0|+k|\mathbf{r}_2-\mathbf{r}_1|+k|\mathbf{r}-\mathbf{r}_2|)}{|\mathbf{r}_1-\mathbf{r}_0||\mathbf{r}_2-\mathbf{r}_1||\mathbf{r}-\mathbf{r}_2|}$$

$$-\frac{k^2}{4\pi^4}\int d\mathbf{r}_1\frac{V(\mathbf{r}_1)V(\mathbf{r})4\pi j_1(k|\mathbf{r}_1-\mathbf{r}_0|+k|\mathbf{r}-\mathbf{r}_1|)}{|\mathbf{r}_1-\mathbf{r}_0||\mathbf{r}-\mathbf{r}_1|}. \qquad (2.8)$$

Evaluating (2.6), (2.7) and (2.8) at $\mathbf{r}_0=\mathbf{r}$, and inserting the results in (2.5) we find

$$T = \frac{k^5}{10\pi^2}\int d\mathbf{r} - \frac{k^3}{2\pi^2}\int d\mathbf{r}_1 V(\mathbf{r}_1)$$

$$+\frac{k^4}{2\pi^3}\iint d\mathbf{r}_1 d\mathbf{r}_2 V(\mathbf{r}_1)V(\mathbf{r}_2)\left[\frac{j_1'(2k|\mathbf{r}_2-\mathbf{r}_1|)}{2k|\mathbf{r}_2-\mathbf{r}_1|}+\frac{4j_1(2k|\mathbf{r}_2-\mathbf{r}_1|)}{(2k|\mathbf{r}_2-\mathbf{r}_1|)^2}\right]. \qquad (2.9)$$

At this stage we make use of the relation that to $0(V^2)$ the displaced charge $\rho-\rho_0$ is given by

$$\rho-\rho_0 = -\frac{k^2}{2\pi^3}\int d\mathbf{r}_1 V(\mathbf{r}_1)\frac{j_1(2k|\mathbf{r}-\mathbf{r}_1|)}{|\mathbf{r}-\mathbf{r}_1|^2}$$

$$+\frac{k^2}{4\pi^4}\iint d\mathbf{r}_1 d\mathbf{r}_2\frac{V(\mathbf{r}_1)V(\mathbf{r}_2)j_1(k|\mathbf{r}_1-\mathbf{r}|+k|\mathbf{r}_2-\mathbf{r}_1|+k|\mathbf{r}-\mathbf{r}_2|)}{|\mathbf{r}_1-\mathbf{r}||\mathbf{r}_2-\mathbf{r}_1||\mathbf{r}-\mathbf{r}_2|} \qquad (2.10)$$

and integrating (2.10) with respect to $\mathbf{r}$ we find

$$\int d\mathbf{r}(\rho-\rho_0) = -\frac{k}{\pi^2}\int d\mathbf{r}_1 V(\mathbf{r}_1)$$

$$+\frac{k}{2\pi^3}\iint d\mathbf{r}_1 d\mathbf{r}_2\frac{V(\mathbf{r}_1)V(\mathbf{r}_2)j_0(2k|\mathbf{r}_2-\mathbf{r}_1|)}{|\mathbf{r}_2-\mathbf{r}_1|}+0(V^3). \qquad (2.11)$$

Thus, using (2.11) we may rewrite (2.9) in the form

$$T = \frac{k^5}{10\pi^2}\int d\mathbf{r} + \frac{k^2}{2}\int d\mathbf{r}(\rho-\rho_0) + \frac{k^4}{\pi^3}\iint d\mathbf{r}_1 d\mathbf{r}_2\frac{V(\mathbf{r}_1)V(\mathbf{r}_2)j_1(2k|\mathbf{r}_2-\mathbf{r}_1|)}{(2k|\mathbf{r}_2-\mathbf{r}_1|)^2}. \qquad (2.12)$$

§ 3. Interaction Energies between Defects

We turn now to deal with the energies of interaction between two defects, carrying charges Z_1 and Z_2. We obtain the interaction energy by considering the changes in total energy as we bring the imperfections from infinity to the near-neighbour distance a.

The basic equation of the perturbation theory of March and Murray (1960) for the perturbing potential V, to first-order in V, is

$$\nabla^2 V(\mathbf{r}) = \frac{2k^2}{\pi^2}\int d\mathbf{r}_1 V(\mathbf{r}_1)\frac{j_1(2k|\mathbf{r}-\mathbf{r}_1|)}{|\mathbf{r}-\mathbf{r}_1|^2} \quad . \quad . \quad . \quad . \quad (3.1)$$

where k is the magnitude of the wave vector at the Fermi surface. This equation reduces to the Mott form

$$\nabla^2 V = q^2 V, \quad q^2 = 4k/\pi \quad . \quad . \quad . \quad . \quad (3.2)$$

when V is slowly varying in space. It will be seen that the superposition property of (3.2), exploited by Alfred and March (1957) in their two-centre work, is also valid for (3.1), and thus, within the first-order approximation, no new calculation of the density and potential for the defect complex under consideration is required. Thus, if charges Z_1 and Z_2 are situated at $\mathbf{a}_1$ and $\mathbf{a}_2$ respectively, and if V_{Z1} and V_{Z2} represent the one-centre potentials as given by March and Murray (1961), then the two-centre potential may be written as

$$V(\mathbf{r}) = V_{Z1}(\mathbf{r}-\mathbf{a}_1) + V_{Z2}(\mathbf{r}-\mathbf{a}_2). \quad . \quad . \quad . \quad . \quad (3.3)$$

We now proceed to obtain the desired kinetic energy change from (2.12), and in doing so we note that if we measure changes from the unperturbed gas state and use charge normalization requirements, only the third term on the right-hand side of (2.12) needs to be considered. We have then, for the complex Z_1Z_2 at separation $a = |\mathbf{a}_2 - \mathbf{a}_1|$, the contribution from this term as

$$\frac{k^4}{\pi^3}\int\int d\mathbf{r}_1 d\mathbf{r}_2 [V_{Z1}(\mathbf{r}_1-\mathbf{a}_1) + V_{Z2}(\mathbf{r}_1-\mathbf{a}_2)][V_{Z1}(\mathbf{r}_2-\mathbf{a}_1) + V_{Z2}(\mathbf{r}_2-\mathbf{a}_2)] \times j_1(2k|\mathbf{r}_2-\mathbf{r}_1|)^2. \quad (3.4)$$

The one-centre terms are now seen to correspond to the kinetic energy changes at infinite separation of Z_1 and Z_2 and thus the kinetic energy change ΔT is given by

$$\Delta T = \frac{k^4}{\pi^3}\int\int d\mathbf{r}_1 d\mathbf{r}_2 \frac{V_{Z1}(\mathbf{r}_1-\mathbf{a}_1)V_{Z2}(\mathbf{r}_2-\mathbf{a}_2)j_1(2k|\mathbf{r}_2-\mathbf{r}_1|)}{(2k|\mathbf{r}_2-\mathbf{r}_1|)^2} + \frac{k^4}{\pi^3}\int\int d\mathbf{r}_1 d\mathbf{r}_2 \frac{V_{Z2}(\mathbf{r}_1-\mathbf{a}_2)V_{Z1}(\mathbf{r}_2-\mathbf{a}_1)j_1(2k|\mathbf{r}_2-\mathbf{r}_1|)}{(2k|\mathbf{r}_2-\mathbf{r}_1|)^2}. \quad . \quad . \quad . \quad (3.5)$$

We turn next to the evaluation of the potential energy change ΔU in this approximation. As we are working in a Hartree framework, we have that the potential energy of the gas with charges Z_1 and Z_2 at $\mathbf{a}_1$ and $\mathbf{a}_2$ is

$$\int d\mathbf{r}\rho(\mathbf{r})V_0(\mathbf{r}) + \tfrac{1}{2}\int d\mathbf{r}_1 d\mathbf{r}_2 \frac{\rho(\mathbf{r}_1)\rho(\mathbf{r}_2)}{|\mathbf{r}_2-\mathbf{r}_1|} + \frac{Z_1Z_2}{|\mathbf{a}_2-\mathbf{a}_1|} \quad (3.6)$$

where

$$V_0(\mathbf{r}) = -\frac{Z_1}{|\mathbf{r}-\mathbf{a}_1|} - \frac{Z_2}{|\mathbf{r}-\mathbf{a}_2|} = V_0(\mathbf{r}-\mathbf{a}_1) + V_0(\mathbf{r}-\mathbf{a}_2)$$

and

$$V(\mathbf{r}_1) = V_0(\mathbf{r}_1) + \int d\mathbf{r}_2 \frac{\rho(\mathbf{r}_2)}{|\mathbf{r}_2-\mathbf{r}_1|}.$$

Clearly (3.6) may now be expressed as

$$\tfrac{1}{2}\int d\mathbf{r}\rho(\mathbf{r})V(\mathbf{r})+\tfrac{1}{2}\int d\mathbf{r}\rho(\mathbf{r})V_0(\mathbf{r})+\frac{Z_1Z_2}{|\mathbf{a}_2-\mathbf{a}_1|}\cdot \qquad (3.7)$$

Using (2.1) for $\rho(\mathbf{r})$, adopting the superposition potential energy (3.3), and omitting unperturbed and one-centre terms we immediately find to second order in V,

$$\Delta U=\tfrac{1}{2}\int d\mathbf{r}\rho_{1Z1}(\mathbf{r}-\mathbf{a}_1)V_{Z2}(\mathbf{r}-\mathbf{a}_2)+\tfrac{1}{2}\int d\mathbf{r}\rho_{1Z2}(\mathbf{r}-\mathbf{a}_2)V_{Z1}(\mathbf{r}-\mathbf{a}_1)$$

$$+\frac{Z_1Z_2}{|\mathbf{a}_2-\mathbf{a}_1|}-\frac{Z_1}{2}\int d\mathbf{r}\frac{\rho_{1Z1}(\mathbf{r}-\mathbf{a}_1)}{|\mathbf{r}-\mathbf{a}_2|}-\frac{Z_2}{2}\int d\mathbf{r}\frac{\rho_{1Z2}(\mathbf{r}-\mathbf{a}_2)}{|\mathbf{r}-\mathbf{a}_1|}, \quad . \quad . \quad (3.8)$$

where ρ_{1Z1} and ρ_{1Z2} are related to V_{Z1} and V_{Z2} by the Poisson equation

$$\nabla^2V_{Zi}(\mathbf{r}-\mathbf{a}_i)=-4\pi\rho_{1Zi}(\mathbf{r}-\mathbf{a}_i), \quad i=1,2. \qquad (3.9)$$

Using (3.9) is it easily shown that (3.8) reduces to the form

$$\Delta U=-\frac{Z_1}{2}V_{Z2}(|\mathbf{a}_1-\mathbf{a}_2|)-\frac{Z_2}{2}V_{Z1}(|\mathbf{a}_2-\mathbf{a}_1|)$$

$$+\tfrac{1}{2}\int d\mathbf{r}\rho_{1Z1}(\mathbf{r}-\mathbf{a}_1)V_{Z2}(\mathbf{r}-\mathbf{a}_2)+\tfrac{1}{2}\int d\mathbf{r}\rho_{1Z2}(\mathbf{r}-\mathbf{a}_2)V_{Z1}(\mathbf{r}-\mathbf{a}_1). \quad (3.10)$$

But from (2.3) we may write

$$\rho_{1Zi}(\mathbf{r}-\mathbf{a}_i)=-\frac{k^2}{2\pi^3}\int d\mathbf{r}_1\frac{V_{Zi}(\mathbf{r}_1-\mathbf{a}_i)j_1(2k|\mathbf{r}_1-\mathbf{r}|)}{|\mathbf{r}_1-\mathbf{r}|^2} \qquad (3.11)$$

and it may then be shown readily that (3.5) has the form

$$\Delta T=-\tfrac{1}{2}\int d\mathbf{r}\,\rho_{1Z1}(\mathbf{r}-\mathbf{a}_1)V_{Z2}(\mathbf{r}-\mathbf{a}_2)$$

$$-\tfrac{1}{2}\int d\mathbf{r}\,\rho_{1Z2}(\mathbf{r}-\mathbf{a}_2)V_{Z1}(\mathbf{r}-\mathbf{a}_1). \quad . \quad . \quad . \quad (3.12)$$

Thus, adding (3.10) and (3.12) we are left with the final simple result

$$\Delta E=\Delta T+\Delta U=-\frac{Z_1}{2}V_{Z2}(|\mathbf{a}_1-\mathbf{a}_2|)-\frac{Z_2}{2}V_{Z1}(|\mathbf{a}_2-\mathbf{a}_1|). \qquad (3.13)$$

We now demonstrate directly the connection with the earlier treatment given by Alfred and March (1957). As pointed out in § 3, the basic eqn. (3.1) of the first-order treatment reduces to the Mott eqn. (3.2) in the slowly varying potential limit, and the appropriate one-centre solution is

$$V_{Zi}(r)=-\frac{Z_i}{r}\exp(-qr). \quad . \quad . \quad . \quad . \quad (3.14)$$

Thus, from (3.13) we then find

$$\Delta E=\frac{Z_1Z_2\exp(-qa)}{a} \quad . \quad . \quad . \quad . \quad . \quad . \quad (3.15)$$

which is just Lazarus's formula. A numerical comparison of the predictions of (3.13) and (3.15) will be made later.

§ 4. Impurity Diffusion in Metals

In the Lazarus theory, as modified by Le Claire, the energy of movement is calculated using a model for the saddle-point configuration which was originated by Huntington (1942). The model consists of a linear three-centre configuration of point charges, $-\frac{1}{2}$, Z, $-\frac{1}{2}$, the half-vacancies being separated from the impurity by distances $(11/16)a$ in the Huntington approximation (see Le Claire 1961 for detailed reasons for this choice). The M-centre calculation of Appendix 1 immediately gives the desired result, which, in Le Claire's notation, may be written quantitatively as

$$\Delta H_2 + \Delta E = -\tfrac{1}{2} V_Z(11a/16) + Z V_{-1/2}(11a/16). \quad . \quad . \quad (4.1)$$

Substituting $V_Z(r)$ as given by the slowly varying potential approximation in (4.1), we find

$$\Delta H_2 + \Delta E = -\frac{Z}{(11/16)a} \exp(-11qa/16). \quad . \quad . \quad (4.2)$$

This differs from Le Claire's result only by a factor α of the order of unity, α being Z dependent (see Alfred and March 1956). The introduction of α may be thought of as an attempt to introduce higher order perturbation corrections (see also the remarks of § 6) and is necessary if curvature in the activation energy versus Z plot is to be explained. In our approximation, α is simply unity.

We record in the table results for divalent impurities in copper, as calculated from (3.13) and (3.15) for ΔE, and from (4.1) and (4.2) for $\Delta H_2 + \Delta E$. It will be seen that the results of our treatment provide a fair numerical justification of the formulae used previously. In view of this, and the crude approximation involved in the choice of the saddle-point model, it does not seem necessary at this stage to rework the predictions of the theory of impurity diffusion as presented by Le Claire (1961), although eventually numerical revision may become desirable.

Results for various energy terms for divalent impurities in copper

Impurity–vacancy interaction energy ΔE ($a = 4{\cdot}8$ atomic units)

(1) From eqn. (3.13): $\Delta E = -0{\cdot}0028$ ($-0{\cdot}076$ ev)

(2) From Lazarus formula (3.15): $\Delta E = -0{\cdot}0021$ ($-0{\cdot}057$ ev)

Energy $\Delta H_2 + \Delta E$, *for saddle-point configuration described by Huntington's model* (see § 4)

(1) From eqn. (4.1): $\Delta H_2 + \Delta E = -0{\cdot}0095$ ($-0{\cdot}26$ ev).

(2) From eqn. (4.2): $\Delta H_2 + \Delta E = -0{\cdot}0128$ ($-0{\cdot}35$ ev).

§ 5. Stability of Di-vacancy: Comparison with the Calculation of Seeger and Bross

The existence of the di-vacancy is important in a number of fields (see, for example, Lomer 1959). As Lomer points out, the stability of the di-vacancy does not seem established beyond doubt, although recent work (Weizer and Girifalco 1960) suggests on both theoretical and experimental grounds that it should be stable.

The only calculation which considers the electronic distribution round the di-vacancy, prior to our work, is that of Seeger and Bross (1956). However, their work on the di-vacancy is incomplete, in the same sense that the work of Fumi (1955) on the formation energy of a vacancy was incomplete. While some justification can be given for the partial cancellation of the terms in the energy omitted by Fumi (Lidiard and Tharmalingham, private communication) for the single-centre problem, no consideration has as yet been given to this point for the di-vacancy.

In order to throw light on this point, we shall now attempt to cast our treatment into a form which parallels the calculation of Seeger and Bross. We note first that these workers, following Fumi, focus attention only on changes in eigenvalue sums for model potentials chosen to satisfy the Friedel sum rule. It is clear then that their calculations are not complete, in that Coulomb corrections are not properly accounted for. In our work, on the other hand, we have used the full energy expressions within a Hartree framework.

We consider then, at this stage, the change in eigenvalue sum using our model of the di-vacancy. The potential energy term in the eigenvalue sum for a complex Z_1Z_2 is

$$\int d\mathbf{r}\,\rho(\mathbf{r})V(\mathbf{r}) \quad . \quad . \quad . \quad . \quad . \quad . \quad . \quad . \quad (5.1)$$

and we now invoke the superposition theorem to obtain

$$\int d\mathbf{r}[\rho_{1Z1}(\mathbf{r}-\mathbf{a}_1)+\rho_{1Z2}(\mathbf{r}-\mathbf{a}_2)-\rho_0][V_{Z1}(\mathbf{r}-\mathbf{a}_1)+V_{Z2}(\mathbf{r}-\mathbf{a}_2)]. \qquad (5.2)$$

Again, separating out unperturbed, one-centre, and two-centre terms, the potential energy contribution to the change in eigenvalue sum is clearly

$$\Delta U_{\text{eigenvalue}}=\int d\mathbf{r}[\rho_{1Z1}(\mathbf{r}-\mathbf{a}_1)V_{Z2}(\mathbf{r}-\mathbf{a}_2)+\rho_{1Z2}(\mathbf{r}-\mathbf{a}_2)V_{Z1}(\mathbf{r}-\mathbf{a}_1)]. \quad (5.3)$$

From the Mott approximate solution (3.14) we obtain the result, for $Z_1=Z_2=-1$

$$-q\exp(-qa). \quad . \quad . \quad . \quad . \quad . \quad . \quad . \quad . \quad (5.4)$$

(5.3) has been obtained numerically, as it seemed of interest to know how the energy change was built up from potential and kinetic contributions. We obtain

$$\Delta U_{\text{eigenvalue}}=-0{\cdot}0074 \text{ atomic units}=-0{\cdot}20\,\text{ev},$$

whereas (5.4) yields $-0{\cdot}26$ ev.

Also, from (3.12) it follows that

$$\Delta T=-\tfrac{1}{2}\Delta U_{\text{eigenvalue}}=0{\cdot}10\,\text{ev},$$

and hence

$$\Delta E_{\text{eigenvalue}}=-0{\cdot}10\,\text{ev},$$

whereas using the approximate solution (3.14) the result is $-0\cdot13$ ev. Since $\Delta E_{\text{eigenvalue}}$ is just the quantity which Seeger and Bross take as a measure of the binding energy, these calculations would then indicate stability of the di-vacancy, and our result of 0·1 ev is to be compared with their value of 0·3 ev. Unfortunately, our estimate of the total energy change ΔE is given by (3.13) and is $\Delta E=0\cdot076$ ev. This indicates repulsion and we see that in the present framework even the sign of the interaction is changed when we use the eigenvalue sum as a measure of the binding energy. Our work therefore must cast serious doubt on the validity of the Seeger–Bross energy; as even if we adopt our Coulomb correction of 0·18 ev and their eigenvalue sum, their binding energy is reduced to 0·12 ev.

§ 6. Higher Approximations

There is little doubt that, even within the framework of the present model, higher approximations should eventually be considered. Thus, the one-centre calculations of March and Murray (1960, 1961) show that the electron density given by the first-order eqn. (3.1), while containing all the essential features of the problem, is not capable of complete quantitative accuracy. The framework developed by these workers is general enough however so that the second-order corrections can be estimated in principle, and this is under discussion for the one-centre case (March and Murray, to be published). The calculations for the present problem are, unfortunately, very formidable indeed, the main obstacle being the loss of the superposition property when terms of $0(V^2)$ are included in (3.1). Thus, entirely new calculations of the two-centre potential and density would be required. However, such a calculation was in fact carried through completely by Alfred and March (1957) for a divalent impurity and a vacancy in Cu, in the simpler case when the potential is assumed slowly varying, and this enables us to draw some tentative conclusions.

Thus, the full computations of Alfred and March referred to above showed that while a result was obtained in quite reasonable numerical agreement with the first-order interaction energy, the inclusion of exchange was important. What emerged was that while the kinetic energy was given reasonably satisfactorily by the first-order treatment, the summation of the slowly varying potential series to infinite order in V led to a rather different potential energy, but the inclusion of exchange brought the result back to approximately the first-order value. Also, as pointed out by Alfred and March (1957), inclusion of exchange in first order completely changes the results, and appears to lead to unreasonable conclusions†. We infer then that, at least for the impurity–vacancy problem, the full calculation of Alfred and March is most probably much better than might

† Nevertheless, for completeness it seemed worthwhile to include in Appendix 2 a brief derivation of the expression for the exchange energy using the density matrix formulation employed here.

have been supposed from the apparently severe approximation of slowly varying potentials.

§ 7. Conclusion

Within a Hartree approximation and assuming the validity of low-order perturbation theory, we have shown how a rather general treatment of interaction energies between point charges in an electron gas may be developed. In particular, the so-called 'electrostatic' model of interactions between defects in metals follows as a direct consequence of our perturbation approach. However, in view of the long-range oscillations in the electrostatic potential around a defect, it follows that the interaction energy also oscillates with distance.

It turns out that in the theory of impurity diffusion, proposed by Lazarus and recently modified by Le Claire, these oscillatory effects should not be very important, and in particular the major term, the energy of movement, should not be greatly affected. Thus, even though detailed numerical revision will eventually be called for, our work seems to substantiate further the current theory. On the other hand, we feel that stability of the di-vacancy cannot be explained in terms of the point charge model employed here. The work of Seeger and Bross (1956) has been shown to be open to criticism on the grounds that changes in sums of eigenvalues are not a proper measure of the interaction energy. Indeed, in our calculations, even the sign is changed by the terms Seeger and Bross omit.

Undoubtedly, in the applications referred to above, reservations must be made concerning the limitations of the model. No doubt relaxation effects around vacancies and impurities will eventually have to be included. Within the uniform gas approximation, a crude account could perhaps be taken of nearest-neighbour displacements using a model like that of Lee and March (1960).

Finally we might briefly refer to some recent measurements of Rayne (1961) on optical absorption in Cu–Zn and Cu–Ge alloys. These experiments may be thought of as measuring the change in the Fermi level with impurity concentration, and are in serious conflict with Friedel's earlier theory (Friedel 1954). However, since the change in the Fermi level is dependent on the interaction between the impurities, that is on the overlap of the screening distributions, it seems very likely that the present treatment, with long-range oscillatory terms, will lead to results markedly different from those based on essentially exponential screening. It is hoped to report on a quantitative treatment of this problem at a later stage.

Acknowledgments

It is a pleasure to thank Dr. A. M. Murray for valuable discussions and material help with some aspects of the work, and Dr. W. H. Young for reading a preliminary manuscript and, by means of numerous suggestions, considerably influencing the final presentation of the theory.

The work reported here was performed under contract with the United Kingdom Atomic Energy Authority, A.E.R.E., Harwell. One of us (N. H. M.) would particularly like to thank Mr. A. D. Le Claire and the other members of his group for the stimulus and encouragement which he derived from a brief visit as a Vacation Consultant in the summer of 1959, during which period the need for the present investigation became clear.

APPENDIX 1

Interaction Energy in Multi-centre Problem

We outline here the generalization of the result of § 3 for two centres to M-centres. We now write

$$V_0(\mathbf{r}) = -\frac{Z_1}{|\mathbf{r}-\mathbf{a}_1|} - \frac{Z_2}{|\mathbf{r}-\mathbf{a}_2|} - \ldots = V_{0Z1} + V_{0Z2} + \ldots. \qquad \text{(A 1.1)}$$

Obvious generalization of the method of § 3 then gives for the potential energy

$$PE = \tfrac{1}{2}\int d\mathbf{r}[\rho_0+\rho_1+\ldots][V_{Z1}(\mathbf{r}-\mathbf{a}_1)+V_{Z2}(\mathbf{r}-\mathbf{a}_2)+\ldots$$
$$+V_{0Z1}(\mathbf{r}-\mathbf{a}_1)+V_{0Z2}(\mathbf{r}-\mathbf{a}_2)+\ldots] + \sum_{i>j}\frac{Z_iZ_j}{|\mathbf{a}_i-\mathbf{a}_j|}$$

and again omitting one-centre and unperturbed terms we have for the potential energy contribution to the interaction energy

$$\Delta U = -\tfrac{1}{2}\sum_{i\neq j} Z_i \int d\mathbf{r}\frac{\rho_{1Zj}(\mathbf{r}-\mathbf{a}_j)}{|\mathbf{r}-\mathbf{a}_i|} + \tfrac{1}{2}\sum_{i\neq j}\frac{Z_iZ_j}{|\mathbf{a}_i-\mathbf{a}_j|} + \tfrac{1}{2}\sum_{i\neq j}\int d\mathbf{r}\,\rho_{1Zj}V_{Zi}(\mathbf{r}-\mathbf{a}_i)$$
$$= -\tfrac{1}{2}\sum_{i\neq j} Z_iV_{Zj}(\mathbf{a}_i-\mathbf{a}_j) + \tfrac{1}{2}\sum_{i\neq j}\int d\mathbf{r}\,\rho_{1Zj}V_{Zi}(\mathbf{r}-\mathbf{a}_i). \qquad \text{(A 1.2)}$$

Similarly, we have for the kinetic energy contribution to the interaction energy

$$\frac{k^4}{\pi^3}\sum_{i\neq j}\int\int d\mathbf{r}_1 d\mathbf{r}_2 \frac{V_{Zi}(\mathbf{r}_1-\mathbf{a}_i)V_{Zj}(\mathbf{r}_2-\mathbf{a}_j)j_1(2k|\mathbf{r}_1-\mathbf{r}_2|)}{(2k|\mathbf{r}_1-\mathbf{r}_2|)^2}$$

and using the result that

$$\rho_{1Zj} = -\frac{k^2}{2\pi^3}\int d\mathbf{r}_1 \frac{V_{Zj}(\mathbf{r}_1-\mathbf{a}_j)j_1(2k|\mathbf{r}_1-\mathbf{r}_2|)}{|\mathbf{r}_1-\mathbf{r}_2|^2}$$

we find

$$\Delta T = -\tfrac{1}{2}\sum_{i\neq j}\int d\mathbf{r}\,\rho_{1Zj}V_{Zi}(\mathbf{r}-\mathbf{a}_i). \qquad \text{(A 1.3)}$$

Thus from (A 1.2) and (A 1.3) we have

$$\Delta E = \Delta T + \Delta U = -\tfrac{1}{2}\sum_{i\neq j} Z_iV_{Zj}(\mathbf{a}_i-\mathbf{a}_j), \qquad \text{(A 1.4)}$$

which is the desired generalization of (3.13).

APPENDIX 2

Exchange Energy in Perturbation Approach based on the Density Matrix

In terms of the Dirac density matrix $\rho(\mathbf{r}, \mathbf{r}_0, k)$ we may write for the exchange energy

$$A = -\tfrac{1}{4}\int\int d\mathbf{r}\, d\mathbf{r}_0 \frac{[\rho(\mathbf{r}, \mathbf{r}_0, k)]^2}{|\mathbf{r}-\mathbf{r}_0|}. \quad . \quad . \quad . \quad (\text{A } 2.1)$$

We now insert the perturbation expansion (2.1) for ρ in (A 2.1) and we find

$$A = -\tfrac{1}{4}\int\int d\mathbf{r}\, d\mathbf{r}_0 \frac{[\rho_0{}^2 + 2\rho_0\rho_1 + \rho_1{}^2 + 2\rho_0\rho_2]}{|\mathbf{r}-\mathbf{r}_0|} \quad (\text{A } 2.2)$$

where we retain only second-order terms in V. We may then show that

$$-\tfrac{1}{4}\int\int d\mathbf{r}\, d\mathbf{r}_0 \frac{\rho_0{}^2}{|\mathbf{r}-\mathbf{r}_0|} = -\frac{3}{4}\left(\frac{3}{\pi}\right)^{1/3}\int d\mathbf{r}[\rho_0(\mathbf{r},\mathbf{r})]^{4/3}, \quad (\text{A } 2.3)$$

which is the usual result for free electrons. For the next term in (A 2.2) we may write almost immediately

$$-\tfrac{1}{2}\int\int d\mathbf{r}\, d\mathbf{r}_0 \frac{\rho_0\rho_1}{|\mathbf{r}-\mathbf{r}_0|} = -\frac{k}{\pi}\int d\mathbf{r}[\rho(\mathbf{r},\mathbf{r}) - \rho_0(\mathbf{r},\mathbf{r})] + 0(V^2). \quad (\text{A } 2.4)$$

There are three contributions to the term of $0(V^2)$, as listed below:

(i) $$\frac{k^2}{2\pi^4}\int\int d\mathbf{r}_1\, d\mathbf{r}_2 \frac{V(\mathbf{r}_1)V(\mathbf{r}_2)}{|\mathbf{r}_1-\mathbf{r}_2|} j_0(2k|\mathbf{r}_2-\mathbf{r}_1|),$$

(ii) $$-\frac{k^4}{16\pi^6}\int\int \frac{d\mathbf{r}\, d\mathbf{r}_0}{|\mathbf{r}-\mathbf{r}_0|}\left[\int d\mathbf{r}_1 \frac{V(\mathbf{r}_1) j_1(k|\mathbf{r}_1-\mathbf{r}_0| + k|\mathbf{r}-\mathbf{r}_1|)}{|\mathbf{r}_1-\mathbf{r}_0||\mathbf{r}-\mathbf{r}_1|}\right]^2,$$

(iii) $$-\frac{k^4}{8\pi^6}\int\int d\mathbf{r}\, d\mathbf{r}_0 \frac{j_1(k|\mathbf{r}-\mathbf{r}_0|)}{|\mathbf{r}-\mathbf{r}_0|^2}\int\int d\mathbf{r}_1\, d\mathbf{r}_2 V(\mathbf{r}_1)V(\mathbf{r}_2)$$
$$\times \frac{j_1(k|\mathbf{r}_1-\mathbf{r}_0| + k|\mathbf{r}_2-\mathbf{r}_1| + k|\mathbf{r}-\mathbf{r}_2|)}{|\mathbf{r}_1-\mathbf{r}_0||\mathbf{r}_2-\mathbf{r}_1||\mathbf{r}-\mathbf{r}_2|}.$$

In the case when V is constant, it is of interest to note that the sum of the above second-order terms reduces to

$$-\frac{V^2}{2\pi^3}\int d\mathbf{r}$$

or in terms of the density difference

$$-\frac{1}{6}\left(\frac{3}{\pi\rho_0{}^2}\right)^{1/3}\int d\mathbf{r}(\rho-\rho_0)^2,$$

a result which is easily shown to follow from the Dirac expression (A 2.3) for the exchange energy. This, in fact, is also true for the contribution (A 2.4).

1296 *On Electron Theory of Interaction between Point Defects in Metals*

References

ALFRED, L. C. R., and MARCH, N. H., 1956, *Phys. Rev.*, **103,** 877; 1957, *Phil. Mag.*, **2,** 985.

BARDEEN, J., and HERRING, C., 1952, *Imperfections in Nearly Perfect Crystals* (New York: Wiley), p. 261.

BIONDI, M. A., and RAYNE, J. A., 1959, *Phys. Rev.*, **115,** 1522.

FRIEDEL, J., 1954, *Advanc. Phys.*, **3,** 44.

FUMI, F. G., 1955, *Phil. Mag.*, **46,** 1007.

HUNTINGTON, H. B., 1942, *Phys. Rev.*, **61,** 325.

LAZARUS, D., 1954, *Phys. Rev.*, **93,** 973; 1960, *Advances in Solid State Physics*, Vol. 10 (New York: Academic Press Inc.).

LE CLAIRE, A. D., 1961, *Phil. Mag.* (to be published).

LEE, P. M., and MARCH, N. H., 1960, *Phys. Rev.*, **118,** 138.

LOMER, W. M., 1959, *Progress in Metal Physics*, Vol. 8 (Pergamon Press), p. 255.

MANNING, J. R., 1958, *Phys. Rev. Letters*, **1,** 365.

MARCH, N. H., and MURRAY, A. M., 1960, *Phys. Rev.*, **120,** 830; 1961, *Proc. roy. Soc.* A, **261,** 119.

RAYNE, J. A., 1961, *Phys. Rev.*, **121,** 456.

SEEGER, A., and BROSS, H., 1956, *Z. Phys.*, **145,** 161.

WEIZER, V. G., and GIRIFALCO, L. A., 1960, *Phys. Rev.*, **120,** 837.

Exact Thomas–Fermi method in perturbation theory

By J. C. Stoddart and N. H. March
Department of Physics, The University, Sheffield

(*Communicated by C. A. Coulson, F.R.S.—Received* 21 *November* 1966)

A variational principle is constructed which leads directly to the March–Murray perturbation series relating the particle density $\rho(\mathbf{r})$ with the one-body potential $V(\mathbf{r})$ in an electron gas. An explicit, though perturbative, form for the Hohenberg–Kohn universal functional, which is essentially the kinetic energy density, is thereby obtained.

When the spatial variations of density and potential are slow, the usual Thomas–Fermi $\rho^{\frac{5}{3}}$ relation between kinetic energy density and particle density is regained. The present approach also leads to a systematic formulation of the gradient expansion of the kinetic energy density.

1. Introduction

Many attempts have been made to generalize the simple Thomas–Fermi theory, which basically relates the particle density $\rho(\mathbf{r})$ to the one-body potential $V(\mathbf{r})$ in which all the particles are assumed to move, through the local relation

$$\rho(\mathbf{r}) = (1/6\pi^2)\,[\{k^2 - 2V(\mathbf{r})\}]^{\frac{3}{2}}, \tag{1}$$

where k is the Fermi momentum. Clearly, this relation is oversimplified when V varies rapidly in space. For example, when $V(\mathbf{r}) \alpha r^{-1}$ for small r, (1) predicts a density which becomes singular as $r^{-\frac{3}{2}}$, whereas it is well known that s-wave functions, and hence electron densities, in an atom are finite at the origin. The singular behaviour of ρ is, of course, a serious defect of the Thomas–Fermi atom.

It is part of the attractiveness of the Thomas–Fermi theory that (1) arises from a variational principle, which for independent particles moving in a prescribed potential $V(\mathbf{r})$, may be written

$$\delta\epsilon(\rho) = 0,$$

where
$$\epsilon(\rho) = \tfrac{3}{10}\,(6\pi^2)^{\frac{2}{3}} \int \{\rho(\mathbf{r})\}^{\frac{5}{3}}\,\mathrm{d}\mathbf{r} + \int \rho(\mathbf{r})\,V(\mathbf{r})\,\mathrm{d}\mathbf{r}. \tag{2}$$

The density variation is carried out subject only to the normalization condition

$$\int \rho(\mathbf{r})\,\mathrm{d}\mathbf{r} = N, \tag{3}$$

where N is the number of particles. Equation (3) essentially defines the Fermi energy or momentum appearing in (1).

For our purposes here, we define a generalized Thomas–Fermi method as one which directly connects $\rho(\mathbf{r})$ and $V(\mathbf{r})$, though normally through a non-local relation, and corresponding to which is a variational principle $\delta\epsilon = 0$, yielding this relationship. The object of the paper is to discuss the generalized $\rho(V)$ relation, which has previously been given by March & Murray (1960, 1961) and to derive its associated variational principle, which is the direct generalization of (2).

2. The density–potential relation

A perturbative generalization of (1) was given some years ago by March & Murray (1961) in the course of a study of the Dirac density matrix. The March–Murray relation between $\rho(\mathbf{r})$ and $V(\mathbf{r})$ may be written

$$\rho(\mathbf{r}) = \sum_{j=0}^{\infty} \rho_j(\mathbf{r}), \tag{4}$$

where

$$\rho_j(\mathbf{r}) = \frac{k^2}{2\pi^2}\left(\frac{-1}{2\pi}\right)^j \int \dots \int \mathrm{d}\mathbf{r}_1\,\mathrm{d}\mathbf{r}_2 \dots \mathrm{d}\mathbf{r}_j\, V(\mathbf{r}_1)\, V(\mathbf{r}_2) \dots V(\mathbf{r}_j) \frac{j_1\left(k\sum_{l=1}^{j+1} s_l\right)}{\prod_{l=1}^{j+1} s_l}. \tag{5}$$

In (5), $j_1(z)$ is the usual first order spherical Bessel function defined by

$$j_1(z) = \frac{\sin z - z\cos z}{z^2},$$

and $s_l = |\mathbf{r}_l - \mathbf{r}_{l-1}|$ with $\mathbf{r}_{j+1} = \mathbf{r}$ and $\mathbf{r}_0 = \mathbf{r}$. This perturbation expansion (4) was shown to sum to (1) in the limit of slowly varying potentials, but it has, so far, not proved possible to obtain a result in closed form, apart from this special case.

Since (4) and (5) constitute the exact perturbative generalization of (1), it is important to inquire what is the exact generalization of the variational principle of (2). At this stage, two procedures are possible. The first and most direct is to use the full expansion of the Dirac density matrix $\rho(\mathbf{r}\mathbf{r}_0)$ in off-diagonal form (March & Murray 1961) to calculate $-\frac{1}{2}[\nabla_\mathbf{r}^2 \rho(\mathbf{r}\mathbf{r}_0)]_{\mathbf{r}_0=\mathbf{r}}$, the kinetic energy density. The second is to work directly from (4) and (5) and construct the variational principle without recourse to the off-diagonal elements of the Dirac matrix. Since this latter procedure is in the spirit of the interesting approach of Hohenberg & Kohn (1964), we shall adopt it here. Needless to say, both procedures described above lead to the same result.

3. The generalized variational principle

We require a generalized form of the variational principle (2), which leads to the exact March–Murray relation (4) between density and potential. To achieve this, as discussed above, we shall follow Hohenberg & Kohn and write the total energy ϵ in terms of a universal functional $T[\rho]$ of the density,

$$\epsilon(\rho) = \int T[\rho(\mathbf{r})]\,\mathrm{d}\mathbf{r} + \int V(\mathbf{r})\,\rho(\mathbf{r})\,\mathrm{d}(\mathbf{r}). \tag{6}$$

In our perturbative approach, we define the displaced charge $\Delta(\mathbf{r})$ through

$$\rho(\mathbf{r}) = \rho_0 + \Delta(\mathbf{r}), \tag{7}$$

where $\rho_0 = k^3/6\pi^2$ is the density of the free Fermi gas. Hence we can write (6) as

$$\epsilon(\rho) = \epsilon_0 + \epsilon(\Delta), \tag{8}$$

and our generalization of the variational principle (2) is therefore sought as the minimization of $\epsilon(\Delta)$ with respect to density variations:

$$\delta\epsilon(\Delta) = \int \mathrm{d}\mathbf{r}\,\delta F[\Delta(\mathbf{r})] + \int V(\mathbf{r})\,\delta\Delta(\mathbf{r})\,\mathrm{d}\mathbf{r} = 0. \tag{9}$$

Here $F[\Delta(\mathbf{r})]$ is a universal functional of the displaced charge density and is evidently intimately related to the kinetic energy density $T[\rho]$ of (6). The precise nature of this relationship is discussed in §5. Equation (9) is equivalent to the exact March–Murray relation between density and potential, namely (4) and (5), and we show below that these equations will give the functional $F(\Delta)$ in our perturbational framework. Knowledge of this functional then completes the generalization of the variational principle (2).

We derive first the potential $V(\mathbf{r})$ in terms of the displaced charge density. From (4) and (5) we have

$$\Delta(\mathbf{r}) = \frac{-k^2}{2\pi^2}\int d\mathbf{r}_1 \frac{V(\mathbf{r}_1)}{2\pi}\frac{j_1(2k|\mathbf{r}-\mathbf{r}_1|)}{|\mathbf{r}-\mathbf{r}_1|^2} + \sum_{j=2}^{\infty}\rho_j(\mathbf{r}). \tag{10}$$

The Fourier transform of (10) gives

$$\tilde{\Delta}(\mathbf{q}) = \frac{-k^2}{4\pi^3}\tilde{V}(\mathbf{q})\,J(k,\mathbf{q}) + \sum_{j=2}^{\infty}\tilde{\rho}_j(\mathbf{q}), \tag{11}$$

where
$$J(k,q) = \frac{4\pi^2}{k}\left[\frac{1}{2}+\frac{k}{2q}\left(1-\frac{q^2}{4k^2}\right)\ln\left|\frac{q+2k}{q-2k}\right|\right] \tag{12}$$

and $\tilde{V}(\mathbf{q})$ is the Fourier transform of the potential. Equation (11) then yields the desired equation for the potential

$$\tilde{V}(\mathbf{q}) = \frac{-4\pi^3}{k^2}\frac{\tilde{\Delta}(\mathbf{q})}{J(k,q)} + \frac{4\pi^3}{k^2}\sum_{j=2}^{\infty}\frac{\tilde{\rho}_j(\mathbf{q})}{J(k,q)}. \tag{13}$$

Hence we can take the inverse transform of (13) to obtain

$$V(\mathbf{r}) = \int d\mathbf{s}K(\mathbf{r}-\mathbf{s})\,\Delta(\mathbf{s}) - \sum_{j=2}^{\infty}\int d\mathbf{s}K(\mathbf{r}-\mathbf{s})\,\rho_j(\mathbf{s}), \tag{14}$$

where
$$K(\mathbf{r}-\mathbf{s}) = \frac{-4\pi^3}{k^2}\int d\mathbf{q}\frac{e^{i\mathbf{q}\cdot(\mathbf{r}-\mathbf{s})}}{J(k,q)}. \tag{15}$$

Using (5) for $\rho_j(\mathbf{s}), j = 2,\dots,\infty$, we can express the potential as an infinite series in terms of Δ, by iteration of (14). The leading terms in this expansion are

$$V(\mathbf{r}) = \int d\mathbf{s}K(\mathbf{r}-\mathbf{s})\,\Delta(\mathbf{s}) - \int d\mathbf{s}\,d\mathbf{r}_1\,d\mathbf{r}_2\,d\mathbf{s}_1\,d\mathbf{s}_2\,K(\mathbf{r}-\mathbf{s})\,K(\mathbf{r}_1-\mathbf{s}_1) \times K(\mathbf{r}_2-\mathbf{s}_2)\,\Delta(\mathbf{s}_1)\,\Delta(\mathbf{s}_2)\,J_2(k,\mathbf{s},\mathbf{r}_1,\mathbf{r}_2) + O(\Delta^3). \tag{16}$$

In general, in the jth order term, we encounter the quantity

$$J_j(k,\mathbf{s},\mathbf{r}_1,\mathbf{r}_2\dots\mathbf{r}_j) = \frac{(-1)^j k^2}{2\pi^2(2\pi)^j}\frac{j_1(k\{|\mathbf{s}-\mathbf{r}_1|+|\mathbf{r}_1-\mathbf{r}_2|+\dots+|\mathbf{r}_j-\mathbf{s}|\})}{|\mathbf{s}-\mathbf{r}_1|\,|\mathbf{r}_1-\mathbf{r}_2|\dots|\mathbf{r}_j-\mathbf{s}|}. \tag{17}$$

We shall use (16) to generate the kinetic energy functional $F[\Delta]$.

We wish then to solve for this functional $F[\Delta]$, such that the variational principle (9) gives equation (16) for the potential. Solution of these equations is not straightforward, but demands an examination of the equations for each successive term in (16) and a final summation of the results for each term. We give below the results of such a procedure and show formally, by direct substitution into (9), that the March–Murray relations are recovered. The solution obtained for $F(\Delta)$ is

$$F[\Delta] = -\sum_{j=1}^{\infty} \frac{j}{1+j}\rho_j(\mathbf{r})\,V(\mathbf{r}) \tag{18}$$

$$= -\sum_{j=1}^{\infty} \frac{j}{1+j}\int d\mathbf{r}_1 d\mathbf{r}_2 \ldots d\mathbf{r}_j\, V(\mathbf{r})\,V(\mathbf{r}_1)\ldots V(\mathbf{r}_j)\,J_j(k,\mathbf{r},\mathbf{r}_1\ldots\mathbf{r}_j), \tag{19}$$

where J_j is given by (17). We have written the right-hand side in terms of the potential merely for compactness. It should be understood that $F[\Delta]$ is expressed explicitly as a functional of Δ by substitution into (19) of the infinite series (16) for the potential in terms of Δ.

We now calculate

$$\int \delta F[\Delta]\,d\mathbf{r} = -\sum_{j=1}^{\infty}\frac{j}{1+j}\int d\mathbf{r}\,d\mathbf{r}_1\ldots d\mathbf{r}_j\,V(\mathbf{r})\,\delta[V(\mathbf{r}_1)\ldots V(\mathbf{r}_j)]\,J_j(k,\mathbf{r},\mathbf{r}_1\ldots\mathbf{r}_j)$$

$$-\sum_{j=1}^{\infty}\frac{j}{j+1}\int d\mathbf{r}\,d\mathbf{r}_1\ldots d\mathbf{r}_j[\delta V(\mathbf{r})]\,V(\mathbf{r}_1)\ldots V(\mathbf{r}_j)\,J_j(k,\mathbf{r},\mathbf{r}_1\ldots\mathbf{r}_j). \tag{20}$$

Considering the last term, it can be easily shown that because of the symmetry of the functions $J_j(k,\mathbf{r},\mathbf{r}_1,\ldots,\mathbf{r}_j)$, the variational operator δ can be moved to any other $V(\mathbf{r}_i)$ in the integrand. Thus we write

$$[\delta V(\mathbf{r})]\,V(\mathbf{r}_1)\,V(\mathbf{r}_2)\ldots V(\mathbf{r}_j) = j^{-1}V(\mathbf{r})\,\delta[V(\mathbf{r}_1)\,V(\mathbf{r}_2)\ldots V(\mathbf{r}_j)]$$

and (20) becomes

$$\int \delta F[\Delta]\,d\mathbf{r} = -\sum_{j=1}^{\infty}\int d\mathbf{r}\,d\mathbf{r}_1\ldots d\mathbf{r}_j\,V(\mathbf{r})\,\delta[V(\mathbf{r}_1)\,V(\mathbf{r}_2)\ldots V(\mathbf{r}_j)]\,J_j(k,\mathbf{r},\mathbf{r}_1,\ldots,\mathbf{r}_j)$$

$$= -\int d\mathbf{r}\,V(\mathbf{r})\,\delta\Delta(\mathbf{r}). \tag{21}$$

We have used (10) and (5) to obtain the last result. Remembering that $V(\mathbf{r})$ on the right-hand side of (21) is expressed as a functional of $\Delta(\mathbf{r})$ from (16), we see that the variational equation (9) has exactly reproduced the March–Murray series, (16). Equivalently we could say that the variational principle (9) is satisfied with respect to potential variations of a functional $F(V)$ given by (19).

This completes the proof of a variational principle leading to the exact March–Murray relation between density and potential, and the set of (4), (5), (8), (9) and (19) represent the exact Thomas–Fermi method in perturbation theory.

4. The kinetic energy functional $F[\Delta]$

The first terms in the series (19) are given by

$$F[\Delta] = -\tfrac{1}{2}\Delta(\mathbf{r})\int d\mathbf{s}K(\mathbf{r}-\mathbf{s})\,\Delta(\mathbf{s}) + \tfrac{1}{2}\Delta(\mathbf{r})\int d\mathbf{s}\,d\mathbf{r}_1\,d\mathbf{r}_2\,d\mathbf{s}_1\,d\mathbf{s}_2\,K(\mathbf{r}-\mathbf{s})\,K(\mathbf{r}_1-\mathbf{s}_1)$$
$$\times K(\mathbf{r}_2-\mathbf{s}_2)\,\Delta(\mathbf{s}_1)\,\Delta(\mathbf{s}_2)\,J_2(k,\mathbf{s},\mathbf{r}_1,\mathbf{r}_2) - \frac{1}{6}\int d\mathbf{s}\,d\mathbf{r}_1\,d\mathbf{r}_2\,d\mathbf{s}_1\,d\mathbf{s}_2\,K(\mathbf{r}-\mathbf{s})$$
$$\times K(\mathbf{r}_1-\mathbf{s}_1)\,K(\mathbf{r}_2-\mathbf{s}_2)\,\Delta(\mathbf{s})\,\Delta(\mathbf{s}_1)\,\Delta(\mathbf{s}_2)\,J_2(k,\mathbf{r},\mathbf{r}_1,\mathbf{r}_2) + O(\Delta^4). \tag{22}$$

It is of interest to express this series in terms of the Fourier transformed density functions $\tilde{\Delta}(\mathbf{q})$ as it then takes a rather simpler form. Defining

$$\tilde{\Delta}(\mathbf{q}) = \int \mathrm{d}\mathbf{r}\, \mathrm{e}^{\mathrm{i}\mathbf{q}\cdot\mathbf{r}}\, \Delta(\mathbf{r}) \tag{23}$$

and

$$\tilde{J}_j(k, \mathbf{q}, \mathbf{k}_1, \mathbf{k}_2 \ldots \mathbf{k}_j) = \int \mathrm{d}\mathbf{r}\, \mathrm{d}\mathbf{r}_1 \ldots \mathrm{d}\mathbf{r}_j \exp\{\mathrm{i}\mathbf{q}\,.\,\mathbf{r} + \mathrm{i}\mathbf{k}_1\,.\,\mathbf{r}_1 + \ldots + \mathrm{i}\mathbf{k}_j\,.\,\mathbf{r}_j\}\, J_j(k, \mathbf{r}, \mathbf{r}_1, \ldots, \mathbf{r}_j), \tag{24}$$

we obtain

$$\begin{aligned} F[\Delta] = &-\tfrac{1}{2}\Delta(\mathbf{r}) \int \mathrm{d}\mathbf{q}\, \mathrm{e}^{\mathrm{i}\mathbf{q}\cdot\mathbf{r}} \frac{\tilde{\Delta}(\mathbf{q})}{J(k,q)} \\ &+ \tfrac{1}{2}\Delta(\mathbf{r}) \int \mathrm{d}\mathbf{k}_1\, \mathrm{d}\mathbf{k}_2\, \mathrm{d}\mathbf{q}\, \mathrm{e}^{\mathrm{i}\mathbf{q}\cdot\mathbf{r}} \frac{\tilde{\Delta}(\mathbf{k}_1)}{J(k,k_1)} \frac{\tilde{\Delta}(\mathbf{k}_2)}{J(k,k_2)} \frac{\tilde{J}_2(k,\mathbf{q},\mathbf{k}_1,\mathbf{k}_2)}{J(k,q)} \\ &- \frac{1}{6} \int \mathrm{d}\mathbf{k}_1\, \mathrm{d}\mathbf{k}_2\, \mathrm{d}\mathbf{k}_3\, \mathrm{d}\mathbf{q} \exp\{\mathrm{i}\mathbf{r}\,.\,(\mathbf{k}_1 - \mathbf{q})\} \frac{\tilde{\Delta}(\mathbf{k}_1)}{J(k,k_1)} \frac{\tilde{\Delta}(\mathbf{k}_2)}{J(k,k_2)} \frac{\tilde{\Delta}(\mathbf{k}_3)}{J(k,k_3)} \\ &\times \tilde{J}_2(k, \mathbf{q}, \mathbf{k}_2, \mathbf{k}_3) + O(\Delta^4) \end{aligned} \tag{25}$$

It can be seen from (25) that the integrations occurring in the functional $F[\Delta]$ essentially depend on $\tilde{\Delta}(\mathbf{q})/J(k,q)$, a property which is valid for all terms in the above series. Attempts at a direct summation of this series have been hampered by the fact that we have not been successful in obtaining any useful closed form for $\tilde{J}_j(k, \mathbf{q}, \mathbf{k}_1, \ldots, \mathbf{k}_j)$ except for $j = 1$. Nevertheless, the series (25) gives an explicit form of the Hohenberg–Kohn functional for the first time.

5. The constant density limit

In spite of the fact that no summation of (22) or (25) has been achieved in the general case, it is of considerable interest to show that these equations sum to the usual Thomas–Fermi result when spatial variations of density and potential are slow.

We consider then (16) and (19) which define $F[\Delta]$, in the case of almost constant density. In (16) we have

$$\int \mathrm{d}\mathbf{s}\, K(\mathbf{r}-\mathbf{s})\, \Delta(\mathbf{s}) = \frac{-4\pi^3}{k^2} \int \mathrm{d}\mathbf{q}\, \mathrm{d}\mathbf{s}\, \mathrm{e}^{\mathrm{i}\mathbf{q}\cdot(\mathbf{r}-\mathbf{s})} \frac{\Delta(\mathbf{s})}{J(k,0)}, \times [1 + c_2 q^2 + c_4 q^4 + \ldots] \simeq -\frac{2\pi^2}{k}\, \Delta(\mathbf{r}). \tag{26}$$

In deriving (26), we have performed a small $|\mathbf{q}|$ expansion of $1/J(k,q)$ inside the integrand and have neglected terms which involve derivatives of the displaced charge density. Equation (26) corresponds essentially to making the long wavelength approximation $J(k,q) \simeq J(k,0)$ in the above integral. Similarly, the higher terms in the series (16) may be obtained by using the result (26). Thus, the second term in (16)

$$\approx \frac{4\pi^4}{k^2} \int \mathrm{d}\mathbf{s}\, K(\mathbf{r}-\mathbf{s}) \int \mathrm{d}\mathbf{r}\, \mathrm{d}\mathbf{r}_2\, \Delta(\mathbf{r}_1)\, \Delta(\mathbf{r}_2)\, J_2(k, \mathbf{s}, \mathbf{r}_1, \mathbf{r}_2).$$

The integration over $\mathbf{r}_1$ and $\mathbf{r}_2$ can be performed approximately for almost constant density functions by using the analysis given by March & Murray (1961). We then obtain

$$\approx \frac{\pi^2}{k^3}\int \mathrm{d}\mathbf{s}\, K(\mathbf{r}-\mathbf{s})\,\Delta^2(\mathbf{s}).$$

Again, the long wavelength approximation $J(k,q) \simeq J(k,0)$ inside the integral must be made, and yields the value $+2\pi^4\Delta^2(\mathbf{r})/k^4$ for the second term in (16). In a similar way, all higher terms may be calculated and we find

$$V(\mathbf{r}) \simeq -\frac{2\pi^2\Delta(\mathbf{r})}{k}+\frac{2\pi^4\Delta^2(\mathbf{r})}{k^4}-\frac{16\pi^6\Delta^3(\mathbf{r})}{3k^7}+\dots \tag{27}$$

In a completely analogous manner, we find that (19) reduces to

$$F[\Delta] = \frac{kV^2(\mathbf{r})}{4\pi^2}-\frac{V^3(\mathbf{r})}{6\pi^2 k}-\frac{V^4(\mathbf{r})}{16\pi^2k^3}+\dots$$

Substitution of $V(\mathbf{r})$ from (28) then yields

$$F[\Delta] = \frac{\pi^2}{k}\,\Delta^2(\mathbf{r})-\frac{2\pi^4}{3k^4}\,\Delta^3(\mathbf{r})+\frac{4\pi^6}{3k^7}\,\Delta^4(\mathbf{r})+\dots \tag{28}$$

Recalling from § 3 that $F[\Delta]$ is intimately related to the kinetic energy $T[\rho]$, which for a free Fermi gas with Fermi momentum k is given by

$$\frac{3(6\pi^2)^{\frac{2}{3}}}{10}\left(\frac{k^3}{6\pi^2}\right)^{\frac{5}{3}},$$

$F[\Delta]$ is readily summed to give the result

$$F[\Delta] = \frac{3(6\pi^2)^{\frac{2}{3}}}{10}\left[\frac{k^3}{6\pi^2}+\Delta\right]^{\frac{5}{3}}-\frac{3(6\pi^2)^{\frac{2}{3}}}{10}\left(\frac{k^3}{6\pi^2}\right)^{\frac{5}{3}}-\frac{k^2}{2}\,\Delta. \tag{29}$$

This equation immediately shows that $F[\Delta]$ is essentially the kinetic energy density $T[\rho]$ measured relative to the free Fermi gas, the first term on the right-hand side of (29) being the usual Thomas–Fermi approximation $\propto \rho^{\frac{5}{3}}$. The third term on the right-hand side of (29) is simply the energy associated with adding extra particles at the Fermi surface, the number of particles being $\int \Delta\,\mathrm{d}\mathbf{r}$. If we had used the result

$$T[\rho] = -\tfrac{1}{2}[\nabla^2\rho(\mathbf{r}\mathbf{r}_0]_{\mathbf{r}_0=\mathbf{r}}$$

referred to above, this term would have appeared directly (cf. the appendix). Thus our exact equations lead to the known Thomas–Fermi kinetic energy for slowly varying particle density.

6. The gradient expansion

We comment briefly here on the gradient expansion, the leading terms of which were considered by Hohenberg & Kohn (1964), as well as by earlier workers referred to in their paper. The derivation of such an expansion is closely connected with the long wavelength approximation to $J(k,q)$ used in the Thomas–Fermi argument of § 5. Now $J(k,q)$ is directly related to the dielectric function of the electron gas, which produces Friedel oscillations; the March–Murray (1960) first-order equation leading directly to this function on Fourier transforming. By the gradient expansion, one is obviously missing out the pathological behaviour of $J(k,q)$ around $q = 2k$, as indeed Hohenberg & Kohn emphasize. It seems to us that it is the full perturbation expansion of the present theory, and not a gradient expansion, which forms a framework for including all effects of Friedel oscillations, shell structure in atoms, etc.

Briefly then, by means of the small q expansion of $J(k,q)$ we find

$$\int \mathrm{d}\mathbf{s}\,\Delta(\mathbf{s})\,K(\mathbf{r}-\mathbf{s}) = \frac{-2\pi^2}{k}\{\Delta(\mathbf{r}) - c_2\nabla_{\mathbf{r}}^2\Delta(\mathbf{r}) + c_4\nabla_{\mathbf{r}}^2\nabla_{\mathbf{r}}^2\Delta(\mathbf{r}) + \dots\}. \tag{30}$$

Substitution of this result into (22) gives us the gradient expansion in the perturbational framework

$$\begin{aligned} F[\Delta] = {} & \frac{-\pi^2}{k}\Delta(\mathbf{r})\,[\Delta(\mathbf{r}) - c_2\nabla_{\mathbf{r}}^2\Delta(\mathbf{r}) + c_4\nabla_{\mathbf{r}}^2.\nabla_{\mathbf{r}}^2\Delta(\mathbf{r}) + \dots] \\ & + \frac{2\pi^4}{k^2}\Delta(\mathbf{r})\int \mathrm{d}\mathbf{s}\,\mathrm{d}\mathbf{r}_1\,\mathrm{d}\mathbf{r}_2\,K(\mathbf{r}-\mathbf{s})\,J_2(k,\mathbf{s},\mathbf{r}_1,\mathbf{r}_2)\,[\Delta(\mathbf{r}_1) - c_2\nabla_{\mathbf{r}_1}^2\Delta(\mathbf{r}_1) + \dots] \\ & \times [\Delta(\mathbf{r}_2) - c_2\nabla_{\mathbf{r}_2}^2\Delta(\mathbf{r}_2) + \dots] + \frac{4\pi^6}{k^3}[\Delta(\mathbf{r}) - c_2\nabla_{\mathbf{r}}^2\Delta(\mathbf{r}) + \dots]\iint \mathrm{d}\mathbf{r}_1\,\mathrm{d}\mathbf{r}_2\,J_2(k,\mathbf{r},\mathbf{r}_1,\mathbf{r}_2) \\ & \times [\Delta(\mathbf{r}_1) - c_2\nabla_{\mathbf{r}_1}^2\Delta(\mathbf{r}_1) + \dots]\,[\Delta(\mathbf{r}_2) - c_2\nabla_{\mathbf{r}_2}^2\Delta(\mathbf{r}_2) + \dots] + O(\Delta^4). \end{aligned} \tag{31}$$

It can be seen from (31) that, even after the gradient expansion has been carried out, terms remain which are such that $F[\Delta]$ is a functional of the density and gradients of the density. If we now tackle these terms, by expanding the kernels as in (30), we can systematically generate an explicit gradient expansion; in principle to all orders in density derivatives. However, in view of its limited validity discussed above, we shall not give the further details.

Appendix

The equivalence of the kinetic energy density (18) with the usual forms of energy density can be seen by defining the potential as $\lambda V(\mathbf{r})$ in terms of a coupling constant λ. Then writing in explicitly the dependence of the density on λ it can be easily shown that

$$F[\Delta] = -\lambda V(\mathbf{r})\,\Delta(\mathbf{r}\lambda) + V(\mathbf{r})\int_0^\lambda \mathrm{d}\lambda'\,\Delta(\mathbf{r}\lambda'). \tag{A 1}$$

The first term is obviously the negative of the potential energy density of the displaced charge. The second term can be easily shown by elementary energy arguments to be the total energy density of the displaced charge, being just the energy density required to bring the charge density $\Delta(\mathbf{r}\lambda)$ from infinity to the potential $\lambda V(\mathbf{r})$. These terms may be compared with the energy densities obtained from the off-diagonal March–Murray relations, as discussed previously. From these we readily obtain,

$$T[\rho]-T_0[\rho_0] = \frac{-k^3}{6\pi^2}\,\lambda V(\mathbf{r})-\lambda V(\mathbf{r})\,\Delta(\mathbf{r}\lambda)-\int \mathrm{d}z\,\frac{\partial^2 j_1(kz)}{\partial z^2}\,f(z,\mathbf{r},\lambda), \tag{A 2}$$

where
$$f(z,\mathbf{r},\lambda) = \frac{1}{2\pi^2}\sum_{j=1}^{\infty}\frac{(-1)^j\lambda^j}{(2\pi)^j}\int \mathrm{d}\mathbf{r}_1\ldots\mathrm{d}\mathbf{r}_j\,V(\mathbf{r}_1)\ldots V(\mathbf{r}_j) \times\frac{\delta(z-|\mathbf{r}-\mathbf{r}_1|-|\mathbf{r}_1-\mathbf{r}_2|\ldots-|\mathbf{r}_j-\mathbf{r}|)}{|\mathbf{r}-\mathbf{r}_1|\,|\mathbf{r}_1-\mathbf{r}_2|\ldots|\mathbf{r}_j-\mathbf{r}|}. \tag{A 3}$$

The two expressions given above for kinetic energy density are equivalent, and differ only by terms which integrate to zero when the total energy is calculated.

One of us (N. H. M.) wishes to acknowledge that his own work on this problem began as a result of a short visit to Brookhaven National Laboratory in the summer of 1966. Particular thanks are due to Dr V. J. Emery, both for his generous hospitality and for his interest and encouragement during the early stages of the work. Thanks are also due to the Science Research Council for a Senior Research Award to one of us (J. C. S.).

References

Hohenberg, P. & Kohn, W. 1964 *Phys. Rev.* **136**, 3B, 864.
March, N. H. & Murray, A. M. 1960 *Phys. Rev.* **120**, 830.
March, N. H. & Murray, A. M. 1961 *Proc. Roy. Soc.* A **261**, 119.

Volume 113A, number 2 PHYSICS LETTERS 2 December 1985

DIFFERENTIAL EQUATION FOR THE GROUND-STATE DENSITY IN FINITE AND EXTENDED INHOMOGENEOUS ELECTRON GASES

N.H. MARCH

Theoretical Chemistry Department, University of Oxford, 1 South Parks Road, Oxford OX1 3TG, UK

Received 25 September 1985; accepted for publication 7 October 1985

The square root $\rho^{1/2}$ of the ground-state electron density can be determined from a Schrödinger equation with a local potential energy. Here, some properties required of this potential are exposed for (a) atoms, by recourse to both the asymptotic density far from the nucleus and to Kato's theorem and (b) a weak perturbation in an electron gas.

Recently we have considered from the standpoint of density functional theory the form of the local potential $\mathcal{V}(\boldsymbol{r})$ which must be inserted in Schrödinger's equation

$$\nabla^2\psi + 2\{\epsilon - [\mathcal{V}(\boldsymbol{r}) - \mathcal{V}(\infty)]\}\psi = 0 \tag{1}$$

to allow the square root $\rho^{1/2} = \psi$ of the ground-state electron density in atoms and molecules to be calculated [1]. In the present work, we shall expose some properties required of $\mathcal{V}(\boldsymbol{r})$ in the special cases of (a) atoms and (b) weak perturbations in an extended system represented by an originally unperturbed uniform but interacting electron gas.

Starting with atoms, two properties of the electron density are of particular importance as they follow from first principles:

(i) Far from the nucleus, the ground-state electron density has the asymptotic form $\rho(r) = Ar^n \times \exp[-2(2I)^{1/2}r]$, I being the ionization potential and atomic units being used throughout. In turn, this will allow the identification of $|\epsilon|$ in eq. (1) with I [‡1] [2–4].

(ii) At the nucleus itself, for an atom with nuclear charge Ze,

$$(\partial\rho/\partial r)_{r=0} = -2Z\rho(0)\,. \tag{2}$$

Evidently, from eq. (1), we can write apart from an additive constant that

$$\mathcal{V}(r) = \tfrac{1}{2}\psi^{-1}\nabla^2\psi \tag{3}$$

and inserting the asymptotic density quoted in (i) above yields the form of $\mathcal{V}(r)$ at large r as [2]

$$\mathcal{V}(r) = -(1+\tfrac{1}{2}n)[(2I)^{1/2}/r - n/4r^2]\,. \tag{4}$$

This then is the first basic property of $\mathcal{V}(r)$ that any quantitative treatment of the local potential $\mathcal{V}(r)$ for atoms must embody.

Turning to the consequence of Kato's theorem given by eq. (2) above, it will be useful to work in Fourier transform or $\boldsymbol{k}$ space. The Fourier transform of the electron density $\rho(r)$ is, of course, the atomic scattering factor $f(k)$, but here we work with the square root of $\rho(r)$, with Fourier transform $\phi(k)$ defined by

$$\phi(k) = (2\pi)^{-3/2}\int \psi(r)\exp(-i\boldsymbol{k}\cdot\boldsymbol{r})\,d\boldsymbol{r}\,. \tag{5}$$

Inserting $\rho = \psi^2$ into the exact relation (2), and using arguments paralleling those of Goscinski and Lindner [5] for the scattering factor $f(k)$, one readily obtains the consequence of eq. (2) as

$$\phi(k) = 2^{3/2}\pi^{-1/2}Z\rho^{1/2}(0)k^{-4}\,, \quad \text{as} \quad k\to\infty\,. \tag{6}$$

It is worth noting that $f(k)$ and $\phi(k)$ are, in fact, related by the convolution property

$$f(k) = \text{const} \times \int \phi(k')\phi(|\boldsymbol{k}-\boldsymbol{k}'|)\,d\boldsymbol{k}'\,, \tag{7}$$

[‡1] See the closely related discussion of the asymptotic form of the first-order density matrix in ref. [2].

Volume 113A, number 2 PHYSICS LETTERS 2 December 1985

which, by putting $k = 0$, relates the normalization integral $\int \phi(k)^2 \, \mathrm{d}\boldsymbol{k}$ to the atomic number Z for a neutral atom.

It is also of interest to note that the Lippmann–Schwinger bound-state equation corresponding to eq. (1) reads

$$\psi(r) = -(2\pi)^{-1} \times \int \frac{\exp[-(2|\epsilon|)^{1/2}|\boldsymbol{r}-\boldsymbol{r}'|]}{|\boldsymbol{r}-\boldsymbol{r}'|} \mathcal{V}(\boldsymbol{r}')\psi(\boldsymbol{r}')\,\mathrm{d}\boldsymbol{r}' , \quad (8)$$

and letting $\boldsymbol{r}$ become large it is apparent then that $|\epsilon|$ must be identified with the ionization potential I from property (i) above. The Fourier transform of this equation reads

$$\phi(k) = P(k)/(2|\epsilon| + k^2) , \quad (9)$$

where, apart from unimportant constants, $P(\boldsymbol{k})$ denotes the Fourier transform of the product $\mathcal{V}(\boldsymbol{r})\psi(\boldsymbol{r})$. Again using the convolution property one has

$$P(k) = \text{const} \times \int \mathcal{V}(\boldsymbol{k}')\phi(\boldsymbol{k}-\boldsymbol{k}')\,\mathrm{d}\boldsymbol{k}' . \quad (10)$$

Clearly, from eqs. (6) and (9), apart from a multiplying constant, $P(k) = Z\rho^{1/2}(0)k^{-2}$ at large k which evidently provides a further important constraint on the Fourier transform of the local potential $\mathcal{V}(\boldsymbol{r})$ through eq. (10). Imposing this constraint will ensure, we reiterate, that Kato's theorem is fulfilled.

From this discussion of finite, atomic, systems, we turn next to discuss the case of extended systems in which a weak perturbing potential energy $V(\boldsymbol{r})$ is introduced into an initially uniform but coulombically interacting electron gas. It will be assumed below that $V(\boldsymbol{r})$ is either repulsive, as for, say, a vacancy in a close-packed simple s–p metal, or if attractive that it is not strong enough to yield any bound states. Then the density perturbation theory of March and Murray [6,7] may be employed. For simplicity, we shall quote results only to first order in V below, but higher order terms can be calculated by the same method should they be required later.

Then for a gas of unperturbed electron density ρ_0 and corresponding Fermi wave number k_f, we can write the explicit form [6]

$$\rho(\boldsymbol{r}) = \rho_0 - \frac{k_\mathrm{f}^2}{2\pi^3} \int \frac{j_1(2k_\mathrm{f}|\boldsymbol{r}-\boldsymbol{r}'|)}{|\boldsymbol{r}-\boldsymbol{r}'|^2} V(\boldsymbol{r}')\,\mathrm{d}\boldsymbol{r}' , \quad (11)$$

where $j_1(x)$ is the first-order spherical Bessel function given by

$$j_1(x) = x^{-2}[\sin x - x \cos x] . \quad (12)$$

Forming $\nabla^2\psi/\psi$ to first order in V, one finds readily the result for the local potential $\mathcal{V}(\boldsymbol{r})$

$$\mathcal{V}(\boldsymbol{r}) = V(\boldsymbol{r}) - \frac{12k_\mathrm{f}^3}{\pi} \int F(2k_\mathrm{f}|\boldsymbol{r}-\boldsymbol{r}'|)V(\boldsymbol{r}')\,\mathrm{d}\boldsymbol{r}' , \quad (13)$$

where $F(x)$ has the explicit form

$$F(x) = x^{-3}\cos x - 5x^{-4}\sin x - 12x^{-5}\cos x + 12x^{-6}\sin x . \quad (14)$$

Of course, it is clear that this non-local relation between $\mathcal{V}(\boldsymbol{r})$ and $V(\boldsymbol{r})$ is only a first-order theory. Its advantage is that it will retain the Friedel oscillations in the density around the perturbing centre at large distances, which are always lost in semiclassical theory based on Thomas–Fermi-like assumptions plus appropriate gradient corrections. Of course, one might try to modify the expression (13) to account approximately for stronger perturbations, by replacing k_f by $(3\pi^2)^{1/3}\rho^{1/3}$ with ρ now the inhomogeneous local density to be determined. In case it should be thought that this would introduce a basically new complication, it must of course be stressed that it is known from the work of Slater [8] for exchange, of Gombas and others for correlation [9], and from the rigorous formal expression for $V(\boldsymbol{r})$ given by Kohn and Sham [10], that $V = V[\rho]$ in any event. On this point, it is relevant to note that the formally exact density functional expression for the local potential $\mathcal{V}(\boldsymbol{r})$ in eq. (1) can be written [1], again aside from an additive constant:

$$\mathcal{V}[\rho] = V[\rho] + \delta T_\mathrm{s}/\delta\rho(\boldsymbol{r}) + \tfrac{1}{4}\nabla^2\rho/\rho - \tfrac{1}{8}(\nabla\rho)^2/\rho^2 ,$$

where T_s is the single-particle kinetic energy functional ‡2. Clearly, all the specific results of the present work must be contained, in a final theory, within this form of the single-particle kinetic energy functional. While V in eq. (15) is known to contain many-elec-

‡2 For a very recent discussion, see ref. [11].

tron effects through the functional derivative of the exchange and correlation energy [10], the "correction" to V to obtain the desired local potential $\mathcal{V}(\boldsymbol{r})$ contains only the one-body functional T_s, plus explicitly known terms in the gradient and the laplacian of the density itself.

References

[1] N.H. March, to be published.
[2] N.H. March, Phys. Lett. 84A (1981) 319; J. Phys. Chem. 86 (1982) 2262.
[3] E.N. Lassettre, J. Chem. Phys. 83 (1985) 1709.
[4] G. Hunter, Sanibel Symposium (1985); Int. J. Quantum Chem., to be published.
[5] O. Goscinski and P. Lindner, J. Chem. Phys. 52 (1970) 2539.
[6] N.H. March and A.M. Murray, Phys. Rev. 120 (1960) 830.
[7] N.H. March and A.M. Murray, Proc. R. Soc. A261 (1961) 119.
[8] J.C. Slater, Phys. Rev. 81 (1951) 385.
[9] P. Gombas, Die statistische Theorie des Atoms und ihre Andwendungen (Springer, Berlin, 1949).
[10] W. Kohn and L.J. Sham, Phys. Rev. A140 (1965) 1133.
[11] M.L. Plumer and M.J. Stott, J. Phys. C18 (1985) 4143.

PHYSICAL REVIEW A VOLUME 33, NUMBER 1 JANUARY 1986

Spatially dependent generalization of Kato's theorem for atomic closed shells in a bare Coulomb field

N. H. March
Theoretical Chemistry Department, University of Oxford, 1 South Parks Road, Oxford OX1 3TG, England
(Received 24 June 1985)

For a bare Coulomb potential energy $-Ze^2/r$, it is shown that the total electron density $\rho(r)$ for an arbitrary number of closed shells is given by $\rho(r)=(2Z/a_0)\int_r^\infty \rho_s(r)dr$ where ρ_s is the s-state contribution to $\rho(r)$. This yields Kato's theorem $[\partial\rho(r)/\partial r]_{r=0}=(-2Z/a_0)\rho(r=0)$ as a limiting case. That $\partial\rho/\partial r$ is always negative follows, for all distances r. For the nth closed shell, with density $\rho_n(r)$, it is further shown that $R_{n0}(r)=[(-a_0/2Z)\partial\rho_n/\partial r]^{1/2}$, with R_{n0} the s-state radial wave function. This result can be used to construct an explicit differential equation for $\rho_n(r)$.

I. INTRODUCTION

In a recent study,[1] the bound-state Slater sum for electrons moving in a bare Coulomb potential energy $V(r)=-Ze^2/r$ has been related to that for s states, for arbitrary temperature. In the present paper, we consider the purely quantum-mechanical case of an arbitrary number of closed electronic shells in the bare Coulomb field defined above. The focal point is the total electron density $\rho(r)$, built by summing the density $\rho_n(r)$ of the nth closed shell over n.

The outline of the paper is as follows. In Sec. II we utilize a result of Blinder[2] that constitutes a generalization of Unsold's theorem. This allows the density $\rho_n(r)$ to be written in terms of the s-state radial wave function $R_{no}(r)$.[3] This result is used here to relate the total density $\rho(r)$ directly to that part, $\rho_s(r)$, contributed by the s states alone. This relation is shown to contain Kato's theorem[4,5]

$$(\partial\rho/\partial r)_{r=0}=(-2Z/a_0)\rho(r=0),\quad a_0=\hbar^2/me^2\ , \tag{1.1}$$

as a special case, and a spatially dependent generalization of Kato's theorem is thereby established for the Coulomb field. A further result that follows from this generalization is that $\partial\rho/\partial r<0$, proving that $\rho(r)$ is a monotonically decreasing function of r for any arbitrary number of closed shells in the Coulomb-field case.

Section III is then concerned with an individual closed shell. An explicit relation is written for the radial wave function $R_{no}(r)$ for s states in terms of $\rho_n(r)$ and it is pointed out that this can be used to derive a differential equation satisfied by $\rho_n(r)$. In the Appendix, the generalization of Kato's theorem established in the present paper for a Coulomb field is discussed in relation to approximate density-functional theories.

II. GENERALIZATION OF KATO'S THEOREM FOR COULOMB FIELD

Blinder's work [2]leads to the result[3]

$$\rho_n(r)=P_n'^2-P_nP_n''\ , \tag{2.1}$$

where $P_n=rR_{no}$ with R_{no} the s-state radial wave function for the principal quantum number n. P_n satisfies the Schrödinger equation

$$P_n''+\frac{2m}{\hbar^2}\left[\epsilon_n+\frac{Ze^2}{r}\right]P_n=0\ , \tag{2.2}$$

with ϵ_n denoting the usual hydrogenlike levels. However, we shall proceed to eliminate these by dividing both sides of Eq. (2.2) by P_n and then differentiating with respect to r. We thereby obtain

$$P_nP_n'''-P_n'P_n''=\frac{2m}{\hbar^2}\frac{Ze^2}{r^2}P_n^2\ . \tag{2.3}$$

Now the left-hand side of Eq. (2.3) is readily shown from Eq. (2.1) to equal $-\partial\rho_n/\partial r$, which yields

$$\frac{\partial\rho_n}{\partial r}=-\frac{2Z}{a_0}\rho_{no}\ , \tag{2.4}$$

where $\rho_{no}(r)$ denotes the s-state density $R_{no}^2=P_n^2/r^2$.

One can sum Eq. (2.4) over an arbitrary number of closed shells to relate the total density $\rho(r)$ thereby obtained to the s-state contribution to this density, $\rho_s(r)$, by

$$\frac{\partial\rho(r)}{\partial r}=-\frac{2Z}{a_0}\rho_s(r)\ . \tag{2.5}$$

Since $\rho_n(r)$ tends to zero as r tends to infinity, Eq. (2.5) can be integrated to obtain

$$\rho(r)=\frac{2Z}{a_0}\int_r^\infty \rho_s(r)dr\ . \tag{2.6}$$

Equation (2.5), or its integral (2.6), constitutes the desired generalization of Kato's theorem (1.1). Evidently, at $r=0$, the p, d, etc., states give zero contribution to $\rho(r=0)$ which is therefore simply equal to $\rho_s(r=0)$, and obviously Eq. (2.5) then reduces to Eq. (1.1) as $r\to 0$.

A further interesting consequence of Eq. (2.5) is that, since $\rho_s(r)>0$, $\partial\rho(r)/\partial r<0$, showing that the total electron density $\rho(r)$ decreases monotonically from the origin. The writer knows of no exception to this result for real atomic densities, although, of course, the present rigorous proof is restricted to the unscreened Coulomb field.

III. RADIAL s-STATE WAVE FUNCTION FOR nTH CLOSED SHELL RELATED TO ELECTRON DENSITY $\rho_n(r)$

Returning to Eq. (2.4), we can write, since the radial wave functions are real quantities,

$$R_{no}=\left[-\frac{a_0}{2Z}\frac{\partial\rho_n(r)}{\partial r}\right]^{1/2}, \tag{3.1}$$

where we have utilized the fact that $\partial\rho_n(r)/\partial r\le 0$, the equality sign obviously corresponding to the nodes of $R_{no}(r)$. This relation (3.1) is of interest in relation to density-functional theory. In particular, such a functional relation between R_{no} and the electron density $\rho_n(r)$ could be made the basis for the use of the variational principle to determine $\rho_n(r)$. However, the resulting Euler equation for $\rho_n(r)$ can only be equivalent to that obtained by substituting $P_n=rR_{no}$, with R_{no} as in Eq. (3.1), into the Schrödinger equation (2.2). Using

$$P_n^2=-(a_0/2Z)r^2\rho_n' , \tag{3.2}$$

it is a straightforward matter to obtain a third-order nonlinear differential equation satisfied by $\rho_n(r)$. Since one would, of course, work with the linear equation (2.2) for P_n, we merely record this third-order equation in the Appendix.

IV. SUMMARY AND CONCLUSION

The main results of the present work are embodied in Eqs. (2.5) and (2.6) which constitute the desired generalization of Kato's theorem (1.1). These Eqs., valid for all r, yield the bonus that they establish that $\partial\rho/\partial r<0$, which in turn means that $\rho(r)$ for the bare Coulomb field decreases monotonically with increasing r. We have noted that, to date, no exception is known to this inequality applied to real atomic densities, though, of course, this latter result is not proved by the present argument.

In addition, Eq. (3.1) demonstrates explicitly that the s-state radial wave function can be written in terms of the closed-shell electron density $\rho_n(r)$. This means that $\rho_n(r)$, if desired, can be determined from an explicit third-order, but nonlinear, differential equation. We have not, so far, found how to generalize this equation (A1) to apply to the total density $\rho(r)$ for an arbitrary number of closed shells. However, because the results of the model problem solved in this present work have interest for density-functional theory, we have thought it worthwhile in the Appendix to relate Eqs. (2.5) and (2.6) to the simplest form of density-functional theory, namely the Thomas-Fermi method.

APPENDIX: DIFFERENTIAL EQUATION FOR CLOSED-SHELL DENSITY $\rho_n(r)$ AND GENERALIZED KATO'S THEOREM RELATED TO DENSITY-FUNCTIONAL THEORY

We first record the differential equation satisfied by the electron density $\rho_n(r)$ of the nth closed shell. This is found by direct substitution of Eq. (3.1) for $R_{no}=P_n/r$ into Eq. (2.2). Then one can express the result in terms of $\rho_n(r)$ and its derivatives, plus the s-shell density $\rho_{no}(r)$ given by $P_n^2/r^2=\rho_{no}(r)$. The final step is then to use Eq. (2.4) to eliminate $\rho_{no}(r)$. The third-order nonlinear differential equation for the closed-shell density $\rho_n(r)$ is

$$r\rho_n'''+2\rho_n''-\frac{1}{2}r\frac{[\rho_n'']^2}{\rho_n'}+\left[4Z-\frac{2Z^2}{n^2}\frac{r}{a_0}\right]\frac{\rho_n'}{a_0}=0 , \tag{A1}$$

which is readily verified, for instance, to be satisfied by the K-shell density

$$\rho_1(r)=(Z^3/\pi a_0^3)\exp(-2Zr/a_0) .$$

Naturally by using Eq. (2.4) one could write an equivalent second-order differential equation for the s-state density $\rho_{no}(r)$.

In the second part of this Appendix, we shall consider briefly the relation of the generalization of Kato's theorem embodied in Eq. (2.5) to the simplest density-functional theory, namely the Thomas-Fermi method. Again for the Coulomb field, the Thomas-Fermi approximation to the total density $\rho(r)$ is[6]

$$\rho^{\mathrm{TF}}(r)=\frac{4\pi}{3h^3}(2m)^{3/2}\left[\mu+\frac{Ze^2}{r}\right]^{3/2}, \tag{A2}$$

where μ is the chemical potential of the electron cloud. One can write the derivative appearing in Eq. (2.5) as

$$\frac{\partial\rho^{\mathrm{TF}}(r)}{\partial r}=\frac{2\pi}{h^3}(2m)^{3/2}\left[\mu+\frac{Ze^2}{r}\right]^{1/2}\left[-\frac{Ze^2}{r^2}\right]. \tag{A3}$$

Next, one can obtain the s-state density by solving the one-dimensional Schrödinger equation (2.2) by a one-dimensional phase-space argument to obtain

$$r^2\rho_s^{\mathrm{TF}}(r)=\frac{2(2m)^{1/2}}{h}\left[\mu+\frac{Ze^2}{r}\right]^{1/2}. \tag{A4}$$

Substituting for $(\mu+Ze^2/r)^{1/2}$ from Eq. (A4) into Eq. (A3), one does indeed find a relation of the form of the generalization (2.5) of Kato's theorem, apart from the presence of a factor of order unity, from this simple approximate density-functional theory.

[1]N. H. March, Phys. Lett. **111A**, 47 (1985).
[2]S. M. Blinder, Phys. Rev. A **29**, 1674 (1984).
[3]R. Shakeshaft and L. Spruch, J. Phys. B **18**, 1919 (1985).
[4]T. Kato, Commun. Pure Appl. Math. **10**, 151 (1957).
[5]See also, for example, S. Fraga and G. Malli, *Many-Electron Systems: Properties and Interactions* (Saunders, Philadelphia, 1968).
[6]See, for example, N. H. March, *Self-Consistent Fields in Atoms* (Pergamon, Oxford, 1975).

Volume 113A, number 9 PHYSICS LETTERS 20 January 1986

THE LOCAL POTENTIAL DETERMINING THE SQUARE ROOT OF THE GROUND-STATE ELECTRON DENSITY OF ATOMS AND MOLECULES FROM THE SCHRÖDINGER EQUATION

N.H. MARCH

Theoretical Chemistry Department, University of Oxford, 1 South Parks Road, Oxford OX1 3TG, UK

Received 9 November 1985; accepted for publication 15 November 1985

The square root $\rho^{1/2}$ of the electron density satisfies a Schrödinger equation with a local potential energy $\mathcal{V}(\boldsymbol{r})$. This is divided into two parts $V(\boldsymbol{r}) + V_{\text{Pauli}}(\boldsymbol{r})$, where $V(\boldsymbol{r})$ is the usual one-body potential of density functional theory. An expression is given for V_{Pauli}, with examples.

The electron density description of the ground state of heavy atoms, introduced by Thomas and Fermi almost 60 years ago [1] remains of interest, not only in its own right [2,3] but also as the simplest approximate version of modern density functional theory [‡1].

That there is an intimate relation between density matrix and density functional theory has also been apparent for a long time [4–6] and this area too continues to attract some attention [‡2]. In the specific case of atoms, the present writer some time ago [8] argued that the asymptotic form of the first-order density matrix $\gamma(r, r')$ for sufficiently large r';

$$\gamma(r, r') = f(r)\rho^{1/2}(r') , \tag{1}$$

with $\rho(r)$ the ground-state electron density, was such that $f(r)$ satisfied a Schrödinger equation with a local potential energy. It was also pointed out that the function $f(r)$ in eq. (1), while defined for all r, must tend to $\rho^{1/2}(r)$ for sufficiently large r, on grounds of the symmetry of $\gamma(r, r')$. Subsequent workers [9–12] have demonstrated that the square root of the density itself, $\rho^{1/2} \equiv \psi$, satisfies such a Schrödinger equation for all r, with a local potential which we shall denote by $\mathcal{V}(r)$.

‡1 Recent reviews are those of ref. [3].

‡2 A recent example is the work of Ludeña and Sieraalta [7], other references can be found from this paper.

This letter exposes just what it is that determines this local potential, in the language of density functional theory, with a view to attempting to construct judicious approximations to it, not only in atoms but also in molecules with many electrons. For if such approximations to $\mathcal{V}(r)$ could be found, then the techniques already developed in computational quantum chemistry could be applied even to molecules large enough to be of biological interest, because essentially only a "ground-state node-less wavefunction" $\psi = \rho^{1/2}$ need be computed.

To illustrate the above approach, we start from the equation following from von Weizsäcker's correction [13] to the Thomas–Fermi theory. Including this term, the Euler equation reads

$$\tfrac{5}{3}c_k\rho^{2/3} + (\hbar^2/8m)[(\nabla\rho)^2/\rho^2 - 2\nabla^2\rho/\rho] + V(r) = \epsilon , \tag{2}$$

and putting $\psi = \rho^{1/2}$ as above one is led to

$$-(\hbar^2/2m)\nabla^2\psi + [V(r) + \tfrac{5}{3}c_k\psi^{4/3} - \epsilon]\psi = 0 , \tag{3}$$

where $c_k = (3h^2/10m)(3/8\pi)^{2/3}$.

Thus, apart from a possible additive constant, the potential $\mathcal{V}(r)$ in the von Weizsäcker approximation is given by

$$\mathcal{V}^{\text{W}}(r) = V(r) + \tfrac{5}{3}c_k[\rho(r)]^{2/3} . \tag{4}$$

Motivated by this approximate result (4), we now turn to construct a formally exact expression for the

0.375-9601/86/$ 03.50 © Elsevier Science Publishers B.V. (North-Holland Physics Publishing Division)

local potential $\mathcal{V}(r)$. By definition of $\mathcal{V}(r)$, the quantity $\rho^{1/2} = \psi$ satisfies

$$\nabla^2\psi + (2m/\hbar^2)[\epsilon - \mathcal{V}(r)]\,\psi = 0\,, \tag{5}$$

and solving this for $\mathcal{V}(r)$ and then replacing ψ by $\rho^{1/2}$ yields

$$\mathcal{V}(r) = e + (\hbar^2/4m)\nabla^2\rho/\rho - (\hbar^2/8m)(\nabla\rho)^2/\rho^2\,. \tag{6}$$

Of course, this is nothing but the definition of $\mathcal{V}(r)$, but to put content into this equation we invoke the Euler equation of the density functional theory in the form of the expression representing the constancy of the chemical potential μ through the charge cloud of the atom or molecule:

$$\mu = \frac{\delta T_\mathrm{s}}{\delta\rho(r)} + V_\mathrm{n} + e^2\int\frac{\rho(r)}{|r-r'|}\,\mathrm{d}r' + \frac{\delta E_\mathrm{xc}}{\delta\rho(r)}\,, \tag{7}$$

where $T_\mathrm{s}[\rho]$ is the single-particle kinetic energy functional, while all many-electron effects are subsumed into $\delta E_\mathrm{xc}/\delta\rho(r)$. V_n denotes the potential of the nuclear framework.

As was emphasized by Slater [14] with specific reference to exchange, by Gombás [13] and others for local density treatments of correlation, and put on a formally exact footing by Kohn and Sham [15], one can view eq. (7) as reducing the many-electron problem to an equivalent one-body problem from the point of view of calculating the ground-state electron density $\rho(r)$. The associated one-body potential energy will be written

$$V(r) = V_\mathrm{n} + e^2\int\frac{\rho(r')}{|r-r'|}\,\mathrm{d}r' + \frac{\delta E_\mathrm{xc}}{\delta\rho(r)}\,. \tag{8}$$

Evidence has built up that, for quite a few, though by no means all, applications in atoms and molecules, local density approximations to $\delta E_\mathrm{xc}/8\rho(r)$ lead to electron densities of useful accuracy for present purposes.

Our aim below is to relate the desired potential $\mathcal{V}(r)$ to $V(r)$, via a correction term, motivated by the approximate result (4). This correction we will write as $V_\mathrm{Pauli}(r) \equiv V_\mathrm{p}$, this being the contribution which distinguishes the fermion problem, with its associated Exclusion Principle, from the Boson problem in the same potential $V(r)$. Our object below is to expose precisely what determines the correction $\mathcal{V} - V = V_\mathrm{p}$.

To do this, we construct V_p by subtracting $V(r)$ defined in eq. (8) from both sides of eq. (6) to obtain

$$V_\mathrm{p}(r) = \epsilon - V(r) + (\hbar^2/4m)\nabla^2\rho/\rho - (\hbar^2/8m)(\nabla\rho)^2/\rho^2\,. \tag{9}$$

But from eqs. (7) and (8), we can now replace $V(r)$ in favour of $\delta T_\mathrm{s}/\delta\rho(r)$ to find, to within an additive constant

$$V_\mathrm{p}(r) = \delta T_\mathrm{s}/\delta\rho(r) + (\hbar^2/4m)\nabla^2\rho/\rho - (\hbar^2/8m)(\nabla\rho)^2/\rho^2\,. \tag{10}$$

Eq. (10) evidently expresses the Pauli potential V_p in terms of the gradient and laplacian of the density itself, plus a functional of ρ subsumed into the single-particle kinetic energy T_s.

If we make the von Weizsäcker approximation

$$T_\mathrm{s}^\mathrm{W} = c_k\int[\rho(r)]^{5/3}\,\mathrm{d}r + \lambda\,\frac{\hbar^2}{8m}\int\frac{(\nabla\rho)^2}{\rho}\,\mathrm{d}r \tag{11}$$

with $\lambda = 1$, the first term in eq. (11) being the Thomas–Fermi kinetic energy and the second the so-called inhomogeneity correction, then we regain eq. (4) above. If, on the other hand, we adopt the Kirznits [16] value $\lambda = 1/9$, then we find

$$V_\mathrm{p}^\mathrm{K} = \tfrac{5}{3}c_k[\rho(r)]^{2/3} + (1-\lambda)(\hbar^2/4m)[\nabla^2\rho/\rho - \tfrac{1}{2}(\nabla\rho)^2/\rho^2]\,. \tag{12}$$

While eq. (12) might perhaps come into its own for really huge numbers of electrons as in biological molecules, it will be disappointing within the range of the periodic table for atoms and for light molecules, as it will lead, as with eq. (4), to a radial charge density $4\pi r^2\rho(r)$ exhibiting no atomic shell structure.

From this, it is clear that to get answers comparable in accuracy with conventional wavefunction calculations based on the potential (8) plus local density approximations to E_xc, one must transcend Thomas–Fermi plus gradient correction treatments of the single-particle kinetic energy T_s. Nevertheless, it is encouraging that the construction of $V_\mathrm{p}(r)$ requires only the solution of a one-body functional problem, unlike $V(r)$ which involves intrinsically many-body theory.

To press this point, in one very simple example, that of an arbitrary number of independent particles filled in one by one into successive levels of a one-dimensional harmonic oscillator potential, the exact differential equation for ρ is known from the work of Lawes and March [17]. From this, for N particles, the derivative of the local potential $\mathcal{V}$ can be written down in terms of the original harmonic oscillator potential and the density ρ along with its derivatives up to and including third order. For $N = 2$, it is easy to show that $V(x)$ has the explicit form

$$V_{\text{Pauli}}(x) = (1 - 2x^2 - 4x^4)/(1 + 2x^2)^2 , \qquad (13)$$

apart from an additive constant. This is a relatively simple symmetrical function with nodes at $x = \pm\frac{1}{2} \times (5^{1/2} - 1)^{1/2}$, tending to 1 as x tends to zero and to -1 as x tends to infinity.

In summary, the local potential $\mathcal{V}(r)$ determining the square root of ρ in an atom or molecule has been separated into two parts, $V(r)$ and $V_{\text{Pauli}}(r)$. While $V(r)$ involves a Hartree term, it also contains a basically many-electron contribution. In contrast, V_{Pauli} involves only the density itself through terms in $\nabla\rho$ and $\nabla^2\rho$ plus the functional derivative $\delta T_s/\delta\rho(r)$ of the single-particle kinetic energy. Eq. (10) is a formally exact expression for V_{Pauli} as a functional of the electron density.

It would seem worthwhile, for the future, to attempt, say for a closed shell atom such as krypton, to construct an explicit form of $\mathcal{V}(r)$, and then to solve the corresponding Schrödinger equation. One point that is worth adding in this context is that it is known that the electron density at sufficiently large distances from the nucleus has the form $\rho(r) = Ar^n \times \exp[-2(2I)^{1/2}r]$, where I is the ionization potential, and this allows the asymptotic form of $\mathcal{V}(r)$ to be calculated explicitly [8,12]. This form should clearly be built in whenever possible. In addition, at small r, one has also Kato's theorem which imposes another constraint on $\mathcal{V}(r)$ [12].

This work formed part of a programme supported by the National Foundation for Cancer Research, Bethesda and subsequently by the Association for International Cancer Research, London. In connection with the latter support, the writer is indebted to Dr. Colin Thomson of the University of St. Andrews for his interest in this work.

References

[1] S. Lundqvist and N.H. March, eds., Theory of the inhomogeneous electron gas (Plenum, New York, 1983).
[2] B.-G. Englert and J. Schwinger, Phys. Rev. A32 (1985) 26.
[3] A.S. Bamzai and B.M. Deb, Rev. Mod. Phys. 53 (1981) 95, 593;
J. Callaway and N.H. March, Solid State Physics, Vol. 38 (Academic Press, New York, 1984) p. 135.
[4] W. Macke, Phys. Rev. 100 (1955) 992; Ann. Phys. 17 (1955) 1.
[5] N.H. March and W.H. Young, Proc. Phys. Soc. 72 (1958) 182.
[6] J.E. Harriman, Phys. Rev. A24 (1981) 680.
[7] E.V. Ludeña and A. Sieraalta, Phys. Rev. A32 (1985) 19.
[8] N.N. March, Phys. Lett. 84A (1981) 319.
[9] M. Levy, J.P. Perdew and V. Sahni, Phys. Rev. A30 (1984) 2745.
[10] G. Hunter, Sanibel Symposium (1985); Int. J. Quantum Chem., to be published.
[11] E.N. Lassettre, J. Chem. Phys. 83 (1985) 1709.
[12] N.H. March, Phys. Lett. 113A (1985) 66.
[13] P. Gombás, Die statistische theorie des atoms und ihre anwendungen (Springer, Berlin, 1949).
[14] J.C. Slater, Phys. Rev. 81 (1951) 385.
[15] W. Kohn and L.J. Sham, Phys. Rev. 140 (1965) A1133.
[16] D.A. Kirznits, Sov. Phys. JETP 5 (1957) 64.
[17] G.P. Lawes and N.H. March, J. Chem. Phys. 71 (1979) 1007.

Phys. Chem. Liq., 1994, Vol. 27, pp. 65–67
Reprints available directly from the publisher
Photocopying permitted by license only

LETTER

Differential Equation for Slater Sum in an Inhomogeneous Electron Liquid

H. LEHMANN and N. H. MARCH

Theoretical Chemistry Department, University of Oxford, 5, South Parks Road, Oxford OX1 3UB, England

(Received 21 July 1993)

For independent electrons moving in a one-body potential $V(\mathbf{r})$ of arbitrarily low symmetry, a partial differential equation is set up for the Slater sum $Z(\mathbf{r}, \beta)$, where $\beta = (k_B T)^{-1}$. As an example, the case when $V(\mathbf{r})$ is the result of an applied time-independent and spatially constant electric field of arbitrary strength is briefly considered.

KEY WORDS: Bloch equation, Thomas-Fermi theory.

The fact that the canonical density matrix satisfies the Bloch equation has been known for a long time. However, it was subsequently shown by March and Murray[1], specifically for central field problems, that the diagonal element of the canonical density matrix, the so-called Slater sum $Z(\mathbf{r}, \beta)$, itself satisfied a differential equation, though of higher order than the Bloch equation.

In the course of work on atoms and molecules in intense external fields, we have found that such a differential equation can be set up for independent electrons moving in a one-body potential energy $V(\mathbf{r})$ of arbitrarily low symmetry. The purpose of this Letter is to present the derivation of this equation, and to exemplify its use by considering, quite briefly, the case when $V(\mathbf{r})$ arises solely from an applied electric field, constant in space and time, of arbitrary intensity.

Let us begin by summarizing the definition of the Slater sum $Z(\mathbf{r}, \beta)$. Suppose that the eigenfunctions $\psi_i(\mathbf{r})$ and corresponding eigenvalus ε_i are generated by the one-body potential $V(\mathbf{r})$. Then the Slater sum is defined by

$$Z(\mathbf{r},\beta) = \sum \psi_i(\mathbf{r})\psi_i^*(\mathbf{r}) \exp(-\beta\varepsilon_i) \tag{1}$$

where $\beta = (k_B T)^{-1}$.

As a first step, let us form the Thomas-Fermi approximation to Z. This may be written from Eq. (1) as

$$Z_{TF}(\mathbf{r},\beta) = Z_0^{(D)}(\beta)\exp(-\beta V(\mathbf{r})) \tag{2}$$

where

$$Z_0^{(D)} = (2\pi\beta)^{-D/2} \tag{3}$$

with D indicating the dimensions of the free motion on to which it is assumed the potential energy $V(\mathbf{r})$ is switched. Equation (2) follows from Eq. (1) with the (usually drastic) approximation that the wave functions are replaced by plane waves in D dimensions while the energy levels are merely shifted by a 'constant' V.

The next step is to set up the differential equation satisfied by Z_{TF}. To do so, dropping the subscript TF for the present for notational convenience; we write from Eq. (2):

$$\nabla Z = (-\beta \nabla V) Z. \tag{4}$$

Differentiating Eq. (4) partially with respect to β then yields

$$\begin{aligned} \frac{\partial \nabla Z}{\partial \beta} &= -Z\nabla V - \beta \nabla V \frac{\partial Z}{\partial \beta} \\ &= -Z\nabla V + \beta Z V \nabla V + \tfrac{1}{2} D Z \nabla V, \end{aligned} \tag{5}$$

the second line following from the use of the explicit form (3).

Rearranging Eq. (5) allows one to write, for the Thomas-Fermi approximation to the Slater sum, the differential equation

$$V\nabla Z + \frac{\partial}{\partial \beta}(\nabla Z) = \left(\frac{D}{2} - 1\right) Z \nabla V. \tag{6}$$

It is to be expected that Eq. (6) will have omitted higher spatial derivatives than the first.

Next we return to Eq. (1) without approximation, but still separating out the free particle motion as in Eq. (3) by writing

$$Z = Z_0^{(D)} f. \tag{7}$$

After forming the quantity $\nabla(\nabla^2 Z)$ directly from Eq. (1), and employing the Schrödinger equation, we find the result which we wish to employ for a potential $V(\mathbf{r})$ corresponding to a uniform electric field, namely

$$\frac{1}{8}\nabla(\nabla^2 Z) - V\nabla Z - \frac{\partial}{\partial \beta}(\nabla Z) + \left(\frac{D}{2} - 1\right) Z \nabla V = 0. \tag{8}$$

which reduces to Eq. (6) when the first term is neglected.

Let us turn immediately to the admittedly simple example where $V(\mathbf{r})$ is generated solely by switching on a constant field of arbitrary strength $\mathscr{E}$ along the z axis. For this case, the result for Z is known and is explicitly[2]

$$Z = (2\pi\beta)^{-3/2} \exp(\beta \mathscr{E} z + \beta^3 \mathscr{E}^2/24). \tag{9}$$

Evidently $\partial^2 Z/\partial z^2 = (\beta\mathscr{E})^2 Z$, and the desired z component of the vector Eq. (8) becomes, with $V = -\mathscr{E}z$:

$$\text{LHS of Eq. (8)} = \frac{1}{8}(\beta\mathscr{E})^2 \frac{\partial Z}{\partial z} + (\beta\mathscr{E}^2) zZ - \frac{\partial}{\partial\beta}(\beta\mathscr{E}Z) - \frac{Z\mathscr{E}}{2}. \tag{10}$$

Using Eq. (9) again to form the β derivative term in Eq. (10) one finds

$$\frac{\partial}{\partial\beta}(\beta\mathscr{E}Z) = Z\beta\mathscr{E}\left(-\frac{3}{2\beta} + \mathscr{E}z + \frac{1}{8}(\beta^2\mathscr{E}^2)\right) + Z\mathscr{E}. \tag{11}$$

Substituting Eq. (11) into Eq. (10) the quadratic and cubic terms in the field immediately cancell. After a brief calculation, the coefficient of the linear term in the field strength $\mathscr{E}$ is found also to be identically zero, verifying that the exact form (9) satisfies Eq. (8), with $D = 3$.

In conclusion, Eq. (8) is the main result of the present work, motivated by the desire to treat atoms and molecules in intense external fields. Equation (8) is shown to be satisfied for free electrons in a constant electric field of arbitrary strength, for which the Slater sum is known to be given by Eq. (9). The generalization of the method outlined in this letter to treat the level spectrum of a hydrogen-like atom in an intense external field will be reported elsewhere.

Acknowledgment

One of us (H. L) wishes to thank the German Academic Exchange Service for a grant which made possible his stay in Oxford.

References

1. N. H. March and A. M. Murray, *Phys. Rev.* **120**, 830 (1960).
2. R. A. Harris and J. A. Cina, *J. Chem. Phys.* **79**, 1381 (1983); see also A. D. Jannussis, *Phys. Stat. Sol.* **36**, K17 (1969).

Journal of Mathematical Chemistry 16 (1994) 9–18 9

Bound-state plus continuum electron densities, and Slater sum, in a bare Coulomb field

S. Pfalzner [a], H. Lehmann [b] and N.H. March [b]

[a] GSI, Planckstrasse 1, 6100 Darmstadt 11, Germany

[b] Theoretical Chemistry Department, University of Oxford, 5 South Parks Road, Oxford OX1 3UB, UK

Received 23 November 1993; revised 31 March 1994

After reporting numerical studies based solely on s-states of total (s + p + d etc.) bound-state densities, which allow the range of validity of the simplest density functional theory of Thomas and Fermi to be critically assessed, two areas in which analytical progress proves possible are focused on. The first of these is the local density of states in the continuum, for which an exact formula is derived. The second concerns the Slater sum, for which an explicit differential equation is established. Prior to this, only the Bloch equation satisfied by the off-diagonal generalization of the Slater sum, namely the canonical density matrix, was available.

1. Introduction

In the important $1/Z$ expansion of the ground-state energy of an atomic ion of nuclear charge Ze and with N electrons, this energy $E(Z,N)$ is expressed as

$$E(Z,N) = Z^2\left(\epsilon_0(N) + \frac{1}{Z}\epsilon_1(N) + \frac{1}{Z^2}\epsilon_2(N) + \dots\right) \tag{1.1}$$

following Hylleraas [1], Layzer [2] and others [3]. In this expansion, $\epsilon_0(N)$ is completely determined by solution of the bare Coulomb field problem, while the higher terms in the series account for effects of screening, plus exchange and correlation. While, for fixed N, the convergence of this series is guaranteed for sufficiently large Z, for other than the lightest atoms it is essential to sum the series (1.1). As was pointed out by March and White [4] in early work, one such partial sum is afforded by the Thomas–Fermi limit of density functional theory, which leads to the result

$$E_{\mathrm{TF}}(Z,N) = Z^{7/3}f(N/Z)\,, \tag{1.2}$$

where the function $f(x)$ is known from numerical solution of the Thomas–Fermi self-consistent field problem [4]. For neutral atoms, ie $N = Z$, the term $Z^{7/3}$ from eq. (1.2) has to be corrected by a term of $O(Z^2)$ from an inner-electron correction neglected in the Thomas–Fermi (TF) approach [5–7], a Dirac–Slater exchange term [5] of $O(Z^{5/3})$ and a correlation energy of $O(Z)$ (see ref. [8] and other authors cited there). Such a development gives an excellent overall account of numerical Hartree–Fock energies calculated for individual atoms.

Such progress in calculating the ground-state energy has prompted recent parallel studies of the ground-state electron density. Two aspects are important in the present context. First of all, since the leading term in the energy expansion (1.1) is for electrons, taken as non-interacting, moving in a bare Coulomb potential $-Ze^2/r$, the ground-state electron density $\rho(r)$ in this potential is of obvious interest. For closed shells, considerable progress has proved possible. One step, for densities of s + p + d etc. states associated with closed shells, has been the derivation of an exact spatial generalization of Kato's theorem to read [9]

$$\frac{\partial \rho(r)}{\partial r} = -\frac{2Z}{a_0}\rho_s(r), \quad a_0 = \hbar^2/me^2 . \tag{1.3}$$

This has already proved valuable [10]; here we shall use it to construct the total (s + p + d etc.) density for 5 closed shells by computing solely the s-state density ρ_s and then determining $\rho(r)$ by numerical quadrature using eq. (1.3). The result for $\rho(r)$ is then utilized to make a critical assessment of the range of validity of the simplest density functional theory of Thomas and Fermi (TF).

Following this numerical study of bound states, attention will then be focused on two specific areas where analytical progress proves possible. The first of these is the local density of states in the continuum. Here, analytical forms are established, both from the (approximate) TF density functional theory and then exactly via a Green function calculation, using the work of Blinder [11] and of earlier authors. This analytical progress is summarized in section 3.

The second area in which analytical progress proves possible is in treating bound plus continuum states via the so-called Slater sum $S(r,\beta)$, defined explicitly in eq. (4.2) below. The essential result of this part of the present investigation is to establish the differential equation (4.13) for $S(r,\beta)$. Prior to the present work, one had to use the Bloch equation, which is obeyed by the off-diagonal generalization of the Slater sum, namely the canonical density matrix. Clearly it is a major simplification to be able to work directly with the diagonal matrix, or Slater sum. Section 5 constitutes a brief summary.

2. Ground-state density for closed shells in a bare Coulomb field

To obtain the ground-state density $\rho(r)$, the simplest route is to calculate the s-state density $\rho_s(r)$ in eq. (1.3) by summing the squares of the s-state hydrogenic wave functions, and then integrating eq. (1.3) to read

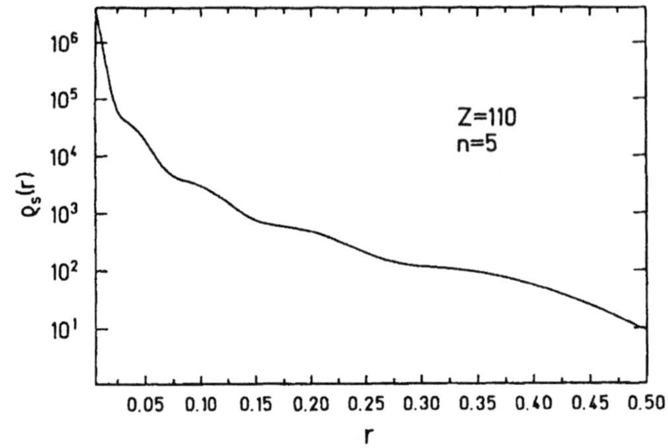

Fig. 1. The s-state density $\rho_s(r)$ calculated for doubly filled levels in a bare Coulomb field. The result is for nuclear charge 110*e* and for 5 closed shells. Atomic units are employed in all figures.

$$\rho(r) = \frac{2Z}{a_0} \int_r^\infty \rho_s(r)\, dr\,. \tag{2.1}$$

As an example, fig. 1 depicts the s-state density for 5 closed shells obtained by summing the squares of the wave functions $\Phi_{n00}(r)$ from $n = 1$ to 5. The plot in fig. 1 has been produced for the specific case $Z = 110$, which corresponds to a "neutral" system in that 5 closed shells in the bare Coulomb field, with $2n^2$ electrons per closed shell, contain 110 electrons. The "ripples" in the s-state density shown become pronounced when $\rho_s(r)$ is multiplied by r^2, as illustrated in fig. 2. All plots in the paper are made in atomic units.

Using the results for $\rho_s(r)$ shown in fig. 1, eq. (2.1) has been utilized to obtain

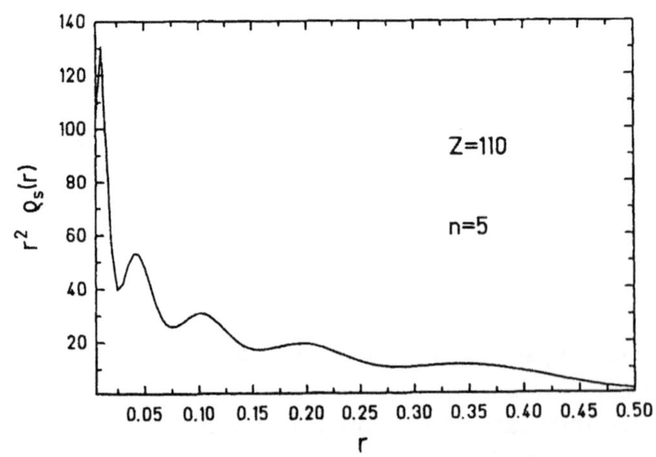

Fig. 2. r^2 times s-state density of fig. 1, to reveal "shells" more clearly.

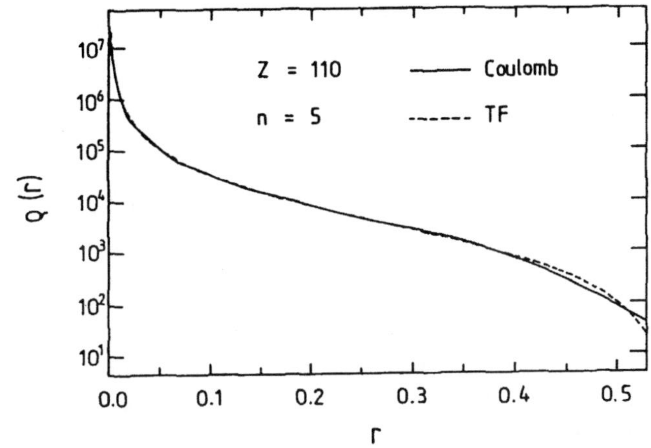

Fig. 3. Solid curve: the exact wave mechanical ground state density $\rho(r)$ calculated directly from the s-state density of fig. 1 by using eq. (2.1). The dashed curve is the result of Thomas–Fermi statistical theory as given in eqs. (2.2) and (2.3). Again $Z = 110$ and $N = Z$.

the total ground-state density $\rho(r)$ for 5 closed shells by numerical quadrature. The result is shown by the solid curve of fig. 3. It is to be stressed that in obtaining this curve, one has avoided recourse to any wave functions having angular dependence, that is p, d, etc., wave functions. For closed shell systems, as exemplified by eqs. (1.3) and (2.1), the s-state density suffices to determine $\rho(r)$. As in fig. 2 for this s-state density, fig. 4 shows $D(r) = 4\pi r^2 \rho(r)$, the radial electron density. In contrast to fig. 2, where the "shells" are pronounced, in the total density $D(r)$ the shells are beginning to be smoothed out, for this example with 110 electrons. Eventually, of course, if one filled a sufficiently large number of closed shells, one must

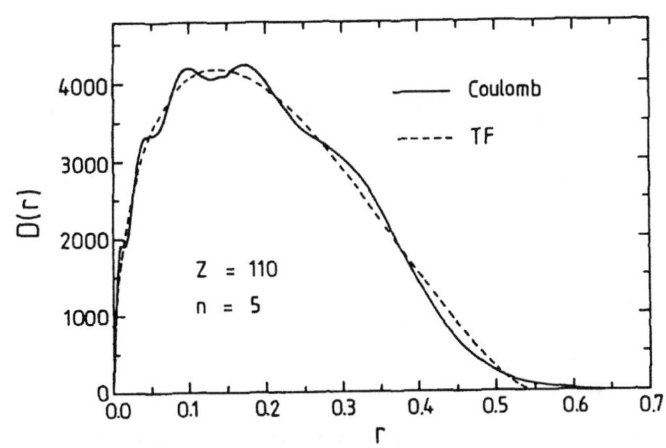

Fig. 4. Radial density $D(r) = 4\pi r^2 \rho(r)$, obtained from $\rho(r)$ in fig. 3.

asymptotically approach the TF semiclassical density. This is given explicitly in the bare Coulomb field problem under discussion by

$$\rho_{\mathrm{TF}}(r) = c_3\left[\mu + \frac{Ze^2}{r}\right]^{3/2}, \quad c_3 = \frac{8\pi}{3h^3}(2m)^{3/2}. \tag{2.2}$$

The chemical potential μ is, of course, to be determined by normalization of the total electron density. In the present case it is given in atomic units by [3]

$$\mu = -\frac{1}{18^{1/3}}Z^2N^{-2/3} \tag{2.3}$$

with $N = \int \rho(r)\mathrm{d}\boldsymbol{r}$ being the total number of electrons as above. For $N = Z = 110$, the result (2.2) is shown by the dotted curve in fig. 3. The corresponding approximation to the radial density $D(r)$ is similarly shown in fig. 4. Clearly, it can be seen that especially from fig. 4 the TF statistical theory is not fully quantitative for the ground-state electron density $\rho(r)$ even for 110 electrons. Furthermore, of course, as follows from eq. (2.2) plus eq. (2.3), it has a classical cutoff radius, r_c say, given by $Ze^2/r_c = |\mu|$, while there is a singularity in the density like $r^{-3/2}$ as one approaches the point nucleus.

Using the present method based on eq. (2.1), it will be a straightforward matter in the future to repeat the calculations shown in figs. 1–4 for a larger number of closed shells. The economy of the present approach is best illustrated by noting that, as already mentioned, there are $2n^2$ electrons in a closed shell of principal quantum number n for the bare Coulomb field, of which only 2 are s electrons.

We turn next to the treatment of continuum states. In the spirit of the TF theory in eq. (2.2), this is moving from the regime of negative chemical potential μ to the case where μ is greater than zero. Here we shall focus all attention on the local density of states, to be defined precisely below.

3. Local density of states in continuum

Let us start out from the TF relation (2.2) and define $\rho(r, E)$ from it by replacing μ by the (now positive) energy E. Since we are in the continuum, let us then differentiate $\rho(r, E)$ with respect to energy E. This defines the local density of states $N(r, E)$, which is evidently given in the TF approximation by

$$N_{\mathrm{TF}}(r, E) = \frac{\partial}{\partial E}\rho_{\mathrm{TF}}(r, E) = \frac{3c_3}{2}\left[E + \frac{Ze^2}{r}\right]^{1/2}. \tag{3.1}$$

If we "switch off" the atomic potential $-Ze^2/r$ in eq. (3.1) by letting Z tend to zero, then we obtain the density of states $N(E)$ of free electrons, the resulting system being described by plane waves and having therefore translational invariance.

The object of the present section is purely analytical; to replace the approximate TF formula (3.1) for the bare Coulomb field by an exact result.

3.1. USE OF COULOMB GREEN FUNCTION

The tool we shall employ here is the Coulomb Green function, the analytical form of which has been studied by many workers [11]. To be precise, we define this Green function by

$$G(\boldsymbol{r}_1, \boldsymbol{r}_2, E) = \sum_{nlm} \frac{\psi^*_{nlm}(\boldsymbol{r}_1)\psi_{nlm}(\boldsymbol{r}_2)}{E - \epsilon_n}, \tag{3.2}$$

where ϵ_n are the exact levels for the Coulomb field, with Im $E>0$ for G^+.

Following Blinder [11], we employ the new independent variables

$$x = r_1 + r_2 + r_{12}, \quad y = r_1 + r_2 - r_{12}, \quad x \geqslant y \geqslant 0, \tag{3.3}$$

where $\boldsymbol{r}_1 - \boldsymbol{r}_2 = \boldsymbol{r}_{12}$ and $r_{12} = |\boldsymbol{r}_{12}|$. Writing $E = k^2/2$, Blinder then finds

$$G^+(x, y, k) = \frac{ik}{2\pi} \int_0^\infty \mathrm{d}s \sinh s \left(\coth \frac{s}{2}\right)^{2i\nu} \exp(ik\xi \cosh s) J_0(k\eta \sinh s) \tag{3.4}$$

with $\xi = (1/2)(x+y), \eta = (xy)^{1/2}$ and $\nu = Z/k$. We note from eq. (3.4) that G^+ is a function of only two space variables x and y. This, in turn, results from the fact that for the bare Coulomb field the Runge–Lenz vector is a constant of motion [12].

As Z tends to zero, ie ν tends to zero, eq. (3.4) reduces to the free-particle Green function, G_0^+ say, namely

$$G_0^+ = \frac{\exp(ik|\boldsymbol{r}_1 - \boldsymbol{r}_2|)}{|\boldsymbol{r}_1 - \boldsymbol{r}_2|}. \tag{3.5}$$

If we take the imaginary part of G_0^+, we evidently find

$$\text{Im } G_0^+(\boldsymbol{r}_1, \boldsymbol{r}_2, E) = \frac{\sin k|\boldsymbol{r}_1 - \boldsymbol{r}_2|}{|\boldsymbol{r}_1 - \boldsymbol{r}_2|}. \tag{3.6}$$

If one now goes on to the diagonal $\boldsymbol{r}_1 = \boldsymbol{r}_2 = \boldsymbol{r}$ say, then eq. (3.6) yields

$$\text{Im } G_0^+(\boldsymbol{r}_1, \boldsymbol{r}_2, E) = k = (2E)^{1/2}, \tag{3.7}$$

which is essentially the free-electron density of states discussed above. In fact, the above correspondence follows from the quite general result for the local density of states $N(\boldsymbol{r}, E)$, namely [13]

$$N(\boldsymbol{r}, E) = \frac{\partial \rho(\boldsymbol{r}, E)}{\partial E} = \frac{1}{\pi^2} \text{Im } G^+(\boldsymbol{r}, \boldsymbol{r}, E). \tag{3.8}$$

The result (3.8), when combined with eq. (3.4) above, provides us with a route to calculate the local density of states in the bare Coulomb field.

3.2. LOCAL DENSITY OF STATES $N(r, E)$

Obviously one wants to take, from eq. (3.8), first the imaginary part of $G^+(x, y, k)$ in eq. (3.4) and secondly to go on to the diagonal $\boldsymbol{r}_1 = \boldsymbol{r}_2 = \boldsymbol{r}$. From the definitions of x and y in eq. (3.3), this second step is equivalent to putting $x = y = 2r$.

To facilitate these two steps, we note with Blinder [11] that the substitution $\sinh s = \operatorname{csch} q$ transforms the integral in eq. (3.4) to give

$$G^{+}(x, y, k) = \frac{ik}{2\pi} \int_0^{\infty} dq \operatorname{csch}^2 q \exp(i[2\nu q + k\xi \coth q]) J_0(k\eta \operatorname{csch} q) . \tag{3.9}$$

Taking the imaginary part of eq. (3.9), one can then put $x = y = 2r$ to find the result

$$N(r, E) = \frac{(2E)^{1/2}}{2\pi^3} \int_0^{\infty} dq \operatorname{csch}^2 q \cos\left[\frac{2Zq}{k} + 2kr \coth q\right] J_0(2kr \operatorname{csch} q) , \tag{3.10}$$

where $k^2 = 2E$. The integral can now be evaluated, with the result

$$N(r, E) = \frac{k}{4\pi^3} \sqrt{\frac{\pi\nu}{\sinh \pi\nu}} \cos\left\{\nu\gamma - \sum_{n=1}^{\infty}\left[\frac{\nu}{n} - \tan^{-1}\left(\frac{\nu}{n}\right)\right]\right\} \times \left[M'_{i\nu,1/2}(ikr)\, W'_{i\nu,1/2}(ikr) - \left(\frac{1}{4} + \frac{\nu}{kr}\right) M_{i\nu,1/2}(ikr)\, W_{i\nu,1/2}(ikr)\right] . \tag{3.11}$$

The functions M and W appearing in eq. (3.11) are Whittaker functions. The quantity γ denotes Euler's constant. It can be shown that the right-hand side of eq. (3.11) is real. The result (3.11) achieves the objective of transcending the semi-classical approximation (3.1) for this case of the bare Coulomb field.

Having discussed the quantum mechanics of bound and continuum electron densities in the Coulomb field case, we shall finally consider the calculation of the Slater sum for this same field. This, of course, involves both bound and continuum states together (see eq. (4.2) below).

4. Differential equation for Slater sum for bare Coulomb field

For a one-electron Hamiltonian H_r of the form

$$H_r = -\frac{1}{2}\nabla_r^2 + V(r) , \tag{4.1}$$

the Slater sum $S(\boldsymbol{r}, \beta)$ is defined by

$$S(\boldsymbol{r}, \beta) = \sum_{\text{all } i} \psi_i^*(\boldsymbol{r})\psi_i(\boldsymbol{r}) \exp(-\beta\epsilon_i), \quad \beta = (k_B T)^{-1} . \tag{4.2}$$

In eq. (4.2), $\psi_i(\boldsymbol{r})$ and ϵ_i represent, respectively, the normalized eigenfunctions and eigenvalues of the hamiltonian (4.1). The Bloch equation

$$H_r C(\boldsymbol{r},\boldsymbol{r}',\beta) = -\frac{\partial C(\boldsymbol{r},\boldsymbol{r}',\beta)}{\partial\beta} \tag{4.3}$$

affords a well-established route to the canonical density matrix $C(\boldsymbol{r},\boldsymbol{r}',\beta)$. $S(\boldsymbol{r},\beta)$ is then the diagonal element $C(\boldsymbol{r},\boldsymbol{r},\beta)$. However March and Murray [14] showed, for central field problems in which $V(\boldsymbol{r}) = V(|\boldsymbol{r}|)$, that a partial wave analysis could lead to $S(r,\beta)$ in the form

$$S(r,\beta) = \sum_{l=0}(2l+1)S_l(r,\beta) \tag{4.4}$$

and that $S_l(r,\beta)$ itself satisfied the differential equation

$$\begin{aligned}\frac{1}{8}\frac{\partial^3}{\partial r^3}(r^2 S_l) &- \frac{l(l+1)}{2r}\frac{\partial}{\partial r}(rS_l) - \frac{1}{2}\frac{\partial V}{\partial r}(r^2 S_l)\\ &- V\frac{\partial}{\partial r}(r^2 S_l) - \frac{\partial^2}{\partial r\,\partial\beta}(r^2 S_l) = 0\,.\end{aligned} \tag{4.5}$$

If one multiplies eq. (4.5) by $(2l+1)$ and sums over all l to form $S(r,\beta)$ in eq. (4.4), the term involving $l(l+1)$ in eq. (4.5) presents a stumbling block, and for a general central field potential $V(r)$ this summation has so far not been achieved. The resulting differential equation therefore relates S and its derivatives to this summation of S_l and its derivative $\partial S_l/\partial r$.

However, for the bare Coulomb field (cf. eq. (1.3) above), Storer [15] and later Pollock [16] have demonstrated that the canonical density matrix $C(\boldsymbol{r},\boldsymbol{r}',\beta)$ entering eq. (4.3) can be directly related to its s-wave component $C_{l=0}(\boldsymbol{r},\boldsymbol{r}',\beta)$. This strongly suggests for this specific model that $S(r,\beta)$ should be determined solely by $S_{l=0}(r,\beta) = S_s(r,\beta)$, the s-wave contribution in eq. (4.4). Storer introduces $P_{ns}(\mathrm{r}) = r\psi_{ns}(\mathrm{r})$ and then shows that

$$S(r,\beta) = -\frac{1}{2}\lim_{s\to 0}\frac{\partial^2}{\partial s^2}\left[\sum_{\mathrm{all}\,n} P^*_{ns}(r+s)P_{ns}(r-s)\ \exp(-\beta\epsilon_n)\right]. \tag{4.6}$$

Expanding the RHS of eq. (4.6) around $s = 0$, one readily finds

$$\begin{aligned}\lim_{s\to 0}\frac{\partial^2}{\partial s^2}\sum_n \exp(-\beta\epsilon_n)\Big[|P_{ns}|^2 &+ s(P^{*\prime}_{ns}P_{ns} - P^*_{ns}P'_{ns})_{s=0}\\ &+\frac{s^2}{2}(P^{*\prime\prime}_{ns}P_{ns} + P^*_{ns}P''_{ns} - 2P^{*\prime}_{ns}P'_{ns})_{s=0} + O(s^3)\Big]\\ = \sum_n \exp(-\beta\epsilon_n)&\left[P^{*\prime\prime}_{ns}P_{ns} + P^*_{ns}P''_{ns} - 2P^{*\prime}_{ns}P'_{ns}\right]_{s=0}\,.\end{aligned} \tag{4.7}$$

From the definition of S_s we find that

$$\frac{\partial^2}{\partial r^2}\left[r^2 S_s(r,\beta)\right] = \sum_n \exp(-\beta\epsilon_n)\left[P^{*\prime\prime}_{ns}P_{ns} + P^*_{ns}P''_{ns} + 2P^{*\prime}_{ns}P'_{ns}\right] \tag{4.8}$$

and

$$-\frac{\partial}{\partial\beta}(r^2 S_s(r,\beta)) = \sum_n \epsilon_n |P_{ns}|^2 \exp(-\beta\epsilon_n). \tag{4.9}$$

Furthermore, the P_{ns} are solutions of the radial Schrödinger equation

$$P''_{ns} + 2\left(\frac{Z}{r} + \epsilon_n\right) P_{ns} = 0. \tag{4.10}$$

Inserting eqs. (4.8)–(4.10) in eq. (4.7), the relation (4.6) finally takes the form

$$S(r,\beta) = \frac{1}{2}\frac{\partial^2}{\partial r^2}(r^2 S_s) - 4\frac{\partial}{\partial\beta}(r^2 S_s) + \frac{4Z}{r}(r^2 S_s). \tag{4.11}$$

Taking the r-derivative of eq. (4.11) and returning to eq. (4.5) for s-states, ie $l = 0$, one obtains

$$\frac{\partial S}{\partial r} = -2ZS_s. \tag{4.12}$$

We stress that eq. (4.12) is the analogue in statistical mechanics of the quantum-mechanical spatial generalization (1.3) of Kato's theorem [9] for the bare Coulomb field. One can now substitute for S_s in eq. (4.11) when the desired differential equation for the Slater sum $S(r,\beta)$ results, namely

$$-2ZS = \frac{r^2 S'''}{2} + 2rS'' + \left[1 + 4r^2\left(\frac{Z}{r} - \frac{\partial}{\partial\beta}\right)\right] S', \tag{4.13}$$

with the primes denoting the derivatives with respect to r.

Equation (4.13) has been shown to hold for the bound part of S:

$$S_b = \sum_{\substack{\text{bound}\\ \text{states}}} \exp(-\beta\epsilon_n)|\psi_{nlm}(r)|^2 \tag{4.14}$$

in earlier work [17]. However the existence of a more general relation seemed to us likely from the infinite series form of S given by Blinder [18]. By means of eq. (4.12), S_s may be calculated from Blinder's result.

5. Summary

The spatial generalization (1.3) of Kato's theorem can be used to provide a direct route to the calculation of the total density from the s-state density (see figs. 1–4 above). The analogue of eq. (1.3) for the Slater sum has been obtained here, the result being exhibited in eq. (4.12).

Specific to continuum states, again for the model of the bare Coulomb field, the local density of states $N(r, E)$ can be obtained. This is given in the form of a one-dimensional quadrature in eq. (3.10). This can be evaluated as an infinite series, as in eq. (3.11).

Finally, by utilizing the result (4.12) relating the Slater sum S to its s-wave component S_s, a third order partial differential equation has been obtained for the Slater sum in a bare Coulomb field. An exact solution of this, again though in the form of an infinite series, is then known from the earlier work of Blinder [18].

Acknowledgement

One of us (H.L.) wishes to thank the German Academic Exchange Service for a grant which made possible his stay in Oxford.

References

[1] E.A. Hylleraas, Z. Phys. 65 (1930) 209.
[2] D. Layzer, Ann. Phys. 8 (1959) 271.
[3] See, e.g., N.H. March, *Electron Density Theory of Atoms and Molecules* (Academic Press, New York, 1993).
[4] N.H. March and R.J. White, J. Phys. B5 (1972) 466.
[5] J.M.C. Scott, Philos. Mag. 43 (1952) 859.
[6] See also B.-G. Englert and J. Schwinger, Phys. Rev. A32 (1985); N.H. March and J.S. Plaskett, Proc. Roy. Soc. A235 (1956) 419.
[7] L. Spruch, Rev. Mod. Phys. 63 (1991) 151.
[8] N.H. March and P. Wind, Mol. Phys. 77 (1992) 791.
[9] N.H. March, Phys. Rev. A33 (1986) 88.
[10] F.J. Galvez and I. Parras, Phys. Rev. A44 (1991) 144; see also P. Csavinsky, Int. J. Quant. Chem. S24 (1990) 557.
[11] S.M. Blinder, J. Math. Phys. 22 (1981) 306.
[12] See, for instance, A. Holas and N.H. March, J. Phys. A: Math. Gen. 23 (1990) 735, where more general central field cases are also treated.
[13] See, e.g., W. Jones and N.H. March, *Theoretical Solid State Physics* (Dover, New York, 1985).
[14] N.H. March and A.M. Murray, Phys. Rev. 120 (1960) 830.
[15] R.G. Storer, J. Math. Phys. 9 (1968) 964.
[16] E.L. Pollock, Comp. Phys. Commun. 52 (1988) 49.
[17] N.H. March, Phys. Lett. A111 (1985) 47.
[18] S.M. Blinder, Phys. Rev. A43 (1991) 13.

PHYSICAL REVIEW A VOLUME 51, NUMBER 3 MARCH 1995

Exact exchange-correlation potential and approximate exchange potential in terms of density matrices

A. Holas
Institute of Physical Chemistry of the Polish Academy of Sciences, 44/52 Kasprzaka, 01-224 Warsaw, Poland

N. H. March
Theoretical Chemistry Department, University of Oxford, 5 South Parks Road, Oxford OX1 3UB, England

(Received 24 August 1994; revised manuscript received 15 November 1994)

An exact expression in terms of density matrices (DMs) is derived for $\delta F[n]/\delta n(\mathbf{r})$, the functional derivative of the Hohenberg-Kohn functional. The derivation starts from the differential form of the virial theorem, obtained here for an electron system with arbitrary interactions, and leads to an expression taking the form of an integral over a path that can be chosen arbitrarily. After applying this approach to the equivalent system of noninteracting electrons (Slater-Kohn-Sham scheme) and combining the corresponding result with the previous one, an exact expression for the exchange-correlation potential $v_{\mathrm{xc}}(\mathbf{r})$ is obtained which is analogous in character to that for $\delta F[n]/\delta n(\mathbf{r})$, but involving, besides the interacting-system DMs, also the noninteracting DMs. Equating the former DMs to the latter ones, we reduce the result for the exact $v_{\mathrm{xc}}(\mathbf{r})$ to that for an approximate exchange-only potential $v_{\mathrm{x}}(\mathbf{r})$. This leads naturally to the Harbola-Sahni exchange-only potential.

PACS number(s): 31.15.Ew, 71.10.+x, 31.25.−v, 71.45.Gm

I. INTRODUCTION

Starting with the Thomas-Fermi theory [1–3], extended later by Dirac [4], in which the ground-state energy E of a heavy atom of atomic number Z can be written quite explicitly in terms of the electron density $n(\mathbf{r})$ and the nuclear potential energy $v(\mathbf{r}) = -Z/r$ (we use atomic units throughout), much attention has been devoted to the functional

$$\widetilde{E}[n] = F[n] + \int d^3r\, n(\mathbf{r})v(\mathbf{r})\,, \qquad (1.1)$$

following the formal existence theorem of Hohenberg and Kohn [5]. In Eq. (1.1), $v(\mathbf{r})$ denotes the external potential acting on an electron (e.g., due to all nuclei of the system). Exact knowledge of $F[n]$ is clearly equivalent to the complete solution of the many-body problem, which still seems a long way off. But if we accept such (formal) knowledge, then the ground-state density itself is to be determined from the minimization of $\widetilde{E}[n]$ with respect to the electron density $n(\mathbf{r})$ normalized to the total number N of electrons in the system under discussion, which yields

$$\mu = \frac{\delta F[n]}{\delta n(\mathbf{r})} + v(\mathbf{r})\,. \qquad (1.2)$$

Though, as already stressed, $F[n]$ is not presently known, many approximations have been proposed: see, for example, the book by Parr and Yang [6]. However, the exact expression for $\delta F[n]/\delta n(\mathbf{r})$ (or, at least, a good approximation to it) rather than $F[n]$ itself is necessary to solve Eq. (1.2).

Usually, in order to provide a high accuracy for the dominant contribution to $F[n]$ — the kinetic energy — the Slater-Kohn-Sham (SKS) scheme is applied [7,8]. In essence then, $F[n]$ is decomposed as

$$F[n] = T_s[n] + E_{\mathrm{es}}[n] + E_{\mathrm{xc}}[n]\,, \qquad (1.3)$$

where $T_s[n]$ is the kinetic-energy functional of a noninteracting-electron system, while $E_{\mathrm{es}}[n]$ is the classical electrostatic energy of the electronic cloud of density $n(\mathbf{r})$. What remains on the right-hand side of Eq. (1.3) is the so-called exchange-correlation energy functional $E_{\mathrm{xc}}[n]$. In this case, the single Eq. (1.2) is replaced by a system of N single-particle Schrödinger equations: the SKS equations

$$\{-\tfrac{1}{2}\nabla^2 + v_s(\mathbf{r})\}\,\phi_j(\mathbf{r},s) = \varepsilon_j\phi_j(\mathbf{r},s), \quad j = 1,\ldots,N\,, \qquad (1.4)$$

in which the effective one-body potential is

$$v_s(\mathbf{r}) = v(\mathbf{r}) + v_{\mathrm{es}}(\mathbf{r}) + v_{\mathrm{xc}}(\mathbf{r})\,. \qquad (1.5)$$

Here, the electrostatic potential is given by

$$v_{\mathrm{es}}(\mathbf{r};[n]) = \frac{\delta E_{\mathrm{es}}[n]}{\delta n(\mathbf{r})} = \int d^3r'\, \frac{n(\mathbf{r}')}{|\mathbf{r}-\mathbf{r}'|}\,, \qquad (1.6)$$

and the exchange-correlation potential by

$$v_{\mathrm{xc}}(\mathbf{r};[n]) = \frac{\delta E_{\mathrm{xc}}[n]}{\delta n(\mathbf{r})}\,. \qquad (1.7)$$

Therefore, in this approach, one requires, ideally, exact expressions for, or in practice good approximations to, E_{xc} and v_{xc} (see, for instance, Parr and Yang [6]; Harbola and Sahni [9]). It is noteworthy in this context that a number of exact relations concerning E_{xc} and v_{xc}

1050-2947/95/51(3)/2040(9)/$06.00 51 2040

have been established [10–12]. These can be very useful for testing approximations, and constructing refinements where necessary, for these quantities E_{xc} and v_{xc}.

In the present paper, which is also a contribution in the field of the investigations referred to immediately above, the following aims are realized:

(i) An exact expression for $\delta F[n]/\delta n(\mathbf{r})$ is obtained in terms of first- and second-order density matrices, by employing the differential virial theorem for interacting electrons (the derivation being given in Appendix A).

(ii) $v_{xc}(\mathbf{r})$ is then obtained in exact form by applying the result (i) above together with the analogous result for the equivalent noninteracting system (SKS scheme).

The results (i) and (ii) are somewhat formal, but they will become useful in practice if some controlled approximations for the necessary density matrices can be inserted. We shall also employ the results (i) and (ii) to obtain an approximation for the exchange-only potential, replacing density matrices of the interacting system by the corresponding matrices of the equivalent noninteracting system in the exact relation for the exchange-correlation potential. This substitution not only simplifies the theory significantly, but also removes difficulties with matrices of higher than first order, since all other matrices can then be expressed in terms of the first-order density matrix. In this way an approximation to $v_x(\mathbf{r})$ is obtained from (ii).

The outline of the present paper is then as follows. In Sec. II, immediately below, the aims (i) and (ii) are realized and followed by approximations. In Sec. III some properties of the approximate $v_x(\mathbf{r})$ and Harbola-Sahni conjecture on $v_{xc}(\mathbf{r})$ are studied, while Sec. IV constitutes a summary, together with proposals for possible future studies. In Appendix A the derivation of the differential virial theorem is given.

II. POTENTIALS VIA DIFFERENTIAL VIRIAL THEOREM

A. Density matrices and the kinetic-energy density tensor

We consider an N-electron system characterized by the Hamiltonian

$$\widehat{\mathcal{H}} = \widehat{\mathcal{T}} + \widehat{\mathcal{U}} + \widehat{\mathcal{V}}, \tag{2.1}$$

with the kinetic energy, electron-electron (e-e) repulsion, and electron-nucleus (e-n) attraction energy operators, respectively, defined as

$$\widehat{\mathcal{T}} = \sum_{i=1}^{N} -\tfrac{1}{2}\nabla_i^2, \tag{2.2}$$

$$\widehat{\mathcal{U}} = \sum_{i=1}^{N-1}\sum_{j=i+1}^{N} u(\mathbf{r}_i,\mathbf{r}_j), \quad u(\mathbf{r}_j,\mathbf{r}_i) = u(\mathbf{r}_i,\mathbf{r}_j), \tag{2.3}$$

$$\widehat{\mathcal{V}} = \sum_{i=1}^{N} v(\mathbf{r}_i). \tag{2.4}$$

In application to real atoms and molecules we have

$$u(\mathbf{r},\mathbf{r}') = 1/|\mathbf{r}-\mathbf{r}'|, \tag{2.5}$$

$$v(\mathbf{r}) = \sum_{J=1}^{M} -Z_J \Big/ |\mathbf{r}-\mathbf{R}_J| \tag{2.6}$$

(for a molecule characterized by M nuclei of charge Z_J, placed at fixed positions $\mathbf{R}_J$). But for all general considerations we do not assume these narrowing specifications (2.5) and (2.6).

Let $\Psi(\mathbf{x}_1,\ldots,\mathbf{x}_N)$ be a normalized eigenfunction (ground- or excited-state one) of the Schrödinger equation

$$\widehat{\mathcal{H}}\Psi = E\Psi, \tag{2.7}$$

where the notation $\mathbf{x}_i = (\mathbf{r}_i, s_i)$ for ith space and spin coordinate is adopted. Then the Nth-order density matrix (DM) generated by Ψ is just the following product:

$$\gamma_N(\mathbf{x}_1,\ldots,\mathbf{x}_N;\mathbf{x}'_1,\ldots,\mathbf{x}'_N) = \Psi(\mathbf{x}_1,\ldots,\mathbf{x}_N)\,\Psi^*(\mathbf{x}'_1,\ldots,\mathbf{x}'_N), \tag{2.8}$$

while the pth-order reduced DM (for $p < N$) is defined in terms of γ_N by integrating out the $(N-p)$ coordinates (see, e.g., [6])

$$\gamma_p(\mathbf{x}_1,\ldots,\mathbf{x}_p;\mathbf{x}'_1,\ldots,\mathbf{x}'_p) = \binom{N}{p}\int d\mathbf{x}_{p+1}\cdots d\mathbf{x}_N\,\gamma_N\,. \tag{2.9}$$

Here $\int d\mathbf{x}_i$ means integration $\int d^3r_i$ and summation over s_i together with the replacement of $\mathbf{x}'_i$ by $\mathbf{x}_i$ in the integrand. The corresponding spinless DM is

$$\rho_p(\mathbf{r}_1,\ldots,\mathbf{r}_p;\mathbf{r}'_1,\ldots,\mathbf{r}'_p) = \sum_{s_1,\ldots,s_p} \gamma_p(\mathbf{x}_1,\ldots,\mathbf{x}_p;\mathbf{x}'_1,\ldots,\mathbf{x}'_p)\Big|_{s'_i=s_i}, \tag{2.10}$$

while the diagonal elements of spinless DMs are denoted

$$n_p(\mathbf{r}_1,\ldots,\mathbf{r}_p) = \rho_p(\mathbf{r}_1,\ldots,\mathbf{r}_p;\mathbf{r}_1,\ldots,\mathbf{r}_p) \geq 0\,. \tag{2.11}$$

The subscript "1" can be omitted. The basic quantity — electron number density — is thus $n(\mathbf{r}) = n_1(\mathbf{r}) = \rho_1(\mathbf{r};\mathbf{r})$.

We define the kinetic-energy density tensor in terms of the DM ρ as

$$t_{\alpha\beta}(\mathbf{r};[\rho]) = \frac{1}{4}\left(\frac{\partial^2}{\partial r'_\alpha \partial r''_\beta} + \frac{\partial^2}{\partial r'_\beta \partial r''_\alpha}\right)\rho(\mathbf{r}';\mathbf{r}'')\Big|_{\mathbf{r}'=\mathbf{r}''=\mathbf{r}}. \tag{2.12}$$

This is a real, symmetric tensor, the trace of which is the non-negative kinetic-energy density (scalar)

$$t(\mathbf{r}) = \sum_{\alpha} t_{\alpha\alpha}(\mathbf{r}) \geq 0\,, \tag{2.13}$$

leading to the global kinetic energy

$$T = \int d^3r\, t(\mathbf{r}) = \langle\Psi|\widehat{\mathcal{T}}|\Psi\rangle\,. \tag{2.14}$$

B. Exact expression for $\delta F[n]/\delta n(\mathbf{r})$

1. *Exact expression for the gradient of the external potential*

The differential virial theorem derived in Appendix A, Eq. (A7), can be rewritten as

$$\boldsymbol{\nabla} v(\mathbf{r}) = -\mathbf{f}(\mathbf{r}\,;[u,n,\rho,n_2])\,, \tag{2.15}$$

where

$$\mathbf{f}(\mathbf{r}\,;[u,n,\rho,n_2]) = \left\{-\tfrac{1}{4}\boldsymbol{\nabla}\nabla^2 n(\mathbf{r}) + \mathbf{z}(\mathbf{r};[\rho]) + 2\int d^3r'\,[\boldsymbol{\nabla} u(\mathbf{r},\mathbf{r}')]\,n_2(\mathbf{r},\mathbf{r}')\right\}\Big/ n(\mathbf{r})\,. \tag{2.16}$$

The vector field $\mathbf{z}$ introduced above, related to the kinetic-energy density tensor via

$$z_\alpha(\mathbf{r}\,;[\rho]) = 2\sum_\beta \frac{\partial}{\partial r_\beta} t_{\alpha\beta}(\mathbf{r}\,;[\rho])\,, \tag{2.17}$$

is a combination of derivatives of DM $\rho(\mathbf{r}';\mathbf{r}'')$ [see Eq. (2.12)] evaluated at the diagonal $\mathbf{r}'=\mathbf{r}''=\mathbf{r}$, so $\mathbf{z}$ can be called a "local" functional of ρ.

Equation (2.15) represents an exact relation between the gradient of the *e-n* potential $v(\mathbf{r})$, the *e-e* interaction potential $u(\mathbf{r},\mathbf{r}')$, and the DMs $n(\mathbf{r})$, $\rho(\mathbf{r}\,;\mathbf{r}')$, $n_2(\mathbf{r},\mathbf{r}')$, generated by some eigenfunction Ψ of the Hamiltonian $\widehat{\mathcal{H}}$ constructed of these potentials; see Eqs. (2.1)–(2.4). It should be noted that no assumption concerning the degeneracy of the system state Ψ is involved in obtaining Eq. (2.15). Also, the actual number of electrons N does not appear explicitly in Eqs. (2.15) and (2.16). The equation obtained from Eq. (2.15) by multiplying both sides by $n(\mathbf{r})$, depends on DMs linearly. Therefore the relation (2.15) remains true if the DM ρ_2 (from which n, ρ, and n_2 are derivable) is replaced by a "mixture" of pure-state matrices $\rho_2^{(i)}$:

$$\rho_2 = \sum_i p_i \rho_2^{(i)}\,, \tag{2.18}$$

where the probabilities p_i satisfy conditions

$$p_i \geq 0\,,\quad \sum_i p_i = 1\,. \tag{2.19}$$

But ρ_2 mentioned above can be obtained by reduction [via Eqs. (2.9) and (2.10)] of the so-called ensemble-state density matrix Γ_N, describing a system in a mixed state, rather than by reduction of γ_N, Eq. (2.8), describing a system in the pure state (see, e.g., Parr and Yang [6]). This Γ_N is defined by

$$\Gamma_N = \sum_i p_i \gamma_N^{(i)}\,, \tag{2.20}$$

where $\gamma_N^{(i)}$ corresponds to $\Psi^{(i)}$, according to Eq. (2.8).

To summarize the above, the exact relation (2.15) holds not only for pure-state density matrices, but also for ensemble-state matrices.

2. *Path-integral form of the external potential $v(\mathbf{r})$*

Equation (2.15) may be viewed as a differential equation for the potential $v(\mathbf{r})$. Because $\mathbf{f}(\mathbf{r}) = -\boldsymbol{\nabla} v(\mathbf{r})$, the force field $\mathbf{f}(\mathbf{r})$ is conservative. Therefore, it follows that the potential at point $\mathbf{r}_0$, say, is the work done in bringing an electron from infinity to $\mathbf{r}_0$ against the force field $\mathbf{f}(\mathbf{r})$:

$$v(\mathbf{r}_0) = -\int_\infty^{\mathbf{r}_0} d\mathbf{r}\cdot\mathbf{f}(\mathbf{r})\,. \tag{2.21}$$

Since $\mathbf{f}(\mathbf{r})$ is conservative, the value of the line integral in Eq. (2.21) does not depend on the path of integration chosen. Note that Eq. (2.21) has been written such that $v(\infty)=0$: a standard choice of gauge for the potential.

Finally, using the Euler-Lagrange equation (1.2) of the density-functional theory, we can reinterpret Eq. (2.21) as

$$\frac{\delta F[n]}{\delta n(\mathbf{r}_0)} = \mu + \int_\infty^{\mathbf{r}_0} d\mathbf{r}\cdot\mathbf{f}(\mathbf{r}\,;[u,n,\rho,n_2])\,. \tag{2.22}$$

The presence of an arbitrary constant μ is connected with the fact that the functional derivative on the left-hand side of Eq. (2.22) is defined only for variations conserving the total number of electrons.

Equation (2.22) together with Eq. (2.16) represents an exact expression for $\delta F/\delta n$ in terms of the *e-e* interaction potential $u(\mathbf{r},\mathbf{r}')$ and three objects n, ρ, n_2 derivable from the DM ρ_2. All remarks on the possible origin of ρ_2 or its construction, made in the preceding section, apply also to the relation (2.22).

C. Exact expression for $v_{xc}(\mathbf{r})$

In this section, in order to have the exchange-correlation potential uniquely defined, we assume that: (i) The original N-interacting electron system is in its ground state. (ii) The electron density of this state possesses the property that it is noninteracting v-representable (see, e.g., Ref. [6]). (iii) The ground state of the equivalent SKS scheme [a system of N noninteracting electrons moving in the external potential $v_s(\mathbf{r})$, Eq.

(1.5)], is nondegenerate and has the same density $n(\mathbf{r})$ as the system in (i). Then the wave function of equivalent system in (iii), $\Phi_s(\mathbf{x}_1,\ldots,\mathbf{x}_N)$, is a Slater determinant constructed of the N lowest-energy orbitals $\phi_i(\mathbf{x})$, the eigenfunctions of the SKS Eq. (1.4). The DMs and their diagonal elements, generated from Φ_s, will be distinguished (from DMs generated by Ψ) by a superscript "s"; e.g., n_2^s, ρ^s, etc.

We can apply the result (2.15) to case (iii) and obtain obviously

$$\nabla v_s(\mathbf{r}) = -\mathbf{f}_s(\mathbf{r}\,;[n,\rho^s])\,, \tag{2.23}$$

where

$$\mathbf{f}_s(\mathbf{r}\,;[n,\rho^s]) = \{-\tfrac{1}{4}\nabla\nabla^2 n(\mathbf{r}) + \mathbf{z}(\mathbf{r}\,;[\rho^s])\}/n(\mathbf{r})\,. \tag{2.24}$$

Now, keeping in mind the decomposition (1.5) of v_s, we subtract Eq. (2.15) from Eq. (2.23) to obtain

$$\nabla(v_{\mathrm{es}}(\mathbf{r}) + v_{\mathrm{xc}}(\mathbf{r})) = \mathbf{f}(\mathbf{r}\,;[u,n,\rho,n_2]) - \mathbf{f}_s(\mathbf{r}\,;[n,\rho^s])\,. \tag{2.25}$$

But, using an analogue of Eq. (1.6) in which the Coulombic interaction, Eq. (2.5), is replaced by a general u, we have

$$\nabla v_{\mathrm{es}}(\mathbf{r}) = \int d^3r'\, \nabla u(\mathbf{r},\mathbf{r}')n(\mathbf{r}')\,. \tag{2.26}$$

After subtracting Eq. (2.26) from Eq. (2.25) we arrive finally at the result

$$\nabla v_{\mathrm{xc}}(\mathbf{r}) = -\mathbf{f}_{\mathrm{xc}}(\mathbf{r}\,;[u,n,\rho^s,\rho,n_2]), \tag{2.27}$$

with

$$\mathbf{f}_{\mathrm{xc}}(\mathbf{r}\,;[u,n,\rho^s,\rho,n_2]) = \left\{\mathbf{z}(\mathbf{r}\,;[\rho^s]) - \mathbf{z}(\mathbf{r}\,;[\rho]) + \int d^3r'\,[\nabla u(\mathbf{r},\mathbf{r}')][n(\mathbf{r})n(\mathbf{r}') - 2n_2(\mathbf{r},\mathbf{r}')]\right\}\Big/ n(\mathbf{r}) \tag{2.28}$$

[see Eq. (2.17) for the definition of $\mathbf{z}$]. The force field $\mathbf{f}_{\mathrm{xc}}(\mathbf{r})$ is conservative because it stems from the potential $v_{\mathrm{xc}}(\mathbf{r})$, i.e., $\mathbf{f}_{\mathrm{xc}}(\mathbf{r}) = -\nabla v_{\mathrm{xc}}(\mathbf{r})$. Therefore, in complete analogy with Eq. (2.21) we have

$$v_{\mathrm{xc}}(\mathbf{r}_0) = -\int_{\infty}^{\mathbf{r}_0} d\mathbf{r}\cdot \mathbf{f}_{\mathrm{xc}}(\mathbf{r}\,;[u,n,\rho^s,\rho,n_2])\,, \tag{2.29}$$

where again the above line integral is independent of the particular path chosen for integration.

Equation (2.29) represents an exact expression for $v_{\mathrm{xc}}(\mathbf{r})$ in terms of objects written explicitly as arguments of $\mathbf{f}_{\mathrm{xc}}$. Contrary to the results (2.21) and (2.22), the expression for v_{xc}, Eq. (2.29), is not valid for ensemble-state DMs (because the equation for $n\nabla v_{\mathrm{xc}}$ is not linear in DMs).

D. Approximate expression for $v_{\mathrm{x}}(\mathbf{r})$

Our further aim is to find, from the exact expression for $v_{\mathrm{xc}}(\mathbf{r})$ in Eq. (2.29), an approximate expression for $v_{\mathrm{x}}(\mathbf{r})$ alone. As a guide to this task let us analyze the separation of the exchange energy

$$E_{\mathrm{x}} = \langle\Phi_s|\widehat{\mathcal{U}}|\Phi_s\rangle - E_{\mathrm{es}}[n] \tag{2.30}$$

from the exchange-correlation energy

$$E_{\mathrm{xc}} = \langle\Psi|\widehat{\mathcal{T}}|\Psi\rangle - \langle\Phi_s|\widehat{\mathcal{T}}|\Phi_s\rangle + \langle\Psi|\widehat{\mathcal{U}}|\Psi\rangle - E_{\mathrm{es}}[n]\,. \tag{2.31}$$

The above equations represent the standard definitions (see, e.g., Parr and Yang [6]). Therefore, combining Eqs. (2.30) and (2.31) we have

$$E_{\mathrm{xc}} = E_{\mathrm{x}} + (T_{\mathrm{c}} + U_{\mathrm{c}})\,, \tag{2.32}$$

where

$$\begin{aligned} T_{\mathrm{c}} &= \langle\Psi|\widehat{\mathcal{T}}|\Psi\rangle - \langle\Phi_s|\widehat{\mathcal{T}}|\Phi_s\rangle \\ &= T - T_s \\ &= \int d^3r \sum_\alpha t_{\alpha\alpha}(\mathbf{r};[\rho-\rho^s])\,, \end{aligned} \tag{2.33}$$

$$\begin{aligned} U_{\mathrm{c}} &= \langle\Psi|\widehat{\mathcal{U}}|\Psi\rangle - \langle\Phi_s|\widehat{\mathcal{U}}|\Phi_s\rangle \\ &= \int d^3r\, d^3r'\, u(\mathbf{r},\mathbf{r}')\{n_2(\mathbf{r},\mathbf{r}') - n_2^s(\mathbf{r},\mathbf{r}')\}\,. \end{aligned} \tag{2.34}$$

We see that, in order to obtain E_{x} from the expression for E_{xc}, it is enough to neglect the differences between the interacting-system DMs and corresponding SKS-system DMs, i.e., to neglect $(\rho-\rho^s)$ in the expression for T_{c}, Eq. (2.33), and $(n_2-n_2^s)$ in the expression for U_{c}, Eq. (2.34).

Now we adopt the above observation in order to split the exact expression for v_{xc} in Eq. (2.29) into the sum of $\widetilde{v}_{\mathrm{x}}$ and $\widetilde{v}_{\mathrm{c}}$; interpreting these terms as approximate exchange potential and approximate correlation potential, respectively. Thus

$$\begin{aligned} v_{\mathrm{xc}}(\mathbf{r}_0) &= \widetilde{v}_{\mathrm{x}}(\mathbf{r}_0) + \widetilde{v}_{\mathrm{c}}(\mathbf{r}_0) \\ &= -\int_{\infty}^{\mathbf{r}_0} d\mathbf{r}\cdot\widetilde{\mathbf{f}}_{\mathrm{x}}(\mathbf{r}) - \int_{\infty}^{\mathbf{r}_0} d\mathbf{r}\cdot\widetilde{\mathbf{f}}_{\mathrm{c}}(\mathbf{r})\,, \end{aligned} \tag{2.35}$$

where the force fields are given by

$$\widetilde{\mathbf{f}}_{\mathrm{x}}(\mathbf{r}) = -\int d^3r'\,[\nabla u(\mathbf{r},\mathbf{r}')][\,2n_2^s(\mathbf{r},\mathbf{r}') - n(\mathbf{r})n(\mathbf{r}')]\Big/ n(\mathbf{r}) \tag{2.36}$$

$$\widetilde{\mathbf{f}}_c(\mathbf{r}) = -\left\{ \mathbf{z}(\mathbf{r};[\rho-\rho^s]) + 2\int d^3r' [\nabla u(\mathbf{r},\mathbf{r}')][n_2(\mathbf{r},\mathbf{r}') - n_2^s(\mathbf{r},\mathbf{r}')] \right\} \Big/ n(\mathbf{r}). \tag{2.37}$$

While we know that the total force field $\mathbf{f}_{xc}$ is conservative, we lack such knowledge concerning the separate pieces $\widetilde{\mathbf{f}}_x$ or $\widetilde{\mathbf{f}}_c$ (examples of the nonconservative character of $\widetilde{\mathbf{f}}_x$ can be constructed). Therefore some specific path of integration must be chosen to complete the definition (2.35). Seemingly the most natural path is along the radius on which the point $\mathbf{r}_0$ lies. As the center of the coordinate system, a position of the nucleus is chosen in the case of a single-ion system, the center of symmetry for a symmetrical molecule, and some "inner" point for a molecule or cluster with lower symmetry. Then

$$\widetilde{v}_x(\mathbf{r}_0) = \int_{r_0}^{\infty} dr\, \mathbf{e}\cdot\widetilde{\mathbf{f}}_x(r\mathbf{e}), \quad \mathbf{e} = \mathbf{r}_0/r_0. \tag{2.38}$$

The same path must be used for the calculation of $\widetilde{v}_c$ in order to leave the sum v_{xc} unchanged.

Using the separation of the pair-distribution function n_2^s into its uncorrelated and correlated parts (see, e.g., Parr and Yang [6])

$$n_2^s(\mathbf{r}_1,\mathbf{r}_2) = \tfrac{1}{2}n(\mathbf{r}_1)n(\mathbf{r}_2)\{1 + h_2^s(\mathbf{r}_1,\mathbf{r}_2)\}, \tag{2.39}$$

and noting that the combination

$$\rho_x(\mathbf{r}_1,\mathbf{r}_2) = h_2^s(\mathbf{r}_1,\mathbf{r}_2)\, n(\mathbf{r}_2) \tag{2.40}$$

is customarily called the exchange hole at $\mathbf{r}_2$ of an electron at $\mathbf{r}_1$, we rewrite Eq. (2.36) as

$$\widetilde{\mathbf{f}}_x(\mathbf{r}) = -\int d^3r'\, \rho_x(\mathbf{r},\mathbf{r}')\nabla u(\mathbf{r},\mathbf{r}'). \tag{2.41}$$

For the case of Coulombic $u(\mathbf{r},\mathbf{r}')$, Eq. (2.5), Eq. (2.41) leads to

$$\widetilde{\mathbf{f}}_x(\mathbf{r}) = \int d^3r'\, \rho_x(\mathbf{r},\mathbf{r}')\frac{(\mathbf{r}-\mathbf{r}')}{|\mathbf{r}-\mathbf{r}'|^3}. \tag{2.42}$$

We recognize $\widetilde{\mathbf{f}}_x(\mathbf{r})$, given by Eq. (2.42), to be identical with the force field $\mathcal{E}_x(\mathbf{r})$ proposed by Harbola and Sahni [9] (in their exchange-only version). This means that our approximation $\widetilde{v}_x(\mathbf{r})$ to the exact $v_x(\mathbf{r})$ coincides with the Harbola-Sahni approximation $\widetilde{v}_x^{HS}(\mathbf{r})$. Since these two approximations were derived by using completely different reasonings, their coincidence strongly enhances the expectation that this function is very close to the exact $v_x(\mathbf{r})$.

III. STUDY OF THE APPROXIMATE EXCHANGE POTENTIAL $\widetilde{v}_x(\mathbf{r})$ AND VALIDITY OF HARBOLA-SAHNI CONJECTURE ON $v_{xc}(\mathbf{r})$

A. Path dependence of exchange-only potential

While in their pioneering work Harbola and Sahni [9] did not discuss the issue of path dependence, this was analyzed by Ou-Yang and Levy [13]. These authors give an example of $\mathcal{E}_x(\mathbf{r})$ which is not conservative; i.e., for which $\nabla\times\mathcal{E}_x(\mathbf{r}) \neq \mathbf{0}$. In their subsequent paper [14], they demonstrate that $\widetilde{v}_x(\mathbf{r})$, given by Eqs. (2.38) and (2.42) [in their notation $h_x^0(\mathbf{r})$], satisfies two necessary conditions for $\widetilde{v}_x$ to be identical with v_x, namely

$$E_x = -\int d^3r\, n(\mathbf{r})\,\mathbf{r}\cdot\nabla\widetilde{v}_x(\mathbf{r}) \tag{3.1}$$

and

$$\widetilde{v}_x(\mathbf{r};[n_\lambda]) = \lambda\widetilde{v}_x(\lambda\mathbf{r};[n]), \tag{3.2}$$

where the scaled density is defined as $n_\lambda(\mathbf{r}) = \lambda^3 n(\lambda\mathbf{r})$. For the proof that $\widetilde{v}_x(\mathbf{r})$ satisfies Eq. (3.1) it is essential that the path, chosen for integration, is that adopted in the definition Eq. (2.38); i.e., along a radius. As Ou-Yang and Levy [14] prove in order to have $\widetilde{v}_x(\mathbf{r}) \equiv v_x(\mathbf{r})$, besides the two conditions (3.1) and (3.2), a third condition must be satisfied by $\widetilde{v}_x$, namely that $\widetilde{v}_x(\mathbf{r})$ be a functional derivative of some functional of n. Unfortunately, this last property has not been shown yet for $\widetilde{v}_x$: therefore there is no general proof that $\widetilde{v}_x(\mathbf{r})$ is identical to $v_x(\mathbf{r})$. Evidently, for systems in which the relation

$$\nabla\times\widetilde{\mathbf{f}}_x(\mathbf{r}) = \mathbf{0} \tag{3.3}$$

is violated in some region, we can conclude that the potential $\widetilde{v}_x$ does not coincide with v_x. But, for systems where Eq. (3.3) is satisfied everywhere, it is not sufficient to claim that the field $\widetilde{\mathbf{f}}_x(\mathbf{r})$ is conservative (see discussion in Wang *et al.* [15]). As dictated by symmetry, $\widetilde{v}_x$ defined by Eq. (2.35) is path independent for closed-subshell atoms, thus uniquely reducing to the form (2.38). Numerical investigation by Li *et al.* [16] of these systems carried out with the exchange-only Harbola-Sahni potential (i.e., $\widetilde{v}_x$) leads to results which demonstrate the high accuracy of this approximate exchange potential

B. Large-r behavior of exchange-only potential

In order to see the large-r behavior of $\widetilde{v}_x(\mathbf{r})$, let us first rewrite the force $\widetilde{\mathbf{f}}_x$, Eq. (2.41), as

$$\widetilde{\mathbf{f}}_x(\mathbf{r}) = -\nabla\int d^3r'\, u(\mathbf{r},\mathbf{r}')\rho_x(\mathbf{r},\mathbf{r}') + \int d^3r'\, u(\mathbf{r},\mathbf{r}')\,\nabla\rho_x(\mathbf{r},\mathbf{r}'). \tag{3.4}$$

For Coulombic interaction this means a separation into the "Slater" and "non-Slater" contributions:

$$\widetilde{\mathbf{f}}_{\mathrm{x}}(\mathbf{r}) = -\nabla v_{\mathrm{Sl}}(\mathbf{r}) + \widetilde{\mathbf{f}}_{\mathrm{nSl}}(\mathbf{r}) , \tag{3.5}$$

with the Slater [7] potential defined as

$$v_{\mathrm{Sl}}(\mathbf{r}) = \int d^3r' \, \frac{\rho_{\mathrm{x}}(\mathbf{r},\mathbf{r}')}{|\mathbf{r}-\mathbf{r}'|} , \tag{3.6}$$

and the non-Slater force (in the present approximation) given by

$$\widetilde{\mathbf{f}}_{\mathrm{nSl}}(\mathbf{r}) = \int d^3r' \, \frac{\nabla \rho_{\mathrm{x}}(\mathbf{r},\mathbf{r}')}{|\mathbf{r}-\mathbf{r}'|} . \tag{3.7}$$

Thus the total approximate exchange potential can be written as

$$\widetilde{v}_{\mathrm{x}}(\mathbf{r}) = v_{\mathrm{Sl}}(\mathbf{r}) + \widetilde{v}_{\mathrm{nSl}}(\mathbf{r}) , \tag{3.8}$$

with

$$\widetilde{v}_{\mathrm{nSl}}(\mathbf{r}_0) = \int_{r_0}^{\infty} dr \, \mathbf{e} \cdot \widetilde{\mathbf{f}}_{\mathrm{nSl}}(r\mathbf{e}), \quad \mathbf{e} = \mathbf{r}_0/r_0 . \tag{3.9}$$

The above decomposition (3.8) for the Harbola-Sahni exchange potential was performed earlier by Ou-Yang and Levy [14].

In the asymptotic large-r region the exchange hole can be shown to behave as

$$\rho_{\mathrm{x}}(\mathbf{r},\mathbf{r}') = \rho_{\mathrm{x}}(\infty,\mathbf{r}')\{1 + O(e^{-br})\}, \quad b > 0 . \tag{3.10}$$

Therefore the non-Slater components, $\widetilde{\mathbf{f}}_{\mathrm{nSl}}(\mathbf{r})$, Eq. (3.7), of the force, and, subsequently, $\widetilde{v}_{\mathrm{nSl}}(\mathbf{r})$, Eq. (3.9), of the potential, are exponentially small, while the Slater potential $v_{\mathrm{Sl}}(\mathbf{r})$, Eq. (3.6), tends to $-1/r$. This provides the correct behavior of $\widetilde{v}_{\mathrm{x}}(\mathbf{r})$, Eq. (3.8) (see, e.g., the figures of $\widetilde{v}_{\mathrm{x}}^{\mathrm{HS}}(\mathbf{r})$ in Li *et al.* [16]).

C. Degree of validity of the Harbola-Sahni conjecture on $v_{\mathrm{xc}}(\mathbf{r})$

Under the assumptions stated in Sec. II C, it is known that the exchange-correlation energy can be written (see, e.g., Parr and Yang [6]) as

$$E_{\mathrm{xc}} = \frac{1}{2} \int d^3r \, d^3r' \, \frac{1}{|\mathbf{r}-\mathbf{r}'|} n(\mathbf{r}) \, \overline{\rho}_{\mathrm{xc}}(\mathbf{r},\mathbf{r}') , \tag{3.11}$$

where

$$\begin{aligned} \overline{\rho}_{\mathrm{xc}}(\mathbf{r},\mathbf{r}') &= \overline{h}_2(\mathbf{r},\mathbf{r}') \, n(\mathbf{r}') \\ &= \{2\overline{n}_2(\mathbf{r},\mathbf{r}') - n(\mathbf{r})n(\mathbf{r}')\}/n(\mathbf{r}) \end{aligned} \tag{3.12}$$

[compare Eqs. (2.40) and (2.39)]. Here the bar over a quantity denotes the coupling-constant averaged value

$$\overline{n}_2(\mathbf{r},\mathbf{r}') = \int_0^1 d\lambda \, n_2^{(\lambda)}(\mathbf{r},\mathbf{r}') , \tag{3.13}$$

when the original system ($\lambda = 1$) and the equivalent SKS system ($\lambda = 0$) are related by means of adiabatic connection [6]. An alternative expression for E_{xc}, already presented in Eqs. (2.31)–(2.34), is

$$E_{\mathrm{xc}} = T - T_s + \tfrac{1}{2} \int d^3r \, d^3r' \, \frac{1}{|\mathbf{r}-\mathbf{r}'|} n(\mathbf{r}) \rho_{\mathrm{xc}}(\mathbf{r},\mathbf{r}') , \tag{3.14}$$

where ρ_{xc} is defined similarly as $\overline{\rho}_{\mathrm{xc}}$ in Eq. (3.12), but with $\overline{n}_2$ replaced by $n_2^{(1)} \equiv n_2$.

Using the notation of Eq. (3.11), the Harbola-Sahni [9] conjecture concerning the exchange-correlation potential means that it is given by

$$\widetilde{v}_{\mathrm{xc}}^{\mathrm{HS}}(\mathbf{r}_0) = -\int_{\infty}^{\mathbf{r}_0} d\mathbf{r} \cdot \widetilde{\mathbf{f}}_{\mathrm{xc}}^{\mathrm{HS}}(\mathbf{r}) , \tag{3.15}$$

where

$$\begin{aligned} \widetilde{\mathbf{f}}_{\mathrm{xc}}^{\mathrm{HS}}(\mathbf{r}) &= \mathcal{E}_{xc}(\mathbf{r}) \\ &= \int d^3r' \, \overline{\rho}_{\mathrm{xc}}(\mathbf{r},\mathbf{r}') \frac{(\mathbf{r}-\mathbf{r}')}{|\mathbf{r}-\mathbf{r}'|^3} \\ &= -\int d^3r' \, \overline{\rho}_{\mathrm{xc}}(\mathbf{r},\mathbf{r}') \, \nabla \left(\frac{1}{|\mathbf{r}-\mathbf{r}'|} \right) \end{aligned} \tag{3.16}$$

[compare Eqs. (1), (4), and (5) of Harbola and Sahni [9]].

Below we shall demonstrate that $\widetilde{v}_{\mathrm{xc}}^{\mathrm{HS}}(\mathbf{r})$ is not identical to the exact $v_{\mathrm{xc}}(\mathbf{r})$, by employing *reductio ad absurdum*. Let us assume that $\widetilde{v}_{\mathrm{xc}}^{\mathrm{HS}}$ is a legitimate potential. Then Eq. (3.15) implies that the gradient of this potential is given by

$$\nabla \widetilde{v}_{\mathrm{xc}}^{\mathrm{HS}}(\mathbf{r}) = -\widetilde{\mathbf{f}}_{\mathrm{xc}}^{\mathrm{HS}}(\mathbf{r}) . \tag{3.17}$$

Now calculate the following energy [in analogy with Eq. (3.1)]

$$\begin{aligned} E_{\mathrm{dif}} &= \int d^3r \, n(\mathbf{r}) \, \mathbf{r} \cdot \{\nabla \widetilde{v}_{\mathrm{xc}}^{\mathrm{HS}}(\mathbf{r}) - \nabla v_{\mathrm{xc}}(\mathbf{r})\} \\ &= \int d^3r \, n(\mathbf{r}) \, \mathbf{r} \cdot \{\mathbf{f}_{\mathrm{xc}}(\mathbf{r}) - \widetilde{\mathbf{f}}_{\mathrm{xc}}^{\mathrm{HS}}(\mathbf{r})\} , \end{aligned} \tag{3.18}$$

showing an (eventual) difference between the Harbola-Sahni expression (3.17) and the exact one, Eq. (2.27), for the gradient of the potential. Using Eqs. (2.28) and (3.16) we evaluate E_{dif} to be

$$E_{\mathrm{dif}} = E_{\mathrm{dif}}^T + E_{\mathrm{dif}}^U , \tag{3.19}$$

where

$$\begin{aligned} E_{\mathrm{dif}}^T &= \int d^3r \, \mathbf{r} \cdot \{\mathbf{z}(\mathbf{r},[\rho^s]) - \mathbf{z}(\mathbf{r};[\rho])\} \\ &= 2\sum_{\alpha\beta} \int d^3r \, r_\alpha \, \frac{\partial}{\partial r_\beta} \{t_{s,\alpha\beta}(\mathbf{r}) - t_{\alpha\beta}(\mathbf{r})\} \\ &= 2(T - T_s) \end{aligned} \tag{3.20}$$

[see Eq. (2.17) and compare with Eqs. (2.13) and (2.14)], and

$$E_{\rm dif}^{U} = \int d^3r\, d^3r' \frac{\mathbf{r}\cdot(\mathbf{r}-\mathbf{r}')}{|\mathbf{r}-\mathbf{r}'|^3} n(\mathbf{r})\{\rho_{\rm xc}(\mathbf{r},\mathbf{r}') - \overline{\rho}_{\rm xc}(\mathbf{r},\mathbf{r}')\}$$
$$= \int d^3r\, d^3r' \frac{\mathbf{r}\cdot(\mathbf{r}-\mathbf{r}')}{|\mathbf{r}-\mathbf{r}'|^3} 2\{n_2(\mathbf{r},\mathbf{r}') - \overline{n}_2(\mathbf{r},\mathbf{r}')\}\,. \tag{3.21}$$

Using symmetry of n_2 and $\overline{n}_2$ we transform Eq. (3.21) to obtain

$$E_{\rm dif}^{U} = \int d^3r\, d^3r' \frac{1}{|\mathbf{r}-\mathbf{r}'|}\{n_2(\mathbf{r},\mathbf{r}') - \overline{n}_2(\mathbf{r},\mathbf{r}')\}$$
$$= \frac{1}{2}\int d^3r\, d^3r' \frac{n(\mathbf{r})}{|\mathbf{r}-\mathbf{r}'|}\{\rho_{\rm xc}(\mathbf{r},\mathbf{r}') - \overline{\rho}_{\rm xc}(\mathbf{r},\mathbf{r}')\}$$
$$= [E_{\rm xc} - (T - T_s)] - E_{\rm xc}\,, \tag{3.22}$$

where Eqs. (3.14) and (3.11) were used in the last step. So we get finally from Eq. (3.19)

$$E_{\rm dif} = T - T_s\,. \tag{3.23}$$

But, as it is known (Levy and Perdew [10]),

$$T_c = T - T_s > 0\,. \tag{3.24}$$

Thus we have arrived at a contradiction and therefore $\widetilde{v}_{\rm xc}^{\rm HS}(\mathbf{r}) \neq v_{\rm xc}(\mathbf{r})$. The values of the correlation kinetic energy T_c, Eq. (3.24), are known to be rather small compared with $E_{\rm xc}$ (see, e.g., Zhao *et al.* [17]); so, on average, $\widetilde{v}_{\rm xc}^{\rm HS}(\mathbf{r})$ may be quite close to $v_{\rm xc}(\mathbf{r})$. But one has also to bear in mind that there is no proof that the definition of $\widetilde{v}_{\rm xc}^{\rm HS}(\mathbf{r})$ in Eq. (3.15) is path independent. Therefore some particular path must be chosen to make this definition complete. It seems to us that the path used in Eq. (2.38) is the best choice, because it proved to be essential for satisfying the requirement Eq. (3.1) by $\widetilde{v}_{\rm x} = \widetilde{v}_{\rm x}^{\rm HS}$.

IV. SUMMARY AND DISCUSSION

The major result of the present study is the path integral formula (2.29) together with Eq. (2.28) for the exchange-correlation potential $v_{\rm xc}(\mathbf{r})$ in terms of quantities directly derivable from the second-order density matrix of the interacting system, of the first-order density matrix of the noninteracting (SKS) system, and of the electron-electron interaction potential.

The simplest possible practical result to obtain from the exact formula (2.29) can be summarized in terms of the following steps:

(a) Replace any interacting density matrix entering the theory by the independent-particle SKS equivalent.

(b) Write the second-order density matrix in terms of the SKS first-order density matrix according to the rules given, e.g., in Parr and Yang [6] for DMs generated from determinantal wave function.

Assumption (a) above is equivalent to the neglect of correlation and hence leads to approximation to the exchange-only potential $\widetilde{v}_{\rm x}(\mathbf{r})$ [given in Eqs. (2.38) and (2.41)]. For Coulombic *e-e* interaction this potential is then found to coincide with the Harbola-Sahni (HS) exchange-only potential. Existing numerical calculations on the latter potential are very encouraging and appear to approximate very accurately (except in the region between atomic shells) the correct exchange potential of the closed-shell atoms. For such atoms, the HS approximation has no path arbitrariness, but for other systems like molecules and clusters, path dependence of their formula will need careful study.

To conclude, we wish to stress a direction which looks promising for future work. It is important to have a careful approximation to the interacting second-order density matrix entering the theory, in the presence of electron correlations (ultimately, as a functional of electron density). In this general context, the pioneering work of Gutzwiller [18] should be referred to. With his correlated wave function, and by judicious approximation, he was able to construct low-order density matrices transcending these from a single Slater determinant.

ACKNOWLEDGMENTS

This collaboration was brought to fruition by the presence of both authors at the Research Workshop on Condensed Matter Physics held at the International Centre for Theoretical Physics, Trieste, Italy. They wish to thank ICTP for hospitality.

APPENDIX: DIFFERENTIAL VIRIAL THEOREM

This local relation between the density, potentials, and the kinetic-energy density tensor can be obtained by a procedure similar to that we employed earlier [19] for noninteracting electrons. The starting point is the Schrödinger equation (2.7). Because $\widehat{\mathcal{H}}$ is real, this equation is satisfied separately by the real and imaginary part of the wave function $\Psi = \Psi^{\rm Re} + i\Psi^{\rm Im}$, i.e.,

$$E = \frac{\widehat{\mathcal{H}}\Psi^{\rm Re}}{\Psi^{\rm Re}} = \frac{\widehat{\mathcal{H}}\Psi^{\rm Im}}{\Psi^{\rm Im}}\,. \tag{A1}$$

Using the definitions (2.1)–(2.4) we first transform (A1) into

$$\widehat{\mathcal{V}} + \widehat{\mathcal{U}} - E = -\frac{\widehat{\mathcal{T}}\Psi^{\rm Re}}{\Psi^{\rm Re}} = -\frac{\widehat{\mathcal{T}}\Psi^{\rm Im}}{\Psi^{\rm Im}}\,, \tag{A2}$$

and next, by differentiating both sides of the left part of Eq. (A2) with respect to $r_{1\alpha}$, we obtain

$$v_{/1\alpha}(\mathbf{r}_1) + \sum_{j=2}^{N} u_{/1\alpha}(\mathbf{r}_1,\mathbf{r}_j) = \tfrac{1}{2}(\Psi^{\rm Re})^{-1}\sum_{i=1}^{N}\sum_{\beta}\Psi^{\rm Re}_{/1\alpha/i\beta/i\beta} - \tfrac{1}{2}(\Psi^{\rm Re})^{-2}\,\Psi^{\rm Re}_{/1\alpha}\sum_{i=1}^{N}\sum_{\beta}\Psi^{\rm Re}_{/i\beta/i\beta}\,. \tag{A3}$$

Here the notation for partial derivatives $f_{/i\alpha} = \partial f/\partial r_{i\alpha}$ is adopted. After multiplying both sides of Eq. (A3) by $(\Psi^{\rm Re})^2$ we arrive at the result

$$\left\{ v(\mathbf{r}_1) + \sum_{j=2}^{N} u(\mathbf{r}_1, \mathbf{r}_j) \right\}_{/1\alpha} (\Psi^{\rm Re})^2 = \sum_{i=1}^{N}\sum_{\beta}\left\{ \tfrac{1}{2}\Psi^{\rm Re}\Psi^{\rm Re}_{/1\alpha/i\beta/i\beta} - \tfrac{1}{2}\Psi^{\rm Re}_{/1\alpha}\Psi^{\rm Re}_{/i\beta/i\beta} \right\}$$

$$= \sum_{i=1}^{N}\sum_{\beta}\left\{ \tfrac{1}{4}(\Psi^{\rm Re}\Psi^{\rm Re})_{/1\alpha/i\beta/i\beta} - (\Psi^{\rm Re}_{/1\alpha}\Psi^{\rm Re}_{/i\beta})_{/i\beta} \right\}. \quad \text{(A4)}$$

An equation, analogous to Eq. (A4), with $\Psi^{\rm Re}$ replaced by $\Psi^{\rm Im}$, can also be obtained from Eq. (A2). After adding two such equations we find

$$\left\{ v(\mathbf{r}_1) + \sum_{j=2}^{N} u(\mathbf{r}_1, \mathbf{r}_j) \right\}_{/1\alpha} |\Psi|^2 = \sum_{i=1}^{N}\sum_{\beta}\left\{ \tfrac{1}{4}(|\Psi|^2)_{/1\alpha/i\beta/i\beta} - (\Psi^{\rm Re}_{/1\alpha}\Psi^{\rm Re}_{/i\beta} + \Psi^{\rm Im}_{/1\alpha}\Psi^{\rm Im}_{/i\beta})_{/i\beta} \right\}. \quad \text{(A5)}$$

Next, we multiply both sides by N, sum over s_1, and integrate over $d\mathbf{x}_2 \cdots d\mathbf{x}_N$ to obtain

$$v_{/1\alpha}(\mathbf{r}_1)n(\mathbf{r}_1) + \sum_{j=2}^{N} N \sum_{s_1} \int d\mathbf{x}_2 \cdots d\mathbf{x}_N \, u_{/1\alpha}(\mathbf{r}_1, \mathbf{r}_j)|\Psi|^2$$

$$= \tfrac{1}{4}\nabla_1^2 n_{/1\alpha}(\mathbf{r}_1) - 2\sum_{\beta}\left\{ N \sum_{s_1} \int d\mathbf{x}_2 \cdots d\mathbf{x}_N \, \tfrac{1}{2}(\Psi^{\rm Re}_{/1\alpha}\Psi^{\rm Re}_{/1\beta} + \Psi^{\rm Im}_{/1\alpha}\Psi^{\rm Im}_{/1\beta}) \right\}_{/1\beta}$$

$$+ \sum_{j=2}^{N}\sum_{\beta} N \sum_{s_1} \int d\mathbf{x}_2 \cdots d\mathbf{x}_N \left\{ \tfrac{1}{4}(|\Psi|^2)_{/1\alpha/j\beta} - (\Psi^{\rm Re}_{/1\alpha}\Psi^{\rm Re}_{/j\beta} + \Psi^{\rm Im}_{/1\alpha}\Psi^{\rm Im}_{/j\beta}) \right\}_{/j\beta}. \quad \text{(A6)}$$

The last integral on the right-hand side of Eq. (A6) vanishes after integration over $dr_{j\beta}$ (because, being a solution of the Schrödinger equation, the wave function Ψ and its derivatives vanish for $|\mathbf{r}_j| \to \infty$). The expression in the curly brackets in the other integral can be recognized as the kinetic-energy density tensor $t_{\alpha\beta}(\mathbf{r}_1)$ defined in Eq. (2.12). The term involving $u_{/1\alpha}$ can be rewritten [using the symmetry of Ψ and the definition of n_2, Eqs. (2.8)–(2.11)] as

$$N(N-1)\sum_{s_1} \int d\mathbf{x}_2 \cdots d\mathbf{x}_N \, u_{/1\alpha}(\mathbf{r}_1, \mathbf{r}_2)|\Psi|^2$$

$$= 2\int d^3r_2 \, u_{/1\alpha}(\mathbf{r}_1, \mathbf{r}_2) n_2(\mathbf{r}_1, \mathbf{r}_2).$$

Finally, Eq. (A6) is reduced to the equation

$$v_{/\alpha}(\mathbf{r})n(\mathbf{r}) + 2\int d^3r' u_{/\alpha}(\mathbf{r}, \mathbf{r}') n_2(\mathbf{r}, \mathbf{r}')$$

$$= \tfrac{1}{4}\nabla^2 n_{/\alpha}(\mathbf{r}) - 2\sum_{\beta} t_{\alpha\beta/\beta}(\mathbf{r}), \quad \text{(A7)}$$

which will be termed the differential virial theorem. This is an exact, local (at space point $\mathbf{r}$) relation involving the external potential $v(\mathbf{r})$, the e-e interaction potential $u(\mathbf{r}, \mathbf{r}')$, the diagonal elements of the first- and second-order DMs, $n(\mathbf{r})$ and $n_2(\mathbf{r}, \mathbf{r}')$, and the first-order DM $\rho(\mathbf{r}_1; \mathbf{r}_2)$ "close to diagonal" — see the definition (2.12) of $t_{\alpha\beta}$. When Eq. (A7) is written for the system with Coulomb potentials [i.e., definitions (2.5) and (2.6) are used] and differentiations of these potentials are performed, the equation obtained turns out to be equivalent to the balance equation between the momentum flux and the force density for the electron system, obtained by Ziesche *et al.* [20] in a quite different way.

The reason for calling Eq. (A7) the differential virial theorem is connected with the fact that the (global) virial theorem can be derived from it by applying the operation $\int d^3r \sum_\alpha r_\alpha$ to both sides of Eq. (A7), so it gives

$$2T = \int d^3r \, n(\mathbf{r}) \, \mathbf{r} \cdot \nabla v(\mathbf{r})$$

$$+2\int d^3r \, d^3r' \, n_2(\mathbf{r}, \mathbf{r}') \, \mathbf{r} \cdot \nabla u(\mathbf{r}, \mathbf{r}') \quad \text{(A8)}$$

[Eqs. (2.13) and (2.14) have been used]. Equation (A8) represents the virial theorem in its most general form (see, e.g., in Levy and Perdew [10]). In the case of Coulombic $u(\mathbf{r}, \mathbf{r}')$, Eq. (2.5), the second integral in Eq. (A8) can be evaluated, using the symmetry of n_2, to be $-E_{ee}$. So, in this case, Eq. (A8) gives

$$2T + E_{ee} = \int d^3r \, n(\mathbf{r}) \, \mathbf{r} \cdot \nabla v(\mathbf{r}), \quad \text{(A9)}$$

the familiar virial theorem for a Coulombically interacting system.

[1] L. H. Thomas, Proc. Cambridge Philos. Soc. **23**, 542 (1926).
[2] E. Fermi, Z. Phys. **48**, 73 (1928).
[3] See, for example, *Theory of the Inhomogeneous Electron Gas*, edited by S. Lundqvist and N. H. March (Plenum, New York, 1983).
[4] P. A. M. Dirac, Proc. Cambridge Philos. Soc. **26**, 376 (1930).
[5] P. Hohenberg and W. Kohn, Phys. Rev. **136**, B864 (1964).
[6] R. G. Parr and W. Yang, *Density-Functional Theory of Atoms and Molecules* (Oxford University Press, Oxford, 1989).
[7] J. C. Slater, Phys. Rev. **81**, 385 (1951).
[8] W. Kohn and L. J. Sham, Phys. Rev. **140**, A1133 (1965).
[9] M. K. Harbola and V. Sahni, Phys. Rev. Lett. **62**, 489 (1989).
[10] M. Levy and J. P. Perdew, Phys. Rev. A **32**, 2010 (1985).
[11] M. Levy, Phys. Rev. A **43**, 4637 (1991).
[12] M. Levy and J. P. Perdew, Int. J. Quantum Chem. **49**, 539 (1994).
[13] H. Ou-Yang and M. Levy, Phys. Rev. A **41**, 4038 (1990).
[14] H. Ou-Yang and M. Levy, Phys. Rev. Lett. **65**, 1036 (1990).
[15] Y. Wang, J. P. Perdew, J. A. Chevary, L. D. Macdonald, and S. H. Vosko, Phys. Rev. A **41**, 78 (1990).
[16] Y. Li, M. K. Harbola, J. B. Krieger, and V. Sahni, Phys. Rev. A **40**, 6084 (1989).
[17] Q. Zhao, M. Levy, and R. G. Parr, Phys. Rev. A **47**, 918 (1993).
[18] M. C. Gutzwiller, Phys. Rev. **137**, A1726 (1965).
[19] A. Holas and N. H. March, Int. J. Quantum Chem. (to be published).
[20] P. Ziesche, J. Gräfenstein, and O. H. Nielsen, Phys. Rev. B **37**, 8167 (1988).

PHYSICAL REVIEW A VOLUME 55, NUMBER 3 MARCH 1997

Line-integral formulas for exchange and correlation potentials separately

Mel Levy [1] and Norman H. March [2]

[1] *Department of Chemistry and Quantum Theory Group, Tulane University, New Orleans, Louisiana 70118*

[2] *Inorganic Chemistry Department, University of Oxford, South Parks Road, Oxford OX1 3QR, England*

(Received 4 August 1995; revised manuscript received 27 November 1996)

Formal *separate* expressions for the exact exchange and correlation potentials v_x and v_c are extracted from the formal line-integral expression of Holas and March for the whole exact exchange-correlation potential. Relations for the components of v_c are extracted for each order in the electron-electron repulsion coupling constant, through use of the coupling-constant expansion of Görling and Levy for the external potential. The resultant expressions for v_c and v_x are separately path independent. The difference between v_x and the Harbola-Sahni approximation to it, v_x^{HS}, is identified as arising from a first-order contribution to the kinetic-energy density tensor. It is shown that this small correction to v_x^{HS}, which we express in terms of perturbation theory, would be precisely *zero* if the Kohn-Sham determinant were identical to the Hartree-Fock determinant for the same density. In other words, v_x^{HS} would equal v_x if the optimized effective potential determinant were the same as the Hartree-Fock determinant. This same property is shared by the Slater potential. [S1050-2947(97)06503-7]

PACS number(s): 31.15.Ew, 71.10.−w, 31.25.−v

I. INTRODUCTION

Recently, Holas and March [1] have derived a formal line-integral expression for the exact density-functional exchange-correlation potential $v_{xc}([n];\vec{r})$. For the purpose of its separation into the exchange and correlation contributions, we shall here formulate the Holas-March [1] expression as a function of the electron-electron repulsion coupling constant λ, in order to develop relations which are associated with each order in λ. The first-order one yields a formal expression for the exact exchange potential, $v_x([n];\vec{r})=\delta E_x[n]/\delta n(\vec{r})$. This exchange expression identifies the correction to the Harbola-Sahni [2] approximation to v_x. We shall show that this correction, which is small in atoms, is zero if the Hartree-Fock single determinant, for the given density, were exactly the same as the Kohn-Sham single determinant for the same density. These determinants are known to be generally quite close in atoms [3].

Higher-order terms in λ yield formal expressions for the corresponding parts of the exact correlation potential, $v_c([n];\vec{r})=\delta E_c[n]/\delta n(\vec{r})$. It is especially important to have knowledge of a separate exact expression for v_c when one employs an approximation to v_x which does not include line integrals or if one attaches an approximation for v_c to an exact optimized effective potential (OEP) calculation or to a Hartree-Fock calculation. Consequently, an explicit formal expression for v_c is presented here.

II. DERIVATION OF SEPARATE EXPRESSIONS FOR v_x AND v_c

In accordance with the constrained-search approach, define $\Psi_{[n]}^{\lambda}$ as that antisymmetric wave function which yields the density n and minimizes $\langle T+\lambda\hat{V}_{ee}\rangle$.

Here, for an N-electron system in atomic units,

$$\hat{T}=\sum_{i=1}^{N}-\frac{1}{2}\nabla_i^2$$

and

$$\hat{V}_{ee}=\sum_{i=1}^{N-1}\sum_{j=i+1}^{N}|\vec{r}_i-\vec{r}_j|^{-1}.$$

The exchange energy associated with $\Psi_{[n]}^{\lambda}$ is

$$E_x^{\lambda}[n]\equiv\langle\Psi_{[n]}^{0}|\lambda\hat{V}_{ee}|\Psi_{[n]}^{0}\rangle-\lambda\int\int\frac{n(\vec{r}_1)n(\vec{r}_2)}{|\vec{r}_1-\vec{r}_2|}d^3r_1d^3r_2$$
$$=\lambda E_x[n], \tag{1}$$

while the corresponding correlation energy is

$$E_c^{\lambda}[n]=\langle\Psi_{[n]}^{\lambda}|\hat{T}+\lambda\hat{V}_{ee}|\Psi_{[n]}^{\lambda}\rangle-\langle\Psi_{[n]}^{0}|\hat{T}+\lambda\hat{V}_{ee}|\Psi_{[n]}^{0}\rangle. \tag{2}$$

Next, define v_{xc}^{λ} through

$$v_{xc}^{\lambda}([n];\vec{r})=v_x^{\lambda}([n];\vec{r})+v_c^{\lambda}([n];\vec{r}), \tag{3}$$

where $v_x^{\lambda}([n];\vec{r})=\delta E_x^{\lambda}[n]/\delta n(\vec{r})$ and $v_c^{\lambda}([n];\vec{r})=\delta E_c^{\lambda}[n]/\delta n(\vec{r})$. From Eq. (1), observe that

$$v_x^{\lambda}([n];\vec{r})=\lambda v_x([n];\vec{r}) \tag{4}$$

and from Görling and Levy [4,5], note that

$$v_c^{\lambda}([n];\vec{r})=\lambda^2 v_{c,2}([n];\vec{r})+\lambda^3 v_{c,3}([n];\vec{r})+\cdots. \tag{5}$$

The essence of the development which follows depends on the fact that v_c^{λ} commences in *second order*, while v_x^{λ} commences in *first order*.

We now observe that Eq. (2.29) of Holas and March [1] can be viewed as an expression for v_{xc}^{λ} with $\lambda=1$. In other words, we view λ as simply determining the square of the charge on the electron. It follows that Eq. (2.29), for all $\lambda\geqslant 0$, becomes

$$v_{xc}^{\lambda}([n];\vec{r}_0)=-\int_{\infty}^{\vec{r}_0}d\vec{r}\cdot\vec{f}_{xc}([u^{\lambda},n,\rho^0,\rho^{\lambda},\Gamma^{\lambda}];\vec{r}), \tag{6}$$

where, using a slightly different notation than in Ref. [1] for simplicity of exposition,

$$\vec{f}_{xc}([u^\lambda,n,\rho^0,\rho^\lambda,\Gamma^\lambda];\vec{r})=\Big\{\vec{z}([\rho^0];\vec{r})-\vec{z}([\rho^\lambda];\vec{r}) + \int d^3r'[\vec{\nabla}u^\lambda(\vec{r},\vec{r}')][n(\vec{r})n(\vec{r}') - 2\Gamma^\lambda(\vec{r},\vec{r}')]\Big\}n^{-1}(\vec{r}). \qquad (7)$$

In expressions (6) and (7),

$$u^\lambda(\vec{r},\vec{r}')=\lambda|\vec{r}-\vec{r}'|^{-1}, \qquad (8)$$

$\Gamma^\lambda_{[n]}(\vec{r},\vec{r}')$ is the diagonal part of the second-order density matrix of $\Psi^\lambda_{[n]}$, and $\rho^\lambda_{[n]}(\vec{r},\vec{r}')$ is its first-order density matrix. Here,

$$\rho^\lambda_{[n]}(\vec{r},\vec{r}')=N\int\cdots\int \Psi^\lambda_{[n]}(\vec{r}\sigma,\vec{x}_2,\ldots,\vec{x}_N)\Psi^{\lambda *}_{[n]}(\vec{r}'\sigma,\vec{x}_2,\ldots,\vec{x}_N)d\sigma d\vec{x}_2\cdots d\vec{x}_N$$

and

$$\Gamma^\lambda_{[n]}(\vec{r},\vec{r}')=\frac{N(N-1)}{2}\int\cdots\int \Psi^\lambda_{[n]}(\vec{r}\sigma,\vec{r}'\sigma',\vec{x}_3,\ldots\vec{x}_N)\Psi^{\lambda *}_{[n]}(\vec{r}\sigma,\vec{r}'\sigma',\vec{x}_3,\ldots,\vec{x}_N)d\sigma d\sigma' d\vec{x}_2\cdots d\vec{x}_N .$$

Hence $\rho^0_{[n]}(\vec{r},\vec{r}')$ is the first-order density matrix of the noninteracting (Kohn-Sham) [6] single determinant $\Psi^0_{[n]}$. Further, the vector field $\vec{z}$ is now given in terms of ρ^λ. In other words,

$$z_\alpha([\rho^\lambda];\vec{r})=2\sum_\beta \frac{\partial}{\partial r_\beta}t_{\alpha\beta}([\rho^\lambda];\vec{r}), \qquad (9)$$

with

$$t_{\alpha\beta}([\rho^\lambda];\vec{r})=\frac{1}{4}\left(\frac{\partial^2}{\partial r'_\alpha \partial r''_\beta}+\frac{\partial^2}{\partial r'_\beta \partial r''_\alpha}\right)\rho^\lambda(\vec{r}',\vec{r}'')\bigg|_{\vec{r}'=\vec{r}''=\vec{r}} . \qquad (10)$$

To arrive at our first desired result, in Eqs. (6) and (7) employ relations (3)–(5) and (8), together with

$$\Gamma^\lambda_{[n]}=\Gamma^0_{[n]}+\lambda\Gamma^c_{1,[n]}+\lambda^2\Gamma^c_{2,[n]}+\cdots \qquad (11)$$

and

$$\rho^\lambda_{[n]}=\rho^0_{[n]}+\lambda\rho^c_{1,[n]}+\lambda^2\rho^c_{2,[n]}+\cdots. \qquad (12)$$

Couple the linear terms in λ to obtain

$$v_x([n];\vec{r}_0)=-\int_\infty^{\vec{r}_0} d\vec{r}\cdot\vec{f}_x, \qquad (13)$$

where

$$\vec{f}_x=\Big\{\int d^3r'[\vec{\nabla}(|\vec{r}-\vec{r}'|^{-1})][n(\vec{r})n(\vec{r}')-2\Gamma^0(\vec{r},\vec{r}')] - \vec{z}([\rho^c_1];\vec{r})\Big\}n^{-1}(\vec{r}) \qquad (14a)$$

or

$$v_x([n];\vec{r}_0)=v_x^{\rm HS}([n];\vec{r}_0)+\int_\infty^{\vec{r}_0} d\vec{r}\cdot\vec{z}([\rho^c_1];\vec{r})n^{-1}(\vec{r}). \qquad (14b)$$

The existence of $\vec{z}([\rho^c_1];\vec{r})$ is responsible for the difference between the exact v_x and $v_x^{\rm HS}$ the Harbola-Sahni approximation to v_x. Hence

$$g([n];\vec{r}_0)=\int_\infty^{\vec{r}_0} d\vec{r}\cdot\vec{z}([\rho^c_1];\vec{r})n^{-1}(\vec{r}) \qquad (15)$$

is the correction to the Harbola-Sahni expression. Since Eq. (14) represents an exact expression for v_x, its value does not vary with the particular path chosen for the integration, while $v_x^{\rm HS}$ and its correction in Eq. (15), taken separately, may, depending upon the situation, vary with the integration path.

To determine $v_x^{\rm HS}$ one requires knowledge of the off-diagonal elements, as a functional of its diagonal density $n(\vec{r})$, of the noninteracting first-order density matrix $\rho^0_{[n]}$ in order to construct $\Gamma^0_{[n]}$. While this information is available [7,8] to all orders in a perturbation theory based on a one-body potential $v_s(\vec{r})$, we have not, to date, been able to sum up the resultant infinite series except in the limit when $v_s(\vec{r})$ varies slowly in space [7,8]. In Eq. (12), the term in $\rho^\lambda(\vec{r},\vec{r}')$, which is of order 1 in λ, is required to calculate the kinetic tensor correction to $v_x^{\rm HS}$. This term, as already noted, suffices, at least in principle, to eliminate any path dependence from the resultant exchange-only potential in density-functional theory.

For the purpose of helping to approximate them, terms in higher orders of λ can be coupled. For instance, for the correlation potential, the coupling of second-order terms gives

$$v_{c,2}([n];\vec{r}_0)=-\int_\infty^{\vec{r}_0} d\vec{r}\cdot\vec{f}_{c,2}, \qquad (16)$$

where

$$\vec{f}_{c,2}=-2\left\{\int d^3r'[\vec{\nabla}|\vec{r}-\vec{r}'|^{-1}]\Gamma_1^c(\vec{r},\vec{r}')+\vec{z}([\rho_2^c];\vec{r})\right\}n^{-1}(\vec{r}). \quad (17)$$

[Note that $v_{c,2}$ is [4,5] the functional derivative of $\lim_{\gamma\to\infty}E_c[n_\gamma]$, where $n_\gamma(x,y,z)=\gamma^3 n(\gamma x,\gamma y,\gamma z)$.] For this reason, $\int n(\vec{r})\vec{r}\cdot\vec{\nabla}v_{c,2}([n];\vec{r})d^3r=0$.

The *whole* correlation component v_c^λ of v_{xc}^λ is obtained by subtracting expression (13) from expression (6). At $\lambda=1$, this gives

$$v_c([n];\vec{r}_0)=-\int_\infty^{\vec{r}_0}d\vec{r}\cdot\vec{f}_c, \quad (18)$$

where

$$\vec{f}_c=\vec{z}([\rho^0+\rho_1^c-\rho^1];\vec{r})n^{-1}(\vec{r})+2\left[\int d^3\vec{r}'[\vec{\nabla}|\vec{r}-\vec{r}'|^{-1}][\Gamma^0(\vec{r},\vec{r}')-\Gamma^1(\vec{r},\vec{r})]\right]n^{-1}(\vec{r}) \quad (19)$$

and where $\vec{z}[\rho^0+\rho_1^c-\rho^1]=\vec{z}[\rho^0]+\vec{z}[\rho_1^c]-\vec{z}[\rho^1]$. In expression (19), ρ^1 and Γ^1 refer to the density matrices at full charge, $\lambda=1$. [Observe that since both v_x and v_c are exact, it follows that the right-hand sides are path independent in both expressions (13) and (18).]

III. REQUIREMENTS FOR ρ^λ AND Γ^λ

We here list several requirements for approximating ρ^λ and Γ^λ. First, path independence is certainly a requisite with either expression (13) or (18). Next, constraints immediately follow from the fact that the density of ρ^λ and Γ^λ is n and therefore independent of λ. Consequently, in expression (19), $\int d^3r'[\Gamma^0(\vec{r},\vec{r}')-\Gamma^1(\vec{r},\vec{r}')]=0$, and, in expressions (11) and (12), we note that $\rho_i^c(\vec{r},\vec{r})=0$ and $\int\Gamma_i^c(\vec{r},\vec{r}')d^3r'=0$.

We now observe that previous work [5] on the behavior of the kinetic energy at very small λ, which utilizes the fact that $\rho^0(\vec{r},\vec{r}')$ is that first-order density matrix which minimizes the kinetic energy at fixed n, dictates that $-\frac{1}{2}\int\nabla^2\rho_{1,[n]}^c(\vec{r},\vec{r}')d^3r=0$ (or $\int d^3r\Sigma_\beta t_{\beta\beta}([\rho_{1,[n]}^c];\vec{r})=0$). Further, the fact [5] that $\lim_{\gamma\to\infty}E_c[n_\gamma]>-\infty$ implies, by the Ghosh-Parr relation [9], that $\int\nabla^2\rho_{2,[n]}^c(\vec{r},\vec{r}')=\int\int\Gamma_{1,[n]}^c(\vec{r},\vec{r}')|\vec{r}-\vec{r}'|^{-1}d^3r\,d^3r'$. Moreover, the required uniform scaling property for g in Eq. (15) is [9,10]

$$g([n_\gamma];\vec{r}_0)=\gamma g([n];\gamma\vec{r}_0),$$

and [9–12] g also satisfies $\int d^3r\,n(\vec{r})\vec{r}\cdot\vec{\nabla}g([n];\vec{r})=0$.

It should prove useful to study $\rho_{[n]}^\lambda(\vec{r},\vec{r}')$ in terms of its natural orbital expansion [13], which is

$$\rho^\lambda(\vec{r},\vec{r}')=\sum_i d_i^\lambda\chi_i^{\lambda*}(\vec{r}')\chi_i^\lambda(\vec{r}), \quad (20)$$

where the d_i^λ are the occupation numbers and the χ_i^λ are the natural orbitals. In Eq. (20),

$$d_i^\lambda=1+\lambda d_{i,1}+\cdots,\quad i\leq N,$$

$$d_i^\lambda=\lambda d_{i,1}+\cdots,\quad i>N,$$

and for $i\leq N$, $\chi_i(\vec{r})=\varphi_i(\vec{r})+\lambda L_{i,1}(\vec{r})+\cdots$, where the φ_i are the occupied Kohn-Sham orbitals of n.

We now find an explicit expression for ρ_i^c in Eqs. (15) and (19), in terms of Kohn-Sham orbitals. For simplicity of notation, let us assume that everything is real and now expand $\Psi_{[n]}^\lambda$ as $\Psi_{[n]}^\lambda=\Psi_{[n]}^0+\lambda\Psi_{1,[n]}+\lambda^2\Psi_{2,[n]}+\cdots$, where $\rho_{1,[n]}^c$ in Eq. (12) is the transition first-order density matrix of $\Psi_{[n]}^0\Psi_{1,[n]}+\Psi_{1,[n]}\Psi_{[n]}^0$ etc. From previous studies [4], we know that $\Psi_{1,[n]}$, the first-order wave function, is given by

$$\Psi_{1,[n]}=-\sum_{k>0}\frac{\langle\Phi_k|\hat{V}_{ee}-\hat{V}_{xu}|\Phi_0\rangle}{E_k-E_0}\Phi_k, \quad (21)$$

where

$$\hat{V}_{XU}=\sum_{i=1}^N v_x([n];\vec{r}_i)+\sum_{i=1}^N\int\frac{n(\vec{r})}{|\vec{r}_i-\vec{r}|}d^3r.$$

Also, the Φ_k are the single-determinant eigenstates of that noninteracting Kohn-Sham Hamiltonian for which $\Phi_0=\Psi_{[n]}^0$ is the ground-state single determinant. The E's are sums of Kohn-Sham orbital energies. It follows that

$$\rho_{1,[n]}^c(\vec{r},\vec{r}')=-\sum_{k>0}\frac{\langle\Phi_k|\hat{V}_{ee}-\hat{V}_{xu}|\Phi_0\rangle}{(E_k-E_0)}\times[\gamma_{k,0}(\vec{r},\vec{r}')+\gamma_{k,0}(\vec{r}',\vec{r})], \quad (22)$$

where

$$\gamma_{k,0}(\vec{r}|\vec{r}')=N\int\cdots\int\Phi_k(\vec{r}\sigma,x_2,\ldots,x_N)\Phi_0(\vec{r}'\sigma,x_2,\ldots,x_N)d\sigma dx_2\cdots dx_N.$$

Now, let Φ_0 be the energy minimizing OEP single determinant [14,15] for some Hamiltonian

$$\hat{H}=\hat{T}+\hat{V}_{ee}+\sum_{i=1}^{N} v(\vec{r}_i) \tag{23}$$

and, in expression (22), add $\hat{T}+\Sigma_{i=1}^{N}v(\vec{r}_i)$ to $\hat{V}_{ee}$ and subtract $\hat{T}+\Sigma_i v(\vec{r}_i)$ from $-\hat{V}_{xu}$. We obtain

$$\rho^c_{1,[n]}(\vec{r},\vec{r}\,')=-\sum_{k>0}\frac{\langle\Phi_k|\hat{H}|\Phi_0\rangle}{(E_k-E_0)}[\gamma_{k,0}(\vec{r},\vec{r}\,')+\gamma_{k,0}(\vec{r}\,',\vec{r})], \tag{24}$$

where, in Eq. (24), each Φ_k pertains to a singly excited determinant with respect to Φ_0, because otherwise $\gamma_{k,0}$ vanishes. Next, *if* Φ_0 is simultaneously the Hartree-Fock determinant of the $\hat{H}$ in Eqs. (23) and (24), then $\rho^c_{1,[n]}(\vec{r},\vec{r}\,')$ would vanish, because, by Brillouin's theorem, $\langle\Phi_k|\hat{H}|\Phi_0\rangle=0$, for all $k>0$. Consequently, $v_x^{\rm HS}$ would equal v_x *if* Φ_0 were a Hartree-Fock determinant. (Of course, it should be noted that we observe that the Slater exchange potential [16] would also equal v_x if Φ_0 were a Hartree-Fock determinant, because the Slater potential [or [17] averaged Fock approximation (AFA)] is invariant to a unitary transformation among the orbitals of which it is composed. However, the Slater potential does not satisfy the Levy-Perdew viral relation [12] while $v_x^{\rm HS}$ does [2]. Incidentally, it has recently been shown by Kleinman [17] that the Slater potential may be obtained as a partial functional derivative of the Hartree-Fock exchange expression.)

Since the Φ_k in Eq. (22) are singly excited determinants, it can be shown that it follows that

$$\rho^c_{1,[n]}(\vec{r},\vec{r}\,')=-\sum_{\ell>N}\sum_{i=1}^{N}\frac{\langle\varphi_\ell|\hat{v}_x^{\rm HF}-v_x|\varphi_i\rangle}{(\varepsilon_\ell-\varepsilon_i)}\times[\varphi_\ell(\vec{r}\,')\varphi_i(\vec{r})+\varphi_\ell(\vec{r})\varphi_i(\vec{r}\,')], \tag{25}$$

where the φ's are Kohn-Sham orbitals, the ε's are the corresponding orbital energies, and $\hat{v}^{\rm HF}$ is the familiar nonlocal Fock potential, except that it is composed of Kohn-Sham orbitals.

Consider $\hat{H}_\lambda=\hat{T}+\lambda\hat{V}_{ee}+\Sigma_{i=1}^{N}v_\lambda([n];\vec{r}_i)$, where v_λ is such that n, the Hartree-Fock density of $\hat{H}_\lambda$, is constrained to be independent of λ. Then it can be shown through a perturbation analysis of Eq. (22) in Ref. [27] that the $\rho^c_{1,[n]}$ in Eqs. (22) and (25), of the present paper, is the difference, to first order in λ, between the Hartree-Fock $\rho^\lambda(\vec{r},\vec{r}\,')$ of $\hat{H}_\lambda$ and the Hartree-Fock (or Kohn-Sham) $\rho^0(\vec{r},\vec{r}\,')$ of $\hat{H}_0$. Note that $\rho^0(\vec{r},\vec{r}\,')$ consists of Kohn-Sham orbitals. Hence, the Harbola-Sahni exchange potential $v_x^{\rm HS}$ would equal the exact exchange potential v_x if $\partial\rho^\lambda/\partial\lambda|_{\lambda=0}=0$, where ρ^λ is the Hartree-Fock $\rho^\lambda(\vec{r},\vec{r}\,')$.

Finally, in calculations, perhaps one can approximate v_x in Eq. (25) for use in the z in Eq. (14). Approximations for the v_x in Eq. (25) might include $v_x^{\rm HS}$ itself, the potentials of Krieger, Li, and Iafrate [18,19] or, even, the Slater potential [20].

IV. CLOSING REMARKS CONCERNING THE CORRELATION POTENTIAL

As implied in the Introduction, knowledge of a *separate* formal exact expression for v_c is important when one wishes to focus upon an approximation for v_c independently, and one employs the exact OEP for v_x, or if one wishes to employ one of the highly encouraging approximations for v_x as presented, or discussed, for instance, in the works in Refs. [2] and [17–22].

In expressions (18) and (19), one would model the correlating second-order density matrix (nondiagonal as well as diagonal elements) as a functional of the density to determine an approximate v_c as a functional of the density (the component ρ_1^c can be extracted through coordinate scaling [5]). The accuracy of this modeled correlating second-order density matrix would be tested by first forming the corresponding correlation energy functional with it through that coupling-constant formula obtained [5,23] by combining the adiabatic connection formula [24,25] with coordinate scaling [5] and then by taking the functional derivative of this correlation functional. For the modeled correlating second-order density matrix to be accurate, the resultant functional derivative would have to agree closely with the approximation for v_c which is obtained through expressions (18) and (19).

Finally, the constraints in Secs. III and IV of van Leeuwen and Baerends [26] should be tested on an approximation to $v_c([n];\vec{r})$, which is obtained through expressions (18) and (19). Also, one could generate an approximation to E_c by performing a functional integration with the approximate $v_c([n];\vec{r})$ by utilizing one of the paths of van Leeuwen and Baerends [26]. As a necessary requisite for accuracy, the resultant approximation to E_c would have to agree closely with that approximation to E_c which is obtained through the coupling-constant formula described in the previous paragraph. The present work concerns line integrals for obtaining an approximation to v_c, while Ref. (26) concerns line integrals for obtaining an approximation to E_c from an approximation to v_c.

ACKNOWLEDGMENTS

We wish to acknowledge that the collaboration resulted from our attendance at the Caracas (1995) Workshop on Condensed Matter Theories, and we are most grateful to Professor E. V. Ludeña for making this possible. One of us wishes to acknowledge that a question on scaling of the electron-electron interaction by Professor N. C. Handy was important in connection with the contribution of N.H.M. to the present study. Also, M. L. wishes to acknowledge his recent coupling-constant collaborations with A. Görling. This research was supported, in part, by NIST. Finally, we thank Professor A. Holas for valuable suggestions.

[1] A. Holas and N. H. March, Phys. Rev. A **51**, 2040 (1995).
[2] M. K. Harbola and V. Sahni, Phys. Rev. Lett. **62**, 489 (1989).
[3] A. Görling and M. Ernzerhof, Phys. Rev. A **51**, 4501 (1995).
[4] A. Görling and M. Levy, Phys. Rev. B **48**, 11 638 (1993).
[5] M. Levy, Phys. Rev. A **43**, 4637 (1991).
[6] W. Kohn and L. J. Sham, Phys. Rev. **140**, A1133 (1965).
[7] N. H. March and A. M. Murray, Proc. R. Soc. London Ser. A **261**, 119 (1961).
[8] N. H. March and A. M. Murray, Phys. Rev. **120**, A830 (1960).
[9] S. K. Ghosh and R. G. Parr, J. Chem. Phys. **82**, 3307 (1985).
[10] H. Ou-Yang and M. Levy, Phys. Rev. Lett. **65**, 1036 (1990).
[11] H. Ou-Yang and M. Levy, Phys. Rev. A **44**, 54 (1991).
[12] M. Levy and J. P. Perdew, Phys. Rev. A **32**, 2010 (1985).
[13] P.-O. Löwdin, Phys. Rev. **97**, 1474 (1955).
[14] R. T. Sharp and G. K. Horton, Phys. Rev. **90**, 317 (1953).
[15] J. D. Talman and W. F. Shadwick, Phys. Rev. A **14**, 36 (1976).
[16] J. C. Slater, Phys. Rev. **81**, 385 (1951).
[17] L. Kleinman, Phys. Rev. B **49**, 14 197 (1994).
[18] J. B. Krieger, Y. Li, and G. J. Iafrate, Phys. Rev. A **45**, 101 (1992).
[19] Y. Li, J. B. Krieger, and G. J. Iafrate, Phys. Rev. A **47**, 165 (1993).
[20] D. M. Bylander and L. Kleinman, Phys. Rev. Lett. **74**, 3660 (1995).
[21] R. van Leeuwen and E. J. Baerends, Phys. A **49**, 2421 (1994).
[22] M. D. Glossman, L. C. Balbás, A. Rubio, and J. A. Alonso, Int. J. Quantum Chem. **49**, 171 (1994).
[23] Q. Zhao, M. Levy, and R. G. Parr, Phys. Rev. A **47**, 918 (1993).
[24] D. C. Langreth and J. P. Perdew, Solid State Commun. **17**, 1425 (1975); Phys. Rev. B **13** 4274 (1976).
[25] O. Gunnarson and B. I. Lundqvist, Phys. Rev. B **13**, 4274 (1976).
[26] R. van Leeuwen and E. J. Baerends, Phys. Rev. A **51**, 170 (1995).
[27] M. Levy, M. Ernzerhof, and A. Görling, Phys. Rev. A **53**, 3963 (1996).

PHYSICAL REVIEW B VOLUME 55, NUMBER 15 15 APRIL 1997-I

Potential-locality constraint in determining an idempotent density matrix from diffraction experiment

A. Holas
Institute of Physical Chemistry of the Polish Academy of Sciences, 44/52 Kasprzaka, 01-224 Warsaw, Poland

N. H. March
University of Oxford, Oxford, England

(Received 2 August 1996; revised manuscript received 8 November 1996)

The successful calculational procedure of Howard *et al.* [Phys. Rev. B **49**, 7124 (1994)] fits an idempotent density matrix (DM) (expanded in a finite basis set of atomic orbitals) to x-ray structure factors in order to determine the electron density. We find that the physical meaning of the fitted DM is that it corresponds to a model noninteracting-electron system, having, in general, a nonlocal one-body potential. It would be desirable to select from these model systems the Kohn-Sham (KS) reference system of the density-functional theory, because of its physical significance. Its unique property — the locality of its potential — is used here to discriminate the KS system among others during fitting. With the help of the equation of motion for the DM we construct a measure of potential nonlocality in terms of the DM, in a form suitable for use as the constraint. We then propose a modification of the Howard *et al.* procedure consisting in enriching the minimized function by adding to their mean squared error a new term — the mean squared departure from potential locality. Thus, in addition to determining the electron density of the system under consideration, this modified procedure will result in the DM of the KS reference system (both quantities optimized for the given basis set and the experimental data set). Using again the equation of motion, this DM can allow the exchange-correlation potential to be directly extracted. When fitting is applied to a given theoretical density (rather than to diffraction data), the discussed procedure can determine the exchange-correlation potential corresponding to this density. Our fitting procedure (applied either to diffraction data or to a given density) can determine the DM and one-body potential of generalized KS reference systems too. An extension of the Howard *et al.* procedure is proposed also, which widens its applicability from the spin-compensated systems to systems of arbitrary spin structure. This opens the possibility of extracting from experiments the spin density in addition to the charge density (and corresponding DM's), especially when polarized neutron diffraction data are available paralleling the x-ray data. [S0163-1829(97)00915-6]

I. INTRODUCTION

X-ray diffraction experiments can provide, for the system under investigation, extensive information about the electron density $n(\mathbf{r})$ — the basic object of the density-functional theory (DFT), which determines, in principle, all properties of the system (see, e.g., Ref. 1). In order to obtain in practice the density from an experiment, $n(\mathbf{r})$ must be somehow parametrized. We are going to discuss in the present paper in detail problems connected with representing the density as the diagonal element,

$$n(\mathbf{r})=\rho(\mathbf{r},\mathbf{r}), \qquad (1.1)$$

of the one-particle density matrix (1DM) $\rho(\mathbf{r},\mathbf{r}')$, which is expanded in a finite basis of K orbitals $\chi_i(\mathbf{r})$ as

$$\rho(\mathbf{r},\mathbf{r}')=\sum_{ij}^{K} P_{ij}\,\chi_i(\mathbf{r})\chi_j(\mathbf{r}'). \qquad (1.2)$$

The 1DM and the orbitals are chosen to be real, which is always possible in the absence of a magnetic field. The real, symmetric matrix P_{ij} represents the above mentioned parameters to be fitted from the experiment.

In a recent paper, Howard *et al.*[2] presented a calculational method in which the matrix P_{ij} may be determined from the elastic x-ray scattering intensities. The constraints of idempotency and the trace value for 1DM, adopted by them from investigation by Pecora,[3] play the crucial role in their procedure. We refer the reader to Howard *et al.*[2] for a review of the historical development of fitting methods for diffraction experiments and an extensive discussion of such practical problems as estimating the number of independent parameters among P_{ij} for a given K and the electron number N, and requirements for a number of experimental data necessary to obtain a unique result. As noticed already by Pecora,[3] if the independent-particle approximation is adopted to describe theoretically the system, this leads to the N-electron wave function in the form of a single Slater determinant of the orthonormal set of molecular orbitals $\psi_j(\mathbf{r})$, and, therefore, to the 1DM in a Dirac form

$$\rho(\mathbf{r},\mathbf{r}')=2\sum_{j=1}^{N/2} \psi_j(\mathbf{r})\psi_j(\mathbf{r}'). \qquad (1.3)$$

Here, for simplicity, a spin-compensated N-electron system is considered (N even) with doubly occupied orbitals, while

0163-1829/97/55(15)/9422(10)/$10.00

the general case is discussed in Appendix A. Such 1DM possesses the two following properties: its trace gives the number of electrons

$$\mathrm{Tr}\rho \equiv \int d^3r \rho(\mathbf{r},\mathbf{r}) = N, \tag{1.4a}$$

and the matrix $\frac{1}{2}\rho$ is idempotent, $(\frac{1}{2}\rho)^2 = \frac{1}{2}\rho$, i.e.,

$$\int d^3r'' \tfrac{1}{2}\rho(\mathbf{r},\mathbf{r}'')\tfrac{1}{2}\rho(\mathbf{r}'',\mathbf{r}') = \tfrac{1}{2}\rho(\mathbf{r},\mathbf{r}') \tag{1.4b}$$

(the factor $\frac{1}{2}$ is present because of double occupancy). The properties (1.4) of $\rho(\mathbf{r},\mathbf{r}')$ lead to the following properties of its basis representation P_{ij}, Eq. (1.2):

$$\sum_i P_{ii} = N, \tag{1.5a}$$

$$\sum_j \tfrac{1}{2}P_{ij}\,\tfrac{1}{2}P_{jk} = \tfrac{1}{2}P_{ik}. \tag{1.5b}$$

To prove this, the basis $\{\chi_j(\mathbf{r})\}$ must be assumed orthonormal. But, with some modification of formulas, nonorthonormal bases can be easily handled too, see Ref. 2. During the fitting process Howard *et al.* force the matrix P_{ij} to satisfy the constraints (1.5) by applying an iterative procedure due to McWeeny.[4] This is the key feature of their fitting, because they consider the "imposition of an idempotency constraint on the density matrix should lead to a unique fit." The last statement may be true for a given, finite basis set and an appropriate collection of experimental data. Determined in this way the density $n(\mathbf{r})$ may represent the best fit (in the sense of minimized the mean squared error χ^2). But we would like to address the question of what is the physical meaning of the fitting by-product — the 1DM $\rho(\mathbf{r},\mathbf{r}')$ [given by Eq. (1.2) from the determined parameters P_{ij}].

The technical problems connected with the role of basis finiteness and with ambiguities arising if some products of basis functions are linearly dependent, have been adequately discussed by earlier workers (see Refs. 5,2). Thus we shall assume for the considerations in Secs. II and III the theoretical ideal situation of infinite, complete, orthonormal basis set. Actually, we are going to use the usual coordinate-$\mathbf{r}$-space representation which is, naturally, equivalent to any of the above mentioned infinite basis set representations. We will then return to a finite basis set in Sec. IV.

II. UNIQUENESS PROBLEM OF FITTING THE IDEMPOTENT 1DM TO DENSITY AND THE DENSITY-FUNCTIONAL THEORY VIEWPOINT

Is the idempotent 1DM $\rho(\mathbf{r},\mathbf{r}')$ unique for a given accurate density $n(\mathbf{r})$ in the situation of a complete, infinite basis set? We can easily give a counter example. During our investigation of the problem of posing the Hartree-Fock approach as a DFT, we obtained[6] the same density $n(\mathbf{r})$ stemming from two different single-determinantal wave functions — the Hartree-Fock one and the Kohn-Sham-like one. The corresponding orbitals were shown explicitly to be different on the example of Be atom.

We are going to generalize now this observation by considering a model noninteracting-N-electron system described by a Hamiltonian

$$\hat{\mathcal{H}}_s = \sum_{i=1}^{N} \hat{h}_s(\mathbf{r}_i), \tag{2.1}$$

with the one-electron Hamiltonian in the form

$$\hat{h}_s(\mathbf{r}) = \hat{t}(\mathbf{r}) + \hat{v}_s(\mathbf{r}), \tag{2.2}$$

where the kinetic energy operator is the usual differential operator

$$\hat{t}(\mathbf{r})\phi(\mathbf{r}) = -\frac{\hbar^2}{2m}\nabla^2(\mathbf{r})\phi(\mathbf{r}) = -\frac{\hbar^2}{2m}\left(\frac{\partial}{\partial\mathbf{r}}\right)^2\phi(\mathbf{r}), \tag{2.3}$$

while the spin-independent potential energy $\hat{v}_s(\mathbf{r})$ is a Hermitian nonlocal operator — an integral operator, in general, with a real, symmetric kernel $\tilde{v}_s(\mathbf{r},\mathbf{r}')$ such that

$$\hat{v}_s(\mathbf{r})\phi(\mathbf{r}) = \int d^3r' \tilde{v}_s(\mathbf{r},\mathbf{r}')\phi(\mathbf{r}'). \tag{2.4}$$

Because of lack of electron-electron interactions in $\hat{\mathcal{H}}_s$, Eq. (2.1), any eigenfunction Φ_s of $\hat{\mathcal{H}}_s$ is a single Slater determinant constructed of a set of $N/2$ doubly occupied orthonormal orbitals $\phi_j(\mathbf{r})$ — the solutions of the one-electron Schrödinger equation

$$\hat{h}_s(\mathbf{r})\phi_j(\mathbf{r}) = \epsilon_j\phi_j(\mathbf{r}). \tag{2.5}$$

(As previously, a spin-compensated system is considered.) So the 1DM of this system (in coordinate-space representation)

$$\rho(\mathbf{r},\mathbf{r}') = 2\sum_{j=1}^{N/2} \phi_j(\mathbf{r})\phi_j(\mathbf{r}') \tag{2.6}$$

obeys relations (1.4). Thus a whole family of model N-electron systems, characterized by various nonlocal potentials $\hat{v}_s$, Eq. (2.4), may have 1DM's which are idempotent. Among them, only one system, with a purely local potential, has a physical meaning for a particular many-electron assembly (such as an atom, a molecule, a cluster, or a crystal) namely, introduced within the DFT, the equivalent reference Kohn-Sham (KS) system, which, by definition, exhibits the same density $n(\mathbf{r})$ as in the original, interacting electron system (see, e.g., Ref. 1). Although the wave function of this KS reference system (and, therefore, the 1DM) is known to be an approximation to the exact wave function of the interacting system, with the exception of giving correctly the density $n(\mathbf{r})$, nevertheless, as practice shows, many properties calculated from the KS wave function are sufficiently accurate to be interesting for the interpretation of various experiments. This includes also properties expressible in terms of excited KS orbitals, which can be obtained if the local KS potential is known. Therefore we propose to specify further the fitting procedure of Howard *et al.*[2] by adding to the existing constraints a new one — the constraint of potential locality, to direct precisely the fitting process towards the determination of the 1DM corresponding to the KS reference system of the many-electron assembly under investigation.

III. ELABORATION OF THE POTENTIAL-LOCALITY CONSTRAINT

In order to convert the qualitative statement, that the 1DM thereby determined should correspond to a model with a local one-body potential, into a mathematical expression suitable for imposing on the fitted 1DM a corresponding constraint, we invoke a notion of the equation of motion (EOM) for 1DM. Actually, EOM's are well known for some time, see, e.g., Dawson and March,[7] not only for noninteracting systems but also for interacting ones, for which the 1DM is coupled with the 2DM. But for a noninteracting case the EOM derivation is so simple and direct that we can give it here. Let us multiply both sides of Eq. (2.5) by $\phi_j(\mathbf{r}')$. Then, from the resulting equation $\phi_j(\mathbf{r}')\hat{h}_s(\mathbf{r})\phi_j(\mathbf{r}) = \epsilon_j\phi_j(\mathbf{r}')\phi_j(\mathbf{r})$ another equation is derived by relabeling variables $\mathbf{r}\leftrightarrow\mathbf{r}'$. Next, after subtracting the second equation from the first, the difference is summed over j running from 1 to $N/2$ and multiplied by the occupancy factor 2. After applying the definition (2.6) of ρ the following EOM is established:

$$[\hat{h}_s(\mathbf{r})-\hat{h}_s(\mathbf{r}')]\rho(\mathbf{r},\mathbf{r}')=0\,. \tag{3.1}$$

According to Eq. (2.2), it can be transformed into

$$[\hat{v}_s(\mathbf{r})-\hat{v}_s(\mathbf{r}')]\rho(\mathbf{r},\mathbf{r}')=-[\hat{t}(\mathbf{r})-\hat{t}(\mathbf{r}')]\rho(\mathbf{r},\mathbf{r}'). \tag{3.2}$$

It will be convenient to have the following functional of 1DM defined:

$$\Gamma(\mathbf{r},\mathbf{r}';[\rho])=-\frac{[\hat{t}(\mathbf{r})-\hat{t}(\mathbf{r}')]\rho(\mathbf{r},\mathbf{r}')}{\rho(\mathbf{r},\mathbf{r}')}. \tag{3.3}$$

It is evident from Eqs. (3.2) and (3.3), that if $\hat{v}_s(\mathbf{r})$ happens to be *local*, $\hat{v}_s(\mathbf{r})=v_{sl}(\mathbf{r})$, i.e., $\tilde{v}_s(\mathbf{r},\mathbf{r}')=\delta(\mathbf{r}-\mathbf{r}')v_{sl}(\mathbf{r})$, then this potential can be directly calculated from ρ in terms of Γ as

$$v_{sl}(\mathbf{r})=v_{sl}(\mathbf{r}')+\Gamma(\mathbf{r},\mathbf{r}';[\rho]). \tag{3.4}$$

The position vector $\mathbf{r}'$ is fixed, although it can be chosen arbitrarily. The presence of the constant $v_{sl}(\mathbf{r}')$, defining the value of the potential at the reference point $\mathbf{r}'$, reflects the fact that potentials can be known only with an accuracy to an additive constant: potentials, which differ by a constant, lead to the same wave functions, and, therefore, the same 1DM.

Now we define the nonlocality indicator Λ, a functional of ρ, to be

$$\Lambda(\mathbf{r}_1,\mathbf{r}_1',\mathbf{r}_2,\mathbf{r}_2';[\rho])=\Gamma(\mathbf{r}_1,\mathbf{r}_2;[\rho])+\Gamma(\mathbf{r}_1',\mathbf{r}_2';[\rho]) -\Gamma(\mathbf{r}_1,\mathbf{r}_2';[\rho])-\Gamma(\mathbf{r}_1',\mathbf{r}_2;[\rho]). \tag{3.5}$$

If the system potential is local, $\hat{v}_s(\mathbf{r})=v_{sl}(\mathbf{r})$, then from Eq. (3.4) it follows immediately that the 1DM of this system satisfies

$$\Lambda(\mathbf{r}_1,\mathbf{r}_1',\mathbf{r}_2,\mathbf{r}_2';[\rho])=0 \quad \text{for any } \mathbf{r}_1,\mathbf{r}_1',\mathbf{r}_2,\mathbf{r}_2'. \tag{3.6}$$

It is important that the reciprocal theorem is also true, namely: if 1DM $\rho(\mathbf{r},\mathbf{r}')$ of the noninteracting model system satisfies Eq. (3.6), then the potential $\hat{v}_s(\mathbf{r})$ of this system is local. The proof is given in Appendix B.

Having in mind an application to a calculational procedure, it is convenient to transform the requirement (3.6) into the following equivalent form:

$$Z_1[\rho]=0, \tag{3.7a}$$

with

$$Z_1[\rho]=\int d^3r_1\, d^3r_1'\, d^3r_2\, d^3r_2'\, w_1(\mathbf{r}_1,\mathbf{r}_1',\mathbf{r}_2,\mathbf{r}_2') \times[\Lambda(\mathbf{r}_1,\mathbf{r}_1',\mathbf{r}_2,\mathbf{r}_2';[\rho])]^2, \tag{3.7b}$$

where the weighting function $w_1>0$ is aimed at making the integrations convergent, thus confining the relevant volume of integration to the "interior" of the system.

Yet another nonlocality indicator, a vector $\boldsymbol{\lambda}$, can be constructed with the help of the functional Γ, namely,

$$\boldsymbol{\lambda}(\mathbf{r}_1,\mathbf{r}_2,\mathbf{r}_2';[\rho])=\boldsymbol{\nabla}(\mathbf{r}_1)[\Gamma(\mathbf{r}_1,\mathbf{r}_2;[\rho])-\Gamma(\mathbf{r}_1,\mathbf{r}_2';[\rho])]. \tag{3.8}$$

If the system potential is local, then we see by appealing to Eq. (3.4) that the 1DM ρ satisfies

$$\boldsymbol{\lambda}(\mathbf{r}_1,\mathbf{r}_2,\mathbf{r}_2';[\rho])=\mathbf{0} \quad \text{for any } \mathbf{r}_1,\mathbf{r}_2,\mathbf{r}_2'. \tag{3.9}$$

As previously, the reciprocal theorem is also true, as it is proven in Appendix B. An equivalent form of the requirement (3.9), more convenient for applications, can be written as

$$Z_2[\rho]=0, \tag{3.10a}$$

with

$$Z_2[\rho]=\int d^3r_1 d^3r_2 d^3r_2'\, w_2(\mathbf{r}_1,\mathbf{r}_2,\mathbf{r}_2')|\boldsymbol{\lambda}(\mathbf{r}_1,\mathbf{r}_2,\mathbf{r}_2';[\rho])|^2, \tag{3.10b}$$

where the weighting function $w_2>0$ should play an analogous role to that played by w_1 in Eq. (3.7b). For obvious reasons the quantities $Z_\nu[\rho]$ will be called mean squared departures from potential locality.

IV. UPGRADING THE FITTING PROCEDURE

As follows from the considerations presented in preceding sections, the 1DM $\rho(\mathbf{r},\mathbf{r}')$, determined from experimental data by the Howard *et al.*[2] procedure, may correspond to some model noninteracting-N-electron system with a nonlocal potential. Its uniqueness and properties (like degree of nonlocality) may strongly depend on such technical details, as properties of particularly chosen atomic orbitals of the basis set, the dimension K of the basis set, etc. In other words, any modification of the basis set may result in a new fitted model, with significantly different $\hat{v}_s$, but giving almost the same $n(\mathbf{r})$ and the same error χ^2. We expect that after including into the fitting procedure the constraint of the potential locality, the 1DM's resulting from fittings performed with various basis sets will be rather close, as corresponding to essentially the same local KS potential, although with some inevitable but small admixture of nonlocality due to finiteness of the bases used.

As shown in Sec. III, both Z_1 Eq. (3.7b), as well as Z_2, Eq. (3.10b), can serve as indicators of the potential nonlocality. Freedom of choice may be resolved by a criterion of calculational convenience (see Appendix C). Dependence of Z_ν on the fitting parameters P_{ij}, $Z_\nu = Z_\nu(P)$, enters Eqs. (3.7b) and (3.10b) via $\rho(\mathbf{r},\mathbf{r}')$, which is expressed in terms of P_{ij} according to Eq. (1.2).

Our proposed modification of the Howard *et al.*[2] fitting procedure therefore consists of a replacement of their minimized function $\chi^2(P')$ by

$$\Xi(P') = \chi^2(P') + \kappa Z_\nu(P'). \tag{4.1}$$

The constant $\kappa > 0$ should be chosen in such a way that both terms in $\Xi(P')$ become similarly sensitive to variations of P' during minimization. It should be noted that for the case of an extremely accurate experiment and of an infinite, complete basis set used in the fitting, the function $\Xi(P')$ can reach the absolute minimum equal zero. The corresponding solution $P' = P_{\min}$ is unique if the ground state of the system is nondegenerate, which follows from the DFT. The use of a finite basis set makes impossible the zero value of Ξ, and the experimental errors increase further the value of χ^2.

Now let us turn to computational details for $Z_\nu(P)$. Since all integrations in Eq. (3.7b) and Eq. (3.10b) must be performed numerically [because of $\rho(\mathbf{r},\mathbf{r}')$ in the denominator, Eq. (3.3)], each integral is to be replaced by a summation over a set of M points, say $\{\mathbf{r}_{0k}\}_{k=1}^{M}$, chosen to be representative for the system [nuclear positions should be avoided, because $\rho(\mathbf{r},\mathbf{r}')$ is nonanalytical there — note that the "thorn" of the density cusp is located there]. The weights pertaining to the numerical integration procedure can be combined with (quite arbitrary) weights w_ν belonging to the integrands, and finally, reduced all together to a common weight equal 1, leading to

$$Z_2 = \sum_{k=1}^{M} \sum_{i=1}^{M} \sum_{j=1}^{M} |\boldsymbol{\lambda}(\mathbf{r}_{0k}, \mathbf{r}_{0i}, \mathbf{r}_{0j})|^2, \tag{4.2}$$

with an analogous expression for Z_1. The number of terms in these summations may be reduced by noting that $|\boldsymbol{\lambda}|^2$ and Λ^2 exhibit some symmetry properties. Details are given in Appendix C.

The minimization of $\Xi(P)$, Eq. (4.1), is aimed to produce all theoretical structure factors as close as possible to experimental ones (via χ^2) and, simultaneously, to make the nonlocal admixtures to the underlying effective local KS potential as small as possible (via κZ_ν). If we want these two aims to be of comparable relevance, the number of (symmetry) independent terms in the summation for Z_2 in Eq. (4.2) (or in its analog for Z_1), i.e., the number (MM' or M'' — see Appendix C) of departure-from-locality events should be chosen of the same order of magnitude as the number of experimental observations included in χ^2. Since, as shown in Appendix C, the computation of a separate event in Z_ν is quite simple and, surely, consuming less time than the computation of a structure factor for an "event" in χ^2, the efficiency of our modified fitting procedure should be similar to that of the original procedure of Howard *et al.*[2]

V. DETERMINING THE POTENTIAL. FITTING TO A GIVEN DENSITY

Having determined the 1DM $\rho(\mathbf{r},\mathbf{r}')$ of the KS reference system, we can immediately obtain the local KS effective potential using Eq. (3.4). Since, as was mentioned, the result of fitting with a finite basis set contains inevitably some admixture of nonlocality, the role of the latter may be diminished by averaging the expression (3.4) over $\mathbf{r}'$. The set $\{\mathbf{r}_{0k}\}$ may be used again for this operation:

$$v_{\text{KS}}(\mathbf{r}) = c_{\text{KS}} + \frac{1}{M} \sum_{k=1}^{M} \Gamma(\mathbf{r}, \mathbf{r}_{0k}; [\rho]). \tag{5.1}$$

For finite systems the constant c_{KS} is traditionally fixed by requiring that $v_{\text{KS}}(\mathbf{r}) \to 0$ for $|\mathbf{r}| \to \infty$. Then the exchange-correlation potential for the system under investigation is determined as

$$v_{\text{xc}}(\mathbf{r}) = v_{\text{KS}}(\mathbf{r}) - v(\mathbf{r}) - v_{\text{es}}(\mathbf{r}; [n]), \tag{5.2}$$

where $v(\mathbf{r})$ is the external (electron-nuclei interaction) potential, and $v_{\text{es}}(\mathbf{r};[n])$ is the classical electrostatic potential calculated with the determined density $n(\mathbf{r})$.

Thus our fitting procedure may be viewed as a route leading to the determination of the exchange-correlation potential — an object of a great theoretical interest for a given system, starting from known x-ray diffraction data (which, in fact, represent the Fourier transform of the electron density). But, very often, a shorter route is even more interesting — namely, that staring just from a given density $n_0(\mathbf{r})$, known from some accurate calculations, e.g., configuration interaction ones (see Leeuwen *et al.*[8] for a recent review of this subject).

Our modification of the Howard *et al.*[2] procedure is, in fact, well suited to perform this task. It is enough to replace in Eq. (4.1) the least-squares sum $\chi^2(P)$ of diffraction intensity errors by an appropriate analog dealing with density errors. The simplest form (used successfully for similar purposes by Engel *et al.*[9]) is

$$\chi^2(P) = \int d^3r \, [\rho(\mathbf{r},\mathbf{r};[P]) - n_0(\mathbf{r})]^2, \tag{5.3a}$$

where the dependence of ρ on P is given in Eq. (1.2). This χ^2 minimizes the squared absolute error, averaged uniformly. Depending on the accuracy distribution of the available input $n_0(\mathbf{r})$, one may also minimize the squared relative error with some weighting function which favors regions of higher accuracy or importance,

$$\chi^2(P) = \int d^3r w(\mathbf{r}) \left(\frac{\rho(\mathbf{r},\mathbf{r};[P]) - n_0(\mathbf{r})}{n_0(\mathbf{r})} \right)^2. \tag{5.3b}$$

The integration may be replaced by the summation over a set of representative points, similarly as in Eq. (4.2).

Although there exists, as reviewed in Ref. 8, a pretty large number of distinct methods allowing the construction of

$v_{xc}(\mathbf{r})$ from a given $n_0(\mathbf{r})$, they almost all have one common feature: the N lowest-energy solutions of the KS Eq. (2.5) must be calculated for many times during iterative processes in which the approximate potential and density are improved. What discerns various methods is the parametrization of the potential and/or construction of the potential for the next iteration. In the simplest case, a few Fourier components of a periodic potential in a solid can serve as fitting parameters,[9] while a potential for atoms and molecules may be expanded in some functional basis or simply represented by its values at a large number of space points. A quite different approach is connected with local-scaling transformations (see Ludeña *et al.*[10] and references therein). Determination of KS orbitals is carried out there via a minimization of the global kinetic energy, during which trial orbitals are modified and scaled in order to fulfill a density constraint. A differential equation, determining the vector scaling function, must be solved numerically at each step of an iterative minimization procedure. Finally, having noncanonical KS orbitals determined, Ludeña *et al.*[10] find a unitary transformation to obtain the orbitals in the canonical form, and next, using the transformed orbitals, they find iteratively the KS eigenenergies and the effective KS potential. It is worth commenting, in connection with this final procedure, that our Eq. (3.4) or (5.1) offers a much simpler and direct method to obtain $v_{\mathrm{KS}}(\mathbf{r})$ from the KS orbitals, because the corresponding 1DM $\rho(\mathbf{r},\mathbf{r}')$, Eq. (2.6), is insensitive to any unitary transformation applied to the orbitals (also the KS eigenenergies are not involved in this case).

Our method of determining $v_{\mathrm{KS}}(\mathbf{r})$ from a given $n_0(\mathbf{r})$ represents a quite different approach: (i) instead of modeling the potential or KS orbitals, parametrization is applied to the 1DM; a convenient orbital basis set $\{\chi_j\}$ and an initial idempotent DM P_{ij} can be taken from the Hartree-Fock calculations; (ii) there is no need to solve the KS equations or the equation for the scaling function; (iii) the evaluation of the minimized function $\Xi(P)$, Eq. (4.1) with (5.3) and (C3) or (C2), for a given set of parameters $\{P_{ij}\}$, involves linear matrix algebra operations only, Eqs. (C6), (C10), (C13); (iv) the minimization of $\Xi(P)$ with respect to $\{P_{ij}\}$ is unconstrained, because the requirements (1.5) are imposed on P_{ij} at each minimization step by means of the McWeeny[4] "purification" procedure.

While numerical tests in the future will be necessary to assess a practical value of the proposed method, we expect it to be more efficient than other methods in the case of large-N systems: the scaling method, as yet, is tested on very small systems only, while all other methods involve frequent solution of an eigenvalue problem, which is a time consuming procedure at large N. On the other hand, the performance of the matrix algebra routines used in our method can be accelerated for large matrices by the use of parallel processors.

VI. DISCUSSION AND CONCLUSIONS

The DFT approach to the problem of fitting the x-ray diffraction data was discussed earlier by Levy and Goldstein.[5] Besides addressing the problem of the relation between the number of data points, the number of electrons, and the number of basis set orbitals in obtaining a unique result, these authors suggest, for the cases when the fitting is not unique, that "the Kohn-Sham determinant, Φ_{KS}, may be approximated via the basis by finding that determinant, which minimizes the kinetic energy expectation value, at fixed n." In the notation of the present paper their prescription for a constraint leading to the results representing the KS system, is the following. Having defined the kinetic energy as a function of P to be

$$T_s(P)=\int d^3r\hat{t}(\mathbf{r})\rho(\mathbf{r},\mathbf{r}';P)|_{\mathbf{r}'=\mathbf{r}}, \tag{6.1}$$

and, similarly, the electron density to be

$$n(\mathbf{r};P)=\rho(\mathbf{r},\mathbf{r};P), \tag{6.2}$$

where ρ depends on P according to Eq. (1.2), let us minimize $\chi^2(P)$ in such a subspace $\{\check{P}\}$ of the space $\{P\}$ of all DM's, fulfilling the conditions (1.5) (as connected with a determinantal N-electron function) that the minimum property

$$T_s(\check{P})\leq T_s(P') \tag{6.3a}$$

is satisfied for all such $P'\in\{P\}$, for which

$$n(\mathbf{r};P')=n(\mathbf{r};\check{P}). \tag{6.3b}$$

Although the Levy and Goldstein[5] constraint (6.3) plays the same role as our potential-locality constraint, there is no obvious way to incorporate it into the fitting procedure. The function $T_s(P)$ cannot be simply added to minimized $\chi^2(P)$, similarly as $Z_\nu(P)$ is in Eq. (4.1), because, besides T_s minimization, Eq. (6.3a), the condition (6.3b) of fixed density must be simultaneously satisfied too. Actually, this idea of minimization (6.3) is implemented in scaling methods[10] (discussed in Sec. V).

However, once the 1DM ρ of the KS system is determined by our method, the value of the noninteracting kinetic energy T_s can be calculated according to Eq. (6.1). This is an important ingredient in various expressions for the approximate total ground-state energy extracted from x-ray data, which are proposed by Levy and Goldstein.[5]

It should be noted that the method discussed in the present paper to obtain n,ρ, and v_{xc} from a diffraction experiment is not confined to finite systems (such as atoms or molecules). All quantities used in these considerations can be well defined for extended systems by the standard device of imposing periodic boundary conditions. In particular, as follows from its derivation in Sec. III, the equation of motion, Eq. (3.1) is valid also for such systems, being a very convenient relation at a local position pair $(\mathbf{r},\mathbf{r}')$. Therefore the potential-locality constraints, obtained from this equation, are applicable to quite general systems.

The analysis given in Appendix A shows that the method can be easily extended and applied to systems having some spin structure. Then two idempotent DM's $\rho_\uparrow$ and $\rho_\downarrow$ are to be fitted, for two subsystems of opposite spins. When results are analyzed in terms of the usual DFT, then one, common KS potential is obtained from Eq. (5.1) for $\rho=\rho_\uparrow+\rho_\downarrow$. But, alternatively, the spin-density-functional theory (SDFT) can be applied, resulting in two KS potentials $v_{\mathrm{KS}\uparrow}(\mathbf{r})$, $v_{\mathrm{KS}\downarrow}(\mathbf{r})$, obtained from Eq. (5.1) after substituting ρ by $\rho_\uparrow$ or $\rho_\downarrow$, respectively. It is worth mentioning that, having separate DM's $\rho_\uparrow$ and $\rho_\downarrow$, the spin density $m(\mathbf{r})$, Eq. (A11), besides the electron number density $n(\mathbf{r})$, Eq. (A9), is available. Taking into account that polarized neutron diffraction data are analyzed in terms of $m(\mathbf{r})$, analogous to x-ray diffraction in terms of $n(\mathbf{r})$, both experiments (if performed simultaneously) can be analyzed by our method together via minimizing Ξ of Eq. (4.1), taken as the sum of the x-ray $\chi^2(n)$, the neutron $\chi^2(m)$ and the appropriate κZ_ν term(s) (one for DFT, two for SDFT). This would lead to experimental determination (for the first time, to the best of our knowledge) of $v_{\mathrm{xc}\uparrow}$ and $v_{\mathrm{xc}\downarrow}$ separately. Although, for any spin-uncompensated system, fitting to x-ray data alone also gives $\rho_\uparrow$ and $\rho_\downarrow$, nevertheless the spin density $m(\mathbf{r})$ obtained in this way may be much less accurate than one obtained from x-ray and neutron experiments analyzed together.

Some wider application of our method is discussed in Appendix D. It is shown there that the role of the fitted noninteracting reference system may be played not only by the most popular KS system, but also by modern so-called generalized KS systems, which allow for some admixture of a well-defined nonlocality to the one-body potential.

As a concluding comment, earlier analyses of diffraction data, such as those of Howard *et al.*[2] and other references cited there, make it clear that imposition of idempotency on a density matrix is both practical and truly helpful in extracting the electron density $n(\mathbf{r})$. We have emphasized, without of course criticism in any way of such analyses, that the density matrix corresponding to the KS reference system can, in principle, be extracted by adding the proposed potential-locality constraint to their procedure. Once this matrix is known, the one-body local potential $v_{\mathrm{KS}}(\mathbf{r})$ can be obtained from Eq. (3.4) with (3.3), which stems from the equation of motion Eq. (3.2) [the kinetic energy operator is defined in a standard way, Eq. (2.3)]. Since the Hartree part of $v_{\mathrm{KS}}(\mathbf{r})$ is determined uniquely by the nuclear framework plus $n(\mathbf{r})$, the many-body exchange correlation contribution $v_{\mathrm{xc}}(\mathbf{r})$, Eq. (5.2), is itself accessible from such diffraction experiments.

When the minimized function $\chi^2(P)$ — the sum of least-squares errors of diffraction intensities — is replaced by an analog concerning errors of the fitted density with respect to a given theoretical density $n_0(\mathbf{r})$, Eq. (5.3), our method can perform the task of constructing the effective KS one-electron Hamiltonian (with its exchange-correlation potential) whose eigenfunctions generate this $n_0(\mathbf{r})$, as it is discussed in Sec. V.

ACKNOWLEDGMENTS

The authors wish to thank International Centre for Theoretical Physics, Trieste, Italy, for hospitality. N.H.M. acknowledges the Leverhulme Trust (U.K.) for supporting his work on density matrices and density-functional theory.

APPENDIX A: SPIN-STRUCTURED SYSTEMS

In the main text of the paper we eliminate electron spin coordinates by considering spin-compensated systems only. Here we remove this limitation. The absence of an external magnetic field is still assumed, so that the real operator $\hat{h}_s$ in Eq. (2.2) acts (depends) on the space coordinate $\mathbf{r}$, but is independent of the spin coordinate s. The spin-dependent 1DM, denoted by γ, of a noninteracting-N-electron system is of Dirac form

$$\gamma(\mathbf{r}s,\mathbf{r}'s')=\sum_{j=1}^{N}\varphi_j(\mathbf{r}s)\varphi_j(\mathbf{r}'s'), \tag{A1}$$

where the set $\{\varphi_j(\mathbf{r}s)\}$ of spin orbitals — eigenfunctions of $\hat{h}_s$, is orthonormal

$$\sum_s \int d^3r\,\varphi_i(\mathbf{r}s)\varphi_j(\mathbf{r}s)=\delta_{ij}. \tag{A2}$$

Therefore γ is idempotent, $\gamma^2=\gamma$, in the space of $(\mathbf{r}s)$ coordinates, i.e.,

$$\sum_{s''}\int d^3r''\,\gamma(\mathbf{r}s,\mathbf{r}''s'')\gamma(\mathbf{r}''s'',\mathbf{r}'s')=\gamma(\mathbf{r}s,\mathbf{r}'s'). \tag{A3}$$

The spinless 1DM $\rho(\mathbf{r},\mathbf{r}')$ is defined by spin summation of γ:

$$\rho(\mathbf{r},\mathbf{r}')=\sum_s \gamma(\mathbf{r}s,\mathbf{r}'s). \tag{A4}$$

Since $\hat{h}_s$ is spin independent, two eigenfunctions can be connected with each eigenenergy ϵ_j, namely, the spin orbital $\varphi_{i\uparrow}(\mathbf{r}s)=\phi_{i\uparrow}(\mathbf{r})\alpha(s)=\phi_j(\mathbf{r})\alpha(s)$ and the spin orbital $\varphi_{\ell\downarrow}(\mathbf{r}s)=\phi_{\ell\downarrow}(\mathbf{r})\beta(s)=\phi_j(\mathbf{r})\beta(s)$.

Therefore the 1DM γ, Eq. (A1), can be written, in general, in the form

$$\gamma(\mathbf{r}s,\mathbf{r}'s')=\rho_\uparrow(\mathbf{r},\mathbf{r}')\alpha(s)\alpha(s')+\rho_\downarrow(\mathbf{r},\mathbf{r}')\beta(s)\beta(s'), \tag{A5a}$$

where

$$\rho_\sigma(\mathbf{r},\mathbf{r}')=\sum_{i=1}^{N_\sigma}\phi_{i\sigma}(\mathbf{r})\phi_{i\sigma}(\mathbf{r}'),\quad \sigma=\uparrow,\downarrow. \tag{A5b}$$

So, according to Eq. (A4), from Eq. (A5a) follows

$$\rho(\mathbf{r},\mathbf{r}')=\rho_\uparrow(\mathbf{r},\mathbf{r}')+\rho_\downarrow(\mathbf{r},\mathbf{r}'). \tag{A6}$$

Of the N electrons in the system, $N_\uparrow$ electrons occupy orbitals of the spin-up type [with spin function $\alpha(s)$], while $N_\downarrow$ electrons are of the spin-down type [with $\beta(s)$], their numbers satisfying

$$N_\uparrow+N_\downarrow=N. \tag{A7}$$

The finite set $\{\phi_{i\sigma}(\mathbf{r})\}_1^{N_\sigma}$ represents some subset of the infinite set of space orbitals $\{\phi_j(\mathbf{r})\}_1^\infty$ — the eigenfunctions of $\hat{h}_s$ in Eq. (2.5). Therefore each of the matrices $\rho_\sigma(\mathbf{r},\mathbf{r}')$, $\sigma=\uparrow,\downarrow$, is idempotent in the space of $\mathbf{r}$ coordinates

$$\int d^3r''\rho_\sigma(\mathbf{r},\mathbf{r}'')\rho_\sigma(\mathbf{r}'',\mathbf{r}')=\rho_\sigma(\mathbf{r},\mathbf{r}'), \tag{A8a}$$

and its trace value is

$$\mathrm{Tr}\rho_\sigma = \int d^3 r \rho_\sigma(\mathbf{r},\mathbf{r}) = N_\sigma . \tag{A8b}$$

If for some system $N=2N_\uparrow=2N_\downarrow$ and the subsets $\{\phi_{i\uparrow}(\mathbf{r})\}_1^{N/2}$ and $\{\phi_{i\downarrow}(\mathbf{r})\}_1^{N/2}$ are identical [i.e., consist of the same $N/2$ orbitals $\phi_j(\mathbf{r})$], then this system is called *spin compensated.* We see that Eqs. (1.3) and (1.4) are in agreement with above definitions and can be viewed as stemming from $\rho_\uparrow=\rho_\downarrow=\frac{1}{2}\rho$. Among other physical systems it is worth mentioning a typical *open-shell* system (for odd N) with $N_\uparrow=N_\downarrow+1$, having $N_\downarrow$ orbitals doubly occupied and one orbital singly occupied. But even a system having $N=2N_\uparrow=2N_\downarrow$ may happen to be locally spin uncompensated, if its $\rho_\uparrow\neq\rho_\downarrow$, i.e., when some orbital belonging to $\rho_\uparrow$ does not belong to $\rho_\downarrow$. However, Eqs. (A5)–(A8) pertain to systems with all possible spin structures.

In connection with Eq. (A8) we see that the fitting procedure, discussed in Sec. IV, must be modified in the case of spin-uncompensated systems by introducing, in the place of the previous set $\{P_{ij}\}$, two sets of fitting parameters: $\{P_{\uparrow ij}\}$ and $\{P_{\downarrow ij}\}$, corresponding to expansions of $\rho_\uparrow(\mathbf{r},\mathbf{r}')$ and $\rho_\downarrow(\mathbf{r},\mathbf{r}')$, respectively, in a finite basis set, in analogy to Eq. (1.2). The expansions may involve the same basis set, but this is not obligatory. Consequently, the idempotency and the trace-value constraints must be imposed in the fitting procedure on each of the matrices $\{P_{\uparrow ij}\}$ and $\{P_{\downarrow ij}\}$ separately.

Applying the relation (1.1) to ρ given in Eq. (A6) we see that the total electron number density of the system is separated as

$$n(\mathbf{r})=n_\uparrow(\mathbf{r})+n_\downarrow(\mathbf{r}), \tag{A9}$$

with

$$n_\sigma(\mathbf{r})=\rho_\sigma(\mathbf{r},\mathbf{r}). \tag{A10}$$

The spin density, defined as

$$m(\mathbf{r})=n_\uparrow(\mathbf{r})-n_\downarrow(\mathbf{r}), \tag{A11}$$

leads to the electron magnetization density equal to $-\mu_B m(\mathbf{r})$ (μ_B is the Bohr magneton). Any system can be characterized either by a pair $\{n_\uparrow(\mathbf{r}),n_\downarrow(\mathbf{r})\}$, or $\{n(\mathbf{r}),m(\mathbf{r})\}$. A spin-compensated system exhibits $m(\mathbf{r})=0$ for any $\mathbf{r}$. It is worth mentioning that the spin density $m(\mathbf{r})$ can be measured by means of polarized neutron diffraction, paralleling the measurement of the total density $n(\mathbf{r})$ by x-ray diffraction.

The equation of motion for the 1DM γ (valid for systems with arbitrary spin structure) can be shown (on the same way as discussed in Sec. III for ρ) to be

$$[\hat{h}_s(\mathbf{r})-\hat{h}_s(\mathbf{r}')]\gamma(\mathbf{r}s,\mathbf{r}'s')=0 . \tag{A12}$$

After summing it over spin coordinate, Eq. (A4), the previous EOM, Eq. (3.1), is reproduced. This means that in the fitting procedure one potential-locality constraint only, either Eq. (3.7) or Eq. (3.10) [both have been derived with the help of Eq. (3.1)] should be imposed. They are expressed in terms of the summary 1DM $\rho=\rho_\uparrow+\rho_\downarrow$. This corresponds to the fact, that the common local potential (the effective KS potential) determines space orbitals for both $\rho_\uparrow$ and $\rho_\downarrow$.

However, besides the traditional DFT, having $n(\mathbf{r})$ as an independent variable, there is known a modified version — the so-called spin-density-functional theory (SDFT) having two independent variables: $n_\uparrow(\mathbf{r})$ and $n_\downarrow(\mathbf{r})$. While in the presence of an external magnetic field the SDFT must be applied, because the DFT is not valid in this case, in the limit of vanishing magnetic field both approaches hold and, in principle, should lead to identical results. The above mentioned equivalent reference noninteracting-N-electron system of DFT, characterized by the effective, local, one-body KS potential $v_{KS}(\mathbf{r})$ and showing the same density $n(\mathbf{r})$ as the original, interacting system, is replaced in the SDFT by two such reference systems, characterized by the local potentials $v_{KS\sigma}(\mathbf{r})$ and showing the same polarized densities $n_\sigma(\mathbf{r})$ as the original, interacting system (for $\sigma=\uparrow,\downarrow$). In consequence, in the SDFT there are two EOM's such as Eq. (3.1), one for $\rho_\uparrow$ and another for $\rho_\downarrow$. Because there are two independent potentials involved, *two* potential-locality *constraints* must be imposed in the fitting procedure, either Eq. (3.7) in terms of $\rho_\uparrow$ and $\rho_\downarrow$, separately, or Eq. (3.10) with the same use. As in the DFT case, two sets of fitting parameters, $P_{\uparrow ij}$ and $P_{\downarrow ij}$ are to be used with the idempotency and trace-value constraints imposed on $P_{\uparrow ij}$ and $P_{\downarrow ij}$ separately. More details about spin-structure aspects of the DFT and the SDFT can be found in Ref. 1.

APPENDIX B: SUFFICIENT CONDITIONS FOR POTENTIAL LOCALITY

We are going to prove that *if the condition* (3.6) *is satisfied* for the 1DM $\rho(\mathbf{r},\mathbf{r}')$ which is connected with some model system, as given in Eqs. (2.1)–(2.6), *then* the one-body *potential* $\hat{v}_s(\mathbf{r})$ of this system, Eq. (2.4), *is local*, i.e., the kernel in Eq. (2.4) has the form

$$\tilde{v}_s(\mathbf{r},\mathbf{r}')=\delta(\mathbf{r}-\mathbf{r}')v_{sl}(\mathbf{r}). \tag{B1}$$

For the purpose of the proof we express the locality property (B1) in the following, equivalent form:

$$\frac{\hat{v}_s(\mathbf{r})f(\mathbf{r})}{f(\mathbf{r})}=\frac{\hat{v}_s(\mathbf{r})g(\mathbf{r})}{g(\mathbf{r})}=\cdots=v_{sl}(\mathbf{r}), \tag{B2}$$

which must hold for each function $f(\mathbf{r}),g(\mathbf{r}),\cdots$. Now, in order to prove by "reductio ad absurdum," we assume that $\hat{v}_s(\mathbf{r})$ is not local. This nonlocality, to be detected in Eq. (3.6), must manifest itself with respect to functions entering $\rho(\mathbf{r},\mathbf{r}')$, Eq. (2.6). Thus, in negation to Eq. (B2), we suppose that there exists an orbital, say $\phi_k(\mathbf{r})$, for which

$$v_{sl}(\mathbf{r})\neq\frac{\hat{v}_s(\mathbf{r})\phi_k(\mathbf{r})}{\phi_k(\mathbf{r})}=v_{sl}(\mathbf{r})+a(\mathbf{r}), \tag{B3}$$

where $a(\mathbf{r})$ is not a constant [see discussion below Eq. (3.4)] and $\nabla a(\mathbf{r})\neq\mathbf{0}$ in some region around $\mathbf{r}_0$, of finite radius $\eta>0$: $|\mathbf{r}-\mathbf{r}_0|<\eta$ (thus we exclude unphysical models with potentials defined on zero-measure sets of points).

Now we calculate Λ in Eq. (3.5) using Γ, Eq. (3.3), expressed in terms of $\hat{v}_s$ and ρ due to Eq. (3.2). After taking into account Eq. (B3) we obtain for Γ,

$$\Gamma(\mathbf{r}_1,\mathbf{r}_2)=v_{\rm sl}(\mathbf{r}_1)-v_{\rm sl}(\mathbf{r}_2)+\Gamma_{\rm nl}(\mathbf{r}_1,\mathbf{r}_2)=v_{\rm sl}(\mathbf{r}_1)-v_{\rm sl}(\mathbf{r}_2)+[a(\mathbf{r}_1)-a(\mathbf{r}_2)]G(\mathbf{r}_1,\mathbf{r}_2), \quad \text{(B4)}$$

where

$$G(\mathbf{r}_1,\mathbf{r}_2)=G(\mathbf{r}_2,\mathbf{r}_1)=\frac{\phi_k(\mathbf{r}_1)\phi_k(\mathbf{r}_2)}{\rho(\mathbf{r}_1,\mathbf{r}_2)}. \quad \text{(B5)}$$

By combining functions Γ of various arguments to form Λ, Eq. (3.5), we observe cancellation of all $v_{\rm sl}$ terms, so what remains is

$$\Lambda(\mathbf{r}_1,\mathbf{r}_1',\mathbf{r}_2,\mathbf{r}_2')=\Gamma_{\rm nl}(\mathbf{r}_1,\mathbf{r}_2)+\Gamma_{\rm nl}(\mathbf{r}_1',\mathbf{r}_2')-\Gamma_{\rm nl}(\mathbf{r}_1,\mathbf{r}_2')-\Gamma_{\rm nl}(\mathbf{r}_1',\mathbf{r}_2). \quad \text{(B6)}$$

Finally, we expand Λ around the point $\mathbf{r}_0$ with respect to all variables to obtain

$$\Lambda(\mathbf{r}_0+\mathbf{h}_1,\mathbf{r}_0+\mathbf{h}_1',\mathbf{r}_0+\mathbf{h}_2,\mathbf{r}_0+\mathbf{h}_2')=[(\mathbf{h}_1-\mathbf{h}_1')\times(\mathbf{h}_2-\mathbf{h}_2')]\cdot[(\nabla a)\times(\nabla_1 G)]+O(h^3), \quad \text{(B7)}$$

with

$$\nabla a=\nabla(\mathbf{r})a(\mathbf{r})|_{\mathbf{r}=\mathbf{r}_0}; \quad \nabla_1 G=\nabla(\mathbf{r}_1)G(\mathbf{r}_1,\mathbf{r}_2)|_{\mathbf{r}_1=\mathbf{r}_2=\mathbf{r}_0}. \quad \text{(B8)}$$

According to our assumption, $\nabla a\neq\mathbf{0}$. We may consider also that $\nabla_1 G\neq\mathbf{0}$ and $\nabla_1 G\nparallel\nabla a$ (because even if this fails exactly at $\mathbf{r}_0$, it should occur at some nearby point $\mathbf{r}_0'$, which then can serve as a new center of expansion). Therefore we see that $\Lambda\neq 0$ for all its arguments varying in some vicinity of $\mathbf{r}_0$. This contradiction with Eq. (3.6) completes the proof.

The second proof, that also the condition Eq. (3.9) is sufficient for potential locality, goes in a similar way as the preceding proof. The following expansion is obtained for $\boldsymbol{\lambda}$, Eq. (3.8):

$$\boldsymbol{\lambda}(\mathbf{r}_0+\mathbf{h}_1,\mathbf{r}_0+\mathbf{h}_2,\mathbf{r}_0+\mathbf{h}_2')=[(\mathbf{h}_2-\mathbf{h}_2')\cdot(\nabla_1 G)]\nabla a-[(\mathbf{h}_2-\mathbf{h}_2')\cdot(\nabla a)]\nabla_1 G+O(h^2). \quad \text{(B9)}$$

This is a linear combination of two nonzero vectors ∇a and $\nabla_1 G$ [see the discussion below Eq. (B8)], with nonzero coefficients in some volume of $(\mathbf{h}_2-\mathbf{h}_2')$ space, so we have $\boldsymbol{\lambda}\neq\mathbf{0}$ in this volume. This contradiction with Eq. (3.9) completes the second proof.

APPENDIX C: SYMMETRIZED SUMMATIONS IN EVALUATING $Z_\nu(P)$

As follows from Eqs. (3.5) and (3.3) [note $\Gamma(\mathbf{r}_2,\mathbf{r}_1)=-\Gamma(\mathbf{r}_1,\mathbf{r}_2)$], the nonlocality measure $[\Lambda(\mathbf{r}_1,\mathbf{r}_1',\mathbf{r}_2,\mathbf{r}_2')]^2$ is invariant with respect to operations $\mathbf{r}_1\leftrightarrow\mathbf{r}_1'$, $\mathbf{r}_2\leftrightarrow\mathbf{r}_2'$, $(\mathbf{r}_1,\mathbf{r}_1')\leftrightarrow(\mathbf{r}_2,\mathbf{r}_2')$, and vanishes on the "diagonals" $\mathbf{r}_1=\mathbf{r}_1'$, $\mathbf{r}_2=\mathbf{r}_2'$, $(\mathbf{r}_1,\mathbf{r}_1')=(\mathbf{r}_2,\mathbf{r}_2')$. Similarly, from Eq. (3.8) it follows that $|\boldsymbol{\lambda}(\mathbf{r}_1,\mathbf{r}_2,\mathbf{r}_2')|^2$ is invariant for $\mathbf{r}_2\leftrightarrow\mathbf{r}_2'$, and vanishes at $\mathbf{r}_2=\mathbf{r}_2'$. In order to take these properties into account in evaluating Z_2 in Eq. (4.2) and its analog for Z_1, let us define a function $\ell=L(i,j)$ numbering consecutively by ℓ the off-diagonal elements of any square matrix A_{ij} in the triangle $i<j$: $L(i,j)=i+(j-1)(j-2)/2$. The pair of functions giving the reciprocal relation, $i=I(\ell)$ and $j=J(\ell)$, can be easily programmed or just stored as arrays. For a matrix of the dimension $M\times M$, the index ℓ varies in the range $1\leqslant\ell\leqslant M'$, where

$$M'=M(M-1)/2. \quad \text{(C1)}$$

So Z_2 in Eq. (4.2) can be transformed into

$$Z_2=2\sum_{k=1}^{M}\sum_{\ell=1}^{M'}|\boldsymbol{\lambda}(k,I(\ell),J(\ell))|^2. \quad \text{(C2)}$$

A convention for the function arguments has been adopted to represent the vector $\mathbf{r}_{0k}$ by its subscript k. The quantity Z_1 is summed in the following way:

$$Z_1=8\sum_{\ell=1}^{M''}\{\Lambda[I(I(\ell)),J(I(\ell)),I(J(\ell)),J(J(\ell))]\}^2, \quad \text{(C3a)}$$

where the number of terms in this one-dimensional summation is

$$M''=(M')'=\frac{1}{2}\frac{M(M-1)}{2}\left[\frac{M(M-1)}{2}-1\right]. \quad \text{(C3b)}$$

Quite a small number M of representative positions $\mathbf{r}_{0k}$ produces a large number (MM') of departure-from-locality events in Z_2, or a number M'' in Z_1 — see a few examples of $(M,\ MM',\ M'')$ triads: (10, 450, 990), (20, 3 800, 17 955), (50, 61 250, 749 700). Therefore it seems feasible, that after performing the time-consuming evaluations of basis functions and their derivatives in advance, these results can be stored in a reasonable volume of a computer memory, ready for use during minimization, as it is shown below.

Because of $P_{ji}=P_{ij}$ symmetry, the number of independent matrix elements of P is

$$K_{\rm ind}=K(K+1)/2, \quad \text{(C4)}$$

and Eq. (1.2) can be rearranged as

$$\rho(\mathbf{r}_1,\mathbf{r}_2)=\sum_{j=2}^{K}\sum_{i=1}^{j-1}P_{ij}[\chi_i(\mathbf{r}_1)\chi_j(\mathbf{r}_2)+\chi_j(\mathbf{r}_1)\chi_i(\mathbf{r}_2)]+\sum_{j=1}^{K}P_{jj}[\chi_j(\mathbf{r}_1)\chi_j(\mathbf{r}_2)]. \quad \text{(C5)}$$

It will be convenient to view this form of ρ as a scalar product of two $K_{\rm ind}$-dimensional vectors:

$$\rho(\mathbf{r}_1,\mathbf{r}_2)=\sum_{a=1}^{K_{\rm ind}}\tilde{P}_a\tilde{\Omega}_a(\mathbf{r}_1,\mathbf{r}_2)=\tilde{\boldsymbol{P}}\cdot\tilde{\boldsymbol{\Omega}}(\mathbf{r}_1,\mathbf{r}_2), \quad \text{(C6)}$$

where

$$\tilde{P}_a = \begin{cases} P_{ij}, & i=I(a),\ j=J(a) \quad \text{for } 1 \leq a \leq K' = K(K-1)/2 \\ P_{jj}, & j=a-K' \quad \text{for } K'+1 \leq a \leq K_{\text{ind}}, \end{cases} \tag{C7}$$

$$\tilde{\Omega}_a(\mathbf{r}_1,\mathbf{r}_2) = \begin{cases} \chi_i(\mathbf{r}_1)\chi_j(\mathbf{r}_2)+\chi_j(\mathbf{r}_1)\chi_i(\mathbf{r}_2), & i=I(a),\ j=J(a) \quad \text{for } 1 \leq a \leq K' \\ \chi_j(\mathbf{r}_1)\chi_j(\mathbf{r}_2), & j=a-K' \quad \text{for } K'+1 \leq a \leq K_{\text{ind}}. \end{cases} \tag{C8}$$

If we define another vector function

$$\tilde{\mathbf{\Theta}}(\mathbf{r}_1,\mathbf{r}_2) = -[\hat{t}(\mathbf{r}_1)-\hat{t}(\mathbf{r}_2)]\tilde{\mathbf{\Omega}}(\mathbf{r}_1,\mathbf{r}_2), \tag{C9}$$

[note $\tilde{\mathbf{\Omega}}(\mathbf{r}_2,\mathbf{r}_1)=\tilde{\mathbf{\Omega}}(\mathbf{r}_1,\mathbf{r}_2)$ and $\tilde{\mathbf{\Theta}}(\mathbf{r}_2,\mathbf{r}_1)=-\tilde{\mathbf{\Theta}}(\mathbf{r}_1,\mathbf{r}_2)$] and evaluate $\tilde{\Omega}_a(\mathbf{r}_{0k},\mathbf{r}_{0\ell})$ and $\tilde{\Theta}_a(\mathbf{r}_{0k},\mathbf{r}_{0\ell})$ for $k,\ell=1,\ldots,M$, we need to store $2\,K_{\text{ind}}M^2$ numbers. Then a single "departure-from-locality" event in terms of Λ is evaluated during minimization by applying arithmetic operations only, namely,

$$[\Lambda(ijk\ell)]^2 = \left[\frac{\tilde{P}\cdot\tilde{\Theta}(ik)}{\tilde{P}\cdot\tilde{\Omega}(ik)} + \frac{\tilde{P}\cdot\tilde{\Theta}(j\ell)}{\tilde{P}\cdot\tilde{\Omega}(j\ell)} - \frac{\tilde{P}\cdot\tilde{\Theta}(i\ell)}{\tilde{P}\cdot\tilde{\Omega}(i\ell)} - \frac{\tilde{P}\cdot\tilde{\Theta}(jk)}{\tilde{P}\cdot\tilde{\Omega}(jk)} \right]^2. \tag{C10}$$

A sum of M'' such events in Eq. (C3a) gives $Z_1(P)$ for actual P.

For the calculation of $\boldsymbol{\lambda}$ one needs two additional functions

$$\tilde{\boldsymbol{\varpi}}(\alpha,\mathbf{r}_1,\mathbf{r}_2) = \nabla_\alpha(\mathbf{r}_1)\tilde{\mathbf{\Omega}}(\mathbf{r}_1,\mathbf{r}_2) \tag{C11}$$

and

$$\tilde{\boldsymbol{\vartheta}}(\alpha,\mathbf{r}_1,\mathbf{r}_2) = \nabla_\alpha(\mathbf{r}_1)\tilde{\mathbf{\Theta}}(\mathbf{r}_1,\mathbf{r}_2) \tag{C12}$$

($\alpha=1,2,3$) to be evaluated and stored (as $2\,K_{\text{ind}}3\,M^2$ numbers). Then a single event in terms of $\boldsymbol{\lambda}$ is

$$|\boldsymbol{\lambda}(kij)|^2 = \sum_{\alpha=1}^{3} \left(\left\{ \frac{[\tilde{P}\cdot\tilde{\vartheta}(\alpha ki)][\tilde{P}\cdot\tilde{\Omega}(ki)] - [\tilde{P}\cdot\tilde{\Theta}(ki)][\tilde{P}\cdot\tilde{\varpi}(\alpha ki)]}{[\tilde{P}\cdot\tilde{\Omega}(ki)]^2} \right\} - \{i\rightarrow j\} \right)^2. \tag{C13}$$

A sum of (MM') such events in Eq. (C2) gives $Z_2(P)$ for actual P. Within the method described above and for fixed K and M, the nonlocality measure Z_1 includes more events than Z_2 does (by a factor of about $M/4$) and requires 1/4 of the memory capacity used by Z_2, so it is much more favorable on both accounts.

APPENDIX D: GENERALIZED KOHN-SHAM SCHEMES

The Kohn-Sham scheme represents an exact, in general, implementation of the DFT. However, first-principle calculations based on it are limited in accuracy, because approximate only forms are available for its important ingredients — the exchange-correlation energy and potential as explicit functionals of the density. Introduced recently generalized Kohn-Sham (GKS) schemes,[11] belong also to exact realizations of the DFT. In these approaches, due to inclusion of a portion of the electron-electron interaction energy into the exactly treated part of a model system, more accurate results can be obtained, because the remaining portion of the interaction, which is treated approximately, plays a less important role. Especially significant improvement in this way is achieved for band-gap results in semiconductors. Therefore it would be of interest to extend our fitting procedure to many-electron systems described in terms of GKS approaches. We are going to show now that important class of GKS schemes, successful in above mentioned calculations, can be treated in our approach in the same way as the KS scheme.

As discussed by Seidl *et al.*,[11] a GKS scheme, similarly as the KS one, introduces a model noninteracting system having the same density $n(\mathbf{r})$ as the original, interacting system (see Sec. II). While, in the case of traditional KS system, the potential occurring in the one-electron Hamiltonian (2.2) is a local operator $\hat{v}_s(\mathbf{r})=v_{\text{KS}}(\mathbf{r})$, that for GKS system is a sum of nonlocal and local contributions,

$$\hat{v}_s(\mathbf{r}) = \hat{v}_{\text{nl}}(\mathbf{r};[\{\phi_j\}]) + v_{\text{eff}}(\mathbf{r};[n]), \tag{D1}$$

functionals of the set of occupied orbitals $\{\phi_j(\mathbf{r})\}$ [solutions of Eq. (2.5)], and the density $n(\mathbf{r})$, respectively. Various GKS schemes differ in a definition of the kernel $\tilde{v}_{\text{nl}}(\mathbf{r},\mathbf{r}')$ [comp. Eq. (2.4)], as an explicit expression in terms of $\{\phi_j\}$ and in an approximation for the corresponding $v_{\text{eff}}(\mathbf{r})$ as a (usually local) functional of the density. Since our approach is formulated in terms of the 1DM — a combination (2.6) of orbitals, therefore we are limited in application to such a class of GKS schemes, for which the nonlocal part of the potential is expressible in terms of 1DM: $\hat{v}_{\text{nl}}=\hat{v}_{\text{nl}}(\mathbf{r};[\rho])$. Then it is obvious that after modifying Eqs. (3.2) and (3.3) with replacements

$$\hat{v}_s(\mathbf{r}) \rightarrow v_{\text{eff}}(\mathbf{r}) \tag{D2a}$$

and

$$\hat{t}(\mathbf{r}) \rightarrow \hat{t}(\mathbf{r}) + \hat{v}_{\text{nl}}(\mathbf{r};[\rho]), \tag{D2b}$$

all previous considerations hold for GKS schemes too, because $v_{\rm eff}(\mathbf{r})$ is local, similarly as $v_{\rm KS}(\mathbf{r})$ does. As the consequence of the replacement (D2a) applied to Eq. (3.4), we see that Eq. (5.1) allows us to obtain now $v_{\rm eff}(\mathbf{r})$ of the GKS system instead of $v_{\rm KS}(\mathbf{r})$ of the KS system.

As an example, it is worth considering in detail the screened nonlocal-exchange scheme, introduced, denoted as sX-LDA, and implemented by Seidl *et al.*[11] The kernel of the nonlocal potential defined there,

$$\tilde{v}_{\rm nl}(\mathbf{r},\mathbf{r}')=\tilde{v}_{\rm x}^{\rm sX}(\mathbf{r},\mathbf{r}';[\rho])=\exp(-k_{\rm TF}|\mathbf{r}-\mathbf{r}'|)\tilde{v}_{\rm x}^{\rm F}(\mathbf{r},\mathbf{r}';[\rho]), \tag{D3a}$$

with

$$\tilde{v}_{\rm x}^{\rm F}(\mathbf{r},\mathbf{r}';[\rho])=-\frac{\rho(\mathbf{r},\mathbf{r}')}{2|\mathbf{r}-\mathbf{r}'|}, \tag{D3b}$$

is a screened kernel of the Fock exchange integral operator $\hat{v}_{\rm x}^{\rm F}$ (occurring in the Hartree-Fock equations for a spin-compensated system). The Thomas-Fermi screening constant $k_{\rm TF}$ is to be taken for the average valence electron density of a considered semiconductor. The local partner in Eq. (D1) of this nonlocal potential is

$$v_{\rm eff}(\mathbf{r};[n])=v(\mathbf{r})+v_{\rm es}(\mathbf{r};[n])+v_{\rm xc}^{\rm sX}(\mathbf{r};[n]). \tag{D4}$$

Besides the external and electrostatic contributions it contains a specific exchange-correlation potential of sX-LDA scheme, which is approximated in Ref. 11 by some function of $n(\mathbf{r})$ and of the ratio $k_{\rm TF}/\bar{k}_{\rm F}$, where $\bar{k}_{\rm F}$ is the Fermi wave vector corresponding to the average density. This sX-LDA scheme put the earlier procedure of Bylander and Kleinman[12] on a solid basis within DFT.

It should be noted that the replacement (D2b) must be applied to Eq. (C9) too. In the case of the sX-LDA nonlocal potential (D3) it leads in Eq. (C10) to the replacement

$$\tilde{\boldsymbol{P}}\cdot\tilde{\boldsymbol{\Theta}}(\mathbf{r}_1,\mathbf{r}_2)\rightarrow\tilde{\boldsymbol{P}}\cdot\tilde{\boldsymbol{\Theta}}(\mathbf{r}_1,\mathbf{r}_2)+\tilde{\boldsymbol{P}}\cdot\tilde{\boldsymbol{\Pi}}(\mathbf{r}_1,\mathbf{r}_2)\cdot\tilde{\boldsymbol{P}}, \tag{D5a}$$

with the following components of the tensor $\tilde{\boldsymbol{\Pi}}$:

$$\tilde{\Pi}_{ab}(\mathbf{r}_1,\mathbf{r}_2)=\int d^3r_3\tilde{\Omega}_a(\mathbf{r}_1,\mathbf{r}_3)\left[\frac{\exp(-k_{\rm TF}|\mathbf{r}_1-\mathbf{r}_3|)}{2|\mathbf{r}_1-\mathbf{r}_3|}-\frac{\exp(-k_{\rm TF}|\mathbf{r}_2-\mathbf{r}_3|)}{2|\mathbf{r}_2-\mathbf{r}_3|}\right]\tilde{\Omega}_b(\mathbf{r}_3,\mathbf{r}_2), \tag{D5b}$$

and to corresponding, obvious modifications in Eq. (C13), which include also the gradient of Eq. (D5a).

[1] R. G. Parr and W. Yang, *Density-Functional Theory of Atoms and Molecules* (Oxford University Press, New York, 1989).
[2] S. T. Howard, J. P. Huke, P. R. Mallinson, and C. S. Frampton, Phys. Rev. B **49**, 7124 (1994).
[3] L. M. Pecora, Phys. Rev. B **33**, 5987 (1986).
[4] R. McWeeny, Rev. Mod. Phys. **32**, 335 (1960).
[5] M. Levy and J. A. Goldstein, Phys. Rev. B **35**, 7887 (1987).
[6] A. Holas, N. H. March, Y. Takahashi, and C. Zhang, Phys. Rev. A **48**, 2708 (1993).
[7] K. A. Dawson and N. H. March, J. Chem. Phys. **81**, 5850 (1984).
[8] R. van Leeuwen, O. V. Gritsenko, and E. J. Baerends, in *Density Functional Theory I*, edited by R. F. Nalewajski, Topics in Current Chemistry Vol. 180 (Springer-Verlag, Berlin, 1996), p. 107.
[9] G. E. Engel, Y. Kwon, and R. M. Martin, Phys. Rev. B **51**, 13 538 (1995).
[10] E. V. Ludeña, J. Maldonado, R. López–Boada, T. Koga, and E. S. Kryachko, J. Chem. Phys. **102**, 318 (1995).
[11] A. Seidl, A. Görling, P. Vogl, J. A. Majewski, and M. Levy, Phys. Rev. B **53**, 3764 (1996).
[12] B. M. Bylander and L. Kleinman, Phys. Rev. B **41**, 7868 (1990).

PHYSICAL REVIEW A VOLUME 60, NUMBER 4 OCTOBER 1999

Field dependence of the energy of a molecule in a magnetic field

A. Holas
Institute of Physical Chemistry of the Polish Academy of Sciences, 44/52 Kasprzaka, 01-224 Warsaw, Poland

N. H. March
University of Oxford, Oxford, England
and Department of Physics (RUCA), University of Antwerp, 2020 Antwerp, Belgium

(Received 10 June 1999)

By combining the variational principle and the Hellmann-Feynman theorem, various forms of virial equations are obtained: for a many-electron system in arbitrary scalar and vector potential fields, for a neutral or charged molecule in an arbitrary vector potential field, for a molecule and an atom in a homogeneous magnetic field. These virial equations and their integrals represent exact results for the field-dependence of the ground-state and excited-state energies of the considered systems. Their validity is demonstrated not only for the exact eigenenergies, but also for a wide class of approximate energies, including the Hartree-Fock approximation and the intense-field limit of the statistical Thomas-Fermi model. [S1050-2947(99)07710-0]

PACS number(s): 31.10.+z, 33.15.−e, 95.30.Ft, 31.15.Pf

I. INTRODUCTION

A good deal of interest now centers on the area of atoms and molecules in intense external magnetic fields (MFs) [1–3]. Part of the motivation, at least, has come from the huge MFs that exist at the surfaces of neutron stars [4]. The problem of stability of negative molecular ions [5,6] has also generated a lot of interest, which has provided additional motivation for the present study embracing molecular geometries for both neutral and charged species.

In the present investigation, a major theme is to explore the knowledge of the field-dependence of the ground-state (GS) energy of a many-electron system (in particular, a molecule) immersed in an arbitrary MF. The principal tools employed below are (a) the variational principle and (b) the Hellmann-Feynman theorem. The outline of the paper is then as follows. In Sec. II below, the Hamiltonian used throughout is defined in terms of arbitrary potential fields, as well as its expectation value in terms of low-order density matrices (DMs). This is followed in turn by a discussion of our principal tools, which leads to the virial equation (VE). Section III specializes this result to a molecule with clamped nuclei in an arbitrary MF, while Sec. IV specializes further to a homogeneous MF of arbitrary strength B. In that context, some attention is given to "uniform scaling," i.e., scaling all bond lengths in the molecular species by the same factor R. This is followed by a derivation of the GS molecular energy as a function of R and B in terms of the first-order DM (1DM). An alternative form for this same energy is derived in terms of the pair density (the diagonal form of 2DM). As a special, and very significant case, the field-dependence of the molecular energy is obtained at a fixed molecular geometry. The simplifications to be achieved for a single atom in a homogeneous MF are then stressed in Sec. V. To this point, all results are exact, but inevitably formal because of the appearance of fully interacting 1DMs and 2DMs in the presence of B. Therefore, it is of importance that for some approximate solutions of the Schrödinger equation these formally exact relations continue to be valid, as discussed at some length in Sec. VI, where, in addition to the Hartree-Fock method [7,8], the Thomas-Fermi statistical model [9], generalized to intense MFs [10–13], is among the approximations treated. The validity of the results for the field-dependence of the GS energy is shown to hold also for the excited-state energies. Section VII constitutes a summary of the main achievements of the present study. In Appendix A, some attention is given to questions involving gauge invariance. In connection with the magnetic-Thomas-Fermi model, it is verified in Appendix B that the known expression [12] for the approximate GS energy of an atom (as a function of the electron number N, the atomic number Z, and the MF B), satisfies the VE and the related to it extrapolating expression.

In view of the importance nowadays of the density-functional theory (DFT) (see, e.g., Ref. [14]), we would like to comment on DFT aspects of VEs. Since all DFT calculations of the GS molecular energy in MF bear some error due to approximate (by necessity) exchange-correlation scalar and vector potentials, the discrepancy in satisfaction of VEs and other relationships derived here may help to determine the best approximants. Our previous work [15] shows that the *differential* VE (which leads to the present "global" VE by means of integration) can be used to reveal properties of the field-dependent exchange-correlation potential of the DFT.

II. ARBITRARY POTENTIAL FIELDS

A. The Hamiltonian and its expectation value in terms of density matrices

The nonrelativistic Hamiltonian for a system of N interacting electrons moving in the fields of the external scalar potential $v(\mathbf{r})$, and of the vector potential $\mathbf{A}(\mathbf{r})$, which defines the MF (the magnetic flux density) via

$$\mathbf{B}(\mathbf{r};[\mathbf{A}])=\boldsymbol{\nabla}(\mathbf{r})\times\mathbf{A}(\mathbf{r}), \tag{2.1}$$

is given by

$$\hat{H}[v,\mathbf{A}]=\sum_{j=1}^{N}\left\{\frac{1}{2m}(-i\hbar\boldsymbol{\nabla}(\mathbf{r}_j)+e\,\mathbf{A}(\mathbf{r}_j))^2+\frac{ge\hbar}{2m}\hat{\mathbf{s}}(\sigma_j)\cdot\mathbf{B}(\mathbf{r}_j;[\mathbf{A}])+v(\mathbf{r}_j)\right\}+\sum_{i<j}^{N}u(\mathbf{r}_i,\mathbf{r}_j). \tag{2.2}$$

Here $\boldsymbol{\nabla}(\mathbf{r})=\partial/\partial\mathbf{r}$, $\mathbf{x}_j\equiv\{\mathbf{r}_j,\sigma_j\}$ denotes the space-spin coordinate of the jth electron, $\{m,-e\}$ are its mass and charge, $\hat{\mathbf{s}}(\sigma)$ is the spin angular momentum operator (its z component having the eigenvalues $\pm\frac{1}{2}$), $g=2.0023$ is the electron Landé factor, and

$$u(\mathbf{r}_i,\mathbf{r}_j)=\frac{e^2}{4\pi\varepsilon_0}\frac{1}{|\mathbf{r}_i-\mathbf{r}_j|} \tag{2.3}$$

is the Coulombic electron-electron interaction potential. To simplify the notation, we suppress the dependence on fixed N in writing the arguments of $\hat{H}$, and, next, of all related quantities. The SI convention for units and electromagnetic relations is used.

The expectation value of $\hat{H}$ with respect to $\Psi(\mathbf{x}_1,\mathbf{x}_2,\ldots,\mathbf{x}_N)$ — an antisymmetric, normalized wave function $\langle\Psi|\Psi\rangle=1$ — can be written in terms of low-order DMs as

$$\langle\Psi|\hat{H}[v,\mathbf{A}]|\Psi\rangle=\tilde{T}_0[\rho_1]+\tilde{T}_1[\rho_1,\mathbf{A}]+\tilde{T}_2[n,\mathbf{A}^2]+\tilde{T}_{1'}[\gamma_1,\mathbf{A}]+\tilde{V}[n,v]+\tilde{E}_{ee}[n_2,u], \tag{2.4}$$

where four kinetic-energy terms $\tilde{T}_0$, $\tilde{T}_1$, $\tilde{T}_2$, and $\tilde{T}_{1'}$ due to the Hamiltonian terms — free of **A**, linear in **A**, quadratic in **A**, and spin-MF interaction, respectively — are functionals of the 1DM:

$$\tilde{T}_0[\rho_1]=\int d^3r\,t(\mathbf{r};[\rho_1])>0, \tag{2.5}$$

$$\tilde{T}_1[\rho_1,\mathbf{A}]=\int d^3r\,e\,\mathbf{j}_p(\mathbf{r};[\rho_1])\cdot\mathbf{A}(\mathbf{r}), \tag{2.6}$$

$$\tilde{T}_2[n,\mathbf{A}^2]=\int d^3r\,n(\mathbf{r})\frac{e^2}{2m}\mathbf{A}^2(\mathbf{r})>0, \tag{2.7}$$

$$\tilde{T}_{1'}[\gamma_1,\mathbf{A}]=-\int d^3r\,\mathbf{m}(\mathbf{r};[\gamma_1])\cdot\mathbf{B}(\mathbf{r},[\mathbf{A}])=-\int d^3r\,\mathbf{A}(\mathbf{r})\cdot\{\boldsymbol{\nabla}(\mathbf{r})\times\mathbf{m}(\mathbf{r};[\gamma_1])\}. \tag{2.8}$$

The terms $\tilde{V}$ and $\tilde{E}_{ee}$ represent the electron-(external-scalar-potential) interaction energy and the electron-electron interaction energy:

$$\tilde{V}[n,v]=\int d^3r\,n(\mathbf{r})v(\mathbf{r}), \tag{2.9}$$

$$\tilde{E}_{ee}[n_2,u]=\int d^3r_1\,d^3r_2\,n_2(\mathbf{r}_1,\mathbf{r}_2)u(\mathbf{r}_1,\mathbf{r}_2)>0. \tag{2.10}$$

Vanishing of the DMs and their derivatives at infinite distances has been used.

The functionals of the one-body DMs, involved in $\tilde{T}_0$, $\tilde{T}_1$, and $\tilde{T}_{1'}$, are defined

$$t(\mathbf{r};[\rho_1])=\frac{\hbar^2}{2m}\boldsymbol{\nabla}(\mathbf{r}')\cdot\boldsymbol{\nabla}(\mathbf{r}'')\rho_1(\mathbf{r}+\mathbf{r}';\mathbf{r}+\mathbf{r}'')|_{\mathbf{r}'=\mathbf{r}''=\mathbf{0}}\geqslant 0 \tag{2.11}$$

for the conventional kinetic energy density,

$$\mathbf{j}_p(\mathbf{r};[\rho_1])=-\frac{i\hbar}{2m}\{\boldsymbol{\nabla}(\mathbf{r}')-\boldsymbol{\nabla}(\mathbf{r}'')\}\rho_1(\mathbf{r}+\mathbf{r}';\mathbf{r}+\mathbf{r}'')|_{\mathbf{r}'=\mathbf{r}''=\mathbf{0}}=\frac{\hbar}{m}\boldsymbol{\nabla}(\mathbf{r}')\,\mathrm{Im}\,\rho_1(\mathbf{r}+\mathbf{r}';\mathbf{r})|_{\mathbf{r}'=\mathbf{0}} \tag{2.12}$$

for the paramagnetic-current density vector, and

$$\mathbf{m}(\mathbf{r};[\gamma_1])=-\frac{ge\hbar}{2m}\sum_{\sigma}\hat{\mathbf{s}}(\sigma)\gamma_1(\mathbf{r},\sigma;\mathbf{r},\sigma')|_{\sigma'=\sigma} \tag{2.13}$$

for the magnetization density vector. The DMs are defined as usual (see, e.g., Refs. [14] and [15]):

$$\gamma_2(\mathbf{x}_1,\mathbf{x}_2;\mathbf{x}_1',\mathbf{x}_2')=\frac{(N-1)N}{2}\int d\mathbf{x}_3\cdots d\mathbf{x}_N\Psi(\mathbf{x}_1,\mathbf{x}_2,\mathbf{x}_3,\ldots,\mathbf{x}_N)\times\Psi^*(\mathbf{x}_1',\mathbf{x}_2',\mathbf{x}_3,\ldots,\mathbf{x}_N), \tag{2.14}$$

$$n_2(\mathbf{r}_1,\mathbf{r}_2)=\sum_{\sigma_1,\sigma_2}\gamma_2(\mathbf{r}_1,\sigma_1,\mathbf{r}_2,\sigma_2;\mathbf{r}_1,\sigma_1,\mathbf{r}_2,\sigma_2)\geqslant 0, \tag{2.15}$$

$$\gamma_1(\mathbf{x}_1;\mathbf{x}_1')=\frac{2}{N-1}\int d\mathbf{x}_2\,\gamma_2(\mathbf{x}_1,\mathbf{x}_2;\mathbf{x}_1',\mathbf{x}_2), \tag{2.16}$$

$$\rho_1(\mathbf{r}_1;\mathbf{r}_1')=\sum_{\sigma_1}\gamma_1(\mathbf{r}_1,\sigma_1;\mathbf{r}_1',\sigma_1), \tag{2.17}$$

$$n(\mathbf{r}_1)=\rho_1(\mathbf{r}_1;\mathbf{r}_1)\geqslant 0, \tag{2.18}$$

the last being the electron-number density.

B. Variational principle

When $\Psi=\Psi(\mathbf{x}_1,\mathbf{x}_2,\ldots,\mathbf{x}_N;[v,\mathbf{A}])$ is the GS wave function of the Hamiltonian $\hat{H}$, Eq. (2.2),

$$\hat{H}[v,\mathbf{A}]\Psi[v,\mathbf{A}]=E[v,\mathbf{A}]\Psi[v,\mathbf{A}], \tag{2.19}$$

then Eq. (2.4) for $\langle\Psi|\hat{H}|\Psi\rangle$, with all DMs being functionals of external fields {e.g., $\rho_1=\rho_1[v,\mathbf{A}]$ via Eqs. (2.14)–(2.18)}, represents an expression for the GS energy of the system:

$$E[v,\mathbf{A}]=T_0[v,\mathbf{A}]+T_1[v,\mathbf{A}]+T_2[v,\mathbf{A}]+T_{1'}[v,\mathbf{A}]+V[v,\mathbf{A}]+E_{ee}[v,\mathbf{A}], \quad (2.20)$$

where $T_1[v,\mathbf{A}]=\tilde{T}_1[\rho_1[v,\mathbf{A}],\mathbf{A}]$ and similarly for other terms.

On the way to obtain the VE let us define the following scaled function:

$$\Psi_\xi(\mathbf{r}_1,\sigma_1,\dots,\mathbf{r}_N,\sigma_N)=\xi^{3N/2}\Psi(\xi\mathbf{r}_1,\sigma_1,\dots,\xi\mathbf{r}_N,\sigma_N;[v,\mathbf{A}]). \quad (2.21)$$

Since Ψ_ξ is a normalized wave function and for $\xi=1$ it coincides with the GS function of the original system, Eq. (2.19), it can be used as a trial function for determination of the GS energy of this system by means of the Rayleigh-Ritz variational principle:

$$E[v,\mathbf{A}]=\min_\xi\langle\Psi_\xi|\hat{H}[v,\mathbf{A}]|\Psi_\xi\rangle=\langle\cdots\rangle|_{\xi=1}. \quad (2.22)$$

Therefore, the necessary condition for this minimum must hold:

$$0=\frac{\partial}{\partial\xi}\langle\Psi_\xi|\hat{H}[v,\mathbf{A}]|\Psi_\xi\rangle|_{\xi=1}. \quad (2.23)$$

In terms of $\rho_{\xi,1},\dots,n_{\xi,2}$ — the DMs derived, according to Eqs. (2.14)–(2.18), from the wave function Ψ_ξ — Eq. (2.21), the last equation (2.23) is rewritten with the help of Eq. (2.4) as

$$0=\frac{\partial}{\partial\xi}\{\tilde{T}_0[\rho_{\xi,1}]+\tilde{T}_1[\rho_{\xi,1},\mathbf{A}]+\tilde{T}_2[n_\xi,\mathbf{A}^2]+\tilde{T}_{1'}[\gamma_{\xi,1},\mathbf{A}]+\tilde{V}[n_\xi,v]+\tilde{E}_{ee}[n_{\xi,2},u]\}_{\xi=1}. \quad (2.24)$$

Then, by taking explicit form of ξ-dependent DMs and their functionals, e.g., $\mathbf{j}_p(\mathbf{r};[\rho_{\xi,1}])=\xi^4\mathbf{j}_p(\xi\mathbf{r};[\rho_1[v,\mathbf{A}]])$, and, next, by renaming variables of integration $\xi\mathbf{r}\to\mathbf{r}$, we find

$$0=\frac{\partial}{\partial\xi}\{\xi^2\tilde{T}_0[\rho_1[v,\mathbf{A}]]+\xi\tilde{T}_1[\rho_1[v,\mathbf{A}],\mathbf{A}_{1/\xi}]+\tilde{T}_2[n[v,\mathbf{A}],\mathbf{A}^2_{1/\xi}]+\xi\tilde{T}_{1'}[\gamma_1[v,\mathbf{A}],\mathbf{A}_{1/\xi}]+\tilde{V}[n[v,\mathbf{A}],v_{1/\xi}]+\xi\tilde{E}_{ee}[n_2[v,\mathbf{A}],u]\}_{\xi=1}, \quad (2.25)$$

where the modified fields are defined by

$$v_\lambda(\mathbf{r})=v(\lambda\mathbf{r}),\quad \mathbf{A}_\lambda(\mathbf{r})=\mathbf{A}(\lambda\mathbf{r}). \quad (2.26)$$

The scaling $u(\xi^{-1}\mathbf{r}_1,\xi^{-1}\mathbf{r}_2)=\xi u(\mathbf{r}_1,\mathbf{r}_2)$ of the Coulombic potential, Eq. (2.3), was taken into account. After performing differentiations in Eq. (2.25) and changing the variable $\xi\to\lambda=1/\xi$, we obtain

$$\frac{\partial}{\partial\lambda}\{\tilde{T}_1[\rho_1[v,\mathbf{A}],\mathbf{A}_\lambda]+\tilde{T}_2[n[v,\mathbf{A}],\mathbf{A}^2_\lambda]+\tilde{T}_{1'}[\gamma_1[v,\mathbf{A}],\mathbf{A}_\lambda]+\tilde{V}[n[v,\mathbf{A}],v_\lambda]\}_{\lambda=1}=2T_0[v,\mathbf{A}]+T_1[v,\mathbf{A}]+T_{1'}[v,\mathbf{A}]+E_{ee}[v,\mathbf{A}]. \quad (2.27)$$

Using Eq. (2.4), we can rewrite the left-hand side of Eq. (2.27) as a derivative of the expectation value of a modified Hamiltonian,

$$\frac{\partial}{\partial\lambda}\langle\Psi[v,\mathbf{A}]|\hat{H}[v_\lambda,\mathbf{A}_\lambda]|\Psi[v,\mathbf{A}]\rangle|_{\lambda=1}=2T_0[v,\mathbf{A}]+T_1[v,\mathbf{A}]+T_{1'}[v,\mathbf{A}]+E_{ee}[v,\mathbf{A}]. \quad (2.28)$$

C. Hellmann-Feynman theorem

By virtue of the Hellmann-Feynman theorem, we have

$$\frac{\partial}{\partial\lambda}E[v_\lambda,\mathbf{A}_\lambda]|_{\lambda=1}=\langle\Psi[v_1,\mathbf{A}_1]|\frac{\partial}{\partial\lambda}\hat{H}[v_\lambda,\mathbf{A}_\lambda]_{\lambda=1}|\Psi[v_1,\mathbf{A}_1]\rangle. \quad (2.29)$$

So, using this identity and also Eq. (2.20) we obtain from Eq. (2.28) the following VE:

$$\frac{\partial}{\partial\lambda}E[v_\lambda,\mathbf{A}_\lambda]|_{\lambda=1}=E[v,\mathbf{A}]+T_0[v,\mathbf{A}]-T_2[v,\mathbf{A}]-V[v,\mathbf{A}]. \quad (2.30)$$

The VE (2.30) is gauge invariant, because this property holds obviously for the total energy E and the external scalar potential energy V, while for the combination T_0-T_2 it is proved in Appendix A.

The left-hand side of Eq. (2.30) — the change of the total energy due to the uniform scaling of the external fields, Eq. (2.26) — can be evaluated by means of functional differentiations, leading to another form of the VE:

$$\int d^3r\left\{\frac{\tilde{\delta}E[v,\mathbf{A}]}{\tilde{\delta}v(\mathbf{r})}v'(\mathbf{r})+\frac{\tilde{\delta}E[v,\mathbf{A}]}{\tilde{\delta}\mathbf{A}(\mathbf{r})}\cdot\mathbf{A}'(\mathbf{r})\right\}=E+T_0-T_2-V, \quad (2.31)$$

where $\tilde{\delta}$ indicates the *partial* functional differentiation, and v' and $\mathbf{A}'$ are defined by [see Eq. (2.26)]

$$f'(\mathbf{r})=\frac{\partial}{\partial\lambda}f(\lambda\mathbf{r})|_{\lambda=1}=\mathbf{r}\cdot\nabla(\mathbf{r})f(\mathbf{r})=r\frac{\partial f(r,\theta,\varphi)}{\partial r}, \quad (2.32)$$

with the last form written in the spherical coordinates.

To explore further the VE (2.31), we need to know the functional derivatives $\tilde{\delta}E/\tilde{\delta}v$ and $\tilde{\delta}E/\tilde{\delta}\mathbf{A}$. The way leading to explicit expressions for them will be demonstrated in detail in the example of the simpler case of δv. By using Eq. (2.4) and the Hellmann-Feynman theorem, we have

$$\int d^3r \frac{\delta E[v,\mathbf{A}]}{\delta v(\mathbf{r})}\delta v(\mathbf{r})$$
$$=\frac{d}{d\varepsilon}E[v+\varepsilon\,\delta v,\mathbf{A}]_{\varepsilon=0}$$
$$=\langle\Psi[v,\mathbf{A}]|\frac{d}{d\varepsilon}\hat{H}[v+\varepsilon\,\delta v,\mathbf{A}]_{\varepsilon=0}|\Psi[v,\mathbf{A}]\rangle$$
$$=\frac{d}{d\varepsilon}\tilde{V}[n,v+\varepsilon\,\delta v]_{\varepsilon=0}=\tilde{V}[n,\delta v] \tag{2.33}$$

[see, e.g., Eq. (A.2) of Ref. [14] for the definition of a functional derivative by means of first equality here], so

$$\frac{\delta E[v,\mathbf{A}]}{\delta v(\mathbf{r})}=n(\mathbf{r};[v,\mathbf{A}]). \tag{2.34}$$

Similarly,

$$\frac{d}{d\varepsilon}E[v,\mathbf{A}+\varepsilon\,\delta\mathbf{A}]_{\varepsilon=0}=\tilde{T}_1[\rho_1,\delta\mathbf{A}]+\tilde{T}_2[n,2\mathbf{A}\cdot\delta\mathbf{A}]$$
$$+\tilde{T}_{1'}[\gamma_1,\delta\mathbf{A}]. \tag{2.35}$$

Therefore

$$\frac{\delta E[v,\mathbf{A}]}{\delta\mathbf{A}(\mathbf{r})}=e\mathbf{j}_p(\mathbf{r};[\rho_1])+2n(\mathbf{r})\frac{e^2}{2m}\mathbf{A}(\mathbf{r})$$
$$-\{\boldsymbol{\nabla}(\mathbf{r})\times\mathbf{m}(\mathbf{r};[\gamma_1])\}. \tag{2.36}$$

After inserting the results (2.34), (2.36), and (2.32) into Eq. (2.31), and using again Eqs. (2.4)–(2.8), we find the VE in the following form:

$$2T_0+E_{ee}-\int d^3r\,n(\mathbf{r})\{\mathbf{r}\cdot\boldsymbol{\nabla}(\mathbf{r})\}\left\{v(\mathbf{r})+\frac{e^2}{2m}\mathbf{A}^2(\mathbf{r})\right\}$$
$$-\int d^3r\,e\,\mathbf{j}_p(\mathbf{r};[\rho_1])\cdot\{\mathbf{r}\cdot\boldsymbol{\nabla}(\mathbf{r})-1\}\mathbf{A}(\mathbf{r})$$
$$+\int d^3r\,\mathbf{m}(\mathbf{r};[\gamma_1])\cdot\{\mathbf{r}\cdot\boldsymbol{\nabla}(\mathbf{r})\}\mathbf{B}(\mathbf{r},[\mathbf{A}])=0. \tag{2.37}$$

The following identity was helpful for obtaining the **B**-dependent term:

$$\int d^3r\{(\mathbf{r}\cdot\boldsymbol{\nabla}(\mathbf{r}))\mathbf{A}(\mathbf{r})\}\cdot\{\boldsymbol{\nabla}(\mathbf{r})\times\mathbf{m}(\mathbf{r})\}$$
$$=\int d^3r\,\mathbf{m}(\mathbf{r})\cdot\{1+\mathbf{r}\cdot\boldsymbol{\nabla}(\mathbf{r})\}\mathbf{B}(\mathbf{r},[\mathbf{A}]). \tag{2.38}$$

If the **B**-dependent term is neglected, Eq. (2.37) happens to be the same as the VE (36) in Ref. [16], obtained by Erhard and Gross, and as Eq. (2.20) in Ref. [15], obtained by us in a different way. The present VE (2.37) extends the above-mentioned investigations by including the **B**-dependent term, which stems from the corresponding term (in its the most general form) in the Hamiltonian (2.2) considered here. It is worth noting that the **B**-dependent term in Eq. (2.37) vanishes for a homogeneous MF (**B** independent of **r**).

III. MOLECULE IN AN ARBITRARY MAGNETIC FIELD

We are going now to specify the external scalar potential field to be created solely by M nuclei held clamped at the positions $\mathbf{R}_l$ and having atomic numbers Z_l:

$$v(\mathbf{r})=v_{en}(\mathbf{r};\mathbf{Z},\mathbf{R})=-\frac{e^2}{4\pi\varepsilon_0}\sum_{l=1}^{M}\frac{Z_l}{|\mathbf{r}-\mathbf{R}_l|}, \tag{3.1}$$

where the arrays of the potential characteristics are denoted as

$$\mathbf{R}=\{\mathbf{R}_1,\mathbf{R}_2,\ldots,\mathbf{R}_M\}=(R_{1x},R_{1y},\ldots,R_{My},R_{Mz}), \tag{3.2}$$

$$\mathbf{Z}=\{Z_1,Z_2,\ldots,Z_M\}. \tag{3.3}$$

So, the N-electron system now represents a *molecule*, which is neutral if $N=\Sigma_l Z_l$, or ionized, if N is different.

From Eq. (3.1) it is easy to evaluate the derivative of the scaled potential [see Eq. (2.32)]:

$$v'_{en}(\mathbf{r};\mathbf{Z},\mathbf{R})=\frac{\partial}{\partial\lambda}\{\lambda^{-1}v_{en}(\mathbf{r};\mathbf{Z},\lambda^{-1}\mathbf{R})\}_{\lambda=1}$$
$$=-\left(1+\mathbf{R}\cdot\frac{\partial}{\partial\mathbf{R}}\right)v_{en}(\mathbf{r};\mathbf{Z},\mathbf{R}). \tag{3.4}$$

So, with the help of Eq. (2.34), the corresponding term in Eq. (2.31) in the case of a molecule can be written in the following form:

$$\int d^3r\frac{\delta E[v,\mathbf{A}]}{\delta v(\mathbf{r})}v'(\mathbf{r})=-E_{en}-\mathbf{R}\cdot\frac{\partial}{\partial\mathbf{R}}E[v_{en}(.;\mathbf{Z},\mathbf{R}),\mathbf{A}], \tag{3.5}$$

where the notation $f(\ )$ represents a function — the mapping $\mathbf{r}\rightarrow f(\mathbf{r})$ — and

$$E_{en}=\int d^3r\,n(\mathbf{r})\,v_{en}(\mathbf{r}) \tag{3.6}$$

denotes the electron-nuclei interaction energy (it replaces the former term V). The Hellmann-Feynman theorem was used in the last step to transform the term involving $\partial v_{en}/\partial\mathbf{R}$ into $\langle\partial\hat{H}/\partial\mathbf{R}\rangle=\partial E/\partial\mathbf{R}$. Thus Eq. (3.5) transforms Eq. (2.31) into the following VE for a molecule:

$$-\mathbf{R}\cdot\frac{\partial}{\partial\mathbf{R}}E+\int d^3r\left\{\frac{\delta E[v_{en},\mathbf{A}]}{\delta\mathbf{A}(\mathbf{r})}\cdot\mathbf{A}'(\mathbf{r})\right\}=E+T_0-T_2. \tag{3.7}$$

It is convenient to consider the total energy of a molecule, E_{mol}, as a sum of the nucleus-nucleus interaction energy

$$E_{\rm nn}(\mathbf{Z},\mathbf{R})=\frac{e^2}{4\pi\varepsilon_0}\sum_{l<k}^{M}\frac{Z_lZ_k}{|\mathbf{R}_l-\mathbf{R}_k|}, \tag{3.8}$$

and the N-electron system energy E

$$E_{\rm mol}=E_{\rm nn}+E=T+E_{\rm C}, \tag{3.9}$$

where

$$T=T_0+T_1+T_2+T_{1'} \tag{3.10a}$$

is the sum of all kinetic energy terms, and

$$E_{\rm C}=E_{\rm en}+E_{\rm ee}+E_{\rm nn} \tag{3.10b}$$

is the sum of all Coulombic (C) energy contributions. Due to the Coulombic character of $E_{\rm nn}$, the following identity holds

$$-\mathbf{R}\cdot\frac{\partial}{\partial\mathbf{R}}E_{\rm nn}(\mathbf{Z},\mathbf{R})=-\frac{\partial}{\partial\lambda}E_{\rm nn}(\mathbf{Z},\lambda\mathbf{R})|_{\lambda=1}=+E_{\rm nn}(\mathbf{Z},\mathbf{R}). \tag{3.11}$$

So, the sum of Eqs. (3.11) and (3.7) gives the following VE (note that $E_{\rm nn}$ is independent of $\mathbf{A}$):

$$-\mathbf{R}\cdot\frac{\partial}{\partial\mathbf{R}}E_{\rm mol}+\int d^3r\left\{\frac{\tilde{\delta}E_{\rm mol}[v_{\rm en},\mathbf{A}]}{\tilde{\delta}\mathbf{A}(\mathbf{r})}\cdot\mathbf{A}'(\mathbf{r})\right\}$$
$$=E_{\rm mol}+T_0-T_2, \tag{3.12}$$

with $\tilde{\delta}E_{\rm mol}/\tilde{\delta}\mathbf{A}\equiv\tilde{\delta}E/\tilde{\delta}\mathbf{A}$ given in Eq. (2.36). The limits $\mathbf{A}'\to\mathbf{0}$ and $T_1,T_2,T_{1'}\to 0$ reduce this equation to the traditional VE for a molecule in the absence of a MF [see, e.g., Eq. (1.6.11) of Ref. [14]].

IV. MOLECULE IN A HOMOGENEOUS MAGNETIC FIELD

A. Virial equations

Vector potentials for the homogeneous MF, $\mathbf{B}=$const., are well known. Let us choose the following:

$$\mathbf{A}(\mathbf{r})=\tfrac{1}{2}\mathbf{B}\times\mathbf{r}=(-\tfrac{1}{2}y,\tfrac{1}{2}x,0)B,\quad \mathbf{B}=(0,0,B), \tag{4.1}$$

where, without loss of generality, for the second form the z axis of the coordinate system is chosen along the direction of $\mathbf{B}$. The derivative, Eq. (2.32), of the scaled vector potential can be found immediately from Eq. (4.1) to be

$$\mathbf{A}'(\mathbf{r})=\frac{\partial}{\partial\lambda}\mathbf{A}(\lambda\mathbf{r})=\mathbf{A}(\mathbf{r}). \tag{4.2}$$

Therefore, with the help of Eq. (2.36), in the case of a homogeneous MF the corresponding term in Eq. (3.12) can be written in the following form [see Eqs. (2.6)–(2.8)]:

$$\int d^3r\left\{\frac{\tilde{\delta}E_{\rm mol}[v_{\rm en},\mathbf{A}]}{\tilde{\delta}\mathbf{A}(\mathbf{r})}\cdot\mathbf{A}'(\mathbf{r})\right\}=T_1+2T_2+T_{1'}. \tag{4.3}$$

By inserting v in the form (3.1) and $\mathbf{A}$ in the form (4.1) into Eq. (2.2) we obtain the Hamiltonian $\hat{H}$ depending explicitly on the arrays $\mathbf{Z},\mathbf{R}$ and the scalar B; therefore, the molecular energy and all its components, Eq. (3.9), depend on the same arguments:

$$E_{\rm mol}=E_{\rm mol}(\mathbf{Z},\mathbf{R},B)=T_0(\mathbf{Z},\mathbf{R},B)+\cdots, \tag{4.4}$$

while the Hellmann-Feynman theorem in connection with such a Hamiltonian gives immediately

$$B\frac{\partial}{\partial B}E_{\rm mol}(\mathbf{Z},\mathbf{R},B)=T_1+2T_2+T_{1'}. \tag{4.5}$$

Thus, due to Eq. (4.3) and (4.5), the VE (3.12) can be rewritten as

$$\left(-\mathbf{R}\cdot\frac{\partial}{\partial\mathbf{R}}+B\frac{\partial}{\partial B}-1\right)E_{\rm mol}=T_0-T_2. \tag{4.6}$$

But, from Eqs. (4.5) and (3.9) we have

$$\left(B\frac{\partial}{\partial B}-1\right)E_{\rm mol}=T_2-T_0-E_{\rm C}. \tag{4.7}$$

The sum of Eq. (4.6) and (4.7) leads to the VE in an alternative form

$$\left(-\mathbf{R}\cdot\frac{\partial}{\partial\mathbf{R}}+2B\frac{\partial}{\partial B}-2\right)E_{\rm mol}=-E_{\rm C}, \tag{4.8a}$$

and, next, using Eq. (3.9), to a form

$$\left(-\mathbf{R}\cdot\frac{\partial}{\partial\mathbf{R}}+2B\frac{\partial}{\partial B}\right)E_{\rm mol}=2T+E_{\rm C}. \tag{4.8b}$$

B. Uniform scaling of a molecule

Although the term $\mathbf{R}\cdot\partial E_{\rm mol}/\partial\mathbf{R}$ in the VEs (4.6) and (4.8) involves $3M$ coordinates of the nuclear skeleton of a molecule, Eq. (3.2), it can be, in fact, expressed as a derivative with respect to a single parameter [compare Eq. (2.32)]:

$$\mathbf{R}\cdot\frac{\partial}{\partial\mathbf{R}}E_{\rm mol}(\mathbf{Z},\mathbf{R},B)=\frac{\partial}{\partial\lambda}E_{\rm mol}(\mathbf{Z},\lambda\mathbf{R},B)|_{\lambda=1}. \tag{4.9}$$

This property suggests the formulation of a problem of the uniformly scaled molecule: we choose and fix some configuration of nuclear positions $\mathbf{R}^0$ [this includes also a particular orientation of a molecule with respect to $\mathbf{B}$, according to the convention in Eq. (4.1)] and, next, we consider a set of configurations $\mathbf{R}$ obtained from $\mathbf{R}^0$ by the uniform scaling with a dimensionless parameter R:

$$\mathbf{R}=R\,\mathbf{R}^0,\quad 0<R\leqslant\infty. \tag{4.10}$$

We will call R the size parameter of a molecule. It should be noted that each bond length is scaled in the same way, $|\mathbf{R}_l-\mathbf{R}_{l'}|=R|\mathbf{R}_l^0-\mathbf{R}_{l'}^0|$, while bond angles and orientation angles remain unchanged during scaling. We are going to find the field dependence of the GS energy $E_{\rm mol}$ of the molecule characterized by $\mathbf{Z}$, Eq. (3.3), and by $\mathbf{R}$, Eq. (3.2), in the form Eq. (4.10). The energy of a highly squeezed molecule (of a very small size R) is very large due to the $E_{\rm nn}\propto 1/R$ term in Eq. (3.9), while the term E tends to the finite energy of the "united atom" having $Z=\Sigma_l Z_l$. At infinite size parameter, the molecule dissociates into atoms. If the initial molecule was ionized, some of these atoms are ionized. Knowledge of the atomic energies together with the ionization potentials (electron affinities) of all atoms would be sufficient to determine the species which are the anions (cations), by looking for the minimum of the total energy of the system of atoms and ions.

From Eq. (4.9) we find

$$\mathbf{R}\cdot\frac{\partial}{\partial\mathbf{R}}E_{\rm mol}(\mathbf{Z},\mathbf{R},B)|_{\mathbf{R}=R\,\mathbf{R}^0}=R\frac{\partial}{\partial R}E_{\rm mol}(\mathbf{Z},R\mathbf{R}^0,B). \tag{4.11}$$

Since now we simplify the notation: $E_{\rm mol}(\mathbf{Z},R\mathbf{R}^0,B)\rightarrow E_{\rm mol}(R,B)$, and similarly for all components of $E_{\rm mol}$, Eq. (3.9), and for DMs, by suppressing the dependence on the fixed $\mathbf{Z}$ and $\mathbf{R}^0$. After inserting the result (4.11), the VEs (4.6) and (4.8) represent partial differential equations for $E_{\rm mol}$ as a function of two independent variables R,B:

$$\left(-R\frac{\partial}{\partial R}+B\frac{\partial}{\partial B}-1\right)E_{\rm mol}(R,B)=T_{0\bar{2}}(R,B), \tag{4.12}$$

with $T_{0\bar{2}}=T_0-T_2$, and

$$\left(-\frac{1}{2}R\frac{\partial}{\partial R}+B\frac{\partial}{\partial B}-1\right)E_{\rm mol}(R,B)=-\frac{1}{2}E_{\rm C}(R,B). \tag{4.13a}$$

$$\left(-R\frac{\partial}{\partial R}+2B\frac{\partial}{\partial B}\right)E_{\rm mol}(R,B)=2T(R,B)+E_{\rm C}(R,B). \tag{4.13b}$$

The difference between Eqs. (4.12) and (4.13a) leads to the expression for the derivative of the molecular energy with respect to the size parameter:

$$R\frac{\partial}{\partial R}E_{\rm mol}(R,B)=-\{E_{C}(R,B)+2T_{0\bar{2}}(R,B)\}, \tag{4.14}$$

a counterpart of Eq. (4.5). At the optimum size $R_{\rm opt}(B)$ for a given B, for which $E_{\rm mol}(R,B)$ reaches minimum, so the equation

$$\frac{\partial}{\partial R}E_{\rm mol}(R,B)|_{R=R_{\rm opt}(B)}=0 \tag{4.15}$$

holds, we see from Eq. (4.14) that the following relation between the Coulombic and kinetic energy contributions must obey

$$E_{\rm C}(R_{\rm opt}(B),B)+2\,T_{0\bar{2}}(R_{\rm opt}(B),B)=0, \tag{4.16}$$

valid at arbitrary B. It should be stressed that during minimization over R, Eq. (4.15), the remaining parameters of the molecule geometry, like ratios of bond lengths, bond angles, orientation angles, all are held fixed; therefore, they may not be at the optimum.

C. (R,B)-dependent molecular energy in terms of 1DM

With the help of Eqs. (2.5), (2.7), and (4.1), the right-hand side of Eq. (4.12) for the homogeneous MF can be written as

$$T_{0\bar{2}}(R,B)=T_0(R,B)-T_2(R,B)$$
$$=\int d^3r\left\{t(\mathbf{r};R,B)-\frac{e^2}{8m}n(\mathbf{r};R,B)(x^2+y^2)B^2\right\}, \tag{4.17}$$

i.e., in terms of the conventional kinetic-energy density $t(\mathbf{r})$ and the electron-number density $n(\mathbf{r})$, both derivable from the 1DM ρ_1, Eqs. (2.11) and (2.18). For small B, $T_{0\bar{2}}(R,B)$ is positive, see Eq. (2.11). It should be recalled here that the separate terms T_0 and T_2 at $B\neq 0$ have no physical meaning as depending on a gauge; however, the combination T_0-T_2 is gauge invariant (see Appendix A).

Let us suppose that the dependence of ρ_1 on R and B is known. So $T_{0\bar{2}}(R,B)$ is available and $E_{\rm mol}(R,B)$ can be found by solving Eq. (4.12). It will be convenient to transform this equation by subtracting from it the $B\rightarrow 0$ limiting equation

$$\left(-R\frac{\partial}{\partial R}+B\frac{\partial}{\partial B}-1\right)\delta E_{\rm mol}(R,B)=\delta T_{0\bar{2}}(R,B), \tag{4.18a}$$

where

$$\delta F(R,B)=F(R,B)-F(R,0). \tag{4.18b}$$

The unique solution $\delta E_{\rm mol}$ of Eq. (4.18a), passing through the initial condition line $\delta E_{\rm mol}^{\rm ini}(R,B_0)$ (assumed to be a known function of R for some fixed B_0) can be written in the form

$$\delta E_{\rm mol}(R,B)=(B/B_0)\,\delta E_{\rm mol}^{\rm ini}(RB/B_0,B_0)$$
$$+\int_{B_0/B}^{1}d\zeta\,\zeta^{-2}\delta T_{0\bar{2}}(R/\zeta,B\zeta). \tag{4.19}$$

It is easy to check directly that this solution satisfies the initial condition and Eq. (4.18a).

But, in fact, we would like to have for the initial condition the GS energy of a molecule in the absence of a MF. For small B_0, we can expand $E_{\rm mol}^{\rm ini}(R',B_0)$ in the Taylor series with respect to B_0, so the corresponding term of Eq. (4.19) is

$$\frac{B}{B_0}\,\delta E_{\rm mol}^{\rm ini}(RB/B_0,B_0)$$
$$=B\left\{\frac{\partial E_{\rm mol}^{\rm ini}(R',B')}{\partial B'}+\frac{1}{2}B_0\frac{\partial^2 E_{\rm mol}^{\rm ini}(R',B')}{(\partial B')^2}+\cdots\right\}_{B'=0,\;R'=RB/B_0}. \tag{4.20}$$

From Eq. (4.5), using Eqs. (4.1) and (2.6)–(2.8) we find

$$\frac{\partial E_{\rm mol}^{\rm ini}(R',B')}{\partial B'}=\int d^3r\left\{\frac{1}{2}(\mathbf{r}\times\mathbf{j}_{\rm p}(\mathbf{r};R',B'))_z+\frac{e^2}{4m}B'\,n(\mathbf{r};R',B')(x^2+y^2)-m_z(\mathbf{r};R',B')\right\}. \tag{4.21}$$

Let us note that at $B=0$ (and $\mathbf{A}=\mathbf{0}$ chosen), the Hamiltonian (2.2) is real, therefore its eigenfunction can be chosen real too, what leads to real 1DM ρ_1 and then to $\mathbf{j}_{\rm p}(\mathbf{r})=\mathbf{0}$, according to Eq. (2.12), second line. Thus only the third term of Eq. (4.21) enters Eq. (4.20) at $B'=0$. Finally, the limiting form of Eq. (4.19) for $B_0\to 0$ is

$$E_{\rm mol}(R,B)=E_{\rm mol}(R,0)-B\,\mathcal{M}_z+\int_0^1 d\zeta\,\zeta^{-2}\{T_{0\bar{2}}(R/\zeta,B\zeta)-T_{0\bar{2}}(R/\zeta,0)\}, \tag{4.22}$$

where, according to Eq. (2.13), the z component of the magnetization vector for a molecule dissociated into atoms at $B=0$ is

$$\mathcal{M}_z=\lim_{R'\to\infty}\int d^3r\,m_z(\mathbf{r};R',0)=-\frac{ge\hbar}{4m}\sum_{l=1}^{M}\{N_\uparrow(Z_l)-N_\downarrow(Z_l)\}. \tag{4.23}$$

Here $N_\sigma(Z_l)$, $\sigma=\uparrow,\downarrow$, is the number of electrons with the spin σ in the atom (ion) characterized by Z_l. The integral in Eq. (4.22) must be convergent at lower limit, because all remaining terms of this equation are finite.

D. (R,B)-dependent molecular energy in terms of pair density

The total Coulombic energy $E_C(R,B)$, involved in the VE (4.13), can be written with the help of Eqs. (3.10b), (3.6), (3.1), (2.10), and (3.8), as

$$E_C(R,B)=\frac{e^2}{4\pi\varepsilon_0}\left\{-\sum_{l=1}^{M}Z_l\int d^3r\frac{n(\mathbf{r};R,B)}{|\mathbf{r}-R\mathbf{R}_l^0|}+\int d^3r_1\,d^3r_2\frac{n_2(\mathbf{r}_1,\mathbf{r}_2;R,B)}{|\mathbf{r}_1-\mathbf{r}_2|}+\frac{1}{R}\sum_{l<k}^{M}\frac{Z_lZ_k}{|\mathbf{R}_l^0-\mathbf{R}_k^0|}\right\}, \tag{4.24}$$

i.e., in terms of the electron-pair density $n_2(\mathbf{r}_1,\mathbf{r}_2;R,B)$ and the electron-number density $n(\mathbf{r};R,B)$. The last can be obtained from the former [see Eqs. (2.15)–(2.18)]:

$$n(\mathbf{r};R,B)=\frac{2}{N-1}\int d^3r'\,n_2(\mathbf{r},\mathbf{r}';R,B). \tag{4.25}$$

Let us suppose that the dependence of n_2 on R and B is known, so the function $E_C(R,B)$ is available for solving Eq. (4.13a). By applying steps analogous to these in Sec. IV C, we have the partial differential equation

$$\left(-\frac{1}{2}R\frac{\partial}{\partial R}+B\frac{\partial}{\partial B}-1\right)\delta E_{\rm mol}(R,B)=-\frac{1}{2}\delta E_C(R,B) \tag{4.26}$$

and its solution

$$\delta E_{\rm mol}(R,B)=(B/B_0)\,\delta E_{\rm mol}^{\rm ini}(R(B/B_0)^{1/2},B_0)-(1/2)\int_{B_0/B}^{1}d\zeta\,\zeta^{-2}\,\delta E_C(R/\zeta^{1/2},B\zeta), \tag{4.27}$$

passing through the initial condition line $\delta E_{\rm mol}^{\rm ini}(R,B_0)$. The $B_0\to 0$ limit of this result is obtained in the same way as previously:

$$E_{\rm mol}(R,B)=E_{\rm mol}(R,0)-B\,\mathcal{M}_z-\tfrac{1}{2}\int_0^1 d\zeta\,\zeta^{-2}\{E_C(R/\zeta^{1/2},B\zeta)-E_C(R/\zeta^{1/2},0)\}, \tag{4.28}$$

where $\mathcal{M}_z$ is given in Eq. (4.23). The integral must be convergent because all remaining terms in the equation are finite.

By equating two solutions (4.22) and (4.28) we find the following relation, valid at arbitrary R and B:

$$\int_0^1 d\zeta\,\zeta^{-2}\{[E_C(R/\zeta^{1/2},B\zeta)-E_C(R/\zeta^{1/2},0)]+2[T_{0\bar{2}}(R/\zeta,B\zeta)-T_{0\bar{2}}(R/\zeta,0)]\}=0, \tag{4.29}$$

which generalizes the relation (4.16) valid at the optimum size parameter.

E. B-dependent molecular energy at fixed geometry

Equation (4.7) can be considered a differential equation for B dependence of $E_{\rm mol}$ when the geometry of a molecule, $\mathbf{R}=R\,\mathbf{R}^0$, is kept fixed. After subtracting from it the $B\to 0$ limiting equation we have

$$\left(B\frac{\partial}{\partial B}-1\right)\delta E_{\rm mol}(R,B)=\delta E_{\rm fd}(R,B), \tag{4.30}$$

where we denoted by $E_{\rm fd}$ the combination of energy terms determining the field dependence (fd):

$$E_{\rm fd}=T_2-T_0-E_C=-(T_{0\bar{2}}+E_C), \tag{4.31}$$

see Eqs. (4.17) and (4.24) for the explicit form. The solution of Eq. (4.30) passing through the initial condition $\delta E_{\rm mol}^{\rm ini}(R,B_0)$ is

$$\delta E_{\rm mol}(R,B)=(B/B_0)\,\delta E_{\rm mol}^{\rm ini}(R,B_0)+\int_{B_0/B}^{1}d\zeta\,\zeta^{-2}\delta E_{\rm fd}(R,B\zeta). \quad (4.32)$$

Using Eqs. (4.20) and (4.21), we find the $B_0\to 0$ limit of the solution

$$E_{\rm mol}(R,B)=E_{\rm mol}(R,0)-B\,\mathcal{M}_z(R)+\int_0^1 d\zeta\,\zeta^{-2}\{E_{\rm fd}(R,B\zeta)-E_{\rm fd}(R,0)\}, \quad (4.33)$$

where $\mathcal{M}_z(R)$ is the magnetization vector of a molecule at $B=0$:

$$\mathcal{M}_z(R)=-\frac{ge\hbar}{4m}[N_\uparrow(\mathbf{Z},R\mathbf{R}^0)-N_\downarrow(\mathbf{Z},R\mathbf{R}^0)]. \quad (4.34)$$

Here $N_\sigma(\mathbf{Z},R\mathbf{R}^0)$ is the number of electrons with the spin σ in the molecule characterized by $\mathbf{Z},R\mathbf{R}^0$ [compare Eq. (4.23)].

F. Extrapolating field dependence

The expressions for $E_{\rm mol}(R,B)$ obtained in the previous Secs. IV C, IV D, and IV E involve, in addition to the information at $B=0$, the functions $T_{0\bar{2}}$, $E_{\rm C}$, and $E_{\rm fd}$ from a wide range of arguments R' and B'. Such information may be difficult to find; therefore, it may prove useful to have expressions extrapolating to the field B the results known at the field B_0 by using the mentioned functions from a narrower range of arguments. Such expressions are easy to obtain from Eqs. (4.18a), (4.26), and (4.30) when δ is omitted both in the equations and in their solutions, and the substitution $E_{\rm mol}^{\rm ini}\to E_{\rm mol}$ is applied:

$$E_{\rm mol}(R,B)=(B/B_0)E_{\rm mol}(RB/B_0,B_0)+\int_{B_0/B}^{1}d\zeta\,\zeta^{-2}T_{0\bar{2}}(R/\zeta,B\zeta), \quad (4.35)$$

$$E_{\rm mol}(R,B)=(B/B_0)E_{\rm mol}(R(B/B_0)^{1/2},B_0)-(1/2)\int_{B_0/B}^{1}d\zeta\,\zeta^{-2}E_{\rm C}(R/\zeta^{1/2},B\zeta), \quad (4.36)$$

$$E_{\rm mol}(R,B)=(B/B_0)E_{\rm mol}(R,B_0)+\int_{B_0/B}^{1}d\zeta\,\zeta^{-2}E_{\rm fd}(R,B\zeta). \quad (4.37)$$

For the solution in terms of $T_{0\bar{2}}$ or $E_{\rm C}$, separately, the extrapolation starts from the molecular energy at the field B_0 and at the size $R_0=RB/B_0$ or $R_0=R(B/B_0)^{1/2}$, and involves functions with the arguments in the range $R\leqslant R'\leqslant R_0$ and $B_0\leqslant B'\leqslant B$, while for the solution in terms of $E_{\rm fd}$, there is $R=R'=R_0$.

V. ATOM IN A HOMOGENEOUS MAGNETIC FIELD

For an atom, the electron-nuclei potential $v_{\rm en}$, Eq. (3.1), reduces to one term ($M=1$, $Z_1=Z$):

$$v_{\rm en}(\mathbf{r};Z)=-\frac{e^2}{4\pi\varepsilon_0}\frac{Z}{|\mathbf{r}|}. \quad (5.1)$$

We have chosen for its position vector $\mathbf{R}_1=\mathbf{0}$, because in a homogeneous MF the atomic energy can not depend on the position of an atom. It depends, in fact, on three scalar parameters only, $E=E(N,Z,B)$. The dependence on Z enters the expectation value (2.4) of the atomic Hamiltonian via the term connected with $v_{\rm en}$, Eq. (5.1),

$$\tilde{V}[n,v]=\tilde{E}_{\rm en}[n,Z]=\int d^3r\,n(\mathbf{r})\,v_{\rm en}(\mathbf{r};Z). \quad (5.2)$$

So, by applying the Hellmann-Feynman identity we can find immediately for the Z dependence

$$Z\frac{\partial}{\partial Z}E(N,Z,B)=Z\frac{\partial}{\partial Z'}\tilde{E}_{\rm en}[n(.;N,Z,B),Z']|_{Z'=Z}=E_{\rm en}(N,Z,B). \quad (5.3)$$

When adapting molecular results of the preceding section to the atomic case, we replace $E_{\rm mol}$ by E, because $E_{\rm nn}$ is absent, Eqs. (3.8), (3.9), and we remove any reference to R. We also simplify the notation: $E(N,Z,B)\to E(B)$, and similarly for all components of E, by suppressing the dependence on N and Z, which are kept fixed for the relations written below. Thus from Eq. (4.5) we have

$$B\frac{\partial}{\partial B}E(B)=T_1+2T_2+T_{1'} \quad (5.4)$$

for the derivative with respect to B, and from Eqs. (4.12) and (4.13)

$$\left(B\frac{\partial}{\partial B}-1\right)E(B)=T_{0\bar{2}}(B), \quad (5.5)$$

$$\left(B\frac{\partial}{\partial B}-1\right)E(B)=-\frac{1}{2}E_{\rm C}(B), \quad (5.6a)$$

$$2B\frac{\partial}{\partial B}E(B)=2T(B)+E_{\rm C}(B) \quad (5.6b)$$

for the VEs. The Coulombic energy for an atom is limited to $E_{\rm C}=E_{\rm ee}+E_{\rm en}$. The difference of Eqs. (5.5) and (5.6a) gives the relation

$$E_{\rm C}(B)+2\,T_{0\bar{2}}(B)=0, \quad (5.7)$$

valid for arbitrary B [the analogous relation for a molecule, Eq. (4.16), holds at the equilibrium size]. The $B\to 0$ limit of Eq. (5.7) or (5.6b) is equivalent to the well known relation for an atom in the absence of a MF (see, e.g., Ref. [14])

$$\langle\hat{V}_{\rm en}+\hat{V}_{\rm ee}\rangle+2\langle\hat{T}\rangle=0. \quad (5.8)$$

The solutions (4.22) and (4.28) of VEs for molecules, adapted now for atoms, are

$$E(B)=E(0)-B\,\mathcal{M}_z+\int_0^1 d\zeta\,\zeta^{-2}\{T_{0\bar{2}}(B\zeta)-T_{0\bar{2}}(0)\},\tag{5.9}$$

$$E(B)=E(0)-B\,\mathcal{M}_z-\tfrac{1}{2}\int_0^1 d\zeta\,\zeta^{-2}\{E_{\mathrm{C}}(B\zeta)-E_{\mathrm{C}}(0)\},\tag{5.10}$$

where $\mathcal{M}_z$ is given by Eq. (4.23) limited to one term only. Equivalence of the two forms of the atomic energy, Eqs. (5.9) and (5.10), is obvious from Eq. (5.7). The convergence of the integral can be proven from the VE (5.5): after inserting there the series for $E(B)$ and $T_{0\bar{2}}(B)$ expanded in powers of B, we find from the linear terms that $\{\partial T_{0\bar{2}}(B)/\partial B\}_{B=0}=0$, so for $\zeta\rightarrow 0$ the integrand tends to the finite value $\frac{1}{2}B^2\{\partial^2 T_{0\bar{2}}(B')/(\partial B')^2\}_{B'=0}$.

All three extrapolating expressions (4.35)–(4.37) for a molecule give one expression when adapted for an atom:

$$E(B)=(B/B_0)E(B_0)+\int_{B_0/B}^1 d\zeta\,\zeta^{-2}T_{0\bar{2}}(B\zeta).\tag{5.11}$$

VI. FIELD DEPENDENCE OF APPROXIMATE GROUND-STATE SOLUTIONS AND OF EXCITED-STATE SOLUTIONS

A. Variational solutions

All results, obtained in the previous sections, concern the exact GS energy and DMs, as corresponding to the GS solution of the Schrödinger equation (2.19). No approximations were involved during the derivations. It is interesting that these results hold also for some approximate solutions. We discuss now the solutions obtained from the variational principle

$$E[v,\mathbf{A}]\leq E^{\Omega}[v,\mathbf{A}]=\min_{\Psi\in\Omega}\langle\Psi|\hat{H}[v,\mathbf{A}]|\Psi\rangle =\langle\cdots\rangle|_{\Psi=\Psi^{\Omega}[v,\mathbf{A}]}.\tag{6.1}$$

Here we denoted by Ω a particular set of normalized N-electron trial wave functions. Used as the superscript, Ω indicates the approximate GS solution defined by Eq. (6.1). Thus $\{E^{\Omega}[v,\mathbf{A}],\Psi^{\Omega}[v,\mathbf{A}]\}$ of Eq. (6.1) will play the role analogous to $\{E[v,\mathbf{A}],\Psi[v,\mathbf{A}]\}$, the solution of Eq. (2.19). The most important example of Ω is a set of Slater determinants, leading, via Eq. (6.1), to the unrestricted Hartree-Fock results. A wider set may define each trial wave function to be a linear combination of a few Slater determinants, exhibiting some particular (e.g., spin) symmetry. There is no need to give further examples. The only requirement for Ω to be applicable for our purposes is that the relation

$$\Psi_{\xi}^{\Omega}\in\Omega\tag{6.2}$$

must be true. Here Ψ_{ξ}^{Ω} denotes the solution $\Psi^{\Omega}[v,\mathbf{A}]$ transformed by means of the scaling defined in Eq. (2.21). The property (6.2) is fulfilled by a general Hartree-Fock set and by a general multideterminantal set (however, not in the restricted cases for which any molecular orbital of a Slater determinant is represented as linear combination of a *finite* number of *fixed* basis functions with the coefficients playing the role of variational parameters). When Eq. (6.2) is true, Eqs. (2.22)–(2.28) hold for $\{E^{\Omega},\Psi^{\Omega}\}$.

The Hellmann-Feynman theorem was invoked many times in the preceding sections. It is easy to verify that it holds not only for the solution $\{E[v,\mathbf{A}],\Psi[v,\mathbf{A}]\}$ of the Schrödinger equation (2.19), but also for the variational solution $\{E^{\Omega}[v,\mathbf{A}],\Psi^{\Omega}[v,\mathbf{A}]\}$ of Eq. (6.1). To see this, let us consider the Hamiltonian $\hat{H}(\epsilon)$ which depends on a parameter ϵ, so the variational solution, corresponding to it, depends on ϵ, too:

$$E^{\Omega}(\epsilon)=\min_{\Psi\in\Omega}\langle\Psi|\hat{H}(\epsilon)|\Psi\rangle=\langle\cdots\rangle|_{\Psi=\Psi^{\Omega}(\epsilon)}.\tag{6.3}$$

Now, applying the rules of differentiation to this solution, we find

$$\begin{aligned}\frac{d}{d\epsilon}E^{\Omega}(\epsilon)&=\frac{d}{d\epsilon}\langle\Psi^{\Omega}(\epsilon)|\hat{H}(\epsilon)|\Psi^{\Omega}(\epsilon)\rangle\\ &=\frac{d}{d\epsilon'}\langle\Psi^{\Omega}(\epsilon)|\hat{H}(\epsilon')|\Psi^{\Omega}(\epsilon)\rangle|_{\epsilon'=\epsilon}\\ &\quad+\frac{d}{d\epsilon'}\langle\Psi^{\Omega}(\epsilon')|\hat{H}(\epsilon)|\Psi^{\Omega}(\epsilon')\rangle|_{\epsilon'=\epsilon}.\end{aligned}\tag{6.4}$$

But the second term of the last result vanishes. To prove this we use the function $\Psi^{\Omega}(\epsilon')\in\Omega$ as a trial function in the minimization (6.3): the minimum with respect to ϵ' is reached at $\epsilon'=\epsilon$. So, finally, Eq. (6.4) can be rewritten as

$$\frac{d}{d\epsilon}E^{\Omega}(\epsilon)=\langle\Psi^{\Omega}(\epsilon)|\frac{d\hat{H}(\epsilon)}{d\epsilon}|\Psi^{\Omega}(\epsilon)\rangle,\tag{6.5}$$

i.e., in the traditional form of the Hellmann-Feynman identity.

Thus all results obtained in Secs. II–V for the exact GS solution hold also for all those variational solutions $E^{\Omega}[v,\mathbf{A}]$, $\Psi^{\Omega}[v,\mathbf{A}]$, for which the set Ω of trial functions shows the property (6.2). In particular, the discrepancy in the satisfaction of the VE (5.7) for atoms and Eqs. (4.14) or (4.16) for molecules may be used for checking the numerical accuracy (convergence) of the Hartree-Fock calculations, similarly as the relation (5.8) is widely used for this purpose for atoms in the absence of a MF.

B. Excited-state solutions

One can define the set Ω of trial functions used in Eq. (6.1) as the subspace of the Hilbert space of normalized N-electron functions, which consists of all functions having particular symmetry properties (i.e., which transform as rows of a particular irreducible representation of the group of the Schrödinger equation for the Hamiltonian $\hat{H}$). If this symmetry is different from the symmetry of the GS, the solution E^{Ω} of Eq. (6.1) represents the *exact excited* energy, the lowest one of this symmetry. However, when the definition of Ω imposes some restrictions (model forms) on trial functions [keeping the relation (6.2) true] in addition to the symmetry requirements, the solution E^{Ω} *approximates* only this *excited* energy. As follows from Sec. VI A, all results obtained in Secs. II–V must hold also for these excited (exact or approximate) states.

It should be noted that in terms of the above-mentioned symmetry-restricted variational problem, the GS energy is also the lowest energy of some particular symmetry, but, in addition, it represents the absolute minimum when all possible symmetries are considered. It may happen for some systems, however, that the symmetry of the GS may be different in various regions of the (R,B) plane. In such a case the results (4.22), (4.28), and (4.33) and also Eqs. (4.35)–(4.37), should be used with caution. As obtained by integration of partial differential equations, they are correct under assumption that $T_{0\bar{2}}$, $E_{\rm C}$, and $E_{\rm fd}$, respectively, are continuous functions of R',B'. But these functions may show discontinuity at the border between regions of different symmetries for the GS solution. To satisfy the mentioned assumption, one should take these functions selected by the criterion of the symmetry of the solution (in the whole range of integration over R',B') rather than by the GS criterion.

Inspired by the very recent Görling paper [17], we are going to argue that our results for the field dependence hold also for *arbitrary excited-state* exact energies. A convenient tool (invoked by Görling) — the basic stationarity principle of quantum mechanics — will play for excited states the same role that the variational principle (2.22) played for the GS. According to this principle, the eigensolution $\{E^{\kappa}[v,\mathbf{A}],\Psi^{\kappa}[v,\mathbf{A}]\}$ of the Schrödinger equation

$$\hat{H}[v,\mathbf{A}]\Psi^{\kappa}[v,\mathbf{A}]=E^{\kappa}[v,\mathbf{A}]\Psi^{\kappa}[v,\mathbf{A}] \tag{6.6}$$

leads to the stationary expectation value of the Hamiltonian

$$\langle\Psi(\varepsilon)|\hat{H}|\Psi(\varepsilon)\rangle=E^{\kappa}+O(\varepsilon^2) \tag{6.7}$$

$$\text{for}\quad \Psi(\varepsilon)\in\Omega,\quad \Psi(\varepsilon)=\Psi^{\kappa}+\varepsilon\delta\Psi+O(\varepsilon^2)$$

(vanishing first variation). The set Ω is the same as that defined at the beginning of Sec. VI B. Thus the composite label κ includes the label of the irreducible representation of Ω and the "principal" quantum number which orders the eigenvalues of the given symmetry into an increasing sequence. It should be noted that the symmetry group of the Hamiltonian of a molecule in homogeneous MF remains the same for all $\{R,B\}$ from the region $0<R<\infty$, $0<B<\infty$.

Similarly as for the GS case, we define the scaled function Ψ^{κ}_{ξ} by applying Eq. (2.21) to Ψ^{κ}. According to Eq. (6.7), the analog of Eq. (2.23) is true

$$0=\frac{\partial}{\partial\xi}\langle\Psi^{\kappa}_{\xi}|\hat{H}[v,\mathbf{A}]|\Psi^{\kappa}_{\xi}\rangle|_{\xi=1}, \tag{6.8}$$

so all remaining equations of Sec. II B hold for the state κ.

In order to have the Hellmann-Feynman theorem

$$\frac{d}{d\epsilon}E^{\kappa}(\epsilon)=\langle\Psi^{\kappa}(\epsilon)|\frac{d\hat{H}(\epsilon)}{d\epsilon}|\Psi^{\kappa}(\epsilon)\rangle \tag{6.9}$$

(and, therefore, the results of Sec. II C) extended to the excited states, we need the eigensolution $\{E^{\kappa}(\epsilon'),\Psi^{\kappa}(\epsilon')\}$ of the Hamiltonian $\hat{H}(\epsilon')$ to be labeled by the same κ as the eigensolution of the close Hamiltonian $\hat{H}(\epsilon)$, i.e., for $|\epsilon'-\epsilon|\ll 1$, the eigenenergy $E^{\kappa}(\epsilon')$ should be close to $E^{\kappa}(\epsilon)$, but far from $E^{\kappa'}(\epsilon)$ for $\kappa'\neq\kappa$ (the separability property). Then the proof of Eq. (6.9) goes similarly as in Eq. (6.4), but now $d\langle\Psi^{\kappa}(\epsilon')|\hat{H}(\epsilon)|\Psi^{\kappa}(\epsilon')\rangle/d\epsilon'=0$ at $\epsilon'=\epsilon$ takes place due to the stationarity principle (6.7).

Thus all results concerning the VEs, obtained for the GS, hold also for excited states. However, in connection with the separability, these results are not available for such states κ at a particular point $\{R,B\}$, for which an accidental degeneracy occurs, $E^{\kappa}(R,B)=E^{\kappa'}(R,B)$.

C. Magnetic Thomas-Fermi solutions

The statistical model of an atom in the absence of a magnetic field, known as the Thomas-Fermi approximation (see, e.g., in Refs. [9,14]), has been extended to an intense magnetic field by Kadomtsev [10] (see also Mueller, Rau, and Spruch [11]). It will be called here the magnetic Thomas-Fermi (MTF) approximation. For a molecule characterized by the electron number N and the external potential $v_{\rm en}(\mathbf{r};\mathbf{Z},\mathbf{R})$ of its nuclear skeleton, Eq. (3.1), immersed in a homogeneous MF B, the GS energy $E^{\rm MTF}(N,\mathbf{Z},\mathbf{R},B)$ and the electron-number density $n^{\rm MTF}(\mathbf{r};N,\mathbf{Z},\mathbf{R},B)$ in the MTF approximation represent the solution of the variational problem

$$E^{\rm MTF}(N,\mathbf{Z},\mathbf{R},B)=\min_{n\to N}\tilde{E}^{\rm MTF}[n,\mathbf{Z},\mathbf{R},B]$$
$$=\tilde{E}^{\rm MTF}[n^{\rm MTF}(.;N,\mathbf{Z},\mathbf{R},B),\mathbf{Z},\mathbf{R},B], \tag{6.10a}$$

where the variational density $n(\mathbf{r})\geqslant 0$ is constrained by the requirement

$$\int d^3r\, n(\mathbf{r})=N. \tag{6.10b}$$

The MTF energy functional is a sum

$$\tilde{E}^{\rm MTF}[n,\mathbf{Z},\mathbf{R},B]=\tilde{E}_{\rm kin}[n,B]+\tilde{E}_{\rm en}[n,\mathbf{Z},\mathbf{R}]+\tilde{E}_{\rm es}[n] \tag{6.10c}$$

of the kinetic energy functional

$$\tilde{E}_{\rm kin}[n,B]=c_{\rm MTF}B^{-2}\int d^3r\, n^3(\mathbf{r}), \tag{6.10d}$$

where $c_{\rm MTF}=2\pi^4\hbar^4/(3me^2)$, which replaces the sum $(\tilde{T}_0+\tilde{T}_1+\tilde{T}_2+\tilde{T}_{1'})$ of the exact theory terms, Eq. (2.4); the electron-nuclei energy functional

$$\tilde{E}_{\rm en}[n,\mathbf{Z},\mathbf{R}]=\int d^3r\, n(\mathbf{r})\,v_{\rm en}(\mathbf{r};\mathbf{Z},\mathbf{R}), \tag{6.10e}$$

which is the same as in the exact theory; and the classical electrostatic energy of electrons

$$\tilde{E}_{\text{es}}[n]=\frac{e^2}{4\pi\varepsilon_0}\int d^3r_1\,d^3r_2\frac{n(\mathbf{r}_1)n(\mathbf{r}_2)}{2\,|\mathbf{r}_1-\mathbf{r}_2|}, \tag{6.10f}$$

which approximates the $\tilde{E}_{\text{ee}}$ term of the exact theory, Eq. (2.10) with (2.3). The solution of Eq. (6.10a) is known to exist [18] for $N\leqslant Z$, where $Z=\Sigma_l Z_l$. The MTF model is valid [10,11] for MFs B satisfying

$$Z^{4/3}\ll B/B_{\text{au}}\ll Z^3, \tag{6.10g}$$

where $B_{\text{au}}=\hbar/(ea_{\text{B}}^2)=2.3506\times10^5$ T is the atomic unit of the field (1 T$=10^4$ G).

In analogy with Eq. (3.9a) we define the total energy of a molecule in the MTF approximation as

$$\begin{aligned}E_{\text{mol}}^{\text{MTF}}(N,\mathbf{Z},\mathbf{R},B)&=E^{\text{MTF}}(N,\mathbf{Z},\mathbf{R},B)+E_{\text{nn}}(\mathbf{Z},\mathbf{R})\\&=E_{\text{kin}}(N,\mathbf{Z},\mathbf{R},B)+E_{\text{C}}^{\text{MTF}}(N,\mathbf{Z},\mathbf{R},B),\end{aligned} \tag{6.11}$$

where E_{nn} is given in Eq. (3.8) and the sum of all Coulombic terms is approximated now by

$$E_{\text{C}}\rightarrow E_{\text{C}}^{\text{MTF}}=E_{\text{en}}+E_{\text{es}}+E_{\text{nn}} \tag{6.12}$$

[compare Eq. (3.10b)]. Here $E_{\text{kin}}(N,\mathbf{Z},\mathbf{R},B)=\tilde{E}_{\text{kin}}[n^{\text{MTF}}(.;N,\mathbf{Z},\mathbf{R},B),B]$, and similarly for other terms. As in Sec. IV B, we are going to find the virial relations for a uniformly scaled molecule, Eq. (4.10), which is characterized by the size parameter R. Now we simplify the notation $E_{\text{mol}}^{\text{MTF}}(N,\mathbf{Z},R\mathbf{R}^0,B)\rightarrow E_{\text{mol}}^{\text{MTF}}(N,R,B)$, and similarly for all components of the total energy and for the density, by suppressing the dependence on the fixed $\mathbf{Z}$ and $\mathbf{R}^0$.

By applying the variational Hellmann-Feynman identity (6.5) to the variational MTF solution (6.10a) (here in terms of the density rather than the wave function) we find

$$\frac{\partial}{\partial B}E^{\text{MTF}}(N,R,B)=\frac{\partial}{\partial B'}\tilde{E}_{\text{kin}}[n^{\text{MTF}}(.;N,R,B),B']|_{B'=B}, \tag{6.13a}$$

$$\frac{\partial}{\partial R}E^{\text{MTF}}(N,R,B)=\frac{\partial}{\partial R'}\tilde{E}_{\text{en}}[n^{\text{MTF}}(.;N,R,B),R']|_{R'=R}. \tag{6.13b}$$

After performing differentiation of $\tilde{E}_{\text{kin}}$ given in the form (6.10d), and taking into account that E_{nn} is independent of B, we obtain for the field dependence of the total molecular energy (6.11) the following result:

$$B\frac{\partial}{\partial B}E_{\text{mol}}^{\text{MTF}}(N,R,B)=-2\,E_{\text{kin}}(N,R,B). \tag{6.14}$$

On the way to obtain the VE we define the following scaled density:

$$n_\xi(\mathbf{r})=\xi^3\,n^{\text{MTF}}(\xi\mathbf{r};N,R,B), \tag{6.15}$$

Since this density satisfies the constraint (6.10b) and for $\xi=1$ coincides with the solution of Eq. (6.10a), it can serve as a trial density for an alternative determination of the MTF GS energy (6.10a) by the following minimization:

$$E^{\text{MTF}}(N,R,B)=\min_\xi\tilde{E}^{\text{MTF}}[n_\xi,R,B]=\tilde{E}^{\text{MTF}}[n_{\xi=1},R,B]. \tag{6.16}$$

Therefore, the necessary condition

$$0=\frac{\partial}{\partial\xi}\tilde{E}^{\text{MTF}}[n_\xi,R,B]|_{\xi=1} \tag{6.17}$$

must hold. Using Eq. (6.15) and the definitions (6.10d)–(6.10f), we find the explicit ξ dependence of various terms in Eq. (6.16) with (6.10c):

$$\begin{aligned}\tilde{E}_{\text{kin}}[n_\xi,B]&=c_{\text{MTF}}B^{-2}\xi^6\int d^3r'[n^{\text{MTF}}(\mathbf{r}';R,B)]^3\\&=\xi^6\,\tilde{E}_{\text{kin}}[n^{\text{MTF}}(.;R,B),B],\end{aligned} \tag{6.18}$$

$$\begin{aligned}\tilde{E}_{\text{en}}[n_\xi,R]&=\xi\int d^3r'\,n^{\text{MTF}}(\mathbf{r}';R,B)\,v_{\text{en}}(\mathbf{r}';\mathbf{Z},\xi R\mathbf{R}^0)\\&=\xi\,\tilde{E}_{\text{en}}[n^{\text{MTF}}(.;R,B),\xi R],\end{aligned} \tag{6.19}$$

$$\tilde{E}_{\text{es}}[n_\xi]=\xi\,\tilde{E}_{\text{es}}[n^{\text{MTF}}(.;R,B)]. \tag{6.20}$$

So the differentiation in Eq. (6.17) leads to

$$0=6E_{\text{kin}}+E_{\text{en}}+E_{\text{es}}+\frac{\partial}{\partial\xi}\tilde{E}_{\text{en}}[n^{\text{MTF}}(.;R,B),\xi R]|_{\xi=1}. \tag{6.21}$$

Finally, using Eqs. (6.14), (6.13b), and (6.10c), the VE is obtained in the form

$$0=3E_{\text{kin}}(N,R,B)+\left(1+R\frac{\partial}{\partial R}-B\frac{\partial}{\partial B}\right)E^{\text{MTF}}(N,R,B), \tag{6.22}$$

or, using Eqs. (6.11) and (3.11), in the form

$$\left(-R\frac{\partial}{\partial R}+B\frac{\partial}{\partial B}-1\right)E_{\text{mol}}^{\text{MTF}}(N,R,B)=3E_{\text{kin}}(N,R,B). \tag{6.23}$$

We see that this VE for the MTF approximation is the same in form as for the exact GS solution if the replacement

$$T_{0\bar{z}}\rightarrow3E_{\text{kin}} \tag{6.24}$$

is performed in Eq. (4.12).

By subtracting Eq. (6.11) from Eq. (6.14) we have

$$\left(B\frac{\partial}{\partial B}-1\right)E_{\text{mol}}^{\text{MTF}}(N,R,B)=-3E_{\text{kin}}(N,R,B)-E_{\text{C}}^{\text{MTF}}(N,R,B). \tag{6.25}$$

Thus the combination of Eq. (6.23) and (6.25) leads to an alternative form of the VE

$$\left(-\frac{1}{2}R\frac{\partial}{\partial R}+B\frac{\partial}{\partial B}-1\right)E_{\mathrm{mol}}^{\mathrm{MTF}}(N,R,B)=-\frac{1}{2}E_{\mathrm{C}}^{\mathrm{MTF}}(N,R,B), \tag{6.26}$$

analogous to the VE (4.13) for the exact solution, and to the size dependence

$$R\frac{\partial}{\partial R}E_{\mathrm{mol}}^{\mathrm{MTF}}(N,R,B)=-\{E_{\mathrm{C}}^{\mathrm{MTF}}(N,R,B)+2[3E_{\mathrm{kin}}(N,R,B)]\}, \tag{6.27}$$

analogous to Eq. (4.14) with replacements (6.12) and (6.24).

It is interesting that in the MTF approximation the derivatives of E^{MTF} can be determined not only with respect to B and R, Eqs. (6.14) and (6.27), but also with respect to N. This is possible due to the fact that the dependence on N enters into Eq. (6.10a) only via the constraint (6.10b) and that it remains meaningful for any real $N>0$, not only for integers. With the help of a Lagrange multiplier μ^{MTF} the minimization (6.10a), constrained by Eq. (6.10b), leads to

$$\mu^{\mathrm{MTF}}=\mathrm{const}=\left.\frac{\delta\tilde{E}^{\mathrm{MTF}}[n,R,B]}{\delta n(\mathbf{r})}\right|_{n=n^{\mathrm{MTF}}}. \tag{6.28}$$

Next, differentiating Eq. (6.10a) and using the result (6.28) we find

$$\begin{aligned}\frac{\partial}{\partial N}E^{\mathrm{MTF}}(N,R,B)&=\int d^3r\left.\frac{\delta\tilde{E}^{\mathrm{MTF}}[n,R,B]}{\delta n(\mathbf{r})}\right|_{n=n^{\mathrm{MTF}}}\times\frac{\partial n^{\mathrm{MTF}}(\mathbf{r};N,R,B)}{\partial N}\\&=\mu^{\mathrm{MTF}}\frac{\partial}{\partial N}\int d^3r\,n^{\mathrm{MTF}}(\mathbf{r};N,R,B)=\mu^{\mathrm{MTF}}.\end{aligned} \tag{6.29}$$

Knowing the explicit form of the functional dependence of $\tilde{E}^{\mathrm{MTF}}$ on $n(\mathbf{r})$, Eqs. (6.10c)–(6.10f), we can find easily the functional derivative $\delta\tilde{E}^{\mathrm{MTF}}/\delta n(\mathbf{r})$. Now, using this result, we multiply both sides of Eq. (6.28) by $n^{\mathrm{MTF}}(\mathbf{r};N,R,B)$, integrate over $\mathbf{r}$, and substitute for μ^{MTF} the result (6.29), to obtain

$$N\frac{\partial}{\partial N}E^{\mathrm{MTF}}(N,R,B)=3E_{\mathrm{kin}}+E_{\mathrm{en}}+2E_{\mathrm{es}}. \tag{6.30}$$

Since the MTF approximation is valid for a sufficiently strong field only, Eq. (6.10g), we cannot obtain analogs of the results from Secs. IV C–IV E, because the $B\rightarrow0$ limit is involved there. However, the results obtained in Sec. IV F are valid for the MTF approximation, provided B_0 is in the range (6.10g) and the replacements (6.12) and (6.24) are applied in Eq. (4.35)–(4.37).

Similarly as for molecules, all results of Sec. V for atoms [except Eqs. (5.9) and (5.10)] are valid for the MTF approximation when the replacements (6.24) and the atomic analog of Eq. (6.12), $E_{\mathrm{C}}\rightarrow E_{\mathrm{C}}^{\mathrm{MTF}}=E_{\mathrm{en}}+E_{\mathrm{es}}$, are applied. In particular, for the analog of Eq. (5.7) we have

$$E_{\mathrm{en}}(N,Z,B)+E_{\mathrm{es}}(N,Z,B)+6E_{\mathrm{kin}}(N,Z,B)=0, \tag{6.31}$$

the VE indicated already by Mueller *et al.* [11]. In Appendix B we show that this VE and also the extrapolating expression (5.11) are satisfied by the known solution of Eq. (6.10) for an atom.

VII. SUMMARY

In conclusion, the main results are summed up in the following.

(i) The virial equation for a *many-electron* system in the fields of *arbitrary scalar* $v(\mathbf{r})$ and *vector* $\mathbf{A}(\mathbf{r})$ potentials is obtained in three equivalent forms (2.30), (2.31), and (2.37). The last one represents a generalization of the result of Erhard and Gross [16] by including the **B**-dependent term.

(ii) For a *molecule* (neutral or ionized) in the field of an *arbitrary vector* potential $\mathbf{A}(\mathbf{r})$ the obtained virial equation (3.12) [with Eq. (2.36)] generalizes the well known VE for a molecule in the absence of a MF.

(iii) For a *molecule* in a *homogeneous* MF, $\mathbf{B}=\mathrm{const}$, the result (4.5) gives the field derivative of the molecular energy, while Eqs. (4.6) and (4.8) provide three forms of the VE. After introducing a uniformly scaled (squeezed/stretched) molecule, characterized by the size parameter R, an expression for the derivative of the molecular energy $E_{\mathrm{mol}}(R,B)$ with respect to R is obtained in Eq. (4.14), while the two virial equations represent partial differential equations (4.12) and (4.13a) for this energy, having the solutions (4.22) and (4.28): the molecular energy for the given size R and the MF B is obtained in terms of the molecular energy and the magnetization vectors (of separate atoms), both in the absence of a MF, and a specific integral involving either the kinetic term (T_0-T_2), or the total Coulombic contribution E_{C}, integrated over sizes larger than R and fields smaller than B. It is worth noting that the T_1 term, involving the paramagnetic-current density vector (which is characteristic for the density-functional theory generalized to the MF case, see, e.g. Refs. [16,15]), and the $T_{1'}$ term, involving the magnetization density, both are absent in these solutions. When the geometry of a molecule for the given R is kept fixed, the field-depending energy is given by Eq. (4.33): it involves the molecular energy and magnetization vector, both in the absence of a MF, and the sum $(T_0-T_2+E_{\mathrm{C}})$, a function of R and B', integrated over B' smaller than B. Eqs. (4.35)–(4.37) provide an extrapolation of results known at the field B_0 to the field B.

(iv) Analogous results for an *atom* in a *homogeneous* MF are given in Eqs. (5.3)–(5.11).

It is interesting to see how the well known virial relation for an atom in the absence of a MF, Eq. (5.8), is modified to include B, Eq. (5.7), and next, when generalized for a molecule, it remains in the same form at the equilibrium size, Eq. (4.16), while at an arbitrary size it holds in an "averaged" sense, Eq. (4.29). It should be stressed that all results for the field-dependence concern not only the exact GS energy, but also exact excited-state energies, and a wide class of *approximate* (GS and some excited-state) energies, resulting from variational solutions, Eq. (6.1), among them, the Hartree-Fock approximation. Even the results for the field dependence of the magnetic Thomas-Fermi energy, Eq. (6.10), show the same form, provided the substitutions (6.12) and (6.24) are applied.

ACKNOWLEDGMENTS

One of us (N.H.M.) acknowledges partial financial support for this work from the ONR. Dr. P. Schmidt of that office is to be thanked for his continuing interest and constant encouragement. Professor Yu Lu and his colleagues are to be thanked by both of us for a stimulating environment at ICTP, Trieste, in which much of the present work was carried out, as well as for their hospitality in Trieste.

APPENDIX A: GAUGE INVARIANCE

The transformation of the vector potential $\mathbf{A}(\mathbf{r})$, defined by

$$\mathbf{A}(\mathbf{r})\rightarrow\mathbf{A}^{\text{new}}(\mathbf{r})=\mathbf{A}(\mathbf{r})-\nabla\Lambda(\mathbf{r}), \tag{A1}$$

where $\Lambda(\mathbf{r})$ is an arbitrary scalar function, does not change a physical quantity — the magnetic field $\mathbf{B}(\mathbf{r})$, Eq. (2.1). It is known that the eigenfunction Ψ^{new} of the Schrödinger equation (2.19) in which $\hat{H}$ is replaced by $\hat{H}^{\text{new}}$;

$$\hat{H}[v,\mathbf{A}]\rightarrow\hat{H}^{\text{new}}\equiv\hat{H}[v,\mathbf{A}^{\text{new}}], \tag{A2}$$

differs from the eigenfunction $\Psi=\Psi[v,\mathbf{A}]$ of the original Schrödinger equation by a Λ-dependent phase factor only

$$\Psi\rightarrow\Psi^{\text{new}}(\mathbf{x}_1,\ldots,\mathbf{x}_N)$$
$$=\exp\left(i\frac{e}{\hbar}[\Lambda(\mathbf{r}_1)+\cdots+\Lambda(\mathbf{r}_N)]\right)\Psi(\mathbf{x}_1,\ldots,\mathbf{x}_N), \tag{A3}$$

while the eigenenergy is independent of the unphysical gauge function $\Lambda(\mathbf{r})$

$$E^{\text{new}}\equiv\langle\Psi^{\text{new}}|\hat{H}^{\text{new}}|\Psi^{\text{new}}\rangle=\langle\Psi|\hat{H}|\Psi\rangle\equiv E. \tag{A4}$$

The transformed DMs are defined by Eqs. (2.14)–(2.18) applied to the transformed wave function Ψ^{new}. One sees immediately that the diagonal (in spatial indices) DMs are gauge invariant, $n^{\text{new}}=n$, $n_2^{\text{new}}=n_2$, $\mathbf{m}^{\text{new}}=\mathbf{m}[\gamma_1^{\text{new}}]=\mathbf{m}[\gamma_1]$, Eq. (2.13). Therefore, from Eqs. (2.8)–(2.10) we find

$$T_{1'}^{\text{new}}=T_{1'},\quad V^{\text{new}}=V,\quad E_{\text{ee}}^{\text{new}}=E_{\text{ee}}. \tag{A5}$$

The transformation of the 1DM, induced by Eq. (A3),

$$\rho_1^{\text{new}}(\mathbf{r}_1;\mathbf{r}_1')=\exp\left(i\frac{e}{\hbar}[\Lambda(\mathbf{r}_1)-\Lambda(\mathbf{r}_1')]\right)\rho_1(\mathbf{r}_1;\mathbf{r}_1') \tag{A6}$$

leads, according to Eqs. (2.11) and (2.12), to

$$t^{\text{new}}(\mathbf{r})\equiv t(\mathbf{r};[\rho_1^{\text{new}}])=t(\mathbf{r})+e\,\mathbf{j}_p(\mathbf{r})\cdot\nabla\Lambda(\mathbf{r})+\frac{e^2}{2m}n(\mathbf{r})\times[\nabla\Lambda(\mathbf{r})]^2, \tag{A7}$$

$$\mathbf{j}_{\text{p}}^{\text{new}}(\mathbf{r})\equiv\mathbf{j}_{\text{p}}(\mathbf{r};[\rho_1^{\text{new}}])=\mathbf{j}_{\text{p}}(\mathbf{r})+\frac{e}{m}n(\mathbf{r})\nabla\Lambda(\mathbf{r}). \tag{A8}$$

Using these results in Eqs. (2.5)–(2.7), one verifies immediately by simple algebra that

$$T_0^{\text{new}}+T_1^{\text{new}}+T_2^{\text{new}}\equiv\int d^3r\Big\{t^{\text{new}}(\mathbf{r})+e\,\mathbf{j}_{\text{p}}^{\text{new}}(\mathbf{r})\cdot[\mathbf{A}(\mathbf{r})-\nabla\Lambda(\mathbf{r})]+\frac{e^2}{2m}n(\mathbf{r})[\mathbf{A}(\mathbf{r})-\nabla\Lambda(\mathbf{r})]^2\Big\}$$
$$=\int d^3r\Big\{t(\mathbf{r})+e\,\mathbf{j}_{\text{p}}(\mathbf{r})\cdot\mathbf{A}(\mathbf{r})+\frac{e^2}{2m}n(\mathbf{r})\mathbf{A}^2(\mathbf{r})\Big\}\equiv T_0+T_1+T_2. \tag{A9}$$

Thus combining Eqs. (A5) and (A9) we check that the total energy, Eq. (2.20), is gauge invariant, $E^{\text{new}}=E$, as announced in Eq. (A4).

Let us find now the gauge transformation of the kinetic energy combination entering the VE

$$T_0^{\text{new}}-T_2^{\text{new}}\equiv\int d^3r\Big\{t^{\text{new}}(\mathbf{r})-\frac{e^2}{2m}n(\mathbf{r})[\mathbf{A}(\mathbf{r})-\nabla\Lambda(\mathbf{r})]^2\Big\}$$
$$=T_0-T_2+\int d^3r\,e\Big\{\mathbf{j}_{\text{p}}(\mathbf{r})+\frac{e}{m}n(\mathbf{r})\mathbf{A}(\mathbf{r})\Big\}\cdot\nabla\Lambda(\mathbf{r})$$
$$=T_0-T_2-\int d^3r\,\Lambda(\mathbf{r})e\,\nabla\cdot\Big\{\mathbf{j}_{\text{p}}(\mathbf{r})+\frac{e}{m}n(\mathbf{r})\mathbf{A}(\mathbf{r})\Big\}. \tag{A10}$$

(Vanishing of the DMs and their derivatives at infinite distances has been used in the last step.) The vector represented by the combination in the curly braces — the physical current $\mathbf{j}(\mathbf{r})$ — is known to satisfy the continuity equation (for stationary states)

$$\nabla\cdot\mathbf{j}(\mathbf{r})\equiv\nabla\cdot\Big\{\mathbf{j}_{\text{p}}(\mathbf{r})+\frac{e}{m}n(\mathbf{r})\mathbf{A}(\mathbf{r})\Big\}=0. \tag{A11}$$

Thus the last integral in Eq. (A10) vanishes, so (T_0-T_2) is gauge invariant.

APPENDIX B: EXAMPLE OF THE MTF ATOM

As shown by Banerjee, Constantinescu, and Rehák [12], the GS energy of an *atom* in the MTF approximation, the solution of Eq. (6.10), can be written as

$$E^{\text{MTF}}(N,Z,B)=Z^{9/5}B^{2/5}\epsilon(N/Z), \tag{B1}$$

where $\epsilon(\nu)$ is a universal function defined for $0<\nu\leqslant1$, available numerically [the authors provide results for $\nu=1,(1-1/Z),(1-2/Z)$ at $Z=5,6,\ldots,100$]. For small ν this function was shown by March and Tomishima [13] to have an expansion

$$\epsilon(\nu)=\text{const}\;\nu^{3/5}[1+O(\nu)]. \tag{B2}$$

We are going to verify now that this solution (B1) satisfies the VE (6.31) and the extrapolating expression (5.11). It will be convenient to introduce an auxiliary function

$$\varphi(\nu)=\frac{d\ln(\epsilon(\nu))}{d\ln(\nu)},\quad \text{so}\ \varphi(\nu)=\frac{3}{5}+O(\nu)\ \text{ for }\ \nu\ll 1, \tag{B3}$$

in terms of which the derivatives of E^{MTF}, Eq. (B1), can be written

$$N\frac{\partial}{\partial N}E^{\mathrm{MTF}}(N,Z,B)=\varphi(N/Z)E^{\mathrm{MTF}}(N,Z,B), \tag{B4}$$

$$Z\frac{\partial}{\partial Z}E^{\mathrm{MTF}}(N,Z,B)=[(9/5)-\varphi(N/Z)]E^{\mathrm{MTF}}(N,Z,B), \tag{B5}$$

$$B\frac{\partial}{\partial B}E^{\mathrm{MTF}}(N,Z,B)=(2/5)E^{\mathrm{MTF}}(N,Z,B). \tag{B6}$$

But the same derivatives are known to be linear combinations of E_{kin}, E_{en}, and E_{es}, Eqs. (6.30), (5.3), and (6.14). The solution of such a system of 3 linear equations can be found immediately to be

$$E_{\mathrm{kin}}=-\tfrac{1}{5}E^{\mathrm{MTF}}, \tag{B7}$$

$$E_{\mathrm{en}}=(\tfrac{9}{5}-\varphi)E^{\mathrm{MTF}}, \tag{B8}$$

$$E_{\mathrm{es}}=(\varphi-\tfrac{3}{5})E^{\mathrm{MTF}}. \tag{B9}$$

After substituting these values into Eq. (6.31) we see that this VE is satisfied identically. No explicit form of the function $\epsilon(\nu)$ was involved, therefore the VE (6.31) is not helpful in checking the numerical accuracy of the solution (B1).

In order to verify that the solution (B1) satisfies the extrapolating expression (5.11) for $E(B)\rightarrow E^{\mathrm{MTF}}(N,Z,B)$, we need to perform the replacement (6.24) using Eq. (B7):

$$\begin{aligned}E^{\mathrm{MTF}}(N,Z,B)&=(B/B_0)E^{\mathrm{MTF}}(N,Z,B_0)+\int_{B_0/B}^{1}d\zeta\,\zeta^{-2}(-3/5)E^{\mathrm{MTF}}(N,Z,B\zeta)\\&=Z^{9/5}\epsilon(N/Z)\left\{B\,B_0^{-3/5}+B^{2/5}\int_{B_0/B}^{1}d\zeta(-3/5)\zeta^{-8/5}\right\}\\&=Z^{9/5}\epsilon(N/Z)\{B\,B_0^{-3/5}+B^{2/5}(1-(B_0/B)^{-3/5})\}\\&=Z^{9/5}\epsilon(N/Z)B^{2/5}.\end{aligned} \tag{B10}$$

So, really, the solution (B1) is recovered by means of the extrapolation.

[1] *Atoms and Molecules in Intense Fields*, edited by L. S. Cederbaum, K. C. Kulander, and N. H. March, Vol. 86 of Structure and Bonding Series (Springer-Verlag, Berlin, 1997).

[2] H. Ruder, G. Wunner, H. Herold, and F. Geyer, *Atoms in Strong Magnetic Fields* (Springer-Verlag, Berlin, 1994).

[3] *Atoms in Strong Fields*, edited by M. Gavrila (Academic, Boston, 1992).

[4] G. Chanmugam, Annu. Rev. Astron. Astrophys. **30**, 143 (1992).

[5] G. R. Freeman and N. H. March, J. Phys. Chem. **100**, 4331 (1996).

[6] M. K. Scheller, R. N. Compton, and L. S. Cederbaum, Science **270**, 1160 (1995).

[7] M. D. Jones, G. Ortiz, and D. M. Ceperley, Phys. Rev. A **54**, 219 (1996).

[8] M. Demeur, P.-H. Heenen, and M. Godefroid, Phys. Rev. A **49**, 176 (1994).

[9] N. H. March, *Electron Density Theory of Atoms and Molecules* (Academic, London, 1992).

[10] B. B. Kadomtsev, Zh. Éksp. Teor. Fiz. **58**, 1765 (1970) [Sov. Phys. JETP **31**, 945 (1970)].

[11] R. O. Mueller, A. R. P. Rau, and L. Spruch, Phys. Rev. Lett. **26**, 1136 (1971).

[12] B. Banerjee, D. H. Constantinescu, and P. Rehák, Phys. Rev. D **10**, 2384 (1974).

[13] N. H. March and Y. Tomishima, Phys. Rev. D **19**, 449 (1979).

[14] R. G. Parr and W. Yang, *Density-Functional Theory of Atoms and Molecules* (Oxford University Press, Oxford, 1989).

[15] A. Holas and N. H. March, Phys. Rev. A **56**, 4595 (1997).

[16] S. Erhard and E. K. U. Gross, Phys. Rev. A **53**, R5 (1996).

[17] A. Görling, Phys. Rev. A **59**, 3359 (1999).

[18] E. H. Lieb, J. P. Solovej, and J. Yngvason, Commun. Math. Phys. **161**, 77 (1994).

PHYSICAL REVIEW A, VOLUME 62, 062512

Nonrelativistic exchange-energy density and exchange potential in the lowest order of the 1/Z expansion for ten-electron atomic ions

I. A. Howard,[1] N. H. March,[2] P. Senet,[1] and V. E. Van Doren[1]
[1]Department of Physics, University of Antwerp (RUCA), Groenenborgerlaan 171, B-2020 Antwerp, Belgium
[2]Oxford University, Oxford, England

(Received 23 June 2000; published 15 November 2000)

The Dirac exchange energy E_x has a density $\varepsilon_x(r)$ in a spherically symmetric ten-electron atomic ion that is determined by the idempotent first-order density matrix (1 DM). In turn, this 1 DM has as its leading term in the $1/Z$ expansion, with Z the atomic number, the bare Coulomb result. This latter quantity is known analytically from the study of March and Santamaria [Phys. Rev. A **38**, 5002 (1988)]. Here $\varepsilon_x(r)$ is calculated analytically, and presented graphically for $Z=92$ in this large-Z limit. The Slater approximation to the exchange potential $V_x(\mathbf{r})$, namely, $V_{\rm Sla}(\mathbf{r})=2\varepsilon_x(r)/\rho(r)$ with $\rho(r)$ the ground-state electron density, is also plotted for $Z=92$. In the large-Z limit, $V_x(\mathbf{r})$ can be obtained by functional differentiation of the resultant exchange energy, and is expressed in terms of electron and kinetic-energy densities. Numerical calculations, based on $V_{\rm Sla}(\mathbf{r})$ plus approximate corrections, are also presented.

PACS number(s): 31.15.Ew

I. BACKGROUND AND OUTLINE

Though density-functional theory (DFT) is now a widely used technique in the many-electron problem encountered in atoms, molecules, and clusters, important problems remain in constructing the exchange-correlation potential $V_{\rm xc}(\mathbf{r})$ (see, for example, Ref. [1]). Therefore, Holas and March [2] gave a complete theory of $V_{\rm xc}(\mathbf{r})$ based on low-order density matrices. This has moved attention to the construction of such density matrices, though of course it is recognized that eventually, to complete the DFT approach, one will require these low-order density matrices as functionals of the "diagonal" ground-state density $\rho(\mathbf{r})$. In the present paper, attention will be focused entirely on the first-order density matrix (1 DM) $\gamma(\mathbf{r},\mathbf{r}')$. This is the natural starting point for the calculation of kinetic-energy density, $t(\mathbf{r})$, for instance, since the total kinetic energy T can be written as

$$T=(\hbar^2/2m)\int \nabla_r\cdot\nabla_{r'}\gamma(\mathbf{r},\mathbf{r}')|_{r'=r}d\mathbf{r}. \quad (1.1)$$

Though $t(\mathbf{r})$ is not defined uniquely by the result (1.1), one acceptable definition is plainly

$$t(\mathbf{r})=(\hbar^2/2m)\nabla_r\cdot\nabla_{r'}\gamma(\mathbf{r},\mathbf{r}')|_{r'=r}. \quad (1.2)$$

What is important for the present paper is that, in the framework provided in the early work of Dirac [3], the exchange energy E_x is also determined by the (now idempotent) first-order density matrix $\gamma(\mathbf{r},\mathbf{r}')$. Thus, in a spin-compensated system on which all attention will be focused below, the idempotency condition reads [for the diagonal density $\rho(\mathbf{r})$]

$$\rho(\mathbf{r})/2=\int \{\gamma(\mathbf{r},\mathbf{r}')/2\}^2 d\mathbf{r}', \quad (1.3)$$

Naturally, however, this idempotency condition precludes inclusion of electron correlation, for which the equality (1.3) has to be replaced by a (much weaker) condition in the form of an inequality.

Even at the idempotent level, however, the Dirac exchange energy [3]

$$E_x=-(e^2/4)\int \frac{\gamma^2(\mathbf{r},\mathbf{r}')}{|\mathbf{r}-\mathbf{r}'|}d\mathbf{r}d\mathbf{r}' \quad (1.4)$$

is not easy to turn into DFT form for the reason already stated above, namely, that one needs the first-order idempotent density matrix $\gamma(\mathbf{r},\mathbf{r}')$ as a functional of the diagonal ground-state density $\rho(\mathbf{r})$. Therefore, in the present paper, we shall combine the Dirac result (1.4) with the important $1/Z$ expansion of atomic theory, which was first connected fundamentally with DFT in the work of March and White [4]. These authors were concerned with the nonrelativistic ground-state energy of a heavy atomic ion having nuclear charge Ze and N electrons, which takes the form (going back to Hylleraas [5] and especially developed later by Layzer and co-workers [6]):

$$E(Z,N)=Z^2[\varepsilon_0(N)+(1/Z)\varepsilon_1(N)+\cdots+(1/Z^n)\varepsilon_n(N)+\cdots]. \quad (1.5)$$

The main result of March and White [4] was to demonstrate, by bringing Eq. (1.5) into contact with the original form of DFT valid for large Z and N (namely, the Thomas-Fermi statistical theory [7]), that the asymptotic form of the coefficients $\varepsilon_n(N)$ for large N was given by

$$\varepsilon_n(N)\approx N^{n+1/3} \quad (1.6)$$

a result subsequently confirmed in a more rigorous study of Tal and Levy [8].

However, the point to be exploited in the present paper is that the leading term in the $1/Z$ expansion of the total energy $E(Z,N)$ of an atomic ion displayed in Eq. (1.5) is completely determined by the properties of the bare Coulomb potential $(-Ze^2/r)$. And for a closed-shell atom with filled k and l shells, March and Santamaria [9] have obtained the first-order idempotent density matrix in this bare Coulomb limit in closed form as

$$\gamma(\mathbf{r},\mathbf{r}')=\rho\left(\frac{r+r'}{2}\right)+|\mathbf{r}-\mathbf{r}'|^2F\left(\frac{r+r'}{2}\right). \tag{1.7}$$

Here the authors show that F is related to the kinetic-energy density $t(r)$ by

$$F=\rho''/24-mt(r)/4\hbar^2. \tag{1.8}$$

We record the explicit forms of $\rho(r)$ and $F(r)$ in Appendix A. Clearly, therefore, for this ten-electron problem, i.e., a nonrelativistic Ne-like atomic ion in the limit of large Z, one can evaluate the Dirac exchange energy E_x from Eq. (1.4) in this large-Z limit by inserting the Coulomb density matrix (1.7).

The outline of the paper is then as follows. In Sec. II immediately below, it will be shown that the exchange energy density $\varepsilon_x(r)$ can be evaluated in closed form using Eqs. (1.4) and (1.7) in this large-Z limit for ten-electron atomic ions, the natural definition of $\varepsilon_x(r)$ following from Eq. (1.4) as

$$\varepsilon_x(\mathbf{r})=-(e^2/4)\int\frac{\gamma^2(\mathbf{r},\mathbf{r}')}{|\mathbf{r}-\mathbf{r}'|}d\mathbf{r}'. \tag{1.9}$$

The resulting (spherically symmetric) exchange energy density will also be presented graphically there for the case $Z\approx 92$, though, of course, it is recognized that for any subsequent contact of all the results with experiment, relativistic corrections will be required for such a uranium highly charged positive ion. Section III then returns to one focal point of current DFT, namely, the potential $V_x(\mathbf{r})$ representing exchange. Of course, in our use of Dirac's result (1.4), we are restricted to idempotent 1 DM's, and therefore correlation is neglected. It is known, however, that exchange enters before correlation, from the work of Scott for exchange [10] and various later workers for correlation, in the spirit of the $1/Z$ expansion of the total energy in Eq. (1.5), but now with partially summed subseries as in the Thomas-Fermi-Dirac (TFD) statistical theory (March and White [4]; see also March and Parr [11]). For example, for neutral atoms, Scott showed that $E_x^{\mathrm{TFD}}\propto Z^{5/3}$, whereas there is agreement that $E_{\mathrm{correlation}}$ depends on a lower power of Z; proposals have been made for Z^n with $n=1$ or 4/3 (March and Wind [12], Kais, Sung, and Herschbach [13]). In Sec. III, an exact functional differentiation of the exchange energy in the large-Z limit is reported, the answer for $V_x(\mathbf{r})$ involving both electron and kinetic-energy densities. Then in Sec. IV, the approximation for $V_x(\mathbf{r})$ proposed by Slater is examined analytically and is also displayed numerically for $Z=92$. It will be argued that the Slater potential is already a useful approximation to the full exchange potential, though it is not quantitative near the atomic nucleus. Corrections are exhibited from the early work of Harbola and Sahni [14], modifications to which have been treated by Holas and March [2] (see also Levy and March [15]). Section V constitutes a summary with some proposals for future work. Finally, in Appendix B, Scott's calculation of the neutral atom exchange energy in the TFD theory is complemented by treating the highly ionized limit corresponding in the present study to $N=10$ and large Z.

II. FORM OF THE EXCHANGE ENERGY DENSITY $\varepsilon_x(r)$ FOR TEN-ELECTRON ATOMIC IONS IN THE (BARE COULOMB) LIMIT OF LARGE ATOMIC NUMBER Z

The reduced first-order density matrix of March and Santamaria [9] [Eq. (1.7)] can be used directly in Eq. (1.9) to evaluate the exchange energy density. Since the functions $\rho((r+r')/2)$ and $F((r+r')/2)$ in Eq. (1.7) are spherically symmetric, the angular integrations in Eq. (1.9) can easily be done, taking $\mathbf{r}$ to lie along the z axis, so that

$$\begin{aligned}\varepsilon_x(r)=-(\pi e^2/2r)\Bigg\{&\int\rho^2((r+r')/2)[(r+r')-|r-r'|]dr'\\&+(2/3)\int r'\rho((r+r')/2)F((r+r')/2)[(r+r')^3\\&-|r-r'|^3]dr'+(1/5)\int r'F^2((r+r')/2)[(r+r')^5\\&-|r-r'|^5]dr'\Bigg\}.\end{aligned} \tag{2.1}$$

We shall return to the functional properties of Eq. (2.1) in Sec. III below. For the present, using the explicit forms of F and ρ given in Eqs. (A1) and (A2), respectively, we note that after radial integration, the resulting total exchange energy density $\varepsilon_x(r)$ is found to be

$$\begin{aligned}\varepsilon_x(r)=&(1/15\,552)(e^2\alpha^3/\pi r)\exp(-4\alpha r)\{15\,552(1+\alpha r)+\exp(\alpha r)[4096+3840(\alpha r)+2304(\alpha r)^2+3456(\alpha r)^3]\\&+\exp(2\alpha r)[-13\,608+2430(\alpha r)+2916(\alpha r)^2+1944(\alpha r)^3+486(\alpha r)^4+243(\alpha r)^5]-4096\exp(5\alpha r/2)\\&+\exp(3\alpha r)[-1944-972\alpha r-972(\alpha r)^2]\},\end{aligned} \tag{2.2}$$

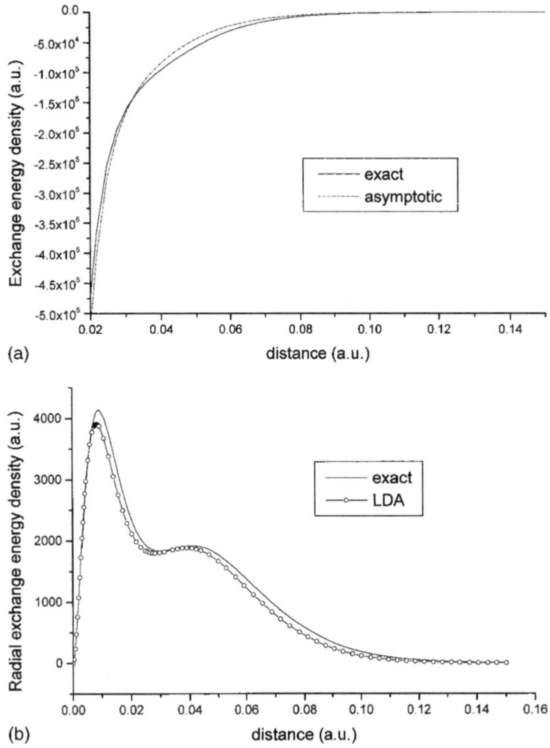

FIG. 1. (a) Exchange energy density $\varepsilon_x(r)$ defined by Eqs. (1.7) and (1.9), for $Z=92$. For comparison, the asymptotic form $\varepsilon_x^{\mathrm{AS}}(r) = -e^2\rho(r)/r$ of the exchange energy density is also plotted. (b) Radial exchange energy density $4\pi r^2\varepsilon_x(r)$, compared to that for the LDA, with $\varepsilon_x^{\mathrm{LDA}}(r) = -c_x\{\rho(r)\}^{4/3}$, $c_x=(3/4)(3/\pi)^{1/3}e^2$, and $\rho(r)$ given by Eq. (A2). The total exchange energies in the limit of $Z=92$ are $E_x=-160.805$ and $E_x^{\mathrm{LDA}}=-147.969$ a.u.

where $\alpha=(Z/a_0)$. In spite of the e^2/r factor outside the curly brackets, it is readily verified that $\varepsilon_x(0)$ is finite. This expression (2.2) is then readily integrated over $\mathbf{r}$ to yield the total exchange energy,

$$E_x = -1.747\,88 Z e^2/a_0 \tag{2.3}$$

(see also Appendix C).

For our case of the $Z=92$, ten-electron ion, the exchange energy density $\varepsilon_x(r)$ is plotted in Fig. 1(a); for comparison we show, in Fig. 1(a), the asymptotic form, $\varepsilon_x^{\mathrm{AS}}(r) = -(1/2)e^2\rho(r)/r$, to which $\varepsilon_x(r)$ tends at large r. It can be seen from Fig. 1 that the asymptotic form takes over when $r \simeq 0.08$ a.u. In Fig. 1b the radial exchange energy density $4\pi r^2\varepsilon_x(r)$ is plotted for $Z=92$ again and compared with the local-density approximation (LDA). The point-point agreement between exact and LDA in Fig. 1b is excellent. But even so, the total exchange energy E_x from exact and LDA calculations is -160.805 and -147.969 a.u., respectively.

III. FUNCTIONAL DIFFERENTIATION OF EXCHANGE ENERGY: THE EXCHANGE POTENTIAL FOR TEN-ELECTRON ATOMIC IONS IN THE LARGE-Z LIMIT

The purpose of this section is to show that an exact expression for the functional derivative $\delta E_x/\delta\rho(\mathbf{r})$, namely, the exchange potential $V_x(\mathbf{r})$, can be obtained from the calculation of the exact energy E_x. To do this it is useful to divide E_x into the sum of three contributions. Returning to Eq. (2.1), we first note that the total exchange energy E_x follows as the sum of three terms:

$$E_x = E_x^{(1)} + E_x^{(2)} + E_x^{(3)}. \tag{3.1}$$

These terms are evaluated in Appendix C.

From the above, it follows that the total exchange energy E_x takes the form

$$E_x = -e^2\left[A_1\int_0^\infty ds\,\rho^2(s)s^4 + A_2\int_0^\infty ds\,\rho(s)F(s)s^6 + A_3\int_0^\infty ds\,F^2(s)s^8\right] \tag{3.2}$$

with $A_1=20\pi^2/3$, $A_2=224\pi^2/9$, and $A_3=32\pi^2$. But from Ref. [9], $F(r)$ can be related to $\rho(r)$ and the kinetic-energy density $t(r)$ by Eq. (1.8) already quoted above. Substitution of this result for $F(s)$ in Eq. (3.2) gives the total exchange energy E_x solely in terms of $\rho(r)$ and its second derivative, plus the kinetic-energy density $t(r)$ (for details see Appendix C). Naturally, this is a result exact only in the bare Coulomb limit of the $1/Z$ expansion considered throughout this paper. In another model, of a metal surface, a related finding has recently been reported [16]. It was stressed by March and Santamaria [17] that it was to be expected that (nonlocal generalizations of) exchange and kinetic-energy densities should be intimately related (see also Lee, Lee, and Parr [18], and a very recent study by Chan and Handy [19]).

To proceed further, let us write in a little more detail the results obtained for the functional derivatives of $E_x^{(i)}$, $i=1-3$. Thus,

$$\delta E_x^{(1)}/\delta\rho(\mathbf{r}) = -(5\pi e^2/3)\int_0^{2\pi}d\phi\int_0^\pi d\theta\sin\theta\int_0^\infty ds\,s^4\partial/\partial\alpha[\rho(s) + \alpha\delta(\phi-\phi')\delta(\cos\theta-\cos\theta')\delta(s-r)/s^2]^2. \tag{3.3}$$

After some manipulation, one reaches the desired result

$$\delta E_x^{(1)}/\delta\rho(\mathbf{r}) = -(10\pi e^2/3)r^2\rho(r). \tag{3.4}$$

Proceeding to treat $E_x^{(2)}$ similarly, one finds

$$\delta E_x^{(2)}/\delta\rho(\mathbf{r})=(56\pi e^2/9)r^4F(r)+(224\pi^2e^2/9)\int_0^\infty ds\,\rho(s)s^6\delta F(s)/\delta\rho(\mathbf{r}). \tag{3.5}$$

After further manipulation and introduction of the relation between F, ρ'', and kinetic-energy density $t(\mathbf{r})$ in Eq. (1.8), one reaches the result

$$\delta E_x^{(2)}/\delta\rho(\mathbf{r})=(56\pi e^2/9)r^4F(r)-(224\pi^2e^2m/27\hbar^2)\times\int_0^\infty ds\,\rho(s)s^6[\delta t(s)/\delta\rho(\mathbf{r})]+[(70\pi e^2/9)r^2\rho(r)+(28\pi e^2/9)r^3\rho'(r)+(7\pi e^2/27)r^4\rho''(r)]. \tag{3.6}$$

Thus this part of the exchange potential can also be obtained formally exactly, but now the functional derivative of the kinetic-energy density appears. To date, we have not succeeded in calculating this term explicitly, since $T[\rho]$ is not yet known in this large-Z limit. One part of $T[\rho]$, that from the $s(l=0)$ states, has been given earlier by one of us [20] as

$$T_s[\rho]=(1/4)\int(\rho/r^2)d\mathbf{r} \tag{3.7}$$

and hence

$$\delta t_s(r)/\delta\rho(\mathbf{r}')=4\pi\delta(\mathbf{r}-\mathbf{r}')/r^2. \tag{3.8}$$

However, no such closed result has been found for the p term.

The third contribution $\delta E_x^{(3)}/\delta\rho(\mathbf{r})$ to the exchange potential can be written by similar arguments as

$$\delta E_x^{(3)}/\delta\rho(\mathbf{r})=-64\pi^2e^2\int_0^\infty ds\,s^8F(s)\delta F/\delta\rho(\mathbf{r}). \tag{3.9}$$

We are then led to the final form

$$\delta E_x^{(3)}/\delta\rho(\mathbf{r})=(64\pi^2e^2m/3\hbar^2)\int_0^\infty ds\,s^8F(s)[\delta t(s)/\delta\rho(\mathbf{r})]-(112\pi e^2/3)r^4F(r)-(32\pi e^2/3)r^5F'(r)-(2\pi e^2/3)r^6F''(r). \tag{3.10}$$

Thus, the exchange potential for this ten-electron system in the large-Z limit can be obtained analytically by direct functional differentiation of the exchange energy E_x, the answer being in terms of the ground-state density $\rho(r)$ and its derivatives, plus the functional derivative $\delta t(s)/\delta\rho(\mathbf{r})$ of the kinetic-energy density. This is therefore a formally complete density-functional theory of the exchange potential in this closed-$(k+l)$ shell atomic ion in the large-Z limit.

Because of the absence, to date, of a calculable form of $\delta t(s)/\delta\rho(\mathbf{r})$, we follow this exact theoretical analysis by

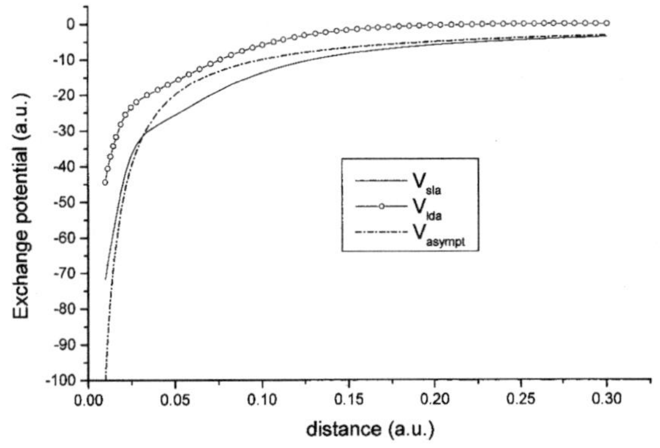

FIG. 2. Slater potential $V_{\rm Sla}(r)=2\varepsilon_x(r)/\rho(r)$ constructed from $\varepsilon_x(r)$ as in Fig. 1 and $\rho(r)$ in Eq. (A2). $V_{\rm Sla}(r)$ has the exact asymptotic form $-e^2/r$ of the exchange potential $V_x(r)$ at sufficiently large r. The LDA potential, $V_{\rm LDA}(r)=-(3/\pi)^{1/3}e^2\rho^{1/3}(r)$, is also shown for comparison.

reporting some approximate numerical results for $V_x(\mathbf{r})$, which we conveniently build around the Slater potential $V_{\rm Sla}(\mathbf{r})$ defined immediately below.

IV. APPROXIMATE NUMERICAL RESULTS FOR EXCHANGE POTENTIAL $V_x(\mathbf{r})$ AND FORCE

Given Eq. (2.2) above, one can immediately find the Slater approximation to the exchange potential, $V_{\rm Sla}(\mathbf{r})=2\varepsilon_x(r)/\rho(r)$, with $\rho(r)=\gamma(\mathbf{r},\mathbf{r}')|_{r'=r}$. Figure 2 shows $V_{\rm Sla}(\mathbf{r})$ for the present $Z=92$, ten-electron case as it approaches the asymptotic limit, $V_x^{\rm AS}(\mathbf{r})=-e^2/r$ as $r\to\infty$.

In the present case of a noninteracting electronic system, the Harbola-Sahni exchange potential [14] can be expressed as the sum of the Slater potential plus a "correction" term:

$$V_{\rm HS}(r)=V_{\rm Sla}(r)-\int_{r''=\infty}^{r}\int\frac{[\partial\rho_x/\partial r'']}{|\mathbf{r}''-\mathbf{r}'|}d^3r'dr'', \tag{4.1}$$

where $\rho_x(r',r'')=\gamma^2(r',r'')/2\rho(r'')$ is the exchange hole density. The correction term has been evaluated numerically for our case, and the resulting exchange potential $V_{\rm HS}(r)$ plotted in Fig. 3. For comparison, the Slater potential $V_{\rm Sla}(r)$ is also shown. As can be seen, the difference between the exact and the Slater potentials decreases rapidly with r. We mention in this context the observation of Kleinman [21] that the Slater potential is a partial functional derivative of the exchange energy with respect to charge density with wave function phases held constant. The phases must contribute to the full functional derivative near the nucleus; this contribution being short ranged, since the Slater potential has the correct asymptotic form.

It is instructive to return to the original form [14] of the Harbola-Sahni potential,

$$V_{\rm HS}(\mathbf{r})=-\int_{r''=\infty}^{r}\int\frac{\gamma^2(\mathbf{r}'',\mathbf{r}')(\mathbf{r}''-\mathbf{r}')}{2\rho(\mathbf{r}'')|\mathbf{r}''-\mathbf{r}'|^3}d^3r'\,d\mathbf{r}'' \tag{4.2}$$

and calculate the corresponding forces for our case. In particular, one can calculate $[\rho(r)r\partial V_{HS}(r)/\partial r]$ and verify that the (negative of the) integral over all space recovers the total exchange energy [the Levy-Perdew equality of Eq. (A5)]. Expressing $\gamma(\mathbf{r},\mathbf{r}')$ in terms of ρ and F and performing the angular integrations, one finds that

$$\rho(r)r\frac{\partial V_{HS}}{\partial r}=(\pi/r)\int_{r'=0}^{\infty}\rho^2((r+r')/2)\left[\frac{(r-r')}{|r-r'|}+1\right]r'^2dr'$$
$$+(2\pi/3r)\int_{r'=0}^{\infty}\rho((r+r')/2)\,F((r+r')/2)[-|r-r'|(2r^2-r'^2-rr')+(r+r')(2r^2-r'^2+rr')]r'dr'$$
$$+(\pi/15r)\int_{r'=0}^{\infty}F^2((r+r')/2)[-|r-r'|^3(4r^2-r'^2-3rr')+(r+r')^3(4r^2-r'^2+3rr')]r'dr'. \quad (4.3)$$

Volume integration then gives

$$-E_x=\int\rho(r)r\frac{\partial V_{HS}}{\partial r}d^3r$$
$$=\int_0^{\infty}4\pi^2rdr\int_{r'=0}^{\infty}\rho^2((r+r')/2)\times\left[\frac{(r-r')}{|r-r'|}+1\right]r'^2dr'$$
$$+[\text{the last two terms from Eq.(4.3)}], \quad (4.4)$$

where $E_x=-160.805$ a.u. as before. Thus, $V_{HS}(r)$ satisfies one of the conditions on the exact exchange potential, although this is not a proof that $V_{HS}=V_{exact}$.

The exchange force resulting from $V_{HS}(r)$ is purely radial, $F_{HS}(r)=-\partial V_{HS}(r)/\partial r$; using Eq. (4.3), one can find analytically the force plotted in Fig. 4, where $F_{Sla}(r)=-\partial V_{Sla}(r)/\partial r$ is shown for comparison. Note that both $F_{Sla}(r)$ and $F_{HS}(r)$ go to zero at the nucleus $r=0$.

V. SUMMARY AND PROPOSED FUTURE DIRECTIONS

The main achievements of the present paper are as follows:

(i) In the spirit of the $1/Z$ expansion written explicitly for atomic ions in Eq. (1.5), the limiting form of the exchange energy density $\varepsilon_x(r)$ at large Z has been obtained (Eq. 2.2) in closed analytic form for nonrelativistic ten-electron ions.

(ii) By integration, the total exchange energy E_x has been shown for these ten-electron ions to be proportional to Z as $Z\to\infty$, with the proportionality coefficient $-1.747\,88(e^2/a_0)$ [see Eq. (2.3) and Appendix C].

(iii) The Slater approximation to the exchange potential $V_x(r)$, namely, $2\varepsilon_x(r)/\rho(r)$, has hence been found analytically and is plotted for $Z=92$ in Fig. 2.

(iv) The Harbola-Sahni exchange potential is also plotted in Fig. 3.

(v) In Sec. III and Appendix C, $\varepsilon_x(r)$ has been constructed in terms of (integrals on) $\rho(r)$ plus its derivatives and the kinetic-energy density $t(r)$. Reference may be made here to a recent condensed matter analogue, namely, the infinite barrier model of a metal surface (March [16]), where again $\varepsilon_x(r)$ is related to $t(r)$.

(vi) Scott's semiclassical calculation of the total exchange energy $\propto Z^{5/3}$ for heavy nonrelativistic neutral atoms is complemented by a study of the strong ionization limit $(N/Z)\ll 1$ in Appendix B. The result has the form $E_x^{\text{semiclassical}}\propto Z^{5/3}(N/Z)^{2/3}$, which becomes proportional to Z for fixed (and still large) N. This then makes contact with

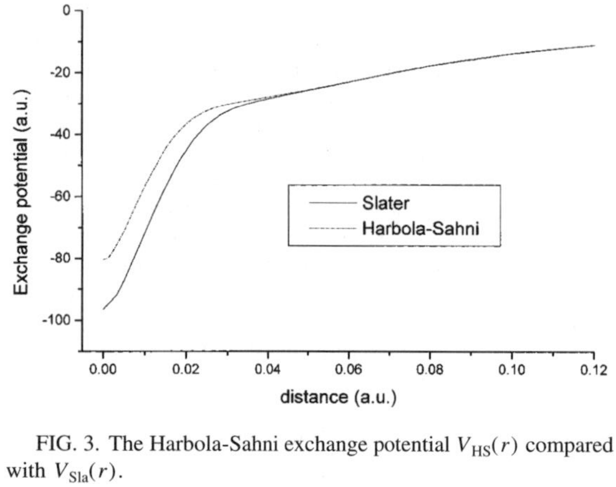

FIG. 3. The Harbola-Sahni exchange potential $V_{HS}(r)$ compared with $V_{Sla}(r)$.

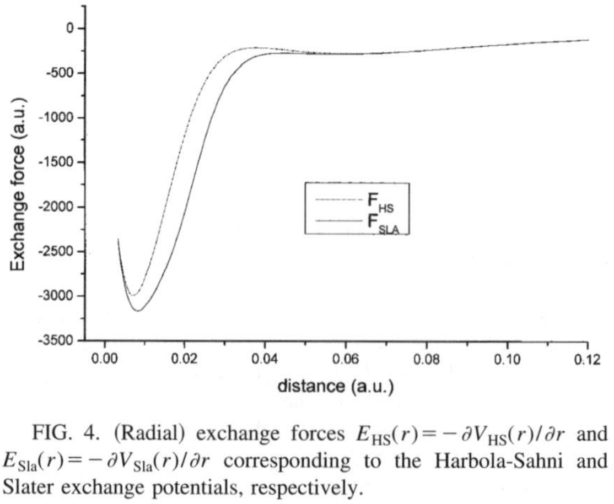

FIG. 4. (Radial) exchange forces $E_{HS}(r)=-\partial V_{HS}(r)/\partial r$ and $E_{Sla}(r)=-\partial V_{Sla}(r)/\partial r$ corresponding to the Harbola-Sahni and Slater exchange potentials, respectively.

item (ii) above in the present section.

As to future directions, it would be of interest to extend the above to homonuclear diatomic molecules with a fixed number of electrons. However, unlike the bare Coulomb one-center problem treated here, the idempotent bare Coulomb density matrix for the two-center problem can no longer be obtained in closed analytic form (except near the united-atom limit for the present ten-electron one-center problem). Nevertheless, a study of the exchange energy $E_x(Z, N, R)$ for fixed (and small) N is called for, at a number of values of the internuclear distance R.

ACKNOWLEDGMENTS

One of us (I.A.H.) wishes to acknowledge support from the Flemish Science Foundation (FWO) under Grant No. G0347.97. We also thank the University of Antwerp (RUCA) for its support. P.S. thanks the FWO for financial support. This work is also supported by the Concerted Action Program of the University of Antwerp.

APPENDIX A: ANALYTIC FORMS OF ELECTRONIC, KINETIC-ENERGY AND EXCHANGE ENERGY DENSITIES FOR TEN-ELECTRON NONRELATIVISTIC ATOMIC IONS IN THE LIMIT OF LARGE Z

From the study of March and Santamaria [9], who worked however with singly occupied states throughout, we find the electron density for the ten-electron atomic ions in the large-Z limit as the diagonal element of the 1 DM $\gamma(r,r')$ given by Eq. (1.7) with

$$F(r)=-(1/32\pi)(Z/a_0)^5\exp(-Zr/a_0) \tag{A1}$$

and

$$\rho(r)=(2/\pi)(Z/a_0)^3\exp(-2Zr/a_0)+(1/4\pi)(Z/a_0)^3\exp(-Zr/a_0)[1-Zr/a_0+1/2(Zr/a_0)^2]. \tag{A2}$$

The corresponding kinetic-energy density $t(r)$ is then readily obtained from Eq. (1.2) as

$$t(r)=(\hbar^2/2\pi m)(Z/a_0)^5\{2\exp(-2Zr/a_0)+(1/16)\exp(-Zr/a_0)[7-3Zr/a_0+(1/2)(Zr/a_0)^2]\}. \tag{A3}$$

It is of interest to note that differentiating (A3), one can verify the equivalence with the result of Amovilli and March [22] that

$$t'(r)=(1/8)\rho'''-(3/4r^2)\rho'-(3Z/2r^2)\rho. \tag{A4}$$

The basically new result of the present investigation is the exchange energy density $\varepsilon_x(r)$ calculated from Eq. (1.9) using the bare Coulomb density matrix (1.7), with the result in Eq. (2.2).

To conclude this appendix, we note that the total exchange energy E_x and the force $-\nabla V_x(\mathbf{r})$ associated with the exchange potential $V_x(\mathbf{r})$ are connected by

$$E_x=-\int \rho(\mathbf{r})\mathbf{r}\cdot\nabla V_x(\mathbf{r})\,d^3r \tag{A5}$$

a result that goes back to Levy and Perdew [23]. Equation (A5) is to be contrasted with the result for the exact E_x in terms of the Slater potential, namely,

$$E_x=\int \varepsilon_x(\mathbf{r})d^3r=(1/2)\int \rho(\mathbf{r})V_{\mathrm{Sla}}(\mathbf{r})d^3r. \tag{A6}$$

We note also that there is ambiguity in defining an exchange energy density. We have taken the definition in Eq. (1.9), which leads to $\varepsilon_x(r)$ in Eqs. (2.1) and (2.2). But one could adopt the (negative of) the integrand in Eq. (A5),

$$4\pi r^2\varepsilon_x^{\mathrm{virial}}(r)=-\rho(r)r\frac{\partial V_x}{\partial r}4\pi r^2. \tag{A7}$$

The area under this curve, namely, the total exchange energy E_x, is, of course, the same as that for $(4\pi r^2)\varepsilon_x(r)$, with $\varepsilon_x(r)$ defined directly from the Dirac expression for E_x [3].

APPENDIX B: SEMICLASSICAL EVALUATION OF EXCHANGE ENERGY E_x FOR HIGHLY CHARGED POSITIVE ATOMIC IONS IN THE LIMIT WHEN THE NUMBER OF ELECTRONS N AND THE NUCLEAR CHARGE Z ARE BOTH LARGE, BUT $N/Z\ll 1$

The objective of this appendix is to evaluate the semiclassical approximation $E_x^{\mathrm{semiclassical}}\equiv E_x^{\mathrm{scl}}$ to the Dirac exchange energy [3]. This will be written explicitly as

$$E_x^{\mathrm{scl}}=-c_x\int \{\rho_{\mathrm{TF}}(r)\}^{4/3}d^3r,\quad c_x=(3/4)(3/\pi)^{1/3}e^2. \tag{B1}$$

The Thomas-Fermi semiclassical density $\rho_{\mathrm{TF}}(r)$ for a bare Coulomb potential is immediately written down from phase-space considerations as

$$\rho_{\mathrm{TF}}(r)=(8\pi/3h^3)(2m)^{3/2}(\mu+Ze^2/r)^{3/2}, \tag{B2}$$

where the chemical potential μ is found from

$$\int \rho_{\mathrm{TF}}(r)d^3r=N \tag{B3}$$

to be

$$\mu=-18^{-1/3}(Z^2/N^{2/3})(e^2/a_0) \tag{B4}$$

with $a_0=\hbar^2/me^2$. Evidently Eq. (B2) is to be used from $r=0$ out to the semiclassical radius r_0 given by

$$\mu=-Ze^2/r_0 \tag{B5}$$

and then one finds from Eqs. (B1), (B2), and (B5) the result

$$E_x^{\rm scl}=-c_x(8\pi/3h^3)(2m)^2\int_0^{r_0}(Ze^2)^2[1/r-1/r_0]^2 4\pi r^2 dr$$

$$=-c_x(8\pi/3h^3)(2m)^2(Ze^2)^2(4\pi r_0/3). \quad \text{(B6)}$$

Substitution for c_x from Eq. (B1) and for r_0 from Eqs. (B4) and (B5) then yields the desired result

$$E_x^{\rm scl}=-(4/3\pi^2)(18N^2)^{1/3}Ze^2/a_0. \quad \text{(B7)}$$

As evidenced in the title of this appendix, Eq. (B7) is physically valid when both Z and N are large, but $N/Z\ll 1$.

In the text, we have noted the "exact" evaluation of the LDA exchange in Eq. (B1) with the correct electron density for ten electrons in a bare Coulomb field. As in Eq. (B7), $E_x^{\rm LDA}\propto Z$, but, of course, $N=10$. Then Eq. (B7) is being used outside its proper range of validity. Thus, substituting $N=10$ in Eq. (B7) gives $-1.6433\,Z(e^2/a_0)$, to be compared with $E_x^{\rm LDA}=-1.6084\,Z(e^2/a_0)$.

It is worth noting also that Eq. (B7) complements the exchange energy calculated by Scott [10] for neutral atoms; he found $-0.221\,Z^{5/3}(e^2/a_0)$ with the self-consistent Thomas-Fermi neutral atom density; Eq. (B7) would give $\approx -0.354\,Z^{5/3}(e^2/a_0)$ due to neglect of screening, which is permissible only for $(N/Z)\ll 1$, with N and Z large.

APPENDIX C: EXCHANGE ENERGY IN TERMS OF ELECTRON AND KINETIC-ENERGY DENSITIES

The total exchange energy $E_x=E_x^{(1)}+E_x^{(2)}+E_x^{(3)}$, where the simplest term $E_x^{(1)}$ is given explicitly by

$$E_x^{(1)}=-(2\pi^2e^2)\int_0^\infty dr\, r\int_0^\infty dr'\, r'$$

$$\times[(r+r')-|r-r'|]\rho^2((r+r')/2). \quad \text{(C1)}$$

The change of variable defined by

$$s=(r+r')/2 \quad \text{(C2)}$$

then allows $E_x^{(1)}$ to be rewritten as

$$E_x^{(1)}=-4\pi^2e^2\int_0^\infty dr\, r\left[\int_{r/2}^{r}2ds(2s-r)^2\rho^2(s)\right.$$

$$\left.+\int_r^\infty 2ds\, r(2s-r)\rho^2(s)\right]. \quad \text{(C3)}$$

Introducing the Heaviside function $H(a,b)$ defined by

$$H(a,b)=1,\quad b>a$$

$$=0,\quad 0<b<a \quad \text{(C4)}$$

one can interchange the order of integration in Eq. (C3) to obtain

$$E_x^{(1)}=-8\pi^2e^2\left\{\int_0^\infty ds\,\rho^2(s)\int_0^\infty dr\, r(2s-r)^2H(r/2,s)\right.$$

$$\times[1-H(r,s)]+\int_{s=0}^\infty ds\,\rho^2(s)\int_{r=0}^\infty dr\, r^2$$

$$\left.\times(2s-r)H(r,s)\right\}. \quad \text{(C5)}$$

The r integration is readily performed to find

$$E_x^{(1)}=-(20\pi^2e^2/3)\int_0^\infty ds\, s^4\rho^2(s). \quad \text{(C6)}$$

By an analogous procedure, $E_x^{(2)}$ and $E_x^{(3)}$ may be calculated, with the results

$$E_x^{(2)}=-(224\pi^2e^2/9)\int_0^\infty ds\,\rho(s)F(s)s^6 \quad \text{(C7)}$$

and

$$E_x^{(3)}=-(32\pi^2e^2)\int_0^\infty dsF^2(s)s^8. \quad \text{(C8)}$$

Inserting the forms (A2) and (A1) for ρ and F, respectively, into the definite integrals in Eqs. (C6)–(C8) yields the results

$$E_x^{(1)}=-(18\,355/3888)Ze^2/a_0, \quad \text{(C9)}$$

$$E_x^{(2)}=(760\,585/139\,968)Ze^2/a_0, \quad \text{(C10)}$$

and

$$E_x^{(3)}=-(315/128)Ze^2/a_0, \quad \text{(C11)}$$

with the final result for the total exchange energy E_x given in Eq. (2.3).

[1] R. G. Parr and W. Yang, *Density Functional Theory of Atoms and Molecules* (Oxford University Press, Oxford, 1989).
[2] A. Holas and N. H. March, Phys. Rev. A **51**, 2040 (1995).
[3] P. A. M. Dirac, Proc. Cambridge Philos. Soc. **26**, 376 (1930).
[4] N. H. March and R. J. White, J. Phys. B **5**, 466 (1972).
[5] E. A. Hylleraas, Z. Phys. **65**, 209 (1930).
[6] See, for example, D. Layzer, Ann. Phys. (N.Y.) **8**, 271 (1959); or the later review of M. Cohen, Adv. At. Mol. Phys. **25**, 195 (1988).
[7] See, e.g., the review by N. H. March, Adv. Phys. **6**, 1 (1957); N. H. March, *Self-Consistent Fields in Atoms* (Pergamon, Oxford, 1975).
[8] Y. Tal and M. Levy, Phys. Rev. A **23**, 408 (1981).
[9] N. H. March and R. Santamaria, Phys. Rev. A **38**, 5002 (1988).
[10] J. M. C. Scott, Philos. Mag. **43**, 859 (1951).

[11] N. H. March and R. G. Parr, Proc. Natl. Acad. Sci. USA **77**, 6285 (1980).
[12] N. H. March and P. Wind, Mol. Phys. **77**, 791 (1992).
[13] S. Kais, S. M. Sung, and D. R. Herschbach, Int. J. Quantum Chem. **49**, 657 (1994).
[14] M. K. Harbola and V. Sahni, Phys. Rev. Lett. **62**, 489 (1989).
[15] M. Levy and N. H. March, Phys. Rev. A **55**, 1885 (1997).
[16] N. H. March, Phys. Rev. B **60**, 5011 (2000).
[17] N. H. March and R. Santamaria, Int. J. Quantum Chem. **39**, 585 (1991).
[18] H. Lee, C. Lee, and R. G. Parr, Phys. Rev. A **44**, 768 (1991).
[19] G. K-L. Chan and N. C. Handy, Phys. Rev. A **59**, 2670 (1999).
[20] N. H. March, Phys. Lett. A **114**, 301 (1986); see also A. Nagy and N. H. March, Chem. Phys. Lett. **181**, 279 (1991).
[21] L. Kleinman, Phys. Rev. B **49**, 14 197 (1994).
[22] C. Amovilli and N. H. March, Phys. Chem. Liq. **38**, 505 (2000).
[23] M. Levy and J. Perdew, Phys. Rev. A **32**, 2010 (1985).

PHYSICAL REVIEW A, VOLUME 63, 012520

Nuclear cusp conditions for components of the molecular energy density relevant for density-functional theory

N. H. March,[1,2] I. A. Howard,[1] A. Holas,[3] P. Senet,[1] and V. E. Van Doren[1]
[1]*Department of Physics, University of Antwerp (RUCA), Groenenborgerlaan 171, B-2020 Antwerp, Belgium*
[2]*University of Oxford, Oxford, England*
[3]*Institute of Physical Chemistry of the Polish Academy of Sciences, 44/52 Kasprzaka, 01-224 Warsaw, Poland*
(Received 29 June 2000; published 14 December 2000)

Nuclear cusp conditions are obtained for the full electron-electron interaction energy density as well as for the exchange and correlation energy densities of density-functional theory. Their form is the same as the form of the well known Kato cusp condition for the electron-number density. All these cusp conditions are valid for both the ground and excited states of a molecule or solid.

DOI: 10.1103/PhysRevA.63.012520 PACS number(s): 31.15.Ew, 31.10.+z, 71.15.Mb, 71.45.Gm

I. INTRODUCTION

A singularity at the nuclear position $\mathbf{R}_A$ of the electron-nucleus interaction potential

$$v_A(\mathbf{r}) = -\frac{Z_A}{|\mathbf{r}-\mathbf{R}_A|} \tag{1.1}$$

causes the N-electron wave function of a molecule to be a nonanalytic function of each electron-position variable $\mathbf{r}_i$, $i=1,2,\ldots,N$, at $\mathbf{r}_i=\mathbf{R}_A$, but the function remains bounded (finite) everywhere. The character of this non-analyticity, known as the nuclear cusp condition, was determined for an atom by Kato [1]. His result was extended to a molecule by Bingel [2], and generalized by Pack and Brown [3] to the case where nuclear motion is included. For a molecule with fixed nuclear positions, as considered in the present investigation, Kato's cusp condition for the wave function leads (see Steiner [4]) to the following condition for the electron-number density $n_1(\mathbf{r})$:

$$\left.\frac{d\overline{n_1}(r)}{dr}\right|_{r=0} = -2\,Z_A n_1(\mathbf{0}), \tag{1.2}$$

where Z_A is the nuclear charge, while the center of the coordinate system is placed at the Ath nucleus position, $\mathbf{R}_A=\mathbf{0}$. The angular average $\bar{F}(r)$ of a function $F(\mathbf{r})$ is defined by

$$\bar{F}(r) = \int \frac{d^2\Omega}{4\pi} F(r\,\mathbf{e}(\Omega)), \tag{1.3}$$

where, in spherical coordinates, $\mathbf{r}=r\,\mathbf{e}(\Omega)$, $|\mathbf{e}|=1$.

We are going to show that the nuclear cusp condition is satisfied not only by the electron-number density, Eq. (1.2), but also by other densities used for the calculation of the energy of a molecule. Such results may be relevant for density-functional theory (DFT) applications to both the ground and excited states (see Nagy [5]).

Consider a molecule as an N-electron system described by the Hamiltonian

$$\hat{H} = \hat{T} + \hat{V} + \hat{U}, \tag{1.4}$$

where $\hat{T}=\Sigma_{i=1}^N \hat{t}(\mathbf{r}_i)$, $\hat{V}=\Sigma_{i=1}^N v(\mathbf{r}_i)$, $\hat{U}=\Sigma_{1\leq i<j\leq N}\, u(\mathbf{r}_i,\mathbf{r}_j)$. Here $\mathbf{x}_i\equiv\{\mathbf{r}_i,s_i\}$ denotes the space-spin coordinate of the ith electron, $\hat{t}(\mathbf{r})=-\frac{1}{2}\nabla^2(\mathbf{r})$ is the kinetic energy operator, $v(\mathbf{r})=\Sigma_A v_A(\mathbf{r})$ [Eq. (1.1)] is the external potential (due to clamped nuclei), and $u(\mathbf{r}_i,\mathbf{r}_j)=1/|\mathbf{r}_i-\mathbf{r}_j|$ is the electron-electron interaction potential. Atomic units are used throughout. Let the N-electron, antisymmetric, normalized ($\langle\Psi|\Psi\rangle=1$) wave function $\Psi(\mathbf{x}_1,\ldots,\mathbf{x}_N)$ be some eigenfunction of the Schrödinger equation

$$\hat{H}\Psi = E\Psi. \tag{1.5}$$

Discussing cusp conditions due to the particular Ath nucleus, we choose its position to be at $\mathbf{R}_A=\mathbf{0}$. Of all (equivalent) arguments of Ψ we focus our attention on $\mathbf{r}_1=r_1\,\mathbf{e}(\Omega_1)$.

The presentation of the nuclear cusp problem given by Davidson [6] serves us now as a starting point of our investigation. Since Ψ is finite everywhere, we have from Eq. (1.5)

$$\lim_{r_1\to 0} r_1 \hat{H}\Psi = 0. \tag{1.6}$$

This means that $\Psi(\mathbf{r}_1,s_1,\mathbf{x}_2,\ldots,\mathbf{x}_N)$ is also the eigenfunction, with zero eigenvalue, of the operator

$$\hat{b}(\mathbf{r}_1) = \lim_{r_1\to 0} r_1\hat{T} + \gamma = \lim_{r_1\to 0} r_1\hat{t}(\mathbf{r}_1) + \gamma, \tag{1.7}$$

where

$$\gamma = \lim_{r_1\to 0} r_1(\hat{V}+\hat{U}) = \lim_{r_1\to 0} r_1 v_A(\mathbf{r}_1) = -Z_A. \tag{1.8}$$

For the wave function $\Psi(\mathbf{r}_1,\mathbf{X})$ where $\mathbf{X}\equiv\{s_1,\mathbf{x}_2,\ldots,\mathbf{x}_N\}$, expanded in spherical harmonics

$$\Psi(\mathbf{r}_1,\mathbf{X}) = \sum_{l=0}^{\infty}\sum_{m=-l}^{m=l} Y_{lm}(\Omega_1)\, r_1^l f_{lm}(r_1,\mathbf{X}) \tag{1.9}$$

with $f_{lm}(r_1,\mathbf{X})$ an analytic function of r_1, the eigenequation mentioned above,

$$\hat{b}(\mathbf{r}_1)\Psi=0, \tag{1.10}$$

leads to the following conditions [3,6] on the partial waves amplitudes:

$$\left.\frac{\partial f_{lm}(r_1,\mathbf{X})}{\partial r_1}\right|_{r_1=0}=\frac{\gamma}{l+1}f_{lm}(0,\mathbf{X}). \tag{1.11}$$

Equations (1.8)–(1.11) are sufficient for obtaining all interesting cusp conditions. These for the system described by the Hamiltonian in Eq. (1.4) will be considered in Sec. II, and those for systems of adiabatic connection (a method that leads to the construction of DFT objects) in Sec. III. The implications of our results will be discussed in Sec. IV.

II. RESULTS FOR THE FULLY INTERACTING SYSTEM

As is well known (see, e.g., Davidson [6]) the total energy $E=\langle\Psi|\hat{H}|\Psi\rangle$—the eigenenergy if Ψ is the solution of the Schrödinger equation (1.5)—can be written in terms of the first and second order density matrix (1DM, 2DM) derived from Ψ, namely,

$$E=T+V+E_{\text{ee}}=\int d^3r_1\{\varepsilon_{\text{kin}}(\mathbf{r}_1;[\rho_1]) +v(\mathbf{r}_1)n_1(\mathbf{r}_1)+\varepsilon_{\text{ee}}(\mathbf{r}_1;[n_2])\}, \tag{2.1}$$

where

$$\varepsilon_{\text{kin}}(\mathbf{r}_1;[\rho_1])=\hat{t}(\mathbf{r}_1)\rho_1(\mathbf{r}_1;\mathbf{r}_1')|_{\mathbf{r}_1'=\mathbf{r}_1} \tag{2.2}$$

is the kinetic energy density, $v(\mathbf{r}_1)n_1(\mathbf{r}_1)$ is the electron-nucleus interaction energy density, and

$$\varepsilon_{\text{ee}}(\mathbf{r}_1;[n_2])=\int d^3r_2\, u(\mathbf{r}_1,\mathbf{r}_2)\, n_2(\mathbf{r}_1,\mathbf{r}_2) \tag{2.3}$$

is the electron-electron interaction energy density. The DMs ρ_1, n_2, and n_1 can be obtained from the 2DM

$$\rho_2(\mathbf{r}_1,\mathbf{r}_2;\mathbf{r}_1',\mathbf{r}_2')=\frac{N(N-1)}{2}\int d\mathbf{Q}\,\Psi(\mathbf{r}_1,\mathbf{r}_2,\mathbf{Q}) \times\Psi^*(\mathbf{r}_1',\mathbf{r}_2',\mathbf{Q}). \tag{2.4}$$

Here the notation $\mathbf{Q}\equiv\{s_1,s_2,\mathbf{x}_3,\ldots,\mathbf{x}_N\}$ is introduced to split the previous $\mathbf{X}$ as $\mathbf{X}=\{s_1,\mathbf{r}_2,s_2,\mathbf{x}_3,\ldots,\mathbf{x}_N\}=\{\mathbf{r}_2,\mathbf{Q}\}$, and $\int d\mathbf{Q}$ means summation over s_i and integration over d^3r_i for all variables included in $\mathbf{Q}$. Thus, the electron-pair density is the diagonal element of the 2DM,

$$n_2(\mathbf{r}_1,\mathbf{r}_2)=\rho_2(\mathbf{r}_1,\mathbf{r}_2;\mathbf{r}_1,\mathbf{r}_2), \tag{2.5}$$

the 1DM is the reduced 2DM,

$$\rho_1(\mathbf{r}_1;\mathbf{r}_1')=\frac{2}{N-1}\int d^3r_2\,\rho_2(\mathbf{r}_1,\mathbf{r}_2;\mathbf{r}_1',\mathbf{r}_2), \tag{2.6}$$

and the electron-number density is the diagonal element of the 1DM,

$$n_1(\mathbf{r}_1)=\rho_1(\mathbf{r}_1;\mathbf{r}_1). \tag{2.7}$$

In order to establish the cusp condition for $n_2(\mathbf{r}_1,\mathbf{r}_2)$, let us calculate this DM using the expansion (1.9) for Ψ and Ψ^* in Eqs. (2.4),(2.5):

$$n_2(\mathbf{r}_1,\mathbf{r}_2)=\frac{N(N-1)}{2}\int d\mathbf{Q}\sum_{lml'm'}Y_{lm}(\Omega_1)Y^*_{l'm'}(\Omega_1) \times r_1^{l+l'}f_{lm}(r_1,\mathbf{r}_2,\mathbf{Q})f^*_{l'm'}(r_1,\mathbf{r}_2,\mathbf{Q}). \tag{2.8}$$

According to Eq. (1.3), its angular average is

$$\overline{n_2}(r_1,\mathbf{r}_2)=\frac{1}{4\pi}\int d^2\Omega_1\, n_2(r_1\mathbf{e}(\Omega_1),\mathbf{r}_2) =\frac{N(N-1)}{2}\int d\mathbf{Q}\frac{1}{4\pi}\sum_{lm}r_1^{2l}|f_{lm}(r_1,\mathbf{r}_2,\mathbf{Q})|^2. \tag{2.9}$$

The derivative of this function,

$$\frac{\partial\overline{n_2}(r_1,\mathbf{r}_2)}{\partial r_1}=\frac{N(N-1)}{2}\int d\mathbf{Q}\frac{1}{4\pi} \times\left\{\sum_{l=1}^{\infty}2l\,r_1^{2l-1}\sum_m|f_{lm}(r_1,\mathbf{r}_2,\mathbf{Q})|^2 +\sum_{l=0}^{\infty}r_1^{2l}\sum_m\frac{\partial|f_{lm}(r_1,\mathbf{r}_2,\mathbf{Q})|^2}{\partial r_1}\right\}, \tag{2.10}$$

together with the result

$$\left.\frac{\partial|f_{lm}(r_1,\mathbf{r}_2,\mathbf{Q})|^2}{\partial r_1}\right|_{r_1=0}=\frac{2\gamma}{l+1}|f_{lm}(0,\mathbf{r}_2,\mathbf{Q})|^2 \tag{2.11}$$

obtained from Eq. (1.11), leads, in the limit $r_1\to0$, to the cusp condition for the electron-pair density

$$\left.\frac{\partial\overline{n_2}(r_1,\mathbf{r}_2)}{\partial r_1}\right|_{r_1=0}=-2Z_A n_2(\mathbf{0},\mathbf{r}_2) \tag{2.12}$$

[using Eq. (1.8)]. We note that its form is the same as in the case of the electron-number density, Eq. (1.2). In fact, we can confirm the validity of Eq. (1.2) by integrating both sides of Eq. (2.12) over d^3r_2 and recalling the definition of n_1 in Eqs. (2.7), (2.6).

To find the cusp condition for $\varepsilon_{\text{ee}}(\mathbf{r}_1;[n_2])$, Eq. (2.3), we calculate the derivative of its angular average,

$$\left.\frac{d\,\overline{\varepsilon_{ee}}(r_1;[n_2])}{dr_1}\right|_{r_1=0}=\int\frac{d^2\Omega_1}{4\pi}\int d^3r_2\frac{\partial}{\partial r_1}(u(r_1\mathbf{e}(\Omega_1),\mathbf{r}_2)\times n_2(r_1\mathbf{e}(\Omega_1),\mathbf{r}_2))_{r_1=0}$$
$$=\int d^3r_2\int\frac{d^2\Omega_1}{4\pi}\times\left(n_2(\mathbf{0},\mathbf{r}_2)\frac{\partial u(r_1\mathbf{e}(\Omega_1),\mathbf{r}_2)}{\partial r_1}+u(\mathbf{0},\mathbf{r}_2)\frac{\partial n_2(r_1\mathbf{e}(\Omega_1),\mathbf{r}_2)}{\partial r_1}\right)_{r_1=0}. \tag{2.13}$$

But, for the Coulombic interaction,

$$\left.\frac{\partial u(r_1\mathbf{e}(\Omega_1),r_2\mathbf{e}(\Omega_2))}{\partial r_1}\right|_{r_1=0}=\frac{\mathbf{e}(\Omega_1)\cdot\mathbf{e}(\Omega_2)}{r_2^2}. \tag{2.14}$$

So

$$\left.\frac{d\,\overline{\varepsilon_{ee}}(r_1;[n_2])}{dr_1}\right|_{r_1=0}=\int d^3r_2\left(\frac{n_2(\mathbf{0},\mathbf{r}_2)}{r_2^2}\mathbf{e}(\Omega_2)\cdot\overline{\mathbf{e}}+u(\mathbf{0},\mathbf{r}_2)\left.\frac{\partial\overline{n_2}(r_1,\mathbf{r}_2)}{\partial r_1}\right|_{r_1=0}\right). \tag{2.15}$$

The first term vanishes because the angular average of the unit vector is zero and for the second term the result (2.12) should be used, so, finally,

$$\left.\frac{d\,\overline{\varepsilon_{ee}}(r_1;[n_2])}{dr_1}\right|_{r_1=0}=-2Z_A\varepsilon_{ee}(\mathbf{0};[n_2]). \tag{2.16}$$

The remaining two densities in Eq. (2.1) satisfy the equation

$$\lim_{r_1\to0}r_1(\varepsilon_{\rm kin}(\mathbf{r}_1;[\rho_1])+v(\mathbf{r}_1)\,n_1(\mathbf{r}_1))=\lim_{r_1\to0}r_1\varepsilon_{\rm kin}(\mathbf{r}_1;[\rho_1])-Z_A\,n_1(\mathbf{0})=0, \tag{2.17}$$

which follows directly from the eigenequation (1.10), when the definitions (1.7), (1.8), and (2.2) are taken into account. The result (2.17) demonstrates that, at the nuclear position $\mathbf{R}_A=\mathbf{0}$, the singularity of the kinetic energy density cancels exactly the singularity of the electron-nucleus interaction energy density.

For DFT applications the classical electrostatic energy of electrons is of interest:

$$E_{\rm es}[n_1]=\int d^3r_1\,\varepsilon_{\rm es}(\mathbf{r}_1;[n_1]), \tag{2.18}$$

where the energy density $\varepsilon_{\rm es}$, defined in analogy with $\varepsilon_{\rm ee}$ [Eq. (2.3)], is

$$\varepsilon_{\rm es}(\mathbf{r}_1;[n_1])=\int d^3r_2\,u(\mathbf{r}_1,\mathbf{r}_2)\,\tfrac{1}{2}n_1(\mathbf{r}_1)\,n_1(\mathbf{r}_2). \tag{2.19}$$

By applying the same steps as in Eqs. (2.13)–(2.15) with $n_2(\mathbf{r}_1,\mathbf{r}_2)$ replaced by $\frac{1}{2}n_1(\mathbf{r}_1)\,n_1(\mathbf{r}_2)$, and, next, using the condition (1.2), we arrive at the cusp condition

$$\left.\frac{d\,\overline{\varepsilon_{es}}(r_1;[n_1])}{dr_1}\right|_{r_1=0}=-2Z_A\varepsilon_{\rm es}(\mathbf{0};[n_1]) \tag{2.20}$$

in the same form as for the densities already considered.

III. RESULTS FOR THE ADIABATIC CONNECTION TO THE NONINTERACTING EFFECTIVE SYSTEM

Adiabatic connection, developed by Harris and Jones [7] (see also Parr and Yang [8], Sec. 8.5, and references therein), is a very convenient tool of DFT for obtaining characteristics of the noninteracting effective Kohn-Sham (KS) system by linking it with the fully interacting system via intermediate systems. For that purpose, a Hamiltonian depending on a parameter $\alpha\in[0,1]$ is introduced,

$$\hat{H}^\alpha=\hat{T}+\hat{V}^\alpha+\alpha\hat{U}, \tag{3.1}$$

where $\hat{V}^\alpha=\Sigma_{i=1}^N v^\alpha(\mathbf{r}_i)$. This $\hat{H}^\alpha$, for $\alpha=1$, due to the requirement

$$\hat{V}^1=\hat{V},\quad v^1(\mathbf{r})=v(\mathbf{r}), \tag{3.2}$$

reduces to the original Hamiltonian $\hat{H}^1=\hat{H}$ [Eq. (1.4)], while for $\alpha=0$ it represents the noninteracting system ($\hat{U}$ is switched off) characterized by the KS potential

$$v^0(\mathbf{r})=v_{\rm KS}(\mathbf{r}). \tag{3.3}$$

From the eigenfunction Ψ^α of the Schrödinger equation

$$\hat{H}^\alpha\Psi^\alpha=E^\alpha\Psi^\alpha, \tag{3.4}$$

the electron density $n_1^\alpha(\mathbf{r})$ can be calculated like $n_1(\mathbf{r})$ in Eqs. (2.4)–(2.7) from Ψ. The effective one-body potential $v^\alpha(\mathbf{r})$ is defined by the requirement that the electron density of the "intermediate" system remains independent of α:

$$n_1^\alpha(\mathbf{r})=n_1^1(\mathbf{r})=n_1(\mathbf{r}). \tag{3.5}$$

In order to investigate the cusp conditions for the intermediate system, let us suppose that the properties of $v^\alpha(\mathbf{r})$ (of unknown explicit form) allow for a finite value of the limit

$$\gamma^\alpha=\lim_{r_1\to0}r_1(\hat{V}^\alpha+\alpha\hat{U})=\lim_{r_1\to0}r_1v^\alpha(\mathbf{r}_1), \tag{3.6}$$

the analog of Eq. (1.8). Then, with Ψ, f_{lm}, γ replaced by $\Psi^\alpha, f_{lm}^\alpha, \gamma^\alpha$, Eqs. (1.9)–(1.11) hold, and, therefore, all previous cusp conditions can be written for α-dependent densities in terms of γ^α, in particular,

$$\left.\frac{d\,\overline{n_1^\alpha}(r)}{dr}\right|_{r=0} = 2\,\gamma^\alpha n_1^\alpha(\mathbf{0}), \tag{3.7}$$

the analog of Eq. (1.2). In view of the requirement (3.5), we conclude from Eqs. (3.7) and (1.2) that

$$\gamma^\alpha = \gamma = -Z_A, \tag{3.8}$$

what justifies *a posteriori* our supposition in Eq. (3.6). Therefore all cusp conditions obtained at $\alpha=1$, Eqs. (1.2), (2.12), (2.16), (2.20), hold for n_1^α, n_2^α, $\varepsilon_{\rm ee}^\alpha$, $\varepsilon_{\rm es}^\alpha$, at any $\alpha \in [0,1]$.

Let us represent the intermediate-system potential as a sum of the original potential due to clamped nuclei and an additional potential that takes into account effective interactions (int) between electrons

$$v^\alpha(\mathbf{r}) = v(\mathbf{r}) + v_{\rm int}^\alpha(\mathbf{r}). \tag{3.9}$$

Then from Eqs. (3.8), (3.6), and (1.8) it follows that the singularity of $v_{\rm int}^\alpha(\mathbf{r})$ at $\mathbf{R}_A=\mathbf{0}$ (if it occurs at all) must be weaker than the Coulombic singularity:

$$\lim_{r_1\to 0} r_1 v_{\rm int}^\alpha(\mathbf{r}_1) = 0. \tag{3.10}$$

It should be noted that the $v_{\rm int}^0$ part of the KS potential, Eqs. (3.3),(3.9), is traditionally split into the electrostatic (Hartree) and exchange-correlation contributions:

$$v_{\rm KS}(\mathbf{r}) = v(\mathbf{r}) + v_{\rm int}^0(\mathbf{r}) = v(\mathbf{r}) + v_{\rm es}(\mathbf{r}) + v_{\rm xc}(\mathbf{r}), \tag{3.11}$$

where $v_{\rm es}(\mathbf{r};[n_1]) = \int d^3r'\, u(\mathbf{r},\mathbf{r}')\, n_1(\mathbf{r}')$. Evidently, the electrostatic potential $v_{\rm es}(\mathbf{r};[n_1])$ is finite everywhere. Therefore, Eq. (3.10) at $\alpha=0$ leads to

$$\lim_{r_1\to 0} r_1\, v_{\rm xc}(\mathbf{r}_1) = 0 \tag{3.12}$$

—an exact property of the exchange-correlation potential at the nucleus position $\mathbf{R}_A=\mathbf{0}$.

In the KS approach to DFT, the total molecular energy, Eq. (2.1), is rewritten as

$$E = T_{\rm s} + V + E_{\rm es} + E_{\rm x} + E_{\rm c} \tag{3.13}$$

(see, e.g., Parr and Yang [8]), where $T_{\rm s} = T^0 = \int d^3r\, \varepsilon_{\rm kin}^0(\mathbf{r})$ is the kinetic energy of the noninteracting ($\alpha=0$) system, $E_{\rm es}$ is given in Eqs. (2.18) and (2.19), the exchange energy $E_{\rm x}$ is defined as

$$E_{\rm x} = E_{\rm ee}^0 - E_{\rm es} = \int d^3r\,[\varepsilon_{\rm ee}^0(\mathbf{r}) - \varepsilon_{\rm es}(\mathbf{r})] = \int d^3r\, \varepsilon_{\rm x}(\mathbf{r}), \tag{3.14}$$

and the correlation energy, given by

$$E_{\rm c} = E_{\rm ee}^1 - E_{\rm ee}^0 + T^1 - T^0, \tag{3.15}$$

can also be written in terms of the adiabatic-connection average (aa)

$$E_{\rm c} = \int d^3r\, \varepsilon_{\rm c}(\mathbf{r}) = \int d^3r\,[\varepsilon_{\rm ee}^{\rm aa}(\mathbf{r}) - \varepsilon_{\rm ee}^0(\mathbf{r})], \tag{3.16}$$

where

$$\varepsilon_{\rm ee}^{\rm aa}(\mathbf{r}) = \int_0^1 d\alpha\, \varepsilon_{\rm ee}^\alpha(\mathbf{r}). \tag{3.17}$$

Here $\varepsilon_{\rm ee}^\alpha(\mathbf{r}) = \varepsilon_{\rm ee}(\mathbf{r};[n_2^\alpha])$, Eq. (2.3). Knowing that the cusp conditions hold for $\varepsilon_{\rm ee}^\alpha(\mathbf{r})$ at any α, and for $\varepsilon_{\rm es}(\mathbf{r})$, Eqs. (2.16),(2.20), we can immediately obtain the cusp condition for the exchange energy density [see Eq. (3.14)],

$$\left.\frac{d\,\overline{\varepsilon_{\rm x}}(r_1)}{dr_1}\right|_{r_1=0} = -2Z_A \varepsilon_{\rm x}(\mathbf{0}), \tag{3.18}$$

and for the correlation energy density [see Eqs. (3.16),(3.17)],

$$\left.\frac{d\,\overline{\varepsilon_{\rm c}}(r_1)}{dr_1}\right|_{r_1=0} = -2Z_A \varepsilon_{\rm c}(\mathbf{0}). \tag{3.19}$$

IV. DISCUSSION AND CONCLUSIONS

It is worth stressing that all results for the cusp conditions are obtained from an arbitrary eigenfunction of $\hat{H}$, Eq. (1.4), or $\hat{H}^\alpha$, Eq. (3.1); therefore they are valid not only for the ground state (GS) of the system, but also for an arbitrary excited state. While the DFT was formulated originally for the GS, various generalizations for excited states are available now too. Thus some of our results such as Eqs. (3.12), (3.18), and (3.19) may prove to be useful for development of approximate density functionals for both the GS and excited state versions of DFT: the quality of an approximation may be checked against fulfillment of these results, or these results may be imposed as constraints during construction of approximate functionals.

The DFT approach to a *single* excited state proposed recently by Nagy [5] is based on the arguments that the cusp condition holds for the density n_1 corresponding to an excited state and that it remains valid also during the adiabatic connection to the effective KS system that describes this excited state. However, these arguments were criticized recently by Moiseyev [9]. Actually, he quotes reasoning from Gross [10] that Nagy's arguments cannot be true for the effective KS system, because the KS one-electron potential does not belong to the class of bare nuclear Coulomb potentials (Nagy's argument is valid only for that class, according to his opinion). Our reasoning, given in Eqs. (3.6)–(3.8), supports Nagy's arguments and shows that there is no foundation for Gross' reservation.

From the fact that the results for the cusp conditions are valid for an arbitrary eigenstate of the system, it follows also that they remain valid if the pure-system densities are replaced by ensemble densities. Ensemble DFT is used for describing a system at a finite temperature (see, e.g., [8]), for obtaining excited state energies [11], or for handling degeneracy of either the GS or excited states [5], and so on. An ensemble 2DM is defined as

$$\rho_2=\sum_{\Lambda} p_\Lambda \rho_{\Lambda,2}, \quad p_\Lambda \geqslant 0, \quad \sum_{\Lambda} p_\Lambda = 1, \tag{4.1}$$

where Λ denotes a set of quantum numbers of a particular eigenfunction Ψ_Λ, from which $\rho_{\Lambda,2}$ is derived according to Eq. (2.4). From this ensemble 2DM other ensemble DMs are obtained according to Eqs. (2.5)–(2.7), and they satisfy the cusp conditions (1.2) and (2.12). The same is true for the ensemble α-dependent densities. Since all considered energy densities depend on DMs linearly, the densities $\varepsilon_{\mathrm{ee}}$, $\varepsilon_{\mathrm{ee}}^{\alpha}$, ε_{c} for ensembles satisfy the cusp conditions (2.16),(3.19) and the density $\varepsilon_{\mathrm{kin}}+v\, n_1$ satisfies the relation (2.17). After calculating $\varepsilon_{\mathrm{es}}$ in terms of the ensemble n_1, Eq. (2.19), the exchange energy density for an ensemble $\varepsilon_{\mathrm{x}}=\varepsilon_{\mathrm{ee}}^{0}-\varepsilon_{\mathrm{es}}$ also satisfies the cusp condition (3.18).

The cusp conditions obtained in the present paper are satisfied not only by exact solutions of many-electron problems, but also by approximate solutions that are derived from an approximate Hamiltonian, provided its potential energy term satisfies Eq. (1.8). The Hartree-Fock approximation evidently belongs to this class. Finally, since solutions for solids can be viewed as obtained by considering solutions for larger and larger clusters, all derived cusp conditions (being local properites) remain valid for condensed matter.

In concluding, our paper demonstrates that nuclear cusp conditions are satisfied by various terms of the energy density that occur in quantum chemistry in general and in a multitude of DFT approaches.

ACKNOWLEDGMENTS

The Flemish Science Foundation (FWO) is acknowledged by I.A.H. for support under Grant No. G.0347.97 and also by P.S. for fiancial support. The University of Antwerp (RUCA) is thanked by I.A.H. for its support. This work was also supported by the Concerted Action Program of the University of Antwerp.

[1] T. Kato, Commun. Pure Appl. Math. **10**, 151 (1957).
[2] W. A. Bingel, Z. Naturforsch. A **18**, 1249 (1963).
[3] R. T. Pack and W. B. Brown, J. Chem. Phys. **45**, 556 (1966).
[4] E. Steiner, J. Chem. Phys. **39**, 2365 (1963).
[5] Á. Nagy, Int. J. Quantum Chem. **70**, 681 (1998).
[6] E. R. Davidson, *Reduced Density Matrices in Quantum Chemistry* (Academic Press, New York, 1976).
[7] J. Harris and R. O. Jones, J. Phys. F: Met. Phys. **4**, 1170 (1974).
[8] R. G. Parr and W. Yang, *Density-Functional Theory of Atoms and Molecules* (Oxford University Press, New York, 1989).
[9] N. Moiseyev, Chem. Phys. Lett. **321**, 469 (2000).
[10] E. K. U. Gross (unpublished), quoted in [9].
[11] E. K. U. Gross, L. N. Oliveira, and W. Kohn, Phys. Rev. A **37**, 2805 (1988); **37**, 2809 (1988).

PHYSICAL REVIEW A **68**, 044502 (2003)

Nonrelativistic variationally optimized exchange potentials for Ne-like atomic ions having large atomic number

I. A. Howard,[1] N. H. March,[1,2] and J. D. Talman[3]
[1]*Department of Physics, University of Antwerp, Groenenborgerlaan 171, Antwerp, Belgium*
[2]*Oxford University, Oxford, England*
[3]*Department of Applied Mathematics, University of Western Ontario, London, Ontario, Canada*
(Received 8 April 2003; revised manuscript received 17 July 2003; published 20 October 2003)

In an earlier study, the exchange energy density $\epsilon_x(r)$ has been calculated analytically for Ne-like atomic ions in the nonrelativistic limit of large atomic number Z. Here, though now by numerical methods based on the optimized exchange potential approximation, corresponding results are obtained for $V_x(r)$ for such ten-electron ions at large Z. The approach to the bare Coulomb limiting form of the 1/Z expansion is considered in some detail, as are corrections to the Slater potential $V_x^{Sl}(r)$, given in terms of the exchange energy density and ground-state density $\rho(r)$ by $2\epsilon_x(r)/\rho(r)$.

DOI: 10.1103/PhysRevA.68.044502 PACS number(s): 31.15.Ew

I. BACKGROUND AND OUTLINE

In recent work, Howard *et al.* [1] have calculated analytically the nonrelativistic limiting form of the exchange energy density for the Ne-like series of positive atomic ions in the limit of large atomic number. These authors defined the exchange energy density $\epsilon_x(r)$ from the Dirac [2] result

$$E_x = -\frac{e^2}{4}\int \frac{\gamma^2(\mathbf{r},\mathbf{r}')}{|\mathbf{r}-\mathbf{r}'|}d\mathbf{r}\, d\mathbf{r}', \tag{1.1}$$

where E_x is the total exchange energy and $\gamma(\mathbf{r},\mathbf{r}')/2$ is the idempotent form of the Dirac density matrix. It then seems natural, though not of course unique [3], to define the exchange energy density $\epsilon_x(r)$ as

$$\epsilon_x(r) = -\frac{e^2}{4}\int \frac{\gamma^2(\mathbf{r},\mathbf{r}')}{|\mathbf{r}-\mathbf{r}'|}d\mathbf{r}'. \tag{1.2}$$

Evidently, from Eqs. (1.1) and (1.2),

$$E_x = \int \epsilon_x(r)d\mathbf{r}. \tag{1.3}$$

In Ref. [1], the exchange energy density $\epsilon_x(r)$ is given analytically for the ten-electron Ne-like series of positive atomic ions in the nonrelativistic limit of large atomic number Z. In particular, it was shown in Ref. [1] that E_x is the sum of three terms

$$E_x = E_x^{(1)} + E_x^{(2)} + E_x^{(3)}, \tag{1.4}$$

but only one of these, $E_x^{(1)}$, is known explicitly in terms of the electron density $\rho(r)$. This has the form

$$E_x^{(1)} = -\frac{5\pi}{3}e^2\int r^2\rho^2(r)d\mathbf{r}. \tag{1.5}$$

Both $E_x^{(2)}$ and $E_x^{(3)}$ are, from the theorems of density functional theory [4], functionals of the ground-state density $\rho(r)$ but contain the kinetic-energy density $t(r)$. This presents problems when one attempts to construct the exchange potential $V_x(r)$ as the functional derivative

$$V_x(r) = \frac{\delta E_x[\rho]}{\delta\rho}. \tag{1.6}$$

Therefore, in Ref. [1], though $\epsilon_x(r)$ is given explicitly, $V_x(r)$ is approximated by (i) the Slater [5] potential $V_x^{Sl}(r)$ and (ii) the Harbola-Sahni [6] potential $V_x^{HS}(r)$. The former potential, following Slater, is given in terms of $\epsilon_x(r)$ and $\rho(r)$ by

$$V_x^{Sl}(r) = 2\frac{\epsilon_x(r)}{\rho(r)}, \tag{1.7}$$

and hence in this approximation the functional differentiation in Eq. (1.6) is bypassed. Given $\gamma(\mathbf{r},\mathbf{r}')$, the Harbola-Sahni [6] potential is known explicitly and again no functional differentiation is involved. Using the limiting large Z form of Ref. [1] for γ (bare Coulomb form, without electron-electron interaction "corrections") both the potentials $V_x^{Sl}(r)$ and $V_x^{HS}(r)$ were plotted as functions of r for Z=92 in Ref. [1]. One had no check on the errors of these two approximations, nor on whether at Z=92 the limiting form of large Z was already quantitative. These two aspects are now addressed in this Brief Report.

The outline of the present study is then as follows. In Sec. II, the investigations of Sharp and Horton [7] and Talman and Shadwick [8] are first discussed. Then, using the so-called optimized exchange potential, the program of Talman [9] is used to obtain a very accurate approximation to $V_x(r)$ for the ten-electron atomic ions. The bare Coulomb limit of this variationally optimized method is also extracted, as well as results for $V_x(r)$ for such ions with Z=50 and Z=92. Then in Sec. III, some consequences of these exchange potentials are summarized, as well as the forces $-\partial V_x/\partial r$ in these cases and in the limiting large Z form. In this section, the virial-like relation of Levy and Perdew [10] is tested on the optimized exchange potentials. Finally in Sec. IV, the "closed-shell" Ar-like atomic ions are discussed more briefly, and the range of validity of the large Z limit is the

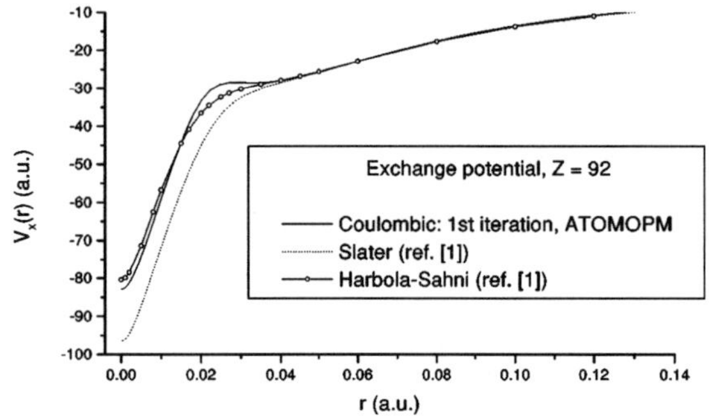

FIG. 1. Limiting form of the optimized exchange potential $V_x(r)$ calculated from Eq. (2.1) in the large atomic number Z limit. Curves in this figure are for $Z=92$ and 10 electrons. The result obtained here is compared with the Slater approximation $V_x^{Sl}(r)$ given in Eq. (1.7) and with the Harbola-Sahni potential $V_x^{HS}(r)$. These latter two potentials are taken from Ref. [1].

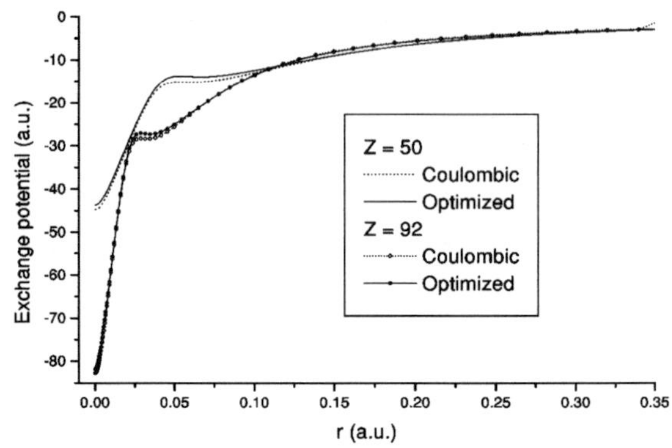

FIG. 2. Illustration of the way $V_x(r)$ is obtained by a numerical iterative solution of Eq. (2.1). The fully converged solutions for the optimized exchange potential for nonrelativistic positive atomic ions with ten electrons are shown for both $Z=50$ and $Z=92$. For comparison, the lowest order iteration using the bare Coulomb potential as the starting point is also plotted.

prime focus here. Section V constitutes a summary, plus some suggestions for future work.

II. OPTIMIZED EXCHANGE POTENTIALS FOR TEN-ELECTRON ATOMIC IONS FOR LARGE VALUES OF Z

The purpose of this section is to obtain numerical results for the optimized exchange potential for (i) the limit of large Z, studied in Ref. [1], and (ii) two positive atomic ions with ten electrons and atomic number Z chosen as 50 and 92. To achieve such results, we made use of the program ATOMOPM of Ref. [9]. The integral equation for the optimized exchange potential $V_x(r)$ which we briefly set out below is solved iteratively, starting with bare Coulomb functions. The equation has the form (Eq. (10) of Ref. [9])

$$\int K(\mathbf{r},\mathbf{r}')V_x(\mathbf{r}')d\mathbf{r}'=Q(\mathbf{r}). \quad (2.1)$$

Here K involves the one-electron orbitals ϕ_i, and the Green function $G_i(\mathbf{r},\mathbf{r}')$ for the single-particle Hamiltonian. Each side of Eq. (2.1) is orthogonal to a constant; therefore its solution is arbitrary up to an additive constant, which is determined by the behavior of $V_x(\mathbf{r})$ at large r. The complete definitions of $K(\mathbf{r},\mathbf{r}')$ and $Q(\mathbf{r})$ are given in Ref. [9].

TABLE I. Data from the first step in the iteration of Eq. (2.1) compared with exact Coulomb data for $Z=92$ and 50, ten-electron ions (in a.u.).

$Z=92$	$Z=50$
First step, one-electron eigenvalues	
$E(1s)=-4232.004$	$E(1s)=-1250.001$
$E(2s)=-1058.003$	$E(2s)=-312.496$
$E(2p)=-1058.001$	$E(2p)=-312.499$
First step, exchange energy	
$E_x=-160.807$	$E_x=-87.400$
Fully converged values, one-electron eigenvalues	
$E(1s)=-3998.307$	$E(1s)=-1125.651$
$E(2s)=-928.580$	$E(2s)=-243.385$
$E(2p)=-911.984$	$E(2p)=-234.858$
Fully converged values, exchange energy	
$E_x=-155.643$	$E_x=-82.230$
Large-Z limit, one-electron eigenvalues	
$E(1s)=-4232$	$E(1s)=-1250$
$E(2s,2p)=-1058$	$E(2s,2p)=-312.500$
Large-Z limit, exchange energy	
$E_x=-160.805$	$E_x=-87.394$

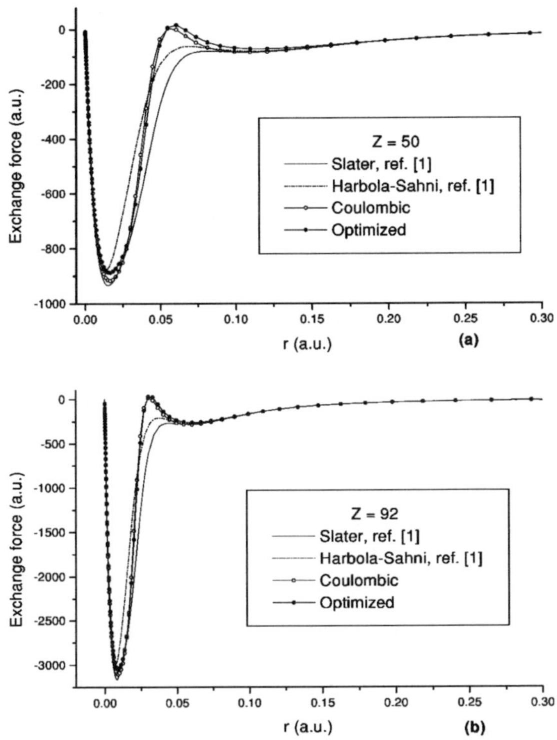

FIG. 3. Limiting large Z (bare Coulomb) form of force $F_x(r)$ in Eq. (3.2) for ten electrons. Converged forms of solution of Eq. (2.1) for (a) $Z=50$ and (b) $Z=92$ are shown for comparison. Slater and Harbola-Sahni curves [6] are also plotted.

In the iterative scheme of Ref. [9], it is very convenient that the first step can be taken with the bare Coulomb potential [1]. Of course, to display the result one must choose a specific atomic number. For the choice $Z=92$, we show the result of the program in Ref. [9] for the optimized exchange potential in Fig. 1. For comparison, the results for $V_x^{Sl}(r)$ and $V_x^{HS}(r)$ given in Ref. [1] are reproduced also in Fig. 1. Additionally, we include in Table I some facts pertaining to the first step, to show the accuracy of the numerical program compared with exact Coulomb values.

We expect that $V_x(r)$ will be a very accurate numerical approximation to the functional derivative result Eq. (1.6) and thus the appropriate curve in Fig. 1 complements the analytic form of the exchange energy $\epsilon_x(r)$ given in Ref. [1] in the limiting bare Coulomb form of the $1/Z$ expansion. But also using the ATOMOPM program, one can answer for the ten-electron system how large Z must be to validate the bare Coulomb limiting form. This is illustrated in Fig. 2, where the large-Z limiting form shown in Fig. 1 is compared with the fully converged iterative solution of Eq. (2.1) for two values of Z, namely, 50 and 92. It can be seen for the ten-electron case that $Z=92$ is already well approximated by the first step in the iteration of Eq. (2.1), namely, the bare Coulomb limit.

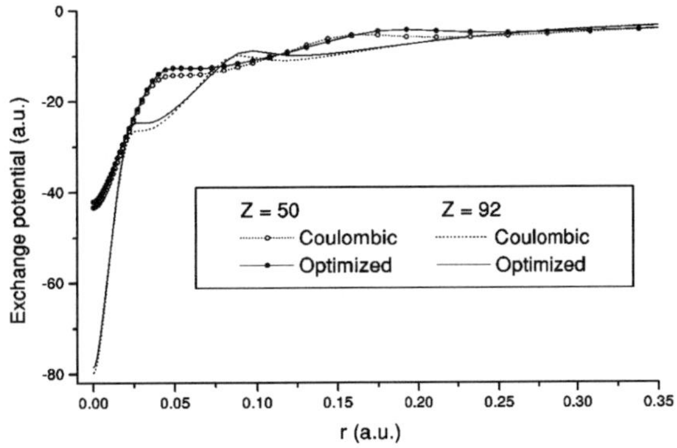

FIG. 4. Fully converged optimized exchange potential $V_x(r)$ for the Ar-like series. This is compared with the first iteration using the bare Coulomb potential. Note that the $Z=92$ curves now differ appreciably, in contrast to the close agreement for the ten-electron ions shown in Fig. 2.

III. SOME OTHER EXCHANGE PROPERTIES: BOTH INTEGRAL AND DIFFERENTIAL

Turning to other exchange properties, we begin with the virial-like relation of Levy and Perdew [10], which reads

$$E_x = -\int \rho(\mathbf{r})\mathbf{r}\cdot\nabla V_x(\mathbf{r})d\mathbf{r}. \tag{3.1}$$

This relation must be satisfied exactly if V_x is the functional derivative of E_x given in Eq. (1.6). Therefore a check of the accuracy of Eq. (3.1) is one way to establish the numerical utility of the optimized exchange potential satisfying Eq. (2.1). Since Eq. (3.1) is virial-like, it involves the force $\mathbf{F}_x$ associated with the exchange potential V_x, namely,

$$\mathbf{F}_x = -\nabla V_x(\mathbf{r}). \tag{3.2}$$

The limiting bare Coulomb form of Eq. (3.2) obtained from the first step in the iteration of Eq. (2.1) is compared with the approximations $\mathbf{F}_x^{Sl}$ and $\mathbf{F}_x^{HS}$ obtained in Ref. [1] in the present Fig. 3 for $Z=50$ and 92. We note that these three force approximations all represent zero "exchange force" at the nucleus. All three approximations have the correct large-r asymptotic forms

$$F_x = -\frac{e^2}{r^2}, \quad r\to\infty. \tag{3.3}$$

Our numerical studies for ten electrons show that the Levy-Perdew relation Eq. (3.1) is satisfied essentially exactly in both the first iteration (i.e., Coulombic) and in the fully optimized calculation. A related, but somewhat less accurate numerical check was also made in the earlier study of Wang *et al.* [11].

In concluding this section, we note that the Harbola-Sahni potential $V_x^{HS}(r)$ discussed in Ref. [1] (see also Fig. 1) obeys Eq. (3.1) exactly [6] even though, as demonstrated by Levy

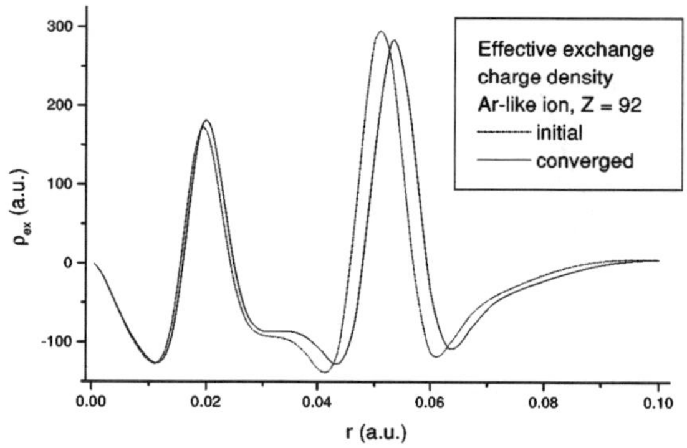

FIG. 5. Radial distribution $r^2\rho_x(r)$ of an effective exchange charge density $\rho_x(r) = -\nabla^2 V_x(r)$ for the Ar-like ion with $Z=92$ and 18 electrons. Results for the initial iteration and for the fully converged optimized exchange potential are compared.

and March [12], "corrections" which are "kinetic" in character are required to convert $V_x^{HS}(r)$ to the exact exchange-only potential.

IV. Ar-LIKE SERIES: BRIEF SUMMARY

Though, in principle, the work of Ref. [1] in the bare Coulomb limit of large Z could be extended analytically to the 18-electron, Ar-like, series of positive atomic ions, the exchange energy density $\epsilon_x(r)$ would be the result, and not $V_x(r)$ as explained above for the Ne-like series. Therefore, here we confine ourselves to showing in Fig. 4 the limiting large-Z form, again from the first step in the iteration of Eq. (2.1), for $Z=50$ and 92. The finally converged curve for the same Z is shown for comparison, again for 18 electrons, in Fig. 4.

Finally, again in response to a suggestion of the referee, we show in Fig. 5 a somewhat different representation of the exchange potential for $Z=92$ and 18 electrons, namely, its Laplacian, related to a charge density $\rho_{ex}(r)$.

V. SUMMARY AND FUTURE DIRECTIONS

The main results of the present study are as follows.

(i) The limiting (Coulomb) forms of the first iteration of the optimized effective potential approximation to the exchange potential in the nonrelativistic theory for large atomic number Z for (a) the ten-electron Ne-like positive atomic ions and (b) the 18-electron Ar-like series.

(ii) The converged iterative solutions of Eq. (2.1) for the ten-electron atomic ions for $Z=50$ and $Z=92$. It would, we believe, in the future, be of interest to attempt to use existing theory [13] on the $1/Z$ expansion for homonuclear diatomic molecules to calculate the limiting forms for some positive molecular ions of the optimized exchange potential $V_x(\mathbf{r})$. Possibly the extension to almost spherical C and B cages [14,15] might also prove practicable.

Finally, the referee has drawn our attention to the work of Nagy [16]. She has kindly informed us that if one knows the exact density (say from diffraction or quantum simulation), one can calculate the exchange-correlation potential either with the Zhao-Morrison-Parr [17] method, or other methods, such as her own. In this sense, as the referee pointed out to us, the Zhao-Morrison-Parr approach can be said to be more accurate than the optimized effective potential method. However, the optimized effective potential can be regarded as the exact exchange potential of density-functional theory in the exchange-only approximation [16,18].

ACKNOWLEDGMENTS

Research by J.D.T. was supported by the Natural Sciences and Engineering Research Council of Canada. I.A.H. acknowledges support from the IWT—Flemish region. N.H.M. wishes to acknowledge partial financial support from the Office of Naval Research for work on density-functional theory. Special thanks are due to Dr. P. Schmidt of that office for much motivation and encouragement.

[1] I.A. Howard, N.H. March, P. Senet, and V.E. Van Doren, Phys. Rev. A **62**, 062512 (2000).
[2] P.A.M. Dirac, Proc. Cambridge Philos. Soc. **26**, 376 (1930).
[3] See, for example, K. Burke, F.G. Cruz, and K.-C. Lam, J. Chem. Phys. **109**, 8161 (1998).
[4] P.C. Hohenberg and W. Kohn, Phys. Rev. **136**, B864 (1964).
[5] J.C. Slater, Phys. Rev. **81**, 385 (1951).
[6] M.K. Harbola and V. Sahni, Phys. Rev. Lett. **62**, 489 (1989).
[7] R.T. Sharp and G.K. Horton, Phys. Rev. **90**, 316 (1955).
[8] J.D. Talman and W.F. Shadwick, Phys. Rev. A **14**, 36 (1976).
[9] J.D. Talman, Comput. Phys. Commun. **54**, 85 (1989).
[10] M. Levy and J.P. Perdew, Phys. Rev. A **32**, 2010 (1985).
[11] Y. Wang, J.P. Perdew, J.A. Chevary, L.D. MacDonald, and S.H. Vosko, Phys. Rev. A **41**, 78 (1990).
[12] M. Levy and N.H. March, Phys. Rev. A **55**, 1885 (1997).
[13] N.H. March and R.G. Parr, Proc. Natl. Acad. Sci. U.S.A. **77**, 6285 (1980).
[14] C. Amovilli, I.A. Howard, D.J. Klein, and N.H. March, Phys. Rev. A **66**, 013210 (2002).
[15] C. Amovilli and N.H. March, Chem. Phys. Lett. **347**, 459 (2001).
[16] A. Nagy, Phys. Rep. **298**, 1 (1998); A. Nagy, J. Phys. B **26**, 43 (1993); A. Nagy, Philos. Mag. B **69**, 779 (1994).
[17] Q. Zhao, R.C. Morrison, and R.G. Parr, Phys. Rev. A **50**, 2138 (1994).
[18] V. Sahne, J. Gruenebaum, and J.P. Perdew, Phys. Rev. B **26**, 4371 (1982).

Recent Progress in Constructing Nonlocal Energy Density Functionals

N. H. MARCH,[1,2] I. A. HOWARD,[1] V. E. VAN DOREN[1]

[1]*Department of Physics, University of Antwerp (RUCA), Groenenborgerlaan 171, B-2020 Antwerp, Belgium*

[2]*Oxford University, Oxford, England*

Received 15 January 2002; accepted 10 March 2002

DOI 10.1002/qua.10506

ABSTRACT: Both model results and some exact relations concerning kinetic, exchange, and correlation energy density functionals will be presented. As to single-particle kinetic energy, a model of harmonic confinement, motivated by recent experimental activity on Fermion vapors, of an arbitrary number of Fermions will be used to construct exact nonlocal kinetic energy density functionals in both one and three dimensions. Then, some further exact results concerning nuclear cusp conditions for exchange only, and for exchange plus correlation, will be discussed. Finally, in Appendices, the one-shell harmonic confinement is generalized to include Coulomb repulsion in the two-electron Hookean atom and some results on the Pauli potentials for many-Fermion systems are obtained. © 2003 Wiley Periodicals, Inc. Int J Quantum Chem 92: 192–204, 2003

Key words: density; kinetic functional; exchange

1. Background and Outline

A long-term aim of what is now called density functional theory (DFT) [1] has been to obtain a differential equation for the Fermion ground-state density $\rho(\mathbf{r})$. Early work in this area was that of Lawes and March [2], who drew on the results of Husimi [3] plus the subsequent study of March and Young [4] on the differential form of the virial theorem [compare Eq. (2.5) below] to obtain such a differential equation for N 1-D Fermions subject to harmonic confinement. Interest in such an approach has been revived because the experimental techniques that led to the achievement of Bose–Einstein condensation in vapors of bosonic atoms [5] are at present being utilized to trap and cool dilute gases of Fermionic alkali atoms [6]. This experimental activity prompted Vignolo et al. [7] to construct a Green's function operator approach to obtain by numerical methods the Fermion particle and kinetic energy densities. Their study has subsequently been related to the theory of Lawes and March [2] by March and Nieto [8], who calculated thereby the particle density, and also utilized the

Correspondence to: I. A. Howard; e-mail: howard@ruca.ua.ac.be study of March et al. [9] to obtain the kinetic energy density.

The outline of this article is as follows. In Section 2, nonlocal single-particle kinetic energy functionals referred to above are explicitly constructed for Fermion harmonic confinement. The relevance for DFT is the continuing interest in so-called "orbital-free" DFT, which of course goes back to the origins of electron density theory through the work of Thomas [10], Fermi [11], and Dirac [12]. Whereas, beginning with Slater [13] (see also Gáspár [14]), whose investigations were formally completed by Kohn and Sham [15], much emphasis moved away from the original approach in terms solely of the density $\rho(\mathbf{r})$ and back to solving (of course, self-consistency for atoms, molecules, and condensed phases) one-particle Schrödinger equations (the Slater–Kohn–Sham equations), interest in orbital-free DFT has continued [16, 17].

However, even within such orbital-free theory, one must still have good approximations, to treat electrons in molecules and clusters, to the exchange and correlation energy density $\epsilon_{xc}(\mathbf{r})$. This topic is therefore taken up in the later sections of this article, with particular emphasis placed on nuclear cusp conditions for exchange and correlation energy densities [18]. In the Appendices, harmonic confinement theory in Section 2 is first related to the two-electron Hookean atom [19], while in Appendix B the concept of the Pauli potential [20, 21] is illustrated numerically for the same harmonic confinement.

2. Fermion Confinement in One and Three Dimensions: Model Single-Particle Kinetic Energy Functionals

2.1. THOMAS–FERMI AND VON WEIZSÄCKER BUILDING BLOCKS

March and Young [4], in early work in which a synthesis was made of the electron density method with Dirac density matrix theory [12], recognized that the single-particle kinetic energy T_s would contain both the Thomas–Fermi kinetic energy T_{TF}, which in d dimensions reads [22]

$$T_{\mathrm{TF}} = c_k^{(d)} \rho^{1+2/d}, \tag{2.1}$$

where $c_k^{(1)} = (\pi^2\hbar^2/24\mathrm{m})$ and $c_k^{(3)} = (3/10)(3\pi^2)^{2/3}(\hbar^2/\mathrm{m})$, and the von Weizsäcker inhomogeneity kinetic energy [23], defined by

$$T_{\mathrm{W}} = \frac{\hbar^2}{8m} \int \frac{(\nabla\rho)^2}{\rho}\, d\mathbf{r}. \tag{2.2}$$

For a 1-D model of harmonically confined Fermions, it will be set out below, following March et al. [9], that the single-particle kinetic energy functional $T_s[\rho]$, although fundamentally nonlocal, contains the building block (2.1) for $d = 1$ and the 1-D form $(\hbar^2/8\mathrm{m}) \int [(\partial\rho/\partial x)^2/\rho]dx$ of T_{W} in Eq. (2.2) as fundamental entities.

2.2. $T_s[\rho]$ FOR 1-D HARMONIC CONFINEMENT WITH AN ARBITRARY NUMBER OF FERMIONS

The chemical potential μ in DFT may be written, in one dimension, as [1, 22]

$$\mu = \frac{\delta T_s[\rho]}{\delta\rho(x)} + V(x). \tag{2.3}$$

The important property of Eq. (2.3) is that while the chemical potential μ must be constant at every point x in the inhomogeneous electronic cloud of density $\rho(x)$, it is made up of the sum of two terms, a kinetic contribution $\delta T_s[\rho]/\delta\rho(x)$, which is the dominant interest in this section, and a potential energy $V(x)$, which in the harmonic model is simply, in suitable units, given by

$$V(x) = \frac{1}{2}x^2. \tag{2.4}$$

Evidently, although their sum is constant, the two contributions of kinetic and potential terms vary from point to point in the electronic distribution.

The second important relation in deriving the single-particle kinetic energy functional for this model is the so-called differential viral theorem derived by March and Young [4], namely,

$$\frac{\partial t(x)}{\partial x} = -\frac{1}{2}\rho(x)\frac{\partial V(x)}{\partial x} - \frac{1}{8}\rho'''(x). \tag{2.5}$$

Equation (2.5) is in fact valid for any confining potential $V(x)$ for Fermions; however, we shall employ it below only for the harmonic form (2.4). Third, Lawes and March [2] derived a differential equation for the ground-state electron density in this harmonic model for an arbitrary number of Fermions as

$$-\frac{1}{2}\rho(x)\frac{\partial V(x)}{\partial x} - \frac{1}{8}\rho'''(x) = (N - V(x))\rho'(x). \quad (2.6)$$

In Eq. (2.6), the lowest level only being occupied corresponds to $N = 1$. Utilizing Eq. (2.5) in the left side of Eq. (2.6) yields immediately the relation

$$\frac{t'(x)}{\rho'(x)} = N - V(x), \quad (2.7)$$

and after differentiating Eq. (2.7) with respect to x and using the chemical potential Eq. (2.3) one is led to the result

$$\left[\frac{\delta T}{\delta \rho(x)}\right]' = \left[\frac{t'(x)}{\rho'(x)}\right]'. \quad (2.8)$$

Writing the kinetic energy density $t(x)$, which is such that

$$T_s = \int_{-\infty}^{\infty} t(x)dx \quad (2.9)$$

as

$$t(x) = t_W(x) + \xi(x)t_{TF}(x), \quad (2.10)$$

March et al. [9] showed that the quantity $\xi(x)$ is given by

$$\xi(x) = \xi(0) + \frac{4}{3}\int_0^x dx\left[\frac{t_W'(x)}{t_{TF}(x)} - \left\{\frac{t_W(x)}{t_{TF}(x)}\right\}'\right]. \quad (2.11)$$

The quantity $\xi(0)$, in turn, is given by

$$\xi(0) = \frac{t(0) - t_W(0)}{t_{TF}(0)}, \quad (2.12)$$

where

$$t(0) = \frac{N\rho(0)}{3} - \frac{\rho''(0)}{12}. \quad (2.13)$$

In Figure 1 we show $\xi(x)$ for the cases $N = 20$ and 50. Equations (2.10)–(2.13) show that indeed the building blocks for this nonlocal kinetic energy density functional are the Thomas–Fermi kinetic energy density $t_{TF}(x)$ and the von Weizsäcker term $t_W(x)$ (Fig. 2). In the early (and of course approximate) density matrix treatment of March and Young [24], the form (2.10) was already obtained (see also Refs. [25]–[27]), but with $\xi(x)$ independent of x and only dependent on N.

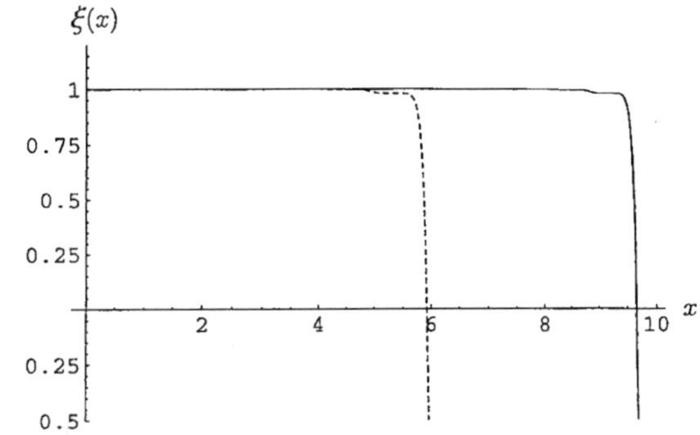

FIGURE 1. Plot of $\xi(x)$ for 1-D harmonic confinement as given by Eqs. (2.10) and (2.11). This quantity describes the (local) weight of the Thomas–Fermi kinetic energy density $t_{TF}(x)$ to be added to the von Weizsäcker form $t_W(x)$. The broken curve is for $N = 20$; the solid curve for $N = 50$. (Results follow from March and Nieto [8].)

The final nonlocal single-particle kinetic energy $T_s[\rho]$ for this 1-D model of harmonic confinement is given by

$$T_s[\rho] = \int_{-\infty}^{\infty} t(x)dx$$

$$= T_W[\rho] + \int_{-\infty}^{\infty} dx\xi(x)t_{TF}(x), \quad (2.14)$$

where the von Weizsäcker kinetic energy is explicitly [cf. Eq. (2.2)]

$$T_W[\rho] = \frac{1}{8}\int_{-\infty}^{\infty} dx\frac{\{\rho'(x)\}^2}{\rho(x)}. \quad (2.15)$$

2.3. ISOTROPIC HARMONIC CONFINEMENT IN THREE DIMENSIONS

Minguzzi et al. [28] recently generalized the differential equation of Lawes and March in one di-

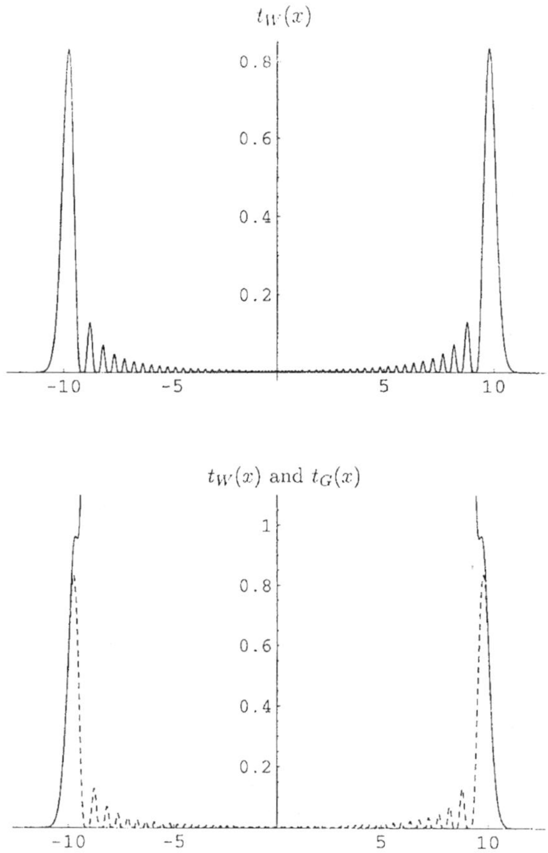

FIGURE 2. Top: Plot of von Weizsäcker kinetic energy $t_W(x)$ for 1-D harmonic confinement for $N = 50$. Bottom: Compares $t_W(x)$ with the (positive definite) form $t_G(x)$ of the kinetic energy density, again for $N = 50$. The remarkable behavior seen is that, rapidly, outside the classically allowed region $t_W(x)$ is equal to $t_G(x)$ to high accuracy. (Results obtained by March and Nieto [8].)

mension to apply to $(M + 1)$ closed shells with isotropic harmonic confinement in three dimensions.

Their equation reads

$$\frac{\hbar^2}{8m}\frac{\partial}{\partial r}(\nabla^2\rho) + [(M+2)\hbar\omega - V(r)]\rho'(r) + \frac{3}{2}\frac{\partial V}{\partial r}\rho(r) = 0, \quad (2.16)$$

with

$$V(r) = \frac{1}{2}m\omega^2 r^2.$$

This Eq. (2.16) has an exact solution given by Howard et al. [29] as

$$\rho(r) = \mathcal{A}\exp(-\omega r^2)\sum_{n=0}^{M} a(n)(\omega r^2)^n, \quad (2.17)$$

where the normalization constant $\mathcal{A}$ is given by

$$\mathcal{A} = \left[\frac{\sqrt{\pi}}{2}\left(\frac{\omega}{\pi}\right)^{3/2} N\right] \Big/ \sum_{n=0}^{M} a(n)\Gamma(n + 3/2) \quad (2.18)$$

for N Fermions in $M + 1$ filled shells [note that $N = (M + 1)(M + 2)(M + 3)/6$] and the coefficients $a(n)$ are related by the recursion relation

$$0 = a(n+2)\left[\frac{(n+2)(2n+5)}{2}\right] + a(n+1)[2(M+1) - 3(n+1)] + a(n)\left[\frac{2(n-M)}{(n+1)}\right]. \quad (2.19)$$

In the work of Minguzzi et al. [28] leading to Eq. (2.16), the kinetic energy density $t(r)$ was also obtained explicitly as

$$t(r) = \frac{t_W(r)}{3} + \left\{C + \frac{\hbar^2}{3m}\int_0^r ds\,\frac{[\rho'(s)]^2}{\rho^{8/3}(s)} \times \left[\frac{1}{2s} + \frac{\rho'(s)}{3\rho(s)}\right]\right\}\rho^{5/3}(r). \quad (2.20)$$

Here, C is a known (M-dependent) constant.

It is important to emphasize first that in Eq. (2.20), the full von Weizsäcker piece

$$t_W(r) = \frac{\hbar^2}{8m}\frac{(\nabla\rho)^2}{\rho} \quad (2.21)$$

does not separate out, as it does in one dimension, but is divided by the dimensionality d [=3 in Eq. (2.20)]. Second, as in one dimension, $t_W(r)$ given above and $t_{TF}(r)$ given by

$$t_{\mathrm{TF}}(r) = c_k\rho^{5/3}(r);\ c_k = \frac{3h^2}{10m}\left(\frac{3}{8\pi}\right)^{2/3} \tag{2.22}$$

figure prominently as building blocks in the fully nonlocal form (2.20) for the kinetic energy density $t(r)$.

It is to be stressed at this point that for harmonic confinement, such as in recent experimental activity on dilute Fermion vapors, a complete DFT is possible. In three dimensions, this constitutes (1) the differential Eq. (2.16) for the Fermion density $\rho(r)$, and its solution (2.17), and (2) the kinetic energy functional $T[\rho] = \int t(r)d\mathbf{r}$, where $t(r)$ is given by Eq. (2.20).

Although this section is dominantly about harmonic confinement, we want to add here that a nonlocal kinetic energy functional has recently been constructed [30] for the model of an arbitrary number of closed shells in a bare Coulomb field. As Amovilli and March [31] have recently shown, for this model one can write, with kinetic energy density $t_G(r)$ now defined from the $(\nabla\psi)^2$ form, which differs from the $\psi\nabla^2\psi$ expression by $1/4\ \nabla^2\rho(\mathbf{r})$, with $\rho(\mathbf{r})$ as usual the electron density,

$$\frac{\partial t_G}{\partial r} = \frac{\rho'''}{8} - \frac{3\rho'}{4r^2} - \frac{3Z}{2r^2}\rho, \tag{2.23}$$

where Z is the atomic number. Forming the Laplacian of $t_G(r)$ from Eq. (2.23), one finds [30]

$$\nabla^2 t_G(r) = \left[\frac{\rho^{IV}}{8} + \frac{\rho'''}{4r} - \frac{3\rho''}{4r^2} - \frac{3Z\rho'}{2r^2}\right]. \tag{2.24}$$

Hence, $t_G(r)$ can be written in the nonlocal form

$$t_G(r') = -\frac{1}{4\pi}\int \frac{\left[\frac{\rho^{IV}}{8} + \frac{\rho'''}{4r} - \frac{3\rho''}{4r^2} - \frac{3Z\rho'}{2r^2}\right]}{|\mathbf{r}-\mathbf{r}'|}d\mathbf{r}. \tag{2.25}$$

Because in this case Kato's theorem reads $Z = -1/2\,\rho'(0)/\rho(0)$, Eq. (2.25) can be viewed as giving $t_G(r)$ as a functional of $\rho(r)$.

The further implications of the form (2.25) will be discussed elsewhere not only for DFT but also for Hartree–Fock theory, and in particular for the two-level Be atom [32] (see also the example in Appendix C). After this discussion of explicitly nonlocal forms of the single-particle kinetic energy, we turn to consider some general properties of exchange only as well as exchange plus correlation energy densities [18].

3. Nuclear Cusp Conditions on Kinetic and Exchange Energy Densities in Atomic Ions

3.1. SIMPLE EXAMPLE OF HYDROGEN-LIKE SYSTEMS

Let us introduce this section on nuclear cusp conditions with the hydrogenic ion having atomic number Z and a single electron. The normalized ground-state wave function $\psi_{1s}(r)$ is given by

$$\psi_{1s}(r) = \left(\frac{Z^3}{\pi a_o^3}\right)^{1/2}\exp\left(-\frac{Zr}{a_o}\right);\ a_o = \frac{\hbar^2}{me^2}. \tag{3.1}$$

Hence, the corresponding ground-state electron density $\rho(r)$ is given by

$$\rho(r) = \frac{Z^3}{\pi a_o^3}\exp\left(-\frac{2Zr}{a_o}\right). \tag{3.2}$$

Forming the ratio $(\partial\rho(r)/\partial r)/\rho(r)$ plainly yields

$$\frac{\partial\rho(r)}{\partial r} = -\frac{2Z}{a_o}\rho(r). \tag{3.3}$$

The limit $r \to 0$ gives immediately

$$\left(\frac{\partial\rho(r)}{\partial r}\bigg/\rho(r)\right)_{r\to 0} = -\frac{2Z}{a_o}, \tag{3.4}$$

which is a special case of a theorem derived originally by Kato [33], valid for (nonrelativistic) atomic ions with an arbitrary number N of electrons, and also at each and every nucleus of multicenter molecules and clusters (see below).

Before turning to exchange and correlation, let us show that there is a corresponding nuclear cusp condition to Eq. (3.4) for the single-particle kinetic energy density $t(r)$ for hydrogenic systems. Again, let us take the hydrogenic density (3.2) as a starting point. Now, $t_G(r)$ above is equal to $t_W(r)$ introduced in Section 2. But, using Eq. (3.3),

$$t_W(r) = \frac{1}{8}\frac{\rho'^2(r)}{\rho(r)} = \frac{Z^2}{a_o^2}\rho(r) \tag{3.5}$$

and hence

$$t'_W(r) = \frac{Z^2}{a_o^2}\rho'(r). \tag{3.6}$$

Thus, by dividing Eq. (3.6) by Eq. (3.5) and using Kato's result (3.4) we obtain

$$\left.\frac{t'_W(r)}{t_W(r)}\right|_{r=0} = \left.\frac{\rho'(r)}{\rho(r)}\right|_{r=0} = -\frac{2Z}{a_o}. \tag{3.7}$$

As with Kato's theorem for the electron density, March et al. [18] demonstrated that analogous conditions hold at every nucleus in multicenter problems. In refining approximate functionals for exchange-only energy $E_x[\rho]$ and exchange plus correlation energy $E_{xc}[\rho]$, it is of obvious importance for the future to build in precisely the nuclear cusp conditions [18]. Let us deal first with spherical atomic ions and then in Section 4 effect the generalization to multicenter problems.

3.2. EXCHANGE-ONLY THEORY FOR FOUR-ELECTRON SYSTEMS (E.G., THE Be ATOM)

In earlier work, one of us [34] discussed the asymptotic form of the exchange energy far from the nucleus of a closed-shell atomic ion. Writing the Dirac [12] exchange energy E_x in terms of the first-order density matrix $\gamma(\mathbf{r}, \mathbf{r}')$ as

$$E_x = -\frac{e^2}{4}\int \frac{\gamma^2(\mathbf{r}, \mathbf{r}')}{|\mathbf{r} - \mathbf{r}'|} d\mathbf{r}d\mathbf{r}', \tag{3.8}$$

one chooses the definition of the exchange energy density $\epsilon_x(\mathbf{r})$ as

$$\epsilon_x(\mathbf{r}) = -\frac{e^2}{4}\int \frac{\gamma^2(\mathbf{r}, \mathbf{r}')}{|\mathbf{r} - \mathbf{r}'|} d\mathbf{r}'. \tag{3.9}$$

In the exchange-only theory of this subsection (correlation will be embodied in Section 4 below), the first-order density matrix satisfies the (idempotency) condition for spin-compensated systems:

$$\frac{\rho(r)}{2} = \int \left\{\frac{\gamma(\mathbf{r}, \mathbf{r}')}{2}\right\}^2 d\mathbf{r}'. \tag{3.10}$$

In Ref. [34], it was shown, from Eqs. (3.9) and (3.10) that the large-r limit of $\epsilon_x(\mathbf{r})$ in a closed-shell atomic ion is given by

$$\epsilon_x(\mathbf{r})|_{r\to\infty} \to -\frac{e^2}{4r}\int \gamma^2(\mathbf{r}, \mathbf{r}')d\mathbf{r}'$$

$$= -\frac{1}{2}\frac{e^2}{r}\rho(r). \tag{3.11}$$

The purpose of this subsection is to obtain as a complementary limit to the large-r form (3.11) the small-r behavior of $\epsilon_x(\mathbf{r})$ for closed-shell atomic ions [18].

Let us start with the specific example of Be-like atomic ions with electronic structure $(1s)^2(2s)^2$. Using the so-called Dawson–March transformation [35]

$$\psi_{1s}(r) = \left\{\frac{\rho(r)}{2}\right\}^{1/2}\cos(\theta(r)) \tag{3.12}$$

and

$$\psi_{2s}(r) = \left\{\frac{\rho(r)}{2}\right\}^{1/2}\sin(\theta(r)), \tag{3.13}$$

it is a straightforward matter to obtain $\gamma(\mathbf{r}, \mathbf{r}')$ as

$$\gamma(\mathbf{r}, \mathbf{r}') = \rho(r)^{1/2}\rho(r')^{1/2}\cos(\theta(r) - \theta(r')) \tag{3.14}$$

in terms of the density amplitude $\rho^{1/2}(r)$ and the phase $\theta(r)$. Inserting Eq. (3.14) into Eq. (3.9), dividing both sides of the resulting equation by $\rho(r)$, and then differentiating the relation thus obtained with respect to r gives almost immediately

$$\frac{d}{dr}\left\{\frac{\epsilon_x(\mathbf{r})}{\rho(r)}\right\} = -\frac{e^2}{4}\int \left[-2\left|\mathbf{r} - \mathbf{r}'\right|\cos(\theta(r) - \theta(r')) \times \sin(\theta(r) - \theta(r'))\frac{\partial\theta(r)}{\partial r} - \cos^2(\theta(r) - \theta(r'))\frac{\mathbf{r}}{r}\cdot\frac{\mathbf{r} - \mathbf{r}'}{|\mathbf{r} - \mathbf{r}'|}\right]\rho(\mathbf{r}')d\mathbf{r}'. \tag{3.15}$$

Nagy and March [36] solved for the phase $\theta(r)$ in the Be atom using the Hartree–Fock approximation for the density $\rho(r)$ and demonstrated that $\partial\theta(r)/\partial r|_{r=0} = 0$ and hence the first term in the square bracket of Eq. (3.15) vanishes at $r = 0$. Further, the angular integration in the term $\mathbf{r} \cdot (\mathbf{r} - \mathbf{r}')$ leads to zero for the integral as $r \to 0$. Hence, for the specific example of this two-level Be atom one finds

$$\frac{d}{dr}\left\{\frac{\epsilon_x(\mathbf{r})}{\rho(r)}\right\}\Bigg|_{r=0} = 0. \tag{3.16}$$

This is readily combined with Kato's theorem (3.4) to yield the nuclear cusp condition for the exchange energy density:

$$\left[\frac{1}{\epsilon_x(r)}\frac{d\epsilon_x(r)}{dr}\right]_{r=0} = -\frac{2Z}{a_o}. \tag{3.17}$$

In Section 4, a generalization of this result (3.17) will be given, following Ref. [18], that (1) incorporates electron correlation and (2) generalizes the above argument for a simple closed-shell atomic ion to multicenter molecules and clusters.

4. Nuclear Cusp Conditions, Including Correlation, and Generalized to Molecules and Clusters

Following the exact results on nonlocal energy densities for models in Sections 2 and 3, the purpose of this section is to demonstrate that the result (3.17) above—a nuclear cusp condition—can be generalized to (1) include correlation and (2) apply to multicenter problems such as molecules and clusters. The argument below follows closely that given by March et al. [18]. We refer the reader also to discussions of cusp conditions in, for instance, Refs. [37] and [38].

4.1. WAVE-MECHANICAL PRINCIPLES EMPLOYED

It has been known since the pioneering work of Kato [33] that the singularity, at a nucleus A say with charge $Z_A e$, of the electron–nuclear interaction

$$V_A(\mathbf{r}) = -\frac{Ze^2}{|\mathbf{r} - \mathbf{R}_A|}, \tag{4.1}$$

has a consequence nonanalyticity of the N-electron wave function where each electronic position vector $\mathbf{r}_i$, $i = 1, 2, \ldots, N$, becomes equal to $\mathbf{R}_A$. Yet, as is clear on physical grounds, the wave function (in nonrelativistic theory, considered exclusively below) remains finite throughout the whole of space. Characterization of such nonanalyticity was first achieved by Kato [33], Eq. (3.4) above reflecting this result, even though this equation is written in electron density form. Bingel [39] subsequently extended the derivation to molecules. As in Ref. [18], the presentation employed by Davidson [40] will supply the background required to present explicit results relevant to the focus here on nonlocal energy density functionals.

With H the total Hamiltonian, let the N-electron antisymmetric normalized wave function $\Psi(\mathbf{x}_i \ldots \mathbf{x}_N)$ be an eigenfunction of the Schrödinger equation

$$H\Psi = E\Psi, \tag{4.2}$$

where $\mathbf{x}_i \equiv \{\mathbf{r}_i, s_i\}$ specifies the space and spin coordinates of the ith electron. In treating cusp conditions due to, say, the Ath nucleus, let us select its position to be at $\mathbf{R}_A = 0$. As Ψ is everywhere finite, it follows from Eq. (4.2) that, if one focuses on the argument $\mathbf{r}_1$ of Ψ,

$$\lim_{r_1 \to 0}[r_1 H\Psi] = 0. \tag{4.3}$$

This condition implies that $\Psi(\mathbf{r}_1, s_1, \mathbf{x}_2 \ldots \mathbf{x}_N)$ is also an eigenfunction, with zero eigenvalue of the operator

$$b(\mathbf{r}_1) = \lim_{r_1 \to 0}[r_1 T] = \lim_{r_1 \to 0}[r_1 t(\mathbf{r}_1)] - Z_A, \tag{4.4}$$

where $T = \Sigma_{i=1}^N t(\mathbf{r}_i)$ denotes the kinetic energy operator. For the wave function $\Psi(\mathbf{r}_1 \ldots \mathbf{X})$ where $\mathbf{X} \equiv \{s_1, \mathbf{x}_2 \ldots \mathbf{x}_N\}$ is expanded in spherical harmonics

$$\Psi(\mathbf{r}_1 \ldots \mathbf{X}) = \sum_{l=0}^{\infty}\sum_{m=-l}^{l} Y_{lm}(\Omega_1) r_1^l f_{lm}(r_1, \mathbf{X}), \tag{4.5}$$

with $f_{lm}(r_1, \mathbf{X})$ an analytic function of r_1, the eigenequation $b(\mathbf{r}_1)\Psi = 0$ yields the following conditions on the partial wave amplitudes:

$$\frac{\partial f_{lm}(r_1, \mathbf{X})}{\partial r_1}\Bigg|_{r_1=0} = -\frac{Z_A}{l+1} f_{lm}(0, \mathbf{X}). \tag{4.6}$$

These Eqs. (4.3)–(4.6) are employed by March et al. [18] to derive nuclear cusp conditions for the exchange and correlation energy densities of DFT.

Before turning to the DFT energy densities $\epsilon_x(\mathbf{r})$ (touched on in Section 3 above) and the corresponding correlation (c) energy density $\epsilon_c(\mathbf{r})$, for DFT applications the classic electrostatic energy of electrons is of interest, namely,

$$E_{es}[\rho] = \int d\mathbf{r}_1 \epsilon_{es}(\mathbf{r}_1), \tag{4.7}$$

where

$$\epsilon_{es}(\mathbf{r}_1) = \frac{1}{2}\int d\mathbf{r}_2 \frac{e^2}{|\mathbf{r}_1 - \mathbf{r}_2|}\rho(\mathbf{r}_1)\rho(\mathbf{r}_2). \tag{4.8}$$

March et al. [18] showed that

$$\left.\frac{\partial \bar{\epsilon}_{es}(r_1)}{\partial r_1}\right|_{r_1=0} = -2Z\epsilon_{es}(0), \tag{4.9}$$

where the bar on $\epsilon_{es}(\mathbf{r}_1)$ denotes an angular average. This is a simple analog for the classic electrostatic energy density of the exchange (x) energy density derived from the bare Coulomb limit system for a 10-electron atomic ion by Howard et al. [41]. This leads us to sketch the generalization derived for a model closed-shell atomic ion to the fully interacting electron problem for atoms, molecules, and solids. The derivation in Ref. [18] appeals to the adiabatic connection, developed by Harris and Jones [42]. This connection links the fully interacting system treated above to the SKS noninteracting reference system with the same (i.e., formally exact) electron density via intermediate systems. For this purpose, a Hamiltonian depending on a parameter α such that $0 \leq \alpha \leq 1$ is defined as

$$H^\alpha = T + V^\alpha + \alpha U, \tag{4.10}$$

where $V^\alpha = \Sigma_{i=1}^N V^\alpha(\mathbf{r}_i)$. This Hamiltonian reduces to the original Hamiltonian for $\alpha = 1$, while for $\alpha = 0$ the electron–electron piece U is switched off, while $V^0(\mathbf{r}) = V_{\mathrm{SKS}}(\mathbf{r})$.

To use this approach to study the cusp conditions for the intermediate system, March et al. [18] assumed (to be checked below) that the properties of $V^\alpha(\mathbf{r})$, whose explicit form remains unknown, permit a finite value for the limit $\gamma^\alpha = \lim_{r_1\to 0}[r_1 V^\alpha(\mathbf{r}_1)]$. Then, the early arguments carry through and one obtains

$$\frac{\partial}{\partial r}\bar{\rho}^\alpha(r)|_{r=0} = 2\gamma^\alpha\rho^\alpha(0). \tag{4.11}$$

But, the one-body potential $V^\alpha(\mathbf{r})$ is defined by requiring that the electron density $\rho^\alpha(\mathbf{r})$ of the intermediate system is independent of α. Hence, one concludes that $\gamma^\alpha = -Z_A$, which justifies *a posteriori* the assumption that $\lim_{r_1\to 0}[r_1 V^\alpha(\mathbf{r}_1)]$ is finite.

March et al. [18] showed that for the exchange correlation potential $V_{xc}(\mathbf{r})$

$$\lim_{r_1\to 0}[r_1 V_{xc}(\mathbf{r}_1)] = 0. \tag{4.12}$$

Without essentially new arguments beyond those presented above, the model result (3.7) for the exchange energy density ϵ_x generalizes to

$$\frac{\partial}{\partial r_1}\bar{\epsilon}_x(r_1)|_{r_1=0} = -2Z_A\epsilon_x(0), \tag{4.13}$$

where the angular average on the left side is not required for closed-shell atomic ions. The final result of March et al. [18] is a similar nuclear cusp condition for correlation energy density $\epsilon_c(\mathbf{r})$, namely,

$$\frac{\partial}{\partial r_1}\bar{\epsilon}_c(r_1)|_{r_1=0} = -2Z_A\epsilon_c(0). \tag{4.14}$$

These relations (4.13) and (4.14) should evidently be built into approximate functionals used in DFT in the future.

5. Summary and Proposed Future Directions

Recent progress in constructing explicitly nonlocal energy density functionals has stemmed from the use of exactly soluble models. Two of these, namely, the Bardeen model [43] of a metal surface and the harmonic confinement model of 1-D Fermions, have led to progress in writing kinetic energy densities explicitly in terms of the ground-state density ρ and its low-order derivatives {see Eqs. (2.10) and (2.11) for harmonic confinement and Ref. [44] for the Bardeen model}. The Bardeen model has also allowed a nonlocal relation to be constructed between kinetic and exchange energy densities [44]. Exchange has also been studied in the large atomic number limit of 10-electron atomic ions, a fully nonlocal calculation being presented in this Coulomb limit system in Ref. [41]; see also the article by Howard et al. [45]. Some attention has then been focused here on exact nuclear cusp conditions on energy densities, studied fully in Ref. [18].

It seems clear that refinements of (inevitably) approximate density functionals for the exchange plus correlation should, in the future, attempt to embody the nuclear cusp conditions set out in Section 4 and Ref. [18]. Also, without necessarily invoking correlation energy, some pointers indicate that there will be merit in studying together the sum $t(\mathbf{r}) + \epsilon_x(\mathbf{r})$ of the kinetic and exchange energy densities, and work is in progress by the writers on this topic. Clearly, for the future, it remains of great interest to transcend LDA (+GGA), and the present study focuses therefore on examples where this can be achieved. The final comment, encouraged by recent progress on the single-particle kinetic energy functional $T_s[\rho]$, some of which is summarized here, is that while it is inevitable, for a long time to come, to approximate the exchange plus correlation energy density functional, the problem of constructing $T_s[\rho]$ may be approaching its final phase.

Appendix A: Hookean Atom

Kestner and Sinanoglu [19] studied the case of two electrons, harmonically confined and interacting Coulombically, and therefore described by the Hamiltonian

$$H = -\frac{1}{2}\nabla_1^2 - \frac{1}{2}\nabla_2^2 + \frac{1}{2}kr_1^2 + \frac{1}{2}kr_2^2 + \frac{e^2}{r_{12}}. \qquad \text{(A1)}$$

The Schrödinger equation for this system can be solved analytically for certain discrete values of the "spring constant" k. For one such value, $k = 1/4$, March et al. [46] derived the analog of Eq. (2.16) of Minguzzi et al. [28], i.e., the differential equation satisfied by the density (given here in a.u.):

$$0 = (-r^2 + r^4)\rho'''(r) + (-2r - 4r^3 + 4r^5)\rho''(r) + (2 - 13r^2 - 4r^4 + 5r^6)\rho'(r) + (-10r^3 + 2r^7)\rho(r). \qquad \text{(A2)}$$

As for the noninteracting case, this is a third-order, linear, homogeneous differential equation, and in fact if the electron–electron interaction is "turned off" we recover Eq. (2.16) with $M = 0$ and $V = r^2/8$.

In addition, March et al. [47] have shown that for this $k = 1/4$ case the electron density is determined by the pair correlation function at coincidence. The ground-state wave function $\Psi(\mathbf{r}_1, \mathbf{r}_2)$ is given by

$$\Psi(\mathbf{r}_1, \mathbf{r}_2) = N_o\left(1 + \frac{1}{2}|\mathbf{r}_1 - \mathbf{r}_2|\right)\exp\left[-\frac{1}{4}(r_1^2 + r_2^2)\right] \qquad \text{(A3)}$$

and the density $\rho(\mathbf{r}_1)$ by

$$\rho(\mathbf{r}_1) = 2\int |\Psi(\mathbf{r}_1, \mathbf{r}_2)|^2 d\mathbf{r}_2, \qquad \text{(A4)}$$

so we can write

$$\rho(\mathbf{r}_1) = 2N_o^2\exp(-r_1^2/2) \times \int\left(1 + \frac{1}{2}\left|\mathbf{r}_1 - \mathbf{r}_2\right|\right)^2 \exp\left[-\frac{r_2^2}{2}\right]d\mathbf{r}_2. \qquad \text{(A5)}$$

It follows then from Eq. (A.3) evaluated on the diagonal that

$$\rho(\mathbf{r}) = 2\Psi(\mathbf{r}, \mathbf{r})\int\left(1 + \frac{1}{2}\left|\mathbf{r}' - \mathbf{r}\right|\right)^2 \Psi(\mathbf{r}', \mathbf{r}')d\mathbf{r}'. \qquad \text{(A6)}$$

Because the pair correlation function $n_2(\mathbf{r}_1, \mathbf{r}_2) = \Psi^2(\mathbf{r}_1, \mathbf{r}_2)$, we see that the pair correlation function at coincidence determines $\rho(\mathbf{r})$.

Appendix B: Pauli Potential for Harmonic Confinement

In early work, Lawes and March [2] derived the differential Eq. (2.6) for the Fermion density $\rho(x)$ for harmonic confinement with potential energy $V(x)$ given by

$$V(x) = \frac{1}{2}x^2, \qquad \text{(B1)}$$

where N in Eq. (2.6) is the number of Fermions, the lowest state corresponding to $N = 1$. Here, instead of considering this equation in the manner proposed by Lawes and March, namely, as a third-order linear homogeneous differential equation to be solved for the Fermion density $\rho(x)$, given $V(x)$ in Eq. (B1) and N the total number of Fermions, we shall view Eq. (2.6) as a first-order differential equation to be integrated for $V(x)$ [of course given by Eq. (B1)!].

Using an integrating factor, $V(x)$ is readily written in the form

$$V(x) = -\rho^2(x)\int_0^x \left[\frac{\rho'''(s)}{4\rho^3} + \frac{2N\rho'(s)}{\rho^3}\right] ds. \quad \text{(B2)}$$

But, as already set out above [1],

$$\frac{\delta T}{\delta\rho(x)} = \mu - V(x), \quad \text{(B3)}$$

where the chemical potential μ is constant throughout the entire inhomogeneous particle density $\rho(x)$. But, the Pauli potential $V_P(x)$ is immediately given in terms of $\delta T/\delta\rho(x)$ and a further functional derivative involving the von Weizsäcker kinetic energy T_W defined by Eq. (2.15). Then (see, e.g., Ref. [48])

$$V_P(x) = \frac{\delta T}{\delta\rho(x)} - \frac{\delta T_W}{\delta\rho(x)}. \quad \text{(B4)}$$

To within an additive constant, $V_P(x)$ is then determined by Eqs. (B4), (B3), and (B2) plus the known functional derivative of Eq. (2.15), namely,

$$\frac{\delta T_W}{\delta\rho(x)} = \frac{1}{8}\left[\frac{\rho'(x)}{\rho(x)}\right]^2 - \frac{1}{4}\frac{\rho''(x)}{\rho(x)}. \quad \text{(B5)}$$

Hence, omitting any additive constant, one obtains the desired result

$$V_P(x) = \rho^2(x)\int_0^x \left[\frac{\rho'''(s)}{4\rho^3} + \frac{2N\rho'(s)}{\rho^3}\right] ds + \frac{1}{4}\frac{\rho''(x)}{\rho(x)} - \frac{1}{8}\left[\frac{\rho'(x)}{\rho(x)}\right]^2. \quad \text{(B6)}$$

This result could alternately be obtained by integrating Eq. (18) of the work of March et al. [9].

We have calculated $V_P(x)$ from Eq. (B6) for $N = 10$ and 20; results are shown in Figure 3. For the $N = 10$ case, we have added a plot of $\rho^2(x)$ (shifted by a constant). For comparison, the form

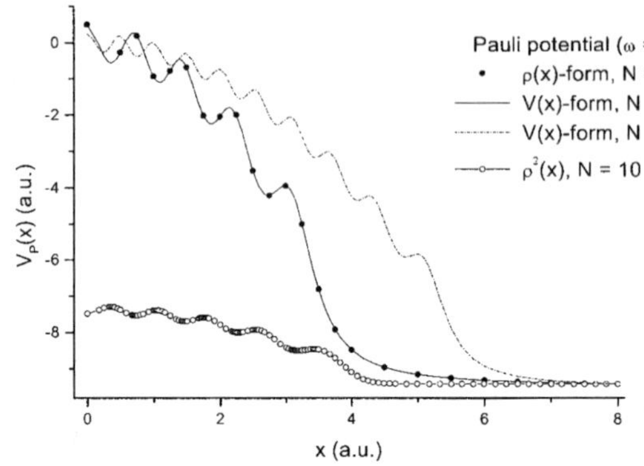

FIGURE 3. Pauli potential $V_P(x)$ for 1-D harmonic confinement for Fermion numbers $N = 10$ and $N = 20$ and for classic angular frequency $\omega = 1$. For comparison, $\rho^2(x)$ is plotted for $N = 10$ [cf. Eq. (B8) with $\lambda = 0$]. The structure in $V_P(x)$ in this case is clearly related to the structure in the square of the Fermion particle density, but with a substantial phase shift.

$$T_{\text{TFWK}} = c_k^{(1)}\int_{-\infty}^{\infty} \rho^3 dx + \frac{\lambda}{8}\int_{-\infty}^{\infty} \frac{\rho'^2}{\rho} dx; \; c_k^{(1)} = \frac{\pi^2}{6} \quad \text{(B7)}$$

may be utilized, the first term of the right side of Eq. (B7) being the usual Thomas–Fermi (TF) kinetic energy while the second term is the von Weizsäcker (W) result, putting $\lambda = 1$, and the Kirznits (K) value λ is taken to be $1/9$ [49]. Using the exact density $\rho(x)$ discussed by Lawes and March, one can then write

$$V_P^{\text{TFWK}} = 3c_k^{(1)}\rho^2(x) + (\lambda - 1)\frac{\delta T_W}{\delta\rho}. \quad \text{(B8)}$$

Minguzzi et al. [28] generalized the low-dimensional arguments on harmonic confinement (see body of text for 3-D result). They demonstrate, for isotropic potential energy $V(r)$, that the differential equation for the (spherical) density $\rho(r)$ for $(M + 1)$ closed shells is Eq. (2.16). This can be viewed, in parallel with the 1-D argument, as a first-order differential equation to solve for $V(r)$:

$$\frac{\partial V}{\partial r} - \frac{2}{3}V(r)\frac{\rho'}{\rho} = -\frac{\hbar^2}{8m}\frac{1}{\rho}\frac{\partial}{\partial r}[\nabla^2\rho(r)] - (M+2)\hbar\omega\frac{\rho'}{\rho}, \quad \text{(B9)}$$

so that

$$\rho^{2/3}\frac{\partial(\rho^{-2/3}V)}{\partial r} = -\frac{2}{3}\frac{\rho'}{\rho}V + \frac{\partial V}{\partial r} \quad \text{(B10)}$$

or

$$\rho^{-2/3}V = -\frac{\hbar^2}{8m}\int_0^r \rho^{-5/3}\frac{\partial}{\partial r}[\nabla^2\rho(r)]dr - (M+2)\hbar\omega\int_0^r \frac{\rho'}{\rho^{5/3}}dr. \quad \text{(B11)}$$

Hence, if $T[\rho]$ denotes the kinetic energy functional,

$$\frac{\delta T}{\delta \rho} = \mu - V(r) \quad \text{(B12)}$$

with μ as usual the chemical potential, which is constant throughout the entire inhomogeneous Fermi density $\rho(r)$.

But, as before,

$$V_{\mathrm{P}} = \frac{\delta T}{\delta\rho(r)} - \frac{\delta T_{\mathrm{W}}}{\delta\rho(r)} \quad \text{(B13)}$$

and because $T_{\mathrm{W}}[\rho]$ is given explicitly by Eq. (2.2), Eqs. (B9)–(B12) determine $V_{\mathrm{P}}(r)$, to within an additive constant, as a functional of the density $\rho(r)$. From Eq. (B5), we have the explicit form

$$\frac{\delta T_{\mathrm{W}}}{\delta\rho(r)} = \frac{\hbar^2}{8m}\left(\frac{\nabla\rho}{\rho}\right)^2 - \frac{\hbar^2}{4m}\frac{\nabla^2\rho}{\rho}. \quad \text{(B14)}$$

Hence, Eqs. (B11)–(B14) suffice to determine the Pauli potential $V_{\mathrm{P}}(r)$, a plot being shown in Figure 4.

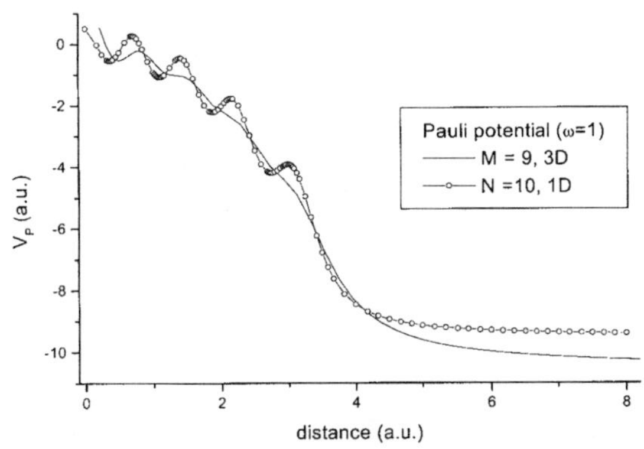

FIGURE 4. Similar to Figure 3 for $N = 10$ but now including the comparable example for 3-D harmonic confinement.

Appendix C: Electrostatic Interpretation of an Electron Density Associated with Spherical Exchange Correlation Potential $V_{XC}(R)$ in the Be Atom

The exchange correlation potential $V_{xc}(\mathbf{r})$ lies at the heart of current applications of DFT. While in the approach of Kohn and Sham, whose work represents the formal completion of the work begun by Slater, this potential is to be obtained by functional differentiation of the as yet unknown exchange correlation energy $E_{xc}[\rho]$:

$$V_{xc}(\mathbf{r}) = \frac{\delta E_{xc}[\rho]}{\delta\rho(\mathbf{r})}, \quad \text{(C1)}$$

the study of Holas and March [50], based on the exact differential virial theorem, expresses $V_{xc}(\mathbf{r})$ explicitly in terms of first- and second-order density matrices of the fully interacting system, plus the SKS one-electron orbitals found from the Schrödinger equation with one-body potential

$$V(\mathbf{r}) = V_n(\mathbf{r}) + V_{es}(\mathbf{r}) + V_{xc}(\mathbf{r}). \quad \text{(C2)}$$

Here, $V_{es}(\mathbf{r})$ is generated by the ground-state density $\rho(\mathbf{r})$, these two quantities being related by Poisson's equation

$$\nabla^2 V_{es}(\mathbf{r}) = -4\pi e^2\rho(\mathbf{r}). \quad \text{(C3)}$$

Equation (C3) will now be used to motivate the central result of this Appendix (see also Ref. [51]).

Let us restrict ourselves to spherically symmetrical atoms, He, Be, Ne, Ar, etc., to be definite. Then, Eq. (C3) can be rewritten as

$$\frac{\partial}{\partial r}\left(r^2\frac{\partial V_{es}}{\partial r}\right) = -4\pi r^2 e^2\rho(r) \quad \text{(C4)}$$

or

$$\frac{\partial V_{es}}{\partial r} = -\frac{Q(r)}{r^2}, \quad \text{(C5)}$$

where the total number of electrons enclosed by a sphere of radius r centered at the nucleus, $Q(r)$, is defined by

$$Q(r) = e^2 \int_0^r 4\pi r^2 \rho(r) dr. \quad (C6)$$

Evidently, for atomic number Z,

$$Q(r)_{r\to\infty} \to Z. \quad (C7)$$

Now, we use the above, elementary, electrostatics to motivate the introduction of an integrated radial density $Q_{xc}(r)$, related to $V_{xc}(r)$ by

$$Q_{xc}(r) = -r^2 \frac{\partial V_{xc}}{\partial r}. \quad (C8)$$

We next note from the work of Almbladh and von Barth [52] that the asymptotic form of $V_{xc}(r)$ in neutral atoms is given by (now in a.u.)

$$V_{xc}(r) = -\frac{1}{r} - \frac{\alpha}{2r^4}, \quad (C9)$$

where α is the dipole polarizability of the singly charged positive ion. Hence, from the definition (C8) of $Q_{xc}(r)$ we find

$$|Q_{xc}(r)| = 1 + \frac{2\alpha}{r^5} + O\left(\frac{1}{r^6}\right) \quad (C10)$$

at sufficiently large r.

Because $Q_{xc}(r) \to 0$ as $r \to 0$, the asymptotic result (C10) shows that, because $|Q_{xc}(r)|$ exceeds unity at large r, it must have, at least, one maximum. In addition, $|Q_{xc}(r)|$ as r is reduced from the position r_m of the above maximum must intersect the value unity at least once. The point to be emphasized here is that there are characteristic features associated with the (at present necessary) modeling of $Q_{xc}(r)$, namely, (1) the dipole polarizability α, (2) the position r_m and height $|Q_{xc}(r)|$ of the above maximum, (3) the position, r_1 say, where $|Q_{xc}(r)| = 1$, and (4) the limit $Q_{xc}(r)$ tends to zero as r tends to zero.

We turn to the example of the Be atom. We bypass the need for fully interacting low-order density matrices by assuming the ground-state density $\rho(r)$ is known, e.g., from X-ray or electron diffraction experiments or from quantal Monte Carlo studies. The idempotent first-order density matrix $\gamma(\mathbf{r}, \mathbf{r}')$ generated by the one-body potential $V(r)$ in Eq. (C2) then has the form given in Eq. (3.14), where the phase $\theta(r)$ is characterized solely by $\nabla\rho/\rho$ and physical boundary conditions through the eigenvalue equation

$$\nabla^2\theta + \frac{\nabla\rho}{\rho}\cdot\nabla\theta + \lambda\sin(2\theta) = 0. \quad (C11)$$

Hence, the eigenvalue λ and the eigenfunction θ are uniquely determined by knowledge of the (assumed exact) ground-state electron density $\rho(r)$ of the Be atom.

Going back to the equation of motion of the density matrix $\gamma(\mathbf{r}, \mathbf{r}')$, namely,

$$\nabla_{\mathbf{r}}^2\gamma - \nabla_{\mathbf{r}'}^2\gamma = 2[V(r) - V(r')]\gamma, \quad (C12)$$

it is a straightforward matter to extract the radial derivative $\partial V/\partial r$ of the one-body potential (C2) in terms of the electron density $\rho(r)$ and the phase $\theta(r)$ to be obtained from the $\nabla\rho/\rho$ by solving Eq. (C11). The result is given by (Ref. [51])

$$\begin{aligned}2\frac{\partial V}{\partial r} &= \frac{\partial}{\partial r}\left\{\frac{\rho''}{\rho} - \frac{1}{2}\left(\frac{\rho'}{\rho}\right)^2\right\} - 4\theta'\theta'' - \frac{1}{r^2}\frac{\rho'}{\rho} \\ &+ \frac{1}{r}\left[\frac{\rho\rho'' - \rho'^2}{\rho^2}\right] + \left[\theta'\left\{\frac{\rho\rho'' - \rho'^2}{\rho^2}\right\} + \theta''\frac{\rho'}{\rho}\right. \\ &\left. + \theta''' - \frac{2}{r^2}\theta' + \frac{2}{r}\theta''\right](\cot\theta - \tan\theta) \\ &+ 2\left[\rho'\frac{\theta'}{\rho} + \theta'' + \frac{2}{r}\theta'\right]\frac{\partial}{\partial r}(\cot 2\theta). \end{aligned} \quad (C13)$$

But, from Eq. (C2)

$$r^2\frac{\partial V_{xc}(r)}{\partial r} = r^2\frac{\partial V}{\partial r} - 4 - r^2\frac{\partial V_{es}(r)}{\partial r} \quad (C14)$$

because for Be $V_n = -4/r$. The last term on the right side of Eq. (C14) is given in terms of $\rho(r)$ by Eqs. (C5) and (C6) and, hence, using the definition (C8)

$$-Q_{xc}(r) = r^2\frac{\partial V}{\partial r} - 4 + Q(r). \quad (C15)$$

Everything on the right side of Eq. (C15) can be determined from the (assumed exact) ground-state density $\rho(r)$ and hence the exchange correlation integrated electron density $Q_{xc}(r)$ can be found.

To close this Appendix, the definition of the exchange correlation integrated density $Q_{xc}(r)$ is proposed as in Eq. (C8), by analogy with the classic electrostatic result in Eq. (C5). For large r, Eq. (C10) gives the asymptotic form of $Q_{xc}(r)$, which also tends to zero as $r \to 0$. The form (C10) shows that $|Q_{xc}(r)|$ must have at least one maximum and must intersect the value unity at least once for finite r. For Be, Eqs. (C15) and (C13) give $|Q_{xc}(r)|$ in a form characterized solely by the ground-state electron density $\rho(r)$, the phase $\theta(r)$ being determined by solution of Eq. (C11), into which $\nabla\rho/\rho$ is the only input information.

ACKNOWLEDGMENTS

The authors acknowledge invaluable discussions on the general area embraced by the present study with Dr. T. Gál, Professors A. Holas and Á. Nagy, and Dr. P. Senet. N.H.M. acknowledges some financial support from the Francqui Foundation. Thanks are extended especially to the Executive Director, Professor L. Eyckmans, for his help and encouragement. I.A.H. acknowledges support from the Flemish Science Foundation (FWO) under Grant G.0347.97. The authors also thank the University of Antwerp (RUCA) for its support in the framework of the Visiting Professors Programme. This area is also supported by the Concerted Action Programme of the University of Antwerp.

References

1. Parr, R. G.; Yang, W. Density Functional Theory of Atoms and Molecules; Oxford University Press: Oxford, UK, 1989.
2. Lawes, G. P.; March, N. H. J Chem Phys 1979, 71, 1007.
3. Husimi, K. Proc Phys Math Soc Jpn 1940, 22, 264.
4. March, N. H.; Young, W. H. Nucl Phys 1959, 12, 237.
5. Anderson, M. H.; Ensher, J. R.; Matthews, M. R.; Wieman, C. E.; Cornell, E. A. Science 1995, 269, 198.
6. DeMarco, B.; Jin, D. S. Science 1999, 285, 1703.
7. Vignolo, P.; Minguzzi, A.; Tosi, M. P. Phys Rev Lett 2000, 85, 2850.
8. March, N. H.; Nieto, L. M. Phys Rev A 2001, 63, 44502.
9. March, N. H.; Senet, P.; Van Doren, V. E. Phys Lett A 2000, 270, 88.
10. Thomas, L. H. Proc Camb Phil Soc 1926, 23, 542.
11. Fermi, E. Z Phys 1928, 48, 73.
12. Dirac, P. A. M. Proc Camb Phil Soc 1930, 26, 376.
13. Slater, J. C. Phys Rev 1951, 81, 385.
14. Gáspár, R. Acta Phys Acad Sci Hung 1954, 3, 26; English translation: J Mol Struct Theochem 2000, 501/502, 1.
15. Kohn, W.; Sham, L. J. Phys Rev 1965, 140A, 1133.
16. Wang, L. W.; Teter, M. P. Phys Rev B 1992, 45, 13196.
17. Pearson, M.; Smargiassi, E.; Madden, P. A. J Phys Condens Matter 1993, 5, 3221.
18. March, N. H.; Howard, I. A.; Holas, A.; Senet, P.; Van Doren, V. E. Phys Rev A 2001, 63, 12520.
19. Kestner, N. R.; Sinanoglu, O. Phys Rev 1962, 128, 2687; see also Kais, S.; Herschbach, D. R.; Handy, N. C.; Murray, C. W.; Laming, G. J. Chem Phys 1993, 99, 417, and references therein.
20. March, N. H. Phys Lett A 1985, 113, 66; March, N. H. Phys Lett A 1986, 113, 476.
21. Levy, M.; Görling, A. Phil Mag B 1994, 69, 763.
22. March, N. H. Electron Density Theory of Atoms and Molecules; Academic: New York, 1992; p 173.
23. Von Weizsäcker, C. F. Z Phys 1935, 96, 431.
24. March, N. H.; Young, W. H. Proc Phys Soc 1958, 72, 182.
25. Gombás, P. Phys Lett A 1969, 28, 585.
26. Goodisman, J. Phys Rev A 1970, 1, 1574.
27. Deb, B. M.; Ghosh, S. K. Int J Quantum Chem 1983, 13, 1.
28. Minguzzi, A.; March, N. H.; Tosi, M. P. Phys Lett A 2001, 281, 192.
29. Howard, I. A.; March, N. H. J Phys A Math Gen 2001, 34, L497.
30. Howard, I. A.; March, N. H.; Van Doren, V. E. Phys Rev A 2001, 63, 62501.
31. Amovilli, C.; March, N. H. Phys Chem Liquids 2000, 38, 505.
32. Howard, I. A.; March, N. H.; Alonso, J. A.; Cordero, N. A.; Van Doren, V. E. Chem Phys Lett 2001, 343, 166.
33. Kato, T. Commun Pure Appl Math 1957, 10, 151.
34. March, N. H. Phys Rev A 1987, 36, 5077.
35. Dawson, K. A.; March, N. H. J Chem Phys 1984, 81, 5.
36. Nagy, Á.; March, N. H. Phys Rev A 1984, 39, 5512.
37. Rassolov, V. A.; Chipman, D. M. J Chem Phys 1996, 104, 9908, and references therein.
38. Nagy, Á.; Sen, K. D. Chem Phys Lett 2000, 332, 154; Nagy, Á.; Sen, K. D. J Phys B 2000, 33, 1745.
39. Bingel, W. A. Naturforsch 1963, 189, 1249.
40. Davidson, E. R. Reduced Density Matrices in Quantum Chemistry; Academic Press: New York, 1976.
41. Howard, I. A.; March, N. H.; Senet, P.; Van Doren, V. E. Phys Rev A 2000, 62, 62512.
42. Harris, J.; Jones, R. O. J Phys 1974, F4, 1170.
43. Bardeen, J. Phys Rev 1936, 49, 653.
44. March, N. H. Phys Rev B 2000, 61, 5011.
45. Howard, I. A.; March, N. H.; Van Doren, V. E. Phys Rev A 2001, 64, 042509.
46. March, N. H.; Gál, T.; Nagy, Á. Chem Phys Lett 1998, 292, 384.
47. March, N. H.; Amovilli, C.; Klein, D. J. Chem Phys Lett 2000, 325, 645.
48. Holas, A.; March, N. H. Phys Rev A 1991, 44, 5521.
49. Kirznits, D. A. Sov Phys JETP 1957, 5, 64.
50. Holas, A.; March, N. H. Phys Rev A 1995, 51, 2040.
51. March, N. H. Phys Rev A 2002, 65, 034507.
52. Almbladh, C.-O.; von Barth, U. Phys Rev A 1985, 31, 3231.

phys. stat. sol. (b) **237**, No. 1, 265–273 (2003) / **DOI** 10.1002/pssb.200301779

Propagator and Slater sum in one-body potential theory

N. H. March[*,1,2] and **I. A. Howard**[1]

[1] Department of Physics, University of Antwerp (RUCA), Groenenborgerlaan 171, 2020 Antwerp, Belgium
[2] University of Oxford, Oxford, England

Received 27 September 2002, accepted 28 November 2002
Published online 30 April 2003

Dedicated to Professor Jozef T. Devreese on the occasion of his 65th birthday

PACS 71.15.Mb

For a one-body potential $V(r)$ generating eigenfunctions $\psi_i(\mathbf{r})$ and corresponding eigenvalues ϵ_i, the Feynman propagator $K(\mathbf{r},\mathbf{r}',t)$ is simply related to the canonical density matrix $C(\mathbf{r},\mathbf{r}',\beta)$ by $\beta \to it$. The diagonal element $S(\mathbf{r},\mathbf{r},\beta)$ of C is the so-called Slater sum of statistical mechanics. Differential equations for the Slater sum are first briefly reviewed, a quite general equation being available for a one-dimensional potential $V(x)$. This equation can be solved for a sech^2 potential, and some physical properties of interest such as the local density of states are derived by way of illustration. Then, the Coulomb potential $-Ze^2/r$ is next considered, and it is shown that what is essentially the inverse Laplace transform of $S(\mathbf{r},\beta)/\beta$ can be calculated for an arbitrary number of closed shells. Blinder has earlier determined the Feynman propagator in terms of Whittaker functions and contact is here established with his work. The currently topical case of Fermion vapours which are harmonically confined is then treated, for both two and three dimensions. Finally, in an Appendix, a perturbation series for the Slater sum is briefly summarized, to all orders in the one-body potential $V(\mathbf{r})$. The corresponding kinetic energy is thereby accessible.

1 Introduction and outline It is a pleasure to be invited to contribute to this Festschrift honouring Professor Devreese. He has, of course, made many important contributions in the area of Feynman propagator theory [1]. It seemed therefore appropriate to summarize recent progress on the calculation of the Feynman propagator, and the closely related Slater sum, within the framework of one-body potential theory. The appeal to such a one-body potential energy $V(\mathbf{r})$ is, of course, central to the current usage in density-functional theory [2], though this potential has an exchange (x) – correlation (c) contribution $V_{xc}(\mathbf{r})$ which is presently unknown. However, it will prove fruitful to appeal to model potentials $V(\mathbf{r})$, such as Coulomb and harmonic oscillator, for which numerous analytical results have been recently established for an arbitrary number of closed shells. Especially the one-dimensional case proves fruitful from the point of view of analytical work, and this is therefore the subject of Section 2 below. Section 3 then considers the bare Coulomb potential $-Ze^2/r$, while Section 4 is devoted to harmonic confinement in both two and three dimensions. Section 5 constitutes a summary, with some proposals made for possible future studies. In an Appendix, a perturbation series for the Slater sum, defined almost immediately below, is given to all orders in the potential $V(\mathbf{r})$, together with the corresponding result for the kinetic energy density.

* Corresponding author: e-mail: howard@ruca.ua.ac.be

Before turning to Section 2, it will be useful to define explicitly the quantities on which most attention is focussed below. These are, in turn, the Slater sum $S(\boldsymbol{r},\beta)$, the canonical density matrix $C(\boldsymbol{r},\boldsymbol{r}',\beta)$ and the Feynman propagator $K(\boldsymbol{r},\boldsymbol{r}',t)$. Throughout this paper we shall be concerned only with the normalized eigenfunctions $\psi_i(\boldsymbol{r})$ and the corresponding eigenvalues ϵ_i generated by the one-body Hamiltonian H_r:

$$H_r = -\tfrac{1}{2}\,\nabla_r^2 + V(\boldsymbol{r}). \tag{1}$$

Then the so-called Slater sum of statistical mechanics, denoted by $S(\boldsymbol{r},\beta)$ below, is defined by [3]

$$S(\boldsymbol{r},\beta) = \sum_{\text{all } i} \psi_i(\boldsymbol{r})\,\psi_i^*(\boldsymbol{r})\,\exp\,(-\beta\epsilon_i) : \beta = (k_B T)^{-1} \tag{2}$$

where k_B is Boltzmann's constant and T the absolute temperature. The integral over all space of the Slater sum evidently yields the partition function pf:

$$\text{pf} = \sum_{\text{all } i} \exp\,(-\beta\epsilon_i)\,. \tag{3}$$

The off-diagonal generalization of the Slater sum is the canonical density matrix $C(\boldsymbol{r},\boldsymbol{r}',\beta)$ defined by [3]

$$C(\boldsymbol{r},\boldsymbol{r}',\beta) = \sum_{\text{all } i} \psi_i(\boldsymbol{r})\,\psi_i^*(\boldsymbol{r}')\,\exp\,(-\beta\epsilon_i) \tag{4}$$

which is related to the Feynman propagator $K(\boldsymbol{r},\boldsymbol{r}',t)$, where t is the time, by the transformation in $C(\boldsymbol{r},\boldsymbol{r}',\beta)$ that

$$\beta \to it\,. \tag{5}$$

With this background, we turn immediately to the discussion of the Slater sum generated by the one-body potential $V(x)$.

2 Differential equation satisfied by the Slater sum $S(x,\beta)$ for a one-dimensional one-body potential $V(x)$ In the early work of March and Murray [4], the Bloch equation

$$H_r C(\boldsymbol{r},\boldsymbol{r}',\beta) = -\frac{\partial C(\boldsymbol{r},\boldsymbol{r}',\beta)}{\partial\beta} \tag{6}$$

was shown to lead, for central fields, to an ordinary differential equation for the s-states. This is, of course, equivalent to a one-dimensional Schrödinger problem in wave mechanics and their Eq. (4.7) leads also immediately to a differential equation for $S(x,\beta)$ for a one-body potential $V(x)$, namely

$$\frac{\hbar^2}{8m}\frac{\partial^3 S}{\partial x^3} = \frac{\partial^2 S}{\partial x\partial\beta} + V(x)\,\frac{\partial S}{\partial x} + \frac{1}{2}\,\frac{\partial V}{\partial x}\,S\,. \tag{7}$$

We shall take as an immediate example the case of a sech^2 potential, for which the bound states are discussed by Landau and Lifshitz [5]. Montroll [6], in discussing the continuum states of this potential, classifies the wave functions into symmetrical forms $\psi_s(x)$ and antisymmetrical forms $\psi_a(x)$. One can use his results to form explicitly $\psi_s^2(x) + \psi_a^2(x)$, which evidently appears in the Slater sum from the definition of $C(\boldsymbol{r},\boldsymbol{r}',\beta)$ with $\boldsymbol{r}'$ put equal to $\boldsymbol{r}$, and motion restricted to the x-axis only. The general solution of the Schrödinger equation in the continuum is given by

$$\psi(x) = \mathcal{N}[\cos\,(kx+\delta) - (\gamma/k)\,\sin\,(kx+\delta)\,\tanh\,(\gamma x)]\,, \tag{8}$$

where $E = k^2/2$ is the energy of the state.

Choosing $\delta = 0$ in Eq. (8) yields the symmetric wave functions $\psi_s(x)$ while $\delta = \pi/2$ generates the antisymmetric forms; $\psi_s^2 + \psi_a^2$ then takes the form

$$\psi_s^2 + \psi_a^2 = a^2(\gamma,k)\,[1 + (\gamma/k)^2\,\tanh^2(\gamma x)]\,. \tag{9}$$

Inserting Eq. (9) into the definition of the Slater sum $S(x, \beta)$ set out above, one finds

$$S_c(x,\beta) = \sum_{\text{all } k>0} a^2(\gamma,k)\,[1 + (\gamma/k)^2 \tanh^2(\gamma x)] \exp\left(-\frac{\beta k^2}{2}\right), \tag{10}$$

where the subscript c on $S(x, \beta)$ indicates that one is summing only over continuum (c) states. This form (10) is important, in that it tells us that $S_c(x, \beta)$ has the general 'shape'

$$\begin{aligned} S_c(x,\beta) &= F(\gamma,\beta) + \tanh^2(\gamma x) \sum_{\text{all } k>0} a^2(\gamma,k)\,(\gamma/k)^2 \exp\left(-\frac{\beta k^2}{2}\right) \\ &\equiv F(\gamma,\beta) + G(\gamma,\beta) \tanh^2(\gamma x). \end{aligned} \tag{11}$$

One sees here the appearance of the sech^2 potential, since

$$\tanh^2(\gamma x) = 1 - \mathrm{sech}^2(\gamma x) \tag{12}$$

and hence Eq. (11) can be rewritten as

$$S_c(x,\beta) = [F(\gamma,\beta) + G(\gamma,\beta)] + \frac{G(\gamma,\beta)}{\gamma^2}\, V(x)\,. \tag{13}$$

Since $V(x) \to 0$ as $x \to \pm\infty$, we can see on physical grounds in Eq. (13) that $S_c(x, \beta)$ must tend to the well-known result for the partition function per unit length (in one dimension) for free electrons. This, in atomic units ($\hbar = m = 1$), is

$$\lim_{x\to\pm\infty} S_c(x,\beta) = (2\pi\beta)^{-1/2}, \tag{14}$$

which shows from Eq. (13) that the γ dependence appearing in F and G separately must cancel in the sum $F + G = (2\pi\beta)^{-1/2}$. This simplification will be utilized below.

Having established the 'shape' of the continuum Slater sum $S_c(x, \beta)$ as

$$S_c(x,\beta) = (2\pi\beta)^{-1/2} + \frac{G(\gamma,\beta)}{\gamma^2}\, V(x) \tag{15}$$

the question that remains is to find $G(\gamma, \beta)$ if possible by direct substitution of Eq. (15) into the general partial differential Eq. (7). One finds straightforwardly that (in a.u.)

$$\frac{1}{8}\frac{G}{\gamma^2}\, V'''(x) - \frac{1}{\gamma^2}\frac{\partial G}{\partial\beta}\, V'(x) - \frac{3}{2}\frac{G}{\gamma 2}\, V(x)\, V'(x) - \frac{1}{2}\,(2\pi\beta)^{-1/2}\, V'(x) = 0\,. \tag{16}$$

On rearrangement, this reads

$$\frac{1}{\gamma^2}\frac{\partial G}{\partial\beta} - \frac{G(\gamma,\beta)}{\gamma^2}\left[\frac{V'''/V'}{8} - \frac{3}{2}\, V\right] + \frac{1}{2}\,(2\pi\beta)^{-1/2} = 0\,. \tag{17}$$

The square bracket in Eq. (17) gives us $\gamma^2/2$, independent of x. The resulting first-order ordinary differential equation may be integrated to yield

$$\begin{aligned} G(\gamma,\beta) &= -\frac{\gamma^2}{2} \exp\left(\frac{\gamma^2\beta}{2}\right)\left[\int_0^\beta \frac{1}{(2\pi\beta)^{1/2}} \exp\left(-\frac{\gamma^2\beta}{2}\right) \mathrm{d}\beta + f(\gamma)\right] \\ &= \left[-\frac{\gamma}{2}\, \mathrm{erf}\left(\frac{\gamma}{2}\sqrt{2\beta}\right) - \frac{\gamma^2 f(\gamma)}{2}\right] \exp\left(\frac{\gamma^2\beta}{2}\right), \end{aligned} \tag{18}$$

where $f(\gamma)$ is a constant of integration, and thus from (15),

$$S_c(x,\beta) = (2\pi\beta)^{-1/2} - \left[\frac{f(\gamma)}{2} + \frac{1}{2\gamma}\, \mathrm{erf}\left(\frac{\gamma}{2}\sqrt{2\beta}\right)\right] \exp\left(\frac{\gamma^2\beta}{2}\right) V(x) \tag{19}$$

Contact may be made with a direct calculation of the local density of states in the continuum at $x = 0$, namely $\partial\rho_c(0,E)/\partial E$ [7]. Using the Laplace transform relation

$$S_c(x,\beta) = \int_0^\infty \frac{\partial\rho_c(x,E)}{\partial E} \exp(-\beta E)\, dE \tag{20}$$

and inserting $\partial\rho_c(0,E)/\partial E$ into the right-hand side of (20) yields the result (15) with $G(\gamma,\beta)$ given in Eq. (18) and $f(\gamma) = -1/\gamma$, so that we can write finally

$$S_c(x,\beta) = (2\pi\beta)^{-1/2} - \frac{1}{2\gamma} \operatorname{erfc}\left(\frac{\gamma}{2}\sqrt{2\beta}\right) \exp\left(\frac{\gamma^2\beta}{2}\right) V(x)\,. \tag{21}$$

It is worth adding, but without elaboration, that the one-dimensional potential for harmonic confinement

$$V(x) = \tfrac{1}{2}\, kx^2$$

can also be readily solved (for the three-dimensional case see also Sondheimer and Wilson [8] and Section 4 below) and the result for the Slater sum is

$$S(x,\beta) = \left(\frac{m}{2\pi\hbar^2\beta}\right)^{1/2} \left(\frac{\beta\hbar\omega}{\sinh(\beta\hbar\omega)}\right)^{1/2} \exp\left(-\frac{m\omega x^2}{\hbar} \tanh\left(\frac{\beta\hbar\omega}{2}\right)\right). \tag{22}$$

3 Bare Coulomb potential energy $-Ze^2/r$ The work of Blinder [9] gives an analytic form for the nonrelativistic Coulomb propagator, involving Whittaker functions. To date, it has not proved easy to use numerically, but Blinder does give a relatively simple result, for what amounts to the Slater sum at the origin $\boldsymbol{r} = 0$. His result may be written, with one-body potential

$$V(r) = -Z/r \tag{23}$$

as

$$S(0,\beta) = (2\pi\beta)^{-3/2} + \frac{Z}{2\pi\beta} + (2\beta)^{-1/2}\pi^{-3/2} \sum_{n=1}^{\infty} Z_n^2 + \frac{1}{2\pi} \sum_{n=1}^{\infty} Z_n^3 \exp(\beta Z_n^2/2) \operatorname{erfc}\left(-Z_n\left(\frac{\beta}{2}\right)^{1/2}\right), \tag{24}$$

with $Z_n = Z/n$. Blinder rearranges this form into a series involving the Riemann ζ function. Writing $S_0(\beta) = (2\pi\beta)^{-3/2}$ his further result is

$$\frac{S(0,\beta)}{S_0(\beta)} = 4\pi^{1/2} \sum_{n=0}^{\infty} \frac{\zeta(n)}{\Gamma\left(\frac{n-1}{2}\right)} \left(\frac{\beta Z^2}{2}\right)^{n/2}, \tag{25}$$

a plot of which is shown in Fig. 1 for the case of $Z = 1$. In view of the complexity of Blinder's result applied to the full Slater sum $S(r,\beta)$, it is of interest to note that Pfalzner et al [10] and also Cooper [11] have derived, by utilizing the spatial generalization of Kato's theorem for the bare Coulomb potential [12], the following differential equation for the Slater sum $S(r,\beta)$:

$$\frac{1}{8}\frac{\partial^3 S}{\partial r^3} + \frac{1}{2r}\frac{\partial^2 S}{\partial r^2} + \left(\frac{1}{4r^2} - V - \frac{\partial}{\partial\beta}\right)\frac{\partial S}{\partial r} + \frac{1}{2}\frac{\partial V}{\partial r} S = 0 \tag{26}$$

One can regard Blinder's result, involving Whittaker, Laguerre, Hermite and error functions (his Eq. (3) in Ref. [9]), as an exact solution of Eq. (26), but it may be, in the future, that a simpler form of solution of Eq. (26) will emerge.

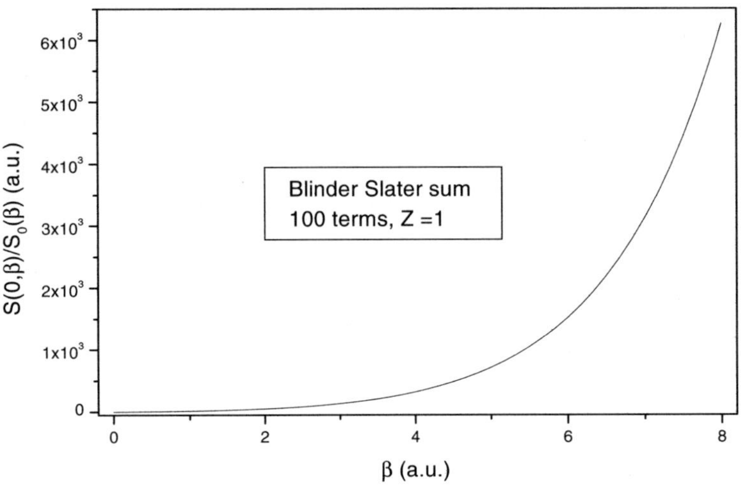

Fig. 1 Plot of Eq. (25) for $Z = 1$ as a function of $\beta = (k_B T)^{-1}$ for the Slater sum $S(\mathbf{r}, \beta)$ of the bare Coulomb field $-Ze^2/r$ at the nucleus $r = 0$.

It is relevant here to add that, using the early work of March and Murray [4], analytical progress can be made in deriving the local density of states $N(r, E)$ in the continuum for the bare Coulomb potential. For its s-wave component $N_s(r, E)$, the result is simple in terms of the Whittaker function $\mathcal{M}(-iZ/k, 1/2, 2ikr)$ with energy $E = k^2/2$, as shown by Howard et al [13], the r-dependence of N_s having the form

$$N_s(r, E) \propto \frac{\mathcal{M}^2(-iZ/k, 1/2, 2ikr)}{r^2} \to \left(\frac{2}{Z}\right)^{1/2} \left(\frac{2E}{\pi}\right) \frac{\cos^2\left(2(2Zr)^{1/2} + \pi/4\right)}{Zr^{3/2}} \tag{27}$$

using an asymptotic form of the Whittaker function [14] which is valid for large Z/k. A plot of this limiting case is shown in Fig. 2. With this fairly brief summary of the Coulomb case, we turn to harmonic confinement.

4 Slater sum for three-dimensional harmonic oscillator Amovilli and March [15] derived the partial differential equation for the Slater sum $S(r, \beta)$ for three-dimensional isotropic harmonic confinement. Their result reads

$$\frac{\hbar^2}{m}\left[\frac{1}{8}\frac{\partial^3 S}{\partial r^3} + \frac{1}{4r}\frac{\partial^2 S}{\partial r^2}\right] - \left(\frac{\hbar^2}{4mr^2} + V + \frac{\partial}{\partial \beta}\right)\frac{\partial S}{\partial r} + \frac{1}{2}\frac{\partial V}{\partial r}\, S = 0 \tag{28}$$

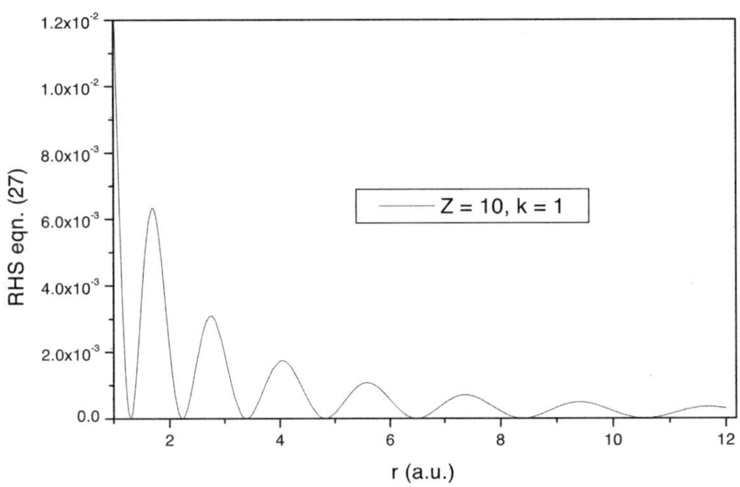

Fig. 2 Plot of limiting form of the s-states local density of states $N_s(r, E)$ for large Z/k, given in Eq. (27).

with $V = (1/2)kr^2 \equiv (1/2)m\omega^2 r^2$, k being the force constant. It is a straightforward, if a little lengthy, calculation to verify that the desired solution of Eq. (28) is

$$S(r,\beta) = \left(\frac{m}{2\pi\hbar}\right)^{3/2} \left(\frac{\omega}{\sinh(\beta\hbar\omega)}\right)^{3/2} \exp\left(-\frac{m\omega r^2}{\hbar} \tanh\left(\frac{\beta\hbar\omega}{2}\right)\right) \tag{29}$$

which certainly can be traced back at least to the study of Sondheimer and Wilson [8] and represents essentially the three-dimensional generalization of Eq. (22).

The corresponding kinetic energy density $t(r,\beta)$ is given by

$$t(r,\beta) = -\frac{\partial S}{\partial \beta} - V(r)\, S(r,\beta) \tag{30}$$

and inserting Eq. (29) into Eq. (30) yields

$$t(r,\beta) = -S(r,\beta) \left[V(r) \tanh^2\left(\frac{\beta\hbar\omega}{2}\right) - \frac{3\hbar\omega}{2} \coth(\beta\hbar\omega)\right] \tag{31}$$

It is of interest here, for this harmonic potential, to note that the ground-state density $\rho(r)$ satisfies, as well as the Slater sum, a third-order ordinary differential equation for this case, which applies to any arbitrary number of closed shells [16, 17]. Essentially the density $\rho(r)$ is related to $S(r,\beta)$ by an inverse Laplace transform on β. The equation given by Minguzzi et al. [16] reads

$$\frac{\hbar^2}{8m} \frac{\partial}{\partial r} [\nabla^2 \rho(r)] + [(M+2)\,\hbar\omega - V(r)] \frac{\partial \rho}{\partial r} + \frac{3}{2} \frac{\partial V}{\partial r} \rho = 0 \tag{32}$$

where $V = (1/2)\, m\omega^2 r^2$ and $(M+1)$ closed shells are occupied in forming the Fermion density $\rho(r)$. Howard and March [17] have presented a series solution of Eq. (32) as (in a.u.)

$$\rho(r) = C \exp(-\omega r^2) \sum_{n=0}^{M} a(n)\, (\omega r^2)^n\,. \tag{33}$$

Here, the normalization constant C is given by

$$C = \left[\frac{\sqrt{\pi}}{2} \left(\frac{\omega}{\pi}\right)^{3/2} N\right] \Big/ \sum_{n=0}^{M} a(n)\, \Gamma(n+3/2)\,, \tag{34}$$

with N the total Fermion number for $(M+1)$ filled shells, which is readily obtained from the degeneracy of the 3D oscillator levels as

$$N = (M+1)\,(M+2)\,(M+3)/6\,. \tag{35}$$

The coefficients $a(n)$ in Eq. (33), which depend on the number of closed shells, are related by the recursion relation

$$0 = a(n+2) \left[\frac{(n+2)\,(2n+5)}{2}\right] + a(n+1)\,[2(M+1) - 3(n+1)] + a(n) \left[\frac{2(n-M)}{(n+1)}\right]. \tag{36}$$

This yields a convenient numerical procedure for calculating the particle density $\rho(r)$ for a substantial number of closed shells.

4.1 Two-dimensional isotropic harmonic confinement Minguzzi et al. [18] have also considered the case of two-dimensional harmonic confinement, their motivation stemming from the current considerable activity in the area of harmonically confined Fermion vapours [19–22]. The equation corresponding to Eq. (32) for the particle density $\rho(r)$ is

$$\frac{\hbar^2}{8m} \frac{\partial}{\partial r} [\nabla^2 \rho(r)] + \left[\left(M + \frac{3}{2}\right) \hbar\omega - V(r)\right] \frac{\partial \rho}{\partial r} + \frac{\partial V}{\partial r} \rho = 0 \tag{37}$$

Also, the Slater sum $S(r,\beta)$ satisfies the partial differential equation

$$\frac{\hbar^2}{8m}\frac{\partial}{\partial r}\left[\nabla^2 S(r,\beta)\right] - \left[V(r) + \frac{\partial}{\partial \beta}\right]\frac{\partial S}{\partial r} = 0 \tag{38}$$

with known solution (see Ref. [23]; also Ref. [24])

$$S(r,\beta) = \left(\frac{m}{2\pi\hbar}\right)\left(\frac{\omega}{\sinh(\beta\hbar\omega)}\right)\exp\left(-\frac{m\omega r^2}{\hbar}\tanh\left(\frac{\beta\hbar\omega}{2}\right)\right). \tag{39}$$

The D-dimensional generalization for isotropic harmonic confinement is given by

$$\frac{\hbar^2}{8m}\frac{\partial}{\partial r}\left[\nabla^2 S(r,\beta)\right] - \left[V(r) + \frac{\partial}{\partial \beta}\right]\frac{\partial S}{\partial r} - \left(1 - \frac{D}{2}\right)\frac{\partial V}{\partial r}\,S = 0\,. \tag{40}$$

This embraces the two- and three-dimensional forms set out in Eqs. (38) and (28). The one-dimensional Eq. (7), for the special case $V(x) = (1/2)\,m\omega^2 x^2$, is also equivalent to Eq. (40).

5 Summary and future directions A survey has been given of progress on propagators, and the related Slater sum on which particular attention has focussed, within the framework of one-body potential theory, central to current usage of density functional theory in molecules [2] and solids [25]. After a brief introduction relating the Feynman propagator $K(\boldsymbol{r}, \boldsymbol{r}', t)$ to the statistical mechanical canonical density matrix $C(\boldsymbol{r},\boldsymbol{r}',\beta)$, through the transformation $\beta \to it$, with $\beta = (k_B T)^{-1}$ and t the time, the Slater sum $S(\boldsymbol{r},\beta) \equiv C(\boldsymbol{r}, \boldsymbol{r}, \beta)$ has been considered for a number of model potentials and in dimensionality D, where $D = 1, 2$ and 3. Quite generally in one dimension, the Slater sum $S(x,\beta)$ satisfies the partial differential Eq. (7) for an arbitrary one-body potential $V(x)$. Though no such general equation is known at the time of writing in higher dimensions, for specific choices of $V(\boldsymbol{r})$, and in particular for the bare Coulomb potential with $D = 3$ and for harmonic confinement with both $D = 2$ and 3, such differential equations for the Slater sum are known (e.g., Eq. (26) for $V = -Ze^2/r$, and Eqs. (28) and (38) for harmonic confinement in $D = 3$ and $D = 2$ respectively). Corresponding equations to (28) and (38) for harmonic potentials for the ground-state density $\rho(r)$ are now also available (see Eqs. (32) and (37) for $D = 3$ and 2 respectively), and finite series solutions are known.

As to the future, simplification of the Blinder [9] Coulomb propagator remains an important matter and the answer may lie in gaining further insight into the types of solution of the established Eq. (26) for the Slater sum, one important 'boundary' condition established through Blinder's study [9] being given in Eq. (25). More study of the nature of long-range oscillations induced in D-dimensional Fermi gases by various types of scattering potentials, such as sech^2 for $D = 1$ and Coulomb for $D = 3$, seems called for. The perturbation theory in the Appendix may provide a way forward here.

Appendix: Perturbation theory to all orders in the one-body potential *V(r)* for the Slater sum *S(r, β)* in three dimensions March and Murray [26] (see also [27]) developed a perturbation theory based on plane waves for an arbitrary potential $V(\boldsymbol{r})$ for the Slater sum $S(\boldsymbol{r},\beta)$, their theory being given to all orders in $V(\boldsymbol{r})$ in three dimensions. Their essential step was to convert the Bloch equation to an integral equation and then to iterate based on the plane wave solution C_0 for the canonical density matrix C in Eq. (4). For $S(\boldsymbol{r},\beta)$ itself the integral equation before iteration then reads

$$S(\boldsymbol{r},\beta) = S_0(\beta) - \int d\boldsymbol{r}_1 \int_0^\beta d\beta\, C_0(\boldsymbol{r},\boldsymbol{r}_1,\beta-\beta_1)\, V(\boldsymbol{r}_1)\, C(\boldsymbol{r}_1,\boldsymbol{r},\beta_1)\,. \tag{A.1}$$

Replacing C inside the integral by C_0, with C_0 given by

$$C_0(\boldsymbol{r},\boldsymbol{r}',\beta) = (2\pi\beta)^{-3/2}\exp\left(-\frac{|\boldsymbol{r}-\boldsymbol{r}'|^2}{2\beta}\right) \tag{A.2}$$

then generates the entire series for $S(\boldsymbol{r},\beta)$ to all orders in $V(\boldsymbol{r})$ (see Ref. [3]). Explicitly, one can write

$$C(\boldsymbol{r},\boldsymbol{r}_0,\beta)=\sum_{j=0}^{\infty} C_j(\boldsymbol{r},\boldsymbol{r}_0,\beta)\,, \tag{A.3}$$

where

$$C_j(\boldsymbol{r},\boldsymbol{r}_0,\beta)=(2\pi\beta)^{-3/2}\int\prod_{l=1}^{j}\left[\frac{-\mathrm{d}\boldsymbol{r}_l\,V(\boldsymbol{r}_l)}{2\pi}\right]\exp\left(-\frac{1}{2\beta}\left(\sum_{l=1}^{j+1}s_l\right)^2\right)\sum_{l=1}^{j+1}s_l\Big/\prod_{l=1}^{j+1}s_l \tag{A.4}$$

and

$$s_l=|\boldsymbol{r}_l-\boldsymbol{r}_{l-1}|\,,\qquad \boldsymbol{r}_{j+1}=\boldsymbol{r}\,. \tag{A.5}$$

In particular,

$$C_1(\boldsymbol{r},\boldsymbol{r}_0,\beta)=(2\pi\beta)^{-3/2}\int\left[\frac{-\mathrm{d}\boldsymbol{r}_1\,V(\boldsymbol{r}_1)}{2\pi}\right][|\boldsymbol{r}_1-\boldsymbol{r}_0|+|\boldsymbol{r}-\boldsymbol{r}_1|]\times\frac{\exp\left(-\{|\boldsymbol{r}_1-\boldsymbol{r}_0|+|\boldsymbol{r}-\boldsymbol{r}_1|\}^2/2\beta\right)}{|\boldsymbol{r}_1-\boldsymbol{r}_0|\,|\boldsymbol{r}-\boldsymbol{r}_1|}\,, \tag{A.6}$$

or

$$S_1(\boldsymbol{r},\beta)=(2\pi\beta)^{-3/2}\int\left[\frac{-\mathrm{d}\boldsymbol{r}_1\,V(\boldsymbol{r}_1)}{2\pi}\right]\frac{2}{|\boldsymbol{r}_1-\boldsymbol{r}|}\exp\left(-2|\boldsymbol{r}_1-\boldsymbol{r}|^2/\beta\right). \tag{A.7}$$

The kinetic energy density $t(\boldsymbol{r},\beta)$ defined in Eq. (30) can thereby be generated to all orders in $V(\boldsymbol{r})$ as

$$t_j(\boldsymbol{r},\beta)=\left[V(\boldsymbol{r})+\frac{3}{2\beta}\left(1-\left(\sum_{l=1}^{j+1}s_l\right)^2\Big/3\beta\right)\right]S_j(\boldsymbol{r},\beta)\,, \tag{A.8}$$

where $t(\boldsymbol{r},\beta)=\sum_{j=0}^{\infty}t_j(\boldsymbol{r},\beta)$. Hence the kinetic energy density is directly determined solely in terms of the one-body potential.

Acknowledgement NHM wishes to acknowledge partial financial support for work on density functional theory from the Office of Naval Research. Especial thanks are due to Dr. P. Schmidt of that Office for continuing encouragement and motivation. IAH acknowledges support from the IWT-Flemish region.

References

[1] F. Brosens and J. T. Devreese, J. Math. Phys. **25**, 1752 (1984), and in: Path Integrals and their Applications in Quantum, Statistical and Solid State Physics, edited by G. J. Papadopoulos and J. T. Devreese (Plenum, New York, 1978).
[2] R. G. Parr and W.-T. Yang, Density Functional Theory of Atoms and Molecules (Oxford University Press, Oxford, 1989).
[3] N. H. March, W. H. Young, and S. Sampanthar, The Many-Body Problem in Quantum Mechanics (Dover, New York, 1995).
[4] N. H. March and A. M. Murray, Phys. Rev. **120**, 830 (1960).
[5] L. D. Landau and E. M. Lifshitz, Quantum Mechanics: non-relativistic theory, 2nd edition (Pergamon, Oxford, 1965).
[6] E. W. Montroll, J. Math. Phys. **11**, 635 (1970).
[7] I. A. Howard and N. H. March, submitted for publication.
[8] E. H. Sondheimer and A. H. Wilson, Proc. Roy. Soc. A **210**, 173 (1951).
[9] S. M. Blinder, Phys. Rev. A **43**, 13 (1991).
[10] S. Pfalzner, H. Lehmann, and N. H. March, J. Math. Chem. **16**, 9 (1994).
[11] I. L. Cooper, Phys. Rev. A **50**, 1040 (1994).

[12] N. H. March, Phys. Rev. A **33**, 88 (1986).
[13] I. A. Howard, N. H. March, and Á. Nagy, submitted for publication.
[14] I. S. Gradshteyn and I. M. Ryzhik, Table of Integrals, Series and Products, 9.228 (Academic Press, New York, 1965).
[15] C. Amovilli and N. H. March, Phys. Chem. Liq. **30**, 135 (1995).
[16] A. Minguzzi, N. H. March, and M. P. Tosi, Phys. Lett. A **281**, 192 (2001).
[17] I. A. Howard and N. H. March, J. Phys. A **34**, L491 (2001).
[18] A. Minguzzi, N. H. March, and M. P. Tosi, Eur. Phys. J. D **15**, 315 (2001).
[19] B. DeMarco and D. S. Jin, Science **285**, 1703 (1999).
[20] M. J. Holland, B. DeMarco, and D. S. Jin, Phys. Rev. A **61**, 053610 (2000).
[21] M. O. Mewes, G. Ferrari, F. Schreck, A. Sinatra, and C. Salomon, Phys. Rev. A **61**, 011403(R) (2000).
[22] F. Schreck, G. Ferrari, K. L. Corwin, G. Cubizolles, L. Khyakovich, M. O. Mewes, and C. Salomon, cond-mat/0011291.
[23] M. J. Stephen and K. Zalewski, Proc. Roy. Soc. A **270**, 435 (1962).
[24] M. Brack and B. P. van Zyl, Phys. Rev. Lett. **86**, 1574 (2001).
[25] J. Callaway and N. H. March, Solid State Phys. **38**, 136 (1984).
[26] N. H. March and A. M. Murray, Proc. Roy. Soc. A **261**, 119 (1961).
[27] F. Despa and R. S. Berry, Phys. Chem. Chem. Phys. **4**, 3774 (2002).

JOURNAL OF CHEMICAL PHYSICS VOLUME 119, NUMBER 12 22 SEPTEMBER 2003

Corrections to Slater exchange potential in terms of Dirac idempotent density matrix: With an approximate application to Be-like positive atomic ions for large atomic number

I. A. Howard
Department of Physics, University of Antwerp, Groenenborgerlaan 171, B-2020 Antwerp, Belgium

N. H. March
Department of Physics, University of Antwerp, Groenenborgerlaan 171, B-2020 Antwerp, Belgium and Oxford University, Oxford, England

(Received 24 April 2003; accepted 3 July 2003)

In earlier studies, we have considered the exchange energy density $\epsilon_x(r)$ in terms of the Dirac density matrix $\rho_1(\mathbf{r},\mathbf{r}')$ for the nonrelativistic limit of large atomic number Z in (i) the Be-like series with configuration $(1s)^2(2s)^2$ and (ii) the Ne-like series with closed $K+L$ shells. Subsequently the work of Della Sala and Görling [J. Chem. Phys. **115**, 5718 (2001)] has appeared, in which an integral equation for the exchange potential $v_x(r)$ is given in terms of the idempotent Dirac density matrix, based on the admittedly drastic approximation that the Hartree–Fock and the Kohn–Sham determinants are equal. Here a formally exact generalization of the integral equation is set up and an approximate solution is presented for the Be series at large Z. [DOI: 10.1063/1.1603711]

I. INTRODUCTION

In density functional theory[1] it is by now widely recognized that, in finite systems, and in particular in atomic ions, it is quite essential to treat the exchange potential, $v_x(r)$ say, in a fully quantitative way. Here, our purpose is to build on our earlier work on two classes of spherically symmetric atomic ions,[2,3] treated in their ground state by nonrelativistic quantum mechanics. The present contribution has been motivated by the recent study in this journal of Della Sala and Görling,[4] who base their treatment of $v_x(r)$ on localized Hartree–Fock methods. Their integral Eq. (11) is based on the, admittedly drastic, approximation that the Hartree–Fock (HF) and exchange-only Slater–Kohn–Sham (SKS) determinants are identical. Then their Eq. (11) is an explicit, but approximate, integral equation for $v_x(r)$ to be solved for corrections to the Slater exchange potential[5] $v_x(\mathbf{r})$, given the determinantal (and therefore idempotent) first-order Dirac density matrix, which we denote by $\rho_1(\mathbf{r},\mathbf{r}')$. It is important to note at the outset that Della Sala and Görling define their density matrix $\rho(\mathbf{r},\mathbf{r}')$ entering their Eq. (11) as having diagonal element $n(\mathbf{r})/2=\rho(\mathbf{r},\mathbf{r}')|_{\mathbf{r}'=\mathbf{r}}$, with $n(\mathbf{r})$ the ground-state electron density.

The outline of the present study is then as follows. In Sec. II immediately below, the integral equation of Della Sala and Görling is set out, and a favorable integral property demonstrated. This is followed in Sec. III by an outline of the way their integral equation can be solved for the Be-like series of atomic ions. Section IV gives a formally exact generalization of this integral equation [see Eq. (2.1) below] and solves it approximately for the same series in the (nonrelativistic) limit in which the atomic number Z becomes very large. Section V consists of a discussion plus a brief summary.

II. DELLA SALA–GÖRLING INTEGRAL EQUATION FOR EXCHANGE POTENTIAL $V_X(R)$

Della Sala and Görling[4] have recently proposed an integral equation of the exchange potential $v_x(\mathbf{r})$. Though it is based on the somewhat drastic approximation that the Hartree–Fock and the exchange-only Kohn–Sham determinants are identical, we shall show below that it has some properties that make it of interest in its own right. Bearing in mind that their density matrix $\rho(\mathbf{r},\mathbf{r}')=\rho_1(\mathbf{r},\mathbf{r}')/2$ defined above, the Della Sala–Görling integral equation reads (with their degeneracy n_s set equal to 2):

$$v_x(\mathbf{r})=v_x^{\mathrm{Sl}}(\mathbf{r})+\frac{1}{2n(\mathbf{r})}\int d\mathbf{r}'\,\rho_1(\mathbf{r}',\mathbf{r})\rho_1(\mathbf{r},\mathbf{r}')v_x(\mathbf{r}')+\frac{1}{4n(\mathbf{r})}\int\int d\mathbf{r}'\,d\mathbf{r}''\frac{\rho_1(\mathbf{r}'',\mathbf{r})\rho_1(\mathbf{r},\mathbf{r}')\rho_1(\mathbf{r}',\mathbf{r}'')}{|\mathbf{r}'-\mathbf{r}''|},\tag{2.1}$$

where $v_x^{\mathrm{Sl}}(\mathbf{r})$ denotes the Slater approximation[1] to the exchange potential. We choose first to multiply Eq. (2.1) by the density $n(\mathbf{r})$ and then to integrate through the whole of space to find

$$\int n(\mathbf{r})v_x(\mathbf{r})d\mathbf{r}=2E_x+\frac{1}{2}\int d\mathbf{r}\int d\mathbf{r}'\,\rho_1(\mathbf{r}',\mathbf{r})\rho_1(\mathbf{r},\mathbf{r}')v_x(\mathbf{r}')+\frac{1}{4}\int\int\int d\mathbf{r}\,d\mathbf{r}'\,d\mathbf{r}''\frac{\rho_1(\mathbf{r}'',\mathbf{r})\rho_1(\mathbf{r},\mathbf{r}')\rho_1(\mathbf{r}',\mathbf{r}'')}{|\mathbf{r}'-\mathbf{r}''|}.\tag{2.2}$$

But the idempotency of our first-order density matrix with the exact exchange-only Kohn–Sham density as its diagonal element, i.e., $n_{HF}(\mathbf{r})=n(\mathbf{r})$ for notational simplicity, reads

$$\frac{\rho_1(\mathbf{r},\mathbf{r}')}{2}=\int d\mathbf{r}''\frac{\rho_1(\mathbf{r},\mathbf{r}'')}{2}\frac{\rho_1(\mathbf{r}'',\mathbf{r}')}{2}. \tag{2.3}$$

Using this result (2.3) in the second and third terms on the right-hand side (RHS) of Eq. (2.2) allows the integrations over $\mathbf{r}$ to be performed, to yield

$$\int n(\mathbf{r})v_x(\mathbf{r})d\mathbf{r}=2E_x+\int n(\mathbf{r}')v_x(\mathbf{r}')d\mathbf{r}' +\frac{1}{2}\int d\mathbf{r}'\,d\mathbf{r}''\frac{\rho_1^2(\mathbf{r}',\mathbf{r}'')}{|\mathbf{r}'-\mathbf{r}''|}. \tag{2.4}$$

But the Dirac[6] formula for the exchange energy E_x is, in atomic units as employed above,

$$E_x=-\frac{1}{4}\int d\mathbf{r}'\,d\mathbf{r}''\frac{\rho_1^2(\mathbf{r}',\mathbf{r}'')}{|\mathbf{r}'-\mathbf{r}''|} \tag{2.5}$$

and so we conclude first that, although the derivation of Eq. (2.1) given by Della Sala and Görling is obviously approximate, this equation yields a precise identity when multiplied by $n(\mathbf{r})$ and integrated over the whole of space.

III. PARAMETRIZATION OF SOLUTION OF INTEGRAL EQ. (2.1) OF DELLA SALA AND GÖRLING FOR BE-LIKE ATOMIC IONS

We express the idempotent density matrix $\rho_1(r,r')$ for the Be-like atomic ions with configuration $(1s)^2(2s)^2$ in terms of $1s$ and $2s$ normalized wave functions denoted by $\psi_1(r)$ and $\psi_2(r)$, respectively, as

$$\rho_1(r,r')=2[\psi_1(r)\psi_1(r')+\psi_2(r)\psi_2(r')]. \tag{3.1}$$

Denoting the integral term in Eq. (2.1) involving the exchange potential $v_x(r')$ by $I(r')$, we can then write

$$\frac{n(r)I(r)}{2}=\psi_1^2(r)\int\psi_1^2(r')v_x(r')d\mathbf{r}' +\psi_2^2(r)\int\psi_2^2(r')v_x(r')d\mathbf{r}' +2\psi_1(r)\psi_2(r)\int\psi_1(r')\psi_2(r')v_x(r')d\mathbf{r}'. \tag{3.2}$$

Since $v_x(r')$ entering the integrals in Eq. (3.2) is unknown at the outset, we shall regard the three integrals in Eq. (3.2) as parameters, denoted λ_i below, so that

$$\frac{n(r)I(r)}{2}=\lambda_1\psi_1^2(r)+\lambda_2\psi_2^2(r)+2\lambda_3\psi_1(r)\psi_2(r). \tag{3.3}$$

We next turn to treat the final term $F(r)$ in the form

$$v_x(r)=\frac{2\epsilon_x(r)}{n(r)}+I(r,\lambda_1,\lambda_2,\lambda_3)+F(r). \tag{3.4}$$

Since the Slater potential in Eq. (2.1) is given in terms of $\epsilon_x(r)$, which is again determined solely by $\rho_1(r,r')$ in Eq. (3.1), as

$$v_x^{\mathrm{Sl}}(r)=\frac{2\epsilon_x(r)}{n(r)},\quad E_x=\int\epsilon_x(r)d\mathbf{r}, \tag{3.5}$$

with

$$\epsilon_x(r)=-\frac{1}{4}\int d\mathbf{r}'\frac{\rho_1^2(r',r)}{|\mathbf{r}-\mathbf{r}'|}, \tag{3.6}$$

one can rewrite Eq. (2.1) as follows. We shall see that, since $\epsilon_x(r)$ and $F(r)$ are determined solely by $\rho_1(r',r)$ in Eq. (3.1), one can converge on the parameters λ_i, $i=1-3$, by iteration (see also Appendix A below).

We have

$$n(r)F(r)=\int d\mathbf{r}'\,d\mathbf{r}''\frac{\rho_1(\mathbf{r}',\mathbf{r}'')}{|\mathbf{r}'-\mathbf{r}''|} \times[\psi_1(r'')\psi_1(r)+\psi_2(r'')\psi_2(r)] \times[\psi_1(r)\psi_1(r')+\psi_2(r)\psi_2(r')], \tag{3.7}$$

or

$$\begin{aligned}n(r)F(r)&=\psi_1^2(r)\int d\mathbf{r}'\,d\mathbf{r}''\frac{\rho_1(\mathbf{r}',\mathbf{r}'')\psi_1(r')\psi_1(r'')}{|\mathbf{r}'-\mathbf{r}''|}\\&+\psi_2^2(r)\int d\mathbf{r}'\,d\mathbf{r}''\frac{\rho_1(\mathbf{r}',\mathbf{r}'')\psi_2(r')\psi_2(r'')}{|\mathbf{r}'-\mathbf{r}''|}\\&+\psi_1(r)\psi_2(r)\int d\mathbf{r}'\,d\mathbf{r}''\frac{\rho_1(\mathbf{r}',\mathbf{r}'')}{|\mathbf{r}'-\mathbf{r}''|}\\&\times[\psi_2(r')\psi_1(r'')+\psi_2(r'')\psi_1(r')]\\&\equiv f_1\psi_1^2(r)+f_2\psi_2^2(r)+f_3\psi_1(r)\psi_2(r).\end{aligned} \tag{3.8}$$

Thus the shape of v_x is given by

$$v_x(r)=2\frac{\epsilon_x(r)}{n(r)}+\frac{2}{n(r)}[\lambda_1\psi_1^2(r)+\lambda_2\psi_2^2(r) +2\lambda_3\psi_1(r)\psi_2(r)]+\frac{1}{n(r)}[f_1\psi_1^2(r) +f_2\psi_2^2(r)+f_3\psi_1(r)\psi_2(r)], \tag{3.9}$$

which is formally equivalent to

$$v_x(r)=v_x^{\mathrm{Sl}}(r)+\frac{1}{n(r)}[p_1\psi_1^2(r)+p_2\psi_2^2(r) +p_3\psi_1(r)\psi_2(r)]. \tag{3.10}$$

The requirement that $v_x(r)$ goes to zero as $r\rightarrow\infty$ changes Eq. (3.10) by $p_2/2$, i.e., a shift of the zero of potential, to yield

$$v_x(r)=v_x^{\rm Sl}(r)+\frac{1}{n(r)}\left[p_1\psi_1^2(r)-\frac{1}{2}p_2\frac{2(\psi_1^2(r)+\psi_2^2(r))}{n(r)}\right.$$
$$\left.+p_2\psi_2^2(r)+p_3\psi_1(r)\psi_2(r)\right]$$
$$\equiv v_x^{\rm Sl}(r)+\frac{1}{n(r)}[a\psi_1^2(r)+p_3\psi_1(r)\psi_2(r)], \quad (3.11)$$

where the parameter a has been written for p_1-p_2. This expression for $v_x(r)$ now goes to zero as $r\rightarrow\infty$. There is no conflict with the integral Eq. (2.1), since, as Della Sala and Görling stress, if $v_x(r)$ is one solution, then also $v_x(r)$ + constant satisfies the integral equation. Equation (3.11) is the central result for the present example of the nonrelativistic exchange potential $v_x(r)$ in the Be-like series of atomic ions, in the Della Sala–Görling approximation.

IV. FORMALLY EXACT GENERALIZATION OF DELLA SALA–GÖRLING INTEGRAL EQUATION, WITH AN APPROXIMATE APPLICATION TO BE-LIKE SERIES FOR LARGE ATOMIC NUMBER

As discussed in Sec. II above, one of the attractions of the Della Sala–Görling integral Eq. (2.1) for the exchange potential $v_x(r)$ is the integral identity established there. Below, we give a formally exact generalization of Eq. (2.1), plus an approximate application.

A. Some consequences of idempotency of the density matrix $\rho_1(\mathbf{r},\mathbf{r}')$

Motivated by Sec. II, we first rewrite $\int nv_x\,d\mathbf{r}$ using the idempotency of $\rho_1(\mathbf{r},\mathbf{r}')/2$ to find

$$\int n(\mathbf{r})v_x(\mathbf{r})d\mathbf{r}=\frac{1}{2}\int\int\rho_1^2(\mathbf{r},\mathbf{r}')v_x(\mathbf{r}')d\mathbf{r}\,d\mathbf{r}'. \quad (4.1)$$

Also, as already noted in Sec. II, again after using idempotency,

$$E_x=-\frac{1}{8}\int\int\int d\mathbf{r}\,d\mathbf{r}'\,d\mathbf{r}''\frac{\rho_1(\mathbf{r}'',\mathbf{r})\rho_1(\mathbf{r},\mathbf{r}')\rho_1(\mathbf{r}',\mathbf{r}'')}{|\mathbf{r}'-\mathbf{r}''|}. \quad (4.2)$$

Writing

$$E_x=\int\epsilon_x(\mathbf{r})d\mathbf{r}=\frac{1}{2}\int n(\mathbf{r})v_x^{\rm Sl}(\mathbf{r})d\mathbf{r}, \quad (4.3)$$

we have the differential form of the addition of the two identities, after dividing by $n(\mathbf{r})$, as

$$v_x(\mathbf{r})=v_x^{\rm Sl}(\mathbf{r})+\frac{1}{2n(\mathbf{r})}\int d\mathbf{r}'\,\rho_1(\mathbf{r}',\mathbf{r})\rho_1(\mathbf{r},\mathbf{r}')v_x(\mathbf{r}')$$
$$+\frac{1}{4n(\mathbf{r})}\int\int d\mathbf{r}'\,d\mathbf{r}''\frac{\rho_1(\mathbf{r}'',\mathbf{r})\rho_1(\mathbf{r},\mathbf{r}')\rho_1(\mathbf{r}',\mathbf{r}'')}{|\mathbf{r}'-\mathbf{r}''|}$$
$$+\frac{1}{n(\mathbf{r})}P(\mathbf{r}), \quad (4.4)$$

where $P(\mathbf{r})$ satisfies the sum rule

$$\int P(\mathbf{r})d\mathbf{r}=0. \quad (4.5)$$

What, to us, is remarkable, is that the arguments of Della Sala and Görling demonstrate that Eq. (4.5) is satisfied by the choice $P(\mathbf{r})=0$ if the equality of SKS and HF determinants is assumed at the outset.

Though we do not have a general proposal for closure of Eq. (4.4), we show by returning to the Be-like series of atomic ions below that a natural (though presumably approximate) closure suggests itself there.

B. Approximate application to Be-like series

Returning to Eq. (3.11) we note again the important orthogonality property $\int\psi_1(r)\psi_2(r)d\mathbf{r}=0$. Comparing this with the sum rule for $P(\mathbf{r})$, we are motivated to propose the closure

$$P(\mathbf{r})=\text{constant}\times\psi_1(r)\psi_2(r). \quad (4.6)$$

Then, using Eqs. (3.11), (4.4), and (4.6) we can write

$$v_x(r)=v_x^{\rm Sl}(r)+\frac{1}{n(r)}[a\psi_1^2(r)+b\psi_1(r)\psi_2(r)], \quad (4.7)$$

where b has now subsumed in it the constant in Eq. (4.6). With the generalized form (4.7) we note that, by multiplying through by $n(r)$ and performing a volume integration

$$a=\int n(r)v_x(r)d\mathbf{r}-2E_x, \quad (4.8)$$

which gives one interpretation of the parameter a (see also Sec. V below). This parameter a is decoupled from the quantity b, subsuming the unknown constant in the assumption (4.6), because of the orthogonality of $\psi_1(r)$ and $\psi_2(r)$.

But we have additionally the Levy–Perdew[8] virial-like relation

$$E_x=-\int rn(r)\frac{\partial v_x(r)}{\partial r}d\mathbf{r}, \quad (4.9)$$

which we can further use to relate the two parameters a and b in Eq. (4.7). From Eq. (4.7), we form $\partial v_x(r)/\partial r$ as

$$\frac{\partial v_x(r)}{\partial r}=\frac{\partial v_x^{\rm Sl}(r)}{\partial r}+a\frac{\partial}{\partial r}\left[\frac{\psi_1^2(r)}{n(r)}\right]+b\frac{\partial}{\partial r}\left[\frac{\psi_1(r)\psi_2(r)}{n(r)}\right]. \quad (4.10)$$

Inserting Eq. (4.10) into Eq. (4.9), and after integrating by parts we find

$$E_x=-\int rn(r)\frac{\partial}{\partial r}\left[\frac{2\epsilon_x(r)}{n(r)}\right]d\mathbf{r}$$
$$+4\pi a\int_0^\infty\psi_1^2(r)\left[3r^2+r^3\frac{n'}{n}\right]dr$$
$$+4\pi b\int_0^\infty\psi_1(r)\psi_2(r)\left[3r^2+r^3\frac{n'}{n}\right]dr. \quad (4.11)$$

Below we evaluate the radial integrals appearing in Eq. (4.11) numerically for the Be-like series of atomic ions in the large-Z limit.

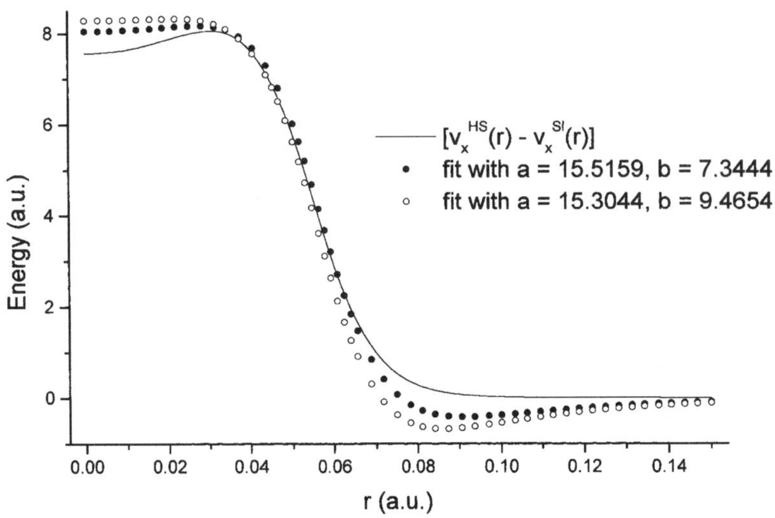

FIG. 1. Shows our earlier results for $v_x^{\mathrm{HS}}(r)-v_x^{\mathrm{SI}}(r)$ for large-Z limit (plot is for $Z=50$). Curves are not very sensitive to small changes in parameters a and b. Even $b=0$ is rather similar, but $b\neq 0$ is needed to impose Levy–Perdew relation (4.9), in addition to the condition (4.8). We have assumed for this Be-like series that, as for the Ne-like analogue, $v_x^{\mathrm{HS}}(r)$ satisfies Eq. (4.9) exactly.

C. Numerical results for Be-like atomic ions at large Z

Given Eq. (4.11), we can evaluate E_x directly, since we know the form of the $1s$ and $2s$ hydrogenic orbitals, for Be in the large-Z limit, to find $E_x=-0.819\,286Z$. Equation (4.11) can then be written in the form

$$E_x=(I_1+3)a+bI_2+I_3, \tag{4.12}$$

where

$$I_1=4\pi\int_0^\infty r^3\psi_1^2(r)\frac{n'}{n}dr, \tag{4.13}$$

$$I_2=4\pi\int_0^\infty r^3\psi_1(r)\psi_2(r)\frac{n'}{n}dr, \tag{4.14}$$

and

$$I_3=-4\pi\int_0^\infty r^3 n\frac{\partial v_x^{\mathrm{SI}}(r)}{\partial r}dr. \tag{4.15}$$

While I_1 and I_2 are independent of Z, I_3 is linearly proportional to Z, and numerical evaluation gives $I_1=-2.536\,99$ a.u., $I_2=0.046\,15$ a.u., and $I_3=-0.969\,75\,Z$ a.u.

For the Be-like ion with $Z=50$, we have numerically evaluated the Harbola–Sahni exchange potential $v_x^{\mathrm{HS}}(r)$[10] and used it to find $a=15.5159$ from Eq. (4.8). Subsequent use of Eq. (4.12) then gives $b=7.3444$. In Fig. 1 we compare this fit to the numerically evaluated Harbola–Sahni potential. An alternative way to evaluate a and b is to note that $\psi_2(r)$ has a node at $r\equiv r_2=2/Z$ a.u. Equation (4.7) at $r=r_2$ can then be used to find a [taking the approximation $v_x(r)=v_x^{\mathrm{HS}}(r)$], and Eq. (4.12) subsequently used to find b. For $Z=50$, we find then $a=15.3044$ and $b=9.4654$. This fit is also shown in Fig. 1. We note that the exact exchange energy for this case is $E_x=-40.964\,32$ a.u.; use of either of the fits just described in the Levy–Perdew virial-like theorem (4.9) gives the same answer.

V. DISCUSSION AND SUMMARY

We have given here a formally exact generalization (4.4) of the Della Sala–Görling integral Eq. (2.1). We stress that the two parameter form in Eq. (4.7) is the basic result of the present application of Eq. (4.4) to the Be-like series at large Z. It is to be noted that in the earlier study of Nagy,[11] an approximate expression for the exchange potential $v_x(\mathbf{r})$ designed to reproduce the HF density has been proposed. Her result, $v_x^N(\mathbf{r})$ say, is given by

$$v_x^N(\mathbf{r})=v_x^{\mathrm{SI}}(\mathbf{r})+\sum_{\text{occupied } i}(\epsilon_i^{\mathrm{KS}}-\epsilon_i^{\mathrm{HF}})\frac{\psi_i^2}{n(\mathbf{r})}. \tag{5.1}$$

Here ϵ_i^{KS} and ϵ_i^{HF} denote the KS and HF eigenvalues. We note that, applied to Be, this appears to be a two-parameter form. But if we require $v_x^N(\mathbf{r})|_{r\to\infty}\to 0$, which affects eigenvalues but not $n(\mathbf{r})$, then for Be this becomes

$$v_x^N(\mathbf{r})=v_x^{\mathrm{SI}}(\mathbf{r})+(\Delta\epsilon_1-\Delta\epsilon_2)\frac{\psi_1^2(\mathbf{r})}{n(\mathbf{r})}, \tag{5.2}$$

where $\Delta\epsilon_i=\epsilon_i^{\mathrm{KS}}-\epsilon_i^{\mathrm{HF}}$. The present approach proposes therefore, by comparison of Eq. (4.7) with Eq. (5.2), the addition of the cross term $\psi_1(\mathbf{r})\psi_2(\mathbf{r})/n(\mathbf{r})$, suitably weighted, to Eq. (5.2). This cross term does not affect the integral $\int n(\mathbf{r})v_x^N(\mathbf{r})d\mathbf{r}$ due to orthogonality of $\psi_1(\mathbf{r})$ and $\psi_2(\mathbf{r})$.

Thus, the Nagy form (5.2) would relate a in Eq. (4.7) to differences between KS and HF eigenvalues of the $1s$ and $2s$ orbitals. While, in the Be-like series at large Z, the term $a\psi_1^2(\mathbf{r})/n(\mathbf{r})$ dominates the cross term until well beyond the radial node of $\psi_2(\mathbf{r})$, a fundamental expression for b is only available to us presently by imposing the Levy–Perdew virial-like result.

ACKNOWLEDGMENTS

The authors are much indebted to Professor A. Holas for drawing our attention to the work of Della Sala and Görling and for many most valuable discussions. I.A.H. would like to acknowledge support from the IWT-Flemish region. Finally N.H.M. wishes to acknowledge partial financial support for work on density functional theory from the ONR. Dr. P. Schmidt of that Office is especially thanked for much motivation and his continuing support.

APPENDIX A: EXPLICIT FORMS FOR PARAMETERS OF BE EXCHANGE POTENTIAL

Let us write Eq. (2.1) in the form

$$v_x(r)=v_x^{\mathrm{SI}}(r)+I(r,\lambda_1,\lambda_2,\lambda_3)+F(r). \tag{A1}$$

Then we have, using Eq. (3.3),

$$\lambda_1=\int \psi_1^2(r)v_x^{\mathrm{SI}}(r)d\mathbf{r}+\int \psi_1^2(r)F(r)d\mathbf{r} \times 2\int \frac{\psi_1^2(r)}{n(r)}[\lambda_1\psi_1^2(r)+\lambda_2\psi_2^2(r)+2\lambda_3\psi_1(r)\psi_2(r)]d\mathbf{r} \tag{A2}$$

or

$$\lambda_1\left[1-2\int \frac{\psi_1^4(r)}{n(r)}d\mathbf{r}\right] =\int \psi_1^2(r)v_x^{\mathrm{SI}}(r)d\mathbf{r}+\int \psi_1^2(r)F(r)d\mathbf{r} +2\lambda_2\int \frac{\psi_1^2(r)\psi_2^2(r)}{n(r)}d\mathbf{r}+4\lambda_3\int \frac{\psi_1^3(r)\psi_2(r)}{n(r)}d\mathbf{r}, \tag{A3}$$

where the first two terms on the right-hand side are known. Likewise, for λ_2 we have

$$\lambda_2\left[1-2\int \frac{\psi_2^4(r)}{n(r)}d\mathbf{r}\right] =\int \psi_2^2(r)v_x^{\mathrm{SI}}(r)d\mathbf{r}+\int \psi_2^2(r)F(r)d\mathbf{r} +2\lambda_1\int \frac{\psi_1^2(r)\psi_2^2(r)}{n(r)}d\mathbf{r}+4\lambda_3\int \frac{\psi_2^3(r)\psi_1(r)}{n(r)}d\mathbf{r}, \tag{A4}$$

and for λ_3,

$$\lambda_3\left[1-4\int \frac{\psi_1^2(r)\psi_2^2(r)}{n(r)}d\mathbf{r}\right] =\int \psi_1(r)\psi_2(r)v_x^{\mathrm{SI}}(r)d\mathbf{r}+\int \psi_2(r)\psi_2(r)F(r)d\mathbf{r} +2\lambda_1\int \frac{\psi_1^3(r)\psi_2(r)}{n(r)}d\mathbf{r}+2\lambda_2\int \frac{\psi_2^3(r)\psi_1(r)}{n(r)}d\mathbf{r}. \tag{A5}$$

Thus, we require $\int\psi_1^4(r)/n(r)d\mathbf{r}$, $\int\psi_2^4(r)/n(r)d\mathbf{r}$, $\int\psi_1^2(r)\psi_2^2(r)/n(r)d\mathbf{r}$, $\int\psi_1(r)\psi_2^3(r)/n(r)d\mathbf{r}$, and $\int\psi_1^3(r)\psi_2(r)/n(r)d\mathbf{r}$.

APPENDIX B: KINETIC CORRECTIONS TO HARTREE–FOCK THEORY FOR BE ATOM

The purpose of this further Appendix is to relate Slater–Kohn–Sham (SKS) and Hartree–Fock (HF) theory, and in particular to exhibit kinetic corrections. These are set to zero when the assumption of Della Sala and Görling[4] that the SKS and HF determinants are identical is imposed from the outset. Defining amplitude and phase from the two sets of wave functions by

$$\psi_1^{\mathrm{HF}}(r)=\left(\frac{n(r)}{2}\right)^{1/2}\cos\theta(r) \tag{B1}$$

and

$$\psi_2^{\mathrm{HF}}(r)=\left(\frac{n(r)}{2}\right)^{1/2}\sin\theta(r), \tag{B2}$$

where $n(r)$ is the HF density, the SKS wave functions corresponding to this same HF density are written as[7]

$$\psi_1^{\mathrm{SKS}}(r)=\left(\frac{n(r)}{2}\right)^{1/2}\cos\theta_S(r) \tag{B3}$$

and

$$\psi_2^{\mathrm{SKS}}(r)=\left(\frac{n(r)}{2}\right)^{1/2}\sin\theta_S(r). \tag{B4}$$

This is the frequently used density amplitude and phase representation for the two-level system, this Appendix being specifically concerned with the neutral Be atom with configuration $(1s)^2(2s)^2$.

In Ref. 7, Eq. (19) gives a correction to the Harbola–Sahni exchange potential $v_x^{\mathrm{HS}}(r)$ via

$$\frac{\partial}{\partial r}[v_x(r)-v_x^{\mathrm{HS}}(r)]=\frac{1}{r^2n(r)}\frac{\partial}{\partial r}[r^2n(r)\{\theta'^2-\theta_s'^2\}]. \tag{B5}$$

But

$$t_{\mathrm{HF}}-t_{\mathrm{SKS}}=\frac{n(r)}{2}[\theta'^2-\theta_s'^2]; \tag{B6}$$

therefore

$$\frac{\partial}{\partial r}[v_x(r)-v_x^{\mathrm{HS}}(r)]=\frac{1}{r^2n(r)}\frac{\partial}{\partial r}[r^2\{2t_{\mathrm{HF}}-2t_{\mathrm{SKS}}\}] =\frac{2}{n(r)}\frac{\partial}{\partial r}\{t_{\mathrm{HF}}-t_{\mathrm{SKS}}\} +\frac{2}{r}\frac{[t_{\mathrm{HF}}-t_{\mathrm{SKS}}]}{n(r)}. \tag{B7}$$

Multiplying Eq. (B7) through by $rn(r)$, and using the Levy–Perdew virial relation

$$E_x=-\int n(r)r\frac{\partial v_x}{\partial r}d\mathbf{r} \tag{B8}$$

we find

$$-E_x-\int n(r)r\frac{\partial v_x^{\mathrm{HS}}}{\partial r}d\mathbf{r}=2\int r\frac{\partial}{\partial r}\{t_{\mathrm{HF}}-t_{\mathrm{SKS}}\}d\mathbf{r}+2[T_{\mathrm{HF}}-T_{\mathrm{SKS}}]. \quad \text{(B9)}$$

Hence, as Levy and March[9] showed, corrections to $v_x^{\mathrm{HS}}(r)$ to get $v_x(r)$ are all about kinetic energy differences.

To use Eq. (B7) one would, of course, have to proceed iteratively. While $t_{\mathrm{HF}}(r)$ and $n(r)\equiv n_{\mathrm{HF}}(r)$ are assumed known at the outset, $t_{\mathrm{SKS}}(r)$ depends on the SKS potential, and hence on the potential $v_x(r)$ to be determined. But the work of Levy and March[9] suggests that kinetic corrections to $v_x^{\mathrm{HS}}(r)$ are small. Thus, one could

(i) use HF orbitals ψ_1 and ψ_2 to calculate $v_x^{\mathrm{HS}(1)}(r)$;

(ii) hence calculate $v_x^{\mathrm{HS}(1)}(r)$ for $v_x(r)$ in the SKS potential; calculate $t_{\mathrm{SKS}}^{(1)}(r)$;

(iii) then use Eq. (B7) with $t_{\mathrm{SKS}}(r)=t_{\mathrm{SKS}}^{(1)}(r)$ to calculate an approximation to $\partial[v_x(r)-v_x^{\mathrm{HS}}(r)]/\partial r$;

(iv) repeat the procedure to self-consistency.

[1] R. G. Parr and W. Yang, *Density Functional Theory of Atoms and Molecules* (Oxford University Press, Oxford, 1989).
[2] I. A. Howard, N. H. March, P. Senet, and V. E. Van Doren, Phys. Rev. A **62**, 062512 (2000).
[3] N. H. March and I. A. Howard, Acta Univ. Debrecen, ser. Physica et Chimica XXXIV–XXXV, 147 (2002).
[4] F. Della Sala and A. Görling, J. Chem. Phys. **115**, 5718 (2001).
[5] A. Holas and N. H. March, Phys. Rev. B **55**, 1295 (1997).
[6] P. A. M. Dirac, Proc. Cambridge Philos. Soc. **26**, 376 (1930).
[7] I. A. Howard, N. H. March, J. A. Alonso, N. A. Cordero, and V. E. Van Doren, Chem. Phys. Lett. **343**, 166 (2001).
[8] M. Levy and J. P. Perdew, Phys. Rev. A **32**, 2010 (1985).
[9] M. Levy and N. H. March, Phys. Rev. A **55**, 1885 (1997).
[10] M. K. Harbola and V. Sahni, Phys. Rev. Lett. **62**, 489 (1989).
[11] Á. Nagy, Phys. Rev. A **55**, 3465 (1997); see also Á. Nagy and N. H. March, Phys. Chem. Liq. **37**, 671 (1999).

Available online at www.sciencedirect.com
SCIENCE DIRECT®
ELSEVIER
Chemical Physics Letters 385 (2004) 231–232
CHEMICAL PHYSICS LETTERS
www.elsevier.com/locate/cplett

Idempotent density matrix derived from a local potential $V(\mathbf{r})$ in terms of HOMO and LUMO properties

I.A. Howard [a,*], N.H. March [a,b], P.W. Ayers [c]

[a] *Department of Physics, University of Antwerp, Groenenborgerlaan 171, B-2020 Antwerp, Belgium*
[b] *Oxford University, Oxford OX1 3UB, UK*
[c] *Department of Chemistry, McMaster University, Hamilton, Ont., Canada L8S 4M1*

Received 17 October 2003; in final form 11 December 2003
Published online: 19 January 2004

Abstract

In a recent paper one of us has exposed a link between the idempotent Dirac density matrix built from occupied Kohn–Sham orbitals and the frontier orbitals. An alternative argument is here presented, from the equation of motion of the density matrix generated by any local potential $V(\mathbf{r})$.

One of us [1], in a recent paper, has established a link between the Dirac idempotent density matrix built from occupied Kohn–Sham orbitals and the frontier orbitals. The purpose of this Letter is to present an alternative argument, from the equation of motion of the Dirac density matrix generated by any local potential $V(\mathbf{r})$.

Following the route used by March and Young [2], let us start from the equation of motion of the density matrix $\gamma(\mathbf{r},\mathbf{r}')$, defined from the occupied orbitals $\psi_i(\mathbf{r})$ generated by a local potential energy $V(\mathbf{r})$. This density matrix satisfies

$$\nabla^2_{\mathbf{r}}\gamma(\mathbf{r},\mathbf{r}') - \nabla^2_{\mathbf{r}'}\gamma(\mathbf{r},\mathbf{r}') = \frac{2m}{\hbar^2}[V(\mathbf{r}) - V(\mathbf{r}')]\gamma(\mathbf{r},\mathbf{r}'). \quad (1)$$

Let us now assume, for simplicity, non-degenerate frontier orbitals $\psi_{\mathrm{HOMO}}(\mathbf{r}) \equiv \psi_{\mathrm{H}}(\mathbf{r})$ and $\psi_{\mathrm{LUMO}}(\mathbf{r}) \equiv \psi_{\mathrm{L}}(\mathbf{r})$, with corresponding one-electron eigenvalues ϵ_{H} and ϵ_{L}, respectively, though non-degeneracy is not an essential requirement. Then we can evidently write from Schrödinger's equation

$$\nabla^2_{\mathbf{r}}\psi_{\mathrm{H}}(\mathbf{r}) + \frac{2m}{\hbar^2}[\epsilon_{\mathrm{H}} - V(\mathbf{r})]\psi_{\mathrm{H}}(\mathbf{r}) = 0 \quad (2)$$

and

$$\nabla^2_{\mathbf{r}'}\psi_{\mathrm{L}}(\mathbf{r}') + \frac{2m}{\hbar^2}[\epsilon_{\mathrm{L}} - V(\mathbf{r}')]\psi_{\mathrm{L}}(\mathbf{r}') = 0. \quad (3)$$

Evidently, Eqs. (2) and (3) can now be solved for $V(\mathbf{r})$ and $V(\mathbf{r}')$, respectively, which can be inserted in Eq. (1). The conclusion is therefore that for any local potential $V(\mathbf{r})$ generating non-degenerate HOMO and LUMO orbitals,

$$\gamma(\mathbf{r},\mathbf{r}') = \gamma[\mathbf{r},\mathbf{r}';\psi_{\mathrm{H}}(\mathbf{r}),\psi_{\mathrm{L}}(\mathbf{r}'),(\epsilon_{\mathrm{H}} - \epsilon_{\mathrm{L}})]. \quad (4)$$

While Eq. (4) is the main result of this Letter, we must emphasize that a solution of (1) is always required which yields $\gamma(\mathbf{r},\mathbf{r}')$ as an idempotent matrix:

$$\gamma(\mathbf{r},\mathbf{r}') = \int \gamma(\mathbf{r},\mathbf{r}'')\gamma(\mathbf{r}'',\mathbf{r}')\,\mathrm{d}\mathbf{r}''. \quad (5)$$

We turn now to consider a simple example, where the one-body potential $V(\mathbf{r})$ in Eq. (1) has the Coulomb form

$$V(\mathbf{r}) = -Z/r. \quad (6)$$

For K plus L shells closed, March and Santamaria [3] effectively solved Eq. (1) for an idempotent $\gamma(\mathbf{r},\mathbf{r}')$. This takes the form

$$\gamma = \gamma_{\mathrm{s}}\left(\frac{x}{2},\frac{y}{2}\right) + \frac{xy}{128\pi}\left(\frac{Z}{a_0}\right)^5 \exp\left(-\frac{Z}{4a_0}(x+y)\right), \quad (7)$$

where $\gamma_{\mathrm{s}}\left(\frac{x}{2},\frac{y}{2}\right)$ is the s-orbital only density matrix. Now we express the two terms on the right-hand side

* Corresponding author.
E-mail address: iris.howard@ua.ac.be (I.A. Howard).

0009-2614/$ - see front matter © 2004 Elsevier B.V. All rights reserved.
doi:10.1016/j.cplett.2003.12.094

implicitly in terms of the HOMO orbital with radial wave function $R_{2p} \equiv R_{21}(r)$ and the LUMO 3s orbital ψ_{3s}, together with the energy difference $\epsilon_H - \epsilon_L$. First we note that, given

$$\psi_{1s}(r) = \left(\frac{Z^3}{\pi}\right)^{1/2} \exp(-Zr), \tag{8}$$

$$\psi_{2s}(r) = \frac{1}{2}\left(\frac{Z^3}{2\pi}\right)^{1/2}\left(1-\frac{Zr}{2}\right)\exp\left(-\frac{Zr}{2}\right) \tag{9}$$

and

$$\psi_{3s}(r) = \frac{1}{3}\left(\frac{Z^3}{3\pi}\right)^{1/2}\left(1-\frac{2Zr}{3}+\frac{2}{27}(Zr)^2\right)\exp\left(-\frac{Zr}{3}\right), \tag{10}$$

we can write

$$\psi_{3s}(r) = \frac{Z}{3^{3/2}}\left(\frac{\psi_{1s}}{\pi}\right)^{1/3}\left[1-\frac{4}{3}\left(1-(8\pi)^{1/4}\frac{\psi_{2s}}{\sqrt{\psi_{1s}}}\right) + \left(\frac{2}{3}\right)^3\left(1-(8\pi)^{1/4}\frac{\psi_{2s}}{\sqrt{\psi_{1s}}}\right)^2\right]. \tag{11}$$

Next we note that, according to Eq. (A.5) of Ref. [3]

$$\frac{3}{4\pi}R_{21}^2 = \frac{r^2}{2}\left[\frac{(\rho_{20}')^2}{\rho_{20}} - \rho_{20}''\right], \tag{12}$$

where

$$\rho_{20} = \frac{(r\psi_{2s})''}{4\pi r^2}. \tag{13}$$

Thus we can write

$$\psi_{3s} \equiv f(Z, \psi_{1s}, R_{21}, \epsilon_H - \epsilon_L) \tag{14}$$

and by inversion (in principle)

$$\psi_{1s} \equiv g(Z, R_{21}, \psi_{3s}), \tag{15}$$

since we can also make use of the fact that

$$\epsilon_H - \epsilon_L = -\frac{Z^2}{8} + \frac{Z^2}{18} = -\frac{5Z^2}{72}. \tag{16}$$

It is interesting here to add that one could determine the atomic number directly from the frontier orbitals (by inverting the Schrödinger equation) rather than through the band gap. Likewise, given Eqs. (12) and (13), we can write in principle

$$\psi_{2s} = h(R_{21}). \tag{17}$$

The s-density matrix γ_s is then just

$$\begin{aligned}\gamma_s\left(\frac{x}{2},\frac{y}{2}\right) &= \psi_{1s}\left(\frac{x}{2}\right)\psi_{1s}\left(\frac{y}{2}\right) + \psi_{2s}\left(\frac{x}{2}\right)\psi_{2s}\left(\frac{y}{2}\right) \\ &= g\left(Z, R_{21}, \psi_{3s}, \frac{x}{2}\right)g\left(Z, R_{21}, \psi_{3s}, \frac{y}{2}\right) \\ &\quad + h\left(R_{21}, \frac{x}{2}\right)h\left(R_{21}, \frac{y}{2}\right).\end{aligned} \tag{18}$$

The second term in Eq. (7) for γ can also be expressed in terms of R_{21}, so we can finally write

$$\begin{aligned}\gamma &= g\left(Z, R_{21}, \psi_{3s}, \frac{x}{2}\right)g\left(Z, R_{21}, \psi_{3s}, \frac{y}{2}\right) \\ &\quad + h\left(R_{21}, \frac{x}{2}\right)h\left(R_{21}, \frac{y}{2}\right) + \frac{3}{4\pi}R_{21}\left(\frac{x}{2}\right)R_{21}\left(\frac{y}{2}\right).\end{aligned} \tag{19}$$

The essence of the above argument may be summarized by saying that Eq. (1) allows the idempotent density matrix $\gamma(\mathbf{r}, \mathbf{r}')$ to be expressed in terms of frontier orbitals plus the HOMO–LUMO gap as key variables. One will then be employing quantities of chemical interest directly in computation, which seems desirable. The use of the 'band gap' in this respect adds to the discussion in [1].

The referee has asked us to add comments on (a) the merits of working with frontier-orbital properties and (b) the influence of long-range exchange-correlation potentials especially on the HOMO orbital. As to (a), one of us [1] has suggested that an ideal Kohn–Sham scheme would be competitive with simple diagonalization for small systems, competitive with linear scaling techniques for large systems, and would provide access to the chemically significant frontier molecular orbitals throughout the entire calculation. The approach in the present Letter seems a step along such a path. Such an approach will be of special computational and conceptual utility when the purpose of a calculation is to elucidate a reagent's chemical activity from the frontier orbitals using, e.g., the Woodward–Hoffmann rules [4] or Fukui's analysis [5]. As to (b), the non-uniform asymptotic decay of the exchange-correlation potential is indeed related to the behaviour of the HOMO orbital and its asymptotic decay. The study of Della Sala and Görling [6] especially should be noted in this context, and also the later investigation of Wu et al. [7] in which the nodal surface problem for formally exact density functional theory was fully analyzed.

References

[1] P.W. Ayers, Theor. Chem. Accounts 110 (2003) 267.

[2] N.H. March, W.H. Young, Nucl. Phys. 12 (1959) 237.

[3] N.H. March, R. Santamaria, Phys. Rev. A 38 (1988) 5002.

[4] R.B. Woodward, R. Hoffmann, J. Am. Chem. Soc. 87 (1965) 2511.

[5] K. Fukui, T. Yonezawa, C. Nagata, J. Chem. Phys. 21 (1953) 174;
K. Fukui, Science 218 (1987) 747.

[6] F. Della Sala, A. Görling, J. Chem. Phys. 115 (2001) 5718.

[7] Q. Wu, P.W. Ayers, W. Yang, J. Chem. Phys. 119 (2003) 2978.

Available online at www.sciencedirect.com
SCIENCE @ DIRECT®
Physics Letters A 324 (2004) 42–45
PHYSICS LETTERS A
www.elsevier.com/locate/pla

Electrostatic interpretation of the force $-\partial V_{xc}/\partial r$ connected with the exchange-correlation potential: direct relation to single-particle kinetic energy density in Be-atom

N.H. March [a,b], P. Geerlings [c,*], K.D. Sen [d]

[a] *Oxford University, Oxford, UK*
[b] *Department of Physics, University of Antwerp (RUCA), B-2020 Antwerp, Belgium*
[c] *Department of Chemistry (ALGC), Free University of Brussels (VUB), B-1050 Brussels, Belgium*
[d] *School of Chemistry, University of Hyderabad, Hyderabad 500 046, India*

Received 27 January 2004; accepted 10 February 2004

Communicated by V.M. Agranovich

Abstract

In a recent study by one of us [Phys. Rev. A 65 (2002) 034501], an electrostatic interpretation has been proposed for the force $-\partial V_{xc}/\partial r$ associated with the exchange-correlation potential energy $V_{xc}(r)$ in a spherical atom such as Be or Ne. Here, this proposal is employed to relate $-\partial V_{xc}/\partial r$ directly to (a) the exact ground-state density $\rho(r)$ (e.g., from diffraction experiments or quantum Monte Carlo simulation) and its low-order derivatives, and (b) the single-particle kinetic energy density $t_s(r)$, calculated again from the exact $\rho(r)$, for the example of the Be-atom. It would, of course, be important if this type of relation could be extended beyond a $(1s)^2(2s)^2$ two-level single-particle configuration.

PACS: 31.15.Ew; 31.10.+z

Keywords: Exchange correlation force; Exchange correlation potential; Single-particle kinetic energy; Be-atom

In a recent study, one of us (NHM) has proposed an electrostatic interpretation for the force $-\partial V_{xc}/\partial r$ associated with the exchange-correlation potential $V_{xc}(r)$ in a spherical atom such as Be or Ne [1]. A specific application to the former case was proposed, given an "exact" ground-state density $\rho(r)$ obtained either from diffraction experiments or from a quantum Monte Carlo simulation.

The purpose of this Letter is to show that analytical progress can be made from the Be-atom example in Ref. [1], and in particular the force $-\partial V_{xc}/\partial r$ can be related directly to (i) $\rho(r)$ and its low-order derivatives and (ii) the single-particle kinetic energy density $t_s(r)$, to be calculated, of course, with the exact ground state density $\rho(r)$.

As in Ref. [1], the starting point for Be is the idempotent first order density matrix $\gamma(\mathbf{r}, \mathbf{r}')$ generated by the one-body potential $V(r)$:

$$V(r) = V_n(r) + V_{es}(r) + V_{xc}(r), \tag{1}$$

* Corresponding author.
E-mail address: pgeerlin@vub.ac.be (P. Geerlings).

0375-9601/$ – see front matter
doi:10.1016/j.physleta.2004.02.040

where $V_n(r) = -4/r$ for Be, while $V_{es}(r)$ is the electrostatic potential (energy) generated by the ground-state density $\rho(r)$. This density matrix may be written for Be in terms of the "density amplitude" $\{\rho(r)\}^{1/2}$ and the phase $\theta(r)$ as [2,3]

$$\gamma(\mathbf{r}, \mathbf{r}') = \rho(r)^{1/2}\rho(r')^{1/2}\cos[\theta(r) - \theta(r')], \quad (2)$$

where the phase $\theta(r)$ is characterized by $(\nabla\rho/\rho)$ plus physical boundary conditions through the eigenvalue equation [2]

$$\nabla^2\theta + \frac{\nabla\rho}{\rho}\cdot\nabla\theta + \lambda\sin 2\theta = 0. \quad (3)$$

It is convenient to rewrite this Eq. (3) in terms of $r^2\rho(r)$ since a central quantity in Ref. [1] was the total number of electrons, say $Q(r)$, enclosed by a sphere of radius r, centred on the atomic nucleus:

$$Q(r) = \int_0^r 4\pi r^2\rho(r)\,dr \equiv \int_0^r D(r)\,dr. \quad (4)$$

For the spherical Be atom, Eq. (3) is then readily rewritten in terms of $D(r)$ as

$$\theta'' + d(r)\theta' + \lambda\sin 2\theta = 0, \quad (5)$$

where $d(r)$ is defined by

$$d(r) = D'(r)/D(r). \quad (6)$$

Using the work of Refs. [3–5] it is straightforward then to show that the force $-\partial V/\partial r$ connected with the one-body potential in Eq. (1) is given by

$$-\frac{\partial V}{\partial r} = \frac{1}{4}d^3 - \frac{1}{2}dd' + d\theta'^2 + \frac{1}{4}\frac{d'''}{d} + 2\lambda\theta'\sin 2\theta. \quad (7)$$

One next eliminates the $\lambda\sin 2\theta$ term in Eq. (7) using Eq. (5) to find

$$-\frac{\partial V}{\partial r} = \frac{d^3}{4} - \frac{dd'}{2} + \frac{d'''}{4d} - \frac{\partial}{\partial r}(\theta'^2) - d\theta'^2. \quad (8)$$

But as shown in Ref. [2], the single-particle kinetic energy density $t_s(r)$ (defined as positive definite via the wave function Ψ form $\frac{1}{2}(\nabla\Psi)^2$) is given in terms of θ' by

$$t_s(r) = t_W(r) + \frac{1}{2}\rho\theta'^2, \quad (9)$$

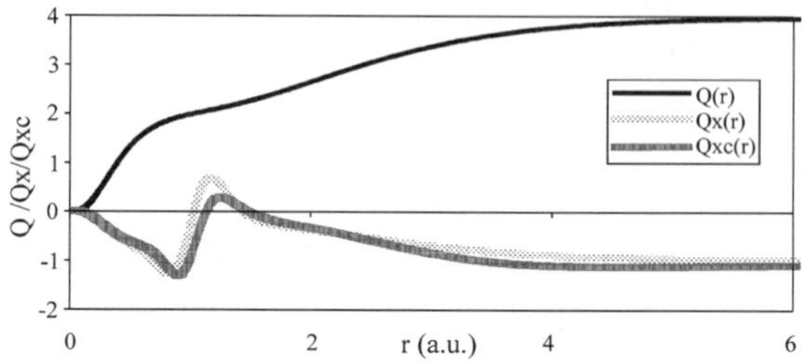

Fig. 1. Total number of electrons $Q(r)$ defined in Eq. (4), calculated from quantum Monte Carlo (QMC) density $\rho(r)$, supplied by C. Umrigar. The $Q_x(r)$ and $Q_{xc}(r)$ contributions to $Q(r)$ as defined in the text are shown for comparison.

where $t_W(r)$ is the von Weizsäcker inhomogeneity kinetic energy density $\rho'^2/8\rho$. Since d and its derivatives in Eq. (8) are determined in terms of ρ and its derivatives by Eqs. (6) and (4), it follows by substitution of θ'^2 from Eq. (9) into Eq. (8) that the force $-\partial V/\partial r$ is determined by $\rho(r)$ and its low-order derivatives, plus the single-particle kinetic energy density $t_s(r)$ and its first derivative. Fortunately, having established this, one can simplify the procedure by invoking a very recent result of Howard et al. [6, see Eq. (2.2)], which is readily rewritten in the form of a force equation analogous to Eqs. (8) plus (9) above, namely

$$-\frac{\partial V}{\partial r} = \frac{\rho'}{2r^2\rho} - \frac{1}{4r^2\rho}\frac{\partial}{\partial r}(r^2\rho'') + \frac{2}{r^2\rho}\frac{\partial}{\partial r}[r^2 t_s(r)]. \quad (10)$$

Combining this result (10) for the force associated with the one-body potential (1) with Eqs. (16) and (8) of Ref. [1], we reach the desired result for the exchange-correlation force $-\partial V_{xc}/\partial r$, namely

$$-\frac{\partial V_{xc}}{\partial r} = -\frac{\partial V}{\partial r} + \frac{4 - Q(r)}{r^2}. \quad (11)$$

Eqs. (10) and (11) evidently relate the exchange-correlation force for the ground state of the Be atom to the exact many-body electron density $\rho(r)$ plus its low-order derivatives together with the single-particle kinetic energy density $t_s(r)$ plus its first derivative.

Using the quantum Monte Carlo density $\rho(r)$ for Be [7–9], Fig. 1 shows $Q(r)$ versus r calculated from Eq. (4). In it we have also presented $r^2\partial V_x(r)/\partial r$ and $r^2\partial V_{xc}(r)/\partial r$ denoted, respectively, by $Q_x(r)$ and

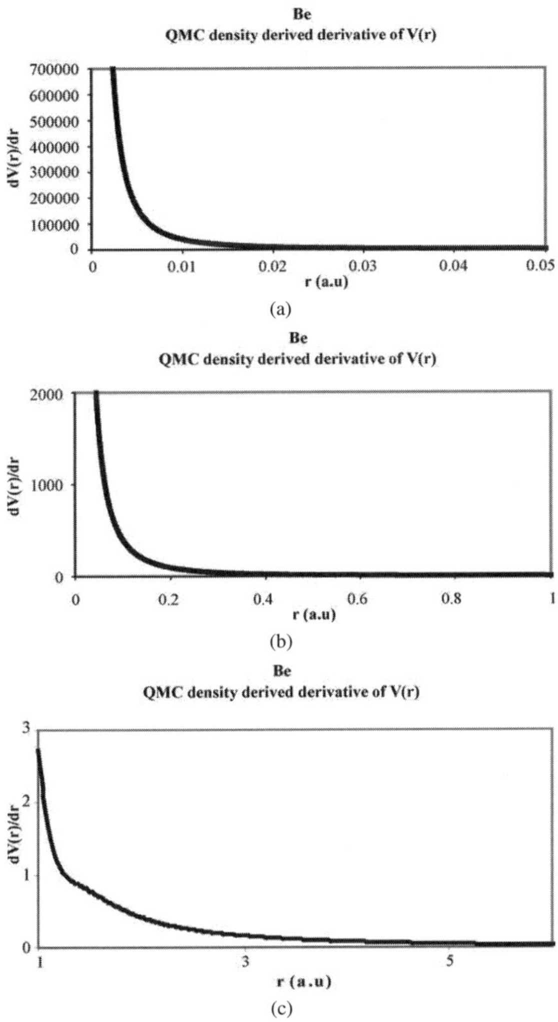

Fig. 2. Derivative of effective one-body potential energy $V(r)$ defined in Eq. (1), as obtained again from QMC density for Be. Parts (a)–(c) correspond to different scales of r.

$Q_{xc}(r)$ for Be using the quantum Monte Carlo density. Fig. 2(a)–(c) depicts $\partial V/\partial r$ derived from this density, respectively, at small, intermediate and large radial distances, while the final Fig. 3 displays the desired (negative) exchange-correlation force $\partial V_{xc}(r)/\partial r$ versus r for the ground state of the Be atom.

In summary, the many-body problem involved in directly calculating $-\partial V_{xc}(r)/\partial r$ has therefore, presently for Be only, been (at least partially!) by-

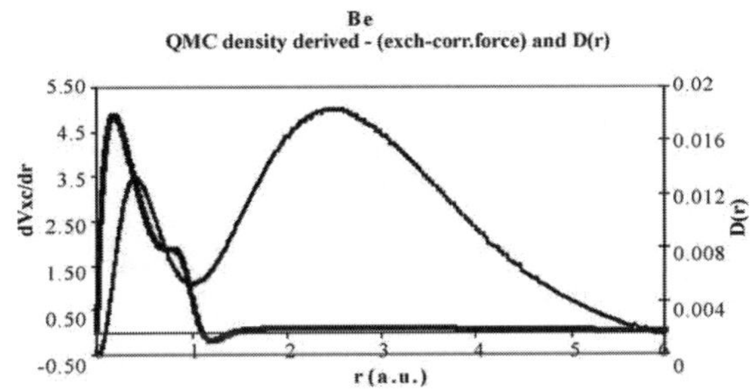

Fig. 3. Negative of exchange-correlation force $-\partial V_{xc}/\partial r$ calculated from Eq. (11) using data displayed in Figs. 1, 2. The principal peak in this figure corresponds to the "K-Shell" of the radial density $D(r)$ defined in Eq. (4). The subsequent "shoulder" reflects the $(2s)^2$ contribution to the density.

passed, since all one requires here to construct this force is (a) an exact (many-electron) density $\rho(r)$ and (b) a single-particle kinetic energy density $t_s(r)$ consistent with the exact density $\rho(r)$. It is tempting to speculate that generalizations of Eqs. (10) and (11) to other spherical atoms might exist, in view of the nature of density functional theory [10], but that remains a (possible important) research project for the future.

Acknowledgements

K.D.S. thanks Professor Umrigar for sending the QMC density data on Be. One of us (N.H.M.) wishes to acknowledge partial financial support from the Office of Naval Research. Especial thanks are due to Dr. P. Schmidt of that Office for much encouragement and motivation. K.D.S. thankfully acknowledges the Fund of Scientific Research Flanders (FWO-Vlaanderen) under the form of a Scientific Research Community Grant which supports this work.

References

[1] N.H. March, Phys. Rev. A 65 (2002) 034501.
[2] A. Holas, N.H. March, Y. Takahashi, C. Zhang, Phys. Rev. A 48 (1993) 2708.
[3] K.A. Dawson, N.H. March, J. Chem. Phys. 81 (1984) 5850.
[4] N.H. March, R.F. Nalewajski, Phys. Rev. A 35 (1987) 525.
[5] A. Nagy, N.H. March, Phys. Rev. A 39 (1989) 5512.
[6] I.A. Howard, N.H. March, V.E. Van Doren, Phys. Rev. A 63 (2001) 062501.
[7] C.J. Umrigar, X. Gonze, Phys. Rev. 50 (1994) 3827.

[8] C. Filippi, C.J. Umrigar, X. Gonze, Phys. Rev. A 54 (1996) 4810.
[9] C. Filippi, X. Gonze, C.J. Umrigar, in: J. Seminario (Ed.), Recent Developments in Applications of Density Functional Theory, Elsevier, Amsterdam, 1996.
[10] See for example, R.G. Parr, W. Yang, Density Functional Theory of Atoms and Molecules, Oxford Univ. Press, New York, 1989.

PHYSICAL REVIEW A **69**, 064101 (2004)

Exchange potential via functional differentiation of the Dirac idempotent density matrix

N. H. March[1,2] and I. A. Howard[1]
[1]*Department of Physics, University of Antwerp, Groenenborgerlaan 171, B-2020 Antwerp, Belgium*
[2]*Oxford University, Oxford, England*
(Received 8 December 2003; revised manuscript received 11 February 2004; published 2 June 2004)

Following two illustrative examples, a quite general form for the functional derivative of the Dirac density matrix $\gamma(\mathbf{r}_1,\mathbf{r}_2)$ with respect to the electron density $\rho(\mathbf{r})$ is proposed. In turn, this functional derivative is related to a further three-point object, which reduces to the linear response function $\chi_0(\mathbf{r}_1,\mathbf{r})$ when $\mathbf{r}_2$ tends to $\mathbf{r}_1$. The main application is then to derive the exchange-only potential of density functional theory. To avoid heavy numerical work in molecules and clusters, an approximation is suggested, with major calculational simplifications as a probable consequence.

DOI: 10.1103/PhysRevA.69.064101 PACS number(s): 03.65.−w, 71.15.Mb

Considerable current interest is being shown in the first-principles treatment of the exchange-only potential $V_x(\mathbf{r})$ of density functional theory (DFT). Thus, the work of Della Sala and Görling [1–3], that of the present writers [4], and numerical calculations on atomic ions [5] using the optimized effective potential (OEP) of Talman and Shadwick [6] may be cited in the above context. In this Brief Report, we shall take as the starting point the Dirac exchange energy [7]—E_x, say—in terms of the idempotent Dirac density matrix $\gamma(\mathbf{r},\mathbf{r}')$. Dirac's result is

$$E_x=-\frac{e^2}{4}\int\frac{\gamma(\mathbf{r}_1,\mathbf{r}_2)^2}{|\mathbf{r}_1-\mathbf{r}_2|}d\mathbf{r}_1 d\mathbf{r}_2. \tag{1}$$

To gain insight into what follows, let us first consider a two-electron spin-compensated system—say, He or H_2—in its ground state with electron density $\rho(\mathbf{r})$.

Then the Dirac density matrix takes the elementary form

$$\gamma(\mathbf{r}_1,\mathbf{r}_2)=\rho(\mathbf{r}_1)^{1/2}\rho(\mathbf{r}_2)^{1/2}, \tag{2}$$

and we can immediately functionally differentiate to obtain a prime focus of this article: namely, the functional derivative $\delta\gamma(\mathbf{r}_1,\mathbf{r}_2)/\delta\rho(\mathbf{r})$. The result following by an elementary calculation from Eq. (2) is

$$\frac{\delta\gamma(\mathbf{r}_1,\mathbf{r}_2)}{\delta\rho(\mathbf{r})}=\frac{1}{2}\left\{\frac{\rho(\mathbf{r}_2)}{\rho(\mathbf{r}_1)}\right\}^{1/2}\delta(\mathbf{r}_1-\mathbf{r})+\frac{1}{2}\left\{\frac{\rho(\mathbf{r}_1)}{\rho(\mathbf{r}_2)}\right\}^{1/2}\delta(\mathbf{r}_2-\mathbf{r}). \tag{3}$$

If we construct from Eq. (1) the exchange potential $V_x(\mathbf{r})$ as the functional derivative of E_x [8]—namely,

$$V_x(\mathbf{r})=\frac{\delta E_x}{\delta\rho(\mathbf{r})}, \tag{4}$$

then we find

$$V_x(\mathbf{r})=-\frac{e^2}{2}\int\frac{\gamma(\mathbf{r}_1,\mathbf{r}_2)}{|\mathbf{r}_1-\mathbf{r}_2|}\frac{\delta\gamma(\mathbf{r}_1,\mathbf{r}_2)}{\delta\rho(\mathbf{r})}d\mathbf{r}_1 d\mathbf{r}_2. \tag{5}$$

As a first, and elementary, application of this result (5), we insert the two-electron result (3) to find

$$V_x(\mathbf{r})=-\frac{e^2}{4}\int\frac{\gamma(\mathbf{r},\mathbf{r}_2)}{|\mathbf{r}-\mathbf{r}_2|}\left\{\frac{\rho(\mathbf{r}_2)}{\rho(\mathbf{r})}\right\}^{1/2}d\mathbf{r}_2-\frac{e^2}{4}\int\frac{\gamma(\mathbf{r}_1,\mathbf{r})}{|\mathbf{r}-\mathbf{r}_1|}\left\{\frac{\rho(\mathbf{r}_1)}{\rho(\mathbf{r})}\right\}^{1/2}d\mathbf{r}_1. \tag{6}$$

Using again γ from Eq. (2) we find, almost immediately,

$$V_x(\mathbf{r})=-\frac{e^2}{4}\int\frac{\rho(\mathbf{r}_2)}{|\mathbf{r}-\mathbf{r}_2|}d\mathbf{r}_2-\frac{e^2}{4}\int\frac{\rho(\mathbf{r}_1)}{|\mathbf{r}-\mathbf{r}_1|}d\mathbf{r}_1=-\frac{1}{2}V_{el}(\mathbf{r}). \tag{7}$$

A special case of this result is already given for the two-electron Hookean atom by Kais *et al.* [9], who, however, bypassed the functional derivative route emphasized here.

Turning to our second example—the Be atom or the H_2 dimer—we follow Dawson and March [10] in writing $\gamma(\mathbf{r}_1,\mathbf{r}_2)$ in terms of the density amplitude $\rho^{1/2}$ again, but now including the phase $\theta(\mathbf{r})$, which yields, for the density matrix,

$$\gamma(\mathbf{r}_1,\mathbf{r}_2)=\rho(\mathbf{r}_1)^{1/2}\rho(\mathbf{r}_2)^{1/2}\cos[\theta(\mathbf{r}_1)-\theta(\mathbf{r}_2)]. \tag{8}$$

Employing the equation of motion for the density matrix, it readily follows that the force $-\partial V/\partial r$ corresponding to the one-body potential $V(r)$ is related to $\gamma(\mathbf{r},\mathbf{r}')$ by

$$\frac{\partial V}{\partial r}=\frac{\partial}{\partial r}\left[\frac{\nabla_r^2\gamma-\nabla_{r'}^2\gamma}{2\gamma}\right]=-2\theta'\theta''-\left[\frac{2}{r}+\frac{\rho'}{\rho}\right]\theta'^2+\frac{\rho'''}{4\rho}+\frac{\rho''}{2r\rho}-\frac{\rho'\rho''}{2\rho^2}-\frac{\rho'}{2r^2\rho}-\frac{\rho'^2}{2r\rho^2}+\frac{\rho'^3}{4\rho^3} \tag{9}$$

or

$$\frac{\partial V}{\partial r}=-\frac{\partial}{\partial r}\left[\frac{\delta T_W}{\delta\rho(r)}\right]+\frac{\partial V_\Delta}{\partial r}, \tag{10}$$

where T_W is the von Weizsäcker [11] inhomogeneity kinetic energy $\int(\nabla\rho)^2/(8\rho)d\mathbf{r}$ and V_Δ is the "correction" to the potential contribution $-\delta T_W/\delta\rho(r)$. Then it follows from Eqs. (9) and (10) that

$$\frac{\partial V_\Delta}{\partial r} = -2\theta'\theta'' - \left[\frac{2}{r} + \frac{\rho'}{\rho}\right]\theta'^2. \quad (11)$$

But following Dawson and March [10] one has the pendulum equation relating density ρ and phase θ:

$$\theta'' + \frac{(r^2\rho)'}{r^2\rho}\theta' = 2\xi \sin(2\theta), \quad (12)$$

where $\xi = (\epsilon_1 - \epsilon_2)/2$, with ϵ_1 and ϵ_2 the $1s$ and $2s$ eigenvalues generated by $V(r)$. Integration of Eq. (11) and use of the pendulum equation (12) plus its first derivative, then yields

$$V_\Delta(r) = -\frac{\theta'^2}{2} + \frac{\theta'''}{4\theta'} + \frac{\theta'}{4\theta'}\left(\frac{2}{r} + \frac{\rho'}{\rho}\right) - \frac{1}{4}\left[\frac{2}{r^2} - \frac{\rho''}{\rho} + \frac{(\rho')^2}{\rho^2}\right]. \quad (13)$$

We note that the above allows an exact functional differentiation of the single-particle kinetic energy, since $V = -\delta T_s / \delta\rho(r)$ to within an additive constant.

In connection with the study of Kleinman [12] on Slater's nonlocal potential [13]—namely,

$$V_x^{Sl}(\mathbf{r}) = -\frac{e^2}{2\rho(\mathbf{r})}\int \frac{\gamma(\mathbf{r},\mathbf{r}_2)^2}{|\mathbf{r}-\mathbf{r}_2|}d\mathbf{r}_2, \quad (14)$$

Holas and March [14] derived the exchange potential in terms of $\rho(\mathbf{r})$ and $\theta(\mathbf{r})$, their result correcting the Slater potential by a term

$$V_x(\mathbf{r}) - V_x^{Sl}(\mathbf{r}) = F[\theta(\mathbf{r})], \quad (15)$$

without functional derivatives entering, which parallels the above example of functional differentiation of $T_s[\rho]$ for $E_x[\rho]$. For the two-level atomic ion with atomic number Z, we may write

$$V_x = V + \frac{Z}{r} - V_{es}, \quad (16)$$

where the electrostatic potential V_{es} is given by

$$V_{es} = -\int^r \frac{Q(r')}{r'^2}dr' : Q(r) = 4\pi\int_0^r r^2\rho(r)dr, \quad (17)$$

so that the exchange potential may be determined as

$$(rV_x)' = \frac{\partial}{\partial r}\left[r\left(V_\Delta - \frac{\delta T_W}{\delta\rho}\right)\right] + \frac{\partial}{\partial r}\left[r\int^r \frac{Q(r')}{r'^2}dr'\right]. \quad (18)$$

Note, however, that we have here bypassed the functional derivatives (4) and (5) by appealing directly to Eqs. (8) and (9), a procedure so far carried out only for a two-level atomic ion. We shall return, albeit briefly, to the two-level example based on the density matrix (8) below.

To obtain a general formula for the functional derivatives in Eqs. (4) and (5), following the above specific examples, we next appeal to the exchange potential derived by Shaginyan [15]. His starting point is an alternative to the Dirac form of E_x, given for example by Pines and Nozieres [16]. Their result involves the frequency-dependent linear response function $\chi_0(\mathbf{r}_1,\mathbf{r}_2,\omega)$ and reads

$$E_x = -\frac{1}{2}\int [\mathrm{Im}\chi_0(\mathbf{r}_1,\mathbf{r}_2,\omega) + \pi\rho(\mathbf{r}_1)\delta(\mathbf{r}_1-\mathbf{r}_2)\delta(\omega)] \times \frac{e^2}{|\mathbf{r}_1-\mathbf{r}_2|}d\mathbf{r}_1 d\mathbf{r}_2 \frac{d\omega}{\pi}. \quad (19)$$

Whereas it is natural from Eq. (1), but not of course unique, to define an exchange energy density $\epsilon_x(\mathbf{r})$ as the negative definite quantity

$$\epsilon_x(\mathbf{r}) = -\frac{e^2}{4}\int \frac{\gamma(\mathbf{r},\mathbf{r}')^2}{|\mathbf{r}-\mathbf{r}'|}d\mathbf{r}', \quad (20)$$

a possible definition from Eq. (19) would be

$$\epsilon_x^{Sh}(\mathbf{r}) = -\frac{1}{2}\int [\mathrm{Im}\chi_0(\mathbf{r},\mathbf{r}_2,\omega) + \pi\rho(\mathbf{r})\delta(\mathbf{r}-\mathbf{r}_2)\delta(\omega)] \times \frac{e^2}{|\mathbf{r}-\mathbf{r}_2|}d\mathbf{r}_2 \frac{d\omega}{\pi}. \quad (21)$$

It is then clear that $\epsilon_x^{Sh}(\mathbf{r})$ following from the expression of Shaginyan [15] taken as starting point can differ from the definition of $\epsilon_x(\mathbf{r})$ in Eq. (20) only by a function—say, $\Delta(\mathbf{r}) = \epsilon_x(\mathbf{r}) - \epsilon_x^{Sh}(\mathbf{r})$—which satisfies

$$\int \Delta(\mathbf{r})d\mathbf{r} = 0, \quad (22)$$

since Eqs. (20) and (21) are both constructed to yield, by volume integration, the same total exchange energy E_x. The Slater potential given in Eq. (14) is equivalently written using Eq. (20) as

$$V_x^{Sl}(\mathbf{r}) = \frac{2\epsilon_x(\mathbf{r})}{\rho(\mathbf{r})}, \quad (23)$$

or as

$$V_x^{Sl}(\mathbf{r}) = \frac{2\epsilon_x^{Sh}(\mathbf{r})}{\rho(\mathbf{r})} + \frac{2\Delta(\mathbf{r})}{\rho(\mathbf{r})}. \quad (24)$$

While $V_x^{Sl}(\mathbf{r})$ given by Eq. (23) is clearly finite everywhere, it may be that singularities appear in the two pieces separately in Eq. (24), though then these must of course cancel in the sum. The merit of Eq. (21) is, as shown by Shaginyan [15], that the functional differentiation involved in Eq. (4) can be carried out using Eq. (21).

Let us then return to the functional derivative $\delta\gamma(\mathbf{r}_1,\mathbf{r}_2)/\delta_\rho(\mathbf{r})$ focused on earlier in Eqs. (3) and (5). We can extract a proposed form for this from Eq. (9) of the study of Shaginyan [15]. This may be expressed as

$$\frac{\delta\gamma(\mathbf{r}_1,\mathbf{r}_2)}{\delta\rho(\mathbf{r})}=\int d\mathbf{r}'\chi_0^{-1}(\mathbf{r},\mathbf{r}')\sum_i[\phi_i(\mathbf{r}')\varphi_i^*(\mathbf{r}_1)G(\mathbf{r}_2,\mathbf{r}',\epsilon_i)+\phi_i^*(\mathbf{r}')\phi_i(\mathbf{r}_1)G^*(\mathbf{r}_2,\mathbf{r}',\epsilon_i)]=\int d\mathbf{r}'\chi_0^{-1}(\mathbf{r},\mathbf{r}')F(\mathbf{r}',\mathbf{r}_1,\mathbf{r}_2), \quad (25)$$

where $\phi_i(\mathbf{r})$ and ϵ_i denote Kohn-Sham orbitals and eigenvalues while G is the single-particle Green function defined by $G(\mathbf{r},\mathbf{r}')=\Sigma\phi_i(\mathbf{r})\phi_i^*(\mathbf{r}')/(\epsilon-\epsilon_i+i\delta)$ [15]. Together with Eq. (5), this equation (25) is one of the central results of this Brief Report. A further important result follows almost immediately from the definition of the function $F(\mathbf{r}',\mathbf{r}_1,\mathbf{r}_2)$ in Eq. (25). Taking the limit $\mathbf{r}_2\to\mathbf{r}_1$, it is easy to show that F has the attractive property that

$$F(\mathbf{r}',\mathbf{r}_1,\mathbf{r}_2)_{\mathbf{r}_2=\mathbf{r}_1}=\chi_0(\mathbf{r}',\mathbf{r}_1), \quad (26)$$

where we recall that χ_0 is the linear response function. This is especially significant because (a) $\chi_0^{-1}(\mathbf{r},\mathbf{r}')$ also enters Eq. (25) and (b) one has the property [15]

$$\int d\mathbf{r}'\chi_0^{-1}(\mathbf{r},\mathbf{r}')d\mathbf{r}'\chi_0(\mathbf{r}',\mathbf{r}_1)=\delta(\mathbf{r}-\mathbf{r}_1). \quad (27)$$

These results (25)–(27) lead us to what seems a natural enough step: namely, to the approximate factorization embodied by writing

$$F(\mathbf{r}',\mathbf{r}_1,\mathbf{r}_2)=\chi_0(\mathbf{r}',\mathbf{r}_1)f(\mathbf{r}_1,\mathbf{r}_2), \quad (28)$$

where

$$f(\mathbf{r}_1,\mathbf{r}_1)=1. \quad (29)$$

Inserting Eq. (28) into Eq. (25) allows integration over $\mathbf{r}'$ to be accomplished using Eq. (27) to yield

$$\frac{\delta\gamma(\mathbf{r}_1,\mathbf{r}_2)}{\delta\rho(\mathbf{r})}=\delta(\mathbf{r}-\mathbf{r}_1)f(\mathbf{r}_1,\mathbf{r}_2). \quad (30)$$

However, there exists one defect of the factorization (28). If one takes the functional derivative of the Dirac density matrix with respect to the density on both sides of the idempotency identity, everything is of course valid still if one uses the exact result (25). In contrast, if one employs Eq. (30), the precise identity is sacrificed. Notwithstanding this, though $f(\mathbf{r}_1,\mathbf{r}_2)$ should eventually be chosen to minimize departures from idempotency requirements, we deem it still worthwhile, motivated by the first example, particularly in Eq. (3), to "symmetrize" Eq. (30) to propose the following form for the approximate functional derivative:

$$\frac{\delta\gamma(\mathbf{r}_1,\mathbf{r}_2)}{\delta\rho(\mathbf{r})}=\frac{1}{2}\delta(\mathbf{r}-\mathbf{r}_1)f(\mathbf{r}_1,\mathbf{r}_2)+\frac{1}{2}\delta(\mathbf{r}-\mathbf{r}_2)f(\mathbf{r}_2,\mathbf{r}_1). \quad (31)$$

It is tempting, from the two-electron example in Eq. (3), to assume that $f(\mathbf{r}_1,\mathbf{r}_2)$ may be modelled in terms of $\gamma(\mathbf{r}_1,\mathbf{r}_2)$ and $\rho(\mathbf{r}_1)$. Thus Eq. (3) is readily rewritten by employing Eq. (2) as

$$\frac{\delta\gamma(\mathbf{r}_1,\mathbf{r}_2)}{\delta\rho(\mathbf{r})}=\frac{1}{2}\frac{\gamma(\mathbf{r}_1,\mathbf{r}_2)}{\rho(\mathbf{r}_1)}\delta(\mathbf{r}_1-\mathbf{r})+\frac{1}{2}\frac{\gamma(\mathbf{r}_2,\mathbf{r}_1)}{\rho(\mathbf{r}_2)}\delta(\mathbf{r}_2-\mathbf{r}), \quad (32)$$

and evidently in this example therefore Eq. (31) corresponds to the exact form (32) if $f(\mathbf{r}_1,\mathbf{r}_2)=\gamma(\mathbf{r}_1,\mathbf{r}_2)/\rho(\mathbf{r}_1)$.

In fact, this form of $f(\mathbf{r}_1,\mathbf{r}_2)$ leads back quite generally to the Slater potential $V_x^{SI}(\mathbf{r})$ in Eq. (14), which Kleinman has stressed is a "partial" functional derivative of E_x in Eq. (1) with respect to $\rho(\mathbf{r})$. As Holas and March [14] point out, in the two-level case one must include in the functional derivative of Eq. (1) the phase angle $\theta(\mathbf{r})$, which then yields calculable corrections to the Slater potential, which is, however, already a very useful starting approximation for $V_x(r)$ for the two-level case of the Be atom. A "correction" to the approximation $f(\mathbf{r}_1,\mathbf{r}_2)\sim\gamma(\mathbf{r}_1,\mathbf{r}_2)/\rho(\mathbf{r}_1)$ which preserves Eq. (29) and has the form $\lambda(N)|\mathbf{r}_1-\mathbf{r}_2|V_x(\mathbf{r}_1)\gamma(\mathbf{r}_1,\mathbf{r}_2)/\rho(\mathbf{r}_1)$ suggests itself; this can be shown to establish then approximate contact with Refs. [1–4]. However, $\lambda(N)$, with N the number of occupied levels, would need to be found by some "least squares" minimization of the departure from idempotency requirements, but it would take us too far from our main theme to go into further details along these lines.

In summary, the main results of this Brief Report are embodied in Eq. (25) for the functional derivative of the Dirac density matrix $\gamma(\mathbf{r}_1,\mathbf{r}_2)$ with respect to the electron density $\rho(\mathbf{r})$ and Eq. (5) for the exchange potential. The function $F(\mathbf{r}',\mathbf{r}_1,\mathbf{r}_2)$ thereby introduced reduces on the diagonal $\mathbf{r}_2=\mathbf{r}_1$ to the linear response function $\chi_0(\mathbf{r}',\mathbf{r}_1)$, which is appealing since its inverse in the form $\chi_0^{-1}(\mathbf{r},\mathbf{r}')$ also enters Eq. (25). Further work on the off-diagonal properties of the three-point function F may be instructive for the future, by both analytical and numerical routes. It could be especially important if useful approximations to F could be found which involved only occupied Kohn-Sham orbitals in contrast to the present form defined in Eq. (25).

I.A.H. acknowledges support by the IWT-Flemish region under Grant No. IWT-161.

[1] F. Della Sala and A. Görling, J. Chem. Phys. **115**, 5718 (2001).
[2] O. V. Gritsenko and E. J. Baerends, Phys. Rev. A **64**, 042506 (2001).
[3] See also M. Grüning, O. V. Gritsenko, and E. J. Baerends, J. Chem. Phys. **116**, 6435 (2002).
[4] I. A. Howard and N. H. March, J. Chem. Phys. **119**, 5789 (2003).
[5] I. A. Howard, N. H. March, and J. D. Talman, Phys. Rev. A **68**, 044502 (2003).

[6] J. D. Talman and W. F. Shadwick, Phys. Rev. A **14**, 36 (1976).
[7] P. A.M. Dirac, Proc. Cambridge Philos. Soc. **26**, 376 (1930).
[8] See, for example, R. G. Parr and W. Yang, *Density Functional Theory of Atoms and Molecules* (Oxford University Press, Oxford, 1989).
[9] S. Kais, D. R. Herschbach, and R. D. Levine, J. Chem. Phys. **91**, 7791 (1989).
[10] K. A. Dawson and N. H. March, J. Chem. Phys. **81**, 5850 (1984).
[11] C. F. Weizsäcker, Z. Phys. **96**, 341 (1935).
[12] L. Kleinman, Phys. Rev. B **49**, 14197 (1994).
[13] J. C. Slater, Phys. Rev. **81**, 385 (1951).
[14] A. Holas and N. H. March, Phys. Rev. B **55**, 1295 (1997).
[15] V. R. Shaginyan, Phys. Rev. A **47**, 1507 (1993).
[16] D. Pines and P. Nozieres, *The Theory of Quantum Liquids* (Benjamin, New York, 1966), Vol. 1.

Available online at www.sciencedirect.com
SCIENCE DIRECT®
Chemical Physics Letters 385 (2004) 231–232
CHEMICAL PHYSICS LETTERS
www.elsevier.com/locate/cplett

Idempotent density matrix derived from a local potential $V(\mathbf{r})$ in terms of HOMO and LUMO properties

I.A. Howard [a,*], N.H. March [a,b], P.W. Ayers [c]

[a] Department of Physics, University of Antwerp, Groenenborgerlaan 171, B-2020 Antwerp, Belgium
[b] Oxford University, Oxford OX1 3UB, UK
[c] Department of Chemistry, McMaster University, Hamilton, Ont., Canada L8S 4M1

Received 17 October 2003; in final form 11 December 2003
Published online: 19 January 2004

Abstract

In a recent paper one of us has exposed a link between the idempotent Dirac density matrix built from occupied Kohn–Sham orbitals and the frontier orbitals. An alternative argument is here presented, from the equation of motion of the density matrix generated by any local potential $V(\mathbf{r})$.

One of us [1], in a recent paper, has established a link between the Dirac idempotent density matrix built from occupied Kohn–Sham orbitals and the frontier orbitals. The purpose of this Letter is to present an alternative argument, from the equation of motion of the Dirac density matrix generated by any local potential $V(\mathbf{r})$.

Following the route used by March and Young [2], let us start from the equation of motion of the density matrix $\gamma(\mathbf{r},\mathbf{r}')$, defined from the occupied orbitals $\psi_i(\mathbf{r})$ generated by a local potential energy $V(\mathbf{r})$. This density matrix satisfies

$$\nabla_{\mathbf{r}}^2\gamma(\mathbf{r},\mathbf{r}') - \nabla_{\mathbf{r}'}^2\gamma(\mathbf{r},\mathbf{r}') = \frac{2m}{\hbar^2}[V(\mathbf{r}) - V(\mathbf{r}')]\gamma(\mathbf{r},\mathbf{r}'). \quad (1)$$

Let us now assume, for simplicity, non-degenerate frontier orbitals $\psi_{\mathrm{HOMO}}(\mathbf{r}) \equiv \psi_{\mathrm{H}}(\mathbf{r})$ and $\psi_{\mathrm{LUMO}}(\mathbf{r}) \equiv \psi_{\mathrm{L}}(\mathbf{r})$, with corresponding one-electron eigenvalues ϵ_{H} and ϵ_{L}, respectively, though non-degeneracy is not an essential requirement. Then we can evidently write from Schrödinger's equation

$$\nabla_{\mathbf{r}}^2\psi_{\mathrm{H}}(\mathbf{r}) + \frac{2m}{\hbar^2}[\epsilon_{\mathrm{H}} - V(\mathbf{r})]\psi_{\mathrm{H}}(\mathbf{r}) = 0 \quad (2)$$

and

$$\nabla_{\mathbf{r}'}^2\psi_{\mathrm{L}}(\mathbf{r}') + \frac{2m}{\hbar^2}[\epsilon_{\mathrm{L}} - V(\mathbf{r}')]\psi_{\mathrm{L}}(\mathbf{r}') = 0. \quad (3)$$

Evidently, Eqs. (2) and (3) can now be solved for $V(\mathbf{r})$ and $V(\mathbf{r}')$, respectively, which can be inserted in Eq. (1). The conclusion is therefore that for any local potential $V(\mathbf{r})$ generating non-degenerate HOMO and LUMO orbitals,

$$\gamma(\mathbf{r},\mathbf{r}') = \gamma[\mathbf{r},\mathbf{r}';\psi_{\mathrm{H}}(\mathbf{r}),\psi_{\mathrm{L}}(\mathbf{r}'),(\epsilon_{\mathrm{H}} - \epsilon_{\mathrm{L}})]. \quad (4)$$

While Eq. (4) is the main result of this Letter, we must emphasize that a solution of (1) is always required which yields $\gamma(\mathbf{r},\mathbf{r}')$ as an idempotent matrix:

$$\gamma(\mathbf{r},\mathbf{r}') = \int \gamma(\mathbf{r},\mathbf{r}'')\gamma(\mathbf{r}'',\mathbf{r}')\,\mathrm{d}\mathbf{r}''. \quad (5)$$

We turn now to consider a simple example, where the one-body potential $V(\mathbf{r})$ in Eq. (1) has the Coulomb form

$$V(\mathbf{r}) = -Z/r. \quad (6)$$

For K plus L shells closed, March and Santamaria [3] effectively solved Eq. (1) for an idempotent $\gamma(\mathbf{r},\mathbf{r}')$. This takes the form

$$\gamma = \gamma_{\mathrm{s}}\left(\frac{x}{2},\frac{y}{2}\right) + \frac{xy}{128\pi}\left(\frac{Z}{a_0}\right)^5 \exp\left(-\frac{Z}{4a_0}(x+y)\right), \quad (7)$$

where $\gamma_{\mathrm{s}}\left(\frac{x}{2},\frac{y}{2}\right)$ is the s-orbital only density matrix. Now we express the two terms on the right-hand side

* Corresponding author.
E-mail address: iris.howard@ua.ac.be (I.A. Howard).

0009-2614/$ - see front matter
doi:10.1016/j.cplett.2003.12.094

implicitly in terms of the HOMO orbital with radial wave function $R_{2p} \equiv R_{21}(r)$ and the LUMO 3s orbital ψ_{3s}, together with the energy difference $\epsilon_H - \epsilon_L$. First we note that, given

$$\psi_{1s}(r) = \left(\frac{Z^3}{\pi}\right)^{1/2} \exp(-Zr), \tag{8}$$

$$\psi_{2s}(r) = \frac{1}{2}\left(\frac{Z^3}{2\pi}\right)^{1/2}\left(1 - \frac{Zr}{2}\right)\exp\left(-\frac{Zr}{2}\right) \tag{9}$$

and

$$\psi_{3s}(r) = \frac{1}{3}\left(\frac{Z^3}{3\pi}\right)^{1/2}\left(1 - \frac{2Zr}{3} + \frac{2}{27}(Zr)^2\right)\exp\left(-\frac{Zr}{3}\right), \tag{10}$$

we can write

$$\psi_{3s}(r) = \frac{Z}{3^{3/2}}\left(\frac{\psi_{1s}}{\pi}\right)^{1/3}\left[1 - \frac{4}{3}\left(1 - (8\pi)^{1/4}\frac{\psi_{2s}}{\sqrt{\psi_{1s}}}\right) + \left(\frac{2}{3}\right)^3\left(1 - (8\pi)^{1/4}\frac{\psi_{2s}}{\sqrt{\psi_{1s}}}\right)^2\right]. \tag{11}$$

Next we note that, according to Eq. (A.5) of Ref. [3]

$$\frac{3}{4\pi}R_{21}^2 = \frac{r^2}{2}\left[\frac{(\rho_{20}')^2}{\rho_{20}} - \rho_{20}''\right], \tag{12}$$

where

$$\rho_{20} = \frac{(r\psi_{2s})''}{4\pi r^2}. \tag{13}$$

Thus we can write

$$\psi_{3s} \equiv f(Z, \psi_{1s}, R_{21}, \epsilon_H - \epsilon_L) \tag{14}$$

and by inversion (in principle)

$$\psi_{1s} \equiv g(Z, R_{21}, \psi_{3s}), \tag{15}$$

since we can also make use of the fact that

$$\epsilon_H - \epsilon_L = -\frac{Z^2}{8} + \frac{Z^2}{18} = -\frac{5Z^2}{72}. \tag{16}$$

It is interesting here to add that one could determine the atomic number directly from the frontier orbitals (by inverting the Schrödinger equation) rather than through the band gap. Likewise, given Eqs. (12) and (13), we can write in principle

$$\psi_{2s} = h(R_{21}). \tag{17}$$

The s-density matrix γ_s is then just

$$\begin{aligned}\gamma_s\left(\frac{x}{2},\frac{y}{2}\right) &= \psi_{1s}\left(\frac{x}{2}\right)\psi_{1s}\left(\frac{y}{2}\right) + \psi_{2s}\left(\frac{x}{2}\right)\psi_{2s}\left(\frac{y}{2}\right)\\ &= g\left(Z, R_{21}, \psi_{3s}, \frac{x}{2}\right)g\left(Z, R_{21}, \psi_{3s}, \frac{y}{2}\right)\\ &\quad + h\left(R_{21}, \frac{x}{2}\right)h\left(R_{21}, \frac{y}{2}\right).\end{aligned} \tag{18}$$

The second term in Eq. (7) for γ can also be expressed in terms of R_{21}, so we can finally write

$$\begin{aligned}\gamma &= g\left(Z, R_{21}, \psi_{3s}, \frac{x}{2}\right)g\left(Z, R_{21}, \psi_{3s}, \frac{y}{2}\right)\\ &\quad + h\left(R_{21}, \frac{x}{2}\right)h\left(R_{21}, \frac{y}{2}\right) + \frac{3}{4\pi}R_{21}\left(\frac{x}{2}\right)R_{21}\left(\frac{y}{2}\right).\end{aligned} \tag{19}$$

The essence of the above argument may be summarized by saying that Eq. (1) allows the idempotent density matrix $\gamma(\mathbf{r}, \mathbf{r}')$ to be expressed in terms of frontier orbitals plus the HOMO–LUMO gap as key variables. One will then be employing quantities of chemical interest directly in computation, which seems desirable. The use of the 'band gap' in this respect adds to the discussion in [1].

The referee has asked us to add comments on (a) the merits of working with frontier-orbital properties and (b) the influence of long-range exchange-correlation potentials especially on the HOMO orbital. As to (a), one of us [1] has suggested that an ideal Kohn–Sham scheme would be competitive with simple diagonalization for small systems, competitive with linear scaling techniques for large systems, and would provide access to the chemically significant frontier molecular orbitals throughout the entire calculation. The approach in the present Letter seems a step along such a path. Such an approach will be of special computational and conceptual utility when the purpose of a calculation is to elucidate a reagent's chemical activity from the frontier orbitals using, e.g., the Woodward–Hoffmann rules [4] or Fukui's analysis [5]. As to (b), the non-uniform asymptotic decay of the exchange-correlation potential is indeed related to the behaviour of the HOMO orbital and its asymptotic decay. The study of Della Sala and Görling [6] especially should be noted in this context, and also the later investigation of Wu et al. [7] in which the nodal surface problem for formally exact density functional theory was fully analyzed.

References

[1] P.W. Ayers, Theor. Chem. Accounts 110 (2003) 267.
[2] N.H. March, W.H. Young, Nucl. Phys. 12 (1959) 237.
[3] N.H. March, R. Santamaria, Phys. Rev. A 38 (1988) 5002.
[4] R.B. Woodward, R. Hoffmann, J. Am. Chem. Soc. 87 (1965) 2511.
[5] K. Fukui, T. Yonezawa, C. Nagata, J. Chem. Phys. 21 (1953) 174;
K. Fukui, Science 218 (1987) 747.
[6] F. Della Sala, A. Görling, J. Chem. Phys. 115 (2001) 5718.
[7] Q. Wu, P.W. Ayers, W. Yang, J. Chem. Phys. 119 (2003) 2978.

Available online at www.sciencedirect.com
SCIENCE DIRECT®
Chemical Physics Letters 402 (2005) 1–3
CHEMICAL PHYSICS LETTERS
www.elsevier.com/locate/cplett

Can the exchange-correlation potential of density functional theory be expressed solely in terms of HOMO and LUMO properties?

I.A. Howard [a,*], N.H. March [a,b]

[a] Department of Physics, University of Antwerp, Groenenborgerlaan 171, B-2020 Antwerp, Belgium
[b] Oxford University, Oxford, England, UK

Received 30 November 2004; in final form 1 December 2004
Available online 18 December 2004

Abstract

In a recent Letter with Ayers, we have demonstrated that the idempotent Dirac density matrix generated by any local potential $V(\mathbf{r})$ can be derived in terms of HOMO and LUMO properties. Subsequently we have pointed out that the exchange-only potential $V_x(\mathbf{r})$ of density functional theory can be built from the Dirac matrix $\gamma(\mathbf{r}_1, \mathbf{r}_2)$ and its functional derivative $\delta\gamma(\mathbf{r}_1, \mathbf{r}_2)/\delta\rho(\mathbf{r})$. By utilizing further the equation of motion of the density matrix, we propose an answer in the affirmative to the question posed in the title of this Letter.

1. Introduction

In a recent study [1], Ayers has established a link between the Dirac idempotent density matrix constructed from Kohn–Sham (KS) orbitals and the frontier orbitals, generalizing thereby an earlier proposal made by Dawson and March [2] in a more limited context. This has led to a demonstration [3] that the idempotent density matrix derived from a local potential $V(\mathbf{r})$ can be constructed entirely from HOMO–LUMO properties. As to the merits of using frontier-orbital properties, in [3] it was stressed that such an approach should be of special computational and conceptual ability when the purpose of a calculation is to elucidate a reagent's chemical activity from the frontier orbitals, using, for instance, either the Woodward–Hoffmann rules [4] or the analysis of Fukui et al. [5].

Subsequent to the study reported in [3], the present authors [6] have stressed the merit for density functional theory (DFT) of writing the exchange-only potential $V_x(\mathbf{r})$ in terms of the functional derivative of the idempotent Dirac density matrix, denoted below by $\gamma(\mathbf{r}_1, \mathbf{r}_2)$. This is readily achieved from the DFT definition of $V_x(\mathbf{r})$, namely [7]

$$V_x(\mathbf{r}) = \frac{\delta E_x}{\delta\rho(\mathbf{r})}, \tag{1}$$

where $\rho(\mathbf{r})$ is the usual electron density while E_x is the total exchange energy given by the Dirac expression

$$E_x = -\frac{e^2}{4}\int \frac{\gamma^2(\mathbf{r}, \mathbf{r}')}{|\mathbf{r} - \mathbf{r}'|}\,d\mathbf{r}\,d\mathbf{r}'. \tag{2}$$

The result of comparing Eqs. (1) and (2) is that [6]

$$V_x(\mathbf{r}) = -\frac{e^2}{2}\int \frac{\gamma(\mathbf{r}_1, \mathbf{r}_2)}{|\mathbf{r}_1 - \mathbf{r}_2|}\frac{\delta\gamma(\mathbf{r}_1, \mathbf{r}_2)}{\delta\rho(\mathbf{r})}\,d\mathbf{r}_1\,d\mathbf{r}_2. \tag{3}$$

In [6], using earlier work referenced there, the present writers give a formally exact expression for the functional derivative $\delta\gamma(\mathbf{r}_1, \mathbf{r}_2)/\delta\rho(\mathbf{r})$ in terms of (a) the linear response function and (b) an appropriate Green function (see also the subsequent study of Liu and Ayers [8]). However, in at least conventional forms of (a) and (b), unoccupied KS orbitals are involved, the

* Corresponding author.
E-mail address: iris.howard@ua.ac.be (I.A. Howard).

0009-2614/$ - see front matter
doi:10.1016/j.cplett.2004.12.002

occupied orbitals only, of course, being required to construct the electron density $\rho(\mathbf{r})$, the central tool of DFT.

We turn, therefore, in Section 2, to point out that the recent work of Della Sala and Görling [9] leads to $V_x(\mathbf{r})$ in a form requiring only the occupied KS orbitals. Related studies of Baerends and coworkers are also referred to in Section 2, while correlation is considered in Section 3. A summary, together with possible directions for future study which may prove fruitful, constitute Section 4.

2. Approximate theories of $V_x(\mathbf{r})$ requiring only knowledge of the Dirac density matrix, and therefore solely of occupied KS orbitals

The earliest theory of which we are aware that expresses the exchange-only potential $V_x(\mathbf{r})$ in terms of the idempotent Dirac density matrix $\gamma(\mathbf{r}_1, \mathbf{r}_2)$ goes back to Slater [10], and was subsequently re-opened by Kleinman [11,12]. Slater's proposal may be written explicitly as

$$V_x^{\mathrm{Sl}}(\mathbf{r}) = -\frac{e^2}{2\rho(\mathbf{r})}\int \frac{\gamma^2(\mathbf{r},\mathbf{r}_2)}{|\mathbf{r}-\mathbf{r}_2|}\mathrm{d}\mathbf{r}_2, \tag{4}$$

which, since only $\gamma(\mathbf{r}, \mathbf{r}_2)$ and its diagonal form $\rho(\mathbf{r}) = \gamma(\mathbf{r}, \mathbf{r}_2)|_{\mathbf{r}_2=\mathbf{r}}$ are involved, is clearly answering 'yes' to the question posed in the title if correlation is omitted. However, Holas and March [12] exhibited exact corrections to the Slater form (4) and so far it has not been shown that such additions to $V_x^{\mathrm{Sl}}(\mathbf{r})$ can be posed in terms of occupied orbitals alone.

Progress beyond the Slater form (4) has come from the much more recent study of Della Sala and Görling [9], followed by the formally exact integral equation theory of the present writers [13]. Because [13] still requires a 'closure' approximation to go from 'formal exactness' to 'potentially useful', let us focus below on the Della Sala–Görling approximation to $V_x(\mathbf{r})$. This is to equate the KS and Hartree–Fock determinants (their 'closure' approximation in the above terminology), which then leads to (cf. also [13])

$$V_x(\mathbf{r}) = V_x^{\mathrm{Sl}}(\mathbf{r}) + \frac{1}{2\rho(\mathbf{r})}\int \gamma(\mathbf{r}',\mathbf{r})\gamma(\mathbf{r},\mathbf{r}')V_x(\mathbf{r}')\mathrm{d}\mathbf{r}' + \frac{1}{4\rho(\mathbf{r})}\int\int \frac{\gamma(\mathbf{r}'',\mathbf{r})\gamma(\mathbf{r},\mathbf{r}')\gamma(\mathbf{r}',\mathbf{r}'')}{|\mathbf{r}'-\mathbf{r}''|}\mathrm{d}\mathbf{r}'\,\mathrm{d}\mathbf{r}'', \tag{5}$$

$V_x^{\mathrm{Sl}}(\mathbf{r})$ being given in terms of γ by Eq. (4). Thus, with this assumed 'closure', $V_x(\mathbf{r})$ is clearly determined solely by $\gamma(\mathbf{r}, \mathbf{r}')$ and therefore by the KS occupied orbitals. Similarly, the 'denominator' approximation used in [14,15] by Baerends and coworkers leads to the same conclusions as for the Della Sala–Görling treatment.

3. Formal inclusion of correlation

Though in Section 4 we comment a little further on the exchange-only case, let us turn to the exact exchange-correlation potential $V_{xc}(\mathbf{r})$ of DFT. One of us [16] has quite recently shown, using electrostatics, how to write the corresponding force $-\partial V_{xc}(\mathbf{r})/\partial\mathbf{r}$ in terms of the (force associated with the) exact one-body potential $V(\mathbf{r})$ of DFT, which in turn can be written as

$$V(\mathbf{r}) = V_{\mathrm{ext}}(\mathbf{r}) + V_{\mathrm{es}}(\mathbf{r}) + V_{\mathrm{xc}}(\mathbf{r}) \tag{6}$$

with, as usual, V_{ext} being the potential of the nuclear framework in the molecule or cluster under consideration, while V_{es} is the electrostatic potential generated by the ground-state electron density $\rho(\mathbf{r})$.

Using next the equation of motion of the Dirac density matrix generated by the exact DFT potential (6), namely [3]

$$\nabla_{\mathbf{r}}^2\gamma(\mathbf{r},\mathbf{r}') - \nabla_{\mathbf{r}'}^2\gamma(\mathbf{r},\mathbf{r}') = \frac{2m}{\hbar^2}[V(\mathbf{r}) - V(\mathbf{r}')]\gamma(\mathbf{r},\mathbf{r}'), \tag{7}$$

we can obtain the force $-\partial V(\mathbf{r})/\partial\mathbf{r}$ explicitly from knowledge of $\gamma(\mathbf{r}, \mathbf{r}')$ by dividing Eq. (7) by γ and taking the gradient of the resulting equation. This establishes, from [3], that $V(\mathbf{r})$ is determined solely by HOMO–LUMO properties. But from Eq. (6), this means that $V_{xc}(\mathbf{r})$ is formally determined by these properties since $V_{es}(\mathbf{r})$ is determined by $\rho(\mathbf{r}) = \gamma(\mathbf{r}, \mathbf{r}')|_{\mathbf{r}'=\mathbf{r}}$, while $V_{\mathrm{ext}}(\mathbf{r}) = V_{\mathrm{ext}}[\rho]$ from a theorem of DFT [7]. Here, though the above argument is of course formal, and without any explicit procedure to construct V_{xc} as a functional of HOMO and LUMO properties, it leads us to propose an affirmative answer to the question posed in the title.

4. Summary and possible future directions

While Eq. (3) is a formally exact expression for the exchange-only potential of DFT in terms of the Dirac idempotent density matrix $\gamma(\mathbf{r}, \mathbf{r}')$, it clearly involves knowledge of γ as a functional of its diagonal density ρ. While such knowledge is explicitly available, the latter functional involves the linear response function and a Green function. Conventionally written, these two objects involve unoccupied KS orbitals, as indeed is the case with the optimized effective potential (OEP) of Talman and Shadwick [17], as well as with the formal integral equation theory of $V_x(\mathbf{r})$ given by the present authors [13].

The approximate expressions for $V_x(\mathbf{r})$ by Slater [10], by Della Sala and Görling [9], and by Baerends and coworkers [14,15] all yield an affirmative answer to the question posed in the title of this Letter, since they are determined solely by $\gamma(\mathbf{r}, \mathbf{r}')$, and therefore from [1,3] entirely by HOMO–LUMO properties. Fruitful directions for the future seem to centre around the 'closure' function $P(\mathbf{r})$ in [13] (see especially Eq. (4.4)), which has been so far conjectured to depend only on HOMO and LUMO properties. If it were true, then the linear response function [6] in combination with the Green function (see again [6,8]) would have to be expressible together in terms of frontier orbital properties. We conjecture that this will be possible, and therefore that the answer to the title question is 'yes'. To prove rigorously the formal generalization set out in Section 3 for the exchange-correlation potential seems important for DFT.

Acknowledgements

I.A.H. acknowledges support from the IWT – Flemish region under Grant No. IWT-161. N.H.M. wishes to acknowledge that his contribution was brought to fruition during a visit to the Abdus Salam International Centre for Theoretical Physics in 2004. He thanks Prof. V.E. Kravtsov for the stimulating atmosphere provided and for generous hospitality.

References

[1] P.W. Ayers, Theor. Chem. Acc. 110 (2003) 267.
[2] K.A. Dawson, N.H. March, Phys. Lett. A 106 (1984) 161.
[3] I.A. Howard, N.H. March, P.W. Ayers, Chem. Phys. Lett. 385 (2004) 231.
[4] R.B. Woodward, R. Hoffmann, J. Am. Chem. Soc. 87 (1965) 2511.
[5] K. Fukui, T. Yonezawa, C. Nagata, J. Chem. Phys. 21 (1953) 74; K. Fukui, Science 218 (1987) 747.
[6] N.H. March, I.A. Howard, Phys. Rev. A 69 (2004) 064101.
[7] R.G. Parr, W. Yang, Density Functional Theory of Atoms and Molecules, Oxford University Press, Oxford, UK, 1989.
[8] S. Liu, P.W. Ayers, Phys. Rev. A 70 (2004) 022501.
[9] F. Della Sala, A. Görling, J. Chem. Phys. 115 (2001) 5718.
[10] J.C. Slater, Phys. Rev. 81 (1951) 385.
[11] L. Kleinman, Phys. Rev. B 49 (1994) 14197.
[12] A. Holas, N.H. March, Phys. Rev. B 55 (1997) 1295.
[13] I.A. Howard, N.H. March, J. Chem. Phys. 119 (2003) 5789.
[14] O.V. Gritsenko, E.J. Baerends, Phys. Rev. A 64 (2001) 042506.
[15] See also M. Grüning, O.V. Gritsenko, E.J. Baerends, J. Chem. Phys. 116 (2002) 6435.
[16] N.H. March, Phys. Rev. A 65 (2002) 034501.
[17] J.D. Talman, W.E. Shadwick, Phys. Rev. A 14 (1976) 36.

THE JOURNAL OF CHEMICAL PHYSICS **123**, 194104 (2005)

Differential virial theorem in relation to a sum rule for the exchange-correlation force in density-functional theory

A. Holas[a)]
Institute of Physical Chemistry, Polish Academy of Sciences, 44/52 Kasprzaka, 01-224 Warsaw, Poland

N. H. March
Donastia International Physics Center, E-20018 San Sebastián, Spain and Oxford University, Oxford OX1 3QR, England

Angel Rubio
Institut für Theoretische Physik, Freie Universität Berlin, Arnimallee 14, D-14 195 Berlin, Germany, Departamento de Física de Materiales, Facultad de Ciencias Químicas, Universidad del País Vasco/Euskal Herriko Unibertsitatea (UPV/EHU) and Centro Mixto Consejo Superior de Investigaciones Científicas-Universidad del País Vasco/Euskal Herriko Unibertsitatea (CSIC-UPV/EHU), E-20018 San Sebastián, Spain, and Donastia International Physics Center, E-20018 San Sebastián, Spain

(Received 7 September 2005; accepted 15 September 2005; published online 11 November 2005)

Holas and March [Phys. Rev. A. **51**, 2040 (1995)] gave a formally exact theory for the exchange-correlation (xc) force $\boldsymbol{F}_{\text{xc}}(\boldsymbol{r})=-\boldsymbol{\nabla}v_{\text{xc}}(\boldsymbol{r})$ associated with the xc potential $v_{\text{xc}}(\boldsymbol{r})$ of the density-functional theory in terms of low-order density matrices. This is shown in the present study to lead, rather directly, to the determination of a sum rule $\langle n\boldsymbol{F}_{\text{xc}}\rangle=0$ relating the xc force with the ground-state density $n(\boldsymbol{r})$. Some connection is also made with an earlier result relating to the external potential by Levy and Perdew [Phys. Rev. A. **32**, 2010 (1985)] and with the quite recent study of Joubert [J. Chem. Phys. **119**, 1916 (2003)] relating to the separation of the exchange and correlation contributions. © *2005 American Institute of Physics*. [DOI: 10.1063/1.2114848]

I. BACKGROUND AND OUTLINE

While density-functional theory (DFT) continues to be widely applied to a variety of problems embracing both chemical physics and biology, evidence is building up that existing approximate functionals are in urgent need of refinement. One very recent example of this is the study by Wanko *et al.*[1] of absorption properties of retinal proteins, for which presently available functionals lead to a disagreement with experiment.

In the search for such refined exchange-correlation (xc) energy functionals, exact results, though generally of integral rather than local form, involving the xc potential $v_{\text{xc}}(\boldsymbol{r})$, can be expected to play an increasingly important role in testing the quantitative virtues of proposed refinements. The present study lies in this area but focuses on the xc force $\boldsymbol{F}_{\text{xc}}(\boldsymbol{r})$ defined by

$$\boldsymbol{F}_{\text{xc}}(\boldsymbol{r}) = -\boldsymbol{\nabla}v_{\text{xc}}(\boldsymbol{r}). \tag{1.1}$$

This quantity was treated exactly by Holas and March[2] in terms of low-order density matrices (DMs) via the so-called differential form of the virial theorem. It should be stressed that in the present paper we will deal exclusively with finite systems, that is, with molecules and clusters.

With this background, the outline of the present article is as follows. In Sec. II, after recalling the March-Young[3] result as a background, we first set out the Holas-March[2] result based on the differential virial theorem. As a consequence, we are led to Eq. (2.7) for the average force $-\boldsymbol{\nabla}v_{\text{ext}}(\boldsymbol{r})$ associated with the external potential. This involves the vector field $z(\boldsymbol{r})$ of the kinetic origin (embracing the kinetic correlation energy) as well as the electron-electron interaction term. A connection is made with an early result of Levy and Perdew[4] for this average force. Throughout Sec. II, we work with the exact exchange-correlation potential without needing any separation into exchange and correlation contributions. In Sec. III, however, we perform such separation and make connections with an earlier study of Joubert.[5] Section IV concludes with a brief summary of the main findings of the present article.

II. EXACT SUM RULES OBTAINED VIA THE DIFFERENTIAL VIRIAL THEOREM

A. Sum rules in a one-dimensional noninteracting system

As a background to the differential virial theorem in this context, we start from the one-dimensional form derived by March and Young.[3] They introduced the kinetic energy per unit length $t(x)$ and considered an arbitrary number of noninteracting fermions moving in a common one-dimensional potential field $v(x)$. By appealing to the equation of motion for the Dirac density matrix [single particle (s) and therefore idempotent] $\rho^s(x_1;x_2)$, expanding it near the diagonal and utilizing the fermion density $n(x)=\rho^s(x;x)$, March and Young found

[a)]Electronic mail: holas@ichf.edu.pl

$$\frac{dt(x)}{dx} = -\frac{1}{2}n(x)\frac{dv(x)}{dx} + \frac{1}{8}\frac{d^3n(x)}{dx^3} \tag{2.1}$$

[rewritten here using an alternative definition of the kinetic-energy density as $t(x)=\frac{1}{2}\partial^2\rho^s(x';x'')/\partial x'\partial x''|_{x'=x''=x}$; atomic units are used throughout the paper]. Forming the virial of Clausius $\boldsymbol{r}\cdot\boldsymbol{F}(\boldsymbol{r})$ which in one dimension is evidently $xF(x)=-xdv(x)/dx$, they stressed that Eq. (2.1), multiplied by $2x$ and integrated, gave back the usual integral virial theorem

$$2T = -\int_{-\infty}^{\infty} dxn(x)xF(x), \tag{2.2}$$

where $T=\int_{-\infty}^{\infty}dxt(x)$. Thus Eq. (2.1) was termed the differential form of the virial theorem. This equation is readily rewritten as a "force-balance" equation, which then reads

$$F(x) = \frac{dv(x)}{dx} = \left[2\frac{dt(x)}{dx} - \frac{1}{4}\frac{d^3n(x)}{dx^3}\right]\frac{1}{n(x)}. \tag{2.3}$$

The achievement of Holas and March[2] was to effect the exact generalization of the independent-fermion result (2.3) to three dimensions and, of course, to include also a full account of electron-electron interactions.

B. Sum rule for the external force in a three-dimensional interacting system

Let us start from the Holas-March[2] result for the external force stemming from the differential virial theorem for the interacting many-electron system, which reads [their Eq. (2.16)]

$$\boldsymbol{F}_{\text{ext}}(\boldsymbol{r}) = -\nabla v_{\text{ext}}(\boldsymbol{r}) = \left\{ z(\boldsymbol{r};[\rho]) - \frac{1}{4}\nabla\nabla^2 n(\boldsymbol{r}) + 2\int d^3r' n_2(\boldsymbol{r},\boldsymbol{r}')\nabla_r(|\boldsymbol{r}-\boldsymbol{r}'|^{-1})\right\} \Big/ n(\boldsymbol{r}). \tag{2.4}$$

Here $n_2(\boldsymbol{r},\boldsymbol{r}')$ denotes the pair density, which is the diagonal element of the second-order density matrix (2DM). The vector field $z(\boldsymbol{r};[\rho])$ is defined solely in terms of the fully interacting first-order density matrix (1DM) $\rho(\boldsymbol{r}_1;\boldsymbol{r}_2)$ via the kinetic-energy density tensor $t_{\alpha\beta}$ as

$$z_\alpha(\boldsymbol{r};[\rho]) = 2\sum_\beta \frac{\partial}{\partial r_\beta} t_{\alpha\beta}(\boldsymbol{r};[\rho]), \tag{2.5}$$

the precise definition of this tensor employed in Ref. 2 being

$$t_{\alpha\beta}(\boldsymbol{r};[\rho]) = \frac{1}{4}\left(\frac{\partial^2}{\partial r'_\alpha \partial r''_\beta} + \frac{\partial^2}{\partial r'_\beta \partial r''_\alpha}\right)\rho(\boldsymbol{r}';\boldsymbol{r}'')_{\boldsymbol{r}'=\boldsymbol{r}''=\boldsymbol{r}}. \tag{2.6}$$

Using the notation $\langle f\rangle=\int d^3rf(\boldsymbol{r})$, we find from Eq. (2.4) the result

$$\langle n\boldsymbol{F}_{\text{ext}}\rangle = \int d^3rz(\boldsymbol{r};[\rho]) - \frac{1}{4}\int d^3r\,\nabla\nabla^2 n(\boldsymbol{r}) + 2\int d^3rd^3r' n_2(\boldsymbol{r},\boldsymbol{r}')\nabla_r(|\boldsymbol{r}-\boldsymbol{r}'|^{-1}). \tag{2.7}$$

C. Sum rule for the external force in a noninteracting system

Adapting the result (2.4) to a system of noninteracting fermions moving in a common potential field $v_s(\boldsymbol{r})$, we omit the term due to the electron-electron interaction (the integral) and replace the interacting 1DM $\rho(\boldsymbol{r}_1;\boldsymbol{r}_2)$ by its single-particle counterpart $\rho^s(\boldsymbol{r}_1;\boldsymbol{r}_2)$. The external force in this system [Eq. (2.24) in Ref. 2] then reads

$$\boldsymbol{F}_s(\boldsymbol{r}) = -\nabla v_s(\boldsymbol{r}) = \{z(\boldsymbol{r};[\rho^s]) - \tfrac{1}{4}\nabla\nabla^2 n(\boldsymbol{r})\}/n(\boldsymbol{r}). \tag{2.8}$$

This leads to a sum rule analogous to Eq. (2.7)

$$\langle n\boldsymbol{F}_s\rangle = \int d^3rz(\boldsymbol{r};[\rho^s]) - \frac{1}{4}\int d^3r\,\nabla\nabla^2 n(\boldsymbol{r}). \tag{2.9}$$

D. Sum rules for forces in the Kohn-Sham system

Considering now the noninteracting system of Sec. II C to be the Kohn-Sham (KS) system of DFT (see, e.g., Ref. 6) equivalent to the original, interacting system of Sec. II B and, therefore, having the same ground-state density $n(\boldsymbol{r})$, we identify its external potential with the formally exact KS potential

$$v_s(\boldsymbol{r}) \equiv v_{\text{KS}}(\boldsymbol{r}) = v_{\text{ext}}(\boldsymbol{r}) + v_{\text{esxc}}(\boldsymbol{r}), \tag{2.10}$$

where $v_{\text{esxc}}(\boldsymbol{r})$ denotes the electrostatic-plus-exchange-correlation potential. The difference between Eqs. (2.7) and (2.9) with the partition (2.10) taken into account leads immediately to the sum rule for the esxc force $\boldsymbol{F}_{\text{esxc}}(\boldsymbol{r})=-\nabla v_{\text{esxc}}(\boldsymbol{r})$, namely,

$$\langle n\boldsymbol{F}_{\text{esxc}}\rangle = \int d^3rz(\boldsymbol{r};[\rho^s-\rho]) - 2\int d^3rd^3r' n_2(\boldsymbol{r},\boldsymbol{r}')\nabla_r(|\boldsymbol{r}-\boldsymbol{r}'|^{-1}). \tag{2.11}$$

It can be reduced next to

$$\langle n\boldsymbol{F}_{\text{esxc}}\rangle = \boldsymbol{0}. \tag{2.12}$$

The vanishing in Eq. (2.11) of the electron-electron interaction contribution is guaranteed by the symmetry of the integrand: the factor $\nabla_r(|\boldsymbol{r}-\boldsymbol{r}'|^{-1})=-(\boldsymbol{r}-\boldsymbol{r}')/|\boldsymbol{r}-\boldsymbol{r}'|^3$ is antisymmetric in $\{\boldsymbol{r},\boldsymbol{r}'\}$, while the remaining factor is symmetric. The vanishing of the kinetic-energy contribution (depending on $\rho^s-\rho$) is due to the asymptotic large-r exponential decay of 1DMs: the long-range behavior of each natural orbital of ρ and of the most extended occupied KS orbital of ρ^s is proportional to $\exp[-(2I_{\min})^{-1/2}r]$, where $I_{\min}$ denotes the ionization potential of the system (see, e.g., Refs. 7 and 6). This applies to finite systems only which are considered in the present paper.

The same arguments lead to the vanishing of all terms in both Eqs. (2.7) and (2.9), so the identities

$$\langle n\boldsymbol{F}_{\text{ext}}\rangle = 0, \quad \langle n\boldsymbol{F}_s\rangle = \boldsymbol{0} \tag{2.13}$$

follow. They reveal the homogeneity of the space as shown by Levy and Perdew.[4]

The identities (2.7) and (2.9) (representing relations between external potentials, interaction potential, and DMs) depend linearly on DMs derived from any eigenfunction of an interacting system (n,ρ,n_2) and of a noninteracting one (n,ρ^s), respectively. Therefore these relations remain true if the pure-state DMs are replaced by the corresponding ensemble DMs. While the DMs generated from the ground-state eigenfunctions are used for application in the basic DFT, those generated from the excited-state eigenfunctions can be applied in the corresponding DFT extension (see, e.g., Levy and Nagy[8]). All above-mentioned remarks concern also Eq. (2.11), which represents the difference of Eqs. (2.7) and (2.9).

Finally, by taking $v_{xc}(\boldsymbol{r})=v_{esxc}(\boldsymbol{r})-v_{es}(\boldsymbol{r})$, the sum rule for the xc force

$$\langle n\boldsymbol{F}_{xc}\rangle=\boldsymbol{0} \tag{2.14}$$

can be obtained from Eq. (2.12) by subtracting the contribution due to the electrostatic force $\boldsymbol{F}_{es}(\boldsymbol{r})=-\boldsymbol{\nabla}v_{es}(\boldsymbol{r})$. Here, the electrostatic potential is defined as usually by

$$v_{es}(\boldsymbol{r};[n])=\int d^3r'n(\boldsymbol{r}')/|\boldsymbol{r}-\boldsymbol{r}'|. \tag{2.15}$$

The identity, which is applied for obtaining Eq. (2.14),

$$\langle n\boldsymbol{F}_{es}\rangle=-\int d^3rd^3r'n(\boldsymbol{r})n(\boldsymbol{r}')\boldsymbol{\nabla}_r(|\boldsymbol{r}-\boldsymbol{r}'|^{-1})=\boldsymbol{0}, \tag{2.16}$$

holds due to the symmetry of the integrand, as discussed just below Eq. (2.12). In the case of the ensemble DFT, the ensemble density $n(\boldsymbol{r})$ is to be used in Eq. (2.15).

III. SEPARATE SUM RULES FOR THE EXCHANGE AND CORRELATION FORCES

As shown by Levy and March,[9] the separate expressions for the exact exchange and correlation forces can be extracted from the Holas-March[2] result for the esxc force $\boldsymbol{F}_{esxc}$ through the use of the coupling-constant adiabatic-connection method.[10] Görling and Levy[10] proposed to link the interacting system and the equivalent noninteracting KS system by a family of intermediate systems with the electron-electron interaction $\lambda/|\boldsymbol{r}-\boldsymbol{r}'|$, i.e., scaled by the coupling parameter $\lambda\in[0,1]$; all systems are required to have the same density $n(\boldsymbol{r})$. Properties of the fully interacting system (at $\lambda=1$) are viewed as derived from the KS system solution (at $\lambda=0$) with the help of the perturbation theory with respect to λ. In this approach, the expanded esxc force in the intermediate system

$$\boldsymbol{F}^{\lambda}_{esxc}(\boldsymbol{r})=\sum_{l=1}^{\infty}\lambda^l\boldsymbol{F}^{\{l\}}_{esxc}(\boldsymbol{r}) \tag{3.1}$$

gives the electrostatic-plus-exchange force $\boldsymbol{F}_{esx}$ as its leading term, and the correlation force $\boldsymbol{F}_c$ as the sum of the remaining terms:

$$\boldsymbol{F}_{esx}(\boldsymbol{r})=\boldsymbol{F}^{\{1\}}_{esxc}(\boldsymbol{r}), \tag{3.2a}$$

$$\boldsymbol{F}_c(\boldsymbol{r})=\boldsymbol{F}_{esxc}(\boldsymbol{r})-\boldsymbol{F}_{esx}(\boldsymbol{r})=\sum_{l=2}^{\infty}\boldsymbol{F}^{\{l\}}_c(\boldsymbol{r}), \tag{3.2b}$$

$$\forall\ l\geqslant 2,\quad \boldsymbol{F}^{\{l\}}_c(\boldsymbol{r})=\boldsymbol{F}^{\{l\}}_{esxc}(\boldsymbol{r}). \tag{3.2c}$$

The sum rule for the esxc force $\boldsymbol{F}^{\lambda}_{esxc}$ in the intermediate system is given by Eq. (2.11) adapted to this system, i.e., with the DMs ρ,n_2 replaced by $\rho^{\lambda},n_2^{\lambda}$ and with the interaction potential scaled by λ:

$$\langle n\boldsymbol{F}^{\lambda}_{esxc}\rangle=\int d^3r\boldsymbol{z}(\boldsymbol{r};[\rho^s-\rho^{\lambda}]) - 2\int d^3rd^3r'n_2^{\lambda}(\boldsymbol{r},\boldsymbol{r}')\boldsymbol{\nabla}_r(\lambda|\boldsymbol{r}-\boldsymbol{r}'|^{-1}). \tag{3.3}$$

This again can be simplified to

$$\langle n\boldsymbol{F}^{\lambda}_{esxc}\rangle=0,\quad \forall\ \lambda\in[\boldsymbol{0},1] \tag{3.4}$$

by applying the symmetry argument to the interaction contribution and by noting that the asymptotic exponential decay of natural orbitals of ρ^{λ} should be of the same form at any strength λ of the interaction because $n^{\lambda}(\boldsymbol{r})=n(\boldsymbol{r})$. The sum rule (3.4) expanded according to Eq. (3.1) results in

$$\langle n\boldsymbol{F}^{\{l\}}_{esxc}\rangle=\boldsymbol{0}\quad \text{for } l=1,2,\ldots. \tag{3.5}$$

The following sum rules follow from Eq. (3.5) and the assignments (3.2a), (3.2b), and (3.2c):

$$\langle n\boldsymbol{F}_{esx}\rangle=\boldsymbol{0}, \tag{3.6a}$$

$$\langle n\boldsymbol{F}_c\rangle=\boldsymbol{0}, \tag{3.6b}$$

$$\langle n\boldsymbol{F}^{\{l\}}_c\rangle=\boldsymbol{0},\quad \text{for } l=2,3,\ldots. \tag{3.6c}$$

The results (3.6a), (3.6b), and (3.6c) are identical with the equations derived by Joubert[5] with the help of different methods.

Finally, by taking $\boldsymbol{F}_x=\boldsymbol{F}_{esx}-\boldsymbol{F}_{es}$ and using the results (2.16) and (3.6a), the separate sum rule for the exchange force is obtained:

$$\langle n\boldsymbol{F}_x\rangle=\boldsymbol{0}. \tag{3.7}$$

IV. DISCUSSION AND SUMMARY

Everything calculated in the present study in Sec. II stems from the differential virial theorem in terms of low-order density matrices. The most important result is the sum rule (2.14) which represents the identity satisfied by the exchange-correlation force $\boldsymbol{F}_{xc}(\boldsymbol{r})=-\boldsymbol{\nabla}v_{xc}(\boldsymbol{r})$ associated with the exact exchange-correlation potential $v_{xc}(\boldsymbol{r})$. Both objects are functionals of the density $n(\boldsymbol{r})$, which occurs also explicitly in the sum rule. To obtain in Sec. III the sum rules for the exchange and correlation forces separately, Eqs. (3.6b) and (3.7), the adiabatic-connection method also needs to be involved.

It should be mentioned that any of the obtained sum rules for a force can be transformed into an equivalent sum rule for the associated potential. Namely, for $\boldsymbol{F}_A(\boldsymbol{r})=-\boldsymbol{\nabla}v_A(\boldsymbol{r})$, the equation $\langle n\boldsymbol{F}_A\rangle=\boldsymbol{0}$ is equivalent to $\langle v_A\boldsymbol{\nabla}n\rangle$

$=\mathbf{0}$. For some systems these rules may be satisfied due to symmetry only, e.g., when the density and potential are spherically symmetric.

Evidently, in the continuing search for refined functionals, as motivated in part by critical comments of Wanko *et al.*[1] for biological applications, it is important to test how well the exact sum rules, and especially Eq. (2.14) involving the exchange-correlation potential, are obeyed by refined proposals in the future.

ACKNOWLEDGMENTS

N.H.M. wishes to acknowledge that his contribution to the present article was brought to fruition during visits to the Donastia International Physics Center, San Sebastian, Spain, and to the Institute of Physical Chemistry of PAS, Warsaw, Poland. He especially thanks Professor P. M. Echenique and Professor A. Rubio in San Sebastian and Professor A. Holas in Warsaw for their most generous hospitality. A.R. was partially supported by the EC Sixth Framework Network of Excellence NANOQUANTA (NMP4-CT-2004-500198), the Spanish MCyT, and the Humboldt Foundation under the Bessel research award (2005).

[1] M. Wanko, M. Hoffman, P. Strodet, A. Koslowski, W. Thiel, F. Neese, T. Frauenheim, and M. Elstner, J. Phys. Chem. B **109**, 3606 (2005).
[2] A. Holas and N. H. March, Phys. Rev. A **51**, 2040 (1995).
[3] N. H. March and W. H. Young, Nucl. Phys. **12**, 237 (1959).
[4] M. Levy and J. P. Perdew, Phys. Rev. A **32**, 2010 (1985).
[5] D. P. Joubert, J. Chem. Phys. **119**, 1916 (2003).
[6] R. G. Parr and W. Yang, *Density-Functional Theory of Atoms and Molecules* (Oxford University Press, Oxford, 1989).
[7] M. M. Morell, R. G. Parr, and M. Levy, J. Chem. Phys. **62**, 549 (1975).
[8] M. Levy and Á. Nagy, Phys. Rev. Lett. **83**, 4361 (1999);Á. Nagy and M. Levy, Phys. Rev. A **63**, 052502 (2001).
[9] M. Levy and N. H. March, Phys. Rev. A **55**, 1885 (1997).
[10] A. Görling and M. Levy, Phys. Rev. B **48**, 11638 (1993).

Available online at www.sciencedirect.com
SCIENCE @ DIRECT®
ELSEVIER
Chemical Physics Letters 423 (2006) 94–97
CHEMICAL PHYSICS LETTERS
www.elsevier.com/locate/cplett

Exchange-correlation potential in terms of the idempotent Dirac density matrix of DFT

C. Amovilli [a,*], N.H. March [b,c]

[a] Dipartimento di Chimica e Chimica Industriale, Università di Pisa, Via Risorgimento 35, 56126 Pisa, Italy
[b] The Abdus Salam International Center for Theoretical Physics, Strada Costiera 11, 34014 Trieste, Italy
[c] Oxford University, Oxford, United Kingdom

Received 8 September 2005; in final form 16 March 2006
Available online 22 March 2006

Abstract

A formal proof has recently been proposed [I.A. Howard, N.H. March, Chem. Phys. Lett. 402 (2005) 1.] to show that the exchange-correlation potential $V_{xc}(\mathbf{r})$ of density functional theory (DFT) is, in principle, determined by the idempotent Dirac density matrix $\gamma(\mathbf{r},\mathbf{r}')$ generated by the one-body potential $V(\mathbf{r}) = V_{ext}(\mathbf{r}) + V_{Hartree}(\mathbf{r}) + V_{xc}(\mathbf{r})$. Here we turn this 'existence' theorem into a concrete result by expressing the Laplacian $\nabla^2 V_{xc}(\mathbf{r})$ solely in terms of $\gamma(\mathbf{r},\mathbf{r}')$, which, although idempotent and therefore not many-electron in character, has the exact ground state density $\rho(\mathbf{r})$ on its diagonal $\gamma(\mathbf{r},\mathbf{r})$. As a specific illustration, we calculate $\rho(\mathbf{r})$ by diffusion QMC and hence $\gamma(\mathbf{r},\mathbf{r}')$ for the Be atom ground-state. $V_{xc}(\mathbf{r})$ is thereby obtained with satisfactory accuracy.

In a recent study in this Journal [1], the importance for DFT of the idempotent Dirac density matrix $\gamma(\mathbf{r},\mathbf{r}')$ generated by the one-body potential

$$V(\mathbf{r}) = V_{\mathrm{ext}}(\mathbf{r}) + V_{\mathrm{Hartree}}(\mathbf{r}) + V_{\mathrm{xc}}(\mathbf{r}) \tag{1}$$

has been stressed. In Eq. (1), $V_{ext}(\mathbf{r})$ represents the electron–nuclear potential energy in the molecule or cluster under consideration, while $V_{Hartree}(\mathbf{r})$ is the electrostatic potential generated by the exact ground-state density $\rho(\mathbf{r})$. Though the Dirac matrix $\gamma(\mathbf{r},\mathbf{r}')$ is idempotent, and therefore does not have many-electron character, its diagonal element, by construction, is the exact density $\rho(\mathbf{r})$. The theorems of DFT [2] tell us that the, as yet unknown, exchange-correlation potential $V_{xc}(\mathbf{r})$ is a unique functional of the electron density $\rho(\mathbf{r})$.

Here, the results of [1] have motivated us to re-open the density matrix theory of the exchange-correlation force, defined as

$$\vec{F}_{\mathrm{xc}}(\mathbf{r}) = -\vec{\nabla} V_{\mathrm{xc}}(\mathbf{r}), \tag{2}$$

given by Holas and March [3] and reviewed, for example, in the recent book by Sahni [4].

This theory was posed in terms of low-order density matrices, including the correlated and formally exact one-particle density matrix plus the pair density of the many-electron system. The studies in [1] now lead us to present here a theory of the Laplacian $\nabla^2 V_{xc}(\mathbf{r})$ which bypasses these unknown many-electron density matrices. The result of Holas and March [3] on which we focus below is for the force $\vec{F}$ defined from the one-body potential $V(\mathbf{r})$ as

$$\vec{F} = -\vec{\nabla} V(\mathbf{r}). \tag{3}$$

In one dimension, March and Young [5] in an early study derived from the equation of motion of the Dirac density matrix $\gamma(x,x')$ a differential form of the virial theorem, namely

$$\frac{\partial t}{\partial x} = -\frac{1}{2}\rho(x)\frac{\partial V}{\partial x} + \frac{1}{8}\rho'''(x) \tag{4}$$

where $t(x)$ is the kinetic energy per unit length in this one-dimensional example. Multiplying both sides by x and integrating from $-\infty$ to ∞ recovers the usual integral virial theorem. Rearranging Eq. (4) yields

* Corresponding author. Fax: +39 050 2219260.
E-mail address: amovilli@dcci.unipi.it (C. Amovilli).

0009-2614/$ - see front matter © 2006 Elsevier B.V. All rights reserved.
doi:10.1016/j.cplett.2006.03.038

$$F = -\frac{\partial V}{\partial x} = \frac{2}{\rho(x)}\frac{\partial t}{\partial x} - \frac{\rho'''(x)}{4\rho(x)}. \tag{5}$$

Of course, though $t \equiv t[\rho]$, the natural tool to calculate this kinetic energy term is the Dirac density matrix $\gamma(x, x')$.

Holas and March [3] gave the exact generalization of Eq. (5), to read in three dimensions

$$\vec{F} = -\vec{\nabla}V(\mathbf{r}) = \vec{F}_K - \frac{1}{4\rho(\mathbf{r})}\vec{\nabla}\nabla^2\rho(\mathbf{r}) \tag{6}$$

where the kinetic contribution to the 'force-balance' equation (6) is determined explicitly by the idempotent matrix $\gamma(\mathbf{r},\mathbf{r}')$. We give the form of $\vec{F}_K$ below, but what we want to emphasize next is the use of the Laplacian operator acting on Eq. (1). This yields, when combined with Poisson's equation for the electrostatic term V_{es} in (1)

$$\nabla^2 V = 4\pi\sum_{i=1}^{n} Z_i\delta(\mathbf{r} - \mathbf{R}_i) - 4\pi\rho(\mathbf{r}) + \nabla^2 V_{xc}. \tag{7}$$

The position of the n nuclei in the cluster under consideration are denoted by $\mathbf{R}_i$ and since this term involving the delta functions is identically zero everywhere except actually on nuclear positions, we shall write Eq. (7) in the form

$$\nabla^2 V_{xc}(\mathbf{r}) = 4\pi\rho(\mathbf{r}) + \nabla^2 V(\mathbf{r}). \tag{8}$$

But we can now utilize Eq. (6) to calculate $\nabla^2 V$ appearing in Eq. (8) since

$$\vec{\nabla}\cdot\vec{F} = -\nabla^2 V = \vec{\nabla}\cdot\vec{F}_K + \frac{1}{4\rho^2}\vec{\nabla}\rho\cdot\vec{\nabla}\nabla^2\rho - \frac{1}{4\rho}\nabla^4\rho. \tag{9}$$

Hence, combining Eqs. (8) and (9) we find

$$\nabla^2 V_{xc} = 4\pi\rho - \frac{1}{4\rho^2}\vec{\nabla}\rho\cdot\vec{\nabla}\nabla^2\rho + \frac{1}{4\rho}\nabla^4\rho - \vec{\nabla}\cdot\vec{F}_K. \tag{10}$$

All the terms on the right-hand-side (RHS) of Eq. (10) are determined explicitly by the ground-state density $\rho(\mathbf{r})$, except the final term $-\vec{\nabla}\cdot\vec{F}_K$. But from the work of Holas and March [3], $\vec{F}_K$ is determined from the idempotent Dirac density matrix $\gamma(\mathbf{r},\mathbf{r}')$ via the so-called kinetic energy tensor, also discussed in full detail in the book by Sahni [4]. Following the work of Holas and March [3] we can construct the force $\vec{F}_K$ by computing first the kinetic energy tensor

$$t_{ij} = \frac{1}{4}\left[\left(\frac{\partial^2}{\partial r_i \partial r'_j} + \frac{\partial^2}{\partial r_j \partial r'_i}\right)\gamma(\mathbf{r},\mathbf{r}')\right]_{\mathbf{r}'=\mathbf{r}}, \tag{11}$$

which leads to the expression

$$(F_K)_i = \frac{2}{\rho(\mathbf{r})}\sum_{j=1}^{3}\frac{\partial t_{ij}}{\partial r_j}. \tag{12}$$

Of course, one can integrate Eq. (10) using the Green function $-1/4\pi|\mathbf{r}-\mathbf{r}'|$ to obtain

$$V_{xc}(\mathbf{r}) = -V_{es}(\mathbf{r}) + \frac{1}{16\pi}\int\frac{\left[\vec{\nabla}\rho\cdot\vec{\nabla}\nabla^2\rho\right](\mathbf{r}')}{\rho^2(\mathbf{r}')|\mathbf{r}-\mathbf{r}'|}\mathrm{d}\mathbf{r}' - \frac{1}{16\pi}\int\frac{\nabla^4\rho(\mathbf{r}')}{\rho(\mathbf{r}')|\mathbf{r}-\mathbf{r}'|}\mathrm{d}\mathbf{r}' + \frac{1}{4\pi}\times\int\frac{\vec{\nabla}\cdot\vec{F}_K(\mathbf{r}')}{|\mathbf{r}-\mathbf{r}'|}\mathrm{d}\mathbf{r}'. \tag{13}$$

Again, everything on the RHS of Eq. (13) is determined by $\rho(\mathbf{r})$ except the final term. This is known, quite explicitly, in terms of the idempotent Dirac density matrix $\gamma(\mathbf{r},\mathbf{r}')$.

To exemplify the use of Eq. (13), let us treat the ground-state of the Be atom. Here we use the density obtained by a computer simulation, more precisely a Diffusion Quantum Monte Carlo calculation performed using the CHAMP package [7]. Writing the desired idempotent density matrix γ in this case in the Dawson and March form [8] involving density amplitude $\rho(r)^{1/2}$ and phase $\theta(r)$

$$\gamma(r,r') = \rho(r)^{1/2}\rho(r')^{1/2}\cos[\theta(r) - \theta(r')], \tag{14}$$

we note that $\theta(r)$ is determined uniquely by the QMC density. This is because $\theta(r)$ satisfies the non-linear pendulum equation [8]

$$\nabla^2\theta + \frac{\vec{\nabla}\rho}{\rho}\cdot\vec{\nabla}\theta + \lambda\sin 2\theta = 0 \tag{15}$$

which has eigenvalue λ. $\theta(r)$ obtained by solving Eq. (15) is plotted in Fig. 1.

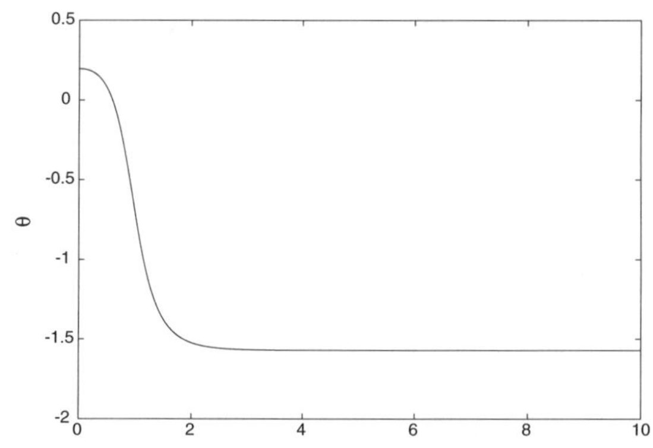

Fig. 1. Phase function $\theta(r)$ solving Eq. (15) with Be atom QMC density.

Since $\vec{F}_K$ in our basic equation (13) is determined through Eqs. (11) and (12) by $\gamma(r,r')$, $V_{xc}(r)$ is determined by the QMC computer simulated experiment.

In a little more detail, we found it best to express Eq. (13) in the more compact form

$$V_{xc}(\mathbf{r}) = -V_{es}(\mathbf{r}) + \frac{1}{4\pi}\int\frac{\vec{\nabla}\cdot\vec{F}(\mathbf{r}')}{|\mathbf{r}-\mathbf{r}'|}\mathrm{d}\mathbf{r}' \tag{16}$$

where the vector field $\vec{F}(\mathbf{r})$ is given in Eq. (6).

While Eq. (6) applies also to molecules and clusters, for Be $\vec{F}$ is along the unit radial vector. We can then express $F(r)$ in magnitude solely in terms of $\rho(r)$ and the single particle kinetic energy

96 C. Amovilli, N.H. March / Chemical Physics Letters 423 (2006) 94–97

$$t_s(r) = t_w(r) + \frac{1}{2}\rho(r)\theta'^2(r), \tag{17}$$

where $\theta(r)$ is given in Fig. 1 and $t_w(r)$ is the von Weizsäcker quantity $\rho'^2/8\rho$. Hence $F(r)$ is determined from the density $\rho(r)$ for Be which satisfies: (i) Kato's electron–nuclear cusp condition, (ii) correct asymptotic form in terms of the ionization potential I (see Eq. (18)) and (iii) good low order moments of $\rho(r)$. As to the asymptotic form, $V_{xc} \to -1/r$ at large r relates via $F(r)$ the exponent n in the large r form of $\rho(r)$, namely

$$\rho(r) = Ar^n \exp\left(-2\sqrt{2I}r\right), \tag{18}$$

to I via

$$-\lim_{r\to\infty} rV_{xc} = 2\sqrt{2I}\left[\frac{1}{2} + \frac{n}{4}\right], \tag{19}$$

the amplitude A in (18) not being involved. V_{xc} and rV_{xc} are given in Figs. 2 and 3, respectively. They have been compared with earlier works of Umrigar et al. [9], Al-Sharif et al. [10] and van Leeuwen and Baerends [11].

We believe our present result is among the most accurate forms of V_{xc} presently available for Be. We have only a remark about Fig. 2. The divergence of the vector field (6), which appears in (16), results from a delicate balance of large contributions with opposite signs coming from the two terms in the RHS of Eq. (6). Such balance, involving null divergence terms, has some relation with the nuclear potential which is dominant at the nuclei positions. V_{xc} at the nuclei is finite but the corresponding value, computed according to the present work, could be influenced by some lack in numerical accuracy related to the discussion above. Fig. 2, at $r = 0$, seems to reflect some effect due to the present numerical treatment but to quantify this uncertainty is a difficult task.

As to future directions, we wish to add two points, both of which are closely related to the use of diffraction experiments: X-ray or electron, in bringing the present result (13) for $V_{xc}(\mathbf{r})$ to full fruition. First, and exemplified below by rare gas elements like Ne or Ar, to evaluate the high derivatives of $\rho(\mathbf{r})$ entering Eq. (13), it will be advantageous to work with the atomic scattering factor $f(\mathbf{k})$ defined by

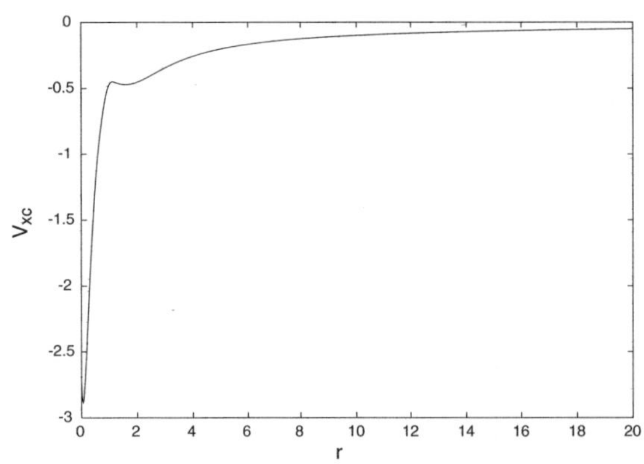

Fig. 2. Exchange-correlation for Be atom calculated using Eq. (13) from diffusion QMC density.

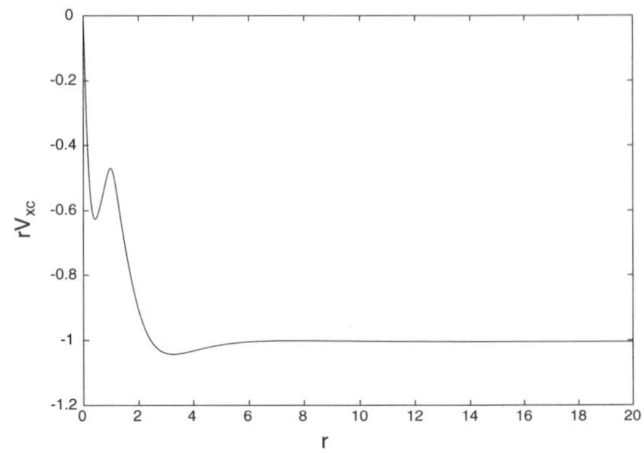

Fig. 3. Plot of rV_{xc} as a function of the radius r using our calculated V_{xc} for Be atom.

$$f(\mathbf{k}) = \int \rho(\mathbf{r}) \exp(i\mathbf{k}\cdot\mathbf{r})d\mathbf{r}. \tag{20}$$

Secondly, it will be essential to make contact with diffraction experiments via the field that is often referred to now as quantum crystallography. To take a specific, and fairly recent, example, we cite the work of Howard et al [6] on formamide. These workers use quantum chemical basis sets [12] (i) to handle the representation of their diffraction data giving $\rho(\mathbf{r})$ and (ii) to construct an idempotent matrix akin to $\gamma(\mathbf{r},\mathbf{r}')$ used here with diagonal elements $\gamma(\mathbf{r},\mathbf{r}) = \rho(\mathbf{r})$. However, as Holas and March [13] have emphasized, the γ obtained by Howard et al. [6] is not necessarily the DFT form of γ required, for that must be calculated from a local potential $V(\mathbf{r})$. Therefore Holas and March [13] have proposed a theoretical approach to a potential locality constraint on γ (of course, with experimental data, only approximately). Thus, by combining theory with diffraction experiments, further important progress on the exchange-correlation potential $V_{xc}(\mathbf{r})$ will be possible.

Acknowledgements

N.H.M. thanks Professor V.E. Kravtsov for making possible his stay at ICTP during 2005, during which his contribution to this study was brought to fruition, and for generous hospitality. N.H.M also acknowledges invaluable discussions on the general area embraced by this article with Professors A. Holas, I.A. Howard and W.H. Young. The authors thank Dr. Claudia Filippi for providing the CHAMP package for QMC calculations.

References

[1] I.A. Howard, N.H. March, Chem. Phys. Lett. 402 (2005) 1.
[2] See, for example, R.G. Parr, W. Yang, Density Functional Theory of Atoms and Molecules, Oxford University Press, New York, 1989.
[3] A. Holas, N.H. March, Phys. Rev. A 51 (1995) 2040.
[4] V. Sahni, Quantal Density Functional Theory, Springer-Verlag, Berlin, 2004.
[5] N.H. March, W.H. Young, Nucl. Phys. 12 (1959) 237.
[6] S.T. Howard, J.P. Huke, P.R. Mallison, C.S. Frampton, Phys. Rev. B 49 (1994) 7124.
[7] C. Filippi, C.J. Umrigar, J. Chem. Phys. 105 (1996) 213.
[8] K.A. Dawson, N.H. March, J. Chem. Phys. 81 (1984) 5850.
[9] C.J. Umrigar, A. Savin, X. Gonze, in: J.F. Dobson, G. Vignale, M.P. Das (Eds.), Electronic Density Functional Theory: Recent Progress and New Directions, Plenum, New York, 1997.
[10] A.I. Al-Sharif, A. Qteish, R. Resta, Phys. Rev. A 60 (1999) 3541.
[11] R. van Leeuwen, E.J. Baerends, Phys. Rev. A 49 (1994) 2421.
[12] R. McWeeny, Rev. Mod. Phys. 32 (1960) 335.
[13] A. Holas, N.H. March, Phys. Rev. B 55 (1997) 9422.

Molecular Physics, Vol. 103, No. 9, 10 May 2005, 1261–1270

Taylor & Francis
Taylor & Francis Group

Integral equation theory of the exchange potential, HOMO–LUMO properties, and sum rules for the exchange-correlation force

I.A. HOWARD†* and N.H. MARCH†‡

†Department of Physics, University of Antwerp, Groenenborgerlaan, 171, B-2020 Antwerp, Belgium
‡Oxford University, Oxford, UK

(Received 3 August 2004; in final form 29 October 2004)

Recently, a formally exact integral equation for the exchange potential $V_x(\mathbf{r})$ has been presented by the authors. In the admittedly simplistic limit in which Slater–Kohn–Sham and Hartree–Fock determinants become equal, this integral equation reduces to that given by Della Sala and Görling. Here, a proposal is made to relate, but now approximately, the formally exact equation for $V_x(\mathbf{r})$ to HOMO–LUMO properties. The addition of a correlation contribution $V_c(\mathbf{r})$ to $V_x(\mathbf{r})$, the sum being the exchange-correlation potential $V_{xc}(\mathbf{r})$, is finally considered, some exact properties and especially sum rules for the force $-\partial V_{xc}/\partial \mathbf{r}$ being the focus.

1. Introduction

It is, of course, a privilege to contribute to this Conference Proceedings honouring Professor Handy. One of us (NHM) was, for a number of years, chairman of the hugely successful sets of lectures which Professor Handy gave as the 'anchor man' of the Coulson Summer Schools in Theoretical Chemistry in Oxford on many-electron theory. The present article is close to a whole programme of work carried out by Professor Handy and his colleagues on density functional theory (DFT). The central equation of that theory [1] is for the constancy of the electron chemical potential μ throughout the entire inhomogeneous electron density $\rho(\mathbf{r})$ in the molecules/clusters under discussion. This equation, in the accepted terminology of DFT, reads

$$\mu = \frac{\delta T_s}{\delta \rho(\mathbf{r})} + V_{ext}(\mathbf{r}) + V_{el}(\mathbf{r}) + V_{xc}(\mathbf{r}). \tag{1}$$

Here, very briefly, $T_s[\rho]$ is the, as yet unknown, single-particle kinetic energy functional, $V_{ext}(\mathbf{r})$ the external potential (equal to $-Ze^2/r$ in an atom having nuclear charge Ze), $V_{el}(\mathbf{r})$ is the electrostatic potential energy created by the charge density $\rho(\mathbf{r})$, while $V_{xc}(\mathbf{r})$ is the exchange-correlation potential. This is given by the functional derivative [1]

$$V_{xc}(\mathbf{r}) = \frac{\delta E_{xc}[\rho]}{\delta \rho(\mathbf{r})}, \tag{2}$$

where $E_{xc}[\rho]$ is the, again unknown, exchange (x)-correlation (c) energy functional.

It is well known that $V_{xc}(\mathbf{r})$ can be separated, formally but exactly, into a sum of two parts [2]

$$V_{xc}(\mathbf{r}) = V_x(\mathbf{r}) + V_c(\mathbf{r}), \tag{3}$$

where $V_x(\mathbf{r})$ is the exchange-only potential while $V_c(\mathbf{r})$ is the correlation contribution to $V_{xc}(\mathbf{r})$. One route to $V_{xc}(\mathbf{r})$ is via adiabatic connection, starting from the work of Harris and Jones [3], and used, for example, by Levy *et al.* [4].

With this as background, the outline of the present contribution is as follows. In section 2 immediately below, the essential content of a recently derived, and formally exact, integral equation [5] for $V_x(\mathbf{r})$ is first summarized. This integral equation theory is then compared and contrasted with earlier approximate studies, and especially those of Gritsenko and Baerends [6] (see also [7]) and of Della Sala and Görling [8]. Following this, some recent results on exchange potentials in closed shell atomic ions are presented in section 3. The content of section 4 is to introduce HOMO and LUMO properties into the now-approximate theory of the exchange potential $V_x(\mathbf{r})$. Section 5 considers some exact sum rules which

* Corresponding author. e-mail: iris.howard@ua.ac.be

Molecular Physics
ISSN 0026–8976 print/ISSN 1362–3028 online
http://www.tandf.co.uk/journals
DOI: 10.1080/00268970412331332970

result when the correlation potential $V_c(\mathbf{r})$ is added to $V_x(\mathbf{r})$ as in equation (3), the particular focus being the force $-\partial V_{xc}(\mathbf{r})/\partial\mathbf{r}$ [9]. Nuclear cusp conditions on exchange and correlation energy densities are set out in section 6, while section 7 presents a brief summary plus some suggestions for further work which may prove fruitful.

2. Formally exact integral equation theory for exchange potential $V_x(\mathbf{r})$

While, of course, DFT theory has now moved well beyond the local density approximation (LDA) in which $V_x(\mathbf{r})$ is taken to be proportional to $\rho(\mathbf{r})^{1/3}$ [10–12], there was already in Slater's 1951 paper [10] a proposed 'non-local' approximation to $V_{xc}(\mathbf{r})$. That this approximation was related to functional differentiation of the exchange-energy functional $E_x[\rho]$ was emphasized especially in an important contribution by Kleinman [13]. His finding was that the Slater approximation, denoted by $V_x^{Sl}(\mathbf{r})$ below, to the exchange potential $V_x(\mathbf{r})$ was a 'partial' functional derivative of the Dirac [14] exchange energy given in terms of the idempotent Dirac density matrix $\gamma(\mathbf{r},\mathbf{r}')$, where $\rho(\mathbf{r}) = \gamma(\mathbf{r},\mathbf{r})$, by

$$E_x = -\frac{e^2}{4}\int\int \mathrm{d}\mathbf{r}\,\mathrm{d}\mathbf{r}'\frac{[\gamma(\mathbf{r},\mathbf{r}')]^2}{|\mathbf{r}-\mathbf{r}'|}. \tag{4}$$

Holas and March [15] completed Kleinman's study by writing the one-electron Slater–Kohn–Sham wave functions $\psi_i(\mathbf{r})$ entering $\gamma(\mathbf{r},\mathbf{r}')$ in equation (4) in terms of a density amplitude $\rho(\mathbf{r})^{1/2}$ and phases and performed the functional differentiation paralleling equation (2) by including these phases beyond the Kleinman 'partial' functional derivative. If we write from equation (4) a definition of the exchange-energy density $\epsilon_x(\mathbf{r})$ (not, of course, unique [16]) as

$$\epsilon_x(\mathbf{r}) = -\frac{e^2}{4}\int \mathrm{d}\mathbf{r}'\frac{[\gamma(\mathbf{r},\mathbf{r}')]^2}{|\mathbf{r}-\mathbf{r}'|} \tag{5}$$

then the form of potential proposed by Slater [10] is

$$V_x^{Sl}(\mathbf{r}) = \frac{2\epsilon_x(\mathbf{r})}{\rho(\mathbf{r})}. \tag{6}$$

Utilizing the asymptotic result of one of us for $\epsilon_x(\mathbf{r})$ for a closed shell atom like Ar as [17]

$$\epsilon_x(\mathbf{r}) \to -\frac{e^2}{2r}\rho(\mathbf{r}) : \mathbf{r}\to\infty, \tag{7}$$

one recovers the correct asymptotic form $V_x(\mathbf{r}) \to -e^2/\mathbf{r}$ at large $\mathbf{r}$ from the Slater form (6) by inserting the limiting form (7) of $\epsilon_x(\mathbf{r})$.

Thus, one major advantage of the non-local Slater potential $V_x^{Sl}(\mathbf{r})$ defined in equation (6) is its long-range behaviour, in contrast to the exponential decay of $V_x^{LDA} \propto \rho^{1/3}$ at large distances from the Ar nucleus in the example cited above.

2.1. *Exact integral equation theory for $V_x(\mathbf{r})$ transcending the Slater potential $V_x^{Sl}(\mathbf{r})$*

Howard and March [5] focus on an exact integral identity for $\int \rho V_x(\mathbf{r})\,\mathrm{d}\mathbf{r}$, which 'corrects' the Slater potential form $\int \rho V_x^{Sl}(\mathbf{r})\,\mathrm{d}\mathbf{r}$. This latter integral was used in Slater's physical arguments resulting in the approximate proposal (6), from which it follows that the Dirac exchange energy is given by

$$E_x = \frac{1}{2}\int \rho V_x^{Sl}(\mathbf{r})\,\mathrm{d}\mathbf{r}. \tag{8}$$

Now we can write therefore that

$$\int \rho V_x(\mathbf{r})\,\mathrm{d}\mathbf{r} - \int \rho V_x^{Sl}(\mathbf{r})\,\mathrm{d}\mathbf{r} = \int \rho V_x(\mathbf{r})\,\mathrm{d}\mathbf{r} - 2E_x. \tag{9}$$

The next step in the argument is to invoke idempotency and since we shall consider throughout spin-compensated systems this reads

$$\frac{\gamma(\mathbf{r},\mathbf{r}')}{2} = \int\left(\frac{\gamma(\mathbf{r},\mathbf{r}'')}{2}\right)\left(\frac{\gamma(\mathbf{r}'',\mathbf{r}')}{2}\right)\mathrm{d}\mathbf{r}''. \tag{10}$$

Using this condition at $\mathbf{r}' = \mathbf{r}$ for $\rho(\mathbf{r})$ on the right-hand side of equation (9) yields then the identity

$$\int \rho V_x(\mathbf{r})\mathrm{d}\mathbf{r} - \int \rho V_x^{Sl}(\mathbf{r})\mathrm{d}\mathbf{r} = \frac{1}{2}\int \gamma(\mathbf{r},\mathbf{r}'')^2 V_x(\mathbf{r})\mathrm{d}\mathbf{r}\mathrm{d}\mathbf{r}'' - 2E_x. \tag{11}$$

But inserting the Dirac form (4) for E_x in the right-hand side of equation (11), and then using 'idempotency' in equation (10) in the resulting term one reaches the integral identity [5]

$$\int \rho V_x(\mathbf{r})\mathrm{d}\mathbf{r} = \int \rho V_x^{Sl}(\mathbf{r})\mathrm{d}\mathbf{r} + \frac{1}{2}\int \gamma(\mathbf{r},\mathbf{r}'')^2 V_x(\mathbf{r})\mathrm{d}\mathbf{r}\mathrm{d}\mathbf{r}'' + \frac{e^2}{4}\int\int\int \mathrm{d}\mathbf{r}\mathrm{d}\mathbf{r}'\mathrm{d}\mathbf{r}''\frac{[\gamma(\mathbf{r},\mathbf{r}')\gamma(\mathbf{r},\mathbf{r}'')\gamma(\mathbf{r}'',\mathbf{r}')]}{|\mathbf{r}-\mathbf{r}'|}. \tag{12}$$

Thus one can construct, but now formally, a differential form of the exact identity (12) which reads

$$V_x(\mathbf{r}) = V_x^{\mathrm{Sl}}(\mathbf{r}) + \frac{1}{2\rho(\mathbf{r})}\int d\mathbf{r}'\gamma(\mathbf{r},\mathbf{r}')^2 V_x(\mathbf{r}') + \frac{e^2}{4\rho(\mathbf{r})}\int\int d\mathbf{r}'\,d\mathbf{r}''\frac{[\gamma(\mathbf{r},\mathbf{r}')\gamma(\mathbf{r},\mathbf{r}'')\gamma(\mathbf{r}'',\mathbf{r}')]}{|\mathbf{r}'-\mathbf{r}''|} + \frac{P(\mathbf{r})}{\rho(\mathbf{r})}, \tag{13}$$

where $P(\mathbf{r})$ must then satisfy the sum rule

$$\int P(\mathbf{r})\,d\mathbf{r} = 0. \tag{14}$$

Besides this result (14), the fact that in closed-shell atomic ions $V_x^{\mathrm{Sl}}(r) \to -e^2/r$ at large r suggests that $P(\mathbf{r})$ will be 'short-ranged'.

Of course, the 'formally exact' integral equation (13) has to be subject, at least to date, to 'closure approximations' since $P(\mathbf{r})$ is only partially specified as discussed above†. The present writers [5] have already noted that, remarkably, equation (13) is 'closed' in the study of Della Sala and Görling (DSG) [8] whose work is based on the admittedly drastic approximation that the Slater determinant built from Slater–Kohn–Sham orbitals $\psi_i(\mathbf{r})$ equals the Hartree–Fock determinant, by the choice

$$P_{\mathrm{DSG}}(\mathbf{r}) = 0. \tag{15}$$

Subsequently, our attention has been drawn to the earlier work of Gritsenko and Baerends [6], who, by again a rather drastic-looking 'energy denominator' approximation, arrive at the same result (15). In the example of the Be atom, the present writers satisfy the sum rule (14) by writing [5]

$$P_{\mathrm{Be}}(\mathbf{r}) \propto \psi_1(\mathbf{r})\psi_2(\mathbf{r}), \tag{16}$$

where ψ_1 and ψ_2 are the orthogonal $1s$ and $2s$ orbitals. Their results for the exchange potential in the neutral Be atom, as corrected from the Slater potential, are reproduced in figure 1.

In section 4 below, we shall propose an alternative approach to approximating $P(\mathbf{r})$ in equation (13) using HOMO and LUMO orbitals.

3. Results for exchange potential in heavier closed-shell atomic ions

In earlier work, the writers used the so-called $1/Z$ expansion of atomic theory [18, 19] which was related to the simplest form of DFT in the early work of March and White [20], to construct an analytic form of the exchange energy density $\epsilon_x(\mathbf{r})$ for an atomic 'Ne-like' ion with nuclear charge Ze in the limit of large Z. For this example, $\gamma(\mathbf{r},\mathbf{r}')$ is known analytically from the study of March and Santamaria [21] and hence $\epsilon_x(\mathbf{r})$ was also obtained analytically. Since $\rho(\mathbf{r})$ is evidently thereby known from $\gamma(\mathbf{r},\mathbf{r}')$, the Slater potential $V_x^{\mathrm{Sl}}(\mathbf{r})$ is also obtained in analytical form. Figure 2 shows $4\pi r^2\epsilon_x(\mathbf{r})$ for the choice $Z = 92$.

For comparison with the approximation $V_x^{\mathrm{Sl}}(\mathbf{r})$, the optimized effective potential (OEP) method of Talman and Shadwick [23] has been used for a bare Coulomb potential [24] and this OEP result is plotted in figure 3 for comparison with the non-local Slater form $V_x^{\mathrm{Sl}}(\mathbf{r})$.

The quantitative corrections to $V_x^{\mathrm{Sl}}(\mathbf{r})$ are obviously of some importance, though $V_x^{\mathrm{Sl}}(\mathbf{r})$ is plainly a very useful 'zero-order' approximation: and is already a major improvement on the LDA result $\propto \rho(\mathbf{r})^{1/3}$.

3.1. *Example of (non-relativistic) four-electron atomic ions in the limit of large atomic number Z*

As an explicit example of equation (13) on which some analytical progress proves possible, we consider the Be-like sequence of atomic ions in the limit of large Z. Then the Dirac density matrix $\gamma(\mathbf{r},\mathbf{r}')$ can immediately be constructed from normalized hydrogenic wave functions ψ_{1s} and ψ_{2s}. It turns out that the term of $\mathcal{O}(\gamma^3)$ in equation (13) can be evaluated analytically, with the result

$$\frac{1}{\rho(r)}\int\int\frac{\gamma(\mathbf{r},\mathbf{r}')\gamma(\mathbf{r}',\mathbf{r}'')\gamma(\mathbf{r},\mathbf{r}'')}{|\mathbf{r}'-\mathbf{r}''|}\,d\mathbf{r}'\,d\mathbf{r}'' = \frac{Z\exp(-Zr)}{2.800\,5264\times 10^{12}}[(482\,638\,515\,625(Z^2r^2+4(1-Zr)) \times \exp(Zr) + 5.797\,742\,72\times 10^{13} + 3103\,059\,345\,408 \times (2-Zr)\exp(Zr/2))]/(32\exp(-Zr)+4-4Zr + Z^2r^2). \tag{17}$$

†The exact $P(\mathbf{r})$ can be expected to relate to the OEP method and will then involve also unoccupied orbitals. See equation (21) below for a proposal for 'closure' involving only unoccupied state ψ_{LUMO}.

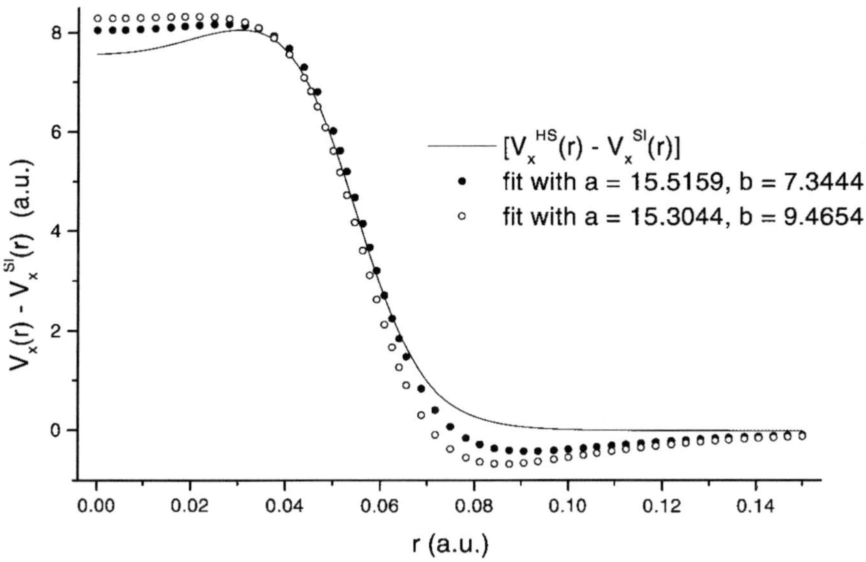

Figure 1. Illustrates results for $V_x(\mathbf{r}) - V_x^{SI}(\mathbf{r})$ for the Be atom using the integral equation (13) with the choice of $P(r) = const. \times \psi_1(r)\psi_2(r)$, with $1 \equiv 1s$ and $2 \equiv 2s$, for the configuration $(1s)^2(2s)^2$. The difference between the Harbola–Sahni exchange potential $V_x^{HS}(\mathbf{r})$ and the non-local Slater potential $V_x^{SI}(\mathbf{r})$ is shown for comparison (see also section 4). The two choices of 'weighting parameters' a and b in [5] are designed to satisfy the virial relation. [Redrawn from [5].]

One interesting consequence of equation (17) is that as $r \to \infty$, with the explicit form of $\rho(r)$ given by

$$\rho(r) = \frac{Z^3 \exp(-Zr)(32\exp(-Zr) + 4 - 4Zr + Z^2r^2)}{16\pi} \quad (18)$$

the ratio shown on the right-hand side of equation (17) tends to $\sim 0.172\,338\,5Z$.

Turning to the term involving $\gamma^2 V_x$ in equation (13), we have evaluated this with the approximation that $V_x = V_x^{SI}$. The latter can again be evaluated analytically, with the result

$$\begin{aligned} V_x^{SI} = {} & \frac{1}{[216r(32\exp(-Zr) + 4 - 4Zr + Z^2r^2)]} \\ & \times [(54Z^3r^3 - 54Z^4r^4 - 216Z^2r^2 - 6048 - 216Zr \\ & + 27Z^5r^5)\exp(-Zr) + (-1024 - 1024Zr \\ & + 768Z^2r^2)Zr\exp(-2Zr) + 6912(1 + Zr) \\ & \times \exp(-3Zr) + 864Zr - 864 - 216Z^2r^2], \end{aligned} \quad (19)$$

which tends correctly to $-1/r$ in the limit as $r \to \infty$. The shapes of the terms (17) and the $\gamma^2 V_x^{SI}$ term when suitably scaled, as in the caption, are shown in figure 4. At large r, the $\gamma^2 V_x^{SI}$ term tends to $\sim -0.186\,5409Z$.

The terms accurately compensate, leaving the Slater potential V_x^{SI} term as an excellent zero-order approximation to V_x in the method of Della Sala and Görling corresponding to equation (15).

4. Dirac density matrix characterized by HOMO and LUMO properties

That the Dirac density matrix constructed from a given potential energy $V(\mathbf{r})$ is a functional of frontier orbitals plus the HOMO–LUMO energy gap has recently been proved by Howard *et al.* [25], following earlier work by Ayers [26] (see also [27]), the latter work using Christoffel–Darboux type relations. In the study of [25], the approach going back to March and Young [28] was adopted, namely starting from the equation of motion for the Dirac matrix $\gamma(\mathbf{r}, \mathbf{r}')$:

$$\nabla_{\mathbf{r}}^2\gamma - \nabla_{\mathbf{r}'}^2\gamma = \frac{2m}{\hbar^2}[V(\mathbf{r}) - V(\mathbf{r}')]\gamma. \quad (20)$$

But for the simplest case of non-degenerate frontier orbitals (an inessential constraint, however) one can construct $V(\mathbf{r})$ and $V(\mathbf{r}')$ from the Schrödinger equations satisfied by $\psi_{HOMO}(\mathbf{r})$ and $\psi_{LUMO}(\mathbf{r}')$ respectively.

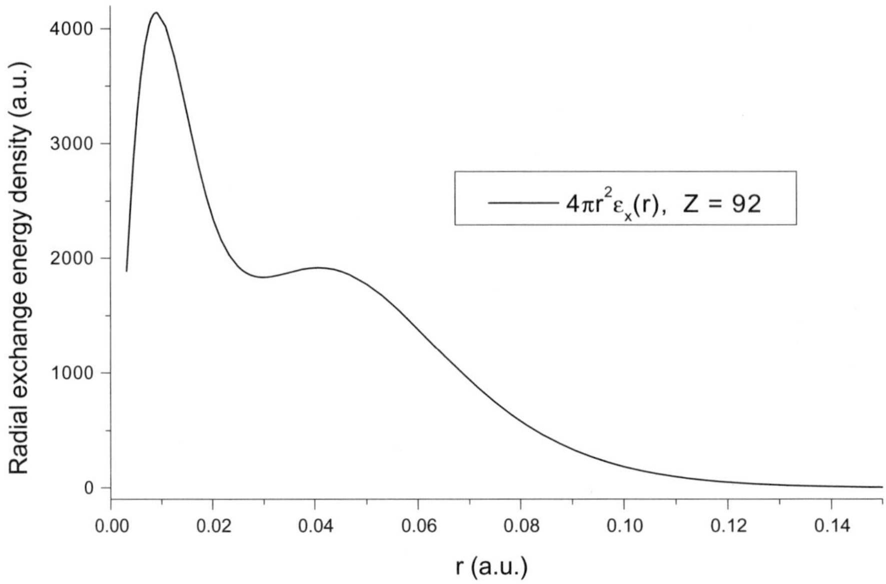

Figure 2. Radial exchange energy density $4\pi r^2 \epsilon_x(r)$ for the 10-electron closed-shell ion with $Z = 92$. [Redrawn from [22].]

$V(\mathbf{r}) - V(\mathbf{r}')$ is then determined uniquely in equation (20) by $\psi_{\text{HOMO}}(\mathbf{r})$ and $\psi_{\text{LUMO}}(\mathbf{r}')$ and their corresponding one-electron eigenvalue difference $\epsilon_{\text{H}} - \epsilon_{\text{L}}$. Of course, one must find an idempotent solution of equation (20), and a model example based on a bare Coulomb potential $V(\mathbf{r}) = -Ze^2/r$ was worked out in [25].

Since $\gamma = \gamma[\psi_{\text{HOMO}}, \psi_{\text{LUMO}}, \epsilon_{\text{H}} - \epsilon_{\text{L}}]$ and $\epsilon_x(\mathbf{r})$ is determined by $\gamma(\mathbf{r}, \mathbf{r}')$, it follows first that the Slater potential $2\epsilon_x(\mathbf{r})/\rho(\mathbf{r})$ in equation (6) is determined already by the frontier orbitals plus the energy separation $\epsilon_{\text{H}} - \epsilon_{\text{L}}$. But since $V_x = \delta E_x[\rho]/\delta\rho(\mathbf{r})$ and E_x is determined solely by $\gamma(\mathbf{r}, \mathbf{r}')$, $V_x = V_x[\psi_{\text{HOMO}}, \psi_{\text{LUMO}}, \epsilon_{\text{H}} - \epsilon_{\text{L}}]$. It seems an interesting direction for the future to use available numerical results (say for positive atomic ions from the OEP method) to see if any features of $V_x(\mathbf{r})$ can be characterized by frontier orbitals plus the HOMO–LUMO energy gap. As to the 'closure' approximation to $P(\mathbf{r})$ entering the integral equation (13), the above argument motivates the proposal

$$P(\mathbf{r}) \propto \psi_{\text{HOMO}}(\mathbf{r})\psi_{\text{LUMO}}(\mathbf{r}), \tag{21}$$

where the orthogonality of the HOMO and LUMO orbitals immediately imposes the sum rule (14). The proportionality constant in the one-parameter approximation (21) could then be determined by imposing the Levy–Perdew virial relation.

5. Exact sum rules for the force $-\partial V_{xc}(\mathbf{r})/\partial\mathbf{r}$; i.e. effect of the correlation potential $V_c(\mathbf{r})$

Following the early work of March and Young [28], who established a differential form of the virial theorem for independent fermions moving in a one-dimensional potential energy $V(x)$, Holas and March [9] generalized their result to three dimensions and to interacting fermions. They thereby obtained a formally exact result for the force $\mathbf{F}_{xc}(\mathbf{r})$ defined from the exchange-correlation potential $V_{xc}(\mathbf{r})$ of density-functional theory [1] by

$$\mathbf{F}_{xc}(\mathbf{r}) = -\frac{\partial V_{xc}(\mathbf{r})}{\partial \mathbf{r}}. \tag{22}$$

The central result of the Holas–March study [9] can be written (their equation (2.28)) as

$$\rho(\mathbf{r})\mathbf{F}_{xc}(\mathbf{r}) = \{\mathbf{z}_s - \mathbf{z}\} + \int \mathrm{d}\mathbf{r}' \nabla u(\mathbf{r}, \mathbf{r}')[\rho(\mathbf{r})\rho(\mathbf{r}') - 2n_2(\mathbf{r}, \mathbf{r}')], \tag{23}$$

where $\rho(\mathbf{r})$ is the exact ground-state density, the vector field $\mathbf{z}$ is related to the kinetic-energy density tensor by equation (2.17) of [9] and the pair correlation function $n_2(\mathbf{r}, \mathbf{r}')$ is the diagonal element of the second-order density matrix [1]. In turn, $\mathbf{z}$ is

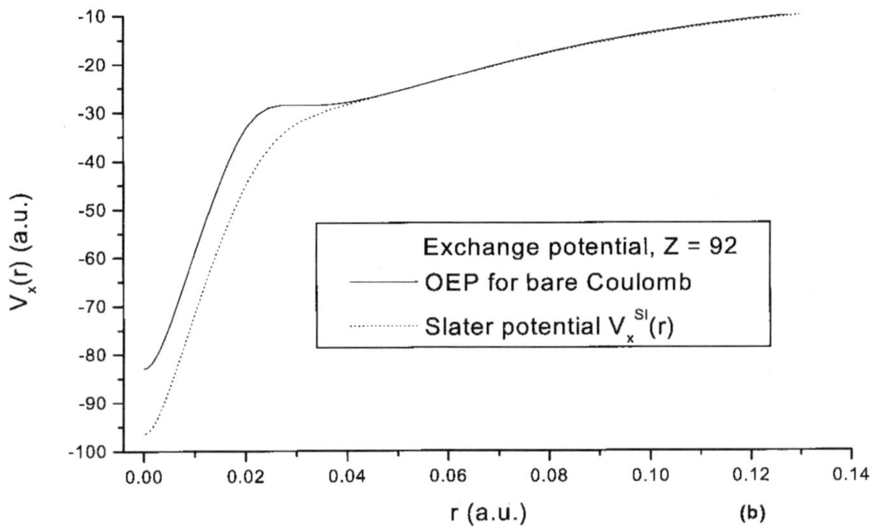

Figure 3. OEP results for the exchange potential with the bare Coulomb potential, compared with $V_x^{SI}(\mathbf{r})$, again for $Z = 92$. [Redrawn from [24].]

determined by the near-diagonal behaviour of the first-order density matrix $\gamma(\mathbf{r}, \mathbf{r}')$ while $\mathbf{z}_s$ is its single-particle counterpart, obtained by replacing the exact matrix γ satisfying the inequality $\gamma^2 < \gamma$ by the idempotent matrix $\gamma_s(\mathbf{r}, \mathbf{r}')$ obtained from a single Slater determinant of Slater–Kohn–Sham (SKS) [10, 12] orbitals yielding the exact density $\rho(\mathbf{r})$ as the sum of the squares of the SKS orbitals over occupied states. Finally, in equation (23) $u(\mathbf{r}, \mathbf{r}')$ represents the interparticle interaction, which in atoms and molecules is given (in au) by

$$u(\mathbf{r}, \mathbf{r}') = \frac{1}{|\mathbf{r} - \mathbf{r}'|}. \tag{24}$$

We now proceed to integrate equation (23) over the whole of space to obtain readily the result

$$\mathbf{I}_{xc} = \int \rho(\mathbf{r})\mathbf{F}_{xc}(\mathbf{r})\,d\mathbf{r} = \int (\mathbf{z}_s - \mathbf{z})\,d\mathbf{r} + \int\int d\mathbf{r}\,d\mathbf{r}'[\nabla u(\mathbf{r}, \mathbf{r}')]\rho(\mathbf{r})\rho(\mathbf{r}') - 2\int\int d\mathbf{r}\,d\mathbf{r}'[\nabla u(\mathbf{r}, \mathbf{r}')]n_2(\mathbf{r}, \mathbf{r}'). \tag{25}$$

The formally exact sum rule (25) will first be illustrated by noting, with Holas and March [9], that the zero-order approximation to the above theory of $\mathbf{F}_{xc}(\mathbf{r})$ is reached by assuming: (i) that $\mathbf{z} = \mathbf{z}_s$ and (ii) that $n_2(\mathbf{r}, \mathbf{r}')$ can be replaced by its form for a single Slater determinant, namely

$$n_2^s(\mathbf{r}, \mathbf{r}') = \rho(\mathbf{r})\rho(\mathbf{r}') - \frac{1}{2}[\gamma_s(\mathbf{r}, \mathbf{r}')]^2. \tag{26}$$

$\mathbf{F}_{xc}(\mathbf{r})$ then corresponds to an exchange-only potential which Holas and March [9] prove is the Harbola–Sahni (HS) potential [29] $V_{xc}^{HS}(\mathbf{r})$. This has difficulties of path dependence, which have been resolved in the subsequent investigation of Levy and March [2]. These authors exhibited 'kinetic' corrections to the HS potential which removed the path dependence of the exchange-only potential $V_x(\mathbf{r})$.

We note here that if we define the integral $\mathbf{I}_x^{HS}$ corresponding to the exact result (25) then

$$\mathbf{I}_x^{HS} = \int \rho(\mathbf{r})\mathbf{F}_x^{HS}(\mathbf{r})\,d\mathbf{r} : \mathbf{F}_x^{HS} = -\frac{\partial V_x^{HS}(\mathbf{r})}{\partial \mathbf{r}} \tag{27}$$

and we have

$$\mathbf{I}_x^{HS} = -\frac{4}{e^2}\int_r \nabla\epsilon_x(\mathbf{r})\,d\mathbf{r}, \tag{28}$$

where γ_s is the idempotent first-order density matrix corresponding to a total wave function which has the form of a single Slater determinant. This result shows

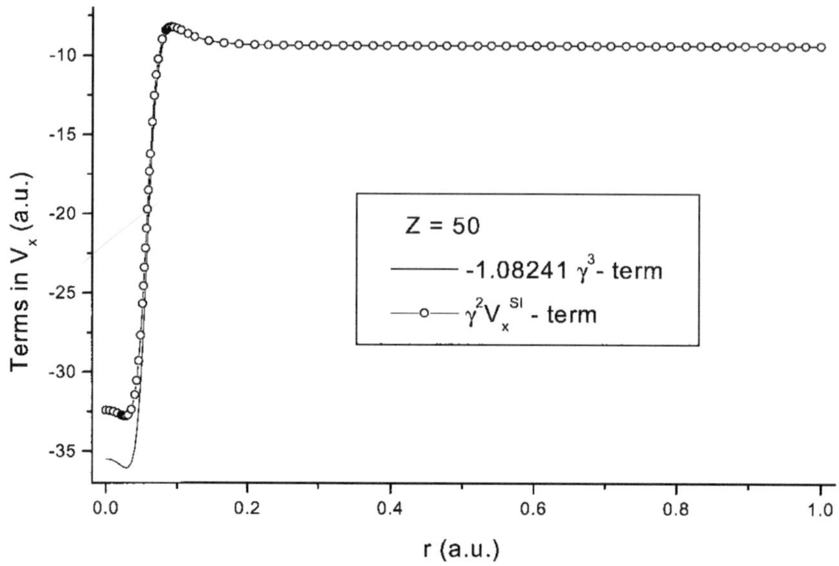

Figure 4. Shows contributions to V_x displayed in equation (13) for four-electron atomic ions for atomic number $Z = 50$. After scaling shown in the figure, remarkable similarity of shape is found between the term of order γ^3 and that involving the product $\gamma^2 V_x$. However, while the γ^3 term is obtained exactly in the limit of large Z, as recorded in equation (17), the $\gamma^2 V_x$ term has been approximated by replacing V_x by the Slater potential V_x^{SI} given, again analytically, in equation (19). This figure makes it clear in this large Z limit for Be-like atomic ions that the correction to the Slater potential is likely to be significant only near the nucleus. But in this region, the nuclear contribution $-Ze^2/r$ is dominant for large Z.

that $\mathbf{I}_x^{HS}$ is intimately related to the Dirac exchange energy

$$E_x^{Dirac} = -\frac{1}{4}\int\int \mathrm{d}\mathbf{r}\,\mathrm{d}\mathbf{r}' u(\mathbf{r},\mathbf{r}')\gamma_s^2(\mathbf{r},\mathbf{r}'). \tag{29}$$

Using the Slater potential (6), the corresponding vector force is

$$\mathbf{F}_x^{SI}(\mathbf{r}) = -\frac{\partial V_x^{SI}(\mathbf{r})}{\partial \mathbf{r}} = \frac{2}{\rho^2(\mathbf{r})}\frac{\partial \rho}{\partial \mathbf{r}}\epsilon_x(\mathbf{r}) - \frac{2}{\rho(\mathbf{r})}\frac{\partial \epsilon_x}{\partial \mathbf{r}} \tag{30}$$

or, with spherical symmetry now assumed,

$$|\mathbf{I}_x^{SI}| = \left|\int \rho(\mathbf{r})\mathbf{F}_x^{SI}(\mathbf{r})\,\mathrm{d}\mathbf{r}\right| = -2\int \frac{\rho'(r)}{\rho(r)}\epsilon_x(r)\,\mathrm{d}\mathbf{r} + 2\int \frac{\partial \epsilon_x}{\partial r}\,\mathrm{d}\mathbf{r}. \tag{31}$$

This integral $\mathbf{I}_x^{SI}$ can be evaluated for 10-electron and 18-electron large-Z ions using the well-known forms of $\gamma(\mathbf{r},\mathbf{r}')$; the results being $|\mathbf{I}_x^{SI}| = 0.968\,735Z^2$ for 10-electron ions and $|\mathbf{I}_x^{SI}| = 1.129\,747Z^2$ for 18-electron ions.

We next go to spin-compensated two-electron systems. For simplicity, let us begin with the ground state of the He atom itself, with normalized spatial wave function $\Psi(\mathbf{r},\mathbf{r}')$. Then $n_2(\mathbf{r},\mathbf{r}')$ in equation (23) is given by

$$n_2(\mathbf{r},\mathbf{r}') = \Psi^2(\mathbf{r},\mathbf{r}')\,, \tag{32}$$

while

$$\rho(\mathbf{r}) = 2\int \Psi^2(\mathbf{r},\mathbf{r}')\,\mathrm{d}\mathbf{r}'. \tag{33}$$

Hence we find from equations (23) and (25) that

$$\begin{aligned}\mathbf{I}_{xc}^{N=2} &= \int \rho(\mathbf{r})\mathbf{F}_{xc}(\mathbf{r})\,\mathrm{d}\mathbf{r} = \int (\mathbf{z}_s - \mathbf{z})\,\mathrm{d}\mathbf{r} \\ &+ \int\int \mathrm{d}\mathbf{r}\,\mathrm{d}\mathbf{r}'[\nabla u(\mathbf{r},\mathbf{r}')]\rho(\mathbf{r})\rho(\mathbf{r}') \\ &- 2\int\int \mathrm{d}\mathbf{r}\,\mathrm{d}\mathbf{r}'[\nabla u(\mathbf{r},\mathbf{r}')]\Psi^2(\mathbf{r},\mathbf{r}').\end{aligned} \tag{34}$$

We next illustrate the exact result (34) for two spin-compensated electrons by referring to the Hookean atom [30–33]. There, the two electrons are attached to the origin by springs (of force constant k) but still interact with Coulombic repulsion, so $u(\mathbf{r},\mathbf{r}') = |\mathbf{r}-\mathbf{r}'|^{-1}$.

From [33], with ϵ_{KS} the Kohn–Sham eigenvalue and $V_H(\mathbf{r})$ denoting the Hartree potential, we can write for the Hookean atom the exchange-correlation potential as

$$V_{xc}(\mathbf{r}) = \epsilon_{KS} - V_{ext}(\mathbf{r}) - 2V_H(\mathbf{r}) - V_{KE}(\mathbf{r}) ; \quad (35)$$

and therefore the corresponding force $-\partial V_{xc}(\mathbf{r})/\partial \mathbf{r}$ is given by

$$\frac{\partial V_{xc}(\mathbf{r})}{\partial \mathbf{r}} = -\frac{\partial V_{ext}(\mathbf{r})}{\partial \mathbf{r}} - 2\frac{\partial V_H(\mathbf{r})}{\partial \mathbf{r}} - \frac{\partial V_{KE}(\mathbf{r})}{\partial \mathbf{r}}. \quad (36)$$

Thus, it follows immediately from equation (36) that

$$\int \rho(\mathbf{r})\frac{\partial V_{xc}}{\partial r}\,d\mathbf{r} = -\int \rho(\mathbf{r})\frac{\partial V_{ext}}{\partial r}\,d\mathbf{r} - 2\int \rho(\mathbf{r})\frac{\partial V_H}{\partial r}\,d\mathbf{r} - \int \rho(\mathbf{r})\frac{\partial V_{KE}}{\partial r}\,d\mathbf{r}, \quad (37)$$

where

$$V_{KE} = -\frac{\delta T_W}{\delta \rho(\mathbf{r})}, \quad (38)$$

with T_W the von Weizsäcker [34] inhomogeneity kinetic energy given by

$$T_W = \frac{1}{8}\int \frac{(\nabla\rho)^2}{\rho}\,d\mathbf{r}. \quad (39)$$

And therefore, returning to equation (34),

$$\begin{aligned}|\mathbf{I}_{xc}^{\text{Hookean }(N=2)}| &= -\int \rho(\mathbf{r})\frac{\partial V_{xc}}{\partial r}\,d\mathbf{r} \\ &= \int (kr)\rho(\mathbf{r})\,d\mathbf{r} + \int \rho(\mathbf{r})\frac{\partial V_{el}}{\partial r}\,d\mathbf{r} \\ &\quad - \int \rho(\mathbf{r})\frac{\partial}{\partial r}\frac{\delta T_W}{\delta\rho(\mathbf{r})}\,d\mathbf{r}, \end{aligned} \quad (40)$$

where

$$V_{electrostatic} \equiv V_{el}(r) = \int \frac{\rho(\mathbf{r}')}{|\mathbf{r}-\mathbf{r}'|}\,d\mathbf{r}'. \quad (41)$$

But now

$$\frac{\partial V_{el}}{\partial r} = \frac{Q(r)}{r^2}, \quad (42)$$

where in this spherically symmetric 'atom', $Q(r)$ denotes the number of electrons enclosed within a sphere of radius r centred on the 'nucleus'. Therefore

$$\int \rho(\mathbf{r})\frac{\partial V_{el}}{\partial r}\,d\mathbf{r} = 4\pi\int_0^\infty \rho(r)Q(r)\,dr. \quad (43)$$

Qian and Sahni [35] give in their equation (A26)

$$\begin{aligned}\int \rho(r)r\,d\mathbf{r} &= 2(2\pi C)^2\left[4\pi + 11(2\pi)^{1/2} + 12\right] \\ &= 3.489\,025\,\text{au}\end{aligned} \quad (44)$$

and so for $k = 1/4$ the 'external potential' contribution to $\mathbf{I}_{xc}^{\text{Hookean }(N=2,k=1/4)}$ in equation (40) is explicitly known.

Since

$$\frac{\delta T_W}{\delta\rho(\mathbf{r})} = \frac{1}{8}\left(\frac{\nabla\rho}{\rho}\right)^2 - \frac{1}{4}\frac{\nabla^2\rho}{\rho} \quad (45)$$

the remaining term in equation (38) can be obtained from the known $\rho(r)$ given in [31] (or [35], equation (A1), with C given immediately below equation (13) of [35]). Hence $\mathbf{I}_{xc}^{\text{Hookean }(N=2)}$ can be obtained exactly, as

$$\begin{aligned}|\mathbf{I}_{xc}^{\text{Hookean }(N=2)}| &= (3.489\,025/4 + 0.600\,801 + 0.561\,209)\,\text{au} \\ &= 2.034\,266\,\text{au},\end{aligned} \quad (46)$$

which constitutes an exact model example to illustrate the sum rule (34).

6. Nuclear cusp conditions on exchange and correlation energy densities

Following the above discussion of integral relations involving $\partial V_{xc}/\partial\mathbf{r}$, we summarize below some additional exact results involving electron correlation in DFT. The discussion below involves both $V_{xc}(\mathbf{r})$ as nuclei in a molecule or cluster are approached, and also exchange and correlation energy densities.

Kato [36] in pioneering studies first characterized non-analyticity of the N-electron wave function as nuclei are approached. Bingel [37] extended this work to multicentre systems but below we shall follow closely the presentation of Davidson [38]. With H denoting the total Hamiltonian, let the N-electron antisymmetric normalized wave function $\Psi(\mathbf{x}_i - \mathbf{x}_N)$ be an eigenfunction of the Schrödinger equation

$$H\Psi = E\Psi, \quad (47)$$

where $\mathbf{x}_i \equiv \{\mathbf{r}_i, s_i\}$ refers to the space and spin coordinates of the ith electron. In treating cusp conditions due to the specified Ath nucleus, let us choose its position to be at $\mathbf{R}_A = 0$. As Ψ is everywhere finite, it is a consequence of equation (47) if we focus on the argument $\mathbf{r}_1$ of Ψ that

$$\lim_{r_1\to 0} r_1 H\Psi = 0. \quad (48)$$

It follows that $\Psi(\mathbf{r}_1, s_1, \mathbf{x}_2 \ldots \mathbf{x}_N)$ is also an eigenfunction, with zero eigenvalue, of the operator [38]

$$b(\mathbf{r}_1) = \lim_{r_1\to 0} r_1 T = \lim_{r_1\to 0} r_1 t(\mathbf{r}_1) - Z_A, \quad (49)$$

with T the kinetic energy operator. For the wave function $\Psi(\mathbf{r}_1, \mathbf{x})$, where $\mathbf{x} \equiv \{s_1, \mathbf{x}_2 \ldots \mathbf{x}_N\}$ expanded in spherical harmonics:

$$\Psi(\mathbf{r}_1, \mathbf{x}) = \sum_{l=0}^{\infty} \sum_{m=-l}^{l} y_{lm}(\Omega_1) r_1^l f_{lm}(r_1, \mathbf{x}) \quad (50)$$

with $f_{lm}(r_1, \mathbf{x})$ an analytic function of r, the eigenequation $b(\mathbf{r}_1)\Psi = 0$ leads to the conditions on the partial wave amplitudes

$$\left.\frac{\partial f_{lm}(r_1, \mathbf{x})}{\partial r_1}\right|_{r_1=0} = -\frac{Z_A}{l+1} f_{lm}(0, \mathbf{x}). \quad (51)$$

These equations (48–51) have been used by March *et al.* [39] to derive nuclear cusp conditions for the exchange and correlation energy densities of DFT.

Before turning to the DFT energy densities $\epsilon_x(\mathbf{r})$ and the corresponding correlation (c) energy density $\epsilon_c(\mathbf{r})$, for DFT applications the classical electrostatic energy of electrons is also of interest, namely

$$E_{es}[\rho] = \int d\mathbf{r}_1 \epsilon_{es}(\mathbf{r}_1), \quad (52)$$

where

$$\epsilon_{es}(\mathbf{r}_1) = \frac{1}{2} \int d\mathbf{r}_2 \frac{e^2}{|\mathbf{r}_1 - \mathbf{r}_2|} \rho(\mathbf{r}_1)\rho(\mathbf{r}_2). \quad (53)$$

March *et al.* [39] then prove that

$$\left.\frac{\partial \overline{\epsilon}_{es}(\mathbf{r}_1)}{\partial r_1}\right|_{r_1=0} = -2Z_A \epsilon_{es}(0), \quad (54)$$

where the bar on $\epsilon_{es}(\mathbf{r}_1)$ denotes an angular average. In the problem for atoms, molecules and solids, the derivation in [39] appeals to the adiabatic connection, developed by Harris and Jones [3]. This connection links the fully interacting system treated above to the SKS non-interacting reference system with the same (i.e. formally exact) electron density via intermediate systems. For this purpose, a Hamiltonian depending on a parameter α such that $0 \le \alpha \le 1$, is introduced:

$$H^\alpha = T + V^\alpha + \alpha U, \quad (55)$$

where $V^\alpha = \sum_{i=1}^N V^\alpha(\mathbf{r}_i)$. For $\alpha = 1$ this Hamiltonian reduces to the original Hamiltonian while for $\alpha = 0$ the electron–electron piece U is switched off, with $V^0(\mathbf{r}) = V_{SKS}(\mathbf{r})$.

To use the above approach to investigate the cusp conditions for the intermediate system, March *et al.* [39] assume (to be checked later) that the properties of $V^\alpha(\mathbf{r})$, whose explicit form remains unknown, allow a finite value for the limit $\gamma^\alpha = \lim_{r_1\to 0} r_1 V^\alpha(\mathbf{r}_1)$. Then one finds

$$\left.\frac{\partial}{\partial r}\overline{\rho}^\alpha(r)\right|_{r=0} = 2\gamma^\alpha \rho^\alpha(0). \quad (56)$$

But the one-body potential $V^\alpha(\mathbf{r})$ is defined by requiring that the electron density $\rho^\alpha(\mathbf{r})$ of the intermediate system is independent of α. Hence one can conclude that $\gamma^\alpha = -Z_A$, which justifies *a posteriori* the assumption that $\lim_{r_1\to 0} r_1 V^\alpha(\mathbf{r}_1)$ is finite.

March *et al.* [39] show that for the exchange-correlation potential $V_{xc}(\mathbf{r})$,

$$\lim_{r_1\to 0} r_1 V_{xc}(\mathbf{r}_1) = 0. \quad (57)$$

Without essentially new arguments beyond those presented above, the result for the exchange energy density ϵ_x takes the form

$$\frac{\partial}{\partial r}\overline{\epsilon_x}(\mathbf{r}_1)|_{r_1=0} = -2Z_A \epsilon_x(0), \quad (58)$$

the angular average on the left-hand side not being required for closed shell atomic ions. The final exact result of [39] is an analogous nuclear cusp condition for correlation energy density $\epsilon_c(\mathbf{r})$, namely

$$\frac{\partial}{\partial r}\overline{\epsilon_c}(\mathbf{r}_1)|_{r_1=0} = -2Z_A \epsilon_c(0). \quad (59)$$

These exact results (58) and (59) should evidently be built into approximate DFT theory in the future.

7. Summary and future directions

The theory of the exchange potential $V_x(\mathbf{r})$ has been focused here on the formally exact integral equation (13) with the 'closure' function $P(\mathbf{r})$ to be constructed to satisfy the sum rule (14). Two approximate closures have been considered, for the example of the Be atom in equation (16) (see also figure 1) and for a general, non-spherical electron density distribution as in molecules and clusters, in equation (21), where the HOMO and LUMO orbitals are directly invoked. In section 3, some analytic progress has been shown to be possible in evaluating terms on the right-hand side of equation (13) in the large Z limit of the $1/Z$ expansion of atomic theory. One interesting consequence is then that whereas $V_x^{Sl}(\mathbf{r}) \to -1/r$ as $r \to \infty$ (see equation (19)) the second and third terms on the right-hand side of equation (13) tend to finite (and almost compensating!) values.

As to future directions, it will be very important to transcend the approximations (16) and (21) proposed here for the 'closure' function $P(\mathbf{r})$. It is possible, in fact, to express $P(\mathbf{r})$ exactly, though not as yet compactly, by invoking (a) the single-particle Green function and (b) the linear response function [40]. But these involve all Slater–Kohn–Sham orbitals generated by the DFT one-body potential $V(\mathbf{r})$; not simply the occupied orbitals entering the idempotent Dirac density matrix $\gamma(\mathbf{r},\mathbf{r}')$ which is the primary focus of the present study. If one had to invoke only HOMO and LUMO orbitals in constructing both $\gamma(\mathbf{r},\mathbf{r}')$ and $P(\mathbf{r})$, as set out in section 4, that would have obvious advantages, both computational and chemical. But it has only been proved that this is exact (in principle) for $\gamma(\mathbf{r},\mathbf{r}')$, as set out in section 4 and [25], whereas equation (21) has the status presently of an approximate proposal.

Finally, the most difficult area of DFT remains, of course, the correlation contribution to the exchange-correlation potential $V_{xc}(\mathbf{r})$. Some exact constraints on the force $-\partial V_x(\mathbf{r})/\partial \mathbf{r}$ corresponding to this potential have therefore been set out in sections 5 and 6. Also, given the quantum Monte Carlo ground-state density for spherically symmetric atoms like Be and Ne, the phases referred to above equation (5) in section 2 can be calculated and $-\partial V_x(\mathbf{r})/\partial \mathbf{r}$ thereby obtained without approximation (see, e.g. [41] for the Be atom). But for molecules and clusters, such an approach still presents formidable computational problems.

We wish to thank Professor P.W. Ayers for making available to us [26] before publication. IAH acknowledges support from the IWT–Flemish region under grant no. IWT-161.

References

[1] R.G. Parr, W. Yang, *Density Functional Theory of Atoms and Molecules*, Oxford University Press, New York (1989).
[2] M. Levy, N.H. March, *Phys. Rev.* A, **55,** 1885 (1997).
[3] J. Harris, R.O. Jones, *J. Phys.* F, **4,** 1170 (1974).
[4] M. Levy, N.H. March, N.C. Handy, *J. Chem. Phys.*, **104,** 1989 (1996).
[5] I.A. Howard, N.H. March, *J. Chem. Phys.*, **119,** 5789 (2003).
[6] O.V. Gritsenko, E.J. Baerends, *Phys. Rev.* A, **64,** 042506-1 (2001).
[7] M. Grüning, O.V. Gritsenko, E.J. Baerends, *J. Chem. Phys.*, **116,** 6435 (2002).
[8] F. Della Sala, A. Görling, *J. Chem. Phys.*, **115,** 5718 (2001).
[9] A. Holas, N.H. March, *Phys. Rev.* A, **51,** 2040 (1995).
[10] J.C. Slater, *Phys. Rev.*, **81,** 385 (1951).
[11] See, for example, A. Nagy, I.G. Csizmadia (Eds), *J. Molec. Struct.* (*Theochem*), 501–502 (2000).
[12] W. Kohn, L.J. Sham, *Phys. Rev.*, **140,** A1133 (1965).
[13] L. Kleinman, *Phys. Rev.* B, **49,** 14197 (1994).
[14] P.A.M. Dirac, *Proc. Camb. Phil. Soc.*, **26,** 376 (1930).
[15] A. Holas, N.H. March, *Phys. Rev.* B, **55,** 1295 (1997).
[16] K. Burke, F.G. Cruz, K.-C. Lam, *J. Chem. Phys.*, **109,** 8161 (1998).
[17] N.H. March, *Phys. Rev.* A, **36,** 5077 (1987).
[18] D. Layzer, *Ann. Phys.* (New York), **8,** 271 (1959).
[19] M. Cohen, *Adv. At. Molec. Phys.*, **25,** 195 (1988).
[20] N.H. March, R.J. White, *J. Phys.* B, **5,** 466 (1972).
[21] N.H. March, R. Santamaria, *Phys. Rev.* A, **38,** 5002 (1988).
[22] I.A. Howard, N.H. March, P. Senet, V.E. Van Doren, *Phys. Rev.* A, **62,** 062512-1 (2000).
[23] J.D. Talman, W.F. Shadwick, *Phys. Rev.* A, **14,** 36 (1976).
[24] I.A. Howard, N.H. March, J.D. Talman, *Phys. Rev.* A, **68,** 044502-1 (2003).
[25] I.A. Howard, N.H. March, P.W. Ayers, *Chem. Phys. Lett.*, **385,** 231 (2004).
[26] P.W. Ayers, *Theor. Chem. Acc.*, **110,** 267 (2003).
[27] K.A. Dawson, N.H. March, *Phys. Lett.* A, **106,** 161 (1984).
[28] N.H. March, W.H. Young, *Nucl. Phys.*, **12,** 237 (1959).
[29] M.K. Harbola, V. Sahni, *Phys. Rev. Lett.*, **62,** 489 (1989).
[30] N.R. Kestner, O. Sinanoḡlu, *Phys. Rev.*, **128,** 2687 (1962).
[31] S. Kais, D.R. Herschbach, R.D. Levine, *J. Chem. Phys.*, **91,** 7791 (1989).
[32] S. Kais, S.M. Sung, D.R. Herschbach, *Int. J. Quantum Chem.*, **49,** 657 (1994).
[33] S. Kais, D.R. Herschbach, N.C. Handy, C.W. Murray, G.J. Laming, *J. Chem. Phys.*, **99,** 417 (1993).
[34] C.F. von Weizsäcker, *Zeits. Phys.*, **96,** 431 (1935).
[35] Z. Qian, V. Sahni, *Phys. Rev.* A, **57,** 2527 (1998).
[36] T. Kato, *Commun. Pure Appl. Math.*, **10,** 151 (1957).
[37] W.A. Bingel, *Z. Naturforsch.* A, **18,** 1249 (1963).
[38] E.R. Davidson, *Reduced Density Matrices in Quantum Chemistry*, Academic Press, New York (1976).
[39] N.H. March, I.A. Howard, A. Holas, P. Senet, V.E. Van Doren, *Phys. Rev.* A, **63,** 012520-1 (2001).
[40] I.A. Howard, N.H. March, *Phys. Rev.* A, **69,** 064101-1 (2004).
[41] See, for example, N.H. March, P. Geerlings, K.D. Sen, *Phys. Lett.* A, **324,** 42 (2004).

Available online at www.sciencedirect.com
SCIENCE DIRECT
ELSEVIER
Physics Letters A 348 (2006) 374–378
PHYSICS LETTERS A
www.elsevier.com/locate/pla

Formally exact integral equation theory of the exchange-only potential in density functional theory: Refined closure approximation

N.H. March [a], Á. Nagy [b,c,*]

[a] *Oxford University, Oxford, England, UK*
[b] *Department of Theoretical Physics, University of Debrecen, H-4010 Debrecen, Hungary*
[c] *Atomic and Molecular Physics Research Group, Hungarian Academy of Sciences, H-4010 Debrecen, Hungary*

Received 9 August 2005; accepted 15 August 2005

Available online 22 August 2005

Communicated by V.M. Agranovich

Abstract

A formally exact integral equation theory for the exchange-only potential $V_x(\mathbf{r})$ in density functional theory was recently set up by Howard and March [I.A. Howard, N.H. March, J. Chem. Phys. 119 (2003) 5789]. It involved a 'closure' function $P(\mathbf{r})$ satisfying the exact sum rule $\int P(\mathbf{r})\,d\mathbf{r} = 0$. The simplest choice $P(\mathbf{r}) = 0$ recovers then the approximation proposed by Della Sala and Görling [F. Della Sala, A. Görling, J. Chem. Phys. 115 (2001) 5718] and by Gritsenko and Baerends [O.V. Gritsenko, E.J. Baerends, Phys. Rev. A 64 (2001) 042506].

Here, refined choices of $P(\mathbf{r})$ are proposed, the most direct being based on the KLI (Krieger–Li–Iafrate) approximation. A further choice given some attention is where $P(\mathbf{r})$ involves frontier orbital properties. In particular, the introduction of the LUMO (lowest unoccupied molecular) orbital, along with the energy separation between HOMO (highest occupied molecular orbital) and LUMO levels, should prove a significant step beyond current approximations to the optimized potential method, all of which involve only single-particle occupied orbitals.

1. Introduction

It is now widely recognized that in applying density functional theory to molecules and clusters [1], the exchange-only potential is the major component of the exchange-correlation potential $V_{xc}(\mathbf{r})$, which is central

* Corresponding author.
E-mail address: anagy@madget.atomki.hu (Á. Nagy).

0375-9601/$ – see front matter
doi:10.1016/j.physleta.2005.08.035

to all such applications. Therefore, a great deal of attention has focussed on obtaining accurate approximations to the exchange-only component $V_x(\mathbf{r})$ of the, as yet unknown, $V_x(\mathbf{r})$.

In a very recent contribution, Howard and March [2] have set up formally exact integral equation theory for $V_x(\mathbf{r})$. Their treatment involved the 'closure' function they denoted by $P(\mathbf{r})$ (see Eq. (8) below) satisfying the exact sum rule $\int P(\mathbf{r})\,d\mathbf{r} = 0$. The simplest choice $P(\mathbf{r}) = 0$ recovers then the approximate method of calculating $V_x(\mathbf{r})$ proposed independently by Della Sala and Görling [3] and Baerends and coworkers [4,5].

The purpose of the present study is to discuss refined choices of the closure function $P(\mathbf{r})$. The most direct approach presented here is based on the OPM [12] and KLI [6,7] approximations to $V_x(\mathbf{r})$. This is set out in some detail in Section 3, following a brief summary of the Howard–March scheme in Section 2. Recently, attention has been drawn to the interest chemically in relating $V_{xc}(\mathbf{r})$ to the frontier orbital properties [8–11] and therefore Section 4, after making a brief summary, makes some proposals, particularly focussed on the introduction of the LUMO orbital, and the HOMO–LUMO energy separation $\varepsilon_{\mathrm{HL}}$, into approximation to the closure function $P(\mathbf{r})$.

2. Summary of formally exact integral equation theory of exchange-only potential

In [2], Howard and March first construct the difference between integrals $\int \varrho(\mathbf{r})V_x(\mathbf{r})\,d\mathbf{r}$ and $\int \varrho(\mathbf{r})V_x^{\mathrm{Sl}}(\mathbf{r})\,d\mathbf{r}$. Here $\varrho(\mathbf{r})$ is the ground-state electron density of the molecule or cluster under consideration, $V_x(\mathbf{r})$ is the sought-after exchange-only potential while $V_x^{\mathrm{Sl}}(\mathbf{r})$ is the Slater approximation [12] to $V_x(\mathbf{r})$. The Slater form may be written in terms of the exchange energy density $\epsilon_x(\mathbf{r})$ which will be defined precisely below in terms of the idempotent Dirac density matrix [13]

$$\gamma(\mathbf{r},\mathbf{r}') = \sum_{\text{occupied}\, i} \phi_i(\mathbf{r})\phi_i^*(\mathbf{r}'), \tag{1}$$

which $\phi_i(\mathbf{r})$ are the Kohn–Sham (KS) orbitals at the exchange-only level. Then we have

$$V_x^{\mathrm{Sl}}(\mathbf{r}) = 2\frac{\epsilon_x(\mathbf{r})}{\varrho(\mathbf{r})}, \tag{2}$$

and hence the integral involving $V_x^{\mathrm{Sl}}(\mathbf{r})$ introduced above is such that the total exchange energy E_x is given by

$$E_x = \int \epsilon_x(\mathbf{r})\,d\mathbf{r} = \frac{1}{2}\int \varrho(\mathbf{r})V_x^{\mathrm{Sl}}(\mathbf{r}). \tag{3}$$

Indeed, Eq. (3) was a pillar supporting Slater's original construction of $V_x^{\mathrm{Sl}}(\mathbf{r})$ as an approximation to the exchange-only potential $V_x(\mathbf{r})$. To complete the above definitions, we write $\epsilon_x(\mathbf{r})$ in terms of the Dirac matrix $\gamma(\mathbf{r},\mathbf{r}')$ as [13]

$$\epsilon_x(\mathbf{r}) = -\frac{1}{4}\int \frac{\gamma^2(\mathbf{r},\mathbf{r}')}{|\mathbf{r}-\mathbf{r}'|}\,d\mathbf{r}'. \tag{4}$$

Then the difference between $\int \varrho(\mathbf{r})V_x(\mathbf{r})\,d\mathbf{r}$ and $\int \varrho(\mathbf{r})V_x^{\mathrm{Sl}}(\mathbf{r})\,d\mathbf{r}$ is immediately given by

$$\int \varrho(\mathbf{r})V_x(\mathbf{r})\,d\mathbf{r} - \int \varrho(\mathbf{r})V_x^{\mathrm{Sl}}(\mathbf{r})\,d\mathbf{r} = \frac{1}{2}\int \frac{\gamma^2(\mathbf{r},\mathbf{r}')}{|\mathbf{r}-\mathbf{r}'|}V_x(\mathbf{r})\,d\mathbf{r}\,d\mathbf{r}' - 2E_x. \tag{5}$$

But from Eqs. (3) and (4) and the idempotency condition for spin-compensated levels considered throughout this Letter:

$$\frac{1}{2}\gamma(\mathbf{r},\mathbf{r}') = \int \frac{\gamma(\mathbf{r},\mathbf{r}'')\gamma(\mathbf{r}'',\mathbf{r}')}{4}\,d\mathbf{r}'' \tag{6}$$

one reaches the exact integral result, already set out in [2]:

$$\int \varrho(\mathbf{r}) V_x(\mathbf{r})\, d\mathbf{r} - \int \varrho(\mathbf{r}) V_x^{\mathrm{SI}}(\mathbf{r})\, d\mathbf{r} = \frac{1}{2} \int \frac{\gamma^2(\mathbf{r},\mathbf{r}')}{|\mathbf{r}-\mathbf{r}'|} V_x(\mathbf{r})\, d\mathbf{r}\, d\mathbf{r}' + \frac{1}{4} \int \frac{\gamma(\mathbf{r},\mathbf{r}')\gamma(\mathbf{r}',\mathbf{r}'')\gamma(\mathbf{r}'',\mathbf{r})}{|\mathbf{r}-\mathbf{r}'|} d\mathbf{r}\, d\mathbf{r}'\, d\mathbf{r}''. \tag{7}$$

Howard and March now write the differential form of the integral identity (7), and after division by $\varrho(\mathbf{r})$ they were led to the formally exact integral equation for the exchange-only potential

$$V_x(\mathbf{r}) = V_x^{\mathrm{SI}} + \frac{1}{2\varrho(\mathbf{r})} \int \gamma^2(\mathbf{r},\mathbf{r}') V_x(\mathbf{r}')\, d\mathbf{r}' + \frac{1}{4\varrho(\mathbf{r})} \int \frac{\gamma(\mathbf{r},\mathbf{r}')\gamma(\mathbf{r}',\mathbf{r}'')\gamma(\mathbf{r}'',\mathbf{r})}{|\mathbf{r}-\mathbf{r}'|} d\mathbf{r}'\, d\mathbf{r}'' + \frac{P(\mathbf{r})}{\varrho(\mathbf{r})}, \tag{8}$$

where $P(\mathbf{r})$ is the so-called 'closure' function, which must, to regain the exact integral identity (7), satisfy

$$\int P(\mathbf{r})\, d\mathbf{r} = 0, \tag{9}$$

as already referred to above.

Of course, the simplest choice of $P(\mathbf{r})$ to satisfy (9) is to put it equal to zero. Then, one regains the method of Della Sala and Görling [3], which makes it quite clear that $P^{\mathrm{DSG}}(\mathbf{r}) = 0$, rests on the (somewhat drastic, but numerically useful) assumption that the Kohn–Sham (KS) determinant of occupied orbitals $\phi_i(\mathbf{r})$ equals the corresponding Hartree–Fock determinant.

Although the studies of Baerends and coworkers [4,5] seem rather different in philosophy, then (also somewhat drastic) denominator approximation leads back to (8) for their approximate $V_x(\mathbf{r})$, after again setting $P(\mathbf{r}) = 0$.

With this as background to the formally exact integral equation theory of $V_x(\mathbf{r})$, let us turn immediately to the main purpose of the present study, namely, to propose refined approximation to the closure function $P(\mathbf{r})$ entering the basic equation (8).

3. Closure function $P(\mathbf{r})$ in the OPM and KLI approximations to the exchange potential

The optimized exchange-only potential can be expressed in the compact form [6,12]

$$V_x^{\mathrm{OPM}}(\mathbf{r}) = V_x^{\mathrm{SI}} + \frac{1}{\varrho(\mathbf{r})} \sum_{\text{occupied}\, i} |\phi_i(\mathbf{r})|^2 \big(\bar{V}_{xi}^{\mathrm{OPM}} - \bar{u}_{xi}\big) - \frac{1}{\varrho(\mathbf{r})} \sum_{\text{occupied}\, i} \nabla\big(\psi_i^* \nabla \phi_i\big), \tag{10}$$

where the function

$$\psi_i^*(\mathbf{r}) = \int d\mathbf{r}'\, \phi_i^*(\mathbf{r}') \big(\bar{V}_{xi}^{\mathrm{OPM}} - \bar{u}_{xi}\big) G_i(\mathbf{r}',\mathbf{r}) \tag{11}$$

satisfies the OPM integral equation

$$\sum_{\text{occupied}\, i} \psi_i^*(\mathbf{r}) \phi_i(\mathbf{r}) + \text{c.c.} = 0. \tag{12}$$

$G_i(\mathbf{r},\mathbf{r}')$ is the Green function that can be expressed with the one-particle properties, i.e., the Kohn–Sham orbitals $\phi_i(\mathbf{r})$ and eigenvalues ε_i as

$$G_i(\mathbf{r},\mathbf{r}') = \sum_{\substack{k=1 \\ k\neq i}}^{\infty} \frac{\phi_k(\mathbf{r}')\phi_k^*(\mathbf{r})}{\varepsilon_i - \varepsilon_k}. \tag{13}$$

Note that this is an infinite sum, not only for the occupied orbitals. $u_{xi}(\mathbf{r})$ is defined via the functional derivative of the exchange energy E_x with respect to the orbital $\phi_i(\mathbf{r})$

$$u_{xi}(\mathbf{r}) = \frac{1}{\phi_i^*(\mathbf{r})} \frac{\delta E_{xc}}{\delta \phi_i(\mathbf{r})}. \tag{14}$$

The averages $\bar{u}_{xi}$ and $\bar{V}_{xi}$ in Eq. (11) denote the following integrals:

$$\bar{u}_{xi} = \int \phi_i^*(\mathbf{r}) u_{xi}(\mathbf{r}) \phi_i(\mathbf{r})\, d\mathbf{r}, \tag{15}$$

$$\bar{V}_{xi} = \int \phi_i^*(\mathbf{r}) V_x(\mathbf{r}) \phi_i(\mathbf{r})\, d\mathbf{r}. \tag{16}$$

Clearly we can again write, using Eq. (5) that

$$V_x^{\mathrm{OPM}}(\mathbf{r}) = V_x^{\mathrm{SI}}(\mathbf{r}) + \frac{1}{2\varrho(\mathbf{r})} \int \gamma^2(\mathbf{r},\mathbf{r}') V_x^{\mathrm{OPM}}(\mathbf{r}')\, d\mathbf{r}' - \frac{2\epsilon_x(\mathbf{r})}{\varrho(\mathbf{r})} + \frac{P(\mathbf{r})}{\varrho(\mathbf{r})}. \tag{17}$$

Hence we reach the conclusion that the closure approximation of the OPM method is such that

$$P^{\mathrm{OPM}}(\mathbf{r}) = 2\epsilon_x(\mathbf{r}) - \frac{1}{2} \int \gamma^2(\mathbf{r},\mathbf{r}') V_x^{\mathrm{OPM}}(\mathbf{r}')\, d\mathbf{r}' + \sum_{\text{occupied } i} |\phi_i(\mathbf{r})|^2 \big(\bar{V}_{xi}^{\mathrm{OPM}} - \bar{u}_{xi}\big) - \sum_{\text{occupied } i} \nabla\big(\psi_i^* \nabla \phi_i\big). \tag{18}$$

To check whether Eq. (18) satisfies the exact sum rule (9), we note that

$$\int P^{\mathrm{OPM}}(\mathbf{r})\, d\mathbf{r} = 2E_x - \int \varrho(\mathbf{r}') V_x^{\mathrm{OPM}}(\mathbf{r}')\, d\mathbf{r}' + \sum_{\text{occupied } i} |\phi_i(\mathbf{r})|^2 \big(\bar{V}_{xi}^{\mathrm{OPM}} - \bar{u}_{xi}\big) \int |\phi_i(\mathbf{r})|^2\, d\mathbf{r} - \sum_{\text{occupied } i} \int \nabla\big(\psi_i^* \nabla \phi_i\big)\, d\mathbf{r}. \tag{19}$$

Finally, we substitute for V_x^{OPM} in Eq. (19) from Eq. (10) and making use of Eq. (17) we find that $\int P^{\mathrm{OPM}}(\mathbf{r})\, d\mathbf{r} = 0$, confirming that the closure in Eq. (18) exactly satisfies the required sum rule (9).

The KLI (Krieger–Li–Iafrate) approximation [6] is obtained by neglecting the last term in the OPM expression (10) for the exchange potential:

$$V_x^{\mathrm{KLI}}(\mathbf{r}) = V_x^{\mathrm{SI}} + \frac{1}{\varrho(\mathbf{r})} \sum_{\text{occupied } i} |\phi_i(\mathbf{r})|^2 \big(\bar{V}_{xi}^{\mathrm{KLI}} - \bar{u}_{xi}\big). \tag{20}$$

Then we are immediately led to the closure approximation of the KLI method

$$P^{\mathrm{KLI}}(\mathbf{r}) = 2\epsilon_x(\mathbf{r}) - \frac{1}{2} \int \gamma^2(\mathbf{r},\mathbf{r}') V_x^{\mathrm{KLI}}(\mathbf{r}')\, d\mathbf{r}' + \sum_{\text{occupied } i} |\phi_i(\mathbf{r})|^2 \big(\bar{V}_{xi}^{\mathrm{KLI}} - \bar{u}_{xi}\big). \tag{21}$$

Applying Eq. (21), as an immediate example, to the Be atom with configuration $(1\mathrm{s})^2(2\mathrm{s})^2$, the requirement that $\varrho(\mathbf{r}) = \varrho_{\mathrm{HF}}(\mathbf{r})$ means that $\varepsilon_{2\mathrm{s}}^{\mathrm{KS}} = \varepsilon_{2\mathrm{s}}^{\mathrm{HF}}$ and hence

$$P^{\mathrm{KLI}}(\mathbf{r}) = 2\epsilon_x(\mathbf{r}) - \frac{1}{2} \int \gamma^2(\mathbf{r},\mathbf{r}') V_x^{\mathrm{KLI}}(\mathbf{r}')\, d\mathbf{r}' + |\phi_{1\mathrm{s}}(\mathbf{r})|^2 \big(\bar{V}_{x1\mathrm{s}}^{\mathrm{KLI}} - \bar{u}_{x1\mathrm{s}}\big). \tag{22}$$

Of course, it has not been proved, to date, that for this example of the Be atom the choice (22) for the closure function $P(\mathbf{r})$ is a better approximation than the Della Sala and Görling result that $P^{\mathrm{DSG}}(\mathbf{r}) = 0$, though we anticipate that thus will prove to be so.

4. Summary and future directions

The formally exact integral equation theory for the exchange-only potential $V_x(\mathbf{r})$ embodied in Eq. (8) presently requires an approximation to the closure function $P(\mathbf{r})$.

In the present Letter, we have attempted the transcend the Della Sala and Görling approximation $P(\mathbf{r}) = 0$, which is evidently a consequence of their starting assumption that the KS and HF determinants are equal, by appeal to the OPM and KLI approximations. This leads to the closure functions P^{OPM} and P^{KLI} set out in Eqs. (18) and (21), respectively. The closure function P^{KLI} has the merit that it involves only the occupied KS orbitals $\phi_i(\mathbf{r})$ already needed to determine the ground-state density $\varrho(\mathbf{r})$.

However, it is well known that the KLI method is an approximation to the OPM method, going back to Horton and Sharp [12]. This method, as is especially clear from the work of Talman and Shadwick [12], involves in its original form, the complete set of KS orbitals that is all unoccupied orbitals as well as those used to determine $\varrho(\mathbf{r})$. It is this circumstance which prompts us to suggest that it will be worthwhile in the future to explore whether the inclusion of a single KS unoccupied wave function, the LUMO orbital, together with the energy separation $\varepsilon_{\mathrm{HL}}$ between the HOMO and LUMO states, can lead to the substantial refinement of the closure function $P(\mathbf{r})$ beyond the KLI approximation of Eq. (21).

Acknowledgements

Thanks are due especially to Dr. C. Amovilli, Professors P. Ayers and I.A. Howard for much valuable help with the general area embraced by the present study. N.H.M.'s contribution to this Letter was brought to fruition during a visit to the University of Debrecen in 2005. He wishes to thank Professor Á. Nagy for generous hospitality. Grant OTKA No. T029469 is gratefully acknowledged.

References

[1] R.G. Parr, W. Yang, Density Functional Theory of Atoms and Molecules, Oxford Univ. Press, New York, 1989.
[2] I.A. Howard, N.H. March, J. Chem. Phys. 119 (2003) 5789.
[3] F. Della Sala, A. Görling, J. Chem. Phys. 115 (2001) 5718.
[4] O.V. Gritsenko, E.J. Baerends, Phys. Rev. A 64 (2001) 042506.
[5] M. Grüning, O.V. Gritsenko, E.J. Baerends, J. Chem. Phys. 116 (2002) 6435.
[6] J.B. Krieger, Y. Li, G.J. Iafrate, Phys. Lett. A 146 (1990) 256;
J.B. Krieger, Y. Li, G.J. Iafrate, Phys. Rev. A 45 (1992) 101.
[7] Á. Nagy, Phys. Rev. A 55 (1997) 3465.
[8] P.W. Ayers, Theor. Chem. Accounts 110 (2003) 267.
[9] K.A. Dawson, N.H. March, Phys. Lett. A 106 (1984) 161.
[10] I.A. Howard, N.H. March, P.W. Ayers, Chem. Phys. Lett. 385 (2004) 231.
[11] N.H. March, I.A. Howard, Chem. Phys. Lett. 402 (2005) 1.
[12] R.T. Sharp, G.K. Horton, Phys. Rev. A 30 (1953) 317;
J.D. Talman, W.F. Shadwick, Phys. Rev. A 14 (1976) 36.
[13] P.A.M. Dirac, Proc. Cambridge Philos. Soc. 26 (1930) 376.

Electronic wave functions round a vacancy in a metal

By N. H. March and A. M. Murray
Department of Physics, The University, Sheffield

(*Communicated by C. A. Coulson, F.R.S.—Received* 14 *December* 1959—
Revised 29 *January* 1960)

A calculation of the wave functions round a vacancy in a metal has been carried out for a simple model. This corresponds to the case of a finite number of particles in a free-electron metal, with the boundary condition that the wave functions vanish on the surface of the spherical metal. Calculations are reported for particle numbers of approximately 2×10^2, 2×10^3 and 10^6. The density fluctuations in the unperturbed metal are found to be serious for the first two cases, but by consideration of the difference between the electronic density distributions in the perturbed and the unperturbed cases, extrapolation from the case of a relatively small number of particles to obtain results valid for very large numbers is shown to be feasible in the region surrounding the vacancy. The possibility of obtaining an accurate density difference by consideration only of states of low orbital angular momentum is exploited. Comparison is made between the finite metal results, existing Thomas–Fermi calculations, and a self-consistent von Weizsäcker calculation which is also reported here. By examination of the solutions of Poisson's equation, it is shown that the total charge density obtained by summing the squares of the normalized wave functions for the assumed Thomas–Fermi potential in the 10^6 particle case is such as to make this potential self-consistent to a good approximation. Our calculations also reveal long-range oscillations in the total perturbed electron density for finite metals. Although the amplitude of these oscillations decreases with increasing volume, extrapolation to the case of an infinite metal does not seem justified here. However, our calculations should provide an upper bound to the amplitude for the assumed potential, and the oscillations are small. Finally, brief reference is made to applications of the present results.

1. Introduction

The theoretical study of the influence of impurities and imperfections on the properties of metal crystals is now a matter of prime importance. Particularly from the standpoint of interpreting experiments on materials damaged by radiation, accurate theoretical computations on vacancies and interstitials seem very well worth undertaking. The related problem of substitutional impurities in metals is also of considerable importance, and the results impinge on such fields of experimental interest as impurity diffusion in metals, Knight shifts in dilute alloys, resistivity and other electrical properties.

Before the present investigation, almost all treatments of defects which make any pretence to self-consistency have employed some form of Thomas–Fermi approximation. It seemed to us desirable to attempt a more accurate computation, and we chose originally to consider the vacancy problem, in order to explore the errors which must be implicit in the earlier treatments.

Unfortunately, it appears essential, for accurate work, to proceed by almost entirely numerical methods for even the very simplest model. Furthermore, the periodic boundary conditions, so useful in the perfect lattice, seem inconvenient in the perturbed lattice, and this creates other problems. We have therefore explored whether, by starting from a finite metal model, and considering the variation of the wave functions as the metal is expanded and particles introduced to keep a constant mean density, results may be obtained which will be valid for very large numbers of particles.

We shall present in this paper the results of three computations, corresponding to different numbers of particles N (actually 186, 1734 and approximately 10^6) but the same mean density, and the results will be analyzed from the above point of view. We should emphasize that all the work reported here is based on the free-electron approximation and no attempt is made to introduce correlation effects at this stage (see, however, the later remarks in § 12).

2. Model employed and earlier approximations

As a simple model for a vacancy in a monovalent metal, we consider a single negative charge embedded in a free-electron gas. The redistribution of electrons and the corresponding screened potential have been fully investigated in the Thomas–Fermi approximation (Alfred & March 1957), and detailed numerical results are available. The potential thus obtained (see curve 1 of figure 6) is employed in the calculations for finite metals described below.

3. Perturbed wave functions in a finite metal

We have now studied the possibility of obtaining results of significance by using a finite metal, with, at first, a relatively small number of particles. In this way, with the aid of a Pegasus electronic computer, we have been able to calculate explicit electronic wave functions, based on the boundary condition that the wave functions vanish on a sphere of radius a. The number of particles N and the radius a have been chosen with a view to yielding a mean electron density equal to that of the conduction electrons in copper (0·0127 atomic unit). Clearly, in such a calculation, the aim must be to find a means of extrapolating the results for N and a finite (and, in practice, quite small) to the limiting case when both N and a become very large.

3·1. *Preliminary calculation for* $N = 186$

We began with an exploratory calculation with a very small number of particles, but the results are somewhat instructive and we give them in some detail. For $N = 186$, for an average density appropriate for copper metal we find $a = 15{\cdot}2$.

The wave-mechanical problem is then that of solving the Schrödinger equations:

$$\nabla^2\psi + \frac{2m}{\hbar^2}E\psi = 0, \quad \psi(a) = 0, \tag{3·1}$$

and

$$\nabla^2\psi_p + \frac{2m}{\hbar^2}[E_p - V_p]\,\psi_p = 0, \quad \psi_p(a) = 0, \tag{3·2}$$

for the unperturbed and perturbed metals, respectively. Here V_p is the perturbing potential energy which we have taken from the numerical calculations of Alfred & March (1957). We should comment here on the fact that we ought really, when we remove an atom from the interior of the metal and place it on the surface, to take account of the expansion of the perturbed lattice. We shall discuss this effect later (see § 8 below), although its influence on the wave functions in regions far from the surface must clearly become less important as the number of particles is increased.

The eigenfunctions of (1) and (2) for the spherically symmetrical perturbing potential $V_p(r)$ are evidently spherical harmonics multiplied by purely radial functions, and thus the problem reduces to integrating the radial Schrödinger equation for both perturbed and unperturbed cases. In the unperturbed metal, the eigenfunctions can immediately be written in terms of Bessel functions and the nodes of these functions yield the eigenvalues. In practice it turned out to be convenient to use the unperturbed eigenvalues to generate the correct Bessel functions by numerical integration.

TABLE 1. EIGENVALUES IN ATOMIC UNITS FOR $a = 15{\cdot}2$

state	unperturbed eigenvalues	perturbed eigenvalues
$1s$	0·02136	0·02568
$1p$	0·04370	0·04450
$1d$	0·07189	0·07209
$2s$	0·08544	0·10037
$1f$	0·10568	0·10575
$2p$	0·12915	0·13324
$1g$	0·14490	0·14493
$2d$	0·17901	0·18037
$1h$	0·18943	0·18945
$3s$	0·19223	0·22055
$2f$	0·23484	0·23539
$1i$	0·23918	0·23919
$3p$	0·25731	0·26692
$1j$	0·29408	0·29408
$2g$	0·29650	0·29676

To give an indication of the eigenvalue changes, and in view of the fact that the later calculations for larger N would involve extensive tabulation, we show in table 1 the eigenvalues for the perturbed and unperturbed cases, the perturbed values being obtained by successive approximation to the boundary condition $\psi_p(a) = 0$. The normalized wave functions were obtained directly from the Pegasus computer, and the form of the total density calculated by summing the squares of the wave functions in the perturbed and unperturbed cases was of obvious interest. The results thus obtained are shown in figure 1. It is then immediately clear that there are very large fluctuations about the mean density in the unperturbed lattice (the maximum density is 0·0195, to be compared with the mean value of 0·0127), evidently a consequence of the small number of particles used. However, the significant quantity on which we shall focus attention is the difference between the densities in the perturbed and unperturbed cases, denoted by $n(r)$ and $n_0(r)$, respectively, and we show $4\pi r^2(n_0 - n)$ against r out to approximately 5 atomic units in curve 1 of figure 2.

In view of the large fluctuations in the unperturbed density it seemed likely that the results reported so far must depend to some extent on the detailed configuration we have chosen. We therefore carried out a calculation in which the radius a was approximately doubled, keeping the density constant (actually $N = 1734, a = 31{\cdot}96$), the results of which we shall now briefly discuss.

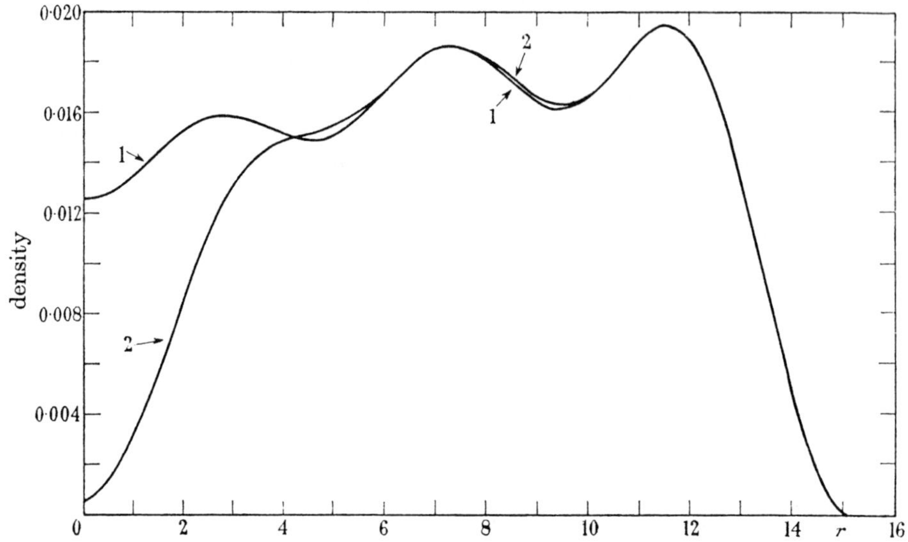

FIGURE 1. Electron densities corresponding to 186 particles. 1, Unperturbed case; 2, perturbed case.

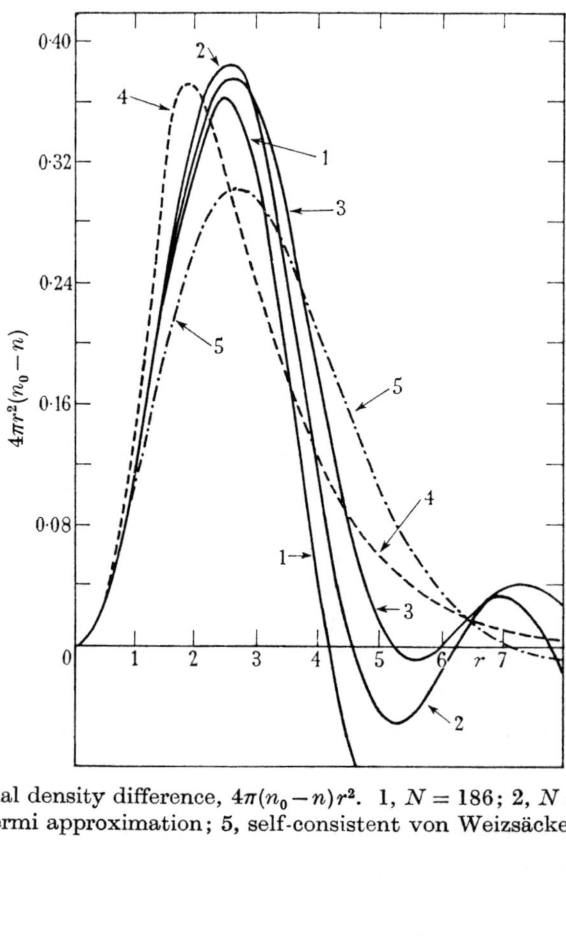

FIGURE 2. Radial density difference, $4\pi(n_0-n)r^2$. 1, $N = 186$; 2, $N = 1734$; 3, $N \sim 10^6$; 4, Thomas–Fermi approximation; 5, self-consistent von Weizsäcker approximation.

3·2. *Summary of results for $N = 1734$*

The labour goes up very considerably, 67 states now being needed in both the perturbed and the unperturbed metal. The most serious point which emerged from the analysis was that marked density fluctuations about the mean still occur in the unperturbed metal for this much larger value of N.

To summarize the results, we show the radial density difference $4\pi(n_0-n)r^2$ in curve 2 of figure 2, and it is encouraging to note that the results are in fair agreement with those for $N = 186$ out to the first node.

4. Estimate of fluctuations for large N

It is clear from the above results that some criterion giving the number of states we should fill in order to restrict fluctuations of the unperturbed density, away from the surface region, to some reasonable value (say 5 %) is highly desirable.

In this connexion it is fortunate that for s states in the unperturbed lattice, we can proceed analytically and can display in closed form the s density as a function of the number of occupied s states N_s say. This is a valuable result because all wave functions corresponding to $l \neq 0$ contribute nothing to the density at the origin.

We show now how we can estimate N_s such that filling the next highest s state changes the density at the origin by 5 %. If we denote the radial wave function for s states by $R(r)$, then

$$R_n^2(r) = \frac{1}{2\pi a r^2}\sin^2\frac{n\pi r}{a}, \tag{4·1}$$

where $n = 1, 2, 3, \ldots$. Here R_n is normalized such that

$$\int_0^a R_n^2\, 4\pi r^2\, \mathrm{d}r = 1. \tag{4·2}$$

Then the total s density corresponding to N_s doubly filled states is given by

$$\begin{aligned}\rho^{N_s}(r) &= \frac{1}{\pi a r^2}\sum_1^{N_s}\sin^2\frac{n\pi r}{a}\\ &= \frac{1}{2\pi a r^2}\left[N_s+\tfrac{1}{2}-\frac{\sin\{\pi(2N_s+1)\,r/a\}}{2\sin\{\pi r/a\}}\right].\end{aligned} \tag{4·3}$$

In particular, the density at the origin is given by

$$\rho^{N_s}(0) = \frac{N_s(N_s+1)\,(2N_s+1)\,\pi}{6a^3}. \tag{4·4}$$

If we doubly fill the (N_s+1)th s state, this contributes a density at the origin given by

$$\rho_{N_s+1}(0) = (N_s+1)^2\,\pi/a^3 \tag{4·5}$$

and therefore we have

$$\frac{\rho_{N_s+1}(0)}{\rho^{N_s}(0)} = \frac{6(N_s+1)}{N_s(2N_s+1)} \tag{4·6}$$

$$\doteqdot 3/N_s, \quad \text{for large } N_s. \tag{4·7}$$

Thus, if we require that the filling of the next s state causes a density increase of 5 %, we find $N_s = 60$.

With the available computational facilities, it appears an enormous task to include all states having energies lower than the 60th s state (actually we chose the 60th p state as the highest occupied state). However, an analysis of the way in which the density difference $n_0 - n$ arises, shows that the states with low l values mainly contribute. Thus we report below calculations carried out for s, p, d, f, g and h states for the present case. While therefore, we have no exact figure for the number of particles N in this case, by equating $\rho^{N_s}(0)$ given by (4·4) to $3N/4\pi a^3$, we immediately obtain $N \simeq 10^6$, for $N_s = 60$.

5. Results for 60 occupied states

We shall not discuss in detail the very extensive numerical computations involved, but merely remark that it was convenient to apply a scale transformation to retain a as 31·96 (the value taken for $N = 1734$), and to solve the whole problem in these units.

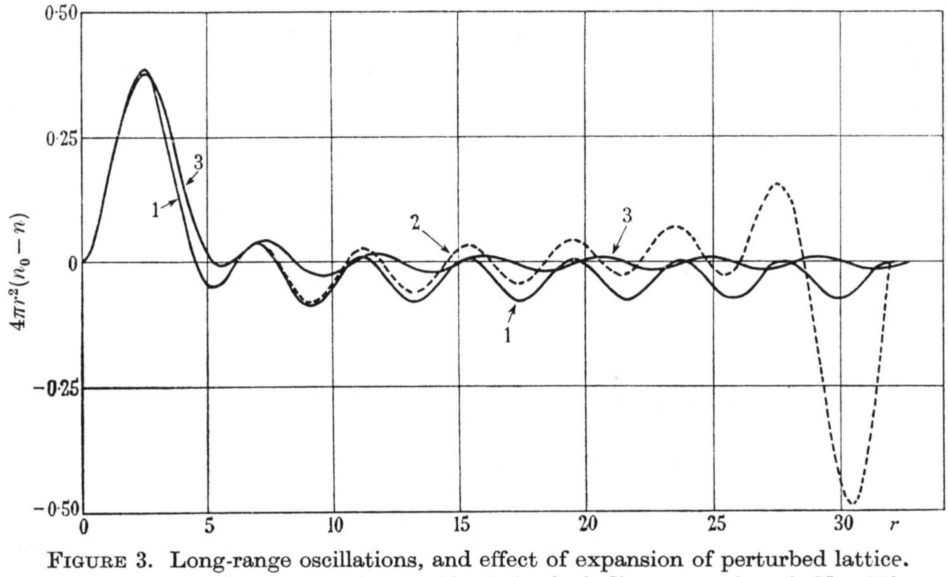

Figure 3. Long-range oscillations, and effect of expansion of perturbed lattice. 1, $N = 1734$, without expansion; 2, $N = 1734$, including expansion; 3, $N \sim 10^6$.

Figure 4 shows the way in which the density difference is built up as we add successive orbital angular momentum states. It will be seen that the convergence is very good (in fact, in a plot of $n_0 - n$, only s, p and d states contribute to graphical accuracy). The final results for the radial density difference are shown in curve 3 of figure 2, for comparison with the cases corresponding to smaller values of N. We thought it also of interest to show in figure 5 the way in which the unperturbed density n_0 is built up from the various states. This figure also well illustrates that the density fluctuations have now largely disappeared in the region where, in the unperturbed metal, the potential due to the imperfection will have an appreciable magnitude. It seems clear therefore that these results should be very much more realistic than those corresponding to the smaller numbers of particles.

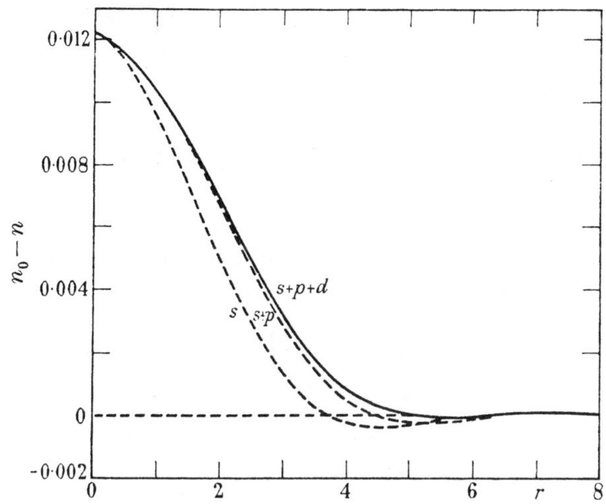

FIGURE 4. Contributions of successive orbital angular momentum states to density difference n_0-n for $N \sim 10^6$.

6. VON WEIZSÄCKER DENSITY ROUND A VACANCY

It appears that the Thomas–Fermi radial density difference plotted in curve 4 of figure 2 is a fair approximation to the results obtained by our finite-metal computations. In view therefore of the practical merits of this approximation in problems such as impurity-vacancy interaction energies (see Alfred & March 1957) which are important in theories of solute diffusion in metals, it seemed of definite interest to inquire whether a correction proposed by von Weizsäcker to the original Thomas–Fermi method would lead to results in improved agreement with our direct computations using wave functions. This class of method has the further advantage that the limiting case when the number of particles N and the volume V tend to infinity such that $N/V \rightarrow n_0$ can be directly examined, with the normalization of n_0-n still under control.

Calculations using this method, which has been shown recently to have a rigorous wave-mechanical basis on the variational theorem (March & Young 1958) have been made previously by Donovan & March (1959) for an attractive potential, in connexion with positron lifetimes in metals. We show here how a transformation of the von Weizsäcker equation allows some analytical progress to be made, and throws light on the general nature of its solutions. The basic equation may be written in the form

$$\frac{3^{\frac{2}{3}}\pi^{\frac{4}{3}}n^{\frac{2}{3}}}{2}+\frac{1}{8}\left[\frac{(\nabla n)^2}{n^2}-\frac{2\nabla^2 n}{n}\right]+V=E_m, \tag{6·1}$$

where E_m is the Fermi energy. If we omit terms involving derivatives of the density, (6·1) reduces to the usual Thomas–Fermi relation between density and potential. For self-consistency, the perturbing potential energy V is related to the density change $n-n_0$ through the Poisson equation

$$\nabla^2 V = 4\pi(n_0-n). \tag{6·2}$$

The transformation

$$\psi = n^{\frac{1}{2}} \tag{6·3}$$

now enables the von Weizsäcker equation (6·1) to be placed in the form of a Schrödinger equation with an effective potential. It is easy to show that this may be written

$$\nabla^2\psi + 2[E_m - V - \tfrac{1}{2}(3^{\frac{2}{3}}\pi^{\frac{4}{3}})\,\psi^{\frac{4}{3}}]\psi = 0. \tag{6·4}$$

The asymptotic form of the solution of (6·4) may now be obtained by writing

$$\psi = n_0^{\frac{1}{2}} + \phi \tag{6·5}$$

and considering the approximation appropriate when $\phi \ll n_0^{\frac{1}{2}}$. It is then shown in appendix 1 that the asymptotic solution is given by

$$r\phi = A\exp\{-\surd[\tfrac{1}{2}(a+b)]\,r\}\sin\{\surd[\tfrac{1}{2}(b-a)]\,r+\eta\}, \tag{6·6}$$

where $$a = \tfrac{4}{3}E_m \quad \text{and} \quad b = 4\pi^{\frac{1}{2}}n_0^{\frac{1}{2}}.$$

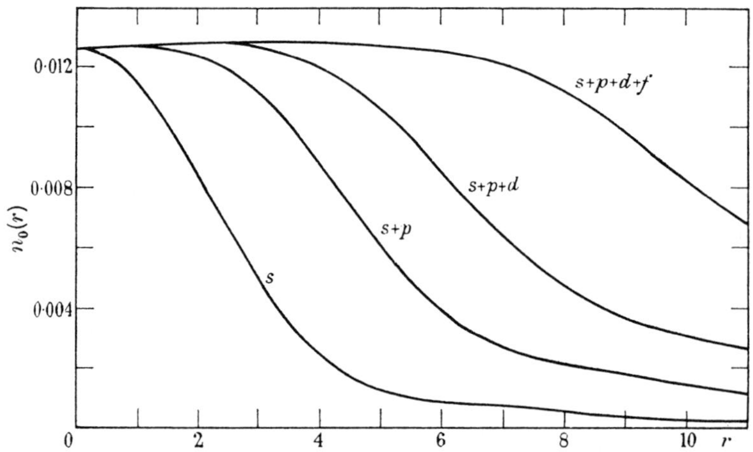

FIGURE 5. Contributions to unperturbed density n_0 of successive orbital angular momentum states for $N \sim 10^6$.

The amplitude A and the phase η must be determined by matching this form to the solution of the full von Weizsäcker equation, which must be obtained numerically. Details of the somewhat involved numerical technique necessary to deal with the simultaneous differential equations (6·1) and (6·2) are given in appendix 2 and the self-consistent results for the radial density difference are shown in curve 5 of figure 2. The values found for A and η are $-0·38$ and $-0·23$, respectively.

7. DISCUSSION OF RESULTS

We will focus attention for the moment on the localized screening charge, and will leave until later the discussion of the long-range oscillations, which can be seen particularly clearly in figure 3. Examination of figure 2 then reveals that with the assumed Thomas–Fermi potential the finite metal computations all agree well concerning the position and height of the principal maximum in the radial density difference. In all three cases the maximum has moved out to around 2·5 or 2·6 atomic units (nearly equal to the atomic sphere radius $r_s = 2·66$), from the Thomas–Fermi value of about 1·8. The maximum as given by the von Weizsäcker approximation is at the correct position, but is rather seriously in error in magnitude

($\sim$ 20 % too low). It may be said therefore that the von Weizsäcker method appears to spread out the screening charge too much, and in fact over-corrects the defects of the original Thomas–Fermi approximation. It will be noticed from figures 2 and 3 that the effect of increasing the number of particles is to spread the screening charge somewhat, the first node in $n - n_0$ moving out to $\sim$ 5·2 from 4·4 in going from the 1734 to the 10^6 particle case. Nevertheless, it seems clear from our results that the von Weizsäcker density difference is much too diffuse. Some exploratory computations we made with the Thomas–Fermi potential in the von Weizsäcker equation (6·1) confirm this by yielding an even more diffuse distribution.

It appears therefore that with a rapidly varying potential of the kind under discussion here, the variational functions on which the von Weizsäcker method is based (transformed plane waves; cf. March & Young 1958) are not sufficiently flexible to yield results of high accuracy. In agreement with conclusions drawn earlier by Berg & Wilets (1955) in a different class of problem, figure 2 suggests that reduction of the coefficient $\frac{1}{8}$ multiplying the density derivatives in (6·1) would probably lead to improved results. Certainly in the vacancy problem, the von Weizsäcker approach has the merit over the Thomas–Fermi scheme that it allows the wave functions to penetrate into the classically forbidden region, but unfortunately the penetration found is now too great. This is well illustrated by noting that the total electronic charge in the vacancy cell (taken to be a sphere of radius $r_s = 2·66$) is $-0·44e$ from the Thomas–Fermi calculation, $-0·56e$ from the self-consistent von Weizsäcker results, and $-0·47e$ from the 10^6 particle case.

8. Effect of expanding perturbed lattice

Before we discuss the long-range oscillations revealed in figures 2 and 3, it is of interest to consider the effect of accounting directly for the expansion of the perturbed lattice which takes place when a vacancy is formed. We have investigated this point in the case $N = 186$ and 1734, and as the results agree qualitatively, those for the latter case only are plotted in figure 3. It will be seen that apart from essential modifications near the boundary, which can be thought of again as another manifestation of localized screening of a perturbation, the main effect is to cause the oscillations to occur symmetrically about the line $n = n_0$. Even in the 1734 particle case, the influence of expansion on $n - n_0$ is seen from figure 3 to be appreciable, but the symmetry of the oscillations for $N \sim 10^6$ seems good evidence that the modification is practically negligible in this case.

9. Discussion of long-range oscillations

Turning now to discuss the long-range oscillations in detail, we note first that Blandin, Daniel & Friedel (1959) have pointed out the existence of such effects due to point imperfections in metals, and have thereby developed a theory of Knight shifts in dilute alloys. These workers refer also to unpublished investigations of Brooks, as also do Machlup & Ogata (1959) in a short abstract which has appeared since our computations were completed. The results given by Blandin *et al.* concerning Knight shifts refer primarily to the fluctuations in the electron density for wave functions corresponding to states at the surface of the Fermi distribution, whereas the present results refer to the electron density obtained by summing over all states, as do those of Machlup & Ogata. The fluctuations induced in this total density by imperfections will also affect the Knight shift, if the nucleus has an electric quadrupole moment.*

It is clear from our results, in spite of the fact that, as shown in § 8, effects of expansion of the lattice bring the $N = 186$ and 1734 particle cases more into line qualitatively with the 10^6 particle case, that the amplitude of these oscillations is decreasing very significantly as the number of particles increases. Although it is not possible to predict with certainty the limiting behaviour as $N \to \infty$ on the basis of these results, we seem able, nevertheless, to draw certain useful conclusions.

First, the wavelength of the oscillations revealed for finite N is well defined, and numerically $\sim 4{\cdot}3$ atomic units. This is in good accord with the result π/k_F, k_F being the magnitude of the wave vector at the Fermi surface, given by Machlup & Ogata (1959), which they obtain from a phase shift argument in the limiting case when N and V tend to infinity. Secondly, the amplitude of the oscillations is very small in the vacancy problem with 10^6 particles. It seems clear from figure 3 that the amplitude in this case will provide an upper bound to that for the infinite metal with the localized potential we have used. In the light of our investigations, we feel that care is needed in discussions of these long-range effects. In particular, careful treatment of normalization is required and this is perhaps a difficulty in existing work on the infinite metal case, where terms of $O(1/N)$ do not seem to be directly under control.

It should perhaps be emphasized here that in view of the long-range density oscillations, the potential we have used at large distances from the centre cannot be entirely self-consistent. However, the results we discuss in § 10 lead us to believe that the potential outside ~ 6 atomic units is so small in magnitude that its detail becomes relatively unimportant, but further work on this aspect is probably called for in order to reach a completely final conclusion concerning its effect on the long-range density fluctuations.

We should note in passing that the von Weizsäcker method predicts oscillations in the density and the potential, but the wavelength of these is about three times too large in the present problem.

10. Investigation of departures from self-consistency

To obtain some quantitative indication of likely changes in potential from the Thomas–Fermi form originally assumed, we have proceeded as follows. Using the density difference $n - n_0$ corresponding to the 10^6 particle case, we have integrated the Poisson equation (6·2) for a charge distribution which we have truncated at 5·9 atomic units, at which point the displaced charge normalizes accurately to unity. We show such results in figure 6, together with the original Thomas–Fermi potential and the von Weizsäcker potential. It will be seen that the differences from the Thomas–Fermi values are nowhere large, and suggest that only minor modifications

* Thanks are due to the referee for pointing this out to us, and for other useful comments on the presentation of this work.

of our results are likely to occur when complete self-consistency is achieved. The von Weizsäcker potential seems less accurate for small r, but around the near-neighbour distance of 4·8 atomic units may have some merit over the Thomas–Fermi result.

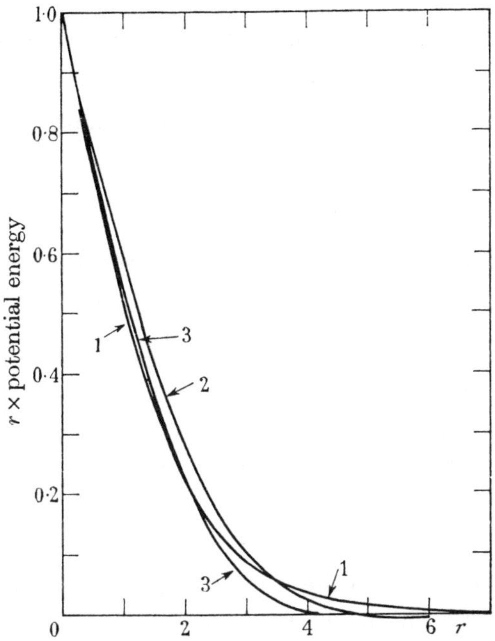

FIGURE 6. Potentials given by: 1, Thomas–Fermi approximation; 2, self-consistent von Weizsäcker calculation; 3, $n_0 - n$ from $N \sim 10^6$ case, truncated at 5·9 atomic units, where localized screening charge is unity.

11. DEPENDENCE OF CHANGES IN EIGENVALUE SUMS ON N

As we remarked earlier, it is not practicable to present here the extensive tables of eigenvalues for all cases considered. However, we shall briefly discuss the change in the eigenvalue sums as this is relevant to the problem of the energy of formation of a vacancy, various aspects of which are at present under consideration (A. B. Lidiard & K. Tharmalingham, private communication). We have recorded in table 2 these changes for the three different values of N we have considered. We give values both for the cases when the unperturbed and the perturbed lattices have the same radius, and when the perturbed lattice is expanded by one atomic volume. It can be seen from the results that no definite indication of the limiting value is given, though we do not believe that the value for $N \sim 10^6$ should be very much higher than the limit. However, if accurate results are to be forthcoming from a finite-metal argument, it seems necessary to discuss analytically the dependence of the eigenvalue change on the particle number N for large N.

We might note, however, that our present results enable us to conclude from rough exploratory calculations, that the Thomas–Fermi approximation to the change in the sum of the eigenvalues is rather seriously in error. We feel therefore that in a limited range of problems—for example, to understand the main features in the redistribution of electrons when two vacancies are brought together to form a divacancy, and the corresponding contributions to the binding energy—the von Weizsäcker energy may reveal the general nature of the limitations of the Thomas–Fermi results, and would not therefore be without interest.

TABLE 2. CHANGES IN SUM OF EIGENVALUES FOR $N = 186$, 1734 AND $\sim 10^6$

Perturbed lattice not expanded

N	ΔE (atomic units)
186	0·212
1734	0·201
$\sim 10^6$	0·170

Perturbed lattice expanded by one atomic volume

N	ΔE (atomic units)
186	0·078
1734	0·083
$\sim 10^6$	0·066*

* In this case, the correction for expansion was not computed directly, but taken as two-fifths of the Fermi energy (see Fumi 1955).

12. CONCLUSION

The present approach, using a finite metal, appears to be successful in yielding the detail of the screening charge distribution round a vacancy, in the one-electron approximation. Our main conclusion is that the total charge density obtained by summing the squares of the wave functions for an assumed Thomas–Fermi potential energy V_p is such as to make V_p self-consistent to a good approximation. The one reservation we would place on this conclusion is that our calculations do not reveal the nature of any small long-range contribution to the final self-consistent potential. Concerning long-range oscillations in the total charge density our calculations with the assumed localized potential show that the amplitude of the oscillations decreases with increasing volume of the metal. However, it does not seem possible to obtain the amplitude in the infinite metal case by extrapolation, but our results should afford an upper bound, which is already quite small.

While it is outside the scope of the present paper to consider detailed physical applications of these results, our findings suggest that the use of the Thomas–Fermi perturbing potential in resistivity calculations should be quite a good approximation, and results based on this potential, modified to take account of relaxation of the lattice around a vacancy which we have neglected in the present work, will be reported elsewhere by Dr P. M. Lee and one of us (N. H. M.). We might remark too that the present findings are highly relevant to the problem of electron correlation in a uniform gas. Whereas the Thomas–Fermi results of Alfred & March (1957) on the model of the vacancy discussed in § 2 were essentially those on which the treatment of electron correlation by Cowan & Kirkwood (1958) was based, the present results remove some of the defects of that treatment, when suitably modified to take account of spin. This application will be discussed in a later paper by Dr W. H. Young and the writers.

We wish to acknowledge that this work was performed under partial contractual support from the U.S. Army. Thanks are due to Mr P. H. Blundell for enabling us to make extensive use of the Ferranti Pegasus computer which is available to the University Computation Laboratory.

Appendix 1

We shall discuss here the asymptotic form of the solutions of (6·4) by writing

$$\psi = n_0^{\frac{1}{2}} + \phi \tag{A 1·1}$$

and considering the approximation appropriate when $\phi \ll n_0^{\frac{1}{2}}$.

Retaining only the leading terms, we then find

$$\nabla^2\phi - \tfrac{8}{3}E_m\phi - 2Vn_0^{\frac{1}{2}} = 0 \tag{A 1·2}$$

and

$$\nabla^2 V = -8\pi n_0^{\frac{1}{2}}\phi. \tag{A 1·3}$$

Introducing

$$a = \tfrac{4}{3}E_m, \quad b = 4\pi^{\frac{1}{2}}n_0^{\frac{1}{2}}, \tag{A 1·4}$$

we may rewrite (A 1·2) and (A 1·3) in the forms

$$\frac{d^2}{dr^2}(r\phi) - 2a(r\phi) - \frac{1}{2\pi^{\frac{1}{2}}}b(rV) = 0 \tag{A 1·5}$$

and

$$\frac{d^2}{dr^2}(rV) = -2\pi^{\frac{1}{2}}b(r\phi). \tag{A 1·6}$$

Eliminating rV between (A 1·5) and (A 1·6), we have

$$\frac{d^4}{dr^4}(r\phi) - 2a\frac{d^2}{dr^2}(r\phi) + b^2(r\phi) = 0. \tag{A 1·7}$$

The solution of (A 1·7) may now be written as a linear combination of terms of the form $e^{\alpha r}$ where α is a solution of

$$\alpha^4 - 2a\alpha^2 + b^2 = 0, \tag{A 1·8}$$

or

$$\alpha^2 = a \pm \sqrt{(a^2 - b^2)}. \tag{A 1·9}$$

We may therefore write

$$\alpha = \pm\sqrt{[\tfrac{1}{2}(a+b)]} \pm \sqrt{[\tfrac{1}{2}(a-b)]} \tag{A 1·10}$$

and since $r\phi \to 0$ as $r \to \infty$, the first term must be negative. Thus we have

$$\alpha = -\sqrt{[\tfrac{1}{2}(a+b)]} \pm \sqrt{[\tfrac{1}{2}(a-b)]} \tag{A 1·11}$$

$$= -\sqrt{(\tfrac{2}{3}E_m + 2\pi^{\frac{1}{2}}n_0^{\frac{1}{2}})} \pm \sqrt{(\tfrac{2}{3}E_m - 2\pi^{\frac{1}{2}}n_0^{\frac{1}{2}})} \quad \text{if} \quad E_m \geqslant 3\pi^{\frac{1}{2}}n_0^{\frac{1}{2}} \tag{A 1·12}$$

$$= -\sqrt{(\tfrac{2}{3}E_m + 2\pi^{\frac{1}{2}}n_0^{\frac{1}{2}})} \pm i\sqrt{(2\pi^{\frac{1}{2}}n_0^{\frac{1}{2}} - \tfrac{2}{3}E_m)} \quad \text{if} \quad E_m \leqslant 3\pi^{\frac{1}{2}}n_0^{\frac{1}{2}}. \tag{A 1·13}$$

The high-density form of solution given in (A 1·12) has been obtained earlier by Donovan & March (1959). However, the present argument indicates that as the density is lowered a point is reached at which the exponential solutions go over to oscillatory solutions, and the transition according to (A 1·12) and (A 1·13) will occur at a density given by

$$E_m = 3\pi^{\frac{1}{2}}n_0^{\frac{1}{2}}. \tag{A 1·14}$$

Introducing for convenience the quantity r_s, defined as the radius of a sphere containing one electron, (A 1·14) becomes

$$r_s = \frac{\pi^{\frac{4}{3}}}{3^{\frac{1}{3}}4^{\frac{4}{3}}} = 0{\cdot}502. \qquad \text{(A 1·15)}$$

For metals under ordinary conditions r_s lies in the range 2 to 5·5 and we see that the asymptotic form including an oscillatory factor will be the appropriate one in all cases. The final result may then be written immediately in the form (6·6).

Appendix 2

The numerical solutions of the radial Schrödinger equation, von Weizsäcker's equation and Poisson's equation were obtained by means of programs written for use with the Pegasus computer.

(*a*) *Radial Schrödinger equation*

The program obtains solutions of

$$\frac{d^2}{dr^2}(r\psi) = \left\{\frac{l(l+1)}{r^2} + 2(V-E)\right\} r\psi \qquad \text{(A 2·1)}$$

which are regular at the origin. V is supplied in tables of rV and E and l are specified for each integration. The mathematical method used is described by Fox & Goodwin (1949; method VII). If (A 2·1) is written as

$$d^2y/dr^2 = fy \qquad \text{(A 2·2)}$$

then a recurrence relation is obtained in the form

$$(1-\tfrac{1}{12}h^2f_1)\,y_1 = (2+\tfrac{5}{6}h^2f_0)\,y_0 - (1-\tfrac{1}{12}h^2f_{-1})\,y_{-1} + \Delta,$$
$$\Delta = (-\tfrac{1}{240}\delta^6 + \ldots)\,y_0, \qquad \text{(A 2·3)}$$

where $y_0 = y(r)$, $y\pm 1 = y(r\pm h)$. The interval h is chosen small enough to eliminate the need for the difference correction Δ and yet not so small as to make the term $-\frac{1}{12}h^2f$ in the expression $(1-\frac{1}{12}h^2f)$ inaccurate because of rounding-off errors in the computer. It is found that the accuracy of oscillating solutions is greatest when h is chosen such that

$$0{\cdot}001 \leqslant |l(l+1)/r^2 + 2(V-E)|\,h^2 \leqslant 0{\cdot}004. \qquad \text{(A 2·4)}$$

This interval corresponds to between 50 and 100 points per node. Provision is made for doubling the interval where necessary to maintain condition (A 2·4). When (A 2·4) is satisfied the accuracy of the solutions is of the order of 1 in 10^6.

Formula (A 2·3) cannot be used at the origin and the first $l+2$ values of $r\psi$ are obtained by a power series method. The ratio of successive values of r is then approximately e and the step-by-step integration proceeds from this point. The values of rV required for the step-by-step integration are first obtained by interpolation and these values may be used for any number of values of E and l.

The integration cycle of 50 orders is looped in about 0·05 s. Thus a solution with six nodes requires about 15 s for integration. Sums and differences of charge densities are obtained by operating on the squares of normalized wave functions within the computer.

(b) *von Weizsäcker equation*

The program described above was amended to obtain solutions of

$$d^2(r\psi)/dr^2 = \{3^{\frac{2}{3}}\pi^{\frac{4}{3}}\psi^{\frac{4}{3}} + 2(V-E)\}\, r\psi. \tag{A 2·5}$$

Since the equation is non-linear the value of ψ at the origin must now also be specified. If we write (A 2·5) as (A 2·2), f is a function of y and in applying (A 2·3) an iteration is performed on y at each step. f_1 is first evaluated using $y_1 = 2y_0 - y_{-1}$. With this value of f_1 an improved estimate of y_1 is obtained. This is repeated until two successive values of y_1 are identical to within 1 in 10^9. For intervals satisfying the relation corresponding to (A 2·4) either three or four iterations are required.

(c) *Self-consistent von Weizsäcker equation*

Simultaneous solutions of (A 2·5) and Poisson's equation

$$d^2(rV)/dr^2 = 4\pi r(n_0 - \psi^2) \tag{A 2·6}$$

were obtained by a second-order Runge–Kutta method (Kamke 1943) which takes advantage of the absence of the first derivative. If we write the equations as

$$y'' = f(x, y), \tag{A 2·7}$$

the values of the functions and gradients at successive steps ν and $\nu+1$ are related by

$$\left.\begin{aligned} y_{\nu+1} &= y_\nu + hy'_\nu + \tfrac{1}{3}(k_1 + 2k_2), \\ \tfrac{1}{2}hy'_{\nu+1} &= \tfrac{1}{2}hy'_\nu + \tfrac{1}{6}(k_1 + 4k_2 + k_3), \\ \text{where} \qquad k_1 &= \tfrac{1}{2}h^2 f(x_\nu, y_\nu), \\ k_2 &= \tfrac{1}{2}h^2 f(x_\nu + \tfrac{1}{2}h,\, y_\nu + \tfrac{1}{2}hy'_\nu + \tfrac{1}{4}k_1), \\ k_3 &= \tfrac{1}{2}h^2 f(x_\nu + h,\, y_\nu + hy'_\nu + k_2), \end{aligned}\right\} \tag{A 2·8}$$

h being the step length.

The solutions of the equations which are well behaved for large r involve two arbitrary constants at the origin. We suppose that the required solution is one for which $\psi \to \psi_\infty$ and $V \to 0$ as $r \to \infty$. A measure of the accuracy of any pair of arbitrary constants (a, b) is obtained by defining a function $R(a, b)$ such that R is the value of r at which either $|r^2(\psi - \psi_\infty)|$ or $|r^2 V|$ exceeds a chosen bound. The problem then reduces to that of finding the maximum of $R(a, b)$ (theoretically infinite). Given three pairs of numbers (a, b) the program is designed to converge on this maximum. In practice it is found that starting with random values of a and b, the program converges on a secondary maximum of R which is quite small. The values of a and b are then correct to about two significant figures. A few values of R obtained by systematic mapping of the (a, b) plane indicates the position of the main maximum. The remaining digits are then obtained by a further convergence.

References

Alfred, L. C. R. & March, N. H. 1957 *Phil. Mag.* **2**, 985.
Berg, R. & Wilets, L. 1955 *Proc. Phys. Soc.* A, **68**, 229.
Blandin, A., Daniel, E. & Friedel, J. 1959 *Phil. Mag.* **4**, 180.
Cowan, R. D. & Kirkwood, J. G. 1958 *Phys. Rev.* **111**, 1460.
Donovan, B. & March, N. H. 1959 *J. Phys. Chem. Solids*, **11**, 68.
Fox, L. & Goodwin, E. T. 1949 *Proc. Camb. Phil. Soc.* **45**, 373.
Fumi, F. G. 1955 *Phil. Mag.* **46**, 1007.
Kamke, E. 1943 *Differentialgleichungen, Lösungsmethoden und Lösungen*, vol. 1, 149.
Machlup, S. & Ogata, H. 1959 *Bull. Amer. Phys. Soc.* **4**, 422.
March, N. H. & Young, W. H. 1958 *Proc. Phys. Soc.* A, **72**, 182.

Self-consistent perturbation treatment of impurities and imperfections in metals

By N. H. March and A. M. Murray
Department of Physics, The University, Sheffield

(*Communicated by C. A. Coulson, F.R.S.—Received* 7 *November* 1960)

A full perturbation treatment of the Dirac density matrix is developed as a basis for self-consistent calculations in free-electron metals containing localized defects. The present perturbation series is shown to sum to the well-known result for the case of slowly varying potentials. To first order in perturbation theory, exact self-consistent results for the radial density of displaced charge and the Hartree potential in the presence of point singularities have been obtained over a density range sufficient to cover all metals under normal conditions. The basic limitations of Mott's first-order method, based on the assumption of slowly varying potentials, are shown to be completely removed and the self-consistent density and potential display long-range oscillations. Finally, the application of the present approach to Bloch wave functions rather than plane waves is briefly considered. Friedel's generalized first-order method, applicable to a band structure, may be obtained from the present theory for sufficiently slowly varying potentials. Unfortunately, such an assumption is seen by comparison with the free-electron findings to lead to serious errors.

1. Introduction

In an earlier study of electronic wave functions round a vacancy in a metal (March & Murray 1960*a*; referred to subsequently as I), calculations were carried out based on the use of a finite metal with box boundary conditions. While it emerged clearly from this work that self-consistency could be established to a reasonable degree of approximation in the immediate neighbourhood of a vacancy, two limitations of the approach were particularly evident:

(i) The long-range oscillations revealed in the density were obtained from a potential which, at large distances, was exponentially decaying, and hence there was no self-consistency at large distances from the vacant site;

(ii) it was not possible to extrapolate with certainty to the desired limiting case of an infinite crystal, even though a monovalent metal with 10^6 atoms was considered.

It seemed clear therefore that to answer the questions left open by our previous work as well as to obtain results of greater generality a different approach was essential. Thus, instead of considering electronic wave functions themselves, it seemed to us very desirable to work with the Dirac density matrix, the diagonal element of which gives the electron density immediately. Pursuing this aim we began in a previous paper (March & Murray 1960*b*) to develop a perturbation theory for the Dirac matrix which should be applicable to localized defects in an infinite metal. In the previous work, we obtained only the first-order result for the diagonal element and no application of the theory was attempted. Here, we show first that the full Dirac matrix may be generated by perturbation theory to infinite order, and demonstrate that for the case of slowly varying potentials the series may be summed to yield the usual result. We then begin the somewhat troublesome numerical application of the perturbation theory by working out self-consistently the density and potential distribution round an impurity to first order in perturbation theory. All the essential features of the problem are thereby revealed, although by comparing the results of I with the present findings, it is clear that eventually it will be very desirable to undertake the extensive computations involved in estimating the second-order correction. The results are sufficiently encouraging, however, to lead us to believe that the perturbation series should converge fairly rapidly in such problems, and that a complete solution for the Hartree potential and density is now within our reach.

2. Perturbation theory for the Dirac density matrix

We begin by showing how the previous argument of March & Murray may be generalized to give the full density matrix to infinite order. Our treatment starts from a calculation of the quantum-mechanical partition function carried out by Green (1952), but since his method is very involved we shall first present a simplified derivation of his results in a form suitable for our purpose.

We consider a Hamiltonian H which we write in the form

$$H = H_0 + V(\mathbf{r}), \tag{2·1}$$

where H_0 is the unperturbed Hamiltonian and $V(\mathbf{r})$ is the perturbing potential energy. If the energy levels and wave functions of H are denoted by E_i and ψ_i then we focus attention first on the canonical density matrix C defined by

$$C(\mathbf{r}, \mathbf{r}_0, \beta) = \sum_i \psi_i^*(\mathbf{r})\, \psi_i(\mathbf{r}_0) \exp(-E_i\beta), \tag{2·2}$$

where $\beta = (kT)^{-1}$. The Bloch equation satisfied by (2·2) is given by

$$\partial C/\partial\beta = -HC, \tag{2·3}$$

with the initial condition $$C(\mathbf{r}, \mathbf{r}_0, 0) = \delta(\mathbf{r}-\mathbf{r}_0). \tag{2·4}$$

In order to develop an iterative method of solution, it is convenient at this stage to convert (2·3) to an integral equation. The result we require is

$$C(\mathbf{r}, \mathbf{r}_0, \beta) = C_0(\mathbf{r}, \mathbf{r}_0, \beta) - \int d\mathbf{r}_1 \int_0^\beta d\beta_1\, C_0(\mathbf{r}, \mathbf{r}_1, \beta-\beta_1)\, V(\mathbf{r}_1)\, C(\mathbf{r}_1, \mathbf{r}_0, \beta_1), \tag{2·5}$$

as may be verified readily by direct substitution in the Bloch equation (2·3), when use is made of the initial condition (2·4). Here, of course, C_0 is the unperturbed canonical matrix. If we now assume that C_0 is the free electron matrix given by

$$C_0(\mathbf{r}, \mathbf{r}_0, \beta) = (2\pi\beta)^{-\frac{3}{2}} \exp(-|\mathbf{r}-\mathbf{r}_0|^2/2\beta) \tag{2·6}$$

then (2·5) may be used iteratively, beginning with the replacement of C by C_0 inside the integral. The basic problem is then to perform the integral over β_1, that is to obtain

$$I(\beta) = \int_0^\beta d\beta_1\, C_0(\mathbf{r}, \mathbf{r}_1, \beta-\beta_1)\, C_0(\mathbf{r}_1, \mathbf{r}_0, \beta_1). \tag{2·7}$$

This may be evaluated either by lengthy elementary methods or, more elegantly, by using a theorem on the Laplace transform of a convolution (Margenau & Murphy 1956, p. 261), as is shown below. Denoting the Laplace transform of f by $\mathscr{L}f$ we have

$$\mathscr{L}[I(\beta)] = \mathscr{L}[C_0(\mathbf{r}, \mathbf{r}_1, \beta)]\,\mathscr{L}[C_0(\mathbf{r}_1, \mathbf{r}_0, \beta)]. \tag{2·8}$$

By using the identity

$$(2\pi)^{-1}\exp\left(-t\sqrt{(2E)}\right) = \int_0^\infty d\beta \exp(-\beta E)\,(2\pi\beta)^{-\frac{3}{2}}\,t\exp(-t^2/2\beta) \tag{2·9}$$

in conjunction with (2·6), it follows that

$$\mathscr{L}[C_0(\mathbf{r}, \mathbf{r}_1, \beta)] = (2\pi|\mathbf{r}-\mathbf{r}_1|)^{-1}\exp\left(-|\mathbf{r}-\mathbf{r}_1|\sqrt{(2E)}\right). \tag{2·10}$$

Thus, from (2·8),

$$\mathscr{L}[I(\beta)] = \exp\left(-\{|\mathbf{r}-\mathbf{r}_1|+|\mathbf{r}_1-\mathbf{r}_0|\}\sqrt{(2E)}\right) \div 4\pi^2\,|\mathbf{r}-\mathbf{r}_1|\,|\mathbf{r}_1-\mathbf{r}_0| \tag{2·11}$$

and hence

$$I(\beta) = (2\pi\beta)^{-\frac{3}{2}}\left(|\mathbf{r}-\mathbf{r}_1|+|\mathbf{r}_1-\mathbf{r}_0|\right)\exp\left\{-\frac{1}{2\beta}\left(|\mathbf{r}-\mathbf{r}_1|+|\mathbf{r}_1-\mathbf{r}_0|\right)^2\right\} \div 2\pi\,|\mathbf{r}-\mathbf{r}_1|\,|\mathbf{r}_1-\mathbf{r}_0|. \tag{2·12}$$

Then from (2·5) we have, to first-order in V,

$$C(\mathbf{r}, \mathbf{r}_0, \beta) = C_0(\mathbf{r}, \mathbf{r}_0, \beta) - (2\pi\beta)^{-\frac{3}{2}}\int d\mathbf{r}_1\,V(\mathbf{r}_1)\left(|\mathbf{r}-\mathbf{r}_1|+|\mathbf{r}_1-\mathbf{r}_0|\right)\exp\left\{-\frac{1}{2\beta}\left(|\mathbf{r}-\mathbf{r}_1|+|\mathbf{r}_1-\mathbf{r}_0|\right)^2\right\} \div 2\pi\,|\mathbf{r}-\mathbf{r}_1|\,|\mathbf{r}_1-\mathbf{r}_0|. \tag{2·13}$$

Clearly we may now insert (2·13) into (2·5) to obtain C to second order, and as can be seen by inspection of (2·13), the basic integral over β_1 is of the same form as in the first-order case, and can therefore be obtained directly from (2·7) and (2·12). Thus, in a straightforward way, we obtain the perturbation series of Green and the result may be written

$$C(\mathbf{r}, \mathbf{r}_0, \beta) = \sum_{j=0}^{\infty} C_j(\mathbf{r}, \mathbf{r}_0, \beta), \tag{2·14}$$

where

$$C_j(\mathbf{r}, \mathbf{r}_0, \beta) = (2\pi\beta)^{-\frac{3}{2}}\int^j \prod_{l=1}^{j}\left\{\frac{-d\mathbf{r}_l\,V(\mathbf{r}_l)}{2\pi}\right\}\left(\sum_{l=1}^{j+1} s_l\right)\exp\left\{-\frac{1}{2\beta}\left(\sum_{l=1}^{j+1} s_l\right)^2\right\} \div \prod_{l=1}^{j+1} s_l,$$

and $s_l = |\mathbf{r}_l - \mathbf{r}_{l-1}|$, $\mathbf{r}_{j+1} = \mathbf{r}$.

To obtain the Dirac matrix ρ, which is our main concern here, we now exploit the relation between C and ρ which was established by March & Murray (1960*b*), namely

$$C(\mathbf{r}, \mathbf{r}_0, \beta) = \beta\int_0^\infty dE\exp(-\beta E)\,\rho(\mathbf{r}, \mathbf{r}_0, E). \tag{2·15}$$

Introducing the spherical Bessel function $j_1(\rho) = (\sin\rho - \rho\cos\rho)/\rho^2$ we have the identity

$$\frac{1}{(2\pi)^{\frac{3}{2}}\beta^{\frac{5}{2}}}\exp\left(\frac{-s^2}{2\beta}\right) = \frac{1}{\pi^2}\int_0^\infty dE\exp(-\beta E)\frac{E}{s}j_1[s\sqrt{(2E)}] \tag{2·16}$$

and hence, setting $E = \frac{1}{2}k^2$ we have

$$\rho(\mathbf{r},\mathbf{r}_0,k) = \sum_{j=0}^{\infty} \rho_j(\mathbf{r},\mathbf{r}_0,k), \tag{2·17}$$

where $$\rho_j(\mathbf{r},\mathbf{r}_0,k) = \frac{k^2}{2\pi^2}\int^j \prod_{l=1}^{j}\left\{\frac{-\mathrm{d}\mathbf{r}_l\, V(\mathbf{r}_l)}{2\pi}\right\} j_1\left(k\sum_{l=1}^{j+1} s_l\right) \div \prod_{l=1}^{j+1} s_l.$$

In particular $$\rho_0(\mathbf{r},\mathbf{r}_0,k) = \frac{k^2}{2\pi^2}\frac{j_1(k\,|\mathbf{r}-\mathbf{r}_0|)}{|\mathbf{r}-\mathbf{r}_0|} \tag{2·18}$$

and $$\rho_1(\mathbf{r},\mathbf{r}_0,k) = \frac{-k^2}{2\pi^2}\int \mathrm{d}\mathbf{r}_1 \frac{V(\mathbf{r}_1)}{2\pi}\frac{j_1(k\,|\mathbf{r}-\mathbf{r}_1|+k\,|\mathbf{r}_1-\mathbf{r}_0|)}{|\mathbf{r}-\mathbf{r}_1|\,|\mathbf{r}_1-\mathbf{r}_0|}. \tag{2·19}$$

Equation (2·17) forms then the basis of the proposed treatment of localized defects in a free-electron metal and in view of its importance in our work we thought it worthwhile to present a quite different derivation in appendix 1.

3. Case of slowly varying potential

We consider the form

$$\rho_{1\epsilon}(\mathbf{r}_2,\mathbf{r}_0,k) = -\frac{k^2}{2\pi}\int \frac{V(\mathbf{r}_1)\,\mathrm{d}\mathbf{r}_1}{2\pi}\frac{j_1(k\,|\mathbf{r}_2-\mathbf{r}_1|+k\,|\mathbf{r}_1-\mathbf{r}_0|+k\epsilon)}{|\mathbf{r}_2-\mathbf{r}_1|\,|\mathbf{r}_1-\mathbf{r}_0|},$$

which reduces to ρ_1 above by setting $\epsilon = 0$. The form of $\rho_{1\epsilon}$ is that required for obtaining ρ_2, ρ_3, etc. We now take the transformations shown in figure 1 and write

$$\alpha = |\mathbf{r}_2-\mathbf{r}_1|, \quad \beta = |\mathbf{r}_1-\mathbf{r}_0|, \quad \gamma = \alpha+\beta.$$

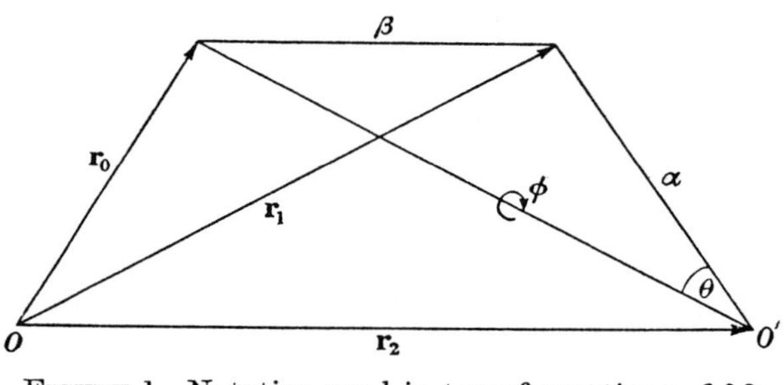

FIGURE 1. Notation used in transformations of § 3.

Moving the origin from O to O' and writing

$$\mathrm{d}\mathbf{r}_1 = \alpha^2\,\mathrm{d}\alpha\sin\theta\,\mathrm{d}\theta\,\mathrm{d}\phi,$$

then, since

$$\beta^2 = \alpha^2+|\mathbf{r}_2-\mathbf{r}_0|^2-2\alpha\,|\mathbf{r}_2-\mathbf{r}_0|\cos\theta, \quad \beta\,\mathrm{d}\beta = \alpha\,|\mathbf{r}_2-\mathbf{r}_0|\sin\theta\,\mathrm{d}\theta,$$

we have $$\mathrm{d}\mathbf{r}_1 = \frac{\alpha\beta}{|\mathbf{r}_2-\mathbf{r}_0|}\mathrm{d}\alpha\,\mathrm{d}\beta\,\mathrm{d}\phi = \frac{\alpha\beta}{|\mathbf{r}_2-\mathbf{r}_0|}\mathrm{d}\alpha\,\mathrm{d}\gamma\,\mathrm{d}\phi.$$

Therefore $$\rho_{1\epsilon}(\mathbf{r}_2,\mathbf{r}_0,k) = -\frac{k^2V}{2\pi^2}\int_{|\mathbf{r}_2-\mathbf{r}_0|}^{\infty}\mathrm{d}\gamma\int_{\frac{1}{2}(\gamma-|\mathbf{r}_2-\mathbf{r}_0|)}^{\frac{1}{2}(\gamma+|\mathbf{r}_2-\mathbf{r}_0|)}\mathrm{d}\alpha\frac{j_1(k\gamma+k\epsilon)}{|\mathbf{r}_2-\mathbf{r}_0|},$$

where V is to be evaluated at the mean of $\mathbf{r}_2$ and $\mathbf{r}_0$, and the assumption of slow variations in V has been made. We have then explicitly that

$$\rho_{1\epsilon}(\mathbf{r}_2,\mathbf{r}_0,k) = -\frac{V}{2\pi^2}\frac{\sin(k\,|\mathbf{r}_2-\mathbf{r}_0|+k\epsilon)}{|\mathbf{r}_2-\mathbf{r}_0|+\epsilon}.$$

Setting $\epsilon = 0$ we have

$$\rho_1(\mathbf{r}, \mathbf{r}_0, k) = -\frac{V[\frac{1}{2}(\mathbf{r}+\mathbf{r}_0)]}{2\pi^2}\frac{\sin k|\mathbf{r}-\mathbf{r}_0|}{k|\mathbf{r}-\mathbf{r}_0|},$$

and in particular

$$\rho_1(\mathbf{r}, \mathbf{r}, k) = -kV(\mathbf{r})/2\pi^2.$$

Similarly ρ_2 may be found by replacing ϵ by $|\mathbf{r}_2-\mathbf{r}|$ and performing the integration over $\mathbf{r}_2$. The diagonal element is then

$$\rho_2(\mathbf{r}, \mathbf{r}, k) = V^2/4\pi^2 k.$$

Likewise, we find

$$\rho_3(\mathbf{r}, \mathbf{r}, k) = V^3/12\pi^2 k^3, \quad \text{etc.}$$

Thus we may write

$$\rho(\mathbf{r}, \mathbf{r}, k) = \frac{k^3}{6\pi^2}\left(1-\frac{3V}{k^2}+\frac{3V^2}{2k^4}+\frac{V^3}{2k^6}+\dots\right)$$

$$= \frac{1}{6\pi^2}(k^2-2V(\mathbf{r}))^{\frac{3}{2}}, \tag{3·1}$$

which is the usual result for slowly varying potentials.

4. Self-consistent first-order perturbation theory

We now begin the detailed application of the theory by giving the exact self-consistent solution to first-order in perturbation theory for a point charge embedded in a free electron gas. It follows from (2·18) and (2·19) that, for doubly filled levels

$$\rho(\mathbf{r}, \mathbf{r}, k) = \frac{k^3}{3\pi^2}-\frac{k^2}{2\pi^3}\int d\mathbf{r}_1\, V(\mathbf{r}_1)\frac{j_1(2k|\mathbf{r}-\mathbf{r}_1|)}{|\mathbf{r}-\mathbf{r}_1|^2}. \tag{4·1}$$

The solution of Poisson's equation for a density $(\rho_0-\rho)$ may be written

$$V(\mathbf{r}) = -\frac{Z}{r}-\frac{k^2}{2\pi^3}\int d\mathbf{r}_0\int d\mathbf{r}_1\frac{V(\mathbf{r}_1)j_1(2k|\mathbf{r}_0-\mathbf{r}_1|)}{|\mathbf{r}_0-\mathbf{r}_1|^2|\mathbf{r}-\mathbf{r}_0|}. \tag{4·2}$$

Now let $|\mathbf{r}_1-\mathbf{r}_0| = s$, $|\mathbf{r}_1| = t$, $|\mathbf{r}_1-\mathbf{r}| = u$, $|\mathbf{r}_0-\mathbf{r}| = v$, and then it follows that

$$d\mathbf{r}_0 d\mathbf{r}_1 = \frac{s\,ds\,v\,dv\,d\phi}{u}\frac{t\,dt\,u\,du\,d\psi}{r}$$

$$= \frac{stv}{r}ds\,dv\,dt\,du\,d\phi\,d\psi. \tag{4·3}$$

The integrations over ϕ and ψ (see figure 2) can be carried out at once, and we then obtain

$$rV = -Z-\frac{2k^2}{\pi}\int_0^\infty dt\int_{|r-t|}^{r+t}du\int_0^\infty ds\int_{|u-s|}^{u+s}dv\,t\,V(t)\frac{j_1(2ks)}{s}$$

$$= -Z-\frac{2k^2}{\pi}\int_0^\infty dt\,t\,V(t)\int_{|r-t|}^{r+t}du\left\{\int_0^u ds\,2j_1(2ks)+\int_u^\infty ds\,2u\frac{j_1(2ks)}{s}\right\}$$

$$= -Z-\frac{1}{\pi}\int_0^\infty dt\,t\,V(t)\int_{2k|r-t|}^{2k(r+t)}du\left\{\int_0^u ds\,j_1(s)+u\int_u^\infty ds\frac{j_1(s)}{s}\right\}. \tag{4·4}$$

Now
$$\int_0^u \mathrm{d}s\, j_1(s) + u\int_u^\infty \mathrm{d}s\frac{j_1(s)}{s}$$
$$= 1-\frac{\sin u}{u}+\frac{u}{2}\left(\int_u^\infty \mathrm{d}s\,\frac{\sin s}{s}+\frac{\sin u}{u^2}-\frac{\cos u}{u}\right)$$
$$= \frac{1}{2}\left(2-\cos u-\frac{\sin u}{u}+u\int_u^\infty \mathrm{d}s\,\frac{\sin s}{s}\right)$$
$$= \frac{1}{4}\frac{\mathrm{d}}{\mathrm{d}u}\left(4u-u\cos u-\sin u+(u^2+2)\int_u^\infty \mathrm{d}s\,\frac{\sin s}{s}\right).$$

Therefore
$$rV(r) = -Z-\int_0^\infty \mathrm{d}t\,tV(t)\,\{F(2k\{r+t\})-F(2k\,|r-t|)\}, \tag{4·5}$$

where
$$F(u) = \frac{1}{4\pi}\left\{4u-u\cos u-\sin u+(u^2+2)\int_u^\infty \mathrm{d}s\,\frac{\sin s}{s}\right\}.$$

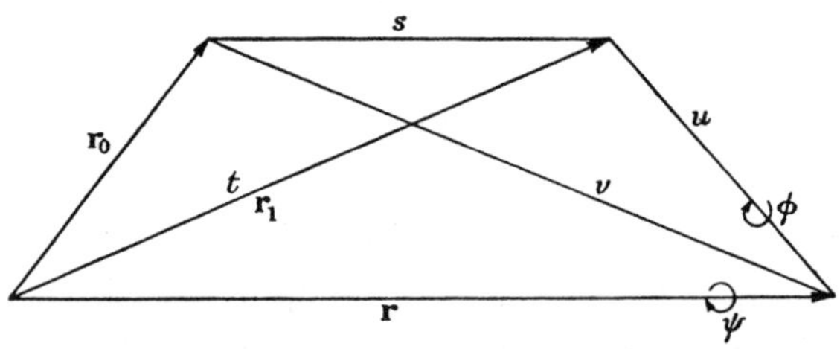

FIGURE 2. Notation used in transformations of § 4.

The forms of $F(u)$ for small and large u are readily obtained and are

$$\text{small } u: \quad F(u) = \frac{1}{4}+\frac{u^2}{8}-\frac{1}{\pi}\left(\frac{u^3}{18}-\frac{u^5}{1800}+\ldots\right);$$

$$\text{large } u: \quad F(u) = \frac{u}{\pi}+\frac{1}{4\pi}\left(\cos u\left\{\frac{20}{u^3}-\frac{672}{u^5}+\ldots\right\}-\sin u\left\{\frac{4}{u^2}-\frac{108}{u^4}+\ldots\right\}\right).$$

The equation corresponding to (4·5) for the radial density of displaced charge $4\pi r^2\rho_1(r)$ is readily shown to be

$$4\pi r^2\rho_1(r) = 4krZF'(2kr)-r\int_0^\infty \mathrm{d}t\,4\pi t\rho_1(t)\,\{F(2k\{r+t\})-F(2k\,|r-t|)\}. \tag{4·6}$$

The integral equations (4·5) and (4·6) may now be solved by replacing them by sets of simultaneous equations, the order of the sets being given by the number of arguments at which V or $4\pi r^2\rho_1(r)$ are being calculated. Details of the method employed are given in appendix 2.

5. Discussion of self-consistent densities

Results for various values of k, the magnitude of the wave vector at the Fermi surface, are presented in table 1 for both the Hartree potential $V(r)$ and the radial density of the displaced charge. The results are tabulated for $Z = -1$ but all cases are thereby included since V and ρ_1 are proportional to Z. Recalling that k and r_s, the radius of a sphere containing one electron, are related by

$$kr_s = (9\pi/4)^{\frac{1}{3}} = 1{\cdot}92, \tag{5·1}$$

and that $2 < r_s < 6$ covers the range of real metallic densities, it will be seen that the results of table 1 cover the interesting cases.

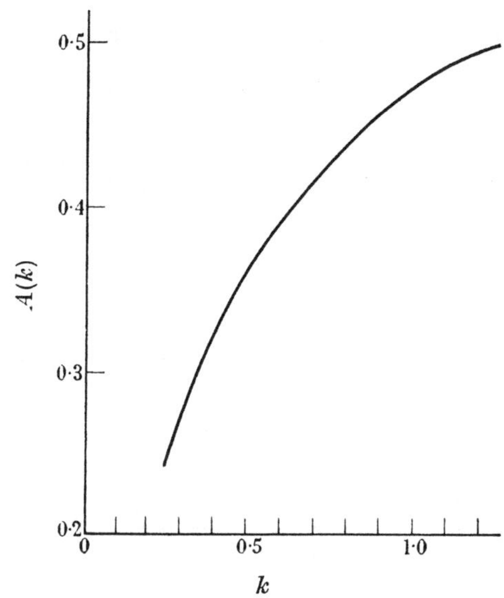

FIGURE 3. Amplitude $A(k)$ in asymptotic equation $4\pi r^2\rho_1(r) = ZA(k)\cos 2kr/2kr$.

We have also thought it worth while to determine the asymptotic form of the density from our numerical solutions. While we have not, as yet, obtained an analytical argument based on (4·1) to justify our form rigorously, the numerical results are fitted to high accuracy by the form

$$4\pi r^2\rho_1(r) = ZA(k)\cos 2kr/2kr, \tag{5·2}$$

where the amplitude $A(k)$ is plotted as a function of k in figure 3. Actually, very recent work by Kohn & Vosko (1960; see also Langer & Vosko 1959) has yielded a form which may be compared with the results of (5·2), and the agreement with their work is excellent for the amplitude, although the phase is somewhat sensitive to the approximations used, as we mention again below.

In addition to the results presented in table 1, we have also carried out calculations for $k = 0{\cdot}72$ ($r_s = 2{\cdot}66$) appropriate to copper metal. We have plotted the radial charge density in figure 4, for comparison with our earlier results on the vacancy problem ($Z = -1$) for a finite monovalent metal containing 10^6 atoms. Also shown in figure 4 are results we have subsequently obtained for $Z = +1$ by a similar finite metal calculation to that described in I, in which approximate self-consistency in the neighbourhood of the impurity has been achieved. However, in view of the similarity between the principal maxima of the radial density for particle numbers of ~ 2000 and 10^6 as seen in figure 3 of I, the results plotted for

TABLE 1. FIRST-ORDER SELF-CONSISTENT SOLUTIONS FOR rV AND $4\pi r^2\rho_1$ FOR $Z = -1$

	rV					$-4\pi r^2\rho_1$					
kr	$k=0\cdot25$	0·5	0·75	1·0	1·25	$k=0\cdot25$	0·5	0·75	1·0	1·25	kr
0·0	+1·0000	+1·0000	+1·0000	+1·0000	+1·0000	+0·0000	+0·0000	+0·0000	+0·0000	+0·0000	0·0
0·2	+0·7688	+0·8098	+0·8322	+0·8473	+0·8585	+0·0258	+0·0321	+0·0360	+0·0388	+0·0410	0·2
0·4	+0·5578	+0·6321	+0·6738	+0·7022	+0·7234	+0·0808	+0·1047	+0·1197	+0·1306	+0·1390	0·4
0·6	+0·3778	+0·4747	+0·5308	+0·5697	+0·5991	+0·1391	+0·1886	+0·2203	+0·2433	+0·2612	0·6
0·8	+0·2354	+0·3427	+0·4077	+0·4537	+0·4889	+0·1845	+0·2630	+0·3143	+0·3520	+0·3816	0·8
1·0	+0·1290	+0·2364	+0·3050	+0·3548	+0·3936	+0·2080	+0·3143	+0·3856	+0·4388	+0·4809	1·0
1·2	+0·0558	+0·1553	+0·2229	+0·2736	+0·3138	+0·2087	+0·3371	+0·4260	+0·4937	+0·5479	1·2
1·4	+0·0103	+0·0965	+0·1597	+0·2087	+0·2484	+0·1888	+0·3305	+0·4329	+0·5125	+0·5773	1·4
1·6	−0·0135	+0·0566	+0·1129	+0·1584	+0·1963	+0·1538	+0·2988	+0·4090	+0·4971	+0·5700	1·6
1·8	−0·0220	+0·0316	+0·0797	+0·1205	+0·1555	+0·1103	+0·2486	+0·3605	+0·4528	+0·5310	1·8
2·0	−0·0206	+0·0177	+0·0572	+0·0927	+0·1242	+0·0650	+0·1879	+0·2957	+0·3882	+0·4685	2·0
2·2	−0·0139	+0·0113	+0·0426	+0·0727	+0·1003	+0·0232	+0·1246	+0·2234	+0·3125	+0·3920	2·2
2·4	−0·0054	+0·0096	+0·0335	+0·0584	+0·0822	−0·0108	+0·0662	+0·1526	+0·2352	+0·3115	2·4
2·6	+0·0025	+0·0101	+0·0278	+0·0480	+0·0683	−0·0344	+0·0178	+0·0899	+0·1640	+0·2354	2·6
2·8	+0·0083	+0·0112	+0·0241	+0·0402	+0·0573	−0·0471	−0·0171	+0·0405	+0·1053	+0·1705	2·8
3·0	+0·0115	+0·0119	+0·0211	+0·0340	+0·0483	−0·0492	−0·0371	+0·0070	+0·0624	+0·1211	3·0
3·2	+0·0120	+0·0116	+0·0182	+0·0286	+0·0405	−0·0428	−0·0430	−0·0102	+0·0364	+0·0886	3·2
3·4	+0·0105	+0·0102	+0·0152	+0·0236	+0·0337	−0·0303	−0·0371	−0·0130	+0·0262	+0·0724	3·4
3·6	+0·0076	+0·0080	+0·0121	+0·0190	+0·0276	−0·0147	−0·0229	−0·0047	+0·0285	+0·0696	3·6
3·8	+0·0039	+0·0053	+0·0088	+0·0147	+0·0221	+0·0014	−0·0043	+0·0106	+0·0393	+0·0760	3·8
4·0	+0·0004	+0·0025	+0·0057	+0·0108	+0·0173	+0·0153	+0·0147	+0·0283	+0·0538	+0·0869	4·0
4·2	−0·0026	−0·0000	+0·0030	+0·0075	+0·0131	+0·0252	+0·0307	+0·0443	+0·0675	+0·0978	4·2
4·4	−0·0046	−0·0020	+0·0009	+0·0048	+0·0097	+0·0304	+0·0414	+0·0555	+0·0771	+0·1049	4·4
4·6	−0·0056	−0·0032	−0·0006	+0·0027	+0·0070	+0·0304	+0·0453	+0·0599	+0·0800	+0·1056	4·6
4·8	−0·0055	−0·0036	−0·0015	+0·0014	+0·0051	+0·0259	+0·0425	+0·0568	+0·0754	+0·0987	4·8

TABLE 1 (*cont.*)

kr	rV					$-4\pi r^2\rho_1$					kr
	$k=0{\cdot}25$	0·5	0·75	1·0	1·25	$k=0{\cdot}25$	0·5	0·75	1·0	1·25	
5·0	−0·0045	−0·0033	−0·0016	+0·0007	+0·0038	+0·0179	+0·0337	+0·0470	+0·0637	+0·0846	5·0
5·2	−0·0030	−0·0025	−0·0013	+0·0005	+0·0031	+0·0080	+0·0207	+0·0321	+0·0465	+0·0650	5·2
5·4	−0·0012	−0·0014	−0·0007	+0·0007	+0·0027	−0·0022	+0·0058	+0·0144	+0·0263	+0·0421	5·4
5·6	+0·0005	−0·0002	+0·0001	+0·0011	+0·0026	−0·0112	−0·0086	−0·0032	+0·0059	+0·0191	5·6
5·8	+0·0019	+0·0009	+0·0009	+0·0015	+0·0027	−0·0178	−0·0205	−0·0183	−0·0119	−0·0014	5·8
6·0	+0·0028	+0·0018	+0·0015	+0·0018	+0·0027	−0·0212	−0·0283	−0·0289	−0·0250	−0·0169	6·0
6·2	+0·0032	+0·0022	+0·0018	+0·0020	+0·0026	−0·0212	−0·0312	−0·0339	−0·0320	−0·0259	6·2
6·4	+0·0030	+0·0023	+0·0019	+0·0019	+0·0024	−0·0180	−0·0290	−0·0330	−0·0324	−0·0279	6·4
6·6	+0·0024	+0·0019	+0·0016	+0·0017	+0·0021	−0·0123	−0·0224	−0·0267	−0·0268	−0·0234	6·6
6·8	+0·0015	+0·0014	+0·0012	+0·0013	+0·0016	−0·0051	−0·0128	−0·0164	−0·0167	−0·0138	6·8
7·0	+0·0004	+0·0006	+0·0007	+0·0008	+0·0011	+0·0023	−0·0018	−0·0040	−0·0039	−0·0012	7·0
7·2	−0·0006	−0·0001	+0·0001	+0·0003	+0·0006	+0·0089	+0·0088	+0·0084	+0·0092	+0·0120	7·2
7·4	−0·0014	−0·0008	−0·0005	−0·0002	+0·0002	+0·0138	+0·0175	+0·0189	+0·0206	+0·0236	7·4
7·6	−0·0019	−0·0012	−0·0009	−0·0006	−0·0002	+0·0163	+0·0230	+0·0260	+0·0285	+0·0317	7·6
7·8	−0·0020	−0·0015	−0·0011	−0·0008	−0·0004	+0·0162	+0·0246	+0·0287	+0·0318	+0·0352	7·8
8·0	−0·0019	−0·0014	−0·0011	−0·0008	−0·0005	+0·0136	+0·0224	+0·0269	+0·0302	+0·0335	8·0
8·2	−0·0014	−0·0012	−0·0009	−0·0007	−0·0005	+0·0091	+0·0168	+0·0211	+0·0241	+0·0272	8·2
8·4	−0·0008	−0·0008	−0·0007	−0·0005	−0·0003	+0·0035	+0·0090	+0·0123	+0·0148	+0·0174	8·4
8·6	−0·0001	−0·0003	−0·0003	−0·0002	−0·0001	−0·0023	+0·0001	+0·0020	+0·0037	+0·0056	8·6
8·8	+0·0005	+0·0002	+0·0001	+0·0001	+0·0001	−0·0075	−0·0083	−0·0080	−0·0074	−0·0062	8·8
9·0	+0·0010	+0·0006	+0·0004	+0·0003	+0·0003	−0·0113	−0·0151	−0·0164	−0·0167	−0·0163	9·0
9·2	+0·0013	+0·0009	+0·0007	+0·0005	+0·0005	−0·0132	−0·0192	−0·0218	−0·0230	−0·0232	9·2
9·4	+0·0014	+0·0010	+0·0008	+0·0006	+0·0006	−0·0130	−0·0202	−0·0236	−0·0253	−0·0260	9·4
9·6	+0·0012	+0·0010	+0·0008	+0·0006	+0·0006	−0·0109	−0·0180	−0·0216	−0·0236	−0·0245	9·6
9·8	+0·0009	+0·0008	+0·0006	+0·0005	+0·0005	−0·0072	−0·0132	−0·0164	−0·0182	−0·0191	9·8
10·0	+0·0005	+0·0005	+0·0004	+0·0003	+0·0003	−0·0026	−0·0066	−0·0089	−0·0102	−0·0109	10·0

$Z = +1$ are for $\sim$ 2000 atoms. We would draw attention particularly to the following points:

(i) The present results show all the features of the finite metal computations of I. In particular, the long-range oscillations are revealed by a completely self-consistent calculation appropriate to an infinite metal.

(ii) The oscillations agree quite remarkably in amplitude and form with the finite metal computation for the vacancy, although there is a significant phase shift. This shows that with the vacancy calculations of I we had virtually reached the limiting form appropriate to an infinite crystal.

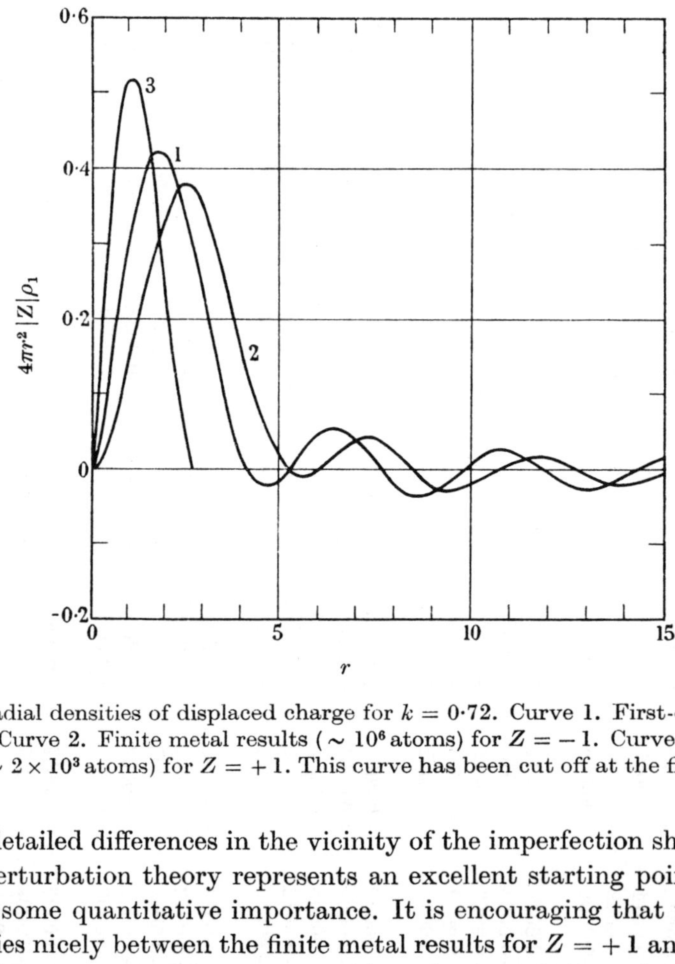

FIGURE 4. Radial densities of displaced charge for $k = 0{\cdot}72$. Curve 1. First-order results for $Z = -1$. Curve 2. Finite metal results ($\sim 10^6$ atoms) for $Z = -1$. Curve 3. Finite metal results ($\sim 2 \times 10^3$ atoms) for $Z = +1$. This curve has been cut off at the first node.

(iii) The detailed differences in the vicinity of the imperfection show that while first-order perturbation theory represents an excellent starting point, the higher terms are of some quantitative importance. It is encouraging that the first-order calculation lies nicely between the finite metal results for $Z = +1$ and -1.

(iv) The Hartree potentials given in table 1 show, we believe for the first time, the correct self-consistent oscillations outside the nearest-neighbour distance. It is easily shown that for Cu there is excellent agreement with Mott's first-order potential $\exp(-qr)/r$, with $q^2 = 4k/\pi$, out to the near-neighbour distance (Mott 1936; see also Mott & Jones 1936, p. 86). This is encouraging in the light of current theories of impurity diffusion in metals (Lazarus 1949; see also Alfred & March 1957 and Le Claire 1961), and Mr G. K. Corless and one of us (N. H. M.) are now engaged in calculating impurity–vacancy interaction energies on the basis of equation (4·1).

One further point seems to deserve brief comment. It is possible to derive an expression for $\rho_1(0)/\rho_0$ from (4·1) in the high density limit ($r_s \to 0$), and the results of this Hartree calculation are compared with previous approximations in appendix 3.

6. Generalization to the case of a band structure

So far, all our considerations have been based on the use of plane waves for describing the electronic structure of the unperturbed metal. It is, of course, very desirable that Bloch waves be inserted eventually into the theory. Unfortunately, however, at present we do not know the canonical density matrix or the Dirac matrix in any periodic potential problem except the free electron case.

However, it seemed of interest finally to point out that some progress might be possible along the present lines for the case of a band structure. In particular, we show that subject to sufficiently slow variations in the perturbing potential, Friedel's generalized first-order approximation may be obtained (Friedel 1954). The argument consists in returning to the basic integral equation (2·4) for the canonical matrix, and taking C_0 now as an unperturbed solution based on Bloch waves. If we then consider the slowly varying potential case we may write, inserting C_0 for C,

$$C(\mathbf{r},\mathbf{r},\beta) = C_0(\mathbf{r},\mathbf{r},\beta) - V(\mathbf{r})\int d\mathbf{r}_1 \int_0^\beta d\beta_1 C_0(\mathbf{r},\mathbf{r}_1,\beta-\beta_1)\,C_0(\mathbf{r}_1,\mathbf{r},\beta_1). \tag{6·1}$$

But it is now easily shown that

$$\int d\mathbf{r}_1 C_0(\mathbf{r},\mathbf{r}_1,\beta-\beta_1)\,C_0(\mathbf{r}_1,\mathbf{r},\beta_1) = C_0(\mathbf{r},\mathbf{r},\beta) \tag{6·2}$$

and hence we may write

$$\begin{aligned} C(\mathbf{r},\mathbf{r},\beta) &= C_0(\mathbf{r},\mathbf{r},\beta) - V(\mathbf{r})\int_0^\beta d\beta_1 C_0(\mathbf{r},\mathbf{r},\beta) \\ &= C_0(\mathbf{r},\mathbf{r},\beta) - V(\mathbf{r})\,\beta C_0(\mathbf{r},\mathbf{r},\beta). \end{aligned} \tag{6·3}$$

Using the inverse Laplace transform relation, we then have

$$\begin{aligned} \rho(\mathbf{r},\mathbf{r},E) &= \frac{1}{2\pi i}\int_{\sigma-i\infty}^{\sigma+i\infty} d\beta \frac{e^{\beta E}}{\beta} C(\mathbf{r},\mathbf{r},\beta) \\ &= \frac{1}{2\pi i}\int_{\sigma-i\infty}^{\sigma+i\infty} d\beta \frac{e^{\beta E}}{\beta} C_0(\mathbf{r},\mathbf{r},\beta) - \frac{V(\mathbf{r})}{2\pi i}\int_{\sigma-i\infty}^{\sigma+i\infty} d\beta\, e^{\beta E} C_0(\mathbf{r},\mathbf{r},\beta). \end{aligned} \tag{6·4}$$

To obtain the second term in (6·4) we return to (2·15), integrate by parts, and invert the resulting Laplace transform and it then follows that

$$\frac{\partial \rho_0(\mathbf{r},\mathbf{r},E)}{\partial E} = \frac{1}{2\pi i}\int_{\sigma-i\infty}^{\sigma+i\infty} d\beta\, e^{\beta E} C_0(\mathbf{r},\mathbf{r},\beta). \tag{6·5}$$

Thus, using (6·5) in (6·4), we have

$$\rho(\mathbf{r},\mathbf{r},E) = \rho_0(\mathbf{r},\mathbf{r},E) - V(\mathbf{r})\,\partial\rho_0(\mathbf{r},\mathbf{r},E)/\partial E, \tag{6·6}$$

which is just Friedel's generalized first-order result. Usually the further replacement of $\partial\rho_0/\partial E$ by the density of states $N(E)$ is made, before attempting to use (6·6) in Poisson's equation to calculate V.

7. Conclusions

The present work lays the foundations for the accurate quantitative calculation of Dirac's density matrix in a form suitable to deal with localized defects in metals. It is not our purpose here to consider detailed applications of our results. Nevertheless, we might point out that in addition to the diffusion of impurities in metals to which we have already made reference, in such problems as residual resistivity, Knight shifts in dilute alloys and the influence of electric quadrupole coupling on nuclear magnetic resonance intensity (Rowland 1960; Kohn & Vosko 1960; Flynn & Seymour 1960) the present approach will undoubtedly be of practical utility. However, we are now proceeding to more detailed calculations, both on the Bloch problem, and on the second-order corrections, before embarking on these applications.

We wish to acknowledge that this work was performed under partial contractual support from the U.S. Army through its European Research Office. Thanks are due to Mr P. R. Becker for helpful discussions of the numerical methods employed and to Mr P. H. Blundell for enabling us to make use of the Ferranti Pegasus computer which is available to the University Computation Laboratory.

References

Donovan, B. & March, N. H. 1959 *J. Phys. Chem. Solids*, **11**, 68.
Flynn, C. P. & Seymour, E. F. W. 1960 *Proc. Phys. Soc.* **76**, 526.
Friedel, J. 1954 *Advanc. Phys.* **3**, 44.
Green, H. S. 1952 *J. Chem. Phys.* **20**, 1274.
Kohn, W. & Vosko, S. H. 1960 *Phys. Rev.* **119**, 912.
Langer, J. S. & Vosko, S. H. 1959 *J. Phys. Chem. Solids*, **12**, 196.
Lazarus, D. 1954 *Phys. Rev.* **93**, 973.
Le Claire, A. D. 1961 *Phil. Mag.* (to be published).
March, N. H. & Murray, A. M. 1960*a* *Proc. Roy. Soc.* A, **256**, 400.
March, N. H. & Murray, A. M. 1960*b* *Phys. Rev.* **120**, 830.
Margenau, H. & Murphy, G. M. 1956 *The mathematics of physics and chemistry*. New York: (Van Nostrand).
Mott, N. F. 1936 *Proc. Camb. Phil. Soc.* **32**, 281.
Mott, N. F. & Jones, H. 1936 *The theory of the properties of metals and alloys*: Oxford University Press.
Quinn, J. J. & Ferrell, R. A. 1958 *Bull. Amer. Phys. Soc.* **3**, 53.
Rowland, T. J. 1960 *Phys. Rev.* **119**, 900.

Appendix 1. Derivation of expression for $\rho(\mathbf{r}, \mathbf{r}_0, k)$ from Schrödinger's equation

Setting $2E = k^2$ in Schrödinger's equation we have

$$\nabla^2\psi + k^2\psi = 2V\psi. \tag{A 1·1}$$

The solution of (A 1·1) with the right-hand side zero which is everywhere finite is

$$\psi(\mathbf{r}) = A\frac{\sin k|\mathbf{r}_1-\mathbf{r}|}{|\mathbf{r}_1-\mathbf{r}|},$$

or any function of the form

$$\psi(\mathbf{r}) = \int d\mathbf{r}_1 A(\mathbf{r}_1)\frac{\sin k|\mathbf{r}_1-\mathbf{r}|}{|\mathbf{r}_1-\mathbf{r}|}.$$

A particular integral of (A 1·1) which may be obtained by using the Green function $\cos kr/r$ is

$$\psi(\mathbf{r}) = -\int d\mathbf{r}_1 2V(\mathbf{r}_1)\psi(\mathbf{r}_1)\frac{\cos k|\mathbf{r}_1-\mathbf{r}|}{4\pi|\mathbf{r}_1-\mathbf{r}|}.$$

Hence (A 1·1) may be written in the form

$$\psi(\mathbf{r}) = -\int d\mathbf{r}_1 V(\mathbf{r}_1)\psi(\mathbf{r}_1)\frac{\cos k|\mathbf{r}_1-\mathbf{r}|}{2\pi|\mathbf{r}_1-\mathbf{r}|} + \int d\mathbf{r}_1 A(\mathbf{r}_1)\frac{\sin k|\mathbf{r}_1-\mathbf{r}|}{|\mathbf{r}_1-\mathbf{r}|}, \tag{A 1·2}$$

where $A(\mathbf{r}_1)$ may be a function of V to be determined by the boundary conditions. We now introduce the quantity $\sigma(\mathbf{r}, \mathbf{r}_0)$ defined by

$$\sigma(\mathbf{r}, \mathbf{r}_0) = \frac{2\pi^2}{k}\frac{\partial\rho(\mathbf{r}, \mathbf{r}_0, k)}{\partial k}, \tag{A 1·3}$$

from which it is clear that

$$\rho(\mathbf{r}, \mathbf{r}_0, k) = \int_0^k \frac{k}{2\pi^2}\sigma(\mathbf{r}, \mathbf{r}_0)\,dk. \tag{A 1·4}$$

It may be shown that $\sigma(\mathbf{r}, \mathbf{r}_0)$ is related to the average wave function product for energy equal to $\frac{1}{2}k^2$ and hence satisfies (A 1·1). For free electrons (singly filled states)

$$\rho_0(\mathbf{r}, \mathbf{r}_0, k) = \frac{k^3}{2\pi^2}\frac{j_1(k|\mathbf{r}-\mathbf{r}_0|)}{k|\mathbf{r}-\mathbf{r}_0|},$$

and therefore from (A 1·3)

$$\sigma_0(\mathbf{r}, \mathbf{r}_0) = \frac{\sin k|\mathbf{r}-\mathbf{r}_0|}{|\mathbf{r}-\mathbf{r}_0|}. \tag{A 1·5}$$

Replacing $\psi(\mathbf{r})$ in (A 1·2) by $\sigma(\mathbf{r}, \mathbf{r}_0)$, we have the integral equation

$$\sigma(\mathbf{r}, \mathbf{r}_0) = -\int d\mathbf{r}_1 V(\mathbf{r}_1)\sigma(\mathbf{r}_1, \mathbf{r}_0)\frac{\cos k|\mathbf{r}_1-\mathbf{r}|}{2\pi|\mathbf{r}_1-\mathbf{r}|} + \int d\mathbf{r}_1 A(\mathbf{r}_1, \mathbf{r}_0)\frac{\sin k|\mathbf{r}_1-\mathbf{r}|}{|\mathbf{r}_1-\mathbf{r}|}, \tag{A 1·6}$$

where $A(\mathbf{r}_1, \mathbf{r}_0)$ is to be determined by the condition that $\sigma(\mathbf{r}, \mathbf{r}_0)$ is symmetrical.

Equation (A 1·6) may now be solved by iteration in powers of V. Thus, we write

$$\sigma(\mathbf{r}, \mathbf{r}_0) = \sum_{j=0}^{\infty}\sigma_j(\mathbf{r}, \mathbf{r}_0),$$

where $\sigma_0(\mathbf{r},\mathbf{r}_0)$ is given by (A 1·5). Inserting $\sigma_0(\mathbf{r},\mathbf{r}_0)$ in the right-hand side of (A 1·6) we have

$$\sigma_1(\mathbf{r},\mathbf{r}_0) = -\int d\mathbf{r}_1 \frac{V(\mathbf{r}_1)}{2\pi}\frac{\sin k\,|\mathbf{r}_0-\mathbf{r}_1|}{|\mathbf{r}_0-\mathbf{r}_1|}\frac{\cos k\,|\mathbf{r}_1-\mathbf{r}|}{|\mathbf{r}_1-\mathbf{r}|} - \int d\mathbf{r}_1 \frac{V(\mathbf{r}_1)}{2\pi}\frac{\cos k\,|\mathbf{r}_0-\mathbf{r}_1|}{|\mathbf{r}_0-\mathbf{r}_1|}\frac{\sin k\,|\mathbf{r}_1-\mathbf{r}|}{|\mathbf{r}_1-\mathbf{r}|},$$

the form of the second integral being fixed by the requirements of symmetry. Thus

$$\sigma_1(\mathbf{r},\mathbf{r}_0) = -\int d\mathbf{r}_1 \frac{V(\mathbf{r}_1)}{2\pi}\frac{\sin k(|\mathbf{r}_0-\mathbf{r}_1|+|\mathbf{r}_1-\mathbf{r}|)}{|\mathbf{r}_0-\mathbf{r}_1|\,|\mathbf{r}_1-\mathbf{r}|}.$$

The other terms may be obtained similarly and we find that

$$\sigma_j(\mathbf{r},\mathbf{r}_0) = \int^j \prod_{l=1}^{j}\left\{-d\mathbf{r}_l\frac{V(\mathbf{r}_l)}{2\pi}\right\}\frac{\sin k\left(\sum_{l=1}^{j+1} s_l\right)}{\prod_{l=1}^{j+1} s_l}, \tag{A 1·7}$$

where $s_l = |\mathbf{r}_l-\mathbf{r}_{l-1}|$ with $\mathbf{r}_{j+1} \equiv \mathbf{r}$.

Inserting (A 1·7) into (A 1·4) we have

$$\rho_j(\mathbf{r},\mathbf{r}_0,k) = \frac{1}{2\pi^2}\int^j \prod_{l=1}^{j}\left\{-d\mathbf{r}_l\frac{V(\mathbf{r}_l)}{2\pi}\right\}\frac{1}{\prod_{l=1}^{j+1} s_l}\int_0^k dk\,k\sin\left(k\sum_{l=1}^{j+1} s_l\right)$$

$$= \frac{k^2}{2\pi^2}\int^j \prod_{l=1}^{j}\left\{-d\mathbf{r}_l\frac{V(\mathbf{r}_l)}{2\pi}\right\}\frac{1}{\prod_{l=1}^{j+1} s_l}\,j_1\left(k\sum_{l=1}^{j+1} s_l\right), \tag{A 1·8}$$

which is the expression we obtained previously.

Appendix 2. Numerical methods of solution of equations (4·5) and (4·6)

The integral equations (4·5) and (4·6) cannot be solved by the usual iteration methods since the kernel $\{F(2k(r+t))-F(2k\,|r-t|)\}$ is not bounded, tending to $4kt/\pi$ for large r. It is easily shown that the solution of (4·5) must satisfy the relation

$$\int_0^\infty dt\,t^2 V(t) = -\pi Z/4k. \tag{A 2·1}$$

Any trial function not satisfying (A 2·1) leads to divergence of V for large r. A starting function in an iteration method which satisfies (A 2·1) does not, in general give a second approximation which satisfies the relation.

A first approximation to the solution of (4·5) was obtained by terminating the integral at a finite distance from the origin. The range of integration was divided into 56 equal intervals and the Newton–Cotes integration formula with nine ordinates applied seven times. The resulting set of 56 simultaneous linear equations (the equation for $r = 0$ being trivial) was solved to give rV. It was found more convenient to obtain $4\pi r^2\rho_1$ by using (4·6) and inserting a second set of right-hand sides than by double differentiation. The number of equations was the largest which could conveniently be handled in the 4096 word store of the Pegasus computer.

The asymptotic forms of the solutions were obtained by fitting curves of the form $\cos 2kr/(2kr)^n$ and $\sin 2kr/(2kr)^n$ along part of the range by a least squares method. The leading terms in the asymptotic expansion of rV and $4\pi r^2\rho_1$ were found to be const. $\times \cos 2kr/(2kr)^2$ and const. $\times \cos 2kr/2kr$, respectively, the sine terms being appreciable only with higher powers of r in the denominator. Substituting these leading terms into the part of the integrals previously neglected, corrections to the equations were obtained analytically. The corrections to the solutions were found to be significant only in the immediate neighbourhood of the point at which the integrals were terminated.

The function $F(u)$ was tabulated at interval 0·2, being computed for large u from the asymptotic expansion, while for small u tables of the sine integral were used. Thus the integral could be terminated at kr equal to any multiples of 5·6. Programs were written for the solution of the integral equation (including the above corrections) with the integral terminated at $kr = 11{\cdot}2$, $22{\cdot}4$ and $28{\cdot}0$. The results of table 1 were obtained with the first of these, while the second and third programs were used to obtain the long-range oscillations and gave a useful check on the accuracy of the solutions, particularly with regard to the end corrections. The accuracy of solutions obtained with the first program should be much greater than that of the others. The results obtained agreed to within about 1 % in the oscillations, the agreement increasing to about 0·1 % nearer the origin.

Appendix 3. Electron density at origin in high-density limit

In view of two formulae in the literature which give the density of electrons at the origin in the one-electron approximation for high densities (Quinn & Ferrell 1958, abstract only; Donovan & March 1959), we obtain here the result to which the first-order equation (4·1) leads.

To do this, it is only necessary to note that we may immediately obtain the leading term by inserting $-Z/r$ for $V(\mathbf{r})$ in (4·1). Then we may write

$$\rho_1(0) = \frac{2k^2 Z}{\pi^2}\int_0^\infty \frac{j_1(2kr_1)}{r_1}\,dr_1. \tag{A 2·1}$$

The integral is easily evaluated and the result is

$$\rho_1(0) = k^2 Z/2\pi. \tag{A 2·2}$$

Using (5·1) and recalling that $\rho_0 = 3/4\pi r_s^3$, we find

$$\rho_1(0)/\rho_0 = \frac{2}{3}\left(\frac{9\pi}{4}\right)^{\frac{2}{3}} Zr_s$$
$$= 2{\cdot}46 Zr_s. \tag{A 2·3}$$

This is to be compared with the result quoted by Quinn & Ferrell (1958) of $2{\cdot}96r_s$ for $Z = 1$ and with $1{\cdot}81r_s$ given by Donovan & March (1959). Owing to lack of detail, we are not able to comment on the reason for the discrepancy between our result and that of Quinn & Ferrell, although it remains somewhat puzzling since these workers apparently use a Hartree method. On the other hand, the lower result of Donovan & March (1959) was based on the von Weizsäcker approximation, and is certainly less accurate than the present calculation.

Proc. Roy. Soc. Lond. A. **317**, 359–366 (1970)
Printed in Great Britain

Lattice dynamics, X-ray scattering and one-body potential theory

BY W. JONES AND N. H. MARCH
Department of Physics, The University, Sheffield

(*Communicated by J. M. Ziman, F.R.S.—Received* 24 *November* 1969)

One-body potential theory, which includes the effect of exchange and correlation forces, is used to calculate the change in the electron density due to small displacements of the ions. The final result contains a Dirac density matrix for the perfect crystal, the diagonal element being the exact ground state density $\rho_0(\boldsymbol{r})$.

The basic quantity $\boldsymbol{R}(\boldsymbol{r})$ determining the electronic contribution to the dynamical matrix is such that the gradient of $\rho_0(\boldsymbol{r})$ is obtained by superposition of $\boldsymbol{R}(\boldsymbol{r}-\boldsymbol{l})$ on each lattice site $\boldsymbol{l}$. An integral equation is obtained which gives $\boldsymbol{R}(\boldsymbol{r})$ uniquely once the exchange and correlation energy is known.

The Fourier transform $\boldsymbol{R}_k$ of $\boldsymbol{R}(\boldsymbol{r})$ is given in terms of the Fourier components ρ_{K_n} of the charge density, which are known from X-ray scattering, by $\boldsymbol{R}_{K_n} = \mathrm{i}\rho_{K_n}\boldsymbol{K}_n$ at the reciprocal lattice vectors $\boldsymbol{K}_n$. This is the same result as the rigid-ion model at the $\boldsymbol{K}_n$'s, which makes the assumption that this is true for all $\boldsymbol{k}$.

Deviations from rigid ions can be evaluated quantitatively from the integral equation obtained here. Such deviations reflect the role of many-body forces in lattice dynamics and the present theory provides a systematic basis for their calculation.

1. INTRODUCTION

Progress in calculating the electron-phonon interaction in metals stemmed from treatments involving the concept of a rigid ion model. Even within more sophisticated modern treatments of lattice dynamics (see, for example, Johnson 1969), the reduction to calculable terms of the dynamical matrix still involves such an approximation. The electronic contribution to the dynamical matrix will be calculated here from one-body potential theory, which can, in principle, take exact account of exchange and correlation forces. An intimate connexion with rigid ions will be established.

As is well known (cf. Vosko, Taylor & Keech 1965) the phonon frequencies ω are determined by the eigenvalue equation

$$\sum_{\beta} D_{\alpha\beta}(\boldsymbol{k})\,\epsilon_k^{\beta} = \omega^2 \epsilon_k^{\alpha}, \tag{1.1}$$

where $\boldsymbol{\epsilon}_k$ is a polarization vector and D the dynamical matrix given in the harmonic approximation by

$$D_{\alpha\beta}(\boldsymbol{k}) = \frac{1}{M}\sum_{l}{}' \Phi_{\alpha\beta}(\boldsymbol{l})\,[\mathrm{e}^{-\mathrm{i}\boldsymbol{k}.\boldsymbol{l}} - 1], \tag{1.2}$$

where

$$\Phi_{\alpha\beta}(\boldsymbol{l}' - \boldsymbol{l}) = \left(\frac{\partial^2 \Phi}{\partial u_{l'}^{\alpha}\, \partial u_{l}^{\beta}}\right)_0, \tag{1.3}$$

Φ being the total potential energy governing the motion of the ions. The vectors $\boldsymbol{l}$ denote the equilibrium sites and the derivative in (1.3) is evaluated at the equilibrium positions.

The electron contribution to $\Phi_{\alpha\beta}$ can be calculated from Feynman's theorem relating the Hamiltonian H to the energy E when a parameter λ is present in H by

$$\frac{\partial E}{\partial \lambda} = \left\langle \Psi_\lambda \left| \frac{\partial H}{\partial \lambda} \right| \Psi_\lambda \right\rangle. \tag{1.4}$$

If $V(r)$ is the potential due to an ion, this equation yields

$$\frac{\partial E}{\partial u_{l'}^{\beta}} = \int \rho(\boldsymbol{r}) \frac{\partial V}{\partial u_l^{\beta}} (\boldsymbol{r} - \boldsymbol{l}' - \boldsymbol{u}_{l'}) \,\mathrm{d}\boldsymbol{r}, \tag{1.5}$$

and therefore

$$\frac{\partial^2 E}{\partial u_l^{\alpha} \,\partial u_{l'}^{\beta}} = \int \frac{\partial \rho(\boldsymbol{r})}{\partial u_l^{\alpha}} \frac{\partial V}{\partial u_{l'}^{\beta}} (\boldsymbol{r} - \boldsymbol{l}' - \boldsymbol{u}_{l'}) \,\mathrm{d}\boldsymbol{r} + \delta_{l'l} \int \rho(\boldsymbol{r}) \frac{\partial^2 V}{\partial u_l^{\alpha} \,\partial u_l^{\beta}} (\boldsymbol{r} - \boldsymbol{l} - \boldsymbol{u}_l) \,\mathrm{d}\boldsymbol{r}. \tag{1.6}$$

The second term of the right-hand side, involving the second derivative of the ionic potential, and representing intrinsic two-phonon processes, is not required for the calculation of the phonon frequencies, as equation (1.2) shows.

The essential point governing our analysis is that the theory of Hohenberg & Kohn (1964; see also Kohn & Sham 1965) shows that there exists a one-body potential which yields the density involved in equation (1.6) exactly.

2. Integral equation for deviations from a rigid ion model

To calculate $\partial \rho / \partial u_l^{\alpha}$ we note that the first-order density change

$$\rho_1(\boldsymbol{r}) = \sum u_l \cdot \frac{\partial \rho(\boldsymbol{r})}{\partial \boldsymbol{u}_l} \tag{2.1}$$

is given by the formal expression

$$\rho_1(\boldsymbol{r}E) = \int \Delta V^{(1)}(\boldsymbol{r}') F(\boldsymbol{r}\boldsymbol{r}'E) \,\mathrm{d}\boldsymbol{r}', \tag{2.2}$$

where (see, for example, Stoddart, March & Stott 1969) F can be expressed in the form

$$\frac{\partial F}{\partial E} = 2\,\mathrm{Re}\left[G_0(\boldsymbol{r}\boldsymbol{r}_1 E_+) \frac{\partial \rho_0}{\partial E} (\boldsymbol{r}\boldsymbol{r}_1 E) \right], \tag{2.3}$$

where G_0 is the perfect lattice Green function and $\rho_0(\boldsymbol{r}\boldsymbol{r}'E)$ is a Dirac density matrix whose diagonal element $\rho_0(\boldsymbol{r})$ is the *exact* crystal density.† The potential $\Delta V^{(1)}(\boldsymbol{r})$ is the *change* occasioned by ionic displacements in the one-body potential incorporating electron–electron interactions. Since we can always write, for a given ionic configuration

$$V'(\boldsymbol{r}) = V_{\text{electrostatic}} + V_{\text{exchange}} + V_{\text{correlation}}, \tag{2.4}$$

where these are *solely* functionals of the electron density, we can express $\Delta V^{(1)}$ to first order in ρ_1 as

$$\Delta V^{(1)} = \Delta V^{(1)}_{\text{electrostatic}} + \int U(\boldsymbol{r}\boldsymbol{r}') \rho_1(\boldsymbol{r}') \,\mathrm{d}\boldsymbol{r}', \tag{2.5}$$

† We emphasize that F is determined solely by one-body and *not* by many-body theory. The local one-body potential which generates the exact density completely determines F.

where $$\Delta V^{(1)}_{\text{electrostatic}} = \int \frac{\rho_1(\boldsymbol{r}')\,\mathrm{d}\boldsymbol{r}'}{|\boldsymbol{r}'-\boldsymbol{r}|} + \sum_l \left[\frac{Ze}{|\boldsymbol{r}-\boldsymbol{l}-\boldsymbol{u}_l|} - \frac{Ze}{|\boldsymbol{r}-\boldsymbol{l}|}\right], \tag{2.6}$$

and Ze represents the charge on each ion. More generally, we could take a potential describing the nucleus plus core electrons, instead of Ze/r. In order to obtain an approximation to U, we can take V_{exchange} and $V_{\text{correlation}}$ to be, for example, the Dirac–Slater and Wigner forms

$$V_{\text{exchange}} = -\left(\frac{3}{\pi}\right)^{\frac{1}{3}} \rho^{\frac{1}{3}}; \quad V_{\text{Wigner}} = -\frac{0.056\rho^{\frac{2}{3}} + 0.0059\rho^{\frac{1}{3}}}{(0.079+\rho^{\frac{1}{3}})^2} \tag{2.7}$$

corrected, when necessary, by $(\nabla\rho)^2$ terms (see, for example, Herman *et al.* 1969).

Equations (2.2) and (2.5) are now to be solved simultaneously and to do so we first write (cf. Johnson 1969)

$$\rho_1(\boldsymbol{r}) = \sum_l \boldsymbol{u}_l \frac{\partial\rho(\boldsymbol{r})}{\partial\boldsymbol{u}_l} = \sum_l \boldsymbol{u}_l \,.\, \boldsymbol{R}_l(\boldsymbol{r}). \tag{2.8}$$

Now if $\boldsymbol{u}_0 . \boldsymbol{R}(\boldsymbol{r})$ represents the density change resulting from a displacement $\boldsymbol{u}_0$ of the ion at the origin say, the physical equivalence of every site implies that the displacement of the ion at $\boldsymbol{l}$ must result in the same density change referred to $\boldsymbol{l}$ as a new origin, i.e. $\boldsymbol{R}_l(\boldsymbol{r}) = \boldsymbol{R}(\boldsymbol{r}-\boldsymbol{l})$. Using this result in (2.8), we can write from (2.5) and (2.6)

$$\Delta V^{(1)}(\boldsymbol{r}) = \sum_l \int \frac{\boldsymbol{u}_l . \boldsymbol{R}(\boldsymbol{r}'-\boldsymbol{l})\,\mathrm{d}\boldsymbol{r}'}{|\boldsymbol{r}'-\boldsymbol{r}|} + \sum_l \left[\frac{Ze}{|\boldsymbol{r}-\boldsymbol{l}-\boldsymbol{u}_l|} - \frac{Ze}{|\boldsymbol{r}-\boldsymbol{l}|}\right]$$
$$+ \sum_l \boldsymbol{u}_l \int U(\boldsymbol{r}\boldsymbol{r}')\,\boldsymbol{R}(\boldsymbol{r}'-\boldsymbol{l})\,\mathrm{d}\boldsymbol{r}'. \tag{2.9}$$

We wish now to show (when ionic terms are expanded to order $\boldsymbol{u}_l$) that this may be expressed in a form analogous to (2.8), namely

$$\Delta V^{(1)}(\boldsymbol{r}) = \sum_l \boldsymbol{u}_l . \boldsymbol{P}(\boldsymbol{r}-\boldsymbol{l}). \tag{2.10}$$

The first two terms on the right-hand side of (2.9) present no difficulty. As for the exchange and correlation terms, we prove in appendix 1 that

$$U(\boldsymbol{r}+\boldsymbol{l}, \boldsymbol{r}'+\boldsymbol{l}) = U(\boldsymbol{r}\boldsymbol{r}'). \tag{2.11}$$

Hence $$\int U(\boldsymbol{r}\boldsymbol{r}')\,\boldsymbol{R}(\boldsymbol{r}'-\boldsymbol{l})\,\mathrm{d}\boldsymbol{r}' = \int U(\boldsymbol{r}\boldsymbol{r}'+\boldsymbol{l})\,\boldsymbol{R}(\boldsymbol{r}')\,\mathrm{d}\boldsymbol{r}'$$
$$= \int U(\boldsymbol{r}-\boldsymbol{l}, \boldsymbol{r}')\,\boldsymbol{R}(\boldsymbol{r}')\,\mathrm{d}\boldsymbol{r}'. \tag{2.12}$$

Thus equation (2.10) follows, with $\boldsymbol{P}(\boldsymbol{r})$ given by

$$\boldsymbol{P}(\boldsymbol{r}) = \int \frac{\boldsymbol{R}(\boldsymbol{r}')\,\mathrm{d}\boldsymbol{r}'}{|\boldsymbol{r}-\boldsymbol{r}'|} - \frac{Ze\,\boldsymbol{r}}{r^2} + \int U(\boldsymbol{r}\boldsymbol{r}')\,\boldsymbol{R}(\boldsymbol{r}')\,\mathrm{d}\boldsymbol{r}'. \tag{2.13}$$

As remarked above, we can generalize (2.13) if we wish by replacing $Ze\,\boldsymbol{r}/r^2$ by the gradient of an appropriate ionic potential $V_b(\boldsymbol{r})$ (for example, a Herman–Skillman (1963) potential).

But we also prove in appendix 1 that

$$F(\boldsymbol{r}+\boldsymbol{l}, \boldsymbol{r}'+\boldsymbol{l}) = F(\boldsymbol{r}\boldsymbol{r}'), \tag{2.14}$$

and hence we find when we insert equation (2.10) into equation (2.2) that

$$\begin{aligned} \rho_1(\boldsymbol{r}) &= \Sigma \boldsymbol{u}_l \cdot \int \boldsymbol{P}(\boldsymbol{r}'-\boldsymbol{l}) F(\boldsymbol{r}\boldsymbol{r}')\,\mathrm{d}\boldsymbol{r}' \\ &= \Sigma \boldsymbol{u}_l \cdot \int \boldsymbol{P}(\boldsymbol{r}') F(\boldsymbol{r}\boldsymbol{r}'+\boldsymbol{l})\,\mathrm{d}\boldsymbol{r}' \\ &= \Sigma \boldsymbol{u}_l \cdot \int \boldsymbol{P}(\boldsymbol{r}') F(\boldsymbol{r}-\boldsymbol{l}, \boldsymbol{r}')\,\mathrm{d}\boldsymbol{r}'. \end{aligned} \tag{2.15}$$

Comparing this equation with equation (2.8) we obtain the basic integral equation of the theory

$$\boldsymbol{R}(\boldsymbol{r}) = \int \boldsymbol{P}(\boldsymbol{r}') F(\boldsymbol{r}\boldsymbol{r}')\,\mathrm{d}\boldsymbol{r}' \tag{2.16}$$

as a consequence of equations (2.11) and (2.14). If the exchange and correlation function $U(\boldsymbol{r}\boldsymbol{r}')$ is known, then equations (2.13) and (2.16) can be solved iteratively to yield $\boldsymbol{R}(\boldsymbol{r})$.

It may in some cases be advantageous to rewrite equations (2.16) and (2.13) in $\boldsymbol{k}$ space and this can be done as follows. From (2.16) we have

$$\begin{aligned} \boldsymbol{R}(\boldsymbol{k}) &= \int F(\boldsymbol{k}\boldsymbol{r}')\,\boldsymbol{P}(\boldsymbol{r}')\,\mathrm{d}\boldsymbol{r}' \\ &= \int F(\boldsymbol{k}\boldsymbol{q})\,\boldsymbol{P}(\boldsymbol{q})\,\mathrm{d}\boldsymbol{q}, \end{aligned} \tag{2.17}$$

$$= \sum_{\boldsymbol{K}} F(\boldsymbol{k}, \boldsymbol{k}+\boldsymbol{K})\,\boldsymbol{P}(\boldsymbol{k}+\boldsymbol{K}). \tag{2.18}$$

Similarly

$$\boldsymbol{P}(\boldsymbol{k}) = \sum_{\boldsymbol{K}'} U'(\boldsymbol{k}, \boldsymbol{k}+\boldsymbol{K}')\,\boldsymbol{R}(\boldsymbol{k}+\boldsymbol{K}') + \mathrm{i}\boldsymbol{k}V_b(\boldsymbol{k}), \tag{2.19}$$

where U' now includes the electrostatic interaction between the electrons. Hence we find

$$\begin{aligned} \boldsymbol{R}(\boldsymbol{k}) = {} & \sum_{\boldsymbol{K}'} \sum_{\boldsymbol{K}} F(\boldsymbol{k}, \boldsymbol{k}+\boldsymbol{K}')\,U'(\boldsymbol{k}+\boldsymbol{K}, \boldsymbol{k}+\boldsymbol{K}'+\boldsymbol{K})\,\boldsymbol{R}(\boldsymbol{k}+\boldsymbol{K}'') \\ & + \mathrm{i} \sum_{\boldsymbol{K}} F(\boldsymbol{k}, \boldsymbol{k}+\boldsymbol{K})\,V_b(\boldsymbol{k}+\boldsymbol{K})\,(\boldsymbol{k}+\boldsymbol{K}). \end{aligned} \tag{2.20}$$

Clearly, given information about the response function F and the correlation function involved in U', we can in principle solve (2.20) for $\boldsymbol{R}(\boldsymbol{k})$.†

† The general response function which modifies F by the interaction function U is discussed in appendix 2.

3. Relation of $\boldsymbol{R}$ to X-ray scattering

It is known that the rigid-ion model follows unambiguously from the assumption of pair forces in lattice dynamics (Johnson 1969). We therefore write

$$\boldsymbol{R}(\boldsymbol{r}) = \nabla\sigma(\boldsymbol{r}) + \boldsymbol{\Delta}(\boldsymbol{r}), \tag{3.1}$$

where the localized density $\sigma(\boldsymbol{r})$ is such that

$$\rho_0(\boldsymbol{r}) = \sum_l \sigma(\boldsymbol{r}-\boldsymbol{l}) \tag{3.2}$$

and curl $\boldsymbol{\Delta} \neq 0$. If we equate all the $\boldsymbol{u}_l$'s in equation (2.8), we must obtain the change in electron density on uniform translation of the lattice through $\boldsymbol{u}_l$ and hence

$$\nabla\rho_0(\boldsymbol{r}) = \sum_l \boldsymbol{R}(\boldsymbol{r}-\boldsymbol{l}). \tag{3.3}$$

Thus it follows that $\boldsymbol{\Delta}(\boldsymbol{r})$ is such that

$$\sum_l \boldsymbol{\Delta}(\boldsymbol{r}-\boldsymbol{l}) = 0. \tag{3.4}$$

In Fourier transform, this equation becomes

$$\boldsymbol{\Delta}_{\boldsymbol{K}} = 0, \tag{3.5}$$

where $\boldsymbol{K}$ is any reciprocal lattice vector, while

$$\boldsymbol{R}_{\boldsymbol{K}} = \mathrm{i}\boldsymbol{K}\sigma(\boldsymbol{K}) = \mathrm{i}\boldsymbol{K}\rho_{\boldsymbol{K}}. \tag{3.6}$$

This is an exact result and the Fourier coefficients of the periodic density are, in principle, known from X-ray scattering experiments (see, for example, Jones, March & Tucker 1965). Clearly, the result (3.6) must be contained in equations (2.13) and (2.16) and hence can be used to check choices of the exchange and correlation function U.

The rigid-ion model agrees *exactly* with the result (3.6) at the reciprocal lattice vectors, but is equivalent to assuming $\boldsymbol{\Delta}(\boldsymbol{r}) = 0$ in (3.1), or that (3.6) is true, for all $\boldsymbol{k}$. In general, equation (3.6) must be replaced by

$$\boldsymbol{R}_k = \mathrm{i}\boldsymbol{\rho}_k \boldsymbol{k}, \tag{3.7}$$

where $\boldsymbol{\rho}_k$ is the Fourier transform of a 'tensor charge density', which reduces exactly to scalar form only when $\boldsymbol{k} = \boldsymbol{K}$.

Deviations of $\boldsymbol{\rho}$ from a scalar quantity when $\boldsymbol{k} \neq \boldsymbol{K}$ imply $\boldsymbol{\Delta}(\boldsymbol{r}) \neq 0$ and so reflect the existence of many-body forces in the ion–ion interaction.† The theory presented here affords a systematic basis for their inclusion in lattice dynamics.

† The elementary results for the pair interaction are shown to follow from the present theory when the response function F is replaced by its free electron approximation, in appendix 3.

Appendix 1

Equations (2.11) and (2.14) can be proved in the following way. On changing the density by $\delta\rho(\boldsymbol{r})$, the resultant change in the parts of the one-body potential incorporating exchange and correlation terms is

$$\delta V_{\mathrm{ex}}(\boldsymbol{r})+\delta V_{\mathrm{corr}}(\boldsymbol{r}) = \int U(\boldsymbol{r}\boldsymbol{r}')\,\delta\rho(\boldsymbol{r}')\,\mathrm{d}\boldsymbol{r}'. \tag{A 1.1}$$

The physical equivalence of every site in the crystal implies that if the density disturbance $\delta\rho(\boldsymbol{r})$ is translated through a lattice vector $\boldsymbol{l}$ to become $\delta\rho(\boldsymbol{r}+\boldsymbol{l})$, the potential $\delta V_{\mathrm{ex}}+\delta V_{\mathrm{corr}}$ will be the original potential also translated through $\boldsymbol{l}$. Thus

$$\delta V_{\mathrm{ex}}(\boldsymbol{r}+\boldsymbol{l})+\delta V_{\mathrm{corr}}(\boldsymbol{r}+\boldsymbol{l}) = \int U(\boldsymbol{r}+\boldsymbol{l},\,\boldsymbol{r}')\,\delta\rho(\boldsymbol{r}')\,\mathrm{d}\boldsymbol{r}' \tag{A 1.2}$$

is also given by

$$\delta V_{\mathrm{ex}}(\boldsymbol{r}+\boldsymbol{l})+\delta V_{\mathrm{corr}}(\boldsymbol{r}+\boldsymbol{l}) = \int U(\boldsymbol{r}\boldsymbol{r}')\,\delta\rho(\boldsymbol{r}'+\boldsymbol{l})\,\mathrm{d}\boldsymbol{r}'. \tag{A 1.3}$$

On change of variable, this implies

$$\int U(\boldsymbol{r}+\boldsymbol{l},\,\boldsymbol{r}')\,\delta\rho(\boldsymbol{r}')\,\mathrm{d}\boldsymbol{r}' = \int U(\boldsymbol{r},\,\boldsymbol{r}'-\boldsymbol{l})\,\delta\rho(\boldsymbol{r}')\,\mathrm{d}\boldsymbol{r}', \tag{A 1.4}$$

and since $\delta\rho(\boldsymbol{r})$ is arbitrary (apart from integrating to zero),

$$U(\boldsymbol{r}+\boldsymbol{l},\,\boldsymbol{r}') = U(\boldsymbol{r},\,\boldsymbol{r}'-\boldsymbol{l}), \tag{A 1.5}$$

which is equivalent to equation (2.11).

From equation (2.2) we may prove

$$F(\boldsymbol{r}+\boldsymbol{l},\,\boldsymbol{r}') = F(\boldsymbol{r},\,\boldsymbol{r}'-\boldsymbol{l}) \tag{A 1.6}$$

in entirely similar fashion.

Appendix 2. General response function

One can write the response to an external perturbation δV_{ext}

$$\delta\rho(\boldsymbol{r}) = \int \mathscr{F}(\boldsymbol{r}\boldsymbol{r}')\,\delta V_{\mathrm{ext}}(\boldsymbol{r}')\,\mathrm{d}\boldsymbol{r}'. \tag{A 2.1}$$

The function F is, of course, the response function for a system of noninteracting particles, with the same unperturbed density. A relation between $\mathscr{F}$ and F is obtainable as follows.

Since
$$\delta V(\boldsymbol{r}) = \delta V_{\mathrm{ext}}(\boldsymbol{r})+\int U'(\boldsymbol{r}\boldsymbol{r}')\,\delta\rho(\boldsymbol{r}')\,\mathrm{d}\boldsymbol{r}', \tag{A 2.2}$$

we also have, by (A 2.1),

$$\delta V(\boldsymbol{r}) = \delta V_{\mathrm{ext}}(\boldsymbol{r})+\int U'(\boldsymbol{r}\boldsymbol{r}')\,\mathscr{F}(\boldsymbol{r}'\boldsymbol{r}'')\,\delta V_{\mathrm{ext}}(\boldsymbol{r}'')\,\mathrm{d}\boldsymbol{r}'\,\mathrm{d}\boldsymbol{r}''. \tag{A 2.3}$$

Inserting this into (2.2) gives, using (A 2.1),

$$\int \mathscr{F}(\boldsymbol{r}\boldsymbol{r}')\,\delta V_{\text{ext}}(\boldsymbol{r}')\,\mathrm{d}\boldsymbol{r}' = \int F(\boldsymbol{r}\boldsymbol{r}')\,\delta V_{\text{ext}}(\boldsymbol{r}')\,\mathrm{d}\boldsymbol{r}' + \int F(\boldsymbol{r}\boldsymbol{r}''')\,U'(\boldsymbol{r}'''\boldsymbol{r}'')\,\mathscr{F}(\boldsymbol{r}''\boldsymbol{r}')\,\delta V_{\text{ext}}(\boldsymbol{r}')\,\mathrm{d}\boldsymbol{r}'\,\mathrm{d}\boldsymbol{r}''\,\mathrm{d}\boldsymbol{r}''', \tag{A 2.4}$$

or, since $\delta V_{\text{ext}}(\boldsymbol{r})$ is arbitrary,

$$\mathscr{F}(\boldsymbol{r}\boldsymbol{r}') = F(\boldsymbol{r}\boldsymbol{r}') + \int F(\boldsymbol{r}\boldsymbol{r}''')\,U'(\boldsymbol{r}'''\boldsymbol{r}'')\,\mathscr{F}(\boldsymbol{r}''\boldsymbol{r}')\,\mathrm{d}\boldsymbol{r}''\,\mathrm{d}\boldsymbol{r}'''. \tag{A 2.5}$$

The corresponding Fourier transform is

$$\mathscr{F}(\boldsymbol{k},\boldsymbol{k}+\boldsymbol{K}) = F(\boldsymbol{k},\boldsymbol{k}+\boldsymbol{K}) + \sum_{\boldsymbol{K}'\boldsymbol{K}''} F(\boldsymbol{k},\boldsymbol{k}+\boldsymbol{K}')\,U'(\boldsymbol{k}+\boldsymbol{K}',\boldsymbol{k}+\boldsymbol{K}'')\,\mathscr{F}(\boldsymbol{k}+\boldsymbol{K}'',\boldsymbol{k}+\boldsymbol{K}). \tag{A 2.6}$$

Instead of solving the integral equation for ρ with explicit and particular form for δV_{ext}, one may obtain a general solution for the static response function by solution of (A 2.5) or (A 2.6).

Appendix 3. Response function F calculated for free electrons and corresponding pair forces

If we define $\sigma(\boldsymbol{r})$ through

$$\phi(\boldsymbol{X}) = \int \sigma(\boldsymbol{r})\,V_b(\boldsymbol{r}-\boldsymbol{X})\,\mathrm{d}\boldsymbol{r}, \tag{A 3.1}$$

then we find

$$\frac{\partial^2\phi(\boldsymbol{X})}{\partial X_\alpha\,\partial X_\beta} = \int \frac{\partial\sigma(\boldsymbol{r})}{\partial x_\alpha}\,\frac{\partial V_b\,(\boldsymbol{r}-\boldsymbol{X})}{\partial X_\beta}\,\mathrm{d}\boldsymbol{r} \tag{A 3.2}$$

and it is evident that

$$R_\alpha(\boldsymbol{r}) = \partial\sigma/\partial x_\alpha, \tag{A 3.3}$$

which is equivalent to the assumption of rigid ions.

For a uniform electron gas, we have

$$F(\boldsymbol{r}\boldsymbol{r}') \equiv F(\boldsymbol{r}-\boldsymbol{r}') \tag{A 3.4}$$

and

$$U(\boldsymbol{r}\boldsymbol{r}') = U(\boldsymbol{r}-\boldsymbol{r}'). \tag{A 3.5}$$

Hence

$$V = \int U(\boldsymbol{r}-\boldsymbol{r}')\,\Delta\rho(\boldsymbol{r}')\,\mathrm{d}\boldsymbol{r}' + V_b(\boldsymbol{r}) + \int\frac{\Delta\rho(\boldsymbol{r}')\,\mathrm{d}\boldsymbol{r}'}{|\boldsymbol{r}-\boldsymbol{r}'|} \tag{A 3.6}$$

where $V_b(\boldsymbol{r})$ is the bare ion potential. Thus we have in Fourier transform

$$V(\boldsymbol{k}) = U(\boldsymbol{k})\,\Delta\sigma_k + V_b(\boldsymbol{k}) + \Delta\sigma_k\,k^{-2} = U'(k)\,\Delta\sigma_k + V_b(\boldsymbol{k}), \tag{A 3.7}$$

where we have written U' for $U+k^{-2}$.

Now we have also that

$$\Delta\sigma_k = F(\boldsymbol{k})\,V(\boldsymbol{k}), \tag{A 3.8}$$

which becomes

$$\Delta\sigma_k = F(\boldsymbol{k})\,U'(\boldsymbol{k})\,\Delta\sigma_k + F(\boldsymbol{k})\,V_b(\boldsymbol{k}). \tag{A 3.9}$$

when we use (A 3.7). Representing the relation between $\Delta\sigma$ and V_b as usual by a dielectric function $\epsilon(\boldsymbol{k})$ we have

$$\Delta\sigma_k = k^2 V_b(\boldsymbol{k})\,[\epsilon^{-1}(\boldsymbol{k}) - 1] \tag{A 3.10}$$

where, by comparing (A 3.9) and (A 3.10)

$$\epsilon^{-1}(k) = \frac{1 - F(k)U(k)}{1 - F(k)U'(k)}. \tag{A 3.11}$$

Explicitly

$$F(\boldsymbol{r}\boldsymbol{r}'E) = -\frac{k^2}{2\pi^3}\frac{j_1(2k\,|\boldsymbol{r}-\boldsymbol{r}'|)}{|\boldsymbol{r}-\boldsymbol{r}'|^2}; \quad E = \frac{k^2}{2}. \tag{A 3.12}$$

Fourier transforming this, we regain the Lindhard theory for a uniform electron gas when $U'(k)$ is replaced by k^{-2}.

From (A 3.1) we may write, adding the ion–ion interaction ϕ_{ii}

$$\begin{aligned}\phi_{\text{total}}(\boldsymbol{X}) &= \iint \Delta\sigma_k\, e^{i\boldsymbol{k}.\boldsymbol{r}}\, d\boldsymbol{k}\, V_b(\boldsymbol{r}-\boldsymbol{X})\, d\boldsymbol{r} + \phi_{ii} \\ &= \int \Delta\sigma_k\, V_b(\boldsymbol{k})\, e^{i\boldsymbol{k}.\boldsymbol{X}}\, d\boldsymbol{k} + \phi_{ii} \\ &= \int \frac{k^2 V_b^2(\boldsymbol{k})}{\epsilon(\boldsymbol{k})}\, e^{i\boldsymbol{k}.\boldsymbol{X}}\, d\boldsymbol{k}. \end{aligned} \tag{A 3.13}$$

This reduces to the $\boldsymbol{r}$ space result of Corless & March (1961) for point ions, whereas for ions with structure the result is the same as that given, for example, by Ziman (1964), though in the present case the dielectric function includes exchange and correlation effects through (A 3.11).

References

Corless, G. K. & March, N. H. 1961 *Phil. Mag.* **6**, 1285.
Herman, F., Van Dyke, J. P. & Ortenburger, I. B. 1969 *Phys. Rev. Lett.* **22**, 807.
Herman, F. & Skillman, S. 1963 *Atomic structure calculations*. New Jersey: Prentice-Hall.
Hohenberg, P. & Kohn, W. 1964 *Phys. Rev.* **136**, B 864.
Jones, W., March, N. H. & Tucker, J. W. 1965 *Proc. Roy. Soc. Lond.* A **284**, 289.
Johnson, F. A. 1969 *Proc. Roy. Soc. Lond.* A **310**, 101, and two preceding papers.
Kohn, W. & Sham, L. J. 1965 *Phys. Rev.* **140**, A 1133.
Stoddart, J. C., March, N. H. & Stott, M. J. 1969 *Phys. Rev.* **186**, 683.
Vosko, S. H., Taylor, R. & Keech, G. H. 1965 *Can. J. Phys.* **43**, 1187.
Ziman, J. M. 1964 *Adv. Phys.* **13**, 89.

ANNALS OF PHYSICS: **64**, 174–210 (1971)

Density-Functional Theory of Magnetic Instabilities in Metals

J. C. STODDART AND N. H. MARCH

Department of Physics, The University of Sheffield, Sheffield, England

Received January 7, 1970

A density-functional formulation of magnetic states in metals is set up in which, initially, a spin-dependent external potential is switched on to the metal. The usual Hartree–Fock treatment of strong short-range correlations in an s-band is shown to follow very simply from the theory.

Building up this Hartree–Fock theory for the lattice by switching on the interactions one site at a time suggests an approximate way of incorporating correlation effects into the density-functional theory. This is consistent with the philosophy that electron correlation effects are best treated via a picture in which the electron density, and, in magnetic metals, the spin density distributions are thought of as localized round atomic sites. In keeping with this philosophy, a new general magnetic instability condition for the lattice is proposed, which is essentially that for the one-site model and involves the density of states through the band, and not simply at the Fermi level, as with the Hartree–Fock condition.

The properties of the Green function associated with this new instability condition are considered. While it is not possible, within the present framework, to resolve completely some ambiguity in higher-order terms of the Green function, nevertheless it is possible to set up a Green function which includes the band and the atomic limits correctly, has required symmetry properties, and exactly conserves states in the band. The density of states in this model does not show simple exchange in the magnetic metal but corresponds to deformation of the magnetic sub-bands by the electron interactions. It seems possible still that the inclusion of the long-range Coulomb correlations may, even in the non-magnetic case, after the basic character of the bands from that of a "cell" model.

1. INTRODUCTION

A great deal of work has been done on the problem of magnetic states in metals over the last decade or so and an extensive review has been given by Herring [1]. Much of this work has centered round model Hamiltonians, of which the investigations of Hubbard [2] are the most notable.

Less attention has been given to an approach, which was historically the earliest, in which a knowledge of the energy of the metal as a function(al) of the spin densities is taken as starting point. This method was used, for example, by Bloch [3] in his paper on the magnetism of a uniform electron gas in the Hartree–Fock approximation and it seems to have fallen into disrepute because of the widespread belief that it is inappropriate for other than a homogeneous electron gas.

The reason for this misunderstanding is also in part due to the fact that this density-functional approach has been identified with the Thomas–Fermi method, with its later refinements due to Dirac and a number of other workers. Early workers had explicitly introduced correlation effects into such Thomas–Fermi schemes, but this was open to criticism in view of uncontrolled approximations made in the one-body terms in the theory. However, the perturbation theory of March and Murray [4] showed that, in principle, there is an exact form of the Thomas–Fermi method in a one-body framework, and so the above specific criticism could be fully met. But a fundamental justification for the introduction of correlations into a method in which the many-body system is characterized by the particle density was given only recently by Hohenberg and Kohn [5] and this has stimulated us to enquire whether there is a useful alternative here to the method of model Hamiltonians. We conclude that this is indeed the case and we present in this paper a new general magnetic instability condition for the lattice, which is based on combining the density-functional approach with the philosophy that correlation effects are best introduced in a framework in which the charge and spin density is thought of as localized round atomic sites [6]. In developing this point of view, we shall see that the emphasis of the theory is shifted from model Hamiltonians to model energy functionals, with, we believe, some definite advantages for building in some physical intuition relating to the role of electron correlations.

To establish the density-functional approach as a useful one in magnetic problems, we shall first set out the general formulation in Section 2 below and then in Section 3 we shall work out the Hartree–Fock treatment of a model of short-range correlations in a narrow s-band. Then, in Section 4, we discuss at some length the introduction of correlation effects into the model, via the ideas of electron localization referred to above. The new instability condition we propose can be incorporated into a Green function, whose properties are analyzed. We shall see that some further motivation for our approach was afforded by interesting features in the treatment of localized moments round impurities in metals by Wolff [7]. In Section 5 we shall establish what to us is a very surprising relation with the instability condition of Richmond and Rickayzen [8], which appeared after much of our work was completed. Their results, as we shall see, are obtained readily as a consequence of the density-functional formalism given here. Finally, in Appendix III, we demonstrate that some of Hubbard's results can also be utilized in a straightforward manner in the density-functional scheme.

2. Variational Principle: General Formulation

The idea behind the density-functional approach is, of course, to write down the total energy of an inhomogeneous electron gas as a functional of its density. Let us suppose that we have spin densities ρ_+ and ρ_- for the upward and downward spin electrons, respectively, and we denote by $\rho(\mathbf{r})$ the total electron density.

We consider first the general case of electrons moving in a spin-dependent external potential $V_{\text{ext}}^{\sigma}(\mathbf{r})$, $\sigma = \pm$. For this system, we write the total energy as a functional of the spin densities ρ_+ and ρ_- as

$$\begin{aligned} E[\rho_+, \rho_-] &= T_s[\rho_+] + T_s[\rho_-] + \frac{1}{2} \iint d\mathbf{r}\, d\mathbf{r}' \frac{\rho(\mathbf{r})\, \rho(\mathbf{r}')}{|\mathbf{r} - \mathbf{r}'|} \\ &+ \int d\mathbf{r}\, [V_{\text{ext}}^{+}(\mathbf{r})\, \rho_+(\mathbf{r}) + V_{\text{ext}}^{-}(\mathbf{r})\, \rho_-(\mathbf{r})] \\ &+ E_x[\rho_+] + E_x[\rho_-] + E_c[\rho_+, \rho_-]. \end{aligned} \tag{2.1}$$

In (2.1), $T_s[\rho_\sigma]$ is the kinetic energy of a noninteracting gas of density ρ_σ and E_x and E_c are the exchange and correlation contributions to the total energy.

Essentially, the spin-dependent external potential V_{ext}^{σ} is introduced, even though we can switch off the spin dependence of the external potential later, to break the directional symmetry of the problem. For example, the external disturbance could be a magnetic field, applied say in the z direction, which would have an energy of interaction with the magnetic moment density $m(\mathbf{r}) = -(1/2c)[\rho_+(\mathbf{r}) - \rho_-(\mathbf{r})]$. Then the total energy may be written as

$$\begin{aligned} E[\rho_+, \rho_-] &\equiv E[\rho, m] \\ &= G[\rho(\mathbf{r}), m(\mathbf{r})] + \frac{1}{2} \int d\mathbf{r}\, d\mathbf{r}' \frac{\rho(\mathbf{r})\, \rho(\mathbf{r}')}{|\mathbf{r} - \mathbf{r}'|} + \int d\mathbf{r}\, [V_{\text{ext}}(\mathbf{r})\, \rho(\mathbf{r})] + E_{\text{field}}, \end{aligned} \tag{2.2}$$

where the last term is simply the energy of interaction with the magnetic field and where $G[\rho, m]$ denotes the total kinetic exchange and correlation energy as a functional of ρ and m, or alternatively, as a functional of ρ_+ and ρ_-.

To be more specific, if we did not introduce spin-dependent external potentials, then it would appear from the rotational symmetry of the problem that only nonmagnetic solutions would be possible, that is $\rho_+ = \rho_-$. Again, from considerations of translational symmetry, it might seem at first sight that only ferromagnetic states would have any significance, whereas, of course, more general magnetic states, such as antiferromagnetic ordering, occur in practice. It should be added that, in spite of the above arguments which necessitate the introduction of a spin-dependent potential into the formalism, when we eventually switch off the spin dependence we regain all the results one would obtain without having recourse to such spin dependence.

We require that $E[\rho_+, \rho_-]$ be stationary with respect to arbitrary variations in $\rho_+(\mathbf{r})$ and $\rho_-(\mathbf{r})$ subject to conservation of the total number of electrons. If we define $t_{\mathbf{r}}[\rho_\sigma]$ by

$$\delta T_s[\rho_\sigma] = \int d\mathbf{r}\ \delta\rho_\sigma(\mathbf{r})\ t_{\mathbf{r}}[\rho_\sigma], \tag{2.3}$$

then the resulting Euler equations have the form

$$t_{\mathbf{r}}[\rho_\sigma] + V_\sigma(\mathbf{r}) = 0, \tag{2.4}$$

where

$$V_\sigma(\mathbf{r}) = V_{\text{ext}}^\sigma(\mathbf{r}) + \int d\mathbf{r}' \frac{\rho(\mathbf{r}')}{|\mathbf{r} - \mathbf{r}'|} + V_{xc}(\mathbf{r}). \tag{2.5}$$

Comparison of (2.4) with the Euler equations obtained when we consider noninteracting electrons moving in external spin dependent potentials $V_{\text{ext}}^\sigma(\mathbf{r})$ shows that we may solve (2.4) by considering spin σ electrons to move in the one-body potential $V_\sigma(\mathbf{r})$, which will give us, via the Schrödinger equation, eigenfunctions $\phi_{\mathbf{k}\sigma}(\mathbf{r})$ and eigenvalues $\epsilon_\sigma(\mathbf{k})$. The density of spin σ electrons is then given by

$$\rho_\sigma(\mathbf{r}) = \sum_{\mathbf{k}(\text{occ})} |\phi_{\mathbf{k}\sigma}(\mathbf{r})|^2 \tag{2.6}$$

with

$$\int d\mathbf{r}[\rho_+(\mathbf{r}) + \rho_-(\mathbf{r})] = N_e = N_+ + N_-. \tag{2.7}$$

The lowest N_σ states are occupied and the numbers of spin σ electrons N_+ and N_- are obtained by filling up the relevant states $\phi_{\mathbf{k}\sigma}(\mathbf{r})$ to the same level μ'. As the one-body potential depends on the density, an iterative procedure is clearly required to solve (2.4), (2.5), and (2.6) for the electron densities.

In an actual crystal lattice, we evidently do not have a spin-dependent external potential, as the external potential arises from the static array of ions. We must therefore let the spin dependence of $V_{\text{ext}}^\sigma(\mathbf{r})$ vanish. However, this would lead us back to complete rotational and translational symmetry and, to retain a preferred direction, we can think of a very small magnetic field remaining, which is so weak that its contribution to the total energy is quite negligible. This device is necessary to give a meaning to the energy functionals $E[\rho_+, \rho_-]$ and $E_c[\rho_+, \rho_-]$, since otherwise the densities ρ_+ and ρ_- cannot be defined: That is they are not independent variables.

We may, therefore, formally modify Eq. (2.1) and let both up and down spin electrons move in the same spin independent external static potential due to the lattice of ions $V_{\text{ext}}(\mathbf{r})$. Then, Eqs. (2.1)–(2.7) constitute the basis for the variational theory which we shall employ. Due to special properties of the one-body potential, which result from our model of electron interactions, much of the complication of the iterative procedure referred to above can be avoided and closed expressions can be obtained for the spin densities. Of course, more sophisticated energy functionals and spin densities can be built into this approach and this will naturally become necessary as the method is refined beyond the fairly primitive starting points of this paper.

A. *Green function for spin-dependent one-body potential*

We first define the Bloch eigenfunctions $\psi_{\mathbf{k}}(\mathbf{r})$ and eigenvalues $\epsilon(\mathbf{k})$ given by the spin-independent average periodic potential $\bar{V}(\mathbf{r})$, which is defined precisely below. This includes the external potential $V_{\text{ext}}(\mathbf{r})$ due to the periodic ionic array and the average effect of electron interactions. This periodic potential defines for us a nonmagnetic lattice and we shall then consider explicitly the many-body interactions between the electrons which can lead to the possibility of a magnetic state. It is clear that these many-body interactions occur only between the spin-density fluctuations from the nonmagnetic state and so, in the density-functional language, we are concerned with that part of the total energy which is a functional of these density fluctuations.

In this model, we can write the total energy (2.1) in the alternative form

$$E[\rho_+, \rho_-] = T_s[\rho_+] + T_s[\rho_-] + \int d\mathbf{r}\ \bar{V}(\mathbf{r})[\rho_+(\mathbf{r}) + \rho_-(\mathbf{r})] + E_I[\rho_+, \rho_-], \quad (2.8)$$

where E_I is thus defined as the many-body interaction energy between the spin-density deviations $\Delta_\sigma(\mathbf{r}) = \rho_\sigma(\mathbf{r}) - \rho_P(\mathbf{r})$ from the nonmagnetic state which has electron density $\rho_P(\mathbf{r})$ for both $\pm$ spin electrons.

This model of the electron interactions requires therefore that the one-body potential has the form

$$V_\sigma(\mathbf{r}) = \bar{V}(\mathbf{r}) + v_\sigma(r), \quad (2.9)$$

where $v_\sigma(\mathbf{r})$ is a functional of $\Delta_\sigma(\mathbf{r})$. The eigenfunctions $\phi_{\mathbf{k}\sigma}(\mathbf{r})$ are defined by

$$[-\tfrac{1}{2}\nabla_{\mathbf{r}}^2 + \bar{V}(\mathbf{r}) + v_\sigma(\mathbf{r})]\ \phi_{\mathbf{k}\sigma}(\mathbf{r}) = \epsilon_\sigma(\mathbf{k})\ \phi_{\mathbf{k}\sigma}(\mathbf{r}) \quad (2.10)$$

and we assume that the $\phi_{\mathbf{k}\sigma}$ can be expanded in terms of the $\psi_{\mathbf{k}}(\mathbf{r})$, or the corresponding Wannier functions $w(\mathbf{r})$. Thus we can write explicitly

$$\phi_{\mathbf{k}\sigma}(\mathbf{r}) = \sum_n b_\sigma(\mathbf{k}, \mathbf{t}_n)\ w(\mathbf{r} - \mathbf{t}_n). \quad (2.11)$$

Instead of calculating the electron density from (2.6), it will be more convenient to obtain it from the Green function $G_\sigma(\mathbf{r}\mathbf{r}_0E)$ defined by the spin-dependent potential $V_\sigma(\mathbf{r})$ given by (2.9). This Green function satisfies the usual integral equation

$$G_\sigma(\mathbf{r}\mathbf{r}_0E) = G_p(\mathbf{r}\mathbf{r}_0E) + \int d\mathbf{r}_1\, G_p(\mathbf{r}\mathbf{r}_1E)\, v_\sigma(\mathbf{r}_1)\, G_\sigma(\mathbf{r}_1\mathbf{r}_0E), \tag{2.12}$$

where G_p is evidently the nonmagnetic lattice Green function. This is given explicitly by

$$G_p(\mathbf{r}\mathbf{r}_0E) = \sum_{\mathbf{k}\gamma} \psi_{\mathbf{k}\gamma}(\mathbf{r})\, \psi_{\mathbf{k}\gamma}(\mathbf{r}_0) \frac{1}{E - \epsilon_\gamma(\mathbf{k}) + i\eta}, \tag{2.13}$$

where γ is a band index, while the electron density is given in terms of the Green function by

$$\rho_\sigma(\mathbf{r}) = -\frac{1}{\pi} \operatorname{Im} \int_{-\infty}^{\mu'} dE\, G_\sigma(\mathbf{r}\mathbf{r}E). \tag{2.14}$$

For future reference, it is convenient to record here that the density of pseudoparticle states for spin σ electrons is given by

$$n_\sigma(E) = -\frac{1}{\pi} \operatorname{Im} \int d\mathbf{r}\, G_\sigma(\mathbf{r}\mathbf{r}E). \tag{2.15}$$

Also, the Fermi energy μ' is defined by

$$\sum_\sigma \int_{-\infty}^{\mu'} dE\, n_\sigma(E) = 2 \int_{-\infty}^{\mu} dE\, n_\gamma(E) = N_e\,, \tag{2.16}$$

where $n_\gamma(E)$ is the nonmagnetic lattice density of states and μ is the corresponding Fermi energy.

3. Hartree–Fock Density Functionals

We turn immediately to discuss the Hartree–Fock approximation in the density-functional approach to magnetic instability. We take the interaction energy E_I to be the Hartree–Fock exchange energy, together with the classical Coulomb energy

$$E_I = E_x[\rho_+] + E_x[\rho_-] + \frac{1}{2} \iint d\mathbf{r}\, d\mathbf{r}' \frac{\rho(\mathbf{r})\, \rho(\mathbf{r}')}{|\mathbf{r} - \mathbf{r}'|} \tag{3.1}$$

with

$$E_x[\rho_\sigma] = -\frac{1}{2} \iint d\mathbf{r}\, d\mathbf{r}' \frac{[\rho_\sigma(\mathbf{r}\mathbf{r}'\mu')]^2}{|\mathbf{r}-\mathbf{r}'|}, \tag{3.2}$$

ρ_σ being the Dirac density matrix

$$\rho_\sigma(\mathbf{r}\mathbf{r}'\mu') = \sum_{\mathbf{k}} \phi^*_{\mathbf{k}\sigma}(\mathbf{r})\, \phi_{\mathbf{k}\sigma}(\mathbf{r}')\, \theta(\mu' - \epsilon_\sigma(\mathbf{k})). \tag{3.3}$$

It will be helpful to work at first with the total spin density as in (3.3), though later we shall change for practical application to the spin-density deviation from the nonmagnetic state, as in (3.13) below.

The exchange energy (3.2) may now be simplified greatly if we assume that the wavefunctions $\phi_{\mathbf{k}\sigma}(\mathbf{r})$ can be expanded in Wannier functions as in (2.11) and that these Wannier functions are sufficiently localized that any overlap between Wannier orbitals on different sites can be neglected. Then it is clear that the spin density $\rho_\sigma(\mathbf{r})$ can be uniquely expressed as a sum of contributions from each lattice site $\mathbf{t}_n$, namely

$$\rho_\sigma(\mathbf{r}) = \sum_n \rho_\sigma^{(n)}(\mathbf{r} - \mathbf{t}_n). \tag{3.4}$$

From (3.2) and (3.4) we then find that, when we consistently drop overlaps,

$$E_x[\rho_\sigma] = -\frac{1}{2} \sum_n \iint d\mathbf{r}\, d\mathbf{r}' \frac{\rho_\sigma^{(n)}(\mathbf{r} - \mathbf{t}_n)\, \rho_\sigma^{(n)}(\mathbf{r}' - \mathbf{t}_n)}{|\mathbf{r}-\mathbf{r}'|}. \tag{3.5}$$

A. *One-body spin-dependent potential*

We shall use the above treatment of exchange to set up an explicit expression for the total one-body potential seen by electrons with spin σ $(= \pm)$. To do so, we note that, if the spin densities overlap only negligibly, an arbitrary variation $\delta\rho_\sigma(\mathbf{r})$ in the total spin density can be built up from arbitrary *independent* variations in $\rho^{(n)}(\mathbf{r} - \mathbf{t}_n)$ in each unit cell, that is

$$\delta\rho_\sigma(\mathbf{r}) \equiv \delta\rho_\sigma^{(n)}(\mathbf{r} - \mathbf{t}_n) \tag{3.6}$$

if the total density is given an arbitrary variation in the cell around $\mathbf{t}_n$.

Defining a spin-dependent exchange potential $v_x^\sigma(\mathbf{r})$ by

$$\delta E_x[\rho_\sigma] = \int d\mathbf{r}\, v_x^\sigma(\mathbf{r})\, \delta\rho_\sigma(\mathbf{r}), \tag{3.7}$$

we obtain from (3.5)

$$\delta E_x{}^\sigma = -\frac{1}{2}\sum_n \iint d\mathbf{r}\, d\mathbf{r}' \left[\frac{\delta\rho_\sigma^{(n)}(\mathbf{r}-\mathbf{t}_n)\,\rho_\sigma^{(n)}(\mathbf{r}'-\mathbf{t}_n)}{|\mathbf{r}-\mathbf{r}'|} + \frac{\rho_\sigma^{(n)}(\mathbf{r}-\mathbf{t}_n)\,\delta\rho_\sigma^{(n)}(\mathbf{r}'-\mathbf{t}_n)}{|\mathbf{r}-\mathbf{r}'|}\right] \tag{3.8}$$

and so

$$v_x{}^\sigma(\mathbf{r}) = -\sum_n \int d\mathbf{r}' \frac{\rho_\sigma^{(n)}(\mathbf{r}'-\mathbf{t}_n)}{|\mathbf{r}-\mathbf{r}'|}. \tag{3.9}$$

Similarly, using (3.4) in the Coulomb energy, we have

$$\delta E_C[\rho_+ + \rho_-] = \int d\mathbf{r}\, V_C(\mathbf{r})\, \delta\rho_\sigma(\mathbf{r}) \tag{3.10}$$

with

$$V_C(\mathbf{r}) = \sum_n \int d\mathbf{r}' \frac{[\rho_+^{(n)}(\mathbf{r}'-\mathbf{t}_n) + \rho_-{}^n(\mathbf{r}'-\mathbf{t}_n)]}{|\mathbf{r}-\mathbf{r}'|}. \tag{3.11}$$

Thus, taking into account the electron interactions in the present model, we find that the total one-body potential felt by spin σ electrons is, in the Hartree–Fock approximation,

$$V_\sigma(\mathbf{r}) = V_{\text{ext}}(\mathbf{r}) + \sum_n \int d\mathbf{r}' \frac{\rho_{-\sigma}^{(n)}(\mathbf{r}'-\mathbf{t}_n)}{|\mathbf{r}-\mathbf{r}'|}. \tag{3.12}$$

B. *Spin-density deviations from nonmagnetic state*

The analysis above has taken two electrons to interact only when they are on the same site and we see that in this approximation we are led to a well-defined interaction energy functional and one-body potential.

It is useful at this stage to reexpress these results in terms of the spin-density deviations from the nonmagnetic state. Defining

$$\Delta_\sigma(\mathbf{r}) = \rho_\sigma(\mathbf{r}) - \rho_p(\mathbf{r}), \tag{3.13}$$

we find that the one-body potential is now given by

$$\begin{aligned} V_\sigma(\mathbf{r}) &= \bar{V}(\mathbf{r}) + \sum_n \int \frac{d\mathbf{r}'\, \Delta_{-\sigma}^{(n)}(\mathbf{r}'-\mathbf{t}_n)}{|\mathbf{r}-\mathbf{r}'|} \\ &= \bar{V}(\mathbf{r}) + v_\sigma^{\text{HF}}(\mathbf{r}) \end{aligned} \tag{3.14}$$

where $\bar{V}(r)$ is the average potential generating the nonmagnetic lattice. It is clear that, in (3.14), $\Delta_{-\sigma}^{(n)}(\mathbf{r}' - \mathbf{t}_n)$ is the density deviation of spin $-\sigma$ electrons on site n from the value in the nonmagnetic state. It is again evident that the spin-dependent part of the one-body potential in the Hartree–Fock approximation is simply a sum of contributions from each atomic site. Each of these contributions involves solely the density deviation of opposite spin electrons on that particular site.

The Hartree–Fock (HF) potential v_σ^{HF} is most usefully discussed in terms of its matrix elements between Wannier states. It is then clear, since the model adopted neglects off-diagonal terms, that v_σ^{HF} has the well-known property of a Koster–Slater impurity potential, namely

$$\int w(\mathbf{r} - \mathbf{t}_n)\, v_\sigma^{\mathrm{HF}}(\mathbf{r})\, w(\mathbf{r} - \mathbf{t}_m)\, d\mathbf{r} = I\delta n_{-\sigma}^{(n)}\, \Delta_{nm} \equiv \lambda_{\sigma\mathrm{HF}}^{nm} \tag{3.15}$$

with

$$I = \iint d\mathbf{r}\, d\mathbf{r}'\, \frac{|\, w(\mathbf{r} - \mathbf{t}_n)|^2\, |\, w(\mathbf{r}' - \mathbf{t}_n)|^2}{|\, \mathbf{r} - \mathbf{r}'\, |}, \tag{3.16}$$

and where Δ_{nm} is the notation used throughout for the Kronecker δ function. $\delta n_\sigma^{(n)}$ is the change in the number of $+\sigma$ electrons on site n from the nonmagnetic state.

C. *Hartree–Fock Green function, density of states and total energy*

First, we write down the general Bloch wave expansion of $G_\sigma(\mathbf{r}\mathbf{r}_0 E)$ in (2.12) as

$$G_\sigma(\mathbf{r}\mathbf{r}_0 E) = \sum_{\mathbf{k}_1\mathbf{k}_2} \psi_{\mathbf{k}_1}(\mathbf{r})\, \psi_{\mathbf{k}_2}(\mathbf{r}_0)\, \tilde{G}_\sigma(\mathbf{k}_1\mathbf{k}_2 E). \tag{3.17}$$

Then the integral equation (2.12) takes the form

$$\tilde{G}_\sigma(\mathbf{q}_1\mathbf{q}_2 E) = \frac{\Delta_{\mathbf{q}_1\mathbf{q}_2}}{E - \epsilon(\mathbf{q}_1) + i\eta} + \sum_{\mathbf{k}} \frac{\alpha_\sigma(\mathbf{q}_1\mathbf{k})\, \tilde{G}_\sigma(\mathbf{k}\mathbf{q}_2 E)}{E - \epsilon(\mathbf{q}_1) + i\eta}, \tag{3.18}$$

where

$$\begin{aligned}\alpha_\sigma(\mathbf{q}\mathbf{k}) &= \int d\mathbf{r}\, \psi_{\mathbf{q}}^*(\mathbf{r})\, v_\sigma(\mathbf{r})\, \psi_{\mathbf{k}}(\mathbf{r}) \\ &= N^{-1} \sum_{nm} e^{-i\mathbf{q}\cdot\mathbf{t}_n + i\mathbf{k}\cdot\mathbf{t}_m} \lambda_\sigma^{nm}\end{aligned} \tag{3.19}$$

and

$$\lambda_\sigma^{nm} = \int d\mathbf{r}\, w^*(\mathbf{r} - \mathbf{t}_n)\, v_\sigma(\mathbf{r})\, w(\mathbf{r} - \mathbf{t}_m). \tag{3.20}$$

Using the form (3.15) for the matrix elements of the spin-dependent potential, it is readily shown that the Hartree–Fock result for $\tilde{G}_\sigma(\mathbf{k}_1\mathbf{k}_2E)$ in the special case in which all sites are equivalent is

$$\tilde{G}_\sigma(\mathbf{k}_1\mathbf{k}_2E) = \Delta_{\mathbf{k}_1\mathbf{k}_2}/(E - \epsilon(\mathbf{q}_1) + i\eta - I\delta n_{-\sigma}). \tag{3.21}$$

Thus, with equivalent sites, the energy levels available to spin σ electrons are shifted by $I\delta n_{-\sigma}$. With $\rho_\gamma(E)$ as the density of states per atom in the nonmagnetic state, the densities of states per atom for electrons described by (3.21) are

$$\rho_\sigma(E) = \rho_\gamma(E - I\delta n_{-\sigma}). \tag{3.22}$$

We can write for the number of spin σ electrons per atom

$$n_\sigma = \int_{-\infty}^{\mu'} dE\, \rho_\gamma(E + I\delta n_\sigma) \tag{3.23}$$

and with $\delta n_\sigma = n_\sigma - \frac{1}{2}n$, the equations must be solved for n_+, n_-, μ'. The condition for a ferromagnetic state to appear may be obtained by expanding (3.23) in a power series in δn_σ. Then we obtain immediately the condition

$$I\rho_\gamma(\mu) = 1, \tag{3.24}$$

which is the Hartree–Fock result obtained by Hubbard, with the energy scale shifted by $\frac{1}{2}In$.

Using the one-body potential theory underlying these equations, it is a straightforward matter to show that the total energy can be written in terms of the Green function as

$$E_{\mathrm{HF}} = -\frac{1}{2\pi}\,\mathrm{Im}\sum_\sigma \int_{-\infty}^{\mu'_{\mathrm{HF}}} dE \sum_{\mathbf{k}} (E + \epsilon(\mathbf{k}))\,\tilde{G}_\sigma(\mathbf{kk}E), \tag{3.25}$$

where μ' is the Fermi energy as usual. Using the Green function solution (3.21), it is readily shown that the total energy is given by

$$E_{\mathrm{HF}} = \sum_\sigma \int_{-\infty}^{\mu'_{\mathrm{HF}}} dE\, En_\gamma(E - I\delta n_{-\sigma}) + IN(\delta n)^2 \tag{3.26}$$

with μ'_{HF} given by

$$\sum_\sigma \int_{-\infty}^{\mu'_{\mathrm{HF}}} dE\, n_\gamma(E - I\delta n_{-\sigma}) = 2\int_{-\infty}^{\mu} dE\, n_\gamma(E). \tag{3.27}$$

We finally consider the minimization of E_{HF} given by (3.26) with respect to variations in the parameter δn_σ. In the equivalent site case, there is, of course, only one independent parameter $\delta n_+ = -\delta n_- = \delta n$. We obtain from (3.26)

$$\frac{\partial E_{\mathrm{HF}}(\delta n)}{\partial(\delta n)} = 2IN\delta n - I\int_{-\infty}^{\mu'_{\mathrm{HF}}} dE\, n_\gamma(E + I\delta n) + I\int_{-\infty}^{\mu'_{\mathrm{HF}}} dE\, n_\gamma(E - I\delta n). \tag{3.28}$$

Equating $\partial E_{\mathrm{HF}}/\partial(\delta n)$ to zero gives the Euler equation in the Hartree–Fock approximation for the equivalent site case and solving this will, of course, give the electron densities.

4. Introduction of Correlation Effects

In order to improve on the Hartree–Fock theory presented in the previous section, we need some information about the functional $E_c[\rho_+, \rho_-]$ or, alternatively, the correlation contribution to $E_I[\Delta_\sigma]$ in Eq. (2.8). We do not, of course, have very much first principle knowledge of correlation effects in a highly inhomogeneous system.

Therefore, it will prove helpful to write down the density functional approach in a different way, which uses, as we shall see below, occupation numbers of Wannier orbitals to parametrize the charge and spin densities.

In introducing the method, it will be helpful to consider the nonmagnetic case.

A. *Occupancy of Wannier states in nonmagnetic case*

When we have a perfectly paired spin situation, we can write the total energy in the form

$$E[\rho] = T_s[\rho] + \int d\mathbf{r}\, V_{\mathrm{ext}}\rho + E_I[\rho]. \tag{4.1}$$

We now introduce a complete orthonormal set of functions which we shall assume are Wannier functions $w(\mathbf{r} - \mathbf{t}_i)$, defined by V_{ext}, or some other periodic potential, eg., $\bar{V}(\mathbf{r})$. We may then expand the charge density $\rho(\mathbf{r})$ in the form

$$\rho(\mathbf{r}) = \sum_{ij} n_{ij} w^*(\mathbf{r} - \mathbf{t}_i)\, w(\mathbf{r} - \mathbf{t}_j) \tag{4.2}$$

in terms of parameters n_{ij} which satisfy

$$n_{ij}^* = n_{ji} \tag{4.3}$$

and the normalization condition

$$\sum_i n_{ii} = N_e . \tag{4.4}$$

Substituting (4.2) into (4.1) gives

$$E = E(n_{ij}) = T_s(n_{ij}) + \sum_{ij} n_{ij} \int d\mathbf{r}\, w^*(\mathbf{r} - \mathbf{t}_i)\, V_{\text{ext}}(\mathbf{r})\, w(\mathbf{r} - \mathbf{t}_j) + E_I(n_{ij}). \tag{4.5}$$

We may now restate the variational principle as a minimization of $E(n_{ij})$ with respect to the parameters n_{ij}, subject to (4.4). Then the Euler equations are found to be

$$\frac{\partial E(n_{ij})}{\partial n_{ij}} = 0 = \frac{\partial T_s(n_{ij})}{\partial n_{ij}} + \int d\mathbf{r}\, w^*(\mathbf{r} - \mathbf{t}_i)\, V_{\text{ext}}(\mathbf{r})\, \mathbf{w}(\mathbf{r} - \mathbf{t}_j) + \frac{\partial E_I}{\partial n_{ij}} + \mu \Delta_{ij} . \tag{4.6}$$

These equations must be solved for the parameters n_{ij} for a given external potential. However, we may note that these Euler equations are in just the form we would obtain if we considered a system of electrons moving in an external static potential $V(\mathbf{r})$ which has matrix elements between the Wannier functions w given by

$$\begin{aligned} \lambda_{ij} &= \int d\mathbf{r}\, w^*(\mathbf{r} - \mathbf{t}_i)\, V(\mathbf{r})\, w(\mathbf{r} - \mathbf{t}_j) \\ &= \int d\mathbf{r}\, w^*(\mathbf{r} - \mathbf{t}_i)\, V_{\text{ext}} w(\mathbf{r} - \mathbf{t}_j) + \frac{\partial E_I}{\partial n_{ij}} . \end{aligned} \tag{4.7}$$

Thus we have the result that a parametrization of the electron density has led us to interpret the derivatives of the interaction energy as matrix elements of a one-body potential. These matrix elements must be used to calculate the single-particle eigenfunctions $\phi_{\mathbf{k}}(\mathbf{r})$ which give the density $\rho(\mathbf{r})$ according to (2.6). As before, an iterative procedure is therefore necessary in order to obtain a solution for n_{ij}.

1. *Generalization to spin densities*

The extension of these ideas to the magnetic situation with spin densities ρ_+ and ρ_- is straightforward. Our model of electron interactions gives the total energy in the form

$$E = T_s[\rho_+] + T_s[\rho_-] + \int \bar{v}(\mathbf{r})[\rho_+(\mathbf{r}) + \rho_-(\mathbf{r})] + E_I[\Delta_\sigma]. \tag{4.8}$$

As expansion functions for the densities we now take the Wannier functions $w(r)$ associated with $\bar{V}(r)$, that is with the nonmagnetic lattice.

We then expand

$$\rho_\sigma(\mathbf{r}) = \sum_{ij} n_{ij}^\sigma w^*(\mathbf{r} - \mathbf{t}_i)\, w(\mathbf{r} - \mathbf{t}_j) \tag{4.9}$$

with, just as before,

$$n_{ij}^{\sigma *} = n_{ji}^\sigma \tag{4.10}$$

and

$$\sum_i [n_{ii}^+ + n_{ii}^-] = N_e\,. \tag{4.11}$$

We now put

$$n_{ij}^\sigma = n_{ij}^0 + \delta n_{ij}^\sigma\,, \tag{4.12}$$

where

$$n_{ij}^0 = \sum_{\mathbf{k}(\text{occ})} e^{i\mathbf{k}\cdot(\mathbf{t}_i - \mathbf{t}_j)} \tag{4.13}$$

is the value of n_{ij} in the nonmagnetic lattice. Thus, we have

$$\rho_\sigma(\mathbf{r}) = \rho_p(\mathbf{r}) + \sum_{ij} \delta n_{ij}^\sigma\, w^*(\mathbf{r} - \mathbf{t}_i)\, w(\mathbf{r} - \mathbf{t}_j). \tag{4.14}$$

Substituting (4.9) into (4.8), we find

$$E = T_s(n_{ij}^+) + T_s(n_{ij}^-) + \sum_{ij}\sum_\sigma n_{ij}^\sigma \int d\mathbf{r}\, w^*(\mathbf{r} - \mathbf{t}_i)\, \bar{V}(\mathbf{r})\, w(\mathbf{r} - \mathbf{t}_j) + E_I(n_{ij}^\sigma). \tag{4.15}$$

Again we wish to minimize E with respect to the parameters n_{ij}^σ, and since n_{ij}^0 are fixed parameters, this is the same as minimizing with respect to δn_{ij}^σ. The Euler equations are found to be

$$\frac{\partial T_s}{\partial n_{ij}^\sigma} + \int d\mathbf{r}\, w^*(\mathbf{r} - \mathbf{t}_i)\, \bar{V}(\mathbf{r})\, w(\mathbf{r} - \mathbf{t}_j) + \frac{\partial E_I}{\partial \delta n_{ij}^\sigma} = 0. \tag{4.16}$$

Once more, we see that we can interpret

$$\frac{\partial E_I}{\partial \delta n_{ij}^\sigma} = \int d\mathbf{r}\, w^*(\mathbf{r} - \mathbf{t}_i)\, v_\sigma(\mathbf{r})\, w(\mathbf{r} - \mathbf{t}_j) \tag{4.17}$$

as the matrix element of $v_\sigma(\mathbf{r})$ between the expansions functions $w(\mathbf{r} - \mathbf{t}_i)$ and $w(\mathbf{r} - \mathbf{t}_j)$.

To make contact with the Hartree–Fock theory given in the previous section, we can write

$$E_I^{\mathrm{HF}} = I \sum_i \delta n_{ii}^{+} \, \delta n_{ii}^{-} \tag{4.18}$$

and we see that this is a function only of the diagonal parameters δn_{ii}^{σ}. In this case it follows immediately that

$$\partial E_I^{\mathrm{HF}} / \partial \delta n_{ii}^{\sigma} = I \, \delta n_{ii}^{-\sigma} \tag{4.19}$$

and

$$\partial E_I^{\mathrm{HF}} / \partial \delta n_{ij}^{\sigma} = 0, \qquad i \neq j. \tag{4.20}$$

These eqs. (4.19) and (4.20) express the Koster–Slater properties of the Hartree–Fock potential $v_\sigma^{\mathrm{HF}}(\mathbf{r})$. Thus, these equations tell us that v_σ^{HF} has matrix elements which are nonvanishing only between Wannier functions on the same site, a property which was used in Section 3 to derive the standard Hartree–Fock results in narrow energy-band theory.

In the general case, we may separate $v_\sigma(\mathbf{r})$ into

$$v_\sigma(\mathbf{r}) = v_\sigma^{\mathrm{HF}}(\mathbf{r}) + v_\sigma^{c}(\mathbf{r}) \tag{4.21}$$

and evidently the matrix elements of the correlation contribution to v_σ are given by

$$\frac{\partial E_I^{c}}{\partial \delta n_{ij}^{\sigma}} = \int d\mathbf{r} \, w^*(\mathbf{r} - \mathbf{t}_i) \, v_\sigma^{c}(\mathbf{r}) \, w(\mathbf{r} - \mathbf{t}_j). \tag{4.22}$$

Then the potential (4.21), with matrix elements (4.19) and (4.22) will give the one-particle Green function according to (2.12) and in the Wannier representation used in this section we may usefully expand G_σ in the form

$$G_\sigma(\mathbf{r}\mathbf{r}_0 E) = \sum_{ij} G_\sigma(ijE) \, w^*(\mathbf{r} - \mathbf{t}_i) \, w(\mathbf{r}_0 - \mathbf{t}_j). \tag{4.23}$$

Substitution of (4.23) into (2.12) gives the following equation for $G_\sigma(ijE)$, when we use (4.17):

$$G_\sigma(ijE) = F(ijE) + \sum_{lr} F(ilE) \left[\frac{\partial E_I}{\partial \delta n_{lr}^{\sigma}} \right] G_\sigma(rjE), \tag{4.24}$$

where

$$F(ijE) = \frac{1}{N} \sum_{\mathbf{k}} \frac{e^{i\mathbf{k}\cdot(\mathbf{t}_i - \mathbf{t}_j)}}{E - \epsilon(\mathbf{k}) + i\eta}. \tag{4.25}$$

Then from (2.14) and (4.24) we obtain the self-consistency equations which determine the parameters δn_{ij}'as

$$\delta n_{ij}^{\sigma} = \frac{-1}{2\pi i} \int_{-\infty}^{\mu'} dE\, [G_{\sigma}(ijE+) - G_{\sigma}(ijE-)] - \frac{1}{2} n. \tag{4.26}$$

As we have already noted, the dependence of E_I^{HF} solely on the diagonal parameters $\delta n_{ii}^{\sigma} = \delta n_{\sigma}^{(i)}$ contains the Koster–Slater property of the Hartree–Fock potential $v_{\sigma}^{\text{HF}}(\mathbf{r})$. Indeed, any theory which is based on E_I, which is a function only of the diagonal parameters δn_{ii}^{σ}, implies a Koster–Slater like one-body potential, whereas there is no reason for the correlation contribution to the spin dependent potential $v_{\sigma}^{c}(\mathbf{r})$ to have matrix elements of such a form. We expect therefore, in a final theory, that the dependence of E_I on the off-diagonal parameters δn_{ij}^{σ}, $i \neq j$, will play a significant role.

Although it is possible to write down at this stage various model correlation energy functions $E_I^{c}(\delta n_{ij}^{\sigma})$, very little is established firmly about E_I^{c} in the case when electrons are localized and therefore, in the following section, we shall consider an alternative way of introducing correlation effects.

However, we wish to stress here that the formulation based on (4.24) has a number of merits which we believe will prove important. First of all, we can build the detailed band structure into the one-body Green function through (4.25). This is, in fact, a fairly standard calculation, once a potential is assumed. Secondly, there is hope that by separating out one-body effects, into which the energy band dispersion relations are built, and many-body effects which are included in $\partial E_I / \partial \delta n_{lr}^{\sigma}$ in (4.24), we can make more drastic assumptions about the many-body term, without such serious consequences as in conventional formulations. This should be possible because, already, some account of electron interactions can be included in the one-body potential. Finally, the important parameters δn_{ij}^{σ} describing the magnetic state can be calculated from the one-body single particle Green function using (4.26), which is in principle exact. We want to stress, of course, that the δn_{ij}^{σ} can also, and more usually, be calculated from a Green function derived from the many-body problem, and it is of considerable interest that we can circumvent this, and, in principle, work with a genuine one-body formulation.

It is scarcely necessary to point out that exact knowledge of $\partial E_I / \partial \delta n_{lr}^{\sigma}$ in (4.24) would be equivalent to solving the many-body problem. However, given a model function $E_I(\delta n_{ij}^{\sigma})$ it is a relatively straightforward matter, for simple magnetic states, to solve the model through the matrix elements (4.17) and the iterative procedure given above.

B. *One-site model and electron localization*

Having given a basic formulation in terms of one-body potential theory, we shall now look more closely at the Hartree–Fock problem in order to arrive at a description of correlation effects. In particular, we shall think in terms of a 'probe' inserted into a nonmagnetic metal, which is such as to scatter + and − spins selectively.

It is helpful in motivating the model we shall use, to refer again to the treatment of localized moments in metals by Wolff. In his treatment, there turns out to be a nontrivial solution when we work out self-consistently the response of electrons to an impurity center and then, at the end of the calculation, we switch off the spin-independent part of the impurity potential. The fact that, under some circumstances, the original paramagnetic state is unstable to the formation of a 'localized imbalance' of spin density suggested strongly that we might have here a useful starting point for a proper treatment of magnetic instabilities and, in particular, for a description of the instability condition.

Bearing in mind the observations above, we shall now enquire whether it is possible to build up the complete Hartree–Fock picture of the lattice by switching on the Hartree–Fock interactions one site at a time. We expect to answer this question in the affirmative since the Hartree–Fock one-body potential is simply the sum of contributions from each atomic site. However, it will turn out that an examination of the Hartree–Fock interactions 'switched on' at one and then at two sites will be sufficient to suggest a method whereby correlation effects can be introduced into the one-body potential. These correlation effects will be introduced as a 'screening' of the Hartree–Fock correlations between density fluctuations on different atomic sites. Such screening, as we shall see, will lead to a simple instability condition, which we expect to transcend the Hartree–Fock condition given earlier.

We start by finding the eigenfunctions and the Green function when the Hartree–Fock interactions are switched on at one site, $\mathbf{t}_a$ say. With the one-site assumption, it is clear from (3.19) that we must write

$$\alpha_\sigma(\mathbf{qk}) = N^{-1}e^{i\mathbf{t}_a\cdot(\mathbf{k}-\mathbf{q})}I\,\delta n_{-\sigma}^{(a)}\,, \tag{4.23}$$

since

$$\lambda_\sigma^{ml} = I\,\delta n_{-\sigma}^{(a)}\,\Delta_{ma}\,\Delta_{la}\,. \tag{4.24}$$

Then, using the expansion (2.8) in terms of Wannier functions, we find

$$b_\sigma(\mathbf{k}\mathbf{t}_n) = e^{i\mathbf{k}\cdot\mathbf{t}_n} + \sum_{m,l} F(nmE)\,\lambda_\sigma^{ml} b_\sigma(\mathbf{k},\,\mathbf{t}_l) \tag{4.25}$$

with

$$F(nmE) = N^{-1} \sum_{\mathbf{q}} \frac{e^{i\mathbf{q}\cdot(\mathbf{t}_n - \mathbf{t}_m)}}{E - \epsilon(\mathbf{q}) + i\eta}. \tag{4.26}$$

In particular

$$\begin{aligned} F(nnE) &= F(E) \\ &= N^{-1} \sum_{\mathbf{q}} \frac{1}{E - \epsilon(\mathbf{q}) + i\eta} \\ &= \int dE' \frac{\rho_\gamma(E')}{E - E' + i\eta} \\ &= P \int dE' \frac{\rho_\gamma(E')}{E - E'} - i\pi\rho_\gamma(E) \\ &= P_\gamma(E) - i\pi\rho_\gamma(E) \end{aligned} \tag{4.27}$$

which is the usual Koster–Slater function, P denoting the principal part, and $\rho_\gamma(E)$ the density of states per atom. Using (4.24) we have

$$b_\sigma(\mathbf{k}, \mathbf{t}_n) = e^{i\mathbf{k}\cdot\mathbf{t}_n} + I\, \delta n_{-\sigma}^{(a)} F(naE)\, b_\sigma(\mathbf{k}\mathbf{t}_a) \tag{4.28}$$

and for $n = a$, we obtain

$$b_\sigma(\mathbf{k}, \mathbf{t}_a) = \frac{e^{i\mathbf{k}\cdot\mathbf{t}_a}}{1 - I\, \delta n_{-\sigma}^{(a)} F(E)} \tag{4.29}$$

which, together with (4.28) gives all the coefficients b_σ. Thus we have a solution for a model in which there may be fluctuations from the nonmagnetic state on all sites in the lattice, but the electrons interact through $v_\sigma(\mathbf{r})$ only through the spin-density deviations $\delta n_\sigma^{(a)}$ on one chosen site $\mathbf{t}_a$.

We record in Appendix I the corresponding Green function and density of states. The latter quantity, for small $\delta n_\sigma^{(a)}$ may be expanded in the form

$$n_\sigma(E) = n_\gamma(E) - I\, \delta n_{-\sigma}^{(a)}\, \rho_\gamma'(E) - (I\, \delta n_{-\sigma}^{(a)})^2\, P_\gamma(E)\, \rho_\gamma'(E) + \cdots \tag{4.30}$$

Here $\rho_\gamma = n_\gamma/N$, $P_\gamma(E)$ is given by (4.27) and primes denote derivatives.

At this stage, it is important to investigate the condition under which the Hartree–Fock interactions on one site can sustain a magnetic deviation from the paired spin state.

In terms of the eigenfunctions, the change in the number of σ electrons on site a in going from the nonmagnetic to the magnetic state is given by

$$\delta n_\sigma^{(a)} = N^{-1} \int_{-\infty}^{\mu'} dE\, n_\sigma(E)\, |\, b_\sigma(E, \mathbf{t}_a)|^2 - N^{-1} \int_{-\infty}^{\mu} dE\, n_\gamma(E). \tag{4.31}$$

We have used here the fact that $b_\sigma(\mathbf{kt}_n) \equiv b_\sigma(\epsilon(\mathbf{k}), \mathbf{t}_n) \equiv b_\sigma(E, \mathbf{t}_n)$ and μ' is defined by (2.16) with the one-site density of states. As n_σ and b_σ depend only on the parameters $\delta n_\sigma^{(a)}$ which represent the density dependence, then (4.31) is essentially the fundamental Euler equation in the density-functional formalism. To investigate whether a magnetic solution corresponding to $\delta n_\sigma^{(a)} \neq 0$ is possible, we use the expansions of $| b_\sigma |^2$ and n_σ for small $\delta n_\sigma^{(a)}$. In the one-site model, $\mu' - \mu = 0(N^{-1})$ and we readily obtain, with $\lambda_\sigma{}^a = I\, \delta n_{-\sigma}^{(a)}$,

$$\lambda_\sigma{}^a = 2If(\mu)\, \lambda_{-\sigma}^a + IE(\mu)\, \lambda_{-\sigma}^{a^2} + \cdots, \tag{4.32}$$

where

$$f(\mu) = \int_{-\infty}^{\mu} dE\; P_\gamma(E)\, \rho_\gamma(E) \tag{4.33}$$

and

$$E(\mu) = \int_{-\infty}^{\mu} dE\; [3P_\gamma{}^2(E) - \pi^2 \rho_\gamma(E)]\, \rho_\gamma(E). \tag{4.34}$$

Thus a magnetic state is permitted when $[2If(\mu)]^2 = 1$, or, as $P_\gamma(E)$ is negative over most of the band, when

$$2I \int_{-\infty}^{\mu} dE\; P_\gamma(E)\, \rho_\gamma(E) = -1. \tag{4.35}$$

Because of the philosophy that we can take some account of correlation effects by emphasizing electron localization round sites, we show in Appendix II that the model in which interaction occurs through the Hartree–Fock potential only when electrons are together on one of the two sites at $\mathbf{t}_a$ and $\mathbf{t}_b$ can also be solved. This helps to envisage the entire Hartree–Fock solution for the lattice, which we can construct (by a very laborious route, however) by continuing with the three-site etc. models.

C. *Properties of Green function with electron localization*

In order to construct a Green function to represent the lattice, we have shown in Section 4A that we must know the interaction energy E_I in terms of the independent parameters δn_{ij}^σ, including, in general, the off-diagonal case $i \neq j$.

As all the parameters δn_{ij}^σ are determined uniquely by the static potential $\overline{V}(\mathbf{r})$, we may assume that the off-diagonal quantities δn_{ij}^σ, $i \neq j$, are functions of the diagonal parameters and functionals of $\overline{V}(\mathbf{r})$. Then, in general, we must have

$$\lambda_\sigma(nm) = \lambda_\sigma^{nm}([\overline{V}], \delta n_{-\sigma}^{(1)}, \ldots, \delta n_{-\sigma}^{(N)}), \tag{4.36}$$

where it is clear that we are considering many-body interactions between only opposite spin electrons in the density fluctuations $\Delta_\sigma(\mathbf{r})$.

With the framework expressed through (4.36), we shall now proceed by noting first that the Hartree–Fock interactions, with which we have dealt so far, overemphasize the formation of a magnetic state, due to the Hartree–Fock correlations between spin-density deviations on different sites. Thus, we shall screen out these correlations and, by this means, we shall make the condition for a magnetic state to occur more restrictive, which certainly agrees qualitatively with the effect we know the electron correlations must have. Thus we turn immediately to construct a Green function for the lattice which includes such a screening effect.

1. *Model in which Hartree–Fock correlations are screened out*

Using the theory of Appendix II, we see that switching the Hartree–Fock interactions on site-by-site leads to different instability conditions from the Hartree–Fock lattice condition. We shall now introduce correlations by requiring that the Hartree–Fock interactions are screened out in such a way that the instability condition for the two-site model is the same as for the one-site case. The same ideas carry over to the N-site model.*

To formulate this 'extreme correlation' model, let us consider the change in the number of electrons on site i from the nonmagnetic situation, when the magnetic state is formed. This may be written as

$$\delta n_\sigma^{(i)} = -\frac{1}{\pi}\,\mathrm{Im}\int_{-\infty}^{\mu'} dE\; G_\sigma(iiE) - \frac{1}{2}\,n. \tag{4.37}$$

We may now construct the Green function for the 'extreme correlation' model by taking for the diagonal component of the lattice Green function in the Wannier representation $G_\sigma(iiE)$ just the component which we obtained when the interactions were switched on at site $\mathbf{t}_i$. Thus, we have

$$G_\sigma(iiE) = F(E)/(1 - I\,\delta n_{-\sigma}^{(i)}\,F(E)). \tag{4.38}$$

It is easy to check the consistency of this solution for the diagonal components by

* *Note added in Proof.* Since our work was completed, a paper by Kim [9] has appeared in which a local moment condition is shown to be relevant to the occurrence of ferromagnetism (or antiferromagnetism) in alloys at finite concentration. We are currently studying a parallel treatment using the Wolff picture for the local moment; Kim's treatment being based on the Anderson model. We have also shown elsewhere, in confunction with D. Wiid, that the density functional method can describe the local moment situation.

substitution of (4.38) into (4.37). The density of states for this model involves only the diagonal components (4.38) and (2.15) gives

$$n_\sigma(E) = \sum_n \frac{\rho_\gamma(E)}{[1 - I\,\delta n_{-\sigma}^{(n)}\,P(E)]^2 + [\pi I\,\delta n_{-\sigma}^{(n)}\,\rho_\gamma(E)]^2}\,. \tag{4.39}$$

Then μ' is given by (2.16) in terms of the density of states (4.39). In general $\mu' - \mu = 0(\delta n^2)$ and (4.37) gives the 'one-site' condition (4.36) as the condition for the appearance of a general magnetic state.

To complete the solution for this model, we need the off-diagonal elements $G_\sigma(nmE)$, $n \neq m$ and to obtain these we must have more information about the potential matrix elements which include, of course, correlation contributions. We may split these matrix elements into two parts by separating out the Hartree–Fock contribution. Thus, we write

$$\lambda_\sigma(lr) = I\,\delta n_{-\sigma}^{(l)}\,\Delta_{lr} + \lambda_\sigma^{c}(lr) \tag{4.40}$$

and Eq. (2.12) for the Green function becomes, in the Wannier representation,

$$G_\sigma(nmE) = F(nmE) + I\sum_l F(nlE)\,\delta n_{-\sigma}^{(l)}\,G_\sigma(lmE) + \sum_{l,r} F(nlE)\,\lambda_\sigma^{c}(lr)\,G_\sigma(rmE). \tag{4.41}$$

The matrix elements λ_σ^{c} must be found so that the diagonal elements $G_\sigma(nnE)$ are given by (4.38). However, the model, as it stands, is not detailed enough to yield completely unambiguous results for the off-diagonal components. From general considerations we may deduce the following approximate solution which has the required diagonal properties:

$$G_\sigma(nmE) = F(nmE) + \frac{1}{2}\,IF(E)\,F(nmE)\left[\frac{\delta n_{-\sigma}^{(n)}}{1 - I\,\delta n_{-\sigma}^{(n)}\,F(E)} + \frac{\delta n_{-\sigma}^{(m)}}{1 - I\,\delta n_{-\sigma}^{(m)}\,F(E)}\right]. \tag{4.42}$$

This Green function has then the following properties, in addition to giving the desired diagonal element. First, the band limit $I \to 0$ is correctly included, coming simply from the first term in (4.42). Secondly, (4.42) has the correct symmetry

$$G_\sigma(nmE + i\eta) = G_\sigma^{*}(mnE - i\eta). \tag{4.43}$$

The atomic limit, discussed in Appendix III, is also correctly given as may be seen by letting $\epsilon(\mathbf{k}) \to T_0$, when we obtain

$$F(nmE) \to \frac{\Delta_{nm}}{E - T_0} \tag{4.44}$$

and

$$G_\sigma(nmE) \rightarrow \frac{\Delta_{nm}}{E - T_0 - I\,\delta n_{-\sigma}^{(n)}}, \tag{4.45}$$

which is the exact one-body Green function in the narrow band limit.

Though the lattice Green function (4.42) has these favorable properties, it may be shown that it does not count correctly the states in the band. In many cases, this should not be a very serious error, but it focusses attention on what, in our case, amounts to some uncertainty in the higher-order terms in the Green function. To illustrate this, it will be useful to discuss a related, but nevertheless different, approach to the construction of a lattice Green function. In this way we shall be able to avoid the difficulty of counting the states, while retaining the other favourable properties.

2. *Model built from superposition of displaced spin densities*

The Green function in the one-site model is given in Appendix I and from (A.2) we have

$$G_\sigma^{(a)}(nmE) = F(nmE) + \Delta G_\sigma^{(a)}(nmE) \tag{4.46}$$

with

$$\Delta G_\sigma^{(a)}(nmE) = I\,\delta n_{-\sigma}^{(a)} \frac{F(naE)\,F(amE)}{1 - I\,\delta n_{-\sigma}^{(a)}\,F(E)}, \tag{4.47}$$

where the chosen interaction site is labeled by $\mathbf{t}_a$. We now define the superposition model as the sum over sites of contributions (4.47) from each site $\mathbf{t}_a$. Thus, we obtain the lattice Green function

$$G_\sigma(nmE) = F(nmE) + I \sum_a \delta n_{-\sigma}^{(a)} \frac{F(naE)\,F(amE)}{1 - I\,\delta n_{-\sigma}^{(a)}\,F(E)}. \tag{4.48}$$

This model Green function differs from (4.42), which on the diagonal $n = m$ corresponds to keeping only the term $a = n$ in the summation in (4.48). The Hartree–Fock correlations discussed previously are obviously not screened completely out of the instability condition as they are in (4.42).

Substitution of (4.48) into (4.36) gives the Euler equations for the $\delta n_\sigma^{(i)}$ and we obtain

$$\delta n_\sigma^{(i)} = \int_\mu^{\mu'} dE\,\rho_\gamma(E) - \frac{1}{\pi}\,\mathrm{Im} \int_{-\infty}^{\mu'} dE \sum_a I\,\delta n_{-\sigma}^{(a)} \frac{F(iaE)\,F(aiE)}{1 - I\,\delta n_{-\sigma}^{(a)}\,F(E)}. \tag{4.49}$$

The density of states may be obtained directly from (2.15) and we find

$$n_\sigma(E) = -\frac{1}{\pi}\,\mathrm{Im}\sum_n G_\sigma(nmE)$$

$$= n_\gamma(E) - \frac{1}{\pi}\,\mathrm{Im}\sum_a I\,\delta n_{-\sigma}^{(a)}\,\frac{(\partial F(E)/\partial E)}{1 - I\,\delta n_{-\sigma}^{(a)}\,F(E)}\,. \qquad (4.50)$$

From (4.50) and (2.16) we obtain μ' as a function of the $\delta n_\sigma^{(i)}$ to complete the Euler equation. It is straightforward to verify that the density of states (4.50) may be rewritten in the more convenient form

$$n_\sigma(E) = n_\gamma(E) - \frac{1}{\pi}\sum_i \frac{d}{dE}\,(\tan^{-1} K_\sigma{}^i(E)), \qquad (4.51)$$

where

$$K_\sigma{}^i(E) = \frac{\pi I\,\delta n_{-\sigma}^{(i)}\,\rho_\gamma(E)}{1 - I\,\delta n_{-\sigma}^{(i)}\,P(E)}\,. \qquad (4.52)$$

The Euler equation (4.49) is a relation which involves all the $\delta n_\sigma^{(i)}$ and is not easy to solve except in the case when all sites are equivalent. In this case, the condition for the appearance of a ferromagnetic state is $I\rho_\gamma(\mu) = 1$, which is the same as the Hartree–Fock condition obtained earlier. The general solutions of (4.49) will not, of course, be the Hartree–Fock solutions.

We can now see that it is the form of the term of $0(\delta n_\sigma^{(i)})$ in $G_\sigma(nnE)$ which determines the instability condition. We have considered the "one-site" instability condition (4.35) to transcend the Hartree–Fock condition and it is straightforward to show that we may build this condition into the superposition Green function (4.48) by modifying the term of $0(\delta n_\sigma^{(i)})$. We write for the diagonal element of the modified Green function

$$G_\sigma{}^m(nnE) = F(E) + I\,\delta n_{-\sigma}^{(n)}\,[F(E)]^2 + \text{higher-order terms in } (I\,\delta n_{-\sigma}) \text{ as in (4.48).} \qquad (4.53)$$

This procedure leads, of course, to a treatment of the first-order term which is on a different footing from the higher-order terms. However, the general form of the density of states is not appreciably altered by this modification and is similar to that obtained from the model defined in Section 4A1.

Using the off-diagonal element of (4.42) to first order in δn_σ, we may sum the series (4.53) to define the model Green function which is based on the superposition principle, but contains the instability condition (4.35). We then obtain

$$G_\sigma{}^m(nmE) = \frac{1}{2} IF(E)\, F(nmE)[\delta n_{-\sigma}^{(n)} + \delta n_{-\sigma}^{(m)}] + \sum_a \frac{(I\,\delta n_{-\sigma}^{(a)})^2\, F(naE)\, F(amE)}{1 - I\,\delta n_{-\sigma}^{(a)}\, F(E)} + F(nmE). \tag{4.54}$$

The corresponding density of states may be written as

$$n_\sigma(E) = n_\gamma(E) - \frac{1}{\pi}\sum_i \frac{d}{dE}(\tan^{-1} K_\sigma^{(i)}(E)) + \Big[I\frac{\partial \rho_\gamma(E)}{\partial E} + 2IP_\gamma(E)\,\rho_\gamma(E)\Big]\sum_a \delta n_{-\sigma}^{(a)}. \tag{4.55}$$

D. *Density of states*

For Green function (4.42) the density of states for spin σ is given by

$$n_\sigma(E) = \sum_i \frac{\rho_\gamma(E)}{D_\sigma^{(i)}(E)} \tag{4.56}$$

with

$$D_\sigma^{(i)}(E) = (1 - I\,\delta n_\sigma^{(i)}\, P_\gamma(E))^2 + (\pi I\,\delta n_{-\sigma}^{(i)}\, \rho_\gamma(E))^2. \tag{4.57}$$

We consider, for $\delta n_\sigma^{(i)} \neq 0$, the contribution from one atomic site n to the sum in (4.56). Then the factor $\rho_\gamma(E)$ may be enhanced or reduced in different parts of the band, depending on whether $\delta n_\sigma^{(n)}$ is positive or negative. In Fig. 1 we show a schematic plot of this situation, assuming that $\delta n_+^{(n)}$ is positive and $\delta n_-^{(n)}$ is negative. The peaking of the density of states factor arises when the part $(1 - I\,\delta n_{-\sigma}^{(n)} P_\gamma)^2$ of $D_\sigma^{(n)}$ has a minimum and for large enough I this factor may vanish.

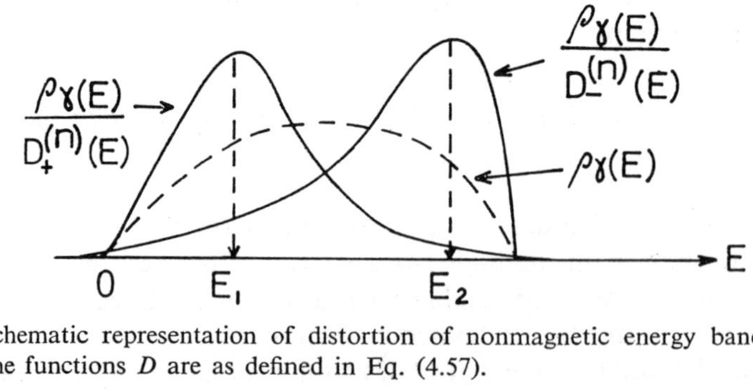

FIG. 1. Schematic representation of distortion of nonmagnetic energy band by magnetic transition. The functions D are as defined in Eq. (4.57).

Thus, adding all site contributions, the density of states for spin σ electrons consists of a lower part or "band" associated with sites upon which we have $<\frac{1}{2}n$ electrons of opposite spin and an upper "band" associated with sites which have $>\frac{1}{2}n$ electrons of opposite spin. These upper and lower "bands" overlap and in this model are not separated by simple exchange splitting in the magnetic state but are substantially deformed from the original band shape.

For the Green function (4.48), the density of states is given by (4.51). Again, taking one factor from site n, we obtain a similar modification of the nonmagnetic density of states per atom $\rho_\gamma(E)$. If $\delta n_+^{(n)}$ is positive, and $\delta n_-^{(n)}$ is negative, then $n_+(E)$ is enhanced in the lower part of the band and $n_-(E)$ in the upper part, and so qualitatively the picture is again as indicated in Fig. 1. We obtain once more the division of the band for spin σ electrons into two parts, associated with sites for which we have $<\frac{1}{2}n$ or $>\frac{1}{2}n$ electrons of opposite spin. Again the two parts of the band overlap and the bands are substantially deformed.

Using $\tan^{-1} K_\sigma{}^a(E) = 0$ at the band edges, we may verify from (4.51) that each spin band contains one state per atom.

E. *Instability condition* $2If(\mu) = -1$

We have regarded as fundamental the general instability condition derived from the one-site interaction model in Section 4B. The condition that the Euler equations may just have a magnetic solution is

$$2If(\mu) = 2I \int_{-\infty}^{\mu} dE \, P_\gamma(E) \, \rho_\gamma(E) = -1 \tag{4.58}$$

with

$$P_\gamma(E) = P \int dx \, \frac{\rho_\gamma(x)}{E - x}, \tag{4.59}$$

where P denotes the principal part of the integral. This condition obviously depends in an important way on the density of states $\rho_\gamma(E)$ through the whole band and not, as in the Hartree–Fock condition for a ferromagnetic state, $I\rho_\gamma(\mu) = 1$, on the density of states at the Fermi level.

In Fig. 2, we give the schematic form for the integral $P_\gamma(E)$ for a highly symmetric density of states $\rho_\gamma(E)$. As $P_\gamma(E)$ is negative in the lower part of the band, the condition (4.58) is more likely to be satisfied for this particular case if μ is near the middle of the band. Also, if $\rho_\gamma(\mu)$ is large, then as in the Hartree–Fock condition, (4.58) is more likely to be satisfied.

We may calculate the condition explicitly for a simple rectangular density of states

$$\rho_\gamma(E) = \Delta^{-1} \qquad \text{for } 0 \leqslant E \leqslant \Delta \tag{4.60}$$

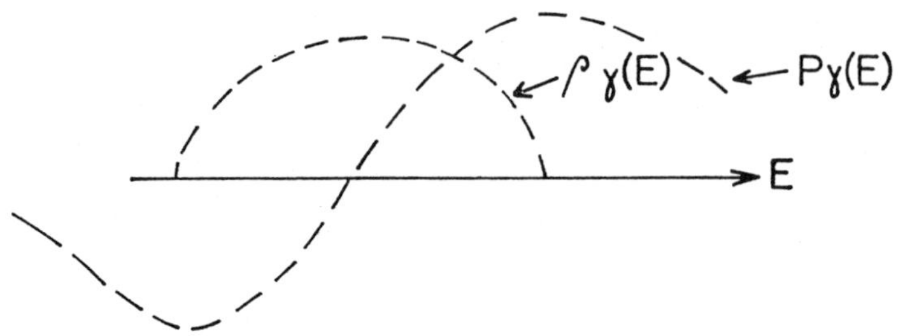

FIG. 2. Schematic form of integral $P_\gamma(E)$ given in Eq. (4.27) for the highly symmetric density of states shown.

and zero otherwise. We obtain

$$2I \int_{-\infty}^{\mu} dE\, \rho_\gamma(E)\, P_\gamma(E) = 2\left(\frac{I}{\Delta}\right)\left[\left(1 - \frac{\mu}{\Delta}\right) \ln \left(1 - \frac{\mu}{\Delta}\right) + \frac{\mu}{\Delta} \ln \left(\frac{\mu}{\Delta}\right)\right]. \tag{4.61}$$

For $\mu = \frac{1}{2}\Delta$ in the middle of the band, a magnetic state is possible at a value of $I/\Delta = (2 \ln 2)^{-1}$. For any other μ, we need a larger value of I to obtain a magnetic solution.

The theory of Appendix II shows that, for a given magnetic state, the condition (4.58) which we propose is always harder to satisfy than that predicted by the Hartree–Fock theory. As, however, (4.58) gives us the condition for the appearance of a general magnetic state, it may well be less stringent than the Hartree–Fock condition for ferromagnetism $I\rho_\gamma(\mu) = 1$, as in the case of the rectangular density of states above. In this case, the predicted instability would be towards another magnetic state.

5. Relation to Instability Condition of Richmond and Rickayzen

Our work above has been concerned with the possibility of a magnetic state arising as the interaction strength is increased, from a nonmagnetic state. We want to conclude the present discussion of the density-functional theory as applied to magnetic problems by establishing a connection with the work of Richmond and Rickayzen [8]. This connection was, initially, very surprising to us, because their approach, based on a variational calculation from the Hubbard Hamiltonian, seemed at first sight to be totally different from ours. However, we shall show below that their work can be posed in a very direct way in the language of density-functional theory.

The above authors consider the energies of a ferromagnetic state, with less that 1 electron per atom, and a state in which one electron with — spin is localized in a Wannier state centred on the origin, the rest of the electrons being + spin electrons. The Hubbard model of electron interactions is assumed, which means that, in the ferromagnetic state, the electrons move in an average one-body potential $\bar{V}(\mathbf{r})$. For the state with an antiparallel spin electron, the trial state $|R\rangle = |\theta\rangle|\chi\rangle$ is assumed, $|\theta\rangle$ representing the — spin electron localized near the origin and $|\chi\rangle$, of course, the + spin electrons. Minimization of the total energy E_+ of the spin + electrons gives

$$E_+ = \sum_{\mathbf{k}(\mathrm{occ})} \epsilon_+(\mathbf{k}), \tag{5.1}$$

where the $\epsilon_+(\mathbf{k})$ are the eigenvalues of the functions $b_+(\mathbf{k}, \mathbf{t}_n)$ which satisfy

$$\sum_s T_{rs} b_+(\mathbf{k}, \mathbf{t}_s) + I\,\Delta_{r0} b_+(\mathbf{k}, 0) = \epsilon_+(\mathbf{k})\, b_+(\mathbf{k}, \mathbf{t}_r), \tag{5.2}$$

$$T_{rs} = N^{-1} \sum_{\mathbf{q}} e^{i\mathbf{q}\cdot(\mathbf{t}_r - \mathbf{t}_s)} \epsilon(\mathbf{q}) \tag{5.3}$$

and

$$\frac{1}{N} \sum_n b_+{}^*(\mathbf{k}, \mathbf{t}_n)\, b_+(\mathbf{k}', \mathbf{t}_n) = \Delta_{\mathbf{k}\mathbf{k}'}\,, \tag{5.4}$$

$$\frac{1}{N} \sum_{\mathbf{k}} b_+{}^*(\mathbf{k}, \mathbf{t}_a)\, b_+(\mathbf{k}, \mathbf{t}_b) = \Delta_{ab}\,. \tag{5.5}$$

In (5.1), the total energy E_+ is calculated by summing the lowest $(N_e - 1)$ eigenvalues and so, in the analysis of Richmond and Rickayzen, the + spin electrons effectively occupy the lowest $N_e - 1$ states, described by $b_+(\mathbf{k}, \mathbf{t}_n)$, which are solutions of (5.2).

This procedure looks very like density-functional theory in which one fills up the lowest eigenstates defined by a one-body potential which depends on the density. In fact, as we shall now show, the same basic equations can be derived from the density-functional theory, using the ideas of the one-site model described in Section 4B.

The treatment of the ferromagnetic state for less than one electron per atom is identical in the density functional treatment, as all + spin electrons are again assumed to move in the average potential $\bar{V}(r)$. When we localize a — spin electron in a Wannier state $w(r)$, then we can obtain directly the density-dependent one-body potential seen by the + spin electrons. It is, in fact, just the Hartree–Fock potential discussed in Section 3 and is given by

$$V_+(\mathbf{r}) = \bar{V}(\mathbf{r}) + \int \frac{d\mathbf{r}' \mid w(\mathbf{r}')\mid^2}{\mid \mathbf{r} - \mathbf{r}' \mid} \tag{5.6}$$

$$= \bar{V}(\mathbf{r}) + v_+(\mathbf{r}) \tag{5.7}$$

and

$$\int d\mathbf{r}\, w^*(\mathbf{r} - \mathbf{t}_a)\, v_+(\mathbf{r})\, w(\mathbf{r} - \mathbf{t}_b) = I\, \Delta_{a0}\, \Delta_{b0} \tag{5.8}$$

represents the Koster–Slater property of the Hartree–Fock potential.

Given the potential, we may now obtain the eigenfunctions $\phi_{\mathbf{k}+}(\mathbf{r})$ which are filled by the + spin electrons. We write

$$\phi_{\mathbf{k}+}(\mathbf{r}) = \sum_n b_+(\mathbf{k}\mathbf{t}_n)\, w(\mathbf{r} - \mathbf{t}_n) \tag{5.9}$$

and from (2.10) we may obtain, after some algebra, the equation for the coefficients b_+, which is found to be identical to the Eq. (5.2) obtained by Richmond and Rickayzen and with a corresponding eigenvalue ϵ_+ which is just the eigenvalue of the one-body state $\phi_{\mathbf{k}+}(\mathbf{r})$ defined through (2.10). It is also easy to verify that the normalization and completeness of the eigenfunctions $\phi_{\mathbf{k}+}(\mathbf{r})$ leads to the relations (5.4) and (5.5) between the coefficients b_σ. This concludes the proof that the eigenfunctions of Richmond and Rickayzen, which diagonalize the Hubbard Hamiltonian for a state in which one has a single − spin electron localized at a lattice site, are essentially the one-body eigenfunctions of the density-functional theory.

We have yet to find an expression for the total energy, within the density-functional theory, to compare with (5.1) and we shall now obtain this as follows. The energy of the + spin electrons may be written as a functional of the density

$$E_+ = T_s[\rho_+] + \int \bar{V}(\mathbf{r})\, \rho_+(\mathbf{r})\, d\mathbf{r} + \iint d\mathbf{r}\, d\mathbf{r}' \frac{\rho_+{}^0(\mathbf{r})\, \rho_-{}^0(\mathbf{r}')}{\mid \mathbf{r} - \mathbf{r}' \mid}. \tag{5.10}$$

In (5.10), $\rho_\pm{}^0(\mathbf{r})$ is the density of ± spin electrons in the cell centered around the origin and incorporated in (5.10) is the model of electrons interacting when opposite spins are found on the same site. The variational principle applied to (5.10) gives the one-body potential (5.6) since $\rho_-{}^0(\mathbf{r}) = \mid w_\gamma(\mathbf{r}')\mid^2$.

We have, in addition,

$$T_s[\rho_+] = -\tfrac{1}{2} \int d\mathbf{r} \sum_{\mathbf{k}(\mathrm{occ})} \phi^*_{\mathbf{k}+}(\mathbf{r})\, \nabla_{\mathbf{r}}{}^2 \phi_{\mathbf{k}+}(\mathbf{r}) \tag{5.11}$$

and with

$$-\tfrac{1}{2}\nabla_{\mathbf{r}}{}^2\phi_{\mathbf{k}+}(\mathbf{r}) = \epsilon_+(\mathbf{k})\, \phi_{\mathbf{k}+}(\mathbf{r}) - [\bar{V}(\mathbf{r}) + v_+(\mathbf{r})]\, \phi_{\mathbf{k}+}(\mathbf{r}), \tag{5.12}$$

it follows that

$$T_s[\rho_+] = \sum_{\mathbf{k}(\text{occ})} \epsilon_+(\mathbf{k}) - \int d\mathbf{r}\, \bar{V}(\mathbf{r})\, \rho_+(\mathbf{r}) - \iint \frac{d\mathbf{r}\, d\mathbf{r}'\, \rho_+{}^0(\mathbf{r})\, \rho_-{}^0(\mathbf{r}')}{|\mathbf{r} - \mathbf{r}'|}. \quad (5.13)$$

Substitution of (5.13) into (5.10) gives

$$E_+ = \sum_{\mathbf{k}(\text{occ})} \epsilon_+(\mathbf{k}), \quad (5.14)$$

where the lowest $(N_e - 1)$ eigenstates are occupied. This is the same result as (5.1) and shows that the density-functional theory produces exactly the same results as the conventional theory when the variational principle is introduced and solved exactly. For this special magnetic system of electrons, the total energy is just the sum of the lowest $N_e - 1$ eigenvalues of the one-body potential and this simple result stems from the fact that the $-$ spin is localized and acts effectively as an external static potential.

Having proved the equivalence of the two approaches discussed above, the total energy of the $+$ spin electrons is easily calculated. From (5.14) we have

$$E_+ = \int_{-\infty}^{\mu'} dE\, En_+(E). \quad (5.15)$$

The density of states $n_+(E)$ calculated from the density-functional theory may be easily shown to be identical to that obtained in the Richmond and Rickayzen theory, being given by (A4) with $\delta n_- = 1$, and so the proof of the equivalence is complete. Calculating E_+ from (5.15) and (A4) and comparing this with the total energy of the ferromagnetic state gives directly the condition for the instability of this state obtained by the above authors.

For large I, we may obtain the same result as Richmond and Rickayzen that if $\rho_\gamma(E)$ is nonzero in the range $E_b \leqslant E \leqslant E_t$, then the ferromagnetic state is unstable if

$$\int_{E_b}^{E_t} dE\, E^2 \rho_\gamma(E) \geqslant I(E_t - \mu) \quad (5.16)$$

which may always be satisfied provided $E_t - \mu$ is sufficiently small.

6. Summary

This paper lays the foundations for the use of the density-functional approach in magnetic problems. It not only shows how the Hartree–Fock results, and the later work of Richmond and Rickayzen, follow from this approach in a very direct manner, but, by shifting attention from model Hamiltonians to model energy functionals, it allows some account of correlation to be added to the Hartree–Fock instability condition. The new lattice stability condition proposed is for a general magnetic instability and is frequently more stringent than the Hartree–Fock condition for ferromagnetism for realistic bands. Unlike the Hartree–Fock condition, which depends only on the density of states at the Fermi level, the new condition involves the density of states through the band.

To attempt to make the model by which correlation effects are included more precise, we have set up lattice Green functions incorporating the new instability condition. These Green functions then agree to first order in $\delta n_\sigma^{(i)}$, but differ somewhat in higher orders. Fortunately, the main features, such as the deformation of the bands by the exchange splitting, are the same in general terms, in spite of these differences in the higher-order terms.

In the model we have set up, we do not include band splitting or a Mott transition. In this early stage of development of the density functional theory, it would be premature to speculate whether a Mott transition could be included within this framework. It is still possible that the answer to the question about the existence of a Mott transition could depend sensitively on the way the interactions are treated. if they are assumed short range. It seems feasible to us that the inclusion of the long-range Coulomb correlations may, in fact, be necessary for this transition to occur, as Mott originally proposed.

It will certainly be of considerable interest to refine the treatment of correlation effects presented here, and to see whether such refinement could lead to band splitting in the non-magnetic state at a critical interaction strength, as a basis for discussing magnetic states around the metal-insulation transition.

APPENDIX I: Green Function for One-Site Interaction Model

The equation for the Green function in the Wannier representation may be obtained for this model by writing

$$G_\sigma(\mathbf{r}\mathbf{r}_0E) = \sum_{nm} w(\mathbf{r} - \mathbf{t}_n)\, w^*(\mathbf{r}_0 - \mathbf{t}_n)\, G_\sigma(nmE). \tag{A.1}$$

Then using the matrix elements $\alpha_\sigma(\mathbf{qk})$ given by (4.23), one can readily verify that the solution for $G_\sigma(nmE)$ is

$$G_\sigma(nmE) = F_\gamma(nmE) + I\, \delta n_{-\sigma}^{(a)} \frac{F_\gamma(naE)\, F_\gamma(amE)}{1 - I\, \delta n_{-\sigma}^{(a)}\, F_\gamma(E)}\,. \tag{A.2}$$

The corresponding density of states for spin σ electrons is given, from (2.15), as

$$n_\sigma(E) = n_\gamma(E) + \frac{1}{\pi} I\, \delta n_{-\sigma}^{(a)}\, \mathrm{Im}\left\{\frac{\partial F_\gamma(E)/\partial E}{1 - I\,\delta n_{-\sigma}^{(a)}\, F_\gamma(E)}\right\} \tag{A.3}$$

and using the fact that $F_\gamma(E) = P_\gamma(E) - i\pi\rho_\gamma(E)$, we obtain

$$n_\sigma(E) = n_\gamma(E) - I\,\delta n_{-\sigma}^{(a)} \left\{\frac{\rho_\gamma'(E)[1 - I\,\delta n_{-\sigma}^{(a)}\, P_\gamma(E)] + I\,\delta n_{-\sigma}^{(a)}\, \rho_\gamma(E)\, P_\gamma'(E)}{(1 - I\,\delta n_{-\sigma}^{(a)}\, P_\gamma(E))^2 + (\pi I\,\delta n_{-\sigma}^{(a)}\, \rho_\gamma(E))^2}\right\}, \tag{A.4}$$

where primes denote derivatives.

In the Bloch wave representation, we can readily obtain the solution for $\tilde{G}_\sigma(\mathbf{q}_1\mathbf{q}_2E)$ as

$$\tilde{G}_\sigma(\mathbf{q}_1\mathbf{q}_2E) = \frac{\Delta_{\mathbf{q}_1\mathbf{q}_2}}{E - \epsilon(\mathbf{q}_1)} + \frac{I\,\delta n_{-\sigma}^{(a)}}{N}\, \frac{e^{i\mathbf{t}_a\cdot(\mathbf{q}_2 - \mathbf{q}_1)}}{[E - \epsilon(\mathbf{q}_1)][E - \epsilon(\mathbf{q}_2)][1 - I\,\delta n_{-\sigma}^{(a)}\, F_\gamma(E)]}. \tag{A.5}$$

APPENDIX II: Two-Site Model with Hartree-Fock Interactions

We assume now that electrons interact through the Hartree–Fock potential when they are together on one of two sites at $\mathbf{t}_a$ and $\mathbf{t}_b$. We have then two parameters in the theory: $\lambda_\sigma^{a,b} = I\,\delta n_{-\sigma}^{a,b}$.

Thus we can write

$$\lambda_\sigma^{ml} = \lambda_\sigma{}^a \Delta_{ml}\Delta_{ma} + \lambda_\sigma{}^b \Delta_{mb}\Delta_{ml} \tag{B.1}$$

and solution of (2.10) and (2.11) gives directly the coefficients b_σ of interest

$$b_\sigma(\mathbf{k}, \mathbf{t}_a) = \frac{e^{i\mathbf{k}\cdot\mathbf{t}_a}[1 - \lambda_\sigma{}^b F(E)] + \lambda_\sigma{}^b e^{i\mathbf{k}\cdot\mathbf{t}_b} F(ab, E)}{[1 - \lambda_\sigma{}^a F(E)][1 - \lambda_\sigma{}^b F(E)] - \lambda_\sigma{}^a \lambda_\sigma{}^b F(ab, E)\, F(ba, E)}, \tag{B.2}$$

with $b_\sigma(\mathbf{k}, \mathbf{t}_b)$ obtained by interchanging a and b. The corresponding Green function may be found from (2.12) but we shall use the eigenfunctions here.

As in the one-site case, the Euler equation in the density functional formalism is

$$\delta n_\sigma^{(i)} = \frac{1}{N}\int_{-\infty}^{\mu'} dE\, n_\sigma(E)\, |\, b_\sigma(E, \mathbf{t}_i)|^2 - \frac{1}{N}\int_{-\infty}^{\mu} dE\, n_\gamma(E); \qquad i = a, b. \tag{B.3}$$

In (B.3), μ' is given by (2.16), with the appropriate two-site density of states. We may obtain, as in the one-site case, the expansion of the right-hand side of (B.3) to $0(\delta n)$, namely

$$\lambda_\sigma{}^a = 2If(\mu)\,\lambda_{-\sigma}^a + I\lambda_{-\sigma}^b\{A(ab,\mu) + B(ab,\mu)\} + \cdots \tag{B.4}$$

with $\lambda_\sigma{}^b$ obtained by interchange of a and b. Explicitly, we have

$$A(ab,\mu) = \frac{1}{N}\sum_{\mathbf{k}} F_R(ab,\epsilon(\mathbf{k}))\cos(\mathbf{k}\cdot\mathbf{t})\,\theta(\mu - \epsilon(\mathbf{k})), \tag{B.5}$$

$$B(ab,\mu) = \frac{1}{N}\sum_{\mathbf{k}} F_I(ab,\epsilon(\mathbf{k}))\sin(\mathbf{k}\cdot\mathbf{t})\,\theta(\mu - \epsilon(\mathbf{k})) \tag{B.6}$$

with

$$F(ab,E) = F_R(ab,E) + iF_I(ab,E) \tag{B.7}$$

and $\mathbf{t} = \mathbf{t}_a - \mathbf{t}_b$.

The general condition for a magnetic solution of the Euler equation is readily obtained from (B.4) but instead of the general case we shall examine two limiting cases, the ferromagnetic (FM) and antiferromagnetic (AFM) solutions, which are characterized by the following:

(i) For FM, $\lambda_\sigma{}^a = \lambda_\sigma{}^b$, i.e., $\delta n_\sigma^{(a)} = \delta n_\sigma^{(b)}$

with the same spin density deviations on both sites.

(ii) For AFM, $\lambda_\sigma{}^a = \lambda_{-\sigma}^b$, i.e., $\delta n_\sigma^{(a)} = \delta n_{-\sigma}^{(b)}$

giving equal and opposite changes on the two sites.

For a general instability, $\lambda_\sigma^{a,b}$ may all be different, but qualitatively will lie between the above limits.

The conditions under which a FM or AFM solution may appear can be found from (B.4). For simplicity, we assume that the energy band is spherical, in which case $B = 0$. The condition for a FM solution is then obtained as

$$2If(\mu) + IA(ab,\mu) = -1 \tag{B.8}$$

and that for the appearance of an AFM solution is

$$2If(\mu) - IA(ab,\mu) = -1. \tag{B.9}$$

Comparison with (4.35) shows that these conditions are just modifications of the single-site condition in Section 4. However, for our picture of correlation effects, it is essential to consider the effect of the extra terms in (B.8) and (B.9). We have first from (B.5) and (4.33), considering $A(ab,\mu) = A(|\,\mathbf{t}\,|,\mu)$ with $\mathbf{t} = \mathbf{t}_a - \mathbf{t}_b$, that $A(0,\mu) = f(\mu)$ which is negative. As $|\,t\,|$ increases, $A(t,\mu)$ may remain negative or may oscillate about zero, with $A(t,\mu) \to 0$ as $t \to \infty$.

For given paramagnetic lattice parameters, we can now illustrate how the conditions for a FM or AFM state are satisfied as I is increased from zero and we can compare this with the one-site instability condition. Let us consider two sites distance t apart. Then two possible cases can occur. First, we consider $A(t, \mu)$ positive for given t and from (B.8), remembering that $f(\mu)$ is negative, it can be seen that an AFM solution is possible at a lower value of I than a FM solution, and also at a lower value of I than for a general magnetic state in the one-site model. For the case of $A(t, \mu)$ negative, the reverse situation holds and a FM solution becomes possible at a lower I value than either an AFM solution or a general magnetic state in the one-site model.

The important point is that, in the two-site interaction case, we have always the possibility of some magnetic state, either FM, AFM or some intermediate state, before a magnetic state becomes possible in the one-site case; that is at a lower value of I. The reason for this is evidently due to the correlations between spin-density deviations on different sites a and b which arise in the Hartree–Fock theory. This can be seen by going back to the one-site interaction model. We can then calculate the spin fluctuations on sites other than the one on which we have allowed the electrons to interact, which we call $\mathbf{t}_a$. These fluctuations can be obtained directly from (4.28) and we obtain, keeping only the first-order term under the assumption of small $\delta n_\sigma^{(i)}$

$$\delta n_0^{(b)} = 2I\, \delta n_{-\sigma}^{(a)}\, A(t, \mu) + \cdots . \tag{B.10}$$

Thus, from (B.10), we see that switching on the interactions on site a sets up spin fluctuations on other sites b, distance t from a, which depend on the form of the function $A(t, \mu)$.

To see the connection and relevance to the two-site interaction model, we consider a chosen site b, distance t from a. If $A(t, \mu)$ is positive for this chosen t, then (B.10) predicts that $\delta n_\sigma^{(b)}$ will have the same sign as $\delta n_{-\sigma}^{(a)}$. If we now switch on the Hartree–Fock interactions on site b, then we know from previous analysis that (for $A(t, \mu)$ positive) the AFM solution is the first possible magnetic state, that is $\delta n_\sigma^{(b)} = \delta n_{-\sigma}^{(a)}$. This is evidently due, as can be seen from the above discussion, to the Hartree–Fock correlations between spin fluctuations on different atomic sites.

APPENDIX III: Atomic Limit of One-Body Green Function

In this Appendix, we investigate the solution of (2.12) in the atomic limit, i.e., with the atoms at infinite separation. What we show is in fact that the exact solution of the Hubbard Hamiltonian in the atomic limit has a correction with the Hartree–Fock theory of Section 3.

We start from (3.17) and define

$$G_\sigma^{ij}(E) = \frac{1}{N} \sum_{\mathbf{q}} e^{i\mathbf{q}\cdot(\mathbf{t}_i - \mathbf{t}_j)} \tilde{G}_\sigma(\mathbf{qq}E). \tag{C.1}$$

From (2.12) we obtain directly an equation for G_σ^{ij},

$$EG_\sigma^{ij}(E) = \Delta_{ij} + \sum_l T_{il} G_\sigma^{lj}(E) + A_\sigma^{ij}(E), \tag{C.2}$$

where

$$A_\sigma^{ij}(E) = \frac{1}{N} \sum_{\mathbf{q}} e^{i\mathbf{q}\cdot(\mathbf{t}_i - \mathbf{t}_j)} A_\sigma(\mathbf{qq}E) \tag{C.3}$$

and

$$A_\sigma(\mathbf{q}_1\mathbf{q}_2E) = \sum_{\mathbf{k}} \alpha_\sigma(\mathbf{qk})\, \tilde{G}_\sigma(\mathbf{kq}_2E). \tag{C.4}$$

We have also

$$T_{ij} = \frac{1}{N} \sum_{\mathbf{q}} e^{i\mathbf{q}\cdot(\mathbf{t}_i - \mathbf{t}_j)} \epsilon(\mathbf{q}). \tag{C.5}$$

The equation for the Wannier "Green function" A_σ^{ij} may be found from (2.12) and (C.4) giving

$$\begin{aligned} EA_\sigma^{ij}(E) = {} & \frac{1}{N} \sum_{\mathbf{q}} \alpha_\sigma(\mathbf{qq})\, e^{i\mathbf{q}\cdot(\mathbf{t}_i - \mathbf{t}_j)} + \sum_l T_{il} A_\sigma^{lj}(E) \\ & + \frac{1}{N} \sum_{\mathbf{qk}} e^{i\mathbf{q}\cdot(\mathbf{t}_i - \mathbf{t}_j)} \alpha_\sigma(\mathbf{qk})\, A_\sigma(\mathbf{kq}E) \\ & - \frac{1}{N} \sum_{\mathbf{qk}} e^{i\mathbf{q}\cdot(\mathbf{t}_i - \mathbf{t}_j)} [\epsilon(\mathbf{q}) - \epsilon(\mathbf{k})] \frac{\alpha_\sigma(\mathbf{qk})\, A_\sigma(\mathbf{kq}E)}{E - \epsilon(\mathbf{k})}. \end{aligned} \tag{C.6}$$

A_σ^{ij} satisfies as usual an integral equation (C.6) and at some point an approximate procedure must be introduced. We show now that, in the atomic limit and with the Hubbard model of interactions, this integral equation can be solved, with no further approximation, leading to an exact solution (C.13) below in this limit.

In the atomic limit, $\epsilon(\mathbf{q}) \to T_0$ and the last term in (C.6) vanishes identically. Also, we have $T_{ij} = T_0\, \Delta_{ij}$ and (C.6) becomes

$$EA_\sigma^{ij}(E) = \frac{1}{N} \sum_{\mathbf{q}} \alpha_\sigma(\mathbf{qq})\, e^{i\mathbf{q}\cdot(\mathbf{t}_i - \mathbf{t}_j)} + T_0 A_\sigma^{ij}(E) + \frac{1}{N} \sum_{\mathbf{qk}} e^{i\mathbf{q}\cdot(\mathbf{t}_i - \mathbf{t}_j)} \alpha_\sigma(\mathbf{qk})\, A_\sigma(\mathbf{kq}E). \tag{C.7}$$

At this point we must specify the one-body potential matrix elements $\alpha_\sigma(\mathbf{qk})$ in the atomic limit and with the Hubbard model of electron interactions. We shall assume that, in the atomic limit, each electron moves in the Hartree–Fock density-dependent potential discussed in Section 3. We have then, from (3.19)

$$\alpha_\sigma(\mathbf{qk}) = \frac{I}{N}\sum_n e^{i(\mathbf{k}-\mathbf{q})\cdot\mathbf{t}_n} n_{-\sigma}^{(n)} - \frac{1}{2}\, In\Delta_{\mathbf{kq}}\,, \tag{C.8}$$

where we have written $\delta n_\sigma^{(n)} = n_\sigma^{(n)} - \frac{1}{2}n$ and in the atomic limit $n_\sigma^{(n)} = 0, 1$. We consider the two parts of α_σ separately. The second term in (C.8) is taken into account in a straightforward way, as we know from the theory of Section 3 that such a form for α_σ simply shifts all energy levels for both $\pm$ spin electrons by $-\frac{1}{2}In$ and so we can neglect this term provided we shift all levels in the final result for the Green function. Thus we have (C.7) with α_σ given by the first term in (C.8).

An exact solution is possible of (C.7) because of the following property of α_σ in the atomic limit. From (C.8)

$$\sum_{\mathbf{p}} \alpha_\sigma(\mathbf{kp})\,\alpha_\sigma(\mathbf{pq}) = I\alpha_\sigma(\mathbf{kq}), \tag{C.9}$$

which follows directly using $(n_\sigma^{(i)})^2 = n_\sigma^{(i)}$ and holds only in the atomic limit. We can now rewrite the last term in (C.7) in the form

$$\begin{aligned}\frac{1}{N}\sum_{\mathbf{qk}} e^{i\mathbf{q}\cdot(\mathbf{t}_i-\mathbf{t}_j)}\alpha_\sigma(\mathbf{qk})\,A_\sigma(\mathbf{kq}E) &= \frac{1}{N}\sum_{\mathbf{qp}}\Big\{\sum_{\mathbf{k}}\alpha_\sigma(\mathbf{qk})\,\alpha_\sigma(\mathbf{kp})\Big\}\,e^{i\mathbf{q}\cdot(\mathbf{t}_i-\mathbf{t}_j)}G_\sigma(\mathbf{pq}E)\\ &= \frac{I}{N}\sum_{\mathbf{qp}} e^{i\mathbf{q}\cdot(\mathbf{t}_i-\mathbf{t}_j)}\alpha_\sigma(\mathbf{qp})\,G_\sigma(\mathbf{pq}E)\\ &= IA_\sigma^{ij}(E).\end{aligned} \tag{C.10}$$

We also have from the first term in (C.7), using (C.8)

$$\frac{1}{N}\sum_{\mathbf{q}} \alpha_\sigma(\mathbf{qq})\,e^{i\mathbf{q}\cdot(\mathbf{t}_i-\mathbf{t}_j)} = In_{-\sigma}\Delta_{ij}\,, \tag{C.11}$$

where

$$n_{-\sigma} = \frac{1}{N}\sum_i n_{-\sigma}^{(i)} \tag{C.12}$$

is the mean number of $-\sigma$ spin electrons per atom.

Thus, collecting all results, Eq. (C.7) is easily solved to give

$$\{A_\sigma^{ij}(E)\}_{\text{atomic limit}} = \frac{In_{-\sigma}\Delta_{ij}}{E - T_0 - I} \tag{C.13}$$

and substitution into (C.2), using (C.5), gives

$$[G_\sigma^{ij}(E)]_{\text{atomic limit}} = \Delta_{ij}\left\{\frac{1-n_{-\sigma}}{E-T_0} + \frac{n_{-\sigma}}{E-T_0-I}\right\}. \tag{C.14}$$

This result is the exact atomic limit for the one-body Green function. Our method of deriving it from the density-functional formalism accords with a physical picture based on the Hartree–Fock approximation. The result implies that all correlation contributions to the one-body density-dependent potential vanish as the atomic separation becomes infinite and emphasizes that these correlation effects are between spin-density fluctuations on different sites. We note finally that to obtain a Green function solution which contains correctly this atomic limit, all we must do is to treat the Hartree–Fock interactions of Section 3 consistently.

To make contact with the rather different notation used for Green-function solutions in the text, we give here an alternative form of the atomic limit solution.

We use (A.1) in the Wannier representation, together with the Hartree–Fock form of the matrix elements

$$\lambda_\sigma^{lr} = I\,\delta n_{-\sigma}^{(l)}\,\Delta_{lr} \tag{C.15}$$

and using the limit as $\epsilon(\mathbf{q}) \to T_0$,

$$F_\gamma(nmE) \to \frac{\Delta_{nm}}{E-T_0}, \tag{C.16}$$

we obtain

$$\{G_\sigma(nmE)\}_{\text{atomic limit}} = \frac{\Delta_{mn}}{E-T_0-I\,\delta n_{-\sigma}^{(n)}}. \tag{C.17}$$

Using (C.1), (A.1) and (3.17), and $\delta n_{-\sigma}^{(n)} = n_{-\sigma}^{(n)} - \frac{1}{2}n$ together with $(n_\sigma^{(i)})^2 = n_\sigma^{(i)}$ in the atomic limit, it is easily verified that (C.17) is the correct atomic limit solution for the one-body Green function.

A. *Inclusion of correlation for finite separation of atoms*

It is worth pointing out, finally, that a method of including correlation effects in the Green function which parallels Hubbard's treatment can be given. We consider a method of decoupling the basic Eqs. (C.2) and (C.6) so that terms which arise in the band and atomic limits are treated exactly and other terms approximated by using the Hartree–Fock value for the potential matrix elements. We assume again that all sites are equivalent.

We may rewrite the terms on the right-hand side of (C.6) as follows:

$$(1)\quad \sum_{l} T_{il} A_{\sigma}^{lj}(E) = T_0 A_{\sigma}^{ij}(E) + \sum_{l\neq i} T_{il} \frac{1}{N} \sum_{\mathbf{q}} e^{i\mathbf{q}\cdot(\mathbf{t}_l-\mathbf{t}_j)} \sum_{\mathbf{p}} \alpha_{\sigma}(\mathbf{qp})\, \tilde{G}_{\sigma}(\mathbf{pq}E)$$

$$\simeq T_0 A_{\sigma}^{ij}(E) + In_{-\sigma} \sum_{l\neq i} T_{il} \mathscr{G}_{\sigma}^{lj}(E) \tag{C.18}$$

using the Hartree–Fock result $\alpha_{\sigma}(\mathbf{qp}) = In_{-\sigma}\, \Delta_{\mathbf{qp}}$ in the second term;

$$(2)\qquad \frac{1}{N} \sum_{\mathbf{qk}} e^{i\mathbf{q}\cdot(\mathbf{t}_i-\mathbf{t}_j)} \alpha_{\sigma}(\mathbf{qk})\, A_{\sigma}(\mathbf{kq}E) = IA_{\sigma}^{ij}(E) \tag{C.19}$$

using the atomic limit (or Hartree–Fock) property of α_{σ} ;

$$(3)\qquad \frac{1}{N} \sum_{\mathbf{q}} \alpha_{\sigma}(\mathbf{qq})\, e^{i\mathbf{q}\cdot(\mathbf{t}_i-\mathbf{t}_j)} = In_{-\sigma} \Delta_{ij} \tag{C.20}$$

as in the atomic limit.

These approximations are consistent with the philosophy which takes all terms which arise in the atomic limit exactly into account and approximates the others by using the Hartree–Fock potential matrix elements. As the matrix elements in the atomic limit for all sites equivalent are identical in form to the Hartree–Fock results, we can make the formal approximations as above.

Collecting together these various results, (C.6) now becomes

$$EA_{\sigma}^{ij}(E) = In_{-\sigma} \Delta_{ij} + T_0 A_{\sigma}^{ij}(E) + In_{-\sigma} \sum_{k\neq i} T_{ik} G_{\sigma}^{kj}(E) + IA_{\sigma}^{ij}(E). \tag{C.21}$$

Equations (C.2) and (C.21) may be solved for G_{σ}^{ij} and A_{σ}^{ij} in an identical manner to that used by Hubbard in his many-body analysis:

$$G_{\sigma}^{ij}(E) = \frac{1}{N} \sum_{\mathbf{q}} e^{i\mathbf{q}\cdot(\mathbf{t}_i-\mathbf{t}_j)} \tilde{G}_{\sigma}(\mathbf{qq}E) \tag{C.22}$$

with

$$\tilde{G}_{\sigma}(\mathbf{qq}E) = \frac{1}{f_{\sigma}(E) - [\epsilon(\mathbf{q}) - T_0]}, \tag{C.23}$$

$$f_{\sigma}(E) = \frac{(E - T_0)(E - T_0 - I)}{E - T_0 n_{-\sigma} - (1 - n_{-\sigma})(T_0 + I)}. \tag{C.24}$$

In the Wannier representation, this solution of (C.23) becomes

$$G_\sigma(nmE) = F_\gamma(nm, E - g_\sigma(E)),$$
$$g_\sigma(E) = E - T_0 - f_\sigma(E) \tag{C.25}$$

and is obviously of a quite different character from the solutions considered in Section 4, which are both based upon a density superposition principle.

Acknowledgments

One of us (N.H.M) wishes to acknowledge a useful discussion with Professor Jim Krumhansl at an early stage of this work. We are also very indebted to the referee, who pointed out to us the need to formulate the basic theory of Section 2 in terms of a spin-dependent external potential, and has helped us to improve the presentation of the paper throughout. We are grateful to Dr. D. M. Edwards for some critical comments on an earlier draft of this work, which have helped us to remove several errors.

References

1. C. Herring, "Magnetism," (G. Rado and H. Suhl, Eds.,) Vol. 4, Academic Press, New York, 1966.
2. J. Hubbard, *Proc. Roy. Soc. Ser. A* **276** (1963), 238; *Proc. Roy. Soc. Ser. A.* **277** (1964), 237.
3. F. Bloch, *Z. Phys.* **57** (1929), 545.
4. N. H. March and A. M. Murray, *Proc. Roy. Soc. Ser. A* **261** (1961), 119. See also J. C. Stoddart and N. H. March, *Proc. Roy. Soc. Ser. A* **299** (1967), 279.
5. P. C. Hohenberg and W. Kohn, *Phys. Rev.* **B136** (1964), 864.
6. N. H. March and J. C. Stoddart, *Repts. Progr. Phys.* **31** (1968), 533.
7. P. A. Wolff, *Phys. Rev.* **124** (1961), 1030.
8. P. Richmond and G. Rickayzen, *J. Phys.* **62** (1969), 528.
9. D. J. Kim, *Phys. Rev.* **B1** (1970), 3725.

Proc. R. Soc. Lond. A. **326**, 97–116 (1971)
Printed in Great Britain

Exchange energy as functional of electronic density from Hartree–Fock theory of inhomogeneous electron gas

BY A. M. BEATTIE, J. C. STODDART AND N. H. MARCH
Department of Physics, The University, Sheffield

(*Communicated by C. A. Coulson, F.R.S.—Received* 5 *July* 1971)

In a previous paper, the validity of gradient corrections to the Dirac–Slater exchange energy was discussed within the density functional theory, for a local potential. This local potential, by definition, generated the exact charge density in the (infinite) system considered.

The present work considers the full Hartree–Fock theory within the same density functional framework. In particular, using a perturbation expansion of the Dirac density matrix, we obtain the exchange energy as an expansion in the displaced charge

$$\Delta_{\text{H.F.}}(\boldsymbol{r}) = \rho_{\text{H.F.}}(\boldsymbol{r}) - \rho_0,$$

ρ_0 being the average electron density and $\rho_{\text{H.F.}}(\boldsymbol{r})$ the Hartree–Fock spatially dependent density.

In contrast to the local potential theory, a basic integral equation must be solved to express the above relation in explicit form. No exact solution of this equation exists to date, but using the work of Kleinman and Langreth on the vertex function we are able to find an explicit approximation for the exchange energy to second-order in $\Delta_{\text{H.F.}}$.

The difficulties encountered in defining gradient series in a local potential theory are not removed in the full Hartree–Fock theory. However, it is possible to sum sub-series of gradient terms to infinite order to give a useful generalization of Dirac–Slater theory.

Although it is known in metals that correlation effects have a major influence on the Hartree–Fock results, the results obtained here may well be useful in atomic and molecular problems with a sufficiently large number of electrons, as well as in non-metallic crystals. The influence of electronic screening in the case of metals is briefly referred to.

1. INTRODUCTION

There is considerable interest currently in the density functional theory of many-body systems. This approach was pioneered by Thomas and Fermi, while later Jensen and others set up approximate energy functionals within a one-body framework. While such a theory can be put on an exact footing to all orders in perturbation theory (March & Murray 1961; Stoddart & March 1967), it was not initially clear that the approach was applicable to a genuinely many-body problem. However, doubts about this were removed by the work of Hohenberg & Kohn (1964), who showed that the philosophy of Thomas and Fermi is indeed correct; namely that the ground state of a many-body system can be characterized by its particle density $\rho(\boldsymbol{r})$.

Thus, in the refinements of the Thomas–Fermi theory which is based on a kinetic energy density proportional to $\{\rho(\boldsymbol{r})\}^{\frac{5}{3}}$ and, in its generalization due to Dirac on an exchange energy proportional to $\{\rho(\boldsymbol{r})\}^{\frac{4}{3}}$, various workers starting with von Weizsäcker (1935) have proposed the addition of gradient corrections. Thus, von Weizsäcker would add a term $(\nabla_r\rho)^2/\rho$ to the kinetic energy, while Herman, Van Dyke & Ortenburger (1969) corrected for exchange by including a term $(\nabla_r\rho)^2\rho^{-\frac{4}{3}}$. In each case, the choice of the multiplying coefficient needs care and in the work of Herman *et al.* on atoms it was treated as a semi-empirical parameter. It came out, though, to be approximately constant for a range of atomic systems.

In addition to atomic applications, it would be very tempting to try to treat large molecules by density functional theory and we feel that the theory developed in this paper will be highly relevant to that problem. The earlier attempts to treat molecules by the Thomas–Fermi theory were reviewed by one of us (March 1957) but the theory was in that approximation too crude to deal quantitatively with the chemical bond. Therefore, orbital theories have largely dominated, but even with major electronic computers these get difficult for molecules having more than say 20 electrons. The degree of difficulty of a density functional theory ought not to grow greatly with the molecular size, though naturally molecular geometry is an important feature via the boundary conditions which must be imposed.

In an earlier paper (Stoddart, Beattie & March 1971, referred to throughout as S.B.M.) we gave a theory based on a local one-body potential which reproduced the exact charge density in the many-electron system. This theory was developed with specific reference to metals, in which case it is known that correlation and exchange effects must be treated together, because of short-range electron screening.

In atoms and molecules, the situation is different and Hartree–Fock theory forms a really excellent starting point. Naturally collective effects such as discussed by Brandt & Lundqvist (1963) cannot be included in such a theory, but it is for many purposes an excellent approximation.

In § 2 we therefore set up the formalism to be used in terms of an integral equation for the canonical density matrix, while in § 3 we develop a technique for expanding the density matrices in powers of a suitable local potential. This expansion is then utilized in § 4 to derive a perturbation expansion of the Dirac density matrix in terms of the Hartree–Fock density from which we obtain the exchange energy in § 5. We deal with the connexion with the Hartree–Fock vertex function in an appendix.

2. The canonical and Dirac matrices in Hartree–Fock theory

In this section we define the basic dynamical equations for the canonical and Dirac density matrices which will be used in later sections.

In Hartree–Fock theory we define the Dirac density matrix by

$$\rho_{\mathrm{H.F.}}(\boldsymbol{r}\sigma, \boldsymbol{r}_0\sigma_0, E) = \sum_{i=1}^{N} \phi_i^*(\boldsymbol{r}\sigma)\,\phi_i(\boldsymbol{r}_0\sigma_0), \tag{2.1}$$

where σ and σ_0 are spin variables. The ϕ_i are the one-particle states which occur in the determinantal wave function, and which satisfy the usual Hartree–Fock equations.

The canonical density matrix is defined in terms of $\rho_{\mathrm{H.F.}}$ by the Laplace transform

$$C_{\mathrm{H.F.}}(\boldsymbol{r}\sigma, \boldsymbol{r}_0\sigma_0, \beta) = \beta \int_0^\infty \mathrm{d}E\, \mathrm{e}^{-\beta E} \rho_{\mathrm{H.F.}}(\boldsymbol{r}\sigma, \boldsymbol{r}_0\sigma_0, E), \tag{2.2}$$

where $\beta = (k_B T)^{-1}$. We may also note for future reference the inverse relation to (2.2), given by

$$\rho_{\text{H.F.}} = \frac{1}{2\pi i}\int_{c-i\infty}^{c+i\infty} d\beta \frac{e^{\beta E}}{\beta} C_{\text{H.F.}}(\boldsymbol{r}\sigma, \boldsymbol{r}_0\sigma_0, \beta). \tag{2.3}$$

From the usual Hartree–Fock equations for the one-particle wave-functions ϕ_i we may obtain in a straightforward manner the dynamical equation for the canonical matrix as

$$\left(\tfrac{1}{2}\nabla_r^2 - \frac{\partial}{\partial\beta}\right) C_{\text{H.F.}}(\boldsymbol{r}\sigma, \boldsymbol{r}_0\sigma_0, \beta) = \left[V_{\text{ext}}(\boldsymbol{r}) + e^2\int \frac{d\boldsymbol{r}_1}{|\boldsymbol{r}-\boldsymbol{r}_1|}\rho_{\text{H.F.}}(\boldsymbol{r}_1 E_f)\right] C_{\text{H.F.}}(\boldsymbol{r}\sigma, \boldsymbol{r}_0\sigma_0, \beta)$$
$$- \sum_{\sigma_1}\int d\boldsymbol{r}_1 \rho_{\text{H.F.}}(\boldsymbol{r}\sigma, \boldsymbol{r}_1\sigma_1, E_f)\frac{e^2}{|\boldsymbol{r}-\boldsymbol{r}_1|} C_{\text{H.F.}}(\boldsymbol{r}_1\sigma_1, \boldsymbol{r}_0\sigma_0, \beta). \tag{2.4}$$

Here $V_{\text{ext}}(\boldsymbol{r})$ is the external potential and

$$\rho_{\text{H.F.}}(\boldsymbol{r}, E_f) = \sum_{i=1}^{N}\sum_{\sigma}|\phi_i(\boldsymbol{r}\sigma)|^2 \tag{2.5}$$

is the electron density in Hartree–Fock theory, with Fermi energy E_f.

Equation (2.4) may be converted into a pure integral equation by use of the Green function $G_0(\boldsymbol{r}-\boldsymbol{r}_1, \beta)$ which satisfies

$$\left[\tfrac{1}{2}\nabla_r^2 - \frac{\partial}{\partial\beta}\right] G_0(\boldsymbol{r}-\boldsymbol{r}_1, \beta) = \delta(\boldsymbol{r}-\boldsymbol{r}_1)\,\delta(\beta). \tag{2.6}$$

From the elementary theory of the Bloch density matrix (see, for example, March, Young & Sampanthar 1967) we obtain

$$G_0(\boldsymbol{r}, \beta) = \theta(\beta)\, C_0(\boldsymbol{r}, \beta) \tag{2.7}$$

where

$$C_0(\boldsymbol{r}, \beta) = (2\pi\beta)^{-\frac{3}{2}}\exp\{-r^2/2\beta\} \tag{2.8}$$

is the free particle Bloch matrix.

From equation (2.4) using equation (2.7) we now obtain the integral equation

$$C_{\text{H.F.}}(\boldsymbol{r}\sigma, \boldsymbol{r}_0\sigma_0, \beta) = C_0(\boldsymbol{r}\sigma, \boldsymbol{r}_0\sigma_0, \beta) - \int d\boldsymbol{r}_1\int_0^\beta d\beta_1 C_0(\boldsymbol{r}-\boldsymbol{r}_1, \beta-\beta_1)$$
$$\times\left[V_{\text{H}}(\boldsymbol{r}_1)\, C_{\text{H.F.}}(\boldsymbol{r}_1\sigma, \boldsymbol{r}_0\sigma_0, \beta) - \sum_{\sigma_2}\int d\boldsymbol{r}_2 \rho_{\text{H.F.}}(\boldsymbol{r}_1\sigma, \boldsymbol{r}_2\sigma_2, E_f)\right.$$
$$\left.\times\frac{e^2}{|\boldsymbol{r}_1-\boldsymbol{r}_2|} C_{\text{H.F.}}(\boldsymbol{r}_2\sigma_2, \boldsymbol{r}_0\sigma_0, \beta_1)\right], \tag{2.9}$$

where we define V_{H} by

$$V_{\text{H}}(\boldsymbol{r}) = V_{\text{ext}}(\boldsymbol{r}) + e^2\int d\boldsymbol{r}_1 \frac{\rho_{\text{H.F.}}(\boldsymbol{r}_1 E_f)}{|\boldsymbol{r}-\boldsymbol{r}_1|} \tag{2.10}$$

and C_0 is the free particle matrix.

We have already assumed that the external potential is spin-independent. If we also assume that the two spin states are equally populated, then we can obtain an equation for the spinless form of the density matrix by summing over spin variables in equation (2.9). The result is

$$C_{\mathrm{H.F.}}(\boldsymbol{r}\boldsymbol{r}_0\beta) = C_0(\boldsymbol{r}\boldsymbol{r}_0\beta) - \int \mathrm{d}\boldsymbol{r}_1 \int_0^\beta \mathrm{d}\beta_1 C_0(\boldsymbol{r}-\boldsymbol{r}_1, \beta-\beta_1)\Big[V_{\mathrm{H}}(\boldsymbol{r}_1) C_{\mathrm{H.F.}}(\boldsymbol{r}_1\boldsymbol{r}_0\beta_1) - \int \mathrm{d}\boldsymbol{r}_2 \rho_{\mathrm{H.F.}}(\boldsymbol{r}_1\boldsymbol{r}_2 E_{\mathrm{f}}) \frac{e^2}{|\boldsymbol{r}_1-\boldsymbol{r}_2|} C_{\mathrm{H.F.}}(\boldsymbol{r}_2\boldsymbol{r}_0\beta_1)\Big]. \quad (2.11)$$

Equation (2.11) is the starting point for this analysis of Hartree–Fock theory in the density functional approach.

3. Expansion of density matrices in powers of local potential

In this section we attempt to follow the procedure set out in previous papers for a one-body potential theory (Stoddart and March 1967; S.B.M. 1971). In one-body potential theory the solutions for the Dirac matrix obtained by March & Murray (1961) in perturbation theory were inverted to give the potential as a functional of the displaced charge $\Delta(\boldsymbol{r}) = \rho(\boldsymbol{r}) - \rho_0$ in an expansion in powers of Δ. Direct substitution back into the series of March & Murray then gave the Dirac matrix in terms of its diagonal component $\Delta(\boldsymbol{r})$. This perturbation expansion and the corresponding gradient expansion have been very useful in the discussion of energy functionals, and the density dependent one-body potential (see S.B.M.).

There are certain difficulties which stand in the way of a direct inversion of (2.11) to give $C_{\mathrm{H.F.}}$ or $\rho_{\mathrm{H.F.}}$ in terms of the Hartree–Fock electron density. These are principally due to the last term in (2.11), which makes the equation nonlinear. The procedure we shall use here is to invert (2.11) to obtain the potential V_{H} in terms of the displaced charge $\Delta_{\mathrm{H.F.}}(\boldsymbol{r})$, which may then be resubstituted into the series for the Dirac matrix to give the desired result. We require therefore a solution of (2.11) in powers of the potential V_{H}. It is easy to see that the existence of a term in (2.11) which is independent of V_{H} does not allow a straightforward iterative solution to be made.

To circumvent these difficulties we proceed as follows. We can rewrite (2.11) formally as

$$C = C_0 + \hat{K}_{\mathrm{H}}.C + \hat{K}_{\mathrm{X}}.C, \quad (3.1)$$

where the integral operators $\hat{K}_{\mathrm{H}}$ and $\hat{K}_{\mathrm{X}}$ are defined by

$$[\hat{K}_{\mathrm{H}}.f](\boldsymbol{r},\beta) = \int \mathrm{d}\boldsymbol{r}_1 \int \mathrm{d}\beta_1 K_{\mathrm{H}}(\boldsymbol{r}\boldsymbol{r}_1\beta\beta_1) f(\boldsymbol{r}_1,\beta_1) \quad (3.2)$$

with
$$K_{\mathrm{H}}(\boldsymbol{r}\boldsymbol{r}_1\beta\beta_1) = -\theta(\beta-\beta_1) C_0(\boldsymbol{r}-\boldsymbol{r}_1, \beta-\beta_1) V_{\mathrm{H}}(\boldsymbol{r}_1) \quad (3.3)$$

and
$$[\hat{K}_{\mathrm{X}}.f](\boldsymbol{r},\beta) = \int \mathrm{d}\boldsymbol{r}_1 \int \mathrm{d}\beta_1 K_{\mathrm{X}}(\boldsymbol{r}\boldsymbol{r}_1\beta\beta_1) f(\boldsymbol{r}_1\beta_1) \quad (3.4)$$

with
$$K_{\mathrm{X}}(\boldsymbol{r}\boldsymbol{r}_1\beta\beta_1) = \theta(\beta-\beta_1) \int \mathrm{d}\boldsymbol{r}_2 C_0(\boldsymbol{r}-\boldsymbol{r}_2, \beta-\beta_1) \frac{e^2}{|\boldsymbol{r}_1-\boldsymbol{r}_2|} \rho_{\mathrm{H.F.}}(\boldsymbol{r}_2\boldsymbol{r}_1 E_{\mathrm{f}}). \quad (3.5)$$

The formal solution of (3.1) is then

$$C = [1 - \hat{K}_{\mathrm{X}} - \hat{K}_{\mathrm{H}}]^{-1}.C_0. \tag{3.6}$$

Using the operator identity

$$\frac{1}{\hat{A}+\hat{B}} = \frac{1}{\hat{A}} - \frac{1}{\hat{A}}.\hat{B}.\frac{1}{\hat{A}} + \ldots + (-1)^n \left[\frac{1}{\hat{A}}\hat{B}\right]^n \frac{1}{\hat{A}} + \ldots \tag{3.7}$$

with
$$\hat{A} = 1 - \hat{K}_{\mathrm{X}}, \quad \hat{B} = \hat{K}_{\mathrm{H}} \tag{3.8}$$

and
$$\hat{\Sigma} = 1/\hat{A}, \tag{3.9}$$

we obtain (3.6) in the form

$$C = \hat{\Sigma}.C_0 + [\hat{\Sigma}.\hat{K}_{\mathrm{H}}.\hat{\Sigma}].C_0 + [\hat{\Sigma}.\hat{K}_{\mathrm{H}}.\hat{\Sigma}.\hat{K}_{\mathrm{H}}.\hat{\Sigma}].C_0 + \ldots. \tag{3.10}$$

However, it can be seen directly that the right-hand side can be summed to give

$$C = [1 - \hat{\Sigma}.\hat{K}_{\mathrm{H}}]^{-1}.(\hat{\Sigma}.C_0). \tag{3.11}$$

The integral equation equivalent to (3.11) is obtained by multiplying by $(1 - \hat{\Sigma}.\hat{K}_{\mathrm{H}})$, and using (3.2), (3.3) we obtain

$$\begin{aligned} C_{\mathrm{H.F.}}(\boldsymbol{r}\boldsymbol{r}_0\beta) &= \int \mathrm{d}\boldsymbol{r}_1 \int \mathrm{d}\beta_1 \Sigma(\boldsymbol{r}\boldsymbol{r}_1\beta\beta_1) C_0(\boldsymbol{r}_1\boldsymbol{r}_0\beta_1) \\ &- \int \mathrm{d}\boldsymbol{r}_1 \int \mathrm{d}\beta_1 \Sigma(\boldsymbol{r}\boldsymbol{r}_1\beta\beta_1) \int \mathrm{d}\boldsymbol{r}_2 \int_0^{\beta_1} \mathrm{d}\beta_2 C_0(\boldsymbol{r}_1 - \boldsymbol{r}_2, \beta_1 - \beta_2) V_{\mathrm{H}}(\boldsymbol{r}_2) C_{\mathrm{H.F.}}(\boldsymbol{r}_2\boldsymbol{r}_0\beta_2). \end{aligned} \tag{3.12}$$

We may write (3.12) in a more convenient way be defining

$$S_0(\boldsymbol{r}\boldsymbol{r}_0\beta) = \int \mathrm{d}\boldsymbol{r}_1 \int \mathrm{d}\beta_1 \Sigma(\boldsymbol{r}\boldsymbol{r}_1\beta\beta_1) C_0(\boldsymbol{r}_1\boldsymbol{r}_0\beta_1) \tag{3.13}$$

and
$$S(\boldsymbol{r}\boldsymbol{r}_1\beta\beta_1) = \int \mathrm{d}\boldsymbol{r}_2 \int \mathrm{d}\beta_2 \Sigma(\boldsymbol{r}\boldsymbol{r}_2\beta\beta_2) C_0(\boldsymbol{r}_2 - \boldsymbol{r}_1, \beta_2 - \beta_1)\,\theta(\beta_2 - \beta_1). \tag{3.14}$$

Then (3.12) becomes, since $S_0(\boldsymbol{r}\boldsymbol{r}_0\beta) = S(\boldsymbol{r}\boldsymbol{r}_0\beta 0)$,

$$C_{\mathrm{H.F.}}(\boldsymbol{r}\boldsymbol{r}_0\beta) = S(\boldsymbol{r}\boldsymbol{r}_0\beta 0) - \int \mathrm{d}\boldsymbol{r}_1 \int \mathrm{d}\beta_1 S(\boldsymbol{r}\boldsymbol{r}_1\beta\beta_1) V_{\mathrm{H}}(\boldsymbol{r}_1) C_{\mathrm{H.F.}}(\boldsymbol{r}_1\boldsymbol{r}_0\beta_1). \tag{3.15}$$

We have reached now the point at which the problem of solving the integral equation (2.11) in a power series in V_{H} falls into two parts. First the equation (3.15) can be solved iteratively to give $C_{\mathrm{H.F.}}$ as a series in V_{H}, S_0 and S (or Σ). Secondly we require the expansion of S in powers of V_{H}.

We can obtain the expansion of S, defined by (3.14), as follows. First we formally expand the exchange operator $\hat{K}_{\mathrm{X}}$ in powers of V_{H}:

$$\hat{K}_{\mathrm{X}} = \hat{K}_{\mathrm{X}}^{(0)} + \hat{K}_{\mathrm{X}}^{(1)} + \hat{K}_{\mathrm{X}}^{(2)} + \ldots \tag{3.16}$$

with $\hat{K}_{\mathrm{X}}^{(n)}$ of order n in V_{H}. From (3.8) and (3.9) we obtain the corresponding series for Σ. Use of the operator identity (3.7) gives

$$\hat{\Sigma} = \hat{\Sigma}^{(0)} + \hat{\Sigma}^{(0)}.\left[\sum_{r=1}^{\infty} \hat{K}_{\mathrm{X}}^{(r)}\right].\hat{\Sigma}^{(0)} + \ldots, \tag{3.17}$$

where
$$\hat{\Sigma}^{(0)} = [1-\hat{K}_X^{(0)}]^{-1}. \tag{3.18}$$

We can write (3.14) formally as
$$S = \hat{\Sigma}.C_0\theta \tag{3.19}$$

and then substitution of (3.17) into (3.19) gives

$$S = \hat{\Sigma}^{(0)}.C_0\theta+\hat{\Sigma}^{(0)}.\left[\sum_{r=1}^{\infty}\hat{K}_X^{(r)}\right].\hat{\Sigma}^{(0)}.C_0\theta+\ldots \tag{3.20}$$

from which we can obtain the expansion of S

$$S = S^{(0)}+S^{(1)}+\ldots \tag{3.21}$$

in powers of V_H, but in terms of the exchange kernel K_X, which itself depends on S.

The exchange kernel K_X is given by (3.5), which we can rewrite in nth order in V_H, in the form

$$K_X^{(n)}(\boldsymbol{r}\boldsymbol{r}_1\beta\beta_1) = \theta(\beta-\beta_1)\int \mathrm{d}\boldsymbol{r}_2\, C_0(\boldsymbol{r}-\boldsymbol{r}_2, \beta-\beta_1)\, L_X^{(n)}(\boldsymbol{r}_2, \boldsymbol{r}_1) \tag{3.22}$$

with
$$L_X(\boldsymbol{r}_2, \boldsymbol{r}_1) = \frac{e^2}{|\boldsymbol{r}_1-\boldsymbol{r}_2|}\rho_{\mathrm{H.F.}}(\boldsymbol{r}_2\boldsymbol{r}_1E_f). \tag{3.23}$$

From (2.3), (3.15), (3.22) and (3.23) we can now obtain the exchange kernel in terms of S. In nth order we have

$$L_X^{(n)}(\boldsymbol{r}\boldsymbol{r}_0) = \frac{e^2}{|\boldsymbol{r}-\boldsymbol{r}_0|}\int_{c-\mathrm{i}\infty}^{c+\mathrm{i}\infty}\mathrm{d}\beta\,\frac{\mathrm{e}^{\beta E_f}}{2\pi\mathrm{i}\beta}C_{\mathrm{H.F.}}^{(n)}(\boldsymbol{r}\boldsymbol{r}_0\beta) \tag{3.24}$$

together with

$$C_{\mathrm{H.F.}}^{(n)}(\boldsymbol{r}\boldsymbol{r}_0\beta) = S^{(n)}(\boldsymbol{r}\boldsymbol{r}_0\beta 0) - \sum_{\substack{l,m\\(l+m=n-1)}}\int \mathrm{d}\boldsymbol{r}_1\int \mathrm{d}\beta_1\, S^{(l)}(\boldsymbol{r}\boldsymbol{r}_1\beta\beta_1)\, V_H(\boldsymbol{r}_1)\, C_{\mathrm{H.F.}}^{(m)}(\boldsymbol{r}_1\boldsymbol{r}_0\beta_1). \tag{3.25}$$

Thus (3.24) and (3.25) can be substituted into (3.22) to give the exchange kernel expansion in terms of S. Equating the terms of order n in V_H on both sides of (3.20) then provides a scheme for determining successively the required terms $S^{(1)}, S^{(2)}, \ldots$ in the expansion for S. Once $S^{(0)}$ is known then this procedure gives for any $S^{(n)}$ a *linear* integral equation of the form

$$S^{(n)} = A[S^{(n)}]+B[S^{(n-1)}, S^{(n-2)}, \ldots, S^{(0)}, V_H]. \tag{3.26}$$

Then direct iteration of (3.15) yields an expansion in V_H with coefficients determined by the linear equations (3.26).

We examine now how these equations give the first three terms in the expansion of $C_{\mathrm{H.F.}}$ or $\rho_{\mathrm{H.F.}}$ in powers of V_H.

3.1. *Lowest order approximation*

In lowest order $V_{\mathrm{ext}} = 0$ and we consider the electrons to be moving in a uniform positive background of charge which cancels exactly their mutual Coulomb repulsion. Then $V_H = 0$, and the one-particle states are the usual free particle states and are filled up to Fermi-wave number k_f. The canonical matrix $C_{\mathrm{H.F.}}^{(0)}(\boldsymbol{r}\boldsymbol{r}_0\beta)$ in this limit is well known, and can be easily shown to come from our basic equations in the following way.

From (3.25) with $V_{\mathrm{H}} = 0$ we have

$$C^{0}_{\mathrm{H.F.}}(\boldsymbol{r}\boldsymbol{r}_0\beta) = S^{(0)}(\boldsymbol{r}\boldsymbol{r}_0\beta 0), \tag{3.27}$$

or

$$C^{(0)}_{\mathrm{H.F.}}(\boldsymbol{r}\boldsymbol{r}_0\beta) = \int \mathrm{d}\boldsymbol{r}_1 \int \beta_1 \Sigma^{(0)}(\boldsymbol{r}-\boldsymbol{r}_1, \beta-\beta_1)\, C_0(\boldsymbol{r}_1\boldsymbol{r}_0\beta_1) \tag{3.28}$$

using (3.13). From (3.18) we have, by expansion of the right hand side, for any function $f(\boldsymbol{r},\beta)$

$$[\hat{\Sigma}^{(0)}.f]\,(\boldsymbol{r},\beta) \equiv \int \mathrm{d}\boldsymbol{r}_1 \int \mathrm{d}\beta_1\, \Sigma^{(0)}(\boldsymbol{r}-\boldsymbol{r}_1, \beta-\beta_1) f(\boldsymbol{r}_1,\beta_1)$$

$$= f(\boldsymbol{r},\beta) + \sum_{n=1}^{\infty}\left\{\prod_{j=1}^{n}\int \mathrm{d}\boldsymbol{r}_j\, \mathrm{d}\boldsymbol{s}_j \int \mathrm{d}\beta_j\, C_0(\boldsymbol{r}_{j-1}-\boldsymbol{s}_j, \beta_{j-1}-\beta_j)\, e^2 \frac{\rho^{(0)}_{\mathrm{H.F.}}(\boldsymbol{s}_j,\boldsymbol{r}_j,E_{\mathrm{f}})}{|\boldsymbol{s}_j-\boldsymbol{r}_j|}\right\} f(\boldsymbol{r}_n,\beta_n). \tag{3.29}$$

In (3.29) we have $\boldsymbol{r}_0 \equiv \boldsymbol{r}$ and $\beta_0 \equiv \beta$.

We denote by $F_k[f(\boldsymbol{r},\beta)]$ the Fourier transform of f with respect to $\boldsymbol{r}$, and $\mathscr{L}_p[f(\boldsymbol{r},\beta)]$ denotes the Laplace transform with respect to β, with $\boldsymbol{k}$ and p being the transform variables. From (3.29) we obtain

$$F_k\mathscr{L}_p\{[\hat{\Sigma}^{(0)}.f]\,(\boldsymbol{r},\beta)\} = F_k\mathscr{L}_p[\Sigma^{(0)}(\boldsymbol{r},\beta)]\,.\,F_k\mathscr{L}_p[f(\boldsymbol{r},\beta)]$$

$$= \left[1 - F_k\mathscr{L}_p(C_0(\boldsymbol{r},\beta))\,.\,F_k\left(e^2\frac{\rho^{(0)}_{\mathrm{H.F.}}(\boldsymbol{r}E_{\mathrm{f}})}{r}\right)\right]^{-1} F_k\mathscr{L}_p[f(\boldsymbol{r},\beta)] \tag{3.30}$$

with

$$\rho^{(0)}_{\mathrm{H.F.}}(\boldsymbol{r},E_{\mathrm{f}}) \equiv \rho^{(0)}_{\mathrm{H.F.}}(\boldsymbol{x},\boldsymbol{y},E_{\mathrm{f}})|_{x-y=r}.$$

Using

$$F_k\mathscr{L}_p(C_0(\boldsymbol{r},\beta)) = (p+\tfrac{1}{2}k^2)^{-1} \tag{3.31}$$

and

$$f(k) = F_k\left[\frac{e^2}{r}\rho^{(0)}_{\mathrm{H.F.}}(\boldsymbol{r},E_{\mathrm{f}})\right] = \frac{e^2k_{\mathrm{f}}}{\pi}\left[1+\frac{k_{\mathrm{f}}^2-k^2}{2kk_{\mathrm{f}}}\ln\left|\frac{k_{\mathrm{f}}+k}{k_{\mathrm{f}}-k}\right|\right] \tag{3.32}$$

and taking for $\rho^{(0)}_{\mathrm{H.F.}}$ the free particle form, with states filled up to k_{f}, we can obtain $F_k\mathscr{L}_p[\Sigma^{(0)}(\boldsymbol{r},\beta)]$ from (3.30) and upon performing the inverse transforms we find

$$\Sigma^{(0)}(\boldsymbol{r}\boldsymbol{r}_0\beta\beta_0) = \frac{1}{(2\pi)^3}\int \mathrm{d}\boldsymbol{k}\, \mathrm{e}^{-i\boldsymbol{k}.(\boldsymbol{r}-\boldsymbol{r}_0)}\,[\theta(\beta-\beta_0)\,(\tfrac{1}{2}k^2-\epsilon_{\mathrm{H.F.}}(k))\,\mathrm{e}^{-(\beta-\beta_0)\epsilon_{\mathrm{H.F.}}(k)} + \delta(\beta-\beta_0)], \tag{3.33}$$

where

$$\epsilon_{\mathrm{H.F.}}(k) = \tfrac{1}{2}k^2 - f(k)$$

is just the well-known Hartree–Fock single particle energy. From (3.28) we have directly, using (3.33),

$$C^{(0)}_{\mathrm{H.F.}}(\boldsymbol{r}\boldsymbol{r}_0\beta) = S^{(0)}(\boldsymbol{r}\boldsymbol{r}_0\beta 0) = \frac{1}{(2\pi)^3}\int d\boldsymbol{k}\, \mathrm{e}^{-i\boldsymbol{k}.(\boldsymbol{r}-\boldsymbol{r}_0)}\,\mathrm{e}^{-\beta\epsilon_{\mathrm{H.F.}}(k)} \tag{3.34}$$

which is the usual Hartree–Fock canonical matrix in the present model. From (2.3) we can obtain the usual Dirac matrix

$$\rho^{(0)}_{\mathrm{H.F.}}(\boldsymbol{r}\boldsymbol{r}_0E_{\mathrm{f}}) = \frac{1}{(2\pi)^3}\int \mathrm{d}\boldsymbol{k}\, \mathrm{e}^{-i\boldsymbol{k}.(\boldsymbol{r}-\boldsymbol{r}_0)}\,\theta(E_{\mathrm{f}}-\epsilon_{\mathrm{H.F.}}(k)), \tag{3.35}$$

where

$$E_{\mathrm{f}} \equiv \epsilon_{\mathrm{H.F.}}(k_{\mathrm{f}}) = \tfrac{1}{2}k_{\mathrm{f}}^2 - \frac{e^2k_{\mathrm{f}}}{\pi}.$$

3.2. *Higher order approximations*

We define first the Fourier transforms of the matrices $C_{\mathrm{H.F.}}$, $\rho_{\mathrm{H.F.}}$, L_{X} and S by

$$\tilde{Z}(\boldsymbol{k},\boldsymbol{k}_0) = \frac{1}{(2\pi)^3}\int \mathrm{d}\boldsymbol{r}\,\mathrm{e}^{\mathrm{i}\boldsymbol{k}.\boldsymbol{r}}\int \mathrm{d}\boldsymbol{r}_0\,\mathrm{e}^{-\mathrm{i}\boldsymbol{k}.\boldsymbol{r}_0}Z(\boldsymbol{r}\boldsymbol{r}_0), \tag{3.36}$$

where Z represents any one of the relevant matrices. We also have

$$\tilde{V}_{\mathrm{H}}(\boldsymbol{k}) = \int \mathrm{d}\boldsymbol{r}\,\mathrm{e}^{\mathrm{i}\boldsymbol{k}.\boldsymbol{r}}\,V_{\mathrm{H}}(\boldsymbol{r}). \tag{3.37}$$

We have then from (3.34) that

$$\tilde{S}^{(0)}(\boldsymbol{k},\boldsymbol{k}_0,\beta) = \delta(\boldsymbol{k}-\boldsymbol{k}_0)\,\mathrm{e}^{-\beta\epsilon_{\mathrm{H.F.}}(k)} \tag{3.38}$$

and from (3.20) and (3.22) we find, after some algebra,

$$\tilde{S}^{(1)}(\boldsymbol{k},\boldsymbol{k}_0,\beta,\beta_0) = \int \mathrm{d}\beta_1\,\tilde{S}^{(0)}(\boldsymbol{k},\beta-\beta_1)\,\tilde{L}^{(1)}_{\mathrm{X}}(\boldsymbol{k},\boldsymbol{k}_0)\,\tilde{S}^{(0)}(\boldsymbol{k}_0,\beta_1-\beta_0) \tag{3.39}$$

and

$$\begin{aligned}\tilde{S}^{(2)}(\boldsymbol{k},\boldsymbol{k}_0,\beta,\beta_0) &= \int \mathrm{d}\beta_1\,\tilde{S}^{(0)}(\boldsymbol{k},\beta-\beta_1)\,\tilde{L}^{(2)}_{\mathrm{X}}(\boldsymbol{k},\boldsymbol{k}_0)\,\tilde{S}^{(0)}(\boldsymbol{k}_0,\beta_1-\beta_0)\\ &+\int \mathrm{d}\beta_1\,\mathrm{d}\beta_2\int \mathrm{d}\boldsymbol{p}\,\tilde{S}^{(0)}(\boldsymbol{k},\beta-\beta_1)\,\tilde{L}^{(1)}_{\mathrm{X}}(\boldsymbol{k},\boldsymbol{p})\,\tilde{S}^{(0)}(\boldsymbol{p},\beta_1-\beta_2)\,\tilde{L}^{(1)}_{\mathrm{X}}(\boldsymbol{p},\boldsymbol{k}_0)\,\tilde{S}^{(0)}(\boldsymbol{k}_0,\beta_2-\beta_0).\end{aligned} \tag{3.40}$$

Upon Fourier transform of (3.24) we have

$$\tilde{L}^{(n)}_{\mathrm{X}}(\boldsymbol{k},\boldsymbol{k}_0) = \frac{e^2}{2\pi^2}\int_{c-\mathrm{i}\infty}^{c+\mathrm{i}\infty}\mathrm{d}\beta\,\frac{\mathrm{e}^{\beta E_{\mathrm{f}}}}{2\pi\mathrm{i}\beta}\int\frac{\mathrm{d}\boldsymbol{p}}{p^2}\,\tilde{C}^{(n)}_{\mathrm{H.F.}}(\boldsymbol{k}-\boldsymbol{p},\boldsymbol{k}_0-\boldsymbol{p}_0,\beta) \tag{3.41}$$

and from (3.25)

$$\begin{aligned}\tilde{C}^{(n)}_{\mathrm{H.F.}}(\boldsymbol{k},\boldsymbol{k}_0,\beta) = \tilde{S}^{(n)}(\boldsymbol{k},\boldsymbol{k}_0,\beta) - \sum_{\substack{l,m\\(l+m=n-1)}}\int \mathrm{d}\boldsymbol{q}\int \mathrm{d}\beta_1\int \mathrm{d}\boldsymbol{p}\,\tilde{S}^{(l)}(\boldsymbol{k},\boldsymbol{q},\beta-\beta_1)\\ \times\tilde{V}_{\mathrm{H}}(\boldsymbol{p})\,\tilde{C}^{(m)}_{\mathrm{H.F.}}(\boldsymbol{q}-\boldsymbol{p},\boldsymbol{k}_0,\beta_1).\end{aligned} \tag{3.42}$$

To obtain the first-order term $S^{(1)}$ or $L^{(1)}_{\mathrm{X}}$ we take (3.42) with $n = 1$, and using (3.38) and (3.39) we find

$$\tilde{C}^{(1)}_{\mathrm{H.F.}}(\boldsymbol{k},\boldsymbol{k}_0,\beta) = \tilde{K}(\boldsymbol{k},\boldsymbol{k}_0,\beta)\,[\tilde{L}^{(1)}_{\mathrm{X}}(\boldsymbol{k},\boldsymbol{k}_0)-\tilde{V}_{\mathrm{H}}(\boldsymbol{k}-\boldsymbol{k}_0)], \tag{3.43}$$

where

$$\tilde{K}(\boldsymbol{k},\boldsymbol{k}_0,\beta) = \frac{\mathrm{e}^{-\beta\epsilon_{\mathrm{H.F.}}(k)}-\mathrm{e}^{-\beta\epsilon_{\mathrm{H.F.}}(k_0)}}{\epsilon_{\mathrm{H.F.}}(k_0)-\epsilon_{\mathrm{H.F.}}(k)}. \tag{3.44}$$

Substitution of (3.43) into (3.41) for $n = 1$ gives immediately an integral equation for $\tilde{L}^{(1)}_{\mathrm{X}}(\boldsymbol{k},\boldsymbol{k}_0)$, namely

$$\tilde{L}^{(1)}_{\mathrm{X}}(\boldsymbol{k},\boldsymbol{k}_0) = \frac{e^2}{2\pi^2}\int\frac{\mathrm{d}\boldsymbol{q}}{q^2}\,K(\boldsymbol{k}-\boldsymbol{q},\boldsymbol{k}_0-\boldsymbol{q},E_{\mathrm{f}})\,[\tilde{L}^{(1)}_{\mathrm{X}}(\boldsymbol{k}-\boldsymbol{q},\boldsymbol{k}_0-\boldsymbol{q})-\tilde{V}_{\mathrm{H}}(\boldsymbol{k}-\boldsymbol{k}_0)], \tag{3.45}$$

where

$$K(\boldsymbol{k},\boldsymbol{k}_0,E_{\mathrm{f}}) = \frac{\theta(E_{\mathrm{f}}-\epsilon_{\mathrm{H.F.}}(k))-\theta(E_{\mathrm{f}}-\epsilon_{\mathrm{H.F.}}(k_0))}{\epsilon_{\mathrm{H.F.}}(k_0)-\epsilon_{\mathrm{H.F.}}(k)}. \tag{3.46}$$

But from (2.3) and (3.43) we have to first order

$$\tilde{\rho}^{(1)}_{\mathrm{H.F.}}(\boldsymbol{k},\boldsymbol{k}_0,E_{\mathrm{f}}) = K(\boldsymbol{k},\boldsymbol{k}_0,E_{\mathrm{f}})\,[\tilde{L}^{(1)}_{\mathrm{X}}(\boldsymbol{k},\boldsymbol{k}_0)-\tilde{V}(\boldsymbol{k}-\boldsymbol{k}_0)] \tag{3.47}$$

and combining (3.45) and (3.47) we can obtain an equation for the first order response of the Dirac matrix to V_{H}. If we use the representation

$$\tilde{\rho}^{(1)}_{\mathrm{H.F.}}(\boldsymbol{k},\boldsymbol{k}_0,E_{\mathrm{f}}) = -\tilde{V}_{\mathrm{H}}(\boldsymbol{k}-\boldsymbol{k}_0)\,H(\boldsymbol{k},\boldsymbol{k}_0) \tag{3.48}$$

then we obtain a linear equation for the coefficient H, which is independent of V_{H}

$$H(\boldsymbol{k},\boldsymbol{k}_0) = K(\boldsymbol{k},\boldsymbol{k}_0,E_{\mathrm{f}})\left[1+\frac{e^2}{2\pi^2}\int\frac{\mathrm{d}\boldsymbol{q}}{q^2}H(\boldsymbol{k}-\boldsymbol{q},\boldsymbol{k}_0-\boldsymbol{q})\right]. \tag{3.49}$$

To obtain the linear response then (3.49) must be solved. In appendix 1 we show that H is essentially the vertex function $\Lambda_k(\boldsymbol{p})$ and (3.49) corresponds to the usual integral equation for $\tilde{\Lambda}_k(\boldsymbol{p})$. We may then use the approximate solutions for the vertex function which have been obtained by various authors to construct the linear response. In particular in § 5, we shall use an approximate solution due to Kleinman to construct a non-perturbative density functional form for the exchange energy.

To obtain the Dirac matrix to second order in V_{H} we start from (3.40), and using (3.38) and performing the β integrals we obtain

$$\tilde{S}^{(2)}(\boldsymbol{k},\boldsymbol{k}_0,\beta-\beta_0) = \tilde{L}^{(2)}_{\mathrm{X}}(\boldsymbol{k},\boldsymbol{k}_0)\,\tilde{K}(\boldsymbol{k},\boldsymbol{k}_0,\beta-\beta_0)+\int\mathrm{d}\boldsymbol{p}\,\tilde{L}^{(1)}_{\mathrm{X}}(\boldsymbol{k},\boldsymbol{p}) \times\tilde{L}^{(1)}_{\mathrm{X}}(\boldsymbol{p},\boldsymbol{k}_0)\,\tilde{K}^{(2)}(\boldsymbol{k},\boldsymbol{p},\boldsymbol{k}_0,\beta-\beta_0), \tag{3.50}$$

where $K(\boldsymbol{k}\boldsymbol{k}_0\beta)$ is given by (3.44) and

$$\begin{aligned}\tilde{K}^{(2)}(\boldsymbol{k},\boldsymbol{p},\boldsymbol{k}_0,\beta) &= \mathrm{e}^{-\beta\epsilon_{\mathrm{H.F.}}(k)}[\epsilon_{\mathrm{H.F.}}(k)-\epsilon_{\mathrm{H.F.}}(p)]^{-1}[\epsilon_{\mathrm{H.F.}}(k)-\epsilon_{\mathrm{H.F.}}(k_0)]^{-1}\\ &\quad+\mathrm{e}^{-\beta\epsilon_{\mathrm{H.F.}}(p)}[\epsilon_{\mathrm{H.F.}}(p)-\epsilon_{\mathrm{H.F.}}(k)]^{-1}[\epsilon_{\mathrm{H.F.}}(p)-\epsilon_{\mathrm{H.F.}}(k_0)]^{-1}\\ &\quad+\mathrm{e}^{-\beta\epsilon_{\mathrm{H.F.}}(k_0)}[\epsilon_{\mathrm{H.F}}(k_0)-\epsilon_{\mathrm{H.F.}}(k)]^{-1}[\epsilon_{\mathrm{H.F.}}(k_0)-\epsilon_{\mathrm{H.F.}}(p)]^{-1}.\end{aligned} \tag{3.51}$$

Now we take (3.42) with $n = 2$ and using (3.39) and (3.50) in the right-hand side we obtain

$$\begin{aligned}\tilde{C}^{(2)}_{\mathrm{H.F.}}(\boldsymbol{k}\boldsymbol{k}_0\beta) &= \tilde{L}^{(2)}_{\mathrm{X}}(\boldsymbol{k}\boldsymbol{k}_0)\,\tilde{K}(\boldsymbol{k}\boldsymbol{k}_0\beta)+\int\mathrm{d}\boldsymbol{p}\,\tilde{L}^{(1)}_{\mathrm{X}}(\boldsymbol{k},\boldsymbol{p})\,\tilde{L}^{(1)}_{\mathrm{X}}(\boldsymbol{p},\boldsymbol{k}_0)\tilde{K}^{(2)}(\boldsymbol{k},\boldsymbol{p},\boldsymbol{k}_0,\beta)\\ &\quad-\int\mathrm{d}\boldsymbol{p}\,\tilde{L}^{(1)}_{\mathrm{X}}(\boldsymbol{k},\boldsymbol{k}_0+\boldsymbol{p})\tilde{V}_{\mathrm{H}}(\boldsymbol{p})[\tilde{K}(\boldsymbol{k}_0+\boldsymbol{p},\boldsymbol{k}_0,\beta)-\tilde{K}(\boldsymbol{k}\boldsymbol{k}_0\beta)]\,[\epsilon_{\mathrm{H.F.}}(k)-\epsilon_{\mathrm{H.F.}}(\boldsymbol{k}_0+\boldsymbol{p})]^{-1}\\ &\quad-\int\mathrm{d}\boldsymbol{p}\,\tilde{V}_{\mathrm{H}}(\boldsymbol{p})\int_0^{\beta}\mathrm{d}\beta_1\,\mathrm{e}^{-(\beta-\beta_1)\epsilon_{\mathrm{H.F.}}(k)}\,\tilde{C}^{(1)}_{\mathrm{H.F.}}(\boldsymbol{k}-\boldsymbol{p},\boldsymbol{k}_0,\beta_1)\end{aligned} \tag{3.52}$$

with $\tilde{C}^{(1)}_{\mathrm{H.F.}}$ given by (3.43).

We may now substitute (3.52) into the right-hand side of (3.41) for $n = 2$ to obtain an integral equation for $\tilde{L}^{(2)}_{\mathrm{X}}$. Performing the transform (2.3) on (3.52) we can obtain $\tilde{\rho}^{(2)}_{\mathrm{H.F.}}$ in terms of $\tilde{L}^{(2)}_{\mathrm{H.F.}}$ and so obtain an integral equation for the second order response $\tilde{\rho}^{(2)}_{\mathrm{H.F.}}$. If we use the representation

$$\tilde{\rho}^{(2)}_{\mathrm{H.F.}}(\boldsymbol{k}\boldsymbol{k}_0E_{\mathrm{f}}) = \int\mathrm{d}\boldsymbol{p}\,\tilde{V}_{\mathrm{H}}(\boldsymbol{p})\tilde{V}_{\mathrm{H}}(\boldsymbol{k}-\boldsymbol{k}_0-\boldsymbol{p})\,H^{(2)}(\boldsymbol{k},\boldsymbol{k}-\boldsymbol{p},\boldsymbol{k}_0) \tag{3.53}$$

then we find that the coefficient $H^{(2)}$ satisfies

$$H^{(2)}(\boldsymbol{k},\boldsymbol{k}_0,\boldsymbol{p}) = M(\boldsymbol{k},\boldsymbol{k}_0,\boldsymbol{p})+\frac{e^2}{2\pi^2}K(\boldsymbol{k},\boldsymbol{k}_0,E_{\mathrm{f}})\int\frac{\mathrm{d}\boldsymbol{q}}{q^2}H^{(2)}(\boldsymbol{k}-\boldsymbol{q},\boldsymbol{k}_0-\boldsymbol{q},\boldsymbol{p}), \tag{3.54}$$

where

$$M(\boldsymbol{k}\boldsymbol{k}_0\boldsymbol{p}) = \frac{K(\boldsymbol{k}-\boldsymbol{p},\boldsymbol{k},E_f)-K(\boldsymbol{k},\boldsymbol{k}_0,E_f)}{\epsilon_{\text{H.F.}}(\boldsymbol{k}_0)-\epsilon_{\text{H.F.}}(\boldsymbol{k}-\boldsymbol{p})}+\frac{K(\boldsymbol{k}_0+\boldsymbol{p},\boldsymbol{k}_0,E_f)-K(\boldsymbol{k},\boldsymbol{k}_0,E_f)}{\epsilon_{\text{H.F.}}(\boldsymbol{k}_0)-\epsilon_{\text{H.F.}}(\boldsymbol{k}-\boldsymbol{p})}$$
$$\times\left[\frac{H(\boldsymbol{k}-\boldsymbol{p},\boldsymbol{k}_0)}{K(\boldsymbol{k}-\boldsymbol{p},\boldsymbol{k}_0,E_f)}-1\right]+K^{(2)}(\boldsymbol{k},\boldsymbol{k}-\boldsymbol{p},\boldsymbol{k}_0,E_f)\left[\frac{H(\boldsymbol{k},\boldsymbol{k}_0+\boldsymbol{p})}{K(\boldsymbol{k},\boldsymbol{k}_0+\boldsymbol{p},E_f)}-1\right]$$
$$\times\left[\frac{H(\boldsymbol{k}-\boldsymbol{p},\boldsymbol{k}_0)}{K(\boldsymbol{k}-\boldsymbol{p},\boldsymbol{k}_0,E_f)}-1\right]. \tag{3.55}$$

In (3.55), $K^{(2)}(\boldsymbol{k},\boldsymbol{p},\boldsymbol{k}_0,E_f)$ is obtained from (3.51) by using the replacement

$$e^{-\beta\epsilon_{\text{H.F.}}(k)}\to\theta(E_f-\epsilon_{\text{H.F.}}(k))$$

throughout the expression.

4. Perturbation expansion for Dirac matrix in terms of Hartree–Fock density

Collecting together the results of the previous section, we have then the following expression for the Dirac density matrix to second order in the potential V_{H}:

$$\Delta\tilde{\rho}_{\text{H.F.}}(\boldsymbol{k},\boldsymbol{k}_0,E_f)=\tilde{\rho}_{\text{H.F.}}(\boldsymbol{k},\boldsymbol{k}_0,E_f)-\tilde{\rho}^{(0)}_{\text{H.F.}}(\boldsymbol{k},\boldsymbol{k}_0,E_f)=-\tilde{V}_{\text{H}}(\boldsymbol{k}-\boldsymbol{k}_0)H(\boldsymbol{k},\boldsymbol{k}_0)$$
$$+\int d\boldsymbol{p}\,\tilde{V}_{\text{H}}(\boldsymbol{p})\tilde{V}_{\text{H}}(\boldsymbol{k}-\boldsymbol{k}_0-\boldsymbol{p})H^{(2)}(\boldsymbol{k},\boldsymbol{k}-\boldsymbol{p},\boldsymbol{k}_0). \tag{4.1}$$

The Fourier transform of the Hartree–Fock charge density is given by

$$\tilde{\rho}_{\text{H.F.}}(\boldsymbol{k})=\int d\boldsymbol{r}\,\rho_{\text{H.F.}}(\boldsymbol{r}E_f)\,e^{-i\boldsymbol{k}.\boldsymbol{r}}$$
$$=\int d\boldsymbol{q}\,\rho_{\text{H.F.}}(\boldsymbol{q},\boldsymbol{q}-\boldsymbol{k},E_f). \tag{4.2}$$

From (4.1) we then find

$$\tilde{\Delta}_{\text{H.F.}}(\boldsymbol{k})=\tilde{\rho}_{\text{H.F.}}(\boldsymbol{k})-\tilde{\rho}^{(0)}_{\text{H.F.}}(\boldsymbol{k})=\tilde{V}_{\text{H}}(\boldsymbol{k})J(\boldsymbol{k})+\int d\boldsymbol{p}\,\tilde{V}_{\text{H}}(\boldsymbol{p})\tilde{V}_{\text{H}}(\boldsymbol{k}-\boldsymbol{p})J^{(2)}(\boldsymbol{k},\boldsymbol{p})+O(V_{\text{H}}^3), \tag{4.3}$$

where

$$J(\boldsymbol{k})=-\int d\boldsymbol{q}\,H(\boldsymbol{q},\boldsymbol{q}-\boldsymbol{k}) \tag{4.4}$$

and

$$J^{(2)}(\boldsymbol{k},\boldsymbol{p})=\int d\boldsymbol{q}\,H^{(2)}(\boldsymbol{q},\boldsymbol{q}-\boldsymbol{k},\boldsymbol{p}). \tag{4.5}$$

Equation (4.3) can now be inverted directly to give the potential V_{H} as a power series in the displaced charge $\Delta_{\text{H.F.}}$. We obtain to second order

$$\tilde{V}_{\text{H}}(\boldsymbol{k})=\frac{\tilde{\Delta}_{\text{H.F.}}(\boldsymbol{k})}{J(\boldsymbol{k})}+\int d\boldsymbol{p}\frac{J^{(2)}(\boldsymbol{k},\boldsymbol{p})}{J(\boldsymbol{k})}\frac{\tilde{\Delta}_{\text{H.F.}}(\boldsymbol{p})\tilde{\Delta}_{\text{H.F.}}(\boldsymbol{k}-\boldsymbol{p})}{J(\boldsymbol{p})J(\boldsymbol{k}-\boldsymbol{p})}+O(\Delta^3). \tag{4.6}$$

This result (4.6), which can in principle be carried to any order in $\Delta_{\text{H.F.}}$, allows us to express the canonical density matrix and the Dirac density matrix in terms of the electron density. Substitution of (4.6) into (4.1) gives one of the central results of this paper, namely the full Dirac matrix in terms of the H.F. charge density. We obtain to second order

$$\Delta\tilde{\rho}_{\text{H.F.}}(\boldsymbol{k},\boldsymbol{k}_0,E_{\text{f}}) = -\frac{H(\boldsymbol{k},\boldsymbol{k}_0)}{J(\boldsymbol{k}-\boldsymbol{k}_0)}\tilde{\Delta}_{\text{H.F.}}(\boldsymbol{k}-\boldsymbol{k}_0)+\int \mathrm{d}\boldsymbol{p}\,\frac{\tilde{\Delta}_{\text{H.F.}}(\boldsymbol{p})\,\tilde{\Delta}_{\text{H.F.}}(\boldsymbol{k}-\boldsymbol{k}_0-\boldsymbol{p})}{J(\boldsymbol{p})\,J(\boldsymbol{k}-\boldsymbol{k}_0-\boldsymbol{p})}$$

$$\times\left[H^{(2)}(\boldsymbol{k},\boldsymbol{k}_0,\boldsymbol{p})-\frac{H(\boldsymbol{k},\boldsymbol{k}_0)\,J^{(2)}(\boldsymbol{k}-\boldsymbol{k}_0,\boldsymbol{p})}{J(\boldsymbol{k}-\boldsymbol{k}_0)}\right]+O(\Delta^3). \qquad (4.7)$$

We must of course obtain H and $H^{(2)}$ by solving (3.49) and (3.54). In the next section we use (4.7) to construct the exchange energy in the density functional theory.

5. The exchange energy

We can obtain the exchange energy directly from the usual expression

$$E_{\text{X}} = -\tfrac{1}{4}e^2\int\!\!\int \mathrm{d}\boldsymbol{r}\,\mathrm{d}\boldsymbol{r}_0\,\frac{\rho_{\text{H.F.}}(\boldsymbol{r}\boldsymbol{r}_0E_{\text{f}})\,\rho_{\text{H.F.}}(\boldsymbol{r}_0\boldsymbol{r}E_{\text{f}})}{|\boldsymbol{r}-\boldsymbol{r}_0|}. \qquad (5.1)$$

Substitution of the response equation (4.7) gives the exchange energy as a perturbation series in the displaced charge $\Delta_{\text{H.F.}}$. The result to second order takes the form, writing $E_x^{(0)}$ for the homogeneous gas exchange energy,

$$\Delta E_{\text{X}} = E_{\text{X}} - E_{\text{X}}^{(0)} = \int\!\!\int \mathrm{d}\boldsymbol{r}\,\mathrm{d}\boldsymbol{s}\,B(\boldsymbol{r}-\boldsymbol{s})\,\Delta_{\text{H.F.}}(\boldsymbol{r})\,\Delta_{\text{H.F.}}(\boldsymbol{s})+O(\Delta^3) \qquad (5.2)$$

and the kernel $B(\boldsymbol{r})$ may be obtained directly by using (4.7) in (5.1), in terms of the functions H and $H^{(2)}$, which are required to calculate the off-diagonal elements of the density matrix.

Alternatively we show in appendix 2 that a simple expression can be obtained for B solely in terms of $H(\boldsymbol{k},\boldsymbol{p})$. If we define

$$B(\boldsymbol{r}) = \frac{1}{(2\pi)^3}\int \mathrm{d}\boldsymbol{q}\,\mathrm{e}^{\mathrm{i}\boldsymbol{q}.\boldsymbol{r}}\,\tilde{B}(q,k_{\text{f}}) \qquad (5.3)$$

then we obtain

$$\tilde{B}(q) = \tfrac{1}{4}\tilde{K}(q)+\tfrac{1}{2}\left[\int \mathrm{d}\boldsymbol{k}\,H(\boldsymbol{k},\boldsymbol{k}-\boldsymbol{q})\right]^{-1}, \qquad (5.4)$$

where $\tilde{K}(q)$ is defined by equation (A 2.3) in appendix 2.

To obtain an explicit response kernel $\tilde{B}(q)$ we must solve the integral equation (3.49). However it is shown in appendix 1 that the function H is given directly by the proper vertex part $\tilde{\Lambda}_k(\boldsymbol{p})$ and in appendix 2 it is shown that $\tilde{B}(q)$ is given by the dielectric function $\epsilon(q)$ due to the usual relation between the vertex function and $\epsilon(q)$.

We may therefore use at once the extensive work which has been done on the vertex function $\Lambda_k(\boldsymbol{p})$ and the dielectric function (Kleinman 1967; Langreth 1969). Equations (A 1.11) and (A 1.12) give the relation between H and $\tilde{\Lambda}_k(\boldsymbol{p})$ and the usual Hartree–Fock equation for $\tilde{\Lambda}_k(\boldsymbol{p})$. Langreth has given a variational solution of (A 1.12) in which $\Lambda_k(\boldsymbol{p})$ is independent of $\boldsymbol{p}$ but which has the correct large and small k limits, and in fact gives the dielectric function $\epsilon(k)$ exactly in these limits. Unfortunately his solution is somewhat complicated and we can see no way of using it in a useful way, although it obviously gives as yet the best possible kernel $\tilde{B}(q)$ and may need to be considered if refinements of the present theory are found to be necessary later.

After certain approximations, it is shown by Langreth that the variational solution reduces to a much simpler form in the static limit, corresponding to a solution obtained by Kleinman. This solution is

$$\tilde{\Lambda}_k = \left[1+\frac{e^2}{2\pi^2}F(k)\left(\frac{1}{k^2+k_f^2+k_s^2}+\frac{1}{k_f^2+k_s^2}\right)\right]^{-1}, \tag{5.5}$$

where

$$F(k) = -\tfrac{1}{2}\int \mathrm{d}\boldsymbol{q}\, K(\boldsymbol{q}, \boldsymbol{k}-\boldsymbol{q}, E_f) \tag{5.6}$$

with $K(\boldsymbol{k}, \boldsymbol{p}, E_f)$ given by (3.46), and k_s is an inverse screening length† ($k_s \approx k_{\mathrm{T.F.}}$, the inverse Thomas–Fermi screening length). If we approximate $\epsilon_{\mathrm{H.F.}}(p)$ in (3.46) by $p^2/2$ then (5.6) gives

$$F(k) = -\frac{\pi^2 k_{\mathrm{T.F.}}^2}{2e^2}\left[\frac{1}{2}+\frac{k_f}{2k}\left(1-\frac{k^2}{4k_f^2}\right)\ln\left|\frac{k+2k_f}{k-2k_f}\right|\right]. \tag{5.7}$$

From (5.5) and (A 1.11) we can construct the corresponding $H(\boldsymbol{p}, \boldsymbol{k})$ and so obtain explicitly the first order term in (4.7) for the Dirac matrix. However to go beyond first order the function $H^{(2)}$ must be obtained by solving (3.54).

Later in this section we require $\tilde{B}(q, k_f)$ which depends on H only through the dielectric function (A 2.14). From (A 1.13) we obtain

$$\epsilon(k) = 1-\frac{(2e^2/\pi^2k^2)\,F(k)}{1+(2e^2/\pi^2k^2)\,F(k)f(k)}, \tag{5.8}$$

where

$$f(k) = \tfrac{1}{4}k^2\left(\frac{1}{k^2+k_f^2+k_s^2}+\frac{1}{k_f^2+k_s^2}\right) \tag{5.9}$$

and we take the approximate $F(k)$ given by (5.7). From (A 2.16) we then have the approximate result

$$\tilde{B}(k, k_f) = \frac{e^2F(k)}{k_f^2 k_{\mathrm{T.F.}}^2} - \pi^3\frac{(1+(2e^2/\pi^2k^2)\,F(k)f(k))}{e^2F(k)}. \tag{5.10}$$

In the next section this expression is used to go beyond the second order perturbation result, to sum to infinite order a sub-series of gradient terms.

5.1. *Gradient series and a one-body exchange potential*

The result (5.4) or (5.10) for the exchange energy kernel is an alternative to the one proposed in a one-body theory by S.B.M. The method proposed here has the advantage of corresponding to a well defined (Hartree–Fock) model of electron

† Actually, in a formally exact H.F. theory, we would have $k_s = 0$. This solution (5.5) of Kleinman has taken some account of correlation effects by introducing a screening factor of the form $\exp(-k_s r)$ into the Coulomb interaction.

interactions in contrast to the result of S.B.M. which included some undefined account of correlation effects. The corrections which must be applied to the present model are therefore relatively well defined.

In S.B.M. the principle aim was to obtain a density dependent one-body potential, in the manner prescribed by Kohn & Sham (1965), which would include electron interactions, and in particular exchange interactions. Gradient series for the exchange part of the one-body potential were investigated, and it was shown that the exchange kernel B could be used to sum an infinite subset of terms in both the series for the exchange energy and the exchange potential. A similar procedure can be followed in the present theory and we give below the main results. As discussed above these results will be of considerable practical interest because they give useful expressions for a well defined part of the Dirac–Kohn–Sham one-body potential.

It was shown in S.B.M. that the singularities of the kernel $\tilde{B}(q, k_f)$ invalidated a normal (local) gradient series for the exchange energy density, and so for the one-body exchange potential. The arguments used were quite general and carry straight over into the present Hartree-Fock theory. We find here the same singularity structure in the Hartree–Fock $\tilde{B}(q, k_f)$ of (5.10), as was found in the corresponding one-body kernel. In particular it may be shown that the exact $\tilde{B}(q, k_f)$ given by (5.4) is singular at $q = 0$, having a term like $q^2 \ln q$ in a low q expansion. There is a singularity also at $q = 2k_f$ of the form $(q-2k_f)^2 \ln(q-2k_f)$. It is these singularities which invalidate a gradient expansion of the exchange energy, which can be seen as follows.

We define the exchange energy density $\epsilon_X(\boldsymbol{r})$ by

$$E_X[\rho] = \int d\boldsymbol{r}\, \epsilon_X(\boldsymbol{r}) \tag{5.11}$$

and we look for an expansion for ϵ_X in the form

$$\epsilon_X(\boldsymbol{r}) = \epsilon_X^{(0)}(\rho(\boldsymbol{r})) + \epsilon_2^{(2)}(\rho(\boldsymbol{r}))\,(\nabla_r \rho(\boldsymbol{r}))^2 + \dots \tag{5.12}$$

where

$$\epsilon_X^{(0)}(\rho(\boldsymbol{r})) = -\frac{3}{4}\left(\frac{3}{\pi}\right)^{\frac{1}{3}} \rho^{\frac{4}{3}}(\boldsymbol{r}) \tag{5.13}$$

is the Dirac result and where higher terms involve higher powers of the gradient operator. The coefficients $\epsilon_X^{(0)}$, $\epsilon_2^{(2)}$, etc. are functions of the density. It was shown in S.B.M. that the coefficient $\epsilon_2^{(2)}$ of the first gradient correction to the Dirac term $\epsilon_X^{(0)}$ is given by the kernel $\tilde{B}(qk_f)$ in the perturbation series, the result being

$$\epsilon_2^{(2)}(\rho(\boldsymbol{r})) = -\tfrac{1}{6}\int d\boldsymbol{s}\, s^2 B(s, k_f)\Big|_{k_f = \{3\pi^2\rho(\boldsymbol{r})\}^{\frac{1}{3}}}. \tag{5.14}$$

Obviously $B(\boldsymbol{s})$ must converge sufficiently fast as $s \to \infty$ for the integral (5.14) to exist. In fact the contribution of the singularity at $q = 0$ of $\tilde{B}(q)$ to the asymptotic behaviour of $B(\boldsymbol{s})$ is

$$B(\boldsymbol{s}) \underset{s\to\infty}{\to} s^{-5}$$

which is not fast enough. The other singularities give a similar behaviour. Hence the coefficient $\epsilon_2^{(2)}$ will not exist and the same will be true for higher order correction terms. This aspect of gradient series is essentially the same as in S.B.M.

However, one way of getting round the difficulty of the non-existence of local gradient series was discussed in S.B.M. The method involved summing an infinite subset of terms in a gradient expansion for the exchange energy density so that all integrals were well defined. The method uses the perturbation series to second order and the result is

$$\epsilon_X[\rho] \doteqdot -\frac{3}{4}\left(\frac{3}{\pi}\right)^{\frac{1}{3}}\rho^{\frac{4}{3}}(\boldsymbol{r}) - \tfrac{1}{2}\int \mathrm{d}\boldsymbol{r}'\, B(\boldsymbol{r}',\rho(\boldsymbol{r}))\,[\rho(\boldsymbol{r}+\tfrac{1}{2}\boldsymbol{r}')-\rho(\boldsymbol{r}-\tfrac{1}{2}\boldsymbol{r}')]^2. \quad (5.15)$$

The first term in (5.15) is the usual Dirac result and

$$B(\boldsymbol{r}',\rho(\boldsymbol{r})) \equiv B(\boldsymbol{r}',k_f)\big|_{k_f=\{3\pi^2\rho(\boldsymbol{r})\}^{\frac{1}{3}}}. \quad (5.16)$$

Equation (5.15) directly generalizes the Dirac–Slater theory in a way which avoids the difficulties of local gradient expansions. The non-local character of (5.15) seems to be an essential requirement.

The exchange contribution V_X to the Dirac–Kohn–Sham one-body potential is defined by

$$\delta E_X[\rho] = \int \mathrm{d}\boldsymbol{r}\, V_X(\boldsymbol{r})\,\delta\rho(\boldsymbol{r}) \quad (5.17)$$

and from (5.15) we obtain

$$V_X(\boldsymbol{r}) = -\left(\frac{3}{\pi}\right)^{\frac{1}{3}}\rho^{\frac{1}{3}}(\boldsymbol{r}) - \tfrac{1}{2}\int \mathrm{d}\boldsymbol{r}'\,\frac{\partial B(\boldsymbol{r}',\rho(\boldsymbol{r}))}{\partial\rho(\boldsymbol{r})}[\rho(\boldsymbol{r}+\tfrac{1}{2}\boldsymbol{r}')-\rho(\boldsymbol{r}-\tfrac{1}{2}\boldsymbol{r}')]^2$$
$$-2\int \mathrm{d}\boldsymbol{r}'\, B(\boldsymbol{r}-\boldsymbol{r}',\rho(\tfrac{1}{2}(\boldsymbol{r}+\boldsymbol{r}')))\,[\rho(\boldsymbol{r})-\rho(\boldsymbol{r}')]. \quad (5.18)$$

Although in Hartree–Fock theory, in contrast to the one-body potential theory of S.B.M., we cannot at present get the kernel B exactly, (5.18) is of considerable interest because it represents a well defined part of the total density dependent one-body potential. The techniques of Ma & Brueckner (1968) will certainly be useful to discuss corrections due to correlation effects, and this point is under investigation at the present time.

5.2. *Relation to Thomas–Fermi–Dirac theory*

We would like to comment briefly on the relation of the present approach to the Thomas–Fermi–Dirac (T.F.D.) theory. This theory originated with the famous paper by Dirac (1930) who showed how exchange could be incorporated into the Thomas–Fermi atom.

It is of interest here that the second order expression (5.2) for the exchange energy gives back directly the linearized form of the T.F.D. equation, which is obtained on the assumption that the potential V_H given by (2.10) varies slowly in space.

The Euler equation in the linearized T.F.D. theory in fact reduces to an equation determining self-consistently the potential $V_H(r)$, which takes the form

$$\nabla_r^2 V_s(\boldsymbol{r}) = q^2 V_H(\boldsymbol{r}), \tag{5.19}$$

where $V_s(\boldsymbol{r}) = V_H(\boldsymbol{r}) - V_{ext}(\boldsymbol{r})$. If we assume in (5.2) that Δ is slowly varying then we can approximate E_X by the local expression

$$E_X \simeq E_X^{(0)} + \int d\boldsymbol{r}\, \Delta^2(\boldsymbol{r})\, \tilde{B}(0). \tag{5.20}$$

It may then be shown that the 'screening constant' q^2 is given by

$$q^2 = 4\pi[\pi^2/k_f + 2\tilde{B}(0)]^{-1}. \tag{5.21}$$

From the exact theory (5.4) and (3.49) it can be shown that $\tilde{B}(0)$ has the same value as in the one-body theory of S.B.M.

$$\tilde{B}(0) = -\pi/2k_f^2 \tag{5.22}$$

and (5.22) with (5.21) gives then the usual expression in T.F.D. theory. With $\tilde{B}(0) = 0$ of course we obtain the usual Thomas–Fermi screening radius.

It is well known that the T.F.D. theory has some basic difficulties associated with it, especially a finite atomic radius at which the density falls suddenly to zero. The present theory, in principle, is free from such limitations, but in practice it does not seem feasible to sum directly the perturbation series. We have earlier obtained forms for the exchange energy in (5.1) by summing, to infinite order, sub-series of gradient terms which provides a useful generalization of the simple T.F.D. theory.

6. Definition of one-body potential

Throughout this paper we have assumed that, provided we could obtain the total energy as a functional of the density, then we could, through the variational principle, define a density dependent one-body potential, according to the prescription of Kohn & Sham (1965), following the pioneering work of Dirac (1930).

If $E[\rho]$ is the total energy then minimization gives

$$\delta E[\rho] = \int d\boldsymbol{r}\, \epsilon_r[\rho]\, \delta\rho(\boldsymbol{r}) = 0 \tag{6.1}$$

or

$$\epsilon_r[\rho] = 0. \tag{6.2}$$

If we introduce the total kinetic energy $T_s[\rho]$ of a system of non-interacting electrons of density ρ (which is a universal functional of ρ) and write

$$\delta T_s[\rho] = \int d\boldsymbol{r}\, t_r[\rho]\, \delta\rho(\boldsymbol{r}) \tag{6.3}$$

then we may express (6.2) as

$$t_r[\rho] + V_r[\rho] = 0, \tag{6.4}$$

where

$$V_r[\rho] = \epsilon_r[\rho] - t_r[\rho] \tag{6.5}$$

is the density dependent one-body potential.

The kinetic energy functional T_s was studied by Stoddart & March (1967). From its definition, together with (6.4), we have an operational method for defining a one-body local potential which will generate a given density $\rho(\boldsymbol{r})$ as the sum of the squares of N one-body wavefunctions $\psi_i(\boldsymbol{r})$. This one-body potential is just of course $V(\boldsymbol{r}) = -t_r[\rho_1]$ if the given density is $\rho_1(\boldsymbol{r})$.

It is of considerable interest to enquire under what conditions we could obtain the density-dependent one-body potential given by (6.5). If we write the total energy in the form

$$E[\rho] = T_s[\rho] + \tfrac{1}{2}\int \mathrm{d}\boldsymbol{r}\,\mathrm{d}\boldsymbol{r}'\frac{\rho(\boldsymbol{r})\rho(\boldsymbol{r}')}{|\boldsymbol{r}-\boldsymbol{r}'|} + E_{\mathrm{XC}}[\rho] + \int \mathrm{d}\boldsymbol{r}\,\rho(\boldsymbol{r})\,V_{\mathrm{ext}}(\boldsymbol{r}) \tag{6.6}$$

then, if we can perform the functional differentiation, the one-body potential is

$$V_r[\rho] = V_{\mathrm{ext}}(\boldsymbol{r}) + \int \frac{\mathrm{d}\boldsymbol{r}'\,\rho(\boldsymbol{r}')}{|\boldsymbol{r}-\boldsymbol{r}'|} + V_{\mathrm{XC}}(\boldsymbol{r}) \tag{6.7}$$

with the variation in exchange plus correlation energy $E_{\mathrm{XC}}[\rho]$ given by

$$\delta E_{\mathrm{XC}}[\rho] = \int \mathrm{d}\boldsymbol{r}\,V_{\mathrm{XC}}(\boldsymbol{r})\,\delta\rho(\boldsymbol{r}). \tag{6.8}$$

The only terms which may present some difficulty are those arising from the exchange and correlation contribution to the total energy. We have shown in section (5) that the exchange contribution can be written in the form of a perturbation expansion in the displaced charge $\Delta_{\mathrm{H.F.}}(\boldsymbol{r})$. Equation (5.2) gives this expansion to second order but in principle this series can be carried to any order. It is possible to show that the existence of such a perturbation series is a sufficient (but not necessary) condition for the existence of the functional derivative and so the exchange part of the one-body potential. Thus to first order we obtain

$$V_{\mathrm{X}}(\boldsymbol{r}) = -\left(\frac{3}{\pi}\right)^{\frac{1}{3}}\rho_0^{\frac{1}{3}} + 2\int \mathrm{d}\boldsymbol{s}\,B(\boldsymbol{r}-\boldsymbol{s})\,\Delta(\boldsymbol{s}) + O(\Delta^2) \tag{6.9}$$

and all higher terms can be dealt with in the same way.

While we cannot at the present time say much about the correlation contribution E_{C} to the total energy, if this also could be expressed in a perturbation expansion then the same conclusion would result. The work of Ma & Brueckner (1968) is certainly in accord with this conclusion.

We want to stress though that the one-body potential discussed here does not give the density of states through a band. It does give the correct Hartree–Fock charge density at the Fermi level, and is sufficient, with knowledge of the energy functionals, to give the exact H.F. ground state energy.

In the Hartree–Fock theory discussed in this paper, this one-body potential $V_r[\rho]$ is of interest, since of course one generally thinks of this theory as based upon an energy-dependent potential.

7. Conclusion

It would clearly be of major interest if the exchange energy functional given by equation (5.15) provided a rapidly convergent way of calculating the H.F. exchange energy in an atom or molecule. That this is a very real possibility is evident from the success of Dirac–Slater exchange theory and its first gradient correction by Herman *et al.* The form (5.15), of course, sums an infinite sub-series of gradient terms.

The aim of the density functional theory, however, must eventually be to transcend the H.F. approximation and to incorporate an explicit account of correlation. In this sense, it may be possible to regard (5.15) as a basic starting point by:

(i) Replacing the first term proportional to $\rho^{\frac{4}{3}}$ by the exchange and correlation functional $E_0[\rho]$ of a homogeneous electron gas, as has been suggested by many earlier workers.

(ii) Regarding the kernel B as being modified by correlation, either from first principles, which at present seems an extremely difficult task, or semi-empirically.

In this connexion, an interesting question is then, of course, how in atoms and molecules the theory of collective effects (Brandt & Lundqvist 1963; see also Wendin 1971 for much more refined calculations) can be approximately incorporated into the density functional scheme. We know, of course, in the uniform electron gas that the plasmon contributes in an important way to the ground-state energy.

Furthermore, as was stressed in S.B.M., it is very well known in metals that one must screen the Coulomb interaction, as in the approximate solution of Kleinman, for example, referred to above. Two different approaches were shown in S.B.M. to lead to very similar results, the first being to use a screened Coulomb interaction, while the second employed a cutoff when the gradient expansion for the density matrix was used in equation (5.1), to avoid trouble far from the diagonal.

Some such method *must* be employed for metals. But, as we remarked at the outset, the application of the theory developed here to large molecules, for which a direct attack via the customary H.F. wavefunctions is still very difficult, would be of prime interest.

One of us (A.M.B.) wishes to acknowledge support from the Science Research Council during the course of this work.

Appendix I. The Hartree–Fock vertex function

We show in this appendix the relation between the theory of the response of the Dirac matrix to the potential V_{H}, given in section (3), and the vertex function $\Lambda_k(\boldsymbol{p})$ which gives the matrix element for the scattering of an electron by a unit external potential from initial state $\boldsymbol{p}$ to final state $\boldsymbol{p}+\boldsymbol{k}$.

In the Hartree–Fock approximation the one-electron potential can be written formally as,

$$V_{\text{H.F.}} = V_{\text{ext}} + V_{\text{Coul}} + V_{\text{exch}}, \tag{A 1.1}$$

where

$$V_{\text{H}} = V_{\text{ext}} + V_{\text{Coul}} \tag{A 1.2}$$

and V_{Coul} is the classical Coulomb potential.

We define the matrix elements $\tilde{V}(\boldsymbol{k}+\boldsymbol{p},\boldsymbol{p})$ of any potential between initial state $\boldsymbol{p}$ and final state $\boldsymbol{p}+\boldsymbol{k}$. Then the vertex function is given by,

$$\tilde{V}_{\text{H.F.}}(\boldsymbol{p}+\boldsymbol{k},\boldsymbol{p}) = \Lambda_k(\boldsymbol{p})\,\tilde{V}_{\text{ext}}(\boldsymbol{k}). \tag{A 1.3}$$

We may introduce the proper vertex part $\tilde{\Lambda}_k(\boldsymbol{p})$ defined by

$$\tilde{\Lambda}_k(\boldsymbol{p}) = \epsilon(k)\,\Lambda_k(\boldsymbol{p}), \tag{A 1.4}$$

where $\epsilon(k)$ is the static dielectric function, which relates the external potential to V_{H} through

$$\tilde{V}_{\text{H}}(\boldsymbol{p}+\boldsymbol{k},\boldsymbol{p}) = \tilde{V}_{\text{ext}}(\boldsymbol{k})/\epsilon(k). \tag{A 1.5}$$

Taking matrix elements of (A 1.1) and using (A 1.3), (A 1.4) and (A 1.5) we obtain

$$\tilde{V}_{\text{exch}}(\boldsymbol{p}+\boldsymbol{k},\boldsymbol{p}) = \frac{\tilde{V}_{\text{ext}}(\boldsymbol{k})}{\epsilon(k)}[\tilde{\Lambda}_k(\boldsymbol{p})-1]. \tag{A 1.6}$$

However we have

$$\tilde{V}_{\text{exch}}(\boldsymbol{p}+\boldsymbol{k},\boldsymbol{p}) = -\frac{1}{(2\pi)^3}\int\!\!\int \mathrm{d}\boldsymbol{r}_1\,\mathrm{d}\boldsymbol{r}_2\,\mathrm{e}^{-\mathrm{i}(\boldsymbol{p}+\boldsymbol{k})\cdot\boldsymbol{r}_1+\mathrm{i}\boldsymbol{p}\cdot\boldsymbol{r}_2}\frac{e^2}{|\boldsymbol{r}_1-\boldsymbol{r}_2|}\rho_{\text{H.F.}}(\boldsymbol{r}_1\boldsymbol{r}_2E_{\text{f}}) \tag{A 1.7}$$

or

$$\tilde{V}_{\text{exch}}(\boldsymbol{p}+\boldsymbol{k},\boldsymbol{p}) = -\int\frac{\mathrm{d}\boldsymbol{q}}{(2\pi)^3}\frac{4\pi e^2}{q^2}\tilde{\rho}_{\text{H.F.}}(\boldsymbol{p}+\boldsymbol{k}-\boldsymbol{q},\boldsymbol{p}-\boldsymbol{q},E_{\text{f}}) \tag{A 1.8}$$

where the Fourier transform $\tilde{\rho}_{\text{H.F.}}$ is given by (3.36). Using the first order response equation (3.48) for $\tilde{\rho}_{\text{H.F.}}$ we obtain,

$$\tilde{V}_{\text{exch}}(\boldsymbol{p}+\boldsymbol{k},\boldsymbol{p}) = \frac{e^2}{2\pi^2}\int\frac{\mathrm{d}\boldsymbol{q}}{q^2}\tilde{V}_{\text{H}}(\boldsymbol{k})\,H(\boldsymbol{p}+\boldsymbol{k}-\boldsymbol{q},\boldsymbol{p}-\boldsymbol{q}). \tag{A 1.9}$$

Using (3.49) and (A 1.5) we finally obtain,

$$\tilde{V}_{\text{exch}}(\boldsymbol{p}+\boldsymbol{k},\boldsymbol{p}) = \frac{\tilde{V}_{\text{ext}}(\boldsymbol{k})}{\epsilon(k)}\left[\frac{H(\boldsymbol{p}+\boldsymbol{k},\boldsymbol{p})}{K(\boldsymbol{p}+\boldsymbol{k},\boldsymbol{p},E_{\text{f}})}-1\right]. \tag{A 1.10}$$

From (A 1.6) and (A 1.10) we have therefore that,

$$H(\boldsymbol{p}+\boldsymbol{k},\boldsymbol{p}) = K(\boldsymbol{p}+\boldsymbol{k},\boldsymbol{p},E_{\text{f}})\,\tilde{\Lambda}_k(\boldsymbol{p}) \tag{A 1.11}$$

which relates directly the vertex function to the response of the Dirac matrix.

Substitution of (A 1.11) into (3.49) gives directly the usual integral equation in Hartree–Fock theory for the reduced vertex part, which is,

$$\tilde{\Lambda}_k(\boldsymbol{p}) = 1+\frac{e^2}{2\pi^2}\int \mathrm{d}\boldsymbol{p}'\,\frac{1}{(\boldsymbol{p}-\boldsymbol{p}')^2}K(\boldsymbol{k}+\boldsymbol{p}',\boldsymbol{p}',E_{\text{f}})\,\tilde{\Lambda}_k(\boldsymbol{p}'). \tag{A 1.12}$$

The solution of (A 1.12) has been discussed in detail by Kleinman (1967) and in particular by Langreth (1969), and we may use at once the methods of solution for

$\tilde{\Lambda}_k(\boldsymbol{p})$ developed by these authors to obtain the corresponding solutions for H, as discussed in §5.

The dielectric function, which is required in §5, is given in terms of $\tilde{\Lambda}_k(\boldsymbol{p})$by the usual expression,

$$\epsilon(k) = 1+\frac{e^2}{\pi^2 k^2}\int \mathrm{d}\boldsymbol{p}'\, K(\boldsymbol{k}+\boldsymbol{p}', \boldsymbol{p}', E_{\mathrm{f}})\, \tilde{\Lambda}_k(\boldsymbol{p}'). \tag{A 1.13}$$

Appendix 2. The dielectric function

In this appendix we show how the theory of the dielectric function can be used to simplify the expression for the exchange energy to second order in the displaced charge $\Delta_{\mathrm{H.F.}}(\boldsymbol{r})$.

We start with the total energy in the Hartree–Fock approximation written as a functional of the electron density in the form

$$E_{\mathrm{H.F.}}[\rho] = T_s[\rho]+\tfrac{1}{2}\int \mathrm{d}\boldsymbol{r}\,\mathrm{d}\boldsymbol{r}'\,\frac{\rho_{\mathrm{H.F.}}(\boldsymbol{r})\,\rho_{\mathrm{H.F.}}(\boldsymbol{r}')}{|\boldsymbol{r}-\boldsymbol{r}'|}+E_{\mathrm{X}}[\rho]+\int \mathrm{d}\boldsymbol{r}\, V_{\mathrm{ext}}(\boldsymbol{r})\,\rho(\boldsymbol{r}), \tag{A 2.1}$$

where $T_s[\rho]$ is the total kinetic energy of a system of non-interacting electrons of density ρ, and $E_{\mathrm{X}}[\rho]$ is the functional representing the exchange energy.

To second order in the displaced charge Δ we have

$$T_s[\rho] = T_s^{(0)}[\rho_0]-\tfrac{1}{4}\int \mathrm{d}\boldsymbol{r}\,\mathrm{d}\boldsymbol{s}\, K(\boldsymbol{r}-\boldsymbol{s})\,\Delta(\boldsymbol{r})\,\Delta(\boldsymbol{s}), \tag{A 2.2}$$

where

$$K(\boldsymbol{r}) = \frac{1}{(2\pi)^3}\int \mathrm{d}\boldsymbol{q}\, \mathrm{e}^{\mathrm{i}\boldsymbol{q}\cdot\boldsymbol{r}}\,\tilde{K}(q) \tag{A 2.3}$$

and

$$\tilde{K}(q) = -\frac{2\pi^2}{k_{\mathrm{f}}}\left[\frac{1}{2}+\frac{k_{\mathrm{f}}}{2q}\left(1-\frac{q^2}{4k_{\mathrm{f}}^2}\right)\ln\left|\frac{2k_{\mathrm{f}}+q}{2k_{\mathrm{f}}-q}\right|\right]^{-1}. \tag{A 2.4}$$

We also have, from (5.2) to second order,

$$E_{\mathrm{X}}[\rho] = E_{\mathrm{X}}^{(0)}[\rho_0]+\int \mathrm{d}\boldsymbol{r}\,\mathrm{d}\boldsymbol{s}\, B(\boldsymbol{r}-\boldsymbol{s})\,\Delta(\boldsymbol{r})\,\Delta(\boldsymbol{s}). \tag{A 2.5}$$

The variational principle $\delta E_{\mathrm{H.F.}}[\rho] = 0$ applied to (A 2.1) gives, upon performing a Fourier transform,

$$-\tfrac{1}{2}\tilde{K}(q)\,\tilde{\Delta}(q)+(4\pi/q^2)\,\tilde{\Delta}(q)+2\tilde{B}(q)\,\tilde{\Delta}(q)+\tilde{V}_{\mathrm{ext}}(q) = 0. \tag{A 2.6}$$

If we write

$$\tilde{V}_{\mathrm{ext}}(q) = -(4\pi/q^2)\,\tilde{\rho}_{\mathrm{ext}}(q) \tag{A 2.7}$$

then the polarizability $\alpha(q)$ is given by

$$\alpha(q) = \frac{\tilde{\Delta}(q)}{\tilde{\rho}_{\mathrm{ext}}(q)} \tag{A 2.8}$$

and from (A 2.6) we obtain

$$\alpha(q) = \left[1-\frac{q^2}{8\pi}\tilde{K}(q)+\frac{q^2}{2\pi}\tilde{B}(q)\right]^{-1}. \tag{A 2.9}$$

The dielectric function is given in terms of $\alpha(q)$ by

$$\epsilon(q) = (1-\alpha(q))^{-1}. \tag{A 2.10}$$

A second expression for the dielectric function can be obtained from the first order term in (4.3).

We can write the potential $V_{\mathrm{H}}(\boldsymbol{r})$ as

$$V_{\mathrm{H}}(\boldsymbol{r}) = V_{\mathrm{ext}}(\boldsymbol{r}) + \int \mathrm{d}\boldsymbol{r}_1 \frac{\Delta(\boldsymbol{r}_1)}{|\boldsymbol{r}-\boldsymbol{r}_1|} \tag{A 2.11}$$

and the displaced charge Δ caused by this potential is given to first order by (4.3) as

$$\tilde{\Delta}(q) = J(q)\tilde{V}_{\mathrm{H}}(q). \tag{A 2.12}$$

However, Fourier transforming (A 2.11) and using (A 2.12) we obtain directly

$$\tilde{V}_{\mathrm{H}}(\boldsymbol{q}) = \frac{V_{\mathrm{ext}}(\boldsymbol{q})}{[1-(4\pi/q^2)J(q)]}. \tag{A 2.13}$$

Hence the dielectric function is, using (4.4) for $J(q)$,

$$\epsilon(q) = 1 + (4\pi/q^2)\int \mathrm{d}\boldsymbol{k}\, H(\boldsymbol{k}, \boldsymbol{k}-\boldsymbol{q}). \tag{A 2.14}$$

This is essentially the same result as (A 1.13) combined with (A 1.11).

Equating (A 2.14) to (A 2.10) gives directly

$$\tilde{B}(q) = \tfrac{1}{4}\tilde{K}(q) - \frac{1}{2J(q)} \tag{A 2.15}$$

or in terms of the dielectric constant

$$\tilde{B}(q) = \tfrac{1}{4}\tilde{K}(q) - \frac{(2\pi/q^2)}{1-\epsilon(q)}. \tag{A 2.16}$$

Using (4.4), (A 2.15) relates $\tilde{B}(q)$ to $H(\boldsymbol{k},\boldsymbol{p})$, and through (A 1.11) to the proper vertex part. The expression (A 2.15) is simpler than the one obtained by use of (4.1) in (5.1) and indicates an inherent integral relationship between H and $H^{(2)}$.

References

Brandt, W. & Lundqvist, S. 1963 *Phys. Rev.* **132**, 2135.
Dirac, P. A. M. 1930 *Proc. Camb. Phil. Soc.* **26**, 376.
Herman, F., Van Dyke, J. P. & Ortenburger, I. B. 1969 *Phys. Rev. Lett.* **22**, 807.
Hohenberg, P. & Kohn, W. 1964 *Phys. Rev.* **136**, B864.
Kleinman, L. 1967 *Phys. Rev.* **160**, 585.
Kohn, W. & Sham, L. J. 1965 *Phys. Rev.* **140**, A1133.
Langreth, D. C. 1969 *Phys. Rev.* **181**, 753.
Ma, S. & Brueckner, K. A. 1968 *Phys. Rev.* **165**, 18.
March, N. H., Young, W. H. & Sampanthar, S. 1967 *The Many-Body Problem in Quantum Mechanics*. Cambridge University Press.
March, N. H. & Murray, A. M. 1961 *Proc. R. Soc. Lond.* A **261**, 119.
March, N. H. 1957 *Adv. Phys.* **6**, 1.
Stoddart, J. C., Beattie, A. M. & March, N. H. 1971 *Inter J. Quantum Chem.* **4**, 35.
Stoddart, J. C. & March, N. H. 1967 *Proc. R. Soc. Lond.* A **299**, 279.
Von Weizsäcker, C. F. 1935 *Z. Phys.* **96**, 431.
Wendin, G. 1970 *J. Phys.* B, **4**, 1080.

Foundations of Walsh's rules for molecular shape

N. H. March

Theoretical Chemistry Department, University of Oxford, 1 South Parks Road, Oxford OX1 3TG, England
(Received 29 January 1980; accepted 28 February 1980)

Combining density functional theory with the form of the virial theorem established by Nelander for polyatomic molecules, the basic foundation of Walsh's rules for molecular shape is clarified.

The rules formulated by Walsh[1] for predicting the shapes of molecules remain of considerable interest, and are supported by a variety of data. Their foundation in first principles theory, however, has been lacking. In particular, the assumption was, at the very least, implicit in Walsh's arguments that the total energy was the sum of one-electron orbital energies. Then, in say HAH, the whole argument depended on following the eigenvalue or orbital energy changes as one went across a correlation diagram from the 90° angle to the linear molecule. As Walsh emphasized, in this example, everything then hinged on the number of valence electrons. If the lowest two levels only were occupied, which required four valence electrons, then the sum of the one-electron energies would be lower for the linear configuration than for the smaller angles; but if there were more than four electrons, then the bent molecule would have a lower orbital energy sum than the linear molecule.

A number of workers pointed out in the early development of Walsh's rules that the total energy could not be the sum of the orbital energies. This is clear because in self-consistent field theories, the electrostatic self-energy of the electronic charge cloud of the molecule is counted twice in the sum of orbital energies; this point is discussed in some detail by Goodisman.[2]

The purpose of the present note is to clarify the first-principles foundations of Walsh's rules. All that is necessary, of course, is that the total energy follows the orbital energy sum; they need not be equal; indeed, as remarked above, they are not! Work of the writer and Plaskett[3,4] showed, for a heavy neutral atom, that the total energy E was 3/2 of the sum of the orbital energies E_s say. The interest in such a relation for molecules at equilibrium was stimulated by Ruedenberg,[5] who proposed a similar relation. This was shown to follow[6,7] from the simplest electron density description of molecules at equilibrium.

Since the changes of energy with molecular geometry are usually small in comparison with total energies, it seemed important to clarify fully what basis there was for Walsh's rules in a many-electron framework. This is the purpose of the argument given below. As discussed elsewhere,[8,9] the basic Euler equation of the electron density theory may be written

$$\mu = \frac{\delta t_r}{\delta \rho} + V_N + V_e + \frac{\delta \epsilon_{XC}}{\delta \rho} , \qquad (1)$$

where μ is the chemical potential, V_N is the potential energy of the nuclear framework, and V_e is the potential energy created by the electrostatic charge density $\rho(\mathbf{r})$. Finally, t_r is the single-particle kinetic energy density as a functional of ρ, while ϵ_{XC} is the exchange and correlation energy density functional. By multiplying Eq. (1) by the density ρ and integrating through the whole of space, one obtains

$$N\mu = \int \rho \frac{\delta t_r}{\delta \rho} d\tau + V_{en} + 2V_{ee} + \int \rho \frac{\delta \epsilon_{XC}}{\delta \rho} d\tau , \qquad (2)$$

where N is the total number of electrons given by $\int \rho(\mathbf{r}) d\tau = N$, while V_{en} and V_{ee} are the electrostatic electron–nuclear and electron–electron potential energy contributions, respectively.

Next we turn to the sum of the orbital energies. Since, in Eq. (1), t_r is defined as the single-particle kinetic energy density, it is clear that one has a one-body potential $V(\mathbf{r})$, with orbital energies ϵ_i, say[7]

$$V(\mathbf{r}) = V_N(\mathbf{r}) + V_e(\mathbf{r}) + \frac{\delta \epsilon_{XC}}{\delta \rho} . \qquad (3)$$

The orbital energy sum $E_s = \sum_{\text{occupied levels}} \epsilon_i$ is clearly, using Eq. (3),

$$\begin{aligned} E_s &= \int t_r d\tau + \int \rho V(\mathbf{r}) d\tau \\ &= \int t_r d\tau + V_{en} + 2V_{ee} + \int \rho \frac{\delta \epsilon_{XC}}{\delta \rho} d\tau , \end{aligned} \qquad (4)$$

which, if we neglect the many-electron contribution ϵ_{XC} for a moment, shows immediately that, as remarked above, the term V_{ee} in the total energy is included twice in the eigenvalue sum.

In the full many-electron theory set out above, we find by subtraction

$$N\mu - E_s = \int \left(\frac{\rho \delta t_r}{\delta \rho} - t_r \right) d\tau . \qquad (5)$$

If we add and subtract 2/3 of the single-particle kinetic energy from the right-hand side of this equation, we find

$$N\mu - E_s = \int \left(\rho \frac{\delta t_r}{\delta \rho} - \tfrac{5}{3} t_r \right) d\tau + \frac{2}{3} \int t_r d\tau . \qquad (6)$$

Now we would like to employ the virial theorem $T = -E$ at equilibrium. Unfortunately, t_r is the single-particle kinetic energy density and its integral $T_s = \int t_r d\tau$ is therefore not equal to T, the exact total kinetic energy, which appears in the virial theorem. It is therefore convenient to define the correlation kinetic energy ΔT say by

$$\Delta T = T - T_s . \qquad (7)$$

Hence, the sum of the orbital energies E_s and the total energy E in the full many-electron theory are related at equilibrium by

$$E=\tfrac{3}{2}E_s-\tfrac{3}{2}N\mu+\frac{3}{2}\int\left(\rho\,\frac{\delta t_r}{\delta\rho}-\tfrac{5}{3}t_r\right)d\tau-\Delta T\,. \quad (8)$$

In the simplest density description,[6,7] the chemical potential of neutral molecules, which is the only case considered in the present work, is zero, as is the integral $\int[(\rho\delta t_r/\delta\rho)-\frac{5}{3}t_r]\,d\tau$. Similarly, as already remarked, ΔT is the correlation kinetic energy which is therefore zero in a one-electron theory. Very recent work by Mehrotra and Hoffman,[10] also on Walsh's rules, considers the relation between total energy and orbital energy sum but does not have electron correlation included explicitly.

What has been demonstrated is that, at equilibrium, there is a fundamental relation (8) between the one-electron orbital energy sum E_s and total energy E. The great successes of one-electron theories of molecules show that many-electron effects are not dominant, but they do appear explicitly through the correlation kinetic energy ΔT in Eq. (8). It is worthy of note that the deviations from the relation $E=(3/2)E_s$ displayed in Eq. (8) are qualitatively of the form that the chemical potential correction term in Eq. (8) is positive, since μ is negative for bound molecules, while both the kinetic energy contributions are expected to be negative. This latter point is supported by using a low order density gradient expansion of t_r as in Ref. 11, the lowest order result being

$$\int\left(\rho\,\frac{\delta t_r}{\delta\rho}-\tfrac{5}{3}t_r\right)d\tau=-\frac{1}{108}\int\frac{(\nabla\rho)^2}{\rho}\,d\tau\,, \quad (9)$$

which is evidently negative to this lowest order. As for the correlation kinetic energy ΔT, this is expected to be positive, because, in forming a wave function from, say, the single-particle basis generated by the one-body potential $V(\mathbf{r})$ defined in Eq. (3) above, one will expect the correlated wave function to involve creation of particles and holes and hence to increase the total kinetic energy above its single-particle value T_s. In general, the success of one-electron theories of molecules shows that ΔT will be normally a very small fraction of the total kinetic energy.

Thus, the first part of our argument in displaying the first-principles foundations of Walsh's rules is achieved in Eq. (8) at equilibrium, for as we have emphasized above the corrections to $E=(3/2)E_s$ are identically zero in the simplest density description and are, on the general grounds discussed above, expected to be small in a full many-body treatment. Furthermore, the chemical potential and kinetic energy corrections at least partly cancel, as argued above.

However, there is still a lacuna in the argument needed for Walsh's rules. Evidently, for the HAH example referred to in the introductory comments, there is one equilibrium geometry. Away from that, if we take an arbitrary geometry, we cannot use the virial theorem in the form $T=-E$. To be explicit, consider the simpler case of a diatomic molecule, with internuclear separation R. Then if U is the total potential energy, such that $T+U=E$, the virial theorem for arbitrary R reads

$$2T+U=-\frac{R\,dE}{dR}\,, \quad (10)$$

where the term on the right-hand side is the virial of the forces required to hold fast the nuclei at separation R. However, this is clearly only zero, for a diatomic molecule, at one configuration, the equilibrium separation R_e say at which dE/dR is zero. Thus, only for this one case in a diatomic molecule can we use $T=-E$ which was, of course, employed to reach the basic equation (8).

However, now the message for the discussion of molecular shapes following Walsh is clear, if we note, from the generalization of Eq. (10) for polyatomics, the importance, in HAH, say, of minimizing with respect to bond lengths at each angular configuration chosen. Provided this is done, in the example HAH, for each angle considered, then as Nelander[12] has proved, for every set of angles for which the energy is minimized with respect to bond distances, the virial theorem holds in its simple form $T=-E$.

In summary, Eqs. (2) and (4) follow, quite generally, from density functional theory, at all arbitrary molecular geometries. However, to reach Eq. (8), required as a basis for Walsh's rules, one must add to these the virial theorem in the simple form $T=-E$. We have emphasized that while this certainly does not hold for arbitrary geometries, provided we regard the correlation diagram for, say, HAH, as plotted at each HAH angle at the bond length which minimizes the energy at that angle, then Eq. (8) holds. Thus, Walsh's arguments go through as before. We stress that, if and when the need arises to discuss possible deviations from Walsh's rules, the present argument points the directions in which such a discussion would go; they reside in chemical potential and kinetic energy corrections spelled out below Eq. (8).

[1] A. D. Walsh, J. Chem. Soc. **1953**, 2260.
[2] J. Goodisman, J. Am. Chem. Soc. **91**, 6552 (1969).
[3] N. H. March and J. S. Plaskett, Proc. R. Soc. London Ser. A **235**, 419 (1956).
[4] N. H. March, Adv. Phys. **6**, 94 (1957).
[5] K. Ruedenberg, J. Chem. Phys. **66**, 375 (1977).
[6] N. H. March, J. Chem. Phys. **67**, 4618 (1977).
[7] J. F. Mucci and N. H. March, J. Chem. Phys. **71**, 5270 (1979); **71**, 1495 (1979).
[8] W. Kohn and L. J. Sham, Phys. Rev. Sect. A **140**, 1133 (1965).
[9] N. H. March, in *Orbital Theories of Molecules and Solids* (Clarendon, Oxford, England, 1974), p. 95.
[10] P. K. Mehrotra and R. Hoffmann, Theor. Chim. Acta **48**, 301 (1978).
[11] N. H. March, J. Chem. Phys. **71**, 1004 (1979). There is a misprint in the sign of the lowest order gradient correction in $\delta t_r/\delta\rho$ in this paper. I am grateful to Professor Parr and his colleagues for drawing my attention to this.
[12] B. Nelander, J. Chem. Phys. **51**, 469 (1969).

The density matrix, density, and Fermi hole in Hartree–Fock theory

K. A. Dawson and N. H. March
Theoretical Chemistry Department, University of Oxford, 1 South Parks Road, Oxford OX1 3TG, England

(Received 4 November 1982; accepted 10 July 1984)

In density matrix and density functional theory, one is concerned with methods that bypass explicit calculation of antisymmetric wave functions but nevertheless take full account of the Fermionic nature of the problem. The purpose of this paper is to examine two quite different approaches which achieve this aim. The first of these involves general many-electron theory and subsequent closure yielding an integrodifferential equation for the density matrix. Despite the interesting nature of this equation, analytical progress is hampered by its nonlinearity. Hence we have considered an alternative more practical variational procedure which, for some simpler systems, has the additional merit of yielding closed differential equations for the density.

I. INTRODUCTION

The methods based on the reduced density matrix, introduced by Löwdin,[1] for treating the many-electron problem are, in principle, extremely attractive but in practice have not been wholly successful. The most obvious difficulty with such non-wave function methods is the task of ensuring that the chosen forms of the low order density matrices have an associated antisymmetric wave function. This is the well known N-representability problem.[2] However, it can be shown that density functional theory is free from these difficulties. Even so, practical progress remains somewhat slow because, although bounded unique functionals of the density are known to exist,[3,4] they have proved difficult to compute. However, the work of Freed and Levy[5] provides both a specific formal algorithm for the exact functional and a functional Taylor expansion of this about a reference density.

It may surprise the reader that these problems remain unresolved even within the traditional Hartree–Fock framework. Based on this approximation, Dirac's early work[6] showed that the single-particle density matrix determines all higher orders of density matrix and, in consequence, the entire wave function. We thus call the single-particle density matrix

$$\gamma(1,2) = \sum_{\substack{\text{occupied}\\ \text{levels}}} \phi_i^*(1)\phi_i(2) \tag{1.1}$$

simply the density matrix. It is well known[7] that the two-particle density matrix has the form

$$\Gamma(12;1'2') = \gamma(11')\gamma(22') - \gamma(12')\gamma(21'). \tag{1.2}$$

A similar closure expression applies for the three-particle density matrix $\Gamma^{(3)}$.

II. THE LOCAL ENERGY METHOD

The basic equations of the local energy method may be derived from the Schrödinger equation.[8] However, there is an interesting and more general procedure based on the field theory commutator. Details of this are given in the Appendix. It is important to note that the same method can be used to generate determining equations for the electron pair function or structure factor.

Denoting the external potential by $U(\mathbf{r})$ and the electron–electron interaction potential by $e^2/r_{12} = V(r_{12})$ we find[8]

$$\begin{aligned} E\gamma(\mathbf{rr}') = &-\frac{\hbar^2}{2m}\nabla_r^2\gamma(\mathbf{rr}') + U(\mathbf{r})\gamma(\mathbf{rr}') \\ &+ \int V(|\mathbf{r}-\mathbf{r}_2|)\Gamma(\mathbf{rr}_2;\mathbf{r}'\mathbf{r}_2)d\mathbf{r}_2 \\ &+ \int U(\mathbf{r}_2)\Gamma(\mathbf{rr}_2;\mathbf{r}'\mathbf{r}_2)d\mathbf{r}_2 \\ &- \frac{\hbar^2}{2m}\int [\nabla_{r_2}^2\Gamma(\mathbf{rr}_2;\mathbf{r}'\mathbf{r}_2')]_{\mathbf{r}_2'=\mathbf{r}_2}d\mathbf{r}_2 \\ &+ \frac{1}{2}\int V(|\mathbf{r}_3-\mathbf{r}_2|)\Gamma^{(3)}(\mathbf{rr}_2\mathbf{r}_3;\mathbf{r}'\mathbf{r}_2\mathbf{r}_3)d\mathbf{r}_2\,d\mathbf{r}_3. \end{aligned} \tag{2.1}$$

This is readily obtained by performing a summation of spin on Eq. (A13). In the Hartree–Fock regime this reduces[9] to

$$\begin{aligned} &-\frac{\hbar^2}{2m}\nabla_r^2\gamma(\mathbf{rr}') + U(\mathbf{r})\gamma(\mathbf{rr}') + \int V(|\mathbf{r}-\mathbf{r}_2|)\rho(\mathbf{r}_2)d\mathbf{r}_2\,\gamma(\mathbf{rr}') \\ &- \int V(|\mathbf{r}-\mathbf{r}_2|)\gamma(\mathbf{r}_2\mathbf{r}')\gamma(\mathbf{rr}_2)d\mathbf{r}_2 \\ &= -\frac{\hbar^2}{2m}\int\gamma(\mathbf{rr}_2)\nabla_{r_2}^2\gamma(\mathbf{r}_2\mathbf{r}')d\mathbf{r}_2 + \int U(\mathbf{r}_2)\gamma(\mathbf{rr}_2)\gamma(\mathbf{r}_2\mathbf{r}')d\mathbf{r}_2 \\ &- \int V(|\mathbf{r}_3-\mathbf{r}_2|)\gamma(\mathbf{rr}_2)\gamma(\mathbf{r}_2\mathbf{r}_3)\gamma(\mathbf{r}_3\mathbf{r}')d\mathbf{r}_2\,d\mathbf{r}_3 \\ &+ \int V(|\mathbf{r}_3-\mathbf{r}_2|)\gamma(\mathbf{rr}_2)\gamma(\mathbf{r}_2\mathbf{r}')\rho(\mathbf{r}_3)d\mathbf{r}_2\,d\mathbf{r}_3, \end{aligned} \tag{2.2}$$

where $\gamma(\mathbf{rr}) = \rho(\mathbf{r})$.

It is interesting to note the absence of the energy E from the above equation. This arises (see the Appendix) by a complicated series of cancellations in the exact equations. It is readily verified that the zero field version of Eq. (2.2) admits the solution

$$\gamma_f = \frac{k_f^3}{\pi^2}\frac{j_1(k_f|\mathbf{r}-\mathbf{r}'|)}{k_f|\mathbf{r}-\mathbf{r}'|}; \tag{2.3}$$

this being the well known free-particle density matrix.

It is also a simple matter to obtain the equation of motion, for the case $V=0$, as

$$\nabla_r^2\gamma(\mathbf{rr}') - \nabla_{r'}^2\gamma(\mathbf{rr}') = \frac{-2m}{\hbar^2}[U(\mathbf{r}') - U(\mathbf{r})]\gamma(\mathbf{rr}') \tag{2.4a}$$

by subtraction of the permuted form of Eq. (2.2) from itself. The exact many-body version of Eq. (2.4a) is derived in the Appendix.

Finally we turn to the equation satisfied by the Fermi hole

$$H_f(\mathbf{r}_1\mathbf{r}_2) = \rho(\mathbf{r}_1)\rho(\mathbf{r}_2) - \Gamma(\mathbf{r}_1\mathbf{r}_2;\mathbf{r}_1\mathbf{r}_2). \tag{2.4b}$$

This function has recently been studied using wave function methods.[10] Substitution of Eq. (2.4b) into Eq. (2.2) yields the exact Hartree–Fock version

$$\frac{\hbar^2}{2m}\frac{[\nabla H_f(\mathbf{rr}')]^2}{H_f(\mathbf{rr}')} - \frac{\hbar^2}{4m}\nabla_r^2 H_f(\mathbf{rr}') + U(\mathbf{r})H_f + \int V(|\mathbf{r}-\mathbf{r}_2|)\rho(\mathbf{r}_2)d\mathbf{r}_2\, H_f - \int V(|\mathbf{r}-\mathbf{r}_2|)[H_f(\mathbf{rr}_2)H_f(\mathbf{r}_2\mathbf{r}')]^{1/2}d\mathbf{r}_2\, H_f^{1/2} = C(\mathbf{rr}'), \tag{2.5}$$

where C is just the right-hand side of Eq. (2.4a) written in terms of H_f. Though anticipating somewhat the discussion of the variational approach in Sec. III below, it is interesting to reflect on the variational origin of Eq. (2.5). In fact, the first two terms arise from the single particle kinetic energy given by

$$T[H_f] = \frac{\hbar^2}{8m}\int \frac{[\nabla_r H_f(\mathbf{rr}')]^2}{H_f(\mathbf{rr}')}\,d\mathbf{r}\,d\mathbf{r}'. \tag{2.6}$$

We see that this has the form of the von Weizsäcker term in density functional theory. There, no exact form of the single-particle kinetic energy exists in terms of the electron density.

Finally, Eq. (2.2) can be used to obtain the asymptotic form of the electron density in an atomic ion. Since the Hartree density matrix separates for large arguments r and r' we obtain

$$\frac{\hbar^2}{8m}\frac{(\nabla_r\rho)^2}{\rho} - \frac{\hbar^2}{4m}\nabla_r^2\rho + [U(r)\rho - (N-1)U(\mathbf{r})\rho]_{r\to\infty} = \lambda\rho(\mathbf{r}), \gamma(\mathbf{rr}') \simeq \rho^{1/2}(\mathbf{r})\rho^{1/2}(\mathbf{r}'). \tag{2.7}$$

From internal considerations λ may be identified as ϵ_{HOMO}, the energy of the highest occupied orbital. The dominant term in the decay of ρ is then just

$$\rho \sim_{r\to\infty} A\exp\left(-2\left\{\frac{2m}{\hbar^2}|\epsilon_{\text{HOMO}}|\right\}^{1/2} r\right). \tag{2.8}$$

We now turn to the second method which is based on the variational approach.

III. VARIATIONAL APPROACHES TO THE CALCULATION OF THE HARTREE–FOCK DENSITY MATRIX

The exact ground state energy of a system of electrons in an external potential U is given by

$$E = -\frac{\hbar^2}{2m}\int \delta(1-2)\nabla_1^2\gamma(12)d\,1\,d\,2 + \int U(1)\gamma(12)\delta(1-2)d\,1\,d\,2 + \frac{1}{2}\int \Gamma(12;12)V(1-2)d\,1\,d\,2, \tag{3.1}$$

where

$$(\mathbf{r}_1,\mathbf{r}_2) \equiv (1,2).$$

Employing the Hartree–Fock closure (1.2) we obtain

$$E^{\text{HF}}[\tilde\gamma] = \int\Big\{\tilde\gamma^*(12)\Big[-\frac{\hbar^2}{2m}\nabla_1^2 + U(1)\Big]\tilde\gamma(12) + \frac{1}{2}\int V(1-2)|\tilde\gamma(13)|^2|\tilde\gamma(24)|^2 d\,3\,d\,4 - \frac{1}{2}\int V(1-2)\tilde\gamma(13)\tilde\gamma^*(23)\tilde\gamma(24) \times\tilde\gamma^*(14)d\,3\,d\,4\Big\}d\,1\,d\,4, \tag{3.2}$$

where $\tilde\gamma$ is any idempotent density matrix. The exact Hartree–Fock kernel is written γ_0, so that

$$\delta E^{\text{HF}}[\gamma_0] = \int \delta\gamma^*(12)\Big\{\Big[-\frac{\hbar^2}{2m}\nabla_1^2 + U(1) + \int V(1-3)\rho_0(3)d\,3\Big]\gamma_0(12) - \int V(1-3)\gamma_0(13)\gamma_0(32)d\,3\Big\}d\,1\,d\,2 + \text{complex conjugate}. \tag{3.3}$$

We now treat the constraint of idempotency by a generalization of the Lagrange method. The full Hartree–Fock variational procedure is given by

$$\delta E^{\text{HF}}[\gamma_0] = 0, \tag{3.4}$$

where the trial $\tilde\gamma$ belongs to the set of all normalized idempotent density matrices. One standard way of applying this constraint is to use a Lagrange multiplier which[1] effectively limits the search in Eq. (3.4) to ensure that one of the solutions of the resulting Euler–Lagrange equation has the required properties. Thus,

$$\delta\{E^{\text{HF}}[\tilde\gamma] - \int[\lambda(21)\tilde\gamma(12) + \lambda(12)\tilde\gamma^*(12)]d\,1\,d\,2\} = 0$$
$$\tilde\gamma = \gamma_0. \tag{3.5}$$

It is easy to show that λ has a diagonal expansion in a basis of the Hartree–Fock subspace and this fact is used below. Using Eq. (3.3) we obtain

$$-\frac{\hbar^2}{2m}\nabla_1^2\gamma_0(11') + U(1)\gamma_0(11') + \int V(1-2)[\gamma_0(11')\rho_0(2) - \gamma_0(12)\gamma_0(21')]d\,2 = \lambda(11'). \tag{3.6}$$

It may be helpful to think of λ as a continuous Lagrange multiplier. In many respects it is the quantal analogue of the forces of constraint which occur in classical mechanics. In the present case though, it has its origins in the Pauli Exclusion Principle. Note that multiplication and integration of Eq. (3.6) yield the Hartree–Fock equations once more,

$$-\frac{\hbar^2}{2m}\nabla_1^2\chi_i(1) + U(1)\chi_i + \int V(1-2)[\rho_0(2)\chi_i(1) - \gamma_0(12)\chi_i(2)]d\,2 = \epsilon_i\chi_i(1),$$
$$\lambda(1,1') = \sum_{\substack{\text{occupied}\\ \text{levels}}} \epsilon_i\phi_i^*(1)\phi_i(1'). \tag{3.7}$$

By comparison of Eq. (3.6) with Eq. (2.2) it is now possible to write λ as a function of γ_0.

Clearly the idempotency condition on the density matrix (the Hartree–Fock manifestation of N-representability requirements) is the constraint which is most difficult to handle in a practical way. In a recent article Lieb[11] has shown that any admissible first order density matrix ($0\leqslant\gamma\leqslant 1$; $\mathrm{Tr}\,\gamma = N$] yields a rigorous upper bound to the exact Hartree–Fock energy when Eq. (3.1) is used with Eq. (1.2). This interesting result may provide a practical alternative to the problem of idempotency constraints discussed above. In what follows, we propose another method that renders this problem tractable. For simplicity we do not deal with problems of degeneracy.

We define

$$T_S^N[\rho] = \min\{\langle\Phi_N|T_N|\Phi_N\rangle = T_s,\quad \Phi_N\to\rho,\quad \Phi_N \text{ is a determinant},\ \langle\Phi_N|\Phi_N\rangle = 1, T_s<\infty\}, \tag{3.8}$$

where T_N is the kinetic part of the Hamiltonian.[12] Levy[4] has used the generalized version of Eq. (3.8) to give an alternative to the Hohenberg–Kohn formulation. It is now a simple matter to show that for a trial $\tilde\rho$ and

$$E_N^{\rm HF}[\tilde\rho] = T_S^N[\tilde\rho] + G_S[\tilde\rho] + \int U(\mathbf r)\tilde\rho(\mathbf r)d\mathbf r + \frac{1}{2}e^2\int\frac{\tilde\rho(\mathbf r)\tilde\rho(\mathbf r')}{|\mathbf r-\mathbf r'|}d\mathbf r\, d\mathbf r',$$

we have

$$E_N^{\rm HF}[\tilde\rho]\geqslant E_N^{\rm HF}[\rho]\geqslant E_N^0;\quad \int\tilde\rho d\mathbf r = N, \tag{3.9}$$

where ρ is the exact Hartree–Fock density, E_N^0 the exact many-body ground state energy, and G_S^N the exchange functional defined as in Eq. (3.8). Now the density obtained from a single determinant is [Eq. (1.1)]

$$\rho(\mathbf r) = \sum_{\substack{\text{occupied}\\ \text{levels}}}\phi_i^2(\mathbf r) \tag{3.10}$$

and if three levels in a six-particle system are doubly occupied we have

$$\frac{\rho}{2} = \phi_1^2+\phi_2^2+\phi_3^2;\quad \langle\phi_i|\phi_j\rangle = \delta_{ij}. \tag{3.11}$$

Clearly the situation in function space is analogous to the equation of a sphere in ordinary Euclidian space. Thus, the density plays the role of the square of the radius and ϕ_1,ϕ_2,ϕ_3 have the same properties as x,y,z. Now the statement that the variational functional depends only on ρ implies the presence of spherical symmetry in function space. The canonical coordinate frame is thus defined in the normal way,

$$\begin{aligned}\sqrt2\phi_1 &= \rho^{1/2}(\mathbf r)\sin\theta(\mathbf r)\cos\phi(\mathbf r),\\ \sqrt2\phi_2 &= \rho^{1/2}(\mathbf r)\sin\theta(\mathbf r)\sin\phi(\mathbf r)\\ \sqrt2\phi_3 &= \rho^{1/2}(\mathbf r)\cos\theta(\mathbf r)\end{aligned} \tag{3.12}$$

so that

$$\rho = 2\phi_1^2+2\phi_2^2+2\phi_3^2,$$

$$\gamma(\mathbf r'\mathbf r) = \rho^{\frac12}(\mathbf r')\rho^{\frac12}(\mathbf r)[\sin\theta(\mathbf r')\sin\theta(\mathbf r)\cos\phi(\mathbf r')\cos\phi(\mathbf r) + \sin\theta(\mathbf r)\sin\theta(\mathbf r')\sin\phi(\mathbf r')\sin\phi(\mathbf r) + \cos\theta(\mathbf r')\cos\theta(\mathbf r)].$$

We next make use of Eqs. (3.1) and (1.2) to obtain $E_3^{\rm HF} = \langle\Phi|H_3|\Phi\rangle \equiv E_3^{\rm HF}[\tilde\rho,\tilde\theta,\tilde\phi]$. Now applying Eq. (3.8) by fixing $\tilde\rho$ and varying $\tilde\theta$ and $\tilde\phi$ subject to the Cartesian-type conditions of orthonormality

$$\begin{aligned}2\int\tilde\phi_i^2\,d\mathbf r = 2 &= \int\tilde\rho\sin^2\tilde\theta\cos^2\tilde\phi\,d\mathbf r\\ &= \int\tilde\rho\sin^2\tilde\theta\sin^2\tilde\phi\,d\mathbf r\\ &= \int\tilde\rho\cos^2\tilde\theta\,d\mathbf r,\ \text{etc.}\end{aligned} \tag{3.13}$$

results in the required functional.

In all, we need six conventional Lagrange multipliers and obtain two equations determining θ and ϕ as functions of $\tilde\rho$. When these expressions are inserted into $E_3^{\rm HF}[\tilde\rho,\theta,\phi]$ we obtain the $E_3^{\rm HF}[\tilde\rho]$ of Eq. (3.9). A similar procedure defines the N-particle Hartree–Fock energy $E_N^{\rm HF}$. Of course the functional relationships involving $\tilde\rho$ and the angle may not have simple analytical forms. However, by varying $E_3^{\rm HF}[\tilde\rho,\theta,\phi]$ with respect to $\tilde\rho$ we obtain three coupled equations. Their solution is equivalent to the density functional method. Nevertheless, we have managed to characterize, in a practical way, the functional defined by Eq. (3.9). Arising from this there are a number of interesting general features to be discussed. However, we shall first illustrate the method by application to a two level, one dimensional problem.[13,14]

Using polar coordinates

$$\begin{aligned}\sqrt2\phi_1(x) &= \sqrt\rho\cos\theta(x),\\ \sqrt2\phi_2(x) &= \sqrt\rho\sin\theta(x),\end{aligned} \tag{3.14}$$

so $\rho = 2\phi_1^2+2\phi_2^2$; $\theta = \tan^{-1}[\phi_2/\phi_1]$,

$$\gamma(x'x) = \sqrt{\rho(x')}\sqrt{\rho(x)}\cos[\theta(x')-\theta(x)].$$

For given exact, one-body orbitals ϕ_1 and ϕ_2, the form (3.14) is also exact, as shown by the explicit inversion for θ given in terms of ϕ_2/ϕ_1. Of course, Eqs. (3.14) avoid assumptions about the equality of the probability densities of the two orbitals, which have often been made hitherto. Thus for a system of four particles moving in external potential V we have

$$E_2^{\rm HF}[\tilde\rho,\tilde\theta] = \int\left\{-\frac18\frac{(d\rho/dx)^2}{\rho} - \frac12\left(\frac{d\tilde\theta}{dx}\right)^2\tilde\rho + V\tilde\rho\right\}dx \tag{3.15}$$

subject to the conditions

$$\begin{aligned}2 &= \int\rho\cos^2\theta\,dx,\\ &= \int\rho\sin^2\theta\,dx,\\ 0 &= \int\rho\sin\theta\cos\theta\,dx.\end{aligned} \tag{3.16}$$

In this way we obtain the equation

$$\frac12\theta'' + \left(\frac{\chi'}{\chi}\right)\theta' = \xi\sin2\theta - \eta\cos2\theta;\quad \psi^2 = \rho. \tag{3.17}$$

It is simpler to work with the two coupled equations in this case. Performing the variation with respect to $\tilde\rho = \tilde\chi^2$ we obtain

$$-\frac{1}{2}\chi'' + \left[\frac{1}{2}(\theta') + V\right]\chi$$
$$= (\lambda + \xi \cos 2\theta + \eta \sin 2\theta)\chi. \quad (3.18)$$

Now differentiation of Eq. (3.18) and substitution of Eq. (3.17) yields

$$\theta'' + \frac{\chi'}{\chi}\theta' + \frac{1}{2\theta'}\left(\frac{\chi'\chi''}{2\chi^2} - \frac{\chi'''}{2\chi} + V'\right) = 0, \quad (3.19)$$

which is just a solvable Bernoulli equation in $(\theta')^2 = u$. Thus,

$$u' + 2\frac{\chi'}{\chi}u = -\left(\frac{\chi'\chi''}{2\chi^2} - \frac{\chi'''}{2\chi} + V'\right) \quad (3.20)$$

a simple linear equation in u with integrating factor

$$I = \exp\int^x \frac{2\chi'}{\chi}\,dt = \chi^2$$

and general solution

$$u = \sqrt{\theta'} = \frac{C_1}{\chi^2} - \frac{1}{\chi^2}\int^x\left(\frac{\chi'\chi''}{2\chi^2} - \frac{\chi'''}{2\chi} + V'\right)\chi^2\,dt. \quad (3.21)$$

After application of the associated boundary conditions our task is, in principle, completed. The θ obtained from Eq. (3.21), being given in terms only of χ and V, is now resubstituted into Eq. (3.18) and an equation relating $\rho = \chi^2$ and V results. This is the final objective of density functional theory and Eqs. (3.21) and (3.18) represent the closed and exact Euler–Lagrange equation of the theory for the above simple example. Whenever interelectronic repulsions are considered we obtain exchange terms. The method is applied in exactly the same way, however. Although the density from a local potential may be the same as the density of the corresponding HF problem, it is easy to see that this is not the case for orbitals by consideration of their asymptotic forms. Note that when V is arbitrary the final equation is already quite complicated. Evidently the correlated theory can only yield equations of greater complexity.

The density matrix given by Eq. (3.14) is now fully determined. It is of considerable interest that θ is fixed by choice of an input χ, providing an explicit map $\rho \to \gamma \to \Phi$ [see Eq. (3.8)]. This is a general feature of the method for arbitrary (integer) N and accomplishes one of the tasks set in density matrix theory. There are a number of other properties of this problem which turn out to be quite general. By examination of Eqs. (3.8) and (3.9) and the arguments following Eq. (3.11) it should be clear that a different functional $E_N^{\rm HF}$ is obtained for each N. This is also true for the correlated theory but is rarely stressed in the literature and confusion on this point has led to confusion about the meaning of the various Lagrange multipliers. For example, the definition (3.8) does not imply any boundedness about noninteger N functionals. Thus, if a density integrating to noninteger N is inserted in $T_S^N[\rho]$ we do not necessarily obtain a trial energy for this type of system. It is then not clear what dE/dN would mean in this case. Perdew *et al.*[15] have tackled this problem and have thereby extended Hohenberg–Kohn theory to fractional particle number.

It is now a relatively simple matter to convince oneself that the first, von Weizsäcker term in Eq. (3.15) occurs in $E_N^{\rm HF}$ for all N, confirming the results of intuitive arguments. Finally, the equivalent of Eq. (3.18) for arbitrary N has the form

$$-\frac{1}{2}\chi'' + [V + f(\theta,\phi...)]\chi = \lambda\chi; \quad f = f(N \text{ angles}) \quad (3.22)$$

emphasizing that the root of the density satisfies a Schrödinger equation with a self-generated potential; again resulting from the requirement of Fermi statistics. It is, in fact, this new term in the potential which provides the "shell structure" of atomic physics.

By setting $\{\psi_i\}$ = natural orbitals, $C_1\psi_1 = \rho^{1/2}\sin\theta\cos\phi$, $C_2\psi_2 = \rho^{1/2}\sin\theta\sin\phi$, $C_3\psi_3 = \rho^{1/2}\cos\theta$, $\rho = C_1^2\phi_1^2 + C_2^2\phi_2^2 + C_3^2\phi_3^2$, $\gamma = C_1^2\psi_1(\mathbf{r})\psi_1(\mathbf{r}') + C_2^2\psi_2(\mathbf{r})\psi_2(\mathbf{r}') + C_3^2\psi_3(\mathbf{r})\psi_3(\mathbf{r}')$ along with the associated constraints it is clear that a similar treatment can be given for the correlated theory though ρ is now described as an ellipsoid in function space.

IV. SUMMARY

The main results of the paper are embodied in the Hartree–Fock form (2.2) of the local energy method, the equivalent form (3.6) and finally, in the relationships between density matrix and density functional theory detailed in Sec. III.

ACKNOWLEDGMENTS

One of us (KAD) wishes to acknowledge the award of a Research Scholarship from the Department of Education for Northern Ireland. Thanks are due to the referee for numerous constructive comments and for drawing our attention to a number of references which are now incorporated.

APPENDIX: DERIVATION OF MANY-ELECTRON FORM OF LOCAL ENERGY METHOD

To retain full generality, we define the density matrices in the language of second quantization as

$$\gamma(1,1') = \langle N|\psi^\dagger(1')\psi(1)|N\rangle \equiv \langle\psi^\dagger(1')\psi(1)\rangle \quad (A1)$$

and the two-particle density matrix as

$$\Gamma(12;1'2') = \langle N|\psi^\dagger(1')\psi(2')\psi(2)\psi(1)|N\rangle$$
$$\equiv \langle\psi^\dagger(1')\psi(2')\psi(2)\psi(1)\rangle. \quad (A2)$$

The three-particle density matrix, which is also required below, is an obvious generalization of the above definitions. It should be noted that the arguments above (1,1′ etc.) characterize both space and spin variables; $i = (\mathbf{r}_i,\sigma_i)$. Thus integrations will normally imply summation over spin variables as well as integration over position vectors. As indicated explicitly in Eqs. (A1) and (A2), all expectation values are taken with respect to the N-body vector in Fock space.

Let us now define the time-dependent operator

$$\psi(1,t) = e^{iHt/\hbar}\psi(1)e^{-iHt/\hbar} \quad (A3)$$

with equation of motion

$$i\hbar\frac{\partial\psi(1,t)}{\partial t} = [\psi(1,t),H], \quad (A4)$$

where H is the conventional second quantized Hamiltonian for interacting electrons moving in an external field:

$$H=\int\psi^\dagger(1)\Big(-\frac{\hbar^2}{2m}\nabla_1^2\Big)\psi(1)d\,1+\int U(1)\psi^\dagger(1)\psi(1)d\,1+\frac{1}{2}\int\int\psi^\dagger(1)\psi^\dagger(2)V(1-2)\psi(2)\psi(1)d\,1\,d\,2. \quad (A5)$$

Using Eqs. (A4) and (A5) one finds

$$i\hbar\frac{\partial\psi}{\partial t}(1,t)=-\frac{\hbar^2}{2m}\nabla_1^2\psi(1t)+U(1)\psi(1t)+\int V(1-2)\psi^\dagger(2t)\psi(2t)\psi(1t)d\,2. \quad (A6)$$

It can readily be shown that

$$\gamma(1t,1't)\equiv\langle\psi^\dagger(1't)\psi(1t)\rangle \quad (A7)$$

is time independent for a Hamiltonian of the form (A5). Thus its time derivative may be set equal to zero:

$$i\hbar\frac{\partial}{\partial t}\gamma(1t,1't)=i\hbar\langle\dot\psi^\dagger(1't)\psi(1t)\rangle+i\hbar\langle\psi^\dagger(1't)\dot\psi(1t)\rangle=0, \quad (A8)$$

where the dot denotes a time derivative. Insertion of Eq. (A6) into Eq. (A8) and use of definitions (A1) and (A2) yields

$$-\frac{\hbar^2}{2m}(\nabla_1^2-\nabla_{1'}^2)\gamma(1,1')+[U(1)-U(1')]\gamma(1,1')+\int[V(1-2)-V(1'-2)]\Gamma(12;1'2)d\,2=0. \quad (A9)$$

This represents the exact generalization of the well-known operator equation

$$[h,\gamma]=0. \quad (A10)$$

It is an elementary matter to evaluate the quantity

$$\Big[i\hbar\frac{\partial}{\partial t}\langle\psi^\dagger(1')\psi(1t)\rangle\Big]_{t=0}=\langle\psi^\dagger(1')\psi(1)H\rangle-\langle\psi^\dagger(1')H\psi(1)\rangle. \quad (A11)$$

After explicit insertion of Eq. (A6) into the left-hand side of Eq. (A11) one finds

$$E\langle\psi^\dagger(1')\psi(1)\rangle-\langle\psi^\dagger(1')H\psi(1)\rangle=-\frac{\hbar^2}{2m}\nabla_1^2\gamma(11')+U(1)\gamma(11')+\int V(1-2)\Gamma(12;1'2)d\,2, \quad (A12)$$

where use has been made of the fact that the N-body ket is an eigenfunction of the Hamiltonian. Use of Eq. (A5) and the basic definitions then yields the exact equation.

$$E\gamma(11')=-\frac{\hbar^2}{2m}\nabla_1^2\gamma(11')+U(1)\gamma(11')+\int V(1-2)\Gamma(12;1'2)d\,2-\frac{\hbar^2}{2m}\int[\nabla_2^2\Gamma(12;1'2')]_{2'=2}d\,2+\int U(2)\Gamma(12;1'2)d\,2+\frac{1}{2}\int V(3-2)\Gamma^{(3)}(123;1'23)d\,2\,d\,3. \quad (A13)$$

Since none of the potentials under consideration are spin dependent, we can evaluate the equation for the spinless situation, to yield, after summation over spin variables, Eq. (2.1) of the main text, which of course now involves spinless density matrices.

Equation (A13) is the exact equation obtained by Cohen and Frishberg by making direct use of the Schrödinger equation. However, it is worth emphasizing that the above route to their equation is very general, allowing parallel treatments of occupation numbers, pair functions, hole functions or even structure factors. Evidently, the above choice of the basic quantity to differentiate with respect to time determines the type of equation to be obtained. Thus, it is an elementary matter to obtain the discrete representation coefficients in the expansion of the exact density matrix. There is one case where this is particularly important, namely interacting electrons moving in a uniform background of neutralizing positive charge; i.e., jellium. The method used above then yields an integral equation for the natural orbital occupation numbers. We intend to take this example further at a later date.

Finally, insertion of a single Slater determinant into Eqs. (A1) and (A2) yields Eqs. (1.1) and (1.2). This is obvious from Wick's theorem or by expansion of ψ in the same basis as used for the determinantal wave function. By applying the same reasoning to Eqs. (A11) and (A6) followed by summation on spins we obtain Eq. (2.2) of the main text.

[1] P. O. Löwdin, Phys. Rev. **97**, 1474 (1955).
[2] A. J. Coleman, Rev. Mod. Phys. **35**, 668 (1963).
[3] J. K. Percus, Int. J. Quantum Chem. **13**, 89 (1978).
[4] M. Levy, Proc. Natl. Acad. Sci. U. S. A. **76**, 6062 (1979); Phys. Rev. A **26**, 1200 (1982).
[5] K. F. Freed and M. Levy, J. Chem. Phys. **77**, 396 (1982).
[6] P. A. M. Dirac, Proc. Cambridge Philos. Soc. **26**, 376 (1930).
[7] N. H. March, W. H. Young, and S. Sampanthar, *The Many-Body Problem in Quantum Mechanics* (Cambridge University, Cambridge, 1967).
[8] L. Cohen and C. Frishberg, Phys. Rev. A **13**, 927 (1976).
[9] L. Cohen and C. Frishberg, J. Chem. Phys. **65**, 4234 (1976).
[10] M. Berrondo and O. Goscinski, Chem. Phys. Lett. **81**, 75 (1981).
[11] E. H. Lieb, Phys. Rev. Lett. **46**, 457 (1981); **47**, 69(E) (1981).
[12] See also Appendix A on kinetic energy in J. P. Perdew and A. Zunger, Phys. Rev. B **23**, 5048 (1981).
[13] For this specific example with $N=4$, the procedure used in the present paper bears considerable resemblance to that of M. R. Nyden and R. G. Parr, J. Chem. Phys. **78**, 4044 (1983); see also M. R. Nyden, *ibid.* **78**, 4048 (1983).
[14] The work of G. Zumbach and K. Maschke, Phys. Rev. A **28**, 544 (1983) and that of J. E. Harriman, *ibid.* A **24**, 680 (1981) also has some similarity to our work in the limited sense that these authors use orbitals which are functions of the density. In each case, however, the orbitals are approximate.
[15] J. P. Perdew, R. G. Parr, M. Levy, and J. L. Balduz, Phys. Rev. Lett. **49**, 1691 (1982).

PHYSICAL REVIEW A VOLUME 36, NUMBER 10 NOVEMBER 15, 1987

Asymptotic formula far from nucleus for exchange energy density in Hartree-Fock theory of closed-shell atoms

N. H. March
Theoretical Chemistry Department, University of Oxford, 1 South Parks Road, Oxford OX1 3TG, England
(Received 24 June 1987)

In Hartree-Fock theory, the exchange energy density can be expressed solely in terms of the first-order density matrix. Far from the nucleus of a closed-shell atom, idempotency of the density matrix yields the exchange energy density as the magnitude of the Coulomb energy e^2/r times the electron density ρ. Thus two lengths enter the asymptotic form in contrast to $\rho^{-1/3}$ alone of local-density theory.

Because of the interest for density-functional theory[1,2] of Slater's simplification of the Hartree-Fock method,[3] in which the exchange energy A is approximated by

$$A_{\rm HFS}=-c_e\int\rho^{4/3}d\mathbf{r},\quad c_e=\frac{3e^2}{4}\left[\frac{3}{\pi}\right]^{1/3} \tag{1}$$

with $\rho(\mathbf{r})$ the electron density, it remains of significance to study the exact Hartree-Fock expression for the exchange energy, which can be traced back to the work of Dirac.[4] In terms of the density matrix $\rho(\mathbf{r},\mathbf{r}_0)$, this takes the form

$$A=\frac{-e^2}{4}\int\frac{\rho(\mathbf{r}\mathbf{r}_0)\rho(\mathbf{r}_0 r)}{|\mathbf{r}-\mathbf{r}_0|}d\mathbf{r}\,d\mathbf{r}_0 . \tag{2}$$

From the expression (2) one can write for the exchange energy density $\epsilon_x(\mathbf{r})$, related to A by

$$A=\int\epsilon_x(\mathbf{r})d\mathbf{r} , \tag{3}$$

the exact expression in the Hartree-Fock theory

$$\epsilon_x(\mathbf{r})=\frac{-e^2}{4}\int\frac{\rho(\mathbf{r}\mathbf{r}_0)\rho(\mathbf{r}_0\mathbf{r})}{|\mathbf{r}-\mathbf{r}_0|}d\mathbf{r}_0 . \tag{4}$$

This expression (4) can be evaluated in two model problems. These are (i) the uniform electron gas in which the density matrix has the closed form

$$\rho_0(\mathbf{r}\mathbf{r}_0)=\frac{k_f^3}{\pi^2}\frac{j_1(k_f|\mathbf{r}-\mathbf{r}_0|)}{k_f|\mathbf{r}-\mathbf{r}_0|} , \tag{5}$$

where $j_1(x)$ is the first-order spherical Bessel function $(\sin x - x\cos x)/x^2$, while k_f is the Fermi wave number related to the uniform density $\rho_0(\mathbf{r}\mathbf{r})\equiv\rho_0$ by the usual formula

$$\rho_0=\frac{k_f^3}{3\pi^2} \tag{6}$$

and (ii) the Bardeen model of a metal surface.[5] In this model, the density matrix has been evaluated analytically by Moore and March,[6] and their result has been utilized by Miglio *et al.*[7] to calculate the exchange energy density.

However, while no such completely analytical results seem presently possible for the density matrix of closed-shell atoms, we can utilize the exponential decay of the Hartree-Fock wave functions far from the nucleus to make an approximate evaluation of Eq. (4) in the asymptotic limit $r\to\infty$, far from the atomic nucleus. The key result in permitting this asymptotic evaluation is the idempotency of the Dirac density matrix $\rho(\mathbf{r}\mathbf{r}_0)$, following directly from the orthonormality of the exact one-electron Hartree-Fock wave functions. Then, for sufficiently large r, the exponential decay of the wave functions ensures that the integral (4) is not dominated by the point $\mathbf{r}_0=\mathbf{r}$ and one can write

$$\lim_{r\to\infty}\epsilon_x(\mathbf{r})\simeq\frac{-e^2}{4r}\int\rho(\mathbf{r}\mathbf{r}_0)\rho(\mathbf{r}_0\mathbf{r})d\mathbf{r}_0 . \tag{7}$$

But the idempotency of $\rho(\mathbf{r}\mathbf{r}_0)$ yields, for paired spins,

$$2\rho(\mathbf{r})=\int\rho(\mathbf{r}\mathbf{r}_0)\rho(\mathbf{r}_0\mathbf{r})d\mathbf{r}_0 , \tag{8}$$

which is, in fact, true for all $\mathbf{r}$. Substitution of Eq. (8) into Eq. (7) yields then

$$\lim_{r\to\infty}\epsilon_x(\mathbf{r})\simeq\frac{-e^2}{2r}\rho(\mathbf{r}) . \tag{9}$$

Hence, in this asymptotic limit, the exchange energy density is determined in absolute magnitude by the product of the Coulomb potential energy e^2/r and the electron density $\rho(\mathbf{r})$.

Clearly, Eq. (1) would lead to the different result $-c_e\rho^{4/3}(\mathbf{r})$, and it should occasion no surprise that Slater's simplification of the Hartree-Fock method, though still valuable for estimating the total exchange energy through $A_{\rm HFS}$ in Eq. (1), is not appropriate, based as it is on the uniform electron gas [example (i) above], in regions far from the nucleus of an atom where the density $\rho(\mathbf{r})$ is decaying exponentially with increasing distance.

In summary, the asymptotic form embodied in Eq. (9) is the main result of this work. If one wishes, one can utilize in Eq. (9) the asymptotic formula of Pucci and

March,[8] that in Hartree-Fock theory the ratio of $\rho(r)$ to its s-state component $\rho_s(r)$ is given by

$$\rho_s(r) \sim \left[\frac{A_s}{A}\right] r^{-4} a_0^4 \rho(r), \quad a_0 = \frac{\hbar^2}{me^2} \tag{10}$$

where A_s and A are measures of the amplitudes of the asymptotic decay of ρ_s and ρ, respectively, at large r. Therefore, combining Eqs. (9) and (10), one may conclude that the r space dependence of $\epsilon_x(\mathbf{r})$ at large r could be simulated by a local-density dependence of the form $\rho(r)^{3/4}\,\rho_s(r)^{1/4}$. However, there is, of course, no assurance that the factor multiplying such a local electron density dependence will not vary strongly from atom to atom. The result (9) appears therefore to be the physically significant asymptotic form to emphasize. This plainly involves two lengths, r itself as well as $[\rho(\mathbf{r})]^{-1/3}$, in contrast to Dirac-Slater local-density theory involving only the second of these lengths. It is worth adding that at the nucleus, $\epsilon_x(0)$ is determined solely by $\rho_s(0)$ in the Dirac-Slater local-density approximation, whereas the Hartree-Fock $\epsilon_x(r=0)$, though involving again only the s-state wave functions, requires nonlocal information also about these.

It is a pleasure to acknowledge valuable discussions on Hartree-Fock theory with Dr. G. Senatore of the International Center for Theoretical Physics, Trieste. Some of this work was carried out while the author was visiting the University of California at Santa Barbara, which was supported in part by the Leverhulme Trustees, to whom the author's thanks are due.

[1] *Theory of the Inhomogeneous Electron Gas,* edited by S. Lundqvist and N. H. March (Plenum, New York, 1983).

[2] J. Callaway and N. H. March, *Solid State Physics,* edited by H. Ehrenreich and D. Turnhill (Academic, New York, 1984).

[3] J. C. Slater, Phys. Rev. **81**, 385 (1951).

[4] P. A. M. Dirac, Proc. Cambridge Philos. Soc. **26**, 376 (1930).

[5] J. Bardeen, Phys. Rev. **49**, 653 (1936).

[6] I. D. Moore and N. H. March, Ann. Phys. (NY) **97**, 136 (1976).

[7] L. Miglio, M. P. Tosi, and N. H. March, Surf. Sci. **111**, 119 (1981).

[8] R. Pucci and N. H. March, Phys. Rev. A **35**, 491 (1987).

Phys. Chem. Liq., 1995, Vol. 30, pp. 135–139
Reprints available directly from the publisher
Photocopying permitted by license only

SLATER SUM AND KINETIC ENERGY TENSOR IN SOME SIMPLE INHOMOGENEOUS ELECTRON LIQUIDS

C. AMOVILLI[a] and N. H. MARCH[b]

[a]*Dipartimento di Chimica e Chimica Industriale, Università di Pisa, Via Risorgimento 35, 56126 Pisa, Italy.*
[b]*Inorganic Chemistry Department, University of Oxford, South Parks Road, Oxford OX1 3QR, England*

(Received 22 February 1995)

Starting from the Bloch equation for the canonical density matrix, the differential equation satisfied by the Slater sum is derived for some simple inhomogeneous electron liquids. The relation of this derivation to the differential form of the virial theorem, and in particular to the kinetic energy tensor, is finally discussed.

KEY WORDS: Bloch equation, slater sum, kinetic energy tensor.

1. INTRODUCTION

Since the work of March and Murray[1], specifically for central field problems, it has been known that the Bloch equation for the canonical density matrix contains within itself a differential equation for the Slater sum $P(\mathbf{r},\beta)$. Here, for independent electrons moving in a common potential energy $V(\mathbf{r})$, $P(\mathbf{r},\beta)$ is defined in terms of the one-electron eigenvalues ε_i and the corresponding one-electron wave functions $\psi_i(\mathbf{r})$ by

$$P(\mathbf{r},\beta)=\sum_{\text{all } i}\psi_i(\mathbf{r})\psi_i^*(\mathbf{r})e^{-\beta\epsilon_i} \quad : \quad \beta=1/k_BT \tag{1.1}$$

where k_BT is the thermal energy. The canonical density matrix $C(\mathbf{r}_1,\mathbf{r}_2,\beta)$ is the direct generalization of eqn (1.1) to the off-diagonal form

$$C(\mathbf{r}_1,\mathbf{r}_2,\beta)=\sum_{\text{all } i}\psi_i(\mathbf{r}_1)\psi_i^*(\mathbf{r}_2)e^{-\beta\epsilon_i} \tag{1.2}$$

If $\hat{H}$ denotes the independent Hamiltonian of the inhomogeneous electron liquid defined by the application of the one-body potential energy $V(\mathbf{r})$:

$$\hat{H}(\mathbf{r})=-\frac{1}{2}\nabla_{\mathbf{r}}^2+V(\mathbf{r}) \tag{1.3}$$

then one will take us starting point the appropriate Bloch equations (see eqns (2.1) and (2.2) below). The differential equation for $P(\mathbf{r},\beta)$ for some simple cases will then be derived by direct expansion of the canonical density matrix about its diagonal. After this development, the connection with the differential form of the virial theorem[2], and in particular with the kinetic energy tensor, will be considered in section 3. Section 4 constitutes a brief summary, with some proposals for future work.

2. DIFFERENTIAL EQUATION FOR SLATER SUM $P(\mathbf{r},\beta)$ IN SOME SIMPLE CASES

The starting point of the present derivation will be the two forms of the Bloch equation for C defined in eqn (1.2):

$$\left[-\frac{1}{2}\nabla_1^2 + V(\mathbf{r}_1)\right] C(\mathbf{r}_1,\mathbf{r}_2,\beta) = -\frac{\partial}{\partial\beta} C(\mathbf{r}_1,\mathbf{r}_2,\beta) \tag{2.1}$$

$$\left[-\frac{1}{2}\nabla_2^2 + V(\mathbf{r}_2)\right] C(\mathbf{r}_1,\mathbf{r}_2,\beta) = -\frac{\partial}{\partial\beta} C(\mathbf{r}_1,\mathbf{r}_2,\beta) \tag{2.2}$$

Introducing new independent variables as $\mathbf{R} = \frac{1}{2}(\mathbf{r}_1 + \mathbf{r}_2)$ and $\mathbf{r}_{12} = \mathbf{r}_1 - \mathbf{r}_2$, summing and subtracting eqns (2.1) and (2.2), one finds

$$\left[-\frac{1}{4}\nabla_R^2 - \nabla_{12}^2 + 2V(\mathbf{R})\right] C(\mathbf{r}_1,\mathbf{r}_2,\beta) = -2\frac{\partial}{\partial\beta} C(\mathbf{r}_1,\mathbf{r}_2,\beta) \tag{2.3}$$

$$\vec{\nabla}_R \cdot \vec{\nabla}_{12} C(\mathbf{r}_1,\mathbf{r}_2,\beta) = \mathbf{r}_{12} \cdot \vec{\nabla} V(\mathbf{R}) C(\mathbf{r}_1,\mathbf{r}_2,\beta) \tag{2.4}$$

in the limit $\mathbf{r}_{12} \to 0$. Now

$$\nabla_R^2 = \frac{1}{R^2}\frac{\partial}{\partial R} R^2 \frac{\partial}{\partial R}$$

and

$$\vec{\nabla}_{12} = \frac{\mathbf{r}_{12}}{r_{12}}\frac{\partial}{\partial r_{12}} + \frac{1}{r_{12}}\left(\frac{\mathbf{R}}{R} - \mu\frac{\mathbf{r}_{12}}{r_{12}}\right)\frac{\partial}{\partial\mu}$$

$$\nabla_{12}^2 = \frac{1}{r_{12}^2}\frac{\partial}{\partial r_{12}} r_{12}^2 \frac{\partial}{\partial r_{12}} + \frac{1}{r_{12}^2}\frac{\partial}{\partial\mu}(1-\mu^2)\frac{\partial}{\partial\mu}$$

with

$$\mu = \frac{\mathbf{R}\cdot\mathbf{r}_{12}}{R r_{12}}$$

Taking as the first example the case of a bare Coulomb field, i.e. $V(R) = -Z/r$, Blinder's arguments[3] lead to

$$C(\mathbf{r}_1,\mathbf{r}_2,\beta) = P\left[\frac{1}{2}(r_1+r_2),\beta\right] + \Delta\left[\frac{1}{2}(r_1+r_2),\beta\right] r_{12}^2 + o(r_{12}^4)$$

$$= P(R,\beta) + \left[\frac{(1-\mu^2)}{8R}\frac{\partial P}{\partial R} + \Delta(R,\beta)\right] r_{12}^2 + o(r_{12}^4) \tag{2.5}$$

while eqns (2.3) and (2.4) become, respectively,

$$\frac{1}{8}P'' + \frac{1}{2R}P' - \left(V + \frac{\partial}{\partial\beta}\right)P + 3\Delta = 0 \tag{2.6}$$

and

$$\Delta' = \frac{1}{2}V'P + \frac{1}{4R^2}P'. \tag{2.7}$$

One then finds, differentiating eqn (2.6) and inserting Δ' from eqn (2.7), for the Slater sum $P(R,\beta)$

$$\frac{1}{8}P''' + \frac{1}{2R}P'' + \left(\frac{1}{4R^2} - V - \frac{\partial}{\partial\beta}\right)P' + \frac{1}{2}V'P = 0. \tag{2.8}$$

This equation is solved by $P(R,\beta\to\infty) = e^{-2ZR+\beta Z^2/2}$.

Using similar arguments and puting (see Sondheimer and Wilson[4])

$$C(\mathbf{r}_1,\mathbf{r}_2,\beta) = P(R,\beta) + \Delta(R,\beta)r_{12}^2 + o(r_{12}^4) \tag{2.9}$$

for the harmonic potential in three dimensions, the differential equation for the Slater sum is

$$\frac{1}{8}P''' + \frac{1}{4R}P'' - \left(\frac{1}{4R^2} + V + \frac{\partial}{\partial\beta}\right)P' + \frac{1}{2}V'P = 0. \tag{2.10}$$

Finally we have considered also the case of a uniform electric field of arbitrary strength along x-axis in three dimensions. In this case the near-diagonal form for C (see Harris and Cina[5], Jannussis[6])

$$C(\mathbf{r}_1,\mathbf{r}_2,\beta) = P(X,\beta) + \Delta(X,\beta)r_{12}^2 + o(r_{12}^4) \tag{2.11}$$

leads to the following differential equation

$$\frac{1}{8}P''' - \left(V + \frac{\partial}{\partial\beta}\right)P' + \frac{1}{2}V'P = 0. \tag{2.12}$$

It is evident, comparing these last two results with the corresponding one dimensional equations, that the dimensionality D gives the coefficient $(D-2)/2$ to the term $V'P$. In a general central field the independent variables are r_1, r_2 and r_{12} giving a greater freedom to the form (2.5) that now can be written

$$C(\mathbf{r}_1,\mathbf{r}_2,\beta) = P(R,\beta) + [\Delta_1(R,\beta) + \mu^2\Delta_2(R,\beta)]r_{12}^2 + o(r_{12}^4) \tag{2.13}$$

and eqns (2.3) and (2.4) are probably not sufficient to give separate equations for P, Δ_1 and Δ_2. The additional condition

$$C(\mathbf{r}_1,\mathbf{r}_2,\beta+\beta') = \int C(\mathbf{r}_1,\mathbf{r},\beta)\,C(\mathbf{r},\mathbf{r}_2,\beta')\,d\mathbf{r}$$

for β' small leads back to the Bloch equation and does not give further information.

3. DIFFERENTIAL FORM OF THE VIRIAL THEOREM

We did some analyses of our previous results and we arrived at the conclusion that the differential form of the virial theorem is already included in eqns. (2.3) and (2.4). The function Δ should be also related directly to the trace of the kinetic energy tensor. The form (2.5) in this context can be read as

$$C(\mathbf{r}_1, \mathbf{r}_2, \beta) = P(R, \beta) + \frac{1}{2}\mathbf{r}_{12} \cdot \mathbf{T}(R, \beta) \cdot \mathbf{r}_{12} + \dots \tag{3.1}$$

where

$$T_{ij} = \lim_{r_{12} \to 0} \frac{1}{2}\left(\frac{\partial^2 C}{\partial x_{i1} \partial x_{j2}} + \frac{\partial^2 C}{\partial x_{j1} \partial x_{i2}}\right) \tag{3.2}$$

is, by definition, a kinetic energy tensor. This tensor must satisfy the differential form of the virial theorem that, for a general central field, could be expressed as a relation of the following type[2]

$$R_i f[V, P] = \sum_{j=1}^{3} \frac{\partial T_{ij}}{\partial R_j} \quad (i = 1, 2, 3). \tag{3.3}$$

Again, from Blinder's arguments, in the Coulomb field the kinetic energy tensor can be written in terms of the trace t, being

$$T_{ij} = t\frac{R_i R_j}{R^2} \tag{3.4}$$

and eqn (3.3) reduces to the following scalar relation

$$f[V, P] = \frac{1}{R}t + t' \tag{3.5}$$

which can be combined with the definition of the kinetic energy density to give an equation for the Slater sum. In a general central field the matrix elements of $\mathbf{T}$ are not related simply to the trace. After some derivation the result was the following. Writing[1]

$$C(\mathbf{r}_1, \mathbf{r}_2, \beta) = \sum_{nl} \chi_{nl}(r_1)\chi_{nl}(r_2)a_{nl}(\cos\theta, \beta) \tag{3.6}$$

one has

$$T_{ij} = \frac{1}{2}\left[\delta_{ij} - \frac{R_i R_j}{R^2}\right]t(R, \beta) + \frac{1}{4}\left[\frac{3R_i R_j}{R^2} - \delta_{ij}\right]s(R, \beta) \tag{3.7}$$

where

$$s(R, \beta) = \sum_{nl} \chi'_{nl}(R)\chi'_{nl}(R)a_{nl}(1, \beta) \tag{3.8}$$

Now eqn (3.7) should be the most general form in central field; one has a new function $s(R, \beta)$ and precisely the same condition as in eqn (2.13).

4. SUMMARY AND FUTURE DIRECTIONS

It has been shown, by a direct argument from the Bloch equations (2.1) and (2.2) for the canonical density matrix, that the differential equations (2.8) and (2.10) for the Coulomb field and the harmonic oscillator respectively can be simply obtained. The differential equation (2.8) for the Coulomb field has also been recently obtained by Pfalzner *et al.*[7] and Cooper[8] by means of the spatial generalization of Kato's theorem[9]. Again, a different derivation has been given from that of Lehmann and March[10] for initially free electrons in a uniform electric field of arbitrary strength. The connection of the arguments of section 2 with the differential form of the virial theorem is set out in section 3. For the specific case of the Coulomb field, this is directly connected with a relation, eqn. (3.4), between the kinetic energy tensor (3.2) and the trace t, of central importance in density functional theory.

We are currently working on the generalization of these arguments to treat the case of a hydrogen-like atom in an electric field of arbitrary strength. Directions in which this is being tackled are (i) semiclassical approaches (see Hill[11] and Blinder[12]) and (ii) fully quantal approaches using the separability in parabolic coordinates. Undoubtedly, in the Coulomb problem in zero electric field, the separability, of the Schrödinger wave equation in both spherical polar coordinates and also parabolic coordinates can be subsumed into the existence of the Runge-Lenz vector as a constant of motion additional to the conventional ones in central fields. Generalizations are being sought presently for non-zero electric field.

Acknowledgments

One of us (N.H.M.) wishes to acknowledge partial financial support from the Office of Naval Research (USA) for the work reported here.

References

1. N. H. March and A. M. Murray, *Phys. Rev.*, **120**, 830 (1960).
2. See for example, A. Holas and N. H. March, *J. Mol. Struct. (Theochem)* **315**, 239 (1994).
3. S. M. Blinder, *Phys. Rev.*, **A43**, 13 (1991).
4. E. H. Sondheimer and A. H. Wilson, *Proc. Roy. Soc.*, **A210**, 173 (1951).
5. R. A. Harris and J. A. Cina, *J. Chem. Phys.*, **79**, 1381 (1983).
6. A. D. Jannussis, *Phys. Status Solidi* **36**, K17 (1969).
7. S. Pfalzer, H. Lehmann and N. H. March, *J. Math. Chem.*, **16**, 9 (1994).
8. I. L. Cooper, *Phys. Rev.*, **A50**, 1040 (1994).
9. N. H. March, *Phys. Rev.*, **A33**, 88 (1986)
10. H. Lehmann and N. H. March, *Phys. Chem. Liquids*, **27**, 65 (1994).
11. R. N. Hill, *J. Math. Phys.*, **9**, 1534 (1968).
12. S. M. Blinder, *Phys. Rev. Lett.*, **52**, 1771 (1984).

PHYSICAL REVIEW B VOLUME 55, NUMBER 3 15 JANUARY 1997-I

Correction to Slater exchange potential to yield exact Kohn-Sham potential generating the Hartree-Fock density

A. Holas and N. H. March
Institute of Physical Chemistry of the Polish Academy of Sciences, 44/52 Kasprzaka, 01-224 Warsaw, Poland
(Received 19 January 1996; revised manuscript received 11 October 1996)

Kleinman has recently pointed out that the Slater, or average Fock approximation, to the exchange potential can be written as the partial functional derivative of an exchange energy functional. Here we demonstrate for an arbitrary N-electron system the explicit corrections to the Slater potential, which yield the Kohn-Sham-like exchange potential generating the exact Hartree-Fock density. The example of the four-electron spin-compensated system is worked out in detail. [S0163-1829(97)08604-9]

In a recent interesting paper,[1] Kleinman has pointed out that the Slater potential $v_{\rm SI}(\mathbf{r})\equiv\mu_x^{\rm AFA}(\mathbf{r})$, or, as he terms it, the average Fock approximation, can be constructed as a partial functional derivative of the Hartree-Fock (HF) exchange energy. Submitted nearly at the same time as Kleinman's paper, the work of Holas *et al.*[2] appeared, in which the exchange-only potential in a density-functional formalism of HF theory was considered. Below, we relate this work to Kleinman's paper by exhibiting explicitly the corrections to the Slater potential $v_{\rm SI}(\mathbf{r})$, constructed to yield the exact Kohn-Sham-like exchange potential generating the HF density, denoted by $\rho(\mathbf{r})$ throughout.

The HF exchange energy $E_x^{\rm HF}$ can be written in terms of the HF first-order idempotent density matrix $\rho_1(\mathbf{r};\mathbf{r}')$ constructed from HF orbitals as

$$E_x^{\rm HF}=-\frac{1}{4}\int d^3r d^3r'\frac{|\rho_1(\mathbf{r};\mathbf{r}')|^2}{|\mathbf{r}-\mathbf{r}'|} \tag{1}$$

(atomic units are used throughout, and, for simplicity, the version for spin-compensated systems only is considered first). Writing this in terms of the Slater potential $v_{\rm SI}(\mathbf{r})$ one has

$$E_x^{\rm HF}=\frac{1}{2}\int d^3r\rho(\mathbf{r})v_{\rm SI}(\mathbf{r}), \tag{2}$$

where

$$v_{\rm SI}(\mathbf{r})=-\frac{1}{2}\int d^3r'\frac{|\rho_1(\mathbf{r};\mathbf{r}')|^2}{\rho(\mathbf{r})|\mathbf{r}-\mathbf{r}'|}. \tag{3}$$

The main object of this Brief Report is to present the results for the exchange potential $v_x^{\rm KSHF}(\mathbf{r})$ [in Holas *et al.*[2] this is termed $v_{\rm xc}^{\rm HF}(\mathbf{r})$], defined similarly to Kleinman[1] as the functional derivative of

$$E_x^{\rm KSHF}[\rho]=E_x^{\rm HF}[\rho]+T^{\rm HF}[\rho]-T_s[\rho] \tag{4}$$

[see Eq. (17) of Ref. 1] with respect to the density:

$$v_x^{\rm KSHF}(\mathbf{r};[\rho])=\frac{\delta E_x^{\rm KSHF}[\rho]}{\delta\rho(\mathbf{r})}. \tag{5}$$

This potential, when added to the external and electrostatic potentials, forms the effective potential occurring in the Kohn-Sham-HF equations yielding the same density as that obtained from the solution of the HF equations for a given system. We stress, as Kleinman does, that this is close to, but not the same as, the "exchange-only" version of the exact Kohn-Sham (KS) equations, due to the presence of the kinetic energy difference $T^{\rm HF}-T_s$ appearing in Eq. (4), and the exchange energy difference $E_x^{\rm HF}-E_x$. Here the KS exchange energy functional $E_x[\rho]\equiv E_x^{\rm KS}[\rho]$ is defined by Eq. (1) but with ρ_1 replaced by ρ_1^s — the density matrix constructed out of KS-HF orbitals.

We concentrate first on the example of the four-electron spin-compensated system, such as the Be atom or He_2 molecule. Following the usage in Kleinman's paper,[1] we represent the HF space orbitals $\psi_1(\mathbf{r})$ and $\psi_2(\mathbf{r})$ for the ground state of the system as constructed from the density amplitude $\rho^{1/2}(\mathbf{r})$ and the phase $\theta(\mathbf{r})$ as[3]

$$\psi_1(\mathbf{r})=(\tfrac{1}{2}\rho(\mathbf{r}))^{1/2}\cos(\theta(\mathbf{r})) \tag{6a}$$

and

$$\psi_2(\mathbf{r})=(\tfrac{1}{2}\rho(\mathbf{r}))^{1/2}\sin(\theta(\mathbf{r})), \tag{6b}$$

which evidently yields $\rho=2(\psi_1^2+\psi_2^2)$. From Eq. (6), the HF density matrix ρ_1 takes the form

$$\rho_1(\mathbf{r};\mathbf{r}')=(\rho(\mathbf{r})\rho(\mathbf{r}'))^{1/2}\cos(\theta(\mathbf{r})-\theta(\mathbf{r}')). \tag{7}$$

Besides the Hartree-Fock form, Eq. (7), to be inserted into Eqs. (1) and (3), we also define another idempotent density matrix $\rho_1^s(\mathbf{r};\mathbf{r}')$, with the same diagonal density $\rho(\mathbf{r})$ imposed as in Eq. (7), but with $\theta(\mathbf{r})$ of HF theory replaced by $\theta_s(\mathbf{r})$ generated from the orbitals which are solutions of the above-mentioned KS-HF equations. The work of Holas *et al.*[2] now allows the exchange potential $v_x^{\rm KSHF}(\mathbf{r})$ for the four-electron spin-compensated system to be related to the Slater potential $v_{\rm SI}(\mathbf{r})$ in Eq. (3), via [see Eq. (3.33) of Ref. 2]

$$v_x^{\mathrm{KSHF}}(\mathbf{r})=v_{\mathrm{SI}}(\mathbf{r})+\tfrac{1}{2}|\nabla\theta(\mathbf{r})|^2-\tfrac{1}{2}|\nabla\theta_s(\mathbf{r})|^2 -\lambda\sin^2(\theta(\mathbf{r}))+\lambda_s\sin^2(\theta_s(\mathbf{r})). \quad (8)$$

Here λ and $\theta(\mathbf{r})$ are the eigenvalue and eigenfunction of the equation [Eq. (3.9) of Ref. 2 after a misprint correction]

$$\nabla^2\theta(\mathbf{r})+\frac{(\nabla\rho(\mathbf{r}))(\nabla\theta(\mathbf{r}))}{\rho(\mathbf{r})} -\frac{1}{2}\int d^3r'\frac{\rho(\mathbf{r}')\sin(2\theta(\mathbf{r})-2\theta(\mathbf{r}'))}{|\mathbf{r}-\mathbf{r}'|} +\lambda\sin(2\theta(\mathbf{r}))=0, \quad (9)$$

thus they are functionals of the density $\rho(\mathbf{r})$. The eigenpair λ_s and $\theta_s(\mathbf{r})$ is the solution of an analogous equation, but with the interaction term (the integral) omitted.

It is known for Be (see, for example, Nagy and March[4]) that the phases tend fairly rapidly to a constant for large r, and therefore that the "correction" terms to $v_{\mathrm{SI}}(\mathbf{r})$ in Eq. (8) are short range. In contrast, as Kleinman[1] emphasizes, $v_{\mathrm{SI}}(\mathbf{r})$ tends to the value of $-1/r$ at large r. That this is the correct behavior of $v_x(\mathbf{r})$ itself at large r is well known.

In the context of the difference between $v_x^{\mathrm{KSHF}}(\mathbf{r})$ and $v_{\mathrm{SI}}(\mathbf{r})$ for atoms, the studies of Gritsenko *et al.*[5,6] are noteworthy. Plots of $v_x(\mathbf{r})$ and $v_{\mathrm{SI}}(\mathbf{r})$ using the optimized potential method (OPM) (i.e., the "exchange-only" version of the exact KS equations, in which the contributions due to the difference between the HF and KS-HF functionals are absent, and where the Slater potential is constructed out of OPM orbitals), for the atoms Kr, Cd, and Mg, in particular, demonstrate well the accord between $v_x(\mathbf{r})$ and $v_{\mathrm{SI}}(\mathbf{r})$ at large r. However, substantial differences occur well inside the atoms. For example, in Mg graphic differences are apparent inside two atomic units. At small distances $\lesssim 0.2$a.u., differences between $v_x(\mathbf{r})$ and $v_{\mathrm{SI}}(\mathbf{r})$ can be between 1 and 2 a.u. [i.e., about 30% of $v_x(\mathbf{r})$].

For systems more complicated than four-electron spin-compensated ones, it is possible to obtain[2] for $v_x^{\mathrm{KSHF}}(\mathbf{r})$ an expression analogous to Eq. (8), but containing more "corrections" to the Slater potential. Their number is $N-1$ if the HF (and KS-HF) system is described by N orbitals, with any separate correction expressed in terms of the HF phase $\theta_i(\mathbf{r})$, its KS-HF counterpart $\theta_{s,i}(\mathbf{r})$, and the corresponding eigenvalues. The system of N-1 eigenequations for phases $\theta_i(\mathbf{r})$, and one (Schrödinger-like) equation for the density amplitude $\rho^{1/2}(\mathbf{r})$, is equivalent to the system of N HF equations [or KS-HF equations in the case of $\theta_{s,i}(\mathbf{r})$]. Methods developed by us in Ref. 7 would be helpful in obtaining the explicit form of the above-mentioned equations.

To be more specific, we consider now a general case of N-electron system treated by the unrestricted HF theory, in which the determinantal wave function is constructed out of $N_\uparrow$ spin orbitals $\psi_i(\mathbf{r}_1)\alpha(s_1)$ and $N_\downarrow=N-N_\uparrow$ spin orbitals $\psi_j(\mathbf{r}_1)\beta(s_1)$, thus leading to the spin density matrix

$$\gamma_1(\mathbf{r}_1s_1;\mathbf{r}_1's_1')=\rho_{1\uparrow}(\mathbf{r}_1;\mathbf{r}_1')\alpha(s_1)\alpha(s_1') +\rho_{1\downarrow}(\mathbf{r}_1;\mathbf{r}_1')\beta(s_1)\beta(s_1'), \quad (10)$$

and the reduced one $\rho_1(\mathbf{r}_1;\mathbf{r}_1')=\Sigma_\sigma\rho_{1\sigma}(\mathbf{r}_1;\mathbf{r}_1')$, where

$$\rho_{1\sigma}(\mathbf{r}_1;\mathbf{r}_1')=\sum_{\ell=i_\sigma}^{f_\sigma}\psi_\ell(\mathbf{r}_1)\psi_\ell^*(\mathbf{r}_1'), \quad (11)$$

$$(i_\uparrow,f_\uparrow)=(1,N_\uparrow),\quad (i_\downarrow,f_\downarrow)=(N_\uparrow+1,N).$$

Due to the real nature of the system Hamiltonian, it is sufficient to consider only real orbitals, $\psi_i^*(\mathbf{r})=\psi_i(\mathbf{r})$. If each space orbital is factorized as

$$\psi_i(\mathbf{r})=\rho^{1/2}(\mathbf{r})K_i(\mathbf{r}), \quad (12)$$

where functions $K_i(\mathbf{r})$ satisfy

$$\sum_{i=1}^{N}K_i^2(\mathbf{r})=1, \quad (13)$$

then $\rho(\mathbf{r})$ represents the electron density corresponding to the considered HF wave function. The constraint (13) may be satisfied by representing the unit vector $[K_1,\ldots,K_N]$ in an N-dimensional space as a function of $N-1$ angles $\vec{\theta}(\mathbf{r})=[\theta_1(\mathbf{r}),\ldots,\theta_{N-1}(\mathbf{r})]$ in hyperspherical coordinate system

$$K_\ell(\mathbf{r})=\tilde{K}_\ell(\vec{\theta}(\mathbf{r}))=\left\{\prod_{i=1}^{\ell-1}\sin(\theta_i(\mathbf{r}))\right\}\cos(\theta_\ell(\mathbf{r})), \quad (14)$$

$$\theta_N(\mathbf{r})\equiv 0.$$

This allows us to write the HF energy in terms of N independent functions $\{\rho(\mathbf{r}),\vec{\theta}(\mathbf{r})\}$ as

$$\tilde{E}^{\mathrm{HF}}[\rho,\vec{\theta}]=T_{\mathrm{W}}[\rho]+\tilde{E}_{\mathrm{P}}[\rho,\vec{\theta}]+\int d^3rv(\mathbf{r})\rho(\mathbf{r})+E_{\mathrm{es}}[\rho] +\tilde{E}_x[\rho,\vec{\theta}], \quad (15)$$

where three terms depending solely on $\rho(\mathbf{r})$, namely the Weizsäcker kinetic energy T_{W}, the external potential energy $\int[v\rho]$, and the electrostatic energy E_{es}, have traditional forms,[7] while the Pauli kinetic energy[7] and exchange energy[2] functionals are

$$\tilde{E}_{\mathrm{P}}[\rho,\vec{\theta}]=\int d^3r\rho(\mathbf{r})\Gamma(\mathbf{r};[\vec{\theta}]), \quad (16)$$

$$\tilde{E}_x[\rho,\vec{\theta}]=-\frac{1}{2}\int d^3r_1d^3r_2\frac{\rho(\mathbf{r}_1)\rho(\mathbf{r}_2)}{|\mathbf{r}_1-\mathbf{r}_2|}\Xi(\mathbf{r}_1,\mathbf{r}_2;[\vec{\theta}]), \quad (17)$$

with $\vec{\theta}$-dependent kernels

$$\Gamma(\mathbf{r};[\vec{\theta}])=\tfrac{1}{2}\nabla_1\nabla_2\left(\sum_{\sigma=\uparrow,\downarrow}\kappa_\sigma(\mathbf{r}_1,\mathbf{r}_2;[\vec{\theta}])\right) \quad (18)$$

taken at $\mathbf{r}_1=\mathbf{r}_2=\mathbf{r}$,

$$\Xi(\mathbf{r}_1,\mathbf{r}_2;[\vec{\theta}])=\sum_{\sigma=\uparrow,\downarrow}\kappa_\sigma^2(\mathbf{r}_1,\mathbf{r}_2;[\vec{\theta}]), \quad (19)$$

in terms of [compare Eq. (11)]

$$\kappa_\sigma(\mathbf{r}_1,\mathbf{r}_2;[\vec{\theta}])=\sum_{\ell=i_\sigma}^{f_\sigma}\widetilde{K}_\ell(\vec{\theta}(\mathbf{r}_1))\widetilde{K}_\ell(\vec{\theta}(\mathbf{r}_2)). \quad (20)$$

When posing HF method as a density-functional theory[2] by means of the constrained-search method, $\widetilde{E}^{\rm HF}[\rho,\vec{\theta}]$ is minimized first, at fixed $\rho(\mathbf{r})$, with respect to variations of all $N-1$ phases $\vec{\theta}(\mathbf{r})$, under the normalization constraints for orbitals (12):

$$\widetilde{\mathcal{N}}_\ell[\rho,\vec{\theta}]\equiv\int d^3r\rho(\mathbf{r})\widetilde{K}_\ell^2(\vec{\theta}(\mathbf{r}))=1\,. \quad (21)$$

This leads to a system of $N-1$ equations

$$\frac{\widetilde{\delta}}{\widetilde{\delta}\theta_i(\mathbf{r})}\left(\widetilde{E}_{\rm P}[\rho,\vec{\theta}]+\widetilde{E}_x[\rho,\vec{\theta}]-\sum_{\ell=2}^{N}\lambda_\ell\widetilde{\mathcal{N}}_\ell[\rho,\vec{\theta}]\right)=0\,, \quad (22)$$

where $\widetilde{\delta}$ denotes a partial functional differentiation. Here λ_ℓ is a Lagrange multiplier connected with the constraint (21). It is sufficient to include $N-1$ constraints only, because the Nth one

$$\int d^3r\rho(\mathbf{r})=N, \quad (23)$$

which is fixed during the considered minimization, is imposed later in connection with the final minimization with respect to $\rho(\mathbf{r})$. Equations (22) taken for $i=1,\ldots,N-1$ represent a system of second-order-differential nonlinear coupled equations to determine $N-1$ eigenfunctions $\vec{\theta}(\mathbf{r})$ and eigenvalues $\{\lambda_\ell\}_2^N$. The solution is a functional of $\rho(\mathbf{r})$, because this function occurs explicitly in Eq. (22).

An equivalent KS-HF system of N noninteracting electrons can be treated in an analogous way. Denoting its phase functions by $\vec{\theta}_s(\mathbf{r})$ and the effective external potential by $v_s(\mathbf{r})$, we obtain the following expression for the system energy functional:

$$\widetilde{E}^s[\rho,\vec{\theta}_s]=T_{\rm W}[\rho]+\widetilde{E}_{\rm P}[\rho,\vec{\theta}_s]+\int d^3rv_s(\mathbf{r})\rho(\mathbf{r}), \quad (24)$$

and eigenequations for phase functions

$$\frac{\widetilde{\delta}}{\widetilde{\delta}\theta_{s,i}(\mathbf{r})}\left(\widetilde{E}_{\rm P}[\rho,\vec{\theta}_s]-\sum_{\ell=2}^{N}\lambda_{s,\ell}\widetilde{\mathcal{N}}_\ell[\rho,\vec{\theta}_s]\right)=0\,. \quad (25)$$

The effective exchange energy of the KS-HF theory, Eq. (4), is then

$$E_x^{\rm KSHF}[\rho]=\widetilde{E}_x[\rho,\vec{\theta}[\rho]]+\widetilde{E}_{\rm P}[\rho,\vec{\theta}[\rho]]-\widetilde{E}_{\rm P}[\rho,\vec{\theta}_s[\rho]] \quad (26)$$

(note cancellation of the Weizsäcker terms), so the corresponding potential is

$$v_x^{\rm KSHF}(\mathbf{r};[\rho])=\frac{\delta E_x^{\rm KSHF}[\rho]}{\delta\rho(\mathbf{r})}=v_{\rm Sl}(\mathbf{r};[\rho])+v_{\rm nSl}(\mathbf{r};[\rho]), \quad (27)$$

with its Slater term

$$v_{\rm Sl}(\mathbf{r};[\rho])=\frac{\widetilde{\delta}\widetilde{E}_x[\rho,\vec{\theta}]}{\widetilde{\delta}\rho(\mathbf{r})}=-\int d^3r'\frac{\rho(\mathbf{r}')}{|\mathbf{r}-\mathbf{r}'|}\Xi(\mathbf{r},\mathbf{r}';[\vec{\theta}[\rho]]), \quad (28)$$

and the non-Slater term

$$\begin{aligned}v_{\rm nSl}(\mathbf{r};[\rho])=&\frac{\widetilde{\delta}}{\widetilde{\delta}\rho(\mathbf{r})}(\widetilde{E}_{\rm P}[\rho,\vec{\theta}]-\widetilde{E}_{\rm P}[\rho,\vec{\theta}_s])+\text{const}\\&+\sum_{i=1}^{N-1}\int d^3r'\left\{\frac{\widetilde{\delta}}{\widetilde{\delta}\theta_i(\mathbf{r}')}(\widetilde{E}_{\rm P}[\rho,\vec{\theta}]\right.\\&+\widetilde{E}_x[\rho,\vec{\theta}])\frac{\delta\theta_i(\mathbf{r}';[\rho])}{\delta\rho(\mathbf{r})}\\&\left.-\frac{\widetilde{\delta}\widetilde{E}_{\rm P}[\rho,\vec{\theta}_s]}{\widetilde{\delta}\theta_{s,i}(\mathbf{r}')}\frac{\delta\theta_{s,i}(\mathbf{r}';[\rho])}{\delta\rho(\mathbf{r})}\right\}. \end{aligned}\quad (29)$$

Equation (17) can be rewritten now in terms of the Slater potential (28) as

$$E_x^{\rm HF}[\rho]=\widetilde{E}_x[\rho,\vec{\theta}[\rho]]=\frac{1}{2}\int d^3r\rho(\mathbf{r})v_{\rm Sl}(\mathbf{r};[\rho]), \quad (30)$$

which is a generalization of Eq. (2).

Functional derivatives in Eq. (29) with respect to $\widetilde{\delta}\rho(\mathbf{r})$ can be easily obtained from Eq. (16), while those with respect to $\widetilde{\delta}\theta_i(\mathbf{r}')$ and $\widetilde{\delta}\theta_{s,i}(\mathbf{r}')$ can be taken directly from Eqs. (22) and (25), thus leading to

$$\begin{aligned}v_{\rm nSl}(\mathbf{r};[\rho])=&\Gamma(\mathbf{r};[\vec{\theta}[\rho]])-\Gamma(\mathbf{r};[\vec{\theta}_s[\rho]])+\text{const}\\&+\sum_{i=1}^{N-1}\sum_{\ell=2}^{N}\int d^3r'\left\{\lambda_\ell\frac{\widetilde{\delta}\widetilde{\mathcal{N}}_\ell[\rho,\vec{\theta}]}{\widetilde{\delta}\theta_i(\mathbf{r}')}\right.\\&\times\frac{\delta\theta_i(\mathbf{r}';[\rho])}{\delta\rho(\mathbf{r})}\\&\left.-\lambda_{s,\ell}\frac{\widetilde{\delta}\widetilde{\mathcal{N}}_\ell[\rho,\vec{\theta}_s]}{\widetilde{\delta}\theta_{s,i}(\mathbf{r}')}\frac{\delta\theta_{s,i}(\mathbf{r}';[\rho])}{\delta\rho(\mathbf{r})}\right\}. \end{aligned}\quad (31)$$

Next, by inserting the result of the total functional differentiation of the constraint (21),

$$\begin{aligned}0=&\frac{\delta\widetilde{\mathcal{N}}_\ell[\rho,\vec{\theta}[\rho]]}{\delta\rho(\mathbf{r})}=\widetilde{K}_\ell^2(\vec{\theta}(\mathbf{r}))\\&+\sum_{i=1}^{N-1}\int d^3r'\frac{\widetilde{\delta}\widetilde{\mathcal{N}}_\ell[\rho,\vec{\theta}]}{\widetilde{\delta}\theta_i(\mathbf{r}')}\frac{\delta\theta_i(\mathbf{r}';[\rho])}{\delta\rho(\mathbf{r})}, \end{aligned}\quad (32)$$

and its analog for the noninteracting system, we obtain the final result for the non-Slater term of $v_x^{\rm KSHF}$, free remarkably of functional derivatives, namely,

$$\begin{aligned}v_{\rm nSl}(\mathbf{r};[\rho])=&\Gamma(\mathbf{r};[\vec{\theta}[\rho]])-\Gamma(\mathbf{r};[\vec{\theta}_s[\rho]])-\overline{\lambda}(\mathbf{r};[\rho])\\&+\overline{\lambda_s}(\mathbf{r};[\rho])+\text{const}, \end{aligned}\quad (33)$$

where we introduced local average eigenvalues

$$\overline{\lambda}(\mathbf{r};[\rho])=\sum_{\ell=2}^{N}\lambda_\ell[\rho]\widetilde{K}_\ell^2(\vec{\theta}(\mathbf{r};[\rho])), \quad (34)$$

$$\overline{\lambda_s}(\mathbf{r};[\rho])=\sum_{\ell=2}^{N}\lambda_{s,\ell}[\rho]\widetilde{K}_\ell^2(\vec{\theta}_s(\mathbf{r};[\rho])). \quad (35)$$

The procedure outlined above can be applied, with minor modifications, to the restricted HF theory, where for an even number $N=2N_\uparrow$ of electrons, only $N_\uparrow$ space orbitals are independent and doubly occupied. This results in replacing Eq. (12) by $\psi_i(\mathbf{r})=\{\frac{1}{2}\rho(\mathbf{r})\}^{1/2}K_i(\mathbf{r})$, N by $N/2$ in Eq. (14) and in all summations (so $\vec{\theta}$ having $\frac{1}{2}N-1$ components), Eqs. (18) and (19) by $\Gamma=\frac{1}{2}\nabla_1\nabla_2\kappa_\uparrow$, $\Xi=\frac{1}{2}\kappa_\uparrow^2$, and in the value 2 instead of 1 for the constraint (21). Then all remaining equations hold for this case. In particular, for a system of $N=4$, Eq. (9) is obtained from Eq. (22) by performing differentiation and using the notation $\theta(\mathbf{r})=\theta_1(\mathbf{r})$ and $\lambda=\lambda_2$, while Eq. (8) is an explicit form of Eq. (27) with Eq. (33) for this case.

The procedure described earlier in this paper allows a construction of all terms of $v_x^{\mathrm{KSHF}}(\mathbf{r})$ in Eq. (27) based solely on the knowledge of $\rho(\mathbf{r})$, which demonstrates the density-functional origin of this solution. In practice, for any particular system, however, wider information is available—the HF orbitals $\psi_\ell(\mathbf{r})$ and the corresponding orbital energies ϵ_ℓ, the electron density $\rho(\mathbf{r})$, and [after applying a procedure yielding the effective KS potential $v_s(\mathbf{r})$ from a given density $\rho(\mathbf{r})$, e.g., Ref. 8]—the KS-HF orbitals $\psi_{s,\ell}(\mathbf{r})$ and energies $\epsilon_{s,\ell}$. Then one finds immediately $\widetilde{K}_\ell(\vec{\theta}(\mathbf{r}))=K_\ell(\mathbf{r})=\psi_\ell(\mathbf{r})/\rho^{1/2}(\mathbf{r})$, and, analogously, $\widetilde{K}_\ell(\vec{\theta}_s(\mathbf{r}))$ from the KS-HF orbitals (for spin-compensated systems replace ρ by $\frac{1}{2}\rho$), omitting thus the need for $\vec{\theta}(\mathbf{r})$ and $\vec{\theta}_s(\mathbf{r})$ as solutions of Eqs. (22) and (25). Moreover, also the need for eigenvalues of these equations can be bypassed, because, as was shown in Ref. 7, $\lambda_{s,\ell}=\epsilon_{s,\ell}-\epsilon_{s,1}$, and, as can be easily proven using the same reasoning, $\lambda_\ell=\epsilon_\ell-\epsilon_1$. Thus, Eqs. (34) and (35) can be rewritten [noting $\lambda_1=0$ and using Eq. (13)] as

$$\overline{\lambda}(\mathbf{r})=\overline{\epsilon}(\mathbf{r})-\epsilon_1;\quad \overline{\lambda_s}(\mathbf{r})=\overline{\epsilon_s}(\mathbf{r})-\epsilon_{s,1}, \quad (36)$$

with

$$\overline{\epsilon}(\mathbf{r})=\sum_{\ell=1}^{N}\epsilon_\ell K_\ell^2(\mathbf{r}) \quad (37)$$

and analogous expression for $\overline{\epsilon_s}(\mathbf{r})$. In this way $v_{\mathrm{nSl}}(\mathbf{r})$ can be calculated for any particular system using the information about its orbitals and orbital energies of the HF and KS-HF solutions.

To conclude, our approach to the N-electron HF and KS-HF problems, realized explicitly in the language of density and phases, as in Kleinman's work, allows the functional dependence of phases on $\rho(\mathbf{r})$ to be explicitly constructed. As Kleinman recognized, the final result for the exchange potential $v_x^{\mathrm{KSHF}}(\mathbf{r})$ is the partial functional derivative of the HF exchange energy with respect to $\rho(\mathbf{r})$, Eq. (28)—namely the Slater potential exhibited in Ref. 1—corrected by the partial functional derivative of the kinetic energy difference with respect to $\rho(\mathbf{r})$, and of all terms with respect to phases, Eq. (29). As was already mentioned, whereas in Ref. 2 the procedure invoked was to solve for $\lambda,\theta(\mathbf{r}),\lambda_s$, and $\theta_s(\mathbf{r})$ given $\rho(\mathbf{r})$, Kleinman's philosophy to express $v_x^{\mathrm{KSHF}}(\mathbf{r})$ directly in terms of orbitals clearly emerges as an alternative approach, discussed in detail in the preceding paragraph.

N.H.M. wishes to acknowledge that his contribution to this study was brought to fruition during a stay in the Institute of Physical Chemistry of PAS in Warsaw. It is a pleasure to thank Professor S. Olszewski for generous hospitality. Partial financial support from the Leverhulme Trust is also acknowledged.

[1] L. Kleinman, Phys. Rev. B **49**, 14 197 (1994).
[2] A. Holas, N. H. March, Y. Takahashi, and C. Zhang, Phys. Rev. A **48**, 2708 (1993).
[3] K. A. Dawson and N. H. March, J. Chem. Phys. **81**, 5850 (1984).
[4] A. Nagy and N. H. March, Phys. Rev. A **39**, 5512 (1984).
[5] O. Gritsenko, R. van Leeuwen, and E. J. Baerends, J. Chem. Phys. **101**, 8955 (1994).
[6] O. Gritsenko, R. van Leeuwen, E. van Lenthe, and E. J. Baerends, Phys. Rev. A **51**, 1944 (1995).
[7] A. Holas and N. H. March, Phys. Rev. A **44**, 5521 (1991).
[8] Yue Wang and R. G. Parr, Phys. Rev. A **47**, 1591 (1993).

26 October 2001

Chemical Physics Letters 347 (2001) 459–464

CHEMICAL PHYSICS LETTERS

www.elsevier.com/locate/cplett

The March model applied to boron cages

C. Amovilli [a,*], N.H. March [b,c]

[a] *Dipartimento di Chimica e Chimica Industriale, Università di Pisa, Via Risorgimento 35, 56126 Pisa, Italy*
[b] *University of Oxford, Oxford, UK*
[c] *Department of Physics, University of Antwerpen, RUCA, Groenenborgerlaan 171, Antwerpen, Belgium*

Received 9 July 2001; in final form 7 September 2001

Abstract

The so-called March model of fullerene, in which a self-consistent spherical distribution of π electrons is combined with the proper nuclear–nuclear potential energy for the correct structure, is here extended to boron cages. The Thomas–Fermi approximation of the initial studies is here replaced by Hartree–Fock calculations. Explicit results for B_{2k} and B^+_{2k+1}, with k ranging from 15 to 27, are discussed and compared with calculations on similar clusters found in the literature.

1. Introduction

In this work we present an ab initio study on 25 spheroidal boron cages with formula B_{2k} and B^+_{2k+1}.

We start from the basic ideas of the March model [1] developed by one of us in 1952. In early work the above model was applied on tetrahedral and octahedral molecules. In GeH_4, this implied smearing the four protons into a uniform surface charge density over a sphere of radius equal to the Ge–H bond length and centred on the Ge nucleus. The same model was used for C_{60} subsequently and termed there by Clougherty and Zhu [2] and Despa [3] the 'March model' (see also Siringo et al. [4]). Though the electron density in this approach is spherically symmetrical, the nuclear–nuclear potential energy is calculated for the correct cage geometry.

Here we apply similar considerations on the less well-studied case of boron cages. The discovery of boron clusters B_n with n up to 52 was realized 10 years ago through laser ablation of hexagonal boron nitride by La Placa et al. [5]. The authors analysed the cluster size by mass spectroscopy but the experimental geometries are still unknown.

The present work is motivated by the close relation existing between the equilibrium geometry of a system of equal point charges (the nuclei), which can move only over the surface of a sphere, and the so-called 'Aufbau principle' [6] used in constructing boron clusters [7]. In fact, when the number of charges smeared over the sphere surface is getting large, the optimized geometry corresponds to a polyhedron with triangular faces and essentially penta- and hexa-coordinated vertices while, following the Aufbau principle proposed for boron compounds, the geometrical forms can be constructed as a combination of two basic units only, namely a pentagonal pyramidal unit B_6 and a hexagonal pyramidal unit B_7 [8].

* Corresponding author. Fax: +39-050-918260.
E-mail address: amovilli@dcci.unipi.it (C. Amovilli).

0009-2614/01/$ - see front matter
PII: S0009-2614(01)01071-5

Using this Aufbau principle several convex, quasi-planar, tubular and spherical boron clusters have been studied at ab initio level [7,9]. The main difficulty in such calculations arises from the consideration of potential energy hypersurfaces which present many local minima and that require heavy computations owing to the number of atoms for large size clusters.

In this Letter the only geometrical parameter optimized at the ab initio level is the radius of the sphere, being the position of nuclei determined separately by minimization of the Coulombic repulsion energy among themselves, once their motion has been constrained over the surface of the sphere. It is important to remark that here we make use of the basic ideas of the March model, but all the final results refer to standard Hartree–Fock calculations performed with the correct electron–nuclei interactions.

2. Computational details and discussion of results

As anticipated in the introduction, the nuclear configuration for the 25 selected spherical cages corresponds to the equilibrium geometry of a system of equal charges smeared over the surface of a sphere. The repulsion energy $E_{\rm rep}$, of a system of n particles with charge Z displaced over the surface of a sphere with radius R, is given by

$$E_{\rm rep} = A_n \frac{(Zn)^2}{R}, \tag{1}$$

where the factor A_n depends on the geometry but does not depend on both the radius R and the charge Z. A_n is similar to the Madelung constant in crystal structures. In the limit of a uniform spherically symmetrical surface charge distribution A_∞ is known to be 0.5. We have minimized A_n for fixed Z and R by means of a simple Monte Carlo procedure and the resulting geometries have been subsequently scaled, by changing the radius R, in order to find the best size of the 25 selected boron clusters. The radius optimization has been performed at the Hartree–Fock level of ab initio calculation. Because the above Monte Carlo step has been performed without preserving any symmetry constraint we have assumed a singlet ground state in all cases. For the SCF calculation we have constructed a basis set slightly better than the standard 3-21G used for large boron clusters. The contraction scheme is (9s,4p) → [2s,2p] and the contracted orbitals are better described in the core region in order to prevent the results from too large basis set superposition errors. All the details of this basis set are shown in Table 1.

In Table 2 we present the main results of our calculations, namely the optimized radii (R_e) and the cohesive or bond energies (E_b), together with the values of the factor A_n defined above. The cohesive energy is given by

$$E_b = \frac{E_n(\infty) - E_n(R_e)}{n}, \tag{2}$$

where $E_n(\infty)$ in terms of the energies of B and B^+ is

$$\begin{aligned} E_n(\infty) &= nE(\mathrm{B}) \quad (n = 2k) \\ &= (n-1)E(\mathrm{B}) + E(\mathrm{B}^+) \quad (n = 2k+1) \end{aligned} \tag{3}$$

and where $E_n(R_e)$ is the SCF energy evaluated at the equilibrium radius. It is important to remark that Eq. (2) is valid to determine the bond energy but $E_n(\infty)$, as defined in Eq. (3), is not the SCF energy at infinite radius; it is well known that the Hartree–Fock method does not work properly in

Table 1
Gaussian contraction scheme used for boron in this work

Type	Exponent	Coefficient[a]
s	2400.000	0.00291
	400.0	0.02128
	90.0	0.10042
	30.0	0.26432
	10.0	0.80193
	3.0	1.00000
s	1.0	−0.16867
	0.3	1.00000
	0.1	0.78505
p	3.0	0.15952
	0.9	0.47478
	0.3	1.00000
p	0.1	1.00000

[a] The orbitals are not normalized, 1.0 is in all cases the biggest coefficient.

Table 2
Equilibrium radius (Å), cohesive energy[a] (eV) and A_n factor for the B_{2k} and B^+_{2k+1} spheroidal cages studied in this work

n	R_e	E_b	A_n
30	2.61	3.03	0.39956
31	2.60	3.10	0.40119
32	2.63	3.06	0.40260
33	2.73	3.05	0.40423
34	2.79	3.04	0.40564
35	2.80	3.06	0.40700
36	2.83	3.00	0.40828
37	2.84	3.10	0.40952
38	2.88	3.12	0.41071
39	2.95	3.09	0.41187
40	2.94	3.14	0.41296
41	3.01	3.18	0.41404
42	3.04	3.14	0.41510
43	3.06	3.25	0.41601
44	3.09	3.17	0.41694
45	3.13	3.25	0.41788
46	3.20	3.16	0.41882
47	3.23	3.24	0.41969
48	3.27	3.09	0.42045
49	3.32	3.23	0.42133
50	3.36	3.14	0.42208
51	3.34	3.15	0.42292
52	3.41	3.14	0.42361
53	3.48	3.16	0.42438
54	3.46	3.11	0.42506

[a] $E(B^+) = -24.16352$ hartree; $E(B) = -24.47151$ hartree.

case of homolytic dissociations, leading to very large errors in such limits.

From the values of A_n listed in Table 2 a simple fitting gives the function

$$A_n = \frac{1}{2} - \frac{0.5505}{\sqrt{n}}, \tag{4}$$

which is valid for n ranging from 30 to 54 with an error smaller than 0.00005. This expression (4) could be of some help in the interpretation of numerical results and in development of simple models for general spherical clusters. Works along these lines are currently in progress [10].

In Fig. 1 we show instead the plot of equilibrium radii obtained from ab initio calculations against the square root of n. Looking at this figure a linear dependence can be supposed; in such case the numeral surface density

$$\sigma = \frac{n}{4\pi R_e^2}, \tag{5}$$

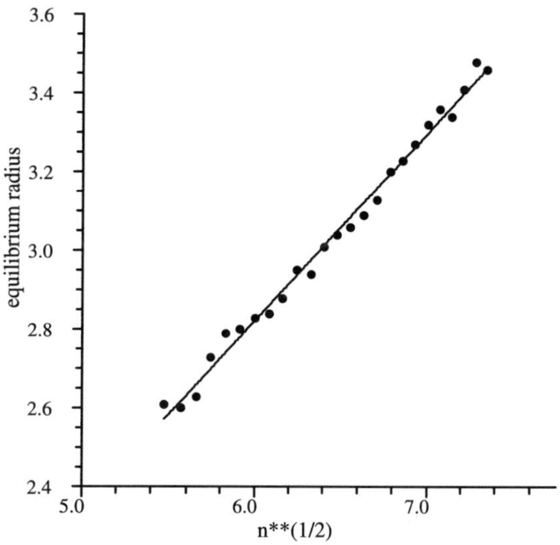

Fig. 1. Plot of the equilibrium radius (Å) of the spheroidal boron cages studied in this work against $\sqrt{n}$. The filled circles refer to ab initio computed values.

is constant for all spherical boron clusters. From data of Table 2 the linear fitting gives the relation

$$R_e = 0.471\sqrt{n}, \tag{6}$$

where the radius is given in Å. This condition allows the average distance between the neighbour atoms to be close to the best boron–boron bond length in all cases. From published calculations on various boron clusters [7] the optimal B–B length is found to be 1.65 Å. In Fig. 2 we show the distribution of the distances between neighbour atoms in the 25 clusters studied in this work; all distances fall in the range 1.50–2.25 Å with a peak at 1.65 Å. The average B–B length is instead 1.72 Å with a standard deviation 0.13 Å. A further analysis made on coordination numbers has shown that coordination six augments with the size of the cluster. The variation of the number of penta- and hexa-coordinate boron atoms with n is plotted in Fig. 3.

Finally the plot of the cohesive energy against n for the clusters B_{2k} and B^+_{2k+1} is shown in Fig. 4. The most stable cluster appears to be B^+_{45} which is schematically viewed in Fig. 5. In this figure it is easy to recognize the characteristic pentagonal and hexagonal subunits.

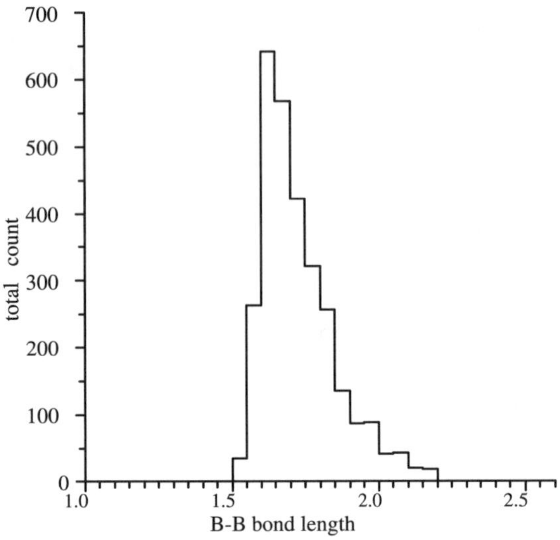

Fig. 2. Shows the distribution of the B–B bond lengths for all the 25 selected spherical cages of this work.

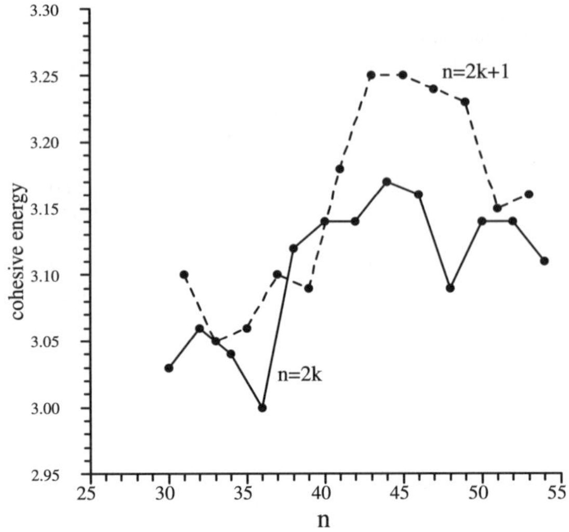

Fig. 4. Plot of the cohesive energy (eV) against the number of atoms for the clusters B_{2k} and B^{+}_{2k+1} studied in this work.

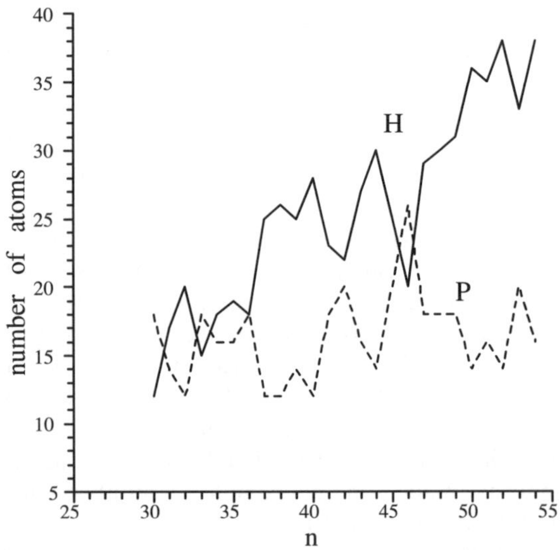

Fig. 3. Variation of the number of pentacoordinate (P) and hexacoordinate (H) atoms with the size of the spherical cages of this work.

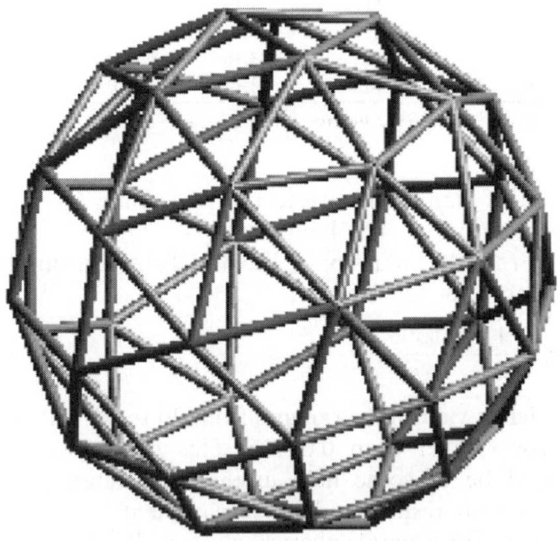

Fig. 5. Schematic view of the B^{+}_{45} boron cage.

The cohesive energy of the solid boron is about 6 eV [11] while most of the Hartree–Fock calculations, performed using the 3-21G basis set on various clusters with different sizes and shapes, gave values ranging from 3 to 3.5 eV [7–9], in substantial agreement with our results of Table 2. The more stable clusters for large n are the tubular isomers [12,13] while in the literature the spherical clusters treated at ab initio level with n ranging from 30 to 54 are B_{32} and B_{42}. B_{32} has been studied in various quasi-spherical structures with the 3-21G basis set [7]; the more stable configuration found was a D_{4h} geometry composed of 12 not regular hexagonal faces and six square faces, with bond length values 1.66 and 1.72 Å, average radius 3.00 Å and cohesive energy 3.10 eV. B_{42}, instead, has been studied resorting to a STO-3G basis set and to a D_{5h} geometry [9]; the bond lengths resulted in the region 1.68–1.72 Å, the radius about 2.99 Å and the cohesive energy 4.56 eV. Here it is very important to remark that the STO-3G basis set overestimates significantly the cohesive energies [9].

The reader must remind that the geometries in the present Letter come from a model and, consequently, they do not correspond necessarily to local minima of the real potential energy surface. Nevertheless, the resulting binding energies tell us that the present approach is an interesting tool for the study of boron spherical cages.

A final remark concerns the effect of electron correlation. Looking at the DFT results of Boustani on small boron clusters [8], one expects higher cohesive energies. On the other hand, correlation lowers $E_n(R_e)$ in Eq. (2) while $E_n(\infty)$ defined by Eq. (3) is already a good estimate to the energy at dissociation. We do not expect other significant effects on the results presented in this work apart from a small change in the equilibrium radii. Of course electron correlation could be important in determining the relative stability of different isomers but this aspect goes outside the aim of the present Letter.

The study of correlation in spherical cages is actually under our attention and it will be considered in future works.

3. Conclusion

In this Letter we have presented a new strategy to define reliable geometries for spherical boron cages without resorting to complicated and heavy ab initio analyses. The method reflects the Aufbau principle proposed for boron clusters but it is not restricted to a series of magic numbers. It is always possible, in fact, to recover a sphere with distorted pentagons and hexagons not necessarily displaced in an ordered manner. This situation coincides with the projection, over the surface of the sphere, of the faces of a polyhedron generated by a system of equal charges, with the motion constrained over the sphere surface itself, in their more stable configuration. Such geometries can be easily generated by a simple Monte Carlo sampling in few minutes for a given numbers n of boron atoms. The radius, R_e, of the sphere is finally found by imposing a constant numeral surface density $n/4\pi R_e^2$.

It is important for the future to investigate the effect of electron correlation in the B_n clusters studied here. The recent study of Flocke, Schmalz and Klein [14] is noteworthy here. These authors investigated the effect of electron correlation specifically in C_{60}, and not only for the ground state but also in excited states, using the Hubbard model to include correlation. Of course, Møller–Plesset perturbation theory would provide an alternative route, considering that Hartree–Fock solutions underly the discussion of the present approach.

Acknowledgements

N.H.M. wishes to acknowledge financial support during the course of this work from the Francqui Foundation, Brussels, Belgium and especially the encouragement and help given by the Executive Director, Professor L. Eyckmans.

References

[1] N.H. March, Proc. Cambridge Philos. Soc. 48 (1952) 665.
[2] D.P. Clougherty, X. Zhu, Phys. Rev. A 56 (1997) 632.
[3] F. Despa, Phys. Rev. B 57 (1998) 7335.
[4] F. Siringo, G. Piccitto, R. Pucci, Phys. Rev. A 46 (1992) 4048.
[5] S.J. La Placa, P.A. Roland, J. Wynne, Chem. Phys. Lett. 190 (1992) 163.
[6] I. Boustani, Phys. Rev. B 55 (1997) 1.
[7] I. Boustani, A. Rubio, J.A. Alonso, Chem. Phys. Lett. 311 (1999) 21.
[8] I. Boustani, Chem. Phys. Lett. 240 (1995) 135.
[9] I. Boustani, J. Solid State Chem. 133 (1997) 182.
[10] I.A. Howard, D.J. Klein, C. Amovilli, N.H. March, in preparation.

[11] R.C. West (Ed.), CRC Handbook of Chemistry and Physics, CRC Press, Boca Raton, FL, 1988.
[12] I. Boustani, A. Quandt, A. Rubio, J. Solid State Chem. 154 (2000) 269.
[13] I. Boustani, A. Quandt, E. Hernández, A. Rubio, J. Chem. Phys. 110 (1999) 3176.
[14] N. Flocke, T.G. Schmalz, D.J. Klein, J. Chem. Phys. 112 (2000) 8233.

PHYSICAL REVIEW A **66**, 013210 (2002)

Dependence of the π-electron eigenvalue sum on the number of atoms in almost spherical C cages

C. Amovilli,[1] I. A. Howard,[2] D. J. Klein,[3] and N. H. March[2,4]
[1]*Dipartimento di Chimica e Chimica Industriale, Università di Pisa, Via Risorgimento 35, 56126 Pisa, Italy*
[2]*Department of Physics, University of Antwerp (RUCA), Groenenborgerlaan 171, B-2020 Antwerp, Belgium*
[3]*Department of Marine Sciences, Texas A&M University, Galveston, Texas 77553*
[4]*Oxford University, Oxford, England*
(Received 6 March 2002; published 30 July 2002)

A free-electron-like model is first shown to lead to the eigenvalue sum of the π electrons in almost spherical C cages, being proportional to N, where N is the number of C atoms. The model is also closely related to a tight-binding model, stemming from the Hückel theory of π electrons. The equilibrium radius R_e of the C cages is assumed to be determined by the "rule" of constant surface area per C atom, and hence $R_e \propto N^{1/2}$. We then compare these models with a series of Hartree-Fock calculations on the fullerenes C_{50}, C_{60}, C_{70}, and C_{84}, from which more accurate energy scaling relations are obtained. Finally, the relation between the total energy of the four C cages at equilibrium and the all-electron eigenvalue sum is confirmed by the Hartree-Fock studies.

DOI: 10.1103/PhysRevA.66.013210 PACS number(s): 36.40.Cg, 73.61.Wp, 71.20.Tx, 61.48.+c

I. BACKGROUND AND OUTLINE

We have been independently examining quite different models for treating the π electrons in buckminsterfullerene and related almost spherical C cages [1,2]. In the first of these references, the so-called March model of fullerene [3–5] was employed; however, it was employed in its simplest form, in which self-consistency was not imposed. Therefore, here we shall turn to the second model of a quite different kind, namely, that of free π electrons confined to the surface of a sphere [6–10].

The π levels, in zeroth order, are those of a rigid rotator, having degeneracy $(2j+1)$ and energies $\hbar^2 j(j+1)/2I$ with I the moment of inertia. With double occupancy up through the nth shell, one then finds a total energy proportional to

$$2\sum_{j=0}^{n} j(j+1)(2j+1)=n(n+1)^2(n+2), \tag{1}$$

which is easily confirmed in the limit of large N by replacing the summations by integration. With the identification of I as mR^2,

$$E_\pi=\frac{n(n+1)^2(n+2)\hbar^2}{2mR^2}. \tag{2}$$

Below, we shall insert an (approximate) equilibrium cage radius [see Eqs. (4) and (5) below] and, recognizing the total number of electrons as $2\Sigma_{j=0}^{n}(2j+1)=2(n+1)^2$, we find

$$E_\pi=\frac{\hbar^2 N}{8mk_e{}^2}-\frac{\hbar^2}{4mk_e{}^2}, \tag{3}$$

which is plainly the correct general form for a graphene sheet. A similar lead N dependence is also obtained for the simple Hückel tight-binding model [11], wherein there is a resonance integral β between each site and its three nearest neighbors. Indeed, by the Gerschgorn circle theorem, all the MO eigenvalues must be bound between $-3|\beta|$ and $+3|\beta|$ so that the total N electron ground-state energy must be bounded between $-3|\beta|N$ and $+3|\beta|N$. Moreover, the MO levels are anticipated to be not too nonuniformly distributed so that the ground state is $\cong -CN$ (with $C\neq 0$ and probably not too different from $3|\beta|/2$, as indicated in Ref. [12]). To reach Eq. (3), we have used an "equilibrium" radius R_e determined by the "rule" that the surface area per C atom remains constant as the cage size becomes larger through an increase in the number of C atoms N. This means that

$$\frac{4\pi R_e^2}{N}=\text{const} \tag{4}$$

or

$$R_e=k_e N^{1/2}, \tag{5}$$

where k_e is a constant, which can evidently be fitted to the experimental value of R_e for buckminsterfullerene itself with $N=60$ (see also Table I below). Substituting Eq. (5) into Eq. (2) then yields Eq. (3) above.

An alternative argument can be given from two-dimensional Thomas-Fermi theory. Suppose the π electrons are modeled as a uniform distribution spread over the almost spherical cage. Then the kinetic energy T is given by [13]

TABLE I. Ground-state energy E and sum of eigenvalues for π electrons E_π in atomic units, and equilibrium radii R_e in Å for the four cages for which Hartree-Fock data has been obtained. N is the number of C atoms in the cage.

N	50	60	70	84
E/N	−37.5884	−37.6029	−37.5926	−37.5874
E_π/N	−0.2232	−0.2275	−0.2234	−0.2223
R_e	3.278	3.582	3.865	4.256
$R_e/\sqrt{N}$	0.463	0.462	0.462	0.464

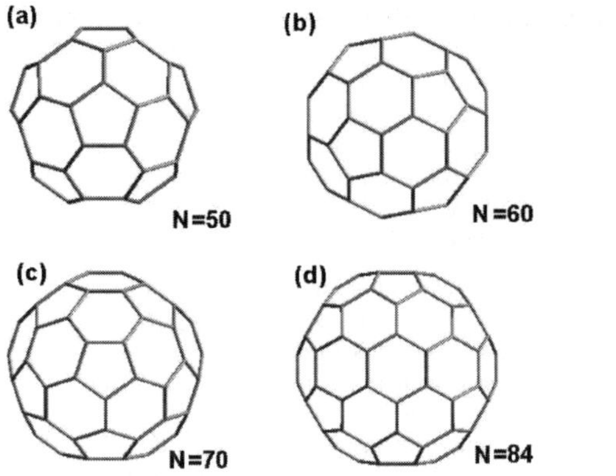

FIG. 1. Structure of the buckminsterfullerene (b), and also the (assumed) high symmetry structures of C_{50} (a), C_{70} (c), and C_{84} (d) used here.

$$T = A c_k^{(2)} n^2, \quad c_k^{(2)} = \frac{\pi \hbar^2}{2m}, \tag{6}$$

where A is the area and n is the number of π electrons per unit area, i.e.,

$$n = \frac{N}{4\pi R^2}. \tag{7}$$

Thus, the total kinetic energy is given by

$$T = (4\pi R^2) \frac{c_k^{(2)} N^2}{(4\pi R^2)^2} = \frac{c_k^{(2)} N^2}{4\pi R^2} = \frac{\hbar^2 N^2}{8 m R^2}, \tag{8}$$

and by again using Eq. (5), we find $T \propto N$. Since in the "rigid-rotator-like" model all the energy is kinetic, we confirm Eq. (3) to leading order by the simplest form of density functional theory.

This is the point at which we shall bring these simple model predictions into contact with quantitative Hartree-Fock (HF) calculations on four C cages, namely, C_{50}, C_{60}, C_{70}, and C_{84}. This is done in Sec. II below. Section III presents a relationship between total energy E and eigenvalue sum E_s. Section IV constitutes a summary with some suggestions for possible future work.

II. HARTREE-FOCK STRUCTURES OF SOME C CAGES

A. Structures taken for four C cages and computational details

For C_{60} it is, of course, well established that the lowest isomer is the European football configuration. For C_{84}, we have used the work of Manolopoulos and Fowler [14], in which a whole variety of isomers were classified. The one we have chosen (albeit with some inevitable arbitrariness) is that of the highest symmetry. The nuclear framework is depicted in Fig. 1(d). For C_{70}, similar symmetry considerations led us to study the nuclear structure in Fig. 1(c), while the framework adopted for C_{50} is depicted in Fig. 1(a) (see also Schmalz *et al.* [15]).

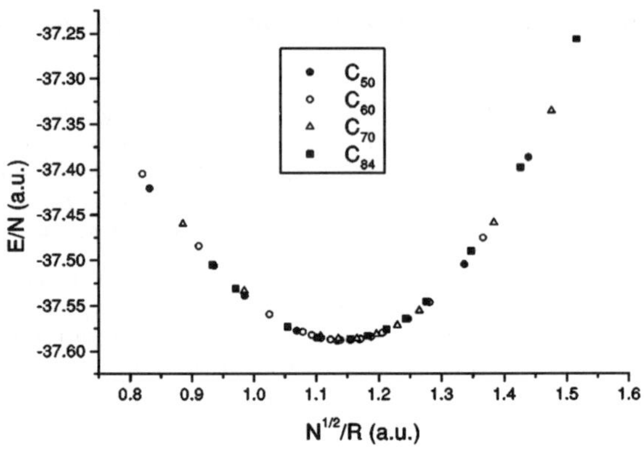

FIG. 2. Total HF energy E/N plotted vs $N^{1/2}/R$ for all four cages. The curves for the individual cages have been shifted by a constant to bring them into coincidence.

As a first step toward understanding overall structural effects of fullerene cages, we report Hartree-Fock energetics of these four C cages constrained to a spherical shape—i.e., with all the nuclei at a common distance R from the center. The Hartree-Fock energies have been obtained using the MINI-I basis set. We discuss our findings below. While our Hartree-Fock calculations involve all electrons, they will yield further insight into this field of C clusters to also extract information directly concerned with the π electrons.

B. Scaling of total energy curves for the four C cages

It turns out that the Hartree-Fock total energy $E(N,R)$, where N denotes the number of C atoms and R denotes the cage radius, can be well represented by the form

$$\frac{E(N,R)}{N} = a + b\frac{N^{1/2}}{R} + c\frac{N}{R^2}, \tag{9}$$

as shown in Fig. 2 for the four cages. In this figure, the four curves have been shifted by a small constant to bring them into coincidence. At the equilibrium radii, precise numerical values of E/N for the four cages are collected in Table I. Naturally, there is chemistry in the small differences between the total energy per carbon atom E/N versus N in the higher decimal places of the HF results. The "law" of constant surface area per C atom for the equilibrium radius embodied in Eq. (4) is already clear from Fig. 2 and is confirmed in the final row of Table I. The sum of occupied orbital energies E_s is plotted in Fig. 3, where it is shown that there is remarkable scaling of the four cages; the solid curve corresponds to the same Eq. (9) with parameters recorded in Fig. 3.

C. π-electron energies

Next, we have extracted the π-eigenvalue sum for the C cages. The reader will note that, owing to the curvature, in spherical clusters it is not possible to immediately identify the π orbitals among the canonical HF orbitals. Our π-eigenvalue sum, for a given C_n cage, corresponds to the

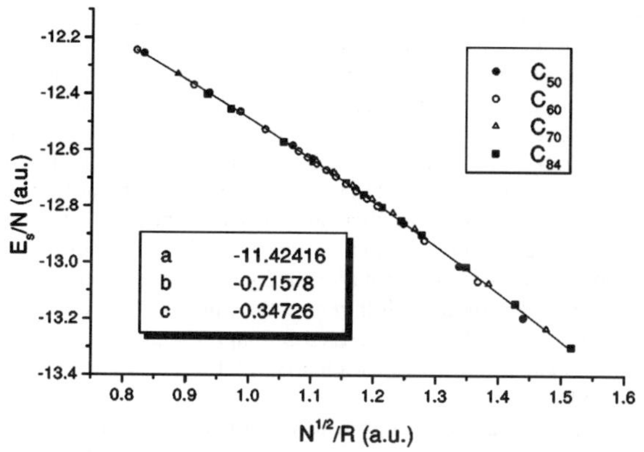

FIG. 3. Eigenvalue sums E_s/N plotted vs $N^{1/2}/R$ for C_{50}, C_{60}, C_{70}, and C_{84}. Values for the parameters a, b, and c representing the best fit to Eq. (9) for the composite curve are recorded. The curves have been shifted by a constant as in Fig. 2.

sum over the n canonical orbitals, which show the highest projection over a set of n p-like atomic functions orthogonal to the spherical cage surface. The four values of E_π/N at equilibrium are also recorded in Table I and show remarkable constancy. Presumably, the constant number already reflects the value anticipated to hold for a single graphene sheet, although, of course, the largest value of N in Table I is only 84. Figure 4 shows the π-electron eigenvalue sum for the four cages.

III. RELATION BETWEEN TOTAL ENERGY AND EIGENVALUE SUM OF ALL ELECTRONS: HF RESULTS AT THE EQUILIBRIUM CAGE RADII

To conclude this discussion of the HF results, we have examined the relation between total energy $E(N,R_e)$ and eigenvalue sum $E_s(N,R_e)$ (now of all electrons) for the four C cages on which we have HF data. The motivation for this goes back to March and Plaskett's [16] study on atoms. They

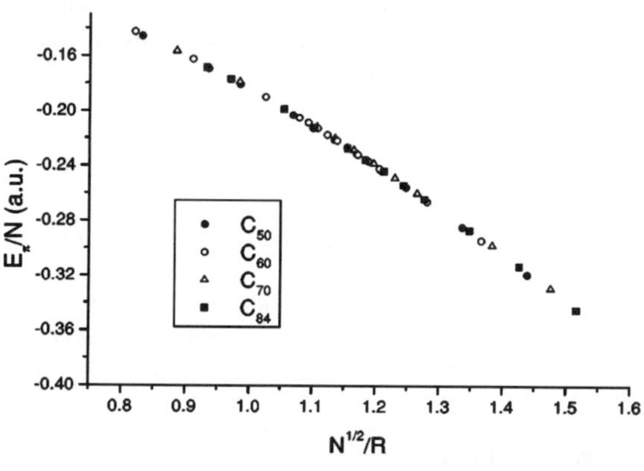

FIG. 4. π-electron eigenvalue sums E_π/N vs $N^{1/2}/R$ for C_{50}, C_{60}, C_{70}, and C_{84}. The curves have been shifted by a constant as in Fig. 2.

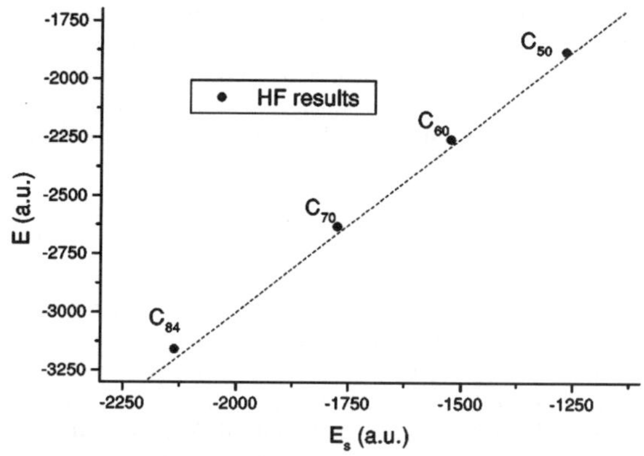

FIG. 5. Total HF energy E vs eigenvalue sum E_s at the equilibrium radius R_e for C_{50}, C_{60}, C_{70}, and C_{84}. The dashed line represents $E=(3/2)E_s$.

showed, using the simplest form of density functional theory, namely, the Thomas-Fermi statistical method, that

$$E=\tfrac{3}{2}E_s. \tag{10}$$

This relation was later shown by Ruedenberg [17] (see also March [18]) to be well obeyed by self-consistent calculations on a variety of molecules at their equilibrium geometry. Therefore, in Fig. 5 we have plotted $E(N,R_e)$ versus the all-electron eigenvalue sum $E_s(N,R_e)$. There is indeed a linear relation, and the dashed line is drawn with slope 3/2 for comparison. The fit is seen to be quantitative.

IV. SUMMARY AND POSSIBLE DIRECTIONS FOR FURTHER STUDY

The main achievements of this study of spherical cages can be summarized as follows. First, the total energy, the eigenvalue sum, and the π-electron eigenvalue sum as a function of N and R can be fitted accurately by the form (9). As seen from Fig. 3, for the total eigenvalue sum E_s there is quite remarkably accurate scaling relating all four cages. We believe this scaling should be robust against later changes in isomer structures for C_{50}, C_{70}, and C_{84} when the lowest energy structure is experimentally established. Second, we have extracted the π-electron eigenvalue sum at the equilibrium cage radii (see Table I). These, in the form of E_π/N, are already remarkably constant and, even though N is small, we expect constancy to have already been closely approached. Third, the relation (10) between the total energy and the eigenvalue sum, for all electrons, calculated at the equilibrium radii, is demonstrated to hold almost quantitatively for the four cages studied.

Typically, there will be notable anisotropic curvature for fullerenes at equilibrium geometry. To see this, first consider a best candidate for uniform curvature: this is a fullerene with the pentagonal "defects" spread as uniformly as possible, that is, an icosahedral-symmetry fullerene. For such a fullerene (which we imagine for the large-N limit), the pentagons may be viewed to be at the corners of an icosahedronlike superstructure, with all other rings being hexagons. Then consider a graphitic region corresponding to one of the triangular faces of the icosahedronlike superstructure. The area of this face (there being 20 such faces), as measured on the surface of a (uniformly curved) spherical surface of radius R, is

$$A_{\mathrm{curved}}=\frac{4\pi R^2}{20}. \tag{11}$$

The length of the side of such a spherical triangle is $l=\theta R$, where θ is the angle subtended by the edge as viewed from the center of the sphere. The area of a planar triangle with the same length edges is $A_{\mathrm{flat}}=(l/2)(\sqrt{3/4}l)$. But A_{flat} (as would correspond to an unstrained section of the graphitic lattice) and A_{curved} do not agree; there is a strain per carbon atom of

$$\sigma\sim\sqrt{A_{\mathrm{flat}}/(N/20)}-\sqrt{A_{\mathrm{curved}}/(N/20)}. \tag{12}$$

This strain occurs for each bond in the graphitic portion of the network. So with about $3N/2$ such bonds, the total stress is $E_{\mathrm{stress}}\approx(3N/2)1/2k\sigma^2$ with k as some suitable force constant (for C—C aromatic bonds). Thus,

$$E_{\mathrm{stress}}\sim 15k\left(\frac{3^{1/4}}{2}\theta-\sqrt{\frac{\pi}{5}}\right)^2 R^2. \tag{13}$$

But $R^2\propto N$, so that stress is quite important, and can even give a correction beyond that of bulk graphite if unrelieved, as with the spherically curved surface. If a structure other than that of icosahedral symmetry for the arrangement of the pentagons is presumed, then the geometric factor above is modified, but still nonzero.

The question of relief of curvature strain then arises (still in the limit of large N). Again as a representative example, the icosahedral symmetry arrangement may be considered. One simply imagines that the structure deforms to look like an icosahedron superstructure with the pentagons at the apices. That is, the triangular graphitic regions of the preceding paragraph are imagined to be changed from the spherically curved form of the preceding paragraph to near planar regions with anisotropic curvature at the edges connecting each triangular region to adjacent ones. Then, within the planar triangular regions there is no strain and no stress; instead, it all occurs at the edges of the triangles. Clearly, the total amount of such edge is $\approx 30R\sim N^{1/2}$ so that now we have a structure with less stress, $E'_{\mathrm{stress}}\sim N^{1/2}$. Again, the proportionality depends on the particular manner of dispersal of the pentagons, the overall conclusion being that geometric Gaussian curvature is [2] preferably localized in the region of the pentagons with then anisotropic curvature (without Gaussian curvature) mediating between such neighboring regions. Notably for computations [19] on large non-open-shell icosahedral-symmetry fullerenes, the suggested polyhedralization can be clearly perceived in diagrams of the geometry optimized structures (most clearly for the largest treated, at $N=540$).

We have already referred to the urgent need to settle the question of the lowest isomer structures for C_{50}, C_{70}, and C_{84}. This presently appears to be a task for experimentation (see [20]). Current theoretical approaches lack the power to decisively distinguish between isomer structures that will differ very little in energy. A final comment concerns connectivity theory. Beyond the constancy of the values of E_π/N in Table I, we conclude that a correction term of $\mathcal{O}(1/\sqrt{N})$ could exist. However, using the HF values, the behavior is not monotonic. For several approximations within the connectivity (or Hückel) theory, a term of $\mathcal{O}(1/\sqrt{N})$ is present in the study by Gutman and Soldatović [21], though their approximations are gauged against a set of (~100) benzenoids (without pentagonal rings).

ACKNOWLEDGMENTS

C.A. wishes to acknowledge the hospitality at the RUCA campus of the University of Antwerp during the period in which this investigation was completed. I.A.H. wishes to acknowledge support from the Flemish Science Foundation (FWO) under Grant No. G.0347.97. We also thank the University of Antwerp (RUCA) for its support in the framework of the Visiting Professors Program. This work is also supported by the Concerted Action Program of the University of Antwerp. N.H.M. acknowledges financial support from the Francqui Foundation. Thanks are especially due to the Executive Director, Professor L. Eyckmans, for his help and encouragement. D.J.K. acknowledges the support of the Welch Foundation of Houston, Texas.

[1] I. A. Howard and N. H. March, in Reviews in Modern Quantum Chemistry: A Celebration of the Contributions of R. G. Parr, edited by K. D. Sen (World Scientific, Singapore, in press).
[2] D. J. Klein and H.-Y. Zhu, in *From Chemical Topology to Three-Dimensional Topology*, edited by A. T. Balaban (Plenum, New York, 1997), pp. 297–341.
[3] D. P. Clougherty and X. Zhu, Phys. Rev. A **56**, 632 (1997).
[4] F. Despa, Phys. Rev. B **57**, 7335 (1998).
[5] F. Siringo, G. Piccitto, and R. Pucci, Phys. Rev. A **46**, 4048 (1992).
[6] R. C. Haddon, L. E. Brus, and K. Raghavachari, Chem. Phys. Lett. **125**, 459 (1985).
[7] M. Ozaki and A. Takahashi, Chem. Phys. Lett. **127**, 242 (1986).
[8] E. Brensdal, S. J. Cyrin, B. N. Cyrin, J. Brunvoll, D. J. Klein, and W. A. Seitz, in *Quasicrystals, Networks, and Molecules of Five-Fold Symmetry*, edited by I. Hargittai (VCH, New York, 1990), pp. 257–276.
[9] R. Saito, G. Dresselhaus, and M. S. Dresselhaus, Phys. Rev. B **46**, 9906 (1992).
[10] M. R. Savina, L. L. Lohr, and A. H. Francis, Chem. Phys. Lett. **205**, 200 (1993).
[11] See, for instance, C. A. Coulson, R. B. Mallion, and B.

O'Leary, *Hückel Theory for Organic Chemists* (Academic, London, 1978).
[12] I. Gutman, A. Nikolić, and Z. Tomović, Chem. Phys. Lett. **349**, 95 (2001).
[13] N. H. March, *Electron Density Theory for Atoms and Molecules* (Academic, New York, 1992), p. 173.
[14] D. E. Manolopoulos and P. W. Fowler, J. Chem. Phys. **96**, 7603 (1992).
[15] T. G. Schmalz, W. A. Seitz, D. J. Klein, and G. E. Hite, J. Am. Chem. Soc. **110**, 1113 (1988).
[16] N. H. March and J. S. Plaskett, Proc. R. Soc. London, Ser. A **235**, 419 (1956).
[17] K. Ruedenberg, J. Chem. Phys. **66**, 375 (1977).
[18] See also N. H. March, J. Chem. Phys. **67**, 4618 (1977).
[19] D. Bakowies and W. Thiel, J. Am. Chem. Soc. **113**, 3704 (1991).
[20] S. J. LaPlaca, P. A. Roland, and J. J. Wynne, Chem. Phys. Lett. **190**, 163 (1992).
[21] I. Gutman and T. Soldatović, Commun. Math. Chem. **44**, 169 (2001).

Nonlocal Energy Density Functionals: Models Plus Some Exact General Results

N. H. MARCH
Oxford University, Oxford, England, Department of Physics, University of Antwerp (RUCA), Antwerp, Belgium, and The Abdus Salam International Centre for Theoretical Physics, Trieste, Italy

Received 19 February 2001; accepted 28 February 2001
Published online 6 January 2003 in Wiley InterScience (www.interscience.wiley.com).
DOI 10.1002/qua.10130

ABSTRACT: Holas and March (Phys Rev 1995 A51, 2040) gave a formally exact expression for the force $-\partial V_{xc}(\mathbf{r})/\partial \mathbf{r}$ associated with the exchange-correlation potential $V_{xc}(\mathbf{r})$ of density functional theory. This forged a precise link between first- and second-order density matrices and $V_{xc}(\mathbf{r})$. Here models are presented in which these low-order matrices can be related to the ground-state electron density. This allows nonlocal energy density functionals to be constructed within the framework of such models. Finally, results emerging from these models have led to the derivation of some exact "nuclear cusp" relations for exchange and correlation energy densities in molecules, clusters, and condensed phases. Int J Quantum Chem 92: 1–10, 2003

Key words: nonlocal energy density functionals; exchange energy density; nuclear cusp relations

1. Background and Outline

In their early attempts to transcend the Thomas–Fermi (TF) approximation, March and Young (MY) [1] recognized that electron density theory should be regarded as a branch of density matrix theory. Their early studies led to a single-particle kinetic energy functional $T_s[\rho]$ of the form

$$T_s[\rho] = T_W[\rho] + \Gamma(N)T_{TF}[\rho], \qquad (1.1)$$

where $T_{TF}[\rho]$ and $T_W[\rho]$ were the TF kinetic energy and von Weizsäcker (W) inhomogeneity kinetic energy, respectively. In particular, the formula of MY for $\Gamma(N)$ led to this TF contribution correctly vanishing in the case of a single-level occupancy only. Subsequently, Herring and coworkers [2, 3] gave stronger arguments as to why in $T_s[\rho]$ it was physically appropriate to subtract out the full W contribution $T_W[\rho]$ in constructing the single-particle kinetic energy functional $T_s[\rho]$.

Of course, the most common route to implement density functional theory (DFT) has been since Slater's work [4] to return to orbitals from which to obtain the single-particle kinetic energy by stan-

This article was presented at the 41st Annual Sanibel Symposium, February 24–March 2, 2001.

dard quantum mechanics. The only question remaining then concerns the one-body potential $V(\mathbf{r})$, now including exchange and correlation interactions, to insert into the Schrödinger equation (the so-called Slater–Kohn–Sham (SKS) equations [4, 5]). However, within the last decade a major effort has been focussed on orbital-free DFT [6–10]. This search for (presently approximations to) the single-particle kinetic energy functional $T_s[\rho]$ is then, of course, additional to the construction of $V(\mathbf{r})$, given formally by

$$V(\mathbf{r}) = V_{\text{Hartree}}(\mathbf{r}) + V_{xc}(\mathbf{r}). \tag{1.2}$$

One direction of progress on the force $-\partial V_{xc}/\partial \mathbf{r}$ associated with $V_{xc}(\mathbf{r})$ was pointed out by Holas and March [11, 12], who showed, via a many-electron form of the differential virial theorem [13], that the force $-\partial V_{xc}(\mathbf{r})/\partial \mathbf{r}$ could be written, explicitly, in terms of first- and second-order density matrices (1 and 2DMs) of the many-electron system under discussion, plus, of course, information contained in the SKS determinant of the "reference" system with the same ground-state density as the correlated electron assembly under discussion. The problem of finding $V_{xc}(\mathbf{r})$ is central, whether one works with SKS orbitals to obtain T_s or with the single-particle kinetic energy functional $T_s[\rho]$ in orbital-free theory because the Euler equation of DFT reads [14]

$$\mu = \frac{\delta T_s}{\delta \rho(\mathbf{r})} + V_{\text{Hartree}}(\mathbf{r}) + V_{xc}(\mathbf{r}), \tag{1.3}$$

where μ is the chemical potential (constant throughout the entire charge cloud) of the molecule, cluster, or condensed phase being considered.

The outline of the present work is as follows. In Section 2, a 1-D model of harmonic confinement of independent fermions will be utilised to expose a nonlocal form of the single-particle kinetic energy functional $T_s[\rho]$ that is exact in this model [15]. Although correlation kinetic energy is "moved" into the exchange-correlation potential $V_{xc}(\mathbf{r})$ in writing the chemical potential Eq. (1.3), it will be convenient for presentation to close Section 2 with a short discussion of an explicit correlation kinetic energy functional, $T_c[\rho]$ say, for two-electron He-like atomic ions in the nonrelativistic limit of the large atomic number Z. This use of the so-called $1/Z$ expansion in atomic theory, related to the TF statistical theory explicitly by March and White [16], leads naturally into Section 3, where the focus is on the large Z limit of atomic ions in which both K and L shells are now closed, i.e., one has moved from He-like ions in Section 2 to Ne-like systems. In the limit of the large atomic number Z, the exchange energy density $\epsilon_x(r)$ [defined precisely in Eq. (3.5)] has been calculated exactly in the recent work of Howard et al. [17]. A nuclear cusp condition is already exposed by the result for $\epsilon_x(r)$ in this limit system. Section 4 is devoted to the demonstration that this limiting result for ϵ_x is exactly true, not only in atomic ions of an arbitrary atomic number but also for multicentre problems of molecular and solid-state theory [18]. The same nuclear cusp condition is shown to embrace the correlation energy, also in Section 4. After a brief summary in Section 5, some proposals are made for future studies.

2. Nonlocal Kinetic Energy Density Functional

In this section, some recent results on a model in which a nonlocal kinetic energy functional can be obtained will first be summarized. However, this is a model of independent Fermions, so it is only relevant to the single-particle kinetic energy functional $T_s[\rho]$. At the end of this section, some results will also be reported on correlation kinetic energy, but now in the restricted case of He-like ions having a large atomic number Z. Throughout the article, the considerations are restricted to nonrelativistic quantum mechanics.

2.1. ARBITRARY NUMBER OF FERMIONS IN A 1-D HARMONIC OSCILLATOR POTENTIAL

In early work, Lawes and March [19] studied the admittedly specialized model of N independent Fermions filling singly the levels of the linear harmonic oscillator potential $V(x) = \frac{1}{2}x^2$. For this model, they already achieved a long-term aim of DFT in that they derived an exact differential equation for the ground-state density $\rho(x)$, namely

$$\frac{\rho'''(x)}{8} + \frac{1}{2}V'(x)\rho(x) + (N - V(x))\rho'(x) = 0. \tag{2.1}$$

The lowest level only being occupied here corresponds to putting $N = 1$ in Eq. (2.1).

Still in one dimension, March and Young [13] earlier derived the so-called differential virial theorem. Defining the (single-particle) kinetic energy per unit length $t_s(x)$ from the one-body wave functions $\psi_i(x)$ generated by the potential energy $V(x)$ as

$$t_s(x) = -\frac{1}{2}\sum_{i=1}^{N}\psi_i^*(x)\frac{\partial^2\psi_i(x)}{\partial x^2}, \tag{2.2}$$

these authors [13] showed, but now for a general potential $V(x)$ that allowed normalisation of $\rho(x)$ to N particles, that

$$t_s'(x) = -\frac{1}{2}\rho V'(x) - \frac{\rho'''(x)}{8}. \tag{2.3}$$

Applying this general 1-D result (2.3) to the above model of harmonic confinement, one can immediately rewrite Eq. (2.1) as [15]

$$\frac{t_s'(x)}{\rho'(x)} = \left(N - \frac{1}{2}x^2\right). \tag{2.4}$$

Differentiating Eq. (2.4) with respect to x yields

$$\left[\frac{t_s'(x)}{\rho'(x)}\right]' = -x \equiv -\frac{\partial V}{\partial x} \tag{2.5}$$

and substituting this harmonic oscillator result for $\partial V/\partial x$ into Eq. (2.3) gives

$$t_s'(x) = +\frac{1}{2}\rho\left[\frac{t_s'(x)}{\rho'(x)}\right]' - \frac{1}{8}\rho'''(x). \tag{2.6}$$

In Ref. [15], Eq. (2.6) is integrated to obtain, after some manipulation,

$$t_s(x) = -\frac{\rho''(x)}{12} + \left[\frac{N}{3\rho^2(0)} + \frac{1}{12}\int_0^x ds\,\frac{\rho'''(s)}{\rho^2(s)}\right]\rho^3(x).^* \tag{2.7}$$

Integrating further, Eq. (2.7) can be expressed as

$$t_s(x) = t_W(x) + \xi(x)t_{TF}(x), \tag{2.8}$$

where the TF and W kinetic energy densities [see also Eq. (1.1)] are defined in one dimension as

$$t_{TF}(x) = c_k\rho^3(x) : c_k = \pi^2/24 \tag{2.9}$$

and

$$t_W(x) = \frac{1}{8}\frac{\rho'^2(x)}{\rho(x)}. \tag{2.10}$$

What emerges is that, while the form (2.8) is not, of itself, new [20–22], $\xi(x)$ in this model can be obtained exactly as

$$\xi(x) = \xi(0) + \frac{4}{3}\int_0^x dx\left[\frac{t_W'(x)}{t_{TF}(x)} - \left(\frac{t_W(x)}{t_{TF}(x)}\right)'\right]. \tag{2.11}$$

Thus, $\xi(x) - \xi(0)$ can itself be expressed solely in terms of $t_{TF}(x)$ and $t_W(x)$ as building blocks. The quantity $\xi(0)$ is evidently from Eq. (2.8) given by

$$\xi(0) = \frac{t_s(0) - t_W(0)}{t_{TF}(0)} \tag{2.12}$$

and $t_s(0)$ from Eq. (2.7) has the value

$$t_s(0) = \frac{N\rho(0)}{3} - \frac{\rho''(0)}{12}. \tag{2.13}$$

Hence, $\xi(0)$ itself is known in terms of ρ, ρ', and ρ'', all evaluated at $x = 0$, using Eqs. (2.9) and (2.10) and the number of Fermions N given by

$$N = \int_{-\infty}^{\infty}\rho(x)\,dx. \tag{2.14}$$

Thus, with complete knowledge of $\xi(x)$ in terms of $\rho(x)$ and its low-order derivatives in this model, the single-particle density is obtained in a fully nonlocal form via Eqs. (2.8), (2.11), and (2.12)–(2.14). Evidently, again of course for this model, the (single-particle) kinetic energy functional T_s follows from Eq. (2.8) as [15]

**Note added after completion of article:* This result (2.7) has now been extended to 3-D isotropic harmonic confinement by A. Minguzzi, N. H. March, and M. P. Tosi (to be published).

$$T_s = \int_{-\infty}^{\infty} t_s(x)\,dx$$
$$= \int_{-\infty}^{\infty} \frac{1}{8}\frac{\rho'(x)^2}{\rho}\,dx + \int_{-\infty}^{\infty} \xi(x)t_{\mathrm{TF}}(x)\,dx$$
$$= T_{\mathrm{W}} + \int_{-\infty}^{\infty} \xi(x)t_{\mathrm{TF}}(x)\,dx. \tag{2.15}$$

This then is the nonlocal kinetic energy functional in this model of harmonic confinement in one dimension.

A byproduct of the above argument [15] is the relation

$$\frac{t_s'(x)}{\rho'(x)} = \frac{3t_s(x)}{\rho(x)} + \frac{\rho''(x)}{4\rho(x)} \tag{2.16}$$

and then it follows that

$$\frac{\delta T_s}{\delta\rho(x)} = \frac{3t_s(x)}{\rho(x)} + \frac{\rho''(x)}{4\rho(x)} + C(N), \tag{2.17}$$

where C is a constant, depending, however, on the number of Fermions N. It is, of course, this functional derivative that is needed to solve the Euler Eq. (1.3) of DFT.

To close this section on the kinetic energy functional, we turn to the explicit example of two (antiparallel) spin electrons repelling Coulombically.

2.2. CORRELATION EFFECTS WITH TWO (ANTIPARALLEL) SPIN ELECTRONS REPELLING COULOMBICALLY

It is the aim of this section to treat two cases: (1) When the interacting electrons are again harmonically confined—the so-called Hookean atom going back to Kestner and Sinanoglu [23, 24]; and (2) He-like ions in the limit of large Z.

Because the differential Eq. (2.1) of Lawes and March for (noninteracting) harmonic confinement was a key result in obtaining the single-particle kinetic energy functional (2.15), we note that in case (1) above the study of March et al. [25] yields the corresponding equation, now for two electrons only but incorporating Coulomb repulsion exactly: for force constant $k = 1/4$

$$(-r^2 + r^4)\rho'''(r) + (-2r - 4r^3 + 4r^5)\rho''(r) + (2 - 13r^2 - 4r^4 + 5r^6)\rho'(r) + (-10r^3 + 2r^7)\rho(r) = 0. \tag{2.18}$$

There is substantial similarity between Eq. (2.18) for two electrons and Eq. (2.1) for an arbitrary number of harmonically confined electrons. In particular, both are homogeneous linear third-order differential equations.

In view of the work of Holas and March [11] on the exchange-correlation potential in relation to 1 and 2DMs, it is noteworthy that the same Hookean atom model with $k = \frac{1}{4}$ also relates 1 and 2DMs through [26]

$$\left(1 + \frac{1}{2}|\mathbf{r} - \mathbf{r}'|\right)\gamma(\mathbf{r}, \mathbf{r}') = \{n_2(\mathbf{r}, \mathbf{r}')\}^{1/2} \int d\mathbf{r}_1\left[1 + \frac{1}{2}|\mathbf{r} - \mathbf{r}_1|\right] \times \left[1 + \frac{1}{2}|\mathbf{r}_1 - \mathbf{r}'|\right]\{n_2(\mathbf{r}_1, \mathbf{r}_1)\}^{1/2}. \tag{2.19}$$

Here $\gamma^2 < \gamma$ because of the full treatment of interactions, and hence this 1DM (2.19) includes correlation kinetic energy. The pair function $n_2(\mathbf{r}, \mathbf{r}')$ is the diagonal element of the 2DM. Having focussed on two of the principle analytic results involving antiparallel spin electron correlation, namely, the differential Eq. (2.18) for $\rho(r)$ and the relation (2.19) between the 1DM and the (diagonal) pair function of the 2DM with harmonic confinement, let us turn to case (2) introduced above.

2.3. He-LIKE ATOMIC IONS WITH A LARGE ATOMIC NUMBER

Gál et al. [27] subsequently obtained a differential equation analogous to Eqs. (2.1) and (2.18) for He-like atomic ions in the (nonrelativistic) limit of the large atomic number Z. The ground-state density $\rho(r)$ again satisfies a differential equation that is linear, third-order, and homogeneous, namely

$$\sum_{i=0}^{3} P_i(r)\rho^{(i)}(r) = 0, \tag{2.20}$$

where $\rho^{(i)}(r)$ denotes the ith derivative of $\rho(r)$: with $\rho^{(0)}(r) \equiv \rho(r)$. The quantities $P_i(r)$ are also polynomials as in the Hookean Eq. (2.18) and are recorded explicitly by Gál et al. [27].

Turning to the 1DM, $\gamma(\mathbf{r}, \mathbf{r}')$ the work of Hall et al. [28] was used by Gál et al. [27] to write, in the limit (two-electron) system with large atomic number Z:

$$\gamma(\mathbf{r}, \mathbf{r}') = \frac{\rho(r')}{2}\left[\frac{\rho_0(r)}{\rho_0(r')}\right]^{1/2} + \frac{\rho(r)}{2}\left[\frac{\rho_0(r')}{\rho_0(r)}\right]^{1/2}, \quad (2.21)$$

where $\rho_0(r)$ is the bare Coulombic density given by

$$\rho_0(r) = \frac{Z^3}{\pi}\exp(-2Zr). \quad (2.22)$$

This result for γ is readily used to obtain the correlation kinetic energy $T_c[\rho]$ in this limit system (two electrons: large Z) as

$$T_c[\rho] = -T_W[\rho] - \frac{1}{2}\left[\frac{\rho'(r)}{\rho(r)}\right]_{r=0}\int\frac{\rho(r)}{r}\,d\mathbf{r} - \frac{1}{8}\left\{\left[\frac{\rho'(r)}{\rho(r)}\right]_{r=0}\right\}^2\int\rho(r)\,d\mathbf{r}, \quad (2.23)$$

where Kato's theorem [see Eq. (3.8) below] has been employed to remove the atomic number Z. Here $T_W[\rho]$ is the (now, of course, 3-D) W inhomogeneity kinetic energy in Eq. (1.1), namely

$$T_W[\rho] = \frac{1}{8}\int\frac{(\nabla\rho)^2}{\rho}\,d\mathbf{r}. \quad (2.24)$$

This discussion of electron–electron interaction in two-electron atomic ions in the limit of large Z leads us naturally to a discussion of exchange in such a limit system, but now with K plus L closed shells, i.e., a 10-electron case.

3. Exchange Energy Density: Model of 10-Electron Atomic Ions in the Limit of Large Atomic Number Z

In early work, March and Santamaria [29] obtained the Dirac density matrix $\gamma(\mathbf{r}, \mathbf{r}')$ for a hydrogenic system with filled K and L shells. The shape of their result is that γ is related to the diagonal density ρ, which is spherically symmetrical in this closed shell example, by

$$\gamma(\mathbf{r}, \mathbf{r}') = \rho\left(\frac{r + r'}{2}\right) + |\mathbf{r} - \mathbf{r}'|^2 F\left(\frac{r + r'}{2}\right). \quad (3.1)$$

The function F in turn is related to ρ itself through

$$F(r) = \frac{1}{24}\rho''(r) - \frac{2m}{\hbar^2}t_s(r), \quad (3.2)$$

because, as discussed in Section 2, t_S is itself determined by ρ. In this bare Coulomb field model, $F(r)$ has the compact form

$$F(r) = -\frac{1}{64\pi}\left[\frac{Z}{a_0}\right]^5\exp\left(-\frac{Zr}{a_0}\right) : a_0 = \frac{\hbar^2}{me^2}. \quad (3.3)$$

Going back to the study of Dirac [30] the exchange energy E_x of a system with idempotent 1DM γ is given by

$$E_x = -\frac{1}{4}e^2\int\frac{\gamma^2(\mathbf{r}, \mathbf{r}')}{|\mathbf{r} - \mathbf{r}'|}\,d\mathbf{r}d\mathbf{r}' \quad (3.4)$$

and from this one can define an exchange energy density $\epsilon_x(\mathbf{r})$ that is everywhere less than zero as

$$\epsilon_x(r) = -\frac{1}{4}e^2\int\frac{\gamma^2(\mathbf{r}, \mathbf{r}')}{|\mathbf{r} - \mathbf{r}'|}\,d\mathbf{r}'. \quad (3.5)$$

In the study of Howard et al. [17], $\epsilon_x(r)$ has been obtained in the limit system defined by the bare Coulomb density matrix (3.1). As noted in Ref. [17], this is the limit of large Z in the spirit of the $1/Z$ expansion [16] for 10-electron atomic ions. The quantity $\epsilon_x(r)$ is given in closed form, but we wish to note here mainly two analytic limits that flow from that analysis, before presenting results in graphic form.

First, as the present writer [18] showed elsewhere,*

$$\epsilon_x(r)|_{r\to\infty} \to -\frac{e^2}{2r}\rho(r) \quad (3.6)$$

and this is shown to be quickly useful away from the nucleus in reproducing the exact results in Ref. [17].

Second, and more importantly in the present context, a direct plot of $\epsilon_x(r)$ vs. r shows that ϵ_x is

*Note added in proof: See also ref [31] below.

coming away from the nucleus ($r = 0$) with a finite slope. This can be found from the analytic formula for $\epsilon_x(r)$ [17] and the slope is such that

$$\frac{\epsilon_x'(r)|_{r=0}}{\epsilon_x(r)|_{r=0}} = -\frac{2Z}{a_0}, \tag{3.7}$$

which has analogy with Kato's theorem for the electron density in a closed-shell atomic ion of nuclear charge Ze:

$$\left.\frac{\rho'(r)}{\rho(r)}\right|_{r=0} = -\frac{2Z}{a_0}. \tag{3.8}$$

While Eq. (3.7) is obtained, at this point, only for the bare Coulomb model density matrix given by Eqs. (3.1) and (3.3), it will be discussed further for "real" atoms, molecules, and solids in Section 4.

However, while Kato's result (3.8) is already applicable to "real" closed-shell atomic ions, we next note, in the present context, that for closed shells (of an arbitrary number, in fact) in the bare Coulomb limit the spatial generalisation of Eq. (3.8) exists, namely

$$\frac{\partial \rho(r)}{\partial r} = -\frac{2Z}{a_0}\rho_{l=0}(r), \tag{3.9}$$

i.e., the total density $\rho(r)$ can be derived by integration of Eq. (3.9) from the density of $s(l = 0)$ electrons alone [32].

In the framework of the present article, which is focussed on relating low-order density matrix to the electron density ρ, it is also noteworthy that Theophilou and March [33] have shown that the diagonal relation between total and $s(l = 0)$ densities in this bare Coulomb limit has a counterpart with density matrices. Their relation is in fact

$$\gamma(\mathbf{r}, \mathbf{r}') = -(x-y)^{-1}\left[\frac{\partial}{\partial x} - \frac{\partial}{\partial y}\right]\left[xy\gamma_{l=0}\left(\frac{x}{2}, \frac{y}{2}\right)\right], \tag{3.10}$$

where

$$x = r + r' + |\mathbf{r} - \mathbf{r}'|, \quad y = r + r' - |\mathbf{r} - \mathbf{r}'|. \tag{3.11}$$

The variables x and y in Eq. (3.11) are in fact motivated by the extra constant of motion in the bare Coulomb limit system, namely the Runge–Lenz vector [34]. Again, as with Eq. (3.9), the result (3.10) is valid for an arbitrary number of closed shells in this model.

Eqs. (3.5) and (3.10) clearly imply that the exchange energy density $\epsilon_x(r)$ calculated in Ref. [17] is determined solely by the s electron density matrix $\gamma_{l=0}(r, r')$, even though it was in fact obtained by direct insertion of Eqs. (3.1) and (3.2) into the definition (3.5).

Howard et al. [17] also discuss the exchange potential $V_x(\mathbf{r})$ defined by

$$V_x(\mathbf{r}) = \frac{\delta E_x}{\delta \rho(\mathbf{r})} \tag{3.12}$$

but because of the results in Eqs. (3.1), (3.2), and (3.4) this still involves a functional derivative with respect to $\rho(r)$ of t_s, e.g., $\delta t_s(r)/\delta\rho(\mathbf{r}')$, and at present this is not known in this model in analytic form. However, approximations to $V_x(\mathbf{r})$ due to (1) Slater [35] and (2) Harbola–Sahni [36] were investigated numerically using the exact density in this Coulomb limit system, the Slater approximation $V_x^{Sl}(\mathbf{r})$ being particularly simple because

$$V_x^{Sl}(\mathbf{r}) = \frac{2\epsilon_x(\mathbf{r})}{\rho(\mathbf{r})}. \tag{3.13}$$

This, using Eq. (3.6), tends to $-e^2/r$ at large r, which is correct.

Returning to $\epsilon_x(r)$ itself, Howard et al. [17] plotted the "radial" exchange energy density $4\pi r^2\epsilon_x(r)$ vs. r in their work, and this is reproduced in Figure 1. On the same plot, the local density approximation (LDA)

$$\epsilon_x^{\mathrm{LDA}}(r) = -c_x\rho(\mathbf{r})^{4/3} : c_x = \frac{3}{4}e^2\left(\frac{3}{\pi}\right)^{1/3} \tag{3.14}$$

going back to Dirac [30] and utilized especially by Slater [4] is given, but we stress with the exact density $\rho(r)$ used. Even though the point–point agreement is remarkable, the total exchange energy E_x is in error by some 8% when the exchange $\epsilon_x(r)$ is replaced by the LDA form (3.14).

In closing this section on exchange, we note qualitatively similar results to the above for a quite different model, namely, that of electron confinement by a metal surface. This has been discussed recently by this author [37]. What we wish to stress here is that in the (infinite barrier) model of a surface the 1DM $\gamma(\mathbf{r}, \mathbf{r}')$ can be directly related to its diagonal density. This then allows a direct nonlocal

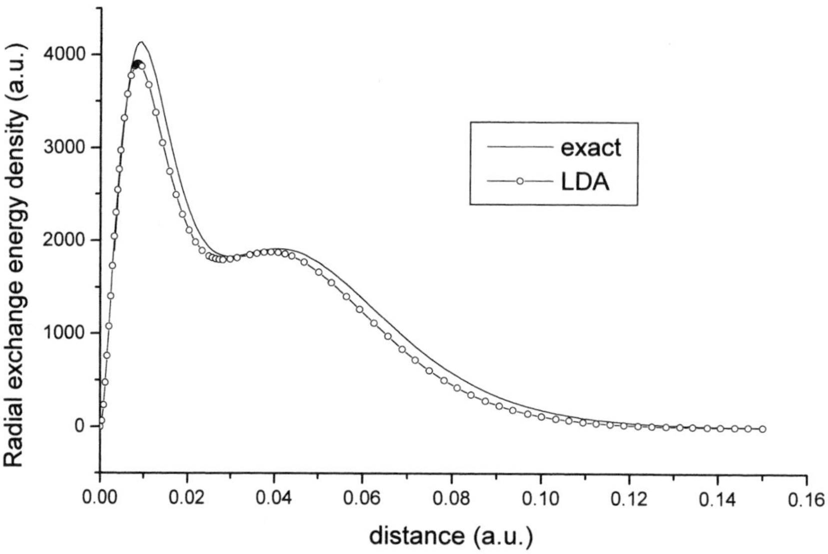

FIGURE 1. Plot of exchange energy density $\epsilon_x(r) \times 4\pi r^2$ obtained analytically by Howard et al. [17] in the 10-electron bare Coulomb limit system. The large Z necessary is chosen in the figure to be 92. The circled line (○) shows the Dirac–Slater local density approximation (3.14), calculated with the exact density $\rho(r)$. (Redrawn from Howard et al. [17].)

relation to be established between $\epsilon_x(\mathbf{r})$ and the kinetic energy density $t_s(\mathbf{r})$ [37].

4. Some Exact General Results: Nuclear Cusp Conditions on Exchange and Correlation Energy Densities

Sections 2 and 3 comprise exact density functional results from models. While such exactitude is important for DFT, "universality" of the functionals obtained cannot be claimed. However, the purpose of this section is to demonstrate that the Coulomb model result (3.7) is not only true in "real" closed-shell atomic ions but also in multicentre problems of molecules, clusters, and condensed matter.

The detailed argument will be presented elsewhere [18]; here we shall restrict ourselves to (1) the principles underlying the derivation and (2) the exact results that flow from these.

4.1. UNDERLYING PRINCIPLES EMPLOYED

The singularity of the nucleus A, with charge $Z_A e$, of the electron–nucleus interaction $V_A(\mathbf{r}) = -Z_A e^2/|\mathbf{r} - \mathbf{R}_A|$ results in nonanalyticity of the N-electron wave function when each electronic coordinate $\mathbf{r}_i$, $i = 1, 2 \ldots N$ becomes equal to $\mathbf{R}_A$, yet the function remains finite everywhere. The characterisation of this analyticity was first revealed in the pioneering work of Kato [38], Eq. (3.8) of the previous section being an example of his findings, for the case of a closed-shell atomic ion. Bingel [39] extended the treatment to a molecule, but here the presentation used by Davidson [40] will provide the basic background needed to present the results relevant to nonlocal energy density functionals.

With H the total Hamiltonian, let the N-electron symmetrical normalised wave function $\Psi(\mathbf{x}_i - \mathbf{x}_N)$ be an eigenfunction of the Schrödinger equation

$$H\Psi = E\Psi, \tag{4.1}$$

where $\mathbf{x}_i \equiv \{\mathbf{r}_i, s_i\}$ denotes the space and spin coordinates of the ith electron. In treating cusp conditions due to the specified Ath nucleus, let us choose its position to be at $\mathbf{R}_A = 0$. As Ψ is everywhere finite, it follows from Eq. (4.1) if we focus on the argument $\mathbf{r}_1$ of Ψ that

$$\lim_{r_1 \to 0} r_1 H\Psi = 0. \tag{4.2}$$

This implies that $\Psi(\mathbf{r}_1, s_i, \mathbf{x}_2 \ldots \mathbf{x}_N)$ is also an eigenfunction, with zero eigenvalue of the operator [40]

$$b(\mathbf{r}_1) = \lim_{r_1 \to 0} r_1 T = \lim_{r_1 \to 0} r_1 t(\mathbf{r}_1) - Z_A, \tag{4.3}$$

where $T = \sum_{i=1}^{N} t(\mathbf{r}_i)$ is the kinetic energy operator. For the wave function $\Psi(\mathbf{r}_1, \mathbf{x}_1)$ where $\mathbf{x} \equiv \{s_i, \mathbf{x}_2 \ldots \mathbf{x}_N\}$ expanded in spherical harmonics:

$$\Psi(\mathbf{r}_1, \mathbf{x}) = \sum_{l=0}^{\infty} \sum_{m=-l}^{m=l} y_{em}(\Omega_1) r_1^l f_{lm}(r_1, \mathbf{x}) \tag{4.4}$$

with $f_{lm}(r_1, x)$ an analytic function of r, the eigenequation $b(\mathbf{r}_1)\Psi = 0$ yields the following conditions on the partial wave amplitudes [40]:

$$\left. \frac{\partial f_{lm}(r_1, \mathbf{x})}{\partial r_1} \right|_{r_1=0} = -\frac{Z_A}{l+1} f_{lm}(0, \mathbf{x}). \tag{4.5}$$

Eqs. (4.3)–(4.5) are used by March et al. [18] to obtain nuclear cusp conditions for the exchange and correlation energy densities of DFT.

Before turning to the DFT energy densities $\epsilon_x(\mathbf{r})$ (discussed in Section 3 for the case of a 10-electron closed-shell atomic ion in the limit of large Z) and the corresponding correlation (c) energy density $\epsilon_c(\mathbf{r})$, for DFT applications the classic electrostatic energy of electrons is of interest, namely

$$E_{es}[\rho] = \int d\mathbf{r}_1 \, \epsilon_{es}(\mathbf{r}_1), \tag{4.6}$$

where

$$\epsilon_{es}(\mathbf{r}_1) = \int d\mathbf{r}_2 \frac{e^2}{|\mathbf{r}_1 - \mathbf{r}_2|} \frac{1}{2} \rho(\mathbf{r}_1)\rho(\mathbf{r}_2). \tag{4.7}$$

March et al. [18] show that

$$\left. \frac{\partial \overline{\epsilon_{es}(r)}}{\partial r_1} \right|_{r_1=0} = -2Z_A \epsilon_{es}(0), \tag{4.8}$$

where the bar on $\epsilon_{es}(\mathbf{r}_1)$ denotes an angular average. This is a simple analogue for the classic electrostatic energy density of Eq. (3.7) for exchange (x) energy density derived from the bare Coulomb limit system (for a 10-electron atomic ion). This leads us to a sketch of the generalization of Eq. (3.7) derived for a model closed-shell atomic ion to the fully interacting electron problem for atoms, molecules, and solids [18]. The derivation in Ref. [18] appeals to the adiabatic connection developed by Harris and Jones [41, 42]. This connection links the fully interacting system treated above to the SKS noninteracting reference system with the same (i.e., formally exact) electron density via intermediate systems. For this purpose, a Hamiltonian depending on a parameter α such that $0 \leq \alpha \leq 1$ is introduced to write

$$H^\alpha = T + V^\alpha + \alpha U, \tag{4.9}$$

where $V^\alpha = \sum_{i=1}^{N} V^\alpha(\mathbf{r}_i)$. This Hamiltonian for $\alpha = 1$ reduces to the original Hamiltonian while for $\alpha = 0$ the electron–electron piece U is switched off, while $V^0(\mathbf{r}) = V_{SKS}(\mathbf{r})$.

To use the above approach to investigate the cusp conditions for the intermediate system, March et al. [18] assume (to be checked later) that the properties of $V^\alpha(\mathbf{r})$, whose explicit form remains unknown, permit a finite value for the limit $\gamma^\alpha = \lim_{r_1 \to 0} r_1 V^\alpha(\mathbf{r}_1)$. Then the early arguments carry through and one obtains

$$\left. \frac{\partial}{\partial r} \bar{\rho}^\alpha(r) \right|_{r=0} = 2\gamma^\alpha \rho^\alpha(0), \tag{4.10}$$

which is the analogue of Eq. (3.8). But the one-body potential $V^\alpha(\mathbf{r})$ is defined by requiring that the electron density $\rho^\alpha(\mathbf{r})$ of the intermediate system is independent of α. Hence we conclude that $\gamma^\alpha = -Z_A$, which justifies *a posteriori* the assumption that $\lim_{r_1 \to 0} r_1 V^\alpha(r_1)$ is finite.

March et al. [18] show that for the exchange-correlation potential $V_{xc}(\mathbf{r})$

$$\lim_{r_1 \to 0} r_1 V_{xc}(\mathbf{r}_1) = 0. \tag{4.11}$$

Without essentially new arguments beyond those presented above, the model result (3.7) for the exchange energy density ϵ_x generalises to

$$\left. \frac{\partial \overline{\epsilon_x(r_1)}}{\partial r_1} \right|_{r_1=0} = -2Z_A \epsilon_x(0), \tag{4.12}$$

where the angular average on the LHS is not required for closed-shell atomic ions. The final exact result of March et al. [18] is an analogous nuclear cusp condition for correlation energy density $\epsilon_c(\mathbf{r})$, namely

$$\left.\frac{\partial \overline{\epsilon_c(r_1)}}{\partial r_1}\right|_{r_1=0} = -2Z_A\epsilon_c(0). \tag{4.13}$$

These exact results (4.12) and (4.13) should evidently be built into approximate DFT theory in the future.

5. Summary

The early connection between density matrix and density functional theories stressed by March and Young [1] (DFT was not used then, but rather electron density theory) in deriving the approximate single-particle kinetic energy functional (1.1) has been greatly strengthened, especially during the last decade. Examples are now available in which an idempotent 1DM can be expressed, either directly or via the single-particle kinetic energy, in terms of the diagonal electron density [15, 17, 37]. Correlation can be added in two-electron atomic ions at a large atomic number to allow the 1DM to embrace correlation kinetic energy [27].

As to exchange energy, an idempotent closed-shell 10-electron 1DM for the Coulomb limit system allows explicit analytic calculation of the exchange energy density $\epsilon_x(r)$ and this has been compared with LDA (but using the exact Coulomb ground-state electron density) in Figure 1. While point-to-point agreement between the two is excellent over the range where $4\pi r^2\epsilon_x(r)$ is substantial, there are errors in LDA both near the nucleus and at large r. In the latter case, the exact asymptotic form is known [compare Eq. (3.6)] while at small r the nuclear cusp condition (3.7) emerges from the model. But, in this case the model result proves generalisable not only to "real" atomic ions but also to molecules, clusters, and condensed phases (see Ref. [18], which is briefly summarised in Section 4). An exact parallel for the correlation energy $\epsilon_c(\mathbf{r})$ follows.

As to future directions, it remains of central interest to improve available forms of the exchange-correlation potential $V_{xc}(\mathbf{r})$. While it has been known since the work of Kohn and Sham [5] that one formal definition is

$$V_{xc}(\mathbf{r}) = \frac{\delta E_{xc}[\rho]}{\delta\rho(\mathbf{r})}, \tag{5.1}$$

already approximate forms available for the exchange-correlation energy functional $E_{xc}(\rho)$ have their errors exacerbated by the functional differentiation in Eq. (5.1). Again it seems that one can look to the further synthesis of density matrix and density functional theory via the many-electron force-balance equation derived by Holas and March [11], namely

$$-\frac{\partial V_{\text{ext}}(\mathbf{r})}{\partial \mathbf{r}} = \mathbf{F}_{\text{kinetic}} + \mathbf{F}_{\text{Laplacian}} + \mathbf{F}_{\text{el-el}} \tag{5.2}$$

and a corresponding (one-body) equation for $V(\mathbf{r})$ in Eq. (1.2) in terms of SKS orbitals, or noninteracting 1 and 2DMs built from a single determinant of these orbitals. In Eq. (5.2) also, $\mathbf{F}_{\text{kinetic}}$ is determined by the fully interacting 1DM [11] while $\mathbf{F}_{\text{el-el}}$ requires only knowledge of the electronic pair correlation function $n_2(\mathbf{r}, \mathbf{r}')$, which is the diagonal element of the (fully interacting) 2DM. It is encouraging that these two many-electron terms on the RHS of the force-balance Eq. (5.2) [$\mathbf{F}_{\text{Laplacian}}$ is given directly in terms of the gradient of the Laplacian of the density $\rho(\mathbf{r})$] are related in the two-electron Hookean atom model discussed in Section 2.2 above by the exact result (2.19) and generalisations (albeit approximate) of this to N electrons are of some urgency.

On a related point, future nonlocal functions of the exchange-correlation energy density $\epsilon_{xc}(\mathbf{r})$ should clearly be constructed with the nuclear-cusp conditions very much in mind, even if, in the end, they can only be embraced semiquantitatively.

ACKNOWLEDGMENTS

This article could not have come to fruition without invaluable collaboration with Professors A. Holas and I. A. Howard. The author is most grateful to them both not only for the joint contributions in Refs. [18] and [17] but also for many most stimulating and informative discussions. Continuous encouragement and help from Dr. P. Senet and Professor D. J. Klein is also gratefully acknowledged.

References

1. March, N. H.; Young, W. H. Proc Phys Soc 1958, 72, 182–192.
2. Herring, C. Phys Rev 1986, A34, 2614–2631.
3. Herring, C.; Chopra, M. Phys Rev 1988, A37, 31–42.
4. Slater, J. C. Phys Rev 1951, 81, 385–390.
5. Kohn, W.; Sham, L. J. Phys Rev 1965, 140, A1133–1138.
6. Wang, L. W.; Teter, M. P. Phys Rev 1992, B45, 13196–13220.

7. Pearson, M.; Smargiassi, E.; Madden, P. A. J Phys Condens Matter 1993, 5, 3221–3240.
8. Perrot, F. J Phys Condens Matter 1994, 6, 431–446.
9. Jesson, B. J.; Foley, M.; Madden, P. A. Phys Rev 1997, B55, 4941–4946.
10. Garciá-González, P.; Alvarellos, J. E.; Chacón, E. Phys Rev 1998, B57, 4857–4861.
11. Holas, A.; March, N. H. Phys Rev 1995, A51, 2040–2048.
12. Levy, M.; March, N. H. Phys Rev 1997, A55, 1885–1889.
13. March, N. H.; Young, W. H. Nucl Phys 1959, 12, 237–240.
14. Parr, R. G.; Yang, W. Density Functional Theory of Atoms and Molecules; Oxford: Oxford University Press; 1989.
15. March, N. H.; Senet, P.; Van Doren, V. E. Phys Lett 2000, A270, 88–92.
16. March, N. H.; White, R. J. J Phys 1972, B5, 466–475.
17. Howard, I. A.; March, N. H.; Senet, P.; Van Doren, V. E. Phys Rev 2000 A, 62, 062512.
18. March, N. H.; Howard, I. A.; Holas, A.; Senet, P.; Van Doren, V. E. Phys Rev 2001 A, 63, 012520.
19. Lawes, G. P.; March, N. H. J Chem Phys 1979, 71, 1007–1009.
20. Gombás, P. Phys Lett 1969, A28, 585–586.
21. Goodisman, J. Phys Rev 1970, A1, 1574–1576.
22. Deb, B. M.; Ghosh, S. K. Int J Quantum Chem 1983, 23, 1–10.
23. Kestner, N. R.; Sinanoğlu, O. Phys Rev 1962, 128, 2687–2696.
24. Kais, S.; Herschbach, D. R.; Handy, N. C.; Murray, C. W.; Laming, G. J. J Chem Phys 1993, 99, 417–425 and earlier references therein.
25. March, N. H.; Gál, T.; Nagy, Á. Chem Phys Lett 1998, 292, 384–386.
26. March, N. H.; Amovilli, C.; Klein, D. J. Chem Phys Lett 2000, 325, 645–647.
27. Gál, T.; March, N. H.; Nagy, Á. Chem Phys Lett 1999, 305, 429–432.
28. Hall, G. G.; Jones, L. L.; Rees, D. Proc Roy Soc 1965, A283, 194–202.
29. March, N. H.; Santamaria, R. Phys Rev 1988, A38, 5002–5006.
30. Dirac, P. A. M. Proc Camb Phil Soc 1930, 26, 376–391.
31. March, N. H. Phys Rev 1987, A36, 5077–5078.
32. March, N. H. Phys Rev 1986, A33, 88–89.
33. Theophilou, A. K.; March, N. H. Phys Rev 1986, A34, 3630–3633.
34. Holas, A.; March, N. H. J Phys A Math Gen 1990, 23, 735–749.
35. Kleinman, L. Phys Rev 1994, B49, 14197–14201.
36. Harbola, M. K.; Sahni, V. Phys Rev Lett 1989, 62, 489–492.
37. March, N. H. Phys Rev 2000, B61, 5011–5012.
38. Kato, T. Commun Pure Appl Math 1957, 10, 151–177; Fraga, S.; Malli, G. Many-Electron Systems: Properties and Interactions; Philadelphia: Saunders; 1968.
39. Bingel, W. A. Z Naturforsch 1963, 18a, 1249–1258.
40. Davidson, E. R. Reduced Density Matrices in Quantum Mechanics; New York: Academic Press; 1976.
41. Harris, J.; Jones, R. O. J Phys 1974, F4, 1170–1186.
42. Levy, M.; March, N. H.; Handy, N. C. J Chem Phys 1996, 104, 1989–1992.

Ground-State Geometry and Electronic Structure of Light Atom Clusters, Especially H Isotopes, Li, B, and C

N. H. MARCH

Department of Physics, University of Antwerp, Groenenborgerlaan 171, B-2020 Antwerp, Oxford University, Oxford, England

Received 29 February 2004; accepted 29 February 2004
Published online 6 August 2004 in Wiley InterScience (www.interscience.wiley.com).
DOI 10.1002/qua.20150

ABSTRACT: The ground-state geometry and electronic structure of light clusters remains an area of considerable current interest for both many-body theory and experiment. We cite here recent experimental studies of H isotopes that may have important implications for energy production. Therefore a summary will first be given of quantum-chemical results on the singly charged cationic clusters $H_3^+(H_2)_m$ for some values of $m \leq 14$, studied by Gobet et al. (2002). More recent theoretical work on doubly charged D clusters is also reported. Some attention is then given to results on small Li clusters, isolated in argon matrices by Kornath et al. (2003), in relation to many-body theory. This is followed by a study of almost-spherical cages of B and of C atoms: both neutral and charged clusters again being referred to in the case of boron. Finally, some mixtures of light atoms are considered, with special focus on BN cages and cationic hydrocarbons. © 2004 Wiley Periodicals, Inc. Int J Quantum Chem 100: 877–886, 2004

Key words: charged D clusters; cages of B and C

1. Background and Outline

The geometry and electronic structure of neutral and charged clusters of light elements in their ground states provide a current focus of much effort, both experimental and theoretical. In the present study we first consider H isotopes in cationic clusters. We cite, in particular, the experimental study by Gobet et al. [1] of $H_3^+(H_2)_m$ clusters with $m \leq 14$ and of multiply charged deuterium clusters used in the experiment of Zweiback et al. [2] related to fusion yield. Theoretical studies on the former species for some values of $m \leq 14$ will be briefly summarized in the above context [3]. Very recent theoretical work on Coulomb explosion in small doubly charged deuterium clusters will also be mentioned [4]. In turn, small Li clusters, almost-spherical B and C cages, and two particular mixtures of light atoms, namely BN cages and cationic hydrocarbons, will then be considered.

A brief outline of this article is as follows. Current experimental interest in charged clusters of hydrogen isotopes [1, 2] is first summarized quite briefly in Section 2 (see also Appendix). There also some theoretical results are presented on singly and doubly charged cationic clusters of such isotopes. Section 3 is motivated by the very recent Raman spectroscopic investigation of small Li clusters isolated in argon matrices by Kornath et al. [5]. Some findings from quantum-chemical studies [6] are discussed in relation to their experimental study. Then in Section 4, quite recent theoretical results on almost-spherical B cages are referred to, both neutral and charged clusters again being treated [7]. Section 5 reports model and Hartree–Fock [8] studies on the fullerenes C_{50}, C_{60}, C_{70}, and C_{84}. In Section 6, mixtures of light atoms are considered, focusing on BN cages and cationic hydrocarbons. Section 7 constitutes a summary plus suggestions for future work that should prove fruitful.

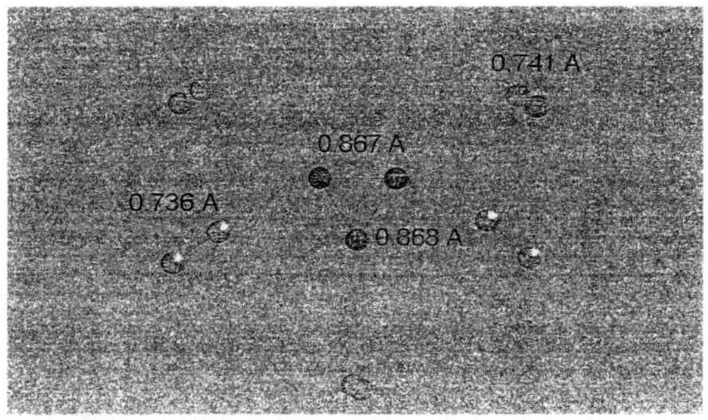

FIGURE 2. Optimized geometry of cation cluster H_{13}^+ at MP2 level. Note that trimer has equilateral geometry. (Redrawn from Ref. [3].)

2. Hydrogen Isotopes in Cationic Clusters

Let us begin with the cationic species $H_3^+(H_2)_m$ focused on in the liquid–vapor transition experiment of Gobet et al. [1]. Howard et al. [3] have reported Hartree–Fock plus Møller–Plesset second-order perturbation theory for a few selected values of m including 14, the largest value considered in Ref. [1].

Hartree–Fock (HF) calculations, supplemented by second-order Møller–Plesset (MP2) perturbation theory, have been carried out [3] for several values of m, including $m = 14$. The results confirm the "chemical" formula $H_3^+(H_2)_m$ in that the net positive charge for each value of m considered resides dominantly on the "central" triangle H_3. Depending on m, this triangle can be either isosceles or equilateral. Figures 1–3 show the structures obtained, together with relevant bond lengths. It is, naturally enough, the intermolecular bond lengths that change significantly when MP2 corrections are added to the HF results. In later work, beyond that reported in Ref. [3], MP4 was added to check convergence of these intermolecular distances, with satisfactory findings.

2.1. MULTIPLY CHARGED DEUTERIUM CLUSTERS

We turn now to work motivated by the experimental study of Zweiback et al. [2]. Using a 35-fs laser pulse to drive explosions in D clusters, these

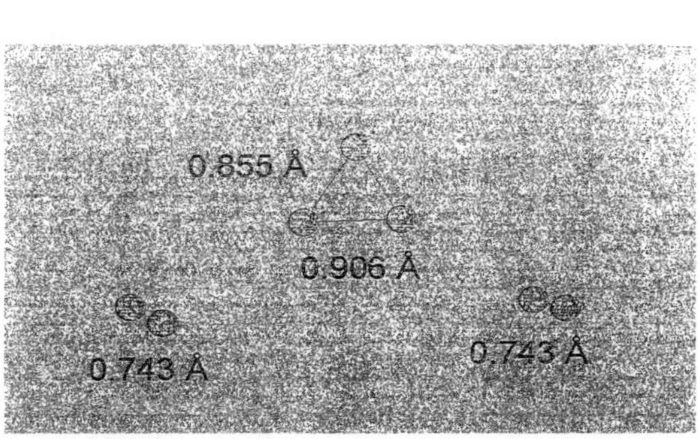

FIGURE 1. Optimized geometry for cationic cluster H_7^+ with an isosceles triangle as the trimer geometry. (Redrawn from Ref. [3].)

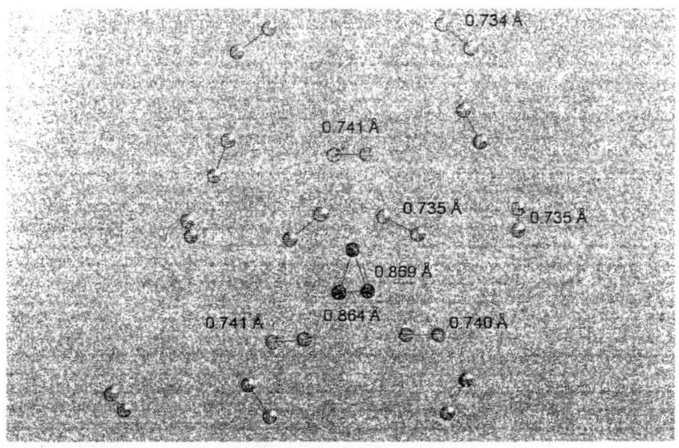

FIGURE 3. Optimized geometry of H_{31}^+ at MP2 level. Note that the central trimer is almost equilateral for this cationic cluster. (Redrawn from Ref. [3].)

FIGURE 4. Coulomb fragmentation of the doubly charged deuterium cluster D_{31}^{2+}. The breaking off of a singly charged cationic dimer from the original charged cation is illustrated. The large fragment has charge residing dominantly on the singly charged cationic trimer D_3^+, which is found to be an almost equilateral triangle with bond length ~0.86 Å. (Redrawn from Ref. [4].)

authors have accelerated ions to sufficiently large kinetic energy to produce DD nuclear fusion. The fusion yield from such exploding clusters is shown to vary strongly with cluster size, which Zweiback et al. [2] conclude is consistent with the acceleration of deuterons via Coulomb explosion forces.

These important findings in Ref. [2] have prompted a theoretical study of multiply charged cationic D clusters [4]. This was carried out, first, for doubly charged clusters, and for a number of clusters static HF calculations supplemented with MP2 and MP4 perturbation corrections showed quite clearly that Coulomb explosion was indeed occurring. The second main finding from this quantal chemical study of doubly charged D clusters was that the fragmentation was usually highly asymmetric, small fragments being broken off and carrying charge. Figure 4 shows a typical case illustrating the main findings recorded above.

We shall return to doubly charged cationic clusters in Section 7, but first consider mixtures, and in particular hydrocarbons.

3. Lithium Clusters: Isolation in Argon Matrices

The main focus of this section is the fact that the small lithium clusters Li_{2n}, with $n = 1, 2$, and 4, have recently been isolated in argon matrices at 15 K by Kornath et al. [5]. These workers have characterized these clusters by Raman spectroscopy. A second point relevant to the study reported below is the tendency of Li clusters to approach "metallic-like" behavior for hundreds of atoms, i.e., $n \sim 50$ or 100. This is in marked contrast to a divalent atom like Hg, where thousands of atoms must be involved in clusters before any "metallic" trends are observable.

3.1. EARLIER EXPERIMENTAL STUDIES

As noted by Kornath et al. [5], small Li clusters have been investigated in the gas phases and embedded in cryogenic matrices by a number of different experimental techniques including ESR, laser-induced fluorescence, depletion spectroscopy, photoionization, cluster beam detection, and optical resonance. Kornath et al. note that most of these methods are difficult to correlate with cluster geometries—a major focus of the present article. However, from earlier experimental studies using Raman spectroscopy on small K and Na clusters in noble gas matrices [9, 10], Kornath et al. have shown that Raman spectroscopy, linked with quantum-chemical calculations, affords a reliable route to the determination of cluster geometries. Below our focus is solely on the lightest alkali, Li.

The conclusion of Kornath et al. [5] is then that the Li_4 cluster had a rhombic structure (D_{2h}). For Li_8, a hypertetrahedral structure (T_d) is predicted, which Kornath et al. [5] take to be an example of a matrix-isolated magic number cluster.

3.2. QUANTUM-CHEMICAL STUDIES

The rhombic structure for tetralithium discussed above agrees with the early prediction from the quantum-chemical studies of Bonacic-Koutecky et al. [11]. It is noted in Ref. [5] that higher symmetry structures, namely, a tetrahedron or a square, have two unpaired electrons. Then Jahn–Teller distortion will occur into the D_{2h} structure, with formation of electron pairing into a singlet ground state.

In Ref. [5], the analysis of the absorption spectrum of Li_8 in the gas phase was combined with quantum-chemical studies to exclude the D_{4d} structure for the Li_8 cluster. When Li_8 had a T_d or D_{2d} symmetry could not be answered purely on the basis of available experimental data at that time. But quantum chemistry predicted that the hypertetrahedron was the most stable geometry, and the Raman spectra [5] have now confirmed this. If the structure had been D_{2d} rather than the above T_d symmetry, Kornath et al. [5] note that a much more complex Raman spectrum than the form they measured would have resulted.

Further quantum-chemical studies on Li_{2n} clusters are in progress [6], both for $n = 3$ and 4, and for some larger n values. Preliminary results indicate that Li_6 is only very weakly bound, and that electron correlation is then the determining factor for the lowest-lying isomer [6]. It is relevant to conclude this section by noting that in Ref. [6] regular trends through the Li_{2n} clusters for n small are exhibited for a variety of properties including (i) LUMO-HOMO gap versus n, (ii) relation between total ground-state energy E and one-electron eigenvalue sum E_s [compare Eq. (5.7) below] and (iii) dissociation energy versus density gradient kinetic energy. The reader interested in further details of these regularities should consult Ref. [6].

4. Almost-Spherical Boron Cages

This section considers a set of almost-spherical B cages. Both neutral cages with formula B_{2k} and cationic clusters B^+_{2k+1} will be considered.

We note first that it is now more than a decade since La Placa et al. [12], through laser ablation of hexagonal boron nitride, discovered B_n clusters with n up to 52. These authors already analyzed the cluster size by mass spectrometry but the experimental geometries remained unknown.

Following Amovilli and March [7], who presented an ab initio study of B_{2k} and B^+_{2k+1} cages referred to above on 25 boron cages, we take as starting point the model used in the early work of one of us on tetrahedral and octahedral molecules. In a molecule like GeH_4, this model implied smearing the four protons into a uniform surface charge density over a sphere of radius equal to the Ge—H bond length and centered on the Ge nucleus. The same model was subsequently applied to C_{60} and termed there by Clougherty and Zhu [13] and Despa [14] as the "March model" (see also Siringo et al. [15]). Although, as explained above in relation to GeH_4, the electron density (say, of the π electrons in C_{60}) is spherically symmetrical, the electrostatic nuclear–nuclear potential energy is calculated for the correct cage geometry. However, in Amovilli and March's study [7] on B cages, the self-consistent Thomas–Fermi (local density) approximation used in the early model [16] was replaced by Hartree–Fock calculations.

Amovilli and March [7] stressed the close relation between the equilibrium geometry of a system of equal point charges (the B nuclei), which are constrained to move only over the surface of a sphere, and the so-called "Aufbau principle" invoked by Boustani [17], (see also Ref. [18]) in constructing B clusters. In reality, when the number of charges smeared over the sphere surface becomes sufficiently large, the optimized geometry corresponds to a polyhedron with triangular faces and essentially penta (P)- and hexa (H)-coordinated vertices while, as emphasized by Amovilli and March [7], following the Aufbau principle proposed earlier for B compounds, the geometric forms can be built as a combination of two basic units only: (i) a pentagonal pyramid B_6 and (ii) a hexagonal pyramidal unit B_7 (see also Boustani [19]).

In the study by Amovilli and March [7], the only geometrical parameter optimized at an ab initio level is the sphere radius. As these workers discuss, the electrostatic repulsion energy E_{rep} of a system of n particles with charge Z constrained on the surface of a sphere of radius R is of the form

$$E_{rep} = A_n(Zn)^2/R, \tag{4.1}$$

where A_n is tabulated by Amovilli and March [7] and is quantitatively represented by

$$A_n = \frac{1}{2} - \frac{0.05505}{n^{1/2}}. \tag{4.2}$$

The Hartree–Fock results for the equilibrium B-cage radius, R_e say, against $n^{1/2}$ [7] are [compare Eq. (4.2)] well fitted by the relationship

$$R_e = 0.471 \quad n^{1/2}, \tag{4.3}$$

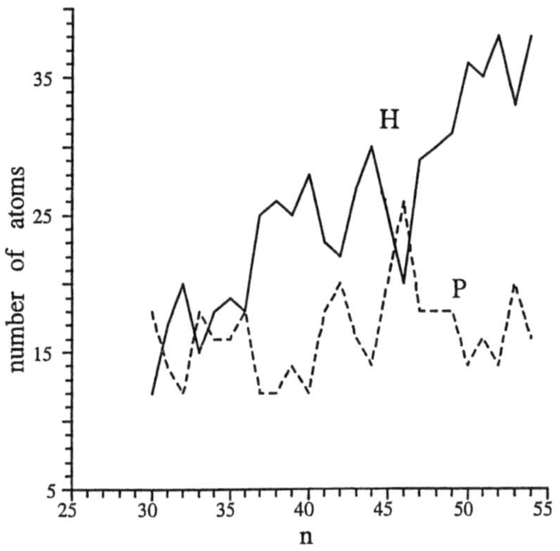

FIGURE 5. Variation of the number of penta- (P) and hexa (H)-coordinated vertices with the size of the spherical B cages considered. (Redrawn from Ref. [7].)

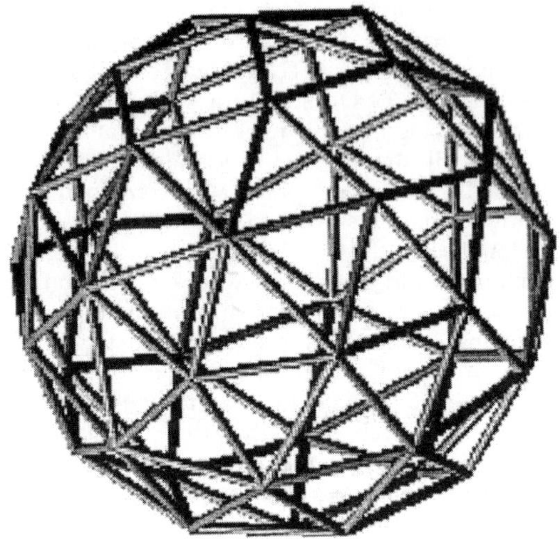

FIGURE 7. Schematic view of the cationic B_{45}^{+} boron cage, following Amovilli and March [7].

where the radius is then in Å. This condition allows the average distance between neighboring atoms to be near to the best B—B bond length in all cages.

Figure 5, following Amovilli and March [7], shows the variation of the number of penta- and hexa-coordinated vertices with cluster size; Figure 6 is a plot of cohesive energy versus number of atoms n for both neutral and singly-charged cationic clusters. The most stable cluster appears to be B_{45}^{+}, with the schematic form in Figure 7 having easily recognizable pentagonal and hexagonal subunits.

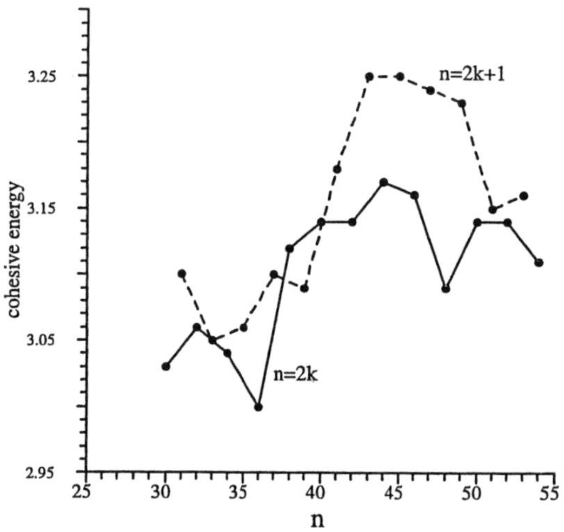

FIGURE 6. Cohesive energy (eV) against the number of atoms n for both neutral clusters B_{2k} and cationic clusters B_{2k+1}^{+}. (Redrawn from Ref. [7].)

We turn from almost-spherical B cages to consider some specific, again highly symmetrical, fullerene structures, in the following section.

5. Regularities in Geometries and π-Electron Eigenvalues for Almost-Spherical C Cages

To introduce recent Hartree–Fock studies [8] on the fullerenes C_{50}, C_{60}, C_{70}, and C_{84}, it will be instructive to refer to a model in which free π electrons are confined to the surface of a sphere [20].

The π eigenvalues, in zeroth order, then correspond to those of a rigid rotator, with degeneracy $(2j + 1)$ and energies $\hbar^2 j(j + 1)/2I$ with I denoting the moment of inertia. Taking double occupancy up through the nth shell, the π-electron eigenvalue sum is then proportional to

$$2 \sum_{j=0}^{n} (2j + 1)j(j + 1) = n(n + 1)^2(n + 2), \quad (5.1)$$

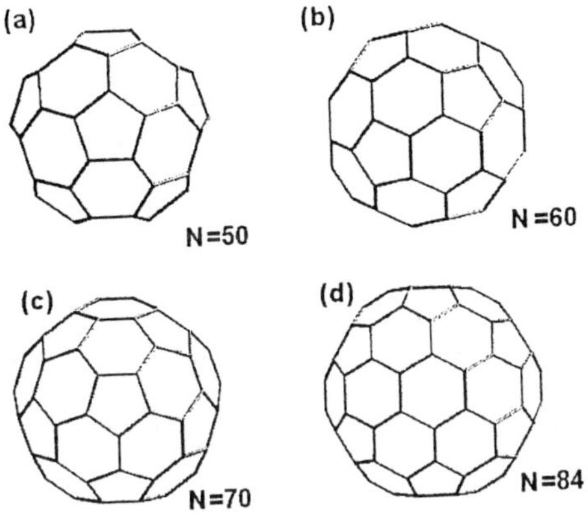

FIGURE 8. Structure of buckyball C_{60} (b) and also the (assumed) high-symmetry geometries of C_{50} (a), C_{70} (c), and C_{84} (d). (Redrawn from Ref. [8].)

which is readily verified in the limit of large N by replacing the summation by an integration. Identifying I with mR^2, where R is the equilibrium radius of the C cages under discussion, one finds the eigenvalue sum E_π to be given by

$$E_\pi = \frac{n(n+1)^2(n+2)\hbar^2}{2mR^2}. \qquad (5.2)$$

We now invoke a "rule" (to be confirmed later) that the equilibrium cage radius R_e is such that the surface area per C atom remains constant as the cage size becomes larger due to increasing the number N of C atoms [8]. This evidently implies that

$$\frac{4\pi R_e^2}{N} = \text{const} \qquad (5.3)$$

or equivalently

$$R_e = k_e N^{1/2}, \qquad (5.4)$$

where $k_e = 0.463$ Å (see Table 1 of Ref. [8]).

Returning to Eqs. (5.1) and (5.2), and recognizing that the number of electrons N is related to n in Eq. (5.2) by

$$N = 2\sum_{j=0}^{n}(2j+1) = 2(n+1)^2, \qquad (5.5)$$

we find, after using Eq. (5.4) that [8]

$$E_\pi = \frac{\hbar^2 N}{8mk_e^2} - \frac{\hbar^2}{4mk_e^2}, \qquad (5.6)$$

which is clearly of the correct general form of a graphene sheet.

Pressing related scaling properties for all electron Hartree–Fock (HF) calculations [8], Figure 8 shows the four C cages considered while Figure 9 plots the total HF energy/C atom versus $N^{1/2}/R$. The curves for the individual cages have only been shifted by a constant in order to bring them into such remarkable coincidence.

5.1. RELATION BETWEEN TOTAL ENERGY AND EIGENVALUE SUM: ALL ELECTRON RESULTS FOR FOUR C CAGES

In early work on the Thomas–Fermi electron density theory [21], March and Plaskett [22] showed that for neutral atoms the total energy E was simply related to the eigenvalue sum, say E_s, by

$$E = \tfrac{3}{2} E_s. \qquad (5.7)$$

This relation was subsequently shown by Ruedenberg [23] (see also Ref. [24]) to be well obeyed by self-consistent field calculations on a variety of molecules at their equilibrium geometry. In Figure 10, we show the results for $E(N, R_e)$ versus the all-electron eigenvalue sum $E_s(N, R_e)$. There is indeed a remarkably linear relation for the four C_n cages,

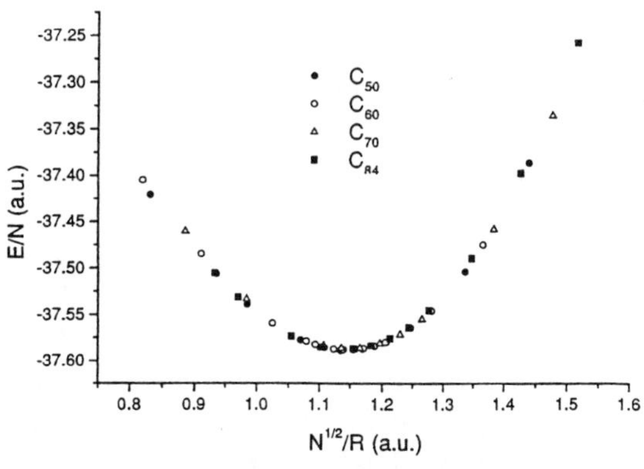

FIGURE 9. Scaling properties of total Hartree–Fock energy per C atom (E/N) vs. $N^{1/2}/R$ where N is the number of C atoms and R is the C cage radius. (Redrawn from Ref. [8].)

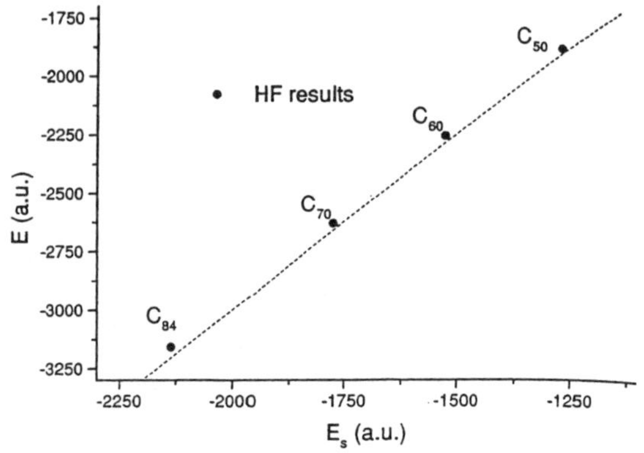

FIGURE 10. Relation between total Hartree–Fock (HF) energy E and eigenvalue sum E_s at the equilibrium radii. The solid circles show the HF results and the dashed line is the theoretical prediction (5.7). (Redrawn from Ref. [8].)

with $n = 50, 60, 70$, and 84. The dashed line shown in Figure 10 represents Eq. (5.7) and the agreement of the theoretical prediction from electron density arguments with the HF data is quantitative.

We close this section with some comments about many-body effects beyond Hartree–Fock or related single determinant theory. The comments can be viewed as applicable to both B_n cages treated in Section 5 and the C cages considered in the present section.

It is a matter of importance, for the future, to study the influence of electron correlation on the B_n clusters discussed in Section 4, as well as in the fullerene cages referred to immediately above. In the latter context, the work of Flocke et al. [25] is noteworthy. These authors focused on C_{60} itself, and not only for the ground state but also in excited states, using the Hubbard Hamiltonian to model correlation effects. For C_{60}, their major finding is that the general description using one-electron theory remains essentially intact, but they warn against drawing the conclusion that this will apply equally well to other fullerenes.

6. Mixtures of Light Atoms: BN Cages and Cationic Hydrocarbons

In this section we want to refer to two specific mixtures of light atoms; namely, some BN cages and a species of cationic hydrocarbons.

6.1. BORON NITRIDE CAGES

Early calculations on BN using a tight-binding-like (Hückel) approximation for the π electrons were reported by Zhu et al. [26]. This and other investigations (see Ref. [27]) have dealt with inorganic analogues of fullerene cages (compare Section 5) that are constructed entirely of BN pairs.

However, Seifert et al. [28] emphasize that a direct geometrical correspondence between BN and C polyhedra is not to be anticipated. By way of example, these authors note that although the truncated icosahedron is an ideal shape for C_{60}, it would be unlikely to yield a stable $B_{30}N_{30}$ structure. Its fivefold rings preclude the possibility of a fully alternating BNBN sequence of atoms and would require a BB or NN bond in each pentagon. Seifert et al. [28] suggest that, although weaker than the heteronuclear bonds, the BB and NN bonds might result in a local minimum for the truncated octahedron.

Fowler et al. [29] carried out a systematic survey of all possible square + hexagon trivalent $(BN)_x$ cages with $4 \leq x \leq 30$. Later work [28] reports in more detail the stable $(BN)_x$ isomers singled out in Ref. [29], discussing geometries and electronic structures as a function of size. Their results for the calculated binding energies per atom of $(BN)_{n/2}$ as a function of the number of atoms n are depicted in Figure 11. This also shows that the structures of especially stable clusters occurring at $n = 24, 32, 44$, and 56.

In addition, Seifert et al. [28] analyzed the HOMO and LUMO energies of the most stable

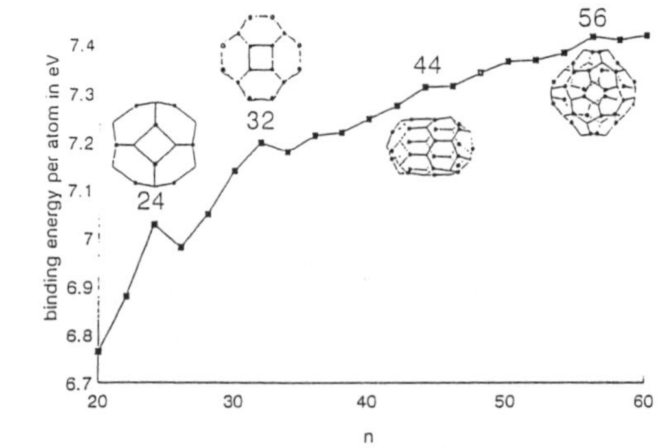

FIGURE 11. Theoretically calculated binding energies per atom in eV of $(BN)_{n/2}$ as a function of the number of atoms n. Note the especially stable clusters at n = 24, 32, 44, and 56. (Redrawn from Ref. [28].)

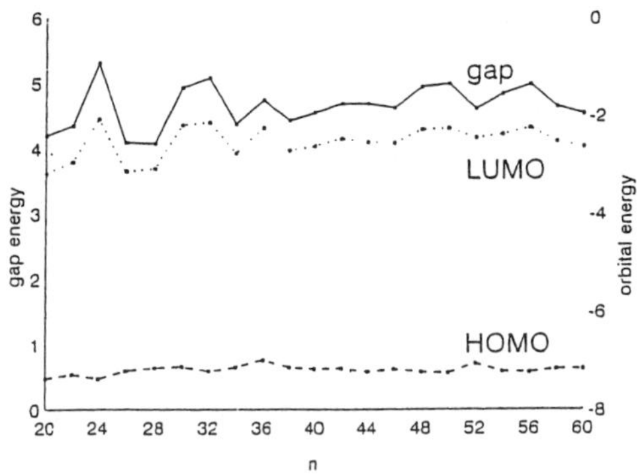

FIGURE 12. HOMO and LUMO energies in eV corresponding to the most stable conformations at each cluster size as a function of the number of atoms *n* in $(BN)_{n/2}$ isomers, as calculated by Seifert et al. [28]. The HOMO-LUMO energy gap is also shown. It is noteworthy [28] that larger energy gaps occur for clusters which have high relative stability. (Redrawn from Ref. [28].)

geometries at each cluster size and their results are depicted in Figure 12, together with the HOMO-LUMO energy gaps. They note that although the simple Hückel model such as considered in Ref. [26] would give little variation of the gap, their more refined treatment leads to larger gaps occurring for those clusters with high relative stability. (see also Ref. [30], especially Fig. 1).

We turn from this brief summary of mixed (BN) cages to a second mixture: cationic hydrocarbons.

6.2. CATIONIC HYDROCARBONS

This very brief subsection has been motivated by Ref. [31], in which the hydrocarbon ion $(C_3H_5)^{2+}$ radical has been considered. In this article, a structure has been proposed. But using Hartree–Fock theory as starting point, refined then by adding second-order Møller-Plesset perturbation corrections, Howard et al. [32] have located a lower-energy stable structure than that proposed in Ref. [31]. The energy difference is 1.05 eV for this 2+ hydrocarbon cation. The new structure found [32] is both nonlinear and nonplanar. To a useful approximation, it can be considered as $(CH_2—CH—CH_2)^{2+}$ with an angle of some 120° between the C—C bonds (see Fig. 13). The vibrational frequencies have also been calculated and, as required for stability of the structure, all turn out to be real.

Our motivation in the above study was linked to the Coulomb explosion in the multiply charged deuterium clusters discussed earlier in Section 2. There, for doubly charged cationic deuterium, Coulomb explosion of the clusters occurred. Here there is no such explosion for the doubly charged hydrocarbon cluster discussed above. Our finding is that most of the positive charge in $(CH_2—CH—CH_2)^{2+}$ is distributed over the hydrogen nuclei.

7. Summary and Future Directions

We have covered four light elements here; hydrogen isotopes, lithium, boron, and carbon. Both neutral and cationic clusters have been referred to, the latter for hydrogen isotopes. In one of the two examples of mixed clusters, namely hydrocarbons, the focus has been on multiply charged cations.

It seems clear that the problem of predicting the lowest isomers of clusters of light elements is a most stringent challenge for present many-electron theories. The best approach currently is therefore to combine experiment and theory, as has been done successfully in the example of the Li_8 cluster referred to in Section 3.

In the writer's opinion, Coulomb explosion of multiply charged clusters is an area well worth fuller development, especially on the theoretical side. Thus, for the examples of H isotopes treated in Section 2, and especially related to DD fusion, experiment is currently well ahead of theory and it seems important to narrow that gap and thereby gain increasing insight into the basic physical processes underlying the dependence say, of neutron yield on cluster size. To further compare and con-

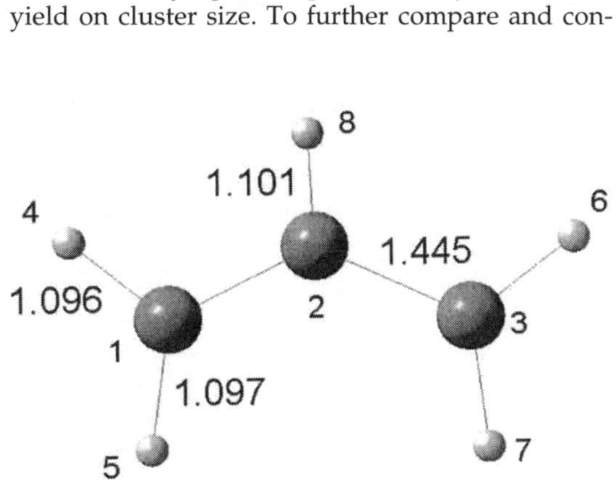

FIGURE 13. Structure of cationic hydrocarbon $(C_3H_5)^{++}$. (Redrawn from Ref. [32].)

trast Coulomb explosion in multiply charged cationic hydrocarbons with that of the hydrogen isotopes seems another task for quantum chemical studies which should bear fruit.

Naturally, because of its practicability it is especially important in the entire area covered in the present article that current fundamental progress in constructing truly ab initio "nonlocal" exchange potentials (i.e., transcending LDA and GGA) [33, 34] is rapidly embedded in density functional programs. Of course, though a final treatment of the exchange potential $V_x(\mathbf{r})$ is now within our grasp, it will be a long time before an equally fundamental approach to the (fortunately generally much smaller) correlation contribution to the effective exchange-correlation potential $V_{xc}(\mathbf{r})$ is designed. That is why, at the present time, it is so important, in the area of the chemical physics of clusters of light elements, that there is the strongest possible interplay between experiment and many-body theory.

ACKNOWLEDGMENTS

It is a pleasure to acknowledge many valuable discussions on the general area embraced by this article with Drs. C. Amovilli and G. G. N. Angilella, Professors I. A. Howard and D. J. Klein, and Dr. F. E. Leys. Partial financial support from the Office of Naval Research for work on many-electron theory is acknowledged. Special thanks go to Dr. P. Schmidt of that Office for much motivation and his continuing support.

References

1. Gobet, F.; Farizon, B.; Farizon, M.; Gaillard, M. J.; Buchet, J. P.; Carré, M.; Scheier, P.; Märk, T. D. Phys Rev Lett 2002, 89, 183403.
2. Zweiback, J.; Smith, R. A.; Cowan, T. E.; Hays, G.; Wharton, K. B.; Yanovsky, V. P.; Ditmire, T. Phys Rev Lett 2000, 84, 2634.
3. Howard, I. A.; Leys, F. E.; March, N. H.; Van Alsenoy, C.; Alonso, J. A.; Rubio, A. In Nanotechnology; Vajtai, R., Aymerich, X., Kish, L. B., Rubio, A., Eds.; Proceedings SPIE, 5118; SPIE: Bellingham, WA, 2003; p 331.
4. Howard, I. A.; Alonso, J. A.; March, N. H.; Rubio, A.; Van Alsenoy, C. Phys Rev A, in press.
5. Kornath, A.; Kaufmann, A.; Zoermer, A.; Ludwig, R. J Chem Phys 2003, 118, 6957.
6. Grassi, A.; Lombardo, G. M.; Angilella, G. G. N.; March, N. H.; Pucci, R. J Chem Phys, in press.
7. Amovilli, C.; March, N. H. Chem Phys Lett 2001, 347, 459.
8. Amovilli, C.; Howard, I. A.; Klein, D. J.; March, N. H. Phys Rev 2002, A66, 013210.
9. Kornath, A.; Ludwig, R.; Zoermer, A. Angew Chem 1998, 110, 1620.
10. Kornath, A.; Kaufmann, A.; Zoermer, A.; Ludwig, R. High Temp Mater Sci 2000, 15, 345.
11. Bonacic-Koutecky, V.; Fantucci, P.; Koutecky, J. J Mol Struct 1988, 166, 221.
12. La Placa, S. J.; Roland, P. A.; Wynne, J. Chem Phys Lett 1992, 190, 163.
13. Clougherty, D. P.; Zhu, X. Phys Rev 1997, A56, 632.
14. Despa, F. Phys Rev 1998, B57, 7335.
15. Siringo, F.; Piccitto, G.; Pucci, R. Phys Rev 1992, A46, 4048.
16. March, N. H. Proc Cambridge Philos Soc 1952, 48, 665.
17. Boustani, I. Phys Rev 1997, B55, 1.
18. Boustani, I.; Rubio, A.; Alonso, J. A. Chem Phys Lett 1999, 311, 21.
19. Boustani, I. Chem Phys Lett 1995, 240, 135.
20. See e.g., Savina, M. R.; Lohr, L. L.; Francis, A. H. Chem Phys Lett 1993, 205, 200.
21. See e.g., March, N. H. Electron Density Theory of Atoms and Molecules, Academic: New York, 1992.
22. March, N. H.; Plaskett, J. S. Proc R Soc London, Ser A 1956, 235, 419.
23. Ruedenberg, K. J Chem Phys 1977, 66, 375.
24. March, N. H. J Chem Phys 1977, 67, 4618.
25. Flocke, N.; Schmalz, T. G.; Klein, D. J. J Chem Phys 2000, 112, 8233.
26. Zhu, H.-Y.; Klein, D. J.; Seitz, W. A.; March, N. H. Inorg Chem 1995, 34, 1377.
27. Sun, M.-L.; Slanina, Z.; Lee, S.-L. Chem Phys Lett 1995, 233, 279.
28. Seifert, G.; Fowler, P. W.; Mitchell, D.; Porezag, D.; Frauenheim, Th. Chem Phys Lett 1997, 268, 352.
29. Fowler, P. W.; Heine, T.; Mitchell, D.; Schmidt, R.; Seifert, G. J Chem Soc, Faraday Trans 1996, 92, 2197.
30. Fowler, P. W.; Rogers, K. M.; Seifert, G.; Terrones, M.; Terrones, H. Chem Phys Lett 1999, 299, 359.
31. Gluch, K.; Fedor, S.; Matt-Leubner, O.; Echt, A.; Stamatovic, M.; Probst, M.; Scheier, P.; Märk, T. D. J Chem Phys 2003, 118, 3090.
32. Howard, I. A.; March, N. H.; Van Alsenoy, C. Internet Electron. J Mol Design 2004, 3, 110.
33. Della Sala, F.; Görling, A. J Chem Phys 2001, 115, 5718.
34. Howard, I. A.; March, N. H. J Chem Phys 2003, 119, 5789.
35. Zweiback, J.; Ditmire, T.; Perry, M. D. Phys Rev 1999, A59, R3166.
36. Ditmire, T.; Donnelly, T.; Rubenchik, A. M.; Falcone, R. W.; Perry, M. D. Phys Rev 1996, A53, 3379.

Appendix: Simple Model for Fusion Yield versus Cluster Size

It is known that large clusters built from atoms having high atomic number Z generally are characterized by "plasmalike" behavior. This is evident from the studies of Zweiback et al. [35] and Ditmire et al. [36], and is a result of space charge force confinement of many photoionized electrons to the body of the cluster. But Zweiback et al. [2] argue that, with the provisos that the laser has a sufficiently fast rise time and a high enough ponderomotive potential, clusters of deuterium can be stripped almost entirely of their electrons before explosion occurs.

Then Zweiback et al. use a simple model of the time needed for a uniformly charged sphere expanding under Coulomb forces to grow, say, to roughly twice the initial radius to estimate the rise time of the laser pulse which is required. For a sphere of ionized deuterons, they estimate this time, τ, say, to be in order of magnitude

$$\tau \sim \left\{\frac{4\pi\epsilon_0 M}{ne^2}\right\}^{1/2}, \tag{A1}$$

M being the mass of deuterium, n is the initial ion density ($\sim 3 \times 10^{22}$ cm^{-3} for D_2), and e is the charge of the electron. From Eq. (A1), $\tau \sim 20$ fs for deuterium clusters.

Although the initial radius of the cluster, say R_i, does not enter this estimate of τ, Zweiback et al. [2] argue that R_i determines the maximum energy E_{max} produced in the explosion. These workers then give, for a uniform density deuterium cluster:

$$E_{max} = \frac{q_{av} n e^2 R_i^2}{3\epsilon_0}, \tag{A2}$$

q_{av} denoting the average charge state in the cluster. Zweiback et al. conclude from this that fully stripped deuterium clusters with $R_i > 25$ Å will suffice to produce the multi-keV ions needed for fusion.

It is important to note the major conclusion of Zweiback et al. [2]; namely, that cluster formation in a deuterium gas jet can lead to nuclear fusion when irradiated by an intense femtosecond laser pulse. These authors uncover a marked dependence of neutron yield on cluster size, which is controlled through reservoir temperature. However (see Fig. 6 of Ref. [2]), there is a maximum neutron yield at a cluster diameter of 60 Å, the neutron yield versus cluster diameter appearing roughly symmetrical about its maximum. Zweiback et al. [2] are optimistic that their measured neutron yield can be increased.

PHYSICAL REVIEW B **72**, 113414 (2005)

BeB_2 nanostructures: A density functional study

L. M. Molina,[1] M. J. López,[1] I. Cabria,[1] J. A. Alonso,[1,2] and N. H. March[1,2,3]
[1]*Departamento de Física Teórica, Universidad de Valladolid, E-47011 Valladolid, Spain*
[2]*Donostia International Physics Center, 20018 San Sebastian, Spain*
[3]*Oxford University, Oxford, England*

(Received 19 May 2005; revised manuscript received 5 August 2005; published 27 September 2005)

Small $(BeB_2)_n$ clusters ($n=1-6$ and 8) as well as BeB_2 nanotubes and cages $(BeB_2)_n$, with $n=10$, 12, 15, 18, and 30, have been studied using density functional theory. These nanostructures strongly resemble the ones formed by carbon, with ringlike geometries for very small clusters and closed cages for medium-sized clusters. The energetics for distortion and closure of planar sheets into cages and nanotubes prove favorable in comparison with carbon systems.

DOI: 10.1103/PhysRevB.72.113414 PACS number(s): 61.46.+w, 61.48.+c, 71.15.Mb

It is well established that carbon graphitic sheets can be rolled into carbon nanotubes[1] and fullerenes.[2] The high energy of dangling bonds at peripheral sites causes an easy spontaneus closing into cage or tubelike structures, that can be further enhanced by employing suitable catalysts.[3] More recently, interest has arisen concerning the possibility of obtaining similar cage or tubular structures based on other materials than carbon. Onionlike cages and multiwalled nanotubes have been synthetized for compounds as various as boron nitride (BN), metal dichalcogenides (MX_2, $M=\text{Mo,W}$ and $X=\text{S,Se}$) and V_2O_5.[4,5] The common feature of all these systems, which rationalizes their ability to form curved nanostructures, is their layered character in the bulk. In all cases, as in graphite, bonding is highly directional, with strong covalent interactions within each layer and weak van der Waals interactions between them. Therefore, most materials with such features are susceptible of forming closed curved nanostructures. These new systems are interesting because of the features introduced by the different electronic structure or topology (for instance, different bonding schemes can lead to variations on the type of defects needed for the closing of the cages[6]).

Very interesting materials for fullerene and nanotube formation, quite unexplored, are alkaline earth diborides (MgB_2 and BeB_2). Their structure is based on a stacking of sheets where B atoms form a hexagonal graphenelike network, with Be (Mg) atoms intercalated midway between the sheets and placed above the center of each boron hexagon. These compounds are also interesting for another reason: the unexpected high-T superconductive behavior found for MgB_2.[7,8] It is also worth mentioning that pure boron clusters are likely to form planar structures (nanotubes and rings), as calculations by Boustani *et al.* have shown.[9,10] Recently, Zhang and Crespi have studied the electronic structure of BeB_2 single-walled nanotubes, finding a metallic behavior. They also find a small interlayer interaction (0.1 eV/atom) between BeB_2 sheets in the bulk.[11]

Using density functional theory (DFT), we study in this paper the structural and electronic properties of small BeB_2 clusters, cagelike structures, and nanotubes. We find that, apart from the very small clusters, most of the systems are metallic [that is, they lack having a highest occupied–lowest unoccupied molecular orbital (HOMO-LUMO) or band gap]. Also, there are close resemblances with the case of carbon nanostructures: in the case of very small clusters, structures based on boron chains are preferred, while, beginning at $Be_{10}B_{20}$, cagelike structures begin to form. Finally, all these systems are found, in principle, easy to form, since the energy cost required to distort and close planar sheets into closed cages and nanotubes is much lower than for analogous carbon systems.

The DFT simulations were performed using a supercell method and a plane wave basis set[12] (expanded up to 25 Ry) and ultrasoft pseudopotentials,[13] as implemented in the DACAPO code.[14] The local density approximation LDA functional of Vosko, Wilk, and Nusair[15] was used for treating exchange-correlation effects. It was checked that non-self-consistent generalized gradient approximation GGA calculations (with the PW91 functional[16]) deliver similar results (for example, the rolling energies reported below are nearly the same). For small clusters and cages, a large unit cell is employed in order to avoid interactions between clusters in neighboring cells. The cell dimensions are chosen depending on the size and shape of each cluster, to ensure a minimum distance between clusters of at least 9 Å. Only one k point (Γ) is used, due to the large size of the supercell. For single-walled nanotubes, large unit cells are used on the x-y plane (the tube axis is along the z direction). The lattice constant along the tube axis was optimized to find the one that minimizes the total energy. A $(1\times1\times4)$ Monkhorst-Pack k-point sampling is used in this case. Finally, all the geometries were fully relaxed using the Broyden-Fletcher-Goldfarb-Shanno algorithm.[17]

Figure 1 shows the most stable relaxed structures for small $(BeB_2)_n$ clusters ($n=1-6$ and 8), and in some cases a few competing isomers, together with the relative energies. In each case, a large number of initial trial structures, with various geometries, were relaxed, in order to arrive at the global minimum with some confidence. Overall, although the structures are complex, the clusters show a marked preference for a chainlike arrangement of the boron atoms. The beryllium atoms form a weblike structure attached from above and below to the boron chain, in the smallest clusters ($n=1-6$). However, the two piled up B chains of Be_8B_{16} are

1098-0121/2005/72(11)/113414(4)/$23.00

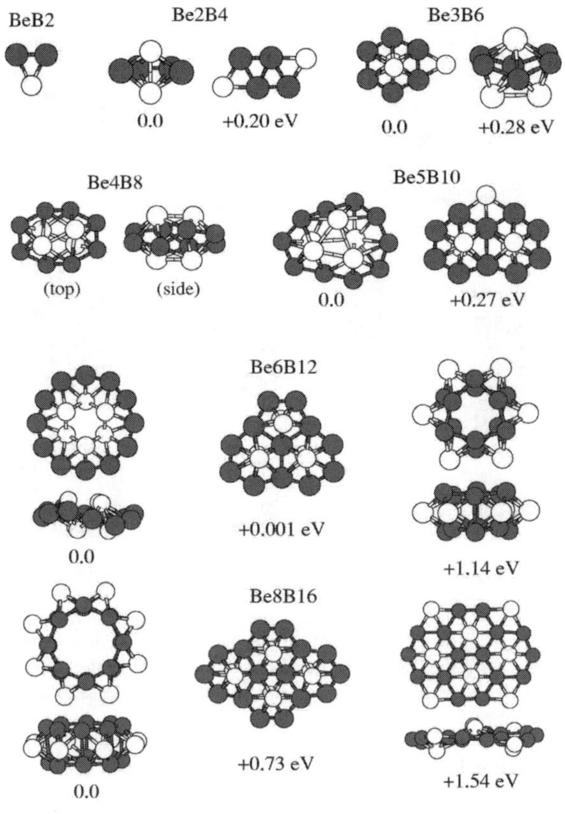

FIG. 1. (Color online) Lowest-energy structures for small $(BeB_2)_n$ clusters ($n=1-6$ and 8). For some of them, competing isomers, and their relative energies, are shown.

fastened with an external ring of Be atoms. These structures compete with, and frequently are more stable than, the ones obtained by cutting pieces of a BeB_2 sheet (one of those pieces is shown for Be_8B_{16}, lying 1.54 eV above the ground state). It is worth noting the close resemblance of these closed chainlike boron structures with the most stable arrangements for small carbon clusters up to $n \approx 20$, that are linear chains. If we consider, given the electronegativity difference, that charge is transferred from beryllium to boron, then the latter can attain an electronic configuration analogous to that of carbon. We have checked the Be-B charge transfer for the ground state structure of the Be_6B_{12} cluster: in Fig. 2 we plot isosurfaces for the induced charge density calculated as

$$\rho_{\text{induced}} = \rho(Be_6B_{12}) - \rho(B_{12}) - \rho(Be_6), \quad (1)$$

where $\rho(B_{12})$ and $\rho(Be_6)$ are the charge densities of the boron chain and of the Be atoms respectively. In the figure, it can be seen that ρ_{induced} becomes negative at the central part of the cluster, where the Be atoms are located, indicating a charge depletion of that region. The transferred charge is accumulated above and below the boron atoms in the chain, indicating an enhancement of π bonding between them. Finally, a small depletion of the σ bonds between boron atoms is observed.

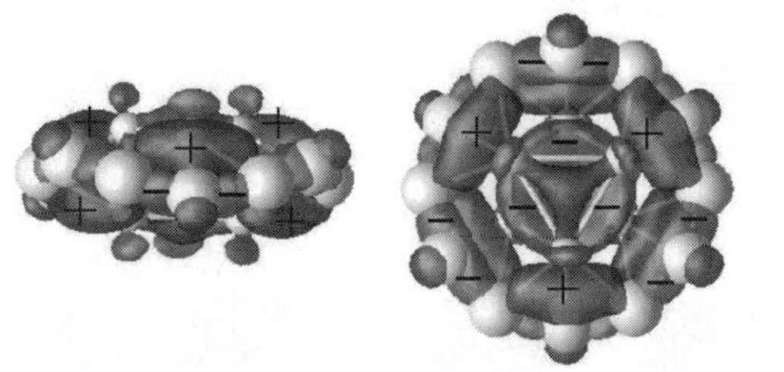

FIG. 2. (Color online) Side and top views of the Be_6B_{12} cluster. Small and large spheres represent Be and B atoms, respectively. Superimposed positive and negative isofurfaces corresponding to $+0.03$ and -0.03 $e^-/\text{Å}^3$ values of the induced charge density ρ_{induced} defined in the text. Plus and minus signs label positive and negative isosurfaces.

The B-B bond distances for small $(BeB_2)_n$ clusters (around 1.55–1.60 Å) are considerably shortened with respect to the B-B distances for either bulk BeB_2 (1.68 Å) or an isolated BeB_2 sheet (1.74 Å). This indicates a somewhat different type of bonding. Let us note that for Be_8B_{16}, the interchain B-B distances (2.06 Å) are much longer than the 1.62 Å distances along each individual chain. Therefore, the structure of Be_8B_{16} can be best viewed as two parallel circular boron chains coupled by the Be capping atoms.

For larger clusters, we find that cagelike structures based on the rolling of BeB_2 sheets are very stable. Figure 3 shows the relaxed structures for $Be_{10}B_{20}$, $Be_{12}B_{24}$, $Be_{15}B_{30}$ $Be_{18}B_{36}$, and $Be_{30}B_{60}$ (this one being an analog of the C_{60} fullerene). Let us note that, for BeB_2 cages maintaining the 1:2 Be to B stoichiometry, two faces of the boron polyhedron are not capped by Be atoms. In order to check the stability of these cages we have performed an extensive search for $Be_{12}B_{24}$, relaxing different types of structures (cage, planar, ring, etc.). Two practically degenerate cagelike isomers, shown in Fig. 3, are found to be much more stable than any other alternative structures. They are 8–10 eV lower in energy than planar structures, showing that it is energetically feasible to form closed cages in this range of sizes. Also, cages are much more stable than the ring geometry shown in the figure (a larger version of the most stable isomer for Be_8B_{16}).

It is interesting to discuss in detail the structure of the two lowest-energy $Be_{12}B_{24}$ isomers; the initial trial structure for the first one (slightly more stable) was a fullerenelike cage where the two noncapped Be pentagons are not facing each other. It is the lower symmetry of this structure that apparently causes, upon relaxation, the opening of some windows in the B network which are capped by two Be atoms. This, effectively, leads to an structure that can be viewed as a "network" of boron chains, the most stable structural motif for smaller clusters. On the other hand, the competing isomer has a more symmetrical cage structure, that can be constructed by cutting a piece of a (4,0) BeB_2 nanotube (see below).

We have to keep in mind that it is not assured that the 1:2 stoichiometry should be maintained upon formation of these

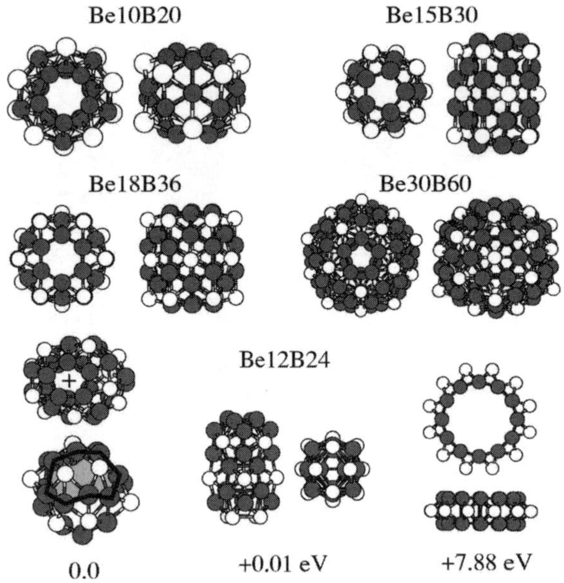

FIG. 3. (Color online) Equilibrium structures of some $(BeB_2)_n$ cages. In the case of $Be_{12}B_{24}$, three isomers are shown. A plus sign indicates the location of the two non-Be-capped B pentagons for the most stable isomer. In the second view, one of the B windows is highlighted.

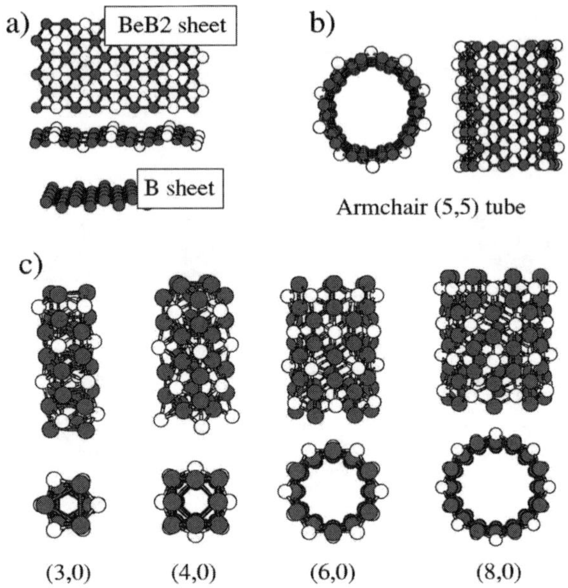

FIG. 4. (Color online) (a) Top and side views of the most stable structure for an isolated BeB_2 sheet, and side view for a pure B sheet. (b) Equilibrium structure of an armchair (5,5) BeB_2 nanotube. (c) Structures of zigzag (3,0), (4,0), (6,0), and (8,0) BeB_2 nanotubes.

clusters, and that other Be concentrations could be reached. To study this point we have optimized the structures of several Be_xB_{24} cagelike isomers, to check if there is any clear preference for the 1:2 stoichiometry, that is, if the cluster with $x=12$ is relatively more stable than clusters with $x=11$, 13, or 14. This is indeed the case, since the energy required to evaporate a Be atom from Be_xB_{24} reaches a local maximum of 5.3 eV at $x=12$.

Finally, the properties of some BeB_2 single-walled nanotubes, constructed by rolling a BeB_2 sheet, were studied. Figure 4 shows the equilibrium structure of the relaxed BeB_2 sheet, and (5,5), (3,0), (4,0), (6,0), and (8,0) nanotubes. It is worth mentioning that for the isolated BeB_2 sheet, several different conformations are possible, depending on whether all the Be atoms are placed on one side of the sheet or if they occupy both sides. Comparing different possibilities we find that the conformer shown in the figure is the most stable, being 0.12 eV more stable (per BeB_2 unit) than the conformer with all Be atoms on one side. Such a buckled structure closely resembles the preferred conformation for a pure boron sheet (also shown in the figure). However, tests for nanotubes of these diameters indicate the preference of the Be atoms for sites on the external surface.

In Fig. 5, we show the energetic cost of rolling a BeB_2 sheet into nanotubes of various diameters, and compare those rolling energies with the analogous ones for carbon nanotubes and for a pure (4,4) boron nanotube. The rolling energy is obtained by subtracting from the energy per C (or B) atom of the nanotube the energy per atom of the planar graphene (or BeB_2) sheet. We have also included in the plot results corresponding to the various $(BeB_2)_n$ cages studied, as well as a few C_n cage structures. It is clear that the BeB_2 material is easier to bend and roll than carbon, making the synthesis of those BeB_2 nanostructures probably feasible. In particular,

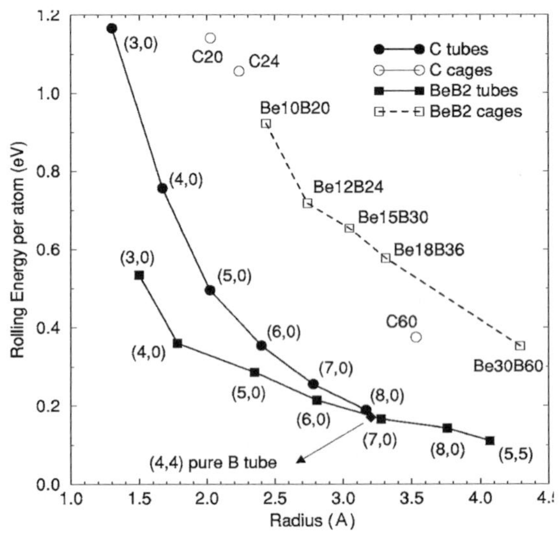

FIG. 5. Rolling energies (see text for definition) per C (or B) atom for various cagelike and tubular carbon and BeB_2 structures, as a function of the radius. The rolling energy for a pure boron (4,4) nanotube is also included.

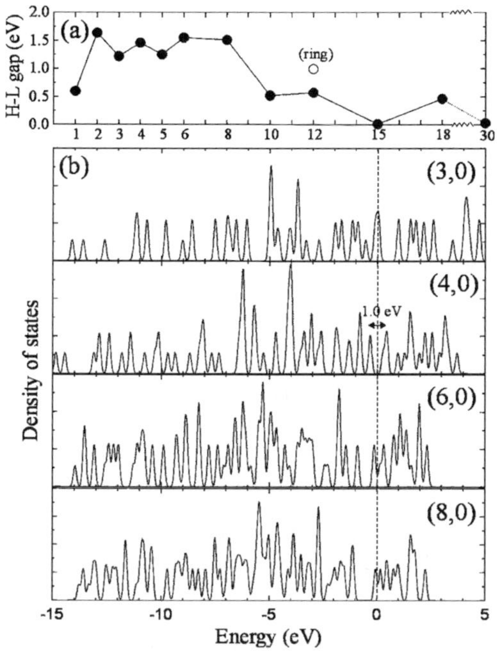

FIG. 6. (a) HOMO-LUMO gap (in eV) as a function of n for $(BeB_2)_n$ small clusters and cagelike structures. In each case, the gap corresponding to the lowest-energy isomer is plotted (for Be_8B_{16}, we show as well the gap for the less stable ring structure). (b) Total densities of states (DOS) for (3,0), (4,0), (6,0), and (8,0) BeB_2 nanotubes. Energies are referred to the Fermi level (0.0 eV), its position highlighted by the vertical dotted line.

it is interesting to realize that the formation of very thin (3,0) or (4,0) BeB_2 nanotubes costs much less energy as compared with analogous thin carbon nanotubes, which have been found only confined in the inner regions of thick multiwalled carbon nanotubes.[18] As for the electronic structure of these tubes, in agreement with Zhang and Crespi,[11] we find them preferentially metallic. Figure 6(a) shows that only small $(BeB_2)_n$ clusters with ringlike structures (up to $n=8$) have a sizable HOMO-LUMO gap; as soon as the bonding character changes to sp^2 in cagelike isomers, the gap nearly closes. The densities of states for the BeB_2 tubes shown in Fig. 6(b) indicate a metallic behavior, with the exception of the (4,0) tube, slightly semiconducting with a gap of around 1 eV. The (6,0) and (8,0) tubes have residual gaps of less than 0.5 eV.

In summary, DFT calculations, with the LDA exchange-correlation potential, have been performed for small $(BeB_2)_n$ clusters, as well as for BeB_2 nanotubes and some closed cages. With the exception of very small clusters, most of the systems studied are found to have metallic character. One of the findings reported here is that the energetics for distortion and closure of planar sheets into cages and nanotubes prove favorable in comparison with the corresponding carbon atom assemblies. As to the LDA approximation dominantly employed in the present study, it has been checked that non-self-consistent GGA calculations[16] lead to nearly the same rolling energies reported here for the LDA studies. Obviously, however, because of the relatively tiny energy differences between competing cluster isomers, some of which are displayed in Fig. 1, it would be of considerable interest if ab initio quantum chemical calculations, using for instance coupled-cluster electronic structure theory, could be carried out in the future, for comparison with the DFT predictions presented here.

ACKNOWLEDGMENTS

This work was supported by the Spanish Ministry of Science and Technology (MCYT) through Grant No. MAT2002-04499. L.M.M. also acknowledges MCYT for supporting his work through the Ramon y Cajal program. N.H.M. wishes to acknowledge that his contribution to this study began during a visit to the University of Valladolid. Thanks are due to J. A. Alonso and L. M. Nieto for very kind hospitality.

[1] S. Iijima, Nature (London) **354**, 56 (1991).

[2] H. W. Kroto, J. R. Heath, S. C. O'Brien, R. F. Curl, and R. E. Smalley, Nature (London) **318**, 162 (1985).

[3] D. S. Bethune, C. H. Kiang, M. S. de Vries, G. Gorman, R. Savoy, J. Vazquez, and R. Beyers, Nature (London) **363**, 605 (1993).

[4] R. Tenne, M. Homyonfer, and Y. Feldman, Chem. Mater. **10**, 3225 (1998).

[5] R. Tenne, Prog. Inorg. Chem. **50**, 269 (2001).

[6] J. A. Ascencio, M. Perez-Alvarez, L. M. Molina, P. Santiago, and M. José-Yacaman, Surf. Sci. **526**, 243 (2003).

[7] J. Nagamatsu, N. Nakagawa, T. Muranaka, Y. Zenitani, and J. Akimitsu, Nature (London) **410**, 63 (2001).

[8] C. Buzea and T. Yamashita, Semicond. Sci. Technol. **14**, R115 (2001).

[9] I. Boustani, Phys. Rev. B **55**, 16426 (1997).

[10] I. Boustani, A. Quandt, E. Hernandez, and A. Rubio, J. Chem. Phys. **110**, 3176 (1999).

[11] P. Zhang and V. H. Crespi, Phys. Rev. Lett. **89**, 056403 (2002).

[12] M. C. Payne, M. P. Teter, D. C. Allan, T. A. Arias, and J. D. Joannopoulos, Rev. Mod. Phys. **64**, 1045 (1992).

[13] D. Vanderbilt, Phys. Rev. B **41**, R7892 (1990).

[14] http://www. fysik.dtu.dk.

[15] S. H. Vosko, K. Wilk, and N. Nusair, Can. J. Phys. **58**, 1200 (1980).

[16] J. P. Perdew, J. A. Chevary, S. H. Vosko, K. A. Jackson, M. R. Pederson, D. J. Singh, and C. Fiolhais, Phys. Rev. B **46**, 6671 (1992).

[17] D. C. Liu and J. Nocedal, Math. Program. **45**, 503 (1989).

[18] L.-C. Qin, X. Zhao, K. Hirahara, Y. Miyamoto, Y. Ando, and S. Iijima, Nature (London) **408**, 50 (2000).

PHYSICAL REVIEW A **73**, 063205 (2006)

Two-dimensional electrostatic analog of the March model of C_{60} with a semiquantitative application to planar ring clusters

Claudio Amovilli*
Dipartimento di Chimica e Chimica Industriale, Università di Pisa, Via Risorgimento 35, I-56126 Pisa, Italy

Norman H. March†
Department of Physics, University of Antwerp, Groenenborgerlaan 171, B-2020 Antwerpen, Belgium
(Received 20 March 2006; published 22 June 2006)

The so-called March model of fullerene uses the one-center approximation of quantum chemistry to calculate the spherical π-electron density ρ by the Thomas-Fermi (TF) statistical method. Then the variational principle of the TF theory is employed by combining the electronic energy determined entirely by ρ with the discrete C_{60} nuclear potential energy to predict, semiquantitatively, the equilibrium radius. Here the same approach is worked out, but now largely analytically, in two dimensions, as a possible model of planar ring clusters. Again the equilibrium radii are calculated and consistency with realistic clusters of hydrogen and carbon rings is discussed in detail. Finally, an estimate of the "breathing" frequency of C_{14}, C_{18}, and C_{22} is given.

DOI: 10.1103/PhysRevA.73.063205 PACS number(s): 36.40.−c, 31.15.Bs

I. INTRODUCTION

The so-called March model [1,2] of fullerene approximates the 60 π electrons peaked around the surface of the buckyball by a spherical distribution of electron density ρ. This then leads, by a variational calculation based on the Thomas-Fermi (TF) energy functional $E_{TF}[\rho]$, but retaining the nuclear-nuclear potential energy for the correct discrete C nuclei positions [1], to an equilibrium radius of the buckminster fullerene somewhat larger than experiment, but already useful. Similar considerations, but with Hartree-Fock method instead of TF, have been successfully used in quantitative calculations on spherical boron cages [3].

The present study has been motivated by the cluster studies on C reported by Jones and Seifert [4]. In particular, they report by the density functional study (DFT), that the likely lowest isomers of C_N with N between 14 and 22 are planar rings with "cumulenic" spacing between the nuclei on the rings. This has prompted us to consider the analog of the three-dimensional March model of C_{60} to apply to such quasi two-dimensional planar ring clusters.

The outline of the paper is then as follows. In Sec. II below, for completeness, we briefly summarize the TF statistical model for such a two-dimensional (2D) planar electronic assembly. Then we focus on the use of electrostatics and charge interactions satisfying the corresponding 2D Poisson equation. Analogous to the March model of C_{60}, ρ, now the number of electrons per unit area, is taken to have circular symmetry. This is, in the spirit of the one-center expansion, equivalent to smearing N point charges f into a uniform line charge with total charge Nf. Whereas in C_{60}, Nf was, naturally enough, taken to be 60, i.e., one π electron per C atom [5,6], in cluster rings we shall study, now largely analytically, the equilibrium radius R_e of the rings as a function of N. The details of this are set out in Sec. III. In Sec. IV we analyze the consistency of the present model with real *ab initio* and DFT calculations on ring clusters. The paper concludes with a summary plus some proposals directions for further study in Sec. V.

II. BRIEF SUMMARY OF THE TWO-DIMENSIONAL THOMAS-FERMI THEORY

The semiclassical TF statistical method is based on two assumptions: (i) a chemical potential equation for the fastest electron with Fermi momentum $p_F(\mathbf{r})$ and (ii) a phase-space argument relating $p_F(\mathbf{r})$ to the electron density $\rho(\mathbf{r})$. Assumption (i) corresponding to writing the chemical potential μ as the sum of the kinetic term $p_F^2/2m$, with m as the electron mass, and a one-body potential energy denoted below by $V(\mathbf{r})$ yields explicitly

$$\mu = \frac{p_F^2}{2m} + V(\mathbf{r}) \tag{1}$$

which is valid for general dimensionality D. However, D enters the phase-space considerations in (ii) above, since a cell of volume h^D can hold two electrons with opposed spin in the spin compensated ground state we consider throughout the present study. Thus in 2D, one has for an area Ω of coordinate space and a momentum space area πp_F^2, that the number of electrons per unit area, $\rho(\mathbf{r})$, is given by

$$\rho(\mathbf{r}) = \left(\frac{n_{el}}{\Omega}\right)_{\mathbf{r}} = \frac{2\pi p_F^2(\mathbf{r})}{h^2}, \tag{2}$$

the numerator of the final term being the spin occupancy two times the area of occupied momentum space, while the denominator is the size of the basic phase-space cell in 2D. Removing $p_F^2(\mathbf{r})$ from Eq. (1), in which μ is constant throughout the whole of the inhomogeneous electron distribution under discussion, using Eq. (2) yields

*Electronic address: amovilli@dcci.unipi.it
†Also at: Oxford University, Oxford, England.

$$\mu = \frac{h^2}{4\pi m}\rho(\mathbf{r}) + V(\mathbf{r}). \quad (3)$$

This is the basic TF equation yielding the areal density $\rho(\mathbf{r})$ in 2D from a given one body potential $V(\mathbf{r})$. As usual μ is determined by the normalization condition on $\rho(\mathbf{r})$.

We turn in the following section to apply this Euler Eq. (3) of the TF statistical method to planar ring clusters.

III. SELF CONSISTENT FIELD AND ITS CONSEQUENCES IN QUASI-2D RING CLUSTERS

As in the TF neutral atom, and in C_{60} in the March model [1,2,5], the chemical potential μ can be put to zero in the neutral ring clusters under discussion. Equation (3) can be combined with the 2D form of Poisson's equation of electrostatics to yield, in atomic units ($\hbar=1$, $m=1$, $q_e=-1$, and $4\pi\epsilon_0=1$), for the electrostatic potential V_{es},

$$V_{es}(r) = -\pi\rho(r), \quad (4)$$

where V_{es} satisfies

$$\frac{\partial^2 V_{es}}{\partial r^2} + \frac{1}{r}\frac{\partial V_{es}}{\partial r} = 4V_{es}. \quad (5)$$

Scaling r to s, via $s=2r$, the electrostatic potential can be obtained by solution of Eq. (5), for a line charge of radius R, and with $t=2R$ as

$$V_{es}(r) = -\lambda[K_0(t)\Theta(t-s)I_0(s) + I_0(t)\Theta(s-t)K_0(s)], \quad (6)$$

where the quantity $\lambda(>0)$ is to be obtained from the density normalization condition ($n_{el}=Nf$),

$$Nf = 2\pi\int_0^\infty \rho(r)rdr, \quad (7)$$

while $\Theta(x)$ is the Heaviside step function and $K_0(x)$ and $I_0(x)$ are modified Bessel functions [7]. From the Gauss theorem, we have additionally that

$$q(s) = \frac{s}{2}V'_{es}(s), \quad (8)$$

which means that $q(s)$, the total charge, must tend to 0 as $s\to\infty$ for a neutral system.

A basic function of the present self consistent TF method for 2D ring clusters is then defined by $Q(s)=2q(s)/\lambda$. The merit of this definition is that $Q(s)$ no longer depends on normalization. It can be written, using Eqs. (6) and (8) as

$$Q(s) = -s[K_0(t)\Theta(t-s)I'_0(s) + I_0(t)\Theta(s-t)K'_0(s)]. \quad (9)$$

The jump (discontinuity due to line charge) at $s=t$ of Q, say ΔQ, is then given by

$$\Delta Q = t|I_0(t)K'_0(t) - K_0(t)I'_0(t)|, \quad (10)$$

where the right-hand side (rhs) of Eq. (10) contains the modulus of the Wronskian evaluated at t. Owing to the special form of the second order differential equation (5), the Wronskian is proportional to t^{-1} and, hence, ΔQ is constant. From (10) one finds $\Delta Q=1$. By the definition of Q this result leads to $\lambda=2Nf$, the jump of the function $q(s)$ being related to the total ring charge Nf. It is important to remark that, when t tends to 0, one recovers the two-dimensional TF model atom already discussed by Kventsel and Katriel [8].

Plots of scaled potential $v(s)=V_{es}(s)/\lambda$ and total charge $Q(s)$ are presented in Fig. 1 for various choices of scaled ring radii t.

A. Variational TF calculation of ring radii as a function of N

As already mentioned we follow the March model of C_{60} by using 2D TF theory variationally, based on the "line charge" circular density $\rho(r)$ given by Eq. (4). Then the energy $E_{TF}[\rho]$ in 2D is explicitly

$$E_{TF}[\rho] = \pi^2\int_0^\infty \rho(r)^2 rdr + 2\pi\int_0^\infty \rho(r)V_{lc}(r)rdr - \int \rho(r_1)\rho(r_2)\ln r_{12}d\mathbf{r}_1 d\mathbf{r}_2 + U_{nn}. \quad (11)$$

Here, V_{lc} is the confining potential due to the nuclear effective charges smeared on a circumference of radius R, namely

$$V_{lc}(r) = \begin{cases} 2Nf\ln R & (r\leq R) \\ 2Nf\ln r & (r>R) \end{cases}. \quad (12)$$

As already emphasized, U_{nn} is to be treated using the discrete nuclei on rings, in order to calculate the equilibrium radius R_e given by the equilibrium condition

$$\left(\frac{\partial E_{TF}}{\partial R}\right)_{R_e} = 0. \quad (13)$$

Following the Hellmann-Feynman theorem, Eq. (13) is equivalent to the balance of the forces acting on nuclei if the electron density entering the energy has been obtained variationally. Using the TF density coming from the potential given by Eq. (6), the radial component of the electric field due to electrons and acting on the ring is

$$\mathcal{E}_e = -\frac{2Nf\chi(2R)}{R}, \quad (14)$$

where

$$\chi(t) = tK_0(t)I'_0(t) \quad (15)$$

is the fraction of electrons inside the circle of radius $t=2R$. Then, for a set of N equispaced point charges f lying on a circumference of radius R, the resultant electric field acting on one of such charges and due to the other $N-1$ has the radial component (still in 2D)

$$\mathcal{E}_n = \frac{(N-1)f}{R} \quad (16)$$

which can be used to establish the force balance equation

$$\mathcal{E}_e + \mathcal{E}_n = 0. \quad (17)$$

Combining Eqs. (14), (16), and (17) we get

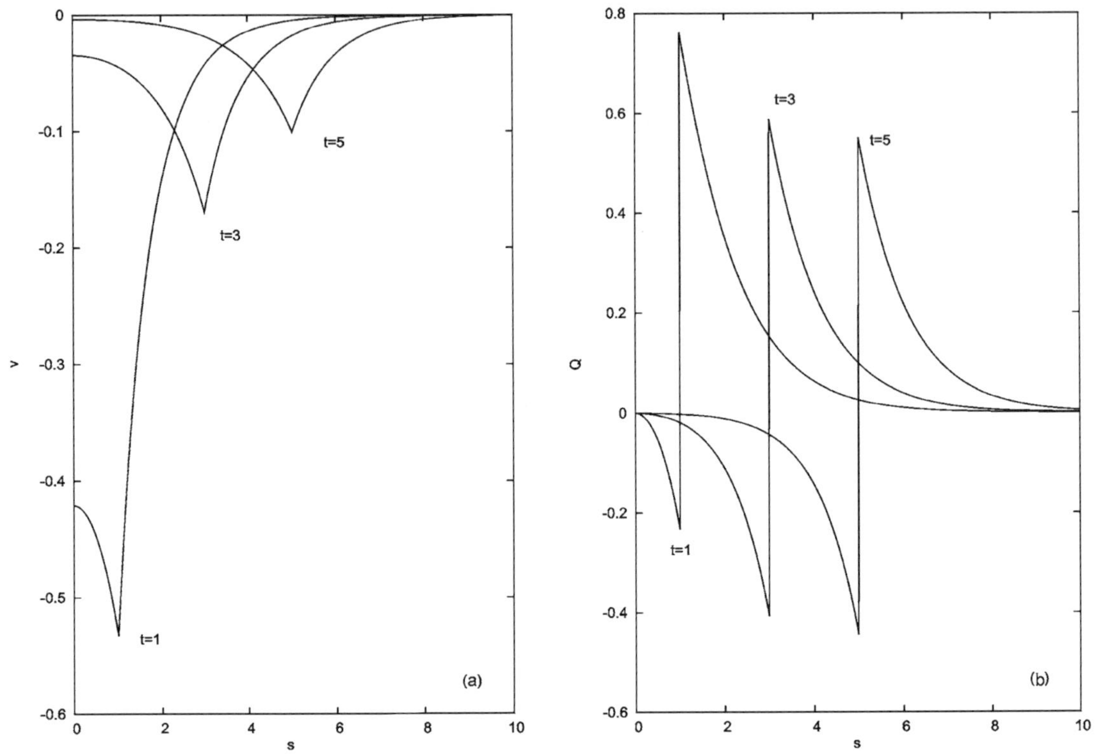

FIG. 1. Thomas-Fermi self consistent reduced potential (a) and charge (b) for a two-dimensional electron system, confined by a positive ring line charge, as a function of a properly scaled distance from the centre. Lengths (s and t) are in units of $2a_0$. Q is in units of Nf, the number of electrons in the ring, and v is in hartrees divided by $2Nf$ (see text for more details).

$$\chi(2R_e) = \frac{1}{2} - \frac{1}{2N}, \tag{18}$$

which can be solved to obtain R_e, the equilibrium radius in this model, as a function only of N, the number of atoms of the ring cluster.

In Fig. 2 we show the variation of the equilibrium radius with the number of ring atoms as it results from Eq. (18). From this plot one can see that R_e, in this model, tends rapidly to the limiting behavior, determined by the asymptotic forms of the modified Bessel functions, of about $N/4$ in atomic units. In the same figure we compare the equilibrium radii of this 2D model with realistic radii obtained from the DFT calculation on almost 2D ring clusters. This comparison will be discussed in the next section.

IV. CONSISTENCY OF THE 2D MARCH MODEL WITH REALISTIC PLANAR RING CLUSTERS

It is of interest now to compare the information that can be obtained from the model developed in the previous section with that arising from *ab initio* or DFT calculations on realistic planar ring clusters.

The prototype of such clusters could be rings of equispaced hydrogen atoms, $f=1$, constrained to lie on a circumference, while, in 2D, the analog of three-dimensional (3D) fullerenes are planar rings of N carbon atoms with N ranging from 14 to 22. The C rings of this size have been really produced in molecular beam experiments [9] and their stability has been discussed by Jones and Seifert [4]. Instead, the rings of hydrogen atoms considered here are only a theoretical construction that can be studied by standard methods for the electronic structure calculation.

In Fig. 2 we report the equilibrium radii of H and C clusters obtained from calculations performed at the B3LYP DFT level with basis sets flexible enough to give good constrained (circular) equilibrium geometries. For H rings we used even numbers of atoms ranging from 6 to 18, while for C rings we followed the study of Jones and Seifert [4] limiting attention to the three clusters C_{14}, C_{18}, and C_{22}. For C rings we made calculations on singlet states. For H clusters we considered singlet ground states when $N=4n+2$, n integer, and triplet states in the other cases.

From Fig. 2 is quite evident that in 2D the agreement between model equilibrium radii and radii of C ring clusters is not good as, instead, it happens in 3D for fullerenes. Some better consistency is instead observed between the model and our hypothetical H rings.

The existence of σ bonds in C rings modifies substantially the effective nuclear interaction from that expected in two dimensions. In order to overcome this inconsistency one can

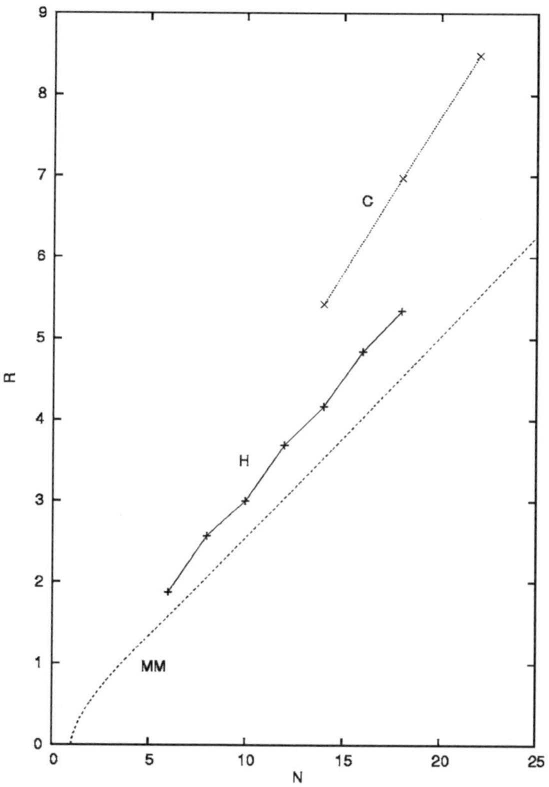

FIG. 2. Equilibrium radius as a function of the number of atoms of a ring cluster for the two dimensional March model of this work (MM) and for realistic H-ring (H) and C-ring (C) planar clusters. The radii are in atomic units.

modify the U_{nn} term in the energy functional (11). Equation (11) with the variational density (4) becomes

$$E = \pi \int_0^\infty \rho(r) V_{lc}(r) r dr + U_{nn} \tag{19}$$

in which we have

$$\pi \int_0^\infty \rho(r) V_{lc}(r) r dr = N^2 f^2 \ln R + N^2 f^2 K_0(2R) I_0(2R) \tag{20}$$

and

$$U_{nn} = -N(N-1) f^2 \ln R + \frac{N^2 f^2}{2} A_N, \tag{21}$$

where $A_N = (1/N) \Sigma_{j=2}^{N} \ln 2(1 - \cos \theta_{1j})$, θ_{1j} being the angle between nucleus 1, the ring centre and nucleus j. A_N does not depend on R. Now, to introduce some effect due to σ bonds, we propose to modify U_{nn} in the terms of order N, N being also the number of such bonds in C rings. σ bonds can be considered by including appropriate short ranged interactions acting between pairs of neighbor C atoms. Thus, we can write

$$U_{nn}^{(+\sigma)} = U_{nn} + \sum_{i=1}^{N} u(d_{i,i+1}) \tag{22}$$

where $u(d_{i,i+1})$ is the bond pair potential which depends on the distance $d_{i,i+1}$ between centres i and $i+1$ ($d_{N,N+1} = d_{N,1}$). In cumulenelike C rings the bond distances are all the same and, with the assumption $d_{i,i+1} \approx 2\pi R/N$, we have

$$U_{nn}^{(+\sigma)} = U_{nn} + N u\left(\frac{2\pi R}{N}\right). \tag{23}$$

At this point, by writing

$$U_{nn}^{(+\sigma)} = -N^2 f^2 \ln R + N g(R) \tag{24}$$

in which

$$g(R) = f^2 \ln R + \frac{f^2}{2} \sum_{j=2}^{N} \ln 2(1 - \cos \theta_{1j}) + u\left(\frac{2\pi R}{N}\right), \tag{25}$$

the energy E takes the simple form

$$E = N^2 f^2 K_0(2R) I_0(2R) + N g(R). \tag{26}$$

In the range of radii of C_N rings considered in this work, the modified Bessel function K_0 and I_0 can be substituted by their asymptotic expression. The product $K_0(t) I_0(t)$ in this regime is simply $1/2t$ with $t = 2R$ in the present case. This fact, combined with the observation that the equilibrium radii are proportional to the number of atoms in the ring, suggests to maintain the $\ln R$ functional dependence in $g(R)$ of Eq. (26) in the region of the equilibrium geometries. Under these considerations we have

$$E(R) = \frac{N^2 f^2}{4R} + N(\alpha \ln R + \beta_N), \tag{27}$$

where β_N is a constant introduced to bring $E(R_{eq})$ to coincidence with that of calculated minimum. Here, we remark that the second term in the rhs of Eq. (27) corresponds to $N g(R)$ of Eqs. (24) and (26). Equation (27) suggests that E/N could be written as a function of R/N in the more general form

$$\frac{E(R)}{N} = a \frac{N}{R} + b \ln\left(\frac{R}{N}\right) + c, \tag{28}$$

with a, b, and c related to f, α, and β_N through Eq. (27) above. Equation (28) is the analog of Eq. (9) of our previous work [6] on fullerenes in 3D.

In Fig. 3 we show results for fitted scaled energies E/N obtained from Hartree-Fock (HF) and B3LYP DFT calculations performed on C_{14}, C_{18}, and C_{22} ring clusters. The three curves, in both the two plots of Fig. 3, look roughly identical apart for a small energy shift. Fitted parameters a, b, and c are given in Table I. These numerical results confirm some universality of Eq. (28). The value of a can also be used to estimate the number of electrons of each atom, namely f, involved in the TF model density (4). From Eq. (27) we can

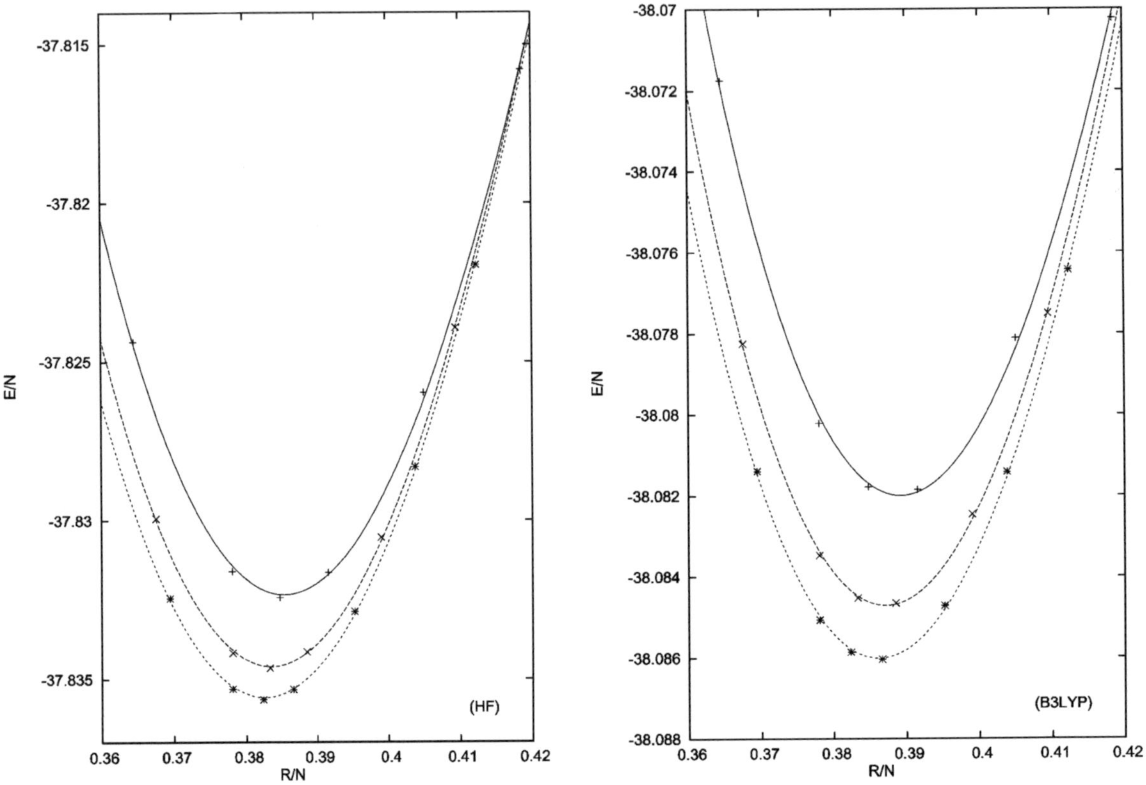

FIG. 3. Curves of scaled energies E/N against the scaled radius R/N, obtained at HF and DFT (B3LYP) level of calculation, for the three carbon rings C_{14} (upper curves), C_{18} (middle), and C_{22} (lower curves). Data are in atomic units.

write $f=2\sqrt{a}$, which tells us that this number is about 2.77 if the HF data are used or about 2.68 with B3LYP results.

Finally, with the parameters of Table I, we can calculate the frequency of the "breathing" symmetrical vibrational mode. The reduced mass for this motion is M/N where M is the mass of the carbon nucleus. By definition, the square of such angular frequency ω is given by

$$\omega^2 = \frac{N}{M}\left(\frac{\partial^2 E}{\partial R^2}\right)_{R_{eq}}, \tag{29}$$

where the energy E is the same as in Eq. (28) and plays the role of potential energy in the harmonic approximation for the study of the motion of nuclei in this case [10,11]. Combining Eqs. (28) and (29), we obtain

$$\omega = \sqrt{\frac{b^3}{a^2 M}}, \tag{30}$$

which is essentially a constant not depending on N. From the data of Table I we obtain $\bar{\nu}=\omega/2\pi c=1163\ \mathrm{cm}^{-1}$ with B3LYP DFT and 1378 cm^{-1} at the HF level.

We conclude this section on connection with realistic systems by mentioning quantum rings. Quantum rings are artificial systems of confined electrons observed at the nanoscale in semiconductors. They originate from 2D quantum dots by

TABLE I. Parameters entering the definition of the scaled energy defined in Eq. (28) obtained from fitting of HF and B3LYP results of calculations performed on C_{14}, C_{18} and C_{22} ring clusters.

Method	Parameter	C_{14}	C_{18}	C_{22}
HF	a	1.93±0.03	1.94±0.02	1.90±0.02
	b	5.02±0.08	5.06±0.06	4.97±0.07
	c	−38.065±0.005	−38.045±0.004	−38.030±0.004
B3LYP	a	1.80±0.03	1.81±0.02	1.77±0.03
	b	4.62±0.08	4.67±0.07	4.57±0.07
	c	−38.343±0.005	−38.323±0.004	−38.308±0.004

depleting the central region of the disk and thereby forming a ring (see, for example, Ref. [12]). We do not enter into the details of the methods used to study such systems, but we say that the model developed in the present work in Secs. II and III can also be formulated in a context of a quantum ring. In order to do this, we need to introduce the effective mass of electrons and the relative dielectric constant of the medium in all expressions of the model. The effective mass m^* enters through the definition of the Fermi energy, while the medium dielectric constant ϵ through the Poisson equation. For this case, we get the following coupled equations in atomic units:

$$\mu = V + \frac{\pi\rho}{m^*} \qquad -\nabla^2 V = \frac{4\pi}{\epsilon}\rho. \tag{31}$$

Of course, some attention will need to be paid to the confinement. The choice of a background positive line charge potential is still valid but the total charge and the radius should be differently defined in terms of some other external parameter. One more important point concerns the effect of an applied magnetic field, but this needs an extension of the present model.

V. SUMMARY AND PROPOSED FUTURE DIRECTIONS

The selfconsistent potential $V_{es}(r)$ in the analog of the March model in 2D planar ring clusters has first been calculated analytically. It is given by Eq. (6) in terms of modified Bessel functions. The 2D TF energy functional is then shown to lead to the potential energy function in Eq. (28), where the energy per atom of the ring cluster of radius R depends on R/N. This is the 2D counterpart of our early scaling of potential energy in the 3D fullerenes. The breathing frequency of the planar ring clusters C_{14}, C_{18}, and C_{22} is thereby estimated.

As to future directions, we already referred at the end of Sec. IV to a possible application of the present model to nanoscale quantum rings. Therefore, to conclude, we want to sketch briefly below how to eventually refine the 2D March model discussed fully above to take account of exchange, in what is essentially a local density (Dirac-Slater like) approximation. In 3D, the Slater $X\alpha$ method works as follows:

$$U_x = \frac{1}{2}\int d\mathbf{r}_1 \int d\mathbf{r}_2 \frac{\rho(r_1)\rho(r_2)f_x(r_{12})}{r_{12}}$$

$$-1 = \int d\mathbf{r}_2 \rho(r_2) f_x(r_{12}),$$

$$f_x(r_{12}) \approx -1(r_{12} \leqslant R) \approx 0(r_{12} > R),$$

$$R = \left(\frac{3}{4\pi\rho}\right)^{1/3},$$

$$U_x \approx -2\pi \int d\mathbf{r}_1 \rho(r_1)^2 \int_0^R r_{12} dr_{12}$$

$$= -\pi \int d\mathbf{r}_1 \rho(r_1)^2 R^2$$

$$\approx -\left(\frac{3\pi^{1/2}}{4}\right)^{2/3} \int d\mathbf{r} \rho(r)^{4/3}. \tag{32}$$

The counterpart of the above argument in 2D goes as follows:

$$U_x = -\int d\mathbf{r}_1 \int d\mathbf{r}_2 \rho(r_1)\rho(r_2) f_x(r_{12}) \ln r_{12},$$

$$R = \left(\frac{1}{\pi\rho}\right)^{1/2},$$

$$U_x \approx 2\pi \int d\mathbf{r}_1 \rho(r_1)^2 \int_0^R r_{12} \ln r_{12} dr_{12}$$

$$= \frac{\pi}{2} \int d\mathbf{r}_1 \rho(r_1)^2 R^2 (2\ln R - 1)$$

$$\approx -\frac{1}{2} \int d\mathbf{r} \rho(r)[1 + \ln(\pi\rho)]. \tag{33}$$

The final line of Eq. (33) prompt us to add that the exchange potential energy U_x in 2D contains a term which is reminiscent of the Shannon entropy, the latter having an integrand of the form $\rho \ln \rho$.

ACKNOWLEDGMENTS

C.A. acknowledges financial support from MIUR (PRIN 2004) and from the University of Pisa (Fondi di Ateneo 2005). N.H.M. wishes to acknowledge that his contribution to this work was made during a visit to Scuola Normale Superiore (Pisa) in 2006. He thanks Professor M. P. Tosi for very generous hospitality.

[1] D. P. Clougherty and X. Zhu, Phys. Rev. A **56**, 632 (1997).
[2] F. Despa, Phys. Rev. B **57**, 7335 (1998).
[3] C. Amovilli and N. H. March, Chem. Phys. Lett. **347**, 459 (2001).
[4] R. O. Jones and G. Seifert, Phys. Rev. Lett. **79**, 443 (1997).
[5] F. Siringo, G. Piccitto, and R. Pucci, Phys. Rev. A **46**, 4048 (1992).
[6] C. Amovilli, I. A. Howard, D. J. Klein, and N. H. March, Phys. Rev. A **66**, 013210 (2002).
[7] M. Abramowitz and I. A. Stegun, *Handbook of Mathematical Functions* (National Bureau of Standards, Washington D.C., 1964).
[8] G. F. Kventsel and J. Katriel, Phys. Rev. A **24**, 2299 (1981).
[9] S. J. La Placa, P. A. Roland, and J. Wynne, Chem. Phys. Lett. **190**, 163 (1992).
[10] N. H. March, Proc. Cambridge Philos. Soc. **43**, 665 (1952).
[11] H. C. Longuet-Higgins and D. A. Brown, J. Inorg. Nucl. Chem. **1**, 60 (1955).
[12] B. C. Lee and C. P. Lee, Nanotechnology **15**, 848 (2004).

Available online at www.sciencedirect.com

ScienceDirect

PHYSICS LETTERS A

ELSEVIER

Physics Letters A 372 (2008) 3253–3255

www.elsevier.com/locate/pla

Molecules in clusters: The case of planar LiBeBCNOF built from a triangular form LiOB and a linear four-center species FBeCN

G. Forte [a], A. Grassi [a], G.M. Lombardo [a], G.G.N. Angilella [b,*], N.H. March [d,c], R. Pucci [b]

[a] *Dipartimento di Scienze Chimiche, Facoltà di Farmacia, Università di Catania, Viale A. Doria, 6, I-95126 Catania, Italy*
[b] *Dipartimento di Fisica e Astronomia, Università di Catania, and CNISM, UdR Catania, and INFN, Sez. Catania, 64, Via S. Sofia, I-95123 Catania, Italy*
[c] *Department of Physics, University of Antwerp, Groenenborgerlaan 171, B-2020 Antwerp, Belgium*
[d] *Oxford University, Oxford, UK*

Received 14 January 2008; accepted 21 January 2008

Available online 1 February 2008

Communicated by V.M. Agranovich

Abstract

Krüger some years ago proposed a cluster LiBeBCNOF, now called periodane. His ground-state isomer proposal has recently been refined by Bera et al. using DFT. Here, we take the approach of molecules in such a cluster as starting point. We first study therefore the triangular molecule LiOB by coupled cluster theory (CCSD) and thereby specify accurately its equilibrium geometry in free space. The second fragment we consider is FBeCN, but treated now by restricted Hartree–Fock (RHF) theory. This four-center species is found to be linear, and the bond lengths are obtained from both RHF and CCSD calculations. Finally, we bring these two entities together and find that while LiOB remains largely intact, FBeCN becomes bent by the interaction with LiOB. Hartree–Fock and CCSD theories then predict precisely the same lowest isomer found by Bera et al. solely on the basis of DFT.

PACS: 31.15.Ne; 36.40.Qv

1. Introduction

Though traditional chemical thinking in which 'atoms in molecules' was a prime focus goes back many decades (see e.g. the early work of Moffitt [1]), more recently the idea has been championed most notably by Bader and coworkers [2]. Here, we have been motivated by the proposal of Krüger [3] on LiBeBCNOF, termed periodane, to study a 'molecules in clusters' approach to this species. This cluster, treated subsequently, in a more refined quantum chemical manner than in [3], by Bera et al. [4], is found to be essentially planar. One molecular grouping which then seemed apparent was LiOB, with the strong bond being B–O. This left Be, C, and N, which would form a radical and would be spin-compensated in the ground state by adding F.

Below, calculations are reported on the ground-state isomer of periodane, by (a) both restricted Hartree–Fock (RHF) and coupled cluster (CCSD) theory, and (b) for LiOB by the coupled cluster singles and doubles (CCSD) methods. Thereby we can make a quantitative comparison with Bera et al. [4] of both bond lengths and angles in the final geometry of the predicted lowest isomer from DFT.

2. Quantum chemical methodology

Several isomers of periodane were considered and, for each of them, a geometry optimization has been performed by using a 6-311(d) standard basis set [5] for all the atoms at Hartree–Fock level. The geometry of the most stable isomer, thermodynamically speaking, was further optimized at coupled cluster

* Corresponding author.
E-mail address: giuseppe.angilella@ct.infn.it (G.G.N. Angilella).

0375-9601/$ – see front matter © 2008 Elsevier B.V. All rights reserved.
doi:10.1016/j.physleta.2008.01.046

with single and double excitations (CCSD) level [6–8]. All the calculations have been performed by using the G03 package [9].

3. Molecules in clusters

The triatomic molecule LiOB has then been studied specifically and accurately by CCSD (see e.g. Ref. [10] for a review). With the B–O strong bond, the geometry predicted by CCSD is shown in Fig. 1, the ground-state energy being −107.422807 Hartree. With the DFT functional of [4], a similar geometry was found with a lower energy of about 0.25 Hartree: It is not clear to us that the DFT variational value lies above the exact ground-state energy because of approximations in the energy functional that are needed to date.

For the spin-compensated four-center molecule FBeCN, a linear structure was obtained as shown in Fig. 2, where the structural parameters are recorded in Table 1.

Fig. 3 shows schematically the way the two isolated molecules, with the individual geometries cited above, are somewhat modified as they are brought together into what we predict, as do Bera et al. [4] by purely DFT, as the lowest isomer of periodane. The four-center molecule is clearly distorted from linearity, the bond lengths and angles being recorded in Table 1. The change in the triatomic LiOB is seen to be much smaller than in the four-center case. Table 2 reports the sum of the RHF eigenvalues for the occupied orbitals for (i) FBeCN as in Fig. 2, and for (ii) isolated FBeCN, but with all constituent atoms held rigid at the HF geometry in Fig. 3 for periodane. The HF eigenvalue sums are seen to be quite close for linear and bent geometries and hence somewhat subtle corrections to Walsh's rules [11] discussed in [12] are required to determine the relative stability between linear and bent forms of FBeCN.

4. Summary

The structure of the lowest isomer as predicted by Hartree–Fock and CCSD theory is shown in Fig. 3. As stated above, our 'molecules in clusters' approach has led to an identical structure reached by Bera et al. [4] on the basis of DFT alone. We have argued that it is useful to be viewed as having building

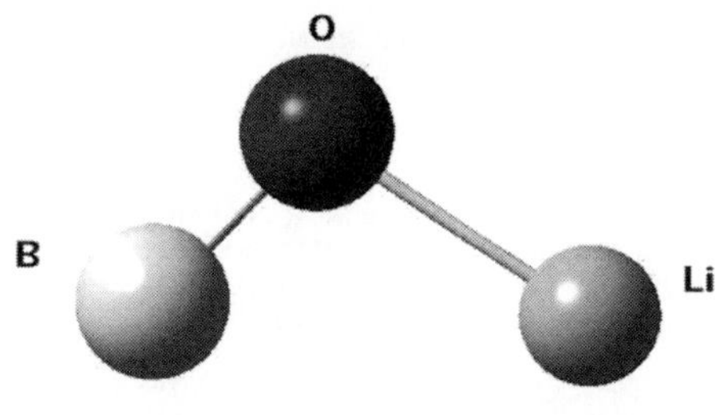

Fig. 1. (Color online.) Shows predicted geometry of molecule LiOB as obtained from a CCSD calculation. The bond lengths, the angle, and the energy are given in Table 1.

Fig. 2. (Color online.) Depicts linear geometry found for FBeCN four-center molecule. Structural parameters are listed in Table 1.

Table 1
Bond lengths (in Å), angles (in degrees), and energies (in Hartree) of the lowest isomers of LiOB (Fig. 1), FBeCN (Fig. 2), and periodane (LiBeBCNOF, Fig. 3), within RHF and CCSD methods

	LiOB	FBeCN		LiBeBCNOF	
	CCSD	RHF	CCSD	RHF	CCSD
Distances (Å)					
O–B	1.26			1.21	1.23
Li–O	1.75			1.86	1.88
C–N		1.13	1.17	1.14	1.17
Be–C		1.67	1.66	1.83	1.81
F–Be		1.36	1.37	1.39	1.40
N–Li				2.00	1.98
B–Be				1.96	1.94
Angles (°)					
B–O–Li	101			105.3	102.9
Be–C–N				155.3	154.9
C–N–Li				101.6	100.1
N–Li–O				120.2	123.0
O–B–Be				146.9	147.3
B–Be–C				90.6	91.8
Energy (Hartree)	−107.423	−206.545	−207.176	−313.705	−315.362

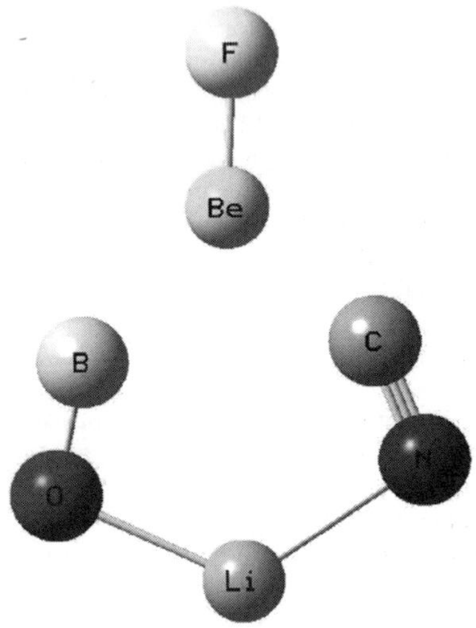

Fig. 3. (Color online.) Shows predicted lowest isomer for periodane from RHF and CCSD theories (with 6-311(d) basis set). Bond lengths and angles (Table 1) agree excellently with DFT predictions [4]. The two strongest bonds are seen to be B–O and N–C, with Be–F also short.

Table 2
Reports the sum of the RHF eigenvalues (in Hartree) for the occupied orbitals for (i) FBeCN as in Fig. 2, and for (ii) isolated FBeCN, but with all constituent atoms held rigid at the HF geometry in Fig. 3 for periodane

(i) FBeCN		(ii) FBeCN	
HF	CCSD	HF	CCSD
−129.992	−129.972	−129.640	−129.642

Table 3
Reports energy differences (in Hartree) between the given clusters and their constituent isolated atoms

	HF	CCSD
LiOB	−0.306	−0.414
FBeCN	−0.489	−0.643

blocks of (a) the bent triatomic molecule LiOB, and (b) the linear four-center molecule formed from BeCN plus F. Both these molecules are stable against dissociation into their isolated neutral atoms (cf. Table 3, reporting the difference between the energies of the clusters LiOB amd FBeCN and their isolated constituent atoms, respectively).

Of course, as pointed out very specifically in [4], it is never possible to exclude the possibility of a (slightly) lower energy isomer than that shown in Fig. 3. Nevertheless, we believe that the present picture of bringing together two molecules LiOB and the linear four-center system BeCN plus F is a favourable way of approaching the final structure of the lowest isomer.

Acknowledgements

N.H.M. made his contribution to the present study during a visit to the University of Catania. He wishes to thank Professors Porto and Pucci for very generous hospitality.

References

[1] W. Moffitt, Proc. R. Soc. A 210 (1951) 245.
[2] R.F.W. Bader, Atoms in Molecules: A Quantum Theory, Oxford Univ. Press, Oxford, 1994.
[3] T. Krüger, Int. J. Quantum Chem. 106 (2006) 1865.
[4] P.P. Bera, P. von R. Schleyer, H.F. Schaefer III, Int. J. Quantum Chem. 107 (2007) 2220.
[5] A.D. McLean, G.S. Chandler, J. Chem. Phys. 72 (1980) 5639.
[6] J. Cizek, Adv. Chem. Phys. 14 (1969) 35.
[7] G.D. Purvis III, R.J. Bartlett, J. Chem. Phys. 76 (1982) 1910.
[8] G.E. Scuseria, C.L. Janssen, H.F. Schaefer III, J. Chem. Phys. 89 (1988) 7382.
[9] M.J. Frisch, G.W. Trucks, H.B. Schlegel, G.E. Scuseria, M.A. Robb, J.R. Cheeseman, J.A. Montgomery Jr., T. Vreven, K.N. Kudin, J.C. Burant, J.M. Millam, S.S. Iyengar, J. Tomasi, V. Barone, B. Mennucci, M. Cossi, G. Scalmani, N. Rega, G.A. Petersson, H. Nakatsuji, M. Hada, M. Ehara, K. Toyota, R. Fukuda, J. Hasegawa, M. Ishida, T. Nakajima, Y. Honda, O. Kitao, H. Nakai, M. Klene, X. Li, J.E. Knox, H.P. Hratchian, J.B. Cross, V. Bakken, C. Adamo, J. Jaramillo, R. Gomperts, R.E. Stratmann, O. Yazyev, A.J. Austin, R. Cammi, C. Pomelli, J.W. Ochterski, P.Y. Ayala, K. Morokuma, G.A. Voth, P. Salvador, J.J. Dannenberg, V.G. Zakrzewski, S. Dapprich, A.D. Daniels, M.C. Strain, O. Farkas, D.K. Malick, A.D. Rabuck, K. Raghavachari, J.B. Foresman, J.V. Ortiz, Q. Cui, A.G. Baboul, S. Clifford, J. Cioslowski, B.B. Stefanov, G. Liu, A. Liashenko, P. Piskorz, I. Komaromi, R.L. Martin, D.J. Fox, T. Keith, M.A. Al-Laham, C.Y. Peng, A. Nanayakkara, M. Challacombe, P.M.W. Gill, B. Johnson, W. Chen, M.W. Wong, C. Gonzalez, J.A. Pople, Gaussian 03, Revision C.02, Gaussian Inc., Wallingford, CT, 2004.
[10] R.J. Bartlett, in: D.R. Yarkony (Ed.), Modern Electronic Structure Theory, World Scientific, Singapore, 1995, p. 1047.
[11] A.D. Walsh, J. Chem. Soc. 2260 (1953).
[12] N.H. March, J. Chem. Phys. 74 (1981) 2973.